A TEXTBOOK OF
ENGINEERING CHEMISTRY

By

Dr. Sunita Rattan

M.Sc, M.Phil, Ph.D
HoD, Department of Chemistry
Amity School of Engineering and Technology, New Delhi
Amity University Uttar Pradesh, Noida

S.K. KATARIA & SONS®

Publisher of Engineering Books
4885/109, Prakash Mahal, Dr. Subhash Bhargav Lane,
Opposite Delhi Medical Association, Daryaganj, New Delhi-110002 (INDIA)
Phones: +91-11-43551243, Mobile: +91-9871775858
e-mail: katariabook@gmail.com; katariabooks@yahoo.com
Website: www.skkatariaandsons.com
(e-books, subscriptions and bulk buying also available on our website)

Published by:

S.K. Kataria & Sons®

4885/109, Prakash Mahal, Dr. Subhash Bhargav Lane,
Opposite Delhi Medical Association, Daryaganj, New Delhi-110002
Phones: +91-11-43551243, Mobile: +91-9871775858
e-mail: katariabook@gmail.com, katariabooks@yahoo.com
Website: www.skkatariaandsons.com

First Edition: 2012; Reprint: 2024

Laser Type Setting by:
SARA ASSIGNMENTS, Patparganj, Delhi-110092.

Printed at:
A.S. OFFSET PRINTERS, Delhi-110051.

A TEXTBOOK OF

ENGINEERING CHEMISTRY

Dr. Sunita Rattan

PREFACE

Learning is not attained by chance; it must be sought for with ardor and attended to with diligence.

—**Abigail Adams**

I am deeply indebted to the faculty and students of various Universities for their enthusiastic response to my previous works. The *Chemistry for Engineering Students* continues with the same spirit: to provide a firm foundation in chemical concepts and principles while presenting a broad range of topics in a clear, concise manner. The book caters to the needs of engineering students of various Indian Universities.

My goal has always been to instill an appreciation of the role of chemistry in many areas of engineering and technology and make the students understand the significance of chemistry in variety of modern technologies. The book is intended to provide an in-depth coverage of a wide range of topics in, keeping in view the applications of chemistry in various fields of engineering. The edition is balanced to work well for the students of engineering and technology. Everyday applications of chemistry that should be of interest to the students appear throughout the text. An integral part of the present text is to develop students' problem-solving and critical thinking skills. At the end of the chapters, short answer questions are given to help the students to grasp the fundamental concepts.

The text presented has been reviewed by the colleagues at other colleges and universities. I profoundly thank the reviewers for their valuable comments which have been used to polish the book in its present form. I am extremely grateful to my colleagues at various institutions for their suggestions and criticisms which have been invaluable in polishing this book in its present form.

I welcome any suggestions for improvements in the organization, scope, and content of the handbook, as well as notification of any errors.

SUNITA RATTAN

Acknowledgements

Gratitude unlocks the fullness of life;
It turns what we have into enough, and more;
It turns denial into acceptance; Chaos into order;
Confusion into clearity;
Gratitude makes sense of our past,
Brings peace for today,
And creates vision for tomorrow.

"Melody Beattie"

I take this opportunity to acknowledge the assistance and contributions of a large number of individual to this effort.

My special thanks are due to my revered teacher, Professor B.N. Misra, Emeritus Fellow, Department of Chemistry, H.P. University, Shimla. The legacy of scholarly and scientific inquiry and the temperament which I have inherited from him, will remain with me as invaluable treasure to be invested in everything, I do and seek in life.

I will be failing in my duties if I do not acknowledge Dr (Mrs) I. K. Mehta (Himachal Pradesh University) whose affectionate advice and moral support have left a deep impression on my life.

It gives me great pleasure to place on record my profound gratitude and indebtness to our esteemed president Dr. Ashok Kumar Chauhan, Founder President RBEF, whose inspiring words, stimulating encouragement and unflinching support have been a source of great strength in accomplishing this assignment.

I express my sincere gratitude to Mr. Atul Chauhan,Chancellor ,Amity University, Uttar Pradesh for his continued encouragement.

My heartfelt thanks are due to Dr. Balvinder Shukla , Pro-Vice Chancellor , Amity University , Uttar Pradesh for her encouragement and boosting my confidence in such type of ventures. I would like to place on record my sincere gratitude to Prof. A.L Verma, Head, Institute of Applied Sciences, and Amity University for providing the valuable outputs.

It is also a pleasure to acknowledge the enthusiastic support and assistance given to me by my colleagues at other institutions. Their constructive comments and suggestions during informal discussions are responsible for smooth completion of the book.

My sincere thanks are due to all my colleagues in Amity University, Uttar Pradesh, for their suggestions and valuable inputs. They have given me both, the invaluable advice on many aspects of the work and encouragement to see the project through to the end.

I also record my sense of appreciation to my students whose inquisitive questions during my laboratory classes provided valuable input into this effort.

Words elude me in expressing my sincere thanks to Praneesh who offered every possible help and supported the long hours of preparation that went into this effort. I feel greatly indebted to my kids, Tanishq and Tanvi without whose patience and sacrifice it was rather impossible to bring out the book in its present form.

My immeasurable gratitude and thanks are due to my family members who have been a driving force behind all my achievements. Their affection and benign blessings will ever remain to be cherished.

Finally, I am thankful to Mr. S.K. Kataria Publications and his entire staff for their efforts without which the book could not have seen the light of the day.

Mr. Dinesh, Mr.Sachin deserve a special thanks for typing this text as error free as possible.

SUNITA RATTAN

CONTENTS

1 WATER TECHNOLOGY

"Science cannot solve the mystery of Nature. It is because in the analysis we ourselves are part of the mystery we are trying to solve"

— ***Max Plank***

CHAPTER OUTLINE

The planet earth is abundantly rich in water, which constitute 75% of its surface. Besides, being an essential ingredient of animal and plant life, water forms the basis of human life. It constitutes about 65% of human body. It is so common that its importance is overlooked many a times. Ben Franklin has rightly said, *"When the well is dry, we know the worth of water."* Water is essential not only for survival of human beings, animals, plants and all other living beings but is also utilized in a large amount in industries. Out of the total water consumed by human beings, more than 50% of it is consumed for industrial activity and only a small proportion is used for drinking purposes.

Most of the water which is available is located in oceans (97.2%). The high concentration of salts make it virtually unusable as a source of water for municipal, agricultural or most industrial needs. However, we do use the oceans for thermal cooling of power plants and as a sink for much of our pollution. Hence we are left with only 2.98% of water, out of which, considerable amount of water is present in the form of polar ice caps and may remain locked as such for thousands of years. A small portion of it is also present in rock crevices and minerals, lying 800 meters deep below the earth's crust, which is again very expensive to pump out. Thus, the conclusion is that, only 0.3 percent of world's water resources is available for possible human use.

Quantity of water is not the only concern. The second major problem is the *quality* of water. Overuse has resulted in the progressive deterioration of water quality. Our traditional confidence in the quality of our drinking water is lost. Even the water which is used for industries has to be treated prior to its use. The treatment to which it is subjected, depends upon the purpose, for which it would be used. Thus, both the problems, *water quantity* and *quality* have to be treated with equal importance.

The impurities present in water vary greatly from one place to another. Treatment of water obtained from *different sources* is therefore essential so that it may be used safely for industrial or municipal purposes.

Maintaining an adequate supply of water for municipal or industrial purposes requires the efforts of scientists and engineers. The following aspects are taken into consideration for treatment of water for various uses.

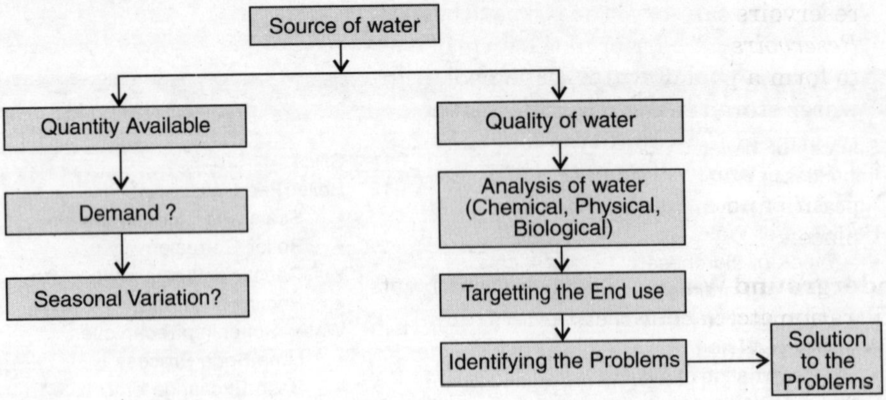

1.1 SOURCES OF WATER

The various sources of water available can be classified into the following.

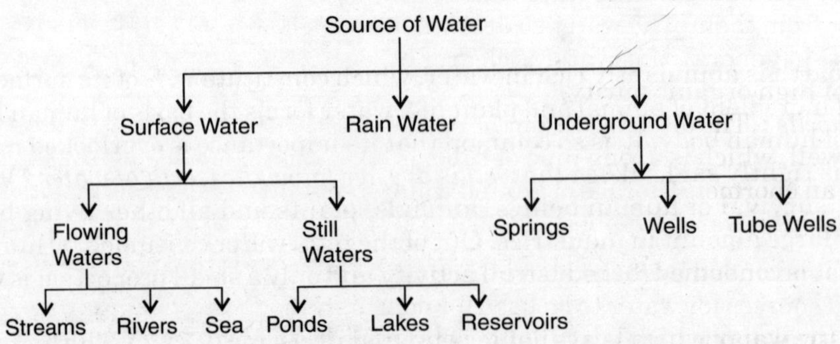

1. **Surface Water :** Surface sources are those sources of water in which the water flows over the surface of the earth, and is directly available for water supplies. The important surface water sources are:

(a) **Flowing waters :** These include streams and rivers. Small stream channels feed their waters to the lakes or rivers.

Streams : Streams are not suitable for water supply schemes, because the quantity of water available in them is generally very small and they may sometimes go dry.

Rivers : Rivers are the most important sources of water for water supply schemes. River water has fairly constant composition. However it is slightly acidic due to the presence of dissolved carbon dioxide and weak organic acids. The disposal of the untreated sewage into the rivers is further liable to contaminate their waters. It also contains some strains of iron, bacteria and small particles of sand and rock in suspension. The river waters must, therefore, be properly analyzed and well treated before use.

Sea Water: Sea waters are too saline and contaminated for industrial use, except cooling.

(b) **Still Waters :** A water supply scheme using river as water source may fail to meet the demands during extremely low or high flows. During high flow it may become difficult to carry out its operations. Thus, the water collected in ponds, lakes or reservoirs can serve the purpose best.

Reservoirs : A barrier in the form of a dam may be constructed across the river, so as to form a pool of water. This pool or artificial lake is called storage reservoir. The water stored in the reservoir can be used easily for different purposes after proper treatment. The quality is just the same as that of river water.

Lakes/Ponds : The lake water has organic impurities due to the decomposition of plants and animals. Water from this source vary in its composition from place to place.

2. **Underground Water :** The water which gets stored in the ground water reservoir through infiltration etc. is known as underground water. This water is generally pure, because it undergoes natural filtration during the percolation through the soil pores. Moreover, these waters are less likely to be contaminated by bacteria. However these are rich in dissolved salts, minerals, gases etc.

Springs : Sometimes, the ground water is brought to the surface by some natural process like springs. The extent of salts and minerals present in the springs depends upon the downward journey of water during which it comes in contact with a number of mineral salts present in the soil.

Wells : The underground water can be drained artificially by constructing wells and lifting water from them. The water so drained, may then be used to fulfill irrigation or municipal needs. In general, this water is clearer but contains more of the dissolved salts. However, it is of high organic purity.

Tubewells : The discharge from the open well is generally limited. To obtain large discharge, tubewell, which is a long pipe of tube, is bored or drilled deep into the ground. Since, we have an enormous storage of ground water in India, the tubewells provide excellent method of obtaining water supplies.

It may be concluded from the above discussion that surface waters and underground waters, both are good sources for water to be used for domestic and industrial purposes. From the point of view of industrial applications, it is not recommended to use sea water and rain water. Sea water is too impure and saline to be used whereas rain water is irregular in supply.

A comparison of surface sources and ground water sources is given in the tabulated form in Table 1.1.

Table 1.1

	Surface Water Sources	*Ground Water Sources*
Forms in which available	Rivers, streams, Lakes, storage reservoirs	Springs, wells, tubewells etc.
Quantity	Huge amounts are available in rainy seasons, but the available quantity is considerably reduced during summers. Storage reservoirs are constructed to over come this problem. These are useful for big cities and industrial towns.	The available yields are comparatively smaller and are useful for towns and villages only.
Quality	Contain huge amount of suspended impurities. Most of the rivers carry huge amounts of inorganic matter, dissolved salts, fine mud suspensions. Lots of organic impurities are also present. Hardness is usually high and hence cause scale formation in boilers. Rivers and streams may get contaminated by industrial waste.	Contain lesser amounts of suspended impurities. Traces of manganese as well as H_2S may also be present. When freshly drawn, this is usually colourless, clear and devoid of finely divided suspensions and hence sparkling. On exposure to air may develop a brown colour due to the presence of small amounts of ferrous ions. The chances of contamination by industrial waste are very less.
Treatment	The water from these source requires lot of treatment before use.	The water from these sources require minor treatments before being supplied.

1.2 IMPURITIES IN WATER

Water from a water supply scheme may contain a number of impurities which it takes up from

- the ground or soil with which it comes in contact. For e.g. soil particles, minerals, garbage etc.
- the sewage and industrial waste. For eg. chemical wastes, acids, colours, heavy metals etc.
- the decomposition of plants and animals.
- the growth of bacteria, algae, viruses etc.
- the atmosphere in the form of dissolved gases.

Types of Impurities: The impurities present in water may be broadly classified as given below in Table 1.2.

Table 1.2

	Impurities
Dissolved Impurities	(a) Salts : Carbonates, bicarbonates, sulphates and chlorides of calcium, magnesium, iron, sodium and potassium. (b) Gases : CO_2, O_2, N_2, H_2S, NH_3 etc.
Suspended Impurities	(a) Inorganic – sand, clay (b) Organic – Vegetable and animal matter
Colloidal Impurities	Finely divided silica and clay, organic waste products, colouring matter etc.
Micro Organisms	Bacteria, Algae, Fungi, Virus etc.

Effect of Impurities: The various types of impurities present in water impart certain properties to water. Colour, decaying aquatic organisms and certain chemicals may impart odour, suspended matter may impart turbidity to the water, presence of dissolved salts make the water hard, whereas excess quantities of metals and dissolved gases makes it corrosive. Bacteriological impurities due to pathogenic bacteria make water unfit for drinking or domestic purposes.

Hence, the water from any source has to be treated before it is used. The treatment to which it is subjected depends upon its use. For e.g., if it has to be used for drinking purposes, the treatment would include removal of objectionable colour, taste and pathogenic microorganisms, where as the water for industrial use require removal of dissolved salts if it is used for steam generation, removal of Ca, Mg, Fe and Mn salts for its use in laundries. Similarly, the water used in alcoholic breweries requires the high purity and the presence of a few micro organisms.

From the foregoing discussion, it is clear that each industry has its special water treatment requirement. Moreover, the treatment for municipal water, used for drinking and domestic purposes is different from the treatment to which water for industrial use is subjected. Subsequent sections give an account of the specifications and treatment of water for its use in industries and for drinking purposes.

WATER FOR INDUSTRIES

Specifications for Industrial Water

The quality of water required for industries depends upon the requirements of a particular industry. But in general, it can be stated that about 80 percent industrial water are used for cooling purposes and not necessarily be of high quality. But the water required for boiler feed purposes must be of high quality. Water required for processing industry must be treated according to its end use. Hence each industry has its own specifications for water which is then treated accordingly. Table 1.3 gives the specifications for water used in some industries.

Table 1.3: *Specifications of Water for Industrial use*

S.No.	*Purpose*	*Specifications of water*	*Remarks.*
1.	Boiler Feed Water	Very soft, should contain least amount of nitrate and organic matter.	Hard water form encrustations or scale on boiler walls, which reduce the efficiency of the boiler. Hardwater may cause priming and foaming.
		Least amount of dissolved solids. Free from alkalies.	On precipitation at high concentrations, these may choke pipes, alkaliesmay cause caustic embrittlement.
2.	Cooling Systems	Must not promote the growth of slime and algae slime and algae.	The spray nozzles of cooling systems or circulating pumps get clogged by
		Should have least amount of dissolved gases.	Dissolve gases may accelerate corrosion.
3.	Alcoholic Distilleries	Should be very pure, free from pathogenic bacteria. Should not be alkaline	Should confirm to the standards of drinking water. Alkalinity tends to neutralize the fruit acids and modifies the taste.

S.No.	Purpose	Specifications of water	Remarks.
4.	Confectionary Dairies and Allied Industries	Free from pathogenic bacteria colourless, odourless.	
5.	Laundries	Should be as soft as possible.	Hard water consumes lot of soaps and detergents.
		Free from Mn and Fe salts.	May cause staining on the cloth due to the formation of iron soap.
6.	Paper Mills	Free from Hardness and alkalinity.	Hardness producing ions may react with the chemicals used for paper.
		Free from SiO_2 Free from lime and magnesia	SiO_2 may produce cracks in paper. These may decompose the resin soaps.
7.	Textiles	Free from hardness causing substances. Free from Fe and Mn salts	Hardwater precipitates basic dyes and decrease the solubility of acidic dyes. These salts produce insoluble precipitates which stain the cloth and cause uneven dyeing.
8.	Sugar Industries	Free from pathogenic bacteria, should confirm to the standards of potable water.	
		Free from sulphates, carbonates and nitrates.	In the presence of these anions the crystallization becomes difficult and sugar obtained is deliquescent.
		Free from micro organisms.	The micro organisms may decompose the sugar partly.
9.	Pharmaceutical Industries	Very pure, free from any pathogenic bacteria, confirming to the standards of potable water. Very soft water.	Salts present in hard water may react with the chemicals to form undesirable products.

1.3 ANALYSIS OF WATER

Analysis of water includes the determination of different constituents in water to ascertain the suitability of water for a specific use, and the treatment to which it should be subjected prior to its use. The following characteristics of water are usually determined :

- Hardness
- Alkalinity
- Dissolved Oxygen (BOD, COD)
- Dissolved Carbondioxide
- Free Chlorine

- Chlorides
- Silica
- Total Solids.

Each of these is discussed in the foregoing sections.

1.4 HARDNESS OF WATER

Hardness is one of the most important characteristic of water. This term was applied to the waters which make cleansing action of the soap difficult. The water which **does not form lather with soap** and is **hard** to wash in, is termed as hard water. Hardness of a water sample is a measure of its capacity to precipitate soap or its soap consuming capacity.

On the other hand, the water which forms copius lather with soap is termed as soft water.

Cause of Hardness : Hardness of water is due to the presence of calcium and magnesium salts. Generally, soaps consist of sodium salts of higher fatty acids such as stearic, oleic, palmitic etc. which are highly soluble in water and thus exert their cleansing action. If calcium and magnesium salts are present in water, these react with soluble sodium soap to form insoluble salts of calcium and magnesium. For example,

$$\underset{\text{(Sodium Stearate)}}{2C_{17}H_{35}COONa} + CaSO_4 \longrightarrow \underset{\text{(Insoluble salt)}}{(C_{17}H_{35}COO)_2\,Ca} + Na_2SO_4$$

$$2C_{17}H_{35}COONa + MgCl_2 \longrightarrow (C_{17}H_{35}COO)_2\,Mg + 2NaCl$$

Thus, a white scum or precipitate of insoluble soap of calcium and magnesium is formed. Lather is not produced untill the cations, Ca^{2+} and Mg^{2+} are completely precipitated out in the form of insoluble salts. As a result, a large quantity of soap is required to produce lather with hard water. Thus, hardness is rightly defined as *soap consuming capacity of water.*

1.4.1 Types of Hardness

Hardness of water is of two types

(a) Temporary Hardness
(b) Permanent Hardness

(a) **Temporary Hardness :** Hardness due to the presence of bicarbonate ions of calcium and magnesium is called temporary hardness or *carbonate hardness.* This is also known as alkaline hardness. This hardness can be removed by mere boiling of water, which converts the bicarbonates into insoluble carbonates or hydroxides. The insoluble carbonates or hydroxides can be removed by filteration.

$$Ca(HCO_3)_2 \xrightarrow{\text{boil}} CaCO_3\downarrow + H_2O + CO_2$$

$$Mg\,(HCO_3)_2 \xrightarrow{\text{boil}} MgCO_3 + H_2O + CO_2.$$

$$Mg\,CO_3 + H_2O \longrightarrow Mg(OH)_2\downarrow + CO_2.$$

Temporary hardness can also be removed by addition of calculated amount of lime ($Ca(OH)_2$)

$$Ca(HCO_3)_2 + Ca(OH)_2 \longrightarrow 2CaCO_3\downarrow + 2H_2O$$

$$Mg\,(HCO_3)_2 + Ca(OH)_2 \longrightarrow MgCO_3\downarrow + CaCO_3\downarrow + 2H_2O$$

(b) **Permanent Hardness :** This type of hardness is also known as *non-carbonate* hardness. It is due to the presence of sulphates and chlorides of calcium and magnesium. Permanent hardness cannot be removed by boiling. Special methods (discussed later) are used for removal of permanent hardness.

1.4.2 Hardness in terms of $CaCO_3$ equivalents

From the foregoing discussion, it is clear that a number of salts can cause hardness such as bicarbonates, sulphates, chlorides of calcium and magnesium. To compare the hardness of different samples of water, it would be easier, if the hardness caused by different salts can be expressed in terms of single salt.

It is customary to express hardness in terms of equivalents of $CaCO_3$. The reason for choosing $CaCO_3$ as the standard for calculating hardness of water is due to–

- its molecular weight is exactly 100, which makes mathematical calculations easier.
- it is the most insoluble salt, thus can be easily precipitated in water treatment processes.

Now 1 mole of $CaCO_3$ (or 100 parts by weight) would react with the same amount of soap as 1 mole of $CaCl_2$ (111 parts by weight), or 1 mole of $MgCl_2$ (95 parts by weight) because 1 mole of all these salts will produce same number of calcium or magnesium ions (Avogadro's number).

Thus $CaCO_3$ equivalents for various salts are as follows:

100 g of $CaCO_3 \equiv$ 111 g of $CaCl_2 \equiv$ 136 g of $CaSO_4 \equiv$ 95 g of $MgCl_2 \equiv$ 120 g of $MgSO_4 \equiv$ 162 g of $Ca(HCO_3)_2 \equiv$ 146 g of $Mg(HCO_3)_2 \equiv$ 164 g of $Ca(NO_3)_2 \equiv$ 44 g of $CO_2 \equiv$ 148 g of $Mg(NO_3)_2$

If x g of $CaCl_2$ is present in a water sample, then the amount of $CaCl_2$ present in terms of its $CaCO_3$ equivalent will be –

1 g mole of $CaCl_2 \equiv$ 1 g mole of $CaCO_3$

111 g of $CaCl_2 \equiv$ 100 g of $CaCO_3$

55.5 g of $CaCl_2$ = 50 g of $CaCO_3$ ($\because$ Eq. wt of $CaCO_3$ = 50 g Eq. wt of $CaCl_2$ = 55.5 g)

$$x \text{ g of } CaCl_2 = \frac{50}{55.5} \times x\,g \text{ of } CaCO_3.$$

Thus to convert various hardness causing salts into $CaCO_3$ equivalents we may use–

$$\text{Equivalents of } CaCO_3 = \frac{\text{Mass of the substance}}{\text{Equivalent weight of the substance}} \times \text{Equivalent weight of } CaCO_3$$

As equivalent weight of most of the hardness causing substances and that of $CaCO_3$ = Molecular weight/2, hence to make mathematical calculations more easier, the above formula can be written in terms of molecular weights i.e.

$$\text{Equivalents of } CaCO_3 = \frac{\text{Mass of the substance}}{\text{Molecular mass of the substance}/2} \times 100/2$$

or
$$\text{Equivalents of } CaCO_3 = \frac{\text{Mass of the substance}}{\text{Molecular mass}} \times 100$$

In otherwords, the weight of the substance has to be multiplied by a certain fraction i.e. 100 / *Molecular Mass*, to obtain equivalents of $CaCO_3$ in case of *divalent salts*. Similarly the multiplication factors for other hardness causing substances can be found out. (Table 1.4)

Table – 1.4

	Salt	*Molar Mass*	*Multiplication Factor*
Divalent Salts	$Ca(HCO_3)_2$	162	100/162
	$Mg(HCO_3)_2$	146	100/146
	$CaCO_3$	100	100/100
	$CaSO_4$	136	100/136
	$CaCl_2$	111	100/111
	$MgCO_3$	84	100/84
	$MgSO_4$	120	100/120
	$MgCl_2$	95	100/95
	$Mg(NO_3)_2$	148	100/148
*	$Al_2(SO_4)_3$	342	100/114
	$FeSO_4.7H_2O$	278	100/278
**	$NaAlO_2$	82	100/82×2
	HCO_3^-	61	100/61×2
	CO_3^{2-}	60	100/60
	OH^-	17	100/17×2
	H^+	1	100/1×2

Notes: *The multiplication factor for $Al_2(SO_4)_3$ is given by:

$$\frac{\text{Equivalent weight of } CaCO_3}{\text{Equivatent weight of } Al_2(SO_4)_3} = \frac{100/2}{342/6} = \frac{50}{57}$$

To maintain the uniformity of the figure 100 in the numerator, as in the case of other multiplication factors, we may take multiplication factor as,

$$\frac{2 \times \text{Equivalent weight of } CaCO_3}{2 \times \text{Equivalent weight of } Al_2(SO_4)_3} = \frac{100}{114}$$

** Similarly for monovalent salts also, we take multiplication factor as:

$$\frac{2 \times \text{Equivalent weight of } CaCO_3}{2 \times \text{Equivalent weight of monovalent substance}}$$

Now, Equivalent mass of monovalent substance = Molar mass.

$$\therefore \qquad \text{Multiplication factor} = \frac{100}{2 \times \text{Molar mass}}$$

1.4.3 Units of Hardness

The results of the analysis of hard water are expressed in one of the following units :

1. **Parts per million (ppm) :** ppm is the parts of calcium carbonate equivalent hardness per 10^6 parts of water.
2. **Milligrams per litre (mg/L) :** It is the number of milligrams of $CaCO_3$ equivalent hardness present per litre of water.

1 mg/L = 1mg of $CaCO_3$ equivalent hardness per L of water.

But 1L of water weighs = 1000 gm.

= 1000 × 1000 mg

1mg/L = 1mg/10^6 mg = 1 ppm.

3. **Clarke's degree (°Cl) :** It is the number of grains (1 grain = $\frac{1}{7,000}$ *lb*) of $CaCO_3$ equivalent hardness per gallon (10 lb) of water. Or simply, it is the parts of $CaCO_3$ equivalent hardness per 70,000 parts of water.

1^0 Clarke = 1 part of $CaCO_3$ equivalent hardness per 70,000 parts of water.

4. **Degree French (°Fr) :** It is the parts of $CaCO_3$ equivalent hardness per 10^5 parts of water.

Relationship between various units of hardness :

1 ppm = 1mg/L = 0.1°Fr = 0.07°Cl

1°Fr = 10 ppm = 10 mg/L = 0.7°Cl

1°Cl = 14.3 ppm = 14.3 mg/L = 1.43°Fr

NUMERICALS BASED ON INTERCONVERSION OF UNITS

Example 1.1: A water sample has the following analysis.

Mg $(HCO_3)_2$ = 83 mg/L; $Ca(HCO_3)_2$ = 134 mg/L; $CaSO_4$ = 124 mg/L; $MgCl_2$ = 84 mg/L; $CaCl_2$ = 94 mg/L; NaCl = 50 mg/L. Calculate the temporary and permanent hardness in degree Clark.

Solution: Calculation of $CaCO_3$ equivalents

Constituent	*Quantity mg/L*	*Multiplication Factor*	*$CaCO_3$ Equivalents (mg/L)*
Mg $(HCO_3)_2$	83	100/146	56.8
$Ca(HCO_3)_2$	134	100/162	82.7
$CaSO_4$	124	100/136	91.2
$MgCl_2$	84	100/95	88.4
$CaCl_2$	94	100/111	84.7
NaCl	50	Does not produce hardness	—

Temporary hardness = hardness due to $(Mg(HCO_3)_2 + Ca(HCO_3)_2$

= 139.5 mg/L = 139.5 ppm

= 139.5 × 0.07°Cl = 9.77°Cl

Permanent hardness = Hardness due to $\{CaSO_4 + MgCl_2 + CaCl_2\}$

= 264.3 mg/L = 264.3 ppm

= 264.3 × 0.07°Cl = 18.50°Cl

Example 1.2: Calculate the total hardness of a water sample having following composition in ppm.

$CaCO_3$ = 20°Cl; $MgSO_4$ = 4° Cl, $CaSO_4$ = 2° Cl; Fe_2O_3 = 0.04° Cl, $MgCl_2$ = 4° Cl ; SiO_2 = 0.09° Cl

Solution:

Constituent	*Amount (°Cl)*	*Multiplication Factor*	*$CaCO_3$ equivalents (°Cl)*
$CaCO_3$	20	100/100	20
$CaSO_4$	2	100/136	1.47
$MgCl_2$	4	100/95	4.21
$MgSO_4$	4	100/120	3.33
Fe_2O_3	0.04	—	—
SiO_2	0.09	—	—

Total hardness of water sample = 29.01°Cl

= 29.01 × 14.3 ppm = 414.84 ppm

Example 1.3: Calculate the total hardness of water in °Fr and °Cl from the following analysis of water sample.

$Mg(HCO_3)_2$ = 20.8 mg/L; $Mg(NO_3)_2$ = 28.8 mg/L; $MgCl_2$ = 21.0 mg/L; $MgSO_4$ = 20.0 mg/L; $CaCO_3$ = 21.2 mg/L; KCl= 73.2 mg/L

Solution: Calculation of $CaCO_3$ equivalents.

Salt	*Amount mg/L*	*Multiplication Factor*	*Equivalents of $CaCO_3$ mg/L*
$Mg(HCO_3)_2$	20.8	100/146	14.2
$Mg(NO_3)_2$	28.8	100/148	19.45
$MgCl_2$	21.0	100/95	22.10
$MgSO_4$	20.0	100/120	16.67
$CaCO_3$	21.2	100/100	21.2
KCl	73.2	—	—
			Total Hardness = 93.62 mg/L

Total Hardness = 93.62 mg/L = 93.62 × 0.1°Fr = 9.362°Fr

Total hardness in °Cl = 93.62 × 0.07°Cl = 6.55°Cl

1.4.4 Determination of Hardness

Hardness of water can be determined by a number of methods :

(i) O.Hehner's Method
(ii) Soap Titration Method
(iii) EDTA Method

Out of the above three methods, the EDTA method is more accurate and is used widely for the estimation of hardness in water. This method is discussed in detail in the following section :

EDTA Method

This method makes use of complexometric titration of water sample with EDTA using Eriochrome Black *T* as the indicator.

EDTA, Ethylene diamine tetra acetic acid is a strong complexing agent having the structure,

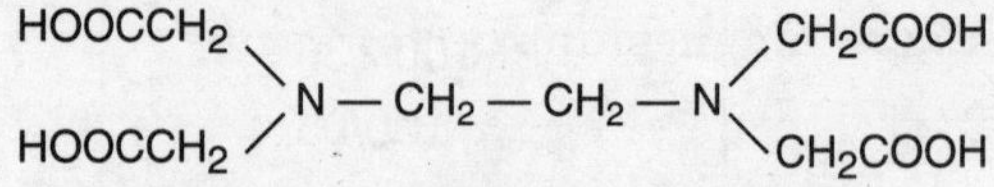

Ethylene diamine tetra acetic acid (EDTA)

When EDTA is dissolved in water, EDTA forms a zwitterion which has the structure,

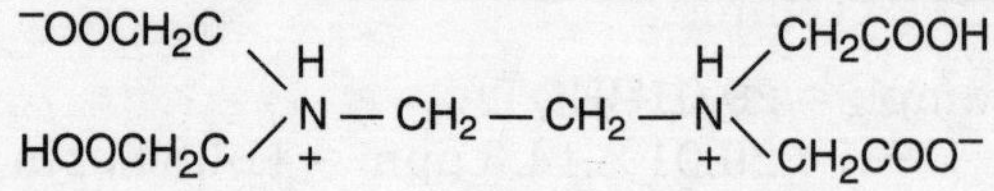

The various EDTA species are abbreviated as H_4Y, H_3Y^-, H_2Y^{2-}, HY^{3-} and Y^{4-} where H represent the four ionizable hydrogens and the rest of the remaining structure is represented by Y. The relative amounts of these species is a function of pH.

EDTA is an hexadentate ligand and can form 4 or 6 dative bonds with Ca^{2+} and Mg^{2+}. As such, it is not very soluble in water and it is used as its disodium salt which is quite soluble in water. It forms a very stable, colourless, 1:1 complex with calcium or magnesium ions. The complex formed is generally very soluble in water (Fig. 1.1).

Eriochrome Black T: Eriochrome black T is an azo dye which is capable of forming soluble wine-red coloured complexes with Ca^{2+} and Mg^{2+}.

The Eriochrome Black T is black in unionized form whereas its ionized form is red in colour. (Fig. 1.2)

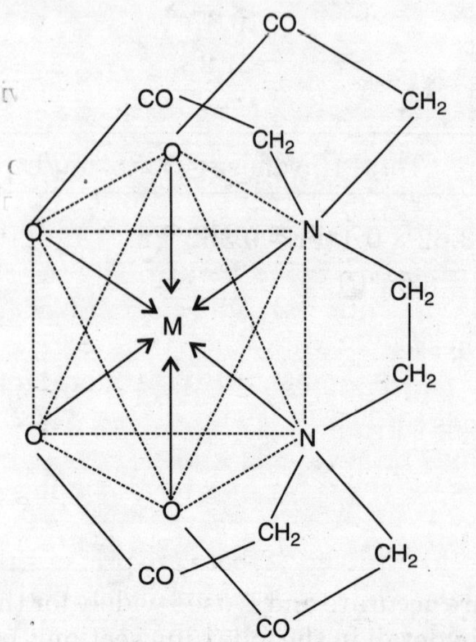

***Fig. 1.1** EDTA Complex*

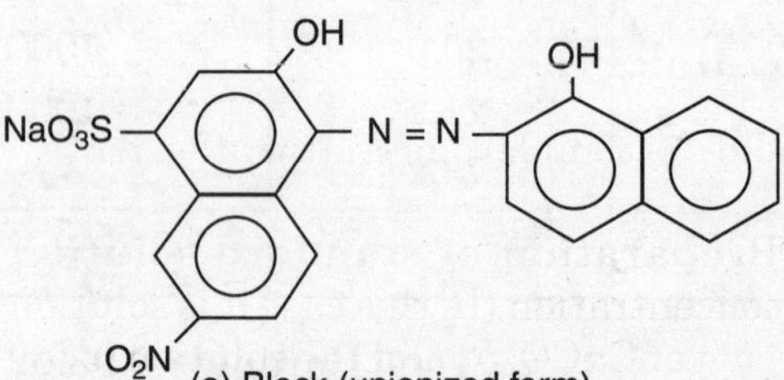

(a) Block (unionized form)

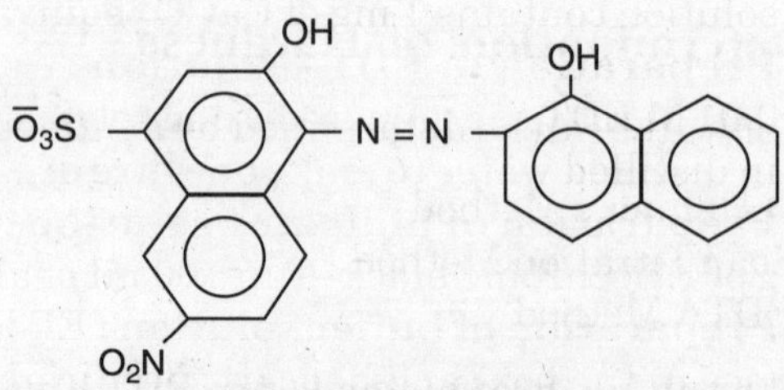

(b) Red (ionized form in aqueous medium)

***Fig. 1.2** Eriochrome Black T (sodium 1 – (1–hydroxy–2–naphthylazo) 6 – nitro –2 – naphthol – 4 – sulphonate*

However, the phenolic groups on EBT may also get ionized at higher pH value (see box)

Principle of Complexometric Titration: Estimation of hardness by EDTA method is based on the following principle –

(i) First, the indicator Eriochrome Black-T, which is a blue coloured dye, forms an unstable complex with calcium or magnesium ions in water at a pH of 9 to 10. The complex is wine red in colour.

$$Mg^{2+}/Ca^{2+} + \text{Eriochrome Black T} \longrightarrow \underset{\text{Unstable Complex (Wine red)}}{[\text{Mg/Ca–Eriochrome Black T}]}$$

(ii) As this solution is titrated against EDTA, the free Ca^{2+} and Mg^{2+} ions in water form stable metal –ion EDTA complex

$$Mg^{2+}/Ca^{2+} + \text{EDTA} \longrightarrow \underset{\substack{\text{Stable Complex} \\ \text{(Colourless)}}}{[\text{Mg/Ca – EDTA}]}$$

Once the free metal ions are complexed, the EDTA replaces Ca and Mg ions from the unstable indicator complex also, to form a stable complex, with the result, the indicator is set free. Since the free indicator is blue in colour at the above mentioned pH, the end point is the appearance of blue colour.

$$\underset{\text{Wine red}}{[\text{Mg/Ca–Eriochrome Black T}]} + \text{EDTA} \longrightarrow [\text{Mg/Ca – EDTA}] + \underset{\text{(Blue)}}{\text{Eriochrome Black T}}$$

Thus the amount of EDTA used corresponds to the hardness of water.

The temporary hardness is removed by boiling and after the removal of precipitate by filteration, the permanent hardness in the filterate is determined by titration with EDTA as above.

Temporary hardness will be given by the difference of total hardness and permanent hardness.

Experimental Procedure

The procedure of complexometric titration for the determination of hardness involves the following steps:

(i) **Preparation of standard solution :** Standard solution is the solution of known concentration (In this case it is solution of known hardness) A standard solution having 1mg of $CaCO_3$/mL of the solution is prepared. As $CaCO_3$ is insoluble in water 1g of it is first dissolved in dil. HCl and then evaporated to dryness on water bath. The residue which is left, is dissolved in distilled water to make one litre solution. Each mL of this solution contains 1 mg of $CaCO_3$ equivalent hardness. (0.01M)

(ii) **Preparation of EDTA solution :** As EDTA forms 1:1 complex with the metal ions, 0.01 M EDTA is prepared by dissolving 3.7 g of its disodium salt (Molar mass = 372.25g) in distilled water to make one litre of EDTA solution.

(iii) **Preparation of indicator solution :** Indicator solution is prepared by dissolving 0.5 g of Eriochrome black–T in 100 ml of alcohol.

(iv) **Preparation of buffer solution :** EDTA and Eriochrome black T, both are pH sensitive (see in the box) and hence a specific pH of the solution has to be maintained. A buffer solution is prepared by dissolving 70g of NH_4Cl in 570 mL of conc. NH_3 and the solution is diluted to one litre with distilled water.

Ethylene diamine tetraacetic acid is highly pH sensitive ligand. Since it is tetraprotic acid, it can be written as H_4Y, where H represent the four ionizable hydrogens. Around

pH > 12 most of the EDTA exists as tetranion, Y^{4-}.

At pH ≥ 8, the EDTA exists mostly as HY^{3-}, whereas

At pH ≈ 5, the predominant specie is H_2Y^{2-}.

The reaction with Ca^{2+} or Mg^{2+} ions can be represented as

$$M^{2+} + H_2Y^{2-} \rightleftharpoons \underset{\text{Metal EDTA Complex}}{MY^{2-}} + 2H^+$$

$$M^{2+} + HY^{3-} \rightleftharpoons MY + HY^-$$

depending upon the pH of the solution. The protons liberated in these reactions would considerably alter the pH of the solution and may cause dissociation of metal–EDTA complex. However this is prevented by *adding an excess of buffer solution* to the titration system.

Erichrome Black T : The ionized form of Eriochrome black T is red in colour (Fig.1.2) The phenolic group on Eriochrome black T may also get ionized at higher pH values. If indicator is represented as H_3In, its dissociations will be represented as –

$$\underset{\text{Black}}{H_3In} + H_2O \longrightarrow H_3O^+ + \underset{\text{Red}}{H_2In^-}$$

$$\underset{\text{Red}}{H_2In^-} + H_2O \rightleftharpoons H_3O^+ + \underset{\text{Blue}}{H\,In^{2-}}$$

$$\underset{\text{Blue}}{H\,In^{2-}} + H_2O \longrightarrow H_3O^+ + \underset{\text{orange}}{In^{3-}}$$

Hence, it follows from the above discussion that the indicator would exhibit different colours, depending upon its dissociation, which in turn depends upon the pH of the solution i.e.

at pH < 6 the indicator exists as H_2In^-(red)

pH > 7 the indicator exists as HIn^{2-}(blue)

Thus, the pH of the titrating solution has to be maintained by adding a suitable buffer solution.

The EDTA titrations are performed in buffers of pH 8 – 10, the range in which the predominant form of the indictor is blue (HIn^{2-}). When metallic salts (Ca^{2+} or Mg^{2+}) are added, the colour of the indicator changes from blue to red as indicator complexes with metal ion. At the end point, the indicator is set free because of the formation of a stable metal EDTA complex and the colour of the solution changes to blue from wine red.

$$\underset{\text{(Indicator)}}{H\,In^{2-}} + M^{2+} \longrightarrow \underset{\text{Wine red complex}}{M\,In^-} + H^+$$

$$M\,In^- + \underset{\text{(EDTA)}}{H_2Y^{2-}} \rightleftharpoons \underset{\text{(Colourless)}}{MY^{2-}} + \underset{\text{(Blue)}}{H\,In^{2-}} + H^+$$

(v) **Standardization of EDTA solution :** To find out the exact strength of EDTA solution, it is titrated against standard hard water prepared in step (i), using Eriochrome black T as indicator and in the presence of 10–12ml of buffer solution. The end point being wine red to blue. EDTA is taken in the burette and water is taken in the titration flask.

(vi) **Titration of the sample water :** In a similar way, the standard hard water is titrated against EDTA solution using Eriochrome black-T as indicator, in the presence of buffer solution.

(vii) **Titration of water having permanent hardness :** 250 ml of the water sample is taken in a 500 ml beaker and is boiled till the volume reduces to about 50 ml. Any temporary hardness present will thereby would be precipitated out as $CaCO_3$ or $MgCO_3$. The precipitates are filtered off and the volume of the filterate is made to 250 ml by adding distilled water. This is now titrated against EDTA as in step (v) and (vi).

Observations : Let volume of the water taken for each titration = 50mL

Volume of EDTA used when titrated against standard hard water = V_1mL

Volume of EDTA used when titrated against sample hard water = V_2mL

Volume of EDTA used when titrated against water having permanent hardness = V_3mL

Calculations :

(a) Strength of EDTA solution:

1 ml of standard Hard water (SHW) contains = 1 mg of $CaCO_3$

50 ml of SHW contains = 50 mg of $CaCO_3$

Volume of EDTA consumed for 50 ml of SHW = V_1 ml

$\therefore$ V_1 ml of EDTA is used for = 50 mg of $CaCO_3$

$$1 \text{ ml of EDTA is used for} = \frac{50}{V_1} \text{ mg of } CaCO_3$$

$$\text{or strength of EDTA solution} = \frac{50}{V_1} \text{ mg/ml of EDTA}$$

(b) Total Hardness :

Volume of EDTA used for 50 ml of sample hard water = V_2ml

$$\text{Since 1 ml of EDTA consumes} = \frac{50}{V_1} \text{ mg of } CaCO_3$$

$$V_2 \text{ ml of EDTA will consume} = \frac{50}{V_1} \times V_2 \text{ mg of } CaCO_3.$$

$$\text{Hence, 50 ml of sample water} = \frac{50}{V_1} \times V_2 \text{ mg of } CaCO_3.$$

$$\text{1000 ml of sample water} = \frac{50 \times V_2}{V_1} \times \frac{1000}{50} \text{ mg of } CaCO_3.$$

$$= \frac{V_2 \times 1000}{V_1} \text{ mg}$$

$$\therefore \quad \text{Total hardness} = \frac{V_2 \times 1000}{V_1} \text{ ppm.}$$

(c) Permanent Hardness:

Volume of EDTA used for 50ml of water, containing permanent hardness = V_3 mL

$$V_3 \text{ ml of EDTA will consume } = \frac{50}{V_1} \times V_3 \text{ mg of } CaCO_3.$$

$$\text{Hence, 50 ml of water contains pemanent hardness } = \frac{50}{V_1} \times V_3 \text{ mg of } CaCO_3.$$

$$\text{1000 ml of water contains permanent hardness } = \frac{50}{V_1} \times \frac{V_3}{50} \times 1000 \text{ mg of } CaCO_3.$$

$$= \frac{V_3 \times 1000}{V_1} \text{ mg}$$

$$\text{Permanent Hardness } = \frac{V_3}{V_1} \times 1000 \text{ ppm.}$$

(d) Temporary Hardness : Total hardness—Permanent hardness.

Table 1.5: *General Scale of Hardness*

Hardness of Water (mg/L)	*Description*
0–17	Soft
17–60	Slightly hard
60–120	Moderately hard
120–180	Hard
> 180	Very hard

Temporary hardness due to the presence of carbonates and bicarbonates is referred to as the carbonate hardness and the amount of hardness in excess carbonate (*i.e.*, permanent hardness) is referred to as the non-carbonate hardness.

∴ Total hardness = Carbonate + Non-carbonate hardness.

NUMERICALS BASED ON HARDNESS OF WATER

Example 1.4: One gram of $CaCO_3$ was dissolved in dil. HCl and the solution diluted to one litre. 100 mL of this solution required 90 ml of EDTA solution, while 100 mL of the sample water required 40 mL of EDTA. On the other hand, 100mL of the boiled sample water when titrated against EDTA consumed 20mL of solution. Calculate each type of hardness in ppm.

Solution: Strength of standard hard water = 1 mg/mL = 1g/L

Strength of EDTA

90 ml of EDTA is required for = 100ml of standard hard water

= 100 mg of $CaCO_3$

∴ 1 ml of EDTA is required for = $\frac{100}{90}$ mg of $CaCO_3$

Total hardness:

100 ml of sample required = 40 ml of EDTA

Since 1ml of EDTA is required for = 10/9mg of $CaCO_3$.

∴ 40ml of EDTA is required for = 10/9 × 40mg of $CaCO_3$.

Hence 100 ml of sample water contains 400/9 mg of $CaCO_3$.

1000ml of sample water contains $= \frac{400}{9} \times \frac{1000}{100}$ mg of $CaCO_3$ = 444.4mg/L

Total hardness = 444.4ppm.

Temporary hardness:

After boiling 100 ml of sample, water requires = 20 ml of EDTA

100 ml of sample water contains $= 20 \times \frac{10}{9} = \frac{200}{9}$ mg of $CaCO_3$.

1000 ml of sample water contains $= \frac{200}{9} \times \frac{1000}{100}$ mg of $CaCO_3$.

Permanent hardness = 222.2mg/L
= 222.2ppm

Temporary hardness = 444.4–222.2ppm = 222.2ppm.

Example 1.5: 100ml of a sample of water required 20ml of 0.01M EDTA for titration using Eriochrome T as indicator. After boiling 100ml of the same sample, required 10ml of 0.01M EDTA. Calculate

(i) the total hardness
(ii) permanent hardness
(iii) and temporary hardness.

Solution: Since EDTA forms 1 : 1 complex with the metal ions hence,

1 molar solution of EDTA reacts with 1mole of $CaCO_3$.

or 1000ml of 1M EDTA reacts with = 100g of $CaCO_3$.

1 ml of 1M EDTA reacts with $= \frac{100}{1000} \times 1000$ mg of $CaCO_3$.
= 100mg of $CaCO_3$.

1ml of .01M EDTA reacts with $= \frac{100}{100}$ = 1mg of $CaCO_3$.

Strength of the given EDTA solution = 1mg/mL.

(i) Total hardness:

Since 1ml of EDTA is required for = 1 mg of $CaCO_3$.
20ml of EDTA is required for = 20 mg of $CaCO_3$.
Hence 100ml of sample hard water contains = 20 mg of $CaCO_3$.

1000ml of sample hard water contains $= \frac{20}{100} \times 1000$ = 200mg of $CaCO_3$.

Total hardness = 200mg/L = 200ppm.

(ii) Permanent hardness:

10ml of EDTA is required for = 1 × 10mg of $CaCO_3$.
Hence 100ml of boiled sample hard water contains = 10mg of $CaCO_3$.
1000 ml of boiled sample hard water contains = 100mg of $CaCO_3$.
Permanent Hardness = 100mg/L = 100ppm.

(iii) Temporary hardness = Total hardness – permanent hardness
= 200 – 100 = 100ppm.

Example 1.6: 500 ml of a sample of water on EDTA titration with Eriochrome Black-T as indicator consumed 18ml of 0.045M EDTA till end point is reached. Calculate the hardness of water.

Solution: (1ml of 0.01M EDTA = 1mg of $CaCO_3$)

$$18\text{ml of } 0.045\text{M EDTA} = \frac{1\times 18}{.01}\times 0.45 \text{ mg of } CaCO_3.$$

$$= 81\text{mg of } CaCO_3.$$

500 ml of water contains 81mg of $CaCO_3$, so hardness present in 1 litre

$$= \frac{81}{500}\times 1000 \text{ mg/L}$$

$$= 162 \text{ mg/L} = 162\text{ppm}.$$

Example 1.7: 0.50g of $CaCO_3$ was dissolved in dil. HCl and diluted to 500 ml. 100ml of this solution required 20ml of EDTA solution for titration. 100ml of a hard water sample required 30ml of the same EDTA solution for titration. On boiling, titration of 100 ml of the same water sample required 10ml of EDTA. Calculate total, permanent and temporary hardness. (1ml of 0.01M EDTA = 1mg of $CaCO_3$)

Solution: 500ml of $CaCl_2$ solution contains 0.50g of $CaCO_3$. eq. hardness

$$= 500\text{mg of } CaCO_3 \text{ eq. hardness}$$

$$1\text{ml of } CaCl_2 \text{ solution} = 1\text{mg of } CaCO_3 \text{ eq. hardness}$$

Strength of EDTA:

$$100\text{ml of } CaCl_2 \text{ solution} = 100\text{mg of } CaCO_3.$$

20ml of EDTA solution is required for 100ml of $CaCl_2$ solution.

$$= 100\text{mg of } CaCO_3$$

1ml of EDTA solution is required for 100/20 mg of $CaCO_3$

$$= 5\text{mg of } CaCO_3$$

Total hardness:

$$100\text{ml of hard water} = 30\text{ml of EDTA solution}$$

$$= 30 \times 5\text{mg of } CaCO_3$$

$$= 150\text{mg of } CaCO_3.$$

$$1000\text{ml of hard water} = 1500\text{mg of } CaCO_3.$$

$$\text{Total hardness} = 1500 \text{ ppm}$$

Permanent hardness:

$$100 \text{ ml of permanent hard water} = 10 \text{ ml of EDTA solution}$$

$$= 5 \times 10\text{mg of } CaCO_3.$$

$$1000\text{ml of permanent hard water} = 500\text{mg of } CaCO_3$$

$$\text{Temporary hardness} = \text{Total hardness} - \text{Permanent hardness}$$

$$= 1500 - 500 = 1000\text{mg of } CaCO_3$$

$$\text{Temporary hardness} = 1000 \text{ ppm}$$

Potential Health Effects of Hard Water

Hard water is not a health hazard. In fact, National Academy of Sciences states that hard drinking water generally contributes a small amount toward calcium and magnesium human dietary needs. They further state that in some instances, where dissolved calcium and magnesium are very high, water could be a major contributor of calcium and magnesium to the diet.

However, hard water has some advantages also.

Advantages of hard water:

1. It tastes better due to the presence of calcium and magnesium ions.
2. It is thought to reduce the number of heart illnesses.
3. It provides useful calcium ions for the healthy growth of bones and teeth.
4. The formation of lime scale in pipes due to hard water causes the inside of the pipe to be covered with insoluble carbonates. This layer of carbonate prevents the water in the pipe from coming into contact with the metal of the pipe and so prevents pipe corrosion and poisonous metal salts becoming dissolved in the water.

1.5 ALKALINITY OF WATER

Another important characteristic of water is its alkalinity, which is a measure of the ability of water to neutralize the acids. The alkalinity of water is normally due to the presence of bicarbonates, carbonates and hydroxides of sodium, potassium, calcium and magnesium. Some of the salts, not all, which cause alkalinity also cause hardness. However, presence of other bases such as borates, silicates, phosphates can also contribute to the total alkalinity, though to a small extent.

Depending upon the anion present, the alkalinity can be classified as:

(i) Caustic alkalinity (due to OH^- and CO_3^{2-} ions) ($\because$ CO_3^{2-} also produce OH^- on hydrolysis)

(ii) Bicarbonate alkalinity (due to HCO_3^- ions)

The alkalinity in water can be present due to the presence of the above mentioned ions present alone or in combination as

- hydroxide only (OH^-)
- Carbonates only (CO_3^{2-})
- Bicarbonates only (HCO_3^-)
- Hydroxides and Carbonates ($OH^- + CO_3^{2-}$)
- Carbonate and bicarbonates ($CO_3^{2-} + HCO_3^-$)

The presence of hydroxide and bicarbonates existing together is ruled out because they undergo the following reaction to form CO_3^{2-} ions.

$$\overline{O}H + HCO_3^- \longrightarrow CO_3^{2-} + H_2O$$

And hence the possibility of all the three ions existing together is also ruled out.

1.5.1 Determination of Alkalinity

Principle: Determination of alkalinity due to different ions is based on the titration of the water sample against a standard acid making *selective use of indicators*. The indicators used are - phenolphthalein and methyl orange. The reactions taking place may be represented by the following equations:

(a) $\overline{O}H + H^+ \longrightarrow H_2O$

(b) $CO_3^{2-} + H^+ \longrightarrow HCO_3^-$

(c) $HCO_3^- + H^+ \longrightarrow H_2CO_3$

[(a) and (b): P; (a), (b) and (c): M]

- The volume of the acid used upto phenolphthalein end point corresponds to the reaction (a) and (b) i.e. complete neutralization of $\overline{O}H$ ions and neutralization of CO_3^{2-} ions upto HCO_3^- stage (half neutralization of CO_3^{2-} ions).

- The volume of the acid used upto methyl orange end point corresponds to the reaction (a), (b) and (c) i.e. complete neutralization of $\bar{O}H$, CO_3^{2-} and HCO_3^- ions.

Thus from the volume of the acid used, the respective strengths of various ion can be determined as discussed in the following section.

Procedure: A known volume of the sample water is titrated against a standard acid, using phenolphthalein as indicator. The end point is disappearance of pink colour. To the same solution is added the indicator, methyl orange.

This solution is then titrated with the standard acid solution till end point, from yellow to pink is reached.

Observations:

Let the volume of the water sample taken for each titration = 100ml

Volume of the acid used upto phenolphthalein end point = V_1 ml

Extra volume of the acid used upto methyl orange end point = V_2ml

Total volume of the acid used upto methyl orange endpoint = $(V_1 + V_2)$ml

Normality of the acid used for titration = N/50

Calculations:

(i) Phenolphthalein alkalinity in terms of $CaCO_3$ equivalents

$$\underset{\text{(Acid)}}{N_1 V_1} = \underset{\text{(Water sample)}}{N_2 \times 100}$$

$$\frac{1}{50} \times V_1 = N_2 \times 100$$

$$N_2 = \frac{1}{50} \times \frac{V_1}{100} = \frac{V_1}{5000}$$

Strength in terms of $CaCO_3$ equivalents = $N_2 \times$ Eq.wt.of $CaCO_3 = N_2 \times 50$g/L

$$= \frac{V_1}{5000} \text{ g/L}$$

Phenolphthalein alkalinity, P = $\frac{V_1}{100} \times 1000$ mg/L = $10V_1$ ppm

(ii) Methyl orange alkalinity in terms of $CaCO_3$ equivalents

$$\underset{\text{(Acid)}}{N_1(V_1 + V_2)} = \underset{\text{(Water sample)}}{N_2 \times 100}$$

$$\frac{1}{50} \times (V_1 + V_2) = N_2' \times 100$$

$$N_2' = \frac{1}{50} \times \frac{(V_1 + V_2)}{100}$$

Strength in terms of $CaCO_3$ equivalent = $N_2' \times$ Eq.wt. of $CaCO_3$

$$= \frac{V_1 + V_2}{5000} \times 50 \text{ g/L}$$

Methyl orange alkalinity, M = $10\ (V_1 + V_2)$ mg/L = $10\ (V_1 + V_2)$ ppm

On the basis of the analysis of water with respect to phenolphthalein alkalinity and methyl orange alkalinity, the alkalinity due to differents ions can be calculated. The results are summarized in the following table.

Table 1.6: *Computing the alkalinity due to different ions*

Alkalinity	*$\bar{O}H$ (ppm)*	*CO_3^{2-} (ppm)*	*HCO_3^- (ppm)*
P = O	NIL	NIL	M
$P = \frac{1}{2}M$	NIL	2P	NIL
$P < \frac{1}{2}M$	NIL	2P	(M-2P)
$P > \frac{1}{2}M$	(2P–M)	2 (M–P)	NIL
P = M	P = M	NIL	NIL

Discussion:

Case (i): When phenolphthalein alkalinity = 0, this implies that volume of the acid used till phenolphthalein end point is 0; This can only happen if CO_3^{2-} and $\bar{O}H$ both are absent. Whatever alkalinity is present is due to HCO_3^- ions and can be detected using methyl orange as indicator.

Case (ii): When $P = \frac{1}{2}M$, this implies that only CO_3^{2-} ions are present. Neutralization using phenolphthalein as indicator reaches upto HCO_3^- stage. Same amount of acid will be further used when methyl orange is used as indicator to neutralize HCO_3^- to H_2O and CO_2 as represented in Fig. 1.3 (a) The strength of CO_3^{2-} will be given by M or 2P

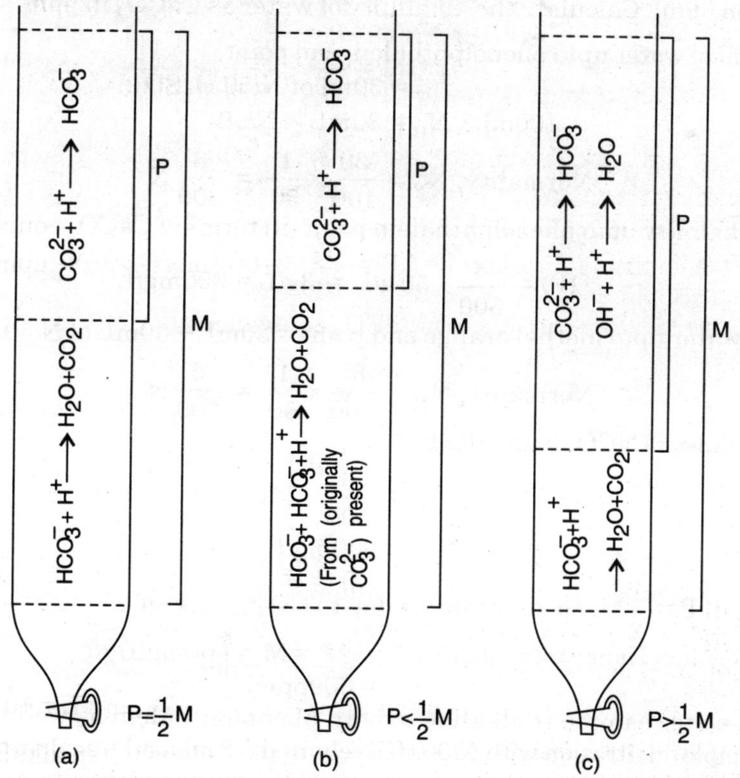

***Fig. 1.3** Relative amounts of Acid using phenolphthalein and methyl orange as indicators*

Case (iii): When $P < \frac{1}{2}M$, this implies that besides CO_3^{2-} ions, HCO_3^- ions are also present. The acid used uptill phenolphthalein end point corresponds to half neutralization of CO_3^{2-} ions. Further titration using methyl orange as indicator corresponds to neutralization of HCO_3^- obtained from CO_3^{2-} ions and HCO_3^- originally present in the reaction mixture as shown in Fig. 1.3(b)

Alkalinity due to $CO_3^{2-} = 2P$

Alkalinity due to $HCO_3^- = (M-2P)$

Case (iv) : When $P > \frac{1}{2}M$. This implies that besides CO_3^{2-} ions, $\overline{O}H$ ions are also present. $\overline{O}H$ will be completely neturalized by the acid whereas CO_3^{2-} will be neutralized up to the HCO_3^- stage, using phenolphthalein as indicator. On further titration with the acid using methyl orange as indicator, the neutralization of HCO_3^- takes place as shown in Fig. 1.3 (c).

Alkalinity due to $CO_3^{2-} = 2\ (M-P)$.

Alkalinity due to $\overline{O}H = M - 2\ (M-P) = (2P-M)$

Case (v) : When $P = M$ implies that only $\overline{O}H$ ions are present

Alkalinity due to $\overline{O}H = P = M$.

SOLVED NUMERICALS BASED ON ALKALINITY OF WATER

Example 1.8: 100ml of a water sample required 30ml of N/50 H_2SO_4 for neutralization to phenolphthalein end point. After this, methyl orange indicator was added to this and further acid required was again 30ml. Calculate the alkalinity of water as $CaCO_3$ in ppm.

Solution: (i) 100ml of water upto phenolphthalein end point

$$= 30\text{ml of N/50 } H_2SO_4.$$

$$100\text{ml} \times N_P = 30\text{mL} \times N/50$$

$$\text{Normality, } N_P = \frac{30}{100} \times \frac{1}{50} = \frac{3}{500}\text{N}$$

Strength of alkalinity upto phenolphthalein point (in terms of $CaCO_3$ equiv.)

$$(P) = \frac{3}{500} \times 50\text{ g/L} = .3\text{g/L} = 300\text{mg/L}$$

(ii) 100ml of water upto methyl orange end point = 30mL + 30mL of N/50 H_2SO_4.

$$\text{Normality, } N_M = \frac{60}{100} \times \frac{1}{50} = \frac{6}{500}\text{N}$$

Strength in terms of $CaCO_3$ equivalents

$$(M) = \frac{6}{500} \times 50\text{g/L g/L} = .6\text{g/L}$$

$$= 600\text{mg/L}$$

From (i) and (ii) $P = \frac{1}{2}M$, so alkalinity is due to CO_3^{2-} ions only.

Hence total alkalinity = 2P = M = 600mg/L

= 600ppm.

Example 1.9: A water sample is alkaline to both phenolphthalein as well as methyl orange. 100ml of water sample on titration with N/50 HCl required 6.8 mL of the acid up to phenolphthalein end point. When a few drops of methyl orange are added to the same solution and the titration further continued, the yellow colour of the solution just turned red after the addition of another 12.6mL of the acid solution. Elucidate the type and extent of alkalinity present in water sample.

Solution: (i) 100ml of water upto phenolphthalein end point = 6.8 mL of N/50 HCl

$$\text{Normality, } N_P = \frac{6.8}{100} \times \frac{1}{50} = .00136\text{N}$$

Strength in terms of $CaCO_3$ eq.

$$(P) = 0.00136 \times 50\text{g/L} = .068\text{g/L} = 68\text{mg/L} = 68\text{ppm}$$

(ii) 100 ml of water upto methyl orange end point = 6.8 mL + 12.6 mL of N/50 HCl

$$\text{Normality, } N_M = \frac{19.4}{100} \times \frac{1}{50} = .0039\text{ N}$$

Strength in terms of $CaCO_3$ eq.

$$(M) = 0.0039 \times 50\text{g/L} = .195\text{g/L} = 195\text{mg/L} = 195\text{ppm}$$

From (i) and (ii), $P < \frac{1}{2}M$, therefore the alkalinity is due to CO_3^{2-} and HCO_3^-

$$\text{Alkalinity due to } CO_3^{2-} = 2P = 2\times 68\text{ppm} = 136\text{ppm}$$
$$\text{Alkalinity due to } HCO_3^- = (M\text{–}2P) = 195\text{– }136\text{ppm} = 59\text{ppm}$$

Example 1.10: A sample of water was alkaline both to phenolphthalein and methyl orange. 100ml of this water sample required 20mL of N/50 H_2SO_4 for phenolphthalein end point and another 15mL for complete neutralization. Calculate the type of alkalinity in ppm.

Solution: (i) 100mL of water upto phenolphthalein end point = 20mL of N/50 H_2SO_4

$$\text{Normality } N_P = \frac{20}{100} \times \frac{1}{50} = .004\text{N}$$

Strength in terms of $CaCO_3$ eq.

$$(P) = 0.004 \times 50\text{g/L}$$

(ii) 100mL of water upto methyl orange end point

$$\text{Normality, } N_M \equiv 20\text{ml} + 15\text{mL of N/50 } H_2SO_4$$

$$\text{Normality, } N_M = \frac{35}{100} \times \frac{1}{50} = .007\text{N}$$

Strength in terms of $CaCO_3$ eq.

$$(M) = 0.007 \times 50\text{ g/L} = 0.35\text{g/L} = 350\text{mg/L}$$

From (i) and (ii), $P > \frac{1}{2}M$, therefore the alkalinity is due to $\overline{O}H$ and CO_3^{2-} ions.

$$\text{Alkalinity due to } CO_3^{2-} \text{ ions} = 2\,(M - P) = 2\,(350\text{–}200)\text{mg/L} = 300\text{mg/L} = 300\text{ppm}$$

$$\text{Alkalinity due to } \overline{O}H \text{ ions} = (2P - M) = [2\,(200)\text{–}350]\text{mg/L} = 50\text{mg/L} = 50\text{ppm}$$

1.6 DISSOLVED OXYGEN

Oxygen is poorly soluble in water. The amount of oxygen in water depends on physical, chemical and biological activities taking place in water. The solubility of dissolved oxygen in water at saturation at any temperature and pressure is given by Henry's law.

Determination: Dissolved oxygen is determined by Winkler's method or Iodometric titration.

Principle: The determination of oxygen by iodometric titration is based on the fact that dissolved oxygen in the water oxidizes potassium iodide and an equivalent amount of iodine is liberated. The liberated iodine is titrated against a standard hypo solution using starch as indicator. However since dissolved oxygen in water is in molecular state and is not capable of reacting with KI, therefore an oxygen carrier such as manganese hydroxide is used. $Mn(OH)_2$ in turn is prepared by the reaction of $MnSO_4$ with KOH.

$$MnSO_4 + 2KOH \longrightarrow Mn(OH)_2 + K_2SO_4$$
$$2Mn(OH)_2 + O_2 \longrightarrow \underset{\text{Basic manganic oxide}}{2MnO(OH)_2}$$
$$2MnO(OH)_2 + H_2SO_4 \longrightarrow MnSO_4 + 2H_2O + [O]$$
$$2KI + H_2SO_4 + [O] \longrightarrow K_2SO_4 + H_2O + I_2$$

Thus the iodine which is liberated in equivalent amount is titrated against hypo solution.

$$2Na_2S_2O_3 + I_2 \longrightarrow Na_2S_4O_6 + 2NaI$$

Some sources of water contain nitrates and sulphates, which can also oxidize potassium iodide and hence their presence can give incorrect results. Thus, before the addition of KI, these have to be eliminated or decomposed. Commonly, sodium azide is used to decompose the nitrites or sulphites.

$$2NaN_3 + H_2SO_4 \longrightarrow Na_2SO_4 + 2HN_3$$
$$HNO_2 + HN_3 \longrightarrow N_2O + N_2 + H_2O$$

Procedure:

(i) **Preparation of Mangenese sulphate solution:** 400 g of $MnSO_4$ is dissolved in 1L of distilled water.

(ii) **Preparation of alkaline iodide solution:** This is prepared by dissolving 150g KI + 500g NaOH and 20g of NaN_3 in one litre of distilled water.

(iii) **Preparation of hypo solution:** 2.48gm of $Na_2S_2O_3$ is dissolved in 1 litre of distilled water to prepare N/100 hypo solution.

(iv) **Titration of the sample water:** A known amount of sample water (say 250mL) is taken in a stoppered bottle avoiding contact with air. 2mL of $MnSO_4$ solution is added to it by means of a pipette, dipping the end well below the surface. Similarly 2mL of alkaline iodine solution is also added to it. The bottle is stoppered and shaken thoroughly. The brown precipitates formed are dissolved in minimum amount of HCl 100mL of this clear solution is pipetted and titrated against N/100 hypo solution using starch as indicator, the end point being, disappearance of blue colour. Let V ml be the volume of the hypo solution used.

Calculations:

$$\begin{array}{cc}\text{Hypo} & O_2 \text{ in water}\end{array}$$
$$N_1V_1 = N_2V_2$$
$$\frac{1}{100} \times V_1 = N_2 \times 100, \quad \therefore N_2 = \frac{V_1}{100}$$

$$\text{Strength of dissolved oxygen} = N_2 \times \text{Eq. of oxygen}$$
$$= \frac{V_1}{10000} \times 8 \text{ g/L}$$
$$= \frac{V_1}{10000} \times 8 \times 1000 \text{ mg/L} = .8V_1 \text{ ppm}$$

Biochemical Oxygen Demand

When biodegradable organic matter is released into water, the micro organisms feed on the wastes and break it down into simpler organic or inorganic substances. If the decomposition takes place in the **presence of oxygen** i.e aerobically, **the non objectionable**, stable end products are formed.

$$\text{Organic Matter} + O_2 \xrightarrow{\text{MICRO ORGANISMS}} CO_2 + H_2O + \text{Other Stable Products.}$$

On the other hand if **insufficient oxygen** is available, the decomposition takes place anaerobically. The micro organisms causing the decomposition of the organic matter without the presence of oxygen are entirely different from aerobic bacteria and produce **highly objectionable** end products including H_2S, NH_3 and CH_4.

$$\text{Organic Matter} \xrightarrow{\text{MICRO ORGANISMS}} CO_2 + CH_4 + \text{Unstable Products}$$

The amount of oxygen required by micro-organisms to oxidize organic wastes aerobically is called biochemical oxygen demand (BOD). The BOD of raw water will indicate the extent of organic matter present, thus indicating the extent of treatment required to make it safe for use. The BOD of treated water should be nil, so as to make it free from any organic matter. If BOD is high, the dissolved oxygen becomes low and this results in greater pollution.

Determination: Polluted waters will continue to absorb oxygen for many months, till the oxidation gets completed and hence it is not practical to determine this ultimate oxygen demand which would require an extended period of time. As a result, it has become standard practice simply to measure the oxygen over a shorter period of 5 days. This is known as 5-day BOD test. **BOD_5 is the total amount of oxygen consumed by micro organisms during the first five days of biodegradation.**

In its simplest form BOD_5 would involve diluting a known volume of sample water with a known volume of pure water, whose oxygen content is already known. This is kept for 5 days at 20°C in a stoppered bottle, away from light. The dissolved oxygen is measured after the period of incubation. The difference between the original oxygen content and residual oxygen content will indicate the oxygen consumed by the water sample in five days in oxidizing the organic matter present in water sample. Thus BOD_5 is given by:

$$BOD_5 = \text{Loss of oxygen in mg/L} \times \text{Dilution factor}$$

Chemical Oxygen Demand

Although BOD test is applicable to organic wastes, there are many drawbacks. Certain organic materials are not biodegredable and hence can give wrong conclusions that less organic matter is present because BOD of such water samples will be low. In this case COD, Chemical Oxygen demand reveals the real organic content present. Here the oxidation of organic substances present in water is done chemically.

COD, is the amount of oxygen required by organic matter in a sample of water for its oxidation by a strong chemical oxidizing agent such as $K_2Cr_2O_7$.

Determination: A known amount of $K_2Cr_2O_7$ is added to a measured amount of the sample and the mixture is boiled with concentrated H_2SO_4. After boiling, the amount of unreacted $K_2Cr_2O_7$ is determined by titrating against a standard Mohr's salt solution using ferroin as indicator.

The difference between the dichromate originally present and the dichromate remaining unreacted gives the amount of dichromate used for oxidation of organic matter.

1.7 DISSOLVED CO_2

Varying amounts of CO_2 are present in water. CO_2 can be picked up from the atmosphere or it is generated at the bottom of the lake due to the decay or organic matter. Rivers receiving acid wastes may also show high CO_2 content.

Determination: CO_2 is determined by titrating the sample water with N/10 sodium carbonate solution using phenolphthalein as indicator. The end point is appearance of pink colour.

If 100ml of sample water is taken for each titration and Vml is the volume of Na_2CO_3 used during titration, then

$$\text{Normality of water w.r.t } CO_2 = \frac{V}{1000}$$

$$\text{Strength of dissolved } CO_2 = \frac{V}{1000} \times 22 \text{ g/L (Equiv. weight of } CO_2 = 22\text{g)}$$

$$\text{Strength of dissolved } CO_2 = 22V \text{ mg/L} = 22V\text{ppm}.$$

1.8 FREE CHLORINE

Chlorine determination is necessary in those samples of water that have been treated with chlorine or hypochlorites.

Determination: The estimation of free chlorine is based on oxidation of KI by free chlorine i.e by iodometric titration.

Principle : When the water sample containing free chlorine is treated with KI, free chlorine oxidizes KI and liberates I_2 in equivalent amount. The liberated I_2 is titrated against hypo solution using starch as indicator.

$$2KI + Cl_2 \longrightarrow 2KCl + I_2$$

$$I_2 + 2Na_2S_2O_3 \longrightarrow 2NaI + Na_2S_4O_6$$

$$I_2 + \text{Starch} \longrightarrow \text{Deep blue complex.}$$

The end point is the disappearance of the blue colour.

Procedure: A known volume of sample water is pipetted out in a titration flask (say 50ml). To it is added about 10ml of KI solution. The flask is kept under dark for 5 minutes so that all the I_2 will be liberated. Then it is titrated against N/10 hypo solution using starch as indicator. Let V be the volume of hypo solution used for titration.

Calculations:

Water sample   hypo solution

$$50 \times \text{Normality} = \frac{N}{50} \times V$$

$$\text{Normality} = \frac{N}{50 \times 50} \times V$$

$$\text{Strength of chlorine} = \frac{V}{2500} \times 35.5 \text{ g/L}$$

$$= \frac{V}{2500} \times 35.5 \times 1000 \text{ mg/L} = 14.2V \text{ ppm.}$$

1.9 DISSOLVED CHLORIDES

The presence of chlorides in natural waters is due to the discharges of effluents from the industries, irrigation drainage, presence of marine sedimentary deposits etc.

Determination: Chlorides ions can be determined by titration of the water sample against a standard solution of $AgNO_3$ using potassium chromate as indicator. (Argentometric titration).

Principle: This determination is based on precipitation titrations. When $AgNO_3$ solution is added to the water sample in presence of K_2CrO_4, the chlorides present in it are precipitated as AgCl. As soon as all the chlorides are precipitated out, even a drop of $AgNO_3$ added in excess gives a red precipitate of silver chromate. This indicates the end point.

$$AgNO_3 + NaCl \rightarrow AgCl\downarrow + NaNO_3.$$

$$AgNO_3 + K_2CrO_4 \rightarrow \underset{\text{Red or Pink.}}{Ag_2CrO_4\downarrow} + 2KNO_3$$

Procedures: 100 ml of water sample is taken in the titration flask. To it are added 2–3 drops of K_2CrO_4.

The solution is titrated against N/50 $AgNO_3$ solution till the appearance of brick-red colour. Let the volume of $AgNO_3$ used be V ml.

Calculations

Water Sample	$AgNO_3$ Solution

$$N_{chloride} \times 100 = \frac{1}{50} \times V$$

$$N_{chloride} = \frac{1}{50} \times \frac{V}{100} = \frac{V}{5000}$$

$$\text{Strength of chloride ions} = \frac{V}{5000} \times 35.5 \text{ g/L}$$

$$= \frac{V}{5000} \times 35.5 \times 1000 \text{ mg/L} = 7.1V\text{ppm}.$$

1.10 TOTAL DISSOLVED SOLIDS (TDS)

Dissolved solids denote mainly the various kinds of minerals present in water. However, some organic substances present in water also contribute to the dissolved solids.

Determination: The amount of total solids in a water sample can be determined very easily as the residue left after evaporation of the filtered sample.

Procedure: Weigh accurately a thoroughly cleaned and completely dry evaporating dish. Take 100ml of water sample in it and evaporate the contents carefully to dryness. Place the dish in an oven at about 110-120°C, for $\frac{1}{2}$ an hour. Cool it in a dessicator and weigh.

Calculations:

Total dissolved solids present in 100 ml of water = weight of dish + residue –weight of dish = Wg.

$$\text{TDS in ppm} = \frac{W}{100} \times 10^6 = 10^4 W\text{ppm}.$$

1.11 SILICA

Determination of silica content in water is very significant in boiler feed waters as silica forms very hard scales in boilers.

Determination: About 500 ml of water is taken in a porcelein dish and to it is added 5–10 mL of conc. HCl.

It is evaporated to dryness and placed in an oven at 110–130°C for about one hour. After one hour it is taken out and cooled in a dessicator and again 10mL of conc.HCl along with 50ml of distilled water is added. It is boiled for half a minute to dissolve all the impurities except SiO_2. The solution is filtered through Whatmann filter paper No. 44. The precipitates and the filter paper are ignited in a weighed crucible. The weight of the residue obtained is the amount of silica present in 500 mL of water. Let it be W.

Calculations: Amount of silica present in 500 mL of Water = W

$$\text{Amount of silica in ppm} = \frac{W}{100} \times 10^6 = 2 \times 10^3 \text{ W ppm}$$

1.12 BOILER FEED WATER

In industries, water is mainly required for cooling and steam generation purposes. The water used for cooling purposes need not to be of high quality and hence requires very less treatment. Sometimes supply of water for cooling purposes are used without any treatment.

On the other hand water required for boiler feed purposes i.e., for steam generation should be of very high quality and thus requires a lot of treatment. However in treatment of boiler feed waters, total elimination of all the impurities is generally not required. Only those impurities which lead to certain problems are eliminated or kept within the prescribed range.

Untreated waters, containing impurities may lead to the following problems in boilers:

1. Scale and sludge formation.
2. Boiler Corrosion
3. Caustic Embrittlement
4. Priming and Foaming

All these problems are discussed below in detail-

1. Scale and Sludge Formation

In boilers water is converted into steam. When water containing impurities is used for generating steam, the concentration of the dissolved salts increases progressively because of evaporation of water. When the concentration reaches beyond the saturation point, the dissolved solids start precipitating out in order of their solubilities, the salt which is least soluble separating out first.

When precipitation of the dissolved salts takes place in the form of loose, non adherent, slimy precipitates, it is called **sludge**. On the otherhand if the precipitates form hard, adhering coating on the inner walls of the boilers, it is called **Scales**.

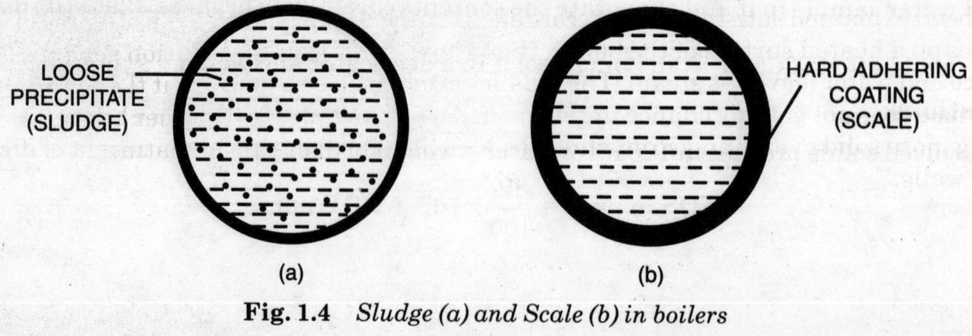

Fig. 1.4 *Sludge (a) and Scale (b) in boilers*

For the deposition of scale and sludge,

- the ionic product of the salt exceeds their solubility product and hence precipitation occurs.
- the increase in temperature leads to reactions that result in the formation of insoluble substances.

The salts responsible for the formation of scales and sludges are $CaCO_3$, $CaSO_4$, Mg $(OH)_2$ $MgCl_2$, and SiO_2 etc. The substances like $CaSO_4$, the solubility of which decrease with temperature forms scales on the hot surface and sludge in the cooler regions. Whereas if the solubility increases with increase in temperature, the material will form scale on the cooler surfaces and sludges in the hotter regions.

Though, formation of scales and sludges both is undersirable but the formation of scales is far more dangerous as compared to sludges as the latter can be removed easily.

Sludge Formation

Sludges are formed by the substances having greater solubilities in hot water than in the cold water e.g $MgCO_3$, $MgCl_2$, $CaCl_2$, $MgSO_4$ etc.

Disadvantages of Sludge Formation

- Wastage of Fuel: Sludges are poor conductor of heat so they tend to waste a lot of amount of heat. The boilers will have to be heated to higher temperatures in order to maintain the supply of heat.
- Decreased efficiency of the boilers: Sludge formation decrease the efficiency of the boilers.
- Choking of the pipes: The formation of sludges slows down the water circulation as it settles in the regions such as pipe connections, plug openings etc. Excessive formation may even cause choking of pipes.

Prevention of Sludge Formation

- Sludge formation can be prevented by using soft water for boiler feed purposes.
- By frequently carrying out the cleaning operations i.e. removal of a portion of concentrated water and replacing it by fresh water. This is called "*blow down operation*" and the soft water which is added to make up the volume is called "*make up water*".

Removal of Sludge

- Sludges can easily be removed by scrapping off with a brush.

Scale Formation

Scales stick very firmly to the inner walls of the boilers and hence are very difficult to remove. Some of them are so hard that they cannot be removed even with the help of hammer and chisel.

The reason for the deposition of some substances as scales and some substances as sludges have not been explained satisfactorily. According to Partridge, when a bubble of water is formed and strikes on a heated surface of the boiler, the salts contained in the solution get deposited on the surface and water leaves as steam. The bubbles are formed so rapidly that the solid deposited at the surface does not get any chance to get washed away and instead another bubble strikes it, depositing more solid over it and so on and subsquently a thick adhering coating is deposited at the inner walls.

Scales are formed through the

- *Decomposition of bicarbonates:* When water containing bicarbonates is heated in the boiler, the bicarbonates present are converted into insoluble $CaCO_3$.

$$Ca\,(HCO)_3 \longrightarrow CaCO_3 + CO_2 + H_2O.$$

- *Hydrolysis of Mg Salts:* Dissolved magnesium salts undergo hydrolysis forming precipitates of $Mg\,(OH)_2$.

$$MgCl_2 + 2H_2O \longrightarrow \underset{\text{Soft Scale}}{Mg\,(OH)_2} + 2HCl.$$

- *Calcium sulphate* is soluble in cold water but the solubility decreases with rise of temperature. Consequently, $CaSO_4$ gets precipitated as hard scale on the heated portions of the boiler. $CaSO_4$ is the main cause of scale formation in high-pressure boilers.
- *Presence of Silica:* Salts like calcium silicate, magnesium silicate are sparingly soluble in cold water, but are completely insoluble in hot water. These deposits stick very firmly on the inner side of the boiler surface and are very difficult to remove.

Disadvantages of Scale Formation

- *Wastage of Fuel :* Because of low thermal conductivity of the scales, the heat supplied to the boiler is not efficiently transferred to water. Hence more heat has to be supplied to maintain the supply of steam leading to wastage of fuels.
- *Bagging:* The distortion of boiler material is known as bagging. The superheating of the boiler leads to the distortion of the boiler material. Due to overheating, rapid reaction between water and iron occurs at high temperature, causing thinning of the boiler material.

$$3Fe + 4H_2O \rightarrow Fe_3O_4 + 4H_2$$

- *Danger of Explosion:* The uneven expansion of the scale may lead to the cracking of the scale. The super-heated steel plates come in contact with boiler water. This causes formation of a large amount of steam suddenly, creating high pressure inside the boiler. When the boiler material is not strong enough to withstand the preasure, it bursts.

Removal of Scales

- Soft scales are removed with the help of scrapper or wire brush.
- Brittle or hard scales can be removed by giving thermal shocks i.e. heating the boiler and suddenly cooling with cold water.
- Scales can also be dissolved in certain chemicals and hence can be removed along with water. For e.g $CaCO_3$ scales can be dissolved by using 5-10% HCl. Similarly EDTA can also be used to dissolve the salts of Ca, with which it forms soluble complexes.

Prevention of Scale Formation

The scale formation can be minimized by subjecting the boiler feed water to certain treatments which may be

(i) **External treatment**. which includes the water softening techniques. (discussed in the next section).

(ii) **Internal Treatment:** The internal treatment of boiler feed waters refers to the treatment of water in the boiler itself (discussed in the next section).

2. Boiler Corrosion

Boiler Corrosion is the decay of boiler material due to the attack of certain chemicals on its surface, resulting into the formation of compounds such as oxides, sulphides etc. The material of the boiler thus gets dissolved and rusted, thereby reducing the life of the boiler.

In general, corrosion occurs, if Fe of the boiler material enters water as Fe^{2+} ions and combines with the water molecules to form ferrous hydroxide, which reacts with oxygen to form Fe $(OH)_3$, or rust.

$$Fe^{2+} + 2H_2O \longrightarrow Fe(OH)_2 + 2H^+$$

$$4Fe(OH)_2 + O_2 \longrightarrow \underset{\text{Rust}}{2[Fe_2O_3 . 2H_2O]}$$

Main reasons for boiler corrosions are-

(a) Presence of dissolved oxygen.
(b) Presence of dissolved carbon dioxide.
(c) Presence of Mineral Acids.

(a) **Presence of dissolved oxygen:** Dissolved oxygen is the main cause of boiler corrosion. At room temperature about 8ml of dissolved oxygen is present in water. Oxygen enters the boiler through raw make up water. When water containing dissolved oxygen is heated in the boiler, the free gas is evolved under prevailing high temperature of boiler and attacks the boiler material.

$$2Fe + 2H_2O + O_2 \longrightarrow 2Fe(OH)_2$$

$$4Fe(OH)_2 + O_2 + 2H_2O \longrightarrow 2[Fe_2O_3 . 2H_2O]$$

Removal of dissolved Oxygen: Oxygen can be removed by chemical or mechanical means.

Chemical Means: Oxygen can be removed by adding calculated amounts of certain chemicals such as Na_2SO_3, Na_2S and N_2H_4.

$$2Na_2SO_3 + O_2 \longrightarrow 2Na_2SO_4$$

$$Na_2S + O_2 \longrightarrow Na_2SO_4.$$

However, the sodium sulphate so formed may decompose in high pressure to form sulphurous acid. Sulphurous acid is corrosive in nature. Therefore a very low concentration of 5 – 10 ppm of Na_2SO_3 is maintained in the boiler water.

Hydrazine is found to be an ideal compound for removing dissolved oxygen since the products are water and nitrogen gas which does not form harmful products.

$$\underset{\text{Hydrazine}}{N_2H_4} + O_2 \longrightarrow N_2 + 2H_2O$$

Thus hydrazine is now extensively used to remove dissolved oxygen in high pressure boilers. Though pure hydrazine is an explosive, inflammable liquid, but 40% aqueous solution is quite safe to handle and is used in water treatment. Calculated amount of hydrazine should be added. Any excess of it decompose to give NH_3, which may cause corrosion of some alloys like copper etc used in condensor tubes.

$$3N_2H_4 \longrightarrow 4NH_3 + N_2.$$

Mechanical Means: Mechanical means for the removal of oxygen make use of Henry's law and

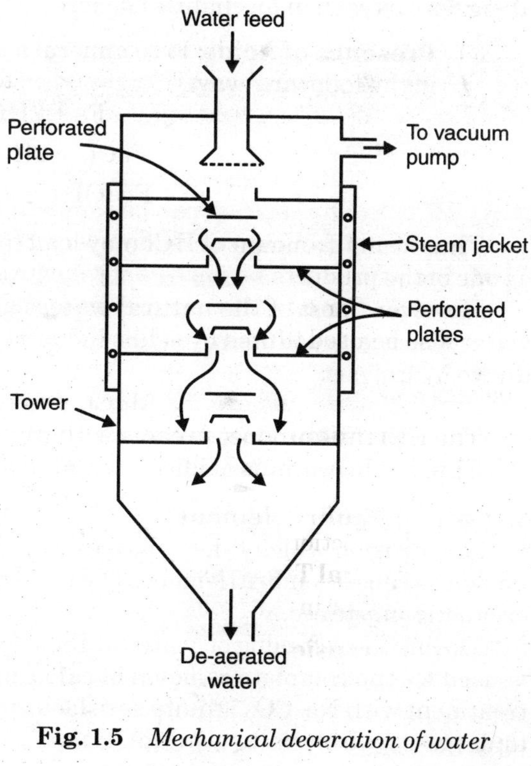

Fig. 1.5 *Mechanical deaeration of water*

Dalton's Law i.e solubility of a gas is directly proportional to pressure and inversely proportional to temperature.

Mechanical deaeration involves, injecting hot feed-water, as a fine spray, into a vaccum chamber heated externally by steam. The chamber is fitted with perforated plates to provide large surface area. Large exposed surface, high temperature, and low pressure reduces the solubility of the gas, the water free of dissolved oxygen is collected at the bottom. (Fig. 1.5)

(b) **Presence of dissolved CO_2:** Carbon dioxide is dissolved in raw water and also it is formed by the decomposition of dissolved bicarbonates under high temperature of boiler.

$$Mg(HCO_3)_2 \longrightarrow MgCO_3 + CO_2 + H_2O.$$

Carbondioxide forms carbonic acid in presence of water which has slow corrosive effect on boiler material.

$$CO_2 + H_2O \longrightarrow \underset{\text{Carbonic acid}}{H_2CO_3}.$$

Removal of CO_2 : Removal of CO_2 also involves chemical or mechanical means.

Chemical means: Chemical means for removal of CO_2 includes addition of calculated amount of NH_4OH.

$$2NH_4OH + CO_2 \longrightarrow (NH_4)_2CO_3 + H_2O.$$

However, if NH_4OH used is more than the calculated amount, it may attack the condeser tubes made of copper.

(ii) Filtering water through lime stone also removes CO_2

$$CaCO_3 + H_2O + CO_2 \longrightarrow Ca(HCO_3)_2$$

But this method leads ot increase in hardness content due to the formation of Ca $(HCO_3)_2$.

Mechanical means: Process is similar as discussed in removal of oxygen. CO_2 is also removed along with oxygen in mechanical deaeration.

(c) **Presence of Acids:** Free mineral acids, present in boiler water, reacts with iron material in the following way

$$Fe + 2HCl \longrightarrow FeCl_2 + H_2$$

$$FeCl_2 + 2H_2O \longrightarrow Fe(OH)_2 + 2HCl$$

$$Fe(OH)_2 + O_2 \longrightarrow 2[Fe_2O_3.2H_2O]$$

Thus a small amount of HCl may lead to corrosion to a great extent as HCl is further formed as one of the products.

However, most of the natural waters are alkaline excepting waters from mining areas or water which gets polluted by acidic industrial wastes. But some salts like $MgCl_2$ etc liberate HCl due to hydrolysis.

$$MgCl_2 + 2H_2O \longrightarrow Mg(OH)_2 + 2HCl.$$

The HCl thus produced reacts with iron and hence corrodes the boiler material.

Thus, if the water is acidic, calculated amount of alkali is added to neutralize it.

3. Caustic Embrittlement

Boiler corrosion which takes place due to the presence of highly alkaline water in the boiler is known as caustic embrittlement. The material of the boiler becomes brittle due to accumulation of caustic substance.

During the softening of water by lime-soda process (discussed later) a slight excess of Na_2CO_3 is used for the complete removal of calcium and magnesium salts. In low pressure boilers, the treatment with Na_2CO_3 is quite satisfactory but in high pressure boilers the free Na_2CO_3 present undergoes hydrolysis to give NaOH as-

$$Na_2CO_3 + H_2O \longrightarrow 2NaOH + CO_2.$$

NaOH so formed along with the natural alkalinity of the water make the boiler water caustic. The NaOH travels to the minute hair cracks, riveted seams, bends and joints. Water evaporates and the concentration of NaOH at these points keep on increasing with time due to poor circulation of water at these points. This caustic soda attacks the surrounding area e.g the intergranular cracks within the cracks, thereby corroding the boiler material or dissolving the iron of the boiler. This causes embrittlement of the boiler parts, particularly the stressed parts. The embrittlement appears to be favoured under the following conditions:

- Materials subjected to great stresses beyond their yield points.
- High conc. of NaOH.
- Presence of compounds such as sodium silicate etc.

The caustic embrittlement can be explained by the formation of the concentration cell as

Iron at Rivets	Conc. NaOH	Dil. NaOH	Iron at plane surfaces

The iron surrounded by concentrated NaOH becomes anodic side, which is corroded whereas the iron surrounded by dil. NaOH becomes cathodic side. (Fig. 1.6)

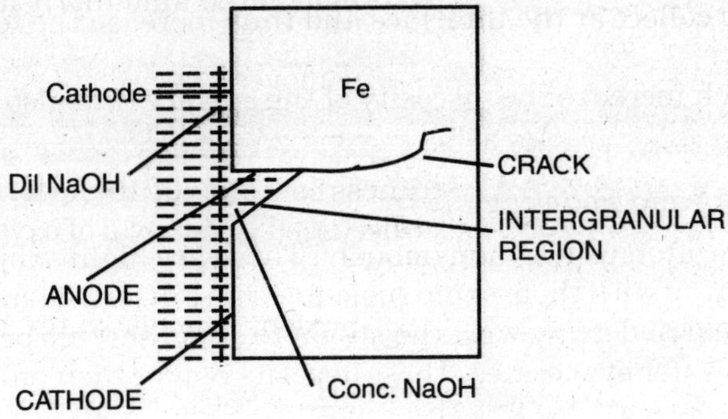

Fig 1.6 *Caustic Embrittlement at the Cracks*

The concentrated alkali dissolves the iron as sodium ferroate ($Na_2 FeO_2$). $Na_2 FeO_2$ decomposes a short distance away from its place of formation according to the reaction.

$$3Na_2FeO_2 + 4H_2O \longrightarrow 6NaOH + Fe_3O_4 + H_2.$$

Further dissolution of iron takes place because of the regeneration of NaOH and precipitation of Fe_3O_4 according to the above reaction.

Prevention of Caustic Embrittlement: The caustic embrittlement can be prevented by

- Using sodium phosphate as softening reagent instead of sodium carbonate.
- By adding liquids or tannin to boiler water. These block up the hair cracks thus prevent the infiltration of sodium hydroxide.
- It is observed that boiler waters containing sodium sulphate or sodium phosphate inhibit the caustic embrittlement by blocking the capillaries, thereby preventing the infiltration of caustic soda solution to these. The concentration of Na_2SO_4 : NaOH are maintained at 1 : 1, 2 : 1 or 3 : 1 for operating pressures of 10, 20 and > 20 atmospheres respectively to check the caustic embrittlement.

However, presence of Na_2SO_4 may lead to the formation of $CaSO_4$ scales. Thus it is preferable to use Na_3PO_4 instead of Na_2SO_4. It is over 300 times as effective as Na_2SO_4 in suppressing embrittlement.

4. Priming and Foaming

Due to the extremely rapid boiling of water in the boiler, the steam formed may be associated with small droplets of water. These droplets of water are carried away along with the steam. *The phenomenon of formation of wet steam i.e. steam associated with liquid water droplets is called priming.*

The liquid contamination in the steam is expressed in percentage by weight of steam. For e.g if steam contains 0.5% moisture, its steam quality will be reported as 99.5%.

Cause: Priming is caused by-

- The high steam velocities.
- Very high water level in the boiler.
- Sudden steam demands leading to sudden drop of pressure in the steam line.
- Faulty boiler design.
- Presence of suspended and dissolved impurities in water.

Foaming is the formation of small persistent bubbles at the surface of water in the boiler. Bubbles are formed when there is a difference in concentration of solute or suspended matter between the surface film and the bulk of the liquid. Any material which lowers the surface tension of the water will collect at the interface and thus increase the foaming tendency of the liquid e.g oil or grease.

The substances which increase the viscosity of the surface film also increase foam forming tendency.

Disadvantages of Priming and Foaming

Priming and foaming are often used synonymously. These are highly objectionable because the droplets of water carry away with them some suspended and dissolved impurities present in the boiler water. These are carried away with the steam to superheaters or turbine blades, where they get deposited as the water evaporates. These deposits reduce their efficiency by hindering the flow of steam.

- A part of the dried salts may be carried along with the steam farther and deposit on the engine valves thereby decreasing their life.
- Presence of water droplets in the steam may lead to corrosion in the steam-inlet ends of superheaters.
- Height of the water column cannot be judged properly due to foaming thereby making the maintenance of the boiler pressure difficult.

Prevention of Priming: Priming can be avoided by-

- Keeping the water level lower.
- Reducing steam velocities also decreases the moisture content of steam because of greater chances of condensing available for the drop.
- Efficient softening and filteration of boiler feed waters so as to minimize the dissolved and suspended impurities.
- Good boiler design fitted with mechanical steam purifiers.

Prevention of Foaming: Foaming can be avoided by-

- Adding antifoaming chemicals: Antifoaming chemicals act by counteracting the reduction in surface tension. For example polyamide antifoamers alters the surface tension and lead to the formation of only large unstable bubbles. Antifoaming agents may reduce foaming by simple mechanical action. For example, castor oil spreads on the surface of water and prevents foaming.

- Removing oil from boiler water also prevents foaming. Oil and grease can be removed by addition of coagulants such as sodium aluminate, ferrous sulphate etc. The coagulants like sodium aluminate etc form hydroxides [Al $(OH)_3$] which entraps oil drops.

5. Removal of Silica

Presence of silica in boiler feed water causes number of boiler problems such as-

- Silica forms thermally resistant scales with calcium and magnesium. With alumina it forms sodium alumino silicate or silica may form scale consisting of silica only.
- In high pressure boilers silica is carried over with steam and deposits on turbine blades.

Silica is removed from the boiler feed water by the following methods:

- By the addition of Magnesia (MgO) to the water, after the removal of temporary hardness. Magnesia acts as magnesium hydroxide. Silica is removed by magnesia by adsorption phenomenon or by the formation of magnesium silicate.
- By the use of ferrous sulphate or sodium aluminate as coagulants. They act as $Fe(OH)_2$ or $Mg(OH)_2$ and $Al(OH)_3$, which enmesh or trap finely suspended and colloidal impurities including oil and silica.
- By the use of highly basic anion exchanger. Water is first dimineralized by passing through anion and cation exchangers. After which it is again passed through a strongly basic special anion exchanger, which removes silica as H_2SiO_3. The silica content is usually reduced to a fraction of a ppm.
- Silicic acid (H_2SiO_3) is very feebly ionized and not removed by passing over weakly basic anion exchanger. Instead flourosilicic acid (H_2SiF_6) is strongly ionized and can be removed by passing over weakly basic anion exchanger. So silica is first converted into flourosilicic acid and then it is removed by passing it over weakly anion exchanger.

Silica in water can be converted into flourosilicic acid by adding sodium flouride and then passing through hydrogen-cation exchanger.

$$SiO_2 + 6NaF + 3H_2R \longrightarrow 3Na_2R + H_2SiF_6 + 2H_2O.$$

For each ppm of SiO_2, 4.2 ppm of NaF is used. Now finally when water is passed through weakly basic anion exchanger, silica is removed.

Silica can also be removed by distillation.

1.13 WATER SOFTENING TECHNIQUES

Internal Treatment of Boiler Feed Water

This involves adding of certain chemicals directly to the water to remove the scale forming salts, either by precipitation of these salts in some or the other form or by converting these salts into soluble complexes, which remain dissolved in water and thus do not form scales. The internal treatment is often referred to as "Sequestration". *Sequestration is a process in which an ion is prohibited to exhibit its usual properties, by either converting it to some other form or by forming a complex with some other compound.* The chemical which is added to bring about sequestration is called sequestering agent.

The methods used for internal treatment of water are-

(a) Colloidal Conditioning.
(b) Carbonate Conditioning.
(c) Phosphate Conditioning.
(d) Calgon Conditioning.

(a) *Colloidal Conditioning:* Scale formation can be reduced by introducing colloidal substances like kerosene, tannin, glue, agar etc. to the boiler water. These substances act as protective coatings. They surround the minute particles of scale forming salts and prevent their coalescence and coagulation. Hence, the salt remains as loose, non sticky deposits in the form of sludges which can be removed by simple blow down operation. This process is suitable for low pressure boilers.

(b) *Carbonate Conditioning:* For the precipitation of a salt, the ionic product of its ions should increase the solubility product. Thus, for the precipitation of $CaCO_3$, the product of its ions Ca^{2+} and CO_3^{2-}, must exceed the solubility product. Similarly, for the precipitation of $CaSO_4$, the ionic product of Ca^{2+} and SO_4^{2-} must exceed the solubility product of $CaSO_4$. The precipitation of $CaSO_4$ thus can be prevented by removing the Ca^{2+} from the solution so that ionic product does not exceed the solubility product. One way of doing it is, by carbonate conditioning, in which Na_2CO_3 is added to the boiler water. The ionization of $CaSO_4$ can be represented as:

$$CaSO_4 \rightleftharpoons Ca^{2+} + SO_4^{2-}$$

By the addition of Na_2CO_3, the concentration ratio of carbonate ions to sulphate ions is increased than the ratio of solubility product of $CaCO_3$ to that of solubility product of $CaSO_4$, i.e.

$$\frac{[CO_3^{\ 2-}]}{[SO_4^{\ 2-}]} > \frac{K_{CaCO_3}}{K_{CaSO_4}} = K'$$

where K_{CaCO_3} and K_{CaCO_4} represent the solubility product of $CaCO_3$ and $CaSO_4$ respectively. From the above discussion, it becomes clear that if $\left[CO_3^{2-}\right] > K'\left[SO_4^{2-}\right]$, then, only $CaCO_3$ will be precipitated in preference to $CaSO_4$, as solubility product of $CaSO_4$ will not be reached under these conditions. Thus the deposition of $CaSO_4$ scales is prevented. The sludge which is formed due to $CaCO_3$ can be removed by blow down operation.

$$Na_2CO_3 + CaSO_4 \rightleftharpoons Na_2SO_4 + CaCO_3$$

(c) *Phosphate Conditioning:* In low pressure boilers, the carbonate conditioning by adding Na_2CO_3 is applicable, while at high pressures, the CO_3^{2-} undergoes hydrolysis to form $\overline{O}H$

$$CO_3^{2-} + H_2O \rightleftharpoons HCO_3^- + \overline{O}H$$

$$HCO_3^- + H_2O \rightleftharpoons H_2O + CO_2 + \overline{O}H$$

Thus the concentration of CO_3^{2-} is decreased and $CaCO_3$ is not precipitated out. Moreover, this is likely to lead to caustic embrittlement. In such cases, it is preferable to precipitate Ca as *the most insoluble compounds of* Ca *i.e* $Ca_3(PO_4)_2$. This is done by the addition of Na_3PO_4

$$3CaSO_4 + 2Na_3PO_4 \rightleftharpoons Ca_3(PO_4)_2 + 3Na_2SO_4.$$

Here also, the excess of PO_4^{3-} is maintained, over that of SO_4^{2-}, to favour the forward reaction i.e the concentration ratio of phosphate ions to sulphate ions is increased than the solubility product of $Ca_3(PO_4)_2$ to $CaSO_4$ by the addition of sodium phosphate.

$$\frac{[PO_4{}^{3-}]^{2/3}}{[SO_4{}^{2-}]} > \frac{[K_{Ca_3(PO_4)_2}]^{2/3}}{K_{CaSO_4}} = K'[SO_4{}^{2-}]$$

or

$$[PO_4{}^{3-}]^{2/3} = K'[SO_4{}^{2-}]$$

Therefore, as long as $[PO_4{}^{3-}]^{2/3} > K'[SO_4{}^{2-}]$, only $Ca_3(PO_4)_2$ will be precipitated because the solubility of $CaSO_4$ cannot be exceeded by the ionic product under these conditions. The sludge, $Ca_3(PO_4)_2$ is removed by blow down operation.

Further, calcium can only be properly precipitated at pH of 9.5 or above. The main phosphates used for colloidal conditioning are-

(a) NaH_2PO_4 - Sodium dihydrogen phosphate (acidic)
(b) Na_2HPO_4 - Disodium hydrogen phosphate (weakly alkaline)
(c) Na_3PO_4 - Trisodium phosphate (alkaline)

The exact choice of phosphate depends upon the alkalinity of the boiler feed water. For instance, if the alkalinity of boiler water is low, trisodium phosphate is used as it is most alkaline. But if the alkalinity is too high then sodium dihydrogen phosphate is used.

(d) *Calgon Conditioning:* Calgon is the commercial name of sodium hexametaphosphate. It forms the complexes with the scale forming salt in which Ca^{2+} is trapped by the phosphate, even if present in very small amounts. The complex which is formed, is soluble in water and hence remains in the dissolved form without causing any harm to the boiler.

$$Na_2[Na_4(PO_3)_6] \longrightarrow 2Na^+ + [Na_4P_6O_{18}]^{2-}$$

$$2CaSO_4 + [Na_4P_6O_{18}]^{2-} \longrightarrow [Ca_2P_6O_{18}]^{2-} + 2Na_2SO_4$$

However, this treatment is not capable of preventing the formation of iron oxide and copper depositions, which can be prevented by adding EDTA or its sodium salt (trylon B) to the boiler water. EDTA binds the cations of scale forming salts into soluble complexes.

External Treatment of Boiler Feed Water

The process of removing the hardness causing salts from water is called softening of water. As is discussed earlier, water used for industrial purposes (boiler feed water, water for cooling purposes) must be sufficiently softened before put to use. The three most common methods adopted for softening of water are-

1. Lime Soda Process
2. Base Exchange or Zeolite Process
3. Demineralization or Ion Exchange Process.

1. Lime Soda Process

Lime soda process is one of the most important method of water softening. The principle involved is to chemically convert all the soluble hardness causing substances into insoluble precipitates which are then removed by settling and filteration.In this process, calculated amount of lime [$Ca(OH)_2$] and soda [Na_2CO_3)] are added to raw water, which react with the calcium and magnesium salts, so as to form insoluble precipitates of calcium carbonate and magnesium hydroxide. The chemical reactions involved are-

Reactions with Lime

(a) Removal of temporary hardness

$$Ca(HCO_3)_2 + Ca(OH)_2 \longrightarrow 2CaCO_3 \downarrow + 2H_2O$$
$$Mg(HCO_3)_2 + Ca(OH)_2 \longrightarrow MgCO_3 + CaCO_3 \downarrow + 2H_2O$$
$$MgCO_3 + Ca(OH)_2 \longrightarrow Mg(OH)_2 \downarrow + CaCO_3 \downarrow$$

(b) Removal of permanent Mg hardness

$$MgCl_2 + Ca(OH)_2 \longrightarrow Mg(OH)_2 \downarrow + CaCl_2$$
$$MgSO_4 + Ca(OH)_2 \longrightarrow Mg(OH)_2 \downarrow + CaSO_4$$

(c) Removal of dissolved iron and aluminium salts

$$FeSO_4 + Ca(OH)_2 \longrightarrow Fe(OH)_2 \downarrow + CaSO_4$$
$$2Fe(OH)_2 + H_2O + \frac{1}{2}O_2 \longrightarrow 2Fe(OH)_3 \downarrow$$
$$Al_2(SO_4)_3 + 3Ca(OH)_2 \longrightarrow 2Al(OH)_3 \downarrow + 3CaSO_4$$

(d) Removal of dissolved CO_2 and H_2S

$$CO_2 + Ca(OH)_2 \longrightarrow CaCO_3 \downarrow + H_2O$$
$$H_2S + Ca(OH)_2 \longrightarrow CaS \downarrow + 2H_2O$$

(e) Removal of free mineral acids

$$2HCl + Ca(OH)_2 \longrightarrow CaCl_2 + 2H_2O$$
$$H_2SO_4 + Ca(OH)_2 \longrightarrow CaSO_4 + 2H_2O.$$

(f) Reaction with $NaHCO_3$

$$2NaHCO_3 + Ca(OH)_2 \longrightarrow Na_2CO_3 + CaCO_3 \downarrow + 2H_2O$$

From the above reactions it can be seen

- that $Ca(OH)_2$ removes temporary hardness due to $Ca(HCO_3)_2$ and $Mg(HCO_3)_2$ completely. Whereas one mole of $Ca(OH)_2$ is required by 1mole of $Ca(HCO_3)_2$, $Mg(HCO_3)_2$ requires two moles of $Ca(OH)_2$ to form insoluble precipitates of $Mg(OH)_2$.
- $Ca(OH)_2$ converts the permanent magnesium hardness to permanent calcium hardness (equations b)
- similarly, reaction of $Ca(OH)_2$ with $FeSO_4$, $Al_2(SO_4)_3$ and free mineral acids also generates an equivalent amount of calcium permanent hardness. (equations c and e)
- One mole of $Ca(OH)_2$ is used to react with $NaHCO_3$, but generates one mole of Na_2CO_3, also. The permanent calcium hardness (original calcium hardness + hardness introduced during the reactions of lime with Mg^{2+}, Fe^{2+}, Al^{3+}, HCl and H_2SO_4) is finally removed by Na_2CO_3.

Reactions of Soda

(a) $CaCl_2 + Na_2CO_3 \longrightarrow 2NaCl + \downarrow CaCO_3$

(b) $CaSO_4 + Na_2CO_3 \longrightarrow Na_2SO_4 + \downarrow CaCO_3$

It is evident from the above equations that the soluble calcium and magnesium salts are precipitated as insoluble calcium salts and removed by filteration. However, the chemical reactions taking place during lime soda treatment are very slow. Moreover, the precipitates formed are very fine and tend to remain in the solution even after filteration. These may cause troubles by getting deposited on the pipes and boiler tubes later. This problem can be solved by adding coagulants like $Al_2(SO_4)_3$ or $NaAlO_2$ to the treated water. These coagulants accelerate the coagulation of the fine particles by entrapping them, which subsequently become heavy and settle down.

Calculation of the amount of lime and soda required

From the above discussion it becomes clear that only required amount of lime and soda are to be added. The calculations of amount of lime and soda required, are based on the amounts of salts present in water. On the basis of the chemical reactions taking place in lime soda process, the following inferences can be drawn.

(i) One equivalent of $Ca(HCO_3)_2$ or calcium temporary hardness requires one equivalent of lime (or 1 mole of $Ca(HCO_3)_2$ requires one mole of $Ca(OH)_2$)

$$Ca(HCO_3)_2 + Ca(OH)_2 \longrightarrow 2\,CaCO_3 + 2\,H_2O$$

Thus 162 parts by weight of $Ca(HCO_3)_2$ requires

$\equiv$ 74 parts by weight of $Ca(OH)_2$

But 162 parts by weight of $Ca(HCO_3)_2 \equiv$ 100 parts by weight of $CaCO_3$

$\therefore$ 100 parts by weight of $CaCO_3$ equivalent hardness requires $\equiv$ 74 parts by weight of $Ca(OH)_2$

1 part by weight of $CaCO_3$ equivalent hardness requires $\equiv \frac{74}{100}$ parts by weight of $Ca(OH)_2$

Hence for x g of $Ca(HCO_3)_2$ present, the lime requirement will be

$$= \frac{74}{100} \text{ (Amount of } Ca(HCO_3)_2 \text{ in terms of } CaCO_3 \text{ equivalents)}$$

$$= \frac{74}{100}\left(x \times \frac{100}{162}\right)$$

(ii) One equivalent of magnesium temporary hardness requires two equivalents of lime

$$Mg(HCO_3)_2 + Ca(OH)_2 \longrightarrow MgCO_3 + CaCO_3 \downarrow + 2H_2O$$

$$MgCO_3 + Ca(OH)_2 \longrightarrow Mg(OH)_2 \downarrow + CaCO_3 \downarrow$$

or

$$\underset{\text{2 equiv.}}{Mg(HCO_3)_2} + \underset{\text{2 (2 equiv.)}}{2Ca(OH_2)} \longrightarrow Mg(OH)_2 \downarrow + 2\,CaCO_3 + 2H_2O$$

Thus lime requirement will be

$$= \frac{74}{100} \text{ } (2 \times Mg(HCO_3)_2 \text{ in terms of } CaCO_3 \text{ equivalents)}$$

(iii) One equivalent of calcium permanent hardness requires one equivalent of soda

$$CaSO_4 + Na_2CO_3 \longrightarrow CaCO_3 + Na_2SO_4$$

$$CaCl_2 + Na_2CO_3 \longrightarrow CaCO_3 + 2\,NaCl$$

In other words,

$$136\,g \text{ of } CaSO_4 \equiv 111g \text{ of } CaCl_2 \equiv 100\,g \text{ of } CaCO_3$$
$$\equiv 106 \text{ g of } Na_2CO_3$$

Thus in terms of $CaCO_3$ equivalents, the soda requirement for permanent calcium hardness

$$= \frac{100}{106} \text{ [(} CaSO_4 + CaCl_2 \text{) in terms of } CaCO_3 \text{ equivalent hardness)]}$$

(iv) One equivalent of magnesium permanent hardness requires one equivalent of lime and one equivalent of soda.

$$MgCl_2 + Ca(OH)_2 \longrightarrow Mg(OH)_2 \downarrow + CaCl_2$$
$$MgSO_4 + Ca(OH)_2 \longrightarrow Mg(OH)_2 \downarrow + CaSO_4$$

$\therefore$ lime Requirement $= \frac{74}{100}$ ($MgSO_4 + MgCl_2$ in terms of $CaCO_3$ equivalents)

The $CaCl_2$ and $CaSO_4$ so generated, react with soda

$$CaCl_2 + Na_2CO_3 \longrightarrow CaCO_3 \downarrow + 2\,NaCl$$
$$CaSO_4 + Na_2CO_3 \longrightarrow CaCO_3 \downarrow + Na_2SO_4$$

$\therefore$ Soda Requirement $= \frac{106}{100}$ ($MgSO_4 + MgCl_2$ in terms of $CaCO_3$ equivalents)

(v) $Al_2(SO_4)_3$ and $FeSO_4$ require equivalent amounts of lime and soda.

$$\underset{\text{(6 equiv.)}}{Al_2(SO_4)_3} + \underset{\text{(6 equiv.)}}{3\,Ca(OH)_2} \longrightarrow 2Al(OH)_3 \downarrow + 3CaSO_4$$

$$FeSO_4 + Ca(OH)_2 \longrightarrow Fe(OH)_2 \downarrow + CaSO_4$$

Lime Requirement $= \frac{74}{100}$ ($Al_2(SO_4)_3 + FeSO_4$ in terms of $CaCO_3$ equiv)

Soda Requirement $= \frac{106}{100}$ ($Al_2(SO_4)_3 + FeSO_4$ in terms of $CaCO_3$ equiv.)

(vi) Lime reacts with HCl, H_2SO_4, CO_2, H_2S etc and hence as in the above cases their respective equivalents must be considered for calculating lime and soda requirement.

(vii) Two equivalents of HCO^-_3 react with two equivalents of lime as:

$$\underset{\text{2equiv}}{2NaHCO_3} + \underset{\text{2equiv}}{Ca(OH)_2} \longrightarrow CaCO_3 + 2H_2O + \underset{\text{2 equiv}}{Na_2CO_3}$$

Thus for equivalents of HCO_3^- ($NaHCO_3$) two equivalents of CO_3^{2-} are generated. Hence this amount in terms of $CaCO_3$ equivalents is *deducted* from soda requirement.

(viii) Presence of $NaAlO_2$ as coagulant neither require lime or soda but forms NaOH on reacting with water-

$$NaAlO_2 + 2H_2O \longrightarrow NaOH + Al(OH)_3$$

1 equiv. of NaOH can be considered to be 1 equiv.of lime and hence corresponding amount should be subtracted from lime requirement.

Hence Total Lime Requirement =

$$\frac{74}{100}\left[\begin{array}{l}2 \times \text{Temp. Mg} + \text{Temp Ca} + \text{Perm Mg} + (\text{Salts of Fe} + \text{Al}) + CO_2 \\ + HCl + H_2SO_4 + HCO_3^- - NaAlO_2 \text{ all in terms of } CaCO_3 \text{ equivalents.}\end{array}\right]$$

Similarly soda is consumed by $CaCl_2$ and $CaSO_4$ originally present and by the equivalent amount of these salts, formed by treating the water with $Ca(OH)_2$ having $MgCl_2$, $MgSO_4$, $Al_2(SO_4)_3$, $FeSO_4$, HCl, H_2SO_4 etc. Now,

100 parts of $CaCO_3$ equiv. hardness requires 106 parts of Na_2CO_3.

$\therefore$ **Total Soda Requirement**

$$= \frac{106}{100}[\text{Perm. Ca} + \text{Perm. Mg} + \text{Salts of Fe} + \text{Salts of Al} + HCl + H_2SO_4 - HCO_3^- \text{ all in terms of } CaCO_3 \text{ equivalents}]$$

Notes: From the above discussion it follows

- The substances like NaCl, KCl, Na_2SO_4, SiO_2, Fe_2O_3 etc. do not impart any hardness to water and these do not consume any lime or soda.
- Temporary hardness due to Ca or Mg salts do not require soda as no corresponding permanent Ca hardness is produced.
- Permanent Ca hardness does not require any lime but only soda.
- Permanent Mg hardness require lime as well as soda. This is also true for the Fe, Al salts and free acids.
- Since the treated water is shown to contain $\overline{O}H$ and CO_3^{2-}. These are formed from excess of $Ca(OH)_2$ and Na_2CO_3 respectively. So during calculations the corresponding amount to these should be added. In case of excess of hydroxide ion, the equivalent amount has to be added in lime as well as soda. This is because it should have been supplied by excess of $Ca(OH)_2$ which introduces subsequently calcium hardness which require soda. However for excess of CO_3^{2-} ions, the equivalent amount has to be added to soda requirement because CO_3^{2-} comes from excess of Na_2CO_3.

Types of Lime Soda Processes

The lime soda process can be carried out either in hot or cold.

1. Cold Lime Soda Process

In this method, calculated amount of lime and soda are mixed with water at room temperature. There are two types of softeners used for softening water by this process.

(a) **Intermittent Type (Batch Process):** Intermittent softener consists of a pair of tanks. Each tank is used in turn for softening of water. Each tank is provided with inlets for raw water and chemicals, outlets for softened water and sludge, and a mechanical stirrer.

Raw water and the calculated quantities of lime and soda are allowed to run in simultaneously and stirring is started. During stirring, some precipitate from the previous treatment is also added in order to form nucleus for the fresh precipitation. By the time tank fills up, the reaction is more or less complete. Stirring is stopped and the precipitate formed is allowed to settle down. At room temperature, the precipitates formed are finely divided, so they do not settle down easily and hence it is essential to add small amounts of coagulants like $Al_2(SO_4)_3$, $NaAlO_2$ etc. The coagulants form gelatinous precipitates of $Al(OH)_3$ and entraps the fine precipitates. Use of sodium aluminate as coagulant also helps the removal of silica as well as oil, if present in water.

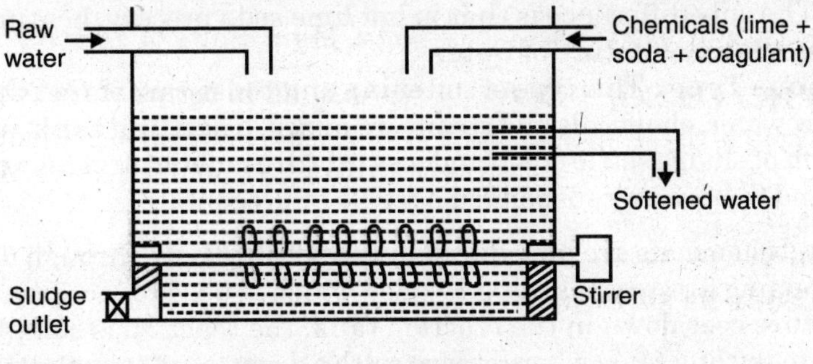

Fig. 1.7 *Batch Process*

The clear softened water is collected through a pipe and sent to the filtering unit. The sludge formed in the tank is removed through the sludge outlet. When one tank is used for reaction and settling, the other tank supplies soft water. Arrangement is such that there is continuous supply of soft water from these tanks (Fig. 1.7).

(b) **Continuous Softener:** In continuous softener, there is continuous feed of raw water and chemicals and uninterrupted outflow of soft water is obtained.

The continuous softener consists of a big steel tank with two chambers. The inner chamber is provided with a strirrer, whereas the outer chamber is fitted with a filtering media (usually made of wood fibres). Raw water and calculated amount of lime, soda and coagulant are fed through two opposite inlets into the inner chamber at room temperature. As raw water and chemicals flow down stirring and mixing takes place. Softening of water takes place and the sludge formed settles down at the bottom of the outer chamber. The softened water rises upwards, passes through the wood filters to ensure complete removal of sludge. Filtered water finally comes out continuously through the outlet provided for the purpose (Fig. 1.8).

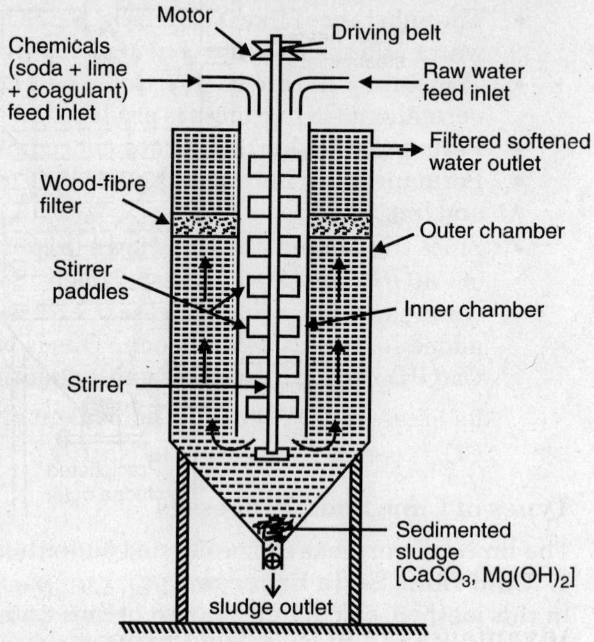

Fig. 1.8 Continuous cold lime-soda softener

2. Hot Lime Soda Process

This method is exclusively used for the boiler feed waters. This process is much more efficient than cold lime soda process. The softening reactions are made to occur almost at the boiling point of water. The high temperature :

- increases the rate of precipitation reactions.
- decreases the viscosity of water thereby making settling of the precipitate and filtering easier and rapid.
- eliminates the need of adding coagulants as precipitates settle down quickly.

Two types of softeners are used for hot lime soda process.

(a) **Intermittent Type:** This is exactly similar to what is described under cold lime soda process. The only difference is that in hot lime soda process, heating coils are fitted in it for raising the temperature of water.

(b) **Continuous Type:** This type of softening unit consists of three parts-**a reaction tank**, where raw water, chemicals and steam are mixed, **a settling tank**, where the precipitates in the form of sludge settle down and a **sand filter** made up of fine sand and coarse sand, to filter and remove any suspended particle or precipitate.

Raw water and chemicals are introduced in the settling tank through different inlets.

A separate opening for steam is also provided which is used to increase the temperature. By the time the mixture goes down in the reaction tank, the reaction is complete. From this tank, the water is fed into settling tank, where sludge settles down. Now a days settling tank is combined with reaction tank as shown in Fig. 1.9. Clear and soften water from the settling tank passes to filter tank where remaining precipitates or suspended particles are removed.

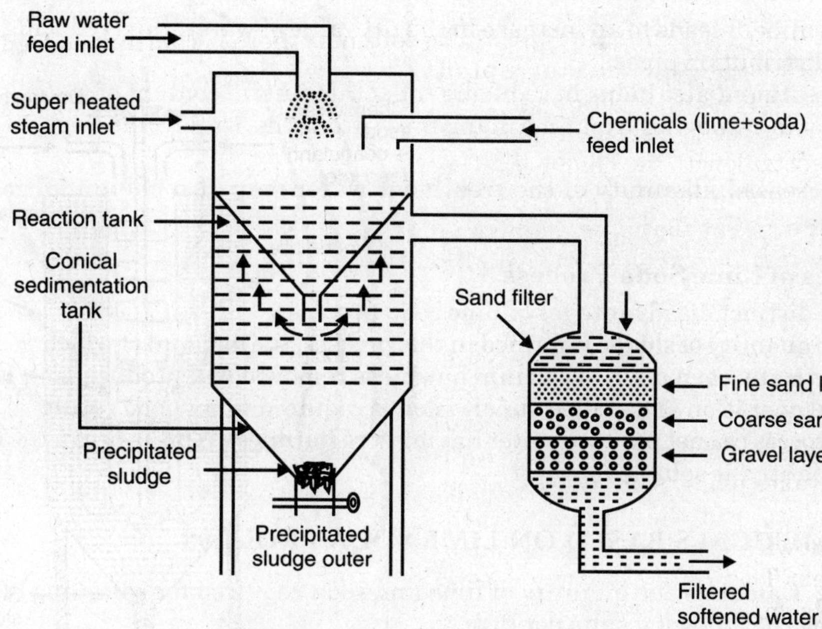

Fig. 1.9 *Hot lime-soda process*

Advantages of hot lime Soda process

(i) The reaction time is greatly reduced as the precipitation takes place rapidly. It takes about 15 minutes whereas the cold process takes several hours for completion.
(ii) Residual hardness is much less than cold lime soda process.
(iii) No coagulant is required because the precipitate and sludge formed setlle down quickly.
(iv) Most of the dissolved gases like CO_2 and O_2 are also removed.
(v) Less chemicals are required as the softening capacity is high.
(vi) Filteration becomes much easier as the viscosity of the water becomes low.

Table 1.7 *Comparison between Cold Lime Soda process and Hot lime Soda process*

Cold Lime Soda Process	*Hot Lime Soda Process*
1. Reactions take place at slower rate.	1. Reactions take place at faster rate.
2. Coagulants are added to settle down the precipitates.	2. Coagulants are not required as the precipitates settle down quickly due to decreased viscosity of water.
3. No fuel consumption.	3. As steam is used, so there is fuel consumption.
4. Temporary hardness is removed by adding lime.	4. Temporary hardness is removed by boiling and hence lime consumption is less.
5. Time taken for softening is longer (few hours).	5. Time taken for softening is less (few minutes)
6. The residual hardness is about 50-60 ppm.	6. The residual hardness is about 17-34 ppm.

Advantages of Lime Soda Process

The lime soda process is widely used because of the following advantages-

(i) The process is highly economical.

(ii) The treatment leads to an increase in the pH value of water, thus reducing the corrosion of the distribution pipes.
(iii) This treatment also helps in reducing the total mineral content of water.
(iv) It helps in removing iron and manganese from the water, although to a very small extent.
(v) The increased alkalinity of the treatment water may also help in killing pathogenic bacteria.

Disadvantages of Lime Soda Process

There are a few distinct disadvantages of lime soda process.

(i) A large quantity of sludge is formed in this process, the disposal of which poses a problem. Theoretically each mg/L of calcium hardness removed will produce 1mg/L of sludge.
(ii) Careful operation and skilled supervision is required to get good results.
(iii) This process cannot produce water having zero hardness. This is because $CaCO_3$, though very slightly, is soluble in water.

SOLVED NUMERICALS BASED ON LIME SODA PROCESS

Example 1.11: Calculate the quantity of lime and soda required for softening 50,000 litres of water containing the following salts per litre.

$Ca(HCO_3)_2$ = 9.2 mg; $Mg(HCO_3)_2$ = 7.9 mg; $CaSO_4$ = 15.3 mg; $MgSO_4$ = 15.0 mg; $MgCl_2$ = 3.0 mg and NaCl = 4.3 mg.

Solution:

Constituent	*Amount mg/L*	*Multiplication Factor*	*$CaCO_3$ Equivalent mg/L*
$Ca(HCO_3)_2$	9.2	100/162	5.68
$Mg(HCO_3)_2$	7.9	100/146	5.41
$CaSO_4$	15.3	100/136	11.25
$Mg SO_4$	15.0	100/120	12.5
$Mg Cl_2$	3.0	100/95	3.16
NaCl	4.3	Neglected	

$$\therefore \text{Lime Requirement} = \frac{74}{100}[Ca(HCO_3)_2 + 2Mg(HCO_3)_2]$$
$$[+MgSO_4 + MgCl_2 \text{ in terms of } CaCO_3 \text{ equiv.}]$$

$$= \frac{74}{100}[5.68 + 2(5.41) + 12.5 + 3.16] \text{ mg/L} = 23.39 \text{ mg/L}$$

Lime required for softening

50,000 litres = 23.40×50,000 mg = 1170 g = 1.17 kg.

$$\text{Soda Requirement} = \frac{106}{100}[CaSO_4 + MgSO_4 + MgCl_2 \text{ as } CaCO_3 \text{ equiv.}]$$

$$= \frac{106}{100}[11.25 + 12.5 + 3.16] = 28.5 \text{ mg/L}$$

Soda required for softening

50,000 litres = 28.52×50,000 mg

= 14.26g = 1.426 kg

Example 1.12: Calculate the amount of lime and soda required per litre for chemical treatment of water containing: Ca^{2+} = 80 ppm; Mg^{2+} = 32 ppm; HCO_3^- = 195 ppm; Fe $SO_4.7H_2O$ added as coagulants = 73.5 ppm.

Solution: Conversion to $CaCO_3$ equiv.

Constituent	*Amount (ppm)*	*Multiplication Factor*	*$CaCO_3$ Equivalents (ppm)*
Ca^{2+}	80	100/40	200
Mg^{2+}	32	100/24	133.3
HCO_3^-	195	100/122	159.8
$FeSO_4.7H_2O$	73.5	100/278	26.44

$$\text{Lime Requirement} = \frac{75}{100}[Mg^2 + HCO_3^- + FeSO_4 \times 7H_2O \text{ as } CaCO_3 \text{ eq}]$$

$$= \frac{74}{100}[133.3 + 159.8 + 26.44]\ ppm = 236.45\ ppm$$

$$\text{Soda Requirement} = \frac{106}{100}[Ca^{2+} + Mg^{2+} + FeSO_4 \times 7H_2O - HCO_3^- \text{ as } CaCO_3 \text{ eq}]$$

$$= \frac{106}{100}[200 + 133.3 + 26.44 - 159.8]ppm = 211.94\ ppm$$

Note: If analytical report shows the quantities of Ca^{2+} and Mg^{2+}, then 1 eq. of lime and 1 eq. of soda is required for Mg^{2+}, whereas 1 eq. of soda is required for Ca^{2+}. In otherwords, the ions Ca^{2+} and Mg^{2+} are treated as permanent hardness due to calcium and magnesium.

Example 1.13: Calculate the amount of lime (92% pure) and soda (98% pure) required for treatment of 30,000 litres of water whose analysis is as follows:

$Ca(HCO_3)_2$ = 40.5 ppm; $Mg(HCO_3)_2$ = 36.5 ppm; $MgSO_4$ = 30.0 ppm; $CaSO_4$ = 34.0 ppm; $CaCl_2$ = 27.75 ppm; NaCl = 10.0 ppm.

Solution: Conversion to $CaCO_3$ eq.

Constituent	*Amount (ppm)*	*Multiplication Factor*	*$CaCO_3$ eq. (ppm)*
$Ca(HCO_3)_2$	40.5	100/162	25
$Mg(HCO_3)_2$	36.5	100/146	25
$MgSO_4$	30.0	100/120	25
$CaSO_4$	34.0	100/136	25
$CaCl_2$	27.75	100/111	25
NaCl	—		

Lime Requirement

$$= \frac{74}{100}\left[Ca(HCO_3)_2 + 2 \times Mg(HCO_3)_2 + MgSO_4 \text{ in term } CaCO_3 \text{ eq.} \times \text{volume of water}\right]$$

$$= \frac{74}{100}[25 + (2 \times 25) + 25]\text{mg/L} \times 30{,}000\text{L} = 74 \times 30{,}000 \text{ mg} = 2220 \text{ g}$$

Since lime is 92% pure, therefore lime required will be

$$= \frac{100}{92} \times 2220\,g = 2.413\,kg$$

Soda Requirement

$$= \frac{106}{100}[MgSO_4 + CaSO_4 + CaCl_2 \text{ in term of } CaCO_3 \text{ eq.}]$$

$$= \frac{106}{100}[25 + 25 + 25]\,mg/L \times 30{,}000\,L = 2385\,g$$

98% pure soda, required for softening 30,000 L

$$= \frac{100}{98} \times 2385 = 2433\text{ g} = 2.433\text{ kg}$$

Note: If the lime and soda used are impure, then the actual requirement is calculated accordingly. In the above example as lime and soda are 92% and 98% pure respectively, therefore the values obtained in each case are multiplied by purity factor $\left(\frac{100}{92} \text{ for lime; } \frac{100}{98} \text{ for soda}\right)$.

Example 1.14: Calculate the amount of lime (74% pure) and soda (92% pure) required for softening 20,000 litres of hard water containing, Mg CO_3 = 84 mg/L; $CaCO_3$ = 40 mg/L; $MgCl_2$ =95 mg/L; $CaCl_2$ = 111mg/L; Mg $(NO_3)_2$ = 37 mg/L; KCl = 30 mg/L.

Solution: Conversion to $CaCO_3$ equivalents.

Contituent	*Amount mg/L*	*Multiplication Factor*	*$CaCO_3$ eq. Mg/L*
Mg CO_3	84	100/84	100
Ca CO_3	40	100/100	40
$MgCl_2$	95	100/95	100
$CaCl_2$	111	100/111	100
Mg $(NO_3)_2$	37	100/148	25
KCl	—	—	-

Lime Requirement

$$= \frac{74}{100}[2 \times MgCO_3 + CaCO_3 + MgCl_2 + Mg(NO_3)_2 \text{ in term of } CaCO_3 \text{ equiv}] \times \text{Volume of water} \times \text{purity factor}$$

$$= \frac{74}{100}[2 \times 100 + 40 + 100 + 25]\,mg/L \times 20{,}000L \times \frac{100}{74}$$

$= 7300$ g $= 7.3$ kg

Soda Requirement

$$= \frac{106}{100}\left[MgCl_2 + CaCl_2 + Mg(NO_3)_2 \text{ in terms of } CaCO_3 \text{ equiv.}\right] \times \text{Volume of water} \times \text{purity factor}$$

$$= \frac{106}{100}[100 + 100 + 25]\,mg/L \times 20{,}000 \times \frac{100}{92}$$

$= 5184$ g $= 5.2$ kg

Note: $CaCO_3$ and $MgCO_3$ are regarded as being present in the form of their bicarbonates, only their weights have been expressed in terms of $CaCO_3$ and $MgCO_3$. So the lime requirement for $MgCO_3$ is 2 equiv. as in the case of Mg $(HCO_3)_2$ whereas for $CaCO_3$ it is one equiv. No soda is required for either Ca or Mg if they are given as $CaCO_3$ and $MgCO_3$.

Example 1.15: Calculate the quantities of lime and soda required for softening of 1,00,000 litres of water using 139 ppm of $FeSO_4.7H_2O$ as a coagulant. The results of the analysis of raw water and softened water are as follows:

Analysis of raw water : $Ca^{2+} = 160$ ppm; $Mg^{2+} = 288$ ppm; $HCO_3^- = 1464$ppm; dissolved $CO_2 = 20$ ppm.

Analysis of treated water; $CO_3^{2-} = 60$ ppm; $\bar{O}H = 34$ ppm.

Solution : Conversion to $CaCO_3$ equiv.

Constituent	*Amount (ppm)*	*Multiplication Factor*	*$CaCO_3$ Equiv. (ppm)*
Raw Water			
Ca^{2+}	160	100/40	400
Mg^{2+}	288	100/24	1200
HCO_3^-	1464	100/122	1200
CO_2	22	100/44	50
$FeSO_4.7H_2O$	139	100/278	50
Treated Water			
CO_3^{2-}	60	100/60	100
$\bar{O}H$	34	100/34	100

Lime Required

$$= \frac{74}{100}\left[Mg^{2+} + HCO_3^- + CO_2 + FeSO_4.7H_2O + \bar{O}H \text{ all in term of } CaCO_3 \text{ equiv.}\right] \times \text{Volume of water}$$

$$= \frac{74}{100}[1200 + 1200 + 50 + 50 + 100]\text{mg/L } L \times 1{,}00{,}000\ L] = 192.4 \text{ kg}$$

Soda Required

$$= \frac{106}{100}\left[Ca^{2+} + Mg^{2+} + FeSO_4.7H_2O + CO_3^{2-} + \bar{O}H - HCO_3^- \text{ in } CaCO_3 \text{ equiv.}\right] \times \text{Volume of water}$$

$$= \frac{106}{100}[400 + 1200 + 50 + 100 + 100 - 1200]\text{mg/L} \times 1{,}00{,}000 \text{ L} = 68.90 \text{ kg}$$

Notes: Since the treated water is shown to contain $\bar{O}H$ and CO_3^{2-}, the excess $\bar{O}H$ in treated water should have been supplied by its equivalent amount of lime used in excess.

$$Ca(OH)_2 \longrightarrow Ca^{2+} + 2\bar{O}H$$

But the corresponding amount of Ca^{2+} so incorporated should have been removed by adding equivalent amount of soda.

$$Ca^{2+} + Na_2CO_3 \longrightarrow CaCO_3 \downarrow + 2Na^+$$

Therefore, in the above example $CaCO_3$ equiv. of $\bar{O}H$ are added in lime as well as soda requirement. The excess CO_3^{2-} required to be present in the treated water must have been supplied by equivalent amount of $Na_2 CO_3$ used in excess.

$$Na_2CO_3 \longrightarrow 2Na^+ + CO_3^{2-}$$

Therefore, $CaCO_3$ equiv of CO_3^{2-} in added is soda requirement only.

2. Base Exchange or Zeolite Process

A reversible exchange of ions between a liquid phase and a solid phase is termed as ion exchange. A certain class of naturally occuring aluminosilicate minerals called zeolites act as solid phase, capable of exchanging its cations.

The name zeolite is derived from two greek words, Zein+lithos-which mean boiling stone. The term was first coined by Cronstedt, a Swedish geologist in 1756 for a group of minerals which released their water of hydration in the form of steam. The chemical structure of sodium zeolite may be represented as:

$$Na_2O.Al_2O_3.xSiO_2.yH_2O \text{ where } x = 2-10 \text{ and } y = 2-6,.$$

It may be regarded as hydrated sodium alumino-silicate, which is capable of exchanging its sodium ions for divalent ions of alkaline earth group and for the multivalent ions of some of the metals in water. Zeolites find application in softening of water for domestic and industrial purpose. Two types of zeolites in use are-

(i) **Natural Zeolites:** The naturally available zeolite is green in colour and is referred to as green sand. The common ones being Thomsonite (Na_2O, $Al_2O_3.3SiO_2.2H_2O$).

(ii) **Synthetic Zeolites:** Synthetic zeolites are porous and posses a gel structure. They can be prepared by heating together solutions of sodium silicate, aluminium sulphate and sodium aluminate. Synthetic Zeolites possess higher exchange capacity compared to that of natural zeolites. The most common artificial zeolite is a white coloured substance prepared from feldspar, kaolin, clay and soda and is called permutit. Hence Zeolite process is sometimes referred to as permutit process.

Structure: A zeolite crystal can be considered to be formed from the linking of several SiO_4 tetrahedra. In it, each oxygen of a tetrahedra is being shared with an adjacent one. The empirical formula is thus $(SiO_2)_n$. However some of the Si^{4+} ions are replaced by Al^{3+} ions. In order to balance the charges, extra positive ion Na^+ or K^+ must be incorporated for every Al^{3+} introduced. Thus the linking of tetrahedra results in an open structure with cavities in which Na^+ or K^+ ions are loosely held. (Fig. 1.10)

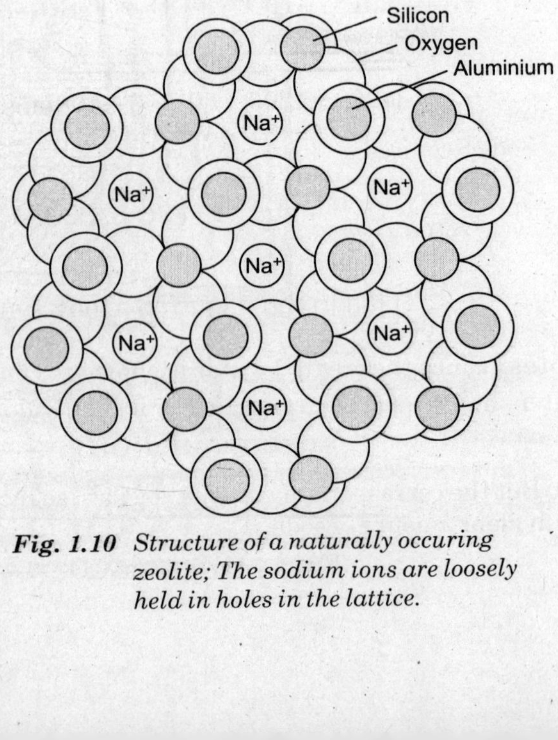

***Fig. 1.10** Structure of a naturally occuring zeolite; The sodium ions are loosely held in holes in the lattice.*

Principle of Water Softening

Zeolites can be represented as Na_2Z, where Z is the insoluble radical framework. When hard water is passed through a bed of active granular Na_2Z, the Ca^{2+} and Mg^{2+} ions of the hard water are exchanged for Na^+ ions of zeolite. The various reactions taking place may be indicated as follows:

$$Ca(HCO_3)_2 + Na_2Z \longrightarrow CaZ + 2NaHCO_3$$
$$Mg(HCO_3)_2 + Na_2Z \longrightarrow MgZ + 2NaHCO_3$$
$$CaSO_4 + Na_2Z \longrightarrow CaZ + Na_2SO_4$$
$$MgSO_4 + Na_2Z \longrightarrow MgZ + Na_2SO_4$$
$$CaCl_2 + Na_2Z \longrightarrow CaZ + 2NaCl$$
$$MgCl_2 + Na_2Z \longrightarrow MgZ + 2NaCl.$$

It is seen from the above reactions that sodium zeolite is converted to calcium and magnesium zeolites. This exchange takes place because the extent of exchange increases with increasing valency of the exchanging ions i.e monovalent ions are exchanged for divalent ions.

Regeneration: When the zeolite bed gets exhausted i.e nearly all its Na^+ are exchanged for Ca^{2+} and Mg^{2+} ions, it loses its sodium exchange capacity. The zeolite bed is regenerated by first backwashing it and then by passing a concentrated NaCl solution through it. Though other solutions like $NaNO_2$, Na_2SO_4, KCl etc can also be used, but brine solution (NaCl solution) is used because it is cheap and also because the products formed by regeneration reactions are $CaCl_2$ and $MgCl_2$ which are highly soluble and can be readily rinsed out from the zeolite bed.

The following reactions take place during regeneration process :

$$\underset{\text{Exhausted zeolite}}{CaZ + 2NaCl} \rightleftharpoons \underset{\text{Regenerated zeolite}}{Na_2Z + CaCl_2}$$
$$MgZ + 2NaCl \rightleftharpoons NaZ + MgCl$$

Process: A zeolite softener consists of a steel tank packed with a thick layer of permutit or zeolite. The hard water is percolated at a specified rate through the bed of zeolite. Calcium and Magnesium ions present in it are retained by the zeolite in the form of CaZ and MgZ respectively. The outgoing water becomes rich in sodium salts. The softened water is collected from the bottom. (Fig. 1.11)

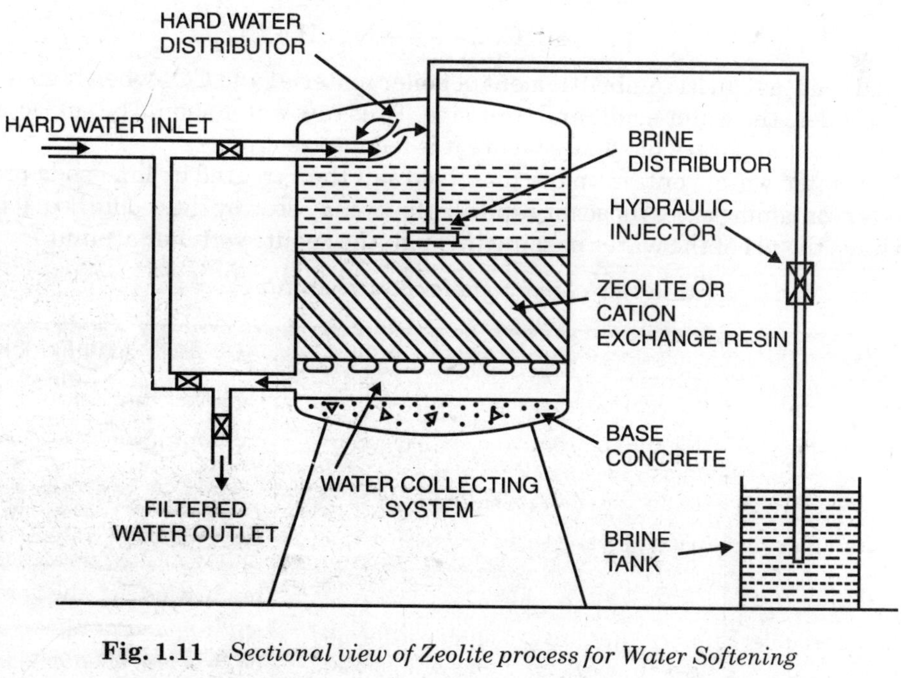

Fig. 1.11 *Sectional view of Zeolite process for Water Softening*

When a significant portion of the sodium in the zeolite has been replaced by calcium and magnesium, it is regenerated by first washing it with water by reversing the flow and then treating it with conc. NaCl solution. The soluble $CaCl_2$ and $MgCl_2$ so formed in regeneration process are washed with soft water and washings are led to sink. The zeolite bed is ready for use again for softening purpose. Theoretically each mg/L of $CaCO_3$ hardness requires 1.17 mg/L of NaCl solution. But thrice the amount of theoretically required amount of NaCl is used to make regeneration reaction more efficient.

Advantages of Zeolite Process

(i) Water having very low hardness of about 10ppm hardness is produced. Even water having zero hardness can be produced.
(ii) The plant is compact and occupies less space.
(iii) No sludge is formed and hence there is no problem of sludge disposal.
(iv) The running, maintenance and operation cost is quite less.
(v) The plant can be installed in the water supply line itself, avoiding double pumping.
(vi) It requires less skill for maintenance as well as operation.

Disadvantages

(i) This process is not suitable for treating highly turbid waters because the turbidity will clog the pores of zeolite bed and thus make zeolite less efficient. Therefore the turbidity must be removed before the water is fed into zeolite.
(ii) Water containing large amounts of Fe and Mn salts when passed through zeolite bed are converted to iron zeolite and manganese zeolite which cannot be regenerated easily. The zeolite is thus wasted.
(iii) The process leaves sodium bicarbonate in water, which causes priming and foaming in industrial or boiler feed waters. Moreover in boilers $NaHCO_3$ dissociates forming NaOH and CO_2.

$$NaHCO_3 \longrightarrow NaOH + CO_2.$$

NaOH causes caustic embrittlement of boiler material and CO_2 when it goes with steam the renders the water acidic and corrosive. Thus raw water should be subjected to boiling to remove temporary hardness before it is fed into zeolite.
(iv) The treated water contains more dissolved salt than treated by lime-soda process.
(v) Water containing excess of acidity and alkalinity may destroy the zeolite bed. It is preferable to have the pH of the water passing through the zeolite softener around 7.

Table 1.8: *Lime Soda Process Vs Zeolite Process*

S.No.	*Property*	*Lime-Soda Process*	*Zeolite Process*
1.	Size of plant	Bulky and Large	Compact and small.
2.	Supervision	Requires careful and skilled supervision because lime and soda have to be added in calculated quantities.	Automatic and easy to operate.
3.	Pretreatment	Turbid waters can be treated.	Highly turbid waters are difficult to treat because the suspended impurities clog the pores of zeolite.
4.	Sludge trouble	Large quantity of sludge is formed which poses problems of disposal.	No sludge is formed and therefore no problem of sludge disposal.
5.	Post treatment if needed.	Coagulation, settling filteration and, removal of precipitates involved.	No such post treatment is required.

S.No.	Property	Lime-Soda Process	Zeolite Process
6.	Dissolved solids	Treated water contains more amount of dissolved sodium salts.	Treated water contain lesser amount of sodium salts.
7.	Results obtained	This can produce waters of hardness not less than about 50 ppm. Therefore, useful for public supplies only.	Water of zero hardness can be obtained. Therefore, useful for industrial supplies also.
8.	Removal of colour due to iron and manganese.	Can remove the colour due to iron and manganese, but only to a very small extent.	Can remove the colour due to iron and manganese, although very costly for treating such waters, because the exhausted Mn and Fe zeolites cannot be regenerated.
9.	The pH of the treated water.	Increases the pH value of water, which reduces the corrosion of distribution pipes.	The pH value of the water is not affected.
10.	Acidic Water	Acidic water can also be treated.	Acidic water cannot be treated since acidic water attack zeolite.
11.	Care needed in handling materials.	Careful handling is required because the materials i.e. (lime and soda) are corrosive.	Brine solution is required to be handled, which does not require much care.
12.	Economy	Process is economical and can be easily combined with usual water treatment methods at slight extra cost.	Process is costlier.

NUMERICAL BASED ON ZEOLITE PROCESS

Example 1.16: A zeolite softener was 95% exhausted, when 10,000 L of hard water was passed through it. The softener required 150 L of NaCl solution of strength 50 g NaCl/L of solution to regenerate. What is the hardness of water.

Solution:

$$
\begin{aligned}
10{,}000 \text{ L of hard water} &= 150 \text{ L of NaCl solution} \\
&= 150 \text{ L} \times 50 \text{ g/L of NaCl} \\
&= 7500 \text{ g of NaCl.} \\
&= 7500 \times \frac{100}{58.5 \times 2} \text{ of } CaCO_3 \text{ eq.} \\
&= 6410.26 \text{ g of } CaCO_3 \text{ eq.} \\
1 \text{ L of hard water} &= 6410.26 \text{ g of } CaCO_3 \text{ eq/10,000.} \\
&= 0.6410 \text{ g of } CaCO_3 \text{ eq.} \\
&= 641 \text{ mg of } CaCO_3 \text{ eq.} \\
\text{Hence hardness of water} &= 641 \text{ mg/L.}
\end{aligned}
$$

Example 1.17: A Zeolite softener was completely exhausted and then regenerated by passing 200 litres of NaCl solution containing 100 g per litre of NaCl. How many litres of a sample of water of hardness 500 ppm can be softened by the softener.

Solution:

$$200 \text{ litres of NaCl} = 100 \text{ L} \times 200 \text{ g/L} = 20000 \text{ g of NaCl}$$

$$= 20000 \times \frac{100}{58.5 \times 2} \text{ g of } CaCO_3 \text{ eq.}$$

$$= 17094 \text{ g of } CaCO_3 \text{ eq.}$$

Now 500 mg of hardness is present in = 1 litre of water

$$17094 \times 10^3 \text{ mg of } CaCO_3 \text{ eq. hardness is present in } = \frac{1}{500} \times 17094 \times 10^3 \text{ L}$$

$$= 34188 \text{ litres of hard water.}$$

Hence, the softener can soften 34188 litres of hard water.

3. Ion Exchange or Deionization Process (Demineralization of Water)

In Zeolite process for softening of water, the anions viz; SO_4^{2-} and HCO_3^- remain in the softened water combined with Na^+. The total content of dissolved solids in water remains the same. However with the development of organic synthetic cation exchangers the removal of even anions has been made possible.

The process of complete removal of all ions present in water is called dimineralization. Demineralization of water is achieved by using ion-exchange resins.

Ion Exchange Resins: Ion exchange resins consist of cross-linked, long chain organic polymers with a microporous structure. The functional groups attached to the polymeric chains are responsible for ion exchange.

Resins of suitable chemical compositions and physical properties may be synthesized for specific ion-exchange applications. The resins usually consist of styrene-divinylbenzene copolymer. Other polymers like methacrylic acid-divinylbenzene or phenol formaldehyde polymers are also used.

The functional groups attached to the resin may be acidic or basic and hence resins can be classified as:

(a) **Cation exchange Resins:** Resins containing acidic groups like –COOH or $-SO_3H$ group are termed as cation exchange resins. These are usually styrene divinyl benzene copolymers which on sulphonation or carboxylation become capable of exchanging their H^+ ions with the cations of the solution. A number of synthetic-resin cation exchangers which contain sulphonated phenolic or aromatic hydrocarbons have been prepared. These resins are called H-form cation exchangers. It is also possible to convert the H-form cation exchanger to Na-form exchanger by using NaCl. Schematically a cation exchanger can be represented as:

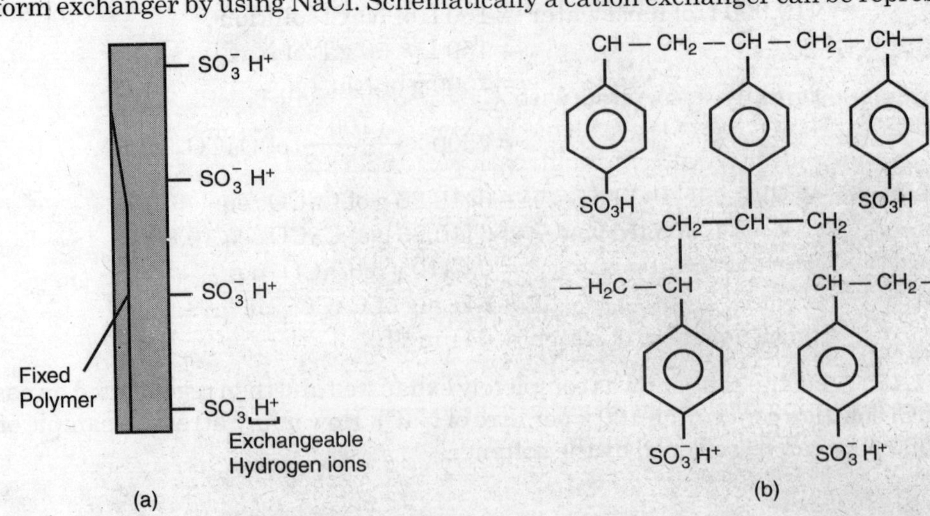

Fig. 1.12 *Representation of Cation Exchanger*

Resins having SO_3H group are termed as *strongly acidic resins* whereas having -COOH group are termed as *weakly acidic resins*. The cation exchange resins can be represented as RH^+.

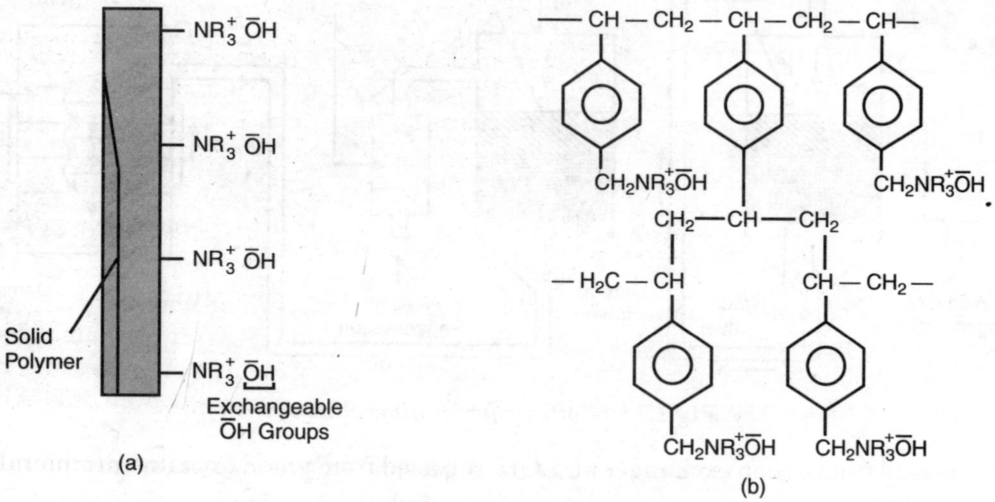

Fig. 1.13 *Representation of Anion Exchanger*

(b) **Anion exchange Resins:** Resin containing basic functional groups like amine, substituted amine or quaternary ammonium groups as their hydroxide salts are termed as anion exchange resins. There are styrene divinyl benzene complexes which, because of their basic functional groups become capable of exchanging their anions with those present in the solution.

Resins having $-NH_2$, = NH group are *weakly basic* whereas those having quaternary ammonium salts are *strongly basic*. Anion exchange resins are represented as $R\bar{O}H$.

Process: The ion exchange unit consists of two tanks. Cation exchange resins and anion exchange resins are placed in first and second tank respectively. First, hard water is made to pass through cation exchange resin which exchanges its H^+ ions for Ca^{2+} and Mg^{2+} ions in water.

$$2RH^+ + CaSO_4 \longrightarrow R_2Ca^{2+} + H_2SO_4$$
$$2RH^+ + MgSO_4 \longrightarrow R_2Mg^{2+} + H_2SO_4$$
$$2RH^+ + CaCl_2 \longrightarrow R_2Ca^{2+} + 2HCl$$
$$2RH^+ + MgCl_2 \longrightarrow R_2Mg^{2+} + 2HCl.$$

The anions originally associated with Ca^{2+} or Mg^{2+} will be associated with H^+ ions. In otherwords HCl, H_2SO_4, H_2CO_3 are formed in amounts equivalent to the anions present in raw water. Thus water after passing through cation exchanger is free of Ca^{2+} and Mg^{2+} ions. This water is then passed through another column having anion exchange resin. Here the anions like SO_4^{2-}, Cl^- etc present in water are exchanged for $\bar{O}H$ ions of the resin.

$$R\bar{O}H + Cl^- \longrightarrow R\bar{Cl} + \bar{O}H$$
$$2R\bar{O}H + SO_4^{2-} \longrightarrow R_2SO_4^{2-} + 2\bar{O}H$$
$$2ROH + CO_3^{2-} \longrightarrow R_2CO_3^{2-} + 2\bar{O}H$$

The water coming out of the anion exchanger is completly free from cations and anions responsible for hardness. It is known as deionized water or demineralized water. It is as pure as distilled water.

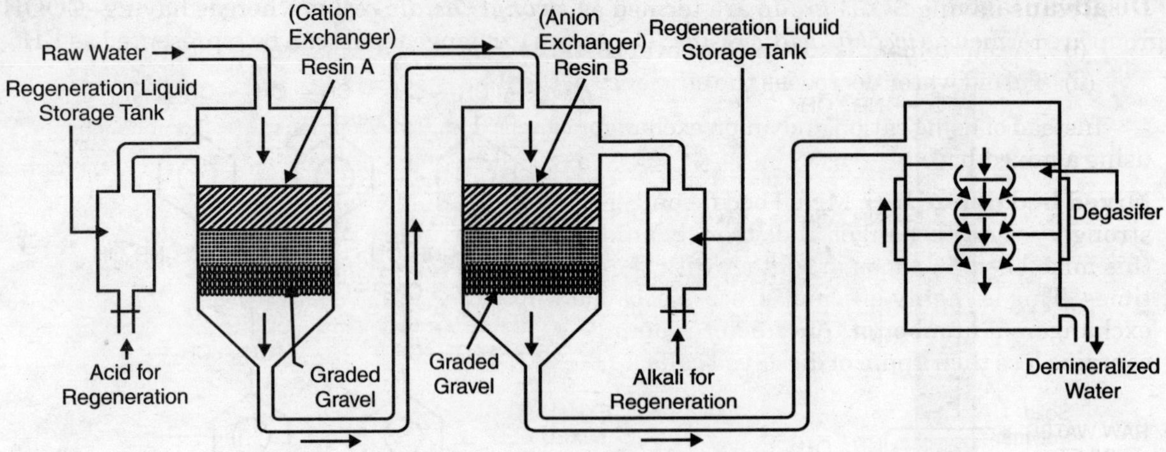

Fig.1.14 *Softener for Demineralization*

H^+ released from cation exchanger and $\bar{O}H$ released from anion exchanger combine to form unionized water.

$$H^+ + \bar{O}H \longrightarrow H_2O.$$

Regeneration: After some time the cation and the anion exchangers get exhausted and stop working.

The cation exchanger is regenerated by first back washing and then passing a 2% H_2SO_4 solution through it. The regeneration reactions can be represented as:

$$R_2Ca^{2+} + H_2SO_4 \rightarrow 2RH^+ + CaSO_4$$
$$R_2Mg^{2+} + H_2SO_4 \rightarrow 2RH^+ + MgSO_4.$$

The column is then rinsed with soft water to remove $CaSO_4$ and $MgSO_4$. However if HCl is used for regeneration, $MgCl_2$ and $CaCl_2$ are formed.

$$R_2Ca^{2+} + 2HCl \longrightarrow 2RH^+ + CaCl_2$$
$$R2Mg^{2+} + 2HCl \longrightarrow 2RH^+ + MgCl_2.$$

HCl is ideally suited for regeneration since $CaCl_2$ and $MgCl_2$ are more soluble.

The anion exchanger is regenerated by back washing and then passing a solution of NaOH. The Na_2SO_4 or NaCl so formed, are rinsed with soft water and washing are led to sink.

$$R_2SO_4{}^{2-} + 2NaOH \longrightarrow 2R\bar{O}H + Na_2SO_4$$
$$R_2Cl^- + NaOH \longrightarrow R\bar{O}H + NaCl.$$

The regenerated exchanger is ready for further use.

Ion exchanger do not remove CO_2 which is removed in degasifier in the deionization process.

Advantages

(i) Highly acidic or alkaline water can be softened.
(ii) Water of very low hardness is produced (about 2ppm)
(iii) Anions as well as cations are removed thereby problems like caustic embrittlement and corrosion are reduced when treated water is used for boiler feed purposes.

Disadvantages

(i) The equipment is costly. Moreover costly chemicals are required for regeneration.
(ii) Turbid water decreases the efficiency of the process.

Instead of using cation and anion exchanger separately, the deminsalization can be carried by using a mixed-bed.

Mixed bed deionizer: Mixed bed deionizer consists of a single column containing a mixture of strongly cation exchanger and strongly anion exchanger. When hard water is passed through this mixed bed, it comes in contact with the two types of ion exchangers alternatively, several times. This is highly efficient process because water comes in contact with cation and anion exchangers a number of times before coming out. The water softened by mixed bed deionizer contains less then 1ppm of dissolved salts.

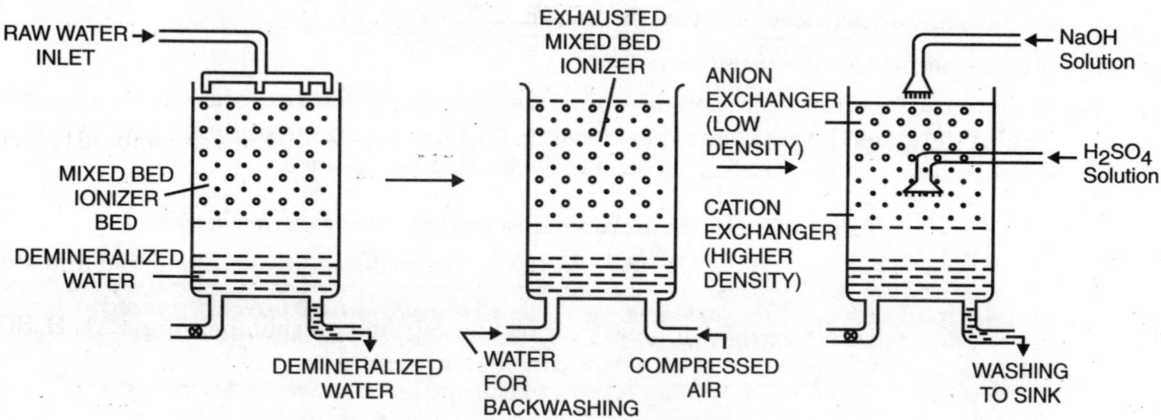

Fig. 1.15 *Regeneration of Mixed-Bed Ion Exchanger*

Regeneration: When the resins are exhausted, the bed is back washed. During washing the anion exchanger being lighter gets displaced above the cation exchanger. This hydraulic separation is highly useful as it makes the regeneration process much easier. The anion exchanger is regenerated by letting in caustic soda solution (NaOH) from the top. The cation exchange resin is regenerated by H_2SO_4 solution. The two beds are then rinsed with soft water. The beds are mixed again by sending a stream of compressed air from below and are used again.

1.14 WATER FOR DOMESTIC PURPOSES (Municipal Water/Drinking Water)

One of the basic question a person asks on arriving at a place is, whether the water available, is safe to drink or not. About 80% of all sickness in the world is attributed to the lack of safe drinking water. It is unfortunate that a large portion of world's population lack a safe supply of water for drinking, washing and other domestic puposes.

As water flows over the surface of earth it picks up a number of impurities in the form of dissolved particles of soil, garbage, sewage, pesticides and other human, animal or chemical wastes. Some of these impurities may make water more useful and potable (water which is safe to drink) and some of these render it harmful and unfit. For example, certain minerals such as iron, calcium, magnesium, flourine etc may be useful in small quantities. But when these materials are present in large amounts or in combination with other impurities, the water may become unfit for human consumption. For example, the water may contain toxic or poisonous substances such as arsenic, barium, cadmium, lead etc which are very harmful even if present in low quantities.

It is clear from the above discussion that water has to be treated to remove various impurities which are objectionable, to make it safe for drinking purposes. Municipal water treatment aim at maintaining an adequate supply of domestic water, confirming to all the required specifications of drinking water.

Specification of Drinking Water

Drinking water standards fall into two categories:

(a) **Primary Standards:** Primary standards specify the maximum contaminant levels (MCL) of various dissolved minerals based on their effect on human health.

(b) **Secondary Standards:** Secondary standards are the standards, which may vary from place to place depending upon the taste, odour, colour and hardness of water etc and do not have any anticipated effect on the health.

The common standards for drinking water are

(i) It should be free from objectionable minerals which have deterimental effect on health. Table (1.9) gives the maximum permissible limit i.e. maximum contaminant levels for certain chemicals and their ill effect if these limits are exceeded.

Table 1.9: *Maximum contaminant levels for certain chemicals and their effect on human health*

Contaminant	*Permissible Limit (mg/L)*	*Principle health effect, if exceeded beyond permissible limit*
Arsenic	0.050	Dermal and nervous system toxicity effects.
Barium	1.000	Circulatory system effects.
Cadmium	0.010	Kidney is effected.
Chromium	0.050	Liver and kidney are effected
Flouride	1.000	Flourosis.
Lead	0.050	Central and peripheral nervous system damage, kidney effects, highly toxic to infants.
Mercury	0.001	Central nervous system disorders.
Nitrite and Nitrate	45.000	Methaemoglobinemia (blue baby syndrome)
Selenium	0.010	Gastrointestinal effects.
Silver	0.050	Skin discolouration.
Sulphate	200.000	Gastrointestinal irritation.

(ii) pH Value: It should have pH value ranging from 6.5 to 8.5. Beyond this range, the water will effect the mucous membrane.

(iii) Total Dissolved Solids: Total dissolved solids should be less than 500ppm.

(iv) Pathogenic bacteria: It should be free from disease causing bacteria. Coliform bacteria (Escherichia coli) are used as indicator organisms whose presence suggests that the water is contaminated. Water should contain one coliform per 100ml.

(v) It should be free of iron and manganese, the permissible limit being .3 mg/L and 0.05mg/L respectively. They change the taste of water and also are responsible for producing stains on clothes in laundry.

(vi) It should be non corrosive.

(vii) It should be free from objectionable dissolved gases like H_2S.

(viii) It should be sparkling clear, odourless and should have pleasant taste.

(ix) It should not be turbid. The permissible turbidity is less than 10 ppm.

WATER TREATMENT FOR DOMESTIC/MUNICIPAL PURPOSE

The treatment of water for domestic supply will depend to a large extent on the quantity of the source water. A typical treatment plant for surface water might include the following sequence of steps:

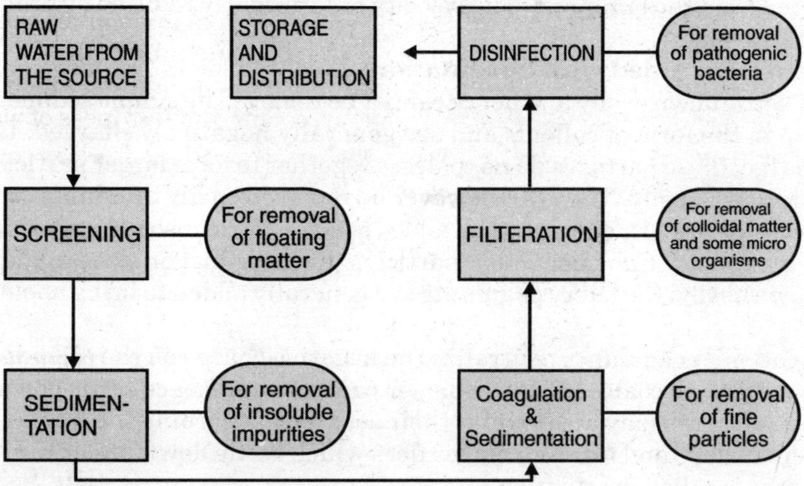

Fig. 1.16 *Schematic Representation of a Typical Water Treatment Plant for Surface Water*

1. **Screening:** Screening is the process by which the floating material like wood pieces, leaves etc are removed. Raw water is made to pass through large screens having number of holes which hold back the floating matter and allow the water to pass.

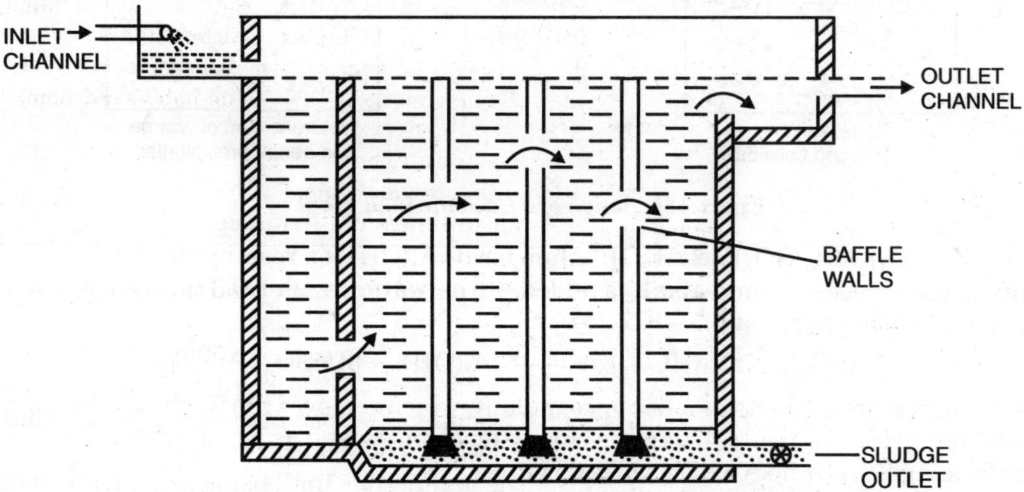

Fig. 1.17 *Horizontal Flow Sedimentation*

2. **Sedimentation:** Sedimentation is the process of allowing water to stand undisturbed in big tanks for sufficient period of time, so that the suspended particles settle down at the bottom. The detention period may vary from 2 to 8 hours. The process of sedimentation is

carried out in continous flow type tanks, in which the flow of the water is slowed down, so that substances held up by turbulence of fast moving water, settle at the bottom under the force of gravity. The basin in which the flow of water is retarded is called the settling tank or sedimentation tank. Thus in such tanks, the flow of water is reduced and water is not brought to complete rest. Fig (1.17) shows a horizontal flow sedimentation tank. It consists of a rectangular tank. It is provided with baffle walls to reduce the velocity of the incoming water. The settled sludge is taken out through the outlet at the bottom.

The working of continous flow type of tanks is simple. Water enters from one end and is collected from the other end. The velocity is sufficiently reduced by providing sufficient length of travel.

3. **Sedimentation Aided with Coagulation:** Finely divided silica, clay and organic matter does not settle down easily and hence cannot be removed by simple sedimentation. Most of these are in the form of colloids and are generally negatively charged. Because of their same charge, these particles do no coalesce together to form larger particles and hence do not settle down easily. They can however be removed easily by adding certain chemicals called **coagulants**. Coagulants when added to water, form insoluble gelatinous precipitates which entrap very fine suspended particles, forming bigger sized agglomerates which settle down easily. Cationic coagulants are generally added to bring about coagulation in water.

The cationic coagulants neutralize the negative charge on the colloids which there by come together and coalesce to form bigger particles and hence settle down.

The coagulants may also exhibit their action by adsorbing the fine colloidal particles over their surface and forming bigger flocs which settle down under the force of gravity. The process is called flocculation.

The most commonly used coagulants in water treatment plants are Alum, sodium aluminate and ferrous sulphate.

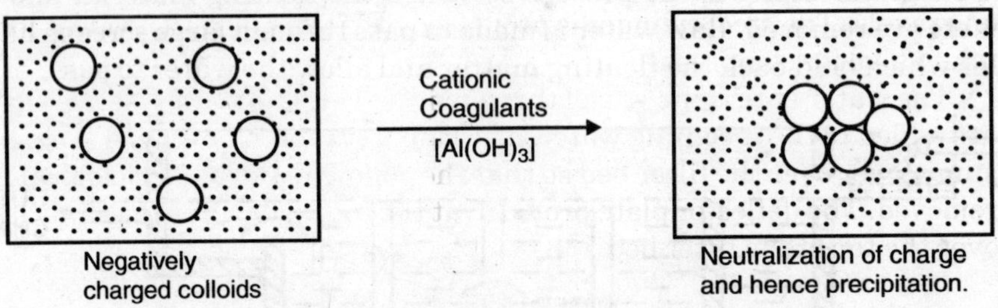

Fig. 1.18 *Action of a Cationic Coagulant*

Alum [$K_2(SO_4)_3$. $Al_2(SO_4)_3$. $24H_2O$]: Alum forms $Al(OH)_3$ on reacting with water in which alkalinity is also present. Some alkali is added, if the water is not sufficiently alkaline. It is mostly used in the pH range of 6.5 – 8.5

$$Al_2(SO_4)_3 + 3Ca(HCO_3)_2 \longrightarrow 2Al(OH)_3 + 3CaSO_4\downarrow + 6CO_2$$

The $Al(OH)_3$ so formed acts as a floc or coagulant, which removes the finely divided particles or colloidal impurities by neutralizing the charge on them as well as by adsorbing the colloidal particles forming heavier flocs which settle down.

Sodium Aluminate [$NaAlO_2$]: For treament of acidic waters sodium aluminate can be used. It works in the range of 5.5 – 8.0. It is costlier than alum. The reactions taking place are-

$$NaAlO_2 + 2H_2O \longrightarrow Al(OH)_3\downarrow + NaOH.$$

$Al(OH)_3$ acts in the same way as discussed above and NaOH react with $MgSO_4$ forming Mg $(OH)_2$.

$$2NaOH + MgSO_4 \longrightarrow Mg(OH)_2\downarrow + Na_2SO_4.$$

Ferrous Sulphate: [$FeSO_4$. $7H_2O$] This is used when water is slightly alkaline, in the pH range > 8.5. If water is not alkaline, the pH must be raised by adding [$Ca(OH)_2$].

$$FeSO_4 + Mg(HCO_3)_2 \longrightarrow Fe(OH)_2 + MgCO_3 + CO_2 + H_2O$$
$$4Fe(OH)_2 + O_2 + 2H_2O \longrightarrow 4Fe(OH)_3.$$

The raw water is thoroughly and rapidly mixed with the required chemicals. It is then led to the flocculation basin where it is gently agitated. During this, the precipitating Al $(OH)_3$ attracts colloidal particles, forming flocs.

After flocculation, the water flows through a sedimentation basin. Solids that collect on the bottom of the tank may be removed manually by periodically shutting down the tank and washing out the collected sludge or it may be mechanically cleaned using a bottom scraper.

4. **Filteration:** The effluent from the tank is then filtered. Filteration is the process of removing insoluble, colloidal and bacterial impurities from water by means of filters. One of the most widely used filteration units is the rapid-sand filter. Though slow-sand filters are also used but have become obsolete these days.

   Rapid sand filters are of two types-

   (a) Rapid gravity filters

   (b) Rapid presssure filters.

   (a) **Rapid Gravity Filters:** It consists of a large rectangular tank made up of concrete. The filtering medium is a layer of coarser sand which is supported by a bed of graded gravel. A layer of fine sand is carefully seived at the top of coarser sand. To collect the filtered water, steel cylinders are used as pressure vessel, containing filtering media consisting of the layers of fine sand followed by coarse sand, which inturn is supported on gravels. The steel cylinder is provided with inlet and outlet valves for water and an inlet valve for compressed air.

   The water enters the vessel through inlet valve, under pressure. Often coagulants are added to the incoming water, after it enters the filter vessel. This helps to form a slimy layer over the filter bed so that the colloidal and bacteriological impurities are removed. The deflector plate provided at the top distributes the slimy layer uniformly over the top of the filter bed.

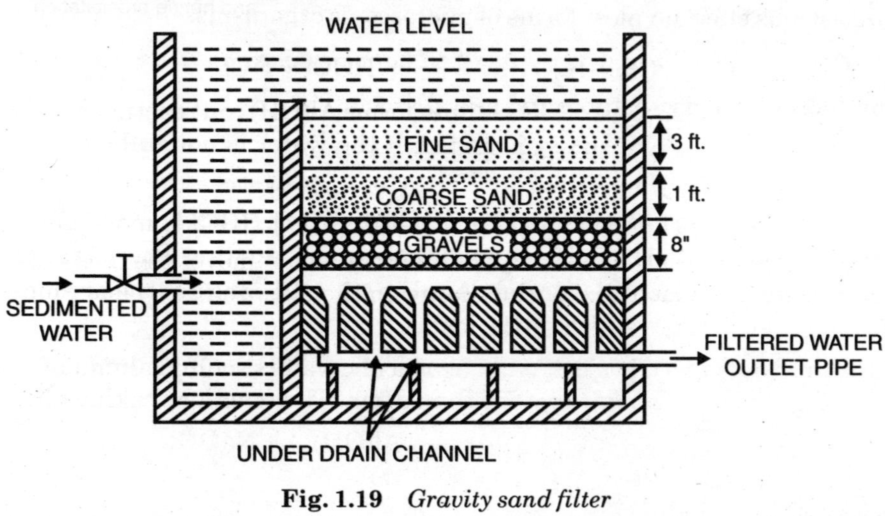

Fig. 1.19 *Gravity sand filter*

The cleaning of the filter is carried out by back washing as is done in gravity filter. Before backwashing, compressed air is passed to agitate the sand. Compact automatic pressure filters, in which cleaning is done at a specific interval of time are available now a days. Water, drains are provided at the bottom. (Fig. 1.19)

Sedimented water entering the sand filter is uniformly distributed over the entire fine sand bed. After passing through the filter beds, the filtered water is taken out from the drain system provided at the bottom. When the filter bed is clogged by the particles removed from the water, it has to be cleaned. When the rate of filteration becomes slow, the working of the filter is stopped. The filter bed is cleaned by backwashing i.e by reversing the flow of water through the filter. The top most layer of sand is scrapped off and replaced with clean sand. The rapid gravity filters get clogged very frequently and have to be washed every 24 to 48 hours.

(b) **Rapid Pressure Filters:** Pressure filters are similar to rapid gravity filters, placed in closed vessels. Water is passed through pressure filters at a pressure greater than the atmospheric pressure, it is necessary that these filters are located in air tight vessels. The raw water is pumped into the vessels by means of pumps.

5. **Disinfection:** During coagulation, settling and filteration, practically all of the suspended solids, most of the colour, and a few percent of the bacteria are removed. The final step is disinfection to kill the remaining pathogenic organisms, (disease causing organisms).

The process of destroying these pathogenic bacteria and micro organisms is known as sterilization or disinfection. The chemicals and the substances used for disinfecting are known as disinfectants.

The process of purification is the most important, because the presence of pathogenic bacteria may lead to spread of various diseases and their epidemics, thus causing disasters to public life.

The disinfection of water can be carried out by the following methods:

(i) **Chlorination:** Chlorination in its various forms is almost universally used for disinfecting domestic water supplies. This is because it is cheap, reliable and easy to handle. Moreover it is capable of providing residual disinfecting effects for long periods and thus prevents future recontamination of water.

Disinfecting Action of Chlorine

When chlorine is added to water, it forms hypochlorous acid or hypochlorite ions which have immediate disastrous effect on most forms of microscopic organisms.

$$Cl_2 + H_2O \xrightarrow{pH > 8} HOCl + HCl$$

The hypochlorous acid is unstable and may dissociate as

$$HOCl \underset{pH > 7}{\overset{pH > 8}{\rightleftharpoons}} H^+ + \overline{O}Cl$$

Hypochlorite ions

The above reaction is pH dependent. The dissociation of HOCl is most effective at higher pH values and vice versa. At pH < 7, HOCl exist without dissociating whereas at pH<5, Cl_2 does not react and remain as elemental chlorine. At pH > 10, $\overline{O}Cl$ ions are the predominant specie. (Table 1.20)

Table 1.20

pH	*Predominant specie*
< 5	Cl_2
> 5 < 7	HOCl
> 10	$\bar{O}Cl$

All the three forms viz. HOCl, $\bar{O}Cl$, Cl_2 existing in a sample water are termed as **free chlorine.** Out of these three forms the HOCl is found to be the most destructive according to Enzymatic hypothesis given by Gleen and stumpt.

For this reason, *the* pH *value of water during chlorination is generally maintained slightly less than* 7 *so as to prevent the dissociation of* HOCl.

According to the enzymatic hypothesis the HOCl reacts with the enzymes in the cells of the bacteria which are essential for their growth and other metabolic processes. Thus due to deactivation of enzymes, in the cells of organisms by HOCl, the bacteria get killed and no multiplication of micro organisms takes place. HOCl is about 80 times more destructive to bacteria than $\bar{O}Cl$ ions.

Moreover the HOCl will immediately react with ammonia likely to be present in water to form various chloramines as

$$NH_3 + HOCl \longrightarrow NH_2Cl + H_2O (pH > 7.5)$$
$$NH_2Cl + HOCl \longrightarrow NHCl_2 + H_2O (pH = 5 \text{ to } 6.5)$$
$$NHCl_2 + HOCl \longrightarrow NCl_3 + H_2O (pH < 4.5)$$

In the usual chlorination process, the pH is kept slightly less than 7 i.e. around 6.5 and hence dichloramine is the predominant specie.

These chloramines so formed are stable and are found to possess disinfecting properties. They act as chlorine reserves. The chlorine in this form is called combined chlorine. As compared to the free chlorine, combined chlorine is less effective.

Break Point Chlorination: The amount of chlorine required for disinfecting water depends upon the inorganic and organic impurities present in water. When chlorine is added to a sample of water and its amount is estimated after a few minutes, it is found that the available chlorine is not equal to the amount of chlorine added. A relationship between the amount of chlorine added to water and the free residual chlorine (determined experimentally) is shown in Fig.1.20. The curve shown in the fig can be divided into four stages.

I stage: It is seen from the figure that for lower doses of Cl_2 all the chlorine added is consumed and there is no residual chlorine. This is due to the fact that initially Cl_2 reacts with the inorganic impurities present in water forming chlorides. These chlorides do not have any residual oxidizing power.

II stage: As the amount of chlorine is increased, there is steady increase in the amount of residual chlorine. This stage

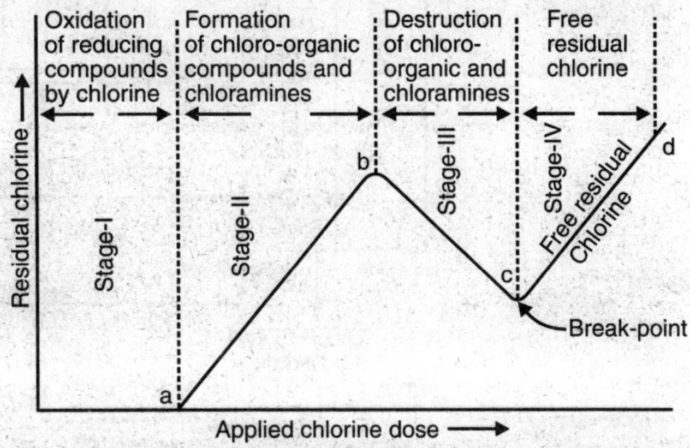

***Fig. 1.20** Break-point chlorination curve*

corresponds to the formation of chloramines i.e chlorine is consumed by NH_3. But chloramines respond to the test for estimation of chlorine in the same way as free chlorine and hence curve represents the chlorine available from combined residuals.

Chloramines remain in water body for longer periods which help to kill the micro organisms even in a distribution system. Simultaneously this stage also corresponds to the formation of chloro organic compounds without oxidizing them. These are responsible for bad odour and unpleasant taste in water.

III stage: **When the dosage of chlorine is further increased, the complete oxidation of chloramines and chloro-organic compounds take place utilizing the chlorine and accordingly free residual chlorine also decreases and reaches a dip when the oxidative destruction is complete. This dip is known as break point. The chlorination upto breakpoint ensures complete destruction of organic compounds which give unpleasant taste and bad odour.**

IV stage: **After breakpoint, the amount of chlorine added is not used in any reaction and the residual chlorine agrees fairly well with the quantity of chlorine added. Thus free chlorine is available only if the demand for chlorine by other reacting substances present in water is met with.**

Thus break point chlorination of water may be defined as the chlorination to such an extent that micro organisms as well as bad tastes and odours present in water are also destroyed.

Chlorination after break point increases the free residual chlorine (Cl_2, HOCl, $\overline{O}$Cl). The free chlorine as well as the combined chlorine (chloramines) will cause germicidal action on micro organisms. Free chlorine will instantaneously kill the pathogens, while the combined chlorine will provide long term germicidal effect. Hence to use chlorine as an effective disinfectant, the chlorine dosage has to be slightly more than the break point. In general, water is satisfactorily disinfected if the free chlorine residual is about 0.2ppm i.e. the chlorine added in the dosage upto break point + 0.2 ppm. Amount greater than 0.2ppm of free residual chlorine causes bad taste in water and is harmful as it irritates the membrane of intestine.

Various forms in which Chlorine can be Applied: Chlorine is generally added to water in the following forms:

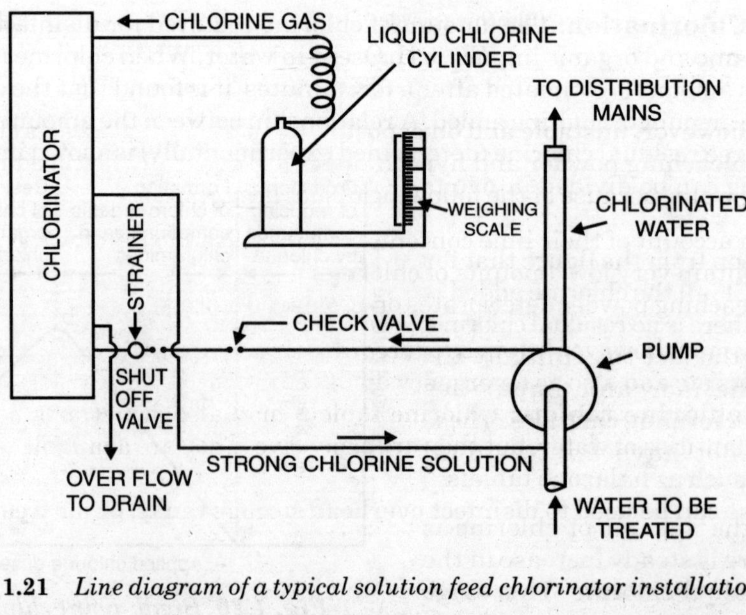

Fig. 1.21 *Line diagram of a typical solution feed chlorinator installation*

(a) Chlorine gas/liquid: The molecular chlorine can be applied to the water in liquid or gaseous form, though liquid form is generally preferred these days. The liquid chlorine is 1½ times heavier than water and is generally kept in steel cylinder called chlorinator. Chlorinators using liquid chlorine, take the chlorine form the container, mix it with a small amount of water to form a strong chlorine solution. The chlorine solution is fed into distribution system by means of a water injector (Fig. 1.21).

Advantages: The liquid chlorine is now-a-days invariably and universally used because of the following advantages-

(i) Can be easily stored for long periods, without any risk of deterioration.
(ii) Cheap and easily available.
(iii) Occupies less space for storage.
(iv) No sludge is formed on its application.
(v) Very powerful disinfectant and may remain in water as residual for enough time, especially if NH_3 is also present in water.

(b) Bleaching powder/Hypochlorites: Calcium and Magnesium hypochlorites may be used for chlorinating small public supplies. When hypochlorite is dissolved in water, it forms hypochlorite ions

$$\underset{\text{Calcium hypochlorite}}{Ca(OCl)_2} \xrightleftharpoons{H_2O} Ca^{2+} + 2OCl^-$$

OCl^- may combine with hydrogen ions present in water.

$$OCl^- + H^+ \underset{pH > 8}{\overset{pH > 7}{\rightleftharpoons}} \underset{\text{(Germicide)}}{HOCl.}$$

HOCl as well as OCl^- cause disinfectioning of water. Commercially available calcium hypochlorite called Hig Test Chlorite or Perchloron, contains 60% to 70% of available chlorine.

Another compound used for chlorination is bleaching powder. When freshly prepared, it contains 30% of available chlorine.

$$\underset{\text{(Bleaching powder)}}{CaOCl_2 + H_2O} \longrightarrow Ca(OH)_2 + Cl_2.$$

$$Cl_2 + H_2O \longrightarrow \underset{\text{(Germicide)}}{HOCl + HCl}$$

It is however, unstable and on exposure to air and light loses its chloride content.

But bleaching powder and hypochlorites are generally not used in modern days for treating water because of the following *disadvantages*;

(i) On account of their lime content, they increase the pH of water and make it hard.
(ii) Contain very low amounts of chlorine.
(iii) Bleaching powder deteriorates on prolonged storage.

They may, however be used for treating small supplies for swimming pools, storage reservoirs etc and also as emergency disinfectants.

(c) Use of chlorine tablets: Chlorine tablets may also sometimes be used to disinfect small quantities of water, but they are expensive. They are available under various trade names, such as halazone tablets.

These can be used to disinfect overhead storage tanks, or for treating ground water supplies.

(d) Use of Chloramines: Chloramines are quite stable. When chlorine and NH_3 are mixed in the ratio of 2:1 by volume, chloromines are formed.

$$Cl_2 + NH_3 \longrightarrow ClNH_2 + HCl.$$

Due to their stability, they remain in water as residuals for a sufficient time. (compared to chlorine which evaporates after some time). They thus provide a greater safeguard against recontamination. However, they are weaker disinfectants compared to free chlorine.

The **advantages** of using chloramines are-

(i) They do not cause bad odour and bad tastes even used in excess.

(ii) Very useful when phenols are present. Free chlorine reacts with phenol to form chlorophenols which impart unpleasant taste to water. But chloramines do not form chlorophenols and hence no such tastes are developed.

(e) Use of ClO_2 Gas: It is very effective and powerful disinfectant, about 2½ times stonger than chlorine. It is produced by passing chlorine gas through sodium chlorite.

$$2NaClO_2 + Cl_2 \longrightarrow 2NaCl + 2ClO_2.$$

It is costly and unstable and has to be used immediately after its production. However, since the action of ClO_2 remains unaffected by high pH values (6 to 10), it can be used for treating highly alkaline waters.

Dechlorination: Overchlorination i.e > .2ppm after the break point produces unpleasant odour and tastes in water. Dechlorination means removing the chlorine from water. The dechlorination process may either be carried out to such on extent that sufficient residual chlorine .1 to .2ppm do remain in water. Otherwise additional chlorine will generally be added to maintain such residues.

Dechlorination is done by using dechlorinating agents such as SO_2 gas, $Na_2S_2O_3$, Na_2SO_3 etc. The reactions are-

$$SO_2 + Cl_2 + 2H_2O \longrightarrow H_2SO_4 + 2HCl$$

$$Na_2SO_3 + Cl_2 + H_2O \longrightarrow Na_2SO_4 + 2HCl.$$

Dechlorination can also be brought about by passing the treated water through bed of molecular carbon or by adding activated carbon to water which is then removed by filteration.

(ii) **Disinfection by Ozone:** Ozone gas is an excellent disinfectant. It can be produced by passing a high tension electric current through a stream of air in a closed chamber.

$$3O_2 \longrightarrow 2O_3$$

It is unstable and readily decomposes to give nascent oxygen.

$$O_3 \longrightarrow O_2 + [O]$$

The nascent oxygen so produced is a powerful oxidizing agent and removes the organic matter and kills the bacteria.

During the treatment, ozone gas is manufactured and bubbled through water contained in a sterilizing chamber having inlets and outlets for water.

Separate inlet for ozone is also provided. The sterilized water is removed from the top.

Ozone and chlorine are quite competative disinfectants, each having its advantages and disadvantages. Chlorine is however preferred over ozone as it is cost effective. Ozone is quite expensive but still it is used in some countries.

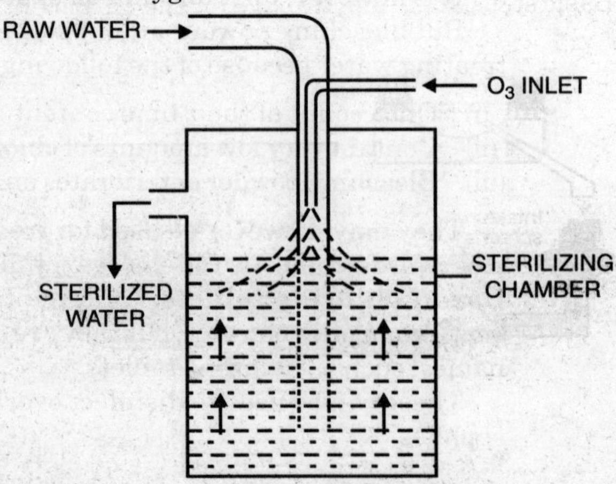

Fig. 1.22 *Ozone Sterilizer*

Advantages of using Ozone:

- Ozone removes the colour, taste and odour from water, in addition to killing bacteria.
- Ozone being unstable does not leave any residue.
- The excess of ozone is not harmful as it decomposes into oxygen.

Disadvantages

- It is very expensive and hence cannot be used for disinfecting municipal water supply.
- No residuals can be maintained and hence no safety against recontaminations.
- Cannot be supplied in cylinders but require, an apparatus ozoniser, to be installed at the treatment plants.

(iii) **Ultra-Violet Rays:** Ultraviolet rays (uv) are found in sun light but can be produced by passing electric current through mercury enclosed in quartz bulbs. Mercury vapour lamps enclosed in quartz bulbs can therefore be used as a good source of ultraviolet rays. These rays are highly effective in killing all types of bacteria. For treatment, the water is passed several times around the quartz bulbs emitting uv rays. These rays can penetrate upto 10cm of distance and hence the depth of water over the bulb should not exeed 10cm. Disinfection with ultra-violet rays does not impart any additional taste or odour to the water as no chemicals are added. But, this is very costly method and is unsuitable for treating large supplies. However, this method is used for treating small quantities of water at places where cost is a minor factor.

(iv) **Potassium Permanganate:** This is used as a disinfectant where water is contaminated with lesser amount of micro organisms. It is added in small doses such as 0.05 to 0.10ppm even to filtered and chlorinated water. It is a popular disinfectant for disinfecting well water in villages. The addition of $KMnO_4$ produces pink colour. However, if pink colour disappears, it shows that more of organic matter is present in water and more quantity of $KMnO_4$ should be added.

$KMnO_4$ is cheap and quite easy to use. But it cannot remove all the bacteria from water. This method of disinfection is also, therefore not recommended except for rural areas, where the water is contaminated with lesser amount of bacteria.

(v) **Boiling:** Water for domestic purposes on a smaller scale may be sterilized by boiling the filtered water for about 20 minutes. This method is very costly and hence can be used only in individaul cases.

Apart from the above methods iodine and bromine tablets are also added to water to kill bacteria. Basic steps, mentioned above for the treatment of municipal water are shown in Fig. (1.23).

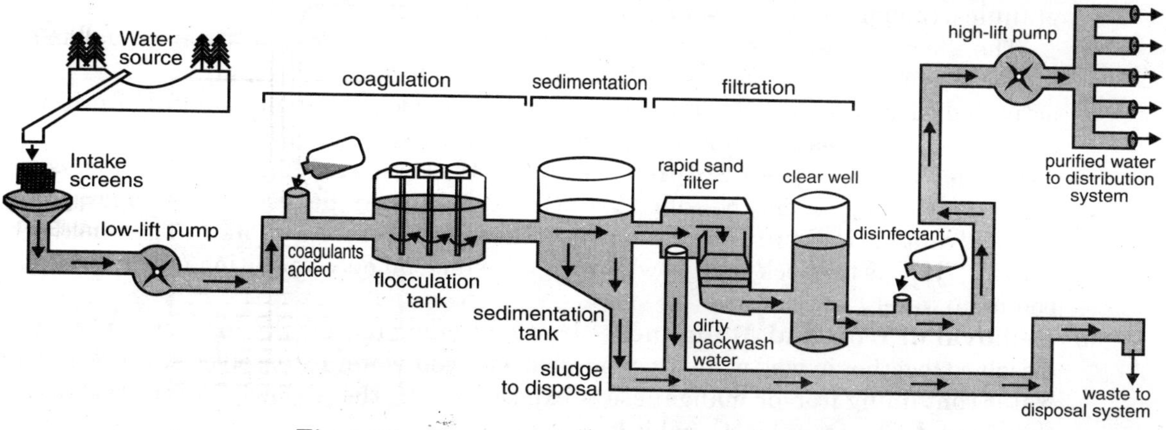

Fig. 1.23 *Treatment of water for drinking purpose*

Besides these normal treatment methods, certain other special treatments are sometimes required in order to remove the minerals, tastes odour and colours etc from water. Some of them are briefly discussed below:

(a) **Aeration:** Under the process of aeration, water is brought in intimate contact with air so as to absorb oxygen and to remove CO_2. It also helps in killing bacteria to a certain extent. The aeration of water can be carried out with the help of spray nozzles. In this method, water is sprinkled in air or atomosphere through special nozzles which breaks the water into droplets, thus permitting the escape of dissolved gases and absorption of oxygen.

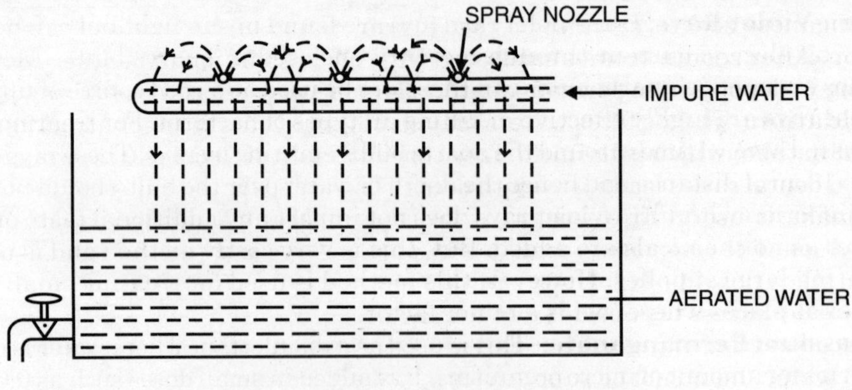

Fig. 1.24 *Aeration through Spray Nozzles*

(b) **Treatment with Activated Carbon:** Activated carbon is specially treated carbon, which possess the property of absorbing and attracting impurities, such as gases, liquids and finely divided solids. It is widely used for removing tastes and odours from public supplies.

Activated carbon is prepared by heating or charring wood or saw dust at about 500°C in a closed vessel and then slowly burning (oxidizing) it under controlled conditions at 800°C. Thus all the hydro carbons, which may interfere with the absorption of organic matter are removed and the residue is called the activated carbon. It is highly porous and absorptive. The activated carbon is mostly used in the powdered form and is added to water before the filteration.

With the passage of time, carbon looses its absorptive power and has to be rejuvenated. During regeneration, the contaminants are usually burned from the surface of the carbon granules, though in some cases a solvent is used to remove them.

The effectiveness of activated carbon is directly related to the surface area available to attract the molecules or particles of contaminants. The most commonly used is granular activated carbon, which has an enormous surface area. A single handful of granular activated carbon has an internal surface area of about one acre.

(c) **Treatment with Copper Sulphate:** $CuSO_4 . 7H_2O$ helps to remove colours, tastes and odours from water. It may be added to the treated water in the distribution pipes, but its main advantage is obtained by adding it to the open reservoirs or lakes. The function of $CuSO_4 . 7H_2O$ is to check the growth of algae, which on decomposition may impart bad odours to water.

(d) **Removal of Iron and Manganese:** Iron and manganese salts are generally found dissolved together in well water or anaerobic reservoir water in dissolved state. When the water containing iron or manganese is exposed to air, these slowly are converted into their oxides. They are objectionable because-

- They cause formation of stains on cloths washed with such waters due to the deposition of red/brown coloured oxides of iron/manganese.
- They cause incrustation of water mains due to deposition of their oxides.
- They impart metallic taste to water.

If present in small quantities, they can be removed during aeration, followed by coagulation, sedimentation and filteration. The soluble ferrous and manganous compounds present in water get oxidized to ferric and manganic compounds which are insoluble and hence can be sedimented out easily.

Softening by ion exchanges removes iron but is not recommended because if Fe is oxidized, it poisons the resins.

Manganese zeolite, can also be used for removing soluble iron and manganese from solution. After zeolite bcomes saturated with metal ions, it can be regenerated by back washing with potassium permanganate.

(e) **Flouridation:** The surfaces waters may sometimes contain lesser amounts of flourides present in them which is harmful. Low concentration of flouride (less than 1mg/L) may result in dental caries leading to tooth decay. Additional flouride may be added to water so as to make it useful for teeth. This process of adding flouride compounds is called flouridation. The compounds used to bring about flouridation are sodium flouride (NaF), sodium silica flouride (Na_2SiF_6), hydroflousilicic acid (H_2SiF_6) etc.

However, excess of flouride (>1.5 to 2mg/L) is also harmful because it causes spotting and discolouration of teeth. Therefore the dose has to be carfully worked out.

Soft water should not be supplied for drinking purposes. Softened water increases the potential for leaching metal from pipes, solder and plumbing fixtures.

$$2Pb + 2H_2O + O_2 \rightleftharpoons 2Pb(OH)_2$$

soft water attacks the lead used in plumbing.

The solubility of lead in water is diminished in the presence of HCO_3^- ions. If water used for drinking purposes contains about 85 ppm of $CaCO_3$ equivalent hardness.

1.15 DESALINATION OF BRACKISH WATER

Water containing high concentrations of dissolved solids is known as brackish water. Since brackish water have a peculiar saltly taste, it is therefore completely unfit for drinking. Depending upon the amount of dissolved solids, water is graded as

(i) Fresh water (having < 1000 ppm of dissolved solids)
(ii) Brackish water (having > 1000 ppm but < 3500ppm of dissolved solids)
(iii) Sea Water (having > 35000 ppm of dissolved solids)

The process of removing dissolved solids (NaCl) from water is known as desalination process. It is well known fact that with the development of civilization, our fresh water demand is increasing day by day, while the supply remains almost the same. Therefore, it becomes evident that the available fresh water is becoming scarce. This problem, can be solved by converting sea water into fresh water through desalination.

Desalination is a costly process. The water obtained by desalination proves to be much costlier than the naturally available treated water. Inspite of high cost, desalination has been used at various places around the world.

Methods of Desalination: The various methods generally adopted for the conversion of salt water into fresh water are-

1. Distillation 2. Freezing
3. Electro dialysis 4. Reverse osmosis.

1. **Distillation:** Brackish water is boiled in huge evaporators and the vapours are condensed to get pure water. This method of obtaining pure water is expensive for common practice.
2. **Freezing:** This method is based on the principle that when salt water freezes, the ice formed in the beginning is almost free from salt. The ice crystals are separated, washed and melted to get pure water. The quantity of water obtained is satisfactory, but the cost of production is high.
3. **Electrodialysis Method:** This is one of the important method of desalination of water. It is based on a standard technique of removal of ions from solution which may be applied at room temperatures. This is based on the fact that the ions present in saline water migrate towards oppositely charged electrodes through ion selective membranes under the influence of an applied electric field.

A schematic diagram of a selective anion exchange membrane is shown in Fig. 1.25:

The electrodialysis cell consists of alternate cation and anion-permeable membranes ie the membranes which allow to pass either positive or negative ions but not both. In an anion exchange membrane anions are exchanged for Cl^- ions in solution but the membrane is not permeable to Na^+ ions as these are repelled by the fixed positive charges inside the membrane (Fig. 1.25). In electrodialysis cell cathode is placed near cation-permeable membrane and anode is placed near anion-permeable membrane. When a strong electric field is applied to the saline water, the sodium ions (Na^+) start moving towards cathode, and chloride ions (Cl^-) start moving towards anode through the membranes. As a result, the concentration of the ions decreases in compartments 3,5 and it increases in the compartments 2,4,6 as shown in the figure 1.26. The desalinated water, free of ions is collected separately whereas the concentrated water in the compartments 1,3,5 etc is replaced from time to time by next lot of saline water.

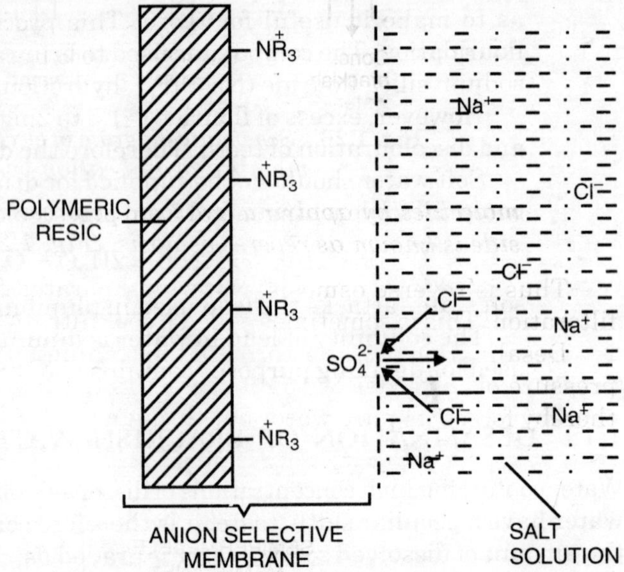

Fig. 1.25 *Schematic Diagram of Anion Exchange Membrane*

Advantages

(i) When very high purification is not required this method is best suited.

(ii) It is a compact unit which is easy to operate.

(iii) The cost of installation of plant is economical.

4. **Reverse Osmosis:** When two solutions having different concentrations are separated by a semipermeable membrane, flow of solvent molecules from the region of low concentration to high concentration takes place, untill the concentration becomes equal on the both sides. The phenomenon is called osmosis. Osmosis is thus the movement of water molecules from pure water to the solution side. The driving force in this phenomenon is called osmotic pressure. Water can also be made to flow in the reverse direction i.e from solution side to the pure water by applying a hydrostatic pressure in excess of osmotic pressure, on the concentrated or solution side. *This process of reversing the flow of solvent*

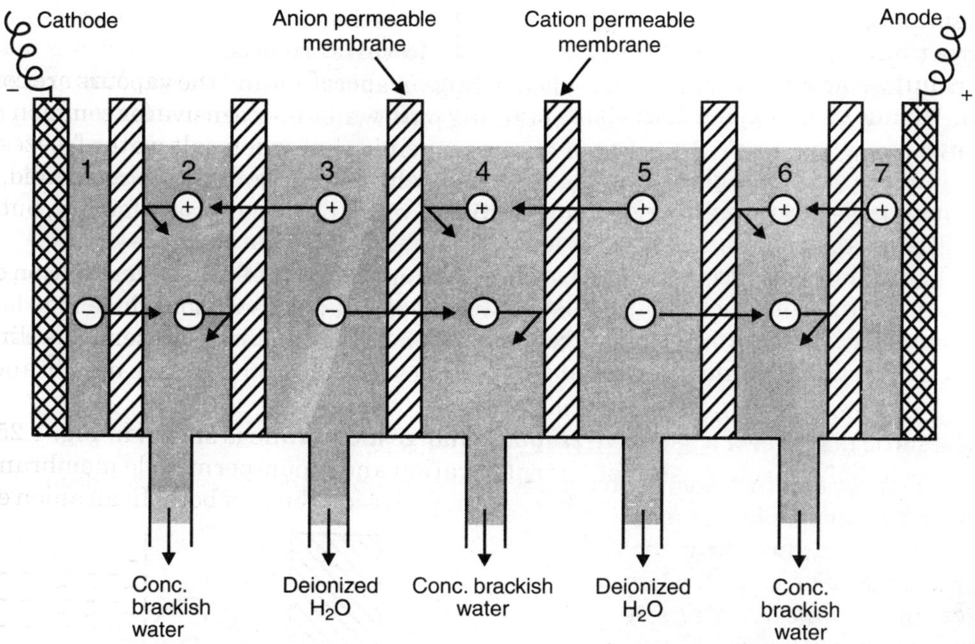

Fig. 1.26 *Schematic diagram of electrodialysis cell: anion permeable membrane and cation permeable membrane.*

molecules by applying hydrostatic pressure in excess of osmotic pressure, on solution side is known as reverse osmosis. (Fig. 1.27 b).

Thus is reverse osmosis, water is separated from the dissolved salts through membrane filteration. This is sometimes called hyper filtration or super filtration.

Desalination of water through the technique of reverse osmosis involves the application of pressure of the order of 15-40 $kgcm^{-2}$ to the impure water (sea water/brackish water) which thereby forces its pure water out through the semipermeable membrance. A major problem in

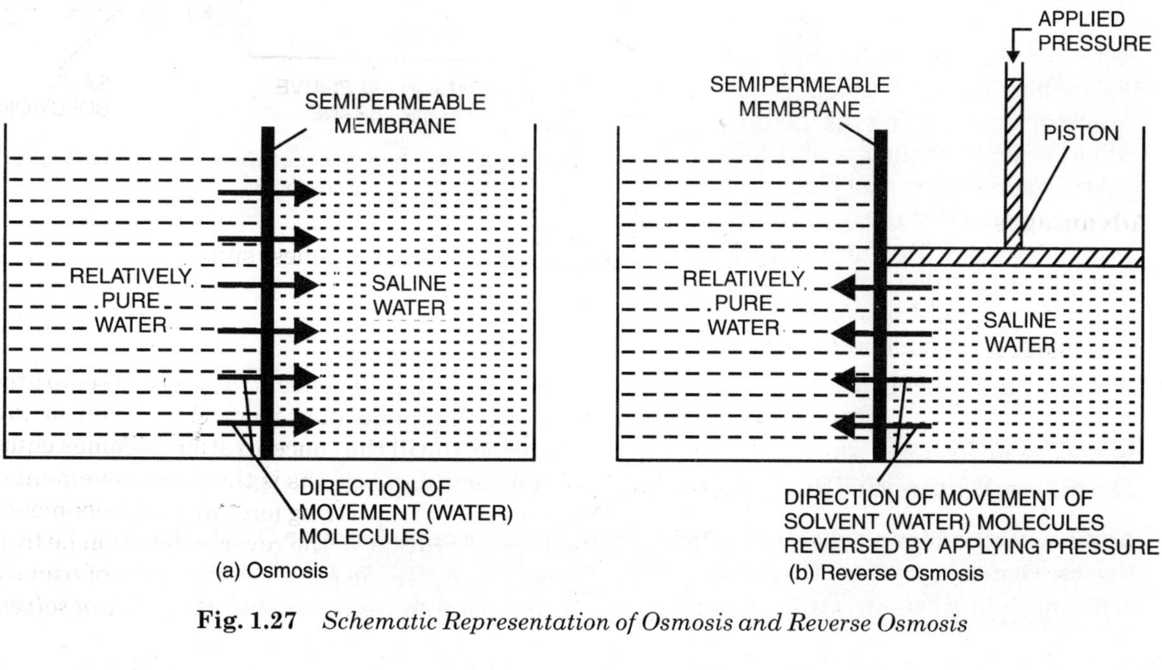

Fig. 1.27 *Schematic Representation of Osmosis and Reverse Osmosis*

reverse osmosis is to find membranes strong enough to withstand the high pressure required and at the same time impermeable to the solute particles dissolved. Synthetic membranes made up of nylon or cellulose acetate have been developed to meet these requirements. More recently the membranes made up of polymethacrylates and polyamide polymers are being used.

Advantages

(i) Purification through reverse osmosis removes ionic as well as non ionic, colloidal and high molecular weight organic matter.

(ii) The energy requirement is 30% lower than that of distillation process.

(iii) The maintenance cost includes mainly the replacement of semipermeable membrane the life time of which is quite high, extending to about 2 years.

Thus purification of water through reverse osmosis is gaining importance because of its simplicity, high reliability and low capital cost.

DOMESTIC ULTRA VIOLET PURIFICATION SYSTEM

The use of ultra-violet (uv) rays is now a days, a common method adopted for disinfection of drinking water at homes, offices, schools and various organizations. Excellent sterilization occurs when uv rays of sufficient intensity irradiate a thin film of water. UV rays sterilization unit is installed immediately before the point of use. A common uv sterilization system is shown in Fig. 1.28. Usually it contains the following units-

(i) A filter Candle.

(ii) Activated Carbon Chamber

(iii) Ultraviolet sterilization chamber

Water is first passed through the filter candle, which has multilayered filtering system to strain out physical impurities present water. Then it is passed through the activated carbon chamber which removes colour, odour, organic impurities and free gases. And finally it is subjected to u.v chamber where u.v rays kill the micro organisms. The rate of flow of water is controlled through the U-tube to ensure complete disinfection.

An electronic monitoring system monitors the purification process.

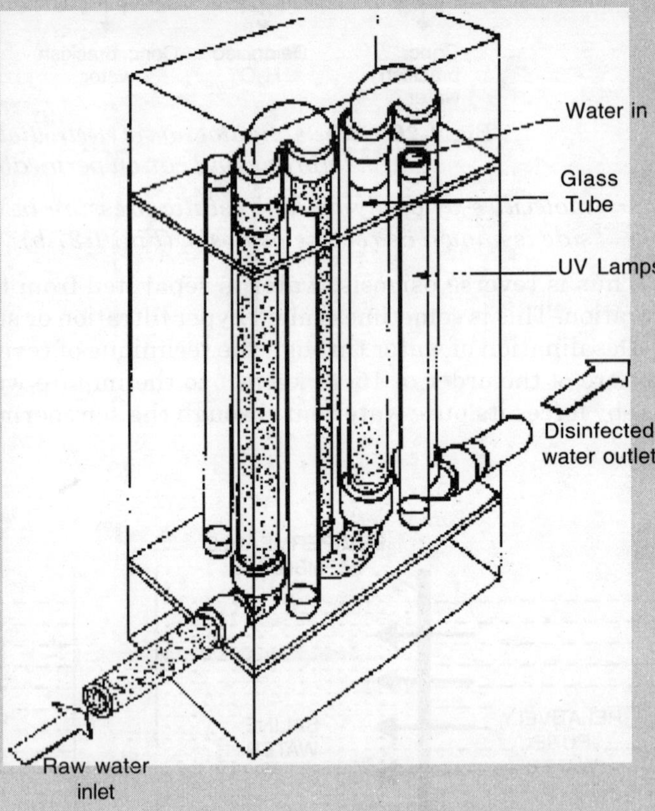

Fig. 1.28 U.V. Sterilizer

RO SYSTEMS

Drinking Water Purification through Reverse Osmosis Units

Another method for purification of drinking water is through Reverse Osmosis Units: The reverse Osmosis process is highly effective in removing a wide range of contaminants from water including heavy metals, chemicals, dissolved salts and micro-organisms. The water is forced through an extremely fine semi-permeable membrane rejecting the contaminated water and leaving only pure, clean and healthy water. Reverse Osmosis is the only effective way to remove fluroride from water.

Such systems typically include a number of steps:

- a sediment filter to remove particulate contaminants such as dirt, rust, slime and sand.
- an activated carbon filter to remove organic chemicals which can give your water its objectionable taste, smell and odour. It also removes chlorine which will attack and degrade reverse osmosis membranes.
- a reverse osmosis (RO) filter which is a thin film composite membrane.
- optionally an ultra-violet lamp is used for disinfection of any microbes that may escape filtering by the reverse osmosis membrane.

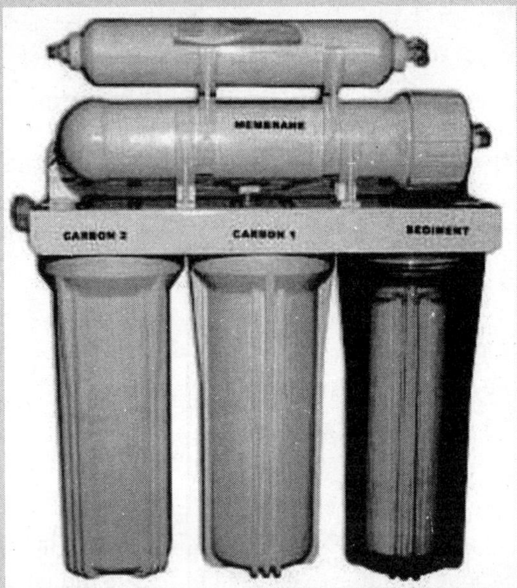

(a) RO Unit

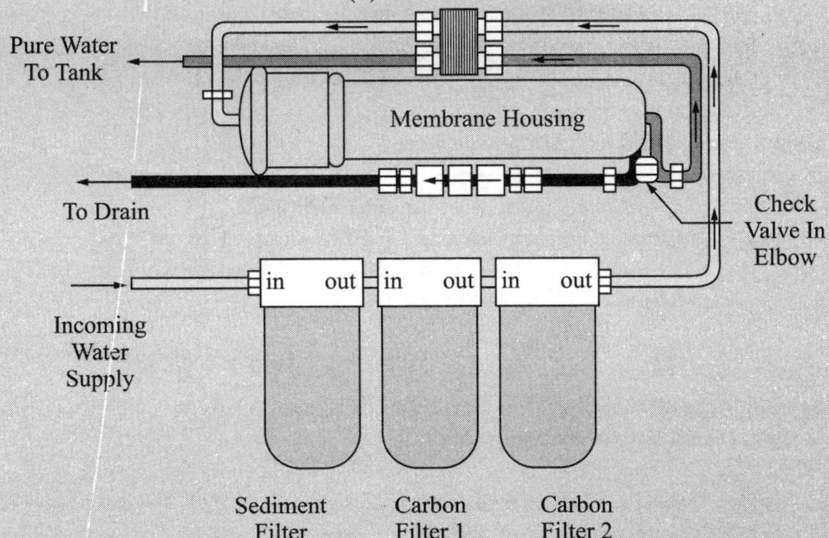

(b) Direction of Flow of Water

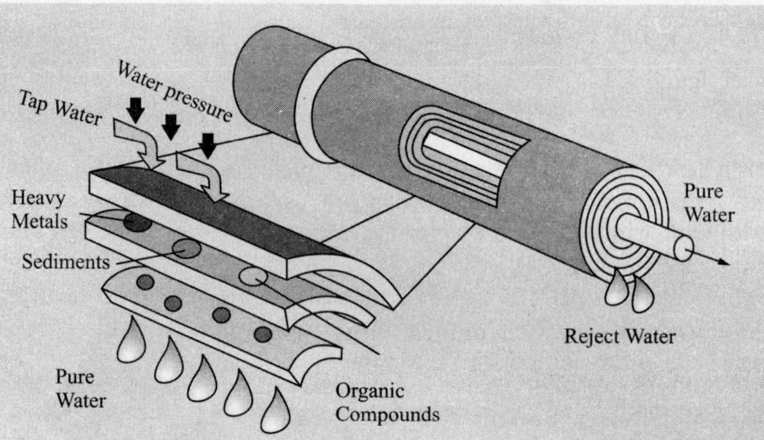

(c) The Membrane

Fig. 1.29 *(a) Reverse Osmosis Unit (b) Direction of Flow of Water in RO (c) Internal View of the Membrane*

QUESTIONS

SHORT ANSWERS QUESTIONS
(Refer to the Respective Sections for Answers)

Section 1.1, 1.2

1. What are the major sources of water
2. What type of dissolved impurities are present in water.
3. What are the specifications of water to be used in Alcoholic Distilleries, and confectionaries.
4. What are the specifications of water to be used in textile industries.
5. Why is rain water a purest form of natural water ?

   Ans. Rain water is obtained through a process similar to distillation hence it is the purest from of natural waters.

Section 1.3, 1.4a, 1.46

6. What do you understand by hardness of water ?
7. What is the cause of hardness of water ?
8. Distinguish between temporary and permanent hardness
9. Why hardness is expressed in terms of $CaCO_3$ equivalents. How is it calculated ?
10. Determine the $CaCO_3$ equiv. of 83 mg of $Mg(HCO_3)_2$ (Molecular weight = 146)

    Ans. $CaCO_3$ equivalents of 83 mg of Mg $(HCO_3)_2$

$$= 83 \times \frac{100}{146} = 56.8 \text{ mg}$$

11. What are the units of hardness?
12. What is the relation between these units?

Section 1.4c

13. What is the principle of hardness of water?
14. What is the structure of Eriochrome Black T.

15. Why $NH_3 - NH_4Cl$ buffer solution is added in determination of hardness of water by EDTA method.
16. What type of ligand is EDTA?
17. Why the complex of Ca^{2+} or Mg^{2+} with EDTA is more stable than the complex of Ca^{2+} or Mg^{2+} with EBT?

    Ans: EDTA is a hexadentate ligand. It forms 4 or 6 bonds with the divalent cations (Ca^{2+} or Mg^{2+}) thus entrapping it like a claw of a crab. Such type of compounds are known as chelate compounds which are very stable.

Section 1.5

18. What is the cause of alkalinity in natural waters?
19. Why alkalinity of water cannot be due to the simultaneous presence of $\overline{O}H$, CO_3^{2-} and HCO_3^-.
20. What type of alkalinity is present if phenolphthalein alkalinity is more than half the methyl orange alkalinity $\left(P > \frac{1}{2}M\right)$.
21. During titration of a water sample with N/10 H_2SO_4 using phenolphthalein indicator zero ml of H_2SO_4 was used. What do you inter from this.

    Ans: Absence of CO_3^{2-} and $\overline{O}H$. Only HCO_3 may be present.

Section 1.12

22. What are the disadvantages of using hard water in boiler operations?
23. Differentiate between scale and sludge?
24. What are the salts responsible for the formation of scales?

    Ans: $Mg(OH)_2$ and $CaSO_4$
25. Why should natural water not be fed to boiler?
26. What is conditioning or internal treatment of hard water ?
27. What is carbonate conditioning?
28. Why phosphate conditioning is better than carbonate conditioning?
29. Why is calgon conditioning is better than phosphate conditioning?

    Ans: In calgon conditioning soluble complexes of Ca^{2+} and Mg^{2+} are formed whereas in phosphate conditioning precipitates of calcium or magnesium phosphate are formed which give rise to sludge which has to be removed. No such problem of sludge removal present in calgon conditioning.
30. What are the reasons of boiler corrosion?
31. How dissolved oxygen can be removed by chemical and mechanical means?
32. What do you mean by caustic embrittlement?
33. How can you prevent caustic embrittlement?
34. How caustic embrittlement is controlled by adding sodium sulphate?
35. What is priming and foaming ?
36. How silica can be removed from water?

Section 1.13

37. What are the various techniques employed for treating water externally ?
38. What is the basic principle applied to remove the hardness of water by Lime Soda process?

    Ans: Calcium and magnesium salts are removed by precipitating them as $CaCO_3$ and $Mg(OH)_2$ respectively.

39. Why does $Mg(HCO_3)_2$ requires double amount of lime for softening?
Ans: From the equation,

$$\underset{(2\text{Eqiv.})}{Mg(HCO_3)_2} + \underset{(4\text{ Equiv.})}{2Ca(OH)_2} \longrightarrow 2CaCO_3 + Mg(OH)_2 + 2H_2O$$

40. Why does $Mg(HCO_3)_2$ requires double amount of
we find that for 1 mole of $Mg(HCO_3)_2$,
2 moles of $Ca(OH)_2$ are required or 1 Equiv of $Mg(HCO_3)_2$ requires 2 equiv. of $Ca(OH)_2$
41. Explain 1 equiv of $Al_2(SO_4)_3$ requires 1 equiv of $Ca(OH)_2$ for softening purposes.
Ans: $\underset{6\text{ Equiv}}{Al_2(SO_4)_3} + \underset{6\text{ Equiv}}{3Ca(OH)_2} \rightarrow 2Al(OH)_3\downarrow + 3CaSO_4$

$$\left[\begin{array}{l}\because 1 \text{ mole of } Al_2(SO_4 = 6\text{equiv. of } Al_2(SO_4)_3 \\ 1 \text{ mole of } Ca(OH)_2 = 2\text{equiv of } Ca(OH)_2\end{array}\right]$$

Thus 6 equiv of $Al_2(SO_4)_3$ requires 6 equiv. of $Ca(OH)_2$ and hence the ratio is 1 : 1
42. Why is the presence of $NaAlO_2$ in water is equivalent to presence of equivalent of $Ca(OH)_2$?
Ans: $\underset{1\text{eq}}{NaAlO_2} + H_20 \rightarrow Al(OH)_3^{-} + \underset{1\text{eq.of } Ca(OH)_2}{NaOH}$

43. Why in the general formulae for lime and soda requirement , $CaCO_3$ equiv of HCO_3^- is added in lime but subtracted in soda requirement.

Ans: HCO_3^- in the form of $Na_2CO_3^-$, requires lime but generate Na_2CO_3 in equivalent amount and hence must be subtracted from and soda requirement.
44. What are the advantages of lime soda process? Why are coagulants used in lime soda process?
45. What do you mean by zeolite?
46. Why zeolite process is sometimes referred to cation-exhange process?
47. Discuss regeneration of zeolite?
48. What are the disadvantages of zeolite process.
49. What are Ion Exchange Resins.
50. What do you mean by cation and Anion Exchangers.
51. How are exhausted ion-exhange resins are regenerated?
52. Compare zeolite and ion-exchange process.
53. What are the advantages of ion-exchange process.
54. In ion exchange process, water is usually passed through cation exchanger and then through anion exchanger and not vice versa. Explain.
Ans: Cation exchangers are attacked by alkalies. Thus if water is passed through anion exchanger first, the alkalies so produced will damage the cation exchanger.

Section 1.14

55. What are the specifications of water for drinking purposes?
56. Why soft water cannot be used for drinking purposes?
Ans: For drinking purposes, some amount of $CaCO_3$ (about 85 ppm) should be there otherwise the soft water reacts with the lead used in plumbing.

$$2Pb + 2H_2O + O_2 \rightarrow 2Pb(OH)_2$$

Presence of HCO_3^- or CO_3^- diminishes the dissolution of lead in drinking water.
57. What is sedimentation and coagulation?
58. Distinguish between primary and secondary standards of drinking water?

59. What are the coagulants used for treating water for drinking purposes.
60. What do you mean by the term disinfection?
61. Why is boiled water is not 100% safe drinking purposes?
    Ans: Boiling kills germs but do not provide protection against contamination during storage etc.
62. Why chloramine is preferred over bleaching powder during chlorination?
63. What is break point chlorination?
64. What is the process of dechlorination?
65. How water is disinfected using $KMnO_4$.

Section 1.15

66. What do you mean by the term desalination?
67. What are the advantages.
68. What is reverse osmosis as applied to water purification ?
69. What is the main advantage of reverse osmosis over ion exchange process?
    Ans: Ion exchange process removes ions whereas reverse osmosis also removes colloidal and high molecular weight organic matter along with ions.

LONG ANSWER QUESTIONS

1. What do you understand by hardness of water ?
   What is its cause? Distinguish between temporary and permanent hardness.
2. What are the specifications of water for industrial use?
3. Give the chemical reactions involved in determination of hardness of water by EDTA titration.
4. What are the disadvantages of hard water ? How is the hardness of water removed by ion-exchange method ?
5. How is water softened by lime-soda process ? Describe its types and suitable chemical reactions.
6. (a) What are the external and internal treatment required for prevention of scale formation in boiler.
   (b) Comment on the statement that internal treatment is only a corrective or supplementary method of treatment for boiler feed water.
7. (a) Give the chemical reaction involved during :
      (i) softening of water by ion-exchange resins, and
      (ii) regeneration of exhausted ion-exchange resins.
   (b) Mention advantages and disadvantages of ion-exchange process of softening water over lime-soda process.
8. (a) What are the disadvantages of using hard water in boiler operation ? Disuss any four methods of disinfection of water.
   (b) Specifications of domestic and industrial water are different. Explain.
   (c) Mention the requisites of water used for paper mills, textile industry and sugar industry.
9. (a) What is meant by phosphate conditioning ?
   (b) What is calgon? What is its application in water treatment?
   (c) What is the role of $NaAlO_2$ in internal treatment of boiler water?
10. (a) What is the ideal pH range for phosphate conditioning?
    (b) Identify from the following compounds, those suitable for treating acidic, neutral and highly alkaline water: NaH_2PO_4, Na_2HPO_4 and Na_3PO_4.
11. (a) What is caustic embrittlement? What are the contributing factors for caustic embrittlement?
    (b) How can the caustic embrittlement be prevented?
12. (a) What are the different factors contributing to boiler corrosion? How can the boiler corrosion be minimised?

(b) What is meant by deaeration? How can it be effected mechanically?

(c) Suggest some chemical reagents for the removal of dissolved oxygen. Which one is the best? why?

(d) Suggest some feasible chemical reagents for the removal of dissolved CO_2 in water.

13. (a) What is priming? How is it caused? What are its disadvantages? How can it be minimised ?

(b) What do you mean by foaming. How it can be minimised?

14. (a) What are cation and anion exchangers?

(b) What are the merits of ion exchange process over the other water softening methods?

(c) What is meant by exhaustion of cation and anion exchangers? How can they be regenerated?

15. Describe the principle and procedure involved in the zeolite process for the treatment of water. What are the limitations, advantages and disadvantages of the process ?

16. Why is demineralization process preferred over zeolite process for softening of water for use in boilers.

17. (a) Write, in brief, how is water purified for drinking purpose by municipalities ?

(b) What is break-point chlorination? What are its advantages?

18. (a) Define the term disa[illegible]. With neat diagram, describe desalination by electrodialysis process.

(b) What do you mean by reverse osmosis? How it help in desalination of water? What are its advantages?

19. Distinguish between :-

(a) Temporary hardness and permanent hardness.

(b) Purification and softening of water.

(c) Boiler scales and sludges

(d) Cold lime soda process and hot lime soda process

(e) Softening and demineralization of water

20. (a) What are the requisites of drinking water?

(b) How is natural water sterlized by : (i) Chlorine, (ii) bleaching powder, (iii) chloramine, and (iv) ozone?

21. (a) State two harmful effects of silica present in water? How it can be removed.

(b) What is the basis of chlorination of water explain.

22. Explain the following steps involved in purification of water (a) Sedimentation (b) Coagulation (c) Filteration

23. Explain the terms : blow down operation, make up water, priming and foaming.

24. (a) Give chemical reactions involved during the softening of water by lime-soda process

(b) Why is hot lime-soda process better than the cold process? Why is sodium aluminate added in cold process? Disuss its function.

25. How is water analyzed for alkalinity. How the alkalinity due to various ions can be determined?

26. What is meant by disinfection in treating domestic water supply? What is its importance? What are the chemicals which are used as disinfectants and what are their comparative merits and demerits?

27. Describe a method for the removal of suspended impurities and micro organisms from water.

28. Describe the methods of determining total solids, alkalinity and chlorides in sample of water.

PROBLEMS

On Hardness of Water

29. A water sample contains 204 mg of $CaSO_4$ per liter. Calculate the hardness in terms of $CaCO_3$ equivalents. **(150 ppm)**

30. How many grams of $FeSO_4$ dissolved per litre gives 210.5 ppm of hardness? **(0.31996g/L)**

31. A sample of water on analysis was found to contain the following impurities in mg/L :

Impurity	$(Ca(HCO_3)_2$	Mg $(HCO_3)_2$	$CaSO_4$	$MgSO_4$
Quantity	10	8.5	12	14
Mol. Wt.	162	146	136	120

Calculate the temporary, permanent & total hardness of water in mg/L?

(11.99, 20.49 and 32.48 mg/L)

32. A sample of water is found to contain 40.5 mg/L $Ca(HCO_3)_2$, 46.5mg/L $mg(HCO_3)_2$, 27.6 mg/L $MgSO_4$, 32.1 mg/L $CaSO_4$ and 22.45 mg/L $CaCl_2$ Calculate the total hardness of water?

(123.68 ppm)

33. A sample of water on analysis has been found to contain the following in ppm :

$$Ca(HCO_3)_2 = 12.5;\ CaCl_2 = 8.2;\ MgSO_4 = 2.6.$$

Calculate the temporary and permanent hardness in degree French?

(1.504 and 1.507 degree French)

34. 100ml of water sample has a hardness equivalent to 12.5ml of 0.08N $MgSO_4$ What is the hardness in ppm? **(500ppm)**

On Determination of Hardness of Water

35. 0.28 g of $CaCO_3$ was dissolved in HCI and the solution was made to one litre with distilled water. 100 ml of the above solution required 28 ml of EDTA solution on titration. 100 ml of the hard water sample required 33 ml of the same EDTA solution on titration. After boiling 100ml of this water, cooling, filterning and then titration required 10 ml of EDTA solution. Calculate the temporary and permanent hardness of water.

(Hardness ; Total = 330, Perm. = 100, Temp. = 230ppm)

36 A standard hard water contains 15g of $CaCO_3$ per lilter. 20 mL a this required 25mL of EDTA solution. 100 ml of sample water required 18 ml EDTA solution. The same sample after boiling required 12 mL EDTA solution. Calculate the temporary hardness of the given sample of water, in terms of ppm. **(Temporary hardness = 720 ppm)**

37. 50 ml of standard hard water containing 1 mg of pure $CaCO_3$ per ml consumed 20 ml of EDTA. 50 mL of water sample consumed 25 mL of the same EDTA solution, using Eriochrome T as indicator. Calculate the total hardness of water sample in ppm. **(1250 ppm)**

38. Calculate the hardness of a water sample, whose 10 mL required 20 mL EDTA. 20 mL of $CaCl_2$ solution, whose strength is equivalent to 1.5 g of $CaCO_3$ per litre, required 30 ml of EDTA solution. **(2000 ppm)**

39. One gm of $CaCO_3$ was dissolved in dil. HCl and the solution diluted to one litre. 50 mL of this solution required 45 mL of EDTA solution, while 50 ml of the sample water required 18 mL EDTA solution. On the other hand, 50 ml of boiled sample water when titrated against EDTA consumed 9 mL of solution. Calculate each type of hardness in ppm.

(Perm. = 200 ppm; Temp.= 200 ppm)

40. 100 mL of a sample of water required 15mL of 0.01M EDTA for titration using Erichrome T as indicator. In another experiment, 100 mL of the same sample was boiled to remove the temporary hardness. The cold sample (100mL), required 8 mL of 0.01 M EDTA solution for titration. Calculate (i) the total hardness (ii) permanent hardness **(150 ppm ; 80 ppm)**

41. 250 ml of a sample of water on EDTA titration with Erio-T as indicator consumed 13 ml of 0.022 M EDTA till end point is reached. Calculate the hardness of water (1ml of 0.01 M EDTA = 1 mg of $CaCO_3$) **(114.4 ppm)**

On Determination of Alkalinity of Water

42. 200 ml. of a sample of water required 20 ml. N/50 H_2SO_4 using methyl orange as indicator but did not give any colouration with phenolphthalien. What type of alkalinity is present ? Express the same in mg/L. **(HCO_3 = 100 ppm)**
43. 200 ml of a sample required 20 ml of N/50 HCl using methyl orange as indicator. Another 200 ml of the same sample required 8 ml of N/50 HCl using phenolphthalien as indicator. Express the alkalinity in terms of mg of $CaCO_3$ per litre. **[80ppm(CO_3^{2-}) and 20 ppm (HCO_3^-)]**
44. A sample of water was alkaline both to phenolphthalien and methyl orange 0.50 ml of this water sample required 15 mL of N/50 H_2SO_4 for phenolphthalien end point and another 10 mL for complete neutralisation. Calculate the type of alkalinity in ppm. **[400 ppm (CO_3^{2-}) and 100 ppm ($\overline{OH}$)]**
45. A water sample is not alkaline to phenolphthalien. However, 100 ml of the sample water on titration with N/50 HCl, required 16.9 ml to obtain the end point, using methyl orange as indicator. What are the types and the amounts of alkalinity present in the sample. **(HCO_3^- = 169mg/l)**
46. A water sample is alkaline to both phenolphthalien as well as methyl orange. 100 ml of water sample on titration with N/50 HCl required 4.7 ml of the acid to phenolphthalien end point. When a few drops of methyl orange are added to the same solution and the titration further continued, the yellow colour of the solution just turned red after addition of another 10.5 ml of the acid solution. Elucidate on the type and extent of alkalinity present in the water sample. **(CO3^{--} = 94 ppm and HCO_3^- = 58 ppm)**
47. 100 ml of a water sample, on titration with N/50 H_2SO_4 gave a titre value of 5.8 ml to phenolphthalein end point and another 100 ml sample on titration with same acid gave a titre value of 11.6 ml to methyl orange end point. Calculate the alkalinity of the water sample in terms of $CaCO_3$ and comment on the type of alkalinity present. **(CO_3^{2-} = 116 ppm)**
48. 100 ml of water sample, on titration with N/50 H_2SO_4 using phenolphthalein as indicator, gave the end point when 5 ml of acid were run down. Another lot of 100 ml of the sample also required 5 ml of the acid to obtain methyl-orange end point. What type of alkalinity is present in the sample and what is its magnitude? **(OH^- = 50 ppm)**

Based on Lime-Soda Process

49. Calculate the amount of lime and soda required to soften 25,000 litres of water having following analysis.
$Ca(HCO_3)_2$= 4.86 ppm; $Mg(HCO_3)_2$ = 7.3 ppm; $CaSO_4$ = 6.8 ppm; $MgCl_2$ = 5.7 ppm; $MgSO_4$ = 9.0 ppm; SiO_2 = 3.5 ppm; NaCl = 5.85 ppm; **(L = 490 g; S = 490 g)**
50. Calculate the quantity of lime and soda required for softening 50,000 litres of water containing the following salts per litre :
$Ca(HCO_3)_2$ = 8.1 mg; $Mg(HCO_3)_2$ = 7.5 mg; $CaSO_4$ = 13.6 mg; $MgSO_4$ = 12.0 mg; $MgCl_2$ = 2.0 mg; and NaCl = 4.7 mg. **(L = 1.01 kg; S = 1.17kg)**
51. Calculate the amount of lime (84% pure) and soda (92% pure) required for treatment of 20,000 litres of water, whose analysis is as follows:
$Ca(HCO_3)_2$= 40.5 ppm; $Mg(HCO_3)_2$ = 36.5 ppm; $MgSO_4$ = 30.0 ppm; $CaSO_4$ = 34.0 ppm; $CaCl_2$ = 27.75 ppm; and NaCl = 10.00 ppm. Also calculate the temporary & permanent hardness of water sample. **(L = 1.762 Kg; S = 1.728 Kg; temp. = 50 ppm; perm. = 75 ppm)**
52. Calculate the amounts of lime and soda needed for softening 100,000 litres of water containing HCi = 7.3 mg/L; $Al_2(SO_4)_3$ = 34.2 mg/L; $MgCl_2$ = 9.5 mg/L : NaCl 29.25 mg/L (AI = 27, Mg = 24; Na = 23; H = 1; CI = 35.5; S = 32; O = 16.) **(L = 3.7 kg ; S = 5.3 Kg)**
53. Explain with equations and calculate the quantity of quick lime and soda ash required to soften 10,000 litres of water containing:
   (i) 219 ppm of magnesium bicarbonate and 234 ppm of sodium chloride:

(ii) 36 pm of Mg^{2+} and 18.3 of HCO_3^-.

(iii) 1.5 ppm of the free acids, 144 ppm of sulphate ions and 71 ppm of chloride ions.

(L = 4.0 kg.; S = 2.226 Kg)

54. A water sample using $FeSO_4.7H_2O$ as a coagulant at the rate of 278 ppm, gave following data on analysis for raw water:
Ca^{2+} = 240 ppm; Mg^{2+} = 96 ppm; CO_2 = 44 ppm HCO_3 = 732 ppm,
Calculate the lime and soda required to soften 250,000 litres of water.
(L = 222 kg; S= 132.5 kg)

55. Calculate the quantities of lime and soda required for softening 300,000 litres of water, using 20 ppm of sodium aluminate as a coagulant. Impurities in water are as follows :
Ca^{2+} = 160 ppm; Mg^{2+} = 96 ppm; HCO_3^- = 403 ppm, dissolved CO_2 = 34 ppm.
(L = 176.6 kg; S = 149.4 kg.)

56. Calculate the quantities of lime ($Ca(OH)_2$ and soda (anhyd. Na_2CO_3) required for cold softening of 125,000 L of water with the following analysis, using 10 ppm of sodium aluminate as coagulant:
Analysis of raw water: Ca^{2+} = 95 ppm; Mg^{2+} = 36ppm, CO_2 = 66ppm; HCO_3^- = 264 ppm; H^+ = 2ppm
Analysis of treated water: $CO^{2-}{}_3$ = 45 ppm; OH^- = 34 ppm; **(L = 65.7 kg; S= 59.1 kg.)**

57. Calculate the amount of lime and soda required to soften 10,000 litres of water containing the following ions per litre : Mg^{2+} = 4.8 mg; Ca^{2+} = 16.0 mg; HCO_3^- = 73.2 mg. **(L = 592 g; S = Nil)**

58. Calculate the cost of lime and soda required for softening 1 million litres of water containing: $(MgHCO_3)_2$ = 73 mg/L; $MgSO_4$ = 120 mg/L; $CaSO_4$ = 68 mg/L; $CaCl_2$ = 111 mg/L. The cost of lime of 80% purity is Rs. 200 per metric tonne and that of soda of 90% purity is Rs. 12,000 per metric tonne. **(Rs. 3,570)**

Based on Zeolite Process of Water Softening

59. A zeolite softner was completely exhausted and was regenerated by passing 100 litres of sodium chloride solution, containing 120 g/L of NaCl. How many litres of a sample of water of hardness 500 ppm can be softened by this softener? **(20,512 L)**

60. The total hardness of 1,000 litres of water was completely removed by a zeolite softener. The zeolite softener required 30 litres of sodium chloride solution, containing 15 g/litre of NaCl for regeneration. Calculate the hardnes of water. **(384.6 ppm)**

61. 10,000 litres of hard water was softened by zeolite process. The zeolite required a total amount of 8 litres of NaCl solution containing 150 g/L of NaCl for regeneration. Calculate the hardness of water. **(102.5 ppm)**

62. A completely exhausted zeolite softener requires 120 litres of NaCl solution having 100 g/L of NaCl. How many litres of water having hardness 500 ppm can be softened by the zeolite.
(20,518 L)

Previous Examination Questions

63. (a) What is the principle of EDTA titration? How can the permanent hardness of water be determined by this method?
(b) Why is NH_3–HN_4Cl buffer solution is added in determination of hardness of water by EDTA titration? **(G.G.S.I.P.U, Ist Term, B.Tech, Dec. 1999)**

64. What are zeolites? How do they function in removing hardness of water? What are the limitations of this process? **(G.G.S.I.P.U., Ist Term, B.Tech, Dec. 1999)**

65. (a) One gm of $CaCO_3$ was dissolved in dil HCl and solution diluted to one litre. 50 ml of this solution required 45 ml of EDTA solution, while 50 ml of sample water required 18 ml of EDTA solution. On the other hand 50 ml of boiled sample water when titrated against EDTA consumed 9 ml of solution. Calculate each type of hardness in ppm.
(b) Differentiate between scale and sludge. **G.G.S.I.P.U., Ist Term, B.Tech., Dec. 1999)**

66. (a) How will you determine the alkalinity of water sample containing hydroxide and carbonate ion.

(b) A water sample is alkaline to both phenolphthalein as well as methyl orange. 100 ml of water sample on titration with $^{N}/_{50}$ HCl required 4.7 ml of the acid to phenolphthalein end point. When a few drops of methyl orange are added to the same solution and the titration further continued, the yellow colour of the solution just turned red after addition of another 10.5 ml of the acid solution. Elucidate on the type and extent of alkalinity present in the water sample. **(G.G.S.I.P.U., Ist Term, B.Tech., Dec. 1999)**

67. (a) Distinguish between softening and demineralisation of water.

(b) Describe ion-exchange method of demineralisation of water.

(c) A water sample contains : $Mg(HCO_3)_2$ = 7.5 mg/L; $CaCl_2$ = 222 mg/L; $MgSO_4$ = 120 mg/L; $Ca(NO_3)_2$ = 164 mg/L. Calculate the quantity of lime (74% pure) and soda (90% pure) for softening 5,000 L of water. **(G.G.S.I.P.U., End Term, B.Tech., Feb.-March 2000)**

68. (a) What is the principle of EDTA titration? Briefly describe the estimation of hardness of water by EDTA method. What is the role of buffer in EDTA titration?

(b) A sample of water was alkaline to both phenolphthalein and methyl orange. 50 ml of this water sample required 15 mL of N/50 H_2SO_4 for phenolphthalein end-point and an other 10 mL for complete neutralization. Calculate the type and amount of alkalinity in ppm. **(G.G.S.I.P.U., End Term, B.Tech., Feb.-March 2000)**

69. (a) What are boiler scales? What are ill effects of scales? How can scale formation be prevented by: (i) phosphate conditioning, (ii) calgon conditioning?

(b) Write short notes on : (i) Caustic embrittlement (ii) Priming and Foaming **(G.G.S.I.P.U., End Term, B.Tech., Feb.-March 2000)**

70. (a) Describe L-S process for water softening. Give the advantages and disadvantages of this process.

(b) Why is hot L-S process better than the cold process.

(c) What are essential requirements of potable water. **(G.G.S.I.P.U., End Term, B.Tech., Dec. 2000)**

71. (a) Describe the principle involved in determination of hardness by EDTA method.

(b) (i) 50 mL of standard containing 1 mg of $CaCO_3$ per mL consumed 20 mL of EDTA solution.

(ii) 50 ml of water sample consumed 25 mL of EDTA solution.

(iii) After boiling 50 ml of water sample consumed 18 mL of EDTA solution. Calculate the temporary and permanent hardness of the sample. **(G.G.S.I.P.U., End Term, B.Tech., Dec. 2000)**

72. Attempt any seven of the following parts :

(a) What is sterilization of water? Name three substances for sterilization of water.

(b) Why do we add buffer during titration of hard water against EDTA solution. Name acidic and basic buffers.

(c) What is reverse osmosis? What are its applications?

(d) What is Calgon conditioning? How is it better than phosphate conditioning?

(e) What are zeolites? Why is water softened by zeolite process unfit for use in boilers.

(f) What are flocculants? What happens when alum is dissolved in water?

(g) What is boiler corrosion? Why should the presence of CO_2 in boiler feed water be avoided?

(h) What is Lime and Soda? Why does magnesium bicarbonate require double amount of lime for softening of hard water? **(G.G.S.I.P.U, End Term, B.Tech, II Batch, Mar. 2001)**

73. (a) What are scales? What are their disadvantages?

(b) What are internal and external treatments required for prevention of scale formation in boilers?

(c) A sample of water is collected from a tank which gives the following results on analysis :

$$CO_2 = 22 \text{ ppm},\ HCO_3^- = 365 \text{ ppm},\ Ca^{2+} = 40 \text{ ppm};\ Mg^{2+} = 48 \text{ ppm}.$$

Calculate the amount of lime and soda that would be required for softening the water if 139 ppm of $FeSO_4 \cdot 7H_2O$ is used as a coagulant.

(G.G.S.I.P.U, End Term, B.Tech, II Batch, Mar. 2001)

74. What is the principle of EDTA titration? Briefly describe estimation of hardness of water by EDTA method. Explain the role of buffer solution in EDTA titration.

(G.G.S.I.P.U, I Term, B. Tech, 2001)

75. (a) Alkalinity of water cannot be due to the simultaneous presence of $\bar{O}H$, CO_3^2 and HCO_3^-.

(b) A water sample is alkaline to both phenolphthalein and methyl orange. 100 ml of water sample on titration with $\frac{N}{50}$ HCl required 4.7 ml of acid to phenolphthalein end point. When four drops of methyl orange are added to same solution and titration further continued the yellow colour of solution just turned red after addition of another 10.5 ml of acid solution. Describe the type and extent of alkalinity present in water sample.

(G.G.S.I.P.U, I Term, B. Tech, 2001)

76. (a) What are boiler scales?
How can scale formation be prevented by :
(i) Phosphate conditioning.
(ii) Calgon conditioning.

(b) What is priming and foaming? Discuss their disadvantages.

(G.G.S.I.P.U, I Term, B. Tech, 2001)

77. (a) What is boiler corrosion? Why the presence of CO_2 in boiler feed water is disadvantageous? How it can be removed?

(b) One gm of $CaCO_3$ was dissolved in dil. HCl and the solution diluted to one litre. 100 ml of this solution required 90 ml of EDTA solution, while 100 ml of the sample water required 40 ml of EDTA. On the other hand, 100 ml of the boiled sample water when titrated against EDTA consumed 20 ml of solution. Calculate each type of hardness in ppm.

(G.G.S.I.P.U, I Term, B. Tech, 2001)

78. (a) Calculate the amount of lime and soda required for softening a million litres of hard water which analysed as follows:

Substance	*Conc. in mg/l.*
$Ca(HCO_3)_2$	243
$Mg(HCO_3)_2$	73
$CaSO_4$	102
$MgCl_2$	95
NaCl	500
$FeSO_4 \cdot 7H_2O$	139

Purity of lime is 94% and that of soda is 86.0%. **(L = 314.8 kg S = 277.3 kg)**

(b) Why is Calgon conditioning better than phosphate conditioning?

(G.G.S.I.P.U, End Term, B. Tech, Dec. 2001)

(c) What happens when temporary hard water is boiled (give equations)?

79. (a) 200.0 ml of a sample required 20.0 ml of N/50 HCl using methyl orange as indicator. Another 200.0 ml of the same sample required 8.0 ml of N/50 HCl using phenolphthalein as indicator. Express the alkalinities in terms of mg. of $CaCO_3$ per liter.

(b) Why is air saturated with steam passed through a bed of hot coal in the production of producer gas? **(G.G.S.I.P.U, End Term, B. Tech, Dec. 2001)**

80. What are the factors that lead to caustic embrittlement in boilers? How can this be prevented?

81. (a) 100 ml of sample of water required 15.0 ml of .01 M EDTA for titration using Erio-T as indicator. In another experiment, 100 ml of the same sample was boiled to remove the

carbonate hardness, the precipitate was removed and the cold solution required 8.0 ml of 0.01M EDTA using Erio-T as indicator. Calculate in terms of ppm.

(i) The total hardness (ii) Permanent hardness (iii) Carbonate hardness

(G.G.S.I.P.U, End Term, B. Tech, Dec. 2001)

82. (a) What are boiler scale? How the scale formation can be prevented by phosphate conditioning?
(b) Why calgon conditioning is better than phosphate conditioning?
(c) What are the limitation of zeolite process? **(G.G.S.I.P.U, I Term, B.Tech, Sept. 2002)**

83. (a) Standard hard water contains 15 *g*. of $CaCO_3$ per litre. 20 ml. of this required 25 ml. of EDTA solution. 100 ml. of sample water required 18 ml. of EDTA solution. The same sample after boiling required 12 ml. EDTA solution. Calculate the temporary hardness of the given sample of water in terms of ppm.
(b) Explain the principle of Complexometric titration.
(G.G.S.I.P.U, I Term, B.Tech, Sept. 2002)

84. (a) Calculate the quantity of lime and soda required to soften 25000 L of water having following analysis Ca $(HCO_3)_2$ = 4.86 ppm, Mg $(HCO_3)_2$ = 7.3 ppm,
$CaSO_4$ = 6.8 ppm, $MgCl_2$ = 5.7 ppm, $MgSO_4$ = 9 ppm, SiO_2 = 3.5 ppm, NaCl = 5.85 ppm.
(b) What is Carbonate Conditioning? **(G.G.S.I.P.U, I Term, B.Tech, Sept. 2002)**

85. (a) What is demineralization of water? How is it done? Explain with the help of diagram.
(b) Differentiate between boiler scales and sludges? **(G.G.S.I.P.U, I Term, B.Tech, Sept. 2002)**

86. (a)What is hardness of water. Mention various units used for its expression and show their relation also.
(b) Discuss the methods for the determination of following (i) Free Chlorine (ii) Alkalinity
(c) Explain Break point Chlorination **(G.G.S.I.P.U, End-Term, B.Tech, Dec. 2002)**

87. (a) Give details of scale and sludge formation in boilers and write about the methods used for their prevention.
(b) Standard hard water contains 15 *g*. of $CaCO_3$ per litre. 20 ml. of this required 25 ml. of EDTA solution. 100 ml. of sample water required 18 ml. of EDTA solution. The same sample after boiling required 12 ml. EDTA solution. Calculate the temporary hardness of the given sample of water in terms of ppm. **(G.G.S.I.P.U, End-Term, B.Tech, Dec. 2002)**

88. Write short notes on
(a) phosphate conditioning
(b) Ion exchange process **(G.G.S.I.P.U, End-Term, B.Tech, Dec. 2002)**

89. (a) Distinguish between temporary and permanent hardness.
(b) Why $NH_3 - NH_4Cl$ buffer solution is added in determination of hardness of water by EDTA method.
(c) A sample of water is alkaline to both phenolphthalein and methyl orange. 100 mL of this water sample required 15 mL of N/50 H_2SO_4 for phenolphthalein end point and another 10 mL for complete neutralization. Calculate the type of alkalinity present.
(d) What do you mean by the term disinfection? **(G.G.S.I.P.U, I-Term, B.Tech., Nov. 2004)**

90. (a) Differentiate between scale and sludge.
(b) Describe the principle and procedure involved in zeolite process for the treatment of water. What are the limitations of this process.
(c) Write short notes on:
(i) Caustic embrittlement
(ii) Priming and foarming **(G.G.S.I.P.U, I-Term, B.Tech., Nov. 2004)**

91. (a) Match the following:

(i) Purest form of natural water	(i) boiling
(ii) Removal of temporary hardness	(ii) sodium chloride
(iii) Calgon is a trade name	(iii) 0.07°C
(iv) Brackish water contains	(iv) sodium hexametaphosphate
(v) Removal of sludges by	(v) 100 ppm

(vi)	1 mg/l	(vi)	Ion Exchange
(vii)	1°Fr	(vii)	Carbonates and bicarbonates
(viii)	M > 2P	(viii)	Mineral Acids
(ix)	Demineralisation	(ix)	Blow-down operation
(x)	Boiler Corrosion	(x)	Rain Water

(b) What is the principle behind the use of selective indicators in the determination of CO_3^{2-}, HCO_3^- and OH^- ions. Explain.

(c) In water softening methods, name the corrective and preventive methods of treatment.

(d) In low pressure boilers, presence of Calcium bi-carbonates leads to scale formation but in high pressure boilers, no scales are formed. Comment.

(e) What are coagulants? **(G.G.S.I.P.U, I-Term, B.Tech., Sept. 2005)**

92. (a) Differentiate between the two:

(i) Boiler Corrosion and Caustic Embrittlement

(ii) Scales and Sludges

(b) Standard hard water contains 1g of $CaCO_3$ per litre. 100ml of this required 25ml of EDTA solution for titration using EBT as an indicator. Further 100ml of sample hard water required 30ml of the same EDTA solution for titration. The sample water was then boiled. 100ml of the boiled water now required 10ml of the same EDTA solution for titration. Calculate total hardness, permanent hardness as well as temporary hardness in the given sample of hard water. Report the values in terms of ppm and °Cl

(G.G.S.I.P.U, I-Term, B.Tech., Sept. 2005)

93. (a) What are Ion-Exchange Resins?

(b) Explain the working (including regeneration) of an Ion-Exchange resin with a suitable sketch. Also report the reactions involved in each case.

(c) A water sample is alkaline to both Phenolphthalein as well as methyl orange. 100ml of water sample on titration with N/50 HCl required 5.0 ml of the acid upto phenolphthalein end point. Few drops of methyl orange was added to the same solution and titration was further continued. The yellow colour of the solution just turned red after the addition of another 11.0ml of the acid solution. Calculate the type and extent of alkalinity present in the water sample. **(G.G.S.I.P.U, I-Term, B.Tech., Sept. 2005)**

94. (a) Explain Caustic Embrittlement.

(b) Explain why is NH_3–NH_4Cl buffer solution added during determination of hardness of water by EDTA titration. **(G.G.S.I.P.U, I-Term, B.Tech., Sept. 2006)**

95. (a) Describe the demineralization process of softening of hard water. What are its advantages over zeolite process?

(b) Calculate the amount of lime (88.3% pure) and Soda (99.2% pure) required to soften 24000 litres of water per day for a year containing the following:

$CaCO_3$ = 1.85 mg/L **$MgCl_2$ = 0.76mg/L**
$MgSO_4$ = 0.90mg/L **SiO_2 = 2.3mg/L**
$CaSO_4$ = 0.34mg/L
$MgCO_3$ = 0.42mg/L
NaCl = 2.34mg/L **(G.G.S.I.P.U, I-Term, B.Tech., Sept. 2006)**

96. (a) Distinguish between softening and demineralization of water (Give equations also).

(b) (i) 50ml of standard hard water containing 1mg of $CACO_3$ per ml consumed 20ml of EDTA solution.

(ii) 50ml of water sample consumed 25ml of EDTA solution.

(iii) 50ml of water sample consumed 18ml of EDTA solution.

(G.G.S.I.P.U, I-Term, B.Tech., Sept. 2006)

97. (a) How calgon conditioning is better than phosphate conditioning?

(b) Chloramine is preferable to bleaching powder for sterilization of drinking water. Give reason.

(c) What is the compound formed when CO combines with blood?
(d) Distinguish between softening and demineralization of water?
(G.G.S.I.P.U, I-Term, B.Tech., Sept. 2007)

98. (a) What are the factors that lead to caustic embrittlement in boilers? How can these be prevented?
(b) 200 ml of a sample required 20 ml of N/50 HCl using methyl orange as indicator. Another 200 ml of the same sample required 8 ml of N/50 HCl using phenolphthalein as indicator. Express the alkalinities in terms of mg of $CaCO_3$ per litre.
(G.G.S.I.P.U, I-Term, B.Tech., Sept. 2007)

99. (a) What is the principle of EDTA titration? Briefly describe the estimation of hardness of water by EDTA method.
(b) A water sample contains:
$Mg(HCO_3)_2$ = 73mg/L, $CaCl_2$ = 222mg/L,
$MgSO_4$ = 12mg/L, $Ca(NO_3)_2$ = 164mg/L
Calculate the quantity of lime (74% pure) and soda (90% pure) for softening 50,000L of water.
(G.G.S.I.P.U, I-Term, B.Tech., Sept. 2007)

100. (a) Discuss the causes of hardness of water. How is hardness removed for industrial purpose?
(b) Give specifications of water to be used for domestic purpose. Is chemically pure water good as potable water? Comment.
(c) Calculate the total hardness of water if its 100 ml sample is found equivalent to 10.2 ml of centimolar E.D.T.A. solution in a titration of p^H 10.0 using Erichrome Black T as indicator.
(G.G.S.I.P.U, End-Term, B.Tech., Dec. 2007)

101. How is sludge formed in boilders? How can this be removed?

□□□

2

FUELS

"Somewhere, there is something incredible waiting to be known."

— ***Carl Sagan***

CHAPTER OUTLINE

Any source of heat energy is called a fuel. The term fuel is applied to a combustible substance which on burning in presence of oxygen produces a large amount of heat that can be used economically for domestic and industrial purposes.

Nature has given us unlimited stores of the substances which can be used as fuels–*e.g.* wood, charcoal, coal, kerosene etc. Combustion of these fuels is used as a source of heat and power. During combustion of these fuels, a chemical reaction involving breakage of bonds of reactants and formation of new bonds in the products takes place which is accompanied by a large amount of heat. Hence these fuels are termed as **chemical fuels**.

It is now well established that combustion is not necessary for a fuel to produce heat. Nuclear energy is also used as a source of heat. In case of **nuclear fuels**, energy is liberated by fission or fusion of nuclei. This energy is very much greater than in the case of chemical fuels.

Fuels in the broad sense, include the stored fuels that are available in the earth's crust *i.e.* fossil fuels or the fuels which are derived from fossil fuels. The term "fossil fuels" is used because these are derived from the fossilized remains of plants and animals. The energy available from these fuels can be considered to be a part of stored solar energy. The solar energy is trapped by the plants during the process of photosynthesis according to the following reaction:

$$H_2O + CO_2 \xrightarrow[\text{Chlorophyll}]{\text{Sunlight}} \text{Carbohydrates}$$

A part of carbohydrate is used as food by plants and rest is converted into cellulose and other constituents of vegetable matter. The vegetable matter thus becomes the source of high chemical potential energy. When the plant dies, the organic matter oxidizes to form peat and finally into coal. The coal having residual potential energy is used as a fuel.

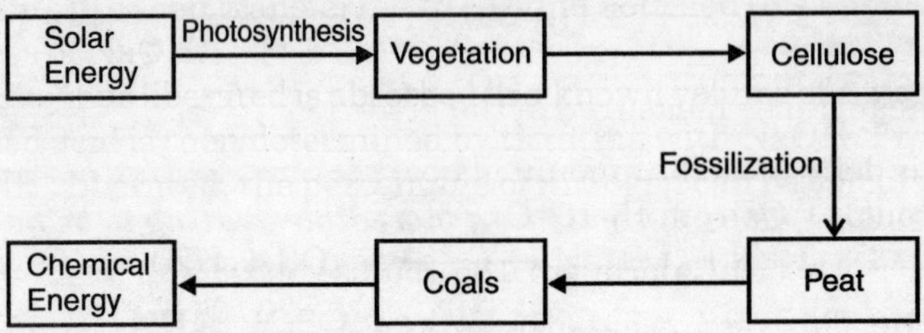

2.1 CLASSIFICATION OF FUELS

Fuels can be classified on the basis of their (i) Occurance and (ii) Physical state.

(i) On the basis of their occurance they are of two types

Primary fuels: Fuels which occur in nature as such are called primary fuels. These can be used either without processing or after little processing which does not change the chemical composition of the fuel. Examples: wood, peat, coal, petroleum and natural gas.

Secondary fuels: The fuels which are derived from the primary fuels by further chemical processing are called secondary fuels. Examples: coke, charcoal, kerosene, coal gas, producer gas etc.

(ii) On the basis of their physical state these may be classified as:

Solid Fuels
Liquid Fuels
Gaseous Fuels
The classification can be summarized as:

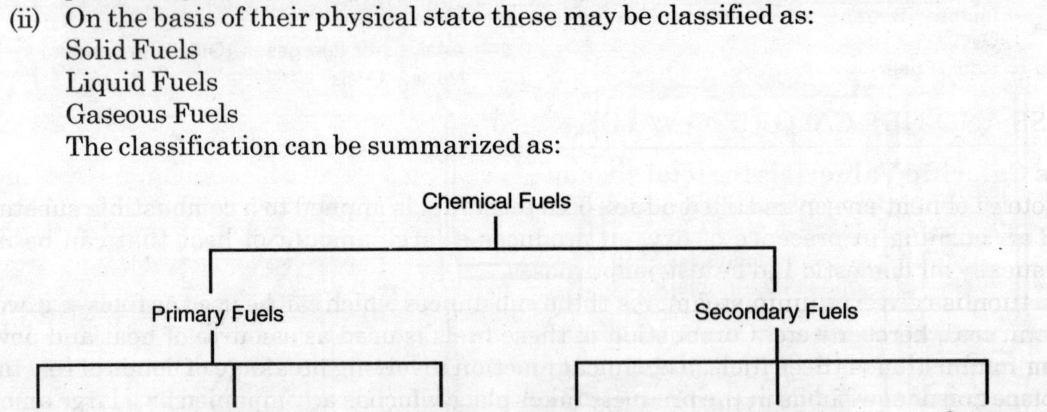

2.2 CALORIFIC VALUE

One of the most important characteristic of a fuel is its calorific value. Calorific value jugdes the efficiency of fuel.

Calorific value is defined as the total quantity of heat liberated when a unit mass of a fuel is burnt completely.

Units of Calorific Value: Units of calorific value for solid, liquid and gaseous fuels are given below.

Units of Calorific Values

System	*Solid/Liquid Fuels*	*Gaseous Fuels*
CGS	Calories/gm	Calories/cm^3
MKS	k cal/kg	k cal/m^3
B.T.U	BTU/1b	BTU/ft^3

The quantity of heat can be measured in the following units:

(i) **Calorie:** It is defined as the amount of heat required to raise the temperature of 1 gram of water through 1°C (from 15-16° C). Also

$$1 \text{ Calorie} = 4.185 \text{ Joules} = 4.185\times10^7 \text{ ergs.}$$

(ii) **Kilo Calorie:** This respresents the unit of heat in MKS system. It is defined as the amount of heat required to raise the temperature of one kilogram of water through 1° C (from 15°–16° C)

$$1 \text{ k cal} = 1000 \text{ cal}$$

(iii) **British thermal unit:** (B. Th. U or B.T.U.) It is defined as the amount of heat required to raise the temperature of 1 pound (lb) of water through 1° F (from 60° F to 61° F)

$$1 \text{ B. Th. u} = 252 \text{ Cal} = 0.252 \text{ k cal}$$
$$= 1{,}054.6 \text{ Joules} = 1{,}054.6\times10^7 \text{ ergs.}$$
$$1 \text{ k cal} = 3.968 \text{ B. Th. U}$$

(iv) **Centigrade heat unit** (C.H.U) It is defined as the amount of heat required to raise the temperature of 1 pound of water through 1° C. (from 15° to 16°C).

$$1 \text{ k cal} = 3.968 \text{ BTU}$$
$$= 2.2 \text{ C.H.U.}$$

GROSS AND NET CALORIFIC VALUE

Gross Calorific Value: It is the total amount of heat generated when a unit quantity of fuel is completely burnt in oxygen and the products of combustion are cooled down to the room temperature (15° C or 60° F).

Usually, all the fuels have sufficient amount of hydrogen. The hydrogen present during combustion is converted into steam. As the products of combustion are cooled down to room temperature, the steam gets condensed into water and latent heat is evolved. Thus in the determination of gross calorific value, this latent heat also gets included in the measured heat. Therefore gross calorific value is also called the *higher calorific value.*

The calorific value which is determined by Bomb calorimeter gives the Gross or Higher Calorific value (HCV).

Net Calorific Value: It is defined as the net heat produced when a unit quantity of fuel is completely burnt and the products of combustion are allowed to escape.

In actual practice when the fuel is burnt, the water vapour, formed due to the presence of hydrogen in the fuel, are not condensed and escape along with hot combustion gases. Hence, a lesser amount than gross calorific value is available. Therefore, this is called net calorific value. It is also known as lower Calorific Value. (LCV)

LCV = HCV–Latent heat of water vapours formed.
Since 1 part by weight of hydrogen gives nine parts by weight of water *i.e.*

$$H_2 + \frac{1}{2}O_2 \rightarrow H_2O$$

$\therefore$ LCV = HCV – Weight of hydrogen × 9 × latent heat of steam
= HCV – Weight of hydrogen × 9 × 587 ($\therefore$ Latent heat of steam = 587 cal/g)

DETERMINATION OF CALORIFIC VALUE

1. *Determination of Calorific Value of Solid and Non volatile Liquid Fuels*: The calorific value of solids and non volatile liquid fuels is determined by **bomb calorimeter.**

Principle: A known amount of the fuel is burnt in excess of oxygen and the heat liberated is transferred to a known amount of water. The calorific value of the fuel is then determined by applying the principle of calorimatery i.e.

Heat gained = Heat lost

Construction: A simple sketch of bomb calorimeter is shown in Fig. 2.1. It consists of the following parts.

(i) *Steel bomb:* Steel bomb consists of a strong cylindrical container made up of stainless steel which is capable of withstanding pressure of atleast 50 atmospheres. The bomb is generally provided with a gas–tight screw cap or lid. The lid in turn is provided with two holes for electrodes and one oxygen inlet valve. A small ring is fitted to one of the electrodes which act as a support for slica crucible.

(ii) The bomb is placed in a *copper calorimeter* containing a known amount of water. The calorimeter is provided with an electrical stirrer and a Beckmann thermometer.

(iii) The copper calorimeter is surrounded by the *air jacket* and *water jacket* to prevent any loss of heat due to radiation.

(iv) A *crucible* made up of nickel, or fused silica which contains the fuel. The crucible is placed inside the ring attached to one of the electrodes.

Working: A known amount of the given fuel (about .5 to 1g) is taken in the silica crucible supported over the ring. A piece of fine magnesium wire is tightly stretched across the electrodes, touching the fuel sample. The lid of the bomb is tightly screwed and bomb filled with oxygen at 25 atmospheric pressure. The bomb is placed inside the copper calorimeter which contains a known amount of water. After thorough stirring, initial temperature of water is noted with the help of Beckmann thermometer, which can read accurately, temperature difference upto 1/100th of a degree. The electrodes are then connected to 6-volt battery and the circuit is completed. The fuel in the crucible burns with the evolution of heat. The heat produced by burning of a fuel is transferred to water, which is stirred throughout the experiment by the electric stirrer. Maximum temperature shown by thermometer is recorded. Time taken to cool the water in the calorimeter from maximum temperature to room temperature is also noted. The calorific value of the fuel is calculated as below:

Calculations

Let

Weight of the fuel sample taken = x g
Weight of water in the calorimeter = Wg

*Water equivalent of the calorimeter, stirrer, bomb, thermometer = w g

Initial temperature of water = t_1°C

Final temperature of water = t_2°C

Higher or gross calorific value = C cal/g

Heat gained by water = W × Δt×specific heat of water

= W ($t_2 - t_1$) × 1 cal.

Heat gained by Calorimeter = $w(t_2 - t_1)$ cal

Heat liberated by the fuel = x C cal

Heat liberated by the fuel = Heat gained by water and calorimeter

$$x\,\text{C} = (W + w)\,(t_2 - t_1)\text{ cal}$$

$$\text{C} = \frac{(W+w)(t_2 - t_1)}{x}\text{ cal/g}$$

Net Calorific Value

Let percentage of hydrogen in the fuel = H

Weight of water produced from 1 gm of the fuel = $\frac{9H}{100}$ gm

(∵ 1 gm of hydrogen produces 9 gm of H_2O)

Heat liberated during condensation of steam = 0.09H×587 cal

Net (Lower) calorific value = GCV – Latent heat of water formed

= C – 0.09H×587 cal/g

Corrections: For obtaining the accurate results the following corrections are also incorporated:

(a) Fuse wire correction: As Mg wire is used for ignition, the heat generated by burning of Mg wire is also included in the gross calorific value as measured above. Hence this amount of heat has to be subtracted from the total value.

---

* Water equivalent of a calorimeter is defined as the number of calories required to heat the calorimeter by 1°C. If M is the mass of the calorimeter and S be its specific heat, then

Water equivalent $w = MS$

If the calorimeter is made of different parts having different specific heats S_1, S_2, S_3 etc. and different masses, M_1, M_2, M_3 etc. then

$$\text{Water equivalent} = M_1S_1 + M_2S_2 + ...$$

The water equivalent of the calorimeter is also determined by burning a fuel of known calorific value and using the equation.

$$\text{GCV} = \frac{(W+w)(t_2 - t_1)}{x};$$

The rise in temperature is determined. As all other values, except w are known in the above formula hence w can be easily calculated. The common fuels used for the purpose of determining water equivalent of the calorimeter are :

Substance	***Calorific Value***
Benzoic acid	6,325 kcal/kg
Naphthalene	9,688 kcal/kg
Camphor	9,292 kcal/kg

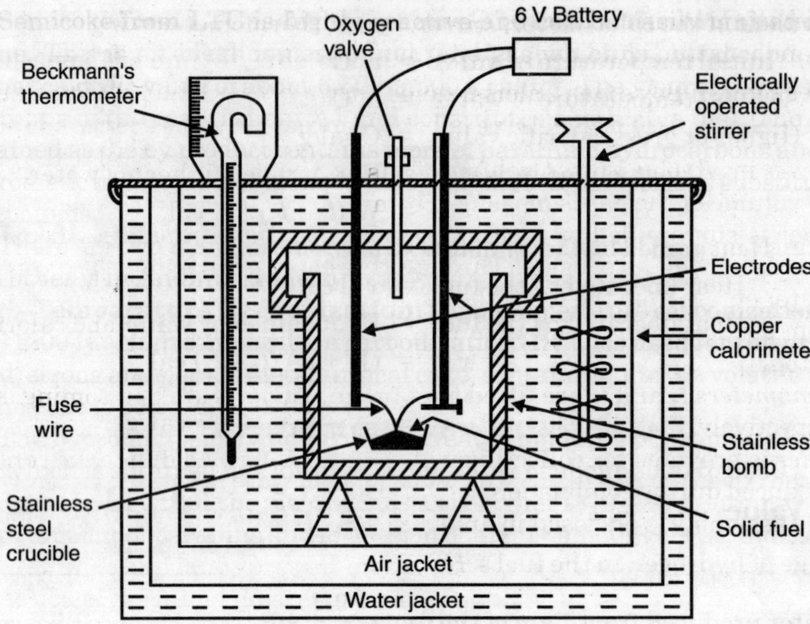

Fig. 2.1 *Bomb calorimeter*

(b) Acid Correction: During combustion, sulphur and nitrogen present in the fuel are oxidized to their corresponding acids under high pressure and temperature.

$$S + O_2 \rightarrow SO_2$$
$$2SO_2 + O_2 + 2H_2O \rightarrow 2H_2SO_4 \qquad \Delta H = -144{,}000 \text{ Calories}$$
$$2N_2 + 5O_2 + 2H_2O \rightarrow 4HNO_3 \qquad \Delta H = -57{,}160 \text{ Calories}$$

The above reactions are exothermic in nature but the heat liberated is not obtained in practical use of coal, because SO_2 and N_2 pass into the atmosphere. Thus corrections must be made for the heat liberated in the bomb by the formation of H_2SO_4 and HNO_3. The amount of H_2SO_4 and HNO_3 is analyzed by washings of the calorimeter.

- For each ml of N/10 H_2SO_4 formed, 3.6 calories should be subtracted.
- H_2SO_4 can also be determined by precipitation as $BaSO_4$ using $BaCl_2$. The correction for 1mg of S formed is 2.25 calories.
- For each ml of N/10 HNO_3 formed, 1.43 calories must be subtracted.

(c) Cooling Correction: Heating and cooling are simultaneous processes. As the temperature rises above the room temperature, the loss of heat does occur due to radiation, and the highest temperature recorded will be slightly less than that obtained, if no loss of heat took place. A temperature correction (cooling correction) is therefore necessary to get the correct rise in temperature.

If the time taken for the water in the calorimeter to cool down from the maximum temperature attained, to the room temperature is x minutes and the rate of cooling is dt/ minute, then the cooling correction $= x \times dt$. This should be added to the observed rise in temperature.

∴ Gross Calorific Value,

$$C = \frac{(W + w)(t_2 - t_1 + \text{Cooling correction}) - [\text{Acid} + \text{fuse corrections}]}{\text{Mass of the fuel}}$$

2. ***Determination of Calorific Value of Gaseous and Volatile Liquid Fuels*:** The calorific value of gaseous and volatile liquid fuels is determined by **Boy's Gas Calorimeter**.

Construction: A sketch of Boy's Gas Calorimeter is shown in Fig. 2.2. It consists of the following parts:

(i) A *gas burner* in which a known volume of gas at a known pressure is burnt at a uniform rate. The volume of the gas is measured by a meter fitted near the inlet of the gas.

(ii) The burner is surrounded by a chimney or *combustion chamber*, which has a copper tubing inside as well as outside. Through this copper coil, water at a constant rate is flown. Water enters from the top of the outer coil, passes through the outer coils, moves to the bottom of the chimney and then moves upwards through the inner coils, and exit from the top.

(iii) The *thermometers* t_1 and t_2 are fitted to note the temperature of incoming and outgoing water respectively.

(iv) A *container* is provided for collecting water which is formed due to the condensation of steam produced during combustion.

(v) The whole assembly is enclosed in an *insulated chamber*.

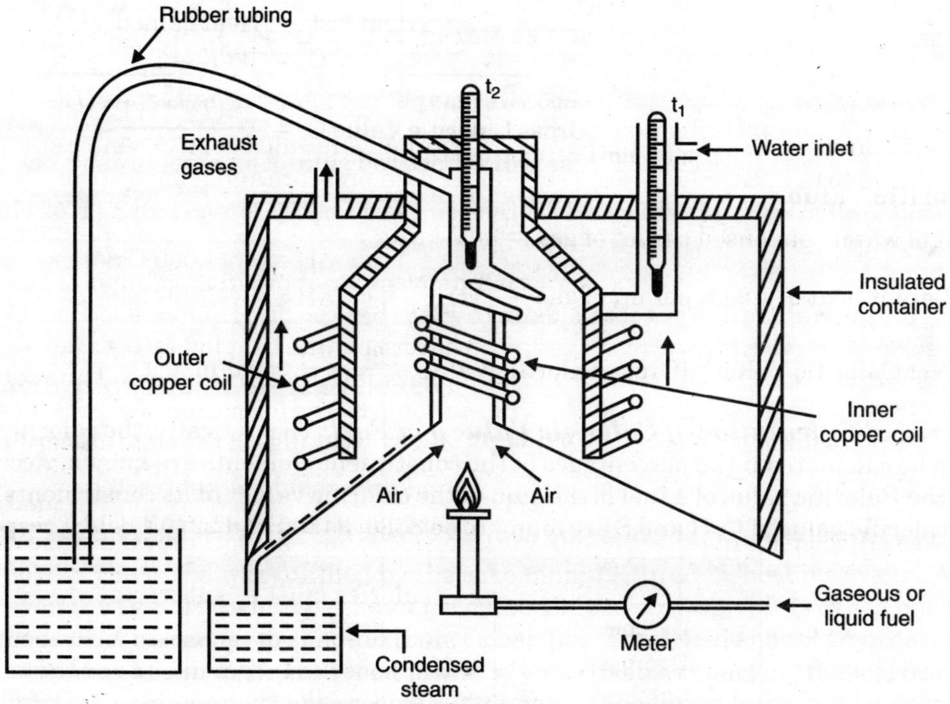

Fig. 2.2 *Boy's gas calorimeter*

Working: Burning of fuel and circulation of water are continued for about 15 minutes for initial warming up period. When the calorimeter gets warmed, the rate of flow is so adjusted that the water leaves the apparatus nearly at atmospheric pressure. Heat produced by burning of gaseous fuels is transferred to water in the copper coil and the steam formed inside the chimney during combustion gets condensed to water which is collected. The following readings are noted, when the conditions become steady:

(i) Volume of gas burnt at a given temperature and pressure in time t. This volume is then converted into STP.
(ii) Weight of water passed through the coil in the same time t.
(iii) Temperature of incoming and outgoing water.
(iv) Weight of water condensed during this time.

Calculations: Let

$$\text{Volume of the gas burnt at STP in a certain time } t = \text{V}m^3$$
$$\text{Weight of water passed through the coil in time } t = W \text{ kg}$$
$$\text{Weight of water condensed during this time} = m$$
$$\text{Temperature of incoming water} = T_1$$
$$\text{Temperature of outgoing water} = T_2$$
$$\text{Rise in temperature} = T_2 - T_1$$
$$\text{Gross Calorific value} = \text{C}$$
$$\text{Heat absorbed by the circulating water} = W(T_2 - T_1)$$
$$\text{Heat produced by combustion of fuel} = \text{V C}$$
$$\text{Heat lost} = \text{Heat gained}$$
$$\text{V C} = W(T_2 - T_1)$$
$$\text{Gross Calorific Value C} = \frac{W(T_2 - T_1)}{V}$$

Net Calorific Value

Weight of water condensed per m^3 of gas = m/V kg

$$\text{Latent heat of steam per } m^3 \text{ of gas} = \frac{m \times 587}{V} \text{ k cal.}$$

$$\therefore \quad \text{Net Calorific Value} = [\text{Gross Calorific Value} - \frac{m \times 587}{V} \text{ k cal/m}^3$$

3. *Theoretical Calculation of Calorific Value of a Fuel:* Theoretically, the calorific value of a fuel can be calculated if the percentages of the constituent elements are known. According to Dulong, the Calorific value of a fuel is the sum of the calorific values of its constituents.

The calorific value of C, H and S are found to be 8080, 34500 and 2240 k cal/kg respectively.

Substrate	Calorific value
Carbon	8080
Hydrogen	34500
Sulphur	2240

If oxygen is also present, it combines with hydrogen to form H_2O. Thus the hydrogen in the combined form is not available for combustion and is called fixed hydrogen.

∴ Amount of hydrogen available for combustion

= Total mass of hydrogen–hydrogen combined with oxygen.

Now, 8 parts by weight of oxygen combines with 1 part by weight of hydrogen to form water *i.e.* for every 8 parts of oxygen, 1 part of hydrogen gets fixed.

$$\underset{1g}{H_2} + \underset{8g}{\tfrac{1}{2}O_2} \longrightarrow \underset{9g}{H_2O}$$

Fixed Hydrogen = Mass of oxygen in the fuel

∴ Amount of hydrogen available for combustion

$$= \text{Total mass of hydrogen} - \frac{1}{8}\text{ mass of oxygen in fuel.}$$

$$= H–O/8$$

Dulong's formula for calculating the calorific value is given as:

Gross calorific Value (HCV) $= \frac{1}{100}\left[8080\,C + 34{,}500(H - \frac{0}{8}) + 2{,}240\,S\right]$ k cal/kg.

where C, H O and S are the percentages of carbon, hydrogen, oxygen and sulphur respectively.

Net Calorific Value (LCV) $= \left[HCV - \frac{9}{100}\, H \times 587\right]$ k cal/kg

$= [HCV - .09H \times 587]$ k cal/kg.

SOLVED NUMERICALS BASED ON DETERMINATION OF CALORIFIC VALUE OF FUELS

Example 2.1. The following data were obtained in a bomb calorimeter experiment.

Weight of Coal burnt = 0.994 g
Weight of water in calorimeter = 2592 g
Weight of bomb, calorimeter etc = 3940 g
Rise in temperature of water = 2.732°C
Mean specific heat of the apparatus = 0.098

Find the gross calorific value of the fuel. If the fuel contains 8% hydrogen, calculate its lower calorific value. (Latent heat of condensation of steam = 587 cal/g).

Solution:

Weight of the fuel, x = 0.994 g
Weight of water, W = 2592 g
Weight of apparatus = 3940 g
Mean specific heat of apparatus = 0.098
∴ Water equivalent of apparatus, w = 3940×0.098 = 386.12
Rise in temperature of eater = 2.732°C

$$\therefore \text{Gross calorific value} = \frac{(W + w)(t_2 - t_1)}{x}$$

$$= \frac{(2592 + 386.12)(2.732)}{0.994} = 8185.3 \text{ cal/g}$$

Net Calorific Value = HCV – 0.09 H × 587 cal/gm
= 8185.3 – 0.09 × 8 × 587 cal/gm
= 8185.3 – 422.64 cal/gm = 7762.6 cal/gm

Example 2.2. A sample of coal contains 92% C, 5% H, and 3% ash. When this coal was tested for its calorific value in the bomb calorimeter, the following results were obtained:

Weight of the coal burnt = 0.95 g
Weight of water taken = 700 g
Water equivalent of bomb and calorimeter = 2000 g
Increase in temperature = 2.48°C

Acid Correction = 60.0 Cal

Cooling Correction = 0.02° C

Fuse wire Correction = 10.0 Cal.

Calculate the net and gross calorific value of coal Given latent heat of condensation of steam = 587 cal/g

Soltuion: $$\text{GCV} = \frac{(W+w)(T_2 - T_1 + \text{Cooling Correction}) - (\text{Acid + Fuse corrections})}{\text{Mass of fuel}}$$

$$= \frac{(2000 + 7000)\ (2.48 + 0.02) - (60 + 10)}{0.95}$$

$$= 7031.6 \text{ cal/g}$$

$$\text{Net Calorific Value} = \text{GCV} - 0.09\text{H} \times 587 \text{ cal/g}$$

$$= 7031.6 - 0.09 \times 5 \times 587$$

$$= 6767.45 \text{ cal/g}$$

Example 2.3. During the determination of calorific value of a gaseous fuel by Boy's calorimeter, the following results were recorded.

Volume of gaseous fuel burnt at N.T.P. = 0.098 m^3

Weight of the water used for cooling the combustion products = 50 kg

Weight of steam condensed = 0.051 kg

Temperature of inlet water = 26.1° C

Temperature of outlet water = 46.5° C

Determine the gross and net calorific values.

Solution: $$\text{Gross calorific value GCV} = \frac{W(t_2 - t_1)}{V}$$

$$= \frac{50 \times (46.5 - 26.1)}{0.098}$$

$$= 10408.2 \text{ k cal/m}^3$$

$$\text{Net calorific value} = \text{Gross calorific value} - \frac{m}{V} \times \text{Latent heat of condensation of steam}$$

$$= 10408.2 - \frac{0.051}{0.098} \times 587$$

$$= 10102.7 \text{ k cal/m}^3$$

Example 2.4. Calculate the gross and net calorific value of a coal sample having the following composition. C: 82 per cent, H_2: 8 per cent, O_2: 5 per cent, S: 2.5 per cent, N_2: 1.4 per cent and ash 2.1 per cent.

Solution: $$\text{GCV} = \frac{1}{100}\left[8080 \times \%\text{C} + 34500\left(\%\text{H} - \frac{\%\text{O}}{8}\right) + 2240 \times \%\text{S}\right]$$

$$= \frac{1}{100}\left[8080 \times 82 + 34500\left(8 - \frac{5}{8}\right) + 2240 \times 2.5\right]$$

$$= \frac{1}{100}[662560 + 254437.5 + 5600] = 9225.9 \text{ k cal/kg}$$

$$\text{Net Calorific Value} = \text{GCV} - 0.09\text{H} \times 587$$
$$= 9225.9 - 0.09 \times 8 \times 587$$
$$= 8803.26 \text{ k cal/kg}$$

Example 2.5. A coal has the following composition by weight: C = 92%, O = 2.0%, S = 0.5%; N = 0.5% and ash = 1.5%. Net Calorific value of the coal was found to be 9,430 k cal/kg. Calculate the percentage of hydrogen and higher calorific value of coal.

Solution:

$$\text{HCV} = \frac{1}{100}\left[8080 \times 92 + 34500\left(\text{H} - \frac{2}{8}\right) + 2{,}240 \times 0.5\right] \text{k cal/g}$$
$$= [7433.60 + 345\text{ H} - 86.25 + 11.2] \text{k cal/kg}$$
$$= [7358.55 + 345\text{ H}] \text{k cal/kg.} \qquad \text{....(i)}$$
$$\text{Also HCV} = (\text{Net Calorific Value} + 0.09\text{H} \times 587) \text{ k cal/kg}$$
$$= 9430 + 0.09\text{ H} \times 587$$
$$= 9430 + 52.83\text{ H} \qquad \text{....(ii)}$$

From (i) and (ii)

$$[7358.55 + 345\text{ H}] = [9430 + 52.83\text{ H}]$$
$$292.17\text{ H} = 2071.45$$
$$\therefore \text{ Percentage of H} = \frac{2071.45}{292.17} = 7.09\%$$
$$\therefore \text{ HCV} = (7358.55 + 345\text{ H}) \text{k cal/kg}$$
$$= 7358.55 + (345 \times 7/09) \text{k cal/kg}$$
$$= 9804.6 \text{ k cal/kg.}$$

2.3 CHARACTERISTICS OF GOOD FUEL

While selecting a fuel for particular purpose, following factors are taken into consideration:

(i) **Suitability:** The fuel selected should be most suitable for the process. For example, coke made out of bituminous coal is most suitable for blast furance.

(ii) **High calorific value:** The fuel should possess high calorific value because calorific value determines the efficiency of the fuel. Higher is the amount of heat per unit mass or volume, better the fuel.

(iii) **Ignition temperature:** Ignition temperature is the lowest temperature to which the fuel must be heated so that it starts burning smoothly. A good fuel should have moderate ignition temperature. If the fuel have low ignition temperature, the storage and transportation become dangerous, since it can cause fire hazards. If, on the other hand the ignition temperature is too high, the transportation and storage become easier but it causes difficulty in kindling.

(iv) **Moisture content:** The moisture content should be low because presence of high moisture lowers the effective calorific value of the fuel.

(v) **Non combustible matter content:** The non combustible matter forms clinkers after combustion. The presence of non combustible matter reduces the heating value and also involves the cost of disposal after the combustion.

(vi) **Velocity of combustion:** Velocity of combustion should be moderate. Too high velocities of combustion are not required, whereas too low velocities cause loss of heat due to radiation.

(vii) **Nature of the products:** The products formed should not be harmful. The gaseous products evolved during combustion should be such that they do not cause unpleasant odours and other harmful effects.

(viii) **Cost of fuel:** The fuel should have a low cost and should be available in large amount at a cheap rate.

(ix) **Smoke:** Formation of smoke should be as low as possible.

(x) **Control of the process:** The fuel should be such that its combustion can easily be controlled *i.e.* it can be easily started or stopped.

A comparison between the solid, liquid and gaseous fuels for the above mentioned characteristics is given in a tabulated form below:

Table: 2.1 *Relative Merits of the Three Physical Forms of Fuel*

	Solid fuels	*Liquid fuels*	*Gaseous fuels*
1.	Relatively cheap and easily available	More costly than solid fuels but cheap in countries of origin	Except natural gas, other gaseous fuels are derived from solid and liquid fuels and hence costly.
2.	Storage, transportation, handling are quite difficult; usually liquefied and gasified products are manufactured very near to the coal site	Can be transported easily from storage tanks through a system of pipes. Greater care must be taken to store them in closed containers. Storage space is less compared to solid and gaseous	Have to be stored in leak proof containers and need considerable storage space; can be distributed through pipelines. Domestic suplies can be arranged either through pipes or by liquefying and fuels storing under pressure in iron cylinders (LPG cylinders)
3.	Combustion is a slow process and once started must be maintained by air. The control and stopping of combustion	Combustion takes place quickly. The process of combustion can be controlled or stopped when needed	More readily miscible with air compared to liquids and the combustion takes place rapidly and more efficiently. Air required is are not easy less compared to solid and liquid fuels. Neutral oxidizing or reducing atmospheres can be maintained by controlling air supply and so a greater flexibility in usage is available.
4.	Invariably some ash is produced and their disposal is a problem.	The burning is clean and no ash is produced but fuels with high carbon and aromatic contents may produce smoke.	Neither ash nor smoke is produced.
5.	Labour is required for	The labour demand is less stocking coal.	The labour demand is less
6.	Not used in internal combustion engine.	Used in internal combustion engines (petrol)	Used in internal combustion engine (CNG)
7.	Have low thermal efficiency.	Their thermal efficiency is higher than that of solid fuels.	Their thermal efficiency is highest.

2.4 SOLID FUELS

Solid fuels that occur naturally (primary fuels) as well as those, that are processed and prepared (secondary) are widely used in domestic and industrial purposes. Wood, and different types of coal constitute the important primary solid fuels. Charcoal, coke etc are the secondary solid fuels derived from the natural varieties.

Wood: Wood has been used as a fuel from ancient times. However, due to large scale deforestation, wood is no longer used except in forest areas where wood is available at a very low cost.

When freshly cut, wood contains 25 to 50 percent moisture. Normally it is used in an air-dried condition with 10 to 15 percent moisture content. Its heating value is directly proportional to the water content. The combustible part of wood is mostly made up of cellulose and lignin having traces of mineral ash. Resins and waxes are also present.

The calorific value of air-dried wood is about 3,500 to 4,500 k cal/kg. It burns with a long and non smoky flame, leaving a small amount of ash. Although ash content is low, the oxygen content is very high. This makes, even dry wood a fuel of low calorific value. Still, its low ignition temperature, non-smoky flame and absence of ash disposal problems are the points of advantage. Wood is largely used as a domestic fuel.

Wood charcoal is obtained by destructive distillation of wood. The major use of wood charcoal is for producing activated carbon which finds extensive application for decolourization in sugar industry.

2.4.1 Coal

Coal is regarded as a fossil fuel produced from the vegetable debris under conditions of high temperature and pressure over million of years. Chemically, coal is highly complex organic matter having varying amount of water, together with nitrogen and sulphur.

The transformation of the vegetable debris to coal takes place in two stages:

(i) **Biochemical or peat stage:** During this stage, the plant materials were at tacked by various micro organisms. The end product of this stage is peat

$$2(C_6H_{10}O_5)_n \longrightarrow \text{Peat}$$

(ii) **Chemical stage or metamorphism:** In this stage, the peat deposit buried under sedimentary deposits lose moisture and volatile components under the effect of high temperature and pressure. The peat gets enriched in carbon whereas its oxygen content decreases. The spongy peat transforms into hard brittle coal gradually. The time required for the formation of young brown coal is of the order of 10^7 years while that of most nature coals is 3×10^8 years.

2.4.2 Classification of Coal

Coals are mainly classified on the basis of their rank, which is defined as the degree of coalification from the parent material, wood. When wood is converted into coal, there is gradual increase in the concentration of carbon and decrease in the percentage of oxygen and nitrogen. The coals are given a ranking depending upon the carbon content of the coal from wood to anthracite.

The ranking can be represented as in table 2.2.

Table: 2.2 *Classification of coal according to rank*

Type of coal	*Percentage (dry, mineral matter free basis)*					*% moisture*	*Calorific value cal g^{-1}*
	C	*H*	*O*	*N*	*VM*		
Wood ↓	40-50	5-6	20-40	0-0.5	–	70-90	4000–4500
Peat ↓	45-60	3.5-6.5	20-45	0.75-3	45-75	70-90	4125–5280
Brown coal ↓	60-75	4.5-5.5	17-35	0.75-2	45-60	30-50	6600–7100
Bituminous Coal ↓	75-90	4.0-5.5	20-30	0.75-2	11-50	10-20	6600–8800
Anthracite	90-95	3-4	2-3	0.5-2	3.8-10	1.5-3.5	8470–8800

Thus it is clear from the table that whereas carbon content and calorific value of coal increases with degree of coalification, hydrogen, oxygen, nitrogen, volatile matter, moisture decreases on increase in degree of coalification.

2.4.3 Analysis of Coal

The composition of coal varies according to the degree of coalification. In order to compare the coals for the use of consumer, the analysis of coal is carried out. The detailed methods of analysis and testing vary from country to country. Coal is analysed in two ways:

1. Proximate Analysis
2. Ultimate Analysis

The results of analysis are generally reported in the following ways:

- As received basis
- Air dried basis
- Moisture free basis (oven dried)
- Moisture and ash free basis

PROXIMATE ANALYSIS

Proximate analysis is an empirical but important analysis. *The data varies with the procedure adopted and hence it is called proximate analysis.* It gives information about the practical utility of coal. Proximate analysis of coal determines the moisture, ash, volatile matter and fixed carbon of coal.

1. **Moisture Content:** Owing to its nature and origin coal is always associated with moisture. When a wet coal is exposed to atmosphere, the external moisture evaporates but the apparently dry coal still contains some moisture. This moisture is referred to air dried moisture.

   Air dried moisture is determined by heating a known amount of coal (air dried) to 105-110°C in an electric hot air oven for about one hour. After one hour, it is taken out from the oven and cooled in a dessicator and weighed. Loss in weight of coal is reported as moisture content on percentage basis.

$$\text{Percentage of moisture} = \frac{\text{Loss in weight}}{\text{Weight of coal taken}} \times 100$$

Significance

- Excess of moisture is undesirable in coal.
- Moisture lowers the heating value of coal and takes away appreciable amount of the liberated heat in the form of latent heat of vapourisation. For every percent of moisture present .1% of heat is lost. Moisture also increases the transport costs.
- Excessive surface moisture may cause difficulty in handling the coal.
- Moreover presence of excessive moisture quenches fire in the furnace.

Hence lesser the moisture content, better is the quality of coal as a fuel. However presence of about 5-10% moisture is desirable as it produces a uniform fuel bed and less of "fly ash".

2. **Volatile Matter:** The volatile matter in coal consists of a complex mixture of gaseous and liquid products resulting from the thermal decomposition of the coal. Volatile matter does not include moisture of coal. It consists mainly of combustible gases such as H_2, CO, CH_4 and other hydrocarbons.
   It is determined by heating a known weight of moisture free coal sample in a covered platinum crucible at 950 ± 20° C for 7 minutes. At this temperature, hydrocarbons and hydrogen are driven off. The crucible is cooled, first in air, then inside a dessicator and weighed. Loss in weight is reported as volatile matter on percentage basis.

$$\text{Percentage of Volatile matter} = \frac{\text{Loss of weight due to removal of volatile matter}}{\text{Weight of coal sample taken}} \times 100$$

Significance

- A high percent of volatile matter indicates that a large proportion of fuel is burnt as a gas or vapour or may escape unburnt.
- The high volatile matter content gives long flames, high smoke and relatively low heating values.
- For efficient use of fuel, the outgoing combustible gases (volatile matter) have to be burnt by supplying secondary air. This requires a larger combustion space. A furnace with a small combustion volume or of short flame is not suitable for burning high volatile coals, since a large proportion of volatile matter will escape unburnt.
- The high volatile coals do not cake well (described later) whereas medium-volatile coals containing 20-30% of volatile matter are capable of forming hard and strong coke on carbonization.
- Coals containing less than 14% of volatile matter do not cake at all and are thus not suitable for manufacturning coke.

However, high volatile matter content is desirable in coal gas manufacture because volatile matter in a coal denotes the proportion of the coal which will be converted into gas and tar products by heat.

3. **Ash:** Coal contains inorganic mineral substances which are converted into ash by chemical reactions during the combustion of coal. Ash usually consists of silica, alumina, iron oxide and small quantities of lime, magnesia, etc.
   Ash content is determined by heating the residue left after the removal of volatile matter at 700 ± 50° C for 1/2 an hour *without* covering. The cooled mass is weighed. From the weight of the residue, the percentage of ash is calculated as.

$$\text{Percentage of ash} = \frac{\text{Weight of the residue left}}{\text{Weight of the coal}} \times 100$$

Ash can be classified as intrinsic ash and extrinisic ash. The mineral matter originally present in vegetable matter from which the coal was formed is called *intrinsic ash*. It consists of oxides of Na, K, Mg, Ca and Si. On the other hand the mineral matter, like clay, gypsum, dirt which gets mixed up during mining and handling of coal constitute the *extrinsic ash* which remains as a residue after the combustion. The extrinsic ash may consist of anhydrous $CaSO_4$, $CaCO_3$, Fe_2O_3 etc.

The minerals responsible for the formation of intrinsic ash cannot be removed by any mechanical means. Fortunately, this mineral matter content is low (about 2 to 3 percent) whereas the extrinsic ash content is relatively higher.

Significance

- The high percentage of ash is undesirable. The ash does not contribute to the calorific value of coal. On the contrary it reduces the calorific value and creates many difficulties in efficient utilization of coal.
- In furnace grate, the ash may restrict the passage of air and lower the rate of combustion.
- High ash leads to large heat losses and cause problems of clinkering (forming of ash lumps). It has been estimated that one percent rise in coal ash is equivalent to 0.3 to 4 percent decrease in boiler efficiency.
- The composition of ash and fusion range also influences the efficiency of coal. Its composition is of considerable importance in metallurgical operations as it affects the slag and metal composition.
- When coal is used in boiler, the fusion temperature of ash is very significant. Generally fusion temperature lies in the range of 1000°–1700° C. Ash having fusion temperature below 1200° C is called *fusible ash* and above 1430°C is called *refractory ash*. If ash fuses at working temperature it leads to clinker formation.
- Apart from loss of efficiency of coal, clinker formation also leads to loss of fuel because some coal particles also get embedded in the clinkers.

However, some ash is desirable since it protects the grate from direct contact with the incandescent coal, which might cause oxidation of the grate bars.

4. **Fixed Carbon:** Fixed carbon content increases from low ranking coals such as lignite to high ranking coals such as anthracite. Higher the percentage of fixed carbon greater is its calorific value and better is the quality of coal. This represents the quantity of carbon that can be burnt by a primary current of air.
The percentage of fixed carbon is given by:

$$\text{Percentage of fixed carbon} = 100 - [\text{\% of Moisture + Volatile matter + ash}]$$

From the above it appears that, what is left after the amount of moisture volatile matter and ash formed is subtracted, corresponds to the fixed carbon. However, the term fixed carbon is misnomer because it contains other non volatile elements also along with carbon.

Significance:

- Higher the percentage of fixed carbon, greater is the calorific value.
- The percentage of fixed carbon helps in designing the furnace and shape of the fire-box because it is the fixed carbon that burns in the solid state.

ULTIMATE ANALYSIS

Ultimate analysis of coal is carried out to ascertain the composition of coal. For the utilization of coal for industrial purpose, it is necessary to determine the percentages of various components so as to classify the coal, to fix its price, along with putting the coal to different uses depending upon its composition. Ultimate analysis includes the estimation of carbon, hydrogen, sulphur, nitrogen and oxygen.

1. Carbon and hydrogen: A known amount of coal is taken in a combustion tube and is burnt in excess of pure oxygen (free from moisture and CO_2). The carbon and hydrogen present in the fuel are converted to CO_2 and H_2O respectively.

$$C + O_2 \longrightarrow CO_2$$

$$H_2 + \frac{1}{2}O_2 \longrightarrow H_2O$$

The gaseous products of combustion are passed through two bulbs, one containing weighed amount of anhydrous $CaCl_2$ which absorbs water, and the other containing weighed amount of KOH which absorbs carbon dioxide. The percentage of carbon and hydrogen are calculated from the increase in weight of the respective absorption tubes.

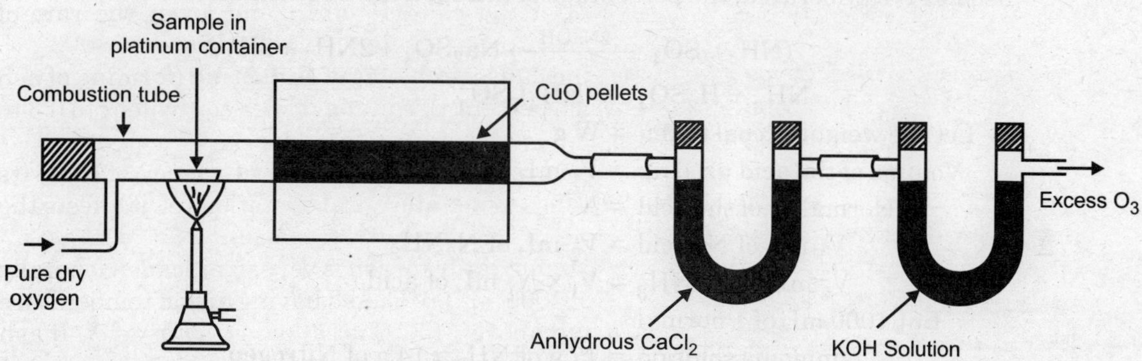

Fig. 2.3 *Estimation of carbon and hydrogen*

$$2KOH + CO_2 \rightarrow K_2CO_3 + H_2O$$

$$CaCl2 + 7H_2O \rightarrow CaCl2\ 7H_2O$$

Let, weight of coal sample taken = x g

Increase in weight of $CaCl_2$ tube = z g

Increase in weight of KOH tube = y g

44 g of CO_2 contain = 12 g of carbon

y g of CO_2 contain = $\frac{12}{44} \times y\ g$ of carbon

$$\textbf{Percentage of carbon} = \frac{12}{44} \times \frac{y \times 100}{\text{weight of coal taken}}$$

18 g of water contain = 2 g of hydrogen

z g of water contain = $\frac{2}{18} \times zg$ of hydrogen

$$\textbf{Percentage of hydrogen} = \frac{2}{18} \times \frac{z \times 100}{\text{weight of coal taken}}$$

Significance

- Calorific value of a fuel is directly related to its carbon content. Hence greater the percentage of carbon, greater is the calorific value of the fuel.
- A higher percentage of carbon also reduces the size of the combustion chamber.
- High percentage of hydrogen also increases the calorific value of coal. The content of hydrogen in coals varies between 4.5 to 6.5 percent from peat to bituminous stage. The hydrogen of coal is responsible for the production of many useful materials, for example, gaseous hydrogen, methane and coal chemicals in carbonization and gastification industries.

2. **Nitrogen:** Nitrogen present in the coal sample can be estimated by Kjeldahl's method.
   - A known weight of the given coal sample along with K_2SO_4+$HgSO_4$ which acts as a catalyst is heated with conc. H_2SO_4. Nitrogen present in coal gets converted to $(NH_4)_2SO_4$, quantitatively.

$$\underset{\text{(from Coal)}}{\text{Nitrogen}} + H_2SO_4 \xrightarrow{\text{Heat}} (NH_4)_2SO_4$$

   - The contents are then transferred to a round bottmed flask and the solution is heated with excess of NaOH.
   - The NH_3 gas thus liberated is absorbed in a known volume of a standard solution of acid.
   - The unused acid is then determined by titrating with NaOH. From the volume of acid used by NH_3 liberated, the percentage of nitrogen can be calculated.

$$(NH_4)_2SO_4 \xrightarrow{2NaOH} Na_2SO_4 + 2NH_3 + 2H_2O$$

$$NH_3 + H_2SO_4 \rightarrow (NH_4)_2SO_4$$

Let the weight of coal taken = W g

Volume of the acid used up = V_1 mL

Normality of the acid = N_1

V_1 mL of N_1 acid = V_2 mL of N_2NH_3

$\therefore$ V_2 mL of IN NH_3 = $V_1 \times N_1$ mL of acid

But 1000 mL of 1 normal ammonia solution ≡ 17 g of NH_3 ≡ 14 g of Nitrogen

$$V_2 \text{ mL of 1 normal ammonia solution} \equiv \frac{14}{1000} \times N_1 \times V_1 \text{ g of Nitrogen} \qquad (\because V_2 = N_1V_1 \text{ ml})$$

$$\therefore \text{Percentage of nitrogen} = \frac{14}{1000} \times \frac{N_1 \times V_1 \times 100}{W} = \frac{1.4 \times N_1 \times V_1}{W}$$

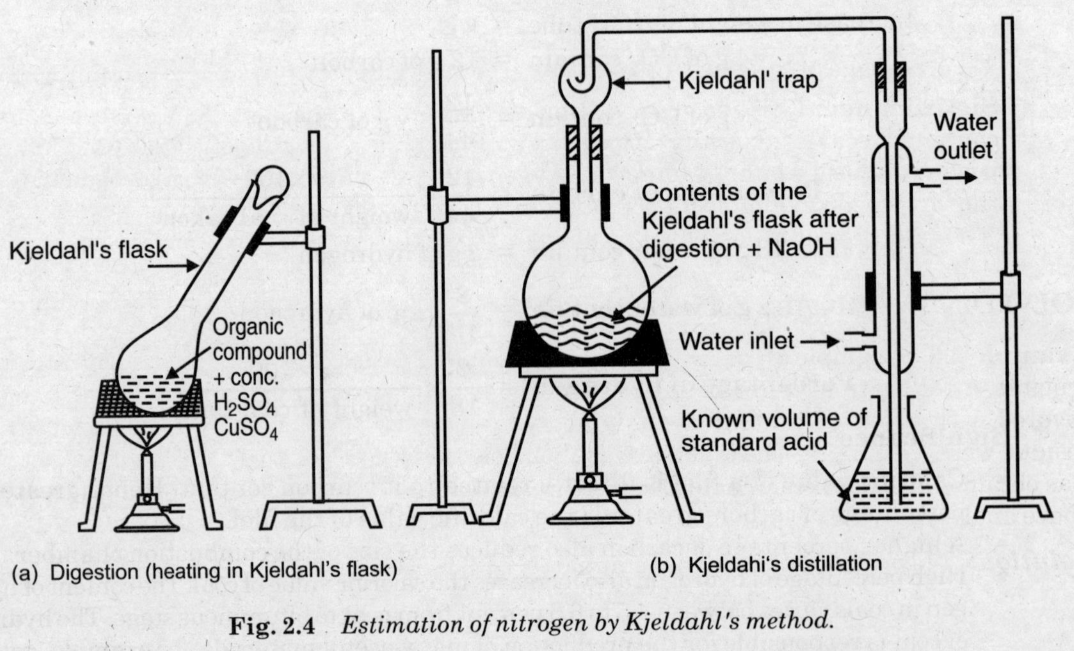

Fig. 2.4 *Estimation of nitrogen by Kjeldahl's method.*

Significance: Nitrogen content in most of coals is between 1 and 2 percent. It has no calorific value and hence its presence is undesirable. However, it is useful in carbonization industries which recover a part of it as valuable by products like NH_3, $(NH_4)_2SO_4$ and pyridine bases.

3. **Sulphur:** Sulphur present in the coal is oxidized by fuming nitric acid into sulphuric acid. The sulphuric acid is precipitated as $BaSO_4$ by the addition of $BaCl_2$, quantitatively. Alternatively, the sulphur can also be determined by the washings obtained from the known mass of coal used in a bomb calorimeter for the determination of its calorific value. The washings contain sulphur in the form of sulphate from which it is precipitated as $BaSO_4$ by the addition of $BaCl_2$.
The precipitate of $BaSO_4$ so obtained from either of the above method is filtered and heated to a constant weight. From the weight of $BaSO_4$ obtained, the sulphur present in the coal is calculated. Let the weight of coal sample = W g

$$\text{the mass of the precipitate be } = W_1 \text{ g}$$

$$\because \quad 233 \text{ g of } BaSO_4 \equiv 32 \text{ g of sulphur}$$

$$W_1 \text{ g of } BaSO_4 \equiv \frac{32}{233} W_1 \text{ g}$$

$$\text{Percentage of sulphur} = \frac{32}{233}\frac{W_1}{W} \times 100$$

Significance: Although sulphur contribute to the calorific value of coal, it is undesirable constituent. It emits corrosive SO_2, SO_3 during combustion and highly objectionable H_2S during carbonization of coal. These gases also cause atmospheric pollution. Sulphur containing coal is not suitable for the preparation of metallurgical coke as it adversely affects the properties of metal. For example, the presence of sulphur in the coke used in iron industry affects the quality and properties of steel.

4. **Ash:** It is determined in the same way as in proximate analysis.
5. **Oxygen:** The oxygen is usually determined by the method of difference. The carbon, hydrogen, nitrogen and sulphur are expressed in percent basis and their sum is subtracted from 100.

Percentage of oxygen = 100 – (% of carbon + hydrogen + Nitrogen + Sulphur + ash)

Significance: Less is the oxygen content, better is the coal. This is because oxygen content of coal is generally associated with moisture, low calorific value and low coking power (discussed later). Moreover, oxygen is in combined form with hydrogen in coal and thus hydrogen available for combustion is lesser than the total hydrogen present.
The calorific value decreases by about 1.7% for every 1% increase in oxygen.

SOLVED NUMERICALS BASED ON PROXIMATE AND ULTIMATE ANALYSIS

Example 2.6. A sample of coal was analyzed as follows: Exactly 2.000 g was weighed into a silica crucible. After heating for one hour at 110° C, the residue weighed 1.975 g. The crucible next was covered with a vented lid and strongly heated for exactly seven minutes at 950 ± 20°C. The residue weighed 1.328 g. The crucible was then heated without the cover, until a constant weight was obtained. The last residue was found to weigh 0.205 g. Calculate the percentage results of the above analysis.

Solution: Mass of moisture in coal sample = 2.000 – 1.975 = 0.025 g

Mass of volatile matter = 1.975 – 1.328 = 0.647 g

Mass of ash = 0.205 g

$$\text{Percent of moisture} = \frac{0.025 \times 100}{2} = 1.25\%$$

$$\text{Percent of volatile matter} = \frac{0.025 \times 100}{2} = 32.35\%$$

$$\text{Percent of ash} = \frac{0.205 \times 100}{2} = 10.25\%$$

$$\text{Percent of fixed carbon} = 100 - (1.25 + 32.35 + 10.25) = 56.15\%$$

Example 2.7. 1.0 g of a sample of coal was used in a bomb calorimeter for the determination of calorific value. Calorific value of coal was found to be 8,800 cal/g. The ash formed in the bomb calorimeter was extracted with acid and the acid extract was heated with barium nitrate solution and a precipitate of barium sulphate was obtained. The precipitate was filtered, dried and weighed. The weight of precipitate was found to be 0.08 g. Calculate the percentage of sulphur in the coal sample.

Solution: Percentage of sulphur in the coal sample.

$$= \frac{\text{Wt. of BaSO}_4\text{ ppt. obtained} \times 32 \times 100}{233 \times \text{Wt. of coal sample taken in bomb}}$$

$$= \frac{0.08\text{g} \times 32 \times 100}{233 \times 1\text{g}} = 1.098\%$$

Example 2.8. 1.56 g of a sample of coal was kjeldahlized and NH_3 gas thus evolved was absorbed in 50 ml of 0.1 N H_2SO_4. After absorption, the excess (residual) acid required 6.25 ml of 0.1N NaOH for exact neutralization. 2.60 g of the coal sample in a quantitative analysis gave 0.1755 g of $BaSO_4$. Calculate the percentage of N and S in the coal sample.

Solution: 6.25 ml of 0.1 N NaOH = 6.25 ml of 0.1 N H_2SO_4

∴ Volume of H_2SO_4 used to neutralize NH_3 evolved = 50 ml of .1 N – 6.25 ml of .1N H_2SO_4

= 43.75 ml of .1 N H_2SO_4

$$\text{Percentage of Nitrogen} = \frac{\text{Vol. of H}_2\text{SO}_4\text{ used} \times \text{Normality} \times 1.4}{\text{Weight of coal sample taken}}$$

$$= \frac{43.75 \times .1 \times 1.4}{1.56} = 3.926\%$$

$$\text{Percentage of sulphur} = \frac{\text{wt. of BaSO}_4\text{ ppt. obtained} \times 32 \times 100}{233 \times \text{wt. of coal sample taken}}$$

$$= \frac{0.1755 \times 32 \times 100}{233 \times 2.60} = 0.927\%$$

2.4.4 Carbonization of Coal (Manufacture of Coke)

Carbonization of coal is the process of heating the coal in absence of air, to a sufficiently high temperature, so that the coal undergoes decomposition and yields a residue which is richer in carbon content than the original fuel. This residue is a lustrous, dense, strong, porous and coherent mass and is known as coke. **In other words the process of converting coal into coke is called carbonization.**

Caking and coking of coals: It must be emphasized here that all types of coal cannot be converted into coke. Some coals have a tendency to soften and swell at higher temperatures, to form a solid coherent mass with porous structure. Such coals are called **caking coals**. The residue so formed is called **coke**. If the coke is hard, porous and strong, than the coal, from which it is formed, it is called **coking coal**. This property is found only in bituminous type of coal and hence only bituminous coal can be coked. Anthracite, sub-bituminous, lignite coals are non-coking coals. All coking coals are caking coals, but only some of the caking coals fall under the category of coking coals.

In general, coals with a high percentage of volatile matter are not fit for 'coking' and are used for gas making. The coals having 20-30% volatile matter are good coking coals.

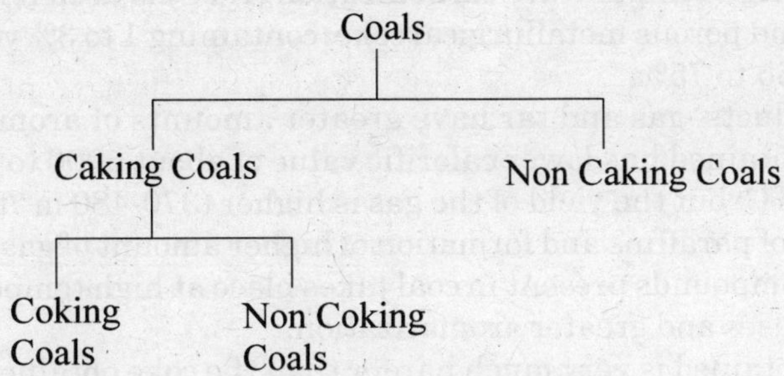

Process of Carbonization:

- First moisture and occluded gases are driven off.
- At about 260° - 270° carbon, water and H_2S, some low molecular alkenes and alkanes are evolved.
- At about 350°C the decomposition of coal is accompanied by evolution of gases and elimination of tarry vapours takes place.
- Around 400°C, caking coals become soft and plastic.
- At about 700°C, hydrogen is evolved.
- Above 800°C, main gaseous products are evolved.
- Gases evolved form the plastic mass, expand it to give faomlike appearance.
- At further high temperatures this foam like mass solidifies to form a solid mass with porous structure called coke.

Types of Carbonization: The properties of the coke produced mainly depend on the type of coal used, temperature of carbonization and the rate of carbonization. Depending upon the temperature, the carbonization is of two types:

(i) Low temperature carbonization.
(ii) High temperature carbonization.

(i) **Low temperature Carbonization:** When the destructive distillation of coal is carried out at temperatures between 500 to 700° C, it is called low temperature carbonization. The low temperature carbonization is practised for the production of semi coke which is also called soft coke or char.

The yield of coke is about 75-80%. The coke thus produced contains 5 to 15% volatile matter and is not sufficiently strong to be used as metallurgical fuel.

The various products of low temperature carbonization are semicoke, low temperature tar, crude low temperature spirit and gas. The quality and yield of these products depend upon the process and the coal used. Low temperature carbonization (LTC) plants normally use low rank coals. These low rank coals produce excessive smoke on burning, but the coke obtained from them is a clean fuel.

Semicoke from LTC is highly reactive and can be easily ignited into a smokeless flame. This is an ideal domestic fuel. The gas which is obtained as a byproduct has higher calorific value of about 6,500-9,500 k cal/m^3. The yield of the gas is about 130-150 m^3/tonne of the coal carbonized and constitutes a valuable fuel. The tar which is also obtained as the by product contains more of paraffinic hydrocarbons and a small amount of aromatic hydrocarbons, along with higher phenols and various substituted nitrogeneous compounds.

There are two main commercial LTC plants in India. One is in Tamil Nadu (Neyveli) and other is in Andhra Pradesh.

(ii) **High-Temperature Carbonization:** High temperature carbonization is carried out at 900-1200° C. High temperature carbonization (HTC) is used for the production of pure, hard, strong and porous metallurgical coke, containing 1 to 3% volatile matter. The yield of the coke is 65 to 75%.

The byproducts–gas and tar have greater amounts of aromatic hydrocarbons. The gas which is obtained has lower calorific value of about 5000 to 6000 k cal/m^3 than that produced in LTC; but the yield of the gas is higher (370-480 m^3/tonne). The formation of lower amount of paraffins and formation of higher amount of gas, indicates that cracking of paraffinic compounds present in coal takes place at high temperature, resulting in the formation of gases and greater aromatization.

The coke obtained is very much harder than the coke obtained from LTC process and hence is called hard coke. This is used as an industrial as well as domestic fuel.

Table 2.3 *Low and High Temperature Carbonizations*

	Characteristic	*Low Temperature Carbonization*	*High Temperature Carbonization*
1.	Carbonization Temperature	500-700° C	1000-1200° C
2.	Yield	75-80%	65-75%
3.	Volatile matter	5-15%	1-3%
4.	Mechanical strength	It is not mechanically strong	It is mechanically strong
5.	Hardness	Soft coke	Hard coke
6.	Use	Used for domestic purposes.	Used for metallurgical purposes.
7.	Yield of the gas	Lower (About 130-150m^3/ tonne)	Higher (About 300-390m^3/ tonne)
8.	Calorific value of the gas	Higher (6500-9,500 k cal/m^3).	Lower (5,400-6000 k cal/m^3)
9.	Percentage of straight chain hydrocarbons in the by-products (tar and gas)	Higher	Lower
10.	Percentage of aromatics in the by-products	Lower	Higher
11.	Smoke produced on burning	Smokeless	Smoky

2.4.5 Metallurgical Coke

The properties of coke depend on porosity, reactivity and the amount of volatile matter retained by coke during carbonization. Coke is mainly used as a heat source and reducing agent in metallurgy. A good coke used in metallurgical process should possess the following requisites:

(i) **Purity:** The metallurgical coke should contain lowest percentages of moisture, ash, sulphur and phosphorus. Moisture reduces the calorific value whereas ash hinders the heating and also restricts the passage of air and lowers the rate of combustion. High percentage of ash also aids in slag formation.
Sulphur and phosphorus if present adversely affect the quality of the metal being produced.

(ii) **Porosity:** The coke should be porous so as to provide contact between carbon and oxygen which ensures efficient and complete combustion of the fuel in the furnace.

(iii) **Strength:** The coke used in metallurgical process should have high strength so as to withstand the weight of the ore, flux etc in the furnaces.

(iv) **Size:** Metallurgical coke should be of medium size–neither too big nor too small. A big size of coke does not give uniform heating whereas if the size is too small, it may cause choking.

(v) **Combustibility:** Coke should burn easily. The combustibility of coke depends on the nature of the coal, carbonization temperature and reaction temperature. Weakly caking or non–caking coals are more combustible than strongly coking coals. Low temperature carbonization produces coke which are easily combustible as compared to those obtained by high temperature carbonization. However, at high combustion temperature (800°-900°C) all cokes have comparable combustibility. At temperature around 1000° C, the rate of combustion depends upon the rate at which air is supplied.

(vi) **Calorific value:** The calorific value should be high.

(vii) **Reactivity:** Reactivity of coke is generally considered as its ability to react with CO_2, steam, air and O_2. The reactivity should not be too high. However the coke used for water gas manufacture must be reactive to steam. The reactivity toward CO_2, represent the reduction of CO_2 according to the equation,

$$CO_{2\,(g)} + C_{(s)} \rightleftharpoons 2\,CO_{(g)}$$

and is significant in producer gas manufacture.

(viii) **Cost:** Coke should be cheap and easily available near the site of the metallurgical plant.

Manufacture of Metallurgical Coke

Coke required for metallurgic purposes can be prepared by two methods which are as follows:

(i) Beehive oven
(ii) Otto Hoffmann oven

(i) **Beehive Oven:** Upto the end of 19th century, the entire requirement of the metallurgical coke of the world was fulfilled by the coke manufactured in 'beehive ovens'. A beehive oven is a fire-brick chamber having a dome-shaped structure. It is called beehive oven because it resembles in shape to that of a beehive. The dimensions of a typical oven are 4 m wide and 2.5m high. The roof is provided with a hole for charging the coal from the top. Another hole, the discharging hole is provided in the circumference of the lower part of the wall. A number of ovens are built in a row with common walls between neighbouring ovens.

The coal is introduced through the top hole in the dome and is spread evenly on the floor to produce a layer of about 60 to 90 cm deep. Some air is supplied initially to ignite the coal. Carbonization starts and volatile matter burns inside the partially closed side door. The carbonization proceeds from the top to bottom of the charge and is completed in two to three days. Heat is supplied by the burning of the volatile matter and hence no by-products are recovered. The exhaust gases are allowed to escape to the atmosphere. At the end of carbonization, the hot coke is quenched with water and is discharged manually through the slide door. The walls and roof of the emptied oven retain enough heat to initiate the carbonization of the next charge. The yield of the coke is about 80% of the coal charged.

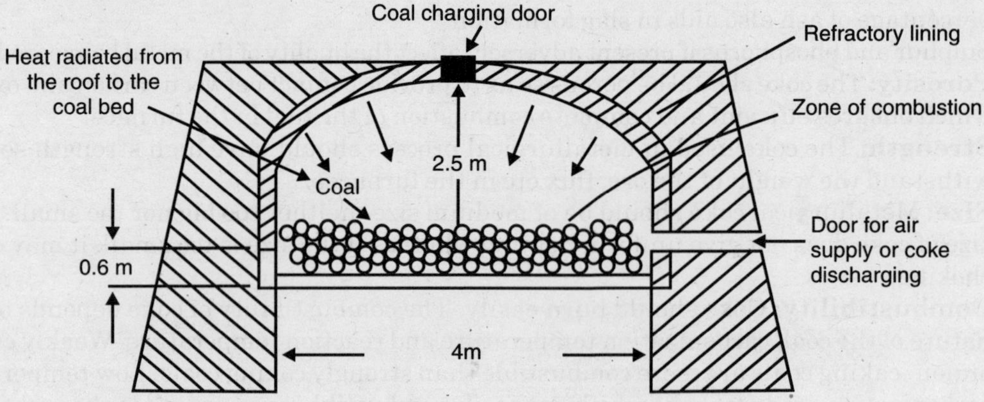

Fig. 2.5 *Beehive Coke Oven*

The process can be made economical by operating a series of ovens in such a way that the hot oven gases escaping to the atmosphere are utilized to ignite the charge in the adjacent oven.

Demerits of Beehive Ovens: The demerits are

- No recovery of by-products, which are useful chemicals and are allowed to escape.
- Lower coke yield due to partial combustion.
- Lack of flexibility of operation.

Inspite of the above demerits, the beehive ovens are still in use because of their low capital and running costs.

(ii) **Otto Hoffmann's Oven or By-product Oven:** The beehive-ovens have been mostly replaced by the chamber ovens which operate on regenerative principle of heat economy *i.e.* utilization of the waste fuel gases for heating of the checker work of bricks. Moreover, all the valuable products are recovered from the outgoing flue gases.

Construction: The Otto Hoffmann's oven consist of a number of narrow rectangular chambers made up of silica bricks. The usual dimensions are 12 to 14m length, 4 to 5 m height and .5 m width. The chambers are tightly closed so that no air can enter. At the top of each chamber there are three holes for charging the coal. Each chamber is further provided with a gas take off and refractory-lined cast iron door at each end for discharging the coke. These carbonization chambers are erected side by side with vertical flues or interspace for combustion in between them. The ovens (chambers) form a sort of battery of 10 to 100 ovens. One single oven is capable of holding 16 to 24 tonnes of coal.

Working: Coal is charged into the chamber and the chambers are closed tightly at both the ends. The coke ovens are heated to 1200° C by burning gaseous fuels (usually coke oven gas).

The process of carbonization takes place layer by layer in the coal charge starting from the two side walls of the oven and moving towards the centre. As the coal adjacent to the oven walls gets heated, a plastic zone is formed which moves away from the walls towards the central zone. This moving plastic zone thus inturn heats up the central zone.

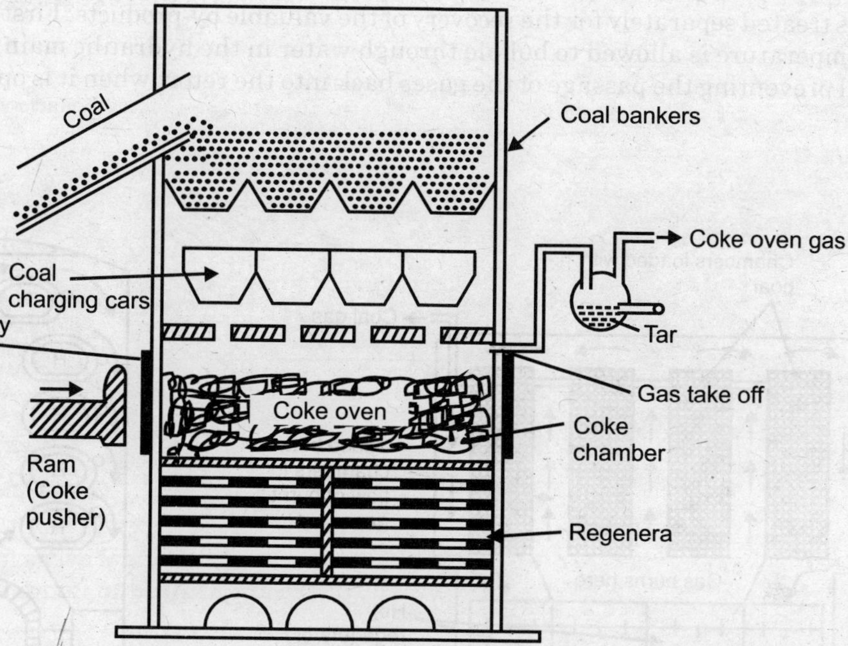

Fig. 2.6 *Schematic representation of a single chamber of Otto Hoffmann's Oven*

As the coal is converted into coke, there is decrease in volume. This is because of the removed of volatile matter in the form of tar and gas at about 500ºC. At further higher temperature (around 750–800ºC), the plastic mass solidifies into hard and porous mass called coke. The hardness is due to the dehydrogenation accompanied by aromatization in the coal.

Regenerative principle is employed to achieve as economical heating as possible. The waste gases produced during combustion (flue gases) are passed through the regenerators and then let off to the atmosphere through the chimney. Regenerators are built underneath the ovens. The flue gases pass their sensible heat to the checker brick work of regenerators until the temperature rises to 1000º C. The flow of heating flue gases is then reversed and the inlet gases (fuel gas and air) used for combustion are passed through the heated checker brickwork of regenerators. Thus inlet gases are preheated before entering into the interspace for combustion. After some time, the flow of gases is again reversed so that the waste combustion gases again heat the checker brick work. Thus regenerators work on the principle of alternate heating and cooling cycles. This is achieved by periodically changing the direction of flow of gases through the vertical flues every 30 minutes or so.

Carbonization of a charge of coal takes about between 11 to 18 hours. After the process is complete, red hot coke is pushed outside by means of a ram which is electrically driven. The coke falls into a quenching car. The yield of the coke is about 75% of coal.

The quenching of red hot coke is achieved by spraying water (wet quenching) or by passing inert gases from boilers (dry quenching). Dry quenching is advantageous over wet quenching because the coke produced is more strong, and dense. Moreover in wet quenching the sensible heat of the coke is wasted whereas in dry quenching the inert gases after quenching the coke become heated up and are then circulated to boilers where they pre-heat the boiler-feed water and then recirculated for cooling fresh charge of coke.

Recovery of by-products: The gases and vapours evolved on carbonization in coke ovens are not allowed to mix with the combustion and are collected separately. The gas known as coke oven gas is treated separately for the recovery of the valuable by-products. First the gas which is at high temperature is allowed to bubble through water in the hydraulic main, which acts as a water seal preventing the passage of the gases back into the retort when it is opened. (Fig. 2.8)

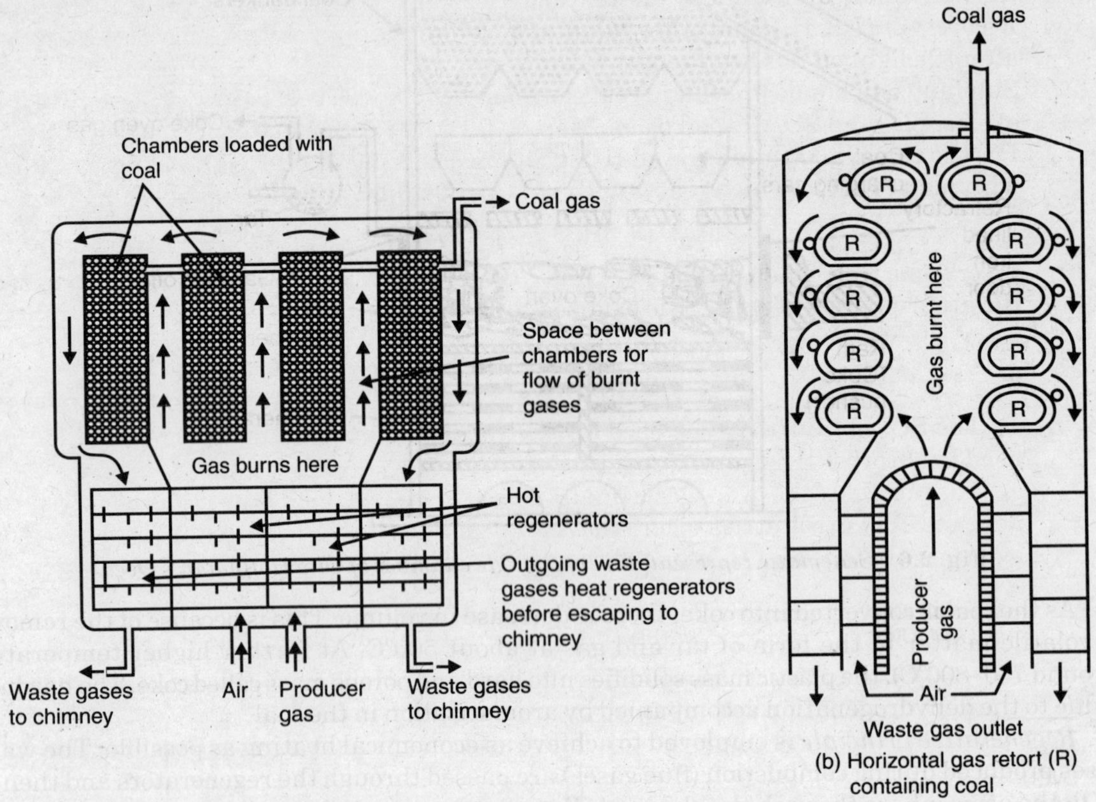

Fig. 2.7 *Otto Hoffmann's Byproduct coke oven*

(i) *Recovery of Tar:* The gas from the coke ovens is passed through a tower in which liquor ammonia is sprayed. Tar and dust get collected in a tank. The tank is provided with heating coils to recover back the ammonia. The ammonia is used again. The 'coal tar' is a valuable source of chemicals. The tar obtained is distilled to give a number of products like crude benzole, phenolic compounds, pyridine bases, heavy naphtha, anthracene oil etc.

(ii) *Recovery of Ammonia:* Next, the gases are then passed through a tower where water is sprayed to recover ammonia. The ammonia can also be removed by dissolving it in H_2SO_4 to form $(NH_4)_2\,SO_4$, which is then used as a fertilizer. The removal of ammonia from the gas is necessary otherwise it may cause blockage of the gas pipes in the form of $(NH_4)_2\,CO_3$.

(iii) *Recovery of Naphthalene:* The gases are passed through a cooling tower, where water at a low temperature is sprayed. The gas is scrubbed with water until its temperature reduces. The naphthalene gets condensed here which is collected along with water and recovered.

(iv) *Recovery of Benzol:* The gases are then introduced into a light oil or benzol scrubber, where benzene along with its homologue is removed and is collected at the bottom. The wash oil along with the benzene homologues is known as the light oil. (Light oil is lighter than water and hence it is called light oil. It is a mixture of low boiling hydrocarbons, benzene, benzene homologues and the solvent used for washing. The solvent is a mixture of ethyl methyl benzene, trimethyl benzene etc.) Separation of different components of the light oil is carried out by distillation. The solvent is returned to the scrubber.

(v) *Recovery of H_2S:* H_2S and other sulphur compounds are removed from the coke oven gas after the light oil has been separated out.

Sulphur compounds present in coal gas must be removed because sulphur along with its oxides, SO_2 and SO_3 aid corrosion. Moreover H_2S has offensive odour.

For the removal of sulphur and H_2S gas, the gas is passed through a chamber, packed with moist Fe_2O_3, which reacts with H_2S to give Fe_2S_3.

$$2Fe_2O_3 + 3H_2S \longrightarrow 2FeO + 3SO_2\downarrow$$

After some time when all the Fe_2O_3 is converted into Fe_2S_3, the purifier is exposed to the atmosphere when Fe_2S_3 is oxidized to the oxide.

$$Fe_2O + 3H_2S \longrightarrow 2FeO + 3SO_2\uparrow$$

$$4FeO + O_2 \longrightarrow 2Fe_2O_3$$

The SO_2 obtained can be used for the manufacturing sulphuric acid, which can be used to absorb NH_3 from the coal gas.

The purified coke oven gas is finally stored over water in holders. It is mostly used as domestic fuel in countries where coal is cheap. The main components of coal gas or coke oven gas are CH_4 and H_2 along with small percentage of CO_2, CO, N_2 and oxygen.

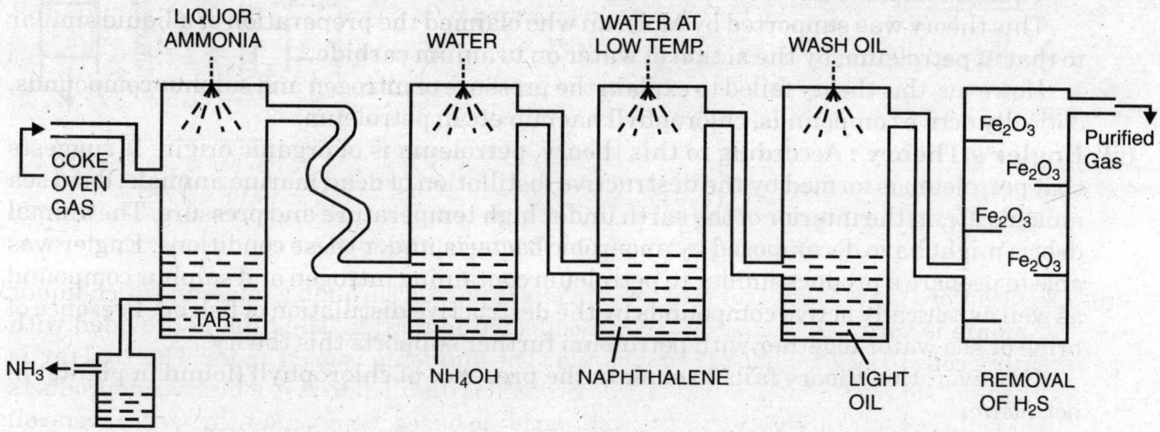

Fig. 2.8 *Schematic Diagram of Coke-Oven Gas Treatment Plant*

2.5 LIQUID FUELS

Liquid fuels are used extensively these days in industrial and domestic fields. Their importance is best understood by the fact that almost all internal combustion engines run on them. The use of liquid fuels in heat generation in ovens and furnaces is also expanding at an increasing rate.

The largest source of liquid fuels is petroleum. The calorific value of petroleum is about 40000 kJ/kg. Though petroleum accounts for the bulk of the liquid fuels, the other sources also supplement the requirement of liquid fuels such as coal tar, crude benzol, synthetic liquid fuels made from coal etc.

Petroleum: The term petroleum means 'rock oil' (in Latin, petra means rock and oleum means oil.) It is also called mineral oil. Petroleum is a complex mixture of paraffinic, olefinic and aromatic hydrocarbons with small quantities of organic compounds containing oxygen, nitrogen and sulphur.

2.5.1 Origin of Petroleum

A number of theories have been put forward to explain the origin of petroleum but none is found satisfactory. Some of these are discussed below briefly:

(i) **Mendeleef Theory :** According to Mendeleef (1876) petroleum is of inorganic origin. It is formed by the action of water on metallic carbides. The metallic carbides are formed by the action of metal and carbon under the conditions of high temperature and pressure, inside the earth.

$$4Al + 3C \rightarrow Al_4C_3$$
$$Al_4C_3 + 12H_2O \rightarrow 4Al(OH)_3 + 3CH_4$$
$$Ca + 2C \rightarrow CaC_2$$
$$CaC_2 + 2H_2O \rightarrow Ca(OH)_2 + C_2H_2$$

The lower hydrocarbons so formed might have undergone hydrogenation or polymerization to form various types of hydrocarbons.

$$3C_2H_2 \xrightarrow{Polymerization} C_6H_6$$

Thus, a complex mixture of paraffinic, olefinic and aromatic hydrocarbons resulted.

This theory was supported by Moissan who claimed the preparation of a liquid similar to that of petroleum, by the action of water on uranium carbide.

However, this theory failed to explain the presence of nitrogen and sulphur compounds, optically active compounds, chlorophyll haemin etc in petroleum.

(ii) **Engler's Theory :** According to this theory, petroleum is of organic origin. It suggests that petroleum is formed by the destructive distillation of dead marine animals, dead sea animals etc, in the interior of the earth under high temperature and pressure. The animal debris might have decomposed by anaerobic bacteria under these conditions. Engler was able to prepare a product similar to petroleum containing nitrogen and sulphur compound as well as optically active compounds by the destructive distillation of fish oil. Presence of brine or sea water together with petroleum further supports this theory.

However, this theory fails to explain the presence of chlorophyll (found in plants) in petroleum.

(iii) **Modern Theory :** According to this theory, it has been suggested that petroleum is formed due to the decomposition of plants as well as dead animals. The dead organic matter (plants as well as animals) settled down to the bottom of shallow seas and lagoons. The settled debris is attacked by anearobic bacteria. Most of the organic compound are destroyed and lost. Unsaturated fatty oils and acids which survive the bacterial attack undergo polymerization. These are buried under the increasing cover of rocks whereas bacteria keep on decomposing them. In the process, the water content reduces from 70% to 10% or even less. As a result of continued bacterial decomposition, various decarboxylation and other resulting processes take place and organic matter eventually changes into petroleum.

The modern theory satisfactorily explains the presence of nitrogen and sulphur compounds as they could only be obtained from proteins. The presence of chlorophyll and haemoglobin in petroleum is explained by the fact that the source of chlorophyll is the plants and the source of heamoglobin present in petroleum, is the sea animals. The theory also explains the presence of optically active compounds. The presence of brine and coal in the vicinity of petroleum further supports the modern theory.

2.5.2 Composition

Crude petroleum obtained from different places have composition which varies within a narrow range. The ultimate analysis shows that the average composition of petroleum is:

Element	Carbon	Hydrogen	Sulphur	Oxygen	Nitrogen
Percentage	80-87	11.1-15.0	0.1-3.5	0.1-0.9	0.4-0.9

The ash of the crude oil is only 0.1% or even less. Practically all metals have been found in the petroleum ash. The most widely occurring metals include silicon, iron, aluminium, calcium, magnesium, nickel and sodium.

Crude petroleum is essentially a mixture of various hydrocarbons. Most of the hydrocarbons are straight chain paraffins and aromatic hydrocarbons *e.g.* benzene, toluene naphthalenes etc. Olefins and acetylenes are absent in crude oil. However, during the processing of petroleum, olefins are formed.

Sulphur is present in the form of derivatives of hydrocarbons such as alkylsulphides, aromatic sulphides etc. Nitrogen is present in the form of pyridine, quinoline derivatives, pyrrole etc. Combined oxygen is present as carboxylic acids (naphthalenic acids), ketones and phenols. The objectionable odour of crude petroleum is due to the presence of sulphur compounds in it.

Classification: The proportions in which the paraffinic, olefinic and aromatic hydrocarbons are present in petroleum, differs from place to place, from where petroleum is obtained. On the basis of the residues left behind on distillation, crude oils can be classified as shown in table 2.4:

Table 2.4 *Classification of Crude Petroleum*

Residue Obtained After Distillation	Name	Contents	Remarks
Paraffin wax	Paraffin base hydrocarbons and small amounts of naphthenes and aromatic hydrocarbons.	Straight chain octane number straight run gasoline.	Furnish low
Asphalt	Asphaltic base naphthenic (cycloparaffins) hydrocarbons.	Aromatic and octane number straight run gasoline.	Furnish high
Paraffin wax and asphalt	Mixed base	Paraffins, naphthenes and aromatic hydrocarbons	Furnish medium grade straight run gasoline.

2.5.3 Processing of Crude Petroleum

Petroleum is found deep below the earth crust. The oil is found floating over salt water or brine. Generally, accumulation of natural gas occurs above the oil.

In order to bring the oil to the surface, (mining of oil) holes are drilled and pipes are sunk right upto the oil-bearing rocks. Oil usually rushes out through these holes because of the pressure of the natural gas. But as the pressure the natural gas subsides, the oil is pumped out with the help of air-lift pumps. The air lift pump consists of two coaxial pipes, lowered into the oil bed. Compressed air is forced through the outer pipe when oil from the inner pipe flows out.

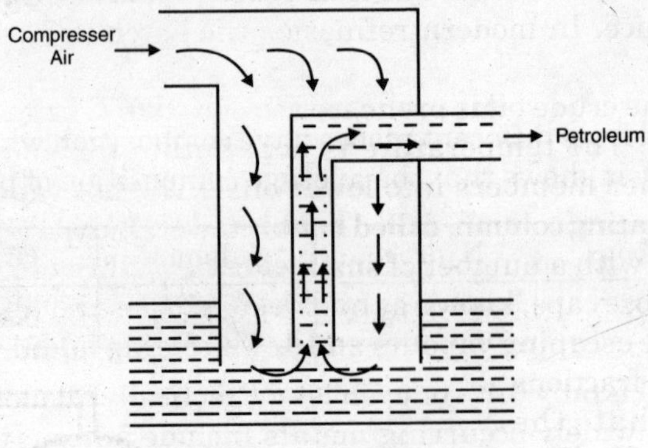

Fig. 2.9 *Pumping of Oil*

2.5.4 Refining of Petroleum

Crude oil reaching the surface, generally consists of a mixture of solid, liquid and gaseous hydrocarbons containing sand and water. Crude oil as such is of little importance. However, after the removal of dirt, water and much of the associated natural gas, the crude oil is separated into a number of useful fractions by fractional distillation. The resultant fractions are then subjected to purification or conversion processes, for the manufacture of various useful petroleum products. All these processes are referred to as **refining** of petroleum and the plant where these processes take place are called **refineries**.

Refining of petroleum includes the following steps

(i) **Demulsification:** The crude oil coming out from the well, is in the form of stable emulsion of oil and salt water, which is yellow to dark brown in colour. The emulsions are stabilized because of the presence of emulsifying agents in the petroleum (*e.g.* sodium salts of naphthenic acids, sulfoacids etc). The emulsions are also stabilized due to the static electricity on the dispersed particles causing mutual repulsion.

The demulsification is achieved by **Cottrell's Process**, in which the water is removed from the oil by electrical process. The crude oil is subjected to an electrical field, when droplets of colloidal water coalesce to form large drops which separate out from the oil.

(ii) **Removal of Harmful Impurities:** Excessive salt content such as NaCl and $MgCl_2$ can corrode the refining equipment. These are removed by washing with water. The objectionable sulphur compound are removed by treating the oil with copper oxide. The copper sulphide so formed is separated by filteration.

(iii) **Fractional Distillation:** The fractional distillation of petroleum is carried out in a specially designed tall fractionating tower or column made up of steel.

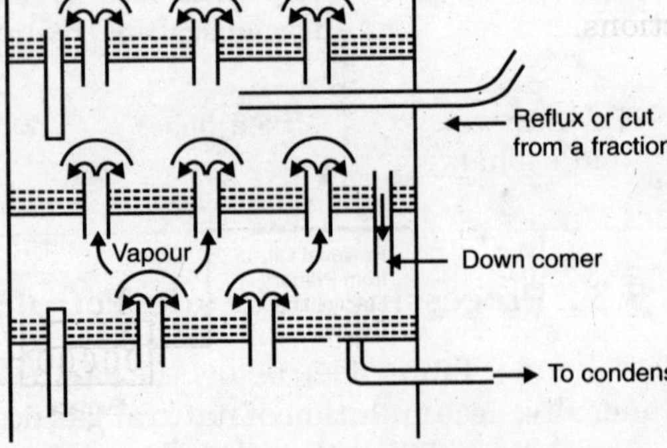

Fig. 2.10 *Simplified representation of a bubble-cap fractionator*

Early distillation processes were discontinuous or "batch processes". In batch process, the crude petroleum was distilled from huge vessels till most of the oil separated into fractions. But this process is obsolete now a days, because due to increased operating temperatures, the cracking of various fractions take place. In modern refineries the batch process has been replaced by continuous process.

In continuous process, the crude oil is preheated to 350-380° C in specially designed tubular furnace known as pipe still. The temperature is kept within the above range so as to avoid thermal degradation of higher members into lower ones. The hot vapours from the crude are passed through a tall fractionating column, called bubble tower. Bubble tower consists of a number of horizontal trays, provided with a number of small chimneys, through which vapours rise. The chimneys are covered with loose caps, known as bubble caps. These bubble caps help to provide an intimate contact between the escaping vapours and down coming liquid.

The separation of various fractions is based on the fact that the hydrocarbons of petroleum boil at different temperatures. The temperature in the fractionating tower decreases gradually on moving upwards (About 300° C at the bottom and decreasing gradually to 125° C at the top). As the vapours of the crude oil go up, they become gradually cooler and fractional condensation takes place at different heights of column. Higher boiling fractions condense first at lower plates whereas lower boiling fractions condense turn by turn at upper plates. Thus, the crude oil gets separated into different fractions in the order of their boiling ranges and are collected at different heights in the columns with the help of condensers (Fig. 2.11).

The various primary fractions collected, their boiling ranges and uses are summarized in Table. 2.5.

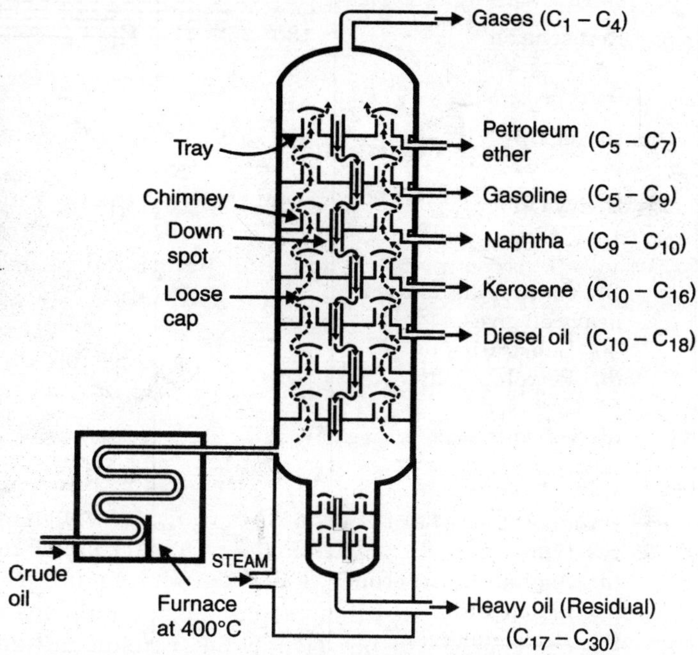

Fig. 2.11 *Fractional distillation of crude petroleum*

The residue from the bottom of the fractionating tower is vacuum distilled to recover various fractions.

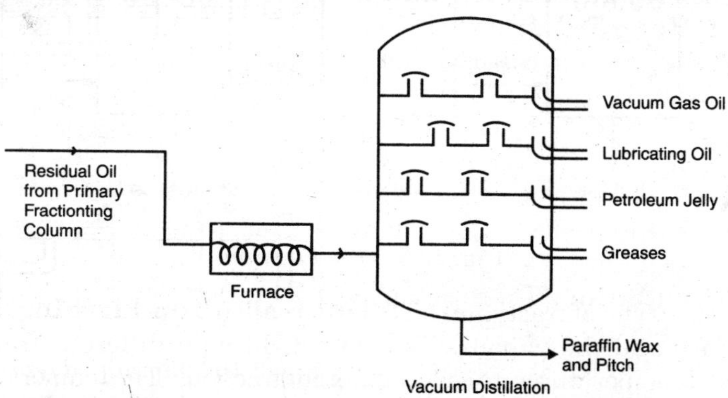

Fig. 2.12 *Vacuum Distillation of Residual Oil*

Table 2.5 *Fractions Obtained on Fractional Distillation of Crude Petroleum*

Fraction	*Boiling Range*	*Composition*	*Uses*
1. Uncondensed fuel gases	<30° C	C_1-C_4	Domestic fuel (LPG), synthesis of organic chemicals.
2. Petroleum ether	30°-70°C	C_5-C_7	Solvent for fats, essential oils. Used in dry cleaning.
3. Gasoline or petrol or motor spirit	40°-120° C	C_5-C_9	As a motor fuel for IC engines, solvent, dry cleaning.
4. Naphtha or solvent spirit.	120°-180° C	C_9-C_{10}	As a solvent and in dry cleaning, feedstock for petrochemicals.
5. Kerosene oil	180°-250° C	C_{10}-C_{16}	Illuminant, fuel for stoves for making oil gas, more volatile portion as jet fuel.
6. Diesel Oil	250°-320° C	C_{10}-C_{18}	Diesel engine fuel, carburetting of water gas.
7. Heavy Oil	320°-400° C	C_{17}-C_{30}	Fuel for ships, metallurgical furnaces, feedstock for cracking processes.
On Vaccum distillation, heavy oil gives:			
(a) Lubricating oil	–	C_{17}-C_{20}	Lubricating oil
(b) Petrolum jelly	–		Used in cosmetics and medicine.
(c) Greases	–	–	As Lubricant.
(d) Paraffin wax	–	C_{20}-C_{28}	Used for preparing candles, wax paper, boot polish, for electrical insulations.
8. Residue: It may be of two types.			
(a) Asphalt (if the crude is asphaltic)	>400° C	–	Used for making roads and water proofing roofs etc.
(b) Petroleum coke	–	–	As a fuel, for moulding electrode rods.

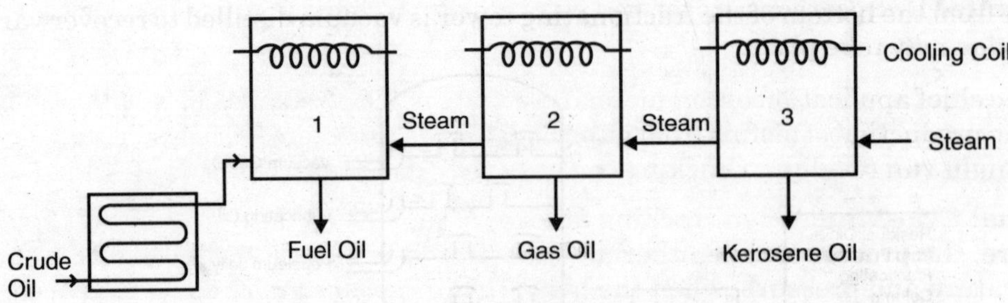

Fig. 2.13 *Top Flashing*

There is yet another type of fractional distillation called **Top-Flashing**. In this, the crude oil is heated in a furnace and the vapours are passed through different fractionating columns maintained at different temperatures, to collect different fractions. The temperatures are maintained with the help of cooling coils. For example when crude oil is made to pass through first column, the temperature of which is so maintained that fuel oil condense at the bottom and the vapour of other components are passed on to second fractionating column. In the second column, the temperature is so maintained that vapours of gas oil condense at the bottom and so on. Steam is injected at the bottom of each fractionating column for its partial pressure effect. Vapour of gasoline finally pass out from the top (Fig. 2.13).

In top flashing, there is better control of product composition, but requires more pumps and instruments and hence is an expensive process.

2.5.5 Cracking

Of all the fractions obtained by fractional distillation of crude petroleum, gasoline is considered to be the most important fraction. With the rapid growth of automobile industry, the consumption of petrol has increased many folds, but the yield of this fraction is only 20% of the crude oil. Also the quality of the gasoline obtained is not good. This gasoline is called straight run gasoline. On the other hand, the yield of heavier petroleum fractions is quite high. Thus, it is necessary to make better use of these heavy fractions, by converting them into more useful fraction, gasoline. This is achieved by a technique called 'cracking.'

'Cracking' is the process by which heavier fractions are converted into lighter fractions by the application of heat, with or without catalyst. Cracking involves the rupture of C—C and C—H bonds in the chains of high molecular weight hydrocarbons. For example-

$$\underset{\substack{\text{Decane} \\ \text{B.pt}=174^\circ\text{ C}}}{C_{10}H_{22}} \xrightarrow{\text{Cracking}} \underbrace{\underset{n\text{-pentane}}{C_5H_{12}} + \underset{\text{Pentene}}{C_5H_{10}}}_{\text{B.pt}=36^\circ\text{ C}}$$

$$C_8H_{18} \xrightarrow{\text{Cracking}} C_5H_{12} + C_3H_6$$

Nearly 50% of today's gasoline is obtained by cracking. It may be emphasized here, that the gasoline obtained by cracking is far more superior to the straight run gasoline. The process of cracking involves the following chemical changes:

- Higher hydrocarbons are converted to lower hydrocarbons by C–C cleavage. The products obtained on cracking have low boiling points than initial reactant.
- Formation of branched chain hydrocarbons takes place from straight chain alkanes.
- Unsaturated hydrocarbons are formed from saturated hydrocarbons.
- Cyclization may take place.

Though, the chief application of commercial cracking is the production of gasolines but it is also used for the production of olefins from naphthas, oil gas from kerosene and the reduction of viscosity of straight run residues. Cracking can be carried out by two methods:

1. **Thermal Cracking:** When cracking takes place simply by the application of heat and pressure, the process is called thermal cracking. The heavy oils are subjected to high temperature and pressure, when the bigger hydrocarbons break down to give smaller molecules of paraffins, olefins etc. The thermal stability among the constituents of petroleum fractions increases as Paraffins < naphthenes (alicyclics) < aromatics. The products obtained, depend upon the operating conditions. Thermal cracking may be carried out either in liquid phase or vapour phase.

   (a) *Liquid phase thermal cracking:* By this method, any type of oil (residues, heavy oil) can be cracked. The charge is kept in the liquid form by applying high pressures of the range 30 kg/cm^2 to 100 kg/cm^2 at a suitable temperature of 476–530° C. The cracked products are separated in a fractionating column. The important fractions are: Cracked gasoline (30-35%); Cracking gases (10-45%); Cracked fuel oil (50-55%).

The cracked gasoline has an octane number of about 70. The cracking gases constitute raw materials for petrochemicals. The cracking residue is used as a fuel.

(b) *Vapour phase thermal cracking:* By this method, only those oils which can vaporize at low temperatures can be cracked. The petroleum fractions of low boiling range like kerosene oil, are heated at a temperature of 670-720° C under low pressure of 10-20 kg/cm^2. The time required for cracking in this process is much less as compared to liquid phase thermal cracking. Petrol obtained from vapour phase cracking has better antiknock properties but poorer stability.

In general, thermal cracking produces petrols which are less stable than straight run petrols due to the presence of highly reactive unsaturated hydrocarbons in them.

Table 2.6: *Comparison of liquid and vapour phase thermal cracking.*

Characteristic	*Liquid Phase-Thermal Cracking*	*Vapour Phase-Thermal Cracking*
Cracking temperature	475 – 530°C	670 – 720°C
Pressure	100 kg/cm^2	10 – 20 kg/cm^2
Octane rating of Petrol	65 – 70	> 70
Type of oil	Any heavy oil can be cracked	Oil should be readily vapourized

Mechanism of thermal cracking: Thermal cracking is believed to follow free radical mechanism. The mechanism can be illustrated by taking the example of n-nonane.

Initiation

$$CH_3(CH_2)_7CH_3 \xrightarrow{\text{heat}} CH_3(CH_2)_3\dot{C}H_2 + \dot{C}H_2(CH_2)_2CH_3$$

Propagation

The free radicals formed are thermally unstable and undergo fission at the β-position to yield a new radical and an olefin.

$$CH_3-CH_2CH_2-CH_2-\dot{C}H_2 \longrightarrow CH_3-CH_2-\dot{C}H_2 + CH_2=CH_2$$

This β scission continues, till a thermally more stable radical is formed.

$$CH_3-CH_2-\dot{C}H_2 \longrightarrow \dot{C}H_3 + CH_2=CH_2$$

$$CH_3(CH_2)_7CH_3 + \dot{C}H_3 \longrightarrow CH_3(CH_2)_5\dot{C}H-CH_2-CH_3 + CH_4$$

The radicals having no C—C bond in the β position give rise to hydrogen free radicals.

$$CH_3\dot{C}H_2 \longrightarrow CH_2=CH_2 + \dot{H}$$

Termination

$$\dot{C}H_3 + \dot{C}H_3 \longrightarrow CH_3-CH_3$$

$$2CH_3-\dot{C}H_2 \longrightarrow CH_3-CH_2-CH_2-CH_3$$

$$CH_3-\dot{C}H_2 + \dot{H} \longrightarrow CH_3-CH_3$$

2. **Catalytic Cracking:** In this method the cracking is brought about in the presence of a catalyst at much lower temperatures and pressures. The catalyst used is usually a mixture of silica and alumina (hydrated aluminium silicate). The most recent catalyst used is zeolite. The quality and yield of gasoline produced by catalytic cracking is greatly improved.

The catalytic cracking has following advantages over the thermal cracking :

(i) High temperatures and pressures are not required in presence of a catalyst and hence there is saving in the production costs.

(ii) The use of catalyst not only acclerates the cracking reactions (decomposition, dehydrogenation, isomerization and polymerization), but also introduces new reactions which considerably modify the yield and the nature of the products.

(iii) The yield of the gasoline is higher. About 70% of raw material is converted into gasoline.

(iv) The gasoline obtained by catalytic cracking contains low percentage of olefins and high percentage of isoparaffin and aromatic hydrocarbons. This increases the antiknock properties of gasoline. Gasoline formed by catalytic cracking has an octane number of 80.

(v) No external fuel is required for cracking, since the heat required is obtained by burning off the coke deposited on the catalyst itself, during regeneration process.

(vi) The process can be better controlled than that of thermal process, so desired products can be obtained.

(vii) The product contains a very little amount of undesirable sulphur, because a major portion of it escapes out as H_2S gas, during cracking.

(viii) The gasoline formed contains much less gum and gum forming compounds.

(ix) Compared with the thermal process, the catalytic cracking yields less coke, less gas and more liquid products.

(x) The evolution of by-product gas can be further minimized, thereby increasing the yield of desired petrol.

(xi) Catalysts are selective in action and hence cracking of only high boiling fractions takes place.

(xii) Coke forming materials are absorbed by the catalysts as soon as they are formed.

Depending upon how the catalyst is used, the catalyst cracking is of two types:

(i) **Fixed bed catalytic Cracking:** In this type of cracking the catalyst is in the form of granules or pellets. These pellets or granules are used in the form of fixed beds in catalyst towers.

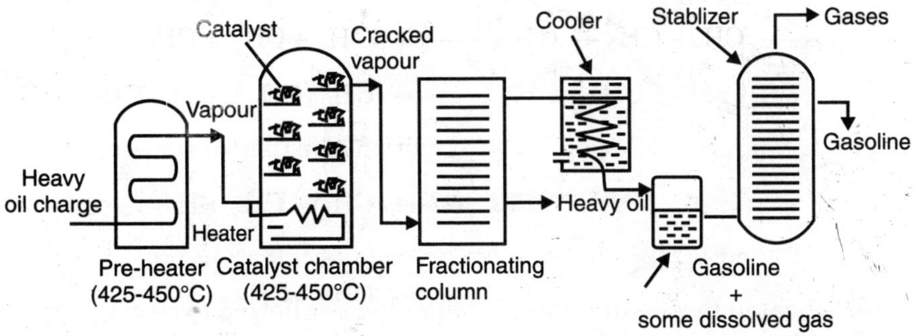

Fig. 2.14 *Fixed-bed type catalytic cracking*

The heavy oil charge is passed through a heater and heated to a cracking temperature (420-450°C). The charge is then passed through a catalyst chamber. The catalyst used is either *silica-alumina gel* (SiO_2. Al_2O_3) *or bauxite, mixed with clay and zirconium oxide.* The catalyst tower is also maintained at 420°-450° C and at a pressure of 1 to 5 kg/cm^2. Cracking of the heavy oil vapours takes place in the catalytic chamber and about 30-40% of the charge is converted into low molecular weight hydrocarbons. The carbon formed (about 4%) during the cracking process gets adsorbed on the catalyst thereby deactivating it. The cracked vapours are passed through the fractionating column, where the gasoline vapours are recovered from the top and the heavy oil gets condensed at the bottom. The gasoline vapours are subjected to a cooler where gasoline along with some other gaseous products gets condensed. Thereafter the gasoline is sent to a stabilizer, where the dissolved gases are removed from the top and pure gasoline is obtained, having the desired boiling range (Fig. 2.14).

During the process, the catalyst stops working after about 10-11 hours, due to the deposition of carbon particles formed during the cracking process. This is reactivated by burning off the carbon in a stream of hot air. During the reactivation of the catalyst, the vapours of heavy oil are diverted through another stand by catalytic chamber so that the reaction proceeds continuously.

As the cracking reactions are *endothermic*, the temperature in the catalytic chamber is maintained by circulating molten salt (mixture of sodium nitrate and sodium nitrite) in pipes passing through the catalyst bed.

(ii) **Moving bed Catalytic Cracking:** This is also known as fluid bed catalytic cracking. In this type of cracking, the catalyst is in the form of fine powder which flows down to the cracking chamber, through a hopper. The catalyst gets mixed with the heavy oil vapours and behave much like a fluid which can be circulated in gas streams. The plant for fluid bed or moving bed catalyst cracking consists of a reactor and a regenerator placed side by side (Fig. 2.15)

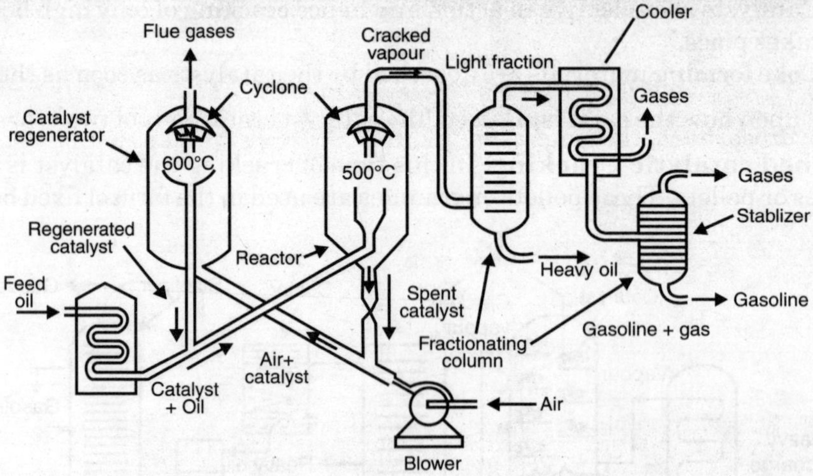

Fig. 2.15 *Moving-bed type catalytic cracking*

The heavy oil is heated in a heater and the vapours are allowed to mix with the hot catalyst, in the form of fine powder, which is introduced from the regenerator. The mixture is forced into a large reactor maintained at 500° C. Due to the high temperature of the catalyst (600° C), the cracking of the heavy oil starts even before its entry into the reactor. This facilitates the cracking of the high boiling fractions also. After cracking of the heavy oil in the reactor, the cracked vapours are allowed to pass on to the fractionating column, through the internal cyclone. The internal cyclone is fixed at the top of the reactor which allows only the cracked oil vapours to pass on to fractionating column, retaining the catalyst powder in the reactor itself through the centrifugal action. The catalyst powder gradually becomes heavier due to the deposition of carbon and settles at the bottom. The spent catalyst is forced by an air blast to the regenerator maintained at 600° C. In regenerator carbon is burnt off and catalyst is reactivated. The reactivated catalyst then flows down through a stand pipe to get mixed with a fresh charge. The cyclone fixed at the top of the regenerator allows only the gases such as CO_2 etc to pass, retaining the catalyst particles. The heat liberated during the regeneration process is used to preheat the catalyst.

Mechanism: Catalytic cracking is believed to involve carbonium-ion intermediates. This can be illustrated by taking the example of n-nonane. First, due to thermal cracking, small amount of alkenes is produced. The alkene produced is thus protonated by the catalyst.

$$C_9H_{20} \xrightarrow{\text{Thermal}} CH_3-CH_2-CH=CH_2+C_5H_{12}$$

$$CH_3-CH_2-CH=CH_2+H^+ \longrightarrow CH_3-CH_2-CH-\overset{\oplus}{C}H_3$$

$$CH_3-(CH_2)_3-CH(H)-(CH_2)_3-CH_3 + CH_3-CH_2-\overset{\oplus}{C}H-CH_3 \longrightarrow CH_3-(CH_2)_3-\overset{\oplus}{C}H-(CH_2)_3-CH_3 + CH_3-CH_2-CH_2-CH_3$$

The nonyl carbonium ion then undergoes β cleavage to yield an alkene and a smaller carbonium ion.

$$CH_3-CH_2-CH_2-CH_2-\overset{\oplus}{C}H-CH_2-CH_2-CH_2-CH_3 \xrightarrow{(1\text{ or }2)} CH_3-CH_2-\overset{\oplus}{C}H_2+CH_3-(CH_2)_3-CH=CH_2$$

The primary carbonium ion rearrange to the more stable secondary carbonium ion by 1,2 hydride shift.

$$CH_3-CH_2-\overset{\oplus}{C}H_2 \longrightarrow CH_3-\overset{\oplus}{C}H-CH_3$$

In aromatic hydrocarbons, the rings remain intact and only the side chains are lost.

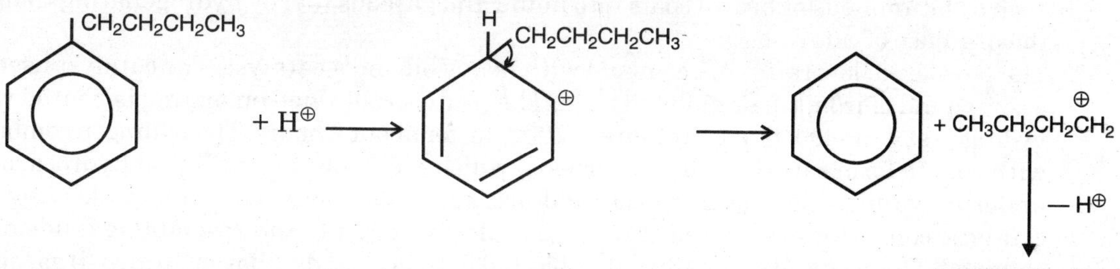

2.5.6 Synthetic Petrol

A great demand of petrol is accomplished by obtaining petrol from cracking and by petroleum distillation process. However, the ever increasing demand of petroleum has necessitated to synthesize petrol from other sources. The chief methods to synthesize petrol are discussed below:

1. **Polymerization:** Polymerization aims at the production of superior gasoline and is complementary to catalytic cracking process. During cracking large quantities of olefins or unsaturated hydrocarbons are obtained as by-product, such as ethene, propene and butenes. When these gases are subjected to high pressure and high temperature in the presence or absence of catalyst, they undergo polymerization to form higher hydrocarbons, resembling gasoline.

$$\underset{\text{Propene}}{CH_3-CH=CH_2}+\underset{\text{1-Butene}}{CH_3-CH_2-CH=CH_2}\rightarrow \underset{\text{5-Methylhexene}}{CH_2=CH-CH_2-CH(CH_3)_2}$$

The gasoline obtained is called polymer gasoline. It has got high octane number (78 to 83).

Similarly, when cracked gases are passed in 70% H_2SO_4 at 10-20° C, two molecules of isobutene polymerize, leaving behind other olefins unaffected. The di-isobutylene is catalytically hydrogenated under pressure over nickel as a catalyst at 160° C to give iso octane. Iso octane in its pure form has octane number of 100.

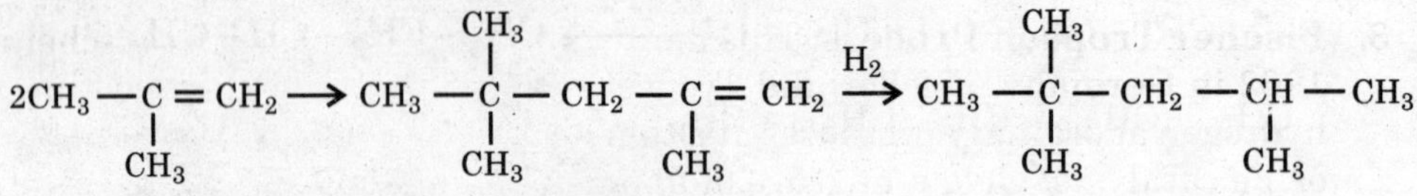

Isobutene Di-isobutylene (isooctene) Iso-octane

Polymerization can be carried out by two methods:

(i) **Thermal polymerization:** This is carried out by subjecting the cracked gases to a temperature of 500-600° C and a pressure of 70-350 kg/cm^2. The products formed are gasoline and gas oil which are separated by fractional distillation.

(ii) **Catalytic polymerization:** The polymerization carried out in presence of a catalyst is known as catalytic polymerization. The catalyst used are phosphoric acid, sulphuric acid etc. The rate of polymerization reaction is increased in presence of a catalyst and the process is carried out at lower temperature (150-200°C). Some molecular combinations which are not possible by thermal polymerisation are possible by catalytic polymerization.

2. **Bergius Process:** The production of gasoline by destructive hydrogenation of coal was developed in Germany by Bergius. The process involves the conversion of low grade coals, such as bituminous or brown coals into liquid and gaseous fuel by hydrogenating them in the presence of catalyst.

The powdered, low rank coal is mixed with heavy oil and a catalyst. The catalyst used, is a compound of iron, tin or molybdenum. The paste so obtained on mixing is heated with hydrogen at 450°-490° C at a pressure of 200 atm. for about 2 hours. The hydrogen combines with coal to form saturated hydrocarbons, which yield low boiling liquid hydrocarbons under prevailing high temperature and pressure (cracking). The vapours leaving the catalytic converter are passed through a cooler, where a liquid resembling crude oil is obtained. The crude oil so obtained is subjected to fractional distillation to give (i) gasoline, (ii) middle oil, (iii) heavy oil. The middle oil is again hydrogenated in the presence of catalyst to give more gasoline. Heavy oil is again used for making a paste with fresh coal dust. The gasoline fraction contains 74% paraffins, 22% aromatics and 4% olefins. One ton of coal is found to yield 140 gallons of gasoline. The hydrogenation of coal is not economical and is adopted by those countries which have no oil resources but have rich coal resources.

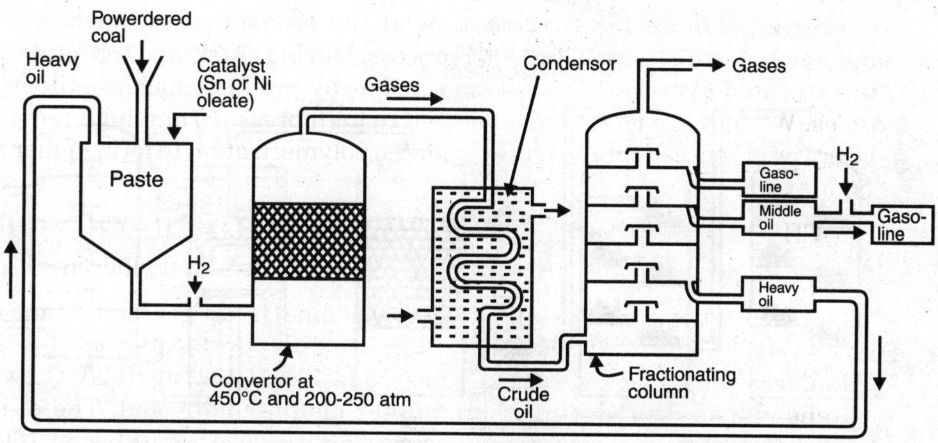

Fig. 2.16 *Bergius process of hydrogenation of coal to gasoline*

3. **Fischer Tropsch Process:** This method was first developed by Fischer and Tropsch in 1962 in Germany. Fischer and Tropsch showed that a mixture of pure water-gas and hydrogen at ordinary pressures at a temperature of 200° C to 300° C in the presence of Co or Fe catalysts, gave a mixture of alkanes and alkenes. From this mixture, a fraction resembling gasoline can be separated by fractional distillation. The nature of the product is determined by temperature, pressure and the nature of the catalyst.

$$nCO + (2n+1)H_2 \longrightarrow C_n H_{2n+2} + nH_2O$$

$$nCO + 2nH_2 \longrightarrow C_n H_{2n+2} + nH_2O$$

Water gas required for the above reaction is obtained from coal. The coal is first converted into coke. The water gas is produced by the action of steam over red hot coke.

$$\underset{\text{Coke Steam}}{C + H_2O} \longrightarrow \underset{\text{Water Gas}}{CO + H_2}$$

Water gas mixed with hydrogen is passed over Fe_2O_3 to remove sulphur and its compounds. The mixture is further purified by passing over Fe_2O_3 mixed with Na_2CO_3 to remove organic sulphur compounds. The pure gas is now subjected into a catalytic converter after being compressed to 5 to 25 atm. The catalyst consists *of cobalt (100 parts), thoria (5 parts), magnesia (8 parts) and kieselguhr (200 parts) maintained at 200-250° C.* The reaction products are predominantly straight chain paraffins and olefins.

The reactions leading to the formation of hydrocarbons are all exothermic. The vapours coming out of the reaction chamber are passed through a cooler where liquid resembling crude oil is obtained. The crude oil is fractionated to get synthetic gasoline, diesel oil and heavy oil. In Fisher-Tropsch process, the gasoline formed is *rich in straight chain hydrocarbons and hence the octane number of gasoline is only 40.* However, the diesel oil obtained is of excellent quality.

From the above discussion it is clear that the catalytic reaction between CO and H_2 is the main basis of Fischer Tropsch process. However its commercial success largely depends upon the method of preparation of water gas and its purification from impurities like sulphur etc which cause catalyst poisoning. About 80% of the operational costs of the process is taken up by the manufacture and purification of synthetic gas.

Mechanism: The reaction is believed to involve the formation of metal carbide (cobalt carbide). This metal carbide on subsequent hydrogenation is cleaved to form methylene radicals and cobalt. The methylene radicals polymerize in presence of hydrogen to form straight chain alkanes and alkenes.

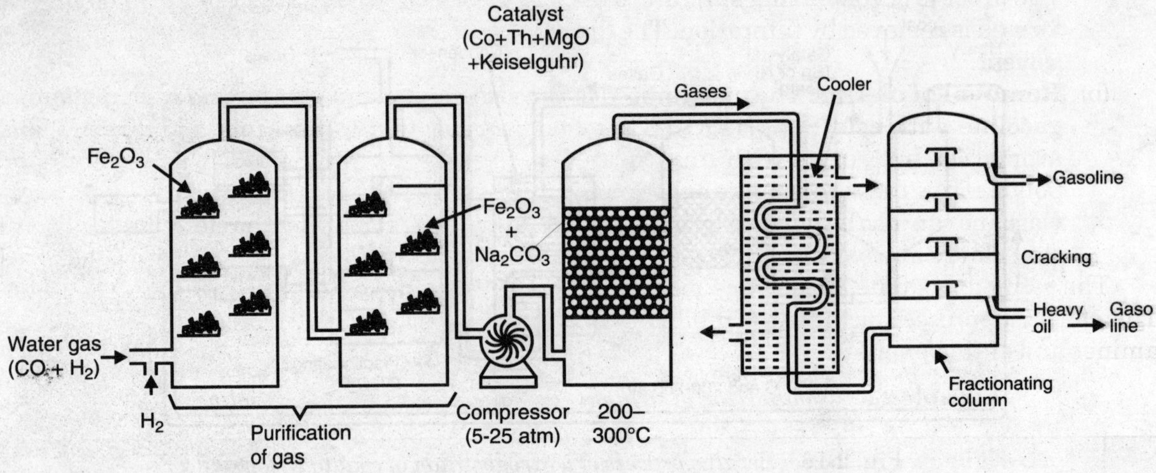

Fig. 2.17 *Fisher-Tropsch method*

The mechanism of the reactions may be represented as follows: (X represents the metal from the catalyst).

$$CO + 2X \longrightarrow \underset{\text{Metal Carbide}}{XC} + \underset{\text{Metal Oxide}}{XO}$$

$$XC + XO + 2H_2 \longrightarrow 2X + \underset{\text{Methylene}}{(CH_2)} + H_2O$$

$$(CH_2) + H_2 \longrightarrow CH_4$$

$$C_nH_{2n} + H_2 \longrightarrow C_n H_{2n+2}$$

According to other view the catalyst seem to function mainly through the formation of metal-oxygen complex. Metal carbide accounts for only 10-15% of the hydrocarbons produced.

2.5.7 Refining of Gasoline

Gasoline obtained from crude oil or by synthetic processes contain some underisable unsaturated olefins and sulphur. Presence of these substances is objectionable because

- The olefins can undergo oxidation and polymerization leading to gum and sludge formation on storage. These deposit on the walls of the carburettor and stop the engine.
- The sulphur compounds can lead to corrosion of internal combustion engines. The compounds of sulphur such as mercaptans or hydrogen sulphide have objectionable odours and are said to make an oil sour.

Thus, the gasoline has to be refined before it reaches the consumer. The removal of these objectionable compounds can be achieved by chemical or physical methods.

(a) **Removal of sulphur compounds:** The sulphur containing gasoline is termed as sour petrol. The sulphur compounds are removed by treating gasoline with sodium plumbite with controlled addition of sulphur.

$$Na_2PbO_2 + 2RSH \longrightarrow \underset{\text{Lead mercaptide}}{Pb(SR)_2} + 2NaOH$$

$$Pb(SR)_2 + S \longrightarrow PbS(s) + \underset{\text{Disulphide}}{RSSR}$$

The process of converting sulphur to disulphides is known as "sweetening". The PbS (s) formed is removed by filteration. The disulphides in gasoline are extracted with a suitable solvent.

(b) **Removal of olefins:** The unsaturated hydrocarbons or olefins can be removed by treating gasoline with cold 80% H_2SO_4. This does not react with paraffins, naphthenes and aromatics, but reacts with unsaturated compounds forming alcohols and esters and polymerizes these to tar-like substances which are removed by filteration.
Gasoline can also be purified by passing through Fuller's earth which preferentially adsorbs the colours and olefins.

The refined gasoline, on storage may still undergo slow oxidation causing gum formation. This can be suppressed by the use of inhibitors which suppress the oxidation processes *e.g.* aromatic amines and amino phenols.

Table 2.6 *Comparison of straight run, cracked and polymer gasolines*

	Straight-run gasoline	*Cracked gasoline*	*Polymer gasoline*
1.	It is obtained from straight distillation of crude petroleum.	It is obtained from heavy oil.	It is obtained from low molecular weight components.
2.	It is physically separated from crude oil	High molecular weight components are broken into low molecular weight components. So, it is a chemical process.	Low molecular weight components are polymerized to give high molecular weight components. So, it is a chemical process.
3.	It is having low octane number	It is having higher octane number.	It is having the highest octane number.
4.	It contains only normal alkanes	It contains more of aromatics, naphthenes and branched chain hydrocarbons.	It contains more of branched chain hydrocarbons.
5.	Composition range is wide (C_5-C_9).	Composition range is narrow (C_6-C_8).	Composition range is very narrow (C_7-C_8).

2.5.8 Knocking

The fuel used for internal combustion engines is gasoline or diesel. In internal combustion engines, a mixture of fuel (gasoline or diesel) is ignited in a cylinder. The ignition is brought about by means of an electric spark (in petrol engines) or by compressing the air (in diesel engines).

The gases evolved during ignition, force the piston down due to high pressure, thus providing the power stroke.

Petrol Engine Fuel

The whole process comprises of four strokes in petrol engine.

(i) The fuel vapours with air in the carburetter is drawn into the cylinder during the induction part and this step is termed as **suction stroke**.
(ii) The fuel air mixture is compressed constituting the *compression stroke.*
(iii) The mixture is ignited by an electric spark. The hot gases produced due to combustion increase the pressure and push the piston down. This is known as the *power stroke.*
(iv) In the last stroke, termed as *'exhaust' stroke*, the piston ascends and expels the exhaust gases from the cylinder.

The next cycle starts again. The fuel air mixture must burn smoothly and rapidly *i.e.* after the reaction is initiated by a spark, a flame should spread rapidly and uniformly through this gas mixture so that the expanding gas drives the piston down the cylinder smoothly.

Figure 2.18 represents the schematic diagram of a four stroke, spark-ignition engine.

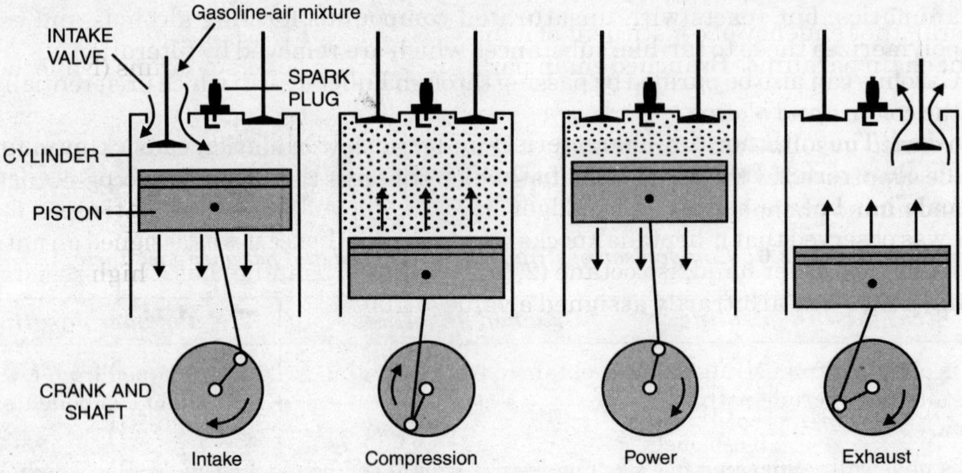

Fig. 2.18 *Schematic diagram of the four stages-intake, compression, power, and exhaust-of an internal combustion engine. This is the type of engine used in practically all automobiles and is described technically as a four-stroke (or the four stages) Otto cycle.*

The ratio of the gaseous volume in the cylinder at the end of the suction-stroke to the volume at the end of compression-stroke of the piston is termed as *compression-ratio.* The efficiency of the engine depends upon this ratio. Higher the compression ratio, higher is the efficiency of the engine. However, if the compression ratio exceeds a certain limit, the fuel-air mixture may get heated to a temperature greater than its ignition temperature and spontaneous combustion occurs even before sparking. This is called *pre-ignition*. Further, it may also happen that the last unburnt portion of the fuel-air mixture after sparking, undergoes spontaneous *self-ignition*, resulting in an explosive violence. This explosion produces a shocking wave, which dissipates its energy by hitting the walls of the cylinder and the piston, producing a characteristic metallic sound termed as *knocking*.

Knocking is thus a sharp metallic sound similar to rattling of hammer which is produced in internal combustion engines due to immature ignition of air gasoline mixture.

Knocking not only decreases the efficiency of the engine but also causes huge loss of energy and damage to pistion and cylinder. The phenomenon of knocking is not yet fully understood but is believed to be closely related with chemical composition of the fuel. The cause of knocking is characteristic of the fuel and engine design.

Chemical structure and knocking: The chemical structure of the fuel used has a close relation with knocking characteristics.

- The knocking tendency decreases with increase in the compactness of the molecules, double bonds and cyclic structures. Thus, olefins of the same carbon-chain length possess better antiknock properties than the corresponding paraffins.
- With normal paraffins, the knocking increases with increase in the length of the hydrocarbon chain. Thus the tendency to knock in straight chain paraffins follow the order:

n–heptane>n–hexane>n–pentane>n–butane

- Branched chain paraffins have better antiknock properties than straight chain paraffins. The resistance to knock increases with the number of branches and their position. Thus the knocking tendency of different isomers of heptane follow the order: n–heptane>2–methyl hexane>2, 2 dimethyl pentane.
- Aromatics such as benzene and toluene have very high antiknock properties.

In general, the tendency of constituent of a fuel to knock is in the following order: Straight chain paraffins>Branched chain paraffins > olefins>Cycloparaffins (Naphthenes)> Aromatics.

Octane Rating: The antiknocking characteristic of gasoline (resistance to knocking) cannot be defined in absolute terms. In order to classify the fuels according to their knocking property, an arbitrary scale has been established by Edger in 1926, known as **Octane rating** or **Octane number.** It was observed that n-heptane knocks very badly and hence it was assigned an antiknock value of zero. On the other hand, iso octane (2, 2, 4 trimethyl pentane) has a high resistance to knocking and hence was arbitrarily assigned a value of 100.

$H_3C–CH_2–CH_2–CH_2–CH_2–CH_2–CH_3$

n-heptane

Antiknock value = 0

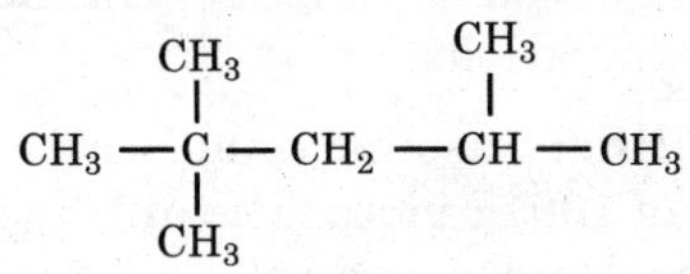

Iso-octane

(2,2,4 trimethylpentane

Autiknock value = 100)

Octane number of a fuel is defined as the percentage of iso octane present in a mixture of iso-octane and n-heptane, which has the same knocking characteristics as that of fuel under examination, under same set of conditions. Thus a gasoline with an octane number of 80, would give the same knocking as a mixture of iso octane and n-heptane containing 80% of iso octane by volume. Greater the octane number, greater is the antiknock property of the fuel.

The octane number of cracked gasoline is higher than straight run gasoline, due to much higher percentage of unsaturated and cyclic hydrocarbons in cracked gasoline. The octane numbers of some common hydrocarbons are given in table 2.8.

Table 2.8 *Octane Numbers of Some Common Hydrocarbons*

Hydrocarbon	*Octane Number*	*Hydrocarbon*	*Octane Number*
Benzene	106	N-pentane	62
Isopentane	90	N-hexane	26
Cyclohexane	77	Alcohol	95
2-methyl pentane	71		

A certain minimum anti-knock quality is required for knock-free operation of a gasoline engine. In USA, the motor gasoline is marketed in two grades, regular having octane number 94 and premium having octane number 100. In India the motor gasoline have specified octane number of 83.

Antiknock Agents: The octane rating of commercial gasolines is not entirely due to the hydrocarbons only. Certain additives are added to the gasoline to improve its antiknock characteristics. These additives are known as antiknock agents. Benzol and alcohols can be used as antiknock agents as these have high antiknock values.

Tetraethyl lead Pb $(C_2 H_5)_4$ is added to gasoline to increase its octane number. TEL is a colourless liquid and is highly poisonous. About 0.5 ml of TEL per litre is added for motor oils and about 1ml of TEL per litre is generally added for aviation petrol. To indicate that petrol is leaded a 2% red dye is also added.

Knocking is believed to be a free radical mechanism leading to a chain reaction which results in an explosion. During combustion Pb $(C_2H_5)_4$ forms Pb and PbO. These act as free radical chain inhibitors and hence the propagation of the chain reaction leading to knocking do not take place.

But lead and lead oxide vapours so formed may contaminate the atmosphere. As soon as explosive combustion occurs, the lead oxide is rapidly reduced to metallic lead which is deposited on the spark plug points, the cylinder walls and the piston head. In order to prevent these problems, ethylene dibromide $(C_2H_4Br_2)$ is also added, so that lead formed is converted into lead bromide.

$$Pb + C_2H_4\,Br_2 \longrightarrow Pb\,Br_2 + CH_2 = CH_2$$

Lead bromide so formed is volatile and escape along with other exhaust gases.

Due to environmental pollution the use of leaded petrol is being minimized now a days.

Other anti-knock agents are tertiary butyl acetate and diethyl telluride, benzol and alcohol. Tricresylphosphate which is capable of suppressing pre-ignition of the fuel can also be used. Benzene is another good antiknock agent. But benzene is carcinogenic and hence it should be used in very low concentrations.

Aviation Gasoline: Gasoline used as a fuel in aeroplanes is known as aviation gasoline. It has octane number of 100 or more. The antiknock quality of the fuel under examination *i.e.* octane number is obtained by comparing it with a mixture of iso octane and a known quantity of tetraethyl lead as reference fuel. For example the octane number of a fuel will be 100+2, if it has the same antiknocking characteristics as a mixture of iso octane and 2 ml of TEL.

Diesel Engine Fuels

(Compression Ignition Engine fuels).

In a diesel engine, the fuel is ignited by the application of heat and pressure and *not by a spark*. In diesel engine, air is first drawn into the cylinder (*suction stroke*) and compressed to a pressure of about 30 to 50 kg/cm^2 (*compression stroke*). Towards the end of the compression stroke, the diesel oil is injected in the form of fine droplets into the very hot compressed air. The fine particles of fuel absorb heat from the compressed air and ignite. The pressure of the gases evolved during the ignition/combustion pushes the piston downwards (*power stroke*). The cycle is completed at the *exhaust stroke* whereby the expulsion of hot gases from the cylinder take place.

Knocking in diesel engine: The main characteristic of diesel fuel is that it should easily ignite below compression temperature. *The interval between the start of fuel injection and its ignition is termed as ignition delay or induction lag.* It is obvious that for efficient functioning of the diesel engine, this induction lag should be as brief as possible.

When a fuel has long induction lag, a large portion of the injected fuel gets accumulated in the cylinder even before ignition and when ignition takes place, the accumulated fuel burn violently with a sudden increase in temperature and pressure. This non-uniform burning of the fuel is known as *diesel knocking*. The greater is the induction lag, higher is the diesel knock.

Chemically, diesel engine fuels consist of straight chain hydrocarbons with minimum amount of branched chain and aromatic hydrocarbons. Thus, in contrast to gasoline, the diesel engine fuels should have low spontaneous ignition temperatures so as to minimize the induction lag.

Cetane Rating: Fuels required for diesel engine are in contrast to petrol engine fuels, hence a separate scale is used to grade the diesel oils as they cannot be graded on octane number scale.

Cetane number is the measure of the ease with which the fuel will ignite under compression. The hydrocarbon cetane (n-hexadecane) has a very short ignition delay as compared to any diesel fuel and is assigned an arbitrary value of 100. Whereas α-methylnaphthalene has a longer ignition delay as compared to any other diesel fuel and hence is assigned a value of 0.

The cetane number of a diesel oil may be defined as the percentage of cetane in a mixture of cetane and α-methyl naphthalene which will have the same ignition characteristics as the fuel under test, under same set of conditions. For example a diesel fuel will be assigned a cetane number 45 if it has the same ignition characteristics as that of a mixture of 45% cetane and 55% α-methyl naphthalene.

$CH_3–(CH_2)_{14}–CH_3$
Cetane
n-hexadecane
(Cetane No. 100)

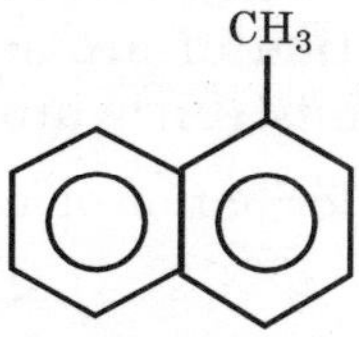

α–Methylnaphthalene
(Cetane No. - 0)

The cetane rating of a fuel depend upon the nature and composition of hydrocarbon. The straight chain hydrocarbons ignite quite readily while aromatics do not ignite easily. Ignition quality order among the constituents of diesel engine fuels in order of decreasing cetane number, is as follows.

n–alkanes > naphthenes > alkenes > branched alkanes > aromatics.

From the above it is clear that the conditions for diesel knock are just the opposite as required for knocking in petrol engines. This is because the cause of knocking is different in the two cases. In petrol engines pre-ignition by compression leads to knocking and hence fuels with higher ignition temperatures are required. On the other hand in diesel engines ignition delay is the cause of knocking and the fuels with low ignition temperatures are required. *Thus, a good petrol engine fuel is a bad fuel for diesel engine and vice versa.*

The cetane number of a diesel oil can be increased by adding certain additives. The commonly used additives are ethyl nitrate, isoamyl nitrate, acetone etc.

Table 2.9 *Comparative account of petrol and diesel oil.*

	Diesel Oil	*Petrol (Gasoline)*
1.	It is a fraction obtained between 250-320° C and is a mixture of $C_{15}H_{32}$–$C_{18}H_{38}$ hydrocarbons.	It is a fraction obtained between 40-120° C and is a mixture of C_5H_{12}–C_9H_{20} hydrocarbons.
2.	It is cheaper.	It is costlier.
3.	Fuel consumption per unit of power produced is less.	Fuel consumption per unit of power produced is more.
4.	Suited for compression engines.	Suited for spark ignition engines.
5.	Knocking is caused by delayed ignition.	Knocking is caused by pre ignition.
6.	Knocking characteristics are expressed in cetane rating.	Knocking characteristics are expressed in octane rating.
7.	Exhaust gases contain lesser amounts of pollutants.	Exhaust gases contain higher amounts of pollutants.
8.	Combustion requires a more expensive fuel injection device.	Combustion requires a simple carburettor.
9.	Alkyl nitrates, acetone etc are used as additives to reduce knocking.	TEL, Diethyl tellurides are used to reduce knocking.

2.5.9 Reforming

In order to increase the anti-knock properties of straight-run gasoline, it is subjected to a special type of treatment known as reforming. Reforming brings about the structural modifications in the components of gasoline. The modifications are essentially a result of rearrangement of molecules without greatly disturbing their average molecular weight. The main reactions in reforming process are (i) the formation of aromatics by dehydrogenation of cylcoalkanes, (ii) the dehydrocyclization of paraffins, (iii) hydro cracking and (iv) isomerization.

(i) Dehydrogenation (formation of aromatics)

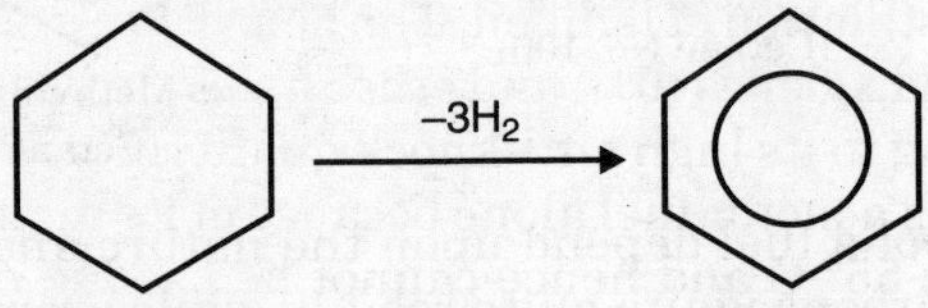

Cyclohexane Benzene

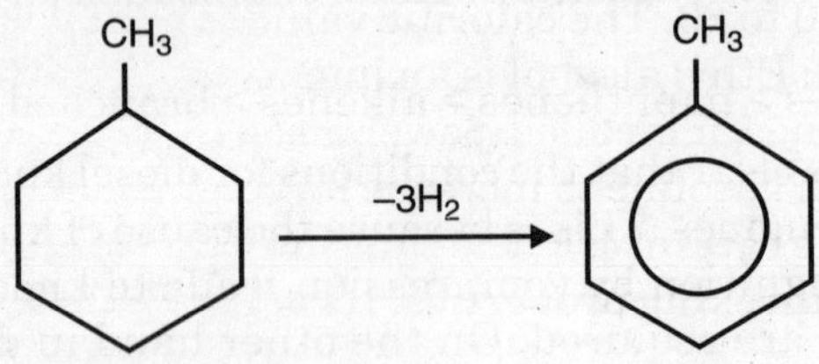

Methylcyclohexane Toluene

(ii) Dehydrocyclization

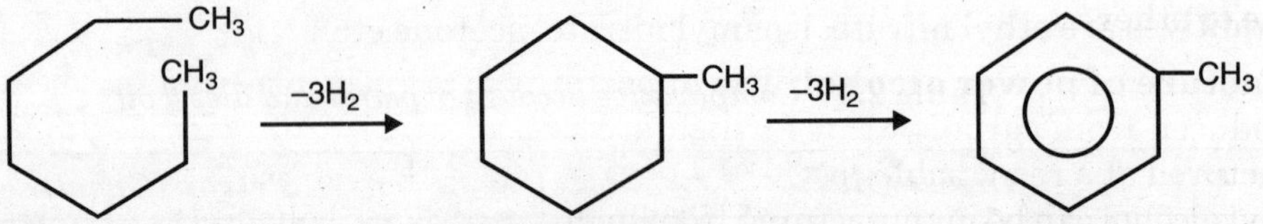

n-Heptane Methyl cyclohexane Toluene

(iii) Hydrocracking:

$$CH_3-(CH_2)_8-CH_3+H_2 \xrightarrow{[Pt]} 2CH_3-(CH_2)_3-CH_3$$

n–pentane

(iv) Isomerization

$$CH_3-CH_2-CH_2-CH_3 \longrightarrow CH_3-\underset{}{CH}(CH_3)-CH_3$$

Isobutane

Reforming can be carried out either thermally or in presence of catalyst.

Thermal reforming: Thermal reforming comprizes of heating the straight run gasoline at about 500-600° C and at a pressure of 85 kg/cm^2. The degree of reforming is controlled by rapid cooling the products by cold-oil spray. The reformed products are then fractionated to remove residual gases. During thermal reforming the alkanes and alkenes formed due to cracking undergo dehydrogenation and cyclization to form naphthenes. N-alkanes may also get converted into branched chained structures contributing higher octane rating to the products.

Catalytic reforming: Catalytic reforming is carried out using platinum supported on alumina as a catalyst. It is carried out at about 460-530° C and a pressure of 35-51 kg/cm^2 pressure.

The catalytic reforming generates a product having superior fuel characteristics as compared to thermal reforming.

The hydrogen evolved in the reforming reactions is used to hydrogenate alkenes, promote cracking of larger alkanes or to remove sulphur in the feedstock in the form of H_2S.

2.5.10 Non-Petroleum Fuels

1. **Benzol:** Benzol is a valuable by-product of coal carbonization. It is also obtained during the fractional distillation of light oil. The benzol is essentially a mixture of benzene (70%), toluene (18%) and xylenes (6%) and other hydrocarbons. It can be used as a component of motor fuels owing to its high anti-knock quality. Its octane rating is about 87-90. Benzol cannot be used as a motor fuel alone because of its high initial boiling point. Moreover its freezing point is 55° C and hence cannot be used in cold climates. It is thus used as a blending agent only. By blending 50% of motor benzol, the octane number of gasoline may be raised from 70 to 80. The calorific value of benzol is about 10,200 k cal/kg.
2. **Power Alcohol:** Ethyl alcohol is an important liquid fuel and when it is used in internal combustion engines, it is called 'power alcohol'. The ethyl alcohol is not used as prime fuel but is mixed with gasoline to increase its octane number. Generally gasoline and alcohol are present in the ratio of 4:1. The calorific value of ethyl alcohol is low (7000 k cal/kg) but it has good antiknocking properties. Its octane number is 90.

Because of the absence of low boiling fractions in ethyl alcohol the starting in the cold engine becomes difficult if ethyl alcohol is used alone as a fuel. Therefore it is always blended with gasoline. It should be emphasized that alcohol mixes well with gasoline only in presence of benzene, tetraline or ether.

Manufacture of power alcohol: The importance of ethyl alcohol as a fuel lies in the fact that it is produced from carbohydrates which are perennially synthesized in nature and hence is manufactured at a reasonable cost.

Ethyl alcohol can be manufactured from hydrocarbons, carbohydrates, starches and molasses, waste sulphite liquor from paper and pulp mills.

In India ethyl alcohol is manufactured from molasses. Molasses is the dark brown mother liquor left after the crystallization of sugar from concentrated sugar cane juice. It contains about 50% uncrystallized sugar and glucose and fructose. The recovery of these sugars from molasses is an expensive process, therefore, molasses is used for the manufacture of alcohol. The formation of alcohol involves the following steps:

(i) **Dilution:** High concentration of sugars in molasses is not suitable for the fermentation reaction and hence molasses are diluted with water to get 8-10% solution of sugars.

(ii) **Addition of dilute H_2SO_4:** For fermentation the pH of the solution should be kept between 4 to 5. Sulphuric acid is added to adjust the pH between 4 to 5. This pH favours the growth of yeast organism which is desirable but checks the growth of any other bacteria which may contaminate the solution.

(iii) **Addition of ammonium salts:** When molasses contain insufficient yeast food, some nutrient substances like ammonium sulphate and ammonium phosphate are added to overcome the deficiency of phosphorus and nitrogen. These act as food for the ferment.

(iv) **Addition of yeast (fermentation):** This is the main step in the formation of alcohol from molasses. Fermentation is a process in which complex organic substances are decomposed into simpler compounds due to the activity of certain micro organisms.

To the above solution a selected strain of yeast is added. The mixture is kept at a temperature of 30° C for 2-3 days. During this period the fermentation takes place. The enzymes invertase and zymase present in the yeast decompose sugars to ethyl alcohol with the evolution of CO_2.

$$\underset{}{C_{12}H_{22}O_{11}} + H_2O \xrightarrow{\text{Invertase}} \underset{\text{Glucose}}{C_5H_{12}O_6} + \underset{\text{Fructose}}{C_6H_{12}O_6}$$

$$C_6H_{12}O_6 \xrightarrow{\text{Zymase}} 2C_2H_5OH + 2CO_2$$

When the evolution of CO_2 ceases, the yeast is filtered off. The fermented solution is filtered. The filterate is called wort which contains 6-10% of ethyl alcohol, 3-5% glycerol, aldehydes, higher alcohols etc. It is distilled to get ethyl alcohol.

(v) **Distillation:** The distillation is carried out in a specially designed plant called Coffey Still. It consists of an analyser and a rectifier, provided with perforated plates with valves opening upwards. Steam and alcohol are made to travel in the opposite directions. (Fig. 2.19)

A current of steam is introduced from the bottom of the analyser and wort/wash (the fiterate obtained in step (iv)) is introduced from the top. The wash is preheated before being introduced to the analyzer. As steam comes in contact with the wash, the alcohol in the wash gets vapourised and is carried to the rectifier along with the steam. In rectifier, the steam gets condensed while the alcohol vapours (b.pt-78.3° C) escape from the rectifier and are condensed seperately. The concentration of alcohol cannot be increased by distillation beyond 97.6% because it forms a constant boiling mixture with water. The constant boiling mixture has lower boiling point than water.

For use in internal combustion engines 100% alcohol (absolute alcohol) should be prepared. This is achieved by azeotropic process.

Fig. 2.19 *Distillation of wash/wort*

(vi) **Azeotropic distillation:** In this process, the alcohol obtained in the above step is mixed with benzene or carbontetrachloride and subjected to distillation. The benzene or carbontetrachloride as the case may be, distills over taking away a portion of alcohol and water, leaving behind absolute alcohol.

The absolute alcohol can also be prepared by digesting the rectified spirit with lime for 2 days and then subjecting it to distillation. The first and last fractions are rejected. The middle fraction is the absolute alcohol.

Advantages of Alcohol blended Gasoline:

1. The octane number of alcohol is quite high (90). Hence on blending with gasoline, it increases the octane number of gasoline. The alcohol blended petrol has better antiknock properties.
2. There is no decrease in power output or increase in specific fuel consumption on blending alcohol with gasoline.
3. The starting difficulties because of high boiling point of alcohol is overcome, when it is used in blended form with gasoline.
4. Alcohol-petrol blend has a tendency to absorb any traces of moisture present.

Disadvantages of Alcohol blended Gasoline:

1. The calorific value of alcohol is low and hence the alcohol lowers the calorific value of petrol.
2. Alcohol is easily oxidized to acids and hence can cause corrosion.
3. Since the air required for complete combustion of blend is less than the air required for petrol, so the air entering the cylinders is to be regulated by modifying the air inlet.
4. The ratio of flow of blend is to be increased to 1.56 times to that of pure petrol by altering the size of carburettor jet.

2.6 GASEOUS FUELS

The remarkable characteristics of gaseous fuels which make them the best fuels are : absence of mineral impurities, consistency in quality, convenience in using, high efficiency, high calorific value, and many more. The popularity of gaseous fuel can be adjudged by the simple fact that the gaseous fuels are adopted on a large scale for domestic as well as industrial use.

Some of the important commercial gaseous fuels are discussed below.

Natural Gas

It is an important primary gaseous fuel associated with petroleum deposits. It is also found in the vicinity of coal mines and oil fields. The natural gas derived from the oil wells may be either dry or wet. When there is no oil but only gas in a petroleum well, the gas is called **dry gas**. In dry gas, lower hydrocarbons such as methane and ethane are present. On the other hand when natural gas occurs along with petroleum in oil wells, it is called **wet gas**. Wet gas consists of higher hydrocarbons along with methane. The calorific value of wet natural gas is higher than that of dry variety because of higher percentage of the heavier unsaturated molecules.

Natural gas is subjected to various treatments before it is used. The various byproducts from the raw natural gas are of great industrial importance and are recovered by various means.

Wet gas gives out very volatile (or low boiling) gasoline which is called casing head gasoline. The gasoline from the gas can be removed by compression, oil scrubbing or adsorption with activated charcoal.

Other valuable byproducts are methane, ethane, propane, butane, which are used as LPG.

In addition to industrially valuable products, natural gas also contains some undesirable components such as water, dust, H_2S and CO_2. Undesirable water is removed by treatment of the gas with drying substance such as alumina, silica gel, sulphuric acid or $CaCl_2$ etc. H_2S is removed by dissolving it in ethanolamine, followed by heating.

$$\underset{\text{Monoethanolamine}}{NH_2CH_2CH_2OH} + H_2S \longrightarrow (NH_2CH_2CH_2OH)H_2S$$

On heating H_2S is liberated. The hydrogen sulphide free gas is called sweet gas. CO_2 is also removed by heating with NH_3

$$2H_3 + CH_2 + H_2O \longrightarrow (NH_4)_2CO_3$$

Composition: The actual composition of natural gas varies widely from field to field, the methane content varying from below 50 to above 90 percent.

The average composition of the gas is; CH_4 = 70-90%; C_2H_6 = 5-10%; H_2 = 3%; CO+CO_2 = rest. Calorific value varies from 12,000 to 14,000 k cal/m^3.

Uses:

(i) Used as domestic and industrial fuel and burns with a hot blue flame. It is delivered to the consumers by a network of pipelines.
(ii) It is used as a raw material for the synthesis of methanol, formaldehyde and other chemical compounds.
(iii) It is widely used as a source of carbon and hydrogen in chemical industries.
(iv) It is also used as a raw material for the manufacture of carbon black.
(v) Microbiological formulation of methane is carried out to obtain synthetic proteins.

Producer Gas

The combustible gaseous mixture obtained by passing air and steam over incandescent coal or coke is known as producer gas. The plant where the reaction takes place is called gas producer or

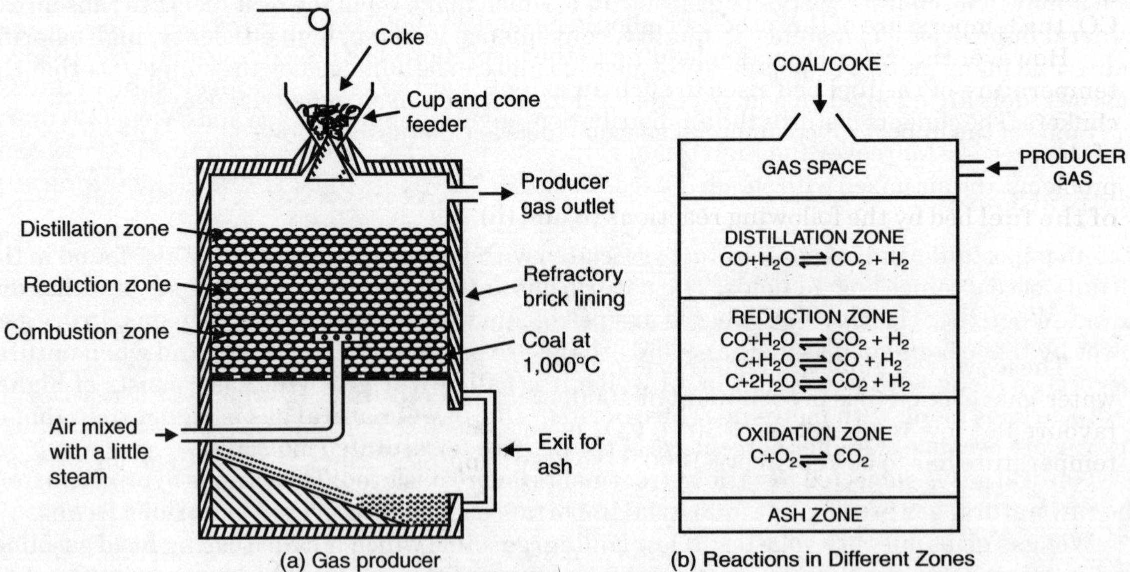

Fig. 2.20 *Manufacture of Producer Gas*

simply producer. The principle components of producer gas are carbon monoxide and nitrogen. It is a cheap industrial fuel which can be obtained from low grade coal. Peat, wood waste, bark and any other type of carbonaceous fuels can be used. The most commonly employed fuels are coal and coke. Though it is considered to be the cheapest gaseous fuel but due to the presence of large proportion of non combustible nitrogen, its calorific value is the lowest (1300 k cal/m^3).

Manufacture: The gas producer consists of a large cylindrical furnace (6-12 feet in internal diameter and 10-15 feet high) lined with fire bricks. It is provided with an air inlet at the base and exit for ash. It is provided with cup and cone feeder at the top and a side opening for the removal of the producer gas formed.

When a mixture of air mixed with steam is passed over red hot coke or coal bed at about 1000° to 1100° C in the producer, the following reactions, which can be divided in four zones, take place.

(a) **Ash Zone :** It is the lowest zone, about 8m thick, which consists of ash formed. The ash protects the grate from the intense heat of combustion. Further, air and steam passing through this zone get heated up.

(b) **Oxidation Zone :** When air comes in contact with solid coke or coal, the carbon of the coal burns in presence of air to form CO_2 and CO. The temperature of this zone is about 1100° C.

$$C_{(S)} + \underbrace{O_{2(g)} + 3.76N_S}_{\text{Air}} \longrightarrow CO_{2(g)} + 3.76N_2 \qquad \Delta H = -97 \text{ kcal}$$

$$C_{(S)} + \underbrace{\tfrac{1}{2}O_2 + 1.88N_2}_{\text{Air}} \longrightarrow CO_{(g)} + 1.88N_2 \qquad \Delta H = -29.5 \text{ kcal}$$

(c) **Reduction Zone :** Carbondioxide formed in the above reaction, undergoes reduction into carbon monoxide by reacting with red hot coke.

$$CO_2 + C \rightleftharpoons 2CO \quad \Delta H = +38{,}270 \text{ kcal}$$

The reduction of CO_2 is an endothermic reaction and is actually the singlemost important reaction which controls the process. The formation of CO is favoured at higher temperatures and at about 1050° C, the mixture contains 99.6% CO and only 0.4% CO_2. Hence for the production of CO, the temperature of the 'producer' should be kept at 1050° C.

However the overall reaction leading to the formation of CO is exothermic. Due to this, the temperature of the fuel bed rises to such an extent that the ash in the fuel bed fuses and form clinkers. The clinkers disturbs the air distribution, engulfs unburnt carbon and lowers the efficiency of the process. Moreover the high temperature may affect the refractory lining. To avoid these problems, the air mixed with steam is used. **Use of steam in air blast reduces the temperature of the fuel** bed by the following reactions (i) and (ii)

$$C_{(S)} + H_2O_{(g)} \rightleftharpoons CO_{(g)} + H_{2(g)} \qquad \Delta H = 28.4 \text{ kcal} \qquad ...(1)$$

$$C_{(S)} + 2H_2O_{(g)} \rightleftharpoons CO_{2(g)} + H_{2(g)} \qquad \Delta H = 18.6 \text{ kcal} \qquad ...(2)$$

These two reactions are endothermic and hence lower the temperature of the fuel bed. The water-gas reaction (i) is predominant at 1000° C and is more desirable whereas the reaction (ii) is favoured at low temperatures. Since CO_2 is an undesired component (non-combustible), the temperature has to be kept above 1000° C so as to suppress the reaction (ii). In actual practice, a balance between the air and steam supply has to be maintained so as to keep the temperature of the 'producer' at 1000° C, which is also essential for the oxidation reaction.

(d) **Distillation zone :** This is the uppermost zone at a temperature of 400-800° C and serves as the preheating zone. In this zone the downcoming coke/coal is heated by the outgoing producer gas. The water vapours and volatile matter of the fuel are added to the escaping gas. Some of the CO may also react with steam in this zone.

$$CO(g) + H2O(g) \rightleftharpoons CO_2 + H_2 \qquad \Delta H = -10 \text{ kcal}$$

This is an undesired reaction which raises the concentration of the inert CO_2 and reduces the calorific value of the product.

Composition: The exact composition of producer gas depends on the type of fuel, the composition of the air + steam and other operation variables. The average compostion of producer gas is CO = 22-30%; H_2 = 8.12%; N_2 = 52.55%; and CO_2 = 3%.

Uses:

(i) It is the cheapest fuel gas, which can be manufactured from waste wood, sowdust, brown coal or peat.

(ii) It is used as a fuel for open hearth-furnaces (in steel and glass manufacture), muffle furnaces, retorts (used in coke and coal manufacture) etc. The hot raw gas is directly fed into the furnace.

(iii) It is used as a reducing agent in metallurgical operations.

Water Gas

Water gas is often called *blue gas* because it burns with the blue flame. It consists mainly of the mixture of carbon monoxide and hydrogen. As both the components are combustible it has higher calorific value (2800 k cal/m^3) than producer gas. It is produced by passing steam over red hot coke at about 1000° C.

$$C + H_2O \longrightarrow \underbrace{CO + H_2}_{\text{Water gas}} \qquad \Delta H = +28 \text{ kcal}$$

The above reaction is endothermic process. Thus large quantity of heat is required to maintain the temperature of the furnace. But supplying the required heat externally is very uneconomical. Therefore, the generation of water gas is accomplished in a cyclic process, comprising two steps. In one of the steps, the heat is generated to raise the temperature of the bed by blowing air and in another step the steam is blown and the formation of water gas takes place with the absorption of heat. These two steps are called *blow* and *run* periods respectively.

Manufacture: The gas generator consists of a steel cylinder lined with refractory lining. It has a cup and cone feeder at the top. Water gas outlet is also provided near the top. The generator has two separate inlets for blowing steam and air. At the bottom it is provided with a exit for ash.

The cyclic operation consists of the following two steps :

(i) Air is passed through the coke bed when carbon of the coke burns to produce CO_2. Since the reaction is exothermic the temperature of the fuel bed increases to about 1000° C.

$$C + O_2 + 3.76\ N_2 \longrightarrow CO_2 + 3.76\ N_2 \qquad \Delta H = -97 \text{ kcal}$$

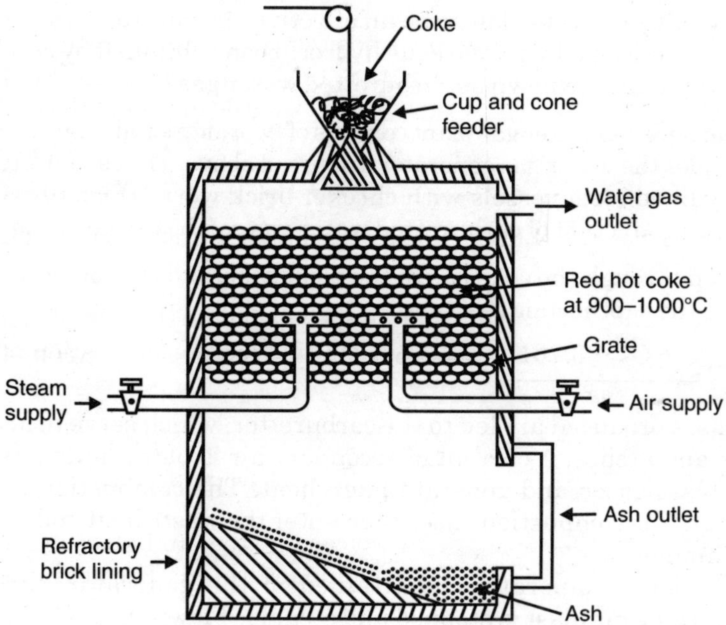

Fig. 2.21 *Water gas production*

(ii) When sufficient high temperature (1000° C) is attained, the air blast is stopped and the steam is blown through the red hot coke to form CO and H_2.

$$C + H_2O \rightleftharpoons CO + H_2 \qquad \Delta H = +28.7 \text{ kcal} \qquad ...(i)$$

$$C + 2H_2O \rightleftharpoons CO_2 + 2H_2 \qquad \Delta H = +18.6 \text{ kcal} \qquad ...(ii)$$

Reaction (i) predominates at higher temperature and is the desirable reaction. Whereas reaction (ii) is predominant at lower temperatures. Due to endothermic nature of these reactions, the temperature of the fuel bed decreases and undesirable reaction (ii) takes place. At this stage, steam supply is cut off and air is again blown to raise the temperature to 1000°–1100° C.

These two steps are thus repeated alternately to maintain proper temperature. The period of steam blow (cold blow) is usually 4 minutes while the period of air blow (hot blow) is about 1-2 minutes.

Composition: The average composition of water gas is : H_2 = 51%; CO = 41%; N_2 = 4%; CO_2 = 4%.

Water gas gives a blue flame owing to high carbon monoxide content, its flame temperature being about 1800° C.

Uses:

1. It is used as a source of hydrogen for the synthesis of ammonia.
2. It is used as a furnace fuel.
3. Mixed with hydrogen it is used in the manufacture of methanol.
4. It is used in Fischer Tropsch synthesis for producing petroleum-like products.
5. Water gas is also used for the preparation of carburetted water gas (discussed below) which can be used for lighting and heating purposes.

Carburetted Water Gas

The calorific value of blue gas is too low to be an effective gaseous fuel. To enhance the calorific value of water gas, it is enriched with gaseous hydrocarbons obtained by cracking of petroleum oils. The enriched water gas is known as carburetted water gas.

Manufacture: A carburetted water gas plant consists of two additional units, namely, a carburetter and superheater besides the water gas generator discussed above. The carburetter and superheater are large refractory lined steel vessels with checker brick work which provide intensive heat transfer surface. The production of carburetted water gas can be explained in three steps-

(i) Air is passed through the coke bed in the generator, and the temperature of the fuel bed rises due to the exothermic reaction.

$$\underbrace{C + O_2 + 3.76N_2}_{\text{air}} \longrightarrow CO_2 + 3.76\,N_2 \qquad \Delta H = -97 \text{ kcal}$$

(ii) The hot gases produced are led to the carburettor, which serve to heat the brick work, side walls and arches. A current of secondary air is blown in the carburettor so as to burn the blast gases and generate more heat. The combustion is completed in the superheater. The combustion gases then enter the waste heat boiler and finally escape through chimney.

(iii) When sufficient temperature is attained in the generator, carburettor and the superheater, the air blast is stopped and steam is blown through the red hot coke. The water gas produced, as described earlier, is led to the carburettor. At the same time the oil is sprayed into the carburettor, which gets vaporized and cracked. The product of cracking mix with the water gas.

(iv) The mixture passes on to superheater where cracking of the gasified oil continues and gets completed. The water gas along with the cracked oil vapours constitute the carburetted water gas which is led through a hydraulic main and a purifier.

The oil used in carburetting is usually a light petroleum fraction consisting mainly of paraffins having boiling temperature of 200° to 300° C. Heavier oils are sometimes used.

Composition: The average composition of carburetted water gas is: CO = 23 to 28%; H_2 = 34 to 38%; Hydrocarbons (saturated and unsaturated) = 30-48%; CO_2 = .2 to 2.5%; N_2 = 2.5 to 5%. Its calorific value is about 3500 to 4450 k cal/m^3.

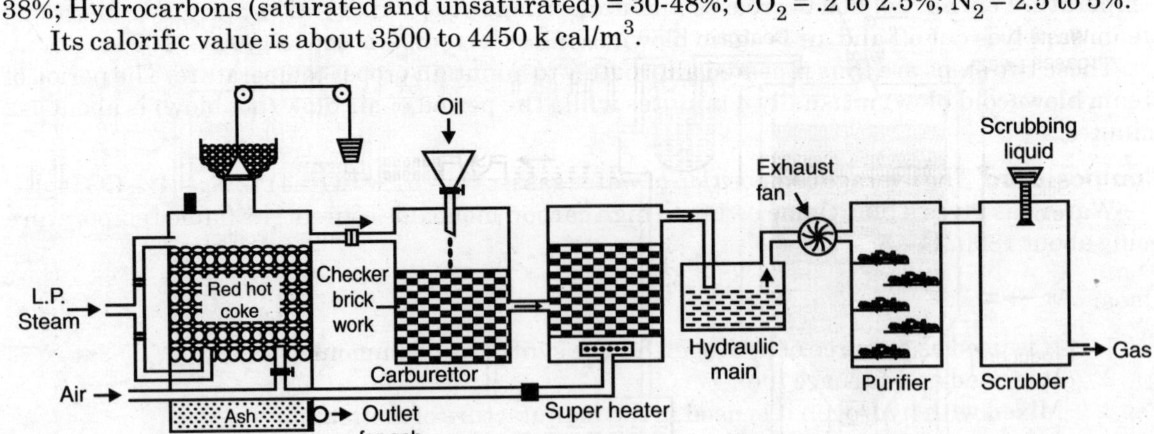

Fig. 2.22 *Carburetted water gas plant.*

Uses:

1. It is used as a source of heat since it posseses higher calorific value.
2. It is employed for illuminating purposes.
3. It is used as a substitute for coal gas to meet the seasonal and peak loads of gas supply.

Coal Gas: (Coke oven gas)

Coal gas and coke oven gas are obtained by the high temperature carbonization of coal. The carbonization of coal leads to the evolution of gaseous products. The gas from gas-making retorts is called **coal gas** while the gas obtained as a by-product from coke-making ovens is called **coke-oven gas**. The two are very similar. The differences are due to the nature of the coal used and also to the different carbonizing conditions.

Coal gas is manufactured by heating coals rich in volatile matter (30-40%) in horizontal or vertical retorts upto 1350° C and the gas is the main product.

$$\text{Coal} \xrightarrow{1300°\text{C}} \text{Coke} + \text{Coal Gas} \uparrow$$

Schematic representation of the process is shown in figure 2.23. The gas evolved contains a large number of valuable impurities which are removed from the coal gas mainly because of their utility value. Further, their removal also improves the quality of the gas.

The gas (at about 700-800° C) is allowed to pass through hydraulic main, where some tar and ammonia get condensed. The hydraulic main, also acts as a water seal preventing the passage of the gases back into the retort.

The gas is led to a large condenser where much of the tar and ammonia are removed and are collected below in two separate layers in a tank. The coal tar from the coal gas is a valuable source of chemicals.

The gases are then passed through scrubbers or washing towers under pressure with the help of an exhaust pump. Here the hot gases are rapidly chilled by spraying with jets of high pressure water when naphthalene is removed.

The cool gases are allowed to pass through another scrubber where they are scrubbed with creosote oil which dissolves benzol, toluene and other aromatic compounds.

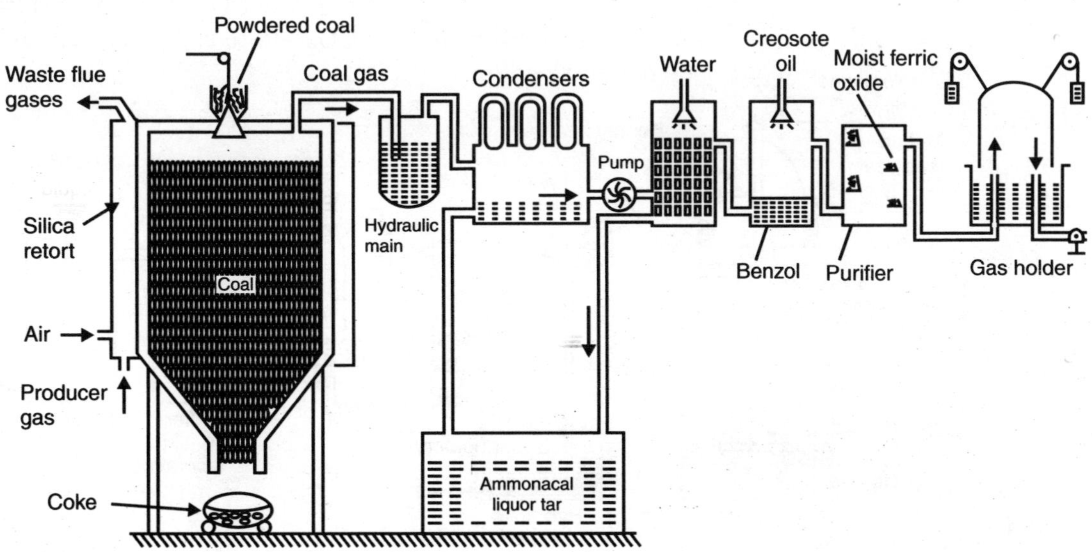

Fig. 2.23 *Manufacture of coal gas*

The gas is purified with the removal of sulphur and hydrogen sulphide by passing it over moist ferric oxide contained in a purifier.

$$Fe_2O_3 + 3H_2S \longrightarrow Fe_2S_3 + 3H_2O$$

When Fe_2O_3 is exhausted it is taken out from the purifier and exposed to air, when it gets oxidised to Fe_2O_3.

$$2Fe_2S_3 + 3O_2 \longrightarrow 3Fe_2O_3 + 6S$$

The coal gas is thus purified and stored over gas holders over water.

Composition: The composition of coal gas is : $H_2 = 40\%$; $CH_4 = 32\%$; $CO = 7\%$; $C_2H_2 = 2\%$; $C_2H_4 = 3\%$; $N_2 = 4\%$; $CO_2 = 1\%$ and rest = 4%. Its calorific value is about 4,900 k cal/m^3.

Coal gas is lighter than air and burns with a smoky flame. The yield of the gas depends on the temperature and nature of coal used.

Uses:

1. Coal gas is mainly used as a fuel.
2. It is also used as an illuminant. Illuminating power of coal gas is due to the presence of unsaturated hydrocarbons.
3. It is also used for providing reducing atmosphere in various metallurgical operations.
4. Owing to high calorific value it is also used in smelting of metals and alloys.
5. Coke oven gas is regarded as a cheapest raw material for ammonia production.

Oil Gas:

Gaseous fuels can also be obtained from petroleum fractions Oil gas consists of mixture of hydrocarbons and is prepared by thermal cracking of kerosene oil.

$$\underset{\text{Kerosene oil}}{C_{12}H_{26}} \xrightarrow{\text{Cracking}} \underset{\substack{\text{Mixture of lower hydrocarbons} \\ \text{(Oil gas)}}}{CH_4 + C_2H_6 + C_2H_4 + C_3H_8 + C_3H_6} + \underset{\text{Tar.}}{C}$$

Manufacture: The plant consists of a strong cast iron retort, which is enclosed in a coal-fired furnace. A bonnet is fitted at the mouth of the retort through a molten lead seal. The bonnet is connected to a hydraulic main through a water seal.

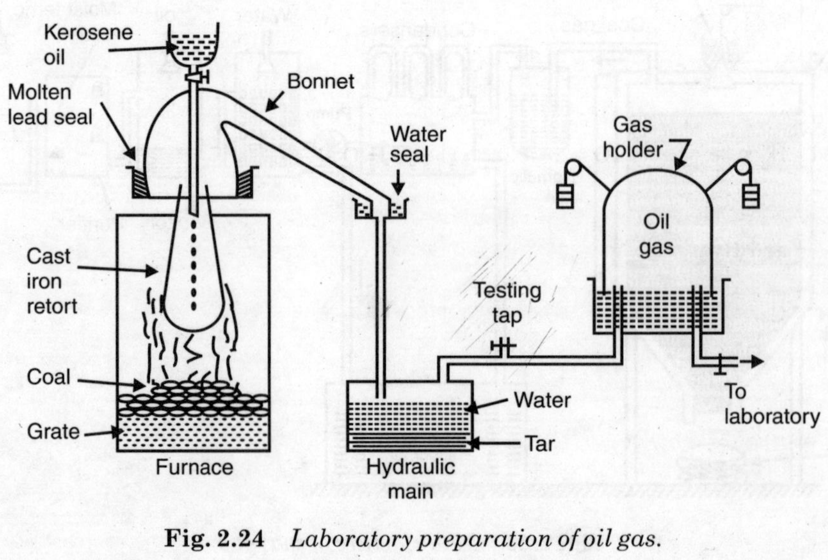

Fig. 2.24 *Laboratory preparation of oil gas.*

A thin stream of kerosene oil is gradually introduced into the red hot iron retort, where it is heated strongly and gets thermally cracked into a mixture of lower hydrocarbons such as methane, ethane, ethylene etc. Tarry matter is also produced and gets mixed up with the gaseous hydrocarbons. The gases so formed passes out through the bonnet cover to a hydraulic main. Tar gets condensed and removed. The extent of completion of cracking process is tested by withdrawing some gas through the testing tap and examining its colour. A good oil gas should appear golden yellow. By adjusting the supply of kerosene and air, the required colour can be obtained. The oil gas is stored over water in gas holders.

Composition: The average composition of the oil gas is CH_4 = 25-30%; H_2 = 50-55%; CO = 11%; CO_2 = 3%.

The calorific value of oil gas is about 6,700 k cal/m^3.

Uses:

1. It is used as a laboratory gas.
2. It is also used to increase the calorific value of water gas (carburetted water gas).

Table 2.8 *Composition and calorific values of gaseous fuels*

	Fuel	*Average Composition*	*Reactions Involved During Manufacture*	*Calorific (Kcal/m³) valve*
1.	Natural Gas	CH_4 = 70 – 90% C_2H_6 = 5 – 10%; H_2 = 3% CO + CO_2 = rest		12,000–14,000
2.	Producer Gas	CO = 22 – 30%; H_2 = 8 – 12% N_2 = 52 – 55%; CO_2 = 3%	Oxidation $C + O_2 + 3.76N_2 \longrightarrow CO_2 + 3.76N_2$ $C + \frac{1}{2}O2 + 1.88\ N_2 \longrightarrow CO + 1.88\ N_2$ Reduction $CO_2 + C \rightleftharpoons 2CO$ $C + H_2O \rightleftharpoons CO + H_2$ $C + H_2O \rightleftharpoons CO_2 + H_2$	1300
3.	Water Gas	H_2 = 51%; CO = 41% N_2 = 4%; CO_2 = 4%	$C + H_2O \longrightarrow CO + H_2$	2800
4.	Carburetted Water Gas	H_2 = 34 – 38%; CO = 23 – 28% N_2 = 2.5 – 5%; CO_2 = 0.2 – 2.5% and Hydrocarbons = 30 – 48%	$C + H_2O \longrightarrow CO + H_2$ $\text{oil} \xrightarrow{\text{cracking}} \text{Hydrocarbons}$	3500–4450
5.	Coal Gas	H_2 = 40%; CH_4 = 32%; CO = 7% C_2H_2 = 2%; C_2H_4 = 3%; N_2 = 4% CO_2 = 1% and rest = 4%	$\text{Coal} \xrightarrow{1300°C} \text{Coke + Coal gas}$	
6.	Oil Gas	CH_4 = 25 – 30%; H_2 = 50 – 55% CO = 11%; CO_2 = 3%	$C_{12}H_{26} \xrightarrow{\text{Cracking}} CH_4 + C_2H_6 +$ $C_2H_4 + C_2H + C_3H_8 \ldots$	6,700

2.7 COMBUSTION

Combustion is a chemical process accompanied by the liberation of heat and light. It is necessarily an exothermic reaction. For example

$$C_{(s)} + O_{2(g)} \longrightarrow CO_{2(g)} \qquad \Delta H = -97 \text{ kcal}$$

To ensure complete combustion, the substance must be brought to its ignition temperature. Ignition temperature is defined as the minimum temperature at which the substance ignites and burns without further addition of heat from outside.

Calculation of Air required for Combustion

The amount of oxygen and air required for complete combustion of a given quantity of the fuel can be calculated on the basis of following principles:

1. Substances always undergo complete combustion by well defined overall chemical reactions *i.e* substances always combine in definite proportions. For *e.g* the combustion of carbon is represented by the following balanced equation:

$$\underset{\substack{\text{(By wt)} \; 12g \\ \text{(By vol)} \; 1 \text{ Vol}}}{C_{(S)}} + \underset{\substack{32g \\ 1Vol}}{O_{2(g)}} \longrightarrow \underset{\substack{44g \\ 1Vol}}{CO_{2(g)}} \qquad \Delta H = -97 \text{ kcal}$$

This equation indicates that: 12 parts by weight of carbon combines with 32 parts by weight of oxygen to produce 44 parts by weight of carbon dioxide or 1 volume of carbon reacts with 1 volume of oxygen to produce 1 volume of carbon dioxide.

Thus from the chemical equation, the amount of oxygen (by weight or by volume) can be calculated easily. The balanced chemical equations for the most common components of any fuel are given below.

(i) $H_2 + \frac{1}{2}O_2 \longrightarrow H_2O$

(ii) $S + O_2 \longrightarrow SO_2$

(iii) $CO + \frac{1}{2}O_2 \longrightarrow CO_2$

(iv) $CH_4 + 2O_2 \longrightarrow CO_2 + 2H_2O$

(v) $2C_2H_6 + 7O_2 \longrightarrow 4CO_2 + 6H_2O$

(vi) $C_2H_4 + 3O_2 \longrightarrow 2CO_2 + 2H_2O$

(vii) $2C_2H_2 + 5O_2 \longrightarrow 4CO_2 + 2H_2O$

Nitrogen or ash present in fuel are non-combustible matters and hence do not consume any oxygen.

2. In actual practice, the combustion is carried out in presence of air. Therefore from the amount of oxygen required (as calculated on the basis of above reactions), amount of air is calculated. The basis of this calculation is the fact that air contains 21% of oxygen by volume and 23% of oxygen by weight, the rest being nitrogen.

Thus $1m^3$ of oxygen is supplied by $1 \times \frac{100}{21} = 4.76 \text{ m}^3$ of air. Similarly 1 kg of oxygen is supplied by $1 \times \frac{100}{23} = 4.35$ kg of air.

3. Mean molecular weight of air is taken as 28.94 g mol^{-1}.

4. The composition of a solid or liquid fuel is usually expressed by weight whereas the composition of gaseous fuel is given by volume. The masses of gaseous substances can be calculated in accordance with gas laws and Avogadro's law:
   (i) Avogadro's Law : 22.4 L of any gas at STP (273 K and 760 mm pressure) has a mass equal to its one mole. Thus at STP, 22.4 L of carbon weighs 12g, which is the mass of one mole of carbon.
   (ii) The mass of any gas can be converted to its volume at certain temperature and pressure by using gas equation PV = nRT, where P is the pressure of the gas, V is the volume of the gas, R is the gas constant and T is the temperature on Kelvin scale.
   (iii) Boyle's law, Charles' law can be used for reducing the volume of a gas at a given temperature and pressure to the corresponding volume at any other specified conditions of temperature and pressure with the help of the equation:
   $$\frac{P_1 V_1}{T_1} = \frac{P_2 V_2}{T_2}$$
   where P_1, V_1, T_1 are the initial pressure, volume and temperature of the gas and P_2, V_2 and T_2 are final pressure, volume and temperature of the gas.
   (iv) The relative weights of the two or more gases is converted to the proportional volume by the relation:
   $$\text{Proportional volume of the gases} = \frac{\text{Weight of the gas A}}{\text{Molar mass of the gas A}} : \frac{\text{Weight of the gas B}}{\text{Molar mass of the gas B}}$$
5. Minimum oxygen required = Theoretical oxygen required $-O_2$ present in the fuel. The minimum oxygen required should be calculated on the basis of complete combustion. In case of incomplete combustion, the products contain CO. In such a case, the excess oxygen is calculated after subtracting the amount of oxygen required to burn CO to CO_2.
6. Combustion seldom takes place efficiently with theoretically calculated amount of air. An excess of air is always used in the furnace.
   $$\%\ \text{Excess Air} = \frac{\text{Actual air used} - \text{Theoretical air required}}{\text{Theoretical air}} \times 100$$
7. Oxygen present in the fuel is present in combination with hydrogen. Thus this amount of hydrogen which is in combination with oxygen present in the fuel, does not take part in combustion. The rest of hydrogen is called *available hydrogen,* only takes part in combustion reaction.

$$\underset{2g}{H_2} + \underset{16g}{\tfrac{1}{2} O_2} \longrightarrow H_2O + \text{Heat}$$

Since 1 part of hydrogen combines with 8 parts of oxygen or in other words for every 8 parts of oxygen present, one part of hydrogen gets fixed and is not available for combustion, therefore available hydrogen

$$= \left[\text{Mass of hydrogen} - \frac{\text{Mass of oxygen}}{8}\right]$$

Hence theoretical amount of oxygen required for complete combustion of 1kg of fuel

$$= \left[\frac{32}{12} \times C + \frac{16}{2}\left[H - \frac{0}{8}\right] + S\right] \text{kg}$$

where C, H, O, and S represent the percentages of carbon, hydrogen, oxygen and sulphur. Consequently theoretical amount of oxygen required for complete combustion of 1 kg of fuel.

$$= \frac{100}{23}\left[\frac{32}{12}\times C+\frac{16}{2}\left[H-\frac{0}{8}\right]+S\right] kg$$

Calculations

Step 1: Write the balanced chemical equations for various components present.

Step 2: Calculate the weight or volume of oxygen required by each one of them.

Step 3: Find the total oxygen required by adding up the oxygen required by individual components and subtracting the amount of oxygen already present.

Step 4: Convert the volume/weight of oxygen into air

$$\text{Weight of air required} = \text{Total } O_2 \times \frac{100}{23} \text{ gm}$$

$$\text{Volume of air required} = \text{Total } O_2 \times \frac{100}{21} \text{ m}^3$$

Step 5: Find out the air actually supplied by taking into consideration excess air.

Step 6: To convert volume into weight use Avogadro's law. Let y litre be the amount of air required.

$$1 \text{ m}^3 = 1000 \text{ L}$$

$$22.4 \text{ L} = 1 \text{ mole} = 28.94 \text{ g of air}$$

$$\therefore \quad y \text{ L} = \frac{28.94}{22.4} \times y \text{ g}$$

Table 2.10 may be helpful for combustion calculations.

Table 2.10

Combustion Equation	*Volume of oxygen needed (m^3) [Let amount of respective constituent = x m^3]*	*Weight of oxygen needed (m^3) [Let amount of respective constituent = x g]*
$C + O_2 \longrightarrow CO_2$	$x \times 1$	$x \times 32/12$
$H_2 + \frac{1}{2}O_2 \longrightarrow H_2O$	$x \times \frac{1}{2}$	$x \times 16/2$
$CO + \frac{1}{2}O_2 \longrightarrow CO_2$	$x \times \frac{1}{2}$	$x \times 16/2$
$S + O_2 \longrightarrow SO_2$	$x \times 1$	$x \times 32/32$
$CH_4 + 2O_2 \longrightarrow CO_2 + 2H_2O$	$x \times 2$	$x \times \frac{32\times 2}{16}$
$C_2H_4 + 3O_2 \longrightarrow 2CO_2 + 2H_2O$	$x \times 3$	$x \times \frac{32\times 2}{28}$
$C_2H_2 + \frac{5}{2}O_2 \longrightarrow 2CO_2 + H_2O$	$x \times \frac{5}{2}$	$x \times \frac{32}{26}\times\frac{5}{2}$
$C_4H_8 + 6O_2 \longrightarrow 4CO_2 + 4H_2O$	$x \times 6$	$x \times \frac{32\times 6}{56}$

SOLVED NUMERICALS BASED ON COMBUSTION

Example 2.9 Calculate the weight and volume of air required for the combustion of 1 kg of carbon.

Solution: Combustion reaction,

$$\underset{12g}{C} + \underset{32g}{O_2} \longrightarrow CO_2$$

Weight of oxygen required to burn 12 kg of C = 32 kg

Weight of oxygen required to burn 1 kg of C = $\frac{32}{12}$ = 2.67 kg.

$\therefore$ Weight of air required = $2.67 \times \frac{100}{23}$ = 11.60 kg = 11600 g

Volume of air required for combustion of 1 kg of C

$$= \frac{1600}{\text{Molar mass of air}} \times 22.4\text{L} = \frac{11600}{28.94} \times 22.4\text{ L}$$

$$= 8978.6 \text{ L} = 8.97 \text{ m}^3$$

Example 2.10 Calculate the minimum amount of air required for the complete combustion of 100 kg of the fuel containing C = 80%, H_2 = 6%; O_2 = 5%; S = 2%; and rest N_2 by weight.

Solution:

Constituent	*Amount (%)*	*Combustion Reaction*	*Weight of O_2 Required*
Carbon	80	$\underset{12g}{C} + \underset{32g}{O_2} \longrightarrow CO_2$	32/12 × 80 = 213.3
Hydrogen	6	$\underset{2g}{H_2} + \underset{16g}{\frac{1}{2}O_2} \rightarrow H_2O$	16/2 × 6 = 48
Sulphur	2	$\underset{32g}{S} + \underset{32g}{O_2} \longrightarrow SO_2$	32/32 × 2 = 2
Oxygen	5	—	—

Nitrogen does not burn and hence does not consume oxygen.

Theoretical oxygen required for 100 kg of fuel = 213.3 + 48 + 2 = 263.3 kg of O_2

Net amount required = O_2 required-O_2 present in fuel = 263.3 – 5 = 258.3 kg of O_2

$\therefore$ Air required = $258.3 \times \frac{100}{23}$ = 1123 kg

Example 2.11 Calculate the volume of air required for complete combustion of 1m^3 of gaseous fuel having the composition: CO = 48%; CH_4 = 8%; H_2 = 40%; C_2H_2 = 2%; N_2 = 1.0 and remaining being ash.

Solution:

Constituent	*Amount*	*Combustion Reaction*	*Volume of O_2 Required*
CO	0.48	$\underset{1\text{mole}}{CO} + \underset{0.5\,\text{mole}}{0.5\ O_2} \longrightarrow CO_2$	$0.48 \times .5 = .24\ m^3$
CH_4	0.08	$\underset{1\text{ mole}}{CH_4} + \underset{2\text{ mole}}{2O_2} \longrightarrow CO_2 + 2H_2O$	$0.08 \times 2 = .16\ m^3$
H_2	0.40	$\underset{1\text{ mole}}{H_2} + \underset{.5\text{ mole}}{0.5\ O_2} \longrightarrow H_2O$	$0.40 \times .5 = .20\ m^3$
C_2H_2	0.02	$\underset{1\text{mole}}{C_2H_2} + \underset{5/2\text{mole}}{\frac{5}{2}O_2} \longrightarrow 2CO_2 + H_2O$	$0.02 \times 5/2 = .05\ m^3$

$\therefore$ Total volume of oxygen required $= .24 + .16 + .20 + .05 = 0.65\ m^3$

$$\therefore \text{Volume of air required} = 0.65 + \frac{100}{21} = 3.095 m^3$$

Example 2.12 A gas has the following composition by volume: H = 22%; CH_4 = 4%; CO = 20%; CO_2 = 6%; O_2 = 3% and N_2 = 45%. If 25% excess air is used, find the weight of air actually supplied per m^3 of this gas.

Solution: Volume of gases in $1m^3$ are: $H_2 = .22m^3$; $CH_4 = .04m^3$; $CO=.20m^3$; $CO_2 = 0.06m^3$. $O_2=.03m^3$; $N_2 = .45m^3$.

Constituent	*Amount*	*Combustion Reaction*	*Volume of O_2 Required*
H_2	.22	$H_2 + \frac{1}{2}O_2 \longrightarrow H_2O$	$.22 \times 0.5 = .11\ m^3$
CH_4	.04	$CH_4 + 2O_2 \longrightarrow CO_2 + 2H_2O$	$.04 \times 2 = .08\ m^3$
CO	.20	$CO + \frac{1}{2}O_2 \longrightarrow CO_2$	$.20 \times .5 = .10\ m^3$
CO_2	.06	—	
O_2	.03	—	
N_2	.45	—	

$$\text{Volume of } O_2 \text{ required} = (.11+.08+.10)-.03=0.26\ m^3$$

$$\text{Volume of air required} = 0.26 \times \frac{100}{21} = 1.238\ m^3$$

$$\text{Volume of air actually supplied} = 1.238 \times \frac{125}{100} = 1.5475\ m^3 = 1547.5\ L$$

$$\text{Weight of air actually supplied} = \frac{1547.5}{22.4} \times \text{molar mass of air}$$

$$= \frac{1547.5}{22.4} \times 28.94 g = 1999.3\ g = 1.999\ kg$$

Example 2.13 The composition by volume of a certain fuel sample is H = 24%; CO = 6%; CO_2 = 8%; CH_4 = 30%; C_2H_6 = 11%; C_2H_4 = 4.5%; C_4H_8 = 25%; O = 2%; N = 12%. What theoretical amount of air would be required at 25° C and 750 mm pressure for complete combustion of 1m³ of the fuel.

Solution:

Constituent	*Amount*	*Combustion Reaction*	*Volume of O_2 (m^3)*
H_2	.24	$H_2 + \frac{1}{2}O_2 \longrightarrow H_2O$	.24 × .05 = .012
CO	.06	$CO + \frac{1}{2}O_2 \longrightarrow CO_2$	.06 × .05 = .003
CO_2	.08	—	—
CH_4	.30	$CH_4 + 2O_2 \longrightarrow CO_2 + 2H_2O$	.30 × 2 = .6
C_2H_6	.11	$C_2H_6 + 3.5O_2 \longrightarrow 2CO_2 + 3H_2O$	.11 × 3.5 = .385
C_2H_4	.045	$C_2H_4 + 3O_2 \longrightarrow 2CO_2 + 2H_2O$	.045 × 3 = .135
C_4H_8	.025	$C_4H_8 + 6O_2 \longrightarrow 4CO_2 + 4H_2O$	.025 × 6 = .15
O_2	.02	—	
N_2	.12	—	

Total volume of oxygen required for combustion

$$= .012 + .003 + .6 + .385 + .135 + .15 - .02$$
$$= 1.40 m^3 \text{ at NTP}$$

Volume of oxygen required at the given temperature and pressure–

$$P_o = 760 \text{ mm} \quad P_1 = 750 \text{ mm}$$
$$V_o = 1.4 \text{ m}^3 \quad V_1 = ?$$
$$T_o = 273 \text{ K} \quad T_1 = 298 \text{ K}$$

$$V_1 = \frac{760 \times 1.4}{273} \times \frac{298}{750} = 1.548 m^3$$

The volume of air required

$$= 1.548 \times \frac{100}{21} = 7.371 m^3$$

Example 2.14 A sample of coal was found to have the following percentage composition: C = 75%; H = 5.2%; O = 12.1%; N = 3.2% and ash = 4.5%.

(i) Calculate the minimum air required for complete combustion of 1 kg of coal.

(ii) Also calculate the higher calorific value and lower calorific value of coal sample. (Gross C.V. in kcal/kg: C= 8,080; H = 34,500; S = 2,240.

Solution:

(i) 1 kg of coal contains: C = 750 g; H = 52g; O = 121 g; N = 32g; ash = 45 g.

Minimum weight of air required for complete combustion of 1 kg of coal

$$= \text{Wt. of } [C \times (32/12) + H \times (16/2) - 0] \times (100/23)$$
$$= [750g \times (32/12) + 52g \times (16/2) - 121 g] \times (100/23)$$
$$= [2{,}000g + 416g - 121g] \times (100/23)$$
$$= 2{,}295g \times (100/23) = 9{,}978 \text{ g or } 9.978 \text{ kg}$$

(ii) $$\text{HCV} = \frac{1}{100}\left[8{,}080\times C + 34{,}500\left(H - \frac{0}{8}\right) + 2{,}240\times S\right] \text{kcal/kg}$$

$$= \frac{1}{100}\left[8{,}080\times 75 + 34{,}500\left(5.2 - \frac{12.1}{8}\right) + 2{,}240\times 0\right] \text{kcal/kg}$$

$$= \frac{1}{100}[606{,}000 + 127{,}219 + 0]\text{kcal/kg}$$

$$= [6{,}060 + 1{,}272] \text{ Kcal/kg} = 7{,}332 \text{ kcal/kg.}$$

$$\text{LCV} = (\text{HCV} - 0.09\text{H}\times 587) \text{ kcal/kg}$$

$$= (7{,}332 - 0.09\times 5.2\times 587) \text{ kcal/kg} = 7{,}057 \text{ kcal/kg.}$$

2.8 ANALYSIS OF FLUE GAS

Flue gas is the gaseous product of combustion of fuels in ovens and furnaces. When combustion is complete, flue gas consists of carbon dioxide, water vapour, nitrogen and excess oxygen. The term flue gas is derived from the flues that lead the gases from the ovens and furnaces to the chimneys.

In order to achieve efficient combustion of a fuel, it is essential that the fuel is brought into intimate contact with sufficient quantity of air, to burn all the combustible matter under appropriate conditions. Analysis of flue gases give an idea about the complete or incomplete combustion process. It is usual to measure the proportions of CO_2, CO and O_2 in the flue gases. Sometimes, N_2, H_2 and hydrocarbons are also determined.

- If the analysis shows the presence of CO, it indicates the incomplete combustion of the fuel or shortage of oxygen. Hence the supply of oxygen is increased.
- If the gas analysis shows the presence of high percentage of oxygen in flue gases then it shows that though the combustion is complete but the supply of air is much in excess. Too much excess of air results in loss of heat. In such cases, the supply of air to fuel is cut short or rate of supply of fuel is increased.
- If the analysis shows the presence of appreciable amounts of oxygen and carbon monoxide, it indicates that the combustion is irregular and non-uniform *i.e.* in some parts of the furnace, there is excess of air and in some other parts the supply of air is insufficient.

The analysis of the gases is carried out with the help of Orsat's apparatus.

Orsat's Apparatus: It consists of a horizontal tube having a three way stop cock at one end. The other end of the horizontal tube is connected to a graduated burette. The burette is surrounded by a water jacket in order to keep it at a uniform temperature. The lower end of burette is further connected to a water reservoir by means of a long rubber tubing. The level of water in the water reservoir can be raised or lowered by raising or lowering the water reservoir.

The horizontal tube is connected in series to a set of three absorption bulbs, A, B and C, each through a stop cock. The absorption bulbs have solutions for absorbing CO_2, O_2 and CO respectively.

- Bulb A contains the KOH solution for absorption of CO_2 (250 g of KOH dissolved in 500 ml of distilled water, freshly boiled to exclude any CO_2 present in it.)
- Bulb B contains a solution of alkaline pyrogallic acid to absorb O_2 (25 g pyrogallic acid +200 g KOH in 500 ml of distilled water.)
- The bulb C contains 'ammoniacal cuprous chloride' solution to absorb CO. (100 g cuprous chloride +125 ml liquor ammonia + 375 ml of water.)

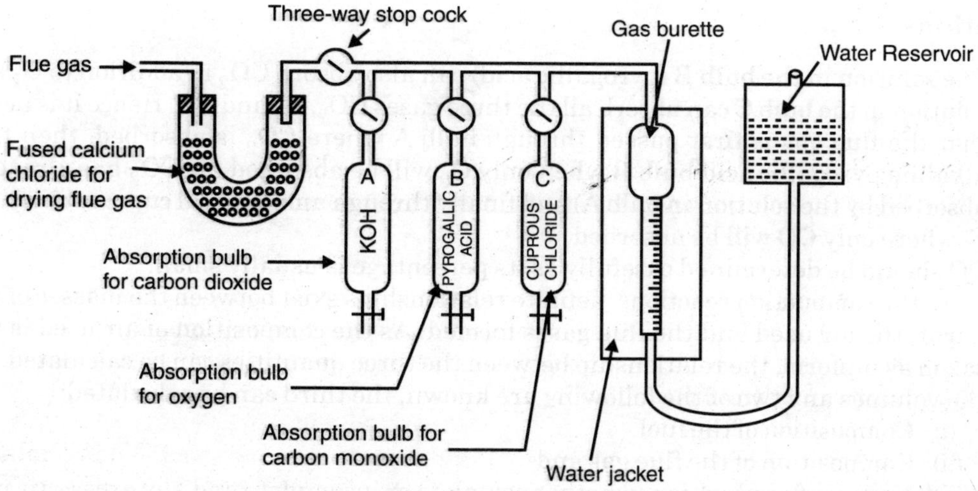

Fig. 2.25 *Orsat's apparatus*

Working

(i) The whole apparatus is cleaned thoroughly. The stoppers are greased and are tested for air tightness. The absorption bulbs are filled with their respective solutions and are closed with their stop cocks. The jacket and the levelling reservoirs are filled with water.

(ii) The three way stop cock is opened to the atmosphere and burette is completely filled with water by raising the water reservoir to remove the air present in the burette.

(iii) The three way stop cock is now connected to the flue gases to be analysed through a $CaCl_2$ U-tube which absorbs the moisture present in the flue gas. The reservoir is lowered to give entry to the flue gas to the burette. However, this flue is expelled to the atmosphere by opening the three-way stopcock and raising the reservior since it is mixed with air present in the apparatus. This process is repeated 3-4 times to ensure complete removal of the air present in the apparatus.

(iv) Finally, gas is sucked in the burette and the volume of the flue gas is adjusted to 100 ml at atmospheric pressure. This is done by adjusting the height of the levelling reservoir. The three way stop cock is opened to the atmosphere and the reservoir is carefully raised, till the level of water in it is the same as in the burette which stands at 100 ml mark. The three way stop cock is then closed.

(v) Now the stopper of bulb A is opened and flue gas is passed through the solution contained in it by raising the reservoir. CO_2 present in the flue gas is absorbed by KOH solution. The gas is again sent to the burette and the process is repeated several times to ensure complete absorption of CO_2 by KOH solution, by raising and lowering of the reservoir. The unabsorbed gases are finally taken back to the burette by raising the liquid in the bulb A till it is completely filled. After that the stopper of the bulb A is closed. The reading of the volume of residual gas is taken in the burette by equalizing the water level in the burette and the water reservoir. The decrease in volume gives the volume of CO_2 in 100 ml of the flue gas sample.

(vi) Now the stop cock of bulb B is opened. The absorption process is repeated as in bulb A. The volume of the gas absorbed gives the volume of O_2 in 100 ml of the flue gas sample. Similarly the flue gas is passed through the bulb C, to determine the amount of O_2 present in the flue gas. The gas remaining in the burette after the absorption of CO_2, O_2 and CO is taken as nitrogen.

Precautions

1. The solution in the bulb B (pyrogallic acid) can also absorb CO_2 in addition to O_2 and the solution in the bulb C can absorb all the three gases CO_2, O_2 and CO. Hence it is necessary that the flue gas is first passed through bulb A where CO_2 is absorbed, then through alkaline pyrogallic acid bulb B when only O_2 will be absorbed (as CO_2 has already been absorbed by the solution in bulb A) and finally through ammoniacal cuprous chloride bulb C, where only CO will be absorbed.
2. CO should be determined carefully as its percentage is usually small.
   In all the combustion reactions, definite relationships exist between the masses of the fuel burnt, the air used and the flue gases formed. As the composition of air used is usually taken as uniform, the relationship between the three quantities can be calculated. Thus if the volumes any two of the following are known, the third can be calculated:
   (i) Composition of the fuel
   (ii) Composition of the flue gas and
   (iii) Nature of combustion (whether complete or incomplete and any excess air used.)

Numerical Problems based on Flue Gas Analysis

The following points should be remembered while solving numericals based on flue gas analysis, in addition to the points discussed in section 2.7.

1. When CO is present in the flue gas, the amount of O_2 necessary to burn it must be deducted from the free oxygen, before calculating the excess air.

$$CO + \frac{1}{2}O_2 \longrightarrow CO_2$$

   For every mole of CO, $\frac{1}{2}$ mole of O_2 is required.
   For example, the flue gas analysis of a fuel gives $CO_2 = 14\%$; $O_2 = 4\%$; $CO = 2\%$; $N_2 = 80\%$; then the total oxygen used can be calculated from the amount of N_2 in the flue gas, i.e
   $80 \times \frac{21}{79} = 26.58$ mols (Air contains 21% of oxygen and 79% of nitrogen by volume). The excess oxygen will be calculated by substracting the oxygen required for combustion of CO, from the amount of oxygen present in the flue gas.

$$\text{Excess oxygen} = 4 - \left(2 \times \frac{1}{2}\right) \text{mols.}$$

$$\text{Excess air} = \frac{3}{(26.58 - 3)} \times 100 = 12.72\%$$

2. The mass of dry flue gases should be calculated by balancing the carbon in the fuel and the carbon present in the flue gases.
3. The composition of flue gases is usually given on dry basis and on volume basis, unless otherwise stated.

Example 2.15 Calculate the minimum weight of air required for the complete combustion of 1 kg of coal having the following percentage compostion by weight. C = 90%; $H_2 = 3.5\%$; $O_2 = 3\%$; S = 0.5%; $N_2 = 0.5\%$ and ash = 1.5%. Calculate the percentage composition by weight of the dry products of combustion.

Solution: 1 kg of coal contains C = .9 kg; H_2 = .035 kg S = 0.05 kg; O_2 = .03 kg; N_2 = 0.005 kg.

Constituent	*Amount in 1 kg of Fuel*	*Combustion Reaction*	*Amount Of O_2 (kg)*	*Amount of Dry (kg) product*
Carbon	.90	$\underset{12g}{C} + \underset{32g}{O_2} \longrightarrow \underset{44g}{CO_2}$	32/12 × .90 = 2.4	44/12 × 0.9 = 3.3
Hydrogen	.035	$\underset{2g}{H_2} + \underset{16g}{\frac{1}{2}O_2} \longrightarrow \underset{18g}{H_2O}$	16/2 × .035 = .28	
Oxygen	.03	—	—	—
Sulphur	.005	$\underset{32g}{S} + \underset{32g}{O_2} \longrightarrow \underset{64g}{SO_2}$	32/32 × .005 = .005	64/32×.005 = .01
Nitrogen	.005			

Oxygen required for burning 1 kg of fuel

$$= (2.4+.28+.005) - .03 = 2.655 \text{ kg}$$

$$\therefore \text{Air required} = 2.655 \times \frac{100}{23} = 11.54 \text{ kg}$$

Amount of dry products:

Amount of CO_2 = 3.3 kg

Amount of SO_2 = 0.01 kg

Amount of N_2 = 77% of the air used + N_2 present in the fuel

$$\therefore \text{Amount of dry products} = 3.3 + .01 + .005 + \left(11.54 + \frac{77}{100}\right)$$

$$= 3.3+.01+8.891=12.2 \text{ kg}$$

Percentage of the dry products:

$$\% \text{ of } CO_2 = \frac{3.3}{12.2} \times 100 = 27.0\%$$

$$\% \text{ of } SO_2 = \frac{0.01}{12.2} \times 100 = 0.08\%$$

$$\% \text{ of } N_2 = \frac{8.891}{12.2} \times 100 = 72.9\%$$

Example 2.16 A sample of coal contains C = 60%; H = 4%; O_2 = 6%; N_2 = 2% and ash = 28%. Calculate the %age composition of the dry products of combustion, assuming that 40% excess air is used.

Solution:

Constituents	*Amount Present in 1 kg of the fuel*	*Combustion Reaction*	*Amount of O_2 (in kg)*	*Amount of Dry products*
Carbon	.60	$C + O_2 \longrightarrow CO_2$	32/12 × .60 = 1.6	44/12 ×. 60 = 2.2
Hydrogen	.04	$H_2 + \frac{1}{2}O_2 \longrightarrow H_2O$	16/2 × .04 = .32	—
Oxygen	.06	—	—	—
N_2	.02	—	—	—

$$\text{Amount of oxygen required} = 1.6 + 0.32 - 0.06 = 1.86 \text{ kg}$$

$$\text{Amount of air} = 1.86 \times \frac{100}{23} = 8.087 \text{ kg}$$

$$\text{Amount of air actually supplied} = 8.087 \times \frac{140}{100} = 11.322 \text{ kg}$$

$$\text{Excess air} = 11.322 - 8.087 = 3.235 \text{ kg}$$

Amount of dry products of combustion

$$\text{Amount of } CO_2 = 2.2$$

Amount of O_2 = Oxygen in the excess air which appears as free state in the flue gas

$$= 3.235 \times \frac{23}{100} = 0.744 \text{ kg}$$

Amount of N_2 = N_2 from the fuel + N_2 from the air used

$$= 0.02 + \left(11.322 \times \frac{77}{100}\right) = 8.738 \, kg$$

$$\text{Total Mass of the dry products} = 2.2 + 0.744 + 8.738 \text{ kg} = 11.682 \text{ kg}$$

$$\% \, CO_2 = \frac{.98}{11.888} \times 100 = 8.24$$

$$\% \, O_2 = \frac{10.03}{11.888} \times 100 = 6.37$$

$$\% \, N_2 = \frac{0.978}{11.888} \times 100 = 74.79$$

Example 2.17: A gaseous fuel has the following composition by volume $H_2 = 20\%$; $CH_4 = 25\%$; $C_2H_6 = 16\%$; $C_2H_4 = 9.5\%$; butene = 2.5%; CO = 4.0%; $CO_2 = 8\%$; $O_2 = 4\%$; and $N_2 = 12\%$. Find the air required for perfect combustion of $1m^3$ of this gas. If 50% of excess air is used, find the volume analysis of the dry products of combustion.

Solution:

Constituent	*Amount in $1m^3$ of Fuel*	*Combustion Reaction*	*Vol. of Oxygen (m^3)*	*Vol. of Dry Products (m^3)*
H_2	*0.20*	$H_2 + \frac{1}{2}O_2 \rightarrow H_2O$	*$0.20 \times 0.5 = 0.1$*	
CH_4	0.25	$CH_4 + 2O_2 \rightarrow 2CO_2 + 2H_2O$	$0.25 \times 2 = 0.50$	$CO_2 = .25 \times 1 = 0.25$
C_2H_6	0.16	$C_2H_6 + 3.5O_2 \rightarrow 2CO_2 + 3H_2O$	$0.16 \times 35 = 0.56$	$CO_2 = .16 \times 2 = 0.32$
C_2H_4	0.095	$C_2H_4 + 3O_2 \rightarrow 2CO_2 + 2H_2O$	$0.095 \times 3 = 0.285$	$CO_2 = .095 \times 2 = 0.19$
C_4H_8	0.025	$C_4H_8 + 6O_2 \rightarrow 4CO_2 + 4H_2O$	$0.025 \times 6 = 0.15$	$CO_2 = .025 \times 4 = 0.1$
CO	0.04	$CO + .5O_2 \rightarrow CO_2$	$0.04 \times 0.5 = 0.2$	$CO_2 = .04 \times 1 = 0.04$
CO_2	0.08	—	—	
O_2	0.04	—	—	
N_2	0.12	—	—	

$$\text{Volume of oxygen required} = 0.1 + 0.5 + 0.56 + 0.285 + 0.15 + 0.2 - 0.04 = 1.755 \text{ m}^3$$

$$\text{Volume of air required} = 1.755 \times \frac{100}{21} = 8.357 \text{ m}^3$$

$$\text{Volume of air actually supplied} = 8.357 \times \frac{150}{100} = 12.54 \text{ m}^3$$

$$\text{Excess air} = 12.54 - 8.357 = 4.183 \text{ m}^3$$

Volume of dry products of combustion:

Volume of CO_2 calculated from the products of combustion = .25 + .32 + .19 + .1 + .04 = .9m^3

Total volume of CO_2 = .9m^3 + CO_2 present in the fuel = .9 + .08 = .98m^3

Volume of N_2 = 79% of air used + N_2 in the fuel

$$= \left(12.54 \times \frac{79}{100}\right) + .12 = 10.03 \text{ m}^3$$

$$\text{Volume of } O_2 = \text{Excess air} \times \frac{21}{100} = 4.183 \times \frac{21}{100} = .878 \text{ m}^3$$

$$\text{Total volume of the dry products} = .98 + 10.03 + .878 = 11.888\text{m}^3$$

$$\%\text{ age of } CO_2 = \frac{.98}{11.888} \times 100 = 8.24$$

$$\%\text{ age of } N_2 = \frac{10.03}{11.888} \times 100 = 84.37$$

$$\%\text{ age of } O_2 = \frac{0.878}{11.888} \times 100 = 7.38$$

Example 2.18 The percentage composition of coal was found to be as: C =54%; H = 6.5%; O = 3.0%; N = 1.8%; moisture = 17.3% and remaining is ash. This coal on combustion with excess of air gave 21.5 kg of dry flue gases per kg of coal burnt. Calculate the percentage of excess air used for combustion.

Solution:

Constituents Kg of the fuel	*Amount per reaction*	*Combustion O_2 (kg)*	*Amount of products (kg)*	*Amount of dry*
Carbon	0.54	$C + O_2 \rightarrow CO_2$	0.54 × 32/12 = 1.44	$0.54 \times \frac{44}{12} = 1.98$
Hydrogen	0.065	$H_2 + \frac{1}{2}O_2 \rightarrow H_2$	0.065 × 16/12 = 0.52	—
Nitrogen	0.018	—	—	—
Oxygen	0.03	—	—	—

$$\text{Amount of oxygen required} = 1.44 + 0.52 - 0.3 = 1.93 \text{ kg}$$

$$\text{Amount of air required} = 1.93 \times \frac{100}{23} = 8.39 \text{ kg}$$

$$\text{Amount of dry products using minimum quantity of air required} = 1.98 + .018 + \left(8.39 \times \frac{77}{100}\right)$$

$$= 8.4583 \text{ kg.}$$

Let the amount of excess air used = x

$$\therefore \quad 21.5\text{kg} = 8.4583 + x\text{kg}$$

$$\therefore \quad x = 21.5 - 8.4583 = 13.042\text{kg}$$

$$\% \text{ of excess air} = \frac{13.042}{8.39} = 155.4$$

Example 2.19 A coal containing 62.4% C; 4.1% H; 6.9% O_2; 1.2% N_2; 0.8% S and 51.1% moisture and 9.7% ash was burnt in such a way that the dry flue gases contained 12.9% CO_2, 0.2% CO; 6.1% O_2, and 80.8% N. Calculate (a) the weight of air theoretically required per kg of coal (b) the weight of air actually used and (c) the weight of dry flue gas produced per kg of coal.

Solution:

Constituent	*Amount in 1kg of fuel*	*Amount of O_2 (kg)*	*Amount of dry flue gases (kg)*
Carbon	0.624	0.624 × 32/12 = 1.664	0.624 × 44/12 = 2.288
Hydrogen	0.041	0.041 × 16/12 = 0.328	
Oxygen	0.069	—	
Nitrogen	0.012	—	
Sulphur	0.008	0.008 × 32/32 = 0.008	0.008 × 64/32 = 0.016
H_2O	0.51	—	

$$\text{Amount of oxygen required} = 1.664 + 0.328 + 0.008 - 0.069 = 1.921 \text{ kg}$$

$$\text{Amount of air required} = 2.003 \times \frac{100}{23} = 8.396 \text{ kg}$$

Now oxygen needed to convert 0.002 kg of CO present in the flue gas, to CO_2

$$= 0.002 \times \frac{16}{28} = 0.00114 \text{ kg}$$

$\therefore$ Excess of oxygen/kg of flue gas =

Amount of oxygen in the flue gas–Amount of oxygen required to convert CO to CO_2

$$= 0.061 - 0.00114 = 0.0598 \text{ kg}$$

Amount of carbon /kg of flue gas

$$\left(\text{wt. of } CO_2 \times \frac{12}{44}\right) + \left(\text{Wt. of CO} \times \frac{12}{28}\right) = \left(0.129 \times \frac{12}{44}\right) + \left(0.002 \times \frac{12}{28}\right)$$

$$= 0.03604 \text{ kg}$$

$\therefore$ 0.03604 kg of carbon is contained in = 1kg of flue gas

$$0.624 \text{ kg of carbon is contained in} = \frac{0.624}{0.03604} = 17.314 \text{ kg of flue gas}$$

(Carbon in the fuel gas and carbon in the flue gas has to be balanced)

$\therefore$ Amount of flue gas per kg of coal burnt = 17.314kg

Weight of excess O_2 in the flue gas = 17.314 × .059 = 1.0319 kg

$$\therefore \quad \text{Excess air} = 1.0319 \times \frac{100}{23} = 4.489 \text{ kg}$$

Hence, weight of air actually supplied = [8.396 (theoretical) + 4.487 (excess)] kg

= 12.885 kg

2.9 NUCLEAR FUELS

So far, we have discussed the chemical fuels which release their chemical energy in the form of heat. These reactions involve the rearrangement of the outer electrons only.

With the rapid increase in power consumption, due to industrialization and an increase in the population, it is estimated that fossil fuels (petroleum, coal) will get exhausted sooner or later. Thus attention has been focussed on the exploitation of other sources to generate power. These alternative sources include nuclear energy, solar energy, wind energy etc.

Nuclear energy is the energy released during a nuclear reaction. The materials which make such energy available are called **nuclear fuels**. The energy released during a nuclear reaction is much greater than that obtained from chemical fuels because the forces involved for binding the nucleons in nucleus are very high. For example, complete fission of 1 kg of uranium (a nuclear fuel) provides the energy equivalent to about 2×10^7 kWh. The same amount of energy requires the burning of 3×10^6 kg of high grade coal. Similarly, a single large nuclear power plant can save over 50,000 barrels of oil per day.

Table 2.11 Difference between Nuclear Fuels and Chemical Fuels

Nuclear Fuels	*Chemical Fuels*
1. In the fission of nuclear fuels, a heavy nucleus, captures a neutron and splits up into nearly equal parts, with the liberation of huge amount of energy.	In the combustion of chemical fuels, the atoms such as carbon, hydrogen etc combine with oxygen with the liberation of heat rapidly. Combustion reactions are always exothermic.
2. In the fission of nuclear fuels, there is a change in the electronic configuration and nucleus of atoms.	In the combustion of chemical fuels, there is a change in the electronic configuration and change in the valence electrons of the atoms involved, but there is no change in their nucleus.
3. Atomic numbers of atoms involved undergo change.	There is no change in the atomic numbers of atoms involved.
4. There is a loss in mass during fission of nuclear fuels and lost mass is converted into energy.	The combustion of chemical fuel involves no change in the total mass.
5. As a result of nuclear reaction, new entities are generally formed.	Combustion of chemical fuels give rise generally to similar products.
6. There is no effect of temperature or pressure on nuclear fission and so it is not possible either to stop or alter the rate of nuclear fission.	Rate of combustion of chemical fuels depends upon temperature and pressure.
7. Sum total of mass and energy remains constant, because matter gets converted into equivalent amount of energy.	During combustion of chemical fuels, conservation of charge and mass energy is maintained.
8. The liberated energy during nuclear fission of nuclear fuels may be million times greater than those during combustion of chemical fuels.	Much smaller energy is liberated as a result of combustion of chemicals fuels.

2.9.1 Nuclear Reactions

The nucleus occupies a very small portion of the total volume of an atom, but it contains most of the mass of the atom because both protons and neutrons (nucleons) reside there.

Nuclear reactions involve the interactions between nucleons of two nuclei. When a bombarding particle (projectile) comes in close contact with the nucleus being hit, the incident particle and the target nucleus form a compound nucleus, and nuclear reaction takes place. In such type of reactions, the nucleus of the target element change into another nucleus, these are called nuclear reactions. As the nuclei undergo change as a result of bombardment by neutrons, protons or other nuclei, this process is called nuclear transmutation.

In nature the nuclei of certain heavy elements undergo nuclear reactions when they spontaneously emit certain type of radiations and the phenomenon is termed as natural radioactivity. The first **artificial nuclear reaction** was carried by Rutherford who bombarded nitrogen with α-particles and the following nuclear reaction took place.

$$^{14}_{7}N + ^{4}_{2}He \longrightarrow ^{17}_{8}O + ^{1}_{1}H$$

An oxygen isotope was produced with the emission of a proton. Nuclear reactions follow laws of conservation as given below:

(i) **Law of conservation of mass and energy :** The total energy of a particle before and after the reaction remains the same.
Since in nuclear reactions, the mass changes into energy, according to the relation,
$E = mc^2$ (Einstein's mass energy equation)
Hence energy and mass are interconvertible quantities inside the nucleus.

(ii) **Law of conservation of mass number :** The total number of nucleons (mass number) before and after the reaction is conserved.

(iii) **Law of conservation of atomic number :** The total charge (atomic number) before and after the reaction is conserved.

Representation of nuclear reactions : The nuclear reactions are represented by the following notation :

[Reactant nucleus (bambarding particle, ejected particle) product nucleus]

For example the reaction,

$$^{14}_{7}N + ^{4}_{2}He \longrightarrow ^{17}_{8}O + ^{1}_{1}H \text{ is represented as } ^{14}_{7}N(\alpha, p)\overset{17}{\underset{8}{O}}$$

Similarly, $Al^{27}(\alpha, p)Si^{30}$ denotes that aluminium-27 is the target nucleus, α-particle is the bombarding particle and proton is emitting particle whereas Si^{30} is the product.

Energy Changes in nuclear reactions : The liberation of energy is explained on the basis that a mass is converted into energy during the nuclear reactions according to the Einstein's relationship. Now,

$$E = mc^2$$

Where m is the mass and c is the velocity of light. For a change of 1 a.m.u, the corresponding energy change can be calculated as follows :

$$1 \text{ a.m.u} = \frac{1}{12}\text{th of the mass of an atom of carbon}$$

$$= \frac{1}{12} \times \frac{12}{6.022 \times 10^{23}} g = 1.6606 \times 10^{-24} \text{ g} = 1.6606 \times 10^{-27} \text{ kg}$$

Taking $c = 2.9979 \times 10^8$ m/s, we have

$$E = mc^2 = (1.6606 \times 10^{-27}) \times (2.9979 \times 10^8)^2 \text{ Joules} = 1.4924 \times 10^{-10} \text{ J}$$

$$= \frac{1.4924 \times 10^{-10}}{1.6022 \times 10^{-10}} \text{eV } (1\text{eV} = 1.6022 \times 10^{-19}\text{J})$$

$$= 931.48 \times 10^6 \text{ eV} = 931.48 \text{ MeV} \approx 931.5 \text{ MeV}$$

$$(\because 1 \text{ MeV} = 10^6 \text{ eV})$$

Thus a mass of 1 amu is equivalent to 931.5 MeV of energy.

Different types of Nuclear Reactions: Many types of nuclear reactions are possible and have been observed. Different bombarding particles can be used as projectiles for bringing about nuclear reactions. Some of the nuclear reactions alongwith the different projectiles are gives below:

(i) Using α-particles ($_2He^4$) as bombarding particles.

$$_2He^4 + {}_7H^{14} \longrightarrow {}_8O^{17} + {}_1H^1 \qquad (\alpha, p \text{ type})$$

$$_2He^4 + 4Be^9 \longrightarrow {}_6C^{12} + {}_0n^1 \qquad (\alpha, n \text{ type})$$

(ii) Using neutrons ($_0n^1$) as the bombarding particles.

$$_0n^1 + {}_{13}Al^{27} \longrightarrow {}_{12}Mg^{27} + {}_1H^1 \qquad (n, p \text{ type})$$

$$_0n^1 + {}_8O^{16} \longrightarrow {}_6C^{13} + {}_2He^4 \qquad (n, \alpha \text{ type})$$

$$_0n^1 + {}_{11}Na^{23} \longrightarrow {}_{11}Na^{24} + \gamma \qquad (n, \gamma \text{ type})$$

$$_0n^1 + {}_{19}K^{39} \longrightarrow {}_{19}K^{38} + 2{}_0n^1 \qquad (n, 2n \text{ type})$$

(iii) Using proton ($_1H^1$) as the bombarding particles :

$$_1H^1 + {}_4Be^9 \longrightarrow {}_3Li^6 + {}_2He^4 \qquad (p, \alpha \text{ type})$$

$$_1H^1 + {}_2Be^9 \longrightarrow {}_3Be^8 + {}_1D^2 \qquad (p, d \text{ type})$$

$$_1H^1 + {}_4Be^9 \longrightarrow {}_5B^{10} + \gamma \qquad (p, \gamma \text{ type})$$

$$_1H^1 + {}_{20}Ca^{14} \longrightarrow {}_{21}Sc^{44} + {}_0n^1 \qquad (p, n \text{ type})$$

(iv) Using deutrons ($_1D^2$) as bombarding particles :

$$_1D^2 + {}_{83}Bi^{209} \longrightarrow {}_{83}Bi^{210} + {}_1H^1 \qquad (d, p \text{ type})$$

$$_1D^2 + {}_{13}Al^{27} \longrightarrow {}_{12}Mg^{25} + {}_2He^4 \qquad (d, \alpha \text{ type})$$

$$_1D^2 + {}_{13}Al^{27} \longrightarrow {}_{14}Si^{28} + {}_0n^1 \qquad (d, n \text{ type})$$

Out of the above mentioned subatomic bombording particles (projectiles), the positively charged particles like α-particles, protons etc. have the limitation that they are repelled by the positively charged nuclei of the atoms being bombarded. The best bombarding particles are neutrons. However the charged particles are also used but they are first accelerated in special types of machines called **particle accelerators**.

Table 2.12 Comparison of Chemical Reactions and Nuclear Reactions

Chemical Reaction	*Nuclear Reaction*
1. Only electron in outer atomic orbitals are involved.	Nuclei of the atoms are involved.
2. Breaking and forming of bonds is involved.	No breaking and forming of the bonds is involved Elements are converted from one to another.
3. Reactions are accompanied by relatively small amounts of energy.	Reactions are accompanied by release of tremendous amounts of energy.
4. Rates of reactions are influenced by temperature, pressure, concentration and catalysts.	Rates of reaction normally are not affected by temperature, pressure and catalysts.

Soddy, Fajans and Russel (1911-1913) observed that when an α-particle is lost, a new element with atomic number less by 2 and mass number less by 4 is formed. When β-particle is lost, a new element with atomic number greater by 1 is obtained. The above results are summarized as "Group Displacement Laws" as follows :

(i) When an α-particle is emitted, the new element formed is displaced two positions to the left in the periodic table than that of the parent element as atomic number decreases by 2. Example,

$$\underset{\text{Parent element}}{{}_{92}U^{238}} \longrightarrow \underset{\text{Daughter (new) element}}{{}_{90}Th^{234}} + {}_{2}H^{4}$$

(ii) When a β particle is emitted, the new element formed is displaced one position to the right in the periodic table than that of parent element because atomic number increase by 1.

$$\underset{\text{Parent element}}{{}_{90}Th^{234}} \longrightarrow \underset{\text{Daughter (new) element}}{{}_{91}Pa^{234}} + {}_{-1}e^{0}$$

2.9.2 Mass Defect and Binding Energy

Mass Defect: The masses of the various atomic nuclei were invariably found to be lesser than the sum of the masses of the corresponding nucleons *i.e.*, the sum of protons and neutrons present in the nucleus. The mass difference is referred to as "Mass Defect". According to Einstein mass-energy relationship ($E=mc^2$), the loss in mass appears as energy given off to the surroundings.

Consider an isotope of an element having atomic number 'Z' and mass number 'A'.

Number of protons in its atom = Z

Number of neutrons in its atom = A–Z

Let m_p m_n and m_e, the masses of proton, neutron and electron respectively, then the mass of this isotope M^1, will be given by

$$M^1 = Zm_p + Zm_e + (A - Z)m_n = Zm_H + (A - Z)\, m_n$$

$$(\because\ M_p + M_e = \text{Mass of hydrogen atom})$$

If M represents the actual atomic mass as determined experimentally, then

$$\text{mass defect} = \Delta m = M^1 - M$$

For example, the ${}_{9}F^{19}$ isotope has an atomic mass of 18.9984 amu. The nucleus has 9 protons and 10 neutrons and therefore a total of 19 nucleons. Using the known mass of ${}_{1}H^{1}$ atom (1.007825 amu) and the neutron (1.008665 amu), we can calculate the mass of the atom as :

$$\begin{aligned}\text{Mass of } {}_{9}F^{19} &= Zm_H + (A - Z)m_n\\ &= (9 \times 1.007825) + (10 \times 1.008665)\\ &= 19.15708 \text{ amu}\end{aligned}$$

The experimentally determined mass of 9F19

$$= 18.9984 \text{ a.m.u}$$

$$\therefore \quad \text{Mass defect, } \Delta m = 19.1570 \text{ a.m.u} - 18.9984 \text{ a.m.u} = 0.1587 \text{ a.m.u}$$

Binding Energy: The energy released when the required number of protons and neutrons coalesce to form the nucleus is known as the binding energy. This also indicates the energy needed to be supplied to break up a nucleus into its component protons and neutrons. This quantity represents the conversion of mass to energy that occurs during a nuclear reaction; *i.e.* the synthesis of a nucleus from the protons and neutrons. Consequently, the greater is the binding energy, the more stable is the nucleus.

Binding energy, $E = (\Delta m) \times c^2$.
Where $\Delta m$is the mass defect.

In the above example, the energy released to the surroundings as a result of the formation of the ${}_9F^{19}$ nucleus can be calculated as,

$$E = (0.1587 \text{ amu})\ (3.00 \times 10^8 \text{m/s})^2$$
$$= 1.43 \times 10^{16} \text{ amu m}^2/\text{s}^2$$

Now 1 kg = 6.022×10^{26} amu

$$1 \text{ J} = 1 \text{ kg m}^2/\text{s}^2$$

$$\therefore \quad E = \left(1.43 \times 10^{16} \frac{\text{amu } m^2}{s^2}\right)\left(\frac{1.00 \text{ kg}}{6.022 \times 10^{26} \text{amu}}\right)$$
$$= 2.37 \times 10^{-11} \text{ kg m}^2/\text{s}^2$$
$$= 2.37 \times 10^{-11} \text{ J}$$

This is the amount of energy released when one flourine-19 nucleus is formed from 9 protons and 10 neutrons. This is known as binding energy. In general, larger is the mass defect, the greater is the binding energy and more stable is the nucleus.

The nuclear binding energy for 1 mole of flourine nuclei

$$= (2.37 \times 10^{-11} \text{ J} \times 6.022 \times 10^{23}/\text{mol})$$
$$= 1.43 \times 10^{13} \text{ J/mol}$$
$$= 1.43 \times 10^{10} \text{ kJ/mol}$$

This is tremendously large quantity if we compare the enthalpies of chemical reactions which are of the order of 200 kJ.

Binding Energy and Nuclear Stability: As is stated earlier that nuclear binding energy is an indication of the stability of the nucleus. For comparing the stability of any two nuclei, we must use the expression, nuclear binding energy per nucleon, as the nuclei of different atoms have different number of nucleons. Thus,

$$\text{Nuclear binding energy per nucleon} = \frac{\text{Nuclear binding energy}}{\text{number of nucleons}}$$

For ${}_9F^{19}$ nucleus,

$$\text{Nuclear binding energy per nucleon} = \frac{2.37 \times 10^{-11} \text{ J}}{19 \text{ nucleons}} = 1.25 \times 10^{-12} \text{ J/nucleon.}$$

The nuclear binding energy per nucleon allows us to compare the stability of all nuclei on common basis. Figure 2.26 shows the variation of nuclear binding energy per nucleon plotted against mass number

It is seen from the graph that the highest binding energies per nucleon belong to the elements with mass numbers in between 40 and 100, and are greatest for elements iron, cobalt and nickel (Group 8B elements). Thus, iron with mass number 56, has much higher binding energy than uranium having mass number 235. It indicates that energy needed to separate nucleons from each other in iron nucleus is higher than needed in uranium nucleus and hence elements of intermediate mass numbers are more stable as compared to both the lighter elements like hydrogen and heavier elements like ${}_{92}U^{235}$.

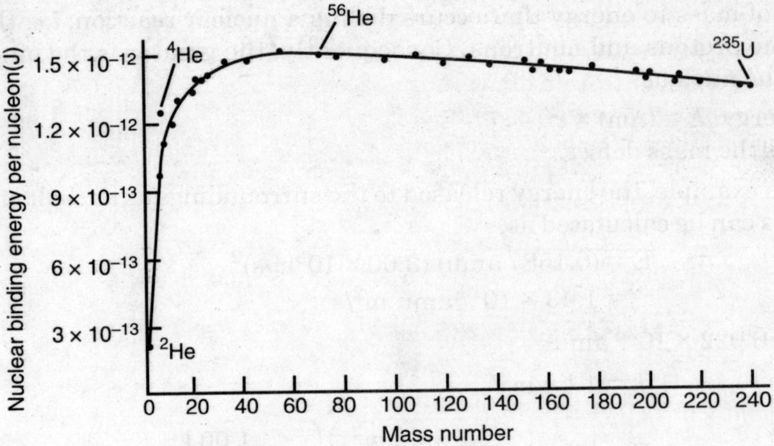

Fig. 2.26 *Binding energy curve.*

When heavier elements (like U–235) are bombarded with fast particles into fragments of lighter nuclei (Nuclear Fisson) a huge amount of energy is released. Similarly when lighter elements (like deuterium) are combined to form heavier nucleus (Nuclear Fusion), it is accompanied by release of tremendous amount of energy.

Because of this tremendous amount of energy which is liberated due to nuclear fission or fusion, these elements are used as **Nuclear Fuels**. At present, three isotopes are known to be suitable as nuclear fuels. These are natural uranium-235 and two artificially prepared isotopes, plutonium-239 and uranium-233. These materials can sustain the fission reactions and are thus termed as nuclear fuels.

2.9.3 Nuclear Fission

Nuclear fission is the process of splitting of a heavy nucleus into two approximately equal parts, with simultaneous liberation of a huge amount of energy. For example-

$$^{235}_{92}U + ^{1}_{0}n \longrightarrow ^{236}_{92}U \longrightarrow \begin{cases} {}_{56}Ba^{140} + {}_{36}Kr^{93} + 3{}_{0}n^{1} + \text{Energy} \\ {}_{54}Xe^{144} + {}_{38}Sr^{90} + 2{}_{0}n^{1} + \text{Energy} \\ {}_{55}Cs^{144} + {}_{37}Rb^{90} + 2{}_{0}n^{1} + \text{Energy} \end{cases}$$

From the above reaction it is clear that the products of the disintegration are not the same in every case. When U-235 atom undergoes fission, it splits up into two unequal fragments accompanied by liberation of 2 or 3 fresh neutrons and energy. Different U-235 atoms split up in many different ways. For example, in addition to the above fission products U-235 can also form,

$$^{235}_{92}U + ^{1}_{0}n \longrightarrow \begin{cases} {}_{37}Rb^{90} + {}_{55}Cs^{144} + 2{}_{0}n^{1} \\ {}_{35}Br^{87} + {}_{57}La^{146} + 3{}_{0}n^{1} \\ {}_{30}Zn^{72} + {}_{62}Sm^{160} + {}_{40}n^{1} \end{cases}$$

In this way fission of U-235, containing billions of atoms, gives a large number of products. Actually, the fission reaction is very complex and more than 35 different elements have been identified among the fission products of U-235. The following types of binary fission are most common :

$${}_{92}U^{235} + {}_{o}n^{1} \rightarrow \left[{}_{92}U^{236}\right]^{*} \rightarrow A^{85-105} + B^{130-150} + 1\,to\,3\,neutrons + energy$$

Calculation of the energy released in nuclear fission reaction

As was discussed in section 2.9a that a mass of 1 amu is equivalent to 931.5 MeV of energy. Now the energy released when one atom of uranium-235 undergoes nuclear fission, according to the following reaction can be calculated as under :

$$_{92}U^{235} + {_0}n^1 \longrightarrow {_{56}}Ba^{144} + {_{36}}Kr^{90} + 2{_0}n^1$$

Actual isotopic mass	235.060	1.009	143.881	89.947	2 × 1.009	a.m.u.
Total	236.069 a.m.u		235.846 a.m.u			

$$\text{Mass Defect} = 236.069 \text{ a.m.u} - 235.846 \text{ a.m.u}$$
$$= 0.223 \text{ a.m.u}$$

∴ Energy released during fission of one U-235 atom

$$= 0.0233 \times 931.5 \text{ MeV} = 208 \text{ MeV}$$

Thus, energy released when 1 g of U-235 undergoes fission

$$= 208 \times \frac{6.022 \times 10^{23}}{235} \text{ MeV} = 5.33 \times 10^{23} \text{ MeV}$$

However, this much of power generation is possible only if fuel were entirely composed of fissionable nuclei and all of them undergo fission. But, in practice, the fuel contains other non-fissionable istopes of uranium, plutonium and thorium.

Nuclear Chain Reaction: The significant feature of uranium-235 fission is not just the enormous amount of energy released, but the fact that more neutrons are produced than are originally used for bombardment. This property makes possible a **nuclear chain reaction**, which is self-sustaining sequence of nuclear fission reactions.

Each fission of U^{235} nucleus is accompanied by 2.5 neutrons on an average. These neutrons act as chain careers and are capable of bringing about a sustaining chain reaction. The neutrons released during the initial stage of fission can induce fission in other U^{235} nuclei, which in turn produces more neutrons and so on. In less than a second the reaction become uncontrollable, liberating tremendous amount of energy to the surroundings.

However, practically some neutrons released escape unused and this lowers the amount of energy released.

Critical Mass: For a chain reaction to occur, enough U^{235} must be present in the sample to capture the neutrons. Otherwise, many of the neutrons will escape from the sample and the chain reaction will not occur.

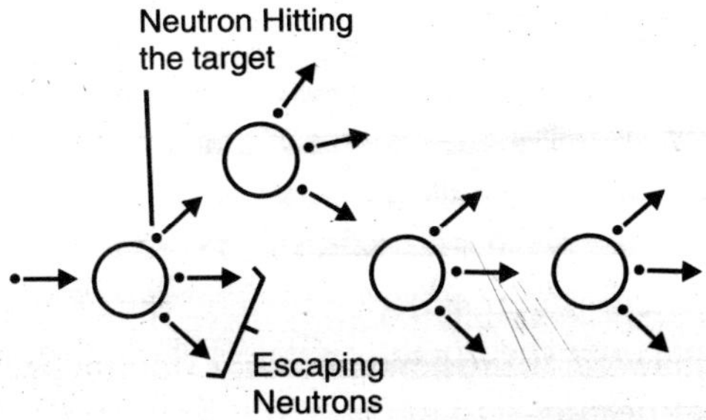

Fig. 2.27

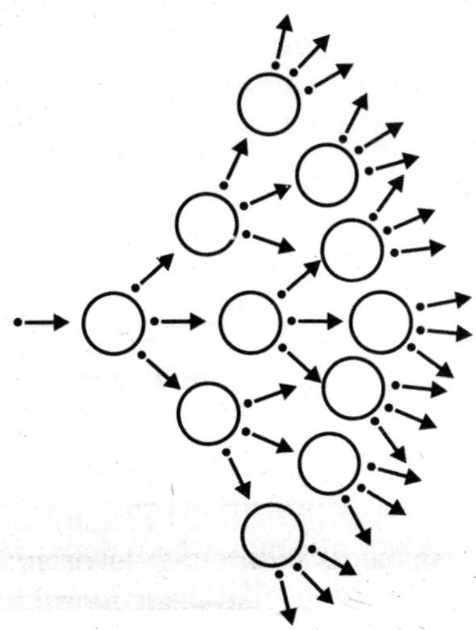

Fig. 2.28 *Chain Reaction*

In this condition, the mass of the sample is said to be subcritical. **The quantity of fissionable material required to initiate the chain reaction is called "critical mass". Critical mass is defined as the minimum mass of fissionable material required to generate a self-sustaining nuclear chain reaction**. Thus if the mass of the fissionable material is less (subcritical) than the critical mass, a number of neutrons escape without causing further fission and the reaction is not explosive. On the other hand if the mass of the fissionable material is more than the critical mass (supercritical), an uncontrollable explosive chain reaction is set up. The critical mass of U-235 is 1 to 100 kg.

However, the chain reaction can be made to proceed in such a way that at least one neutron is available for sustaining the chain reaction in each cycle of fission. This is done by using moderators which absorb the neutrons and hence limit their number available for inducing fission. Under these conditions the controlled fission chain reaction can be used to produce power in nuclear reactor.

The number of neutrons resulting from a single fission, is known as the multiplication factor. When this is less than 1, a chain reaction does not take place. In nuclear reactor, the fissionable material is so arranged that the multiplication factor can be carefully controlled to one.

Fissionable Materials: Only isotopes such as U^{235}, Pu^{239} and U^{233} are fissionable by neutrons of all energies. Hence these are suitable as the nuclear fuels. The isotopes like U^{238}, Th^{232}, Pu^{240} are fissionable only by high energy neutrons. Natural uranium is a mixture of U^{238} (99.3%), U^{235} (0.7%) and traces of U^{234}. Out of these, only U^{235} is fissionable by slow and fast moving neutrons whereas U^{238} absorbs slow moving neutrons and instead of undergoing fission, it yields transuranic elements Np and Pu.

$$_{92}U^{238} + \underset{\text{(slow)}}{_{0}n^{1}} \rightarrow {}_{92}U^{239} \xrightarrow{\beta} {}^{230}_{93}Np \xrightarrow{\beta} {}^{239}_{94}Pu$$

U^{238} is however fissionable by fast neutrons, the fission probability is being rather low.

$$_{92}U^{238} + {}_{0}n^{1} \rightarrow {}_{54}Xe^{143} + {}_{38}Sr^{94} + 2{}_{0}n^{1} + Energy\ (very\ rare\ reaction)$$

Mechanism of Nuclear Reaction: According to Bohr's mechanism of nuclear reaction, the projectile particle enters the nucleus to form a compound nucleus. The compound nucleus possesses a high excitation energy of the order of several MeV. If a particle acquires enough energy to escape out, it will be emitted, otherwise the excess energy is emitted in the form of γ-rays. The compound nucleus so formed disintegrates into smaller fractions. The mode of its disintegration depends upon factors such as its excitation energy, the heights of potential barriers for different reactions and the stabilities of the residual nuclei. In fact, it may be formed in different ways and may break down also in several different ways.

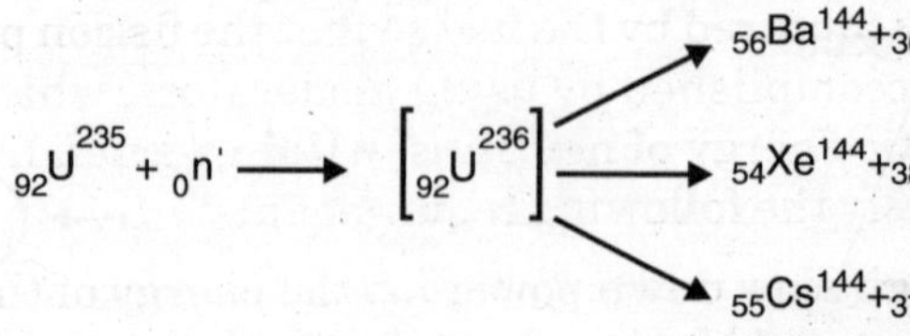

Different modes of disintegration

The probability of occurance of a given nuclear reaction depends upon the cross-section of the target nucleus, which depends upon the property of the nucleus, type of the reaction and energy associated with the incident neutrons. Nuclear cross sections, (σ) having the dimensions of area, are generally of the order of 10^{-24} cm^2 and are expressed in barn (1 barn = 10^{-24} cm^2).

2.9.4 Nuclear Reactor

The most important application of energy released in nuclear fission depends on proper control of the propagation of chain reaction. This is usually carried out in a nuclear reactor. **A nuclear reactor is a device for carrying out fission reaction at a controlled rate to produce energy which can be utilized for peaceful purposes such as generation of electricity.**

Depending on the velocities of the fission inducing neutrons, nuclear reactors may be classified as

(i) Thermal reactors-A reactor where fission is brought about by slow neutrons.
(ii) Fast reactors-A reactor where fission is brought about the fast neutrons.

Classification can also be done on the basis of purpose for which reactor is operated as materials testing reactor, power reactor, breeder reactor etc.

A typical nuclear reactor has the following parts.

(1) Reactor Core

The principle component of any nuclear reactor is Reactor Core. Here the controlled fission reaction is made to occur, liberating a huge amount of energy. The reactor core consists of the following parts:

(i) Fuel : The fissionable material used in the reactor is called the fuel. The fuels used in the reactor generally are-natural uranium (containing 0.7% of the fissile U^{235} isotope) enriched uranium (which contains high percentage of U^{235}), plutonium-239 (obtained from U-238) and thorium-232 in which fissile U^{233} is mixed.

The fuel material is usually made into the form of pellets or rods, shielded by placing them in stainless steel tubes or zirconium aluminium alloy. The metal jacket serves the following purposes :

- It protects the fuel material from chemical reaction with air, water or other materials used as coolant.
- The fission products are not allowed to escape and hence cannot contaminate the coolant.
- The metal jacket act as a structural support for the fuel.

The shielding material should possess good mechanical and corrosion resistance properties. It should have high temperature stability and low neutron absorption cross-section Al or Mg is preferred as shielding material for moderate temperatures whereas stainless steel are used for high temperatures, existing in power reactor.

(ii) Moderator: The neutrons produced in the fission have to be slowed down to speeds at which they are easily captured by the fuel so that the fission process can take place most efficiently. This is accomplished by using moderators, which are the substances that can reduce the kinetic energy of neutrons, within a small fraction of a second. A good moderator must satisfy the following requirements:

- It should have high slow down power *i.e.* the energy of the fast neutrons should be slowed down from 1 or 2 MeV to about 0.025 eV.
- It should be cheap and abundantly available in pure form because a very large quantities are required.
- It should resist conversion to radioactive substances by neutron bombardment.
- It should have good chemical, thermal and radiation stability.
- It should be corrosion resistant (in case it is solid).
- It should have good thermal conductivity.

No substance fits in all these requirements, although water comes closer than many others which can be used as moderator. The commonly used moderators are water, heavy water, graphite, beryllium and some other organic compounds.

Heavy water is considered to be the best moderator. The main advantage of heavy water reactor is that it eliminates the need for using expensive uranium enrichment facilities. Thus natural uranium is used as a fuel if heavy water is used as a moderator. Heavy water slows down the neutrons emerging from the fission reaction less efficiently and hence they travel great distances so the probability that they will strike the proper targets (U^{235}) is correspondingly high. High cost of heavy water (about Rs 500/kg) is the only disadvantage.

Graphite is considered as a good moderator because of its cheapness and can be machined easily into any shape. It is available in pure state, has good thermal conductivity, high temperature resistance and non toxic properties. However, the nuclear properties of graphite are inferior to those of heavy water and beryllium. It breaks easily when struck. It reacts with O_2 and CO_2 at high temperatures. Its porous nature and dimensional instability under radiation are some more limitations for its use as moderator.

Beryllium, because of its excellent chemical corrosion resistance and high radiation stability, is considered a good moderator. However, it is brittle and susceptible to attack by air and water at high temperature and is quite expensive, hence is used only in special cases. Moreover it is toxic also.

Water is also used as moderator in many of the commercial reactors because of its cheapness and availability.

In homogeneous reactors, the fuel and the moderator are mixed to form a uniform mixture (U^{235} and graphite) and is used in the form of rods or plates in the reactor core and in heterogeneous reactors, the fuel is used in the form of rods or plates and moderator surrounds the fuel element.

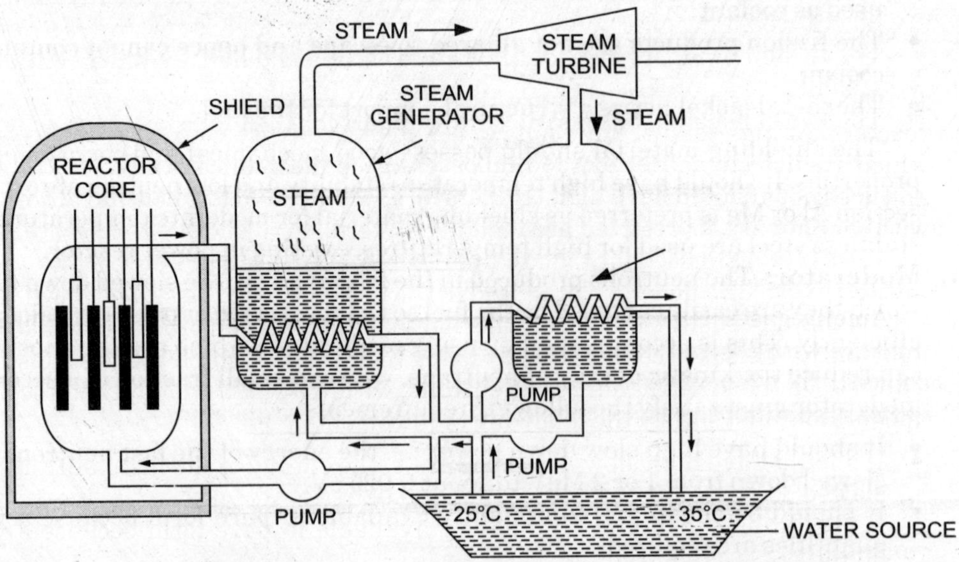

Fig. 2.29 *A complete design of nuclear power plant*

(iii) Coolants: To carry away the heat produced during fission, a liquid is circulated in the reactor core. This liquid is termed as coolant. In case of a power reactor, the heat from the coolant is exchanged to a working fluid to produce steam or a hot gas which is then used to produce enough steam to drive an electric generator. In this respect nuclear power plant is similar to a conventional power plant that burn fossil fuel. Thus most nuclear power plants are built near a river or a lake. In some reactors, water is boiled within the core of the reactor to directly generate steam.

The most important requirements for a coolant are:

- High thermal conductivity and high specific heat capacity.
- Stability to heat and radiation.
- Low cost and low power requirment for pumping.
- Non corrosive and non toxic.
- In case of liquids, high boiling point and low melting point.

Important coolants are: ordinary water and heavy water, liquid metals (Na or K-Na alloy), organic liquids and gases.

Because of their excellent heat exchange properties heavy water and ordinary water are good coolants. The advantage of using D_2O or H_2O is that, they serve as both moderator and coolant. However, their boiling point being low, they are used under pressurization to prevent their boiling in the reactor operating at high temperatures. Moreover, water becomes corrosive at high temperature and hence fuel rods should be shielded in non-corrosive materials.

Liquid metals (Na, K) are excellent coolants and are recommended for use at high temperature fast reactors. Liquid Na has excellent heat exchanging properties. No pressurization is required even at high temperatures. However, Na is very reactive with oxygen and hydrogen. Moreover due to the capture of neutron it also becomes radioactive, when passed through the reactor. It decays with the emission of β-particles and γ-rays which pose potential hazards and hence needs a special shielding. It is usually employed in fast reactors.

Organic compounds such as benzenes and polyphenyls (*e.g.* diphenyls and terphenyls) are also used as coolants. Their heat removing properties are inferior to those of water and liquid sodium. But they are non-corrosive and are fairly stable to heat and radiation.

The gaseous coolants used are CO_2, H_2, He and air. Air is suitable for low power reactors. At high temperatures, it reacts chemically with materials like Al, Be, Mg etc which are used in the reactor. The most commonly gaseous coolant is CO_2. It is cheap and does not attack metals even at fairly high temperatures (up to 500° C). But it reacts with graphite at elevated temperature.

$$CO_2 + C \rightarrow 2CO$$

Among gases H_2 and He may also be used as coolants. But H_2 forms explosive mixture with air and it also renders many metals brittle, thus is not used. The other choice is He which is used as a coolant in some experimental power reactors. But He is very expensive and special precautions have to be taken to minimize the losses due to leakage.

(iv) **Control Rods :** The control system of a reactor consists of materials usually in the form of rods, which can absorb neutrons and thereby prevent the chain reaction becoming violent.

The material should not become radioactive as a result of neutron capture. The commonly used materials for control rods are boron, cadmium, and hafnium.

Earlier cadmium was extensively used as a control material due to its availability and ease of fabrication. However due to its relatively low melting points, cadmium is used only in reactors operating at low temperatures Moreover it has low neutron absorption crosssection.

Boron is the most common material used for reactor control due to its very high melting point and very large neutron absorption cross section over a wide energy range. It is employed in the form of boron incorporated in stainless steel or boron carbide dispersed in Al.

In the nuclear reactor, the overall amount of fuel present is greater than the critical mass, and hence a violent nuclear chain reaction may take place. Unless the growth of neutron is checked at some stage, the reactor is likely to be damaged. The control of the chain reaction is possible either by raising, withdrawing or inserting the control rods. By inserting the control rods deeper into the central core, any amount of excess neutrons can be absorbed. If at some stage, it is desired to increase the power level, the control rods are partially withdrawn. The control rods serve the following purposes:

- For increasing the neutron flux in a thermal reactor, the control rods are lifted up whereas they are pushed deep in the reactor to decrease the neutron flux or fission rate.
- For maintaining a uniform ratio of power production.
- For controlling the nuclear chain reaction so as to avoid explosion.
- For start-up or for shutting the nuclear reactor.

(2) Reflector: Reflector surrounds the reactor core so as to reflect back some of the neutrons that leak out from the surface of the core. Reflectors should have low absorption and high reflection for neutrons, high resistance to oxidation and irradiation and high radiation stability. In thermal reactors, the same material which is used as moderator can be used as reflector (D_2O, H_2O or graphite)

(3) Pressure-vessel: Pressure vessel encloses the reactor core and the reflector. Pressure vessel is designed to withstand pressures of the order of 200 kg/cm^2. Pressure vessels are provided with the holes at the top to insert or pull the control rods. It is also provided with entrance and exit passage for coolant.

(4) Shielding: To prevent the losses of heat and to protect the persons operating the reactor from the radiation and heat, the entire reactor core is enclosed in a heavy steel or concrete dome called shield. Two types of shield are usually provided:

(i) **The thermal shield:** It is very close to the reactor core and is 50 to 60 cm thick made up of iron. It absorbs most of the γ-rays. It prevents the biological shield from damage due to overheating. It is located in such a way that it can be readily cooled by circulation of water.

(ii) **The biological shield:** It surrounds the thermal shield and comprise of a layer of concrete. It absorbs both gamma rays and neutrons escaping from inner thermal-shield.

(5) Heat exchanger: Heat exchanger exchanges the heat liberated in fission process to operate a steam turbine.

(6) Turbine: The steam at high pressure, generated in the heat exchanger is used to operate a steam turbine, which drives a generator to produce electricity.

The steam after driving the turbine has to be condensed. For this purpose additional cooling water is needed. This is obtained from a large source such as river or lake having flowing water.

Breeder Reactor:

The rate at which U^{235} is being used up to produce power, the stocks are likely to exhaust very soon. The more abundant isotope U^{238} is non-fissionable but can be converted to fissionable material Pu^{239}. A breeder reactor is a nuclear reactor designed in such a way that it can generate its own fuel by converting the non fissionable material to a fissionable material, such reactors are also called regenerative reactors or converters.

In these reactors U^{235} is fissioned by slow neutrons and on an average 2.5 neutrons are ejected and .5 neutron escapes out of the reactors. Some of the neutrons are captured by ${}_{92}U^{238}$ present in the parent uranium and following reaction take place.

$$\underset{\text{(Non-fissionable)}}{{}_{92}U^{238}} + {}_0n^1 \longrightarrow {}_{92}U^{239} \longrightarrow {}_{93}^{239}Np + {}_{-1}^{0}B$$

$${}_{93}^{239}Np \longrightarrow {}_{-1}^{0}B + \underset{\text{(fissionable)}}{{}_{94}Pu^{239}}$$

Plutonium-239 is the fissionable isotope. In otherwords ${}_{94}Pu^{239}$ is the man made nuclear fuel and can be termed as secondary nuclear fuel. The ${}_{92}U^{238}$ used to produce fissionable material is termed as fertile isotope.

Anothor fertile isotope is ${}_{90}Th^{238}$. Upon capturing slow neutrons, thorium is transmuted to uranium-233, which is again a fissionable material. The reactions taking place are

$$\underset{\text{(Non-fissionable)}}{{}_{90}Th^{232}} + {}_0n^1 \longrightarrow {}_{90}Th^{233} \xrightarrow{\text{b emission}} {}_{91}Pa^{233}$$

$${}_{91}Pa^{233} \xrightarrow{\text{b emission}} {}_{92}^{233}U$$

Uranium-233 is stable enough for long term storage.

Thus it is evident that the burn up of primary nuclear fuel can be compensated to some extent by generating secondary fuel. The extent of compensation is measured by a factor, called **conversion factor**. It is the ratio of the number of secondary fuel atoms produced to the number of primary fuel consumed. In uranium graphite moderator type of reactors, the conversion factor is generally =.9. In other words the net consumption of the nuclear fuel is just 10%, since 90% is recovered as secondary fuel. However in the reactors where the conversion factor is more than unity are called breeder reactors because they produce more fissionable material than what is consumed.

In a typical breeder reactor, nuclear fuel containing plutonium-239 is mixed with uranium-238 so that the breeding takes place within the core. No moderator is employed to slow the neutrons. During the operation of the breeder reactor, for every Pu^{239} nuclear undergoing fission, more than one neutron is captured by uranium-238 which is turn generate Pu^{239}. Thus the stock of fissionable material is increased steadily as the fissionable material is consumed.

Therefore, the long term future of nuclear energy depends on the efficiency of converter reactors in which Pu-239 is consumed as a fuel and at the same time, is also regenerated from U-238.

2.9.5 Nuclear Fusion

In contrast to the nuclear fission process, nuclear fusion is defined as a process in which the lighter nuclei fuse together to form heavier nucleus releasing a large amount of energy.

Figure 2.26 depicts that for the lightest elements, nuclear stability increases with increasing mass number. This behaviour suggests that if two light nuclei combine or fuse together to form a larger, more stable nucleus, an appreciable amount of energy will be released in the process. Example,

$$_1H^3 + {}_1H^3 \longrightarrow {}_2He^4 + {}_0n^1 + 17.6\text{ MeV}$$
$$_1H^2 + {}_1H^2 \longrightarrow {}_1H^3 + {}_1H^1 + 4\text{MeV}$$

The fusion of two or more nuclei is, however very difficult because all the nuclei are positively charged. The fusion can only take place if the nuclei move with very high velocities *i.e.* their kinetic energies must be increased to a very high values so as to overcome the strong forces of electrostatic repulsion. Since the electrostatic repulsion increases with the atomic number of the nuclei, lighter nuclei like the isotopes of hydrogen and helium are preferred for fusion reactions. Further a fusion reaction takes place only when the combining nuclei are made to approach each other within a distance of the order of 1 fermi.

The kinetic energies of the interacting nuclei can be increased thermally. It has been estimated that the temperature of 10^{6}°C will be required for effecting the fusion. Such a high temperature is obtained by nuclear fission. Nuclear reactions which proceed at very high temperatures, such as nuclear fusion are termed as **thermonuclear reactions.**

In nature, high temperatures of the order of 10^{6}°C are available only in stars including sun. H.A. Bethe (1939) proposed that most of the energy of the sun and the stars is due to thermonuclear reactions, which involve the isotopes of hydrogen. In these thermonuclear reactions hydrogen is converted into He. This conversion from hydrogen to helium may take place either by the proton cycle or by carbon cycle.

Proton Cycle

$$_1H^1 + {}_1H^1 \longrightarrow \underset{\text{Deuterium}}{{}_1H^2} + \underset{\text{positron}}{{}_{+1}e^0}$$
$$_1H^2 + {}_1H^1 \longrightarrow {}_2He^3 + \gamma$$
$$_2He^3 + {}_2He^3 \longrightarrow {}_2He^4 + 2\,{}_1H^1$$

The net effect is conversion of four protons to give one $_2He^4$ nucleus with the emission of γ-rays and two positrons *i.e.*

$$4\,{}_1H^1 \longrightarrow {}_2He^4 + 2\,{}_{+1}e^0 + \gamma$$

Carbon-Nitrogen Cycle:

$$_6C^{12} + {}_1H^1 \longrightarrow {}_7N^{13} + \gamma$$
$$_7N^{13} \longrightarrow {}_6C^{13} + {}_{+1}e^0$$
$$_6C^{13} + {}_1H^1 \longrightarrow {}_7N^{14} + \gamma$$
$$_7N^{14} + {}_1H^1 \longrightarrow {}_8O^{15} + \gamma$$
$$_8O^{15} \longrightarrow {}_7N^{15} + {}_{+1}e^0$$
$$_7N^{15} + {}_1H^1 \longrightarrow {}_2He^4 + {}^{12}_{6}C$$

$$4\,{}_1H^1 \longrightarrow {}_2He^4 + 2\,{}_{+1}e^0 + 26.7\text{ MeV}$$

The net result in both the above reaction sequences is the formation of helium-4 nucleus and the same amount of energy (26.7 MeV) is released. In carbon-nitrogen cycle, carbon acts like a catalyst.

On earth, temperature as high as required for fusion to start can be made available only by nuclear fission. This forms the basis of hydrogen bomb. The temperature attained during the fission of U-235 or Pu-239 in fission reactions, are of the order of the magnitude required to cause thermonuclear fission of $_1H^2$, $_1H^3$ etc. Thus atomic bomb, based on nuclear fission is used to attain the temperature required to cause nuclear fusion in the hydrogen bomb.

Fusion Reactors

One of the major concern in choosing the proper nuclear fusion process for energy production is the high temperature which is required for such reactions to take place. Attempts are still being made to use themonuclear reactions as energy source. Some of the promising reactions are:

(i) $_1H^2 + {}_1H^2 \longrightarrow {}_1H^3 + {}_1H^1 + 4$ MeV
(ii) $_1H^2 + {}_1H^3 \longrightarrow {}_2He^4 + {}_0n^1 + 17.6$ MeV
(iii) $_3Li^6 + {}_1H^2 \longrightarrow 2\,{}_2He^4 + 3.2$ MeV

Out of these, the (i) reaction is particularly of great interest. This is because deuterium is present in all natural waters on this planet and hence would represent an inexhaustible source of energy. Water contains about one deuterium atom for every 6000 hydrogen atoms and the separation of hydrogen isotopes is not a difficult task. The sea water has been estimated to contain 10^{14} kg of deuterium and could be a cheap source for nuclear fuel. The energy equivalent of the deuterium in one gallon of water is calculated to be the same as that obtained from the combustion of 300 gallons of gasoline.

From the above discussion nuclear fusion appears to be a wonderful source of energy. But the technical difficulties to design the nuclear fusion reactor have not yet been solved. Even if the high temperature required for fusion is produced, there is no known material to withstand such a high temperature.

At temperatures of about 10^{6}°C, molecules cannot exist and all the atoms are stripped of their electrons. This state of matter, a gaseous mixture of positive nuclei and electrons is called **plasma**. Now the problem is that, what solid container can hold this plasma. Research work in fusion reactions is directed towards producing plasma and simultaneously to keep the plasma away from the walls of the container for a long enough time for fusion to occur. One approach to solving this problem is to use **magnetic confinement**. Since a plasma consists of charged particles moving at high speeds, a magnetic field will exert force on it. Figure 2.30 shows that the plasma moves through a dough-nut shaped tunnel, confined by a complex magnetic field. Thus the plasma never comes in contact with the walls of the container. Such type of magnetic trap is known as Tokomak (Russian acronym for toroidal cavity with magnetic field.)

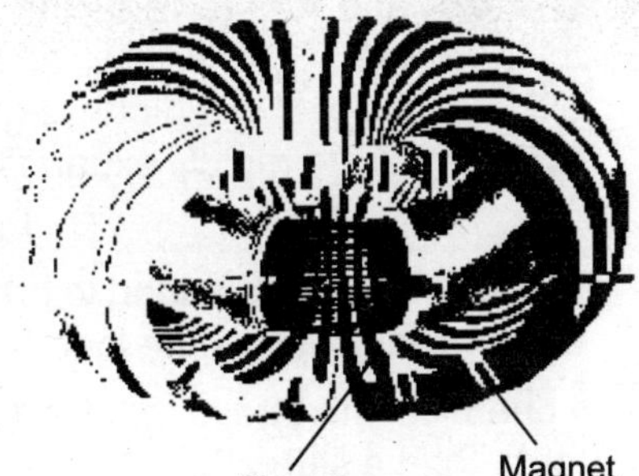

Fig. 2.30 *A Magnetic Confinement Design Called Tokamak*

Another development to carry out fusion reaction is, by using high-power lasers. In test runs, a number of laser beams transfer energy to a small fuel pellet (a solid deuterium tritium pellet) which is dropped into the middle of the chamber, heating it and causing it to compress or collapse inwards from all sides. The hot compressed plasma has a density of about 10^{19} to 10^{22} particles cc^{-1}. Under such conditions, nuclear fusion without problems of container, takes place and energy thus released can be collected. However both the techniques, magnetic confinement and laser fusion presents a number of technical difficulties that still need to be solved before these can be used on a large scale.

It must be emphasized here that for a controlled fusion reaction to provide useful energy, the process must be self-sustaining *i.e* the energy released in the fusion process must be sufficient at least to maintain the high temperature for further reaction. A balance must be maintained between the energy released in the fusion reactions and the energy lost, mostly due to escaping radiations.

Nuclear Fusion Versus Fission

As a fuel or energy source nuclear fusion has several advantages over nuclear fission. Fusion products are stable isotopes such as $_2He^4$ whereas fission products are hazardous radioactive isotopes. The isotopes used for fusion are far more abundant than the heavy isotopes required for fission. A comparsion of the two process is given in the following table :

Nuclear Fission	*Nuclear Fusion*
(1) It is a process of breaking up of heavier nucleus into lighter nuclei.	It involves the union of two or more lighter nuclei to form heavier nucleus.
(2) The neutrons ejected initiate further reaction resulting in a chain reaction.	It is not a chain process.
(3) It is initated by neutrons and does not require high temperatures.	It requires very high temperature of the order of a few million degrees.
(4) This process results in the emission of radioactive isotopes.	It does not form any radioactive isotopes.
(5) Nuclear fission can be performed under control conditions.	Nuclear fusion cannot be performed under control conditions.
(6) It requires minimum size of fissionable material i.e the critical mass.	There is no limit to the size of the fuel for the reaction to start.

HOW DO SPARKLERS WORK (CHEMISTRY OF THE SHOWER OF SPARKS)

All Fireworks are not created equal! For example, there is a difference between a firecracker and a sparkler. The goal of a firecracker is to create a controlled explosion. A sparkler, on the other hand, burns over a long period of time (up to a minute) and produces a brilliant shower of sparks. How do fireworks produce their brilliant colours and loud bangs? Actually, only a handful of different chemicals are responsible for these spectacular effects. These consists of :

- An oxidizer • a fuel • metal powder • A combustible binder
- Oxidizer and fuel (a reducing agent) are used to produce noise and flashes. A common mixture is $KClO_4$ or $KClO_3$ as oxidizer and aluminium and sulphur as fuel. The perchlorate oxidizes the fuel in a highly exothermic reaction, which produces a brilliant flash and a loud sound from the rapidly expanding gases produced.

$$KClO_4 + \text{fuel} \rightarrow \text{Products} + \text{Heat}$$

- Metal powders are added to produce brilliant flashes.
- Binders hold the mixture together. Common binders are dextrin (a sugar) dampened by water.

The chemical mixture is moulded onto a rigid stick or wire. The chemical often are mixed with water to form a slurry that can be coated on a wire (by dipping) or poured into a tube. Once the mixture dries, sparkler is ready. Aluminium, iron, steel, zinc or magnesium dust or flakes may be used to create the bright, shimmering sparks. The metal flakes heat up until they are incandescent and shine brightly or, at a high enough temperature, actually burn. A variety of chemicals can be added to create colours.

Electrons in these substance are excited to high -energy orbitals. The excited atoms then release this excess enegy by emitting light of specific wavelengths in visible region. The energy to excite these electrons comes from combustion reactions.

Common chemical used are

Oxidizes	*Fuels*	*Colouring effects/Special*
Potassium nitrate	Aluminium	Red flame : Strontium nitrate
Potassium chlorate	Magnesium	Blue flame : Copper Carbonate
Barium nitrate	Sulphur	Green Flame : Barium nitrate
Barium chlorate	Antimony sulphide	Yellow flame : Sodium Oxalate
Strontium nitrate	Dextrin	Gold sparks : iron filings, charcoal Whistle effect : potassium benzoate Coloured smoke : mixture of Potassium nitrate and sulphur.

QUESTIONS

SHORT ANSWERS QUESTIONS
(Refer to the Respective Sections for Answers)

Section 2.1, 2.2

1. Define chemical fuel?
2. What are secondary fuels.
3. Name any two primary solid fuels.
4. What is meant by calorific value of a fuel.
5. Distinguish between gross and net calorific value of a fuel.
6. How can you determine the calorific value of solid and volatile liquid fuels.
7. What are the corrections to be made in the calorific value of a fuel, determined by Bomb Calorimeter.
8. What is the Dulong Formula for calculating calorific value of a fuel?

Section 2.3, 2.4a, 2.4b

9. What are the characteristics of good fuel?
10. Why are gaseous fuels are better than solid fuels?
11. What are different varieties of coal formed inside the earth.
12. Why should an ideal fuel have moderate ignition temperatures?
13. Arrange wood, peat, lignite, bituminous coal and anthracite in increasing order of their calorific value
14. Which is the highest ranking coal?
    Ans: Anthracite.

Section 2.4c,

15. What do you mean by proximate analysis of coal. Why is it so called?
16. What is ultimate analysis of coal?
17. What is the significance of volatile matter in coal?
18. What all is determined in proximate analysis ?
19. What is the purpose of determining the fixed carbon in coal?
20. How nitrogen in coal is determined ?
21. What is ash? Why a fuel should have low ash content?

Section 2.4d, 2.4e

22. What do you mean by carbonization of coal?
23. What is the difference between caking and coking coals?
24. All coking coal are caking coals but all caking coals are not coking coals. Explain.
25. What is high temperature carbonization? What is its use?
26. What is low Temperature carbonization of coal?
27. What type of coke is produced from high temperature carbonization and low temperature carbonization.
28. Which type of carbonization is done to produce metallurgical coke?
29. Which variety of coal is used in metallurgy?
    Ans: Anthracite.
30. The reactivity of the coke used for metallurgical purposes should not be too high. Explain.
31. What are the demerits of Beehive oven?
32. What is regenerative principle of heat used in Otto Hoffmann's oven?
33. How the red hot coke is quenched in Otto Hoffmann's process of carbonization.
34. How tar is recovered from outgoing gases in Otto Hoffmann's process?
35. How ammonia is recovered from the outgoing gases. Why it should be removed?
36. How H_2S is removed from the outgoing gases?
37. What are the advantages of Otto Hoffmann's oven?

Section 2.5

38. What is straight run gasoline ?
    Ans: The gasoline obtained directly from the fractionation of crude oil.
39. What is the average composition of petroleum?
40. What are paraffin based crude oils?
41. What is meant by demulsification of crude oil?
42. What are the fractions obtained from the residual oil obtained from primary fractionating column.
43. What is the composition range of gasoline and diesel
    Ans: Gasoline $C_5 — C_9$
    Diesel $C_{10} — C_{18}$
44. What is meant by the term cracking ?
45. What are the advantages of catalytic cracking over thermal cracking ?
46. Which catalyst is generally used for cracking purposes?
47. What is synthetic petrol?
48. Name two processes used to convert coal into petrol.
49. What type of coal is used in Bergius process?
50. What is the catalyst used in Fischer Tropsch process?
51. What is the difference between straight rum and cracked gasoline?
52. What is meant by knocking?
53. What is the reason for knocking in petrol engine and in diesel engine.
54. What is the relation between chemical structure and knocking in petrol engine?
55. Out of straight chain hydrocarbons and aromatics, which one would have high anti knocking characteristics?
56. What is octane rating for petrol?
57. What are the antiknock agents used for petrol engine fuel?
58. Why ethylene bromide is added to the leaded petrol?
59. What is cetane number?
60. Out of aromatics and straight chain hydrocarbons which one would have antiknock characteristics?
    Ans: Straight chain hydrocarbons.

61. A good petrol engine fuel is bad fuel for diesel engine. Comment.
62. What is power alcohol? What its advantages?
63. What are the disadvantages of using power alcohol?
64. What are the advantages and disadvantages of using methanol as a fuel in automobiles?

    Ans: Methanol has very high combustion efficiency and the pollutant emitted are less as compared to gasoline. However, it is corrosive to engine, fuel tanks etc. On partial oxidation it releases formaldehyde which forms ozone in troposphere.
65. Mention a few advantages of diesel over petrol as a fuel?

Section 2.7, 2.8

66. What do you mean by combustion.
67. What type of energy changes take place during combustion?
68. What is flue gas?
69. How the flue gas is analysed ?
70. Why is it necessary that the flue gas is passed through a solution of KOH before it is passed through pyrogallic acid in Orsat's appartus.

Section 2.9

71. Compare solid fuels with gaseous fuels.
72. Distinguish between water gas and producer gas.
73. (a) Water gas is superior to producer gas. Why?

    (b) Distinguish between blue gas and producer gas.
74. What is natural gas? Name the different types of natural gas. Give the uses of natural gas.
75. What is carburetted water gas? Discuss its advantages over water gas.
76. What is the difference in the composition of producer gas, water gas and carburetted water gas ?
77. What are the merits and demerits of solid, liquid and gaseous fuels.
78. Distinguish between nuclear fission and nuclear fusion.
79. Write short notes on :

    (a) Binding energy

    (b) Mass defect
80. Compare chemical fuels with nuclear fuels.
81. Write a detailed note on breeder reactors.
82. Describe briefly, controlled nuclear fission as source of useful energy.
83. Write a short note on nuclear reactor.
84. What do you mean by moderator in nuclear reactor. What is it function?
85. The process of nuclear fission is essential for nuclear fusion. Explain.
86. (a) What is the function of control rods?

    (b) Give a concise account of nuclear power reactor.

LONG ANSWER QUESTIONS

1. What is a fuel? Give the different types of fuels ? What are the requirements of fuel to be used in an industry?
2. What do you understand by the calorific value of a fuel and how is it determined by bomb calorimeter?
3. How would you determine the calorific value of coal by Bomb calorimeter. Explain with the help of a neat diagram.
4. Give an account of the determination of calorific value of a gaseous fuel by Boy's calorimeter.
5. Enumerate the characteristic of a good fuel.
6. How are fuels commonly classified?
7. What do you understand by proximate and ultimate analysis of coal?

8. What is ultimate and proximate analysis of coal? Discuss their significance. How are ultimate and proximate analysis done?
9. Name the various solids fuels. Which of them is best suited for manufacture of metallurgical grade coke and why?
10. (a) What are the requisties of a metallurgical coke?
    (b) Describe the Otto-Hoffman's process for preparing coke. What are its advantages over the earlier methods?
11. (a) Why is coke, and not coal, used as a fuel in metallurgical processes?
    (b) Discuss in detail the by-product coke oven process for the manufacture of coke and mention the by-products recovered in the process.
12. (a) What is the difference between low temperature and high temperature carbonization?
    (b) Why is coke preferred to coal in metallurgical operations?
13. (a) Write briefly on metallurigical coke.
    (b) What is coke? Describe the manufacture of coke by beehive-coke oven method.
14. (a) List out the constituents of petroleum.
    (b) How is crude petroleum processed? How is it refined?
    (c) Indicate the major fractionation products obtainable from crude oil. Mention their uses.
15. With the help of a neat diagram, describe the fractional distillation of crude pertroleum and name the various products obtained.
16. What is thermal cracking and how is it carried out?
17. What is cracking ? Why is cracking considered as an important process?
18. (a) Describe the moving-bed catalytic cracking method to obtain petrol from heavy oil.
    (b) Discuss the merits and demerits of gaseous fuels over liquid and solid fuels.
19. What is synthetic petrol? Describe the Fischer-Tropsch and Bergius processes for producing the same.
20. (a) What is knocking? How is it related to a chemical constitution? Describe the function of TEL in I.C. engine. Name two preignition dopes used in diesel engine.
    (b) What do you understand by cetane and octane number?
21. Describe the fixed-bed catalytic cracking process for obtaining gasoline.
22. Disuss the relation between octane number and chemical constitution of fuel. How can it be improved?
23. What is power alcohol ? Give its merits and demerits. How is it obtained from molasses.
24. What are the merits and demerits of solid, liquid and gaseous fuels.
25. Give the name diagram and working of the apparatus used for the analysis of flue gas. What conclusion will you be able to draw from this analysis?
26. Distinguish between the following:
    (a) Proximate and ultimate analysis
    (b) Coal and coke
    (c) Coking coals and caking coals
    (d) Octane number and cetane number
    (e) Thermal and catalytic cracking.
27. How do you explain knocking in a diesel engine? How can it be controlled? What is cetane number
28. Describe the manufacture of producer gas. Mention its applications.
29. Give only the neat labelled diagram for the manufacture of water gas, giving necessary chemical equations involved in it. Give approximate composition, calorific value and uses of water gas.
30. Discuss giving equations, sketch of the manufacture, composition and uses of :
    (i) Coal gas
    (ii) Oil gas.
31. Give the name diagram and working of the apparatus used for the analysis of flue gas. What conclusion will you be able to draw from this analysis?
32. Write a detailed note on breeder reactors.

PROBLEMS

Based on Calorific Value determination

33. When 0.935 g of a fuel underwent complete combustion in excess of oxygen, the increase of temperature of water in a calorimeter containing 1,365 g of water was 2.40°C. Calculate the higher calorific value of the fuel, if the water equivalent of calorimeter, etc. is 135g.
[HCV = 3850.3 cal/g]

34. Liquid fuel weighing 0.98 g and containing 90% of carbon and 8 % of hydrogen gave the following results in a bomb calorimeter experiment:
Amount of water taken in the calorimeter = 1,450 g; water equivalent of the calorimeter = 450 g; rise in temperature of water = 1.8°C.
If the latent heat of steam is 587 cal/g, calculate the gross and net calorific value of the fuel.
[GCV = 3,490 cal/g; NCV = 3,067 cal/g]

35. (a) Calculate the gross and net calorific value of coal sample having the following composition: C = 80%, H = 7%, O = 3%, S = 3.5%, N = 2.1% and ash = 4.4%.
[GCV = 8,828 kcal/kg; NCV = 8,458 kcal/kg]
(b) The temperature of 950.0 g of water increased from 25.5°C to 28.5°C on burning 0.75 g of a solid fuel in a bomb calorimeter. Water equivalent of calorimeter and latent heat of steam are 400.0 g and 587.0 cal/g respectively. If the fuel contains 0.65% of hydrogen, calculate its net calorific value.
[5,366 cal/g]

36. The determination of the calorific value of a coal sample gave the follwing data:

Weight of coal sample	:	0.9 g
Water equivalent of calorimeter	:	440g
Weight of water	:	2560g
Rise in temperature	:	2.42°C
Cooling correction	:	0.052°C
Fuse wire correction	:	10.0 cal

Calculate the gross calorific value and net calorific value, if the total contains 6% hydrogen and assume latent heat of steam equal to 600 cal/g.
[8228, 79 04 cal/g]

37. Calculate the gross calorific value and net calorific value of coal sample having the following composition: C = 80%, H = 7%, O = 3%, S = 3.5%, N = 2.1%, and ash = 4.4%
[8828, 8458 cal/g]

38. A coal has the following ultimate analysis data: C = 84%, S = 1.5%, N = 0.6%, H = 5.5% and O = 8.4%. Find the gross and net calorific values with the help of Dulong's formula.
[8356, 8065 cal/g]

39. A sample of coal containing 92% C; 5% H; 3% ash. When this coal was tested in the laboratory for its calorific value in the bomb calorimeter, the following data were obtained:

Weight of coal burnt	=	0.95 g
Weight of water taken	=	700g
Water equivalent of bomb and calorimeter	=	2,000 g
Rise in temperature	=	2.48°C
Cooling correction	=	0.02°C
Fuse wire correction	=	10.0 cal
Acid correction	=	60.0 cal.

Calculate the net and gross calorific values of the coal in cal/g. (Assume the latent heat of condensation of steam as 580 cal/g).
[GCV = 7031.6 cal/g; LCV = 6770.6 cal/g]

40. 0.834 g of fuel on complete combustion in excess of oxygen increased the temprature of water in calorimeter from 14.36°C to 18.10°C. The mass of water in calorimeter was found to be 1,365 g. Calculate the higher calorific value of the fuel, if the water equivalent of calorimeter, etc. in 135 g.
[6726.6 cal/g]

Based on Proximate and ultimate analysis

41. A sample of coal was analyed as follows: 1.000g of an air-dired coal sample was weighed in a silica crucible. After heating for 1 hour at 105–110°C, the dry coal residue weighed 0.985 g. The crucible was covered with a vented lid and then heated strongly for exactly 7 minutes at 950 + 20°C. The residue weighed 0.800 g. The crucible was then heated strongly in air, until a constant weight was obtained. The last residue was found to weigh 0.100 g. Calculate the proximate analysis **[Moisture = 1.5; volatile matter-18.5; ash = 10.0; and fixed C = 70.0%]**
42. A sample of the Gondwana Coal of Jharia was analysed as follows: Exactly 2.500 g was weighed into a silica crucible. After heating for 1 hour at 110°C, the residue weighed 2.415 g. The crucible was then covered with a vented lid and strongly heated for exactly 7 minutes at 950 + 20°C. The residue weighed 1.528 g. The crucible was then heated without the cover, until a constant weight was obtained. The last residue was found to weight 0.245 g. (i) Calculate the percentage result of the above analysis. (ii) To which type of analysis does the above description belong? Why is the analysis so-named?
[Moisture = 3.4%; volatile matter = 35.48%; ash = 9.80%; and fixed C = 51.32%]
43. 1.56 g of the coal was kjeldahlized and NH_3 gas evolved was absorbed in 50.0 mL of 0.1 N H_2SO_4. After absorption, the excess (residual) acid-required 6.25 mL of 0.1 N NaOH for exact neutralization. 2.60 g of the coal sample in a quantitaive analysis gave 0.1755 g of $BaSO_4$. Calculate the percentage of N and S in the coal sample.
[N = 3.926 % ; S = 0.927 %]
44. 0.5 g of a sample of coal was used in a bomb calorimeter for the determination of calorific value. Calorific value of coal was found to be 8600 cal / g. The ash formed in the bomb calorimeter was extracted with acid and the acid extract was heated with barium nitrate solution and a precipitate of barium suphate was obtained. The precipitate was filtered, dried and weighed. The weight of precipitate was found to be 0.05 g. Calculate the percentage of sulphur in the coal sample.
[1.3734%]
45. 1.0 g of a sample of coal was used for nitrogen estimation by kjeldahl method. The evolved ammonia was collected in 25 m L (N/10) sulphuric acid. To neutralise excess acid, 15 mL of 0.1 N sodium hydroxide was required. Determine the percentage of nitrogen in the given sample of coal.
[Ans.: 1.4%]
46. A sample of coal was analysed as follows: Exactly 2.500 g was weighed into a silica crucible. After heating for one hour at 110°C. The residue weighed 2.415 g. The crucible next was covered with a vented lid and strongly heated for exactly seven minutes at 950 + 20°C. The residue weighed 1.528 gm. The crucible was heated without the cover, until a constant weight was obtained. The last residue was found to weigh 0.245 g. Calculate the percentage results of the above analysis.
[Moisture = 3.40; volatile matter = 35.48%; ash = 9.80%; and fixed Carbon = 51.32%]

Based on Combustion

47. On analysis, an oil was found to have the following percentage composition: C = 48%, H = 12% and O = 4%. Calculate the weight of air required for the complete burning of kg of this oil.
[9.57 kg]
48. A sample of petrol was found to contain 15.4% hydrogen and 84.6% carbon by weight. Calculate the weight and volume of air required for the complete combustion of kg of this fuel.
[15.17 kg, 11.74 m^3]
49. Calculate the volume of air required for the complete combustion of 1m^3 of the gaseous fuel having the following composition by volume. H_2 = 50%, CH_4 = 36%, N_2 = 1.5%, CO = 6%, C_2H_4 = 4% and water vapour = 2.5%. **[5.33 m^3]**
50. Calculate the volume of air at STP needed for the complete combustion of 1kg of methane.
[13.33 m^3]
51. A sample of coal was found to have the following percentage composition: C = 75%, H = 5.3%, O = 12.8%, S = 1.2%, N = 3.7%, Ash = 2.1%. Calculate the minimum amount of air necessary for complete combustion. **[10.0 kg]**

52. An oil, on analysis, gave the following result by weight: C = 86%, H = 11.75% and O = 2.25%. Find the minimum amount of air required for burning 1 kg of the fuel. **[14.06 kg]**
53. The minimum amount of air required for the complete combustion of IL of water gas is 2.8L at NTP. Assuming that 20% excess air is used for the actual composition, calculate the volume of air used at 740 mm pressure and 25°C. **[Ans 3.77L]**
54. A sample of fuel oil analyses are as follows: 85%C, 6%H, 4.5%S, 2%O and 2.5% ash. This oil was employed for heating in a furnace with 30% excess air. Calculate a) the theoretical amount of air for complete combustion of 1 kg of fuel and b) the actual amount of air used per kg of fuel. **[Ans (a) 12.13 kg (b) 15.77 kg]**
55. Calculate the theoretical quantity of air that will be required for the complete combustion of a liquid fuel weighing 1 kg. It contains 80% by weight benzene and 20% by weight ethanol (Assume the composition of air to be 80% N_2 and 20% O_2 by weight). **[Ans 14.7 kg]**
56. What amount of air would be required for the complete combustion of 10 kg of fuel containing 85% ethanol and 15% methanol? **[Ans 95.6 kg]**
57. Calculate the volume of air required for the complete combustion of 1 litre of CO (Volume % of oxygen in air is 21). **[Ans 2.38 L]**

Based on Flue gas Analysis

58. A sample of coal was found to contain the following: C = 81%, H = 4%, O = 2%, N = 10%, the remaining being ash. Estimate the quantity of minimum air required for complete combustion of 1kg of the sample. Find the composition of dry flue gas by weight, if 40% excess air is supplied. **[10.7 kg, CO_2 = 19.06%, N_2 = 74.65%, O_2 = 6.29%]**
59. A gaseous fuel has the following composition by volume: H_2 = 40%, CH_4 = 40%, C_2H_6 = 6%, CO = 6% and N_2 = 8%. Calculate the volume of air required for the combustion of 1m^3 of the fuel, if 20% excess air is supplied. Find out the percentage composition by volume of the dry products of combustion. **[6.05 m^3, CO_2 = 8.73%, N_2 = 87.45%, O_2 = 3.82%]**
60. Calculate the minimum weight and volume of air required for the complete combustion of 1 kg of coal having the following percentage composition by weight: C = 90%, H = 3.5%, S = 0.5%, N = 0.5% and ash = 2.5%. Also calculate the percentage composition by weight of the dry products of combustion if 25% excess air is supplied. **[11.54 kg, 8.93 m^3, CO_2 = 21.86%, SO_2 = 0.07%, N_2 = 73.67% and O_2 = 4.41%]**
61. 100 kg of a liquid hydrocarbon fuel containing 89.4% carbon by weight is burnt with (i) theoretical amount of air (ii) 20% excess of the theoretically required amount of air. Calculate the volumetric composition of the products in each case. **[1036.52 kg, 1243.5 kg, CO_2 = 17.27%, N_2 = 79.28%, O_2 = 3.45%]**
62. 1000 m^3 of a producer gas having the volumetric composition CO = 22%, CO_2 = 4%, CH_4 = 2%, H_2 = 14%, N_2 = 56%, O_2 = 2% has to be burnt for heating a furnace. (i) Calculate the minimum quantiy of air required for the complete combustion (ii) if 40% excess air is used, what is the volumetric composition of dry flue gas? **[952.38 m^3, CO_2 = 14.19%, N_2 = 81.76%, O_2 = 4.05%]**
63. A gaseous fuel has the following composition: H_2 = 50%, CH_4 = 30%, C_2H_6, CO = 8%, N_2 = 6%. Calculate the volume of air required per cubic meter of the gas. Also find out the percentage composition by volume of the dry products of combusiton, if 20% excess air is supplied. **[5.24 m^3, CO_2 = 8.7%, N_2 = 87.47%, O_2 = 3.83%]**
64. The percentage composition of a sample of anthracite coal was found by weight to be C = 90, H = 3.3, O = 3.0, N_2 = 8.0, S = 0.9 and ash 2.0. Calculate the minimum weight of air required for the complete combustion of 1 kg of this fuel. If 50% excess of air is supplied, find the percentage composition of the dry flue gases by weight. **[11.49 kg, CO_2 = 18.31%, N_2 = 74.12%, O_2 = 7.47%, SO_2 = 0.1%]**
65. A sample of coal was found to contain the following C = 81%, H = 4%, O = 2%, N = 10%, S = 2% and the remaining being ash. Estimate the quantity of minimum air required for the complete combustion of 1 kg of the sample. Find the composition of dry flue gas by weight, if 40% excess air is supplied. **[10.78 kg, CO_2 = 18.89%, N_2 = 74.55%, O_2 = 6.31%, SO_2 = 0.25%]**

66. Analysis of coal used in a boiler is: C = 82%; H = 4.2%; O = 4.8% ash = rest. When this coal was used in a boiler, the analysis of flue gas was CO_2 = 10%; CO = 1.5%; O_2 = 8.0%, N_2 = 80.5%. Determine (i) the amount of excess air per kg of coal burnt. (ii) total mass of air supplied per kg of coal burnt. (iii) percentage of excess air supplied for combustion.

[(i) 5.970 kg (ii) 16.731 kg (iii) 55.5%]

EXAMINATION QUESTIONS

67. (a) What are fuels? How are they classifed?
(b) Describe the methods employed for the ultimate analysis of coal

[G.G.S.I.P.U, I Term, Dec-1999]

68. (a) Describe with a neat diagram, the production of coal gas; Mention its application.
(b) What is catalytic cracking? What are its advantages? **[G.G.S.I.P.U, II Term, Dec-1999]**

69. (a) How is pertrol manufactured artificially by Fischer Tropsch process.
(b) A sample of coal was found to have the following percentage composition: C = 75%, 11 = 5.2%, O = 12.8%, S=1.2%, N=3.7% & ash=2.1% Calculate
(i) The minimum air necessary for complete combustion of 1kg. coal.
(ii) The higher Calorfic value and the lower Calorific Value of the coal sample

OR

What is carbonization? Explain with a neat diagram, the Otto-Hoffman by-product oven method for the manufacture of metallurgical coke. **[G.G.S.I.P.U, II Term, Dec-2000]**

70. (a) Enumerate the characteristics of good fuel.
(b) Differentiate between gross calorific value and net calorfic value.
(c) A coal sample has the following composition by weight C = 90%, O = 3.0%, S = 0.5%, N = 0.5% and ash = 2.5% Net calorific value of the coal found to be 8,490 kcal/kg. Calculate the percentage of hydrogen and gross calorific value of hydrogen.

[G.G.S.I.P.U, End Term, B. Tech, Mar, 2000[

71. (a) What is meant by knocking? How is it related to chemical constitution? Describe the function of TEL. Explain octane number and cetane number.
(b) A coal sample was found to contain: C = 66.2%, H = 4.2%, O = 6.1%, N = 1.4%, S = 2.9% moisture = 9.7% and ash = 9.5% by weight. Calculate the quantity of dry products of combustion, if 1 kg of coal is burnt with 25% excess air

[G.G.S.I.P.U, End Term, B. Tech, Mar, 2000]

72. (a) Describe the manufacture of gasoline by Fisher-Tropsch method.
(b) Differntiate between chemical fuel and nuclear fuel.
(c) Describe the manufacture of producer gas. What is its average composition and calorfic value? Mention its applications. **[G.G.S.I.P.U, End Term, B. Tech, Mar, 2000]**

73. (a) Describe with the help of a neat diagram, how the calorific value of a fuel is determined by the "Bomb-calorimeter"
(b) Explain why the good fuel for petrol engine is a bad fuel for diesel engine.

[G.G.S.I.P.U, II Term, B. Tech, Oct, 2000]

74. (a) Expain with a neat diagram "moving bed catalytic cracking"
(b) Write short notes on:-
(i) Octane Number
(ii) Cetane Number
(iii) Antiknocking agents **[G.G.S.I.P.U, II Term, B. Tech, Oct, 2000]**

75. (a) Describe with the help of a neat diagram the prepartion of coal gas. Give its properties and applications.
(b) A sample of coal was analysed as follows: Exactly 2.500 gm was weighed into a silica crucible. After having heated for one hour at 110°C the residue weighed 2.415 gm. the crucible was

next covered with a lid and strongly heated for exactly 7 minutes at 950 + 20°C. The residue weighed 1.528 gm. Calculate the percentage result of the above analysis.

[G.G.S.I.P.U, II term, B. Tech, Oct, 2000]

76. (a) Describe briefly ultimate analysis of coal.
(b) A sample fo coal was found to have the following percentage composition: C = 70%; H = 6.2%; O = 16.1%; N = 3.2% and Ash 4.5% Calculate (i) the gross and net calorific values (ii) Minimum air required for complete combustion of 1 kg of coal.

[G.G.S.I.P.U, II term, B. Tech, Oct, 2000]

77. (a) Desribe how petrol is synthesized by " Bergius Process"
(b) Which one, coal gas or producer gas has higher calorific value and why?
(c) Why is it necessary to remove sulphur compounds from oil & natural gas.

[G.G.S.I.P.U, II term, B. Tech, Oct, 2000]

78. (a) Writes notes on
(i) Fischer-Tropsch process.
(ii) Octane Number.
(b) State the characteristics of a good fuel.
(c) A sample of coal was analysed as follows: Exactly 2.500 gm, was weighed in silica crucible. After heating for one hour at 110°C, the residue weighed 2.415 gm. The crucribe next was covered with a vented lid and strongly heated for exactly seven minutes at 950° + 20°C. The residue weighed 1.528 gm. The crucible was then heated without the cover, until a constant weight was obtained. The last residue was found to weight 0.245 gm. Calculate the percentage results of the above analysis. **[G.G.S.I.P.U, End term, B. Tech, Dec. 2000]**

79. With the help of a neat diagram explain how the calorfic value of a fuel can be determined by static oxygen bomb calorimeter. What are the fuels used for the determination of water equivalent of bomb calorimeter and why?

80. Distinguish between
(a) Producer and water gas.
(b) Octane and cetane number
(c) Nuclear fission and nuclear fusion
(d) Bituminous coal and anthracite
(e) Gross and Net calorific value
(f) High and low temperature carbanization
(g) Proximate and ultimate analysis. **[G.G.S.I.P.U, End term, (II Batch), Mar. 2001]**

81. (a) Describe the moving bed catalytic cracking method to obtain petrol from crude oil.
(b) A producer gas has the following composition by volume; CO = 30%, H_2 = 12%, CO = 4%, CH_4 = 2% and N_2=52%. What will be the composition of dry flue gas, if 50% excess air was used for burning 100m^3 of the gas? **[G.G.S.I.P.U, End term, (II Batch), Mar. 2001]**

82. (a) Define gross calorific value & net calorific value?
(b) Describe with the help of a neat diagram, how is the calorific value of a fuel determined by a Bomb calorimeter. **[G.G.S.I.P.U, II Term Nov. 2001]**

83. (a) Describe briefly the ultimate analysis of coal.
(b) A samples of coal was found to have the following composition by mass:
C = 75%, H = 5.2%, O = 12.1%, N = 3.2% and ash 4.5% Calculate (i) the minimum air required for the complete combustion of 1kg coal. (ii) Highest and the lowest calorific value of fuel.

[G.G.S.I.P.U, II Term Nov. 2001]

84. (a) Describe with a neat diagram the Fischer-Tropsch process for the manufacture of synthetic petrol.
(b) Calculate the gross and net calorific value of a fuel having the following percentage composition:
C = 80%, H = 9%, S = 2%, N = 3%, Ash - 4%, rest oxygen. The latent heat of condensation of steam being 587 cal. **[G.G.S.I.P.U, II Term Nov. 2001]**

85. (a) What is cracking? Describe fixed-bed catalytic method for cracking of petrol.
(b) 1.56g of a sample of coal was kjeldhalised and NH_3 thus evolved was absorbed in 50ml of 0.1N H_2SO_4. After absorption the residual acid required 8.25 ml of 0.1N NaOH for exact neutralization. 2.60g of caol sample in a quantitative analysis gave 0.1755g of $BaSO_4$. Calculate the % of N & S in coal sample. **[G.G.S.I.P.U, II Term Nov. 2001]**

86. (a) Why and how corrections are made in determination of Calorific value by Bomb's Calorimeter?
(b) The following data were obtained in a bomb calorimeter experiment.
Weight of coal burnt = 0.996g.
Weight of water in calorimeter = 2490g.
Weight of bomb, calorimeter etc. = 3900 g.
Rise in temperature of water = 2.592°C
Mean specific heat of apparatus = 0.099
If the fuel contains 6% hydrogen, calculate its lower calorific value assuming 584.0 cal/g as latent heat of steam.
(c) Name the analysis of coal in which moisture, volatile matter, ash and fixed carbon are determined. Give significance of each component. **[G.G.S.I.P.U, II Term Nov. 2001]**

87. Differentiate between chemical fuel and nuclear fuel. **[G.G.S.I.P.U, II Term Nov. 2001]**

88. (a) A gaseous fuel has the following composition. H_2 = 50.0%, CH_4 = 30.0%, C_2H_6 = 6.0%, CO = 8.0% and N_2 = 6.0%, calculate the volume of air required per cubic meter of gas if 20.0% excess air is supplied for combustion. Also find out the % composition by volume of dry products of combustion. (Assume air to be a mixture of 80.0% N_2 and 20.0% O_2 by volume). **[G.G.S.I.P.U, End Term, B. Tech, Dec. 2001]**

89. Write short notes on:
(a) Carbonisation of coal
(b) Nuclear reactors **[G.G.S.I.P.U, End Term, B. Tech, Dec. 2001]**

90. (a) Explain knocking. How it is related to chemical constitution of fuel?
(b) What are advantages of catalytic cracking over thermal cracking?
(c) Explain cetane No. and octane No. **[G.G.S.I.P.U. II Term, Nov. 2002]**

91. (a) Describe Otto-Hoffmann's process for preparing Metallurgical coke.
(b) 20 g. sample of coal was used for nitrogen estimation by kjeldahl method. The evolved NH_3 was collected in 50 ml (N/10) H_2SO_4. To neutralize excess acid, 25 ml of 0.1 N NaOH was required. Determine the percentage of N_2 in the given sample of coal. **[G.G.S.I.P.U. II Term, Nov. 2002]**

92. (a) How calorific value can be determined by Bomb's calorimeter? Explain with the help of neat diagram.
(b) Differentiate between Net and Gross calorific value. **[G.G.S.I.P.U. II Term, Nov. 2002]**

93. A Coal has following composition by weight. C = 80%, O_2 = 5%, S = 2.5%, N = 1.4%, ash = 2.1%. Net colorific value of the coal was found to be 9430 K cal/kg, calculate the percentage of *hydrogen* and *higher calorific value.* **[G.G.S.I.P.U. II Term, Nov. 2002]**

94. (a) Give the complete classification of chemical fuels and mention the criteria for selecting a good fuel
(b) Explain proximate analysis of coal and report the method of its determination.
(c) Distinguish between gross and net calorific value of fuel. **[G.G.S.I.P.U. End Term, Dec. 2002]**

95. (a) What is metallurgical coke? Give a method for its manufacturing.
(b) A coal containing 62.4%, C; 4.1% H; 6.9% O; 1.2% N; 0.8% S; 15.1% moisture and 9.7% ash was burnt in such a way that dry fuel gas contained 12.9% CO_2; 0.2% CO; 6.1% O_2 and 80.8% N_2. Calculate
(i) The weight of air theoretically required per kg of coal.
(ii) The weight of dry flue gas obtained per kg of fuel.
(iii) The weight of air actually used. **[G.G.S.I.P.U. End Term, Dec. 2002]**

96. Write short notes on:
(a) Catalytic cracking
(b) Nuclear reactor
(c) Octane and Octane number. **[G.G.S.I.P.U. End Term, Dec. 2002]**

97. Differentiate the following:
(i) Gross calorific value and Net calorific value.
(ii) Low temperature Carbonization and High temperature Carbonization.
(iii) Octane Number and Cetane Number.
(iv) Liquid Phase Thermal Cracking and Vapour Phase Thermal Cracking.
[G.G.S.I.P.U. II-Term, B.Tech. Oct. 2003]

98. Give the average composition of Petroleum. Draw the neat and labeled diagram for the refining of Petroleum. Name any four separated components and give their uses.
[G.G.S.I.P.U. II-Term, B.Tech. Oct. 2003]

99. (a) A sample of coal gas was found to have the following percentage composition.

$$C = 71\%, H = 7.8\%, O = 11.2\%, N = 5.5\%, Ash = 4.5\%$$

(i) Calculate the Higher calorific value and Lower calorific value of the coal sample.
(ii) Calculate the minimum air required for complete combustion of 1 Kg of coal.

(b) A gaseous fuel has following composition by volume:

$$H_2 = 36\%, CH_4 = 6\%, N_2 = 38\%, O_2 = 16\%, CO = 4\%.$$

if 30% of excess of air is used, find the weight of air actually supplied per m^3 of the gas.
[G.G.S.I.P.U. II-Term, B.Tech. Nov. 2005]

100. (a) Match the following:

(i)	Petrochemicals	(i)	$C_5 - C_9$
(ii)	Naphtha	(ii)	Presence of oxygen
(iii)	Carbonisation	(iii)	Non-volatile liquids
(iv)	Bomb Calorimeter	(iv)	$C_9 - C_{10}$
(v)	Gasoline	(v)	Aerobic method of decomposition
(vi)	Boy's Calorimeter	(vi)	Burning wastes in large furnaces
(vii)	Proximate Analysis	(vii)	Derived from crude oil and natural gas
(viii)	Compositing	(viii)	Absence of oxygen
(ix)	Incineration	(ix)	Volatile liquids
(x)	Combustion	(x)	Moisture content

(b) Briefly comment on the following:
(i) What is meant by calorific intensity of coal?
(ii) Why should an ideal fuel have moderate ignition temperature?
(iii) Why is the oxidation of fuel a self sustaining process?
(iv) Why a good fuel must have low ash content?
(v) Why are no aromatic compounds are found in gases or gasoline fractions?
[G.G.S.I.P.U. II-Term, B.Tech. Nov. 2005]

101. (a) Differentiate between the two
(i) Fixed bed Catalytic Cracking and Fluidised bed Catalytic Cracking
(ii) Synthetic petrol and Power Alcohol

(b) A sample of coal was tested for its calorific value using Bomb's Calorimeter. Following data was obtained.
Weight of coal burnt = 0.920g
Weight of water taken = 550g
Water equivalent of bomb and calorimeter = 2200g
Rise in temperature = 2.42°C

Fuse wire correction = 10.0 cal

Acid correction = 50.0 cal

Assuming coal to contain C = 93%, H = 6% and ash = 1%, calculate gross and net calorific value of coal. The latent heat of condensation of steam is 580 cal/g.

[G.G.S.I.P.U. II-Term, B.Tech. Nov. 2005]

102. (a) Using a suitable figure, explain the Otto-Hoffmann's by-product oven method for the manufacture of metallurgical coke.

(b) A gaseous fuel undergoes combustion using 20% excess air. Composition of the gaseous fuel (by volume) is as follows, CH_4 = 5%, H_2 = 20%, CO = 25%, CO_2 = 6% and the rest in N_2. Calculate volume of air supplied per m^3 of fuel and composition of dry flue gases.

[G.G.S.I.P.U. II-Term, B.Tech. Nov. 2005]

103. (a) Match the following:

(i) Petrochemicals	(i) Tetra ethyl lead
(ii) Anthracite	(ii) Presence of oxygen
(iii) Carbonisation	(iii) Non volatile liquids
(iv) Bomb Calorimeter	(iv) Aerobic method of decomposition
(v) Anti knocking agent	(v) Highest calorific value
(vi) Boy's Calorimeter	(vi) Burning wastes in large furnaces
(vii) Proximate Analysis	(vii) Derived from crude oil and natural gas
(viii) Compositing	(viii) Absence of oxygen
(ix) Incineration	(ix) Volatile liquids
(x) Combustion	(x) Moisture content

(b) Briefly comment on the following:

(i) What is meant by sweetening of petrol?

(ii) What is meant by wet gas?

(iii) What is meant by compression ratio of an I.C. engine?

(iv) What are Petrochemicals?

(v) What are fossil fuels and how are they formed?

[G.G.S.I.P.U. II-Term, B.Tech. Nov. 2005]

104. (a) Differentiate between the two:

(i) Octane number and Cetane number

(ii) Thermal Cracking and Catalytic cracking.

(b) Calculate the gross and net calorific value of coal having the following compositions, C = 85%, H = 8%, S = 1%, N = 2%, ash = 4%.

Assume the latent heat of steam = 587 cal/g. **[G.G.S.I.P.U. II-Term, B.Tech. Nov. 2005]**

105. (a) Describe the synthesis of petrol by Bergius process.

(b) 4kg of carbon undergoes combustion. Calculate the requirement of the following:

(i) weight of oxygen

(ii) volume of oxygen

(iii) weight of air

(iv) volume of air **[G.G.S.I.P.U. II-Term, B.Tech. Nov. 2005]**

106. Write short notes on the following:

(i) Proximate analysis and ultimate analysis

(ii) Octane rating and cetane rating **[G.G.S.I.P.U. II-Term, B.Tech. Nov. 2005]**

107. (a) Match the following:

(i) Reforming	(i) Ethanol blended Gasoline
(ii) Synthetic Petrol	(ii) Presence of oxygen
(iii) Carbonisation	(iii) Non volatile liquids
(iv) Refining	(iv) Aerobic method of decomposition
(v) Combustion	(v) Removal of olefins and 'S' compounds

(vi) Boy's Cǎlorimeter	(vi) Burning wastes in large furnaces
(vii) Power alcohol	(vii) Structural modification of gasoline
(viii) Compositing	(viii) Absence of oxygen
(ix) Incineration	(ix) Volatile liquids
(x) Bomb Calorimeter	(x) Petrol from non Petroleum source

(b) Briefly comment on the following:
(i) What is the importance of the determination of fixed carbon in coal?
(ii) What is wet gas?
(iii) Why is it necessary to remove 'S' compounds from oil and natural gas?
(iv) Why is coke preferred to coal in metallurgical processes?
(v) Why does coal evolve more heat on burning?
[G.G.S.I.P.U. II-Term, B.Tech. Nov. 2005]

108. (a) Differentiate between the two
(i) Proximate Analysis and Ultimate Analysis
(ii) Caking coal and coking coal

(b) A sample of coal was analysed as follows: Exactly 2.500g was weighed into a silica crucible. After having heated for 1hr at 110°C, the residue weighted 2.41g, the crucible was next covered with a lid and strongly heated for exactly 7min. at 950 ± 20°C. The residue weighed 1.528g. The crucible was then heated with cover until a constant weight was obtained. The last residue weighed 0.245g. Calculate the percentage of the above results.
[G.G.S.I.P.U. II-Term, B.Tech. Nov. 2005]

109. (a) With the help of a suitable figure, represent the different fractions (along with their boiling point, composition range and uses) obtained by fractional distillation of crude petroleum.

(b) Calculate the quantity of dry products of combustion, if 1 kg of coal is burnt with 25% excess air. The coal sample was found to have the following composition (by weight)
C = 66.2%, H = 4.2%, O = 6.1%, N = 1.4%, S = 2.9%, moisture = 9.7% and ash = 9.5%
[G.G.S.I.P.U. II-Term, B.Tech. Nov. 2005]

110. (a) Why is coke and not coal used as a fuel in metallurgical processes?
(b) Why should an ideal fuel have moderate ignition temperature?
(c) What is cetane value of a diesel fuel?
(d) Give the condition when GCV = NCV. **[G.G.S.I.P.U. II-Term, B.Tech. Nov. 2006]**

111. (a) Name the analysis of coal in which moisture, volatile matter, ash and fixed carbon are determined. Give explanation and significance of each component.

(b) The percentage composition of a sample of coal was found to be as under:
C = 75.4% H = 5.3%
O = 12.6% N = 3.2%
S = 1.3% and ash = rest

Calculate the minimum weight of air necessary for complete combustion of 1 Kg of coal and the percentage composition of the dry products of combustion by weight.
[G.G.S.I.P.U. II-Term, B.Tech. Nov. 2006]

112. (a) Why and how corrections are made in determination of calorific value by bomb's calorimeter?

(b) The following data were obtained in a bomb calorimeter experiment:

Weight of coal burnt = 0.996g
Weight of water in calorimeter = 2490g
Weight of bomb calorimeter etc. = 3900g
Rise in temp of water = 2.592°C
Mean specific heat of apparatus = 0.099

If the fuel contain 6% hydrogen, calculate its lower calorific value assuming 587.0 cal/g as latent heat of steam. **[G.G.S.I.P.U. II-Term, B.Tech. Nov. 2006]**

113. (a) Describe the terms: carbonization and cracking.
(b) Hydrocarbons that are poor gasoline fuels are quite good diesel fuels. Explain the term.
[G.G.S.I.P.U. II-Term, B.Tech. Nov. 2007]

114. (a) Describe the method of a carbonization of coal to yield coke.
(b) 5g of a sample of coal when dried at 110°C to constant weight left a moisture free residue of 4.80g. On further heating out of contact with air to about 900°C, in a crucible with a vented lid to derive off all the volatile matter, a residue of 3.38g of coke was obtained. On heating the crucible with a lid open till all the carbon was oxidised, a residue of 0.075g was obtained. What is the proximate analysis of the coal. **[G.G.S.I.P.U. II-Term, B.Tech. Nov. 2007]**

115. What do you understand by destructive distillation of coal? Comment on the statement, "the products obtained depend upon the temperature of carbonization". Give data to support your answer. **[G.G.S.I.P.U. End-Term, B.Tech. Dec. 2007]**

116. (a) What is a fuel? How are fuels classified? Support your answer with examples.
(b) Give an account of proximate analysis of solid fuels. What is the significance of proximate analysis?
(c) 50 ml mixture of dry gases was analyzed using orsat's apparatus. Volume of the gaseous mixture left after passing through tubes containing KOH, alkaline pyroyallol, acidified Cu_2Cl_2 and Br_2 water was 48.7 ml, 48.2 ml, 41.6 ml and 41.4 ml respectively. Calculate the percentage of each component in the mixture. **[G.G.S.I.P.U. End-Term, B.Tech. Dec. 2007]**

117. How does ash left after the coal is burnt differs from the ash originally present in the sample.
[G.G.S.I.P.U. End-Term, B.Tech. Dec. 2007]

3

CHEMICAL BONDING-I

"What we observe is not nature itself, but nature exposed to our method of questioning".

— Werner Heisenberg

CHAPTER OUTLINE

3.1 Types of Chemical Bond
3.2 Electrovalent or Ionic Bond
- Factors Governing The Formation of Ionic Bonds:
- Characteristics of Ionic Compounds
- Determination of Lattice Energy
- Lattice Energy and Solubility of Ionic Compounds

3.3 Covalent Bond
- Valence Bond Theory
- Polar and Non Polar Covalent Bond
- Covalent Character of Ionic Bond (Polarization)
- Coordinate Bond or Dative Bond
- Covalent Bond Characteristics
- Characteristics of
- Covalent Compounds

3.4 Valence Shell Electron Pair Repulsion Theory (VSEPR)
3.5 Hybridization
- Types of Hybridization

3.6 Resonance
3.7 Hydrogen Bond
3.8 Intermolecular Forces - The Vander Waals' Forces

At ordinary temperatures atoms are rarely capable of free existence. Most of the substances exist as molecules like H_2, O_2, P_4, S_8, NH_3, CH_4, etc. But then a number of questions arise –

- Why and how do atoms combine?
- What kind of force holds these atoms together in a molecule?
- How these atoms are arranged in a molecule?
- Why do some atoms combine, while certain others do not. Two chlorine atoms combine to form Cl_2 molecule whereas two helium atoms do not combine to form He_2.

The study of Chemical Bonding attempts to answer all the above questions. **Chemical bond is defined as the force which binds the atoms together within a molecule**. Atoms combine to form molecules which are more stable and have lower energy than the individual atoms. A number of attempts were made to explain the formation of chemical bonds. Some of them are briefly explained below.

1. Electronic Theory of Valency: Kossel and Lewis (1916) put forward the first satisfactory explaination of the formation of chemical bonds. As this explaination was based on the electronic concept of atom, so it is referred to as "Electronic Theory of Valency."

The study of spectra of inert gases of zero group revealed that they are quite stable and do not combine chemically with any other element or with each other. Their electronic configurations are as follows:

Element	*Atomic Number*	*Electronic Configuration*
Helium	2	2
Neon	10	2, (8)
Argon	18	2, 8, (8)
Krypton	36	2, 8, 18, (8)
Xenon	54	2, 8, 18, 18, (8)
Radon	84	2, 8, 18, 32, 18, (8)

The electronic configuration of these gases indicate that each of *s* and *p* orbitals of the outermost energy shell is completely filled with 2 and 6 electrons respectively i.e. a total of 8 electrons. Hence it was suggested that these constitute a structure of maximum stability (s^2p^6) and minimum energy. But in case of all other elements, which show chemical reactivity, the outermost orbitals are found to be incomplete. Therefore, all other elements, other than noble gases, tend to acquire stable gas configuration, by losing, gaining or sharing of electrons with other elements. This led to the conclusion that atoms of different elements participate in bond formation in order to complete their respective octets (i.e 8 electrons in the outer most shell) or duplet, (i.e 2 electrons in the outer most shell) in case of H, Li and Be, to attain stable inert gas configuration. This is the fundamental cause of chemical combination.

Thus in light of electronic theory of valency, it follws that, "*Chemical bonding may be defined as the combination of two or more atoms through rearrangement of electrons in their outer shells, either by sharing of electrons amongst themselves or by the transfer of electrons from one atom to another, so that all the atoms attain the stable configuration of nearest inert gas.*"

Limitations of Electronic Theory: Though octet rule has explained the formation of chemical bond in large number of compounds, still it has the following limitations:

(i) The formation of electron deficient compounds such as $AlCl_3$ or BF_3 where the central atom Al or B do not have their octet completed, cannot be explained on the basis of this theory.

(ii) Similarly the formation of compounds such as SF_6, PCl_5, where central metal atom has more than 8 electrons cannot be explained.

(iii) Noble gases having eight electrons in their valence shell are not expected to form any compounds on the basis of this theory, but compounds like XeF_2, XeF_4, XeF_6 have actually been prepared.

The formation of chemical bond can thus better be explained on the basis of lowering of energy.

2. Lowering of Energy: In the formation of a chemical bond, there is invariably a lowering of potential energy of the system. Energy considerations show that systems associated with less energy are more stable than the systems possessing more energy. Thus atoms combine with each other, to lower their energy and hence gain extra stability. This will be discussed in more detail a little later in this chapter.

3.1 TYPES OF CHEMICAL BOND

As discussed above, atoms combine together in order to complete their respective octets so as to acquire stable inert gas configuration. This can happen in two ways.

1. By complete transference of one or more electrons from one atom to another. This type of bond is called **Ionic or Electrovalent bond.**
2. By sharing of electrons: Sharing can take place in two ways:
   (a) When the shared electrons are contributed equally by the two combining atoms. This type of bond is called **covalent bond.**
   (b) When the shared electrons are contributed by one atom but shared by both, the bond so formed is known as **co-ordinate bond or dative bond.**

Considering the electronegativities of the elements, the following ideal types of bonds are formed.

Electropositive Element + Electronegative Element —**Ionic Bond**
Electronegative Element + Electronegative Element—**Covalent Bond**
Electro positive Element + Electropositive Element—**Metallic Bond**

The above mentioned bonds are extreme forms. In most of the cases the bond type is some where between these extreme forms.

3.2 ELECTROVALENT OR IONIC BOND

When a bond is formed by complete transference of electrons from one atom to another so as to complete their outermost shell by acquiring 8 *electrons* (*or* 2 *electrons in case of hydrogen, lithium etc*), *the bond formed is called ionic or covalent bond.* The atom that loses one or more electrons acquires positive charge and the ion so formed is called a **cation.** The atom that gains one or more electrons acquires negative charge and the ion so formed is called an **anion.** An ionic bond is the result of electrostatic attraction between positively charged cations and negatively charged anions.

Examples:

The most striking example of electrovalent bonds is the combination of sodium and chlorine to form sodium chloride.

Electronic configuration of sodium atom = $1s^2, 2s^2, 3s^1$.
Electronic configuration of chlorine atom= $1s^2\ 2s^2, 2p^6, 3s^2, 3p^5$.

By losing one electron from its outermost shell, sodium acquires the inert gas configuration of neon and changes into cation.

$$\underset{2,\,8,\,1}{Na} \longrightarrow \underset{2,\,8}{Na^+} + e$$

On the other hand, chlorine atom accepts one electron, released by sodium to complete its octet by attaining stable gas configuration of argon. In this process, chlorine is converted into chloride ion.

$$\underset{2,\,8,\,7}{Cl + e^-} \longrightarrow \underset{2,\,8,\,8}{Cl^-}$$

Now we have two oppositely charged ions. As soon as they approach each other, they are held together by strong electrostatic forces of attraction. Thus, formation of sodium chloride takes place-

$$Na^+ + Cl \longrightarrow NaCl$$

These steps can be represented by Lewis structures as follows:

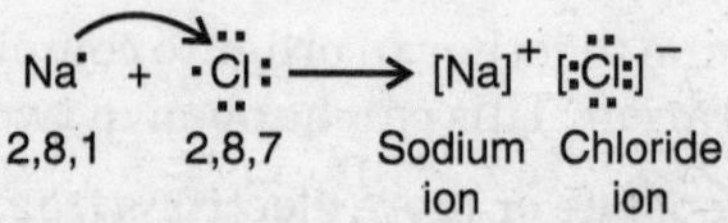

The number of electrons lost or gained during the formation of an electrovalent linkage is termed as the electrovalency of the element. For example, in the above case the electrovalency of sodium as well as chlorine is one. In other words, valency is equal to the charge on the ion.

It is worthwhile to mention here that an ionic compound consists of an aggregate of cations and anions thus having a number of ionic bonds.

X-ray analysis have confirmed that the crystals of NaCl contain charged ions in the form of Na^+ Cl^-. Such solids which consists of ions only are known as **Ionic Solids**. In a similar manner we can explain the formation of ionic bonds in a large number of compounds eg LiF, KCl, CaF_2, $MgCl_2$, MgO, $MgBr_2$, CaS etc. Some of which are depicted below.

Formation of Calcium Flouride CaF_2

Electronic Configuration of Calcium (Z = 20) = 2, 8, 8, 2
Electron Configuration of Flourine (Z = 9) = 2, 7

A calcium atom loses 2 electrons to achieve the stable noble gas configuration of argon and forms a calcium cation with two positive charges. A flourine atom can accept one electron only to form flouride ion. Therefore two flourine atoms are needed to accept the two electrons released by calcium atom. Therefore, the ratio of cations to anions in calcium flouride is 1:2. The oppositely charged ions are then held together by strong electrostatic forces of attraction.

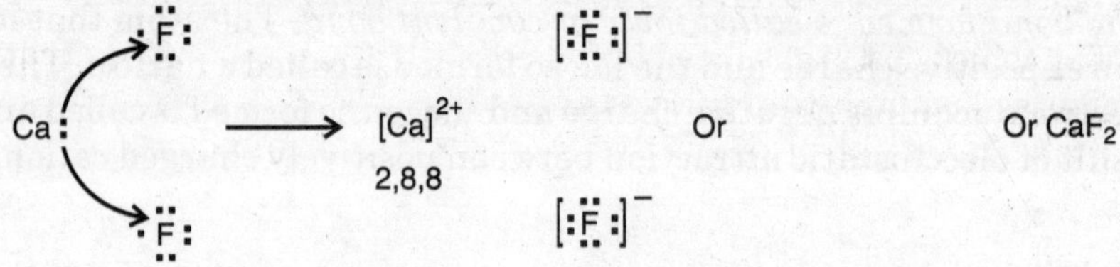

Formation of Magnesium Oxide MgO

Electronic Configuration of Magnesium Atom (Z = 12) = 2, 8, 2
Electronic Configuration of Oxygen atom (Z = 8) = 2, 6.

A magnesium atom loses two electrons and oxygen atom gains two electron to acquire noble gas configuration. The magnesium ions and oxide ions attract each other to form an ionic bond.

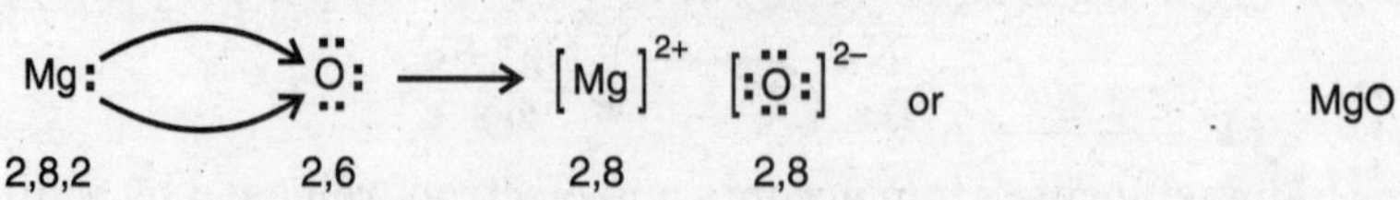

Formation of Sodium sulphide Na_2S

Electronic Configuration of Sodium atom (Z = 11) : 2, 8, 1
Electronic Configuration of Sulphur atom (Z = 16) : 2, 8, 6

A sodium atom loses one electron to form a sodium cation. But a sulphur atom requires two electrons to complete its octet. Thus two sodium atoms should lose one electron each which are taken up by sulphur atom to form sulphide ion.

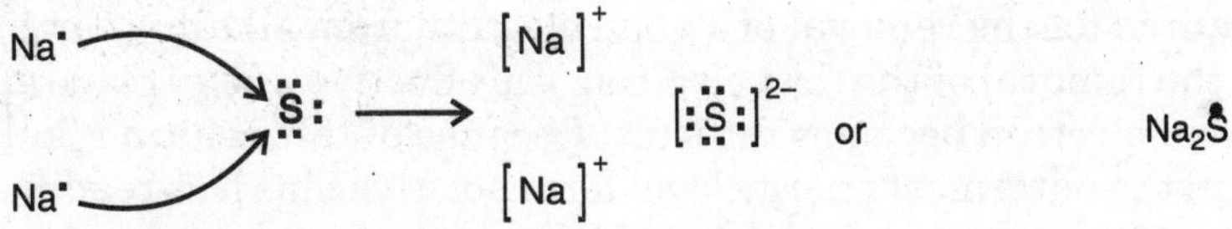

Formation of Aluminium Oxide Al_2O_3

Electronic Configuration of Aluminium Atom (Z = 13) = 2, 8, 3
Electronic Configuration of Oxygen Atom (Z = 8) = 2, 6

An Aluminium atom loses 3 electrons to achieve the stable gas configuration whereas an oxygen atom can accept only 2 electrons to form oxide ion. As a compound is electrically neutral, the positive charges in aluminium oxide compound must be balanced by equal number of negative charges. This is possible when 6 electrons released by 2 aluminium atoms (3 each) are accepted by three oxygen atoms as shown below.

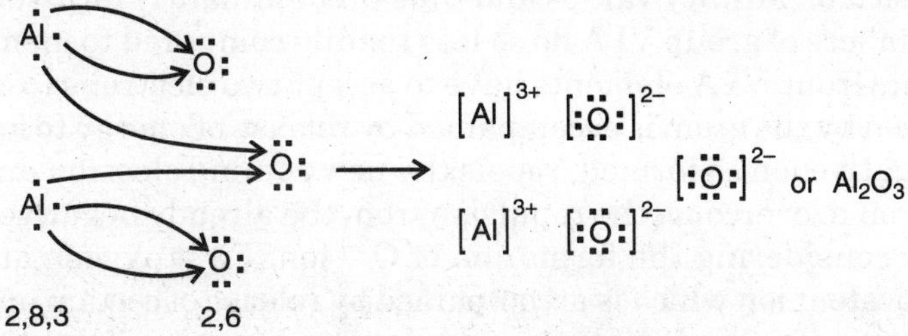

3.2.1 Factors Governing The Formation of Ionic Bonds

The formation of an ionic bond is governed by the ease with which anions and cations are formed. The following factors play important role in the formation of an ionic bond:

1. Ionization Energy.
2. Electron Affinity.
3. Lattice Energy.

Let us discuss these factors in detail.

Ionization Energy: It is defined as the amount of energy required to remove the most loosely bound electron from an isolated gaseous atom of an element. It can easily be understood that, lesser the ionization energy, easier will be the removal of an electron and hence greater are the chances of formation of ionic bond.

$$M \xrightarrow{I.E_1} M^+ + e^- \qquad (I.E_1 = \text{First ionization energy})$$

$$M^+ \xrightarrow{I.E_2} M^{2+} + e^- \qquad (I.E_2 = \text{Second ionization energy})$$

The alkali metals (group I elements) have low ionization energies, hence they have more tendency to form positive ions. On the other hand, alkaline earth metals (group II elements) do so less readily. The reason being that alkaline earth metals have to lose two electrons from their valence shells to acquire stable noble gas configuration.

$$Na \xrightarrow[(I.E_1)]{495\text{ kJ mol}^{-1}} Na^+(g) + e^- \qquad \text{(Alkali metal)}$$

$$Mg(g) \xrightarrow[(I.E_1 + I.E_2)]{735.5 + 1450\text{ kJ mol}^{-1}} Mg^{++}(g) + 2e^- \qquad \text{(Alkaline earth metal)}$$

The energy required for the removal of second electron from alkaline earth metal ion is very high because after the removal of the first electron, the effective nuclear charge increases hence the removal of second electron becomes difficult. Because of this reason alkaline earth metals with two electrons in the outermost energy level form positive ions less readily. Similarly, it can be concluded that the elements of group III A and IV A do not form ionic bonds easily as elements of these groups have to lose three and four electrons respectively and thus would require a huge amount of ionization energy.

2. Electron Affinity: It is defined as the amount of energy released when an electron is added to an isolated gaseous atom of an element. In other words, electron affinity is a measure of the tendency of an atom to change into an anion. Higher the value of electron affinity, the greater is the ease with which the anion is formed.

Among the nonmetals, elements of group VI A and VII A have high electron affinity values and thus readily change into negatively charged ions. The members of halogen family, group VII A have the highest electron affinity values and thus predominantly form ionic compounds with other atoms. The members of group VI A do so less readily compared to member of Group VII A. The reason being that Group VI A elements have to accept two electrons to complete their octet. Addition of one electron by the atom is accompanied by release of energy to form a singly charged negative ion. The negative ion so formed, repels the new coming electron and energy have to be supplied to add electron to overcome the repulsion from the already negatively charged ion. This can be illustrated by considering the formation of O^{2-} ion. First oxygen atom will add up one electron to form monovalent ion which is accompained by release of certain amount of energy. But in the process of addition of second electron to the monovalent oxygen ion, energy is required to overcome the force of repulsion exerted by the negatively charged monovalent ion to the incoming electron.

$$O_{(g)} + e^- \longrightarrow O^- \quad DH = -140.9 \text{ kJ mol}^{-1}$$
$$O^- + e^- \longrightarrow O^{2-} \quad DH = +769.9 \text{ kJ mol}^{-1}$$

Electron affinity for the overall process will be given by-

$$\Delta H = 629 \text{ kJ mol}^{-1}$$

The positive sign indicates that the energy is **required** to form divalent oxide ion. Hence it can be concluded that monovalent anions are formed much more easily than divalent anions.

3. Lattice Energy: The lattice energy of an ionic solid is defined as energy released when one mole of crystal is formed from gaseous ions. Higher the magnitude of lattice energy, greater is the tendency for the formation of ionic bonds.

$$Na^+_{(g)} + Cl^+_{(g)} \longrightarrow Na^+Cl^- + \text{Lattice Energy.}$$

Lattice Energy of NaCl is -782 kJ mol^{-1}

Higher the electrostatic attraction between the constituent ions, higher is the lattice energy. The force of attraction between the two oppositely charged ions in air (vacuum) is given by the well known Coulomb's Law as $F = \frac{q_1 q_2}{r^2}$: where q_1 and q_2 are the respective charges on the ions and r is the distance between them. Thus the magnitude of the lattice energy depends upon the following factors-

(a) **Size of the ions:** Smaller the size of the ions, lesser the internuclear distance and hence higher will be the lattice energy.

(b) **Charge on the ions:** Greater the charge on ions greater will be attractive forces between the ions and hence higher will be the value of Lattice energy.

Example: Lattice energy of Na^+Cl^- is more than lattice energy of Cs^+Cl^- because anion being the same, comparison has to be made between Na^+ and Cs^+ ion. As Na^+ ion is smaller than Cs^+ ion hence lattice energy of Na^+Cl^- is more than that of Cs^+Cl^-.

In another example, let us compare the lattice energies of LiCl and CaO. In case of LiCl the internuclear distance is 241 pm and in case of CaO it is 239 pm. As both the values are similar, thus size has no role to play as far as lattice energy is considered. But as we can see in *CaO* both Ca^{2+} and O^{2-} ions are doubly charged whereas in LiCl, Li^+ and Cl^- ions are singly charged, the lattice energy for CaO must be almost 4 times as large as for LiCl. The observed value is 845 kJ mol^{-1} for LiCl and 3460 kJ mol^{-1} for CaO.

3.2.2 Characteristics of Ionic Compounds

1. **Crystalline Nature:** Ionic compounds are usually crystalline in nature.X-ray diffraction patterns have proved that the constituent units in a crystal are ions and not molecules. These ions are arranged in a regular pattern to form a lattice. The attraction between ions is electrostatic and non-directional. For example, in crystal of *NaCl*, every sodium ion is surrounded by six chloride ions and each chloride ion is surrounded by six sodium ions in a regular fashion, thus giving rise to a three dimensional octahedral crystal structure.
2. **High melting and boiling points:** Ionic compounds have high melting and boiling points due to electronic cation-anion interactions, extending throughout the crystal lattice. Considerable amount of energy is needed to break up their crystal lattices.

   Closer the ions in a crystal, stronger will be the attractive forces and higher will be the melting points. (Table – 3.1)

Table 3.1: Melting Point of Sodium Halides

Compound	*Na F*	*NaCl*	*NaBr*	*NaI*
M. pt (°C)	995	800	750	651

3. **Hardness:** The existance of electrostatic forces of attraction extending throughout the crystal lattice makes ionic compounds hard. Hardness of the ionic crystals increases with decrease in interionic distances and increase in ionic charge. Further it has been found that ions with an inert gas shell configuration octet give harder crystals than ions with pseudo inert gas shell configuration (18 electrons).
4. **Solubility:** In general, ionic compounds are soluble in water but insoluble in benzene or other organic solvents. Water and other polar solvents having high dielectric constants, decrease the electrostatic forces of attraction and the ions get seperated. Further the interaction between the ions and the polar molecules ie ion-solvent dipole interactions, provide considerable energy for breaking up the ionic lattice and thus help in the process of dissolution.
5. **Conductance:** Ionic compounds do not conduct electricity in the solid state. In the solid state, the ions occupy fixed places and hence are not free to move. But the compound conduct electricity when present in fused state or in aqueous solution. This is because, in fused state or in aqueous solution, the ions are free to move and hence are able to conduct electricity.
6. **Ionic Reactions:** Ionic compounds react very rapidly. Chemical reactions of an ionic compound are the reactions of its constituent ions, and as these already exist, they have only to collide with the other type of ion.

   Example

$$Na^+Cl^-_{(aq)} + Ag^+NO^-_{3(aq)} \longrightarrow AgCl_{(s)} + NaNO_{3(aq)}$$

7. **Brittle in solid state:** This is because when a force is applied, the relative positions of cations and anions change with respect to each other. The cation-cation and anion-anion repulsions make the crystal brittle or cause it to break apart.

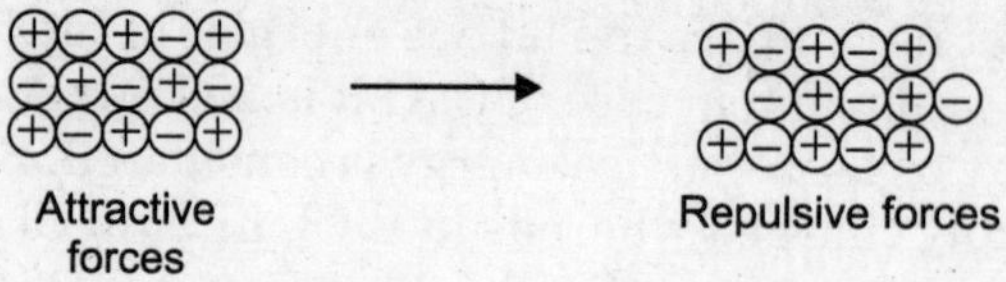

3.2.3 Determination of Lattice Energy

As discussed earlier in the chapter, lattice energy (U) of a crystal is the energy evolved when one mole of crystal is formed from gaseous ions.

Lattice energies can be determined by using:

(a) **Born-Lande Equation:** This equation is used for theoretical calculations of lattice energy.
(b) **Born Haber Cycle:** Experimental values of lattice energy are obtained from thermodynamic data using Born Haber Cycle.

Born-Lande Equation

The lattice energy of ionic solids can be calculated theoretically on the basis of coulombic interactions between the ions present in the crystals.

The interactions are of two types-

(i) Attractive interactions between oppositely charged ions.
(ii) Repulsive interactions due to repulsion of electron clouds of two atoms.

Attractive Interactions: Consider the two oppositely charged ions M^{Z+} and X^{Z-}, as point charges separated by a distance 'r', Z^+ and Z^- are the number of charges on cations and anions respectively.

The electrostatic energy of attraction is given by-

$$E = -\frac{z^+ z^- e^2}{r} \quad ...(1)$$

Where e is the electronic charge (1.6×10^{-19}C)

For more than two ions, the electrostatic energy depends on the number of ions. For one mole the attractive energy is-

$$E = \frac{-N_0 z^+ z^- z^2}{r} \quad ...(2)$$

Where N_0 is the Avogadro's constant. The above equation represents the electrostatic energy between ion pair, whereas within a crystal, the same two ions are subjected to additional coulombic forces from other ions. Furthermore, ionic crystals are not composed of simple ionic pairs but of ionic clusters in which each cation is surrounded by a definite number of anions and viceversa. Therefore, in order to calculate the coulombic energy of any single ion, allowance must be made for the number of, and arrangement of neighbouring ions. *Because of electrostatic interactions with surrounding ions, the coulombic energy of a single ion is greater than indicated by the above equation, by a factor typical of each crystal structure. This factor is designated as A and is known as Madelung Constant.* The Madelung constant is a correction factor to take care of ion pair and is entirely dependent upon the geometry of the crystal. Thus the equation takes the form-

$$E = \frac{-N_0\ A\ z^+ z^- e^2}{r} \quad ...(3)$$

Where A is the Madelung Constant. Interpretation of Madelung constant is given in the next section.

Repulsive Interactions: The equation for the attractive forces between the ions gives a negative value for energy, that is energy is given out when a crystal is formed. It is seen that smaller is the value of r, greater amount of energy will be evolved when crystal is formed and hence more stable the crystal will be.

If $r = 0$, infinite amount of energy should be evolved according to the equation. But this is not so. When the cation and anion approach close to one another repulsive force comes into play due to the interaction between the electron clouds on the two atoms or ions.

The repulsive forces increase rapidly as r decreases. The repulsive force is given by B/r^n where B is a constant termed as repulsion co-efficient and n is a constant called the **Born exponent**. Born exponent is the power to which r must be raised to give the correct repulsive force. The value of n increases with increase in electron charge density around the ions.

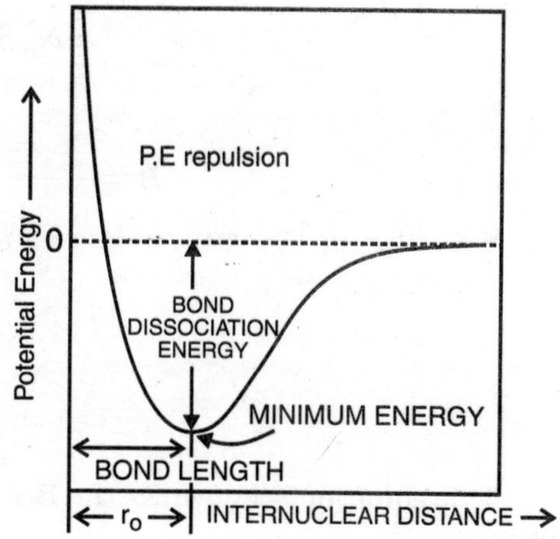

Fig. 3.1 *Variation of Potential energy with nuclear distance giving a minima at r_o.*

$$P.E \text{ (repulsion)} = B/r^n$$

For one mole:

$$P.E = N_0 \, B/r^n$$

Note: In case of ions having Neon configuration the value of n is taken as 7 (eg. Na^+ and F^- ion) while in ions having Ar, Kr and Xe configuration, n is taken as 9, 10 and 12 respectively.

The total energy holding the crystals together is U, the lattice energy. This is the sum of the attractive and the repulsive forces

$$U = \frac{-N_0 \, A z^+ z^- e^2}{r} + \frac{N_0 \, B}{r^n} \qquad ...(4)$$

As the two terms are of opposite signs, the plots of U against r will show a minima at the equilibrium internuclear distance r_0.

At equilibrium distance r_0, the attractive and repulsive forces exactly balance (Fig. 3.1).

It is possible to eliminate 'B' from the expression by considering that, at equilibrium distance U is minimum and hence $\frac{dU}{dr}$.

Differentiating equation (4) with respect to r, we get

$$\frac{dU}{dr} = -N_0 A z^+ z^- e^2 \, (-1) r^{-2} + B N_0 (-n) r^{(-n-1)}$$

$$\left| \quad \frac{dx^n}{dx} = nx^{n-1} \qquad \frac{dx^{-n}}{dx} = -nx^{-n-1} \right.$$

$$= \frac{N_o A \, z^+ z^- e^2}{r^2} - \frac{n N_o B}{r^{n+1}} \qquad ...(5)$$

At equilibrium r is replaced by r_O and $\left[\frac{dU}{dr}\right]_{r=ro} = 0$, thus equation (5) becomes,

$$\left[\frac{dU}{dr}\right]_{r=r_O} = 0 = \frac{N_o A z^+ z^- e^2}{r_o^2} - \frac{nN_o B}{r_o^{n+1}}$$

$$\frac{N_o A z^+ z^- e^2}{r_o^2} = \frac{nN_o B}{r_o^{n+1}}$$

$$B = \frac{A z^+ z^- e^2 r_o^{n-1}}{n} \qquad ...(6)$$

Substituting the value of B in equation (4)

$$U = \frac{-N_o A z^+ z^- e^2}{r_o} + \frac{N_o A z^+ z^- e^2 r_0^{n-1}}{r^n}$$

$$= \frac{-N_o A z^+ z^- e^2}{r^o} + \frac{N_o A z^+ z^- e^2}{nr^o} = \frac{-N_o A z^+ z^- e^2}{r^o}\left[1 - \frac{1}{n}\right]$$

This equation is known as the **Born Lande Equation.** Using SI units the equation takes the form,

$$U = \frac{-N_o A z^+ z^- e^2}{4p\, e_o\, r_o}\left(1 - \frac{1}{n}\right)$$

Where ε_0 is the permittivity of free space = 8.854×10^{-12} Fm^{-1}

Interpretation of Madelung Constant

Madelung constant A, is a correction factor to take care of ion pair and is entirely dependent upon the geometry of the crystal. In a crystal, there is not one cation and one anion but a large number of cations and anions arranged together according to their specific geometry.

A study of geometry of *NaCl* crystal shows that each sodium ion is surrounded by

- 6 chloride ions at a distance r
- 12 sodium ions at a distance $\sqrt{2r}$
- 8 more chloride ions at a distance $\sqrt{3r}$
- 6 more sodium ions at a distance $\sqrt{4r}$
- 24 more chloride ions at a distance $\sqrt{5r}$
- 24 more sodium ions at a distance $\sqrt{6r}$ and so on.

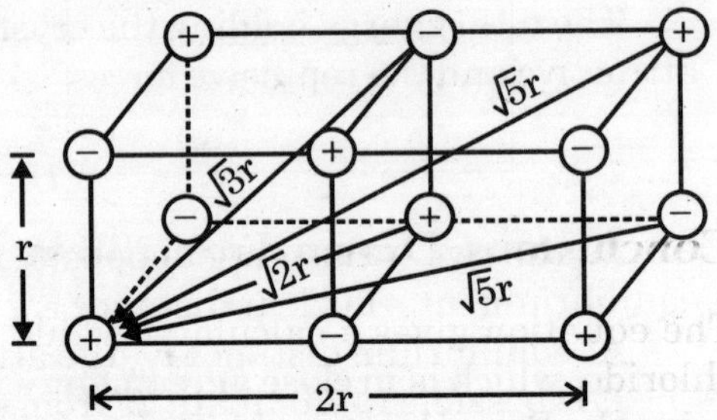

Fig. 3.2 *Interionic distances in NaCl crystal*

Because of electrostatic interactions with surrounding ions, the coulombic energy of single ion is greater than indicated by equation (1) by a factor typical of each of crystal structure called Madelung constant *A*. Thus coulombic energy because of these interactions will be given by a equation similar to equation (1) but which includes the contribution from the surrounding ions. *P.E* due to attraction is given a negative sign while that due to repulsion is given a positive sign.

Thus–

$$E = -\frac{6\, z^+ z^- e^2}{r} + \frac{12 z^+ z^- e^2}{\sqrt{2\, r}} - \frac{8\, z^+ z^- e^2}{\sqrt{3r}} + \frac{6 z^+ z^- e^2}{\sqrt{4r}} - \frac{24\, z^+ z^- e^2}{\sqrt{5r}} + \frac{24 z^+ z^- e^2}{\sqrt{6r}}$$

$$= -\frac{z^+z^-e^2}{r}\left(6-\frac{12}{\sqrt{2}}+\frac{8}{\sqrt{3}}-\frac{6}{\sqrt{4}}+\frac{24}{\sqrt{5}}-\frac{24}{\sqrt{6}}\ldots\ldots\right)$$

$$= -\frac{z^+z^-e^2A}{r}$$

where $A = \left(6-\frac{12}{\sqrt{2}}+\frac{8}{\sqrt{3}}-\frac{6}{\sqrt{4}}+\frac{24}{\sqrt{5}}-\frac{24}{\sqrt{6}}\ldots\ldots\ldots\right)$

$= 6 - 3.486 + 4.649 - 3 + 10.733 - 9.900 + \ldots$

$= 1.7476 + \ldots$ for several terms

The value converges to a single value because the electrostatic interactions at great distances become unimportant.

Thus, Madelung constant can be a sum of inverse distance from an ion to other ions with a plus sign for opposite ion and a minus sign for the same ion.

Any other salt having the same geometrical arrangement like *NaCl* will have the same value of Madelung constant. The values of Madelung constant for some typical crystal structure are given in Table 3.2.

Table 3.2: *Madelung Constant for some Typical Crystal Structures*

Crystal Structure	*Coordination number of Ions*	*Madelung Constant*
NaCl	6 : 6	1.7475
Cs Cl	8 : 8	1.7627
Zn S (zinc blende)	4 : 4	1.6381
Zn S (wurtzite)	4 : 4	1.6413
$Ca\ F_2$	8 : 4	5.0388

Conclusions Drawn From Born Lande's Equation

The equation gives a calculated value of $U = -778$ kJ mol^{-1} for the lattice energy for sodium chloride, which is in close agreement with the experimental value of -775 kJ mol^{-1} obtained by using the Born-Haber cycle. The experimental and theoretical values for the alkali metal halides and the oxides and halides of the alkaline earths, all agree within 3%. However Be show a little more variation. The Born Lande's equation leads to the following conclusions.

1. The lattice energy U_0 depends upon the value of r_o, smaller the value of r_o, more is the value of lattice energy.

   Example:

Ionic Solid	r_o *(A°)*	U_o *K cal/mol*
Li F	2.01	–240
Li Cl	2.57	–203
Li Br	2.75	–197
Li I	3.00	–180

The lattice energy U_0 depends upon the value of r_o, smaller the value of r_o, more is the value of lattice energy.

Example:

2. Higher the charge on cation and anion, greater would be the magnitude of lattice energy and therefore greater would be the stability of the crystal.

Example:

Ionic Solid	*Nature of Ionic Solid*	U_o *k cal/mol*
Li F	uni-univalent	240
Ca F_2	uni-bivalent	632
Mg S	bi-bivalent	777

3. The lattice energy depends on the value of the Madelung constant, which depends upon the coordination number and the geometric pattern of the ions in the crystal lattice of the crystal.

Born-Haber Cycle

Experimental approach to lattice energy have been made through Born Haber thermodynamic cycle. This cycle devised by Born and Haber in 1919 relates the lattice energy of a crystal to other thermochemical data.

As a necessary consequence of the first law of thermodynamics, the enthalpy of a reaction will be the same, whether the reaction takes place in one or several steps. This law of conservation of energy is the Hess's law.

Consider the formation of M^+X^- from M and X_2 (e.g. NaCl from $Na_{(s)}$ and $Cl_{2(g)}$)

$$M_{(s)} + \frac{1}{2}X_{2(g)} \longrightarrow MX_{(s)}$$

The enthalpy change for this reaction is called enthalpy of formation and is denoted as ΔHf.

The formation of $MX_{(s)}$ can be broken down into a series of steps. The various steps are:

1. Conversion of M (solid) into gaseous atoms. The energy involved for this process is the sublimation energy.
   It is represented as S.
   $$M_{(s)} \xrightarrow{\text{Sublimation}} M_{(g)}$$
2. $M_{(g)}$ is converted into the gaseous cation M^+. This involves the ionization energy represented as $I.E.$
   $$M_{(g)} \xrightarrow{\text{I.E}} M^+_{(g)} + e^-$$
3. Diatomic gas X_2 is dissociated into atoms. This step involves the dissociation energy i.e. the amount of energy required to dissociate one mole of gaseous molecules into gaseous atoms. It is represented by D.
   $$X_{2(g)} \xrightarrow{\text{D}} 2X_{(g)}$$
   The energy required to produce one mole of gaseous atoms would be D/2.
4. Gaseous atom X are converted into the anions by releasing the corresponding electron affinity. It is represented by $E.A.$
   $$X_{(g)} + e^- \xrightarrow{-\text{EA}} X^-_{(g)}$$
5. Gaseous cations and anions combine to form crystals of M^+X^- with the release of the lattice energy U_o of the crystal :
   $$M^+_{(g)} + X^-_{(g)} \xrightarrow{-U_o} M^+X^-_{(s)}$$

The various steps can be represented as:

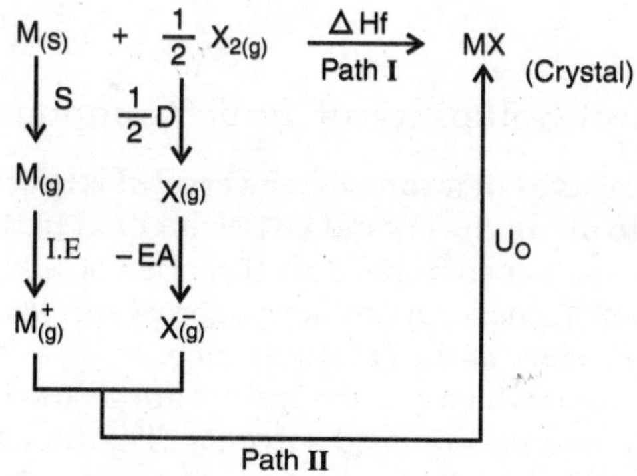

Fig. 3.3 Born Haber Cycle

According to Hess's law, the enthalpy of formation of MX should be the same whether the reaction takes place in one step or several steps.

Hence

$$\Delta H_f = S + \frac{1}{2}D + I.E + EA + U_0.$$

Using positive sign for energy absorbed and negative sign for energy released, we have

$$-\Delta H_f = S + \frac{1}{2}D + I.E - EA - U_0.$$

Applications of Born Haber Cycle

1. **To Calculate electron affinities:** The most important application of Born Haber Cycle is to determine electron affinities which are otherwise difficult to determine by other means. By using known crystal structures, lattice energy is calculated by using Born Lande's equation and hence it is possible to calculate electron affinity. Example for NaCl crystal –

$$-\Delta H_f = S + \frac{1}{2}D + I.E - E.A - U_0$$

$$-381.2 = 108.4 + 495.4 + 120.9 - E.A - 757.3$$

$$E.A = 348.6 \text{ kJ mol}^{-1}$$

Enthalpy of formation of ionic compounds gives us a clear idea about the stability of the compound.

2. **Stability of unknown compounds:** The values of S, I, D and $E.A$ for most of the elements and their ions are available. The ionic radii can be calculated by using Pauling's equation. These are used to find out the interionic distances in a crystal lattice. Thus, lattice energy can be calculated by Born Lande's equation and hence enthalpy of formation of ionic compounds can be calculated with reasonable degree of accuracy.
3. **Variations in stability of various ionic compounds:** Born Haber cycle serves as a valuable tool for correlating the variations in stability of various ionic compounds.

Examples:

MgO is highly stable compound even though considerable energies are required to vaporize $Mg_{(s)}$ and to dissociate $O_2(g)$. Moreover the formation of Mg^{2+} and O^{2-} are again endothermic processes. The stability of MgO is explained on the basis of high lattice energy which is released during the formation of MgO which more than balances out the above opposing factors.

3.2.4 Lattice Energy and Solubility of Ionic Compounds

When an ionic compound dissolves in a solvent, the crystal lattice is broken and ions move far apart. The energy required to break the crystal lattice is equal to the lattice energy. The energy for rupturing the crystal lattice comes from the interaction of solvent and the ions i.e from the solvation of ions. The solvation of ions is an exothermic process releasing a lot of amount of energy which is utilized in causing rupture of the crystal lattice.

It follows therefore that liquids which cause solvation are good solvents and have high dielectric constant. Thus water and ammonia are good solvents. The liquids which have low dielectric constants such as benzene, CCl_4 do not solvate the ions and are therefore not good solvents for ionic compounds. In this case, since no solvation takes place and hence no energy is released to cause rupture of bonds. For example sulphates, oxalates, phosphates of alkaline earth metals are much less soluble than the corresponding alkali metals because former has much higher values of lattice energies.

3.3 COVALENT BOND

Lewis Theory: Lewis (1916) proposed a new concept of bonding among non-metallic elements. When two atoms taking part in a chemical combination are short of electrons than the nearest inert gas configuration, they can share their electrons in order to complete their octets. *The bond formed between the two atoms by mutual sharing of electrons between them, so as to complete their octets is called covalent bond. The number of electrons contributed by each atom is known as covalency.*

The electron pair shared between two atoms for the formation of a covalent bond is called the bonding pair of electrons.

Examples:

Formation of Hydrogen molecule: Hydrogen atom has one electron in its valence shell. In the formation of hydrogen molecule, both the atoms contribute one electron each for mutual sharing and thus acquiring the configuration of helium gas.

$$\text{H}\bullet \; \bullet\text{H} \longrightarrow \text{H : H or H–H}$$

Formation of Chlorine molecule: A chlorine atom has 7 electrons in its valence shell. Each chlorine atom is short of one electron to complete its octet, so by sharing one electron with another chlorine atom both atoms attain noble gas configuration.

$$:\ddot{\underset{\cdot\cdot}{\text{Cl}}}\cdot \quad \cdot\ddot{\underset{\cdot\cdot}{\text{Cl}}}: \longrightarrow :\ddot{\underset{\cdot\cdot}{\text{Cl}}}:\ddot{\underset{\cdot\cdot}{\text{Cl}}}: \text{ or Cl – Cl}$$

Formation of NH_3: Nitrogen atom has five outer electrons and in NH_3, it shares three of these forming three bonds and thus attaining an octet. Hydrogen has only one electron and by sharing it attains a stable configuration of helium.

$$\cdot\ddot{\text{N}}\cdot + 3[\text{H}\cdot] \longrightarrow \text{H}:\ddot{\underset{\text{H}}{\text{N}}}:\text{H}$$

The shared pairs of electrons present between the atoms are called **bond pairs** because they are responsible for bonding between the atoms. On the other hand, the valence electrons not involved in bonding (i.e sharing) are shown as such and are called non-bonding pairs or **lone pairs**. Thus NH_3 has three bond pairs and one lone pair.

Limitations of Lewis Concept: Lewis concept of covalent bond as explained above had the following limitations–

(i) It could not explain how the atoms are held together in the molecules like H_2, Cl_2 etc, as no electrostatic forces are operating between them.
(ii) It could not explain the formation of molecules which violate the octet rule, example PCl_5, SF_6, $AlCl_3$ etc.
(iii) It could not explain the shapes of molecules containing covalent bonds.
(iv) It could not explain the formation of the molecules which have odd number of electrons, such as NO and ClO_2.
(v) It could not explain why oxygen molecule is paramagnetic and have two unpaired electrons.

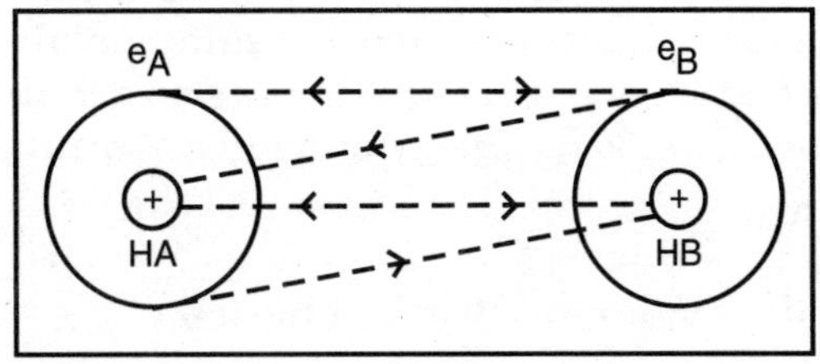

Fig. 3.4 Interactions between two hydrogen atoms as they approach each other.

Lowering of Energy: Formation of a stable covalent bond between two atoms can be explained by using the concepts of electrostatic attraction. The formation of a chemical bond of any type is possible if it results in decrease in energy.

When two atoms approach each other, the new forces come into play viz–

(i) **Forces of attraction:** These forces exist in between the nucleus of one atom and electrons of other atom. These forces tend to decrease the total energy of the system and result in formation of chemical bond.
(ii) **Forces of repulsion:** These forces exist in between the nuclei of the reacting atoms and in between the electrons of the two atoms. These forces tend to increase the energy of the system and no chemical bonding is possible.

Consider the formation of hydrogen molecule. Suppose we have two hydrogen atoms A and B.

(a) When the two atoms are far apart from each other, there are no attractive or repulsive forces between hydrogen atoms.
(b) When the two atoms approach each other, there is appreciable interaction and new forces start-operating.
   - Force of attraction between nucleus of A and electrons of B and between nucleus of B and electrons of A.
   - Force of repulsion between nuclei of two atoms and electrons of two atoms.

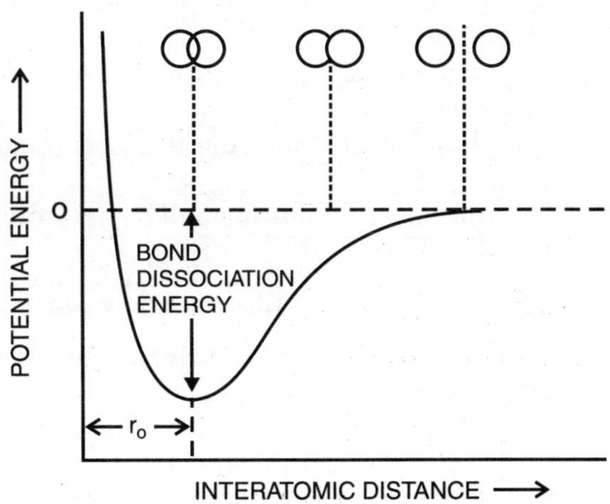

Fig. 3.5 Potential Energy Diagram showing variation of P.E with internuclear distance for two hydrogen atoms.

Graphically, the energy changes taking place during the formation of H_2 molecule can be represented as in Figure 3.5.

When the hydrogen atoms lie far away from each other, by convention the electrostatic potential energy is taken as zero. As the two atoms approach each other, they start interacting with each other and the system starts losing its energy because of attractive forces between the two atoms become more and more dominant. This decrease is continued till a certain minimum value is obtained. After this, if the atoms are brought still closer, the curve show a sharp rise due to the repulsion between the two nuclei at such small internuclear distance. Thus the energy of the system is minimum when the forces of repulsion are just balanced by the forces of attraction. At this stage the two atoms are said to be bonded together to form a stable system i.e hydrogen molecule.

At this stage the distance between the two nuclei is referred to as **Bond Length.**

It can be concluded that a certain amount of energy is released during the formation of covalent bond or chemical bond. More is the amount of energy released stronger is the bond. If we want to dissociate the bond, the same amount of energy has to be supplied. This is known as bond dissociation energy. *Thus, the energy required to dissociate one mole of bonds of the same kind is known as Bond Energy or bond dissociation energy. The bond energy for H-H bond is* 433 *kJ* mol^{-1}.

3.3.1 Valence Bond Theory

To overcome the shortcomings of Lewis concept of the formation of covalent bond Linus Pauling and J.C Slater proposed Valence Bond Theory. Infact this theory was first proposed by W. Heitler and F. London in 1927 and later developed by Pauling and Slater. The various points of the Valence Bond Theory are as follows:

1. An atom must possess one or more unpaired electrons to enter into a chemical combination.
2. Atoms with unpaired electrons tend to combine with other atoms which also have unpaired electrons. In other words a covalent bond is formed by the overlapping of two atomic orbitals belonging to two different atoms, having unpaired electrons. The electrons in the orbitals should have opposite spins.
3. The new orbital which is formed due to overlapping of atomic orbitals is more stable.
4. The strength of the bond is determined by the extent of overlapping of two half-filled orbitals. More is the overlapping, more is the decrease in energy and stronger is the bond formed.
5. The atoms which unite to form a molecule completely retain their identities in the resulting molecule.
6. The shape of the molecule is determined primarily by the directions in which orbitals point.
7. The number of bonds formed by an atom is equal to the number of unpaired electron it has.

The above points can be illustrated by considering a few examples–

Formation of hydrogen molecule, s–s overlap: Hydrogen has 1 electron in its valence shell i.e. 1s′. H_2 molecule is formed when two s-orbitals of two hydrogen atoms overlap along their axis. The electrons shared between the two atoms are located in the region of space between the nuclei of the atoms. Thus 1*s* orbital of the two atoms gets fully occupied. This bond is called *s-s* bond as it is formed by the overlapping of two *s* orbitals. (Fig. 3.6)

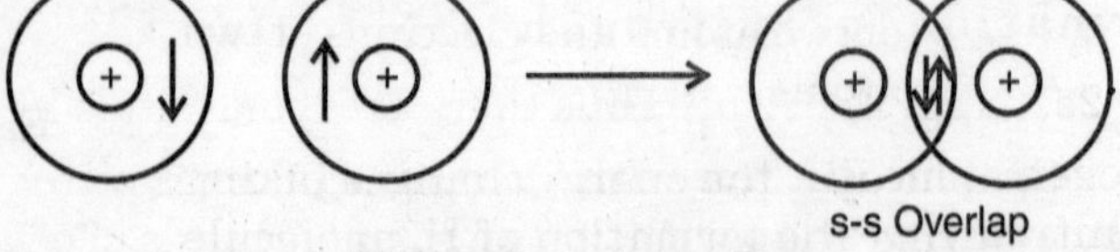

Fig. 3.6 s–s overlap

Formation of flourine molecule, p-p overlap: Electronic configuration of flourine = $1s,^2\ 2s,^2$, $2p_x^2, 2p_y^2, 2p_z^1$ Thus its $2p_z$ orbital having unpaired electron overlaps with the $2p_z$ orbital of other flourine atom having electron with opposite spin. As the overlap is between two p-orbitals, this is called p-p overlapping. (Fig. 3.7)

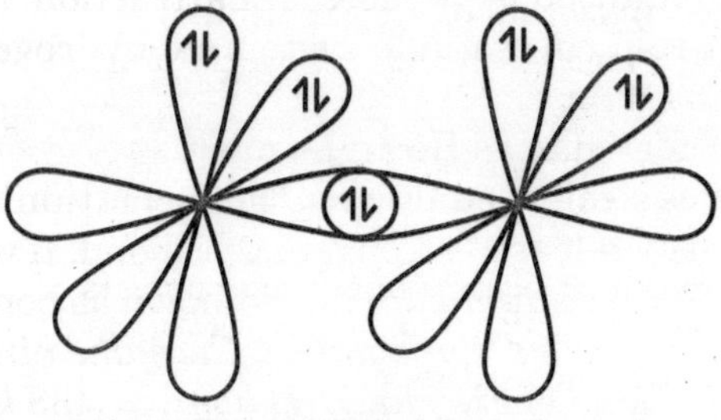

Fig. 3.7 p–p overlap along internuclear axis.

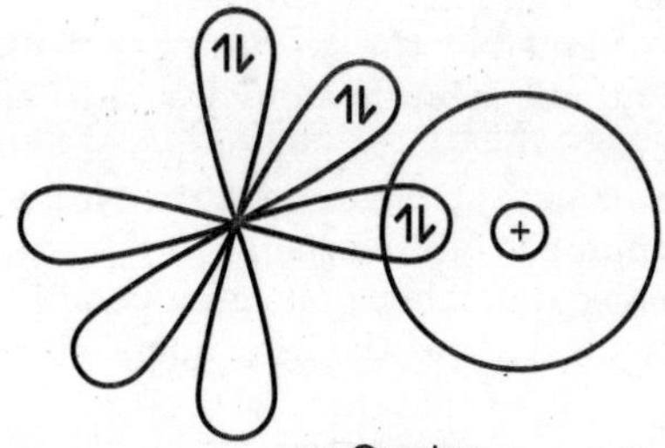

Fig. 3.8 s–p overlap

Formation of hydrogen flouride molecule s-p overlap: In the formation of hydrogen flouride, half filled $2p_z$ orbital of flourine overlaps with half filled $1s$ orbital of hydrogen atom resulting in s-p overlap. (Fig. 3.8)

Sigma and Pi Bonds

As is seen in the above examples the overlap is along the internuclear axis. When a bond is formed by the overlapping of the orbitals along the internuclear axis, it is known as **Sigma Bond**. On the other hand if the bond is formed by the sidewise overlapping or lateral overlapping of the atomic orbitals the bond is termed as **pi (π) Bond**. Formation of pi bond only takes place if sigma bond is already present i.e. it cannot be present alone. Formation of pi-bond is illustrated in the following example.

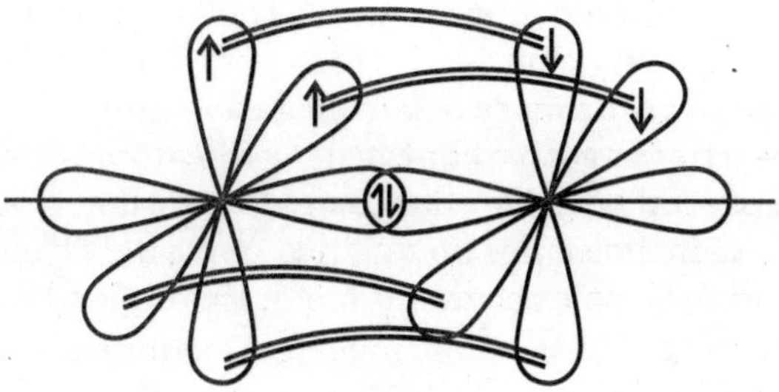

Fig. 3.9 Formation of N_2 molecule.

Formation of nitrogen molecule: Electronic configuration of nitrogen is $1s,^2\ 2s,^2\ 2p_x^1, 2p_y^1, 2p_z^1$. Thus there are 3 p-orbitals having unpaired electrons. $2p_z$ orbital of one nitrogen atom overlaps with $2p_z$ orbital of other nitrogen atom forming a σ bond. $2p_x^1$ and $2py^1$ of one nitrogen overlap with $2p_x^1$ and $2py^1$ orbitals of other nitrogen atom laterally to form two pi-bonds. Nitrogen atom has one sigma bond and two π bonds i.e a triple bond.

Table 3.3: Comparison of Sigma and Pi Bonds

Sigma Bond	*Pi Bond*
1. This bond is formed by the overlap of atomic orbitals along internuclear axis.	This type of bond is formed by the lateral (sideways) overlap of atomic orbitals.
2. This is formed by the overlapping of s-s, s-p, or p-p atomic orbitals along internuclear axis.	This is formed by the lateral overlapping of p-p atomic orbitals.
3. Sigma bond is stronger because extent of overlapping is quite large.	π-bond is weaker because overlapping is to a small extent.
4. The molecular orbital of sigma bond is symmetrical about the internuclear axis.	The molecular orbital of π bond consists of two electron clouds, one above and other below the plane of atomic nuclei.
5. There can be free rotation of the atoms around the sigma bond.	Free rotation about a π-bond is not possible.
6. The bond may be present alone or with π bond.	The bond is always present along with the sigma bond only.

3.3.2 Polar and Non Polar Covalent Bond

A covalent bond may be non-polar or polar depending upon the electronegativity of the bonded atoms. (cf : electronegativity is the tendency of an atom in a molecule to attract the shared pair of electrons towards itself.)

Non-Polar Covalent Bond: If the bond is formed between two identical atoms, the shared pair will be attracted equally by the two atoms and hence the bond formed will be non-polar covalent bond. For example-H_2, Cl_2 etc. In non-polar bond the electron cloud is completely symmetrical and there is no charge separation at all.

H : H — Hydrogen molecule

:Cl : Cl: — Chlorine molecule

Polar Covalent Bond: If the bond is formed between two dissimilar atoms, the shared pair will be attracted more towards the atom having greater electronegativity. The distribution of electrons gets distorted as the electron cloud is displaced towards more electronegative atom. This leads to the development of partial negative charge near the atom which is more electronegative and partial positive charge near the other atom which is less electronegative. Such molecules having two oppositely charged poles are called polar molecules and the bond is said to be polar covalent bond. Examples-HCl, BrCl etc. The nonpolar molecules are called dipoles.

H : Cl: or $\overset{d+}{H} - \overset{d-}{Cl}$

↑ Electron pair lie towards chlorine atom.

Partial Ionic Character of Polar Covalent Bond: The polar covalent bond is said to possess partial ionic character. The ionic character of the covalent bond depends upon the electronegativity difference between the two atoms. Several empirical equations have been proposed to calculate the percent ionic character of the covalent bond from the electronegativities of the bonded atoms. It has been proposed by Pauling that if the difference in electronegativities of the bonded atoms is 1.9 then the bond is 50% covalent and 50% ionic in character. If it is less than 1.9, then bond is predominantly covalent and if it is more than 1.9 then bond is predominantly ionic. He established the following approximate relationship-

Electronegativity Difference	0.2	0.6	0.8	1.2	1.9	2.2	2.8	3.2
Percent Ionic Character	1	10	15	30	50	70	86	92

3.3.3 Covalent Character of Ionic Bond (Polarization) (Fazan's Rules)

A covalent bond acquires some ionic character due to the greater difference in electronegativities of the two atoms, similarly ionic bond has some covalent character due to interaction between cation and anion.

The formation of a bond, intermediate between an ionic and a covalent bond occurs through a phenomenon known as **polarization of ions**.

When the cation approaches the anion, the net positive charge on the cation attracts the electron cloud of the anion and repels the positively charged nucleus of the anion. Therefore the electron cloud of the anion gets distorted or bulged towards the cation. This is called polarization of the anion. The electron charge cloud of the anion no longer remains spherical but gets distorted.

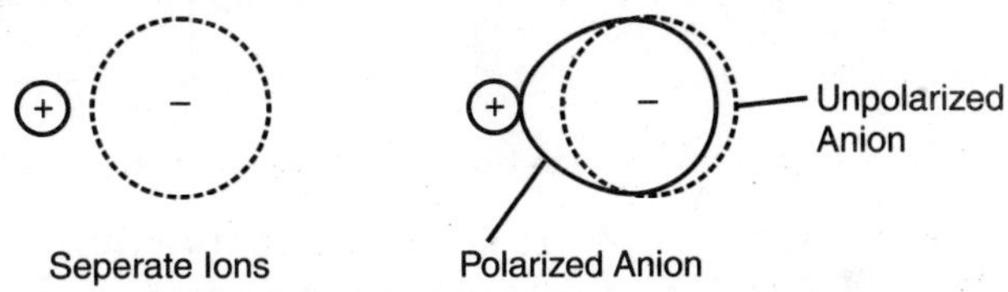

Fig. 3.10 Polarization of the anion

This polarization of ions results in high electron density between the two nuclei and hence appreciable covalent character is introduced. This type of bond is called polar covalent bond. *The power of an ion to distort the other ion is known as its polarizing power and the tendency of an ion to get polarised or distorted is called polarizability.* Fazan gave a number of empirical rules known as Fajan's rules to explain the polarization of ions. According to these rules the increased polarization in ionic compound is favoured by-

1. Small positive ion
2. Large negative ion
3. High charge on the ions.

1. **Small Positive Ions:** Small positive ions have higher polarizing power because positive charge is concentrated over a small surface area and thus it can distort the electron charge cloud of the anion efficiently. Example Li^+ has more polarizing power than Na^+.
2. **Large size of the anion:** The hold of the nucleus on the orbitals of the large anion will be less and hence such anions will get polarized easily.

   Example I: has greater tendency to get polarized than Cl^-.
3. **High Charge:** High charge on the cation has a greater tendency to polarize the anion and high charged anion has a greater tendency to get polarized. Example Mg^{2+} has more polarizing power than Na^+
4. **Electronic configuration of the cation:** Further the molecules having cations with 18 electron valency shell configuration, are more covalent than those having 8 electron valency shell configuration. This is because *d* electrons of the 18 electrons shell shield the nuclear charge less effectively than the *s* and *p* electrons of the 8 electron shell. Hence a cation with 18 electron shell has larger effective nuclear charge than the cations with 8 electrons in the outermost shell, even though the two may have same size and same charge. Example AgCl is more covalent than Na Cl even though the size of Ag^+ and Na^+ is

almost the same. This is due to the fact that Ag^+ has 18 electrons in its outermost shell where Na^+ has 8 electrons.

3.3.4 Coordinate Bond or Dative Bond

It is a type of covalent bond formed by the mutual sharing of electrons between the two atoms but the shared pair of electron is contributed by one of the two atoms only. Coordinate or dative bond is formed between two atoms, one of which has a complete octet and possesses a pair of valency electrons while the other is short of a pair of electrons. The atom which contributes electron pair for sharing is called the **donor** while the other atom which accepts it for sharing is called **acceptor**. A coordinate bond is represented by an arrow head ⟶. Example-NH_4^+ ion.

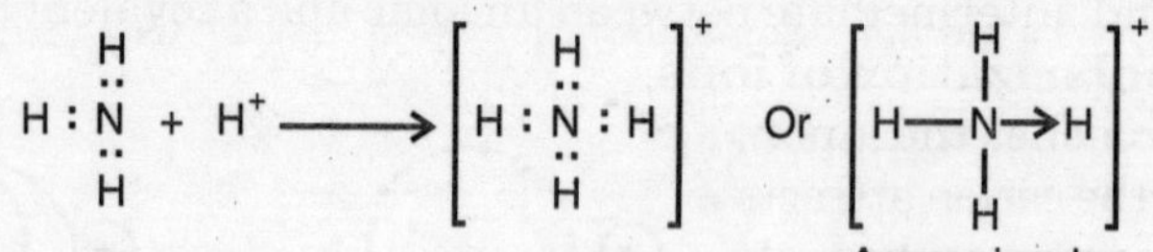

Ammonium Ion

Coordinate bond once formed is indistinguishable from a covalent bond.

Orbital Concept of Formation of Dative Bond: According to this concept a coordinate bond is formed when an orbital containing a lone pair of electrons overlaps with an empty orbital present in the other atom. This can be illustrated by considering the formation of hydronium ion, H_3O^+.

Formation of H_3O^+: The central atom oxygen in water has four orbitals $2s^2$, $2p_x^2$, $2p_y^1$, $2p_z^1$. Two of these are used up in formation of covalent bonds with hydrogen, while the other two contain lone pair of electrons. The H^+ ion has empty 's' orbital which overlaps with one of the p orbital of exygen atom containing lone pair forming hydronium ion, H_3O^+.

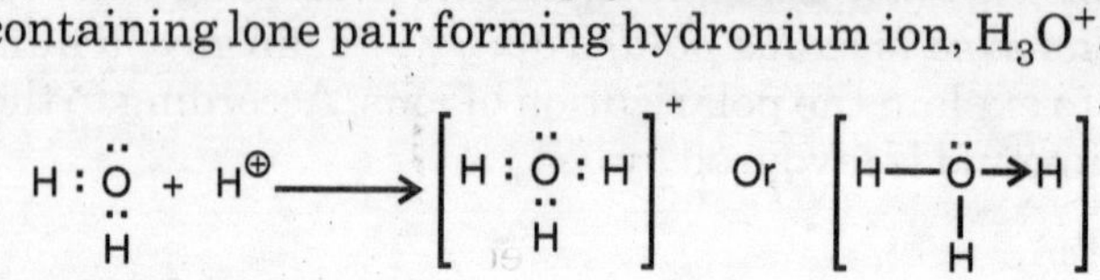

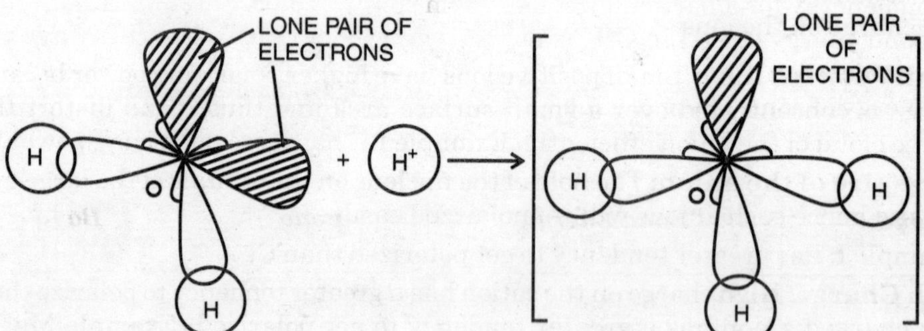

Fig. 3.11 Formation of hydronium ion

3.3.5 Covalent Bond Characteristics

The important characteristics of a covalent bond are:

1. Bond length
2. Bond Angle
3. Bond Energy

1. Bond Length: Bond length may be defined as the average equilibrium distance between the centres of the nuclei of the two bonded atoms. Since bond length is very small it is usually measured in Angstrom units ($1A^0 = 10^{-8}$cm) During the formation of diatomic molecule, let us say H_2, we have already seen that a stage is reached which corresponds to minimum energy. At this stage the atoms get bonded together. The distance between the nuclei of hydrogen atoms is called the bond length of H-H bond.

Bond length can be measured by certain physical methods such as X-ray diffraction in case of crystals.

The bond length of any multiple bond is shorter than that of single bond eg $C-C$ single bond length is 1.54 A° and $C=C$ (double bond) is 1.34 A° and $C\equiv C$ triple bond is 1.20 A°. Some of the common bond lengths are given in Table 3.4:

Table 3.4 Some common Bond lengths (A°)

Bond	*Bond length (A°)*	*Bond*	*Bond length A°*
H-H	0.74	C – C	1.54
Cl-Cl	1.98	C = C	1.34
Br - Br	2.28	C ≡ C	1.29
I-I	2.67	H-F	0.92
C-H	1.00	H-Cl	1.27
O-H	0.96	H-Br	1.41
C=O	1.20	H-I	1.61

2. Bond Angle: Covalent bond is formed by the overlap of atomic orbitals. The direction of overlap gives the direction of bond. Bond angle is defined as the internal angle between the bonded orbitals of the central atom in the molecule. For example the bond angles in case of H_2O, NH_3 and CH_4 are 104.5, 107° 109° 28′ respectively.

3. Bond Energy: The bond energy may be defined as the amount of energy required to break one mole of the bonds of that type in a substance in gaseous state. This energy is the measure of the strength of the bond. More is the energy required to break the bond, stronger is the bond. The value of bond energy for a particular bond is same only when present in similar molecules. For example, bond energy of C-H in CH_4 is not the same as in $CHCl_3$. Hence always average bond energy is considered.

Bond energy of some of the common bonds are given in Table 3.5 :

Table 3.5 Bond Energies

Bond	*Bond Energy kJ mol^{-1}*	*Bond*	*Bond Energy kJ mol^{-1}*
H–H	433	H–F	563
C–C	347	H–Cl	432
C = C	619.0	H–Br	366
C ≡ C	836.8	H–I	292

3.3.6 Characteristics of Covalent Compounds

1. **Physical State:** Generally covalent compounds are gases or liquids at room temperature because of weak intermolecular forces between the molecules. They are solids only when their molecular weights are very high.

2. **Crystal Structure:** A few covalent compounds exist in the solid state. Examples-Iodine, diamond and graphite. Generally crystals of covalent compounds are of three types-
   (i) Molecules in the crystals of inorganic substances like sulphur, iodine etc are held together by weak Vander Waals forces. These compounds are generally soft, easily fusible, and volatile.
   (ii) The second type of crystal structure results when atoms are linked with other atoms by covalent linkages throughout the crystal. For example in diamond, carbon is linked with four neighbouring carbon atoms held at the corners of a regular tetrahedron by covalent bonds. Since these covalent bonds are very strong, the crystals are very hard.
   (iii) The third type of structure includes the crystals which have lattice layers. Example graphite.
3. **Melting and boiling point:** Generally covalent compounds have low melting and boiling points because in majority of these compounds only weak Vander Waals forces are operative. Thus a small amount of energy is required to overcome these weak forces. In few cases such as diamond, silica, the structures are covalent giant lattices instead of discrete molecules and hence a large amount of energy is required to break this lattice. These compounds have high melting points and are very hard.
4. **Solubility:** Covalent compounds are normally insoluble in polar solvents but are soluble in non-polar solvents such as benzene and tetrochloromethane. Covalent compounds having giant molecules are practically insoluble in all solvents.
5. **Conductance:** Covalent compounds contain neither ions nor mobile electrons, so they do not conduct electricity in solid or fused state. On the other hand, substances like graphite which consist of separate layers conduct electricity as it has free electrons.
6. **Reaction Rates:** Reaction rates between covalent molecules are much slow compared to those between ionic compounds. Ionic reaction simply involves regrouping of ions whereas covalent reactions involve breaking a bond and substituting another group.

3.4 VALENCE SHELL ELECTRON PAIR REPULSION THEORY (VSEPR)

In 1940 Sidgwick and Powell reviewed the structures of various molecules known at that time. According to them, the shape of a molecule can be predicted from the number of electron pairs in the outer or valence shells of the central metal atom. The electron pairs arrange themselves around the central metal atom in such a way that there is minimum repulsion between them so that the molecule has minimum energy and maximum stability. Since there can be only one orientation of orbitals corresponding to minimum energy a molecule has invariably a definite shape i.e. a definite geometry.

In Sidgwick and Powell theory all the electron pairs whether bond pairs or lone pairs were considered equivalent. However this theory was further developed by Gillespie and Nyholm in 1957 to predict and explain molecular shapes and bond angles more exactly. This theory was called Valence Shell Electron Pair Repulsion (VSEPR) Theory. The basic points of this theory are as follows:

1. The central metal atom is bonded to the other atoms through electron pairs by single covalent bonds.
2. The electron pairs around the central metal atom are so arranged in space so as to minimize the coulombic repulsions between them, leading to stable arrangement.
3. The geometry of the molecule is determined by the total number of electrons pairs, bond pairs as well as lone pairs, surrounding the central metal atom.
4. When the central atom in a molecule is surrounded by bond pairs only, the molecule will have a regular shape or geometry as given in the following table.

Table 3.6: Molecular shapes predicted by VSEPR theory

No. of electron pairs	*Shape*	*Bond Angle*	*Example*
2	Linear	180°	Be F_2
3	Plane triangular	120°	BF_3
4	Tetrahedral	109° 28'	CH_4
5	Trigonal bipyramid	120° and 90°	PCl_5
6	Octahedral	90°	SF_6
7	Pentagonal biyramid	72° and 90°	IF_7

5. When the central atom in a molecule is surrounded by bond pair as well as lone pair the geometry is distorted or irregular. This is because a lone pair of electrons takes up more space around the central atom than a bond pair as lone pair is attracted to one nucleus whereas bond pair is under the influence of two nuclei. Consequently, the presence of lone pairs will alter the bond angle to a small extent which does not pertain to any regular geometry.
6. It is found that the repulsion between the lone pair electrons is greater than the repulsion between bond pair electrons. The order of decreasing repulsion being
   Lone pair – Lone pair > Lone pair – bond pair > Bond pair-bond pair.
7. Electronegativity difference between the central atom and the other atoms also influence the magnitude of repulsion.

Application of VSEPR Theory: In the light of above mentioned points, the geometry of various molecules can be predicted. Let us discuss a few examples–

1. **Shape of BeF_2 molecules:** In BeF_2, the central atom Be has two valence electrons. It shares three electrons with two flourine atoms i.e. the outer shell of Be has two electron pairs (bond pairs). According to VSEPR theory, for minimum energy and maximum stability, these bond pairs arrange themselves at an angle of 180° and molecule in such case will be linear.

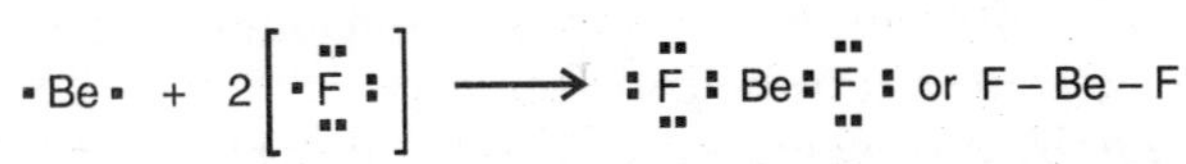

2. **Shape of BF_3:** Boron, the central metal atom has 3 electrons in the valence shell which are used to form covalent bonds with three flourine atoms Hence the central atom has three bond pairs which are directed towards the three corners of a equilateral triangle.

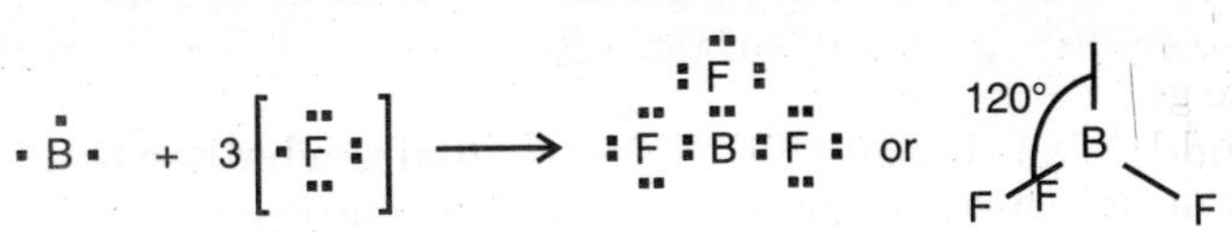

3. **Shape of CH_4:** Central atom carbon has four valence electrons which it shares with four hydrogen atoms. Hence 4 bond pairs are present around central atom giving tetrahedral shape to the molecule with a bond angle of 109°, 28′.

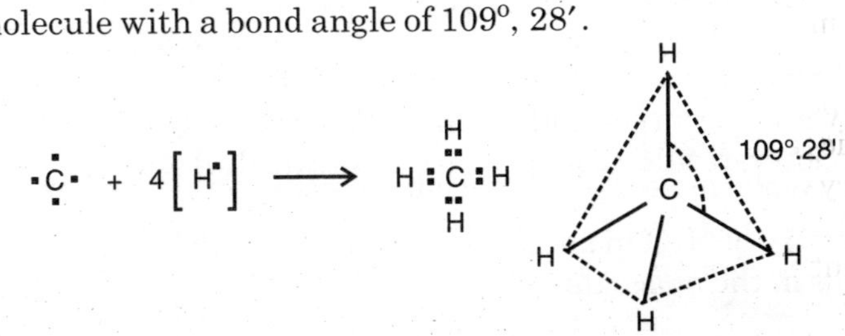

4. **Shape of PCl_5:** Phosphorus has five valence electrons which it shares with five chlorine atoms to form five bond pairs. Hence the geometry of the molecule is triagonal bipyramidal with bond angles of 120° and 90°.

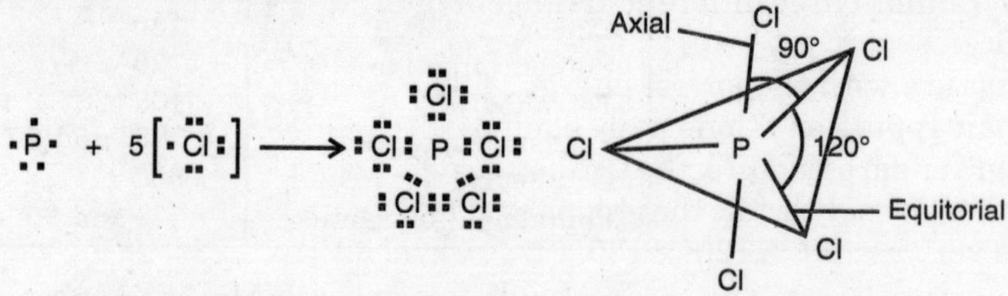

5. **Shape of SF_6:** Sulphur has six valence electrons and forms six bonds with flourine atoms. The geometry of the molecule is octahedral, with a bond angle of 90°.

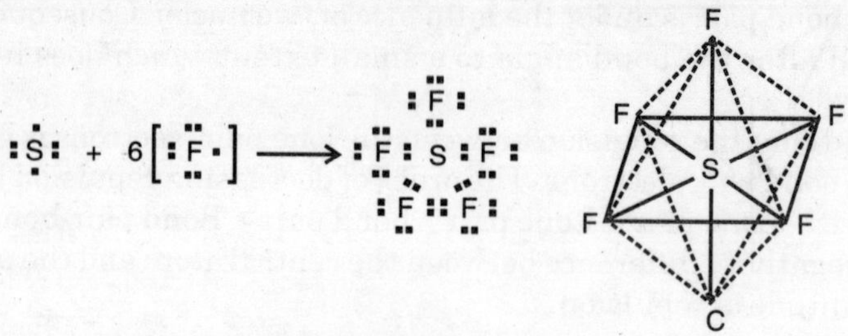

Now we will discuss the geometry of the molecules having bonding as well as lone pairs of electrons.

NH_3 Molecule: In NH_3, N is the central atom and has five valence electrons (Z=7) Three of these are used to form bonds with three hydrogen atoms where as two electrons are not involved in bonding and therefore constitute lone pair. Thus outer shell has 4 electron pairs, i.e three bond pairs and one lone pair. Presence of four electron pair should give the molecule a tetrahedral geometry but due to the presence of one lone pair, the geometry is slightly distorted from tetrahedral. Since lone pair-bond pair repulsion is more than the bond pair bond pair– repulsion, it causes three N–H bonds to move slightly closer to each other and bond angle decreases from 109°.28' to 107°.48' and molecule acquires pyramidal shape.

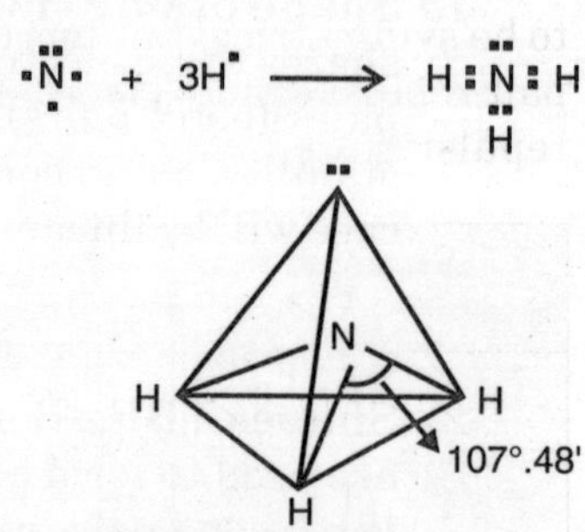

Fig. 3.12

H_2O Molecule: In water oxygen (Z=8) has 6 valence electrons. Two of which are used to form bonds with hydrogen. The other four outer electrons on oxygen are non-bonding i.e lone pairs. Thus, a total of four electron pairs are present in the valence shell out of which two pairs are bond pairs and two are lone pairs. Repulsion between lone pair-lone pair > lone pair-bond pair>bond pair-bond pair, therefore the geometry is distorted or irregular. The two lone pairs in water move away from each other while the two O–H bonds are forced to come closer, resulting in decrease of tetrahedral angle from 109°.28' to 104°.5'.

Thus the geometry of H_2O molecule is V-shaped as there are only two O–H bonds in the molecule.

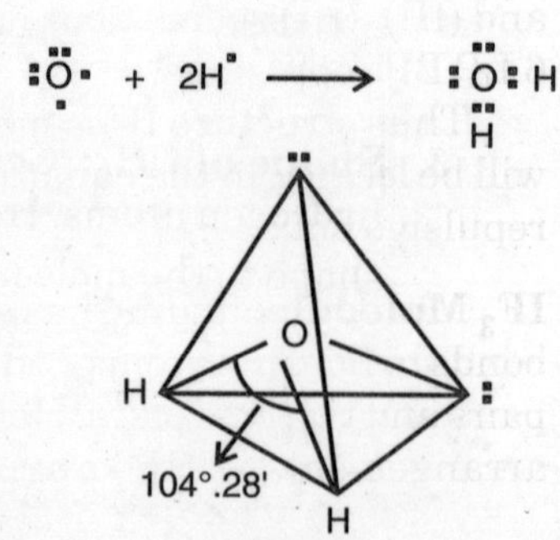

Fig 3.13

BrF_3 Molecule: Bromine (Z = 35) has seven electrons in its valence shell. Three electrons form bond to flourine atoms and four electrons do not take part in bonding. Thus in BrF_3 the bromine atom has five electron pairs i.e three bond pairs and two lone pairs. Hence the structure is trigonal bipyramid. Three different arrangements are theoretically possible as shown in figure 3.14.

The structure with the lowest energy will be the most stable one. Now we know that lone pair– lone pair repulsion > lone pair–bond pair repulsion > bond pair–bond pair repulsion. In trigonal bipyramidal structure, the two bond angles are 120° and 90°. Groups at 90° to each other repel each other strongly, while the groups at 120° repel each other less. The magnitude of repulsive interactions very as : 90° > 120° > 180°.

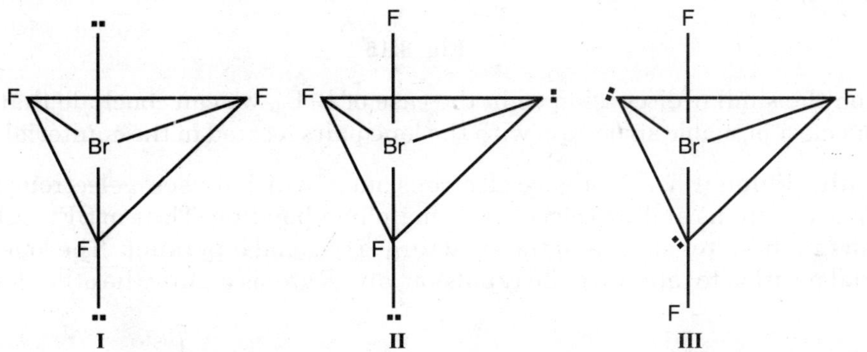

Fig. 3.14

Considering the repulsions at 90°, let us examine all the three structures. Structure (I) seems to be symmetrical but has 6 lone pair-bond pair repulsions. Structure (II) has one lone pair–lone pair repulsion and 3 lone pair–bond pair repulsion at 90°. Structure (III) has 4 lone pair-bond pair repulsions.

Structure	*LP-LP Repulsions at 90°*	*L-P-BP Repulsion at 90°*	*BP BP Repeat 90°*
I	0	6	0
II	1	3	2
III	0	4	2

Structure (II) has one lone pair-lone pair repulsion and hence least stable. Out of structure (I) and (III), (III) seems to be more stable as it has only 4 LP-BP repulsion whereas structure (I) has 6 LP BP repulsions.

Thus structure III is most probable. Generally, if lone pairs occur in trigonal bipyramid they will be located in the equatorial positions rather than the apical positions since this minimize the repulsive force.

IF_3 Molecule: Iodine has seven electrons in its valence shell. Out of which three electrons form bonds to flourine atoms and four electrons do not take part in bonding Thus, there are three bond pairs and two lone pairs. Hence the structure is trigonal bipyramid. Similar to BrF_3, three different arrangements are theoretically possible, as shown in Fig. 3.15

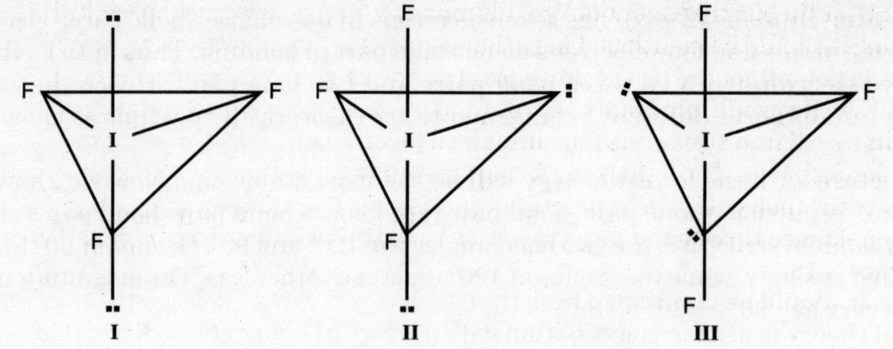

Fig. 3.15

Based on the similar discussion as in the case of BrF_3, we can conclude that structure III would be the most probable structure, with the lone pairs located in the equitorial positions.

SF_4 Molecule: Sulphur has 6 valence electrons out of which it uses 4 electrons to form bonds with four flourine atoms and two electrons remain non-bonding. Thus in SF_4, sulphur has five electron pairs in the outer shell, and the structure is trigonal bipyramid. The lone pair occupies the equitorial position to minimize the repulsion, and F atoms occupy the other four corners.

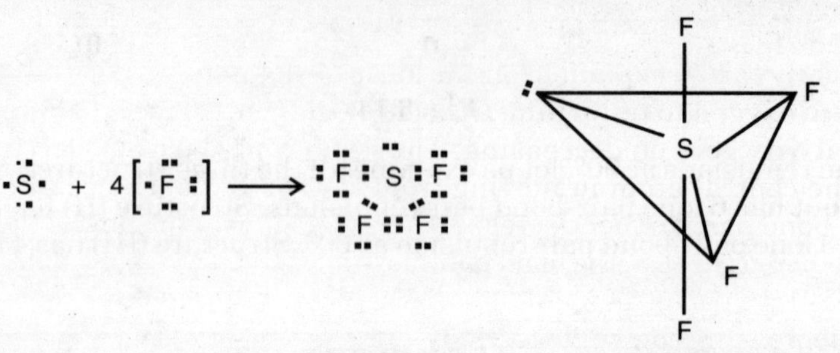

Fig. 3.16

$SnCl_2$ Molecule: The outer electronic configuration of central metal tin is $5s^2$, p_x^1, p_y^1. It has four valence electrons out of which it uses two electrons to form two bonds with chlorine atoms. Thus it has two bond pairs and one lone pair. The geometry should be trigonal planar. But due to the presence of lone pair, the two bonds move closer to each other resulting in a bond angle less than 120° having a V shaped geometry.

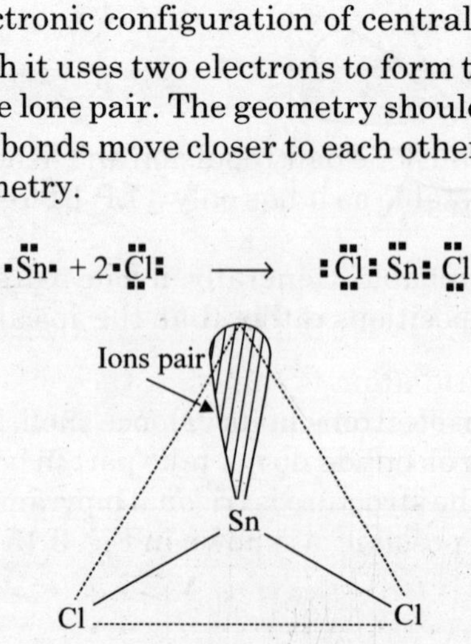

Fig. 3.17

CO_3^{2-} Ion: So far, in our treatment of VSEPR model, we have not considered the molecules with multiple bonds. It should be mentioned here that geometry of the molecule having double bonds, is not influenced by the presence of double bonds. As a general rule, for the VSEPR model, multiple bonds are counted as one effective electron pair.

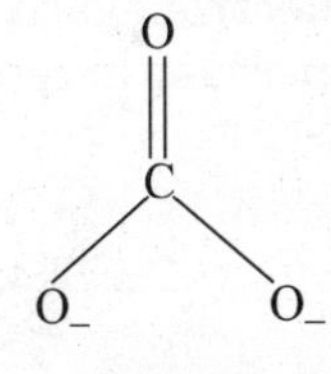

Fig. 3.18

Carbon in its excited state, has four unpaired electrons, three of which are used to form sigma bonds. The fourth one forms a pi bond, but it does not influence the geometry of the molecule. The geometry is decided by the three electron pairs around carbon. Hence three bond pairs give rise to trigonal planar geometry as shown in figure 3.18.

However, it should be mentioned here that structure of CO_3^{2-} ion is satisfactorily explained when VSEPR theory is used in combination with concept of hybridization, which is discussed in the next section.

Effect of Electronegativity on Geometry of the Molecule: In the hydrides of group VI elements i.e H_2O, H_2S, H_2Se and H_2Te, the central atom in all the four molecules is surrounded by four electron pairs two of which are bonding and two are non-bonding (lone pairs) but still the bond angles are different.

H_2O	H_2S	H_2Se	H_2Te
04.5°	92.5°	91.0°	89.5°

The above anomaly can be explained on the basis of electronegativity difference. As we move down the group from oxygen to tellurium, the size of the central metal atom goes on increasing and its electronegativity goes on decreasing. Thus the bonded pair of electron shifts more and more away from the central atom in moving from H_2O to H_2Te. Hence, the force of repulsion between bond pair-bond pair further decrease, thereby reducing the angle. In H_2O, the bond pair lie closer to oxygen whereas in H_2S, the bond pair move slightly away from sulphur (electronegativity of sulphur is less than that of oxygen) and hence the bond pair-bond pair repulsion is less in case of H_2S than H_2O, decreasing the angle to 92.5°. For the same reason, as we move from Se to Te, the angle decreases further.

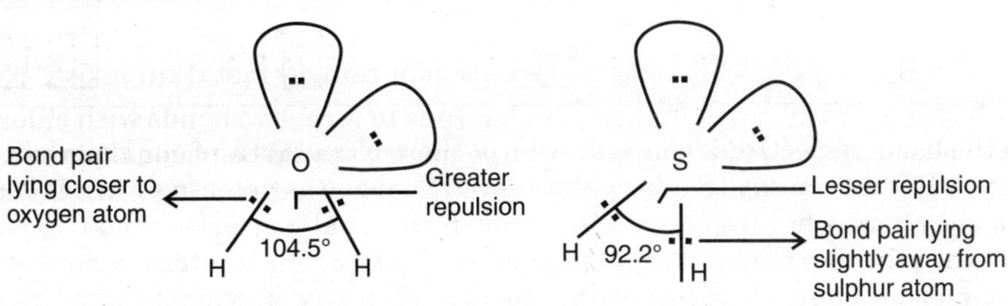

Fig. 3.19

On the contrary when central atom is same but the bonded atoms are different, the bond angle increases with decrease in electronegativity of the bonded atoms. This can be illustrated by considering the case of phosphorus halides.

Compound	PF_3	PCl_3	PBr_3	PI_3
Bond Angle	98°	100°	101.50°	102°

It can be seen that as we move from PF_3 to PI_3, the size of the bonded atoms increases and electronegativity decreases. Thus the bonded pair lie more towards the central atom causing increase in repulsion between bond pairs and hence bond angle increases.

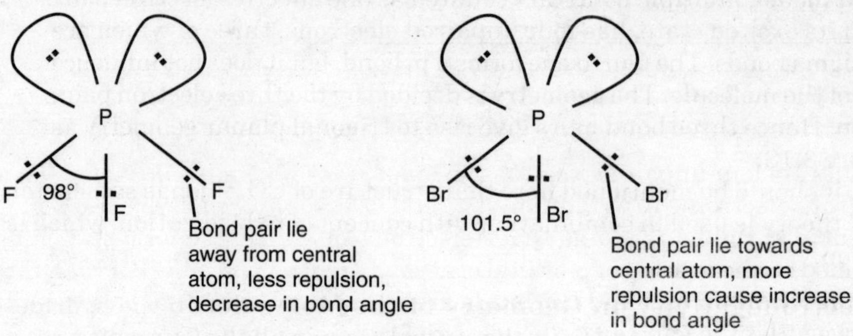

Fig. 3.20

3.5 HYBRIDIZATION

In the theories of covalent bonding, discussed in the previous sections, it is discussed that half filled atomic orbitals undergo bond formation. The number of half filled atomic orbitals of the element is referred to as covalency of the element. However this concept could not explain the formation of beryllium, boron and carbon compounds. The electronic configuration of these elements can be represented schemetically as-

	Z	*1s*	*2s*	*2p*
Beryllium	4	↑↓	↑↓	
Boron	5	↑↓	↑↓	↑ | |
Carbon	6	↑↓	↑↓	↑ | ↑ |

Thus, on the basis of electronic configuration of these elements and our knowledge of orbital overlap theory (VBT) we can say that beryllium should behave as noble gas since it does not have any unpaired electrons, whereas boron and carbon have covalency of one and two respectively. But this is not the case. Beryllium forms compounds like BeF_2, BeH_2 etc. in which it is bivalent. Similarly boron forms a number of compounds like BCl_3, BF_3 etc in which it is trivalent whereas carbon forms a large number of compounds in which it is tetravalent.

To explain this anomaly it was assumed that some of the paired electrons become unpaired before the atoms participate in bonding. These unpaired electrons then become available for the formation of covalent bonds. The energy required for excitation will be available from heat of reaction when covalent bonds are formed. For example in carbon one of the electron from 2s orbital jumps to 2p orbital-

	2s	2p
Carbon (Ground State)	↑↓	↑ ↑
Carbon (Excited State)	↑	↑ ↑ ↑

Hence, a total of four unpaired electrons are available for bonding. But now one more problem arises. With this electronic distribution, it would be expected that the three p electron cloud would be oriented at right angles and that *s*-electron cloud will be spherically symmetrical. Thus the bonds formed by the overlap of p-orbital of carbon atom will be different than the bond formed by the *s* orbital of carbon as *s* and *p* orbitals are of different energies. Experimental evidence, however, showed that the four carbon bonds are of equal strength i.e **they are equivalent in all respects**.

In order to explain the formation of four equivalent bonds, the **concept of hybridization** was introduced. It is assumed that one *s* and 3*p* orbitals of carbon get mixed up and form four equivalent hybrids.

The phenomenon of mixing up of atomic orbitals of similar energies, so that a redistribution of energy takes place between them, resulting in the formation of new orbitals of equal energies and identical shapes is known ashybridization. The new orbitals thus formed are known as hybrid orbitals. The bonds formed by these orbitals are stronger than the bonds formed by pure *s*, *p* or *d* orbitals.

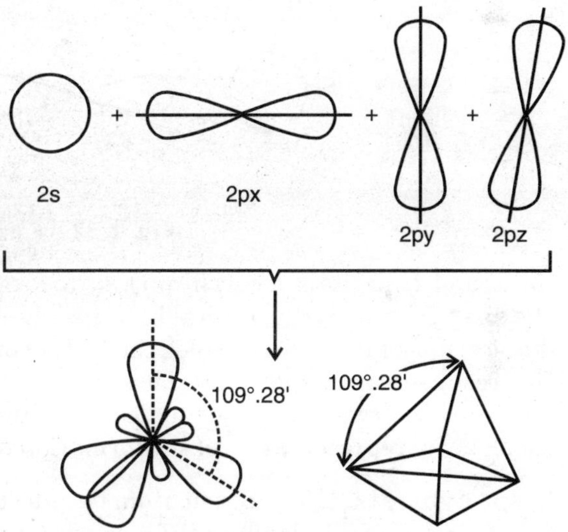

Fig. 3.21

The four new bonds formed in the case of carbon are directed towards the four corners of a regular tetrahedron. Diagrammatically it can be represented as shown in fig. 3.21.

The concept of hybridization is hypothetical but is extremely helpful in predicting the structure of various compounds. For understanding the phenomenon of hybridization, following points should be considered.

1. Hybridization involves the orbitals of similar energies belonging to the same atom or ion.
2. Number of hybrid orbitals produced is equal to the number of orbitals that participate in hybridization process.
3. Hybridization not only involves the half filled orbitals, even completely filled orbitals with slightly different energies can also participate in hybridization.
4. It is not necessary that all the half filled orbitals must hybridize.
5. Hybridization never takes place in isolated atom but it occurs only at the time of bond formation.
6. Type of hybridization indicates the shape of the molecule.

3.5.1 Types of Hybridization

1. sp hybridization: The combination of an s and a *p*-orbital leads to two hybrid orbitals known as *sp* orbitals. *sp* hybridization is well examplified by BeF_2 molecule. The electronic configuration of Be is $1s^2$, $2s^2$. It is assumed that one of the electron from 2*s* orbital jumps to 2*p* orbital in excited state of beryllium.

	1*s*	**2*s***	**2*p***
Be (Ground State)	↑↓	↑↓	
Be (Excited State)	↑↓	↑	↑
Hybridization	↑↓	↑	↑
			sp hybridization

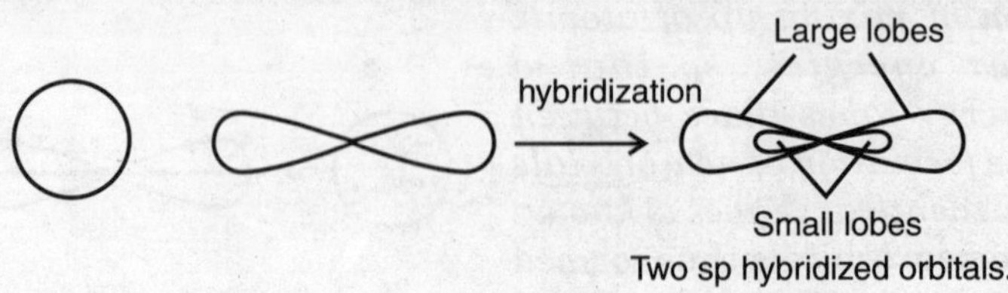

Fig. 3.22 Formation of sp orbitals

2*s* and 2*p* orbitals of beryllium mix to form two *sp* hybridized orbitals having equal energies. The size of the *sp* hybrid orbital has one lobe of larger size and other a smaller one. The broader shape of one of the lobes enables the hybrid orbital to bring about overlapping more effectively. This results in a stronger bond. These hybrid *sp* orbitals overlap with the *2p* orbitals from flourine atoms to form two bonds. The molecule is linear with a bond angle = 180°. Fig. 3.22 represents the pictorial representation of the formation of BeF_2 molecule.

2. sp² hybridization: Boron atom has electronic configuration as $1s^2$, $2s^2$, $2p_x^1$. It is assumed that 1 electron from 2*s* orbital of boron jumps to 2*p* orbital. Thus boron in the excited state has three unpaired electrons. The three orbitals i.e. 1s and two p orbitals hybridize to give three sp^2 hybridized orbitals which are coplanar and are directed at angles 120° to each other. sp^2 hybridized orbitals are able to form stronger bonds than *s* or *p* orbitals because of their larger size, they overlap more effectively.

	1*s*	**2*s***	
Boron atom (Ground state)	↑↓	↑↓	↑
Boron atom (Excited state)	↑↓	↑	↑ ↑
Hybridization	↑↓	↑	↑ ↑
			sp^2 hybridization

The three sp^2 hybridized orbitals overlap with p orbitals from three flourine atoms to form BF_3 molecule.

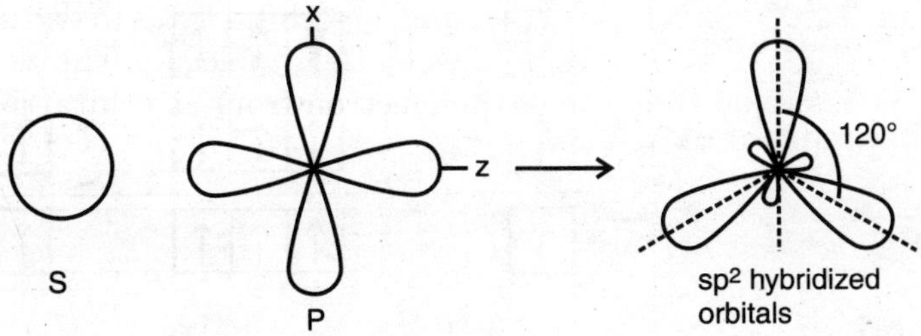

Fig. 3.23

Other examples of sp^2 hybridization are SO_2, CO_3^{2-} etc.

3. sp³ hybridization: The electronic configuration of carbon atom is $1s^2, 2s^2, 2p_x^1, 2p_y^1$. One of the electrons from $2s$ orbital jumps to $2p_z$ orbital thus forming four orbitals having unpaired electrons. These four orbitals, one s and three p orbitals hybridize to form four sp^3 hybridized orbitals of equal energy.

	1s	**2s**	**2p**
C atom (Ground State)	↑↓	↑↓	↑ ↑
C atom (Excited State)	↑↓	↑	↑ ↑ ↑
Hybridization	↑↓	↑	↑ ↑ ↑ (sp^3 hybridization)

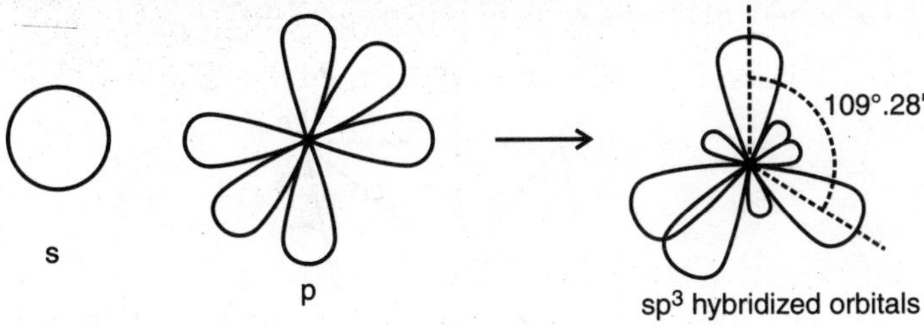

Fig. 3.24

Four sp^3 hybrid orbitals overlap with four $1s$ orbitals of hydrogen atoms to form four bonds in CH_4. The bond angle = 109°28′.

4. sp³ d hybridization: The combination of one s, three p and one d orbital to form five hybrid orbitals of equal energy is known as sp^3d hybridization. Example PCl_5 molecule.

	3s	3p	
Phosphorus (Ground State)	↑↓	↑ | ↑ | ↑	
Phosphorus (Excited State)	↑	↑ | ↑ | ↑	↑ | | | |
Hybridization	↑	↑ | ↑ | ↑	↑ | | | |
	sp^3d		

The shape of the molecule is trigonal bipyramid, the equitorial and axial angles being 120° and 90° respectively. (Fig. 3.25)

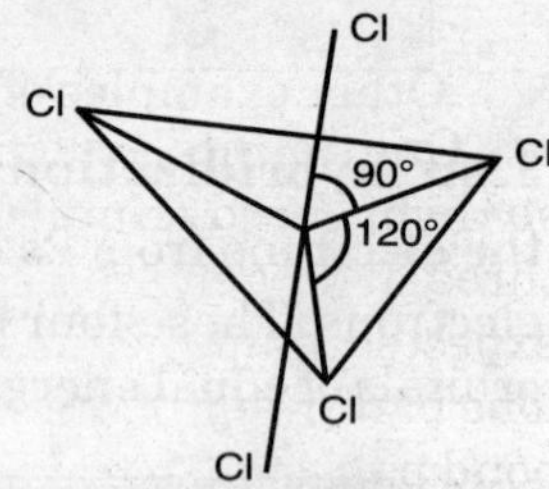

Fig. 3.25

5. sp^3d^2 hybridization: The combination of one *s*, three *p* and two *d* orbitals to form six hybrid orbitals of equal energy is known as sp^3d^2 hybridization. An example, is the structure of SF_6.

S (Ground State)	↑↓	↑↓ | ↑ | ↑	| | | |
S (Excited State)	↑	↑ | ↑ | ↑	↑ | ↑ | |
Hybridization	↑	↑ | ↑ | ↑	↑ | ↑ | |
		sp^3d^2	

The molecule has octahedral shape with bond angle of 90°.

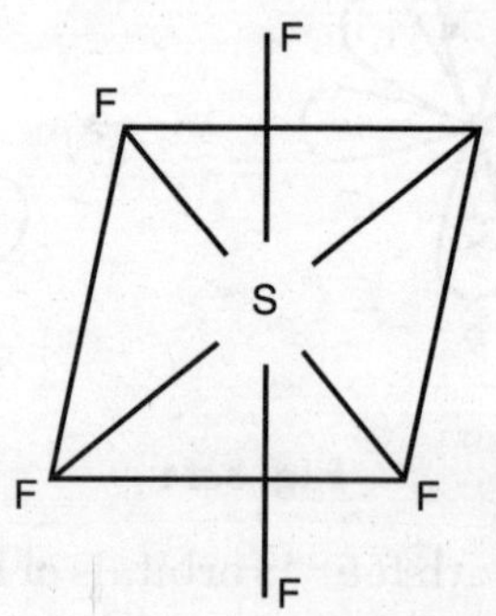

Fig. 3.26

So far we have discussed the molecules in which orbitals having unpaired electrons are involved in hybridization. But hybridization not only involves the half filled orbitals, *even completely filled orbitals with slightly different energies can participate in hybridization.* This can be illustrated by considering the following examples-

Hybridization in NH_3 Molecule

Nitrogen has electronic configuration $1s^2, 2s^2, 2p_x^1, 2p_y^1\ 2p_z^1$

	1s	2s	2p
Nitrogen	↑↓	↑↓	↑ ↑ ↑
Hybridization	↑↓	↑↓	↑ ↑ ↑
		sp^3 hybridization	

One *s* orbital which is completely filled and three half filled *p* orbitals hybridize to form four sp^3 hybridized orbitals. Out of four hybridized orbitals, three orbitals are used for bonding with three hydrogen atoms, while fourth orbital contains a lone pair. Due to sp^3 hybridization the expected shape is tetrahedral, but actual shape is pyramidal. This is because of the presence of lone pair in one of the orbitals. The force of repulsion between lone pair-bond pair is greater than bond pair-bond pair and hence the shape of the molecule is distorted from tetrahedral and bond angle reduces to 109.5° to 107°.

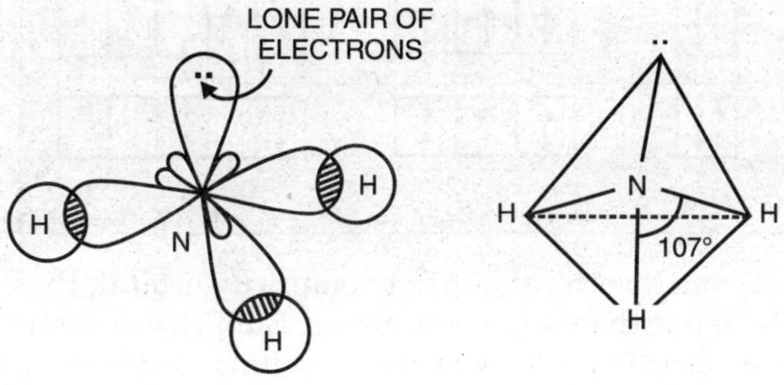

Fig. 3.27 Pyramidal structure of NH_3.

Hybridization in H_2O Molecule

Oxygen has electronic configuration $1s^2, 2s^2, 2p_x^2, 2p_y^1\ 2p_z^1$

	1s	2s	2p
Oxygen atom	↑↓	↑↓	↑↓ ↑ ↑
Hybridization	↑↓	↑↓	↑↓ ↑ ↑
			sp^3 hybridization

One *s* and three *p* orbitals hybridize to form four sp^3 hybridized orbitals. Out of these four hybridized orbitals, two contain lone pairs while the other two are utilized for bond formation with two hydrogen atom. The expected structure is tetrahedral but due to the presence of two lone pairs the geometry is V shaped. Presence of two lone pairs distort the bond angle which reduces to 104.5° from 109.5°.

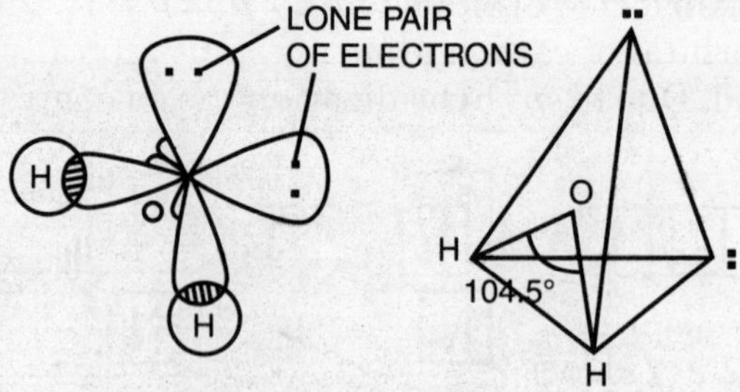

Fig. 3.28 Shape of water molecule

Hybridization in SF_4 molecule: The outer electronic configuration of central atom sulphur is $3s^2$, $3px^2$, $3py^1$, $3pz^1$.

	3s	*3p*	*3d*
S (Ground State)	↑↓	↑↓ ↑ ↑	☐ ☐ ☐ ☐ ☐
S (Excited State)	↑↓	↑ ↑ ↑	↑ ☐ ☐ ☐ ☐
Hybridization	↑↓	↑ ↑ ↑	↑ ☐ ☐ ☐ ☐
		sp^3d	

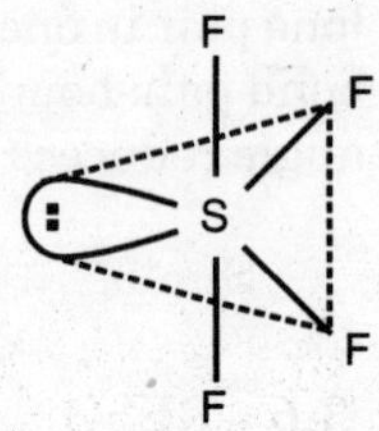

Fig. 3.29

One of the electron from 3*p* orbital jumps to empty 3*d* orbital, thus making available four unpaired orbitals to overlap with four flourine atoms. But prior to bond formation, one *s*, 3*p* and one *d* orbitals hybridize to form five sp^3d hybridized orbitals. Out of these five, one accommodates the lone pair and the other four form bonds with flourine atoms. The shape of the molecule is trigonal bipyramidal. The expected bond angles are 90°, 120° and 180°. But due to the presence of lone pair in one of the hybridized orbital, the angle reduces to 89°, 118° and 177° instead of 90°, 120° and 180° respectively. The lone pair occupies equatorial position.

Hybridization in $SnCl_2$ molecule: The outer electronic configuration of central atom tin is $5s^2$, $p_x^{\ 1}$, $p_y^{\ 1}$.

Sn (Ground State)	↑↓	↑ ↑ ☐
Hybridization	↑↓	↑ ↑ ☐

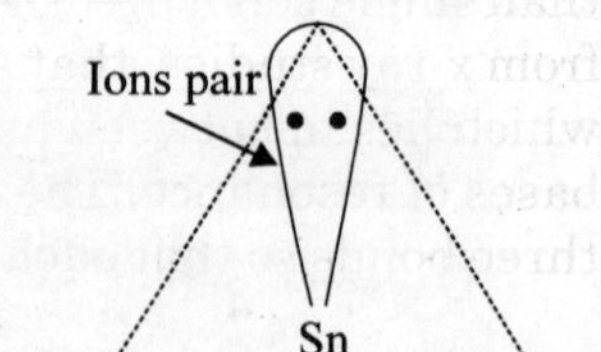

Fig. 3.30

Thus sp^2 hybridizatin takes place. Out of the three hybrid orbitals one accomodates the lone pair and the other two form bonds with chlorine atoms.The shape of the molecule is trigonal planar.The expected angle is 120° but due to the presence of lone pair the angle is reduced and $SnCl_2$ has a V shaped geometry. (Fig. 3.30)

Hybridization in CO_3^{2-} ion: The outer electronic configuration of carbon atom in ground state is $2s^2, 2p_x^1, 2p_y^1$ and in excited state it is $2s^1, 2p_x^1, 2p_y^1, 2p_z^1$. Out of three oxygen atoms of CO_3^{2-}

ion, two carry one additional electron each, so that two of the oxygen atoms contain only one half filled *p* orbital. The third oxygen atom has 2 unpaired electrons. Carbon atom undergoes sp^2 hybridization. The three hybridized orbitals form σ bonds with three oxygen atoms as shown, whereas unhybridized p orbital of carbon atom overlaps laterally with *p* orbital of one of the oxygen atom to form π bond. Due to sp^2 hybridization the geometry of the ion is trigonal planar with bond angle of 120°.

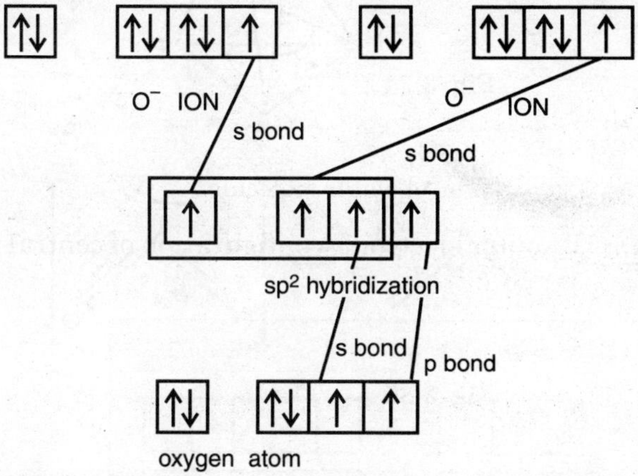

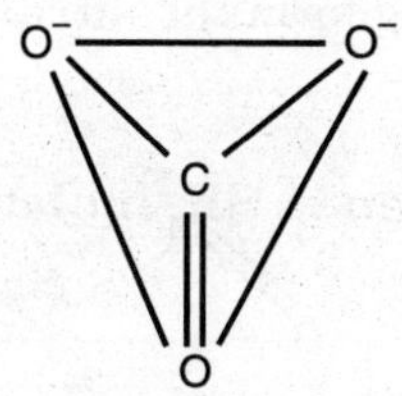

Fig. 3.31

Similarly hybridization of other molecules can be explained.

3.6 RESONANCE

In case of certain molecules, a single structure, as elucidated by various concepts, cannot explain all the properties of the molecule. The molecule is then represented by more than one structures, which explain most of its properties, but none can explain all the properties of the molecule. The actual structure of the molecule is in between of all these contributing structures and is called the resonance hybrid of all the contributing structures. This phenomenon is called **resonance.** The contributing structures are called **resonating structures**. The phenomenon of resonance can be illustrated by taking the example of CO_3^{2-} ion. The structure of this ion is planar triangular with bond angles of 120°. Central atom carbon uses sp^2 orbitals. In the structure of CO_3^{2-}, there are three carbon to oxygen bonds, two of which are single bonds and one is double bond. Evidently as double bond is shorter than single bond, in carbonate ion, two different bond lengths should be observed. But it is known from x-ray studies, that all the three carbon to oxygen bonds in carbonate ion are of equal length which lies in between usual single and double bond lengths. This can only be explained on the bases of resonance. The double bond in CO_3^{2-} ion is not localized but is spread out, over all the three bonds so that each bond has a bond order of 1/2. The carbonate ion :

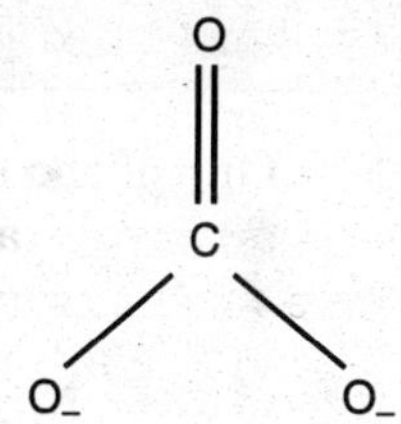

Fig. 3.32

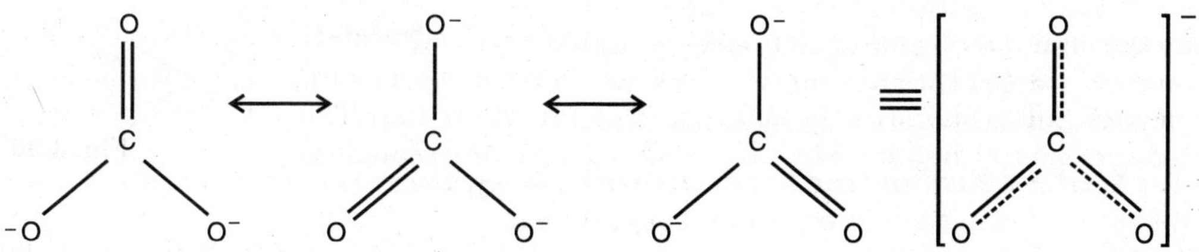

Fig. 3.33

It should be emphasized here that the contributing structures do not actually exist. The true structure is somewhere in between and is called **resonance hybrid.** The true structure has no pictorial representation.

Similarly resonance structure for ozone O_3 can be written as-

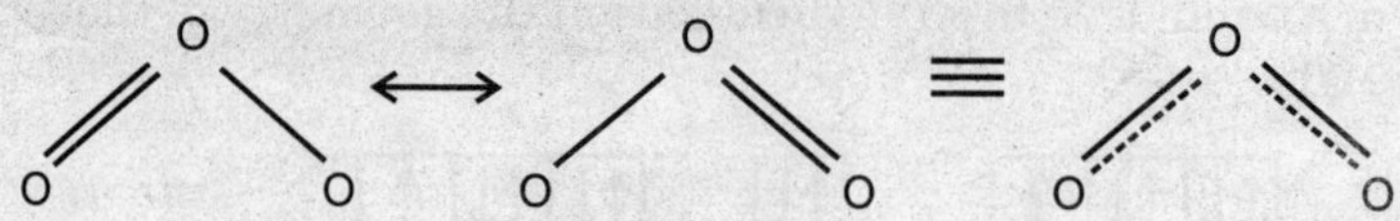

Fig. 3.34

Resonating forms of Nitrite ion

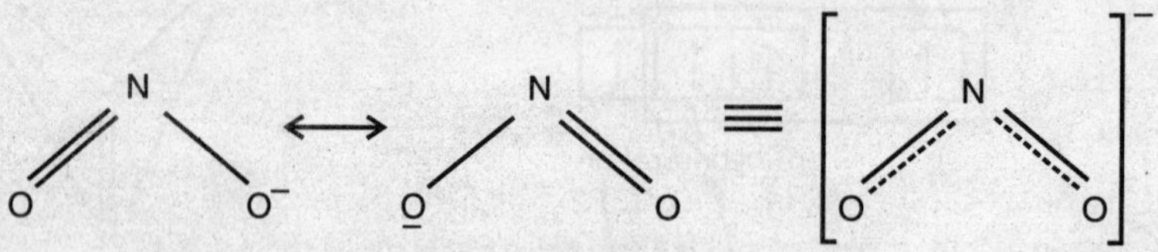

Fig. 3.35

Resonating forms of SO_3

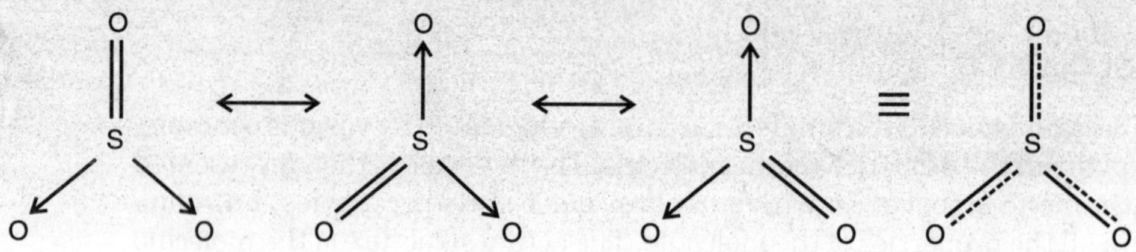

Fig. 3.36

Resonance Energy: It is found that the actual or true structure of the molecule is more stable than any of the contributing structure i.e. the resonance hybrid has lower energy than any of the contributing structures. In otherwords we can say that resonance brings extra stability to the system. *The difference in energy of the resonance hybrid and the most stable contributing structure is called resonance energy*. Resonance energy can be calculated from heat of hydrogenation and heat of combustion values. For example resonance energy of CO_2 is 159.5 $kJmol^{-1}$, it means the actual molecule of CO_2 is more stable than the most stable contributing structure whose energy is calculated theoretically, by 159.5 $kJmol^{-1}$.

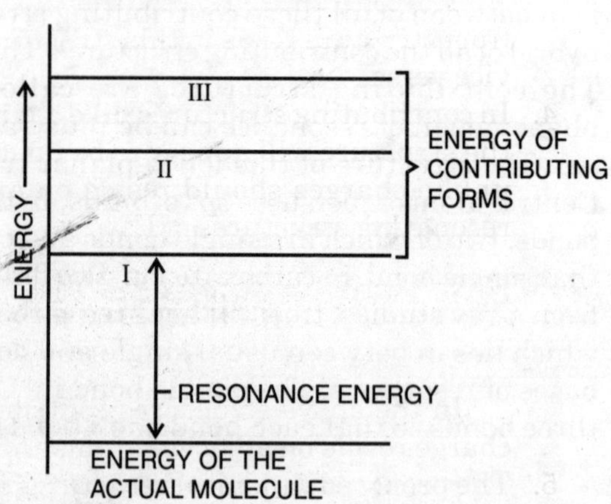

Fig. 3.37 Concept of Resonance Energy

Rules for Writing Resonating Structures: While writing resonance structures for various molecules, the following conditions should be satisfied :

1. The contributing structure should have the same relative positions of atoms. They should differ only in the position of electrons, eg

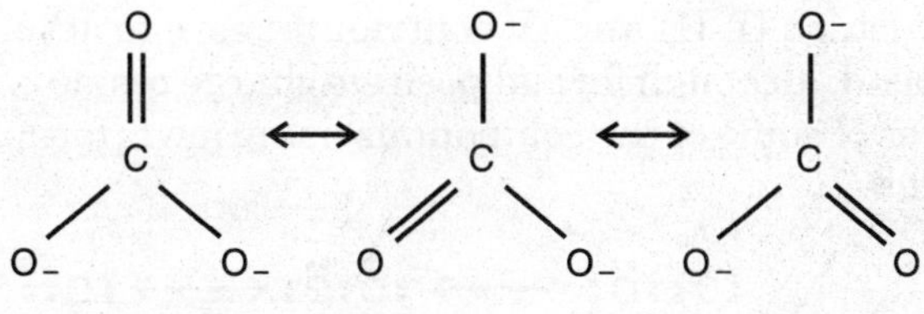

Fig. 3.38

Examples:

Out of the following three structures for NNO, II is not the resonating structure for NNO, because the position of oxygen is changed.

I II III

(Not a resonating Structure)

2. Number of unpaired electrons, if any, in the molecule must remain the same in all the forms. The resonating structure of O_2 cannot be represented as-

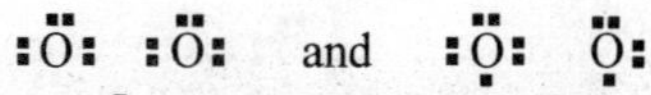

I II

These two structures are not the resonating structures of oxygen because the number of unpaired electrons in the two structures is different.

3. Contributing structures should be such that the negative charge resides on the most electronegative atom and positive charge on the most electropositive atom.eg-

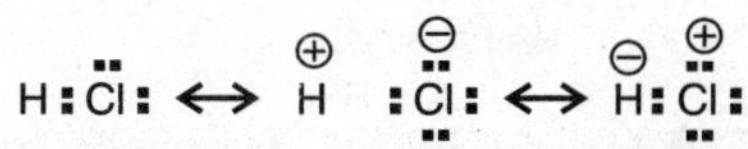

I II III

The contributing structures can be I and II whereas there is practically no contribution from structure III as in this structure positive charge resides on electronegative atom and vice versa.

4. In contributing structures, like charges should not reside on the adjacent atom otherwise the structure will appear to be unstable as a result of electrostatic repulsion. Similarly, unlike charges should reside on atoms close together in contributing structures, e.g- resonating structures of HN_3.

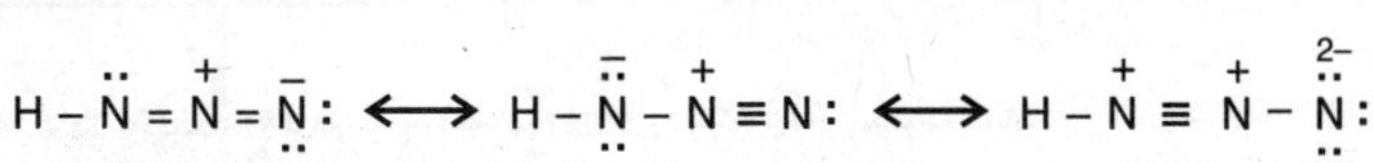

I II III

In the above structures contributing structure III is unimportant because positive charge reside on adjacent atoms.

5. The proposed canonical structures should have maximum number of bonds. For e.g., the contributing structures of BF_3

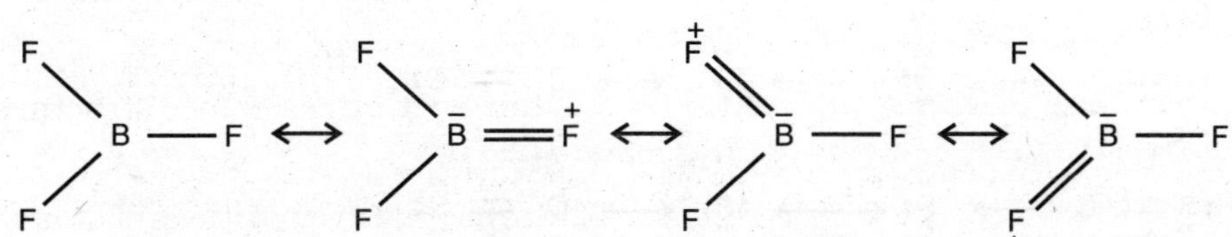

I II III IV

Out of these the structure II, III and IV contribute more to actual structure because these contain one additional bond, although formal positive charge reside on electronegative atom.

Resonance structures of some of the compounds using lewis formulae are:

(i) Carbon monoxide

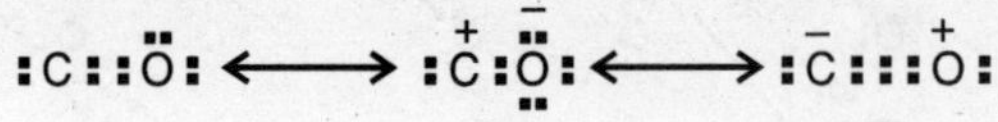

I II III

All the three structures are important.

(ii) Nitrate ion

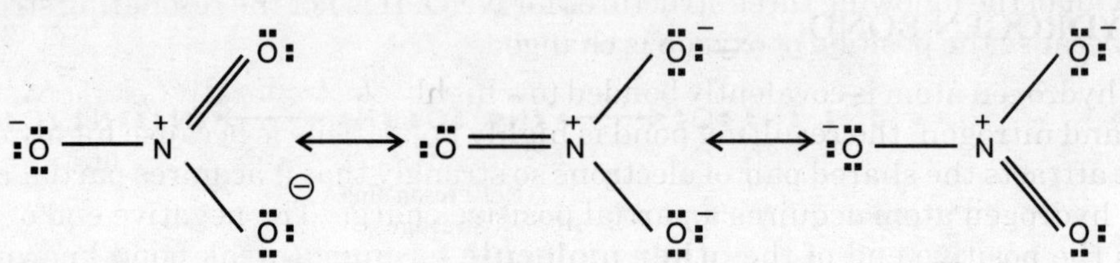

I II III

(iii) Nitric Acid

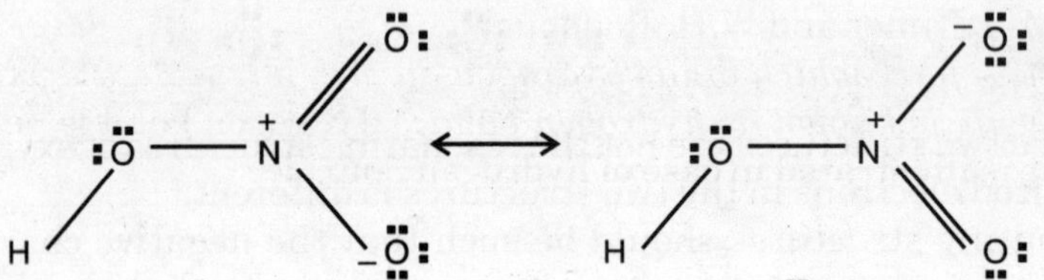

I II

(iv) Carbonyl group

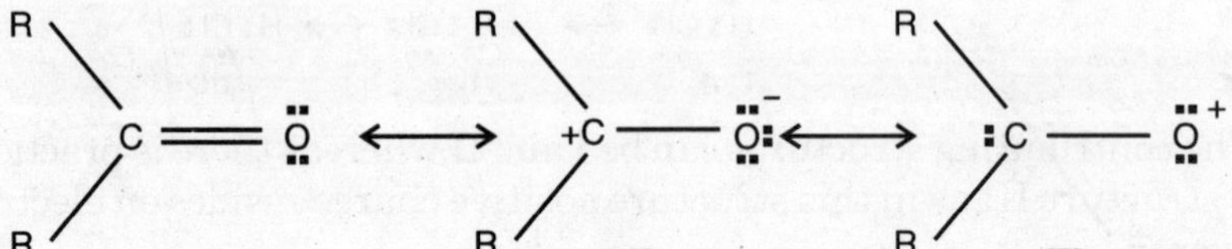

I II III

(v) Azide ion

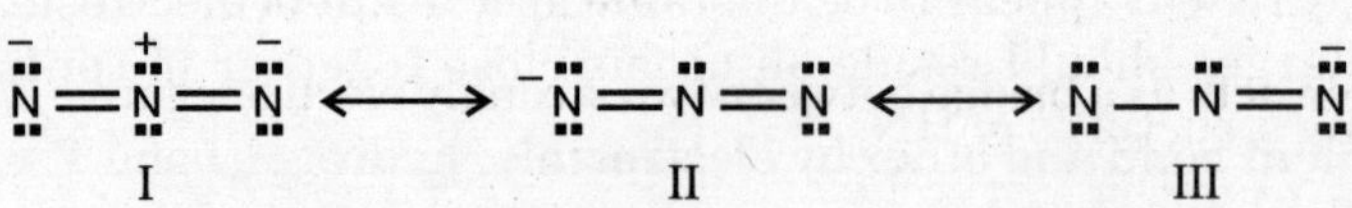

I II III

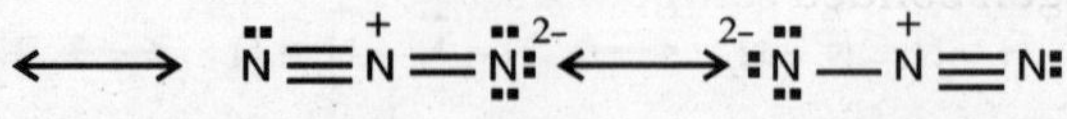

IV V

Structure I is the most stable form.

(vi) SO_2

(vii) CO_2

(viii) CS_2

(ix) $SOCl_2$

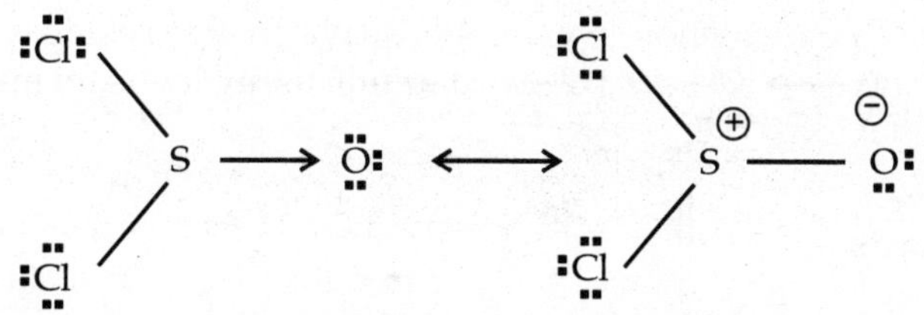

3.7 HYDROGEN BOND

When a hydrogen atom is covalently bonded to a highly electronegative element such as flourine, oxygen and nitrogen, the resulting bond is highly polar. This is because highly electronegative element attracts the shared pair of electrons so strongly that it acquires partial negative charge and the hydrogen atom acquires a partial positive charge. The negative end of **one molecule** attracts the positive end of the **other molecule** forming a weak bond known as **hydrogen bond**. Thus hydrogen atom is able to act as a link between two electronegative atoms. This phenomenon was first considered by T.S. Moore T.F. Winwill and P.Pfeiffer. Its applicability was realized by W.M. Latimer and W.H. Rodebush.

"*The attractive force which binds hydrogen atom of one molecule with electronegative atom of another molecule is known as hydrogen bond.*" Hydrogen bond is generally represented by dotted line. This is illustrated in case of hydrogen flouride.

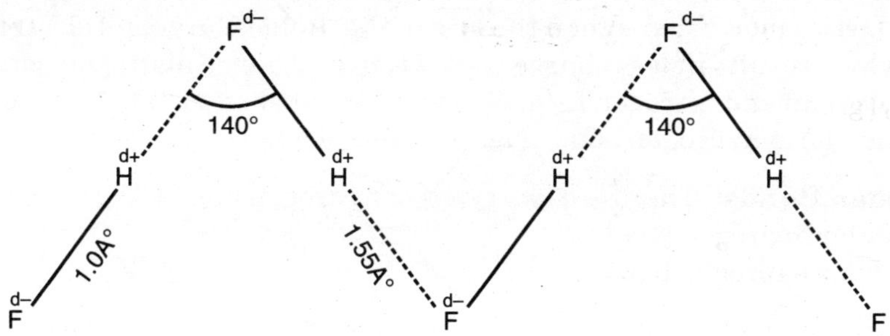

Fig. 3.39

Thus hydrogen acts as a bridge between two strongly electronegative atoms holding one flourine atom by the covalent bond and other by electrostatic hydrogen bond.

Examples of hydrogen bonded compounds:

Ammonia

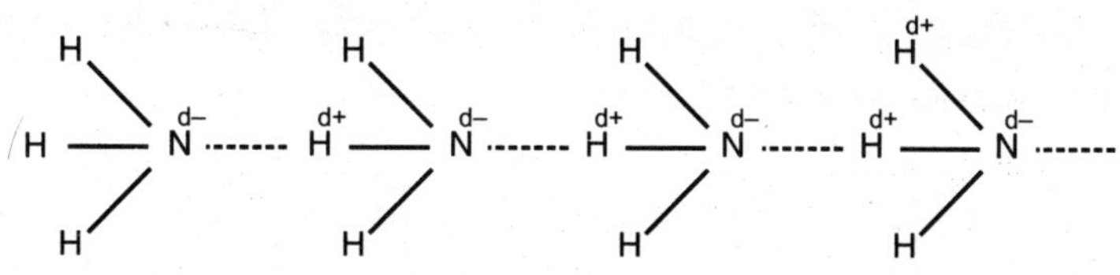

Water

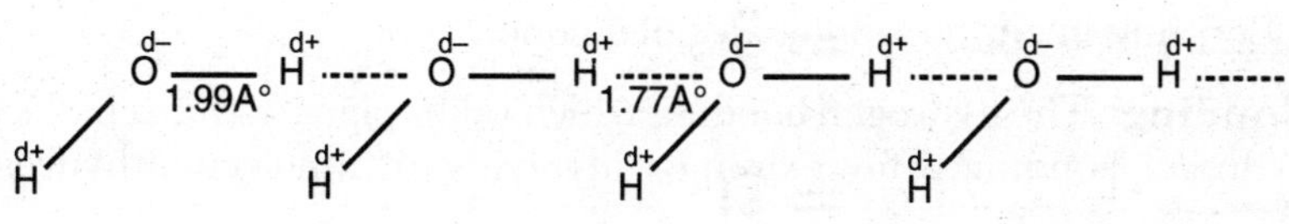

Alcohol

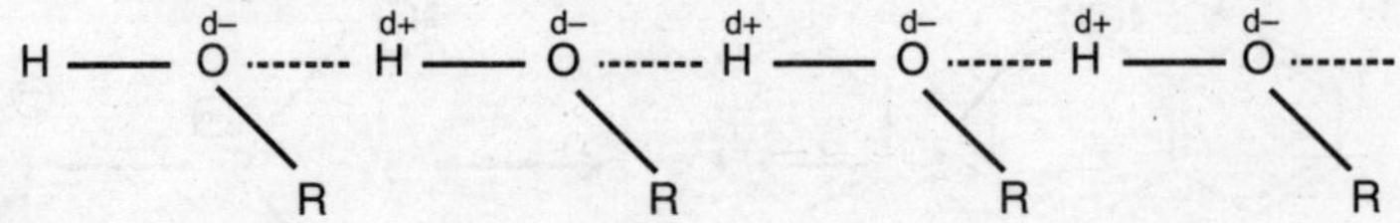

Carboxylic Acids

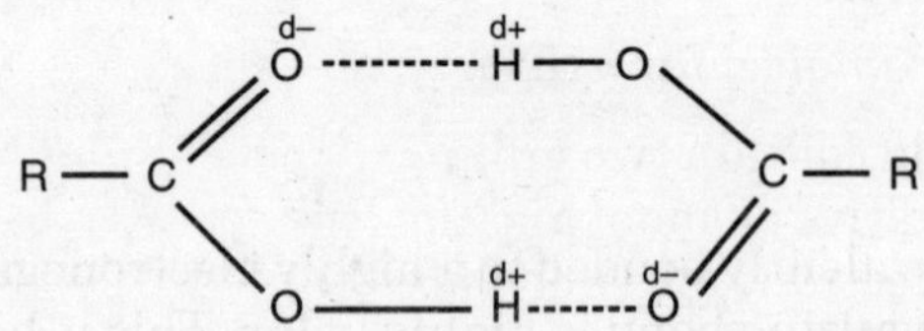

Conditions Necessary for the Formation of Hydrogen Bond: The formation of hydrogen bond is based on the following factors.

1. **High electronegativity:** The electronegativity of the atoms to which hydrogen atom is linked should be high. More is the electronegativity of the element to which hydrogen is linked, greater is the polar character and hence stronger is the hydrogen bond. The electronegativity of flourine is highest among flourine, oxygen and nitrogen. Hence strength of hydrogen bond is highest in HF (41.8 kJ mol^{-1}).
2. **Small size of the atom attached to hydrogen:** The smaller the size, greater will be the polarity of the covalent bond and hence greater will be the electrostatic attraction between the two molecules. When the size of the atom is large, its electron cloud is more diffused which results in less charge separation and hence hydrogen bond is not formed. Flourine, oxygen and nitrogen form strong hydrogen bonds but chlorine due to its large size, does not form hydrogen bond of appreciable strength.

Types of Hydrogen Bonds: There are two types of hydrogen bonds-

1. Intermolecular hydrogen bond.
2. Intramolecular hydrogen bond.

Intermolecular Hydrogen Bonding: When hydrogen bonding takes place between different molecules of the same or different compounds, it is called intermolecular hydrogen bonding. Examples include HF, H_2O and ROH where the hydrogen bonding is in between the molecules of the same compound. Hydrogen bonding in between molecules of different compounds is exhibited by alcohol and water or ammonia and water;

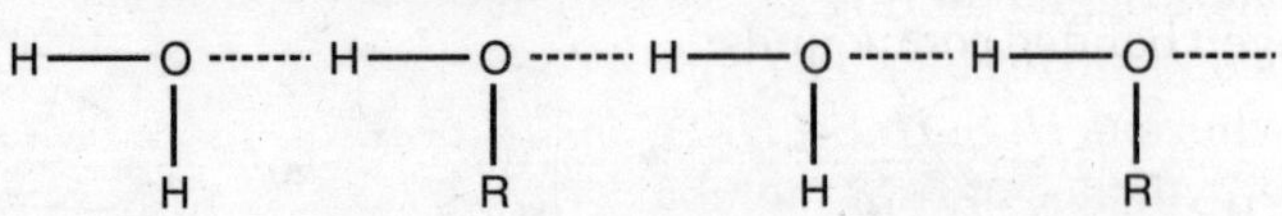

Intermolecular Hydrogen Bonding in water and alcohol molecules.

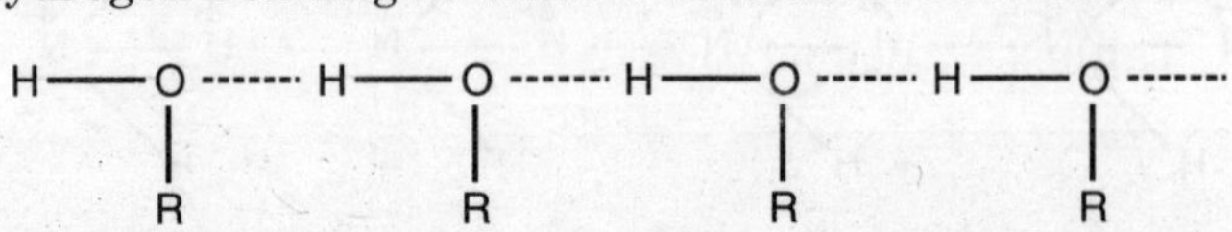

Intermolecular Hydrogen Bonding in similar molecules of alcohol.

Intramolecular Hydrogen Bonding: The hydrogen bonding which takes place within a molecule itself is called intramolecular hydrogen bonding. A few examples involving intramolecular hydrogen bonding include–

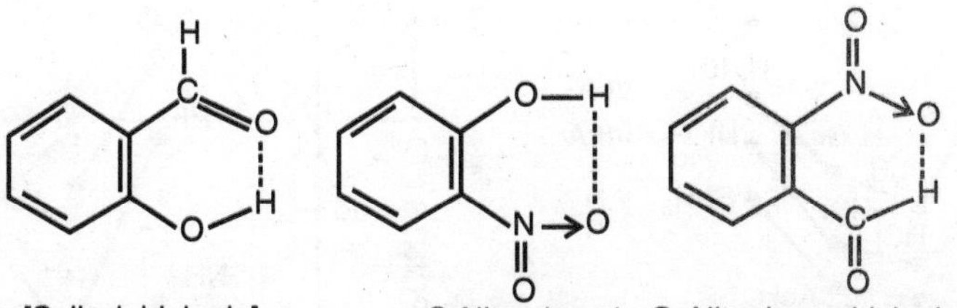

[Salicylaldehyde] O-Nitrophenol O–Nitro benzaldehyde

For the formation of intramolecular hydrogen bonding the following conditions apply–

(i) The molecule should contain two groups such that one group contains hydrogen atom linked to electronegative element and the other group contains electronegative element attached to lesser electronegative element.

(ii) The molecule should be planar.

(iii) The hydrogen bonding should lead to the formation of a six membered ring including hydrogen atom.

Strength of Hydrogen Bond: Hydrogen bond is a weak bond with the strength intermediate between the weak Vander Waals forces and strong covalent bonds. The bond dissociation energy of a covalent bond is 209-418 $kJ\ mol^{-1}$ and that for hydrogen bond it is only 12.6-41.8 $kJ\ mol^{-1}$. The hydrogen bond energies in case of three elements, F, O and N are-

Hydrogen Bond	*Bond Dissociation Energy* $kJ\ mol^{-1}$
H.........F	41.8
H.........O	29.3
H.........N	12.6

Consequences of Hydrogen Bonding

Though hydrogen bond is a weak bond, but still it affects a large number of physical properties of compounds-

1. ***High Melting and Boiling points :*** Substances in which molecules have stronger inter-molecular hydrogen bonding, show abnormally high melting and boiling points. To illustrate the point let us take the example of hydrides of group VI A. As we move down the group (Table 3.7), the boiling point increases from H_2S to H_2Te. But hydride of oxygen i.e H_2O has abnormally high boiling point. This can only be explained on the basis of hydrogen bonding. Due to hydrogen bonding association of water molecules takes place and hence extra energy is needed to break these bonds. For the same reason it can be explained that H_2O is a liquid whereas H_2S, H_2Se, H_2Te are all gases at room temperature.

   Similarly the hydrides of first member of group V A and VI A also have abnormally high melting and boiling points due to hydrogen bonding in these compounds. Fig. 3.35.

Table 3.7 Boiling points of hydrides of *V* group

Formula (°C)	*B. Pt*
H_2O	+100
H_2S	–59.6
H_2Se	–42.0
H_2Te	–1.8

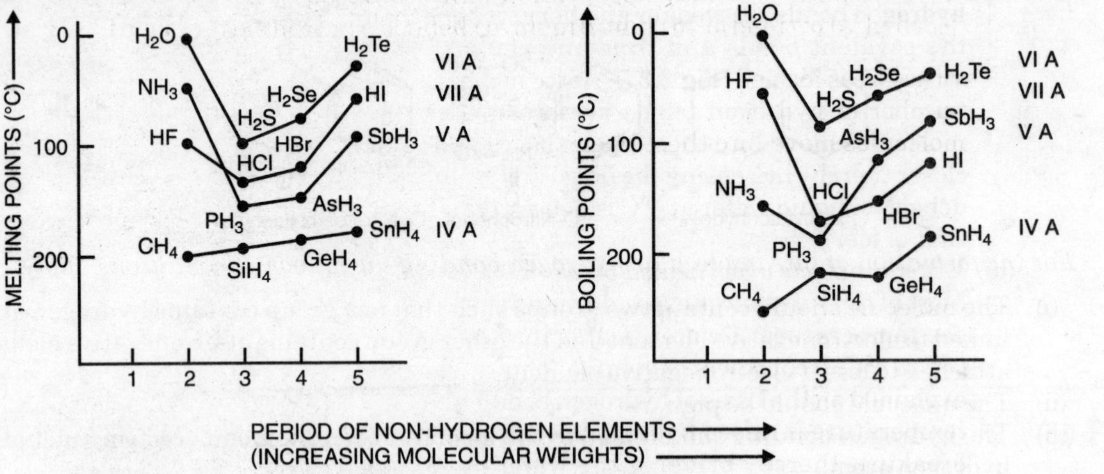

Fig. 3.40 The abnormally high m.p. and b.p. of H_2O, NH_3 and HF.

2. **Solubility:** The compounds which can form hydrogen bonds with the solvent molecules are soluble in such solvents. The organic compounds generally are insoluble in polar solvent, water. But those organic compounds, which can form hydrogen bond with water molecules are soluble. Example, alcohols can form hydrogen bonding with water and hence are soluble.

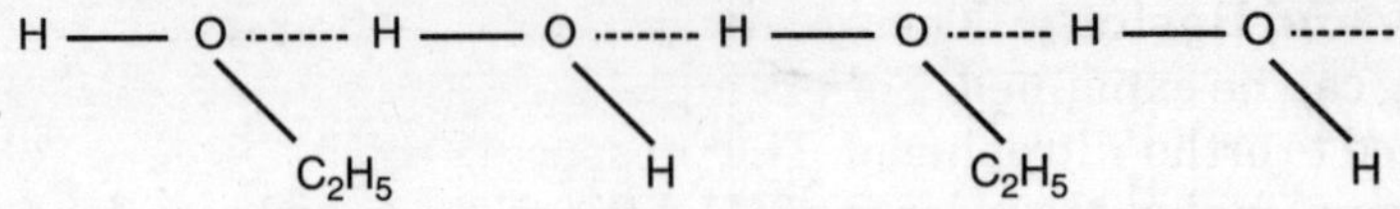

Similarly dimethyl ether is soluble in water because it is capable of forming hydrogen bonds with water.

3. **Viscosity:** Hydrogen bonded molecules have higher viscosity. Example glycerol has three -OH groups per molecule. There is extensive hydrogen bonding in glycerol and hence it is highly viscous.
4. **Association:** Hydrogen bond links up two or more molecules of a compound to form bigger units as in the case of carboxylic acids. Carboxylic acids exist as dimers even in the vapour state due to hydrogen bonding

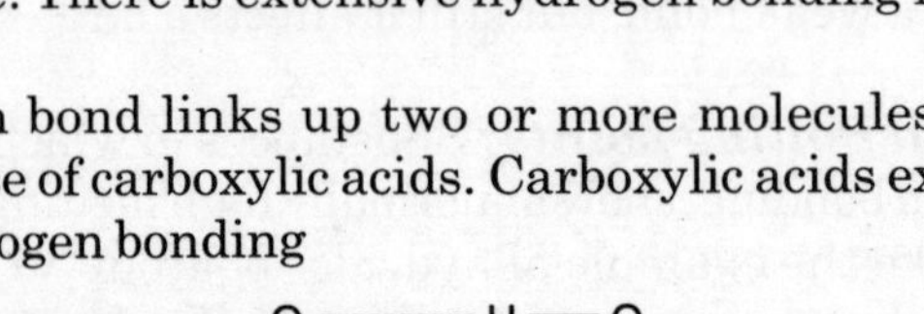

Fig. 3.41

However in aqueous solutions, the molecules of these acids associate with water molecules rather than existing as dimers.

5. **Some unique properties of water and Ice:** Due to hydrogen bonding water has some unique properties.
   (a) *Density of ice is less than that of water* : Gen-erally density of a substance in solid state is more as compared to liquid state. Actually, in ice the hydrogen bonding between water molecule is more extensive than in liquid water. The molecules are tetrahedrally oriented with respect to one another. Oxygen atom is surrounded by four hydrogen atoms, two of these are bonded covalently and the other two by hydrogen bonds. The

hydrogen bonds are weaker and hence longer than the covalent bonds and thus an open cage like structure is formed (Fig.3.37). As ice melts, a large number of hydrogen bonds are broken and the molecules move into these open spaces and come closer thereby increasing the density of water. The density of liquid water at 4°C is hence greater than that of ice.

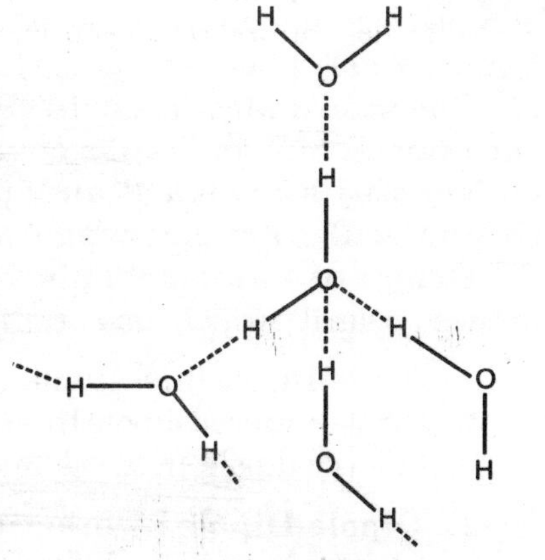

Fig. 3.42

(b) *Water contracts when heated between 0°C and 4°C:* Most of the substances expand when heated but as the temperature of water (ice) is increased from 0°C to 4°C, it contracts in volume. This can also be explained on the basis of hydrogen bonding. As the temperature is raised hydrogen bond in ice starts breaking, thereby bringing the water molecules closer and closer together. This leads to contraction. Above 4°C the normal expansion as in the case of other substances takes place.

6. **Boiling point and intramolecular hydrogen bonding:** Intramolecular hydrogen bonding leads to lower boiling points than expected. Example, in O-nitrophenol, hydrogen bonding prevents the association of molecules and therefore the compound has low boiling point.
7. **Acidity and Basicity:** The acidity and basicity of some compounds exhibiting hydrogen bonding can be explained. For example, para-nitrophenol has higher acidic character as compared to ortho-nitrophenol. The presence of intramolecular hydrogen bonding in ortho nitrophenol makes the release of H^+ difficult decreasing its acidic strength.

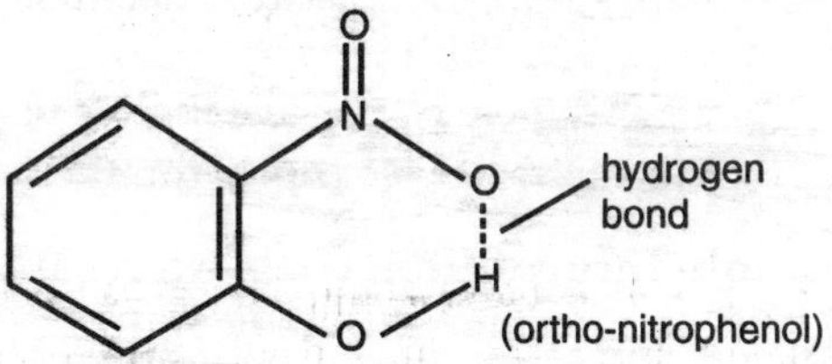

Similarly the basic strength of $(CH_3)_4 N^+OH^-$ is higher as compared to $(CH_3)_3 NH^+ \overline{O}H$. This is again due to the presence of intramolecular hydrogen bonding in $(CH_3)_3 NH^+ \overline{O}H$, which makes the release of $\overline{O}H$ difficult.

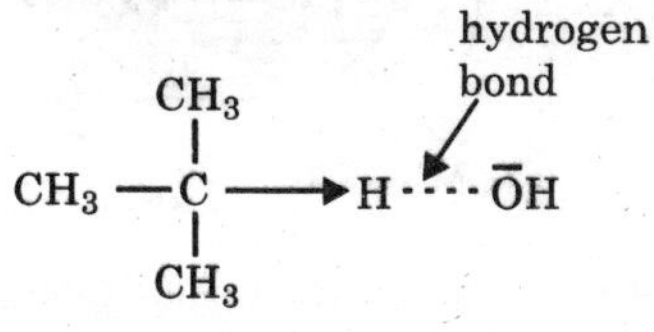

3.8 INTERMOLECULAR FORCES – THE VAN DER WAALS' FORCES

Colvalent molecules like H_2, Cl_2, CH_4 etc cannot form additional bonds because the bonding capacity of the atoms involved in these molecules is saturated. But then the question is, what type of force holds these covalent molecules together? The condensation of covalent substances like hydrogen gas to liquid state or even solid state at low temperatures, implies that attractive forces are present between these molecules. These intermolecular attractions between the molecules are grouped together under the heading of Vander Waals' forces in honour of the Dutch chemist J.H. Vander Waals who first studied their nature.

The intermolecular cohesive forces operative between the molecules, which cause the neutral molecules of a covalent compound to condense into the liquid or solid are collectively known as Vander Waals' forces. These forces are significant only when the molecules are close together.

Origin of Vander Waals' Forces: These forces are electrical in nature and are due to induced polarity which may originate from three types of interactions :

1. Dipole-Dipole Interaction.
2. Dipole-Induced dipole Interaction.
3. Induced dipole-Induced dipole Interaction.

1. **Dipole-Dipole Interaction:** (Keesom forces). This effect was studied by Keesom in 1912 and is also referred to as the orientation effect. Gases such as HF, HCl, NH_3 etc have permanent dipoles. These dipoles orient themselves in such a way that the positive end of the molecule will be attracted to the negative end of the other molecule, when these are in close proximity. As a result, there is appreciable dipole-dipole interactions between the molecules. Because of interactive interactions these gases can easily be liquefied.

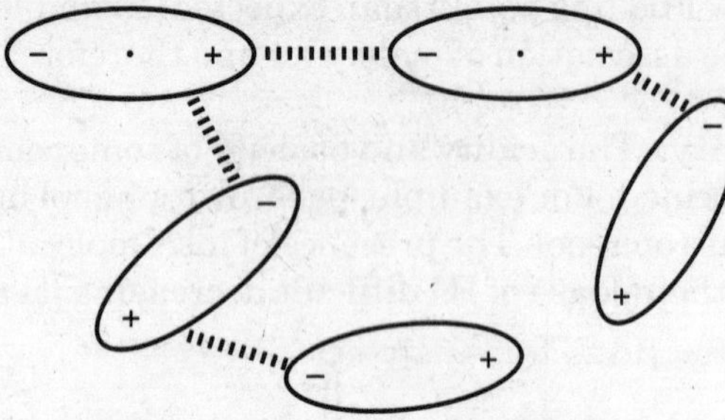

Fig. 3.43 Dipole Dipole Interaction

2. **Dipole-Induced dipole interaction:** (Debye forces) This effect was studied by Debye (1920). A polar molecule or a permanent dipole induces or polarizes a nonpolar molecule, which lies appreciably close to it. The induced dipole interacts with the permanent dipole. The attraction between the two will largely depend upon the polarizing power of the permanent dipole and polarizability of the non-polar molecule.

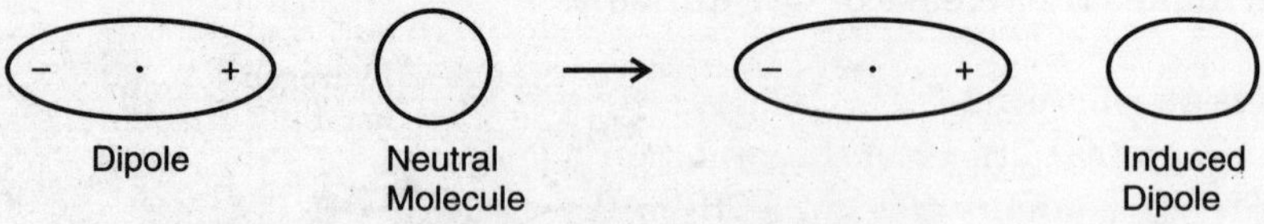

Fig. 3.44 Dipole-Induced dipole interaction

3. ***Induced dipole-Induced dipole Interaction:*** (London forces or Dispersive forces)- Vander Waals' forces exist in non-polar molecules also. F. London (1930) provided a satisfactory explaination for such type of forces. In non-polar substances, such as H_2, O_2, N_2 and also inert gases, there are no initial dipoles. The molecules are electrically neutral. A non-polar molecule may be considered as a positive centre surrounded by symmetrical electron cloud. But electron clouds keep on oscillating with respect to nuclei of the atoms. Therefore at a given instant the electron cloud may be concentrated on one side and a

temporary dipole exists in a molecule i.e a non-polar molecule may become temporarily polarized. Now this self-polarized molecule may induce a dipole moment in a neighbouring molecule and in this way a large number of non-polar molecules become temporarily polar which are mutually attracted by weak forces. These forces are also known as dispersive forces.

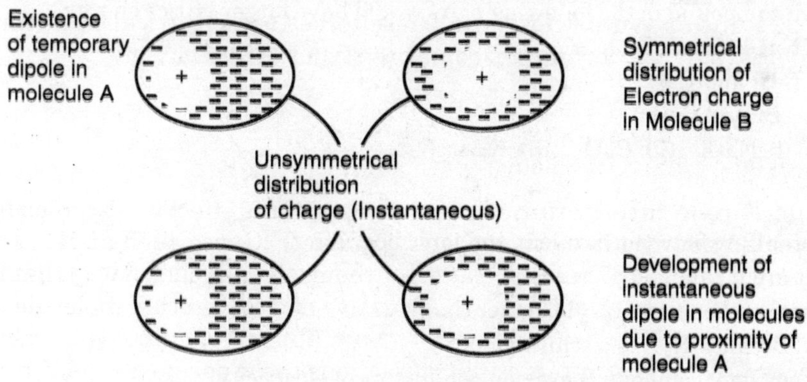

Fig. 3.44 Induced dipole-Induced dipole interaction

These forces though weak, are responsible for liquefaction of non-polar and noble gases.

QUESTIONS

SHORT ANSWERS QUESTIONS
(Refer to the Respective Sections for Answers)

Section 3.1

1. Define chemical Bond? Why is it formed?
2. What type of bond is formed between
   (i) Electropositive element and Electropositive element.
   (ii) Electropositive element and Electronegative element.
3. What type of bond is formed when the shared pair of electrons is contributed by one of the atom taking part is bonding.

Section 3.2

4. What is an ionic bond ?
5. What are the conditions for the formation of ionic bond ?
6. Define electon affinity and ionization energy.
7. Why do ionization energies always have positive value?
   Ans: Energy is always required to remove the electrons which are bound to the nucleus and hence I.E have a positive value always.
8. Why is second ionization energy is greater than first ionization energy?
9. Why alkaline earth metals form positive ions less readily?
10. Why group VI A elements form negative ions less readily ?
11. Why does argon have a higher I.E (1520 KJ mol^{-1}) than potassium (418 KJ mol^{-1})?
    Ans: In argon the electron is removed from a ***stable*** 3p subshell which is quite difficult
12. Why NaCl conducts electricity in molten state and not in solid state.
13. Why ionic solids are brittle.

14. Why oxides of alkaline earth metals are formed despite the unfavourable electron affinities.
15. On the basis of Born Lande's equation, explain how lattice energy depends on the internuclear distance between the molecules and charge on cation and anion.
16. What are the applications of Born Haber Cycle?
17. Why alkaline earth metals are less soluble in water than corresponding alkali metals
18. Which compound in the following pairs would have larger lattice energy?
    (i) NaCl and CsCl
    (ii) LiCl and CaO
    (iii) KBr and CsBr
    Ans: (i) NaCl (ii) CaO (iii) KBr

Section 3.3

19. Differentiate between covalent and ionic bond?
20. What are sigma and pi bonds?
21. Why is bond energy of N_2 is higher than O_2?
    Ans: Because N_2 has a triple bond.
22. Why covalent compounds are non-conductors of electricity?
23. Arranges the covalent bonds C-F, C-Cl and C-I in increasing order of strength, giving reasons.
    Ans: The higher the electronegativity difference, the stronger is the bond. The electronegativity values increases in the order: I < Br < CI < F.
24. How many sigma and pi bonds are there between C-C atoms of $CaCO_2$?
    Ans: One sigma and two π bonds, because the triple C ≡ C bond is a combination of these.
25. Arrange LiCl, $BeCl_2$, BCl_3 and CCl_4 in order of increasing covalent bond character. Give reason.
    Ans: LiCl < $BeCl_2$ < BCl_3 < CCl_4.
    Electronegativity difference in M-Cl bond decreases as we move from Li → Be → B → C and, therefore, covalent character varies as indicated.
26. Arrange the following according to the bond length giving reasons.
    (i) H-F, H-Cl, HBr, HI
    (ii) C-C, C = C, C ≡ C
27. What do you understand by partial ionic character of covalent bonds?
28. Give reason for the following
    (i) Ionic compounds are soluble in water whereas covalent compounds are mostly insoluble in water.
    (ii) Ionic compounds have higher melting point than the covalent compounds.
29. What are the important characteristics of covalent compounds?
30. KI is soluble in alcohol whereas KCl is not.
31. How many sigma and pi bonds are there between C-C atoms of CaC_2
    Ans: are and two sigma bonds

Section 3.4

32. Why is geometry of NH_3 pyramidal ?
33. Predict the shape of H_2O molecule?
34. Based on VSEPR theory explain the order of bond angles in
    (i) H_2O (104.5°) > H_2S (92.50°) > H_2Se (91.0°) > H_2Te (89.5°)
    (ii) PF_3 (98°) < PCl_3 (100°) < PBr_3 (101.5°) < PI_3 (102°)
35. Which of the following posses the largest bond angle?
    NH_3, PH_3, AsH_3, SbH_3

36. Explain why the bond angle of NH_3 (107°) is larger as compared to NF_3 (102°)

Ans: Though both the molecules, NH_3 and NF_3 have 4 valence pairs (one lone pair and three bond pairs), so the geometry should be a distorted tetrahedron. However electronegativity of F is more than H, so electron pair is away from the central atom and hence less repulsion between bond pairs takes place. Thus bond pairs come closer forming an angle of 102°

37. CO_2, N_3^-, NO_2^+, are linear.

Ans: All these have two bond pairs and hence are linear.

Section 3.5

38. What do you mean by the term hybridization?

39. What is the type of hybridization of each carbon atom in the following molecule

$\overset{4}{C}H_3 - \overset{3}{C}H = \overset{2}{C} = \overset{1}{C}H_2$

Ans: (i) sp^2 (ii) sp (iii) sp^2 (iv) sp^3

40. Indicate the type of hybrid orbitals of the bold atom

(i) **Be**Cl_2 (ii) **C**Cl_4 (iii) C_2F_4 (iv) SF6

Ans: (i) sp (ii) sp^3 (iii) sp^2 (iv) sp^3d^2

41. In benzene, what is the valence bond hybridization on each carbon atom?

Ans: sp^2

42. Determine the geometry of each of the following molecules and the hybridization about the central atom.

(i) BeF_2 (ii) NH_3 (iii) $HC \equiv CH$

43. What are the relative energies of sp, sp^2 and sp^3 hybrid orbitals.

Ans: More is s character lower is the energy. so the energies follow the order

$$sp^3 > sp^2 > sp$$

44. Can nitrogen use ground state atomic orbitals to form equivalent bonds.

Ans: Yes. Ground state nitrogen has three half filled atomic orbitals which can overlap with the orbitals of other atoms to form bonds. ($2s^2$, $2p_x^1$, $2p_y^1$, $2p_3^1$)

45. Why do then nitrogen use hybrid orbitals for bonding.

Ans: Hybrid orbitals give stronger bonds than atomic orbitals.

46. Draw outline sketches to show the formation of sp., sp^2 and sp^3 hybrid orbitals.

47. What type of hybridization is associated with the central atom when the atoms attached to it form

(a) an equilateral triangle

(b) a regular tetrahedron ?

48. Discuss the hybridization in PCl_5 and SF_6

49. What is the state of hybridization in CO^{-3} Ion?

50. Indicate the type of hybridization in the following molecules

(i) $CH - CH_3$ (ii) $H_3C - C \equiv C - CH_3$

(iii) $CH_2 = C = CH_2$ (iv) $CH_3 - C \equiv N$

51. What is the state of hybridization in SO_3 molecule.

Ans: sp^2

Section 3.6

52. Differentiate between structural isomers and resonance structures.

Ans: Structural isomers are real molecules whose atoms are linked together in different ways. Whereas resonance structures are not real and are written when a single structure cannot explain all the properties of the molecule

53. Write the contributing structures of the following

(i) N_2O (ii) HCO_2^- (iii) NO_2^-

54. Explain why the following pairs are not resonance structures

(i) $:\ddot{O}: = :\ddot{O}:$ and $:\ddot{O}: - :\ddot{O}:$

(ii) $CH_3 - C(OH) = CH_2$ and $CH_3 - C(=O) - CH_3$

Ans: (i) Contributing structures should have same number of un paired electrons

(ii) The position of all the atoms is not the same, i.e., H atom.

55. Write the resonance structures for

(i) CS_2 (ii) SO_2

56. How bond order is calculated on the basis of resonance.

Ans: $\dfrac{\text{Total no. of bonds between two atoms in all the structures}}{\text{Total no. of resonating structures}}$

e.g (i) In case of carbonate ion

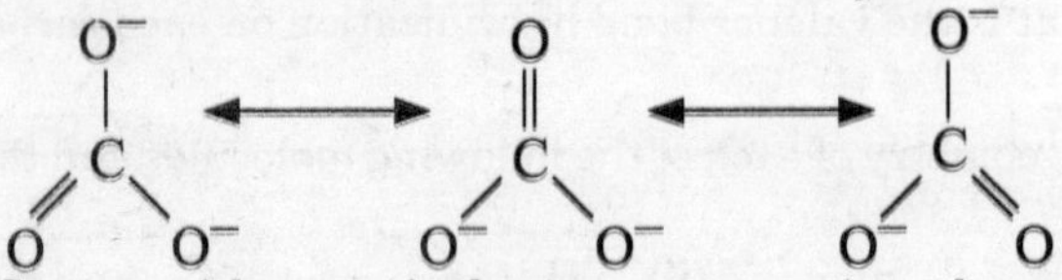

Looking at bonds of C-atom with a particular oxygen atom (say the one linked on the left) the bonds are 2,1 and 1 respectively in the three structures and number of resonating structures is 3,

$$\text{Bond order} = \frac{2+1+1}{3} = \frac{4}{3} = 1.33$$

Section 3.7

57. Which one of the following is expected to show hydrogen bonding?

(i) CH_3CH_2OH and CH_3OCH_3

(ii) CH_3NH_2 and CH_3SH

(iii) CH_3OH and $(CH_3)_3N$

Ans: (i) CH_3CH_2OH (ii) CH_3NH_2 (iii) CH_3OH

58. Why the boiling point of $H_2O > CH_3OH$

Ans: Because extent of hydrogen bonding is more in H_2O

59. Why CH_3OCH_3 is soluble in water and CH_3OCH_3 is not?

60. Why water contracts when heated between 0°C and 4°C ?

61. Why melting or boiling point of para isomer is more than ortho isomer?

62. Why o-nitrophenol can be steam distilled and not p-nitrophenol?

Ans: In p-nitrophenol intermolecular hydrogen bonding is present decreasing its volatility whereas in o-nitrophenol interamolecular hydrogen bonding is present hence the single molecules have high volatility and can be steam distilled.

63. Explain why the acidic strength of 0-nitrophenol < p-nitrophenol

64. Account for the following

(i) Why H_2O is a liquid while H_2S is a gas?

(ii) Why HF has higher boiling point than HCl?

(iii) Why ethyl alcohol is completely miscible with water?

65. Why KHF_2 exists but $KHCl_2$ does not ?

Ans: Due to H-bonding in HF, we have H—F....H—F....H—F..... This can dissociate to give HF_2^- ion and hence KHF_2 exists but there is no H-bonding in H—Cl. So HCl_2^- ion does not exist and hence $KHCl_2$ also does not exist.

66. When we move from HF to HCl, the boiling point drops sharply but on moving further to HBr and HI the boiling point increases. Why? Or Out of HF, HCl, HBr and HI which has lowest boiling point and why?

67. How many H- bonds are formed by each H_2O molecule and how many water molecules are attached to each water molecule and in what direction?

Ans: Each H_2O molecule forms four H-bonds, two with O-atom and two with H-atoms. Further, each H_2O molecule is linked to four H_2O molecules through H-bonds tetrahedrally.

68. Though Cl has nearly same electronegativity as N, yet there is no H-bonding in HCl. Why?

Ans: Chlorine atom has a large size.

69. Why glucose, fructose, sucrose etc. are soluble in water though they are covalent compounds?

Ans: These compounds contain—OH groups which can form H-bonds with water.

70. Viscosity of glycerol > dimethyl ether. Explain.

Section 3.8

71. What are van der Waals forces?

72. Arrange the following interactions in order of increasing stability covalent bond, van der waals forces, hydrogen bond

Ans: Covalent > hydrogen bond > van der waals

73. Why boiling point of halogens follow the order:

$$I_2 > Br_2 > Cl_2 > F_2$$

Ans: More is the molecular mass, more will be the van der waals forces between the molecules.

74. Why the boiling point of following hydrides follow the order:

$$GeH_4 > SiH_4 > CH_4$$

Ans: Strength of the van der waals forces depend upon size. The size decreases from Ge to C and hence the boiling point also decreases in the same order.

75. In which molecule the van der waals forces are likely to be the most important in determining the melting and boiling point : ICl, HCl, Br_2, H_2S, CO

Ans: Br_2, because in Br_2 only van der waals forces are present.

LONG ANSWER QUESTIONS

1. What do you mean by a chemical bond? How do atoms combine? How many types of bonds are there?
2. What are essential conditions for the formation of ionic bond? Explain the formation of an ionic bond between an atom of Na and Cl.
3. Describe various characteristics of electrovalent compound.
4. List the following bond in order of increasing ionic character : LiF, K_2O, N–N, Cl–F_3.
5. How is covalent bond formed? Explain the formation of covalent bond between two hydrogen atoms. Why is such a bond not formed between two He atoms?
6. What do you understand by overlap of orbitals? How does it account for the stability of the bond between the two atoms? Give an example for each of the following types of orbital overlap. Also sketch the corresponding diagram leading to the bond formation.

   (a) s–s overlap,

   (b) s–p overlap

(c) Sideway p–p overlap and
(d) end to end p–p overlap.

7. What are sigma and pi bonds? Explain the different ways of their formation diagrammatically. Which one of them is stronger and why?
8. Explain the terms bond length, bond energy and bond angle.
9. Briefly explain orbital theory of covalent bond formation.
10. (a) What do you understand by polar and non-polar molecules? Is it possible for the non- polar molecules to have polar bonds? Justify your answer with examples.
    (b) Explain why CO_2 and CCl_4 molecules are non-polar while $CHCl_3$ molecule is polar.
    (c) Which is more polar, H_2S or H_2O and why?
11. (a) What is the difference between coordinate bond and covalent bond?
    (b) Under what circumstances can a fully filled orbital participate in bond formation? Discuss with an example.
12. (a) Flourine has seven valence electrons ($2s^22p^5$), so seven covalent bonds in principle could form around the atom. Such a compound might be FH_7 or FCl_7. These compounds have never been prepared. Why?
    (b) Which of the following hydrogen halides has the most polar molecules and why?

    HI, HBr, HCl, HF

    (c) What type of chemical bonds are involved in the formation of the following compounds? Explain the bond formation on the basis of electronic structures of atoms involved in each case :
    (i) NaCl
    (ii) HCl
    (iii) NH_4Cl
    (d) Which one of the following has highest melting point?

    C_3H_7OH, KCl, C_6H_5COOH

13. What is lattice energy? How does the solubility of an ionic solid depend upon its lattice energy?
14. Discuss in detail Born–Haber cycle for calculating lattice energy.
15. Give reason for the following:
    (a) Ionic compounds are soluble in water whereas covalent compounds are mostly insoluble in water.
    (b) Ionic compounds have higher melting points than covalent compounds.
    (c) NaCl solution gives white precipitate with $AgNO_3$ solution but CCl_4 or chloroform does not.
    (d) $AlCl_3$ is largely covalent while AlF_3 is largely ionic.
16. What is VSEPR theory? How does it explain the bond angles observed in CH_4, NH_3 and H_2O molecules?
17. Explain molecular orbital theory and the following observations.
    (i) CCl_4 and $SiCl_4$ are tetrahedral
    (ii) CO_2 molecule is linear and SO_2 is angular.
    (iii) BF_3 is planar but NH_3 is not
    (iv) H–S–H angle in H_2O is closer to 90° then H–O–H angle in H_2O.
18. What is hybridization of orbitals? Draw outline sketches to show the formation of sp, sp2 and sp3 hybrid orbitals.
19. (a) Explain why, carbon has valency of four and not two and why are four C–H bonds in methane are identical.
    (b) What type of hybridization is associated with the central atom when the atoms attached to it form

(i) an equilateral triangle

(ii) a regular tetrahedron.

20. (a) Making use of concept of hybridization, predict the shape of C_2H_2 molecule.

(b) Using the concept of hybridization, predict the geometry of NH_3, H_2O, BeF_2, PCl_5 and SF_6.

21. (a) What is resonance? Discuss by taking example of CO_3^{2-} ion.

(b) What do you understand by the term resonance energy? What are the essential rules for writing resonating structures?

(c) Is it possible to "trap" a resonance structure of a compound for study? Explain.

22. What is meant by hydrogen bonding? Discuss the nature and consequences of hydrogen bonding. Discuss the importance of hydrogen bonding in sustaining life. Why do H_2O and HF have abnormally high boiling points?

23. Account for the following:

(i) Water is a liquid while H_2S is a gas.

(ii) NH_3 has higher boiling point than PH_3.

(iii) Boiling point of HF is lower than that of water.

24. (a) What is hydrogen bonding? In what respects is it different from ionic and covalent bonding?

(b) Explain why ordinary H_2S is a gas while H_2O a liquid even though both S and O are elements of the same group in the periodic table and S has a higher atomic mass.

(c) Water is a liquid and hydrogen sulphide is a gas, although the molecular mass of hydrogen sulphide is almost double than that of water. Explain.

25. (a) What requirement should a molecule fulfil for the formation of a hydrogen bond?

(b) Give one example of intramolecular hydrogen bond.

(c) The boiling and melting points of water are abnormally higher than those of other hydrides of group 16 of the periodic table. Give reasons.

26. Account for the following:

(a) Although o-and p-nitrophenols have the same mol. wt. (134), yet the former boils at a very low temperature as compared to latter which boils at 114°C.

(b) Compared to p-nitrophenol, o-nitrophenol is more soluble in benzene and less in water.

(c) o-Hydroxy acetophenone is steam volatile but its meta and para isomers are not.

27. Write short notes on :

(a) Inter and intramolecular hydrogen bonding.

(b) Vander waals' forces.

(c) Resonance.

(d) Sigma and pi bond.

(e) Born Haber Cycle.

28. How many lone pairs and bonded pairs are present in ClF_3? Write down the shape and the type of hybridization.

29. (a) What is the difference between a coordinate bond and a covalent bond? What are the factors influencing the formation of an ionic bond.

(b) Briefly, discuss the steps involved in the formation of a metal chloride from the metal and chlorine gas in terms of energy changes. **(G.G.S.I.P.U. I Term, May 2000)**

30. Write short notes on the following :

(a) Resonance

(b) Hydrogen bonding

(c) Effective Atomic number rule **(G.G.S.I.P.U, I Term, May 2000)**

31. What do you understand by Hybridization? Explain with diagram the formation of H_2O, SiF_4 and BrF_3. **(G.G.S.I.P.U, I Term, May 2000)**

32. (a) Discuss the factors affecting the formation of an ionic bond.
(b) List the characteristics of covalent compounds.
(c) Illustrate with an example each of the following with respect to coordination compounds:
(i) Ionization Isomerism
(ii) Geometrical isomeriem
(iii) Optical isomerism. **(G.G.S.I.P.U, End Term, July 2000.)**

33. (a) Write a short note on Van der Waals forces.
(b) On the basis of hydrogen bonding, explain the following :
(i) Ice is lighter than water
(ii) o-Nitrophenol can be separated from p-nitrophenol by steam distillation.
(c) Discuss the structures of SiF_6 and BrF_3.
(d) Write the energy level diagram for O_2^+ and calculate the bond order.
(G.G.S.I.P.U. End Term, July 2000.)

34. (a) Explain the term Hybridization. Give an account of the different types of hybridization with suitable examples.
(b) BeF_2 molecule is linear, BF_3 molecule is triangular, while methane molecule is tetrahedral in structure. How would you account. **(G.G.S.I.P.U. Ist Term, Mar 2001)**

35. (a) What are the factors influencing the formation of ionic bond? Give the difference between a co-ordinate and covalent and covalent bond by giving suitable examples.
(G.G.S.I.P.U. Ist Term, Mar. 2001)
(b) Discuss the steps involved in the formation of a metal chloride from the metal and chlorine gas in terms of energy changes.

36. (a) Write resonance structure for
(i) SO_2 (ii) CO (iii) O_3
(iv) OCS (v) $SOCl_2$ (vi) CS_2
(b) Which compound in the following pairs would have larger crystal energy.
(i) NaCl and $MgCl_2$
(ii) SrS and SrSe
(iii) KBr and CsBr
(iv) LiCl and LiI **(G.G.S.I.P.U., End Term, May 2001)**

37. (i) Indicate the type of hybridization of each carbon atom in the following compounds.
(a) $H_3C–CH_3$ (b) $CH_2 = C = CH_2$
(c) $CH_3 - C \equiv C - CH_3$ (d) $CH_3 - C \equiv N$
(ii) How hydrogen bonding effect the boiling point of water and solubility of compounds.
(G.G.S.I.P.U., End-Term Exam, May 2001)

38. (a) Mention in brief 'Werner's Coordination Theory'.
(b) Discuss 'E.A.N.' Rule, giving at least two examples.
(c) How many chloride ions would you expect to be released when each of the following complex.
(i) $CoCl_3.6NH_3$
(ii) $CoCl_3.5NH_3$ individually dissolves in aqueous solution : supposing that Cobalt prefers octachedral coordination? **(G.G.S.I.P.U., End Term, II Batch, B.Tech, July 2001)**

39. (a) What does the VSEPR theory dictate regarding the influence of electronic repulsion on the shapes of molecules? Give your comments on the shapes of NF_3, BF_3 and ClF_3 molecules following VSEPR model.
(b) Forward a proper reasoning and explain.
(i) SF_2 molecule acquires an angular shape while BeF_2 is linear.
(ii) NF_3 molecule shows a net dipole moment, while BF_3 molecule does not.

(iii) Bond angle in H–S–H is shorter that the bond angle in H–O–H.

(G.G.S.I.P.U., End Term, II Batch B.Tech, July 2001)

40. (a) Explain with reason.

(i) O-nitrophenol can be steam distilled while para and meta-nitro phenol can not.

(ii) H_2S is a gas while H_2O is a liquid.

(b) Which of the following compounds has larger lattice energy.

(i) NaCl and $MgCl_2$

(ii) LiCl and Li I **(G.G.S.I.P.U., I Term, March 2002)**

41. (a) What do you understand by hybridization. Give the structure of SF_6 and PCl_5.

(b) Give resonating structure of following.

(i) CO_3^{2-}

(ii) O_3

(iii) NO_2– **(G.G.S.I.P.U., I Term, March 2002)**

42. Describe the structure of H_2O and NH_3 molecules interms of VSEPR theory, explain the decrease in bond angle from NH_3 to H_2O. **(G.G.S.I.P.U., End Term, 2004)**

❑❑❑

4 CHEMICAL BONDING-II

"The most incomprehensible thing about the world is that, it is comprehensible"

—Albert Einstein

CHAPTER OUTLINE

4.1 MOLECULAR ORBITAL THEORY

According to Valence Bond Theory discussed earlier, a covalent bond is formed by the overlapping of atomic orbitals and the bonded atoms in the molecule are held together by one or more pair of electrons. According to this theory, *the atoms retain their identity even after bond formation.*

On the other hand in the molecular orbital method, the valency electrons are considered to be associated with all the nuclei in the molecule i.e. electrons move in the field of more than one nucleus.

The molecular orbital theory given by R.S. Mulliken and F. Hund incorporates the following points.

1. When nuclei of two atoms come close to each other, their atomic orbitals interact, to form **molecular orbitals** in which the identity of both the atomic orbitals is lost.
2. The number of molecular orbitals is equal to the number of atomic orbitals involved in their formation.
3. Just as atomic orbitals in an atom are quantized, similarly molecular orbitals are also believed to be quantized.
4. The electrons in the atomic orbital are influenced by one nucleus, the electrons in the molecular orbitals are influenced by all the nuclei.
5. Electrons are filled in the molecular orbitals in the same way as they are filled in atomic orbitals, following Aufbau principle, Hund's rule of maximum multiplicity.

6. The molecular orbitals are represented by $\sigma\ \sigma^*\ \pi\pi^*$ etc.

A molecular orbital is constructed by combination of two atomic orbitals and this method is known as Linear Combination of Atomic orbitals (LCAO).

Table 4.1 Difference between Valence Bond Method and Molecular Orbital Method

Valence Bond Method	*Molecular Orbital Method*
1. In valence bond method, a molecule is composed of atoms which retain their individual identity even after bond formation.	In molecular orbital method, the combining atoms lose their individual identity.
2. An electron moves in the field of one nucleus only, i.e. atomic orbitals are monocentric.	An electron moves in the field of more than one nucleus i.e. molecular orbitals are polycentric.
3. Valence bond method does not explain the paramagnetism of O_2 successfully.	Molecular orbital theory explains the paramagnetism of oxygen successfully.
4. It does not involve the formation of bonding and antibonding orbitals.	It involves the formation of bonding and antibonding orbitals.
5. Atomic orbitals in valence bond method are represented by *s, p, d* and *f*.	Molecular orbitals in molecular orbital method are represented by $\sigma\ \sigma^*\ \pi\ \pi^*$ etc.

4.2 LINEAR COMBINATION OF ATOMIC ORBITALS

Electrons may be considered either as particle or waves. An electron in an atom may therefore be described by a wave function ψ, which is a solution to Schrodinger wave equation. According to Molecular Orbital Theory the electrons in a molecule are said to occupy molecular orbitals. The wave function describing a molecular orbital may be obtained by the method of **linear combination of atomic orbitals.**

Consider two atoms A and B which have atomic orbitals described by the wave functions Ψ_A and Ψ_B. The electron clouds of these two atoms overlap when the atoms approach each other. The wave function for the molecular orbital Ψ_{MO} can be obtained by a linear combination of the atomic orbitals Ψ_A and Ψ_B.

$$\Psi_{MO} = C_1\Psi_A + C_2\Psi_B$$

Where C_1 and C_2 are the coefficients chosen to give the lowest-energy for the molecular orbital. For homoatomic systems atoms A and B are same, and hence C_1 and C_2 are equal.

Now the probability of finding an electron in a volume of space *dv* is $\Psi^2 dv$, so the probability density for the combination of two atoms is given by:

$$\psi^2_{MO} = (C_1\Psi_A + C_2\Psi_B)^2$$

$$\psi^2_{MO} = (C_1^2\Psi_A^2 + 2C_1C_2\Psi_A\Psi_B + C_2^2\Psi_B^2)$$

The term $C_1^2\Psi_A^2$ gives the probability of finding an electron on atom A, if A is an isolated atom. The term $C_1^2\Psi_B^2$ gives the probability of finding an electron on atom B, if B is an isolated atom. The middle term $2C_1C_2\Psi_A\Psi_B$ is called the *overlap integral* as this term represents the main difference between the electron clouds in individual atoms and in the molecule. The larger this term the stronger the bond.

Overlap has a positive sign when the superimposed regions of the two orbitals have the same sign i.e. both (+) or both (-) (same phase). Overlap has a negative sign when the superimposed regions of the two orbitals have opposite signs (different phase). Zero overlap results when there are equal regions of overlap with opposite signs. Let us discuss this in more detail by considering the overlap of different orbitals.

Combination of s-s orbitals

Consider a simple case of combination of 1*s* atomic orbital of one hydrogen atom (A) with 1*s* atomic orbital of another hydrogen atom (B). Let wave function Ψ_A and Ψ_B represent the 1*s* atomic orbitals of the atoms A and B respectively. There are two possible ways of linearly combining the atomic orbitals.

(i) Additive combination-where the signs of the two wave functions are the same.*

(ii) Subtractive combination-where the signs of the two wave functions are different.* (Fig. 4.1)

Wave functions which have the same sign may be regarded as waves that are in phase, which when combined add up to give a larger resultant wave. Similarly wave functions of different signs correspond to waves that are completely out of phase and which cancel each other by destructive interfer-ence. Thus two combinations are—

$$\Psi_{(g)} = \Psi_A + \Psi_B$$

$$\Psi_{(u)} = \Psi_A - \Psi_B$$

The molecular orbital wave functions are designated by $\Psi_{(g)}$ and $\Psi_{(u)}$: *g* stands for *gerade* (even) and *u* for *ungerade* (odd). Actually the terms g and u refer to the symmetry of the orbital about its centre. If the sign of the lobe remains unchanged when the orbital is rotated about the line joining the two nuclei and then about a line perpendicular to this, then, the orbital is termed as gerade and if the sign changes, the orbital is termed as ungerade.

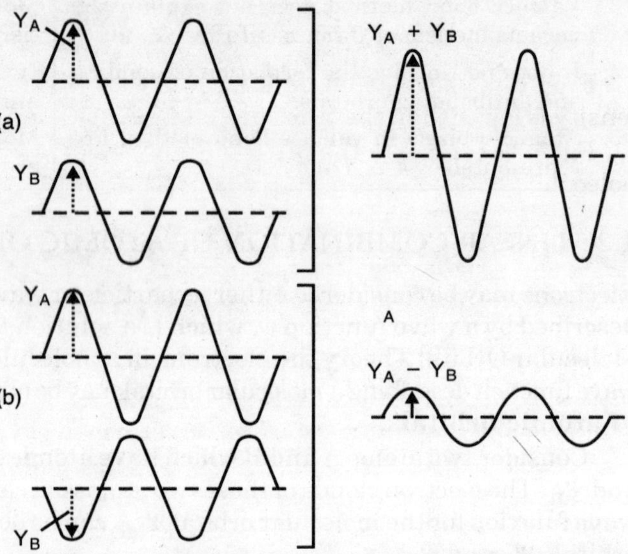

Fig. 4.1 (a) Additive effect of the electron waves in phase.
(b) Subtractive effect of the electron waves out of phase.

Bonding Molecular Orbital

The molecular wave function $\Psi_{(g)}$ leads to increased electron density in between the nuclei and is therefore a *Bonding Molecular Orbital*. The probability density of electron cloud can be obtained by squaring the wave function $\Psi_{(g)}$.

$$\Psi_g^2 = (\Psi_A + \Psi_B)^2$$

$$\Psi_g^2 = \Psi_A^2 + \Psi_B^2 + 2\Psi_A\Psi_B$$

* *One of the wave function Ψ_A is arbitrarily assigned a positive sign, the other may be either +ive or –ive. The signs + and – refer to signs of the wave functions, which determine their symmetry and have nothing to do with electrical charges.*

From the above equation, overlap criteria becomes clear. In a region where two orbitals Ψ_A and Ψ_B have positive overlap, the electron density is higher than the mere sum of the electron densities of the two separate orbitals i.e. $(\Psi_A + \Psi_B)^2$ is greater than $\psi^2_A + \psi^2_B$ by $2\,\psi_A\,\psi_B$. The attraction of both nuclei for these electrons is greater than the mutual repulsion of the nuclei. A net attractive force or bonding interaction results in the formation of stable chemical bond.

Antibonding Molecular Orbial

Conversely $\psi_{(u)}$ results in two lobes of opposite signs giving zero electron density in between the nuclei and hence is called the *antibonding molecular orbital*. The probability density of electron cloud in this case also is obtained by squaring the wave function ψ_u

$$\Psi_u^2 = (\Psi_A - \Psi_B)^2$$

$$= \Psi_A^2 + \Psi_B^2 - 2\Psi_A\Psi_B$$

From the above equation it is clear that if ψ_A and ψ_B have negative overlap, the electron density is lower than the sum of electron densities of the two separate orbitals.

Fig 4.2 represents the pictorial representation of bonding and antibonding orbitals of H_2 molecule.

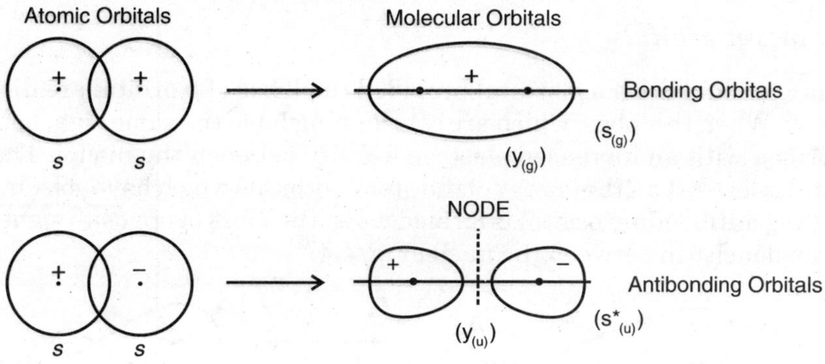

Fig 4.2 *s-s* combination of atomic orbitals

The energy of the bonding molecular orbital is lower than that of the atomic orbital by an amount Δ. This is known as **stabilization energy**. Similarly the energy of the antibonding molecular orbital is increased by Δ. The relative energies of the two molecular orbitals and the atomic orbitals are shown in Fig. 4.3.

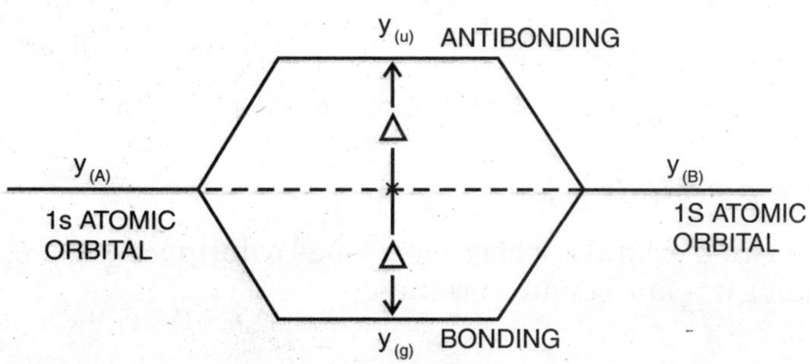

Fig. 4.3 Molecular Orbitals (Bonding and Antibonding)

Further, since the overlap of s-s orbital is along the axis joining the nuclei, the resulting molecular orbitals are designated as σ orbital. As a convention the antibonding molecular orbital is designated with an asterisk σ^*. Thus the two orbitals formed in the case of two combining hydrogen atoms, are designated as σ_{1s} and σ^*_{1s}.

Table 4.2 Difference between Bonding and Antibonding Molecular Orbitals

Bonding Molecular Orbital	*Antibonding Molecular Orbital*
1. They are formed by the additive effect of the atomic orbitals $\Psi_{MO} = \Psi_1 + \Psi_2$	They are formed by the subtractive effect of the atomic orbitals $\Psi_{MO} = \Psi_1 - \Psi_2$
2. It increases the electron density between the nuclei	It decreases the electron density between the nuclei
3. These orbitals have lower energy than the atomic orbitals from which they are formed.	These orbitals have higher energy than the atomic orbitals from which they are formed.
4. Bonding molecular orbitals stabilize the molecule.	Antibonding molecular orbitals destabilize the molecule.
5. These are represented by σ, π etc.	These are represented by σ^*, π^* etc.

Combination of s-p orbitals

An *s* orbital may combine with a *p* orbital provided the lobes of *p* orbital are directed along the internuclear axis. When the lobes, which are overlapping have the same sign, bonding molecular orbitals are formed with an increased electron density between the nuclei. The energy of this molecular orbital is lower than the energy of the atomic orbitals which have gone into its formation. The corresponding antibonding orbital is formed when the lobes of opposite signs overlap, with a reduced electron density in between the nuclei Fig. 4.4.

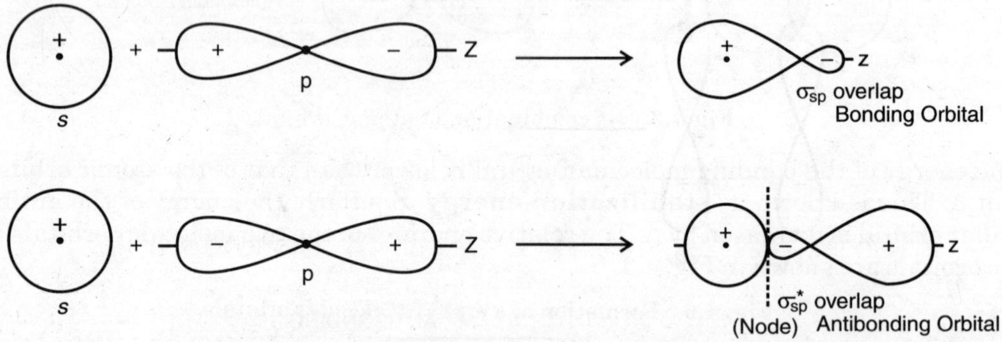

Fig. 4.4 Combination of *s-p* orbitals

Combination of p-p orbitals

The overlap of the two *p* orbitals, which have lobes pointing along the axis joining the nuclei results in σ_{2p} bonding σ^*_{2p} antibonding orbitals.

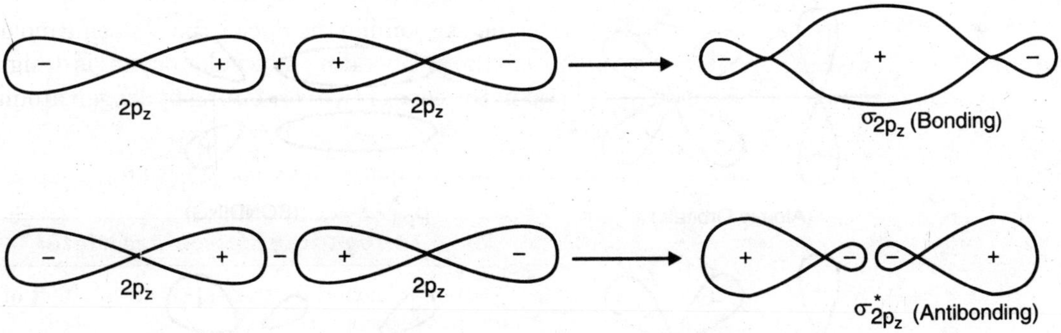

Fig. 4.5 Combination of *p-p* orbitals

Electron charge density between the two nuclei in σ^*_{2pz} is enhanced whereas the electron density in σ^*_{2pz} gets reduced.

The overlap between the two *p* orbitals, which both have lobes perpendicular to the axis joining the nuclei, is lateral. This results in the formation of **π bonding and π* antibonding molecular orbitals**.

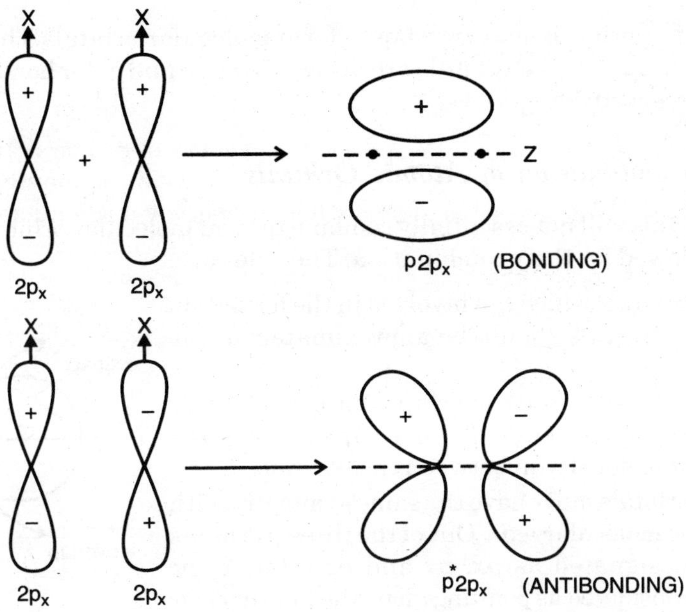

Fig. 4.6 Formation of π and π* molecular orbitals.

It is to be noted that for π molecular orbitals, ψ is zero along the internuclear line and consequently the electron density ψ^2 is also zero. This is in contrast to σ orbitals.

Combination of p-d Orbitals

A *p* orbital on one atom may overlap with a *d* orbital on another atom forming bonding and antibonding orbitals. Since the overlap is lateral and hence molecular orbitals are designated as π orbitals.

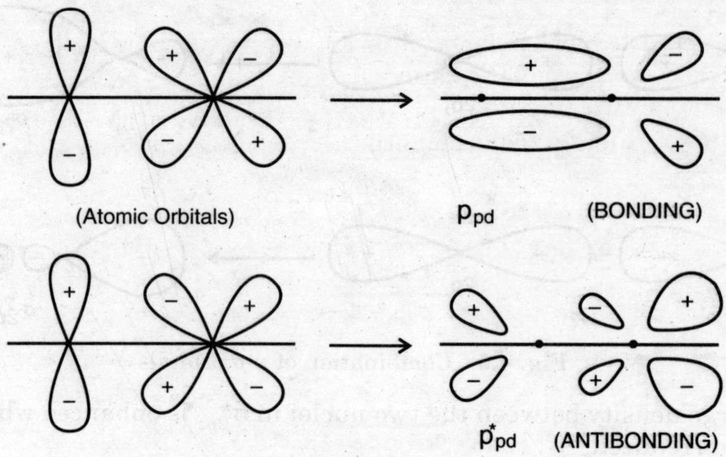

Fig. 4.7 *p-d* combination of atomic orbitals

Combination of d-d Orbitals

Two *dxy* or two *dx*^{2}-*y*^{2} orbitals may overlap to form molecular orbitals.The molecular orbitals formed by the combination of *d-d* orbitals are designated as δ and δ* for bonding and antibonding molecular orbitals respectively.

Criteria for the Combination of Atomic Orbitals

Any two atomic orbitals will not essentially combine to form molecular orbitals. There are certain rules which are followed for their combination. The rules are:

1. The atomic orbitals which are involved in the formation of molecular orbitals should be approximately of the same energy.
2. The atomic orbitals must overlap to a considerable extent to form stable bonds. The overlap increases the electron charge density in between the two nuclei.
3. The atomic orbitals must have the same symmetry with respect to the molecular axis. Out of the three p orbitals which are designated as *px*, *py* and *pz* orbitals, *pz* orbitals are considered as pointing along the internuclear axis by convention. Thus *px* or *py* orbital of one atom does not combine with *s* orbital of the other atom. In fact ++ overlap is cancelled by +– overlap.

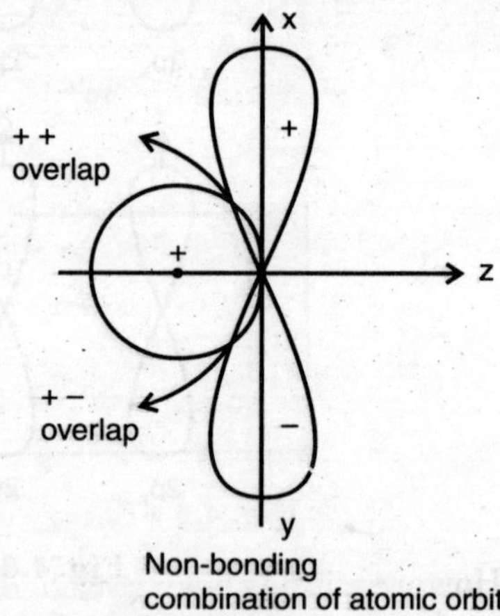

Fig. 4.8

4.3 ENERGY LEVEL DIAGRAM FOR MOLECULAR ORBITALS

The order of energy of molecular orbitals has been determined from spectroscopic data. The energy generally increases in the following order as we move from left to light –

$$\sigma_{1s},\ \overset{*}{\sigma}_{1s},\ \sigma_{2s},\ \overset{*}{\sigma}_{2s},\ \sigma_{2pz},\ \begin{Bmatrix}\pi_{2px}\\ \pi_{2py}\end{Bmatrix}\ \begin{Bmatrix}\overset{*}{\pi}_{2px}\\ \overset{*}{\pi}_{2py}\end{Bmatrix}\ \overset{*}{\sigma}_{2pz}$$

increasing energy →

The bonding π_{2px} and π_{2py} molecular orbitals have exactly the same energy and are said to be double degenerate. In a similar way the antibonding π^*_{2px} and π^*_{2py} molecular orbitals have same energy and are also doubly degenerate.

The energy levels of different molecular orbitals are shown diagrammatically in Fig. 4.9.

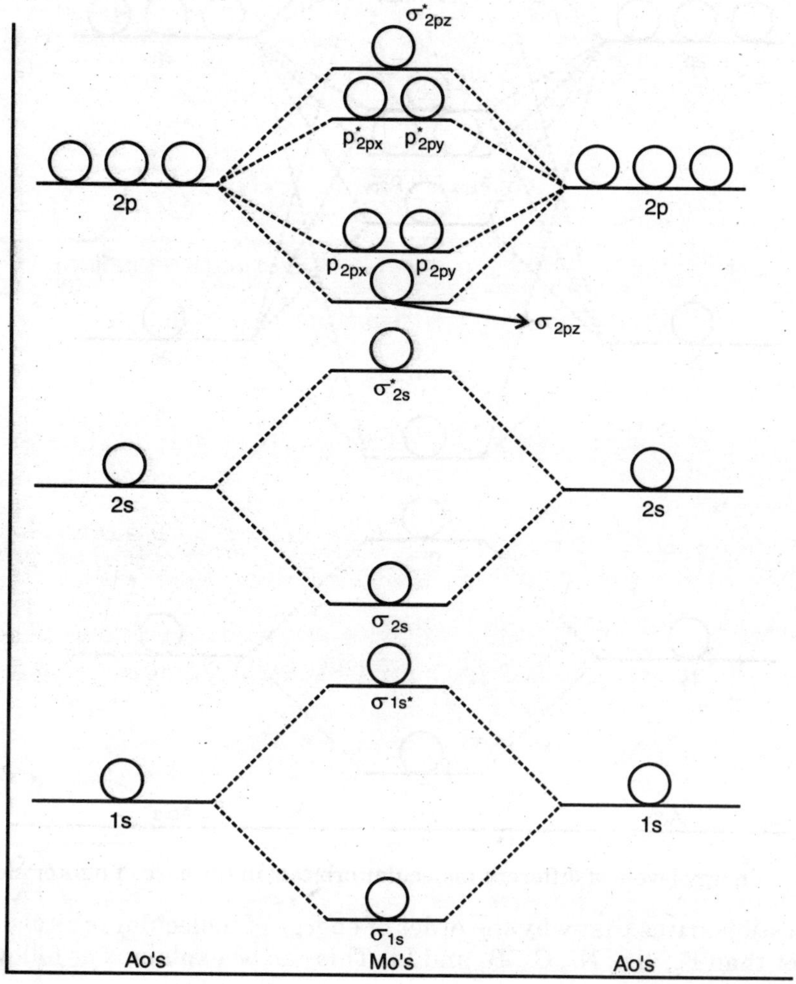

Fig. 4.9 Energy levels of different molecular orbitals of O_2, F_2, Ne_2.

However, it has been revealed by spectroscopic studies that the above mentioned sequence of energy levels is not correct for all molecules. In the case of homonuclear diatomic molecules of second row elements of the periodic table upto N_2 (viz. Li_2, Be_2, B_2, C_2, N_2), the σ_{2pz} orbitals are of higher energy than the π_{2px} and π_{2py} molecular orbitals. For these atoms the order is:

$$\sigma_{1s},\ \overset{*}{\sigma}_{1s},\ \sigma_{2s},\ \overset{*}{\sigma}_{2s},\ \left\{\begin{matrix}\pi_{2px}\\ \pi_{2py}\end{matrix}\right.\ \sigma_{2pz},\ \left\{\begin{matrix}\pi^*_{2px}\\ \pi^*_{2py}\end{matrix}\right.\ \overset{*}{\sigma}_{2pz}$$

$\longrightarrow$ *increasing energy*

The energy level diagram for these molecules is given in Fig. 4.10.

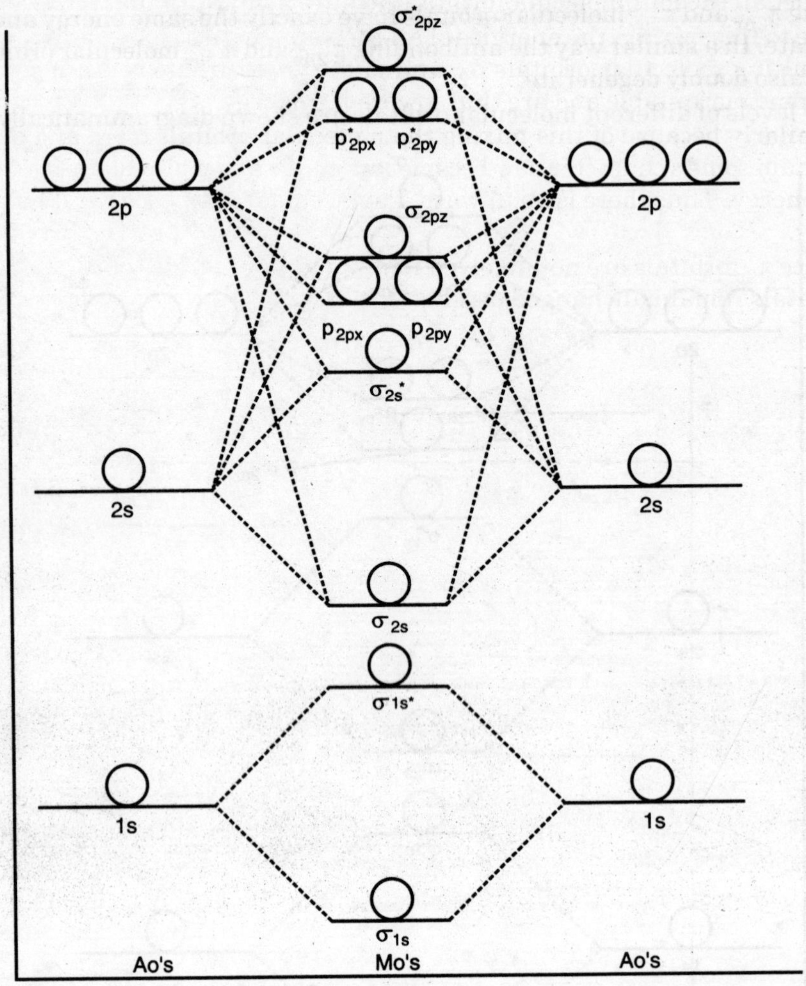

Fig. 4.10 Energy levels of different molecular orbitals in the case of lighter elements.

Now the question arises that why the order of energy of molecular orbitals is different for molecules lighter than F_2, viz., N_2, C_2, B_2 and Li_2 This can be explained as follows:

- In case of homonuclear diatomic molecules of second row elements, if the energy difference between 2s and 2p atomic orbital is **large**, then the 2s orbital of one atom interacts slightly with the 2pz orbital of the other atom. Though the symmetry is proper for overlap, but the energy difference between the two orbitals is so great that the overlap is not effective. Thus σ_s molecular orbitals are composed of the two 2s atomic orbitals and the order of energy levels is as shown in Fig. 4.11 This is true for diatomic molecules of O_2 and F_2.
- In case of elements upto nitrogen, in the second row, the energy difference between 2s and 2p atomic orbitals is **small** i.e. 2s orbital of one atom is close to 2pz orbital of second atom. As the two orbitals are of same symmetry their overlap cannot be ignored. As a result of interaction between *s* orbital of one atom with the *p* orbital of the other and vice versa, σ_{2s} and σ^*_{2s} molecular orbital **do not retain their pure *s* character** and σ_{2pz} and σ^*_{2pz} molecular orbitals **do not retain their pure *p* character**.

— Infact all the four molecular orbitals acquire a mixed *sp* character.

— Due to this mixing, the energies of all the four orbitals change in such a way that the resulting molecular orbitals σ_{2s} and and σ^*_{2s} which also contain some *p* character, become more stable and are thus lowered in energy.

— Similarly because of this mixing the molecular orbitals $\sigma_{(2pz)}$ and $\sigma^*_{(2px)}$ which also contain some *s* character now become less stable and molecular orbital are thus raised in energy. Thus there is an upward displacement of σ_{2pz} so that it lies above π_{2px} and π_{2py}.

— Since π_{2p} orbitals are not involved in mixing, the energies of π_{2px} and σ_{2py} molecular orbitals remain unchanged, as shown in the Fig. 4.11.

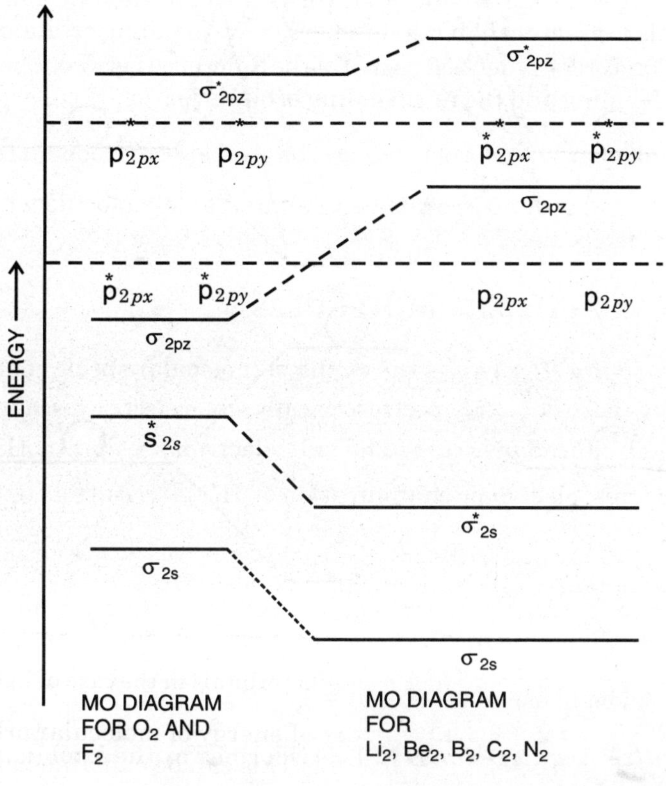

Fig. 4.11 The change in energies of MOs due to mixing of orbitals of the same symmetry.

- Considering the above points it can be concluded that the difference in 2*s* and 2*p* atomic orbitals of atoms up to nitrogen are small and hence these would overlap effectively resulting in the orbitals, the energies of which will be different than pure σ_{2s} and σ_{2pz} molecular orbitals. Same explanation is applicable for antibonding σ_{2s} and σ^*_{2px} orbitals. On the contrary, in the case of O_2 and F_2 the energy difference between 2*s* and 2*p* is quite large and hence no overlapping between these atomic orbitals is possible. Therefore energies of σ_{2s}, σ^*_{2s}, σ_{2pz}, σ^*_{2pz} remain unaltered and hence σ_{2pz} molecular orbital has lower energy than the energies of π_{2px} and π_{2py} molecular orbitals.

4.4 FILLING OF ELECTRONS IN MOLECULAR ORBITALS

For simple homoatomic molecules, the filling of electrons are carried out in a similar fashion as in Atomic Orbitals. Following rules are followed:

1. ***Aufbau Principle:*** The orbital with lowest energy is to be filled first of all i.e. electrons enter various molecular orbitals in the order of their increasing energies.
2. ***Pauli's Exclusion Principle:*** The maxmimum number of electrons in a molecular orbital cannot exceed two and these should have opposite spins.
3. ***Hund's Rule:*** Pairing of electrons takes place only when each molecular orbital of the same energy has at least one electron each.
4. A bond between two atoms is said to be formed when the total number of electrons in bonding orbitals is greater than the numbers in antibonding orbitals.
5. ***Bond order:*** Bond order is defined as half of the difference between the number of electrons present in the bonding and the antibonding orbitals i.e.

$$\text{Bond order} = \frac{1}{2}\text{ (No. of bonding orbitals–No. of antibonding orbitals).}$$

A negative or zero bond order corresponds to an unstable molecule whereas positive bond order indicates a stable molecule.

4.5 HOMONUCLEAR DIATOMIC MOLECULES

1. ***Hydrogen molecule-ion H_2^+:*** This is the simplest molecular specie containing one electron only. Its existence has been detected spectroscopically when electric discharge is passed through hydrogen gas under reduced pressure. The only electron present in H_2^+ ion enters the σ_{1s} molecular orbital. Thus, electronic configuration of H_2^+ is written as $\sigma_{(1s)}{}^1$ and bond order = $\frac{1}{2}(1-0) = \frac{1}{2}$
2. ***Hydrogen molecule H_2:*** Each hydrogen atom contains 1 electron and hence there are two electrons in the molecule, occupying the lowest energy level. Thus, the electronic configuration of H_2 will be σ_{1s}^2 and bond order = $\frac{1}{2}(2-0) = 1$.
3. ***Hydrogen molecular negative ion H_2^-:*** This is formed by the combination of one hydrogen atom having one electron in 1s orbital with a hydrogen ion having 2 electrons. Thus H_2^- ion has 3 electrons. The electronic configuration will thus be $\sigma_{1s}^2\ \overset{*}{\sigma}{}_{1s}^{1}$ and bond order = $\frac{1}{2}(2-1) = \frac{1}{2}$.

Table 4.3 Characteristics of H_2^+, H_2, H_2^-

Specie	*Electronic config.*	*Bond Order*	*Magnetic character*
H_2^+	σ_{1s}^1	$\frac{1}{2}$	Paramagnetic
H_2	σ_{1s}^2	1	Diamagnetic
H_2^-	$\sigma_{1s}^2\ \overset{*}{\sigma}{}_{1s}^{1}$	$\frac{1}{2}$	Paramagnetic

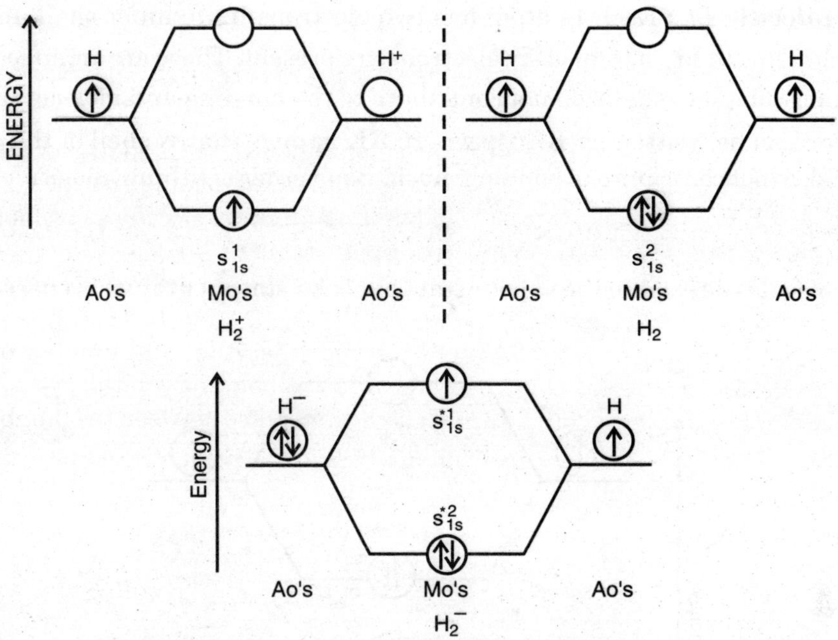

Fig. 4.12 Energy Level Diagrams for H_2^+ H_2 and H_2^-

4. ***Helium Molecule ion*** He_2^+**:** This may be considered to be formed from He atom and a He^+ ion. In the molecular ion there are three electrons which are arranged in the molecular orbitals as σ_{1s}^2, $\overset{*1}{\sigma_{1s}}$. Thus helium molecular ion can exist, but it is not very stable. Bond order = $\frac{1}{2}(2-1)=\frac{1}{2}$.

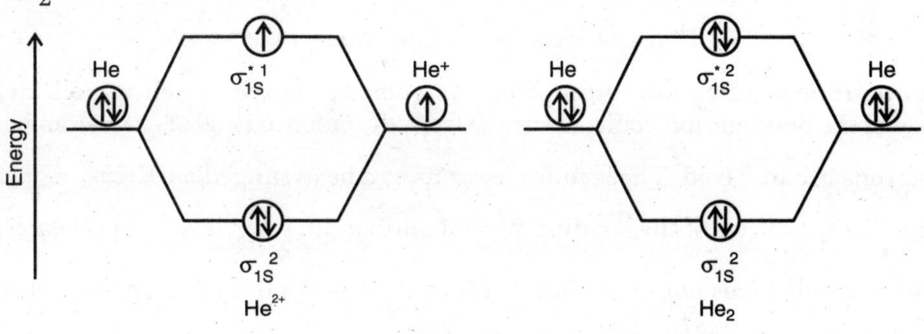

Fig. 4.13

5. ***Helium Molecule,*** He_2**:** The electronic configuration of the atom is $1s^2$. *He* molecule can be considered to form from two atomic orbitals. In the molecular He_2, there are four electrons which are arranged as σ_{1s}^2, $\overset{*2}{\sigma_{1s}}$. Thus, the stabilizing effect of bonding orbitals is cancelled out by the destabilizing effect of antibonding orbitals and the molecule does not exist. Bond order = $\frac{1}{2}(2-2)=0$.

6. ***Lithium Molecule Li_2:*** Each Li atom has two electrons in its inner shell and one in the valence shell. Thus in Li_2, a total of six electrons are present. These are arranged as σ_{1s}^2, $\overset{*}{\sigma}{}_{1s}^2$, σ_{2s}^2. The inner complete shells do not contribute to the bonding and molecular electron i.e. configuration can be written as $KK\sigma_{2s}^2$ where KK implies that K shell of the two atoms is complete and do not contribute to bonding much. Bond order of lithium molecule comes out to be $\frac{1}{2}(4-2)=1$

Lithium molecules do exist in the vapour state but solid state of lithium is energetically more favourable.

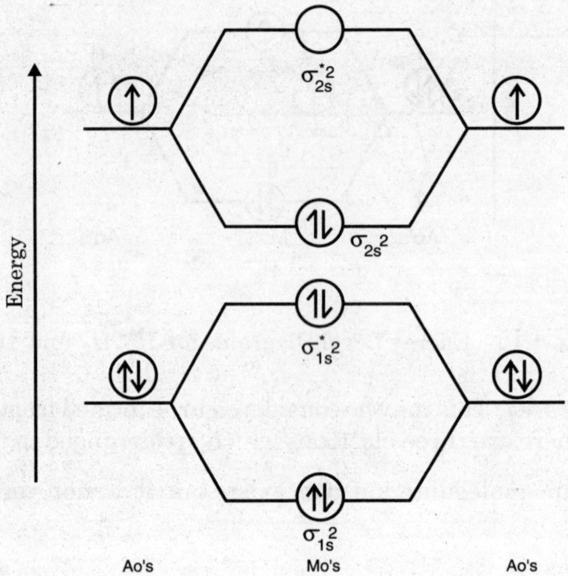

Fig. 4.14 Energy level diagram for Li_2.

7. ***Beryllium Molecule Be_2:*** Beryllium atom has 2 electrons in the valence shell. Hence in the formation of the diatomic molecule, electrons from the valence shell of each atom i.e. a total of four electrons are involved. These four electrons can be arranged as $KK\sigma_{2s}^2\ \overset{*}{\sigma}{}_{2s}^2$. Thus, it is apparent that the effects of the bonding σ_{2s} and antibonding $\overset{*}{\sigma}_{2s}$ levels cancel each other and hence the molecule does not exist. Bond order = $\frac{1}{2}(4-4)=0$. This zero bond order further suggests that existance of Be_2 molecule is not possible.
8. ***Boron Molecule B_2:*** The valence shell of each atom of boron contains 3 electrons (Z = 5). When two atoms of boron combine to form B_2 molecule the molecule contains a total of 10 electrons. These are arranged in the molecular orbitals as- $KK\,\sigma_{2s}^2\ \overset{*}{\sigma}{}_{2s}^2 \begin{cases}\pi 2p1_y \\ \pi 2p1_x\end{cases}$. It may again be pointed out here that boron is a light atom and order of energies of molecular orbits is different from the usual arrangement i.e π_{2px} and $2\pi_{2py}$ orbitals are at lower energy level than the σ_{2pz} orbital for the reason explained in section 4.3. Since π_{2px} and π_{2py} are degenerate,

the electrons are filled according to Hund's rule i.e each of these orbitals is singly occupied. The order of the energies of molecular orbitals as given above is supported by the experimental facts also. Boron molecule is found to be paramagnetic in nature which can be very well explained on the basis of this order, as there are two unpaired electrons one each in π_{2px} and π_{2py} molecular orbitals.

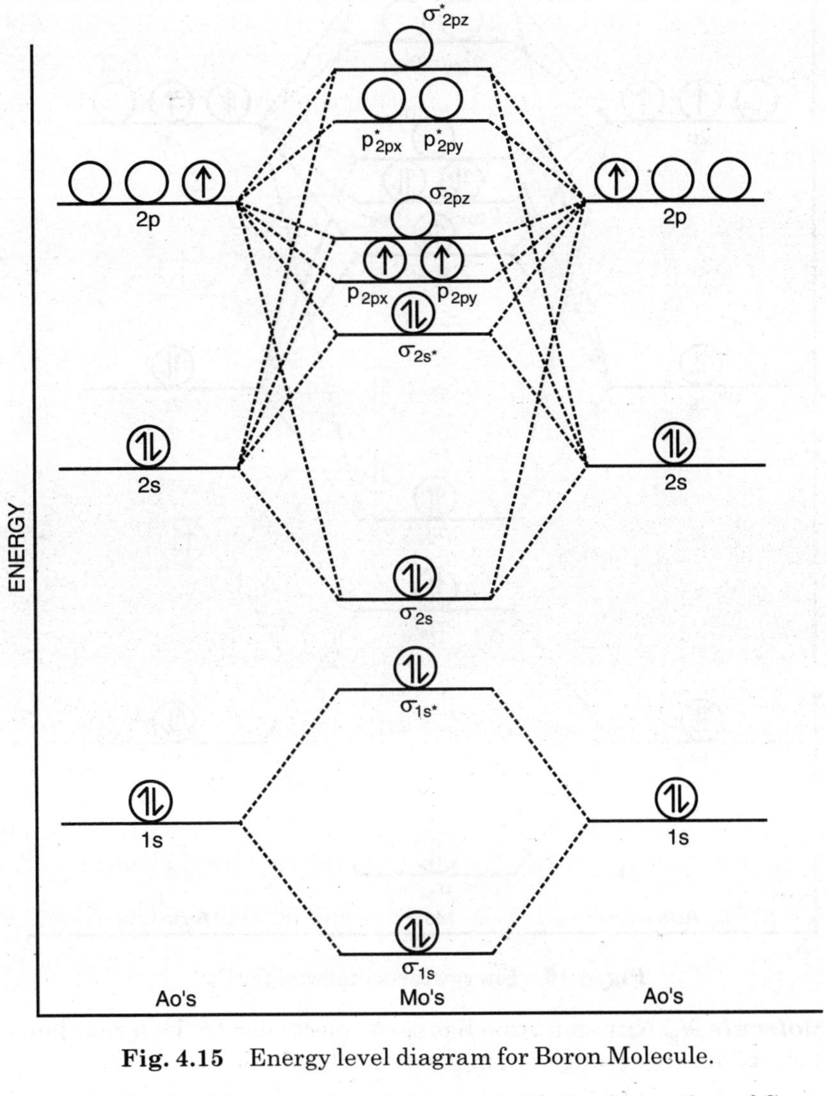

Fig. 4.15 Energy level diagram for Boron Molecule.

9. ***Carbon Molecule C_2:*** Carbon atom has 6 electrons. In the formation of C_2 molecule, a total of 12 electrons have to be accommodated in the molecular orbitals. These are arranged as:

$KK\ \sigma_{2s}^2\ \overset{*}{\sigma}{}_{2s}^{2} \begin{cases} \pi_{2py}^2 \\ \pi_{2px}^2 \end{cases}$. Bond order $= \frac{1}{2}(8 - 4) = 2$. Thus the molecule should be stable and diamagnetic because all the electrons are paired. Similar to the case of B_2 molecule, in C_2 also reversal of energies of σ_{2pz} and π_{2px}, π_{2py} orbitals take place, due to the mixing of the 2s and

2p orbitals. Though it appears that C_2 molecule is stable, but carbon exists as macro molecule in graphite and diamond. These are even more stable than C_2, hence in preference to C_2, diamond and graphite are formed.

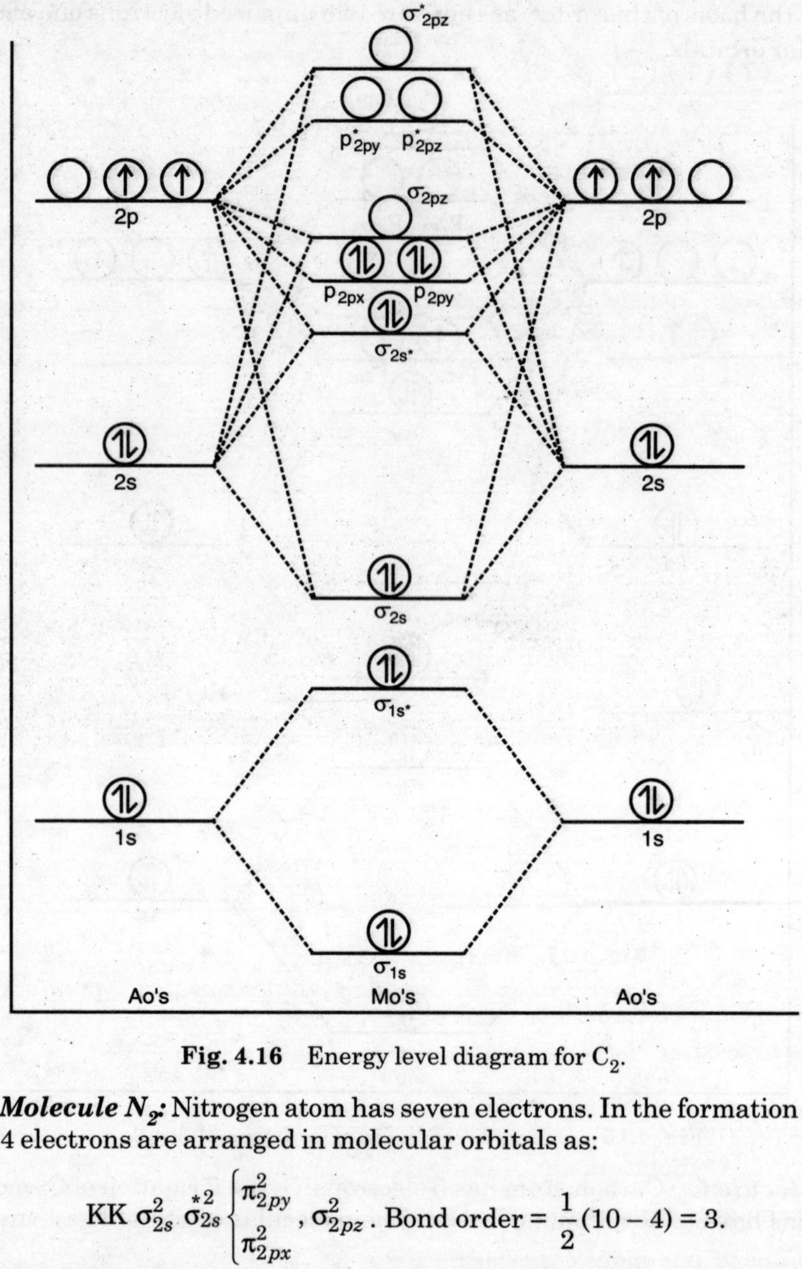

Fig. 4.16 Energy level diagram for C_2.

10. *Nitrogen Molecule N_2:* Nitrogen atom has seven electrons. In the formation of N_2 molecule, a total of 14 electrons are arranged in molecular orbitals as:

$$KK\ \sigma_{2s}^2\ \overset{*2}{\sigma}_{2s} \begin{cases} \pi_{2py}^2 \\ \pi_{2px}^2 \end{cases}, \sigma_{2pz}^2 \text{. Bond order} = \frac{1}{2}(10-4) = 3.$$

Thus nitrogen molecule contains a triple bond and is highly stable therefore nitrogen exist as N_2 molecule. Moreover all the electrons are paired therefore it is diamagnetic.

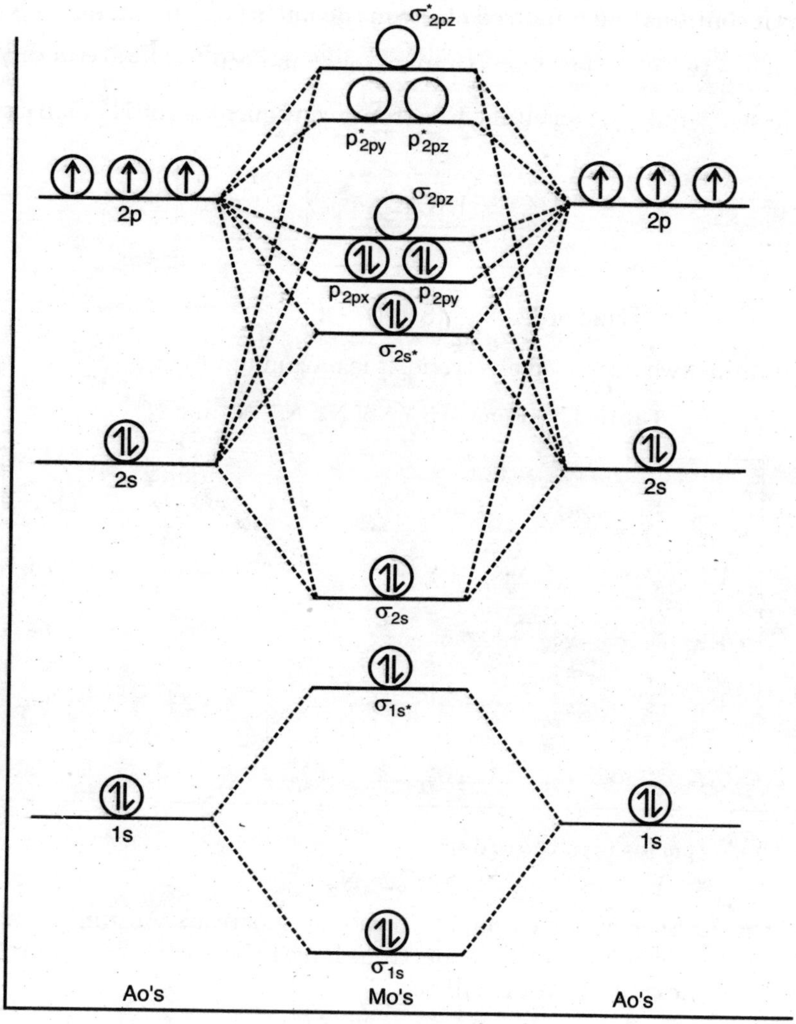

Fig. 4.17 Energy Level diagram for N_2.

N_2^+ ion: N_2^+ ion has one electron less than N_2 molecule. This electron will be lost from σ_{2pz} orbital. Hence the molecular electronic configuration will be –

$$N_2^+ = KK\ \sigma_{2s}^2\ \sigma_{2s}^2\ \pi_{2px}^2 = \pi_{2py}^2\ \sigma_{2pz}^1$$

$$\text{Bond order} = \frac{1}{2}(7-2) = 2\frac{1}{2}$$

The ion is paramagnetic due to the presence of unpaired electron.

N_2^- ion: N_2^+ ion has or one more electron than the nitrogen molecule. This extra electron is added in $\overset{*}{\pi}_{2px}$ $\overset{*}{\pi}_{2py}$ orbital, thus molecular ion electronic configuration will be-

$$N_2^+ = KK\ \sigma_{2s}^2\ \overset{*}{\sigma}{}_{2s}^2\ \pi_{2px}^2 = \pi_{2py}^2\ \sigma_{2pz}^2\ \overset{*}{\pi}{}_{2px}^1$$

$$\text{Bond order} = \frac{1}{2}(8-3) = 2\frac{1}{2}$$

Moreover, as it contains one unpaired electron the ion is paramagnetic.

N_2^{2-} ion: N_2^{2-} ion has two more electrons than nitrogen molecule. These two extra electrons are added one each in $\overset{*}{\pi}_{2px}$ and $\overset{*}{\pi}_{2py}$ orbitals. Electronic configuration of N_2^{2-} ion can be written as:

$$N_2^{2-} = KK\ \sigma_{2s}^2\ \overset{*}{\sigma}{}_{2s}^2 \left\{\begin{matrix}\pi_{2px}^2 \\ \pi_{2py}^2\end{matrix}\right.\ \sigma_{2pz}^2 \left\{\begin{matrix}\overset{*}{\pi}{}_{2px}^1 \\ \overset{*}{\pi}{}_{2py}^1\end{matrix}\right.$$

$$\text{Bond order} = \frac{1}{2}(8-4) = 2$$

Further it contains two unpaired electrons, it is paramagnetic.

Table 4.4 Comparision of N_2, N_2^+, N_2^- and N_2^{2-}

Molecular Specie	*Electronic Configuration*	*Bond Order*	*Magnetic Character*
N_2	$KK\ \sigma_{2s}^2\ \overset{*}{\sigma}{}_{2s}^2\ \pi_{2px}^2\ \pi_{2py}^2\ \sigma_{2pz}^2$	3	Diamagnetic
N_2^+	$KK\ \sigma_{2s}^2\ \overset{*}{\sigma}{}_{2s}^2\ \pi_{2px}^2\ \pi_{2py}^2\ \sigma_{2pz}^1$	$2\frac{1}{2}$	Paramagnetic
N_2^-	$KK\ \sigma_{2s}^2\ \overset{*}{\sigma}{}_{2s}^2\ \pi_{2px}^2\ \pi_{2py}^2\ \sigma_{2pz}^2\ \overset{*}{\pi}{}_{2px}^1$	$2\frac{1}{2}$	Paramagnetic
N_2^{2-}	$KK\ \sigma_{2s}^2\ \overset{*}{\sigma}{}_{2s}^2\ \pi_{2px}^2\ \pi_{2py}^2\ \sigma_{2pz}^2\ \overset{*}{\pi}{}_{2px}^1\ \overset{*}{\pi}{}_{2py}^1$	2	Paramagnetic

Stability of these species is in the order:

$$N_2 > N_2^+ = N_2^- N_2^{2-}$$

Oxygen Molecule O_2: Each oxygen atom has 6 valence electrons. Assuming that inner orbitals do not participate in bonding, altogether 12 electrons have to be accomodated in molecular orbitals. The molecular electronic configuration will be:

$$O_2 = KK\ \sigma_{2s}^2\ \overset{*}{\sigma}{}_{2s}^2\ \sigma_{2pz}^2 \left\{\begin{matrix}\pi_{2py}^2 \\ \pi_{2px}^2\end{matrix}\right. \left\{\begin{matrix}\overset{*}{\pi}{}_{2py}^1 \\ \overset{*}{\pi}{}_{2px}^1\end{matrix}\right.$$

In the case of oxygen, the difference between the energies of 2s and $2p_z$ atomic orbitals is quite large (1430 kJ/mol) and hence interaction between these orbitals and subsequent *changes in order of energies of molecular orbitals do not take place.* Thus, the usual order of energies as given in section 4.3 is followed. As is evident from the energy level diagram that two unpaired electrons are present, thus the molecule is paramagnetic. This is one of the great successes of molecular orbital theory, as it could predict the paramagnetism of oxygen molecule which could not be explained on the basis of valence bond theory.

Bond order = (10-6)/2 = 2. Thus a double bond is present in the molecule.

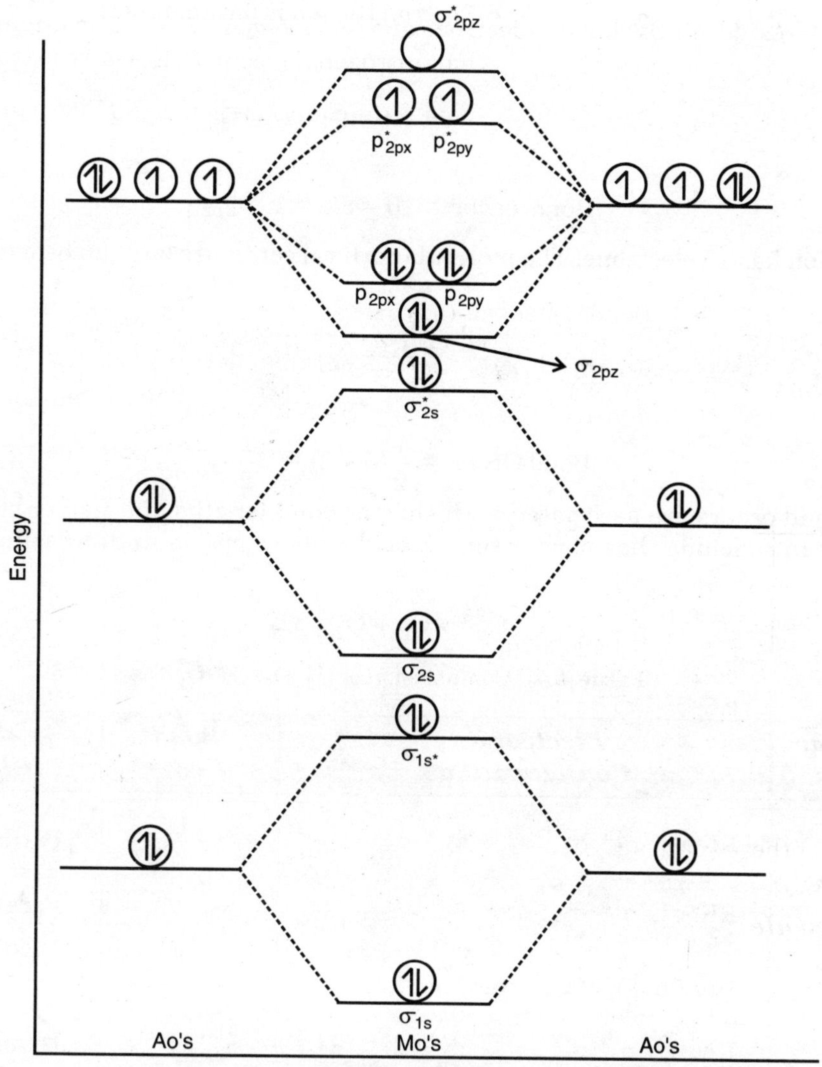

Fig. 4.18 Energy level diagram for O_2 molecule

O_2^- ion: These superoxide ions O_2^- has 17 electrons, i.e., one electron more than O_2 and hence this extra electron enters the $\overset{*}{\pi}_{2px}$ or $\overset{*}{\pi}_{2py}$ orbital. The molecular orbital electron configuration becomes:

$$= KK\ \sigma_{2s}^2\ \sigma_{2s}^{*2}\ \sigma_{2pz}^2 \begin{cases} \pi_{2y}^2 \\ \pi_{2px}^2 \end{cases} \begin{cases} \overset{*1}{\pi}_{2py} \\ \overset{*2}{\pi}_{2px} \end{cases}$$

$$\text{Bond order} = (10 - 7)/2 = 1\frac{1}{2}$$

The superoxide ion is paramagnetic.

O_2^{2-} ion: The peroxide ion O_2^{2-} has 18 electrons and has molecular orbital configuration as:

$$= \text{KK}\ \sigma_{2s}^2\ \overset{*}{\sigma}{}_{2s}^2\ \sigma_{2pz}^2 \begin{cases} \pi_{2py}^2 \\ \pi_{2px}^2 \end{cases} \begin{cases} \overset{*}{\pi}{}_{2py}^2 \\ \overset{*}{\pi}{}_{2px}^2 \end{cases}$$

$$\text{Bond order} = 10 - 8/2 = 1.$$

O_2^+ ion: This ion has 15 electrons. The molecular orbital configuration can be written as:

$$= \text{KK}\ \sigma_{2s}^2\ \overset{*}{\sigma}{}_{2s}^2\ \sigma_{2pz}^2 \begin{cases} \pi_{2py}^2 \\ \pi_{2px}^2 \end{cases} \begin{cases} \overset{*}{\pi}{}_{2py}^0 \\ \overset{*}{\pi}{}_{2px}^1 \end{cases}$$

$$\text{Bond Order} = \frac{1}{2}(8 - 3) = 2\frac{1}{2}.$$

Higher bond orders are associated with shorter bond lengths and higher bond strengths. Threfore we can conclude that bond strength of the above specie in decreasing order can be arranged as:

$$O_2^{2+} > O_2 > O_2^- > O_2^{2-}$$

Table 4.5 Comparision of O_2^+, O_2, O_2^-, O_2^{2-}

Molecular Species	*Electronic Configuration*	*Bond Order*	*Magnetic Character*
O_2^+	KK $\sigma_{2s}^2\ \overset{*}{\sigma}{}_{2s}^2\ \sigma_{2pz}^2\ \pi_{2px}^2\ \pi_{2py}^2\ \overset{*}{\pi}{}_{2px}^1$	$2\frac{1}{2}$	Paramagnatic.
O_2	KK $\sigma_{2s}^2\ \overset{*}{\sigma}{}_{2s}^2\ \sigma_{2pz}^2\ \pi_{2px}^2\ \pi_{2py}^2\ \overset{*}{\pi}{}_{2px}^1\ \overset{*}{\pi}{}_{2py}^1$	2	Paramagnatic.
O_2^-	KK $\sigma_{2s}^2\ \overset{*}{\sigma}{}_{2s}^2\ \sigma_{2pz}^2\ \pi_{2px}^2\ \pi_{2py}^2\ \overset{*}{\pi}{}_{2px}^2\ \overset{*}{\pi}{}_{2py}^1$	$1\frac{1}{2}$	Paramagnatic.
O_2^{2-}	KK $\sigma_{2s}^2\ \overset{*}{\sigma}{}_{2s}^2\ \overset{*}{\sigma}{}_{2pz}^2\ \pi_{2px}^2\ \pi_{2py}^2\ \overset{*}{\pi}{}_{2px}^2\ \overset{*}{\pi}{}_{2py}^2$	1	Diamagnetic

Flourine Molecule F_2: Flourine atom has 7 electrons in the outermost shell and hence 14 electrons are arranged in the molecular orbital as:

$$= \text{KK}\ \sigma_{2s}^2\ \overset{*}{\sigma}{}_{2s}^2\ \sigma_{2pz}^2 \begin{cases} \pi_{2py}^2 \\ \pi_{2px}^2 \end{cases} \begin{cases} \overset{*}{\pi}{}_{2py}^2 \\ \overset{*}{\pi}{}_{2px}^2 \end{cases}$$

$$\text{Bond Order} = \frac{1}{2}(8 - 6) = 1$$

The F-F bond is rather weak. Since there are no unpaired electrons, the molecule is diamagnatic.

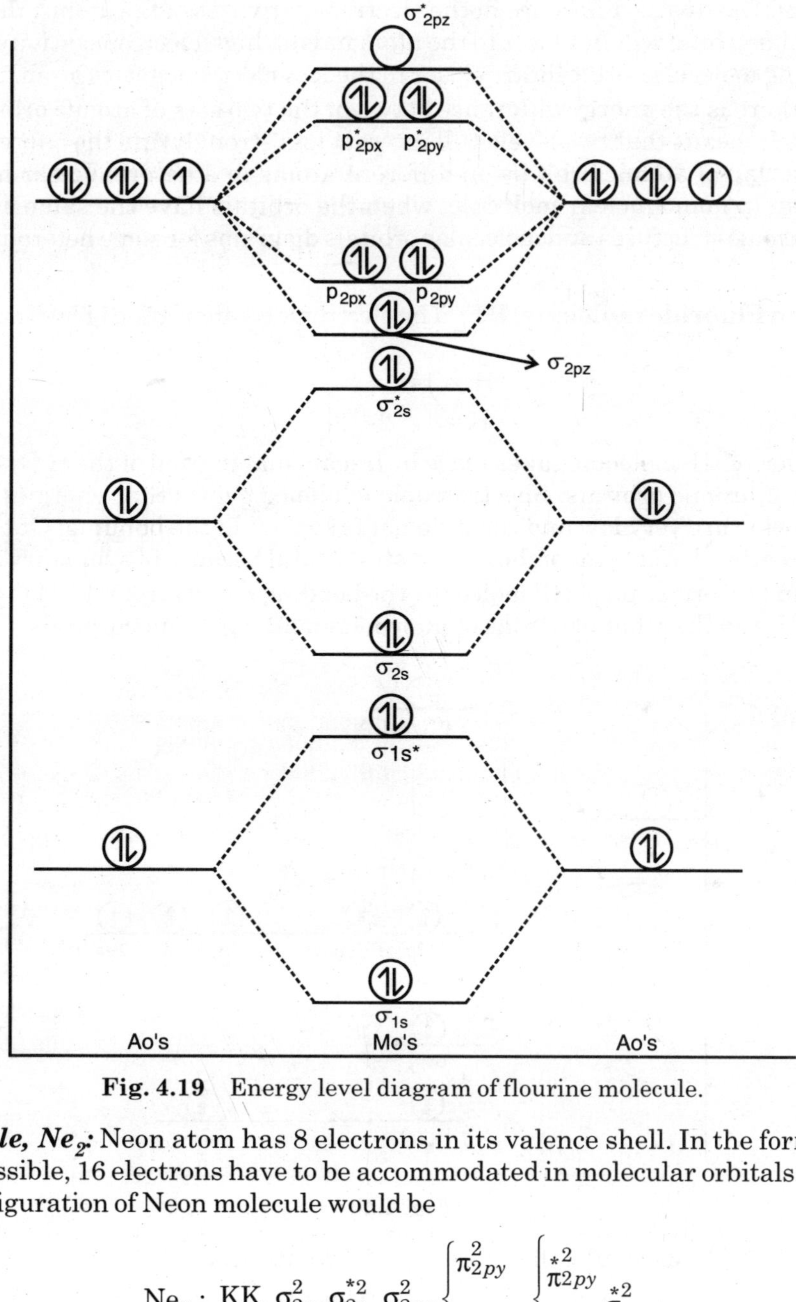

Fig. 4.19 Energy level diagram of flourine molecule.

Neon Molecule, Ne_2: Neon atom has 8 electrons in its valence shell. In the formation of Neon molecule, if possible, 16 electrons have to be accommodated in molecular orbitals. The molecular electronic configuration of Neon molecule would be

$$Ne_2 : KK\ \sigma_{2s}^2\ \sigma_{2s}^{*2}\ \sigma_{2pz}^2 \begin{cases} \pi_{2py}^2 \\ \pi_{2px}^2 \end{cases} \begin{cases} \pi_{2py}^{*2} \\ \pi_{2px}^{*2} \end{cases} \sigma_{2pz}^{*2}$$

It is apparent that the effects of the bonding orbitals and autibonding orbitals cancel, so the molecule does not exist.

4.6 HETERONUCLEAR DIATOMIC MOLECULES

The molecular orbitals of heteronuclear diatomic molecules differ from those of homonuclear diatomic molecules. Since the two atoms are different, the energies of their atomic orbitals are slightly different. Due to the difference in the electronegativities of the atoms, the electrons in a bonding molecular orbital will lie closer to the atom having high electronegativity. The electrons in an antibonding molecular orbitals are closer to the less electronegative atom.

Moreover, there is the energy difference between the two sets of atomic orbitals on the two different atoms. It means that two levels will interact less strongly and the energy lowering as a result of the overlap of atomic orbitals on different atoms in a heteronuclear molecule is less pronounced than in homonuclear molecule, when the orbitals have the same energies. Let us discuss the electronic structures and molecular orbitals diagrams for some heteronuclear diatomic molecules.

1. Hydrogen Fluoride molecule, HF: The electronic configuration of hydrogen and fluorine atoms are

$$H = 1s$$
$$F = 1s^2\ 2s^2\ 2p^5$$

The formation of HF molecule takes place by linear combination of the H (1s) atomic orbital with one of the F atomic orbitals. Spectroscopic evidence shows that energies of $1s^2$ and $2s^2$ electrons of fluorine are very low and these do not take part in the bonding. Out of the three 2p orbitals,only 2pz orbital is able to combine with an s-orbital because of symmetry considerations.

Therefore, in the formation of HF molecule, the bonding results from the H (1s) electrons and $2p_z$ electrons of F and the other electrons of fluorine remains in atomicorbitals.

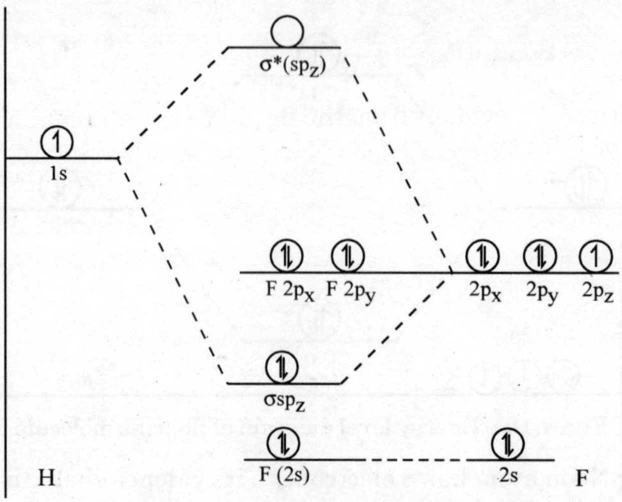

Fig. 4.20

The structure of HF molecule, therefore, may be written as

$$H\ (1s) + F\ (1s\ 2s^2\ 2p^5) \longrightarrow HF\ [K(2s)^2\ (\sigma_{spz})^2\ (2p_x)^2\ (2p_y)^2]$$

The bonding σ_{2pz} is the only molecular orbital formed in HF molecule.

The rest of the electrons remain in their atomic orbitals. The configuration of HF molecule, closely resembles that of the neon atom ($1s^2 2s^2\ 2p^6$). However, in neon, all the orbitals are atomic.

It may be noted that $\sigma_{(sp)}$ is the only MO formed in HF molecule and bonding MO contains two electrons.

In case of heteronuclear HF molecule, the electrons are no longer shared equally between the two nuclei. The F($2p_z$) atomic orbital is contributing more than the H (1s) orbital to the molecular orbitals. As seen in Fig. 4.20, the molecular orbital has an energy which lies closer to that of $2p_z$ orbital of fluorine than to that of 1s-orbital of hydrogen.

Similarly, the other HX molecules have single bond with an unequal sharing of the electrons between the nuclei. The chlorine, bromine and iodine atoms contribute respectively $3p_z$, $4p_z$ and $5p_z$ atomic orbitals in formation of HX molecules.

The molecular orbital diagram for HCl will be similar to HF.

2. Nitric Oxide Molecule (NO): The electronicconfigurations of N and O atoms are:

$$N = 1s^2\ 2s^2\ 2p_x^{\ 1}\ 2p_y^{\ 1}\ 2p_z^{\ 1}$$
$$O = 1s^2\ 2s^2\ 2p_x^{\ 2}\ 2p_y^{\ 1}\ 2p_z^{\ 1}$$

There are five electrons in the outer most shell of N atom and six electrons in outer shell of O atom.Thus, a total of 11 electrons are to be accommodated in the molecular orbitals of NO molecule.The MOdiagram for NO molecule is shown in Fig. 4.21. It is clear from the diagram that the bonding MO's are closer to oxygen and antibonding MO's are closer to N atom because oxygen is more electronegative than nitrogen.

The molecular orbital electronic configuration of NO molecule is:

$$NO = \left[\sigma_{2s}^2 \quad \sigma_{2s}^{*2} \quad \sigma_{2p_z}^2 \quad \begin{matrix} \pi_{2p_x}^2 \\ \pi_{2p_y}^2 \end{matrix} \quad \begin{matrix} \pi_{2p_x}^{*1} \\ \pi_{2p_y}^{*o} \end{matrix} \right]$$

$$\text{Bond order} = \frac{8-3}{2} = 2\frac{1}{2}$$

NO is paramagnetic. As is evident from the fig. 4.21 that it is similar to as that of diatomic molecules such as O_2 or F_2 but energy level of N and O are not the same.

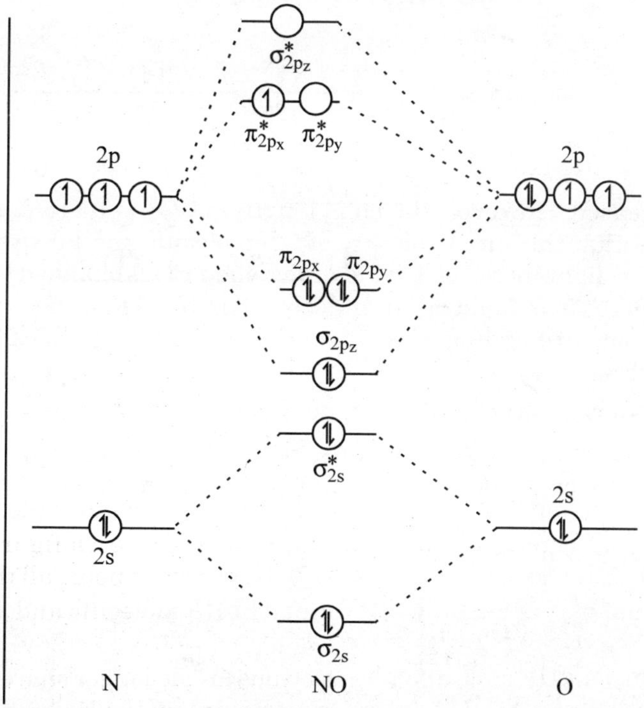

Fig. 4.21 Molecular orbital diagram for NO molecule.

3. Carbon monoxide molecule, CO: The electronic configuration of carbon and oxygen atoms are:

$$C : 1s^2\ 2s^2\ 2p^2 \quad \text{and} \quad O : 1s^2\ 2s^2\ 2p^4$$

There are four electrons in the outermost shell of carbon atom and six electron in the outermost shell of oxygen. Thus, a total of 10 electrons are to be accommodated in the molecular orbitals of CO molecule. The MO diagram for CO molecule may be drawn similar to NO molecule as shown in Fig. 4.21.Then the molecular orbital configuration may be written as:

$$CO : KK \left[\sigma_{2s}^2 \quad \sigma_{2s}^{*2} \quad \sigma_{2p_z}^2 \quad \begin{matrix} \pi_{2p_x}^2 \\ \pi_{2p_y}^2 \end{matrix} \right]$$

$$\text{Bond order} = \frac{8-2}{2} = 3$$

Thus CO molecule has a **triple bond**. Its bond dissociation energy has been found to 1067 kJ mol^{-1} and bond length equal to 1.128 Å. However, ***it may be noted that the molecular orbital diagram or CO is somewhat complicated than NO molecule.***The above configuration is not correct.

Oxygen is more electronegative than carbon and therefore, the atomic orbitals of oxygen are lower than those of carbon. Due to this difference, the $\overset{*}{\sigma}_{2s}$ MO is slightly higher in energy than σ_{2pz}, π_{2px} and π_{2py} MOs. The diagram for CO molecule is shown in 4.22.

The better understanding of the structure is possible by comparison of CO and CO$^+$. CO$^+$ is formed by the loss of one electron from CO molecule. If the electronic coinfiguration for CO molecule similar to NO, is correct, then electronic configuration for CO$^+$ is

$$CO : KK \left[\sigma_{2s}^2 \quad \sigma_{2s}^{*2} \quad \sigma_{2p_z}^2 \quad \begin{matrix} \pi_{2p_x}^2 \\ \pi_{2p_y}^1 \end{matrix} \right]$$

$$\text{Bond order} = \frac{7-2}{2} = 2\frac{1}{2}$$

Thus, when CO is ionized to CO$^+$, then its bond order should be reduced to $2\frac{1}{2}$ and the bond length should be increased. However, the bond length in CO$^+$ is 1.115 Å and is less than in CO (1.128 Å). This means that the simple picture of CO molecule similar to other molecules is not correct. In fact, there is slight increase in bond order when CO is changed toCO$^+$. The most likely explanation for the increase in bond order or decrease in bond length is that the electron must have been removed from antibonding molecular orbital. This has been explained on the basis of energy levels of C and O atoms in Fig. 4.22.

The above figure shows that $\overset{*}{\sigma}_{2s}$ MO is higher in energy than σ_{2pz}, π_{2px} and π_{2py} MO's. Now when an electron is lost, it is removed from $\overset{*}{\sigma}_{2s}$ MO and therefore, bond order increases. As a result, CO$^+$ has slightly strong bond than does CO and bond length in CO$^+$ is less than that in CO molecule.

The molecular orbital configuration may be written as:

$$CO : KK \left[\sigma_{2s}^2 \quad \sigma_{2p_z}^2 \quad \begin{matrix} \pi_{2p_x}^2 \\ \pi_{2p_y}^2 \end{matrix} \quad \sigma_{2s}^{*2} \right]$$

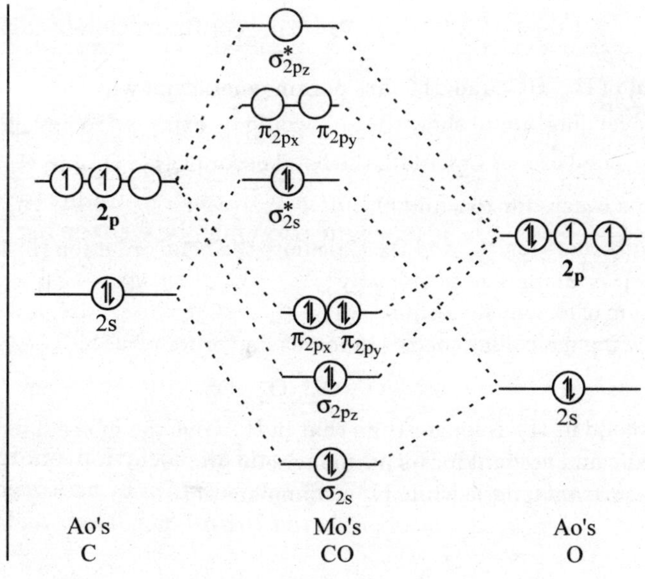

Fig. 4.22 Molecular orbital diagram for CO molecule.

QUESTIONS

SHORT ANSWERS QUESTIONS
(Refer to the Respective Sections for Answers)

Section 4.1

1. What do you understand by molecular orbital?
2. What is the difference between VBT and MOT?
3. What are the main point of MOT?

Section 4.2

4. What do you understand by LCAO method
5. Sketch the shapes of molecular orbitals formed by the overlap of
   (i) Two s-orbitals.
   (ii) End on overlap of two p-orbitals.
   (iii) Side on overlap of two p-orbitals. Name the orbitals formed in each case.
6. What is the maximum number of electrons that can occupy a molecular orbital? How many MO.'s of H_2 originate from the hydrogen 1s atomic orbitals?
7. What is meant by bonding and antibonding molecular orbitals?
8. What are the conditions of combination of atomic orbitals?

Section 4.3

9. Give molecular orbital description of hydrogen molecule and deduce the bond order.
10. Discuss the energy level diagram of F_2, O_2 and Ne_2.
11. What is the order of the energy of the molecular orbitals in case of diatomic molecular of second row elements of the periobic table up to N_2.

Section 4.4, 4.5

12. Explain why Ne_2 does not exist
13. Which species out of H_2, H_2^{2+} and H_2^{2-} are paramagnetic and why?
14. Use the energy level diagram to show that N_2 would be expected to have a triple bond.
15. What is meant by bond order? Calculate the bond order of H_2^+, O_2 and N_2 molecules.
16. Calculate the bond orders for He_2 and O_2^+.
17. Calculate the bond orders for N_2^+ and O_2. Calculate the bond order for H_2^+ ion. Is this ion expected to be paramagnetic or diamagnetic and why?
18. Draw M.O. diagram of oxygen molecule.
19. Arrange the following molecular species in increasing order of stability (giving bond orders):—

$$O_2,\ O_2^+,\ O_2^-,\ O_2^{2-}$$

20. Explain why the bond in H_2^+ is longer than that in H_2. Give the molecular orbital energy diagram for oxygen molecule and account for its paramagnetic property.
21. Explain why O_2 is paramagnetic while N_2 is diamagnetic.

Section 4.6

22. What do you mean by coordination compounds?
23. What is the difference between a double salt and a coordination compound?
24. Explain the terms (i) Central metal atom (ii) Ligand (iii) Coordination sphere.
25. What are polydentate ligands?
26. Name a ligand which forms a chelate?
27. What is meant by unidentate and bidentate ligand?
28. What are the main postulates of Werner's theory of coordination compounds?
29. What is meant by EAN ?
30. It is said EAN is over emphasis on the stability of the inert gas configuration. Discuss?
31. Explain the following terms with respect to metal complex
    (i) Primary valence.
    (ii) Secondary valence.

Section 4.9

32. What is structural isomerism.
33. How structural isomerism differs from stereo isomerism.
34. Illustrate with an example.
    (i) Ionization isomerism
    (ii) Linkage isomerism.
    (iii) Hydrate isomerism.
35. What is geometrical isomerism?
36. Discuss optical isomerism in compounds having coordination number 4.
37. Discuss geometrical isomerism in Ma_2b_2 type of complexes.

LONG ANSWER QUESTIONS

1. (a) What is Molecular Orbital theory?
   (b) Give the electron charge density diagrams for bonding and antibonding molecular orbitals.
2. (a) What do you understand by the terms bonding and antibonding molecular orbitals? Why are they so called? Illustrate your answer by taking into consideration the structure of hydrogen ion H_2^+.

(b) Is B_2 molecule paramagnetic or diamagnetic? Discuss.

3. Explain how the atomic orbitals combine to form bonding and antibonding molecular orbitals. What are the limitations to such combinations? Explain why in a homonuclear molecule, an s orbital of one atom cannot combine with a px or a py orbital of another atom of the same element.
4. With the help of molecular orbital diagrams. explain why

   (a) the bond order in N_2^+ ion is less than that in N_2 molecule whereas the bond order in O_2^+ is greater than the O_2 molecule.

   (b) Oxygen molecule is paramagnetic while nitrogen molecule is diamagentic.

   (c) Hydrogen forms diatomic molecule while helium remains monoatomic.

   (d) The bond order in O_2^- is less than that in O_2 which, in turn, is less than that in O_2^+
5. What is meant by bond order? Calculate the bond order of $He_2{}^+$, O_2, O_2^{2-} molecules.
6. Give the essential features of valence bond and molecular orbital theories.
7. Explain the following on the basis of MO theory

   (a) Ne_2 does not exist.

   (b) Why N_2 has greater bond dissociation energy than N_2^{2+} whereas O_2^{2+} has lesser bond dissociation energy than O_2

   (c) Why the bond in H_2 is shorter than that in H_2^-
8. What is meant by bond order? Calculate the bond order of He_2^+, O_2 and N_2 molecules.
9. Which species out of H_2, H_2^{2+} and H_2^{2-} are paramagnetic and why?
10. Explain LCAO. What are the conditions of combination of atomic orbitals.
11. Arrange the following molecular species in increasing order of their stability: O_2, O_2^+, O_2^-, O_2^{2-}
12. Discuss the shapes of molecular orbitals formed by combination of the following atomic orbitals:

    (a) $2pz + 2pz$ (along internuclear axis)

    (b) $2py + 2py$ (Lateral overlap)

    (c) $2s + 2pz$
13. Draw MO energy level diagram for the F_2 molecule. What is the order of stability among F_2^{2+}, F_2 and F_2^-.
14. State main points of difference between bonding and anti-bonding molecular orbitals.
15. (a) How are the bond order and bond length related? Explain with example.

    (b) Arrange the molecules H_2, O_2, F_2 and N_2 in the order of increasing bond length.
16. Differentiate between:

    (a) Valence bond theory and molecular orbital theory

    (b) Atomic orbital and molecular orbital.
17. Write notes on:

    (a) Hydrogen bond

    (b) Effective atomic number

    (c) Effect of temperature on enzyme catalysis **G.G.S.I.P.U. I Term, Feb. 2001**
18. Discuss briefly the molecular orbital theory. Discuss the formation of H_2 and O_2 molecule on the basis of this theory. **G.G.S.I.P.U, II Term, Apr. 2001**
19. What do you understand by the term bonding and antibonding orbitals? Give the conditions for the combination of atomic orbitals to form molecular orbitals. How does molecular orbital theory help to explain the paramagnetic character of oxygen and diamagnetic character of nitrogen (O = 8, N = 7) **G.G.S.I.P.U. End Term, May 2001**

20. (a) Explain with reasoning
    (i) Hybridization in SF_6 and PCl_5 molecules.
    (ii) Phosphorous forms both PCl_5 and PCl_5 while Nitrogen forms NCl_3 only.
    (iii) Aqueous KCl solution is electrically conducting, but Chlorine water is not.
    (iv) Carbon di-oxide molecule is linear and Sulphur Dioxide molecule is not.
  (b) Arrange the following.
    (i) O_2, O_2^{2-}, O_2^{2+}, O_2^+ in the order of increasing bond strengths.
    (ii) BCl_3, $BeCl_2$, LiCl, CCl_4 in the order of increasing covalent bond character.
    (iii) H_2O, NH_3, HF, H_2S in the order of increasing polar Character.
    (iv) NaCl, LiCl, KCl in the order of increasing melting points.
    (v) CH_4 H_2O, NH_3 in the order of increasing bond angles.

G.G.S.I.P.U. End Term, II Batch, B.Tech, July 2001

21. Discuss briefly werner's theory of co-ordination compounds with special reference to complexes of cobalt. **G.G.S.I.P.U. I Term, March 2002**

22. (a) Discuss the formation of F_2, N_2^+ and H_2^- on the basis of molecular orbital theory.
  (b) Which of the following would have the maximum bond strength

$$O_2,\ O_2^+,\ O_2^-,\ O_2^{-2}$$

G.G.S.I.P.U. II Term, April 2002

23. (a) Explain VSEPR with the examples of CH_4, NH_3 and H_2O.
  (b) What is Werner's Theory, Find out E.A.N. of the following
    (i) $K_3[Fe(CN)_6]$ (ii) $Ni(CO)_4$
    (iii) $[CoF_6]^{3-}$ (iv) $[CuCl_4]^{-2}$

G.G.S.I.P.U. End Term, B. Tech, May 2002

24. (a) Find out hybridizations and structures of any four of the following
    (i) BrF_3 (ii) SF_6 (iii) XeF_4
    (iv) IF_7 (v) SiF_4
  (b) Discuss Born-Haber Cycle with the help of an example of NH_4Cl

G.G.S.I.P.U. End Term, B. Tech, May 2002

25. (a) Give reasons
    (i) O_2 is paramagnetic (ii) He_2 does not exist
    (iii) H_2O is a liquid while H_2S, a gas
  (b) Find out the bond orders for CO, N_2, O_2^{2-}, O_2^{2+} **G.G.S.I.P.U. End Term, B. Tech, May 2002**

26. Which of the following statements given below are correct?
  (a) Valance electrons are usually the electrons in the outermost energy level.
  (b) Greater the overlap of the orbitals involved in bonding, stronger is the resulting bond.
  (c) The minimum in the potential energy curve represents state of a system at which its potential energy is minimum.
  (d) Electron with same spins can also lead to bonding.
  (e) Acid hydrolysis is reversible where as base hydrolysis is irreversible.
  (f) Hybridization in both NH_3 and PH_3 is sp^3.
  (g) Adsorption arises due to unsaturation of valence forces atoms or molecules on the surface.
  (h) The bond angle for atoms where the central atom is sp^3 hybridized is 120º.
  (i) sp hybrid orbitals have equal s and p character.
  (j) An atom with a large electronegativity has a small tendency to attract electron.

G.G.S.I.P.U. II. Term, B.Tech Feb. 2002

27. (a) What are the important postulates of Werner's theory. How was Werner able to show that the only geometry possible for a complex with co-ordination number 6 is octahedral.
  (b) Write resonating structure for SO_3 and NO_2^-. **G.G.S.I.P.U. II. Term, B.Tech Feb. 2002**

28. (a) Illustrate with an example each of the following with respect to co-ordination compounds.
(i) Ionization isomerism (ii) Geometrical isomerism
(iii) Optical isomerism
(b) AB, CD and EF are three compounds. The electro negativity differences between atoms in these compounds are 1.8, 1.5 and 0.1. Which one in these molecules would be expected to be (i) more polar, (ii) less ionic in character, and (iii) more covalent?
G.G.S.I.P.U. II. Term, B.Tech Feb. 2002

29. (a) Write the energy level diagram of a diatomic molecule having bond order three.
(b) Explain why N_2 has greater dissociation energy than N_2^+ whereas O_2 has less dissociation energy than O_2^+.
(c) Discuss the structure of BrF_3 and SiF_4 molecules by VSEPR theory.
G.G.S.I.P.U., II. Term, Apr. 2002

30. Write notes on:
(a) Resonance and Resonance energy.
(b) Application of Born Haber Cycle in calculating lattice energy.
(c) Linkage isomerism. **G.G.S.I.P.U., II. Term, May 2003**

31. Explain the terms bonding and antibonding molecular orbitals. Give the conditions for the combination of atomic orbitals to form molecular orbitals. How does M.O. theory explain the paramagnetic character of oxygen and diamagnetic character of nitrogen.
G.G.S.I.P., II. Term, May 2003

❑❑❑

5 COORDINATION COMPOUNDS

"The most incomprehensible thing about the world is that, it is comprehensible"

—Albert Einstein

CHAPTER OUTLINE

5.1 COORDINATION COMPOUNDS

The term coordination compound is widely applicable to the compounds in which the central metal atom is linked to a number of ions or neutral molecules by coordinate bonds.

When solution containing two or more salts in stoichiometric proportions are allowed to evaporate, crystals of new compounds separate out. These compounds fall under two catogories.

(a) Double Salts

(b) Coordination Compounds.

Double Salts: These are the molecular compounds which exist only in crystal lattices. When they are dissolved in water or any other solvent they break down into their constituent ions. Their physical and chemical properties are essentially the same as their individual components. For example, when solutions of two salts viz. KCl and $MgCl_2$ are mixed together in stoichiometric proportion and are allowed to evaporate, crystals of Carnellite, a double salt, are obtained

$$KCl + MgCl_2 + 6H_2O \longrightarrow \underset{Carnellite}{KCl.MgCl_2.6H_2O}$$

$$KCl.MgCl_2.6H_2O \longrightarrow K^+_{(aq)} + Mg^{2+}_{(aq)} + 3C\ell^-_{(aq)}$$

But when these crystals are dissolved in water, the double salt break down into its constituent ions i.e. the solution shows the properties of K^+, Mg^{2+} and Cl^- ions. Thus double salts have their identity only in the crystalline state. Mohr's salt, potash alum, etc are some more examples of double salts.

$$FeSO_4(NH_4)_2\,SO_4.6H_2O \text{ (Mohr's Salt)}$$

$$K_2SO_4.Al_2(SO_4)_3.24H_2O \text{ (Potash Alum)}$$

Coordination Compounds: Coordination compounds are those molecular compounds which *retain their identity* even when dissolved in water or any other solvent. The properties of coordination or complex compound is entirely different from those of its components. For example, when a solution of ferrous cyanide is mixed with potassium cyanide, potassium ferrocyanide is formed. When this complex compound is dissolved in water it **does not** give the tests of its constituent ions Fe^{2+} and CN^- ions but gives the test of ferrocyanide ion $[Fe(CN)_6]^{4-}$, though a small dissociation does take place.

$$Fe(CN)_2 + 4KCN \longrightarrow K_4[Fe(CN)_6]$$

$$K_4[Fe(CN)_6] \rightleftharpoons 4K^+ + \underset{\text{Complex ion.}}{[Fe(CN)_6]^{4-}}$$

Thus coordination compounds do not lose their identity even when dissolved in water. Similarly tetrammine cupric sulphate when dissolved in water does not give the test of Cu^{2+} or NH_3 but instead gives $[Cu\,(NH_3)_4]^{2+}$ ions.

$$[Cu(NH_3)_4]SO_4 \rightleftharpoons [Cu(NH_3)_4]^{2+} + SO_4^{2-}$$

Terms Used in Coordination Compounds

The terms most frequently used in the study of coordination compounds are-

Central Ion: Central ion is the cation which accepts the electron pairs from the ligands to form coordinate bonds, e.g. in the complex $[Cu\,(NH_3)_4]\,SO_4$, Cu^{2+} is the central ion.

Ligands: The anions or neutral molecules which are attached to the central ion through coordinate bonds are called ligands, e.g. in the above complex $[Cu\,(NH_3)_4]SO_4$, NH_3 molecules act as ligands. The common ligands are NH_3, H_2O, CN^- etc. The ligands should have atleast a lone pair of electron, so that it may act as donor for the formation of covalent bond to the central metal atom. The ligands may have one or more than one point of attachment and hence are named accordingly.

Ligand	*Point of Attachment*	*Example*
Monodentate	1	H_2O, NH_3, CO
Bidentate	2	$C_2O_4^{2-}$, $NH_2\,(CH_2)_2\,NH_2$
Polydentate	2 or more than 2	EDTA

Further, ligands can be neutral, (H_2O, NH_3, ethylene diamine) or anionic ($C_2\,O_4^{2-}$, F^-, Cl^-, Br^-, CN^-, OH^-)

Complex ion: A complex ion is an electrically charged species which consists of central metal ion surrounded by a number of ligands. When dissolved in water the complex ion does not dissociate but retains its identity. For example K $[Ag\,(CN)_2]$ contains complex ion $[Ag\,(CN)_2]^+$ which does not dissociate into Ag^+ and CN^- ions in solution but behaves as a separate entity i.e. $[Ag\,(CN)_2]^+$ ion.

Coordination Number: Coordination number is the number of unidentate ligands attached to the central metal atom or simply it is the number of coordinate bonds formed with the central metal atom. In the complex ions $[Ag (CN)_2]^-$, $[Cu (NH_3)_4]^{2+}$ $[Cr (H_2O)_6]^{3+}$ the coordination number of Ag, Cu and Cr are 1, 2 and 3 respectively. If on the other hand, bidentate ligands are attached to the central metal atom, then the coordination number becomes double the number of bidentate ligands, because each ligand forms two coordinate bonds. For e.g. in the coordinate compound $[Fe (C_2O_4)_3]^{2-}$, the coordination number of Fe is 6 because each $C_2O_4^{2-}$ forms two coordinate bonds.

Oxidation Number: It is the charge which the central atom has or appears to have when attached to the ligands to form coordinate compound. For example in $[Co (NH_3)_6] Cl_3$, cobalt has oxidation number of +3, because charge on the complex ion $[Co (NH_3)_6]^{3+}$ is +3. Now as NH_3 is neutral so cobalt carries a oxidation number of +3.

Coordination Sphere: It consists of central metal atom and the surrounding ligands. It is always enclosed in square brackets. In $[Pt (NH_3)_4] Cl_2$, $[Pt (NH_3)_4]^{2+}$ constitutes the coordination sphere.

Ionization Sphere: The ions outside the coordination sphere constitute what is known as ionization sphere. In the example $[Pt (NH_3)_4] Cl_2$, Cl_2 constitute the ionization sphere.

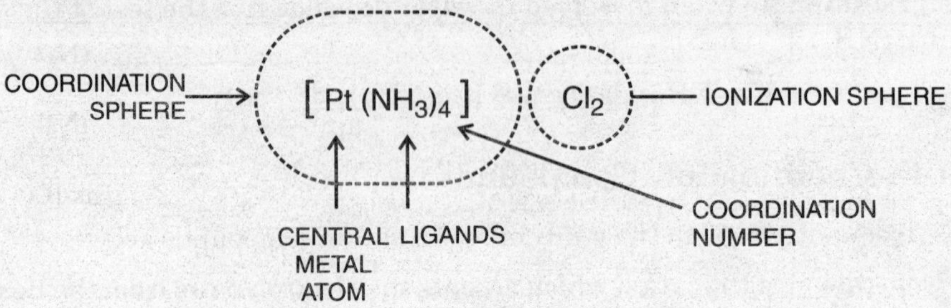

Fig. 5.1

5.2 WERNER'S THEORY OF COORDINATION COMPOUNDS

The fundamental studies in coordination chemistry were carried out by the Danish chemist S.M. Jorgensen and the Swiss chemist Alfred Werner. It was rather surprising that how could a stable molecule like $CoCl_3$ combine with a group of stable molecules like NH_3 to form complex compound like $[(Co (NH_3)_6] Cl_3$ which has entirely new properties. Werner's coordination theory in 1893 provided the first adequate description of the structures of complex ions. At that time, since no modern instrumental techniques were developed, therefore all the studies were carried out using simple reaction chemistry. Werner received the Noble Prize in 1913 for his remarkable work in coordination compounds.

The main postulates of Werner's Coordination Theory are:

1. Most metals possess two types of valencies, (a) Primary Valence and (b) Secondary Valence. The primary Valence is ionizable and corresponds to the oxidation state, whereas secondary valence is nonionizable and corresponds to the coordination number in modern terminology.
2. Every metal atom tends to satisfy both, its primary and secondary valence.
   Primary valence is satisfied by negative ions in simple as well as in complex salts. For example, in simple salt $Cu SO_4$ (Cu^{2+} SO_4^{2-}) the primary valency of Cu is 2 and in complex salt $[Cu (NH_3)_4]Cl_2$ which actually exist as $[Cu (NH_3)_4]^{2+}$ and $2Cl^-$, the primary valence of copper is again 2 as there are 2 ionic bonds.

Secondary valence, the nonionizable linkage is satisfied either by negative ions or neutral molecules. For example in [Cu $(NH_3)_4$] Cl_2, the secondary valency is satisfied by 4 NH_3 molecules whereas in compound K [Ag $(CN)_2$], the secondary valence of Ag, the central metal ion is satisfied by 2 negativly charged CN^- ions.

3. The secondary valencies are directed towards fixed positions in space. (This postulate forms the basis for the stereochemistry of the coordination compounds). Primary valencies are non-directional.

On the basis of above postulates, Werner explained the structures of a number of coordination compounds. Let us discuss the structures of cobalt (III) ammine complexes in the light of above postulates.

When an aqueous solution of $CoCl_3$ is treated with ammonia, not one but a number of compounds result as shown in table 5.1.

Table 5.1 Various Ammines of Co (III)

Complex	*Number of Cl^- ions precipitated*	*Colour*	*Name*	*Present Formulation*
$CoCl_3$. $6NH_3$	3	Yellow Chloride	Luteocobaltic	[Co $(NH_3)_6$] Cl_3
$CoCl_3$. $5NH_3$	2	Purple cobaltic Chloride	Purpureo	[Co $(NH_3)_5$ Cl] Cl_2
$CoCl_3$. $4NH_3$	1	Green Chloride	Praseocobaltic	trans-[Co $(NH_3)_4$ Cl_2] Cl
$CoCl_3$. $4NH_3$	1	Violet Chloride	Violeocobaltic	cis-[Co $(NH_3)_4$ Cl_2] Cl

Werner was able to find out the number of ionizable chloride ions by adding silver nitrate solution to aqueous solutions of the various compounds. The silver chloride thus formed was weighed and the stoichiometry was established on its basis.

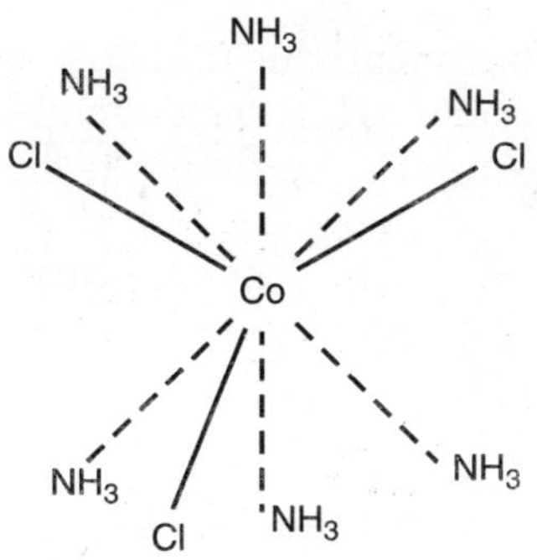

Structure of $CoCl_3.6NH_3$
Number of Cl^- Precipitated = 3
Total Number of ions = 4

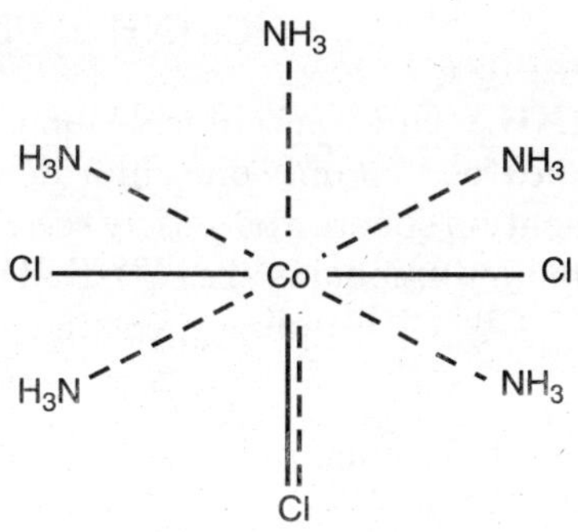

Structure of $CoCl_3.5NH_3$
Number of Cl^- Precipitated = 2
Total Number of ions = 3

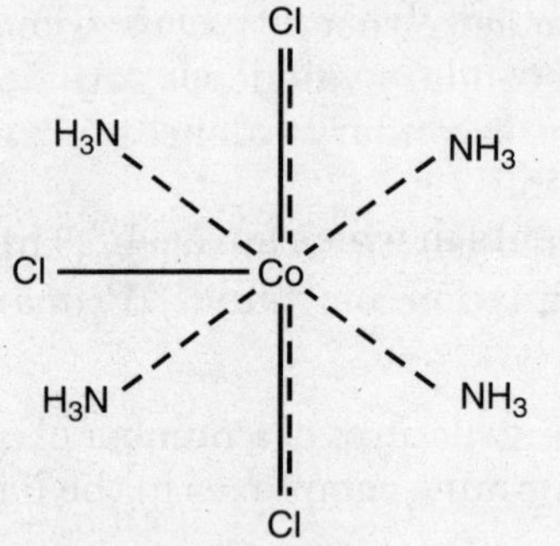

$CoCl_3.4NH_3$

Structure of $CoCl_3.4NH_3$
Number of Cl^- precipitated = 1
Total Number of ions = 2

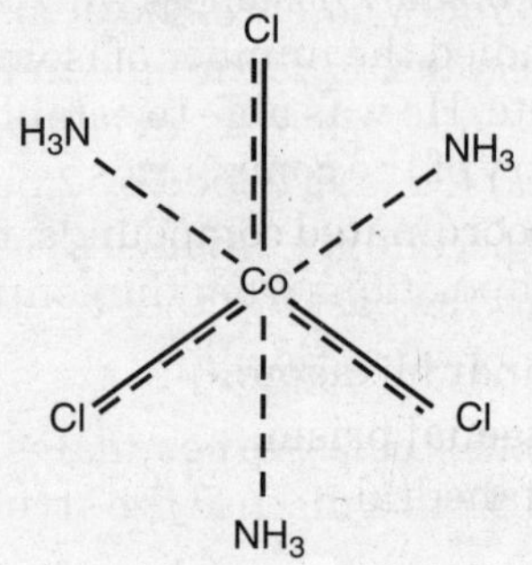

$CoCl_3.3NH_3$

Structure of $CoCl_3.3NH_3$
Number of Cl^- Precipitated = zero
Neutral Molecule

Fig. 5.2

$CoCl_3$. $6NH_3$: The addition of a solution of Ag NO_3 to a freshly prepared solution of $CoCl_3$. $6NH_3$ results in the immediate precipitation of all the three chloride ions Thus, it was deduced that three chloride ions satisfy the primary valence whereas six ammonia molecules satisfy the secondary valence forming a complex $[Co(NH_3)_6]^{3+}$. The primary valence are denoted by solid lines and secondary valencies are denoted by dotted lines in Figure 5.2. The above explaination is further supported by the fact that conductance measurements show the presence of four ions, as is indicated–

$$[Co(NH_3)_6]Cl_3 \longrightarrow [Co(NH_3)_6]^{3+} + 3Cl^-$$

$CoCl_3$. $5NH_3$: This compound was found to have 2 ionizable chloride ions. Now only five ammonia molecules are there to satisfy the secondary valence. As the number of secondary valence remains fixed, (6 in this case) therefore it was assumed that one chloride ion must serve the dual function of satisfying both primary and secondary valence. The chloride which satisfy both the valencies is not precipitated by the addition of $AgNO_3$ solution. Thus, the compound is formulated as $[Co(NH_3)_5Cl]Cl_2$. It gives 3 ions in aqueous solutions.

$$[Co(NH_3)_5Cl]Cl_2 \longrightarrow [Co(NH_3)_5Cl]^{+} + 2Cl^-$$

Co Cl_3. $4NH_3$: Only one chloride ion is precipitated, when this compound is treated with $AgNO_3$ which indicates that only one chloride ion is ionizable. The other two chloride ions are present in the coordination sphere and satisfy secondary valency in addition to primary valency. The remaining four secondary valencies are satisfied by four NH_3 ligands. Thus the compound is formulated as $[Co(NH_3)_4Cl_2]Cl$. Conductance measurement also show the presence of two ions.

$$[Co(NH_3)_4Cl_2]Cl \longrightarrow [Co(NH_3)_4Cl_2]^{+} + Cl^-$$

Two related compounds (Table: 4.6) one violet and other green have been prepared having the same composition. Werner explained this by assuming that the two are different isomers, one is trans and other is cis-type.

Co Cl_3. $3NH_3$: This is the next compound in the series. It does not give any precipitate on addition of $AgNO_3$ solution, which indicates the absence of ionizable chloride ions. Thus, it can be formulated as $[Co(NH_3)_3Cl_3]$. Compound is a nonconducting molecule.

A significant contribution of Werner's theory was the determination of the structure of complexes having coordination number six. As no modern techniques like X-ray or spectroscopy were available at that time, the spatial configuration of molecules was determined by comparing the number of known isomers with the number theoretically possible isomers for several structures. Werner studied the number of isomers of various metals including cobalt, chromium, copper and platinum etc. He was able to establish the existence of definite coordination number, along with the geometry of the compounds.

In six coordinated compounds, the possible arrangements in which six ligands are situated at equidistant positions from the central atom are-

(a) Planar hexagon.
(b) Trigonal prism.
(c) Octahedral

Werner studied a number of compounds of the type $[MX_6]$, $MX_5 Y]$, $[MX_4 Y_2]$, $[MX_3 Y_3]$ and compared the number of known isomers with the number of theoretically predicted structures. On the basis of this comparison, he was able to establish the shape of six coordinated compounds.

To illustrate the above point let us take the example of the complexes of cobalt of the type $[Co (NH_3)_6]^{3+}$, $[Co (NH_3)_5 Cl]^{2+}$, $[Co (NH_3)_4 Cl_2]^+$ and $[Co (NH_3)_3 Cl_3]$. The observed as well as the number of predicted isomers is as given in Table 5.2.

Table 5.2

Complex	*No. of Isomers Observed*	*No. of Isomers Predicted*		
		Planar Hexagon	*Trigonal Prism*	*Octahedral*
$[Co (NH_3)_6]^{3+}$	1	1	1	1
$[Co (NH_3)_5 Cl]^{2+}$	1	1	1	1
$[Co (NH_3)_4 Cl_2]^+$	2	3 (1,2; 1,3; 1,4)*	3 (1,2; 1,4; 1,6)	2 (1,2; 1,6)
$[Co (NH_3)_3 Cl_3]$	2	3 (1, 2, 3; 1, 2, 4; 1, 3, 5)	3 (1, 2, 3; 1, 2, 4; 1, 2, 6)	2 (1, 2, 3; 1, 2, 6)
		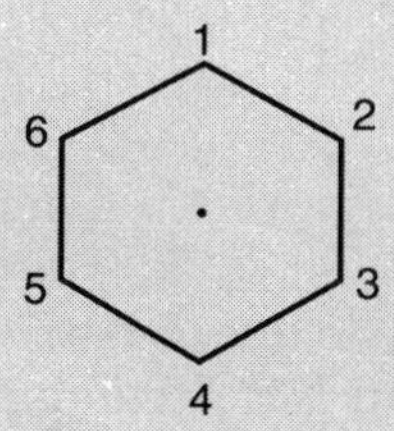	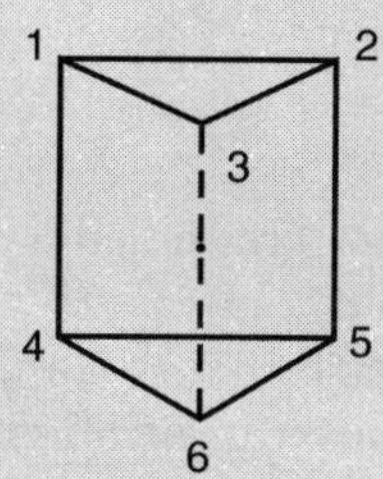	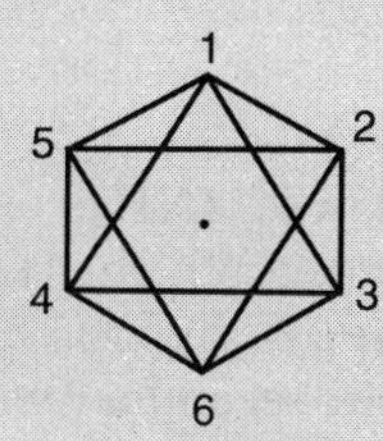

*Number in the brackets indicates the position of negative ligands i.e chloride ions in different isomers.

From comparison of the results from the above table, it is clear that the number of isomers actually isolated for various compounds are in accordance with the octahedral shape. Thus, Werner was able to conclude that in all the *six coordinated compounds the geometry is octahedral.*

It should be emphasized here that the above method of predicting the shape does not furnish a positive proof for the octahedral structure. The other isomers in planar hexagonal or trigonal prism might be more difficult to isolate or may be very unstable to isolate. But fortunately, later on when the structure of six coordinated compounds were determined using X-ray techniques, the results were in accordance with what Werner had established several years ago.

5.3 EFFECTIVE ATOMIC NUMBER

It has been stated earlier that ligands are attached to the central metal atom through coordinate bonds. All neutral molecules or anions which are capable of being coordinated to central metal ion have atoms with unshared lone pair of electrons in their outermost shells. Professor N.V. Sidgwick of Oxford University postulated that the total number of ligands attached to the central metal atom is such that, which makes the total number of electrons around the central metal atom same as that of noble or inert gas. This is also known as **Effective Atomic Number.**

Effective Atomic Number is the total number of electrons possessed by the central metal in the complex including those gained through bonding. It is abbreviated as EAN. Example, EAN of Pt (IV) in the complex $[Pt(NH_3)_6]^{4+}$ is calculated as follows:

Central metal atom Pt

Atomic Number of Pt = 78

Number of electrons in Pt^{4+} = 78–4 = 74

Number of electrons gained through ligands = 2 × 6 = 12

Total Electrons possessed by the central atom = 74 + 12 = 86

Thus the EAN is same as possessed by the noble gas radon.

Similarly EAN determined for a number of complexes show that in most cases the EAN is same as that of atomic number of the next noble gas. EAN for a number of complexes are shown in Table 5.3.

Table 5.3

Complex	*Central Metal ion*	*Atomic Number*	*Electrons on Metal ion*	*Electrons Gained Through Ligands*	*EAN*	*Atomic Number of Next Noble Gas*
$[Pt(NH_3)_6]^{4+}$	Pt^{4+}	78	78 – 4 = 74	2 × 6 = 12	86	86 (Rn)
$[Pt(NH_3)_4Cl_2]^{2+}$	Pt^{4+}	78	78 – 4 = 74	2 × 6 = 12	86	86 (Rn)
$[Fe(CN)_6]^{4-}$	Fe^{2+}	26	26 – 2 = 24	2 × 6 = 12	36	36 (Kr)
$[Co NH_3)_6]^{3+}$	Co^{3+}	27	27 – 3 = 24	2 × 6 = 12	36	36 (Kr)
$[Pt Cl_6]^{2-}$	Pt^{4+}	78	78 – 4 = 74	2 × 6 = 12	86	86 (Rn)
$[Cr(CO)_6]$	Cr	24	24	2 × 6 = 12	36	36 (Kr)
$[Co(NO_2)_6]^{3-}$	Co^{3+}	27	27 – 3 = 24	2 × 6 = 12	36	36 (Kr)
$[Cu(CN)_4]^{3-}$	Cu^{+}	29	29 – 1 = 28	2 × 4 = 8	36	36 (Kr)
$[Cd(NH_3)_4]^{2+}$	Cd^{2+}	48	48 – 2 = 46	2 × 4 = 8	54	54 (Xe)

However there are a few cases where EAN is not equal to the number of electrons in the next noble gas but a few units above or below it. For example the EAN of the central metal atom Fe in the complex $[Fe(CN)_6]^{3-}$ is-

Atomic Number of Fe = 26

Number of electrons in Fe^{3+} = 26–3 = 23

Number of electrons gained through ligands

= 2×6 = 12

EAN = 35

Some more complexes which do not obey EAN rule are given in Table 5.4.

Table 5.4

Complex	Central Metal ion	Atomic Number	Electrons on Metal ion	Electrons Gained Through Ligands	EAN	Atomic Number of Next Noble gas
$[Cr(NH_3)_6]^{3+}$	Cr^{3+}	24	24 – 3 = 21	2 × 6 = 12	33	36 (Kr)
$[Pt(NH_3)_4]^{2+}$	Pt^{2+}	78	78 – 2 = 76	2 × 4 = 08	84	86 (Rn)
$[Ag(NH_3)_2]^{+}$	Ag^{+}	47	47 – 1 = 46	2 × 2 = 04	50	
$[Ni(en)_3]^{2+}$	Ni^{2+}	28	28 – 2 = 26	2 × 6 = 12	38	36 (Kr)
$[Ni(NH_3)_6]^{2+}$	Ni^{2+}	28	28 – 2 = 26	2 × 6 = 12	38	36 (Kr)
$[CuCl_2]^{-}$	Cu^{+}	29	29 – 1 = 28	2 × 2 = 04	33	36 (Kr)

From the above discussion, one can conclude that tendency to attain inert gas configuration is though an important factor but not a necessary condition for complex formation. Moreover, the complexes in which EAN does not correspond to the number of electrons in the next noble gas are equally stable.

5.4 ISOMERISM IN COORDINATION COMPOUNDS

We all are well acquainted with the word "isomers" which can be defined as the compounds having the same molecular formula but different structural arrangements. The phenomenon is known as isomerism. Isomerism is very common in organic compounds but is not encountered that frequently in inorganic chemistry. However coordination compounds represent a class of inorganic chemistry in which overwhelming majority of compounds exhibit isomerism.

Isomerism in co-ordination compounds can be classified into the following two types.

(i) Structural Isomerism

(ii) Stereoisomerism.

These two isomers are further subdivided into different types as discussed below.

5.4.1 Structural Isomerism

Structural isomerism arises due to difference in structure of coordination compounds having same molecular formula. It is further divided into following types-

1. ***Ionization Isomerism:*** The isomers which yield different ions in solution are known as ionization isomers and the phenomenon is known as ionization isomerism. This type of isomerism is due to interchange of a ligand anion within the complex and external anion associated with the complex. A classical example is the violet $[CoBr(NH_3)_5]SO_4$ and red $[CoSO_4(NH_3)_5]Br$. The violet complex gives the test for SO_4^{2-} ions i.e in solution gives a precipitate with $BaCl_2$ confirming the presence of SO_4^{2-} ions. On the other hand, red complex does not give test for sulphate ion but instead gives precipitate of Ag Br when treated with $AgNO_3$ solution.

$$\underset{\text{Violet Complex}}{[CoBr(NH_3)_5]SO_4} \longrightarrow [CoBr(NH_3)_5]^{2+} + SO_4^{2-}$$

$$\underset{\text{Red Complex}}{[CoSO_4(NH_3)_5]Br} \longrightarrow [CoSO_4(NH_3)_5]^{+} + Br^{-}$$

Some more pairs of isomers showing ionization isomerism are-

$[Pt\ Br\ (NH_3)_3]\ NO_2$ and $[Pt\ NO_2\ (NH_3)_3]\ Br$

$[Co\ NO_3\ (NH_3)_5]\ SO_4$ and $[CoSO_4\ (NH_3)_5)\ NO_3$

2. ***Hydrate Isomerism:*** This type of isomerism arises due to the difference in number of water molecules present as ligands. For example, three different isomers with the formula $Cr\ Cl_3.\ 6H_2O$ are known:

$[Cr\ (H_2O)_6]\ Cl_3$	$[Cr\ Cl\ H_2O)_5]\ Cl_2.\ H_2O$	$[Cr\ Cl_2\ (H_2O)_4]\ Cl.\ 2H_2O$
Violet	Light Green	Green

On dehydration over sulphuric acid, the violet compound does not lose any water molecule, light green compound loses one water molecule and green compound loses 2 molecules of water of crystallization. Some other compounds in which hydrate isomerism is observed:

$[Co\ Cl\ (H_2O)\ (NH_3)_4\ Cl_2$ and $[Co\ Cl_2\ (NH_3)_4]\ Cl.\ H_2O$

$[Co\ Cl\ (en)_2\ H_2O]\ Cl_2$ and $[Co\ Cl_2\ (en)_2]\ Cl.H_2O$

3. ***Linkage Isomerism:*** Some of the ligands can coordinate to metals in more than one way. For example, in the case of NO_2^- ion, either nitrogen or an oxygen atom may act as a donor giving rise to linkage isomerism.

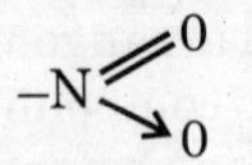

Nitro (Linkage through nitrogen)

–O – N = 0

Nitrito (Linkage through oxygen)

Example of linkage isomerism include the following pairs.

$[Co\ (NO_2)\ (NH_3)_5]\ Cl_2$ and $[Co\ (ONO)\ (NH_3)_5]\ Cl_2$

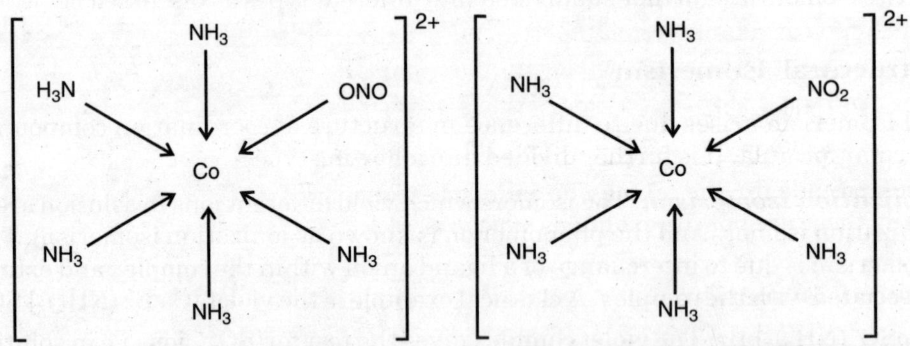

Pentaamminenitrito cobalt (III) ion (Red) | Pentaamminenitro cobalt (III) ion (Yellow)

Fig. 5.3

The red compound undergoes slow transformation into the yellowish one. Besides NO_2^-, other groups which can result into linkage isomerism are SCN^-, $S_2O_3^{2-}$ and CN^-.

The ligands which have more than one atom which can form coordinate bond as above are known as **ambidentate ligands**. Ambidentate ligands attach to central metal atom through one site at a time.

4. ***Coordination Isomerism:*** Compounds which contain both cationic and anionic complexes are capable of forming isomers by exchanging the ligands. Such type of isomers are known as coordination isomers.

Examples:

$$[Co(NH_3)_6][Cr(C_2O_4)_3] \quad \text{and} \quad [Cr(NH_3)_6][Co(C_2O_4)_3]$$
$$[Cu(NH_3)_4][PtCl_4] \quad \text{and} \quad [Pt(NH_3)_4][CuCl_4]$$

5. ***Coordination Position Isomerism:*** In polynuclear complexes, the coordinating groups may be present in the same number but their placement may be different with respect to the different metal ions present, *i.e.* exchange of ligands between different metal nuclei results in coordination position isomerism. Example

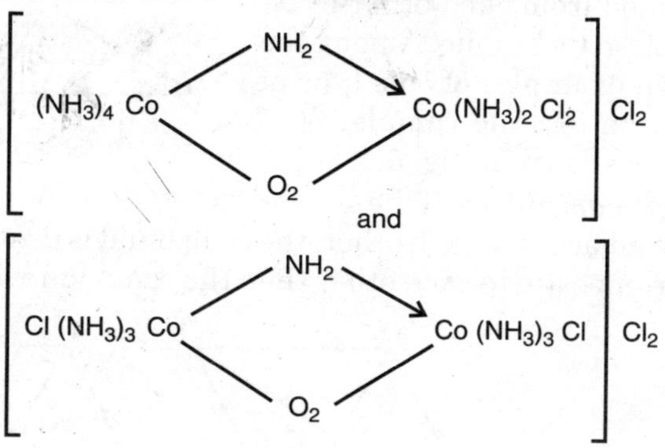

Fig. 5.4

6. ***Polymerization Isomerism:*** This term is actually used to denote compounds of the same empirical formula but different molecular formula, therefore infact, to use the word isomerism for such compounds is wrong. Polymerization isomers are the compounds which have the same stoichiometric composition but whose molecular compositions are the multiples of the empirical formula. Example-

$[Pt(NH_3)_2Cl_2]$; (a) $[Pt(NH_3)_4][PtCl_4]$; (b) $[Pt(NH_3)_3Cl]_2[PtCl_4]$ (c)

Thus compounds (b) and (c) may be called the polymer of (a). But in the strict sense of the word actually these are not the examples of polymerisation.

5.4.2 Stereo Isomerism

Stereoisomerism arises on account of the different positions and arrangements of atoms or groups in space around the central metal atom.

The coordinate bonds around the central metal atom are directional in nature which give rise to stereoisomerism. Coordination compounds exhibit stereoisomerism on account of the different positions and arrangements of ligands in space, around the metal ion. It is of two types.

1. ***Geometrical Isomerism:*** Geometrical isomerism arise due to the different positions of the ligands around the central metal atom. If the two different ligands are adjacent to each other, the isomer is referred to as cis form and if the ligands are opposite to one another the isomer is trans. Geometrical isomerism is not possible for complexes having coordination number 2 and 3, because in this arrangement all positions are adjacent to

one another. However cis-trans isomerism is exhibited by square planar and octahedral complexes.

Square Planar Complexes:

(a) ***$Ma_2\ b_2$ Type Complexes:*** In these compounds M is the central metal atom and *a* and *b* are the ligands. These type of compounds exist in cis and trans isomers as shown in Fig. 5.5.

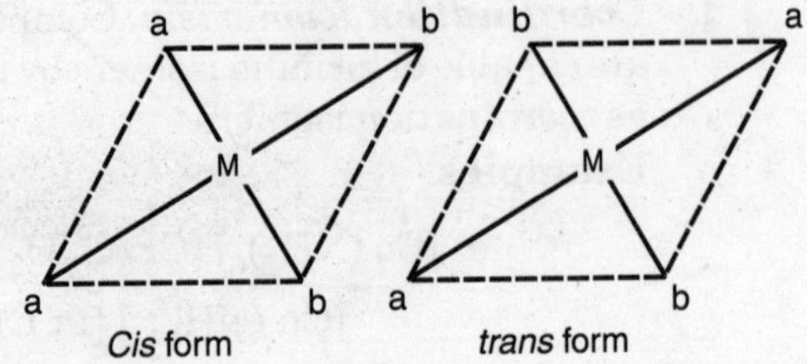

Fig. 5.5

These two isomers can be readily distinguished from each other by X-ray diffraction technique. Among the best known examples of this type of isomerization are the complexes of platinum as shown in Fig. 5.6.

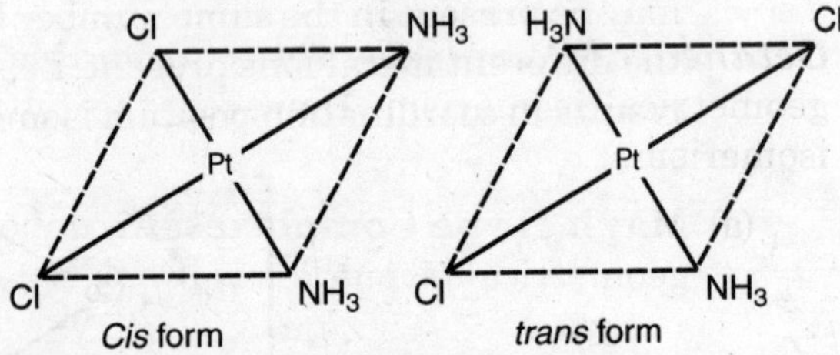

Fig. 5.6

(b) ***$Ma_2\ bc$ type complexes:*** If the similar groups are adjacent to each other, the compound is designated as cis and if the similar groups are opposite to each other then the compound is trans.

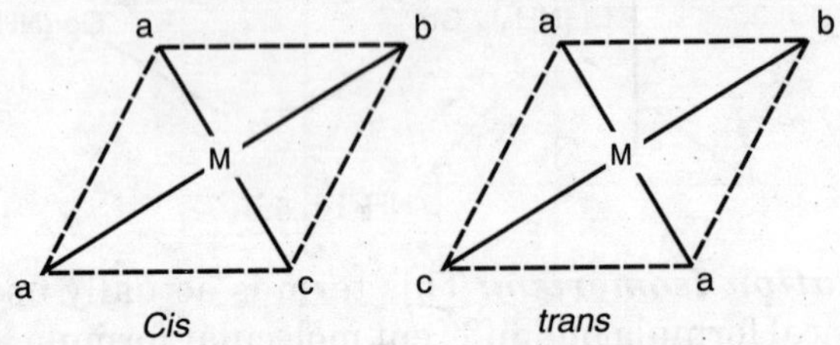

Fig. 5.7 The cis and trans forms of the complex Ma_2bc type.

(c) ***Mabcd type Complexes:*** A few compounds of *Mabcd* type are known. Example [Pt NH_3 (NH_2OH) py $NO_2]^+$. Three possible geometrical forms are:

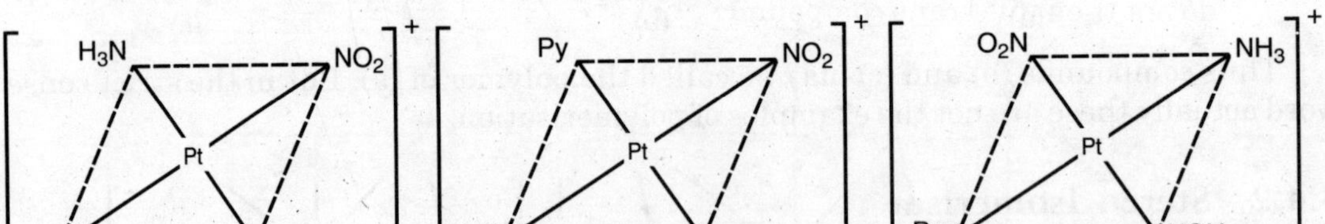

Fig. 5.8 The three isomeric structure of the complex.
[Pt (NH_3) (NH_2OH) py (NO_2)$]^-$

(d) **Ma_3 b or Mab_3 type:** These compounds do not show geometrical isomerism as all the possible spatial arrangements for any of these complexes will be equivalent.

(e) **M $(ab)_2$ type:** This denote square planar complexes containing unsymmetrical bidentate ligands. Example of such type of ligand is $NH_2\ CH_2\ COO^-$. The compound [Pt $(gly)_2$] where gly = $NH_2\ CH_2\ COO^-$ exist in cis and trans form:

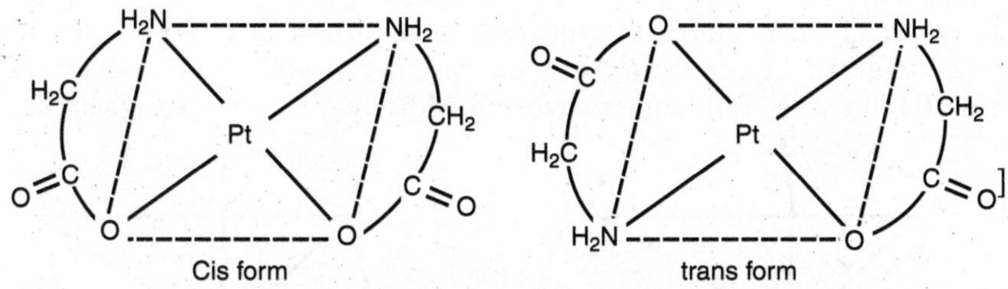

Fig. 5.9 Cis and trans forms of the complex $[Pt(gly)_2]$.

Octahedral Complexes: The octahedral complexes of the type Ma_6 Ma_5b and Mab_5 do not show geometrical isomerism. However, the following octahedral compounds exhibit geometrical isomerism.

(a) **Ma_4 b_2 Type Complexes*:*** An important example of this type of compound exhibiting geometrical isomerism is $[Co\ (NH_3)_4\ Cl_2]^+$, which exist in two forms-

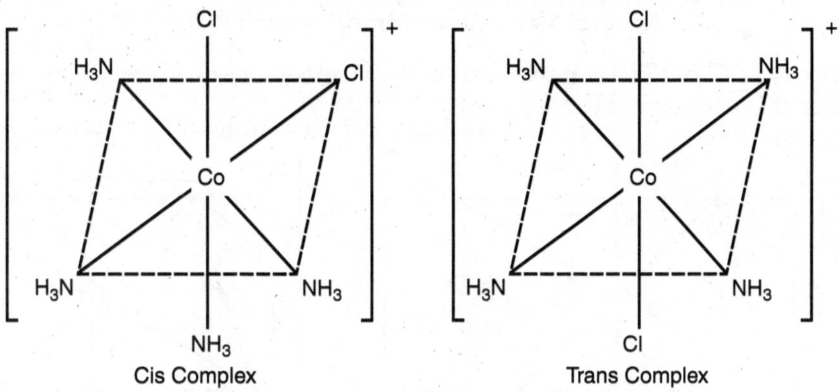

Fig. 5.10 Cis and trans form of the complex $[Co(NH_3)_4\ Cl_2]^+$

(b) ***$Ma_3\ b_3$ Type:*** The isomers of Rh $(py)_3$ Cl_3 have the following structures. If the like groups occupy the corners of one of the octahedral faces the isomer is designated as cis and if they do not then it is trans compound.

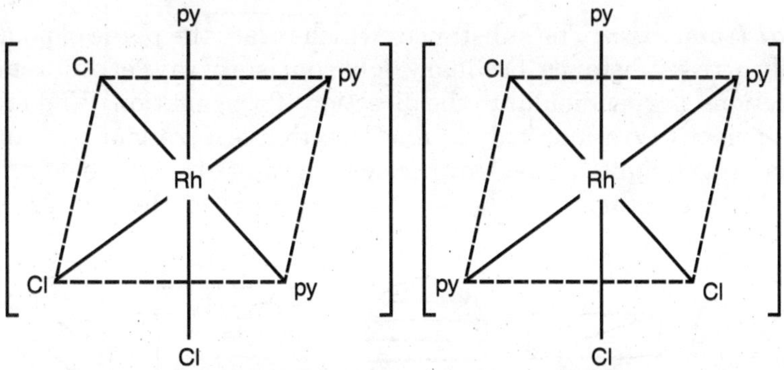

Fig. 5.11 Cis and trans forms of $[RhCl_3(py)_3]$.

(c) ***Mabcdef Type:*** Such a species can exist in 15 different geometrical forms and each form is further capable of showing optical isomerism. The only compound that has been prepared is [Pt (py) (NH_3) (NO_2) (Cl) (Br) (I)]. It was obtained in three different forms, other forms have not been isolated yet.

(d) ***M (aa)$_3$ Type:*** The octahedral complexes having bidentate ligands are also capable of showing geometrical isomerism. For example, the cis-trans isomers of triglycinato chromium (III) have the following two forms. Moreover both the forms are optically active.

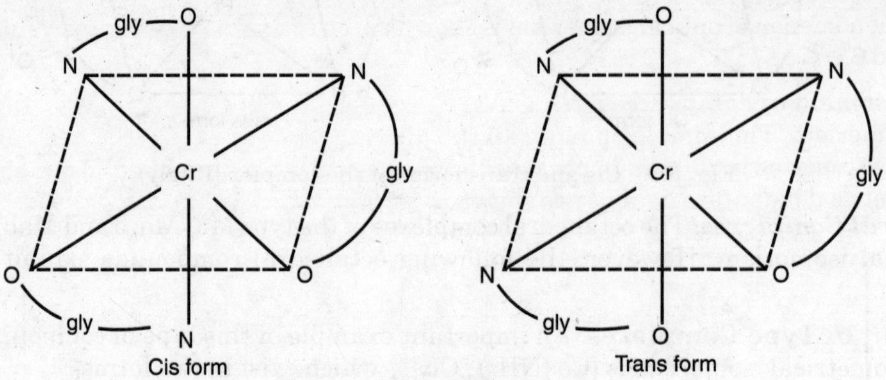

Fig. 5.12 Cis and trans forms of $[Cr(gly)_3]^{3+}$.

(e) ***M (aa)$_2$ b$_2$ Type:*** The common example of the compound showing geometrical isomerism under this category is $[Co\ Cl_2\ (en)_2]^+$.

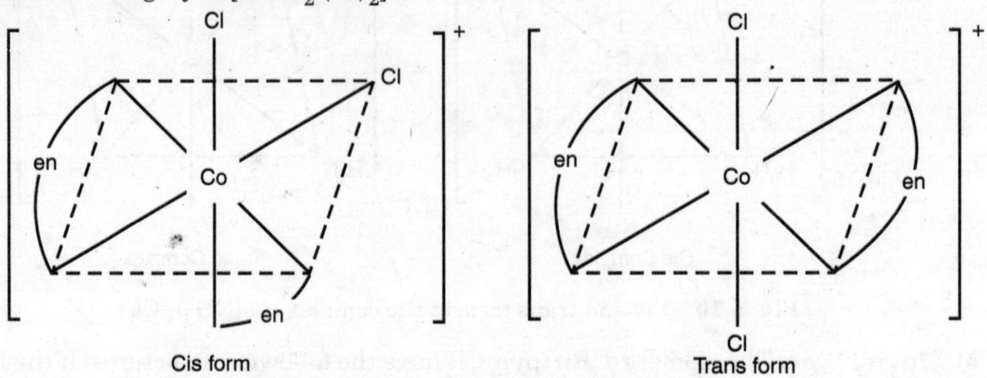

Fig. 5.13 Cis and trans of $[CoCl_2(en)_2]^+$
In this the cis isomer is optically active.

2. ***Optical Isomerism:*** The substances which rotate the plane of polarized light are called optically active substance. Ordinary light consists of rays of different lengths vibrating in all directions, perpendicular to the direction of propagation. When this is passed through a Nicol prism, vibrations only in one direction are transmitted and this light is called plane polarized light. The substances which rotate the plane polarized light to right are called dextrorotatory and those which rolate it to the left are called laevorotatory (Fig. 5.11).

Fig. 5.14

Lack of symmetry in a molecule renders it the property of optical activity. The optically active isomers which rotate the plane of polarized light equally but in opposite directions are called enantiomers.

The phenomenon of optical isomerism is shown in the complexes having coordination number (C.N) 4 and 6.

Optical Isomers of complexes with C.N4: Tetrahedral complexes with C.N 4 should exhibit optical isomerism. The complex having all the different group i.e. [As (CH_3) (C_2H_5) (S) $(C_6H_4COO)]^{2+}$ was reported in 1963 but it could not be resolved. However, tetrahedral complex containing two symmetrical bidentate ligands are shown to exhibit optical isomerism.

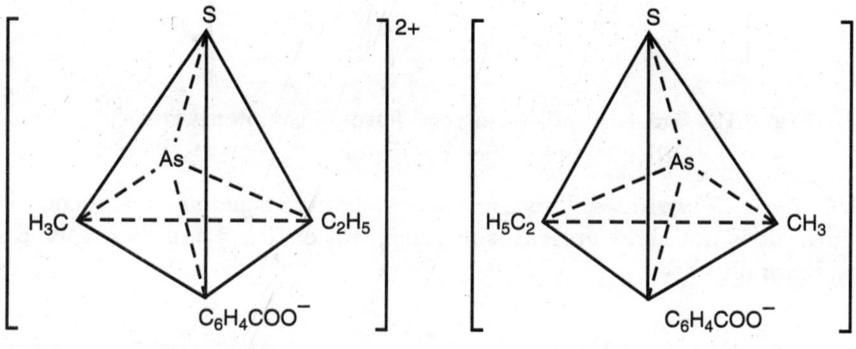

Fig. 5.15 Optical isomers of [As (CH_3) (C_2H_5) (S) $(C_6H_4COO^-]^{2+}$

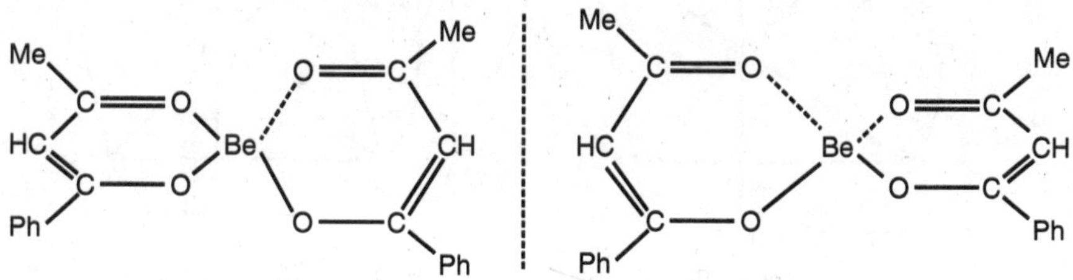

Fig. 5.16 Optical isomers of tetrahedral bis (benzoylacetonato) beryllium (II) Complex.

Square Planar Complexes: As the square planar complexes have all the four ligands and the central metal ion in the same plane and hence optical isomerism is rarely observed in these type of complexes. However Mills and Quibell in 1935 were successful in preparing a four coordinated complex of Pt (II), which could be resolved into two isomeric forms (Fig. 6.14).

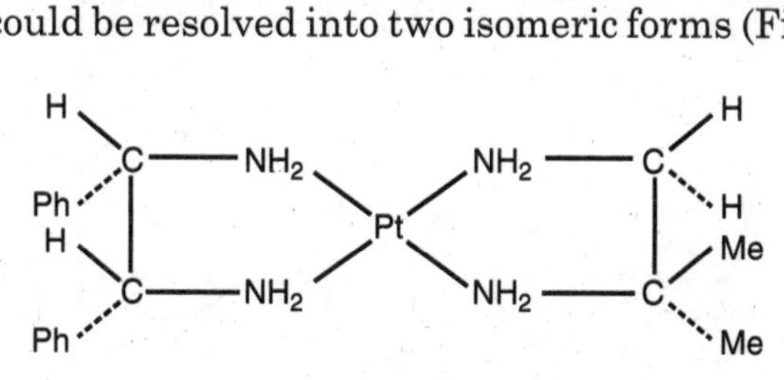

Fig. 5.17

Optical Isomers of complexes with C. N. 6: Optical isomerism is very common in the complexes having C. N 6.

(a) ***M(aa)$_3$ Type:*** Octahedral complex as $[Cr(C_2O_4)_3]^{3-}$ exist in two optically active forms–

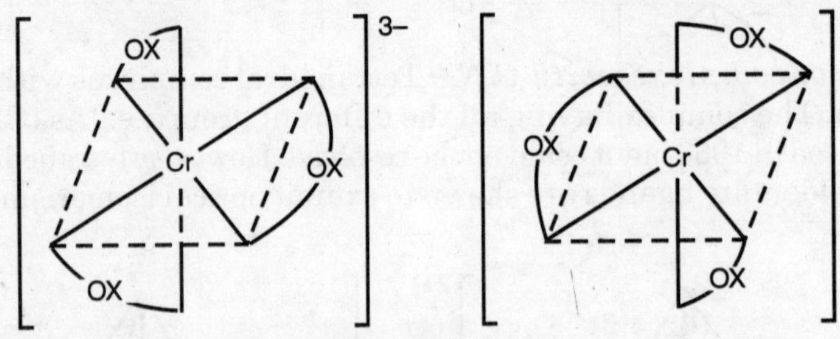

Fig. 5.18 The two optical isomeric forms of complex $[Cr(ox)_3]^{3-}$. Other examples include $[Co\ (en)_3]^{3+}$, $[Pt\ (en)_3)^{4+}$ etc.

(b) ***M(aa)$_2$b$_2$ Type Complexes:*** This complex exists in two geometrical forms cis and trans. Trans form does not show optical isomerism but cis form can be resolved into optical isomers. Example-

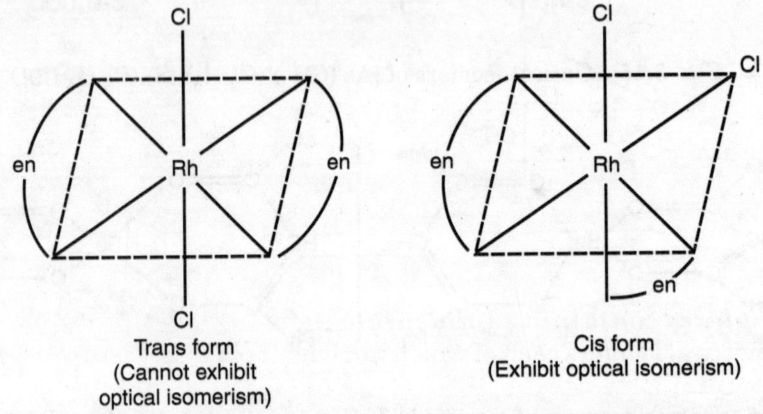

Fig. 5.19 Cis and trans isomers of $[RhCl_2(en)_2]$

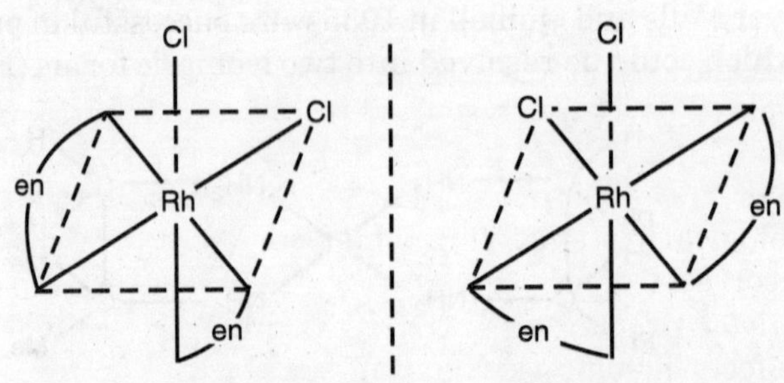

Fig. 5.20 Optically active cis form of $[RhCl_2(en)_2]$

(c) ***M(aa)b$_2$c$_2$ Type:*** Example of such type of optical active compounds are furnished by $[CoCl_2(en)(NH_3)_2]^+$

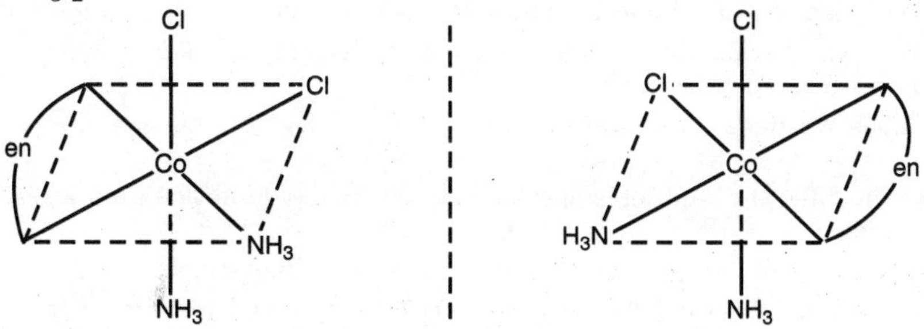

Fig. 5.21 Optical isomers of $[CoCl_2(en)(NH_3)_2]^+$.

(d) ***Complexes containing polydentate ligands:*** Example $[Co(EDTA)]^-$

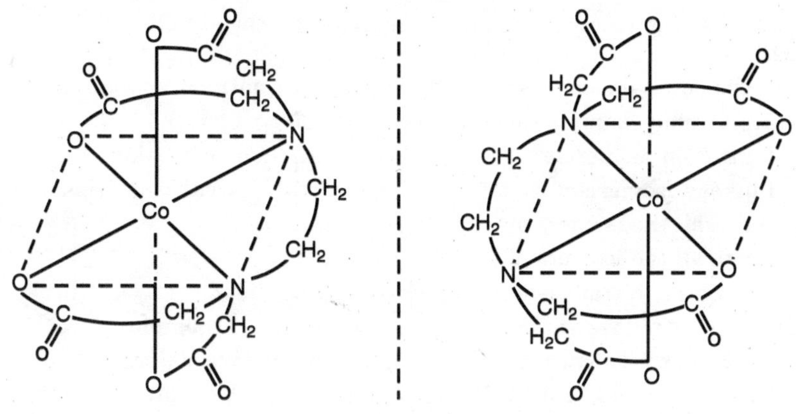

Fig. 5.22

(e) ***Complexes containing Unidentale ligends:*** *Mabcdef* type compounds can show 15 geometrical isomers each of which can show optical isomerism.

SIGNIFICANCE OF GEOMETRICAL ISOMERS IN BIOLOGICAL SYSTEMS

Some of the most important discoveries in science are accidental. One of such accidental discovery is the action of cis —$[Pt(NH_3)_2]Cl_2$ and cis $[Pt(NH_3)_2]Cl_4$. It was found that these two complexes inhibited the cell division in some type of bacteria. The discovery triggered the scientists to investigate the action of these complexes on cancer cells which multiply very fast. Thus these complexes were evaluated as antitumour agents, which inhibit the division of cancer cells. The results showed that cis $[Pt(NH_3)_2]Cl_2$ was active against a wide variety of tumours. However, trans complex had no effect on tumours. This shows the importance of isomerism in biological systems. Thus whenever drugs are synthesized, great care must be taken to get the correct isomer.

Unfortunately cis-isomer, cis-$[Pt(NH_3)_2Cl_2]$ has proved to be effective, but it has got a number of side effects including kidney damage.

QUESTIONS

1. (a) What are the important postulates of Werner's Theory of co-ordination compounds?
   (b) How was Werner able to show that the only geometry possible for a complex with co-ordination number 6 is octahedral?
2. How does Werner's co-ordination Theory account for non-ionic nature of the complex $CoCl_3$ 3 NH_3?
3. Name the different kinds of isomerism possible in co-ordination complexes. Give one example of each kind.
4. Name the type of isomerism exhibited by the following isomers:-
   (i) $[Cr(NH_3)_6]$ $[Cr(CN)_6]$ and $[Cr(CN)_2(NH_3)_4]$ $[Cr(CN)_4(NH_3)_2]$
   (ii) $[(CoCl_2(py)_2(H_2O)_2]$ Cl and $[CoCl_3(py)_2(H_2O)]$ H_2O
   (iii) $[PtBr_2(NH_3)_4]$ Cl_2 and $[PtCl_2(NH_3)_4]$ Br_2
   (iv) $[Co(NO_2)(NH_3)_5]$ Cl_2 and $[Co(ONO)(NH_3)_5]$ Cl_2
5. What is the EAN rule for the co-ordination compounds?
6. It is said that the EAN rule is an overmphasis on the stability of the inert gas configuration. Discuss.
7. Explain geometrical isomerism exhibited by co-ordination compounds.
8. Illustrate with an example each of the following terms:
   (i) Ionization isomerism (ii) Co-ordination isomerism
   (iii) Linkage isomerism (iv) Optical isomerism
9. (i) Write all the isomers of $[Co(NH_3)_6][Cr(C_2O_4)_3]$
   (ii) Write all the isomers of $[Co(NH_3)_5$ SCN] Cl_2
   (iii) Draw cis and trans isomers of $[Co(NH_3)_3Cl_3]^+$ ion.
10. (a) Explain the following terms with respect to a metal complex
    (i) Primary Valence (ii) Secondary Valence.
    (b) What is a complex? How is it different from a double salt.
    (c) The two complexes $PtCl_4.2NH_3$ and $PtCl_4.2KCl$ do not give precipitates of AgCl when treated with $AgNO_3$. The conductance studies indicate zero and three moles of ions per mole of the complex respectively in their solutions. Write the structure of these compounds.
11. Discuss briefly the Werner's theory of Co-ordination compounds with special reference to complexes of cobalt **G.G.S.I.P.U. I Term, Feb. 2001**

❑❑❑

6 ELECTROCHEMISTRY

To accomplish great things, we must not only act, but also dream; not only plan, but also believe.

—Anatole France

CHAPTER OUTLINE

6.1 INTRODUCTION

Electrochemistry is a branch of physical chemistry that studies the relationship between chemical and electrical phenomena and the laws of their interaction.

We know that on passing electric current through an electrolyte solution a chemical reaction takes place. In other words, an electric current can cause a chemical reaction to take place under suitable conditions. The phenomena is called *electrolysis*. The reverse is also true i.e. we can generate electric current by the use of a chemical reaction. Hence, depending upon these two types of electrochemical processes, there are two types of cells such as:

(*i*) Electrolytic cells

(*ii*) Electrochemical cells.

(*i*) Electrolytic Cells: A device in which electrical energy is converted into chemical energy by passing an electric current to produce the desired chemical change is called an electrolytic-cell and the process is called electrolysis.

The refining of metals, electroplating with silver, gold and chromium or production of many chemical substances such as chlorine, sodium hydroxide, hydrogen peroxide etc., are carried out in electrolytic cells. The charging of storage batteries is also based on electrolytic cells.

(*ii*) **Electrochemical Cells:** An electrochemical cell is a device which converts chemical energy into electrical energy. It is also called *galvanic cell*. Daniell cell is a well known example of galvanic cell.

In this chapter, we will limit our discussion to the second type i.e. electrochemical cell and the processes involved therein. But before that, it is worthwhile to discuss some of the terms and concepts used in electrochemistry.

6.2 REDOX REACTION

Various chemical reactions occur through redistribution of electrons among the reacting substances. Any substance that loses electrons is said to be *oxidized* and the one which gains electrons is said to be *reduced*.

However, in a chemical reaction, a substance can lose electrons only if there is present another substance, which can gain electrons. This implies that oxidation can take place only if reduction also occurs at the same time or vice-versa. These reactions in which reduction and oxidation take place are called redox reactions.

$$Zn_{(s)} + Cu^{2+}_{(aq)} \longrightarrow Zn^{2+}_{aq} + Cu_{(s)}$$

In this $Zn_{(s)}$ loses two electrons whereas Cu^{2+} gains two electrons, and hence this is a redox reaction. The redox reactions are of two types:

(*a*) **Direct Redox Reactions:** The redox reactions in which oxidation and reduction take place in the same vessel are called direct redox reactions. Example, displacement of copper from $CuSO_4$ solution when a zinc rod is dipped in it.

$$Zn + CuSO_4 \longrightarrow ZnSO_4 + Cu_{(s)} \quad \text{or} \quad Zn_{(s)} + Cu^{2+} \longrightarrow Zn^{2+} + Cu_{(s)}$$

The above reaction can be split into two halves,

$$Zn \longrightarrow Zn^{2+} + 2e^- \text{ (oxidation half reaction)}$$
$$Cu^{2+} + 2e^- \longrightarrow Cu \text{ (reduction half reaction)}$$

(*b*) **Indirect Redox Reaction:** If redox reaction is allowed to take place in such a way that oxidation reaction takes place in one beaker and the reduction half reaction in another beaker, the electrons given out by former will be taken by latter and a current will flow. These indirect redox reactions form the basis of electro-chemical cells.

6.3 ELECTRODE POTENTIAL

A metal rod or plate in contact with its own ions is called electrode. When a metal is in contact with the solution of its own ions, it can undergo oxidation or reduction. *The tendency of an electrode to lose or gain electrons is called electrode potential*. Since each electrode represents a half cell, therefore, electrode potential is also called potential for half cell.

Consider a copper rod dipped in the solution of its own ions, i.e. Cu^{2+} ions. The copper rod has a tendency to lose electrons to form Cu^{2+} ions which pass into the solution. Thus, copper rod acquires a net negative charge as electrons, which are released, get accumulated around the Cu rod. These electrons will attract Cu^{2+} ions from the solution. This indicates that copper rod will not keep on losing electrons and a equilibrium will soon be established.

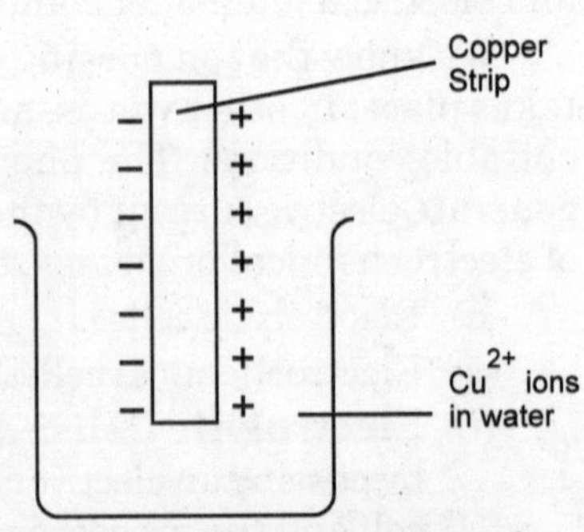

Fig. 6.1(a) Development of a pote-ntial difference between metal and its ions.

$$Cu_{(s)} \rightleftharpoons Cu^{2+}_{(aq)} + 2e^-$$

When such an equilibrium is reached, a separation of positive and negative charges would occur. This separation creates a potential difference between metal rod and the solution. This potential difference is called the *electrode potential.*

Thus, electrode potential of a metal is the measure of tendency of a metallic electrode to lose or gain electrons, when it is in contact with its own salt. It magnitude depends upon:

(*a*) The nature of metal and ions
(*b*) Concentration of ions
(*c*) Temperature.

Electrode potential measured under standard conditions i.e. 1 molar concentration of metal ions and a temperature of 298 K is called *standard electrode potential* and is denoted by E_o.

Moreover, the tendency of an electrode to *lose* electrons is a direct measure of its tendency to get oxidised and this tendency is called *oxidation potential*. Similarly, the tendency of an electrode to *gain* electrons is the measure of its tendency to get reduced, and this tendency is known as *reduction potential*. The value of reduction potential is negative of its oxidation potential.

i.e. $$E_{oxdn} = -E_{red}$$

Generally, the term standard electrode potential refers to reduction potential.

6.4 ELECTROCHEMICAL CELL

Electrochemical cell or the galvanic cell is a device which converts free energy of a physical or chemical process into electrical energy. Electrochemical cell usually consists of two electrodes immersed in one or more suitable electrolytes. When the electrodes are connected externally, a chemical reaction occurs in the cell resulting into oxidation at one electrode and reduction at the other electrode. The electrode at which oxidation occurs is called *cathode* and the electrode at which reduction occurs is called anode.

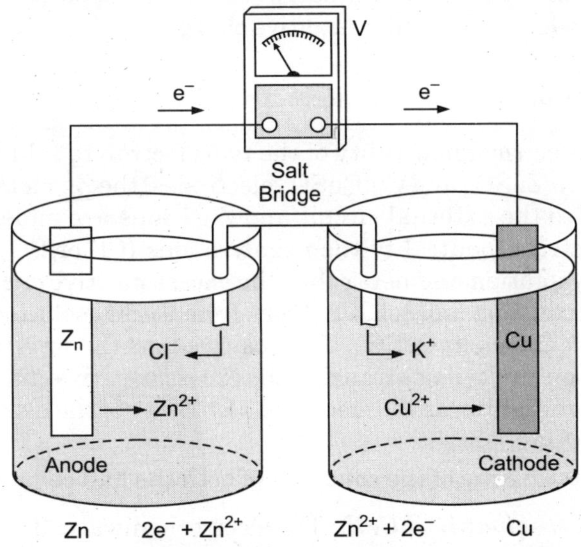

Fig. 6.1(b) Daniell cell

A typical example of galvanic cell is that of Daniell cell (Fig. 6.1(b)). It consists of a zinc rod dipping in $ZnSO_4$ solution, where oxidation takes place and a copper rod dipped in $CuSO_4$ solution, where reduction takes place. In other words, each electrode may be regarded as a half cell.

The metallic rods in the two beakers are connected to an ammeter by means of an insulated wire. The solutions in two beakers are connected by an inverted U-tube known as salt bridge. The tube is filled with a solution of electrolyte such as KNO_3, KCl, NH_4NO_3 or solidified solution of an electrolyte agar-agar and gelatine. The ends of the U-tube are plugged with glass wool or cotton wool. The salt in the bridge is such that it does not chemically react with either of the two solutions connected by it.

As soon as the circuit is closed, i.e. the two electrodes are connected, a current starts flowing which is shown by the deflection in the ammeter. The electric current flows due to the passage of electrons from zinc to copper rod. The electrode reactions involved as:

At anode $\quad Zn \longrightarrow Zn^{2+} + 2e^-$ (Oxidation)

At cathode $\quad Cu^{2+} + 2e^- \longrightarrow Cu$ (reduction)

$$Zn + Cu^{2+} \longrightarrow Cu + Zn^{2+}$$

Since electrons, i.e., negative electricity flows from zinc to the copper rod, the conventional current, i.e., positive electricity flows from copper to zinc rod. Thus zinc rod serves as negative electrode and copper rod serves as positive electrode.

As shown in the figure, in an electrochemical cell the two half reactions take place in two different compartments. The electrode where oxidation takes place (the anode) is assigned the *negative charge* because oxidation is accompanied by liberation of negatively charged electrons. Each time an ion leaves the electrode surface, two electrons are left on the electrode surface making it negatively charged. Similarly, the electrode where reduction takes place (the cathode) is assigned the *positive charge* because the metal cations from the solution tend to withdraw electrons from electrode surface and get deposited as metal atoms on electrode. This process decreases the availability of electrons on the electrode surface and thus leaves the electrode surface positively charged.

Note: However, in an *electrolytic cell*, the situation will be just opposite. The anode would acquire positive charge and cathode would acquire negative charge.

Function of Salt Bridge

(*i*) It maintains the electrical neutrality of the two electrolyte solutions. In the Daniell cell, which consists of $Zn/ZnSO_4$ and $Cu/CuSO_4$ electrodes, the Zn metal undergoes oxidation to release electrons to the external circuit and Zn^{2+} ions are added to $ZnSO_4$ solution. To maintain the electrical neutrality the negative ions (Cl^- or SO_4^{2-}) from the salt bridge migrate to $ZnSO_4$ solution and neutralize the excess positive charge in the oxidation half cell. Similarly in reduction half cell, Cu^{2+} ions from $CuSO_4$ solution are reduced. This leads to an excess of SO_4^{2-} ions in $CuSO_4$. The positive ions (K^+ or NH_4^+) from the salt bridge migrate to reduction half cell neutralizing the excess negative charge in $CuSO_4$ solution. In this way salt bridge maintains the electrical neutrality of the electrolytes.

(*ii*) It also completes the circuit.

(*iii*) It prevents the intermixing of the solutions of both the half cells.

Representation of Electrochemical Cell: The electrochemical cell is represented as

Anode	Salt bridge	Cathode
Zn/Zn^{2+}	KNO_3	Cu^{2+}/Cu

The electrode at which oxidation takes place (anode) is written on the left hand side. The electrode at which reduction takes place (cathode) is written on the right hand side of the salt bridge.

In some galvanic cells, the same electrolyte is used in two half cells. In such cases, no separation is shown between two electrolytes of two electrodes for example

$$\text{Pt, } H_2 \text{ (1 atm); HCl (0.1 M); } Cl_2\text{, Pt}$$

Here HCl is an electrolyte and provides H^+ ions to the hydrogen electrode and Cl^- ion to the chlorine electrode.

Cell Potential (E.M.F. of the Cell): An electrochemical cell is made up of two electrodes, one of these electrodes must have a higher electrode potential (higher tendency to lose electrons) than the other electrode. As a result of this potential difference, the electrons flow from the electrode at higher potential to the electrode at lower potential. A cell that has not reached equilibrium, can do the electrical work till a transfer of electron takes place. *The electron transfer depends upon the potential difference between two electrodes. This potential difference is called the cell potential and* measured in *volts*. This potential is also known as *electromotive force of the cell*. When the cell potential is large, a given number of electrons travelling between the electrodes can do a large amount of electrical work. When the cell potential is small, the work done by the same number of electrons will be small. When the cell potential is zero, the reaction is at equilibrium and no work is done. The total cell potential can be written as the sum of the two electrode potentials. Thus

$$E_{cell} = E_{cathode} + E_{anode}$$

The total cell potential is a measure of the driving force of a chemical reaction whereas $E_{cathode}$ and E_{anode} are representatives of oxidising and reducing capabilities of concerned reactants, respectively. Now, we will explain how to measure the electrode potential of a single electrode.

The maximum electrical work done by a system is given by value of ΔG

$$W_{max} = \Delta G \text{ at the constant temperature and pressure.}$$

Therefore, the e.m.f. of a cell or cell potential is a measure of overall tendency which determines the spontaneity of a reaction to take place.

The various electrodes are discussed below:

1. Reference Electrodes:

(*a*) Standard Hydrogen Electrode: Standard hydrogen electrode contains hydrogen gas in contact with H^+ ions of 1 M concentration at 298 K and at 1 atmospheric pressure (Fig. 6.2).

In hydrogen electrode, hydrogen gas is bubbled through a solution of hydrogen ions. The inert metal (which is often Pt) act as a source or sink of electrons, but takes no other part in the reaction other than acting as a catalyst for it. This electrode is denoted as

$$\text{Pt} \mid H_{2(aq)} \mid H^+(aq)$$

The hydrogen electrode may act as either a cathode or an anode, depending on the other electrode in the cell and the spontaneous direction of the overall reaction. When the electrode is acting as *cathode*, the reaction is

$$2H^+_{(aq)} + 2e^- \longrightarrow H_{2(g)} \quad \text{(reduction)}$$

when the hydrogen electrode is acting as *anode*

$$H_{2(g)} \longrightarrow 2H^+(aq) + 2e \quad \text{(oxidation)}$$

For determining the electrode potential of the given electrode, the electrode is coupled with SHE. The E.M.F. of the cell is then determined with the help of a potentiometer. The electrode potential of SHE is arbitrarily taken as zero, the E.M.F. of cell gives the value of electrode potential of the given electrode.

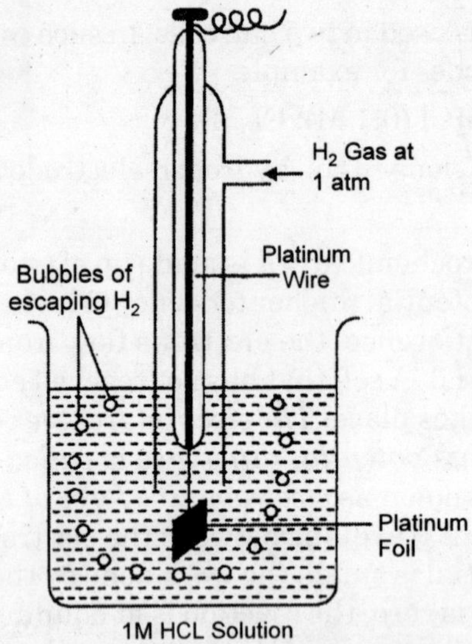

Fig. 6.2 Standard hydrogen electrode.

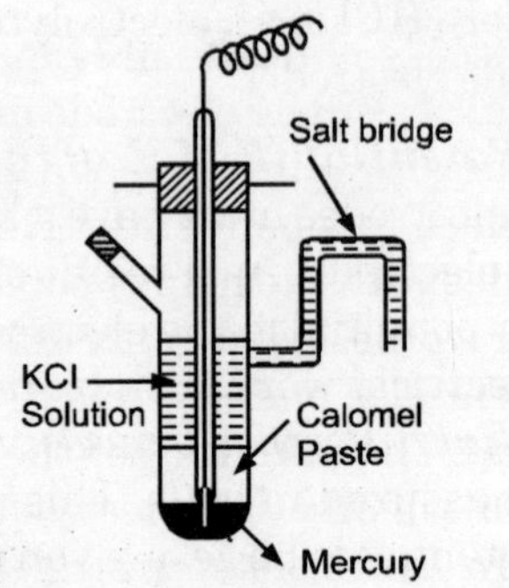

Fig. 6.3 Calomel Electrode.

(b) Calomel Electrode: It is very convenient secondary reference electrode and is used in place of hydrogen electrode which is not easy to prepare. It consists of mercury in contact with a solution of potassium chloride saturated with mercurous chloride (calomel). The electrode is represented as

$$Hg/Hg_2Cl_{2(s)},\ KCl_{(aq)}$$

It must be emphasized here that the essential requirement of a reference electrode is that its potential through the potentiometric measurement must remain constant (Fig. 6.3).

2. *Indicator Electrodes:*

(a) Quinhydrone Electrode: Quinhydrone is the name given to quinone-hydroquinone system involving the reaction,

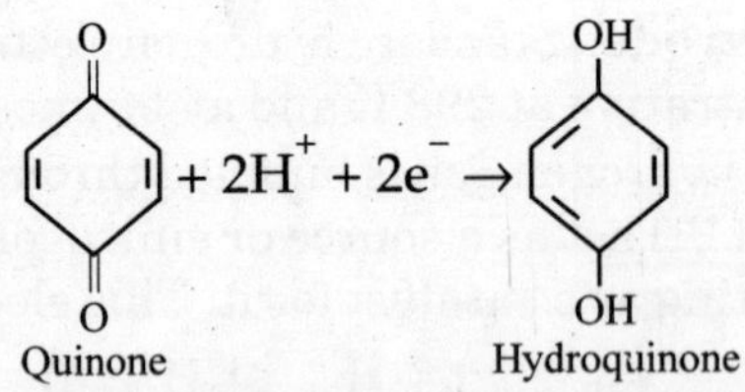

Reversible Cells: A cell is said to work reversibly in thermodynamic sense, when it is generating infinitesimally small current so that the cell reaction always remains virtually in a state of equilibrium.

A reversible cell satisfies the following conditions:

(*i*) If it is connected to an external source of e.m.f. acting in opposite direction, the opposing e.m.f. is exactly equal to the e.m.f. of the cell i.e. no current is given out and no chemical reaction takes place.

(*ii*) If opposing e.m.f. is infinetesimally smaller than that of the cell itself, an extremely small current is given out by the cell and corresspondingly small amount of chemical reaction takes place in cell.

(*iii*) If the opposing e.m.f. is infinitesimally greater than that of cell itself, an extremely small current flows through the cell in opposite direction, and a small amount of the chemical reaction also takes place in reverse direction. A reversible galvanic cell would thus consist of two reversible electrodes, one of which acts as anode and other as cathode.

6.5 MEASUREMENT OF ELECTRODE POTENTIAL

As is discussed in the previous section, the electrochemical cell consists of two half cells. The potential of a single electrode in a half cell is called the single electrode potential. It is impossible to know the absolute value of a single electrode potential because neither oxidation nor reduction takes place independently. However, we can measure the *potential difference* between the two electrodes with the help of potentiometer, by combining the two electrodes to form a complete cell. In other words, we can only determine the relative value of electrode potential, if we fix arbitrarily the potential of one electrode. The electrode, whose potential is arbitrarily fixed, is called *reference electrode*. The potential of the reference electrode is arbitrarily assigned a zero value and the potentials of all other electrodes are then referred to it. The reference electrode so chosen is the *standard hydrogen electrode* (SHE) or *normal hydrogen electrode* (NHE) discussed in previous section.

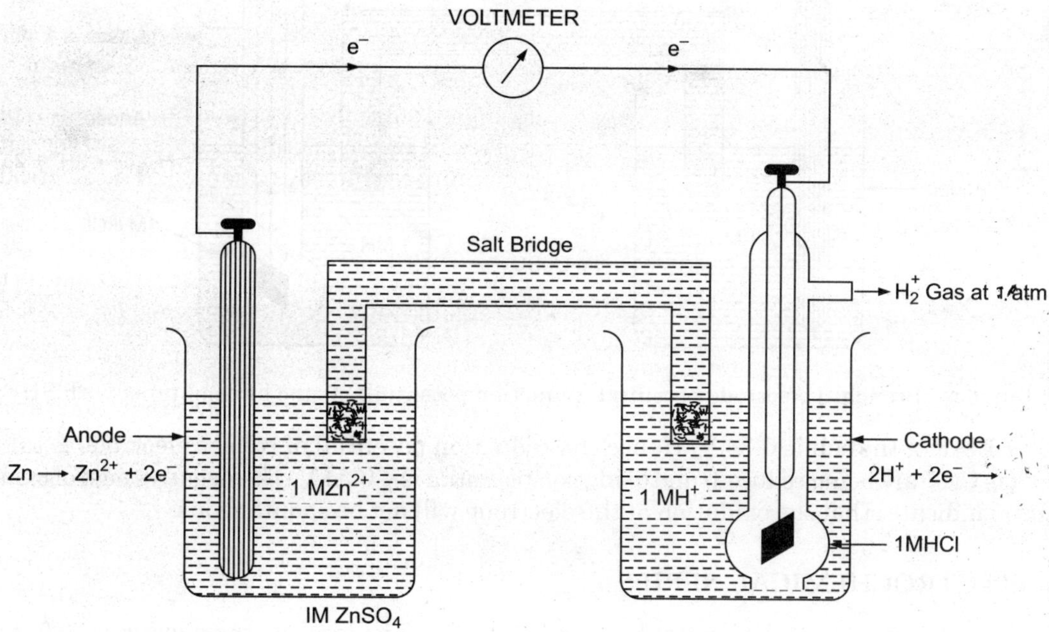

Fig. 6.4 Standard electrode potential (reduction potential) of zinc by combining with SHE

In order to measure the electrode potential of the given electrode, the electrode is coupled with SHE. The e.m.f. of the complete cell is then measured using voltmeter. The measured e.m.f. would give the electrode potential of the given electrode because electrode potential of SHE is arbitrarily fixed as zero. The direction of the flow of current further indicates whether oxidation takes place or reduction takes place at the electrode under investigation with respect to SHE. Accordingly the electrode potential is termed as oxidation or reduction potential.

Let us illustrate this by taking a few examples. Suppose it is desired to measure and determine the electrode potential of zinc electrode, $Zn/ZnSO_4$. The electrode is connected with the SHE as shown in figure 6.4. The two half cell reactions are represented as:

$$Zn_{(s)} \longrightarrow Zn^{2+} + 2e^- \quad \text{(Oxidation)}$$
$$2H^+ + 2e^- \longrightarrow H_2 \quad \text{(Reduction)}$$

The EMF of the cell is found to be 0.76 V. Therefore, the oxidation potential of zinc electrode is taken as 0.76 V and reduction potential is taken as – 0.76 V.

Similarly, the EMF of copper electrode, $Cu^{2+}/CuSO_4$ can be determined by pairing it with SHE (Fig. 6.5) when the cell reaction can be represented as:

$$H_{2(g)} \longrightarrow 2H^+ + 2e^- \quad \text{(Oxidation)}$$
$$Cu^{2+} + 2e^- \longrightarrow Cu_{(s)} \quad \text{(Reduction)}$$

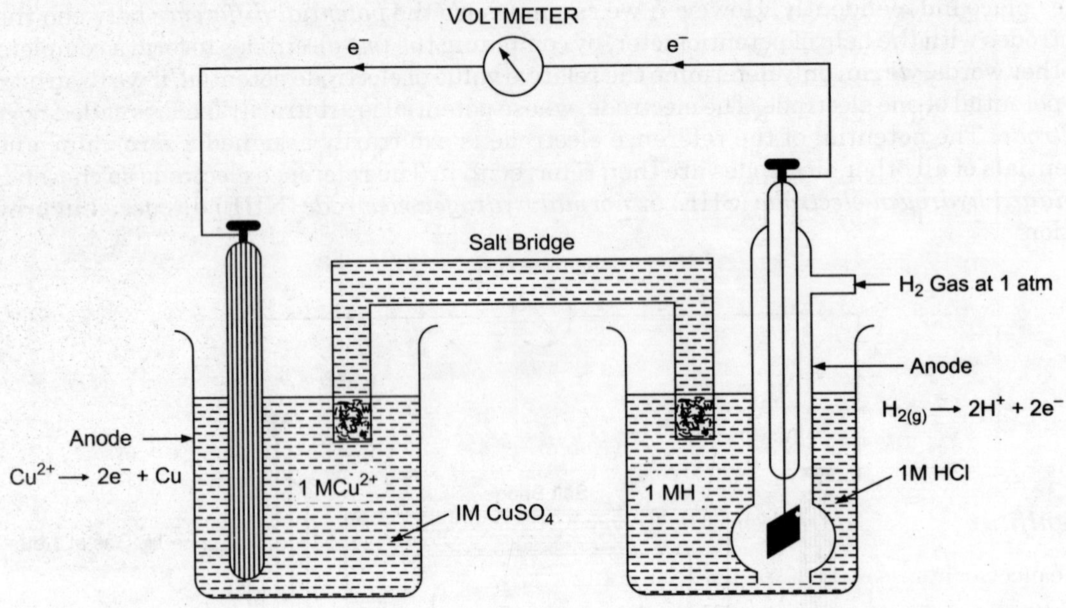

Fig. 6.5 Standard electrode potential (reduction potential) of zinc by combining with SHE

The EMF of the cell is 0.34 V. Hence the reduction potential of copper electrode is taken as 0.34 V (as Cu^{2+} are being reduced) and oxidation potential is – 0.34 V. The negative sign of oxidation potential indicates that the oxidation at the electrode will not be spontaneous.

6.6 ELECTROCHEMICAL SERIES

The measurement of reduction potential can be used to calculate the thermodynamic functions and standard potential of the cell.

Now, we know that the standard potential of the electrode can be measured with reference to standard hydrogen electrode, which has been assigned the value zero. IUPAC recommends the adoption of the reduction potential as electrode potentials of the corresponding electrodes. When electrodes are arranged in increasing order of their standard (reduction) electrode potential in a series, the series is called *electrochemical series*. The electrochemical series is also called *activity series* (Table 6.1).

Table 6.1 Standard Electrode Potentials (Reduction) at 25°C.

Metalion		*Potential in volts*	
$Li^+ + e^- \rightarrow Li$	(base)	−3.05	(Anode)
$K^+ + e^- \rightarrow K$		−2.93	
$Ca^{2+} + 2e^- \rightarrow Ca$		−2.90	
$Na^+ + e^- \rightarrow Na$		−2.71	
$Mg^{2+}\ 2e^- \rightarrow Mg$		−2.37	
$Al^{3+} + 3e^- \rightarrow Al$		−1.66	
$Zn^{2+}\ 2e^- \rightarrow Zn$		−0.76	
$Cr^{3+} + 3e^- \rightarrow Cr$		−0.74	
$Fe^{2+} + 2e^- \rightarrow Fe$		−0.44	
$Ni^{2+} + 2e^- \rightarrow Ni$		−0.23	
$Sn^{2+} + 2e^- \rightarrow Sn$		−0.14	
$Pb^2 + 2e^- \rightarrow Pb$		−0.13	
$F^{3+} + 3e^- \rightarrow Fe$		−0.04	
$H^+ + e^- \rightarrow 1/2H$		0.00	
$Cu^{2+} + 2e^- \rightarrow Cu$		+0.34	
$Ag^+ + e^- \rightarrow Ag$		+0.80	
$Pt^{4+} + 4e^- \rightarrow Pt$		+0.86	
$Au^+ + e^- \rightarrow Au$		+1.69	
$1/2F_2 + e^- \rightarrow F^-$	(Noble)	+2.87	(Cathodic)

Significance of Electrochemical Series

The electrochemical series provide valuable information regarding the following:

(i) **To Calculate Standard Potential of the Cell:** The electrode potential of a cell formed from any two electrodes can be calculated by electrochemical series.

$$E^o_{Cell} = \text{[Reduction potential of reduction half cell} - \text{Reduction potential of oxidation half cell]}$$

In representation of a cell, the electrode at which oxidation takes place is shown on left hand side (L.H.S.) and the electrode at which oxidation takes place is shown on right hand side (R.H.S.) hence the above expression can be written as:

$$E^o_{Cell} = E^o_{(R.H.S.\ Electrode)} - E^o_{(L.H.S.\ Electrode)}$$

$$= E^o_{cathode} - E^o_{anode}$$

(Both the electrode potentials are reduction potentials)

Problem 6.1: Calculate the standard cell potential for the reaction:

$$Cu + Ag^+ \longrightarrow Cu^{2+} + 2Ag$$

For which the half-cell reactions and corresponding electrode potentials are given below:

$$Cu \longrightarrow Cu^{2+} + 2e^- \quad -0.34\ V \quad \text{(oxidation potential)}$$

$$Ag^+ + e^- \longrightarrow Ag \quad 0.80\ V \quad \text{(reduction potential)}$$

Solution: The reduction potentials of the anode and the cathode are + 0.34 V and 0.80 V respectively. Hence

$$E^o_{Cell} = E^o_{(R.H.S.)} - E^o_{(L.H.S.)}$$

$$= E^o_{cathode} - E^o_{anode} = [0.80 - 0.34]\ V = 0.46\ V$$

(ii) **Predicting the Spontaneity or Feasibility of the Cell Reaction:** If the standard EMF of the cell as calculated above is positive, the cell reaction is spontaneous, i.e., feasible. If E^o_{Cell} is negative, the cell reaction would not be feasible.

Problem 6.2: Predict whether the reaction

$$2Ag_{(s)} + Zn^{2+}_{(aq)} \longrightarrow 2Ag^+_{(aq)} + Zn_{(s)}$$

is feasible or not.

Solution: The half reactions are

$$2Ag_{(s)} \longrightarrow 2Ag^+_{(aq)} + 2e^- \qquad E^o = 0.80\ V$$

$$Zn^{2+}_{(aq)} + 2e^- \longrightarrow Zn_{(s)} \qquad E^o = -0.763\ V$$

$$E^o_{Cell} = E^o_{cathode} - E^o_{anode}$$

$$= [-0.763 - 0.80]\ V = -1.563\ V$$

Since E^o_{Cell} is negative, the reaction is not feasible.

Problem 6.3: Using electrochemical series, find out whether zinc and silver would react with dilute H_2SO_4 or not.

Solution: The reaction of zinc with dilute H_2SO_4 is represented as:

$$Zn + H_2SO_4 \rightleftharpoons ZnSO_4 + H_{2(g)}$$

The above reaction is split into two half reactions as

$$Zn_{(s)} \rightleftharpoons Zn^{2+} + 2e^- \qquad \text{(Oxidation)}$$

$$2H^+ + 2e^- \rightleftharpoons H_{2(g)} \qquad \text{(Reduction)}$$

The galvanic cell can be written as

$Zn, Zn^{2+}/H^+, H_{2(g)}: Pt$

$$E^o = E^o_{RHS} - E^o_{LHS}$$

$$= 0 - (-0.76\ V) = +0.76\ V$$

Since E^o is positive, the reaction of zinc with H_2SO_4 is feasible i.e. Zn will react with H_2SO_4. Similarly the reaction of silver with dil. H_2SO_4 is written as

$$2Ag + H_2SO_4 \rightleftharpoons Ag_2SO_4 + H_{2(g)}$$

The above reaction is split into two half cells as:

$$Ag \longrightarrow Ag^+ + e \qquad \text{(oxidation)}$$

$$2H^+ + 2e^- \longrightarrow H_{2(g)} \qquad \text{(reduction)}$$

The cell involving the above reaction is written as:

$Ag, Ag^+ \mid H^+, H_2 : Pt$

$$E^o = E^o_{RHS} - E^o_{LHS}$$

$$= 0 - (0.80\ V) = -0.80\ V$$

Since E^o is negative, hence the reaction of silver with sulphuric acid is not feasible i.e. Ag will not react with H_2SO_4.

(iii) **To Predict the Relative Oxidising and Reducing Powers:** The behaviour of different elements can be interpreted easily by using the electrochemical series. A system with high reduction potential has a great tendency to undergo reduction. In the electrochemical series the substances are arranged in increasing order of their reduction potential. Thus, the element at the bottom of the table having high value or reduction potential has the greatest tendency to get reduced and hence consequently is the strongest oxidising agent. Thus F_2 has the greatest tendency to get reduced and act as strongest oxidising agent. Similarly the substance lying at the top of the table has lower values of reduction potential and, therefore, has least tendency to get reduced. In other words, it has the tendency to get oxidised and hence is strongest reducing agent. Thus Li^+ having least value of reduction potential, is reduced with great difficulty to Li or conversely, Li is easily oxidised to Li^+. Therefore, Li is the strongest reducing agent. In general,

- the species which are easily reduced than hydrogen are placed below it having positive values of E^o. The oxidising agents have positive E^o values.
- higher the positive value, stronger will be the oxidising agent.
- the species which are easily oxidised than hydrogen are placed above it and they have negative E^o values.
- higher the negative value, stronger will be the reducing agent.

(iv) **To Predict Whether a Metal will React with Acid to give H_2 Gas:** The metals which are placed above hydrogen in electrochemical series have great tendency for oxidation and hence can displace hydrogen from acids.

(v) **Replacement Tendency:** A metal having low reduction potential value shows great tendency to assume the oxidised form and hence can reduce the metal lying below it or displaces it from the solution. On the contrary the metals lying below in the series are strong reducing agents and are oxidised to metal ions.

Problem 6.4: What will happen if a strip of zinc is immersed in $CuSO_4$ solution?

Solution: As zinc is placed above in the electrochemical series compared to copper, zinc has a tendency to get oxidised and act as reducing agent. Thus it displaces Cu^{2+} ions from the solution to form Cu and in turn strip of zinc dissolves in solution.

$$Zn + Cu^{2+}_{(aq)} \longrightarrow Cu + Zn^{2+}_{(aq)}$$

NUMERICALS BASED ON ELECTROCHEMICAL SERIES

Problem 6.5: Why copper does not discolve in HCl but dissolves in dilute HNO_3. E_o values of Cu^{2+}/Cu and H^+/H_2 are 0.34 and 0. Also, $E^o_{NO_3^-/NO}$ is 0.96 V.

Solution: The standard reduction potential for

$$Cu^{2+} + 2e^- \longrightarrow Cu$$

is +0.34V as compared to standard hydrogen electrode for which it is 0.0 V. Hence copper is a poorer reducing agent than hydrogen. The copper will only dissolve in acids which contain oxidising anions like NO_3^-. Since $E^o_{NO_3^-/NO}$ is larger than $E^o_{Cu^{2+}/Cu}$, NO_3^- can be reduced by Cu.

Problem 6.6: Predict which of the following reactions will be spontaneous with the help of standard potential values:

(*i*) $Fe + 2H^+ \longrightarrow Fe^{2+} + H_2$ (*ii*) $Cu + 2H^+ \longrightarrow Cu^{2+} + H_2$

(*iii*) $2Fe^{3+} + 2I^- \longrightarrow 2Fe^{2+} + I_2$

Solution:

(*i*) $Fe + 2H^+ \longrightarrow Fe^{2+} + H_2$

$$E^o_{cell} = E^o_{anode} + E^o_{cathode}$$
$$= [-(-0.44) + 0]$$
$$= 0.44V$$

The cell potential is positive and hence reaction is spontaneous

(*ii*) $Cu + 2H^+ \longrightarrow Cu^{2+} + H^2$

$$E^o_{cell} = E^o_{anode} + E^o_{cathode}$$
$$= [-(0.34) + 0]V$$
$$= -0.34\ V$$

(Since for Cu^{2+}/Cu ; $E^0 = +0.34V$ and hence for $Cu/Cu^{2+} = -0.34V$)

Reaction is non-spontaneous because E^o_{cell} is negative.

(iii) $2\ Fe^{3+} + 2I^- \longrightarrow 2Fe^{2+} + I_2$

$$E^o_{cell} = E^o_{anode} + E^o_{cathode}$$
$$= [-(0.54) + 0.77]V = 0.23V$$

Reaction is spontaneous.

6.7 NERNST EQUATION

The concentration of the electrolyte in a half-cell influences its electrode potential. The concentration dependence of EMF can be obtained from thermodynamic considerations.

Consider a general cell reaction:

$$M^{n+}_{(aq)} + ne^- \longrightarrow M_{(s)}$$

The decrease in free energy, $-\Delta G$, accompanying the process is given by the well known thermodynamic equation,

$$\Delta G = \Delta G^o + RT \ln \frac{a_{Products}}{a_{reactants}} \quad ...(1)$$

where ΔG^o refers to the free energy change for the reaction in the standard state i.e. when the activities of the reactants and products are all unity, $a_{product}$ and $a_{reactant}$ represent the activities of the reactants and the products under a given set of conditions, R is the gas constant and T is the temperature.

Since $\Delta G = -nFE$ and $\Delta G^o = -nFE^o$

where n is the number of electrons liberated at one electrode and F is Faraday of electricity ($1F = 96500$ coulombs) and E and E^o are the electrode potential under the operating conditions and standard conditions respectively.

Hence, substituting the value of ΔG in equation (1) we get

$$-nEF = -nE^oF + RT \ln \frac{a_{products}}{a_{reactants}} \quad ...(2)$$

$$E = E^o - \frac{RT}{nF} \ln \frac{a_p}{a_r} \quad ...(3)$$

$$E = E^o - \frac{2.303\,RT}{nF} \log \left(\frac{a_{\text{products}}}{a_{\text{reactants}}} \right) \qquad ...(4)$$

Equation (4) is known as Nernst equation and gives the dependence of the electrode or cell potential in terms of activities of the reactants and products. The quantity E^o is known as standard EMF of the cell and is the cell potential when the activities are all unity.

At $T = 298$ K, equation (4) reduces to:

$$E = E^o - \frac{(8.314)(298)(2.303)}{n(96485)} \log \frac{a_{\text{products}}}{a_{\text{reactants}}}$$

$$\boxed{E = E^o - \frac{0.0591}{n} \log \frac{a_{\text{products}}}{a_{\text{reactants}}}} \qquad ...(5)$$

In dilute solutions the activities may be replaced by molar concentration terms,

$$\therefore \qquad \boxed{E = E^o - \frac{0.0591}{n} \log \frac{[M_{(s)}]}{[M^{n+}_{(aq)}]}} \qquad ...(6)$$

For pure solids, $[M_{(s)}] = 1$

Hence equation (6) can be written as

$$E = E^o - \frac{0.0591}{n} \log \left[\frac{1}{M^{n+}_{(aq)}} \right] \qquad ...(7)$$

Equation (7) is the Nernst equation for pure solids at 298 K. For any other temperature it can be written as

$$E = E^o - \frac{2.303RT}{n} \log \left[\frac{1}{M^{n+}_{(aq)}} \right] \qquad ...(8)$$

For a cell reaction,

$$aA + bB \longrightarrow lL + mM$$

The Nernst equation can be written as

$$E = E^o - \frac{2.303RT}{nF} \log \frac{[L]^l [M]^m}{[A]^a [B]^b} \qquad ...(9)$$

or

$$E = E^o + \frac{2.303RT}{nF} \log \frac{[A]^a [B]^b}{[L]^l [M]^m} \qquad ...(10)$$

It is seen from the Nernst equation that,

- If the product of the molar concentration of the products is greater than the product of the molar concentration of the reactants, the EMF is less than the standard EMF of the cell.
- The electrode potential (E) decreases as temperature is increased.

APPLICATIONS OF NERNST EQUATION

(*i*) **Calculation of the Electrode Potential of the Cell:** Nernst equation is used for the calculation of cell potential under the operating conditions.

Problem 6.7: Calculate the potential of the cell:

$$Cr/Cr^{3+}_{(0.1\,M)} \,//\, Fe^{2+}_{(0.01\,M)}/Fe$$

Given: $E^o_{Cr^{3+}/Cr} = -0.74$ V

$E^o_{Fe^{2+}/Fe} = -0.44$ V

Solution: The half reactions are:

$$Cr \longrightarrow Cr^{3+} + 3e^- \;] \times 2 \quad \text{(Oxidation at anode)}$$

$$Fe^{2+} + 2e^- \longrightarrow Fe \;] \times 3 \quad \text{(Reduction at cathode)}$$

$$2Cr + 3Fe^{2+} \longrightarrow 2Cr^{3+} + 3Fe \quad \text{Cell reaction}$$

$\therefore \quad n = 6$

$$E_{cell} = E^o_{cell} - \frac{0.0591}{6} \log \frac{[Cr^{3+}]^2[Fe]^3}{[Fe^{2+}]^3[Cr]^2}$$

For pure solids the activity or the concentration is taken as unity, so the equation becomes,

$$E_{cell} = E^o_{cell} - \frac{0.0591}{6} \log \frac{[Cr^{3+}]^2}{[Fe^{2+}]^3}$$

$$= (E^o_{cathode} - E^o_{anode}) - \frac{0.0591}{6} \log \frac{[Cr^{3+}]^2}{[Fe^{2+}]^3}$$

$$= (E^o_{Fe^{2+}/Fe} - E^o_{Cr^{3+}/Cr}) - \frac{0.0591}{6} \log \frac{[Cr^{3+}]^2}{[Fe^{2+}]^3}$$

$$= [-0.44\text{ V} - (-0.74\text{ V})] - \frac{0.0591}{6} \log \frac{(0.10\,M)^2}{(0.01\,M)^3}$$

$$= 0.3\text{ V} - \frac{0.0591}{6} \log 10^4 = 0.3\text{ V} - 0.0394\text{ V} = +\,0.2606\text{ V}$$

The cell potential = 0.2606 V.

(*ii*) To Study the Effect of Electrolyte Concentration on Electrode Potential: The concentration of the electrolyte affects the electrode potential of the single electrode. This can be studied by Nernst equation.

For *reduction half cell reaction* of a metal in contact with M^{n+} ions the reaction can be written as:

$$M^{n+} + ne^- \longrightarrow M$$

The Nernst equation takes the form:

$$E = E^o - \frac{0.0591}{n} \log \frac{[M]}{[M^{n+}]}$$

[*M*] being solid, its concentration is taken as unity.

$$E = E^o - \frac{0.0591}{n} \log \frac{1}{[M^{n+}]}$$

Problem 6.8: What is the potential of a half cell consisting of zinc electrode in 0.01 M $ZnSO_4$ solution at 25°C given $E^o = -0.763$ V.

Solution: The half cell reaction is

$$Zn^{2+}_{(aq)} + 2e^- \longrightarrow Zn_{(s)}$$

$$E = E^o - \frac{0.0591}{n} \log \frac{1}{[M^{n+}]}$$

$$= -0.763 - \frac{0.0591}{2} \log \frac{1}{(0.01)} = -0.763 - 0.0591 = -0.8221 \text{ V.}$$

***(iii)* Calculation of Equilibrium Constant:** Nernst equation can be used to find out the equilibrium constant of the given cell.

Consider a reaction at equilibrium

$$aA + bB \rightleftharpoons lL + mM$$

$$K_c = \frac{[L]^l [M]^m}{[A]^a [B]^b}$$

where K_c is the equilibrium constant.

The Nernst equation can be written as:

$$E_{\text{cell}} = E^o_{cell} - \frac{0.0591}{n} \log \frac{[L]^l [M]^m}{[A]^a [B]^b}$$

or

$$E_{\text{cell}} = E^o_{\text{cell}} - \frac{0.0591}{n} \log K_c$$

But at equilibrium, the cell reaction is balanced, i.e. at the equilibrium concentration of the two metal ions, there is no potential difference between the two electrodes and the cell potential becomes zero. The Nernst equation may now be written as:

$$O = E^o_{\text{cell}} - \frac{0.0591}{n} \log K_c$$

$$\text{Log } K_c = \frac{nE^o_{\text{cell}}}{0.0591}$$

Problem 6.9: Calculate the equilibrium constant at 298 K for the reaction taking place between copper sulphate and aluminium. Given $E^o_{\text{cell}} = 2.0$ V.

Solution: The equation for the above reaction is:

$$2Al + 3Cu^{2+} \longrightarrow 2Al^{3+} + 3Cu$$

or Half cell reactions are:

$$Al \longrightarrow Al^{3+}_{(aq)} + 3e^-] \times 2 \quad \text{(oxidation)}$$

$$Cu^{2+}_{(aq)} + 2e^- \longrightarrow Cu_{(s)} \quad] \times 3 \quad \text{(reduction)}$$

$$2Al + 3Cu^{2+} \longrightarrow 2Al^{3+} + 3Cu$$

Thus $n = 6$

According to Nernst Equation,

$$\log K_c = \frac{nE^o_{\text{cell}}}{0.0591}$$

$$\log K_c = \frac{6 \times 2}{0.0591} = 203.39 \quad \text{or} \quad K_c = \text{Antilog } 203.39 = 2.5 \times 10^{203}$$

(iv) To Find Out the Concentration of One of the Ionic Species in a Cell if the Concentration of the other Species is Known.

Problem 6.10: Determine the concentration of Cd^{2+} ions in the following electrochemical cell.

$$Zn/Zn^{2+}_{(0.1\,M)} \;//\; Cd^{2+}_{(xM)}\,/\,Cd$$

Given
$$E^{o}_{Zn^{2+}/Zn} = -0.76\text{ V};$$
$$E^{o}_{Cd^{2+}/Cd} = -0.40\text{ V}$$
$$E_{cell} = 0.3305\text{ V at 298 K}$$

Solution: The cell reaction is

$$Zn_{(s)} + Cd^{2+}_{(aq)} \longrightarrow Zn^{2+}_{(aq)} + Cd_{(s)}$$

$$E^{o}_{cell} = E^{o}_{Cd^{2+}/Cd} - E^{o}_{(Zn^{2+}/Zn)}$$
$$= -0.40 - (-0.76) = 0.36\text{ V}$$

According to Nernst equation,

$$E = E^{o} - \frac{0.0591}{2}\log\frac{[Zn^{2+}]}{[Cd^{2+}]}$$

$$0.3305 = 0.36 - \frac{0.0591}{2}\log\frac{0.1}{x}$$

$$-0.0295 = -0.0295\log\frac{1}{10x}$$

$$\frac{-0.0295}{-0.0295} = \log\frac{1}{10x}$$

$$\log\frac{1}{10x} = 1$$

$$\frac{1}{10x} = \text{antilog } 1 = 10$$

$$\therefore \quad \frac{1}{x} = 100, \quad x = 0.01\text{ M}$$

(v) To Find out the pH of a Solution: Nernst equation is also used to find out the pH of a solution.

Problem 6.11: Calculate the pH of the cell:

$$Pt_{(s)}/H_{2(g,\,1\,atm)}/\,H^{+}_{(aq)} \;//\; Ag^{+}_{(aq,\,1.00\,M)}\,/\,Ag_{(s)}$$

Given
$$E = 0.900\text{ V at 25°C}$$
$$E^{o}_{Ag^{+}/Ag} = +0.80\text{ V}$$

Solution: The cell reaction is:

$$H_{2(g)} \longrightarrow 2H^{+}_{(aq)} + 2e^{-} \qquad \text{(Anodic reaction)}$$
$$Ag^{+} + e \longrightarrow Ag \;\;]\times 2 \qquad \text{(Cathodic reaction)}$$

$$H_2 + 2Ag^{+} \longrightarrow 2H^{+} + 2Ag$$

$\therefore \quad n = 2$

$$E^o_{cell} = 0.80\text{ V} - 0 = 0.80\text{ V}$$

$$E_{cell} = E^o_{cell} - \frac{0.0591}{n}\log\frac{[H^+]^2}{pH_2 + [Ag^+]^2}$$

$$E_{cell} = E^o_{cell} - \frac{0.0591}{n}\log\frac{[H^+]^2}{1\text{ atm.}\times[1\text{ M}]^2}$$

$$= E^o_{cell} - \frac{0.0591}{n}\log[H^+]^2$$

$$\log[H^+]^2 = (E^o_{cell} - E_{cell})\times\frac{n}{0.0591}$$

$$= (0.80\text{ V} - 0.90\text{ V})\times\frac{2}{0.0591} = -0.10\text{ V}\times\frac{2}{0.0591} = -3.381$$

$$2\log[H^+] = -3.381$$

$$\log[H^+] = \frac{-3.381}{2} = -1.695$$

$$pH = -\log[H^+] = 1.695 \approx 1.7$$

NUMERICALS BASED ON NERNST EQUATION

Problem 6.12: Calculate the cell potential at 298 K for the reaction:

$$Al^{3+} + Fe \longrightarrow Al + Fe^{3+}$$

The concentrations of Al^{3+} and Fe^{3+} are 1.2 and 2.5 M

Solution: For the reaction,

$$Al^{3+} + Fe \longrightarrow Al + Fe^{3+}$$

We can write,

$$E_{cell} = E^o_{cell} - \frac{0.059}{n}\log\frac{\left[Fe^{3+}\right]}{\left[Al^{3+}\right]} = -1.62 - \frac{0.059}{3}\log\frac{2.5}{1.2} = -1.63\text{V}$$

Problem 6.13: Calculate the equilibrium constant of the reaction:

$$Fe^{2+} + Ag^+ \longrightarrow Fe^{3+} + Ag_{(s)} \text{ at } 25°C, \text{ if}$$

if $\quad E^0_{Fe^{3+}/Fe^{2+}} = 0.77\text{V}$ and $E^0_{Ag^+/Ag^{2+}} = 0.80$ V.

Solution: $E^0_{cell} = E_{Ag^+/Ag} - E^0_{Fe^{3+}/Fe^{2+}}$

$$= 0.8\text{V} - (-0.77\text{V}) = 1.57\text{ V}$$

$$\therefore \quad \log K_{eq} = \frac{nE^o_{cell}}{0.0592V} = \frac{1\times1.57V}{0.0592V} = 26.52$$

Hence K_{eq} = Antilog 26.52 = 3.31×10^{26}

The pH Meter

- In practice, a glass electrode containing a silver wire coated with silver chloride immersed in a solution of HCl is used as the test electrode; the solution in the cell is separated from the test solution by a thin glass membrane (Fig. 6.6).
- The reference electrode is a calomel (Hg_2Cl_2) electrode in contact with liquid mercury and $KCl_{(aq)}$:

$$Hg_2Cl_2(s) + 2e^- \longrightarrow 2Hg(l) + 2Cl^-_{(aq)};\ E^o = +0.28\ V$$

- A typical pH meter measures the cell potential, and electronically converts the potential to the pH value:

$$pH = \frac{E_{cell} - 0.28\ V}{0.0592\ V}$$

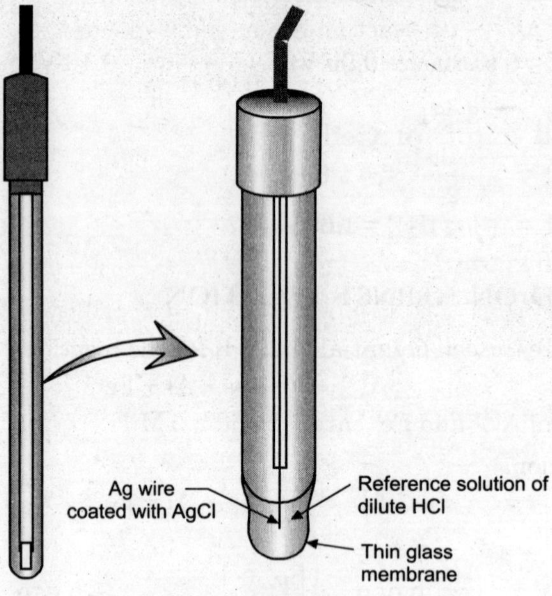

Fig. 6.6 pH meter.

6.8 THERMODYNAMIC FUNCTIONS AND ELECTROCHEMICAL CELLS

The change in thermodynamic properties such as Gibbs function, enthalpy and entropy can be calculated from cell potential measurements.

6.8.1 Gibbs Free Energy and e.m.f. of Cell

The electrical energy produced by a galvanic cell is given by the product of its electromotive force and the quantity of electricity which passes through it. The net electrical work performed by a reaction is the product of electromotive force E and the quantity of electricity Q.

$$\text{Net electrical work done} = QE \qquad ...(11)$$

If EMF is measured in volts and quantity of electricity in coulombs, the electrical energy obtained is in volt-coulombs or joules (volt coulombs = joule).

Also in accordance with Faraday's laws, the quantity of electricity produced is:

$$\text{Quantity of electricity, } Q = nF \qquad ...(12)$$

where n is the number of electrons and $F = 96500$ coulombs.

$$\text{Hence, net electrical work done} = nFE \quad ...(13)$$

It was pointed out by Gibbs and independently by Helmholtz, that the electrical energy of a reversible cell is given by the free energy decrease ($-\Delta G$) of the reaction occuring in the cell.

As electrical Energy produced by the cell = nFE

Hence $$-\Delta G = nFE \quad ...(14)$$

Thus the reversible EMF of any cell is determined by the free energy change of the reaction occurring in the cell.

In the cell reaction, if the substances involved are in their standard states i.e. at one atmosphere pressure and 298 K, then

$$\Delta G^o = -nFE^o \quad ....(15)$$

When $\Delta G = -ve$, or $\Delta E = +ve$, the reaction is spontaneous
$\Delta G = +ive$, or $\Delta E = -ve$, reaction is non-spontaneous
$\Delta G = 0$, or $\Delta E = 0$ the reaction is at equilibrium.

6.8.2 Enthalpy and e.m.f. of Cell

The enthalpy of a reaction is given by:

$$\Delta H = \Delta G + T\Delta S \quad ...(16)$$

According to the well known Gibbs-Helmholtz equation, decrease in free energy, $-\Delta G$, of the cell reaction at constant pressure would be given by the expression,

$$-\Delta G = -\Delta H - T\left(\frac{\partial(\Delta G)}{\partial T}\right)_P \quad ...(17)$$

where $-\Delta H$ is the decrease in enthalpy of the cell reaction at constant pressure. Substituting the value of ΔG from equation (14) we get,

$$nFE = -\Delta H - T\left(\frac{\partial(-nFE)}{\partial T}\right)_P = -\Delta H + TnF\left(\frac{\partial E}{\partial T}\right)_P \quad ...(18)$$

$$E = -\frac{\Delta H}{nF} + T\left(\frac{\partial E}{\partial T}\right)_P \quad ...(19)$$

where n and F are constants and $[\partial E/\partial T]_P$ is the temperature coefficient of the cell. The electrical energy depends upon the temperature coefficient of the e.m.f. of the cell.

- If temperature coefficient is O, the electrical energy is equal to the enthalpy of the cell reaction.
- If temperature coefficient is $+ve$, i.e. the e.m.f. of the cell increases with temperature, the electrical energy will be greater than the enthalpy of the reaction. The additional energy will be supplied by the surroundings. If that does not happen, the temperature of the cell fall during its working.
- If temperature coefficient is $-ve$, the electrical energy will be smaller than the enthalpy of the reaction. Either the heat is given out to the surroundings or the temperature of the cell will rise during its working.

In the case of Daniell cell, temperature coefficient is very small and hence electrical energy is very close to the enthalpy of the cell.

$$\Delta G = \Delta H - nFT\left(\frac{\partial E}{\partial T}\right)_P \quad ...(20)$$

Also $$\Delta G = \Delta H - T\Delta S$$

Thus $$\Delta S = nF\left(\frac{\partial E}{\partial T}\right)_P \quad ....(21)$$

Thus ΔS can be calculated using this equation.

NUMERICALS BASED ON CALCULATION OF ΔS, ΔH AND ΔG

Problem 6.14: For Weston cell, emf is 1.018 V at 293 K. Its temperature coefficient $\left(\frac{\partial E}{\partial T}\right)_P = -4.00 \times 10^{-5}\ VK^{-1}$. Calculate ΔG, ΔS and ΔH for the cell reaction of the cell.

Solution: $\Delta G = -nFE = -2 \times 96{,}500C \times 1.018\ V$

$$= -2 \times 96{,}500 \times 1.018\ J$$
$$= -1.964 \times 10^5 J$$

$$\Delta S = nF\left(\frac{\partial E}{\partial T}\right)_P$$
$$= 2 \times 96{,}500\ C \times (-4.00 \times 10^{-5}\ VK^{-1})$$
$$= -7.72\ JK^{-1}$$

$$\Delta H = -nFE + TnF\left(\frac{\partial E}{\partial T}\right)_P$$
$$= -2 \times 96{,}500\ C \times 1.018\ V + [293K \times 2 \times 96500\ C \times (-4.00 \times 10^{-5}\ VK^{-1})]$$
$$= -1.99 \times 10^5\ J$$

Problem 6.15: Consider the cell

$$Cd/CdCl_2 \cdot 2H_2O/AgCl_{(s)}/Ag$$

Solution: The emf of the cell at 15°C is 0.67531V and temperature coefficient is 0.00065 V deg^{-1}. Calculate the value of ΔG, ΔS and ΔH, if the process is carried reversibly.

$$Cd/CdCl_2 \cdot 2H_2O/AgCl_{(s)}/Ag$$

or $$Cd(s) + 2Cl^- \rightarrow CdCl_{2(s)} + e^-$$

$$2AgCl_{(s)} + 2e^- \rightarrow 2Ag_{(s)} + 2Cl^-$$

$$Cd_{(s)} + 2AgCl_{(s)} \rightarrow CdCl_{2(s)} + 2Ag_{(s)}$$

$$\Delta G = -nFE = -2 \times 96500\ C \times 0.67531V = -130.335\ kJ$$

$$\Delta S = nF\left(\frac{\partial E}{\partial T}\right)_P$$

$$\Delta S = 2 \times 96500 \times 0.00065 = 125.45\ JK^{-1}$$

$$\Delta H = \Delta G + T\Delta S$$
$$= (-130.335\ kJ) + (288\ K) + (125.45\ JK^{-1}) \times 10^{-3} kJ/J$$
$$= -166.465\ kJ$$

Problem 6.16: The voltage of the cell $Pb/PbSO_4/Na_2SO_4 \cdot 10H_2O/Hg_2SO_4/Hg$ is 0.9647 V at 25°C. The temperature coefficient is $1.74 \times 10^{-4}\ VK^{-1}$. Calculate the values of ΔG, ΔS and ΔH.

Solution: The cell reaction is;

$$Pb_{(s)} + Hg_2\,SO_4 \longrightarrow PbSO_4 + 2Hg_{(l)}$$

$$
\begin{aligned}
E^{\circ}_{cell} &= 0.9647 \\
\Delta G &= -nFE^{\circ} = -2 \times 0.9647 \times 96500 \\
&= -1.86 \times 10^{5} = -186 \text{ kJ} \\
\Delta S &= nF\left(\frac{\partial E}{\partial T}\right)_P \\
&= 2 \times 96500 \times 1.74 \times 10^{-4} \text{ VK}^{-1} = 33.58 \text{J/deg} \\
\Delta H &= \Delta G + T\Delta S \\
&= (-186 \text{ kJ}) + 298 \text{ K} + 33.58 \text{ JK}^{-1} \times 10^{-3} \text{ kJ/J} \\
&= -176 \text{ kJ}
\end{aligned}
$$

Problem 6.17: Calculate the standard free energy change of the cell reaction:

$$Fe^{2+} + Ag^{+} \longrightarrow Fe^{3+} + Ag_{(s)}$$

Given: $E^{0}_{Fe^{3+}/Fe^{2+}} = 0.77\text{V}$; $E^{0}_{Ag^{+}/Ag} = 0.80$ V.

Solution:

$$
\begin{aligned}
E^{\circ}_{cell} &= \text{E}_{cathode} - E^{\circ}_{anode} \\
&= E^{0}_{Ag^{+}/Ag} - E^{0}_{Fe^{3+}/Fe^{2+}} \\
&= 0.80\text{V} - 0.77\text{V} = 0.03\text{V} \\
\Delta G &= -nFE^{\circ} = -1\text{mol} \times 96500 \text{ C mol}^{-1} \times 0.03 \text{ V} \\
&= -2895 \text{ J.}
\end{aligned}
$$

6.9 TYPES OF ELECTRODES

A large number of electrodes are known and these can be conveniently divided into several categories depending upon the nature of the half cell reactions. Some common types reversible electrodes are discussed below:

6.9.1 Metal-Metal ion Electrode

Metal-metal ion electrode consists of a metal rod dipping in a solution of its own ions. For example, zinc rod dipping in zinc sulphate solution. If the metal is univalent, the electrode reaction may be represented as:

$$M^{+}_{(aq)} + e^{-} \rightleftharpoons M_{(s)}$$

The electrode is represented as: $M^{+}_{(aq)}/M_{(s)}$

If the metal rod behaves as positive electrode i.e. the reaction at the electrode involves reduction, the concentration of M^{+} ions would decrease. If it behaves as negative electrode i.e. the electrode reaction involves oxidation, the concentration of M^{+} ions in solution will, therefore, increase. Thus the electrode is reversible with respect to M^{+} ions.

6.9.2 Metal-Insoluble Salt Electrode

This electrode consists of a metal in contact with one of its sparingly soluble salt immersed in solution of soluble salt having a anion common with that of sparingly soluble salt. The typical examples of such electrodes are *Calomel electrode* and *silver-silver chloride electrode.*

(*i*) **Calomel Electrode:** Calomel electrode is commonly used as a secondary reference electrode for potential measurements in place of hydrogen electrode, which is inconvenient and difficult to prepare.

It consists of mercury in contact with sparingly soluble salt mercurous chloride Hg_2Cl_2 (Calomel) and a solution of soluble chloride ions such as KCl of known concentration.

It is represented as:

$$Cl^-_{(aq)}/Hg_2Cl_{2(s)}/Hg_{(l)}$$

Mercury is placed at the bottom of the tube and is covered with a paste of solid mercurous chloride, Hg_2Cl_2. A solution of KCl is placed over the paste. Electrical contact is made by a platinum wire, dipping into the mercury layer. The contact with the salt bridge is made through a side tube (Fig. 6.7).

When the electrode acts as a cathode, mercurous chloride is reduced to mercury. The half cell reaction is:

$$Hg_2Cl_{2(s)} + 2e^- \rightleftharpoons 2Hg + 2Cl^-$$

The half-cell potential for saturated calomel electrode is 0.2682 V.

The calomel electrode is reversible with respect to Cl^- ions. When a saturated solution of KCl is used, the electrode is called saturated Calomel electrode (SCE). Normal Calomel Electrode (NCE) has 1.0 N KCl solution whereas Deci Normal Calomel Electrode (DNCE) consists of 0.1 N KCl solution.

The reduction potentials of the three types are:

SCE	0.2422 V
NCE	0.2800 V
DNCE	0.3338 V

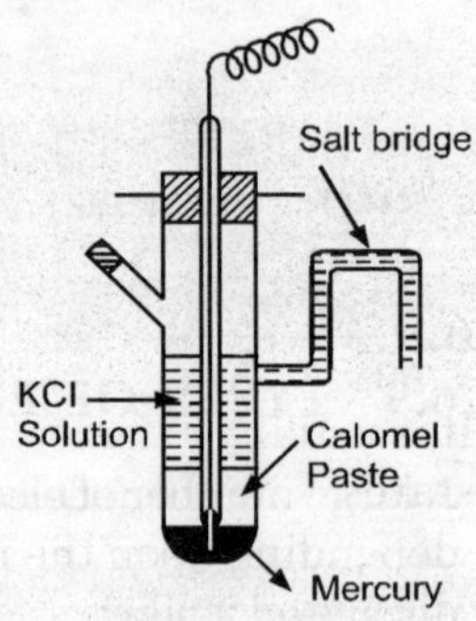

Fig. 6.7

Calomel electrode is preferred to hydrogen electrode. Calomel electrode is convenient to use and its results are reproducible and do not vary with temperature.

(*ii*) **Silver-Silver Chloride Electrode:** It consists of silver metal in contact with solid silver chloride and a solution of soluble chloride ions such as KCl or HCl. It is prepared by depositing a thin layer of AgCl on a silver or platinum wire which is immersed in a solution containing chloride ions. The electrode is represented as:

$$Cl^-_{(aq)}/AgCl_{(s)}/Ag_{(s)}$$

The electrode reaction is: $AgCl_{(s)} + e^- \rightleftharpoons Ag_{(s)} + Cl^-_{(aq)}$

The electrode is reversible with respect to chloride ions (Fig. 6.8).

6.9.3 Gas Electrodes

These electrodes consist of gases which are continuously bubbled through the solution of their ions. The gases are non-conductors of electricity, hence an inert metal (Pt, Au etc.) is employed for establishing electrical contacts.

The inert metal is dipped in a solution containing ions of the gas and the gas is bubbled through it. The metal absorbs the gas and an equilibrium is at once established between the gas and ions in the solution. Some of the examples of gas electrodes are: hydrogen electrode (described in section 6.4), oxygen electrode and chlorine electrode.

(i) Oxygen Electrode: In this type of gas electrode, oxygen gas is bubbled through a solution containing hydroxyl (OH) ions with platinum strip dipped into it. The half cell reaction is:

$$\frac{1}{2}O_{2(g)} + H_2O + 2e^- \rightleftharpoons 2\bar{O}H_{(aq)}$$

It is represented as:

$$\bar{O}H_{(aq)}/O_{2(g)}/Pt$$

It is reversible with respect to $\overline{OH}$ ions. The disadvantage of oxygen electrode is that an oxide layer is formed on the metal and the equilibrium is not established quickly.

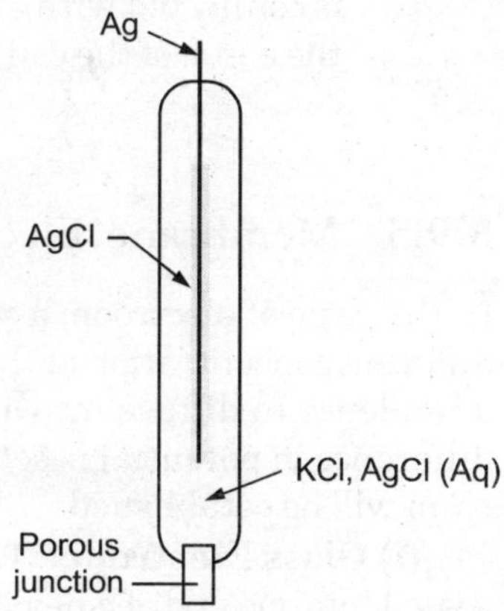

Fig. 6.8 Silver-silver chloride electrode

(ii) Chlorine Electrode: As the name implies, here chlorine gas is bubbled through the solution containing Cl^- ions in HCl solution with an inert metal dipped into it. The electrode is represented as:

$$Cl^-_{(aq)} / Cl_{2(g)} / Pt$$

$$Cl^-_{2(g)} + 2e^- \rightleftharpoons 2Cl^-_{(aq)}$$

The electrode is reversible to Cl^- ions.

(iii) Hydrogen Electrode: Refer to section 6.4.

6.9.4 Redox Electrodes

It is usually referred to as a system in which a species exists in solution in two different oxidation states. The electrodes are non reactive with solution and are just carriers of electrons.

(i) A typical example is Pt/Fe^{3+}, Fe^{2+} electrode which is represented as:

$$Pt/Fe^{3+}_{(a_1)}, Fe^{2+}_{(a_2)}$$

The half cell reaction is represented by:

$$Fe^{3+} + e^- \longrightarrow Fe^{2+}$$

(ii) **Quinhydrone Electrode:** Quinhydrone electrode behaves as a form of hydrogen electrode and is used for measuring pH of solutions as an indicator electrode (Fig. 6.9).

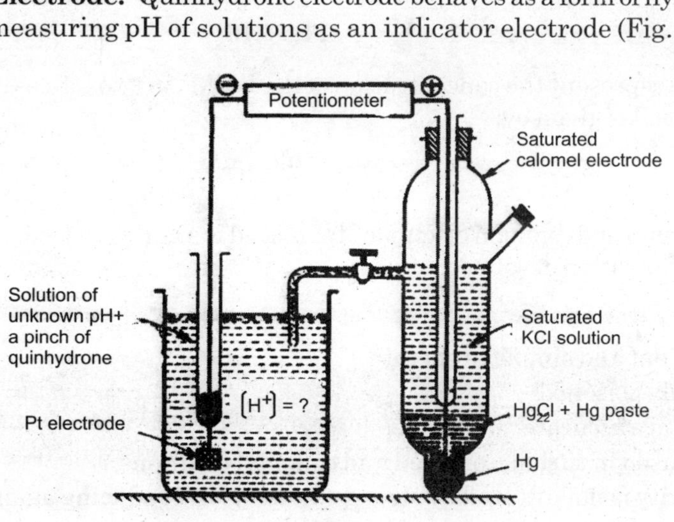

Fig. 6.9 Quinhydrone Electrode Connected to Calomel Electrode.

Quinhydrone is dissolved in the solution whose pH is to be determined until the solution becomes saturated and a platinum wire is dipped into the solution. The half cell so formed is connected with any other reference electrode, usually saturated calomel electrode and the e.m.f. of the cell so formed is determined potentiometrically. The cell is represented as,

$$Pt \mid H_2Q, Q\,H^+_{(unknown)}//KCl_{(satd.)}, Hg_2Cl_{2(s)}/Hg$$

6.9.5 Membrane Electrode

In this type of electrode, a semipermeable membrane, say X separates two solutions containing different concentration of the salt MA and is permeable to one of the ions say M^+. The ion M^+ have a tendency to diffuse into a more dilute solution whereas the anion A^- cannot diffuse. Hence a difference in potential is set up. This will retard the migration of M^+ and after sometime equilibrium will be established.

***(i)* Glass Electrode:** This indicator electrode consists of a thin, pH sensitive glass membrane sealed into one end of a heavy walled glass tube. The glass membrane functions as an ion-exchange resin. A small volume of dilute HCl saturated with AgCl is contained in the tube. A silver wire in this solution forms a silver/silver chloride reference electrode, which in turn is connected to one of the terminals of potentiometer. To the other terminal of potentiometer, calomel electrode is connected. Thus glass electrode system contains two reference electrodes: viz, the external calomel electrode and the internal silver/silver chloride electrode. The glass electrode can be represented as:

$Ag/AgCl_{(s)}$, HCl (0.1 M)/Glass, 0.1 MHCl/Glass

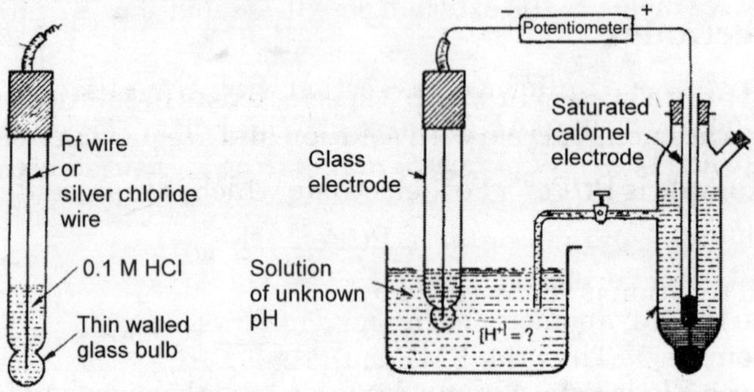

Fig. 6.10

If C_1 and C_2 represent the concentration of the ion M^+ in two halves of the cell respectively, the standard potential is given by:

$$E^o = \frac{RT}{F} \ln \frac{C_1}{C_2}$$

The most useful and common electrode of this kind is a glass electrode. This is used for measuring the H^+ ion concentration of solutions.

Merits of Glass Electrode

- Convenient and simple to operate
- Not easily poisoned
- Results are accurate
- Can be used in turbid, coloured and colloidal solutions.
- Its activity is not affected by strong oxidising and reducing agents.

6.10 MEASUREMENT OF E.M.F. OF A CELL (POTENTIOMETRY)

Potentiometry deals with the measurement of difference in potential between two electrodes which have been combined to form electrochemical cells. The cell potential measured nearly under zero or negligible current flow is called electromotive force (e.m.f.) of the cell. The difference between e.m.f. and potential difference may be sumarized as follows:

EMF	*Potential Difference*
1. EMF is the potential difference of the cell when no current is flowing in the circuit i.e. an open circuit.	Potential difference is the difference between the electrode potentials of two electrodes under any condition.
2. EMF can be measured by potentiometric method only.*	Potential difference can be measured by simple voltmeter.
3. EMF is the maximum voltage obtainable from the cell.	Potential difference is less than the maximum voltage obtainable from the cell.

*The EMF sent by the potentiometric source is equal and opposite to that of the cell and there is no net current flowing in the circuit.

The e.m.f. of a cell is measured with the help of Potentiometer. The experimental set up is as follows:

Potentiometer: In potentiometer, the principle generally used is that of Poggendorff compensation method. This consists in balancing the unknown e.m.f. against a known one in potentiometric circuit. In other words, we measure the external potential difference required to stop the passage of current in the cell.

Potentiometer consists of a uniform cross section wire *AB*. A battery *S* of a constant e.m.f. is connected to the two ends of the wire *AB*. The e.m.f. of this battery or cell should be larger than the cell e.m.f. to be measured. The cell *C* whose e.m.f. is to be measured is connected to A, with the poles in the same direction, as that of the cell *C*. The other terminal of the cell *C* is connected through a galvanometer to a sliding contact *J*, which can be moved along the wire *AB*.

Firstly the sliding contact is moved till at a certain point there is no deflection in the galvanometer i.e. no current flows through it. Say the point is *Y*. Thus the EMF of the cell E_C would be directly proportional to the distance *AY*.

$$E_c \alpha AY$$

Then the cell *C* is replaced by the standard cell *D* and a point (say *X*) is found out when no current flows through the galvanometer. Then, EMF of this standard cell *D*, E_D will be given by

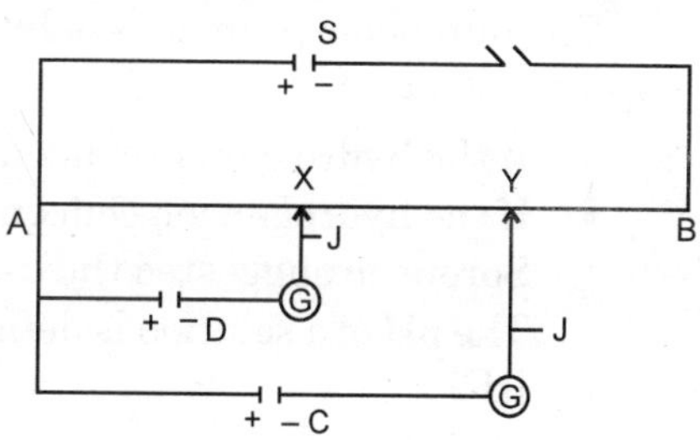

Fig. 6.11 Measurement of emf of cells (potentiometer).

$$E_D \alpha AX$$

Now
$$\frac{E_C}{E_D} = \frac{AY}{AX}$$

Therefore,
$$E_C = \frac{AY}{AX} \times E_D$$

Knowing the value of E_D, E_C can be calculated.

The commonly used standard cell for the above purpose is Western Cadmium cell (Cadmium amalgam cell).

Western cadmium cell consists of an *H* shaped glass vessel with a platinum wire sealed through the bottom of each limb. One limb contains mercury which serves as positive electrode (anode). It is covered with a paste of sparingly soluble mercurous sulphate and cadmium sulphate crystals. The other limb i.e. the negative pole contains a cadmium amalgam containing 12.5 percent cadmium. It is covered with layer of cadmium sulphate crystals ($3CdSO_4 . 8H_2O$). The electrolyte is a saturated solution of cadmium sulphate and Hg_2SO_4.

The cell is represented as

$Cd - Hg/3\,CdSO_4 . 8H_2O\ Hg_2SO_{4(s)}/Hg$

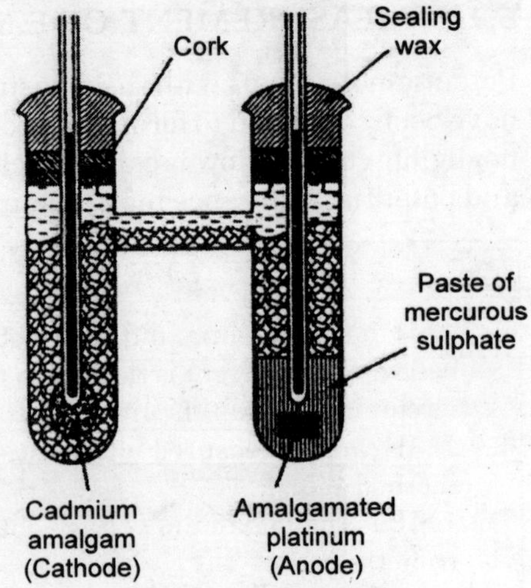

Fig. 6.12 Standard cell.

Direct Reading Potentiometers: Now a days direct potentiometers are available in most of laboratories. In these the meter wire of a simple potentiometer is replaced by a series of coils of standard resistance (usually 2 ohms). Separate terminals are provided for connecting the working battery, standard cell, galvanometer and the EMF to be tested. By means of a selector switch, the potentiometer can be connected separately to the standardization and measurement of unknown e.m.f.

APPLICATION OF EMF MEASUREMENTS

pH Metric Measurements: An important application of potentiometry is the determination of pH of a solution. The technique is known pHmetry.

The term pH was introduced by Sorensen (1909) to express H^+ ion concentration i.e. the degree of acidity or alkalinity of a solution. As is well known that the product of H^+ and $\bar{O}H$ ions in water (ionic product) is constant and is equal to 10^{-14}. In neutral water, therefore, we have

$$[H^+] = [\bar{O}H] = 10^{-7}$$

The number of gram ions of hydrogen present in one litre of the solution is known as hydrogen ion concentration.

- If the hydrogen ion concentration is more than 10^{-7}, solution is acidic.
- If the hydrogen ion concentration is less than 10^{-7}, solution is basic.

  Sorensen suggested the use of pH for expressing the acidity or alkalinity.

  The pH of a solution is defined as the negative logarithm of the hydrogen ion concentration $[H^+]$

$$pH = -\log [H^+]$$

For pure water, $[H^+] = 10^{-7}$

$\therefore$ pH of pure water = 10^7

It is evident that

- For neutral solution pH = 7
- For acidic solution pH < 7
- For basic solution pH > 7

	pH	$[H^+]$ (mol l^{-1})	$[OH^-]$ (mol l^{-1})	
Acidic	0	10^0	10^{-14}	Acidic
	1	10^{-1}	10^{-13}	
	2	10^{-2}	10^{-12}	
	3	10^{-3}	10^{-11}	
	4	10^{-4}	10^{-10}	
	5	10^{-5}	10^{-9}	
	6	10^{-6}	10^{-8}	
Neutral →	7	10^{-7}	10^{-7}	← Neutral

	pH	$[H^+]$ (mol l^{-1})	$[OH^-]$ (mol l^{-1})
Basic	8	10^{-8}	10^{-6}
	9	10^{-9}	10^{-5}
	10	10^{-10}	10^{-4}
	11	10^{-11}	10^{-3}
	12	10^{-12}	10^{-2}
	13	10^{-13}	10^{-1}
	14	10^{-14}	10^{-0}

Thus, lower the pH, more acidic is the solution and higher the pH more basic is the solution.

Determination of pH: For determining the pH of a solution, a number of methods are employed viz., potentiometry, cryoscopy etc. In the potentiometric determination of pH, first a cell is assembled in which the indicator electrode used is reversible to hydrogen ions and dips in the solution whose pH is to be determined. The reference electrode is usually a calomel electrode of known potential. The junction between the two is made through salt bridge or by immersing the reference electrode directly into the solution. By measuring the voltage of the cell so formed, pH of the solution can be determined using Nernst equation. A typical cell may be represented as follows:

$$\text{Electrode reversible to } H^+ \text{ ion} \Big/ \begin{matrix}\text{standard or} \\ \text{unknown solution}\end{matrix} \Big/ \begin{matrix}\text{KCl saturated} \\ Hg_2Cl_{2(s)} \text{ solution}\end{matrix} \Big/ Hg$$

The emf of this cell, containing standard solution will be given by,

$$E = E_{\text{ref}} - \frac{RT}{F} \text{In} [H^+]$$

where R = gas constant, T = Absolute temperature and F = Faraday.

A number of indicator electrodes reversible to hydrogen ion are employed for pH measurements. These include hydrogen electrode, quinhydrone electrode and glass electrode. Standard Calomel Electrode (SCE) is used as reference electrode.

Determination of pH Using Glass Electrode: When placed in a solution, the potential of the glass electrode depends on the H^+ ion concentration of the solution. The potential develops across the glass membrane as a result of a concentration difference of H^+ ions on the two sides of the membrane.

The magnitude of this potential depends on the difference in the pH values of two solutions.

To determine the pH of the unknown solution, a glass electrode is immersed in the solution of unknown pH. It is coupled with a standard calomel electrode (SCE) as shown in Fig. 6.11. The EMF of the cell is given by:

Ag, AgCl / 1M HCl / Glass / Solution of unknown pH | | SCE

and is determined experimentally.

Now the potential of glass electrode, E_G at 25°C is given by:

$$E_G = E_G^o + \frac{2.303\,RT}{F}\,\text{pH}$$

$$= E_G^o + 0.0591\,\text{pH}$$

Now $E_{\text{Cell}} = E_R - E_L$

$$E_{\text{Cell}} = E_{SCE} - E_G$$

$$E_{\text{Cell}} = E_{SCE} - E_G^o - 0.0591\,\text{pH (from eq.)}$$

$$\therefore \quad \text{pH} = \frac{E_{SCE} - E_G^o - E_{\text{Cell}}}{0.0591}$$

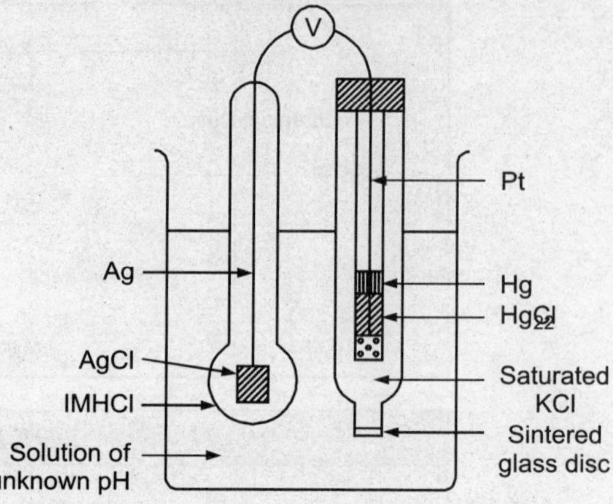

Fig. 6.13 A glass electrode coupled with Standard calomel electrode.

The value of E_{SCE} is known, E_{Cell} can be found experimentally. If E_G^o is known, pH can be found out using the above relation.

6.11 VARIETIES OF CELL

In general, electrochemical cells can be classified into two classes:

(*i*) Chemical Cells
(*ii*) Concentration Cells

(*i*) Chemical Cells: In chemical cells, the e.m.f. arises due to the chemical reaction occurring in the cell. A typical example is Daniell Cell.

If the electrodes of the half cells are immersed in the same electrolyte, so that there is no liquid junction (described later), the cell is known as *cell without transport*. If the electrolytes of the half cells are different or electrolyte is same but its concentration is different in two half cells, the ions of the electrolytes migrate across the liquid junction. Such a cell is called *a cell with transport*.

(*ii*) Concentration Cells: If the chemical reaction in the two half cells is the same, but the concentration of the electrolyte solutions or electrodes materials are different, then the net reaction is the transfer of species from higher concentration to lower concentration. Such cells are known as concentration cells. Here the e.m.f. arises not due to any chemical reaction but due to transfer of matter from one half-cell to the other.

Concentration cells are further classified into two types:

(*i*) Electrode concentration cells
(*ii*) Electrolyte concentration cells

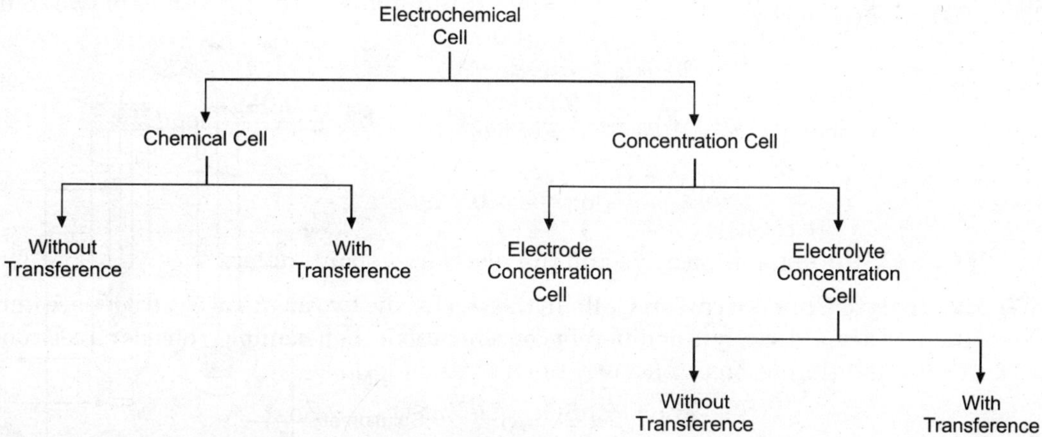

***(i)* Electrode Concentration Cells:** In these cells, two identical electrodes of different concentrations are dipped in the common electrolyte containing ions of the electrode material. Two hydrogen electrodes at unequal gas pressures immersed in the same solution of hydrogen ions constitute an electrode-concentration cell. The electrode with a higher pressure of the gas forms the anode and the cell may be denoted as:

$$Pt/H_2(p_1)/HCl_{(a)}/H_2(p_2)/Pt:$$

where $p_1 > p_2$

The cell reactions are;

$$\begin{array}{rcll} H_2(p_1) & \rightleftharpoons & 2H^+ + 2e^- & \text{(Oxidation)} \\ 2H^+ + 2e^- & \rightleftharpoons & H_2(p_2) & \text{(Reduction)} \\ \hline H_2(p_1) & \rightleftharpoons & H_2(p_2) & \end{array}$$

This reaction is evidently independent of the concentration of the electrolyte. The cell reaction involves the transfer of hydrogen gas from higher pressure (p_1) to lower pressure (p_2).

The e.m.f. of the cell is:

$$E = E^o - \frac{0.0591}{2} \log p_2 / p_1$$

Since E^o in concentration cell = 0, so

$$E = -0.02955 \log p_2/p_1 = 0.02955 \log p_1/p_2$$

When $p_1 > p_2$, EMF is positive and the reaction is spontaneous.

Another example of electrode concentration cell is that of an amalgam electrodes at different concentrations in contact with a solution of an electrolyte containing ions of the metal dissolved in amalgam. Example,

$$Hg - Pb_{(C1)} / PbSO_{4\,(a)} / Hg - Pb_{(C2)}$$

where $C_1 > C_2$

The electrode reactions are:

$$\begin{array}{rcll} Pb_{(C1)} & \rightleftharpoons & Pb^{2+} + 2e^- & \text{(Oxidation)} \\ Pb^{2+} + 2e^- & \rightleftharpoons & Pb_{(C2)} & \text{(Reduction)} \\ \hline Pb_{(C1)} & \rightleftharpoons & Pb_{(C2)} & \end{array}$$

The EMF of the cell is given by:

$$E = E_{cathode} - E_{anode}$$

$$= \left(E^o_{Pb} - \frac{0.0591}{2} \log C_2\right) - \left(E^o_{Pb} - \frac{0.0591}{2} \log C_1\right)$$

$$= \frac{0.0591}{2} \log \frac{C_1}{C_2} = 0.02955 \log \frac{C_1}{C_2}$$

$\therefore$ If $C_1 > C_2$, the e.m.f. is positive and the reaction is spontaneous.

***(ii)* Electrolyte Concentration Cell:** In these cells, the two identical electrodes are immersed in two solutions of same electrolyte at different concentrations. For example, consider a cell consisting of two half-cells at different Zn^{2+} concentrations C_1 and C_2.

$$Zn(s) / ZnSO_{4(C1)} // ZnSO_{4(C2)} / Zn_{(s)}$$

The overall reaction involves the passage of Zn^{2+} from a solution of higher concentration (C_2) to that of lower concentration (C_1).

Cell Reaction:

$$Zn \longrightarrow Zn^{2+}_{(C_2)} + 2e^- \quad \text{(Oxidation)}$$

$$Zn^{2+}_{(C_2)} + 2e^- \longrightarrow Zn^{2+}_{(C_1)} \quad \text{(Reduction)}$$

$$Zn^{2+}_{(C_2)} \longrightarrow Zn^{2+}_{(C_1)}$$

E^o_{Cell} for a concentration cell is zero as shown:

$$E^o_{Cell} = E^o_{cathode} - E^o_{anode} = E^o_{Zn^{2+}/Zn} - E^o_{Zn^{2+}/Zn} = 0$$

The EMF of the cell can be calculated as:

$$E_{Cell} = E_{cathode} - E_{anode}$$

$$= \left\{E^o + \frac{2.303\ RT}{2F} \log C_2\right\} - \left\{E^o + \frac{2.303\ RT}{2F} \log C_1\right\}$$

$$E_{Cell} = \frac{-2.303\ RT}{2F} \log \frac{C_1}{C_2}$$

At 298 K, the expression becomes;

$$E_{Cell} = -\frac{0.0591}{2} \log \frac{C_1}{C_2} = \frac{0.0591}{2} \log \frac{C_2}{C_1}$$

Thus EMF so developed is due to the mere transference of metal ions from the solution of higher concentration C_2 to the solution of lower concentration C_1.

Another similar concentration cell is represented by:

$$Pt, H_{2(g)}, H^+_{(a_1)} // H^+_{(a_2)}, H_{2\,(g)}, Pt$$

In this case, both the electrodes are hydrogen electrodes which are in contact with hydrogen ions of different activities. The two solutions are separated by a salt bridge. Cell reaction

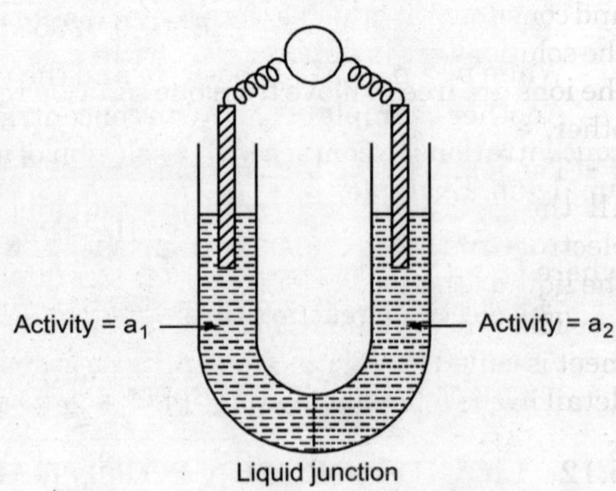

Fig. 6.14 Concentration cell with a liquid junction

$$\frac{1}{2}H_{2(g)} \rightleftharpoons H^+_{(a_1)} + e^- \quad \text{(Oxidation)}$$

$$H^+_{(a_2)} + e^- \rightleftharpoons \frac{1}{2}H_{2(g)} \quad \text{(Reduction)}$$

$$H^+_{(a_2)} \rightleftharpoons H^+_{(a_1)}$$

EMF of cell $E = E_{cathode} - E_{anode}$

At 298 K,

$$E_{cell} = -\frac{0.0591}{2}\log\frac{a_1}{a_2} = \frac{0.0591}{2}\log\frac{a_2}{a_1}$$

Electrolyte concentration cells in which solutions of same electrolyte of different concentrations are used can further be of two types:

(*a*) **Concentration Cells Without Transference:** In these types of cells, the two electrolytic solutions are not in direct contact with each other and transference of ions from one solution to the other does not take place directly. The two solutions are separated from each other by means of a salt bridge or by some other means. The above described electrolyte concentration cells belong to this type.

(*b*) **Concentration Cells with Transference:** In this type, the two solutions are in direct contact with each other. For example,

$$\underset{1\,atm}{Pt, H_{2(g)}}, HCl_{(a_1)} \;:\; HCl_{(a_2)}, \underset{1\,atm}{H_{2(g)}}, Pt$$

$$H^+ \longrightarrow$$

$$\longleftarrow Cl^-$$

Note: The double line has been replaced by a dotted line indicating that the two solutions are in direct contact.

Cell reaction:

$$\frac{1}{2}H_{2(g)} \rightleftharpoons H^+_{(a_1)} + e^- \quad \text{(Oxidation)}$$

$$H^+_{(a_2)} + e^- \rightleftharpoons \frac{1}{2}H_{2(g)} \quad \text{(Reduction)}$$

The H^+ ions are generated at one electrode and consumed at other as the current flows. Since the solutions are in direct contact with each other, the ions are free to move from one solution to the other.

The e.m.f. of the cell corresponds to sum of all the reactions occurring in the cell, i.e., electrode reactions and migration of ions across the liquid junction.

The junction at which the two electrolytes meet is called liquid junction and is described in detail in the following section.

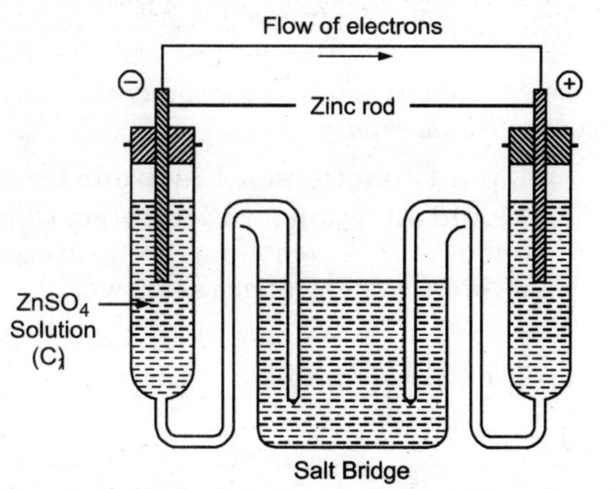

Fig. 6.15 Typical concentration cell where $C_2 > C$

6.12 LIQUID JUNCTION POTENTIAL (L.J.P.)

Whenever two solutions of different concentration are in contact, there develops a potential difference at their interface. This is called liquid junction potential (LJP).

Liquid Junctional Potential (LJP) arise between the two different solutions when the two solutions:

- are in contact
- have ions present at different concentrations, and
- contain ions of different mobilities.

The simplest example is the case of two solutions containing the same salt in different concentrations. The salt will diffuse from higher concentration to lower concentration. However, the diffusion rate of the cations and anions of the salt are not the same. Let us consider the case of two solutions having different concentrations of the electrolyte, say NaCl. Now sodium and chloride have different mobilities, chloride having higher mobility than sodium ions. When the two solutions are in contact, the electrolyte will diffuse from higher concentration to lower concentration. The chloride ions will move faster than sodium ions. As a consequence, an excess negative charge will accumulate on the side having low concentration of electrolyte. Similarly because of slow moving sodium ions, a positive charge will accumulate behind it in the region depleted of chloride ion (Fig. 6.14). Thus a potential difference is set up at the junction of the two solutions. The junction potential slows the movement of chloride ions but accelerates the movement of sodium ions, till they move at the same rate. Thus the potential difference between the solutions remains steady. In steady state conditions, the two ions will move at the same speed and a potential difference is created between the two solutions.

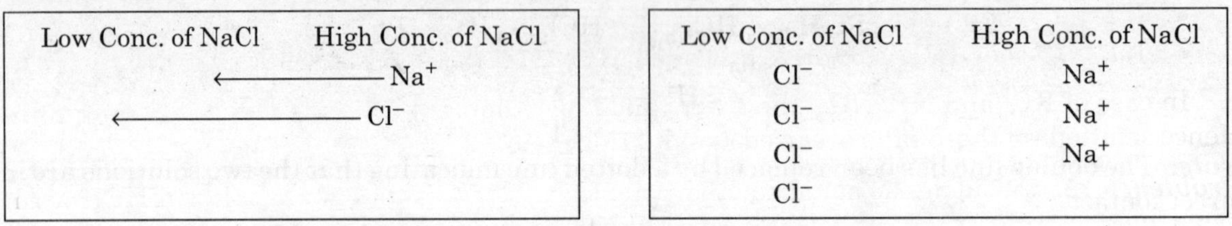

Fig. 6.16

Whenever two solutions of different concentration are in contact, there develops a potential difference between the two surfaces, besides the potential difference in concentrations of the solutions and this should be taken into account while deriving the equations for the EMF of different types of concentration cells:

(*i*) **Liquid Junction for Electrode Concentration Cell:** These cells do not have LJP.

(*ii*) **Liquid Junction for Electrolyte Concentration Cell:** Let us consider a concentration cell formed by combining two hydrogen gas electrodes in contact with HCl solutions of different concentrations as shown:

$$PtH_{2(g)}, HCl_{(a1)} : HCl_{(a2)}, H_{2(g)}; Pt$$

The EMF of the cell with transference ($E_{\omega,t}$) which includes the LJP is given by

$$E_{\omega,t} = t_a\left(\frac{RT}{F}\right)\text{In}\,\frac{a_2}{a_1}$$

where a_1 and a_2 stand for activities of the two half cells and t_a stands for transport number of the Cl^- ions. Transport number of an ion is the fraction of charge carried in electrolyte by that ion. In a simple electrolyte, containing one cation and one anion, the transport numbers are represented as t_+ and t_- with $t_+ + t_- = 1$.

Thus for NaCl, the transport number of Na^+ will be $t_+ = 1 + t_-$, and for Cl^- it will be,

$$t_- = 1 + t_+$$

When the two solutions are not in direct contact with each other, LJP is not involved and e.m.f. of such cell ($E_{\omega \cdot o \cdot t}$) is given by:

$$E_{\omega \cdot o \cdot t} = \frac{RT}{F} \ln \frac{(a\pm)_2}{(a\pm)_1}$$

where $(a\pm)_2$ and $(a\pm)_1$ are mean ionic mobilities of the HCl solutions in two cells. Hence, we can calculate LJP as

$$E_{LJP} = E_{\omega \cdot t} - E_{\omega \cdot o \cdot t}$$

$$= (2t_- - 1)\frac{RT}{F} \ln \frac{(a\pm)_2}{(a\pm)_1}$$

$$= [t_- + (1 + t_+) - 1]\frac{RT}{F} \ln \frac{(a\pm)_2}{(a\pm)_1}$$

$$E_{LJP} = (t_- + t_+)\frac{RT}{F} \ln \frac{(a\pm)_2}{(a\pm)_1}$$

It is evident from the above equation that the sign as well as the magnitude of LJP depends upon the transference number of the anion and cation.

- If $t_- = t_+$, then LJP = 0
- If $t_+ > t_-$, then LJP is negative
- If $t_- > t_+$, then LJP is positive and adds to e.m.f. of the cell.

In case of KCl and NH_4NO_3, the $t_- = t_+$ and hence liquid junction is reduced to minimum. Hence solutions of these salts are used as salt bridge.

Problem 6.18: Calculate the liquid junction potential at 25°C between two solutions of HCl having mean ionic activities of 0.01 and 0.001 respectively. The transference number of H^+ ion (t_+) in HCl may be taken as 0.83.

Solution:

$$t_+ - t_- = 1$$

$$\therefore \quad t_- = 1 - t_+ = 1 - 0.83 = 0.17$$

$$(a\pm)_1 = 0.01; \quad (a\pm)_2 = 0.001$$

$$E_{LJP} = (t_- - t_+)\frac{RT}{F} \ln \frac{(a\pm)_2}{(a\pm)_1}$$

$$= (0.17 - 0.83)\ 0.0591 \log \frac{0.001}{0.01} = 0.0039 \text{ Volts.}$$

6.13 COMMERCIAL CELLS/BATTERIES

As in evident from the previous section, there are many kinds of electrodes. All these electrodes may combine in different ways to give a large number of galvanic cells. Combination of two or more galvanic cells electrically connected to work together, to produce electric energy is called a *battery*. Commercially, batteries are used as a source of electrical energy for various purposes. Various combinations of electrolytes and electrodes are employed to make batteries for specific purposes. Thus commercial batteries can be made from different types of electrochemical cells.

The term battery should not be used while referring to a single cell. The batteries act as a portable source of electrical energy.

There are many different types of batteries ranging from the relatively large 'flash light' batteries to the miniature versions used for wrist watches or calculators. Batteries consist of either of two types of cells:

(*i*) Primary cells
(*ii*) Secondary cells.

6.13.1 Primary Cells

In primary cells, there is fixed amount of energy stored and once that energy has been used up, the cell is considered dead. In other words, the chemical reaction occurs only once and the cell then becomes dead and cannot be used again. Some of the important primary cells are described below:

(a) Dry Cell (Leclanche Cell): This is the most familiar type of dry cell. It consists of a zinc anode which is draped as a container. The cathode is comprised of a graphite rod surrounded by MnO_2 a moist electrolyte paste of NH_4Cl and $ZnCl_2$. Starch is added to the electrolyte (NH_4Cl, $ZnCl_2$ and MnO_2) to make it a thick paste like. To reduce leakage the cell is enclosed in polypropylene cylinder.

The cell reactions are:

$$Zn_{(s)} \longrightarrow Zn^{2+}_{(aq)} + 2e^- \qquad \text{(At Aanode)}$$

$$2MnO_2 + 2NH_4^+ + 2e^- \longrightarrow Mn_2O_3 + 2NH_3 + H_2O \qquad \text{(At Cathode)}$$

Thus, the reduction reaction occurs within the moist paste comprised of NH_4Cl and MnO_2. The NH_3 which is liberated reacts with Zn^{2+} from $ZnCl_2$ to form the complex $[Zn(NH_3)_2]\ Cl_{2(s)}$. The cell can be represented as:

$$Zn/Zn^{2+},\ NH_4^+/MnO_2/C$$

The dry cell 'couple' produces about 1.5 volts.

Dry cells do not have indefinite life. The acidic NH_4Cl corrodes the zinc container even when not in use.

In *alkaline version,* a modified form of dry cell, the NH_4Cl is replaced by KOH or NaOH. Thus corrosion of Zn container is prevented because zinc does not dissolve readily in basic medium.

The half cell reactions of alkaline cell are:

$$Zn + 2\bar{O}H \longrightarrow ZnO + H_2O + 2e^- \qquad \text{(Anode)}$$

$$2MnO_2 + 2e^- + H_2O \longrightarrow Mn_2O_3 + 2\bar{O}H \qquad \text{(Cathode)}$$

The alkaline dry cell lasts much longer compared to Leclanche cell.

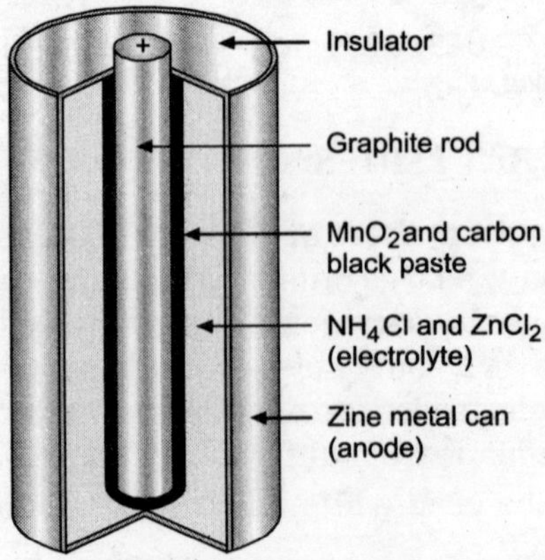

Fig. 6.17 Dry cell.

(b) Mercury Cell: The type of cell commonly used for calculations, hearing aids, digital watches and cameras is the mercury cell. The mercury cell, also known as Ruben Cell was developed in 1947. The anode consists of zinc-mercury amalgam and cathode is comprised of a paste of HgO and carbon. A paste of KOH and ZnO forms the electrolyte.

The cell reaction is:

$$Zn_{(amalgam)} + 2\bar{O}H \longrightarrow ZnO_{(s)} + H_2O + 2e^- \quad \text{(At Anode)}$$

$$HgO_{(s)} + H_2O + 2e^- \longrightarrow Hg(l) + 2\bar{O}H \quad \text{(At Anode)}$$

$$Zn_{(amalgam)} + HgO_{(s)} \longrightarrow ZnO_{(s)} + Hg_{(l)}$$

Mercury cells are much more expensive to manufacture than are dry cells or alkaline cells. This is due to high cost of the mercury (II) oxide they contain. At 25°C the cell voltage is 1.344 V. Since the overall cell reaction does not involve any ion in solution whose concentration can change, the cell shows constant potential throughout its life. This makes it useful in sensitive instruments.

Similarly is *silver button cell* the anodic reaction is the same as mercury cell whereas anodic reaction involves $Ag_2O_{(s)}$.

$$Zn_{(s)} + 2\bar{O}H_{(aq)} \longrightarrow ZnO(s) + H_2O_{(l)} + 2e^-$$

$$Ag_2O_{(s)} + H_2O_{(l)} + 2e^- \longrightarrow 2Ag_{(s)} + 2O\bar{H}_{(aq)}$$

$$E_{Cell} = 1.3 \text{ V (Mercury)}, 1.6 \text{ V (Silver)}$$

Other types of primary cell include lithium cell which has been developed for special uses where long life, small size and very low power are required. Cameras, heart pacemakers and electronic wrist watches are powered by these cells.

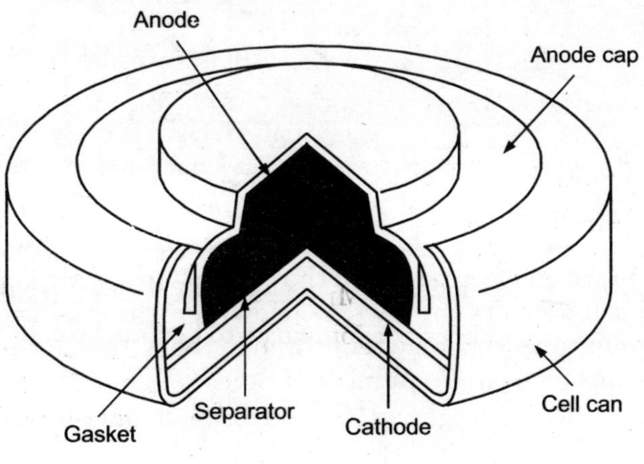

Fig. 6.18 Mercury cell.

A brief account of primary cells is given in Table 6.2

Tabel 6.2: Primary Cells

Type	*Anode (–)*	*Cathode (+)*	*Electrolyte*	*Voltage*	*Uses*
(1) *Leclanche or dry cell* Zn | ($ZnCl_2$), NH_4Cl | MnO_2,C	Zinc case $Zn(s) \rightarrow Zn^{2+}(aq) + 2e^-$	Carbon rod in contact with carbon & MnO_2 $NH_4^+(aq) + MnO_2(s) + e^- \rightarrow NH_3(aq) + MnO(OH)(s)$	Paste of NH_4Cl, $ZnCl_2$ (acid electrolyte)	1.5 volts	Low drain appliances such as radios, torches, liquid crystal calculators, toys
(2) *Alkaline cell* Zn, ZnO | KOH (paste) | MnO_2,C	Brass rod in contact with powdered zinc $Zn(s) + 2OH^-(aq) \rightarrow ZnO(s) + H_2O(l) + 2e^-$	Steel case in contact with carbon & MnO_2 $MnO_2(s) + H_2O(l) + e^- \rightarrow MnO(OH)(s) + OH^-(aq)$	Paste of 7 M KOH (alkaline electrolyte)	1 volt	High current or high total capacity appliances such as electronic phot-graphic flash units, tape recorders, radios, shavers. Can deliver higher current then Leclanche cell without severe volt-age drop.
(3) *Mercury button Cell*	Steel cap in contact with powdered zinc	Steel case in contact with powdered HgO $HgO(s) + H_2O(l) + 2e^- \rightarrow Hg(l) + 2OH^-(aq)$	Paste of KOH (alkaline electrolyte)	1.3 volt	Watches, pacemakers, hearing aids, micro-phones.
(4) *Silver button cell* Zn, ZnO | KOH(paste) | Ag_2O, Ag	Steel cap in contact with powdered zinc $Zn(s) + 2OH^-(aq) \rightarrow ZnO(s) + H_2O(l) + 2e^-$	Steel case in contact with powdered Ag_2O $Ag_2O(s) + H_2O(l) + 2e^- \rightarrow 2Ag(s) + 2OH^-(aq)$	Paste of KOH (alkaline electrolyte)	1.3 volts	Watches, pacemakers, hearing aids. As cell discharges, concentrations do not Change as reactants and products are solids. [OH^-] in electrolyte does not change. So, the voltage of this cell remains constant throughout its lifetime.

6.13.2 Secondary Cells

A secondary cell is one in which electrodes and the electrolyte are altered by the chemical action that takes place when cell delivers the current. However, these cells can be restored to their original condition by forcing an electric current through them. In other words, such batteries can be recharged and the cycle is repeated a number of times.

The automobile battery (lead storage battery) and nickel cadmium cells fall under this category.

***(i)* Lead Storage Cell:** In 1859, the French physicist Gaston Plante (1834-89) invented a device called the lead-acid accumulator, which is still used today as motor car battery with minor changes. It is an example of secondary electrochemical cell, since it can be recharged.

The cell consists of a lead grid filled with a spongy lead as anode and a lead grid packed with lead oxide as cathode. A solution of H_2SO_4 (38% by mass) is used as an electrolyte. This is the battery acid which one can buy at service stations.

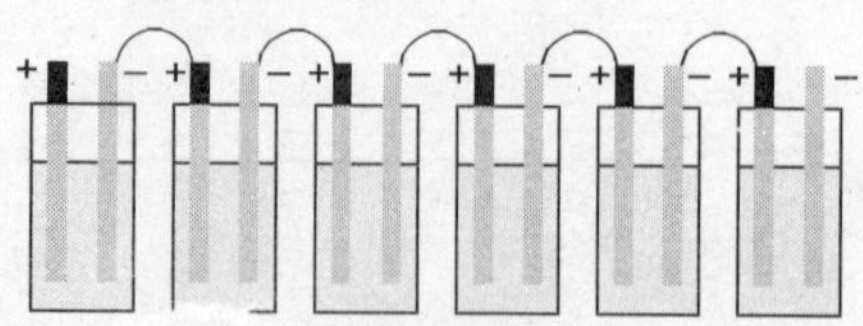

6 × 2 V = 12 V

Fig. 6.19

The battery consists of 6 such cells connected in series, each cell having an emf of about 2 V, giving 12 V as the overall emf of the battery.

When the battery is discharging, i.e., when it is supplying a current, the reactions are

At the anode:

$$Pb(s) + SO_4^{2-}(aq) \longrightarrow PbSO_4(s) + 2e^-$$

At the cathode:

$$PbO_2(s) + 4H^+(aq) + SO_4^{2-}(aq) + 2e^- \longrightarrow PbSO_4(s) + 2H_2O\ (l)$$

The overall reaction (during discharge) is, therefore,

$$Pb(s) + PbO_2(s) + 4H^+\ (aq) + 2\,SO_4^{2-}\ (aq) \longrightarrow PbSO_4(s) + 2H_2O\ (l)$$

Note that the electrolyte is sulphuric acid (H_2SO_4), which is gradually used up. Both electrodes become coated with an insoluble layer of lead sulphate, which, being an insulator, would eventually ruin the battery.

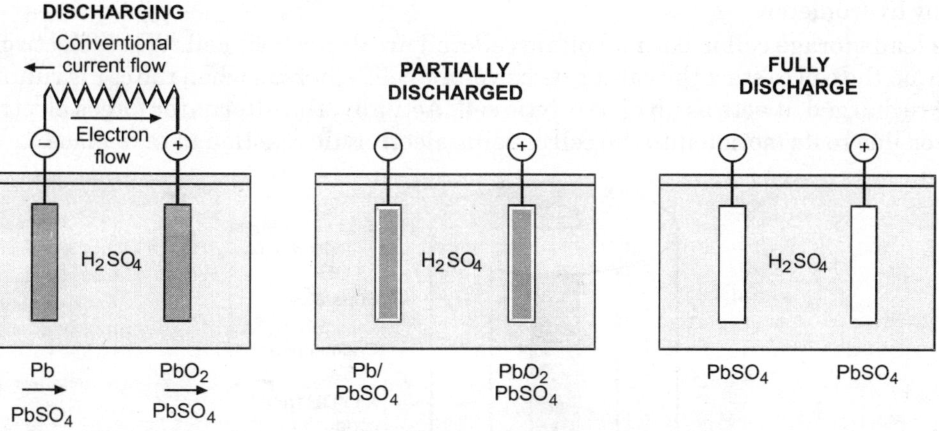

Fig. 6.20

As the battery provides energy, it is discharged and this lead to the gradual formation of lead sulphate at the electrodes, and a steady decrease in the concentration of the sulphuric acid:

If the battery is totally discharged, both electrodes will have been converted to lead sulphate. It will no longer be able to provide a current, and cannot be recharged. It is only good for scrap!

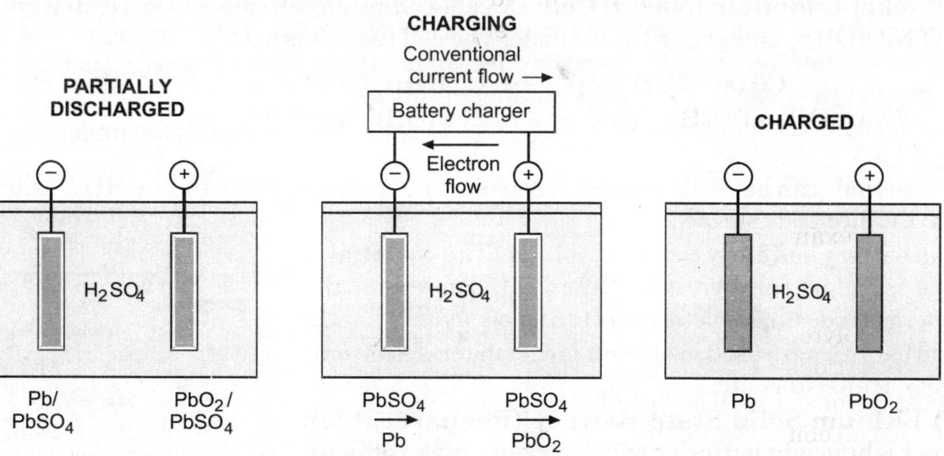

Fig. 6.21

When the battery is being charged, (using a battery charger or the motor car's alternator), electrons are supplied to the anode of the battery and this converts the lead sulphate to lead. At the same time, the lead sulphate at the anode is converted to lead dioxide. That is the cell reactions are reversed and the cell operates like an electrolyte cell.

The reactions are:

$$PbSO_4(s) + 2e^- \longrightarrow Pb(s) + SO_4^{2-}(aq) \quad \text{(at cathode)}$$

$$PbSO_4(s) + 2H_2O(O) \longrightarrow PbO_2(s) + 4H^+(aq) + SO_4^{2-}(aq) + 2e^- \quad \text{(at anode)}$$

$$2PbSO_4 + 2H_2O \longrightarrow Pb(s) + PbO_2(s) + 4H^+(aq) + SO_4^{2-}(aq)$$

The concentration of H_2SO_4 decreases during discharging whereas it increases during charging. Thus the extent to which the battery is discharged can be checked by measuring the density of H_2SO_4 by hydrometer.

The lead storage cell is both a voltaic cell and an electrolytic cell. When electricity is being drawn from the cell to start the car it acts as voltaic cell. Whereas when the car is running, the cell is being recharged, it acts as an electrolytic cell. Actually, the alternator forces electrical energy generated due to its motion into the cell causing electrolytic reaction to take place.

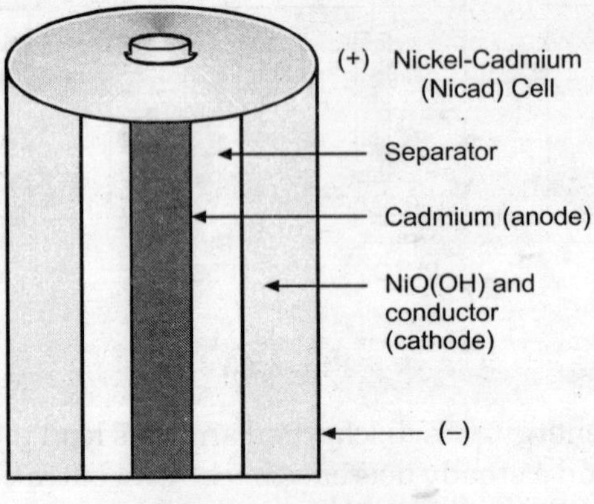

Fig. 6.22 Nicad cell.

***(ii)* Nickel-Cadmium (Nicad) Cell:** Nickel-cadmium cell consists of a cadmium anode and a paste of $NiO(OH)_{(s)}$ comprises the cathode. The cell reactions are:

$$Cd\ (s) + 2OH^-\ (aq) \longrightarrow Cd\ (OH)_2\ (s) + 2e^- \quad \text{(At Anode)}$$

$$2NiO\ (OH)\ (s) + 2H_2O\ (l)\ 2e^- \longrightarrow 2Ni\ (OH)_2\ (s) + 2OH^-(aq) \quad \text{(At Cathode)}$$

$E_{cell} = 1.4$ V

The reactions can be easily reversed, because the reaction products, $Ni(OH)_{2(s)}$ and $Cd(OH)(s)$ adhere to the electrode surface.

Nicad battery has fairly constant voltage. The potential of Cd anode is below the hydrogen potential. Therefore, the cadmium electrode is completely inert to electrolyte.

Nicad batteries are used in electronic calculators, electronic flash units, transistors etc.

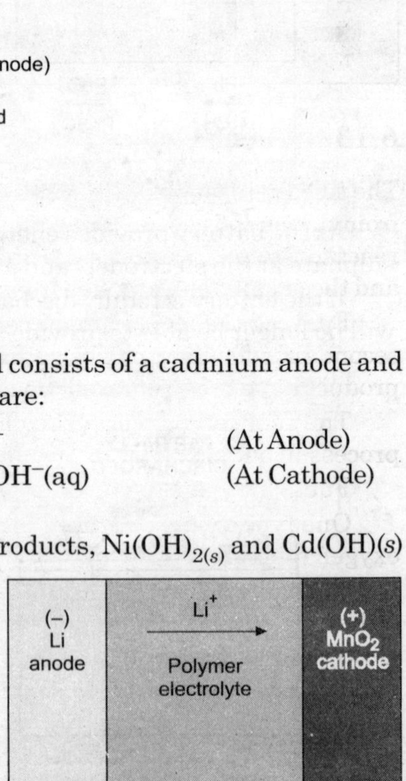

Fig. 6.23 Lithium Solid-state Battery

***(iii)* Lithium Solid-State Battery (Rechargeable):** These are Lightweight batteries which produce high voltages. Lithium solid-state batteries are common in computers, digital cameras, cell phones, etc. The cell reactions are:

$$Li(s) \longrightarrow Li^+ \text{ (in solid electrolyte)} + e^- \quad \text{(Anode)}$$
$$MnO_2(s) \longrightarrow Li^+ + e^- \longrightarrow LiMnO_2(s) \quad \text{(Cathode)}$$

A brief account of secondary batteries is given in Table 6.3.

Table 6.3

Type	*Anode (–)*	*Cathode (+)*	*Electrolyte*	*Voltage*	*Uses*
(1) *Lead-acid battery (or accumulator)* Pb, $PbSO_4$ | $H_2SO_4(aq)$ | PbO_2, Pb	Lead plates $Pb(s) + HSO_4^-(aq) \rightarrow PbSO_4(s) + H^+(aq) + 2e^-$	Lead plates covered with PbO_2 $PbO_2(s) + 3H^+(aq) + HSO_4^-(aq) + 2e^- \rightarrow PbSO_4(s) + 2H_2O(l)$	4M H_2SO_4	2.1 volts per cell (6 cells in series makes up a 12 volt car battery)	Motor vehicle batteries. As cell delivers current, lead sulfate deposits on electrodes and $[H_2SO_4]$ falls. Six 2 V cells connected in series make up a 12 V battery. To obtain high current output from the battery, the surface area of the electrodes in contact with the electrolyte must be large and the plates close together.
(2) *Nickel-cadmium cell (NiCad)* Cd, $Cd(OH)_2$ | KOH | NiO_2, $Ni(OH)_2$, Ni	Grid covered with cadmium $Cd(s) + 2OH^-(aq) \rightarrow Cd(OH)_2(s) + 2e^-$	Grid covered with nickel (III) hydroxide $NiO(OH)(s) + H_2O(l) + e^- \rightarrow Ni(OH)_2(s) + OH^-(aq)$	KOH	1.25 volts	Video cameras, phones, cordless drills, laptop computers

6.13.3 Fuel Cells

The third category of cells are the fuel cells. These are the electrical cells which convert the energy from the combustion of fuels into the electrical energy. It differs from an ordinary battery that the reactants are not contained within the cell, but are continually supplied from outside i.e. the fuel and the oxidant are supplied from outside the cell to the electrodes at which they undergo reactions.

Fuel cells is not a primary cell, since neither the reactants nor products are stored. It is not a secondary cell which can be recharged. Rather, the reactants have to be fed continuously and the product have to be removed. So it is reasonable to keep fuel cell in a separate category.

The fuel cell converts the chemical energy of the fuel directly to electricity. The essential process in a fuel cell is:

Fuel + Oxygen $\longrightarrow$ Oxidation product + Electricity

One type of fuel cell, used in space crafts and advanced non-linear submarines, is hydrogen-oxygen fuel cell.

Hydrogen-Oxygen Fuel Cell: It consists of two inert porous electrodes which are made up of either graphite impregnated with finely divided Pt or a 75/25 alloy of Pd with Ag or Ni. The electrolyte solution is 25% KOH. Hydrogen and oxygen are bubbled through the anode and cathode compartment respectively. The cell reactions are:

$$H_2(g) \longrightarrow 2H^+_{(aq)} + 2e^- \quad \text{(Anode)}$$
$$O_2(g) + 4e^- + 4H^+_{(aq)} \longrightarrow 2H_2O_{(l)} \quad \text{(Cathode)}$$
$$2H_{2(g)} + O_{2(g)} \longrightarrow 2H_2O_{(l)} \quad \text{(Overall reaction)}$$

The theoretical e.m.f. of the cell is 1.23 V

$$E^o = E^o_{oxd.} + E^o_{red} = 0.83\text{ V} + 0.40\text{ V} = 1.23\text{ V}$$

However, in actual practice, the e.m.f. of the cell is about 0.8 V to 1.0 V.

The only by-product formed in fuel cell is water which is used by the spacecraft or submarine crews.

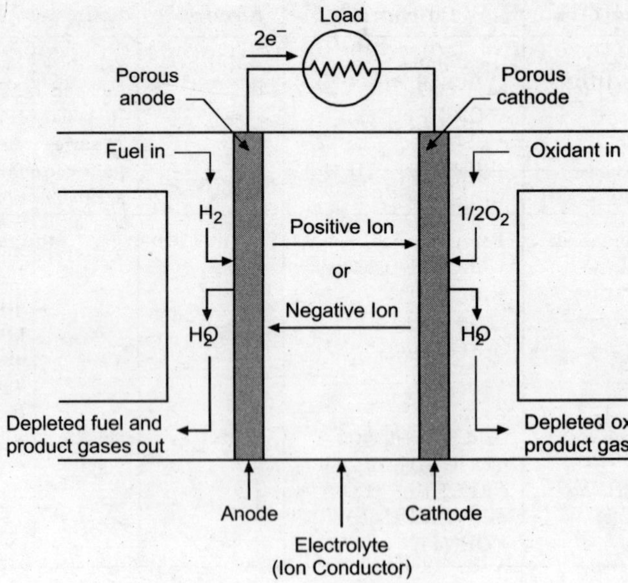

Fig. 6.24 Schematic of an individual fuel cell.

Fig. 6.24 shows a basic design of a fuel cell. However, is practice, it is very complex piece of equipment.

Let us look in more detail at the key cell components:

(a) The Fuel: Oxygen is usually taken from the air (filtered and cleaned before introduction to the cell) and hydrogen is generally made from the decomposition of a hydrocarbon source (such as methane). Any hydrogen rich material can serve as possible fuel source such as methanol, ethanol, propane etc.

(b) The Electrodes: The function of the electrodes is to bring about reaction between the reactant (fuel or oxygen) and the electrolyte, without itself being consumed or corroded. They must also, by definition, be an electronic conductor and bring the three phases (gaseous fuel, liquid electrolyte and electrode itself) to contact. They must be good electron sources or sinks.

There are several methods for gas-liquid interface stabilisation which rely on capillary action. A biporous electrode allows liquid to penetrate the small pores and a gas pressure is applied that drives the liquid out of the large pores. All this is set-up within the structure of the electrode. Then the merging of the two phases may be achieved in several ways. The electrolyte will tend to form a thin wetting film over part of the internal surface of the electrode. The reactant gas, sparingly soluble in the electrolyte, can diffuse through this film

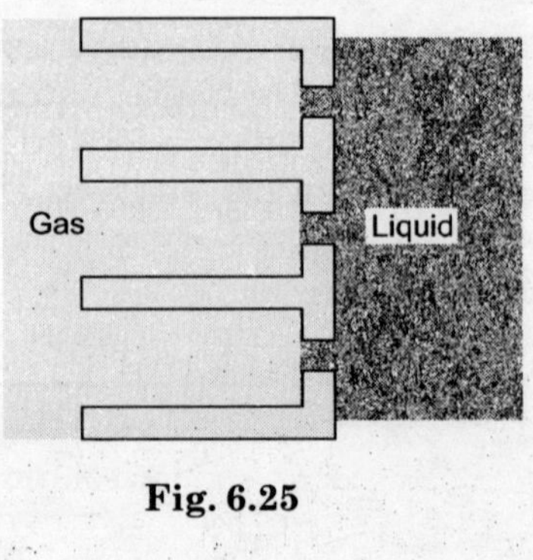

Fig. 6.25

and reach the electrode surface, where a solid-liquid reaction can occur. The electrode structure have to be designed to maximise the area of the wetted film (Fig. 6.25).

(c) *The Electrolyte:* A large range of fuel cells have been developed using a variety of different electrolytes. These developments have been driven by the desire to improve the economics of the cells and the potential applications at a range of different temperatures. In fact fuel cells are usually classified in terms of their electrolyte contents. Electrolytes most often used are aqueous KOH or H_2SO_4 or ion exchange resin saturated with water.

Depending upon the electrolyte used the fuel cell can be of different types. Table 6.4 gives an account on the different types of fuel cells.

Table 6.4 Fuel Cells

Type	*Anode (–)*	*Cathode (+)*	*Electrolyte*	*Voltage*	*Uses*
Alkaline fuel cell	Porous carbon anode with catalyst in contact with $H_2(g)$ $H_2(g) + 2OH^-(aq) \rightarrow 2H_2O(l) + 2e^-$	Porous carbon cathode with catalyst in contact with $O_2(g)$ $O_2(g) + 2H_2O(l) + 4e^- \rightarrow 4OH^-(aq)$	Hot KOH (*aq*)	~ 1 volt	• Promising future • More efficient than coal or gas-fired power stations. • Non-polluting • Porous electrodes lined with catalysts, eg, Ni or Pt, making it expensive
Acid fuel cell	Porous carbon anode with catalyst in contact with $H_2(g)$ $H_2(g) \rightarrow 2H^+(aq) + 2e^-$	Porous carbon cathode with catalyst in contact with $O_2(g)$ $O_2(g) + 4H^+(aq) + 4e^- \rightarrow 2H_2O(l)$	hot H_3PO_4 (*aq*)	~ 1 volt	• Systems required to continuously supply reactants and remove products • CH_4, CH_3OH, C_3H_8 can be used instead of $H_2(g)$

6.14 ACID-BASE TITRATIONS

Acid base titrations are the titrations in which acid is titrated against an alkali or vice versa. These titrations in aqueous solutions are essentially based on the reaction between hydrogen and hydroxyl ions to form unionized water (actually water is ionized to a very small extent). A typical example is the titration of sodium hydroxide against hydrochloric acid. The reaction involved is:

$$NaOH + HCl \longrightarrow NaCl + H_2O$$

or $$Na^+ + \bar{O}H + H^+ + Cl^- \longrightarrow Na^+ + Cl^- + H_2O$$

or $$H^+ + \bar{O}H \longrightarrow H_2O$$

Acid base titrations are also known as acidimetry and alkalimetry.

Acidimetry: To find the strength of an unknown acid with the help of a known base is known as acidimetry. The solution of the given acid is titrated with an alkaline solution and the $\bar{O}H$ ions of the alkali combine with H^+ ions of the acid. The concentration of H^+ decreases gradually and hence the *pH of the solution increases. At a definite pH value, the end point is reached and no more alkali is added.

---

* $pH = -\log [H^+]$. A solution is acidic if $pH < 7$
A solution is basic if $pH > 7$
A solution is neutral if $pH = 7$

Alkalimetry: To find the strength of an unknown base with the help of known acid is known as alkalimetry. The solution of an alkali is titrated with an acid solution and the $\bar{O}H$ ions combine with H^+ ions. The concentration of H^+ ions increases gradually and pH of the solution decreases. At a definite pH value, the end point is reached.

Acid Base Titrations and Indicators

The process of acid-base titrations is accompanied by a change in pH. The equivalence point in neutralization titrations occurs at a definite pH value. Therefore, substances which change colour in accordance with pH (indicators) are used to indicate the end point.

Acid-Base indicators are complex organic compounds which have one colour in acid solution while an altogether different colour in alkaline solution.

Theories of Acid Base Indicators

1. Ostwald's Theory. According to this theory the indicators themselves are weak acids or weak bases. They have different colours in ionized and unionized forms. The behaviour of phenolphthalein and methyl orange, the common acid base indicators is described as follows:

Action of Phenolphthalein. Phenolphthalein is a weak acid and can be represented as, HPh. Its unionized molecules are colourless while on ionization it gives pink coloured Ph^- ions

$$\underset{\substack{\text{(Colourless)}\\\text{(Acid Colour)}}}{HPh} \rightleftharpoons \underset{\text{(Colourless)}}{H^+} + \underset{\substack{\text{(Pink)}\\\text{(Base Colour)}}}{Ph^-} \quad \text{...(i)}$$

In presence of *acid,* due to increase in concentration of common H^+ ions in solution, the dissociation of HPh is suppressed and thus the solution becomes *colourless.* On the other hand, when a base like NaOH is added, the $\bar{O}H$ ions of the base, combine with H^+ of the indicator to form unionized water. Thus, the equilibrium shifts towards the right, producing more of coloured Ph^- ions. The solution therefore appears pink.

Action of Methyl Orange: Methyl orange is a weak base and can be represented as Me. It changes colour from yellow to red depending upon the pH of the solution. The equilibrium can be represented as:

$$\underset{\substack{\text{(Yellow)}\\\text{(Base colour)}}}{Me} + H_2O \rightleftharpoons \underset{\substack{\text{(Red)}\\\text{(Acid colour)}}}{MeH^+ + \bar{O}H}$$

If the solution is acidic, the hydrogen ions furnished by acid combine with $\bar{O}H$ ions and the equilibrium shifts to right and hence the solution appears red. If on the other hand, a base is added which furnishes $\bar{O}H$ ions then the equilibrium shifts to left (common ion effect) and hence the solution appears yellow.

2. Modern Quinoid Theory. The colour change of acid base indicators have also been explained on the basis of quinoid theory. According to this theory, the acid-base indicators can exist in two tautomeric forms. Ordinarily, one form is *benzenoid* and the other is *quinoid* form. The two forms have different colours and usually quinoid form has deeper colour than benzenoid form. Out of these two forms, one form exist in acidic medium while the other in alkaline medium. Change in pH causes the conversion of benzenoid form to quinoid form or vice versa and hence change in colour.

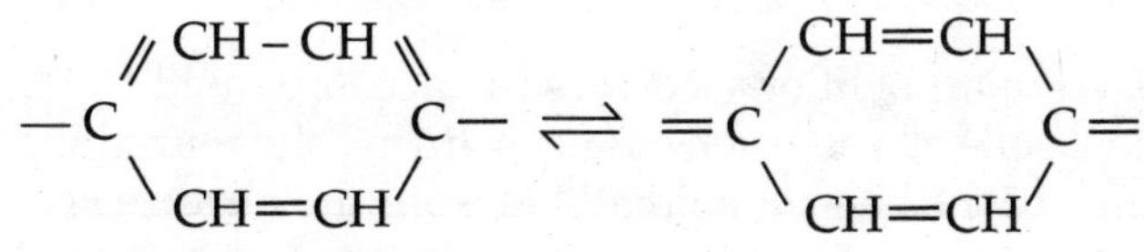

(Benzenoid form) (Quinoid form

The action of the phenolphthalein and methyl orange on the basis of this theory is explained.

Action of Phenolphthalein: Phenolphthalein possesses a benzenoid structure in acidic medium, whereas it changes to pink coloured tautomeric form in alkaline medium having quinoid form.

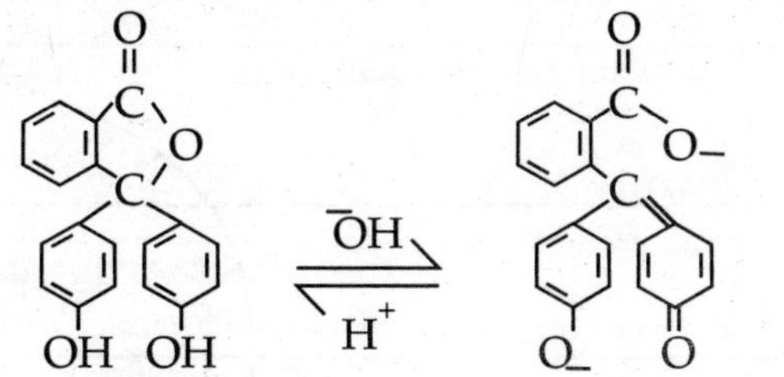

Colourless (Benzenoid form in acid solution) — Red colour (Quinoid form in base)

Action of Methyl Orange. Methyl orange also exists in two tautomeric forms. In alkaline solution it exists in benzenoid form and is yellow in colour. In acidic solution it exists in red colour having quinoid structure.

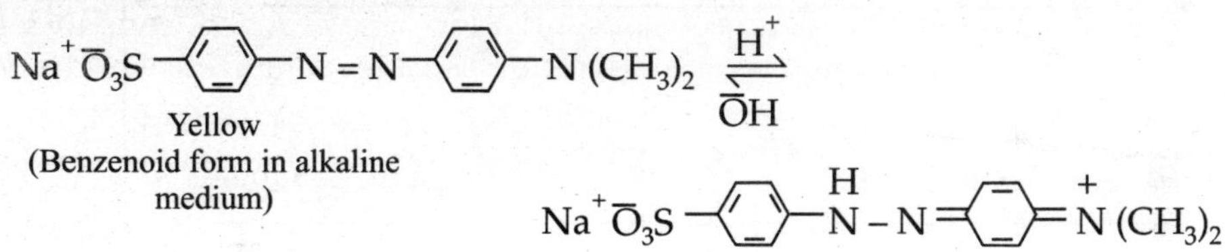

Yellow (Benzenoid form in alkaline medium)

(Quinoid form in acidic medium)

Selection of Proper Indicator

The pH value at the end point in titrations depend upon the relative strengths of acid and base being titrated. Since different indicators have different pH ranges within which they can be used, the selection of a proper indicator for a given titration is very important. As a general rule, one should select an indicator which changes colour at approximately the pH at equivalence point of titration. Some common indicators together with their useful ranges are given in Table 6.5

Table 6.5

Indicator	*pH range*	*Colour in acidic medium*	*Colour in basic medium*
Methyl Orange	3.1 – 4.4	Red	Yellow
Methyl Red	4.2 – 6.3	Red	Yellow
Litmus	4.5 – 8.3	Red	Blue
Phenolphthalein	8.0 – 9.6	Colour less	Pink

Acid Base Titration Curves

The addition of a base to an acid or vice versa is accompanied by the change in hydrogen ion concentration and hence pH of a solution changes during the course of acid base titration. A plot of pH against the volume of the base or acid added from the burette is called pH titration curve or acid base titration curve.

The pH titration curves are very significant in the detection of equivalence point and selection of suitable indicators for an acid base pair. For most of acid-base titrations, we can select an indicator which exhibits a distinct colour change at a pH close to that obtained at the equivalence point. The pH at equivalence point for a given acid base pair is determined experimentally by pH titration curves or by theoretical calculations as is exemplified below.

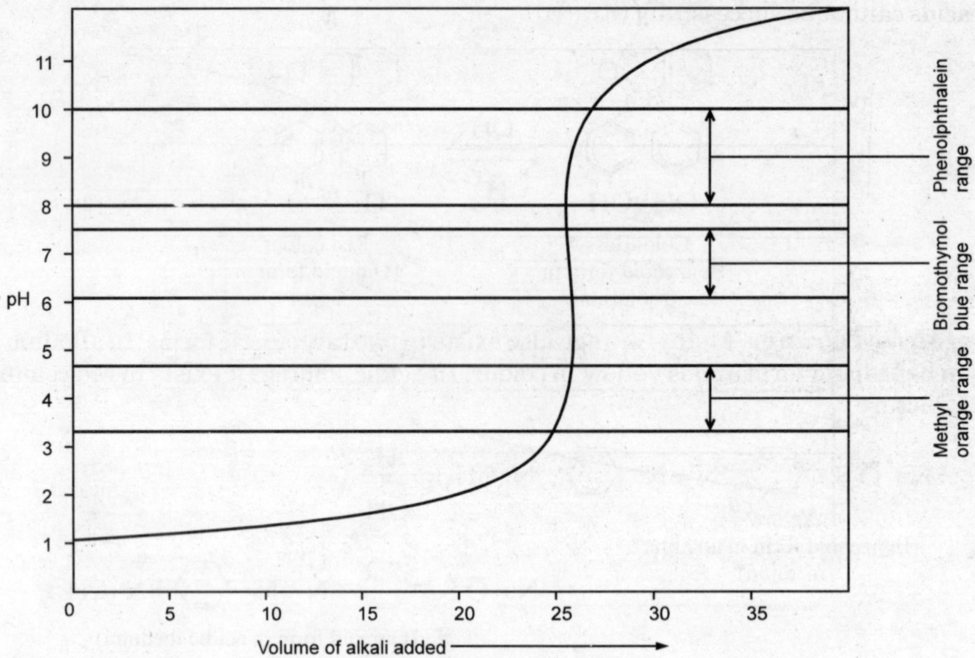

Fig. 6.26 The pH titration curve of strong acid with a strong base

(a) A Titration of a Strong Acid with a Strong Base The change of pH with the volume of the alkali added in case of titration of strong acid with a strong base is depicted in Fig. 6.1(a). It is seen that the pH changes very slowly at first, increases rapidly near the equivalence point. At the equivalence point, there is a sharp increase in the pH of the solution for even a very small amount of alkali added. This sharp and sudden change in pH from 4 to 10 near the equivalence point is very significant because all the indicators which show colour change in the pH range 4 to 10 can be used to detect end point. Thus, methyl orange, methyl red, phenolphthalein all can be used as indicators for detecting the end point in titrations involving strong acid and strong bases. After the equivalence point, on further addition of alkali, the pH again changes slowly.

It may be emphasized here that if the solutions are dilute (~ 0.01 N), the change in pH and equivalence point is not very sharp. The pH range is limited to 5.5 – 8.5 at equivalence point * and hence the indicators such as methyl red or bromothymol would be a suitable choice.

---

* **Note.** The equivalence point in all cases, whether the solution is dilute or concentrated in titration of strong acid with strong base remains the same i.e. 7.

(b) Titration of a Weak Acid with a Strong Base The titration curve of weak acid with a strong base (CH_3COOH Vs NaOH) is depicted in Fig. 6.27. In this case the pH at equivalence point will not be 7, because of the hydrolysis of sodium acetate so formed. As sodium acetate is a salt of a weak acid and a strong base it gives excess of $\bar{O}H$ ions in aqueous solution thereby increasing the pH at equivalence point beyond 7. Thus the indicator used should have pH range slightly on alkaline side. Hence phenolphthalein or thymol blue would be satisfactory indicators.

For titration of acids like boric acid against sodium hydroxide the titration curve does not show any sharp rise in pH. This is because the salts so formed (sodium borate) during the titration undergo hydrolysis to a greater extent resulting in alkaline solution. The pH, therefore goes on increasing gradually without showing a sharp increase. Thus very weak acids cannot be successfully titrated.

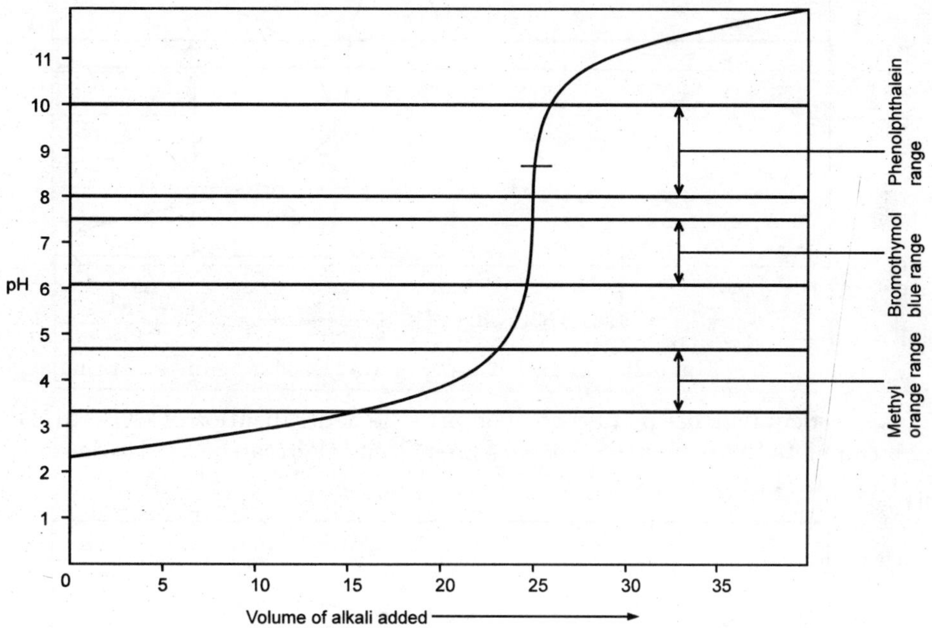

Fig. 6.27 The pH titration curve of weak acid with a strong base

(c) Titration of a Weak Base with a Strong Acid: The titration curve of a weak base against a strong acid (ammonium hydroxide vs hydrochloric acid) is depicted in Fig. 6.28. The pH at equivalence point in this case lies in the acid range i.e. below 7. This is due to the formation of a salt of weak base and a strong acid (NH_4Cl) which undergoes hydrolysis releasing H^+ ions in the solution. Thus the pH change at equivalence lies in the range 6 to 4. Thus the indicator having pH range on acidic side i.e. methyl red, or methyl orange can be used successfully.

(d) Titration of a Weak Acid by a Weak Base: In these types of titrations the pH changes during the titration are continuous i.e. no sharp pH change is observed and hence none of the indicator is suitable. However, a mixed indicator may be used.

(e) Titration of Na_2CO_3 with Hydrochloric Acid: When Na_2CO_3 is titrated with a strong acid, such as hydrochloric acid, the carbonate ions are first converted to bicarbonate ions, and then to carbonic acid.

$$CO_3^{2-} + H^+ \longrightarrow HCO_3^-$$
$$HO_3^- + H^+ \longrightarrow H_2CO_3$$

As the neutralization takes place in two steps, the pH titration curve show two inflection points as shown in Fig. 6.29.

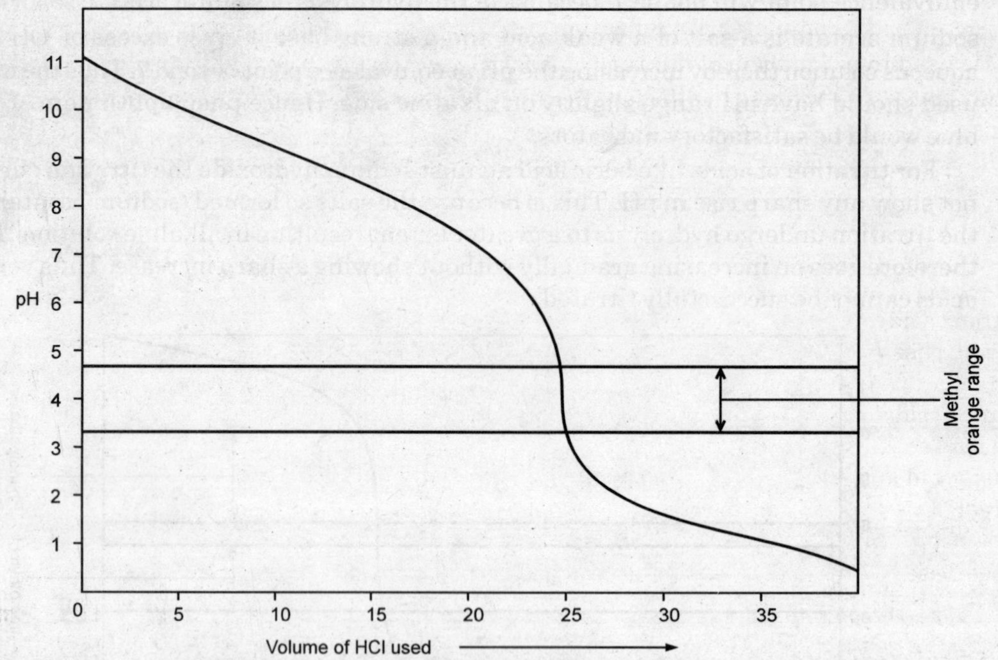

Fig. 6.28 Titration curve of a weak base against a strong acid

- the first equivalence point corresponds to the neutralization of CO_3^{2-} to HCO_3^- stage. This is completed at about pH 8.5. Hence phenolphthalein can be used to detect the conversion of CO_3^{2-} to HCO_3^- stage.

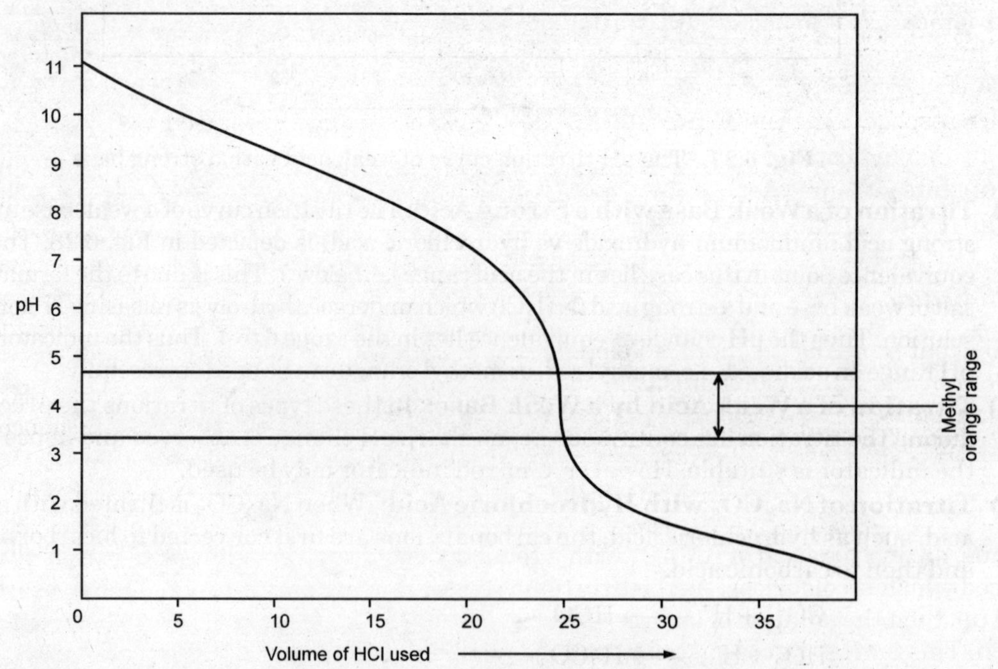

Fig. 6.29 Titration curve of sodium carbonate versus HCl

- the second equivalence point indicates the neutralization of HCO_3^- to H_2CO_3 stage. This reaction gets completed at pH 4.3. Thus, methyl orange can detect the end point in this case i.e. the complete neutralization of CO_3^{2-} ions. This behaviour of Na_2CO_3 and $NaHCO_3$ is used in estimation of mixture of Na_2CO_3 and $NaHCO_3$.

The choice of indicators in different acid-base titrations can be summarized in Table 6.6.

Table 6.6

Titrations	*Approximate pH range at equivalence point*	*Suitable indicators*	*pH at colour change of indicator*
1. Strong acid and strong base	4 to 10	Methyl Orange Litmus Phenolphthalein	3.1 – 4.4 4.5 – 8.3 8.0 – 9.6
2. Strong acid and weak base	3.5 to 7	Methyl Orange Methyl Red	3.1 – 4.4 4.2 – 6.3
3. Weak acid and strong base	6.5 to 10	Phenolphthalein	8.0 – 9.6
4. Weak acid and weak base –	No sharp inflection at equivalence point	None satisfactory	—

6.15 REDOX TITRATIONS AND INDICATORS

A redox indicator is defined as a substance whose oxidized form is of different colour from that of its reduced form. The oxidation and reduction of the indicator is readily reversible. As the change in colour in acid-base indicator is dependent upon the pH of the solution, similarly change in colour of redox indicator is dependent upon the electrode potential of the system. Therefore, while choosing a redox indicator for a particular titration, it is necessary to ensure that its potential lies within that of the system. Moreover the colour change should occur as close to the equivalence point as possible.

With respect to the use of indicators, redox indicators can be classified as:

(i) Self indicator
(ii) Internal indicator
(iii) External indicator

Self Indicator

If the reagent (reacting solution) is strongly coloured and is decolourized or changes into a different colour by the substance that is being determined then it is called self indicator. For example, potassium permanganate is purple in its oxidized form, but its reduced product i.e. manganese (II) ion is almost colourless. In acidic solution, the reaction is,

$$\underset{\text{Pure}}{MnO_4^-} + 8H^+ \rightarrow \underset{\text{Colourless}}{Mn^{2+}} + 4H_2O$$

Thus, during titration of $KMnO_4$ with a reducing agent such as oxalic acid, manganese (II) is produced which is colourless. As the titration proceeds and when whole of reducing agent is consumed up, then the addition of an excess drop of $KMnO_4$ solution gives its own colour (pink) to the solution. This marks the completion of the reaction (end point). Hence $KMnO_4$ acts as self indicator since no other indicator is added to mark the end point.

Internal Indicator

Broadly speaking, internal indicators are the substances which are added to the volumetric flask in which titration is carried out (The acid base indicators like phenolphthalein, methyl orange etc. are all examples of internal indicators). In redox titrations, diphenylamine is one of the first internal indicator discovered by Knop. This is used for the titration of iron (II) with potassium dichromate. This indicator has different colours in oxidized and reduced form. When all the iron (II) ions are oxidized to iron (III) ions by $K_2Cr_2O_7$ then an extra drop of $K_2Cr_2O_7$ added in the solution will oxidize the indicator (diphenylamine) which changes to blue-violet colour and hence marks the completion of reaction.

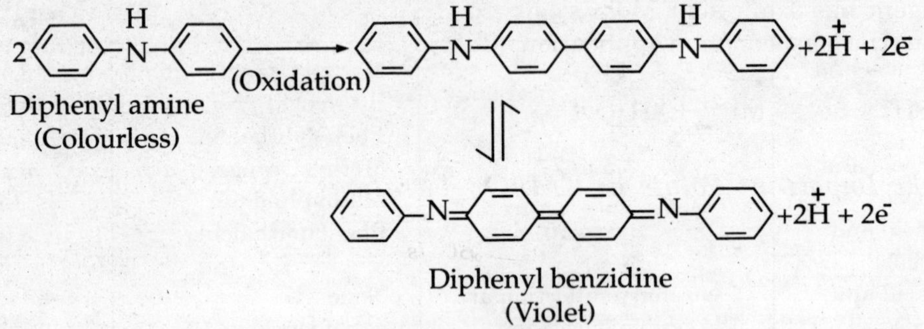

External Indicator

The indicators which cannot be added to the volumetric flask in which the titration is carried out, but are used externally are called external indicators. For example, potassium ferricyanide is used as an external indicator in the titration of $K_2Cr_2O_7$ with ferrous ammonium sulphate. Potassium ferricyanide (K_3 [Fe $(CN)_6$]) gives a blue colour with Fe^{2+} ions due to the formation of the complex ferrous ferricyanide, according to the following equation:

$$2\,K_3\,Fe\,(CN)_6 + 3\,FeSO_4 \longrightarrow \underset{\text{Ferrous ferri cyanide (Dark blue complex)}}{Fe_3\,[Fe\,(CN)_6]_2} + 3\,K_2SO_4$$

Thus, as long as Fe^{2+} ions are present, K_3 [Fe $(CN)_6$] will react with it to give blue colour. When all Fe^{2+} ions are converted to Fe^{3+} ions, blue colour will no longer be obtained.

The drops of external indicator [K_3[Fe$(CN)_6$] are placed on a groove tyle by means of a dropper. During the titration, the drops of the reaction mixture are taken out with the help of glass rod at intervals and tested by mixing with the drops of indicator solution in the groove tile. Appearance of blue colour indicates that the end point has not reached and the titration is continued further, till the reaction mixture gives no blue colour. This marks the end point. This is a rough titration. An almost exact end point is obtained in a subsequent titration in which number of drops taken out from the titration solution are minimum.

Adsorption Indicator

In some cases, though the reagent is coloured but colour change at the end point is not distinct to ensure accuracy. For example, the iodine is brown in solution whereas iodide ion is colourless,

$$\underset{\text{Brown in solution}}{I_2 + 2e^-} \longrightarrow \underset{\text{Colourless}}{2I^-}$$

As the iodine is reduced, the colour fades to a pale yellow. It is difficult to distinguish the pale yellow colour from the colourless iodide ion at the end point. However if starch is used as an indicator, iodine is *adsorbed* at the surface of the starch molecules and produce blue colour. The solution becomes colourless at the end point as the iodine is removed from the solution. Hence starch is used as an indicator in the iodine titrations as it gives distinct end point i.e. from blue to colourless.

II. A Potassium Permanganate Titrations

Potassium permanganate is a very strong oxidizing agent and is employed in the estimation of reducing agent like oxalic acid, Mohr's salt etc. It acts as an oxidizing agent in acidic and basic medium, but the most common application of $KMnO_4$ is in acidic medium according to the equation,

$$Mn + 8H^+ + 5e^- \rightarrow Mn^{2+} + 4H_2O$$

Some of the Important Points in $KMnO_4$ Titrations are

1. *For acidification of $KMnO_4$ solution, only H_2SO_4 is used whereas other mineral acids like HCl and HNO_3 are not used.*
   - *HCl cannot be used because some of $KMnO_4$ will oxidize HCl to Cl_2 gas and thus interferes in the quantitative estimations.*

     $$2Mn + 16H^+ + 10Cl^- \rightarrow 2Mn^{2+} + 8H_2O + 5Cl_2$$

   - *HNO_3 cannot be used because it is itself a strong oxidizing agent and may oxidize the reducing agent, thereby introducing error.*
2. *$KMnO_4$ is not a primary standard.*
   - *It cannot be obtained in perfectly pure state as it always contains traces of manganese dioxide.*
   - *$KMnO_4$ may react with organic matter present in water in which it is dissolved.*
   - *$KMnO_4$ solution slowly decomposes when exposed to bright sun-light, reducing gases etc. Therefore, it is stored in dark coloured bottles.*
   - *$KMnO_4$ solution can be standardized using a suitable primary standard such as oxalic acid or Mohr's salt, which can be directly weighed.*
3. ***$KMnO_4$ is a self indicator.** This point is already discussed under the heading "redox indicators".*

III. Potassium Dichromate Titrations

Potassium dichromate acts as an oxidizing agent in the presence of the acid. The chemical equation in presence of an acid being,

$$K_2Cr_2O_7 + 4H_2SO_4 \rightarrow K_2SO_4 + Cr_2(SO_4)_3 + 4H_2O + 3[O]$$

or Ionic equation $Cr_2O_7^{2-} + 14H^+ + 6e^- \rightarrow 2Cr^{3+} + 7H_2O$

As the above equation involves the transfer of 6 electrons during reduction of $Cr_2O_7^{2-}$ to Cr^{3+}, hence the equivalent weight of $K_2Cr_2O_7$ will be,

$$= \frac{\text{Molar mass of } K_2Cr_2O_7}{6} = \frac{294}{6} = 49$$

Some of the Important Points Involved in $K_2Cr_2O_7$ Titrations are

1. *It is a less powerful oxidizing agent than $KMnO_4$.*
2. *It is a excellent primary standard.*
   - *It is available in pure state.*
   - *It is very stable. Its solutions do not decompose even on long standing.*
   - *Its solutions are stable towards light and are little affected by dust, organic matter or by autocatalytic reactions.*

   Thus, $K_2Cr_2O_7$ has above advantages over $KMnO_4$.
3. *For acidification of $K_2Cr_2O_7$, apart from H_2SO_4, HCl can also be used. $K_2Cr_2O_7$ do not oxidize chlorides at room temperature.*
4. *$K_2Cr_2O_7$ acts as on oxidizing agent only in acidic medium. In presence of an alkali, $K_2Cr_2O_7$ reacts to give chromate salt and cannot act as oxidizing agent.*
5. ***Indicators.** A dilute solution of $K_2Cr_2O_7$ has a faint orange colour and its reduction product chromium (III) is green in colour. Thus, a drop of $K_2Cr_2O_7$ in excess at the end point is not sufficient to give a distinct colour to the solution (unlike $KMnO_4$) and hence it cannot act as self indicator. Therefore a redox indicator should be used in $K_2Cr_2O_7$ titrations. There are two types of indicators used in $K_2Cr_2O_7$ titrations:*

   (a) ***External Indicator.*** *Potassium ferricyanide is used as an external indicator.*

   (b) ***Internal Indicators.*** *Commonly used internal indicators are:*

Indicator	*Colour in reduced state*	*Colour in oxidized state*
• *N Phenylanthranilic acid*	*green*	*violet-red*
• *Diphenylamine in conc. H_2SO_4*	*colourless*	*violet*
• *Diphenyl benzidine in conc. H_2SO_4*	*colourless*	*violet*
• *Ferroin in conc. H_2SO_4*	*Red*	*faint blue*

IV. Iodine Titrations (Iodometry and Iodimetry)

Iodine is a mild oxidizing agent and is used in volumetric estimations. Iodine acts as an oxidizing agent in neutral aqueous medium according to the reaction,

$$I_2 + 2e^- \rightarrow 2I^-$$

On the other hand an oxidizing agent can oxidize I_2^- ions to I_2. Both these reactions are used in iodine titrations.

Some of the Important Points Regarding Iodine Titrations are

1. *Iodine is a much weaker oxidizing agent than potassium permanganate and dichromate. It is used to estimate the amount of various reducing agents like sodium thiosulphate, sodium sulphite, sodium arsenite and antimony salts in their solutions.*
2. *Iodine can also be employed to estimate the strength of oxidizing agents as well.*
3. *Hence, Iodine Titrations are of two types:*

   (a) ***Iodimetry.*** *All such titrations which involve the direct titration of iodine with a reducing agent are grouped under iodimetry. Iodimetry is employed to determine the strength of reducing agents such as sodium thiosulphate, sulphite, arsenite, stannous and ferrous ions in neutral or moderately acidic solutions.*

(b) ***Iodometry:*** *In these titrations, iodine is liberated as a result of chemical reaction involving oxidation of an iodide ion by a strong oxidizing agent in neutral or acidic medium. The liberated iodine is then titrated with a reducing agent. Iodometry is used to estimate the strength of oxidizing agents. For example, the estimation of Cu^{2+} with thiosulphate–*

$$2Cu^{2+} + 4I^- \longrightarrow Cu_2I_2 + 2I_2$$

$$I_2 + 2S_2O_3^{2-} \longrightarrow S_4O_6^{2-} + 2I^-$$

4. ***Indicator:*** *The indicator used in these types of titrations is starch solution* which is used as an internal indicator. Starch forms a dark blue complex with triiodate ions. The blue colour disappears when there is no more of free iodine.*

Iodometric Titrations

As is discussed earlier the estimation of oxidizing agents can be done by iodometric titrations.

Following Points Must be Kept in Mind During Iodometry.

1. *Since iodine is volatile, titrations should be performed in cold. Moreover, the indicator starch also loses its sensitivity at high temperatures.*
2. *Iodometry titrations cannot be performed in strongly alkaline solutions. This is because iodine reacts with alkalies to give hypoiodite which is a strong oxidizing agent and may oxidize thiosulphate ions.*

$$I_2 + 2NaOH \rightarrow NaIO + NaI + H_2O$$

or

$$I_2 + 2\bar{O}H \rightarrow I\bar{O} + I^- + H_2O$$

$$S_2O_3^{2-} + 4I\bar{O} + 2\bar{O}H \rightarrow 4I^- + 2SO_4^{2-} + H_2O$$

3. *Always use excess of KI: This is because the solubility of I_2 in water is very low. But in the presence of KI it forms an unstable complex salt, KI_3. This soluble complex will release iodine slowly during titration due to reversible reaction,*

$$KI + I_2 \rightleftharpoons KI_3$$

$$I^- + I_2 \rightleftharpoons I_3^-$$

When free iodine is used up in the reaction with thiosulphate, the equilibrium shifts to the left releasing more I_2.

4. *Since the rate of reaction between I^- ion and the oxidizing agent is low, therefore the solution is allowed to stand for 2 – 3 minutes before titrating.*
5. *The reaction mixture is kept in the dark before titrating. This is due to the reason that light accelerates the oxidation of iodide ions to iodine by atmospheric oxygen*

$$4I^- + 4H^+ + O_2 \rightarrow 2I_2 + 2H_2O$$

---

* **Note.** The starch indicator is not added in the beginning of the titration but towards the end of the titration. This is because the concentration of iodine is very high in the beginning and it is strongly adsorbed by colloidal starch particles. It is not released even at the end point i.e. at the end point, though the solution becomes colourless but the blue coloured starch flocks make the detection of end point difficult.

The sodium thiosulphate (or the reducing agent) is added to the reaction mixture (which is brown in colour due to free iodine). When this I_2 is used up, the brown colour fades to straw (pale yellow) colour, indicating that only a small amount of iodine is left. Then starch solution is added near the end. The solution acquires the blue colour and the titration is continued till the solution just becomes colourless.

6.16 OVER VOLTAGE

In an electrolysis cell the discharge of an ion on the cathode would occur at the standard half-reaction potential of that ion. Thus, H^+ ions will discharge at the cathode when $E^o = 0.00$ V. This has been found to be nearly true if finelly divided platinum is used as electrode.

However, if other metal electrodes are used, the voltage needed for the discharge of H^+ ions i.e. evolution of hydrogen gas, is higher than its standard reduction potential. *This difference between the potential of the electrode when gas evolution is actually observed and theoretical value for the evolution of gas is called the overvoltage.*

Let us illustrate this point by taking the example of the following cell;

$$Pt/H_{2(1\ atm)}/HCl_{(a\,=\,1)}/Pt/O_{2\ (1\ atm)}$$

Now, the voltage of this cell is 1.12 volts. If the voltage (reversible) greater than 1.12 volts is applied to the cell, the cell reaction get reversed and evolution of H_2 and O_2 takes place. However, in reality, significantly higher voltage than 1.12 volts is required for the discharge of H_2 and O_2. In the above case a voltage of 1.17 volts is required. Thus the difference (1.17 – 1.12 = 0.58 Volts) of 0.58 V is the overvoltage. In other words, the excess of voltage, over the voltage of the cell that is necessary to cause the reverse reaction, is termed as over voltage.

Table 6.7 Gives the Hydrogen Over-Potentials on Some Metal Cathodes in Dilute H_2SO_4

Cathode	*Hydrogen Over potentials*	*Oxygen Over potential*
Mercury	0.78 V	—
Zinc	0.70 V	—
Copper	0.23 V	—
Silver	0.15 V	0.41
Platinum Black	0.00 V	0.25

Over potential exists for any gas at any electrode.

DENTAL GALVANIC CELL

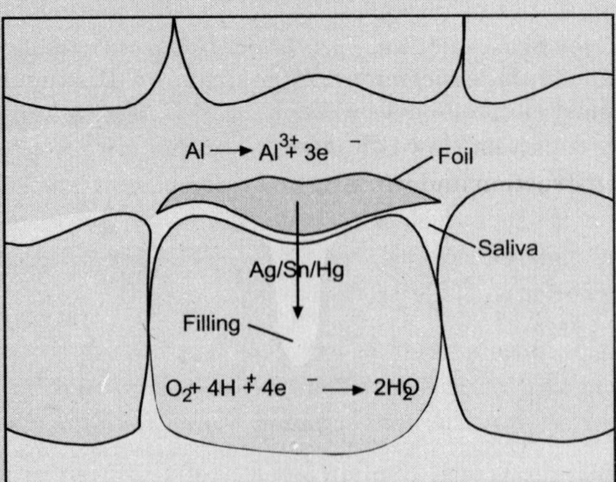

Biting on aluminium foil produces pain if you have fillings in your teeth. Al acts as the anode (–1.66 V), saliva as the electrolyte, and the silver/tin/mercury alloy fillings as an inert cathode. O_2 is reduced to water (1.23 V), and the short circuit between the foil in contact with the filling creates a current that jolts the nerve in the tooth.

QUESTIONS

1. What is a galvanic cell? Describe a Daniell cell.
2. Differentiate between an electro chemical cell and an electrolytic cell
   Give two important differences between e.m.f. and potential difference
3. What is an elecrochemical series ? Discuss its three important applications.
4. Differentiate between
   (i) Galvanic cell and electrolytic cell
   (ii) E.m.f. and potential cell
   (iii) Specific and equivalent conductance.
   What is the effect of dilution on specific and equivalent conductance?
5. A silver rod is dipped in a solution at 25°C which is 0.1 M in Ferric ions. Calculate the equilibrium concentration of all the ions in the solution
   $$E° (Fe^{3+}, Fe^{2+}) = 0.771 \text{ V and}$$
   $$E° (Ag^{+}, Ag) = 0.799 \text{ V}$$
6. Devise a cell for which the cell reaction is $Ag^{+}(aq) + Cl^{-}_{(aq)} = AgCl(s)$
   The standard e.m.f. of this cell is 0.503 V. Calculate the equilibrium constant for the reaction.
7. What do you understand by a reversible electrode? Describe one reversible electrode.
8. (a) Describe the construction and working of the following electrodes:
   (i) Hydrogen electrode (ii) Calomel electrode
   (b) Write down the reaction taking place at the separate electrode and in the complete reversible cell given below: $Pt\ H_2(g) \mid HCl\ (aq) \mid KCl\ (aq) \mid Ag\ Cl\ (s) \mid Ag$
9. (a) Give an example where a metal chloride behaves reversibly with respect to an ion.
   (b) Explain the meaning of an electrode potential being positive or negative with respect to hydrogen electrode.
10. (a) What is a reference electrode? Describe briefly.
    (b) What do you understand by electrochemical series? How does it help in predicting whether a redox reaction is feasible or not
11. Can we store copper sulphate in an iron vessel? Why.
12. Derive a mathematical relation between the electrical energy of a reversible galvanic cell and the change in free energy of the reaction taking place in the cell.
13. (a)What is the source of energy in a galvanic cell? Give the relation between electrical energy and chemical energy.
    (b)How is electrode potential experimentally determined? Distinguish between single electrode potential and standard electrode potential. **(Madras I.B.E. Oct. 1997)**
14. (a) Derive Nernst's equation. How is it useful? **(Bharathiar I B.E., Nov. 1997)**
    (b) Describe the construction and working of a calomel electrode. **(Andhra I B.E., 1997)**
    (c) What is pH? How is pH of an electrolyte determined with hydrogen electrode?
    (d) Give an account of glass electrode. **(Sri Krishnadevaraya I B.E., July 1997)**
15. (a) Derive Nernst's equation of electrode potential.
    (b) Give the constructional details of glass electrode. How can this be used for the determination of pH of a solution? **(Sri Krishnadevaraya I B.E., Oct. 1997)**
16. (a) Derive Nernst's equation.
    (b) Explain the various applications of e.m.f. series.
    (c) What are concentration cells? Explain an electrolyte concentration cell and its applications.
    (d) What is meant by overvoltage and decomposition potential? Explain.
    (e) What is fuel cell? Give detail of one such cell. **(Anna I B.E., July 1997)**
    (f) Describe the construction of a glass electrode. How is it significant in its applications? **(Anna I B.E., May 1996)**

(g) Explain the different applications of e.m.f. series. **(Anna I B.E., May 1995)**

(h) What are reversible and irreversible cells? Give examples. **(Anna I B.E., July 1997)**

17. Write short note on electrode potentials.

(a) What is a reference electrode? Explain why is it introduced with measurement of electrode potentials? Describe its construction.

18. (a) What are secondary cells? Describe the construction of one secondary cell. Write the cell reactions and mention its applications.

(b) Write short note on fuel cells. **(Sri Venkatesware I B.E., Sept. 1997)**

(c) Write short note on fuel cell. How is it different from commercial cell? Mention the advantages of fuel cells. **(VT, Aug. 2K, March 01, Aug. 01, Anna, Nov. 2K)**

(d) Describe the construction of lead acid battery with the reaction occurring during discharge. **(VT, March 99)**

(e) How does a fuel cell differ from a battery? **(VT, Aug. 99)**

19. Define fuel cell. Explain the construction and working of H_2–O_2 fuel cell. What are the advantages and limitation of a fuel cell? Write the reaction involved. Why is water formed in this cell removed continuously? **(VT Marth 99, Aug. 99; Calicut June 01)**

20. Discuss the working principle of primary batteries? **(VT April, 2K)**

(a) Write briefly about (i) Primary cells (ii) Secondary cells **(S.K. Aug. 01)**

(b) Distinguish between a cell and battery. Give the classification of cells with examples.

21. (a) What is alkaline battery? Describe the construction and working of Ni-Cd battery with relevant reactions taking place during discharge. Mention its advantages and applications. **(Calicut, June 01; VT April 2L; Aug. 01)**

(b) How is glass electrode constructed? Write the cell reactions.

(c) What is the effect of electrolyte concentration on electrode potential?

(d) Explain the construction of Ni-Cd cell. **(Kerala, April 97)**

22. (a) The electrode potential of Zn is assigned a negative value whereas that of Cu of is assigned a positive value. Give reasons.

(b) For the Danial cell involving the cell reaction:

$$Zn(s) + Cu^{2+}_{(aq)} \rightleftharpoons Zn^{2+}_{eq} + Cu_{(s)},$$

the standard free energies of formation for $Cu^{2+}_{(aq)}$ and $Zn^{2+}_{(aq)}$ are 15.66 kcals mol^{-1} and – 35.14 kcal mol^{-1} respectively. Calculate E° of the cell.

(c) Calculate the e.m.f. of the Zn–Ag cell at 25°C when $[Zn^{2+}] = 0.01M$ and $[Ag^+] = 10$ M. E°cell at 25°C = 1.56 Volt.

(d) Write a short note on calomel Electrode. **(PTU, Dec. 2002)**

23. (a) Why do electrochemical cells stops functioning after some time?

(b) the EMF of the cell Cd, $CdCl_2$, $2.5H_2O$ (saturated / $AgCl_{(s)}$ Ag in which the cell reaction is: Cd + $2AgCl_{(s)}$ + aq → $CdCl_2$ $2.5H_2O$ (Sat) + 2Ag, is 0.6753 volt at 25°C and 0.6915 volt at 0°C. Calculate ΔG, ΔH and ΔS of the cell at 25°C.

(c) What do you understand by electrochemical series? **(PTU, June 2003)**

24. Write in brief about: (i) polarization, (ii) overvoltage. **(Nagpur I B.E., 1997 Summer)**

(a) Write short note on EMF of a cell. **(Rajasthan I B.E., 1997)**

25. ΔG can be calculated, if half-cell potentials are known. Explain. **(Pb. Tech., Dec. 1997)**

26. (a) Represent standard Weston cell. **(DDIT Nadiad, II B.E., May 1997)**

(b) Give the use of silver-silver chloride electrode.

(c) What is concentration cell? Discuss its applications. **(DDIT Nadiad, II B.E., May 1997)**

27. (a) Explain the construction and working of quinhydrone electrode. Discuss its merits and demerits.

(b) What is a concentration cell? Explain its working with example. Write the electrode and cell reactions.

(c) Explain the construction and working of : (i) dry cell, (ii) hydrogen-oxygen fuel cell.

28. Write a short notes on
    (a) Liquid Junction Potential.
    (b) Overvoltage.

29. What are the indicators used in acid-base titrations.

30. (a) Discuss the types of redox titrations
    (b) What is the difference between iodometery and iodimetry.

31. (a) The electrode potential of Zn is assigned a negative value whereas that of Cu of is assigned a positive value. Give reasons.
    (b) For the Daniel cell involving the cell reaction:

$$Zn(s) + Cu^{2+}_{(aq)} \rightleftharpoons Zn^{2+}_{eq} + Cu_{(s)},$$

the standard free energies of formation of $Cu^{2+}_{(aq)}$ and $Zn^{2+}_{(aq)}$ are 15.66 kcals mol^{-1} and – 35.14 kcal mol^{-1} respectively. Calculate E° of the cell.

(c) Calculate the EMF of the Zn–Ag cell at 25°C when $[Zn^{2+}] = 0.01M$ and $[Ag^{+}] = 10$ M. E°cell at 25°C = 1.56 Volt.

(d) Write a short note on calomel Electrode. **(PTU, Dec. 2002)**

32. (a) Why do electrochemical cells stops functioning after some time?
    (b) The EMF of the cell Cd, $CdCl_2$, $2.5H_2O$ (saturated / $AgCl_{(s)}$ Ag in which the cell reaction is: Cd + $2AgCl_{(s)}$ + aq → $CdCl_2$ $2.5H_2O$ (Sat) + 2Ag, is 0.6753 volt at 25°C and 0.6915 volt at 0°C. Calculate ΔG, ΔH and ΔS of the cell at 25°C.
    (c) What do you understand by electrochemical series? **(PTU, June 2003)**

33. Write down the cell for which the overall reaction is

$$Cd + Cu^{2+}\ (a=1) \rightleftharpoons Cd^{2+}(a=2) + Cu$$

(PTU, Dec. 2003)

34. (a) Derive Nernst equation for the calculation of cell e.m.f.
    (b) The standard electrode potential for Hg_2^{+2}/Hg and Hg^{+2}/Hg and Hg^{+2}/Hg are 0.799V and 0.855V respectively. Calculate at 298 K the equilibrium constant for the reaction. **(PTU, Jun. 2004)**

35. What is the cell reaction and cell e.m.f. at 298 K of the cell?

$$Zn/Zn^{2+}\ (a=1)\ ||\ Pb^{+2}\ (a=1)\ |\ Pb$$

(PTU, June. 2004)

36. Why does a dry cell become dead after a long time, even it has not been used? **(PTU, Dec., 2004)**

37. Describe a typical galvanic cell and show how the chemical energy is converted to electrical energy? What is the relation between free energy, enthalpy, entropy and e.m.f. of a cell. **(PTU, May 2005)**

38. (a) Derive Nernst equation and give its significance.
    (b) Write equations for each half reaction and calculate E_{cell} for the following concentration cell $Zn/Zn^{++}_{(1.0M)}/Zn^{++}_{(0.15M)}/Zn$ as the cell discharges, does the concentration of two solutions becomes smaller or larger. **(PTU, Dec. 2005)**

39. Calculate the ΔG of the following galvanic cell

$$Zn(s)/Zn^{2T}_{(1.0M)}/Cu^{2+}_{(0.175m)}/Cu(s) \text{ at } 25°C.$$

$$E^{\circ}_{Zn^{2+}/Zn} = -0.76\ V,\ E^{\circ}\ Cu^{2+}/Cu = 0.34\ V.$$

(PTU, May, 2006)

40. What is Liquid Junction potential? How is it minimized. **(PTU, May, 2006)**

41. (a) What are concentration cells? Explain with an example.
    (b) Write the electrode reaction and calculate the oxidation potential of each of the following at 298 K
    (i) Ag, Ag Cl(s)/Cl^- (a = 0.001) E° (Ag/Ag^+) = – 0.225 V
    (ii) Pt H_2(1 atm)/H^+ (a = 0.15) **(PTU, May 2006)**

42. For a cell reaction $A(s) + 2B^{+}_{(aq)} \longrightarrow A^{2+}_{(aq)} + 2B(s)$ at 298 K, the equilibrium constant is 1.0×10^{4}. Calculate E°_{cell}. **(PTU, May 2007/Dec. 2006)**

43. (a) What is Nea-rnst Equation? Write its applications.
(b) The emf of cell reaction $3Sn^{4+} + 2Cr \longrightarrow 3Sn^{2+} + 2Cr^{3+}$ is 0.89 V. Calculate the standard free energy change of reaction. **(PTU, May 2007/Dec. 2006)**

44. (a) Sketch the electrochemical cell and write down the cell reaction. If Zn and Ag electrode are dipped in respected solution of their ions. Find out the emf of the cell given that:

$$E^{\circ}_{Zn^{2+}/Zn} = -0.76 \text{ V and}$$

$$E^{\circ}_{Ag^{+}/Ag} = 0.8 \text{ V}$$

Concentration of Zn^{2+} and Ag^{+} ions are 0.2 M and 0.01 M respectively.
(b) What is reference electrode? Describe the concentration of calomel electrode and hydrogen electrode. **(PTU, May 2007 (C))**

45. How is ΔG calculated for Zn – Cu cell if 1M electrolyte are present. **(PTU, May 2007)**

46. Standard EMF of Daniell cell involving cell reaction

$$Zn(s) + Cu^{2+}_{(aq.)} \rightleftharpoons Zn^{2+}_{(aq.)} + Cu(s) \text{ is } 1.10 \text{ V}$$

Calculate the equilibrium constant of cell reaction at 25°C. **(PTU, Jane. 2008)**

47. (a) What are secondary cells? Describe construction of one secondary cell. Write cell reactions and mention its uses.
(b) A cell consist of two hydrogen electrodes negative electrode is in contact with a solution of 10^{-6} M hydrogen ions. The EM of cell is 0.118 V at 250°C. Calculate concentration of hydrogen ions at positive electrode. **(PTU, may 2008)**

PROBLEMS

1. The e.m.f. of the cell

$$Pb \mid PbCl_{2(s)} \mid KCl_{(m)} \mid Hg_2Cl_2 \mid Hg$$

is 0.535 V at 25°C and increases with temperature by 1.45×10^{-4} V/K.

Write the electrode and overall cell reactions and calculate the enthalpy of the reaction taking place in the cell.

2. Calculate the e.m.f. at 298°K of the cell

$$Zn \mid ZnSO_4(m{=}1) \mid \mid CuSO_4(m{=}1) \mid Cu$$

Given
$$E^0\, Zn \mid Zn^{2+} = -0.763 \text{ V}$$
$$E^0\, Cu \mid Cu^{2+} = 0.337 \text{ V}$$
$$\gamma\pm (CuSO_4) = 0.047,$$
$$\gamma\pm (ZnSO_4) = 0.045$$

where γ = mean ionic activity coefficient

Given $\frac{RT}{F} = 25.6$ mV

3. Zinc displaces copper from aqueous solution of copper sulphate but it does not displace magnesium from the corresponding sulphate solution. Given the relative order for the standard redox potentials :

$$E^0\,(Zn^{2+}, Zn); \quad E^0\,(Cu^{2+}, Cu) \text{ and } E^0\,(MG^{2+}, Mg)$$

Give reasons for your answer.

4. Given
$$E^0(Cu^{2+}, Cu) = 0.337 \text{ V}$$
$$E^0(Cu^{+}, Cu) = 0.521 \text{ V}$$

Calculate the equilibrium constant for this reaction

$$2Cu^{+} \Leftrightarrow Cu^{2+} + Cu$$

5. For the cell

$$Zn \mid Zn^{2+} (a = 0.0001) \mid\mid Mg^{2+} (a = 0.001) \mid Mg$$

The standard reduction potential for the Zn and the Mg electrodes are –0.764 and –2.364 volt respectively. Now

(a) Write the half cell reaction.
(b) Write the overall cell reaction.
(c) Calculate E° of the cell.
(d) Calculate the free energy change of the cell reaction.
(e) Is this reaction spontaneous?

6. The e.m.f. of the cell

$$Pt, H_2 \text{ (1 atm)} \mid HCl \text{ (a=1)} \mid AgCl_{(s)} \mid Ag$$

at 25°C is 0.2224 V. Write the spontaneous cell reaction. If the temperature coefficient is -6.45×10^{-4} V/deg, calculate the enthalpy change for the cell reaction :

(F = 96,493 coulomb ; 1 cal = 4.184 Joules)

7. (i) Write the spontaneous reaction occuring in the galvanic cell

$$Zn \mid Zn^{2+} \mid\mid Cu^{2+} \mid Cu$$

(ii) From the following standard electrode potentials.

$$Zn^{2+} + 2e^{-} = Zn, \ E^{0} = -0.763 \text{ V Vs SHE}$$

$$Cu^{2+} + 2e^{-} = Cu, \ E^{0} = +0.33 \text{ V Vs SHE}$$

❑❑❑

7

CORROSION AND ITS CONTROL

"What lies behind us and what lies before us are small matters compared to what lies within us."

Ralph Waldo Emerson

CHAPTER OUTLINE

7.1 CORROSION

Corrosion can be defined as degradation or deterioration of a metal by chemical or electrochemical reaction with its environment (gaseous or liquid medium).

Degradation or deterioration means reduction in the useful properties of the material which include:

- Weakening of the material due to loss of crossectional area.
- Loss of properties such as malleability, ductility.
- Decaying of surfaces of metals.
- Cracking of a polymer due to sunlight.

The materials affected by corrosion are metals, polymers (plastics or rubbers etc.), ceramics (concrete) etc. But metals, the most useful class of materials corrode most frequently and hence this discussion is mainly devoted to corrosion in metals.

Some of the common examples of corrosion in metals are:

Rusting of Iron: When exposed to the atmospheric conditions, a layer of reddish scale and powder of Fe_3O_4 is formed on the surface.

Formation of green film on the surface of copper: A green layer of basic carbonate consisting of $CuCO_3 + Cu(OH)_2$ is formed on the surface of copper when exposed to moist air

Tarnishing of silver: When silver is exposed to the atmosphere, a black coating is formed.

7.2 WHY METALS CORRODE

All metals except gold, platinum and palladium corrode spontaneously when exposed to atmosphere, which contains 21% of oxygen. Metals, with the exception of gold, platinum and palladium are found in nature in combined forms as their oxides, carbonates, sulphides, chlorides or silicates. The metals in the combined form are known as minerals. The natural combined form of the metal i.e., mineral is more stable than the metal. During the extraction of the metal from the mineral, a considerable amount of energy is required. As energy is consumed in the extraction process, it implies that metal has a higher energy state than the mineral or compound of the metal. Thus, the metal which is in higher energy state (unstable) tends to revert back to its stable natural combined form which is of low energy, when it comes in contact with the elements in the environment. *In other words, corrosion process is the return of the refined material to its natural state.*

Greater is the amount of energy required to separate the metal from its mineral or ore, greater will be the tendency of the metal to revert back to its natural state resulting in degradation of the metal.

Metallic Ore (Mineral or metal in combined from) [or Stable State]	Extration of Metal → ← Corrosion (Reaction with atmosphere)	Pure Metal [Unstable State]

7.3 CONSEQUENCES OF CORROSION

Corrosion has a number of economic and social consequences which include:

(i) Enormous waste of machineries and different types of metallic materials.

(ii) It leads to unpredictable machinery failure which may lead into dangerous situations such as loss of life.

(iii) It leads to decrease in efficiency of machine and frequent replacement of corroded equipment, which is an expensive proposition.

(iv) It may cause leakage of inflammable gas from the corroded pipe line resulting into fire hazards and leakage of toxic product.

(v) It causes contamination of potable water.

(vi) It has been estimated that 25% of annual world production of iron is wasted due to corrosion.

From the above discussion it becomes clear that study of corrosion and its prevention is essential to both the engineers and metallurgists. For this, an understanding of the mechanism of corrosion is imperative.

7.4 THEORIES OF CORROSION

A number of different views have been put forward to explain as to how corrosion takes place. Some say, corrosion is oxidation or electrochemical phenomena while others say, it is a chemical attack. Still others have opinion that it is electrical phenomena. Each is partially true. The different theories of corrosions are discussed below.

7.4.1 Chemical or Dry Corrosion

Chemical or dry corrosion is the simplest case of corrosion, where corrosion takes place by direct chemical attack. This type of corrosion occurs mainly when metal surface come in direct contact with the atmospheric gases such as oxygen, halogens, hydrogen sulphide, nitrogen etc. or with anhydrous inorganic liquid. The corrosion product formed can be insoluble, soluble or liquid product.

Insoluble corrosion product implies that a solid film of the corrosion product is formed on the surface of the metal, which goes on thickening as corrosion proceeds. After sometime this layer protects the metal from further corrosion because rate of penetration of the corroding environment decreases and fresh metal is not exposed to it.

When the corrosion product is soluble or liquid, it is removed from the surface as soon as it is formed, exposing fresh metal to environment. This leads to excessive corrosion.

Dry or chemical corrosion is generally of three main types:

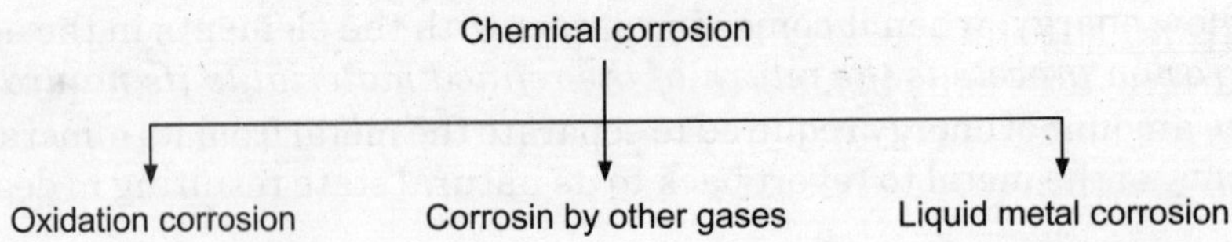

7.4.1.1 Oxidation Corrosion

Oxidative corrosion takes place by direct action of oxygen on metals, at low or high temperature, forming oxides of the metal. It generally takes place in absence of moisture. The reaction of a metal with air or pure oxygen to form the corresponding oxide can take place spontaneously in case of alkali and alkaline earth metals even at ordinary temperatures. However, most of the metals react with oxygen only at high temperatures. Almost all the metals except Ag, Au, and Pt, get oxidized at higher temperatures.

Mechanisms of Dry Corrosion

The growth of the oxide film on metal takes place at the surface as:

- The oxygen is adsorbed at the surface of the metal by physical adsorption.
- When the temperature rises the physical adsorption turns into chemisorption, wherein metal gets oxidized to metal ions ($M \rightarrow M^{n+} + ne$)
- The electrons are taken up by oxygen which gets reduced to oxide ion ($nO_2 + 2ne^- \rightarrow 2nO^2$)
- The M^{n+} O^{2-} combine to form metal oxide (M_2O_n)
- The reactions can be written as

$$2M \longrightarrow 2M^{n+} + 2\,ne^- \quad \text{(Oxidation of metal)}$$
$$n\,O_2 + 2ne^- \longrightarrow 2n\,O_2 \quad \text{(Reduction of oxygen)}$$

$$\underset{\text{Metal}}{2M} + \underset{\text{Oxygen}}{O_2} \longrightarrow \underset{\text{Metal ion}}{2M^{n+}} + \underset{\text{Oxide ion}}{nO^{2-}}$$

$$\text{Metal Oxide } (M_2O_n)$$

- The metal oxide scale so formed forms a barrier between the metal and the oxygen and restricts further oxidation or corrosion of the metal. For oxidation to continue

— either metal must diffuse outwards through the scale to the surface.
— or oxygen must diffuse through the scale to the underlying metal.

- As the metal cation is much smaller in size, the outward diffusion of metal ion is much more rapid than inward diffusion of oxide ion. Hence corrosion continues by rapid diffusion of highly mobile cations. (Fig. 7.1)

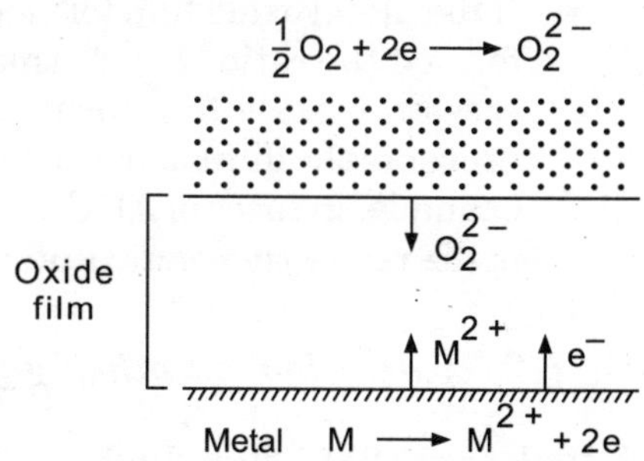

Fig.7.1 Mechanism of Growth of Oxide film

Nature of the oxide film formed plays an important part in corrosion process. If the thickness of the oxide layer is less than 300 Å then such layer is called a **film** and if the thickness of the film exceeds 300 Å than it is known as **scale**. The nature of the film decides the continuation of corrosion process as discussed below:

(i) **Stable film:** The stable oxide film is fine grained in structure and adheres tightly to the metal surface. This can be impervious in nature, thereby preventing the penetration of oxygen to the underlying metal. In otherwords, such a film forms a protective coating over the metal and prevents further oxidation or corrosion. The oxide films on Al, Cr, Cu, Pb, Sn etc. are stable and adhere to the surface. Therefore these metals are employed for alloy formation with other reactive metal to protect them from corrosion.

(ii) *Unstable film:* The oxide film formed on the surface of metal decomposes back into metal and oxygen and hence the film is called unstable film.

$$\text{Metal Oxide} \rightleftharpoons \text{Metal} + \text{Oxygen}$$

Hence in such cases corrosion is not possible. Unstable film is formed by the oxides of Ag, Au and Pt and hence these do not undergo corrosion.

(iii) *Volatile film:* The metal-oxide film volatizes from the surface as soon as it is formed. Hence the underlying metal is exposed to the further attack by oxygen leading to rapid and continuous corrosion. For example, molybdenum oxide (MoO_3) is volatile and hence molybdenum undergoes excessive corrosion.

(iv) *Porous film:* Sometimes the oxide film so formed is porous or have cracks. Hence the diffusion of oxygen takes place through these pores leading to further oxidation or corrosion. The metal oxides of metals like alkali and alkaline earth metals fall under this category.

Pilling-Bedworth Rule

The oxide layer can be protective or non-protective depending upon the ratio of volume of the metal oxide to the volume of the metal consumed. This is known as Pilling-Bedworth ratio.

$$\text{Pilling} - \text{Bedworth ratio} = \frac{\text{Volume of the metal oxide formed}}{\text{Volume of the metal consumed}}$$

According to Pilling-Bedworth's rule,

- If the volume of the oxide film formed is less than the volume of the respective metal from which it is formed, the ratio is less than unity then the oxide film will be non-protective in nature. In such cases the oxide film formed over the surfaces of the metals is porous in nature and oxygen can diffuse through these pores to bring about further attack on the metal. For example, in case of alkali and alkaline earth metals such as Na, Ca, Mg (except

Al) oxide film produced has smaller volume than that of the metal from which it is formed. Consequently, the oxide film develops cracks and pores in its structure.

- If the metal oxide film formed is equal or greater than the respective metal from which it is formed the ratio is more than unity, the metal oxide film is protective in nature. In such cases the oxide film formed is continuous, adhering and non-porous. This type of film prevents the diffusion of oxygen and hence the rate of corrosion by oxidation decreases. For example, in cases of Al, Cr, and W etc., the volume of oxide layer is greater than the volume of the respective metal and hence is protective in nature.

7.4.1.2 *Corrosion by other gases*

In addition to oxygen, a number of other gases such as CO_2, SO_2, Cl_2, H_2S, F_2 etc. also have corrosive effect on metals. The extent of corrosion depends on the chemical affinity between the metal and the gas involved. In this case also the film so formed may be protective or non-protective in nature.

- If the film formed is protective in nature, the extent of attack or corrosion decreases. For example, chlorine gas attacks silver metal surface to form non-porous protective film which prevents further corrosion.
- If the film formed is non-protective in nature then it destroys the whole metal eventually. For example, chlorine attacks Sn to form $SnCl_4$ film which is volatile and hence leaves the metal surface exposing the fresh metal for further attack.
- H_2S attacks steel forming FeS layer which is porous in nature.

7.4.1.3 *Liquid-metal corrosion*

Sometimes the corrosion also results due to the action of flowing liquid metal at high temperatures on solid metals or alloys. Liquid sodium is used as heat transfer media in nuclear reactor. It is shown to cause corrosion in carrier tubes. The corrosion reaction involves either dissolution of a solid metal by a liquid metal or by penetration of the liquid metal into the solid metal causing corrosion in solid metal.

7.4.2 Electrochemical Corrosion or Wet Corrosion

The corrosion of metals in aqueous medium seems to be the electrochemical in nature. This type of corrosion is also known as immersed corrosion and generally takes place when:

(i) a metal is in contact with the moist air or any liquid medium or
(ii) when two different metals or an alloy is in contact with a solution.

The most familiar example of electrochemical corrosion is the *formation of rust on iron*. We all know that iron pipes corrode in wet soil or moist atmosphere. In this type of corrosion the metal is oxidized to a soluble form or ionic state.

Before the corrosive process takes place, the following conditions must be satisfied for the formation of electrochemical cell:

(i) There must be a cathode and an anode.
(ii) There must be an electrical potential between the anode and cathode.
(iii) There must be a metallic path electrically connecting the anode and cathode.
(iv) The anode and cathode must be immersed in an electrically conductive medium.

Once these conditions are satisfied a "corrosion cell" is formed. The corrosion takes place as discussed below:

(a) **Anodic reactions:** At anodic area, oxidation reaction takes place where metal atoms lose their electrons to the environment and pass into the solution.

$$\underset{\text{(Metal)}}{M} \longrightarrow \underset{\text{(Metal ion)}}{M^{n+}} + ne^-$$

Thus metal is destroyed by dissolving or changing into metallic ions. *Hence, corrosion takes place at anodic areas or metal, which forms the anode, is corroded.*

(b) **Cathodic reactions:** The electrons released in the anodic reactions are consumed in the reactions at the cathode or cathodic reactions. Depending upon the nature of the corrosive environment, cathodic reactions may result in:

(i) *Evolution of hydrogen:* In absence of oxygen and in acidic environment evolution of hydrogen takes place. The H^+ ions of acidic solution take up electrons released at anode and are eliminated as hydrogen gas

$$2H^+ + 2e^- \longrightarrow H_2 \uparrow \quad \text{(in acidic medium)}$$

In alkaline or neutral solution in absence of oxygen, the reaction will be:

$$2H_2O + 2e^- \longrightarrow H_2 + 2OH^- \quad \text{(in neutral or alkaline medium)}$$

Example: Rusting of iron in acidic medium in absence of oxygen takes place as follows,

$$Fe \longrightarrow Fe^{2+} + 2e^- \quad \text{(Anode reaction)}$$

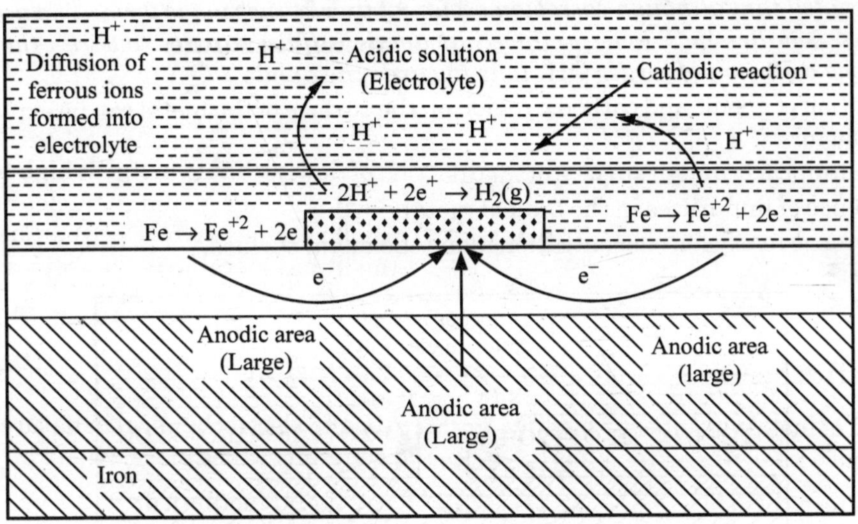

Fig. 7.2 Mechanism of wet Corrosion by Hydrogen Evolution

The electrons that are released, flow through the metal surface (electronic conductor) from anode to cathode. The H^+ ions of **acidic** solution take up these electrons and are eliminated as hydrogen gas.

$$2H^+ + 2e^- \longrightarrow H_2 \uparrow \quad \text{(Cathodic reaction)}$$

The overall reaction is: $Fe + 2H^+ \longrightarrow Fe^{2+} + H_2 \uparrow$

Hence, in this type of corrosion hydrogen ions are displaced from the acidic solution by metal ions.

(ii) *Absorption of oxygen:* This type of corrosion reaction takes place in the presence of atmospheric oxygen. In neutral aqueous solution or weakly alkaline solution, containing oxygen the reaction, $2H^+ + 2e^- \longrightarrow H_2$, does not take place and oxygen gets reduced as, $2H_2O + O_2 + 4e^- \longrightarrow 4\bar{O}H$

Example: Rusting of iron in neutral aqueous solution in presence of oxygen can be explained, with the following reactions.

$$Fe \longrightarrow Fe^{2+} + 2e^- \quad \text{(Anode reaction)}$$

The electrons travel to cathode through the metal surface and react with oxygen,

$$\frac{1}{2}O_2 + H_2O + 2e^- \longrightarrow 2OH^- \quad \text{(Cathodic reaction)}$$

The Fe^{2+} produced at anode travel to the cathodic region through the moisture and react with OH^- ions produced at cathode as follows:

$$Fe^{2+} + 2OH^- \longrightarrow Fe(OH)_2 \downarrow$$

The $Fe(OH)_2$ produced is oxidized as,

$$Fe(OH)_2 + O_2 + 2H_2O \longrightarrow 4Fe(OH)_3 \text{ or } 2Fe_2O_3 \cdot 3H_2O \text{ (Rust)}$$

$Fe(OH)_3$ corresponds to rust, which is hydrated iron (III) oxide of variable composition, Fe_2O_3, nH_2O. The degree of hydration of the iron (III) oxide affects the colour of the rust which may vary from black (Fe_2O_4) to yellow or reddish brown ($Fe_2O_3 \cdot H_2O$).

It may be noted that although corrosion occurs at anode, the rust is formed near the cathode. This is because smaller Fe^{2+} *ions produced at anode diffuse rapidly towards the cathode as compared to the diffusion of larger* OH^- *ions towards the anode.*

The formation of anodic and cathodic region on the surface of the metal can be explained due to the following factors:

(i) Non-uniform or rough surface of metal.
(ii) Presence of impurities in the bulk of metal
(iii) Difference in oxygen concentration along the metal surface.

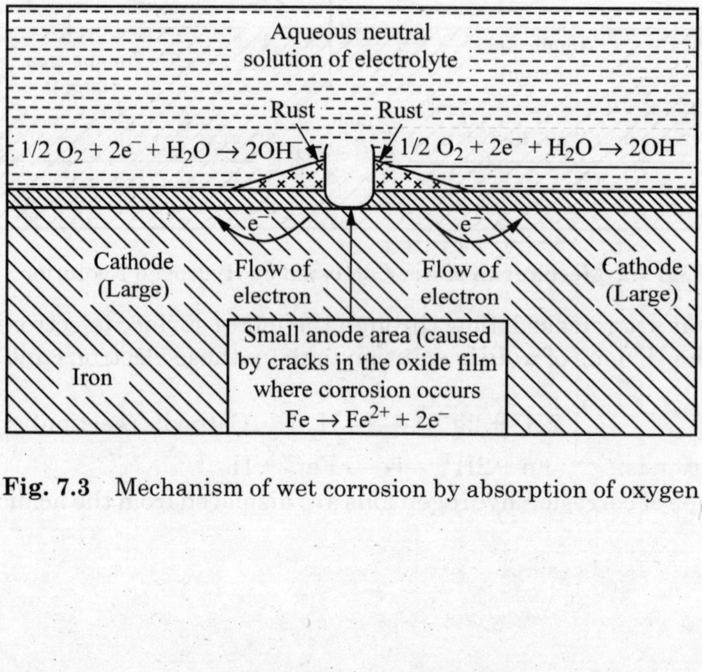

Fig. 7.3 Mechanism of wet corrosion by absorption of oxygen

Difference between Chemical Corrosion and Electrochemical Corrosion

Chemical Corrosion	*Electrochemical Corrosion*
1. It takes place in dry conditions	It takes place in presence of wet conditions such as presence of aqueous solution or electrolytes.
2. It takes place by direct chemical attack of environment on the metal	It takes place through the formation of a series of galvanic cells.
3. It can takes place on homogeneous and heterogeneous surface	It takes place on heterogeneous metal surfaces only.
4. Corrosion product accumulates at the same spot where corrosion has occured	Corrosion product accumulates at the cathode whereas corrosion occurs at the anode.
5. Uniform corrosion takes place	Non uniform corrosion takes place.

7.5 TYPES OF CORROSION

There are a number of methods of classifying corrosion but no universally accepted terminology is in use. In broad sense, corrosion can be classified into following types:

7.5.1 Galvanic Corrosion

Galvanic corrosion is caused when two dissimilar metals are electrically connected with each other and are exposed to an electrolyte. The two metallic parts constitute "a galvanic cell" in which the metal with the lower electrochemical potential (i.e., the more active metal) will get corroded. For example, if zinc and copper are in contact with each other in presence of an electrolyte then zinc, which is higher in electrochemical series, (with lower electrochemical potential) forms the anode and gets dissolved (corroded), while copper which is lower in electrochemical series (with higher electrochemical potential) acts as cathode and is protected.

The use of copper pipes in conjunction with iron pipes in water distribution systems has been the cause of galvanic corrosion. The iron which acts as anode corrodes, whereas the more noble Cu acts as cathode and is protected.

Larger the potential difference between the two metals, greater will be the galvanic corrosion. Galvanic corrosion is also known as differential metal corrosion (Fig. 7.4).

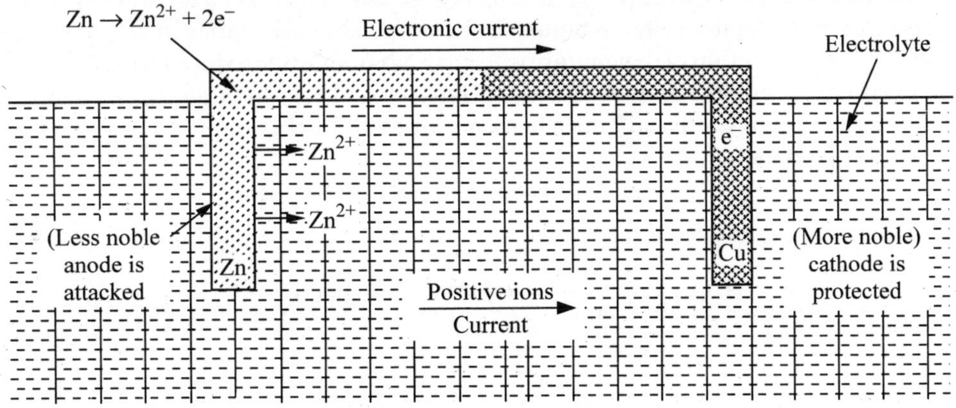

Fig. 7.4 Galvanic cell corrosion

7.5.2 Concentration Cell Corrosion

Concentration cell corrosion takes place when two or more areas of a metal surface are in contact with different concentration of the same electrolyte or are exposed to an electrolyte of varying aeration.

Differential Oxygen Concentration Cell

This is the common type of concentration cell corrosion where corrosion takes place due to "differential aeration." Differential aeration occurs when a metallic surface is partially immersed in an electrolyte and the other part of the metal is exposed to the air i.e., the one part of metal is exposed to a different air concentration than the other. The electrode potential of a metal will be different at these two locations. This is a direct consequence of Nernst law. In such circumstances, those areas which are exposed to greater amount of air or oxygen become cathodic while the areas which are little exposed or not exposed to air or oxygen become anodic. The anodic part suffers the corrosion.

For example, if zinc metal is partially immersed in a dilute solution of NaCl and the solution is not stirred, the parts above and closely adjacent to the waterline are well aerated and act as cathode (Fig. 7.5). Areas deep inside the solution are anodic, since the concentration of oxygen is smaller. Thus a potential is developed which causes a flow of current between two areas of the same metal. Zn dissolve at the anodic area,

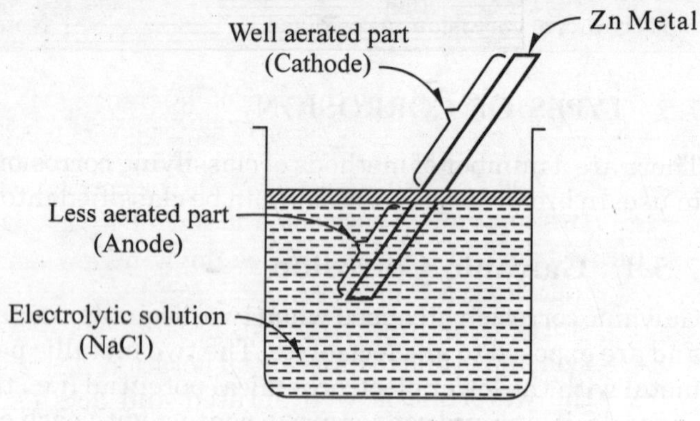

Fig. 7.5 Corrosion by differential Aeration.

$$Zn \longrightarrow Zn^{2+} + 2e^- \quad \text{(At Anode)}$$

At the cathodic area oxygen takes up electrons to form hydroxyl ions:

$$\frac{1}{2}O_2 + H_2O + 2e^- \longrightarrow OH^- \quad \text{(At Cathode)}$$

Other examples of corrosion by differential aeration are:

(i) **Drop corrosion:** It takes place when a drop of on electrolyte solution is in contact with the metal surface. For example, if a drop of electrolyte is placed on an iron surface, rusting occurs in the central region beneath the drop. This is because at the periphery of the drop, the concentration of oxygen/air is higher as compared to the center of the drop. Thus the peripheral region becomes cathodic. The central region beneath the drop has lower concentration of oxygen and hence becomes anodic. At this region, Fe is oxidized to Fe^{2+} i.e., gets corroded.

$$Fe \longrightarrow Fe^{2+} + 2e^- \quad \text{(At anode)}$$

At cathode O_2 is reduced to $\overline{O}H$ ions.

$$\frac{1}{2}O_2 + H_2O + 2e^- \longrightarrow OH^- \quad \text{(At cathode)}$$

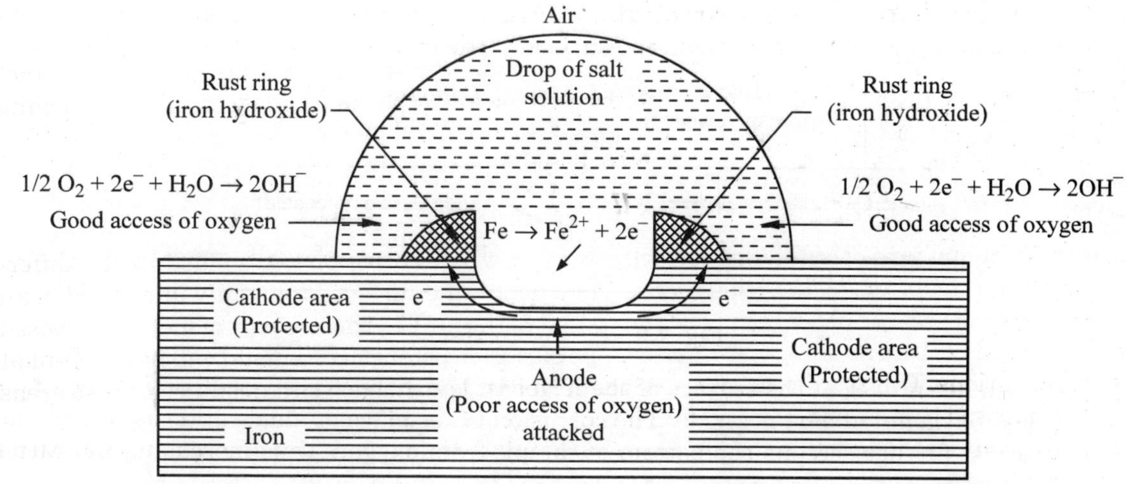

Fig. 7.6 Mechanism of differential aeration caused by the presence of drop of salt solution on the iron surface.

The Fe^{2+} ions move towards the cathode and OH^- ions move towards the anode. These two combine in between and form $Fe(OH)_2$ which subsequently is oxidized to rust.

(ii) **Waterline corrosion:** You might have observed rusting along a line just beneath the level of the water stored in an iron tank. This is called waterline corrosion and is due to differential aeration. The area above the waterline is well aerated and acts as cathode whereas the surface below it acts as anode (less oxygenated/aerated) and hence gets corroded. The reactions at the cathode and anode are same as discussed in drop corrosion.

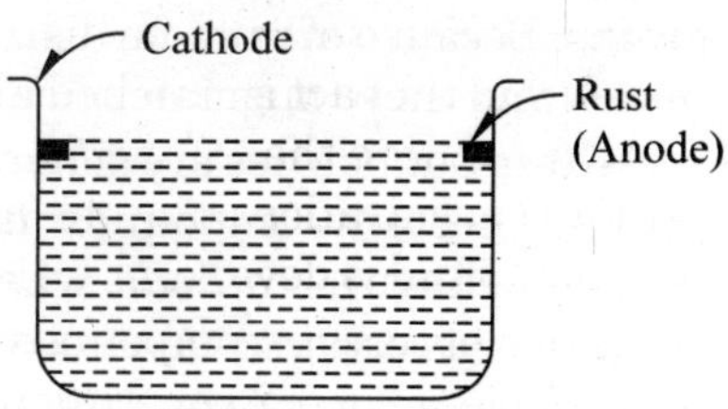

Fig. 7.7 Waterline corrosion.

7.5.3 Crevice Corrosion

Crevice or contact corrosion is the corrosion produced at the region of contact of metals with metals or non-metals. For example, it occurs at the crevice between different metallic objects such as bolts, rivets, washers, nuts etc. When these crevices come in contact with the liquids, these become the anodic region because oxygen supply to these areas is restricted compared to exposed area which becomes the cathode. The corrosion occurs at the crevices.

This type of corrosion is accelerated by the deposition of dirt, debris etc. These materials restrict the supply of oxygen underneath the covered portion which become anodic and hence suffers corrosion. Crevice corrosion can also be attributed to concentration cell effect.

7.5.4 Pitting Corrosion

Pitting corrosion is a localized corrosion resulting in the formation of pits or holes. Pitting usually takes place due is cracking of the protective film on the surface of the metal. An anode is formed where the film has broken, while the unbroken film (or protective layer) acts as a cathode. The formation of small anodic and large ca-

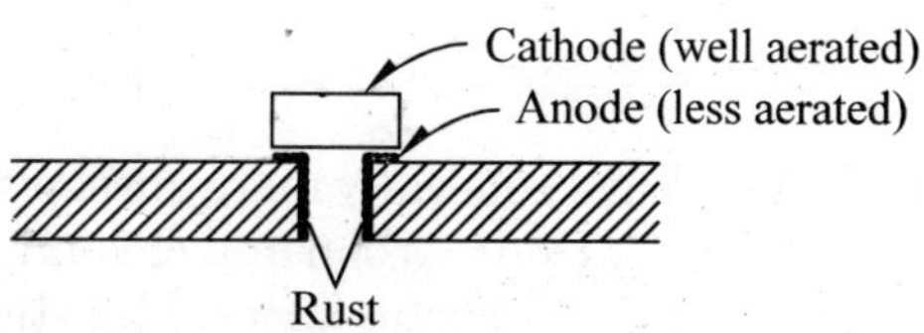

Fig. 7.8 Crevice Corosion

thodic areas results in severe corrosion of the anode region. Because of corrosion of smaller anodic region, small holes or pits are formed on the surface of the metal.

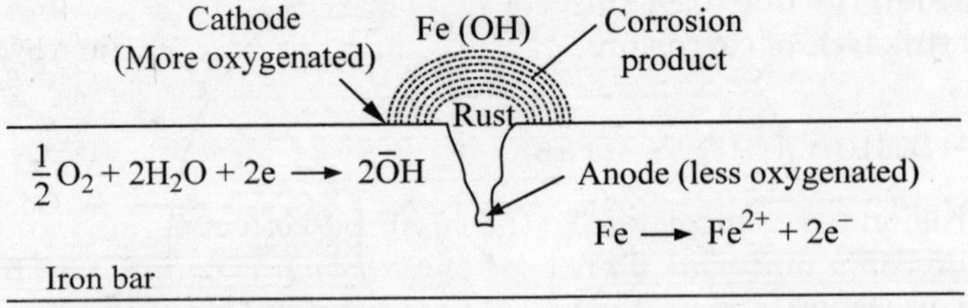

Fig. 7.9 Pitting Corrosion.

As shown in Fig. 7.9, at the surface of the iron bar, the more oxygenated portion is cathode whereas less oxygenated part is anode. The corrosion takes place at anode and rust is accumulated. The rust further screens the bottom of the pit from oxygen, thus increasing the rate of corrosion.

7.5.5 Intergranular Corrosion

When a molten metal is solidified, the solidification starts at number of randomly distributed nuclei within the molten mass. Each of them grow in a regular pattern or atomic array to form grains. Because of random distributions of the nuclei, the planes of atom in adjacent grains do not match and these mismatched area between adjacent grains are known as grain boundaries.

Grain boundaries are generally more prone to corrosive attack. This may be due is the precipitation of certain compounds at grain boundaries. The precipitated compound as well as grain centers behave as *cathode*, whereas grain boundaries behave as *anode*. This is because, the precipitation of certain compounds at the grain boundaries leaves the solid metal solution adjacent to them depleted in one constituent. Thus, a potential difference is created leading to the *corrosion at anodic grain boundaries*.

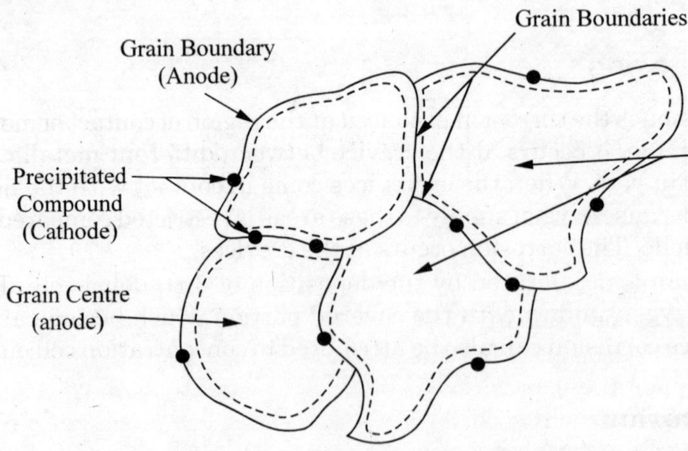

Fig. 7.10 Intergranular Corrosion

Alloys are generally more susceptible to intergranular corrosion. For example, stainless steel is an alloy of iron, carbon and chromium. During its welding the chromium carbide gets precipitated at the grain boundaries. Thus the chromium composition is *less in the region adjacent to the grain boundaries* as compared to the grain center and the precipitate. The chromium depleted region becomes more anodic and suffers corrosion. The corrosion proceeds along the grain boundary and continues until the affected grain is completely dislodged. Thus type of corrosion cause sudden failure of the material due to loss of cohesion between the grains. Aluminium copper alloys are also attacked by this type of corrosion.

7.5.6 Stress Corrosion (Stress Cracking)

The term stress corrosion takes place due to the combined effect of mechanical stresses and the corrosive environment on a material. This type of corrosion is common in fabricated articles of some alloys such as zinc brasses or nickel brasses. During their fabrication, these materials are subjected to different types of uneven stresses such as rolling, drawing or insufficient annealing. In such cases, *the metal under stress become more anodic* as compared to other areas which act as cathodes. In presence of corrosive environment the anode areas suffer corrosion. These areas become very reactive chemically and may be attacked even by traces of corrosive environment, resulting in the formation of cracks.

Stress corrosion (cracking) is usually unpredictable in nature. Even after a few years of trouble free service, a metal can suddenly crack without any warning or previous deformation.

It may be noted that stress corrosion is due to the combined effect of static tensile stresses and the corrosive environment. But every type of stress cannot cause corrosion. The stress should be of sufficient magnitude to produce cracking. Moreover, stress cracking takes place only in highly selective environment i.e., different environments are required for different alloys for stress cracking. For example, corrosion cracking of brass takes place only in presence of traces of ammonia or amines.

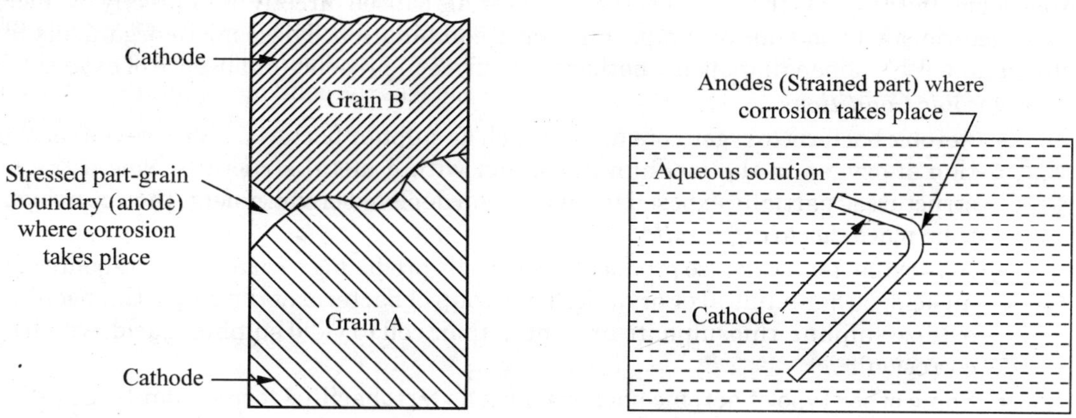

Fig. 7.11 Stress Corrosion

Some examples of stress corrosion are:

(a) *Season cracking:* Season cracking due to stress corrosion is commonly observed in cold-drawn brass (copper alloy) articles. Pure copper is immune to stress corrosion but presence of alloying agents such as Zn, Al, P, Sb etc., make it prone to stress corrosion. In the presence of atmosphere containing traces of ammonia or amines, brass show intergranular cracking. Both copper and zinc present in brass can form stable complex ions, $[Cu(NH_3)_4]^{2+}$ and $[Zn(NH_3)_4]^{2+}$ in ammonia solution and this causes dissolution of brass leading to formation of fissures. Fissures propagates and cause cracking in the presence of high tensile stress.

(b) *Caustic embrittlement of mild steel:* When mild steel is exposed to alkaline solutions at high temperatures and stresses, it causes failure of the material due to caustic embrittlement. This type of corrosion mainly occurs in boilers or heat transfer equipment. Boiler water contains Na_2CO_3 (added for softening purposes) which undergoes decomposition under high pressures

$$Na_2CO_3 + H_2O \longrightarrow 2NaOH + CO_2$$

The alkaline water flows into the hair cracks present at the inner surface of the boiler due to capillary action. Eventually the concentration of NaOH in the cracks/crevices increases. The boiler material is thus under the two different concentration of the electrolyte, higher inside the crack and dilute outside.

Thus a concentration cell sets in.

$$\underset{\text{Cathode}}{\text{Fe | Conc. NaOH}} \;||\; \underset{\text{Anode}}{\text{Dil. NaOH | Fe}}$$

The iron of the main body containing dil. NaOH behaves as cathode whereas the iron of the cracks and crevices surrounded by conc. NaOH behaves as anode. The anodic iron is corroded by the action of NaOH forming sodium ferroate. The sodium ferroate so formed decomposes as,

$$2Na_2FeO_2 + 4H_2O \longrightarrow 6NaOH + Fe_3O_4 + H_2$$

Fe_3O_4 is precipitated and NaOH is regenerated causing further corrosion.

7.5.7 Microbiological Corrosion

Microbiological corrosion is the deterioration of materials caused directly or indirectly by microbes such as bacteria, algae, moulds or fungi, singly or in combination. These micro-organisms can be aerobic or anaerobic depending upon whether they thrive in presence or absence of oxygen.

(i) *Aerobic Conditions*

- Aerobic bacteria (oxygen consuming) such as sulphur bacteria, decrease the concentration of oxygen in the medium in contact with a metal structure. Now if the metal structure is partially exposed to this oxygen depleted environment, then corrosion will take place *due to differential aeration.*
- Aerobic bacteria can also produce corrosion by producing a corrosive metabolite usually an acid, either mineral or organic. The group of bacteria of the genus thiobacillus can oxidize sulphur, thiosulphate present in their cell to yield sulphuric acid, which cause corrosion.
- Some other type of aerobic bacteria such as ferrobacillus ferrooxidants may directly oxidize iron into iron oxide or hydroxides and cause the formation of precipitates of ferric *hydroxides.*

(ii) *Anaerobic conditions*

- Anaerobic bacteria such as sulphate reducing bacteria (desulfovibrio desulfuricans) cause corrosion of iron and steel. The sulphate reducing bacteria will not grow even in traces of oxygen. Various theories have been put forward to explain their action but no uniform theory seems to apply. The main products of the corrosion are iron sulphide and ferrous hydroxide. Anaerobic bacterial corrosion (due to sulphate reducing bacteria) is recognizable on a freshly exposed metal surface by a black corrosion product with the strong rotten-egg smell of hydrogen sulphide. On cast iron, due to corrosion by anerobic bacteria, graphitization occurs forming iron sulphide i.e the surface layer takes the appearance of graphite and can be easily cut with a knife.

The anaerobic bacteria provide the reducible chemicals that cause corrosion. The corrosion that was found on the hull of the Titanic occurred due to anerobic bacteria.

7.5.8 Erosion Corrosion

Erosion corrosion is the result of relative movement between a corrosive fluid and metal surface. Generally due to the high velocity of the turbulent motion of the fluid, the mechanical rubbing of the solid metal takes place thereby removing the protective coating at the point of impingement. This results in the development of differential cells at these points leading to corrosion. Metals that are soft and are readily damaged due to mechanical abrasion are quite susceptible to this type of corrosion.

Many types of corrosive medium moving with high velocities can cause erosion corrosion. These can be gases, aqueous solutions, liquid metals etc. For example, hot gases may oxidize a metal and then at high velocity blow off an protective scale around the metal.

All types of equipment exposed to moving fluids are subjected to erosion corrosion. Examples include, ship propellers, hydraulic turbines, pump impellers and diesel engine cylinder.

7.5.9 Soil Corrosion

The two factors which promote corrosion viz: moisture and dissolved electrolytes are present in soil, making it corrosive. Moreover, presence of micro-organisms in the soil further leads to corrosion of the underground structures and pipelines for oil gas and water. Thus soil corrosion of the structures embedded in soil is due to the combined effect of the presence of

(i) Acidity in the soil
(ii) Moisture and electrolytes
(iii) Micro-organisms
(iv) Differential aeration.

According to the types of soil the corrosion may be galvanic or micro-biological.

Buried pipelines passing from one type of soil to another, (say from highly porous soil to less porous soil), get corroded due to *differential aeration.*

Microbiological corrosion of the buried structures may take place in water-logged soils which give the environment in which bacteria can grow.

7.5.10 Selective Leaching/Corrosion

Selective leaching is the removal of one element from a solid alloy by corrosion process. Dezincification i.e., removal of zinc from zinc copper alloy (brass) is a common example of this type of corrosion. Similar processes occur in other alloy systems in which aluminium, iron, cobalt, chromium and other elements are removed from their alloys.

Dezincification is prevalent in condenser tubes made of brass, where sea water is used as condenser water. Although both zinc and copper go into the solution but only zinc ions remain in solution and copper is redposited as a spongy layer. This is because zinc is quite reactive whereas copper is nobler. Hence dezincification proceeds continuously. The dezincification results into a porous metal with poor mechanical properties.

Similarly, gray cast iron and also shows the effects of selective leaching in mild environments. The cast iron appears to become graphitized. Actually, graphite is present in the gray iron. After selective leaching of the iron, graphite network is left behind. Graphitization does not occur in malleable cast irons because graphite network is not present to hold together the residue. Similarly white iron has no free carbon and is not subjected to graphitization.

7.5.11 Hydrogen Damage

Hydrogen damage is a general term which refers to mechanical damage of a metal caused by the presence of or interaction with hydrogen. Hydrogen damage can be classified into the following types:

(a) *Hydrogen blistering:* It is caused by penetration of hydrogen into metal. Surface bulges appear on the surface of the metal due to hydrogen blistering.

(b) *Hydrogen embrittlement:* Hydrogen embrittlement is also caused by penetration of hydrogen into a metal leading to loss of ductility and tensile strength.

(c) *Decarburization:* Moist hydrogen at high temperatures may cause removal of carbon from steel. This is called decarburization. It lowers the tensile strength of steel.

(d) *Hydrogen attack:* It refers to the interaction between hydrogen and a component of an alloy at high temperature. For example the oxygen-containing copper gets disintegrated in presence of hydrogen.

7.5.12 Stray Current Corrosion

Electric current from electric system can leak into adjacent conducting structures. The metal structure such as water pipes, gas pipes etc. thus get corroded because these become anodic due to leakage from the electric circuits.

7.6 PASSIVITY

In an electrochemical series, metals are arranged from top to bottom in increasing order of their reduction electrode potentials. A metal high in the series is more anodic and undergoes corrosion faster than the metal below it. For example, Zn corrodes faster than Fe which inturn corrodes faster than Sn. This is because in electrochemical series, Zn is placed high above followed by Fe, which inturn is followed by Sn.

However, there are number of *exceptions*. For example, Ti is placed above Ag in electrochemical series, and is expected to corrode faster than Ag. But it is just the reverse i.e., Ti is corroded to a lesser extent than Ag. This can only be explained by passivity of the metals. *A metal is said to be passive in a certain environment if it exhibits much lower corrosion rate than what is expected from its position in the electrochemical series.*

However, on changing the environmental a passive metal can be rendered active.

Passivity of the metal is due to the formation of a highly protective, but very thin film on the surface of the metal under particular environmental conditions. The film so formed at the surface makes the metal or the alloy, act like a noble metal.

For example, metals like chromium, aluminium, nickel, titanium and a wide variety of stainless steel alloys containing chromium, exhibit outstanding corrosion resistance in *oxidizing* environments. These metals and alloys form insoluble oxide films on their surface in presence of oxygen and hence are rendered passive. (Some other metals such as zinc, cadmium, tin, uranium and thorium have also been observed to exhibit passivity effects under limited conditions.)

On changing the environment i.e., *in reducing environment*, all these metals become *chemically active* and undergo rapid corrosion. Fig. 7.12 illustrates the effect of passivity on corrosion of a metal. Fig. 7.12(a) shows the behaviour of the normal metal i.e. metal which do not show possivity effect. For example, metal immersed in an air-free acid solution.

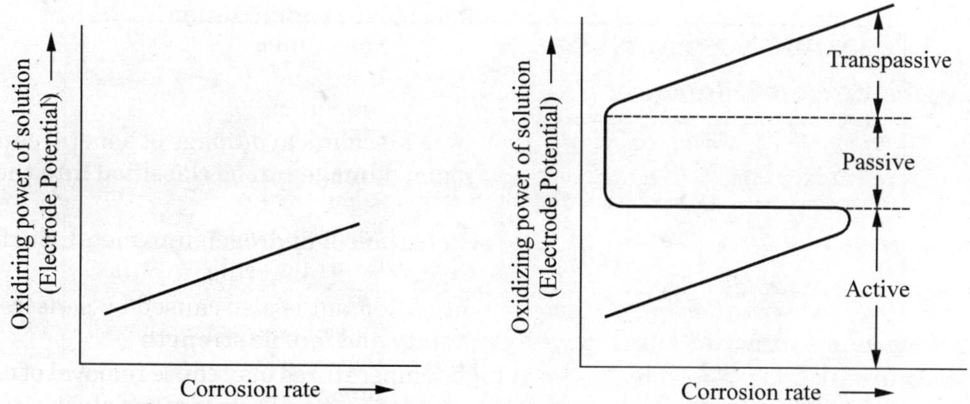

(*a*) Corrosion rate of the normal metal (*b*) Corrosion rate of an active-passive metal

Fig. 7.12

Fig. 7.12 (b) illustrates the typical behaviour of a metal which exhibits passive effect. Initially, the metals shows corrosion rate similar to normal metal. But once the protective oxide film is formed it is rendered passive and corrosion rate shows a sudden decrease. This is the beginning of the passive region. Further increase in oxidizing power produce little, if any, change in corrosion rate of the material. However, at very high concentrations of the oxidizers, the corrosion rate again increases. This region is termed as transpassive region.

Significance: Usually 10^3 to 10^6 times reduction in corrosion rate is observed on passivity. From an engineering standpoint, the metals most susceptible to this kind of behaviour are the common engineering and structural materials, including iron, nickel, silicon, chromium and alloys of these metals. Passivity has been used successfully to develop new methods of preventing corrosion and to predict corrosion resistance. Passive oxide films are artificially produced on certain metals and alloys by treatment in acidic or alkaline medium in presence of strong oxidizing agents. For example, the action of conc. HNO_3 on Al produce a thin protective oxide film, thereby making it passive. Metals which are susceptible to corrosion can also be made passive by alloying with one or more metals which are passive or resist corrosion. For example, iron is rendered passive when alloyed with any of the transition metals such as Cr, Ni and Mo.

Iron is heavily attacked by dilute HNO_3 but in conc. HNO_3, it forms a very thin protective passive film and behaves as a noble metal.

7.7 GALVANIC SERIES

As is discussed in the previous section that electrochemical series can provide the basis for predicting the corrosion of the metals. The metal lying at the top of the series is more anodic and undergoes corrosion faster than the metal below it in electrochemical series. However, due to the formation of protective oxide film at the surface of some metals (passivity), these metals behave exactly opposite to that predicted by the electrochemical series. Hence a more reliable series for predicting corrosion behaviour in different environments is the '**Galvanic Series**'. This series gives the reliable information about the corrosion behaviour of metals and alloys in different environments. The galvanic series for sea water is given in table:

Table 7.1 Galvanic Series in Sea Water

ANODIC/MOST ACTIVE	Magnesium
	Zinc
	Aluminium
	Cadmium
	Duralumin
	Steel
	Lead Tin Solder
	Lead
	Tin
	Copper and copper alloys
	Cupro-nickel
	Bronze
	Nickel (passive)
	Passive stainless steel
	Silver
	Titanium
	Graphite
	Gold
CATHODE OR NOBLE METALS	Platinum

7.8 FACTORS INFLUENCING CORROSION

Corrosion depends upon number of factors which include (i) nature of the metal and (ii) environmental factors

7.8.1 Nature of the Metal

A number of properties of metal affect corrosion rate.

(a) **Position in Galvanic Series:** Extent of corrosion depends upon the position of the metal in galvanic series. When two metals are in electrical contact in presence of an electrolyte, the metal higher up in galvanic series becomes anodic and suffers corrosion. Moreover, greater the difference in their positions in the electrochemical series, faster is the corrosion of the anodic metal.

(b) **Purity of the metal:** The rate of corrosion increases with the extent of impurities. Impurities accelerate the corrosion due to the formation of minute/tiny electrochemical cells because of 'heterogeneity' caused by them. The anodic part gets corroded.

For example, very pure and smooth zinc will not corrode in very pure hydrochloric acid. But zinc metal containing impurities of Pb or Fe corrodes rapidly. The figures below show the effect of impurities on corrosion of aluminum by hydrochloric acid.

% Aluminium	99.998	99.97	99.2
Relative Corrosion rate	1	1000	30,000

However pure metals are expensive and they are usually weak. Therefore, pure metals cannot be used in engineering applications where strength and cost are main considerations.

(c) **Physical state of the metal:** The physical state of the metal such as orientation of grains, grain structures, localized stresses, scratches etc., highly influence the rate of corrosion. For example grain boundaries are attacked more rapidly than the grain faces when exposed to a corrosive. Similarly, smaller the grain-size of the metal or alloy, greater will be its solubility as compared to the macroscopic crystal and hence, greater will be its corrosion.

Further the areas under stress, even in case of pure metals tend to be anodic and undergo corrosion.

(d) **Nature of the oxide film:** It was discussed earlier that in aerated atmosphere, almost all the metals get covered with an oxide film.

- If the specific volume of the metal oxide film formed is lesser than the volume of the metal atom, then the film will be porous. This porosity will easily allow oxygen to diffuse into the metal and cause further corrosion. For example alkaline earth metals such as Be, Mg, Ca etc., form oxides which have smaller volumes than the metals from which they are formed and hence are easily corroded even at low temperatures.
- On the other hand if the specific volume of the metal oxide film formed is less than that of the metal from which it is formed, it will not allow the oxygen to diffuse into it. Thus, this type of oxide film protect the metal from further oxidation or corrosion. For example, Al and heavy metals form the films whose specific volume is more than the metal itself and hence this film protects the metal from further corrosion.

The ratio of the volume of the metal oxide to the metal is known as specific volume ratio. If,

Specific volume ratio > 1, the metal will be protected from corrosion
Specific volume ratio < 1, the metal will corrode further.

(e) **Area Effect:** Another important factor in corrosion is the area effect i.e., relative areas of anode and cathode. If the area of cathode is large as compared to anode then the corrosion of the anode is rapid and severe. This is because, when the cathodic area is large, it will demand more electrons, which can only be met by brisk and rapid corrosion of anodic area, may be 100 or 1000 times greater than if the anodic and cathodic areas were equal in size. For example, in copper plates (cathode) with steel (anode) rivets, the steel rivets are completely corroded because the smaller steel rivets are coupled to large copper cathodic area.

(f) **Solubility of the products of corrosion:** If the corrosion product is soluble in the corroding medium then corrosion proceeds at a faster rate. Whereas insoluble corrosion product suppresses the corrosion as it functions as physical barrier. Further, if the product formed interacts with the medium to form insoluble product then also corrosion is suppressed. For example the corrosion of Pb in H_2SO_4 proceeds at power much lower rate because of the formation of insoluble $PbSO_4$.

(g) **Volatility of corrosion product:** The volatility of the corrosion product has a marked significance on corrosion rate. It the corrosion product is volatile, it leaves the surface of the metal as soon as it is formed. Hence it leaves the underlying fresh metal surface exposed to further attack. This results into excessive corrosion. One of the volatile corrosion product MoO_3 is formed by the oxidation corrosion of molybdenum.

7.8.2 Environmental Factors

The nature and aggressiveness of the environment determine to a large extent the degree of corrosion. Some of the common environment variables are:

(a) **Temperature:** Temperature increases the rate of almost all chemical reactions. Therefore rate of corrosion also increases with increase in temperature.

At higher temperatures even the passive metal can become active and get corroded. In high pressure boilers intergranular corrosion takes place only at high temperatures.

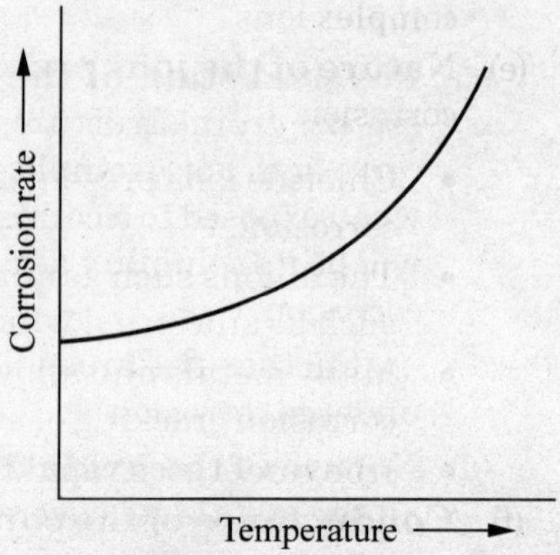

Fig. 7.13 Effect of temperature on corrosion.

(b) **Humidity:** One of the important environment variable in corrosion is relative humidity. Corrosion increases with the increase in humidity of the atmosphere. The effects of humidity on corrosion of enameled steel is apparent in the following table.

Relative Humidity	0 – 65	65 – 80	80 – 90	100
Corrosion (Appearance)	No corrosion	Thin corrosion filaments	Wide corrosion filaments	Blisters due to corrosion

Though presence of moisture increases the corrosion rate, but it should be emphasized that metal does not rust if covered with ice. This is due to the fact that electrochemical attack (corrosion) requires water in the form of liquid.

The relative humidity above which the atmospheric corrosion rate of metal increases sharply is known as **critical humidity.** The increase in corrosion rate due to increase in relative humidity is due to the following reasons:

- Moisture acts as a solvent for oxygen CO_2 etc., present in the atmosphere and helps in setting up the electrochemical corrosion cell.
- Further, the metal oxide film possesses the property of absorbing moisture which can enhance the corrosion. Moreover, rain water may also wash away oxide film from the surface partially, leading to enhanced corrosion.
- Solids like NaCl etc dissolve in water and increase its conductivity and hence increase the rate of cossosion.

(c) **Presence of impurities in atmosphere:** Atmospheric corrosion increases if air is polluted by smoke or gases like H_2S, SO_2, sodium and other chlorides in marine atmosphere.

- NaCl and $(NH_4)SO_4$ if present absorb moisture and act as strong electrolytes.
- Some dust particles containing charcoal are capable of adsorbing SO_2 and so enhance the corrosion process.
- The presence of H_2S, H_2SO_4, HCl near the industrial areas increase the acidity of the liquid adjacent to the metal surface and hence increase its conductivity. This results into increase in corrosion rate.

(d) **Effect of pH:** Lower is the pH, greater is the corrosion. This implies that acidic media having pH less than 7 are more corrosive than alkaline or neutral media. Zinc is easily corroded even in weakly acidic solutions such as carbonic acid.

However, amphoteric metals like Al, Pb etc., go into the solution as complex ions at higher alkalinities. For example, zinc suffers corrosion in acidic medium and corrosion rate can be reduced by increasing the pH, but only up to a certain value. It suffers minimum corrosion at pH 11. On increasing the alkalinity further, it goes into the solution as complex ions.

(e) **Nature of the ions present:** The nature of the ions of the electrolytes greatly affect the corrosion.

- Chloride ions present in the medium destroys the passive film and expose the metal to corrosion.
- The anions such as silicates in the medium lead to the formation of insoluble product such as silicon gel which inhibits the corrosion
- Many metals and alloys can undergo corrosion in the presence of cyanide ions. Stress corrosion cracking of steel is frequently reported in cyanide containing solutions.
- Iron undergoes rapid corrosion in the presence of NH_4^+ salts.

(f) **Conductance of the corroding medium:** Conductance of the corroding medium plays an important role in corrosion. Dry sandy soils have lower conductance as compared to clay or mineralized soils. Hence the metallic structures buried under mineralized or clay soils undergo faster corrosion compared to dry sandy soils.

(g) **Effect of Oxygen and Oxidizers:** The effect of oxygen and oxidizers have been discussed in some details under the heading 'passivity'. Fig. 7.14 illustrates the effect of oxidizers an corrosion rate.

The graph can be divided into three sections. Section 1 is the behavior exhibited by normal metals when they exist in active state. When oxidizer is added, some metals achieve passivity due to the formation of oxide film which is protective in nature, represented by section 2. Section 3 is characteristic of the metals such as copper in acid solution which do not passivate.

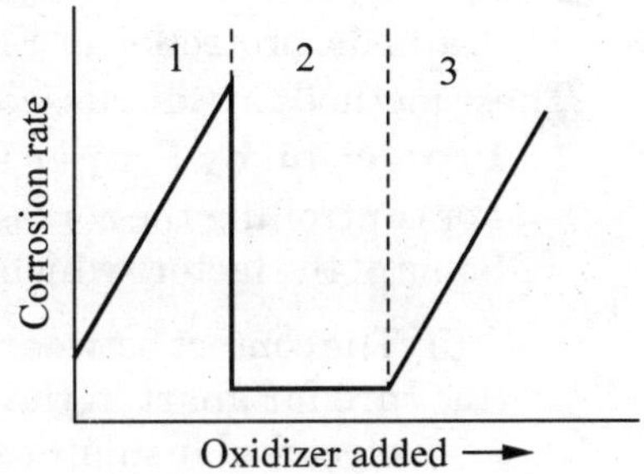

Fig. 7.14 Effect of oxidizers or oxygen on corrosion rate

(h) **Effect of Velocity:** Velocity of the environment plays an important role in erosion corrosion. Increase in velocity generally result in increased corrosion attack. For example, hardened straight chromium stainless steel valves and plugs give excellent service in most steam application. But, if the steam velocity is increased, corrosion increases leading to 'wire drawing.' High velocity may also reduce surface protection thereby increasing the corrosion.

However, velocity can decrease the corrosion attack in some cases by preventing the deposition of dirt which would cause corrosion. Increasing velocity may also sweep away the corrosive environment thus reducing corrosion. For example, under stagnant conditions, steel is attacked by nitric acid autocatalytically, because of the formation of nitrous acid as a cathode reaction product. Increasing velocity sweeps away the nitrous acid and thus removes one of the corrosive agents.

(i) **Formation of Oxygen Concentration Cell:** The rate of corrosion increases due to differential aeration by the formation of concentration cells. The corrosion by differential aeration has already been discussed. The corrosion due to differential aeration or at the surfaces where oxygen cannot diffuse readily is predominant in:

- metal washers where oxygen cannot diffuse
- buried pipe-lines passing from one type of soil to another
- crevice corrosion and waterline corrosion.

7.9 PROTECTION AGAINST CORROSION (CORROSION CONTROL)

The process of corrosion is slow but the losses incurred are enormous. In fact our economy would be drastically changed if there were no corrosion. But while the corrosion is inevitable, its cost can be considerably reduced. The main responsibility of a corrosion engineer is to control corrosion. Corrosion engineering mainly deals in applying scientific principles to prevent or control corrosion damage.

Methods for Controlling Corrosion

The various methods by which corrosion can be controlled and prevented are:

1. Protection by proper designing.
2. Material selection.
3. Modifying the environment.
4. Modification of the properties of the metal.
5. Use of inhibitors.
6. Use of protective coatings.
7. Cathodic protection or Electrochemical protection.

These methods are discussed below in detail.

1. Protection by Proper Designing

For controlling the corrosion, proper designing of the equipment plays a significant role. Some of the factors which should be considered for proper designing are:

(i) The contact between different metals and alloys should not be made, especially if they are far apart in the galvanic series. Where it cannot be avoided delete an efficient electrical insulator should be used. If this is not followed then the more active metal will get corroded causing local corrosion.

Example: Plastic washers are preferred over metal washers because plastic washers act as insulators and prevent corrosion.

(ii) The rate of corrosion also depends upon the relative sizes of the anode and the cathode. As far as possible, if two dissimilar metals are in contact, *the combination of small anode and large cathode should be avoided.* (refer to section 7.9) Screws, bolts, nuts and welds should therefore be made from a more noble material than the materials which are joined through them. Thus the screws nuts and bolts will act as cathodes and because of their small areas (cathodes), a limited cathodic reaction occurs leading to decreased rate of corrosion.

Example: Copper rivets are used to fasten steel sheets and not vice versa because copper is nobler than steel and hence acts as cathode. Due to the small area of the copper rivets, the cathodic reaction is limited and anodic reaction of steel proceeds slowly. Whereas if steel rivets are used for joining copper sheet, the small areas of the rivet anode will keep on furnishing electrons to the large cathodic copper sheet thereby corroding steel rivets completely.

(iii) The presence of crevices between adjacent parts of structure, even in the case of the same metal, should be avoided. The welded butt joints of the same metal/alloy should be used instead of riveted or bolted joints otherwise crevice corrosion may result. *In other words welding is better joining technique than mechanical fastners.*

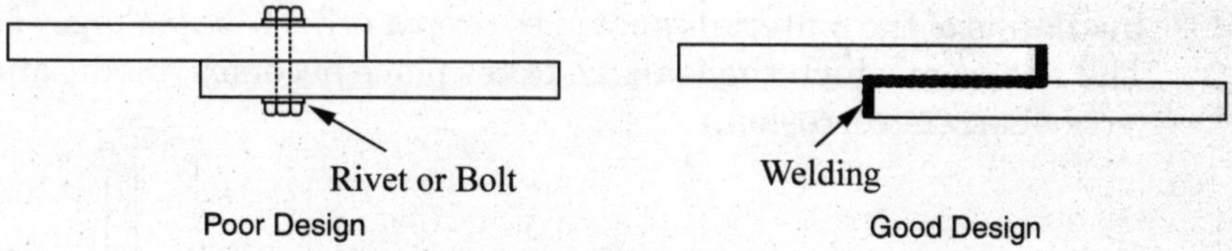

Fig. 7.15 Different techniques of joining the metals/alloys

(iv) As moisture is the main factor governing corrosion, the design should be such that retention of moisture should be as low as possible. Properties should be so designed that water is drained completely. For that, sharp corners should be avoided.

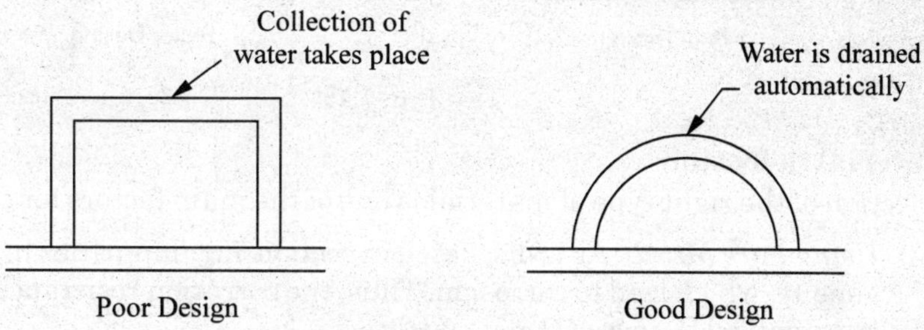

Fig. 7.16 Water stagnation

Storage containers should be designed in such a way that they can be completely drained and cleaned.

(v) Sharp corners and recesses should be avoided because dust or solid particles may accumulate leading to the corrosion by differential aeration.

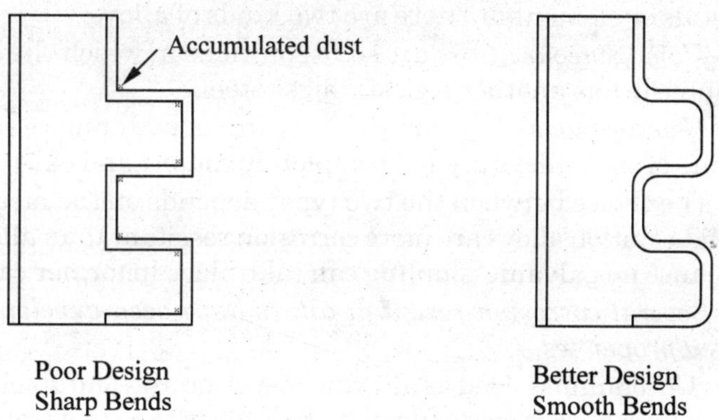

Fig. 7.17

(vi) The anodic parts should be readily replaceable or should be made thicker for longer life.

(vii) Increasing pipe diameters helps from mechanical standpoint, by decreasing velocity of the flowing fluids and ensuring laminar flow (refer to section 7.9 for the effects of velocity) Pipelines should be free from obstruction or crevices, in which water may stay giving rise to corrosion by differential aeration.

(viii) The equipment should be supported on stands to allow circulation of air.

(ix) Insulation of the materials should be proper otherwise leakage of current takes place. The site from which the leakage takes place becomes anodic and leads to corrosion (stray-current corrosion.)

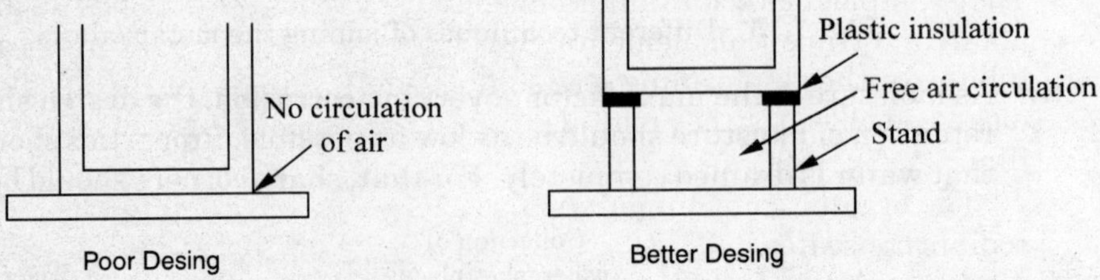

Fig. 7.18

2. Material Selection

Selection of the right type of material is one of the main factors for corrosion control.

(i) *Using Pure Metal:* As is discussed in section 7.8, impurities in a metal cause heterogeneity, which lead to corrosion. Thus, the corrosion resistance of a given metal may be improved by increasing its purity.

However it is not always fruitful to use a metal of high purity because:

(a) Pure metals are expensive

(b) They are usually weak and do not posses the required strength. For example, one would not build a bridge of pure iron.

(ii) *Using metal alloys:* Corrosion resistance of most commercial metals can be increased by alloying them with suitable alloying elements. (Alloys are mixtures of two or more metals or elements.) There are two kinds of alloys:

(a) *Homogeneous alloys are* solid solutions in which the components are completely soluble in one another e.g., stainless steel.

(b) *Heterogenous alloys* are the mixtures of two or more separate phases. The components of such alloys are not completely soluble and exist as separate phases.

The choice between the two types depends on the mechanical properties desired. Solid-solution alloys are more corrosion resistant than alloys with two or more phases because no galvanic coupling can take place in former case.

Several corrosion-resistant alloys have been developed depending upon the desired properties.

(a) Chromium is used as alloying metal for iron and steel as it produces an exceptionally coherent oxide film which protects the steel from further attack.

(b) Silicon iron alloy called (15% Si and 85% Fe) is highly resistant to the acids because of the formation of silicon oxide film on its surface.

(c) Cupro-nickel alloys (70% Cu + 30% Ni) containing 0.2% Fe are used for condenser tubes and for bubble trays used in fractionating columns.

3. Use of Inhibitors

Corrosion inhibitors are the substances which when added in small quantities to the corrosive environments decrease the corrosion rate. Inhibitors are organic as well as inorganic substances that dissolve in the corroding medium.

Inhibitors are generally classified on the basis of electrode at which their action is predominant. Thus we have two types of inhibitors, Cathodic inhibitors and Anodic inhibitors.

(i) *Anodic Inhibitors:* The anodic reaction takes place when the ions from the anodic metal pass into the solution. (For example, $Fe \rightarrow Fe^{2+} + 2e^-$). The anodic inhibitors are the compounds which react with these metal ions to produce insoluble compounds. This insoluble compound gets adsorbed on the metal surface forming a film or a protective barrier thereby reducing the corrosion rate.

The examples of anodic inhibitors include molybdates, phosphates and chromates of alkali and transition elements. Substances like Na_2CrO_4, $NaNO_2$ act as anodic inhibitors either by forming less soluble compounds with the anodic metal ions (as discussed above) or they may also retard the corrosion by oxidative repair of the oxide film.

The organic compounds can also acts as anodic inhibitors and example include sodium benzoate.

Limitations of anodic inhibitors: Anodic inhibitors are not considered satisfactory due to the following reasons:

(a) The inhibitors can prevent corrosion only above a certain critical concentration of solution.

(b) If insufficient concentration is used then the film so formed may not cover all the anodic area and corrosion of small areas left exposed suffer intense corrosion and cause severe pitting. (due to small anodic area and large cathodic area)

(ii) *Cathodic Inhibitors:* The catholic reaction in acidic medium is represented by:

$$2H^+ + 2e^- \longrightarrow H_2$$

Thus corrosion can be controlled by slowing down the diffusion of H^+ ions to the cathode. The diffusion of H^+ ions is considerably reduced by organic inhibitors such as amines, mercaptans, heavy metal soaps, substituted ureas and thioureas. These substances are capable of being adsorbed at the metal surfaces and thus restrict the diffusion of H^+ ions. Antimony or arsenic oxides are used as cathodic inhibitors because of the formation of adherent film on the surface of the metal.

In neutral solutions, the cathodic reaction is

$$H_2O + \frac{1}{2}O_2 + 2e^- \rightleftharpoons 2\bar{O}H$$

In this environment, the corrosion can be controlled by either removing the oxygen from the corroding medium or by not conducting the electrons.

The removal of oxygen is achieved by adding reducing agents such as sodium sulphite or by de-aeration.

$ZnSO_4$ and $MgSO_4$ can also be added which react with $\overline{O}H$ ions and the respective hydroxides precipitated on the cathode surface. The important cathodic inhibitor is $Ca(HCO_3)_2$ which is found in water.

$$HCO_3^- + \bar{O}H \longrightarrow H_2O + CO_3^{2-}$$

Compared to anodic inhibitors cathodic inhibitors are quite safe. This is because, even if they are used in concentrations less than required for complete protection, the partially covered cathodic sites decrease the surface of the cathode whereas anodic area remains practically the same. Thus corrosion decreases due to area effect.

(iii) *Vapour Phase Inhibitors:* These are the inhibitors which readily vapourize and form a protective layer of inhibitor on the metal surface. Vapour phase inhibitors (VPI) are used to prevent corrosion in enclosed spaces and also during storage, packing, shipping etc. Small metals are protected from corrosion by keeping them in the envelopes made from paper impregnated with a suitable vapour phase inhibitor.

Dicyclohexylammonium nitrile and cyclohexyl amine carbonate are amongst the most widely used VPIs.

4. Cathodic Protection or Electrochemical Protection

For the structures immersed in soils or electrically-conducting liquids, the cathodic protection is the best way to control corrosion. Cathodic protection is to force the metal to be protected to behave like a cathode as a result of which it is not corroded.

Cathodic protection can be achieved in two ways:

(i) By using galvanic or sacrificial anode (Galvanic Protection).

(ii) By using impressed current.

(i) *Galvanic Protection:* In galvanic protection or sacrificial anodic protection, the metal structure to be protected is made the cathode by connecting it by a wire to a more anodic material. As a result, the corrosion takes place at anodic material and the metal structure is saved from corrosion. The more active metal used for the purpose is known as *sacrificial anode*. The sacrificial anode is replaced by a fresh one when it gets corroded or consumed from time to time. Most commonly metals used for sacrificial anodic protection are Mg, Zn, Al and their alloys.

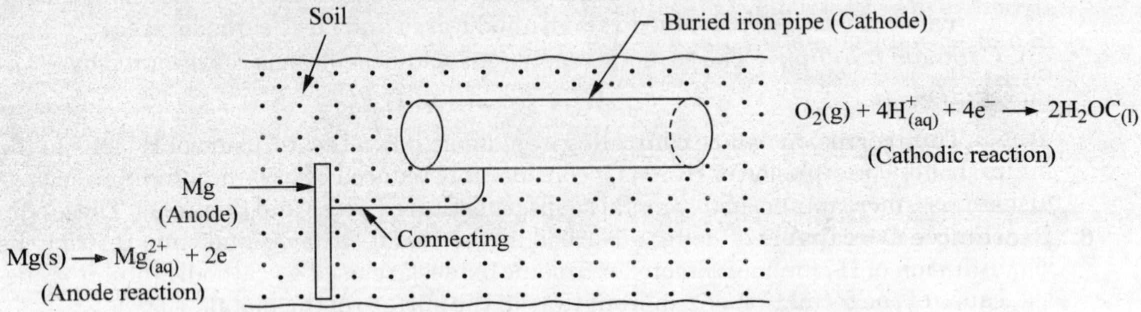

Fig. 7.19 Galvanic Protection using Sacrificial Anode

The application of cathodic protection include:

(a) Protection of buried pipelines, underground cables.

(b) Protection of marine structures, ship hulls etc.

(c) Magnesium rods are inserted into domestic water boilers or tanks to prevent rusting.

(ii) *Impressed Current Cathodic Protection:* This method, known as electrolytic method of cathodic protection, the object to be protected is made cathode by connecting it to negative terminal of DC source. The positive terminal is connected to a insoluble anode like graphite, scrap iron or platinum. The objective is to apply the current from an external source (insoluble anode) in the opposite direction to that of the corrosion current. The insoluble anode is buried in the soil or in the conducting medium adjacent to the metal which is to be protected (Fig. 7.20) These are placed in such a manner that a uniform potential is obtained over the entire metal surface. The electrons flow to the metal structure and as a result it act as cathode and is protected. The anode has to be replaced periodically.

The impressed current cathodic protection has been applied to protect.

(a) Buried pipelines, transmission line towers.

(b) Water pipes, water tanks, marine piers etc.

This kind of technique is particularly useful for large structures which require long term applications.

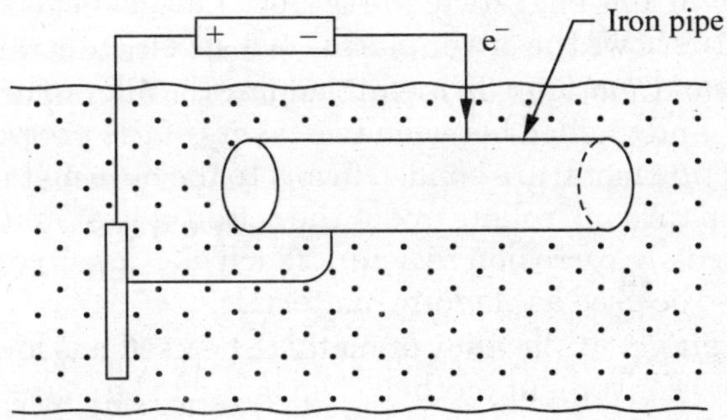

Fig. 7.20 Impressed Current Cathodic Protection

5. Anodic Protection

Anodic protection is applicable to the metals which can be passivated by the deposition of oxide films. In anodic protection, a protective oxide film is grown on the surface of the metal structure to be protected. This is done by electrochemical means i.e., by applying current in a direction that makes the metal structure more anodic. Under these conditions the metal structure is passivated when the required potential is maintained and hence rate of corrosion decreases.

Anodic protection has been applied in case of steel, stainless steel aluminium and chromium. It must be remembered that potential has to be accurately maintained so that it just induces passivity.

6. Protective Coatings

The metal structure to be protected can be coated with metallic or non-metallic coatings. These coatings insolate the insulate from the surrounding corrosive environment. The various types of coatings used for controlling corrosion are:

(i) Metallic Coatings (ii) Non-metallic Coatings

(i) *Metallic Coatings:* The metals used for coatings are Zn, Ni, Sn, Cd and Cr because these are capable of forming their respective oxide films which adhere to the surface of the metal and prevents corrosion. For example, to prevent the rusting of iron it can be coated with Zn or Cr.

Metallic coatings can be applied by using following techniques

(a) *Electroplating:* This is the most widely used method of coating metals to protect them from corrosion. The metal to be plated is made the cathode in a plating bath

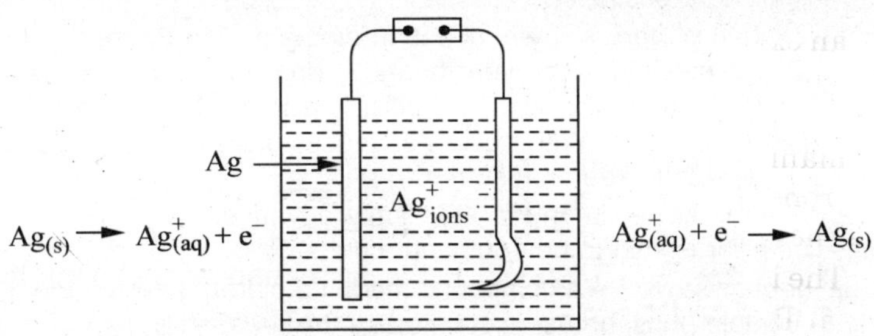

Fig. 7.21 Electroplating

containing the ions of the metal to be deposited. The anode is either an inert material or coating metal to be deposited. When current is passed, the coating metal ions migrate to the cathode and get deposited in the form of thin layer. Fig. 7.16 shows the silver plating of a spoon

(b) *Metal Cladding:* In metal cladding the alloy to be protected, (known as base metal) is sandwitched between two layers of the protecting metal. These layers of protecting metal are bonded firmly to the base metal by pressing the whole combination through rollers under the action of heat and pressure. The cladding material used are corrosion resistant (such as Nickel, copper, lead, silver etc.) Alloys can also be used as cladding materials.

(c) *Dipping:* It the alloy or metal to be used has low melting point then dipping is a very convenient method. The metal to be coated is passed through a bath of the molten coating metal and the thickness of the coating is adjusted by squeezing out the excess of the coating metal with rollers. Coating of zinc on iron by this process is called galvanizing.

(d) *Spraying:* A stream of 'atomized' droplets is blown on to the surface of the metal to be protected. This type of coating can be applied to the finished structures such as tanks, ships and towers etc.

(ii) *Non-Metallic Coatings:* The non-metallic coatings may be inorganic or organic coatings and fall under the following categories.

(a) *Organic Coatings:* Organic coatings are inert organic barriers which provide corrosion resistance as well as decorative value to the structure which is coated. The organic coatings include; paints, varnishes, enamels, and lacquers.

(b) *Chemical Conversion Coatings:* These are inorganic surface-barriers produced by chemical or electrochemical reactions at the surface of the metal. These coatings provide the protection to the base metal from corrosion and also provide a surface ideal for painting or other decorative measures. Examples of chemical coatings include, phosphate coating, chromate coating etc. The phosphate coating is produced by the chemical reaction of base metal with aqueous solution of phosphoric acid and zinc or manganese phosphate. Such coatings are applied to iron, steel and zinc. The chemical reaction between the phosphating solution and the base metal results in the formation of a surface film consisting of crystalline zinc-iron or manganese-iron phosphates.

ASHOKA PILLAR

This pillar was built by samrat Ashoka who ruled around 280 B.C, and is known by his name Ashoka Pillar. It is situated near Qutab Minar in Delhi. This pillar is the testimony of the advanced metallurgical technology ancient India possessed.

The pillar is solid shaft of iron, sixteen inches in diameter and 23 feet high. What is most astounding about it is that it has never rusted even though it has been exposed to wind and electrolysis! The technique used to cast such a gigantic, solid pillar is also a mystery, as it would be difficult to construct another of this size even today. With today's advances only four foundries in the world could make this piece and none were able to keep it rust free.

The pillar stands as mute testimony to the highly advanced scientific knowledge that was known in ancient India. The pillar is said to be 1600 years old.

A protective layer of 'misawite'— a compound made up of iron, oxygen and hydrogen on the steel pillar, which is said to contain phosphorus- is claimed as the reason for the non-corrosive existence.

QUESTIONS

SHORT ANSWERS QUESTIONS

Section 7.1, 7.2, 7.3

1. Why some metal have tendency to get corroded ?
2. What are the consequences of corrosion ?
3. The oxide film formed by alkali and alkaline earth metals such as Na, Ca, Mg etc. develops cracks and pores leading to excessive corrosion whereas the oxide film formed by Cr, W etc. is protective in nature explain.

Section 7.4

4. Discuss mechanism of rusting of iron
5. List differences between chemical corrosion and electrochemical corrosion.
6. What is pilling-bedworth rule ?
7. What are the various factors that contribute to the rusting of iron ?
8. Iron corrodes faster than aluminium, even though iron is placed below aluminium in the electrochemical series.
   Ans: The film formed by aluminium on oxidation is non porous, very thin, tightly adhering which protects the metal from further corrosion, whereas iron forms non-adhering film thus exposing the metal surface to atmospheric attack.
9. Which part of an electrochemical cell formed by two metals, suffer corrosion cathode or anode.
   Ans. Anode

Section 7.5

10. If a drop of an electrolyte is placed on an iron surface, rusting occurs in the central region. Explain.
    Ans: This due to the formation differential oxygen concentration cell. At the periphery of the drop the concentration of oxygen is higher as compared to the centre of the drop. Thus peripheral region becomes cathodic. Central region becomes anodic and get corroded.
11. Wire mesh corrodes factor at joints. Why ?
12. Bolt and nut should be of the same material. Why ?
13. Why does the part of nail inside the wood corrodes easily ?
14. What is the casuse of water line corrosion ?
    Ans. Differential aeration (Refer to 1.7.2)
15. What is intergranular corrosion ?
16. In presence of atmosphere containing traces of ammonia or amines, brass show intergranular cracking. Explain
17. What is the causes of corrosion of buried pipelines passing from one type of soil to another
18. What do you mean by selective leaching?

Section 7.6, 7.7

19. What do you understand by the term-passivity ?
20. Presence of moisture increases the corrosion rate but metals do not rust if covered with ice
21. Steel connected to copper plumbing corrodes faster.

Section 7.8

22. With increase in temperature, the rate of metallic corrosion increases.
    Ans: Because with increase in rate of reaction, the rate of reaction as well as diffusion increases which increase the rate of corrosion.

23. Impure metals corrode faster than pure metal under identical conditions. Give reason.
    Ans: Impurities in a metal cause heterogenity and form minute electrochemical cells at the exposed parts causing anodic part to get corroded.
24. Rusting of iron in saline water is quicker than ordinary water.
    Ans: Presence of NaCl (an electrolyte) increases the conductivity of water and hence speed up rusting.

Section 7.9

25. Copper equipment should not possess steel bolt. Give reason.
    Ans: Iron acts as anode and is corroded. Moreover anodic area (steel bolt) is very small and hence corrosion is more rapid because demand for electrons be the cathodic area can only be fulfilled by the smaller anodic area by undergoing corrosion more rapidly.
26. Why plastic washers are preferred over metal washers.
27. Welding is better joining technique than mechanical fasteners.
    Ans: Welding by the same metal prevents crevice corrosion, which may occur if mechanical fasters are used.
28. Why chromium is used as an alloying metal for steel and iron
29. What is impressed current cathodic protection ?
30. Can we use aluminium in place of zinc for cathodic protection of iron from rusting ?
    Ans: In galvanic series the electrode potential of aluminium is more than that of zinc and hence we can use aluminium.

LONG ANSWER QUESTIONS

1. Why do metals corrode?
2. What are the consequences of corrosion?
3. Discuss the various factors which influence corrosion.
4. What do you mean by 'dry corrosion' and 'wet corrosion' ?
5. What is the difference between rusting and corrosion?
6. Discuss the mechanism of dry corrosion.
7. The nature of the oxide film decides the rate of corrosion. Explain.
8. What is pilling-bedworth rule ?
9. What are the reasons for the formation of anodic and cathodic region on the surface of the metal ?
10. What is the difference between chemical and electrochemical corrosion ?
11. Explain the mechanism of hydrogen evolution and oxygen absorption in electrochemical corrosion ?
12. Write short note on galvanic corrosion.
13. What is concentration cell corrosion? Discuss drop corrosion and waterline corrosion.
14. What is pitting corrosion?
15. Write a short note on microbial corrosion.
16. What is the cause of erosion corrosion? What type of equipments are susceptible to erosion corrosion?
17. What do you mean by selective leaching?
18. What is hydrogen damage? Discuss its various types.
19. What is meant by the term passivity? What is its significance ?
20. Galvanic series is a preferred way of predicting the corrosion of the metals as compared to electrochemical series. Comment.
21. Impurities increase the rate of corrosion. Explain.
22. If the area of the cathode is larger than the anode, the corrosion is severe and very rapid. Explain.
23. How does temperature and humidity affect the corrosion rate ?
24. What are the various factors influencing corrosion ?
25. How does nature of the ions affect the corrosion rate ?

26. Rusting of iron is quicker in saline water than ordinary water. Explain.
27. The welded joints of the same metal/alloy to be joined are preferred over bolted joints-why?
28. How alloying decreases the rate of corrosion?
29. What are corrosion inhibitors ? Discuss cathodic and anodic inhibitors ?
30. What do you mean by vapour phase inhibitors ?
31. Discuss the importance of design and material selection in controlling corrosion.
32. A pure metal rod half immersed vertically in water starts corroding at the bottom. Give reason.
33. What is cathodic protection?
34. A copper equipment should not possess a small steel bolt. Give reason.
35. What is impressed current cathodic protection?
36. How is cathodic protection of iron different from its galvanization ?
37. Why does corrosion occur in steel pipe connected to copper plumbing?
38. Bolt and nut made of same metal is preferred. Explain.
39. Why does part of a nail inside the wood undergoes corrosion easily.
40. Discuss sacrificial anodic protection. What is the condition for a metal to act as a sacrificial anode to iron?
41. How much rust ($Fe_2O_3 \times 3H_2O$) will be formed, when 100 kg of iron have completely rusted away? (Mol.wt of rust = 214 g)
Ans. 191.67 kg].

❑❑❑

8

SPECTROSCOPIC METHOD OF ANALYSIS I

Knowledge is a treasure, but practice is the key to it.

— Thomas Fullen

CHAPTER OUTLINE

Spectroscopy is the branch of science which deals with the determination of the structure of a compound through the interaction of electromagnetic radiation with the matter.

Though, a number of classical methods of structural analysis, such as examination of physical properties (melting point, boiling point, refractive index, optical rotation, solubility etc.) and chemical properties (functional group analysis etc.) are available, but spectroscopy has a number of advantages over these methods. *Some of these advantages are:*

- Spectroscopic methods require much less time as compared to classical methods.
- A very small amount of substance is required for spectroscopic analysis.
- Generally the substance remains unchanged during spectroscopic examination and can be used again for other tests.
- Spectroscopic methods are highly reliable in establishing the identity of two compounds.

Before going into the details of spectroscopy let us discuss the nature of electromagnetic radiations and how these interact with the matter.

8.1 ELECTROMAGNETIC RADIATIONS

An electromagnetic radiation is a form of energy that is transmitted through space at an enormous velocity. The common example of electromagnetic radiation is visible light. The quantum mechanics suggests that electromagnetic radiation has a dual character i.e. exhibiting wave as well as particle behaviour.

According to wave nature of radiations, the electromagnetic waves consist of oscillating electric and magnetic field and hence the name, electromagnetic radiations. The electric and magnetic components are mutually perpendicular to each other and are coplanar. (Fig. 8.1).

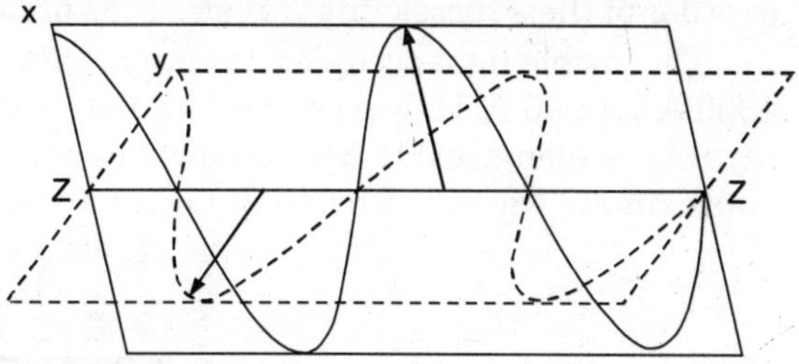

Fig. 8.1 Planes of electromagnetic waves

The electromagnetic radiations are characterized by different parameters, such as

(i) Wavelength: It is the distance between the two adjacent crests (points *C—C* in Fig. 8.2) or troughs (points *T—T* in Fig. 8.2) in particular wave. It is denoted by 'λ'. Mathematically,

$$\lambda = \frac{c}{v}$$

where c is the velocity of light and is equal to 3×10^{10} cm per sec and v is the frequency. Wavelengths are expressed in a number of units such as, cm, nm, Å etc.

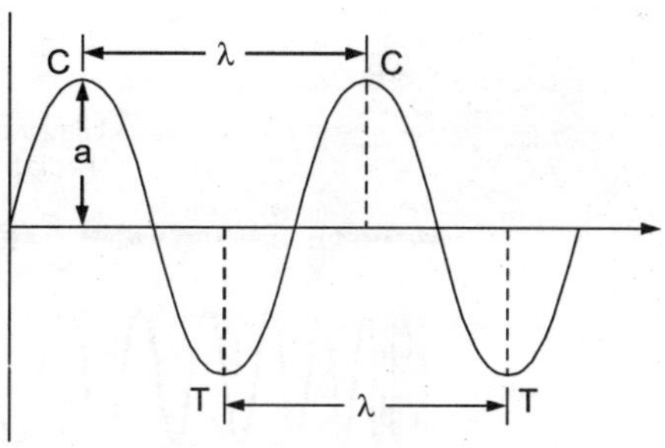

Fig. 8.2 Wave motion

(ii) Wave Number: Wave number is the reciprocal of the wavelength expressed in centimeters. It is the total number of waves which can pass through a space of one cm. It is denoted as $\overline{v}$ and is expressed in cm^{-1}.

$$\overline{v} = \frac{1}{\lambda \text{ in cm}}$$

(iii) Frequency: It is defined as the number of waves which can pass through a point in one second. It is denoted as 'v' and is expressed in cycles per second or in Hertz where 1 Hz = 1 cycle sec^{-1}. Mathematically,

$$v = \frac{c}{\lambda}$$

where c is the velocity of light and λ is the wavelength. As λ and v are inversely related to each other, the shorter the wavelength, higher the frequency and vice versa.

(iv) Energy: According to quantum mechanics, an electromagnetic radiation consists of discrete packets of pure energy, called photons or quantas. Each photon is associated with a definite amount of energy. Energy of a wave of the particular radiation is given by Plank's relation,

$$E = hv$$

where h = Plank's constant and its value is 6.62×10^{-34} J-s and v is frequency of radiation in cycles sec^{-1}.

8.2 ELECTROMAGNETIC SPECTRUM

The radiation used in spectra is electromagnetic radiation. A wide range of electromagnetic radiation stretching from cosmic rays at one end to radiowaves at the other, constitutes the electromagnetic spectrum. The electromagnetic spectrum covers an enormous range of frequencies from radiowaves of lowest frequencies (or highest wavelengths) to cosmic rays of very high frequencies (or low wavelengths). In other words, electromagnetic spectrum is the arrangement of electromagnetic radiations in order of their increasing wavelengths or decreasing frequencies.

The visible light occupies a very small region in the electromagnetic spectrum extending from 4300 Å to 6900 Å. However, the complete spectrum may extend over very wide range from cosmic rays of wavelength .001 Å to radiowaves of wavelength 3×10^{13} Å. Table 8.1 gives an idea about the approximate wavelengths, frequencies and energies of different types of radiations.

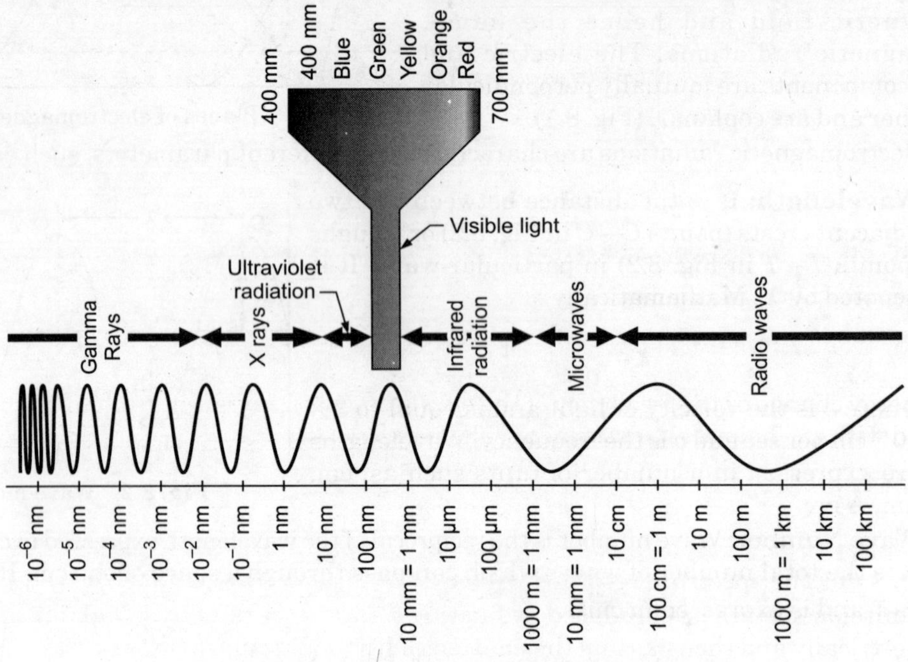

Fig. 8.3 Electromagnetic Spectrum

Table 8.1: The Electromagnetic Spectral Region.

Spectral Region	*Wavelength*	*Frequency, Hz*	*Energy J/Mole*	*Type of quantum change*
γ rays	100 to 1 pm	3×10^{18} to 3×10^{20}	10^9 to 10^{11}	Change of nuclear configuration
X rays	10 nm to 100 pm	3×10^{16} to 3×10^{18}	10^7 to 10^9	Change of electron distribution
U.V. and visible	1000 to 10 nm	3×10^{14} to 3×10^{16}	10^5 to 10^7	Change of electron distribution
I.R.	100 μm to 1000 pm	3×10^{12} to 3×10^{14}	10^3 to 10^5	Change of configuration
Microwave	1 cm to 100 μm	3×10^{10} to 3×10^{12}	10^1 to 10^3	Change of orientation
ESR	100 cm to 1 cm.	3×10^8 to 3×10^{10}	10^{-1} to 10	Change of spin.
NMR	10 m to 100 cm	3×10^6 to 3×10^8	10^{-3} to 10^{-1}	Change of spin

8.3 ORIGIN OF ELECTRONIC SPECTRA (INTRODUCTION OF RADIATION WITH MATTER)

If some light radiation is passed through a simple of an compound, then some of the wavelengths are absorbed while others remain unaffected. A molecule can only absorb radiation of certain frequency, if there exists within a molecule an energy difference of magnitude ΔE which is equal to $h\upsilon$. Suppose E_1 and E_2 are the two possible energy states of a system. The energy difference between the two states is given by ΔE.

$$\Delta E = E_2 - E_1$$

The energy transition from E_1 to E_2 or from E_2 to E_1 corresponds to the absorption or emission of energy exactly equivalent to the energy of the wavelength absorbed or emitted.

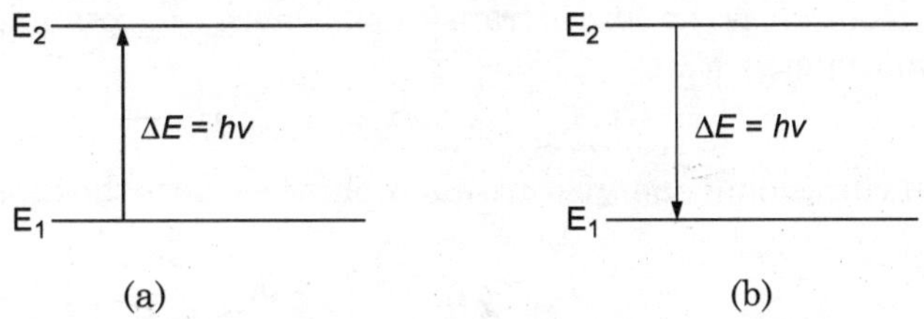

Fig. 8.4 Energy transition for the (a) absorption and (b) emission of electromagnetic radiation

The absorbed or emitted energy is the form of electromagnetic radiation, the frequency of which is given by:

$$\upsilon = \Delta E / h \quad \text{Hz}$$

The energy change or the frequency of the electromagnetic radiation emitted or absorbed can be recorded with the help of an instrument called *spectrometer*.

Thus, spectra so obtained can be broadly classified into two categories:

(i) **Emission Spectrum:** When there is a transition from a state of high energy to a stage of lower energy (Fig. 8.4b), the excess energy is emitted as a photon of energy $h\upsilon$ provided, $\Delta E = h\upsilon$. The spectrum recorded will be *emission spectrum*.
Emission spectra are produced by heating a substance directly in a flame or heating it electrically and then passing the emitted radiation through a prism.

(ii) **Absorption Spectrum:** When there is a transition from a state of lower energy to a state of higher energy (Fig. 8.4a), the energy absorbed will be exactly equal to the energy difference ΔE. The spectrum recorded will be *absorption spectrum*.

Absorption spectrum is produced when the light from a source emitting a continuous spectrum is first passed through an absorbing substance, and then observed through a spectroscope. It will be found that certain wavelengths of light are absorbed by the substance and will leave dark lines or bands at their places giving rise to absorption spectrum.

Emission or absorption spectrum may be classified as Atomic spectra of Molecular spectra.

Atomic Spectra

Atomic spectra is concerned with the interaction of electromagnetic radiation with atoms which are commonly in their ground stage. In atomic spectra, the transition of electrons takes place from one *electronic energy* state to another i.e. between the atomic energy levels. The spectra consists of a series of sharp lines.

Molecular Spectra

Molecular spectroscopy deals with the interaction of electromagnetic radiation with molecules. The spectra by molecules are quite different from the atomic spectra because it involves the transition between *rotational* and *vibrational energy levels* in addition to *electronic transitions.*

The study of molecular spectra has given invaluable information regarding the structure of molecules, incorporating their size and shape, bond distance, bond strength and dissociation energy of the molecules.

In this chapter we will confine our discussion to *Molecular spectroscopy.*

8.4 TYPES OF ENERGY CHANGES

The total energy of the molecule is the sum of transational energy E_t, rotational energy E_r, vibrational energy E_v and electronic energy E_e.

$$E_{\text{total}} = E_t + E_r + E_v + E_r$$

The rotational and vibrational energies are also quantized as is the case with electronic energy. Also,

$$E_e >> E_v >> E_r > E_t$$

The transitional energy which is very small is dissipated as heat and is not involved in the molecular spectra.

The energy levels fall into widely spaced groups which are considered to correspond to different electronic states of the molecule. Associated with each electronic state are a series of vibrational energy levels designated by the vibrational quantum number v. Moreover, each vibrational energy level has a set of rotational sub-levels designated by the rotational quantum number J. The various levels can be represented by Fig. 8.5.

Electronic energy is associated with the transition of an electron from the ground state energy level to an excited state energy level of the molecule due to the absorption of a photon of suitable frequency.

Vibrational energy is due to the to-and-fro motion of the molecule such that centre of gravity does not change.

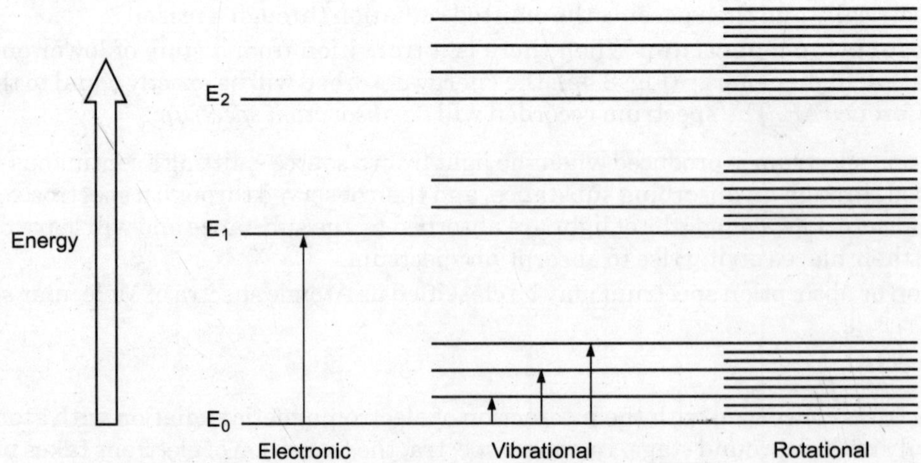

Fig. 8.5 Energy level diagram showing the various molecular energies

Rotational energy arises when the molecule rotates about an axis perpendicular to the internuclear axis and passing through the centre of gravity of the molecules.

Thus energy required for transition will be least for rotational and higher for vibrational and still higher for electronic transitions.

From the above discussion, it is clear that absorption of energy can cause three changes in a molecule.

(i) If the exciting energy is small so as to produce only *rotational transitions* within a vibrational energy level, pure rotation spectra will result. The rotational spectra of a molecule are observed in the *microwave region* or *far infrared region* (> 20 μm).

(ii) If the energy required for exitation are sufficiently large to cause *vibrational transitions* within the same electronic level, the energy changes observed correspond to the change in vibrational quantum numbers. But change in vibrational levels involves changes in rotational levels also, thus the resulting spectrum is called vibrational rotational spectrum. Such type of spectra are generally observed in the *near infrared region* (< 20 μm).

(iii) If the exciting energies are still higher, the *electronic transitions* take place. The electronic transitions are accompanied by a number of vibrational transitions and each vibrational transition in turn is associated with a number of rotational transition. Thus, the resulting spectrum will be complex, consisting of lines due to electronic, vibrational and rotational transitions. The set of rotational lines corresponding to a particular vibrational transition is called a band. Each vibrational transition gives rise to a series of such bands. The entire set of bands for a given electronic transition is known as *band group.* Such type of spectrum is called *electronic band spectra.* These type of transitions take place in *visible and ultraviolet spectral regions,* since considerably higher energies are involved.

8.5 TYPES OF MOLECULAR SPECTRA

Different parts of the electromagnetic spectrum have very different effects upon interaction with matter. Fig. 8.6 gives an idea of the transitions accompanying the absorption/emission of different parts of electromagnetic spectrum.

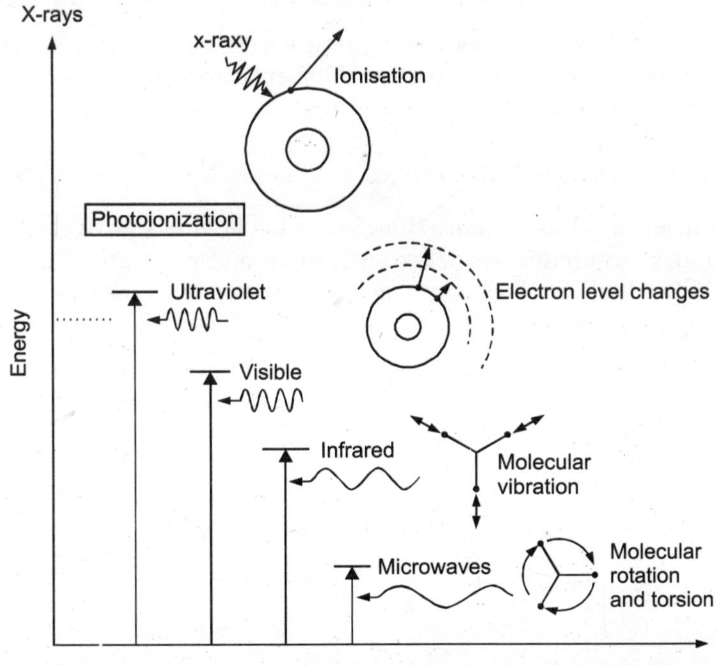

Fig. 8.6 Transitions accompanying the absorption of different parts of electromagnetic spectrum

Energy transitions within molecules are usually studied by the selective absorption of *radiations* passing through them. Molecules spectra are classified according to the region of electromagnetic spectrum in which they are observed or according to the transitions that take place.

Table 8.2 Various Types of Molecular Spectra and their Origin are Summarized

Various Types of Molecular Spectra			
Spectral region	***Wavelength involved***	***Transition***	***Spectroscopy***
1. Microwave	1 – 100 cm^{-1}	Between rotational energy levels	Microwave spectroscopy (Rotational spectroscopy)
2. Infrared	500 – 400 cm^{-1}	Between vibrational energy levels	Infrared spectroscopy (vibrational and vibrational–Rotational spectroscopy)
3. Visible and ultraviolet	12500 – 25000 cm^{-1} (visible) 25000 – 70,00 cm^{-1} (uv)	Between energy levels	Ultraviolet-visible spectroscopy (Electronic spectroscopy)
4. Radio frequency < 1 cm^{-1} region		Between the nuclear spin energy of the molecules in applied magnetic field.	Nuclear Magnetic Resonance spectroscopy
5. Microwave region	1 – 100 cm^{-1} (2 – 9.6 GHz)	Between the electron spin energy of the molecules in applied magnetic field.	Electron Spin Resonance spectroscopy

Although almost all parts of the electromagnetic spectrum are used for studying matter but in this chapter will confine our discussion to ultraviolet spectroscopy, infrared spectroscopy nuclear magnetic resonance spectroscopy and mass spectroscopy.

8.6 GENERAL INSTRUMENTATION

The instrument which is used to measure the absorption or emission of electromagnetic radiation by the substance under examination is called spectrometer. Spectrometers used in various regions of the spectrum differ widely in their construction, but the basic features remain the same. The absorption spectrometers consists of :

(i) **Source of Electromagnetic Radiation:** The incident light of required wavelength is provided by different sources for different spectroscopic techniques. The sources for different spectroscopic techniques are tabulated in Table 8.3.

Table 8.3: Source for Different Spectroscopic Techniques

Spectral Region	*Source*	*Dispersing Elements*	*Detector*
1. Ultraviolet Spectroscopy	Hydrogen discharge lamp Deuterium Discharge lamp	Quartz prism	Photomultiplier tube
2. Visible Spectroscopy	Tungsten lamp	Glass prism	Photomultiplier
3. Infrared Spectroscopy	Electrically heated (~ 1500°C) rods of • Nernst glower (Sintered mixtures of oxides of Zr, Y Er etc.) • Globar (silicon carbide) • Ceramic element	Rock salt prism	Thermocouple
4. Microwave Spectroscopy	Klystron tube (A specially developed electron tube in which electrons generate electromagnetic radiations	—	Quartz crystal detector.

(ii) **Monochromator:** For conversion of polychromatic radiation from source into narrowbands or monochromatic radiations, the devices like filters or monochromators are used.

(iii) **Sample Holder:** The sample is placed in the sample container which are cuvettes or cells. The windows of the sample container must be chemically resistant to the sample. Sample holders for various techniques are given in Table 8.4.

Table 8.4

Technique	*Sample holder*
Visible-ultraviolet	Glass or quartz
Infrared	NaCl and KBr

(iv) **Detector Unit:** The main function of any detector is to absorb the energy of photons and convert it into measurable quantity such as electric current or thermal changes. Generally, the detectors generate an electrical signal which activates some type of meter or recorder. In UV-visible spectroscopy, photo cells or photomultipliers are used as detectors. In the infrared regions, thermocouples are used whereas in microwave region, the detector is a crystal rectifier.

(v) **Amplification and Recording Unit:** The main function of amplifier is to amplify a signal so that it can be measured. An amplifier takes an input signal from circuit and through electronic operations, produce on output signal which is several orders of magnitude than input. The signal is recorded on the graph or on the computer monitor. Most recording spectrophotometers record wavelength versus absorbance. The absorbance A or *optical density* is given by

$$A = \log \frac{I_0}{I}$$

where I_0 is the intensity of incident light and I is the intensity of transmitted light. Fig. 8.7 gives a schematic representation of a spectrometer in general.

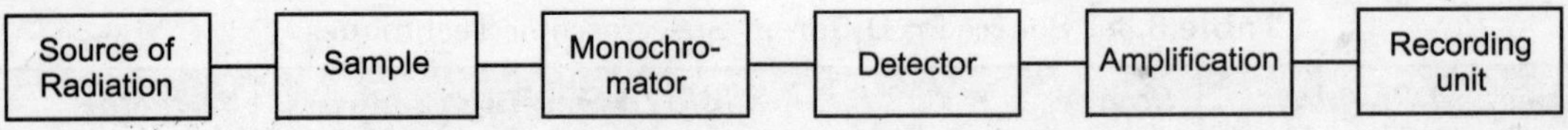

Fig. 8.7 Block Diagram of Spectrometer

Spectrometers for different spectroscopy units are discussed in their respective sections.

8.7 ABSORBANCE AND TRANSMITTANCE

Every molecular specie is capable of absorbing certain characteristic wavelength of electromagnetic radiation. When radiations are passed through a sample of an organic compound some of the frequencies are absorbed while other frequencies are transmitted through the sample, without being absorbed. If the percent Absorbance or percent Transmittance is plotted against frequency, the result is an *absorption spectra. Transmittance of a solution is defined as the ratio of intensity of radiation after passing through the solution to the intensity of intensity light i.e.,*

$$T = \frac{I}{I_0}$$

where I = intensity of light after passing through the solution
I_0 = intensity of the incident light

The transmittance can also be expressed in terms of absorbance of the solution i.e.,

$$A = \log \frac{I_0}{I}$$

where A is absorbance.

Laws of Light Absorbance: (Lambert's and Beer's Law)

The absorbance of radiation is governed by two empirical laws:

(a) **Lambert's law:** *When monochromatic radiation is passed through a solution, the decrease in intensity of radiation, with the thickness of the solution, is directly proportional to the intensity of the incident radiation. Let*

Intensity of incident radiation = I

$\therefore$ Rate of decrease in intensity of intensity radiation with thickness dx of the medium $= -\frac{dI}{dx}$

According to the law, $-\frac{dI}{dx} \propto I$ or $-\frac{dI}{dx} = kI$

where k is a proportionality constant.

$$\text{Further } \frac{-dI}{I} = kdx$$

On integrating the above expression between limits $I = I_0$ at $x = 0$ and $I = I$ at $x = l$, we get

$$In \frac{I}{I_0} = -kl$$

$$2.303 \log \frac{I}{I_0} = -kl.$$

$$\log\frac{I}{I_0} = \frac{-k}{2.303}l$$

$$\log\frac{I_0}{I} = \frac{k}{2.303}l$$

where $$\log\frac{I_0}{I} = A \text{ (Absorbance or optical density)}$$

$$\frac{k}{2.303} = \varepsilon \text{ (absorption coefficient)}$$

$$A = \varepsilon Cl$$

(b) **Beer's Law:** Beer's law also takes into consideration the concentration of solution along with the thickness of the solution. This law states that *when a mono-chromatic light is passed through a solution, the decrease in intensity of radiation with the thickness of the solution is directly proportional to the intensity of intensity light as well as concentration of the solution.* Let a monochromatic light of intensity I pass through a solution of concentration C molar and thickness dx, then rate of decrease of intensity of radiation is given by :

$$-\frac{dI}{dx} \ \alpha\ CI$$

$$-\frac{dI}{dx} = k'CI$$

where k' is the proportionality constant.

$$\frac{dI}{I} = -k'Cdx.$$

On integrating the above equation between $I = I_0$ at $x = 0$ and $I = I$ at $x = l$, we get

$$\int_{I_0}^{I}\frac{dI}{I} = -k'C\int_{0}^{I}dx$$

$$In\frac{I}{I_0} = -k'Cl$$

$$2.303\log\frac{I}{I_0} = -k'Cl$$

$$\log\frac{I}{I_0} = -\frac{-k'Cl}{2.303}$$

$$\log\frac{I_0}{I} = \frac{k'Cl}{2.303}$$

$$A = \varepsilon Cl$$

If the concentration C = moles/cm^3 and length of solution l is in cm, then ε is known as *molar absorptivity* (extinction coefficient). The magnitude of molar absorption coefficient is directly related to probability of a particular transition.

The transition with ε_{max} value more than 10^4 are called *allowed transitions* and ε_{max} value less than 10^4 are called *forbidden transitions.*

Unit of ε = Litre/mole/cm

However, generally the units of ε are omitted.

8.8 LINE WIDTHS

The transitions in molecular spectra are not precisely 'sharp'. There is always a finite width to the observed spectral lines. In otherwords spectral lines are not infinitely narrow but more or less occurs as broad peaks. Three different processes may contribute to the finite width of a spectral line.

(i) Natural broadening
(ii) Doppler broadening
(iii) Interaction with neighbouring particle (collision broadening)

(i) **Natural Broadening:** According to quantum mechanical picture, the absorption or emission consist of transitions between two discrete energy levels. Energy levels are not perfectly sharp and there is a small range of energies allowed for a transition to occur because of the uncertainty principle. It is impossible to specify the energy levels exactly. If on an average a system survives in a state for time τ, then

$$\Delta E \,.\, \Delta\tau \approx h$$

where $\Delta\tau$ is the uncertainty in time associated with finding of the system in that particular state and is measured by mean life time τ of the state. This is also called "*uncertainty broadening*". No excited state has infinite life time. Therefore, all states are subject to some life time broadening, and the shorter the life times of the state involved in a transition, is the broader is the spectral line.

(ii) **Doppler Broadening:** Doppler broadening is the result of well known "Doppler effect" which is the apparent shift in wavelength of the signal from a source moving towards or away from the observer. The first of these motions, towards the observer cause a decrease in wavelength and the second one increase in frequency. In liquid and gases, the molecules are moving with high speed, some molecules approach the observer, some move away. Correspondingly the detected spectral line is the absorption or emission frequencies arising from all the Doppler shifts. Hence the spectral line is broadened. Doppler broadening increases with temperature because the molecules acquire a wider range of speeds.

$$\Delta \upsilon_{\text{Doppler}} = \upsilon \frac{V}{C}$$

where υ = Correct frequency absorbed
V = Velocity of the emitter
C = Velocity of light

(iii) **Interaction with Neighbouring Particles:** Atoms or molecules in liquid and gaseous phases are in continuous motion and collide frequently with each other. The collision can cause some deformation of the particles. In general molecular interaction takes place more in liquids than in gases, and gas phases spectra usually exhibit sharper lines than those of corresponding liquid. In solid phase spectra are often sharp because the motion of particles is more limited.

8.9 INTENSITY OF SPECTRAL LINES

The spectral lines occur with a variety of intensities. Some lines that might be expected to occur do not appear at all. The intensity of spectral lines depend on

(i) the number of molecules in various states – *Boltzmann population.*
(ii) The likelihood of a system in one state changing to another state – *transition probability.*
(iii) The amount of materials present giving rise to the spectrum – *concentration of the sample.*

(i) **Population of States:** Whenever there are transitions among different energy levels, the most spectral line will arise from the level which initially has a greater population. We may use the Boltzmann distribution to write the population difference as

$$N - N' = N(1 - N'/N) = N(1 - e^{-\Delta E/kT})$$

where, ΔE is the energy difference between the ground and excited states, K is Boltzmann constant.

N = Number of molecules at lower level

N' = Number of molecules at higher level

Therefore, the intensity of net absorption is proportional to

(a) population of lower states

(b) difference in population of upper and lower state.

(ii) **Transition Probability:** Transition probability decides whether a particular transition occurs or not, i.e., whether the transition is forbidden or allowed. This process is essentially the deduction of selection rules, which allows us to decide between which levels transitions will gives rise to spectral lines. The selection rules are a set of rules that determines whether a transition is forbidden or allowed. Only allowed transitions occur with more probability. However, it is not unusual to observe forbidden transition as weak transition.

(iii) **Path Length of the Sample:** If a sample is absorbing energy from a beam of radiations, the more sample the beam traverses, the more energy will be absorbed from it. The relationship between concentration (C), path length (l) and the incident and transmitted intensities of radiation (I_0 and I respectively) can be expressed in terms of Beer-Lambert law.

$$I/I_0 = \exp(-KCl)$$

where K is a constant for a particular spectroscopic transition under investigation.

8.10 PURE ROTATIONAL SPECTROSCOPY (MICROWAVE SPECTROSCOPY)

The microwave spectroscopy explores that part of the electromagnetic spectrum which is extending from 100 μm to 1 cm. This region is called microwave region and lies in between far infrared and ratio frequency regions.

This type of spectra results entirely due to the changes in the rotational energy of the molecule i.e. absorption of microwave energy represents changes of the absorbing molecule from one rotational level to another.

Condition: The condition for observing rotational spectrum is that a molecule must *possess permanent dipole moment.* When a molecule having dipole moment rotates, it generates an electric field which can interact with the electric field of microwave radiation. As a result of this interaction, energy can be emitted or absorbed giving rise to rotational spectrum. No spectrum will be obtained if a molecule has no permanent dipole moment. The molecules having permanent dipole moments are *microwave active* and molecules having no dipole moments are *microwave inactive.* Thus, molecules like HCl, CH_3Cl, CO etc. are microwave active and show rotational spectrum whereas molecules like H_2, Cl_2 etc. are microwave inactive and hence rotational spectrum is not observed in their case.

8.10.1 Microwave Spectra of Diatomic Molecules

Pure rotation spectra offers a powerful method for determining molecular parameters of simple molecules in the gaseous phase. It may be emphasized here that rotational spectra are not observed in liquid and solid states because the rotational energy levels are not quantized in these states.

Consider the case of a diatomic molecule having permanent dipole. The rotating diatomic molecule whose nuclei are supposed to be separated by a definite mean distance may be treated as a rigid rotator with free axis of rotation. Let the molecule be represented by two atoms A and B having masses m_1 and m_2 respectively. Let the distance of atom A from the centre of gravity C be r_1 and the distance of atom B be r_2 as shown in Fig. 8.8.

Diatomic Molecule with Molecular Axis along the z Direction

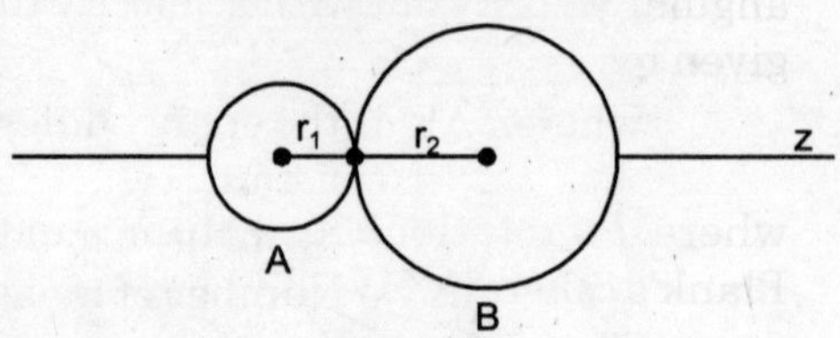

Fig. 8.8

As the system is balanced about its centre of gravity the moments of two atoms are equal around it, i.e.

$$m_1 r_1 = m_2 r_2 \quad \text{...(1)}$$

Also $$r = r_1 + r_2 \quad \text{...(2)}$$

Hence $$m_1 r_1 = m_2 (r - r_1) \quad \text{...(3)}$$

$$m_1 r_1 + m_2 r_1 = m_2 r$$

$$(m_1 + m_2) r_1 = m_2 r$$

$$r_1 = \frac{m_2}{m_1 + m_2} \quad \text{...(4)}$$

Further, solving in a similar way for r_2, we get

$$r_2 = \frac{m_1}{m_1 + m_2} r \quad \text{...(5)}$$

The moment of inertia of a diatomic molecule will be given as

$$I = m_1 r_1^2 + m_2 r_2^2 \quad \text{...(6)}$$

Putting the values of r_1 and r_2 from equations (4) and (5) we get,

$$I = m_1 \left(\frac{m_2}{m_1 + m_2} r \right)^2 + m_2 \left(\frac{m_1}{m_1 + m_2} r \right)^2 \quad \text{...(7)}$$

$$I = \frac{m_1 m_2^2}{(m_1 + m_2)^2} r^2 + \frac{m_1^2 m_2}{(m_1 + m_2)^2} r^2$$

$$= \frac{m_1 m_2 (m_1 + m_2)}{(m_1 + m_2)^2} r^2$$

$$= \frac{m_1 m_2}{m_1 + m_2} r^2$$

$$\therefore \quad I = \mu r^2 \quad \text{...(8)}$$

where μ is the reduced mass of the diatomic molecule and is given the $\mu = \frac{m_1 m_2}{m_1 + m_2}$. The equation (8) defines the moment of inertia in terms of atomic masses and bond length.

Now, the *angular momentum* L of a rotating molecule is given by

$$L = I_\omega \quad \text{...(9)}$$

where ω is the angular velocity.

Rotational Energy Levels

The angular momentum is quantized and hence the molecule cannot rotate with any arbitrary angular velocity but with a velocity that leads to the quantized values of angular momentum being given by:

$$L = \sqrt{J(J+1)}h/2\pi \qquad ...(10)$$

where J = rotational quantum number and can have any integral values (0, 1, 2, 3, ...) and h is Plank's constant.

The *rotational energy levels* of a rigid diatomic rotating molecules are given by:

$$E_J = \frac{1}{2}I\omega^2 = \frac{(I\omega)^2}{2I} = \frac{L^2}{2I} \qquad ...(11)$$

where I = Moment of Inertia

On substituting equation (10) in equation (11) we get the expression

$$E_J = \frac{h^2}{8\pi^2 I}J(J+1) \text{ Joule} \qquad ...(12)$$

where J = 0, 1, 2, 3 ...

However in rotational region, spectra are usually expressed in terms of wave numbers rather than Joules. Thus equation (12) in terms of wave number is written as

$$F(J) = \frac{E_J}{hc} \quad (\because \ \overline{\nu} = E/hc) \qquad ...(13)$$

$$= \frac{h}{8\pi^2 Ic}J(J+1) \text{ cm}^{-1} \qquad ...(14)$$

$$= B\,J\,(J+1) \text{ cm}^{-1}$$

where $$B = \frac{h}{8\pi^2 Ic} \text{ cm}^{-1}$$

The rotational energy $F(J)$ is called *rotational term* and B is known as *rotational constant.*

Selection Rule

It was discussed earlier that a molecule will give rise to rotational spectrum only if it possess a permanent dipole. However according to quantum theory only those rotational transitions are allowed for which

$$\Delta J = \pm 1$$

In absorption $\Delta J = +1$ while in emission $\Delta J = -1$. This general rule is called selection rule.

Rotational Energy Levels

Putting the values of J = 0, 1, 2, 3 ... in equation (14) we get the rotational energy levels of a diatomic molecules as shown below:

where (cf. equation 14)

$J = 0$ $\quad F(J) = 0$

$J = 1$ $\quad F(J') = 2B \text{ cm}^{-1}$

$J = 2$ $\quad F(J'') = 6B \text{ cm}^{-1}$

$J = 3$ $\quad F(J''') = 12B \text{ cm}^{-1}$

$J = 4$ $\quad F(J'''') = 20B \text{ cm}^{-1}$

Fig. 8.9 Various Rotational Energy levels of a Diatomic Molecule

Rotational Spectra

We known, according to selection rule $\Delta J = \pm 1$. For a transition taking place from J to $J + 1$, the rotational frequency is given by,

$$v_r = F(J+1) - F(J) = B(J+1)(J+2) - BJ(J+1)$$

$$= B(J+1)(J+2-J) = 2B(J+1) \text{ cm}^{-1}$$

where $J = 0, 1, 2 \ldots$

Therefore, the rotational transitions occur at frequencies:

$$v_{r_{0\to1}} = 2B$$

$$v_{r_{1\to2}} = 4B$$

$$v_{r_{2\to3}} = 6B$$

Hence the rotational spectrum of a rigid diatomic molecule consists of a series of lines at $2B$, $4B$, $6B$, etc. Therefore on the frequency scale, the lines are equidistant (Fig. 8.10).

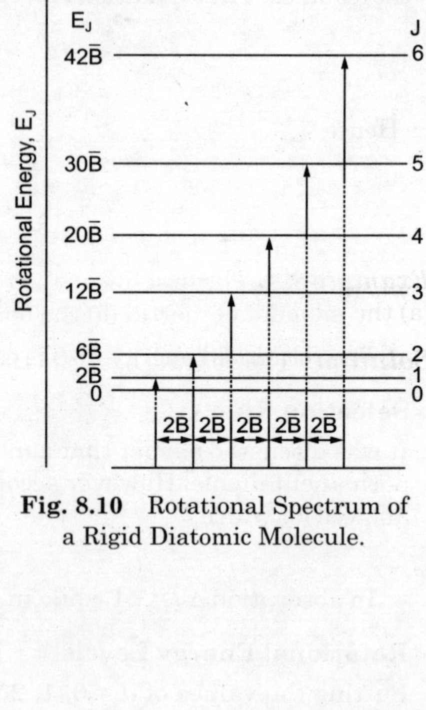

Fig. 8.10 Rotational Spectrum of a Rigid Diatomic Molecule.

Bond Length from Rotational Spectra:

From the study of rotational spectrum the frequency of separation can be calculated as,

$$\bar{v} = 2B(J+1) \qquad \ldots(17)$$

Knowing the value of B, it is possible to calculate the moment of inertia I of the molecule from equation,

$$B = \frac{h}{8\pi^2 Ic} \text{ cm}^{-1}$$

Further, since,

$$I = \mu r^2$$

hence putting the value of masses of atoms present in diatomic molecule, it is also possible to calculate the distance between them (bond length) from equation (18)

$$\therefore \qquad r = \sqrt{\frac{I}{\mu}} \quad \text{where } \mu = \frac{m_1 m_2}{m_1 + m_2} \qquad \ldots(18)$$

SOLVED NUMERCIALS BASED ON DETERMINATION OF BOND LENGTH AND MOMENT OF INERTIA

Example 8.1: **The pure rotational spectrum of gaseous HCl consists of a series of equally spaced lines separated by 20.6 cm^{-1}. Calculate the internuclear distance of the molecule. The molar masses are $H = 1.008 \times 10^{-3}$ kg and Cl = 35.3×10^{-3} kg.**

Solution: The spacing between the lines = $2B = 20.6$ cm^{-1}

$$\therefore \quad B = 10.3 \text{ cm}^{-1} = 10.3 \times 10^2 \text{ m}^{-1}$$

Now
$$B = \frac{h}{8\pi^2 Ic} \text{ cm}^{-1}$$

$$\therefore \quad I = \frac{h}{8\pi^2 Ic} = \frac{6.626\times10^{-34}\text{ kg m}^2\text{s}^{-1}}{8\times\left(\frac{22}{7}\right)^2\times(10.3\times10^2\text{m}^{-1})(3\times10^8\text{ms}^{-1})}$$

$$= 2.655 \times 10^{-47} \text{ kg m}^2$$

Further
$$I = \mu r^2$$

$$r = \sqrt{\frac{I}{\mu}}$$

$$\mu = \frac{m_1 m_2}{m_1 + m_2} = \frac{(1.008\times10^{-3}\text{kg})(35.5\times10^{-3}\text{kg})}{(1.008\times10^{-3}\text{kg} + 35.5\times10^{-3}\text{kg})(6.023\times10^{23}\text{mol}^{-1})}$$

$$= 1.628 \times 10^{-27}\text{kg.}$$

Hence

$$r = \sqrt{\frac{I}{\mu}} = \sqrt{\frac{2.655\times10^{-47}\text{kg m}^2}{1.628\times10^{-27}\text{kg}}} = 1.276 \times 10^{-10} \text{ m}$$

Example 8.2: **The first line in the rotational spectrum of CO appears at 3.84235 cm^{-1}. Determine (a) the moment of inertia (b) the bond length. The molar masses for C = 12 g and 0 = 15.999 g.**

Solution: The spacing between the lines = $2B = 3.84235$ cm^{-1}

$$\therefore \quad B = 1.92118 \text{ cm}^{-1}$$

$$I = \frac{h}{8\pi^2 Bc} = \frac{6.626\times10^{-34}}{8\times\left(\frac{22}{7}\right)^2\times3\times10^{-10}\times1.92118} = 14.56 \times 10^{-47}\text{kg m}^2$$

$$I = \mu r^2$$

$$\mu = \frac{12\times15.999}{12+15.999}\times\frac{1}{6.023\times10^{23}} = 11.3836 \times 10^{-23}\text{g}$$

$$= 11.3836 \times 10^{-27}\text{kg}$$

Now
$$r = \sqrt{\frac{I}{\mu}} = \sqrt{\frac{14.56\times10^{-27}\text{kg m}^2}{11.3836\times10^{-27}\text{kg}}}$$

$$= \sqrt{1.2799\times10^{-20}}\text{ m}^2 = 0.1131 \text{ nm}$$

8.11 INFRARED SPECTROSCOPY

One of the most widely used tools for detection of functional groups in pure compounds and mixtures is Infrared Spectroscopy.

Infrared Spectroscopy, also known as vibrational spectroscopy, involves the transition between the vibrational energy levels or more precisely, the transition between vibrational and rotational energy levels of a molecule on absorption of radiation falling in the spectral range of 650 – 4000 cm^{-1} (infrared region). Infrared radiations do not have sufficient energy to cause the excitation of electrons. However, it causes atoms and groups of atoms of organic compounds to vibrate faster about the covalent bonds connecting them. As these vibrations are quantized and hence the compound absorbs the infrared energy in particular regions of the spectrum. *In other words, an infrared spectrum is obtained when the frequency of molecular vibration corresponds to the frequency of the infrared radiation absorbed.* In addition to the changes in vibrational energy, simultaneous changes in rotational energy can occur and hence some fine structures may be observed on the vibrational band.

Infrared Spectrum

The plot of percentage transmission of IR radiation versus wavelength is known as infrared spectrum. When a beam of IR radiation of varying frequency is passed through sample of a compound, then from time to time the energy of the beam is absorbed. This happens when the energy matches the difference between vibrational energy levels belonging to the bonds. Now the infrared spectrometer is so designed that the pen of the recorder moves downwards if radiation is absorbed, so vibrational spectrum shows a series of dips or troughs. Fig. 8.11 shows a typical example of IR spectrum.

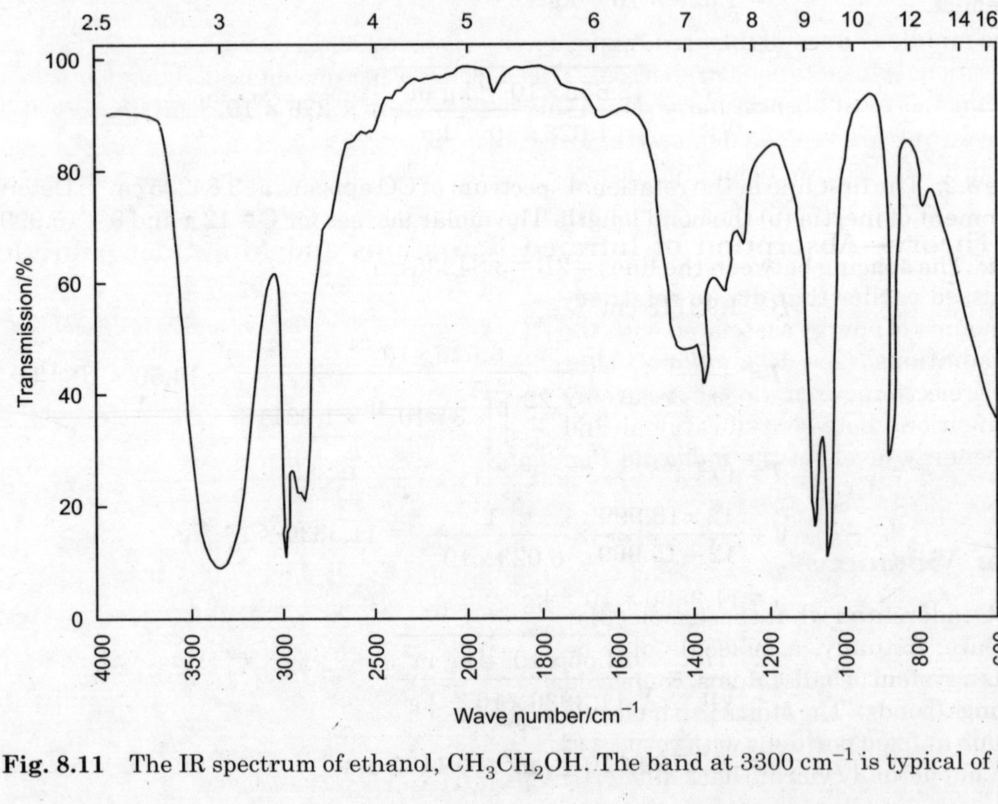

Fig. 8.11 The IR spectrum of ethanol, CH_3CH_2OH. The band at 3300 cm^{-1} is typical of alcohols.

In the Fig. 8.11 the scale on the side shows degree of transmission of the radiation. The higher the transmission, the less is the radiation absorbed and vice-versa. Fig. 8.8 shows a strong dip at around 3300 cm^{-1} and another one at 1050 cm^{-1}. These dips are called *bands*. Both bands are said to be *strong* because of large amount of absorption. The band at 3300 cm^{-1} is known as *broad* because it spreads over quite a wide range of frequencies.

The IR bands can be strong, medium and weak depending upon the amount of absorption. It must be noted again that infrared radiations can be expressed either in terms of wavelength (λ) or in wave number $(\bar{\upsilon})$. The relationship between the two is expressed as:

$$\text{Wave number } \bar{\nu} = \frac{1}{\text{wavelength } (\lambda) \text{ in cm}}$$

Hence the wave number has units of cm^{-1}. Further, IR spectroscopy is divided into three regions as shown below:

Near Infrared Region	Middle infrared (Fundamental vibrational Region)		Far Infrared Region (Pure Rotational Region)
	Functional Group Region	Finger Print Region	

12500 cm^{-1} 4000 cm^{-1} 1400 cm^{-1} 900 cm^{-1} 667 cm^{-1} 50 cm^{-1}

- The near infrared region corresponds to the wave number 12500 cm^{-1} to 4000 cm^{-1}. As there are few absorptions of organic molecules in this range, it is of little use to organic chemist.
- The middle or normal infrared region extends from 4000 cm^{-1} to 667 cm^{-1} and results in vibrational plus rotational transitions. This region is of paramount importance for an organic chemist as most absorptions occur in this region.
- The far infrared region deals with pure rotational motion of the molecules and only few absorptions occur here for organic molecule. It extends from 667 cm^{-1} to 50 cm^{-1}.

8.11.1 Theory—Absorption of Infrared Radiations and Molecular Vibrations

As is discussed earlier that due to relatively smaller amounts of energy associated with the infrared radiations, (2 – 10 k cal/mol) they cannot cause electronic excitations but can only induce transitions between vibrational and rotational energy levels of the molecule Fig. 8.12.

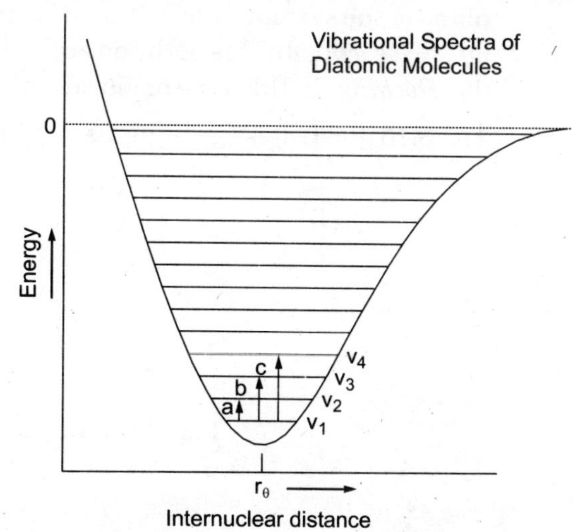

Fig. 8.12 Vibrational levels for vibrating bond.

Molecular Vibrations

Let us first understand what these molecular vibrations are. Actually, a molecule may be regarded as a system of balls (atoms) connected by tiny springs (bonds). The atoms in a molecule do not remain at fixed positions with respect to each other, but actually vibrate back and forth about an average value of interatomic distance (Fig. 8.13)

Now this vibrational motion is quantised as shown in Fig. 8.13. At room temperature most of the molecules in a given sample are in lowest vibrational state. However, when infrared light is passed through it, the molecule absorbs the radiation of appropriate energy and gets excited to higher vibrational level. This results in increasing the amplitude of the vibrational energy. *It may be mentioned here that IR light is absorbed only when oscillating dipole moment due to molecular vibrations interacts with the oscillating electric vector of an infrared beam.*

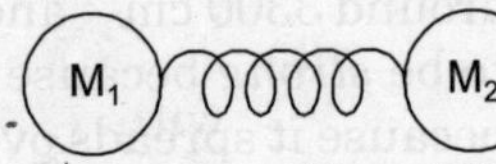

Fig. 8.13 M_1 and M_2 at equilibrium bond length

The simplest vibrational motions in molecules giving rise to absorption of infrared radiation are stretching and bending.

(1) Stretching Vibrations: In this type of vibrations, the distance between atoms increases or decreases but the atoms remain in the same bond axis. It is of two types: *symmetric* and *Antisymmetric*. For the molecule of the type AB_2 the vibrations can be illustrated as in Fig. 8.14.

(a) (b)

Fig. 8.14 Stretching Vibrations.

(i) *Symmetric:* In symmetric stretching the movement of the atoms with respect to the particular atom in a molecule is in the same direction. (Fig. 8.14a)

(ii) *Antisymmetric:* In this, one atom approaches the central atom while the other moves away. (Fig 8.14b)

(2) Bending Vibrations: In this type of vibrations, the positions of the atom change with respect to the original bond axis but the distance between the atoms remains constant.

Bending vibrations are of two types:

(i) *In plane bending vibrations:* In this type atoms remain in the same plane as the nodal plane of the system. It is further of two types:

(a) *Scissoring:* In this both the atoms swing in opposite direction.

(b) *Rocking:* In this type of vibration both the atoms move in the same direction.

Fig. 8.15 illustrates the in-plane bending vibrations.

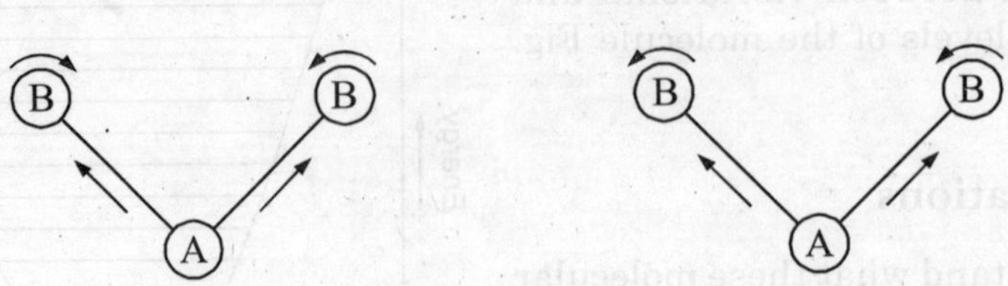

(a) (b)

Fig. 8.15 In-plane bending vibrations

(*ii*) *Out of plane bending vibrations:* Out of plane vibration is of two types:

(a) *Wagging:* In this both the atoms swing up and below the plane with respect to central atom.

(b) *Twisting:* In this type, one atom moves up the plane and other moves down the plane with respect to the central atom.

Fig. 8.16 illustrates the out of plane bending. + and – sings indicate the motion towards and away from reader.

It should be noticed that as more energy is required to stretch a bond than to bend it hence stretching absorption of a band appears at higher frequency as compared to bending vibrations.

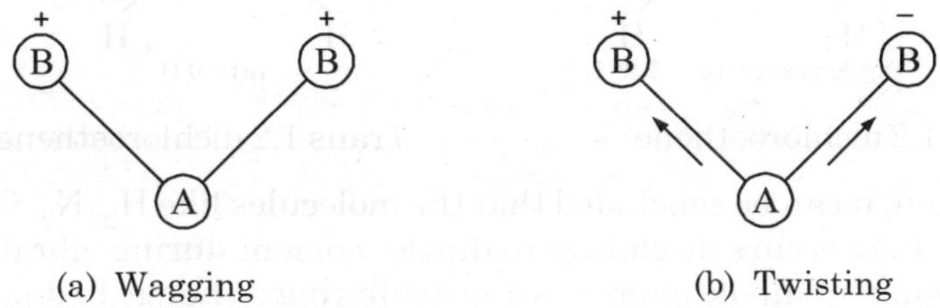

Fig. 8.16 Out-of-plane bending motion.

8.11.2 Dipole Moment and Infrared Spectrum

Generally, the absorption of an infrared radiation of particular frequency by a molecule can occur only if the dipole moment of the molecule is different in the two vibrational levels. As stated earlier, the infrared light is absorbed when the oscillating dipole moment interacts with the oscillating electric vector of an infrared beam. Now, for this interaction to take place the dipole moment of the molecule in the ground state and the vibrational excited state should be different. The greater is the difference in these dipole moments, the more intense the absorption. If no change in dipole moment accompanies the vibrations, then that mode is infrared *inactive* or the molecule is said to be infrared inactive. For example, the symmetrical stretching vibration of — C = C — bond in ethylene can be represented as:

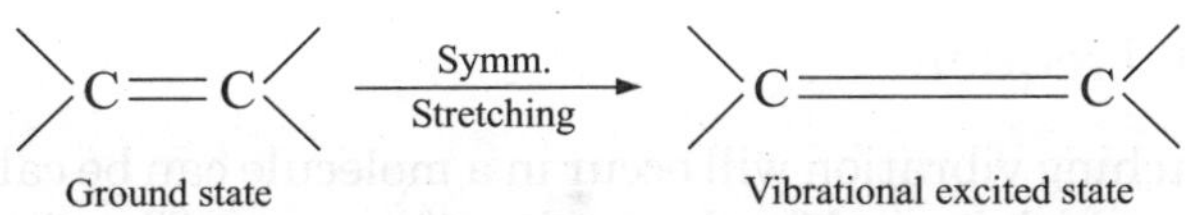

As $>C=C<$ is a nonpolar carbon-carbon double bond, on symmetrical stretching there is little or no change in dipole moment and hence infrared absorption is either absent or very weak. On the other hand the symmetrical stretching of polar carbonyl group can be represented as:

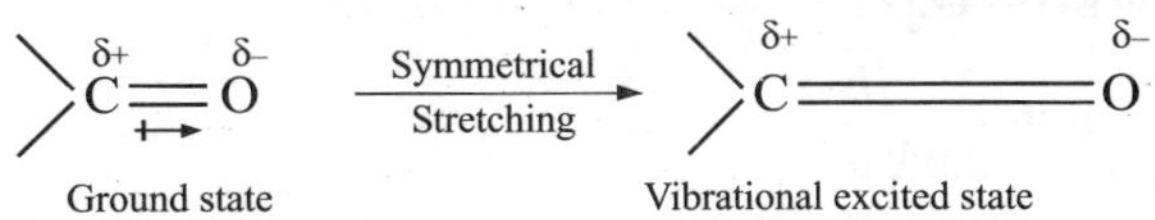

In carbonyl group ($>C=O$), the bond is connected by the atoms having different electronegativities and hence is polar. On stretching, the dipole moment increases considerably or there is a change in dipole moment on symmetrical stretching of the molecule. This change in dipole moment leads to the generation of an oscillating electrical field. If the frequency of this oscillating electrical field is equal to the frequency of fluctuating electric field of infrared radiation, the energy can be absorbed or emitted resulting in absorption infrared spectrum or emission infrared spectra respectively. This $>C=O$ is a infrared active molecule and will show an intense absorption.

Example 8.3: Why trans dichloroethylene shows no C=C stretching absorption while Cis isomer shows this bond.

Solution: The (>C=C<) bond in trans isomer is non-polar having zero dipole moment. On stretching, the dipole moment does not change and hence no absorption of infrared radiation takes place. Whereas Cis isomer has a dipole moment, the magnitude of which increases on stretching and hence C is isomer shows C=C stretching absorption band. However both the isomers show bands for C—H and C—Cl stretchings.

Dipole moment (μ) = 2.95 D (μ) = 0 D

cis 1,2 dichloroethene Trans 1,2 dichloroethene

For the same reason, it can be concluded that the molecules like H_2, N_2, Cl_2, O_2 are *infrared inactive molecules* as there occurs no change in dipole moment during vibrations. Whereas the molecules having permanent dipole moments and exhibiting a change in dipole moment during the vibration are *infrared active* molecules.

However, there are molecules which do not have permanent dipole moment but some of their vibrational modes are accompanied by change in dipole moment making these molecules infrared active. For example, the symmetrical stretching of C=O of CO_2 is infrared inactive while the asymmetrical stretching and bending vibrations are infrared active.

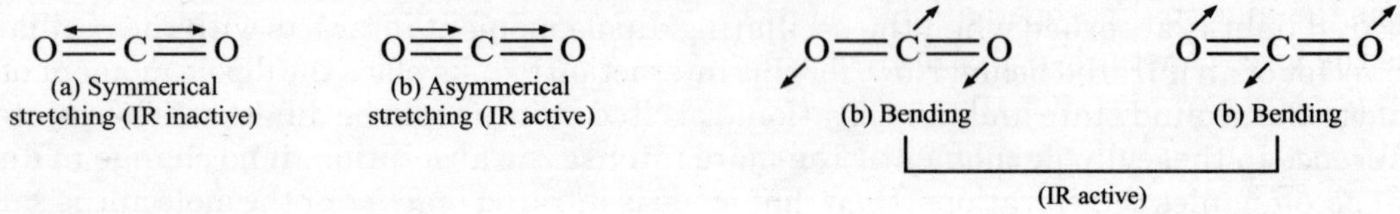

Vibrations of Carbondioxide

8.11.3 Vibrational Frequency

The region where a stretching vibration will occur in a molecule can be calculated fairly accurately by applying *Hooke's law*, which is used for the motion of a spring. The vibrating diatomic molecule with atomic masses m_1 and m_2, joined by a chemical bond vibrates as a one dimensional simple harmonic oscillator and hence according to Hooke's law, frequency at which it is oscillating or vibrational frequency υ is given by:

$$\upsilon = \frac{1}{2\pi}\sqrt{\frac{k}{\mu}} \qquad \text{...(1)}$$

where k is the force constant related to the strength of the spring (the force constant of the bond), μ is the reduced mass of a diatomic molecule.

$\because$ $\mu = m_1 m_2 / m_1 + m_2$

$\therefore$ Equation (1) can be written as:

$$\upsilon = \frac{1}{2\pi}\sqrt{\frac{k}{m_1 m_2 / m_1 + m_2}} \qquad \text{...(2)}$$

From equation (3) it because clear that the value of vibrational frequency of a bond depends upon:

(i) Force constant which is a measure of bond strength. For a single bond, it is approximately 5×10^5 dynes/cm. For a double and triple bond it becomes double and triple respectively. Stronger the bond, greater will be the frequency of absorption.

(ii) Masses of the atoms involved in bonding or reduced mass. Thus, greater the reduced mass, the lower is the vibrational frequency.

Example 8.4: Out of the following pairs which one is expected to absorb at higher frequency for stretching vibration?

(a) C = C and C—C (b) O—H and C—C

Solution: (*a*) C = C is expected to absorb at higher frequency than C—C stretching because of higher bond strength of double bond as compared to single bond.

(*b*) O—H is expected to absorb at higher frequency because it has smaller value of reduced mass as compare to C—C.

Example 8.5: Calculate the approximate frequency of the C—H stretching from the following data: $k = 500\ \text{N m}^{-1}$, mass of carbon atom = 20×10^{-24}g, mass of hydrogen atom = 1.6×10^{-24} g.

Solution: According to Hooke's law:

$$\upsilon = \frac{1}{2\pi}\sqrt{\frac{k}{\mu}} = \frac{1}{2\pi}\sqrt{\frac{k}{m_1 m_2/m_1 + m_2}}\ \frac{1}{2\pi}\sqrt{\frac{k(m_1 + m_2)}{m_1\, m_2}} \quad \text{...(i)}$$

where υ is the vibrational frequency

m_1 is the mass of carbon atom = 20×10^{-24} g

m_2 is the mass of hydrogen atom = 1.6×10^{-24} g

k is the force constant in gs^{-2} or dyne cm^{-2}

$\therefore \quad k = 500\ \text{Nm}^{-1} = 5.0 \times 10^5\ \text{gs}^{-2}$ $\qquad (\because\ 1\text{N} = 10^3\text{g m s}^{-2})$

Putting all the values in equation (*i*) we get

$$\upsilon = \frac{7}{2\times 22}\left[\frac{(5.0\times 10^5\,\text{g s}^{-2})\,(20+1.6)\,10^{-24}\,\text{g}}{(20\times 10^{-24}\text{g})\,(1.6\times 10^{-24}\text{g})}\right]^{1/2} = 9.3 \times 10^{13}\ \text{s}^{-1}$$

The wave number $\bar{\upsilon}$ is given by

$$\bar{\upsilon} = \frac{\upsilon}{c}$$

where c is the velocity of light.

hence $$\bar{\upsilon} = \frac{\upsilon}{c} = \frac{9.3\times 10^{13}\,\text{s}^{-1}}{3.0\times 10^{8}\,\text{ms}^{-1}} = 3.1 \times 10^5\ \text{m}^{-1} = 3100\ \text{cm}^{-1}$$

The stretching frequency of C—H bond is 3100 cm^{-1}.

8.11.4 Fundamental Vibrations and Overtones

The vibrational energy levels are quantized. At room temperature most of the molecules in a given sample are in the lowest vibrational level. For the excitation of a molecule from one vibrational energy level to another, the molecule has to absorb IR radiation of a particular wavelength or frequency. Transition from the ground state to the first excited state i.e. $\upsilon_0 \rightarrow \upsilon_1$, absorb strongly and give rise to intense bands called fundamental bonds or *fundamental vibrations*. Whereas the transition from the ground state to second, third, fourth or any multiple of a given fundamental frequency i.e. $\upsilon_0 \rightarrow \upsilon_2$ or $\upsilon_0 \rightarrow \upsilon_3$, give rise to weak bands called *overtones*.

The transition from v_0 to v_2 is first overtone and transition from v_0 to v_3 is known as second overtone and so on. The frequency required for the first and second overtone will be twice and three times respectively of the frequency of fundamental vibration. For example, the fundamental vibration for carboynl group occurs at $v_0 = 1700 \text{ cm}^{-1}$. Therefore the first overtone peak will appear near 3400 cm^{-1} and second at 5100 cm^{-1} and so on.

The overtones are very weak but these can be important for characterization of certain classes of compounds, particularly benzene derivatives.

8.11.5 Number of Fundamental Vibrations

Polyatomic molecules may exhibit more than one fundamental vibrational absorption bands. The number of these fundamental bands is related to the degrees of freedom in a molecule. (The number of degrees of freedom is equal to the sum of the co-ordinates necessary to locate all atoms of a molecule in space). Each atom has 3 degrees of freedom corresponding to the three Cartesian co-ordinates (x, y and z) which are necessary to describe its position relative to other atoms in the molecule. In an isolated atom only transational motion is there hence it has three transational degrees of freedom.

Now when atoms combine to form molecules then total number of degrees of freedom in the molecule will be given by $3n$, where n is the number of atoms in the molecule. This $3n$ includes transational, vibrational and rotational degrees of freedom as the molecule has vibrational as well as rotational motion

$$\therefore \quad 3n = \text{Transational} + \text{Vibrational} + \text{Rotational degrees of freedom.}$$

But here we are only concerned with *vibrational* degrees of freedom, so let us calculate the number of vibrational degrees of freedom which will inturn gives number of fundamental bands.

Since in all the molecules three co-ordinates are necessary to locate a molecule in space, therefore molecule has always three transational degrees of freedom.

(a) For linear molecule of *n* atoms,

Total degrees of freedom = $3n$

Transational degrees of freedom = 3

Rotational degrees of freedom = 2

($\because$ Rotation of linear molecule about its axis of linearity does not bring about any change in position of atoms while rotation about other two axes do change the position of the atom)

$\therefore$ Vibrational degrees of freedom = $3n - (3 + 2) = 3n - 5$

Each vibrational degree of freedom corresponds to a fundamental mode of vibration which inturn corresponds to a fundamental band. Thus for linear molecule there will be $(3n - 5)$ possible fundamental bands.

Example 8.6: Calculate the possible number of fundamental bands in CO_2 molecule.

Solution: Since CO_2 is a linear molecule hence it can have $(3n - 5)$ fundamental vibrational modes giving rise to fundamental bands.

$\therefore$ Vibrational degrees of freedom = $3n - 5 = 3(3) - 5 = 4$.

Therefore, theoretically 4 fundamental bands are possible in CO_2 molecule.

(b) For Non-linear molecules of *n* atoms,

Total degrees of freedom = $3n$

Transational degrees of freedom = 3

Rotational degrees of freedom = 3

(∵ rotation about all three axis will result in change in position of the atoms)

∴ Vibrational degrees of freedom = $3n - 6$

Example 8.7: Calculate the number of fundamental vibrations in CH_4.

Solution: Since CH_4 is a non-linear molecule,

∴ No. of degrees of freedom = $3n - 6 = 3(5) - 6 = 9$

Thus 9 fundamental vibrational bands are possible.

Similarly in C_2H_6

No. of degrees of freedom = $3n - 6 = 3\,(8) - 6 = 18$

Thus 18 fundamental vibrational bands are possible.

It may be emphasized here that many of these fundamental vibrations occur at the same frequency and are therefore degenerate, so all possible peaks do not appear as independent absorptions.

8.11.6 Combination Bands and Fermi Resonance

The interaction of the vibrational states in polyatomic molecules can produce combination band which is the sum of two or more fundamental vibrational frequencies.

When an overtime or combination band falls near a strong fundamental vibration, the interaction between the two leads to a type of resonance wherein the molecule transfers its energy from fundamental to overtone and back. Thus this resonance mixes their character and both the bonds share their intensity and appear as a pair of equally intense bands in IR spectrum. Such type of resonance is known as Fermi resonance.

8.11.7 Factors Influencing Vibrational Frequencies

As is discussed earlier that vibrational frequency can be calculated using Hooke's law. But the calculated value is never exactly equal to experimental value because vibrations of each group are significantly influenced by the structure of the molecule in immediate neighbourhood of bond. A number of factors are responsible for vibrational shifts from their normal values such as:

(i) **Physical state:** Variations in the physical state influence the infrared frequencies by means of association effects. In varpour state, a substance usually absorbs at higher frequency as compared to, when it is in liquid or solid state. For example, cyclopentanone shows the carbonyl absorption band in vapour phase at 1772 cm^{-1} and in liquid phase at 1746 cm^{-1}.

(ii) **Effect of solvent:** Change of solvent also shifts the IR band. For example, carbonyl stretching frequency of acetone occurs at 1742 cm^{-1} in the vapour phase whereas in the liquid phase dipole-dipole association of the type ($\text{---}\!>\!\overset{\delta^+}{C}=\overset{\delta^-}{O}\text{---}\!>\!\overset{\delta^+}{C}=\overset{\delta^-}{O}\text{---}$) lowers the frequency to 1718 cm^{-1}. In different solvents, the frequency of absorption is different viz.

Solvent	Hexane	Chloroform	Ethanol
Absorption band	1726 cm^{-1}	1713 cm^{-1}	1709 cm^{-1}

(*iii*) **Coupled Vibration and Fermi Resonance:** Vibrational coupling may take place between two bonds, through a common atom, vibrating with similar frequency. When the coupling is between two fundamental bands then it is called **coupled vibration** and when it is between fundamental vibration and overtone it is called **Fermi resonance**. In aldehydes, C—H stretching absorption appears at doublet at 2820 cm^{-1} and 2720 cm^{-1} due to interaction between C—H stretching (fundamental band) and the overtone of C—H bending.

(*iv*) **Electronic Effects:** Under the influence of electronic effects (inductive effect, mesomeric effect) force constant or bond strength changes and the absorption shifts from the normal value. For example the absorption of carbonyl group in the following compounds occur at:

Compound	HCHO	CH_3CHO	CH_3COCH_3
Absorption (cm^{-1})	1750	1745	1715

The introduction of —CH_3 group cause +*I* effect in CH_3 CHO and CH_3 CO CH_3 which results in lengthening or weakening of the bond and hence force constant is lowered and wave number of absorption decreases.

(*v*) **Hydrogen Bonding:** Hydrogen bonding brings about downward frequency shifts. Stronger the hydrogen bonding, greater is the absorption shift towards lower wave number than normal value. Moreover inter and intramolecular hydrogen bonding can be easily distinguished by IR spectroscopy. Generally, intermolecular hydrogen bonds give rise to broad bands whereas intramolecular hydrogen bonds given rise to sharp and well defined bands. Moreover intermolecular hydrogen bonds are concentration dependent and intensities of such bonds decrease and finally disappear on dilution whereas intramolecular hydrogen bonds are independent of concentration.

8.11.8 Applications of IR Spectroscopy

Infrared spectrum of a compound gives the maximum information about its structure. It is highly characteristic property of an organic compound. The applications of IR spectrum are discussed below:

(1) Comparison of Two Compounds—Fingerprint Region

The most powerful function of infrared spectroscopy is establishing conclusively the identity of two samples. The possibility of two different compounds having the same IR spectra is negligible. That is why IR spectra is also called the "fingerprint" of a molecule.

- Thus, if two pure samples have *different* IR spectra, the compounds must be *different*.
- If they have exactly *same* spectra, when run in the same medium under similar conditions, the two samples are of the *same* compound..

Actually, the region from 4000–1500 cm^{-1} in an infrared spectrum, is useful for the identification of functional groups.

This region shows the absorption due to stretching mode, whereas, the region below 1500 cm (1500 – 650 cm^{-1}) contains number of vibrations caused by bending and stretching vibrations. As there are many more bending vibrations in a molecule than stretching vibrations, this region shows a number of vibrations which makes it complex. However, each organic compound has its own unique absorption pattern in this region. This part of spectrum is therefore known as **"Finger Print Region"**.

The molecules containing the same functional group, show similar absorptions above 1500 cm^{-1}, but if the molecules are diferent than the spectra differ in finger print region, on the other hand if along with high frequency region (above 1500 cm^{-1}), the finger print region also matches, the two molecules are of the same compound.

This spectral comparison thus provides confirmation of their identity.

(2) Detection of Impurities in a Compound

The infrared spectrum cannot commonly distinguish a pure sample from an impure sample. However, spectrum of pure sample will have fairly sharp and well-resolved absorption bands. The spectrum of a impure sample that contains many different kinds of molecules will display broad and poorly resolved absorption bands.

(3) Detection of a Functional Group in a Compound

Infrared spectrum helps to reveal the functional group present in an unknown sample. In this technique almost all functional groups have their characteristic absorption bonds. Fig. 8.17 shows some of the most common groups with their absorption wavelength.

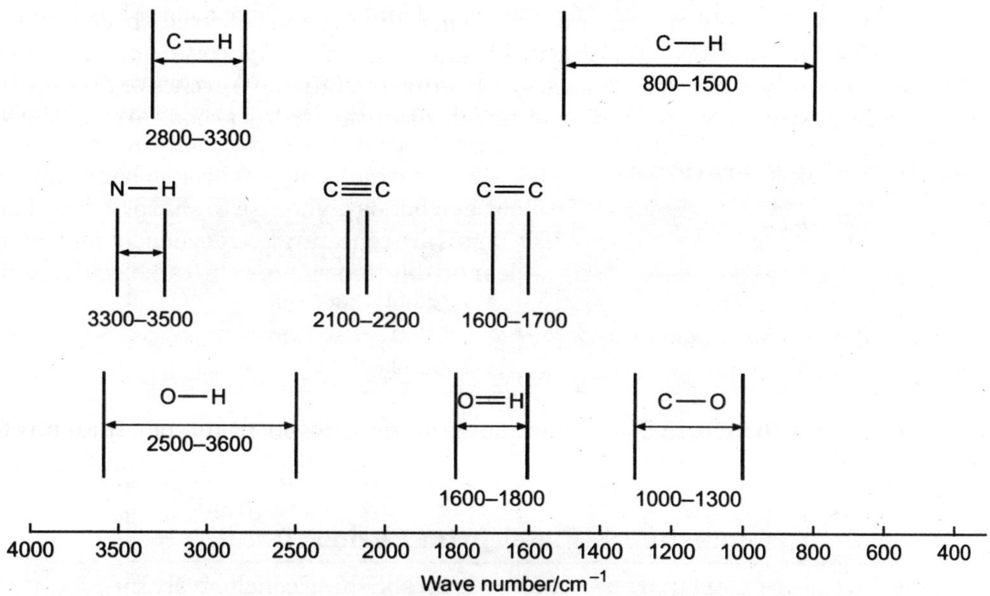

Fig. 8.17 The diagram shows where some of the most common groups appear on an infrared spectrum

For example, if the spectrum contains a strong bond at 1718 cm^{-1}, the compound almost certainly contains a carbonyl group (refer Fig. 8.17) However, the spectrum by itself does not always provide further information as to the nature of the group i.e. whether the compound having the carbonyl group is aldehyde, a ketone, an acid or an amide, all of which have carbonyl group. For this, interpretation of infrared spectra for other diagnostic absorption bands has to be carried out. On the other hand, if the spectrum does not contain an absorption band characteristic of a particular functional group, this suggests the molecule does not contain that functional group. This negative evidence is sometimes more useful in confirming the presence or absence of functional group.

(4) Elucidation of Structure

Structural elucidation of an unknown sample by infrared spectroscopy is one of the most significant application of infrared spectroscopy. When the spectrum of unknown material is obtained, a number of questions arise in the mind of a chemist, such as

- What is the type of carbon skeleton present in the compound?
- Which is the functional group present?
- What other environments are influencing it?

These questions can be answered fairly satisfactorily b;, systematic interpretation of the infrared spectrum. Next section discusses some of the important families giving characteristic absorptions and the environmental effects on them. *However, it is indeed not necessary to memorize all the absorption ranges but only a few values for general position can be remembered to interpret the data.*

8.11.9 Interpretation of Infrared Spectra of Some Representive Compounds

(1) IR Spectra of Hydrocarbons

As hydrocarbons are made up of carbon and hydrogen only, thus in infrared spectrum of these compounds, we shall see absorption bands due to vibrations of *carbon-hydrogen* and *carbon-carbon* bonds. Hydrocarbons may be saturated (alkanes), unsaturated (alkenes) cyclics or aromatics. The characteristic bonds due to stretching and bending vibrations in hydrocarbons are as follows:

(*a*) **C—H stretching Vibrations:**

Groups	***Infrared Absorption Wavelength (cm^{-1})***
Alkanes	2800 – 3000 cm^{-1}
Alkenes and Aromatics	3000 – 3100 cm^{-1}
Alkanes	3300 cm^{-1}

(*b*) C—H bending Vibration: These vibrations occur at lower frequency than stretching vibration.

Groups	***Infrared Absorption Wavelength (cm^{-1})***
Alkanes, C—H group	1340 cm^{-1}
Alkene, —CH_2— (Methylene group)	1485 – 1445 cm^{-1}
Alkane, —CH_3 (Methyl group)	1380 – 1370 cm^{-1} (Symmetrical and bending) and 1470 – 1430 cm^{-1} (Asymmetric bending)
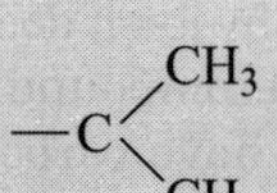 (Isopropyl group)	1385 cm^{-1} and 1370 cm^{-1} (Doublet with equal intensity)
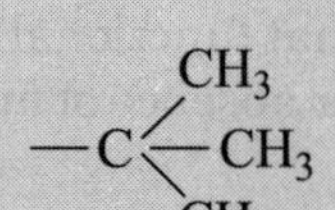 (Tertiary butyl group)	1370 cm^{-1} and 1395 cm^{-1}

Alkenes and Aromatic rings: In these compounds both in-plane and out-of-plane bending is present. However out of plane bending is more useful, which gives strong bonds in 800 –1000 cm^{-1}.

Groups	*IR Absorption*
Alkane, monosubstituted RCH = CH_2	910 – 920 cm^{-1} and 990 – 1000 cm^{-1}
Alkene, disubstituted, Cis RCH = CHR	675 – 730 cm^{-1}
Alkene, disubstituted, trans RCH = RHR	965 – 975 cm^{-1}
Alkyne ≡ C—H	630 cm^{-1}
Aromatic rings Monosubstituted,	730 – 770 cm^{-1}
Aromatic, disubstituted	
Aromatic, disubstituted ortho	735 – 770 cm^{-1} and 690 – 710 cm^{-1}
Aromatic, disubstituted meta	750 – 810 cm^{-1} and 690 – 710 cm^{-1}
Aromatic, disubstituted para	810 – 840 cm^{-1}

(c) C—C stretching Vibration:

Groups	*IR absorption*
Aromatic	1500 – 1600 cm^{-1}
Alkenes, nonconjugated	1680 – 1620 cm^{-1}
Alkene, monosubstituted	1645 cm^{-1}
Alkene, disubstituted, cis	1658 cm^{-1}
alkene, disubstituted, trans	1675 cm^{-1}
Alkyne monosubstituted	2140 – 2100 cm^{-1}

Some of the characteristic bands of hydrocarbon are represented by IR spectrum of 1-octane, oct- 1-ene, oct-1-yne Fig. 8.18 *a*, *b*, *c*

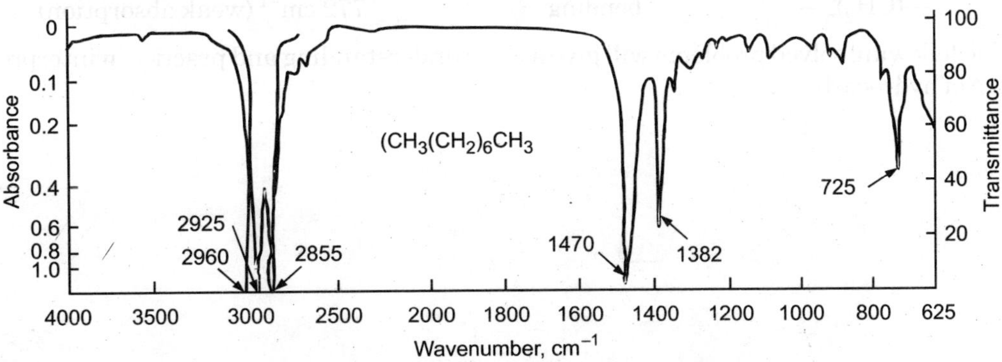

Fig. 8.19(a) Infrared spectrum of *n*-octance.

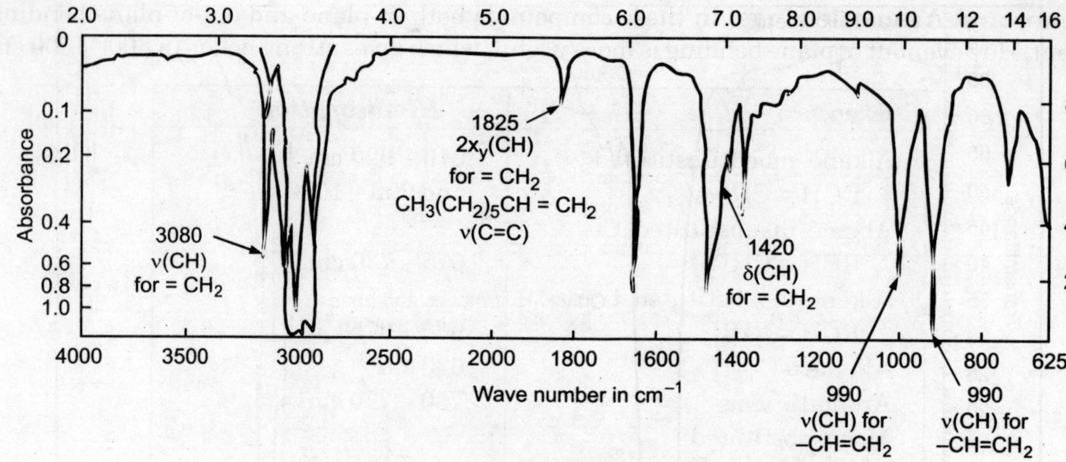

Fig. 8.18(b) Infrared spectrum of oct-1-ene.

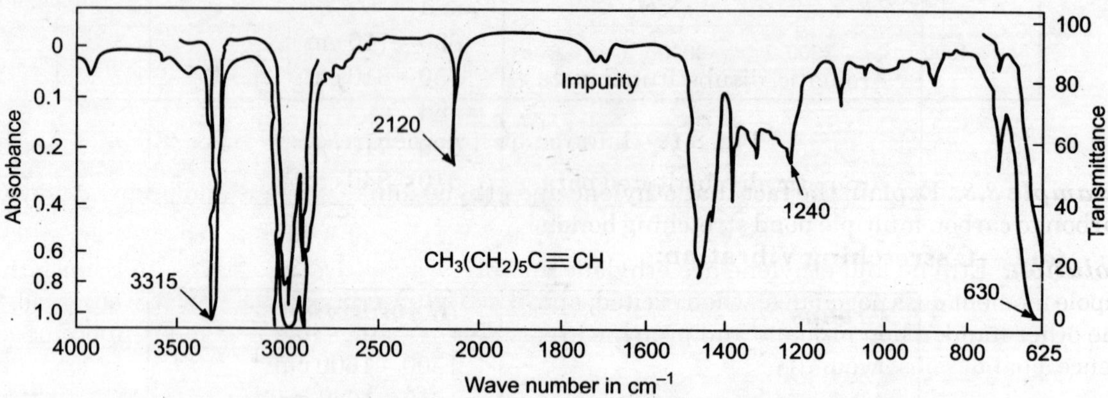

Fig. 8.18(c) Infrared spectrum of oct-1-yne.

The infrared spectrum of nujol (a mixture of saturated hydrocarbons also show a number of absorptions resulting from vibrations typical of groups that are present in such molecules: (Fig. 8.19)

C—H	stretching	2950 and 2820 cm^{-1}
—CH_2—	bending	1485 cm^{-1}
C—CH_3	bending	1458 and 1380 cm^{-1}
—$(CH_2)_n$—	bending	772 cm^{-1} (weak absorption)

The following solved problems will give a clear understanding and practice in interpretation of spectra of hydrocarbons.

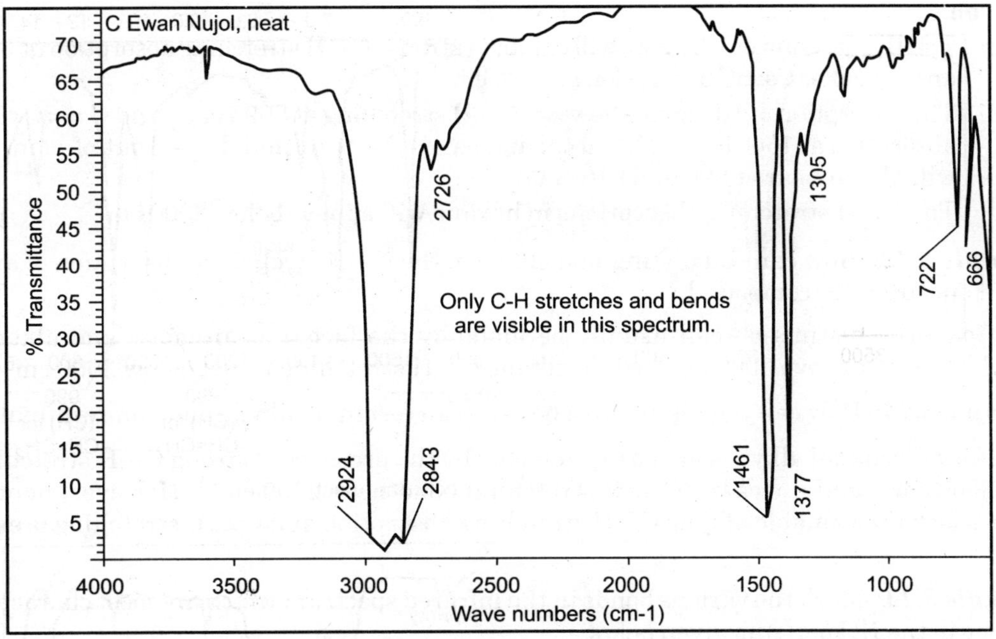

Fig. 8.19 Infrared spectrum of Nujol

Example 8.8: Explain the fact that ethylene and ethyne unlike propene and propyne, have no carbon to carbon multiple bond stretching bonds.

Solution: Ethyne and ethylene are ethylene and are *symmetrical* molecules and hence their dipole moment does not change when excited, and hence no absorption in IR region is observed. On the other hand, when propane and propyne are excited, a change in dipole moment occurs and hence a band is observed.

Example 8.9: The stretching frequencies for C—C, C = C and C ≡ C bonds increase in the order,

C—C	C=C	C≡C
1200 cm^{-1}	1680 – 1620 cm^{-1}	2140 – 2100 cm^{-1}

Explain.

Solution: The frequency of vibration is directly proportional to the force constant of the bond, which increase in the order,

$$C—C < C{=}C\ C < C \equiv C$$

and hence the stretching frequencies increase in the same order.

Example 8.10: How can you distinguish between an aliphatic and aromatic compound?

Solution: For aromatic compounds apart from other absorption peaks the most characteristic C=C stretching bands are at 1600 cm^{-1}, 1580 cm^{-1}, 1500 cm^{-1} and 1450 cm^{-1}. If there is no absorption in this region, it is a fair proof that the compound is not aromatic.

Example 8.11: How can you distinguish among hexane, 1-hexene and 1-hexyne using infrared spectroscopy.

$CH_3\ CH_2\ CH_2\ CH_2\ CH_2\ CH_3$
hexane

$CH_3\ CH_2\ CH_2\ CH_2\ CH = CH_2$
1-hexene

$CH_3\ CH_2\ CH_2\ CH_2\ C\ °\ CH$
1-hexyne

Solution:

(i) The IR spectrum of 1-hexyne will exhibit a strong ≡C—H stretching absorption at 3300 cm^{-1} and a weaker C≡C absorption at 2120 cm^{-1}.

(ii) The absorption of 1-hexene show ═C—H stretching at 3095 cm^{-1} and C═C stretching absorption at 1640 cm^{-1}. The additional bands due to terminal C—H out-of-plane bending will also appear at 990 and 910 cm^{-1}.

(iii) The C—H stretching absorptions in hexane will appear below 3000 cm^{-1}.

Example 8.12: How can you distinguish between the C—H stretching absorption of an alkane and alkene/aromatic compound

Solution: The distinction can usually be made by the fact that aromatic and alkene C—H stretching is just above 3000 cm^{-1} while alkane C—H stretching is just below 3000 cm^{-1}.

Example 8.13: How can you distinguish between an terminal alkyne and non-terminal alkyne.

Solution: Terminal alkynes are easily detected by the presence of strong C—H stretching band near 3300 cm^{-1} and a weaker C≡C stretching bond near 2200 cm^{-1}. However, nonterminal alkynes lack the valuable alkyne C—H stretching absorption and C≡C stretching is extremely weak..

Example 8.14: Mark the various bonds in the infrared spectrum which are most characteristic of benzene in the IR spectrum given below:

Infrared spectrum of benzene:

Solution: Position of characteristic absorptions,

(A) C—H Aromatic stretching vibration	3100 – 3000 cm^{-1}
(B) C═C stretching vibrations	1600, 1500 cm^{-1}
(C) C—H bending vibrations	900 – 700 cm^{-1}

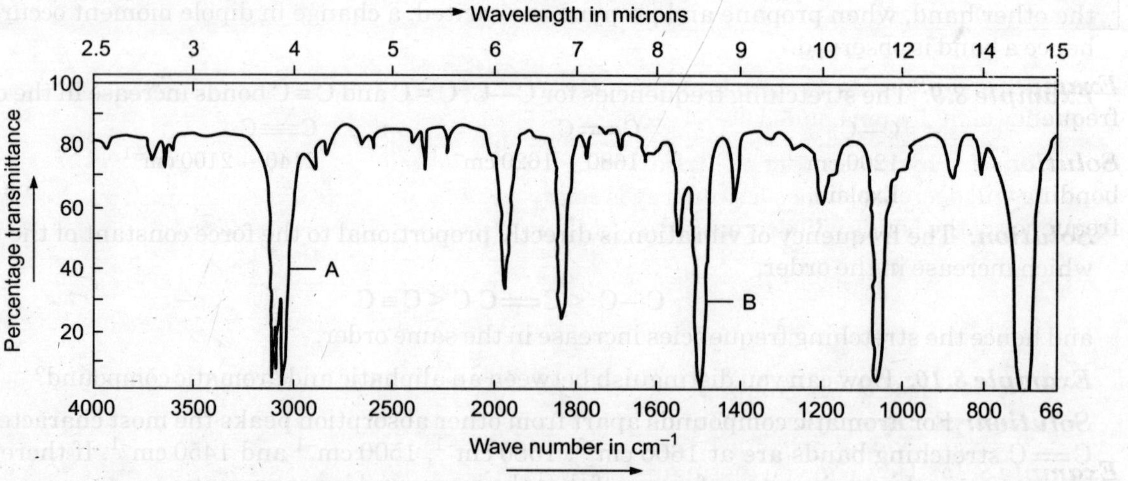

Fig. 8.20 Infra red spectrum of benzene

(2) Infrared spectra of O—H Compounds

Alcohols and phenols exhibit the property of hydrogen bonding and display the following characteristic absorptions due to vibrations and interactions of O—H and C—O bonds.

O—H Stretching vibrations: Alcohols and phenols, in varpour state or in dilute solution in non-polar solvents exhibit a weak O—H stretching absorption due to free OH group. When the spectrum is taken in concentrated solutions, the sharp band disappears and a broad band appears at lower frequencies

Group	*Type of Vibration*	*Region*	*Remarks*
Alcohols and Phenols			
Free OH group	O—H str	3700 – 3500 cm^{-1}	—
Intermolecular hydrogen bonded	O—H str	3400 – 3200 cm^{-1}	Dependent on concentration
Intramolecular hydrogen bonded	O—H str.	3570 – 3450 cm^{-1}	Independent of concentration

C—O stretching vibrations: The C—O stretching vibrations in alcohols and phenols produce a strong band near 1260 – 1000 cm^{-1}. The exact frequency depends on the nature of the alcohol.

Group	*Type of Vibration*	*Region*
Primary alcohols	C—O str. and O—H bending	1050 cm^{-1} and 1350 – 1260 cm^{-1}
Secondary alcohols	C —O str. and O—H bending	1100 cm^{-1} and 1350 – 1260 cm^{-1}
Tertiary alcohols	C—O str and O—H bending	1150 cm^{-1} and 1410 – 1310 cm^{-1}
Phenols	C—O str and O—H bending	1200 cm^{-1} and 1410 – 1310 cm^{-1}

Example 8.15: The O—H stretching frequency of free O—H group is higher than O—H stretching frequency of hydrogen bonded O—H group in alcohols and phenols. Explain.

Solution: In a hydrogen bonded hydroxy compound, the bond order of both the bonds involved in bonding will decrease, hence less energy is required to stretch it and consequently the stretching frequency of the bonds will decrease.

R—C—H

free O—H group
Absorption frequency = 3600 cm^{-1}

R—O—H····O(R)—H

hydrogen bonded O—H group
Absorption frequency = 3400 – 3200 cm^{-1}

Example 8.16: How can you distinguish between intermolecular and intramolecular hydrogen bonding on the basis of infrared spectrum.

Solution: The IR spectra of the compounds in non polar solvents is examined in different concentrations. Intermolecular hydrogen bonded molecules will show absorption shifts on changing concentration whereas for intramolecular hydrogen bonded compounds there is no absorption shift on changing the concentration.

Example 8.17: The solutions of ethanol and 1, 2 – glycol in carbon tetrachloride show a broad O—H str. absorption at 3350 cm^{-1} in their IR spectra. On dilution with CCl_4, the spectrum of 1, 2 glycol does not change but the spectrum of ethanol show a sharp band at 3600 cm^{-1} in addition to the band at 3350 cm^{-1}. Explain.

Solution: No change in the spectra of 1, 2 glycol on dilution indicates the presence of intramolecular hydrogen bonding. Whereas intermolecular hydrogen bonding is present in ethanol and accounts for the appearance of higher frequency band on dilution.

(3) Infrared Spectra of Carbonyl Compounds (Aldehydes and Ketones)

The carbonyl group ($>C=O$) is the most important in infrared spectrum because it absorbs strongly in this region and exhibit high sensitivity towards minor changes in its environment.

The appearance of a strong band due to C═O stretching in the spectrum between **1650 – 1950 cm^{-1}** shows the presence of carbonyl group in the compound. The frequency of absorption due to the carbonyl group depends mainly on the force constant which intrun depends upon—

(*a*) Inductive effect
(*b*) Conjugative effect
(*c*) Ring Strain
(*d*) Field effects.

(*a*) Inductive Effect: Carbonyl group can be written in the form

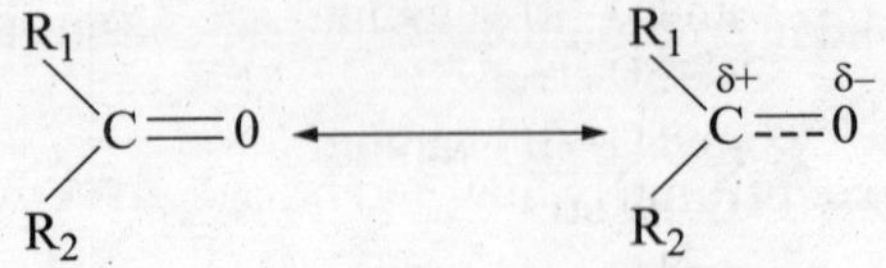

(I) (II)

- If R_1 and R_2 are electron repelling then structure II (polar) is predominant, the force constant for which is lowered as compared to structure I and hence absorption occurs at lower frequency.
- If R_1 and R_2 are electron withdrawing then structure I (double bond structure) is predominant, the force constant is high and hence absorption occurs at high frequency.

Due to large $+I$ effect operating in ketones, the ketones absorb at lower frequency as compared to aldehydes

Aldehyde	HCHO	1750 cm^{-1}
	CH_3CHO	1745 cm^{-1}
Ketone	CH_3COCH_3	1718 cm^{-1}

(*b*) **Conjugative Effect:** Conjugation with an olefinic group or benzene ring results in delocalisation of the Iĺ electrons of both the unsaturated groups thus reducing the double bond character of both. Due to the introduction of single bond character, the carbonyl absorption frequency decreases from 1718 cm^{-1} to about 1690 cm^{-1}.

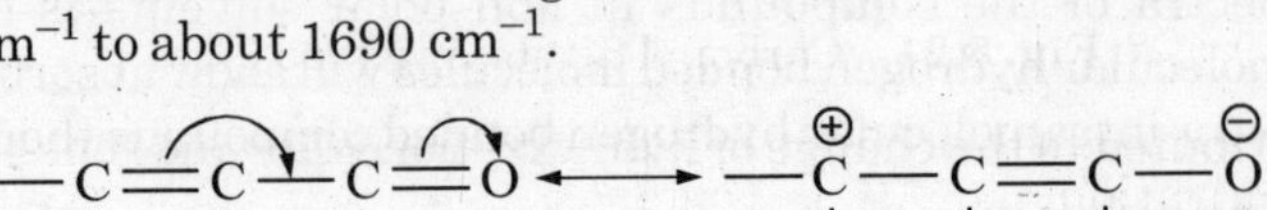

Example 8.18: Compare the C═O stretching absorption in the following compounds:

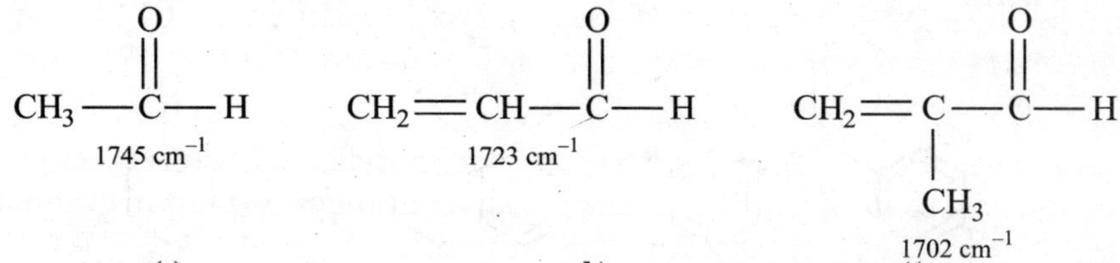

(a) (b) (c)

In compound (*b*) conjugative effect overpowers the –*I* effect and absorption occurs at lower frequency as compared to (*a*) In compound (*c*) both conjugative effect as well as +*I* effect due to CH_3 group lowers the absorption frequency as compared to the compound (*b*).

Example 8.19: Crotonaldehyde exhibit C═O and C═C stretching at 1700 cm^{-1} and 1630 cm^{-1} whereas an aliphatic aldehyde containing unconjugated double bond exhibits C═O and C═C stretching at 1720 –1740 cm^{-1} and near 1650 cm^{-1} respectively. Explain.

Solution: Crotonaldeyhyde can be represented by the following two resonating forms:

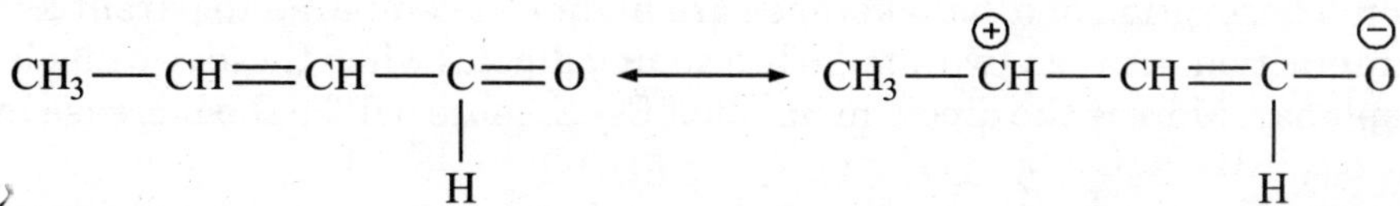

The presence of conjugated bond in crotonaldehyde reduce the stretching frequencies of C═O and C═C bond.

Example 8.20: Explain the exceptionally low carbonyl stretching frequency in 2, 4, 6 cycloheptatrienone.

Solution: Cycloheptatrienone can be represented as:

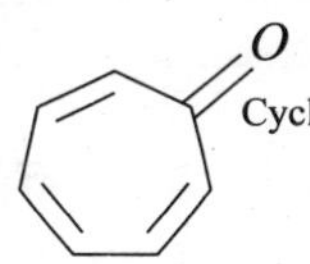

Cycloheptatrienone

The extremely low frequency of cyloheptatrienone is due to the presence of greater extent of conjugation because more is the conjugation more is the lowering of absorption frequency.

(c) Ring Strain: The C═O stretching frequencies for unstrained molecules (planar, trigonal bond angles of 120°) are considered normal. However if the size of the ring decreases i.e. the ring strain increases, it has considerable effect on absorption frequencies. The C═O stretching frequencies in cycloalkanones increase with a decrease in the ring size as shown below:

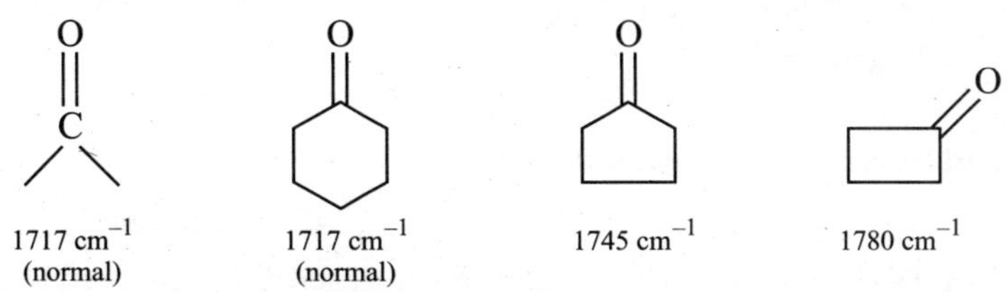

Fig. 8.21 Carbonyl Stretching Frequency

This change is attributed to the change of hybridization state and consequently change in force constant.

This provides a dependable test to distinguish clearly between three, four, five and larger membered ring ketones.

Example 8.41: Arrange the following carbonyl compounds in order of their increasing carbonyl stretching frequencies:

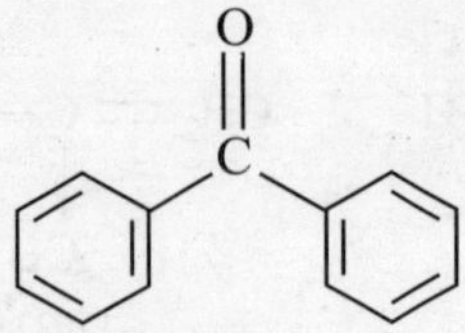

(*a*)

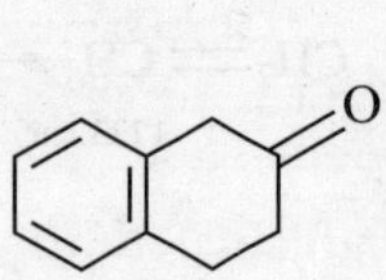

(*b*)

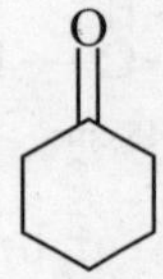

(*c*)

Solution: (*a*) is a conjugated carbonyl and as conjugation lowers the frequency, it will have lower absorption frequency. (*b*) is nonconjugated compound having five membered ring (ring strain) which increases its frequency of absorption. (*c*) is six membered (normal) ketone. Hence the order will be:

$$(a) < (c) < (b)$$

(*d*) **Field Effects:** A polar group can interact with carbonyl group through its dipole if it is oriented properly relative of the carbonyl group.

The carbonyl absorption of α-halo ketones are higher as compared to parent ketones due to electrostatic interaction between two similarly oriented dipoles when these two dipoles (C—X and C═O) are coplanar. More is the dipole moment of C—X, more will be the increase in absorption frequency.

The absorption frequencies due to C═O chromophore are summarized as follows:

Ketone Carbonyl Stretching Vibrations

(*i*) Saturated acyclic ketones — 1725 – 1705 cm^{-1}

(*ii*) Saturated, cyclic,

- 6-membered ring — 1725 – 1705 cm^{-1}
- 5-membered ring — 1750 – 1740 cm^{-1}
- 4-membered ring — ~ 1745 cm^{-1}

(*iii*) α, β unsaturated, acyclic — 1685 – 1665 cm^{-1}

(—C═C—C═O)

(*iv*) α, β unsaturated, cyclic

- 6-membered ring — 1685 – 1665 cm^{-1}

(*v*) Aromatic ketones — 1700 – 1680 cm^{-1}

(C_6H_5—C(═O)—R)

Aldehydes Carbonyl Stretching Vibrations

(*i*) Simple saturated — 1740 – 1720 cm^{-1}

- HCHO — 1750 cm^{-1}
- CH_3CHO — 1745 cm^{-1}

(*ii*) α, β unsaturated — 1750 – 1680 cm^{-1}

—C═C—C(═O)—H

(*iii*) Aryl aldehyde — 1705 – 1695 cm^{-1}

Aldehydes C—H Stretching Vibrations

Aldehydes	2900 – 2820 cm^{-1} and 2775 – 2700 cm^{-1}

Example 8.22: How can you distinguish aldehydes from ketones using IR spectroscopy.

Solution: Aldehydes can be distinguished from ketones by the presence of following bands:

(i) A doublet due to C—H stretching absorption of the CHO group in the regicn 2830—2695 cm^{-1}.

(ii) A weak band due to C—H out-of-plane bending absorption in the region from 975–780 cm^{-1}.

Example 8.23: Which of the following molecules has the highest carbonyl stretching frequency:

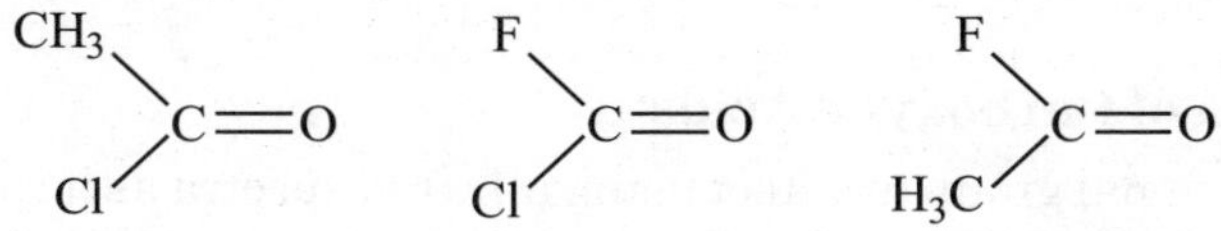

(*a*) (*b*) (*c*)

Solution: The stretching carbonyl frequency will be highest in the compound (*b*) because it involves the largest *–I* effect which withdraws electrons from carbonyl group and hence increases the stretching frequency.

Example 8.24: A compound has molecular formula C_8H_8O and has the following IR spectrum. Identify the compound.

Solution:

(i) The weak absorption at about 3040 cm^{-1} indicates the presence of aromatic C—H stretching.

(ii) The bands at 2980 and 2860 cm^{-1} due to stretching vibration of C—H group, indicate the presence of methyl group.

(iii) Prominent absorption at 1690 cm^{-1} indicates C═O stretching vibration due to an aromatic ketone.

(iv) The peaks at 750 cm^{-1} (aromatic C—H out-of-plane bending) and 680 cm^{-1} (aromatic C—C out of plane bending vibration) further indicate a monosubstituted benzene ring.

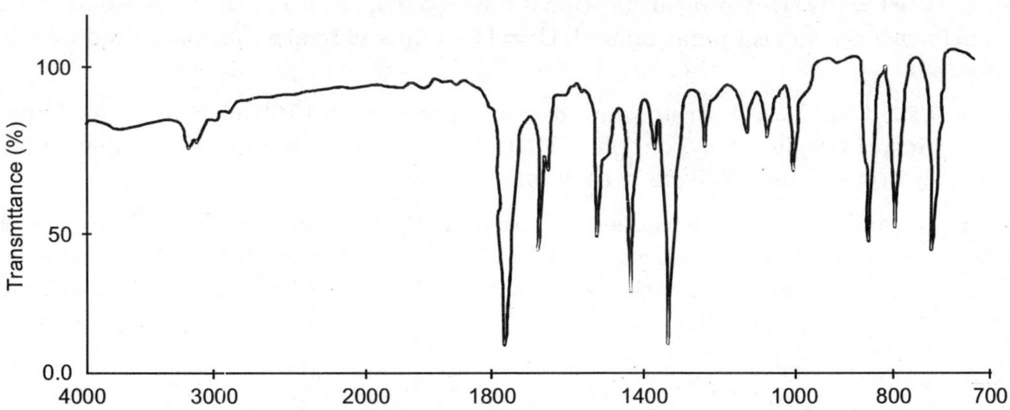

Fig. 8.22 IR spectrum of compound C_8H_8O

(v) Absence of doublet around 2700 cm^{-1} exclude the possibility of the presence of aldehydic group. Hence the structure consistent with the spectrum is

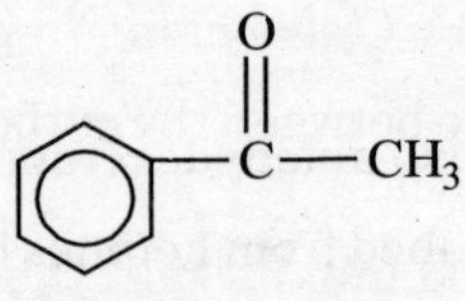

Acetophenone.

Example 8.25: Infrared absorption due to carbonyl stretching occurs at higher frequency than stretching of carbon-carbon double bond. Explain.

Solution: The π bond stretch of C═O bond is considerably higher than the π bond stretch of C═C bond.

(4) Infrared Spectra of Carboxylic Acids

This is the easiest functional group to detect using infrared spectra as it can be considered as being formed from C═O and O—H units. Thus IR spectra of these compounds will show the absorption bands due to both the groups.

(i) Absorption due to *O—H stretching* appears as a broad bond near 3000 cm^{-1}, but extending from 3300 to 2500 cm^{-1}. (The carboxylic acids in liquid or solid state usually exist as dimers and hence the formation of bridges lowers the force constants of O—H and C═O bonds thus absorption is shifted to lower frequency)

(ii) Absorption due to C═O group occurs at the following positions for respective acids:

Aliphatic acid	1725 – 1700 cm^{-1}
Aromatic acid	1715 – 1680 cm^{-1}

or conjugated acids

Example 8.26: An carboxylic acid is formed from an aldehyde on replacing hydrogen atom by OH group. OH has strong –*I* effect and as a result C═O stretching absorption for acids should occur at higher wave number as compared to aldehydes. But actually it occurs at lower wave number. Explain.

Solution: The C═O stretching absorption is lowered in acids because the lone pair on oxygen of OH group forms internal conjugation with C═O group and hence the absorption wave number decreases.

Example 8.27: The C═O stretching absorption occurs at 1760 cm^{-1} in case of monomeric carboxylic acid (acid studied in vapour phase), whereas the saturated aliphatic ketones show C═O stretching absorption near 1720 cm^{-1}. Explain.

Solution: In carboxylic acid, two effects are predominant which work in opposite directions i.e. inductive effect and mesomeric effect, since in carboxylic acids the inductive effect is dominant over mesomeric effect and hence carbonyl stretching frequency in carboxylic acids is higher than ketones.

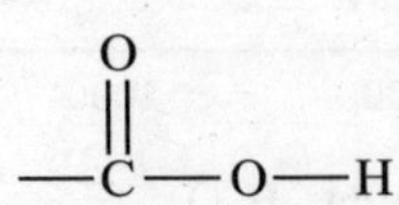

–*I* effect exerted by –OH group, increases absorption frequency

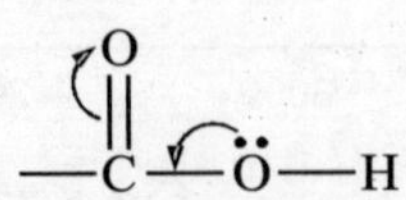

Mesomeric effect causes internal conjugation

Confirmation of the presence of carboxylic acid is done through conversion of acid to its salt. The spectrum of resulting carboxylate anion contains all the absorption pertaining to —COO^- group. Whereas absorptions due to –COOH group will be absent i.e. the absorption due to O—H stretching bond will disappear and C=O stretching frequency will be modified.

Example 8.28: Explain the difference between the carbonyl absorption of carboxylate anion and the carbonyl absorption in carboxylic acid.

Solution: When the acid is converted into its salt, the carboxylate anion formed is resonance hybridised having two CO bonds of equivalent strength, intermediate between C=O and C—O.

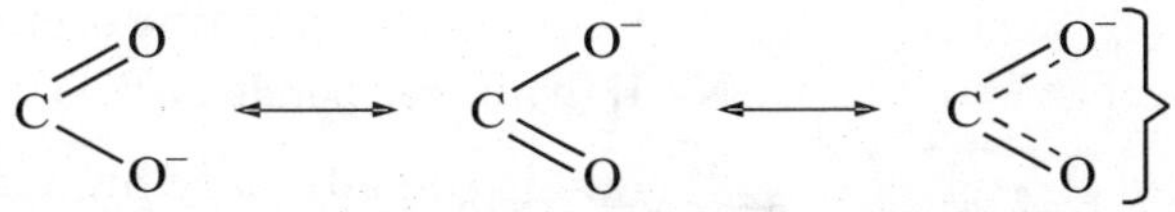

The coupling interaction between two equivalent C=O bonds of carboxylate anion will give rise to two bonds:

a strong asymmetic stretching bond	at 1650 – 1550 cm^{-1}
weak symmetric stretching bond	at 1400 cm^{-1}.

When the salt is converted to the free acid, the C=O stretching absorption appears at higher frequency than C=O stretching of carboxylate anion.

(5) Infrared Spectra of Derivatives of Carboxylic Acids

All the derivatives of carboxylic acids contain acyl group i.e. 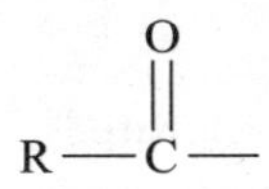. The derivatives are:

Acid halides 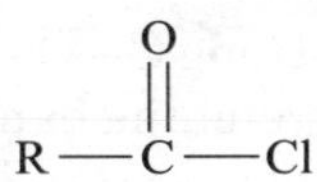

Acid anhydrides 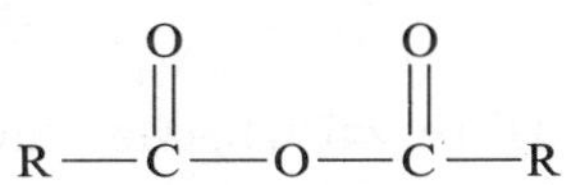

Amides 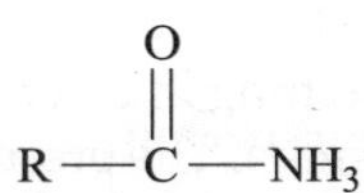

Esters 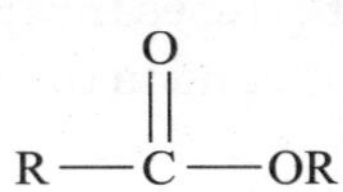

The various stretching absorptions for these derivatives are tabulated below:

(I) Acid Halides

Group	Vibration	Region
Acid chloride	C=O	1795 cm^{-1}
Acid bromide	C=O	1790 cm^{-1}

(II) Acid Anhydride

saturated, acyclic compounds	C=O	1850 – 1800 cm^{-1} and 1790 – 1740 cm^{-1}
$\alpha-\beta$ unsaturated	C=O	1830 – 1780 cm^{-1} 1770 – 1725 cm^{-1}

Note: (i) Acid anhydrides are always characterized by two high frequency bands usually separated by about 60 cm^{-1}.

(ii) All classes of anhydrides show two bands in the region 1300 – 1050 cm^{-1} due to C—O stretching.

(III) Amides

Group	Vibration	Region
Primary amides	C = O str.	1690 cm^{-1}
	N—H (free) str.	3500 cm^{-1}
		3400 cm^{-1}
	N—H (hydrogen bonded) str.	3350 cm^{-1}
		3180 cm^{-1}
N—H deformation	1620 – 1590 cm^{-1}	
Secondary amides	C = O str	1700 – 1670 cm^{-1}
	N—H str (free)	3425 cm^{-1}
	N—H str (hydrogen bonded)	3330 – 3140 cm^{-1}
	N—H deformation	1550 – 1510 cm^{-1}
Tertiary amides	C = 0 str.	1670 – 1630 cm^{-1}

IV Esters

saturated	C═O str.1750 – 1735 cm^{-1}
α – β unsaturated or aryl esters	C═O str.1730 – 1715 cm^{-1}

Notes: All esters exhibit strong C—O stretching absorptions in the region 1300 – 1030 cm^{-1} Aryl esters exhibit two C—O stretching bands near 1280 cm^{-1} and 1120 cm^{-1} caused by asymmetric and symmetric stretching of C—O—C groups. Alkanoate esters show one C—O stretching band only.

Example 8.29: Distinguish between an ester and ketone on the basis of IR spectroscopy.

Solution:

(i) The C═O stretching in esters occurs at higher frequency (~ 1740 cm^{-1}) as compared to ketones (~1720 cm^{-1}). The presence of one more oxygen atom in esters R—COOR compared to ketones (RCOR) raises the frequency of absorption due to –*I* effect.

(ii) A strong C—O str. band in the region 1300—1030 cm^{-1} in case of esters distinguish them from ketones

Example 8.30: Identify a particular acid derivative in each of the following carbonyl stretching frequencies:

(*a*) 1700 – 1815 cm^{-1} (*b*) 1720 – 1750 cm^{-1}

(*c*) 1630 – 1690 cm^{-1} (*d*) 1740 – 1790 cm^{-1} and 1800 – 1850 cm^{-1}

Solution:

(*a*) Acid chloride (*b*) Ester

(*c*) Amide (*d*) Anhydride

Example 8.31: The carboxylic acid and its derivatives have common C═O group. What are the distinctive features of each of the derivatives which can distinguish it from other.

Solution:

(i) Carboxylic acid show a strong and broad O—H str. frequency in the region 2500 – 3000 cm^{-1} which is absent in esters. Further, it shows C—O stretching at 1250 cm^{-1} whereas C—O stretching in esters occurs as two strong bands in the region 1050 – 1300 cm^{-1}.

(ii) Amides show N—H stretching in 3050 – 3540 cm^{-1} and also a strong absorption band due to N—H bending in the region 1645 cm^{-1} Acids and esters do not show these bands.

Example 8.32: Arrange the following non-conjugated carbonyl compounds in order of increasing carbonyl stretching frequency: aldehydes, esters, amides acyclic ketones, acid chlorides.

Solution: The carbonyl stretching frequency depends upon the inductive and resonance effect. The order is

Amides < Acyclic ketones < aldehydes < esters < acid chlorides.

(6) Infrared Spectra of Amines

Amines are characterised by the appearance of the following absorptions in IR spectrum :

Group	*Type of Vibration*	*Region*	*Remarks*
(*a*) Primary amines	N—H str.	3500 – 3300 cm^{-1}	Two sharp bands
Secondary amines	N—H str	3500 – 3300 cm^{-1}	One band
Tertiary amines	—	—	Do not absorb in this region
(*b*) Primary amines	N—H in-plane bending	1650 – 1580 cm^{-1}	—
Secondary amines	N—H in-plane bending	1650 – 1550 cm^{-1}	—
(c) Aromatic primary amines	C—N vibration	1340 – 1250 cm^{-1}	
Aromatic secondary amine	C—N vibration	1350 – 1280 cm^{-1}	
Aromatic tertiary amine	C—N vibration	1360 – 1310 cm^{-1}	
Alphatic amines	C—N vibration	1220 – 1020 cm^{-1}	

Similarities in O—H and N—H Stretching Bands

The typical O—H stretching absorptions in alcohols and phenols occur around 3650 – 3200 cm^{-1} which is also the region in which N—H stretching vibrations of amines occur. Hence O—H str. and N—H str. regions can be sometimes confused with one another. However, since oxygen is more electronegative than nitrogen the O—H ... O hydrogen bonds are stronger than N—H ... N hydrogen bonds, hence frequency shifts due to hydrogen bonding in amines are smaller than alcohols and phenols. The N—H absorptions occur at lower absorption frequencies compared to O—H.

Example 8.33: How can you distinguish between N—H stretching absorption of a primary amine (I) and a secondary amine (II)

$$R\,NH_2, \quad R_2\,NH$$

Solution: These can be distinguished by N—H stretching absorptions. For primary amines N—H str. would appear as doublet (symmetric and antisymmetric stretching). For secondary amines generally a single band will appear in this region.

Example 8.34: How will you distinguish between $(CH_3)_3 N$ and $CH_3 CH_2 CH_2 NH_2$.

Solution: The spectrum of the primary n-propylamine exhibits two N—H stretching bands in the region 3500-3300 cm^{-1}. These N—H stretching bands will be absent in tertiary butyl amine.

(7) Infrared Spectra of Nitro Compounds

In nitro compounds, the two oxygen atoms are equidistant from nitrogen atom and π electrons are distributed between the two N—O bonds:

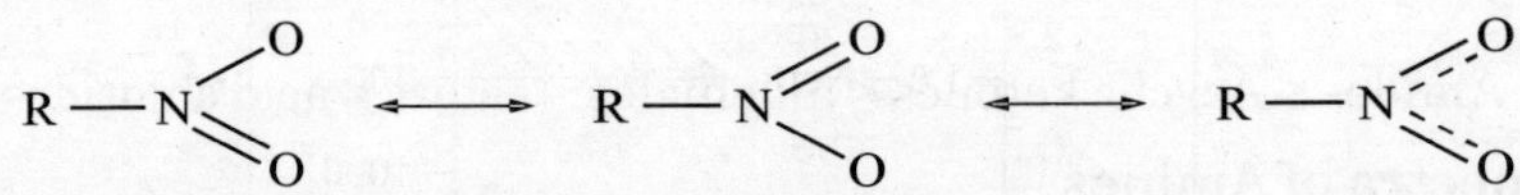

Thus in IR spectrum two strong bands arise from the symmetrical and asymmetrical stretching modes at 1620 – 1535 cm^{-1} and 1375 – 1275 cm^{-1} respectively.

Group	*Type of Vibration*	*Region*	*Remarks*
(i) Aliphatic nitro compounds	N═O str.	1620 – 1535 cm^{-1} 1375 – 1275 cm^{-1}	Symmetric str. Antisymmetric str.
(ii) Aromatic nitro compounds	N═O str	1570 – 1500 cm^{-1} 1370 – 1300 cm^{-1}	Electron Releasing groups at 'o' or *p* positions lower the N═O str. frequencies

(8) Infrared Spectrum of Ethers

Ethers, the derivatives of alcohols show characteristic C—O—C stretching absorptions in the range **1300 – 1000 cm^{-1}**. But a band in this region is also shown by alcohols, aldehydes, ketones and carboxylic acids. Hence the presence of an ether is confirmed only if the compounds shows no absorption band in the hydroxy (3650 – 3200 cm^{-1}) or carbonyl (1870 – 1650 cm^{-1}) region.

In case of dialkyl or diarylethers the symmetric C—O—C stretch is infrared inactive and only antisymmetric stretch is observed (one band). Whereas in case of highly unsymmetrical ethers i.e. alkyl aryl ethers, both the bands appear in infrared spectrum

Example 8.35: How can you distinguish between on ethyl alcohol and diethyl ether.

Solution: The absence of O—H str. band at 3200 – 3600 cm^{-1} in ether and presence of this band in alcohol distinguishes between the two compounds.

8.12 ULTRAVIOLET AND VISIBLE SPECTROSCOPY

Ultraviolet spectroscopy is used to measure the number of conjugated double bonds and also aromatic conjugation within the various molecules.

The electromagnetic radiations in UV-visible region extends from 200 – 750 nm (1 nm = 10^{-7} cm).

The light is of higher frequency and hence of greater energy than infrared light. When UV-visible light is absorbed by a molecule it produces the changes that require greater energy i.e. changes in electronic states. Hence UV-visible spectroscopy involves the transitions of electrons from the ground state to the higher energy states i.e. it leads to transitions among the electronic energy level of the molecule. It is also referred to as *electronic spectroscopy*.

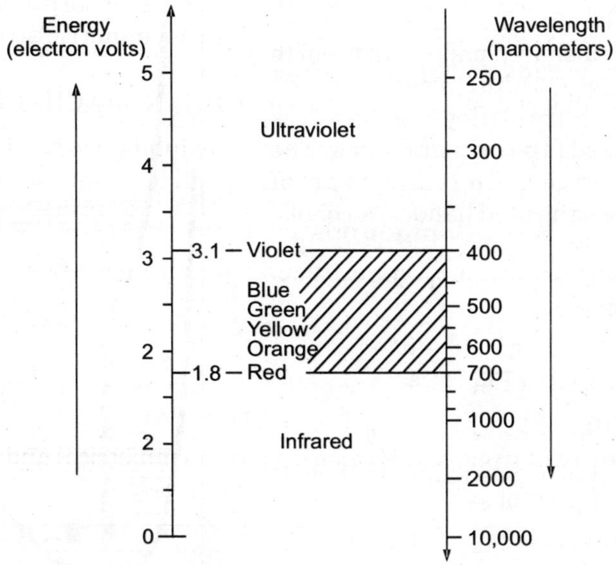

Fig. 8.23

8.12.1 Electronic Transitions: Types of Electronic Transitions

Absorption of ultraviolet radiation by an organic molecule leads to electronic transitions between various energy levels. These transitions consist of an excitation of an electron from an occupied molecular orbital to the next higher orbital.

Now, the electrons in organic molecules may be involved in bonding as strong σ bonds, weak π bonds or present as the non-bonding lone pair. Absorption of UV radiations can bring about the following transitions:

(*i*) **σ – σ* Transition:** The σ electrons are held firmly in the molecule, the transition from σ to σ* require large energies corresponding to absorptions in the far ultraviolet region in the range of 120 – 200 nm.

(*ii*) **π – π* Transitions:** This transition takes place in the compounds containing double or triple bonds. The excitation of π electron requires smaller energy and hence transition occurs at longer wavelengths.

In unconjugated alkenes, absorption bonds appear around 170 – 190 nm.

$$\rangle C :: C \langle \xrightarrow{\pi - \pi^*} \rangle C : . \dot{C} \langle$$

(*iii*) ***n* – π* Transitions:** In this transition, an electron of unshared electron pair on a hetero atom is excited to π* anti-bonding orbital. As these non-bonding electrons are held loosely, the transition requires longer wavelengths. These transitions occur with compound containing double bonds involving hetero atoms having unshared pair of electrons (e.g. $C = O$, $C = S$, and $N = O$ etc.).

For example, aldehydes and ketones exhibit two bands at 180 – 200 nm (π – π* transition) and 280 nm (*n* – π*). *n* – π* transition is always less intense because the electrons in the *n*-orbital are situated perpendicular to the plane of the π bond and hence the probability of the jump of an electron from *n* to π* is low.

(*iv*) **$n-\sigma^*$ Transitions:** The excitation of an electron in an unshared pair on an atom to an antibonding σ orbital is called $n-\sigma^*$ transition. Such transitions require less energy compared to σ to σ* transition. The compounds undergoing this type of transitions are saturated halides, alcohols, ethers etc.

The relative energies required for the various transitions obey the order

$$\sigma-\sigma^* > n-\sigma^* > \pi-\pi^* > n-\pi^*$$

The order is illustrated in Fig. 8.24.

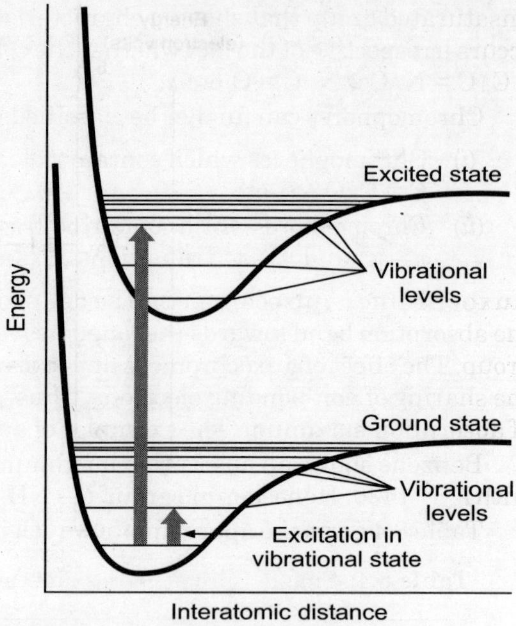

Fig. 8.24 Energy curves for a diatomic molecule

Transition Probability

Depending upon the value of extinction coefficient the transitions can be classified as :

(*a*) Allowed Transitions

(*b*) Forbidden Transitions

The transitions with values of ε_{max} more than 10^4 are usually called *allowed transitions.* They are generally due to $\pi-\pi^*$ transition.

The transitions with values of ε_{max} below 10^4 are *forbidden transitions.* They are generally due to $n-\pi^*$ transitions. For example in benzophenone, the two types of transitions involved are

(*i*) 252 nm ε_{max} 20,000 (Allowed)

(*ii*) 325 nm ε_{max} 180 (Forbidden)

Designation of UV Bands

The UV absorption bands are either designated by using electronic transitions such as ($\pi\rightarrow\pi^*$, $n\rightarrow\pi^*$ etc.) or by letter designation such as:

R Bands: The bands due to $n\rightarrow\pi^*$ transitions are called *R*-Bands (German-Radikalartig). They have low molar absorptivities ($\varepsilon_{max} < 100$) and are called forbidden bands.

K Bands: Such type of bands arise due to π–π* transitions in conjugated π systems (German - Konjugierte). These type of bands are found in dienes, polyenes, enones etc. The ε_{max} is usually more than 10^4.

B Bands: B bonds are characteristic of aromatic and heteroaromatic compounds representing π to π* transitions. The B-band occurs in the region from 250–255 nm as a broad band containing multiplet five structures.

E Bands: These are also characteristic of aromatic and heteroaromatic compounds. The E_1 and E_2 bands of benzene occur near 180 and 200 nm respectively with the ε value between 2000 – 14000.

8.12.2 Chromophores and Auxochromes

Chromophore: Originally, a chromophore was considered any system which is responsible for imparting colour to the compound. (Greek Chromophorus - colour carrier). For example, in nitro compounds the yellow colour is carried by — NO_2 group and hence NO_2 group is a chromophore.

The term has been extended these days and chromophore is defined as a covalently bonded unsaturated group that shows a characteristic absorption in UV or visible region. The absorption occurs irrespective of the fact whether the colour is produced or not. Typical examples are C = C, C ≡ C, C = N, C ≡ N, C = O etc.

Chromophores can further be classified into two types:

(*i*) Chromophores which contain π electrons and they undergo π – π* transitions. Example, C = C, C ≡ C etc.

(*ii*) Chromophores which contain both π electrons as well as non-bonding electrons and undergo π – π* and *n* – π* transitions. Examples are C = O, C ≡ N etc.

Auxochrome: Auxochrome can be defined as any group whose presence brings about a shift of the absorption band towards the longer wavelength. Actually auxochrome is the colour enhancing group. The effect of auxochrome is due to its ability to extend the conjugation of a chromophore by the sharing of non-bonding electrons. Thus, a new chromophore results which has different value of absorption maximum. The examples of auxochrome groups include — OR, — NH_2, — NR_2 etc.

Benzene shows an absorption maximum at 255 nm with ε_{max} 203 whereas aniline at 280 nm with ε_{max} 1430. Hence amino group (— NH_2) is an auxochrome.

Table 8.4 gives the approximate wavelength of absorption of organic chromophores.

Table 8.4: Simple Chromophores and the Approximate Wavelengths at which they Absorb.

Chromophore	*Transition*	*Absorption max. (mμ)*	ε_{max}	*Solvent*
C = C	π – π*	~ 175	~ 15000	Vapour
		(*i*) ~ 175	~ 10000	Hexane
C ≡ C	π – π*	(*ii*) 196	~ 2000	"
		(*iii*) 220	~ 150	"
C=O	*n* → σ*	160	18000	
	π – π*	180	10000	Hexane
	n → π*	285†	15	
R – NO_2	π – π*	~ 200	5000	Methanol
	n → π*	~ 274	15	
—C(OH) = O	*n* → π*	204	60	Methanol
—N = N—	*n* → π*	338	~ 5	Ethanol
— $CONH_2$	π → π*	178	9500	Hexane
	n → π*	220	63	Hexane

8.12.3 Absorption and Intensity Shifts

(*a*) **Bathochromic shift (Reds Shift):** Shift of an absorption maximum to longer wavelength is called Bathochromic shift or Red shift. It may be due to the presence of an auxochrome or change in the polarity of the solvent.

- The *n* – π* transition for carbonyl compounds experiences bathochromic shift when the polarity of the solvent is decreased.
- Conversion of phenol to phenolate anion causes bathochromic shift because the non-bonding electrons become available for interaction with the π-electron system of the ring.

(*b*) **Hypsochromic Shift (Blue Shift):** Shift of absorption maximum to shorter wavelength is called Hypsochromic shift or Blue shift. It may be due to increase in the polarity of the solvent or the removal of conjugation.

- Aniline absorbs at 280 nm because of the pair of electrons on nitrogen is in conjugation with the π bond system of benzene ring. However in its acidic solutions, a blue shift is caused and absorption occurs at ~ 203 nm because 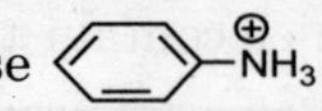 formed in acidic solution, the electron pair on nitrogen is not present and hence the conjugation is removed.

(*c*) **Hyperchromic Shift:** It is the effect leading to increase in intensity of the absorption maximum i.e. ε_{max} increases. The introduction of auxochrome usually increases the intensity of the absorption maximum.

- The B-bands for pyridine at 257 nm with ε_{max} 2750 shifts to 262 nm with ε_{max} 3560 for 2-methyl pyridine.
- Phenol absorbs at 210 nm with ε_{max} 6200 but the value of ε_{max} increases to 9400 in phenolate ion with λ_{max} 210 nm.

(*d*) **Hypochromic Shift:** It is the effect leading to decrease in intensity of the absorption maximum i.e. ε_{max} decreases. The groups which distort the geometry of the molecules cause hypochromic shift.

- Benzene absorbs at 230 nm with ε_{max} 11600 whereas benzoate absorbs at 224 nm with ε_{max} 8700. Hence the ε_{max} is decreased in benzoate ion.

Terminology for Absorption Shift

Nature of Shift	*Descriptive Term*
To Longer Wavelength	Bathochromic
To Shorter Wavelength	Hypsochromic
To Greater Absorbance	Hyperchromic
To Lower Absorbance	Hypochromic

Fig. 8.25(a)

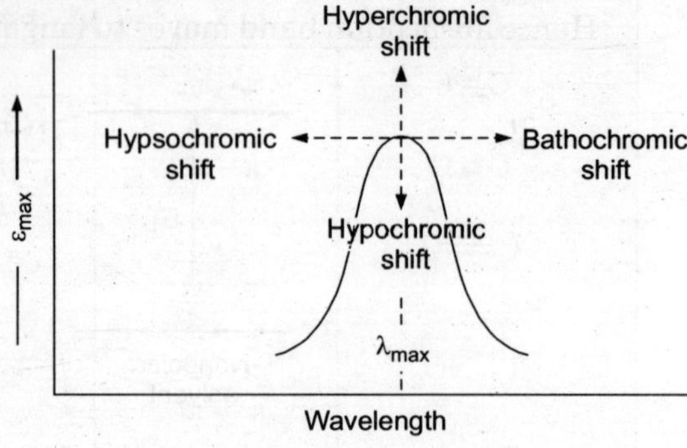

Fig. 8.25(b)

8.12.4 Solvent Effects

The most suitable solvent for UV spectroscopy is the one which does not absorb in the region under investigation. Most commonly useds solvents are ethanol, hexane, methanol, cyclohexane etc. They are transparent (do not absorb) upto above 210 nm.

The position and intensity of absorption bands is greatly affected by the polarity of the solvent used.

Following shifts are absorbed on changing the polarity of the solvents:

- Conjugated dienes and aromatic hydrocarbons experience very little shifts on changing the polarity of solvents.
- The polar compounds show a change in absorption maximum on changing the polarity of the solvent. α, β-unsaturated carbonyl compounds show two different shifts on increasing the polarity of the solvent:

(*a*) **The π-π* band** moves to longer wavelength. This is because the dipole-dipole interactions with the solvent molecules lower the energy of the excited state more than that of the ground state. In other words, π* orbitals are more stabilized by hydrogen bonding with polar solvents because of greater polarity of π* compared to π orbitals. Thus the energy required for π - π* transition is less and absorption moves to longer wavelengths.

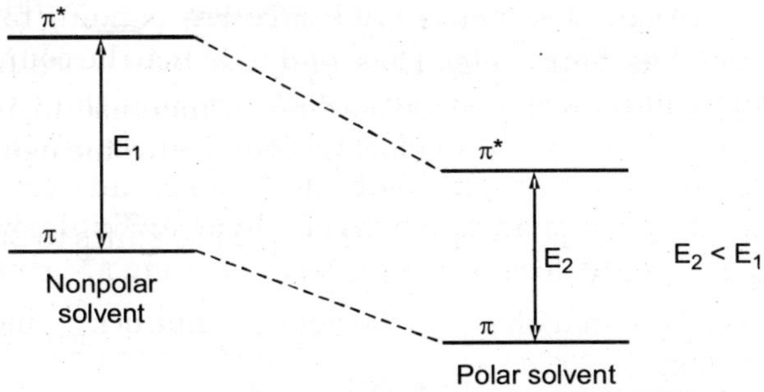

Fig. 8.26 Absorption shift with change in polarity in π-π* band.

(*b*) **The *n*-π* bond** moves to shorter wavelength by increasing the polarity of the solvent. This is because the polar solvent stabilizes the *n* orbitals more than the π* orbitals. This stabilization is more with hydrogen-bonding solvents (H_2O, ethanol etc.) Thus the net energy of *n* orbital is lowered, increasing the energy gap between ground and excited state. Hence absorption band moves to longer wavelengths.

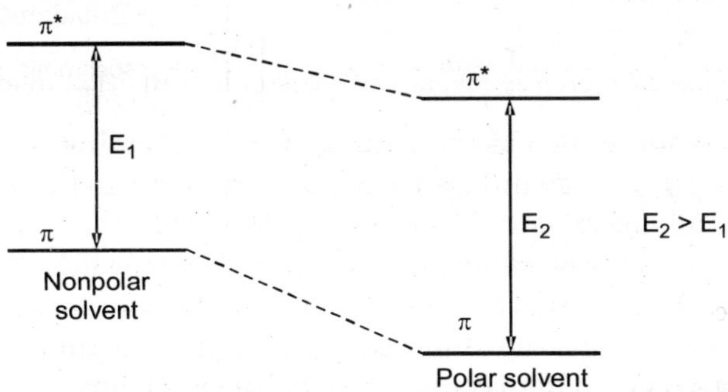

Fig. 8.27 Absorption shift with change in polarity in *n* – π* band

8.13 ULTRA-VIOLET SPECTRA OF ORGANIC COMPOUNDS

Organic applications of UV spectroscopy are mainly concerned with conjugated systems. Some of the most significant applications of UV spectroscopy in organic chemistry are discussed below:

(1) **Extent of Conjugation:** A conjugated system requires lower energy for the π-π* transition than an unconjugated system. Longer the conjugation, longer the absorption wavelength. For example, ethylene has one double band and absorbs at 170 nm due to π-π* transition whereas butadiene, having two double bonds in conjugation absorbs at 217 nm. Some more examples are

Compound	λ_{max}	ε_{max}
Ethylene	170	21,000
2,3 dimethyl butadiene	226	21,400
1,3,5 Hexatriene	254	21,400

How conjugation affects the λ_{max} value can be explained by considering the cases of ethylene and butadiene.

Ethylene has two π orbitals, one ground state π bonding orbital and one excited state π* antibonding orbital. When a molecule absorbs light at its longest-wavelength, an electron is excited from its highest occupied molecular orbital (HOMO) to the lowest unoccupied molecular orbital (LUMO). In ethylene or simple alkenes, the HOMO orbital is π orbital and LUMO is π* orbital. The energy difference between these two is 176 k cal corresponding to a wavelength of 170 nm.

Butadiene, however has four π electrons and has four π orbitals. Two bonding and two antibonding. The two bonding are designated as π_1, π_2 and the two antibonding orbitals are designated as π_1* and π_2*. The π_1 orbital of butadiene is more stable than π orbital of ethylene because the electrons clouds in π_1 orbital butadiene is spread over four carbon atoms. The π_2 orbital of butadiene is less stable than π_1 orbital of ethylene. Similarly the two π_1* and π_2* orbital of butadiene are more stable and less stable respectively than π* orbital of ethylene. (Fig. 8.28)

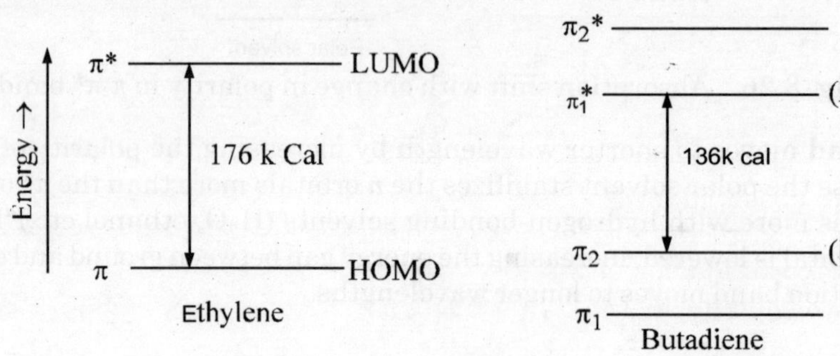

Fig. 8.28 Molecular orbital energy levels and electronic transition in ethylene and 1, 3 butadiene.

On absorption of energy the electron jumps from π_2 bonding → π_1* antibonding (HOMO → LUMO) orbital. Since the energy difference between π_2 to π_1* in butadiene is less as compared to π - π* in ethylene, butadiene absorbs at longer wavelengths.

Hence, greater the number of π orbitals (conjugation) lower will be the energy difference between HOMO and LUMO and longer will be the absorption wavelength.

Sufficient conjugation shifts the absorption to wavelengths that reach the visible region of the spectrum i.e. a compound with high conjugation becomes coloured.

For example, tomatoes are red in colour because of the presence of lycopene, the compound having eleven conjugated double bonds. Thus it absorbs at λ_{max} value of 454 nm which is the wavelength of blue colour - the complementary colour of red and hence the tomatoes appear red.

Similarly dyes owe their colours to conjugated systems.

Example 8.36: Which of the following will exhibit higher value of λ_{max} in the UV spectra ?

(*a*) CH_3COCH_3 and $CH_2=CHCOCH_3$

(*b*) $CH_2=CH-CH=CH-CH_2-CH_3$ and $CH_2=CH-CH_2-CH=CH-CH_3$

Solution:(*a*) $CH_2=CHCOCH_3$

(*b*) $CH_2=CH-CH=CH-CH_2-CH_3$

The UV absorption wavelength in conjugated dienes and trienes can be calculated using *Woodward–Feiser Rules* given in section 8.11:

(2) **Determination of Geometrical Isomers:** It has been found that cis-alkenes absorb at different wavelengths as compared to their corresponding trans isomers. Coplanarity is needed for

the most effective π orbital which eases the π–π* transition. The steric strain in cis-isomer prevents complete coplanarity. Thus in cis-isomer λ_{max} arises at shorter wavelength.

Example 8.37: Cis stilbene absorbs at 274 and trans stilbene absorbs at λ_{max} 294 nm. Explain.

Solution: The steric hindrance in cis-isomer prevent complete coplanarity required for full conjugation and hence it absorbs at shorter wavelength than trans isomer.

(3) **UV Spectra of Carbonyl Compounds:**

(*i*) **Saturated Carbonyl Compounds:** For a carbonyl compound, two types of transitions can take place $\pi \rightarrow \pi^*$ and $n \rightarrow \pi^*$.

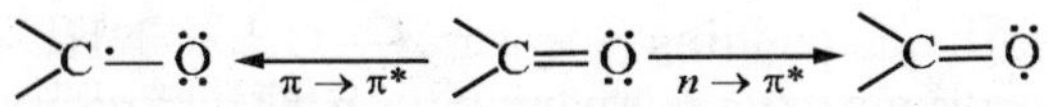

The $\pi \rightarrow \pi^*$ transition is very intense and corresponds to shorter wavelength. $n \rightarrow \pi^*$ transition is less intense and corresponds to longer wavelength. The λ_{max} values for isolated > C = 0 band are:

Transition	λ_{max}	ε_{max}
$\pi \rightarrow \pi^*$	170	10,000
$n \rightarrow \pi^*$	290	~ 20

Thus two bonds are observed for isolated > C = 0 group in UV spectra. When the carbonyl group is substituted by an auxochromic - as in ester, amide or acid, the π* orbital are raised and hence $n - \pi^*$ transitions occur at shorter wavelengths.

The presence of weak band in the 275 – 295 nm region is positive identification of a carbonyl group.

(*ii*) **α–β unsaturated Carbonyl Compounds:** When a carbonyl group is conjugated with an ethylenic linkage as in α – β unsaturated carbonyl compounds, then both the bands undergo bathochromic shift i.e. move towards longer wavelengths in the range of 300 – 350 nm.

The position and intensity is also imfluenced by solvent effects discussed in section 8.12.5. Increasing the polarity of the solvent induces red shift in $\pi - \pi^*$ transition and blue shift in $n - \pi^*$.

Moreover axially substituted isomers absorb at longer wavelengths than equitorially substituted isomers.

Example 8.38: Explain why a polar solvent usually shift the $\pi - \pi^*$ transition to longer wavelength and n - π^* transition to shorter wavelength.

Solution: Refer to section 8.12.5

Example 8.39: Acetone absorbs at 279 nm in hexane whereas the values of λ_{max} in water is 264.5 nm. Explain.

Solution: The shift to shorter wavelength in water is due to the change in polarity of the solvent. Hydrogen bonding of acetone with water molecule lowers the energy of n-orbitals and hence n-π* transition requires more energy i.e. shorter wavelength.

(4) **UV Spectra of Benzene and its Derivatives:** Benzene shows three absorption bonds, all due to π–π* transitions. The three bonds occur at

λ_{max}	ε_{max}	*Remarks*
184 nm	60,000	Allowed transition
204 nm	7900	Forbidden transition
256 nm	200	Forbidden transition

The bond at 256 nm (B-bond) is easily identifiable and is called benzenoid band. It shows a great deal of fine structure in vapour state.

- When benzene is substituted by simple alkyl groups the absorptions are shifted to slightly longer wavelength. This bathochromic shift is due to hyperconjugation in which the σ-electrons of C—H of alkyl substituent takes part in resonance with the ring. For example, B band of benzene appears at 254 nm and is shifted to 261 nm in case of toluene.
- In case of disubstituted benzene the para isomer absorbs at the longest wavelength as compared to meta and ortho. Ortho absorbs at shortest wavelength. This is due to steric interaction between the ortho substituents, which reduce the effect of hyperconjugation.
- The substituents with non bonding electrons (— OH, — NH_2, — OR etc.) present in benzene ring, shift the absorptions to longer wavelengths. The fine structure of the B band is either diminished or eliminated, because of *n*-π conjugation.. Table 8.5 lists some of the absorption values of *K*-band of benzene and its derivations:

Table 8.5 Benzene and its Derivative

Compound	*λ_{max} in water*		*Intensity ε_{max}*	
	K Band	*B-Band*	*K Band*	*B-Band*
Benzene	204 nm	254 nm	7900	204
Chlorobenzene	290 nm	263.5 nm	7400	190
Aniline	230 nm	280 nm	8600	1430
Nitrobenzene	252 nm	280 nm	10,000	1000

- Absorption maximum for polynuclear aromatic hydrocarbons also move to longer wave-lengths.

254 nm
Benzene

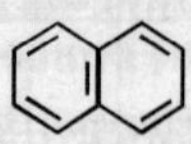
286 nm
Naphthalene

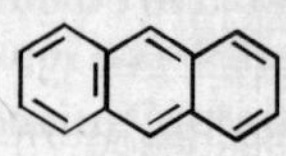
375 nm
Anthracene

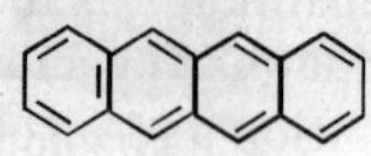
Tetracene

As might be expected from their absorption wavelengths that naphthalenes and anthracene are colourless but tetracene is coloured (orange).

8.14 WOODWARD FIESER RULES

The absorption maxima for a conjugated or substituted diene can be predicted by Woodward Fieser Rules. According to these rules, each type of diene has a certain fixed value and the value of λ_{max} depends upon:

(*i*) the number of alkyl substituents or ring residues on double bond.
(*ii*) the number of double bonds with extended conjugation.
(*iii*) the presence of polar groups such as — Cl, — Br, — OR, —SR etc.

The various rules are summarized in Table 8.6:

Table 8.6 Woodward Fieser Rules for Diene and Trienes

Base value	λ max
Parent acyclic diene	217 nm
*Homoannular diene	253 nm
**Heteroannular diene	214 nm
Addition for each substituent	
— R (alkyl) or ring residue	5 nm
— OR (alkoxy)	6 nm
*** Exocyclic double bond (to one ring)	5 nm
Exocyclic double bond (to two rings)	10 nm
Double bond extending conjugation	30 nm
Auxochrome	
— OR	6 nm
— SR	30 nm
— Cl, — Br	5 nm
— NR_2	60 nm
— $OCOCH_3$	0 nm

λ_{max} (calculated) = Base value + substituent contribution

Let us illustrate these rules by taking a few examples in the following problems.

Example 8.40: Calculate the λ_{max} in UV spectrum of 2, 4 hexadiene.

Solution: $CH_3 — CH = CH — CH = CH — CH_3$

2, 4 hexadiene

Basic unit is CH = CH — CH = CH i.e. butadiene

Substituents are two alkyl groups.

∴ Basic Value (from the table) = 217 nm

2-alkyl substituents (2 × 5 nm) = 10 nm

Calculated λ_{max} 227 nm

Example 8.41: Calculate the λ_{max} value in the following compounds:

(*a*) 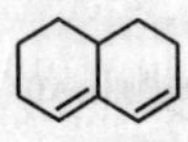 (*b*) 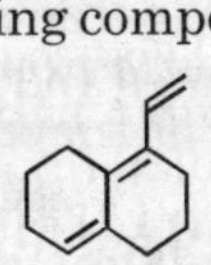

Solution: (*a*) 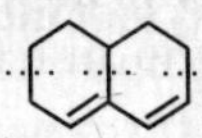

Base value of heteroannular diene = 214 nm

Substituents or ring residues (3 × 5) = 15 nm

Exocyclic double bond (1 × 5) = 0.5 nm

Calculated λ_{max} 234 nm

* If the double-bands are present in the same ring it is called homoannular diene.

** If the conjugated double bands are not present in the same ring then it is called heteroannular diene.

*** 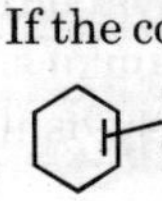 endocyclic double bond 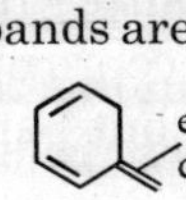 exocyclic double bond

(*b*) 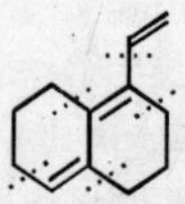

Base value of heteroannular diene	=	214 nm
Double bond extending conjugation	=	30 nm
Ring residues (4 × 5)	=	20 nm
Exocyclic double bond (2 × 5)	=	10 nm
Calculated λ_{maxm}		274 nm

Example 8.42: Calculate λ_{max} for the following compounds.

(*a*) 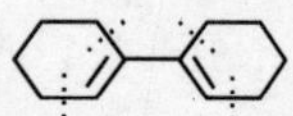 (*b*) 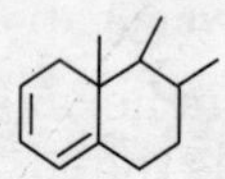

Solution: (*a*) Base value of heteroannular diene = 214 nm

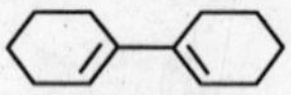 Ring residues (4 × 5) = 20 nm

Calculated λ_{max} 234 nm

(*b*) 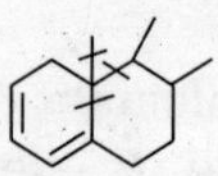

Base value of homoannular diene	=	253 nm
Ring Residues (3 × 5)	=	15 nm
Exocyclic double bond (1 × 5)	=	5 nm
Calculated λ_{max}		273 nm

APPLICATION OF ULTRA-VIOLET SPECTROSCOPY

(1) **Extent of Conjugation:** UV spectroscopy is widely used to determine the extent of conjugation in polymers. The value of λ_{max} increases with the extent of conjugation. (Refer to section - 8.13)

(2) **Distinction between Conjugated and Non-conjugated Compounds:** UV spectroscopy distinguishes between conjugated and non-conjugated compounds. For example in the following compounds,

$(CH_3)_2C=CH-\underset{\underset{O}{\|}}{C}-CH_3$ (*a*)

$CH_2=\underset{\underset{CH_3}{|}}{C}-CH_2-\underset{\underset{O}{\|}}{C}-CH_3$ (*b*)

the compound (*a*) will have longer λ_{max} value because C = O group is in conjugation with double bond.

(3) **Characterization of Aromatic Compounds:** Anomatic compounds can be characterized by their characteristic absorption bands in UV spectroscopy. (Section 8.13).

(4) **Detection of Impurities:** Benzene is present as an impurity in ethanol. Because of toxic nature of benzene, it is important that ethanol for internal use should be freed from benzene. The presence of even minute amounts of benzene can be confirmed by UV spectrum of sample at 280 nm. At this wavelength ethanol is transparent but benzene exhibits an absorption band with $\varepsilon_{max} = 230$.

(5) **Quantitative Analysis:** This analysis is based on Beer-Lambert law.

$$A = \varepsilon cl$$

where A = absorbance at a particular wavelength
c = Concentration of the sample
ε = Constant known as molar absorptivity or extinction coefficient.

The absorbance of standard solution (A_1) and of that of the sample is measured (A_2) in the same sample cell (l = constant), concentration of standard solution, knowing the (c_1) the concentration of the sample (c_2) can be determined as

$$\frac{A_1}{A_2} = \frac{c_1}{c_2}$$

Example 8.43: For a solution of camphor in hexane in a 10 cm cell, the absorbance A was found to be 2.52 at 295 nm with $\varepsilon_{max} = 14$. What is the concentration of camphor.

Solution:

$$A = \varepsilon cl$$
$$A = 2.52, \quad \varepsilon = 14, \quad l = 10 \text{ cm}$$
$$\therefore \quad 2.52 = 14 \times c \times 10$$
$$c = \frac{2.52}{14 \times 10} = 1.8 \times 10^{-2} \text{ moles/litre}$$

(6) **Determination of strength of hydrogen bonding:** On the basis of shifts absorbed in polar solvents, it is possible to deduce the strength of hydrogen bonds. (Section 8.12.5)

(7) **Study of isomerism:** Refer to section 8.13 for geometric isomerism.

SOLVED PROBLEMS BASED ON UV SPECTROSCOPY

Example 8.44: Explain the variation in the value of λ_{max} for the following compounds:

CH_3Cl, λ_{max} = 173 nm
CH_3Br, λ_{max} = 204 nm
CH_3I, λ_{max} = 258 nm

Solution: These compounds undergo $n-\sigma^*$ transitions. As we move down the halogen group, the quantum numbers increase from Cl to I and hence the n electrons lie for away from the nucleus. Thus these are easily excited and λ_{max} occurs at higher values.

Example 8.45: Why λ_{max} for diene (I) is observed at lower nm than (II)

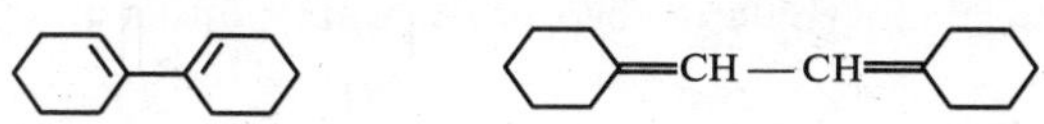

I II

Solution: Though there are four ring residues as substituents in I and II. But in diene (II), the two double bonds are exocyclic and hence according to Wood-Fieser rules, the λ_{max} will be higher by 2 × 10 nm.

Example 8.46: Why PABA (*p*-aminobenzoic acid) is used in sunscreen lotions.

Solution: PABA has λ_{max} at 289 nm with a high extinction coefficient of around 19,000. Due to this property it absorbs UV light and protect our skin from high energy UV radiation.

Example 8.47: β-carotene has eleven conjugated double bonds and is responsible for orange colour of carrots. It absorbs light at 454 nm in the visible region. Explain, why it absorbs in the visible region and what colour is absorbed ?

Solution: As the number of conjugated double bonds increases, the energy difference between HOMO and LUMO decreases. Thus ΔE corresponds to longer wavelengths in the region of visible light. The colour absorbed at 454 nm is blue which is the complementary colour of orange colour.

QUESTIONS

Infra Red Spectroscopy

1. What is the difference between absorption and emission spectra.
2. Distinguish between atomic and molecular spectroscopy.
3. What do you understand by molecular spectroscopy. Name the different types of molecular spectra.
4. What are vibrational and rotational quantum numbers?
5. Explain Beer-Lambert's Law.
6. Why absorption spectroscopy is preferred? Give its experimental set-up.
7. (a) What are the factors that affect the width of spectral line?
   (b) What are the factors which affect the intensity of spectral line?
8. What are electromagnetic radiations? Give the regions of electromagnetic radiations in the increasing order of frequency.
9. Show vibrational, rotational and electronic energy levels on an energy level diagram. To which region of electromagnetic spectrum do rotational, vibrational and electronic spectra are concerned?
10. What type of molecules give pure rotational spectra?
11. Explain the principle and working of spectrophotometer, given its neat diagram.
12. Briefly explain, how a vibrational-rotational spectrum is obtained? What do you understand by fundamental bands and overtone bands?
13. What do you understand by force constant? How is the force constant of a molecule related to its vibrational frequency?
14. What do you understand by modes of vibration? Explain with suitable examples.
15. What type of molecules exhibit vibrational-rotational spectra and why?
16. Distinguish between the following, from their IR-spectra:
    (i) Ethanol and dimethylether (ii) Primary, secondary and tertiary amides.
17. What modes of vibrations are active in I.R. absorption spectra and why? Taking nitro group as an example show the various types of vibrations which will be observed in IR spectra.
18. Three test tubes contain three ketones one of them absorbs near 1785, one near 1750 and one near 1710 cm^{-1}. What types of ketones are they?
19. Explain, from the spectrum of acidified ethanol, it is seen that methyl absorption is split into a triplet (relative areas 1 : 2 : 1) and the methylene absorption is split into a quartet (relative areas 1 : 3 : 3 : 1).
20. (a)How many vibrational modes are there in a linear and a non linear molecules having n atoms? Explain with the help of degrees of freedom. Draw the diagram of vibrational modes for CO_2 molecules.
21. (a) What are the different kinds of electronic transitions.
    (b) Distinguish between the following pairs from their IR spectra.
    (i) 2-propanol and propanone (ii) Ethanol and dimethylether
    (iii) Ethylamine and acetamide

UV – Visible Spectroscopy

1. What is an absorption spectra? Write what do you know about $n \rightarrow \pi^*$ and $n \rightarrow \sigma^*$ transitions? Explain by giving examples.

2. Write short note on the electronic transitions caused by energy absorbed in the U.V. region.
3. Why molecules absorb in UV-visible region? What are the type of electronic transitions that can occur in a molecule? Discuss giving examples.
4. (a) Give one example each of $\sigma \rightarrow \sigma^*$, $n \rightarrow \sigma^*$, $n \rightarrow \sigma^*$, $n \rightarrow \pi$ and $\pi \rightarrow \pi^*$ transitions.
   (b) What is the effect of the presence of alpha-beta unsaturation on the carbonyl absorption in the U.V. spectrum.
5. List all the electronic transitions possible for (a) CH_4, (b) CH_3Cl, (c) $H_2C = 0$.
   (Ans. (a) $\sigma \rightarrow \sigma^*$, (b) $\sigma \rightarrow \sigma^*$ and $n \rightarrow \sigma^*$ (there are no π or π^* MO's), (c) $\sigma \rightarrow \sigma^*$, $\sigma \rightarrow \pi^*$, $\pi \rightarrow \sigma^*$, $n \rightarrow \sigma^*$, $\pi \rightarrow \pi^*$ and $n \rightarrow \pi^*$)
6. The UV spectrum of acetone shows two peaks at λ_{max} = 280 nm. ε_{max} = 15, and λ_{max} = 190 nm, ε_{max} = 100. (a) Identify the electronic transition for each (b) Which is more intense ?
   (Ans. (a) The longer wavelength (280 nm) is associated with smaller energy ($n \rightarrow \pi^*$) transition, $\pi \rightarrow \pi^*$ transition occurs at 190 nm. (b) $\pi \rightarrow \pi^*$ has the larger ε_{max} and is more intense.)
7. Identify the two geometric isomers of stilbene, C_6H_5 CH = CH C_6H_5, from their λ_{max} values, 294 nm and 278 nm.
   (Ans. The higher energy *cis*-isomer has the shorter λ_{max}. Steric strain prevents full coplanarity of the *cis* phenyl groups and the conjugative effect is attenuated.)
8. Predict the kind of electronic transitions in (a) Cl_2 and (b) C = 0 group. Also give their intensity.
   (Ans. (a) $\sigma \rightarrow \sigma^*$ (allowed transition, strong)
   $n \rightarrow \sigma^*$ (forbidden transition, weak)
   (b) $\pi \rightarrow \pi^*$ and $\sigma \rightarrow \sigma^*$ (both allowed, strong)
   $n \rightarrow \pi^*$, and $n \rightarrow \sigma^*$ (both forbidden, weak)
9. (a) How 1, 3-Pentadeiene and 1, 4 Pentadiene can be distinguished by UV spectroscopy.
   (b) Name two molecules which undergo $n \rightarrow \pi^*$ electronic transitions. Discuss the effect of solvent on these transitions.
10. (a) How will you distinguish between benzene and anthrancene by UV spectroscopy?
    (b) Arrange the following compounds in the increasing order of their UV absorption maxima: (a) Ethylene (b) Naphthalene (c) Anthracene (d) 1, 3-Butadiene
11. Write short notes on : (i) Franck condon principle (ii) Chromophores and Auxochromes.
12. Explain the effect of polar solvents on (1) $n \rightarrow \pi^*$ and $\pi \rightarrow \pi^*$ transitions.
13. (a) Define the following terms
    (i) Bathochromic shift (ii) hypochromic shift
    (iii) A chromophore (iv) Hyperchromic effect
14. Explain the various applications of ultra-violet spectroscopy.
15. "Increase in polarity of the solvent shifts $\pi \rightarrow \pi^*$ and $n \rightarrow \sigma^*$ band to shorter wavelength"
16. (a) Write short notes on K-band and R-band.
    (b) Which solvent in generally used in ultra-violet spectroscopy and why?
17. (a) What do you understand by Photoelectron spectroscopy.
    (b) Why does molecular spectroscopy utilise absorption spectra and not emission spectra.
    (c) What do you understand by UV spectroscopy?
18. (a) What do you understand by Electronic spectroscopy?
    (b) What type of Electronic transitions are involved in ultra-violet and visible regions? Explains the significance of one type and give suitable examples.
    (c) Explain; "Doppler broadening of spectral lines".
19. Draw a block diagram of a typical instrument for making UV/Visible absorption measurement. Briefly describe the various components of the instrument.
20. Explain the following in respect of UV-Visible spectroscopy:
    (b) Red shift or bathochromic effect. (b) Blue shift or hypsochromic effect.
    (c) Hyperchromic effect (d) Isobestic point

21. (a) Explain the terms: (i) Molar extinction coefficient (ii) Optical density
    (b) What are charge transfer transitions?
22. (a) Explain Frank Condon principle with examples.
    (b) Describe a UV-Vis double beam spectrophotometer.
23. (a) Explain the following:
        (i) Frank-Condon principle (ii) Chromophores and Auxochromes
    (b)A solution shows a transmittance of 20% when taken in a cell of 2.5 cm thickness. Calculate its concentration if the molar absorption coefficient is 12,000 $dm^3\ mol^{-1}\ cm^{-1}$.
24. (a) Explain photosensitization.
    (b) Give requirement of laser action.
25. What do you understand by quantum yield?
26. What is photosensitization? Write the photosensitized reaction that occurs in nature.
27. Show how does a laser acquire a strong coherent beam.
28. (a) Define quantum yield? Provide an explanation for high and law quantum yield found in certain photochemical reactions.
    (b) In the photochemical reaction of $H_2 + Cl_2 \xrightarrow{hv} 2HCl$ the quantum yield was found to be 1.0×10^6 upon absorption of radiation of wavelength 4800 Å. Calculate the number of moles of HCl(g) produced under conditions for calorie of radiant energy absorbed.
    (c) Using Jablonski diagram, differentiate between phosphorescence and fluorescence.
29. How does temperature affect the rate of photosynthesis in plants.
30. (a) State and explain Einstein law of Photochemical Equivalence.
    (b) Describe Jablonski diagram for depicting various photophysical processes.
    (c) Write a short note on lasers and its uses.
31. Define the term photosensitization.
32. (a) Explain photosensitization and quenching by taking examples.
    (b) What are lasers? Explain laser action with reference to three level and four level laser. Discuss practical uses of lasers.

EXAMINATION QUESTIONS

1. "IR spectra is often characterized as molecular finger-prints". Comment on it.
   [U.P. Tech., B. Tech., 1st Sem., 2000-01]
2. Which of the following molecules will show infrared spectrum and why?
   H_2, HCl, CH_4, CO_2, H_2O *[U.P. Tech., B. Tech., 1st/2nd Sem., 2001-02]*
3. Point out the True and False statements out of the following:
   (i) Ultraviolet technique helps in predicting the number of conjugated double bonds in a molecule.
   (ii) The substance which absorbs in the visible region appears colored to the human eye.
   (iii) Spectroscopy involves the interaction between matter and electromagnetic radiation.
   (iv) In the nuclear magnetic resonance spectrum, radiation of radio frequency range is used.
   (v) In infra red, electron donating groups lower the wave number of absorption while the electron attraction groups raise the wave number. *[RGPV, B.E., Ist Sem., 2001]*
4. How can you distinguish $CH_3CH_2CH_2CHO$ from $(CH_3)_2CHCHO$ by proton NMR spectroscopy?
   [U.P. Tech., B. Tech., 2nd Sem., 2000-01]
5. Write a brief not on IR spectroscopy. *[U.P. Tech., 1st Sem. (C.O.), 2003]*
6. A gaseous hydrocarbon 'A' on passing through a quartz tube heated at 600°C, gave a liquid compound 'B' (Molecular weight : 78 amu). The latter compound was found to undergo electrophilic substitution reactions. It gave the following physical data on analysis. The IR spectrum showed a characteristic absorption band at 3040 cm^{-1} and a UV absorption, due to π–π^* transition, at 204 nm ($\log_e$, 3.84). The ^1H-NMR spectrum displayed a downfield, singlet (6H) at 7.3 τ. Identify the compound 'A' and 'B' and give your reasoning. *[U.P. Tech, 1st Sem., 2004-05]*

7. Define the terms chromophone and auxochrome, in UV spectroscopy. [*U.P. Tech., 2st Sem., 2006-07*]
8. What is Beer-Lambert law in UV-VIS absorption spectrospcopy ? A compound having concentration 10^{-3} g/l resulted absorbance value **0.20** at λ_{max} **510** nm using **1.0** cm cell. Calculate its absorptivity and molar absorptivity values. Molecular weight of compound is 400.

[*U.P. Tech, B.Tech. Ist Sem 2008-09*]

❑❑❑

9 SPECTROSCOPIC METHOD OF ANALYSIS II

The best way to escape from a problem is to solve it.

— Alan Saporta

CHAPTER OUTLINE

Nuclear magnetic resonance spectroscopy involves the transition of a nucleus from one spin state to another with the absorption of electromagnetic radiation by spin active nuclei, when they are placed in a magnetic field. Spin active nuclei are the nuclei having nuclear spin not equal to zero.

The electromagnetic radiations used in NMR spectroscopy have very long wavelengths ($\lambda = 10^7$ μ to 10^8 μ) and have extremely low energy. These radiations are *radio frequency waves* which are *not* energetic to effect the electronic, vibrational or rotational energy levels of the molecules. However they are able to interact with the nuclei of certain atoms which are exposed to a strong magnetic field, hence the name nuclear magnetic spectroscopy.

Before discussing the theory of NMR it is worth to review the magnetic properties of the nuclei of certain atoms.

9.1 MAGNETIC PROPERTIES OF NUCLEI

(*a*) Nucleus—A Tiny Magnet: The nucleus of certain isotopes (such as ^1H) behave as tiny spinning bar magnet because it possess both electric and magnetic spin. Hence like any other spinning charged body, the nucleus of hydrogen atom also generates a magnetic field. Thus the nucleus is equivalent to a minute magnet having magnetic moment μ.

(*b*) Effect of External Magnetic Field: Now if a magnetic nucleus such as ^1H is placed in a uniform external magnetic field, it is found that it can only adopt two orientations with respect to the direction of the applied external magnetic field —

- either aligned with the field (parallel)
- or opposed to the external field (antiparallel)

(c) Precessional Motion : The precessional motion of the nucleus can be understood by taking an example of a spinning top, the spinning axis of which moves slowly around the vertical axis of earth's gravitational field. This waltz like motion of the spinning top around the vertical axis ϕ is known as precessional motion.

Now as the proton is also a spinning magnet, it will also *precess* around the axis of the applied magnetic field. Either it can precess aligned with the field or opposed to the field. The aligned orientation is stable and is of low energy whereas the opposed orientation is unstable and is of higher energy.

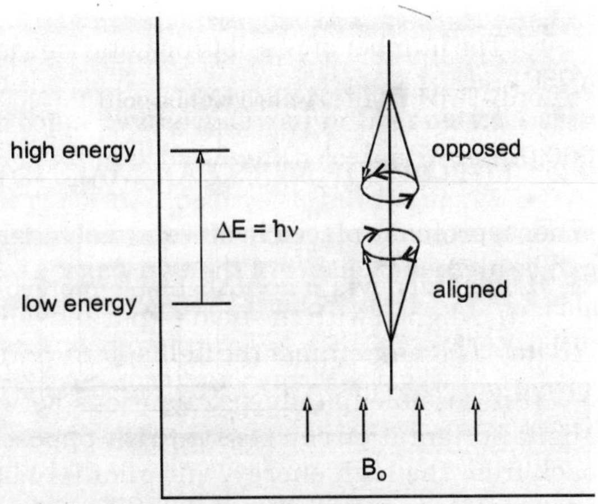

Fig. 9.1 Representation of precessing nuclei, and the DE transition between the aligned and opposed conditions.

The processional frequency υ is directly proportional to the strength of external field B_0 i.e.

$$\upsilon \propto B_0$$

This frequency of precession is extremely important in NMR spectroscopy.

(d) Transitions between the Two States: It is possible to induce transitions between these two transitions i.e. from lower energy state to higher energy state. The frequency of the radiation required to induce the transition is given by,

$$\upsilon = 2\,\mu B_0/h$$

where μ = magnetic moment
B_0 = strength of external magnetic field.
h = Planck's constant.

This frequency υ of electromagnetic radiation, required to induce transition from one nuclear spin state to another, is exactly equal to the processional frequency of the spinning nucleus. In other words, the processing proton will absorb energy from the radio frequency region only if the processing frequency is the same as the frequency of the radio frequency radiation. In such situation, the nucleus and the electromagnetic radiation are said to be in **resonance** and hence the name nuclear magnetic resonance.

(e) Spin-active Nuclei: Nuclei that exhibit the NMR phenomena are known as spin-active nuclei. As is evident from the above discussion that nuclear spin is the fundamental property of the nucleus to be involved in the NMR spectroscopy. Only those nuclei will exhibit the phenomenon of NMR for which the spin quantum number *I is greater than zero.* These nuclei are known as spin-active nuclei. For example, ^{1}H, ^{13}C and ^{19}F etc. ^{12}C, ^{16}O, ^{32}S etc. have $I = 0$ and are *not spin-active.* The spin active nuclei have permanent magnetic moments and quantized nuclear spin states.

The spin quantum number I is associated with mass number and atomic number of the nuclei

Mass Number	*Atomic Number*	*Spin Quantum Number I*
Odd	Odd or even	$\frac{1}{2}, \frac{3}{2}, \frac{5}{2}, \ldots$
Even	Even	0
Even	Odd	1, 2, 3

The number of possible orientations is given by $2I + 1$. In the nuclei with spin $\frac{1}{2}$ i.e. ^{1}H, ^{13}C etc. only two orientations are allowed.

9.2 THEORY OF NUCLEAR MAGNETIC RESONANCE

When a proton is placed in an external magnetic field, its magnetic moment, as discussed above, can be aligned in either of the two ways:

(*i*) Aligned with the field—parallel alignment.
(*ii*) Aligned against the field— antiparallel alignment.

It is possible to induce transitions between these two states i.e. a proton processing in the aligned orientation can pass into the opposed orientation by absorbing energy. Also, it can come back from the high energy, antiparallel alignment by losing energy. This transition from one energy state to the other is called *flipping* of the proton.

Now, how much energy is required to flip the proton from one state to another depends upon the strength of the external magnetic field applied. Stronger the field, greater will be the tendency of the nuclear magnet to remain lined up with it and higher will be the energy required to flip the proton to the higher energy state. If ΔE is the energy difference between the two states, it is related to the external field strength B_0 as,

$$\Delta E = 2\mu B_0 \qquad \text{...(ii)}$$

or

$$\upsilon = \frac{2\mu B_0}{h} \qquad \text{...(iii)}$$

where h is Planck's constant, μ is the magnetic moment.

For example, in the field of 14092 gauss, the energy required to cause flipping of the proton corresponds to the electromagnetic radiation of frequency 60 million cycles per second or 60 MHz. These radiations are in the radio frequency region. Hence, if we irradiate the processing nuclei with a beam of electromagnetic radiations of the required frequency, the flipping of proton from low energy state to higher energy state takes place. It should again be emphasized here that the processing proton will only absorb energy if the processing frequency of the spinning nucleus is same as the frequency of the radio frequency beam. When these occurs, the nucleus and the radio frequency beam are said to be in resonance.

Thus in NMR spectroscopy in principle, the substance can be placed in a magnetic field of constant strength and irradiated with electromagnetic radiations of varying frequencies and the frequency at which absorption occurs is observed. In practice however, it is generally more convenient to *keep the radio frequency constant and the strength of the magnetic field is constantly varied.* Now, at some value of the field strength, the energy required to flip the proton matches the energy of the radiation and absorption takes place. A signal is observed in the spectrophotometer. The spectrum so obtained is called nuclear magnetic resonance or NMR spectrum. When the nucleus involved is proton the spectrum is also called *PMR* or *proton magnetic resonance.*

9.3 RELAXATION PROCESS

Relaxation processes involve some non-radiative transitions by which a nucleus in an upper transition state returns to the lower spin state. Two kinds of relaxation processes are: (i) spin-spin relaxation (ii) Spin-lattice relaxation.

Spin-spin Relaxation: It is due to the mutual exchange of spins by two processing nuclei which are in close proximity to each other. We know that each precessing nucleus is associated with a magnetic vector component rotating in plane perpendicular to the field. If this small rotating magnetic field is the same as in required to induce a transition in the neighbouring proton, then mutual exchange of spin takes place. In other words, it involves the transfer of energy from one nucleus to the other. There is no net loss of energy. The spread of energy among the nuclei concerned results in line broadening which makes *nmr* spectra of solids comparatively more interesting.

Spin-Lattice Relaxation (Longitudinal Relaxation): It involves the transfer of energy from the nucleus in its higher energy state to the molecular lattice. The energy is transferred to the components of the lattice as the additional translational vibrational and rotational energy. The total energy of the system remains the same An efficient relaxation process involves a short time and results in the broadening of absorption peaks. Smaller the line of excited state, greater is the line width.

9.4 THE CHEMICAL SHIFT

If the resonance frequencies for all the protons in a molecule were the same, then all the protons in the molecule would absorb at exactly the same field strength and the spectrum would consist of a single signal. Hence, NMR technique would be of little use to the organic chemist as it would tell us little about the structure of the molecule.

Fortunately, the precessional frequency of all the protons in the same external field is *not the same* and the precise value for any one proton depends upon a number of factors. The frequency at which a proton absorbs depends on the magnetic field which that proton *feels or experiences* and this effective field strength is different from the *applied field* strength. The magnetic field strength which the proton feels depends upon the environment of the proton, electron density at the proton, presence of other nearby protons and a number of other factors. For example, the surrounding electrons may shield the nucleus, so the effective magnetic field felt by the nucleus is not quite the same as the applied field.

Hence, each proton or more precisely, each set of equivalent protons will have a slightly different environment from every other set of protons. Thus each set will require a particular field strength at which absorption takes place. This particular field strength is measured and the result is that the spectrum show many absorption peaks.

Thus, the shift in the position of NMR spectrum as compared to a standard reference, due to chemical environment of the proton is called *Chemical Shift.*

Measurement of Chemical Shift: The measurement of precessional frequency of a group of nuclei in absolute frequency unit is extremely difficult and is not required also. However, relative absorption values can be obtained using a suitable standard. The chemical shift values of the protons in a particular compound are then determined with reference to this standard.

The universally accepted standard is NMR signal of tetramethyl silane (TMS)

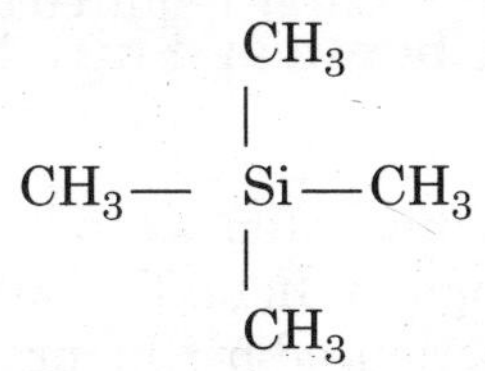

Tetramethyl silane (TMS)

TMS has 12 protons in magnetically equivalent positions and hence a sharp single signal is obtained. Moreover it is chemically inert and soluble in organic solvents used. It can be added to the sample solution as an internal standard.

As TMS is not soluble in water or D_2O, for solutions of the compounds in these solvents the sodium salt of 3-(trimethyl silyl)- propoanesulphonic acid is used.

The NMR signal value for a particular proton in a molecule will appear at different field strengths compared with a signal from TMS. This difference in the absorption position of the proton with respect to TMS signal is called *chemical-shift value.*

Units used in NMR: Chemical shift positions are frequently expressed in δ (delta) units which are defined as *the difference in parts per million from the TMS signal.* The δ unit is proportionality and is thus dimensionless.

The value of δ for a substance with respect to TMS can be obtained by measuring resonance frequency of the sample υ_s and resonance frequency of TMS

$$\delta = \frac{\upsilon_{\text{sample}} - \upsilon_{\text{TMS}}}{\text{operating frequency in magacycles}}$$

In majority of organic compounds, protons resonate at lower field than the protons of tetramethyl silane. Thus δ value of TMS is arbitrary fixed at zero and the protons absorbing at lower field than TMS are given positive value for δ. Most chemical shifts have δ values from 0 to 10.

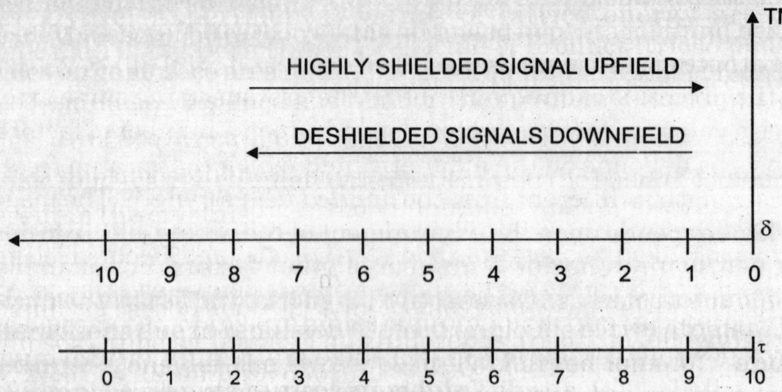

Fig. 9.2

NMR signal is usually plotted with magnetic field strength increasing to the right. Thus, the signal for TMS appears at extreme right of the spectrum with δ = 0 ppm. Greater the deshielding of the protons, larger will be the value of δ (Fig. 9.2). For most of the organic compoundes, signals appear downfield to the left of TMS signal.

Another scale used for the measurement of chemical shift is

τ scale: In τ (tau) scale, TMS is given an arbitrary value of 10 ppm. Thus

$$\tau = 10 - \delta$$

9.4.1 Factors Influencing the Chemical Shift Shielding and Deshielding Effects

In a magnetic field, the electrons around the proton are induced to circulate and thus they produce secondary magnetic field i.e. induced magnetic field. This magnetic field is opposed (diamagnetic) to the applied field. Thus the field experienced by the proton is diminished than that applied externally and the proton is said to be *shielded* from the influence of the applied field. Thus to bring a shielded proton to resonance, greater external field is required. Greater the electron density circulating around the proton, greater would be the shielding and greater the external field required to overcome the shielding effect.

On the other hand, rotation of electrons (π electrons) generates a field that can either oppose or reinforce the applied field at the proton. If the induced field reinforces the applied field the proton experiences a higher field strength as compared to the applied field and is said to be *deshielded*. Thus smaller external field is required to bring the deshielded proton to resonance.

Hence, shielding shifts the absorption peaks towards lower value of δ (higher values of external field B_0) and deshielding shifts the absorption peaks towards higher values of δ (lower value of external field, B_0)

Shielded proton → Higher Field (up field) → lower value of δ

Deshielded proton → lower field (down field) → Higher value of δ

Shielding and deshielding and hence chemical shift depends upon a number of factors discussed below:

(1) *Electronegativity:* There is a reasonable correlation between shielding and electronegativity of the substituent groups in simple saturated molecules. An electronegative group withdraws electron density from the neighbourhood of the proton, so that smaller field is needed to cause resonance. Hence absorption occurs downfield (higher δ values) farthest from TMS signal.

The chemical shift for the protons of halomethane is directly related to electronegativity of halo atom as is illustrated below:

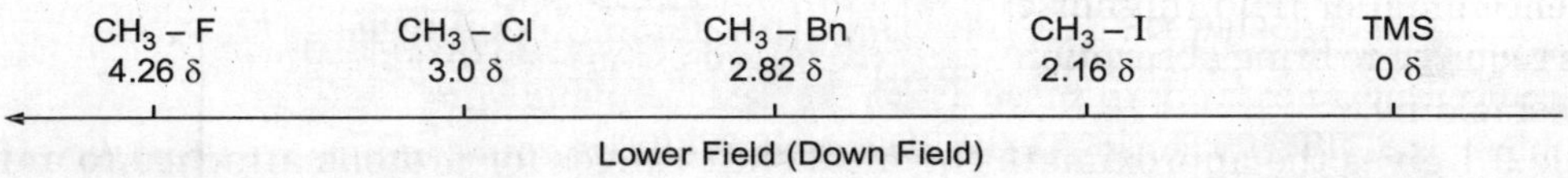

Similarly, silicon is electropositive and the electrons are pushed towards the methyl group. Hence the methyl proton are highly shielded and higher field is needed to bring about the resonance and thus absorption occurs at lower δ value (High field) which is fixed arbitrarily at 0.

(2) *Anisotropic Effects:* It has been found that in unsaturated compounds shielding or deshielding depends upon the manner in which π electrons circulate under the influence of applied field. The effects are diamagnetic (opposed to the field) in certain directions and paramagnetic (aligned with the field) in others, so these effects are called Anisotropic effects.

Alkene: Alkene is so oriented in space that the plane of the double bond is at right angles to the applied field. Circulation of π electrons generates a magnetic field which is diamagnetic around the carbon atoms but paramagnetic in the region of alkene protons. Thus protons experience deshielding and require lower value of external magnetic field to exhibit resonance and hence absorption occurs at lower field (higher δ value) than expected.

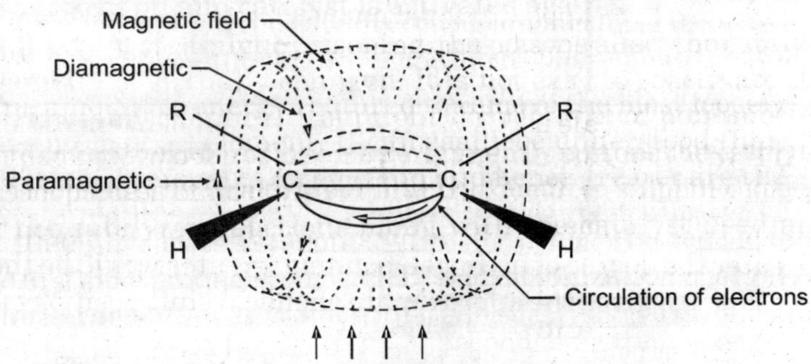

Fig. 9.3 Induced anisotropic magnetic field around an alkene group.

Alkynes: In alkynes, electronic circulation around triple bond takes place in such a manner that the protons experience diamagnetic shielding effect. Thus protons feel smaller field strength and resonance require higher applied field (low δ values).

Aromatic Compounds: The movement of π electrons under the influence of applied magnetic field, around a benzene ring is called a ring current. Aromatic rings display strong magnetic anisotropy. Above and below the ring, there is strong shielding region (i.e. induced magnetic field is in opposition to the applied field) whereas in the plane of the ring there is strong deshielding region (induced magnetic field is in the same direction as the applied field). Consequently, the protons and groups attached to the ring are in the plane of the ring and are thus deshielded and give peaks at down field (higher δ values). The protons above and below the ring are shielded and resonate at low δ values.

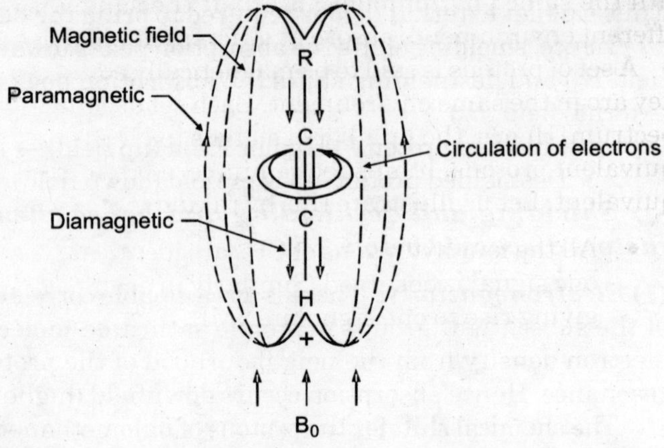

Fig. 9.4 Anisotropic shielding of a proton in an alkyne group.

Carbonyl Compounds: Similar situation arises for carbonyl compounds. The induced magnetic field by the π electrons in the region where protons are located, is oriented in the same direction as the applied field. Hence smaller field (higher δ value) is required to bring absorption due to deshielding.

Table 9.1 gives the approximate chemical shift ranges for protons attached to anisotropic groups.

Table 9.1: Chemical shifts due to anisotropic effects.

Compound	*Effect experienced by protons*	*Absorption*	*δ Value*
Alkene	Deshielding	Lower field	High (5 – 6)
Alkyne	Shielding	Higher field	Low (1.5 – 3.5)
Aromatic	Deshielding (for the protons attached to the ring)	Lower field	High (7 – 8)
Carbonyl	Deshielding	Lower field	High (9.5 – 10.0)

(3) *Hydrogen Bonding:* The hydrogen bonded proton being attached to a highly electronegative atom will have smaller electrons density around it and hence is deshielded. Thus absorption occurs at low field (higher δ). Also the downfield shift depends upon the strength of hydrogen bonding. The greater the degree of hydrogen bonding, greater would be the downfield shift.

Fig. 9.5 gives the approximate chemical shift positions for protons in organic molecules.

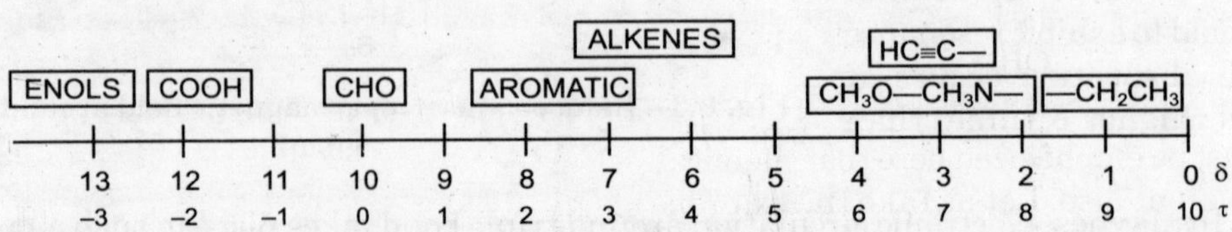

Fig. 9.5 Approximate chemical shift positions for protons in organic molecules.

9.5 NUMBER OF SIGNALS

Equivalent and non-equivalent protons: From the above discussion it becomes clear that protons with the same environment absorb at the same applied field strength, while the protons with different environments absorb at different applied field strength.

A set of protons is said to be magnetically equivalent if they exhibit the same chemical shift i.e. they are in the same environment. Each set of equivalent protons will give rise to a signal in NMR spectrum. Hence, the number of signals in NMR spectrum tells us the number of different sets of equivalent protons. It must be renumbered that magnetically equivalent protons are chemically equivalent. Let us illustrate this by taking few examples.

- All the four hydrogens of CH_4 are identical i.e. are in the same environment and hence only one signal is observed. Similarly in acetone all the six protons are in similar environment giving rise to one signal in NMR.

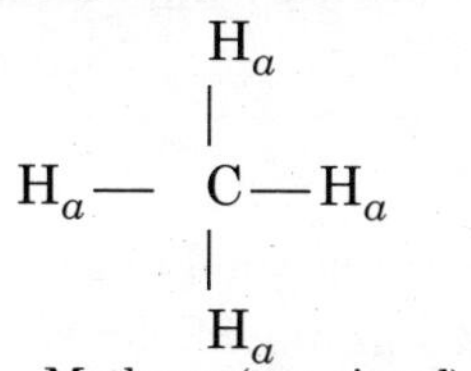

Methane (one signal)

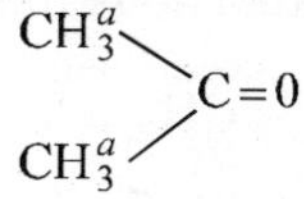

Acetone (One signal)

- In ethyl chloride the protons are in two different environments and hence two signals are observed. Same is true for other molecules such as methanol, propane, isobutylene etc.

$\underset{a}{CH_3}—\underset{b}{CH_2}—Cl$

Ethyl Chloride
(2 NMR signals)

$\underset{a}{CH_3}—\underset{b}{OH}$

Ethyl Chloride
(2 NMR signals)

$\overset{a}{CH_3}—\overset{b}{CH_2}—\overset{a}{CH_3}$

Propane
(2NMR signals)

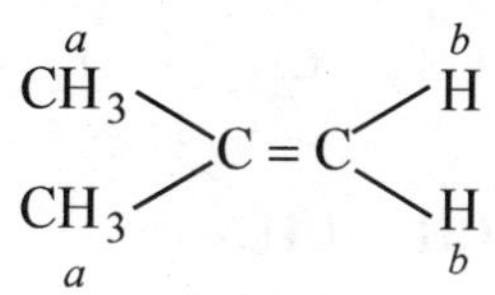

Isobutylene
(2 NMR signals)

- Some of the compounds having 3 NMR signals are as follows:

$\overset{c}{CH_3}—\overset{b}{CH_2}—\overset{a}{CH_2}—Cl$

3 NMR signals

$\overset{c}{CH_3}—\overset{b}{CH_2}—\overset{a}{NH_2}$

3 NMR signals

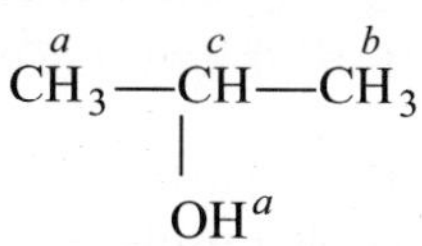

3 NMR signals

- It must be emphasized here that chemically equivalent protons must also be stereochemically equivalent also. Let us take the example of vinyl chloride.

$ClHC = CH_2$
Vinyl Chloride

or

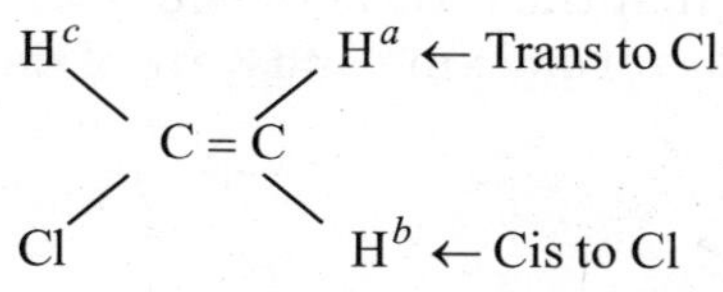

At first sight it appears that it has two sets of equivalent protons and would give rise to 2 NMR signals. But that is not true as one can see that out of two hydrogens attached to one of the carbon atom, one is cis to Cl atom while other is trans to Cl atom and hence are in different environment.

Moreover, these protons are in different environments than the protons on the C—Cl carbon. Thus all the three protons are in different environment, giving rise to 3 NMR signals. Based on similar arguments the number of signals in 2-bromopropene is 3

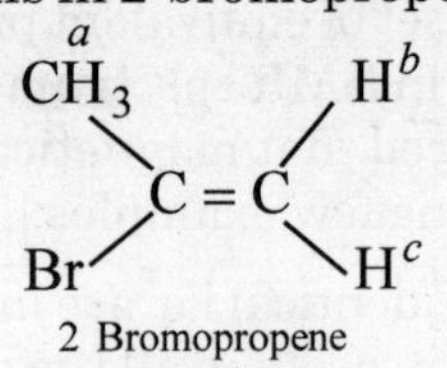

2 Bromopropene
3 NMR signals

Problem: Predict the number of NMR signals in the following compounds: (*a*) Tert-butyl amine (*b*) Benzene (*c*) Propanol (*d*) benzene.

Solution: (*a*) 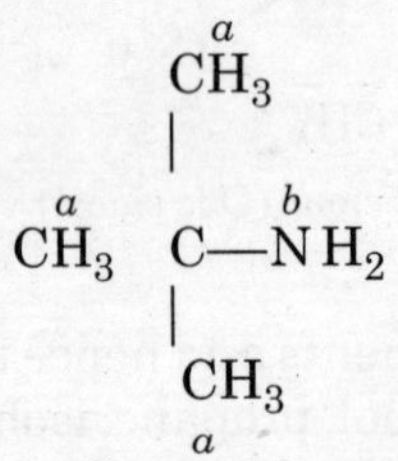 2 NMR signals

Tert–butylamine

(*b*)  1 NMR signal

(*c*) $\overset{d}{CH_3}—\overset{c}{CH_2}—\overset{b}{CH_2}—\overset{a}{OH}$ 4 NMR signals

(*d*) 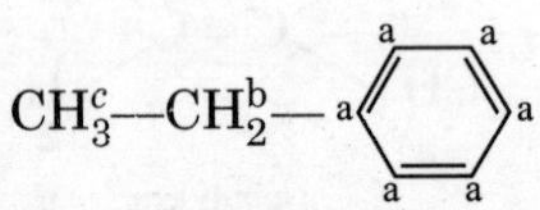 3 NMR signals

9.6 PEAK AREA AND PROTON COUNTING

The number of signals in NMR tells us how many different 'kinds' of protons are there in the molecule. The peak area tells us about *the number of protons in each equivalent set* present in the compound. The area under an NMR signal is directly proportional to the number of protons giving rise to the signal. Greater is the number of protons in a particular set, greater will be the energy absorbed.

In NMR, squares under each peak are counted and from this, the ratio between various kinds of protons is found out and converted into whole numbers. These whole numbers or some multiple of them tell the number of protons represented by each signal. Let us consider the NMR spectrum of toluene.

Two signals indicate two sets of equivalent protons. If the number of squares under each peak is counted, it will be found that the ratio of the number of squares under two peaks will be 5 : 3. Thus in toluene 5 protons are of one kind and 3 are of the another kind.

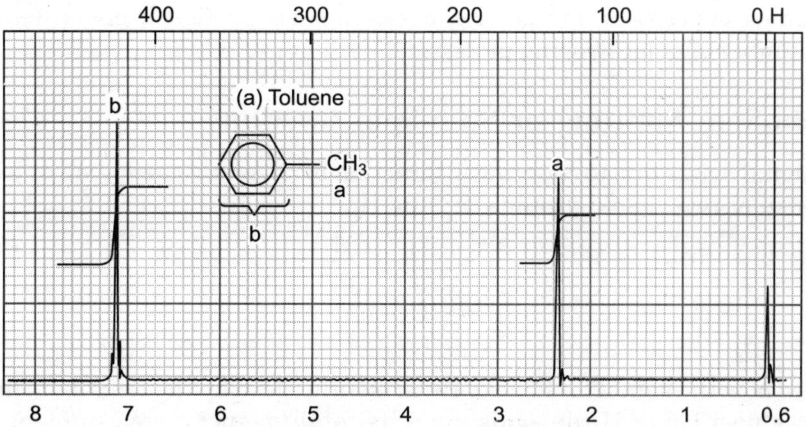

Fig. 9.6 NMR spectra of toluene.

9.7 SPLITTING OF SIGNALS — SPIN-SPIN COUPLING

Each signal in an NMR spectrum represents one kind or one set of protons in a molecule. However, in certain molecules, a single peak (singlet) may not be observed for a particular set of protons but instead the signal may split into a group of peaks (multiplet).

The splitting of a singlet signal into multiplet is due to spin-spin coupling interaction between neighbouring protons. In other words the multiplicity of signal is related to the number of protons in neighbouring groups. Let us illustrate by taking the example of ethyl bromide CH_3CH_2Br.

$\overset{a}{CH_3}\,\overset{b}{CH_2}\,Br$ has two kinds of protons in it and hence two signals are expected in NMR spectra. However it is observed that the signals we expect from each set of equivalent protons do not appear as a single peak but as a *group* of peaks.

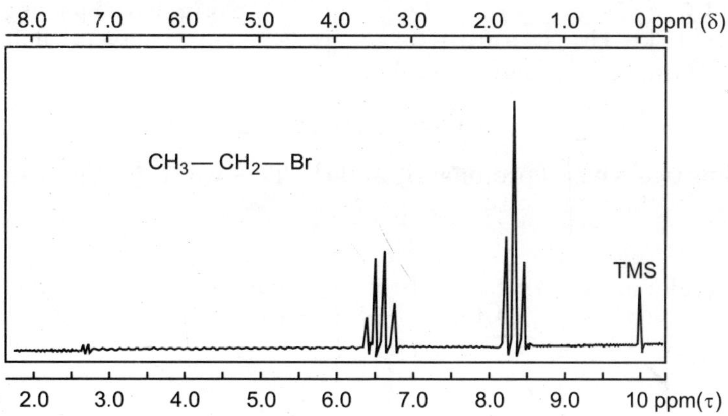

Fig. 9.7 NMR spectrum of ethyl bromide.

It is further observed from the figure that for methyl protons (CH_3—) a group of three peaks is observed (triplet) whereas for — CH_3— protons a group of four peaks (quartet) is observed.

The number of peaks observed for a particular signal is related **to the number of protons in the neighbouring groups.** In ethyl bromide, the protons H_a have two neighbouring protons and the signal appears as three peaks (triplet). The protons H_b have 3 neighbouring protons and the signal appear as four peaks (quartet) in NMR spectrum.

Thus the number of peaks observed for equivalent set of protons is governed by $(n + 1)$ rule where 'n' is the number of neighbouring protons.

Theory of Spin-Spin Splitting or Spin-Spin Coupling

Splitting of the signal arises due to coupling interaction between two neighbouring protons.

(*i*) Let us consider two vicinal protons, H_a and H_b

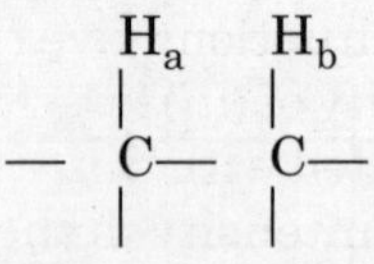

Now let us see how the signal for each of them is influenced by the neighbouring proton.

The resonance position of H_b depends upon its total magnetic environment which is also influenced by the neighbouring proton H_a.

H_b can either have its nuclear magnet aligned with H_a or opposed to H_a. Thus the magnetic field that H_b proton feels at a particular instant is slightly *increased* or *decreased* by the spin of H_a. However it does *both*. The two spin orientations of H_b create two types of magnetic field around H_b and hence it gives rise to a doublet in NMR spectrum. The doublet has equal peak intensities.

Similarly the magnetic field experienced by H_a is slightly increased or decreased by the spin of H_b and hence give rise to two spin orientations. Thus the signal spits into a doublet.

The two peaks of the doublet have the equal intensities.

The spacing between the peaks is the multiplet is called the coupling constant τ.

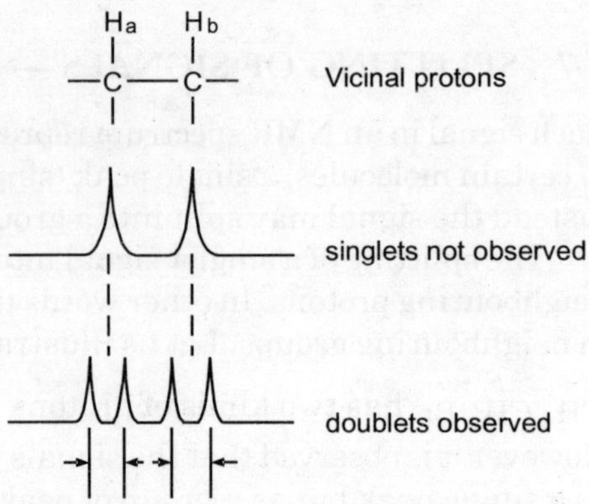

Fig. 9.8 Splitting of the signals of two vicinal protons.

(*ii*) Now let us take the example of 1, 1, 2 trichloroethane to see how a triplet signal arise.

$$\overset{a}{ClCH_2}—\overset{b}{CHCl_2}$$

The compound has two types of protons: H_a and H_b. The two H_b protons can be differentiated as H_b and H_b'.

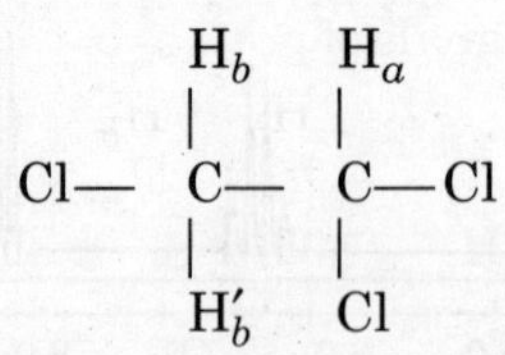

- For proton H_a, magnetic field is influenced by three different combinations of the neighbouring H_b protons.

  (*a*) Both parallel to H_a (*b*) Both antiparallel to H_a

  $\overset{\uparrow}{H_b}\ \overset{\uparrow}{H_b'}$ $\overset{\downarrow}{H_b}\ \overset{\downarrow}{H_b'}$

  (*c*) One is parallel and the other is antiparallel and this can happen in two ways:

- H_b parallel and H_b' antiparallel

$$\overset{\uparrow}{H_b}\ \overset{\downarrow}{H_b'}$$

- H_b' parallel and H_b antiparallel

$$\overset{\uparrow}{H_b}\ \overset{\downarrow}{H_b'}$$

Thus, these three distinct energy situations give rise to a triplet. The probability of (*i*) and (*ii*) energy states is equal but the probability of (*iii*) is twice as it can take place in *two* different ways. Thus the intensity of the signal associated with this state is twice that of the lines associated with the first two states. Thus the relative intensity in the triplet is: 1 : 2 : 1.

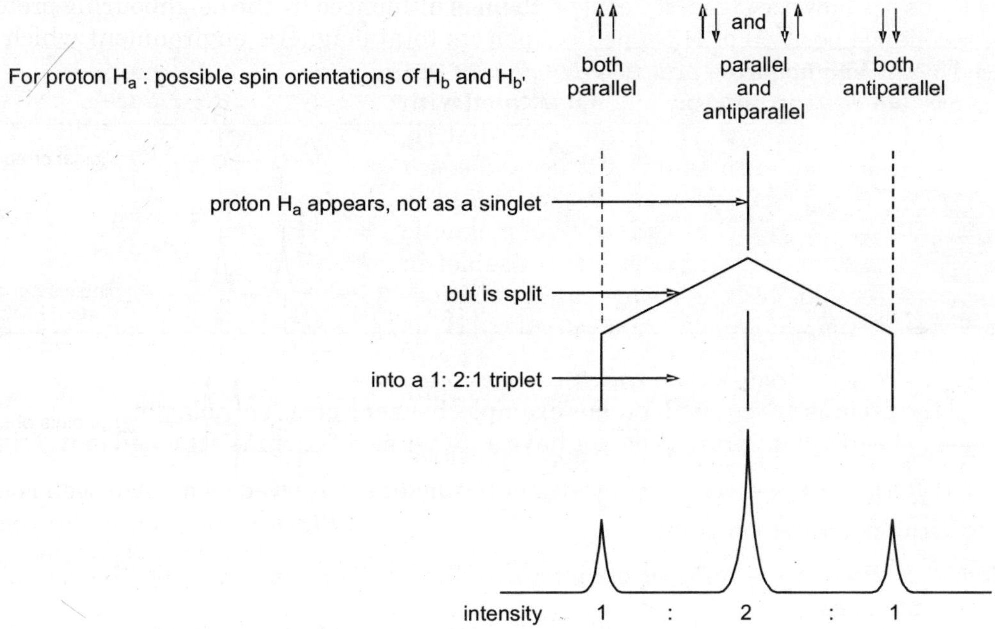

Fig. 9.9 Representation of spin coupling between a proton H_a and two neighbour protons H_b and H_b'.

(*ii*) Now let us explain the splitting of signals of protons of the molecule of ethyl bromide, CH_3CH_2Br

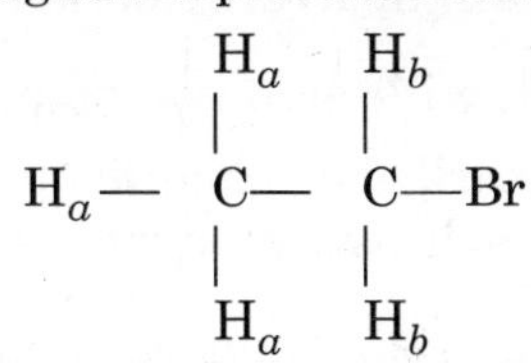

- The signal of the three equivalent H_a proton will be influenced by two H_b protons. Thus H_a can experience three different combinations of the neighbouring H_b protons.

  For proton H_a possible spin orientation of H_b:

↑↑	↓↑ and ↑↓	↓↓
Both H_b protons parallel to H_a	One H_b is parallel and other is anti-parallel to H_a	Both H_b protons are antiparallel

  Thus a triplet of peaks arises with the intensity ratio 1 : 2 : 1.

- Similarly the signal for H_b protons is influenced by three H_a protons which may having following possible combinations:

  For proton H_b possible spin orientations of three H_a protons:

  ↑↑↑ (H_a protons parallel to H_b protons)
  ↑↑↓ ↑↓↑ ↓↑↑ (Two are parallel one is antiparallel)
  ↓↓↑ ↓↑↓ ↑↓↓ (One is parallel two are antiparallel)
  ↓↓↓ (All the three are antiparallel)

Thus, a quartet of peaks result with an intensity ratio of 1 : 3 : 3 : 1.

The relative intensities of different signals can be obtained by the use of Pascal's triangle, where each term coefficient is the sum of the two terms diagonally above it.

Table 9.2

Number of Neighbouring protons	*Number of Peaks (Multiplicity, n + 1)*	*Intensity distribution*
0	1	1
1	2	1 1
2	3	1 2 1
3	4	1 3 3 1
4	5	1 4 6 4 1
5	6	1 5 10 10 5 1

It may be remembered that no splitting takes place if –

- all the protons are equivalent. For example, benzene gives only singlet.
- a set of equivalent protons do not have another set of protons attached to it. Example in $\overset{c}{C}H_3\overset{b}{C}H_2—CO—\overset{a}{C}H_3$, for proton *a*, only singlet is observed as no hydrogen is attached with adjacent carbon atom.

Problem 9.1: Predict the number of signals and multiplicity of the respective signals in the following compounds:

(*a*) CH_3CH_3 (*b*) $Cl—CH_2—CH_2—CH_2—Cl$
(*c*) $Cl_2—CH—CH_2—Cl$

Solution:

(*a*) CH_3CH_3 - only one signal as all the hydrogens are equivalent and it will be a singlet.

(*b*) $Cl—\overset{a}{C}H_2—\overset{b}{C}H_2—\overset{a}{C}H_2—Cl$ willl give two signals as the four protons marked *a* are equivalent and two protons marked *b* are equivalent.

The four protons marked '*a*' will give the signal as a *triplet* as these have two neighbouring protons and hence $(n + 1) = 3$.

The two protons marked *b* will appear as *quintet* (5 peaks) as it has 4 neighbouring protons and hence $(n + 1) = 5$.

(*c*) $Cl_2—\overset{b}{C}H—\overset{a}{C}H_2—Cl$ will give the signal for '*a*' type of proton will appear as *doublet* as it has one proton in its neighbourhood. The signal for '*b*' type will be a *triplet*, as it is having two hydrogens in its neighbourhood.

9.8 COUPLING CONSTANTS

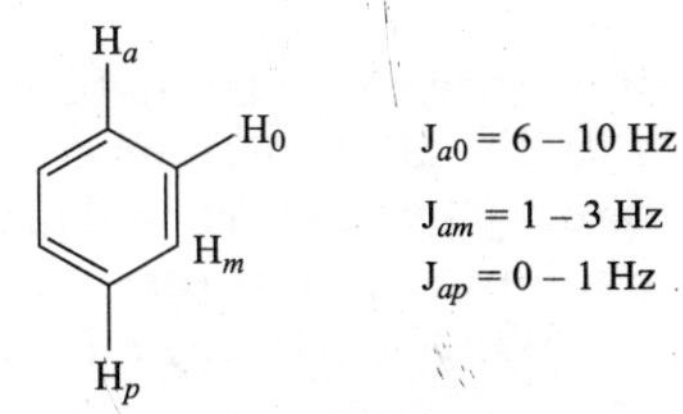

Fig. 9.10

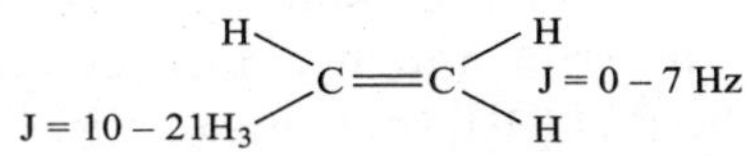

Fig. 9.11

The distance between the centres of two adjacent peaks in a particular multiplet is called the coupling constant *J*. Coupling constant is measured in Hertz (Hz) and is a measure of the effectiveness of spin-spin coupling.

Coupling constant do not depend upon the magnitude of induced magnetic field and remains the same whatever the applied magnetic field may be. Coupling constant depends markedly on the structural relationship between the coupled protons. Some of the factors on which it depends are:

(*i*) *Number of bonds separating the nuclei*: The coupling constants depend upon the number of bonds separating the nuclei. For example $J_{ortho} > J_{meta} > J_{para}$

(*ii*) *Geometry of the molecule:* The magnitude of coupling constants for protons in general lies in the range 5 – 17 Hz and are very sensitive to the geometry of the coupling path. For example in any pair of geometric isomers - *J* is always larger between trans protons than between cis protons.

Problem 9.2: How can you distinguish between two singlets and a doublet in NMR spectroscopy.

Solution: This can be done by simple recording the spectrum at two different radio frequencies. In case of doublet, the value of *J* i.e. the separation between the lines remains constant. On the other hand, if the separation between the lines increases with increasing frequency then this is the case of two signals.

9.9 NMR SPECTRUM AT MORE THAN ONE RADIO-FREQUENCY

It may be noted that the chemical shift positions for the various sets of protons are field dependant. If we work the spectra of a particular compound at different radio-frequencies, then it has been found that the value of coupling constant (measured in cps) remains the same whatever the applied field. Scanning at different radio-frequencies tells clearly whether a particular signal is a multiplet or a few singlets. Let us consider a case in which a doublet is observed and is being suspected as two singlets. For this the spectrum of the same compound is re-run at a higher radio frequency. Now the signal will appear at a different field strength but if the distance between the two peaks (coupling constant J) remains the same, then it is necessarily a doublet. If the distance between the two peaks increase by working at a higher radio-frequency, the formation of two singlets can be safely declared. Sometimes, it happen that at a ceratin radio-frequency, the two signals in the NMR spectrum overlap and thus, the analysis becomes difficult. This technique helps in simplifying such a complex spectrum. The multiplets can be pulled apart by scanning the spectrum of the same compound at a higher radio-frequency. Thus, the complex spectrum is improved.* For example, in the *nmr* spectrum of 4-chlorobutyric acid ($ClC^4H_2C^3H_2C^2H_2C^1OOH$) at 60 mega cycles sec^{-1}, the quintet for the central methylene and the triplet for C_1-methylene overlap and thus complexity arises. But the signals get apart by working the spectrum at 200 mega cycles sec^{-1}. At 200 mega cycles sec^{-1}, two distinct signals (quintet and triplet) are observed with different chemical shifts.

*The degree of improvement of the spectrum depends upon the differences in chemical shift and also upon the coupling constants involved.

9.10 DEUTERIUM EXCHANGE REACTIONS

It has been found that if few drops of deuterium oxide are added in the sample, the D_2O exchanges with the labile protons such as —OH, —NH, —SH and also with the reactive methylene protons flanked by the carbonyl groups. The mechanism involves the same course as in seen in **proton exchange reactions.**

When a little D_2O is added to ROH, then due to rapid exchange, ROH becomes ROD.

$$ROH + D_2O \rightleftharpoons R\text{–}OD + H\text{–}OD$$

Thus, the signal for —OH proton normally observed in ROH will be missing in the PMR spectrum and instead, a signal for proton in H—OD will appear. Similarly, if a little D_2O is added to RCOOH, then due to rapid exchange, it becomes RCOOD.

$$RCOOH + D_2O \rightleftharpoons RCOOD + H\text{—}OD$$

Clearly, the signal for the proton in RCOOH (in the PMR spectrum) which normally appears (in the absence of D_2O) at a negative tau value will be missing or diminished and a signal for proton corresponding to H—OD appears instead. This technique which is employed for detecting the presence of OH, NH groups etc. is called deuteration. For the deuterium exchange technique, two spectra are run.

(*i*) One with the sample dissolved in solvent other than D_2O.

(*ii*) Second spectrum with the sampel dissolved in the same solvent and containing a few drops of D_2O.

On comparing the two spectra, if the peak areas are seen to diminish then the sample may contain —OH, —NH,—SH group in which deuterium exchange in possible.

9.11 DOUBLE RESONANCE (SPIN DECOUPLING)

Double resonance is a technique to simplify the complex spectrum. It is just opposite to the spin-spin coupling and hence is also called spin-decoupling.

This technique involves the irradiation of a proton or a group of equivalent protons with sufficiently intense radio-frequency energy to eliminate completely the observed coupling to the neighbouring protons. We know that a signal for a particular proton or a set of equivalent protons is split up into a multiplet under the influence of the neighbouring proton/protons under different environ-ments. Consider the case of a compound —C—C—, in which H_a and H_b are in different environ-

H_a H_b

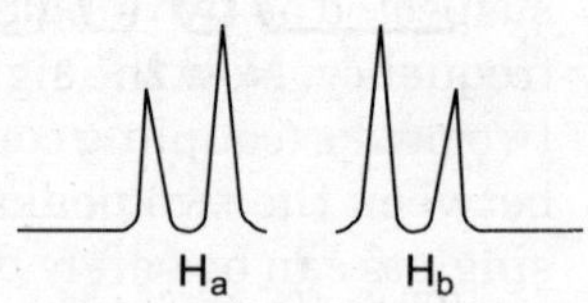

Fig. 9.12

ments. Clearly, in the nmr spectrum, two doublets corresponding to each proton should be observed at different field strengths.

It has been observed that if we irradiate H_a with correct frequency energy (intense) so that the rate of its transitions between the two energy states (spin states) becomes much larger, then the life time of this nucleus in any one spin state will be too short to resolve coupling with H_b. In such a case, H_b protan will have one time average view of H_a and hence H_a will come to resonance only once and H_b will appear as a singlet (not doublet).

In the same way, if we irradiate H_b with sufficiently intense radio frequency energy, then due to its rapid transitions between the two spin states, H_a will have one time average view of H_b and hence H_b will come to resonance only once and H_a will also appear as a singlet.

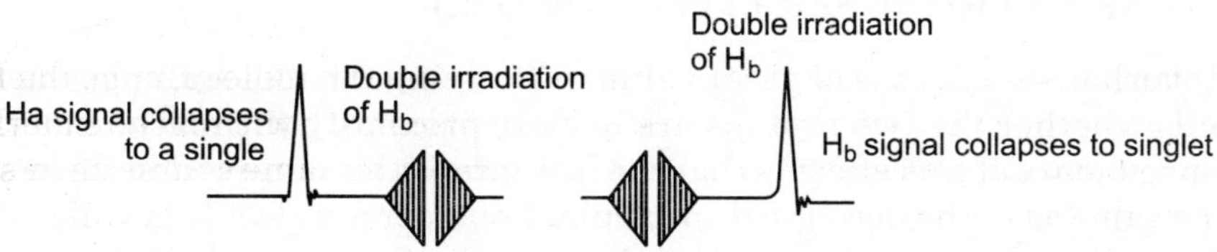

Fig. 9.13

It may be noted that the time (dt) needed to resolve the two lines of a doublet is related to the separation between the lines, *i.e.*, coupling constant J. Thus, the formation of doublet in the above example is possible if each spin state of H_a has a life time greater than dt with double irradiation, the said life time becomes still less and thus, due to the increase in the rate of transition, coupling is not possible and thus, singlets result.

In this technique, we make simultaneous use of two radio frequency sources. In addition to the normal *nmr* instrument, a second tunable radio frequency source is needed to irradiate H_a at the necessary frequency and the recording of the spectrum is done in the same way. It is called double resonance or double irradiation.

Double resonance technique is a powerful tool for simplifying a spectrum and is of great value to organic chemists working with complex molecules. It helps in the identification of coupled protons in spectra that are too complex for detailed analysis. Let us consider the nmr spectrum of Furfural (furan 2-aldehyde).

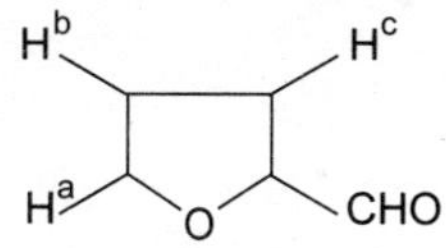

Proton H_a couples with H_b which splits into a doublet but H_a also couples with H_a so that each line of H_a is further split into two giving four lines in all for H_a proton. Similarly, four lines result for each of H^b and H^c. Thus, a spectrum consisting of 12 lines (three multiplets) result for three protons. Each multiplet is found two show two different value of J (Fig. 9.14).

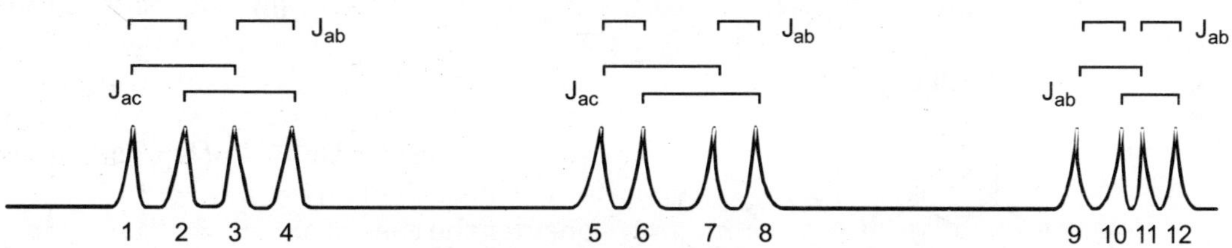

Fig. 9.14

The effect of strong decoupling eliminates the coupling of H_a with H_b and H_c. This results in the collapse of each of the four line multiplet (for H_b and H_c) into the doublet. (Fig. 9.15)

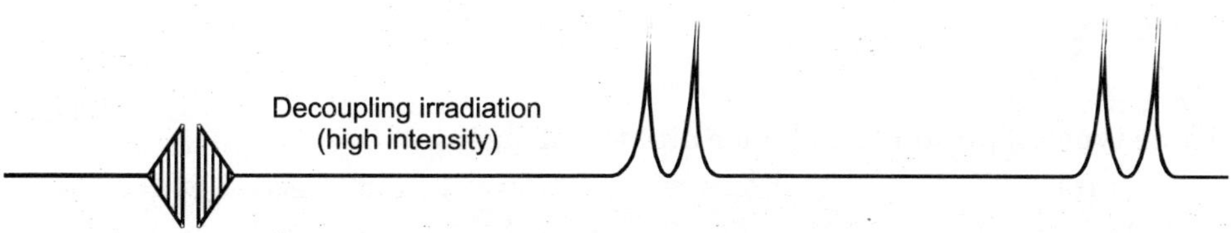

Fig. 9.15

9.12 NUCLEAR OVERHAUSER EFFECT (N.O.E.)

The nuclear overhauser effect is of great value in studying the molecular geometry of the compounds. It tells whether the two protons are in close proximity within the molecule or not. An important consequence of this effect is that the line intensities observed in the normal spectrum may not be the same as in the decoupled spectrum. Consider a molecule in which two protons are close enough to allow through-space interactions of the fluctuating magnetic vector. For this effect, the number of intervening bonds between the two concerned protons have no significance. Consider a hypothetical molecule in which two protons are in close proximity.

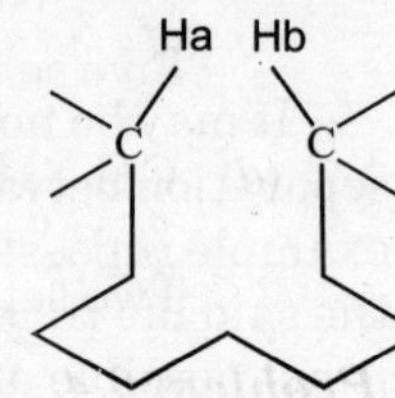

Fig. 9.16

In such a compound, if we double irradiate H_b, then this proton gets stimulated and the stimulation is transferred through space to the relaxation mechanism of H_a. Thus, due to the increase in the spin lattice relaxation of H_a, its signal will appear more intense by 15 — 50%. Thus, we say that if the intensity of absorption of H_a signal is increased by double irradiating H_b, then the protons H_a and H_b must be in close proximity in molecule.

SOLVED PROBLEMS BASED ON NMR SPECTROSCOPY

Problem 9.3: Determine the equivalent and non-equivalent hydrogen in the following compounds and indicate the number of signals expected.

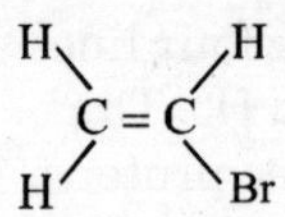

(*a*)

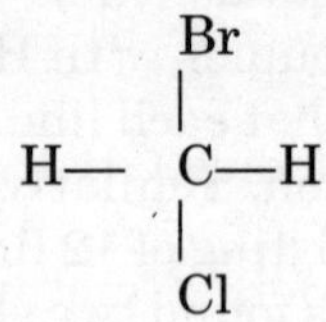

(*b*)

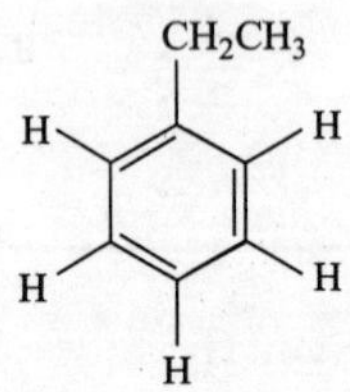

(*c*)

$CH_3CH_2—O\,CH_2CH_3$

(*d*)

Solution:

(*a*)

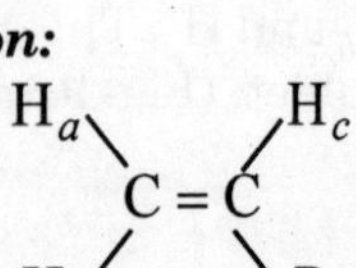

Three types of protons and hence number of signals = 3.

(*b*)

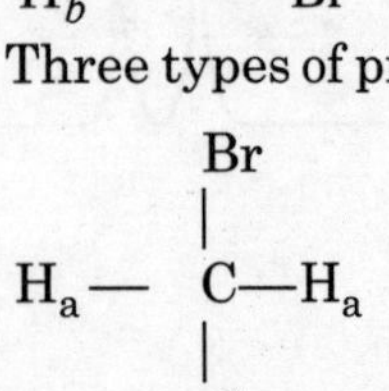

One set of equivalent protons hence number of signals = 1.

(c) 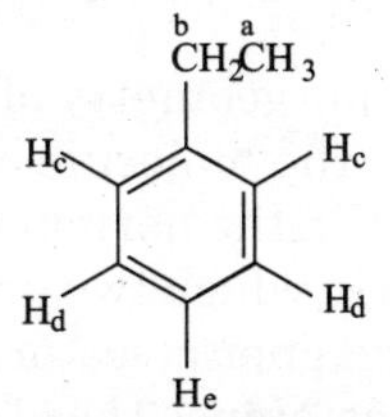

Five sets of equivalent protons and hence the number of signals = 5

(d) $\underset{b}{CH_3}\underset{a}{CH_2}—O—\underset{a}{CH_2}\underset{b}{CH_3}$

Two set of equivalent protons and hence number of signals = 2.

Problem 9.4: Which of the following compounds would be expected to show only a single peak in NMR or PMR spectrum

(*i*) Acetone
(*ii*) Methyl acetate
(*iii*) 1, 2-dibromo methane
(*iv*) Methyl iodide
(*v*) 1-chloro-2-bromoethane

Solution:

(*i*) Acetone $\underset{a}{CH_3}—\underset{\underset{O}{\|}}{C}—\underset{a}{CH_3}$

(*ii*) 1, 2-Dibromomethane $C\overset{a}{H_2}Br_2$

(*iii*) Methyl Iodide $\underset{a}{CH_3}I$

Problem 9.5: Methyl, methylene, methine and quaternary hydrogens resonate in successively lower fields in NMR spectroscopy, explain.

Solution: The replacement of hydrogen by an alkyl group causes a downfield shift because hydrogen is more electropositive than carbon. Thus, methylene methine and quarternary hydrogens resonate at successively lower fields as shown below

	CH_4	CH_3CH_3	$(CH_3)_2CH_2$	$(CH_3)_3CH$
δ	0.23	0.86	1.33	1.68

Problem 9.6: A compound has molecular formula $C_{10}H_{14}$. It gives the following NMR data

(*i*) 0.88 δ (9H, singlet),
(*ii*) 7.28 δ (5H, singlet, aromatic protons)

Assign the structural formula to the compound.

Solution:

(*i*) The nine proton singlet at 0.88 δ indicates the presence of tertiary butyl group.
(*ii*) Five proton singlet at 7.88 δ confirms the phenyl ring.

Hence the structural formula of the compound may be

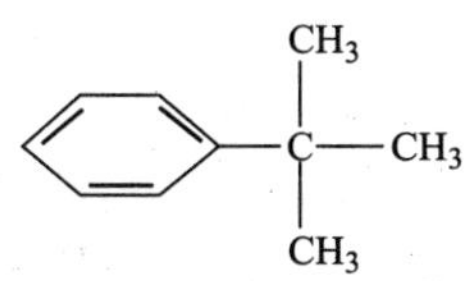

Problem 9.7: The following compounds show only one signal in the NMR spectrum. Write their structural formulae.

(*a*) C_5H_{12} (*b*) C_2H_6O

(*c*) $C_2H_4Br_2$ (*d*) C_4H_6

Solution:

(*a*) **C_5H_{12}:** Single signal means all the 12 hydrogens are equivalent. Hence these can be in four equivalent methyls. The structure is:

$$(CH_3)_4C$$

(*b*) **C_2H_6O:** Two equivalent methyl groups account for the single signal. Hence the structure is:

$$CH_3—O—CH_3$$

(*c*) **$C_2H_4Br_2$:** Two equivalent methylene groups account for the single signal.

$$Br—CH_2—CH_2—Br$$

(*d*) **C_4H_6:** Two equivalent methyl groups, with two carbons having no hydrogens account for the single signal

$$H_3C—C\equiv C—CH_3$$

Problem 9.8: How many NMR signals are expected in each of the following compounds

(*a*) $CH_3OCH_2CH_2CH_3$ (*b*) CH_3OCH_3

(*c*) $CH_3CH_2CH_2CH_2OH$

(*d*) CH_3CHO (*e*) $CH_3—CH(CH_3)—CH_2—Cl$

Solution:

(*a*) $\overset{a}{CH_3}\,O\overset{b}{CH_2}\,\overset{c}{CH_2}\,\overset{d}{CH_3}$ 4 signals (*b*) $\overset{a}{CH_3}O\overset{a}{CH_3}$ 1 signal

(*c*) $\overset{a}{CH_3}\,\overset{b}{CH_2}\,\overset{c}{CH_2}\,\overset{d}{CH_2}OH$ 5 signals (*d*) $\overset{a}{CH_3}\,\overset{b}{CHO}$ 2 signals

(*e*) $\overset{a}{CH_3}—\overset{b}{CH}(\underset{a}{CH_3})—\overset{c}{CH_2}—Cl$ 3 signals

Problem 9.9: How will you distinguish between the following two compounds on the basis of NMR spectroscopy.

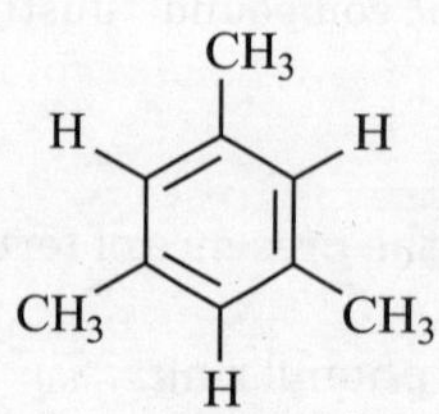

Mesitylene

2 sets of equiv. protons 2 signals

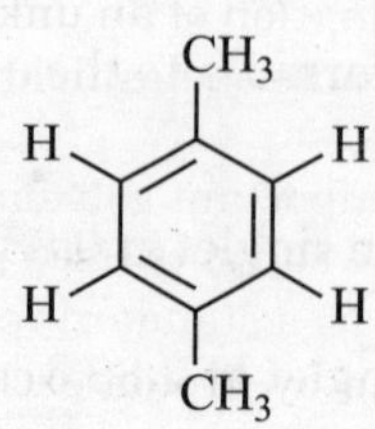

Xylene

2 sets of equiv protons 2 signals

Solution: Both the compounds would give two signals but the intensities of the peaks would be different in two compounds. In mesitylene the ratio of methyl protons to ring protons is 9 : 3 whereas in *p*-xylene it is 6 : 4.

QUESTIONS (NMR)

1. What do you mean by spin Active Nuclei?
2. Which of the following nuclei do not show nuclear magnetic resonance:

$$^{1}H,\ ^{2}H,\ ^{12}C,\ ^{13}C,\ ^{14}N,\ ^{15}N,\ ^{16}O,\ ^{19}F,\ ^{31}P$$

[**Ans.** Nuclei that contain odd number of electrons show NMR spectra. Thus, ^{13}C, ^{16}O, and ^{32}S having even number of electrons will not show NMR spectra.]

3. Describe briefly the theory of *nmr* spectrometry.
4. What do you understand by the positions of the signals in a *nmr* spectrum? How many signals are expected in each of the following compounds?
   (a) Propane (b) Isobutane (c) Ethanol (d) Cyclobutane (e) Ethylmethyl other
   (f) Ethyl acetate (g) Butanal (h) Glycol (i) α-alanine (f) Caprolactum.
5. (a) What is meant by the term chemical shift?
6. Write with examples, the shielding and the deshielding effects involved in *nmr* spectroscopy.
7. Describe with examples the various factors which affect the magnitude of the chemical shift.
8. (a) Briefly describe how *nmr* spectrum of a compound is scanned?
   (b) What do you understand by the term "Splitting of the signals"? Explain with examples.
   (c) Describe briefly the difference in the positions of absorption as noted in the case of acetylene and benzene.
9. (a) Explain clearly the term 'spin-spin coupling'. Why does a peak for a particular set of protons split into a multiplet? Give examples.
   (b) Write a short note on the use of solvents in the *nmr* spectrometry.
10. (a) Write brief notes on the following :
    (i) Chemical shift (ii) Spin-spin coupling
    (iii) Coupling constant.
    (b) Explain why *nmr* spectrum of benzene is observed at a lower field whereas that of acetylene is observed at a higher field strength.
11. How many kinds of H^{1}s are there in :
    (i) $CH_3—CH_2—CH_3$? (ii) $CH_2=CH_2$?
    (iii) $CH_3—CH=CH_2$? (iv) $C_6H_5—CH_3$?
12. Write the possible isomers corresponding to the following molecular formulae and draw their *nmr* structures.
    (i) C_4H_{10} (ii) C_2H_6O
    (iii) C_4H_8O (iv) $C_4H_6O_2$
    (v) $C_4H_8O_2$.
13. "NMR spectroscopy is the most important tool compared to ultra-violet and Infra-red for the structure determination of an unknown organic compound." Justify this statement.
14. (a) Write a short note on deshielding due to hydrogen bonding in NMR spectrum taking at least four examples.
    (b) Name some important solvents, used in *nmr* spectroscopy. What are the important characteristics of the solvents used in this technique?
15. Why is TMS used as a standard reference ?
16. What do you mean by Proton exchange reactions ? How does spin decoupling occurs in certain groups ?
17. (a) Describe clearly what do you know about coupling constant ?
    (b) How will you distinguish between Cis and Trans isomers with the help of *nmr* spectroscopy ?
18. Write short notes on :
    (i) Spin-spin coupling (ii) Areas of the various signals
    (iii) Deshielding due to hydrogen bonding.

19. Write notes on :
    (i) Nuclear overhauser effect. (ii) Deuterium exchange reactions.
20. How many kinds of protons are there in
    (i) CH_3CH_3 (ii) $CH_3CH_2CH_3$ (iii) $(CH_3)_2CHCH_2CH_3$
    (iv) $C_6H_5CH_3$ (v) $C_6H_5CO_2$ (vi) $CH_2{=}CH_2$
    (vii) $CH_3CH{=}CH_2$?
21. Determine which of the following molecules will show spin-spin coupling in their NMR spectra. If the coupling or splitting is observed, give the multiplicity of each kind of proton, *i.e.*, give the number of lines arising from each proton :
    (a) $ClCH_2CH_2Cl$ (b) $ClCH_2CH_2I$ (c) 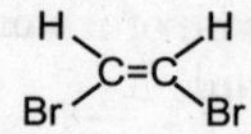
    (d) 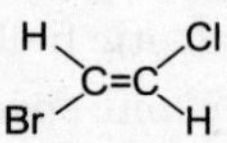 (e) 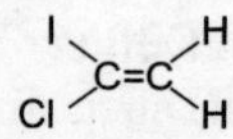 (f) $CH_3—CH{=}0$
    (g) 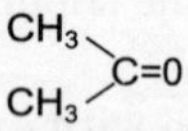 (h) $CH_3—C(Cl_2)—CH_3$
22. Predict the low-resolution NMR spectrum and the high-resolution NMR spectrum of acetone.
23. An organic compound C_3H_6O contains a carbonyl group

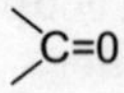

How will its NMR spectrum decide whether it is an aldehyde or a ketone ?
24. Predict the NMR spectrum of the compound

$$CH_3COCH_2C{\equiv}CCH_3$$

25. How do you account for the fact that nuclear spin-spin splitting is observed in 2-methyl propene put not in 1-chloro-2, 2-dimethyl-propane ?
26. How can the NMR spectrum distinguish between the isomers *p*-xylene and ethyl benzene?
27. Explain diagrammatically
    (i) shielding of an acetylenic proton
    (ii) deshielding of an aldehydic proton in terms of simple diamagnetic effect.
28. (a) Why is an internal standard is used in obtaining an NMR Spectrum? Why TMS is a good standard?
    (b) What is the multiplicity of each proton in
        (i) Isobutyl bromide
        (ii) See butyl bromide
29. (a) Indicate diagramatically the splitting of signals in NMR spectra of
        (i) $CH_2Br—CHBr_2$
        (ii) $CH_3—CHBr_2$
        (iii) $CH_3—CH_2Br$
    What is the reason for the multiplicity of peaks?
30. (a) Explain shielding and deshielding effects of protons in benzene and acetylene towards an applied field H_0.
    (b) What are the different kinds of electronic transitions?
31. What type of nuclei will give NMR spectra? Give two examples of each type.
32. Explain the following in respect of NMR spectroscopy:
    (a) Chemical shift (b) Shielding and Deshielding of protons
    (c) Spin-spin interactions
33. Explain the applications of NMR spectroscopy to:
    (a) Structural diagnosis (b) Hydrogen bonding
    (c) Keto-enol tautomerism

34. Give the high resolution HNMR spectrum of Ethanol.
35. (a) State and explain the principle of NMR spectroscopy.
    (b) Explain the term coupling constant with examples.
36. Write short notes on:
    (i) Nuclear Over Hauser Effect (ii) Double Resonance.
37. In relation to NMR spectra, explain any one of the following:
    (i) Chemical shift and TMS (ii) Shielding and Deshielding.
38. What do you understand by NMR spectroscopy?
39. What type of nuclei shows ESR spectra?
40. What type of information is obtained by studying the UV, IR and H-NMR spectra of an organic compound.
41. What is meant by shielding and deshielding of protons in H-NMR?
42. Calculate the frequency required for an ESR experiment employing a magnetic field of 12000G.
43. (a) What is the principle behind (NMR) spectrum
    (b) Write short notes on:
    (i) Spin spin splitting (ii) Chemical shift
    (iii) Nuclear Overhauser effect.
44. How will you verify that a particular signal in NMR spectrum arries from –OH, –NH or –SH groups?
45. (a) How will you distinguish between primary, secondary and tertiary alcohols on the basis of PMR spectroscopy?
    (b) Write short notes:
    (i) Spin-spin coupling or splitting (ii) Chemical shift
    (iii) Coupling constant.
46. How many NMR signals are observed in the spectrum of ethanol, isopropyl alcohol, bulanone and benzene?
47. (a) Write notes on :
    (i) NOE (ii) Spin-spin Coupling
    (iii) Deuterium Exchange.
    (b) What is meant by term chemical shift? How will you distinguish between cis and trans isomer with the help of NMR spectroscopy.
48. Why TMS is used as internal standard in NMR spectroscopy?
49. (a) What are advantages of taking NMR spectrum at more than one radio frequency?
    (b) Write brief notes on:
    (i) Chemical shift (ii) Spin-spin coupling
    (c) How will you verify that a particular signal in NMR arises from – OH, – NH or – SH group.

EXAMINATION QUESTIONS

1. What are the basic principles involved in the Nuclear-Magnetic Resonance (NMR).
   [U.P. Tech., 2nd Sem., 2003-04]
2. (i) How many NMR signals do you expect from each of the following compounds. Indicate also the splitting pattern of the various signals:
   (a) CH_3OCH_3 (b) $CH_3OCH_2CH_3$
   (ii) Define Infra-red Spectroscopy. Describe the various molecular vibrations in the technique.
   [U.P. Tech., 2nd Sem., 2003-04]
3. How will you distinguish vinyl acetate and methyl acrylate with the help of NMR spectroscopy?
   [U.P. Tech., 1st Sem. (C.O.), 2003]
4. What are the basic principles involved in infrared and 1H-NMR spectroscopy?

5. A compound has molecular formula C_2H_3Cl. It can show geometral isomerism. The compound has two NMR-Signal. The splitting, under high resolution of NMR shows one doublet and one triplet. Identify the compound with the help of proper explanation. *[U.P. Tech., 1st Sem., 2006-07]*
6. Give the structure consistent with the following NMR data:
   $C_3H_5Cl_3$ (a) Singlet, d 2.20, 3H
   (b) Singlet, d 4.20, 2H *[U.P. Tech., 2nd Sem., 2006-07]*
7. (a) Write down the absorption positions of common functional groups found in organic compounds.
   (b) What is 'Shielding' and 'Deshielding'? *[U.P. Tech., 1st Sem., 2007-08]*
8. Calculate the NMR signals and splitted signals in the following molecules :
   (a) $CH_3-O-CH_2-CH_3$
   (b) 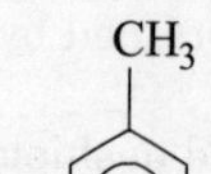
   (c) 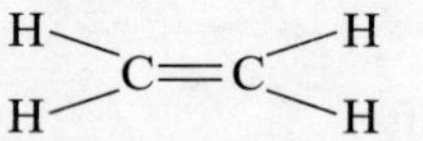
   (d) $CH_3—CH_2—CH_2—Cl$
   (e) CH3—CH—COOH
   with Br on the CH (| Br)
   [U.P. Tech., B.Tech., 1st Sem., 2007-08]
9. (a) The absorbance is directly proportional to :
   (i) Wavelength (ii) Path length
   (iii) Concentration (iv) Concentration
   (b) IR active molecules are those which undergo a net change in ....
   (c) Number of NMR signals obtained in CH_3COCH_3 will be
   (i) 2 (ii) 6
   (iii) 1 (iv) 3 *[U.P. Tech, B. Tech., Ist Sen., 2008-09]*
10. (i) An organic compound having molecular formula C_7H_6O shows absorption peaks at 3010, 2700, 1600, 1580, 1520, 1480 and 1720 cm^{-1} in its IR spectrum Suggest its structure.
    (ii) Define chemical shift Show the expected NMR signals and their splitting in the following compounds

$$CH_3\ CH_2\ CH_2\ OH \text{ and } C_6H_5\ CH_3.$$

[U.P. Tech, B. Tech., Ist Sen., 2008-09]
11. Differential with one suitable example:
    (a) Shielding and deshielding of proton.
    (b) Chromophores and auxochromes
    (c) Bathochromic and hypsochromic shift. *[U.P. Tech, B. Tech., Ist Sen., 2008-09]*

10 CATALYSIS

"God grant me the serenity to accept the things I can not change, the courage to change the things I can, and the wisdom to know the difference."

— Reinhold Niebuhr

CHAPTER OUTLINE

10.1 Concept of Activation energy and catalysis
10.2 Characteristics of Catalysis
10.3 Types of Catalysis
- Homogeneous
- Heterogeneous

10.4 Classification of Catalysis
10.5 Theoris of Catalysis
- Intermediate compound formation theory
- Concept of Adsorption
- Adsorption Theory

10.6 Acid Base Catalysis
- Kinetics

10.7 Enzyme Catalysis
- Mechanism of Enzyme Action
- Characteristics of Enzyme catalysis

10.8 Catalysis by Metal Salts
10.9 Surface Catalysis
10.10 Catalysis in Industry

It is a well known fact that all the chemical reactions take place at a definite rate under given experimental conditions. Some reactions are so rapid that they appear to be instantaneous, e.g. neutralization of strong acids and bases. But some reactions are so slow at ordinary temperatures, that over the years no detectable products are formed, e.g. combination of hydrogen and oxygen. Between these two extremes, lie a number of reactions which proceed with measurable velocities at ordinary temperatures. **The study of rates of reactions proceeding under given conditions of temperature, pressure and concentration is known as chemical kinetics.**

It has been a continuous effort to increase the rate of reaction, by working out different conditions for getting maximum yields of several products. This is done by varying the conditions on which rate of reaction depends. The most important of these include varying the concentration of the reactants, changing the temperature, use of radiation and presence of a catalyst. **The presence of a catalyst is the most wonderful aspect of chemical kinetics**. It has been observed that certain reactions which do not proceed to completion even when carried out for indefinite periods of time, are speeded up and get completed very fast when a small quantity of a catalyst is added.

The term catalyst was coined by Berzelius in 1836. **Catalyst is defined as a substance, which alters the rate of the reaction without undergoing any change in mass and composition at the end of the reaction.** This phenomenon of altering the velocity of a chemical reaction by the presence of catalyst is called **catalysis.**

How do catalysts increase reaction rates? The general answer is that catalysts provide a new path for the reaction to proceed. To understand this more clearly we must review the concept of activation energy and what difference the presence of catalyst make.

10.1 CONCEPT OF ACTIVATION ENERGY AND CATALYSIS

According to the collision theory of reaction rates, for a chemical reaction to occur, collisions between reactant molecules must take place. But only those collisions result in chemical reaction, in which the colliding molecules are associated with a certain minimum amount of energy. The minimum amount of energy which must be associated with molecules, so that the collisions between them result in chemical reaction is called **threshold energy**. The collisions between the molecules which are associated with energy less than the threshold energy do not result into chemical reaction. Most of the molecules, however have much less kinetic energy than the threshold energy. **The excess energy that the reactants molecules must acquire to change into products is known as activation energy.** Thus

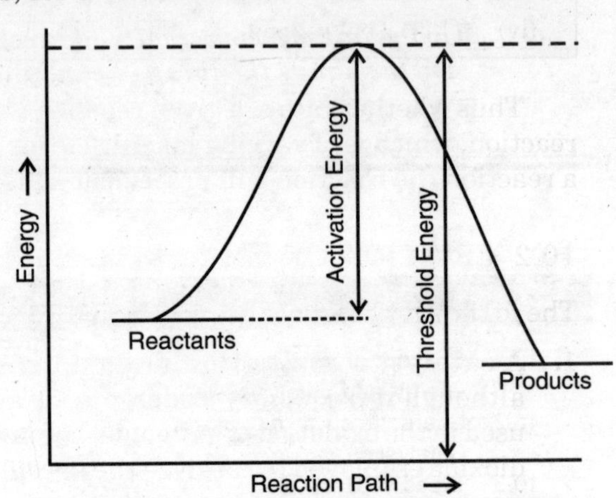

Fig. 10.1 The concept of activation energy

Activation Energy=Threshold Energy–Energy possessed by the molecules.

Activation energy is denoted by E_a.

Effect of Catalyst on activation energy

Catalyst provides an alternative route, with lower activation energy to the reactants to change into the products. In other words, a catalyst lowers the activation energy of the reaction by providing a new pathway. For instance, the activation energy for the decomposition of hydrogen peroxide in solution is 76 kJ mol^{-1} and the reaction is slow at room temperature. When a little iodine is added, the activation energy falls to 57 kJ mol^{-1} and the rate constant increases by a factor of 2000. Similarly enzymes, the biological active catalysts increase the reaction rate upto a factor of 10^{12} at body temperature. For example the enzyme saccharase, reduces the activation energy for acid hydrolysis of sucrose from 107 kJ mol^{-1} to 36 kJ mol^{-1} and hence reaction proceeds much faster at body temperature (310 K). Table (10.1) gives an idea about lowering of activation energy of catalyzed reaction.

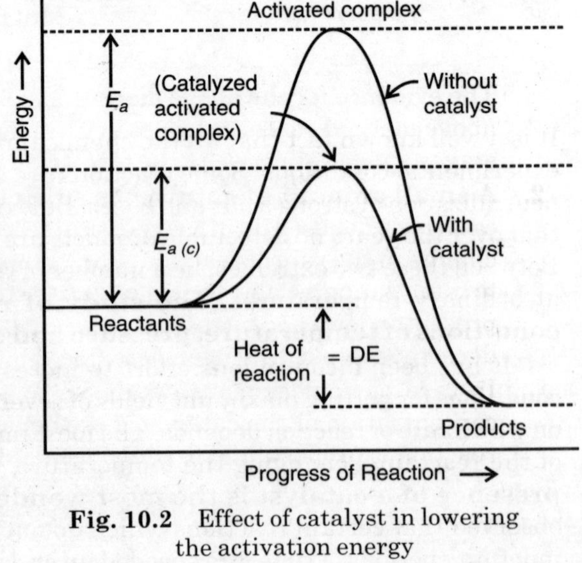

Fig. 10.2 Effect of catalyst in lowering the activation energy

Table 10.1

Reaction	*Catalyst*	*Activation Energy E_a (kJ mol^{-1})*	
		In Absence of Catalyst	*In Presence of Catalyst*
(i) $2HI \rightarrow H_2 + I_2$	$Au_{(s)}$	184	105
	$Pt_{(s)}$	184	59
(ii) $2HN_3 \rightarrow N_2 + 3H_2$	$W_{(s)}$	350	162
(iii) $2N_2O \rightarrow 2N_2 + O_2$	$Au_{(s)}$	245	121
	$Pt_{(s)}$	245	134
(iv) $(C_2H)_2O$ pyrolysis	$I_{2(g)}$	224	144

Thus whether rate of a given reaction is fast or slow depends upon the activation energy of the reaction, which is of varying magnitude for different reactions. If the activation energy is less for a reaction the reaction will proceed at a faster rate and vice versa.

10.2 CRITERIA OR CHARACTERISTICS OF CATALYSIS

The fo1llowing features are common to all catalytic reactions.

1. **A catalyst remains unchanged in mass and composition at the end of the reaction:** although it frequently undergoes a physical change. For example platinum gauze catalyst used in the oxidation of ammonia, becomes quite rough after some weeks. Similarly manganese dioxide employed to catalyze the decomposition of potassium chlorate, is in the form of large crystals in the beginning of the reaction but changes to a fine powder in the end.
   Since the catalyst undergoes a physical change, it implies that it is involved in the mechanism of the reaction but regenerated in the end. The participation of MnO_2 in the decomposition of $KClO_3$ is believed to be as explained below :
   $$2KClO_3 + 2MnO_2 \longrightarrow 2KMnO_4 + Cl_2 + O_2$$
   $$2KMnO_4 \longrightarrow K_2MnO_4 + MnO_2 + O_2$$
   $$K_2MnO_4 + Cl_2 \longrightarrow 2KCl + MnO_2 + O_2$$
   The evidence for the above mechanism is provided by the fact, that oxygen prepared by the above method contains traces of chlorine. Moreover the reaction mixture if cooled at early stages appear pink due to the formation of $KMnO_4$.
2. **A small amount of catalyst is sufficient to alter the rate of the reaction:** Since the catalyst is not used up in the chemical reaction, a small amount is sufficient to bring appreciable change in the rate of reaction. For example, 1 gm of copper (II) ions in 10^9 litres is sufficient to catalyze appreciably the oxidation of sodium sulphite by oxygen. Similarly colloidal platinum, present in low concentration to the extent of 1 mole in 10^8 litres, is enough to accelerate the decomposition of hydrogen peroxide. This is true in the case of heterogeneous catalysis.
   In case of homogeneous catalysis, however, the rate of reaction is proportional to the concentration of the catalyst. For the acid and alkaline hydrolysis of an ester, the rate of reaction is proportional to the concentration of the acid, which acts as an catalyst.
   $$RCOOR' + H_2O \xrightarrow[OH^-]{H^+ \text{ or}} RCOOH + R'OH$$

In Friedel Crafts reaction, anhydrous aluminium chloride function as a catalyst effectively, if present to the extent of 30% of the mass of benzene which is appreciably large amount.

$$C_6H_6 + C_2H_5Cl \xrightarrow{AlCl_3} C_6H_5C_2H_5 + HCl$$

In certain heterogeneous reactions, the rate of the reaction is proportional to the area of the catalytic surface. This explains why the finely divided state of a solid catalyst is much more effective in increasing the rate of reaction.

3. **A catalyst is specific in its action:** A particular substance can act as a catalyst only in particular reaction. For example MnO_2 may catalyze the decomposition of $KClO_3$, but not of $KClO_4$. Highly specific action of a catalyst can be compared to the specific use of a key which can open a particular lock. Enzymes are also highly specific in their action. But on the other hand, the transition metals like iron, cobalt, nickel, platinum and palladium can catalyze various reactions.
4. **Nature of the products is unaltered by the presence of the catalyst:** The combination of N_2 and H_2 under suitable conditions will give NH_3 in presence or absence of catalyst. Similarly SO_2 and O_2 will always combine to form SO_3 whether a catalyst is present or not. However there are some **exceptions** also. For example CO and H_2 combine to give three **different** products when different catalysts are used.

$$CO + 3H_2 \xrightarrow{Ni} CH_4 + H_2O$$
$$CO + H_2 \xrightarrow{Cu} HCHO$$
$$CO + 2H_2 \xrightarrow{ZnO + Cr_2O_3} CH_3OH$$

Similarly chlorination of toluene in the presence of a halogen carrier such as iron in the **absence of sunlight** takes place in the benzene ring.

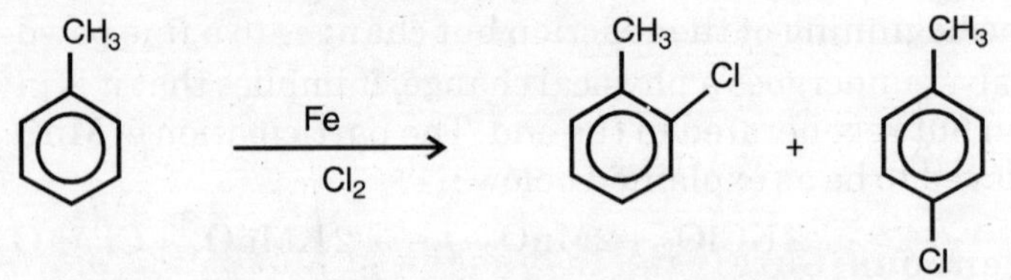

But chlorine is substituted in the side chain in the absence of a catalyst but in **presence of sunlight.**

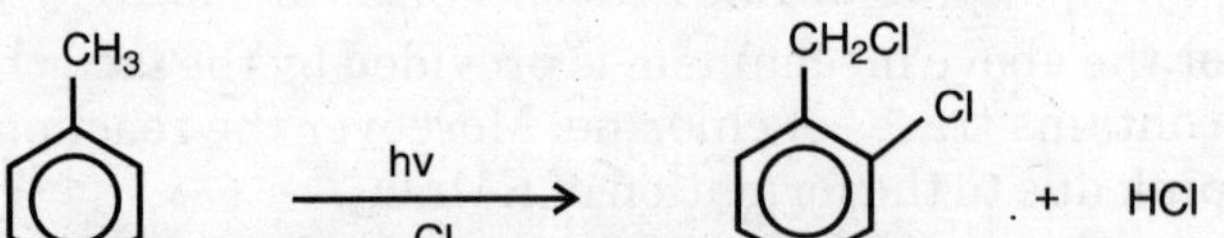

5. **The catalyst does not alter the position of equilibrium in a reversible reaction:** The catalyst helps in attaining the equilibrium more quickly by increasing the rates of both the forward and the reverse reactions to the same extent, but do not alter the concentration of the reactant and the products at equilibrium. Thus the equilibrium constant remains the same. It is also evident from the Fig. 10.2 that, the activation energy for the reverse reaction is also lowered by exactly the same amount as for the forward reaction, by the addition of the catalyst. Thus, if a catalyst doubles the rate of the forward reaction, it also doubles the rate of the backward reaction. In other words, the catalyst for the forward process in a reaction is also a catalyst for the reverse reaction.
6. **A catalyst does not intitiate the reaction:** It can only accelerate the rate of the reaction already taking place, even if very slowly. But this subject has been a matter of great controversy.

According to W. Ostwald (1895) catalyst cannot initiate the reaction but H.W. Armstrong (1885-1903) and T.M. Lowry (1925-1928) have suggested that there are certain reactions which can occur only if a catalyst is present and will not occur in its absence. For example, perfectly dry H_2 and O_2 do not combine to form water, even if they are left in contact for years, but in presence of a little water which act as a catalyst, the reaction proceeds quite rapidly. But still this does not prove that change did not occur at all, but that uncatalyzed combination may be so slow that millions of years would be required for a detectable amount of water to be formed.

7. **Optimum temperature:** There is a particular temperature at which the activity of the catalyst is maximum, this temperature is called optimum temperature.
8. **Catalytic promotors:** The addition of small amounts of foreign substances which themselves are not catalytically active, increases the activity of a catalyst. Such substances are called promoters. In the manufacture of NH_3 by Haber's process, finally divided iron acts as a catalyst, while molybdenum acts as a promotor.
9. **Catalytic poisons:** The activity of a catalyst is inhibited or completely destroyed by the presence of even minute traces of foriegn substances called catalytic poisons or anticatalysts. In the manufacture of H_2SO_4, a trace of As_2 O_3 destroys the catalytic efficiency of spongy platinum.
10. **High surface area:** Catalysts have high surface areas. Even a small lump (1 cm^2) of a catalyst can have a surface area ranging to 100 m^2 because of the presence of pores.

10.3 TYPE OF CATALYSIS

Broadly two types of catalysis are known:

1. Homogeneous Catalysis
2. Heterogeneous Catalysis

Homogeneous Catalysis: In homogeneous catalysis, the catalyst is present in the same phase as the reactants and is evenly distributed throughtout. The homogeneous catalysis can occur in gaseous or liquid phase.

(i) **Homogeneous Catalysis in gaseous phase:** There are only a few cases of homogeneous catalysis in gaseous phase e.g.

(a) Oxidation of sulphur dioxide (SO_2) to sulphur trioxide (SO_3) with nitric oxide as catalyst.

$$\underset{\text{vapour}}{CH_3\,CHO} + \underset{\text{vapour}}{[I_2]} \longrightarrow \underset{\text{gas}}{2SO_3} + \underset{\text{gas}}{[NO]}$$

(b) Decomposition of acetaldehyde with I_2 as catalyst

$$\underset{\text{vapour}}{CH_3\,CHO} + \underset{\text{vapour}}{I_2} \longrightarrow \underset{\text{gas}}{CH_4} + \underset{\text{gas}}{CO} + I_2$$

(c) NO gas also acts as a catalyst in combination of CO with oxygen.

$$\underset{\text{gas}}{2CO} + \underset{\text{gas}}{O_2} \xrightarrow[\text{gas}]{NO} 2CO_2$$

(d) NO gas catalyses the combination of moist sulphur dioxide and oxygen in the lead chamber process for the manufacture of H_2SO_4.

$$\underset{\text{gas}}{2SO_2} + \underset{\text{vap}}{2H_2O} + \underset{\text{gas}}{O_2} \xrightarrow[\text{gas}]{[NO]} 2H_2SO_4$$

(ii) Homogeneous catalysis in liquid phase: Most of the reactions under this category involve the acid or base as the catalyst.

(a) Hydrolysis of an ester in the presence of acid or alkali.

$$CH_3COOC_2H_{5(l)} + H_2O_{(l)} \xrightarrow{H^+ \text{ or } \overline{OH}} CH_3COOH_{(l)} + C_2H_5OH_{(l)}$$

(b) Inversion of sucrose in presence of hydrogen ions in aqueous solution.

$$C_{12}H_{22}O_{11} + H_2O \xrightarrow{H^+} \underset{\text{Glucose}}{C_6H_{12}O_6} + \underset{\text{Fructose}}{C_6H_{12}O_6}$$

(c) The decomposition of hydrogen peroxide is catalyzed in the presence of chloride ions.

$$2H_2O_{2(l)} \xrightarrow{Cl^-_{(l)}} 2H_2O + O_2$$

(d) Conversion of acetone into diacetone alcohol is catalyzed by hydrogen ions

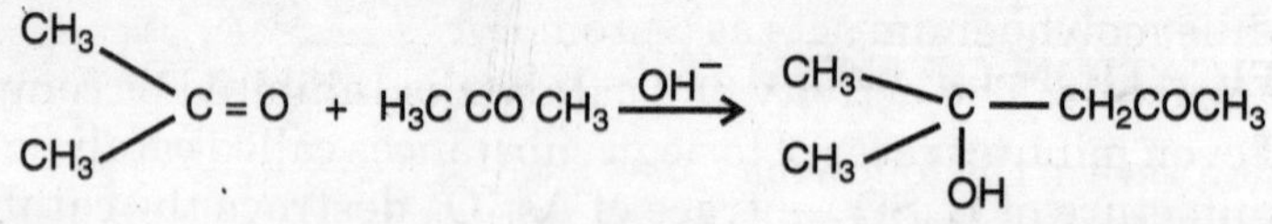

Diacetone alcohol

The mechanism of homogeneous catalysis can be explained on the basis of unstable intermediate compound formation theory. (discussed later).

Heterogeneous Catalysis: In heterogeneous catalysis the catalyst is in different phase from the reactants. In majority of the cases, the catalyst is solid while the reactants are either liquids or gases. Often the term "Contact Catalysis" is used for this type of catalysis as the reaction occurs by the contact of the reactants with the catalyst surface. This form of catalysis has a number of reactions of technical importance. The catalyst is usually a metal or metal oxide. Heterogeneous catalysis involve solid, liquid or gaseous reactants.

(i) **Heterogeneous Catalysis with gaseous reactants:** There are a large number of reactions which can be put under the heading of heterogeneous catalysis. Some of them are–

(a) Combination of SO_2 and oxygen in the presence of finely divided platinum or vanadium penta oxide V_2O_5 (Contact Process).

$$\underset{\text{gas}}{2SO_2} + \underset{\text{gas}}{O_2} \xrightarrow{Pt} \underset{\text{gas}}{2SO_3}$$

(b) Combination of hydrogen and oxygen in presence of finely divided platinum.

$$2H_2 + O_2 \xrightarrow{Pt} 2H_2O$$

(c) Hydrogenation of oils in presence of Nickel

$$\underset{\text{unsaturated oil}}{R-CH=CH-R'} + \underset{\text{gas}}{H_2} \xrightarrow{Ni} R-CH_2-CH_2-R'$$

(d) Synthesis of petroleum by the combination of CO and H_2 in the presence of iron, cobalt or nickel

$$x\,CO + yH_2 \xrightarrow[\text{or Ni}]{\text{Fe, Co}} \text{Petroleum} + H_2O.$$

(e) Manufacture of ammonia by Haber process

$$N_2 + 3H_2 \underset{Mo}{\overset{Fe}{\rightleftharpoons}} 2NH_3$$

(ii) **Heterogeneous catalysis with liquid reactants :** The common example under this category are–

(a) Benzene and ethanoyl chloride react in the presence of anhydrous aluminium chloride to form phenyl methyl ketone ($C_6H_5COCH_3$)

$$C_6H_{6(l)} + CH_3COCl_{(l)} \xrightarrow{AlCl_3} C_6H_5COCH_3 + HCl$$

(b) Decomposition of aqueous solution of hydrogen peroxide (H_2O_2) is catalysed by MnO_2 or platinum in colloidal form.

$$2H_2O_2 \xrightarrow{Pt} 2H_2O + O_2$$

(c) Decomposition of hypochlorites : The decomposition of hypochlorites in aqueous solutions is catalysed by oxides of cobalt and nickel.

$$Ca(ClO)_2 \xrightarrow{NiO} CaCl_2 + O_2$$

10.4 CLASSIFICATION OF CATALYSIS

Catalytic reactions are of the following types–

1. Positive Catalysis
2. Negative Catalysis
3. Auto Catalysis

1. **Positive Catalysis:** When the catalyst used accelerates the speed of a chemical reaction it is known as a positive catalyst. Initially the term 'catalyst' coined by Berzelius was used for the positive catalyst only, but later on this term was used for all the substances which **alter** the rate of reaction, and hence the substances which **increase** the rate of reaction are placed under positive catalysis. All the reactions discussed above are the example of positive catalysis.

2. **Negative Catalysis:** There are a few cases, where a catalyst instead of accelerating a reaction may actually **reduce** the rate of reaction. Such a catalyst used is termed as negative catalyst. The phenomenon is known as negative calatysis or inhibition. Negative Catalyst slows down or stops altogether an unwanted reaction. For example–

(i) $CHCl_3$ on oxidation by air forms carbonyl chloride ($COCl_2$) which is a poisonous substance. The formation of $COCl_2$ is suppressed by the addition of ethanol which acts as a negative catalyst.

(ii) Small amount of glycerol or acetanilide retards the decomposition of hydrogen peroxide.

3. **Auto Catalysis:** There are certain reactions where one of the products of the reaction acts as the catalyst. Such reactions are known as Auto Catalyzed reactions. In auto catalysis the initial rate of the reaction rises as the catalytic product is formed thereby indicating that product accelerates the reaction rate. (Fig. 10.3).

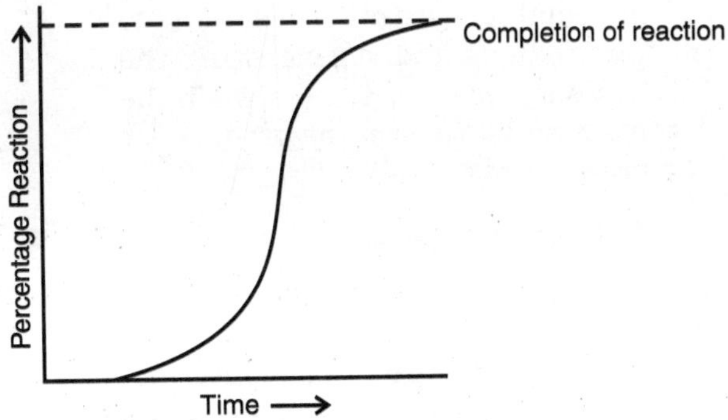

Fig. 10.3 Rise of Rate of reaction with the formation of product

Examples:

(i) During titration of warm solution of oxalic acid by acidified $KMnO_4$ solution, the first two drops take appreciable time to decolourize, since the reaction is initially very slow. But after some time, the decolourization takes place rapidly as Mn^{2+} ions formed during the reaction, catalyze the reaction.

$$2KMnO_4 + 5\begin{array}{c}COOH\\|\\COOH\end{array} + 3H_2SO_4 \longrightarrow \underset{\text{Catalyst}}{\underline{2MnSO_4}} + K_2SO_4 + 8H_2O + 10CO_2$$

(ii) Hydrolysis of an ester is autocatalyzed by the acid which is formed during the reaction.

$$RCOOR' + H_2O \longrightarrow RCOOH + R'OH.$$

(iii) The free arsenic produced by the decomposition of arsine (AsH_3) auto catalyze the reaction.

$$2AsH_3 \longrightarrow \underset{\text{Catalyst}}{2As} + 3H_2$$

(iv) The action of HNO_3 on copper is very slow initially but NO_2^- ions produced during the reaction act as catalyst and the reaction proceeds rapidly.

$$3Cu + 8HNO_3 \longrightarrow 3Cu(NO_3)_2 + 4H_2O + 2NO$$

Nitric oxide formed above dissolves in H_2O to provide NO_2^- ions.

10.5 THEORIES OF CATALYSIS

Different theories have been put forward to explain the homogeneous and heterogeneous catalysis.

Theory of Homogeneous Catalysis (Intermediate Compound Formation Theory)

This theory was put forward by element and Desormes in 1806. This theory explains satisfactorily the mechanism of homogeneous catalysis. According to this theory, the catalyst combines with one of the reactants to form an intermediate compound which is unstable and hence react immediately with other reactants to form the products and regenerating the catalyst. Consider a reaction between the reactants X and Y to form the product XY in the presence of a catalyst C.

$$X + Y \xrightarrow{C} XY$$

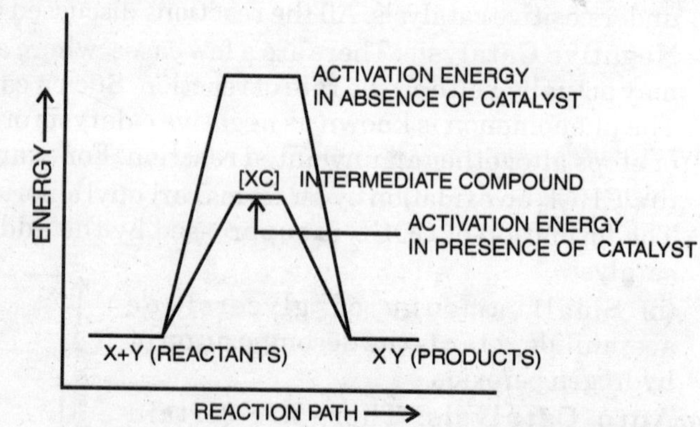

Fig. 10.4 Energy Profile for Homogeneous Catalysis

According to the intermediate compound formation theory, the reaction may be split up into the following two steps:

(i) **First step** – Formation of intermediate

$$X + C \longrightarrow [XC] \text{ intermediate Compound}$$

(ii) **Second step** – Dissociation of intermediate.

$$[XC] + Y \longrightarrow \underset{\text{Product}}{XY} + \underset{\text{Catalyst}}{C}$$

Infact, the formation of intermediate compound provides an alternative reaction path having lower activation energy and hence reaction proceeds at an increased rate.

It is seen from the Fig. 10.4 that the total energies of the reactants *X* and *Y*, and those of the product *XY* are unaffected by the catalyst. The only difference between the two is lowering of the activation energy in the presence of a catalyst. A number of catalytic reactions can be explained on the basis of intermediate compound formation theory. A few of them are discussed below:

(i) Ordinarily, the gas phase reaction of sulphur dioxide with oxygen proceeds slowly:

$$2SO_2 + O_2 \longrightarrow 2SO_3$$

But if the same reaction is carried out in presence of NO as catalyst, the reactions proceeds relatively faster.

$$\underset{\text{Catalyst}}{NO + \frac{1}{2}O_2} \longrightarrow \underset{\text{Intermediate}}{NO_2}$$

$$NO_2 + SO_2 \longrightarrow SO_3 + \underset{\text{Catalyst Regenerated}}{NO}$$

(ii) In Friedel Craft's reaction, anhydrous aluminium chloride is used as a catalyst for the preparation of benzene. The mechanism can be discussed on the basis of intermediate formation theory. For example, benzoylation of benzoyl chloride proceeds as follows–

$$\underset{\text{(Reactant)}}{C_6H_5COCl} + \underset{\text{(Catalyst)}}{[AlCl_3]} \longrightarrow \underset{\text{Intermediate}}{C_6H_5COCl.AlCl_3}$$

$$C_6H_5COCl.AlCl_3 + C_6H_6 \longrightarrow \underset{\text{Product}}{C_6H_5COC_6H_5} + HCl + \underset{\text{Catalyst}}{AlCl_3}$$

(iii) Decomposition of hydrogen peroxide into water and oxygen catalyzed by bromide ions is believed to take place as:

$$H_2O_{2(aq)} + \underset{\text{Catalyst}}{Br^-_{(aq)}} \longrightarrow H_2O_{(l)} + BrO^-_{(aq)}$$

$$H_2O_2 + BrO^- \longrightarrow H_2O_{(l)} + O_{2(g)} + \underset{\text{Catalyst}}{Br^-_{(aq)}}$$

(iv) Reaction for the manufacture of chlorine by the reaction of oxygen and hydrochloric acid in the presence of cuprous chloride as catalyst.

$$\underset{\text{Catalyst}}{4CuCl} + O_2 \longrightarrow \underset{\text{Intermediate}}{2Cu_2OCl_2}$$

$$2Cu_2OCl_2 + 4HCl \longrightarrow 4CuCl_2 + 2H_2O$$

$$4CuCl_2 \longrightarrow \underset{\text{Catalyst}}{4CuCl} + 2Cl_2$$

(v) Formation of ether from excess of ethylalcohol using H_2SO_4 as catalyst.

$$C_2H_5OH + H_2SO_4 \longrightarrow \underset{\text{Intermediate}}{C_2H_5HSO_4} + H_2O$$

$$C_2H_5HSO_4 + C_2H_5OH \longrightarrow \underset{\text{Ethylether}}{C_2H_5OC_2H_5} + \underset{\text{Catalyst}}{H_2SO_4}$$

(vi) An example of homogeneous catalysis from atmospheric chemistry includes, the decomposition of Ozone in the stratosphere by the presence of chlorine atoms, resulting from the photolysis of chloroflourocarbons (CFCs).

$$\underset{\text{Catalyst}}{Cl} + O_3 \longrightarrow \underset{\text{Intermediate compound}}{ClO} + O_2$$

$$ClO + O \longrightarrow \underset{\text{Catalyst}}{Cl} + O_2$$

Further, the two most important reaction under homogeneous catalysis viz. acid base catalysis and enzyme catalysis are dealt in detail in section 10.6 and 10.7.

Limitations

1. This theory fails to explain the action of catalytic promoters and poisons.
2. This is only applicable to homogeneous catalysis. In heterogeneous catalysis,where the catalyst is a solid and reactants are gases the formation of intermediate compound is not possible.

The mechanism of heterogeneous catalysis is explained on the basis of Adsorption Theory. But before we discuss this theory it is important to learn what the term **adsorption means.**

Adsorption

The term refers strictly to the existence of a higher concentration of the molecules of a substance at the surface of a liquid or solid phase than present in the bulk.

The surface of a liquid is in a state of strain due to strong inward pull, due to which the liquid surface tends to contract. This property arises due to the presence of unbalanced forces on the surface. Similar to it, surface of a solid also has unsatisfied valence or residual forces. When a solid surface is exposed to a gas or a liquid, molecules of the gas or solution phase accumulate or adsorbed on the surface. The substance that is adsorbed on the surface is called **Adsorbate** and the solid on whose surface the molecules are adsorbed is called **Adsorbent.**

Adsorption is a surface phenomenon and at a given temperature and pressure depends upon the surface area of the adsorbent. Solids in finely divided state, because of large surface area are good adsorbents.

Adsorption is an exothermic process. The phenomenon of adsorption is thermodynamically a spontaneous process, it should be accompanied by a decrease in free energy of the system. When the gas molecules are adsorbed on the surface of the metal, the process is accompanied with a decrease in entropy. In other words ΔS is negative. Thus, it is clear from the thermodynamic relation,

$$\Delta G = \Delta H - T\Delta S$$

that adsorption is an exothermic process.

There are two categories of adsorption depending upon whether the association between the gas and the solid surface is physical or chemical in character, the two types are–

(a) **Physical adsorption:** Physical adsorption is also referred to as Vander Waals adsorption in which gas molecules or atoms are held to the solid surface by weak Vander Waals forces. It is characterized by low enthalpy of adsorption of the order of 20 kJ mol^{-1}, which is of the same order as the heat of condensation. This small enthalpy change is insufficient to lead to breaking of bonds. Physical adsorption is usually observed at low temperatures or on relatively inert surfaces.

(b) **Chemisorption:** In chemisorption, a term for chemical adsorption, the molecules are adsorbed to the surface by forming chemical bonds. These bonds may be covalent or ionic in nature. The enthalpy of chemisorption is much greater than that of physiosorption of the order of 200 kJ mol^{-1}. The chemisorbed molecules may get completely dissociated into atoms and the existence of these dissociated molecules on the surface is one reason why solid surface catalyze the reactions. Chemisorption of hydrogen ion on nickel can be represented as–

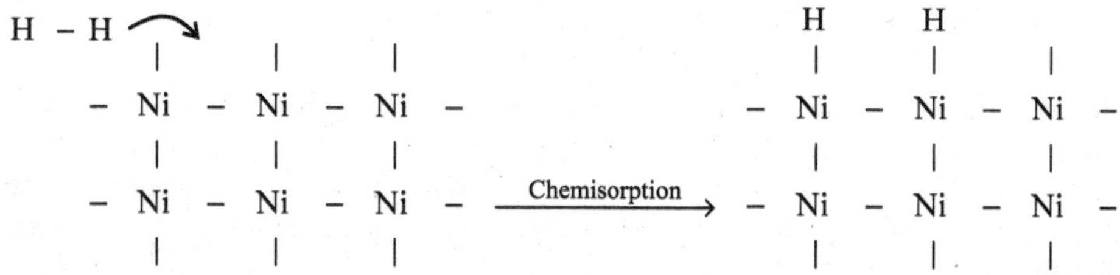

The difference between the physical adsorption and chemical adsorption is given in Table 10.2.

Table 10.2

Physical Adsorption	***Chemisorption***
1. It involves the physical forces of attraction, like van der Waals forces etc.	It involves the chemical forces of attraction like covalent bonds etc.
2. It is characterized by low enthalpy of adsorption of the order of 20 kJ mol^{-1}.	It is characterized by high enthalpy of adsorption of the order of 80-200 kJ mol^{-1}. It is irreversible.
3. The adsorption is reversible.	
4. Physiosorption are non specific. Every gas is adsorbed to more or less extent on all solid surfaces.	Chemisorptions are specific in nature. A gas will be chemisorbed on such solids only with which it can combine chemically.
5. Multilayers may be formed in physical adsorption.	Only monolayer is formed in chemisorption.
6. Physical adsorption occurs rapidly at low temperature and decreases with increasing temperature.	Chemisorption usually occurs at high temperatures.
7. The activation energy of desorption is very low as weak van der Waals forces are involved.	The activation energy of desorption is high as adsorbate molecules are held by comparatively strong valence forces.
8. Some examples of physical adsorption are– (a) Adsorption of various gases on charcoal (b) Adsorption of nitrogen on mica.	Some examples of chemical adsorption are– (a) Adsorption of O_2 on tungsten. (b) Adsorption of H_2 on nickel.

Theory of Heterogeneous Catalysis (Adsorption Theory)

This theory was postulated by Faraday (1883) and later on modified by others. This theory is based on the phenomenon of adsorption.

The mechanism of heterogeneous catalysis can be explained satisfactorily on the basis of adsorption of reactant molecules on the surface of the catalysis. According to this theory,

(i) The surface of the catalyst has certain active centres due to the unsaturation of valencies at the surface.

(ii) The reactant molecules strike the surface and are chemisorbed on it.

(iii) The exothermic nature of the chemisorptive process is able to supply energy of the same order of magnitude as in chemical reactions and hence the dissociation of bonds of adsorbed molecules may take place

(iv) The bonds between adsorbate and adsorbent are usually covalent but sometimes may be ionic also, as in the case of metal oxygen bonds.

(v) Due to the close proximity between adsorbed molecules on the catalyst surface, they undergo interaction to form an activated complex.

(vi) The activated complex breaks to form the products and the free surface of the catalyst is regenerated.

(vii) The molecules of the products diffuse away from the surface of the catalyst.

Hydrogenation of ethene in presence of Nickel can be explained, to illustrate the above points.

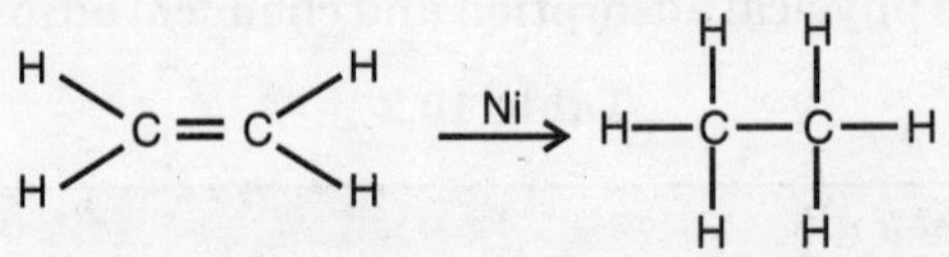

The catalyst operates by the following steps–

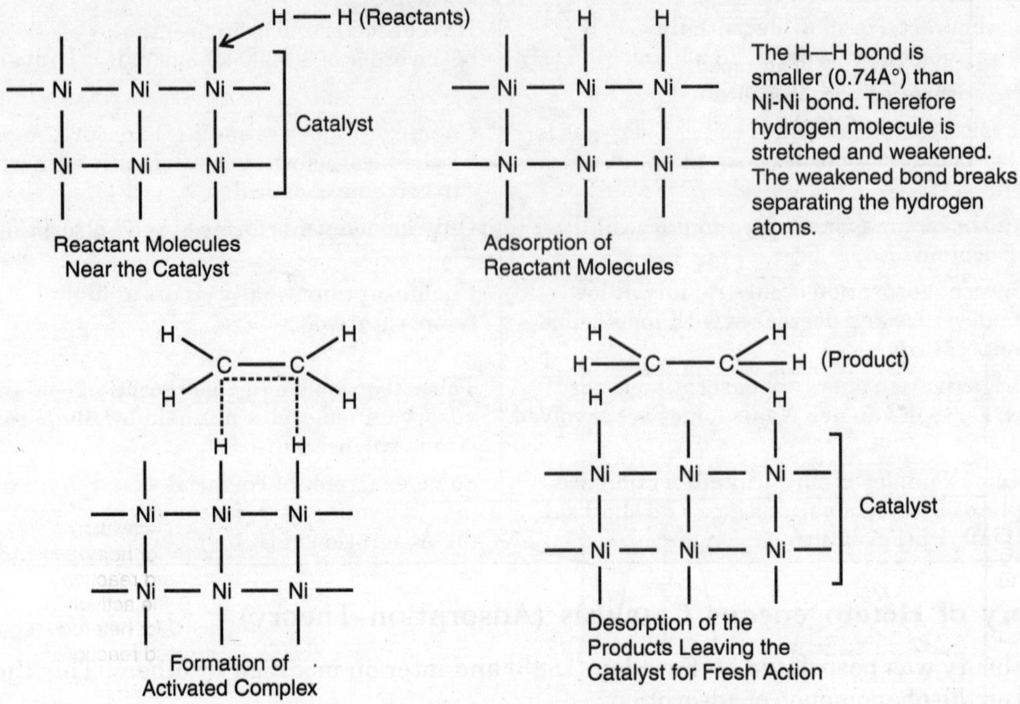

Fig. 10.5 Mechanism of Heterogeneous catalysis.

The process of heterogeneous catalysis also provides an alternative reaction path of much lower activation energy as compared to uncatalysed reaction. Fig. 10.5 give energy curves fon non-catalytic reaction, and heterogeneous catalysis. (Fig. 10.5)

A number of reactions have been studied both as homogeneous and heterogeneous processes. It has been found that lowering of activation energy is more in case of heterogeneous catalysis as compared to homogeneous catalysis (Fig. 10.6).

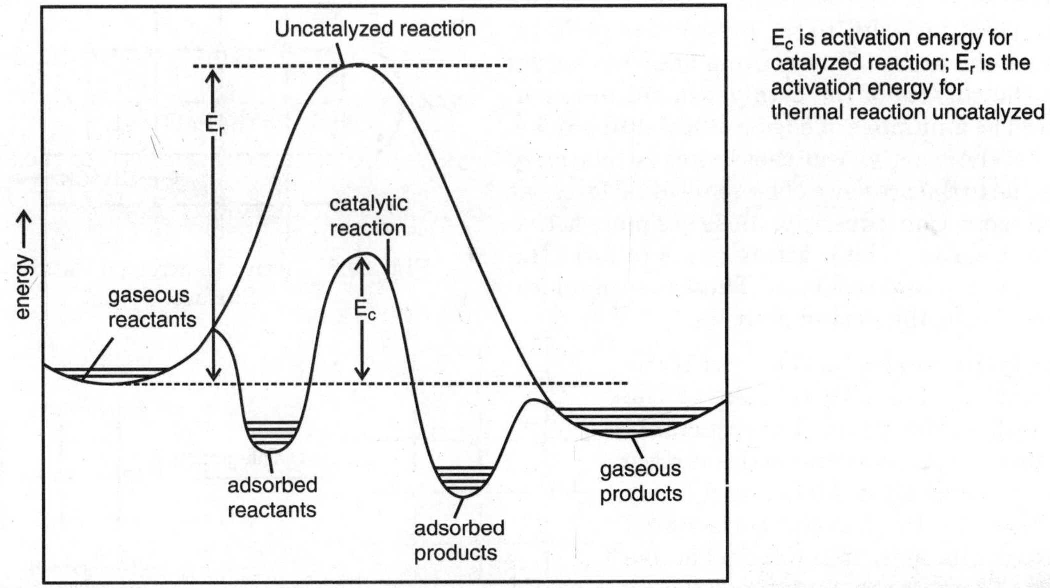

Fig. 10.6 Energy profiles for Heterogeneous catalysis.

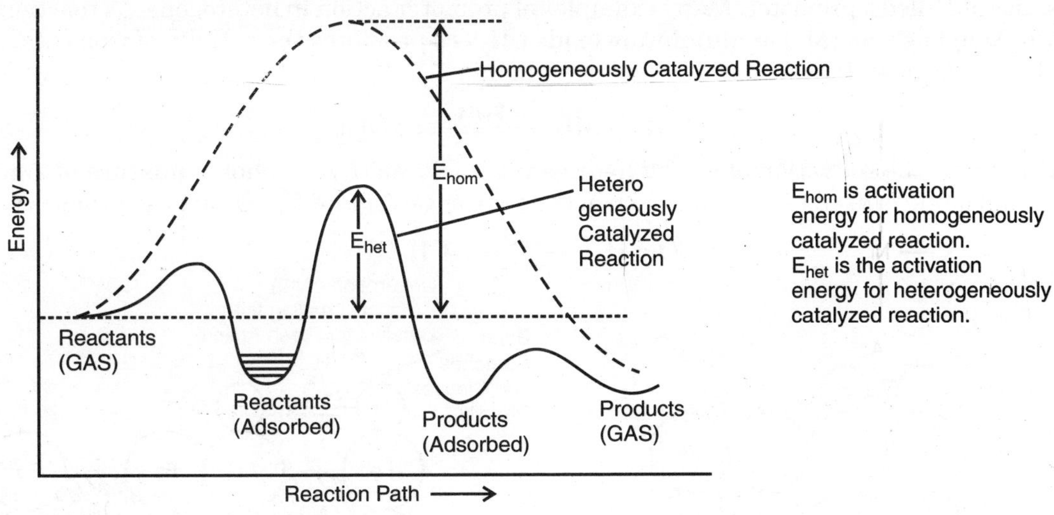

Fig. 10.7 Energy Profile for Homogeneously & Heterlogeneously Catalyzed Reaction

Adsorption theory of heterogeneous catalysis explains a number of the observed facts satisfactorily. Some of there are discussed below.

Active Centers: As discussed earlier, the activity of the catalyst is due to the presence of residual forces at the surface. X-ray examination has shown that the surface of the catalyst is not uniform. There are a number of edges, peaks and cracks. The catalytic activity at these sites is relatively high due to the presence of more residual forces at these spots. Consequently, these are more active in chemisorbing the reacting gases to form the adsorbed activated complex. These are therefore, referred to as the **active centres.**

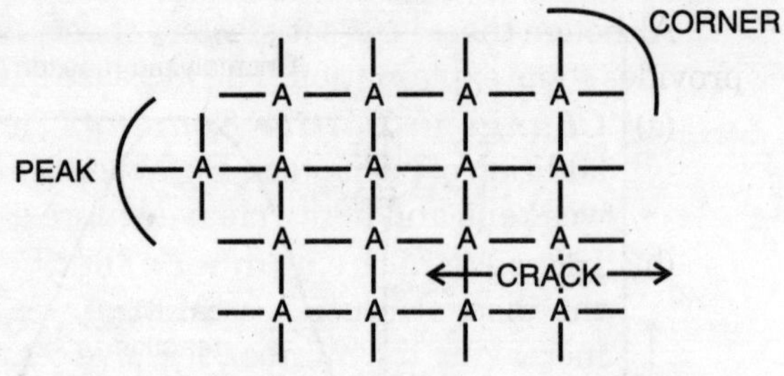

Fig. 10.8 Active Centres on Catalytic Surface.

Catalytic Activity: The catalytic activity is inversely related to the strength of adsorption of the reactants and directly proportional to the surface area of the catalyst. Metals in a finely divided state provides large surface area due to the increase in number of active sites. Thus large surface area is available for adsorption which enhance the activity of the catalyst.

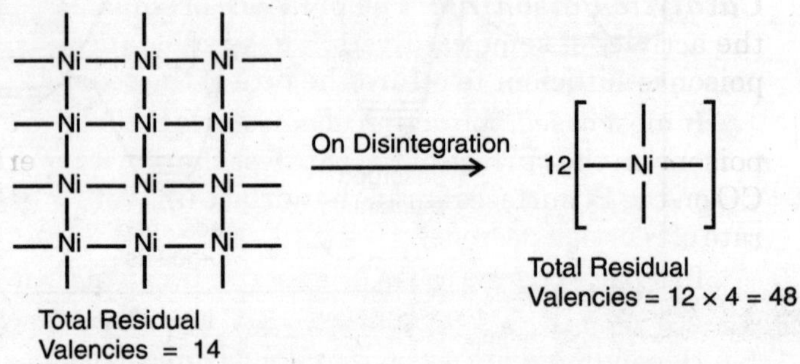

Promoters: It has been shown that the activity of a catalyst can be increased by adding a small amount of another substance, which itself is either a non catalyst or a feeble catalyst. Such a substance is called a promoter. Many examples of promoter action in heterogeneous reactions are known. Molybdenum (Mo) or aluminium oxide (Al_2O_3) promotes the activity of iron catalyst in the Haber synthesis for the manufacture of ammonia.

$$N_2 + 3H_2 \xrightleftharpoons{\text{Fe/Mo}} 2NH_3$$

Similarly, the reduction of CO by hydrogen to yield methyl alcohol, a mixture of zinc and chromium oxide is used as a catalyst. Zn acts as the catalyst and Cr_2O_3 acts as promoter.

$$CO + H_2 \xrightarrow[Cr_2O_3]{ZnO} CH_3OH$$

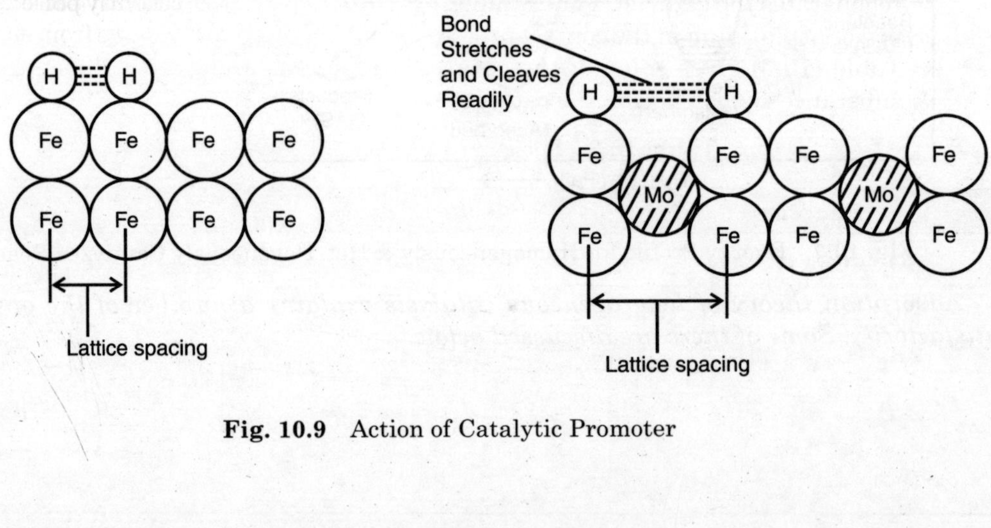

Fig. 10.9 Action of Catalytic Promoter

Although there is no satisfactory theory to explain the action of promoters, but adsorption theory provides some explaination. The increase in the catalytic activity is presumed to be because of–

(a) ***Change in Lattice Spacing:*** The lattice spacing of the catalyst is increased by the addition of promoters. Because of this, the bond between the adsorbed atoms of a molecule weakens and hence cleaved more easily thereby making the reaction go faster.

(b) **Increase in the number of active sites:** It is believed that probably promoters increase the discontinuities (peaks, curves and cracks) at the surface of the catalyst thereby increasing the number of active centres. The increase in the active sites results in greater adsorption of the reactant molecules on the interface between the promoter and the catalyst, where the free valencies are crowded. Increased adsorption results in higher rate of reaction.

Catalytic Poisoning: The presence of small amount of certain substances destroys or lowers the activity of some catalysts. These substances are referred to as poisons. A small amount of poison is sufficient to retard the rate of the reaction.

It most cases, the action of poisons is explained on the basis of preferential adsorption of the poisons on the surface of the catalyst thereby rendering it inactive. For example, a monolayer of CO on the Pt surface marks the surface unavailable for the adsorption of reactants and hence the rate of reaction decreases

In some extreme cases, it may combine with the catalyst. Fe is poisoned by presence of H_2S

$$Fe + H_2S \longrightarrow FeS + H_2$$

Depending upon the extent and strength of adsorption, catalytic poisoning is of two types–

(a) ***Temporary poisoning:*** In this type of poisoning, the activity of the catalyst is restored when the substance responsible for poisoning is removed from the reaction mixture. For example, in the combination of nitrogen and hydrogen, iron is used as a catalyst. Water vapour or oxygen act as temporary poisons by uniting with iron to form an oxide of iron. If the poison (H_2O or O_2) is removed from the gas the iron oxide is reduced to iron by the hydrogen present in the reaction mixture and its catalytic activity is restored.

Similarly, in the catalytic cracking of hydrocarbons, the aluminium silicate is used as a catalyst. A deposit of carbon temporarily poisons the catalyst by filling up the sites on the surface of the catalyst. This is burnt off and catalyst is activated again.

(b) ***Permanent Poisoning:*** In this type of catalytic poisoning, the poisons affect the catalyst to such an extent, that its activity is virtually changed. It is not easy to reactivate the catalyst by simple means. However, chemical treatment for removal of poison often regenerates the catalyst. For example, oxides of arsenic permanently poisons the catalytic power of platinum in the contact process for making sulphur trioxide from sulphur dioxide. Table (10.3) gives some of the catalyzed reactions along with the catalyst used and substances which poison the catalyst.

Table 10.3 Poisons in Various Catalytic Reactions

Reaction	*Catalyst*	*Poison*
$SO_2 + O_2 \rightarrow SO_3$	Pt	As_2O_3 As_2O_3
$N_2 + 3H_2 \rightarrow 2NH_3$	Fe	H_2S
$2H_2 + O_2 \rightarrow 2H_2O$	Pt	CO
$H_2O_2 \rightarrow H_2 + \frac{1}{2}O_2$	Glass	$C_6H_5-NHCOCH_3$
$H_2O_2 \rightarrow H_2 + \frac{1}{2}O_2$	Pt	HCN, H_2S

In some cases one of the products of a reaction is strongly adsorbed by the catalyst surface; thus the rate of the reaction falls considerably and the phenomenon is called **autocatalytic** poisoning. The decomposition of ammonia catalyzed by platinum is retarded after sometime on account of the adsorption of one of the products i.e hydrogen on the surface of platinum which is used as a catalyst.

Poisoning of catalysts sometimes may be done deliberately to improve the selectivity of the catalyst. For example in dehydrocyclization of alkanes to aromatic compounds using chromia-alumina catalysts, potassium oxide, K_2O is used as poison. The function of K_2O is to neutralize the acidic sites, which would otherwise promote cracking.

Specificity or Selectivity: The specific action of the catalyst is due to the fact that the extent of adsorption of the reactant molecules on the catalyst depend upon the chemical affinity of the catalyst for the reactant molecules. In other words, the adsorption depends on the nature of both the adsorbent i.e. catalyst, and adsorbate i.e. reactants. So a catalyst under consideration cannot possess the same affinity for the different reactants. Hence, the action of the catalyst is also specific. **Lock & Key Model** developed by Emil Fischer (1894) explains the specificity of enzymes (biological catalysts). This model is discussed in section (10.7).

Because a particular catalyst increases the rate of a reaction, it may not do so for another reaction. Moreover, it has been found that different catalyst can bring about completely different reactions with the same reactants, as is evident from the following reactions–

$$C_2H_5OH \xrightarrow{Al_2O_3} C_2H_4 + H_2O$$

Thus ethanol when passed over hot Al_2O_3 gives ethene. But with hot copper it gives ethanal (CH_3CHO).

$$C_2H_5OH \xrightarrow{Cu} CH_3CHO + H_2$$

SOME OF THE IMPORTANT CATALYTIC PROCESSES

Development, selection and application of catalysts forms the basis of a major part of chemical industries. To give a more clear idea of the concept, some of the catalysis processes are discussed in detail in the following sections.

10.6 ACID BASE CATALYSIS

The most important category of homogeneously catalyzed reactions in solution is acid-base catalysis.

Early investigations by Ostwald and Arrhenius on the study of acid base catalysis showed that the rate of the reaction was dependent on the concentration of H^+ ions in acid catalyzed reactions and on the concentration of $\bar{O}H$ ions in base catalyzed reactions. Thus H^+ or $\bar{O}H$ were thought to be effective catalytic agents. It was also observed the nature of anion or undissociated acid in acid catalyzed reaction and the nature of cation or undissociated base in base catalysed reaction does not influence the rate of reaction. But later on, the work of Taylor, Lowry and others showed the reactions catalysed by H^+ ions are influenced by the undissociated acids and the reactions catalyzed by $\bar{O}H$ ions are influenced by the undissociated bases. Many observations of similar nature were made by other workers including S. Arrhenius during the course of his work later. Acid-base catalysis can be divided into the following categories–

(a) Specific acid catalysis.
(b) Specific base catalysis.
(c) General acid catalysis.
(d) General base catalysis.

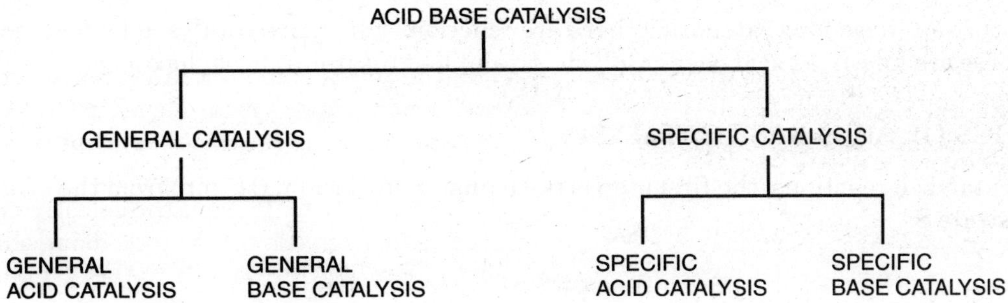

(a) **Specific Acid Catalysis:** These are the reactions which are catalysed by H^+ or H_3O^+ ions **only** or simply these are proton catalyzed reactions. Example – inversion of cane sugar in the presence of acid as a catalyst. The rate is influenced by the concentration of H^+ ions.

$$\underset{}{C_{12}H_{22}O_{11}} + H_2O \xrightarrow{H^+} \underset{\text{Glucose}}{C_6H_{12}O_6} + \underset{\text{Fructose}}{C_6H_{12}O_6}$$

Keto Enol transformation also involves specific acid catalysis.

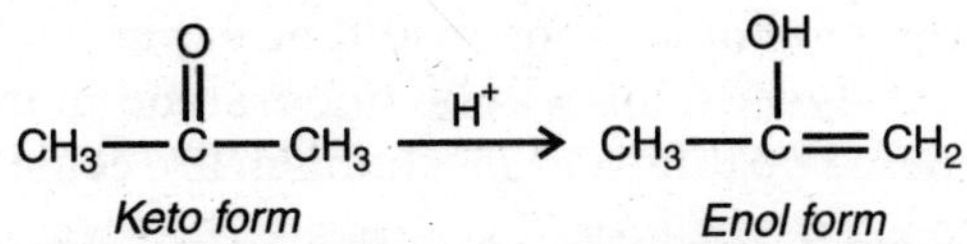

The rate of reaction of these reactions is found to be proportional to the concentration of hydrogen ions present in the solution.

(b) **Specific base Catalysis:** These reactions are catalyzed by $\bar{O}H$ ions only and are not influenced by undissociated bases or proton acceptors. For example conversion of acetone into diacetone alcohol is catalyzed by $\bar{O}H$ ions only

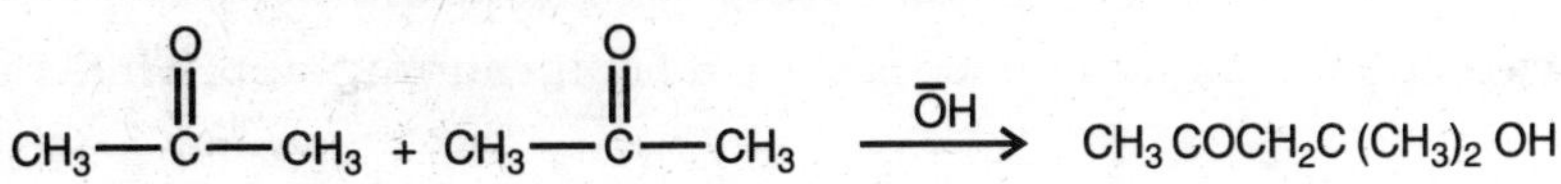

The rate of reaction in this case is dependent upon the concentration of $\bar{O}H$ ions and is independent of any other basic specie present in the system.

(c) **General Acid Catalysis:** In this type of catalysis the rate is affected by not only H^+ ions present in the system, but also undissociated acid or cations of weak bases, influence the rate of reaction. In other words, all the substances which have a tendency to lose a proton (bronsted acid) can function as catalyst. Example–Iodination of acetone catalyzed by monochloroacetic acid and its salt.

$$CH_3COCH_3 + I_2 \xrightarrow[ClCH_2CO\bar{O}]{ClCH_2COOH} CH_3COCH_2I + HI$$

The rate of the reaction of these type of reactions depend upon the concentration of $\bar{O}H$ ions as well as all effective proton donors.

(d) **General Base Catalysis:** These reactions are catalyzed by hydroxyl ions, undissociated bases and even anions of weak acids. In other words, all the substances which have a tendency to gain a proton (bronsted bases) will act as catalyst. Example-decomposition of nitramide is catalyzed by hydroxyl ions, or by acetate ions.

$$NH_2NO_2 \xrightarrow{CH_3COO^-} N_2O + H_2O \text{ (General base catalyzed)}$$

$$NH_2NO_2 \xrightarrow{\bar{O}H} N_2O + H_2O \text{ (Specific base Catalyzed)}$$

Apart from these four categories there are reactions which are catalysed by both acid and base. These are known as acid base catalysis. Example–Mutarotation of glucose.

KINETICS OF ACID BASE CATALYSIS

In acid catalyzed reactions, the first step is the transfer of a proton (H^+ ion) from the acid HA to the substrate S.

$$S + AH^+ \underset{k_2}{\overset{k_1}{\rightleftharpoons}} SH^+ + A^- \quad \text{(fast)} \qquad .....(1)$$

Where SH^+ is the intermediate compound and k_1 and k_2 are the rate constants for the forward and backward reactions.

There are two possibilities with the intermediate compound.

(a) either it reacts with water molecule and transfer its proton with simultaneous formation of product.

(b) Or it reacts with the base A^- and transfer its proton to it with simultaneous formation of product.

Hence addition and removal of protons would be expected to play an essential part in the mechanism of acid-base catalysis. This will be understood more clearly by discussing the two possibilities separately where two different mechanism are considered.

(a) Let us consider the first possibility i.e. in the second step, the intermediate formed in step (1) reacts with H_2O molecule.

$$S + AH^+ \underset{k_2}{\overset{k_1}{\rightleftharpoons}} SH^+ + A^- \quad \text{(fast)}$$

$$SH^+ + H_2O \xrightarrow{k_3} P + H_3O^+ \quad \text{(slow)} \qquad ...(2)$$

This reaction (2) is slow and hence is the rate determining step. Thus rate of the reaction is given by–

$$rate = \frac{-d[S]}{dt} = k_3[SH^+] \qquad ...(3)$$

We are assuming that the concentration of H_2O remains constant in very dilute solutions and hence the term $[H_2O]$ is not included in the rate law equation.

The concentration of the intermediate SH^+ practically does not change with time interval i.e. the rate of formation of the intermediate and its disappearance are equal. The constant concentration of an unstable intermediate species is called **steady state approximation**. Thus the concentration of SH^+ can be obtained by applying the steady state approximation–

$$\frac{d[SH^+]}{dt} = 0 = k_1[S][AH^+] - k_2[SH^+][A^-] - k_3[SH^+] \qquad ...(4)$$

or $$k_2[S][AH^+] = k_2[SH^+][A^-] + k_3[SH^+] = \{k_2[A^-] + k_3\}[SH^+]$$

or $$\frac{k_1[S][AH^+]}{k_2[A^-] + k_3} = [SH^+] \qquad ...(5)$$

Substituting the value of $[SH^+]$ in rate equation (3) the rate of formation of the product is given by–

$$k_3[SH^+] = \frac{k_3 k_1[S][AH^+]}{k_2[A^-] + k_3} \qquad ...(6)$$

From equation (6) two cases may arise–

(i) If $k_3 \gg k_2[A^-]$, then ignoring $k_2\,[A^-]$

$$\text{Rate} = k_1\,[S]\,[AH^+] \quad ...(7)$$

In this case as $[AH^+]$ appears in the derived rate expression i.e. the rate of the reaction is affected by proton donor in general hence the reaction is **general acid catalysis.**

(ii) If $k_2\left[A^-\right] >> k_3$ then, ignoring k_3 in the denominator.

$$\text{Rate} = \frac{k_1\,k_3\,[S]\,[AH^+]}{k_2\,[A^-]} \quad ...(8)$$

But for a weak acid

$$AH^+ \rightleftharpoons A^- + H^+ \quad ...(9)$$

The equilibrium constant for this acid base reaction is

$$K = \frac{[A^-]\,[H^+]}{[AH^+]} \quad ...(10)$$

or

$$[H^+]/K = \frac{[AH^+]}{[A^-]} \quad ...(11)$$

substituting this value in equation (8) we get

$$\text{Rate} = \frac{k_1\,k_3\,[S]\,[H^+]}{k_2\,K} \quad ...(12)$$

Thus only $[H^+]$ appears in the rate equation i.e. H^+ or H_3O^+ is the **only** acid that catalyzes the reaction. The reaction rate is unaffected by the addition of any other potential proton donors which are also acids by definition. Hence the reaction is **specific hydrogen catalyzed.**

(b) In the second possibility, we consider that in the second step, the intermediate SH^+ reacts with a base A^- instead of water molecule.

$$S + AH^+ \underset{k_2}{\overset{k_1}{\rightleftharpoons}} SH^+ + \bar{A} \qquad (fast)$$

$$SH^+ + A^- \xrightarrow{k_3} P + HA \qquad (slow) \quad ...(13)$$

The rate of the reaction is given by

$$Rate = \frac{-d[S]}{dt} = k_3\,[SH^+]\,[A^-] \quad ...(14)$$

Applying steady state approximation,

$$\frac{d\,[SH^+]}{dt} = 0 = k_1\,[S]\left[AH^+\right] - k_2\left[SH^+\right]\left[A^-\right] - k_3\left[SH^+\right]\left[A^-\right] \quad ...(15)$$

$$k_1\,[S]\left[AH^+\right] = k_2\left[SH^+\right]\left[A^-\right] + k_3\left[SH^+\right]\left[A^-\right] \quad ...(16)$$

$$\left[SH^+\right]\left[A^-\right] = \frac{k_1\,[S]\left[AH^+\right]}{k_2 + k_3} \quad ...(17)$$

Substituting in equation (14) we get–

$$Rate = \frac{k_3\,k_1\,[S]\left[AH^+\right]}{k_2 + k_3} \quad ...(18)$$

As $[AH^+]$ appears in the rate equation, hence it is a case of **general acid catalysis.**

While the **acid catalysis** involves the transfer of a proton from the catalyst to the reactant, **base catalysis** involves the transfer of a proton from the reactant to the catalyst forming an intermediate, which decomposes to give the products.

Thus the first step is the transfer of a hydrogen atom from the substrate to the base.

$$\underset{\text{Substrate}}{SH + B} \rightarrow S^- + BH^+ \qquad \text{...(19)}$$

The expressions for the rate for base catalyzed reaction can be obtained in a manner similar to that of acid catalyzed reactions.

EXAMPLES OF ACID BASE CATALYSIS

Specific Acid Catalysis

Hydrolysis of ester, inversion of cane sugar etc are specific acid catalyzed reactions. The mechanism of acid catalyzed hydrolysis can be explained as follows–

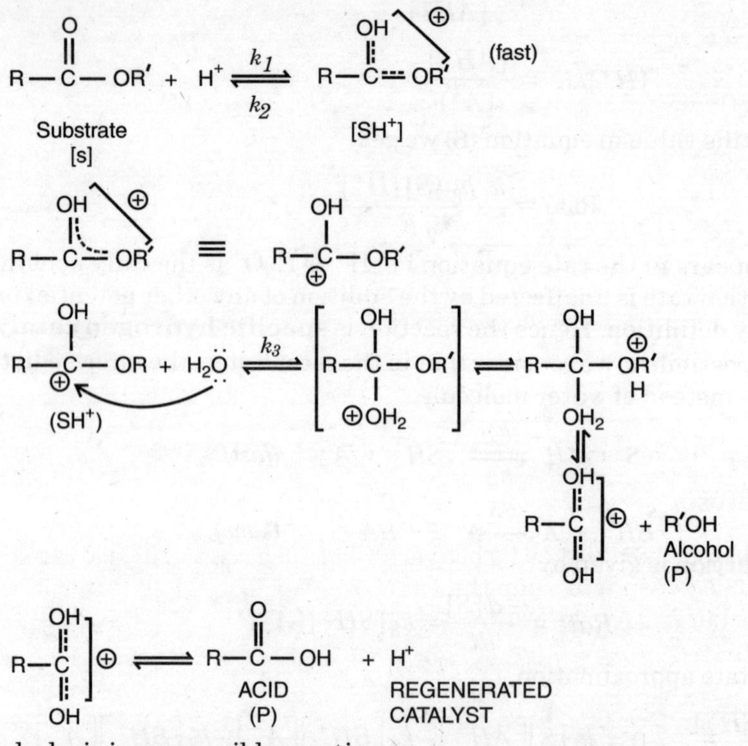

Acidic hydrolysis is a reversible reaction

$$RCOOR' + H_2O \underset{H^+}{\overset{H^+}{\rightleftharpoons}} RCOOH + R'OH$$

Thus in the reverse process i.e. esterification the mechanism is just the mechanism of hydrolysis taken in the opposite directions.

Rate of the reaction is given by–

$$r = k_3 [SH^+]$$

$$\text{or } r = \frac{k_1 k_3 [RCOOR'][H^+]}{k_2 + k_3} \qquad \text{...(20)}$$

as is discussed in kinetics of the reaction. Thus rate is first order with respect to ester as well as H^+. At constant H^+ the reaction is first order with respect to ester, or can be termed as pseudo first order reaction.

Specific Base Catalysis

Hydrolysis of ester may also be accomplished by alkalies. But this is irreversible reaction, since the reverse reaction, esterification, cannot be brought about in the presence of alkalies. This is because carboxylic acid reacts with alkali to form a resonance stabilised carboxylate anion which has very little tendency to react with an alcohol.

Hydrolysis of ester depends upon the concentration of $\bar{O}H$ ion and hence is specific base catalysed reaction. The mechanism can be outlined as follows:

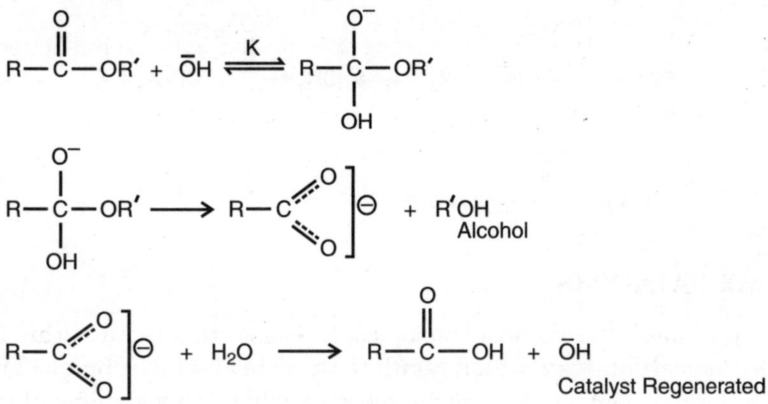

Thus the rate of the reaction depends upon the concentration of the ester as well as hydroxide ion.

$$r = k\,[R\overset{O}{\overset{\|}{C}} - OR]\,[\bar{O}H] \qquad ...(21)$$

General Acid Catalysis

The halogenation of a ketone is catalyzed by both acids as well as bases. In acid catalyzed reaction, the rate of the reaction is dependent upon the anion of the acid and hence this is called general acid catalysis. The mechanism can be outlined as follows:

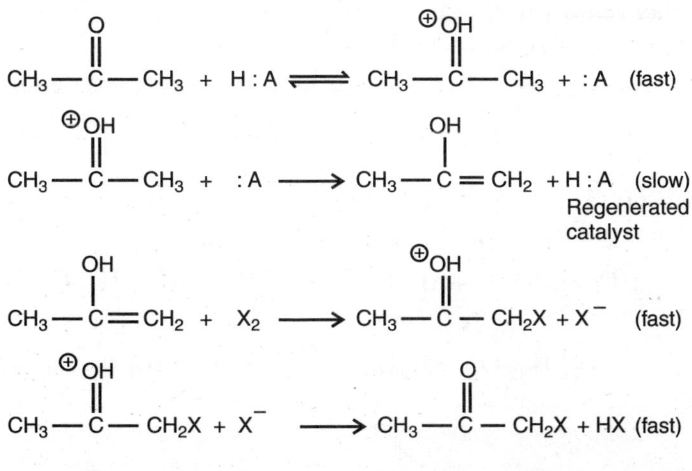

It is seen from the above mechanism that the rate determining step (slowest step) involves the anion :A, the anion of the acid.

General Base Catalysis

The halogenation of a ketone in the presence of a base. The rate determining step involves the bronsted base (the substance which can accept the proton) and not the hydroxide ion and hence this reaction is general base catalysed reaction.

The mechanism can be outlined as follows–

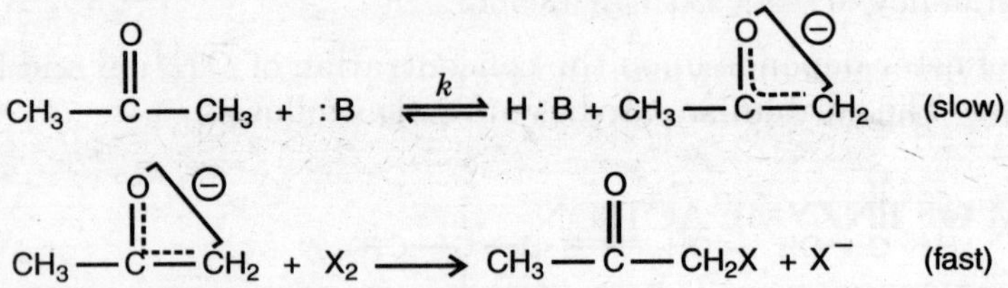

Thus rate of the reaction is given by

$$r = k\,[CH_3\overset{\overset{\displaystyle O}{||}}{C} - CH_3]\ [:B] \qquad \text{...(22)}$$

10.7 ENZYME CATALYSIS

Enzyme catalysis is one of the essential biological processes that occur within the living organism. Enzymes are biological catalysts which regulate the rate at which chemical reactions proceed.

Although a number of enzymes were discovered earlier, but a significant confirmation of their importance in living systems was found in 1897 by the German chemist Eduard Buchner. Since then more than 1000 enzymes have been recognised, each specific to a particular chemical reaction. A number of enzymes have been crystallized and studied in detail. The first enzyme to be crystallized was urease, which was isolated from the jackbean by James Batcheller Sumner in 1926. The enzymes which are isolated and studied have been found to be **proteins** comprising of about 20 natural amino acids.

An enzyme is typically a large protein molecule that contains one or more active sites where interaction with the substrate takes place. The amazing fact about the enzyme is that they can increase the rate of biochemical reactions by factors ranging from 10^6 to 10^{12}. The enzymes are highly specific. It has been estimated that an average living cell may contain some 3000 different enzymes, each of them catalyzing a specific reaction.

Some of the examples of enzyme catalysis are–

(i) Inversion of cane sugar by the enzyme invertase present in yeast.

$$C_{12}H_{22}O_{11} + H_2O \xrightarrow{\text{Invertase}} \underset{\text{glucose}}{C_6H_{12}O_6} + \underset{\text{fructose}}{C_6H_{12}O_6}$$

(ii) Conversion of starch to form maltose, which in turn undergoes hydrolysis to form glucose in presence of maltase.

$$2(C_6H_{10}O_5)_n + nH_2O \xrightarrow{\text{Diastase}} nC_{12}H_{22}O_{11}$$

Maltose

$$C_{12}H_{22}O_{11} + H_2O \xrightarrow{\text{Maltase}} 2C_6H_{12}O_6$$

Glucose.

(iii) Hydrolysis of urea by urease present in soyabean

$$H_2N-\overset{\overset{\large O}{\|}}{C}-NH_2 + H_2 \xrightarrow{\text{Urease}} 2NH_3 + CO_2$$

(iv) Conversion of glucose into ethanol by zymase present in yeast.

$$\underset{\text{Glucose}}{C_6H_{12}O_6} \xrightarrow{\text{Zymase}} \underset{\text{Ethanol}}{2C_2H_5OH} + 2CO_2$$

(v) The enzyme ptyalin present in saliva converts starch into glucose.

$$(C_6H_{10}O_5)_n + nH_2O \xrightarrow{\text{ptyalin}} \underset{\text{Glucose}}{nC_6H_{12}O_6}$$

(vi) The enzymes pepsin, trypsin and erepsin present in gastric juice, pancreatic juice and intestine respectively converts the proteins to amino acids.

MECHANISM OF ENZYME ACTION

Many of the reactions in the cells have very slow rates with very high activation energies. We have seen earlier that a catalyst accelerates the formation of products in a reaction by decreasing the energy of activation. This is the role that enzymes play. They react with the substrate to form an intermediate complex that requires less energy for the reaction to proceed. The unstable intermediate compound quickly breaks down to form reaction products and the enzyme is free to react with other substrate molecules.

Enzymes are proteins which form a type of specific folding pattern depending upon the conformation of protein molecule, in which only a specific substrate molecule can fit, thus making enzymes highly specific. This concept is explained on the basis of Lock and Key model given by Emil Fischer.

Lock & Key Model

Emil Fischer proposed this model in (1894) to explain the specificity of enzymes. The main points of this hypothesis are–

(i) An enzyme is ordinarily a very large protein molecule coiled in a definite pattern which gives the enzyme a specific geometry having cavities on the surface called active sites.

(ii) The reaction with the substrate takes place at their active sites whereas rest of the protein molecule remains undisturbed.

(iii) A substrate molecule which has got complementary structure to the active centre may fit into it. In other words the rigid structure of active sites on protein molecules can be compared to a lock into which the substrate molecule having complementary shape can fit, just as key fits into a lock.

(iv) As the substrate molecules are drawn into the active site, they are activated and undergo the catalysed reaction forming the products.

(v) The product seperate from the enzyme, readying the active sites for fresh action.

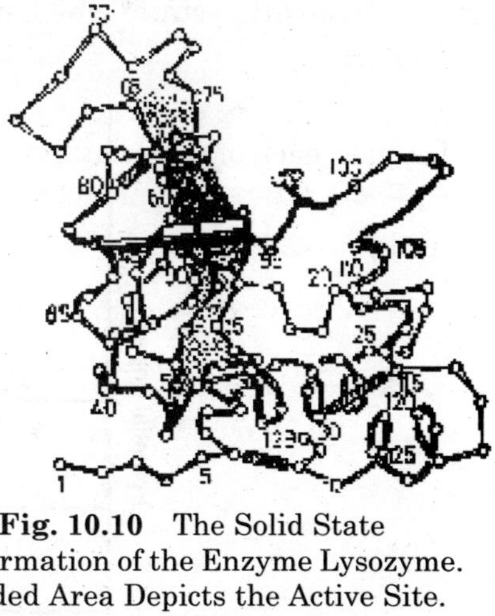

Fig. 10.10 The Solid State Conformation of the Enzyme Lysozyme. Shaded Area Depicts the Active Site.

Fig. 10.11 gives the schematic illustration of lock and key model.

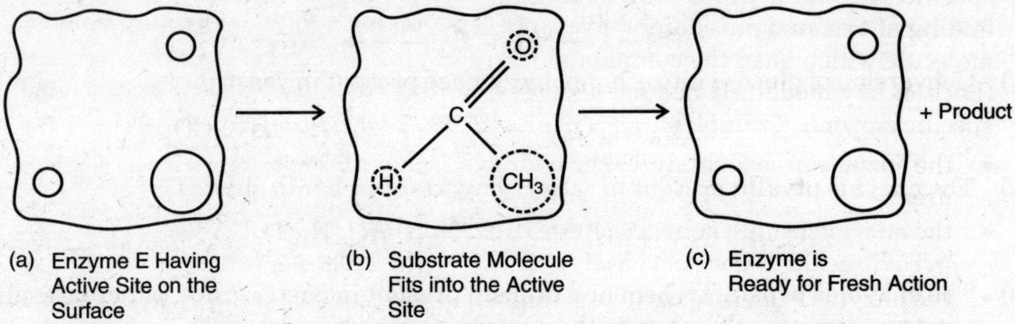

Fig. 10.11 Lock and Key Model

Characteristics of Enzyme Catalysis

1. **Enzymes constitute a class of most efficient catalysis.** They increase the rate biochemical reactions by factors ranging from 10^6 to 10^{12}. For example, starch which we eat as rice or bread, gets broken down in the body cells in about twenty steps to finally form CO_2 and H_2O.

$$\text{Starch} \longrightarrow \text{Glucose} \xrightarrow{O_2} 6CO_2 + 6H_2O$$

The total time taken for all these 20 steps in presence of enzymes as catalyst is less than a hour in the cell. If we were to attempt these reactions in laboratory, this would take months. The catalytic efficiency of enzymes is the highest known in nature. The number of substrate molecules converted in 1 minute by an enzyme are of the order of 10^2-10^6 which is also referred to as turnover number.

2. **Enzymes function as catalysts by lowering down the activation energy of a reaction.** For example, the activation energy for the decomposition of hydrogen peroxide is 18,000 cal mol^{-1}. If we use platinum as a catalyst the activation energy is lowered to 12,000 cal mol^{-1}. But in the presence of enzyme catalyst, the activation energy reduces to 5500 cal mol^{-1}. The enzyme reacts with the substrate to form an intermediate complex–a "transition state" which requires much less energy for the reaction to proceed.

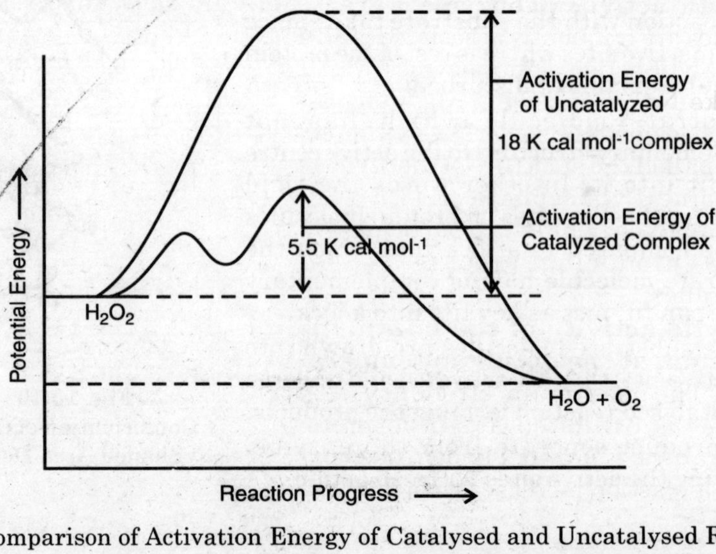

Fig. 10.12 Comparison of Activation Energy of Catalysed and Uncatalysed Reaction

3. **Enzymes are highly specific:** The important characteristic of enzymes is their extreme specificity. This is quite evident from the fact that the conformation or overall chain folding of the enzyme confers a specific conformation to the active site. The substrate molecule which have the complementary shape would fit into this shape otherwise not. (Lock & key model). It has been suggested that each biochemical process has its own specific enzyme. Examples
   - the enzyme urease catalyzes the hydrolysis of urea and no other amide, not even methyl urea.
   - the enzyme lactate dehydrogenase catalyzes the dehydrogenation of lactic acid to yield pyruvic acid but does not catalyze the dehydrogenation of other acids.
   - the enzyme fumerase catalyzes the reaction of fumaric acid with water to form malic acid but the cis isomer of fumaric acid is not catalyzed.
   - Where the compound can exist in optically active isomers, an enzyme can act on one of the isomers and is unable to act on the other. The enzyme present in Pencillium glaucum, ordinary mould, when added to $\pm$ tartaric acid, decomposes the (+) form only.
4. **Enzyme activity can be inhibited by the presence of certain compounds:** Inhibition occurs when molecules, very similar in structure to substrate molecules bind to the enzyme substrate and prevent binding of the actual substrate. The use of penicillin is based on the competitive inhibition. Penicillin blocks the active site of the enzyme which is used by bacteria to construct their cell wall. Fig. (10.13) represents schematically the action of inhibitor.

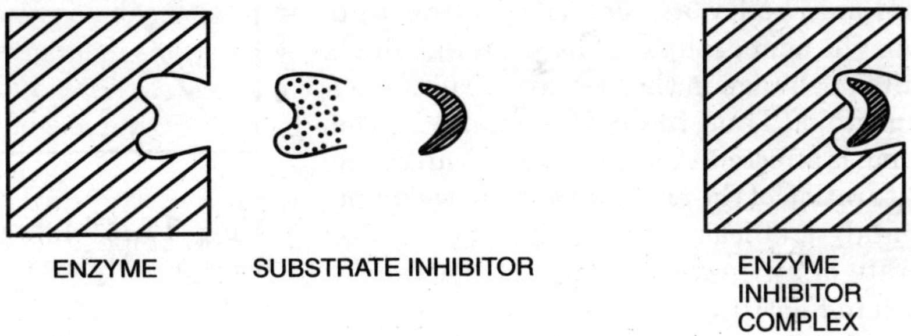

Fig. 10.13 Competitive Inhibition

5. **Catalytic activity of enzyme is greatly enhanced by the presence of Co-enzymes:** Enzymes are frequently associated with activators called coenzymes. Coenzymes promote the catalytic action of the enzymes. Metal ions like Na^+, Co^{2+}, Cu^{2+}, Mg^{2+} etc can act as coenzymes or sometimes vitamins form a part of the structure of coenzymes. Adenosine triphosphate (ATP) is an important coenzyme which participate in energy producing process in cell membranes.

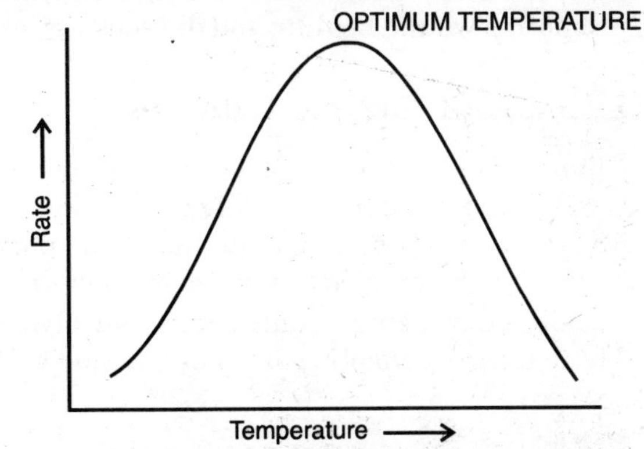

Fig. 10.14 Effect of Temperature on the rate of Reaction of Enzyme Catalyzed Reaction

6. **Catalytic activity of the enzyme is maximum at the optimum temperature**– The maximum efficiency of a catalyst is exhibited at an optimum temperature. It is a well known fact that rate of a reaction increases with increase

in temperature, but this is not the case with enzyme catalyzed reactions. The rate of a enzyme catalyzed reaction is increased upto a certain point with rise in temperature thereafter it decreases and eventually approaches zero. The temperature at which the reaction rate is maximum is called optimum temperature. The effect of temperature on enzyme catalyzed reactions is illustrated in Fig. 10.14.

The above can be explained on the basis of conformational aspects. At high temperatures the degree of folding of the protein is altered and this affects the formation of enzyme substrate complex. In other words the enzyme is denatured at high temperatures and its protein structure is gradually destroyed. At temperatures above 45° C the thermal denaturation is quite significant whereas at 55° C the denaturation of enzyme protein completely destroys the catalytic activity.

The optimum temperatures of enzyme reactions occuring in human body is 37^0 C. If the temperature becomes higher the enzymatic activity is reduced.

7. **Catalytic activity of Enzyme is maximum at optimum pH :** The rate of reaction and hence activity of an enzyme is maximum at a given pH called optimum pH. The enzymes of the body exhibit their maximum activity at pH close to 7.4. Change in conformation of the enzyme with change in external conditions, again explains the dependence of enzyme catalysis on pH. Since enzymes are proteins which are inturn made up of a large number of amino acids. Change in pH markedly affects the amino and carboxylic acid groups of the protein, which results in change in the tertiary structure of the protein. Thus the catalytic site of the enzyme is affected. Also pH change can cause denaturation of the enzyme protein, reducing the catalytic efficiency. The pH dependence of the rate of an enzyme catalyzed reaction is shown in Fig. 10.15.

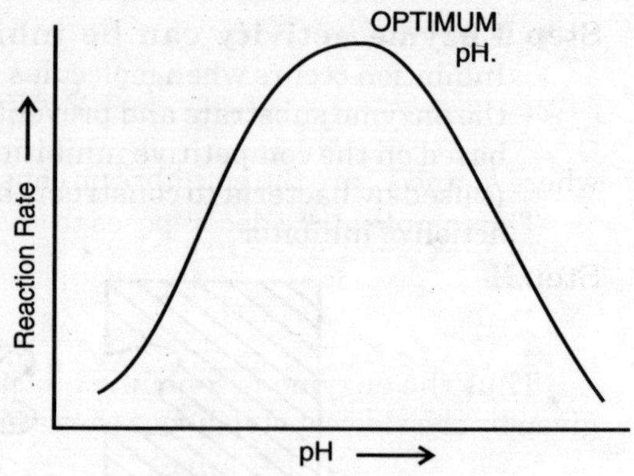

Fig. 10.15 Effect of pH on Rate of Reaction of Enzyme Catalyzed Reaction

Enzymes also have valuable industrial and medical applications. The fermenting of wine, curdling of cheese, brewing of beer have been practised from earlier times. Since then, enzymes have assumed an increasing importance in industrial processes that involve organic chemical reactions. The uses of enzymes in medicine include, killing disease causing micro organisms, promoting wound healing and diagnosing certain diseases.

Example of Enzyme Catalysis

Chymotrypsin is a digestive enzyme which hydrolyse certain peptide links in proteins. Its conformation and sequence of amino acids have been determined. It is a soluble globular protein folded in such a way that its lipophilic parts are turned inwards, away from water, forming maximum intramolecular hydrogen bonding.

Chymotrypsin not only cause hydrolysis of proteins, but also of certain amides and esters. The action of chymotrypsin is illustrated below in the hydrolysis of amide linkage in proteins.

I Step: The enzyme acts as an alcohol and protein undergoes nucleophilic acyl substitution.

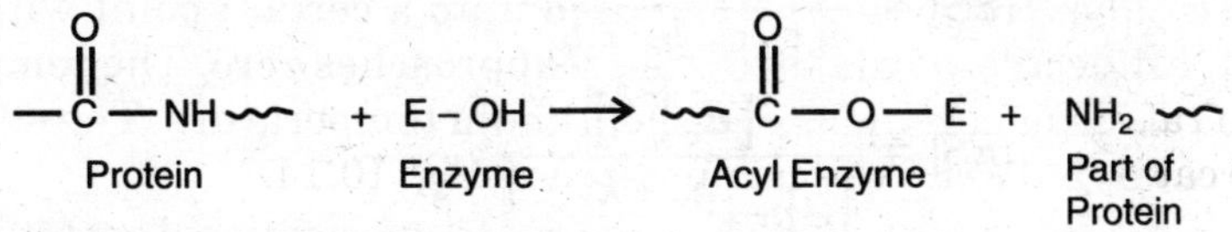

Thus an enzyme ester is formed.

II Step: In this step, the ester is hydrolyzed to form carboxylic acid, the remaining portion of substrate molecule. The enzyme is regenerated.

Kinetics of Enzyme Catalyzed Reactions

The mechanism of enzyme catalyzed reactions was proposed by L. Michaelis and M.L. Menten. This mechanism is known as **Michaelis-Menten mechanism.** The mechanism can be explained as follows:

Let us consider, the substrate [S] forms a complex with the enzyme [E]

Step I

$$E + S \underset{k_2}{\overset{k_1}{\rightleftharpoons}} \underset{(Complex)}{[ES]} \qquad (fast) \qquad ...(23)$$

where k_1 and k_2 are the equilibrium constants of forward and the backward reaction respectively.

The complex [ES] decomposes to give the product [P] and enzyme [E] is regenerated.

Step II

$$\text{ES} \xrightarrow{k_3} \text{P + E} \qquad \text{(slow)} \qquad ...(24)$$

Thus the enzyme is consumed in step I and regenerated in step II. The rate of reaction is given by the slowest step hence the rate of appearance of the product is given by–

$$-\frac{d[S]}{dt} = k_3[ES] \qquad ...(25)$$

Applying the steady state approximation to [ES] we get–

$$\frac{d[ES]}{dt} = 0 = k_1[E][S] - \{k_2[ES] + k_3[ES]\} \qquad ...(26)$$

$$k_1[E][S] = k_2[ES] + k_3[ES]$$

$$[ES] = \frac{k_1[E][S]}{k_2 + k_3} \qquad ...(27)$$

Now [*E*] cannot be measured experimentally. But the total enzyme concentration $[E_O]$ is a measurable quantity and is related to [E] and [ES] as–

$$[E_o] = [E] + [ES] \qquad ...(28)$$

where $[E_o]$ is the initial concentration of the enzyme and [*E*] is the concentration of free enzyme,

$$\therefore \quad [E] = [E_o] + [ES] \qquad ...(29)$$

Substituting the above relation in equation (27)

$$[ES] = \frac{k_1\{[E_o] - [ES]\}[S]}{k_2 + k_3}$$

$$k_1[E_o][S] = \{k_1[S] + k_2 + k_3\}[ES] \qquad ...(30)$$

$$[ES] = \frac{k_1[E_o][S]}{k_1[S] + k_2 + k_3} \quad ...(31)$$

$$\therefore \quad [ES] = \frac{[E_o][S]}{[S] + \left\{ {k_2 + k_3}/{k_1} \right\}}$$

(Dividing the numerator and the denominator by k_1)

$$= \frac{[E_o][S]}{[S]k_m} \quad ...(32)$$

where $k_m = \frac{k_2 + k_3}{k_1}$ and is known as **Michaelis–Menten constant**. The constant k_m and k_3 are characteristic of a particular enzyme catalysed reaction.

Substituting the value of $[ES]$ from equation (32) in rate equation (25) we get–

$$Rate = \frac{k_3[E_o][S]}{[S] + k_m} \quad ...(33)$$

This equation is known as **Michaelis–Menten equation.**

If $[S_o]$ denotes the initial concentration of the substrate then initial rate of the reaction will be given by–

$$r_o = \frac{k_3[E_o][S_o]}{[S_o] + k_m} \quad ...(34)$$

Thus initial rate is dependent upon the concentration of the substrate at a given initial concentration of the enzyme. Two cases can be discussed–

(i) **When the concentration of substrate is high :** If the concentration of substrate is very high, the constant k_m can be ignored in comparison to $[S_o]$ and hence equation (34) reduces to–

$$r_o = k_3 [E_o] \quad ...(35)$$

Thus initial rate of the reaction is independent of the concentration of substrate i.e. it is a zero order reaction with respect to S. In other words equation (35) gives the maximum possible rate for the reaction at a given concentration of enzyme. Hence we can write equation (35) as:

$$r_{max} = k_3 [E_o] \quad ...(36)$$

k_3 is known as the turnover number of the enzyme. It is the number of substrate molecules converted into the products by one molecule of enzyme.

In enzymology r_{max} is denoted by V_{max} so we can write–

$$V_{max} = k_3 [E_o] \quad ...(37)$$

Using equation (37) the Michaelis–Menten equation can be written in the form

$$r = \frac{V_{max}[S]}{k_m + [S]} \quad ...(38)$$

(ii) **When the concentration of the substrate is low :** At low substrate concentration, $[S]$ can be ignored in comparison to k_m and hence equation (33) reduces to–

$$r = \frac{k_3}{k_m}[E_o][S] \quad ...(39)$$

Substituting k_3 $[E_o]$ with V_{max},

$$r = \frac{V_{max}}{k_m}[S] \qquad ...(40)$$

that is, the rate is first order with respect to substrate concentration.

Fig. 10.16 represents the variation of initial rate with the initial substrate concentration. Initially when the concentration of the substrate is low, the reaction follows first order kinetics. At very high concentrations of the substrate, initial rate is zero order with respect to S.

Michaelis Constant: If $k_m = [S]$ then from equation (38)

$$r = \frac{1}{2}V_{max} \qquad ...(41)$$

Thus, Michaelis constant is equal to that concentration of substrate at which the reaction rate is just half the maximum rate obtained at high concentration of S. (Fig. 10.16)

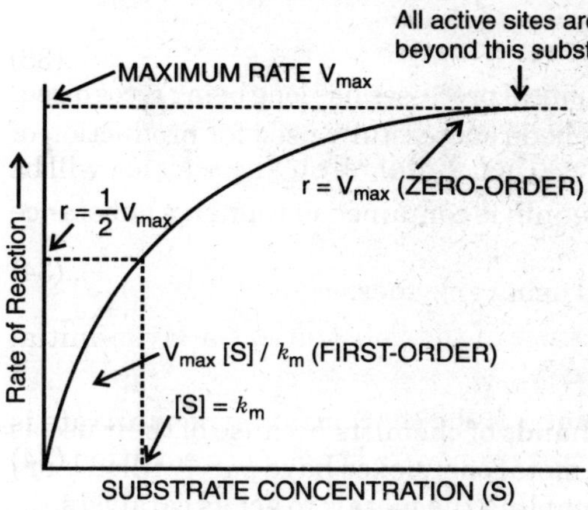

Fig. 10.16 Plot of Rate of Reaction Versus Substrate Concentration in an Enzyme Catalysed Reaction

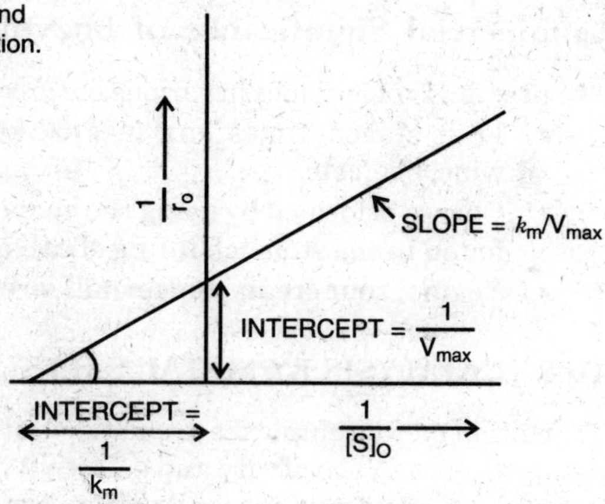

Fig. 10.17 The Lineweaver-Burk method of plotting enzyme kinetic data

From the above discussion it is clear that reaction rate of an enzyme catalyzed reaction changes from first order to zero order with increase in concentration of the substrate. The physical reason lies in the fact that at low concentrations of the substrate molecule, some of the active sites on the surface of the enzyme may remain unoccupied. When the concentration of the substrate is increased the number of occupied active sites also increases, thereby increasing the rate of the reaction. But as all the active sites on the surface of the enzyme are occupied at high concentration of the substrate, therefore no increase in rate of reaction is observed with further increase in the concentration of the reactants.

Another common form of Michaelis-Menten equation which makes the determination of V_{max} and hence k_m easier is obtained by taking the reciprocal of both sides in Michaelis-Menten equation.

$$\frac{1}{r_o} = \frac{[S_o] + k_m}{k_3[E_o][S_o]} \qquad ...(42)$$

$$= \frac{1}{k_3[E_o]} + \frac{k_m}{k_3[E_o][S_o]}$$

or

$$\frac{1}{r_o} = \frac{1}{V_{\max}} + \frac{km}{V_{\max}[S_o]} \quad ...(43)$$

Now, $\because$ slope $= \frac{k_m}{V_{\max}} = k_m \times$ intercept

$\therefore$ $k_m =$ slope/Intercept

A plot of $\frac{1}{r_o}$ against $\frac{1}{[S_o]}$ will be a straight line with slope $k_m/V_{\max}$ and intercept on the x axis is given by $-1/k_m$ whereas on y axis the intercept is equal to $1/V_{\max}$. Thus, from the two intercepts or one intercept and the slope, the constants $V_{\max}$ and k_m can be determined. This method is known as Lineweaver-Burk method of determination of V_{max}. Fig. 10.17.

Commercial Significance of Enzymes

The usefulness of enzymes in various life processes and chemical processes has long being recognized.

- From ancient times, enzymes are known to have been successfully used for production of wine, been etc.
- Cheese is formed by using the enzyme rennin. Rennin is contained in rennet, a substance found in the stomach lining of calves.
- Yogurt, sour cream, buttermilk are all produced using enzymes.

10.8 CATALYSIS BY METAL SALTS

Transition metal complexes are important tools in the hands of chemists because of their use as catalysts of high specificity and selectivity. Transition metal complexes have partly filled *d* or *f* subshells. The vacant subshells act as bonding sites, enabling the metals to act as catalysts.

The catalytic properties of transition metals are probably due to the fact, that these exhibit multiple oxidation states which help them to form various unstable intermediate compounds. In certain other cases, transition metal provides a large surface area with free valencies on which the reactant molecules are adsorbed.

A metal complex is formed by the metal and certain ions or molecules known as ligands. These ligands as such take no direct part in the reaction that is being catalyzed but still these help to determine the course of reaction. They stabilize the complex and make it soluble in organic solvents. In some cases the ligands are substituted by the reactant molecules during the course of the reaction.

Metal brings the reactant molecules in close proximity to each other by forming bonds with them. This is achieved by substituting some of its ligands with the reactant molecules. The reactant molecules then interact with each other to form the products. The product departs and the metal complex is again ready to start the catalytic cycle.

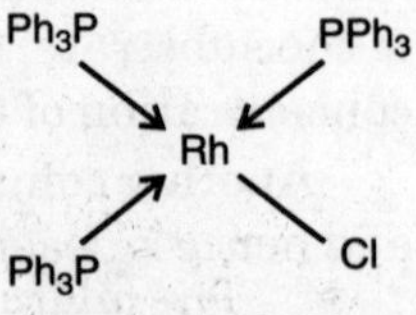

Wilkinson's Catalyst

The mechanism of catalysis by metal salt is well illustrated by **Wilkinson's Catalysis.**

Wilkinson's catalyst is most widely used for homogeneous hydrogenation. This catalyst was discovered by Sir Geoffrey Wilkinson. It is a complex of a transition metal rhodium called tris (triphenylphosphine) chloro rhodium (I) having the formula $RhCl\,(PPh_3)_3$ where Ph stands for phenyl group.

The phosphorus atom of phosphines has an unshared electron pair. It is through this electron pair, that triphenylphosphine is bonded to rhodium.

Rhodium can exhibit variable oxidation state of I and III and hence forms unstable intermediate with the reactant molecules. The mechanism can be explained as follows–

- It is believed that in solution the complex RhCl $(PPh_3)_3$ exchanges one of its PPh_3 for a solvent molecule, reversibly.

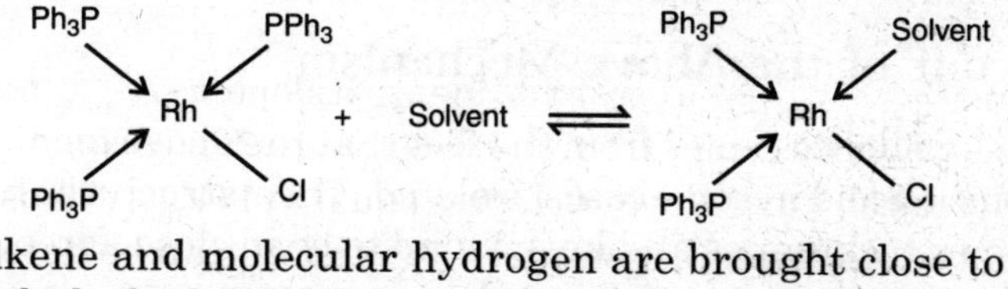

- The reactants, alkene and molecular hydrogen are brought close to the catalyst. The catalyst reacts with the hydrogen to form dihydrido complex RhH_2 Cl $(PPh_3)_2$. The H–H bond breaks and each hydrogen attach seperately to the rhodium.

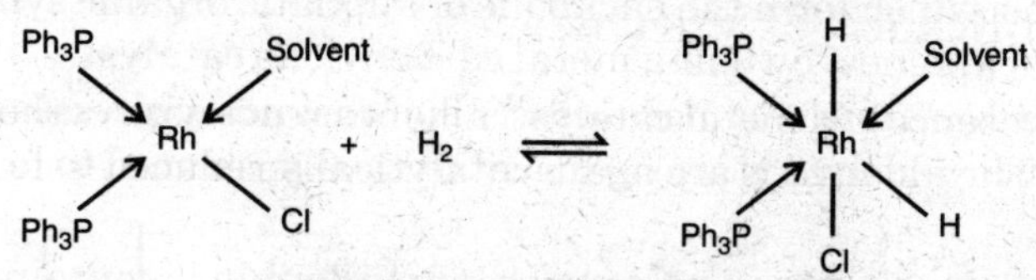

Thus in this step rhodium changes its oxidation state from (I) to (III).

- In the next step, alkene reacts with the catalyst by replacing a solvent molecule. The bond between the alkene and the metal atom rhodium is formed by the overlap of an empty orbital of the metal with the π cloud of alkene resulting in π complex.

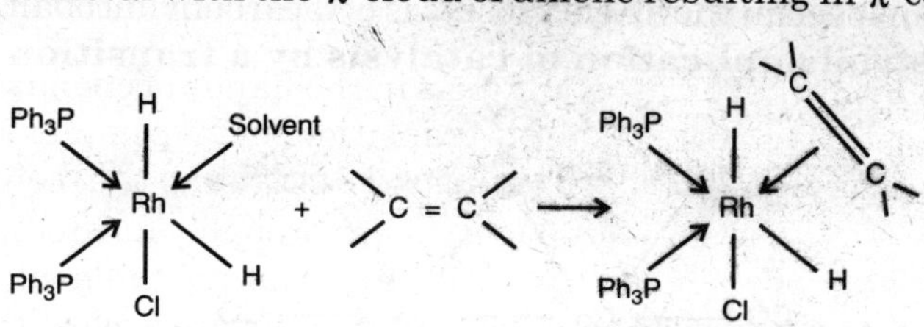

At this point both the reactants are bounded to rhodium and are in close proximity to react.

- A hydrogen atom migrates from the metal atom to one of the carbon atom of alkene whereas the other carbon attaches itself to the metal by a σ bond forming metal alkyl.

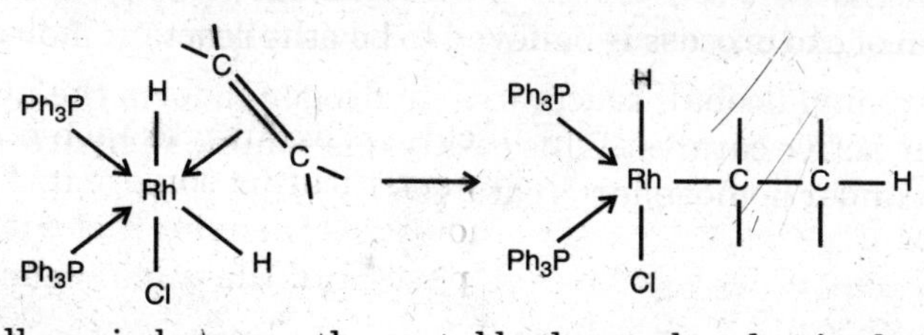

- The migration of alkene in between the metal hydrogen bond as in the above case is often referred to as insertion. Insertion can be defined as a process in which any atom or group is inserted between two other atoms initially bound together.
- Now, the second hydrogen migrates from the metal to carbon, which is still bonded to the metal atom, forming the hydrogenated product. The product detaches itself from the metal atom regenerating the catalyst. The rhodium atom regains its oxidation state of (I).

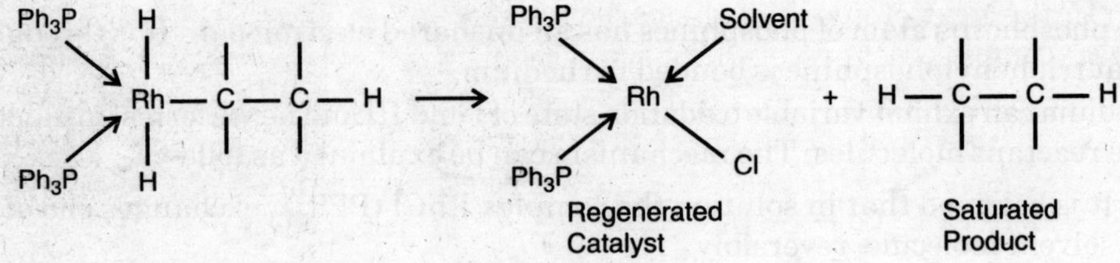

Evidence in Favour of the Above Mechanism

The most significant evidence comes from the fact that the above mentioned intermediates have been detected in solution and in some cases, isolated. Their structures have been well established by using spectroscopic techniques and have found to be in close agreement with the postulated structures.

Metal catalyst makes a significant contribution in industrial applications of catalysts. For example, the alcohols which form the backbone of aliphatic organic synthesis, can be obtained by the oxo process from alkenes, by using metal carbonyl, as catalyst.

Oxo Process: In presence of a metal catalyst, alkenes react with carbon monoxide and hydrogen to yield aldehydes. The aldehydes are again catalytically reduced to form alcohols.

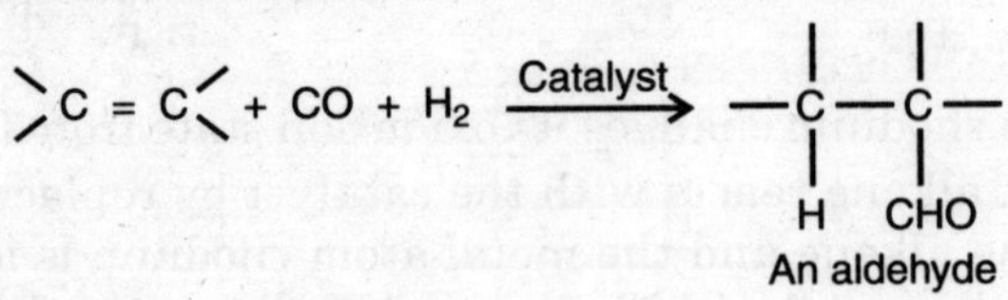

The catalyst employed in the above process is octacarbonyldicobalt, $Co_2(CO)_8$, **This process is the first industrial application of catalysis by a transition metal complex.**

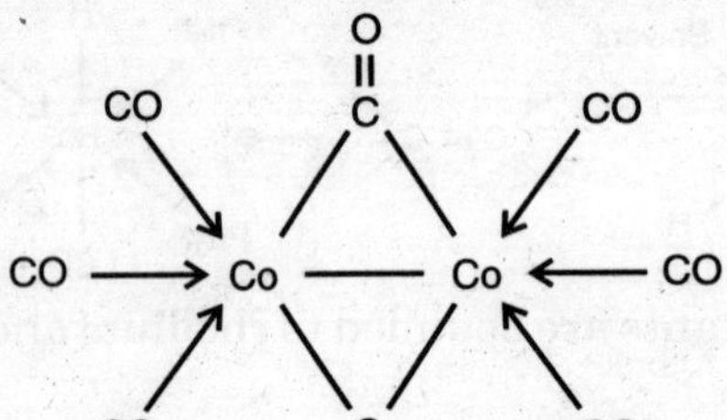

(OCTACARBONYLDICOBALT)

The mechanism of oxo process is believed to be as follows:

- The octacarbonyldicobalt reacts with hydrogen to form the hydrido complex $CoH(CO)_4$ which is the active catalyst. This catalyst is soluble in hydrocarbons, so oxo process can be grouped under homogeneous catalysis.

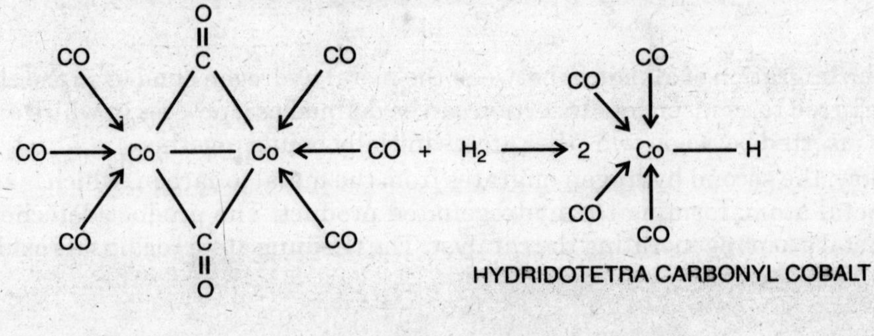

HYDRIDOTETRA CARBONYL COBALT

- In the next step one of the carbonyl group is replaced by alkene molecule to form a π - complex.

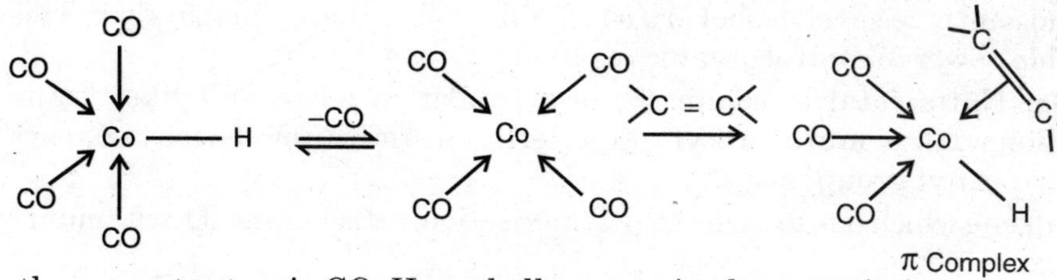

Thus all the three reactants, viz CO, H_2 and alkene are in close proximity.

- The hydrogen migrates from Co to one of the carbon atom of alkene and the other carbon atom of alkene attaches itself to cobalt forming metal alkyl. In other words the alkene is inserted inbetween cobalt and hydrogen atom.

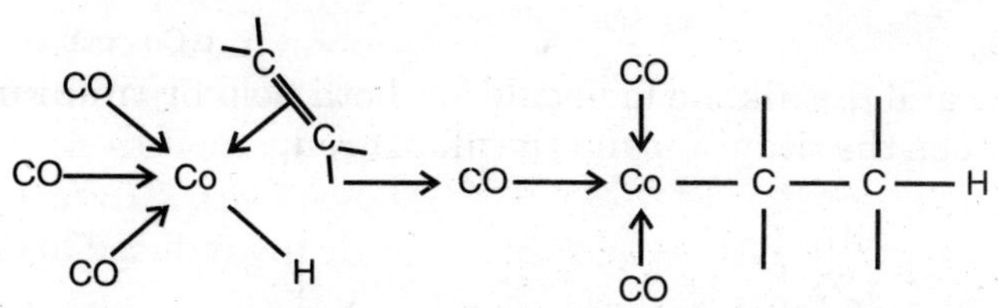

One molecule of CO is further added to the metal alkyl.

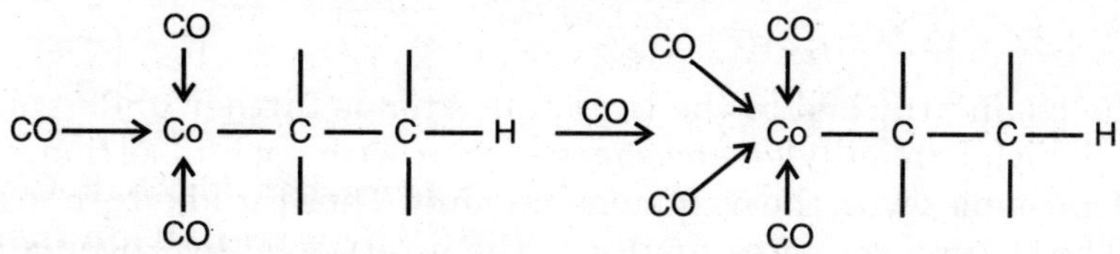

- The newly formed alkyl group then migrates to the carbonyl group, forming carbon carbon bond.

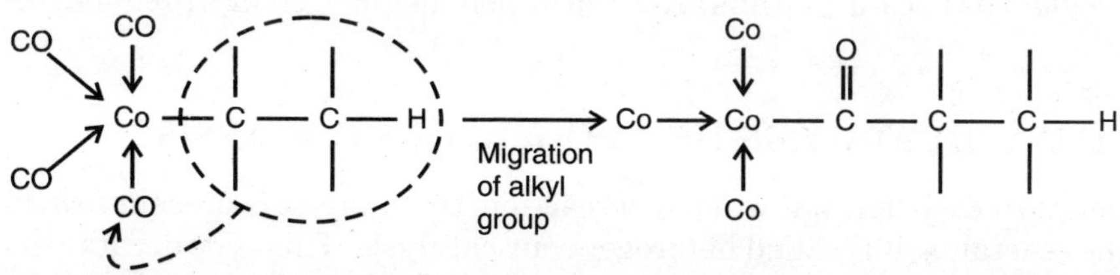

- At this stage hydrogen is absorbed to form dihydrido complex. One of these hydrogens migrates to carbon of the C = O group and aldehyde molecule is formed which detaches itself from the metal complex, which becomes ready for fresh action.

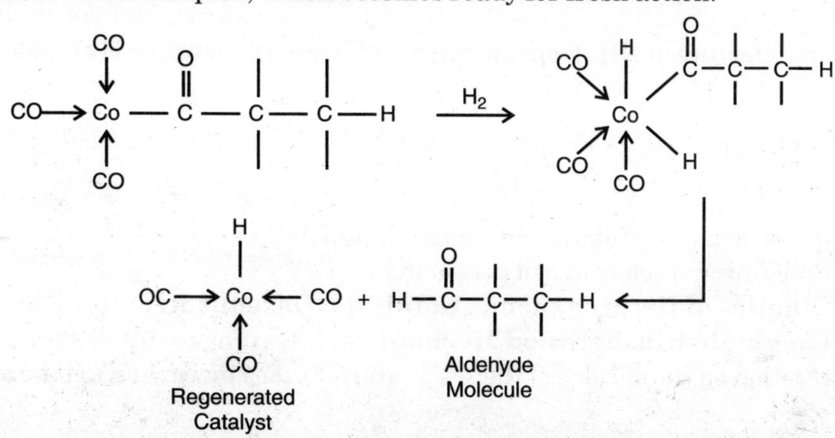

The catalyst which has revolutionised the field of polymer chemistry also belongs to metal salts i.e. the **Ziegler Natta catalyst**. This catalyst was discovered by Karl Ziegler and Giulio Natta who jointly received Nobel prize in 1963 for developing this catalyst. This catalyst exhibits remarkable power to control organic reactions.

Ziegler Natta catalyst comprises of a transition metal salt like titanium trichloride in combination with a metal alkyl like triethyl aluminium. These two react to form a complex having an ethyl group in situ.

The alkene which has to undergo polymerisation, attaches itself to titanium by π bond forming π complex.

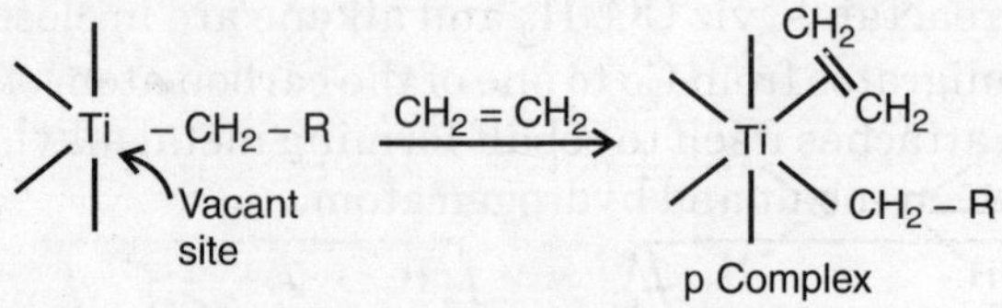

Now the alkyl and the alkene molecule are both held by titanium atom. The ethylene unit inserts itself between the titanium and the alkyl group.

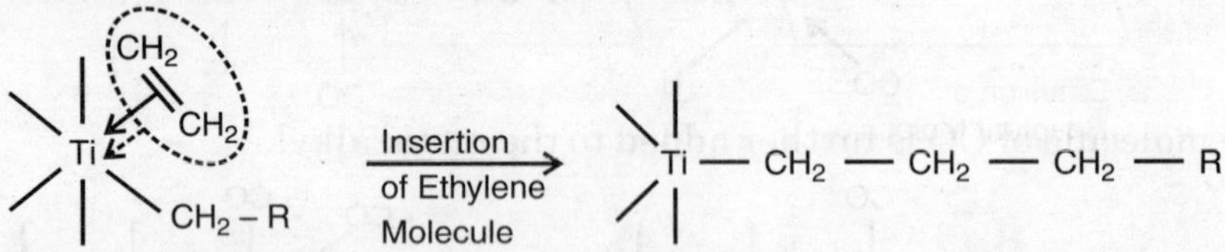

Thus, the chain attached to the titanium extends further undergoing similar steps. The bonding site which held ethylene becomes vacant with each insertion, and is followed by the formation of π complex with the next monomer unit. Thus the alkyl group keeps on growing with each cycle. The chain terminates finally by the insertion of hydrogen atom between titanium atom and the alkyl chain.

Ziegler Natta catalyst thus gives linear polymer molecules a great degree of stereochemical control.

10.9 SURFACE CATALYSIS-HETEROGENEOUS CATALYSIS

It has been stated earlier that when in a reaction, the catalyst is present in a different phase than that of the reactants, it is called heterogeneous catalysis. This type of catalysis generally include gaseous reactants, whereas the catalyst is a solid. The reaction takes place at the surface of the catalyst and hence it is also called surface catalysis. As heterogeneous catalysts can be readily recovered thus they have wide applicability in industry, compared to homogeneous catalysis. Examples:

(i) Manufacture of NH_3 from nitrogen and hydrogen using finely divided iron as catalyst.

$$N_{2(g)} + 3H_{2(g)} \xrightarrow{\text{Fe}} 2NH_{3(g)}$$

(ii) Contact process for manufacture of H_2SO_4 using finely divided V_2O_5 or Pt.

$$2SO_{2(g)} + O_{2(g)} \xrightarrow[\text{Or } V_2O_5]{Pt} 2SO_{3(g)}$$

(iii) Hydrogenation of oils using nickel as catalyst

(iv) Cracking and reforming of petroleum.

Kinetics of Heterogeneous Catalysis (Gaseous Reactants on Solid Surfaces)

Langmuir-Hinshelwood Mechanism: According to this mechanism a gaseous reaction on the surface of the solid catalyst takes place through the following steps:

(i) The reactant molecules are diffused to the surface of the catalyst.
(ii) Reactant molecules are adsorbed on the surface of the catalyst, which results in stretching and weakening of the bonds.
(iii) Chemical reaction of the adsorbed molecules on the surface of the catalyst forming product.
(iv) Product molecules are desorbed from the surface.
(v) The product molecules diffuse away from the surface.

The above sequence is illustated by hydrogenation of ethylene in the presence of hydrogen.

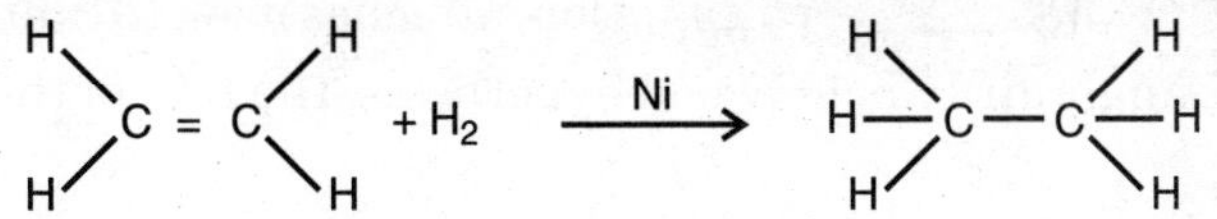

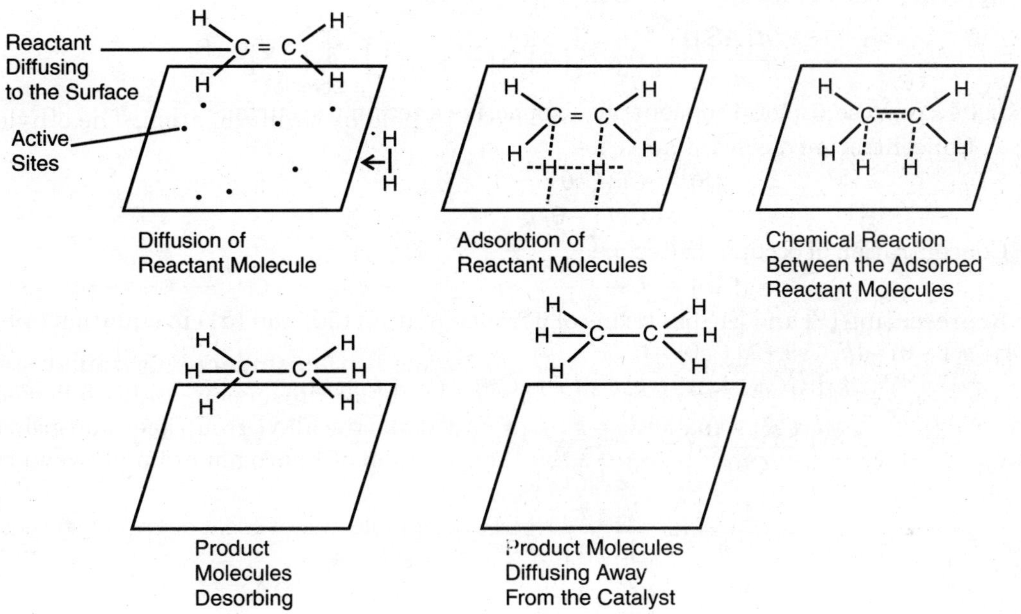

Fig. 10.18 Mechanism of Heterogeneous Catalysis

Any of the steps mentioned above may be slowest and hence becomes rate determining. Diffusion of the reactants towards the surface of the catalyst and diffusion of the products away from the catalyst viz, step (i) and (v) are considered very fast. Adsorption of the reactants and desorption of the products i.e. step (ii) and (iv) are faster than the step (iii) Therefore step (iii)–i.e the chemical reaction at the surface is the slowest and rate determing step.

Thus kinetics of surface reactions is based on the following assumptions–

(1) The chemical reaction between the adsorbed molecules is the rate determining step.
(2) An equilibrium exists between adsorbed and desorbed molecules at the surface.
(3) The rate of reaction per unit surface area is proportional to θ, where θ is the fraction of the surface covered by reactant molecules.

Thus according to Langmuir and Hinshelwood mechanism, the rate of reaction can be expressed as

$$r \propto \theta \qquad \text{..... ..(44)}$$

$$r = k_1\theta \qquad \text{...(45)}$$

where k_1 is the constant of proportionality

$$\theta = \frac{\text{Number of adsorption sites occupied}}{\text{Number of adsorption sites available}}$$

Let us consider a simple unimolecular reaction, where a reactant molecule A reacts with the vacant site on the surface.

$$A + S \underset{k_2}{\overset{k_1}{\rightleftharpoons}} AS \qquad \text{...(46)}$$

$$AS \xrightarrow{k_3} Products \qquad \text{...(47)}$$

Now according to Langmuir Hinshelwood hypothesis, the rate of the reaction is given by–

$$r = k_3\theta \qquad \text{...(48)}$$

Applying seady state approximaion to $[AS]$, we have

$$\frac{d[AS]}{dt} = 0 = k_1[A][S] - \{k_2[AS] + k_3[AS]\} \qquad \text{...(49)}$$

Let Cs represents total concentration of active sites on the surface

∴ Concentration of vacant sites, i.e.

$$[S] = Cs - Cs\theta$$

$$= Cs(1-\theta) \qquad \text{...(50)}$$

Concentration of occupied sites, i.e.

$$[AS] = Cs\theta \qquad \text{...(51)}$$

Representing $[S]$ and $[AS]$ in terms of θ from equation (50) and (51) in equation (49) we have

$$k_1[A]Cs(1-\theta) - [k_2Cs\theta + k_3Cs\theta] = 0 \qquad \text{...(52)}$$

$$k_1[A]Cs(1-\theta) = k_2Cs\theta + k_3Cs\theta = (k_2+k_3)Cs\theta \qquad \text{...(53)}$$

$$\text{or } \frac{Cs(1-\theta)}{Cs\theta} = \frac{k_2+k_3}{k_1[A]}$$

$$\text{or } \frac{1-\theta}{\theta} = \frac{k_2+k_3}{k_1[A]}$$

$$\text{or } \frac{1}{\theta} - 1 = \frac{k_2+k_3}{k_1[A]}$$

$$\text{or } \frac{1}{\theta} = \frac{k_2+k_3}{k_1[A]} + 1 = \frac{k_1[A]+k_2+k_3}{k_1[A]}$$

$$\theta = \frac{k_1[A]}{k_1[A]+k_2+k_3} \qquad \text{...(54)}$$

Substituting the value of θ in equation (48) we get–

$$r = \frac{k_1k_3[A]}{k_1[A]+k_2+k_3} \qquad \text{...(55)}$$

or it can be rewritten in the form,

$$\frac{1}{r} = \frac{k_1[A]+k_2+k_3}{k_1k_3[A]} = \frac{1}{k_3} + \frac{k_2+k_3}{k_1k_3[A]} \qquad \text{...(56)}$$

Thus a plot of $\frac{1}{r}$ versus $\frac{1}{[A]}$ would give a straight line having slope = $k_2 + k_3 / k_1 k_3$ and intercept = $\frac{1}{k_3}$. For the gaseous reactants,-we can use partial pressures instead of the concentration terms and hence equation (55) modifies to–

$$r = \frac{k_1 k_3 p_A}{k_1 p_A + k_2 + k_3} \quad ...(57)$$

$$\text{or } \frac{1}{r} = \frac{1}{k_3} + \frac{k_2 + k_3}{k_1 k_3 p_A} \quad ...(58)$$

In this case plot of $\frac{1}{r}$ versus $\frac{1}{p_A}$ would give a straight line.

Two different cases are discussed below

(i)If the rate of decomposition is very large in comparison with the rate of adsorption or desorption i.e. $k_3 >> k_1 p_A + k_2$, then equation (57) reduces to

$$\boxed{r = k_1 p_A}$$

Thus, the surface reaction is first order with respect to the pressure of the reactant gas.

If the rate of decomposition is very small as compared to the rate of adsorption and desorption, i.e. $k_3 << k_1 p_A + k_2$. Then equation (57) reduces to

$$r = \frac{k_1 k_3 p_A}{k_1 p_A + k_2} \quad ...(59)$$

Dividing the numerator and denominator by k_2 we get–

$$r = \frac{(k_1/k_2) k_3 p_A}{(k_1/k_2) p_A + 1} = \frac{K k_3 p_A}{k p_A + 1} \quad ...(60)$$

K is the adsorption equilibrium constant and is given by, $K = k_1/k_2$.

Under this case, i.e. when the rate of decomposition is very small as compared to the rate of adsorption and desorption, two different situations may arise depending upon the pressur:

(a) **When the pressure is low** : In this condition, the reactant molecules are slightly adsorbed or $\theta \to 0$. Thus $K_{PA} << 1$ and equation (60) reduces to

$$r = k_3 KpA \quad ...(61)$$

as $k_3 K$ is a constant, thus the rate is first order with respect to concentration of reactant A.

(b) **When the pressure is high:** When the pressure is high the reactant molecules almost occupy whole of the surface of the catalyst or $\theta \to 1$. Thus $KpA >> 1$ and equation (60) reduces to

$$r = k_3$$

Thus the rate is zero order with respect to the reactant molecules A.

The above results may be explained by the fact that the decomposition step is the slowest step and hence is rate determining. When the pressure is low only a fraction of the molecules are adsorbed on the surface. Hence on increasing the pressure, more of the reactant molecules will be adsorbed thereby increasing the rate of the reaction.

At high pressures, since all the active sites at the surface are occupied and hence further increase in pressure does not bring any change in the rate of reaction because the concentration of the adsorbed molecules remains constant. Thus the reaction becomes zero order with respect to the adsorbed reactant.

The variation of reaction rate with pressure is shown in Fig. 10.19.

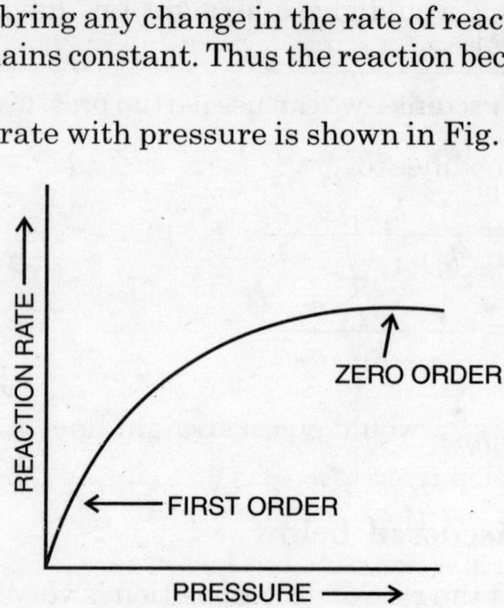

Fig. 10.19 Variation of Reaction Rate with Pressure

10.10 CATALYSTS IN INDUSTRY

From the foregoing discussion, it is apparent that a variety of substances have been used as catalysts. Many industrially important processes involve the use of catalysts. The criteria for choosing a catalyst for industrial applications are as follows:

(i) The catalyst must possess maximum possible activity so as to give maximum yield of the desired product in minimum time.

(ii) The cost of the catalyst should be low.

(iii) The catalyst must be easily available.

(iv) The catalyst should be such that it is regenerated easily without much difficulty.

(v) The catalyst should be highly selective so that the desired product should be formed in major amount.

Many of the primitive methods of preparing certain substances involved the unconscious use of catalysis. For example

- the fermentation of wine to acetic acid.
- the manufacture of soap from fats and alkalies.
- the formation of ether from alcohol and H_2SO_4.

All of these are catalytic reactions. But the application of catalyst to industrial processes was undertaken in 19th century. P. Phillips used platinum to oxidize sulphur dioxide to sulphur trioxide in presence of air. This process became a great success towards the close of 19th century. There after numerous catalysis were discovered and affected the growth and development of industrial processes. Catalyst manufacture is itself a rapidly growing industrial process.

Modern chemical industry depends on the development, selection and application of the catalysts. Some of the important industrial processes along with the corresponding catalyst are tabulated in Table 10.4.

Table 10.4

Process	*Catalyst*
1. Synthesis of NH_3 (Haber's Process) $N_2+3H_2 \rightleftharpoons 2NH_3$	*Fe*
2. Manufacture of sulpuric acid	
(a) Contact Process : $2SO_2 + O_2 \rightleftharpoons 2SO_3$	V_2O_5
(b) Chamber Process : $2SO_2 + O_2 \rightleftharpoons 2SO_3$	NO_2
3. Cracking of Petroleum $C_{10}H_{22} \rightarrow C_5H_{12} + C_5H_{10}$	$Al_2(SiO_3)_3$ Zeolites Silica-Alumina Catalyst.
4. Hydrogenation of Unsaturated Hydrocarbons. $-CH=CH-+H_2 \rightarrow -CH_2-CH_2-$	Ni, Pt
5. Oxidation of Hydrocarbons in Automobile Exhausts	CuO, V_2O_5, Pt, Pd
6. Isomerization of n–butane to isobutane $n-CH_3-CH_2-CH_2-CH_3 \xrightarrow{AlCl_3} CH_3-CH-CH_3$	$AlCl_3$, HCl
7. Catalytic Reforming 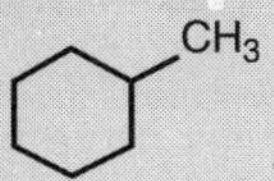 $\longrightarrow C_6H_5CH_3 + 3H_2$ Methylcyclohexane	$Mo_2O_3-Al_2O_3$
8. Polymerization $nCH_2=CH_2 \rightarrow -(CH_2-CH_2)_n$	Ziegler Natta Catalyst $TiCl_3 + (C_2H_5)_3Al$
9. Manufacture of Chlorine (Deacon's process) $4HCl + O_2 \rightleftharpoons 2CH_3COOH$	Cu_2Cl_2
10. Oxidation of NH_3 to NO (Ostwald's Process) $4NH_3 + 5O_2 \rightleftharpoons 4NO + 6H_2O$	Pt Gauze

CATALYST AND AUTOMOBILE INDUSTRY

A recent area of research in the field of industrial catalysis is the development of catalysts which can convert the harmful auto exhaust gases into less harmful ones. For example-Catalytic converters in the cars. One such type of catalytic converter is discussed briefly.

Catalytic Converters

In running cars, the temperature inside the engine becomes high. At this high temperature nitrogen and oxygen gases react to form nitric oxide.

$$N_{2(g)} + O_{2(g)} \rightleftharpoons 2NO_{(g)}$$

$NO_{(g)}$ is released into the atmosphere where it combines with O_2 to form NO_2. Nitrogen dioxide and other gases which are also emitted by the automobile like CO and unburnt hydrocarbons are the major cause of air pollution.

To overcome this difficulty to some extent most of the cars are fitted with the catalytic converters. The catalyst employed are mostly Pt or Pd. Sometimes transition metal oxide such as CuO or Cr_2O_3 are also used.

An efficient catalytic converter serves the following purposes

(i) It oxidizes CO and unburnt hydrocarbons to CO_2 and H_2O.

(ii) It reduces NO and NO_2 to N_2 and O_2.

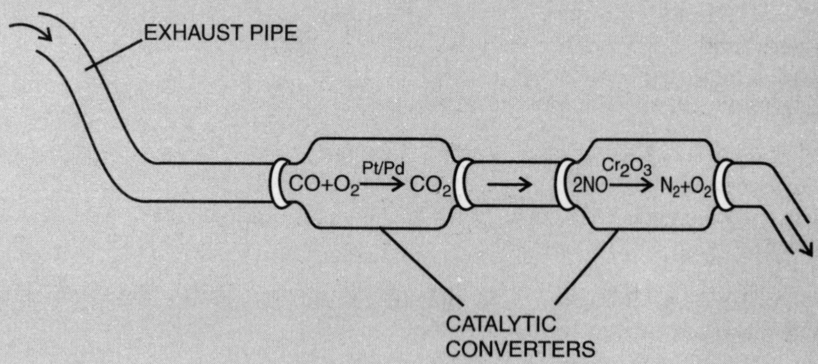

Fig. 10.19 Catalytic Converters for an Automobile.

Fig. 10.20 gives a simplified diagram of catalytic converter. Hot exhaust gases are first mixed with air and are passed through the first chamber where the complete burning of hydrocarbons take place. This chamber has Pt or Pd as catalyst. Then the exhaust gases are passed through the second chamber containing a different catalyst mostly a transition metal oxide (Cr_2O_3 or CuO), which operates at a low temperature. In this chamber NO is dissociated into N_2 and O_2 before the exhaust is allowed to escape through tail pipe.

QUESTIONS

Short Answer Questions

Section 10.1, 10.2

1. What do you mean by activation energy?
2. What is the role of the catalyst?
3. What are negative catalysts?
4. What is the advantage of catalyst over high temperature to promote chemical reactions.
5. What are catalytic promoters?
6. What are catalytic poisons?
7. Light speeds up many reactions and is sometimes spoken of as a catalyst. Criticize the statement.
8. What is the difference between intermediate and a catalyst?

Section 10.3

9. What do you understand by the term homogeneous catalysis and heterogeneous catalysis.
10. Explain induced catalysis?

    Ans. When one reaction influences the speed of the other reaction, the phenomenon is termed as induced catalysis. For example, oxidation of sodium sulphite acts as a catalyst for oxidation of sodium arsenite solution.
11. Give three examples of homogeneous catalysis.
12. Give three examples of heterogeneous catalysis.

Section 10.4, 10.5

13. Define (*i*) Positive Catalyst (*ii*) Negative Catalyst (*iii*) Auto catalyst
14. What are the limitations of intermediate compound formation theory?
15. What is adsorption?
16. Differentiate between physical adsorption and chemisorption.
17. What are active centres.

18. What is catalytic poisoning?
19. Distinguish between temporary and permanent poisoning.
20. In catalytic cracking of hydrocarbons, the catalyst gets exhausted after some time. Explain?

Section 10.6

21. What is general and specific acid catalysis?
22. Give some examples of specific acid catalysis.
23. Give some examples of specific base catalysis.

Section 10.7, 10.8

24. What is lock and key model?
25. Enzymes constitute a class of most efficient catalysis. Explain.
26. Enzymes are highly specific. Comment
27. What are co-enzymes?
28. What do you understand by optimum temperature and optimum pH for Catalyst efficiency?
29. What is Wilkinson's catalysis?

Long Answer Questions

1. What is meant by catalysis and catalyst? Give general characteristic of catalytic reactions.
2. Describe briefly homogeneous and heterogeneous catalysis giving one example of each. Explain Intermediate Compound Formation theory of catalysis.
3. How does a catalyst work?
4. What are promoters and inhibitors? Give suitable examples.
5. (a) What do you understand by homogeneous catalysis? Give three examples each of the homogeneous catalytic reactions taking place in
   (i) gaseous phase (ii) liquid phase.
   (b) Briefly explain the mechanism of homogeneous catalytic reactions taking suitable examples.
   (c) Briefly discuss the kinetics of the decomposition of acetaldehyde in presence of iodine vapours as catalyst.
   (d) Briefly discuss the kinetics of the dehydration of tert-butyl alcohol in the presence of HBr.
6. Explain the difference between homogeneous and heterogeneous catalysis. Discuss the theories of catalysis.
7. Give:
   (a) Mechanism of solid catalyst.
   (b) Action of Promoters and Poisons with examples.
   (c) Applications of catalysts.
8. Derive a general expression for the rate of reaction of an acid-catalysed reaction.
9. Explain with a suitable example that heterogeneous catalysts are highly specific in nature.
10. What are enzymes? Give their classification based on reactions. Describe the effect of the factors that affect the enzyme reactions.
11. Discuss the catalytic hydrogenation of alkenes on metal surface. What may be the order of such reactions?
12. Write down the mechanism of enzyme catalysis suggested by Michaelis and Menten.
13. Explain the following giving one example each:
   (a) Homogeneous catalysis (b) Heterogeneous catalysis
   (c) Auto catalysis (d) Negative catalysis
   (e) Positive catalysis
14. What effect does a catalyst have on an equilibrium system?
15. Write a short note on Activation Energy.
16. Write note on
   (i) Negative Catalysis (ii) Auto Catalysis

17. Discuss the mechanism of enzyme catalysed reactions. Deduce Michaelis Menten equation and give conditions under which an enzyme catalysed reaction changes its order.
18. Briefly describe the mechanism of heterogeneous catalytic reactions, taking suitable examples.
19. A reaction involving only one gaseous reactant decomposes on the surface of a solid catalyst according to the equation

$$A \rightarrow \text{Products}$$

Write expression for the reaction rate. How is the equation modified when:
    (i) the reactant gas is slightly adsorbed?
    (ii) the reactant gas is strongly adsorbed?
20. What is the effect on the rate of a reaction if the product molecules are adsorbed on the surface of a solid catalyst? Give reason for your answer.
21. Under what conditions, an enzyme catalysed reaction follows:
    (i) second order kinetics (ii) first order kinetics
    (iii) zero order kinetics?
22. Most reactions, including enzyme-catalyzed reactions, proceed faster at higher temperatures. However, for a given enzyme the rate drop off abruptly at a certain temperature. Account for this behaviour.
23. The following mechanism has been proposed for the enzyme catalysis :

$$E + S \underset{k_{-1}}{\overset{k_1}{\rightleftharpoons}} ES$$

$$ES \xrightarrow{k_2} P + E$$

using steady state approximation for [ES], show that the reaction rate is given by $r = k_2 [E]_o [S]/km + [S]$, where symbols have their usual meanings.
24. (a) Explain the characteristics of catalysts.
    (b) Define enzyme catalysis. Give the kinetics and mechanism of enzyme catalyzed reaction.
    G.G.S.I.P.U, II Term, B.Tech, June 2000
25. Write short notes on the following:
    (a) Mean free path
    (b) Promotors and Inhibitors.
    (c) Catalysis by metal salts or effect of pH on enzyme catalysis.
    G.G.S.I.P.U, II Term, B. Tech, June 2000
26. Explain the following with examples:
    (i) Auto catalysis (ii) Catalyst poisons
    (iii) Catalyst promoters (iv) Negative catalyst
    (v) shape selective catalyst (vi) Homogeneous catalysis
    (vii) Heterogeneous catalysis. **G.G.S.I.P.U, End-Term, B.Tech, July 2000**
27. (a) What is catalysis and what are the general characteristics of a catalyst.
    (b) Give the mechanism and kinetics of enzyme catalyzed reactions (Michaelis Menton Equation). **G.G.S.I.P.U, I-Term, B. Tech, March 2001**
28. Write notes on:
    (a) Homogeneous catalysis
    (b) Promoters/Effect of temperature on enzyme catalysis.
    G.G.S.I.P.U, I Term, B. Tech, II Batch, May, 2001.
29. (a) What are the criteria of catalysis.
    (b) Give the mechanism of kinetics of enzyme catalyzed reaction (Michaelis-Menton equation).
    G.G.S.I.P.U, I Term, B. Tech, II Batch, May, 2001
30. (a) Explain adsorption theory of catalysis.
    (b) Explain catalysis by metal salts (Wilkinson's catalysis)
    G.G.S.I.P.U, I Term, B. Tech, II Batch, May, 2001
31. Why chloroform used as an anaesthetic is kept in dark coloured bottles and mixed with absolute alcohol?

32. When acidic KMnO4 solution is added to a hot solution of oxalic aicd, the colour is decolorized slowly in the beginning, but after sometime, it disappears rapidly. Why?
33. Why a rough surface piece of platinum metal acts as a good catalyst in comparison to a smooth surface in the manufacture of H2 SO4 by contact process.
34. Acid hydrolysis is a reversible reaction where as base hydrolysis is irreversible reaction. Explain.
35. (a)A particular mass of charcoal absorbs a large volume of ammonia than of hydrogen at a given temperature. Explain.
    (b) Discuss in details various theories of hetrogeneous catalysis.
    (c) Define the term (i) adsorption (ii) adsorbate (iii) adsorbent
    (d) State a condition under which a catalyst looses its influence over a reaction.

G.G.S.I.P.U, End Term, B. Tech, May, 2001

36. (a) Write down the catalysis, conditions, and chemical reactions involved in each of the following industrial process.
    (i) Bosch process (ii) Haber's process
    (iii) Contact process (iv) Ostwald process

$$\left[\textit{Bosch process}: (co + H_2) + H_2O \underset{400°C}{\overset{Fe_2O_3}{\rightleftharpoons}} CO_2 + 2H_2 \right]$$

    (b) What are enzymes? Describe the chemical conversions promoted by the following enzymes: Diastase, Maltase, Invertase, Urease, Ptyalin

G.G.S.I.P.U, End Term, B. Tech, II Batch, July 2001

37. (a) Discuss the kinetics of Acid-Base catalysis.
    (b) Explain the action of catalyst in terms of activation energy.

G.G.S.I.P.U, I Term, B. Tech, March 2002

38. Write short note on.
    (i) Wilkinson's Catalyst
    (ii) Effect of temperature on enzyme catzalysis **G.G.S.I.P.U, I Term, March 2002**
39. Explain why rough surface of a catalyst is more effective than smooth surface.

G.G.S.I.P.U, I Term, March 2002

40. (a) What is catalysis? Discuss homogeneous and heterogeneous catalysis with examples.
    (b) Describe the mechanism of catalysis between a single substrate and a catalyst which is applicable in case of acid and base catalysed as well as enzymatic reactions.

G.G.S.I.P.U, End Term, May 2002

41. (a) Define a general expression for the rate of reaction of an acid – catalyzed reaction
    (b) Write short notes on (i) Promoter. (ii) catalyst poison, (iii) Activation energy

G.G.S.I.P.U, I Term, Feb. 2002

42. (a) Write the effect of temperature on enzyme catayst.
    (b) Write the differences between physical adsorption & chemical adsorption
    (c) What do you understand by homogeneous catalysis? Give three examples.
43. (a) Assuming that chemisorption of nitrogen on iron at 770 K follows longmuir isotherm, pressure in Haber process for the manufacture of ammonia.
    (b) Write notes on following:
    (i) Homogeneous and Heterogeneous catalysis
    (ii) Physical absorption and chemical absorption. **G.G.S.I.P.U, End Term, May 2003.**
44. Explain the action of promoters in catalysis **G.G.S.I.P.U, End Term, May 2003.**
45. (a) What is meant by catalysis? Discuss the general characteristics of a catalyst.
    (b) What is meant by enzyme catalysis? Why are they highly specific in their actions?
    (c) Discuss in detail the mechanism of an enzyme catalysed reaction.

G.G.S.I.P.U, End Term, May 2004.

46. Discuss the role of inhibitors in catalysis. **(G.G.S.I.P.U, End Term, May 2004)**
47. What are the criteria of catalysis? Give the mechanism of kinetics of enzyme catalysed reactions (Michaelis-Meneton equation). **(G.G.S.I.P.U, I-Term, B.Tech., Feb. 2005)**
48. What do you understand by activity and selectivity of a catalyist.

(G.G.S.I.P.U, I-Term, B.Tech., Feb. 2006)

49. What are the various steps involved in a gaseous reaction taking place on the surface of the solid catalyst. Which is the rate limiting step. **(G.G.S.I.P.U, I-Term, B.Tech., Feb. 2006)**
50. Discuss the mechanism of enzyme catalyzed reactions. Deduce Michaelis-Menten equation and give the conditions under which an enzyme catalyzed reaction changes in order. **(G.G.S.I.P.U, I-Term, B.Tech., Feb. 2006)**
51. Explain the action of Catalytic promoters and poison. **(G.G.S.I.P.U, I-Term, B.Tech., Feb. 2007)**
52. Under what conditions an enzyme catalyzed reaction follows:
    (i) First order kinetics (ii) Zero order kinetics
    (G.G.S.I.P.U, I-Term, B.Tech., Feb. 2007)
53. (a) When acidic $KMnO_4$ solution is added to a hot solution of oxalic acid, the color is decolorized slowly in the beginning, but after sometime it disappears rapidly.
    (b) Why chloroform is kept in dark colored bottles and mixed with absolute alcohol?
    (G.G.S.I.P.U, I-Term, B.Tech., Feb. 2007)
54. Discuss the kinetics of an acid catalyzed reaction. **(G.G.S.I.P.U, I-Term, B.Tech., Feb. 2007)**
55. (a) Define the term 'catalyst'. Give various types of catalysis you have studied. Support your answer with suitable examples.
    (b) Oxidation of SO_2 to SO_3 is catalyzed by NO. Explain, giving chemical equations, the various steps involved in this conversion. **(G.G.S.I.P.U, End-Term, B.Tech., Dec. 2007)**

❑❑❑

11 POLYMERS

"Every great advance in science has issued from a new audacity of imagination".

— John Dewey

CHAPTER OUTLINE

The word 'polymer' is derived from the classical Greek words *poly* meaning 'many' and *meres* meaning 'parts'. Simply stated, a polymer is a molecular compound with high molecular mass, ranging into thousands and millions of gram. It is composed of a large number of repeating units of identical structure called *monomers*. For example, polythene is a polymer formed by linking together of a large number of ethene molecules.

$$\underset{\text{(Monomer)}}{\underset{\text{Ethene}}{n\,CH_2=CH_2}} \longrightarrow \underset{\text{(Polymer)}}{\underset{\text{Polyethene}}{-\!\!\left(CH_2-CH_2\right)\!\!-}}$$

Certain polymers, such as proteins, cellulose, and silk are found in nature and hence are called natural polymers, whereas, a number of polymers are produced by synthetic routes and are called synthetic polymers. In some cases, naturally occuring polymers can also be prepared synthetically. For e.g. rubber, which is also known as polyisoprene.

Both natural and synthetic polymers impinge on every aspect of our existence. The idea is often expressed that we live in "a plastic age". The range of polymeric materials is so wide that they have found applications in practically every branch of industry. About 80% of today's chemists, including biochemists, work with polymers.

The number of repeating units in a chain formed in a polymer, is known as the **"degree of polymerization"**. Polymers with high degree of polymerization are termed as "high polymers" and those with low degree of polymerization are called oligopolymers.

11.1 CLASSIFICATION OF POLYMERS

Synthetic polymers have been classified into various groups, in various manners.

Classification of polymers

- (i) On the basis of Chemical structure
  - Homopolymer
  - Copolymer
- (ii) On the basis of Polymeric structure
  - Linear
  - Branched
  - Crosslinked
- (iii) On the basis of arrangement of monomers
  - Block copolymers
  - Graft copolymers
- (iv) On the basis of tacticity
  - Isotactic polymer
  - Syndiotactic polymer
  - Atactic polymer
- (v) On the basis of thermal behaviour
  - Thermoplastic
  - Thermosets
- (vi) On the basis of ultimate form
  - Plastic
  - Elastomers
  - Fibers

(i) On the basis of the chemical structure of their backbone, polymers are of two types.

Homopolymer: If a polymer consists of identical monomer, the polymer is termed as homopolymer.

Copolymer: If the polymer is a mixture of more than one type of monomer it is termed as copolymer.

$$\frown A-A-A-A-A \frown \qquad \frown A-B-A-B-A-B \frown$$

Homopolymer   Copolymer

(ii) On the basis of their polymeric structures, the polymers can be classified as linear, branched or crosslinked.

∽A—A—A—A—A∽

Linear Homopolymer

∽A—B—A—B—A—B∽

Linear Copolymer

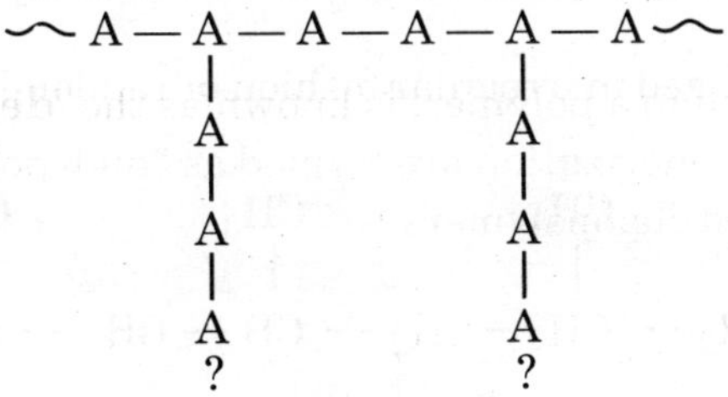

Branched Chain Homopolymer

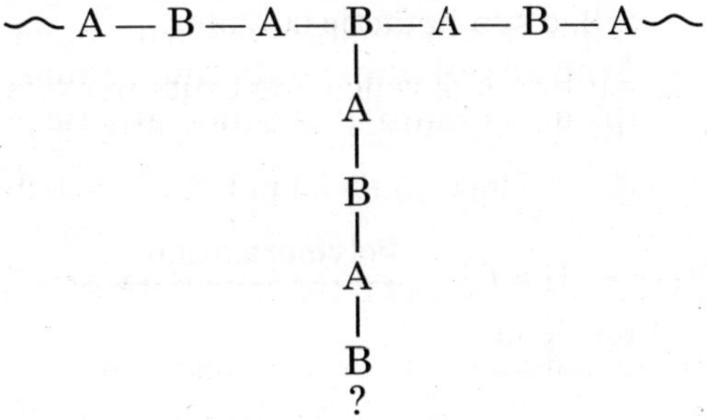

Branched Chain Copolymer

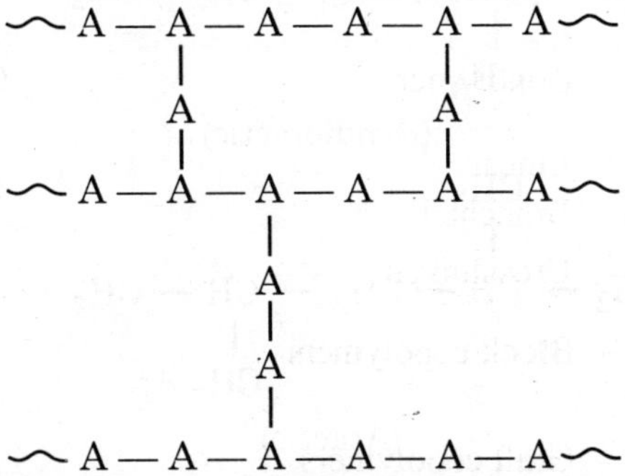

Cross Linked Homopolymer

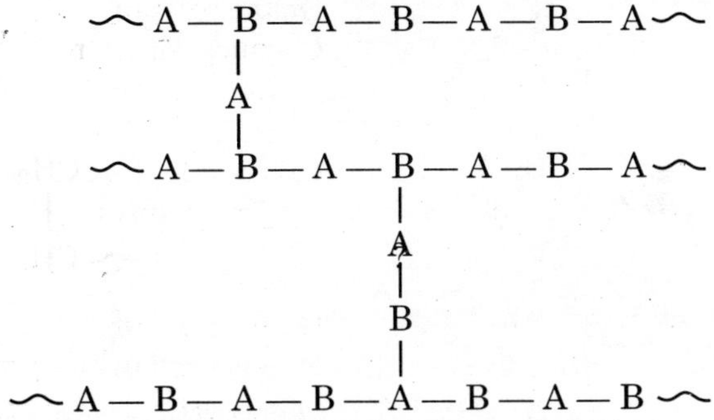

Cross Linked Copolymer

(iii) Depending upon the arrangement of the monomers in the case of copolymers, the polymers are divided into block copolymers and graft copolymers.

Block Copolymers: Linear polymers in which the identical monomeric units occur in relatively long sequences are called block co-polymers.

—A—A—A—B—B—B—A—A—A—B—B—B—

Block Copolymer

Graft Copolymers: Graft copolymers are branched copolymers in which the backbone is formed from one type of monomer and branches are formed of the other, i.e. the monomer segments on the branches and backbone are not the same.

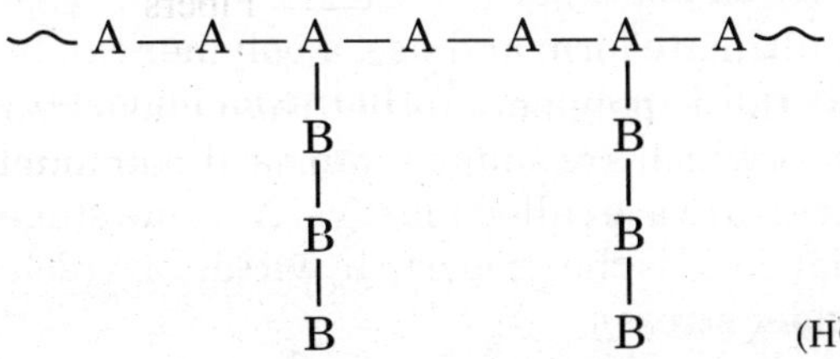

Graft Copolymer

(HOMOCHAIN POLYMER)
(HETROCHAIN POLYMER)

(iv) Depending upon the orientation of monomer units in a polymer molecule: With respect to the main chain the polymers are classified as:

Isotactic polymer: If the side groups of the monomers lie on the same side of the chain, it is called an "isotactic" polymer.

Syndiotactic polymer: If the side groups are arranged in an alternate fashion, it is called syndiotactic polymer.

Atactic polymer: If the side groups are arranged in irregular fashion or randomly around the main chain, it is called atactic polymer.

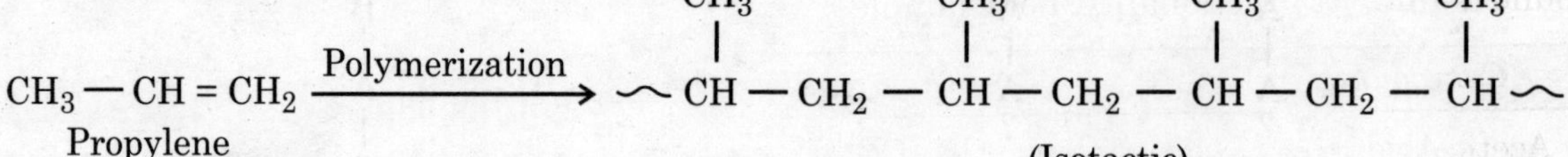

(Isotactic)

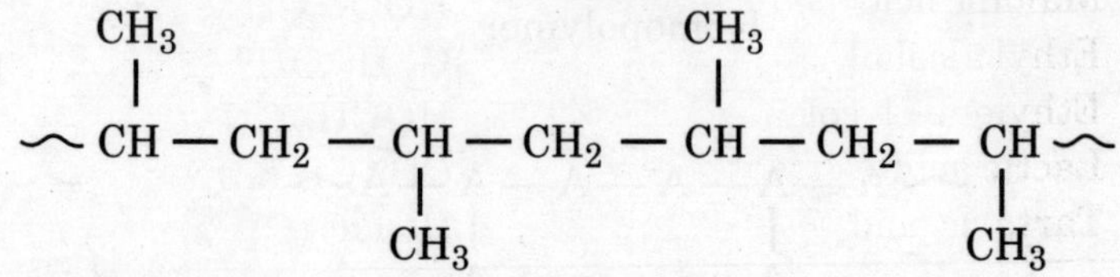

(Syndiotactic)

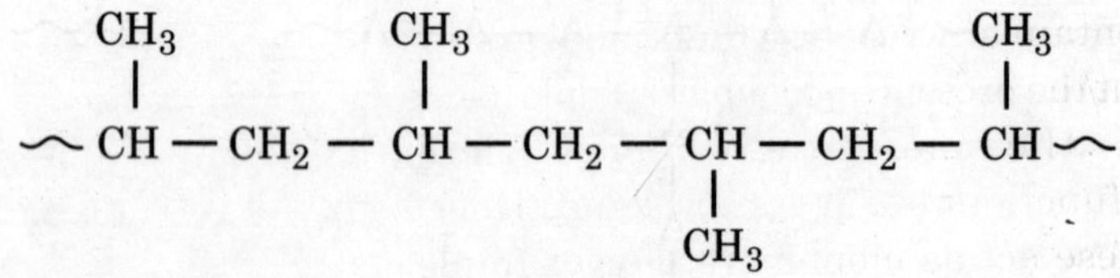

(Atactic)

(v) If the main chain is composed of the atoms of the same specie, the polymer is called homochain and if the main chain is mads up of different atoms then they are called heterochain polymers.

∼C—C—C—C—C—∼

∼C—C O—C—O—C∼

(vi) According to their behaviour on heating, the polymers are divided into two groups.

Thermoplastics: Those polymers which soften on heating and become plastic so that they can be converted to any shape by moulding are known as thermoplastics.

Thermosets: Those polymers which change irreversibly into hard and rigid materials on heating and cannot be reshaped, once they are set are called thermosets. They form infusible and insoluble mass on heating.

(vii) Depending on their ultimate form and use, a polymer can be classified as:

Plastics: All the synthetic polymers (other than elastomers) are usually referred to as plastics. The polymers which are shaped into hard and tough utility articles by the application of heat and pressure are called plastics. At some stage of manufacture, these are in plastic condition (Plasticity is the property by virtue of which a material undergoes permanent deformation under stress).

Elastomers: The polymers which are capable of being stretched rapidly at least 150 percent of their original length without breaking and return to their original shape on release of stress are known as elastomers. Synthetic rubbers constitute important examples of this group.

11.2 FUNCTIONALITY

The functionality of a molecule is the number of reactive sites it has. For a substance to act as a monomer, it must have two reactive sites. In otherwords, they should have a functionality of atleast two. A compound assumes functionality because of the presence of reactive functional groups like –OH, –COOH, $–NH_2$, –SH etc. The number of functional group present in a compound defines its functionality. For example-

Compound	*Chemical Formula*	*Functionality*
Acetic Acid	$CH_3\ COOH$	1
Malonic acid	$HOOC\ CH_2\ COOH$	2
Ethyl alcohol	$C_2\ H_5\ OH$	1
Ethylene glycol	$HOCH_2\ CH_2\ OH$	2
Lactic acid	$CH_3\ CH\ (OH)\ COOH$	2
Tartaric acid	$HOOC\ (CHOH)_2\ COOH$	4

Some compounds, however, do not contain any reactive functional groups but the presence of double or triple bonds in the molecules makes them bifunctional or polyfunctional. Hence these act as monomers. For example, in olefins the double bond can be considered as a site for two free valencies. When a double bond is broken, two single bonds become available for combination Thus,

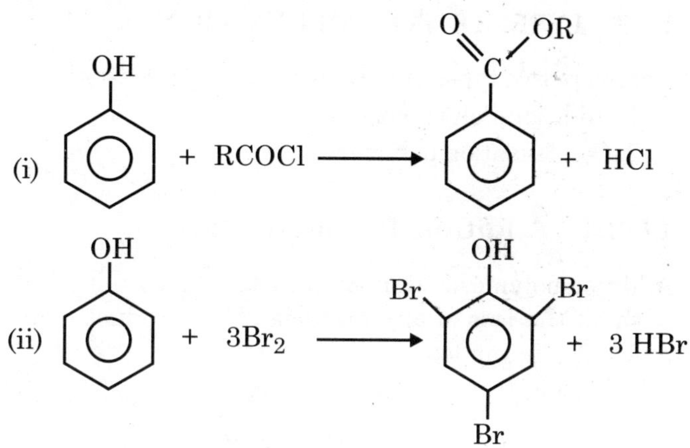

$$CH_2 = CH_2 \longrightarrow —CH_2–CH_2—$$

Acelytene, on the other hand, has a functionality of four.

$$CH \equiv CH \longrightarrow \text{---}CH = CH\text{-----} \longrightarrow \text{---}\ CH — CH\ \text{---}$$

There are some other compounds in which the presence of easily replaceable hydrogen atoms impart functionality. For example, phenol. It has got –OH groups as a functional group but can also undergo substitution reaction at three sites replacing its three hydrogen atoms. For example, in reaction (i), phenol exhibits monofunctionality and in reaction (ii), it exhibits a functionality of three.

Depending on the functionality of the monomers used, we get linear, branched or three dimensional cross-linked polymers.

(i) In case of bifunctional monomers, each molecule can add two more molecules and give a linear polymer as shown:

n •A• ⟶ •A — A — A — A — A•

Bifunctional Monomer — Linear polymer

(• represent functionality)

(ii) In case of trifunctional monomers, we get a cross-linked polymer:

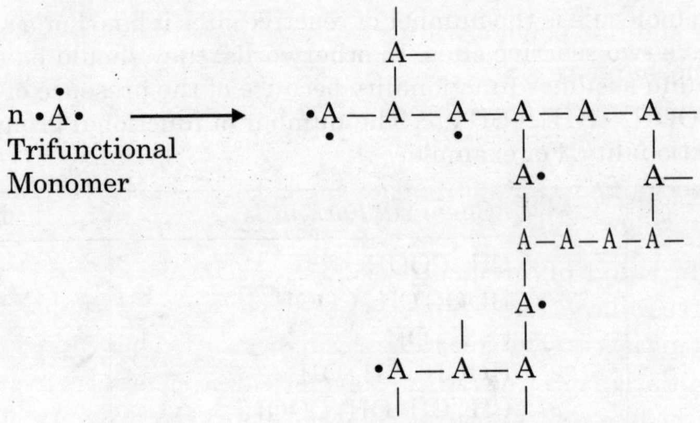

Cross-linked polymer

11.3 TYPES OF POLYMERIZATION

There are two types of polymerization processes:

1. Addition Polymerization,
2. Condensation polymerization.

11.3.1 Addition Polymerization

Addition polymerization is characterized by self-addition of the monomer molecules to each other, without the loss of any material. The product is an exact multiple of the original monomeric molecule. Examples:

(i) Addition by breaking up of multiple bonds:

$$\underset{\text{(ethene)}}{nCH_2 = CH_2} \longrightarrow \underset{\text{(Polyethene)}}{\left(CH_2 - CH_2 \right)_n}$$

(ii) Addition by opening up of ring structures:

$$CH_3OH + (n+1)\ CH_2 - CH_2\ (\text{bridged by } O) \longrightarrow CH_3O \left[CH_2 - CH_2 - O \right]_n CH_2CH_2OH$$

Vinyl compound ($CH_2 = CHX$), allyl compounds ($CH_2 = CH.CH_2X$), olefins ($CH_2 = CHR$) and dienes ($CH_2 = CR–CH = CH_2$) typically undergo addition polymerization. Since a majority of these fall under the 'vinyl' category, addition polymerization is also customarily termed as vinyl polymerization.

MECHANISM OF ADDITION POLYMERIZATION

Addition polymerization is a typical chain reaction and consists of three important steps:

(i) Initiation or the formation of active centres.

$$A \longrightarrow A^* \text{ (Excited momomer)}$$

(ii) Chain propagation

$$A^* + A \longrightarrow A - A^*$$

$$A - A^* + A \longrightarrow A - A - A^* \quad \text{(Activated growing chain)}$$

(iii) Chain termination or removal of active centre.

$$\sim A - A - A^* \longrightarrow \sim A - A - A \text{ (Inactivated chain)}$$

The process of addition polymerization can be brought about by three methods:

1. Free radical mechanism
2. Ionic mechanism (Anionic and Cationic)
3. Coordination mechanism

The three types of mechanisms for addition polymerization, depending upon the active centres formed, are discussed below.

1. **Free Radical Addition Polymerization:** The free radical addition polymerization involves the formation of free radicals in the initiation step, which can be produced by decomposition of compounds called initiators. The free radical polymerization has three principal steps:

(i) **Initiation:** Initiation in a free radical polymerization consists of two steps. The first is the generation of the free radical species, followed by the second step which involves the addition of a single monomer molecule to the free radical.

The generation of free radical takes place as a result of homolytic dissociation of an initiator, which may take place in the presence of heat energy, light energy or catalysts. A number of low molecular weight compounds, comprising mainly of azo compounds, peroxides, hydroperoxides, peracids and peresters are useful as initiators. An important example of a free radical initiator is benzoyl peroxide which dissociates as:

$$C_6H_5\overset{\overset{O}{\|}}{C} - O - O - \overset{\overset{O}{\|}}{C} - C_6H_5 \xrightarrow{\Delta} 2C_6H_5 - \overset{\overset{O}{\|}}{C} - \dot{O}$$

In general, the dissociation of the initiator (R–R) to form free radical specie can be represented as

$$\underset{\text{(Initiator)}}{R - R} \longrightarrow \underset{\text{(Free radicals)}}{2R^*}$$

In the second step of initiation, the free radical so formed gets attached to the monomer molecule (M) to form the chain initiating species

$$\underset{\text{(free radical)}}{R^*} + \underset{\text{Monomer}}{CH_2 = CH_2} \longrightarrow R - CH_2 - \dot{C}H_2$$

(ii) **Propagation:** In the next step, additional monomer units are added to the initiated monomer species, as.

$$R - CH_2 - \underset{\underset{\times}{|}}{\dot{C}H} + \underset{\underset{\times}{|}}{CH_2 = CH} \longrightarrow R - CH_2 - \underset{\underset{\times}{|}}{CH} - CH_2 - \underset{\underset{\times}{|}}{\dot{C}H}$$

$$R - CH_2 - \underset{\underset{\times}{|}}{CH} - CH_2 - \underset{\underset{\times}{|}}{\dot{C}H} \quad \underset{\underset{\times}{|}}{CH_2 = CH} \longrightarrow R - CH2 - \underset{\underset{\times}{|}}{CH} - CH2 - \underset{\underset{\times}{|}}{CH} - CH2 - \underset{\underset{\times}{|}}{\dot{C}H}$$

The propagation step involves a continuing attack on fresh monomer molecules which, in turn, keep successively adding to the growing chain one after another.

(iii) **Termination:** Propagation will continue until some termination process takes place. The propagating polymer chain is terminated by any of the following reactions.

(*a*) *By combination:* Since the decomposition of the initiator produces many free radicals at the same time, each one of them can initiate and propagate the growth of a number of chains simultaneously. The two growing chains may come close and collide with each other and get deactivated.

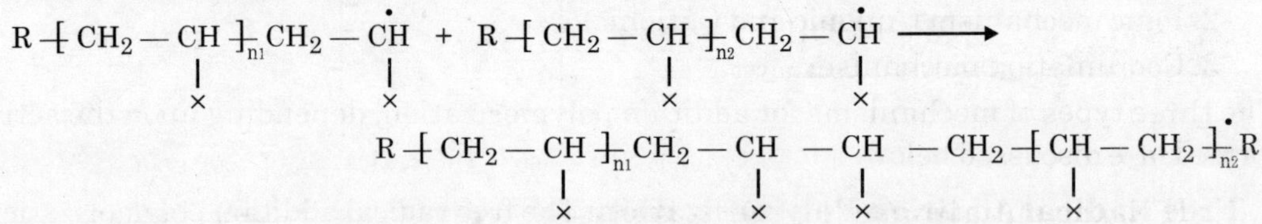

(*b*) *By disproportionation:* In some cases, one hydrogen atom from one growing chain is abstracted by the other growing chain and results in the formation of two polymer molecules, one saturated and other unsaturated. In this case the termination process results in the formation of two polymer molecules of shorter chain length as against a single molecule of a longer chain length obtained by first method.

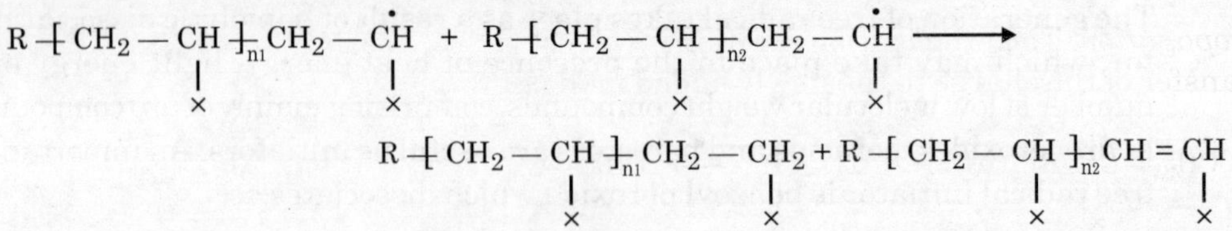

Thus the products so formed do not contain any reactive sites and may be termed as dead polymers as these cannot grow further.

(c) In addition to the above two methods of chain termination, chain transfer by hydrogen abstraction from solvent molecule, may also terminate the growing chain.

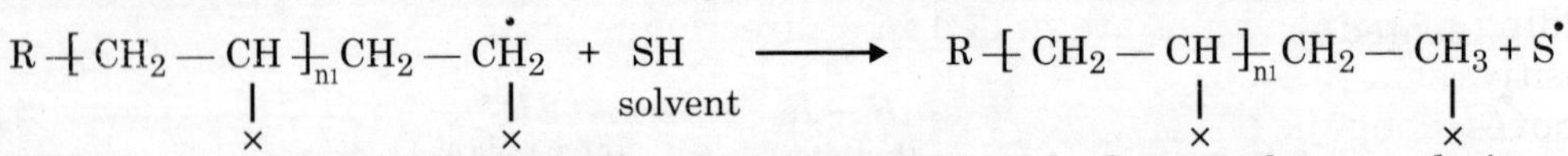

In this reaction the original growing chain is terminated whereas the new chain gets initiated by the solvent free radical, hence this is called termination by chain transfer. Certain chemical compounds which are capable of inhibiting or completely stopping the chain growth process are called radical inhibitors. Phenolic compounds are widely used as inhibitors, because phenolic hydrogen can be easily abstracted by the polymer chain. The phenolic free radical so formed is highly stabilized and no further reaction is possible. Hydroquinone is often added as inhibitor.

Polythene, polypropylene, styrene etc are some of the polymers that can prepared by free radical polymerization.

Branching and crosslinking during the free-radical addition polymerization:

2. **Ionic Addition Polymerization:** Ionic polymerization follow the same basic steps as free-radical addition polymerization i.e initiation, propagation and termination. In this process, the initiation step involves the formation of carbonium ion or the carbanion as the active site and hence the polymerization can be termed as cationic or anionic polymerization respectively.

(i) **Cationic Polymerization**

Cationic polymerization takes place in the presence of catalyst that is regenerated at the end of polymerization (in contrast to free radical and anionic polymerization where the catalyst added become incorporated into the terminated polymer chain) Polymerization of vinyl monomers with an electron-donating group (such as $-OC_2H_5$ $-OCH_3$, $-C_6H_5$ etc.) may take place by a cationic mechanism which involves the following steps.

Initiation: This step involves the formation of a carbonium ion by the attack of a proton on the monomer. Any strong Lewis acid like BF_3, $AlCl_3$ etc can be used as a catalyst. In this case a cocatlyst like water is required to provide the proton source.

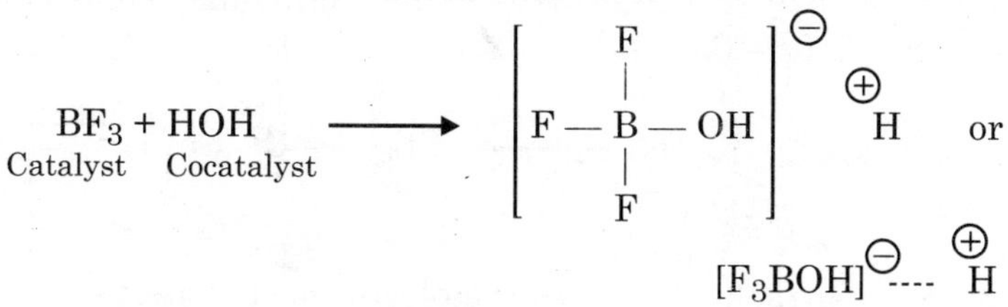

H^+ is the initiator and $[F_3\ BOH]^-$ is the counter ion.

The H^+ now attacks the π electron of the monomer and forms the carbonium ion. The counter ion $[F_3\ BOH]^-$ anion faces the growing chain end.

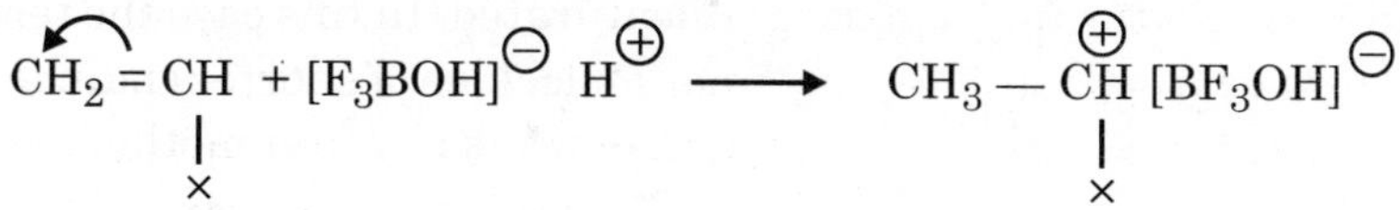

Propagation: The propagation reaction involves the addition of monomer units and simultaneous transfer of the charge to the newly added monomer unit.

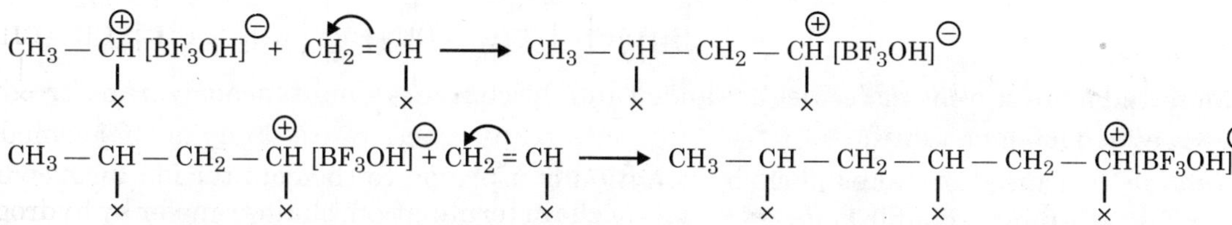

More and more units are added up and the chain keeps on growing. The electron pair of the adding monomer units are pulled in a direction opposite to the growth of the chain and the positive charge keeps on moving in the direction of the chain growth. The counter ion $[BF_3\ OH]^-$, moves along the positive charge all the time.

Termination: Termination takes place when the growing polymer chain having C^+ collides with an anion, which may be added deliberately or with the counter ion. The possible reactions may be:

(a) Donation of a proton to the counter ion by the growing polymer chain: This results in the formation of a double bond at the end of growing polymer molecule.

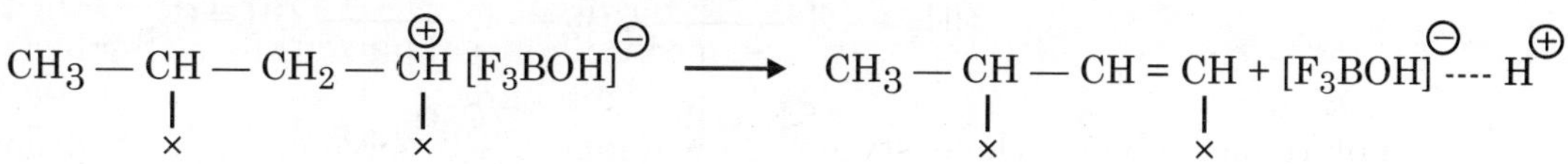

Thus the initiator or catalyst is regenerated.

(b) Formation of covalent bond between the carbonium ion and the counterion (coupling):

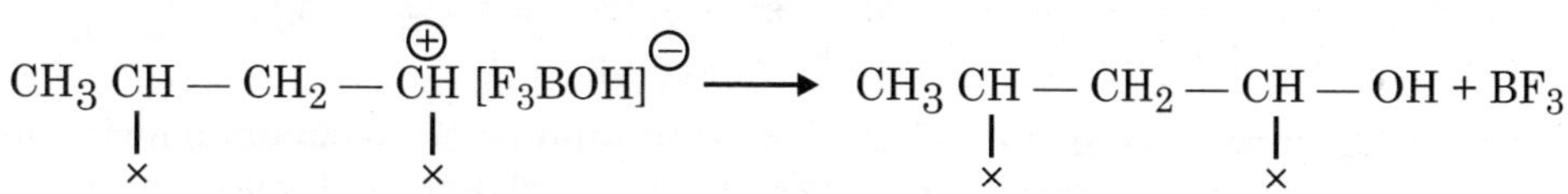

In this case also the regeneration of the initiator takes place.

Examples of the monomers that can undergo cationic polymerization are isobutylene, styrene, methyl styrene and many vinyl ethers. Cationic polymerization proceeds at a very fast rate even at low temperatures. For example, polymerization of isobutylene is completed within a few seconds at $-100°$ C.

Similarly cationic polymerization of styrene in the presence of $SnCl_4$ as the catalyst an be explained as follows:

Initiation: This involves the formation of the carbonium ion.

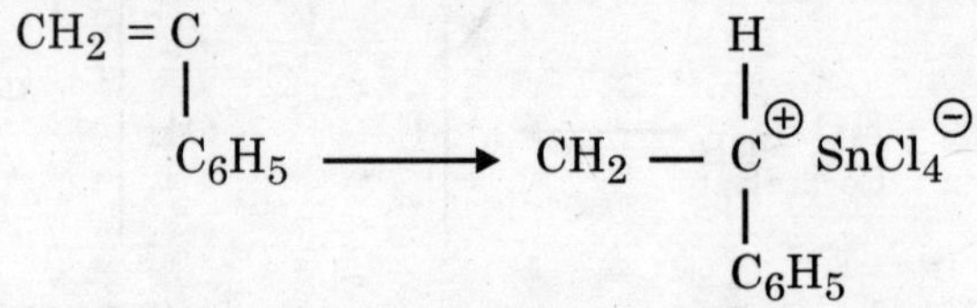

Propagation: Carbonium ion attacks the another molecule of styrene.

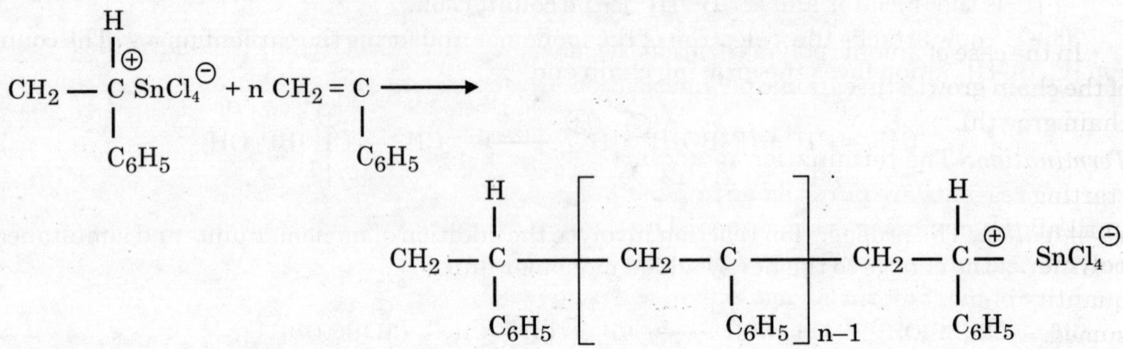

More and more styrene molecules are added and the charge is simultaneously transferred to the newly added monomer unit.

Termination: Termination takes place by donation of a proton to the counter ion by growing polymer chain forming a double bond

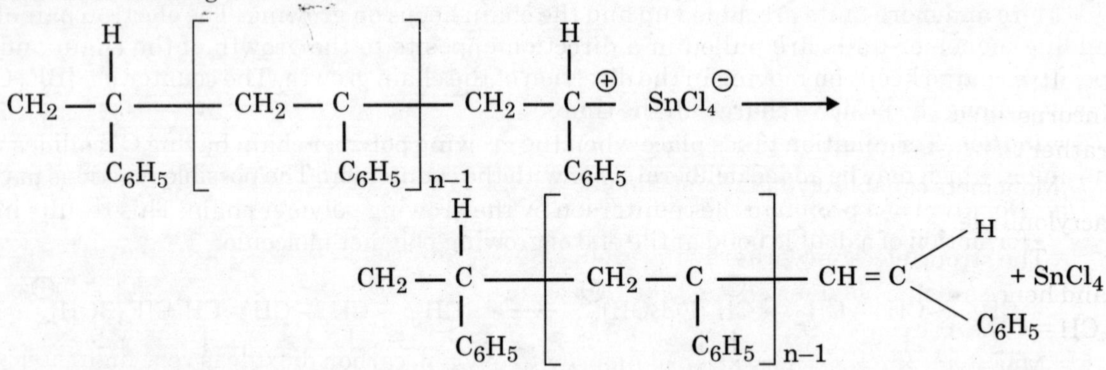

(ii) **Anionic Polymerization**

Anionic polymerization involves the formation of the carbanion in the initiation step. Polymerization of vinyl monomers with an electron withdrawing group can proceed by an anionic mechanism, involving the following steps:

Initiation: The initiators or the catalysts used in anionic addition polymerization are strong bases such as organo alkali compounds (alkyl or aryl derivatives of alkali metals) like n-butyl lithium, ethyl sodium etc. Alkali metal amides, hydroxides etc can also be used.

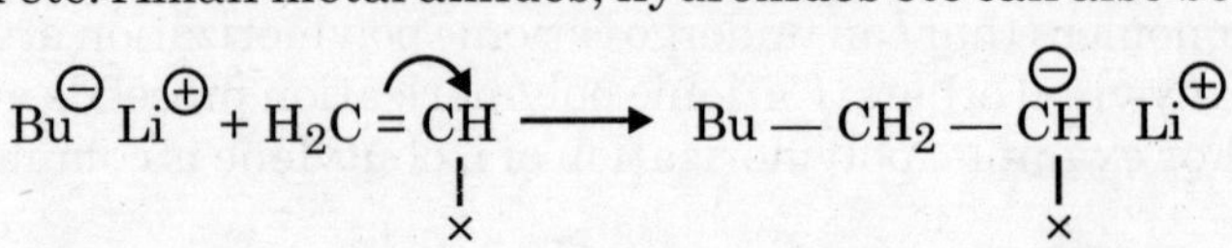

The electron pair in this case is pushed to the end of the molecule forming the carbanion. Li^+ is the counter ion.

Propagation: The carbanion formed, propagates the chain growth by attacking the second monomer unit, pushing its π electron pair further away to the end and forming a sigma bond with the new monomer unit.

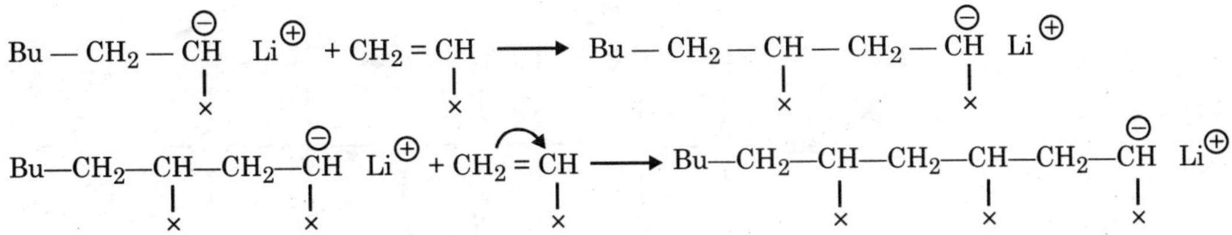

In the case of anionic polymerization the movement of the π electrons is towards the direction of the chain growth (in cationic polymerization, the movement of the π electrons is opposite to the chain growth).

Termination: The termination in anionic polymerization is not a spontaneous process. If the starting reagents are pure and no impurities are present, propagation can proceed indefinitely or until all the monomer is consumed. Thus, anionic polymerization is also referred to as "living polymerization", because the carbanions at the chain ends remain potentially active. If a fresh quantity of monomer is added, polymerization again goes on until the added monomer is consumed.

Termination is generally accomplished by the transfer of negative charge to a species which is not directly involved in the reaction. For example CO_2, methanol, water etc.

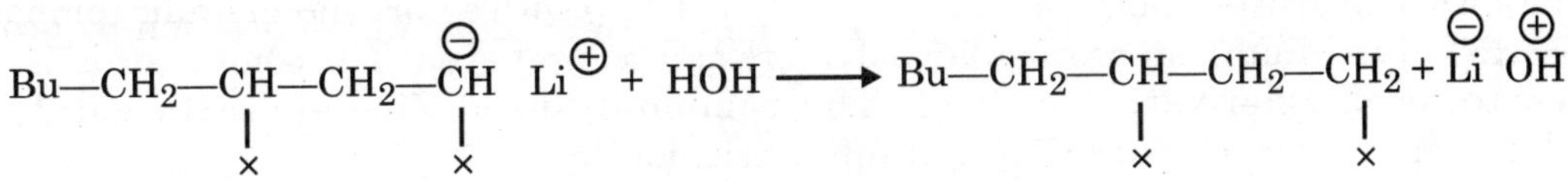

It may be noted that like free radical polymerization, the initiating specie (Bu) has been incorporated as the end group of the terminated polymer. Thus Butyl-lithium is an initiator rather than a catalyst.

Monomers capable of undergoing anionic polymerization include butadiene, isoprene, styrene, acrylonitrile.

The strong electron withdrawing groups have stabilizing effect on the carbanion so formed and hence anionic polymerization takes place with the monomers having – CN, – COOR, and – $CH = CH_2$ groups.

Moreover, the reactivity of carbanions with oxygen, carbon dioxide is very high, hence the system should be free of these impurities.

The mechanism of anionic polymerization can be further illustrated by polymerization of acrylonitrile in the presence of butyl lithium.

Initiation:

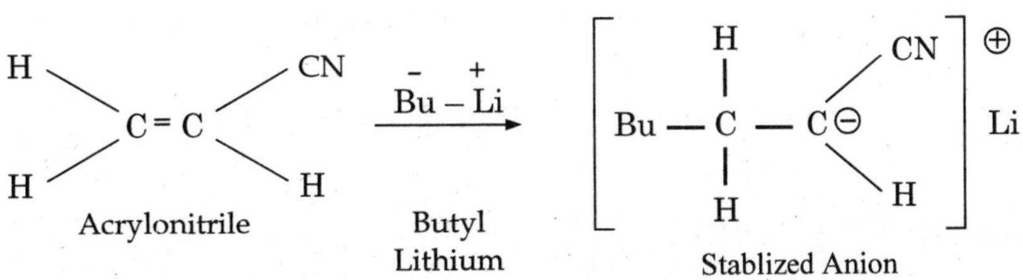

Propagation:

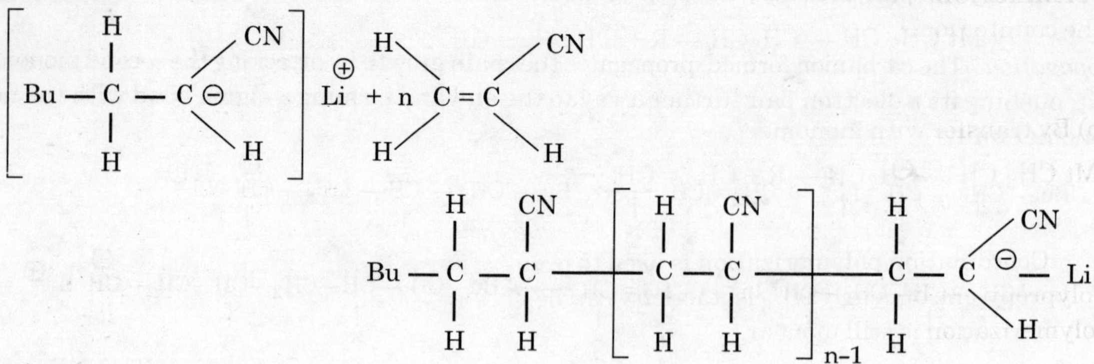

Termination:

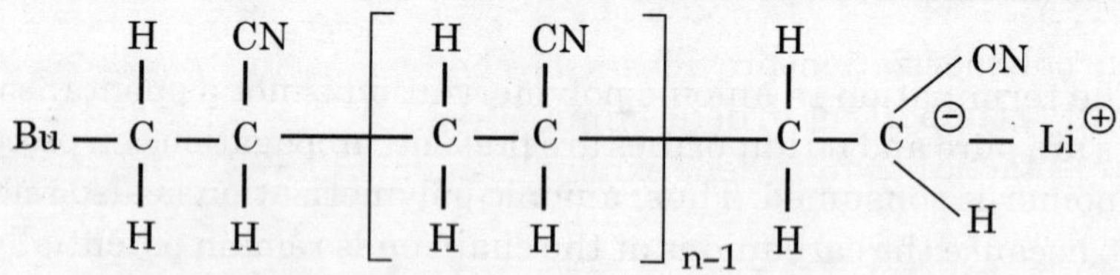

3. **Coordination Polymerization:** Polymerization reactions, taking place in the presence of organo metallic compounds as catalysts, are termed as coordination polymerizations.

Ziegler (1953) and Natta (1955) discovered that in the presence of a catalyst which is a combination of a transition metal halide (like $TiCl_4$ or $TiCl_3$) with an organo-metallic compounds (like triethyl aluminium), stereoregular polymers can be prepared. The combination is often referred to as, ZieglerNatta catalyst. The commonly used Ziegler-Natta catalyst is triethylaluminium in combination with titanium trichloride.

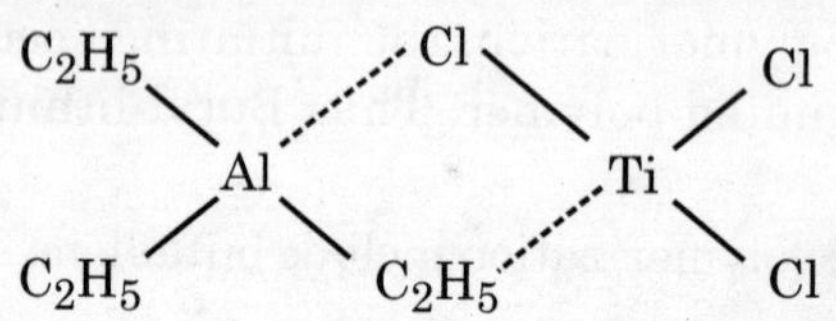
(Ziegler Natta Catalyst)

In the formation of the monomer catalyst complex, a co-ordinate bond is involved between a carbon atom of a monomer and metal atom of the catalyst, that is why, this type of polymerization is known as coordination polymerization. The co-ordinated metal-carbon bond formed in the monomer-catalyst complex acts as an active center from where propagation starts.

Mechanism of coordination polymerisation can be illustrated as follows:

Initiation:

$$Mt - R + CH_2 = \underset{\times}{\underset{|}{CH}} \longrightarrow Mt\,CH_2\underset{\times}{\underset{|}{C}}HR$$

The alkene is inserted in between the metal - carbon bond of the catalyst

Propagation:

$$Mt\,CH_2\,\underset{\times}{\underset{|}{C}}HR + CH_2 = \underset{\times}{\underset{|}{CH}} \longrightarrow Mt\,CH_2\,\underset{\times}{\underset{|}{CH}} - CH_2 - \underset{\times}{\underset{|}{CH}} - R$$

$$Mt\,CH_2\,\underset{\times}{\underset{|}{CH}} - CH_2\underset{\times}{\underset{|}{CH}} - R + CH_2 = \underset{\times}{\underset{|}{CH}} \longrightarrow Mt\,CH_2\,\underset{\times}{\underset{|}{CH}} - CH_2 - \underset{\times}{\underset{|}{CH}} - CH_2 - \underset{\times}{\underset{|}{CH}} - R$$

Termination: (a) By active hydrogen compound

$$\text{Mt}\,CH_2\underset{\displaystyle \times}{\underset{|}{CH}}-CH_2\underset{\displaystyle \times}{\underset{|}{CH}}-R + R'H \longrightarrow \text{Mt}-R' + CH_3\underset{\displaystyle \times}{\underset{|}{CH}}-CH_2\underset{\displaystyle \times}{\underset{|}{CH}}-R$$

(b) By transfer with monomer

$$\text{Mt}\,CH_2\underset{\displaystyle \times}{\underset{|}{CH}}-CH_2\underset{\displaystyle \times}{\underset{|}{CH}}-R + CH_2=\underset{\displaystyle \times}{\underset{|}{CH}} \longrightarrow \text{Mt}-CH_2-\underset{\displaystyle \times}{\underset{|}{CH_2}} + CH_2=\underset{\displaystyle \times}{\underset{|}{CH}}-CH_2\underset{\displaystyle \times}{\underset{|}{CH}}-R$$

Coordination polymerization is used to prepare stereospecific polypropylene, polydiene etc. Polypropylene of nearly 90% isotacticity can be prepared. The exact mechanism of coordination polymerization is still unclear.

Advantages of Coordination Polymerization

1. Coordination polymerization provides a method to produce stereoregular polymers. The stereoregularity is important in determining the properties of the polymer. For example, the stereoregular isotactic polypropylene exhibit better mechanical properties over a wide range of temperatures.
2. The stereoregular control of the coordination polymerization process produce the polymers, which have fewer branches and therefore could be obtained in a higher degree of crystallinity. For example high density polyethylene produced by coordination polymer has higher degree of crystallinity as compared to the one produced by other methods. This is because the polyethylene having less branching pack more efficiently resulting in high degree of crystallinity.

11.3.2 Condensation Polymerization

Condensation polymerization is brought about by monomers containing two or more reactive functional groups with the elimination of small molecules like H_2O, HCl etc. For example,

$$n\ HO-\overset{\displaystyle O}{\overset{\|}{C}}-(CH_2)_4-\overset{\displaystyle O}{\overset{\|}{C}}-OH + n\ H_2N-(CH_2)_6-NH_2$$

(adipic acid) (hexamethylene diamine)

$$HO\left[\overset{\displaystyle O}{\overset{\|}{C}}-(CH_2)_4-\overset{\displaystyle O}{\overset{\|}{C}}-NH-(CH_2)_6-NH\right]_n H + (2n-1)H_2O$$

nylon 6, 6

It is seen from the above reaction that the polymer formed still contains both the reactive functional groups at its chain ends and hence, is 'active' and not 'dead'.

In the above example, the different type of reactive functional groups, viz., $-NH_2$ and $-COOH$ are present on two different monomers. In some cases, both the reactive groups may be present on the same monomer and hence lead to self condensation reaction. For example, polycondensation of amino acids.

$$n\ H_2N\left(CH_2\right)_6 COOH \longrightarrow H\left[HN-(CH_2)_6-CO\right]OH + (n-1)\ H_2O$$

Some polycondensation reactions leading to different types of polymers are:

(i) Polycondensation of dihydric alcohols and dicarboxylic acids leads to the formation of **polyesters.**

$$n\,HO-R-OH + n\,HOOC-R'-COOH \longrightarrow HO\!\left[R-OOC-R'-COO\right]_n\!H + (2n-1)\,H_2O$$

(ii) Polycondensation of dicarboxylic acids and diamines to give **polyamides.**

$$n\,HOOC-R-COOH + n\,H_2N-R-NH_2 \longrightarrow HO\!\left[OC-R-CONH-R-NH\right]_n\!H + (2n-1)\,H_2O$$

(iii) Polycondensation of dihydric alcohols and methyl esters of dicarboxylic acids to give **polyesters.**

$$n\,HO-R-OH + n\,CH_3COO-R'-COOCH_3 \longrightarrow HO\!\left[R-COO-R-COO\right]_n\!CH_3 + (2n-1)\,CH_3OH$$

(iv) Self condensation of ω-hydrogen acids yields **polyesters.**

$$n\,HO-R-COOH \longrightarrow HO\!\left[R-COO\right]_n\!H + (n-1)\,H_2O$$

(v) Self condensation of amides to form **polyamides.**

$$n\,H_2N-R-COOH \longrightarrow H\!\left[NH-R-CO\right]\!OH + (n+1)\,H_2O$$

(vi) Polycondensation of bisphenol A and phosgene to form **polycarbonate.**

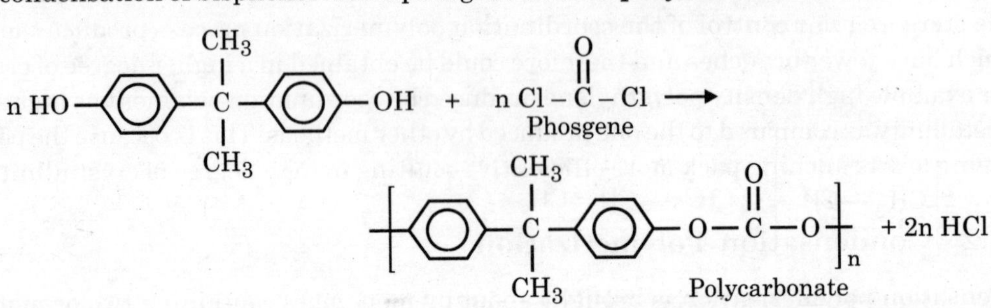

11.4 CHAIN GROWTH POLYMERIZATION AND STEP GROWTH POLYMERIZATION

Recently, another classification scheme based on polymerization kinetics has been adopted over the traditional addition and condensation categories. According to new scheme, all polymerization mechanisms are classified as either chain growth or step growth polymerizations.

Chain Growth Polymerization: During chain growth polymerization, high molecular weight polymer is formed early during polymerization and polymer yield gradually increases with time. Chain growth polymerizations require the presence of an initiating molecule that is used to activate the monomer molecule at the start of the polymerization. Most of the addition polymers are chain growth polymers and hence in most of the books the term chain growth and addition polymerization are used interchangeably.

Step Growth Polymerization: In step growth polymerization, high molecular weight polymer is formed only near the end of the polymerization. Most of the condensation polymers are step growth polymerization, but a number of important exceptions exist. In step growth polymerization, the reaction takes place in stepwise manner and polymer build-up is slow (unlike chain growth, where the polymer build-up is very rapid).

Apart from condensation reactions discussed earlier, step growth polymerization include some addition reactions also. Example, the non-condensation step growth polymerization of a polyurethane prepared by ionic addition of a diol (1, 4 butanediol) to a diisocyanate (1, 6-hexane diisocyanate.)

$$n\,HO-(CH_2)_4-OH \quad + \quad O=C=N-(CH_2)_6-N=C=O$$
1,4 butanediol 1,6 hexane diiso cyanate

$$\left[O-(CH_2)_4-O-\overset{\overset{O}{\|}}{C}-NH-(CH_2)_6-NH-\overset{\overset{O}{\|}}{C}\right]_n$$
Polyurethane

11.5 COPOLYMERIZATION

Commonly available natural and synthetic polymers suffer from many disadvantages and therefore cannot be used for special purposes. However, it is possible to modify a number of natural and commercially available synthetic polymers to give products which can be utilized for performing desired functions.

Copolymerization is the most general and powerful method of affecting modification in polymers and is widely used in the production of commercial polymers. It provides a technique to combine the properties of two or more different polymers into a single polymer. *Copolymerization can be defined as any process whereby two or more monomers are copolymerized to form a single polymer.* The resulting high molecular weight compound is known as copolymer. For example, butadiene and styrene, copolymerize to form butadiene-styrene rubber.

$$n\underset{\underset{C_6H_5}{|}}{CH_2=CH} + nCH_2=CH-CH=CH_2 \xrightarrow{\text{Copolymerization}}$$
Styrene Butadiene

$$\left[CH_2-\underset{\underset{C_6H_5}{|}}{CH}\right]_n (CH_2-CH=CH-CH_2)_n \quad \text{or} \quad \left[CH_2-CH=CH-CH_2-\underset{\underset{C_6H_5}{|}}{CH}\right]_n$$

Styrene butadiene rubber (SBR)

The styrene butadiene rubber so formed has better overall properties than polystyrene or polybutadiene alone.

Similarly, acrylonitrile and vinylchloride are copolymerized to form a copolymer. The copolymer so formed is readily soluble in acetone, whereas polyacrylonitrile and polyvinyl chloride are soluble only in high boiling solvents

$$n\left[\underset{\underset{CN}{|}}{CH_2=CH}\right] + n\left[\underset{\underset{Cl}{|}}{CH_2=CH}\right] \xrightarrow{\text{Copolymerization}} \left[CH_2-\underset{\underset{CN}{|}}{CH}-CH_2-\underset{\underset{Cl}{|}}{CH}\right]_n$$
Acrylonitrile Vinyl chloride Polyacrylonitrile-co-vinyl chloride

A polymer of vinylidene chloride and vinyl chloride is known as saran wrap.

$$n\,CH_2=\underset{\underset{Cl}{|}}{\overset{\overset{Cl}{|}}{C}} + CH_2=CHCl \rightarrow \left[-CH_2-\underset{\underset{Cl}{|}}{\overset{\overset{Cl}{|}}{C}}-CH_2-\underset{\underset{Cl}{|}}{\overset{\overset{Cl}{|}}{C}}-\right]_n$$
Vinylidene Chloride Vinyl Chloride Saran Wrap

Saran wrap has a high tensile strength and excellent sealing properties as compared to the individual polymers involved in its preparation.

It should be clearly understood that copolymer is not a mixture of two or more separate polymers but it consists of the chains which contain both the units.

The composition of the copolymer with respect to the polymer involved in its formation depends upon the following factors:

(i) **Reactivity of the monomers involved:** Different monomers have different tendencies towards undergoing copolymerization. Moreover, the reactivity of the monomers is also governed by the technique used for copolymerization. (free radical, ionic or coordination). For example, in a copolymer of butadiene and styrene the ratio of butadiene to styrene is about 1:3 in the usual products.

(ii) **Amounts of the monomers in the reaction mixture:** The composition of the copolymer is sometimes governed by the relative ratio of the monomers in the reaction mixture. In condensation-copolymerization because the polymer undergo stepwise increase in degree of polymerization therefore amounts of monomers in the reaction mixture has no significance in condensation copolymerization. However, where the copolymerization takes place by addition i.e. in vinyl copolymerization, the composition of the copolymer depends upon the amounts of monomers present in the reaction mixture.

Depending upon the sequence in which different polymeric units are arranged, copolymers are of following types:

(a) Random copolymers.
(b) Alternating copolymers.
(c) Block copolymers.
(d) Graft copolymers.

(a) **Random copolymers:** In random copolymers the two monomers A and B are distributed randomly along the polymer chain.

$$nA + nB \longrightarrow [A-A-B-B-A-B-A-A]_n$$

(b) **Alternating copolymers:** In these copolymers the monomers have regular alternate arrangement along the chain of the copolymer.

$$nA + nB \longrightarrow [A-B-A-B-A-B-A]_n$$

(c) **Block copolymers:** In block copolymers, a sequence or block of one monomer is followed by the block of the other monomer.

$$nA + nB \longrightarrow [\underbrace{A-A-A}_{\text{Block}}-\underbrace{B-B-B}_{\text{Block}}-A-A-A]$$

(d) **Graft copolymer:** Graft copolymers are branched polymers and contain main chain or backbone consisting of one monomer and branches formed from the other monomer:

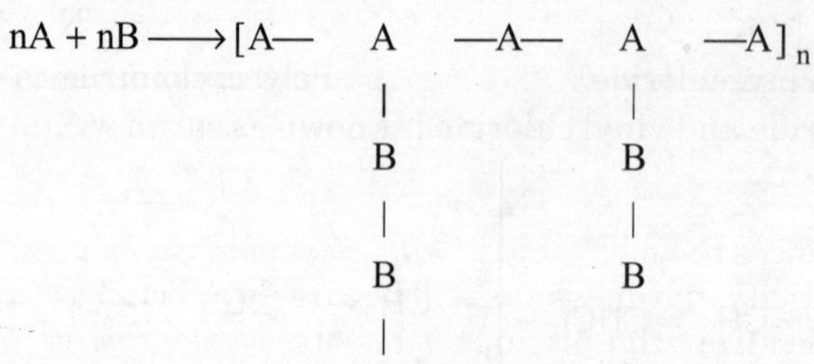

Block and graft copolymerization require special techniques for their formation.

Advantages of Copolymers

Copolymerization provide an excellent way to combine the properties of individual polymers into single polymers. Modification of polymers through graft or block copolymerization impart certain desirable properties which are difficult to achieve otherwise. The tremendous advances in the synthesis of copolymers are an indication of their growth. Some of the advantages can be listed below:

1. **Reinforcement of Natural Rubber:** The natural rubber, graft copolymerized by methyl methacrylate has been used for the manufacture of unbreakable electric plugs and sockets which have many advantages over those prepared from brittle thermoplastics or vulcanized rubber. Natural rubber grafted with styrene has been claimed as a reinforcing resin for natural rubber.
2. **High-Impact-Strength Polymers:** The discrete particles of a different phase dispersed uniformly in glass like polymers, such as polystyrene, increase the impact resistance. Thus polystyrene can be given high-impact properties by blending with it a mixture of polybutadiene or any other rubber i.e., by preparing block copolymer of rubber with polystyrene.
3. **Surface Active Agents**: The basic structure of block copolymers impart exceptional surface active properties to these polymers. This is because the polymeric segments that are water soluble and others that are oil soluble may be built into the same molecule.
4. **Fibers**: Many fibers cannot retain the dyes and hence are difficult to die. This problem can be overcome by grafting or block copolymerizing with a monomer, the polymer of which accepts the dye and hence aids in colouring. Acrylamide or methacrylonitrile is grafted into polyacrylonitrile to improve the dyeing characteristics. The physical properties of cotton i.e., strength, abrasion resistance and crease resistance can be achieved by graft copolymerizing vinyl monomers into it.
5. **Ion-Exchange Membranes**: Styrene is grafted to divinylbenzene, polypropylene, nylon etc. The resulting copolymer can be chemically converted to ion exchange resins.

11.6 STRUCTURE PROPERTY RELATIONSHIPS BETWEEN THE POLYMERS

Polymer properties are significantly influenced by the structure of the polymer. The structural details which affect the properties of a particular polymer include:

- the chemical composition
- sequence of monomer units
- types of bonds present
- nature of packing
- molecular weight

In the light of these structural details we will discuss some of the properties of the polymers.

1. Mechanical Properties

The structure of a polymer influences the mechanical properties of a polymer significantly. These properties such as strength, elasticity, hardness etc. will be largely affected by the *crystallinity* of the polymer, whereas the tensile strength and impact resistance are related to the *molecular weight* of the polymer.

(a) **Strength:** The strength of the polymer depends upon the intermolecular forces of attraction between the molecules and its molecular weight.

- In the case of straight chain and branched chain polymers, the individual chains are held together by weak intermolecular forces of attraction. *The magnitude or strength of these forces increases with the chain length and molecular weight.* Thus, for a polymer to show appreciable mechanical strength it should have chain length upto 150 - 200 units which corresponds to molecular weights of roughly 20000. These high molecular weight polymers are tougher and more heat resistant than the monomers of low molecular weight. On the otherhand, low molecular weight polymers are soft, gummy and brittle. Hence, the strength of the polymer can be varied by controlling their molecular weight or chain length which influence the intermolecular forces between them. Fig. 11.1 exhibits a relation between molecular weight and mechanical properties such as tensile and impact strength etc.
  The nature of the curve is valid for all polymers.

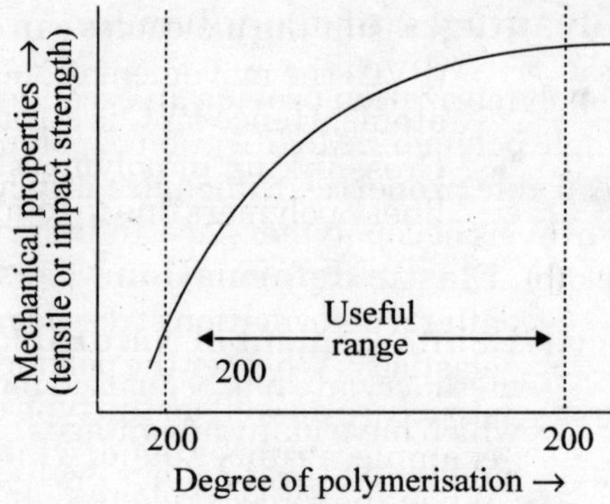

Fig. 11.1

- Every polymer has a threshold value for its degree of polymerization below which the polymer does not possess any strength and exists as powder or a liquid resin.
- Apart from increasing the strength of the polymer by increasing its molecular weight, the same can be done through increasing intermolecular forces of attraction through the introduction of polar groups. The polar groups which may be introduced are —COOH, —OH, —Cl, —F etc which increase the intermolecular forces to a greater extent.
- *Crystallinity of the polymer* is another property which affects the strength of the polymer. For a partially crystalline polymer, its crystalline and amorphous regions will exhibit different properties, even though both the regions are chemically same. For example, the density of the crystalline regions will be higher than the amorphous regions. Similarly, hardness and tensile strength of crystalline polymer will be higher as compared to the amorphous polymer.
- A combined effect of the molecular weight and crystallnity on the physical properties is show in Fig. 11.2.

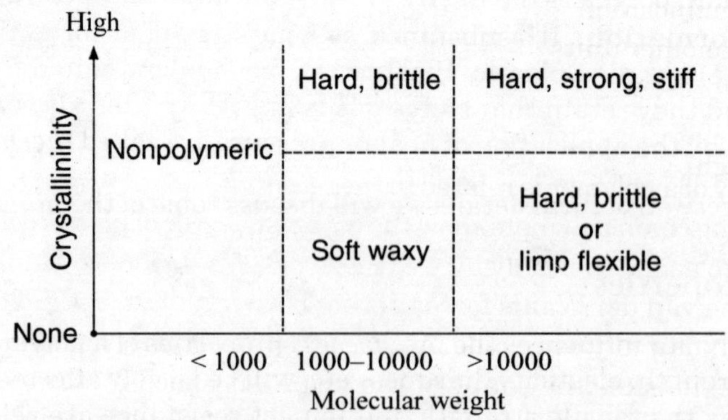

Fig. 11.2 Dependence of Physical properties of polymers on their molecular weight and crystallinity

- Strength of the polymer is also greatly influenced by configuration of the polymer units. For example, polyethylene (PE) molecule has a relatively simple and uniform shape, as a result the molecules can slip past each other. On the other hand, in polyvinyl chloride (PVC) the movement of the molecules is restricted due to the presence of larger chloride atoms. Hence PVC is stronger and tougher than PE.

- Crosslinking in polymers leads to increased strength and toughness as compared to linear polymers due to totally restricted movement of the chains in crosslinked polymers.

(b) **Elastic deformation:** The term 'elasticity' refers to the recoverance of original shape, after a deformation stress is released. We can say, elastomer is a substance which exhibits elasticity. Whether the polymer will behave as elastomer or not depends upon its geometrical structure as well as the type of interchain arrangement within the chain aggregates. For example, a rubber band, which is an elastomer, can be elongated to many times its length. When the force is released, it snaps back to its original length. A substance will behave as an elastomer if it consists of long chain of molecules having free rotating groups. In unstressed or unstretched conditions, these chains have peculiar arrangement of irregularly coiled and entangled snarls. Fig. 11.3.

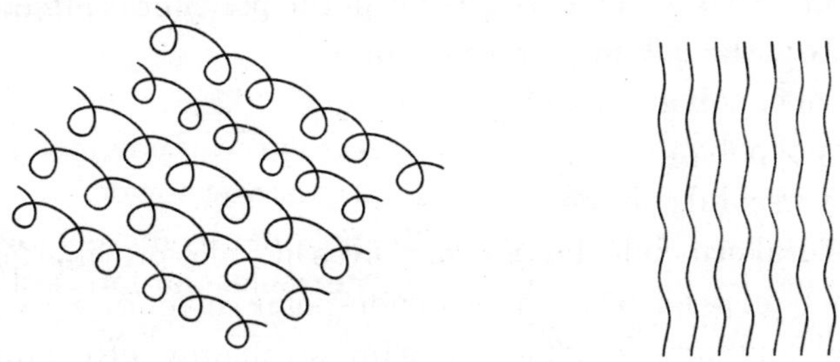

Irregularly Coile chains in unstreched conditions

Uncoiled chains in stretched conditions

Fig. 11.3

When such a polymer is stretched, the chains begin to disentangle and straighten out. This causes crystallisation with the result the intermolecular forces between the chains increase and hence stiffening of the material takes place.

Thus long chain polymers having free rotating groups exhibit elastic properties.

(c) **Plastic Deformation:** If a substance, at some stage of its formation, is capable of flowing (plastic stage) is called plastic. Plastics can be moulded into different shape under flow condition and they retain that shape on cooling The materials which can undergo plastic deformation by the application of heat or pressure are called *thermoplastics*.

The plasticity of a polymer is related to the *secondary bonds* present in the polymer molecules. The linear long chain polymers show the greatest degree of plastic deformation. The molecules are bonded to each other only by weak van der Waals forces. Under high temperatures or pressures the van der Waals forces acting between different molecules become more and more weak and finally overcome. As a result, the polymer become soften and flows. In this heated state, it takes the shape of the mould in which it is injected under pressure. On cooling, the secondary bonds are re-established as a result the material becomes hard in the moulded shape.

On the other hand the *cross linked polymers do not undergo plastic deformation*. In these polymers only primary covalent bonds are present throughout the structure making it hard and rigid. The plastic deformation cannot take place because the molecules cannot slip past each other due to strong directional covalent bonds. These crosslinked polymers are referred to as *thermosets.* If however the strong covalent bonds of thermosets are broken by the application of heat and pressure the result is total destruction of the molecule and not the plasticity.

2. Solubility

The process of dissolution of a polymer in a solvent may be divided into two steps. In the first step the solvent slowly diffuses into the polymer molecule to produce a swollen gel. In the second step the solvent disintegrates the polymer and a solution of the polymer in the solvent is obtained. The process of dissolution of a polymer is slow as compared to that of simple molecules and sometimes it takes weeks to a high polymer to dissolve.

The diffusion of the solvent molecules into the polymer solute structure depends upon polymer-solvent interaction, which leads to unwinding or expanding of the polymer surface and allows the penetration of the solvent molecules. The dissolution of the polymer is affected by:

(i) Chemical nature of the polymer and solvent
(ii) Molecular weight of polymer
(iii) Chain flexibility of polymer
(iv) Packing density in solid polymer

Further, the general principle that like dissolves like also prevails for polymeric solutions.

- Non polar polymers form solutions in non-polar solvents.
- Cross linked polymers do not dissolve but swell only if they interact at all with the solvent.
- Linear polymers are soluble in some solvent. The solvent molecules can get inbetween the chains and tear them apart after which the chains move freely. Hence the substance swells and as more solvent gets into the mass, a visous solution results. This viscous solution becomes thinner as more solvent is added thus forming a solution.
- It may be emphasized here that in polymeric solution, no saturation limits are reached as are observed with low molecular weight compounds. In otherwords at no stage, the unswollen polymer starts separating out of the solution on further addition of polymer molecules.
- Amorphous polymers containing bulky polar substituents readily swell in polar solvents but do not form solutions at room temperature.
- Crystalline polymers with close packing between the chain segments do not easily dissolve and require high temperatures so that crystalline order is broken to facilitate the individual segment-solvent contacts.

3. Chemical Reactivity

The chemical reactivity of a polymer depends upon the chemical nature of monomeric units and their molecular configuration.

- The polymers involving only C—C and C—H bonds such as polyethylene etc. are mostly inert and undergo only substitution reactions. However if branching is there then oxidation may take place at tertiary carbon atoms.

- Polymers containing double bonds usually are vulnerable to oxidation and scission of the chain.
- Esters, amides and carbonate groups are susceptible to hydrolysis. However, if benzene ring is present adjacent to such groups it protects these from hydrolyzing.
- Polymers containing C—F bonds (PTFE) are very stable and very inert.
- Cellulosics polymers have free hydroxyl group which are attacked by number of reagents which incorporates useful modification of properties. For example, reaction with nitric acid produces nitro-cellulose from which gun cotton and celluloid are formed.

4. Electrical properties

In general, polymers are insulators, that is, they are bad conductors of heat and electricity. Polymeric materials especially plastics have been used in many electrical applications which are related to their insulating capacity. Use of polymers as insulators in place of other media such as vacuum is highly preferred. This is because apart from acting as insulator it also provides mechanical support for the field carrying conductors.

The important electrical properties for electrical test purpose are dielectric strength, dielectric constant and dissipation factor. These properties and their dependence on the structure of the polymer are discussed below.

Dielectric strength: To evaluate the insulating property of a material its dielectric strength is determined. The dielectric materials (insulators) have high values of resistivity or very low values of conductivity. Higher the resistance of the material for electrical flow, larger is the insulating capacity or dielectric strength. The dielectric strength of the material is defined in terms of maximum voltage the material withstands and beyond which the dielectric nature of the material is lost. It is thus expressed as voltage per unit thickness

$$\text{Dielectric strength} = \text{V/mm}$$

Other term which is used to determine the insulating or the dielectric capacity of a material is Dielectric constant.

Dielectric constant: The dielectric constant or permittivity (ε) is a dimensionless parameter and describes the capacitance of a dielectric i.e. the ability of a dielectric to store electrically separated charges (electro-static energy). It is defined as the ratio of the capacitance induced by two metallic plates with an insulator sand-witched between them and capacitance induced by the same metallic plates without the material i.e. with vacuum as a medium between them.

Structure of the polymer and electrical properties: The dielectric constant of electrical insulating material depends on the role of polarisation processes that occur in dielectric under the influence of an applied voltage.

- In case of symmetrical molecules, such as benzene C_6H_6, polyethylene, polytetraflouroethylene etc. which are non-polar dielectrics, only electronic polarisation is responsible for insulating or dielectric constant. Thus these materials have low dielectric constants. (usually less than 3).
- In case of polar dielectrics, apart from the electronic polarisation the dipole movement also takes place and results into dipolar polarization. Electronic polarisation involves movement of the electrons whereas a dipole polarisation involves movement of a part or whole of a molecule. Hence the dielectric constant is high and lies between 3 and 7.
- Dielectric constant of some of the polymers are given in Table 11.1

Table 11.1

Polymer	*Dielectric Constant (1000 cycles)*
Nylon 66	3.9 - 4.5
ABS	2.4 - 4.5
Cellulose Acetate	3.4 - 7.0
Poly(methyl meth acrylate)	3.0 - 3.6
Poly(vinylidene flouride)	7.46
Polybutene	2.25

- Also the dielectric properties in a polar material would depend upon the location of the dipole in the structure. For example, if the dipole is located in the main chain then the polarization would depend upon the segmental movement of the chain. The polarization would be low below T_g and material will act as the insulator only at temperature below T_g.
- Electrical properties of a polymer are also affected by humidity. It lowers the resistivity of a material. Moreover it may also cause plasticising effect and thereby raises the value of the dielectric constant.

5. Light Transmission

Polymers differ in their ability to transmit light. Some plastics are ***transparent***, exhibiting optical properties similar to glass. Other plastics are ***opaque***, and do not allow ay light to pass through. ***Translucent*** plastics allow some light to pass through them, but the images are cloudy and unfocused. Lastly, some plastics are ***semi-opaque***, allowing enough light to pass through them to be able to make out shadows and vague outlines. Since the boundaries between these different light transmission values are not clear, there are some rules. If a newspaper is placed on the other side of the plastic, and it can be easily read, the plastic is transparent. If it is cloudy and cannot be read, but general shapes can be distinguished, then the plastic is translucent. If only vague shadows can be seen, the plastic is semi-opaque. If no light is transmitted, the plastic is opaque.

The polymer crystallinity plays a major role in determining the optical properties of plastics.

- **Amorphous polymers**, the acrylic, polycarbonate, and polystyrene do not form crystals. They have constant value of refractive index throughout the material. Thus no scattering of light takes place and material appears transparent. In general amorphous polymers are transparent.
- **In Crystalline polymers** the refractive index arising out of the difference in densities of crystalline regions is different. This will cause scattering of light, which makes the polymer transluscent. If the crystalline structure are approximately the same size as the wavelength of visible light, it will also cause the light to scatter and hence will appear transluscent. In general crystalline polymers are transluscent.
- Sometimes, even crystalline polymers appear transparent. If the crystalline structures such as spherulites, are smaller than the wavelength of light, they do not interfere with the passage of light. As a result polymer appear transparent. Also if the densities of crystalline and amorphous region of a crysalline polymer is same then also the polymer will be transparent as no scattering of light would take place.
- If any additive or impurity is present in amorphous polymer, which otherwise has uniform refractive index, then also amorphous polymer will become transluscent. Additive or impurity results in deviation of the incident light and causes haze.

6. Permeability

In package industry one of the most important property to be considered is permeability. Permeability of a polymer is also dependent upon the crystallinity of the polymer. Permeability in actually the rate of penetration of a liquid or gases through the polymer matrix.

- In crystalline polymers because of close packing, the permeability is quite low. As crystallinity increases permeability decreases.
- In amorphous polymers, above glass transition temperature, there is enough free volume between the molecules for the passage of liquid or gases and hence they are permeable.

7. Degradation of Polymers

Any change in a polymer during its service, whether good or bad, can be technically termed as degradation. For example, if a plastic bucket is left for long in the sun and rain, it loses its lustre and strength. In other words it has undergone degradation. In the process of degradation, the molecular weight of the polymer is reduced.

A polymer can suffer degradation at two stages -

(i) During the fabrication process which involve heating, which a polymer may not withstand and degrades.

(ii) During its usage where the polymer is subjected to mechanical stresses, radiation, atmospheric oxygen, moisture etc.

Types of polymer degradation: Polymer degradation can be of two types:

(i) *Chain end degradatin* - In chain end degreation, the deradation starts from the chain ends resulting in the release of monomer units.

M—M—M—M—M⋮M ⟶ M—M—M—M—M⋮M ⟶ M + M—M—M—M

(ii) *Random degradation:* It takes place at any random point along the polymer chain

M—M—M⋮M—M—M ⟶ M—M—M + M—M⋮M

In this type of degradation, the polymer undergoes degradation into low molecular weight fragments but unlike first case, no monomer units are released.

Factors responsible for degradation: Degradation of polymers can be brought up by -

- physical factors such as heat, light etc.
- chemical agents such as oxygen, ozone, acid or alkalies.

We will be discussing these under the following heads.

(i) **Thermal Degradation:** Degradation under the influence of heat is known as thermal degradation. The thermal stability of the polymer is dependent upon the stability of C—C bond of the backbone. Generally, bulkier substituents decrease the stability of C—C bond. For example,

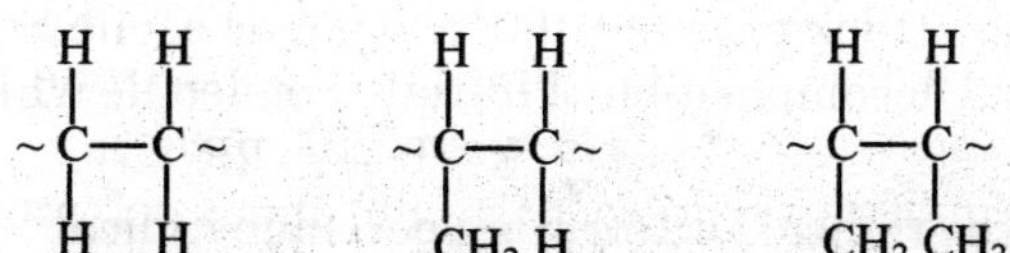

polyethene   polypropylene   polybutylene

in the above polymers, as one moves from polyethene to polybutylene, bulkier substituent (—CH_3) replaces H and hence the stability of C—C bond also decreases. Thus polyisobutylene is least stable whereas polyethene is most stable.

- Also, aromatic groups in polymer backbone increase the thermal stability whereas presence of oxygen atom in the polymer chain decreases the thermal stability.

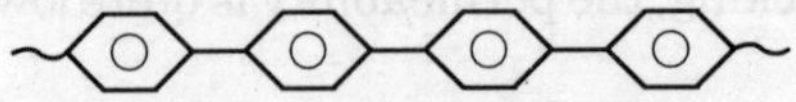

Polyphenylene (Highly thermally stable)

$\sim O—CH_2—CH_2—O—CH_2—CH_2—O—CH_2\sim$

potyethylene oxide (thermally unstable)

- Polymer degradation is also possible because of the breaking of the substituent groups, but not the main chain. Example poly(vinyl chloride) PVC degrades at about 200°C releasing HCl.

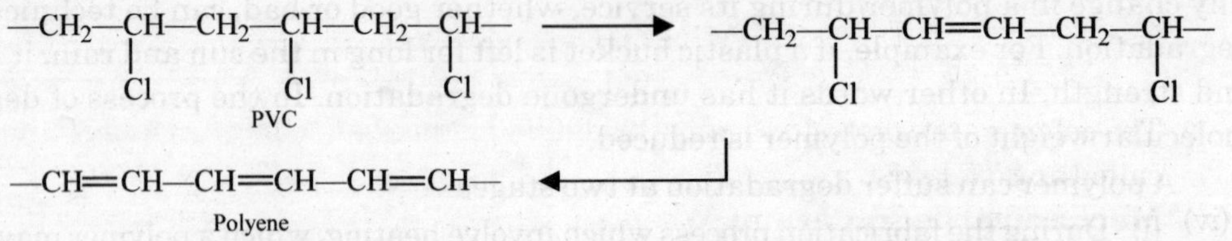

As a result of degradation PVC changes colour from white to black (white to → pale yellow → orange → brown → black) depending upon the number of concentration of conjugated double bonds.

(ii) **Photo degradation:** Photodegradation is brought about the ultra-violet (UV) light. Some of the transparent plastics become yellowish and brittle on storage This is due to photo degradation. The UV light has wavelength less than 400 nm corresponding to energy of 390 kJ/mole. The energy required to break a C—H, C—C and C=C bonds is 99, 83 and 145 kcal/mole respectively. Thus it is evident that homolytic bond fusion will occur readily to give free-radicals. These free radicals can then react with any oxygen present, leading to the oxidation of polymer chain. This is the phenomenon of "ageing or weathering" which is seen quite often.

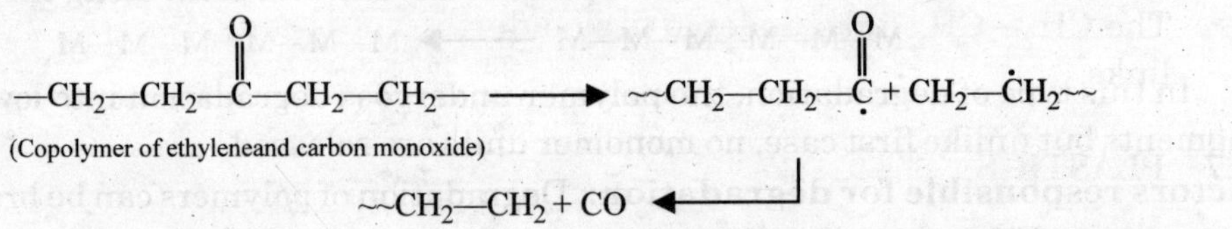

In order to protect the polymers from photodegradation some photo stabilizers are added to the polymer.

(iii) **Oxidative degradation:** Oxidative degradation of a polymer results in hordening, discolouration and surface changes of the polymer. Oxidative degradation of a polymer is dependent upon the structure of the polymer. The double bonds in polymers such as polyisoprene or polybutadiene are easily attacked by oxygen.

In the first stage of oxidative degradation free radicals are formed on the chain backbone of polymer by the attack of molecular oxygen or ozone Sometimes free radicals are also formed by thermal decomposition of initiator molecule which may be present as impurity.

$$R—R \longrightarrow 2R^{\cdot}$$

Oxygen adds on the radical site forming peroxide radical

$$R^{\cdot} + O_2 \longrightarrow ROO^{\cdot}$$

Peroxide radical attacks the polymer chain and generates a new radical site

$$M + ROO^{\cdot} \longrightarrow ROOH + M^{\cdot}$$

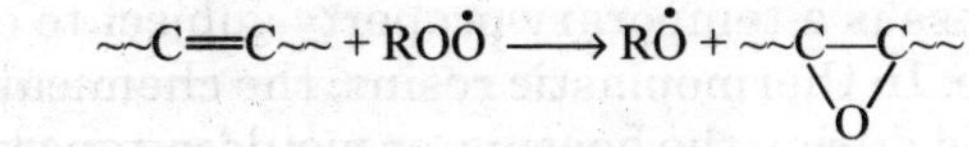

The hydroperoxide can form several free radical sites

$$ROOH \longrightarrow RO^{\cdot} + OH^{\cdot}$$
$$M^{\cdot} + ROOH \longrightarrow MOH + RO^{\cdot}$$
$$RO^{\cdot} + ROOH \longrightarrow ROH + ROO^{\cdot}$$

Recombination reactions of free radicals result in termination of the chain reactions

$$2ROO^{\cdot} \longrightarrow ROOR + O_2$$
$$2RO^{\cdot} \longrightarrow ROOR$$
$$M^{\cdot} + \dot{O}H \longrightarrow MOH$$
$$2M^{\cdot} \longrightarrow M\text{—}M \text{ etc.}$$

The polymers are protected against oxidative degradation by incorporating chemical compound called antioxidants.

(iv) **Mechanical degradation:** When a polymer material is subjected to a mechanical stress, the most vulnerable part of the structure gets ruptured, thus bringing about degradation in the polymer. This type of degradation is called mechanical degradation. For example when polytyrene dissolved in a suitable solvent is subjected to vigorous stirring, it undergoes considerable molecular degradation or fragmentation. In rubber industry rubber is converted into soft, supple mass from hard and tough material through mechanical degradation. The process is called mastication in which the rubber is passed through rotating rollers to decrease its molecular weight. The rubber so obtained is more processable.

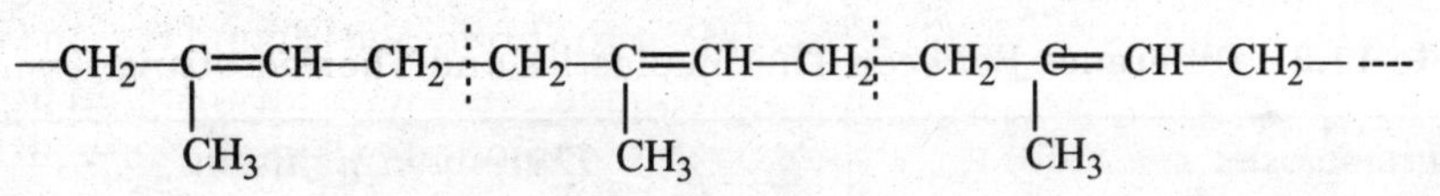

Polyisoprene

The CH_2—CH_2 links undergo fission during mastication because they are the weakest links.

11.7 PLASTICS

As was discussed earlier, plastics are a group of artificially synthesized materials, which can be moulded into any desired form, when subjected to heat and pressure in the presence of a catalyst. At some stage in their manufacture or fabrication, these materials exist in plastic condition and hence their name.

The resins are the basic binding materials which form the major part of the plastics and determine the type of treatment needed in the moulding operations. However, there is no sharp line of demarcation between plastics and resins, and the two terms are considered synonymous. From engineering point of view there are two types of resins-

(i) Thermoplastic resins.
(ii) Thermosetting resins.

Plastic materials or the plastic articles depending on the bases of the type of resins used in their preparation, are called thermoplastics or thermosetting resins/plastics.

(i) **Thermoplastics:** Thermoplastics or resins soften on heating and become plastic so that they can be converted to any desired shape by moulding. On cooling, they become hard and rigid but their hardness is a temporary property subject to change with increase or decrease in temperature. In thermoplastic resins, the chemical structure or the molecular weight are not changed during the heating or moulding operations, because the changes involved are of physical nature. Only the secondary bonds between the individual molecular chains are broken on heating which results in their softening. On cooling, these secondary bonds are again formed and hence they become hard again.

Some common examples of thermoplastics are polyethene, polypropylene, polystyrene, polyvinyl chloride, teflon, polyacrylonitrile, etc.

Plasticizer: Sometimes, the substances called plasticizers are added to improve plasticity and flexibility of the plastics. The plasticizing properties are believed to be mainly due to the neutralization of intermolecular forces of attraction in the resin molecules. A variety of organic materials are used as plasticizers including non-drying vegetable oils, tributyl phosphate, triphenyl phosphate, triacetin, camphor, esters of oleic or stearic acid.

(ii) **Thermosetting resins:** Thermosetting resins are those which upon heating, change irreversibly into hard and rigid materials on heating. After cooling, if the set article is again heated, it will not soften again. The thermosetting resins during moulding acquire three dimensional cross linked structures with predominantly covalent bonds, which retain their strength even on heating. Thus, the three dimensional network structure joined by strong covalent bonds, make the thermosetting resins harder, stronger and more brittle than thermoplastic resins. If heating of a thermoset resin is prolonged, these undergo change in chemical composition to give a hard, infusible and insoluble mass.

Some common examples include phenol formaldehyde, malamine formaldehyde, alkyds, epoxy resins etc.

Table 11.2 Difference Between Thermoplastics and Thermosetting Resins

	Thermoplastic	*Thermosetting Resin*
1.	They are generally long chain linear polymers.	They have three dimensional crosslinked net work structure joined by strong covalent bonds.
2.	On heating, they soften readily because secondary forces between the chain can be easily broken.	They do not soften on heating because cross links and strong covalent bonds retain their strength on heating. They can, however, he charred on strong heating.
3.	They can be softened, reshaped and even reused by reheating to a suitable temperature.	They can not be reshaped and reused because they do not soften on heating and retain their shape and structure.
4.	They are generally weak, soft and less brittle.	They are usually hard, strong and more brittle.
5.	They can be reclaimed from waste.	They can not be reclaimed from waste.
6.	They can dissolve in some organic solvents.	They are insoluble almost in all organic solvents because of the presence of cross linking and strong covalent bonds.

11.8 COMMERCIALLY IMPORTANT THERMOPLASTICS

1. **Polyethene:** Ethylene, one of the most important petrochemicals, may be polymerized by a number of methods to produce thermoplastic polymer, polyethene.

$$\underset{\text{ethene}}{n\,CH_2 = CH_2} \longrightarrow \underset{\text{Polyethene}}{\left(CH_2 - CH_2 \right)_n}$$

The starting material (monomer) is obtained either by dehydration of alcohol or by the cracking of petroleum.

There are two varieties of polyethylenes, viz., low density polyethylene (LDPE) and high density polyethylene (HDPE).

(a) LDPE: Low density polyethylene is prepared by polymerizing ethylene at high pressures (1000-5000 atm) at a temperature of 250° C in the presence of oxygen as a free radical initiator. LDPE consists of highly branched structure which does not allow the polymeric chains to pack close together and hence results in low density (0.91to 0.925 g/cm^3) of the polymer.

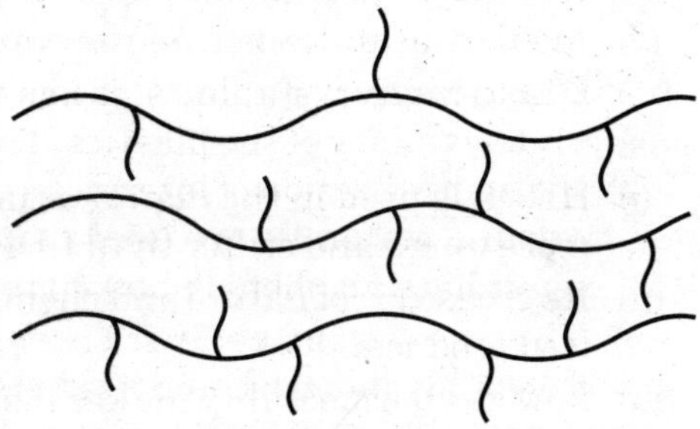

Fig. 11.4 Branched Structure of LDPE.

Properties:

(i) Low density polyethylene melts at 110–125° C and is only 40–50% crystalline.
(ii) It has low density (0.91 g/cm^3).
(iii) It is a transparent polymer having moderate tensile strength and high tonghness.
(iv) It is chemically inert, slightly flexible and poor electrical conductor.
(v) Its elasticity is preserved even at lower temperatures
(vi) It is moisture resistant.

Uses:

(i) LDPE films are mainly used for packing and wrapping frozen food, textile products etc because of their flexibility, chemical and moisture resistance.
(ii) The non-polar nature of the polymer makes it suitable for insulation for electric wires and cables.
(iii) Its inertness to chemicals and resistance to breakage makes it a useful material for the manufacture of 'squeeze bottles' and in many attractive containers.
(iv) It is also used for making pipes for agricultural, irrigation and cold water plumbing and chemical plants.

(b) **HDPE:** High density polyethylene is prepared by polymerizing ethylene at low pressure (1200-2000 atm) and at a temperature of 180–200° C in the presence of metal oxide catalysts. HDPE can also be produced by using Ziegler Natta catalyst by coordination polymerization. In this second method ethylene is heated to 60-200° C under a pressure of 6-7 atmospheres in the presence of a catalyst consisting of triethylaluminium and titanium tetrachloride (Ziegler-Natta catalyst)

$$n\,CH_2 = CH_2 \xrightarrow[\substack{6\text{–}7\text{ atm} \\ (C_2H_5)_3Al + TiCl_4}]{60 - 200°C,} \left(CH_2 - CH_2 \right)_n$$

The polyethene so produced consists of linear chains of polymer molecules which pack close together resulting in high density polyethene (0.97 g/cm^3).

Properties:

(i) High density polyethene has melting point in the range of 145–150° C and is 90% crystalline.
(ii) It possesses density in the range of 0.941 to 0.965 g/cm^3.
(iii) It is much stiffer than LDPE and has higher tensile strength and hardness.
(iv) It is chemical resistant and has significantly lower gas permeability.
(v) Insulating and moisture resistant properties are similar to LDPE.
(vi) Due to high crystallinity, it has low gas and water vapour permeability.

Uses:

(i) HDPE is used in the manufacture of toys, crates closures, caps, buckets etc because it is more tough and stiffer than LDPE.
(ii) Its grease proof nature and chemical inertness makes it suitable for wrapping material for food products.
(iii) Low gas permeability makes it suitable for domestic gas piping.
(iv) It is also used for manufacturing bottles for milk, household chemicals and drug packaging.
(v) It is used in the manufacture of overhead tanks for water storage.

Whenever high tensile strength and stiffness are required HDPE finds better use than LDPE.

2. **Polypropylene:** Polypropylene is probably the lightest known industrial polymer and can be prepared in isotactic, syndiotactic or atactic forms. The monomer propylene is obtained as a byproduct from gasoline refineries. Polypropylene (isotactic) is produced by polymerizing propylene in the presence of Ziegler-Natta catalyst [Al $(C_2H_5)_3$–$TiCl_4$)]

$$n\,\underset{CH_3}{\overset{|}{CH}}=CH_2 \xrightarrow{Al(C_2H_5)_3-TiCl_4} \left(-\underset{}{\overset{CH_3}{\overset{|}{CH}}}-CH_2- \right)_n$$

Properties:

(i) Polypropylene is stereoregular (isotactic) and highly crystalline polymer (slightly lower than HDPE).
(ii) Isotactic polypropylene (IPP) melts at 160-170° C.
(iii) It exhibits high stiffness, hardness and tensile strength.
(iv) Its moisture resistance is comparable to that of polyethylene.
(v) Polypropylene is resistant to many chemicals such as alkalies, acids and oils. But due to presence of methyl groups attached to alternate carbon atoms, it is susceptible to oxidation.
(vi) Its electrical properties are similar to polyethylene.
(vii) Films of polypropylene are impermeable to a number of vapours and gases.

Uses:

(i) The higher melting point of polypropylene allows it to be used in products that must be steam sterilized such as baby bottles, flasks etc.
(ii) Its fibres are extremely strong by weight and hence are used in making ropes and filaments.

(iii) Polypropylene films are used for packaging due to high clarity and their impermeable nature to a number of vapours and gases. It is also used for wrapping confectionary items (bread, biscuits etc) and potato chips.

(iv) Due to high strength, rigidity and temperature and chemical resistance it is used for the production of chemical process equipments such as pipes and tanks.

(v) It is used for the preparation of textile machinery parts (bobbins, spools, spindles etc).

(vi) It has excellent resistance to continued flexing and hence is used for joints in boxes, containers etc.

(vii) When blended with suitable elastomer it is widely used for making bumpers, radiator grills etc.

(viii) It is also used as an insulating material for electrical wires.

(ix) It is also used for the manufacture of drinking glasses, dinner wares etc.

3. **Polystyrene:** Polystyrene is also known as polyvinyl benzene. The monomer styrene is produced from ethylene and benzene. When ethylene is passed into benzene in the presence of $AlCl_3$ catalyst, ethyl benzene is formed, which on passing over iron oxide catalyst at high temperatures is converted into styrene.

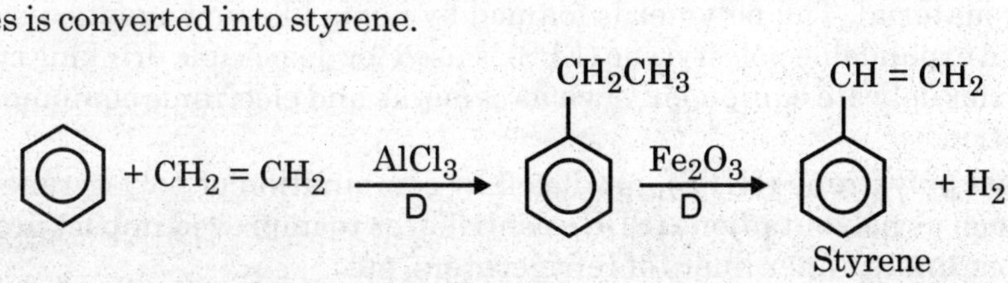

Polystyrene is prepared by polymerization of styrene in the presence of benzoyl peroxide as initiator.

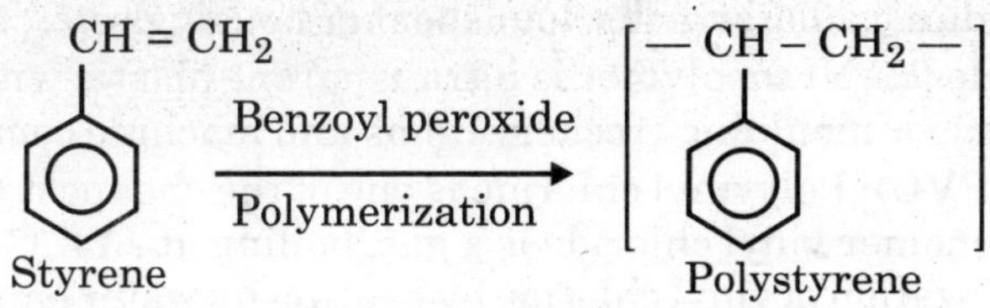

Properties:

(i) Polystyrene is transparent and has high refractive index (1.6) which gives it high brilliance.

(ii) It has got low heat distortion temperature (85° C) and hence articles made of polystyrene cannot be sterilized with steam.

(iii) It consists generally of linear chains and is chemically inert and resistant to most alkalies, oxidising and reducing agents and acids.

(iv) It is susceptible to photooxidative degradation, which results in brittleness and yellowing. Degradation is initiated by the phenyl groups, which absorb UV radiation from sunlight resulting in the formation of free radicals.

(v) It has got excellent moisture resistance.

(vi) It has good electrical insulating properties.

(vii) It can be dyed to different colours.

(viii) It has the unique property of transmitting light of all wavelengths through curved sections.

(ix) Polystyrene is attacked by a number of solvents.

Uses:

(i) Polystyrene is widely used in the manufacture of articles such as lids, jars, bottles, combs, brush handles, radio and television cabinets.

(ii) It is used for preparing synthetic ion exchange resins in combination with p-divinyl benzene. The resulting crosslinked polymer on sulphonation gives cation exchange resin.

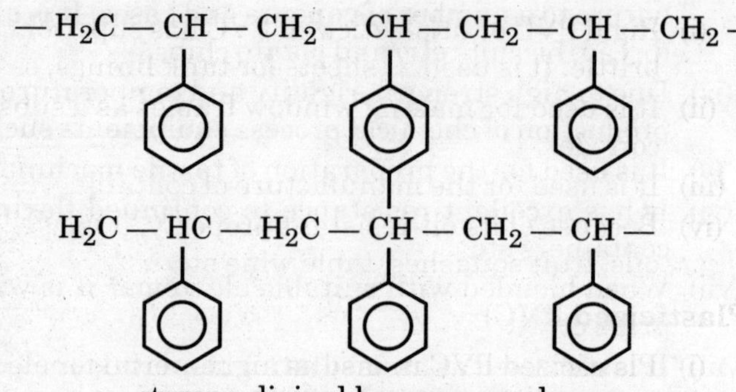

styrene-divinyl benzene copolymer

whereas, when styrene-divinyl benzene copolymer is treated with formaldehyde and HCl in presence $ZnCl_2$ as catalyst, it results in anion exchange resin.

(iii) An important use of polytyrene is the manufacture of foam and bead for insulation and packaging material. The polymer is foamed by using blowing agents, usually pentane. This foamed expandable polystyrene (EPS) is used for disposable drinking cups, cushioned packaging (for delicate equipment such as cameras and electronic equipments), and thermal insulation.

(iv) High-impact polystrene (HIPS), produced by combination of polystyrene with rubbery material such as polybutadiene (SBR), contributes to improved impact properties and is used for door liners, inner liners of refrigerators, etc.

(v) Polystyrene copolymerized with acrylonitrite and butadiene (ABS), due to its high impact resistance and high dimensional stability at elevated temperatures is used in the manufacture of door handles, radiator grills, loudspeaker housings etc.

(vi) Styreneacrylonitrile (SAN) copolymer is a transparent plastic with a very good impact strength and is used for moulding crockery items and machine components.

4. **Polyvinyl Chloride (PVC):** Polyvinyl chloride is one of the cheapest and most widely used plastic globally. The monomer vinyl chloride is a gas, boiling at –14° C. Polyvinyl chloride is prepared by the polymerization of vinyl chloride by heating its water emulsion in presence of a small amount of benzoyl peroxide.

$$\underset{\text{Vinyl chloride}}{CH_2=\overset{\overset{Cl}{|}}{C}H} \xrightarrow{\text{Polymerization}} \underset{\text{Polyvinyl chloride}}{\left[-CH_2-\overset{\overset{Cl}{|}}{C}H-\right]_n}$$

Polyvinyl chloride is hard and horny material owing to the high intermolecular attraction forces, but when plasticised with tricresyl phosphate it becomes very useful product. Thus, PVC is available in two general grades, viz, rigid and flexible.

Properties:

(i) It is colourless, odourless and non inflammable.

(ii) PVC is partially syndiotactic and have low crystallinity.

(iii) PVC is thermally not very stable and degrades above 200° C, with the evolution of HCl.

(iv) It is resistant to atmospheric oxygen, inorganic acids and alkalies.

(v) It has greater abrasion and flex resistance compared to leather.

(vi) It is water proof and washable.

(vii) It is less brittle and lighter when compared to glass.

Uses: (Rigid PVC)

(i) Rigid PVC or unplasticized PVC has superior chemical resistance and high rigidity but is brittle. It is used as sheets for tank linings, safety helmets, mudguards.
(ii) It is used for making window frames as a substitute for wood because the frames do not corrode.
(iii) It is used for the manufacture of containers for detergents, cosmetics, mineral water.
(iv) Because of its oil-resistant property, it is used for storing consumable liquids like edible oils, fruit squashes, table wine etc.

(Plasticized PVC)

(i) Plasticized PVC is used as an insulating material.
(ii) Because it is washable and has greater abrasion resistance therefore it is used for making ladies-handbags, bathroom curtains, kitchen upholstery.
(iii) Used for upholstery for seats in public transport vehicles.
(iv) In the film form it is used in membrane separation technology.

Properties of PVC can be modified through chemical modification or copolymerization. For example chlorinated PVC (CPVC) increases the heat-distortion temperature of PVC which can be used for the manufacture of hot water pipe.

5. Polyvinylacetate (PVA): Polyvinyl acetate is produced by heating vinyl acetate in the presence of small amount of benzoyl peroxide.

$$\underset{\text{Vinyl acetate}}{n\,CH_2 = CH - COOCH_3} \xrightarrow{\text{Polymerization}} \underset{\text{Polyvinyl acetate (PVA)}}{\left(CH_2 - \overset{\overset{\large COOCH_3}{|}}{CH} \right)_n}$$

Properties

(i) Polyvinyl acetate is a solid, clear resin, and has no taste or odour and is soft sticky material.
(ii) It is insoluble in water and is saponified with great difficulty and hence is not absorbed by the digestive system. Moreover if at all saponification takes place, the products polyvinyl alcohol and acetic acid are not harmful. Due to the above stated reasons, polyvinyl acetate is used as a basic material for chewing gums and surgical dressings.
(iii) It is resistant to atmospheric oxygen, water and chemicals.

Uses

(i) Polyvinyl acetate is used for the manufacture of lacquers, paints and adhesives. Polyvinyl acetate adhesives are used for bonding of textile to textile, leather to leather, paper to paper, paper to textile, leather to wood etc.
(ii) It is used as a basic material for chewing gums and surgical dressings.
(iii) It is also used for finishing of textile fabrics.
(iv) On copolymerization with polyvinyl chloride, it is converted into a polymer having varying degree of softness. For example, vinylite which is a copolymer of polyvinyl acetate and polyvinyl chloride, is especially useful as a surface coating material for metals.
(v) It is mainly used for the manufacture of polyvinyl alcohol.

$$\underset{\text{Polyvinyl acetal}}{— CH_2 - \overset{\overset{COOCH_3}{|}}{CH} - CH_2 - \overset{\overset{COOCH_3}{|}}{CH} —} \xrightarrow{\text{Hydrolysis}} \underset{\text{Polyvinyl alcohol}}{— CH_2 - \overset{\overset{OH}{|}}{CH} - CH_2 - \overset{\overset{OH}{|}}{CH} —}$$

Polyvinyl alchohol is used in the preparation of polyvinyl butyral on condensation of polyvinyl alcohol with butyraldehyde.

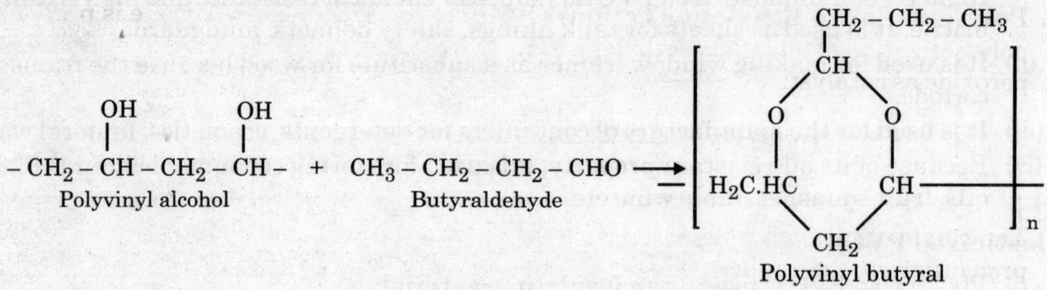

Polyvinyl butyral is tough and elastic and is used for laminating safety glass. It holds the broken pieces of glass in case of accident and thus minimizes the danger from flying glass fragments.

5. **Polymethylmethacrylate (PMMA):** Polymethyl methacrylate (PMMA) is one of the important thermoplastic from commercial point of view, because of its outstanding optical properties. It is prepared by the polymerization of methylmethacrylate in presence of free radical initiator such as peroxides

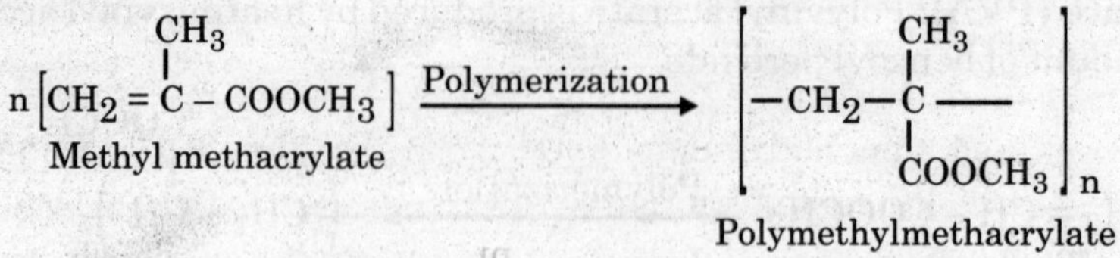

Properties

(i) It is a colourless transparent plastic with an excellent outdoor life period and good strength.
(ii) Owing to the presence of bulky side groups, it is amorphous in nature.
(iii) It can be depolymerized to yield back the monomer.
(iv) It has high softening point of about 130-140° C but it becomes rubber-like at a temperature above 65° C.
(v) The main feature of this plastic is its high optical transparency and capacity of transmitting light accurately, even in curved sections.
(vi) It has good mechanical properties but has poor scratch resistance as compared to glass.
(vii) It has low chemical resistance to hot acids and alkalies.
(viii) It has excellent weathering properties.

Uses

(i) PMMA is the nearest alternative to glass as it does not shatter and can be moulded into any shape. Infact, PMMA sheets known as plexiglass or lucite are used as an alternative to glass.
(ii) PMMA finds major use in automotive industry (e.g. housing for rear lamps, profiles, light fittings, ceiling lighting signal-light etc).
(iii) It is used for making hard contact lenses due to its high transparency.
(iv) It is used as composite materials for kitchen sinks, basins and bathtubs because it provides wide range of colour and is cheap and light.
(v) PMMA is also used for making wind screens T. V. screens etc. The fragments formed after cracking are less sharp and hence less harmful.

(vi) It is also used for decorative purposes such as in jewellery etc.

(vii) It is also used in bone splints, artificial eyes, dentures etc.

7. **Polytetraflouro Ethylene (Teflon) (PTFE):** Polytetra flouroethylene is prepared by the polymerization of tetraflouroethylene ($CF_2 = CF_2$), under high pressure in presence of benzoyl peroxide as catalyst.

$$n\,CF_2 = CF_2 \xrightarrow{\text{Polymerization}} \left(CF_2 - CF_2 \right)_n$$

Polytetraflouroethylene (PTFE)

The polymerization is associated with the release of large quantities of heat and if sufficient precautions are not taken, this can result in violent explosions.

Properties

(i) PTFE has a highly regular structure and is highly crystalline.

(ii) Due to the presence of highly electronegative flourine atoms and regular configuration of PTFE molecule there are very strong attractive forces between different chains, which give the material extreme toughness.

(iii) PTFE has high softening point of about 327° C. Above 400° C it decomposes without changing into vapour.

(iv) It is extremely resistant to attack by chemicals such as strong alkalies and acids. Only hot alkali metals and hot flourine can attack PTFE.

(v) PTFE is highly dense polymer (2.1 to 2.3 g/cm^3), having excellent mechanical properties.

(vi) It can be machined, punched and drilled.

(vii) It has a very low coefficient of friction.

(viii) It has extremely good electrical insulating properties.

(ix) It is slippery and waxy to touch.

(x) It does not melt like other thermoplastics and hence behaves somewhat like a thermosetting polymer.

(xi) It has non-adhesive characteristics.

Uses

(i) Due to its chemical inertness, it is used in chemical carrying pipes, tubings, stop-cock for burettes.

(ii) Its extreme inertness and stability to temperature makes it a suitable material for making gaskets and seals.

(iii) It is used as an insulating material for transformers, cables, wires, fittings etc.

(iv) Because of its low coefficient of friction, it is used in non-lubricating bearings.

(v) Its most important application is, as non-stick coating for pots and pans. PTFE is applied as a coating over the rollers and mixing equipments used for preparing sticky materials such as adhesives and jams.

(vi) Its fibre is used to form belts, filter cloth etc.

8. **Polycarbonates (PC):** The polycarbonates are amorphous polymers with interesting engineering properties. Polycarbonates are the polyesters of phenol and carbonic acid and are prepared by the condensation of diphenoxy methylene derivatives with diphenyl carbonate. Bis phenol-A [2, 2–bis (4–hydroxyphenyl) propane] on interaction with diphenyl carbonate, gives a polycarbonate as follows:

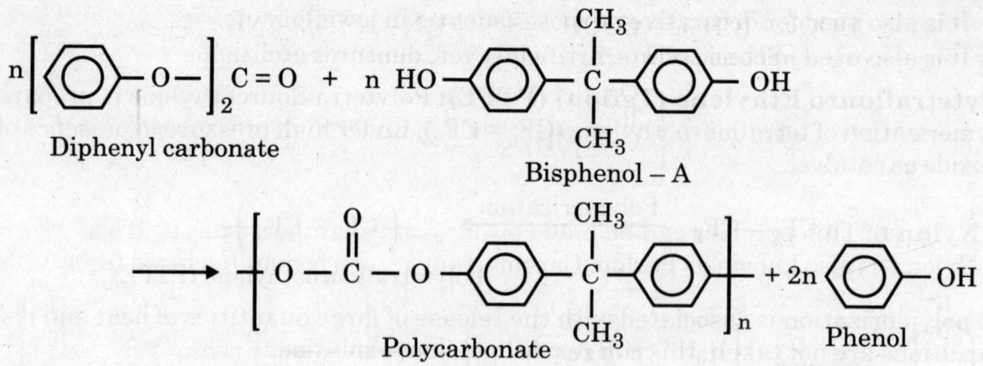

Properties

(i) Polycarbonate is an amorphous polymer.
(ii) It has very high impact strength.
(iii) It is resistant to water.
(iv) It has high light transmittance.
(v) Polycarbonate melts at around 265° C.
(vi) It has good dimensional stability.
(vii) It has good electrical insulation characteristics.

Uses

(i) Polycarbonate, due to its electrical properties and dimensional stability is used for mouldings for computers, calculating machines, for moulding compact disks etc.
(ii) It is used for the manufacture of hair drier, and electric razor housing.
(iii) As it is not stained by the usual domestic beverages and fruit juices, it is used for making food processor bowls, coffee maker and food mixer housings.
(iv) Due to its high temperature stability and toughness it is used for microwave cookware and transparent baby bottles.
(v) It is also used in cameras (lens holders, shutter assembly etc) safety shields, and for the manufacture of unbreakable caps of pens.

(a) **Polyamides:** Polyamides are a group of polymers which contain the amide (–CONH–) linkage in the main polymer chain. Polymers of this type are the synthetic linear aliphatic polyamides which are capable of fibre formation.

Nylon is the generic name for synthetic polyamides capable of forming fibres. Polyamides (nylons) are prepared by the polycondensation between dicarboxylic acids and diamines. Nylons are designated by two numbers, the first representing the diamine. Polyamides of commercial importance are Nylon 6, 6; Nylon 6, 10, Nylon 6 etc.

(i) **Nylon 6,6:** Nylon 6,6 is prepared by the polycondensation of hexamethylene diamine with adipic acid.

$$\underset{\text{Adipic acid}}{HOOC-(CH_2)_4-COOH} + \underset{\text{Hexamethylene diamine}}{H_2N-(CH_2)_6-NH_2} \longrightarrow$$

$$-\overset{\overset{O}{\|}}{C}-(CH_2)_4-\overset{\overset{O}{\|}}{C}-NH-(CH_2)_6-NH-$$

(ii) **Nylon 6, 10:** Nylon 6, 10 is prepared by the condensation between hexamethylene diamine and sebacic acid. Nylon 6, 10 is generally not used as a fibre, but it is suitable for monofilaments.

$$\underset{\text{Sebacic acid}}{HOOC-(CH_2)_8-COOH} + \underset{\text{Hexamethylene diamine}}{H_2N-(CH_2)_6-NH_2} \longrightarrow$$

$$\underset{\text{Nylon 6, 10}}{\left[\overset{O}{\overset{\|}{C}}-(CH_2)_8-\overset{O}{\overset{\|}{C}}-NH-(CH_2)_6-NH \right]_n}$$

(iii) **Nylon 6:** This polymer can be made either from ε-aminocaproic acid or ε-caprolactam. Nylon 6 is also known as Perlon. Caprolactam inturn is manufactured from cyclohexane.

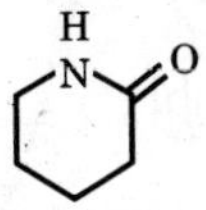

$$\xrightarrow{H_2O,\ D} \left[H_3N^+-(CH_2)_5-COO^-\right] \xrightarrow[\text{Polymerization}]{D} \underset{\text{Nylon}-6}{\left[-NH-(CH_2)_5-\overset{O}{\overset{\|}{C}}-\right]_n}$$

(iv) **Nylon-11:** It is prepared by self condensation of ω-amino undecanoic acid.

$$\underset{-\text{ Aminoundecanoic acid}}{n\left[H_2N-(CH_2)_{10}-COOH\right]} \xrightarrow{\text{Self condensation}} \underset{\text{Nylon-11}}{\left[HN-(CH_2)_{10}-\overset{O}{\overset{\|}{C}}\right]_n}$$

Properties of Nylons

(i) Nylons are characterized by a combination of high strength, elasticity, tonghness and abrasion resistance.
(ii) They have very good moisture resistance.
(iii) They are insoluble in common solvents.
(iv) The molecular chains are held together to each other by hydrogen bonds having linear structure, permitting side-by-side alignment.
(v) They are very flexible and retain their original shape after use.
(vi) Nylons provide excellent resistance to wear and abrasion.
(vii) Due to the presence of hydrogen bonding, all nylons are water sensitive. Water acts as plasticizer, which reduces tensile strength and dimensional stability.

Uses

(i) Nylon 6, 6 is used as a plastic as well as a fibre. Its most important fibre applications include automobile tire cords, ropes, threads, cords having high tenacity and good elasticity.
(ii) Nylon 6,6 is also used to make textile fibres for use in dresses, undergarments, socks etc.
(iii) Nylon 6,6 being a tough plastic, is used as a substitute for metals in gears and bearings etc. Nylon 6,6 can be moulded in one piece whereas similar item in metal require assembling of several parts. Moreover, nylon bearings and gears perform quietly and need little or no lubrication.
(iv) It is also used in making rollers, slides and door latches.
(v) Nylon 6, 10 is not used much as synthetic fibre, but is used to manufacture articles like brushes and bristles.
(vi) Nylon 11, is used as textile fibre.
(vii) Due to its good barrier properties, nylon-6 film is used for packaging some oxygen sensitive food.
(viii) Due to its high tenacity, nylon is suitably used for parachute fabric.
(ix) Nylons are also used in making transmission belting, ribbons for electric-typewriters, spindle bands on textile spinning machines etc because of its high durability.

(x) Nylon is also used for making insect screens because of their biological resistance.

Aromatic polyamides, often called aramids have been produced to improve the heat and flammability resistance of nylons. These polymers have very high melting points (greater than 500° C). One such nylon is Kevlar. It is prepared by polycondensation between aromatic dichloride and aromatic diamines.

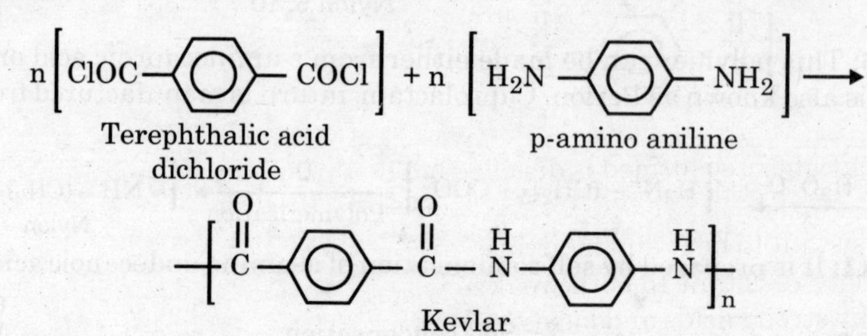

Kevlar is exceptionally strong and decomposes only above 500° C. It is 5 times stronger than steel and it is used as a substitute for steel in belted radial tires as well as in bullet resistant vests. It is also used in making car parts such as tyres, brackes, clutch linings etc.

11.9 COMMERCIALLY IMPORTANT THERMOSETTING RESINS

1. **Polyesters:** Polyester resins are the condensation products of dicarboxylic acid with dihydroxy alcohols. The industrially most important polyster is poly (ethylene terephthalate) (PET), with the trade name Terylene or Dacron. PET can be prepared from terephthalic acid and ethylene glycol, either by direct esterification or by catalyzed ester inter change.

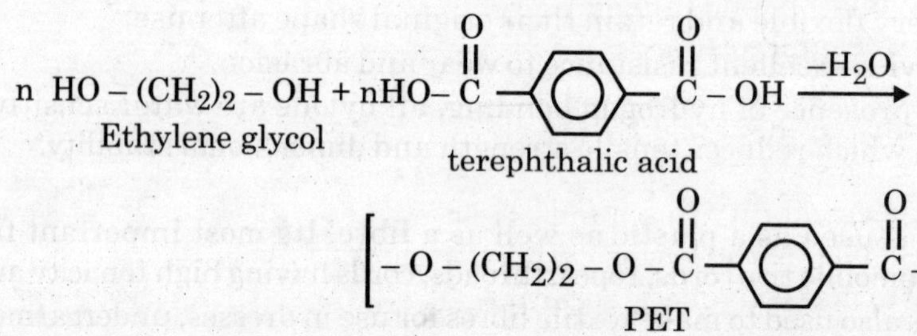

Now a days, dimethyl terephthalate is used instead of terephthalic acid. This is because of the fact that terephthalic acid is nonvolatile and is very slightly soluble in most of the solvents. The reaction takes place in two successive stages:

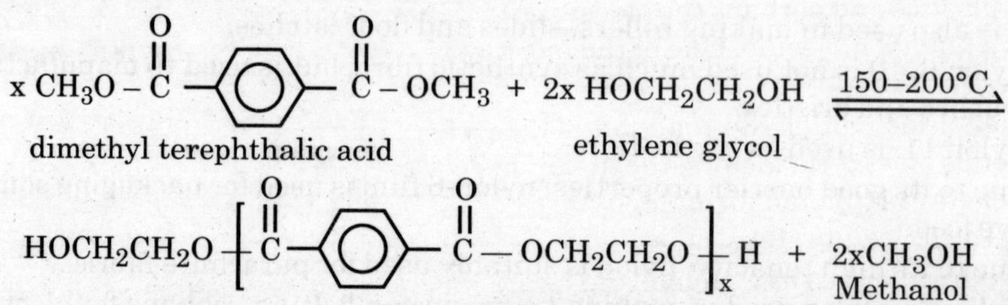

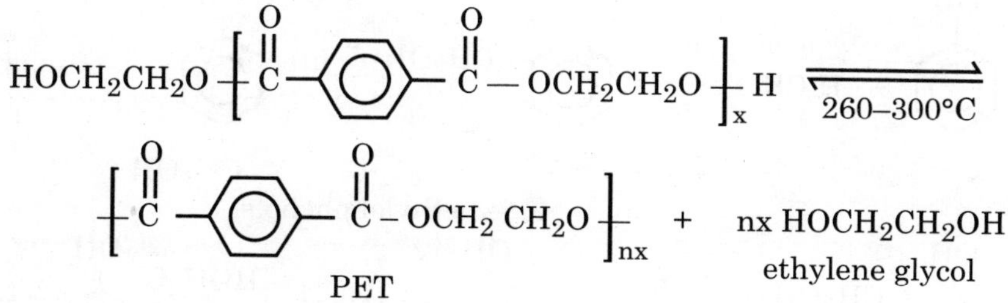

The enthylene glycol so formed is distilled off at such high temperature.

Properties

(i) PET is resistant to heat and moisture.

(ii) It is strongly resistant to both weak acids and bases at room temperature but readily attacked by strong acids and bases.

(iii) Presence of numerous polar groups results in powerful dipole dipole attractions in it.

(iv) It has good mechanical strength upto 175° C.

(v) It is resistant to almost all kinds of insect attacks.

(vi) It can be dyed at 100° C or in presence of a carrier.

(vii) The fibre has good wrinkle resistance due to good elasticity.

(viii) PET fibres are abrasion and oxidation resistant.

Uses:

(i) PET is mostly used for making synthetic fibres like Terylene, Dacron etc. Garments made from its fibres resist the formation of wrinkles and have many desirable properties such as warmth to touch, low moisture absorption (and hence dry rapidly), crease resistance, shape retention, and mildew proof. Moreover, these may be blended with other fibres, such as wool and cotton.

(ii) PET is also useful for making films because of its high strength, resistance to tearing durability, transparency etc. PET is used in making the transparencies of overhead projector.

(iii) Due to its impermeability to a number of organic and inorganic gases it is used in the manufacture of plastic bottles for carbonated beverages (barrier to CO_2 and water)

(iv) It is also used to make magnetic recording tapes.

(v) Due to their high dielectric strength they are used for electrical insulation.

(vi) Glass reinforced PET is used for the manufacture of housings for coffee machines, car heaters, toasters.

2. Phenolic Resins: Phenol formaldehyde resins are the most important and extensively used synthetic plastics and perhaps the oldest synthetic plastics. In these resins one of the monomers is always a phenol molecule. One of the most important member of this class is *phenol formaldehyde resin.*

The phenol formaldehyde reaction is catalyzed by alkalies and acids. The nature of the product depends on several factors, the major one being the nature of the catalyst and the proportion of the reactants (phenol formaldehyde ratio). The formation of phenol formaldehyde resin comprises of the following steps:

(i) **Methylolation:** When phenol and formaldehyde react together, the first step is the entry of methylol (CH_2 OH) groups in ortho and para positions to the hydroxy group. The reaction takes place in the presence of acid or alkali. Depending upon phenol formaldehyde ratio, various phenol-alcohols may be formed.

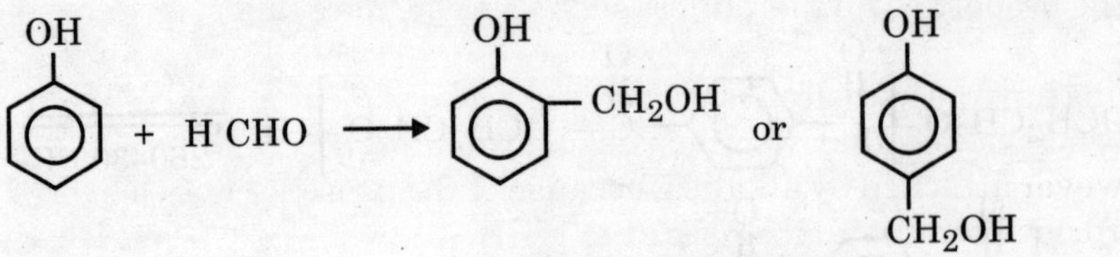

Monomethylol phenols

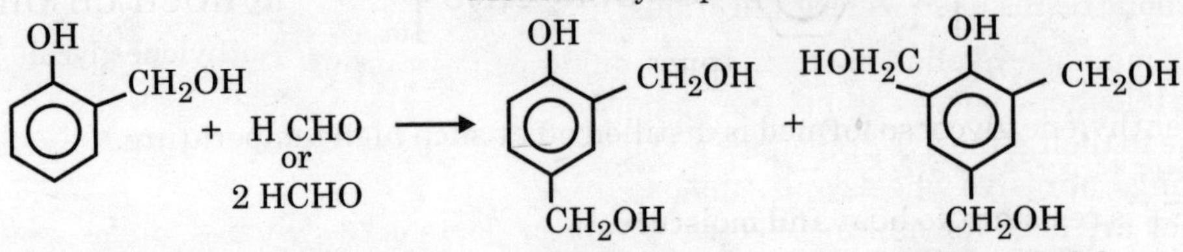

Dimetheylol phenols   Trimethylol phenols

(ii) **Novolac formation:** Depending upon the ratio of phenal formaldehyde (P/F) different resins, namely novolac and resol resins are obtained. In the presence of acid catalyst, when the P/F ratio is greater than unity, the methylol derivatives condense with phenol to form a linear polymer with little methylol groups. The product is thermoplastic in nature and is known as novolac.

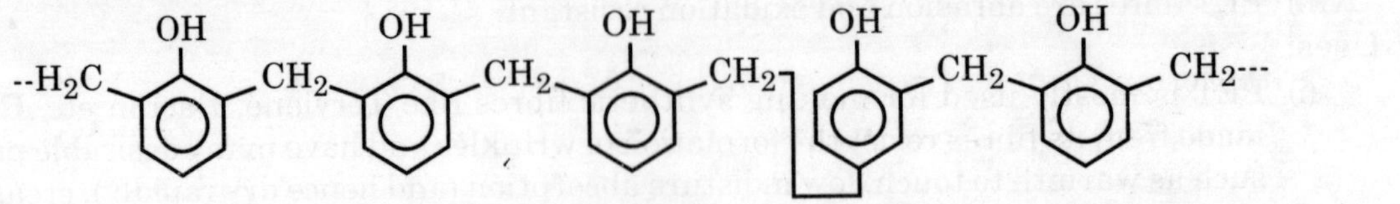

Novolac

(iii) **Resole formation:** In the presence of alkaline catalysts with P/F ratio less than 1, the methylol phenols condense to form linear structure called resoles. The resoles are soluble in the reaction mixture and have excess of methylol groups capable of further reaction during continued heating. Resols, therefore have limited shelf-life. However, the reaction may be stopped at the desired step by cooling.

(iv) **Resite formation:** On further heating, resole is transformed into a three dimensional polymer, called resite, due to the presence of active groups and mobile hydrogens Resite is an infusible and insoluble material.

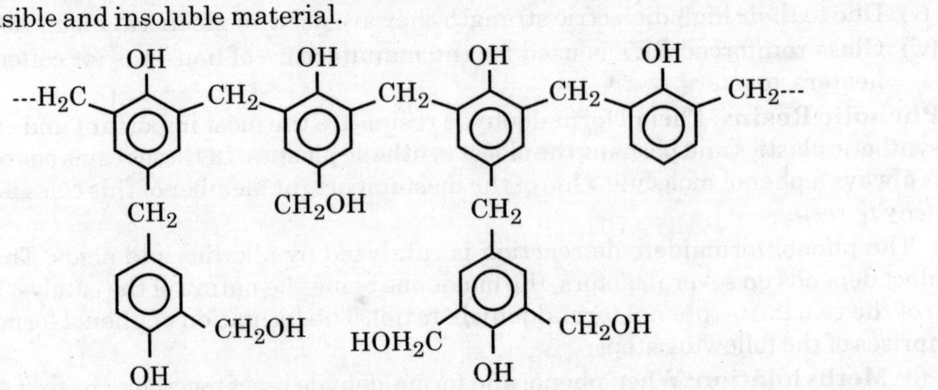

Resite

Properties

(i) Phenol formaldehyde moulding resins have excellent heat resistance.

(ii) These have high dimensional stability.

(iii) To improve the properties further, fillers such as wood, flour, mica, asbestos etc are added to them.
(iv) Phenolic resins have good dielectric properties.
(v) These have remarkable adhesive properties and bonding strength.
(vi) They are however attacked by alkalies, because of the presence of free hydroxy groups.
(vii) They are hard, rigid and scratch-resistant.

Uses Phenolic resins have wide applications.

(i) The main use of phenolic resin is in moulding applications. They are widely used in making telephone parts, cabinets for radio, television and automobile parts
(ii) Due to their adhesive properties and bonding strength, they are used for producing brake linings, abrasive wheels and sand paper.
(iii) They are used for making electric insulator parts like switches, plugs, switch boards, heater-handles etc.
(iv) Phenolic resins are used for impregnating paper, wood and other fillers.
(v) These are also used in varnishes, paints and protective coatings.
(vi) These are used in the production of ion exchange resin for water softening.

3. **Amino Resins:** Amino resins or amino plasts are prepared by the reactions of formaldehyde with nitrogen compounds such as urea, melamine and other amino compounds. Because of their attractive light colours, these resins have great commercial importance.

(a) **Urea formaldehyde resins:** This is prepared by the condensation reaction between urea and formaldehyde in neutral or alkaline conditions. The first products formed during the formation of resin are monomethylol and dimethyol ureas.

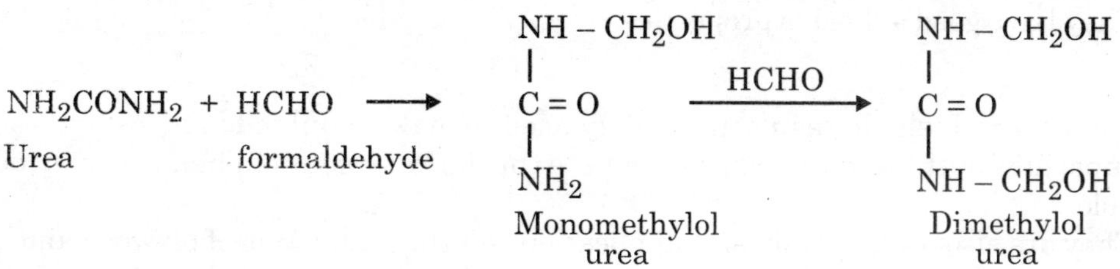

(i) Polymerization can take place from mono or dimethylol urea, or possibly through both, with the formation of long chains.

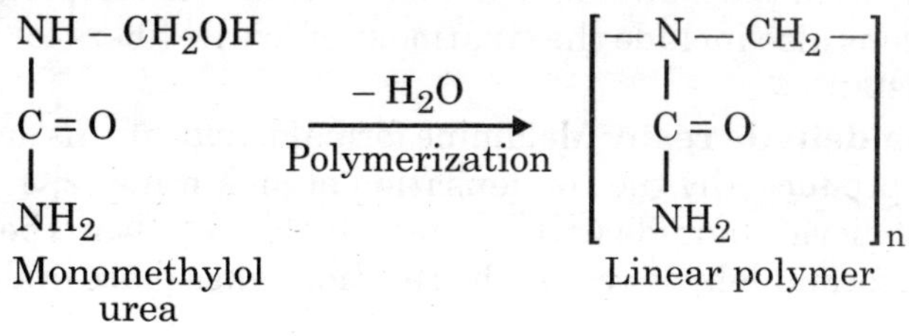

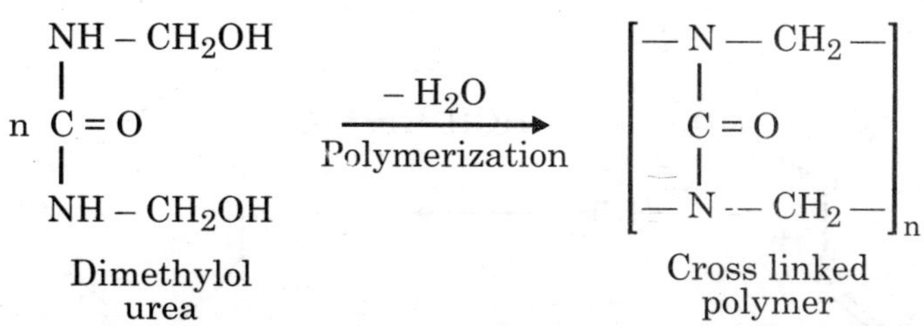

(ii) Thermosetting of the resin takes place during the moulding process. For moulding, the methylol derivatives are compounded with fillers, plasticizers, pigments etc and then are subjected to heat and pressure, resulting into a hard, infusible product. A fully cross linked urea-formaldehyde resin can be represented as

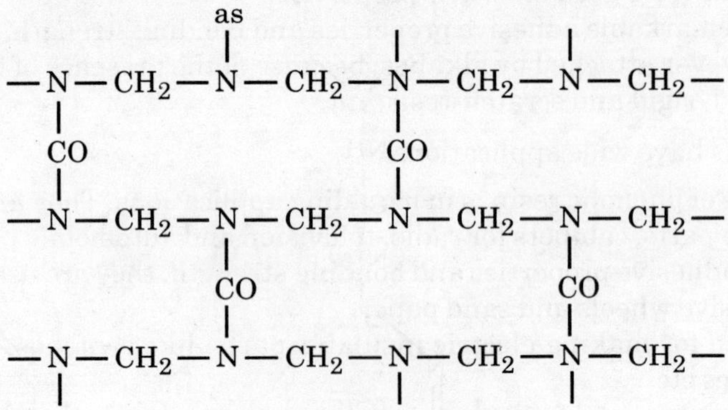

Urea-formaldehyde resin
(Cross linked polymer)

Properties

(i) The urea formaldehyde resins are light in colour and enable the production of light or pastel coloured objects.
(ii) Their heat and moisture resistance are lower than those of phenolic resins.
(iii) They have good electrical insulating properties.
(iv) They are resistant to oxidation to oil, grease and weak acids.
(v) They are hard, resist abrasion and scratching.
(vi) They have good adhesive properties.

Uses

(i) Urea-formaldehyde resins are widely used to make moulded articles like vacuum flask cups, jugs buttons, bottle caps etc where the darker colour of phenolics may be objectionable.
(ii) They are also used for adhesive applications for the production of plywood and laminating.
(iii) They are used for the manufacture of coated abrasive paper and binder for foundry cores.
(iv) They are used for the manufacture of cation exchange resins.
(v) These also find use in the manufacture of electrical switches, plugs and insulating foams.
(vi) Their applications also include the treatment of textile fibers for improving their shrink and crease resistance.

(b) **Melamine-formaldehyde resin:** Melamine formaldehyde resins are another class of amino resins. They are produced by the condensation of melamine (2, 4, 6 triamino - 1, 3, 5 triazine), a trimer of dicyandiamide with formaldehyde, exactly in the same manner as in the formation of urea formaldehyde resin. The reaction takes place under mild alkaline or neutral conditions.

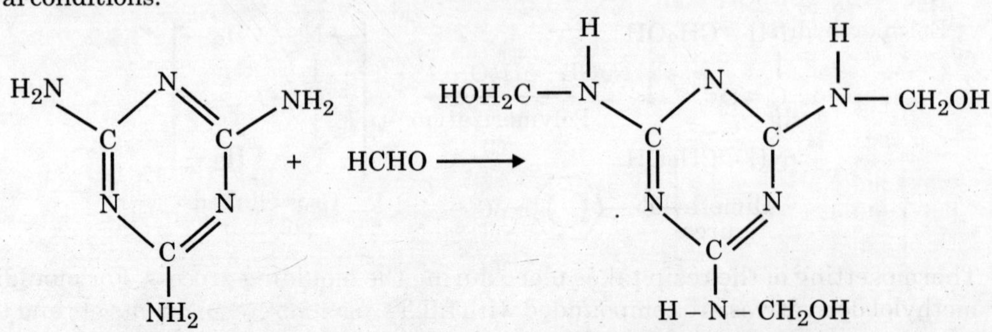

Melamine Trimethylol melamine

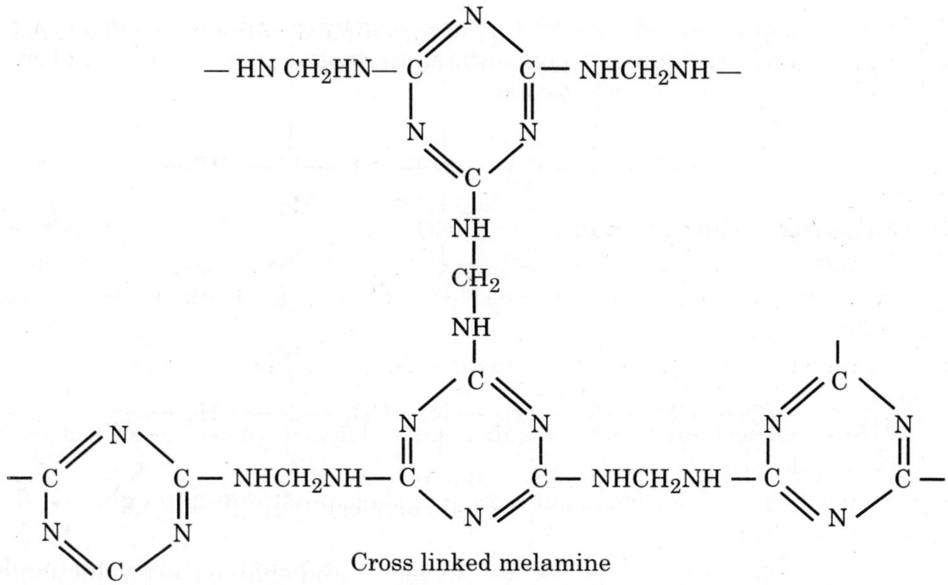

Cross linked melamine

Properties:

(i) The melamine resins have better hardness, heat resistance, moisture resistance than the urea-formaldehyde resins. However these are expensive. Rest of the properties are similar to urea-formaldehyde resins.

Uses

(i) Typical applications of urea formaldehyde resins include the production of decorative plastic ware, laminated worktops.

(ii) They are used in electrical fittings.

(iii) They are used for resin bonded grinding wheels.

(iv) They are used for textile finishing to provide crease resistance, stiffness, water repellency etc.

(v) They are also used in automotive finishes.

(vi) Melamine resin is also used as a tanning material for leather. The presence of small amount of this resin in the tanning material improves the strength of the leather.

(vii) They are also used for the preparation of lacquers.

4. Epoxy Resins: Epoxy resins are so called because their starting material is epichlorohydrin and the polymer before crosslinking contains epoxide groups.

The epoxy resins are formed by the reaction of epichlorohydrin and bisphenol A.

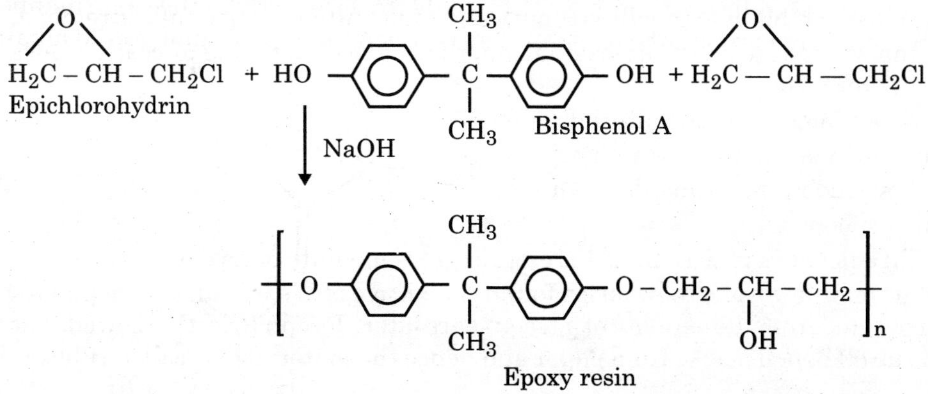

Epoxy resin

The linear epoxy resins are converted into thermosets by reacting with many types of materials including polyamines, polyamides, poly-sufides etc. to give three dimensional crosslinked structures. This is called curing of epoxy resins.

Properties

(i) Epoxy resins have remarkable chemical resistance due to the presence of very stable ether linkages.
(ii) They are tough and heat resistant.
(iii) The polar nature of the molecules produces excellent adhesion quality in them.
(iv) They are highly flexible because the reactive groups (epoxide and hydroxyl groups) are comparatively widely-spaced.
(v) They have good electrical properties and dimensional stability.

Uses

(i) Epoxy resins are excellent structural adhesives and hence are used as adhesives (araldite) for glass, metals etc.
(ii) They are used as surface coatings because of their outstanding toughness, flexibility, adhesion and chemical resistance.
(iii) Their electronic applications include their use as encapsulating material for electronic components.
(iv) Because of their toughness, epoxy resins are used to provide skid-resistant surfaces for highways.
(v) Epoxy resins are applied over cotton, rayon and bleached fabrics to impart crease resistance and shrinkage controls.
(vi) The structural application of epoxy resin is the production of glass fibre reinforced articles with better mechanical strength.

11.10 RUBBERS (ELASTOMERS)

It was discussed in the beginning of the chapter that depending on the ultimate form and use, a polymer can be classified into two broad categories: Plastics and Elastomers (rubbers).

Elastomers are the polymers which are capable of being stretched rapidly at least 150 precent of their original length without breaking and return to their original shape on release of stress. For example, rubber band is a typical elastomeric material. When stretched, it is elongated to many times its length. But when the force is released it gets back to its original length.

The properties of typical elastomers are defined by the following:

- They must stretch rapidly and considerably when stretched, reaching high elongations (500-1000%).
- They must exhibit high tensile strength and high stiffness when fully stretched.
- They must snap back to their original length once the stretching force is withdrawn or the stress is released.

These criteria are met with the materials which:

- have high molecular mass (high polymers).
- are above their glass transition temperature.
- are amorphous.
- contain a network of crosslinks to restrain gross mobility of its chains.

Columbus on this voyage to America found the native of West Indies were playing with the solid mass collected from the exudate of a tree. Years later, Joseph Priestly showed that this solid mass could rub off pencil marks from paper and hence the material was called rubber. The main source of natural rubber is the species of tree known as 'Hevea Basiliensis' even today, which account for multi billion dollar rubber industry.

Rubber or elastomers can be obtained naturally or synthetically. Consequently we have natural rubber and synthetic rubber.

NATURAL RUBBER

Natural rubber of nearly same characteristics occur in the inner bark and to a lesser extent in the leaves and roots of a number of tropical plants. But the latex from the tree 'Hevea Basiliensis' is proved to be the main source of rubber. Latex is an emulsion of poly hydrocarbon droplets in an aqueous solution and looks like milk. The negatively charged particles of polyhydrocarbons, a colloidal dispersion, stabilizes the emulsion.

Processing

The latex is collected by making small incisions in the trees in such a manner so as to allow the latex to accumulate in small cups. The percentage of rubber in it is 25-35%. A fully grown tree gives about 5 gms of rubber in a day.

The average composition of the latex is;

Water = 60%; Rubber = 35%; Proteins, enzymes and nucleic acid = 3%; Fatty acids and esters = 1% and inorganic salts = 0.5%.

In these trees, the rubber is formed from polymerization of isoprene i.e (2-methyl-1,3-butadiene) to produce poly-(cis)isoprene by a biochemical reaction in which a particular type of enzyme acts as a catalyst.

$$CH_2 = \underset{\underset{CH_3}{|}}{C} - CH = CH_2$$

Isoprene (2 – methyl – 1, 3 -- butadiene)

The latex is first sieved to remove the impurities such as leaves, bark etc. From 25-35% dry rubber content the latex is diluted to contain about 15-20% of rubber content. The rubber is then separated from the latex by the process of coagulation. The acetic acid or formic acid is added as coagulants and the rubber is coagulated to soft white mass. About 1 g of acetic acid/formic acid is added per 200 kg of rubber material. This coagulated mass is the crude rubber and consists of 90-95% rubber, 2-4% proteins and 1-2% resins. The crude rubber is wached and treated for producing the following types of rubber depending upon the process.

1. **Crepe Rubber:** Crepe rubber is prepared by adding a small amount of sodium bisulphite ($NaHSO_3$) to the coagulated mass to bleach the rubber. Then the rubber is passed through a creping machine. The creping machine is rolling machine which consists of two rollers having longitudinal grooves. Thus, after passing through it, the spongy rubber is converted into a sheet having a rough surface which resembles the crepe paper and hence the name.
2. **Smoked Rubber:** The smoked rubber is produced by carrying out the coagulation of latex in the tanks provided with vertical grooves on sides and fitted with the metal plates which run across the width of the tank. First the diluted latex is poured into the tanks, with the metal plates removed. Then the coagulant is added (acetic or formic acid) and the plates are inserted. The tanks are kept undisturbed for 15-16 hours. The tough slabs are formed. These slabs are passed through a series of rolling machines and finally rubber sheets having ribbed pattern are obtained. The ribbed surface prevents the sticking of the rubber with each other. These sheets are then dried in smoke houses at about 50°C in the

smoke obtained from burning wood or coconut shells. The smoked rubber so obtained is transluscent and ambre coloured.

GUTTA PERCHA

Gutta percha is another variety of natural rubber. It is obtained from the mature leaves of dichopsis gutta and palagum gutta trees. The hevea rubber and gutta percha, both are composed of isoprene units with structures shown in the figure? But **natural rubber** is *cis-isomer* where a bending back of successive isoprene units take place giving the molecule a coiled structure. The **gutta percha**, on the other hand is a *trans-isomer* which consist of straight chains, giving a rod like structure to the molecule. The rod-like structure results into close packing of adjacent chains. Thus gutta percha is more crystalline.

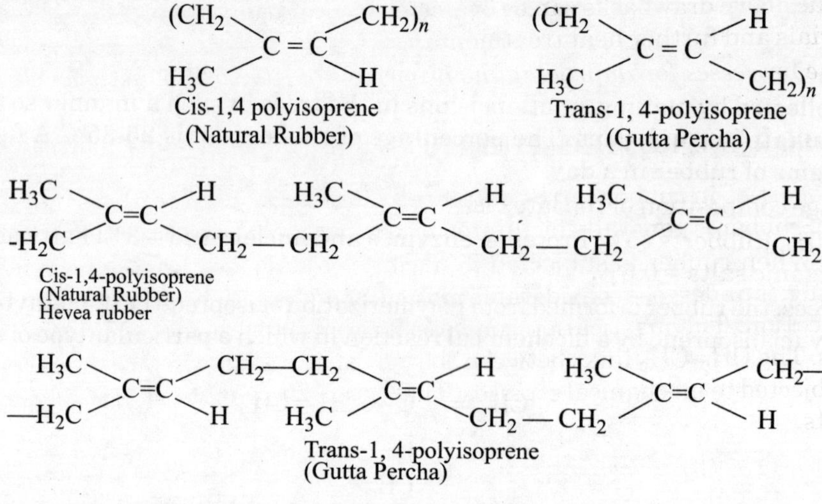

Fig. 11.5 Configuration of isoprene units in natural rubber and gutta-percha.

Processing of Gutta Percha

The mature leaves obtained from the trees Dichopsis gutta or palagum gutta are carefully ground in mills and are then treated with water maintained at 70ºC for about half an hour when the gutta percha is extracted in water. The solution is dropped into cold water and gutta percha floats on the surface. Alternatively, gutta percha can be recovered by solvent extraction.

Properties of natural (Hevea) rubber and gutta percha:

	Natural Rubber (Hevea rubber)	Gutta Percha
1.	Natural rubber is highly soft and elastic material	Gutta percha is a hard thermoplastic material at room temperature but it softens and becomes tacky at about 1000ºC
2.	It is soluble in carbon disuphide and petrol.	It is soluble in petrol.
3.	It is cis polyisioprene	It is transpolyisoprene.

Drawbacks of Raw Rubber: Raw rubber has a number of drawbacks when put to use. Some of the drawbacks are:

(i) Crude rubber is soft at high temperatures but becomes brittle at low temperatures. So it can be used in a limited temperature range.

(ii) It is too weak with a tensile strength of 200 kg/cm^2.

(iii) It is not resistant to mineral oils, organic solvents.

(iv) It has large water absorption capacity.

(v) Due to oxidation in air, it deteriorates and hence is durability its decreased.

(vi) It is also attacked by the oxidizing agents like nitric acid, conc. sulphuric acid, chromic acid etc.

(vii) It undergoes permanent deformation when stretched. This may be due to stretching the individual chains slip past each other and get separated resulting in the breakage of the material.

However the above drawbacks can be overcome by improving the properties by the addition of suitable materials and further heat treatments.

Some of the processes for improving the properties of rubber are discussed below:

PROCESSES FOR IMPROVING THE PROPERTIES OF RUBBER

The desired properties in rubber can be obtained by compounding it with certain chemicals. The compounding of rubber is generally facilitated by a process called mastication.

Mastication: When rubber is subjected to mechanical stresses arising from say, high-speed stirring or milling, it undergoes considerable molecular degradation or fragmentation. The process is called mastication. The mastication converts hard and tough rubber into a soft, supple and semisolid mass. The CH_2-CH_2 links between the isoprene units in the rubber are weak and hence break when subjected to mechanical stresses. The net effect is that the big molecule is broken into small fragments.

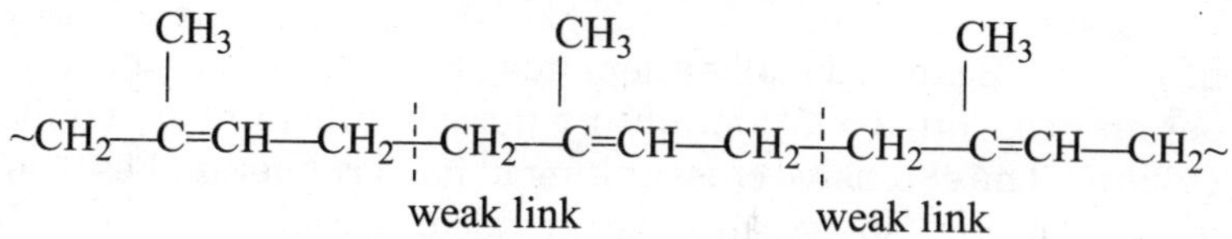

Polyisoprene

After mastication is complete the compounding agents are added to the rubber.

Compounding: The process involving the incorporation of chemicals such as plasticizers, vulcanizing agents, stabilizers, fillers, etc., into the rubber is known as compounding of the rubber. Out of above *vulcanization* is very significant and requires a detailed study.

Vulcanization

The process of vulcanization which revolutionized the rubber industry was discovered by Charles Goodyear in 1839. The process by which a network of crosslinks is introduced into an elastomer is called vulcanization. Charles Goodyear used sulphur to crosslink polyisoprene molecules in natural rubber. The added sulphur combines chemically at the double bonds bringing about changes in its properties profoundly. Vulcanization transforms an elastomer which is a weak thermoplastic mass of no use into a strong, elastic and tough rubber. The tensile strength, stiffness, durability, resistance to changes in temperature, elasticity and chemical resistance of a vulcanized rubber are increased manifolds as compared to unvulcanized rubber.

The introduction of sulphur links in between the polymer chains restrict the intermolecular movement which brings about stiffening of rubber.

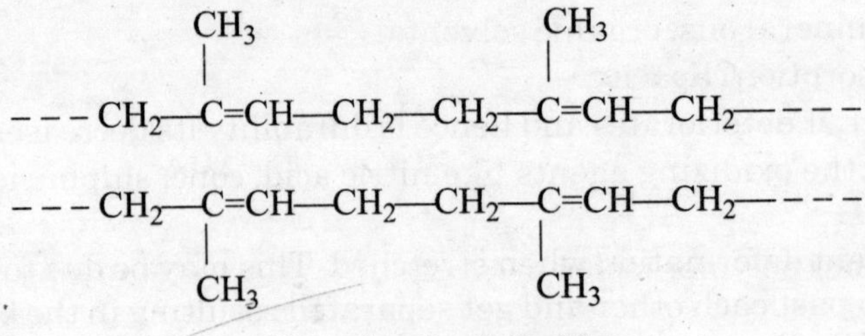

Unvulcanized Rubber

Sulphur
↓

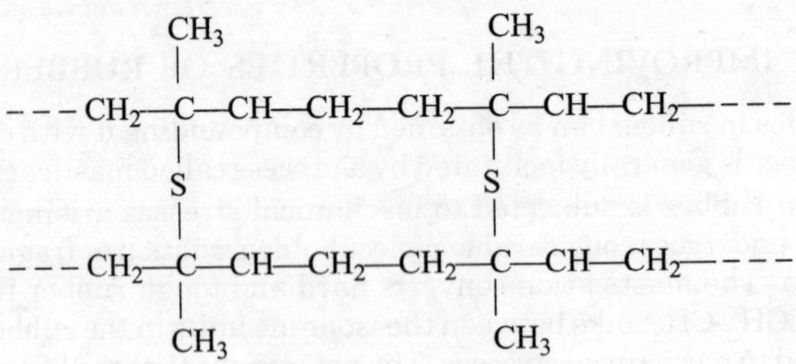

Vulcanized Rubber

The extent of stiffness of vulcanized rubber depends upon the amount of sulphur added. For example, a tyre rubber may contain 3 to 5% sulphur whereas Vulcanite or ebonite (hard rubber) contains about 32% sulphur. The extensive crosslinking in hard rubber makes it so stiff that there is practically no elasticity in it. Fig. shows the effect of vulcanization on rubber. Ordinary rubber molecules are bent and convoluted (a) When stretched the molecules can slip past one another. But vulcanization introduces crosslinks between these and hence on stretching the molecular chains cannot slip past each other.

(a) Unvulcanized Rubber (b) Vulcanized Rubber (Unstretched) (c) Vulcanized Rubber (Stretched)

Chemistry of Vulcanization

The addition of sulphur develops crosslinks, thereby improving the processibility and utility of rubber. But the mechanism of valucanization is quite complex and is still not known in detail. Both free radical and ionic intermediates have been postulated. Sulphur exists as S_8 i.e cyclo octasulphur. The probable mechanism can be:

- Sulphur (S_8) dissociates into S^+ and S^-

$$S_8 \longrightarrow S^+ + S^-$$

where *m* and *n* are the numbers of sulphur atoms.

- The sulphur cation Sm^+ reacts with a diene molecule of rubber to give I.

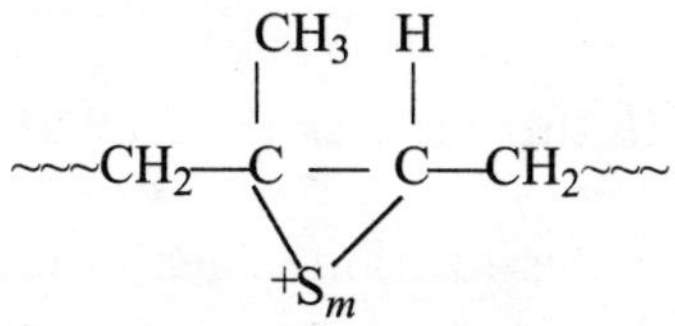

I

- This cation adds to other polymer molecule to give II and carbonium ion III

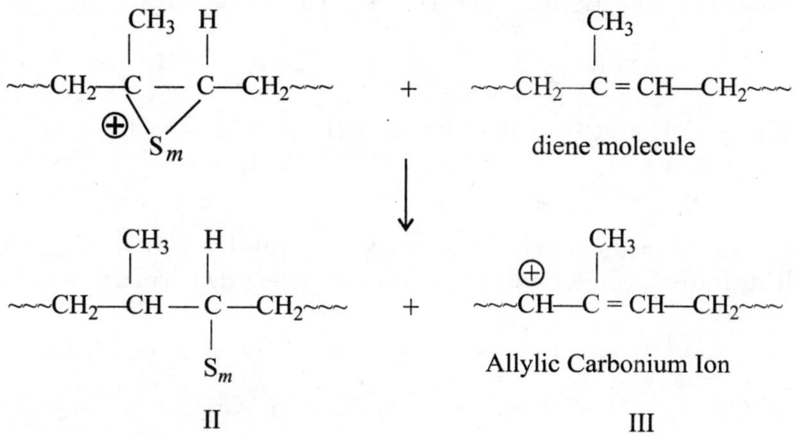

- The allylic carbonium ion reacts with S_8 to form IV which adds to polydiene to form sulphur bridges (V) This can further react with S_8 and the cycle goes on

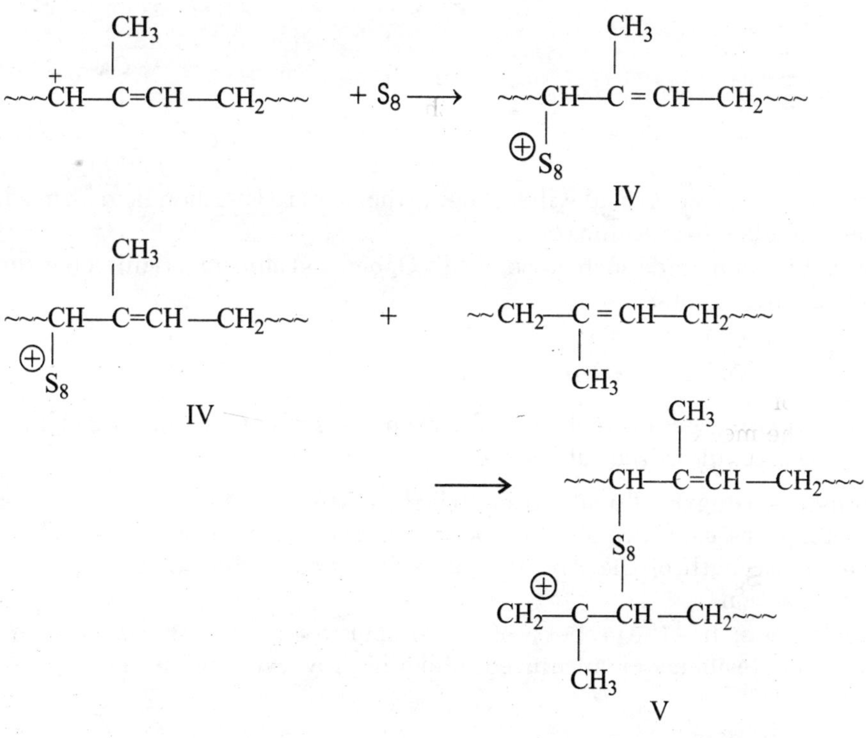

It must be emphasized that neither heat nor sulphur is essential for the vulcanization process. *Vulcanization can even be carried out by number of other substances.* Hence vulcanization can be of two types:

(a) **Sulphur Vulcanization:** As stated earlier, in this process the rubber is heated with sulphur or CS_2.

(b) **Non-sulphur Vulcanization:** Non-sulphur vulcanization is achieved through a free radical.

- Free radical can be generated from a curing agent such as benzoyl peroxide or as a step in the oxidative degradation of rubber.
- This free radical abstracts a hydrogen atom from one of the α-methylene groups.

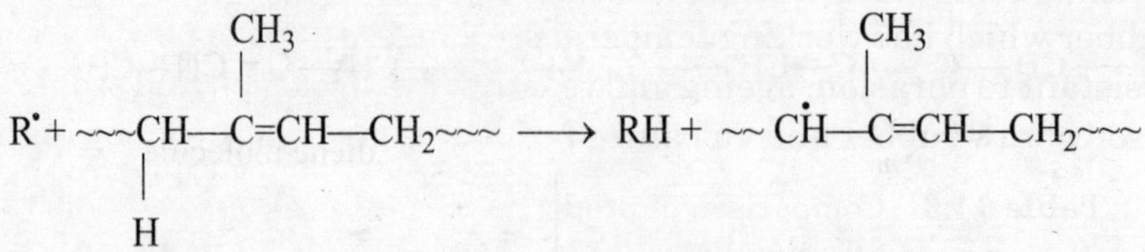

- The rubber free radical so formed attacks a double bond in an adjacent polymer chain resulting into crosslink (vulcanization) and regeneration of a free radical.

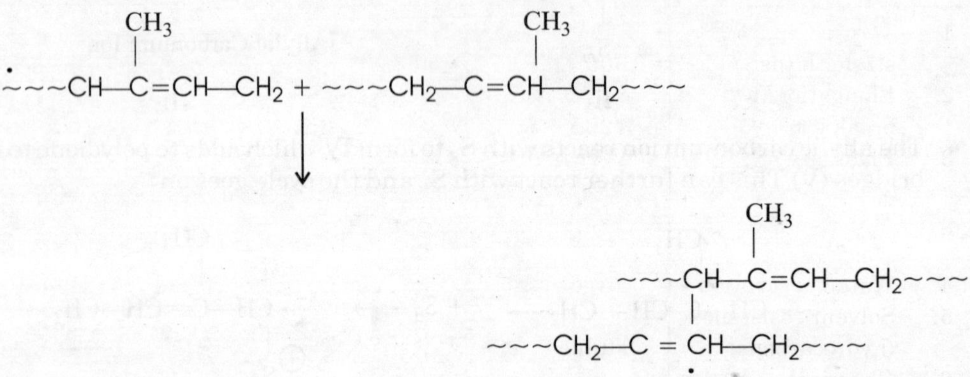

This cross linked free radical will further attacks more C=C bonds to form a large number of cross linkages leading to vulcanization.

Further, the compounds such as Se, Te, ZnO, benzoyl chloride, trinitrobenzene, etc., are also used as vulcanizing agents.

Advantages of Vulcanization

Vulcanized rubber has got ideal elastomeric properties which increase its utility manifolds. The main advantages of vulcanized rubber are:

1. **Tensile strength:** Tensile strength is the measure of the ability of a polymer to withstand pulling stresses. Vulcanized rubber has got good tensile strength and extensibility. The tensile strength of the raw rubber is 20 kg/cm^2 which increases to 2000 kg/cm^2 on vulcanization.
2. **Resilience:** It is the property of a elastomer to return to its original shape on the release of stress. Resilience of vulcanized rubber improves manifolds as compared to raw rubber.

3. **Water Absorption Capacity:** The water absorption capacity of vulcanized rubber is greatly reduced.
4. Vulcanized rubber is resistant to organic solvents such as benzene CCl_4, fats and oils. Though it swells in these solvents but to a lesser extent than raw rubber.
5. **Insulation Properties:** Vulcanized rubber has better insulation properties than raw rubber. Raw rubber vulcanized with 32% sulphur (ebonite) has extremely good insulating properties.
6. **Elasticity:** The elasticity of vulcanized rubber is greatly reduced. The elasticity of the vulcanized rubber depends upon the extent of vulcanization. If 30-50% of sulphur is present in vulcanized rubber then a rigid, non-elastomeric plastic (ebonite or hard rubber) is formed.
7. The working temperature range for vulcanized rubber is –40°C to 100°C as compared to raw rubber which has working temperature range of 10°C to 60°C.
8. It is resistant to abrasion, ageing and reactivity with oxygen and ozone.

A comparison of raw rubber and vulcanized rubber is given in the following Tabl 11.3:

Table 11.3 Comparison of properties of Raw, Vulcanized Rubber, Reinforced Vulcanized Rubber.

	Property	*Raw Rubber*	*Vercanized Rubber*	*Reinforced Valcanized Rubber*
1.	Tensile strength (psi)	300	3000	4500
2.	Elongationat break (%)	1200	800	600
3.	Elasticity	Very high	low, depends upon extent of vulconization	Very low, depends upon extent of vulcanizatin and reinforcement
4.	Water absorption	large	small	small
5.	Solvent resistence (hydrocarbons)	sluble	swells only	swells only
6.	Chemical resistance	Poor	Much better	Much better

Other Compounding Techniques

Though vulcanization improves the properties of natural rubber to a great extent but even vulcanized rubber do not exhibit satisfactory properties for various utility items. These properties can be enhanced by other compounding techniques discussed below:

(i) **Reinforcing Fillers:** To give strength and rigidity to the rubber, fillers are added. Fillers are of two types: (a) **inert fillers,** such as clay which make the rubber mixture easy to handle before vulcanization, but have no effect on the physical properties, and (b) **reinforcing fillers** which greatly improve the properties of rubber or vulcanized rubber.

Common fillers include carbon black, zinc oxide, magnesium carbonate, barium sulphate, calcium carbonate etc. Carbon black is a remarkable reinforcing filler for both natural and synthetic rubbers. Table give the comparison of raw rubber and non-reinforced vulcanized rubber and reinforced vulcanized rubber.

In vulcanization, though the long range movement of the polymer molecules is restricted, still movement of local chains is high. Reinforcement of vulcanized rubber restricts this local freedom of movement, making the polymer stiffer and tougher.

(ii) **Plasticizers or Softeners:** Plasticizers are added to rubber to improve flow and therefore, processability. Plasticizers also reduce the brittleness of the product. The common plasticizers used are hydrocarbon oils, resins, waxes, etc. The oil serves as a plasticizer and a softener by reducing the melt viscosity of the rubber to the levels required in compounding

(iii) **Antioxidants:** Rubbers that retain double bonds in their vulcanized structure are sensitive to heat, light and particularly oxygen. Thus rubber deteriorates in presence of light and air due to oxidation and as a result becomes soft and degrades. Antioxidants are substances that protect the polymer by preventing oxidation. In many cases these also prevent undesirable reactions promoted by oxygen. Common antioxidant are complex organic amines like phenyl β-naphthylamine, phenolic substances and phosphates. The structures of two common antioxidants are given below:

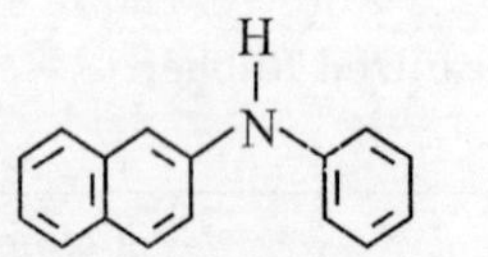

Phenyl β-naphthyl amine
(PBNA)

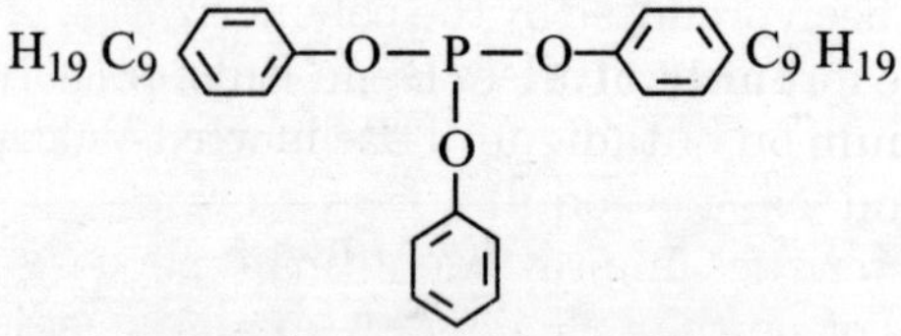

Tris-(nonly phenyl) phosphite

Antioxidants

(iv) **Accelerators:** Vulcanization takes place by heat and in presence of sulphur, but the process is *very slow.* The process can be speeded up by addition of small amounts or organic or inorganic compounds known as accelerators. The *inorganic accelerators* include lime, magnesia and white lead whereas *organic accelerators* are complex organic compounds such as aldehydes, amines, thiocarbonates and 2-mercaptol benzothiozole.

Sometimes accelerators require the presence of still other chemicals known as activators or promoters. These activators are usually metallic oxides such as zinc oxide. The best *combination* for efficient vulcanization include sulphur, an organic accelerator, a metallic oxide and a soap.

(v) **Colouring Agents:** Several dyes and pigments are used to impart colours to elastomers. Common colourants for polymers include:

Appearance	*Chemical used*
White	TiO_2, $BaSO_4$
Yellow	Lead Chromate
Red	Ferric oxide
Crimson	Antimony sulphide
Blue	Ultramarine
Green	Chromium trioxide
Silver metallic effect	Flake Aluminimum
Pearlescence	Lead carborate and mica.

The oil-soluble dyes or which have small particle size and refractive index near that of the plastic (organic pigments such as phthalocyanins) are added to get transparent coloured products from transparent plastics. The inorganic pigments impart opacity to the plastic.

SYNTHETIC RUBBERS

The polymers possessing similar physical properties as that of natural rubber are called synthetic rubbers. As natural rubber suffers from many disadvantages, attempts to find out a substitute for natural rubber began very early.

Greville William (1860) was first to isolate isoprene as a decomposition product by heating rubber to a high temperature. This gave a clue that rubber can be synthesized by the polymerization of monomers like isoprene. By the end of 19th century the fact was confirmed when polymerization of isoprene yielded rubber like product.

Since then attempts were on to prepare synthetic rubber by other monomers as well. Synthetic rubbers have been produced by the polymerization of a large number of conjugated dienes resembling isoprene. The foremost of all is, Buna Rubber (polymer of butadiene) which was prepared by the action of sodium on butadiene. That is why synthetic rubber is known as Buna rubber—'Bu' for Butadiene and 'Na' for sodium. However, the product prepared had poor tensile strength, and ageing characteristics. Improvements in properties were made by developing *copolymers* of butadiene. Several types of synthetic rubber are available today and some of them are described below:

1. SBR or Buna-s Rubber

Styrene-butadiene rubber (SBR) is the first commercially made synthetic rubber. It is a copolymer obtained by the free-radical polymerization of two monomers, styrene and butadiene. The two components are allowed to react in a mixing vessel containing an aqueous solution of an emulsifying agent in presence of free radical initiator such as, cumene hydroperoxide or p-menthane hydroperoxide.

According to the temperature used, two processes have been developed, the hot process (at 50°C) and the cold process. In the cold process the temperature is kept in the range –18°C to 5°C. The rubber is known as cold rubber. Cold process is more efficient than the hot process because chain length can be controlled more efficiently. The SBR copolymer contains about 75% butadiene and 25% styrene and is a random copolymer. Both type of rubbers 'cold' as well as 'warm rubbers' can be vulcanized by conventional methods using sulphur. A reinforcing filler is added to achieve good physical properties.

Properties:

(i) SBR possesses high abrasion loses resistance and high load bearing capacity.
(ii) At elevated temperatures SBR almost 2/3 of their tensile strength whereas natural rubber only loses 1/4th.
(iii) Its resilience is poorer than that of natural rubber which restricts its use in tire industry.
(iv) The weatherability of SBR is better than that of natural rubber.
(v) 'Cold' SBR is superior to the standard product because it contains less chain branching and crosslinking.
(vi) SBR gets easily oxidized in presence of traces of ozone present in atmosphere.
(vii) It swells in oils and solvents.

Uses:

Of all the synthetic elastomers, SBR enjoys the major share of the market. SBR has the following applications.

(i) The principal use of SBR is in the production of tires and tire products which are typically blends of both natural and synthetic rubbers. SBR is used for lighter duty tyres but is inferior to the natural rubber in its use for truck tyres.

(ii) For many mechanical goods, it is superior to natural rubber and is used because of its easier processing.

(iii) It is used for making hose, shoe soles, gaskets, floor tiles, carpet backing, tank linings etc.

(iv) It is widely used for electrical insulation, although its properties are not good as those of butyl rubber.

2. Buna-N or Nitrile Rubber (NBR) or GR-A

The nitrile rubbers are polymers of butadiene and acrylonitrile in the ratio of 75% and 25% respectively. It is prepared by emulsion polymerization using cumene hydroperoxide as initiator. Hydrogen peroxide and ferros sulphate can also be used as initiators. Vulcanization is done by conventional means by adding sulphur.

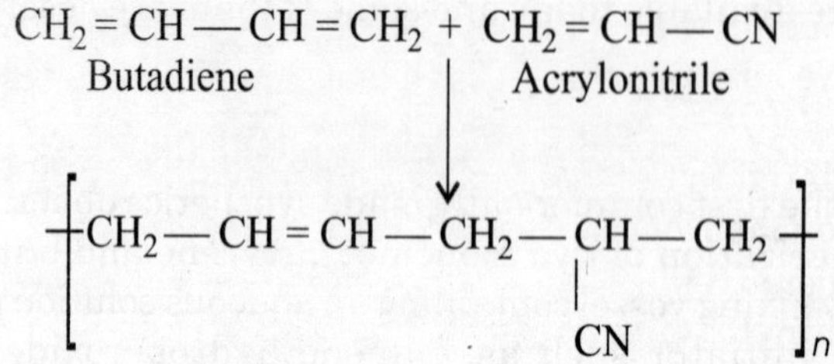

Properties:

(i) Nitrile rubbers are noted for their oil resistance. However the oil resistance varies with composition. If the acrylonitrile content is between 18% then the rubber has fair oil resistance and if it is more than 40% then it is extremely oil resistant.

(ii) It possesses excellent resistance to heat, sunlight, acids and salts.

(iii) Because of the presence of cyano groups (-CN), it is less resistant to alkalies than natural rubber.

(iv) It has good tensile strength and abrasion resistance.

(v) The rubbers are less resilient than natural rubbers.

(vi) It has good heat resistance and if properly protected by antioxidants, it shows resistance to oxidative degradation as well.

Uses:

(i) Nitrile rubber is used primarily for its oil resistance. These are extensively used for fuel tanks, gasoline hoses, gaskets, printing rollers and high attitude aircraft components.

(ii) They can be used in contact with water and antifreeze solutions.

(iii) They are also used in the form of latex for impregnating paper, textile and leather.

(iv) It can also be used as blend component in tyre manufacture.

3. Neoprene or GR-M Rubber

This is also known as polychloroprene or polychlorobutadiene or duprene. Neoprene was the first synthetic rubbers developed in the United States.

Preparation: Neoprene or polychloroprene is prepared by polymerization of chloroprene, which is 2-chloro-1,3 butadiene.

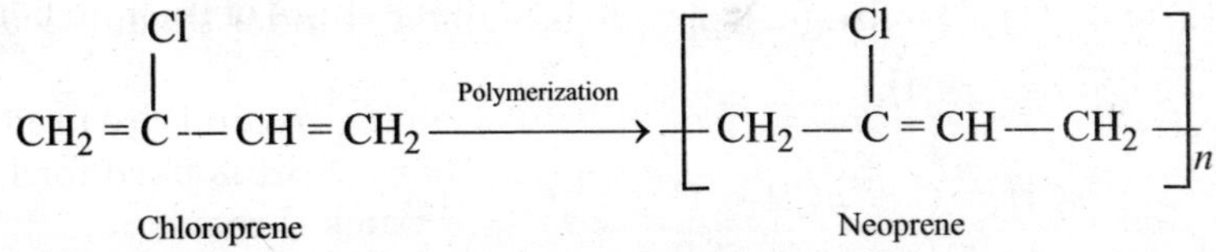

The preferred configuration produced during the polymerization is trans 1-4 polychloroprene. The polymerization of chloroprene takes place through free radical emulsion polymerization.

The starting material, chloroprene is manufactured by the addition of hydrogen chloride to vinyl acetylene, which inturn is prepared by dimerization of acetylene.

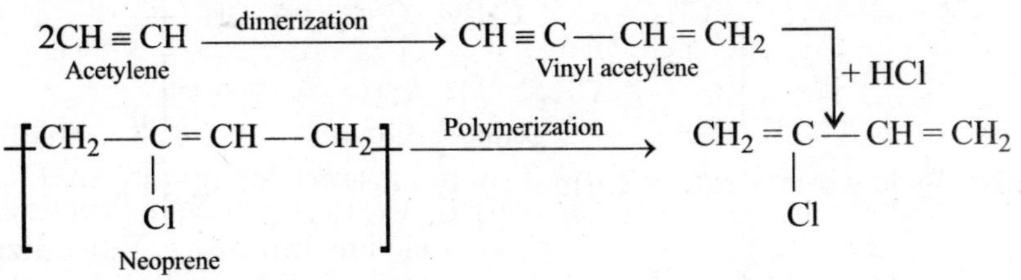

The vulcanization of neoprene is different from that of other elastomers in that it can be vulcanized by heat alone. Although some types are polymerized in presence of sulphur which introduces crosslinks in the polymer. Neoprene can also be vulcanized using zinc oxide or magnesium oxide.

Properties: Neoprene is closely related to the natural rubber and can replace rubber because similarity in the properties of the two. Some of the properties of Neoprene are:

(i) As H atom in the natural rubber is replaced by Cl atom in Neoprene, Neoprene has superior resistance to vegetable and mineral oils.
(ii) It is superior to natural rubber as far as ageing and effects of high temperatures are concerned.
(iii) Neoprene is soluble in polar solvents due to the presence of chloro-group.
(iv) Vulcanized neoprenes have high tensile strength (3500-4000 psi) in the absence of carbon black.
(v) Vulcanized neoprene is extremely resistant to oxidative degradation if suitably protected.
(vi) It has good weathering and ozone resistance.
(vii) It has got excellent oil resistance, though inferior to nitrile rubber but better than natural rubber or SBR.

Uses

Neoprene is used for:

(i) It is used in oil resistant wire and cable coatings, industrial hoses, shoe holes.
(ii) Its latex is used for the manufacture of gloves and coated fabrics.
(iii) It is used for making tubings for carrying corrosive gases and oils.
(iv) It is used for the manufacture of sponges, conveyer belts, linings of reaction vessels etc.

4. Butyl Rubber (GR-I)

Butyl rubber belongs to the class of elastomers which do not have unsaturated chain structure. However, a small amount of diene monomer that provides a site for vulcanization is included by means of copolymerization.

Preparation

Butyl rubber or polyisobutylene are prepared by the copolymerization of isobutylene with a small amount (1.5 to 4.5%) of isoprene. Isoprene is added to render them vulcanizable.

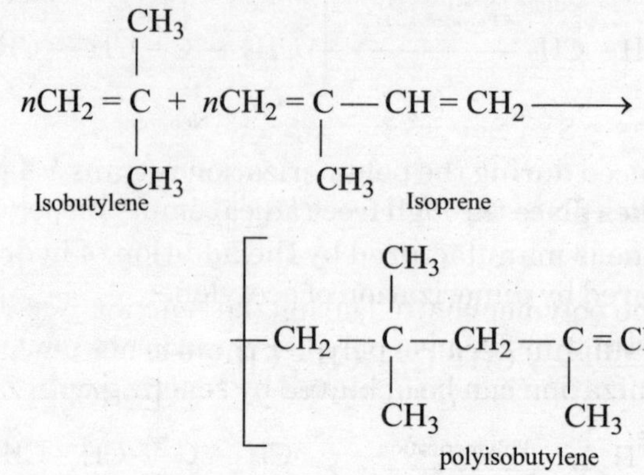

It is prepared by low temperature cationic polymerization using Friedal-crafts catalyst at –80ºC.

Properties

The absence or low concentrations of double bonds in the main chain of butyl rubber results in high resistance to attack by a number of chemicals. It has the following properties.

(i) Polyisobutylene or butyl rubber is amorphous under normal conditions but crystallize on stretching.

(ii) It exhibits outstanding resistance to attack by oxygen and ozone due to low concentration of double bonds. It is less sensitive to oxidative ageing as compared to other elastomers except silicons.

(iii) Butyl rubber becomes soft on oxidative degradation.

(iv) It exhibits low gas permeability due to closepacked linear paraffinic chains.

(v) It has high ozone and acid resistance.

(vi) Due to very low unsaturation, it cannot be hardened much on vulcanization.

(vii) Unstabilized polyisobutylenes are degraded by light or heat to sticky low-molecular weight products. Therefore usual rubber antioxidants should be added to stabilize the polymer.

(viii) It has excellent insulating properties as predicted from its non-polar, saturated nature.

(ix) It is soluble in hydrocarbons such as benzene.

(x) The tear resistance of butyl rubber is quite good and is retained well at high temperature and for long times, in contrast to natural rubber.

Uses

(i) Butyl rubber is used for inner tubes for tires because of its impermeability to gases.

(ii) It is used for wire and cable insulation.

(iii) It is used for linings of tanks, hoses etc.

5. Polysulphide Rubber (Thiocol or GR-P)

Polysulphide Rubbers are the condensation products of ethylene dihalides and alkali sulphides. These are linear condensation polymers.

Preparation

Polysulphide rubber is obtained by the reaction between sodium polysulphide and ethylene dichloride

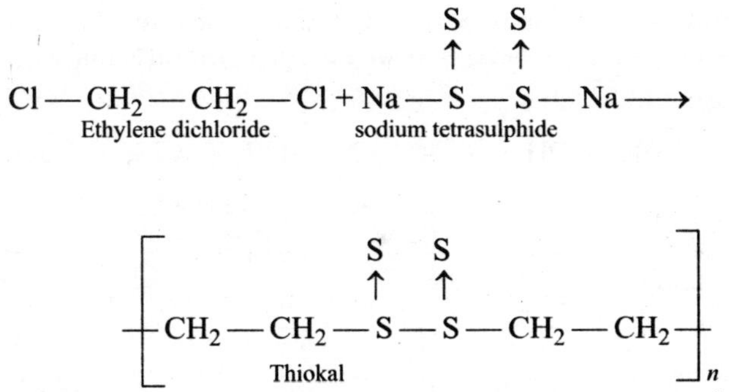

Thus sulphur forms the part of the polymer chain. During the reaction NaCl gets eliminated. Vulcanizaticn cannot be achieved by sulphur because polymer chain is not unsaturated. Hence it is not a hard rubber. However vulcanization can be achieved by reacting with ZnO which reacts with sulphur links forcing crosslinks.

Properties

The properties of the materials depend on the length of the aliphatic groups and the number of sulphur atoms present in it. With four sulphur per monomer, the product is rubbery, whereas if only two sulphur atoms are present per monomer, the polymer do not behave as elastomer. In that case at least four methylene groups are needed in the dihalide to induce elastomeric properties to the products. Some of the important properties are:

(i) Thiokol is resistant to swelling.
(ii) It is resistant to oils and solvents. Fuel oils, lubricating oils, gasoline and kerosene have no effect on thiokol.
(iii) Thiokol films are impermeable to gases to a large extent.
(iv) It cannot be vulcanized and hence does not form hard rubber.

Thiokol rubbers from many disadvantages:

(i) It has disagreeable odour (mercaptan odour).
(ii) It has poor heat resistance, poor abrasion resistance and low tensile strength.
(iii) It tends to lose its shape under continuous pressure.

However, recent developments of Thiokol have overcome these disadvantages.

Uses

(i) Polysulphides or Thiokol is used to make sealants, gaskets, balloons, fabric coating.
(ii) Due to low permeability it is used to make gasoline hoses.
(iii) Liquid Thiokol can be cured at room temperature to tough solvent-resistant rubbers. These liquid compounds are used as gasoline tank sealants and liners for aircraft and in variety of other sealing applications.
(iv) The liquid polymers, when combined with oxidizers burn with great intensity and generate large volume of gas. Thus it can be used in solid-fuel rocket propellents.

6. Polyurethane (PU)

Polyurethanes are polymers containing the group (carbonate group)

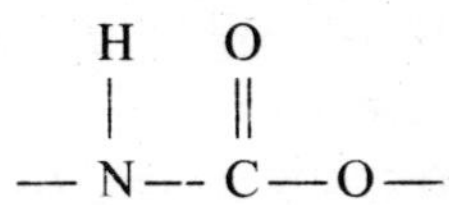

These are known as urethane linkages.

Preparation: The polyurethanes are formed through the reaction of diisocyanate with the compound containing atleast two active hydrogen atoms such as diols or diamines. The reaction of diisocyanate with glycol is represented as:

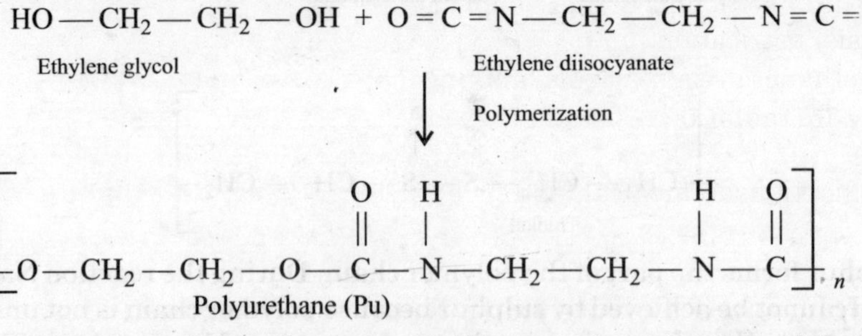

The properties of the final product depend upon the type of monomers and their chain length. A wide variety of polymers can be prepared ranging from rigid to elastic polymers. Crosslinking takes place through isocyanate groups which are capable of reacting with hydrogen atoms attached to the polymers leading to vulcanization. The polyurethanes can be produced in the form of fibers, elastomers or foams.

The foamed plastic can be obtained by blowing air/gas through molten polyurethane. This foamed plastic is also known as *thermocol.*

Note: Thermocol is also prepared by blowing air/gas through polystyrene.

Alternatively, for the production of foamed polyurethane little water is mixed with the reaction mixture, which reacts with isocyante group to evolve CO_2. The CO_2 produced foams the polymer as it sets.

$$O{=}C{=}N-R-N{=}C{=}O + 2H_2O \longrightarrow H_2N-R-NH_2 + 2CO_2$$

Properties

(i) Polyurethanes have high strength, good resistance to gas and aromatic hydrocarbons.

(ii) They have excellent resistance to ozone and oxygen but are susceptible to microbial attack.

(iii) The polyurethanes can be prepared in different forms varying from hard coatings to flexible rubbers and rigid foams, each having different properties:

Coatings: Coatings are hard, glassy and solvent resistant. They have good flexibility and impact resistance.

Elastomers: Elastomers have excellent abrasion resistance and are resistant to oils, greases and solvents.

Fibers: Fibers are obtained by reacting di-isocyanate with long chain glycols which are usually polyesters or polyethers.

Foams: Foams can be flexible or tough depending upon the polymer and the type of crosslinking produced. They are light, tough and resistant to heat.

Uses

(i) The polyurethane foams have found the widest use. The flexible foams are used as upholstery material for furniture, insulation and crash pads.

(ii) The rigid polyurethane foams are used as core material as insulation cover in refrigerated appliances and vehicles.

(iii) Flexible polyurethanes can also be used for coating rubber articles to give them additional resistance to abrasion, such as in gymnasium and dance floors.

(iv) Polyurethane resin is used to insulate the wires polyurethane coated wires can be soldered directly.

(v) Polyurethane fibers with unusually high elasticity are used in foundation garments and swim wear. They have replaced rubber latex thread in this use.

(vi) Flexible urethane foams are used as cushions for furniture and automobiles and has replaced rubber foams in these applications because of improved strength, lower density and easier fabrication.

(vii) Polyurethanes are also used in making matteresses and pillows.

(viii) As they are radiation resistant, they are also used in encapsulation for integrated-circuit devices.

(ix) Its applications in medical field include its use in dental materials, artificial heart and skin.

7. Chlorosulphonated Polyethylene Rubber (Hypalon)

Preparation: When polyethylene is treated with a mixture of chlorine and sulphur dioxide, chlorosulphonated polyethylene is formed.

$$\left(CH_2-CH_2\right)_n + Cl_2 + SO_2 \xrightarrow{-HCl} \left[CH_2-\underset{\underset{Cl}{|}}{CH}-\left(CH_2-CH_2\right)_n-CH_2-\underset{\underset{SO_2Cl}{|}}{CH} \right]$$

The chlorosulphonation can be carried out either on the solid material or in solutions.

Properties: The chlorosulphonylation of ethene provides two modifications:

(a) The chlorine atoms breaks the regularity of polyethene chain structure changing the crystalline polyethylene to a elastomer.

(b) The sulphonyl groups which are introduced, provide the sites for crosslinking.

The chlorosulphonyl ethane has the following properties:

(i) Its resistance to ozone is better than neoprene and butyl rubber.

(ii) It has good heat resistance.

(iii) It has good chemical resistance and flame resistance.

(iv) It is resistant to solvents.

Uses

It is used in chemical industry.

11.11 BIOPOLYMERS/NATURAL POLYMERS

Biopolymers are the polymers found in the living organisms. They are also called natural polymers. They are formed within the organism by complex metabolic processes. The growth and existence of living organisms is attributed to a great extent to the natural polymers or macromolecules. These polymers provide protective surfaces, membranes, skeletal system, energy storage systems and pathways for electrical conduction to the living organisms.

The biological/natural polymers can be studied under the four heads:

(1) Carbohydrates (Polysaccharides)
(2) Proteins (Polyamides)
(3) Nucleic Acids (Polynucleotides)
(4) Polyisoprene (cis-1,4-polyisoprene)

These are discussed below in detail.

I. Carbohydrates (Polysaccharides)

In higher plants, polysaccharides are the most important structural biopolymers. They have macromolecular structure built up of many monosaccharide units (polyhydroxy aldehydes and ketones) joined together with the elimination of water molecule.

Polysaccharides are of almost universal occurrence in living organisms where they perform various useful functions such as:

- In plants, they serve as excellent materials, for skeletal structure and supporting tissues.
- They serve as main source of our food and material for nutrition of cell.
- Apart from that they are used for clothing and providing shelter.

Though cellulose and starch are the best known polysaccharides but a number of other are also present in living organisms and are equally important.

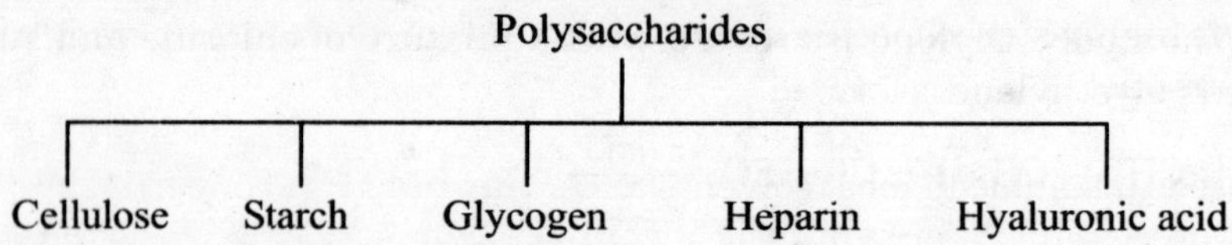

A brief discussion of these is given below:

(1) Cellulose

Cellulose is the primary structural component of plant cell walls. It is a linear polymer consisting of D-Glucose residues linked by β-(1-4) glycoside bonds.

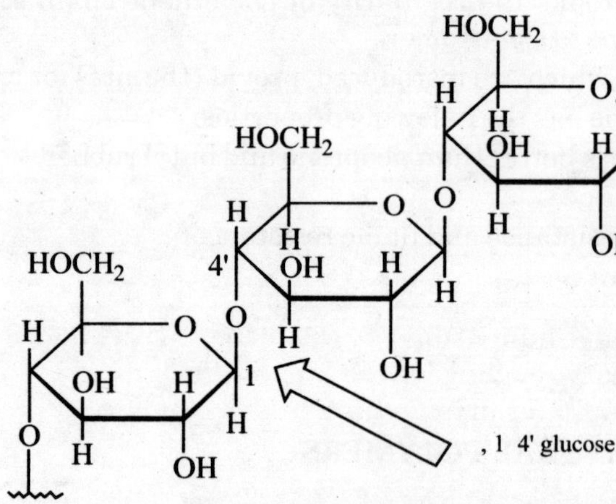

Fig. 11.6 Cellulose

In plants, cellulose is synthesized from the simple monosaccharide-glucose. Glucose is present in the vegetables and also in the blood of animals. Nature converts glucose into cellulose where thousands of glucose molecules are linked together through hydroxyl groups at C_1 and C_4 to form cellulose.

Cotton which contains almost 95% of cellulose is the oldest natural fibre. Filter paper which we use in the lab is nearly pure cellulose. Others fibers like jute, flex, hemp are also very rich in cellulose.

Properties

(i) Cellulose is a linear polymer whose molecular weight varies from 50,000 to several millions.

(ii) It can be easily hydrolysed by acids to glucose.

(iii) The glucose units are linked together by acetal bonds formed between hemiacetal of carbon at C_1 of the cyclic glucose structure in one unit and a hydroxyl group at C_4 in the other unit.

(iv) Cellulose cannot be digested by vertebrates (example-human beings) because they do not posses the enzyme capable of hydrolyzing the β (1-4) linkages of cellulose. However termites as well as symbiotic microorganisms (found in digestive tracts of herbivores) secrete enzymes which can degrade cellulose.

(v) Despite the presence of several hydroxyl groups, cellulose is insoluble in water. This is because the chains are very stiff and hydrogen bonding between two—OH groups on adjacent chains favours close packing of the chains.

(vi) Cellulose is highly stereospecific macromolecule, each glucose unit being at right angle to the previous one.

Uses

(i) As cellulose is insoluble in most solvents and is infusible, it is converted into its derivatives to make it processable. The synthesis of various cellulose derivatives by chemical reactions is known as *regeneration*. The various regenerated forms of cellulose are:

Regenerated forms of Cellulose

Cellulose derivatives	*Uses*
(a) Cellulose Nitrate	• Plastic-grade cellulose nitrate is used for coatings. • Cellulose containing 12.8 to 13.5% nitrogen is used as explosive (Example—Gun cotton)
(b) Cellulose Acetate	• Non-flammable material • Used in photo films, for wrapping small envelopes or bags.
(c) Viscose Rayon	• Used as an apparel fiber. • High tenacity viscose rayon is used as tire cord.
(d) Cellophone	• Used as wrapping and packaging material.
(e) Carboxy methyl cellulose (CMS)	• Used as thickening agent for ice cream • Builder for synthetic detergents.

(ii) Cellulose is of considerable human dietry value. Though we cannot digest it but it provides indigestible but necessary roughage in the human diet.

(iii) Cellulose is used as raw material in textile and paper industry.

(2) Starch

Starch is also made up of glucose units joined together. But it has α-glucose as the repeat unit whereas cellulose has β-glucose as repeat unit. It is the major constituent of human diet in the form of potatoes, grains, rice, corn etc. Starch is a branched polymer having two units amylose and amylopectin.

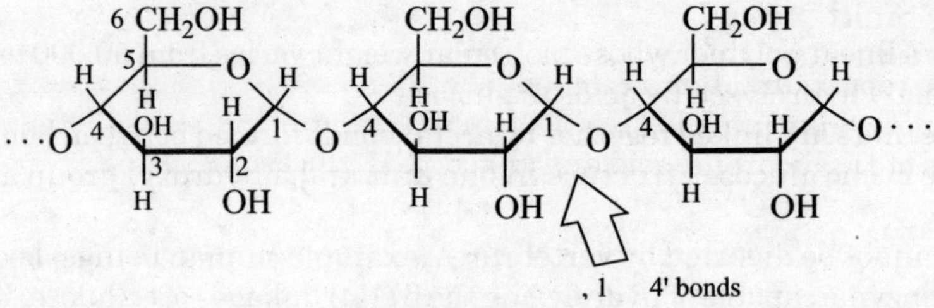

Fig. 11.7 Amylose

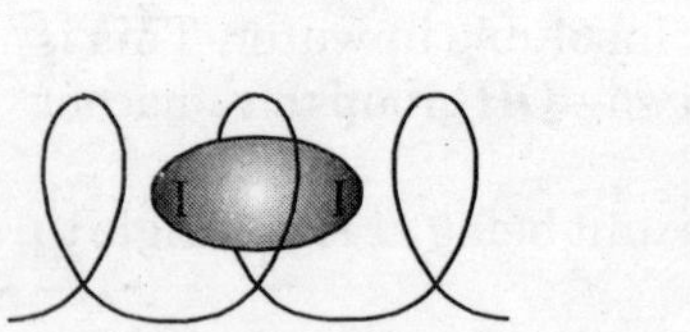

The amylose molecules form helices around I_2 molecules and as a result a deep blue colour arises from electronic interactions between the two. In the iodine test for starch, a solution of iodine is added to an unknown as a test for the presence of starch.

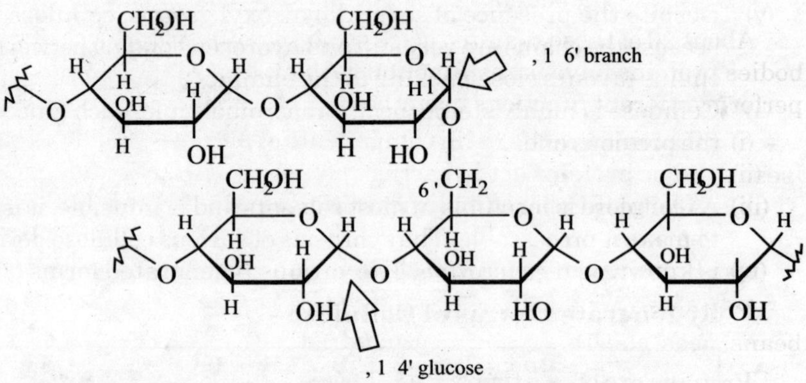

Fig. 11.8 Amylopectin

Properties

(i) Starch can be hydrolyzed by amylases present in the saliva of human beings.
(ii) It is amorphous in nature which makes it available in granular form.
(iii) Water molecules can penetrate between the chains and cause swelling to take place forming paste or hydrated mass.
(iv) Can wet and adhere to the polar surfaces.

Uses

(i) It is used as foodstuff.
(ii) Used in paper sizing and textile industry.
(iii) Crystalline glucose and corn syrup are manufactured from starch.
(iv) Starch acetate is used in making various sweets.

(3) Glycogen

It is also a polysaccharide found in animals, and occurs in liver, muscles and also in some fungi and bacteria. It is a reserve food and releases energy by oxidation to lactic acid.

Glycogen has a structure very similar to starch except that it is highly branched.

(4) Heparin

It is a polysaccharide of animal origin and is found in extra-cellular matrix.

It is used as anticoagulant for blood and antithrombotic agent.

(5) Hyaluronic Acid

It has disaccharide repeat unit. It is a component of synovial fluid (fluid that lubricates the joints), the vitreous humor of eye and umbilical cord. Its molecules are composed of β(1-4)linked disaccharide units that consists of D-glucuronic acid and N-acetyl-D-glucosamine linked by αβ(1-3) bond.

II. Proteins

Proteins are the natural polymers made of α-amino acid units linked through peptide linkages

— CONH —
peptide linkage

About 20 of these α-amino acids occur in nature, of which 8 are essential to sustain life. In our bodies proteins have many functions. About 15% of the human body is protein matter. Proteins perform different *functions* in our body.

(i) Proteins are the main components of our muscles, hair and skin.
(ii) Some proteins act as carriers of inorganic and organic compounds.
(iii) A large number of biological catalysts are proeins. For example, enzymes are proteins by nature.
(iv) Proteins are structural elements and function as protective agents in the living systems.

Proteins are the important component of most vegetables and lean meat. These are found in beans, peas, cheese, soyabean, groundnut, egg yolk and egg white.

Proteins are of two types:

(i) Simple Proteins
(ii) Conjugated Proteins

Simple proteins yield only α-amine acids on hydrolysis whereas conjugated proteins yield amino acids and some other compounds on hydrolysis. The other compound yielded is known as prosthetic *group*.

Depending upon the prosthetic group the conjugated proteins are subdivided into the following:

Conjugated Proteins

- Glycoprotein (Carbohydrate) Prothetic Group
- Lypoprotein (Lipid)
- Nucleoprotein (Nucleic Acid)
- Phosphoprotein (Phosphoric Acid)

Proteins can also be classified depending upon the bonding in between the polymer chains:

(i) **Fibrous Proteins**: Fibrous proteins are intermolecularly hydrogen bonded and are insoluble in water. Examples include:

- keratin (protein of hair and nails)
- collagen (protein of connective tissue)
- elastin (protein of arteries, tendons)
- myosin (the muscle protein)
- fibroin (the silk thread)

(ii) **Globular Proteins:** Globular proteins are intramolecularly hydrogen bonded and exist roughly as spherical entities. They are soluble in water.
Examples include:
- hormones (like insulin and glucogens)
- enzymes (like trypsin, invertase and papain)
- haemoglobin (respiratory protein)
- albumins (egg albumin)

Structure

The structure of proteins is very complex. The proteins have primary, secondary and tertiary structures:

(i) **Primary structure:** Primary structure of protein is represented by the amino acids arranged in a definite sequence through peptite linkage.

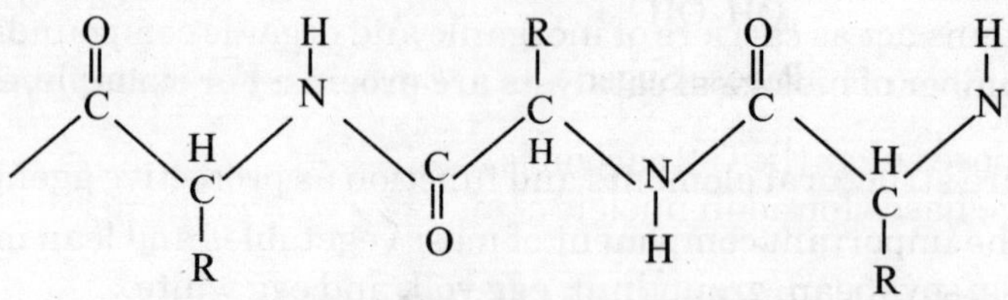

For example, a fibre such a silk is made up of linear protein molecules. The adjacent chains of these linear protein molecules are held together by hydrogen bonds.

(ii) **Secondary structure:** When protein molecules are so arranged that they form right bonded helical arrangement or pleated sheet structure, then they are said to posses secondary structure.

(iii) **Tertiary Structure:** Tertiary structure represents the folding pattern of the proteins which results due to sulphur-sulphur cross links between macromolecular chains. Tertiary structure gives a specific conformation to the particular protein molecule. The specificity of enzymes is due to the tertiary structure of proteins. Moreover proteins lose their physiological activity because of disruption of folded structure (denatured)

III. Nucleic Acids

Nucleic acids are the macromolecules that store genetic information and control protein synthesis. It is rightly said that nucleic acids are the substance of heredity. Nucleic acids are found in the nuclei of all living cells in the form of nucleoproteins i.e., proteins containing nucleic acid as prosthetic group.

There are two types of nucleic acids:
- Ribonucleic acid (RNA
- Deoxyribonucleic acid (DNA)

It is the DNA which carries the genetic information necessary to carry on reproduction, and to sustain the organism. DNA molecules are among the largest molecules known. They have molar masses of up to tens of billions of grams.

The RNA is responsible for controlling the synthesis of proteins.

The monomers of nucleic acids called 'nucleotides' are formed from three distinct parts:

(i) A sugar (pentose or five carbon sugar).
(ii) A nitrogen containing base
(iii) Phosphoric acid

Thus in general nucleotide can be represented as:

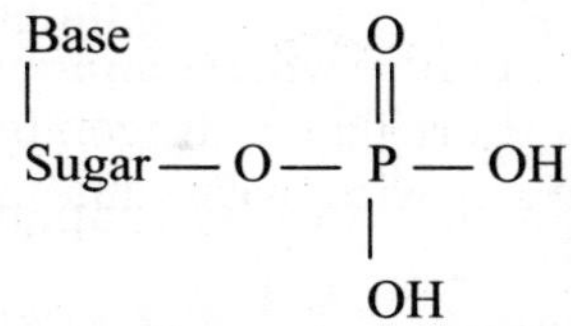

Let us discuss these three parts in little detail:

(i) **Sugar:** The sugar part in nucleotide is a pentose sugar. Two types of pentose sugars are present in nucleotides; deoxyribose in DNA and ribose in RNA.

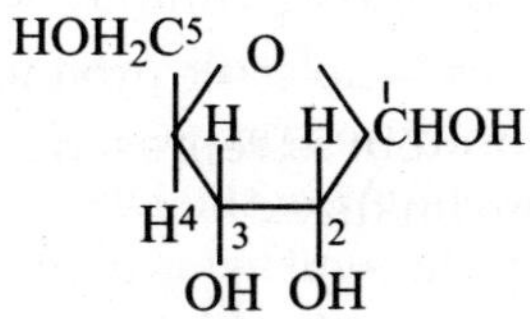

Ribose Sugar

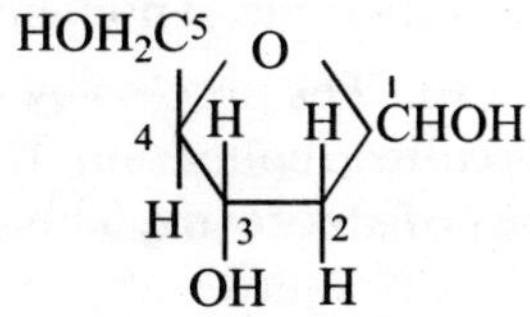

Deoxyribose

In deoxyribose sugar the C-2 carbon is attached to H instead of — OH in ribose sugar.

(ii) **Bases:** The bases found in nucleic acids are heterocyclic nitrogeneous bases and are of two types:
   (a) Purines, which include adenine (A) and Guanine (G)
   (b) Pyrimidines which include, uracil (U), Thymine (T) and Cytosine (C).

The base and the sugar combine to form a unit known as *nucleoside*.

The bases present in RNA are: Uracil, Cytosine, Adenine and Guanine.

The bases present in DNA are: Cytosine, Thymine, Adenine and Guanine.

(iii) **Phosphoric Acid:** The nucleoside (base-sugar unit) combine with phosphoric acid through the esterification of one of the hydroxyl group of sugar forming the *nucleotide*. Thus nucleotides are

Nucleoside mono phosphate

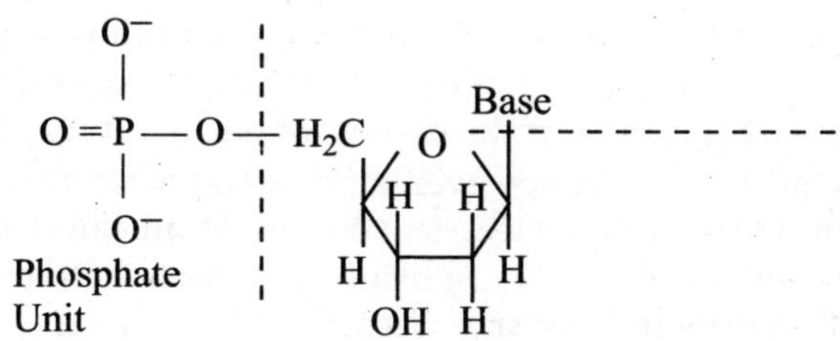

Sugar Unit

Nucleotide

These nucleotides combine to form nucleic acids.

$$\text{Sugar + base} \longrightarrow \text{Nucleoside} \xrightarrow{\text{Phosphoric Acid}} \text{Nucleotide} \xrightarrow{\text{Polymerization}} \text{Nucleic acid}$$

$$\text{Nucleic acid} \xleftarrow{\text{Polymerization}} \text{Nucleotide}$$

Structure

DNA has got a double helical structure with complimentary bases on the two strands. The bases form hydrogen bonds to each other. The hydrogen bonding can exist between specific bases. A purine of one strand always pairs up with pyrimidine of the other strand. The only possible pairings are:

Guanine ≡ Cytosine
Adenine = Thymine
(– denotes hydrogen bonds)

Unlike DNA, RNA has single strand

The important functions of nucleic acids are:

(i) **Replication:** The process by which a single DNA molecule produces two identical copies of itself is called replication. It is an enzyme catalysed process.

(ii) **Synthesis of Proteins:** The genetic information from the DNA is passed to the RNA or messanger RNA (mRNA). This is achieved by the synthesis of mRNA and the process is called transcription. The mRNA so synthesized give the message of the DNA for a specific protein synthesis in the ribosome. The specific ribonucleotide sequence in mRNA determines the order in which the different amino acids have to be joined to form a protein.

Each segment of a DNA molecule which codes for a specific protein is called a *gene* and hence every protein in a cell has a corresponding gene.

11.12 CONDUCTING POLYMERS

In general, polymers are found to be insulators. Infact electricity wires are coated with polymers because of the insulating properties of the latter.

In 1800s, it was observed that the conductivity of natural rubbers-which is normally an excellent insulator-could be significantly increased by adding carbon black which has natural conductivity. In 1930s, natural rubber filled with acetylene black was used as an antistatic device in hospitals where it was necessary to control the danger of sparks due to the building of static electricity.

Earlier (before 1973) the only polymer capable of showing appreciable electrical conductivity ($\sim 10^3$ S cm^{-1}) was polysulphurnitride $(SN)_x$. However this polymer could not generate much interest among scientists because of the explosive nature of its precursor monomer, S_2N_2.

During the past 15 years, however, a new class of organic polymers has been devised with the remarkable ability to conduct electrical current. Some of these conductive polymers are already under development for practical applications, which include rechargeable batteries, electrolytic capacitors and 'smart windows' that absorb sunlight in summer. Thus conductive polymers have tremendous potential for future scientific and technological development.

Now, the major question is that, how the electric current passes through the conducting polymers. A common feature of all conducting polymer is a conjungated π-electron system (consisting of alternating single and double bonds) along the polymer chain backbone or ring structure. Among many polymers known to be conductive, **polyacetylene, polyanaline, polypyrrole, poly (phenylenevinylene)** have been studied the most. However, the conductive polymer that actually launched this new field was polyacetylene.

Prof. Alan J. Heeger, Prof Alam. G. MacDiarmid, and Prof Hideki Shirakawa in 1977 discovered that partial oxidation of polyacetylene with iodine, chlorine or bromine vapours made polyacetylene 10^9 times more conductive. The process of transforming a polymer to its conductive form via chemical oxidation or reduction is called **doping** by analogy to somewhat related doping of semiconductors. Thus, the conductive form of polyacetylene is called "doped polyacetylene." The conductivity of doped polyacetylene was found to be as high as 10^3 S cm^{-1}, which was higher than that of any known polymer. By comparison, silver has conductivity of 10^6 S cm^{-1}. The conductivity of doped polyacetylene is comparable to that of copper on an equal weight basis. The three scientists received the Nobel prize-2000 in chemistry, for their discovery and development of electrically conductive polymers.

In general, conducting polymers can be classified into the following categories:

1. **Conjngated π-electrons conducting polymers:** As is discussed earlier, a key property of a conductive polymer is the presence of conjugated double bonds along the backbone of the polymer. The orbitals of conjngated π-electrons, overlap over the entire backbone of the polymer, resulting into the formation of valence bands as well as conduction bands, extending over the entire polymer molecule.

   The energy spacing between the highest occupied and lowest unoccupied band is called the band gap. The highest occupied band is called the **valence band** and the lowest unoccupied band is the **conduction band**. If the band gap is narrow, the electrons from the valence band are excited to conduction band giving rise to conductivity. If the band gap is too wide, the substance is an insulator.

   In an electric field, conjugated π electrons get excited and electrons have sufficient energy to jump the gap and reach conduction band.

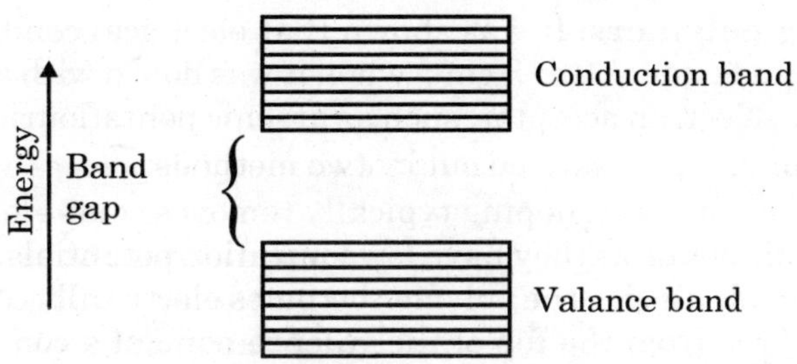

In otherwords electrons thereby can be transported through the solid polymeric material and hence polymer becomes conducting. The most common examples are polyacetylene and polyaniline. However a number of heterocyclic aromatic conducting polymers are also being studied.

Examples:

1.

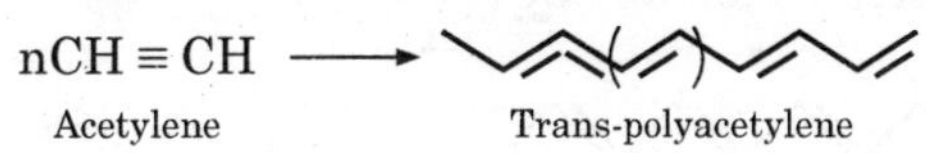

2.

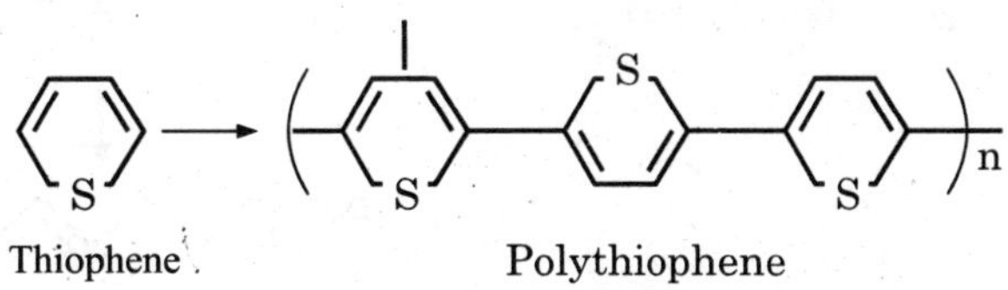

3.

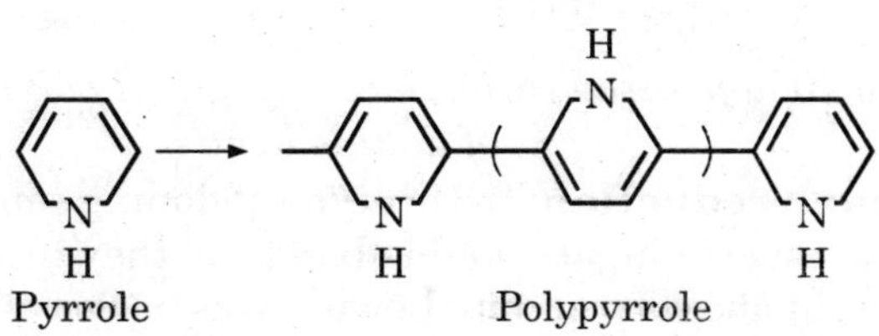

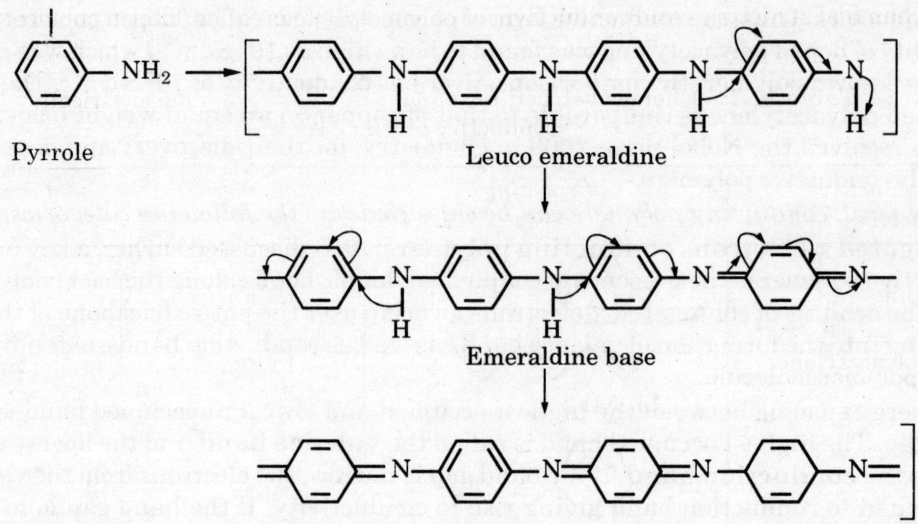

However, conjugation is not enough to make a polymer conductive. The conductivity of polyacetylene is 1.7×10^{-9} S cm^{-1} and that of polyaniline is 10^{-10} S cm^{-1}. The conductivity is greatly enhanced through doping.

2. **Doped conducting polymers:** It was shown that electrical conductivity of polyacetylene could be increased by a factor of 10^{12} S cm^{-1} when it was doped with an electron donor, such as alkali metal ion or an electron acceptor, such as arsenic pentaflouride (AsF_5) or iodine.

   The process of doping can be carried out by two methods:

   (i) **Oxidation**: In this process, doping typically removes some electrons from π bond of the conjugated double bonds as they have low ionization potentials. The holes so created can move along the molecule i.e. the polymer becomes electrically conductive. Thus, when an electron is removed from the top of the valence bond of a conjugated polymer, such as polyacetylene or polypyrrole, a valency (hole or radical cation) is created, causing partial delocalization, extending over several monomeric units. The radical cation is called 'polaron'. It stabilizes itself by polarizing the medium around it and hence the name. The polorons are mobile and can move along the polymer chain by rearrangement of double and single bonds and hence the polymer become conducting. The oxidation process is generally brought about by Lewis acids such as $FeCl_3$ etc.

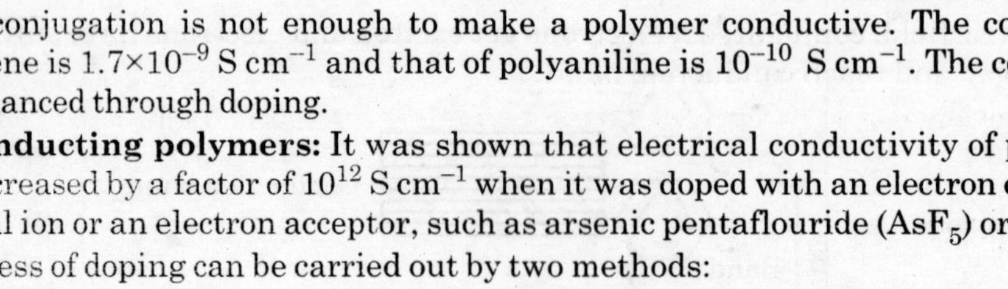

Propagation of a polaron through a conjugated polymer chain by shifting of double bonds.

   (ii) **Doping through reduction**: In this type of doping some electrons are introduced into the polymer having conjugate double bonds as they have high electron affinities. This process is brought about by adding Lewis bases.

The chemical structures and conductivities of some electrically conducting polymers are given in (Table 11.2).

Table 11.2: Chemical structures and conductivities of some electrically conductive polymers

Conductive polymer	*Repeating unit*	*Dopants*	*Conductivity ($S\ cm^{-1}$)*
Polyaniline	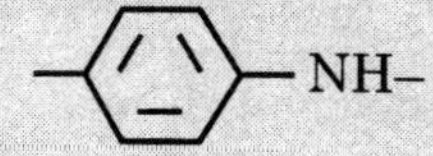	HCl	200
trans-polyacetylene	 Li, AsF_5	I_2, Br_2,	10^4
Polypyrrole	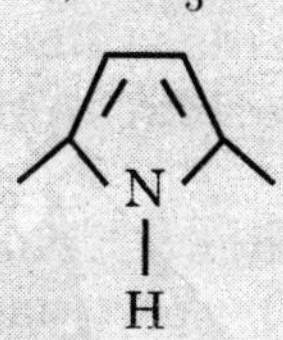	BF_4, ClO_4	500-7500
Poly (p-phenylene)		AsF_5, Li, K	10^3
Poly (p-phenylene vinylene)	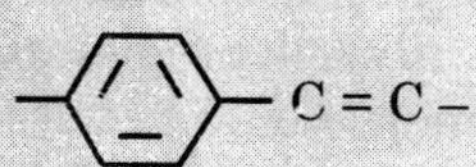	AsF_5	10^4

It should be noted that polyaniline exists in four oxidation states of which, only the emeraldine salt is conducting which require protonic doping of the imine nitrogen.

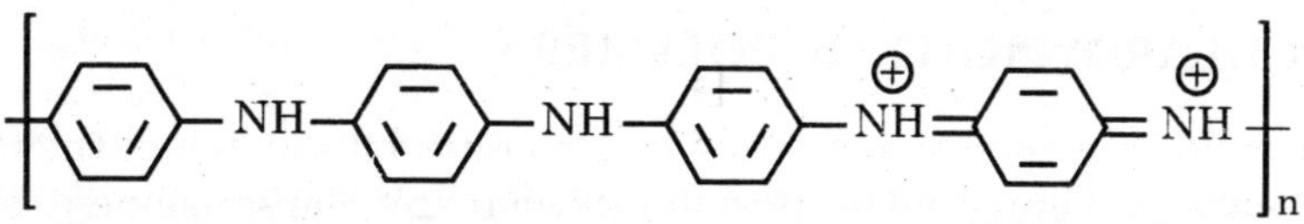

doped polyaniline (emeraldine salt) (Conductivity ~10^3 Scm^{-1})

Because of its conductivity, it is also known as *synthetic metal.*

3. **Conductive element-filled polymers:** These are the polymers which are filled with conducting elements, such as carbon black, metallic fibres, metal oxides etc. The polymer acts as a binder to hold the conducting elements together. These polymers possess reasonable conductivity. Moreover, these are low in cost, light in weight and mechanically durable and strong. Carbon black which is used as a filler is special conducting grade carbon. This carbon has very high surface area and is highly porous.

   However, the addition of the filler affect the properties of the polymer such as tensile strength, and impact strength etc. This problem can be overcome by the use of blended conducting polymers.

4. **Blended conducting polymer:** Conducting polymers can be blended with the conventional polymers to produce blends. These blends so formed possess better physical, chemical and mechanical properties.

ENGINEERING APPLICATIONS OF CONDUCTING POLYMERS

Potential applications of conductive polymers are quite significant. This is due to the fact that these are generally lighter, more flexible and easier to fabricate than many of the materials, which are conventionally used. Some of the promising applications include:

(i) **Rechargeable batteries:** The use of conducting polymers as cathodes and solid electrolytes in batteries for automotive and other applications is highly advantageous as an alternative to lead acid batteries. The polymeric batteries have high reliability, light weight, nonleakage of electrolyte solution and are small in size.

(ii) Some conducting polymers like polyaniline show different colours in different oxidation forms. This electrochromic property can be used to produce "**smart windows,**" and electrochromic displays. (Smart windows are the windows which change colour in response to sunlight or temperature changes).

(iii) These are also used in **electroluminance displays** like in mobile phones etc.

(iv) Other possible uses of conductive polymers include **analytical sensors**, (for pH, SO_2, glucose etc) conductive paints, etc.

(v) These can also be used in **Light Emitting Diodes** (LEDs), electromagnetic shielding and non-linear optical material (Non-linear optical property means that the optical properties varies with the intensity of light).

(vi) Can also be used in photovoltaic devices e.g in Al/polymer/Au photovoltaic cells.

(vii) Their wider commercial utilization awaits further increase in their conductivities and improvements in the mechanical properties and processability. One of the promising near-future applications of conductive polymers would be in the integrated circuits. The integrated circuits have attained the minimum pattern dimension of 200 nm. But the dream is to have the pattern on a single molecule, which can only be realised with the conductive polymers. This will increase the speed and dynamic memory of computers by a factor of 10^8.

11.13 MOLECULAR WEIGHT OF POLYMERS

The molecular weight of polymers is related to the chain length and the extent of crosslinking between different chains. The extent of cross-linking depends on the concentration of the monomer during polymerization. The length of the polymer chain depends on the random encounter between the monomer and the reactive site of the chain. Due to the randomness, some polymeric chains grow longer than the others. Thus, a polymer sample is a mixture of different molecules having different chain lengths. Hence molecular weight of a polymer is always expressed as some sort of average of molecular weight. The molecular weights can be expressed as–

(i) Number Average Molecular Weight M_n:

$$\bar{M}_n = \frac{\text{Total Mass of the polymer sample}}{\text{Number of molecules present in the sample}}.$$

or

$$\bar{M}_n = \frac{n_1 M_1 + n_2 M_2 + n_3 M_3 + \ldots}{n_1 + n_2 + n_3 + \ldots}$$

$$= \frac{\Sigma n_i M_i}{\Sigma n_i}$$

where n_1, n_2, n_3 etc. are the number of molecular species having molecular mass M_1, M_2, M_3 etc ... respectively. Measurement of an appropriate colligative property affords the Number Average Molecular Weight.

(ii) Weight Average Molecular weight Mw:

$$\bar{M}_W = \frac{W_1M_1 + W_2M_2 + W_3\ M_3 + \ldots}{W_1 + W_2 + W_3 + \ldots} = \frac{W_1\ M_1 + W_2M_2 + W_3M_3 + \ldots}{W}$$

where W_1 = Total mass of the species having molecular weight M_1.
$W_1 = n_1M_1$
W = Total weight of the polymer sample

$$\therefore \quad \bar{M}_W = \frac{\sum n_i\ M_i^2}{W}$$

(iii) Viscosity Average Molecular Weight, $\overline{M}_v$:

$$\bar{M}_v = \left[\frac{\sum Ni\ Mi^{(1+a)}}{\sum Ni\ Mi}\right]^{1/a}$$

where Ni is the number of molecules of molecular weight Mi a is the exponent in Mark-Kuhn-Houwink equation i.e. $[\eta] = KM^a$, where η is the intrinsic viscosity and K is a constant.

(iv) Degree of polymerization (*DP*): It is the average number of monomer units in a chain. It can be number average degree of polymerization (DP_n) or weight average degree of polymerization ($\overline{DP}_W$).

$$\overline{DP}_n = \frac{\bar{M}_n}{M_o} \text{ and } \overline{DP}_W = \frac{\bar{M}_W}{M_o}$$

where $\bar{M}_n$, $\bar{M}_W$ and $\bar{M}_o$ are number average, weight average and molecular weight of the monomer respectively.

NUMERICAL PROBLEMS

Example 1. Calculate the number average molecular weight, if two polymers, having mass equal to 100 to 10,000 are mixed.

Solution: $\bar{M}_n = \dfrac{100 + 10{,}000}{2} = 5050.$

Example 2. Calculate the number and weight average molecular wights of a polymer sample containing 20% of polymer *A* and 80% of polymer *B*. The molecular weights of *A* and *B* are 3000 and 30,000 respectively.

Solution: $\bar{M}_w = \dfrac{0.2 \times 3000 + 0.8 \times 30{,}000}{0.2 + 0.8} = 24600.$

For finding out $\bar{M}_n$ we have to calculate the number of molecules of *A* and *B*.
Number of molecules of *A* having

$$\frac{\text{molecular mass 30,000}}{\text{Number of molecules of } B \text{ having molecular mass 3000}} = \frac{80 \times 3000}{20 \times 30{,}000} = \frac{2}{5}$$

Thus for every 2 macromolecules of mass 3000, there are 5 molecules of mass 30,000. Hence,

$$\bar{M}_n = \frac{5 \times 3000 + 2 \times 30{,}000}{(5+2)} = 10714.$$

Example 3. Equal weights of polymer molecules with molecular weights 20,000 g/mol and 200,000 g/mol are mixed. Calculate $\bar{M}_n$ and $\bar{M}_w$.

Solution: Let the weight of one of the polyemer = W_1

Weight of the other polymer = W_2

$$W_1 = W_2$$

Let it $W_1 = W_2 = 400{,}000$ g

Number of molecules of first monomer

$$n_1 = \frac{400{,}000}{20{,}000} = 20$$

Number of molecules of second polymer

$$n_2 = \frac{400{,}000}{202{,}000} = 2$$

$$\bar{M}_n = \frac{n_1M_1 + n_2M_2}{n_1 + n_2} = \frac{20 \times 20{,}000 + 2 \times 200{,}000}{20 + 2}$$

$$\bar{M}_w = \frac{n_1M_1 + n_2M_2}{n_1M_1 + n_2M_2} = \frac{20 \times (20{,}000) + 2 \times (200{,}000)^2}{20 \times 20{,}000 + 2 \times 200{,}000}$$

$$= \frac{20 \times 2 \times 10^8 + 2 \times 2 \times 10^{10}}{8 \times 10^5} = \frac{4 \times 10^9 + 4 \times 10^{10}}{8 \times 10^5} = \frac{4 \times 10^9\ (1 + 10)}{8 \times 10^5}$$

$$= 110{,}000 \text{ g/mol.}$$

Example 4. If 1000 g of a polymer of molecular weight 1000 g/mole is mixed with 1000 g of another polymer of molecular weight 10^6 g/mole, what is the ratio of $\bar{M}_w / \bar{M}_n$.

Solution:

Polymer	*Weight*	*Mol.wt.*	*No. of moles in*
1	1000 g	1000 g/mol	1 (n_1)
2	1000 g	10^6 g/mol	$10^{-3}(n_2)$

$$\bar{M}_n = \frac{n_1M_1 + n_2M_2}{n_1 + n_2} = \frac{(1 \times 1000) + (10^{-3} \times 10^6)}{1 + 10^{-3}} = \frac{2000}{1.001} \approx 2000$$

$$\bar{M}_w = \frac{n_1M_1^2 + n_2M_2^2}{n_1M_1 + n_2M_2} = \frac{1 \times (1000)^2 + 10^{-3} \times (10^6)^2}{(1000 + (10^6 \times 10^{-3}))}$$

$$= \frac{10^6 + 10^9}{1000 + 1000} = \frac{10^6(1 + 10)}{2000} = 5.5 \times 10^5$$

$$\frac{\bar{M}_w}{\bar{M}_n} = \frac{5.5 \times 10^5}{2000} = 2.75 \times 10^2$$

Problem: A polydisperse sample of polystyrene is prepared by mixing three monodisperse samples in the following proportions:

1 g 10,000 molecular weight, 2 g 50,000 molecular weight, 2 g 100,000 molecular weight. Using this information, determine the following:

(a) Number-average
(b) Weight-average.

QUESTIONS

1. Define the terms: Polymer; Monomer; Degree of polymerization; copolymerization.
2. Define functionality of a molecule.
3. How are polymerization reactions classified on the basis of mechanism?
4. What are addition and condensation polymerization processes? Explain with example.
5. Differentiate between thermoplastics resins and thermosetting resins. Classify the following as thermoplastic or thermosetting:
   polyethylene, nylon, polyester, urea-formaldehyde resin, teflon, polystyrene, bakelite.
6. What do you mean by tacticity? Explain an isotactic, a syndiotactic and an atactic polymer.
7. Explain (a) cationic polymerization, (b) anionic polymerization, (c) coordination polymerization.
8. Explain step growth and chain growth polymerization with examples.
9. What are the chemical and structural differences between thermosetting and thermoplastic polymers. Name three important examples of each class.
10. What are the monomers used for each of the following polymers. Give commercial application of each:
    (a) Bakelite (b) Nylon-6
    (c) Nylon 6-6 (d) Terylene
    (e) Teflon (f) PVC
11. (a) List out the difference between LDPE and HDPE.
    (b) What are the materials from which an epoxy adhesive is prepared.
    (c) Name any flourine containing polymer. State whether it is thermoplastic or thermosetting.
12. Give properties and uses of the following polymers:
    (a) Nylon 6-6 (b) PMMA
    (c) PVC (d) Teflon
    (e) Polyester
13. Give main properties and uses of:
    (a) HDPE (b) Plexiglass
    (c) Epoxy resins (d) Polypropylene.
14. Distinguish between nylon-6 and nylon 6,6.
15. What are the commercial uses of plastics.
16. Write a short note on coordination polymerization.
17. Mention the uses of urea-formaldehyde resins and polyurethane.
18. Differentiate between homopolymer and copolymer. Illustrate your point with examples.
19. Explain:
    (a) Why can not thermosetting resins be reused and reshaped?
    (b) Why is bakelite used in electrical appliances.
    (c) Why is teflon highly chemical resistant.
    (d) Why is PVC generally plasticized?
20. Distinguish between the following:
    (a) Addition and condensation polymerization
    (b) Anionic and cationic polymerization.

(c) LDPE and HDPE
(d) Thermoplastic and thermosetting resins.

21. Give properties and uses of the following:
(a) PVC (b) PVA
(c) Polystyrene (d) PMMA
22. Outline briefly on the applications of plastics in engineering and industry.
23. What is polyamide? Discuss various types of nylons.
24. Write short notes on:
(i) Mechanism of cationic polymerization
(ii) Mechanism of anionic polymerization
(iii) Conducting polymers
25. (a) Distinguish between thermoplastic and thermosetting resins.
(b) Give the properties and uses of
(i) Phenolic resins
(ii) Nylon 6:6
(iii) Urea-formaldehyde. **G.G.S.I.P.U, II Term, B.Tech. July 2000**
26. (a) Differentiate between thermoplastic and thermosetting resins.
(b) Write the structural unit and two important applications of each of the following:
(i) PE (ii) PVC
(iii) Nylon 6,6 (iv) PMMA
(v) Bakelite **G.G.S.I.P.U, End Term, B.Tech. July 2000**
27. (a) Classify the following polymers on the basis of action of heat on them
(i) Bakelite (ii) Nylon
(iii) Dacron (iv) Polyethylene
(b) Write a short note on conducting polymers. **G.G.S.I.P.U, End Term, B.Tech. May 2001**
28. What is PET? Discuss its properties and applications. What are the advantages of blending PET with wool and cotton.
29. What is an epoxy resin? Give its properties and uses.
30. Name the polymers used for the following:
(i) Squeeze bottles (ii) Domestic gas piping
(iii) Overhead tanks for water storage. (iv) Baby bottles
(v) Wrapping confectionary items (vi) Bobbins, Spools, Spindles etc
(vii) Radiator grills (viii) Radio and television cabinets
(ix) Ion exchange resins (x) Inner liners for refrigerators
(xi) Loudspeaker housings (xii) Safety helmets
(xiii) Containers for detergents, cosmetics, mineral water
(xiv) Ladies handbags (xv) Adhesives
(xvi) Laminating safety glass (xvii) Alternative to glass
(xviii) Housing for rear lamps, (xix) Gaskets and reals
(xx) Non-stick coating
(xxi) Moulding for computers, calculating machines
(xxii) Microwave cookware
(xxiii) Substitute for metals in gears and bearings.
(xxiv) Bullet resistant vests (xxv) Switch board, heater handles.
(xxvi) Tanning material for leather. (xxvii) Araldite (adhesive)
(xxviii) Skid resistant surface for highways.
31. What are elastomers?
32. Give the structure and uses of:
(i) SBR (ii) Butyl Rubber
(iii) Buna N (iv) Styrene acrylonitrile rubber.

33. Compare the structure of elastomers and plastics.
34. Explain vulcanization of rubber.
35. Explain number average and weight average molecular mass.
36. Weight average molecular weight is higher than number average molecular weight. Explain
37. What do you mean by the term copolymerization. What are its advantages?
38. How structure of a polymer is related to its (i) Mechanical property (ii) Electrical property.
39. (a) What are resins? Describe some distinguished characteristics of 'Thermoplastic' and 'Ther mosetting' Resins.
    (b) Write down the preparation and uses of any two industrially important "Vinyl" based Resins. **G.G.S.I.P.U, End Term, B.Tech. II Batch, July 2001**
40. Define resins. Describe thermoplastic and thermosetting resins, their properties and applications. **G.G.S.I.P.U, End Term, May 2002**
41. Give the preparation, propert ies and uses of following:
    (a) Polyvinyl chloride
    (b) Bakelite
    (c) Polyester **G.G.S.I.P.U., End Term, Apr. 2002**
42. (a) Distinguish between thermosetting and thermoplastics resins.
    (b) What are nylons? How are they distinguished from one another. Explain with the help of suitable example.
    (c) What is meant by functionality of monomer. **G.G.S.I.P.U., End Term, May 2003**
43. Write notes on
    (a) Conducting Polymers
    (b) Fiber reinforced composites.
44. (a) Give the preparation, properties and applications of any one thermoplastic polymer.
    (b) Define functionality of a monomer; classify the following on the basis of their functionality.
        (i) Ethylene Glycol
        (ii) Vinyl Chloride
        (iii) Lactic Acid
    (c) Discuss a technique by which a polymer can be made into a conducting polymer, Mention few applications of conducting polymers. **G.G.S.I.P.U. End Term, May 2004**

❑❑❑

12 Composites

"The whole of science is nothing more than a refinement of everyday thinking".

— Albert Einstein

CHAPTER OUTLINE

Modern technologies require materials which have unusual combinations of properties that cannot be provided by any single material. For example, aircraft engineers would like to have structural materials having low density, stiffness, high strength, impact resistant and which are not easily corroded. Now all these properties can not be met with any single material, because strength is associated with high density and increase in stiffness would mean decrease in impact strength. This is where the concept of composite materials comes into play.

Generally speaking, **a composite material is a combination of two or more materials that exhibit a significant proportion of the properties of all the constituent materials**. A combination of two or more distinct materials can provide better combination of properties such as strength, heat resistance or stiffness etc. Such composite material systems result in a performance which cannot be achieved by any of the individual constituents. Thus, the composite materials offer the great advantage of combining different materials to tailer make the desired properties as per the required specification.

As an illustration, in Fig. 12.1 a comparison of different properties of individual constituents and their composite is made. This figure indicates that composite materials provide great improvement over the properties of individual constituents.

Practically everything in this world is a composite material. For example, wood is a fibrous composite made up of strong and flexible cellulose fibres in a stiffer lignin matrix. Bone is a composite made up of short and soft collagen fibres embedded in a hard and brittle mineral matrix called apatite. However, in the present context, a composite material is the material which meets the following conditions:

(i) It is prepared artificially.

(ii) It consists of two or more physically or chemically distinct phases, which are separated by a distinct interface.

(iii) It has properties which are not depicted by any of its individual components.

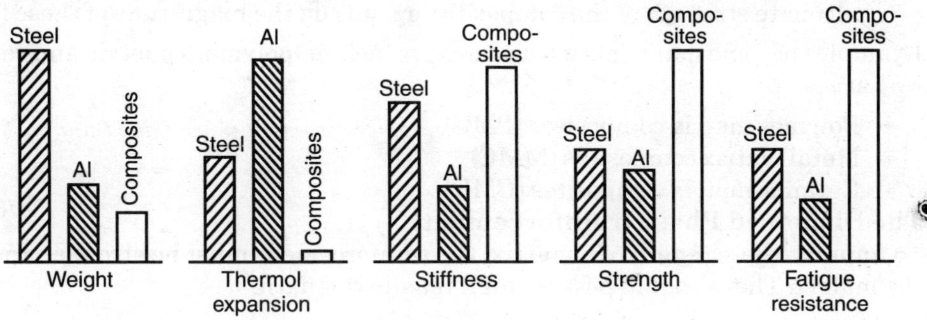

Fig. 12.1 Comparison between conventional materials and composite materials.

12.1 CONSTITUENTS OF COMPOSITES

Composite materials generally comprise of just two phases

(i) The constituent forming the body of the composite, called the **'matrix'** and

(ii) The structural constituent called **'dispersed phase'** or the reinforcing material.

The properties of the composites are determined by the properties of the constituent phases, their relative amounts and their distribution and orientation.

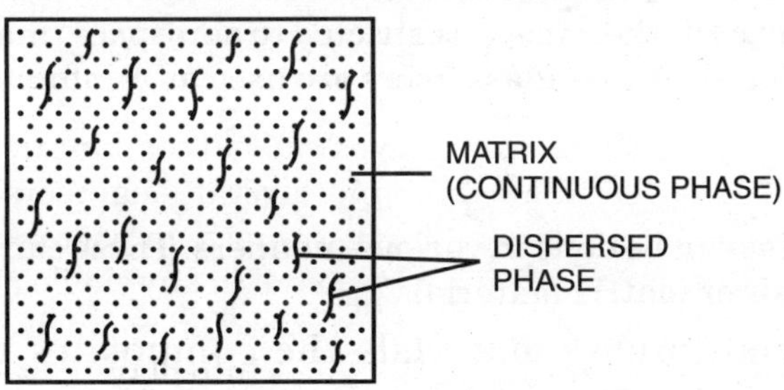

Fig. 12.2 Representation of a composite material

The most common example of a composite material is the **fiber glass reinforced plastic**. The plastic acts as the matrix and fibre glass is the dispersed phase. The plastic alone is relatively weak and has a low elastic modulus, bends and stretches easily. The glass fibers provide the strength and stiffness and their modulus of elasticity may be 50 times greater than that of plastic. Glass fibers take most of the load when the composite is stressed.

(i) **Matrix Phase**

The matrix phase of composites has several important functions like-

- It binds the reinforcing material (dispersed phase) together,
- protects the reinforcing material from surface damage due to mechanical abrasion or chemical reaction.
- It separates the individual particles of the reinforcing materials.

- It acts as a medium by which external load is transmitted and distributed to the dispersed phase.
- It prevents the propagation of brittle cracks because of its plasticity.

The matrix material should meet the following criteria.

- Its elastic modulus should be lower than that of the dispersed phase :
- It should be ductile.
- The adhesive bonding forces between the fiber and the matrix must be high. The ultimate strength of the composite depends on the magnitude of these forces.

Examples of some materials used as matrix include-polymers, metals and ceramics. Thus we have—

- Polymer matrix composites (PMC)
- Metal matrix composites (MMC)
- Ceramic matrix composites (CMC)

(ii) **The Dispersed Phase (reinforcement)**

To improve the strength of a matrix, the reinforcement must be stronger and stiffer than the matrix. These should possess high tensile strength.

Properties of the composites, to a large extent, depend upon the properties of the reinforcing materials, their geometry, which is described by shape, size and size distribution and orientation. They are chosen for their high strength and modulus. The physical form and shape of the reinforcements vary greatly depending on many factors. The most effective reinforcements are *long fibres*, and the composites so formed are called *fibre reinforced composites*. Some of the fibers used as reinforcement are glass fibres, carbon fibers, boron fibers and ceramic fibers. Other reinforcing materials include whiskers, metallic wires, particles, flakes etc. But still, fibers form the most commonly used reinforcing materials because the composites so formed are stiffer and stronger than the composites formed by any other reinforcing material. Another factor which is responsible for the choice of composite material is the cost. For example, carbon fibers have very high modulus values (upto five times that of steel) and therefore, make excellent reinforcement. However, due to their high cost, their use is restricted to aerospace industry. For non-aerospace applications cheaper material such as glass fibers are used as reinforcing materials.

12.2 CLASSIFICATION

Classification of composites can be done in various manners. These can be classified on the basis of the reinforcing materials or matrix materials.

(A) On the basis of reinforcing materials the composites can be classified as:

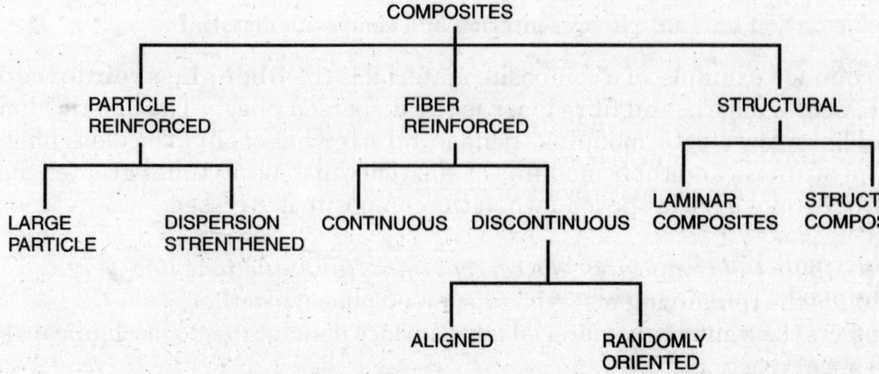

Fig. 12.3 Classification scheme on the basis of reinforcing materials

As in evident from the Fig. 12.3 that reinforcements can be in the form of particles, fibers or structural. Following is the brief description of these materials:

1. **Particle Reinforced Composites**

   When the reinforcing material is in the form of large or small particles suspended in a matrix, the composites are called particle reinforced composites. The particles can be either metallic or non metallic. These are of two types :

   (a) **Large-particle Composites:** The term "Large"is used to indicate that particle-matrix interactions cannot be treated on the atomic and molecular level. For most of these composites the particulate phase is harder and stiffer than the matrix. The reinforcing particles can have quite a variety of geometries, but they should be of approximately the same dimensions in all directions (equiaxed). For effective reinforcement, the particles should be small and evenly distributed throughout the matrix. Common examples of large-particle composites include:

   automobile tyres in which polymer matrix contains spherical particles of carbon with 20-50 mm diameters. The carbon increases the tensile strength, toughness and abrasion resistance.

   Cermet is another important particle reinforced composite which is composed of extremely hard particles of a refractory carbide, such as tungsten carbide or titanium carbide, embedded in a matrix of metal such as cobalt or nickel. These composites are utilized extensively as cutting tools for hardened steels. No single material could possibly provide the combinaiton of properties possessed by a cermet. In these composite materials (cermets) the hard carbide particles provide the cutting surface, but, being extremely brittle, are not themselves capable of withstanding the cutting stresses. The metal matrix thus enhances the toughness, which also isolates the carbide particles from one another and prevents particle to particle crack propagation.

   Another example of large-particle composites is *concrete*, which is composed of cement matrix and particulates of sand and gravel.

   (b) **Dispersion strengthened composites:** In these type of composites, very small particles of the range 10 to 100 nm are used as strengthening material. The particle-matrix interactions occur on the atomic or molecular level. The dispersed phase i.e. the reinforcing particles may be metallic or non-metallic. The strengthening mechanism involves interactions between the particles and dislocations within the matrix. The strengthening is retained at elavated temperatures and for longer time periods because the material for dispersed phase is so chosen, which is unreactive with the matrix phase. While the matrix bears the major part of the applied load, the dispersed particles hinder the motion of dislocations. This helps in restricting the plastic deformation and improving the tensile strength and hardness.

Fine metallic or non-metallic particles are used as dispersed particles for strengthening metals and metal alloys. For example, the high temperature strength of nickel alloy is improved by adding thoria (ThO_3) in the form of finely dispersed particles in 3 volume%.

2. **Fiber-Reinforced Composites :** Fibers constitute an important class of reinforcing materials. These composites provide high strength and stiffness on a weight basis. The use of fiber as reinforcing materials in due to

   - The small diameters of fiber with respect to other microstructural units.
   - The aspect ratio (length/diameter) of fibers is very high and hence, a very large fraction of the applied load is transferred via the matrix to the fibers which are stiff and strong.

- A very high degree of flexibility permits the use of a variety of techniques for the formation of composites, using the fibers.

Two most important characteristics of fiber reinforced composites are Fiber length and Fiber orientation.

(i) **Fiber Length:** The applied load on the composite is transmitted to the fibers by the matrix phase and hence, some critical fiber length is necessary for effective strengthening and stiffening of composites. The critical length l_c is dependent on the fiber *diameter d*, its *tensile strength* σ_f and on the *fiber matrix bond strength* (shear yield strength) Tc as per the following equation:

$$l_c = \frac{\sigma_f d}{T_c}$$

For a number of fibers this critical length is of the order of 1 nm which is, in most of the cases, ranges between 20 to 150 times of the fiber diameter. If the fiber length is greater than the critical length, the fibers are termed as **continuous** and if it is consideraly less than the critical length, then the fibers are termed as **discontinuous** or short fibers.

(ii) **Fiber Orientation :** The strength and other properties of the fiber reinforced composites are significantly influenced by the orientation and arrangement of fibers relative to one another. There are two extreme possibilites with respect to the orientation of fibers:

- A parallel alignment of the longitudinal axis of the fibers in a single direction.
- A totally random alignment.

Continuous fibers are normally aligned whereas discontinuous fibers may be aligned or randomly oriented. Fig. 12.3 represents the three types of fiber composites schematically.

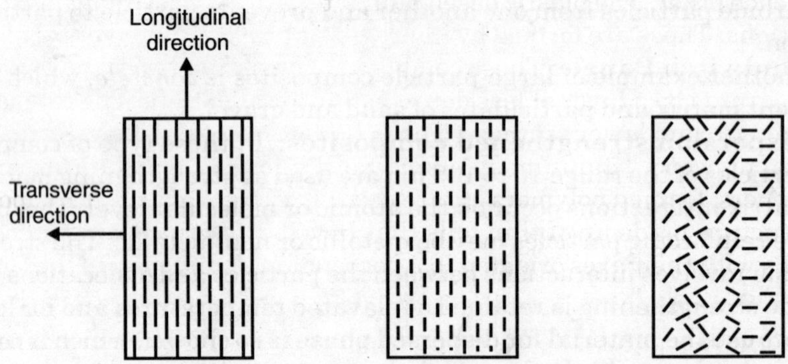

Fig. 12.4 Schematic representations of (a) continuous and aligned (b) discontinous and aligned, and (c) discontinuous and randomly oriented fiber reinforced composites

(a) **Continuous and aligned Fiber Composites:** A uniform fiber distribution results in better overall composite properties. The properties of continuous and aligned composites are highly anisotropic, that is, they depend on the direction in which they are measured. In the longitudinal direction, i.e. along the alignment, maximum strength and reinforcement is achieved. In the transverse direction, as the fiber reinforcement is virtually non-existent, hence strengthening is least.

(b) **Discontinuous Fiber Composites :** The reinforcing property of discontinuous fiber composites (aligned as well as random) is lower than that of continuous fiber composites, but still these are used extensively because of very low cost as compared to continuous fibers. Further, in case of discontinuous (short) fiber composites, intricate shapes can be formed which are rather impossible with continuous fiber reinforcement. Different type of fibers used for reinforcing materials are discussed in the next section (12.3)

3. **Structural reinforced composites :** A structural composite consists of both, the homogeneous material and composite material. These two different materials are bonded together in layers. The properties of these composites are not only governed by the properties of the constituent materials but also on their geometrical arrangements. Broadly structural composites are classified as :
   (i) Laminar composites and
   (ii) Sandwitch composites.

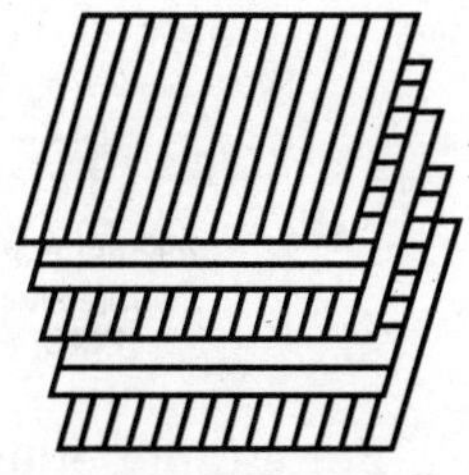

Fig. 12.5 Stacking of successive oriented fiber-reinforced layers for producing a laminar composites

   (i) **Laminar Composites:** In laminar composites, two dimensional layers of two different materials are stacked together to form the composite structure. Each single layer is known as lamina. A single lamina is unsuitable for any useful purpose. The laminae are stacked and glued together such that the orientation of the high strength direction varies with each successive layer as shown in Fig. 12.4. The individual unidirectional lamina piles are oriented in such a way that the resulting structural component has the desired mechanical or physical characateristic in different direction.
   Examples of laminar composites include plywood, cladded metals, plastic based laminates. Adjacent sheets in plywood are aligned with the grain direction at right angles to each other. The plastic based laminates are very useful in high temperature applications. These are formed by impregnating silicons in layers of glass or asbestos.

   (ii) **Sandwitch Panels:** These composites consists of two strong outer sheets or faces, separated by a layer of less dense material called core. The core material is of lower strength and lower stiffness. The face materials are generally fiber reinforced plastics, plywood, titanium, steel and aluminium alloys. For core material, synthetic rubbers, foamed polymers etc are used. The core can also be composed of honey comb structure made up of thin foils that have been formed into interlocking hexagonal cells with their axes oriented in a perpendicular direction to the face planes as shown in Fig. 12.5

The core serve to separate the "faces" and hence resist any deformations perpendicular to the face plane.

Sandwitch panels are used extensively in aircraft for wings. They are also used in roofs, floor and walls of buildings. Use of lighter composites in aircraft structures results in energy savings. For a given aerodynamic configuration of an aircraft, there is a direct correlation between aeroplane weight and fuel consumption. Example, in 1986 aircraft voyager completed non stop flight around the world without refueling because of the use of sandwitch composite in its manufacture, resulting in lower weight and hence, fuel saving.

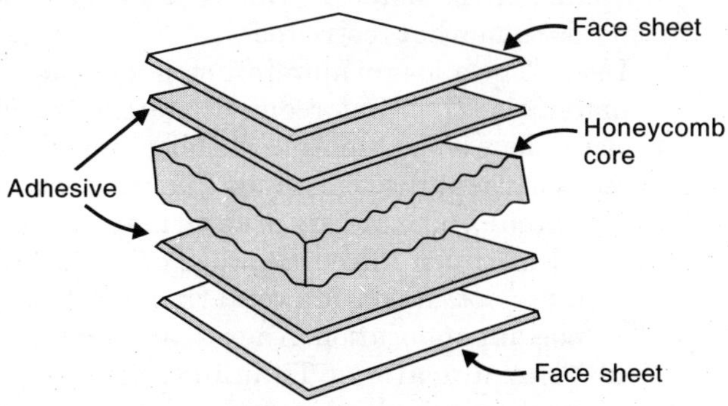

Fig. 12.6 Schematic diagram showing the construction of a honeycomb core sandwitch panel

Sandwitch composites also find uses in safety glasses. Safety glass is a layer of PVB sandwitched between two layers of glass. The glass in the composite protects the plastic from scratching and gives it stiffness. The plastic provides the toughness to the composites.

(B) On the basis of matrix material, composites can be classified as:

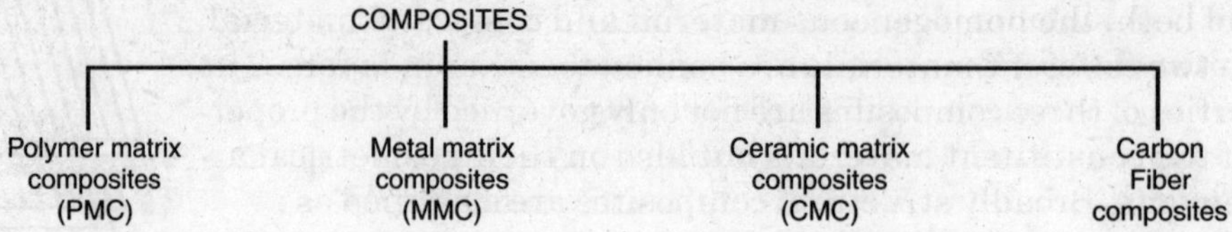

Fig. 12.7 Classification scheme on the basis of matrix material.

1. **Polymer-matrix composites (PMC):** Polymer matrix composites (PMCs) are extensively used as engineering structural materials. The polymers which are used as the matrix material can be thermoplastic or thermoset. The most widely utilized polymer resins are the polyesters and vinyl esters. Epoxies have also been used as matrix resins for advanced composite materials because of ease in processing and excellent mechanical properties. Polyether ketones, polyphenylene sulphide and polyetherimide thermoplastics are used as matrix materials for high temperature applications.
   A variety of reinforcements can be used with both thermoset and thermoplastic PMCs. High performance fibers such as carbon, boron, aramid, glass fiber etc are used as reinforcing materials. Nevertheless, glass fiber reinforced polymers represent the largest class of PMCs.
   Uses: Glass fiber reinforced polymer matrix are used in a wide variety of industries. Tanks and vessels in the chemical process industry, as well as process and effluent pipelines are made of glass fiber reinforced polymer matrix composites. Use of PMCs in military and commercial helicopters has been most significant. The main driving force in such applications is that of weight reduction. Boron/epoxy composites are used in sporting goods, e.g., golf clubs and tennis rackets.
2. **Metal Matrix Composites (MMC) :** Metal matrix composites consist of metal or metal alloy as matrix material. MMCs are of interest these days because they offer the opportunity to combine the desirable properties of two materials which are not available with the single material. The use of metals as matrix materials is due to their high operating temperature, nonflammability, and greater resistance to degradation by organic fluids. Metal matrix composites are costlier than the polymer matrix composites and hence, their use is somewhat restricted.
   Uses: The alloys of aluminium, magnesium, titanium and copper are employed as matrix materials. The reinforcement materials can be particulates, short fibers or whiskers or continuous fibers. Continuous fiber material include carbon, silicon carbide, boron, alumina etc, wheraeas discontinuous fiber material include chopped fibers and alumina and carbon. Some of the metals used as matrix materials are.
   (i) **Aluminum alloy:** Because of its low density, excellent strength and resistance to corrosion makes it a remarkable matrix material. The composites so formed find important application in aerospace field.
   (ii) **Titanium alloys:** Titanium has a relatively high melting point (1672^{0}C) and retains its strength at high temperatures with good oxidation and corrosion resistance. All these properties makes it suitable as matrix material for aerospace applications. Tita-

nium alloys are used in jet engines, fuselage parts etc. However, it is an expensive material.

(iii) **Magnesium alloy:** Magnesium is one of the lightest material and hence, its composites are used in aircraft gearbox housing, and housing for electronic equipments.

(iv) **Copper alloy :** The conductivity of copper makes it a good matrix material where good thermal conductivity is required. The major application of copper in a composite material is in niobium based superconductor.

3. **Ceramic-Matrix Composites:** Ceramic material are excellent matrix materials because of their high strength, high stiffness, chemical inertness and low density. However, ceramics materials lack toughness and hence, are prone to be easily damaged during processing. Thus the main purpose of ceramic matrix composite is to toughen the ceramics by incorporating reinforcing fibers. The toughness of ceramics have been improved significantly by reinforcing ceramics with particulates, whiskers etc. The whiskers of one ceramic material can be incorporated into another ceramic material resulting in toughened composite materials.

   Uses: CMCs are used widely in aerospace vehicles including aircraft, helicopters, missiles etc. Another area of CMCs is in the area of cutting tools. For example silicon carbide whisker reinforced alumina is used as a cutting tool inserts for high-speed cutting of superalloys.

4. **Carbon-Carbon Composites:** One of the most promising engineering materials is the carbon fiber reinforced carbon matrix composite i.e. both reinforcement and the matrix are carbon. Carbon carbon composites can withstand high temperatures in an inert atmosphere. These have high thermal conductivities and low coefficient of thermal expansion. Carbon-Carbon Composites have high tensile strengths, high-tensile moduli which are retained upto service temperature exceeding 1000^0C and even approaching 2200^0C. The major drawback of carbon carbon composites is the formation of gaseous oxides of carbon upon reaction with oxygen. Moreover, these are expensive too.

   Uses: Major applications of carbon carbon composites involve their use at high temperature for example, as heat shields for reentry vehicles, aircraft brakes, hot pressing dies and high temperature parts such as nozzles. Because of their excellent biocompatibility they are also used as implants and internal repair of bone fractures.

12.3 DIFFERENT FIBERS USED AS REINFORCING MATERIALS

Some of the common fibers used as reinforcing materials include glass fiber, carbon fibers, boron fiber, aramid fiber. A brief description of their properties and structures is given below :

1. **Glass Fibers :** Glass is a very extensively used reinforcing material for several reasons such as:
   - It is commonly available and can be economically fabricated into glass reinforced plastic by a wide variety of composite manufacturing techniques.
   - The fibers are very strong. When embedded in a plastic matrix, they produce a composite which have very high specific strength.
   - It is easily drawn into high strength fiber from the molten state and exhibits chemical inertness and is therefore useful in a variety of corrosive environments.

In order to protect glass monofilaments from abrasion, the fibers are usually coated during the drawing process with the material that protects the surface from damage.

Common glass fibers are silica based i.e. 50-60% SiO_2 in combination with a number of other oxides of calcium, boron, sodium, aluminium etc. The different type of glasses having varying proportion of oxides are commonly used and are designated as E Glass, C Glass and S Glass. The composition of these fibers is given in Table 12.1

Table 12.1 Chemical Composition of Some Glass Fibers

Compostion	*E Glass*	*C Glass*	*S Glass*
SiO_2	55.2	65.0	65.0
Al_2O_3	8.0	4.0	25.0
CaO	18.7	14.0	–
MgO	4.6	3.0	10.0
Na_2O	0.3	–	–
K_2O	0.2	–	–
B_2O_3	7.3	5.0	–

The E type glass was originally developed for electrical purposes and is least expensive and most common. C type has a better resistance to chemical corrosion whereas S type stands for high silica content and hence can withstand higher temperatures than other glasses.

The density of glass fibers is quite low whereas the strength is quite high. However, glass fibres suffer from a limitation that Young's modulus of these fibers is not very high. Thus, while strength-to-weight ratio of glass fibers is quite high, the modulus-to-weight ratio is low. However, it is quite cheap and is available in variety of forms (chopped fibers, continuous fibers, woven fabrics and non woven mats etc).

Uses: Glass fibers are used as reinforcing materials in building and construction industry. Commonly used composites containing glass fiber are glass-reinforced plastics or GRP.

- They are used in the manufacture of structural materials like wallpanel, window frames, tanks, bathroom units, pipes etc.
- In chemical industry, glass fiber reinforced materials are used as storage tanks, pipelines etc.
- The other big user of these fibers is rail and road transport industry and the aerospace industry. During world war II radomes of aircraft were manufactured using 'E' glass fibers Reinforcement of rubber tyres, underground petrol storage tank is another area of their use.
- S glass is used for high strength applications like rocket motor cases etc.

2. **Carbon Fiber:** Carbon is a high performance fiber material and is most commonly used reinforcement material now-a-days. The desirable properties which makes it a good reinforcing material are :
   - Carbon is very light element with a density equal to 2.268 g cm^{-3}
   - Carbon fibers have highest specific modulus and specific strength among all reinforcing materials.
   - Unlike glass, these are not affected by moisture at room temperature. Moreover, these are not attacked by the solvents like acids and bases.
   - Like glass fibers, carbon fibers are available in a variety of forms with monofilament diameter in the range of 5 to 10 microns.
   - Carbon fibers of extremely high modulus can be made by carbonization of organic precursor fibers, followed by graphatization. (Graphitization is a thermal treatment that improves the properties of carbon fiber).

Uses:

- Application of carbon fibers in aerospace industry were the major driving force for carbon fiber reinforced composites. The steady decline in the prices of carbon fiber over the years has led carbon fiber to assume an important position as reinforcing materials.
- The aerospace industry makes use of lightweight structures made of carbon fiber reinforced polymer matrix composites.
- In commercial aircrafts, such as Boeing 757, 767 and 777, carbon fiber composites are used in a number of locations. It is claimed that carbon composite helicopter motor blades have lower direct operating costs than aluminium blades.
- In road transport industry carbon fiber epoxy composites have been used in drive shafts and various chassis parts. Conventional two piece drive shaft are being replaced by one-piece design made up of carbon composite, which results in 50-60% weight reduction.
- Blades in turbines, compressors, windmills, flywheels have been made of carbon fiber.
- Other application of carbon fiber composites include cargo shipping containers for rail, sea and land transportation.
- Pressurized tanks, especially CNG (compressed natrual gas) for bus and other vehicles are other uses of carbon fiber composites.

In sports goods industry the items where carbon fiber reinforced resin are used are archery bows, fishing rods, tennis rackets, cricket balls and skies.

A few companies are making musical instruments such as guitars, violins out of carbon fiber reinforced composites.

3. **Boron Fibers :** Boron is a brittle material. Boron fibers are manufactured by chemical vapour depositon of boron on a substrate. Generally, the material is tungsten wire. A carbon substrate can also be used. Thus, boron fiber as produced is itself a composite material. Due to the composite nature of the boron fiber, complex internal stresses and defects such as voids and structural discontinuities result from the presence of a core and the deposition process.

Uses:

- Boron fiber composites have been used in military aircraft components, helicopter rotor blades and some sporting goods.
- They are also used for stiffening golf shafts, tennis rackets and bicycle frames.
- The major limitation to the use of boron fiber is its high cost compared to other fibers.
- The high price is due to the high cost of its tungsten substrate.

4. **Aramid Fibers :** Aramid fiber is a manufactured fiber in which the fiber forming substance is a long-chain synthetic aromatic polyamide. Commercial name of aramid fiber is Kevlar. The basic chemical structure of aramid fibers consists of oriented para-substituted aromatic units, which makes them rigid rod like polymers. Kevlar is p-phenylene terephthalamide. There are several grades of Kevlar (Kevlar 29, 49 and 149) which have different mechanical behaviour.
   - The aramid fibers have higher longitudinal tensile strengths and tensile moduli than other polymeric fiber materials.
   - These are known for their toughness, impact resistance and resistance to creep and fatigue failure.
   - Though aramids are thermoplastics, they are resistant to combustion and stable at relatively high temperatures (2000°C).
   - Aramid fibers are sensitive to ultraviolet (UV) light.
   - When exposed to UV light, aramid fibers discolour from yellow to brown and lose mechanical properties.

Uses:

- Kevlar is mainly used as reinforcing material for tires (belts or radial tires for cars and trucks).
- Kevlar 29 is used for ropes, cables coated fabrics for inflatables and ballistic (bullet proof) protection fabrics.
- Kevlar 49 is used for reinforcement of epoxy, polyester and other resins for use in aero-space, marine and automotive industries.

5. **Other Fiber Reinforcement Materials :** Other fiber materials that are also used as reinforcements, though to a limited extent are silicon carbide, ceramic fibers, aluminium oxide etc.

QUESTIONS

1. What are composite materials? How are they classified on the basis of reinforcing material?
2. Discuss the classification of composite on the basis of matrix material?
3. Explain laminated composites giving their properties.
4. What are fiber reinforced composites?
5. What do you mean by metal matrix composites.
6. Write short notes on :
   (a) Carbon fiber as reinforcing materials.
   (b) Sandwitch Panels
   (c) Polymer matrix composites
   (d) Metal matrix composites
7. What are the uses of Aramid Fiber composites.
8. What do you mean by particle reinforced composites? What are dispersion strengthened composites?
9. Name a few alloys used as matrix materials for composite. Also give reasons for their use as matrix materials.
10. What are the advantages of using glass as reinforcing materials?
11. Why is surface protection of glass important?
12. Briefly discuss structural composites.
13. What are composite materials? Discuss some important types of fibre-reinforced composites.
    G.G.S.I.P.U. End Term, May 2002
14. (a) How is a composite defined? Give a broad classification of composite material
    (b) What are the factors on which the properties of a composite material depend?
    G.G.S.I.P.U. End Term, 2004

❑❑❑

13

INORGANICS POLYMERS

"Everyone takes the limits of his own vision for the limits of the world."

Asthur Schopenhauer

CHAPTER OUTLINE

An outstanding feature of organic polymer chemistry is the ability of carbon atoms to link together to form chains and rings. This property of self linkage is known as catenation which is responsible for one of the rapidly growing branch of chemistry i.e polymer chemistry.

Similarly catenation is an important feature of some of the elements other than carbon, belonging to group 14, 15 and 16, though the extent of catenation is much less as compared to carbon. Therefore, very long chains of homoatomic molecules are seldom formed, but polymers having heteroatomic chains nevertheless constitute by and large, the chemistry of inorganic polymers.

Inorganic polymers are the macromolecules composed of atoms other than carbon having covalent linkages. Among the wide variety of inorganic polymers, the frequently known polymers are–

(1) Phosphorus based polymers e.g. polyphosphazenes, polyphosphates.
(2) Silicon based polymers e.g. silicones.
(3) Sulphur based polymers e.g. polymeric sulphur, polymeric sulphur nitride etc.

General Properties of Inorganic Polymers:

(1) Inorganic polymers have high density of covalent bonds and if cross linking is considerable, they become much stiffer and harder than organic polymers.
(2) Inorganic polymers can be obtained in pure amorphous or pure crystalline forms. In contrast, organic polymers have structures with partly crystalline regions, known as crystallites along with amorphous regions.
(3) Generally, inorganic polymers do not burn. At high temperature they soften or melt.

(4) Inorganic polymers do not allow intercalation of solvent molecules, because chain segments between cross links in polymers are usually short and stiff.

(5) Inorganic polymers cannot be extended much as compared to organic polymers. They are much less ductile.

(6) These are stronger but harder and more brittle than organic polymers.

Some of the important inorganic polymers are discussed below in detail.

13.1 PHOSPHAZENES

Nitrogen and phosphorus have little tendency on their own to undergo catenation. However nitrogen and phosphorus catenate together resulting in a very extensive chemistry of polymeric compounds known as phosphazencs with P–N and P = N bonds. These polymers have the general structure $\left(-\overset{\displaystyle R}{\underset{\displaystyle R}{\overset{|}{\underset{|}{P}}}} = N - \right)_n$ where R can be Cl, OCH_3 and OC_2H_5. Monophosphazene ($R_3P{=}NR'$) diphosphazenes ($R_3P{=}N{-}(PR_2{=}N)_n \ldots PR_2$) and polyphosphazenes ($R_3P{=}N{-}(\text{PR}_2{=}\text{N})_n \ldots PR_2$) are all known. However it is the cyclophosphazenes and long chain polyphosphazenes that currently attract the greatest attention.

The polymers where R is Cl are known as **polyphosphonitrilic chlorides**. If Cl is replaced by OCH_3 or OC_2H_5, the compounds are known as **polydimethoxy** and **polydiethoxy phosphazenes** respectively.

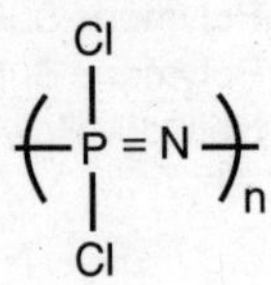

Polyphosphonitrilic chlorides

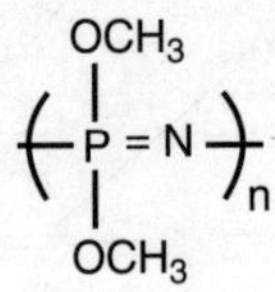

Polydimethoxy phosphazenes

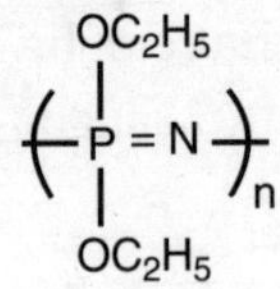

Polydiethoxy phosphazenes

Preparation: When phosphorus pentachloride is heated with ammonium chloride in 1, 1, 2, 2 tetrachloroethane or chloro benzene as solvent, a mixture of colourless solids of formula $(NPCl_2)_n$ is formed where n = 3 or 4.

$$nPCl_5 + nNH_4Cl \xrightarrow[C_2H_2Cl_4]{} (NPCl_2)_3 + (NPCl_2)_4 + [NPCl_2)_n + 4n - 1HCl$$

Depending upon the reaction conditions cyclic phosphazenes or linear polyphosphazenes can be formed. The trimeric or tetrameric compounds are easily seperated by distillation under reduced pressure. The trimer $(NPCl_2)_3$ is an intermediate in the synthesis of many other phosphazenes.

Mechanism: The reaction mechanism is complicated and is believed to take place in the following sequence:

(i) NH_4Cl first dissociates

$$NH_4Cl \rightarrow NH_3 + HCl$$

(ii) PCl_5 reacts in its ionic form $[PCl_4^+][PCl_6]^-$

(iii) Nucleophilic attack by NH_3 on $[PCl_4^+]$ results in the formation of intermediate species (I)

$$NH_3 + [PCl_4] + [PCl_6]^- \rightarrow \underset{\text{I}}{[HN{=}PCl_3]} + PCl_5 + 2HCl$$

(iv) Species I reacts with $[PCl_4]^+ [PCl_6]^-$

$$\underset{\text{I}}{[HN = PCl_3]} + [PCl_4]^+ [PCl_6]^- \rightarrow \underset{\text{II}}{[Cl_3P\,N\,PCl_3]^+ [PCl_6]^-} + HCl$$

(v) (II) reacts with NH_3 to give the following polymer–

$$[Cl_3P\ N\ PCl_3]^+[PCl_6]^- + NH_3 \longrightarrow \underset{\text{Linear Polymer}}{Cl - \overset{Cl}{\underset{Cl}{\overset{|}{\underset{|}{P}}}} = N - \overset{Cl}{\underset{Cl}{\overset{|}{\underset{|}{P}}}} = NH}$$

This process then continues and size of the chain grows as long as PCl_5 is available.

(vi) Cyclization occurs by intramolecular elimination of HCl to form rings.

Similarly $(NPBr_2)_n$ can be prepared by reacting PBr_5 with $NH_4\,Br$.

Cyclophosphazenes: The most common cyclophosphazenes are $(Cl_2P - N)_3$ and $(Cl_2P - N)_4$, which are commercially available. A mixture of these along with linear polymer can be obtained by the following reaction as stated earlier–

$$n\,PCl_5 + nNH_4Cl \xrightarrow[\text{or } C_6H_5Cl]{C_2H_2Cl_4} H\!\left(\overset{Cl}{\underset{Cl}{\overset{|}{\underset{|}{P}}}} = N\right)_{n=7}\!Cl + \underset{\text{Trimer}}{(PCl_2N)_3} + \underset{\text{Tetramer}}{(PCl_2N)_4} + (4n-1)\,HCl$$

It may be noted that hexachloro-cyclotriphosphazenes $(NPCl_2)_3$ is a key intermediate in the synthesis of many other phosphazenes and is manufactured commercially. It is a white crystalline solid (*mp* 113° C) that can be sublimed.

Structure: Structurally cyclophosphazenes can be represented as–

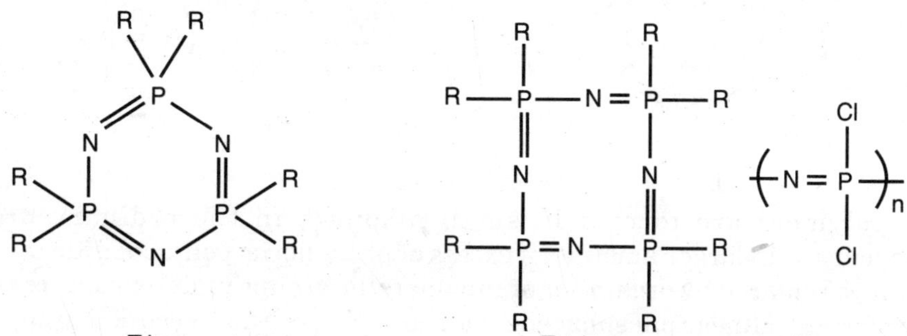

Trimer Tetramer Linear

The six membered rings are planar or nearly so, while the larger rings are nonplanar with *NPN* angles of ~120° and *PNP* angles of ~132°. Fig. 13.1 shows the structure of $(NPCl_2)_3$.

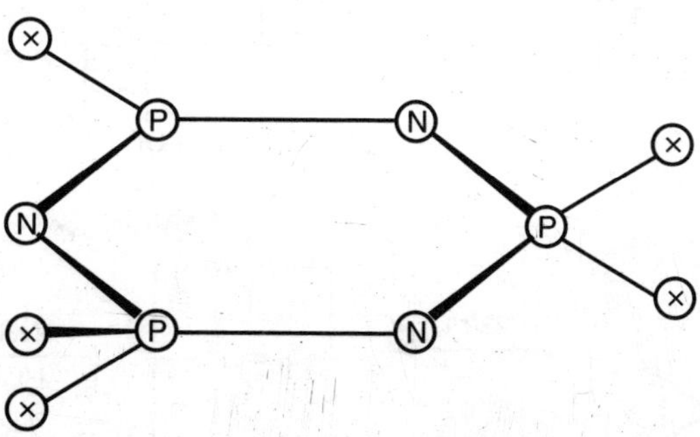

Fig. 13.1 Structure of cyclic trimer $(NPCl_2)_3$

In all of phosphazene compounds the apparent P-N and P=N bonds are equivalent.

Bonding: The P–N distances lie in the range of 1.55 to 1.60 A°. They are thus shorter than the expected single bond length of 1.75 to 1.80 A°. Thus cannot be represented as alternate single or double bonds. Much attention has been paid to understand the nature of P–N π bonding, but the matter is still a subject of controversy. The trimer appear to show marked resemblances to benzene as it is agreed that *N* atom, is sp^2 hybridized. Two of the five valence shell electrons of nitrogen form σ bonds with the phosphorus atoms while lone pair is retained in the third sp^2 hybrid orbital. Out of five valence shell electrons of phosphorus, four are used to form σ bonds with nitrogen and halogen atoms. Compared to benzene the double bonding does not arise from $p\pi - d\pi$ interactions but from $p\pi - d\pi$ interactions i.e $3d$ orbital of phosphorus and 2p orbital of nitrogen. Thus double bond interactions are weaker than that of benzene.

The extent of delocalization is also not clearly understood, that whether there is complete delocalization all around the rings, similar to that of benzene or these are localised. All the phosphazenes have a much lower electrical conductivity than analogous carbon based polyethylenes, suggesting that π bonds are not as delocalised in polyphosphazenes, as are the double bonds in carbon based unsaturated polymers. Further there are large number of orbitals potentially involved and hence σ and π character to individual orbitals cannot be assigned with complete surity.

Linear poly phosphazenes: A great variety of linear phosphazenes are known. Hexachlorocyclotriphosphazene $(NPCl_2)_3$ is the key intermediate to synthesize linear phosphazenes. Very pure $(NPCl_2)_3$,free of PCl_5 has to be used.

When $(NPCl_2)_3$ is heated above 230°C it gives the linear polymer, where degree of polymerization, *n*, can be = 15,000.

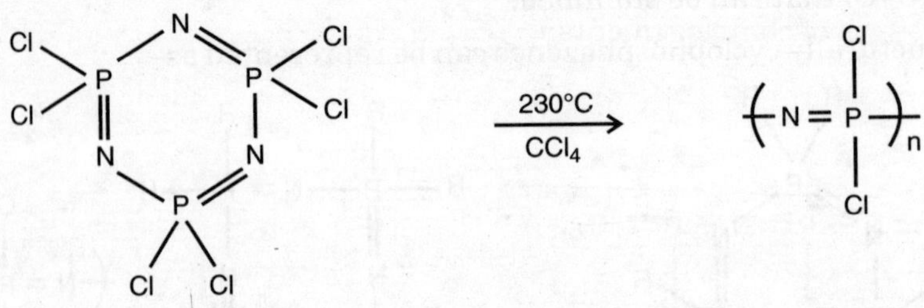

Linear polymers are formed in small amounts in the ordinary preparation of cyclophasphazenes or in larger amounts if excess of phosphorus pentachloride is used. The chlorine atoms can be replaced by organic or organometallic groups under moderate reaction conditions. The properties of linear phosphazenes can be varied by the choice of R groups in the $(R_2\,P\text{–}N)_n$ chain. The scope for the preparation of mixed substituent polymer is very vast.

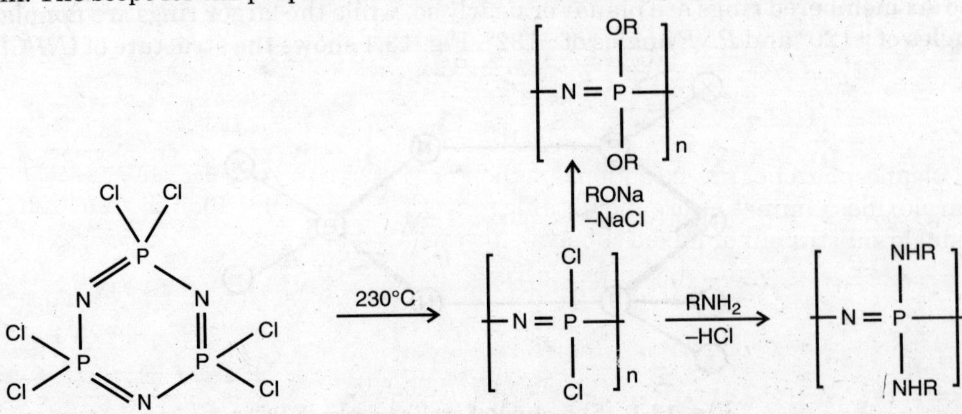

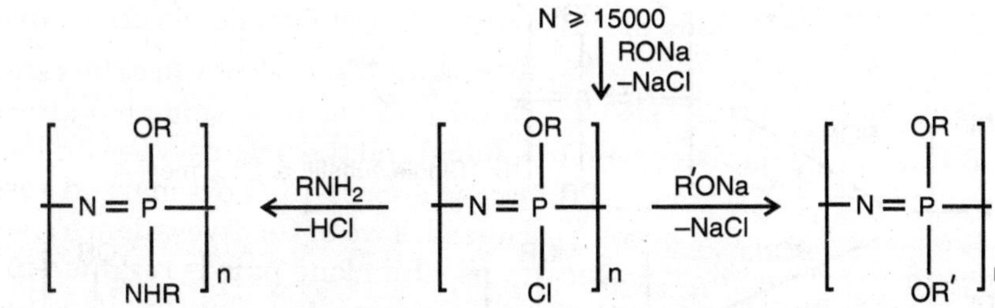

Properties of Polyphosphazenes: The properties of these polymers depend on both, the inorganic backbone and the organic or organometallic side groups.

- The backbone confers flexibility, fire resistance and stability to oxidation.
- The side groups control solubility, hydrolytic stability, surface characteristics, secondary reactivity, thermal stability and a wide range of optical, electrical and biological properties.
- The size and polarity of the side groups also influence material properties such as crystallinity and glass transition temperature. Some of the important properties of these polymers are discussed below–

1. Polyphosphazenes have mechanical properties similar to those of rubber. These exhibit high elasticity and can be stretched many folds with almost complete recovery.
2. Solubility : The freshly prepared polyphosphonitrilic chlorides are soluble in chloroform but insoluble in petroleum ethers.
3. When samples of highly elastic rubbers are kept in absence of air their elastic properties do not change. However, material becomes brittle due to the storage in moist air. This is due to the fact that oxygen bridges are formed from chain to chain so that there is reduction in average length of chain segments.

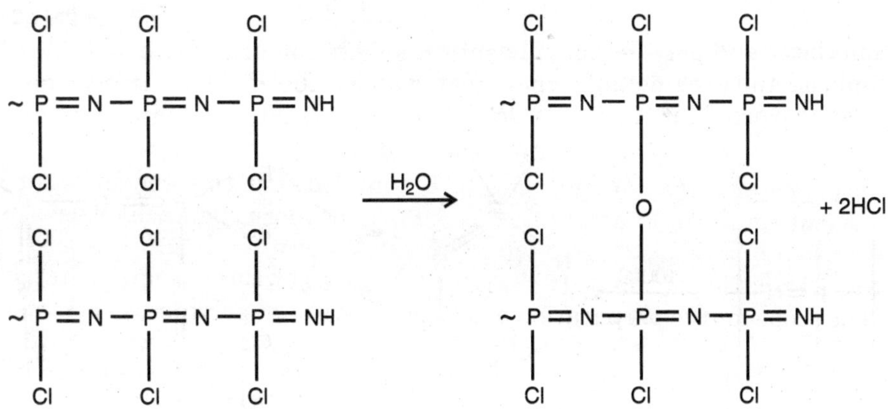

4. Polyphosphazenes are exceptional in their high reactivity to nucleophiles such as alkoxides, aryloxides, amines or organometallic reagents. Depending upon the reaction conditions single substituent or mixed substituents are formed.

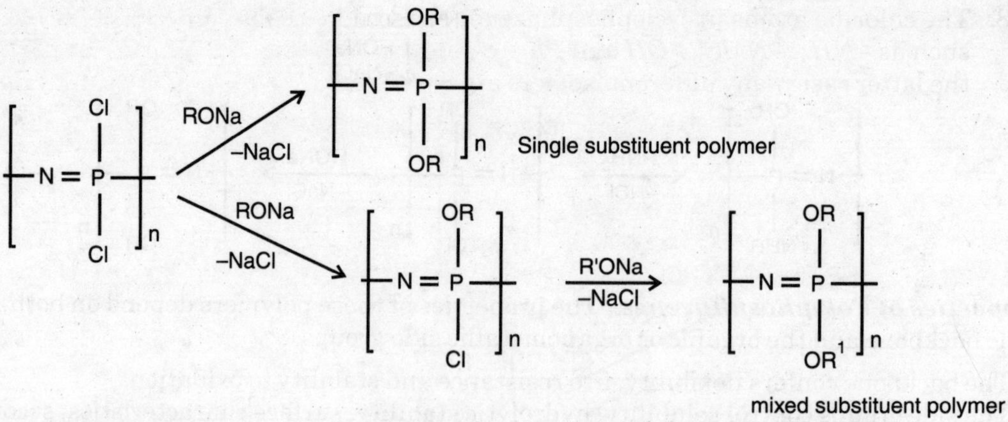

5. When polyphosphazenes react with NH_3 it produces cross linked polymers

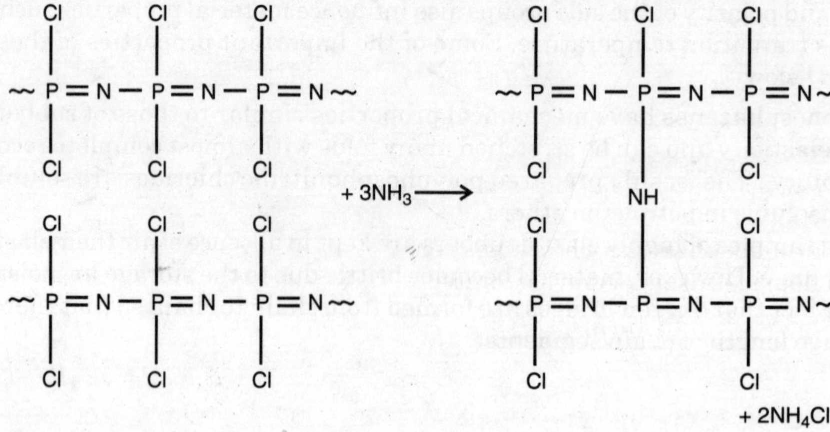

6. Polydimethoxy and polydiethoxyphosphazines are colourless transparent film forming thermoplasts and are unstable when heated above 100°C. Above this temperature, they slowly get converted into cyclic $[P(OR)_2N]_3$ or $[P(OR)_2N]_4$ polymers ($R = CH_3$ or C_2H_5)

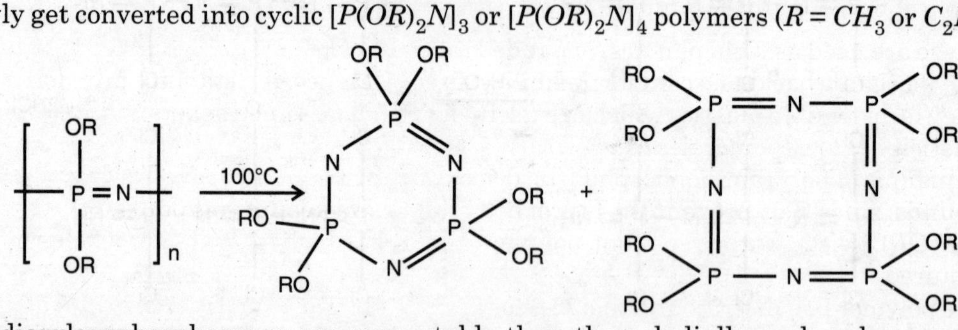

7. Polydiaryloxyphosphazenes are more stable than the polydialkoxyphosphazenes. The polyphenylphenoxy phosphazene 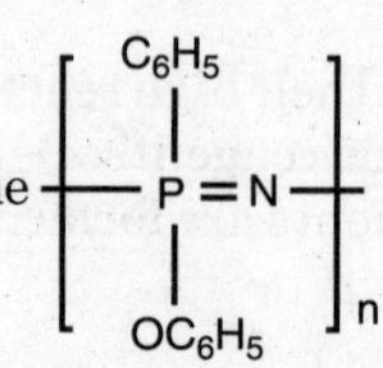 melts at 225°C and can be spun into fibres.

The Youngs modulus of the fibre is similar in magnitude to that of rayon.

8. The chlorine atoms in cyclophosphazenes can also be replaced by different nucleophiles such as $-NH_2$, $-NMe_2$, $-OH$ and Ph etc. Substitution may be complete or partial and in the latter case many different isomers are possible.

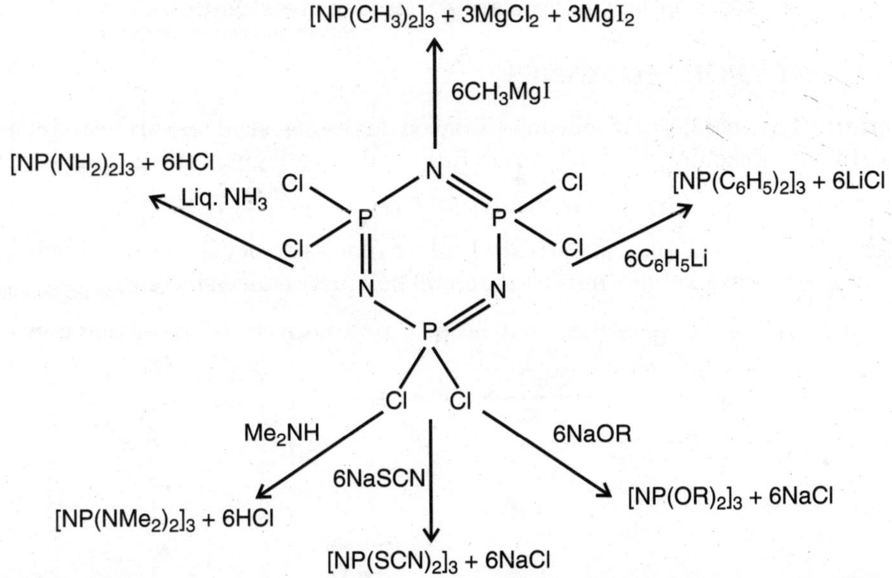

9. In chlorophosphazenes, chlorine is replaced by flourine by heating the chlorides with sodium flouride suspended in acetonitrile or nitrobenzene.

$$[NPCl_2]_3 + 6NaF \xrightarrow{C_6H_5NO2} [NPF_6] + 6NaCl$$

Applications of Polyphosphazenes

There are many uses of polyphosphazenes as they have mechanical properties comparable to those of rubber. Some of the important properties are listed below :

1. These are used as rigid plastics, expanded foam and fibres.
2. The inorganic backbone confers flexibility, fire resistance and stability to oxidation.
3. They form flexible plastics which are useful for fuel hoses and gaskets, since they retain their elasticity at low temperatures.
4. Thin films of poly (aminophosphazenes) are used in hospitals to cover severe burns and other wounds, since they prevent the loss of body fluids and keep germs out.
5. $[PN(OR)_2]_n$ is a water repellent polymer and is used where water proof thermoplasts are required.
6. The polymer polydi (triflouroethoxy) phosphazene is used as high performance elastomer. The polymer is so inert that it can be used for the construction of artificial blood vessels and organs.
7. Polyphosphazenes are also used where good biostability and biocompatibility are necessary, such as in the encapsulation of pancreatic cells.
8. Poly (organophosphazenes) also exhibit extremely high permeability for fixed gases and organic liquids and may find use in membrane applications.
9. The polyphosphazenes having alkyl ether side chain are used as solid solvents for salts such as lithium.

These ion conducting materials can be fabricated into light weight, thin-film rechargeable lithium batteries.

Applications of polyphosphazenes in the areas of membranes, hydrogels, bone-replacement materials, hydrogels and optical devices are still under investigations.

13.2 SILICONES-POLYSILOXANES

***Introduction*:** The chemistry of silicones is based on the classical researches of Frederick Kipping (1901-44) Silicon tetrachloride, $SiCl_4$ when fully hydrolysed gives silica, SiO_2–

$$SiCl_4 + 4H_2O \longrightarrow Si(OH)_4 \xrightarrow{-2H_2O} SiO_2$$

But the hydrolysis of alkyl substituted chlorosilanes do not lead to the expected alkyl silicon compound analogous to a ketone, but to long chain polymers called silicones.

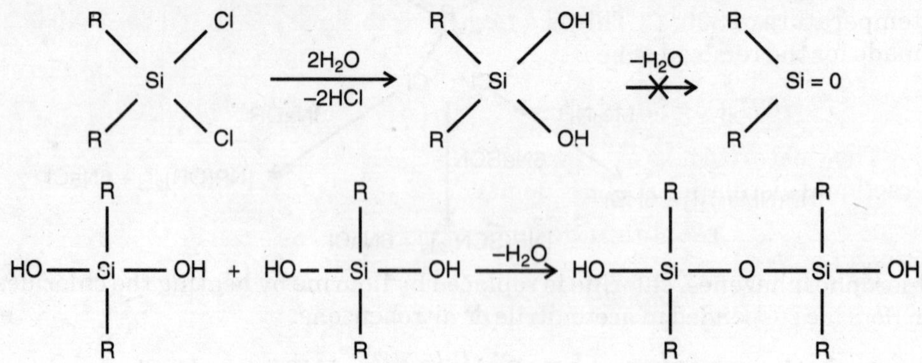

Since an active OH group is left at each end of the chain, polymerization reaction continues and a long chain product is formed.

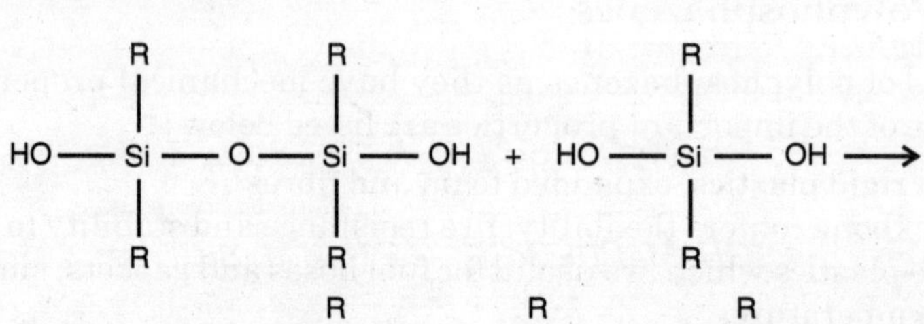

These products are known as silicones or poly siloxanes. Thus silicones are a group of neutral organosilicon polymers, based on network of alternating silicon and oxygen atoms, having the general formula :

$[R_x SiO_{(4-x)/2}]n$ where n is very large while x varies from 0 to 3.

Silicones can be categorized both as organic as well as inorganic polymers. These polymers resemble inorganic polymers in having high percentage of ionic character of Si–0 bonds and they resemble organic polymers in having organic groups on silicon atoms. The silicon products possess certain unusual, but very useful properties such as remarkable stability over wide range of temperature (from –70° C to 250° C), good water repellency, chemical and physiological inertness, good resistance to the effect of weathering, low vapour pressure and desirable low temperature characteristics. The broad spectrum of properties combined with the many forms in which silicones can be formulated (eg. liquids, gums, resins and rubbers) lead to their wide spread application in a broad range of processes and industries.

Preparation: Silicones are generally prepared through the hydrolysis of alkyl or aryl substituted chloro silanes which in turn are obtained through various methods. Thus the preparation of silicones requires two steps–

(a) Preparation of intermediates.
(b) Hydrolysis/polymerization of intermediates.

(a) ***Preparation of Intermediates:*** There are three commercial methods for the manufacture of organochlorosilanes.

(i) ***Direct Method:*** Used for the preparation of methyl chlorosilanes. In this method methylchloride is reacted with silicon in the presence of copper as a catalyst at a temperature of 300° C. This is a highly exothermic reaction and provisions must be made for the removal of heat.

$$2CH_3Cl + Si \xrightarrow[300°C]{Cu} (CH_3)_2\,SiCl_2 + (CH_3)SiCl_3 + (CH_3)_3Si\,Cl$$

The yield of (CH3)2Cl_2 is over 50% and it can be separated from the mixture through fractional distillation. Boiling points of different silanes are given in Table 13.1.

Table 13.1: Boiling Points of Chlorosilanes

Compound	*B.Pts. °C*
$(CH_3)_3SiCl$	87.7
$(CH_3)_2SiCl_2$	69.6
$(CH_3)SiCl_3$	66.9

(ii) ***Use of Grignard Reagent:*** In this method introduced by Kipping, a solution of $SiCl_4$ in dry ether is allowed to react with the Grignard reagent

$$CH_3MgCl + SiCl_4 \longrightarrow \underset{\text{Methyltrichlorosilane}}{CH_3\,SiCl_3} + MgCl_2$$

$$2CH_3MgCl + SiCl_4 \longrightarrow \underset{\text{Dimethyl dichlorosilane}}{(CH_3)_2\,SiCl_2} + 2Mg\,Cl_2$$

$$2CH_3MgCl + SiCl_4 \longrightarrow \underset{\text{Trimethyl chlorosilane}}{(CH_3)_3\,SiCl} + 3MgCl_2$$

Depending upon the amount of Grignard reagent, 1 to 4 organic groups can be added to silicon. After $MgCl_2$ is filtered, the products formed are seperated through fractional distillation.

(iii) ***Aromatic chloro silanes :*** These are prepared by the reaction of benzene and other aromatic hydrocarbons at 230–300° C with the compounds containing *Si–H* bonds, using $AlCl_3$ or BF_3 as catalysts.

$$C_6H_6 + HSiCl_3 \xrightarrow{230° - 300\,C} C_6H_5SiCl_3 + H_2$$

(b) ***Hydrolysis of Intermediates :*** Hydrolysis of different intermediates under controlled conditions yield different products. Since these intermediates are very reactive and inflammable, they are preserved under dry inert gases such as nitrogen.

(i) ***Dimethyldichlorosilane*** $(CH_3)_2SiCl_2$**:** This is mainly a chain building unit. On controlled hydrolysis $(CH_3)_2SiCl_2$ gives rise to straight chain polymers. As an active OH group is left at each end of the chain, polymerization continues and finally a linear thermoplastic polymer is formed.

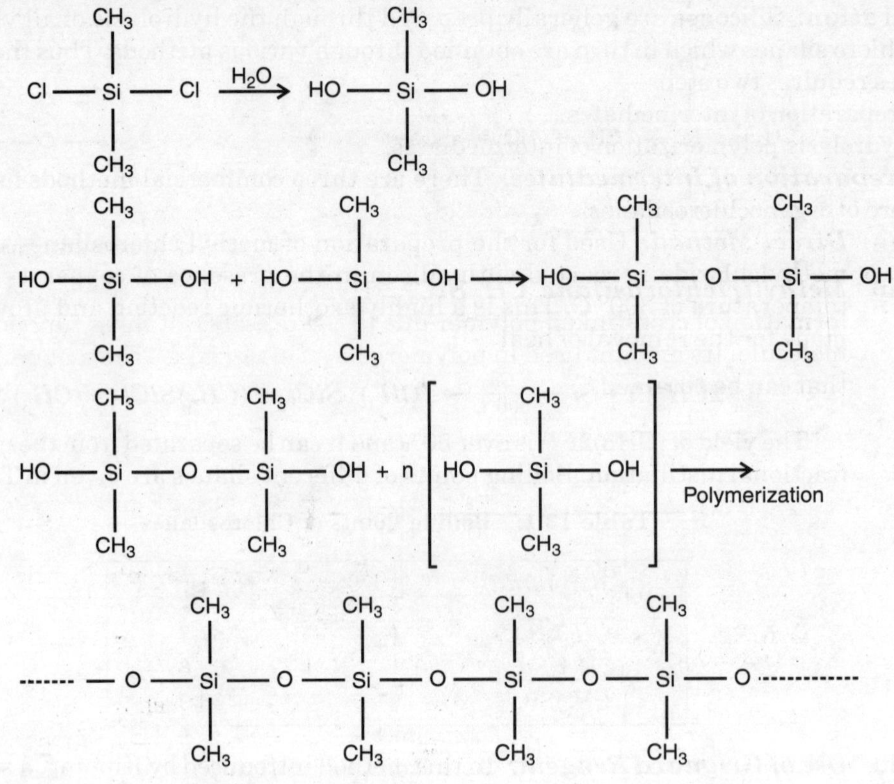

Linear Thermoplastic Polymer

Hydrolysis under carefully controlled conditions can produce cyclic structures, with rings containing three to six silicon atoms.

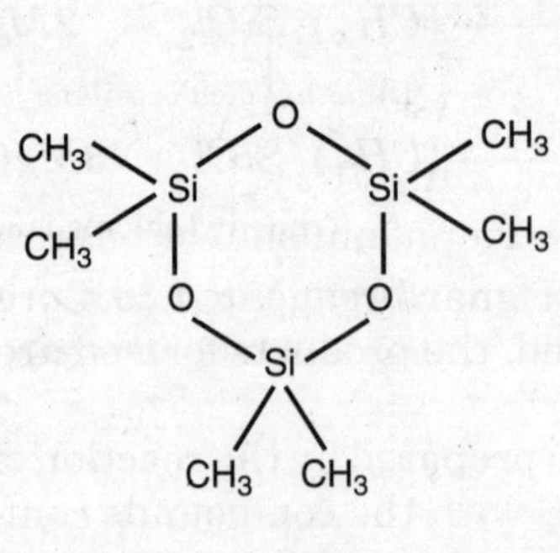

Triscyclo dimethyl Siloxane

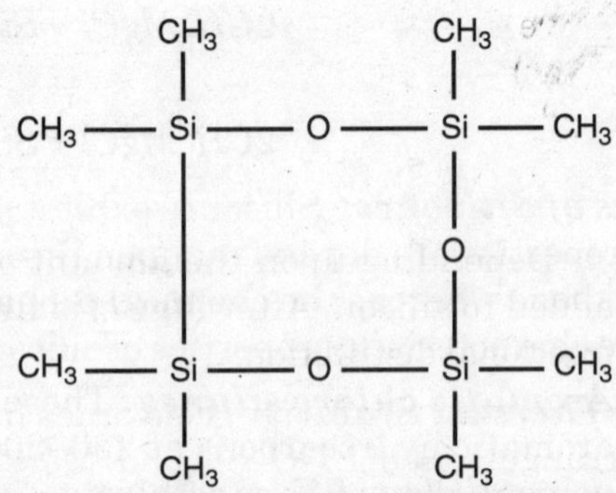

Tetrakis cyclo dimethyl Siloxane

(ii) ***Trimethylmonochlorosilane*** $(CH_3)_3SiCl$**:** Hydrolysis of this compound yields trimethylsilanol as a volatile liquid which condenses to give hexamethyldisiloxane. Since the resulting compound does not have any OH group, it cannot polymerize further. Therefore it is also used as a "chain stopper".

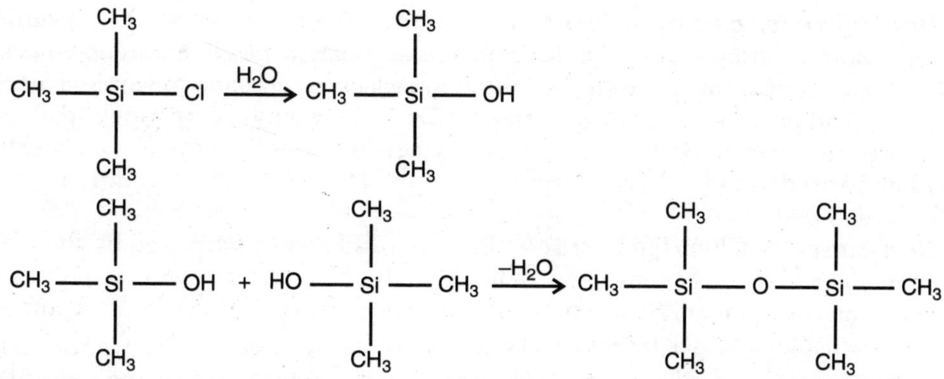

Hexamethyl disiloxane

(iii) ***Methyltrichlorosilane CH_3SiCl_3:*** Hydrolysis of this compound results in the formation of crosslinked polymer due to the presence of three functional groups per molecule. Its amount used in polymerization, determines the amount of crosslinking that can be obtained.

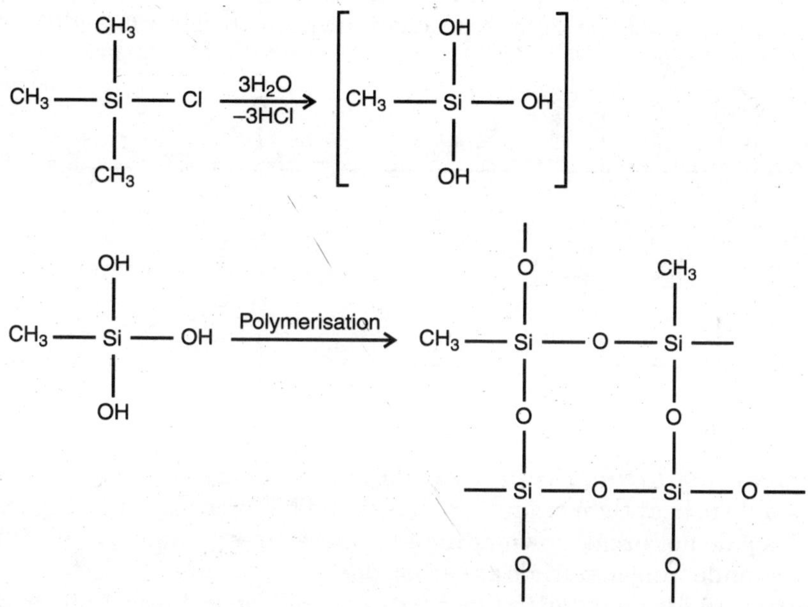

Properties of Silicones: Silicones exhibit a number of unusual but desirable properties. The physical properties are dependent upon the kind and number of organic groups attached to the silicon backbone whereas the chemical properties are governed by *Si–O–Si* bonds in siloxanes. Some of the characteristic properties of silicones are :

(1) ***High Thermal Stability :*** Siloxanes have high thermal stability because of stable silica like arrangement of *Si–O–Si* bond. Moreover *Si–O* bond has very high bond energy. The *Si–O–Si* bond in siloxanes is shorter than the expected *Si–O–Si* bond as calculated from their radii, which indicates that there is some ionic character or partial double bond character in the bond which confers stability to the bond. In addition, they remain flexible in the temperature range of –90° C to 250° C which is a much wider range than for natural rubber. The most heat stable side groups are phenyl groups, followed by methyl, ethyl and propyl groups. Silicones can withstand temperature upto 250-300° C without decomposing. Silicone rubbers withstanding 400° C for 100 hours have been prepared.

(2) ***Resistance to Oxidation :*** Siloxanes are quite resistant to attack by oxygen. On heating in air to 250° C oxidation of organic groups takes place which leads to cross linking. However the presence of aromatic groups on silicon increases the stability towards oxygen. Crosslinking due to oxidation renders the polymer brittle and low molecular weight polymers and cyclic structures are evolved. Organosilicon polymers with methyl groups do not show any change when heated at 200° C even for a very long time, but surface oxidation takes place when heated at 300° C.

(3) ***Resistance to Chemical attack :*** Poly siloxanes are unaffected by the majority of mild chemical reagents such as weak acids and alkalies. However due to polar nature of *Si–O* bond and the ability of *Si* to expand its valency due to the presence of vacant orbitals makes it susceptible to attack by several reagents.

(i) Depolymerization takes place by the attack of strong bases and concentrated acids.
(ii) Cleavage of *Si–O* bond can take place by the attack of electron deficient compounds.
(iii) Chlorination of alkyl group can result by the attack of dry chlorine.

Anyhow silicones are considered to be chemically inert because other common chemicals like H_2O_2, ozone, NH_4OH, metal salts etc have little or no effect on polysiloxanes.

(4) ***Chemical Properties :*** As is discussed above, silicones are generally chemically inert. However, reagents like Grignard reagent, lithium aluminium hydride can cleave the *Si–O* bond. For example–

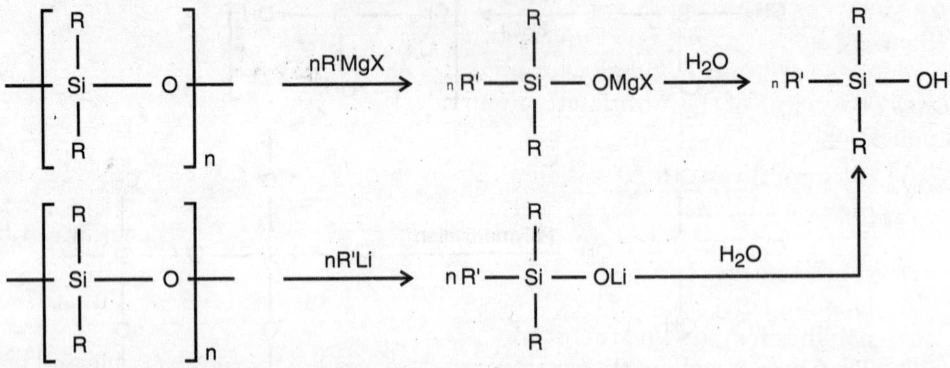

(5) ***Electrical Insulating Properties :*** Polysiloxanes are good electrical insulators. They are more stable to heat than organic polymers and hence are used for high temperature insulation. They do not break down at high temperatures ordinarily but if they do, they do not produce conducting materials as carbon does.

(6) ***Variation in viscosity with temperature :*** Silicon fluids exhibit very small variations in viscosity with temperature. Thus, these can function over a considerably wide range of operating temperatures. This property is attributed to the molecular structure of silicones. Silicones have regularly coiled structure with weak Vander Waals' forces in between the linear chains. At low temperatures the molecules have kinked and coiled structure, which opens up at high temperatures. The opening up causes greater interaction and entanglement of polymer molecules, thus increasing the viscosity. This increase partly compensates the normal decrease in viscosity caused by increase in temperature. Thus, the net result is that the viscosity of silicones does not change much on increasing the temperature i.e silicones show very small variations in viscosity with temperature.

Large size of silicon atom facilitates the coiling which is not possible in its linear organic analogs.

(7) ***Water repellency :*** Silicon resins are highly water repellent. Silicone chain is surrounded by organic groups and thus high content of hydrocarbon is responsible for this property. The high degree of water repellency coupled with the spreading efficiency due to low surface tension allows the deposition of silicon fluid films over plastics, metals, glass etc. When glass is coated with polysiloxanes, the oxygen atom of *Si–O–Si* chain adheres to the glass surface while the organic groups are directed outwards.

(8) ***Specific Gravity :*** The specific gravity for silicone greases and fluid resins lies between 1.03 and 1.09, whereas for silicon rubbers its value lies between 1.7 to 2.1

Types of Silicones And Their Uses

The silicones are generally classified on the basis of their physical state and appearance at room temperature. Depending upon the length of the chain, amount of crosslinking, nature of organic groups attached to the silicon atom, they can be obtained in the following forms–

(1) Liquid silicones or silicone fluids.
(2) Silicone Greases.
(3) Silicone Rubber or Elastomers.
(4) Silicone Resins

(1) ***Silicone Fluids :*** As was discussed in one of the earlier sections that low molar mass straight chain polymers are obtained by the hydrolysis of dichlorosilanes. These polymers are known as silicone fluids.
When low molecular weight products are required, a suitable amount of trimethyl chlorosilane is added to serve as an end blocker.
Uses : By virtue of their unusual properties, the silicone fluids have got a large number of applications.

(i) Silicone fluids are available in wide range of viscosities. They have high viscosity index which makes them useful as hydraulic fluids.
(ii) These are used for water repellent finishes for a large number of natural and synthetic textiles.
(iii) Dust and dirt come off easily from surface coated with silicone fluids, they are used in polishes for cars and furniture.
(iv) They have excellent dielectric properties and are thermally stable and hence are used in capacitors and small transformers.
(v) They are used in high temperature water baths as they can withstand high temperature.
(vi) They are also used as high temperature lubricants. Silicones with phenyl groups are better lubricants.
(vii) Silicone fluids have antifoam properties (e.g. dimethylsilicone fluids) and are used to eliminate foam in petroleum oils.
(viii) Silicon fluids are nontoxic.

2. ***Silicone Greases :*** These are modified silicone fluids and are formed by adding fillers such as silica, carbon black, lithium soap etc to silicone oils. Silicone greases vary in properties depending on the type of fluid and thickner used.
Uses :

(i) Greases have a wide range of service temperature range from –100 to 450° C and hence can be very useful where effective lubrication is required at very high and very low temperatures.
(ii) Greases with lithium soap filler can function from –75° C to 150° C and are used for ball bearings, operating and high speed and high temperatures.

(iii) Greases with carbon black filler are heat resistant and are used in high temperature conveyers and oven doors etc.

(iv) The greases in which silica is used as filler have been used as electrical greases in car ignition systems.

3. ***Silicone Rubbers or Elastomers :*** Silicon rubbers are the polymers containing high molecular weight dimethyl silicon polymer, mixed with a filler, usually finely divided SiO_2 and a peroxide as curing agent. Peroxide helps to form few bridges between the linear chains, SiO_2 is added to reinforce the polysiloxane structure. The mechanism of formation of silicon rubbers from straight chain compound can be depicted as–

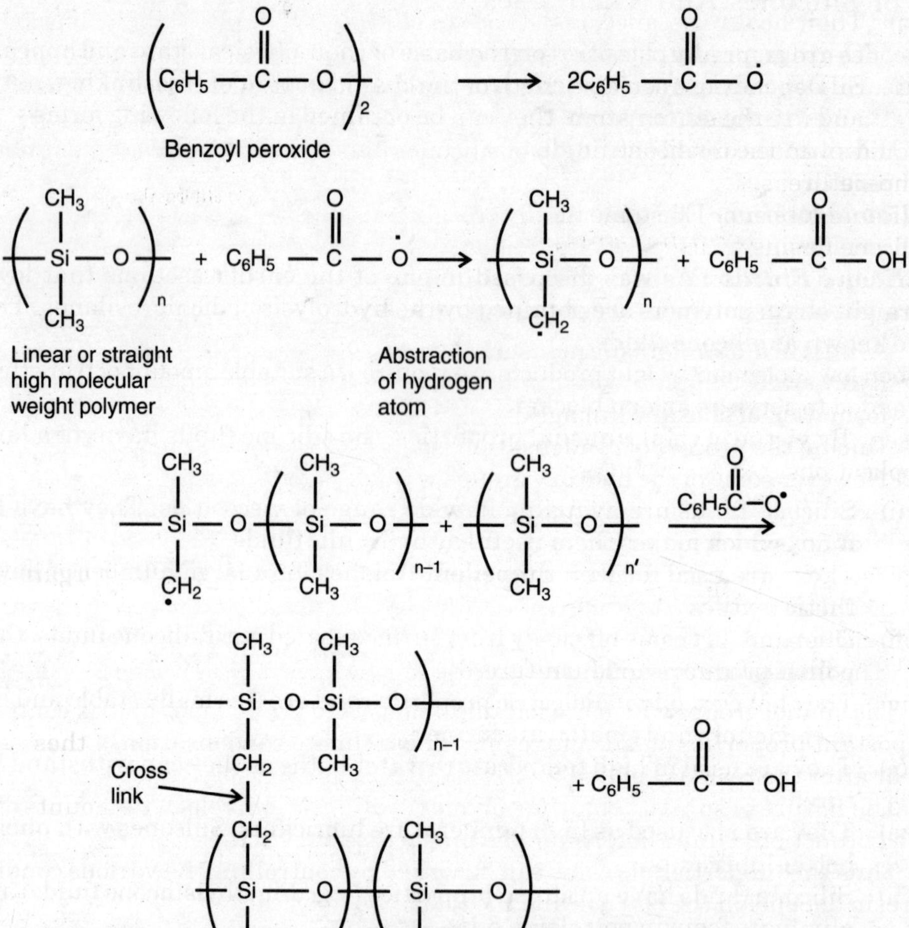

Uses: Silicone rubbers are exceptionally resistant to prolonged exposure to weathering, boiling water, dilute acids and alkalies. They have got remarkable stability at higher temperatures. Some of the uses of silicone rubbers are–

(i) Because of their property to withstand heat, they are used in sealing joints and seams of jet aircrafts. They are also useful for insulating electrical parts.

(ii) Silicone rubbers are very resistant to many chemicals. That is why they are used especially when chemical inertness is required.

(iii) They remain flexible in the temperature range of 90–250° C and hence used in making tyres for fighter aircrafts, because they prevent damage on landing whereas ordinary rubber tyre becomes brittle and undergoes disintegration.

(iv) Silicon rubber is available in variety of forms like thin sheets, mouldable stocks, smooth or heavy pastes. These may be dispersed in organic solvents to give free flowing dispersion which can be applied to surfaces by brushing, spraying or dipping. The pastes are used for coating organic and inorganic fabrics. Coated fabrics are used to make hoses for carrying hot gas or liquid.

(v) Pastes can provide an electrically insulating coating as it can adhere well to ceramic, glass and metalware.

(vi) Their heat resistance, resistance to mineral oils makes them exceptionally useful as O–rings, pressure cooker seals etc.

(vii) Silicon rubbers are also used for making artificial heat valves, transfusion tubings and in implant surgery.

However, the tensile strength of silicone rubbers is inferior to organic rubber at normal temperature.

4. ***Silicone Resins:*** These are highly crosslinked silicone polymers obtained by hydrolysing and condensing mixtures of bifunctional or trifunctional alkyl chlorosilanes under carefully controlled conditions. Their crosslink density is much higher as compared to silicone rubbers. The preparation of silicone resins constitute the following steps –

(a) A mixture of chlorosilanes such as $CH_3\ SiCl_3$, $(CH_2)_2\ SiCl_2$, $(C_6\ H_5)CH_3\ SiCl_2$ is hydrolyzed. Simultaneous condensation, polymerization during hydrolysis leads to the formation of siloxane linkages.

(b) During the process of condensation, some ring closures also take place.

(c) However some of the hydroxy groups which do not condense (about 10%) get fixed in the network because increasing viscosity of the resin makes it difficult for them to approach and get condensed with one another.

(d) The partly polymerized product is washed with organic solvent. Organic layer is separated and washed to remove any traces of HCl in order to check the reversibility of the reaction.

(e) The final product is extensively cross linked.

The silanol groups which are left uncondensed in the polymer matrix contribute towards important properties of silicone resins. The ultimate condensation of these silanol groups occur after the resin is applied.

The nature of substituent, the polymer configuration, relative amounts of bifunctional and trifunctional units determine the properties of the resins.

Thus the properties of resins can be varied by controlling the various conditions. Choice of resin depend on the end use and the method of fabrication.

Uses: Silicon resins are of many types i.e coating resins, laminating resins, release resins, water repellent resins, electrical resins etc.

(i) ***Coating resins :*** As resins are water repellent and resistant to most of the aqueous chemicals they are used as films on exhausts, stoves, boilers, steam pipes etc where conditions are too severe for the use of usual type of protective coatings. As they can withstand the temperature of the order of 500° C they can be applied on high temperature processing equipments.

(ii) ***Laminating resins :*** Resin laminated structures are used in electrical industry in the form of insulated boards and rods. Methyl phenyl silicone resins are used in the manufacture of heat resistant glass cloth laminates, especially for electrical applications.

(iii) ***Release resins :*** These are used in bread pans, candy pans etc.

(iv) ***Water repellent resins :*** These resins are applied as dilute solutions to coat the pores in masonry and concrete. Due to their water repellency they keep the masonry dry.

(v) ***Electrical resins*** : For water proof and high temperature insulation, silicon resins are used. Impregnation of silicone resins are also used in the treatment of capacitor insulation and motor windings.

13.3 SULPHUR BASED POLYMERS

Sulphur, because of its ability for catenation, though much less than that of carbon, forms open and cyclic S_n species. For cyclic compounds, n can have values between 2 to 20 but for linear compounds it may have higher values.

A large number of sulphur based polymers are known. They fall under two categories–

(1) Linear chain polymers which include polymeric sulphur and sulphur nitrides.

(2) Network polymers which include chalcogenide glasses.

Polymeric Sulphur

Polymeric sulphur is prepared by melting the rhombic sulphur (S_8) at 165-180°. When this molten sulphur is poured into ice water, the so called plastic sulphur is obtained. The rhombic S_8 structure opens up to produce free radical which combine with S_8 to produce long chains.

$$S_8 \underset{165\text{–}180°C}{\rightleftharpoons} \cdot S - S_6 - S\cdot$$

$$\cdot S - S_6 - S\cdot + S_8 \rightleftharpoons \cdot S - S_{14} - S\cdot$$

$$\cdot S - S_{14} - S\cdot + S_8 \rightleftharpoons \cdot S - S_{22} - S\cdot \text{ and so on.}$$

Normally plastic sulphur has S_8 inclusions but on washing with CS_2 traces of rhombic sulphur are removed as plastic sulphur is insoluble in CS_2 unlike other allotropes of sulphur. Plastic sulphur can be stretched to as much as 15 times its normal length.

Polymeric sulphur can also be obtained as long fibres by heating S_8 in nitrogen at 300° C for 5 minutes and quenching a thin stream in ice water. These fibers can be stretched under water and appear to contain helical chains of sulphur atoms with ~3.5 atoms per turn.

Polymeric sulphur is unstable below the melting point of sulphur i.e 114° C. Below this temperature, sulphur is unstable and reverts back to rhombic sulphur. Moreover, if traces of S_8 are present, then also transition from polymeric sulphur to S_8 is accelerated.

Polymeric Sulphur Nitride $(SN)_n$

Sulphur and nitrogen have similar electronegativities and strong tendencies to form single and multiple covalent bonds. In 1950s Becke-Goehring and coworkers attempted to synthesize sulphur-nitrogen compounds. Since then a large number of monocyclic and bicyclic ring systems containing only nitrogen and sulphur have been reported.

Among best known sulphur nitrogen compounds, S_4N_4 tetrasulphur tetranitride is studied the most. S_4N_4 is a source compound for the preparation of other important S–N compounds. S_4N_4 is obtained as diamagnetic orange solid by passing ammonia into a warm solution of disulphur dichloride in dry CCl_4.

$$6S_2Cl_2 + 16NH_3 \rightarrow S_4N_4 + 12NH_4Cl + S_8$$

The reaction is complex and is not well understood. S_4N_4 melts at 178° C. It must be handled with care since grinding, friction or rapid heating can cause it to explode.

S_4N_4 is used to prepare polymeric sulphur nitride compounds $(SN)_n$. *Polymeric sulphur nitride* was first prepared in 1910 by Burt. S_4N_4 is converted to S_2N_2 by passing S_4N_4 vapours over silica wool at 525-575 K and 0.1 to 1.0 mm Hg. Rapid cooling through nitrogen cooled trap permits the isolation of S_2N_2. The main purpose of adding silver is to consume sulphur generated by the thermal decomposition by S_4N_4.

$$S_4 N_4 + 8Ag \longrightarrow 4Ag_2 S + 2N_2$$

$$S_4 N_4 \xrightarrow{Ag_2 S} 2S_2 N_2$$

S_2N_2 is a colourless crystalline substance insoluble in water. It decomposes explosively when struck or warmed above the room temperature (298K). It is polymerized to $(SN)_n$ probably in a ring opening system. S_2N_2 at the ordinary temperature changes into fibrous polymer $(SN)_n$.

Structure: According to most recent x-ray diffraction study, there are two S-N bond lengths (1.59 and 1.63 A°) and NSN and SNS bond angles are 106° and 119° respectively. The closest approaches of chains are represented by sulphur-sulphur contacts at 3.5 A°. The structure can be represented approximately as,

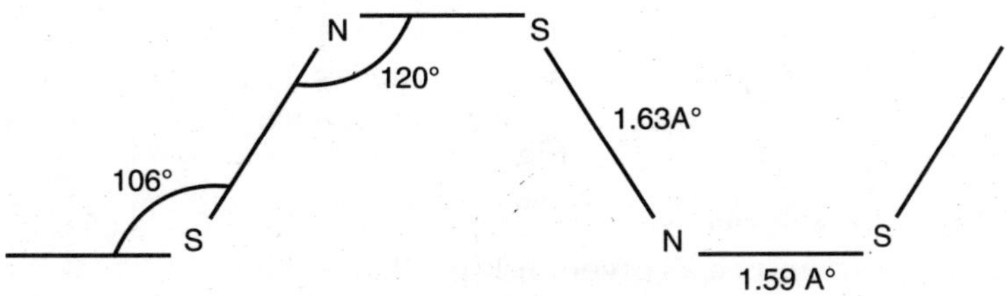

Fig. 13.2

Properties:

(i) Polymeric $(SN)_n$ is golden bronze in colour
(ii) When heated in air it decomposes explosively at about 510 K.
(iii) In vacuum it decomposes at 140° C to give sulphur and nitrogen.
(iv) It is soft and malleable.
(v) It is slowly attacked by moisture when exposed to it for several months.
(vi) One of the remarkable property of polymeric sulphur nitride is that, the material posseses metallic conductivity in the range of 1200–3700 S cm^{-1}, AT 0.26 K it becomes superconductor. Polymeric sulphur nitride is the first example of a non metallic superconductor.

Network Polymers

Sulphur based network polymers are cross linked polymers formed by reacting poly valent elements such as silicon antimony, arsenic, bismuth, germanium, indium etc with chalcogens. The resulting binary compounds are of great variety and complexity. These compounds are known as **chalcogenide glasses**. The nature of the products usually also depends on the ratios of reactants, the temperature of reaction and other conditions. Examples include polymeric silicon disulphide, $(SiS_2)_n$, $(As_2\,S_3)_n$ which are binary compounds. However many multicomponent systems like *Si–Sb–S, Si–P–Te* are also known.

The x-ray studies reveal that these network polymers consist of short chains of chalcogen atoms i.e *S, Se* etc which are linked together with polyvalent atoms. Polyvalent atoms serve as bridges between the chalcogen atoms. For example $(SiS_2)_n$ consists of infinite chains of SiS_4

tetrahedra as shown in Fig.13.3. Sb_2S_3 and Bi_2S_3 chalcogenide form infinite bonds that are then held in parallel strips in the crystal by weak secondary bonds.

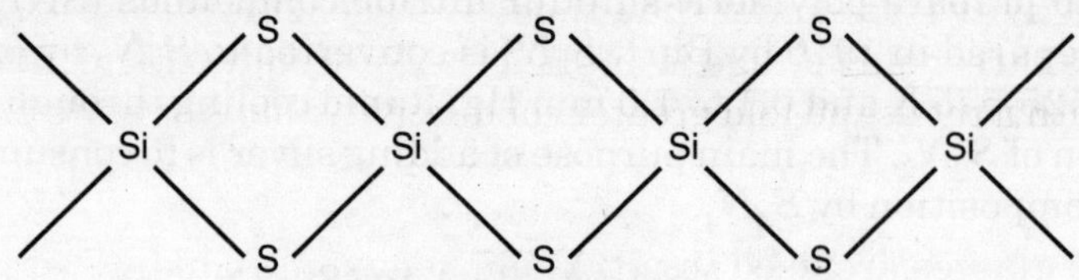

Silicon disulfide

Fig. 13.3

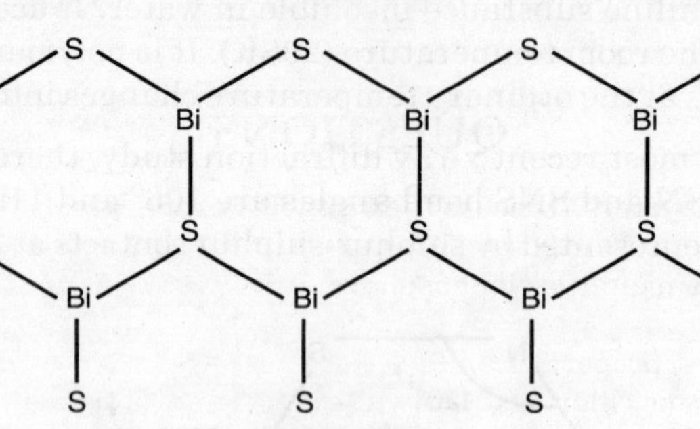

Fig. 13.4

An example of best known two-component chalcogenide glass is $(As_2\ S_3)_n$. A typical *As–S* polymer network has the structure as given in Fig. 13.5

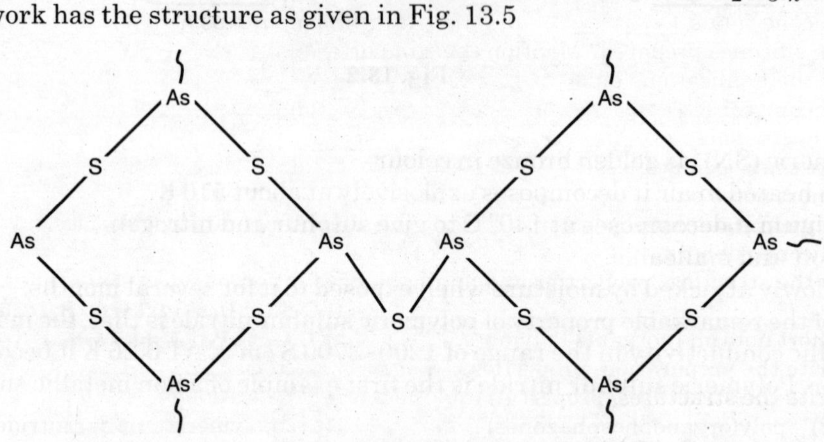

Fig. 13.5

Properties: These polymers resemble organic polymers. Some of the properties of network polymers are–

(i) These polymers have lower softening points and lower tensile strengths as compared to other inorganic polymers such as borophosphate glasses.

(ii) These are stable to acids but are attacked by concentrated alkalies.

(iii) A remarkable property of these polymers is that, these can be distilled in vacuum without decomposition. Actually dissociation of polymeric network into small fragments in vapour phase in vacuum takes place. As the vapours condense, the small fragements recombine spontaneously forming the polymer network again.

(iv) Chalcogenide glasses are deeply coloured.

(v) Unique property of network polymer of sulphur include the change from low conductivity to high conductivity state under the applied voltage. This phenomenon is called **switching**. After the threshold voltage is applied the low conducting polymer becomes highly conducting with a thousand fold greater conductivity than the state prior to voltage application.

Uses:

(i) Chalcogenide glasses are used in ultrasonic delay lines, high energy particle detector, memory devices for computers.

(ii) These are also used for infrared transparent windows because most of them are good infrared transmitting materials. They are used for fabrication of IR windows for army and civil optical devices.

QUESTIONS

1. What are inorganic polymers? Discuss their general properties. How do they differ from organic polymers.
2. Discuss in detail linear and cyclophosphazenes. Give an account of the bonding present in phosphazenes.
3. Discuss the structures, properties and uses of phosphazenes.
4. What are phosphonitrilic chlorides? Draw the structures of trimer and tetramer.
5. What are silicones and how are they classified? Which fundamental property of silicon makes the formation of silicones possible.
6. Give a brief account of the applications of silicones in technology.
7. Write short notes on :
   (a) Silicone fluids (b) Silicone resins
8. How are silicones prepared? Mention their important properties.
9. Discuss the applications of:
   (a) Silicon greases (b) Silicon fluids
   (c) Silicon resins
10. What are chalcogenide glasses? Give an account of the properties of network polymers of sulphur.
11. Wrtie short notes on :
    (a) Polymeric sulphur. (b) Polyhosphonitrilic chlorides.
    (c) Silicone rubbers.
12. Discuss the structure, properties and applications of different types of siloxanes.
    G.G.S.I.P.U, End Term, B.Tech, March, 2000.
13. Write short note on Inorganic Polymers. **G.G.S.I.P.U, End Term, B.Tech, Dec., 2000.**
14. (a) Write the preparation, properties and uses of silicones.
    (b) Write the structures, properties and industrial application of
    (i) poly(organophosphazenes) (ii) polymeric sulphur nitride.
    (iii) chalcogenide glasses
    G.G.S.I.P.U, End Term, B.Tech, II Batch, March, 2001.
15. What are inorganic polymers? How are they distinguished from organic polymers? Give the uses of polyphosphazenes. **G.G.S.I.P.U, I Term, B.Tech, Sept., 2001.**
16. Write short notes on of the following :
    (a) Chalcogenide glasses
    (b) Different types of Siloxanes
    (c) Polyphosphazenes

❑❑❑

14 CHROMATOGRAPHY

Even a mistake may turn out to be the one thing necessary to a worthwhile achievement.

— ***Henry Ford***

CHAPTER OUTLINE

14.1 INTRODUCTION

The term 'Chromatography' was coined by a Russian botanist Tswett in 1906. The term 'Chromatography' is applied to the systems and techniques used for separation, identification, and determination of the chemical components in complex mixtures. No other separation method is as powerful as chromatography.

Twett employed the technique to separate various plant pigments, such as chlorophylls and xanthophylls, by passing their solutions through glass columns which were packed with finely divided calcium carbonate. The separated species appeared as *coloured* bands on the column. Because of the formation of coloured bands the technique was named as chromatography. In Greek *chroma* means colour and *graphein* means to write. Though the advance chromatographic techniques employed today for the separation of complex mixtures frequently involves colourless substances, but the original term *chromatography* is still retained.

All the chromatographic methods have one thing in common i.e. the use of two mutually immiscible phases, stationary phase and a mobile phase.

Stationary Phase: It is the phase that is fixed in a place, either packed in a column or on a fixed planar surface.

Mobile Phase: The mobile phase carriers the mixture of the compounds to be separated along with it and is moved through the stationary phase. The mobile phase can be a liquid or a gas. The components of the mixture are separated depending upon the rates at which they are carried along with the mobile phase through the stationary phase.

Chromatography involves a sample (or sample extract) being dissolved in a *mobile phase* (which may be a gas, a liquid or a supercritical fluid). The mobile phase is then passed through a immobile, immiscible *stationary phase*. The phases are so chosen such that components of the sample have differing solubilities in each phase. A component which is quite soluble in the stationary phase will take longer to travel through it, than a component which is soluble in the mobile phase. As a result of these differences in mobilities, sample components will become separated from each other as they travel through the stationary phase.

Some of the general terms which are used in chromatography are:

(*i*) *Elution:* It is a process in which solutes are carried through a stationary phase by the movement of mobile phase.

(*ii*) *Eluent:* It is a solvent used to carry the components of a mixture through a stationary phase.

14.2 CLASSIFICATION

Chromatographic techniques can be classified on the basis of nature of mobile phase or the stationary phase or the mechanism of separation.

(*i*) ***Based upon the nature of the mobile phase, chromatography is of two types:***

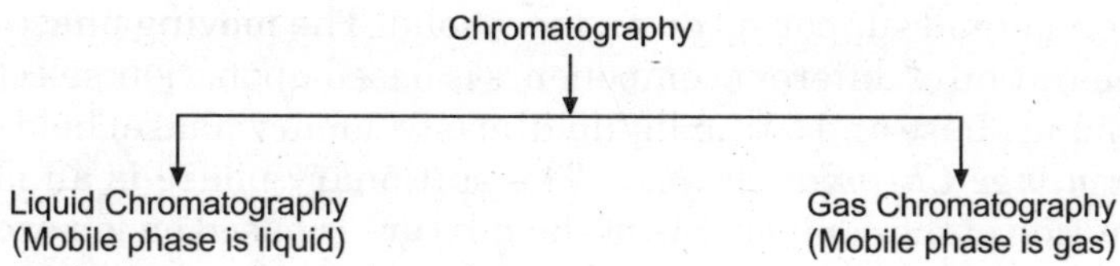

(*ii*) ***Based upon the nature of the stationary phase, these chromatography techniques are further divided into various types:***

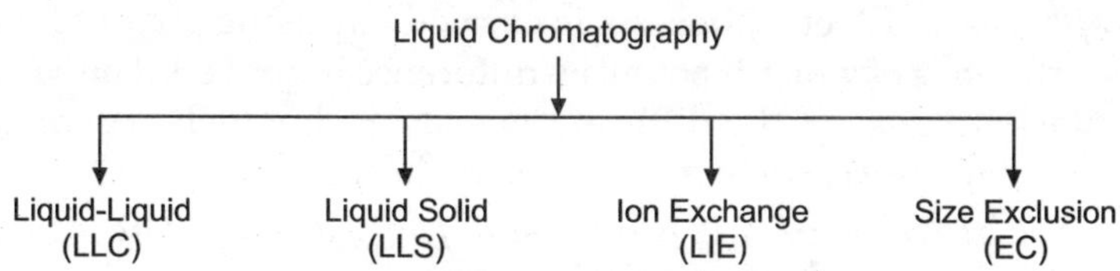

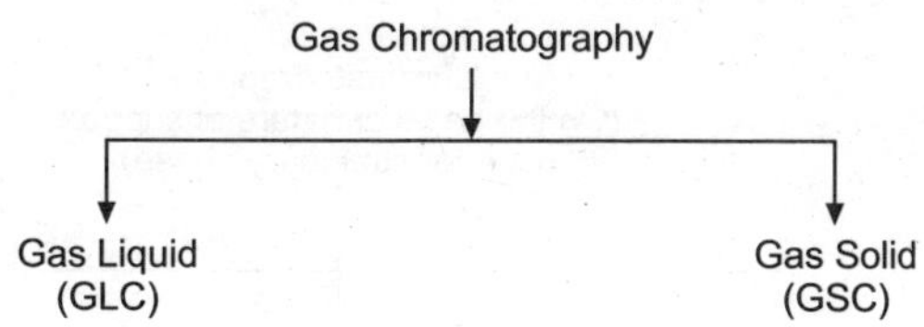

(a) *Liquid-Liquid Chromatography:* In this technique both the phases (stationary and mobile) are in the form of liquids. The liquid which forms the stationary phase is supported over a stationary support such as paper, plate etc.

(b) *Liquid-Solid Chromatography:* The stationary phase is solid (e.g. alumina, silica gel etc.). The solid phase can be placed in a column through which the mobile phase is passed or it can be thin layer supported on a glass plate.

(c) *Ion-Exchange Chromatography:* The ions present in the stationary phase are stoichiometrically exchanged for the ions in the mobile liquid phase.

(d) *Size-Exclusion Chromatography:* Separation is based mainly on exclusion effects, such as differences in molecular size and/or shape or in charge. The term *Size-Exclusion Chromatography* may also be used when separation is based on molecular size. The terms *Gel Filtration* and *Gel-Permeation Chromatography* (GPC) were used earlier to describe this process when the stationary phase is a swollen gel. The term *Ion-Exclusion Chromatography* is specifically used for the separation of ions in an aqueous phase.

(*iii*) ***Based on the mechanism of separation or the interacting phenomenon between the solute molecules and the stationary phase, the various chromatographic techniques have also been classified into the following types:***

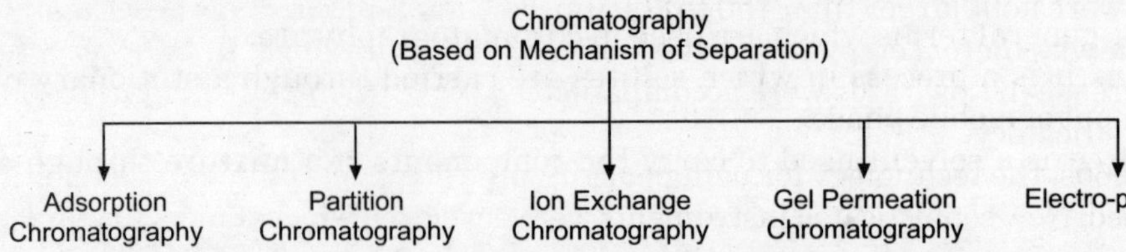

(*a*) *Adsorption Chromatography:* In this type of chromatography, the stationary phase consists of a solid adsorbent and the mobile phase is liquid or gas containing the mixture. The selective adsorption of the components of the mixture occurs on the surface of the solid.

(*b*) *Partition Chromatography:* In this type of chromatography, the stationary phase is a liquid or a liquid supported on an inert solid. The moving phase can be liquid or a gas. The separation of different components is based upon their selective partition between the moving phase and a thin liquid film (stationary phase) held on an inert solid.

(*c*) *Ion Exchange Chromatography:* The stationary phase is an ion exchanger and the distribûtion of the components of the mixture is based on ion exchange principle. This type of chromatography is applicable to ionic species.

(*d*) *Gel Permeation Chromatography:* In gel chromatography, the stationary phase is gel and the separation is based on its sieving action.

(*e*) *Electrophoresis:* Electrophoresis is carried out using strips of moistened filter paper between the ends of which a potential difference is applied. The separation of the mixture takes place because of the different speeds with which organic ions move under the influence of an electric field towards cathode or anode.

(*iv*) ***Based on the nature of support used for holding the stationary phase, the chromatography is broadly classified as:***

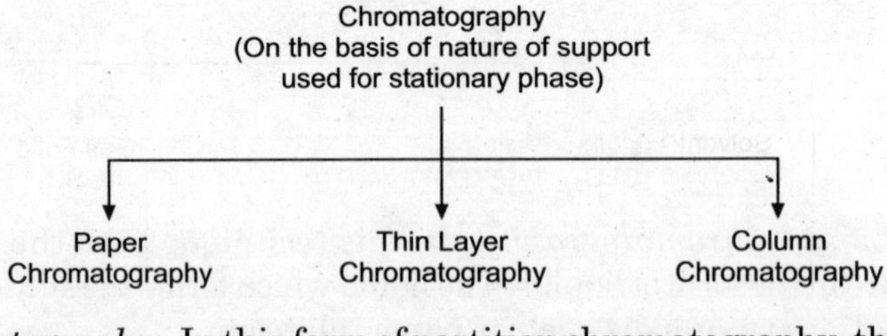

(*a*) *Paper Chromatography:* In this form of partition chromatography, the stationary phase is the absorbed water, present in the filter paper and the moving phase is a solvent saturated with water. Filter paper used is made of highly purified cellulose $(C_6H_{10}O_5)_n$.

(*b*) *Thin Layer Chromatography:* In this, the stationary phase is thin layers of certain fine solid particles, such as silica gel and alumina, supported on glass plates or plastic plates.

(*c*) *Column Chromatography:* In this type of chromatography, the stationary phase is formed into a column in a cylindrical tube.

14.3 LIQUID CHROMATOGRAPHY

As is discussed in section 14.2, that the term 'Liquid chromatography' has been used to cover a range of chromatographic techniques, all of which employ a *mobile liquid phase*.

Earlier liquid chromatography was performed in glass columns having diameters between 10 to 50 mm. These columns were packed with solid particles which constitute the stationary phase and through which the mobile phase percolates under gravity. The particle size in these earlier techniques were kept larger than 150 to 200 μm. The main problem in these techniques was slow speed which was increased by application of vacuum or pressure. But these were not effective because increase in flow rates were accompanied by increase in plate heights and also decreased the column efficiency.

In late 1960s, the technology for using packings with particle diameters as small as 5 to 10 μm was developed. New sophisticated instruments were developed which could withstand high pressures employed to increase the efficiency of the process.

Thus, the name High Performance Liquid Chromatography (HPLC) is often employed for these newer techniques to distinguish them from the older methods.

High Performance Liquid Chromatography (HPLC) is the chromatography is which liquid is the mobile phase and very finely divided particles comprise the stationary phase. To increase the efficiency or to obtain increased flow rates, the liquid is pressurised to several hundred pounds per square inch. HPLC have led to very successful separations of complex mixtures and is more versatile than gas chromatography.

14.4 INSTRUMENTATION FOR HPLC

The instrument used for chromatography is known as chromatograph. The basic HPLC equipment consists of the following components:

(*i*) Solvent reservoir
(*ii*) Pumping systems
(*iii*) Sample injection system
(*iv*) Column
(*v*) Detector
(*vi*) Data control and display

Fig. 14.1 represents the flow diagram of a typical liquid chromatograph.

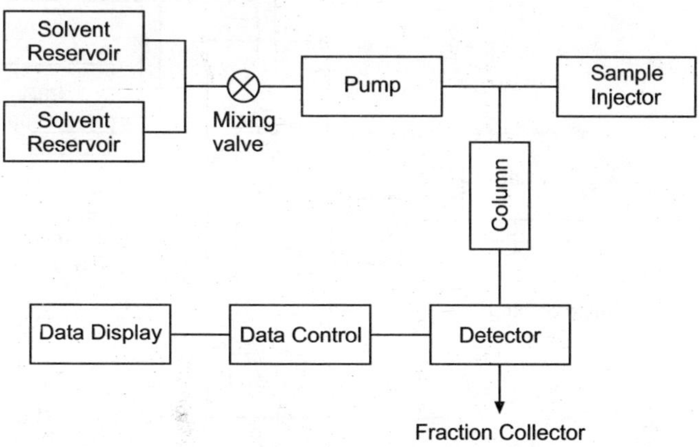

Fig. 14.1 Flow diagram of modern liquid chromatograph.

The solvents from the reservoirs are mixed by proper positioning of mixing valve. This mixed liquid is pumped at high pressure into the column along with the sample from the sample injector. The inlet pressure of the column is monitored with a manometer. After leaving the column, the sample components are monitored by the detector and the data is traced in a recorder.

(*i*) **Solvent Reservoir:** Solvent reservoirs are glass or stainless steel containers which contain 500 ml or more of a solvent. Solvents, as supplied may contain substantial amounts of dissolved air. Formation of air bubbles can seriously interfere with satisfactory separation of the components. Thus the solvent reservoirs are equipped with degassifiers. Degasser may consist of vacuum pumping system, a distillation system, a device for heating, stirring etc.

To ensure consistent performance, the solvent needs to be of high purity as any traces of impurities can seriously affect the column efficiency. HPLC grade solvents are hence expensive.

The process in which solutes are washed by the movement of mobile phase is termed as ***elution***. An elution with the single solvent is known as isocratic. But generally two solvents that differ in polarity are employed and the elution is known as gradient elution. In this, the composition of the solvent is changed continuously by changing the ratio of the two solvents.

(*ii*) **Pumping Systems:** In HPLC, the columns used are quite narrow and are packed with particles of small sizes. Hence there is high resistance to flow of solvent. Therefore, high pressures are required to achieve satisfactory and constant flow rate. The liquid chromatographic pumping systems are of wide varieties. The most popular types is the pump which can provide accurately controlled rates of the flow of solvents (upto 20 ml/min) at pressures upto 300-400 atmospheres.

The pumping systems operate on the principle of either constant pressure or constant displacement. The most commonly used pumps are:

(*a*) *Srew-driven syringe type:* These pumps operate on the principle of solvent displacement by a piston mechanically driven at a constant rate in a chamber of limited capacity (~500 ml). However, it is inconvenient when frequent change of solvents is desired.

(*b*) *Reciprocating pumps:* These pumps are widely used and consist of a small cylindrical chamber that is filled and then emptied by the back and forth movement of a piston. This pump facilitates replacement of solvents but provide pulsating flow. The pulsating flow is subsequently damped by using pump dampness to smoothen the flow.

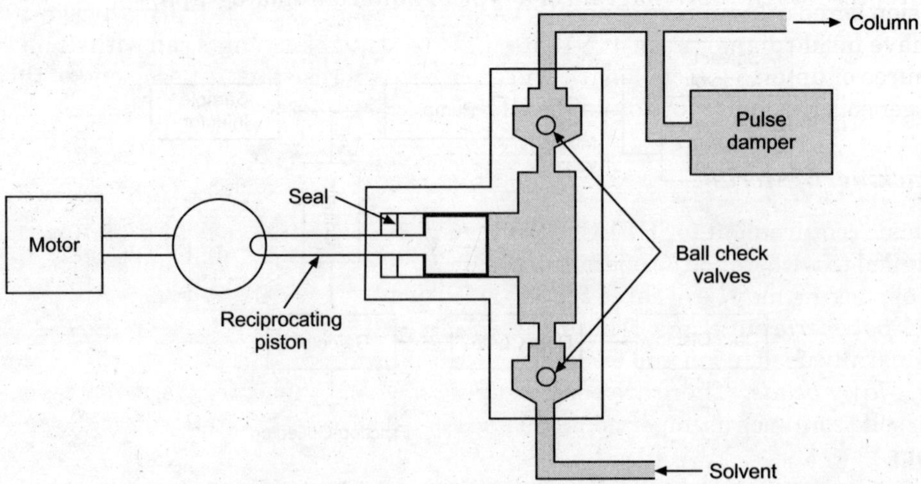

Fig. 14.2 A reciprocating pump for HPLC

(c) *Pneumatic pumps:* This is the simplest type of pump and consists of a collapsible solvent container placed in a vessel that can be pressurized by a compressed gas. These pumps are inexpensive and pulse free but have limited capacity and pressure output. Pumping rates depend upon the solvent viscosity.

(*iii*) **Sample Injection System:** Sample can be introduced into he column either by a syringe injection into the solvent stream or by a sample loop from which it is swept into the column by the solvent.

(a) *Stop flow injection:* In stop-flow injection, the solvent flow is stopped momentarily and sample is injected directly on top of the column by means of a syringe.

(b) *Loop injection:* This is the most widely used method of sample introduction. Here the sample is introduced with the help of a metal loop of fixed volume. The loop is filled with the sample and by approximately adjusting the sample valve the solvent from the pumped is passed through the loop. Thus sample is taken by the solvent through an outlet which opens directly at the top of the column (Fig. 14.3).

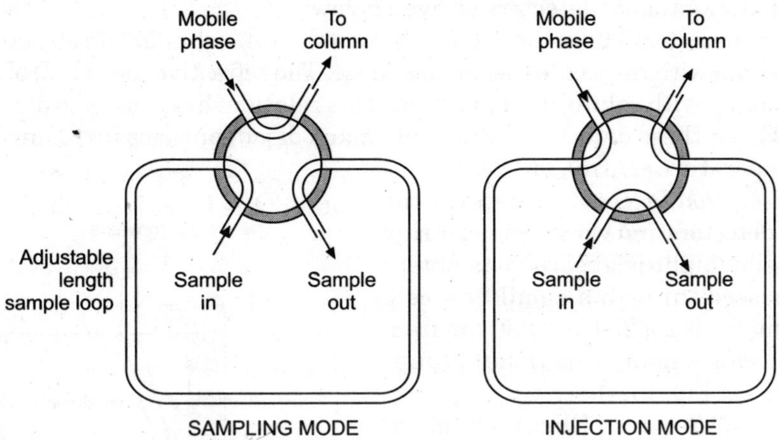

Fig. 14.3 The sample loop injection

(*iv*) ***Column:*** Since glass tubing cannot withstand pressures employed in HPLC, stainless steel precision bored columns are generally used. Most columns range in length from 10 to 30 cm and have inside diameters of 4 to 10 mm. These straight columns can withstand very high pressures of upto 15–50 cm length and are corrosion resistant. At each end of the column, homogeneously porous plugs of stainless steel or teflon are used to retain the packing material.

Column Packing Material

One of the basic requirement for HPLC is that packing material should be pressure stable i.e. it must to designed to withstand the operating pressures applied during separation.

The type of packing may be of three types:

(a) *Porous polymeric materials:* These are based on styrene-divinyl benzene copolymers. These are generally used in ion and exclusion chromatography.

(b) *Porous-layer beads:* There are porous supports where a thin, porous, active layer is coated onto a solid core such as impervious glass beads. The thickness of the porous layer is generally 1–3 μm.

(c) *Totally porous silica particles:* In these packings the silica particles, which are generally 5–10 μm in diameter, form the packing material. These give considerable improvements in column efficiency and speed of analysis.

Column Packing Technique

The column can be packed depending upon the mechanical strength of the packing and its particle size. Particles having diameter more than 20 μm can be dry packed where-as particles having diameters less than 20 μm are packed through preparing slurries.

(*v*) **Detectors:** The function of detector is to monitor the composition of mobile phase as it comes out from the column. Two types of detectors are generally employed:

(*a*) *Bulk property detectors:* These monitor the difference in some physical property of the solute in the mobile phase as compared to pure mobile phase. For example refractive index, conductivity detectors.

(*b*) *Solute property detectors:* These detectors respond to the physical or chemical property of the solute such as fluorescence and electrochemical detectors.

Some commercially available detectors have a number of different detection modes built into a single unit such as UV absorption, fluorescence and conductometric detention. A description of some of the commonly available detectors is given below:

Refractive Index Detector: These detectors are based on the change of refractive index of the eluent from the column with respect to the mobile phase. The refractive index (RI) of dilute solutions changes proportionally with solute concentration. This relationship is used to detection of solutes in the eluate. However, these detectors suffer from many disadvantages including low sensitivity, need for strict temperature control, etc.

Ultra Violet Absorption Detectors: These are the most widely used detectors and work by measuring how much UV/visible light is absorbed as eluent from column is passed through a small flow cell. The UV detector make use of 254 and 280 nm from mercury source because many functional groups absorb in this region. (Fig. 14.4)

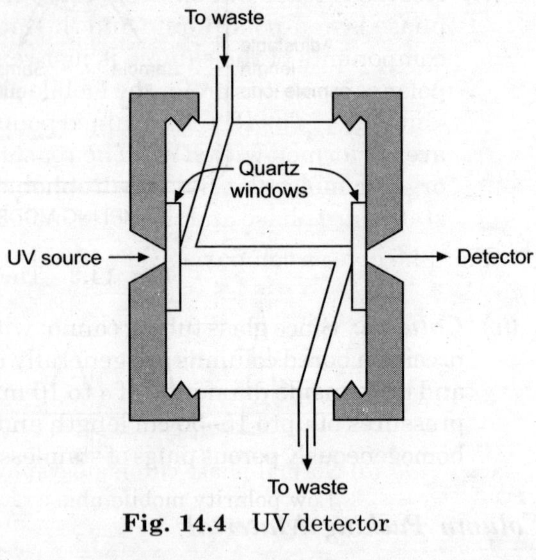

Fig. 14.4 UV detector

Electrochemical Detectors: Electrochemical detectors respond to the change in electrochemical property of the bulk liquid flowing through the measuring cell or change in voltage or current when an analyte is passed through the cell.

Fluorescence Detectors: The fluorescence of a number of compounds is used in conventional fluorescence detectors. Organic and inorganic naturally flourescent compounds, biologically active compounds and environmental contaminants are flourescent and can be detected with high sensitivity by these detectors.

(*vi*) **Recorder:** The signals from the detectors are recorded as deviation from a base line. The peak position along the curve, relative to the starting point, denotes the particular component.

14.5 TYPES OF LIQUID—CHROMATOGRAPHY

The most commonly and widely used types of HPLC techniques are:

(*i*) Partition or liquid-liquid chromatography
(*ii*) Absorption or liquid-solid chromatography
(*iii*) Ion exchange chromatography

(*iv*) Gel permeation chromatography

(*v*) Gel filteration chromatography

It should be emphasized here that in each chromatography technique, though one mechanism predominates but two or more mechanisms may be involved simultaneously.

14.5.1 Partition or Liquid-Liquid Chromatography

Partition chromatography is based upon the distribution of the solute particles between two immiscible liquid phases depending upon their relative solubilities.

One of the liquid is held on a suitable solid support and acts as a stationary 'phase'. The other liquid is used as the mobile phase. The stationary phase may be in the form of a packed column, a thin layer on glass plate or a strip of paper. The two liquids which are used must be immiscible and differ in the degree of polarity.

Based on the relative polarities of the mobile phase liquid and stationary phase liquid, the liquid chromatography can be categorized as given below:

(*i*) ***Normal-Phase Chromatography:*** In normal phase chromatography, the highly polar liquid is made the stationary phase i.e. the inert support (silica, alumina etc.) which holds the polar solvent and the non-polar solvent constitutes the mobile phase. As like dissolves like, the non-polar components of the sample prefer the mobile phase and elute first while polar components prefer the stationary phase and elute later.

(*ii*) ***Reverse Phase Chromatography (RPC):*** In reverse phase chromatography the stationary phase is non-polar liquid and the mobile phase is polar liquid. The order of elution of the components of the sample is just reversed of that normal liquid chromatography i.e. the polar components prefer the mobile phase and elute first, whereas the non polar components elute later. The RPC has gained popularity and about three quarters of HPLC separations are performed with RPC. The reason for its polarity arises due to the fact that almost all organic molecules have hydrophobic regions and, therefore, interact with the non polar stationary phase and the polar molecule will elute first along with polar mobile phase.

Fig. 14.5 gives a comparison of normal and reversed phase chromatography.

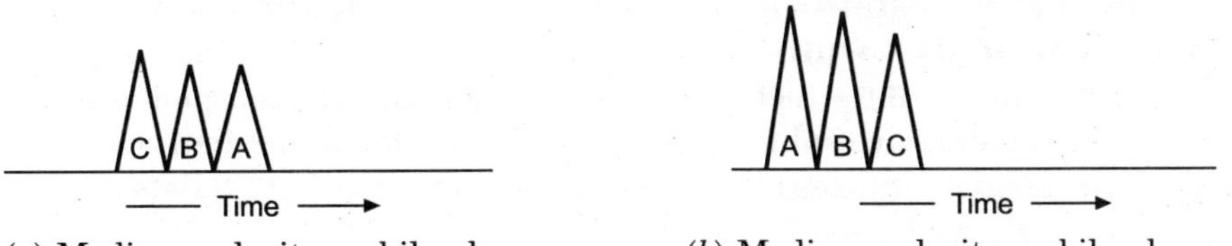

(*a*) Normal-phase chromatography
Low polarity mobile phase

(*b*) Reversed-phase chromatography
High polarity mobile phase

(*a*) Medium polarity mobile phase

(*b*) Medium polarity mobile phase

Solute polarities: A > B > C

Fig. 14.5 Normal (polar column) versus Reversed Phase (non-polar) elution

Stationary Phase Packings

The partition chromatography can be subdivided into liquid-liquid and liquid bonded phase chromatography. In liquid-liquid partition chromatography the stationary phase is the liquid held in place by *adsorption* on the surface of packing particles. In liquid-bonded partition chromatography, the stationary phase is an organic specie that is attached to the surface of support particles by chemical bonds.

The most common bonded hydrocarbon packings are prepared by reaction of an organochlorosilane with the —OH groups formed on the surface of silica particles by hydrolysis. An organosiloxane is formed as the product.

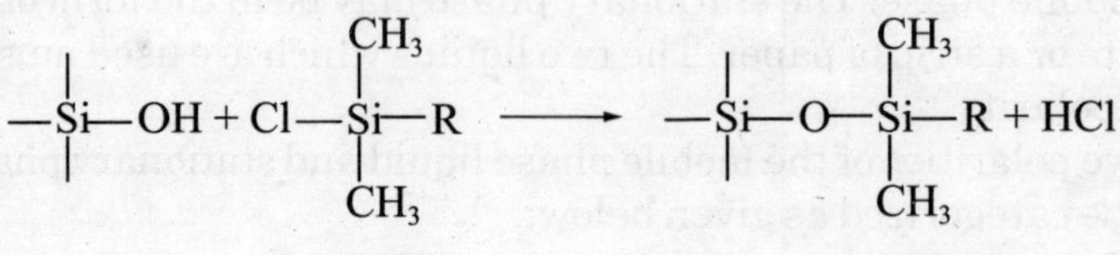

Chlorosilane Organosiloxane

R is often a straight chain octyl or octyldecyl group. Bonded phase packings have greater stabilities compared to physically held liquid phase in LLC. Moreover, in LLC the liquid stationary phase may get dissolved away with the mobile phase. This type of problem is not met within bonded phase packing.

Selection of Solvents for Mobile and Stationary Phase

A successful chromatographic separation depends upon the differences in the interaction of the solutes with the mobile phase and the stationary phase.

Thus the polarity difference between the two liquids is of prime importance in order to obtain efficient separations, and should be properly balanced.

- If the polarity difference between the two phases (stationary and mobile) is too large then the solutes will either be retained too strongly or not be retained.
- If the polarity difference between the two phases is low then the solute will be soluble in both the solvents to some extent and hence separations would not be effective.

Thus the choice of a suitable mobile phase is vital in HPLC. The eluting power of mobile phase is determined by its overall polarity, the polarity of the stationary phase and the nature of the sample components. In general, the polarities of common organic functional groups in increasing order are:

Aliphatic hydrocarbons < olefins < aromatic hydrocarbons < halides < sulphides < ethers < nitro compounds < esters ≈ aldehydes ≈ ketones < alcohols ≈ amines < sulphones < amides < carboxylic acids < water.

Currently there is no one system which adequately ranks all of the common solvents in terms of polarity. However, the solvents are split into eight groups with similar characteristics.

Table 14.1 gives the polarity indices of various solvents and Table 14.2 lists the eight groups of solvents.

Table 14.1: Representative Solvent Polarity Indices

Solvent	*Polarity Index*	*Solvent Groups*
Isooctane	– 0.4	0
Hexane	0.0	0
Carbon tetrachloride	1.7	7
Dibutyl ether	1.3	1
Triethylamine	1.8	1
Toluene	2.3	7
Chlorobenzene	2.7	8
Diphenyl ether	2.8	8
Diethyl ether	2.9	1
Benzene	3.0	7
Ethyl bromide	3.1	6
Methylene chloride	3.4	5
1,2-Dichloroethane	3.7	5
Bis-2-ethoxyethyl ether	3.9	5
n-Propanol	4.1	2
Tetrahydrofuran	4.2	3
Chloroform	4.3	9
2-Butamine	4.5	6
Dioxane	4.8	6
Ethanol	5.2	2
Nitroethane	5.3	7
Pyridine	5.3	3
Acetone	5.4	6
Benzyl alcohol	5.5	4
Methoxyethanol	5.7	4
Acetic acid	6.2	4
Acetonitrile	6.2	6
Dimethylformamide	6.4	3
Dimethyl sulfoxide	6.5	6
Methanol	6.6	2
Nitromethane	6.8	7
Water	9.0	9

Table 14.2: Solvent Classification by Selectivity Groups

Solvent group	*Representative classes or examples of compound*
0	*n*-Alkanes, cyclohexane, saturated fluoro-chlorohydrocarbons
1 (I)	Alphatic ethers
2 (II)	Alphatic alcohols
3 (III)	Pyridine derivatives, tetrahydrofuran, amides
4 (IV)	Glycols, benzyl alcohol, acetic acid
5 (V)	Methylene chloride ethylene chloride, bis-2-ethoxyethyl ether
6 (VIA)	Aliphatic ketones and esters, nitriles, sulfoxides
7 (VIB)	Nitrocompounds, phenyl alkyl ethers, aromatic hydrocarbons, carbon tetrachloride
8 (VII)	Halobenzenes, diphenyl ether
9 (VIII)	Fluoroalkanols, chloroform, water

14.5.2 Adsorption or Liquid Solid Chromatography

In adsorption or liquid solid chromatography the stationary phase is a finely divided solid adsorbent and the mobile phase is the organic liquid. The components of the sample are adsorbed onto the surface of the solid adsorbent. The mobile phase carries the sample through the stationary phase and the component which has greater affinity for stationary phase is retained by the stationary phase and the component with relatively higher affinity of the mobile phase travels at a much faster rate as it is carried along the mobile phase and hence separation results.

The solid adsorbent is packed in a column or spread on a plate. Finely divided silica and alumina are the stationary phases which are used commonly in adsorption chromatography. These solids have high surface areas. Silica is generally preferred for most of the applications because it is available in various useful forms.

It should be noted that highly active adsorbents usually give rise to irreversible solute adsorption. For example, silica gel which is slightly acidic may strongly retain basic components whereas alumina (non acid washed) is basic and, therefore, retains the acidic components strongly. Thus silica should not be used for acid sensitivie compounds and alumina should not be used for base sensitive compounds.

The role of mobile phase (solvent) in adsorption chromatography is very important because it is the only variable that affects the distribution of solute between two phases. This is in contrast to partition chromatography where the polarity of the stationary liquid phase can also be varied. Moreover, solvent molecules also compete with the solute molecules for polar adsorption sites on stationary phase.

Liquid solid HPLC is used for the compounds which are soluble in organic solvents and are non-ionic. The water soluble non-ionic compounds are separated using bonded phase chromatography.

14.5.3 Ion Exchange Chromatography

Ion exchange may be defined as the reversible interchange of like ions between a solid phase (ion-exchange material) and a liquid phase in which there is no permanent change in structure. Thus, Ion Exchange Chromatography may be regarded as a special example of liquid-solid chromatography wherein strong ionic attractions replace relatively weak polar adsorption forces. In this technique, substances to be separated are previously adsorbed on the ion exchangers and then released by eluting or washing the column with a solution containing a solute which is more tightly held by the adsorbent than the component to be separated.

Ion Exchange Resins: Ion exchange resins consist of cross linked, long chain organic polymers with a microporous structure. The functional groups attached to the polymeric chains are responsible for ion exchange.

Resins of suitable chemical compositions and physical properties may be synthesized for specific ion-exchange applications. The resins usually consist of styrene-divinylbenzene copolymer. Other polymers like methacrylic acid-divinylbenzene or phenol formaldehyde polymers are also used.

The functional groups attached to the resin may be acidic or basic and hence resins can be classified as:

(*a*) ***Cation exchange Resins:*** Resins containing acidic groups like –COOH or $-SO_3H$ group are termed as cation exchange resins. These are usually styrene divinyl benzene copolymers which on sulphonation or carboxylation become capable of exchanging their H^+ ions with the cations of the solution. A number of synthetic-resin cation exchangers which contain sulphonated phenolic or aromatic hydrocarbons have been prepared. These resins are called

H-form cation exchangers. It is also possible to convert the H-form cation exchanger to Na-form exchanger by using NaCl. Schematically a cation exchanger can be represented as:

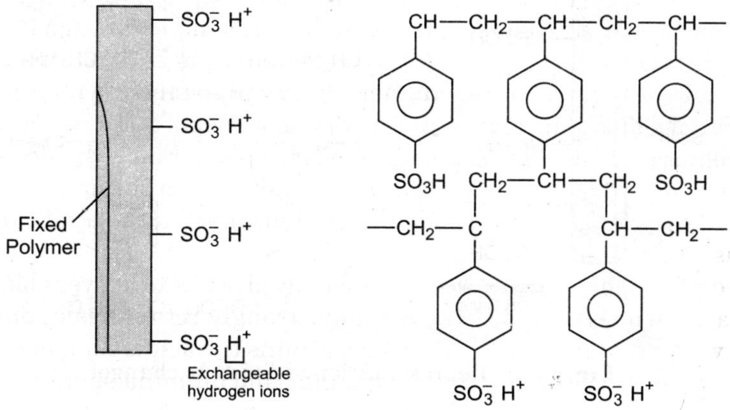

Fig. 14.6 Representation of Cation Exchanger

Resins having SO_3H group are termed as strongly acidic resins whereas having COOH group are termed as weakly acidic resins. The cation exchange resins can be represented as RH^+.

Table 14.3 gives some of the examples of commercial cation Exchange resins.

Table 14.3 Commercial Cation Exchange Resins

Types of Resin	*Functional Group*	*Nature of Resin*	*pH Range*	*Commercial Name*	*Applications*
Strongly acidic	$-CH_2SO_3^-H$	Sulphonated Polystyrene	1–14	Amberlite IR-100 IR-105 IR-109	Inorganic separations, fractionation of cations
	$-SO_3^-H^+$	Sulphonated Polystyrene	1–14	IR-112 IR-120	Peptides, amino acids, vitamin B
Weakly acidic	–COOH	Carboxylic	5–4	Amberlite IR-6 IR-50	Biochemical separations, amino acids, antibiotics
	$-PO(OH)_2$	Phosphonated Polystyrene	5–4	Duolite C-63	Organic bases, transition elements

(*b*) ***Anion exchange Resins:*** Resins containing basic functional groups like amine, substituted amine or quaternary ammonium groups as their hydroxide salts are termed as anion exchange resins. These are styrene divinyl benzene complexes which, because of their basic functional groups become capable of exchanging their anions with those present in the solution.

Resins having $-NH_2$, =NH group are weakly basic whereas those having quaternary ammonium salts are strongly basic. Anion exchange resins are represented as $R\overline{O}H$.

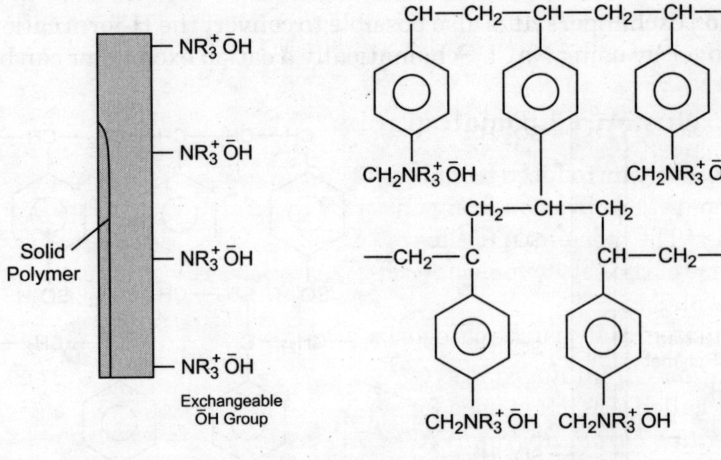

Fig. 14.7 Representation of Anion Exchanger

Table 14.4 gives the examples of some of the commercially available anion exchangers.

Table 14.4 Commercial Anion Exchange Resins

Types of Resin	*Functional Group*	*Nature of Resin*	*pH Range*	*Commercial Name*	*Applications*
Strongly basic	$-CH_2NR_2OH$	Quaternary Polystyrene	0–12	IRA-410	Halogens, fatty acids,
	$-CH_2N$	Ammonium Polystyrene	0–12		Alkaloids, vitamin B complex
	$-(CH_2COOH)_2$	Chelating amino diacetic acid	0–12	DowexA-I	Fractionation of anions
	$-CH_2NH_2$		0–12	Dowex IR-4B	Fatty acids
Weakly basic	$-N(C_2H_5)_2$	Polyamine polystyrene	0–9	De-Acedite -E Amberlite IR-4B	Amino acids, vitamins
	–OH	Phenol formaldehyde	0–9	Duolite A-1 to 7	Fractionation of anionic compounds of metals

Action of Ion Exchangers: The ion exchange resins are insoluble in aqueous media. When a cation exchanger is immersed in an aqueous solution of the sample containing M^{x+} ions, the following exchange equilibrium is quickly established.

$$\underset{\text{Resin}}{xRS_3^-H^+} + \underset{\text{Solution}}{M^{x+}} \rightleftharpoons \underset{\text{Resin}}{(RSO_3^-)_x\, M^{x+}} + \underset{\text{solution}}{xH^+}$$

Similarly the anion exchanger can exchange anions (A^{x-}) in a similar way:

$$\underset{\text{Resin}}{xRN(CH_3)_3^+\,\bar{O}H} + \underset{\text{Solution}}{A^{x-}} \rightleftharpoons \underset{\text{Resin}}{[RN(CH_3)_3^+]_x\, Ax^-} + \underset{\text{Solution}}{x\bar{O}H}$$

The ion exchangers have been found to have greater affinities for polyvalent ions as compared to univalent ions i.e. they retain polyvalent ions more strongly than singly charged ions. For a sulphonated cation exchange resin, the affinity for univalent ions decrease in the order:

$$Ag^+ > Cs^+ > Rb^+ > K^+ > NH_4^+ > Na^+ > H^+ > Li^+$$

For divalent the order is
$Ba^{2+} > Pb^{2+} > Sr^{2+} > Ca^{2+} > Ni^{2+} > Cd^{2+} > Cu^{2+} > Co^{2+} > Zn^{2+} > Mg^{2+} > UO_2^{2+}$

14.5.4 Size Exclusion Chromatography

This is one of the recently introduced technique of chromatography. Size exclusion chromatography is a technique of separation of various components depending upon the difference in their molecular weights and shapes. The mechanism relies only on the relative pore sizes of the stationary phase compared to the size of the solute molecules. There is no chemical interaction between the solute and the stationary phase.

Exclusion chromatography can be categorized as two types:

(i) *Gel filteration chromatography (GFC):* The chromatography based on the hydrophilic packing is called GFC. It is used for separating polar species.

(ii) *Gel permeation chromatography (GPC):* The technique based on hydrophobic packing is called GPC. It is used for separating non-polar species.

The stationary phase in size exclusion chromatography consists of small (~ 10 μm) silica or polymer particles containing a network of uniform pores into which solute and solvent molecules can diffuse. The solute molecules having smaller sizes than the pores are trapped in the pores and are retained before being removed by the mobile phase. The penetration by each solute into the stationary phase is different and depends upon its size and shape. The larger molecules will be effectively excluded from entering these pores and would be first to be eluted with the mobile phase. The retained molecules are eluted at different intervals depending upon their retention times; smaller molecules are more strongly retained and are last to be eluted.

The materials used for size exclusion chromatography are micro particles of styrene-divinylbenzene copolymers, silica etc.

The exclusion chromatography is widely used for separation of large molecules or groups of molecules from smaller ones. This has wide applications in biosciences. Another important application of size exclusion chromatography involves the rapid determination of the molecular weight of polymers.

14.5.5 Planar Chromatography

Planar chromatographic methods include thin-layer chromatography (TLC) and paper chromatography (PC).

(a) Thin Layer Chromatography

Thin layer chromatography (TLC) is an efficient, fast and simple method of separating the mixtures. In this technique a thin layer of silica gel or alumina is spread on a glass or plastic plate and hence the name. The basic principle of separation is the *selective absorption* of the components of the mixture on the stationary phase, when they are carried through it by the flow of mobile phase. The steps involved are as follows:

(i) Preparation of thin layer plates.
(ii) Application of the material on the plate.
(iii) Development of the plate.
(iv) Visualization
(v) Calculation of R_f values.

(*i*) *Preparation of thin layer plates.*

The plates are usually prepared using glass plates or plastic plates of the size 20 × 20 cm or 5 × 20 cm. A thin layer of silica or alumina is spread on the glass or the plastic plate. The prepared thin layer on glass is often called a *chromaplate*. Chromaplates are prepared by applying a uniform layer of silica or alumina in the form of a thin aqueous paste. The solid adsorbent (silica or alumina) is usually mixed with a small amount of 'binder' such as calcium sulphate.

The binder helps in proper adherence of the adsorbent onto the plate. Special applicators are available commercially to spread the adsorbent. The thickness of the adsorbent layer is generally 0.25 mm. The paste or slurry of the adsorbent is prepared in water.

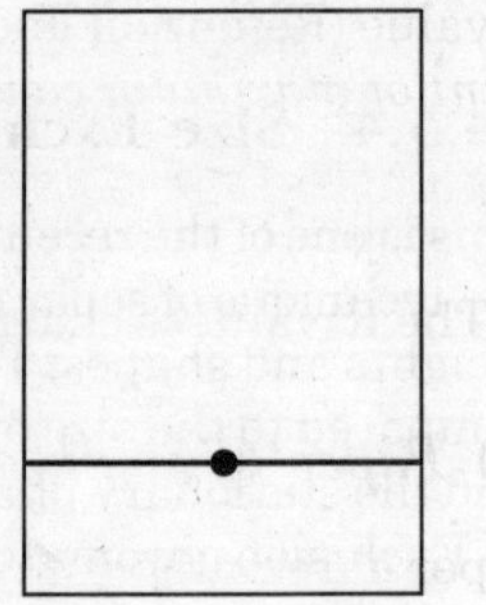

Fig. 14.8 Thin layer plate with the spot of sample

(*ii*) *Application of the material on the plate.*

The material is dissolved in an appropriate solvent and deposited on the thin layer with the help of a capillary. The capillary is first dipped in the sample solution, the solution goes up the capillary. The sample is applied on the plate by touching the capillary to the adsorbent surface at about 2 cm from the edge. Position of spot is marked with a pencil so that the Rf values may be calculated after separation.

(*iii*) *Development of the plates*

When the spot is dry, the chromaplate is placed in a jar (developing chamber) containing small amount of solvent. The jar must be saturated with solvent vapour before the plate is placed in it. This can be achieved by lining the jar with filter paper dipping into the solvent. Wide-mouthed, capped bottles are used as developing chambers in the laboratories. The plate is kept in such a way that lower end of the plate dips 1–2 cm below the surface of the solvent. The spot should be above this level. The top of the jar is firmly closed. The solvent rises through the adsorbent layer, *and the components of the mixture ascend at different rates depending on their affinities for the adsorbent.* When the solvent has nearly reached the top of the adsorbent layer (3/4 of the TLC plate), the components should be well separated. The relative distance travelled by the components can be increased by using a solvent of higher or lower polarity. Typical solvents in order of increasing elutive power:

$$\text{Petroleum} < \text{benzene} < \text{chloroform} < \text{ether} < \text{ethanol}$$

(*iv*) *Visualization*

The plate is removed from the tank, the position of the solvent front is marked with a pencil and the solvert is allowed to evaporate from the plate. If the components of the mixture are coloured, they can be detected by simple visual examination. However, if they are colourless, they are located by

(a) keeping the plate in a closed dry tank containing crystals of iodine. The spots show up as brown spots

(b) spraying the plate with concentrated H_2SO_4 and then heating the plate to about 200ºC. The separated spots of organic compounds will render visible due to highly sensitive charring effect

(c) Irradiation with UV light to locate the substance which give flourescence on exposure to UV.

(v) *Calculation of R_f Values*

R_f value (Retention Factor) is defined as the *ratio of the distance travelled by the component at its point of maximum concentration to the distance travelled by the solvent.*

$$R_f \text{ Value} = \frac{\text{Distance moved by component}}{\text{Distance moved by the solvent front}}$$

The R_f value is characteristic of a substance for a given solvent.

(b) Paper Chromatography

Paper chromatography (PC) is a simple technique to separate complex mixtures of metal ions, amino acids, sugar, dyes and drugs. It is a form of partition chromatography in which the stationary phase is the absorbed water, always present in filter paper. The support is the paper itself and the mobile phase is a solvent previously saturated with water.

Paper chromatography is a form of partition chromatography but adsorption phenomenon is also involved. Filter paper used is made of highly purified cellulose which, being a polyhydroxy compound of high molecular weight, absorbs and retains water molecules. The stationary phase is not the filter paper, but a very thin film of the solvent, usually water adhering to the surface of the filter paper. Thus the paper chromatography will not work in dry climates unless the paper is exposed to air of high humidity, before being placed in the developing solvent.

A drop of a solution of the mixture (or drops of individual components of the mixture) is applied to the paper by means of the capillary tube and the paper dried. The paper is then suspended in a suitable, closed container so that it can be irrigated with an organic liquid or a mixture of liquids. The bottom of the paper should immerse in the solvent. The capillary action of the paper will cause the solvent to rise steadily up the paper. During this process, the solvent which now contains the mixture in solution, is continuously extracted by the retained water molecules in the paper.

The extent to which a particular solute has moved, relative to the extent to which the solvent has moved, known as R_f value, will be a function of the partition coefficient of the solute between the solvent and the film of retained water.

When the ascending solvent has travelled the required distance, the paper is removed, the position of solvent-front marked and the paper is dried. The position of various solutes, if they are coloured, appear as coloured spots. If the spots are not coloured, the location of the spots is determined by (i) spraying the paper with a chemical reagent which produces insoluble coloured derivatives with the solutes (e.g. ninhydrin for amino acids) (ii) or by UV light in which the solute exhibit flourescence. The R_f value is calculated.

$$R_f = \frac{\text{Distance moved by the substance}}{\text{Distance moved by the solvent}}$$

In paper chromatography, two types of techniques are used.

(*i*) *Ascending technique:* In this the solvent ascends through the filter paper as described above.

(*ii*) *Descending technique:* In this the solvent is contained in a small trough at the top of developing tank and travels downwards. The mixture is applied at a point external to the trough.

14.6 GENERAL AND FUNDAMENTAL CONCEPTS OF CHROMATOGRAPHY

Chromatography involves a number of separation methods which have many similarities in experimental technique, basic principles and operating conditions. In all the chromatographic techniques various components in a mixture get separated during their passage through a porous media due to the differences in their migration rates. The separation is due to the difference in their distribution behaviour between stationary and mobile phase.

Fundamental Concepts

Some of the fundamental concepts used in all types of chromatographic techniques are:

(*i*) ***Distribution Ratio:*** This is one of the main fundamental concept in chromatographic technique. It is defined as the ratio of concentration of a solute in two phases (stationary and mobile) at equilibrium.

$$\text{Distribution ratio, } K_D = \frac{[\text{Solute}]_{\text{stationary phase}}}{[\text{Solute}]_{\text{mobile phase}}}$$

The value of K_D depends upon the temperature, pressure and is independent of concentration. If K_D is large the concentration of solute is more in stationary phase i.e. greater is the retention of the solute in stationary phase.

(*ii*) ***Retention Time:*** The retention time t_R is the time between injection of a sample and the appearance of a solute peak at the detector:

$$t_R = \frac{\text{Length of the column packing } (L)}{\text{Average linear rate of solute migration } (\bar{V})}$$

$$t_R = \frac{L}{\bar{V}}$$

$$\therefore \quad \bar{V} = \frac{L}{t_R}$$

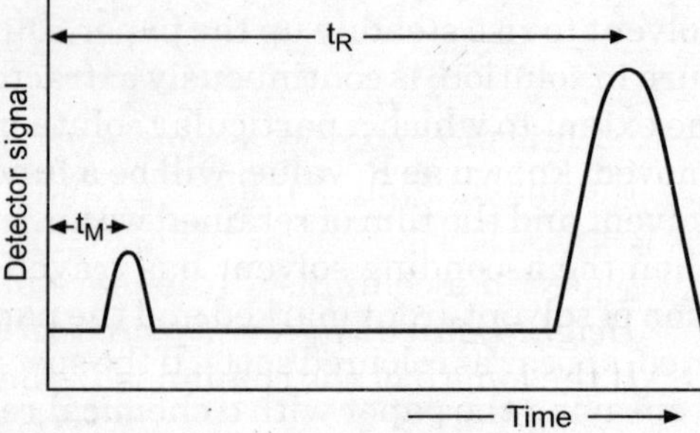

Fig. 14.9

The time taken for mobile phase to pass through the column is called t_m.

A term called the *retention factor*, k', is often used to describe the migration rate of an analyte on a column. It is also called the *capacity factor*. The retention factor for analyte A is defined as:

$$k'_A = \frac{t_R - t_M}{t_M}$$

tR and tM are easily obtained from a chromatogram. When an analytes retention factor is less than one, elution is so fast that accurate determination of the retention time is very difficult. High retention factors (greater than 20) mean that elution takes a very long time. Ideally, the retention factor for an analyte is between one and five.

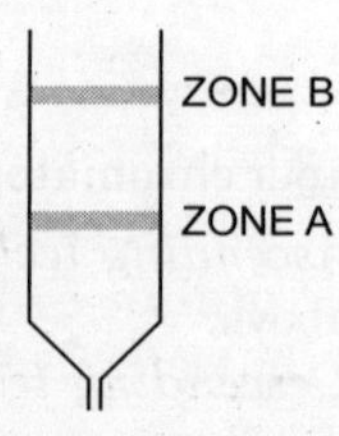

Fig. 14.10 Column showing different zones

(*iii*) ***Retardation Factor (R_f):*** The retention of the solute in partition or adsorption chromatography is described by its migration relative to the solvent of the mobile phase.

$$R_f = \frac{\text{Distance travelled by solute}}{\text{Distance travelled by mobile phase}}$$

The R_f values change with the type of migration, adsorbent and solvent.

(*iv*) ***Resolution:*** The effectiveness of the separation is determined by two general factors, the distance between the separated zones and the compactness of the zones.

As the distance between the zone centres increases, the degree of separation increases. The measure of this difference is called resolution. Sometimes as zones travel with the mobile phase, they tend to spread and broaden and overlapping of the bands take place.

(*v*) ***Selectivity: (Relative migration rates):*** The selectivity factor α for solutes A and B is defined as the ratio of partition coefficient of more strongly solute (say B) to the partition coefficient of the less strongly held solute (say A)

$$\alpha = \frac{K_A}{K_B}$$

Thus α is always greater than unity.

(*vi*) ***Theoretical Plate Theory:*** This theory was given by Martin and Synge who were awarded Nobel prize for their work in chromatography. They treated a chromatographic column as if it were made up of a number of bubble-cap like plates, similar to that of fractionating column. These plates are thus theoretical plates where perfect equilibrium is assumed in each plate for mobile and stationary phase. The plate number and plate height is a useful index of column efficiency to describe the broadening of peaks and zones.

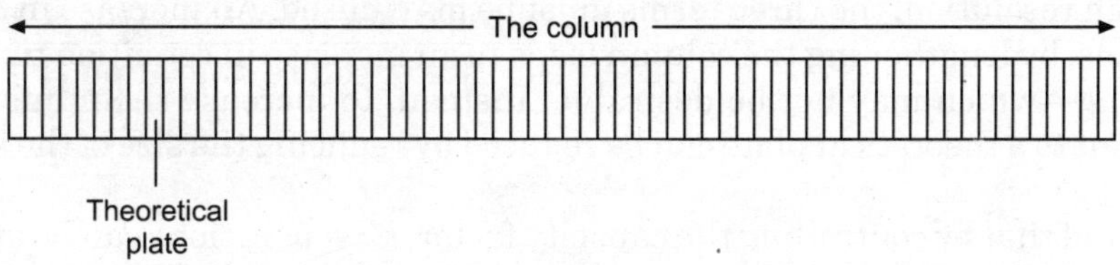

Fig. 14.11

It is important to remember that the plates do not really exist, they are a figment of the imagination that helps us understand the processes at work in the column. They also serve as a way of measuring column efficiency, either by stating the number of theoretical plates in a column, N (the more plates the better), or by stating the plate height, the *Height Equivalent to a Theoretical Plate* (HETP) (Smaller the better).

If the length of the column is L, then the HETP is

$$\text{HETP} = L/N$$

The number of theoretical plates r, is given by

$$r = \left[\frac{4t_r}{y}\right]^2$$

where t_r is retention time

y = time of emergence of 68.3% of vapour.

(*vii*) ***Column Efficiency:*** The efficiency of column can be expressed in terms of number of theoretical plates N and plate height H

$$N = L/H$$

where L is the length of column packing.

The efficiency of a chromatographic column increases as the number of plates become greater and the plate height becomes smaller.

The N is determined from a chromatogram. The retention time of a peak t_r and the width of the peak at its base W (in units of time) are measured and number of plates can be computed by simple relationships:

$$N = 16\left(\frac{t_R}{W}\right)^2$$

Since N is inversely related to H, an increase in N improves the efficiency of resolution.

(viii) ***Resoluton:*** Although the selectivity factor, α, describes the separation of band centres, it does not take into account peak widths. Another measure of how well species have been separated is provided by measurement of the *resolution*. Resolution is the quantitative measure of the ability of chromatographic column to separate solute A and B. The resolution of two species, A and B, is defined as

$$R = \frac{2[(t_R)_B - (t_R)_A]}{W_A + W_B}$$

Baseline resolution is achieved when $R = 1.5$.

It is useful to relate the resolution to the number of plates in the column, the selectivity factor and the retention factors of the two solutes;

$$R = \frac{\sqrt{N}}{4}\left(\frac{\alpha - 1}{\alpha}\right)\left(\frac{1 + k_B}{k_B}\right)$$

To obtain high resolution, the three terms must be maximised. An increase in N, the number of theoretical plates, by lengthening the column leads to an increase in retention time and increased band broadening—which may not be desirable. Instead, to increase the number of plates, the height equivalent to a theoretical plate can be reduced by reducing the size of the stationary phase particles.

It is often found that by controlling the capacity factor, k', separations can be greatly improved. This can be achieved by changing the temperature (in Gas Chromatography) or the composition of the mobile phase (in liquid chromatography).

The selectivity factor, α, can also be manipulated to improve separations. When α is close to unity, optimising k' and increasing N is not sufficient to give good separation in a reasonable time. In these cases, k' is optimised first, and then α is increased by one of the following procedures:

1. Changing mobile phase composition
2. Changing column temperature
3. Changing composition of stationary phase
4. Using special chemical effects (such as incorporating a species which complexes with one of the solutes into the stationary phase).

14.7 CHROMATOGRAPHIC DEVELOPMENTS

Chromatographic development is the process in which a sample is carried through the stationary phase by a mobile phase.

Chromatographic development may be carried out by the following four techniques:

(i) Elution development
(ii) Gradient elution
(iii) Frontal analysis
(iv) Displacement development.

(i) Elution Development

In elution analysis a single portion of the sample dissolved in the mobile phase is introduced at the head of the column. Additional fresh solvent (mobile phase) is then added where upon the components of the sample distribute themselves between the mobile phase and the stationary phase. In moving down the column, the mobile phase carries the components of the sample with it at different migration rates. The components that favour retention by the stationary phase move slowly (component *B* and *C* in Fig. 14.12) whereas the components which are least strongly held travel rapidly through the column with the mobile phase. In this way the different migration rates of the components of the sample cause these to separate into bands or zones along the length of the column. The separated species are then isolated by passing sufficient quantity of mobile phase through the column to cause the individual bands to pass out or eluted from the column.

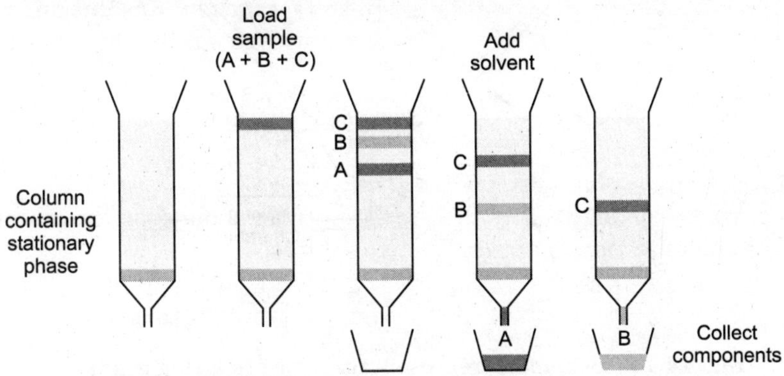

Fig. 14.12 Elution development

In adsorption chromatography, the separation is due to differences in adsorption affinity of solutes. In partition chromatography, separation is due to solute's ability to distribute itself between the two phases. Elution development is the most popular chromatographic technique.

If a detector that responds to solute concentration is placed at the end of the column and it signal is plotted as a function of time or of volume of added mobile phase, a series of peaks are obtained. The plot is called chromatogram. The components of the sample are identified from the peaks of the chromatogram and area under the peaks provide the quantitative measure of the amount of the component. (Fig. 14.13)

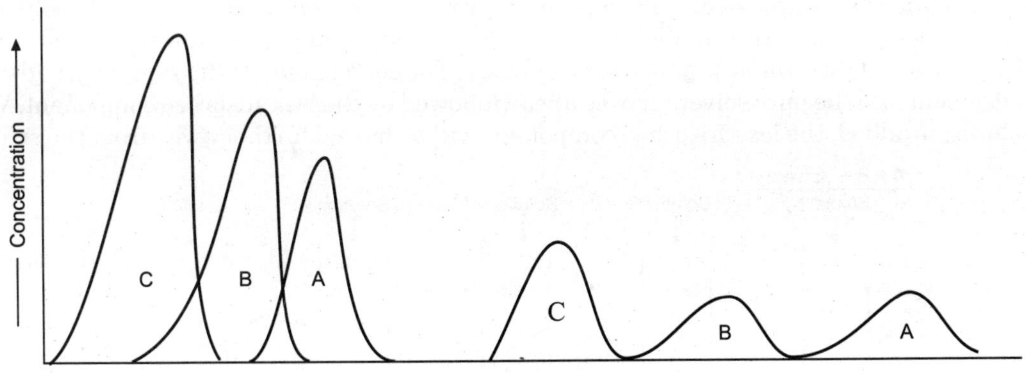

Fig. 14.13 Chromatogram

(ii) Gradient Elution

Gradient elution is significant in separating the components of the sample that cannot be separated by single solvent. An elution with a single solvent of constant composition is termed as isocratic. In gradient elution, two or more solvent systems are employed. It is widely applicable where the sample components differ widely in polarity. The solvent systems that differ significantly in polarity are employed. The ratio of the two solvents is varied in a progressive manner from 0 to 100% of one component, in a series of steps or continuously.

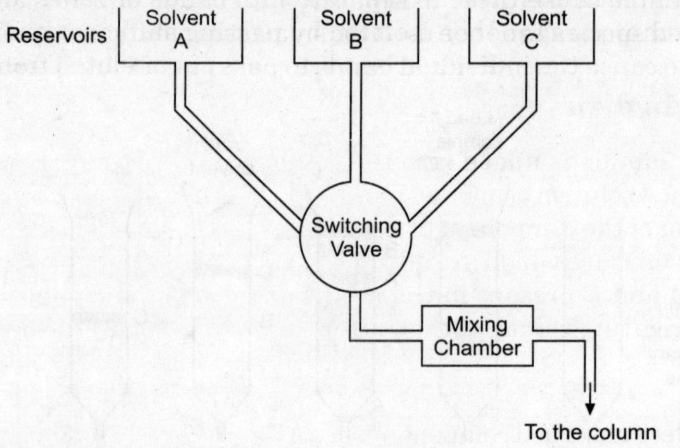

Fig. 14.14 Schematic representation for preparing gradient

The two liquids which are combined may be miscible solvents but are kept in separate reservoirs. For gradient elution using a low pressure mixing system, the solvents from separate reservoirs are fed into a mixing chamber and the mixed solvent is pumped to the column (Fig. 14.14). Modern HPLC instruments are equipped with proportionating valves that introduce liquids from two or more reservoirs at rates that vary continuously.

The change in the composition of the eluent liquid causes a change in the partition coefficients of the components and lead to improved resolution.

(iii) Frontal Analysis

In this technique the sample is dissolved in an appropriate solvent and this sample solution is passed continuously through the adsorbent which may be silica gel or alumina. No solvent is added but solution of the sample is added continuously. The components of the sample get adsorbed on the adsorbent and the pure solvent moves ahead followed by the less adsorbed component. When more solution is added, the less adsorbed component will be moved further away from the strongly

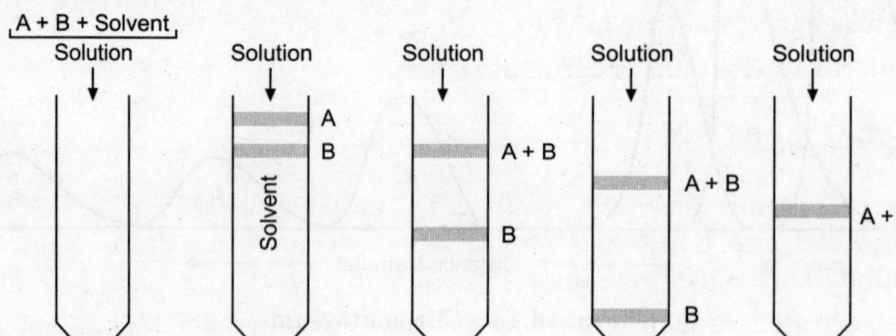

A is strongly adsorbed component
B is less strongly adsorbed component.

Fig. 14.15 Frontal analysis

adsorbed component resulting into separation. The component which is adsorbed least will appear first in front while the strongly adsorbed component will elute last. The technique is called frontal analysis because a series of 'fronts' are formed.

In frontal analysis a great amount of mixture is required as it has to be applied continuously to the column thus the technique cannot be used for small samples.

(iv) Displacement Analysis

In this technique the sample is added into the column and the components of the sample are adsorbed on the column. A solution of another substance, which have a higher preferential adsorption on the column than that of the component of the mixture, is then introduced into the column. Thus the components of the mixture which are adsorbed are displaced by this solution. Suppose there are two components *A* and *B* present in the mixture and *C* is the displacing component in the solution, known as irrigating solution, then after the completion of the separation, the elution order will be

(a) Pure solvent
(b) Solution of the less absorbed component (say *A*)
(c) Solution of *B*
(d) Solution of *C*.

The component of solution *C* is called displacing agent because it displaces the component of the mixture. As the flow of the displacing agent is continued, it causes all other solutes to be forced ahead of the displacer.

14.8 APPLICATIONS OF CHROMATOGRAPHY

In any chemical or bioprocessing industry we need to purify a product from a complex mixture, which is an important step. There are number of techniques available for separating the components of a mixture but chromatography offers a very special technique for number of reasons:

- It can separate complex mixture with great precision.
- It can separate even very similar components. For example, the proteins which differ only in one amino acid can be separated with chromatography.
- The chromatography can purify basically any soluble or volatile substance, if right adsorbent material, mobile phase and operating conditions are employed.
- Since the conditions under which chromatography is performed are not typically severe, it can be used to separate delicate produces.

Some of the general applications of chromatography are:

1. In inorganic Chemistry

HPLC and its various techniques have been applied in inorganic chemistry for separation of a number of complex mixtures. Inorganic compounds can be ionic compounds (composed of ions) or molecular compounds. Different chromatographic techniques have been used for these two classes.

- The ions are separated through ion exchange chromatography or ion exclusion chromatography.
- The non-dissociated molecular compounds are separated by reversed phase liquid chromatography.
  (*i*) ***Separation through ion exchange chromatography:*** In ion exchange chromatography the anions are removed by using the resins of the type – NR^{3+}. Higher is the affinity of the anion to the exchange sites of the ion, higher is the retention of the respective ionic solutes.

The observed order of affinity for groups of ions is:

$$\overline{O}H, F^- < CH_3COO^- < HCOO^- < H_2PO_4^- < HCO_3^- < Cl^- < NO_2^- < HSO_3^- < Br^- < NO_3^- < HSO_4^- < I^-$$

Thus, a mixture of ionic compounds containing anions such as F^-, Cl^-, PO_4^{3-}, Br^- and SO_4^{2-} can be separated easily. Fig. 14.16 shows the separation of these anions.

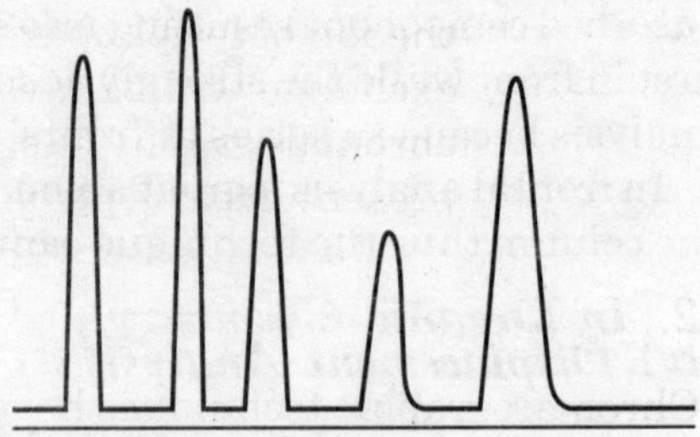

Fig. 14.16 Separation of F^-, Cl^-, PO_4^{3}, Br^- and SO_4^{2-} anions.

For the separation of anions such as I^-, IO_3^-, SCN^-, ClO_4, MoO_4^{2-}, CrO_4^{2-} and VO_4^{2-}, stronger eluents are generally employed.

(*ii*) ***Separation of strong acids from weak acids (organic acids):*** Using the stationary phase containing sulphonate type ion exchanger, strong acids are separated from weak acids. The strong acids are eluted with the mobile phase and weak acids are retained by the ion exchanger. The eluent is generally HCl.

(*iii*) ***Separation of the cations:*** For the separation of cations, sulphonated polymeric resins are used. These are used as thin coatings on the surface of the spherical glass beads (available under the name Zipax). Alkali metals like Li^{2+} to Cs^+ are eluted by using HCl. The exchange of cations depends upon the following factors:

- At low concentrations, the extent of exchange increases with increasing charge of the exchanging ion:

$$Na^+ < Ca^{2+} < Al^{3+} < Th^{4+}$$

- For singly charged ions, the extent of exchange increases with decrease in size of the hydrated cation:

$$Li^+ < Na^+ < NH_4^+ < K^+ < Rb^+ < Cs^+$$

- To exchange an ion of higher charge on the exchanger for one of lower charge in solution, exchange will be favoured by increasing the concentration of the solution. If

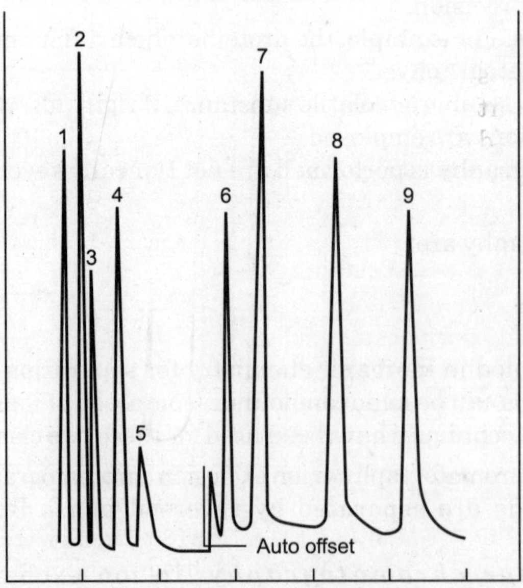

	ppm
1. Li^+	0.5
2. Na^+	2
3. NH_4^+	3
4. K^+	3
5. Morpholine	30
6. Cyclohexylamine	10
7. Mg^{2+}	1
8. Ca^{2+}	2
9. Sr^{2+}	10

Fig. 14.17 Ion chromatogram of a mixture of cations

the ion of lower charge is in exchanger and ion of higher charge is in solution, the exchange will be favoured by high dilutions. Fig. 14.17 shows applications of ion chromatography to separate a mixture of cations. Note that the amount of cations present in parts per million which would have been difficult to separate otherwise.

2. In Organic Chemistry

Chromatographic techniques have wide range of applications in organic chemistry. The range of applications in organic chemistry include separation, preparation, purification and identification of the organic compounds.

(*i*) **Preparative HPLC:** It refers to the process of isolation and purification of the compounds. The compounds are isolated in high purity.

(*ii*) **Chemical Separations:** Almost all fields of organic chemistry utilize the chromatographic methods to separate the complex mixture of compounds due to its simplicity and relatively low cost.

Table 14.5 illustrates some of the fields and typical mixtures where this technique is applied extensively.

Table 14.5 Typical Applications of High-Performance Liquid Chromatography

Field	*Typical Mixtures*
Pharmaceuticals	Antibiotics, sedatives, steroids, analgesics
Biochemicals	Amino acids, proteins, carbohydrates, lipids
Food products	Artificial sweeteners, antioxidants, additives
Industrial chemicals	Condensed aromatics, surfactants, propellants, dyes
Pollutants	Pesticides, herbicides, phenols, PCBs
Forensic science	Drugs, poisons, blood alcohol, narcotics
Clinical Medicine	Bile acids, drug metabolites, urine extracts, estrogens
Petroleum and coal industries	Coal and oil products.

(*iii*) **Purification:** It refers to the process of separating or extracting the target compound from other structurally related compounds or contaminants. Under chromatographic conditions each compound produces a characteristic peak. Depending on the compound which has to be separated and the closely related samples, one can choose the proper conditions (such as mobile phase, stationary phase, etc.) in order to collect the desired compound as it elutes from the stationary phase. For purification, care must be taken to avoid overloading the column so that no contamination of the desired sample (to be collected) takes place. Moreover, the rate of migration of the desired sample and the contaminant should vary considerably.

(*iv*) **Identification:** This is the crucial part of any HPLC. It is widely used for recognizing the presence or absence of the components in mixtures.

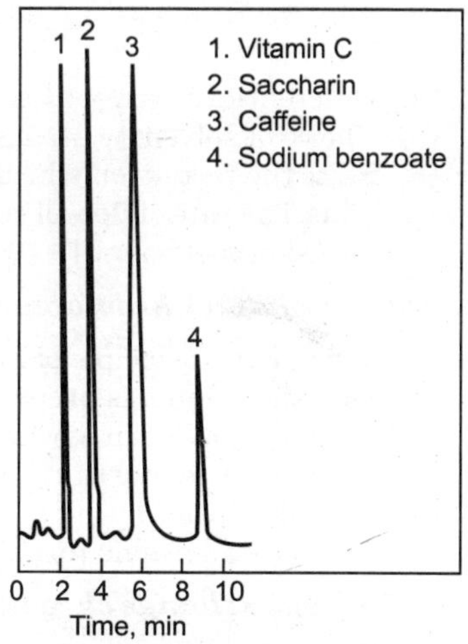

Fig. 14.18 Liquid bonded-phase chromatogram of soft-drink additives

For example, around 30 or more amino acids in a protein hydrolyrate can be detected with a reasonable degree of certainty with a chromatogram. However, the chromatogram may not lead to positive identification of the specie in a sample because it provides only single information i.e. retention time of the specie. But it provides sure evidence of the *absence* of specie in the mixture. The failure of a sample to produce a peak at the same retention time as that obtained for standard under identical condition is strong evidence that the compound is absent.

The technique has wide applicability in identifying the pollutants such as pesticides and carcinogens in air, water and food supplies. It has also been used in identifying the preservatives, antioxidants, additives in food products. Fig. 14.18 illustrates the identification of soft drink additives.

QUESTIONS

1. Define the term chromatography.
2. What are the different types of chromatographic techniques?
3. Define partition chromatography.
   [**Ans.** It is a technique in which mixtures of substances are separated by means of partition between a moving solvent and a stationary liquid which is held on a suitable support. The difference in partition coefficient of individual solutes in mobile and stationary phase is basis of separation.]

Based on thin layer chromatography

4. What is the principle of thin layer chromatography?
5. How the solid layers adhere to the glass plates?
   [**Ans.** With the help of binding agents like $CaSO_4$ etc.]
6. Define chromatogram, developer, adsorbent.
7. What is R_F value? How it is determined?
8. What are different materials used as adsorbent in thin layer chromatography?
9. What is the structure of ninhydrin.
10. What is silica Gel G?
11. In what ways TLC is superior to other chromatographic techniques?
    [**Ans.** It is simple, having short development time and is highly sensitive.]

Based on paper chromatography

12. What is the paper chromatography?
13. What is R_F value?
14. Differentiate between ascending and descending techniques?
15. Choice of solvent is an important criteria for separation in paper chromatography. Explain.
16. State the factors on which the rate of flow of substance depend?
    [**Ans.**The rate of flow of substance depend on the size, nature of the substance, solvent density and diffusion coefficient.]

Based on Column Chromatography

17. What is the principle of column chromatography?
18. How the column is prepared in the laboratory?
19. What do you mean by chromatogram?
20. How the pure constituents are recovered?
21. What is elution?
22. Why it is advisable to start elution with the non polar solvent in the beginning?

Based on ion exchange chromatography

23. Define ion exchange chromatography?
24. What are ion exchangers? How these are classified?
25. What is the principle of ion exchange chromatography?
26. What is ion exchange capacity?

27. Name some of the commercially available ion exchangers.
28. What is regeneration? What chemicals are used for regeneration for cationic and anionic exchangers respectively.

Mixed

29. Define the following:
    (a) Chromatography
    (b) Elution
    (c) Eluent
30. State the principle of column chromatography.
31. How are the different chromatographic techniques classified?
32. Give a brief outline of instrumentation for HPLC.
33. What are the various types of liquid-chromatography?
34. What are the main requirements of solid adsorbents used in column chromatography. Name some adsorbents commonly used.
35. Discuss ion exchange chromatography.
36. What are cation and anion exchange membranes.
37. Why do we change from less polar to more polar eluting solvents and not in the reverse order in column chromatography?
38. What is meant by R_f? What is its significance?
39. What are the main considerations in the selection of eluting solvent in column chromatography?
40. What is TLC? How will you test the purity of an organic compound using TLC?
41. What is paper chromatography? How is it better than column chromatography?
42. Compare HPLC with column chromatography.
43. Write short note on Frontal Analysis.
44. Why chromatography is considered as a unique technique for the separation of substances.
45. Discuss some applications of chromatography.
46. What is R_f value in chromatography? **(PTU, Dec. 2004)**
47. (a) What is the basic principle of Gel chromatographic technique
    (b) Write short notes on:
       (i) liquid Chromatography.
       (ii) Chromatographic development. **(PTU, Dec. 2004)**
48. What do you understand by distribution ratio (D) in chromatography? **(PTU, May, 2005)**
49. How are chromatographic techniques classified. Discuss in brief chromatographic mechanism? **(PTU, May, 2005)**
50. Explain the chromatographic development with suitable example. **(PTU, May, 2005)**
51. How the performance of a particular chromatographic system can be assesed? **(PTU, Dec. 2005)**
52. Write short note on :
    (i) Liquid chromatography.
    (ii) Classification of chromatography methods. **(PTU, Dec. 2005)**
53. What do you mean by Retardation factor (R_f) in chromatography? **(PTU, Dec. 2005)**
54. Mention the advantage of chromatography. **(PTU, May 2006)**
55. (a) How are chromatographic techniques classified?
    (b) Compare
       (i) HPLC with GC and
       (ii) HPLC with column chromatography. **(PTU, May 2006)**
56. What is R_f value in chromatography? **(PTU, Jan. 2008; PTU, May 2007/Dec. 2006)**
57. (a) What are the various classes of chromatography?
    Bring out clearly the principle involved in each case.
    (b) Write short notes on the following:
       (i) Liquid chromatography.
       (ii) Vapour phase chromatography. **(PTU, Jan. 2008; PTU, May 2007/Dec. 2006)**

❑❑❑

15 LUBRICANTS

"What we observe is not nature itself, but nature exposed to our method of questioning."

Werner Heisenberg

CHAPTER OUTLINE

15.1 FRICTIONAL RESISTANCE

When the surface of a metal is allowed to move tangentially with respect to another metal, with which it is in contact, it experiences a resistance to its motion. This resistance is known as frictional resistance.

Actually, all the material surfaces, no matter how smooth they appear, show many irregularities in the form of peaks and valleys (of the order of 1000 A°). When the two metal surfaces come in contact with each other, the area of contact is only a small fraction of the apparent contact area, because the contact is through the peaks as shown in Fig. 15.1.

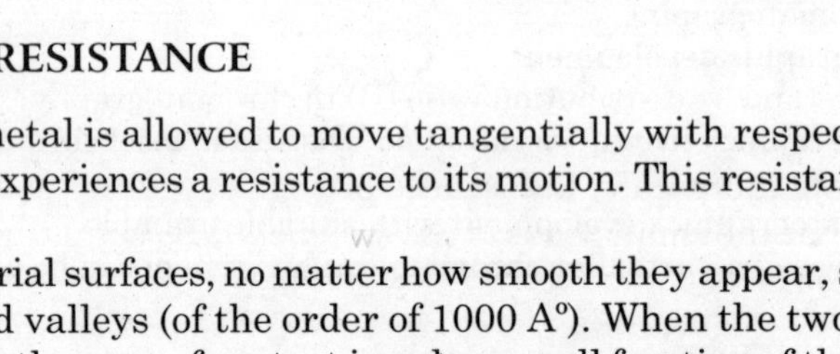

Fig. 15.1 Contact between peaks

These real areas of contact between the two surfaces are called *asperites.* When two metal surfaces slide over one another, the interlocking of the aspirites or peaks cause the resistance to the motion and give rise to frictional resistance. Moreover, due to sliding of one metal over the other, the peaks also get broken, resulting in wear and tear of the surfaces.

Further, the local pressure at the asperites may be sufficiently high to cause fusion of the peaks which leads to the formation of *welded junctions* between asperites. Additional force is then required to break these junctions. This further adds to frictional resistance.

Frictional resistance results in and loss of energy, a lot of wear and tear of surfaces of moving parts thereby causing loss in efficiency of machine. Sometimes, the moving parts get heated up and may also cause welding of two surfaces due to heat.

15.2 LUBRICANTS

In order to reduce frictional resistance and wear and tear of machine parts, a layer of fluid is introduced in between the moving parts. This layer of fluid is known as *lubricant*. The lubricant prevents direct contact of the rubbing surfaces by forming a film in between. Lubricant also acts as a coolant by removing the frictional heat generated due to rubbing of the surfaces.

Thus a lubricant can be defined as any substance introduced between two moving or sliding surfaces which minimizes the frictional resistance. The process of introducing the lubricant is known as lubrication (Fig. 15.2). The study of the phenomena and mechanism of friction, wear and lubrication of surfaces in relative motion is called *Tribology*, the main objective of which is to minimize the wear and tear of moving machine parts by reducing friction.

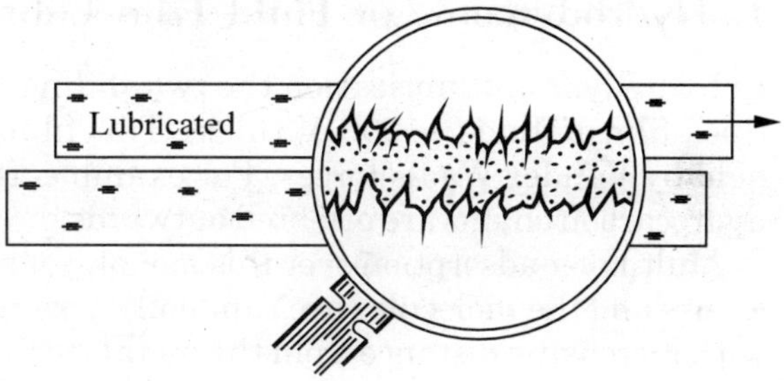

Fig. 15.2 Effect of lubrication

15.3 FUNCTIONS OF LUBRICANTS

(i) *Lubricants reduce the frictional resistance:* To reduce the frictional resistance, is one of the main functions of lubricants. Lubricants prevent the direct contact between the sliding metal surfaces so that the small peaks of the two metals do not interlock and hence frictional resistance is reduced.

(ii) *It reduces the surface deformation, wear and tear:* When two solid surfaces are in contact the real area of contact is very small as compared to the apparent area of contact, because the real contact is only at a limited number of asperites (Fig. 15.1). Even under light loads, high pressures are developed at the real contact areas. As a result the asperites are crushed causing appreciable deformation of the ductile metals. However, through lubrication the direct contact between the metal surfaces is avoided and hence the deformation as well as wear and tear of the machine parts is minimized.

(iii) *It acts as a coolant:* The frictional heat generated at the point of contact between the rubbing parts is localized at the surface peaks. This may raise their temperature to the melting point of the material and may cause welded junctions. The lubricating oil absorbs the heat and hence acts as coolant.

(iv) *It provides protection against corrosion:* The lubricants cover the metal surface and hence the attack of atmospheric corrosive environment on metal surface is prevented. Thus lubricants provide protection against corrosion.

(v) *It acts as a seal in some cases:* The lubricants also act as a seal preventing the leakage of gases. For example, in internal combustion engines the lubricant used between piston and cylinder not only reduces the friction but prevents the leakage of gases from the cylinder.

(vi) *It improves the efficiency of the machine:* Since the use of lubricant minimizes the liberation of frictional heat, it reduces expansion of metal. Moreover it avoids welded joints and hence reduces the maintenance and running cost of the machine.

15.4 MECHANISM OF LUBRICATION

Efficient lubrication plays a vital role in engineering. Depending upon the operating conditions and the lubricant characteristics, lubrication are of the following types:

1. Hydrodynamic or fluid film lubrication.
2. Thin film or boundary lubrication.
3. Extreme pressure lubrication.

1. Hydrodynamic or Fluid Film Lubrication

In hydrodynamic lubrication the two sliding metal surfaces are separated from each other by a thick film of fluid (~ 1000 A° thick). This film of lubricant in adsorbed on the metal surface and held by Vander Waals forces. For example, if hydrocarbon oils are used as lubricant, the long hydrocarbon chains are oriented outwards.

Multilayer adsorption over this monolayer also occurs and the molecules are randomly arranged with increasing distance from the metal surface.

Thus the two metal surfaces are separated by the oil film and direct surface to surface contact between the two metals is prevented and hence the peaks or the asperities do not interlock. This consequently reduces the friction. Whatever friction is there after lubrication, is only due to the internal resistance between the particles of lubricant. In the above example we may say that the friction is only along the tails of hydrocarbon chain. The coefficient of friction in such cases is as low as 0.001 to 0.03 as compared to 0.5 to 1.5 for sliding metal to metal surfaces. The longer the hydrocarbon chain, greater will be the separation between the surfaces and a consequent decreased coefficient of friction.

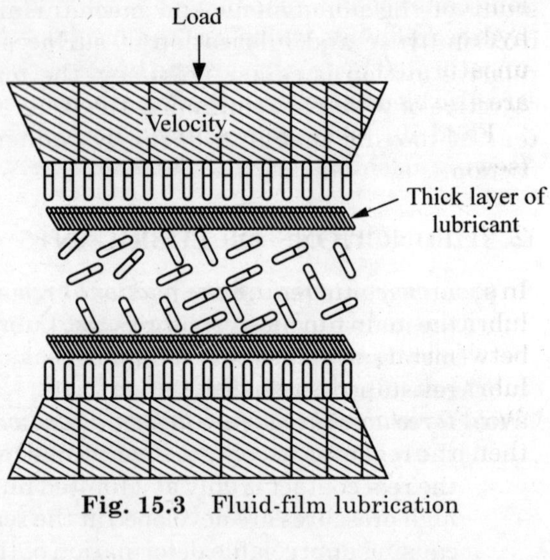

Fig. 15.3 Fluid-film lubrication

In fluid film lubrication, one of the rubbing surfaces is slightly displaced in relation to the other and a wedge-shaped film of the lubricant is formed between the two rubbing surfaces against each other. The pressure developed in the wedge shaped film, supports the load and prevents the direct contact between the rubbing surfaces.

Let us illustrate this by taking the example of a journal bearing. A journal bearing is formed of a shaft (journal) rotating with respect to stationary surface called bearing. The lubricant is introduced in the annular space. Fig 15.4.

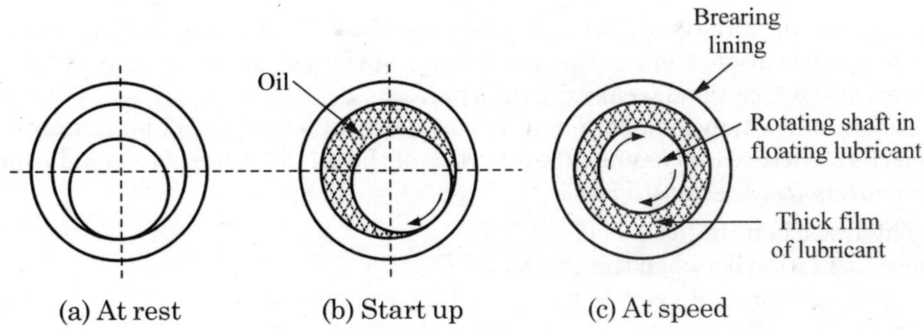

Fig. 15.4 Fluid Film Lumbrication in a Journal bearing.

In rest position, the shaft rests on the bottom of the bearing and all of the lubricant is squeezed out resulting in contact between the two surfaces. When a torque is applied to the shaft, it begins to rotate and tends to roll up the bearing. A *hydrodynamic wedge* is developed as a result. The lubricant is drawn between the bearing and the shaft or into the wedge. As more and more lubricant is drawn the pressure exerted by the lubricant on the journal increases and the bearing is properly lubricated. A fluid film between the shaft and bearing is formed. Fig. 15.4 (c) gives the final running position.

The lubricants used for fluid film lubrication are hydrocarbon oils. However the viscosity of the hydrocarbon oil decreases with rise in temperature, therefore these oils are blended with selected long chain polymers to maintain the viscosity of the oil in all seasons of the year. Moreover, the hydrocarbon oils which are the fractions from petroleum refining contain small quantities of unsaturated hydrocarbons and hence get oxidised forming gummy products. Hence antioxidants are also blended with the hydrocarbons used as lubricants.

Fluid film lubrication is provided in delicate instruments such as watches, clocks, light machines (sewing machines, scientific instruments) etc.

2. Thin Film or Boundary Lubrication

In some cases, under the conditions of high load and low speed or low viscosity of the lubricant, the lubricant tends to be squeezed out of position or a continuous film of lubricant cannot persist between the two surfaces. In such cases, lubrication is maintained by a very thin film of the lubricant adsorbed on the surface *by physical or chemical forces or both.* These adsorbed layers avoid direct metal to metal contact. Though, the moving surfaces are not completely separated but then points of contact are considerably decreased resulting in lowering of frictional resistance.

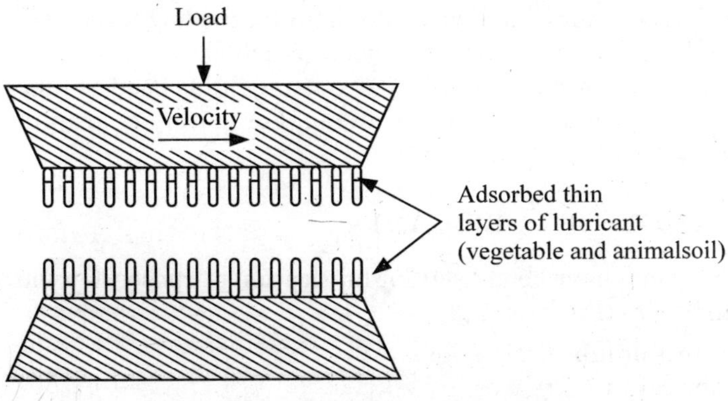

Fig. 15.5 Boundary Lubrication

The thickness of the boundary film may not exceed one or two molecular layers Fig. 15.5.

In this type of lubrication, some of the peaks in the surface of the metal may be higher than the film thickness, and hence some wear may take place.

The coefficient of friction in such a case is generally 0.05 to 0.15. This friction is dependent mainly upon the structure and chemical properties of the oil to be used for boundary lubrication.

The lubricant molecules should have:

(a) Long hydrocarbon chains.
(b) Lateral attraction between the chains.
(c) Polar groups to promote wetting or spreading over the surface.
(d) Active groups to form chemical bonds with the metal surface.
(e) High viscosity index, resistance to heat and oxidation, good oiliness and low pour-point (*discussed in section* 15.8).

These properties are met with the *organic acids* which have both, polar groups as well as hydrocarbon chains. Examples include oleic acid and stearic acid. Other lubricants used in boundary lubricants are:

Vegetable and animal oils: Glycerides of higher fatty acids and their soaps get adsorbed on the metal surfaces either physically or chemically forming a thin film, which acts as lubricant.

Mineral oils: Mineral oils are also used for boundary lubrication. Although fatty acids have more oiliness than mineral oil, fatty acids tend to break down at higher temperatures. Hence these cannot be used in internal combustion engines. Mineral oils can be used, as they are thermally more stable. The oiliness of mineral oils can be increased by adding a small amount of fatty acids.

Graphite and MoS_2 are also used for boundary lubrication. These can bear compression as well as high temperatures. These are solid lubricants and are either used alone or as stable suspension in oil.

3. Extreme Pressure Lubrication

If the sliding surfaces are subjected to very high pressure and speed, excessive frictional heat is generated. Under such conditions, the liquid lubricants fail to stick and may decompose and even vaporize. In order to provide effective lubrication under these conditions, special additives called "extreme pressure additives" are used along with the lubricants. The active chemicals used are compounds of chlorine such as chlorinated esters, sulphur such as sulphurized fats and oils, and phosphorus such as tricresol phosphate. At prevailing high temperatures, these additives form solid surface films of chlorides, sulphides or phosphides as the case may be. These metallic compounds possess high melting points and serve as good lubricants under the conditions of extreme pressure and temperature.

These lubricants have an additional advantage that if the low shear strength films formed are broken by rubbing action, they are immediately replenished.

Extreme pressure additives are also needed in cutting fluids used as lubricants in machining of tough metals. These are also used in wire drawing. For example, graphite is used for drawing wires made up of mild steel.

15.5 CLASSIFICATION OF LUBRICANTS

A wide variety of lubricants have been used for lubrication purposes. On the basis of their physical state lubricants can be classified as:

1. Liquid lubricants or lubricating oils.
2. Semi-solid lubricants or greases
3. Solid lubricants.

15.5.1 Lubricating Oils

The lubricating oils or liquid lubricants are used to reduce the friction and wear between two sliding metallic surfaces. Apart from reducing the friction they also act as cooling medium. Lubricating oils can further be divided into three categories viz;

(i) Vegetable and animal oils.
(ii) Mineral oils or Petroleum oils.
(iii) Blended oils.

A brief discussion of all the three types is given below:

(i) **Animal and Vegetable Oils:** These oils contain glycerides of higher fatty acids. These oils were used as lubricants before the development of petroleum industry, after which they were replaced by mineral oils. Animal and vegetable oils decompose on heating but do not distil and hence are also known as fixed oils. These have the following limitations when used as lubricants:

- They are expensive
- They undergo oxidation and form gummy and acidic products, which are disastrous to the metal parts
- They get thickened on coming in contact with air.
- They hydrolyze in presence of water.

Because of these limitations they are not widely used now a days. However both animal and vegetable oils have a property called "oiliness" by virtue of which the oil sticks to the surface of the metal even under high temperature and high load. Hence they are blended with the mineral oils to improve the oiliness of the latter.

Vegetable oils are obtained by crushing the seeds and washing the crushed seeds with a suitable solvent (solvent extraction). Therefore the washings are distilled to separate oil and the solvent.

Animal oil is obtained by a process called rendering in which the enclosing tissue is broken by a combined action of steam and water.

Both the types of oils, vegetable as well as animal oils require further treatments such as cooling, filteration, neutralization and coagulation.

Some of the important vegetable and animal oils along with their uses are given in Table.

Table. 15.1 Animal and Vegetable Oils

Vegetable Oils	*Uses*	*Remark*
1. Olive Oil	Good lubricant for machine parts working under low pressure and high speed.	Expensive and tends to decompose on storage
2. Castor Oil	Used as a lubricant for bearing and machine parts subjected to low pressure and high speeds.	Possesses excellent oiliness and is soluble in alcohol and not in petroleum. Oxidizes into gummy products.
3. Palm Oil	Lubricant for delicate article such as watches, clocks, etc.	Pleasant smelling oil.
4. Rape seed oil	Lubricant for steam cylinders and delicate apparatus.	Most extensively used fatty oil as lubricant.

(i) **Mineral Oils:** These oils are obtained by fractional distillation of crude petroleum. These are obtained as heavy residual fraction. This heavy residual fraction is subjected to vacuum distillation and the lubricating oil is extracted. The length of the hydrocarbon chain varies

Animal Oils	*Uses*	*Remarks.*
1. Tallow oil	Used as lubricant as such or can be blended with mineral oils.	Obtained by compression of cattle or sheep fat under high pressure.
2. Lard oil	Used as a cutting oil and for lubricating ordinary machine parts.	Obtained from fats, kidneys and intestines of pigs.
3. Neats foot oil	Used for lubricating instruments such as clocks and sewing machines.	Obtained from the hoofs of slaughtered cattle and sheep by extraction process.
4. Whale Oil	Used for lubrication of light machinery	Obtained by different parts of sperm whale flesh. It is relatively a cheap oil.
5. Sperm Oil	Good spindle oil and is suitable for light machinery.	Lightest of fatty oils.

from 12 to 50 carbon atoms. Mineral oils are used most widely as lubricants because they are cheaper and are available in abundance. However they have poor oiliness which can be increased by blending them with higher fatty acids.

The lubricating oils obtained from mineral residues have a large number of impurities such as paraffin waxes, resins, asphalts etc. These impurities are removed by different processes:

(a) *Dewaxing.* Waxes if present in the lubricating oil increase the setting point and also increase the pour point and hence lubricant cannot be used in low temperatures. The wax can be removed by dewaxing, in which oil is mixed with a suitable solvent (propane, trichloroethylene) and is then refrigerated. The wax crystallizes and is removed by filteration. The solvent present in oil is then recovered by distillation.

(b) *Acid Refining.* The asphaltic constituents in mineral oils undergo decomposition at higher temperatures causing carbon deposition and sludge formation. These are removed by agitating the oil with the required amount of H_2SO_4. Some of the unwanted impurities dissolve in acid while others react chemically to form tarry sludge. The oil is separated through filteration and is neutralized with a calculated quantity of ammonia or caustic soda. The oil is decolourized by passing through fuller's earth at 108-230°C.

(c) *Solvent Refining.* The other undesirable impurities such as naphthenic, asphaltic and resinous constituents can be removed by solvent refining. In this refining the mineral oil is mixed with a suitable solvent in which it is immiscible but the impurities are highly soluble. The mixture of solvent and the oil is left undisturbed for some time when two layers are separated-the oil layer free from impurities and the solvent layer containing the dissolved impurities. The oil layer is recovered and distilled to remove the small amount of solvent which may be present. The solvent is recovered from the solvent layer through distillation and the impurities left behind as residue can be used as a fuel oil.

Solvent refining has number of advantages over acid refining as it is cheaper and gives higher yield of refined oil. However solvent refined oils are generally have less oxidation resistance due to removal of natural oxidation inhibitors present in it. Moreover these oils possess lower oiliness. Thus it becomes imperative to add some additives to improve their desirable properties.

(ii) **Blended Oils:** Small quantities of certain substances are added to the lubricating oils to improve their desirable properties. These substances are called additives. The oils to which additives are added are known as blended oils or compounded oils. These blended oils are capable of giving desired lubricating properties required for a particular machinery. Some of the additives along with their different functions are tabulated in table.

Table 15.2: Additives and their functions

Additive Type	*Chemical Used*	*Functions*	*Application*
1. Antioxidant or inhibitor	Aromatic phenolic or amino compounds.	Retard the oxidation of lubricating oils by getting themselves preferentially oxidised.	Added in lubricants used in internal combustion engines.
2. Oiliness-carrier	Vegetable oils like castor oil and fatty oils like stearic acid, palmitic acid	Increase the oiliness of the lubricating oil	Added in lubricants used in boundary lubrication.
3. Viscosity-index improver	Polymers and copolymers of olefins.	Increases the viscosity index	Added in the machines working under broad range of temperature change.
4. Pour-point depressant	Alkylated naphthalene and phenolic polymers, polymethacrylates	Lowers the pour point of the oil	Reduce interlocking
5. Detergents and deflocculants	Basic calcium and barium salts of phosphates and sulphonates	Reduce or prevent deposits	Used in engines operated at high temperatures.
6. Dispersants	High molecular weight amines and amides, polymers	Prevent or retard sludge formation	In lubricating oils used in internal combustion engine.
7. Metal deactivators	Amines, sulphides or phosphides	Stop the catalytic effect of metals on oxidantion rate	Protect the metal surfaces by forming a protective layer over the metal surface
8. Antifoaming agents	Glycol, glycerol, silicon polymers	Prevent the formation of stabel foams.	
9. Corrosion inhibitors	Organo metallic compounds, zinc dithiophosphates,	Protect the metal in contact with the lubricant from corrosion.	In lubricating oils used for bearing etc.
10. Rust inhibitors	Amine phosphates, Alkyd succinic acid, fatty acids.	Protect ferrous metals from rusting.	
11. Extreme Pressure additives	Sulphurized fats, organic phosphorus compounds etc.	React chemically with the metal, forming a surface layer which prevent seizure and welding of the metals.	
12. Thickness	Polystyrene, Polyesters, soaps of lithium, sodium, calcium and barium	Given high viscosity to the lubricating oil.	Used for forming greases.
13. Abrasion inhibitors	Tricresyl phosphate	They provide abrasion resistance.	
14. Emulsifiers	Sodium salts of sulphonic acids and sodium salts of organic acids	Help the formation of emulsions fo lubricating oils with water.	

15.5.2 Semi Solid Lubricants or Greases

In textile, paper and food mills, if liquid lubricants are applied to the metal surfaces of the machinery, there is always a possibility of spilling or spurting of lubricating oil. This is detrimental to the product being manufactured. In these cases the lubricating oil is thickened with the thickening agent and the semi-solid so obtained is known as 'grease'. Greases are ideal in such cases as these do not spill or splash. Greases are also employed in the situation where the compounds have to be sealed against the entry of dirt, water etc.

Greases can be defined as a semisolid combination of a lubricating oil and a soap (of lithium, sodium, calcium or barium etc.) as a thickening agent. Other substances used as thickener include carbon black, asphaltenes, siloxanes and flouro carbons. These substances act as structural elements which trap the oil and give stiffness to the oil so that it clings to the moving parts. The thickening agents are also known as gelling agents. Thus soaps which are used as thickeners or gelling agents give a interconnected structure held together by intermolecular forces. At high temperature, the interconnected forces cease to exist because the soap dissolves in the oil and hence greases get liquified as the temperature rises.

Preparation of Greases

Greases are generally prepared by saponification of fat (such as fallow oil or fatty acids) with alkali ($NaOH$, $Ca(OH)_2$) followed by addition of hot lubricating oils with constant mixing. Sufficient oil is added to obtain the desired consistency. The nature of the soap decides the temperature upto which grease can be used and its consistency.

Some of the important aspects of greases are as follows:

(i) Consistency of the greases can vary from a heavy viscous liquid to that of a solid mass. The consistency changes suddenly with change in temperature because the interaction of gelling agents and the viscosity of the oil changes with increase in temperature.

(ii) Greases have a tendency to separate into oils and soaps on storage.

(iii) Greases are less stable to oxidation than the component lubricating oil present. Thus antioxidants or inhibitors are added to grease.

(iv) Most of the thickening agents are hydrophillic in nature and hence greases absorb water. Thus the greases are poor water resistant.

(v) Greases have higher shear or frictional resistance than oils and therefore can support much heavier loads at lower speeds.

(vi) Greases do not require as much attention as oils and hence are convenient to use.

(vii) Greases cannot dissipate heat from the bearing effectively as compared to oils.

Applications of Greases

Greases are mainly used in the following:

(i) When a machine is worked at slow speeds and high pressures the lubricating oil cannot remain in place and in these situations greases are used for lubrication.

(ii) Greases are used in bearings and gears that work at high temperatures.

(iii) They are used in machines used in the manufacture of paper, textiles or edible articles because unlike oil, there is no danger of spilling of grease.

(iv) They are used in bearing which required to be sealed against entry of dust, dirts, grit or moisture.

Classification of Greases

The consistency, temperature upto which grease can be used are all dependent upon the nature of soap and hence greases are classified on the basis of soap used in their manufacture. Important classes of greases are:

(i) Calcium-based or cup greases.
(ii) Soda-based greases.
(iii) Lithium based greases.
(iv) Axle greases.

The soaps used in their preparation and their uses are given in Table 15.3.

Table 15.3: Types of greases and the thickening agents used

Type of Greasem	*Oil and thickening agent used*	*Remarks*
1. Calcium based greases	Emulsions of petroleum oils with calcium soaps	Cheapest, water resistant and can be used upto 80°C, suitable for lubricating water pumps, tractors etc.
2. Soda based greases	Petroleum oils thickened by mixing sodium soaps.	Not resistant to water and hence cannot be used in the machine parts exposed to wet conditions. Due to high melting points can be used upto 175°C. Suitable for use in ball bearings.
3. Lithium based greases	Petroleum oils thickened by mixing lithium soaps	Have good water resistance and good thermal stability. They are excellent multipurpose greases for a wide range of applications. However, they are expensive.
4. Axle greases	Resin mixed with lime.	Cheapest of all greases, water resistant and find applications in equipment working under high loads and low speeds.
5. Non-soap greases	Oil mixed with carbon black, silica gel etc.	Suitable for high temperature applications. Talc, mica etc. are added as filters.

15.5.3 Solid Lubricants

There are number of situations where lubricating oils and greases cannot be used satisfactorily. For example:

(i) Where the operating temperatures and pressures are too high, the liquid lubricants cannot be used because these are easily combustible.
(ii) Where a liquid or semi solid lubricant cannot be maintained in position.
(iii) Where the contamination of lubricating oil by the entry of dust particles is not acceptable such as in aerospace devices.

Under these conditions, to minimize the friction, *solid lubricants* are used. The solid lubricants are used either in dry powder form or mixed with water or oil.

The most important solid lubricants that have widely been used are:

(i) Graphite
(ii) Molybdenum disulphide

Their lubrication properties are due to their layered structure or laminar structure.

(i) *Graphite:* Graphite consists of a network of hexagonal carbon rings separated from the adjacent layer by a distance of about 3.4 A° (Fig. 15.6). The adjacent layers are held together by weak Vander Waals' forces. Consequently the parallel layers slide over one another easily. Graphite is soapy to touch, non-inflammable and not oxidized in air below 375° C. When graphite is used as a lubricant between uneven surfaces, it fills into the valleys on the surface making the surface smooth. Moreover as there is no strong bonding between the adjacent layers of graphite, thus these layers slide past each making the surface lubricated. Graphite is used either as powder, as suspension, or as grease. Suspension of graphite is prepared in oil or water using *tannin* as emulsifying agent. Its suspension in water is called 'aquadag' and in oil is called 'oildag'. (dag-*deflocculated adhesion graphite*). Oil dag is used in internal combustion engines because it provides a tight fit contact by forming a film between the piston and rings. Aquadag is used where lubricant free from oil is needed. Graphite greases are used for lubrication at very high temperatures.

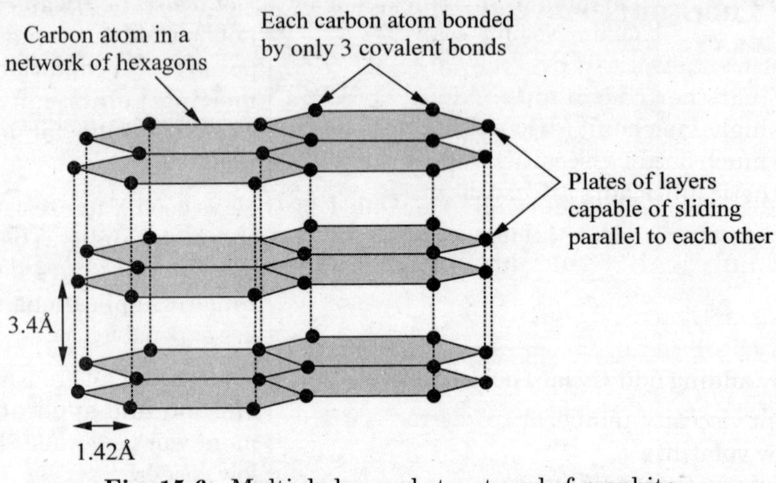

Fig. 15.6 Multiple layered structured of graphite

(ii) *Molybdenum Disulphide:* MoS_2 has a sandwitch like structure in which a layer of Mo atoms lies between two layers of sulphur atoms (Fig. 15.7). It has low coefficient of friction and is thermally stable in air upto 400° C. It is also used as a powder or mixed with solvents or greases. It is used as dispersion in petroleum oils or water. Above 800°C, MoS_2 gets oxidized to MoO_3 which is slightly abrasive. Therefore, for higher temperatures, graphite is preferred.

MoS_2 is used in airframe lubrication and wire-drawing dies. A solid-film lubricating surface useful for space

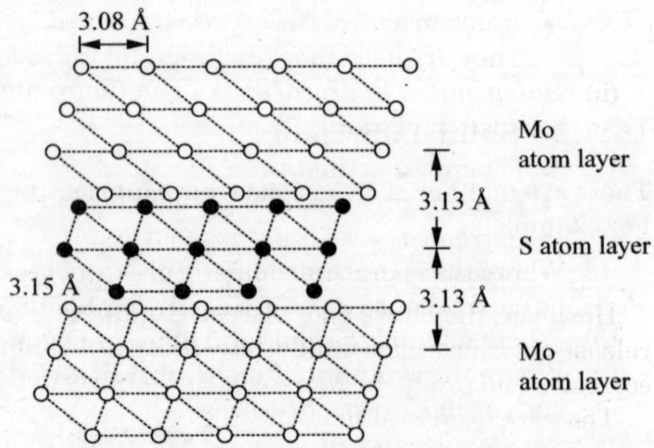

Fig. 15.7 Sandwitch-like structure of molybdenum disulphide.

vehicles is made from 70% MoS_2 + 7% graphite bonded with 23% silicates which can with stand extreme temperatures, low pressures and nuclear radiations

Other solid lurbicants beside graphite and molybdenum disulphide are soapstone, talc, mica etc. Teflon, copper, phthatocyanine are examples of solid organic lubricants.

15.6 SYNTHETIC LUBRICANTS

Mineral lubricants are used extensively and satisfactorily in engineering applications because of their ready availability, desirable viscosity temperature characteristics and lower costs. However, under severe operating conditions, the mineral lubricants, even with suitable additives, have been found to be unsatisfactory. Such operating conditions include extreme temperatures, wide temperature fluctuations in machinery. For example in air craft turbines, which require lubricants to have low viscosities and high working temperature range (– 55° C to + 260° C). These requirements have led to the development of synthetic lubricants.

Synthetic lubricants are the lubricants prepared to meet the special requirements *i.e.* which can work under severe or extreme conditions satisfactorily. However synthetic lubricants are used only for special applications because of their high cost and some of them are toxic and corrosive.

The Synthetic Lubricants Have the Following Characteristics

(i) They perform satisfactorily over a wide range of temperature, say from – 50° C upto 260° C.
(ii) They have high thermal stability, oxidation resistance and resistance to hydrolysis.
(iii) They have high flash point (discussed later) and rust preventive qualities.
(iv) They have much better viscosity temperature characteristics.
(v) Many synthetic lubricants are excellent solvents for important additives.

The Synthetic Lubricants that are Currently in use are Classified as

(i) *Dibasic Esters:* The most widely used dibasic ester is di-2-ethyl hexyl sebacate, which has satisfactory performance from – 50°C to 230°C. The properties of these lubricants can be modified by adding additives. Their advantages are:
- excellent viscosity-temperature characteristics
- very low volatility
- high temperature stability
- non-corrosive, non-toxic
- stable to hydrolysis.

They are used in jet engines and military equipment.

(ii) *Halogenated hydrocarbons:* The flouro and chloro hydrocarbons are used as lubricants because of following advantages:
- low-inflammability
- extreme pressure lubricating properties
- extreme resistance to oxygen and H_2O_2
- resistant to chemicals.

However, they have poor viscosity-temperature characteristics and chloro-hydrocarbons may release chlorine atoms under the conditions of high loading and high temperatures and result into corrosive end products. Moreover they are very costly.

They are used in submarines.

(iii) *Organo-phosphate esters:* Phosphatic esters are used because of their fire-resistant characteristics. They can be used as extreme pressure additives and are also used in air craft hydraulic oils.

(iv) *Polyglycols and their derivatives:* These can be water soluble or water insoluble depending upon the carbon chain length. Higher polyalkylene oxides are water insoluble but can absorb lot of water whereas polythylene glycol, poly-propylene glycol are water soluble and can be used in water diluted lubricants.

The advantages of these lubricants are:

- low pour points
- no effect on rubber and other gasket materials
- comparable to petroleum oils in thermal stability, fire resistance and lubricity
- cheaper.

However, they decompose at high temperatures into volatile parts which undergo oxidation subsequently. Hence they are not useful as high temperature lubricants. They are used for combustion engines, gears, compressors and pumps. They are also used as aircraft turbine lubricants because of their thermal stability.

(v) *Silicones:* Silicones are used as lubricants because of their excellent properties which are as follows: They

- are oxidation resistant upto 200° C.
- possess high viscosity index
- are chemically more inert than other synthetics
- hamve smallest change in viscosity with temperature.

Silicon fluids are used for rubber and plastic surfaces such as slide rules, bushings, bearings etc. They are also used as moisture repellent lubricants for clocks, timers and other electronic devices.

(vi) *Amines, imines and amides:* They have got high viscosity index and low pour point. They can be used in the temperature range – 50°C to 250°C.

15.7 LUBRICATING EMULSIONS

Emulsion is a heterogeneous system consisting of disperse phase and dispersion medium. It consists of two immiscible liquids, the one being dispersed as fine droplets in the other. It is stabilized by adding a third component called emulsifying agent. The liquid which forms the droplets is called *dispersed phase* and the surrounding liquid is known as the continuous or *dispersing phase.* The size of the dispersed droplets can vary from 1 to 6 μ.

There are two types of emulsions which are used as lubricants :

1. **Oil-in water emulsions or cutting emulsions:** These are obtained by adding an oil containing about 3-20 % water soluble emulsifying agent (*e.g.*, water soluble soap) to a suitable quantity of water.

   In various types of machining operation, (milling, boring, turning etc.), the tool employed gets heated to a very high temperature. This may cause oxidation and subsequent corrosion of metal. In order to prevent overheating, efficient cooling and lubrication is provided. This is usually done by employing emulsions of oil-in water which are called *cutting fluids* or cutting emulsions. Oil lubricates whereas water act as a coolant. Fatty oils such as lard oil and sperm oil are very good for cutting oils.

2. **Water-in-oil emulsions or cooling liquids:** These are prepared by mixing together water and oil containing 1 to 10% of water insoluble emulsifiers (*e.g.*, alkaline earth metal soaps). There are very useful as compressors and pneumatic tool lubricants as they give internal cooling effect due to the evaporation of water.

15.8 PROPERTIES OF LUBRICATING OILS

The selection of the lubricant for a particular job depends on a number of factors including design of machine, operating conditions of the machine (temperature, pressure, speed of machine, etc.), and its physical and chemical characteristics.

Some of the physical and chemical properties which are significant in selection of the lubricant are given below:

- Viscosity
- Cloud point and pour point
- saponification value
- Iodine value
- Emulsification
- Aniline point
- Neutralization point
- Flash point and Fire point
- carbon Residue

The physical and chemical properties such as viscosity, specific gravity, flash point, volatility, acid value, iodine value etc., measured in the laboratory do not indicate that how a lubricant will behave under operating conditions. Thus, it becomes imperative that these properties measured in the laboratory should be correlated to the conditions when the machine would be used. Some of the important properties of lubricants which would help an engineer to make a correct choice of the lubricant are discussed below in detail.

15.8.1 Viscosity

Viscosity is the property of a liquid or fluid by virtue of which it offers resistance to its own flow. It is mainly due to the forces of cohesion between the fluid molecules. It is a measure of a fluid's resistance to flow. A liquid, such as gasoline, which flows easily has low viscosity; and liquid, such as tar, which flows slowly has high viscosity.

Absolute viscosity may be defined as the tangential force per unit area required to maintain a unit velocity gradient between two parallel planes in the fluid unit distance apart. The units of absolute viscosity (denoted by η) in C.G.S system are dynes/cm^2 (poise).

Kinematic Viscosity is the ratio of absolute viscosity to density of oil in g/m^3. This is expressed in centistokes (cm^{2}s^{-1}) (1 centistoke = centipoise/density)

$$r = \frac{\eta}{\rho}$$

where r = absolute kinematic viscosity
η = absolute dynamic viscosity
ρ = density of the fluid

Viscosity is one of the most important properties of a lubricating oil. Viscosity of the lubricating oil determines its performance under operating conditions.

- If the viscosity of the liquid is too low, the fluid lubricant film cannot be maintained between the moving surfaces, as a result of which excessive wear may take place.
- If the viscosity of the lubricating oil is to high, excessive friction will take place.

Thus, in thick film lubrication the lubricant selected should be such, that the selected lubricant possess a sufficiently high viscosity so that it adheres to the surface and do not squeeze out under operating conditions. On the other hand the viscosity of the lubricant should not be too high otherwise, it will cause excessive friction.

Measurement of Viscosity of an Oil

Viscosity is usually expressed in centipoise or centistoke, but a more common practical measure of the viscosity of an oil is the *time in seconds for a given quantity of oil to flow through a orifice under the specified set of conditions. The instruments used to measure the viscosity of a liquid are known as viscometers.*

Different viscometers are used in different counteries for the measurement of viscosities in seconds.

Country	*Viscometer Used*	*Units in which viscosity is expressed*
Commonwealth countries	Redwood viscometer	Redwood seconds.
Europe	Engler's viscometer	Engler degrees (Time in sec. of outflow of oil Time in sec. of out flow of water)
U.S.A.	Saybolt's Viscometer	Saybolt universal seconds (SUS).

In these viscometers, a fixed volume of the liquid is allowed to flow through a capillary tube of specified dimensions under given set of conditions and the time of flow is measured at a particular temperature. The results are expressed in terms of the time as seconds taken by the oil to flow through the standard orifice of the apparatus used.

The description of redwood viscometer and saybolt viscometer is given below.

Redwood Viscometers

There are two types of Redwood Viscometers *i.e.* Redwood viscometer No. 1 and Redwood Viscometer No.2. The two viscometers work on same principle and have same shape and method of testing. The only difference lies in the dimensions of the discharge capillary (jet).

Redwood viscometer No.1 is commonly used for determining the viscosities of thin lubricating oils (kerosene, mustard oil etc.) as it has smaller jet diameter of 1.62 mm with a length 10 mm. Redwood viscometer No. 2 is used for determining the viscosities of highly viscous oils (mobile oil, glycerol etc.) and has jet diameter of 3.8 mm and length 50 mm.

Viscometer	*Diameter*	*Length of the Jet*
RW 1	1.62 mm	10 mm
RW 2	3.8 mm	50 mm

Redwood viscometer consists of the following parts:

(i) *Oil cup:* Oil cup is a silver-plated brass cylinder having height of 90 mm and a diameter of 46.5 mm. The bottom of the cup is provided with a central hole in which an agate jet with bore diameter 1.62 mm and length 10 mm is fitted. This jet can be opened or closed by a valve rod. The valve rod comprises of a small silver plated brass ball fixed to a stout wire. To indicate the level to which the cylinder is to be filled with oil, a pointer is fixed on the inner side of the cylinder. The pointer is a stout, tapered, upward-pointing wire. A thermometer is inserted in the cup through the lid of the cup.

(ii) *Heating bath:* A cylindrical copper bath containing water surrounds the oil cup. It is provided with a tap for emptying the water and has a long side tube projecting outwards, for heating the water by means of a gas burner. A thermometer is also inserted in the copper bath to know the temperature of water. Heating bath is also provided with a stirrer for stirring the water to keep the temperature of water uniform throughout. The stirrer is provided with a circular shield to prevent any water splashing into the cylinder.

(iii) *Spirit Level:* A spirit level used for levelling the apparatus vertically is also provided in the lid of the cup. The entire apparatus is supported on three legs, provided at the bottom with levelling screws.

(iv) *Kohlrausch flask:* A specially shaped flask known as Kohlrausch flask is provided to receive the oil from the outlet. The flask is of 50 ml capacity.

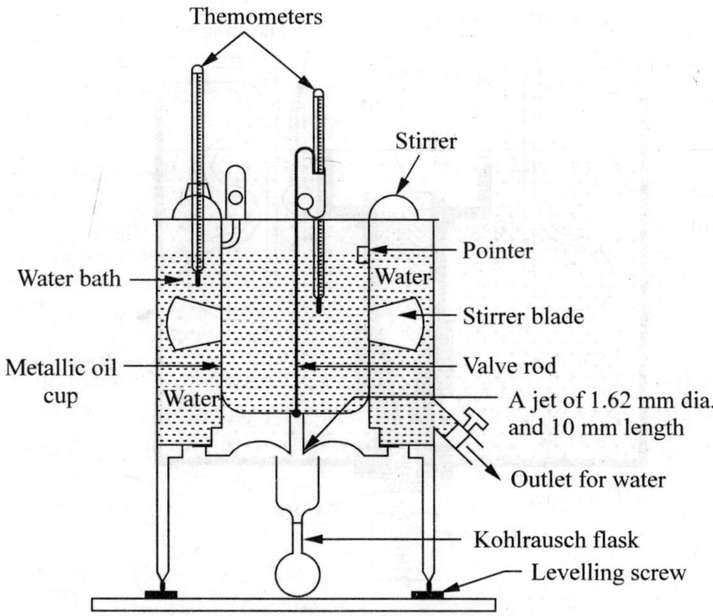

(The apparatus is set over electric heater having temperature regulating device.)

Fig. 15.8 Redwood viscometer No.1

Working: The viscometer is levelled with the help of levelling screws. The water bath is filled with water upto the indicator mark. The oil cup is cleaned and ball of valve rod is placed on the agate jet to close it. The oil under test is poured into the oil cup upto the tip of indicator. An empty Kohlrausch flask is kept just below the jet. The water in the side tube is heated slowly with constant stirring of the bath. When the oil attains the desired temperature, heating is stopped, the ball valve is lifted and suspended from thermometer bracket, and stop clock is started. The time taken for 50 ml of the oil to collect in the flask is noted in seconds and ball valve is replaced in the original position to prevent the outflow of the oil. The result is expressed in Redwood No. 1 .. seconds at a particular temperature.

The Saybolt Viscometer

The saybolt viscometer measures the time required, in seconds, for 60 ml of the test oil at 100°F to pass through a standard orifice.

It consists of the following parts:

(i) *Oil cup:* It is cylindrical brass cup provided with an orifice at the bottom.

(ii) *Heating bath:* A liquid heating bath surrounds the oil cup. A thermometer is also inserted in the liquid bath to know the temperature. Heating units are inserted into the bath to heat the liquid used for heating.

(iii) *Standard flask:* A graduated standard flask is provided to receive the liquid from the orifice. The flask is of 60 ml capacity

Working: The lubricating oil is taken in the oil cup and is heated to the test temperature. It is made to pass through the orifice and the time required to collect 60 ml of the oil in the standard flask is noted in seconds. The result is reported as saybolt universal seconds (SUS) at a particular temperature.

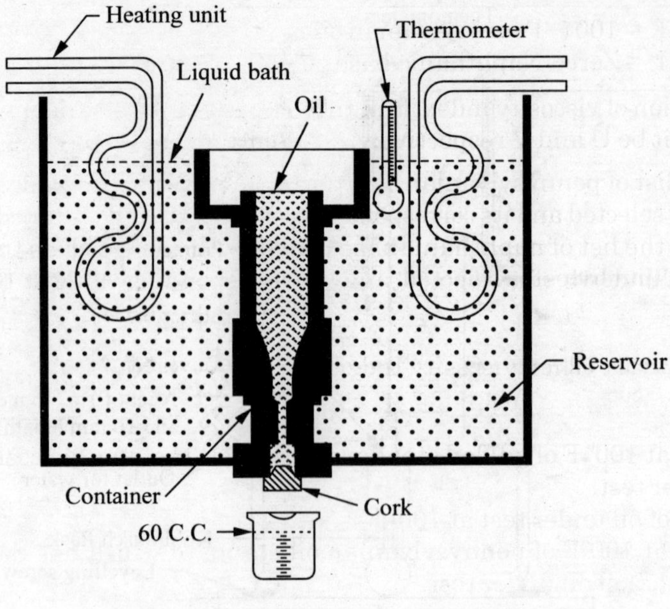

Fig. 15.9 Saybolt vircometer

Change of Viscosity with Temperature (Viscosity-Index)

Viscosity of liquids decreases with increasing temperature so that lubricating oil becomes thinner as the operating temperature is raised; and lubricant can be squ eezed out from the subbing parts. It is therefore necessary that viscosity of the lubricating oil do not change much on changing temperature. Since the viscosity decreases with temperature, it is impossible to select an oil having same viscosity over wide range of temperature. However, an oil whose variation with temperature is minimum can be selected.

The rate at which viscosity of a lubricant changes with temperature is measured by an arbitrary scale known as the **viscosity index** (V.I). V.I represents the average decrease in viscosity of an oil per degree rise in temperature between 100°F and 210°F. Viscosity index is the numerical expression of the average slope of the viscosity-temperature curve of a lubricating oil between 100° F to 210° F.

Determination of Viscosity-Index

For determining viscosity-index, two types of references are used. One of these is **Pennysylvanian oils.** This Pennysylvanian oils consist mainly of paraffins and show a small decrease in the viscosity with increase in temperature. These oils are arbitrary given a value of 100 for V.I. The other reference taken is **naphthanic-base, Gulf oils.** These consist of alicyclic hydrocarbons or naphthenes and exhibit a relatively large variation in viscosity with temperature. These are arbitrarily assigned a V.I. of 0.

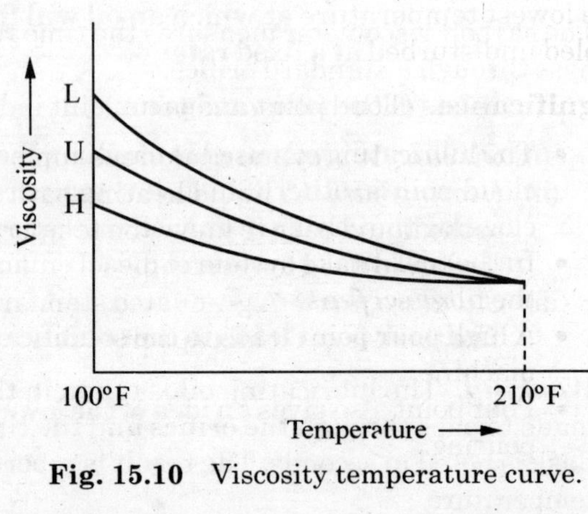

Fig. 15.10 Viscosity temperature curve.

V.I. = 100	Pennysylvanian oils
V.I. = Zero	Naphthanic-base gulf oils

For the determination of viscosity index, first the viscosity of the oil under test is determined at 100°F and 210°F. Let it be U and V respectively.

- Now, from the list of pennysylvanian oils, the oil having the same viscosity at 210°F as the oil under test is selected and its corresponding viscosity at 100°F is noted down. Let it be V_H.
- Similarly from the list of naphthanic base Gulf oils, the oil which as the same viscosity at 210°F as the oil under-test is selected, and its corresponding value at 100°F is noted down. Let it be V_L.

$$\text{Then viscosity Index} = \frac{V_L - U}{V_L - V_H} \times 100$$

where

V_L = Viscosity at 100° F of gulf oil standard which has the same viscosity at 210° F as that of oil under test.

U = Viscosity of oil under test at 100° F

V_H = Viscosity at 100°F of pennysylvanian oil standard which has the same viscosity at 210°F as that of oil under test.

Significance: Higher the V.I, lesser is the variation of viscosity with change in temperature. Thus a good lubricating oil should possess high V.I.

15.8.2 Cloud Point and Pour Point

Cloud Point: Cloud point is the temperature at which the oil becomes cloudy or hazy in appearance when it is cooled in a standard apparatus at a standard rate.

Lubricating oils which are derived from petroleum, mostly contain dissolved paraffin wax and other asphaltic or resinous impurities. The amounts of these impurities may vary depending upon the refining process used. At low temperatures *i.e.* when the oil is cooled down, these impurities tend to separate out of the oil in the form of minute crystals and hence result in cloudiness of the oil.

Pour Point: Pour point is the temperature at which the oil just ceases to flow when cooled at a standard rate in a standard apparatus.

If, after the cloud point, the oil is cooled further, more and more of impurities will crystallize and a stage is reached when the oil solidifies and stops flowing. Thus pour point can also be defined as the lowest temperature at which an oil will flow or pour under prescribed conditions, when it is cooled undisturbed at a fixed rate.

Significance. Cloud point and pour point indicate the suitability of a lubricant in cold conditions.

- The lubricating oils used in capillary feed systems or wicking arrangement should have low cloud points, otherwise at low temperatures the crystals of impurities will separate out clogging the capillary or wick interstices.
- In the fuel intake system of diesel engine, the lubricants having high cloud points may clog the filter screens.
- A high pour point leads to the solidification of the lubricant that may cause jamming of the machine.
- Pour point also gives an idea of the lowest temperature at which an oil can be transferred by pouring.

- Pour point determines the suitability of the lubricant for low temperature installations which may be required to start and operate at sub-zero temperatures. Examples include refrigerator plants and air craft engines.

Determination

The cloud and pour point of a lubricating oil is determined with the help of 'cloud and Pour Point Apparatus'. It consists of the following parts:

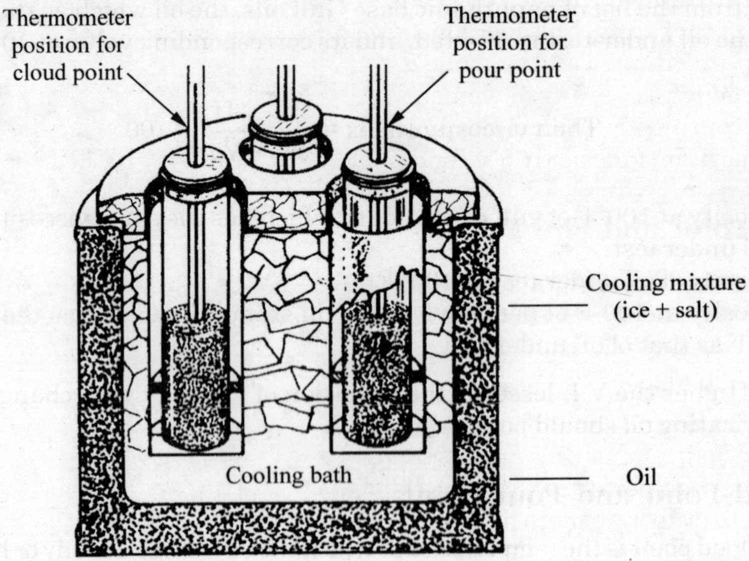

Fig. 15.11(a) Apparatus for cloud and pour points determination

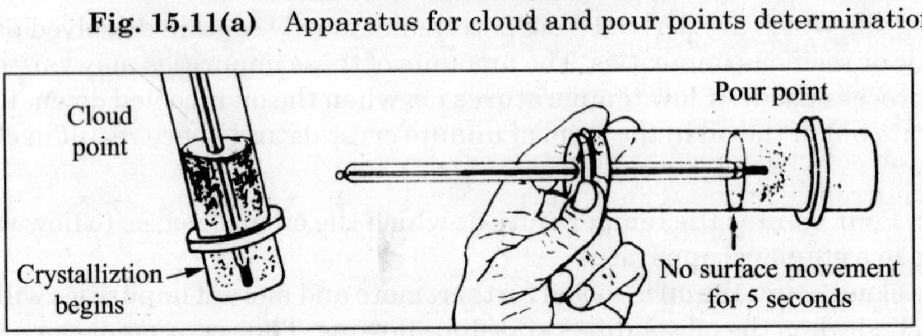

Fig. 15.11(b) Cloud and Pour points.

(i) *Flat bottomed tube:* The flat bottomed tube is about 3 cm in diameter and 12 cm high and is fitted with a thermometer.

(ii) *Air jacket:* The tube is surrounded by the air jacket.

(iii) *Cooling bath:* The jacket is surrounded in turn by a cooling bath containing freezing mixture. The freezing mixture can be crushed ice and salt or ice and $CaCl_2$ or solid CO_2 and petrol, depending upon the cooling temperatures required. A thermometer is inserted in the cooling bath to determine the temperature.

Working

- Lubricating test oil is dried by shaking it with a small amount of anhydrous sodium sulphate and filtered through lintless filter paper.

- The oil is filled upto the mark inside the flat bottomed glass tube and tube is fitted with a cork into which a thermometer is inserted. The bulb of the thermometer should dip inside the oil.
- The tube is kept in the cooling air jacket surrounded by the cooling bath containing the cooling mixture. The temperature of the oil will start falling down.
- After every 1°C fall in temperature of oil, the test jar is taken out from the jacket and inspected for cloudiness. If no cloudiness appears, it is immediately replaced in the jacket.
- The temperature at which such inspection first reveals a distinct cloudiness in the oil near the bottom of the jar, is recorded. This gives the cloud point.
- The cooling is continued and the tube is taken out after every 3° C fall of temperature and tilted just enough to see any movement or flow of oil.
- The test is continued until no movement of oil is observed when the test jar is held in a horizontal position for exactly 5 seconds. The temperature is recorded as pour point.

15.8.3 Flash Point and Fire Point

Flash Point: The flash point of a lubricating oil is defined as the minimum temperature at which the oil gives off enough vapours that ignite for a moment when a tiny flame of standard dimensions is brought near it. The flash immediately disappears for want of more vapours *i.e.* the temperature is not enough for the vapours to be formed at high rate.

Fire Point: The fire point of an lubricating oil is defined as the lowest temperature at which the vapours of oil burn continuously for atleast five seconds, when a tiny flame is brought near it.

At slightly higher temperature than the flash point *i.e.* at fire point the heat from the flash becomes sufficient to evaporate more liquid and maintain combustion. The fire point is usually 5 to 40° C higher than the flash point.

Significance: The flash and fire points do not have any bearing with the lubricating property of the oil, but these are important from the following point of view:

- The knowledge of Flash and Fire point is helpful in providing safeguards against fire hazards during their storage, transportation, handling and use.
- A lubricant selected for a particular job should have a flash point which is reasonably above its working temperature. This ensures safety.
- To ensure safety of illuminating oils like kerosene etc., the flash point of these should be reasonably above the average maximum temperature of a country.
- Liquids having flash point less than 140°F are called flammable and those having flash point above 140°F are called combustible liquids.
- Low flash point indicates the possibility of formation of gas-locks in fuel pipes or Spark Ignition Engines.
- Flash and Fire points have also been used to detect solvent contamination.

Determination

The determination of flash and fire points can be made by using the following apparatus:

(i) Cleverland Flash Point apparatus
(ii) Abel's closed cup apparatus
(iii) Pensky-Marten's apparatus.

The description of Pensky-Marten's apparatus is given below.

Pensky Marten's Flash and Fire Point Apparatus. It consists of the following parts:

(i) *Oil Cup:* The oil cup made up of brass, is about 5 cm in diameter and 5.5.cm deep. It has a marking inside to indicate the level upto which the cup is filled. The lid of the cup is provided with four openings. Through one opening, a thermometer is inserted while a stirrer passes through the other, carrying two brass blades. The third opening is for introducing a test flame and the fourth one is for introducing the air.

(ii) *Shutter:* The lid is provided with a shutter which can be moved to introduce the flame when required.

(iii) *Flame exposure device:* It is a device for introducing a standard flame. It is connected to the shutter in such a way that when the shutter is turned, openings of the test flame and air are opened and the flame exposure device dips into the opening over the surface of the oil.

(iv) *Air bath:* The oil cup is supported by its flange, over air bath which is heated by a gas burner.

(v) *Pilot Burner:* Pilot burner lights the test flame which gets extinguished when it is introduced into the opening for the test. The moment test flame returns to the original position pilot burner lights it.

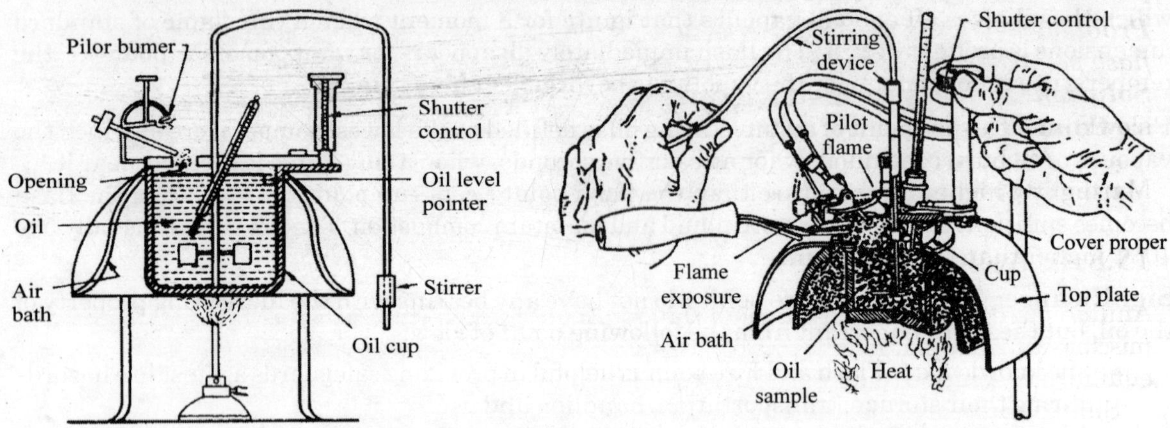

Fig. 15.12 (a) Panskey-Marten's flash and fire point apparatus.
(b) Cutaway view of Panskey-maten's flash point apparatus.

***Working*:** The oil cup is filled with the test oil upto the mark. The lid is fixed on the top of the cup, through which are inserted a thermometer and a stirrer. The flame exposure device is fixed on the top. The test flame is lighted and is adusted to about 4 mm in diameter. The apparatus is heated by a gas burner so that the temperature of the test oil increases by about 5 to 6° per minute, while the stirrer is rotated at approximately 60 revolutions per minute. At every 1° rise of temperature, the test flame is introduced for a moment into the oil vapour. This is done by operating the shutter. On moving the knob of the shutter, the test flame is lowered in the oil vapours through the opening. The moment knob is released, the test flame springs back to its original position and the opening is closed. When the test flame causes a distinct flash in the interior of the cup, the temperature is noted, which represents the Flash point. Further the test oil is heated at the rate of 1° C/minute and the test flame is applied as before. The temperature at which the vapours of the oil give a clear and distinct blue flash for five seconds is recorded as the Fire Point of the oil.

Problem 1. *What should be the flash point of a good lubricant?*
Solution: Flash point must be atleast above the temperature at which the lubricant is to be used, otherwise there is every risk of fire during use.

Problem 2. *In Pensky Marten's apparatus, what is the purpose of the air jacket surrounding the oil cup?*

Solution: Air being a bad conductor of heat ensures very slow and uniform temperature change whether the oil is being heated or cooled.

Problem 3. *What are the factors that affect the flash and fire point.*

Solution:
- Presence of moisture
- Vapour pressure
- Apparatus used.
- Frequency of application of test flame
- Rate of heating the test oil etc.

Problem 4. *How moisture is removed from the lubricating oil.*

Solution: By absorption through dehydrating agent such as Na_2SO_4, anhydrous $CaCl_2$ etc.

Problem 5. *Fatty oils do not have sharp or distinct flash points?*

Solution: These may undergo thermal decomposition with the evolution of vapours of decomposition products which may give irregular flashes (freaky flashes) before the true flash point.

Problem 6. *What is the difference between Abel's Apparatus and Pensky Marten's Apparatus for flash point determination?*

Solution: In Abel's apparatus, heating is done by water bath and the temperature rises very slowly where as in Pensky Marten's Apparatus, the heating is done by electric heater.

Abel's apparatus is employed for oils having flash point upto 120° F whereas Pensky Marten's apparatus is employed for oils having flash point above 120° F.

15.8.4 Aniline Point

Aniline point of an oil is defined as the minimum temperature at which the oil is completely miscible with an equal volume of freshly distilled aniline. Alternatively, Aniline point is the minimum equilibrium solution temperature for equal volumes of aniline and the lubricating sample.

Since like dissolve like therefore aniline is readily soluble in those lubricants which are rich in aromatic and naphthenic compounds. Whereas the lubricants which are rich in paraffins, the dissolution takes place only if the temperature is increased.

Significance

- Aniline point is a measure of aromatic content of the lubricating oil.
- A lubricating oil with low aniline point have a high aromatic content and will attack the rubber seals used in the system. This is because the aromatic compounds have a tendency to dissolve natural and certain types of synthetic rubbers.
- A higher Aniline point means low percentage of hydrocarbons and hence is desirable.
- Thus aniline point is used as an indication of possible deterioration of rubber sealings etc.

Determination

For the determination of Aniline Point, equal volumes of aniline and sample are taken in a test tube. The mixture is heated till a homogeneous solution is obtained. The tube is allowed to cool at a uniform rate. The temperature at which the two phases (oil and aniline) separate out is recorded as the aniline point. The separation of the two phases is indicated by the appearance of cloudiness in the medium. This is noted down as aniline point.

Problem 1. *Is there any relation between Aniline Point and ignition quality of a diesel fuel ?*

Solution: High aniline point of a diesel fuel means low aromatic content. And low aromatic content means less knocking in diesel fuel.

15.8.5 Neutralization Number

The acidity or alkalinity of lubricating oil is determined in terms of neutralization number. Determination of acidity is expressed as Acid Number or Acid Value and determination of basicity is expressed as Base Number or Base Value.

Acid Number or Acid Value. Acid value of a lubricating oil is the number of milligrams of potassium hydroxide required to neutralize all the free acid present in 1 g of the oil.

Base Number or Base Value. Base value of a lubricating oil is the number of milligrams of a titrating acid required to neutralize all the basic constituent present in 1 g of the oil.

Out of the two, determination of acidity is more common. Fatty oils consist mostly of glycerol or other esters of higher fatty acids and in some cases, considerable amounts of free acids are present in the lubricating oil if these are not refined carefully, *i.e.* acids and bases used in refining process have not been completely removed.

Significance

- The small intrinsic acidity may not be harmful in itself but the degree to which it increases in used oils is usually taken as a measure of deterioration of the oil due to oxidation or contamination.
- The acid content of the fatty oils increases with time due to hydrolysis with moisture and is therefore a rough indicator of age of the oil.
- The deterioration in the flavour of edible oils with time is due to the increase in free acid concentration formed by hydrolysis or oxidation.
- Periodic determination of Acid number is thus helpful in deciding the state, where the lubricant needs replacement.

However the presence of free acid in lubricating oils is very rare and indicates contamination. A lubricating oil must possess acid value less than 0.1.

Determination of Acid Value

The empty beaker in which the oil is to be taken is weighed. About 150 drops of the lubricating oil (about 5g) is transferred into the beaker and weighed again. The exact weight of the oil taken is calculated by subtracting the weight of empty beaker from it. Now 50 ml of the solvent (neutral alcohol) is added to the oil and shake vigorously to dissolve the oil. The contents are transferred into a 250 ml conical flask. The flask is heated over a water bath for 30 minutes and cooled to room temperature. phenolphthlein is added as indicator. This solution is titrated with the standard N/10 KOH solution until a faint permanent pink colour appears at end point. The volume of alkali used is noted down. Let the volume used by *y* ml. A blank titration on the same volume of hot solvent is also carried out. Let the volume of KOH used be z ml.

Let Initial weight of the beaker = $x_1$gm
Final weight of the beaker and oil = $x_2$gm
Weight of oil = $x_2 - x_1$gm
Volume of N/10 KOH used in the test oil = *y* ml.
Volume of N/10 KOH used in the blank titration = *z* ml
Therefore volume of N/10 KOH used for titrating the acid in the oil = $(y - z)$ ml.

$$\text{Acid value} = \frac{\text{ml of KOH used} \times \text{Normality} \times 56}{\text{weight of oil taken}}$$

$$= \frac{(y-z)}{\text{weight of oil}} \times \frac{1}{10} \times 56 = \frac{(y-z) \times 5.6}{\text{weight of oil}}$$

where 5.6 represents the amount of KOH in mg present per ml of N/10 KOH solution. (1000 ml of 1 N KOH ≡ 56 g of KOH)

15.8.6 Saponification Value/Number

Saponification value or number of an oil is the number of milligram of potassium hydroxide required to saponify 1 g of oil. (Saponification is the hydrolysis of an ester with KOH to give alcohol and sodium or potassium salt of the acid).

Mineral oils are the mixture of hydrocarbons and hence do not react with KOH and are not saponifiable. On the other hand vegetable and animal oils are the mixtures of glyceryl esters of fatty acids and require large amount of alkali to get hydrolyzed. Their saponification values are very high.

Significance

(i) The determination of saponification value helps us to ascertain whether the oil under reference is mineral or vegetable oil or a compounded oil containing vegetable and mineral oil.

(ii) Each of the oil has its own specific saponification number. Any deviation from this value in a given sample indicates the probability and extent of adulteration.

15.8.7 Iodine Value

Iodine value or iodine number is the number of grams of iodine equivalent to the amount of Iodine Monochloride (ICl) absorbed by 100g of the oil.

As is well known, oils and fats are mixtures of mixed glycerides and can be represented as:

$$\begin{array}{c} CH_2OOCR \\ CHOOCR' \\ CH_2OOCR'' \end{array}$$

These are formed by glycerol and monobasic fatty acids. The acids present in combination with glycerol have all even number of carbon atoms (C_4 to C_{22}) and may be unsaturated or saturated. Example, $C_{15}H_{31}$ COOH (palmitic acid) $C_{17}H_{35}COOH$ (stearic acid), CH_3 $(CH_2)_4$ CH = $HCCH_2CH$ = CH $(CH_2)_7$. COOH (linoleic acid) etc.

The oils having predominance of short chain saturated fatty acids or long chain fatty acids with a considerable degree of unsaturation are liquids at room temperature and are called oils. Others are solid at room temperatures and are called fats.

The degree of unsaturation is reported in terms of Iodine Number.

Significance

(i) As each oil has a specific iodine value so the determination of iodine value can be utilized to detect the extent of contamination in any specific oil

(ii) It can be used to classify oils based on iodine value because unsaturation confers drying characteristics on oils.

Iodine Value	*Type of oil*	*Examples*
> 150	Drying oil	Linseed oil, tung oil
100 to 150	Semidrying oil	Castor oil (dehydrated), soyabean oil
<100	Non Drying oil	Coconut oil, olive oil

(iii) When a thin film of an oil with a high degree of unsaturation (high Iodine Value) is spread on a surface, it takes up oxygen at the double bond sites and gets oxidized and polymerized into a hard, tough, infusible mass. Thus these oils having high I.V. are used in varnishes and paints.

(iv) To use oil as lubricant its Iodine Value should be low so that it does not deteriorate.

15.8.8 Carbon Residue

Lubricating oils have a tendency to form carbon residue on combustion because they have high percentage of carbon in combined form.

Significance

- These carbon deposits are very harmful because excessive build up of carbon deposits in combustion chamber (in internal combustion engine) results in decreased volume of the charge at the end of compression stroke which gives high compression ratio leading to denotation.
- The carbon may also be deposited on cylinder walls or on injector nozzles leading to friction and wears.

Determination

The carbon residue is determined by (*i*) Conradsom Method or (*ii*) Ramsbottom method.

Conradsom Method: In this method a weighed quantity of oil is taken in a silica crucible, which is placed in a skidmore crucible. The skidmore crucible is provided with a lid and a small opening for the escape of volatile matter. It is inturn placed in a wrought iron crucible which has a chimney-shaped iron hood. (Fig. 15.13). The outer iron hood is heated by a Meker burner till flame appears. After about 15 minutes when all the vapours are burnt completely, the apparatus is allowed to cool and the residue left is weighed. The result is reported as percentage of the residue in the weight of oil taken.

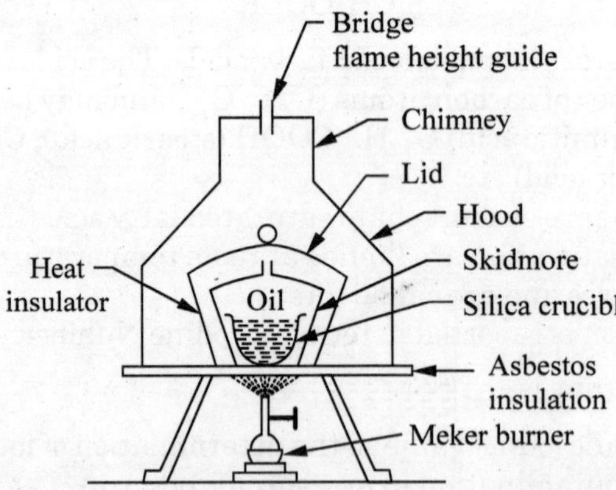

Fig. 15.13 Couradson's apparatus for carbon residue estimation

Ramsbottom Method: In this method, a hard glass bulb is used. Stainless steel bulbs are also used now-a-days. 1-4 g of oil sample is taken in the bulb and weight of the bulb along with oil is determined. The bulb is then is placed in a special furnace (Ramsbottom apparatus) consisting of an iron tube immersed in bath of molten lead. The lead bath is electrically heated to 550° C for 20

---

* Wij's solution:- Wij's solution is available commercially or it can be prepared by dissolving 1.6 g of commercially available iodine monochloride (ICL) in 100 ml of glacial acetic and. Wij's solution is stored in embus coloured stoppered bottles in a dark place below 30°C.

minutes. Bulb containing oil is taken out, cooled and weighed. The result is reported as percentage of carbon residue of the weight of oil taken.

$$\% \text{ carbon residue} = \frac{\text{Weight of residue in bulb}}{\text{Weight of original oil sample}} \times 100$$

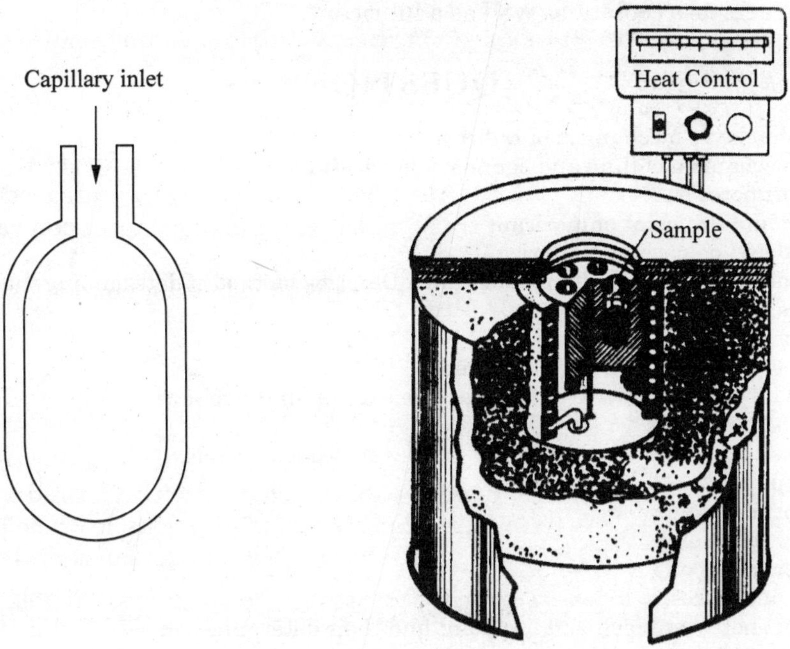

Fig. 15.14 Ramsbottom apparatus for carbon residue test.

15.8.9 Emulsification

Emulsification can be defined as the property of oils to get intimately mixed with water easily. Emulsions can be oil in water emulsion or water in oil emulsion. Certain oils form emulsions with water very readily. These emulsion tend to collect impurities (dirt, foreign matter etc.) and may cause abrasion and wearing out of the part that is lubricated.

Thus a good lubricating oil should form such an emulsion with water which breaks easily (demulsify). This property is called demulsification.

The time in seconds in which a given volume of oil and water emulsion separates out in distinct layers is called steam demulsification number. A good lubricating oil should have lower demulsification number.

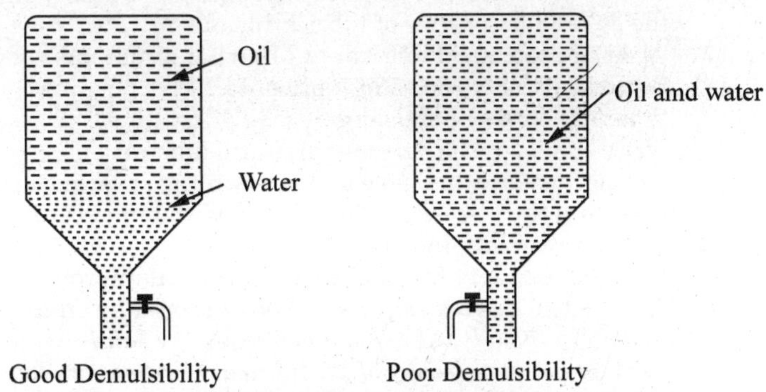

Fig. 15.15 Demulsibility

Significance

- In steam turbines, oils are contaminated with steam and hence quicker the oil separates out from the emulsion formed, better is the lubricating oil. Otherwise oil along with the steam may be carried away and can clog oil lines and pumps.
- In cutting oils the higher the emulsification number, better the oil is. This is because the emulsion acts as a coolant as well as a lubricant.

QUESTIONS

1. Explain thin layer mechanism of Lubrication. **[Raj. Univ., 2004]**
2. Explain mechanism of fluid and boundary lubrication. **[Raj. Univ., 2002]**
3. Write short notes on :
   (a) Flash and fire point of lubricant
   (b) Cloud and pour point of lubricant **[Raj. Univ., 2001]**
4. Discuss the principle underlying lubricants. Describe method of determining the viscosity index of lubricants. **[MREC 2000]**
5. Write notes on : solid lubricants **[Raj. Univ. 1999]**
   Write short notes on flash and fire point. **[Raj. Univ. 1998]**
6. What are lubricants? How are they classified? Explain the following properties of lubricants and give their significance:
   (1) Viscosity (2) Viscosity Index;
   (3) Pour point **[Raj. Univ. 1997]**
7. Write short notes on :
   Cloud and pour point **[Raj. Univ. 1996]**
8. Write short notes on :
   Viscosity and viscosity index **[Raj. Univ. 1996 Back]**
9. Write short notes on Flash and fire point and their determination **[Raj. Univ. 1995]**
10. Define viscosity and viscosity index. How are these properties of lubricating oil improved? **[Raj. Univ. 1994]**
11. What are lubricants and how are these classified? Discuss the function of lubricants.
12. What are solid lubricants? How the chemical structure of lubricants contribute to its functions? Illustrate your answer with two suitable examples.
13. What is viscosity? How it is determined by Red wood viscometer no. – 1? Explain with diagram.
14. What are greases? Mention their some specific uses.
15. Distinguish between fluid film and boundary lubrication.
16. What do you mean by blended or compounded oils? What are the various additives used to improve the desired properties of lubricating oils ?
17. What are semisolid lubricants ? In what situations they are preferred over liquid lubricants ?
18. Write short notes on solid lubricants.
19. What are cutting emulsions ?
20. Write short note on extreme-pressure lubrication.
21. Synthetic lubricants have added advantage over natural lubricants. Explain.
22. How does the viscosity of liquid vary with temperature ?
23. What is Viscosity Index ?
24. How the Viscosity Index of a lubricating oil improved ?
    Ans: *V.I. of lubricating oils can be improved by adding certain polymers, which are only partially soluble in the oils. At low temperatures the solubility of the polymers is low and hence the effect of polymer on the viscosity of the lubricant is also very low. At high temperatures, the solubility of the polymer in lubricating oil is quite high and hence the increased concentration of the polymer raises the viscosity of the oil. This compensates for the decrease in the viscosity of the oil itself.*
25. What is the difference between Redwood Viscometer No. 1 and Redwood Viscometer No. 2 ?
26. Name a few lubricants with high viscosity index ?
    Ans: *Silicones, polyglycol ethers, Diesters or triesters.*

27. How can the pour point of an oil be lowered ?
Ans: *Wax pour point can be lowered by dewaxing or adding suitable pour point depressant. Viscosity pour point can be lowered by lowering the viscosity of the oil which is achieved by removing the viscous constituents of the oil.*
28. What are pour point depressants ?
Ans: *These are the substances which when added to oils containing wax get adsorbed on the surface of wax crystals and thus reduce the size of wax crystals.*
29. Define flash point and fire point of a lubricating oil.
30. What should be the flash point of a good lubricant ?
Ans: *It must be above the temperature at which the lubricant is to be used.*
31. Define Aniline Point. What is the relation between aniline point and aromatic content ?
32. Define Acid number. What do you mean by base value ?
33. What is iodine value?
34. What is steam emulsion number (SEN) what is its significance.
35. What is Wij's solution?
Ans: *Wij's solution is prepared by dissolving 1.6 g of ICl in 100 ml of glacial acetic acid.*

16 CERAMIC, CEMENT, REFRACTORIES AND GLASS

"What we observe is not nature itself, but nature exposed to our method of questioning."

Werner Heisenberg

CHAPTER OUTLINE

16.1 CERAMICS

Ceramics include a variety of materials such as rocks, minerals, glass, metallic compounds, cement, abrasives, refractories, insulators etc. Ceramic materials of wide range applications can be prepared.

The word "ceramics" has derived its name from the greek word "Keramos" which means burnt material i.e. the desired properties are achieved through high temperature heat treatment process called, firing. The science of ceramics is the processing of earthly materials by heat. Early man prepared his crude utensils from clay, which was dried and fired to develop the permanent structure. Now a days more advanced techniques such as injection moulding sintering and hot pressing etc. are used to prepare ceramics of desirable properties.

Chemically speaking, ceramic materials are non-metallic solids, made up of inorganic compounds such as oxides, nitrides, borides, silicides and carbides for which the interatomic bonds are either totally ionic or predominantly ionic having some covalent character.

16.1.1 General Properties

In general, ceramic materials are characterized by the following properties:

(i) The ceramics are hard, strong and dense.
(ii) They have got higher stability and higher melting points due to the presence of ionic bonds.
(iii) These materials have high temperature resistance, good electrical resistance and resistance to chemical attack.

(iv) These materials posses high compressive strength and good tensile strength.
(v) Some of these are insulators like other covalent materials.

16.1.2 Classification

The very existence of ceramic materials having wide range of properties is due to the reason that many combinations of metallic and non-metallic atoms, having a number of different structural arrangements are possible.

A general classification of ceramics is difficult because of the great versatility of these materials. However, the following form the major classes:

Table 16.1

Type	*Properties*	*Examples*	*Uses*
1. Naturally Occuring	Most of these contain silica, alumina and oxides of sodium, calcium, magnesium and iron.	Gravel, sand, rocks, clay etc.	Important constituents of most of ceramic materials because they are available in plenty.
2. Produced from minerals through the action of heat			
(a) Cement	Higher compressive strength.	Portland cement, pozzolanic cement,	Used as structural materials.
(*b*) Glasses	• Contain magnesium or lithium ions. • They are hard and have high impact strength.	Pure silica glass, soda lime glass, lead glass.	Manufacture of utensils, lenses, bulbs, chemical apparatus, table ware etc.
(*c*) Refractories	• Can retain its shape and structure even at very high temperatures. • Does not react chemically with slag or molten charge.	• Acid refractories (consisting of SiO_2 or Al_2O_3) , • Basic refractories (consisting of MgO, CaO etc.). • Neutral refractories (consisting of carborundum, SiC chromite (FeO, Cr_2O_3)	Used in furnace walls, lining for hot metal mixers, crucibles etc.
(*d*) Porcelein	Good chemical resistance even at very high temperatures.		Used in furnace linings, used for protective.
3. Abrasive Materials	These substances rub down the objects, in order to give them desired size, shape or finish.	Diamond, Corrundum, Garnet — Natural Quartz, Flint, Sand Stone — Siliceous abrasives	• Used for making sand papers, srubbing powders etc. • Solt abrasives are used for polishing the metals.

16.2 CEMENTS

The term 'Cement' is used for the materials possessing adhesive and cohesive properties, which make them capable of binding mineral fragments (like bricks, stones tiles etc.) into a compact coherent structures. Cements have the characteristic property to form a paste with water and setting into a hard solid mass in few hours. The hardened mass can have varying degree of strength and bonding properties.

Egyptians were the first to use a cementing material for uniting blocks and slabs of stone. The chemical analysis of this material later on indicated it to be burnt gypsum containing calcium carbonate. Greeks and Romans used a mixture of lime, sand and water as cementing material. Joseph Aspidin (1924) found that strongly heated mixture of limestone and clay, when mixed with a small amount of water, set in a few hours to a hard stone which resembled Portland rock of England. Hence he named it Portland Cement. With time, a number of advancements have been made to improve strength and bonding properties of cements.

16.2.1 Classification

Depending upon the property of hardening in air or water cements can be classified as:

Hydraulic Cements: These are capable of setting and hardening under water. Example the cements used in construction industry such as Portland cement.

Non-hydraulic Cements: These are the cementing material which harden in air and cannot be used under water. For example, ordinary lime. It hardens in air and cannot be used under water.

Hydraulic Cements are most widely used and can be classified as follows:

1. **Natural Cements:** The naturally occurring limestone (argillaceous limestone) contains 20 – 40% clay. It is calcined at high temperature and crushed. The silica and alumina present in the clay combine with lime to form calcium and magnesium silicates.
   Natural cements are quick setting but have low strengths and hence are replaced by other types of cements.
2. **Puzzolana Cement:** Puzzolana cement was used by Greeks and Romans in the form of mortar (a mixture of lime, sand and water). Puzzolana are the deposits of volcanic aeh produced by rapid cooling of lava (mixture of silicates of calcium, iron and aluminium). Puzzolana is mixed with slaked lime. Such mixtures are capable of giving a hard mass with water and are called hydraulic limes.
   Presently the Puzzolana cements are not used alone but are mixed with other types of cements.
3. **High-alumina Cement:** It is prepared by heating a mixture of lime stone and banxite (containing iron oxide, magnesia and silicon). High-alumina cements have high rate of hardening and compression strength.
4. **Slag Cement:** Slag cement is prepared from blast furnace slags to which hydrated lime is added. The blast furnace slags consist of mixture of calcium and aluminium silicates. The rate of setting of slag cement is low. They have lower strength and hence find applications in the areas where strength is not required.
5. **Portland Cement:** The credit for the discovery of cement consumed in largest quantity i.e. Portland cement goes to *William Aspidin*. Portland cement is one of the most important building materials at the present time. It is prepared by grinding and mixing clay and lime bearing minerals in proper proportions and then heating the mixture to 1400°C. The resulting product is then powder and $CaSO_4 \cdot 2H_2O$ (gypsum) is added.

Portland Cement is produced in different varieties. The properties of portland cement, *i.e.* setting time and strength depends upon its composition to a large extent. Portland cement is most widely used cement through out the world in construction industry and hence is discussed in detail in the following sections.

16.2.2 Portland Cement

16.2.2a Manufacture of Cement by Rotary Kiln

As portland cement is formed by mixing clay and lime, the raw materials are the substances which supply these components. Thus the raw materials for the manufacture of Portland cement are:

(i) *Calcareous materials:* These materials supply lime and include lime stone, chalk, cement rock, waste calcium carbonate from industrial processes. Calcareous materials used, should also contain 3 - 4% SiO_2, Fe_2O_3 and Al_2O_3 and < 3.3% MgO.

(ii) *Argillaceous materials:* These supply silica, alumina and iron oxide. These include clay, blast furnace slag, shale, slate, etc. All these materials should have 2.5 to 4 times more of silica than alumina.

The relative proportions of the raw materials must be properly regulated. The small variations in established ratios of the principal components considerably change the characteristics of the mixture. For example, excess of lime reduces the strength of the cement as it makes the cement to expand and disintegrate. On the other hand lesser amount of it than required makes it quick setting and also reduces the strength.

The manufacture of the portland cement involves the mixing of calcareous and argillaceous materials in the required ratio. The mixture is heated to the fusion point (1400°C to 1600°C). The clinker obtained is cooled and mixed with 2 to 3% gypsum which is powdered and stored. The general outline of the various operations is as follows:

1. **Mixing of the raw materials:** The properly ground raw materials are mixed either by: (*a*) dry process or (*b*) wet process.

   (a) *Dry Process:* This process is used when the raw materials are quite hard (cement rock or blast furnace slag). The raw materials are crushed and ground separately to fine powder. Each of these is stored in a separate hopper. The two are then mixed in required proportions to form a raw mixture, which is stored in bins.

   (b) *Wet Process:* In this process, the raw material is crushed and ground to particles of suitable size and are mixed with water (30 – 40%) to get a free flowing slurry. The slurry is thoroughly homogenised with the help of compressed air. The slurry is then led into a correcting basin where its chemical composition is adjusted to a proper ratio. The slurry is stored in storage tanks. This process permits a better control of the cement composition in the slurry stage. However the consumption of fuel is more than dry process since water used for grinding has to be removed.

The comparison between dry and wet process is given in Table 16.2.

Table 16.2 Comparison between dry and wet process of cement manufacture.

	Dry Process	*Wet Process*
1.	It is a slow process.	It is relatively faster process
2.	Cost of production of cement is less, as the fuel consumption is low. Shorter kiln is sufficient.	Cost of production is somewhat higher because of the higher fuel consumption. As longer kiln is needed to drive off the excess water.
3.	The quality of cement produced is inferior.	The quality of cement produced is somewhat superior since more accurate control of composition can be attained.
4.	This process is adopted when the raw material are quite hard.	This process is preferred when the raw materials are soft.
5.	This process is adopted when raw materials do not have any inherent moisture content of 15% or more because it is uneconomical to drive away the excessive quantity of water	This process can be easily adopted in such cases.

The choice between the dry and wet processes largely depends upon:

(i) Physical conditions of raw materials.
(ii) Climatic conditions of factory area.
(iii) Price of the fuel.
(iv) Availability of water of appropriate quality in abundance.
(v) Local demand.
(vi) Transportation facilities,
(vii) Capital investment.

2. **Burning:** The ground mix is burnt to form clinker in a rotary kiln. Rotary kiln consists of a long horizontal steel cylinder, lined with refractory bricks and rotating at a speed of 0.5 to 2 rotations per minute. The dimensions of the steel cylinder vary with the quantity of production. The kiln is slightly inclined at an angle of 3 – 6º (Fig. 16.1). The inclination allows the material fed in the upper end to travel slowly to the other end. The kiln is mounted on roller bearings.

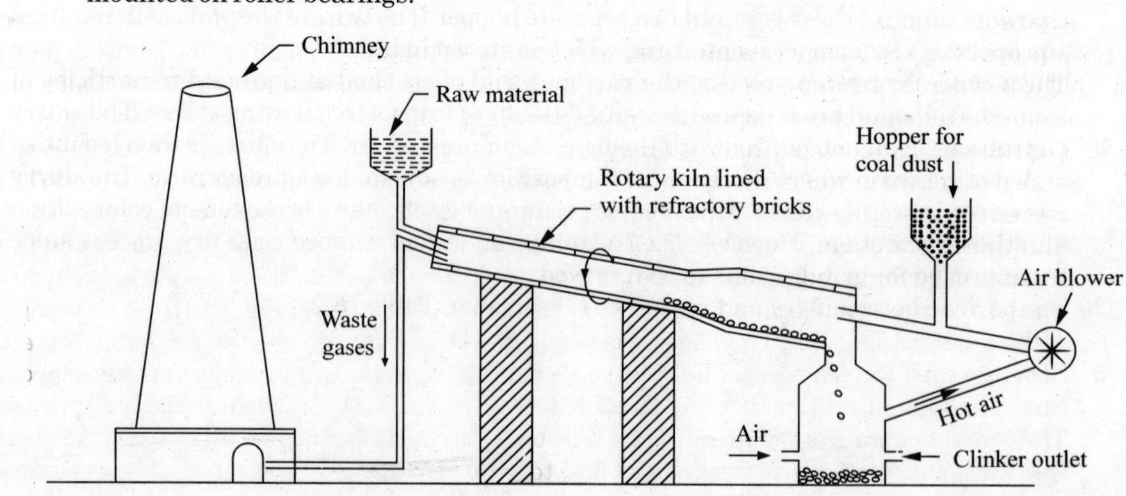

Fig 16.1 Manufacture of portend cement by rotary kiln.

The material is fed from the upper end of the rotary kiln while the burning fuel (coal or oil) and air are injected at the lower end. The hot flames produced in the lower end, heat up the kiln to a temperature of about 1750°C. As the material passes from the upper end towards the lower end due to inclination and rotatory motion of the kiln, it is heated up and it reaches the point of clinckering temperature. The chemical reactions which take place in the kiln are divided into the following zones:

(a) *Drying zone:* In this zone moisture and slurry get evaporated. This is the upper part of kiln having a temperature of about 400°C. The dry material passes down the kiln.

(b) *Calcination zone:* This is the central part of kiln, where the temperature is around 1000°C. The dry material reaching, this point undergoes decomposition of lime stone to form CaO and CO_2

$$\underset{\text{limestone}}{CaCO_3} \rightleftharpoons \underset{\text{Quick lime}}{CaO} + CO_2 \uparrow$$

CO_2 escapes out and material acquire the shape of lumps.

(c) *Clinkering zone:* This is the hottest zone of the kiln having a temperature of about 1400°C to 1600°C. Here the main reaction between the lime and clay takes place forming calcium aluminates and calcium silicates.

$$2CaO + SiO_2 \longrightarrow 2CaO.SiO_2$$

$$3CaO + SiO_2 \longrightarrow \underset{\text{Tricalcium silicate}}{3CaO.SiO_2}$$

$$3CaO + Al_2OI_3 \longrightarrow 3CaOAl_2O_3$$

These calcium aluminates and silicates combine together to form small hard, greyish stones called clinkers. The clinker is actually a mixture of many compounds and cannot be assigned a simple chemical formula. Chemical analysis of the clinker gives only the percentages of CaO, SiO_2, Al_2O_3 Fe_2O_3 etc. Main constituents of clinker are $2CaO.SiO_2$, $3CaO.SiO_2$ and $3CaO.Al_2O_3$, which form 90% of the cement by weight.

3. **Cooling:** The clinker is cooled by a steam of air, either by rotary coolers or by air quench type coolers. The hot air so produced is used for drying the coal before pulverisation.
The quantity of cement produced depends upon the rate of cooling among other factors. If the rate of cooling is *very high* supercooling results in the formation of glassy phase (due to the rapid solidification of alumina and iron into glass). If the rate of cooling is *slow*, dicalcium silicate $2CaO.SiO_2$ converts into powder form (due to allotropic modification of $2CaO.SiO_2$). The powder form has no binding property. Thus cooling is an important process which has to be controlled properly to produce a definite degree of crystallisation of the melted clinker.

4. **Grinding:** The cooled clinker is finely ground with about 2 – 5% of *gypsum*. The gypsum is added to retard or delay the setting of cement when it comes in contact with water. When cement is mixed with water, initial set takes place due to the hydration of tricalcium aluminate. Gypsum, $CaSO_4.2H_2O$, reacts with tricalcium aluminate and retards its hydration forming insoluble calcium sulpho aluminate which does not hydrate quickly.
The size of the ground cement also matters. Finer is the cement, greater will be the strength of the concrete made from it. The powder obtained after grinding is called *Portland Cement*.

5. **Storage and Packaging:** The storage and packaging of portland cement require special care. The cement retains its properties until it comes in contact with the moist air or water. The cement coming out of grinding mill is stored in concrete storage called *silos*. Moisture free air is used to agitate the cement and to keep it free from compaction. The cement is packed in bags by automatic packaging machines.
Fig.16.2 gives the flow diagram for the manufacture of portland cement.

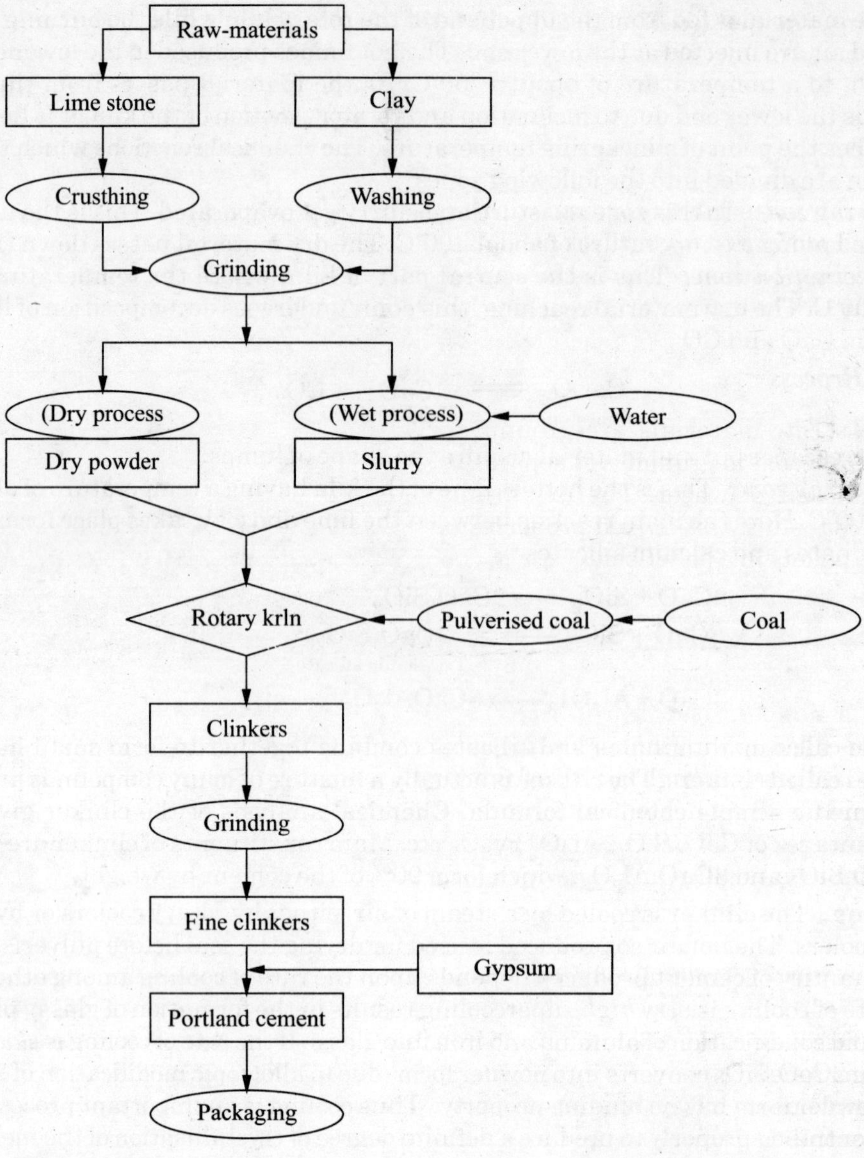

Fig 16.2 Flow diagram for the manufacture of portland cement.

16.2.2b Manufacture of cement by vertical shaft kiln Technology

Large big cement plants with large capital investment use rotary kiln technology but certain mini cement plants use Vertical Shaft kiln instead of long horizontal kilns used in dry and wet process. The vertical shaft kiln (VSK) technology was in use upto 1920, before the advent of rotary kiln technology. The VSK technology was again revived in 1950 with the inventions of pan noduliser (*pelletiser*). In VSK technology, pellets of raw material are made by pan noduliser or pelletiser. Uniform pellets are then fed in to VSK and good quality cement is obtained

Lime (calcareous), clay (argillaceous) and fuel (low volatile) are used as main raw materials in V.S.K process.

In V.S. kilns, the fuel (coal) is mixed with proportioned batch of raw materials either by (i) black meal process or by (ii) fuel slurry process and it is converted to pellets with the help of pelletiser.

Black meal Process

In the process raw materials are ground with coal in the calculated proportion to get 'black meal'. This powdered materials in then converted into pellets by using (about 15%) water.

Fuel Slurry Process

In this process raw materials are ground dry separately. While coal is ground wet separately, with equal amount of water. The dry ground material and coal slurry are mixed properly and converted into pellets in a pelletiser.

Process

Now the ready pellets are fed into the vertical shaft kiln through a hopper. The kiln is lined with refractory and is filled up with clinker leaving about 2 feet. The pellets pass through the kiln from the top to downwards, and get heated and sintered by ascending hot gases. The whole coal charge composed of unburnt nodules and clinker rests on a specially designed grate, rotating slowly at the bottom of the kiln.

The various zones of reaction starting from the top of the kiln are:

(i) Preheating zone
(ii) Calcining zone
(iii) Sintering zone
(iv) Cooling zone

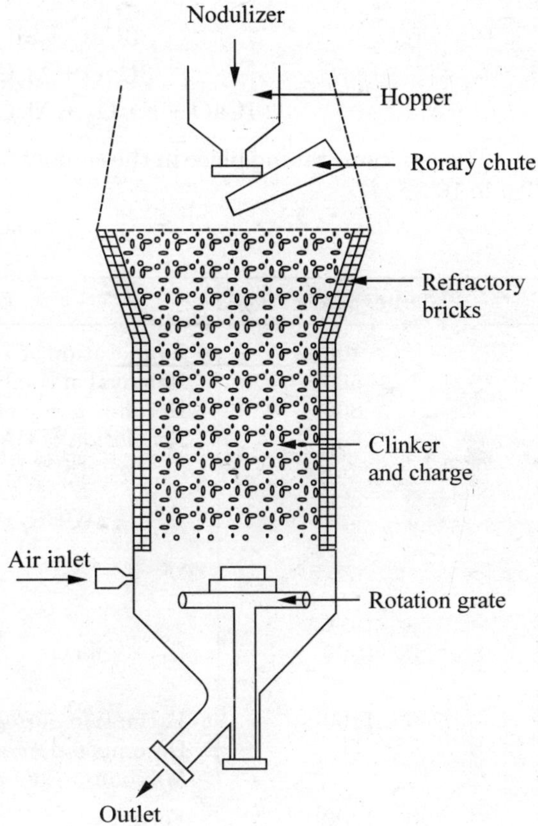

Fig. 16.3 Vertical Shaft Kiln for manufacture of cement

The cold air is blown through the discharge unit to cool the sintered clinkers of cement. The air gets heated in cooling the clinkers which is then utilized in combustion of coal present in pellets. The clinker is finally extracted from kiln bottom through discharge gates and finally ground with 4.5% gypsum before packing.

Advantages of vertical shaft kiln technology

(i) Due to use of pellets and low velocity of exit gases, no dust is discharged into the atmosphere so air pollution is less.
(ii) Control over the shape of pellets reduces the wear of grinding machinery.
(iii) The plant is quite compact since the kiln is usually 10 metres tall.
(iv) It has lower maintenance cost and requires less capital investment.

16.2.3 Reaction Taking Place in the Rotary Kiln

The main chemical reactions taking place can be summarized as follows.

$$CaCO_3 \rightarrow CaO + CO_2$$
$$2CaO + SiO_2 \rightarrow 2CaO \cdot SiO_2$$
$$3CaO + SiO_2 \rightarrow 3CaO \cdot SiO_2$$
$$3CaO + Al_2O_3 \rightarrow 3CaO \cdot Al_2O_3$$
$$4CaO + Fe_2O_3 + Al_2O_3 \rightarrow 4CaO \cdot Al_2O_3 \cdot Fe_2O_3$$

The reactions taking place in the cement kiln at various temperatures are given in the following Table 16.3.

Table 16.3

Temperature (°C)	*Reaction*	*Comments*
> 100	Evaporation of water	
> 500	Removal of combined water from clay	
> 800	Decomposition of $CaCO_3$, $MgCO_3$. Formation of CA, C_2F and C_2S begins	Pure $MgCO_3$ decomposes at 600 – 700°C and pure $CaCO_3$ at 900°C. But in mixture (dolomite) decompositon occurs at lower temperature.
800 – 900	$C_{17}A_{17}$ forms	Decomposes to C_3A at higher temperatpre.
900 – 1100	Formation of C_3A and C_4AF. Decomposition of $CaCO_3$ completed Maximum CaO present.	
1100 – 1200	Formation of C_2S reaches maximum major parts of C_3A and C_4AF are formed.	
1250 – 1280	Formation of liquid phase	C_3A, C_4AF, F, A dissolve and recrystallise on cooling.
1250 – 1450	Formation of C_3S, free lime content decreases. ($A = Al_2O_3$, $C = CaO$, $F = Fe_2O_3$ and $S = SiO_2$)	

16.2.4 Chemical Composition of Cement

The properties of cement depend upon the relative proportions of the components present. The properties of a particular cement can be modified by varying the amounts of various components.

The main constituents of portland cement are:

(i) Tricalcium silicate $3CaO.SiO_2$ (C_3S)

(ii) Dicalcium silicate 2CaO2 CaOiSiO_2SiO_2 (C_2S)

(iii) Tricalcium aluminate $3CaO.Al_2O_3$ (C_3A)
(iv) Tetracalcium alumina $4CaO.Al_2O_3.Fe_2O_3$ (C_4AF)
(v) Calcium sulphate $CaSO_4$

Apart from these some free CaO and MgO may also be present.

The characteristics of the main constituents are:

(i) **Tricalcium silicate (C_3S):** This is responsible for ultimate strength of the cement. Its rate of hydration is medium. It develops its ultimate strength very rapidly.
(ii) **Dicalcium silicate (C_2S):** Its ultimate strength is quite comparable to C_3S but is slightly less. Moreover it does not develop its ultimate strength *initially* and hydrates slowly. Its heat of hydration is 420 J/kg. It is also responsible for imparting strength to cement.
(iii) **Tricalcium aluminate (C_3A):** It undergoes hydration at a very fast rate. The initial set or flash set of cement is due to tricalcium aluminate. However, its early strength is good but ultimate strength is poor.
(iv) **Tetracalcium alumino ferrite:** It has slow rate of hydration. Both its early and ultimate strength are poor and hence do not contribute much to the strength of cement.

Different portland cements contain the four major constituents in different proportions depending on the proportions of raw materials used. For example,

- If clay used is free of Fe, *i.e.* C_4AF is absent, the cement is white.
- If iron is present, the ordinary greenish-black colour of portland cement results.

Different properties such as setting, hardening and strength of cement are also dependent upon the relative proportions of these compounds.

16.2.5 I.S.I. Specifications of Portland Cement

The Indian Standard Institute specifications of Portland cement are as follows:

(i) The ratio of the percentage of lime (CaO) to that of silica (SiO_2), alumina (Al_2O_3) and iron oxide Fe_2O_3 should be more than 1.02 and not less than 0.66, when calculated by the formula:

$$\frac{CaO - 0.7\ SO_3}{2.8\ SiO_2 + 1.2\ Al_2O_3 + 0.65\ Fe_2O_3}$$

Too much lime will result in unsound concrete. Unsound concrete is the one which expands and cracks.

(ii) The ratio of percentage of alumina (Al_2O_3) to that of iron oxide (Fe_2O_3) shall not be less than 0.66.

$$Al_2O_3/Fe_2O_3 \geq 0.66$$

(iii) The weight of insoluble residue should not exceed 2%.
(iv) Total loss on ignition should not be more than 4%.
(v) The sulphur contents shall not be more than 2.75%.
(vi) Weight of MgO should not exceed 6%. The excessive amount of MgO will produce unsound concrete, because a large volume change on hydration is produced if excessive MgO is present resulting into cracking.

16.2.6 Physical Requirements

(i) *Setting Time:* Initial setting time should be not less than 30 minute and final setting time should not be more than 600 minutes.

(ii) *Soundness:* (If very small volume change on hydration is produced then the cement is said to be sound).
Lechatelier's method expresses the expansion of cement set in 24 hours between 25°C and 100°C as soundness. For unaerated cement the soundness should be maximum 10 mm and for aerated cement it should be maximum 5 mm.

(iii) *Compressive strength:* (1 : 3 cement mortar cube of cement and Ennore sand)
3 days : Not less than 1.6 kg/mm^2
7 days : Not less than 2.2 kg/mm^2

(iv) *Fineness:* Not less than 215 m^2/kg or after sieving the residue by weight on B.S. 170-mesh test sieve not to exceed 10%.

(v) *Specific gravity:* should be 3.1 to 3.2.

(vi) *Tensile strength:* Not less than 300 lbs/sq inch after 72 hours.

Indian Portland cement on average had been found to have the following composition:

SiO_2 ... 23%
Fe_2O_3 ... 2.19%
Al_2O_3 ... 5.97%
CaO ... 64.86%
MgO ... 1.00%
SO_3 ... 1.75%
Losses ... 0.74%

16.2.7 Setting and Hardening of Cement

When cement is mixed with water and allowed to stand, it sets to a hard rigid mass. This is known as setting. Thus *setting* can be defined as stiffening of the original plastic mass due to initial gel formation and chemical reactions.

Finally, further solidification of stiffened mass takes place to form a compact rock-like material. This is known as hardening. *Hardening* can be regarded as development of strength due to crystallisation.

Several theories are given to explain the setting and hardening of cement. The setting and hardening of portland cement is attributed to the hydration and hydrolysis of its constituents which are all anhydrous. The hydrated compounds so formed have much less solubility and precipitate out as insoluble gels or crystals. These insoluble gels have the ability to surround the inert materials (like sand or crushed stones) and hence bind them very effectively.

Hardening of the cement takes place due to interlocking of these crystalline products formed during hydration.

Mostly setting takes 24 hours whereas hardening requires 15 – 20 days.

The strength developed by cement paste at any time, depends upon the amount of gel formed and the extent of crystallisation. *The chemical reactions taking place during setting and hardening follow the sequence:*

(i) When the cement is mixed with water, the paste so formed becomes stiff within a short time. This initial set or *flash set* is due to the presence of tricalcium aluminate C_3A which hydrates very rapidly and a considerable heat of hydration is produced

$$3CaO \cdot Al_2O_3 + 6H_2O \rightarrow 3CaO \cdot Al_2O_3 \cdot 6H_2O + 210 \text{ cals/g}$$

$$\text{or} \quad C_3A + 6H_2O \rightarrow \underset{\text{(Crystals)}}{C_3A.6H_2O} + 210 \text{ cals/g}$$

The flash set is not desirable because it prevents the hydration of other constituents of cement. *Thus, gypsum is added to retard the flash set.*

(ii) Gypsum reacts with C_3A to form insoluble complex calcium sulphoaluminate which do not hydrate rapidly.

$$3CaO \,.\, Al_2O_3 + xH_2O + y\, CaSO_4 \,.\, 2H_2O \rightarrow \underset{\text{(Insoluble complex)}}{3CaO.Al_2O_3\, yCaSO_4.zH_2O}$$

or $$C_3A + x\, H_2O + y\, CaSO_4 \,.\, 2H_2O \rightarrow C_3A \,.\, yCaSO_4 \,.\, zH_2O$$

(iii) Tetracalcium alumino ferrite also undergo hydrolysis forming crystalline compound

$$4\, CaO \,.\, Al_2O_3 \,.\, Fe_2O_3 + 7H_2O \rightarrow \underset{\text{(Crystals)}}{3CaO.Al_2O_3.6H_2O + CaO.Fe_2O_3.H_2O}$$

or $$C_4AF + 7H_2O \longrightarrow \underset{\text{(Crystals)}}{C_3A.6H_2O} + \underset{\text{(gel)}}{CF.H_2O}$$

(iv) Next the tricalcium silicate also hydrates within 24 hours and hydration is completed in 7 days.

$$2\,(3\, CaO \,.\, SiO_2 + 6H_2O \rightarrow \underset{\text{Tobermonite gel}}{3CaO.SiO_2.3H_2O} + 3Ca(OH)_2 + 120\, cal/g$$

or $$2C_3S + 6H_2O \rightarrow C_3S_2 \times 3H_2O + 3Ca(OH)_2 + 120\ Cal/g$$

Calcium hydrate gel also known as tobermonite gel, serves as bonding material between grains of aggregates. The strength of cement in early days is due to formation of this gel.

(v) The dicalcium silicate also reacts with water but at a slower rate between 7 to 28 days.

$$2\,(2CaO \,.\, SiO_2) + 4H_2O \rightarrow \underset{\text{Tobermonite gel}}{3CaO.SiO_2.3H_2O} + Ca(OH)_2 + 60\, cal/g$$

(vi) These gels and crystalline products harden gradually to form hardened product.

As hydrolysiss and crystallisation proceed, the strength and hardness of the set mass increase. The process continues for 10 – 28 days.

It should be emphasized here that the engineering properties of concrete is due to tobermonite gel (calcium-silicate hydrate gel) which is the main cementing agent. This gel has an enormous surface area and very strong adhesive strength.

To summarise, the process of hydration and hydrolysis results in the formation of a mixture of gels and crystals in the cement water paste. These crystals adhere firmly to the grains of the aggregate giving the hardening and cementing property to cement.

The whole process can be schemetically represented as:

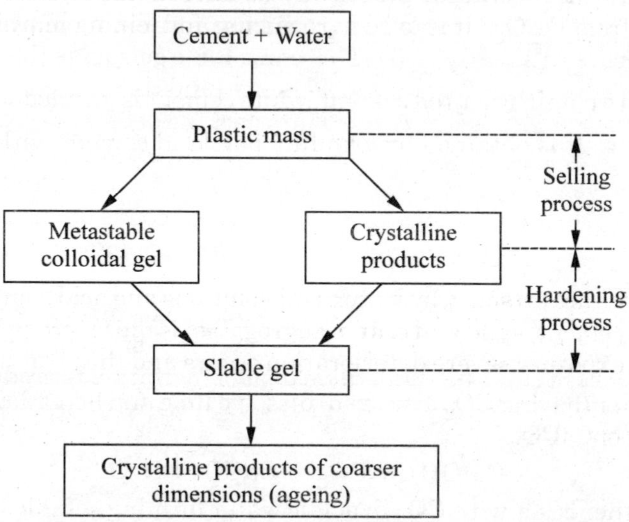

Fig. 16.4 Schematic diagram of setting and hardening of cement

16.2.8 Heat of Hydration of Cement

The process of hydration is exathermic and the heat evolved is the resultant of the heats evolved during the hydration of various constituents. This heat is also known as heat of setting or hardening. The heat of hydration of different constituents are in the following order:

$$C_3A > C_3S > C_4AF > C_2S$$

878   502   418   251 kJ/kg

It is essential that the heat generated during the hydrolysis should be dissipated rapidly otherwise the shrinkage cracks may occur in large concrete constructions. Thus curing plays an important role in the proper hardening

16.2.9 Concrete and Mortar

Cement alone is not used in construction work because it is highly moisture sensitive. Moreover internal stresses can develop in cement leading to cracks. Thus cement is used as a admixture with sand and crushed stones. It may be used in the form mortor or concrete.

Mortor: Mortor is a mixture of portland cement, sand and water. The mortar is used in masonry work for bonding and surface covering. For example, in bonding the bricks together and also for plastering.

Concrete: Concrete is a mixture of gravel or crushed stones, cement, sand and water. Concrete is used in flooring works, roofs, columns, foundations etc. where compressive strength is required. If instead of cement, lime is added then the concrete is known as lime concrete.

Reinforced concrete construction (R.C.C.): The concrete has little ability to withstand tension. Hence reinforced concrete is used in the structures which require to bear high tensile stresses. Reinforced concrete is the concrete reinforced with steel rods or heavy wire mesh giving high tensile strength to the structures. Reinforced concrete is used in floor-beams, girders, slabs bridges etc.

16.2.10 Special Portland Cement

The portland cements used for special purposes are:

White Portland Cement: It contains practically no ferric oxide and is obtained by calcinating the raw material free from Fe_2O_3. It is used in repairing and joining marble pillars, manufacture of tiles and mosaic work.

Coloured Cement: To obtain coloured cement, white cement is grinded with 5 – 10% of pigment.

Water-proof Cement: It is obtained by grinding portland cement with calcium stearate or a non-saponifiable oil.

Decay of Cements

The cement constituents are attacked by water, salt solutions and acid solutions. Although cement concrete (a mixture of cement, sand and coarse aggregates) is quite strong but due to the presence of lime (CaO), on long exposure severe deterioration occurs and the structure weakens.

- The acidic water (having CO_2 dissolved) dissolve lime and hence leaches out free lime with the formation of $CaCO_3$

  $$Ca(OH)_2 + CO_2 \rightarrow CaCO_3 + H_2O$$

  This may further react with CO_2 in acidic water forming soluble Ca $(HCO_3)_2$

  $$CaCO_3 + CO_2 + H_2O \rightarrow Ca\,(HCO_3)_2$$

The process continues till the CO_2 depletes.

- Alkaline waters do not have marked effect on strength of concrete.
- The water can also hydrolyse the silicates and aluminates and dissolve the liberated $Ca(OH)_2$.
- Lime of concrete can also be leached out by sulphates and chlorides present in water. Sulphates also combine with tricalcium aluminate to form sulphoaluminates. Sulphoaluminates occupy more volume causing expansion and cause disruption and weakening of the structure. This can be avoided by increasing the proportion of C_4AF at the expense of C_3A.

16.3 LIME

Another cementing material which is widely used is lime. Lime is essentially calcium oxide and is formed by heating of naturally occurring calcium carbonate, to about 895°C.

When CaO comes in contact with water, it is known as slaking of lime. Slaking is an exothermic process.

$$CaO + H_2O \longrightarrow Ca(OH)_2$$

The slaked lime is obtained as fine powder.

Lime is of various types:

(i) *Fat limes:* Fat limes are high-calcium limes and are produced by heating pure $CaCO_3$. About 95 – 98% CaO and balance SiO_2, Al_2O_3 and MgO is present in fat limes. These limes slake rapidly.

(ii) *Lean limes:* These contain about 5% of MgO and 70 – 80% of CaO. Balance is silica and iron oxide impurities. These slake very slowly.

(iii) *Hydraulic limes:* These contain 5 – 30% clay matter but not more than 2% magnesia along with 70 – 80% CaO. They set into hard structures when mixed with water.

16.3.1 Manufacture of Lime

Raw material used for manufacture of lime is lime stone. The manufacture include two processes:

(i) Calcination (ii) Hydration

(i) **Calcination:** Calcination is carried out at a temperature of 1000 – 1200°C resulting into complete dissociation of $CaCO_3$.

$$CaCO_3 \rightleftharpoons CaO + CO_2$$

The reaction is exothermic and reversible. To ensure that the reaction proceeds in the forward reaction, it is essential that CO_2 so produced is removed and the temperature should not be allowed to decrease below 900°C. $MgCO_3$ present in the lime stone also dissociate into MgO and CO_2.

If temperature employed increases 1200°C, the lime formed will combine with silica, alumina and Fe_2O_3 present and lime will be vitrified outside the lumps. The lime thus becomes slow slaking which is undesirable. The slow slaking lime escapes hydration and expands in the walls causing blisters in the plaster works.

Calcination takes place in kilns either rotary or vertical kiln.

(ii) **Hydration:** The lime obtained after calcination is high-melting white solid. When water is added, it is absorbed through pores.

$$CaO + H_2O \longrightarrow Ca(OH)_2$$

This is known as slaked lime. Slaked lime is fine white powder obtained by adding water to lime in a limited amount. Thus hydrated lime is partially slaked lime.

16.3.2 Setting and Hardening of Lime Mortar

As such the lime paste is not a satisfactory cementing material because it shrinks and cracks. Thus lime mortor, which is a paste of sand (3 parts) and hydrated lime (1 part) is used. The mortar used for building purposes contain a minimum of 75% CaO and 20% MgO rest being Al, Si and Fe. The lime mortar slakes slowly. However of the amount of sand used is less, this will cause shrinkage and cracking. On the other hand if amount of sand is large, a weak mortor is used.

The setting of lime-mortar involves the following processes:

(i) **Dehydration:** First, the water is lost from the lime mortar either due to evaporation or by absorption of water by porous bricks. The mortar becomes stiff but has no strength.

(ii) **Carbonation:** The stiffened mass absorbs CO_2 on exposure to air and a part of $Ca\ (OH)_2$ is converted into $CaCO_3$

$$Ca\ (OH)_2 + CO_2 \longrightarrow CaCO_3 + H_2O$$

$$CaO + CO_2 \longrightarrow CaCO_3$$

The $CaCO_3$ crystals so formed interlock the sand particles and a hard coherent mass is formed. The process of setting involves loss of water, so shrinkage in the volume of mortar takes place which lead to the development of cracks. The sand particles present in the mortar reduce the development and propagation of cracks.

Hardening process is very slow taking months and years together. The crystalline layer formed at the surface prevents the diffusion of air and CO_2 in the interior layers, extending the hardening process to years.

From the above discussion, it becomes clear that ordinary lime hardens in air and hence cannot be used under water. This type of lime is known as *non-hydraulic lime.*

Setting of hydraulic lime takes place in presence of water due to the decomposition of complex silicates of calcium and aluminium into simpler compounds like calcium silicate, calcium aluminate and calcium hydroxide. Calcium silicate and aluminate interlock resulting into the formation of hard coherent mass and $Ca\ (OH)_2$ diffuse to the surface. This $Ca(OH)_2$ by the carbonation of atmospheric CO_2 form $CaCO_3$ which gets hardened. Hence hydraulic limes can be used under water.

16.4 PLASTER OF PARIS

Plaster of Paris $\left(CaSO_4 . \frac{1}{2} H_2O\right)$ is also used as cementing material in making castings for structural tiles or as plaster for walls.

16.4.1 Manufacture

The raw material for plaster of paris is Gypsum. Gypsum is hydrated calcium sulphate $CaSO_4 . 2H_2O$. Plaster of Paris is produced by heating gypsum to a temperature of about 120 – 160ºC.

$$\underset{\text{Gypsum}}{CaSO_4 . 2H_2O} \xrightarrow[150^\circ C]{\text{heat}} \underset{\text{plaster of paris}}{CaSO_4 . \frac{1}{2} H_2O}$$

If plaster of paris is heated further the anhydride is produced

$$CaSO_4 . \frac{1}{2} H_2O \longrightarrow \underset{\text{(anhydrous)}}{CaSO_4}$$

The manufacture of plaster of paris is schemetically represented as:

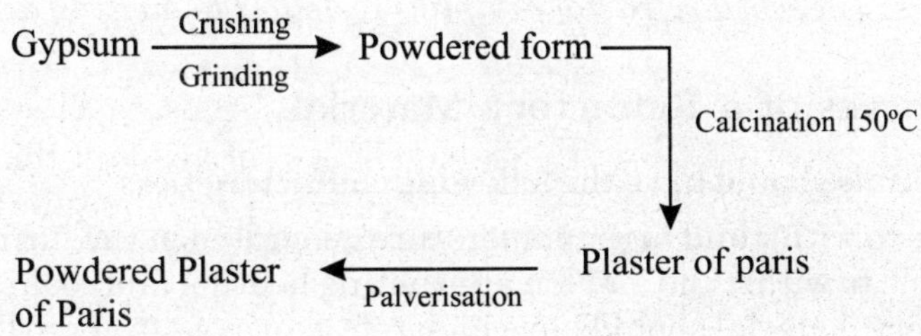

16.4.2 Setting and hardening of Plaster of Paris

Contrary to the setting process of lime, plaster of paris sets and hardens through hydration. It is partially hydrated to gypsum

$$\underset{\text{Plaster of Paris}}{CaSO_4.\frac{1}{2}H_2O} \xrightarrow{\text{hydration}} \underset{\text{Gypsum}}{CaSO_4.2H_2O}$$

Thus on absorption of water, Plaster of Paris expands to form a plastic mass which quickly hardens and converted into a crystalline, closely packed structure of gypsum. In order to increase the rate of setting of Plaster of Paris it is mixed with accelerators such as common salt, K_2SO_4, Na_2SO_4 etc.

Thus hardening of plaster of paris is hydration reaction.

16.4.3 Uses of Plaster of Paris

1. As plaster of Paris expands on setting, due to the absorption of water, it is used for making moulds.
2. It is used in surgery for plastering fractured body parts.
3. It is used as plaster for walls, interior ceilings and decorations etc.

16.5 MAGNESIUM OXYCHLORIDE CEMENT

Magnesium oxychloride cement is made by the addition of a strong solution of magnesium chloride to finely ground and calcined magnesia. The reaction is highly exothermic. It is also known as Sorel Cement.

$$MgCl_2 + 3MgO + 11\ H_2O \rightarrow \underset{\text{Magnesium oxychloride cement}}{3MgO.MgCl_2.11H_2O}$$

The hardening of the cement takes place in three to four hours. This kind of cement is generally prepared 'in situ' or in moulds. The rate of hardening of cement can be increased by addition of calcium chloride or formate.

The magnesium oxychloride cement is used in composite floorings with coloured pigments producing very pleasant surfaces.

16.6 REFRACTORIES

Refractories are inorganic materials that can withstand high temperatures without softening, melting or deformation.

Furnaces used in extraction of metals, cement kins, are subjected to very high temperatures. These should be protected with suitable lining materials which are capable of withstanding high temperatures, corrosive and erosive action of melts, slags and hot gases. These lining materials are called refractories. *The word refractory implies resistant to melting or fusion.*

16.6.1 Requirements of a Refractory Material

Good refractory materials should have the following characteristics:

(i) It must be able to withstand high temperature generated in the furnace.
(ii) It should be able to withstand sudden alternating heating and cooling i.e. thermal shocks.
(iii) A good refractory material should have high refractoriness (refer to the properties).
(iv) The refractory material should be such that it does not chemically react with slag, molten charge and gases like CO, CO_2, SO_2, Cl_2 etc. produced in its immediate contact in furnaces.
(v) It should be able to withstand abrasion action of rapidly moving solid particles of the charge and molten slag moving in the furnace.

16.6.2 Classification

Refractories can be classified on the basis of temperature range of fusion or on the basis of chemical composition.

1. **Classification based on temperature range of fusion:** On this basis the refractories can be classified as:

Refractory	*Fusion Range*
Low refractory	1580 – 1780°C
High highly refractory	1780 – 2000°C
Super refractory	> 2000°C

This class fiction is useful for furnace operation.

2. **Classification based on chemical composition:** On the basis of chemical composition refractories are classified as:

(i) *Acid Refractories:* These refractories are not attacked by acid slags (siliceous slags) or acidic materials in the furnace. Acid refractories consist of acidic oxides like SiO_2 and fire clays containing 30% to 42% Al_2O_3. Silica is used in the form of flint, quartz, sandstone etc.
The acid refractories are readily attacked by the bases or basic slags.

(ii) *Basic Refractories:* Basic refractories are not attacked by basic slags (limy slags) or basic materials. These consist mainly of basic oxides like CaO and MgO without free silica. The most common basic refractories are Magnesite, Dolomite, Lime etc.

(iii) *Neutral Refractories:* These are the refractories which do not combine with either acidic or basic oxides and hence constitute the most satisfactory furnace lining. Example of neutral refractories include graphite, chromite (FeO . Al_2O_3), Zirconia (ZrO_2) and carborundum (SiC).
Generally, several types of refractories are required for the construction of a furnace. This is because no single refractory can withstand all the different conditions which are met in the various parts of the furnace.

3. Classification based on oxide content

(i) Single oxide refractories, e.g. alumina, magnesia and zircon
(ii) Mixed oxide refractories, e.g., spinel, mullite
(iii) Non oxide refractories, e.g., borides, carbides, and silicides

16.6.3 Properties of Refractories

The refractories are selected on the basis of the properties best suited for a particular operation. The following are the important properties of the refractories:

Properties of Refractories

(i) Refractiveness
(ii) Strength
(iii) Thermal Expansion and contraction
(iv) Thermal spelling
(v) Thermal Conductivity
(vi) Porosity
(vii) Permeability
(viii) Abrasion and excision Resistance
(ix) Electrical Conductivity
(x) Heat capacity

1. Refractoriness

When a refractory material is heated to very high temperatures, it may soften and become fluid. As the fluidity increases, the material begins to deform. Refractoriness is the ability of a material to withstand high temperatures without appreciable softening or deformation under working conditions. Refractoriness is usually reported in terms of softening or fusion temperature of the material.

A refractory material should have higher refractoriness i.e. higher softening temperature than the operating furnace temperatures.

Measurement of Refractoriness

Most of the refractories do not exhibit sharp melting points and soften gradually over a wide range of temperatures. This is because they are not pure but composed of several minerals, both crystalline and amorphous in nature. The softening points (fusion point) of some of the refractories are given in Table 16.4.

Table 16.4 Fusion Temperature of some Refractories

Refractory Material	*Fusion Temperature, °C*
Silica (SiO_2)	1710
Silica brick	1700
Fire clay brick	1600 – 1750
Kaolinite ($Al_2O_3 . 2SiO_2 . 2H_2O$)	1785
Bauxite brick	1732 – 1850
High alumina clay brick	1802 – 1880
Alumina (Al_2O_3)	2050
Magnesia brick	2200
Magnesia	2830
Chromite ($FeO . Cr_2O_3$)	1770
Chromite brick	1950 – 2200
Silicon Carbide (SiC)	2700
Zirconia (ZrO_2)	2710
Zirconia brick	2200 – 2700
Zirconium boride (ZrB)	3040
Lime	2570
Carbon (C)	3500

The refractoriness is determined by means of *pyrometric cone equivalent* (PCE) test. The pyrometric cones are small pyramid-shaped 38 mm high cones having triangular base with 19 mm side. The refractory material is moulded into small cones. The softening and bending behaviour of these cones is compared with the behaviour of standard cones of known softening points (Seger Cones). The sager cones are of definite mineral composition and have definite melting or fusing temperatures. These cones are assigned cone numbers based on their fusion temperatures (Table. 16.5).

Table 16.5 Seger cone Number and Fusion Temperature

Seger cone Numbers	*Temperature*
1	1110
2	1120
3	1140
4	1160

For measuring refractoriness of a refractory material, the material is heated at a uniform rate along with seger cones having different cone numbers. The temperature at which the test cone deforms (indicated by its apex touching the base, Fig. 16.5) is noted. Now the number of the *seger cone* (standard cone) which also deforms at that particular temperature along with the test cone is reported as Pyrometric cone equivalent (PCE) of the refractory under that test.

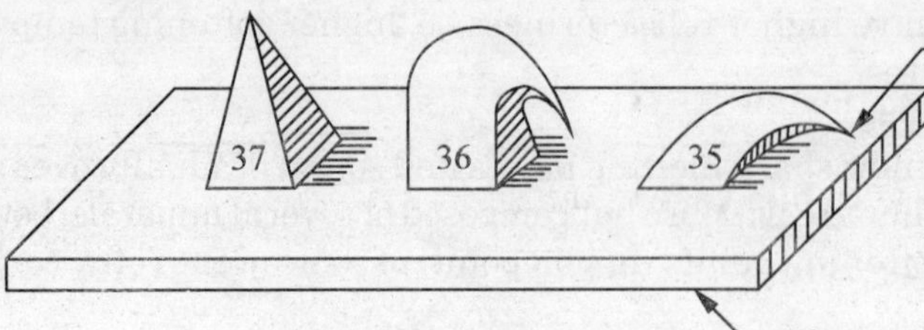

Fig. 16.5 Measurement of refractoriness – Seger cone-test.

2. Strength of Refractory (Refractoriness under load)

The refractories used for lining the furnaces have to withstand heavy loads of charge or the products. Hence refractory materials should possess high mechanical strengths to bear the maximum possible load without deforming at operating temperatures. (The strength at lower temperatures is different from strength at high temperatures).

The high temperature load-bearing characteristics of a refractory material is determined by refractory-under-load test or R.U.L. Test.

In R.U.L. test the specimen of refractory of size 5 cm^2 and height 75 cm under a load of 1.75 kg/cm^2, is heated in a carbon-resistant furnace. The heating rate is maintained at 10°C/minute. The temperature at which 10% deformation occurs on test specimen is reported as R.U.L. The determination of deformation is done by recording the height of the specimen at different temperatures. A plot of deformation versus temperature is plotted and the point at which 10% deformation occurs is reported as R.U.L.

On the basis of R.U.L. test, the refractory materials are classified as:

R.U.L.	*Refractory material*
1350°C	High heat duty brick
1300°C	Intermediate heat duty brick
1100°C	Moderate heat duty brick

3. Thermal Expansion and Contraction (Dimensional stability)

Refractory materials may shrink or expand when heated. For example, fire brick, magnesite, chrome magnesite etc. shrink when in service whereas refractories like siliceous refractories expand when heated.

The resistance of a material to any changes in volume due to exposure to high temperature is known as dimensional stability. The *contraction* of a refractory material may occur due to transformation of one crystalline form of material into another, more dense form. For example,

$$\underset{\substack{\text{Sp. gr} = 3.05 \\ \text{Less dense}}}{\text{MgO (Amorphous)}} \rightleftharpoons \underset{\substack{\text{sp.gr} = 3.54 \\ \text{More dense}}}{\text{Periclase}}$$

Thus, there is a natural shrinkage of the material.

The *expansion* of a refractory material may occur due to phase transformation. For example in silica bricks the transformation of quartz to tridymite and finally to cristobalite is accompanied by a considerable increase in volume.

$$\underset{\text{sp.gr} = 2.65}{\text{Quartz}} \rightleftharpoons \underset{\text{sp.gr} = 2.26}{\text{Tridymite}} \rightleftharpoons \underset{\text{sp–gr} = 2.32}{\text{Cristobalite}}$$

4. Thermal Spalling

The fracturing or flaking off of the refractory material due to temperature changes is known as thermal spalling. Spalling is due to the uneven stresses or strains or compressions. Stresses and strains may occur due to rapid changes in temperature causing uneven expansion and contraction of the refractory material. Bricks with highest expansion and least uniformity are susceptible to thermal spalling.

Spalling can be reduced by

- using bricks having low coefficient of expansion, high uniformity and high porosity.
- avoiding sudden temperature changes
- improving furnace designs.

All the three factors minimize stresses and strains and help in reducing the spalling tendency.

5. Thermal Conductivity

It is an important property of the refractory material as it tells us the amount of heat that flows out through a furnace wall. In industrial operation, refractory materials of both high thermal conductivity and low thermal conductivity are required.

Refractories to be used in muffle furnaces, coke-oven retorts etc. should be capable of transmitting heat and hence should be of high thermal conductivity. On the other hand refractories used in lining the walls of blast furnace, copper hearth furnace etc. should be insulating i.e. have low thermal conductivities because these will not allow much loss of furnace heat and maintain the high temperature inside the furnace.

Thermal conductivity depends upon the porosity and chemical composition. The porous bricks possess low thermal conductivity, since the air in the pores is a poor conductor of heat. The least porous bricks possess the highest thermal conductivity due to the absence of air in the voids.

6. Porosity

Porosity of a refractory material is the ratio of its pores volume to the bulk volume. The brick with the lowest porosity will have high strength, heat capacity and thermal conductivity. However, high porosity increases resistance to thermal spalling.

All refractories are porous to a certain extent. The porosity in the refractory can be the result of either manufacturing methods or by deliberately incorporating combustibles such as raw dust and cork.

7. Permeability

The rate of diffusion of gases and liquids through a refractory is known as its permeability. It depends on the size and number of pores. The permeability of gases and liquids increases on increasing the temperature because the viscosity of these materials decreases.

8. Abrasion and Erosion Resistance

The factors responsible for failure of a refractory lining among others are; abrasion or erosion by descending hard charge, flue gases, carbon particles etc. Slags, molten metals etc. can also cause the corrosion and erosion of refractory lining due to the chemical reaction between the slag and refractory lining.

Thus a refractory material should have erosion and corrosion resistance.

9. Electrical Conductivity

Refractories are poor conductors of electricity except graphite. The refractories used for lining of electrical furnaces should have low electrical conductivity.

10. Heat Capacity

Heat capacity of the refractory material is another property to decide its use in industrial application.

In case of intermittently operated furnaces, light weight bricks having low heat capacity are suitable because the working temperature of the fuel is achieved in lesser time with lesser consumption of fuel. On the other hand, the dense, heavy fire-clay bricks having higher heat capacity are suited for regenerative checker work systems used in coke ovens, stoves for blast furnaces, glass furnaces etc.

16.6.4 Manufacture of Refractories

The manufacture of refractories include a number of physical operations which are almost same for all the materials from which refractory has to be manufactured. A brief outline of the different operations is given below:

1. **Crushing:** The raw materials which are generally in the form of big lumps are crushed to uniform size (about 25 mm).
2. **Grinding:** The crushed material is ground in suitable grinding machines such as hammer mills, ball mills or tube mills.
3. **Screening:** The unwanted material is screened off from the raw materials by the processes like magnetic separation, electrostatic separation and froth floatation.
4. **Blending and Mixing:** The raw materials are then blended with binders which may be organic or inorganic in nature. The organic binders include molasses, starch, shellac, gum, ethylsilicate etc. The inorganic binders include bentonile, sodium silicate, lime, calcined gypsum etc. The type and percentage of the bond additions depend upon the type of refractory to be produced.
   Blending or mixing makes the moulding easier by producing a homogeneous mixture of the batch ingredients. Moreover, it also distribute the plastic material so as to coat the non-plastic constituents thereby developing plasticity.
5. **Moulding:** Moulding may be done either manually or through mechanical means by the application of high pressure. Mechanical moulding is superior to hand moulding because refractories of high density and high strength are produced through the former process.
6. **Drying:** The moisture is removed from refractories in tunnel dries, under selected set of humidity and temperatures, depending on the type of refractory. Rack cars of bricks are placed in a long tunnel and heat is supplied by steam, passing through the tunnel.

7. **Firing or Burning:** The final stage of the manufacture of refractory is firing or burning. The chemical changes that take place include removal of water of hydration, calcination of carbonates and oxidation of iron (II) to iron (III). Firing causes vitrification of the mix which imparts mechanical strength to the refractory. The firing techniques and schedules vary with the type of refractory being manufactured.

16.6.4.1 Acidic Refractories

1. Silica Bricks

Silica brick is also called 'dinas'. It is a refractory material containing more than 93% silica (SiO_2). About 2% lime is added during grinding which helps to bond the brick and also, on firing reacts chemically with silica to impart strength to the finished product.

Manufacture

The raw material used for the manufacture of silica brick is mainly ganister (grained quartzite). Other raw materials include sandstone, quartz, rock crystal etc. Ganister is a hard, dense, fine grained variety having 90 – 96% pure silica. The quartz or quartzite (ganister) is crushed ground and mixed with 2% lime which acts as binder. From this mixture the bricks are moulded manually or by mechanical means. Drying is usually carried out in tunnel driers at 90 – 120°C for about 24 hours until the moisture is reduced to 0.2%.

Firing of silica bricks is usually done in continuous tunnel kilns under a temperature range of 1400°C to 1800°C. Firing is the most vital step in the manufacture of silica bricks. In this step the quartzite is converted into crystobalite. After this, the slow cooling is done which changes crystobatile to tridymite. Thus final bricks is the mixture of crystobalite and tridymite.

$$\text{Quartzite} \rightleftharpoons \text{Crystobalite} \rightleftharpoons \text{Tridymite}$$

The conversion of quartzite to crystobalite is accelerated by lime whereas formation of tridymite is accelerated by Fe_2O_3.

Properties

(i) Silica bricks have homogeneous texture and are free from air pockets.

(ii) The porosity of silica bricks is low (17 to 25%).

(iii) These are yellowish in colour. The colour varies from buff to pale yellow depending upon the raw material used.

(iv) Silica bricks undergo allotropic changes on heating.

$$\underset{\text{(Crystalline silica)}}{\text{Quartzite}} \rightleftharpoons \underset{\alpha\text{-form}}{\text{Tridymite}} \rightleftharpoons \underset{\beta\text{-form}}{\text{Crystobalite}}$$

(v) Silica bricks do not contract but expand when change from quartzite to crystobalite. This change is reversible and the brick returns to original size on cooling.

(vi) During firing it is essential that the silica bricks are heated at 1500°C for 12 hours. This is because otherwise quartzite will not be converted into tridymite and crystobalite and the bricks will undergo a permanent volume change during the operating temperatures of the furnace. This undesirable expansion may damage the refractory lining.

(vii) Silica bricks spall readily when subjected to sudden temperature changes below 500°C. This may be due to stress created when silica brick is cooled, because of inversion of crystobalite. However, they are resistant to thermal spalling below 800°C.

(viii) They can withstand a load of about 3.5 kg/cm^2 upto 1500 – 1600°C.

(ix) They posses good mechanical strength, rigidity and are comparatively light.

Uses

(i) Since they are not deformed under load at high temperatures, these are extensively used in roof arches as in open hearth furnaces and reverberatory furnaces.

(ii) Due to their high thermal conductivity they are used in by-product coke ovens and gas retort settings.

(iii) They are also used in acid converter linings because these bricks are acidic and are not attacked by acids.

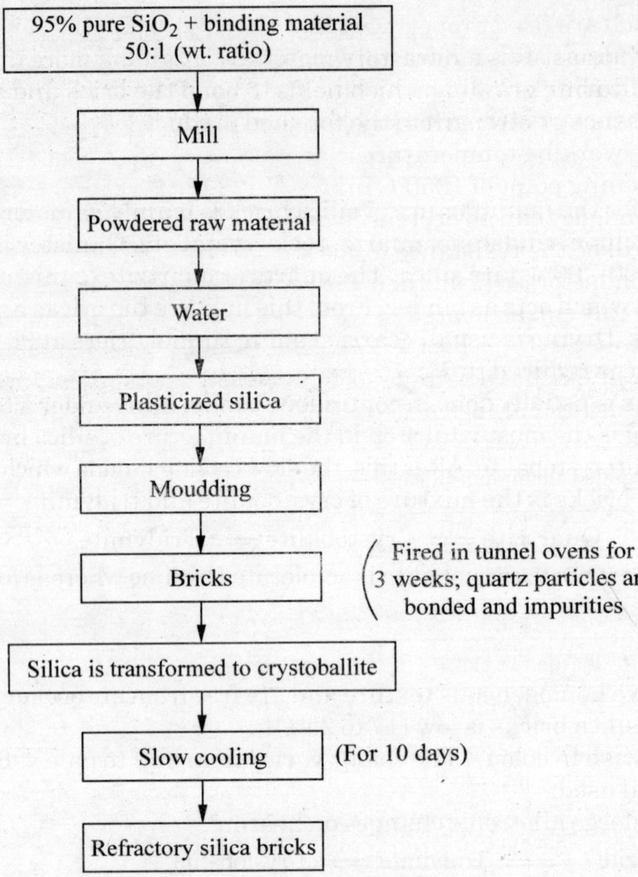

Fig. 16.6 Flow chart for the manufacture of silica bricks

2. Fire Clay Bricks

Manufacture

Fire clays are hydrated aluminium silicates $Al_2O_3.2SiO_2.2H_2O$, with impurities such as alkalies, oxides of iron sulphate and sulphides. At higher temperature the fire clays lose their water of hydration and mainly consist of alumina and silica.

The raw material used for fire clay bricks include raw and calcined fire clay. The calcined fire clay known as *grog* is the non-plastic constituent which is blended with plastic constituent i.e. $Al_2O_3.2SiO_2.2H_2O$ (kaolin). Grog constitutes about 50% more of batch. Fire-clay bricks are manufactured in the same manner as silica-bricks. The exact proportions of the constituents depend on the type of bricks to be made. General composition of different types of fire-clay bricks is:

	SiO_2	Al_2O_3	*Balance*
Feebly Acidic	55%	35%	Clays like
Neutral	55%	40%	K_2O, FeO, CaO, MgO

Properties

(i) The fire clay bicks are pale buff to light brown in colour owing to the varying amount of iron oxide present.

(ii) The bricks which are fired properly are as hard as steel.

(iii) They have low porosity and lower refractoriness, higher resistance to thermal spalling compared to silica bricks.

(iv) Due to the presence of SiO_2, fire clays are acidic in nature and hence is not attacked by acid. Their safe working temperature is about 1545°C.

(v) They have softening point of 1350°C under a load of 2 kg/cm^2.

(vi) The spalling tendency can be decreased by using a coarse grog. Greater the percentage of grog, greater will be the resistance to spalling.

(vii) They are bad conductors of heat, their radiation power is high and hence the fire chambers made up of fire clay can be heated to very high temperatures.

Uses

(i) As the fire clay bricks have better resistance to thermal spalling than silica bricks and hence are suitable for making checker works of regenerative furnaces.

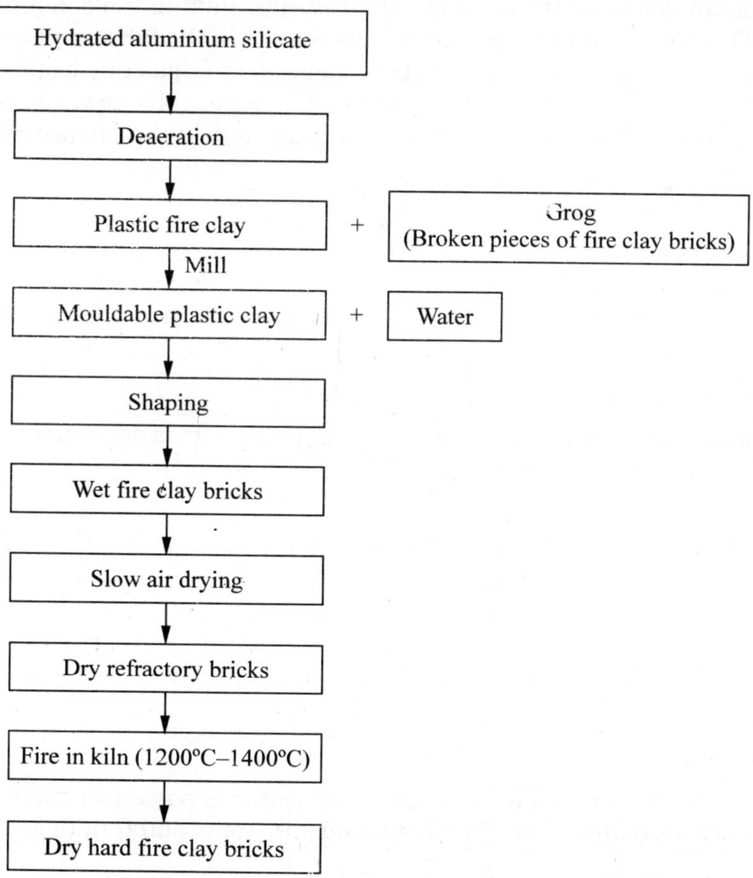

Fig. 16.7 Flow chart for the manufacture of fire clay bricks

(ii) They are also used as linings for blast furnaces, open hearths, stoves, ovens etc. Steel industries are the largest consumers of these refractories for the linings of furnaces.
(iii) Fire clay refractories are also widely used in foundries, lime kilns, regenerators, pottry kilns, glass furnaces, brass and copper furnaces.

16.6.4.2 Basic Refractories

3. Dolomite Bricks

Dolomite bricks are the basic refractories. Dolomite is mixed carbonate of calcium and magnesium, $CaMg(CO_3)_2$.

Manufacture

The dolomite is crushed, washed and calcined when dolomite decompose to give CaO and MgO. Dolomite bricks are made by mixing calcined dolomite (CaO + MgO mixture in equimolecular proportions) with silica which act as binding material. Magnesium silicate, basic slags, clays etc. can also be used as binding material. The dried bricks are fired at 1500°C in a kiln for about 24 hours.

Properties

(i) Dolomite bricks are soft, hygroscopic, having high porosity.
(ii) They are more resistant to slags and spalling than magnesia bricks but are less stronger.
(iii) They can withstand a load of 3.5 kg/cm^2 at 1650°C.
(iv) Properties of dolomite bricks can be improved by mixing the dolomite with serpentine, $MgO.SiO_2$ and calcining the mixture. This results in the formation of di- and tri-calcium silicates. It is then mixed with silicate binder and moulded into bricks, which on firing at 1500°C form stabilized bricks. These bricks are more stable towards basic slags.
(v) Dolomite bricks being soft and porous, wear away quickly and hence are not very useful as refractories.

Uses

(i) Dolomite is rarely used as refractory but is useful as repair material.
(ii) In the granular form, they can be used to pitch the bottom of open-hearth furnace.
(iii) Stabilized dolomite bricks are used in Bessemer convertors, ladle linings and for basic electric furnace linings.

4. Magnesite Refractories

Magnesite refractories are basic type of refractories. Magnesite is naturally occuring $MgCO_3$ and occurs as hydrated carbonate of magnesia.

Preparation

Magnesite ($MgCO_3$) is calcined at a sufficiently high temperature to drive off CO_2 and to form a dense, hard variety of magnesia, called *periclase*. This is also known as dead-burnt magnesite. Calcined magnesite is powdered and mixed with caustic magnesia or iron oxide which act as binders. It is ground with water and moulded into bricks which are fired at 1500°C for about 8 hours. The fired magnesite brick can be given a cladding of mild steel on three sides. In service, steel oxidizes and cements adjacent bricks.

Properties

(i) Magnesite bricks are highly resistant to the action of basic slags and iron oxide.
(ii) They can be used upto 2000°C without load and upto 1500°C under a load of 3.5 kg/cm^2.

(iii) They are not capable of resisting sudden changes of temperature and show a tendency to spall under such conditions.
(iv) They have high electrical resistance and hence are useful in electric furnaces.
(v) Have low resistance to abrasion.

Uses

(i) These refractories are used in electric furnace walls and in the roofs of non-ferrous reverberatory furnaces.
(ii) They are costlier than silica and fireclay bricks and used in the construction of those parts of a furnace which are required to withstand the corrosive action of basic slags.
(iii) Used in the steel industry for the lining of basic converters and open-hearth furnaces.
(iv) Also used in the furnaces used for refining noble metals such as gold, silver and platinum etc.

16.6.4.3 Neutral Refractories

5. Silicon Carbide (Carborundum) Refractories

Silicon carbide is very hard abrasive and have a very high melting point.

Manufacture

It is manufactured by the fusion of sand and coke in an electric furnace together with some saw dust and a little salt. The saw dust helps to increase the porosity whereas salt reacts with iron and impurities, present in the raw materials, forming volatile chlorides. The silicon carbide so formed is in the form of interlocking crystals which are crushed to get a dense packing and mixed with the binding agent. Graphite, ethyl silicate or finely divided SiC itself act as binding agent. The moulded bricks are fired at 1500°C in inert atmosphere.

Self-bonding silicon carbide refractory bricks can also be prepared by mixing SiC particles with a temporary binding agent like glue. The mixture is then fired at 2000°C when inter-crystalline bonds of silicon carbide are formed.

Properties

(i) These bricks have got high thermal conductivity and low thermal expansion.
(ii) Silicon nitride bonded bricks have a high strength and superior thermal shock resistance than clay bonded refractories.
(iii) They have high refractoriness and ability to withstand sudden temperature fluctuations.
(iv) They have high resistance to spalling and abrasion.
(v) Their resistance to acid slag is high whereas to basic slags it is low.
(vi) Self-bonded silicon carbide refractories have superior properties than silicon nitride-bonded bricks which in turn are superior to clay bonded bricks.
(vii) Silicon carbide refractories tend to oxidise to silica when heated in air in the temperature range of 900 to 1000°C.

Uses

(i) These are mainly used in muffles because of their good thermal conductivity.
(ii) They are used in recuperators, partition walls of chamber kilns because of their ability to absorb and release heat rapidly.
(iii) Owing to high electrical conductivity, they are used as heating elements in furnaces in the form of rods and bars known as globars.
(iv) Used in making high conductivity crucibles along with tar.

6. Chromite Bricks

Chromite ($FeO.Cr_2O_3$) bricks constitute neutral type of refractory.

Manufacture

Chromite bricks are made from core ores which mainly contain chromite. Chromite is blended with clay as binding material. The mixture is moulded into bricks, dried and fired at 1500 – 1700ºC.

Properties

(i) They have high resistance to acid as well as basic slags.
(ii) They have moderate resistance to spalling.
(iii) They have crushing strength of about 350 – 550 kg/cm^2.
(iv) They posses high density, moderate spalling resistance and moderate thermal conductivity.
(v) They have refractoriness of 1430ºC under a load of 3.5 kg/cm^2.

Uses

(i) Chrome bricks are used in separating acidic and basic refractory lining to prevent their interaction. For example, they are used between acid-roofs of open-hearth furnaces.
(ii) They are used in steel furnaces, inner lining of acid converters used in the extraction of Cu from low grade ores.
(iii) They are used in bottom of soaking pits, sodium carbonate recovery furnaces etc.

16.7 GLASS

A true solid is crystalline with definite melting point, but amorphous solid is not a true solid. Amorphous solid can be understood as a supercooled liquid with infinite viscosity. Amorphus solid has no definite melting point. Deformity occurs only in the form of softening. On increasing the temperature these amorphous solids first softens and then start flowing. The amorphus solid breaks with conchoidal fracture.

Glass is a fused mixture of silicates for sodium, potassium, calcium and lead which is amorphous, hard, brittle, transparent or translucent supercooled liquid with infinite viscosity.

Within certain limits the glass may be represented by the general formula x R_2O. YMO.6SiO_2 where R is the atom of monovalent alkali metal like sodium and potassium etc. and M is the atom of bivalent metal like calcium, lead and zinc, *x* and *y* are whole number. Thus approximate composition of ordinary glass (soda lime glass) is $Na_2O.CaO.6SiO_2$. In some of the glasses in place of silica, oxides of aluminium, boron and phosphorous (Al_2O_3, B_2O_3, P_2O_5) etc. are used.

16.7.1 Manufacture of Glass

16.7.1a Raw Materials and Their Effect

The main ingredients of glasses are SiO_2, Na_2O, K_2O, B_2O_3, CaO, etc. The raw materials of glass must contain these glass forming oxides. The different raw materials and their effects are as follows.

1. Soda Ash ($NaCO_3$)

It provides Na_2O which imparts softness. Salt cake (Na_2SO_4) and sodium nitrate ($NaNO_3$) may also be used as a source of Na_2O.

2. Lime stone ($CaCO_3$)

It provides CaO. Dolomite (CaO.MgO) may also be used as a substitute of lime stone. The alkaline earth metal oxides like CaO make the glass resistant to water, weathering and chemicals.

3. Sand (SiO_2)

SiO_2 may also be provided by quartz and white sand in pure and iron free condition. It imparts strength and resistance to water, weathering and chemicals.

4. Potash feldspar ($K_2O.Al_2O_3$. $6H_2O$)

It given K_2O. K_2O may also be provided by k_2CO_3 or KNO_3. K_2O imparts hardness and a superior quality of glass is formed.

5. Borax ($Na_2B_4O_7.10H_2O$)

It is the source of B_2O_3. its fluxing capability is good. The borax decreases the expansion coefficient of glass.

6. Litharge or red lead

It is used as a source of PbO. It increases the refractive index and electrical resistance of glass.

7. Phosphorous Penta Oxide (P_2P_5)

It provides brightness in glass used for table wares.

8. Modifiers

Since pure silica glass requires extremely high temperature to melt and work, therefore, modifiers are used which reduce the melting and working temperature. The commonly used modifiers one, the alkalies, the alkaline earths, zinc oxide, lead oxide boron oxide etc.

9. Refining agents

The refining agents are mixed with 'batch' (raw materials) for the removal of small gas bubbles from the glass. Generally used refining agents are Na_2SO_4, $NaNO_3$, NaCl, CaF_2. carbon etc.

10. Colouring agents

Cr_2O_3 – green colour
CrO_3 – yellow green
CoO – blue
CuO – red
Fe_2O_3 – brown
MnO_2 – purple
CdS – yellow
Uranium oxide – fluroescent greenish yellow.
Cryolite (Na_3AlF_6) – Opaque milky white

11. Decolouring agents

These are selinium, cerium oxide, neodymium oxide neodymium oxide etc.

12. Cullets

Cullets are the pieces of broken glass. They increase the fusibility.

16.7.1b Manufacturing Steps

It can be divided into four stages

1. Melting
2. Forming and shaping
3. Annealing
4. Finishing.

1. Melting

The melting of raw materials is carried out in either pot furnace or tank furnace. These furnaces may be of two types.

(i) Regenerative type

(ii) Recuperative type

(i) *Regnerative type:* In this type of furnace heat of waste gases is made available for the heating purpose. In this way a flame of greater intensity is created.

(ii) *Recuperative Type:* The purpose of recuperative furnace in similar to regenerative furnace that is utilization of heat of waste gases and production of high temperature in the furnace. In this type of furnace two set of passages are available. In one set of passage (tube type) gas flows in one direction while in the other set of passage air flows and is preheated as it goes to furnace.

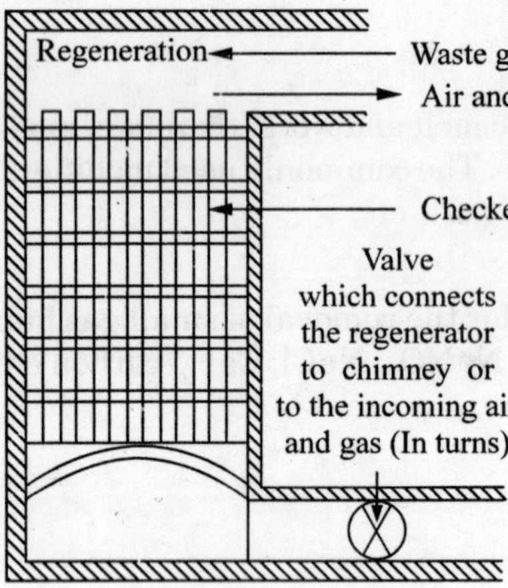

Fig. 16.8 Regenerator

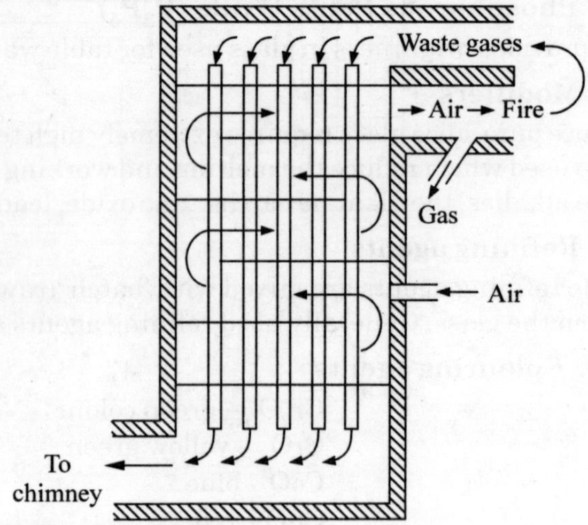

Fig. 16.9 Recuperator

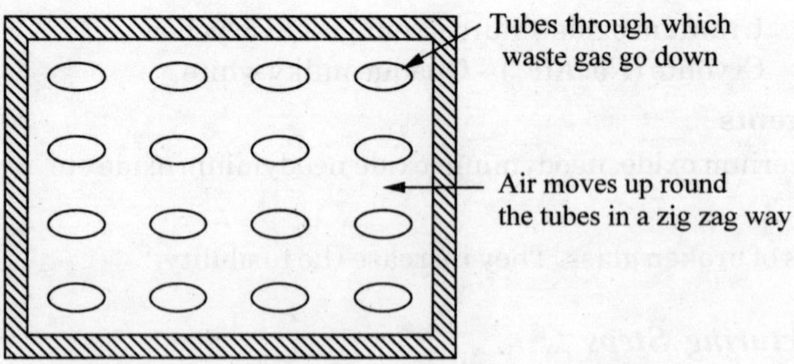

Fig. 16.10 Cross section of Recuperator

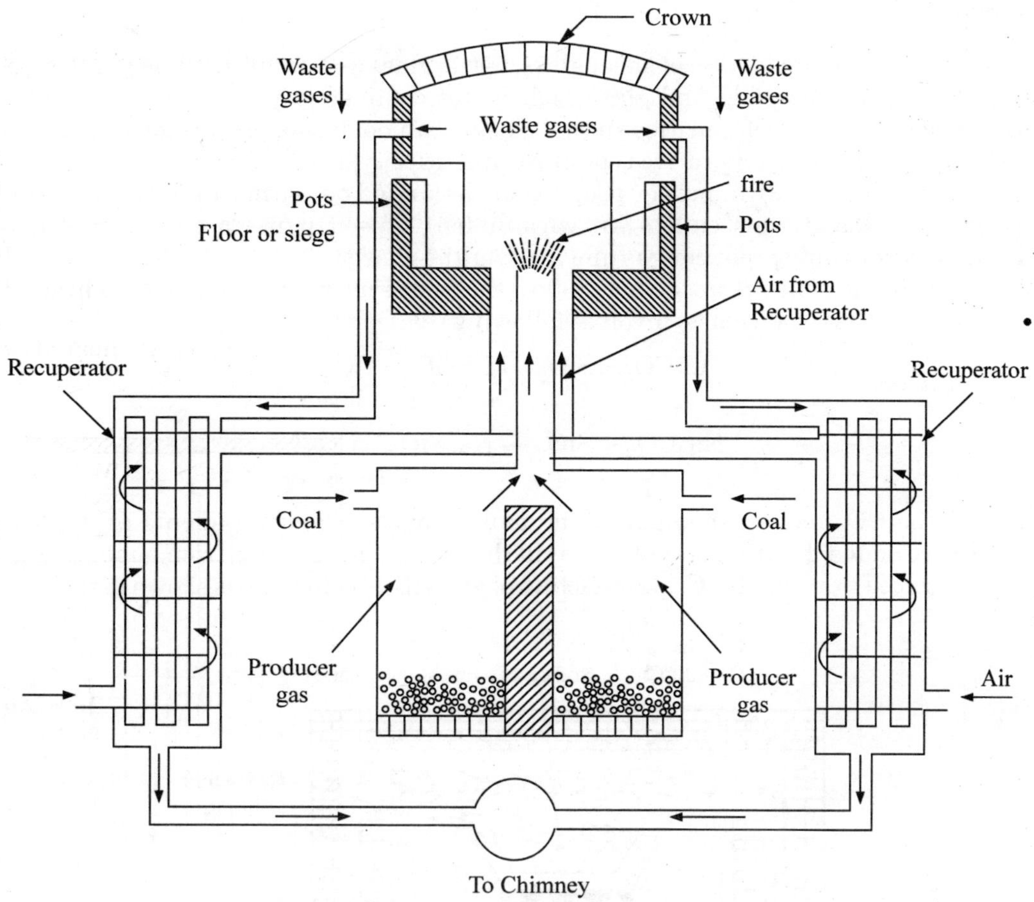

Fig. 16.11 Recuperative Pot Furnace

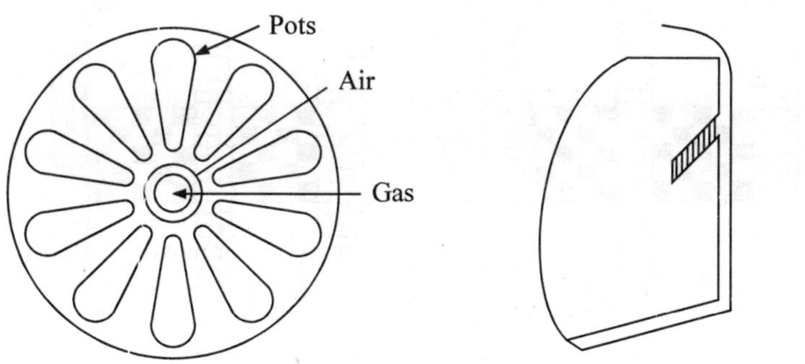

Fig. 16.12 Position of Pots in the pot furnace

Pot Furnace

These furnaces are used to prepare optical glass, and special glass of fine quality. The raw material (Batch) is melted in pots. No of different types of glasses may be melted at one time because many (1 to 20 or more) pots containing similar or different 'batch' may be place at pots floor or siege in the one time. The pots are made of high alumina clay.

Tank Furnace

It is used for large scale production of sheet glass, bottle glass and other common glasses. In tank furnace stirring is not possible and glass melt is not completely protected from combustion impurities. Because of these impurities special glasses can not be made by using tank furnace.

Both furnaces may be regenerative type or recuperative type.

Melting of raw materials in proper proportions is carried out either in pot or tank furnace. Finely powdered raw materials, cullets (known as batch) is fused in fireclay pots or tank in which heating is done by burning producer gas and air over the charge.

At high prevailing temperature (about 1800°C) the charge melts and fuses. The melting process in case of ordinary glass (soda glass) involves following reactions

$$CaCO_3 + SiO_2 \rightarrow \underset{\text{Cal. silicate}}{CaSiO_3} + CO_2\uparrow$$

$$Na_2CO_3 + SiO_2 \rightarrow \underset{\text{Sodium silicate}}{Na_2SiO_3} + CO_2\uparrow$$

When all the CO_2 has escaped out of the molten mass, colouring agents or decolourizer (as required) are added. Heating is continued till the molten mass is free from bubbles and glass balls and then cooled to about 800°C. Manufacture of glass in tank furnace is shown in the following Fig. 16.13.

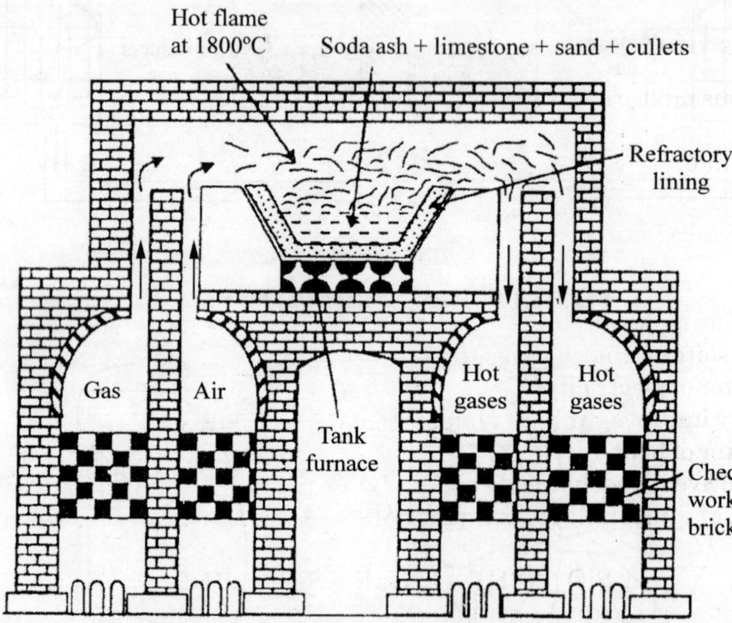

Fig. 16.13 Manufacture of glass in Tank furnace

2. Forming and Shaping

Molten glass is then worked into articles of desired shapes by using techniques like blowing, moulding or pressing between rollers.

3. Annealing

The glass articles are allowed to cool gradually to room temperature by passing through a series of chambers with descending temperature. Thus stepwise or gradual or slow cooling is known as annealing.

If the manufactured glass is cooled suddenly, its upper layer will be cooled down leaving inner layers in a slate of strain. During this unequal expansion, the glass articles are likely to crack. Therefore, annealing is needed.

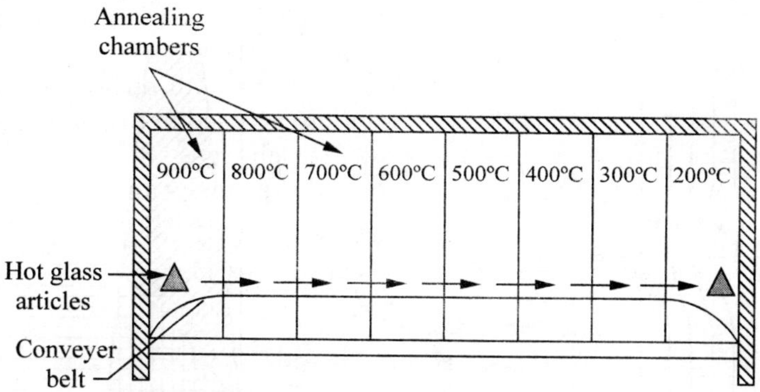

Fig. 16.14 Annealing process

4. Finishing

The decorating or finishing steps include cutting, polishing, exterior colouring, metallic lining etc.

16.7.2 Properties of Glass

(i) Homogeneous mixture
(ii) Amorphus
(iii) Non crystalline
(iv) Transparent
(v) Insoluble in water
(vi) Hard substance
(vii) Brittle
(viii) No sharp melting point
(ix) On heating softens and changes to viscous liquid.
(x) Bad conductor of electricity
(xi) Conductivity improves at high temperature
(xii) Bad conductor of heat
(xiii) Highly sensitive to mechanical shocks
(xiv) No reaction of almost all acids and alkalies. Only HF reacts with silicates present in glass.

$$Na_2SiO_3 + 6HF \rightarrow 2NaF + SiF_4 + 3H_2O$$
$$CaSiO_3 + 6HF \rightarrow CaF_2 + SiF_4 + 3H_2O$$

(xv) *Devitrification:* When glass is softened by heating and is kept as such for long time it appears to loose its transparency due to crystallization of the individual silicates. This property of glass in known as devitrification.

16.7.3 Fracture of Glasses

Glass does not have any crystal lattice structure and is brittle.

Fracture or creak in glass is caused due to small imperfections, irregularity. Most of the cracks occur at the surface, improper annealing in also important cause of fracture.

Flows or irregularities like fine crests (invisible even under microscope) cause concentration of stress which will be intensified to many times greater than the nominal stress at that point under tension. Thus, a crack proceeds and propagates quickly resulting into fracture as shown in the following figures 16.15, 16.16.

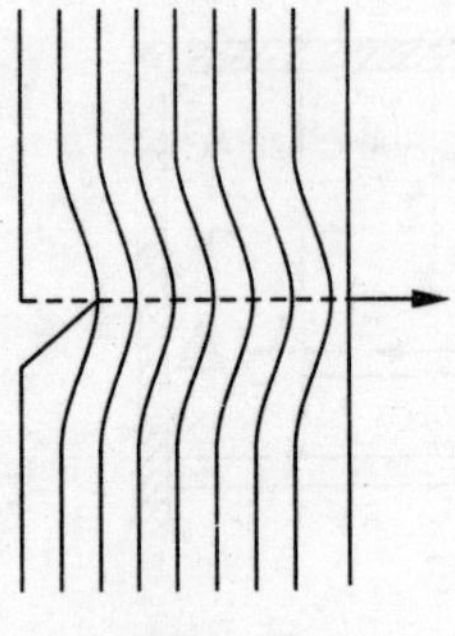

Fig. 16.15 Propagation of crack in glass

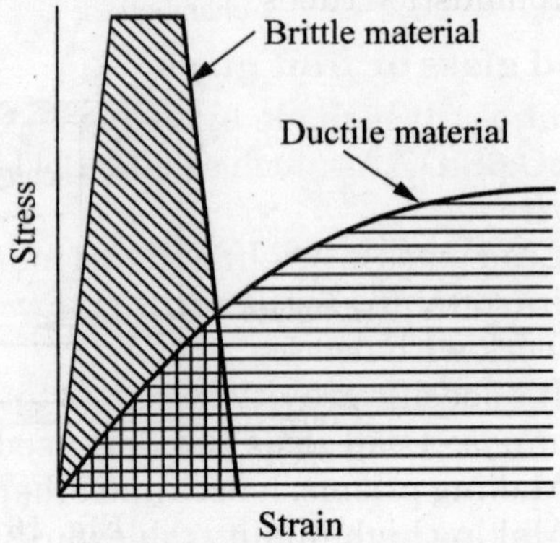

Fig. 16.16 Energy requirements to fracture in ductile and brittle materials

16.7.4 Types of Glasses

Some commercial glasses are discussed below:

1. Soda-lime glass or soft glass

It is the simplest silicate glass in which Na_2O is added to the melt for lowering down the temperature for melting of silica glass

Preparation: It is made by fusing the sodium carbonates, calcium carbonate and SiO_2. It's approximate composition is

$$Na_2O \cdot CaO. 6SiO_2$$

As the carbonates decompose to oxides on heating.

Properties:

(i) Low melting point and thus can be hot-worked easily,
(ii) Resistant to water and devitrification,
(iii) Attacked by common reagents like acids and hence limited chemical stability
(iv) High temperature resistance is Poor.
(v) Low cost

Applications: Soda-lime glass is widely used for

(i) Window glasses
(ii) Cheaper tablewares like bottles, jars etc.
(iii) Cheap laboratory glasswares.

2. Potash glass or Hard glass

Preparation. It is made by fusing K_2CO_3, $CaCO_3$ and SiO_2. Its average composition is $K_2O.CaO.6SiO_2$

Properties: Compared to soda glass, potash glass:

(i) possesses high melting point and can withstand high temperatures.
(ii) is more harder
(iii) fuse with difficulty

(iv) more resistance to acids, alkalies and other solvents.
(v) more costly.

Applications: It is used for heating operations like
(i) chemical apparatus
(ii) combustion tubes

3. Lead glass or flint glass

Preparation: It is made by fusing K_2CO_3, red lead and SiO_2. The approximate composition is $K_2O.PbO.6SiO_2$. As much as 80% PbO is incorporated for dense optical glasses.

Properties
(i) Lead glass is bright and lustrous due to its higher refractive index.
(ii) Its softening temperature is lower than soda-glass and thus it is much easier to shape and work with it.
(iii) It's specific gravity is high (3 to 33).

Applications: Lead glass is widely used for
(i) Making prisms, lenses and other optical devices;
(ii) Making high-quality tablewares and art-objects, because of high lusture;
(iii) Making extra-dense optical glasses (High lead content flint glass is used) for windows and shields to protect personnel from γ-rays and X-rays in medical and atomic energy fields respectively.

4. Borosilicate glass or Pyrex glass or Jena glass

It contains boron trioxide (B_2O_3) and is very rich in silica and little amounts of alumina and alkali metal oxide.

Al_2O_3 and B_2O_3 provide Al^{3+} and B^{3+} ions for pyrex glass. Al^{3+} may be present in structure as a free metal ion, or it may replace Si^{4+} in SiO_4. Similarly, B^{3+} replaces some Si^{4+} in the tetrahedral skeleton.

The percentage composition of different components would be

Component	SiO_2	B_2O_3	Al_2O_3	K_2O	Na_2O
%	80.5	13	3	3	0.5

Properties: Borosilicate glass has
(i) low thermal coefficient of expansion.
(ii) high chemical resistance.
(iii) high softening point.
(iv) excellent shock-resistance (shock-proof).

Applications: Borosilicate glass is widely used in
(i) industry for pipelines for corrosive liquids.
(ii) superior laboratory apparatus like flasks, beakers etc.
(iii) television tubes, electrical insulators etc.
(iv) kitchenware.

5. Alumino silicate glass

Compared to borosilicate glass aluminosilicate glass has higher percentage of Al_2O_3 and lower percentage B_2O_3. The percentage combustion of different components is given below:

Component	SiO_2	Al_2O_3	B_2O_3	MgO	CaO	$Na_2O + K_2O$
%	55	23	7	9	5	1

Property: It pocesses exceptionally high softening temperatures.
Applications: It used for

(i) high-pressure mercury discharge tubes.
(ii) chemical combustion tubes and
(iii) certain domestic equipment, etc.

6. Optical or Crooks Glass

Optical glasses are used in optical instruments because they possess the following important properties

(i) Low viscosity so that it becomes free for bubbles in molten state.
(ii) No devitrification even on long annealing.
(iii) Even on long exposure to the atmosphere, no spoilage of its surface.
(iv) Absence of impurities like iron and thus colourless.
(v) Ability to take up the desired polish on grinding and polishing.

Optical glasses contain phosphorus and lead silicate in which calcium oxide is replaced by oxides of Pb, phosphorus etc. They have high refractive index and used for the manufacture of lenses.

Crooke's glass is a special type of optical glass which contains little cerium oxide which is capable of absorbing ultraviolet light (which is injurious to eyes).

Compared to ordinary glass, optical glasses are relatively soft, possess low melting point, low durability and slow chemical resistance.

7. Safety glass Laminated glass

Preparation. Safety glass is made by placing a thin layer of plastic (either acetyl resin or cellulose accetate) between two sheets of ordinary glass. This sandwiched structure is then heated under slight pressure till the glass layers and plastic layers merge into one another.
Properties:

(i) Safety glass is quite tough
(ii) It is shatter proof.

(When it breaks, it does not fly in to pieces, since the inner plastic layer tends to hold back the broken pieces of the glass. That's why broken pieews does not cause any harm to the people around).
Applications:

(i) There-layer laminated (safety) glass is used as wind-shields in automobile and aeroplane industries.
(ii) Five-layer laminated glass is used in bulking construction.
(iii) Several layered laminated glass is bullet-proof and is used for making automobile wind screens, looking windows etc.

8. Toughened Glass

Preparation. After conversion of glass into desired shape articles, the hot articles are dipped in oil-bath so that some chilling takes place. By doing so the outer layers of the articles acquire a state of compression because of shrinkage; while the inner layers are in a state of tension.
Properties:

(i) Toughened glass in more elastic.
(ii) It is capable of withstanding thermal and mechanical shocks.
(iii) When toughened glass breaks it is reduced to fine powder and the pieces do not fly away.

Applications: Toughned glass is used for making

(i) Window shields of furnaces;
(ii) Window shields of trucks, cars, aeroplanes, and other fast moving vehicles;
(iii) Large show-cases.
(iv) Automatic opening doors

9. Wired glass

Preparation: During casting of glass sheet, wire-mesh is embedded at the centre of it.
Properties:

(i) More fire-resistant than ordinary glass, and
(ii) Shatter-proof

Applications. It is used for making fire-resisting doors, roofs, windows and sky lights.

10. Photochormic Glass

Photochromic glass change colour in the presence of high-energy radiation (usually ultra-violet radiation) but reverts back to their original appearance in the absence of this radiation.

Preparation: With the base composition of borosilicate glass, silver, copper and a metal halide is mixed and heated until the sold melt and about 1200°C. As the glass cools, small crystallites (about 10 nm in diameter and 100 nm apart) are too small to scatter or absorb visible light and thus the glass appears transparent.

11. Glass wool

It is intermingled fine threads or filament so glass.

Preparation: It is obtained by forcing molten mass of completely alkali-free glass through small holes (Average diameter of 0.0005 to 0.007 mm) continuously. The filaments of glass are then thrown over a rapidly revolving drum to get the material in the form of intermingled fine threads of filaments of glass.
Properties: Glass wool is

(i) It is having low thermal and electrical conductivity.
(ii) Resistant to heat, fire and chemicals
(iii) Its density is quite low (about 63 kg/m^3) but tensile strength is about with times that of steel.

Applications:

(i) It is used in domestic and industrial appliances like over, insulation of metal pipelines, motors, vacuum-cleaners, etc. where heat insulation is required.]
(ii) It is also used for manufacturing fibre-glass, by making a composite with pulsate resins;
(iii) It is also used for sound and electrical insulation.
(iv) It is also used for filtration of acidic and other corrosive liquids where resistant to chemicals is required;

QUESTIONS

1. Discuss the chemical reactions involved in formation of clinker (in burning zone) during manufacture of cement. Explain setting and hardening of portland cement. **[MNIT 2003, Raj. Univ. 2003]**
2. Draw labelled diagram of rotary kiln used for the manufacture of portland cement by wet process and state its working. **[Raj. Univ. 2002]**
3. Draw flow diagram to show the various steps involved during portland cement manufacture by rotary kiln technology. Also write the chemical reactions along with temperatures which occur during the process. **[Raj. Univ. 2001]**

4. Write short note on: Portland cement. **[Raj. Univ. 2000]**
5. What are the constituents of cement? Discuss the mechanism of setting and hardening of cement. **[Raj. Univ. 1999 Main]**
6. Describe the manufacture of Portland cement by rotary kiln technology. **[Raj. Univ. 1998]**
7. How is cement manufactured? Explain. **[Raj. Univ. 1997]**
8. What are the constituents of portland cement. **[Raj. Univ. 1996]**
9. Discuss the setting and hardening of portland cement. **[Raj. Univ. 1996 Back]**
10. Explain briefly the chemistry of setting and hardening of cement. **[Raj. Univ. 1994]**
11. What do you understand by the following terms in relation of refractories.
    (i) Dimensional stability
    (ii) Thermal spalling
    (iii) Porosity?
12. What are refractory materials? How are they classified? Explain seger cone test and RUL test of refractory materials. **[Raj. Univ. 1999 Main]**
13. Give the essential requirement of a good refractory. **[Raj. Univ. 1999]**
14. What are refractories? Explain the essential requirements of a good refractory. **[Raj. Univ. 1998]**
15. Write the important properties of good refractories? What are acidic, basic and neutral refractories? Explain. **[Raj. Univ. 1996 Main]**
16. What are refractories? How are they classified? Give examples. **[Raj. Univ. 1996]**
17. What are the refractories? What are the essential conditions for a good refractory? **[Raj. Univ. 1994]**
18. What is glass? Describe the manufacturing process of Borosilicate glass. Discuss the important of silica glass.
19. Give classification of glass with their respective uses. How is glass manufactured? **[Raj. Univ. 2002]**
20. Discuss the following in brief (any three)
    (i) Laminated glass,
    (ii) Borosilicate glass,
    (iii) Optical glass,
    (iv) Glass wool. **[Raj. Univ. 2001]**
21. What is glass? Describe the chemistry involving in formation of glass. **[Raj. Univ. 2000]**
22. What is glassy state of matter? Why is annealing important in manufacturing of glass? Explain the importance of pure silical glass. **[Raj. Univ. 1999 Main]**
23. Write short note on special glasses. **[Raj. Univ. 1999]**
24. What do you understand by glassy state of matter? Give the approximate composition of soda-glass. **[Raj. Univ. 1997]**
25. Write a short note on manufacture of glass. **[Raj. Univ. 1996]**
26. What is glass? Describe briefly the various types of glasses. **[Raj. Univ. 1995]**
27. What is glass? Describe the chemistry involved in formation of glass. **[Raj. Univ. 1994]**
28. Give the compound composition of a typical portland cement sample. Explain the influence of each one in the properties of portland cement. **[MREC 2001]**
29. Name the raw materials used for the manufacture of portland cement. Discuss the various physico-chemical changes taking place in the kiln during manufacture of cement. Explain the role of gypsum in portland cement. **[MREC 2000]**
30. Write short note on:
    Porosity of refractories. **[MREC 2002]**
31. The selection of a refractory for a furnace lining has to be choosen on the bases it its chemical nature. Explain with examples. **[MREC 2001]**
32. What is devitrification of glass? Why annealing is required in the manufacturing of glass? Why borosilicate glass is preferred in chemical laboratory? **[MREC 2002]**
33. Write short note on:
    A comparison between pot and tank furnace. **[MREC 2002]**
34. Write short note on:
    Annealing **[MREC 2001]**

35. Who invented cement? Give chemistry of manufacturing of cement.
36. Draw neat and well labelled diagram of manufacturing of cement by rotary kiln.
37. What is Gypsum? Why is it added to cement?
38. Explain with equations the setting and hardening of cement.
39. What are refractory materials. How are they classified?
40. Give preparation properties and uses of fire clay refractory.
41. How silica bricks are manufactured? Write properties and uses of silica bricks.
42. Write short notes on the following
    (i) RUL Test
    (ii) Seger cone test
43. What is glass? What are the raw materials and their functions used for the manufacturing of glass.
44. Why is curing necessary during construction.
45. Write short notes on toughened glass pullet proof glass, crooks glass and pyrex glass.

❑❑❑

17

GASEOUS STATE

"There comes a time when the mind takes a higher plane of knowledge but can never prove how it got there. All great discoveries have involved such a leap. The important thing is not to stop questioning".

— Albert Einstein

CHAPTER OUTLINE

The three principal states of matter are-Gaseous state, liquid state and solid state. Depending upon the temperature and pressure, a substance can exist in any of these states. Gaseous state is characterized by having very weak forces of attraction between their molecules. Gaseous state is studied most extensively. The properties of gases are well defined, uniform and well understood.

Gases have the following main characteristics:

- Gases have marked sensitivity of volume change with temperature and pressure.
- Gases are most compressible of all the states of matter.
- Gases can diffuse easily and mix evenly with each other.
- Gases have lower densities compared to liquids and solids.
- Gases exert pressure on the walls of the container in which these are enclosed.

Out of the known elements, only eleven elements are gases under normal atmospheric conditions. These are listed in Table 17.1.

Apart from these elements which exist as gases, a number of compounds are also found in gaseous state at ordinary temperature and pressures.

Gases can be described in terms of four parameters (measurable properties)– viz; Volume, pressure, temperature and number of moles. However, it is not necessary to specify each and every parameter since these are interrelated.

Table 17.1 Substances Existing as Gases at 1 atm. and 25° C

Diatomic Gases	*Monoatomic Gases*
H_2 (molecular hydrogen) N_2 (molecular nitrogen) O_2 (molecular oxygen) F_2 (molecular flourine) Cl_2 (molecular chlorine)	He (Helium) Ne (Neon) Ar (Argon) Kr (Krypton) Xe (Xenon) Rn (Radon)

17.1 THE GAS LAWS

The empirical generalizations based on experimental results gave rise to well known gas laws like Boyle's Law, Charles' Law, Graham's Law, Dalton's Law and Avogadros' Law. But before the development of Kinetic Molecular Theory there was no theoretical basis for these laws. All these laws were derived through experiments. Boltzman, Maxwell and Clausius proposed the theoretical basis of these laws and hence related theory to the experimental results.

Boyle's Law: The conclusions reached by R. Boyle (1662) from his experiments on the compressibility of air have been generalized in the statement known as Boyle's Law. **Boyle's law states that at constant temperature, the volume of a definite mass of gas is inversely proportional to its pressure, i.e.**

$$V \, \alpha \frac{1}{P}$$

or

$$PV = \text{Constant} \qquad ...(1)$$

Where P is the pressure and V is the volume of the gas. If V_1 is the volume occupied by the gas at pressure P_1 and V_2 its volume when pressure changes to P_2 then from Boyle's law it follows :

$$P_1 V_1 = P_2 V_2 \qquad ...(2)$$

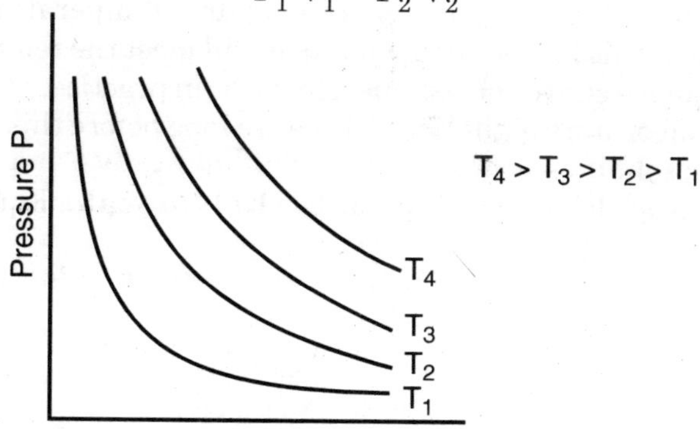

Fig. 17.1 The pressure-volume dependence of a gas for a fixed mass

Charles' Law: The variation of the volume of a gas with temperature at constant pressure is expressed as the generalization often called Gay-Lussac's Law. Actually much before J.L Gay-Lussac proposed this generalization in 1802, the same conclusions had been reached by J.A.C. Charles in 1787, but his work remained unpublished. However, in his honour this law is commonly known as Charles' Law.

Charles' Law states that at constant pressure, the volume of any gas increases or decreases by 1/273.16 of its volume at O°C for every 1° rise or fall of temperature. Mathematically,

$$V_t = V_o + \frac{t}{273.16} V_o \quad ..(3)$$

$$= V_o\left(1 + \frac{t}{273.16}\right) \quad ...(4)$$

$$= V_o\left(\frac{273.16 + t}{273.16}\right) \quad ...(5)$$

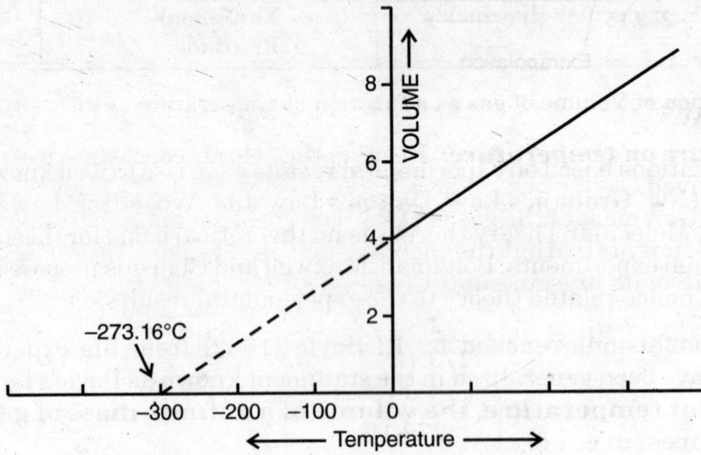

Fig. 17.2 Volume of a gas as a function of temperature in centigrade

It is obvious from the equation (5) that volume of a gas must become zero when the temperature reaches −273.16 °C. This temperature gives us a new temperature scale, the so called **absolute scale**. The temperature on absolute scale is represented by T and is given by T=273.16 + t, where t is the temperature in °C. Therefore, 0°C is given by T_o = 273.16.

If the volume of a certain mass of a gas is plotted against temperature at constant pressure, a straight line is obtained. If this line is extrapolated, it will meet the temperature axis at −273°C. Thus, at −273°C a gas occupies zero or no volume. However in practice, no such phenomenon have been encountered in any laboratory of the world, because long before this temperature is reached, the gas liquefies and then solidifies.

The new temperature scale is also known as the **Kelvin scale**. Equation (5) becomes

$$V_t = V_o \frac{T}{T_o} \quad (\because T = 273.16 + t,\ T_o = 273.16) \quad ...(6)$$

$$\frac{V_t}{V_o} = \frac{T}{T_o} \text{ or } \frac{V_t}{T} = \frac{V_o}{T_o} \quad ...(7)$$

For a given mass of an ideal gas at constant pressure, therefore, the quantity V/T is constant, i.e. at constant pressure the volume is proportional to the absolute temperature. This is an alternative statement of Charles' Law.

$$V \propto T \text{ or } \frac{V}{T} = \text{Constant} \quad ...(8)$$

According to equation 8, the volume of a gas should be a linear function of absolute temperature at constant pressure. Fig. 17.3.

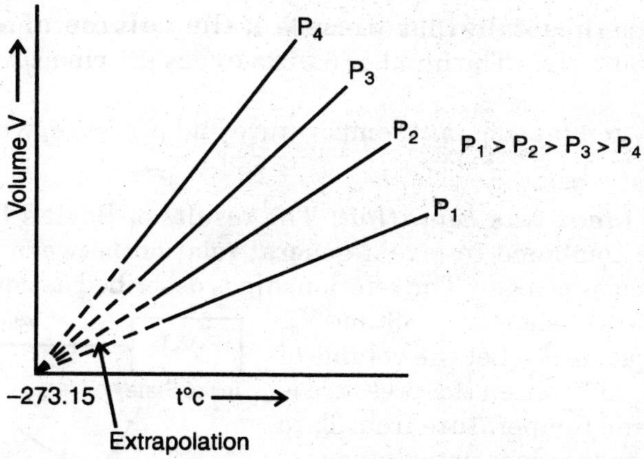

Fig. 17.3 Variation of Volume of gas as a function of temperature at different pressures

Dependence of pressure on temperature: Keeping the volume constant, an expression similar to the above can be derived.

The pressure of a given mass of the gas varies linearly with temperature at constant volume. It has been found experimentally that the increase in pressure per degree rise in temperature comes out to be 1/273.16 of its pressure at 0°C, i.e.

$$P_t = P_o + \frac{t}{273.16} P_o \quad ...(9)$$

$$= P_o\left(1 + \frac{t}{273.16}\right) \quad ...(10)$$

$$= P_o\left(273.16 + \frac{t}{273.16}\right) = \frac{P_o T}{T_o} \quad ...(11)$$

$$\frac{P_t}{P_o} = \frac{P_o}{T_o} \quad ...(12)$$

Where, P_o is the pressure at 0°C.
P_t is the pressure at t°C.

Thus the pressure of a given mass at constant volume is directly proportional to its absolute temperature.

$$P \propto T \quad ...(13)$$

P-T plots for a given amount of the gas are given in Fig. 17.4

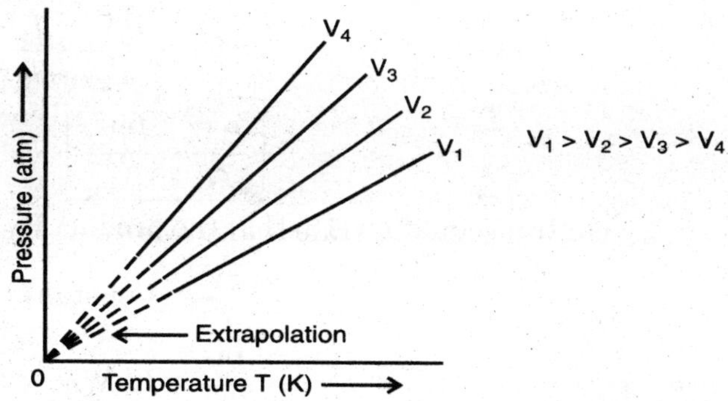

Fig. 17.4 Variation of pressure with temperature

Avogadro's Law: In 1811, **Avogadro suggested that equal volumes of different gases at the same temperature and pressure contain equal number of molecules.**

In otherwords, if temperature and pressure are kept constant, the volume of a gas is directly proportional to the number of molecules (N) present in the gas.

Mathematically

$$V \propto N \text{ (At constant temperature and pressure)} \quad ...(14)$$

It has been found experimentally that one mole of every gas contains 6.023×10^{23} molecules and occupies 22.4 litres at N.T.P. The number 6.023×10^{23} is known as Avogadro's Number and is represented by No.

We can also write, V α n (at constant temperature and pressure, where n is the number of moles) ...(15)

Equation of State or Ideal Gas Equation: The results of Boyle's Law, Charles Law and Avogadro's Law can be combined to give a general relation between pressure, volume and temperature of a given mass of a gas. The relationship is described as equation of state.

Consider a given mass of a gas at a volume V_1, pressure P_1 and temperature T_1. Let the volume of the gas changes from V_1 to V_2 when the pressure is changed from P_1 to P_2 and temperature from T_1 to T_2. Suppose the change takes place in two steps :

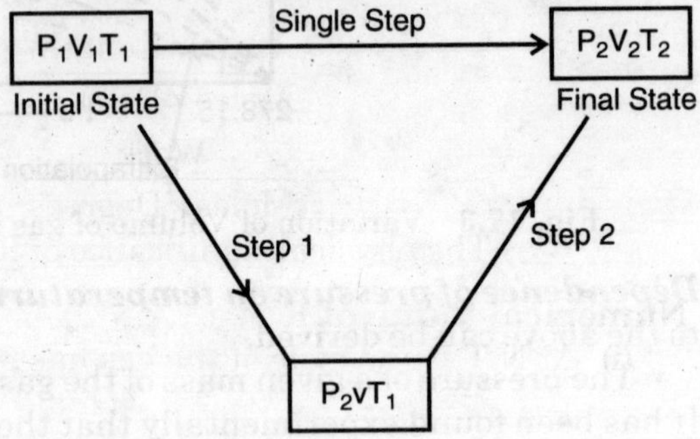

Step 1: In this step suppose the volume of a given mass of a gas changes from V_1 to v, on changing the pressure from P_1 to P_2 at constant temperature T_1. The volume v is given by Boyle's law.

$$P_2 \times v = P_1 \times V_1$$

or
$$v = \frac{P_1V}{P_2} \qquad ...(16)$$

Step 2: Now let volume v change to V_2 when temperature is changed from T_1 to T_2 at constant pressure P_2. The final volume V_2 is given by Charles' Law.

$$\frac{v}{T_1} = \frac{V_2}{V_2}$$

or
$$V_2 = \frac{vT_2}{T_1} \qquad ...(17)$$

Substituting, the value of v from equation (16) in equation (17) we have,

$$V_2 = \frac{P_2V_1}{P_2} \times \frac{T_2}{T_1}$$

$$\boxed{\frac{P_1V}{T_1} = \frac{P_2V_2}{V_2}} \qquad ...(18)$$

It is clear from equation (18) that the product PV divided by T is always constant.

$$\frac{PV}{T} = \text{Constant}$$

or
$$\frac{PV}{T} = \text{K (Where K is any constant)} \qquad ...(19)$$

The value of this constant depends upon the amount of gas in the system since volume is proportional to the amount of the gas present in the system. If n is the number of moles of the gas taken, it is found that

$$K \propto n$$
$$K = n R \qquad ...(20)$$

Where R is constant of proportionality and is known as **universal gas constant**. It is found to be independent of the nature of the gas. Substituting the value of K in equation (19) we get,

$$\frac{PV}{T} = nR$$

$$PV = nRT \qquad ...(21)$$

This is the most common form of ideal gas equation.

Evaluation of Gas constant R: From the ideal gas equation PV=nRT we have-

$$R = \frac{PV}{nT} = \frac{\text{Pressure} \times \text{Volume}}{\text{Moles} \times \text{Degrees}} = \frac{\frac{\text{Force}}{\text{Area}} \times (\text{Length})^3}{\text{Moles} \times \text{Degrees}}$$

$$= \frac{\text{Force} \times \text{Length}}{\text{Moles} \times \text{Degrees}} \qquad (\because \text{ Area} = \text{length}^2)$$

$$= \frac{\text{Work}}{\text{Moles} \times \text{Degrees}}$$

∴ The physical significance of R is work per degree per mole.

Numerical Value of R

(i) At N.T.P one mole of the gas has pressure P = 1 atm;

Volume V = 22.4 litres,

Temperature T = 273 K

$$R = \frac{1\text{atm} \times 22.4\,\text{litres}}{1\text{mole} \times 273.16\,\text{K}}$$

$$= 0.0821 \text{ litres atm } K^{-1} \text{ mol}^{-1}$$

(ii) In the C.G.S units, pressure is expressed in dynes/cm^2 and volume in cubic centimeters.

$$P = hdg = 76 \times 13.6 \times 981 \text{ dynes/cm}^2$$

$$V = 22400 \text{ cm}^3$$

$$T = 273K$$

$$n = 1 \text{ mol}$$

$$R = \frac{(76 \times 13.6 \times 981) \text{ dynes/cm}^2 \times 22400 \text{ cm}^3}{1 \text{ mol} \times 273 \text{ K}}$$

$$= 8.314 \times 10^7 \text{ ergs } K^{-1} \text{ mol}^{-1} \quad (\because \text{dyne cm=erg})$$

(iii) In terms of joules:

$$10^7 \text{ ergs} = 1 \text{ joule.}$$

$$R = \frac{8.314 \times 10^7}{10^7} \text{ joules per kelvin per mol.}$$

$$= 8.314 \text{ J } K^{-1} \text{ mol}^{-1}$$

(iv) In terms of Calories:

$$4.184 \text{ joules} = 1 \text{ calorie}$$

$$4.184 \text{ joules} = \frac{8.314}{4.184} \text{ calories } K^{-1} \text{ mol}^{-1}$$

$$R = 1.987 \text{ calories } K^{-1} \text{ mol}^{-1}$$

Thus R can be expressed in different units. To sum up, different values of R in different units are-

0.0821 L atm K^{-1} mol^{-1}
8.314×10^7 ergs K^{-1} mol^{-1}
8.314 Joules K^{-1} mol^{-1}
1.987 Cal K^{-1} mol^{-1}

NUMERICAL PROBLEMS BASED ON IDEAL GAS EQUATION AND GAS LAWS

Example 17.1. A given mass of a gas occupies a volume of 25 ml at 25°C when the pressure is 1 atm. Calculate the new volume if the pressure is increased to 1.25 atm keeping the temperature constant.

Solution:

$$V_1 = 25 \text{ ml } P_1 = 1 \text{ atm}$$
$$V_2 = ? \; P_2 = 1.25 \text{ atm}$$
$$P_1V_1 = P_2V_2$$
$$\therefore \quad V_2 = \frac{P_1V_1}{P_2} = \frac{(1 \text{ atm})(25 \text{ ml})}{1.25 \text{ atm}} = 20 \text{ ml}$$

Example 17.2. A given amount of a gas occupies a volume of 400 ml at a pressure of 1 atm and at a constant temperature of 298K. Calculate the pressure that should be applied on the gas to lower its volume to 320 ml.

Solution:

$$P_1 = 1 \text{ atm } V_1 = 400 \text{ ml}$$
$$P_2 = ? \; V_2 = 320 \text{ ml}$$
$$P_1V_1 = P_2V_2$$
$$P_2 = \frac{P_1V_1}{P_2} = \frac{(1 \text{ atm})(25 \text{ ml})}{1.25 \text{atm}} = 20 \text{ ml}$$

Example 17.3. 5 g of ethane is confined in a bulb of one-litre capacity. The bulb is so weak that it will burst if the pressure exceeds 10 atm. At what temerature will the pressure of the gas reach the bursting value? (R = 0.821 lit. atm /K/mole)

Solution:

$$PV = n\,RT$$
$$PV = \frac{\text{Amount of } C_2H_6}{\text{Mol. wt. of } C_2H_6}.RT$$
$$\therefore \quad T = \frac{PV}{R} \times \frac{\text{Mol.wt of } C_2H_6}{\text{Amount of } C_2H_6}$$
$$\therefore \quad T = \frac{10 \times 1 \times 30}{0.0821 \times 5}$$
$$= 730.81 \text{ K}$$
$$= (730.81 - 273)°\text{C}$$
$$= 457.81°\text{C}.$$

Example 17.4. A gas cylinder containing cooking gas can withstand a pressure of 14.9 atm. The pressure gauge of cylinder indicates 12 atm at 27°C. Due to sudden fire in the building, its temperature starts rising. At what temperature the cylinder will explode?

Solution: Suppose the cylinder will burst at T_2K.

When the pressure will increase from 12 atm to 14.9 atm we have,

$$\frac{P_1V_1}{T_1} = \frac{P_2V_2}{T_2} \quad P_1 = 12 \text{ atm}, T_1 = (27+273)$$

Here

$$P_2 = 14.9 \text{ atm}, T_2 = ?$$
$$V_1 = V_2 \text{ as the volume does not change}$$
$$\therefore \quad T_2 = \frac{P_2T_1}{P_1} = \frac{14.9 \times 300}{12} = 372.5 \text{ K}.$$

Example 17.5. Oxygen is present in a one-litre flask at a pressure of 7.6×10^{-10} mm of Hg. Calculate the number of oxygen molecules in the flask at 0°C.

Solution:

$$P = 7.6 \times 10^{-10} \text{ mm}$$
$$= 0.01 \times 10^{-10} \text{ atm (760 mm = 1 atm)}$$
$$T = (0 + 273) = 273 \text{ K}$$

We have,

$$PV = nRT$$

$$0.01 \times 10^{-10} \times 1 = n \times 0.0821 \times 273, \quad \therefore n = \frac{10^{-12}}{0.0821 \times 273}$$

$\therefore$ No. of oxygen molecules = Mole × Avogadro's number

$$= \frac{10^{-12}}{0.0821 \times 273} \times 6.022 \times 10^{23} = 2.68 \times 10^{10}$$

Dalton's Law of Partial Pressures: This law deals with the total pressure of a mixture of gases. In nineteenth century John Dalton made some observations and reached a conclusion known as Dalton's law. According to this law, **"if two or more than two gases which do not react chemically are enclosed in a vessel, the total pressure exerted by the gaseous mixture is equal to the sum of all the partial pressures that each gas would exert when present alone in that volume."** If $p_1, p_2, p_3, \ldots p_n$ be the partial pressures of n gases enclosed in a given vessel and P be the total pressure of the gaseous mixture, then according to Dalton's law-

$$P = p_1 + p_2 + p_3 + \ldots + p_n \qquad \ldots(22)$$

The partial pressure of a constituent gas is defined as the pressure it would exert if it is present alone in that volume at the same temperature. This may be calculated using the formula:

$$\text{Partial pressure of the gas} = \frac{\text{Moles of the gas}}{\text{Total Moles of the mixture}} \times \text{Total pressure.}$$

or

$$\text{Partial pressure} = \text{Mole fraction} \times \text{Total pressure.}$$

This can be derived as follows: Let a mixture of gases contains the amount n_1 of the first gas, n_2 of the second, n_3 of the third and so on. Let the corresponding partial pressures be p_1, p_2, p_3........ The total pressure is given by

$$P_{\text{Total}} = p_1 + p_2 + p_3 \ldots\ldots$$

Assume that the gases in the mixture behave ideally, then for each gas we can write

$$p_1 = \frac{n_1RT}{V} \qquad \ldots(23)$$

$$p_2 = \frac{n_2RT}{V} \qquad \ldots(24)$$

$$p_3 = \frac{n_3RT}{V} \qquad \ldots(25)$$

$$\text{hence } (p_1 + p_2 + p_3 \ldots\ldots)V = (n_1 + n_2 + n_3 \ldots\ldots)RT$$

$$P = (n_1 + n_2 + n_3 \ldots\ldots)\frac{RT}{V}$$

$$\therefore \quad P = \frac{nRT}{V} \text{ where } n = n_1 + n_2 + n_3 \ldots\ldots \qquad \ldots(26)$$

Dividing equations (23), (24), (25) by equation (26)

$$p_1 = \frac{n_1 P}{n} \quad ...(27)$$

$$p_2 = \frac{n_2 P}{n} \quad ...(28)$$

$$p_3 = \frac{n_3 P}{n} \quad ...(29)$$

Since the fraction $\frac{n_1}{n}, \frac{n_2}{n}, \frac{n_3}{n}$ ... represents the moles fraction of the constituent 1, 2 and 3... denoted by x_1, x_2 and x_3..... respectively.

$$\therefore \quad p_1 = x_1 P \quad ...(30)$$

$$p_2 = x_2 P \quad ...(31)$$

The mole fraction of a constituent of any mixture is defined as the number of moles of that constituent divided by the total number of moles in the mixture.

Graham's Law of Diffusion: This law was put forward by Thomas Graham in 1833 which states that *under the same conditions of temperature and pressure the rate of diffusion of a gas is inversely proportional to the square root of its density or molar mass.* If r_1 and r_2 are the rates of diffusion of two gases, whose densities are d_1 and d_2 respectively, then from Graham's Law-

$$\frac{r_1}{r_2} = \sqrt{\frac{d_2}{d_1}} \quad ...(32)$$

or

$$\frac{r_1}{r_2} = \sqrt{\frac{M_2}{M_1}} \quad ...(33)$$

Where M_1 and M_2 are the molar masses of the two gases.

NUMERICAL PROBLEMS BASED ON DALTON'S LAW & GRAHAM'S LAW

Example 17.6. What pressure is exerted by a mixture of 2.00 g of H_2 and 8.00 g of N_2 at 273 K in a 10 litre vessel?

Solution:

$$PV = nRT$$

$$P = \frac{nRT}{V}$$

$$\text{No. of moles of } H_2 = \frac{2.00}{2.02} = 0.990$$

$$\text{No. of moles of } N_2 = \frac{8.00}{28} = 0.286$$

$$p_{H_2} = \frac{0.990 \text{ mole} \times 0.0821 \text{ atm. litre K}^{-1} \text{ mol}^{-1} \times 273 \text{ K}}{10.0 \text{ litre}}$$

$$= 2.22 \text{ atm.}$$

$$p_{H_2} = \frac{0.286 \text{ mol} \times 0.0821 \text{ atm.litre K}^{-1} \text{mol}^{-1} \times 273\text{K}}{10.0 \text{ litre}}$$

$$= 0.64 \text{ atm.}$$

$$P_{total} = p_{H_2} + p_{H_2}$$

$$= 2.22 \text{ atm} + 0.64 \text{ atm}$$

$$= 2.86 \text{ atm.}$$

Example 17.7. Small amounts of ammonia (NH_3) and hydrogen chloride (HCl) are released simultaneously at the opposite ends of a 2.0 m long tube. At what point the formation of ammonium chloride (NH_4Cl) would start?

Solution: The formation of NH_4Cl would start when NH_3 and HCl come in contact. Let x be the distance from the HCl and where the formation of the salt NH_4Cl starts. The distance covered by NH_3 in the same time is $2 - x$. The situation may be shown as:

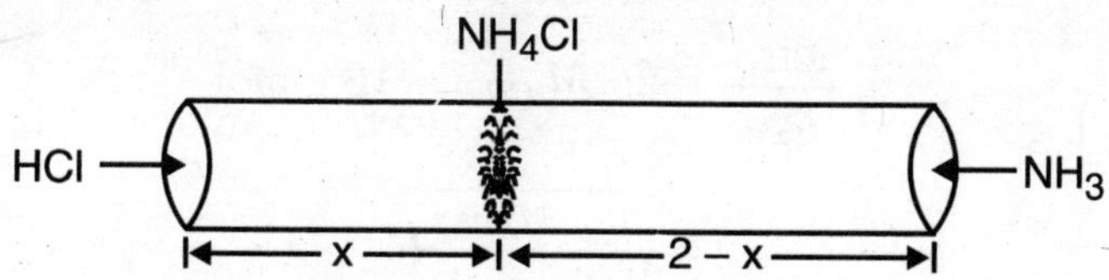

If C(HCl) and C(NH_3) are the molecular speeds of the two gases, then the time taken to cover the respective distances is given by,

$$t = \frac{x}{C(HCl)} = \frac{2.0 - x}{C(NH_3)}$$

or
$$\frac{C(HCl)}{C(NH_3)} = \frac{x}{2.0 - x}$$

Since the molecular speed is inversely proportional to the square root of molar mass, we shall have,

$$\frac{[M(NH_3)]^{1/2}}{[M(HCl)]^{1/2}} = \frac{x}{2.0 - x}$$

But, $\frac{(17.0\,g\,mol^{-1})^{1/2}}{(36.5\,g\,mol^{-1})^{1/2}} = 36.5$ g mol^{-1} and M(NH_3) = 17.0 g mol^{-1}, therefore,

$$\frac{(17.0\,g\,mol^{-1})^{1/2}}{(36.5\,g\,mol^{-1})^{1/2}} = \frac{x}{2.0 - x} = 0.682$$

$\therefore$ $$x = 0.81 \text{ m}$$

The formation of NH_4Cl would start when HCl has moved 0.81m and NH_3 has moved 1.19 m inside the tube.

Example 17.8. A 2L flask contains 1.6 g of methane and 0.5 g of hydrogen at 27°C. Calculate the partial pressure of each gas in the mixture and hence calculate the total pressure.

Solution:

$$\text{No. of moles of } CH_4 = \frac{1.6}{16} = 0.1 \text{ mol}$$

$$\text{Partial pressure } p_{CH_4} = n_{CH_4} \times \frac{RT}{V} = \frac{0.1 \times 0.0821 \times 300}{2} = 1.23 \text{ atm}$$

$$\text{No. of moles of } H_2 = \frac{0.5}{2} \text{ mole} = 0.25 \text{ mol}$$

$$\text{Partial pressure } p_{H_2} = n_{H_2} \times \frac{RT}{V}$$

$$= \frac{0.25 \times 0.0821 \times 300}{2} = 3.079 \text{ atm.}$$

$$\text{Total pressure} = p_{CH_4} + p_{H_2}$$

$$= 1.23 + 3.079 = 4.31 \text{ atm.}$$

Example 17.9. The rate of diffusion of methane at a given temperature is twice that of a gas X. What is the molecular mass of X.

Solution:

$$\frac{r_{(CH_4)}}{r_{(x)}} = \sqrt{\frac{M_{(x)}}{M_{(CH_4)}}}$$

$$\frac{r_{(CH_4)}}{r_{(x)}} = 2; \quad M_{(CH_4)} = 16 \text{ g mol}^{-1}$$

$$\therefore \quad 2 = \sqrt{\frac{M(x)}{16 \text{ g mol}^{-1}}}$$

$$\therefore \quad M(x) = 16 \text{ g mol}^{-1} \times 4 = 64 \text{ g mol}^{-1}$$

17.2 THE KINETIC THEORY OF GASES

All the gas laws discussed above are based on experimental facts. Prior to Kinetic Molecular Theory, there was no theoretical model to explain the gas laws. Several attempts were made to elucidate the behaviour of gases on theoretical grounds. Kinetic theory of gases is the result of all such attempts, which successfully provides theoretical basis for all the observed experimental facts that have led to the enunciation of various laws.

Kinetic theory of gases has developed from the ideas proposed by D. Bernovilli (1738); J.J. Waterson (1845), K.A. Krönig (1856) and R. Clausius (1857). It was further elaborated by J. Clark Maxwell (1860) and L. Boltzmann (1868).

Kinetic theory of gases, includes the following assumptions:

1. A gas consists of a large number of particles (atoms or molecules) that are so small that their actual volume is neglible as compared to the total volume occupied by the gas. This assumption is justified considering the fact that gases are highly compressible.
2. There are no attractive forces between the molecules and the molecules are completely independent of one another.
   This assumption is supported by the fact that a highly compressed gas if allowed to expand, occupies all the available space, suggesting the absence of any intermolecular attractions.
3. The molecules of a gas move randomly in all possible directions. During this movement the molecules collide with one another and also with the walls of the container. The pressure exerted by the gas is due to the bombardment of its molecules on the walls of the container. The movement of the particles cannot be directly observed because molecules are too small to be seen through the microscope. However particles of smoke floating in air can be observed which showed the irregular motion of smoke particles. This phenomenon is called Brownian movement. Though molecules are much smaller than the smoke particles but in absence of attractive forces between them it is reasonable to assume Brownian movement in gas molecules too. The fact that gas fills up a vessel of any size in a very short time, further supports this postulate.
4. The collisions of gas molecules are perfectly elastic, i.e. there is no loss of kinetic energy. This assumption is supported by the fact that if there were loss of kinetic energy as a result of collisions, the motion of the particles after some time would eventually stop, thereby temperature and pressure will become zero. Since this does not happen unless we change the temperature or pressure, one can conclude that collisions are perfectly elastic.

5. At any particular instant, molecules in a gas can have different velocities and hence different kinetic energies, but the average kinetic energy remains the same at a given temperature, i.e. average kinetic energy of the molecules is directly proportional to the absolute temperature.

When the temperature of a gas is raised, the molecules start moving faster, i.e. their individual velocities increase and hence the kinetic energy increases.

17.3 KINETIC GAS EQUATION (Pressure of a gas as derived from Kinetic Theory)

Gas pressure can be calculated on the basis of postulates of kinetic theory by considering the change of momentum resulting from the impacts of the molecules on the walls of the container. Though this is a simple method and give approximately correct pressure but it has a drawback as it gives a wrong estimate of number of collisions with the walls of the vessel.

Imagine a definite mass of a gas contained in a cubical vessel. Let-
Length of the cube edge = l cm.
Number of molecules present in the cube = N
Mass of each molecule = m gm.
Velocity of the molecule under consideration = u cm/s.

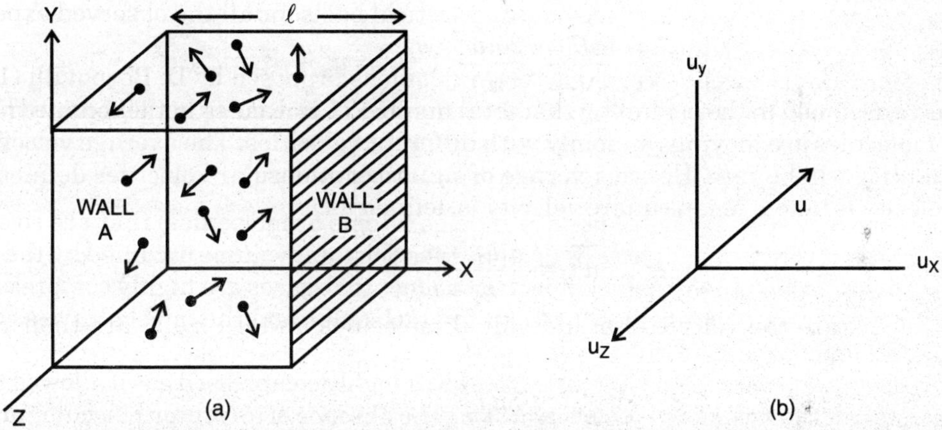

Fig. 17.5 *(a) Gas Molecules Enclosed in a Cubical Container (b) Resolution of Velocity component*

Velocity u of the molecule can be resolved into three components u_x, u_y, u_z which are mutually perpendicular to each other and parallel to the edges of the cube. The components are related to each other by the expression:

$$u^2 = u_x^2 + u_y^2 + u_z^2 \quad ...(34)$$

Each of the component of the velocity can be treated for a single molecule of mass m. Let us consider the velocity component along x-axis i.e. u_x, striking the wall B which is perpendicular to its motion. The molecule strikes the wall and bounce back and collide with the wall A subsequently. It is the total effect of such collisions with the walls of the container that produces the gas pressure.

The momentum of the molecule before collision with the face A

$$= mu_x \quad ...(35)$$

As the collision is perfectly elastic, only the direction is reversed,

∴ the momentum of the molecule after the collision = – mu_x. ...(36)

Change of momentum per collision= $mu_x - (-mu_x) = 2mu_x$. ...(37)

Before the molecule can again strike the same wall B, it has to travel across the cube, strike the wall A and rebound to hit face B. Thus the molecule must travel a distance of $2l$ between two successive collisions with the wall A.

$\therefore$ Number of collisions per second with wall A by the molecule

$$= \frac{u_x}{2l} \quad \text{...(38)}$$

Hence, number of collisions per second with the two opposite walls (A and B)

$$\frac{1}{t} = 2 \times \frac{u_x}{2l} = \frac{u_x}{l}. \quad \text{...(39)}$$

Change in momentum per unit time is given by:

$$\frac{\Delta P_x}{\Delta t} = 2mu_x \times \frac{u_x}{l} = \frac{2mu_x^2}{l} \quad \left(\frac{1}{t} = \frac{u_x}{l}\right) \quad \text{...(40)}$$

Similarly, the total change in momentum per molecule per second due to collisions on all the six walls is given by:

$$\frac{\Delta P}{\Delta t} = \frac{2mu_x^2}{l} + \frac{2mu_y^2}{l} + \frac{2mu_z^2}{l}$$

$$= \frac{2m}{l}(u_x^2 + u_y^2 + u_z^2)$$

$$\frac{\Delta P}{\Delta t} = \frac{2mu^2}{l} \quad (\because u^2 = u_x^2 + u_y^2 + u_x^2) \quad \text{...(41)}$$

We have assumed in the beginning that total number of molecules in the container is N. Now all these molecules are moving randomly with different velocities. The average velocity being a vector quantity will be zero. Hence, average of square of molecular velocities defined as mean square velocity is taken. Mean square velocity is defined as:

$$\overline{u^2} = \frac{n_1u_1^2 + n_2u_2^2 + ...}{n_1 + n_2 + ...} \quad \text{...(42)}$$

where u_1, u_2 .... are the velocities of individual molecules and n_1, n_2 .... are their respective numbers.

$$\overline{u^2} = \frac{\Sigma n_i u_i^2}{\Sigma n_i} \quad \text{...(43)}$$

$$\overline{u^2} = \frac{\Sigma n_i u_i^2}{N} \quad \text{...(44)}$$

The total change in momentum per unit time for all the N molecules in the container is given by considering average velocity of all the molecules. Thus

$$\frac{\Delta P}{\Delta t} = \frac{2m N\overline{u}^2}{l} \quad \text{...(45)}$$

Where u^{-2} is the mean square velocity.

It is known from mechanics that the force acting on a body is equal to the rate of change of momentum. Moreover, pressure is defined as force acting on unit area.

$$\therefore \quad \text{Pressure} = \frac{\text{Force}}{\text{Area}} = \frac{\text{Change of momentum per unit time}}{\text{Area}}$$

$$= \frac{\Delta P/\Delta t}{\text{Area}} \quad \text{...(46)}$$

substituting the value of $\Delta P / \Delta t$, from equation (45),

$$\therefore \quad \text{Pressure} = \frac{2mN\bar{u}^2}{l \times 6l^2} = \frac{1}{3}\frac{mN\bar{u}^2}{V} \quad ...(47)$$

$$\boxed{PV = \frac{1}{3}mN\bar{u}^2} \quad ...(48)$$

This is known as the Kinetic Gas Equation.

17.4 KINETIC GAS EQAUTION AND GAS LAWS

All the gas laws stated earlier can be derived using Kinetic Gas Equation.

Boyle's Law: According to Kinetic Gas Equation (48).

$$PV = \frac{1}{3}mN\bar{u}^2$$

$$= \frac{2}{3}.\frac{1}{2}mN\bar{u}^2$$

$$= \frac{2}{3}.\frac{1}{2}M\bar{u}^2 \quad ...(49)$$

where m N = M, the total mass of the gas. If a definite mass is under consideration then this remains constant.

$$\therefore \quad PV = \frac{2}{3}.K.E \quad \left(\because K.E = \frac{1}{2}M\bar{u}^2\right) \quad ...(50)$$

where KE is the total Kinetic energy of gas molecules.

But according to Kinetic Theory of Gases,

$$K.E \propto T$$

$$K.E = KT$$

where K is the proportionality constant. Thus equation (50) becomes

$$PV = \frac{2}{3}KT \quad ...(51)$$

Now, $\frac{2}{3}$ is a constant quantity, K is also constant, hence PV = Constant, if temperature is kept constant. This is the expression for Boyle's law.

Charles' Law: From equation (51), we have,

$$PV = \frac{2}{3}KT$$

$$\frac{V}{T} = \frac{2}{3}\frac{K}{P} \quad ...(52)$$

$\frac{2}{3}$ is a constant quantity, K is also a constant. Thus if pressure is also kept constant, the equation becomes,

$$\frac{V}{T} = \text{Constant (At constant pressure)} \quad ...(53)$$

or

$$V \propto T \quad ...(54)$$

Thus at constant pressure the volume of a given mass of a gas is directly proportional to its absolute temperature. This is the statement of Charles' Law.

Avogadro's Law: Let us consider two gases 1 and 2. From kinetic gas equation we have,

$$P_1V_1 = \frac{1}{3} m_1 N_1 \bar{u}_1^2 \qquad ...(55)$$

where P_1, V_1, m_1, n_1 and $\bar{u}_1^2$ are the pressure, volume, mass, number of molecules and root mean square velocity of the molecules of the first gas respectively. Also,

$$= \frac{1}{3} m_2 N_2 \bar{u}_2^2 \qquad ...(56)$$

where P_2, V_2, m_2, n_2 and $\bar{u}_2^2$ are the pressure, volume, mass, number of molecules and root mean square velocity of the molecules of the second gas respectively.

When pressure and volumes of the two gases are same i.e $P_1 = P_2$, $V_1 = V_2$, then it follows;

$$\frac{2}{3} \cdot \frac{1}{2} m_1 N_1 \bar{u}_1^2 = \frac{2}{3} \cdot \frac{1}{2} m_2 N_2 \bar{u}_2^2 \qquad ...(57)$$

Also, K.E of a gas α absolute temperature.

If the temperature is kept constant then mean molecular kinetic energy of each gas is the same.

$$\frac{1}{2} m_1 \bar{u}_1^2 = \frac{1}{2} m_2 \bar{u}_2^2 \qquad ...(58)$$

Dividing equation (57) by equation (58) we get,

$$N_1 = N_2 \qquad ...(59)$$

Thus equal volumes of all gases under the same conditions of temperature and pressure contain equal number of molecules. This generalization is known as Avogadro's law.

Graham's Law of Diffusion:m Since from equation (48) we have,

$$PV = \frac{1}{3} m N \bar{u}^2$$

$$\bar{u} = \frac{1}{2} mN\bar{u}^2 \qquad ...(60)$$

$$\bar{u} = \sqrt{\frac{3P}{d}} \qquad ...(61)$$

$$\left(\frac{\text{mN}}{\text{V}} = \frac{\text{Total mass}}{\text{Volume}} = \text{density (d)} \right)$$

Where d represents density

Now, the rate of diffusion (r) of the gas, depends upon the mean velocity of the gas.

$$r \ \alpha \ \bar{u}$$

$$\alpha \ \sqrt{\frac{3P}{d}} \qquad ...(62)$$

At constant pressure

$$r \ \alpha \ \sqrt{\frac{1}{d}} \qquad ...(63)$$

Hence, the rate of diffusion of a gas is inversely proportional to the square root of the density of the gas if pressure is kept constant. This is Graham's law of diffusion.

Dalton's Law of Partial Pressures: Let us again consider two gases 1 and 2. According to Kinetic Gas Equation,

$$PV = \frac{1}{3} mN\bar{u}^2$$

If only the first gas is enclosed in the vessel of volume V, the pressure exerted by the first gas would be,

$$p_1 = \frac{1}{3}\frac{m_1 N_1 \bar{u}_1^2}{V} \quad \text{...(64)}$$

If only the second gas is enclosed in the same vessel the pressure exerted by the second gas would be,

$$p_2 = \frac{1}{3}\frac{m_2 N_2 \bar{u}_2^2}{V} \quad \text{...(65)}$$

Now if both the gases are enclosed in the same vessel then, according to kinetic theory, the molecules behave independently of each other. Hence, the total pressure exerted would be

$$\text{P} = \frac{1}{3}\frac{m_1 N_1 \bar{u}_1^2}{V} + \frac{1}{3}\frac{m_2 N_2 \bar{u}_2^2}{V}$$

$$= p_1 + p_2 \quad \text{...(66)}$$

Similarly if more than two gases are presents.

$$\text{P} = p_1 + p_2 + p_3 \ldots \quad \text{...(67)}$$

The above expression is Dalton's law of partial pressures.

17.5 SOME OTHER EXPRESSIONS DERIVED ON THE BASIS OF KINETIC GAS EQUATION

Root mean square velocity: The term $\bar{u}^2$ in kinetic gas equation represents mean square velocity. It may be recalled that $\bar{u}^2$ is given by,

$$\bar{u}^2 = \frac{\Sigma n_i \mu_i^2}{N}$$

From kinetic gas equation (48)

$$\bar{u}^2 = \frac{1}{3} m N \bar{u}^2$$

$$\bar{u}^2 = \frac{3PV}{mN}$$

$$\bar{u} = \sqrt{\frac{3PV}{mN}} \quad \text{...(68)}$$

$\because \quad mN = M$ total mass of the gas

$$\therefore \quad \boxed{\bar{u} = \sqrt{\frac{3PV}{M}}} \quad \text{...(69)}$$

$\bar{u}$ is known as root mean square velocity.

Further, for 1 mole of the gas

$$PV = RT$$

$$\bar{u} = \sqrt{\frac{3RT}{M}} \qquad ...(70)$$

Thus root mean square velocity is directly proportional to the square root of temperature and inversely proportional to the square root of molar mass.

Average Kinetic Energy: The kinetic gas equation for one mole of a gas can be written as

$$PV = \frac{1}{3} m N_o \bar{u}^2$$

where N_o represents the Avogadro's number. The above equation can be written as.

$$PV = \frac{2}{3} \cdot \frac{1}{2} m N_o \bar{u}^2 = \frac{2}{3} . K.E_t \qquad ...(71)$$

where $K.E_t$ is the transational kinetic energy of one mole of the gas. Also for one mole

$$PV = RT \qquad ...(72)$$

$$\frac{2}{3} KE_t = RT \qquad ...(73)$$

From the above equation it may be concluded that the total translational kinetic energy of the molecule of an ideal gas is directly proportional to absolute temperature. The KE_t is independent of the volume, pressure and molar mass of the gas. It only depends upon the absolute temperature (T).

The average kinetic energy per molecule of an ideal gas will be given by:

$$\bar{e} = \frac{KE_t}{N_o} = \frac{3}{2} \frac{RT}{N_o} \qquad ...(74)$$

$$= \frac{3}{2} kT \qquad ...(75)$$

where $k = \frac{R}{N_o}$ is the Boltzmann constant whose numerical volume is given by:

$$k = \frac{R}{N_o} = \frac{8.314 \times 10^7 \text{ erg K}^{-1} \text{ mol}^{-1}}{6.023 \times 10^{23} \text{ mol}^{-1}}$$

$$= 1.38 \times 10^{-16} \text{ erg K}^{-1}$$

$$= 1.38 \times 10^{-23} \text{ J K}^{-1}$$

17.6 IDEAL AND REAL GASES: DEVIATIONS FROM IDEAL BEHAVIOUR

An ideal gas is the one which obeys the gas laws or the Kinetic Gas Equation under all conditions of temperatures and pressures.

The behaviour of gases has so far been presumed to follow the ideal gas equation, $PV = nRT$. Such behaviour is stated as ideal or perfect. This is the reason that the gas equation is often referred to as ideal gas equation. But it has been found that no real gas obeys the ideal gas equation under all conditions of temperatures and pressures. These gases are thus termed as **nonideal gases**. Deviations from the ideal gas equation are pronounced at high pressure and low temperatures.

The deviations of a real gas from the ideal gas behaviour can be depicted in terms of a new quantity 'Z' called compressibility factor. For an ideal gas:

$$PV = nRT$$

$$\frac{PV}{nRT} = Z$$

or $$\frac{PV_{obs}}{PV_{ideal}} = Z \quad ...(76)$$

The quantity $\frac{PV}{nRT}$ is called compressibility factor. For ideal gas Z = 1 under all conditions of temperatures and pressures. Deviation of compressibility factor from unity determines the departure of a real gas from ideal behaviour.

The quantity Z can exhibit both positive and negative deviations from unity. Some of the compressibility curves exhibiting non ideal behaviour are discussed below.

Effect of Pressure on Deviations: Fig. 17.6 represents the plots of compressibility factor Z, plotted against pressures for different gases at constant temperature. The dotted line represents the behaviour of an ideal gas, for which Z is always = 1.

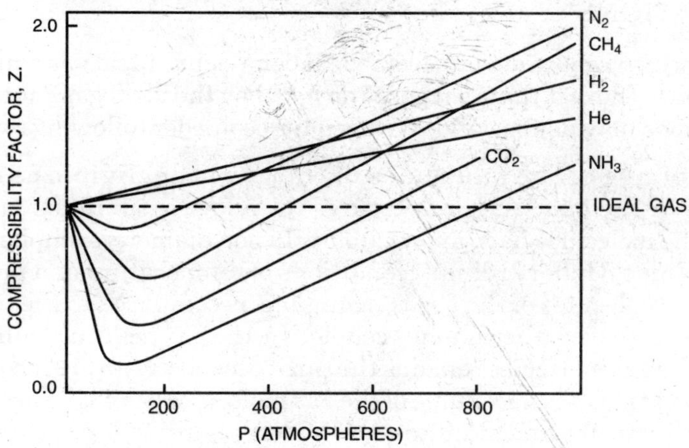

Fig. 17.6 Deviations of real gases from ideal behaviour

It is seen from the graph that at very low pressures, real gases exhibit nearly ideal behaviour. As the pressure is increased, considerable deviations set in.

- For hydrogen and helium Z>1, hence these gases are less compressible than the ideal gas at all pressures.
- For NH_3, CO_2, N_2 Z<1 at low pressures and Z>1 at high pressures. Thus NH_3, CO_2 and N_2 are more compressible at low pressures than expected on the bases of ideal behaviour and are less compressible than ideal gas at high pressures.

Effect of Temperature on deviations: Plots of Z versus pressure are represented in Figure 17.7 for nitrogen gas at different temperatures. From the graph it is seen that the deviations from ideal gas behaviour becomes smaller as the temperature is increased. At lower temperature, the dip in the curve is large. As the temperature is increased the dip becomes smaller and smaller and eventually disappears. At higher temperature of about 323.16 K, the curve seems to be almost horizontal for appreciable range of pressures. At this temperature, compressibility factor is almost unity and product PV remains approximately constant. Thus, Boyle's law is obeyed within this range of pressure, hence this temperature for the gas is called *Boyle temperature*. It is characteristic of each gas. For N_2 it is 332 K.

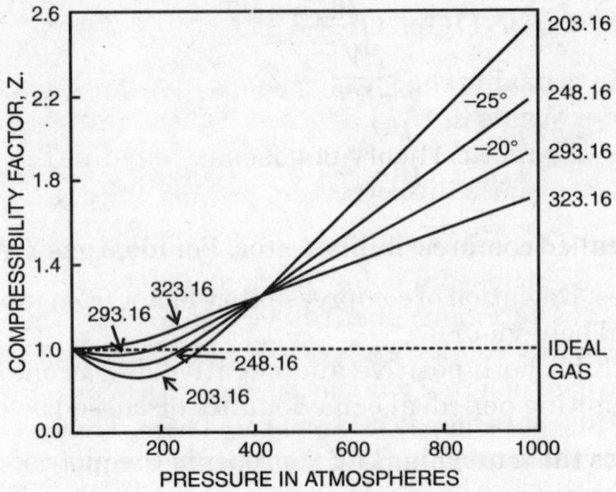

Fig. 17.7 Effect of temperature on deviation of nitrogen from ideal behaviour

CAUSES OF DEVIATIONS

Now the question is, why do gases deviate from ideal behaviour? Ideal gas equation is based on a number of assumptions. The fact that real gases do not obey the ideal gas equation indicates that some of the assumptions may not be valid. We need to reconsider following two assumptions.

(i) The actual volume of the molecules of the gas is negligible as compared to the total volume of the gas: At low pressures the above assumption may appear correct. The fact that the molecules have appreciable collision diameters, implies that they possess an effective volume. Under ordinary conditions of temperature and pressure the volume occupied by the molecules of the gas is negligible as compared to the total volume. But if the pressure is increased, the volume occupied by the gas molecule will decrease whereas the volume of the gas molecules remains the same because these are incompressible. Hence, under these conditions actual volume of the gas molecules cannot be neglected. Further, by increasing the pressure and simultaneously decreasing the temperature, gases can be liquified and eventually solidified. The solids are incompressible which further confirms the fact that molecules of a gas occupy definite volume which corresponds to the volume that these molecules will occupy when it solidifies.

(ii) There are no intermolecular forces of attraction between the molecules: This assumption of kinetic molecular theory is also not correct. The very fact that the gases can be converted to liquids under the conditions of low temperature and high pressures, indicates the presence of attractive forces between the molecules. These attractive forces are short range forces, since at low pressures the distance between the molecules is large enough and hence the real gases approach ideal behaviour. The most convincing experimental evidence for the existence of attractive forces comes from Joule Thomson effect. When a compressed gas is allowed to expand under adiabatic conditions, cooling takes place. This is known as Joule Thomson effect. The fall in temperature is due to the work done by the gas during expansion to **overcome inter molecular forces of attraction.** The energy for this work is taken from the kinetic energy of the molecules. Thus, the decrease in kinetic energy of the molecules results in decrease in temperature.

Hence this assumption also does not hold under condition of high pressure and low temperature.

From the above discussion it is clear that suitable corrections to ideal gas equation has to be made to make it applicable for real gases.

17.7 VAN DER WAALS' EQUATION OF STATE

Several attempts were made to modify the ideal gas equation. An approximate equation of state was suggested by J.H. van der Waals in 1873. The systematic and scientific approach to correct the erroneous assumptions of Kinetic Theory of gases resulted in the famous van der Waals' equation for real gases. Let us discuss the corrections one by one.

(i) **Volume Correction:** The volume of a gas is considered to be equal to the volume of the container in which it is enclosed, i.e. for an ideal gas, volume V represents the volume where the gas molecules can move about freely. In an ideal gas, it is assumed that the molecules do not occupy any volume and entire space in the container is available for their movement. But in the case of a real gas the gas molecules have finite volumes, thus a part of the free space in the container is occupied by them. Hence the compressible or free volume for a real gas is the total volume of the container minus the volume occupied by the gas molecules.

If b represents the effective volume of molecules occupied by 1 mole of a gas then this volume b must be subtracted from the total volume to get the actual volume available for free movement of gas molecules. **Thus the volume in the ideal gas equation is corrected as (V–b)**

For n moles, the corrected volume is $(V-nb)$ (77)

b is the excluded volume of the gas. Its value is constant and characteristic for each gas.

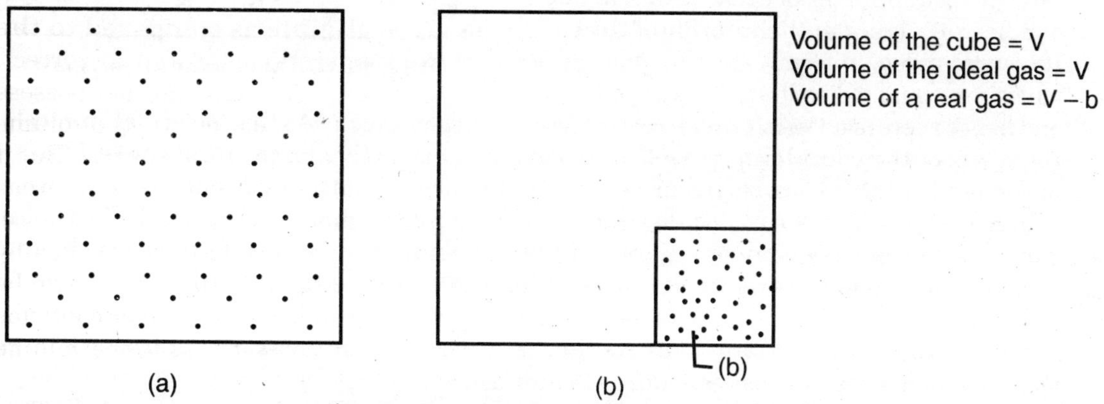

Fig. 17.8 Volume of a Real Gas

It can be shown that excluded volume is not equal to the actual volume of the gas molecules, but four times the actual volume of the gas molecules.

Consider two molecules represented by incompressible spheres of radius r, undergoing bimolecular collisions. Let the collision diameter of the gas molecules represented by σ which is equal to 2r. The space indicated by the dotted sphere will not be available to all other molecules of the gas. In other words, the dotted spherical space is excluded volume per pair of molecules. Thus

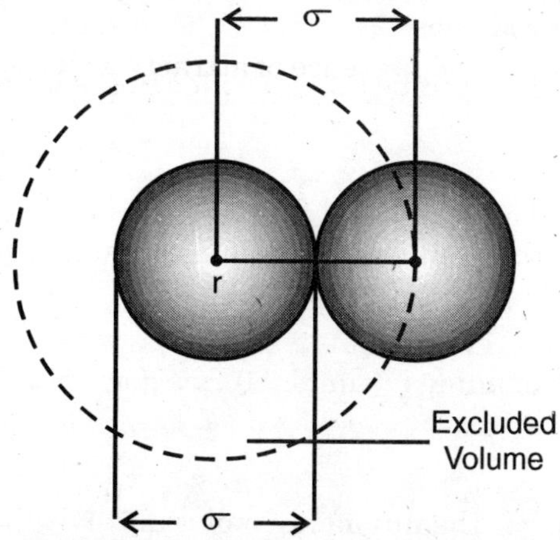

Fig. 17.9 Excluded Volume per pair of molecules

Excluded volume per pair of molecules

$$= \frac{4}{3}\pi\sigma^3 = \frac{4}{3}\pi(2r)^3 = 8\left(\frac{4}{3}\pi r^3\right)$$

Excluded volume per molecule

$$= \frac{1}{2}\times 8\left(\frac{4}{3}\pi r^3\right) = 4\left(\frac{4}{3}\pi r^3\right) \quad ....(78)$$

$$= 4 \text{ (volume occupied by one molecule)}$$

Now b represents total excluded volume per mole of the gas thus,

$$b = N_o\left[4\left(\frac{4}{3}\pi r^3\right)\right]$$

where N_o is the Avogadro's number.

(ii) **Pressure Correction:** As the presence of intermolecular forces is established, allowance should be made for these forces of attraction in the ideal gas equation, which is based on the wrong assumption that there are no intermolecular forces of attraction. In order to take into account the effect of intermolecular forces of attraction, consider a molecule x lying in the interior of the gas. It is surrounded uniformly by other molecules which are equally distributed in all directions and hence there is no resultant attractive force on this molecule. As this molecule approaches the walls of the containing vessel, it experiences attractive forces from the bulk of the molecules behind it because the gas molecules are present on one side only. In other words a force is exerted on the molecule which tends to pull the molecules inward. (Fig. 17.10) Now the molecules which strike the wall of the container contribute towards the total pressure of the gas. Therefore, it appears that the molecule striking the wall, due to the inward pull, will strike it with lower velocity and will exert a lower pressure as it would have, had there been no force of attraction. As the result the measured pressure P is less than the ideal pressure corresponding to an ideal gas. Therefore a certain term 'p' should be added to the measured pressure. Thus ideal pressure will be P + p.

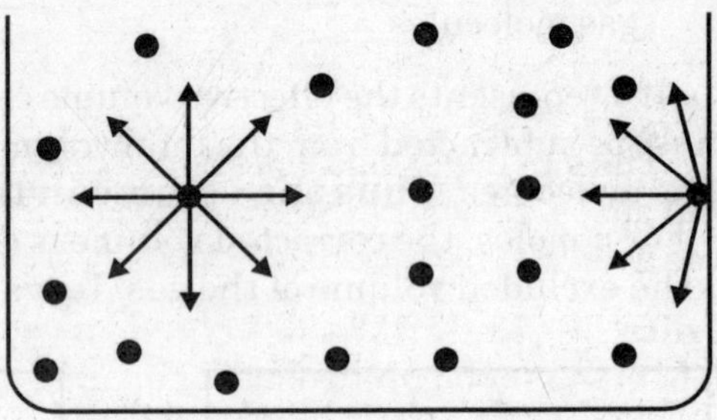

Fig. 17.10 Molecular attraction

The dragging force or inward pull experienced by the molecules about to strike the wall is proportional to:

- The number of molecules in the bulk of the gas or the density of the gas, i.e.

$$p \propto \frac{n}{V} \quad ....(79)$$

- The number of molecules striking the wall per unit area, which inturn is proportional to the number of molecules per unit volume of the gas or its density.

$$p \propto \frac{n}{V} \quad ...(80)$$

Combining (79) and (80), we get,

$$p \propto \frac{n^2}{V^2} \quad ...(81)$$

where n is the number of moles and V is the volume occupied by the gas.

$$\therefore \qquad p = \frac{an^2}{V^2} \qquad ...(82)$$

where a is the constant of proportionality. Therefore the **corrected pressure** is:

$$P_{ideal} = P + \frac{an^2}{V^2} \qquad ...(83)$$

Substituting the values of P and V from equation (83) and (77) in ideal gas equation, PV = n RT we get,

$$\boxed{\left(P + \frac{an^2}{V^2}\right)(V - nb) = nRT} \qquad ...(84)$$

This equation is known as **van der Waals' equation**. The constants a and b are known as vander Waals' constants. For 1 mole of the gas the above equation can be written as:

$$\boxed{\left(P + \frac{a}{V^2}\right)(V - b) = RT} \qquad ...(85)$$

The van der Waals equation expresses the pressure, volume and temperature relationship in real gases. This is one of the best known equations of state for real gases. In 1910 van der Waals was awarded Nobel prize for physics for developing this equation.

Units of vander Waals' constants: The units for the van der Waals constants 'a' and 'b' depend upon the units used for expressing pressure and volume.

The term $\frac{an^2}{V^2}$ is actually the pressure which has to be added to correct the observed pressure.

$$p = \frac{an^2}{V^2}$$

$$\therefore \qquad a = p\frac{V^2}{n^2} \qquad ...(86)$$

If pressure is in atmospheres, volume in litres then a is expressed in litre^2–atm.mol^{-2}.

In S.I units 'a' is expressed as.

$$a = \frac{(\text{Nm}^{-2})(\text{m}^3)^2}{(\text{mol})^2} = \text{Nm}^4\text{mol}^2$$

Relation between the two

$$\text{litre}^2\text{–atm. mol}^{-2} = \frac{(1.013 \times 10^5\ \text{Nm}^{-2})(10^{-3}\ \text{m}^3)^2}{(\text{mol})^2}$$

$$= 1.013 \times 10^{-1}\ \text{Nm}^4\text{mol}^{-2}$$

Similarly, it is evident from van der Waals' equation that b is actually the incompressible volume per mole of a gas. Hence, its units are:

$$b = \frac{\text{Volume}}{\text{mol}} = \text{litre mol}^{-1}$$

In S.I units the units of b are $m^3\ mol^{-1}$.

Relation between the two-

$$1\ \text{litre mol}^{-1} = 10^{-3}\ \text{m}^3\ \text{mol}^{-1}$$

van der Waals constants can be evaluated either from critical constants or from P–V–T relation. The van der Waals constant can also be calculated from the co-efficient of thermal expansion and of compression of the gas.

Validity of van der Waals' Equation: It is worthwhile to consider the behaviour of real gases and explain their behaviour on the basis of van der Waals' equation. It has been observed that results obtained from van der Waals equation are approximately in agreement with the experimental values. The departure of real gases from ideal behaviour at different conditions is explained below with the help of van der Waals equation.

1. **At moderate pressures:** van der Waals' equation for one mole of a gas is given by:

$$\left(P + \frac{a}{V^2}\right)(V - b) = RT$$

Now at moderate pressures, V is sufficiently large and hence the term b can be ignored in comparison and van der Waals' equation assumes the form:

$$\left(P + \frac{a}{V^2}\right)(V) = RT$$

$$PV = RT - \frac{a}{V} \quad \text{...(87)}$$

Hence, under the conditions of moderate pressures PV should be less than RT by a factor $\frac{a}{V}$. Dividing by RT, the compressibility factor Z comes out to be

$$Z = 1 - \frac{a}{VRT} \quad \text{...(88)}$$

From the equation (88) it is clear that when the pressure is low, Z is less than 1. The dip in the curve of PV against P is thus explained on the basis of above equation.

2. **At high pressures:** At high pressure the volume occupied by the gas is considerably small and hence the constant b cannot be ignored in comparison to volume occupied by the gas. But the term $\frac{a}{V^2}$ may be ignored and hence the equation (85) assumes the form:

$$P(V - b) = RT$$

$$PV = RT + Pb$$

$$Z = 1 + \frac{Pb}{RT} \quad \text{...(89)}$$

Now, PV is greater than RT and increases further on increasing pressure and $Z > 1$. Thus equation (89) explains the rise in the curve after passing through a minimum when the pressure is high in Fig. (17.6). This is what is expected also, since the forces of repulsion become appreciable when the molecules are close together at high pressures.

3. **At very low pressures and high temperatures:** At very low pressure the volume is very large and hence the terms $\frac{a}{V^2}$ and b are negligible in comparison to P and V. The equation (85) reduces to

$$PV = RT \quad \text{...(90)}$$

At high temperatures the volume occupied by the gas molecules will be quite large and thus the term a / V^2 will be small enough to be neglected. Moreover in comparison to V, the volume occupied by the gas molecules, the term b can also be neglected and hence equation (85) assumes the form $PV = RT$ i.e. ideal gas equation. Thus real gas tends to approach ideal gas equation at high temperature and low pressures.

4. **Behaviour of hydrogen and helium:** The attractive forces between hydrogen and helium are very small. Thus the correction term a/V^2 which is introduced due to the presence of intermolecular forces can be ignored. The equation (85) assumes the form :

$$P(V - b) = RT$$

$$PV = RT + Pb$$

Dividing by RT

$$Z = 1 + \frac{Pb}{RT} \quad ...(91)$$

It becomes clear from the above equation that Z is always greater than 1 and it increases with increase in pressure. The PV curve for hydrogen shows no dip at ordinary temperature (Fig. 17.6).

Problem: Explain why the value of a is larger for NH_3 as compared to N_2
Solution: The constant 'a' is related to intermolecular forces. In NH_3, due to hydrogen bonding these forces are higher and hence it has larger value for 'a'

Problem: Why the value of b is larger for N_2 as compared to NH_3
Solution: The constant b is related to molecular volume. In NH_3, the hydrogen atoms practically take up no volume and therefore the volume of NH_3 is less than that of N_2. Hence the value of constant b is more for N_2.

17.8 BOYLE TEMPERATURE

While discussing the effect of temperature on deviations from ideal behaviour, it was observed from the plot of Z versus P for nitrogen gas at different temperatures (Fig. 17.11) that, as the temperature is raised, dip in the curve decreases and the position of the minimum moves to left. At a certain temperature the minimum in the curve vanishes and the curve remains horizontal over an appreciable range of pressure. Boyle's law is obeyed within this range.

The temperature at which the real gas obeys Boyle's law over an appreciable range of pressure is called Boyle point or **Boyle temperature**. This temperature is characteristic of each gas.

It is further clear from Fig. 17.11 that below Boyle temperature the value of Z decreases to a minimum with increase in pressure and then increases. Above Boyle temperature the value of Z increases continuously.

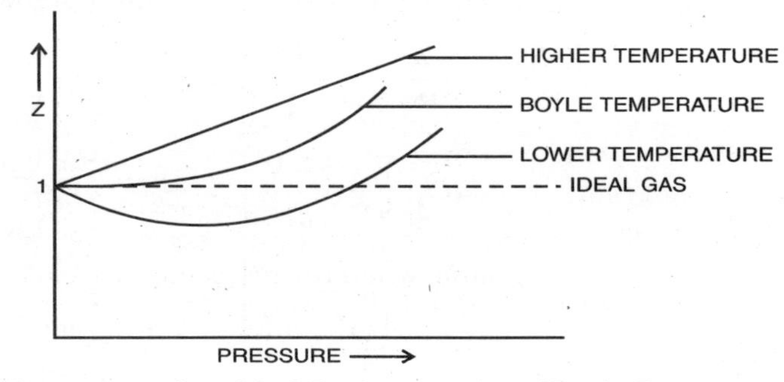

Fig. 17.11 Graphical Representation of Boyle Temperature

It is also observed from the Fig. 17.11 that at Boyle temperature the curve after remaining horizontal for appreciable range of the pressure, rises very slowly above the ideal gas curve.

The Boyle temperatures for hydrogen and helium are very low i.e. –165° and –240°C respectively. Thus they have Z values greater than unity when are studied under ordinary experimental conditions. Nitrogen, carbondioxide etc have comparitively high Boyle temperatures and hence a dip below this temperature is observed at low pressure range.

The Boyle temperature (T_B) of a gas can be readily obtained using van der Waals' equation.

Calculation of Boyle Temperature: It is seen from Fig. 17.11 that minima on the curves showing variation of Z with P lie on a parabolic curve. As the temperature is raised we get a point at which minimum occurs at zero pressure, theoretically. At still higher temperatures no minimum is observed. Above a certain temperature (Boyle temperature) any gas will behave in the same manner as hydrogen and helium do at ordinary temperatures.

The condition for the PV to be minimum when $P \rightarrow O$ i.e. for the Boyle temperature can be obtained from van der Waals equation.

According to van der Waals' equation for 1 mol of a gas:

$$\left(P + \frac{a}{V^2}\right)(V - b) = RT$$

$$P = \frac{RT}{V-b} - \frac{a}{V^2} \quad \text{...(92)}$$

Multiplying throughout by V we get-

$$PV = \frac{RT(V)}{V-b} - \frac{a}{V} \quad \text{...(93)}$$

Differentiating the above equation w.r.t. P at constant temperature, we get-

$$\left[\frac{\partial(PV)}{\partial P}\right]_T = RT\left\{\frac{1}{V-b}\left(\frac{\partial V}{\partial P}\right)_T + V\frac{-1}{(V-b)^2}\left[\frac{\partial V}{\partial P}\right]_T\right\} + \frac{a}{V^2}\left[\frac{\partial V}{\partial P}\right]_T$$

$$= RT\left[\frac{1}{V-b} - \frac{V}{(V-b)^2}\right]\left[\frac{\partial V}{\partial P}\right]_T + \frac{a}{V^2}\left[\frac{\partial V}{\partial P}\right]_T$$

$$= \left[\frac{RT}{V-b} - \frac{RTV}{(V-b)^2} + \frac{a}{V^2}\right]\left(\frac{\partial V}{\partial P}\right)_T \quad \text{...(94)}$$

But the condition for a minimum in the curve of PV against P is that $\partial(PV)/\partial P$ should be equal to zero. $(\partial V/\partial P)_T$ is always negative and hence cannot be equal to zero. Hence the expression within square brackets on right hand side is equal to zero. Thus

$$\left[\frac{RT}{V-b} - \frac{RTV}{(V-b)^2} + \frac{a}{V^2}\right] = 0 \text{ or } RT\left[\frac{1}{V-b} - \frac{V}{(V-b)^2}\right] = \frac{-a}{V^2} \quad \text{...(95)}$$

or

$$RT\left[\frac{V-b-V}{(V-b)^2}\right] = \frac{-a}{V^2}$$

or

$$RT\left[\frac{-b}{(V-b)^2}\right] = \frac{-a}{V^2}$$

$$RT = \frac{a}{b}\left(\frac{V-b}{V}\right)^2 \quad \text{...(96)}$$

For the minimum to occur when P is 0 i.e at Boyle temperature $T = T_B$, V will be infinite and the above expression reduces to

$$RT_B = \frac{a}{b}$$

$$\boxed{T_B = \frac{a}{Rb}}$$

The volume of Boyle temperature T_B calculated from the above formula are in good agreement with experimental values for a number of gases.

Alternatively Vander Waals' equation can be expressed in the Virial form as a power series in $1/Vm$ and an expression for the Boyle Temperature can be obtained there of.

The Virial Equation of State: This is the most general equation of state for real gases and was proposed by Kammerlingh Onnes. This equation expresses *PV* as *a* power series of pressure P at any given temperature and is written as:

$$PV = A + BP + CP^2 + DP^3 \ldots \qquad \ldots(97)$$

Where *P* is the pressure in atmosphere and *V* is the molar volume in litres. The coefficients A, B, C etc are known as first, second, third ...... virial coefficients. The virial coefficients are different for different gases. At very low pressures, the first virial coefficient is significant and is equal to RT, i.e. A = RT and the equation reduces to PV = RT. However at higher pressures the other virial co-efficients also become important and must be considered. The value of virial coefficients change with change of temperature.

The first virial coefficient A is always positive and increases with increase in temperature. Whereas second virial coefficient B is negative at low temperature and hence on increasing the temperature it increases to zero and then become more and more positive. The temperature at which second virial coefficient is zero is called **Boyle temperature**.

The Virial equation can also be expressed by expressing the compressibility factor *Z* as a power series in $\frac{1}{V}$ i.e.

$$Z = \frac{PV}{RT} = 1 + \frac{B}{V} + \frac{C}{V2} + \ldots \qquad \ldots(98)$$

or

$$PV = RT\left[1 + \frac{B}{V} + \frac{C}{V^2} + \ldots\right] \qquad \ldots(99)$$

Any equation can be expressed into virial form.

VAN DER WAALS EQUATION IN VIRIAL FORM AND DETERMINATION OF BOYLE TEMPERATURE.

van der Waals equation for 1 mole of a gas is given by

$$\left(P + \frac{a}{V^2}\right)(V - b) = RT$$

$$P = \left(\frac{RT}{V - b}\right) - \left(\frac{a}{V^2}\right)$$

Multiplying the above equation by V

$$PV = \left(\frac{RTV}{V - b}\right) - \frac{a}{V} \qquad \ldots(100)$$

Dividing by RT we get,

$$\frac{PV}{RT} = \left(\frac{V}{V - b}\right) - \frac{a}{V\,RT}$$

$$Z = \left(1 - \frac{b}{V}\right)^{-1} - \frac{a}{V\,RT} \qquad \ldots(101)$$

At very low pressures V>>b and hence $\frac{b}{V}$ is less than unity, i.e. b/V<<1. Thus the expression $\left(1-\frac{b}{V}\right)^{-1}$ can be expanded to a power series in b/V by making use of binomial theorem :

$$Z = \left[1+\frac{b}{V}+\left(\frac{b}{V}\right)^2+\left(\frac{b}{V}\right)^3+\ldots\ldots\right]-\frac{a}{VRT} \quad \ldots(102)$$

$$= 1+\frac{1}{V}\left(b-\frac{a}{RT}\right)+\left(\frac{b}{V}\right)^2+\left(\frac{b}{V}\right)^3+\ldots\ldots \quad \ldots(103)$$

V may be taken approximately $= \frac{RT}{P}$. Putting the value to the above equation :

$$Z = 1+\frac{1}{RT}\left(b-\frac{a}{RT}\right)P+\left(\frac{b}{RT}\right)^2 P^2+\ldots\ldots \quad \ldots(104)$$

The slope of Z versus P curve is obtained by differentiating the above equation with respect to pressure, keeping the temperature constant, i.e.

$$\left(\frac{\partial Z}{\partial P}\right)_T = \frac{1}{RT}\left(b-\frac{a}{RT}\right)+2P\left(\frac{b}{RT}\right)^2+\ldots \quad \ldots(105)$$

At P = 0 all the higher terms drop out and the derivative reduces to-

$$\left(\frac{dZ}{\partial P}\right)_T = \frac{1}{RT}\left(b-\frac{1}{RT}\right) \quad \ldots(106)$$

This derivative is the initial slope of the plot of X versus P.

Now three cases arise–

Case 1: If $b > \frac{a}{RT}$

In this case the inital slope is positive and the size effect dominates the behaviour of gas.

Case 2: If $b > \frac{a}{RT}$

In this case the initial slope is negative and effect of intermolecular forces dominate the behaviour of gases.

As an interpretation of the (Fig. 17.6) one can say that at 0°C, the effect of attractive forces dominate initially the behaviour of CH_4 and CO_2, size effect dominates the behaviour of hydrogen and helium.

Similarly in the plots of Z versus P of a single gas at various temperatures (Fig. 17.11), one may interpret that is the temperature is low enough, the term a/RT will be larger than b and hence the inital slope will be negative. As the temperature is increased a/RT becomes smaller and eventually becomes less than b. At that temperature the initial slope of Z versus P curve becomes positive.

Case 3:
$$b = \frac{a}{RT_B}$$

or
$$b-\frac{a}{RT_B} = 0$$

At some intermediate temperature T_B, i.e. the Boyle temperature, the two effects, is finite size and the intermolecular attractions roughly compensate each other. At the Boyle temperature, initial slope of Z versus P plot is zero.

$$b = \frac{a}{RT_B}$$

$$T_B = \frac{a}{Rb}$$

Thus at Boyle temperature, the real gas shows ideal behaviour over a wide range of pressure because the effect of finite size of molecules and intermolecular forces roughly compensate each other.

NUMERICAL PROBLEMS BASED ON VAN DER WAAL'S EQUATION

Example 17.10. 2 moles of ammonia occupied a volume of 5 litres at 27° Calculate the pressure if the gas obeys van der Waals equation. (a = 4.17 atm L^2mol^{-2}; b = 0.03771 L mol^{-1})

Solution:

$$n = 2;\ V = 5 \text{ litres};\ T = 273 + 27 = 300 \text{ K}$$

Applying van der Waals' equation for n moles

$$\left(P + \frac{an^2}{V^2}\right)(V - nb) = nRT$$

$$\left(P + \frac{4.17 \times 2^2}{5^2}\right)[5 - (2 \times 0.371)] = 2 \times 0.082 \times 300$$

$$P = 9.33 \text{ atm.}$$

Example 17.11. Calculate the pressure exerted by 1.00 mole of methane (CH_4) in a 250 ml container at 300 K using van der Waal's equation. What pressure will be predicted by ideal gas equation? $a = 2.253$ L^2 atm mol^{-2}; $b = 0.0428$ L mol^{-1}; $R = 0.0821$L atm mol^{-1}K.

Solution:

$$\left(P + \frac{n_2 a}{V^2}\right)(V - nb) = nRT$$

$$P = \frac{nRT}{V - nb} - \frac{n^2 a}{V^2}$$

$$P = \frac{1 \times 0.0821 \times 300}{0.250 - (1 \times 0.0428)} - \frac{1^2 \times 2.253}{(0.250)^2}$$

$$= 82.8 \text{ atm}$$

If the gas obeys ideal gas equation, then the pressure would be –

$$P = \frac{nRT}{V} = \frac{1 \times 0.0821 \times 300}{0.250} = 98.5 \text{ atm.}$$

Example 17.12. van der Waals constant 'b' for a real gas is 0.02788 L mol^{-1}. Calculate the radius of the molecule of the gas.

Solution:

$$\text{Volume of one molecule} = \frac{4}{3}\pi r^3$$

$$\therefore \quad \text{Volume of 1 mole} = \frac{4}{3}\pi r^3 \times \text{Avogadro's No.}$$

$$b = 4\times\frac{4}{3}\pi r^3 \times \text{Avog.No.}$$

$$\text{Now} \quad b = 0.02788 \text{ L mol}^{-1}$$
$$= 27.88 \text{ mL mol}^{-1}$$

$$27.88 = 4\times\frac{4}{3}\times\frac{22}{7}\times r^3\times 6.022\times10^{23}$$

$$r^3 = \frac{27.88}{4}\times\frac{3}{4}\times\frac{7}{22\times6.022\times10^{23}}$$
$$r = 1.4\times10^{-8} \text{ cm.}$$

17.9 MAXWELL'S LAW OF DISTRIBUTION OF VELOCITIES

The molecules of a gas are continuously colliding with each other and with the walls of the container. As the collisions are supposed to be perfectly elastic there is no net loss of kinetic energy. However, as a result of impacts, the molecule of a gas cannot all have the same velocity. The speed of a molecule of a gas also changes continuously. Before a collision, a particular molecule may be travelling rapidly, but after the collision it may be slowed down. Thus it is very difficult to speak of the speed of an individual molecule. But one can only obtain a distribution of molecular speeds at a given temperature.

The manner in which the molecules of a gas are distributed over the possible speed ranges was first investigated by J.C. Maxwell. In 1886 Maxwell pointed out that inspite of the random motion of the molecules, at a particular temperature, the fraction of the molecules possessing particular velocities remain almost constant. Such a state is called **steady state**. The term steady state does not mean that all the velocities of individual molecules remain unchanged. It only implies that if some molecules change their velocities due to collisions, an equal number of other molecules acquire their velocities which they were possessing prior to collisions. Thus the distribution of velocities remains unchanged. James Clark Maxwell calculated the distribution of velocities from the laws of probability. He derived the following equation for the distribution of molecular velocities.

$$\frac{dN_C}{N} = 4\pi\left(\frac{m}{2\pi KT}\right)^{3/2} c^2\, e^{-mc^2/2kT}\, dc \qquad \text{...(107)}$$

where dN_C = Number of molecules having velocities between c and $(c+dc)$

N = total number of molecules

m = mass of a molecule

T = Absolute temperature

k = Boltzman constant.

Consider a molecule with velocity c, having its components along three axes at right angles u, v and w, so that

$$c^2 = u^2 + v^2 + w^2 \qquad \text{...(108)}$$

Let us assume that the values of u, v and w for any molecule are completely independent of each other. Thus the probability that the u component lies between u and $u+du$ is a function of u alone, i.e. $f(u)\,du$. The kinetic energy ε associated with this velocity component is given by $\frac{1}{2}m\mu^2$.

In the 19th century, Boltzmann had shown that the probability for a molecule to have an energy E was proportional to $e^{-\varepsilon/kT}$. We can equate this probability with $f(u)$. Thus

$$f(u) \propto e^{-\varepsilon/kT} \qquad ...(109)$$

$$f(u)du = Ae^{-\varepsilon/kT}du \qquad ...(110)$$

$$= Ae^{-mu^2/2kT}du \quad (\because \varepsilon = \frac{1}{2}mu^2) \qquad ...(111)$$

where A is the proportionality constant.

The constant A can be evaluated. Since the velocity components can have all values between $+\infty$ and $-\infty$, it follows that the probability of a value of u between these limits must be unity. In other words, the integral of $f(u)\,du$ between $+\infty$ and $-\infty$ must be equal to unity, i.e.

$$\int_{-\infty}^{+\infty} f(u)du = \int_{-\infty}^{+\infty} Ae^{-mu^2/2kT}du = 1 \qquad ...(112)$$

$$= A\int_{-\infty}^{+\infty} e^{-mu^2/2kT}du = 1 \qquad ...(113)$$

Put $m/2kT = a$ in equation (113)

$$A\int_{-\infty}^{+\infty} e^{-au^2}\,du = 1 \qquad ...(114)$$

$$A\left(\frac{\pi}{a}\right)^{\frac{1}{2}} = 1 \qquad \left(\because \int_{-\infty}^{+\infty} e^{-au^2}\,du = \left(\frac{\pi}{a}\right)^{\frac{1}{2}}\right) \qquad ...(115)$$

$$A = \left(\frac{a}{\pi}\right)^{\frac{1}{2}} \qquad ...(116)$$

$$= \left(\frac{m}{2\pi kT}\right)^{\frac{1}{2}} \qquad \because a = m/2kT \qquad ...(117)$$

Substituting the value of A from equation (117) in equation (111)

$$\boxed{f(u)du = \left(\frac{m}{2pkT}\right)^{\frac{1}{2}} e^{-mu^2/2kT}du} \qquad ...(118)$$

This is the Maxwell's distribution of molecular velocities in one dimension. Similarly, expression for Maxwells distribution of molecular velocities in three dimensions can also be derived.

The probability that the three components of the molecular velocity shall simultaneously lie between u and $u+du$, between v and $v+dv$, and w and $w+dw$ respectively is given by:

$$f(u)\,f(v)\,f(w)\,du\,dv\,dw = f(u)\,du\,f(v)\,dv\,f(w)\,dw. \qquad ...(119)$$

$$= \left(\frac{m}{2\pi kT}\right)^{3/2} e^{-m(u^2+v^2+w^2)/2kT}du\,dv\,dw$$

$$= \left(\frac{m}{2\pi kT}\right)^{3/2} e^{-mc^2/2kT}du\,dv\,dw \qquad ...(120)$$

The expression on the left hand side of this equation gives the fraction of molecules in the speed range c and $c+dc$ per unit interval of speed and is thus the probability of finding the molecules having the speeds between c and $c+dc$ regardless of direction. These are the molecules having their velocity points lying within a spherical shell of thickness dc at a distance c. The volume elements $du\ dv\ dw$ would then be equal to the volume of the shell having thickness dc. This shell has the volume $4\pi c^2 dc$. Hence, we can write:

$$\text{Volume of outer sphere} = \frac{4}{3}\pi(c+dc)^3$$

$$\text{Volume of inner sphere} = \frac{4}{3}\pi c^3$$

$$\text{Volume of spherical shell} = \frac{4}{3}\pi(c+dc)^3 - \frac{4}{3}\pi c^3$$

$$= \frac{4}{3}\pi[c^3 + (dc)^3 + 3c(dc)^2 + 3c^2(dc)] - \frac{4}{3}\pi c^3$$

$$= 4\pi c^2 dc$$

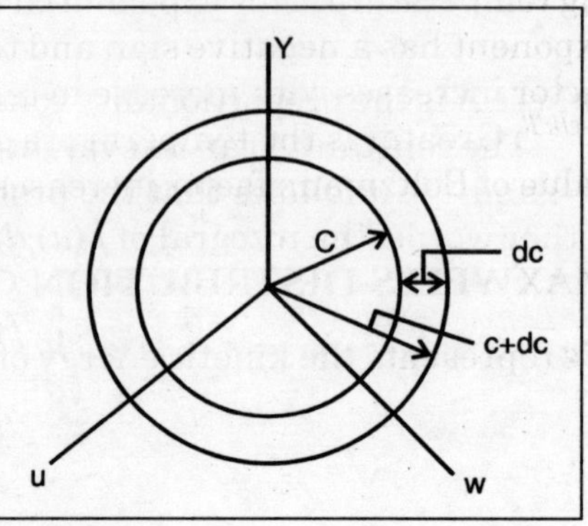

$$f(c)dc = 4\pi\left(\frac{m}{2\pi kT}\right)^{3/2} c^2 e^{-mc^2/2kT} dc \qquad ....(121)$$

This expression is called the Maxwell distribution of speeds. The term $f(c)\,dc$ can also be written as dN_C/N. The quantity dN_c/N gives the fraction of molecules with speeds between c and $c+dc$. Moreover $m = M_m/N_o$ where M_m is the molar mass and N_o is the Avogadro's number. Thus equation (121) can be written as –

$$\frac{dN_c}{N} = 4\pi\left(\frac{M_m}{2\pi RT}\right)^{3/2} c^2\, e^{-M_m c^2/2RT}\, dc \qquad \left(\because k = \frac{R}{No}\right) \qquad ...(122)$$

Dividing equation (121) by dc, we get

$$f(c) = \frac{1}{N}\frac{dN_c}{d_c} = 4\pi\left(\frac{m}{2\pi kT}\right)^{3/2} c^2 e^{-mc^2/2kT} \qquad ....(123)$$

The distribution of molecular speeds is represented graphically in Fig. 17.12. The quantity $\frac{1}{N}\cdot\frac{dN}{dc}$ is plotted as ordinate against various values of c as abscissae.

It is evident from the graph that:

1. The fraction of the molecules with too low or too high velocities is very small.
2. The fraction of the molecules having speeds greater than zero goes on increasing with increasing velocity passes through a maximum and then again falls towards very high velocities.
3. There is a certain velocity for which the fraction of molecules is maximum. This is called the most probable velocity.
4. By altering the temperature, the general shape of the distribution curve remains unchanged but flattening of maximum takes place which occurs at higher velocity if the temperature is increased.

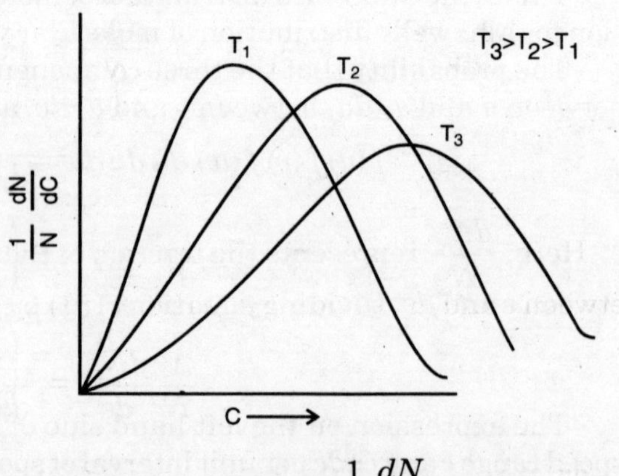

Fig. 17.12 Plots of $\frac{1}{N}\frac{dN}{dc}$ versus c

Infact, on increasing the temperature the entire distribution curve shifts to the right. The fraction of the molecules having high velocities increases with increase in temperature. This effect is mainly due to the presence of the exponential term which increases rapidly with increasing temperature. This exponential term, i.e. $e - mc^2/2kT$ is also known as Boltzmann factor. The exponent has a negative sign and temperature T appears in the denominator. Thus Boltzmann factor increases with increase in temperature. In terms of *K.E*, this factor can be represented as $e^{-\varepsilon/kT}$. Greater is the temperature greater is the value of ε. Hence with increase in temperature value of Boltzmann factor increases remarkably.

MAXWELLS DISTRIBUTION OF MOLECULAR KINETIC ENERGIES

If ε represents the kinetic energy of 1 mole of molecules having the velocity c, then

$$\varepsilon = \frac{1}{2}mc^2 \quad ...(124)$$

$$c^2 = \frac{2\varepsilon}{m} \quad ...(125)$$

Differentiating equation (124) we get,

$$d\varepsilon = \frac{1}{2}m\, 2cdc$$

$$d\varepsilon = mcdc \quad ...(126)$$

Multiplying the above equation by c, we get,

$$cd\varepsilon = mc^2dc \quad ...(127)$$

But $$c = \left(\frac{2\varepsilon}{m}\right)^{1/2} \text{ from equation (125)} \quad ...(128)$$

Substituting the value of c from equation (128) in equation (127), we get,

$$\left(\frac{2\varepsilon}{m}\right)^{1/2} d\varepsilon = mc^2dc \quad ...(129)$$

$$\frac{\sqrt{2e}}{(m)^{3/2}} d\varepsilon = c^2dc \quad ...(130)$$

Substituting the above value of c^2dc in equation (121) we have,

$$\frac{dN_\varepsilon}{N} = 4\pi\left(\frac{m}{2\pi kT}\right)^{3/2} \frac{\sqrt{2\varepsilon}}{m^{3/2}} d\varepsilon\, e^{-\varepsilon/kT}$$

$$= \frac{2\sqrt{\varepsilon}}{\sqrt{\pi}(kT)^{3/2}} e^{-\varepsilon/kT} d\varepsilon. \quad ...(131)$$

Here, $\frac{dN_\varepsilon}{N}$ represents the fraction of the total number of molecules having kinetic energies between ε and $d\varepsilon$. Dividing equation (131) by $d\varepsilon$ we get,

$$\frac{1}{N}\frac{dN_\varepsilon}{d\varepsilon} = \frac{2\sqrt{\varepsilon}}{\sqrt{\pi}(kT)^{3/2}} e^{-\varepsilon/kT} \quad ...(132)$$

The plots of $\frac{1}{N}\frac{dN_\varepsilon}{d\varepsilon}$ versus ε are similar to those for the distribution of velocities.

17.10 TYPES OF MOLECULAR VELOCITIES

The three different types of velocities which are used in the study of gases are:

- Most Probable Velocity
- Average Velocity
- Root mean square Velocity.

The expression for these three types of velocities can be obtained on the bases of Maxwell's Distribution law of velocities.

(a) **Most Probable Velocity:** The most probable velocity is the velocity possessed by maximum number of molecules of a gas at a given temperature. Maxwells Distribution Law is given by equation (123), i.e.

$$\frac{1}{N}\frac{dNc}{dc} = f(c) = 4\pi\left(\frac{m}{2\pi kT}\right)^{3/2} c^2 e^{-mc^2/2kT}$$

Differentiating the right hand side of the above equation with respect to c and then equating the results to zero, conditions for maxima or minima in the distribution curve are obtained. Thus

$$\frac{d}{dc}\left[4\pi\left(\frac{m}{2\pi kT}\right)^{3/2} c^2\, e^{-mc^2/2kT}\right] = 0 \qquad \text{...(133)}$$

$$\because \quad \frac{d}{dx}f(x)g(x) = dfg + dgf$$

$$\therefore \quad \frac{d}{dc}\left[4\pi\left(\frac{m}{2\pi kT}\right)^{3/2} \underset{f}{c^2}\; \underset{g}{e^{-mc^2/2kT}}\right]$$

$$= 4\pi\left(\frac{m}{2\pi kT}\right)^{3/2}\left[\frac{dc^2}{dc}e^{-mc^2/2kT} + \frac{d}{dc}e^{-mc^2/kT}c^2\right]$$

$$= 4\pi\left(\frac{m}{2\pi kT}\right)^{3/2}\left[2ce^{-mc^2 2kT} + c^2\, e^{-mc^2/2kT}\left(\frac{-2mc}{2kT}\right)\right]$$

$$= 4\pi\left(\frac{m}{2\pi kT}\right)^{3/2}\left[2ce^{-mc^2/2kT}\left(1-\frac{mc^2}{2kT}\right)\right] = 0$$

$$= 4\pi\left(\frac{m}{2\pi kT}\right)^{3/2}\left[2ce^{-mc^2/2kT}\left(1-\frac{mc^2}{2kT}\right)\right] = 0 \qquad \text{...(134)}$$

Now, $4\pi\left(\frac{m}{2\pi kT}\right)$ is a constant, different from zero.

$$\therefore \quad \left[2ce^{-mc^2/2kT}\left(1-\frac{mc^2}{2kT}\right)\right] = 0 \qquad \text{...(135)}$$

Now there are three possibilities :

(i) $c = 0$ (ii) $\varepsilon^{-mc^2/2kT} = 0$ when $c = \infty$

(iii) $\left(1-\frac{mc^2}{2kT}\right) = 0$

As is seen from the graph that first two possible solutions corresponding to maxima or minima when $c = 0$ or $c = \infty$ respectively are not possible. This is because there are no molecules having velocity equal to zero or infinity. Thus the solution (iii) is possible and hence,

$$\left(1 - \frac{mc^2}{2kT}\right) = 0 \text{ or } c^2 = \frac{2kT}{m} \qquad ...(136)$$

$$\therefore \qquad c = \sqrt{\frac{2kT}{m}} \qquad ...(137)$$

$$c \equiv c_p$$

$$c_p = \sqrt{\frac{2RT}{M}} \qquad (138)$$

where M is the molecular mass, R is the boltzmann constant

$$k = \frac{R}{mN_o} = \frac{R}{M}$$

(b) **Average Velocity:** The average velocity is given by the arithmatic mean of the different velocities possessed by the molecules of the gas at a given temperature If c_1, c_2, c_3.... are the velocities possessed by n_1, n_2, n_3.... number of molecules respectively, then average velocity will be:

$$\bar{c} = \frac{c_1 + c_2 + c_3 + ... + c_N}{N} = \frac{1}{N}\sum_i c_i \qquad ...(139)$$

The average velocity $\bar{c}$ can also be written in the form:

$$\bar{c} = \frac{1}{N}\int_0^\infty c\,dN_c = \int_o^\infty c\frac{dN_c}{N} \qquad ...(140)$$

where dN_c is the number of molecules having velocity c.

The summation of different velocities is replaced by integration because all types of velocities ranging from zero to infinity are involved.

Substituting the value of $\frac{dNc}{N}$ from equation (122) in equation (140), we get–7

$$\bar{c} = \int_0^\infty 4\pi\left(\frac{m}{2\pi kT}\right)^{3/2} c^3 e^{-mc^2/2kT}\,dc \qquad ...(141)$$

$$= 4\pi\left(\frac{m}{2\pi kT}\right)^{3/2} \int_0^\infty c^3 e^{-mc^2/2kT}\,dc \qquad ...(142)$$

put $$\frac{mc^2}{2kT} = x^2 \qquad ...(143)$$

$$\therefore \qquad c^2 = \frac{x^2 2kT}{m} \qquad ...(144)$$

$$c = \sqrt{\frac{2kT}{m}}\,x \qquad ...(145)$$

or $$c^3 = \left(\frac{2kT}{m}\right)^{3/2} x^3 \qquad ...(146)$$

Differentiating equation (145), with respect to x

$$\frac{d}{dx}c = \sqrt{\frac{2kT}{m}} \quad ...(147)$$

$$dc = \sqrt{\frac{2kT}{m}}dx \quad ...(148)$$

Substituting, the values of c and dc from equation (145) and (148) in equation (142) we get :

$$\overline{c} = 4\pi\left(\frac{m}{2\pi kT}\right)^{3/2} \int_0^{\infty}\left(\frac{2kT}{m}\right)^{3/2} x^3 e^{-x^2}\left(\frac{2kT}{m}\right)^{\frac{1}{2}} dx \quad ...(149)$$

$$= 4\pi\left(\frac{m}{2\pi kT}\right)^{3/2}\left(\frac{2kT}{m}\right)^2\int_o^{\infty} x^3 e^{-x^2}\, dx \quad ...(150)$$

Substituting, the value of $\int_0^{\infty} x^3 e^{-x^2} dx = \frac{1}{2}$, from the table we get,

$$\overline{c} = 4\pi\left(\frac{m}{\pi kT}\right)^{3/2}\left(\frac{2kT}{m}\right)^2\left(\frac{1}{2}\right) \quad ...(151)$$

$$\overline{c} = \sqrt{\frac{8kT}{\pi m}} \quad ...(152)$$

Equation (152) is the expression for the average velocity. It can also be expressed as:

$$\overline{c} = \sqrt{\frac{8RT}{\pi M_m}} \quad ...(153)$$

Table 17.2 Values of Integrals used in Derivation of Different Velocities

	Integral	*Value*
1	$\int_{-\infty}^{\infty} e^{-au^2\, du}$	$\left(\frac{\pi}{a}\right)^{\frac{1}{2}}$
2	$\int_o^{\infty} x^3 e^{-x^2} dx$	1/2
3	$\int_o^{\infty} x^4 e^{-x^2} dx$	$\frac{3}{8}\sqrt{\pi}$

3. **Root Mean Square Velocity:** The root mean square velocity is defined as the square root of the mean of squares of different velocities possessed by molecules of a gas at a given temperature. It can be expressed as:

$$(\overline{c}^2)^{\frac{1}{2}} = \sqrt{\overline{c}^2} = \left[\frac{c_1^2 + c_2^2 + c_3^3 \ldots c_N^2}{N}\right]^{\frac{1}{2}} = \left[\frac{1}{N}\sum_i c_i^2\right]^{\frac{1}{2}} \quad ...(154)$$

This can also be expressed as:

$$\sqrt{\overline{c}^2} = \left[\int_0^{\infty} c^2 \frac{dNc}{N}\right]^{\frac{1}{2}} \quad ...(155)$$

Substituting the value of $\frac{dNc}{N}$ in equation (155) from equation (122) we get the following expression for mean square velocity:

$$\bar{c}^2 = \int_0^\infty 4\pi\left(\frac{m}{2kT}\right)^{3/2} c^4 e^{-mc^2/kT}\, dc \qquad ...(156)$$

$$= 4\pi\left(\frac{m}{2kT}\right)^{3/2} \int_0^\infty c^4 e^{-mc^2/kT}\, dc \qquad ...(157)$$

Let $$\frac{mc^2}{2kT} = x^2 \qquad ...(158)$$

and hence $$c^2 = \frac{2kT}{m}x^2 \text{ or } c^4 = \left(\frac{2kT}{m}\right)x^4 \qquad ...(159)$$

$$c = \left(\frac{2kT}{m}\right)^{\frac{1}{2}} x \qquad ...(160)$$

Differentiating equation (160) we get,

$$dc = \left(\frac{2kT}{m}\right)^{\frac{1}{2}} dx \qquad ...(161)$$

Putting the values of c^4 and dc from equations (159) and (161) in equation (157) we get,

$$\bar{c}^2 = 4\pi\left(\frac{m}{2\pi kT}\right)^{3/2} \int_0^\infty \left(\frac{2kT}{m}\right)^2 x^4 e^{-x^2} \left(\frac{2kT}{m}\right)^{\frac{1}{2}} dx \qquad ...(162)$$

$$= 4\pi\left(\frac{m}{2\pi kT}\right)^{3/2} \left(\frac{2kT}{m}\right)^{3/2} \int_0^\infty x^4 e^{-x^2} dx \qquad ...(163)$$

Substituting the value of $\int_0^\infty x^4 e^{-x^4} dx$ from the table i.e $\frac{3}{8}(\pi)^{\frac{1}{2}}$ we get,

$$\bar{c}^2 = 4\pi\left(\frac{m}{2\pi kT}\right)^{3/2} \left(\frac{2kT}{m}\right)^{3/2} \frac{3}{8}(\pi)^{\frac{1}{2}} \qquad ...(164)$$

$$= \frac{3kT}{m} \qquad ...(165)$$

And hence root mean square velocity will be given by,

$$\boxed{(\bar{c}^2)^{\frac{1}{2}} = \sqrt{\frac{3kT}{m}}} \qquad ...(166)$$

This is the expression for root mean square velocity which can also be expressed as:

$$(\bar{c}^2)^{\frac{1}{2}} = \sqrt{\frac{2RT}{M_m}} \qquad ...(167)$$

It has been found that the ratio of these three types of velocities is as follows :

$$\sqrt{\bar{c}^2} : \bar{c} : c_p = 1.00 : 0.92 : 0.82$$

NUMERICALS BASED ON MOLECULAR VELOCITIES

Example 17.13. Calculate the rms velocity of chlorine molecules at 12°C and 78 cm pressure.
Solution:

At STP	At given conditions
$V_1 = 22400$ mL	$V_2 = ?$
$T_1 = 273$ K	$T_2 = 12 + 273 = 285$ K
$P_1 = 76$ cm	$P_2 = 78$ cm

Now $$\frac{P_1V_1}{T_1} = \frac{P_2V_2}{T_2}$$

$$\therefore \quad V_2 = \frac{P_1V_1T_2}{T_1P_2} = \frac{76\times22400\times285}{273\times78} = 22785 \text{ ml.}$$

$$\text{rms velocity} = \sqrt{\frac{3PV}{M}}$$

$$P = hdg = 78 \times 13.6 \times 981 \text{ dynes cm}^{-2}$$

$$V = 22785 \text{ ml; } M = 71$$

$$\text{rms} = \sqrt{\frac{3\times78\times13.6\times981\times22785}{71}}$$

$$= 31652 \text{ cm sec}^{-1} = 316.52\text{m sec}^{-1}$$

Example 17.14. A gas has a density of 1.2504 gL^{-1} at 0°C and a pressure of 1 atm. Calculate the rms, average and most propable speeds of its molecules at 0°C.
Solution:

$$P = 76 \times 13.6 \times 981 \text{ dynes cm}^{-2}$$

$$d = 1.2504 \text{ gL}^{-1} = 0.0012504 \text{ gcm}^{-1}$$

$$\text{r.m.s. velocity} = \sqrt{\frac{3P}{d}} = \frac{3\times76\times13.6\times981}{0.0012504}$$

$$= 4.93 \times 10^4 \text{ cm sec}^{-1}$$

$$\text{Average velocity} = 0.92 \times 4.93 \times 10^4 \text{ cm sec}^{-1}$$

$$= 4.54 \times 10^4 \text{ cm sec}^{-1}$$

$$\text{Most Probable velocity} = 0.82 \times 4.93 \times 10^4 \text{ cm sec}^{-1}$$

$$= 4.04 \times 10^4 \text{ cm sec}^{-1}$$

Example 17.15. Calculate the temperature at which the root mean square velocity of helium will be the same as that of methane molecules at 27°C.
Solution:

$$\text{r.m.s velocity} = \sqrt{\frac{3RT}{Mm}}$$

For Helium, M = 4T = 300K

For Methane M = 16

$$\text{Now} \quad \sqrt{\frac{3RT}{M_{He}}} = \sqrt{\frac{3RT}{M_{CH4}}}$$

$$\sqrt{\frac{3RT}{4}} = \sqrt{\frac{3R}{16}\times300K}$$

$$T = \frac{4\times300K}{16} = 75\text{K}$$

Example 17.16. Calculate the temperature at which the rms velocity of hydrogen molecules is 10^3m sec^{-1}

Solution:

$$\text{For hydrogen } M = 0.002 \text{ kg mol}^{-1}$$

$$c_{rms} = \sqrt{\frac{3RT}{M}} = 10^3 \text{ m sec}^{-1}$$

$$\frac{3RT}{M} = (10^3 \text{ m sec}^{-1}) = 10^6 \text{ m}^2 \text{ sec}^{-2}$$

$$T = \frac{M \times 10^6 \text{m}^2 \text{sec}^{-2}}{3R}$$

$$= \frac{(0.002 \text{ kg mol}^{-1})(10^6 \text{m}^2 \text{ sec}^{-2})}{3 \times 8.314 \text{ kgm}^2 \text{sec}^{-2} \text{mol}^{-1} \text{ K}^{-1}}$$

$$= 80.2 \text{ K}$$

Example 17.17. Calculate rms speed of oxygen molecules in the lungs at normal body temperature 37°C.

Solution:

$$T = 273 + 37° = 310 \text{ K}$$

$$c_{rms} = \sqrt{\frac{3RT}{M}}$$

$$= \sqrt{\frac{3 \times 8.314 \times 10^7 \times 310}{32}}$$

$$= 4.92 \times 10^4 \text{ cm sec}^{-1}$$

Example 17.18. The average speed at T_1K and the most probable speed at T_2K of CO_2 gas is 9×10^4 cm sec^{-1}. Calculate the value of T_1 and T_2.

Solution:

$$\text{Average speed} = \sqrt{\frac{8RT_1}{\pi Mm}} = 9 \times 10^4 \text{ cm sec}^{-1}$$

$$T_1 = \frac{\pi Mm}{8R} \times (9 \times 10^4)^2$$

$$= \frac{3.142 \times 44 \text{ g mol}^{-1} \times (9 \times 10^4 \text{ cm sec}^{-1})}{8 \times 8.314 \times 10^7 \text{ erg K}^{-1} \text{ mol}^{-1}}$$

$$= 1684 \text{ K}$$

$$\text{Most probable velocity } c_p = \sqrt{\frac{2RT_2}{Mm}} = 9 \times 10^4 \text{ cm sec}^{-1}$$

$$T_2 = \frac{Mm}{2R} \times (9 \times 10^4 \text{cm sec}^{-1})^2$$

$$= \frac{44 \text{ g mol}^{-1} \times (9 \times 10^4 \text{cm sec}^{-1})^2}{2 \times 8.314 \times 10^7 \text{ erg K}^{-1} \text{ mol}^{-1}}$$

$$= 2143\text{K}$$

17.11 COLLISION PROPERTIES

Collision is a term used to describe an encounter between any two particles of the gas. While considering collisions of the molecules we assume the molecules to be rigid, non-interacting spheres.

Collision Diameter: When two gas molecules approach one another, they cannot come closer beyond a certain distance. The closest distance of approach between the two molecules taking part in collision is called the collision diameter. It is denoted by σ. The collision diameter of hydrogen is 2.74 A° and that of oxygen is 3.61 A°.

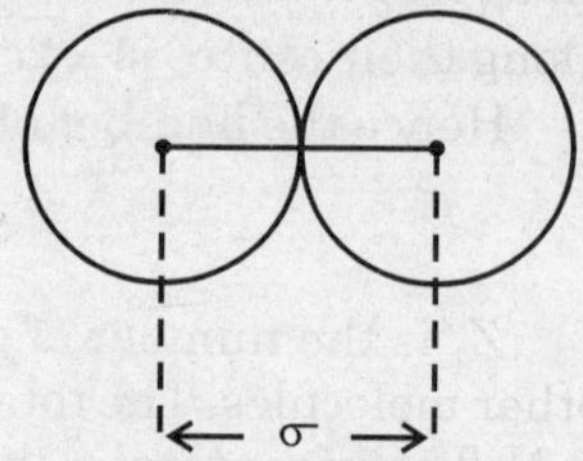

Fig. 17.13 Collision Diameter of Molecules

Collision Frequency (Z_1): Collision frequency is the number of collision per unit time per unit volume of the gas. This is calculated as follows :

Suppose there are *N* molecules of a gas per unit volume, each of which is having the collision diameter σ. Let us assume that only one molecule *x* is moving and others are stationary. Molecule *x* moves with an average velocity of c cm/sec^{-1}. A collision will occur each time when *x* approaches other molecule at a distance σ. In one second molecule *x* will sweep out a cylinder having radius σ and length c. Whenever the centre of another molecule comes within this cylinder, a collision will result.

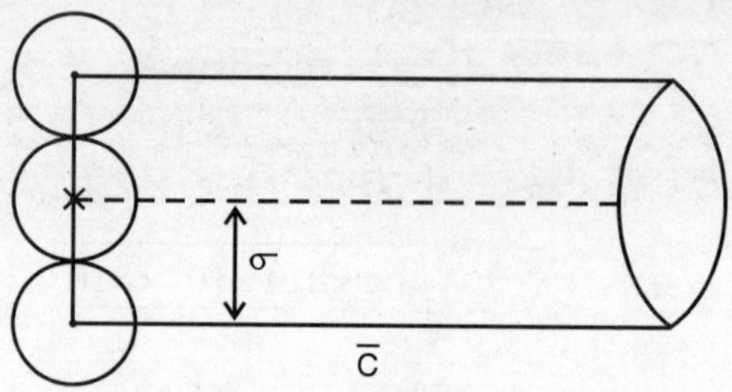

Fig. 17.14 Path swept out by a molecule of a gas in unit time

$$\text{Volume of the collision cylinder} = \pi\sigma^2\bar{c}N \quad ...(168)$$

$$\text{Number of molecules per unit volume in the cylinder} = N$$

$$\text{Number of collision experienced by } x = \pi\bar{c}N \quad ...(169)$$

But other molecules are also not stationary as was assumed earlier. In order to account for the movements of all molecules, one must consider the average velocity along the line of centre of two colliding molecules, instead of the average velocity of a single molecule.

Two molecules may encounter head on collision, or grazing collision or collision at any angle.

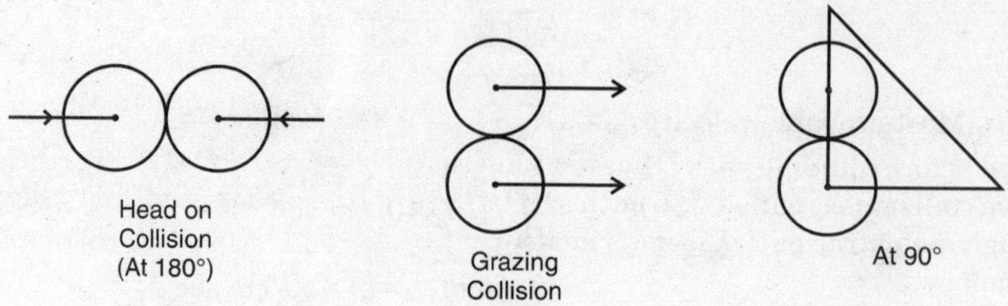

Fig. 17.15 Types of Collisions

All the angles are possible from zero to 180°. If it is assumed that on an average, molecules collide at right angles, then the average velocity along their centre is $\sqrt{2}c$. Fig. (17.16)

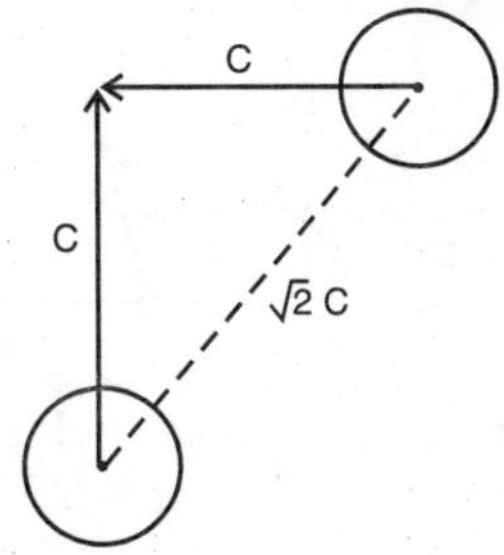

Fig. 17.16

Hence the number of collisions per second is given by :

$$Z_1 = \sqrt{2}\pi\sigma^2\bar{c}N \quad ...(170)$$

Z_1 is the number of collisions which a single molecule makes with other molecules. The total number of molecules present per unit volume is N. Therefore total number of collisions made by all the molecules will be given by

$$Z_{11} = Z_1 \times \frac{1}{2}N \quad ...(171)$$

$$= \frac{1}{2}\sqrt{2}\,\pi\sigma^2\,\bar{c}\,N^2 \quad ...(172)$$

The factor $\frac{1}{2}$ is introduced to ensure that each collision is not counted twice, since a single collision involves two molecules. Hence

$$\boxed{Z_{11} = \frac{1}{\sqrt{2}}\pi\sigma^2\,\bar{c}N^2} \quad ...(173)$$

But $$\bar{c} = \sqrt{\frac{8kT}{\pi m}}$$

Substituting in equation (173)

$$Z_{11} = \frac{1}{\sqrt{2}}\pi\sigma^2\,N^2\sqrt{\frac{8kT}{\pi m}} \quad ...(174)$$

Collision frequency for a mixture of gases : For a mixture of two gases A and B, the number of collisions Z_{AB} per unit volume per unit time is given by

$$Z_{AB} = N_A\,N_B\,\sigma_{AB}^2\,\pi\sqrt{\frac{8kT}{\pi\mu}} \quad ...(175)$$

Where N_A and N_B are the number of molecules per unit volume and σ_{AB} is the collision diameter, μ represents the reduced mass such that

$$\mu = \frac{m_A\,m_B}{m_A + m_B} \quad ...(176)$$

Mean Free Path: According to the kinetic theory, a molecule moves along a straight line with constant speed before colliding with other molecules. The path traversed by any molecule between two successive collisions is called free path and the average of all such paths is called **mean free path**. Although mean free path λ is not a measurable quantity but it can be determined by using collision frequency.

$$\text{Mean free path} = \frac{l_1 + l_2 + l_3....}{\text{No. of collisions}}$$

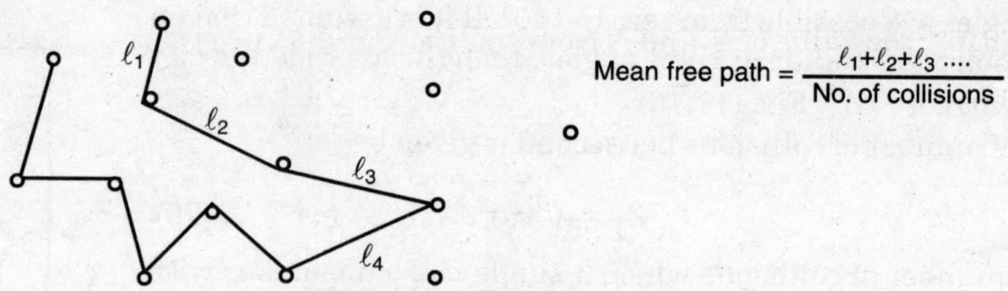

Fig. 17.17 Mean Free Path

As the distance traversed by a molecule in one second is $\bar{c}$, therefore, the mean free path is given by :

$$\lambda = \frac{\text{Average distance travelled by the molecule per unit time}}{\text{Collision frequency}} = \frac{\bar{c}}{Z_1} \quad ...(177)$$

$$= \frac{1}{\sqrt{2}\,\pi\sigma^2 N} \quad ...(178)$$

Mean free path is inversely proportional to the concentration of the gas and molecular size.

Effect of Temperature and Pressure on Collision Frequency and Mean Free Path: According to ideal gas equation,

$$PV = n\,RT$$

$$= \frac{n'}{N_o} RT \quad ...(179)$$

where n' is number of molecules and N_o is the avogadro's number.

or
$$P = \frac{n'}{V}\frac{R}{N_o}T = \frac{n'}{V}kT \quad ...(180)$$

Thus number of molecules per unit volume is given by:

$$\frac{n'}{V} = \frac{P}{kT} \quad ...(181)$$

In the above equations (180), (181), (182) n'/v represents the number of molecules per unit volume and hence is equal to N which also represents the number of molecules per unit volume in equation (179). Thus,

$$N \,\alpha\, \frac{P}{kT} \quad ...(182)$$

Substituting, the value of N in equation (179) we get,

$$\lambda = \frac{kT}{\sqrt{2}\,\pi\sigma^2 P} \quad \text{or} \quad \boxed{\lambda \,\alpha\, \frac{T}{P}} \quad ...(183)$$

Thus mean free path is

(i) directly proportional to temperature and

(ii) inversely proportional to pressure.

Further, the average velocity $\bar{c}$ is given by :

$$\bar{c} = \sqrt{\frac{8RT}{\pi M}} = \sqrt{\frac{8kT}{\pi m}} \quad ...(184)$$

$$\bar{c} \propto \sqrt{T} \quad ...(185)$$

Substituting, the value of $\bar{c}$ and N from equations (185) and (183) in equation (170) we get,

$$Z_1 = \sqrt{2}\,\pi\sigma^2\left(\frac{8kT}{\pi m}\right)^{\frac{1}{2}}\frac{P}{kT} \quad ...(186)$$

$$Z_1 \propto \sqrt{T}\left(\frac{P}{T}\right) \quad ...(187)$$

$$Z_1 \propto \frac{P}{\sqrt{T}} \quad ...(188)$$

Thus collision frequency is

(i) directly proportional to pressure at constant temperature and

(ii) inversely proportional to square root of temperature when pressure is kept constant.

NUMERICALS BASED ON COLLISION FREQUENCY AND MEAN FREE PATH

Example 17.19. At 0°C and one atmospheric pressure the molecular diameter of a gas is 4A°. Calculate the mean free path of its molecule.

Solution:

$$\text{The mean free path } \lambda = \frac{1}{\sqrt{2}\pi\sigma^2 N}$$

Where σ is the molecular diameter and N is the number of molecules per cm^3

$$\sigma = 4 \times 10^{-8} \text{ cm}$$

No. of molecules per unit volume,

$$N = \frac{6.02\times10^{23}}{22400}$$

$$= 2.689\times10^{19} \text{ molecules}$$

$$\lambda = \frac{1}{\sqrt{2}\times3.14\times(4\times10^{-8})\times2.689\times10^{19}}$$

$$= 0.524 \times 10^{-5} \text{ cm}$$

Example 17.20. The mean free path of the molecule of a certain gas at 300 K is 2.6×10^{-5}m. The Collision diameter of the molecule is 0.26 nm. Calculate (i) Pressure of the gas, and (ii) number of molecule per unit volume of the gas.

Solution:

$$\lambda = 2.6 \times 10^{-5} \text{ m}$$

$$\sigma = 0.26 \times 10^{-9} \text{ m}$$

$$T = 300 \text{ K}$$

Now,

$$\lambda = \frac{1}{\sqrt{2}\pi\sigma^2 N}$$

$$N = \frac{1}{\sqrt{2}\pi\sigma^2\lambda} = \frac{1}{1.414\times3.14\times(0.26\times10^{-9}\text{ m})^2\times(2.6\times10^{-5}\text{m})}$$

$$= 1.281 \times 10^{23} \text{ m}^{-3}$$

Also,

$$N = \frac{P}{kT}$$

$$P = NkT$$

$$= (1.281 \times 10^{23} \text{ m}^{-3})\,(1.38066 \times 10^{-23} \text{ JK}^{-1})\,(300\text{K})$$

$$= 5.306 \times 10^2 \text{ Jm}^{-2} = 5.306 \times 10^2 \text{ Pa}$$

Example 17.21. The collision diameter of a nitrogen molecule is 3.74×10^{-10} m. Calculate the number of collisions for each nitrogen molecule in one second at 1 atm and 298 K. Also calculate the total number of collisions in a volume of $1 m^3$ in one second.

Solution: Number of molecules in $1 m^3$ at 298 K and 1 atm is given by:

$$N = \frac{No}{V} = \frac{NoP}{RT} \quad (\therefore PV = RT, V = RT/P)$$

$$= \frac{(6.023\times10^{23} \text{ molecules mol}^{-1})(1\,\text{atm})(10^3 \text{ litre m}^{-3})}{(0.0821 \text{ atm litre K}^{-1} \text{ mol}^{-1})(298\,\text{K})}$$

$$= 2.462 \times 10^{25} \text{ molecules m}^{-3}$$

The average speed, $\bar{c}$ is given by:

$$\bar{c} = \sqrt{\frac{8RT}{\pi M}} = \sqrt{\frac{8(8.314\,\text{kgm}^2 \text{ sec}^{-2} \text{ K}^{-1}\text{mol}^{-1})(298 \text{ K})}{\Pi(28\times10^{-3}\text{kg mol}^{-1})}}$$

$$= 4.75 \times 10^2 \text{ m sec}^{-1}$$

Number of Collision per second,

$$Z_1 = \sqrt{2}\Pi\sigma^2\bar{c}N = \sqrt{2}\times\frac{22}{7}\times(3.74\times10^{-10}\text{ m})^2(4.75\times10^2\text{m sec}^{-1})$$

$$= 7.26 \times 10^9 \text{ collisions sec}^{-1}$$

Total number of collisions per unit volume per second,

$$Z_{11} = \frac{1}{2}NZ_1 = \frac{1}{2}(2.462\times10^{25} \text{ molecules m}^{-3})(7.26\times10^9\text{sec}^{-1})$$

$$= 8.93 \times 10^{34} \text{ collisions m}^{-3} \text{ sec}^{-1}$$

17.12 LIQUEFACTION OF GASES: CRITICAL PHENOMENON

A gas can be liquefied by two ways:

— by lowering the temperature or
— by increasing the pressure.

The molecules of a gas lie far apart from one another and are in continuous random motion. As the temperature of the gas is lowered, the kinetic energy of the gas molecules also decreases. As a result of this, some of the molecules get closer and closer to one another and ultimately the gas changes into the liquid state.

Gases can also be liquefied by the application of pressure. On increasing the pressure the molecules are brought closer and closer and hence permit the attractive forces to be large enough to cause liquefaction.

Hence, a gas can be liquefied by cooling or by the application of pressure or by the combined effect of both.

Gases like NH_3, SO_2, CO_2, etc. could be liquefied by any one of the above mentioned methods. It is possible to liquefy most of the gases at atmospheric pressure, but it is observed that many of the gases could not be liquefied at ordinary temperatures even under pressure of several thousand atmospheres. For example, gases like hydrogen, helium, oxygen, nitrogen, etc. could not be liquefied by the application of pressure alone, howsoever high it may be. Such gases were therefore, referred to permanent gases. But to use the term permanent gases for the above mentioned gases has lost its significance because of Andrews experiments, which have clearly shown that even these gases can be liquefied.

The essential condition for liquefaction was discovered by Andrews in 1863. He established that the effect of temperature is more important than that of the pressure. He demonstrated that **for each gas, there exists a certain temperature, above which it is impossible to liquefy it, no matter how great the pressure applied is. This temperature is called the critical temperature of the gas (*Tc*).** Thus, a gas cannot be liquefied unless it is cooled down to its critical temperature. The critical temperatures for easily liquefiable gases lie in the ordinary temperature range. For example, the critical temperature of CO_2 is 31.1°C and hence it can be liquefied by the application of pressure alone below this temperature. But it is not possible to liquefy CO_2 above 31.1°C no matter how great the pressure applied is.

The pressure required to liquefy a gas at critical temperature is called critical pressure (P_c). The volume occupied by one mole of a gas at critical pressure and temperature is called critical volume (Vc). *Tc*, *Pc*, and *Vc* are known as the critical constants for a gas. For example at critical temperature of CO_2 i.e 31.1°C, the pressure required to convert CO_2 gas into liquid state is 72.9 atm, thus critical pressure of CO_2 is 72.9 atm. Volume occupied by 1 mole of CO_2 at critical temperature and critical pressure is 940 ml per mole.

At critical temperature and critical pressure, the gas becomes identical with its liquid. The two phases can no longer be distinguished. The smooth merging of a gas with its liquid form is called the **critical phenomenon**.

Thus discovery of critical phenomenon by Andrews opened a new approach for liquefaction of gases. If a gas is cooled below its critical temperature, it can be liquefied easily when subjected to adequate pressure.

Various techniques are employed to attain low temperatures. Some of which are discussed below:

1. **Use of freezing mixtures:** A simple method for producing lower temperature than ice is to add various salts to the ice. Addition of salt to ice results in melting of some ice. The melting of ice and dissolution of salt both are endothermic processes and hence the temperature of the mixture is lowered upto a temperature when the solution becomes saturated with respect to both salt and ice.

   Addition of common salt to ice depresses its melting point much below 0°C i.e –22°C. Freezing mixture of ice and $CaCl_2$ produces temperature as low as –55°C. Faraday (1823) succeeded in liquefying a number of gases such as SO_2, CO_2 and Cl_2 by this method.

   He used a *U*-shaped tube, in one arm of which the gas was prepared. The other limb of the gas was immersed in the freezing mixture. When the pressure inside becomes sufficiently high, the gas is compressed under its own pressure and hence is liquefied in the arm immersed in the freezing mixture.

   Thilorier (1834) found that a mixture of solid CO_2 and ether evaporating under reduced pressure could produce a temperature as low as –100°C. This freezing mixture was used under the name Thilorier mixture.

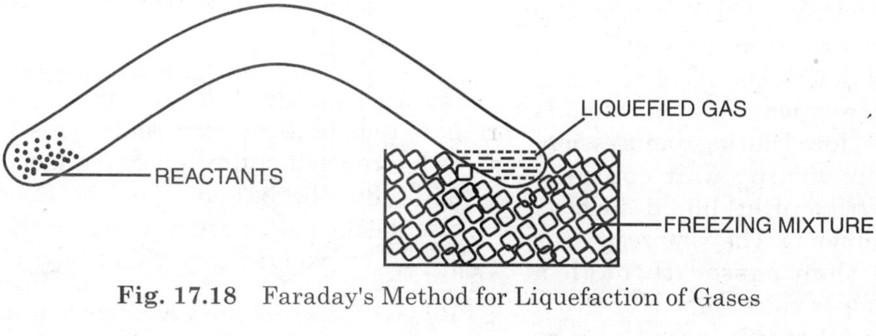

Fig. 17.18 Faraday's Method for Liquefaction of Gases

However gases like H_2, O_2, N_2 and helium which have very low critical temperatures could not be liquefied by simple techniques.

2. **Cooling by rapid evaporation of volatile liquid:** In 1877, Cailletet and Pictet succeeded in liquefying oxygen and air. This method is based upon the fact that when a liquid evaporates, latent heat is required for its evaporation. This latent heat is supplied by the surrounding gas which is kept in contact with the evaporating liquid. As a result, gas cools down. First ammonia refrigerator was made in 1862 based on this principle.
3. **Cooling by Joule Thomson Method:** Linde and Claude were able to liquefy air utilizing the principle based on Joule Thomson effect.
Joule and Thomson (1852-1862) observed that if a compressed gas is allowed to expand into a region of low pressure or vacuum through a fine orifice, under adiabatic conditions, it suffers a fall in temperature. This effect is known as Joule Thomson effect. The reason of cooling is that, when the gas expands from a region of high pressure to the region of low pressure, the molecules fall apart from each other. Therefore, work has to be done in overcoming the forces of attraction between the molecules. This is done at the cost of kinetic energy of the gas, as the system is thermally insulated. Consequently the kinetic energy decreases and cooling takes place.
It may be mentioned here that for each gas there is an inversion temperature, below which the gas experience cooling and above which it experience heating if allowed to expand under Joule Thomson effect. This temperature is known as inversion temperature and is given by

$$Ti = \frac{2a}{Rb} \quad \text{...(190)}$$

Where Ti is the inversion temperature of the gas, a and b are van der Waals' constants. At inversion temperature the gases do not experience Joule Thomson effect, i.e. there is neither fall or rise of temperature if the gas is subjected to expansion from region of high pressure to low pressure.
In most of the gases the inversion temperature lie well within the range of ordinary temperature and hence these gases experience cooling on expansion under Joule Thomson effect. However hydrogen and helium have very low inversion temperature, 193 K (–80°C) and 33 K (–240°C) respectively and hence at ordinary temperatures these gases are warmed up in Joule Thomson expansion. If however these gases are cooled to their inversion temperatures, these will also experience cooling on expansion.
Two methods for liquefaction based on Joule Thomson effect are discussed below:

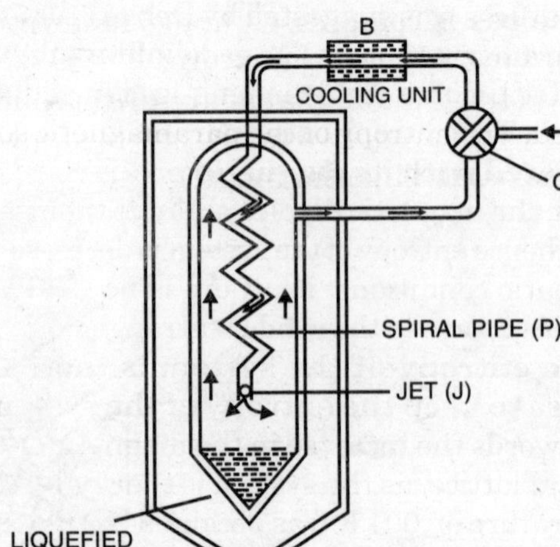

Fig. 17.19 Linde's process for the liquefaction of air

(a) **Linde's Process:** In this process (Fig.17.19) pure dry air is compressed to about 200 atmospheres in the compressor *C*. The compressed air is passed through cooling unit B to remove any heat that is produced during compression. The cooling unit contains refrigerating liquid such as ammonia. The compressed air is then passed through a copper spiral pipe (P) with a jet (J) at the lower end. The free expansion of the air at the jet results in considerable drop of temperature. The cooled air which is at about one atmosphere pressure, moves up the chamber and inturn cools the incoming air further and returns to the compressor again. This process is repeated a number of times till a temperature low enough to liquefy air is reached. The liquefied air collects at the bottom of the chamber.

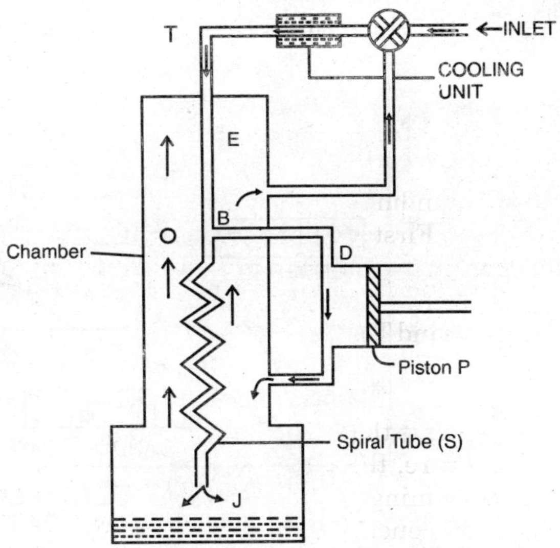

Fig. 17.20 Claude's Process for the liquefaction of air

(b) **Claude's Process:** This is again based on the adiabatic expansion of the compressed gas but it involves mechanical work also. This method is more efficient than Linde's. The gas which is subjected to Joule Thomson expansion is also made to do mechanical work by driving a piston. The energy for it comes from the kinetic energy of the gas which gets cools thereby. Thus in this process the gas is cooled not only by overcoming the intermolecular forces but also by performing some external work. Cooling produced is hence greater. Pure dry air is compressed to about 200 atmospheres (Fig. 17.20) It is passed through a cooling unit to remove any heat produced during compression. The compressed air then passes through the tube T which bifurcates at the point B. A part of it goes to the cylinder, fitted with an airtight piston P. It pushes the piston P and thus some external work is performed at the expense of kinetic energy. The air gets cooled which is then made to enter the chamber E. This cooled air further cools the incoming air in the coil S.The gas in the coil S which is already cooled by the surrounding air is allowed to expand through a jet J and is cooled further. The process is repeated a number of times till the air is cooled to such an extent that it liquefies. The liquefied air collects in the chamber.

4. **Cooling by Adiabatic Demagnetization:** This method for the production of very low temperatures was suggested by Debye (1926) and Giauque (1927) independently.

Certain paramagnetic salts like gadolinium sulphate octahydrate, compounds of cerium, dysprosium, etc. tend to become highly magnetized in a magnetic field but lose their magnetism once the field is removed. The entropy of the paramagnetic solid may be regarded as **partly thermal and partly associated with its magnetic properties**. When paramagnetic substances is placed in a magnetic field, the magnetic dipoles arrange themselves in an orderly manner, i.e. randomness is decreased and hence entropy of the system is decreased. If now the demagnetization is allowed to take place in adiabatic conditions, i.e. there is no net change in the total entropy of the system, the magnetic dipoles return to the random arrangement. In other words on removing the magnetic field **the magnetic entropy of the system is increased and consequently thermal entropy must decrease to keep the entropy of the system constant** and results in the fall of temperature. In other words the increase in the magnetic entropy takes place at the expense of the energy from the crystal lattice, as the system is thermally isolated and hence the temperature of the salt falls. A temperature of .001 K was obtained by this technique by Lerden at al.

The magnetic entropy is only appreciable, in comparison with thermal entropy at low temperatures so the paramagnetic substance is first cooled to 1K by liquid helium.

The temperature entropy curve for a paramagnetic salt is shown in Fig. 17.21.

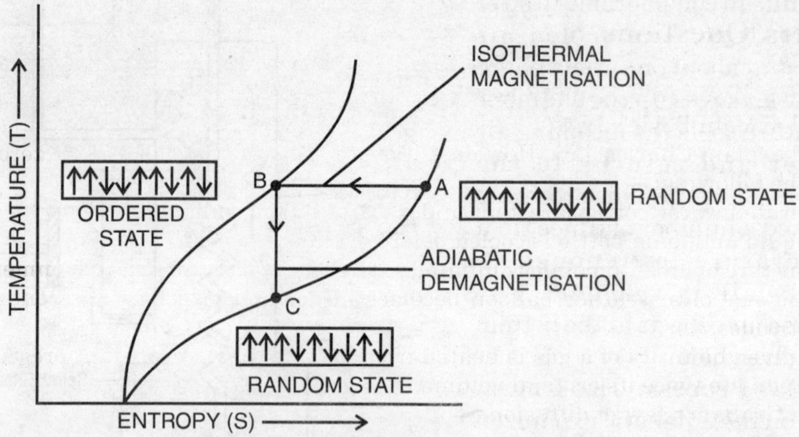

Fig. 17.21 Temperature entropy curve for a paramagnetic substance

In the unmagnetised state the spins of the unpaired electrons of a paramagnetic salt are oriented randomly (A). Now the substance, kept in contact with liquid helium bath at 1K, is **isothermally magnetized** (i.e. T = constant). The spins of the electrons orient themselves in a more orderly state, i.e. the entropy of the system decreases (B). In other words the magnetized substance has lower entropy than unmagnetized substance at a given temperature.

In the second step, the substance is removed and adiabatically demagnetized, as a result the substance is brought back to its disorderly state, i.e. the entropy arising out of magnetic property increases (C). The total entropy of the system remains constant and hence entropy of the system arising out of thermal property decreases causing cooling.

It may be mentioned here that even lower temperature can be reached if nuclear spins are used instead of electron spins. The application of adiabatic demagnetization to nuclear magnetic moments has made it possible to attain a temperature of 10^{-6} K.

PRESSURE OF GASES AND DEEP DIVING

The mole fraction of O_2 in air is 0.21, therefore the partial pressure of O_2 at sea level = 0.21 × 1 atm = 0.21 atm (Dalton's Law).

Now in each normal breath we take in about 0.02 mole of O_2 molecules. Our bodies operate very efficiently under these conditions, but the situation changes if we subject ourselves to greater pressures. For example, by diving in deep water, a diver at a depth of 100 feet experiences approximately an extra 3 atmospheres of pressure, and at 300 feet the pressure is about 10 atmospheres. This increased pressure affects theear canals and squeezes the lungs, but a more serious effect involves the increased partial pressure of oxygen in air breathed at these pressures. Elevated concentrations of oxygen are actually quite harmful for reasons that are not well understood. The symptoms of oxygen poisoning include confusion, impaired vision and nausea. At 300 feet, at a pressure of 10 atmospheres the diver inhales 2 atm. Thus oxygen must be diluted by another gas. The probable candidate appears to be nitrogen. But it cannot be used because at high pressures a large quantity of nitrogen dissolves in the blood causing a condition called 'bends'. The divers joint become locked in a bent position. Helium is often used as diluting agent in deep diving tanks. It has low solubility in blood as compared to oxygen or nitrogen.

However, it has the effect or raising the pitch of the human voice. This occurs because the pitch of the voice depends on the density of the gas surrounding the vocal cords. Lower density produces a higher pitch. Because the mass of a He atom is much smaller than the masses of N_2 and O_2, the density of helium gas is much lower than that of air and hence results in raising the pitch.

QUESTIONS

Short Answers Questions

Section 17.1

1. State and explain Boyle's Law.
2. State and explain Charles' law. How does it lead to the concept of 'absolute zero'?
3. Explain the following:
   (a) Aerated water bottles are kept under water during summer.
   (b) Liquid ammonia bottle is cooled before opening the seal.
   (c) The tyre of an automobile is inflated to slightly lesser pressure in summer than in winter.
   (d) The size of a weather balloon becomes larger and larger as it ascends up into higher altitudes.
   (e) A given quantity of a gas is heated from 20°C to 40°C at constant pressure. The gas will expand to twice its original volume.
4. (a) State Graham's law of diffusion.
   (b) Give reasons :
   (i) Methane diffuses faster than oxygen.
   (ii) Why does CO_2 diffuse faster than SO_2.
5. (a) Discuss the nature of gas constant R. Derive its value when
   (i) P is expressed in atm and V in litres.
   (ii) P is expressed in dynes/cm^2 and V in cm^3.
   (b) Does the value of R depend upon the nature of the gas?
6. Two flasks A and B have equals volume. The flask A contains H_2 and is maintained at 300K, while the flask B contains an equal mass of CH_4 gas and is maintained at 600K. Answer the following questions:
   (a) Which flask contains greater number of molecules? How many times more?
   (b) In which is the pressure greater? How many times as great?
   (c) In which are the molecules moving faster? How many times as fast?
   (d) In which flask are the number of collisions with the wall greater? How many times as great?
7. A container is filled with an ideal gas to a pressure of 40.0 atm at 0°C
   (i) What will be the pressure in the container if it is heated to 45°C?
   (ii) At what temperture would the pressure be 1.50×10^2 atm?
   (iii) At what temperature would the pressure be 25.0 atm.
8. Calculate the volume occupied by 1.00 mol of an ideal gas at STP.
9. A compressed gas cylinder, at 13.7 MPa and 23°C, is in a room where a fire raises the temperature to 450°C. What is the new pressure in the cylinder?
10. A bicycle tire is filled with air to a pressure of 610 kPa, at a temperature of 19°C. Riding the bike a hot day increases the temperature of the tire to 58°C. The volume of the tire increases by 4.0%. What is the new pressure in the bicycle tire?
11. A sample of oxygen gas is collected over water at 25°C and a total pressure of 641 torr. The volume of gas collected is 500.0 mL. What mass of oxygen is collected? (At 25°C the vapor pressure of water is 23.8 torr.)
12. The partial pressure of CH_4(g) is 0.175 atm and O_2(g) is 0.250 atm in a mixture of the two gases.
   (a) What is the mole fraction of each gas in the mixture?
   (b) If the mixture occupies a volume of 10.5 L at 65°C, calculate the total number of moles of gas in the mixture.
   (c) Calculate the number of grams of each gas in the mixture.

Section 17.2, 17.3, 17.4

13. Using the postulates of kinetic theory derive
   (i) Charles law (ii) Dalton's law
   (iii) Boyle's law
14. What are the postulates of kinetic gas theory.

Section 17.6, 17.7

15. What is an ideal gas? Does it exist in nature?
16. What are the factors which led van der Waals to modify the ideal gas equation PV = nRT.
17. What are the units of van der Waals constants (a) and (b).
18. What is the significance of 'a' and 'b'.
19. Distinguish between ideal and real gas.
20. How does van der Waals' equation explain the exceptional behaviour of hydrogen and helium.
21. How does van der Waals' equation explain the behaviour of real gases when:
    (i) Pressure is low
    (ii) Pressure is high
    (iii) Temperature is high
22. What are different types of molecular velocities of gases.
23. What do you mean by 'mean' and r.m.s. velocity?
24. Calculate r.m.s. of CO_2 at 27°C.
25. What is most probable velocity?
26. Starting from Maxwell's Distribution law, derive an expression for root mean square velocity.
27. How the rms velocity varies with temperature.
28. Define the terms:
    (i) Mean free path.
    (ii) Collision frequency
29. Explain the effect of temperature and pressure on mean free path and number of collisions of the gas molecules.

Section 17.12

30. Write short note on liquefaction of gases.
31. What is the principle of Linde's process of liquefaction.
32. What is the difference in Linde's process and Claude's process of liquefaction of gases.

Long Answer Questions

1. (a) Deduce the following gas laws from the kinetic gas equation.
    (i) Boyle's law
    (ii) Charles' law
    (iii) Graham's law

   (b) Why is kinetic theory applicable only to ideal gases?
2. (a) Experimentally, it is known that gas laws are not applicable to real gases under all conditions of temperature and pressure. Under what conditions are these laws obeyed by real gases?

   (b) What is an ideal gas? Does it exist in nature?
3. (a) What are main assumptions of the kinetic theory of gases?

   (b) Derive kinetic gas equation.
4. (a) Explain the factors which led van der Waals to modify the ideal gas equation PV = nRT, and hence derive the van der Waals equation of state

$$\left(P + \frac{n^2 a}{V^2}\right)(V - nb) = nRT$$

   (b) What are the units of van der Waals constants a and b? Do they have the same values for all real gases? What is their significance?
5. (a) It is stated that the excluded volume b is approximately four times the actual volume occupied by the molecules. How would you derive this conclusion?

   (b) From the given value of b for a real gas, how would you estimate the molecular diameter of the molecule of the given gas?

6. (a) Explain how the van der Waals equation of state accounts for the behaviour of real gases.
   (b) Two van der Waals gases have the same value of b but different a values. Which of these would occupy greater volume under identical conditions? If the gases have the same a value but different values of b, which of the two would be more compressible?
   (**Hint :** Out of the gases having same value of b, but different value of a, the gas with a larger value of 'a' will have larger force of attraction between its molecules and hence occupies lesser volume. Out of the gases having same value of a and different values of b, the gas having smaller value of b will be more compressible.)
7. (a) Show that the Boyle temperature as predicted by the van der Waals equation is $Tb = \frac{a}{Rb}$
   (b) Can a van der Waals' gas with a = 0, be liquefied? Explain
8. Explain whether a gas approaches ideal behaviour or deviates from ideal behaviour if it is
   (a) compressed to a smaller volume at constant temperature,
   (b) temperature is raised keeping the volume constant,
   (c) more gas is introduced into the same volume and at the same temperature.
9. Give a brief account of Maxwell's distribution of velocities. Explain with the help of a graph how molecular velocities change with increase in temperature.
10. Define most probable velocity, average velocity and root mean square velocity. Write the expression for each of them and hence calculate the ratio between them.
11. Following the Maxwell distribution of molecular speeds, derive the expressions for:
    (i) average velocity
    (ii) root mean square velocity, and
    (iii) most probable velocity.
    Arrange these three speeds in order of increasing magnitude.
12. (a) With reference to gases explain the following terms:
    (i) Collision frequency
    (ii) Collision diameter
    (iii) Mean free path.
    (b) Explain the effect of temperature and pressure on mean free path and number of collisions of the gas molecules.
13. (a) Define the terms : Critical temperature, critical pressure and critical volume.
    (b) Explain why beyond a certain temperature gases cannot be liquefied, whatever the pressure may be.
14. Discuss Claude's process for liquefaction of gases.
15. Explain, giving reasons:
    (i) The mean free path of gas molecules decreases and number of collisions increases per unit time, with lowering of temperature.
    (ii) The molecular attractions between gas molecules are pronounced at low temperature.
    (iii) The difference between r.m.s. and most probable velocities remains constant with in crease in temperature.
    (iv) An ideal gas is not expected to show any cooling on free expansion.
16. Write short notes on:
    (a) Liquefaction of gases.
    (b) Mean free path.
    (c) Kinetic theory of gases.
    (d) Linde's process for liquefaction.

Problems

17. Calculate the volume of 11g of CO_2 at 760 mm Hg and 273°C, if the same mass of gas occupies 11.21 at 0°C and 380 mm Hg pressure. (11.2 L)
18. A gas occupies a volume of 150 dm^3 at 27°C and 1 atm. At what temperature the same gas will occupy 75 dm^3 at 1 atm? (150 K)

19. An iron cylinder contains helium at a pressure of 250 kPa at 300K. The cylinder can withstand a pressure of 1×10^6 Pa. The room in which cylinder is placed catches fire. Predict whether the cylinder will blow up before it melts or not (melting point of iron = 1800 K) (Cylinder will blow up)
20. A sample of air occupying 4.0 dm^3 at 127°C and 1.5 atm is cooled to 27°C and its pressure is reduced to 1 atm. What is the new volume of the sample? (4.5 L)
21. A gas occupies 75 mL at 25°C and 750 Torr. Calculate its volume at STP. (67.8 mL)
22. What volume would a gas occupy at 2 atm and 227°C if its volume at STP is 2.5 dm^3?
23. As the pressure on 6.20 dm^3 of nitrogen gas is increased from 1.0 atm to 3.1 atm, its temperature rises from 25°C to 50°C. Calculate the new volume of the gas. (2.17 L)
24. Calculate the volume occupied by 5.0 g of acetylene gas at 50°C and 740 mm Hg. (5.23L)
25. Two gases A and B having molecular weights 60 and 45 respectively are enclosed in a vessel. The weight of A is 0.5 g and that of B is 0.2 g. The total pressure of the mixture is 750 mm. Calculate the partial pressure of the two gases. (490 mm; 260 mm)
26. When 2.0 g of a gas A is introduced into an evacuated flask kept at 25°C, the pressure is found to be 1 atm. If 3 g of the another gas B is then added to the same flask, the total pressure becomes 1.5 atm. Assuming ideal gas behaviour, calculate the ratio of molecular weights $M_A : M_B$. (1:3)
27. Calculate the temperature at which the root mean square speed of SO_2 molecules is the same as that of oxygen at 27°C. (600K)
28. Calculate the r.m.s. speed of ozone kept in a closed vessel at 20°C and 82 cm Hg pressure. (3.9×10^4 cm/sec.)
29. A gas has a density of 1.2504 g/l at 0°C and a pressure of 1 atm. Calculate the rms, average and the most probable speeds of its molecules at 0°C. (4.93×10^4; 4.59×10^4; 4.03×10^4)
30. The rms speed of an ideal gas at 600 K is 569 ms^{-1}. What is the molar mass of the gas? (46.2 g mol^{-1})
31. Calculate the r.m.s. velocity of SO_2 molecules kept in a closed container at 47°C and 76 cm Hg pressure. (353.14 ms^{-1})
32. The molar mass of a gas is 16×10^{-3} kg mol^{-1}. What is the speed of its molecules at 298 K? At what temperature will the molecules of an another gas (molar mass 30×10^{-3} kg mol^{-1}) have the same speed? (6.9×102 m s–1, 562.6K.)
33. 1×10^{-2} m^3 of N_2 and 2×10^{-2} m^3 of O_2 at pressure of 700 and 500 mm of Hg respectively are introduced into a 10×10^{-3} m^3 flask. What will be the mole fraction of each gas in the mixture?
34. Calculate the ratio of the r.m.s. velocities of $^{238}UF_6$ and $^{235}UF_6$ molecules at 27°C. (0.996)
35. Calculate the ratio of (a) r.m.s. velocity to most probable velocity (b) average velocity to most probable speed. [(a) 1.22 (b) 1.13]
36. What is the ratio of the average molecular kinetic energy of $^{238}UF_6$ to that of H_2, both at 300 K? (1 : 1)
37. A certain saturated hydrocarbon effuses about half as fast as CH_4 gas. What is the molar mass and the molecular formula of the hydrocarbon? (64 g mol^{-1}; Butane)
38. Calculate the mean free path of O_2 molecules at 25°C and a pressure of 10^{-3} mm Hg, given that the collision diameter is 361 pm. (53 mm)
39. The coefficient of viscosity of O_2 gas is 208 micropoise. Calculate the mean free path and collision diameter of oxygen molecules at N.T.P. (7.17×10^{-8} m; 3.57×10^{-10} m)
40. The molecular diameter of CO is 3.19×10^{-10} m. At 300 K and a pressure of 100 mm Hg, what will be (a) the number of molecules colliding per unit volume, (b) the number of collisions per metre cube per second and (c) the mean free path of the gas molecules?
(a) 3.22×10^{24}, (b) 1.06×10^{22} and (c) 6.87×10^{-5}m.)
41. Calculate the pressure exerted by 22×10^{-2} kg of CO_2 in 0.5×10^{-3} m^3 capacity at 298 K using (a) the van der Waals equation and (b) the ideal gas equation (a and b are 0.0368 N m^4 mol^{-3} and 0.4267×10^3 m^3 mol^{-1} respectively. (a) 21.95 atm. (b) 24.43 atm.
42. The van der Waals constant 'b' for helium is 2.4333×10^{-5} m^3 mol^{-1}, calculate the diameter of helium atom. (2.68×10^{-10}m.)
43. Three different containers of equal volume at 27°C contain H_2, N_2, and O_2, respectively.
(a) Which one of the gases has the maximum average velocity of its molecules?

(b) Which of the following gases will have higher average energy per molecule at the same temperature? Give reasons.
(i) CO_2, (ii) H_2O vapour.

44. Calculate the pressure exerted by 22 g of carbon dioxide in 0.5 dm^2 at 298.15 K using : (a) the ideal gas law and (b) van der Waals equation. Given:
a = 363.76 kPa dm^6 mol^{-2} and b = 42.67 cm^3 mol^{-1} (2.479 × 10^3 kPa), (2225.55 kPa)

45. (a) Calculate the rms velocity of oxygen molecules at 27°C (R = 8.314 JK^{-1} mol^{-1}.
(b) What is meant by mean free path? Calculate the mean free path of oxygen at 25°C and 1 atm.
Collision diameter, σ = 3.61 A°
Boltzman constant k = 1.38 × 10^{-23} JK^{-1}
1 atm = 101325 N
(c) What is Boyle Temperature? Calculate the Boyle temperature T_B for CO_2 gas, assuming it to be a van der Waals' gas (a= 3.59 L^2 atm. mol^{-2}, b = 0.04271 mol^{-1}; R = 0.0821 L atm K^{-1} mol^{-1})
(d) Give suitable reasons, support or contradict the following :
(i) The parameters b in van der Waals' equation is related to intermolecular forces.
(ii) At low pressure and high temperature, a real gas behaves more or less like an ideal gas.
(G.G.S.I.P.U. End Term, B-Tech, July-2000)

46. (a) Discuss the reasons of deviation of gas equation PV = nRT from ideal behaviour. Derive Vander Waals' equation.
(b) Discuss Maxwell's Distribution of molecular velocities with the help of a diagram. Illustrate the effect of temperatures on this distribution. **(G.G.S.I.P.U II Term, B-Tech June-2000)**

47. (a) Derive the expression for root mean square velocity by using Maxwell's distribution equation.
(b) Calculate the temperature at which the root mean square velocity, the average velocity and the most probable velocity of oxygen gas are all equal to 1500 m/sec.
(c) Discuss Linde's process for liquefaction of air. **(G.G.S.I.P.U. II Term, B-Tech. June-2000)**

48. Give an account of kinetic theory of gases. How may the various gas laws be derived from it.
(G.G.S.I.P.U II Term B.Tech, April-2001)

49. (a) Discuss the three types of molecular velocities and derive an expression for root mean square velocity.
(b) A gas has a density of 1.2504 kg/m^3 at 0°C and a pressure of 1 atmosphere. Calculate (a) the root mean square velocity and (b) the average velocity of gas molecule.
(G.G.S.I.P.U. II Term, B.Tech, Apr. 2001)

50. (a) Derive van der Waals' equation of state for real gas. Give the units of van der Waals' constants.
(b) What will be the mean free path of oxygen at a pressure of 10^{-5} mm Hg at 25°C (0 = 3.57 × 10^{-10}m)? **(G.G.S.I.P.U. II Term B.Tech, Apr. 2001)**

51. (a) How the van der Waals' equation explains the behaviour of real gases when the pressure is very high or very low. Also discuss the exceptional behaviour of H_2 and He.
(b) Discuss Clande's process for liquefaction of gases.
(c) 2.0 moles of NH_3 at 300K occupy a volume of 5×10^{-3} m^3. Calculate the pressure using van der waals equation. (a = 0.417 Nm^4 mol^{-2} and b = 0.037 × $10^{-3}$$m^3$/mol)
(G.G.S.I.P.U. End Term, B-Tech May 2001)

52. Discuss the gas equation PV = nRT in relation to the forces responsible for the deviation from its ideal behaviour. **(G.G.S.I.P.U End Term, B-Tech., II Batch, July 2001)**

53. (a) Discuss Maxwell's Distribution of molecular velocities illustrating the effect of temperature on this distribution.
(b) Derive an expression for "Root mean square velocity."
(G.G.S.I.P.U. End Term, B-Tech, II Batch, July 2001)

54. (a) Explain the factors which are responsible for modifications of ideal gas equation. Derive the vander Waal's equation of state.
(b) Explain the behaviour of real gas when the pressure is very high and very low.
(G.G.S.I.P.U. II Term, April 2001)

55. (a) Discuss maxwell's distribution of molecular velocities showing the effect of temp. on its distribution
(b) Derive an expression for average velocity.

(c) 2.0 moles of ammonia at 300K occupy a volume 5×10^{-3} m^3. Calculate the pressure using the van der waal's equation. (a = 0.417 Nm^4 mol^{-2} and b = 0.037 × m^3/mole.)

(G.G.S.I.P.U. II Term, April 2001)

56. Write short notes on
Boyle's temperature
Mean free Path
LCAO theory **(G.G.S.I.P.U. II Term, April 2001)**

57. Explain giving reasons, the following statements
(a) The mean free path of a gas molecules decreases and the number of collisions per unit time increases with the lowering of temperature.
(b) An ideal gas can not be liquefied.
(c) The tyre of an automobile is inflated to lesser pressure in summer then in winter.
(d) The size of the weather balloon becomes larger and larger as it ascends up into higher altitude.
(e) The molecular attractions between gas molecules are pronounced at low temperature.

58. (a) Explain that excluded volume is four times the actual volume of gas
(b) Calculate the average and the root mean square speeds for oxygen molecule at 298 K. At what temperature would hydrogen have the same values of these speeds?

(G.G.S.I.P.U. II Term, Apr. 2002)

(c) Deduce the Charles' Law from the kinetic gas equation.
(d) Show that the most probable speed is given by 2RT/M.
(e) A teacher enters in the classroom from the front door while a student from the back door. There are 13 equidistant rows of benches in the classroom. The teacher releases N_2O, the laughing gas from the first bench while the student releases the weeping gas $C_6H_{11}OBr$ from the last bench. At which row will the student start laughing and weeping simultaneously?

(G.G.S.I.P.U. End Term, Apr. 2002)

59. Discuss the van der Waals equation at
(i) Mild P (ii) High P **(G.G.S.I.P.U. End Term, Mag. 2003)**

60. (a) How do real gases deviate in their behaviour from ideal gases
(b) What are fundamental assumptions of kinetic theory of gases

(G.G.S.I.P.U. II Term, May. 2003)

61. What is inversion temperature. How is it related to van der Waals constant

(G.G.S.I.P.U. End Term, Apr. 2003)

62. (a) Calculate the temperature at which the RMS, the average velocity and the most probable velocityiof oxygen are all equal to 1500 ms^{-1}.
(b) Derive the kinetic gas equation from the 1st principle for an ideal gas.
(c) What do you understand by mean free path. At 273 K and one atm pressure the coefficient of viscosity of H_2 is 8.41×10^{-6} PaS. Its density is 9×10^{-2} kg m^{-3} and average speed is 1.69×10^3 meters/s. Calculate the mean free path. **(G.G.S.I.P.U. End Term, May 2004)**

63. (a) What is the physical significance of constants 'a' and 'b' in van der Waals equation?
(b) Under what conditions a sample of ideal gas not only exhibits a pressure of one atm, but also a concentrationof one mole dm^3.

64. Write short notes on van der Waals forces. **(G.G.S.I.P.U. End Term, May 2004)**

65. (a) Derive various gas laws from kinetic gas equation.
(b) Explain the terms collision diameter and mean free path.
(c) Calculate the Boyle temperature for a gas whose van der Waals constants, a = 7.18 L^2 atm Mol^{-2}, b = 0.854 Mol^{-1} and R = 0.0821 L atm K^{-1} Mol^{-1}. **(G.G.S.I.P.U. II-Term, April 2004)**

66. What factors led to modification of ideal gas equation PV = nRT.
Derive van der Waals equation of state. What is the significance of constants a and b.

(G.G.S.I.P.U. II-Term, April 2004)

67. The pressure exerted by 4g of a gas A in a flask increases from 4 atm to 6 atm when 6g of another gas B is introduced into the flask at 298°K. Calculate the ratio of their molecular weights assuming the gases to be ideal. **(G.G.S.I.P.U. II-Term, April 2004)**

68. How does vanderwaal's equation explain the exceptional behaviour of Hydrogen and Helium.

(G.G.S.I.P.U. II-Term, B.Tech, April 2006)

69. Calculate the mean free path of oxygen at a pressure of 10^{-5} mm Hg at 25°C. (s = 3.57×10^{-10}m).
(G.G.S.I.P.U. II-Term, B.Tech, April 2006)

70. (a) Starting from Maxwell's distribution law derive an expression for root mean square velocity.
(b) A tank holding 56 L of oxygen gas at 5.0 atm. pressure and at 25°C developed at leak. When the leak was repaired, the tank had lost 82.0 g of the gas. Calculate the pressure of the gas remaining in the tank at 25°C.
(c) Gases can not be liquefied merely by the application of pressure above their critical temp. comment. **(G.G.S.I.P.U. II-Term, B.Tech, April 2006)**

71. What do you understand by the statement; "The mean free path of a gas molecule in a gaseous sample at 1 atm pressure and 25°C is 70.2 nm". What is the effect of temperature on the mean free path? **(G.G.S.I.P.U. End-Term, B.Tech, May-June 2006)**

72. Calculate the average kinetic energy of an ideal gas per molecule at 25°C.
(G.G.S.I.P.U. End-Term, B.Tech, May-June 2006)

73. What do you understand by the non-ideal behaviour of real gases? What are its causes? Derive van der Waal's Equation of state for a real gas.
(G.G.S.I.P.U. End-Term, B.Tech, May-June 2006)

74. Calculate the temperature at which the average velocity of O_2 equals that of H_2 at 20K.
(G.G.S.I.P.U. End-Term, B.Tech, May-June 2006)

75. At what temperature does a real gas behave as an ideal gas.
(G.G.S.I.P.U. II-Term, B.Tech, April 2007)

76. Can we liquefy a gas by increasing pressure alone? Why?
(G.G.S.I.P.U. II-Term, B.Tech, April 2007)

77. A fluoride of phosphorus in the gaseous state was found to diffuse more slowly by a factor of 2.12 than Nitrogen under similar conditions. Calculate the molecular weight of this fluoride. If it contains only one phosphorus atom, write its formula.
(G.G.S.I.P.U. II-Term, B.Tech, April 2007)

78. Calculate the r.m.s. velocity of Nitrogen at 300 K and 700 mm pressure. Density of Hg is 13.6 g per cm^3. **(G.G.S.I.P.U. II-Term, B.Tech, April 2007)**

79. (a) Derive the expression for r.m.s. velocity by using Maxwell's distribution equation.
(b) If two gases have same value of '*a*' and different values of '*b*', which gas is more compressible.
(G.G.S.I.P.U. II-Term, B.Tech, April 2007)

80. (a) (i) Why methane diffuses faster than oxygen.
(ii) The size of a weather balloon becomes larger and larger as it ascends up into higher altitudes. Explain.
(b) How does van der Waal's equaiton explain the behaviour of real gases when pressure is high.
(G.G.S.I.P.U. II-Term, B.Tech, Nov. 2007)

81. (a) What is excluded volume? Show that excluded volume is four times the actual volume of gas.
(b) Calculate the r.m.s. velocity and average velocity of nitrogen molecule at 27°C.
(c) Discuss Claude's process for liquefying air. **(G.G.S.I.P.U. II-Term, B.Tech, Nov. 2007)**

82. (a) Differentiate between a 'Real gas' and an 'Ideal gas'. Explain, how real gases deviate from ideal behaviour.
(b) Derive van der Waal's equation for real gases. Mention the significance of constant '*a*' and '*b*'.
(G.G.S.I.P.U. End-Term, B.Tech, Dec. 2007)

83. (a) List various methods used for liquification of gases. Describe any one method in detail.
(G.G.S.I.P.U. End-Term, B.Tech, Dec. 2007)

❑❑❑

18

THERMOCHEMISTRY

"There is a single light of science, and to brighten it anywhere is to brighten it everywhere"

— Issac Asimov

CHAPTER OUTLINE

Thermochemistry is the branch of physical chemistry which deals with the heat changes accompanying chemical reactions or change of phase. It is based on the first law of thermodynamics.

Chemical reactions are almost always accompanied by energy changes. These changes appear in the form of evolution or absorption of heat. If a chemical reaction is accompanied by evolution of heat, it is called exothermic reaction. For example:

$$C_{(s)} + O_{2(g)} \rightarrow CO_{2(g)} + 393.5 \text{ kJ}$$

On the other hand, if the chemical reaction is accompanied by absorption of heat, it is called endothermic reaction. For example :

$$H_{2(g)} + I_{2(g)} \rightarrow 2HI_{(g)} - 52.5 \text{ kJ}$$

But now the question is, how energy change occurs in a chemical reaction? This is because of the fact, that every substance is associated with a definite amount of energy known as its internal energy. Since, the total energy of the products of a chemical reaction may be different from the total energy of the reactants, so it is evident that the process will be accompanied by absorption or evolution of energy. It may be mentioned here that majority of the chemical reactions which go to near completion are accompanied by evolution of heat, i.e. they are exothermic.

So, it is the difference in the energies of reactants and products which are responsible for the energy changes in the chemical reactions. This difference of energies is also attributed to the different bond energies of reactants and products. Chemical reactions involve breaking of the existing bonds between the atoms and formation of new bonds. This change of bond energy that takes place in a chemical reaction appears in the form of heat.

18.1 INTERNAL ENERGY AND CHEMICAL REACTIONS

The energy possessed by a system due to translational, vibrational and rotational motion of the molecules, along with the energy possessed by its electrons, protons, etc. is known as its internal energy. In other words, the internal energy is the energy associated with the matter due to the position and motion of molecules, atoms and their subatomic particles. It is denoted by *E*.

Every substance is associated with a definite amount of internal energy but it is not possible to find out the absolute value of internal energy. This is because, it is impossible to determine accurately most of the quantities which contribute towards internal energy of the system. Fortunately, in thermodynamics the absolute value of internal energy is not required also. What is required, is the change in internal energy when reactants change into the products and this change in internal energy is a measurable quantity and can be determined experimentally.

Let us suppose that a system has internal energy E_1. It undergoes a chemical change and its internal energy changes to E_2. Change in internal energy of the system will be given by:

$$\Delta E = E_2 - E_1 \qquad ...(1)$$

As E_2 and E_1 are definite quantities, therefore, change in internal energy is also a definite quantity. Now the energy change will depend upon whether the reaction is taking place at *constant pressure* or at *constant volume.*

Let us consider that the reaction takes place at **constant volume** at a given temperature. This implies that work of expansion, i.e. $P\Delta V$ is zero. According to first law of thermodynamics

$$\Delta E = q + w \qquad ...(2)$$

$$\Delta E = q_v \qquad (\because w = 0) \qquad ...(3)$$

where q_v represents the heat change at constant volume.

Thus if the chemical reaction takes place at constant volume at a given temperature then the heat change is given by the difference in internal energies of products and reactants. The heat change is known as heat of reaction at constant volume.

But most of the reactants normally occur at atmospheric pressure, i.e. at constant pressure, therefore, heat changes are usually determined at constant pressure. This gives us a new thermodynamic term called **enthalpy** or heat content of the system.

18.2 ENTHALPY

To understand the meaning of the term 'enthalpy' let us consider a change of a system from state 1 to state 2 at constant pressure, at a given temperature. In such a case, change of volume will take place. Let the volume changes from V_1 to V_2, then the work of expansion is given by

$$w = P(V_2 - V_1) \qquad ...(4)$$

According to the first law:

$$\Delta E = q - w \qquad \text{(Assuming that work is done by the system and hence 'w' has negative sign)} \qquad ...(5)$$

Substituting the value of *w* in the equation (5)

$$\Delta E = q - P(V_2 - V_1) \qquad ...(6)$$

$$E_2 - E_1 = q - PV_2 + PV_1 \qquad ...(7)$$

$$(E_2 + PV_2) - (E_1 + PV_1) = q \qquad ...(8)$$

The quantity E + PV is known as the enthalpy of a system and is denoted by H. It represents the total energy stored in the system.

From Equation (8) we have,

$$H_2 - H_1 = \Delta H = q_p; \quad ...(9)$$

where q_p is the heat change at constant pressure

Since, H_2 and H_1 are definite quantities, ΔH is also a definite quantity. Further, it follows from equation (8) that

$$(E_2 - E_1) + P(V_2 - V_1) = q \quad ...(10)$$

$$\Delta E + P\Delta V = q \quad ...(11)$$

or

$$\Delta E + P\Delta V = \Delta H \quad ...(12)$$

Thus, enthalpy of the reaction at a given temperature is given by the difference in the enthalpies of the products and the reactants, the quantities of the reactants and products being the same as represented by the chemical equation.

Sign of ΔH

(i) If the enthalpy of the reactants is greater than that of the products $H_1 > H_2$, the energy will be evolved and given out to the surroundings, i.e. the reaction will be exothermic. In that case ΔH will have negative sign. For example, for the reaction,

$$N_{2(g)} + 3H_{2(g)} \rightarrow 2NH_{3(g)} \qquad \Delta H = -92.0 \text{ kJ}$$

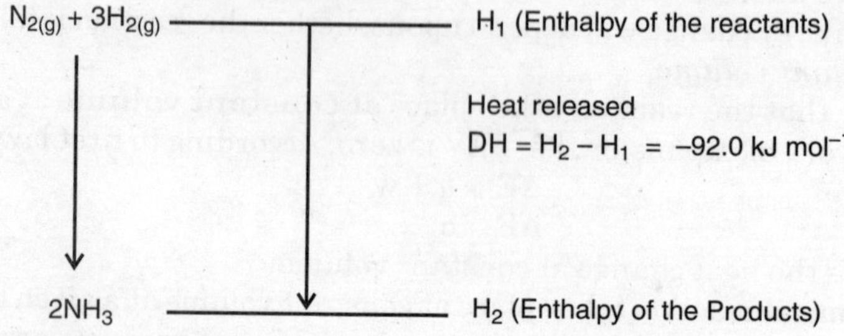

Fig. 18.1 Enthalpy Level Diagram for Exothermic Reaction

(ii) If the enthalpy of the products is greater than that of the reactants $H_1 < H_2$, the energy will be absorbed from the surroundings, i.e. the reaction will be endothermic. Thus in all endothermic reactions, ΔH will have positive sign. For example, for the reaction

$$C_{(graphite)} + 2S_{(rhombic)} \rightarrow CS_{2(1)} \qquad \Delta H = +92.0 \text{ kJ}$$

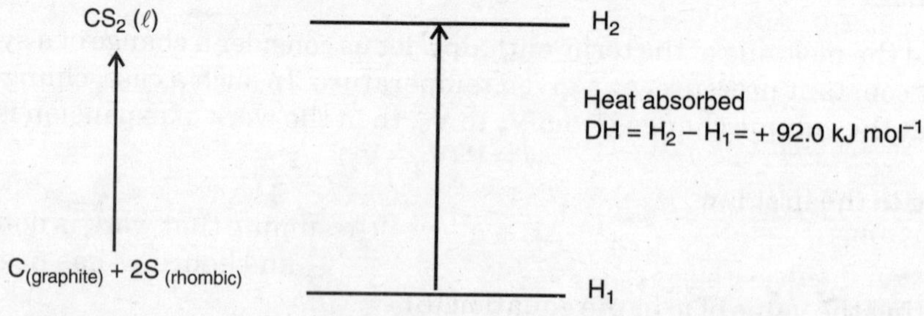

Fig. 18.2 Enthalpy Level Diagram For an Endothermic Reaction

Heat Changes at Constant Volume & Constant Pressure-Relation between the two: The relationship between the heat of reaction at constant volume ΔE and at constant pressure ΔH is given by equation (12) which is written as

$$\Delta E = \Delta H - P\Delta V \quad ...(13)$$

where ΔV is the increase of volume when the reaction occurs at constant pressure.

For a reaction involving *solids* and *liquids* only, the volume change ΔV is usually very small and hence $P\Delta V$ is negligible. Thus in this case $\Delta E = \Delta H$.

But for a reaction involving gases, the change in volume may be quite large. For example:

$$\underset{\text{2 moles}}{2H_{2(g)}O} \rightleftharpoons \underset{\text{2 moles}}{2H_{2(g)}} + \underset{\text{1 mole}}{O_{2(g)}}$$

$$\underset{\text{2 moles}}{2H_{2(g)}} + \underset{\text{1 moles}}{O_{2(g)}} \rightarrow \underset{\text{2 moles}}{2H_{(g)}}$$

If the above two reactions are allowed to take place at constant pressure then in the first case, as the total number of moles of the product is more than the number of moles of reactants and hence the work of expansion is done by the system. Whereas in the second case, the number of moles of the product is less than the total number of the reactants and hence the work of contraction is done on the system.

Thus in the case of gaseous reactions, the change in volume may take place if the reaction is carried out at constant pressure.

It has already been discussed that,

$$q_p = \Delta H \text{ and } q_v = \Delta E$$

Equation (13) can be written as

$$q_p = q_v + P\Delta V \qquad \text{...(14)}$$

If a system undergoes a volume change from V_1 to V_2 then equation (14) becomes

$$q_p = qv + P(V_2 - V_1) \qquad \text{...(15)}$$

$$= q_v + PV_2 - PV_1 \qquad \text{...(16)}$$

But for ideal gas:

$$PV = nRT \qquad \text{...(17)}$$

$\therefore$ For gaseous reactants:

$$PV_1 = n_1RT \qquad \text{...(18)}$$

and for gaseous products

$$PV_2 = n_2RT \qquad \text{...(19)}$$

Where n_1 and n_2 are the number of moles of gaseous reactants and products respectively. Substituting, the values of PV_1 and PV_2 from equation (18) and (19) respectively, in equation (16) we get,

$$q_v = q_p + n_2RT - n_1RT = q_p + (n_2 - n_1) = q_p + \Delta nRT \qquad \text{...(20)}$$

or

$$\Delta H = \Delta E + \Delta n\ RT \qquad \text{...(21)}$$

Conditions at which ΔH = ΔE

(i) When reaction involves only liquids and solids and hence the volume changes of solids and liquids are negligible.

(ii) When reaction involves the gaseous reactants and products which have same number of moles; Example

$$\underset{\text{1 mole}}{H_{2(g)}} + \underset{\text{1 mole}}{I_{2(g)}} \rightarrow \underset{\text{2 moles}}{2HI_{(g)}}$$

No of moles of reactants = 2, No. of moles of products = 2

Numericals Based on Relationship between ΔH and ΔE

Formulae	*Units*
$q_p = q_v + \Delta nRT$ or $\Delta H = \Delta E + \Delta nRT$	If q_p and q_v are in calories, R = 1.987 cal/deg/mol If q_p and q_v are in k calories. R = 0.002 k Cal/deg/mol If q_p and q_v are in joules R = 8.314 J/deg/mol

Example 18.1. The heat of combustion of CO at constant volume and at 17°C is –283.3 kJ. Calculate its heat of combustion at constant pressure.

Solution: The reaction is

$$CO_{(g)} + \frac{1}{2}O_{2(g)} \rightarrow CO_{2(g)}$$

$$1 \text{ mole} \quad \frac{1}{2} \text{ mole} \quad 1 \text{ mole}$$

No. of moles of products = 1; No. of moles of reactants = $1\frac{1}{2}$

$$\Delta n = 1 - 1\frac{1}{2} = -\frac{1}{2}$$

Also,

$$\Delta E = -283.3 \text{ kJ}$$

$$T = 17°C = (273+17)K = 290 \text{ K}$$

$$R = 8.314 \text{ J} = 8.314 \times 10^{-3} \text{ kJ}$$

Now,

$$\Delta H = \Delta E + \Delta n RT$$

$$= -283.3 + \left[-\frac{1}{2} \times (8.314 \times 10^{-3}) \times 290\right] = -284.5 \text{ kJ}$$

∴ Heat of combustion at CO at constant pressure is = –284.5 kJ

Example 18.2. For a reaction ΔH° = – 285.1 kJ mol^{-1} and ΔE = – 280.5 kJ mol^{-1}. What is the value of Δ*n* at 298 K?

Solution:

$$\Delta H = -285.1 \text{ kJ mol}^{-1}$$

$$\Delta E = -280.5 \text{ kJ mol}^{-1}$$

$$T = 298 \text{ K}$$

$$R = 8.314 \times 10^{-3} \text{ kJ mol}^{-1}$$

$$\Delta H = \Delta E + \Delta nRT$$

$$-285.1 \text{ kJ mol}^{-1} = -280.5 \text{ kJ mol}^{-1} + \Delta n 8.314 \times 10^{-3} \text{ kJ mol}^{-1} \times 298 \text{ K}$$

$$= \frac{-285.1 \text{kJ mol}^{-1} + 280.5 \text{kJ mol}^{-1}}{8.314 \times 10^{-3} \text{kJmol}^{-1} \times 298\text{K}} = -2$$

$$\therefore \quad \Delta n = -2$$

Example 18.3. When 1 mole of ice melts at 0°C at a constant pressure of 1 atm, 1440 cal of heat are absorbed by the system. The molar volumes of ice and water 0.0196 and 0.0180 litre respectively. Calculate Δ*H* and Δ*E*.

Solution: $H_2O_{(s)} \rightarrow H_2O_{(l)}$

$$\Delta H = 1440 \text{ cal mol}^{-1}$$

$$\Delta H = q_p = 1440 \text{ cal}$$
$$\Delta H = \Delta E + P\Delta V$$
$$P\Delta V = P(V_2 - V_1)$$
$$= 1\,(.0180 - 0.0196)$$
$$= -\,0.0016 \text{ atm. litre} = -\,0.039 \text{ Calories}$$

Since, $P\Delta V$ is very small compared to ΔE, hence $P\Delta V$ can be neglected and therefore $\Delta H = \Delta E$ = 1440 Calories.

18.3 HEAT CAPACITY

The heat capacity of a substance is defined as the amount of heat required to raise the temperature of the system from lower to higher temperature divided by the temperature difference, between any two temperatures.

If *q* calories is the heat absorbed by a substance having mass m, in raising its temperature from T_1 to T_2, the heat capacity will be given by the expression

$$C = \frac{q}{m \times (T_2 - T_1)} \qquad ...(22)$$

If m = 1gm, T_2–T_1=1K, then C is defined as **the amount of heat required to raise the temperature of 1 gm of a substance through one degree centigrade.** This is sometimes also referred to as **specific heat** of the substance. Its units are *cal deg*$^{-1}$ *g*$^{-1}$ whereas **molar heat capacity is defined as the amount of heat required to raise the temperature of 1 mole of a substance through one degree centigrade.** Its units are cal deg^{-} mol^{-1} or J deg^{-1} mol^{-1}.

Further as the heat capacity varies with temperature it should be defined as the differential equation. If dq is the small amount of heat absorbed by the system producing a small temperature change dT, then heat capacity is given by

$$C = \frac{dq}{dT} \qquad ...(23)$$

Heat capacity is **not a state function** and hence the amount required to produce a given change depends upon the path followed. The following two types of heat capacities are commonly used:

(i) Molar heat capacity at constant volume: As no external work is done by the system or on the system when the volume is kept constant, therefore heat absorbed by the system will be:

$$q_v = \Delta E \qquad ...(24)$$

$$\therefore \qquad C_V = \left[\frac{\Delta E}{T_2 - T_1}\right]_V \quad \text{or} \quad C_V = \left[\frac{\partial E}{\partial T}\right]_V$$

Since, the internal energy of an ideal gas is not a function of volume, for an ideal gas we can write

$$\boxed{C_V = \left[\frac{dE}{dT}\right]_V} \qquad ...(25)$$

where C_v is the heat capacity at constant volume, which is thus defined as rate of change of internal energy with temperature at constant volume.

(ii) Molar heat capacity at constant pressure: At constant pressure there is change of volume and hence some work is always done on the system or by the system. Suppose the work is done by the system, therefore heat absorbed by the system when it changes from temperature T_2 to T_1 will be :

$$q_p = \Delta H \qquad ...(26)$$

$$\therefore \quad C_P = \left[\frac{\Delta H}{T_2 - T_1}\right]_P = \left[\frac{\partial H}{\partial T}\right]_P \qquad ...(27)$$

Since, the enthalpy of an ideal gas is a function of temperature only and is independent of pressure, it follows that

$$\boxed{C_P = \left[\frac{dH}{dT}\right]_P} \qquad ...(28)$$

where C_P is the heat capacity at constant pressure, which is defined as the rate of change of enthalpy with temperature at constant pressure.

Relation Between C_P and C_v: By the definition of enthalpy, we know

$$H = E + PV$$

For ideal gas equation,

$$PV = RT$$

$$\therefore \quad H = E + RT \qquad ...(29)$$

Differentiating w.r.t temperature T we get,

$$\frac{dH}{dT} = \frac{dE}{dT} + R \qquad ...(30)$$

$$C_P = C_V + R \qquad ...(31)$$

$$C_P - C_V = R \qquad ...(32)$$

From equation (31) and (32) it is clear that:

(i) The difference between the molar heat capacity of a gas at constant pressure (C_P) and at constant volume (C_V) is equal to the gas constant R.

(ii) The two heat capacities are not equal and C_P is greater than C_V. This can be explained as follows: When a gas is heated at constant volume, the heat absorbed by the system is used to increase its internal energy as no external work is done. However if the gas is heated at constant pressure, there will be increase in its volume, i.e. work of expansion is done by the system. So some of the supplied heat will be used for performing this work by the system. Thus some extra heat, in addition to the heat required to raise the temperature of the system or to increase the internal energy, must be supplied and hence the system at constant pressure will absorb more heat or its heat capacity will be more than the heat capacity at constant volume. In other words, $C_P - C_V$ gives the work of expansion done by one mole of the gas, when it is heated through 1°C at constant pressure.

18.4 HEAT OF REACTION OR ENTHALPY OF REACTION

Heat of reaction is defined as the amount of heat absorbed or evolved in a reaction when the number of moles of reactants as represented by the balanced chemical equation, are converted into the products at constant temperature and pressure. Since most of the reactions occur at constant pressure, it is usual to express the heat of the reaction as enthalpy change accompanying the reaction.

When 1 mole of gaseous methane react with 2 moles of oxygen at 298 K and 1 atmospheric pressure to form 1 mole of gaseous CO_2 and 2 moles of liquid water at the same temperature and pressure, 890.35 kJ of energy is given out to the surroundings. It can be represented by thermochemical equation as:

$$CH_{4(g)} + 2O_{2(g)} \rightarrow CO_{2(g)} + 2H_2O_{(l)} \qquad \Delta H = -890.35 \text{ kJ}$$

A balanced chemical equation which indicates the amount of heat absorbed or evolved in the reaction process is called thermochemical equation.

Similarly, formation of liquid H_2O from gaseous hydrogen and oxygen can be represented as the following thermochemical equation:

$$H_{2(g)} + \frac{1}{2}O_{2(g)} \rightarrow H_2O_{(l)} \qquad \Delta H = -285.8 \text{ kJ.}$$

Thus heat of reaction or enthalpy of reaction in this case is –285.8 kJ.

Heat of Reaction Depends Upon the Following Factors

(i) **Physical state of the reactants and products:** Physical state of the reactants and products affects the heat of reaction. For example, the heat of reaction when water vapour is formed from gaseous hydrogen and gaseous oxygen, is different from heat of reaction in the formation of liquid CO_2 from the same constituents at same temperature and pressure.

$$H_{2(g)} + \frac{1}{2}O_{2(g)} \rightarrow H_2O_{(l)} \qquad \Delta H = -285.8 \text{ kJ}$$

$$H_{2(g)} + \frac{1}{2}O_{2(g)} \rightarrow H_2O_{(g)} \qquad \Delta H = -248.8 \text{ kJ}$$

This is because the absorption of latent heat of vaporisation is required to convert liquid water to gaseous state.

The physical states of reactants and products are indicated by writing 's', '*l*', g for solid state, liquid state, gaseous state respectively in the parenthesis as shown above.

(ii) **Amounts of reactants involved:** If the amount of the reactants is doubled, the heat of reaction is also doubled. For example, if the quantities of hydrogen and oxygen are doubled in the above reaction, the heat of reaction will also be doubled.

$$2H_{2(g)} + O_{2(g)} \rightarrow 2H_2O_{(l)} \qquad \Delta H = -571.6 \text{ kJ}$$

Heat of reaction is generally reported for 1 mole of the reactant.

(iii) **Whether the reaction is taking place at constant pressure or constant volume:** This factor has already been discussed in section (18.1).

(iv) **Temperature:** The enthalpy change accompanying a chemical reaction varies with temperature. The values of heat of reaction are usually reported at 298 K.

(v) **Concentration of the solution:** If the solutions are involved in the reaction, their concentration may affect the heat of reaction.

18.5 DIFFERENT TYPES OF ENTHALPIES OF REACTIONS

Heat of the reaction is a general term used for heat change or enthalpy change accompanying any reaction. However depending upon the nature of reaction–whether it is neutralisation or combustion etc. the enthalpy of the reaction is named accordingly. Some of the important heats of the reaction are discussed below.

1. Heat of Formation: The heat of formation of a substance is defined as the heat change when 1 mole of the substance is formed from its elements under given conditions of temperature and pressure. It is usually represented as ΔH_f. For example, the heat of formation of CO_2 and SO_2 are –393.5 kJ and –294.5 kJ respectively.

$$C_{(s)} + O_{2(g)} \rightarrow CO_{2(g)} \qquad \Delta H = \Delta H_f = -393.5 \text{ kJ}$$
$$S_{(rhombic)} + O_{2(g)} \rightarrow SO_{2(g)} \qquad \Delta H = \Delta H_f = -297.5 \text{ kJ}$$

Similarly, the reaction between gaseous H_2 and Cl_2 to for $HCl_{(g)}$ can be represented as following thermochemical equation :

$$H_{2(g)} + Cl_{2(g)} \rightarrow 2HCl \qquad \Delta H = -184.6 \text{ kJ}$$

But in this case –184.6 kJ is not the enthalpy of formation of HCl, because heat change accompanying in the above reaction involve the formation of two moles. The heat of formation of hydrogen chloride will be –184.6 kJ/2 = –92.3 kJ and the thermochemical equation will be represented as:

$$\frac{1}{2}H_{2(g)} + \frac{1}{2}Cl_{2(g)} \rightarrow HCl \qquad \Delta H = \Delta H_f - 92.3 \text{ kJ}$$

The formation of 2 moles of NH_3 involve the following thermochemical equation:

$$N_2 + 3H_2 \rightarrow 2NH_3 \qquad \Delta H = -92.0 \text{ kJ}$$

The formation of 1 mole of NH_3 or heat of formation of NH_3 is given by:

$$\frac{1}{2}N_2 + \frac{3}{2}H_2 \rightarrow NH_3 \qquad \Delta H = \Delta H_f = -46.06 \text{ kJ}$$

Standard Heat of Formation

The standard state of any substance is defined as its most stable physical form at one atmospheric pressure at a specified temperature. For making comparisons and for convenience the temperature chosen is 298 K.

Standard heat of formation of a substance is defined as the heat change accompanying the formation of 1 mole of the substance in the standard state from its elements, also taken in the standard state, i.e. at 298 K and 1 atmospheric pressure. It is represented as ΔH_f^o. For example:

$$\frac{1}{2}H_{2(g)} + \frac{1}{2}Br_{2(g)} \rightarrow HBr_{(g)} \qquad \Delta H_f^o = 36.4 \text{ kJ}$$

$$H_{2(g)} + \frac{1}{2}O_{2(g)} \rightarrow H_2O_{(l)} \qquad \Delta H_f^o = -285.8 \text{ kJ}$$

By convention, the enthalpy of every element in its standard state is arbitrarily assigned a zero value. Enthalpy of formation of Na, Mg, Al, H_2, O_2, F_2 etc is 0.00 kJ.

Exothermic and Endothermic Compounds

(i) **Exothermic Compound:** The negative value of enthalpy of formation of a compound indicates that, it has lower energy than the elements from which it is formed, and hence is considered more stable. Such compounds are called **exothermic compounds.** For example:

$$H_{2(g)} + \frac{1}{2}O_{2(g)} \rightarrow H_2O_{(g)} \qquad \Delta H_f^o = -24.8 \text{ kJ}$$

As enthalpy of formation has negative value and thus $H_2O_{(g)}$ is more stable than either a molecule of H_2 or O_2. The chemical bonds in H_2O are stronger than chemical bonds in H_2 and O_2.

(ii) **Endothermic Compounds:** The positive value of enthalpy of formation of a compound indicates that it has higher energy than the elements from which it is formed, and hence considered less stable. Such compounds are called **endothermic compounds** For example:

$$\frac{1}{2}H_{2(g)} + \frac{1}{2}I_{2(g)} \rightarrow HI_{(g)} \qquad \Delta H_f = +25.9\ kJ$$

As enthalpy of formation has positive value and thus $HI_{(g)}$ is less stable than either of the molecules of H_2 or I_2. The chemical bonds in HI are weaker than those in H_2 and I_2.

Significance of Standard Enthalpies of Formation

1. **Calculation of standard heat of reaction (ΔH°) from standard heat of formation (ΔH_f^o):** We can calculate the heat of reaction under standard conditions knowing the standard heat of formation or enthalpy of formation of different compounds involved in a chemical reaction. The standard heat of reaction is equal to standard heat of formation of products minus the standard heat of formation of reactants.

$$\Delta H^o = \begin{bmatrix}\text{Sum of standard heat} \\ \text{of formation of products}\end{bmatrix} - \begin{bmatrix}\text{Sum of standard heat} \\ \text{of formation of reactants}\end{bmatrix}$$

$$\Delta H^o_{reaction} = \Sigma\, \Delta H^o_{f(Products)} - \Sigma\, \Delta^o_{f(Reactants)}$$

For a general reaction,

$$p^P + q^Q \rightarrow x^X + y^Y$$

The standard heat of reaction is given by:

$$\Delta H^o = \Delta H^o_{f(Products)} - \Delta^o_{f(Reactants)}$$
$$= [x \times \Delta H_f^o(X) + y \times \Delta H_f^o(Y)] - [P \times \Delta H_f^o(p) + q \times \Delta H_f^o(Q)]$$

2. **Calculating the standard enthalpies of the compounds:** It was discussed earlier, that every substance is associated with a definite amount of enthalpy, but it is not possible to find out the absolute enthalpy of a substance. But enthalpy of formation enables us to determine the relative enthalpy of a compound. As, enthalpies of elements in their standard state are taken as zero, therefore enthalpies of the compounds so obtained become identical with their standard enthalpies of formation. Consider the formation of $NH_{3(g)}$ from $N_{2(g)}$ and $H_{2(g)}$

$$\frac{1}{2}N_{2(g)} + \frac{3}{2}H_{2(g)} \rightarrow NH_{3(g)} \qquad \Delta H_f^o = -46.0\ kJ$$

Now we know

$$\Delta H_{reaction} = \Sigma\, \Delta H^o_{f(Products)} - \Sigma\, \Delta^o_{f(Reactants)}$$
$$= \text{Enthalpy of } NH_3 - (\text{Enthalpy of } N_2 + \text{Enthalpy of } H_2)$$
$$= \text{Enthalpy of } NH_3\ (\because \text{Enthalpy of elements are taken as zero}).$$

$$\text{But } \Delta H_{reaction} = \Delta H_f^o = -\,46.0\ kJ$$

Thus standard *enthalpy of formation of NH_3* evidently becomes the enthalpy of the compound itself, *i.e. enthalpy of NH_3*.

Standard enthalpy of a compound is often referred to as standard enthalpy of **formation** of compound. The standard enthalpies of formation of some important compounds are given in Table 18.1.

Table 18.1 Standard Enthalpies of Formation of Compounds

Substance	*Standard Enthalpy of Formation, ΔH_f^o (kJ mol^{-1})*	*Substance*	*Standard Enthalpy of Formation, ΔH_f^o (kJ mol^{-1})*
	Inorganic Compounds		
$H_2O(l)$	–286.2	AgCl (s)	–127.0
H_2O (g)	–245.5	$AgNO_3$ (s)	–123.0
HCl (g)	–92.0	Ag_2O (s)	–30.6
HBr (g)	–36.2	CuO (s)	–156.9
HI (g)	+25.9	$CaCO_3$ (s)	–1206.9
HNO_3 (l)	–173.2	FeO(s)	–269.0
H_2SO_4 (l)	–811.3	Fe_2O_3(s)	–822.1
H_2S (g)	–20.2	Fe_3O_4(s)	–1117.1
SO_2 (g)	–297.5	KCl(s)	–435.9
SO_3 (g)	–395.2	KOH(s)	–451.9
NO (g)	+90.4	NaCl(s)	–411.0
NO_2 (g)	+33.8	NaOH(s)	–426.8
NH_3 (g)	–46.0	Na_2SO_4(s)	–140.2
CO (g)	–110.5	Na_2CO_3 (s)	–1129.7
CO_2 (g)	–393.5	$PbSO_4$ (s)	–914.2
	Organic Compounds		
Methane (g)	–74.8	*Acetylene (g)*	+226.7
Ethane (g)	–89.2	*Benzene (l)*	+49.0
Propane (g)	–103.8	*Naphthalene (s)*	+60.2
n-Butane (g)	–124.7	*Methanol (l)*	–238.6
n-Hexane (g)	–167.2	*Ethanol (l)*	–277.6
n-Octane (g)	–208.4	*Ethanol (g)*	–236.4
Ethylene (g)	+52.3	*Acetic acid (l)*	–487.0

Numerical Problems Based On Enthalpy of Formation

Formula

$$\Delta H^o = \Sigma\, \Delta H^o_{f(Products)} - \Sigma\, \Delta^o_{f(Reactants)}$$

Units of ΔH = kJ mol^{-1}

Example 18.4. What is the standard enthalpy change for the reaction:

$$SO_{2(g)} + 2H_2S_{(g)} \rightarrow 3S_{(s)} + 2H_2O_{(l)}$$

$$\Delta H_f \text{ of } SO_2 = -296.9 \text{ kJ mol}^{-1}$$

$$\Delta H_f \text{ of } H_2S = -20.15 \text{ kJ mol}^{-1}$$

$$\Delta H_f \text{ of } H_2O = -285.9 \text{ kJ mol}^{-1}$$

Solution:

$$\Delta H^o = \left[3\Delta H_f^\circ\, S_{(g)} + 2\Delta H_f^\circ\, H_2O_{(l)}\right] - \left[\Delta H_f\, SO_{2(g)} + 2\Delta Hf\, H_2S_{(g)}\right]$$

$$= \left[3\times 0 + 2(-285.9)\text{kJ mol}^{-1}\right] - \left[-296.9 + 2\left(-20.15\text{kJ mol}^{-1}\right)\right]$$

$$= [-571.8\text{kJ}] - [-337.20\text{kJ}]$$

$$= -234.6 \text{ kJ}$$

∴ Standard enthalpy change for the reaction is –234.6 kJ

Example 18.5. Compute the enthalpy of formation of nitric oxide (NO) from the following data:

$$NO_{(g)} + CO_{(g)} \rightarrow \frac{1}{2}N_{2(g)} + CO_{2(g)} \qquad \Delta H^\circ = -372.2 \text{ kJ mol}^{-1}$$

$$\Delta H_f^\circ(CO)_{(g)} = -110.5 \text{ kJ mol}^{-1}$$

$$\Delta H_f^\circ(CO_2)_{(g)} = -393.5 \text{ kJ mol}^{-1}$$

Solution:

$$\Delta H^\circ = \left[\frac{1}{2}\Delta H_f^\circ N_{2(g)} + \Delta H_f^\circ CO_2\right] - \left[\Delta H_f^\circ NO_{(g)} + \Delta H_f^\circ CO_{(g)}\right]$$

$$-372.2 \text{ kJ mol}^{-1} = \left[\frac{1}{2}\times 0 + (-393.5)\text{kJ mol}^{-1}\right] - \left[\Delta H_f^\circ NO_{(g)} - 110.5\text{kJ mol}^{-1}\right]$$

$$\therefore \qquad \Delta H_f^\circ NO = (-393.5 + 110.5 + 372.2)\text{kJ mol}^{-1}$$

$$= 89.2 \text{ kJ mol}^{-1}$$

Example 18.6. Using the standard enthalpy of formation calculate the standard enthalpy change (ΔH°) and standard internal energy change (ΔE°) for the reaction at 298 K.

$$OF_{2(g)} + H_2O_{(g)} \rightarrow 2HF_{(g)} + O_{2(g)}$$

$$\Delta H_f^\circ OF_{2(g)} = 23 \text{ kJ mol}^{-1}$$

$$\Delta H_f^\circ H_2O_{(g)} = -241.8\text{kJ mol}^{-1}$$

$$\Delta H_f^\circ HF_{(g)} = -268.6 \text{ kJ mol}^{-1}$$

Solution:

$$\Delta H^\circ = \left[2\Delta H_f^\circ HF_{(g)} + \Delta Hf\, O_{2(g)}\right] - \left[\Delta H_f^\circ OF_{2(g)} + \Delta H_f^\circ H_2O_{(g)}\right]$$

$$= [-2\times 268.6 + 0]\text{ kJ mol}^{-1} - [23.0 - 242.8] \text{ kJ mol}^{-1}$$

$$= -537.2\text{kJ mol}^{-1} + 218.8\text{kJ mol}^{-1} = -318.4\text{kJ mol}^{-1}$$

Now for the given reaction:

$$\Delta n = \text{Number of moles of products–Number of moles of reactants.}$$

$$= 3 - 2 = 1$$

$$\Delta H^\circ = \Delta E^\circ + \Delta n\, RT$$

$$-318.4 \text{ kJ mol}^{-1} = \Delta E^\circ + 1\times 8.314\times 10^{-3}\text{ kJ mol}^{-1}\text{ K}^{-1}\times 298\text{K}$$

$$\Delta E^\circ = -318.4\text{kJ mol}^{-1} - 2.477\text{kJ mol}^{-1}$$

$$= -320.87 \text{ kJ mol}^{-1} = -320.9 \text{ kJ mol}^{-1}$$

3. **Heat of Combustion:** The heat of combustion of a substance is defined as the heat change when one mole of the substance is completely burnt in excess of air or oxygen. For example, the heat of combustion of methane is –890.4 kJ as shown below:

$$CH_{4(g)} + 2O_{2(g)} \rightarrow CO_{2(g)} + 2H_2O_{(l)} \qquad \Delta H = -890.4 \text{ kJ}$$

It should be noted that in determining heat of combustion of a substance, the substance should be completely oxidised. Let us make this point clear by considering the following example:

$$C_{(s)} + \frac{1}{2}O_{2(g)} \rightarrow CO_{(g)} \qquad \Delta H = -110.5\text{kJ}$$

$$C_{(s)} + O_{2(g)} \rightarrow CO_{2(g)} \qquad \Delta H = -393.5\text{kJ}$$

Now heat of combustion of carbon will be taken as –393.5 kJ and not –110.5 kJ because in first reaction, combustion is not complete, CO can be further oxidised to carbon dioxide.

Heat of combustion of a substance is always negative, i.e. energy is always evolved as a result of combustion.

Significance of Heat of Combustion

1. **Calculation of heat of formation:** Heat of combustion of a number of compounds can be determined easily hence this may be used to calculate the heat of formation of a compound which otherwise may be quite difficult. For example the heat of combustion of ethane is –1559.7 kJ.

$$C_2H_{6(g)} + \frac{7}{2}O_{2(g)} \rightarrow 2CO_{2(g)} + 3H_2O_{(l)} \qquad \Delta H^\circ = -1559.7\,kJ$$

The enthalpy of formation of $C_2H_{6(g)}$ can be calculated if we know the heat of formation of $CO_{2(g)}$ and $H_2O_{(l)}$ as we know:

$$\Delta H^\circ = -1559.7\,kJ = \left[2\times\Delta H_f^\circ\, CO_{2(g)} + 3\times\Delta H_f^\circ\, H_2O_{(l)}\right] - \left[\Delta H_f^\circ\, C_2H_{6(g)} + \frac{7}{2}\Delta H_f^\circ\, O_{2(g)}\right]$$

$$\therefore\ \Delta H_f^\circ\, C_2H_{6(g)} = \left[2\times\Delta H_f^\circ\, CO_{2(g)} + 3\times\Delta H_f^\circ\, H_2O_{(l)} - \frac{7}{2}\Delta Hf^\circ O_{2(g)} + 1559.7\,kJ\right]$$

$$= \left[2\times\Delta H_f^\circ\, CO_{2(g)} + 3\Delta H_f^\circ\, H_2O_{(l)} + 1559.7\,kJ\right]$$

Hence, knowing the values of ΔH_f° of $CO_{2(g)}$ and ΔH_f° of $H_2O_{(l)}$, $\Delta H_f^\circ C_2H_{6(g)}$ can be determined.

2. **Determination of Calorific Values of foods and fuels:** The calorific value is defined as the amount of heat in calories or joules produced when 1 gram of a substance is completely burnt. From heat of combustion of different substances calorific values of these substances can be determined. For example, the heat of combustion of butane the main constituent of LPG is –2878.7 kJ.

$$C_4H_{10}(g) + 6\frac{1}{2}O_2(g) \rightarrow 4CO_2(g) + 5H_2O(\ell) \quad \Delta H = -2878.7\,kJ$$

Thus by complete burning of 1 mole of butane, i.e. 58 g of butane –2878.7 kJ of energy is produced. The heat produced by burning 1 gram of butane will be –2878.7/58 kJ = –49.6 kJ. Thus calorific value of butane is –49.6 kJ g^{-1}.

NUMERICALS BASED ON HEAT OF COMBUSTION

Example 18.7. **The heat of combustion of butane (C_4H_{10}) is –2878.7 kJ. From the given data, find out the standard enthalpy of formation of butane.**

$$\Delta H_f^\circ\, CO_{2(g)} = -393.5\,kJ\,mol^{-1}; \quad \Delta H_f^\circ\, H_2O_{(l)} = -285.8\,kJ\,mol^{-1}$$

Solution. The thermochemical equation of combustion of butane is represented as:

$$C_4H_{10(g)} + 6\frac{1}{2}O_{2(g)} \rightarrow 4CO_{2(g)} + 5H_2O_{(l)} \qquad \Delta H^\circ = -2878.7\,kJ$$

$$\therefore \qquad \Delta H^\circ = \left[4\Delta H_f^\circ\, CO_{2(g)} + 5\Delta H_f^\circ\, H_2O_{(l)}\right]$$

$$-2878.7\,kJ = \left[4\times(-393.5) + 5\times(-285.8)\,kJ\,mol^{-1}\right] - \Delta H_f^\circ\, C_4H_{10(g)} + 0$$

$$\Delta H_f^\circ\, C_4H_{10(g)} = -4\times393.5 - 5\times285.8 + 2878.7\,kJ\,mol^{-1} = -124.3\,kJ\,mol^{-1}$$

Example 18.8. The heat of combustion of methane is –809 kJ. Calculate the calorific value of methane.

Solution: The combustion of methane is represented as :

$$CH_{4(g)} + 2O_{2(g)} \rightarrow CO_{2(g)} + 2H_2O_{(g)} \qquad \Delta H = -809\,kJ$$

$$\text{Molar mass of } CH_4 = 16 \text{ g}$$

Calorific value is heat liberated by burning 1g of the substance hence,

$$\text{Calorifice Value of } CH_4 = \frac{809}{16} = 50.6 \text{ kJ g}^{-1}$$

Example 18.9. The standard heat of formation of $CH_{4(g)}$ $CO_{2(g)}$ and $H_2O_{(g)}$ are –76.2, –394.8 and 241.6 kJ respectively. Calculate the amount of heat evolved by burning 1m^3 of methane measured under normal conditions.

Solution: $CH_{4(g)} + 2O_{2(g)} \rightarrow CO_{2(g)} + 2H_2O_{(l)} \quad \Delta H° = ?$

$$\Delta H° = \Sigma\, \Delta H^o_{f(Products)} - \Sigma\, \Delta^o_{f(Reactants)}$$

$$= \left[\Delta H^\circ_f\, CO_{2(g)} + 2\Delta H^\circ_f\, H_2O_{(l)}\right] - \left[\Delta H^\circ_f\, CH_4 + \Delta H^\circ_f\, O_2\right]$$

$$= \left[-394.8 + 2(-241.6)\right]kJ - \left[-76.2\right]kJ = -801.8\,kJ\,mol^{-1}$$

$$1 \text{ mol} = 22.4 \text{ litres} = 22400 \text{ cm}^3 \text{ at STP}$$

$$1 \text{ m}^3 = (100 \text{ cm})^3 = 10^6 \text{ cm}^3.$$

$$22400 \text{ cm}^3 = 1 \text{ mole of methane}$$

$$10^6 \text{ cm}^3 = \frac{1}{22400} \times 10^6 = 417.64 \text{ mol}$$

Heat evolved on burning 1 mole of CH_4 = 801.8 kJ
Heat evolved on burning 44.64 mole of $CH_4 = 801.8 \times 44.64 = 3.58 \times 10^4$ kJ

3. Heat of Neutralisation: The heat of neutralisation of an acid is defined as **the heat change when one gram equivalent of an acid is neutralized by a base in dilute solutions**. For example the heat of neutralisation of hydrochloric acid by sodium hydroxide in dilute solutions is –57.32 kJ and is represented by:

$$HCl_{(aq)} + NaOH_{(aq)} \rightarrow NaCl_{(aq)} + H_2O \qquad \Delta H° = -57.32\,kJ$$

The symbol (aq.) indicates that the reaction occurs in such a dilute solution that no more heat changes are observed on further dilution.

Similarly, heat of neutralisation of a base is defined **as the heat change when one gram equivalent of a base is neutralised by an acid in dilute solutions**. The heat of neutralisation of by NaOH by HCl is also –57.32 *kJ* and is represented as:

$$HCl_{(aq)} + NaOH_{(aq)} \rightarrow NaCl_{(aq)} + H_2O \qquad \Delta H = -57.32\,kJ$$

Heat of neutralisation of strong acid by strong bases :

The heat of neutralisation of strong acid and strong bases are given in Table 18.2 :

Table 18.2 Enthalpy of Neutralisation of strong acids by strong bases

Chemical Equation	*ΔH(kJ/equiv)*
$HCl_{(aq)} + NaOH_{(aq)} \rightarrow NaCl_{(aq)} + H_2O_{(l)}$	–57.32
$HNO_{3(aq)} + NaOH_{(aq)} \rightarrow NaNO_{3(aq)} + H_2O_{(l)}$	–57.28
$HCl_{(aq)} + KOH_{(aq)} \rightarrow KCl_{(aq)} + H_2O_{(l)}$	–57.45
$HCl_{(aq)} + LiOH_{(aq)} \rightarrow LiCl_{(aq)} + H_2O_{(l)}$	–57.38

From the table, it is evident that enthalpy of neutralisation of strong acid with a strong base is identical in all cases. This can be explained on the basis of theory of ionisation. Strong acid and strong bases are assumed to be completely ionised in dilute aqueous solutions. Moreover, the salts they form on mixing are also completely ionised. The reaction between a strong acid like HCl and a strong base like NaOH can be written as:

$$H^+_{(aq)} + Cl^-_{(aq)} + Na^+_{(aq)} + \bar{O}H_{(aq)} \rightarrow Na^+_{(aq)} + Cl^-_{aq} + H_2O_{(l)} \qquad \Delta H = -57.3\,kJ$$

$$H^+_{(aq)} + \bar{O}H_{(aq)} \rightarrow H_2O_{(l)} \qquad \Delta H = -57.3\,kJ$$

Thus neutralisation is simply a reaction between the H^+ ions given by the acid and $\bar{O}H$ ions given by the base. In case of strong acids and strong bases the number of H^+ ions and $\bar{O}H$ ions produced by one gm equivalent of any strong acid or strong base is always the same. Hence the heat of neutralisation between a strong acid and a strong base is always same.

Heat of Neutralisation if Either the Acid or Base is Weak

The heat of neutralisation of strong acid and weak base or vice-versa is always less than –57.32 kJ, and is different in different cases as shown in Table 18.3.

Table 18.3

Chemical Equation	*ΔH(kJ/equiv)*
$CH_3COOH + NaOH \rightarrow CH_3COONa + H_2O$	–55.23
$HCN + NaOH \rightarrow NaCN + H_2O$	–12.13
$H_2S + NaOH \rightarrow Na_2S + H_2O$	–15.90
$HCl + NH_4OH \rightarrow NH_4Cl + H_2O$	–51.46

In such cases the neutralisation process not only involves the combination of hydrogen and hydroxyl ions, but also the dissociation of weak acid or base. Consider the neutralisation of a weak acid like CH_3COOH with a strong base like NaOH. Acetic acid ionises to a small extent whereas NaOH ionises completely.

$$CH_3COOH \rightleftharpoons CH_3COO^- + H^+ \qquad \text{(i)}$$

$$NaOH \longrightarrow Na^+ + \bar{O}H \qquad \text{(ii)}$$

$$H^+ + \bar{O}H \longrightarrow H_2O \qquad \text{(iii)}$$

As H^+ ions given by the acid combine with $\bar{O}H$ ions given by the base, the equilibrium shifts towards the right resulting in dissociation of more of acetic acid. Both the reactions, (i) and (iii) proceed side by side till acetic acid is completely neutralised. **Thus a part of heat produced during the combination of H^+ and OH^- ions is used up for the complete dissociation of acetic acid**. This heat required for dissociation of acetic acid is called heat of dissociation of acetic acid.

The heat of neutralisation of acetic acid by NaOH has been found to be –55.23 kJ which is less than the average value of –57.32 kJ for the combination of H^+ and $\bar{O}H$ ions. The difference between the two, i.e. +2.09 kJ is the heat required for dissociation of acetic acid and hence is called heat of dissociation of acetic acid.

Similarly, the neutralisation of a weak base like NH_4OH with a strong acid like *HCl* can be represented as:

$$NH_4OH \rightleftharpoons NH_4^+ + \bar{O}H \quad ...(i)$$

$$HCl \longrightarrow H^+ + Cl^- \quad ...(ii)$$

$$H^+ + \bar{O}H \longrightarrow H_2O \quad ...(iii)$$

The heat of neutralisation comes out to be –51.34 kJ which is less than the usual value and hence the enthalpy of dissociation of NH_4OH is 57.32–51.34 = 5.98 kJ.

NUMERICALS BASED ON HEAT OF NEUTRALISATION

Example 18.10. Calculate the amount of heat released when 400 mL of 0.2 M of H_2SO_4 is mixed with 600 mL of 0.1 MNaOH solution. What would be the rise in the temperature of the final solution? Take specific heat of water = 4.18 $JK^{-1}\,g^{-1}$.

Solution: Number of moles of H_2SO_4 in 400 mL of 0.2 M acid $= 0.2 \times \frac{400}{1000} = 0.08$ mol

Number of moles of NaOH in 600 mL of 0.1 M $= \frac{0.1 \times 600}{1000} = 0.06$ mol

Thus 0.06 mol NaOH will neutralise 0.06 mol of H^+ ions and the net reaction is

$$0.06\,\text{mol}\,H^+ + 0.06\,\text{mol}\,\bar{O}H \rightarrow 0.06\,\text{mol}\,H_2O$$

Heat of Neutralisation of strong acid and strong base = 57.32 kJ mol^{-1}

∴ Heat released in the above reaction = 0.06 × 57.32 = 3.439 kJ

Temperature Change

Heat gained = msθ

where m is mass = 1000 gm

(∵ Density of solution is assumed to be = density of water = 1g/mL) s is specific heat θ is temperature change

$$\therefore \quad \theta = \frac{\text{Heat Gained}}{\text{mass} \times \text{s}} = \frac{3.439 \times 1000\,\text{J}}{1000\,\text{g} \times 4.18\,\text{Jk}^{-1}\text{g}^{-1}} = 0.82\text{ K.}$$

So, the temperature of the solution will rise by 0.82 K.

Example 18.11. Calculate the amount of heat evolved when 500 cm^3 of 0.1 M hydrochloric acid is mixed with 200 cm^3 of 0.2 M sodium hydroxide solution. What would be the rise in temperature if the specific heat of water is 4.18 JK^{-1} g^{-1}.

Solution: 500 cm^3 of 0.1 MHCl $= \frac{0.1}{1000} \times 500 = 0.05$ moles of H^+ ions

200 cm^3 of 0.2 M NaOH $= \frac{0.2}{1000} \times 200 = 0.04$ moles of $\overline{O}H$ ions.

Thus the net reaction is– $0.04 H^+ + 0.04 \overline{O}H \rightarrow 0.04 H_2O$

Heat evolved when 1 mole of H^+ combines with 1 mole of $\overline{O}H = 57.32$ kJ.

∴ Heat evolved when 0.04 mole of H^+ combines with 0.04 mole of $\overline{O}H = 57.32 \times 0.04$
$= 2.292$ kJ.

Heat produced = 2.292 kJ.

Density of solution = Density of water = 1g/ml

Total mass of the solution = 500 + 200 = 700 ml = 700 g

specific heat = 4.184 JK^{-1} g^{-1}

$$\text{Heat produced } \theta = ms\theta, \quad \therefore \theta = \frac{Q}{m \times s}$$

$$\theta = \frac{2292}{4.184 \times 700} = 0.78 \text{ K}$$

4. Heat of Solution: The heat of solution of a solute in a particular solvent depends upon the amount of solvent and accordingly it can be defined in two ways:

(i) ***Integral Heat of solution :*** The heat of solution of a solute in a particular solvent is defined **as the heat change when one mole of the solute is dissolved in the pure solvent to give a solution of specified concentration.** Thus for integral heat of solution, it is necessary to specify the number of moles of the solvent used. For example if one mole of *HCl* is dissolved in 25 moles of water, 72.266 *kJ* of heat is evolved. The thermochemical equation for the above dissolution is written as :

$$HCl_{(g)} + 25H_2O_{(l)} \rightarrow HCl(25H_2O) \quad \Delta H = -72.266 \text{ kJ}$$

The integral heat of solution of hydrogen chloride at different dilutions are given in Table 18.4.

Table 18.4 Integral Heats of Solution of HCl in Water at 298 K

Chemical Equation	*ΔH in kJ*
(i) $HCl_{(g)} + 2H_2O_{(l)} \rightarrow HCl(2H_2O)$	–47.99
(ii) $HCl_{(g)} + 5H_2O_{(l)} \rightarrow HCl(5H_2O)$	–63.09
(iii) $HCl_{(g)} + 10H_2O_{(l)} \rightarrow HCl(10H_2O)$	–68.07
(iv) $HCl_{(g)} + 50H_2O_{(l)} \rightarrow HCl(50H_2O)$	–72.00
(v) $HCl_{(g)} + 100H_2O_{(l)} \rightarrow HCl(100H_2O)$	–72.59
(vi) $HCl_{(g)} + 200H_2O_{(l)} \rightarrow HCl(200H_2O)$	–72.93
(vii) $HCl_{(g)} + 400H_2O_{(l)} \rightarrow HCl(400H_2O)$	–72.18
(viii) $HCl_{(g)} + aq \rightarrow HCl(aq)$	–73.35

The limiting value of integral heat of solution will be the total heat of solution which is –73.35 kJ in the above case.

(ii) **Differential heat of solution:** The differential heat of solution is defined as the heat change when 1 mole of solute is dissolved in such a large quantity of the solution that there occurs no appreciable change in concentration of the solution.
Mathematically differential enthalpy of solution can be represented as :

$$\text{Differential enthalpy of solution} = \frac{d\Delta H}{dm}$$

where *d*(Δ*H*)*d*(Δ*H*) is the small change in value of ΔH when a small amount *dm* of solute is added to a solution of definite composition.
To illustrate the differential heat of solution, consider a solution containing one mole of HCl dissolved in 20 moles of water. Now if we take large volume of this solution and dissolve 1 mole more of HCl, the heat change produced will be called the differential heat of solution of HCl. This heat change is produced without changing the concentation of the solution appreciably.

It should be mentioned that;

(i) Salts like $CuSO_4\ 5H_2O$, $CaCl_2.\ 6H_2O$, etc. or in general hydrated salts dissolve with the absorption of heat. For example:

$$CuSO_4.5H_2O + aq \rightarrow CuSO_{4(aq)} \qquad \Delta H = +11.7\ kJ$$

(ii) Salts which do not form hydrates also dissolve with the absorption of heat

$$kCl_{(s)} + aq \rightarrow kCl_{(aq)} \qquad \Delta H = +18.6\ kJ$$

(iii) The anhydrated salts which are capable of forming hydrates, e.g., $CaCl_2$, $CuSO_4$, etc. are accompanied by evolution of heat and as a result the solution heats up.

$$CuSO_{4(s)} + aq \rightarrow CuSO_{4\ (aq)} \qquad \Delta H = -\ 66.5\ kJ$$

5. Heat of Hydration: The heat of hydration is defined as **the heat change when 1 mole of anhydrous substance combines with the requisite amount of water to form a hydrated stable salt**. Hydration of $CuSO_4$ is represented by the following thermochemical equation.

$$CuSO_{4(s)} + 5H_2O_{(l)} \rightarrow CuSO_4\ 5H_2O_{(s)} \qquad \Delta H = -78.868\,kJ$$

Heat of hydration is almost always negative, i.e. heat is always evolved in such reactions. Heat of hydration is always calculated from heat of solutions of hydrated and anhydrous substances. For example heat of hydration of $CuSO_4$ can be calculated from the following thermochemical equations:

(i) $CuSO_{4(s)} + 800\,H_2O_{(l)} \rightarrow CuSO_4\,(800\,H_2O) \qquad \Delta H_1 = -68.74\,kJ$

(ii) $CuSO_4\,.5H_2O + 795\,H_2O_{(l)} \rightarrow CuSO_4\,(800\,H_2O) \quad \Delta H_2 = 10.12\,kJ$

∴ Heat of hydration of $CuSO_4 = \Delta H_1 - \Delta H_2 = -78.86$ kJ

6. Heat of Dilution: Heat of dilution is defined as **the change of enthalpy when a solution containing 1 mole of a solute is diluted from one concentration to another**.

It has been mentioned earlier that heat of solution per mole of a solute varies with the concentration of the solution. (Table : 18.4) This implies that when a given solution is diluted on addition of solvent there must be a change in enthalpy which is called heat of dilution.

The enthalpy of dilution also called integral enthalpy of dilution can be calculated by finding the integral heat of solution at two different concentrations. For example the enthalpy of dilution when a solution is diluted from kCl (20 H_2O) to (kCl 200H_2O) can be found out by following thermochemical equations:

(i) $kCl_{(s)} + 20H_2O \rightarrow kCl(20H_2O)$ $\Delta H_1 = +15.90\,kJ$

(ii) $kCl_{(s)} + 200H_2O \rightarrow kCl(200H_2O)$ $\Delta H_2 = +18.58\,kJ$

Heat of dilution = $\Delta H_2 - \Delta H_1 = 2.68$ kJ

18.6 HESS'S LAW OF CONSTANT HEAT SUMMATION

G.H Hess, a Russian chemist, in 1840, proposed the law of constant heat summation based on his experimental observations. This law states that **the amount of heat-evolved or absorbed in a process is the same whether the process takes place in one step or in number of steps**. In other words, the total amount of heat change depends upon the initial state and final state of the reaction and is independent of the path by which the change has been brought about.

This law is a consequence of first law of thermodynamics. Consider a general reaction where a substance A changes to a product D directly releasing Q Joules of energy

$$A \rightarrow D + Q$$

Now, suppose the same reaction takes place in three different steps and q_1 q_2 and q_3 is the energy released in first, second and third step respectively,

$$A \rightarrow B + q_1$$
$$B \rightarrow C + q_2$$
$$C \rightarrow D + q_3$$

Thus the total heat evolved = $q_1 + q_2 + q_3$ = Q' Joules

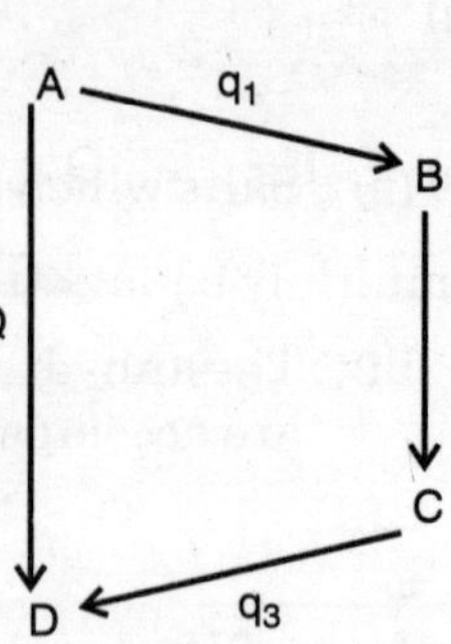

Fig. 18.3

According to Hess's Law:

$$Q = Q'$$

If Hess's law is not correct then, either $Q > Q'$ or $Q < Q'$.

Suppose $Q' > Q$. This means that if we go from A to D in number of steps but return to A directly, then heat evolved will be more than heat absorbed. Thus there is net gain of energy which will be $Q' - Q$. By repeating this process a number of times unlimited energy can be produced. This is however against the law of conservation of energy, i.e. first law. Hence Q must be equal to Q′. Thus Hess's law must be correct. This is verified experimentally also. For example, carbon can be burnt to CO_2 in one step or in number of steps.

1st path

$C_{(s)} + O_{2(g)} \rightarrow CO_{2(g)}$ $\Delta H = -393.7\,kJ$

2nd path

$C_{(s)} + \frac{1}{2}O_{2(g)} \rightarrow CO_{(g)}$ $\Delta H = -110.5\,kJ$

$CO_{(s)} + \frac{1}{2}O_{2(g)} \rightarrow CO_{2(g)}$ $\Delta H = -282.8\,kJ$

$C_{(s)} + O_{2(g)} \rightarrow CO_{2(g)}$ $\Delta H = -393.7\,kJ$

It is evident from the above that the total heat energy evolved is the same in the two paths.

Applications of Hess's Law

Hess's law has been extremely useful in the indirect determination of heats of formation, heats of transition, heats of combustion, etc. which cannot be determined experimentally. One of the consequences of Hess's law which forms the basis of these determinations is that the thermochemical equations can be treated as algebric equations which can be subtracted, added, multiplied or divided. A few important applications of Hess's law are discussed below :

1. **Calculation of heat of formation:** Hess's law has made it possible to calculate heat of formation of many compounds which cannot be measured directly. This is illustrated in the following example. Suppose we want to determine the heat of formation of ethane.

$$2C_{(s)} + 3H_{2(g)} \rightarrow C_2H_{6(g)} \qquad \Delta H = ?$$

This is impossible to determine by direct measurement, because 2 moles of carbon do not combine directly with 3 moles of hydrogen to form ethane. However Hess's law has made it possible to determine the heat of formation of ethane on the basis of certain reactions whose heat of reactions can be determined directly.

$$\text{(i)}\quad C_{(s)} + O_{2(g)} \rightarrow CO_{2(g)} \qquad \Delta H = -393.5\,\text{kJ}$$

$$\text{(ii)}\quad H_{2(g)} + \frac{1}{2}O_{2(g)} \rightarrow H_2O_{(l)} \qquad \Delta H = -285.85\,\text{kJ}$$

$$\text{(iii)}\quad C_2H_{6(g)} + \frac{7}{2}O_{2(g)} \rightarrow 2CO_2 + 3H_2O_{(l)} \qquad \Delta H = -1559.88\,\text{kJ}$$

If we multiply equation (i) by 2 and equation (ii) by 3 and add (i) and (ii), we get

$$2C_{(s)} + 2O_{2(g)} \rightarrow 2CO_{2(g)} \qquad \Delta H = -787.0\,\text{kJ}$$

$$3H_{2(g)} + \frac{3}{2}O_{2(g)} \rightarrow 3H_2O_{(l)} \qquad \Delta H = -857.55\,\text{kJ}$$

$$\text{(iv)}\quad 2C_{(s)} + 3H_2 + \frac{7}{2}O_{2(g)} \rightarrow 2CO_2 + 3H_2O \qquad \Delta H = -1644.55\,\text{kJ}$$

Reversing equation (iii)–

$$\text{(v)}\quad 2CO_{2(g)} + 3H_2O_{(l)} \rightarrow C_2H_{6(g)} + \frac{7}{2}O_{2(g)} \qquad \Delta H = 1559.88\,\text{kJ}$$

Adding equation (iv) and (v) we get–

$$2C_{(s)} + 3H_2 \rightarrow C_2H_{6(g)} \qquad \Delta H = -84.67\,\text{kJ}$$

Thus heat of formation of ethane = –84.67 kJ

2. **Calculation of heat of allotropic transformations:** Several elements exist in allotropic forms. The change of one allotropic form to another involves a very small amount of heat and is a very slow process. Hence it is difficult to determine experimentally. These can be calculated by the application of Hess's law. Example–we can find the heat change accompanying the transformation of C (graphite) to C (diamond) because heats of combustion of graphite and diamond can be determined experimentally.

$$\text{(i)}\quad C_{(graphite)} + O_{2(g)} \rightarrow CO_{2(g)} \qquad \Delta H = -393.5\,\text{kJ}$$

$$\text{(ii)}\quad C_{(diamond)} + O_{2(g)} \rightarrow CO_{2(g)} \qquad \Delta H = -395.4\,\text{kJ}$$

Subtracting equation (ii) from (i) we get,

$$C_{(graphite)} \rightarrow C_{(diamond)} \qquad \Delta H = +1.9\,kJ$$

Hence, heat of transition of $C_{(graphite)}$ to $C_{(diamond)}$ is +1.9 kJ.

3. Calculation of heat of reaction: By using Hess's law we can calculate the heats of many reactions which otherwise cannot be measured directly. Example : from the following thermochemical equations we can find out the heat of dimerisation of NO_2.

$$\text{(i)}\ N_{2(g)} + 2O_{2(g)} \rightarrow 2NO_{2(g)} \qquad \Delta H = 67.9\,kJ$$

$$\text{(ii)}\ N_{2(g)} + 2O_{2(g)} \rightarrow N_2O_{4(g)} \qquad \Delta H = 9.3\,kJ$$

Subtracting equation (i) from (ii) we have

$$2NO_{2(g)} \rightarrow N_2O_{4(g)} \qquad \Delta H = -58.6\,kJ$$

Thus heat of dimerization of NO_2 is –58.6 kJ.

NUMERICAL PROBLEMS BASED ON HESS'S LAW

I. Calculation of Heat of Formation

Example 18.12. Calculate the heat of formation of benzene at 25° C if the heats of combustion of benzene, carbon and hydrogen are –3267.62, –393.5 kJ and –285.85 kJ respectively at 298 K and 1 atm.

Solution: The combustion of benzene, carbon and hydrogen at 298 K can be represented by the following thermochemical equations:

$$\text{(i)}\ C_6H_{6(l)} + 7\frac{1}{2}O_{2(g)} \rightarrow 6CO_{2(g)} + 3H_2O_{(l)} \qquad \Delta H = -3267.62\,kJ$$

$$\text{(ii)}\ C(s) + O_{2(g)} \rightarrow CO_{2(g)} \qquad \Delta H = -393.5\,kJ$$

$$\text{(iii)}\ H_{2(g)} + \frac{1}{2}O_{2(g)} \rightarrow H_2O_{(g)} \qquad \Delta H = -285.85\,kJ$$

$$6C_{(s)} + 3H_{2(g)} \rightarrow C_6H_{6(l)} \qquad \Delta H = ?$$

Multiply equation (ii) by 6 and equation (iii) by 3 and add (ii) and (iii)

$$6C_{(s)} + 6O_{2(g)} \rightarrow 6CO_{2(g)} \qquad \Delta H = -2361.0\,kJ$$

$$3H_{2(g)} + \frac{3}{2}O_{2(g)} \rightarrow 3H_2O_{(\ell)} \qquad \Delta H = -857.55\,kJ$$

$$\text{(iv)}\ 6C_{(s)} + 3H_{2(g)} + 7\frac{1}{2}O_{2(g)} \rightarrow 6CO_{2(g)} + 3H_2O_{(l)} \qquad \Delta H = -3218.55\,kJ$$

Subtract equation (i) from equation (iv)

$$6C_{(s)} + 3H_{2(g)} \rightarrow C_6H_{6(l)} \qquad \Delta H = +49.07\,kJ$$

Thus heat of formation of benzene at 25° C is +49.07 *kJ*.

Example 18.13. Calculate the standard heat of formation of acetylene from the heat of combustion of C_2H_2, graphite and H_2 being as –1300 kJ mol^{-1}, 395 kJ mol^{-1} and –286 kJ mol^{-1} respectively.

Solution: The combustion of C_2H_2, graphite and H_2 can be represented as:

(i) $C_2H_2 + 2.5O_2 \rightarrow 2CO_2 + H_2O$ $\quad \Delta H = -1300\,kJ\,mol^{-1}$

(ii) $H_2 + \frac{1}{2}O_2 \rightarrow H_2O$ $\quad \Delta H = -286\,kJ\,mol^{-1}$

(iii) $C + O_2 \rightarrow CO_2$ $\quad \Delta H = -395\,kJ\,mol^{-1}$

We have to calculate for the following reaction:

$$2C_{(graphite)} + H_{2(g)} \rightarrow C_2H_{2(g)} \qquad \Delta H = ?$$

Multiply equation (iii) by 2 and add (ii) (iii)

$$2C + 2O_2 \rightarrow 2CO_2 \qquad \Delta H = -790\,kJ\,mol^{-1}$$

$$H_2 + \frac{1}{2}O_2 \rightarrow H_2O \qquad \Delta H = -286\,kJ\,mol^{-1}$$

(iv) $2C + H_2 + 2.5O_2 \rightarrow 2CO_2 + H_2O$ $\quad \Delta H = -1076\,kJ\,mol^{-1}$

Subtract equation (i) from equation (iv)

$$2C + H_2 \rightarrow C_2H_2 \qquad \Delta H = 224\,kJ\,mol^{-1}$$

Thus the standard heat of formation of acelytene = 224 kJ mol^{-1}

II Calculation of Heat of Allotropic Transformation

Example 18.14. Calculate the heat of transition of S (rhombic) to S (monoclinic) from the following data.

(i) $S_{(rhombic)} + O_{2(g)} \rightarrow SO_{2(g)}$ $\quad \Delta H = -296.9\,kJ$

(ii) $S_{(monoclinic)} + O_{2(g)} \rightarrow SO_{2(g)}$ $\quad \Delta H = -297.2\,kJ$

Solution:

(i) $S_{(rhombic)} \rightarrow S_{(monoclinic)}$ $\quad \Delta H = ?$

Subtracting equation (i) from equation (ii)

$$S_{(rhombic)} \rightarrow S_{(monoclinic)} \qquad \Delta H = 2.5\ kJ$$

III Calculation of Heat of Reaction

Example 18.15. Calculate the heat of reaction for the hydrogenation of acetylene at constant volume at 25°C given that heat of formation of water = –68.3 k Cal, heat of combustion of acetylene = –310.6 k Cal and heat of combustion of ethylene = –337.2 k Cal.

Solution:

(i) $H_{2(g)} + \frac{1}{2}O_{2(g)} \rightarrow H_2O_{(l)}$ $\quad \Delta H = -68.3\,k\,Cal$

(ii) $C_2H_{2(g)} + \frac{5}{2}O_{2(g)} \rightarrow H_2O_{(l)} + 2CO_{2(g)}$ $\quad \Delta H = -310.6\,k\,Cal$

(iii) $C_2H_{4(g)} + 3O_{2(g)} \rightarrow 2H_2O_{(l)} + 2CO_{2(g)}$ $\quad \Delta H = -337.2\,k\,Cal$

Add equation (i) and (ii)

(iv) $H_{2(g)} + C_2H_{2(g)} + 3O_{2(g)} \rightarrow 2H_2O + 2CO_2$ $\Delta H = -378.9\,k\,Cal$

Subtract equation (iii) from equation (iv)

$C_2H_{2(g)} + H_{2(g)} \rightarrow C_2H_4$ $\Delta H = -41.7\,k\,Cal$

Heat of reaction at constant pressure = –41.7 k Cal.
We know

$$\Delta H = \Delta E + \Delta nRT$$

$$\Delta n = 1 - 2 = -1$$

$$\therefore \quad -41.7\,k\,Cal = \Delta E + [(-1)mol\,(1.987\times10^{-3}\,k\,Cal\,K^{-}\,mol^{-1})\times 298\,K]$$

$$= \Delta E - 0.5921$$

$$\Delta H = -41.7 + 0.5921\,k\,Cal$$

$$= -41.1\,k\,Cal$$

Heat of hydrogenation of acetylene at constant volume = –41.1 k Cal.

18.7 VARIATION OF ENTHALPY OF A REACTION WITH TEMPERATURE-KIRCHHOFF'S EQUATION

The enthalpy of a reaction varies with the temperature of the reaction. The variation of enthalpy with temperature can be derived by the following two methods:

I. Consider a simple hypothetical reaction in which the reactants in a state A at temperature T_1 change to the final state B at the temperature T_2.

$$AT_1 \rightarrow BT_2$$

This change can be brought about by two paths:

(a) The change A to B is made to occur at T_1 at constant pressure and the heat absorbed in the reaction is equal to the heat contents of the final and initial state at the temperature T_1, i.e. $(H_B - H_A)_1 = \Delta H_1$. The temperature of the state B now is raised from T_1 to T_2, thus the heat supplied for this change will be $C_p\,(T_2 - T_1)$ where C_p is the heat capacity of the products. Hence the total heat change will be given by:

$$\Delta H_{(path\,I)} = C_P\,(T_2 - T_1) + \Delta H_1 \quad ...(33)$$

(b) Alternatively, in the second path the temperature of A is first raised from T_1 to T_2. The heat supplied for this change will be $C_P(T_2 - T_1)$ where $C_p'$is the heat capacity of the reactants. Now A at temperature T_2 is allowed to change into product B accompanying an enthalpy change ΔH_2.

Total heat change will be given by :

$$\Delta H_{(path\,II)} = Cp'\,(T_2 - T_1) + DH_2$$

Graphically, it can be represented as Fig (18.4). Now according to first law of thermodynamics:

$$\Delta H_{(Path\,I)} = \Delta H_{(Path\,II)} \quad .....(34)$$

$$\text{or } C_P\,(T_2 - T_1) + \Delta H_1 = C_P'\,(T_2 - T_1) + \Delta H_2$$

$$\frac{\Delta H_2 - \Delta H_1}{(T_2 - T_1)} = C_P - C'_P = \Delta C_P \qquad \text{... (35)}$$

where ΔCp is the difference in the heat capacities of final and initial states. This equation can also be written in the form:

$$\frac{\Delta H_2 - \Delta H_1}{\Delta T} = \Delta CP \qquad \text{... (36)}$$

If ΔT is small increase in temperature then the above equation can be written as:

$$\left(\frac{\partial(\Delta H)}{\partial T}\right)_P = \Delta C_P \qquad \text{... (37)}$$

This relationship, developed by G.R. Kirchhoff (1858) is known as **Kirchhoff equation.**

II. Alternatively the same results can be obtained by utilizing the heat capacities as follows:

Consider a hypothetical reaction.

$$p^P + q^Q \rightarrow x^X + y^Y$$

The enthalpy change for the above reaction is expressed as:

$$\Delta H = \Sigma H_{(Products)} - \Sigma H_{(Reactants)}$$

or

$$\Delta H = ({}_x HX + {}_y HY) - ({}_P HP + {}_q HQ)$$

At constant pressure, its variation with temperature is given as:

$$\left[\frac{\partial(\Delta H)}{\partial T}\right]_P = \left[x\left(\frac{\partial HX}{\partial T}\right)_P + y\left(\frac{\partial HY}{\partial T}\right)_P\right] - \left[p\left(\frac{\partial HP}{\partial T}\right)_P + q\left(\frac{\partial HQ}{\partial T}\right)\right]$$

$$= (x.C_{P,X} + y C_{P,Y}) - ({}_p C_{P,P} + q C_{P,Q})$$

$$= \Delta C_p \qquad \text{... (38)}$$

where ΔC_p = Sum of the heat capacities of the products – sum of the heat capacities of the reactants

Thus

$$\left[\frac{\partial(\Delta H)}{\partial T}\right]_P = \Delta C_p \qquad \text{... (39)}$$

i.e. change in heat of reaction at constant pressure per degree change of temperature is equal to difference in heat capacities of products and reactants at constant pressure. Similarly the Kirchhoff's equation for a reaction **at constant volume** can be deduced. The increase in internal energy will give the heat absorbed. If change in heat capacity at constant volume is represented as ΔC_V then

$$\left[\frac{\partial(\Delta E)}{\partial T}\right]_V = \Delta C_V \qquad \text{......(40)}$$

Two cases may be discussed on the basis of Kirchhoff's relation.

Case 1: If the temperature difference is not too large : In this case it is assumed that heat capacity is independent of temperature. Thus integrating equation (39) between two temperatures T_1 and T_2 gives

$$\int_{T_1}^{T_2} d(\Delta H) = \int_{T_1}^{T_2} \Delta C_P \, dT = \Delta C_P \int_{T_1}^{T_2} dT$$

$$\Delta H_2 - \Delta H_1 = \Delta C_P (T_2 - T_1) \qquad \text{...... (41)}$$

By the use of equation (41) it is possible to calculate the enthalpy change for reactions at different temperatures, if enthalpy change and ΔC_P at one temperature is known.

Case 2: If temperature range is not small : In this case the heat capacities of reactants and products vary with temperature. The heat capacities are expressed as a function of temperature as power series as

$$C_p = a + bT + cT^2 \ldots.$$

where a, b, c are constants for a given substance. Their values are determined from experimental data on heat capacities over a range of temperature. Hence,

$$\Delta C_p = \Delta a + \Delta bT + \Delta cT^2 \ldots. \qquad \text{......(42)}$$

For the reaction

$$_pP + {_q}Q \rightarrow {_x}X + {_y}Y$$

$$\Delta a = (xa_X + ya_Y) - (pa_p + qa_Q)$$

$$\Delta b = (xb_X + yb_Y) - (pb_p + qb_Q)$$

$$\Delta c = (xc_X + yc_Y) - (pc_p + qc_Q)$$

Integrating equation (39) after substituting the value of ΔC_p from equation (37) we get;

$$\int_{T_1}^{T_2} d(\Delta H) = \int_{T_1}^{T_2} \left(\Delta a + \Delta bT + \Delta cT^2 \ldots.\right) dT$$

$$= \int_{T_1}^{T_2} \Delta a\, dT + \int_{T_1}^{T_2} \Delta bT\, dT + \int_{T_1}^{T_2} \Delta cT^2\, dT \ldots.$$

$$= \Delta a (T_2 - T_1) + \Delta b \left(\frac{T_2^2 - T_1^2}{2} \right) + \Delta c \left(\frac{T_2^3 - T_1^3}{3} \right) + \ldots.$$

$$= \Delta a (T_2 - T_1) + \frac{1}{2} \Delta b (T_2^2 - T_1^2) + \frac{1}{3} \Delta c (T_2^3 - T_1^3) + \ldots. \qquad \text{...(43)}$$

Equation (43) is the integrated Kirchhoff's equation.

NUMERICALS BASED ON KIRCHHOFF'S EQAUTION

Example 18.16. Calculate heat of reaction at 358 K for the reaction $Fe_2O_{3(s)} + 3H_{2(g)} \rightarrow 2Fe_{(s)} + 3H_2O_{(l)}$. Given that, heat of reaction at 298 K is –33.29 kJ mol^{-1}. The heat capacities C_P for $Fe_2\,O_{3(s)}$, $Fe_{(s)}$, $H_2O_{(l)}$ and $H_{2(g)}$ are 103.8, 25.1, 75.3 and 28.8 JK^{-1} mol^{-1}.

Solution: $Fe_2O_{3(s)} + 3H_{2(g)} \rightarrow 2Fe_{(s)} + 3H_2O_{(l)}$

$$\Delta C_p = \text{Heat capacities of products} - \text{Heat capacities of reactants.}$$

$$= [2 \times 25.1 + 3 \times 75.3] - [103.8 + 3 \times 28.8] \text{JK}^{-1} \text{ mol}^{-1}$$

$$= 85.9 \text{ JK}^{-1}$$

According to Kirchhoff's equation:

$$\frac{\Delta H_2 - \Delta H_1}{T_2 - T_1} = \Delta C_p$$

Now, $\Delta H_1 = -33.29\,kJ\,mol^{-1} = -33290\,J\,mol^{-1}$

$\Delta H_2 = ?$

$T_1 = 298$ K; $T_2 = 358$ K

Hence, $85.9\ JK^{-1}\ mol^{-1} = \dfrac{\Delta H_2 - (-33290)}{358 - 298}$

$\Delta H_2 = -28136\ J\ mol^{-1} = -28.136\ kJ\ mol^{-1}$

Example 18.17. The heat of reaction of $N_2 + 3H_2 \rightarrow 2NH_3$ at 27°C was found to be –21.976 k Cal. What will be the heat of reaction at 50°C? Given that molar heat capacities at constant pressure and at 27°C for Nitrogen, hydrogen and ammonia are 6.8, 6.77 and 8.86 cal mol^{-1} $degree^{-1}$.

Solution: Here; $\Delta H = -21.976$ k Cal

$T_2 = 273 + 50 = 323\,K$

$T_1 = 273 + 27 = 300\,K$

ΔC_p = Heat capacities of products – Heat capacities of reactants.

$= (2 \times 8.86) - [6.8 + (3 \times 6.77)]$

$= -9.39\ cal = -9.39 \times 10^{-3}$ k Cal

According to Kirchoff's equation,

$$\Delta H_2 = \Delta H_1 + (T_2 - T_1)\Delta C_P$$

$$= -21.976 + [23 \times (-9.39 \times 10^{-3})] = -22.192 \text{ k Cal}$$

Heat of reaction at 50° C = –22.192 k Cal.

Example 18.18. The heat of reaction $2Cu + O_2 \rightarrow 2CuO$ is –3138 kJ at 298 K. Calculate the heat of reaction at 398 K. The molar heat capacities for Cu, O_2 and CuO at constant pressure between 298 and 398 K are 23.05, 35.15 and 45.56 $JK^{-1}\ mol^{-1}$.

Solution: $2Cu + O_2 \rightarrow 2CuO$ $\qquad \Delta H_{298} = -3138\,kJ$

$\Delta C_p = (2 \times 45.56) - [(2 \times 23.05) + 35.15]\,JK^{-1}\,mol^{-1}$

or $9.52\,JK^{-1}\,mol^{-1} = 9.52 \times 10^{-3}\,kJ\,K^{-1}\,mol^{-1}$

$$\Delta H_2 = \Delta H_1 + (T_2 - T_1)\Delta C_P$$

Here, $\Delta H_1 = 3138$; $T_2 = 398$; $T_1 = 298$

Hence, $\Delta H_2 = -3138 + (398–298)9.52 \times 10^{-3}$

$= -3138 + 0.952\ kJ\ K^{-1}\ mol^{-1}$

$= -3137\ kJ\ K^{-1}\ mol^{-1}$

18.8 FLAME AND EXPLOSION TEMPERATURE

Consider a process that takes place at constant pressure under adiabatic conditions, i.e. no heat enters or leaves the system. In such case, temperature of the system would change and products would be at different temperature than the reactants.

The combustion of a gaseous fuel in air takes place so quickly that the entire heat produced during combustion is utilized in heating the products of the combustion, without being dissipated to the surroundings. Thus, the combustion is similar to adiabatic process.

In other words, during the combustion of the fuel, definite amount of products is formed and a definite quantity of heat is liberated. This quantity of heat raises the temperature of the products of combustion sufficiently to produce a flame.

The maximum temperature attained in the flame zone due to the liberation of heat by burning the fuel in excess of oxygen or air under adiabatic conditions is known as **flame temperature.**

For a combustion process taking place at constant pressure, the rise in temperature can be calculated from the Kirchoff's equation. Consider the combustion reaction taking place in two stages. In stage I the reactants are changed into the products at constant pressure with the liberation of heat ΔH_o. In stage II the temperature of the products is raised from T_o to T_f by absorbing the heat ΔH_P which is equal in magnitude to ΔH_o.

$$\text{Stage I Reactants}(T_o,R) \rightarrow \text{Products}(T_o,P) \qquad \Delta H_o$$

$$\text{Stage II Products}(T_o,P) \rightarrow \text{Products}(T_f,P) \qquad \Delta H_P$$

In other words the heat liberated in stage I is absorbed by the products in stage II to raise the temperature of the products from T_o to T_f. So in overall reaction,

$$\text{Reactants}(T_o,R) \rightarrow \text{Products}(T_f,P) \qquad \Delta H$$

the net heat change will be:

$$\Delta H = \Delta H_o + \Delta H_P \qquad ...(44)$$

But
$$\Delta H = 0 \left[\begin{array}{l}\text{Because reaction is carried out}\\ \text{under adiabatic conditons}\end{array}\right]$$

$$\therefore \quad \Delta H_p = \Delta H_o \qquad ...(45)$$

$\therefore -\Delta H_o$ is the heat absorbed in increasing the temperature of the products from T_o to T_f.

The heat involved in the second stage is:

$$\Delta H_P = \Delta H_{(heating)} = -\Delta H_o = \int_{T_o}^{T_f} C_{P(products)}\, dT$$

$$= C_p(T_f - T_o)$$

$$\therefore \quad T_f = \frac{-\Delta H_o}{C_{P(products)}} + T_o \qquad ...(46)$$

It is assumed that ΔC_p remains constant over a given temperature range. But this assumption is not correct since T_f is very large of the order of thousands Kelvin.

One of the hottest flames (500 k) is produced by the reaction,

$$H_2 + F_2 \rightarrow 2HF \qquad \Delta H^\circ = -136 \text{ kCal/mol}$$

Combustion in Air & Flame Temperature: In combustion, excess of air or oxygen is supplied. If combustion takes place in air, which happens in majority of cases, the unreacted nitrogen along with the excess air which is used for combustion, is also heated up along with the other products. Thus combustion in air involving one mole of oxygen will result into the production of four moles of nitrogen (Air contains approximately 80% of N_2 and 20% of oxygen).

Thus combustion of carbon in air can be represented as:

$$C_{(S)} + \underbrace{O_{2(g)} + 4N_{2(g)}}_{\text{Air}} \rightarrow CO_{2(g)} + 4N_{2(g)}$$

Thus the heat capacity of the products is considered to be the sum of heat capacities of the products plus the heat capacities of 4 moles of unreacted N_2 per mole of O_2 used. Hence $C_P = C_{P\,products} + 4\,C_P, N_2$.

∴ C_p in equation (46) is replaced by ΣC_P where ΣC_P is the summation of heat capacities of products of combustion and the unreacted oxygen.

$$T_f = \frac{-\Delta H_o}{\Sigma C_P} + T_o$$

Explosion Temperature: If the reaction takes place in a closed vessel i.e at constant volume under adiabatic conditions, the maximum temperature attained is called maximum explosion temperature. The heat evolved in the reaction, i.e. ΔE is entirely used up in raising the temperature of the products and hence C_P is replaced by C_V The explosion temperature will be given by:

$$T_f = \frac{-\Delta E_o}{\Sigma C_V} + T_o$$

Limitations of Equation

(i) The process is assumed to be adiabalic but in actual practice there is exchange of heat between the system and the surroundings.

(ii) In actual practice, excess air is used for combustion. This excess air along with the unreacted nitrogen also utilizes considerable amount of heat liberated during combustion.

(iii) At high temperature attained in the burning of hydrocarbons the products H_2O and CO_2 may get-dissociated as:

$$H_2O \rightarrow H_2 + \frac{1}{2}O_2$$

$$CO_2 \rightarrow CO + \frac{1}{2}O_2$$

Thus, these reactions involve considerable amount of heat which is not accounted for by theoretical calculations.

Example 18.19: The heat of combustion of CH_4 is 809 kJ mol^{-1} measured at 300 K. Calculate the theoretical temperature which the combustion products would have attained if the experiments were carried adiabatically and oxygen is used for combustion. Given $C_P\,(CO_2) = 30$ JK^{-1} and $C_P\,(H_2O) = 35$ JK^{-1}.

Solution: $CH_4 + 2O_2 \rightarrow CO_2 + 2H_2O$ $\Delta H = -809$ kJ mol^{-1}

Now

$$\Sigma C_P = C_P(CO_2) + 2(C_P\,H_2O)$$

$$= 30 + 2 \times 35 = 100\ \text{JK}^{-1}\ \text{mol}^{-1}$$

$$= 100 \times 10^{-3}\ \text{kJ mol}^{-1}$$

$$T_2 = \frac{\Delta H_{(heating)}}{C_P} + T_1 = \frac{809\,\text{kJ mol}^{-1}}{100 \times 10^{-3}\,\text{kJ K}^{-1}\,\text{mol}^{-1}} + 300\,\text{K}$$

$$= 8090\ \text{K} + 300\ \text{K} = 8390\ \text{K}$$

Explosions can be classified as:

(i) Deflagarations (flames)

(ii) Detonations

Deflagarations: If the speed of the flame is less than the speed of sound then it is known as deflagration. Generally the term flame is used to describe deflagaration. For example, in Bunsen burner, a controlled explosion of hydrocarbon-air mixture takes place and flame travels with a speed of about 100 cm/sec. Thus it is termed as flame or deflagaration.

Detonations: When the speed of the flame exceeds that of the speed of sound then the term detonations is used. These are also known as 'shock waves'. Example, in petrol engine the compressed mixture of hydrocarbons and air is ignited by a spark. Under ordinary conditions, the speed of the flame is about 1000 cm/sec. But when the reaction becomes out-of control (knocking) the flame speeds exceeds to the order of 10^5 cm/sec and are known as detonations.

FIRE WALKING : MAGIC OR SCIENCE

You must have been amazed at the ability of some people to walk across beds of glowing coals with ease. Do you think that these fire workers have supernatural powers? No. There are scientific explanations, of why fire walking is possible. These explanations are based on thermochemistry. Actually the heat capacity of feet is high because human tissue is composed mainly of water which has relatively large specific heat. Thus a large amount of heat must be transferred from the coals to significantly change the temprature of the feet. During the brief contact between feet and coals, there is relatively little time for energy flow so that feet do not reach a high enough temperature to cause damage.

The amount of energy available for heat flow depends upon the quantity of matter at a given temperature. This is the cause that when a tiny sparkler, though has a very high temperature, does not hurt when it hits our hand. The sparkle has so little mass that no significant energy transfer occurs to our hand. Similarly although the surface of the coals has a very high temperature, the red hot layer is very thin and hence no significant heat transfer occurs to the feet.

The third factor which helps fire walkers is the Leidenfrost effect. This is a phenomenon that allows water droplet to skate around on a hot griddle for a long time. Fire walkers first wet their feet before walking on the fire which aids in Leidenfrost effect.

QUESTIONS

Short Answer Questions

Section 18.1, 18.2

1. What is internal energy of a system?
2. Why chemical reactions are accompanied by heat changes?
3. What do you understand by the term enthalpy?
4. What is the relation between enthalpy and internal energy.
5. What are the conditions at which $\Delta H = \Delta E$.
6. Explain the term enthalpy and enthalpy change?
7. What is a thermochemical equation?

Section 18.3

8. Define the term specific heat.
9. What is molar heat capacity?
10. Name the two types of heat capacities.
11. Define C_p and C_v. What is the relation between the two?

Section 18.4, 18.5

12. What do you understand by heat of reaction?
13. What are the factors on which heat of reaction depends?

14. Define enthalpy of formation.
15. What are exothermic and endothermic compounds?
16. What is the significance of standard enthalpy of formation?
17. Why heat of neutralization of strong base and strong acid always comes out to be – 57.3 kJ/mol?
18. What is the difference between heat of hydration and heat of solution?
19. Distinguish between differential and integral heat of solution.
20. Diamond is an elementary substance, yet its standard heat of formation is not taken as zero. Why?
21. Can ΔH combustion have negative value?
22. Heat of solution of $CuSO_4$ is ______. and heat of solution of $CuSO_4 . 5H_2O$ is______.
23. One kg of graphite is burnt in an open vessel. Will the same amount of heat is evolved in the two cases.

    Ans. Yes, because $\Delta n_g = 0$.

Section 18.6, 18.7

24. Define Hess's law of constant heat summation.
25. State some of the applications of Hess's Law?
26. Write Kirchhoff's equation.
27. What form it takes if the temperature difference is not too large.

Section 18.8

28. What is flame temperature.
29. What do you understand by explosion temperature.

Long Answer Questions

1. What do you mean by internal energy change and enthalpy change? What are the conditions when enthalpy change (ΔH) is equal to the energy change (ΔE)?
2. Derive a relation between enthalpy change (ΔH) and internal energy change (ΔE) for a chemical reaction.
3. Define the terms C_P and C_V. What is the relation between the two?
4. Distinguish the terms in the following pairs :
   (i) Exothermic reactions and Endothermic reactions
   (ii) Enthalpy of formation and standard enthalpy of formation.
   (iii) Heat of reaction at constant volume from that at constant pressure.
5. Explain and illustrate the following terms.
   (i) Enthalpy of a reaction;
   (ii) Enthalpy of formation;
   (iii) Enthalpy of combustion;
   (iv) Enthalpy of neutralization
   (v) Integral and differential enthalpies of solutions.
6. State and illustrate Hess's law of thermochemistry. Mention its important applications with suitable examples.
7. Explain. giving reasons the following:
   (i) Enthalpy of neutralization of a strong monobasic acid by a strong base is always equal to – 57.32kJ mol^{-1}.
   (ii) The Hess's law of constant heat summation is a direct consequence of the first law of thermodynamics.
8. Derive Kirchoff's equation. What is the principle or law on which it is based?
9. Explain the terms :
   (i) Adiabatic flame temperature
   (ii) Explosion temperature
10. Derive the expression for the adiabatic flame temperature, if the reaction is carried out under constant volume condition.

11. (a) Derive Kirchoff's equation?
    (b) What is adiabatic flame temperature?
    (c) The following reaction carried out at 25°C and 1 atm. pressure :
    $SO_{2(g)} + \frac{1}{2}O_{2(g)} \rightarrow SO_{3(g)}$ is found to be exothermic. The energy evolved at constant volume being 97.030 kJ. Calculate the enthalpy change for the reaction. (Given R = 8.314 J/K/mol)
    (Ans. = – 98.269 KJ) (G.G.S.I.P.U, I term, May 2000)
12. Derive the following expressions :
    (i) Kirchoff's relation
    (ii) $\Delta H = \Delta E + \Delta n_{(g)}RT$
13. Write short notes on
    (i) Flame temperature
    (ii) Hess's law
    (iii) Endothermic and Exothermic compounds.
14. What is the basic difference between the heat of formation and heat of reaction? Illustrate with suitable examples.
15. Hess's law is just an extension of the law of conservation of energy. Justify this statement.

Problems

16. The heat of combustion of ethene $C_2H_{4(g)}$ at constant volume and 25°C is found to be – 900 kJ mol^{-1}. Calculate the enthalpy of combustion at constant pressure. (R = 8.314 kJ^{-1}mol^{-1}.
    (Ans. = – 904.95kJ mol^{-1})
17. For a reaction $\Delta H° = -285.1$kJ mol^{-1} and $\Delta E° = -280.5$kJ mol–1. What is the value of Δn at 298K.
    (Ans. = –2)
18. Calculate ΔE° for the following reaction at 298 K: $H_{2(g)} + \frac{1}{2}O_{2(g)} \rightarrow H_2O_{(g)}$ $\Delta H° = -241.6$kJ mol^{-1}
19. The heats of combustion of carbon, hydrogen and methane are 94 kCals, 68 kCals and 213 kCals respectively. Calculate the heat of formation of methane.
20. The heats of combustion of carbon, hydrogen and ethylene are 94 kcals, 68 kCals and 373 kCals respectively. Calculate the heat of formation of ethylene.
21. Calculate the heat of formation of ethane from the following data :
    (i) $C_{(s)} + O_2(g) \rightarrow CO_2(g);$ $\Delta H = -94.0$ kCal
    (ii) $H_2(g) + \frac{1}{2}O_2(g) \rightarrow H_2O(l);$ $\Delta H = -68.0$ kCal
    (iii) $C_2H_6(g) + 3\frac{1}{2}O_2(g) \rightarrow 2CO_2(g) + 3H_2O(l);$ $\Delta H = -373$ kCal
    [Ans. –19.0 kcal mol–1]
22. Calculate the enthalpy of formation of H_2SO_4(l) from the following data :
    (i) $S_{(S)} + O_2(g) \rightarrow SO_2(g)$ $\Delta H = -71.0$ kCal
    (ii) $SO_2(g) + \frac{1}{2}O_2(g) \rightarrow SO_3(g)$ $\Delta H = -23.5$ kCal
    (iii) $SO_3(g) + H_2O(l) \rightarrow H_2SO_4(l)$ $\Delta H = -31.2$ kCal
    (iv) $H_2(g) + \frac{1}{2}O_2(g) \rightarrow H_2O(l)$ $\Delta H = -68.5$ kCal
    (Ans. = –194.2 kcal)
23. Standard enthalpy of neutralization for CH_3COOH with NaOH is –51.63 mol^{-1}. Calculate enthalpy of ionization of CH_3COOH.
    (Ans. = 5.27 kJ mol^{-1})
24. When H_2O gas is burnt in O_2 gas, 240.5 kJ heat is liberated per mole of the explosion products at 27°C. Calculate the maximum temperature of explosion. Given $C_V(H_2O,g) = 24.15$ JK^{-1} and the reaction involved is

$$H_2(g) + 1/2\,O_2(g) \rightarrow H_2O(g);\ \Delta E_{300} = -240.5\text{kJ}$$

(Ans. = Approx.10^4K)

25. (a) The heat of formation of methane at constant pressure at 25°C is –74.85kJ. What will be the heat of formation at constant volume?
    (b) The heat of reaction (ΔH) for the formation of NH_3 according to reaction $N_2 + 3H_2 \rightleftharpoons 2NH_3$ at 27°C was found to be –91.94 kJ. What will be ΔH at 50°C? The average molar heat capacities at constant pressure and between 27°C and 50°C for N_2, H_2, and NH_3 are 28.45, 28.32 and 37.07 J, respectively. **(Ans. = –92.85kJ)**
    (c) Explain the following :
    (i) Heat of Dilution.
    (ii) Heat of Neutralization.
    (iii) Flame Temperature **(G.G.S.I.P. Univ, End term, July 2000)**
26. (a) What is heat of reaction and how it is related to temperature. Derive Kirchoff's equation.
    (b) The standard heat of formation at 25°C of liquid water is given
    $H_2(g) + \frac{1}{2}O_2(g) \rightarrow H_2O(l)\ +68.3\,\text{kCal}$. Calculate the standard heat of formation at 20°C and also total change in heat capacities at constant pressure. Take the molar heat capacities at constant pressure in the temperature range 25 to 25°C to be
    (i) $H_2(g) = 6.89$ cal/deg/mol
    (ii) $O_2(g) = 6.97$ cal/deg/mol
    (iii) $H_2O(l) = 18.0$ cal/deg/mol **(G.G.S.I.P.U, End Term, B.Tech, May 2001)**
27. (a) Find the total pressure exerted by 2g of C_2H_6 and 3g of CO_2 contained in a 5l vessel at 50°C.
    (b) Calculate heat of reaction at 25°C for the following reaction from the standard heat of formation of reactants and products.

$$SO_{2(g)} + 2H_2S(g) \rightarrow 3S_{(rhombic)} + 2H_2O(l)$$

$$\Delta H_f^\circ \text{kCal/mol} \quad -70.96 \quad -4.82 \quad 0 \quad 68.32 \text{ respectively}$$

    (c) With the help of suitable example, explain Hess's Law.
    (G.G.S.I.P.U, End Term, B.Tech, May 2001)
28 (a) State Hess's law. Give its applications.
    (b) Calculate the standard heat of formation of acetylene from the heat of combustion of C_2H_2 graphite and H_2 being as -1300kJ mol^{-1}, -395 kJmol^{-1} and -286 kJ mol^{-1} respectively.
    (Ans. = 224 kJ.mol) (G.G.S.I.P.U, II Term, April 2002)
29 (a) Derive Kirchoff's equation.
    (b) The enthalpy of reaction (Δ*H*) for the formation of ammonia according to the reaction $N_2 + 3H_2 \rightarrow 2NH_3$ at 27°C was found to be –91.94 kJ. What will be the enthalpy of reaction (Δ*H*) at 50°C. The molar heat capacities at constant pressure and at 27°C for nitrogen, hydrogen and ammonia are 28.45, 28.32 and 37.07 joules respectively?
    (G.G.S.I.P.U, II Term, April 2002)
30 (a) Explain Hess's law.
    (b) What is meant by Heat of Combustion? The internal energy change accompanying the complete combustion of n-heptane (l) at constant volume and at 25°C is $-1150\text{ kCal/mol}^{-1}$. Calculate the ΔH_f^0 for n-heptane.
    ($R = 1.987\text{ cal deg}^{-1}\text{ mol}^{-1}$, ΔH_f^0 for CO_2 (g) and H_2O (l) are –94 and -68 kCal/mol^{-1})
    (Ans. = –52 kCal/mol)
    (G.G.S.I.P.U, End Term, May 2002)
31. Calculate herat of reaction

$$CO_2 + H_2 \longrightarrow CO + H_2O_{(g)}$$

From the following data:

$C + \frac{1}{2}O_2 \longrightarrow CO_2$ $\quad \Delta H_f^0 = -24.2$ k Cal/mol

$C + O_2 \longrightarrow CO_2$ $\quad \Delta H_f^0 = -94.1$ k Cal/mol

$H_2 + \frac{1}{2}O_2 \longrightarrow H_2O$ $\quad \Delta H_f^0 = -57.8$ k Cal/mol

(G.G.S.I.P.U, End Term, May 2003)

32. State and explain Hess's law. How Hess's law is only special case of first law of thermodynamics. **(G.G.S.I.P.U, End Term, May 2003)**

33. (a) For fiollowing exothermic changes, explain whether the heat evolved at constant pressure is < or > or = the heat evolved at constant volume. Give reason.

$1/2\ H_2(g) + 1/2\ Cl_2(g) \rightleftharpoons HCl(g)$

$SO_2(g) + 1/2\ O(g) \rightleftharpoons SO_3(g)$

$PCl_5(g) \rightleftharpoons PCl_3(g) + Cl_2(g)$

(b) Calculate the enthalpy of reaction.

$CH_3Cl(g) \rightleftharpoons C(g) + 3H(g) + Cl(g)$

The bond energies of C—H and C—Cl bonds are 415.8 kJ^{-1} and 326.35 kJ mol^{-1} respctively.

(c) Derive Kirchoff's equation and show how the heat of reaction at constant volume is related to that at constant. pressure. **(G.G.S.I.P.U, End Term, May 2004)**

34. Find ΔH_f for MgF_2 from the following data;

$\Delta H_{sub} = 1.46$ kJ/mol

$\Delta H_D = 158.9$ kJ/mol

I.E. $(Mg^{+2}) = 2184.0$ kJ

E.A. = – 334.7 kJ

Lattice energy = – 2922.5 kJ/mol. **(G.G.S.I.P.U, End Term, 2004)**

35. (a) Hess's law is just an extension of the law of conservation of energy. Justify.

(b) When a tire is pumped up rapidly, its temp. increases. Predict the behaviour if air was an idea gas. **(G.G.S.I.P.U, II-Term, B.Tech., April 2006)**

36. (a) In a Bunsen burner methane gas is pre mixed with sufficient air to allow complete combustion and at the flame temp. water is converted into steam. Using the following thermochemical data:

$$CH_{4(g)} + 2O_{2(g)} \longrightarrow CO_{2(g)} + 8N_2; \quad \Delta H^0_{298} = -890.36 \text{ kJ}$$

$$2H_2O_{(l)} \longrightarrow 2H_2O_{(g)}; \quad \Delta H^0_{298} = 44.02 \text{ kJ}$$

Calculate the flame temp. given SC_p is 41.78 JK^{-1} mol^{-1} (for products), initial temp. is 25°C. Assume that the gaseous products behave ideally.

(G.G.S.I.P.U, II-Term, B.Tech., April 2006)

37. (a) Calculate the enthalpy of fusion ΔH_f of ice at –10°C from the following data:

$\Delta H_f = 6.02$ $kjmol^{-1}$ at 0°C

C_p(ice) = 37.66 JK^{-1} mol^{-1}

C_p(water) = 75.31 JK^{-1} mol^{-1}

(b) Define heat of neutralisation. Why is heat of neutralisation of a strong acid and strong base always constant? **(G.G.S.I.P.U, II-Term, B.Tech., April 2006)**

38. For the reaction $N_{2(g)} + 3H_{2(g)} \rightarrow 2NH_{3(g)}$, what the relation between ΔH and ΔE.

(G.G.S.I.P.U, II-Term, B.Tech., April 2007)

39. For the hypothetical reaction: $2A_{(g)} \rightarrow A_{2(g)} \Delta C_p = (6.00 + 2.00 \times 10^{-3}T)$J, and $\Delta H^0_{298} = -20$kJ per mol. Estimate the temperature at which $\Delta H^0 = 0$ for the reaction.

(G.G.S.I.P.U, II-Term, B.Tech., April 2007)

40. What is adiabatic flame temperature? **(G.G.S.I.P.U, II-Term, B.Tech., April 2007)**

41. The heat of combustion of methane is 809 kJ per mol measured at 300 K. Calculate the theoretical temperature at which the combustion products would have attained if the experiment were carried adiabatically and oxygen is used for combustion.

Given $C_p(CO_2) = 30JK^{-1}$ and $C_p(H_2O) = 50JK^{-1}$ **(G.G.S.I.P.U, II-Term, B.Tech., April 2007)**

42. What is the relation between heat change of reaction at constant volume and constant pressure?

(G.G.S.I.P.U, II-Term, B.Tech., Nov. 2007)

43. (a) Derive Kirchoff's equation.

(b) Standard enthalpy of neutralization for acetic acid with sodium hydroxide is –51.63 $kJmol^{-1}$. Calculate the enthalpy of ionisation of acetic acid.

(c) What is explosion temperature? **(G.G.S.I.P.U, II-Term, B.Tech., April 2007)**

44. State Hess Law and illustrate its validity choosing a suitable example.

(G.G.S.I.P.U, End-Term, B.Tech., Dec. 2007)

45. The heat of reaction $\frac{1}{2}H_2 + \frac{1}{2}Cl_2 = HCl$ at 27°C is –22.1 kCal. Calculate the heat of this reaction at 77°C. Molar heat capacities for H_2, Cl_2 and HCl at constant pressure and 27°C are 6.82, 7.70 and 6.80 Cal mol^{-1} respectively. **(G.G.S.I.P.U, End-Term, B.Tech., Dec. 2007)**

❑❑❑

19 SOLUTIONS

"Advances are made by answering questions. Discoveries are made by questioning answers."

— ***Bernard Haisch***

CHAPTER OUTLINE

A solution is defined as a homogeneous mixture of two or more nonreacting substances whose composition may be varied upto certain limits.

When only two components are mixed to form the homogeneous mixture, it is called a **binary solution,** which is the most common type of solution we come across. In a binary solution, the component which is present in smaller proportion is termed as **solute** and the component present in larger proportion is termed as the **solvent**. For example, when a small amount of NaCl (Solute) is mixed with water (Solvent) a homogeneous solution is obtained. But the terms solute and solvent are arbitrary and do not signify any essential difference between the two components. For example, a mixture of alcohol and water may be regarded as a solution of alcohol in water or solution of water in alcohol.

However, in the case of solutions of solids in liquids, the terms solute and the solvent are frequently used. The substance that is dissolved is referred to as solute and the substance in which dissolution takes place is referred to as solvent.

19.1 TYPES OF SOLUTIONS

Depending on the physical state of the two components, different types of solutions are distinguished in Table 19.1.

Table 19.1

Solute	*Solvent*	*Examples*
Gas	Gas	Air, Cooking Gas
Gas	Liquid	Aerated waters (CO_2 in water)
Gas	Solid	H_2 Gas in Palladium
Liquid	Gas	Moisture in Air
Liquid	Liquid	Alcohol in water
Liquid	Solid	Mercury in Silver (Silver Amalgam)
Solid	Gas	Smoke (Carbon Particles in Air)
Solid	Liquid	Salt in water
Solid	Solid	Alloys like Brass, Bronze, etc.

Out of various types of solutions, our focus in the present chapter will be on solutions having liquid as the solvent, i.e. gas liquid, liquid-liquid and solid-liquid binary solutions.

Solutions are also characterized by their capacity to dissolve a solute in a given solvent.

Saturated solution: A solution that contains the maximum amount of solute in a given solvent at a given temperature is called a saturated solution.

Unsaturated solution: A solution that contains less solute than its capacity to dissolve the solute at a given temperature is called unsaturated solution.

19.2 FORMATION OF SOLUTIONS

Molecules in a substance are held together by intermolecular forces of attractions. These forces play a vital role in the formation of solutions. When a substance (solute) is dissolved in another substance (solvent), particles of the solute disperse throughout the solvent. The ease with which the solute particles dissolve or the solubility of the solute depends on the relative strengths of the following interactions:

- solvent-solvent interaction
- solute-solute interaction
- solvent-solute interaction

The formation of the solution can be assumed to proceed in three steps :

(i) separation of the solute molecules
(ii) separation of the solvent molecules
(iii) mixing of solute and solvent molecules.

The first two steps require energy to overcome the intermolecular forces hence are endothermic. The third step may be exothermic or endothermic depending upon the interactions between solute and solvent molecules. The heat of the solution is given by:

$$\Delta H_{soln} = \Delta H_1 + \Delta H_2 + \Delta H_3$$

where ΔH_1, ΔH_2, and ΔH_3 are the heat change accompaning (i), (ii) and (iii) step respectively. Two cases can be distinguished :

Case 1: If the solute-solvent attractions are stronger than the solvent-solvent attraction and solute-solute attraction, the solution process is favourable. The ΔH_{soln} will be negative, i.e. the process will be exothermic.

Case 2: If the solute-solvent interaction is weaker than the solvent-solvent and solute-solute interactions, the ΔH_{soln} will be positive, i.e. the process will be endothermic.

Hence, the formation of solution can be exothermic or endothermic. One may wonder-why a solute dissolves in a solvent at all, if the intermolecular attraction between solvent-solute is weaker as compared to solute-solute interaction (case 2). In fact, similar to other chemical processes the solution process is also governed by two factors – first is *enthalpy*, which determines whether a process is exothermic or endothermic and second is *entropy*, i.e. tendency towards disorder. In the formation of solutions, where the reaction is endothermic, the second factor, i.e. entropy or randomness favours the process. In the pure state, the solvent and the solute possess a fair degree of order. This ordered arrangement is destroyed when the solute dissolves in the solvent. Therefore, the solution process is always accompanied by an increase in disorder or randomness.

Thus it is the increase in disorder of the system that favours the solubility of any substance, even if the solution process is endothermic.

Factors Affecting Solubility: The general rule for the formation of the solution is "Like dissolves like"– i.e. polar substances dissolve polar substances and non-polar substances dissolve non-polar substances.

The two substances having similar type and magnitude of intermolecular forces are likely to be soluble in each other. For example, both CCl_4 and benzene are non-polar liquids and have same type of intermolecular forces (dispersion forces). When these liquids are mixed, they readily dissolve in each other as the attraction between CCl_4 and C_6H_6 molecules is comparable in magnitude to the attraction between CCl_4 molecules and attraction between C_6H_6 molecules. These type of solutions, where two liquids are completely soluble in each other in all proportions are called **miscible**.

Alcohols are soluble in water due to the formation of hydrogen bonds with water molecules. Let us consider the solubility of ethanol (C_2H_5OH) and n-hexanol ($CH_3\ CH_2\ CH_2\ CH_2\ CH_2\ CH_2$ OH) in water, a polar solvent. Though both alcohols contain OH groups which can form hydrogen bonding with water, but n-hexanol having a larger non-polar alkyl chain is much less polar then ethanol. Hence, ethanol is much more soluble than hexanol in water due to its polar nature.

19.3 CONCENTRATION OF A SOLUTION

The amount of solute present in a given amount of solution is known as the concentration of the solution. There are various ways of expressing the concentration of a solution. The most common units of concentration are discussed below:

(i) **Mass Percent:** The mass percent (also known as percentage by weight or weight percent) is defined as the amount of the solute in grams present in 100 gms of the solution.

$$\text{Mass percent of solute} = \frac{\text{Weight of solute}}{\text{Weight of solution}} \times 100$$

$$= \frac{\text{Weight of solute}}{\text{Weight of solute} + \text{Weight of solvent}} \times 100$$

For example, a 10% solution of sodium chloride means a solution containing 10 g of sodium chloride in 100 g of the solution.

(ii) **Strength:** The strength of a solution is defined as the number of grams of the solute dissolved per litre of the solution.

$$\text{Strength of a solution} = \frac{\text{Weight of solute in grams}}{\text{Volume of the solution in litres}}$$

(iii) **Molarity:** It is defined as the number of moles of the solute present per litre of the solution.

$$\text{Molarity} = \frac{\text{Number of moles of the solute}}{\text{Volume of the solution in litres}}$$

Molarity is denoted by 'M'. Molarity of a solution can be calculated from the strength of the solution as below:

$$M = \frac{\text{Strength in grams per litre}}{\text{Molar mass of solute}}$$

If 'g' gram of the solute is present in V ml of a given solution, then

$$\text{Molarity, } M = \frac{g}{\text{Molar mass of the solute}} \times \frac{1000}{V}$$

For example the molarity of a solution of NaOH having 4 g of it dissolved in 500 cm^3 of solution will be:

$$M = \frac{4}{40} \times \frac{1000}{500} = 0.2$$

In case of ionic compounds like NaCl, Na_2CO_3, etc., the term *formality* is used in place of molarity. Formality is defined as the number of *formula weight dissolved in one litre of solution. It is usually denoted as F.

(iv) **Molality:** Molality is defined as the number of moles of the solute dissolved in 1 kg (1000 g) of the solvent.

$$\text{Molality} = \frac{\text{Number of moles of the solute}}{\text{Weight of solvent in grams}} \times 1000$$

It is denoted by 'm'. Molality of a solution can also be calculated from the strength of the solution as:

$$\text{Molality} = \frac{\text{Strength per 1000 grams of solvent}}{\text{Molar mass of the solute}}$$

If 'g' grams of the solute is dissolved in 'G' gram of the solvent then

$$\text{Molality 'm'} = \frac{g}{\text{Molar mass of the solute}} \times \frac{1000}{G}$$

For example the molality of a solution of anhydrous sodium carbonate, having 26.5 grams of it dissolved in 500 grams of water will be:

$$M = \frac{26.5}{106} \times \frac{1000}{500} = 0.5$$

(Molar mass of anhyd. sodium carborate = 106)

(v) **Normality:** The normality of a solution is defined as the number of gram equivalents of the solute dissolved per litre of the solution.

$$\text{Normality} = \frac{\text{Number of gram equivalents of the solute}}{\text{Volume of solution in litres}}$$

It is denoted by 'N'. Normality of a solution can also be calculated from the strength.

---

* *The term formula weight is used in place of molecular mass because ionic compounds exist as ions and not as molecules. Formula mass is the sum of the atomic masses of the atoms in the formula of the compound.*

$$\text{Normality N} = \frac{\text{Strength in grams per litre}}{\text{Equivalent weight of solute}}$$

If 'g' gram of the solute is present in Vml of given solution then

$$\text{Normality} = \frac{g}{\text{Equivalent weight of the solute}} \times \frac{1000}{V}$$

For example, normality of a solution of sulphuric acid having 4.9 grams of it dissolved in 500 ml of a solution will be:

$$N = \frac{4.9}{49} \times \frac{1000}{500} = 0.2$$

(Equivalent weight of sulphuric acid = 49)

(vi) **Mole fraction:** The mole fraction of a component in a solution is defined as the number of moles of that component divided by the total number of moles of all the components.
Mol fraction of the solute in the solution

$$= \frac{\text{Number of moles of the solute}}{\text{Number of moles of solute} + \text{Number of moles of solvent}}$$

Mol fraction is denoted by x.
If x_1 is the mol fraction of the solvent
x_2 is the mol fraction of the solute
n_1 is the number of moles of the solvent in the solution
n_2 is the number of moles of the solute in the solution then,

$$\text{Mol fraction of solvent in the solution, } x_1 = \frac{n_1}{n_1 + n_2}$$

$$\text{Mol fraction of solute in the solution, } x_2 = \frac{n_2}{n_1 + n_2}$$

Evidently, $x_1 + x_2 = 1$, i.e. the sum of mol fractions of all the constituents of a solution is always equal to one.
For example, if a solution contains 5 moles of alcohol and 10 moles of water then,

$$\text{Mol fraction of water} = \frac{10}{5+10} = 0.67$$

$$\text{Mol fraction of alcohol} = \frac{5}{5+10} = 0.34$$

Choice of Concentration Units: The choice of concentration unit is based on the purpose of measurement.

- Generally it is easier to measure the volume of a solution using precisely calibrated volumetric flasks, than to weigh the solvent. For this reason **molarity** is preferred over **molality**.
- But if one has to work under different temperature conditions with the same solution then **molality** is the preferred term. This is because in molality the concentration is expressed in number of moles of solute and mass of solvent which remain the same at all temperatures. Whereas, in molarity the concentration is expressed in number of moles and volume of the solution. Now volume of a solution typically increases with increasing temperature. For example, a solution that is 1.0 M at 25°C may become 0.97 M at 45°C because of increase in volume, Thus the dependence of concentration on temperature can affect the accuracy of an experiment.
- Mass percent is also independent of temperature and hence can also be used while working under different temperatures.

19.4 VAPOUR PRESSURE OF SOLUTIONS AND RAOULT'S LAW

VAPOUR PRESSURE: Every liquid has a tendency to escape as vapours in the free space above it. The molecules in the liquid are moving with different kinetic energies. The molecules that possess comparitively high energies than others, can overcome the intermolecular forces that hold them in the liquid phase and hence tend to escape into the vapour phase.

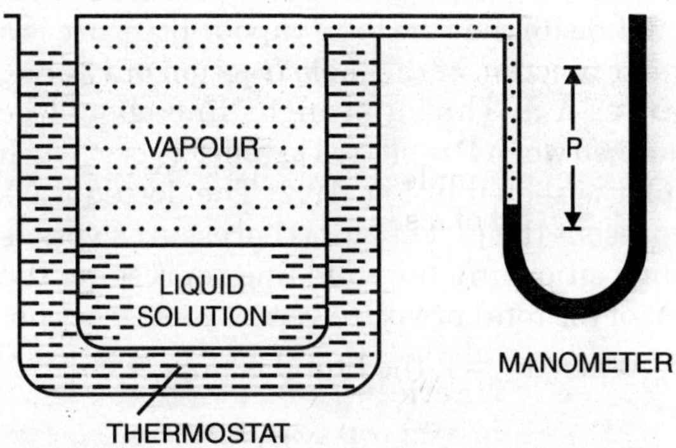

Fig. 19.1 Concept of measurement of vapour pressure of a solution at a particular temperature

At the same time, if the liquid is placed in a closed vessel, some of the escaping molecules tend to condense and find their way back into the liquid phase. If a pure liquid or solution is taken in an evacuated vessel placed in a themostat at a particular temperature and the vessel is connected to a manometer as shown in Fig 19.1, the amount of vapours in the free space goes on increasing and ultimately becomes constant. This is because an equilibrium is established or rate of evaporation becomes equal to rate of condensation. The pressure exerted by the vapours is called the vapour pressure. Hence vapour pressure can be defined as **the pressure exerted by the vapours in equilibrium with the liquid/solution at a particular temperature.**

Raoult's Law

The relationship between the vapour pressure of a liquid and its mole fraction in solution was first studied by the French chemist, Francois Marie Raoult (1830–1901). Based on his studies he gave the following generalization, popularly known as **Raoult's Law.**

The partial vapour-pressure of any volatile component of a solution, at any temperature, is equal to the vapour pressure of the pure component multiplied by the mole fraction of that component in the solution.

Let us assume that a solution contains n_A moles of volatile liquid A and n_B moles of volatile liquid B. Let p_A and p_B represent the partial pressures of liquid A and B respectively and x_A and x_B represent their respective mole fractions. Then according to the Raoult's Law.

$$p_A = x_A \overset{\circ}{p}_A \quad ...(1)$$

$$p_B = x_B \overset{\circ}{p}_B \quad ...(2)$$

Where $\overset{\circ}{p}_A$ and $\overset{\circ}{p}_B$ are the vapour pressures of pure components A and B respectively.

If P is the total vapour pressure and the vapour behaves as an ideal gas, then, according to Dalton's Law of partial pressures,

$$P = p_A + p_B \quad ...(3)$$

$$P = x_A \overset{\circ}{p}_A + x_B \overset{\circ}{p}_B \quad ...(4)$$

$$\text{Since, } x_A + x_B = 1$$

$$\therefore x_A = 1 - x_B \quad ...(5)$$

Equation (4) can be rewritten as:

$$P = (1 - x_B)\overset{\circ}{p}_A + x_B \overset{\circ}{p}_B$$

$$= \overset{\circ}{p}_A - x_B \overset{\circ}{p}_A + x_B \overset{\circ}{p}_B$$

$$\boxed{P = (\overset{\circ}{p}_B - \overset{\circ}{p}_A)x_B + \overset{\circ}{p}_A} \quad ...(6)$$

Since, $\overset{\circ}{p}_A$ and $\overset{\circ}{p}_B$ are constant at a particular temperature, it is evident from equation (6) that the total vapour pressure is a linear function of the mole fraction of x_B or x_A ($\because x_A = 1 - x_B$). Thus a straight line should be obtained when P is plotted against x_A or x_B such a plot is shown in Fig. 19.2. The dotted lines represent the plot of partial pressures versus composition and the solid line represents the plot of the total pressure versus mole fraction.

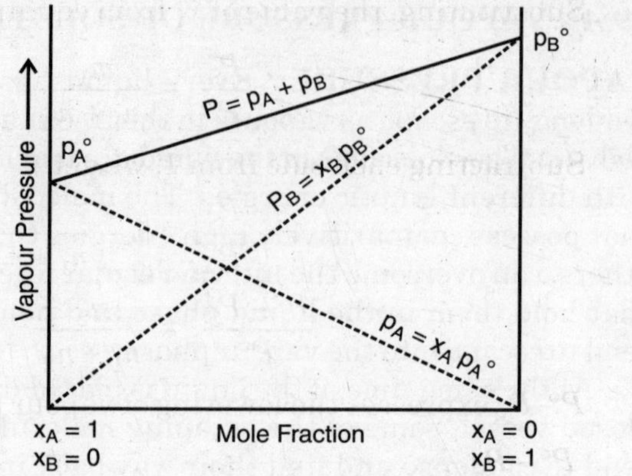

Fig. 19.2 Relationship between vapour pressure and mol fraction of an ideal solution

When $x_A = 1$, the liquid is pure A

$$\therefore P = \overset{\circ}{p}_A$$

When $x_B = 1$, the liquid is pure B

$$P = \overset{\circ}{p}_B$$

At any other value of x_A or x_B the total pressure is given by equation (6).

In the above case (Fig. 19.2), it can be seen that both the components which are considered, are volatile. Therefore the vapour above them will have both the components. However, the vapour will be richer in more volatile component.

Raoult's Law for non-volatile solutes: For a binary solution, vapour pressure of the solvent in the solution is given by Raoult's Law as:

$$\textit{Vapour pressure of the solvent in solution} = \begin{bmatrix}\text{Mole fraction of the} \\ \text{solvent in the solution}\end{bmatrix} \times \begin{bmatrix}\text{Vapour pressure} \\ \text{of the pure solvent}\end{bmatrix} \quad \text{...(7)}$$

But if the solute is non-volatile, it will not contribute to the total vapour pressure of the solution. Thus the vapour pressure of the solution will be the vapour pressure due to the solvent in the solution only. Thus

$$\begin{matrix}\text{Vapour pressure} \\ \text{of the solution}\end{matrix} = \begin{matrix}\text{Vapour pressure of the} \\ \text{solvent in the solution}\end{matrix} \quad \text{...(8)}$$

From (7) and (8), for a solution containing non-volatile solute, the total vapour pressure will be given by

$$\begin{matrix}\text{Vapour pressure} \\ \text{of the solution}\end{matrix} = \begin{bmatrix}\text{Mole fraction of the} \\ \text{solvent in the solution}\end{bmatrix} \times \begin{bmatrix}\text{Vapour pressure} \\ \text{of the pure solvent}\end{bmatrix} \quad \text{... (9)}$$

$$\text{or } P_S = x_1 P^\circ \quad \text{... (10)}$$

Where P_S is the vapour pressure of the solution

x_1 is the mole fraction of the solute

P° is the vapour pressure of pure solvent.

$$\frac{P_S}{P^\circ} = x_1 \quad \text{...(11)}$$

Further, n_2 moles of the solute are mixed in n_1 moles of the solvent, then mole fraction of solvent x_1 is given by;

$$x_1 = \frac{n_1}{n_1 + n_2} \quad \text{... (12)}$$

Substituting, the value of x_1 from equation (12) in equation (11), we have

$$\frac{P_S}{P^\circ} = \frac{n_1}{n_1 + n_2} \quad \text{... (13)}$$

Subtracting each side from 1, we get

$$1 - \frac{P_S}{P_O} = 1 - \frac{n_1}{n_1 + n_2}$$

$$\frac{P^\circ - P_S}{P^\circ} = \frac{n_2}{n_1 + n_2} \quad \text{... (14)}$$

$P^\circ – P_S$ expresses the lowering of vapour pressure,

$\frac{P^\circ - P_S}{P^\circ}$ is relative lowering of vapour pressure

$\frac{n_2}{n_1 + n_2}$ is the mole fraction of solute

Thus, the relative lowering of vapour pressure of a solution containing non volatile solute is equal to the mol fraction of the solute in the solution.

19.5 IDEAL AND NON IDEAL SOLUTIONS

Ideal Solution: An ideal solution is defined as the solution which obeys Raoult's Law at all concentrations at a given temperature. Thus if two liquids A and B having mole fractions x_A and x_B respectively, are mixed together, the solution formed by them will be termed as ideal if:

$$p_A = x_A \overset{\circ}{p}_A \text{ and}$$

$$p_B = x_B \overset{\circ}{p}_B$$

Where p_A and p_B are the partial vapour pressures of A and B respectively and $\overset{\circ}{p}_A$ and $\overset{\circ}{p}_B$ are the vapour pressure of pure A and pure B respectively.

It has been observed that the solution formed by mixing the components A and B is ideal if the interactions between A and B are of the same magnitude as among the molecules of pure A or pure B. This implies that the intermolecular or cohesive forces in these solutions are completely uniform. The component A will have the same tendency to escape into the vapour phase, whether it is surrounded by its own molecules A or by the molecules of B or partly by the molecules A and partly by the molecules B. The escaping tendency of a component of an ideal solution will be the same as that of its escaping tendency when it is in pure state. The only difference is that in solution its proportionality is reduced as a result of its lower mole fraction in solution.

Liquids form an ideal solution only if they have nearly the same molecular size and related constitution. Liquid heptane and liquid hexane form an ideal solution because the interaction between hexane-hexane molecules or between heptane-heptane molecules or between hexane-heptane molecules are all of same magnitude

Further it is observed that the solution formed on mixing the two solutions is ideal if-

(a) There is no volume change on mixing, i.e. $\Delta V_{mix} = 0$

(b) No heat is evolved or absorbed on mixing, i.e. $\Delta H_{mix} = 0$

Practically no solution is ideal. However when concentration of solute is very low, the solution behaves ideally. The following binary mixtures form ideal solutions as they have similar structures and polarities:

(i) Hexane + Heptane
(ii) Carbon tetrachloride+Silicon tetrachloride
(iii) Benzene + Toluene
(iv) Ethyl chloride + Ethyl bromide

The behaviour of ideal solutions with the help of P–*x* plot is already discussed in Fig. 19.2.

Non-Ideal Solution: A solution which does not obey Raoult's law is termed as non-ideal solution. It shows deviations from Raoult's law.

In non-ideal solution formed by mixing of A and B, the forces of interaction between A–B are of different magnitude than the forces of interactions among the molecules of pure A or pure B. For such solutions,

$$\Delta V_{mix} \neq 0$$
$$\Delta H_{mix} \neq 0$$

The escaping tendency of A or B in a solution may be more or less than what it is in their pure states, this may give rise to positive or negative deviations as discussed below.

(a) **Non-ideal solutions showing positive deviations:** Consider a binary solution formed by the mixing of two components A and B. Let us assume that the intermolecular forces between A and B are weaker than those among the molecules of either of the pure components, *i.e.* A–A interactions or B–B interactions. The escaping tendency of the molecules of A and B in the solution will thus be more as compared to pure A or pure B.

A–B Interactions < A–A Interaction or B–B Interactions

In other words the vapour pressure of each component in the binary solution will be considerably greater than expected on the basis of Raoult's law. The total vapour pressure for any composition of the solution is thus greater than correspondng to an ideal solution of the same composition. This is shown in the vapour pressure versus composition diagram in Fig. 19.3. The dotted line represent the ideal behaviour while the solid lines represent the actual vapour pressure of each component and total vapour pressure of the solution. These solutions are said to show positive deviation from ideality or Raoult's law.

For a solution showing positive deviation from ideality, ΔH is positive, i.e. heat is absorbed during the formation of solution. Also as the bonds between solute and solvent molecules are weaker than solvent-solvent or solute-solute bonds, thus mixing is accompanied by increase in volume Thus

$$\Delta H_{mix} = \text{positive}$$
$$\Delta V_{mix} = \text{positive}$$

Examples of binary solutions exhibiting positive deviations include:

(i) ethanal-water
(ii) ether-acetone
(iii) Carbon tetrachloride-methanol
(iv) Cyclohexane ethanol
(v) Carbondisulplide-acetone

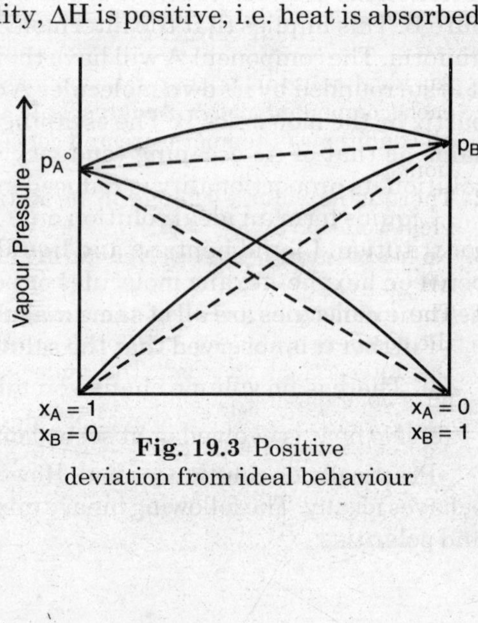

Fig. 19.3 Positive deviation from ideal behaviour

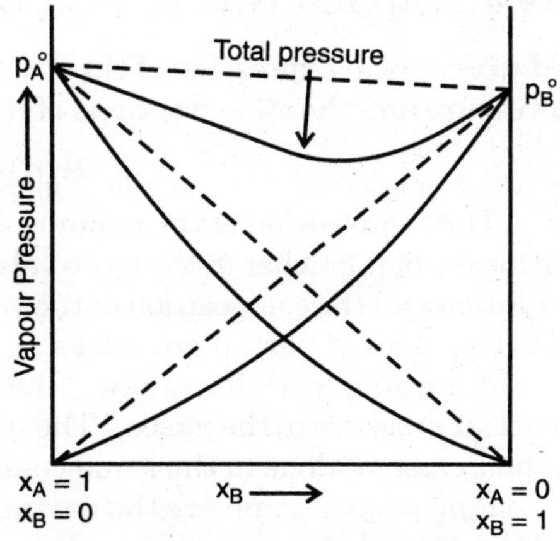

Fig. 19.4 Negative deviation from ideal behaviour

(b) **Non-ideal solutions exhibiting negative deviations:** In a binary solution formed by mixing the two components A and B, let us assume that the intermolecular forces between A and B are stronger than those of among the molecules of either of the individual components, i.e.

A–B Interactions > A–A Interactions or B–B Interactions

In this case the escaping tendency of the molecules of A and B will be less in the solution as compared to the pure components A and B. The vapour pressure of each component will be considerably lower than expected on the basis of Raoult's law. The total vapour pressure for any composition of the solution will be lower than corresponding to an ideal solution of the same composition. This is shown in Fig. 19.4. These solutions are said to show negative deviadation from ideality or Raoult's law.

For a solution showing negative deviation from ideality ΔH is negative, i.e. heat is evolved during the formation of solution. Also the stronger bonds are formed between solvent and solute molecules than either of solute-solute or solvent-solvent bonds and mixing is accompanied by decrease in volume. Thus for a solution showing negative deviation,

$$\Delta H_{mix} = \text{negative}$$
$$\Delta V_{mix} = \text{negative}$$

Examples of binary solutions exhibiting negative deviations include:

(i) Chloroform-benzene
(ii) HCl-water
(iii) HNO_3-water
(iv) Acetone-Aniline
(v) Chloroform-Diethyl ether
(vi) Acetic acid-pyridine

Difference between ideal and non-ideal solutions: The various points of difference may be summed up as:

Ideal Solution	*Non-Ideal Solution*
1. An ideal solution is that solution in which each component obeys Raoult's Law under all conditions of temperature and concentrations.	An non-ideal solution which do not obey Raoult's Law.
2. There is no volume change in formation of ideal solution $\Delta V_{mix} = 0$	Formation of non-ideal solution is accompanied by volume change $\Delta V_{mix} \neq 0$
3. No heat is absorbed or evolved during mixing $\Delta H_{mix} \neq 0$	Heat is evolved or absorbed during mixing $\Delta H_{mix} \neq 0$
4. The interactions between the molecules of the components in the solution is similar to the interactions between the molecules of pure components.	The interaction between the molecules of the components are different from the interactions between the molecules of pure components.

19.6 VAPOUR PRESSURE—COMPOSITION DIAGRAM FOR IDEAL SOLUTION

Mathematical expression of the Raoult's law (Equation 6) gives a relationship between the total pressure and the mole fractions of the components of the *solution*.

$$P = \overset{\circ}{p}_A + x_B(\overset{\circ}{p}_B - \overset{\circ}{p}_A) \qquad \text{(cf equation 6)}$$

The composition of the vapour phase will evidently be different than the composition of the solution, being richer in the more volatile component. The relation between the composition of the solution and the composition of the vapours above the solution can be worked out with the help of Dalton's Law of partial pressures.

According to Dalton's law, "the pressure exerted by a mixture of gases is the sum of the partial pressure of the gases." The partial pressure of a gas is the pressure which it would exert when present alone in the same vessel at the same temperature.

If p_1, p_2, p_3, ...pn are the partial pressure of the individual gases then P, the total pressure of the gaseous mixture is given by:

$$P = p_1 + p_2 + p_3 ..... p_n.$$

or $$P = \frac{n_1 RT}{V} + \frac{n_2 RT}{V} + \frac{n_3 RT}{V} + ... \qquad (\because PV = nRT \text{ and } P = \frac{nRT}{V})$$

$$P = \frac{RT}{V}(n_1 + n_2 + n_3 + ... n_n)$$

where $n_1 + n_2 + n_3 ... n_n$ = total number of moles (n_i)

Hence, $$P = \frac{RT}{V} n_i$$

Now, $$\frac{P_1}{P} = n_1 \frac{RT}{V} \Big/ n_i \frac{RT}{V} = \frac{n_1}{n_i}$$

or $$\frac{P_1}{P} = x_1$$

where x_1 = mole fraction of $n_1 = \frac{n_1}{n_i}$

Similarly,

$$\frac{P_2}{P} = x_2, \quad \frac{P_3}{P} = x_3$$

where x_2 and x_3 are the mole fractions of n_2 and n_3.

In a binary solution having components A and B, let the mole fraction of the component B in the solution be x_B and its mole fraction in the vapour phase be Y_B, then according to Dalton's law of partial pressures.

$$Y_B = \frac{p_B}{P} \qquad \text{...(15)}$$

Where p_B is the partial vapour pressure of B

P is the total vapour pressure of the solution

or $$Y_B = \frac{p_B}{p_A + p_B}$$

Applying Raoult's Law:

$$Y_B = \frac{x_B \overset{\circ}{p}_B}{x_A \overset{\circ}{p}_A + x_B \overset{\circ}{p}_B} = \frac{x_B \overset{\circ}{p}_B}{\overset{\circ}{p}_A + x_B(\overset{\circ}{p}_B - \overset{\circ}{p}_A)} \qquad \text{...(16)}$$

As $\overset{\circ}{p}_A$ and $\overset{\circ}{p}_B$ are constants and are known quantities, hence the composition of the vapour phase, i.e. Y_B can be calculated for different values of x_B. From equation (16) it is clear that x_B and Y_B are not identical.

However, if $\overset{\circ}{p}_A = \overset{\circ}{p}_B$ then $x_B = Y_B$.

To express the total vapour pressure in terms of composition of vapour phase

Rearranging equation (16), we get

$$x_B = \frac{Y_B \overset{\circ}{p}_A}{\overset{\circ}{p}_B + Y_B(\overset{\circ}{p}_A - \overset{\circ}{p}_B)} \quad ...(17)$$

Substituting the value of x_B in equation (6) we get,

$$*P = \overset{\circ}{p}_A + \frac{Y_B \overset{\circ}{p}_A}{\overset{\circ}{p}_B + Y_B(\overset{\circ}{p}_A - \overset{\circ}{p}_B)}(\overset{\circ}{p}_B - \overset{\circ}{p}_A) \quad ...(18)$$

$$\text{or} \quad P = \frac{\overset{\circ}{p}_A \, \overset{\circ}{p}_B}{\overset{\circ}{p}_B + Y_B(\overset{\circ}{p}_A - \overset{\circ}{p}_B)} \quad ...(19)$$

Relation (19) gives the dependence of the total pressure on the composition of the vapour phase. Equation (19) can be written in the linear form by taking its reciprocal, i.e.

$$\frac{1}{P} = \frac{\overset{\circ}{p}_B + Y_B(\overset{\circ}{p}_A - \overset{\circ}{p}_B)}{\overset{\circ}{p}_A \, \overset{\circ}{p}_B}$$

$$= \frac{1}{\overset{\circ}{p}_A} + \left(\frac{1}{\overset{\circ}{p}_B} - \frac{1}{\overset{\circ}{p}_A}\right) Y_B \quad ...(20)$$

Thus the reciprocal of the total pressure varies linearly with Y_B, the mole fraction of the component B in the vapour phase. This is shown in Fig. 19.5.

$$* P = \overset{\circ}{p}_A + \frac{Y_B \overset{\circ}{p}_A}{\overset{\circ}{p}_B + Y_B(\overset{\circ}{p}_A - \overset{\circ}{p}_B)}(\overset{\circ}{p}_B - \overset{\circ}{p}_A)$$

$$= \frac{\overset{\circ}{p}_A[\overset{\circ}{p}_B + Y_B(\overset{\circ}{p}_A - \overset{\circ}{p}_B)] + Y_B \overset{\circ}{p}_A(\overset{\circ}{p}_B - \overset{\circ}{p}_A)}{\overset{\circ}{p}_B + Y_B(\overset{\circ}{p}_A - \overset{\circ}{p}_B)}$$

$$= \frac{\overset{\circ}{p}_A[\overset{\circ}{p}_B + Y_B \overset{\circ}{p}_A - Y_B \overset{\circ}{p}_B] + Y_B \overset{\circ}{p}_A \overset{\circ}{p}_B - Y_B \overset{\circ}{p}_A \overset{\circ}{p}_A)}{\overset{\circ}{p}_B + Y_B(\overset{\circ}{p}_A - \overset{\circ}{p}_B)}$$

$$= \frac{\overset{\circ}{p}_A \overset{\circ}{p}_B + Y_B \overset{\circ}{p}_A \overset{\circ}{p}_A - Y_B \overset{\circ}{p}_A \overset{\circ}{p}_B + Y_B \overset{\circ}{p}_A \overset{\circ}{p}_B - Y_B \overset{\circ}{p}_A \overset{\circ}{p}_A)}{\overset{\circ}{p}_B + Y_B(\overset{\circ}{p}_A - \overset{\circ}{p}_B)}$$

$$= \frac{\overset{\circ}{p}_A \overset{\circ}{p}_B}{\overset{\circ}{p}_B + Y_B(\overset{\circ}{p}_A - \overset{\circ}{p}_B)}$$

Ratio of Mole fraction of B component in vapour phase to its mole fraction in solution:

Now we know,

$$Y_B = \frac{p_B}{p_A + p_B}$$

$$= \frac{\overset{\circ}{p}_B x_B}{\overset{\circ}{p}_A x_A + \overset{\circ}{p}_B x_B}$$

or
$$\frac{Y_B}{x_B} = \frac{\overset{\circ}{p}_B}{\overset{\circ}{p}_A x_A + \overset{\circ}{p}_B x_B} \quad ...(21)$$

Dividing the numerator and denominator by p_B we get,

$$\frac{Y_B}{x_B} = \frac{1}{\frac{\overset{\circ}{p}_A x_A}{\overset{\circ}{p}_B} + x_B} \quad ...(22)$$

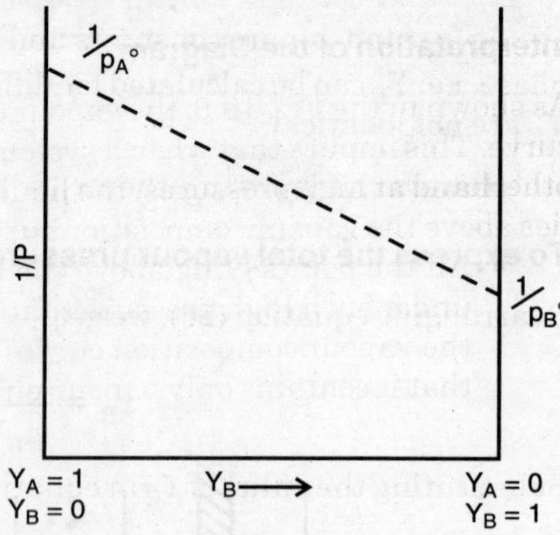

Fig. 19.5 Plot of $1/P$ versus Y_B, the composition of component B in vapour phase

If $P_B^\circ > P_A^\circ$, i.e. the mole fraction of the component B in the vapour phase is larger than component A, the ratio Y_B/x_B would be greater than one. In other words the vapour phase is richer in more volatile component B than the liquid phase. This conclusion is known as the **Konowaloff's rule, which states that the vapour phase is relatively richer in the component, whose addition to the liquid mixture results in an increase of total vapour pressure.**

With the help of equation (16), vapour pressure composition diagram can be obtained which give the plot of vapour pressure versus vapour composition. (The composition diagram discussed earlier were for vapour pressure versus liquid composition). Generally both, the vapour composition and the liquid composition are of equal interest and hence it is more sensible to combine the liquid as well as vapour composition versus mole fraction of one of the components. Such type of graph is shown in Fig. 19.6.

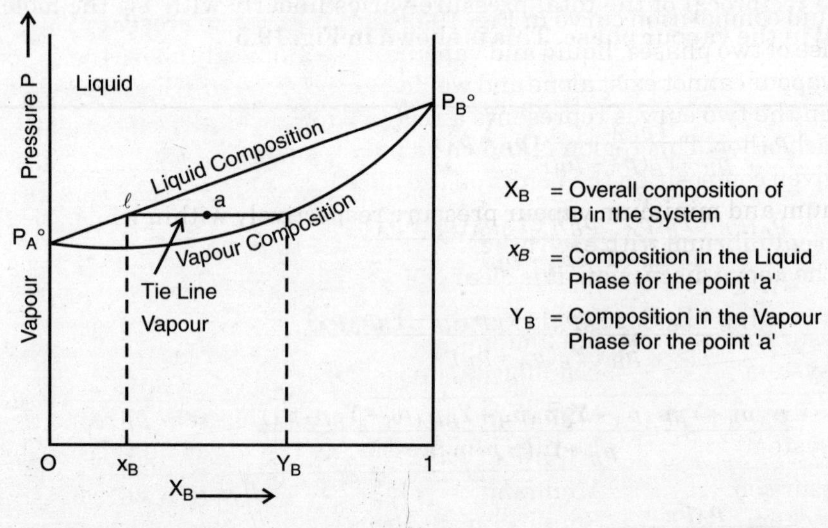

Fig. 19.6 Liquid and vapour composition curves for ideal solutions

Interpretation of the Diagram

As shown in the Fig. 19.6, the vapour composition curve always lies below the liquid composition curve. This implies that when a system is under low pressure vapour phase predominates. On the otherhand at high pressures, the liquid phase predominates. Hence the liquid composition curve lies above the vapour composition curve.

- All the points lying above the liquid composition curve correspond to a system which is under such high pressure that it contains only a liquid phase. Similarly, all points below the vapour composition curve correspond to a system which is under such low pressures that it contains only vapour phase.

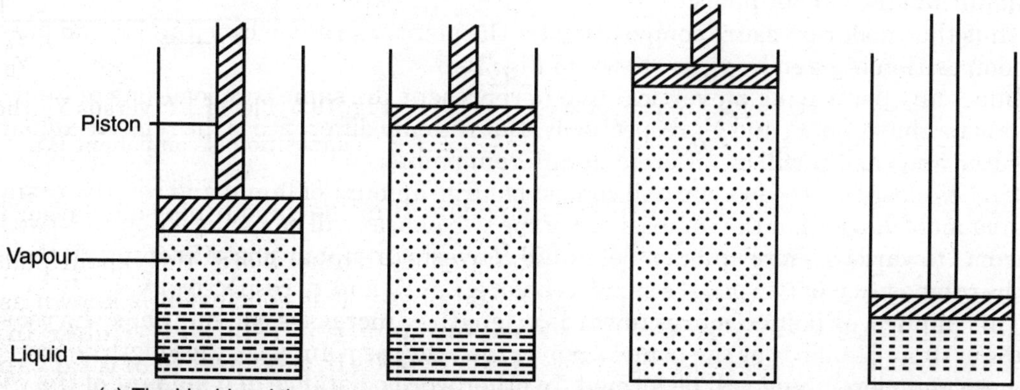

(a) The Liquid and the Vapour Phase are in equilibrium with each other. (b) Change in equilibrium pressure Changes the amounts of Liquid and Vapour Phase (c) System under Low Pressure than the equilibrium vapour pressure contains only Vapour Phase (d) System under High Pressure than the equilibrium vapour pressure contains only Liquid Phase.

Fig. 19.7 Variation in amounts of liquid and vapour with change in pressure

- Points that lie between the liquid composition curve and vapour composition curve correspond to a system where liquid and vapour coexist (Fig. 19.7(a) and (b)). This implies that below the liquid composition curve in Fig. 19.6, liquid cannot be present alone, the system will comprise of two phases, liquid and vapour phase. Similarly above the vapour composition curve vapour cannot exist alone and would be associated with liquid phase. Thus the region between the two curves represents a region where both the phases exist in equilibrium with each other. This region is known as liquid-vapour region.
- For a given composition the points on the liquid and vapour composition curves represent maximum and minimum vapour pressure respectively within which the two phases can exist in equilibrium with each other.
- From the above discussion, it is clear that –

Pressure on the system	$>$	Maximum Equilibrium pressure	System will consist of liquid phase only
Pressure on the system	$<$	Minimum Equilibrium pressure	System will consist of vapour phase only
Pressure on the system	$>$	Minimum Equilibrium pressure	System will consist of liquid and vapour phase in Equilibrium.
	$<$	Maximum Equilibrium pressure	

Relative Amounts of Liquid and Vapour Present in Equilibrium with Each Other

As is discussed above the points within the two curves correspond to the liquid vapour region. But now it is of significant interest to know that what will be the respective amounts of liquid and vapours for a given composition in this region.

Let us reconsider the plot of liquid vapour composition for ideal solution. (Fig. 19.6)

- Consider a point 'a' lying in the liquid-vapour region.
- The composition in liquid phase and vapour phase for this particular point is obtained by drawing a horizontal line lav known as a "tie line", because it connects the composition of liquid and the vapour phase.
- Thus the vapour pressure composition for the point 'a' is given by Y_B and liquid pressure composition is given by x_B as shown in Fig. 19.6.
- Infact any point lying on vertical line lv represent the same compositions of liquid and vapour phase, i.e. x_A and Y_B respectively. But the only difference is the relative amounts of liquid and vapour will be different at different points.
- If a lies close to l, the system will consist of large amount of liquid and relatively smaller amount of vapour but the composition of the two phases will be x_B and Y_B only. As we move from l towards v, more and more of liquid converts to vapour phase keeping all the while the composition of two phases same as before, i.e. Y_B and x_B respectively.
- This implies, as point a moves towards v, length al increases, vapour phase increases in the system. Similarly as the point 'a' moves towards the point l, i.e. the length av increases, more and more liquid will be formed. In other words distance al α amount of the vapour formed distance av α amount of the liquid formed

or
$$\frac{al}{av} = \frac{\textit{Amount of vapour formed}}{\textit{Amount of liquid formed}}$$

This expression is known as the **Lever rule**. Thus the ratio of amount of vapour formed to the amount of liquid formed is given by the ratio of lengths of line segments connecting a to l and a to v.

19.7 VAPOUR PRESSURE—COMPOSITION DIAGRAMS FOR NON-IDEAL SOLUTIONS

As was pointed out earlier that the non-ideal solution deviates from ideal behaviour and exhibits positive and negative deviations. The vapour pressure and composition in liquid phase for non-ideal solutions have already been discussed in section (19.5). In this section we will study vapour pressure versus vapour composition and liquid composition curves simultaneously. For non-ideal solutions the following three types of vapour pressure–compostion (vapour as well as liquid composition) diagrams have been identified.

Type 1: Solutions with very little deviations from Ideality : These are the solutions in which the vapour pressure increases or decreases regularly with the composition of the mixture. The total vapour pressure of the mixture lies between those of pure components. The curves are very similar to those of an ideal solution. (Fig. 19.8)

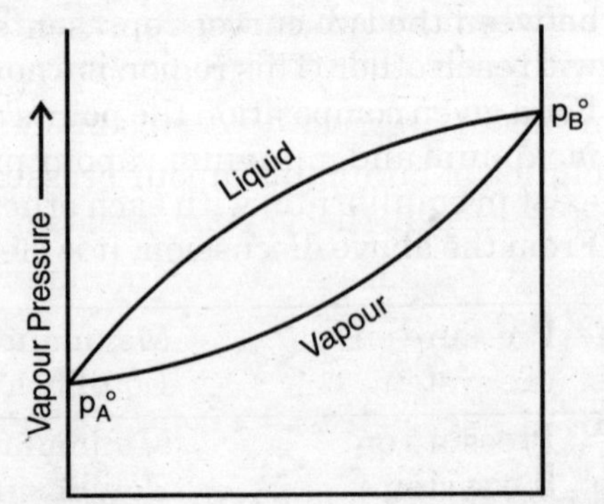

Fig. 19.8 Vapour pressure versus liquid and vapour composition for the solutions of type I

Type II: Solutions with large positive deviation from ideality exihibiting a maximum in the vapour pressure curve : In this category, the solution exhibits maximum in the pressure versus composition curve. It is seen from the graph (Fig. 19.9) that at point O, the composition of the vapour and liquid phase are the same. Points $\overset{\circ}{p}_A$ and $\overset{\circ}{p}_B$ denote the vapour pressures of pure components.

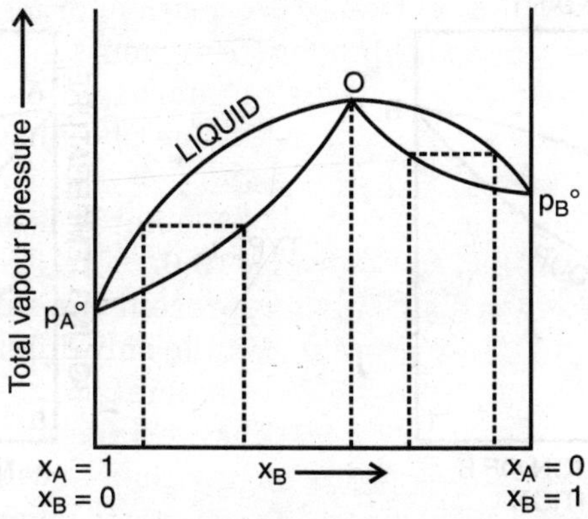

Fig. 19.9 Vapour pressure versus liquid and vapour composition for solutions showing positive deviation from raoult's law

The diagram thus can be demarcated into two portions, one between vapour pressure of pure component A, $\overset{\circ}{p}_A$ and a solution of maximum vapour pressure, O and the other lies between the vapour pressure of pure component B, $\overset{\circ}{p}_B$ and a solution of maximum vapour pressure, O

Type III : Solution with large negative deviations from ideality exhibiting minimum in the vapour pressure composition curve : In this category are the solutions which exhibit the minimum in the pressure versus composition curve. From Fig. 19.10, it is clear that at point O, the composition of the vapour and the liquid phase are the same. The curve may again be divided into two portions, one lying between vapour pressure of pure component A, $\overset{\circ}{p}_A$ and a solution of minimum vapour pressure O and the other between the vapour pressure of pure component B, $\overset{\circ}{p}_B$ and a solution of minimum vapour pressure, O (Fig. 19.10)

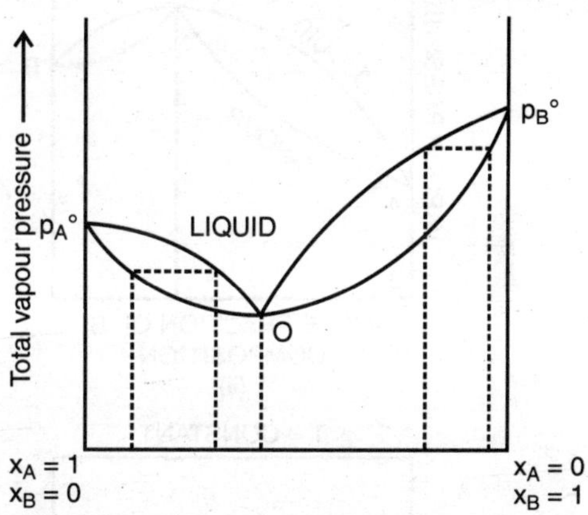

Fig. 19.10 Vapour pressure versus liquid and vapour composition for non ideal solution showing negative deviation from Raoult's Law

19.8 BOILING POINT-COMPOSITION DIAGRAMS

The boiling point of a liquid is the temperature at which the vapour pressure of the liquid becomes equal to the atmospheric pressure.

A liquid which has higher vapour pressure will have a lower boiling point since less heat will have to be supplied to attain the vapour pressure equal to the atmospheric pressure. Similarly a liquid with low vapour pressure will have a high boiling point. Since the solutions of different compositions have different vapour pressures, it follows that these all will boil at different temperatures. This implies that a solution having higher vapour pressure will boil at low temperature and vice-versa.

Thus boiling point is inversely related to vapour pressure. These facts help us to construct the boiling point versus composition diagrams from vapour pressure–composition diagrams. Boiling point composition diagrams corresponding to three types of vapour composition diagram are shown in Fig. 19.11.

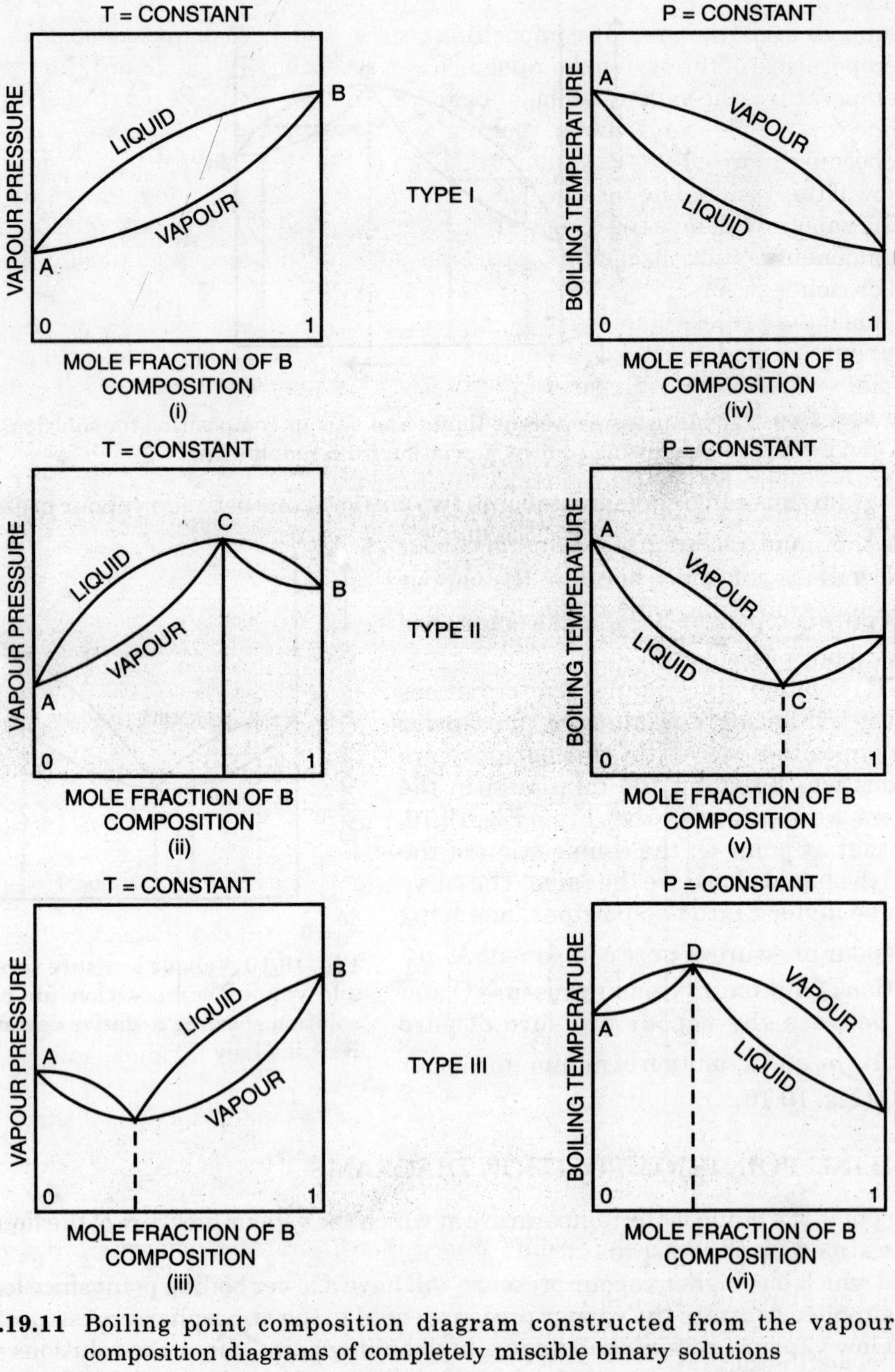

Fig.19.11 Boiling point composition diagram constructed from the vapour pressure composition diagrams of completely miscible binary solutions

19.9 FRACTIONAL DISTILLATION OF BINARY LIQUID SOLUTIONS

The distillation of various solutions can be discussed with the help of boiling point – composition diagram. The distillation of the above mentioned three types of solutions is discussed below.

Distillation of Solution of Type-I : The distillation of solution which show slight deviation from ideality can be explained with the help of boiling point composition diagram as shown in Fig. 19.12.

- Suppose a solution having composition corresponding to point 'a' is to be distilled. Let the temperature of the system be raised. When it reaches T_a, the liquid starts boiling. At this temperature, the vapour which appear will have a composition a_1 (tie line).
- Since a_1, is richer in B than a, the composition of the residual liquid will become richer in A because more of B have vapourised from it. Let this composition be represented by 'b'.
- Now if this residual liquid 'b' is heated, it will start boiling when temperature T_b is reached. The vapour will have the composition corresponding to b_1. Again vapour b_1 is richer in the component B and consequently the composition of the residual liquid will become poorer in B or richer in A.
- Thus if the process of heating the residue is continued with simultaneous removal of the vapours formed, the residual solution will progressivly become richer in A and the boiling point will subsequently rise from T_a to T_A which is the boiling point of the pure liquid. This means that if the process is contined for sufficiently long time, a final residue having pure A can be obtained.
- Further the vapour obtained on heating the solution 'a' i.e. vapours having the composition a_1 are withdrawn and condensed. The liquid composition of these condensed vapours will be the same, i.e. a_1.
- Now if this mixture is distilled again, it will boil when it reaches the temperature Ta_1. The composition of the vapours at this stage is given by a_2 which is richer in B. Evidently if this process of condensing the vapours and redistilling the liquid is continued, the distillate obtained in due course will compose of pure B.

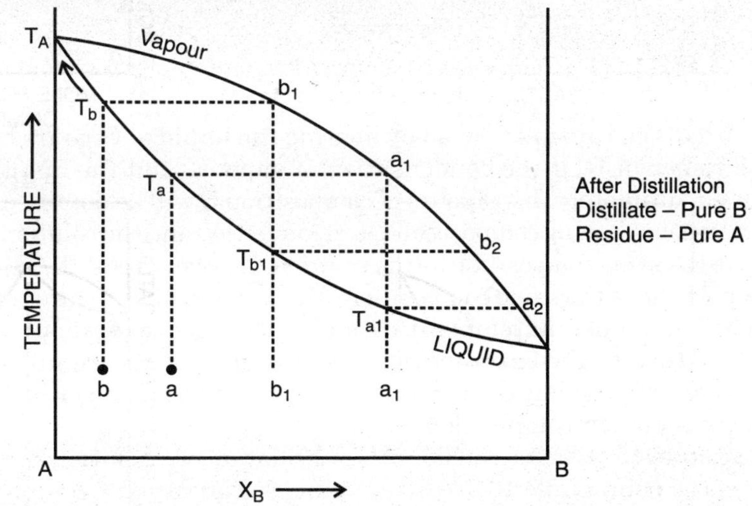

Fig. 19.12 Boiling point – composition diagram for a binary solution

From the above discussion one can conclude that in case of solutions which show slight deviation from ideal solutions or in ideal solutions, a complete seperation of the components is possible by distillation.

Distillation of Solution of Type-II: The distillation of the solutions which show positive deviations in vapour pressure composition curves can be explained by considering their temperature–composition diagram Fig. 19.13.

Consider the distillation of a mixture having composition 'a' lying between the points A and C.

- The liquid mixture will start boiling at temperature T_a and the vapours so formed will have composition a_1.
- The residual liquid will be rich in A and if the vapours formed are removed, the composition of the residual liquid will change. Let the new composition be represented by b.
- If b is heated it will boil at temperature T_b and the vapours so formed will have composition b_1. The residual liquid will become richer in A.
- Thus is this process of heating the residue is continued with simultaneous removal of the vapour formed, the residual solution will become richer in A, thus ultimately the residue having pure A is formed.

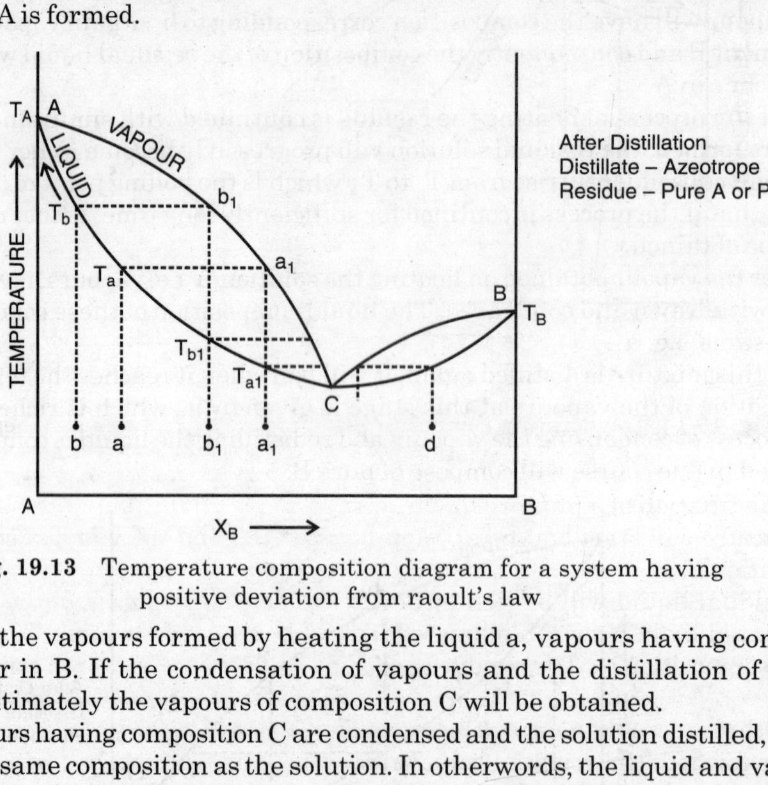

Fig. 19.13 Temperature composition diagram for a system having positive deviation from raoult's law

- As regard to the vapours formed by heating the liquid a, vapours having composition a_1, will be richer in B. If the condensation of vapours and the distillation of the liquid is continued, ultimately the vapours of composition C will be obtained.
- If these vapours having composition C are condensed and the solution distilled, the distillate will have the same composition as the solution. In otherwords, the liquid and vapour phases at this point have the same composition. Thus, the liquid mixture represented by the point C will boil at a constant temperature and will distil over without change of composition. Such a mixture is called **azeotropic mixture**. In this case we will get **minimum boiling azeotrope** as the boiling point of azeotrope is minimum than either of the pure component or any other of their composition.
- If on the other hand let us consider the distillation of a solution having composition lying between the point C and B, let us say point 'd'. The vapours coming off will be richer in A and the residue will be richer in B. As the distillation proceeds, the successive residues will become increasingly rich in pure component B. Whereas the distillate will become progressively rich in the constant boiling mixture. It will never be possible to have pure A in this case.
- The above results can be summarized as follows:

A solution having composition between A and C on fractional distillation will give pure A as residue and azeotrope C as a final distillate. No pure B can be obtained.

Whereas a solution having composition between C and B on fractional distillation will give pure B as the residue and azeotrope C as the final distillate. No pure A can be obtained.

Distillation of Solution of Type III: The temperature – Composition diagram for the solutions showing negative devaitions in vapour pressure-composition is shown in Fig. 19.14.

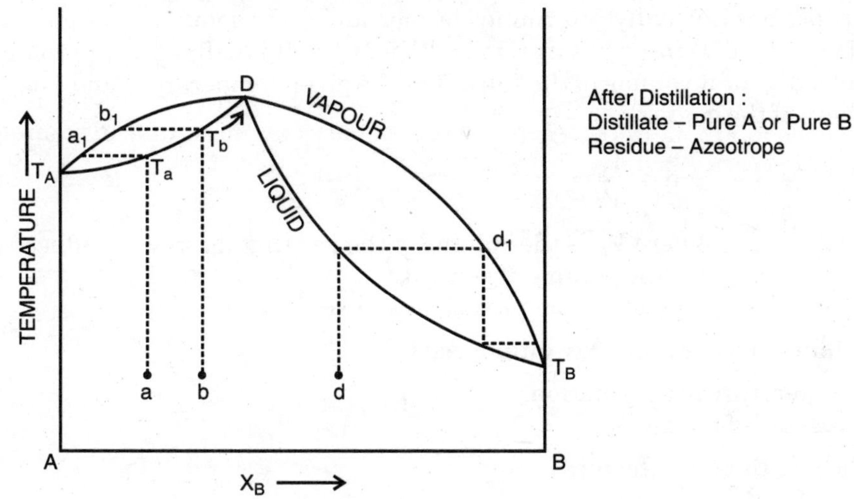

Fig. 19.14 Temperature composition diagram of a system having negative deviation from Raoult's Law

Consider the distillation of a mixture having composition 'a' lying between points A and D.

- The mixture will start boiling at a temperature T_a and the vapours so formed will have composition a_1.
- The residual liquid will be rich in B. Let the new composition of the residual liquid be represented by b. Now if b is heated, it will boil at temperature T_b giving off vapours having composition b_1. The residue so obtained will become richer in Band will shift toward D.
- Thus if this process of heating the residue is continued with simultaneous removal of the vapours formed, the residual solution of composition D will be formed. It will never be possible to have pure B in this case.
- If the solution having composition D is heated, the distillate will have the same composition to that of the residue represented by the point D. Thus, this liquid mixture is the azeotropic mixture which distils over without change of composition. The azeotropic mixture in this case is **maximum boiling azeotrope** as it has maximum boiling point than either of the pure components or for any other of their composition.
- The vapours which are formed, in different stages on subsequent cooling and heating will eventually lead to a solution having pure component A.
- Similarly a mixture of composition lying between B and D say 'd' will on distillation give vapours d_1 which will be rich in B and the residue shifts towards the point D. As the distillation proceeds, the distillate eventually will have a solution having pure B and residue will have constant boiling mixture having composition D.

- Thus, a solution having composition between A and D on fractional distillation will give pure A as distillate and azeotrope D as residue. No pure B is obtained whereas a solution having composition between D and B on fractional distillation will give pure B as distillate and azetrope D as residue. No pure A is obtained in this case.

19.10 SOLUTIONS OF GASES IN LIQUIDS—HENRY'S LAW

Most of the gases are soluble in liquids, i.e. water or any other solvent to more or less some extent. The solubility of a gas is usually determined by measuring the volume of the gas that dissolves. It is expressed in terms of Bunsen absorption coefficient "α". It is defined as the volume of the gas at N.T.P., dissolved by unit volume of the solvent at the given temperature and under a pressure of one atmosphere of the gas *viz.*

$$\alpha = \frac{V_O}{V_P} \quad \text{... (23)}$$

Where V_O is the volume of the gas that dissolves, reduced to STP,
V is the volume of solvent
P is the partial pressure of the gas.

The solubility of a gas is greatly influenced by:

(a) The temperature of the solution.
(b) The pressure of the gas.

Let us discuss these two factors.

(a) **Effect of temperature on solubility:** Dissolution of a gas in a liquid is accompanied by evolution of heat. According to Le-Chatelier principle, the solubility of a gas should decrease with increase of temperature and has been found so. For example, it is a common observation that when water is heated in a beaker, the bubbles of dissolved air start forming on the side of the glass. As the temperature rises, the dissolved air molecules begin to come out of the solution even before the water boils.

(b) **Effect of Pressure on solubility:** Solubilities of gases in liquids vary considerably with external pressure. The quantitative relationship between gas solubility and pressure is given by Henry's law. William Henry in 1803 on the bases of his systematic studies concerning the solubilities of the gases in liquids gave a generalisation known as **Henry's Law**, which can be stated as :

The mass of a gas dissolved per unit volume of a solvent is proportional to the pressure of the gas in equilibrium with the solution at constant temperature.

If m is the mass of the gas dissolved by unit volume of solvent at equilibrium pressure p, then according to Henry's law

$$m \propto p$$
$$m = kp \quad \text{... (24)}$$

Where k is the constant of proportionality.

Henry's law can be understood in terms of Le Chatelier's principle. Suppose we have a gas in dynamic equilibrium with a solution.

$$\text{Gas} \rightleftharpoons \text{Gas in Solution}$$

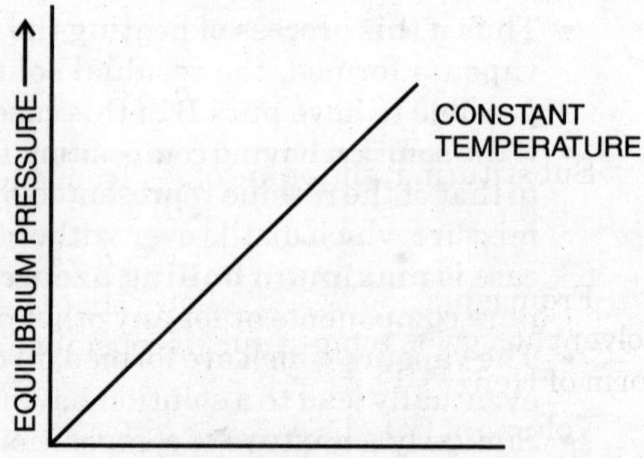

Fig. 19.15 Variation of solubility of a gas with pressure.

If pressure is increased on the system, the equilibrium will move in the direction which will reduce the pressure. The pressure can be reduced if more gas dissolves in the solution.

From equation (24) it is clear that graph obtained by plotting solubility of a gas against the equilibrium pressure of the gas at a constant temperature should be a straight line (Fig. 19.15).

Limitations of Henry's Law: Gases obey Henry's Law provided:

(i) The gases exhibit nearly ideal behaviour.

(ii) The pressure is not too high.

(iii) The temperature is not too low.

(iv) The solubility of the gas in the given solvent is low.

(v) The dissolved gas does not react with the solvent. For example, NH_3 or HCl which react with water do not obey Henry's Law, because these gases react with water.

$$NH_3 + H_2O \rightleftharpoons NH_4^+ + \bar{O}H$$

(vi) The gas does not associate or dissociate on dissolving in the solvent.

Alternative Form of Henry's Law

Henry's law may be stated in terms of volume of the gas dissolved instead of the mass.

Let v be the volume of the gas dissolved, at pressure p, in a given volume of the solvent at temperature T. According to ideal gas law,

$$pv = nRT$$

$$= \frac{m}{M} RT$$

$$v = \left(\frac{m}{p}\right)\frac{RT}{M}$$

$$= k'\frac{m}{p} \qquad ...(25)$$

Where k' is a constant and is equal to $\frac{RT}{M}$

M is the molar mass of the gas.

From Henry's law, stated above we have

$$\frac{m}{p} = k \qquad ...(26)$$

Substituting, the value of $\frac{m}{p}$ from equation (26) in equation (25) we have,

$$v = k \times k' \text{ or } \quad v = \text{Constant} \qquad ...(27)$$

From equation (27) it follows that the volume of the gas dissolved in a given volume of the solvent at a given temperature is constant and is independent of the pressure. This is the alternative form of Henry's Law.

Volume of CO_2 dissolved at different pressures at a given temperatures is given in Table 19.2. It is seen from the table the volume of the gas dissolved remains constant at different pressures verifying the above statement.

Table 19.2: Volume of CO_2 Dissolved at different Pressures.

Pressure (mm of Hg)	*Mass of Gas Dissolved (g per ml of water)*	*Volume of Gas Dissolved (cc per ml of water)*
271	0.270	0.825
495	0.492	0.825
755	0.751	0.826
927	0.922	0.826
1211	1.205	0.825
1350	9.343	0.824

When a mixture of a number of gases are in contact with the solvent, each of the gases dissolves according to its partial pressure, i.e. Henry's Law is applicable to each gas independent of the presence of the other gases.

The effect of partial pressure on solubility is utilised in making carbonated drinks. The so called soda water is prepared by dissolving CO_2 in water at a pressure of about 4 atmospheres. Whenever a soft drink bottle is opened in the air, the partial pressure of CO_2 above the solution decreases and CO_2 bubbles out.

Raoult's Law as a special case of Henry's Law: According to Henry's law,

$$m = kp \qquad \text{(cf eq. 24)}$$

Where m is the mass of the gas dissolved in a unit volume of the solvent and p is the pressure of the gas.

The mass of a gas dissolved in a given volume of the solvent can be expressed in terms of mole fraction of the solute in the solution. Let m' represent the mass of the solvent.

Dividing equation (24) throughout by m' we have,

$$\frac{m}{m'} = \left(\frac{k}{m'}\right)p = k'p \qquad \text{...(28)}$$

Further dividing the two masses by their respective molar masses, so as to get their number of moles, we have –

$$\frac{m/M}{m'/M'} = k'p\frac{M'}{M} \qquad \text{...(29)}$$

$$= \left(k'\frac{M'}{M}\right)p = k''p$$

$$\text{or} \quad \frac{n}{n'} = k''p \qquad \text{...(30)}$$

Where n is the number of moles of the gas

n' is the number of moles of the solvent.

Hence, the mole fraction of the gas above the solution is given by:

$$x = \frac{n}{n+n'} \qquad \text{...(31)}$$

For a dilute solution, the number of moles of the solute are very less as compared to the number of moles of the solvent such that $n' >> n$ and hence in equation (31), the mole fraction can be approximated as:

$$x = \frac{n}{n'} \qquad \text{...(32)}$$

Substituting, this value of $\frac{n}{n'}$ from equation (32) in equation (30) we have

$$x = k'' p \quad \text{...(33)}$$

Thus the solubility of a gas expressed in mole fractions is directly proportional to the pressure of the gas.

Also we can regard p as the vapour pressure of a volatile solute when its mole fraction x is present in the solution.

Let us consider a special case where Henry's Law is applicable over the entire range of concentration starting from $x = 0$ to $x_2 = 1$. The constant k'', when

$x = 1$ is given as

$1 = k'' p^\circ$ (When $x = 1$, p becomes the vapour pressure of the pure solute) ...(34)

$$\text{or } k'' = \frac{1}{p^\circ} \quad \text{...(35)}$$

Substituting, the value of k'' in equation (33), we have

$$x = \frac{1}{p^\circ} \times p \quad \text{...(36)}$$

$$\boxed{\text{or } p = xp^\circ} \quad \text{...(37)}$$

The above expression is identical with that of Raoult's Law for volatile solutes. Hence Raoult's Law can be regarded as special case of Henry's law.

19.11 THE NERNST DISTRIBUTION LAW

When a solute is added to two immiscible solvents in contact with each other, at a constant temperature, the solute distributes itself between the two solvents in such a way that the ratio of its equilibrium concentration in the two solvents is constant. This generalisation is known as distribution law.

For example, when iodine is added to water and carbondisulphide, it distributes itself in such a way that at equilibrium, the ratio of the concentration of iodine in water to that of its concentration in carbon disulphide is constant at any given temperature (Fig. 19.16) *viz.*

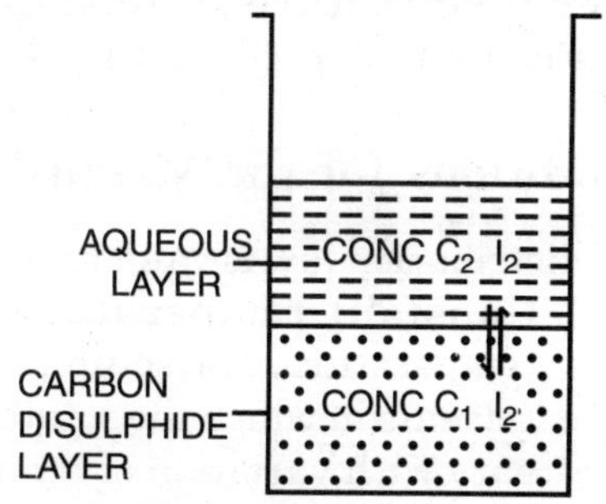

Fig. 19.16 Distribution of Solute I_2 between Solvent water and Carbondisulphide

$$\frac{[I_2]_{H_2O}}{[I_2]_{CS_2}} = \frac{C_1}{C_2} = \text{Constant}$$

Where C_1 is the conc. of I_2 in H_2O

C_2 is the conc. of I_2 in CS_2

$$\frac{C_1}{C_2} = K_D \quad \text{.... (38)}$$

The constant K_D is known as distribution or partition coefficient. The value of constant K_D–

- depends on the temperature of the system
- is independent of the relative amounts of the two layers, i.e. the two solvents.
- is independent of the amount of solute, provided the solution is dilute.

Berthelot and Jungfleish studied the distribution of iodine between carbon disulphide and water at 291K. The results for K_D are given in Table 19.3.

Table 19.3: Distribution of I_2 between CS_2 and H_2O at 291K

Conc. of Iodine In g/ml of Iodine (C_1)	*Conc. of Iodine In g/ml of water (C_2)*	$\frac{C_1}{C_2} = K_D$
0.174	0.00041	424.4
0.121	0.00032	403.1
0.066	0.00016	412.5
0.041	0.00010	410.0
0.0071	0.000017	417.6

Nernst found out that the ratio C_1/C_2 is constant only if the solute is present in the same molecular specie, i.e. it has same molecular conditions. For example if a solute associates to form dimer in one of the solvent and not in the other, the law is not valid and has to be modified. The law will be valid only if the ratio of concentrations of single molecules in the two solvents is considered. The distribution of benzoic acid between water and benzene is one such example. Benzoic acid does not exist in the same molecular condition in the two solvents. In water, it exists mostly as single molecules, whereas in benzene, it exists as dimer $(C_6H_5COOH)_2$ along with a small proportion of single molecules. Thus in this case, the Nernst distribution law is valid only for concentrations of single molecules in the two solvents.

The statement of distribution law is thus modified as follows : **When a solute distributes itself between two immiscible solvents in contact with each other, the ratio of the equilibrium concentration in the two solvents for the same molecular specie at a given temperature is constant.** Thus C_1 and C_2 in equation (38) are the concentrations of same molecular species in the two solvents at a given temperature.

Conditions for the Validity of the Distribution Law

The distribution law is valid under the following conditions:

(i) **Constant temperature:** The solubility of the solute generally changes with the temperature. The solubility of the solute may not change to the same extent for both the solvents in which the solute is distributed. Hence the distribution coefficient is expected to vary with variation of temperature. The variation of partition coefficient for 1°C rise in temperature is known as the temperature coefficient. Thus the value of distribution coefficient is constant, if the temperature is kept constant.

(ii) **Same molecular species of the solute dissolved, in the two solvents:** As is discussed earlier, the solute should be present in the same molecular condition in the two solvents i.e. the molar mass of the solute in the two solvents should be same.

(iii) **The solution should be dilute:** It has been found that the law is applicable when the concentration of the solution is low. At high concentration the departure from distribution law sets in. Higher the concentration, larger is the deviation.

(iv) **The two solvents should be immiscible:** The law is valid when the two solvents involved are mutually insoluble.

Modified Distribution law under Different Conditions

It was pointed out earlier that C_1/C_2 is constant only if the molecular state of the solute remains same in both the solvents. If the solute undergoes association or dissciation in one of the solvents

$C_1/C_2 \neq$ constant or distribution law, as given before is not valid. However the distribution law is applied in these cases with certain modifications as discussed below :

(i) When solute undergoes association in one of the solvents. Consider a solute having the molecular formula *M*, be distributed between two immiscible solvents *A* and *B*. Let us assume that it remains as such in solvent *A* but associates in solvent *B*. *n* molecules of *M* associate to form *(M)n*.

Let the conc. of *M* in solvent *A* = *C*
the total conc. of *M* in solvent $B = C_2$
(Total conc. = Associated molecules + Single molecules)

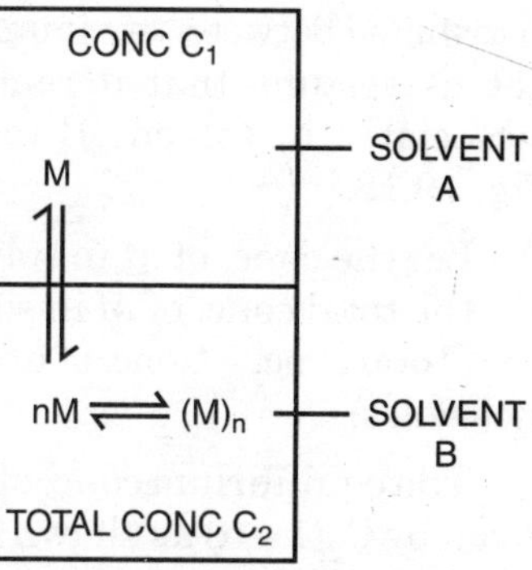

Fig. 19.17 Association of solute in one of the solvents

For applying distribution law we are interested in knowing the concentration of single molecules in solvent *B*.

The equilibrium between associated and single molecules in solvent *B* is given as:

$$\begin{array}{cc} nM & \rightleftharpoons (M)n \\ C_2(1-\alpha) & C_2\left(\frac{\alpha}{n}\right) \end{array}$$

Where α is the degree of association and *n* is the number of solute molecules that combine to form associated molecules.

Applying the law of chemical equilibrium to the above equilibrium we get,

$$K = \frac{C_2\left(\frac{\alpha}{n}\right)}{[C_2(1-\alpha)]^n} \qquad \text{... (39)}$$

$$\text{or } C_2(1-\alpha)^n = C_2\left(\frac{\alpha}{nK}\right) C_2\left(\frac{\alpha}{nK}\right)$$

$$C_2(1-\alpha) = \left[C_2\frac{\alpha}{nK}\right]^{\frac{1}{n}} \qquad \text{... (40)}$$

But $C_2(1-\alpha)$ is the conc. of single molecules in solvent B. Hence the distribution constant is given as:

$$K_D = \frac{C_1}{C_2(1-\alpha)} \qquad \text{... (41)}$$

$$= \frac{C_1}{\left(C_2\frac{\alpha}{nK}\right)^{\frac{1}{n}}} = \frac{C_1}{(C_2)^{\frac{1}{n}}}\left(\frac{nK}{\alpha}\right)^{\frac{1}{n}}$$

At constant temperature $\frac{nK}{\alpha}$ is a constant, thus

$$K_D = \frac{C1}{n\sqrt{C_2}} \cdot \text{Constant}$$

$$K'_D = \frac{C_1}{\sqrt[n]{C_2}} \qquad \text{... (42)}$$

(ii) When solute undergoes dissociation in one of the solvents

Consider a solute having the molecular formula M be distributed between two immiscible solvents A and B as before. Let us assume that it remains as such in solvent A but dissociates in solvent B to species X and Y as shown in Fig. 19.18.

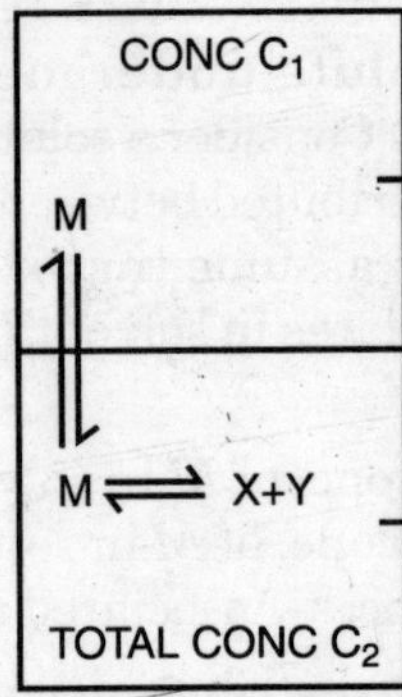

Fig. 19.18 Dissociation of solute in one of the solvents

Let the conc. of M in solvent A = C_1
the total conc. of M in solvent B = C_2
(Total conc. = Conc. of undissociated + conc. of dissociated molecules)

The equilibrium conc. of single molecules in solvent B is given by $C_2(1-\alpha)$ as shown below:

$$\begin{array}{ccc} M & \rightleftharpoons & X + Y \\ C_2(1-\alpha) & & \alpha C_2 \quad \alpha C_2 \end{array}$$

Where α is the degree of dissociation. The distribution constant is given as:

$$K_D = \frac{C_1}{C_2(1-\alpha)} \quad ...(43)$$

(iii) When solute enters into chemical combination with one the solvents

Again, consider a solute having the molecular formula M be distributed between two immiscible solvents A and B. Let us assume that it does not undergo any chemical change in solvent A but it enters into a chemical combination with solvent B.

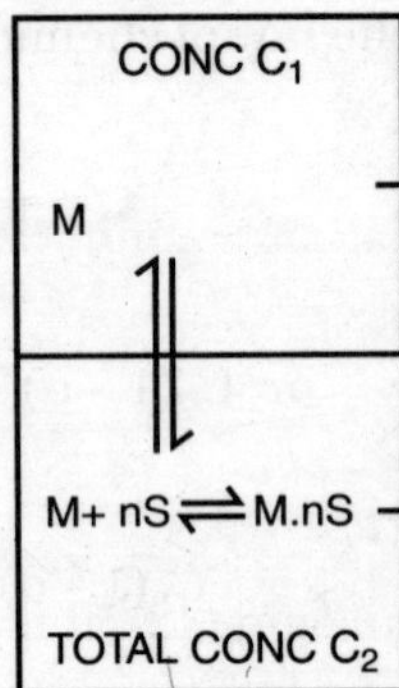

Fig. 19.19 Solute enters in chemical combination with one of the solvents

Let the conc. of M in solvent $A = C_1$
Total conc. of M in solvent $B = C_2$
(Total conc. = Conc. of uncombined molecules + Conc. of combined molecules)

Thus the conc. of uncombined solute molecules will be given by $C_2(1\text{-}\alpha)$.

$$\begin{array}{ccc} M + nS & \rightleftharpoons & M.nS \\ C_2(1-\alpha) & & C_2\alpha \end{array} \quad ...(44)$$

Thus the distribution law will be:

$$K_D = \frac{C_1}{C_2(1-\alpha)} \quad ...(45)$$

However, it can be shown that the combination of the solute with one of the solvents does not change the fundamental equation of distribution law but the numerical value of the distribution coefficient changes. This can be shown as below : Applying the law of chemical equilibrium to the equation (44), we have

$$\text{K} = \frac{C_2\alpha}{C_2(1-\alpha)[S]^n} \quad ...(46)$$

As the solvent is in excess, its concentration may be taken as constant, hence

$$\frac{C_2\alpha}{C_2(1-\alpha)} = \text{Constant} \quad ...(47)$$

Dividing equation (45) by (47) we get,

$$\frac{C_1}{C_2\alpha} = \text{Constant}$$

Now α is a constant at a given temperature, hence

$$\frac{C_1}{C_2} = \alpha.\text{Constant}$$

$$\boxed{\frac{C_1}{C_2} = \text{Constant}} \qquad \dots (48)$$

Thus the fundamental equation remains the same, but the numerical value of the partition coefficient changes.

Applications of Distribution Law

There are numerous applications of distribution law. Some of the important ones are discussed below :

1. **Association of a solute:** When a substance is associated in one of the solvent and exist as singlet molecules in the other, the Distribution law is modified as:

$$\frac{C_1}{n\sqrt{C_2}} = K'_D \qquad \text{(cf eq. 42)}$$

Where n is the number of molecules which combine to form an associated molecule. It is possible to find out the value of n from Distribution law.

Taking log of equation (42),

$$\log C_1 = \log K'_D + \frac{1}{n}\log C_2 \qquad \dots (49)$$

From equation (42), it is clear that a plot of Log C_1 versus Log C_2 will give a straight line with slope equal to $\frac{1}{n}$ and the intercept equal to log K'_D. Hence, the value of K'_D and n can be calculated by simply determining the concentration of the solute in the two solvents.

Example : Study of distribution of acetic acid and benzoic acid between water and benzene showed that these substances exist in benzene as dimers, the value of n being 2.

2. **Determination of degree of dissociation:** As has been discussed earlier, if a solute undergoes disssociation in one of the solvents and remains as such in the other, then dissociation constant is given by:

$$\frac{C_1}{C_2(1-\alpha)} = K_D \qquad \text{(cf Eq. 43)}$$

The value of α can be determined from conductivity measurements and C_1 and C_2 are determined experimentally. Thus value of K_D can be calculated. Once the value of K_D has been calculated, the value of α i.e. degree of dissociation for any other concentrations of the solute can be found out.

3. **Determination of equilibrium constant:** The equilibrium constant can be conveniently determined by using the distribution law. Let us take for example, the determination of equilibrium constant of the reaction,

$$I_2 + KI \rightleftharpoons KI_3$$

or

$$I_2 + I^- \rightleftharpoons I_3^-$$

This can be done as follows:

(a) Determine the value of distribution coefficient of iodine, distributed between H_2O and Carbon disulphide.

$$K_D = \frac{[I_2]_{H_2O}}{[I_2]_{CS_2}}$$

[The concentration of I_2 in the two solvents can be found out by titrating against sodiumthiosulphate, using starch as indicator.]

(b) A solution of KI having conc. 'a' is prepared.

(c) Now dissolve some I_2 in KI solution and shake the solution with carbon disulphide so that I_2 distributes itself between aqueous KI solution and carbon-disulphide layer (Fig. 19.20)

(d) Determine the concentration of I_2 in the two layers.

Let conc. of I_2 in CS_2 Layer = b

Let total conc. of I_2 in water layer = c

(Total Conc. = Free I_2 + KI_3)

But we know from step (a), that

$$K_D = \frac{[I_2]_{H_2O}}{[I_2]_{CS_2}}$$

$\therefore$ Conc. of free I_2 in water $= [I_2]_{CS_2} \times K_D$

$= b \times K_D$.

Hence, Conc. of KI_3 in water = $c - K_D b$

Conc. of KI in water $= a - (c - K_D b)$

$= a - c + K_D b$

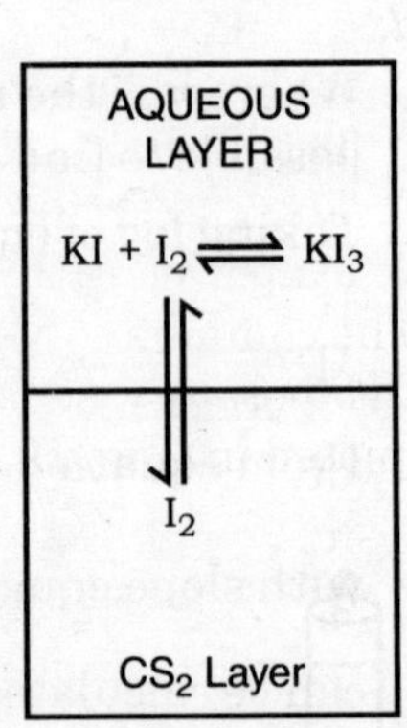

Fig. 19.20 AI_2 distributed between Aq. KI solution and CS_2

Now the equilibrium constant of the reaction,

$KI + I_2 \rightleftharpoons KI_3$ is given by:

$$K_{eq} = \frac{[KI_3]}{[KI][I_2]}$$

$$= \frac{c - K_D b}{(a - c + K_D b)(K_D b)}$$

Since, all the quantities on right hand side are known, the equilibrium constant K_{eq} mcan be calculated.

4. **Distribution Indicators:** The distribution coefficient of I_2 dissolved in water and carbontetrachloride is in favour of carbontetrachloride, i.e. conc. of iodine is more in carbontetrachloride than in water, when it is distributed between the above two solvents. This fact is used in titrating extremely dilute solutions of iodine in water. A few drops of carbontetrachloride are added to the solution. Much of the iodine present in water passes into the organic layer and imparts intense violet colour to it. Disappearance of violet colour marks the end point of the reaction.
5. **Confirmatory test for Bromide and Iodide:** The salt solution having bromide or iodide ions is treated with chlorine water which liberates bromine or iodine, as the case may be. To this solution is added chloroform. The iodine or bromine distributes between the two layers, water and chloroform. The free bromine or iodine being more soluble in chloroform passes into it, making it red in case of bromine and violet for iodine, Thus the above test is used as a confirmatory test for iodine and bromine ions.
6. **Solvent Extraction:** One of the most important application of the distribution law is in the process of solvent extraction. The extraction of an organic substance from an aqueous solution is the most common application of distribution law in the laboratory. The solute is generally extracted from the aqueous solution by shaking the solution with some organic solvent. The organic solvent employed for extraction is known as extracting solvent. For most of the organic substances, the distribution coefficient is in favour of organic solvent and hence most of it passes into the organic layer, leaving a small portion of it in water layer. On standing organic and water layers separate. The organic layer is removed and is transferred to a distillation flask. Organic layer is distilled over while the organic substance is left as a residue in the flask.

The solvent used for extraction should be chosen carefully keeping the following points in mind:

(a) It should be completely immiscible with the aqueous layer.
(b) The partition coefficient should be in favour of organic solvent employed for separation.
(c) The extracting solvent should be volatile so that the solute can be extracted by simply evaporating the organic solvent.

The common solvent employed for extraction are hexane, benzene, chloroform, etc.

In laboratory, separating funnel is used for the purpose of extraction. The process of extraction is depicted in (Fig. 19.21).

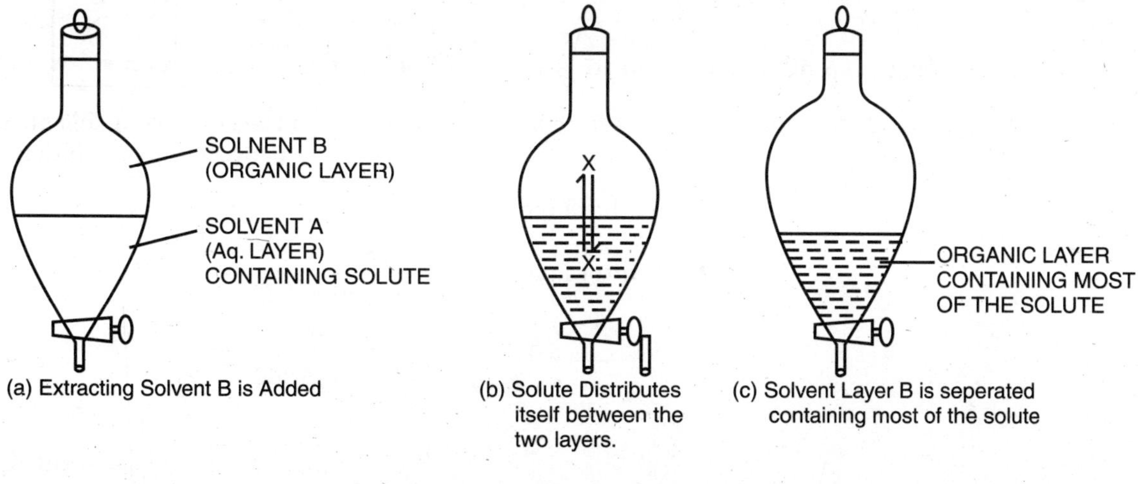

Fig. 19.21 Solvent extraction

The extraction with the extracting solvent can be carried out in two different ways :

(a) By using the entire volume of the extracting solvent in a single step.

(b) By using the extracting solvent in number of instalments, i.e. multiple step extraction.

However, it can be shown that multiple extraction is more advantageous than the single step extraction because in multiple extraction the amount of solute recovered is more as compared to single step extraction. Let us illustrate this by taking the following example :

Let A be the amount of the solute present in 1000 ml of aqueous solution. Let 1000 ml of ether is available as an extracting solvent, the distribution coefficient being twice in favour of ether. We can carry out the process either in a single step or in a multiple step.

(i) **Single step extraction:** If all the ether, i.e. 1000 ml is used in one step, let x g of the solute be extracted or passes into the ethereal layer leaving (A–x) g in the aqueous solution, such that,

Concentration of solute in ether layer = $x/1000$

Concentration of solute in aqueous layer = A-x/1000.

$$K_D = \frac{C_{ether}}{C_{water}} \text{ (Distribution law)}$$

$$2 = \frac{x/1000}{A - x/1000}$$

$$\therefore \quad x = \frac{2}{3} \text{ that is 66\% of the solute is extracted using single step.}$$

(ii) **Multiple extraction:** Let us see what happens if the above process is carried out in two instalments by using 500 ml ether in each instalment. Let x_1 g of solute extracted in the first step with 500 ml of ether.

$$\text{Conc. of ether layer} = x_1/500$$

$$\text{Conc. of water layer} = \frac{A - x_1}{1000}$$

$$K_D = \frac{x_1/500}{A - x_1/1000}$$

$$x_1 = \frac{1}{2} A$$

Thus in the first step 50% of the solute is extracted. Now the solute left in the water layer = $\frac{1}{2} A$. Let x_2 g of the solute is extracted in the second step with the remaining 500 ml of ether.

$$\text{Conc. of ether layer} = x_2/500$$

$$\text{Conc. of water layer} = \frac{1}{2} A - \frac{x_2}{1000}$$

$$K_D = 2 = \frac{x_2/500}{\frac{1}{2} A - x_2/1000}$$

$$\therefore \quad x_2 = \frac{1}{4} A$$

Thus 25% of the solute is extracted in the second step.

Therefore, a total of 75% of solute is extracted by carrying out the process in two steps as against 66% when 1000 ml of ether is used in a single step.

Similarly, it can be shown that if extraction is carried out in four instalments using 250 ml of ether in each step, it is possible to extract 80.2% of substance. In other words the process of extraction becomes more efficient if carried out in number of steps.

However, it is not possible to remove whole of the dissolved substance, however large the number of steps may be. This is because everytime the solute distributes itself between the two solvents and some portion of it, no matter however small must always be present in the aqueous layer.

General Formula for the Amount of Substance Left Unreacted

A general formula which enables us to find out the amount of substance unextracted can be derived as follows :

Let Volume of the solution from which the solute has to be extracted be = V_O ml

Amount of the solute present in the above solution = Wg

Amount of extracting liquid used in each step = V_1 ml

Amount of solute remained unextracted after first step = w_1g

$$K_D = \frac{C_{original\ solvent}}{C_{extracting\ solvent}}$$

$$= \frac{w_1/V_o}{W - w_1/V_1}$$

$$= \frac{w_1 V_1}{V_o (W - w_1)}$$

$$\text{or } w_1 = \frac{K_D V_o W}{V_1 + K_D V_o} \quad \text{... (50)}$$

Similarly, in the second extraction when volume V_1 of extracting solvent is used and if w_2 is the amount that remains unextracted then K_D is given by:

$$K_D = \frac{w_2/V_o}{w_1 - w_2/V_1} \quad \text{... (51)}$$

$$\text{or } w_2 = \frac{K_D V_0 w_1}{V_1 + K_D V_0} \quad \text{... (52)}$$

Putting the value of w_1 from equation (50) into (52)

$$w_2 = \left(W \frac{K_D V_0}{V_1 + K_D V_0} \right) \left(\frac{K_D V_0}{V_1 + K_D V_0} \right) \quad \text{... (53)}$$

$$= W \left(\frac{K_D V_0}{V_1 + K_D V_0} \right)^2 \quad \text{... (54)}$$

If the process is carried out in 'n' steps, then at the end of nth extraction the amount of solute remains unextracted will be given by:

$$w_n = W \left(\frac{K_D V_0}{V_1 + K_D V_0} \right)^n \quad \text{... (55)}$$

Thus it is evident that in order to extract maximum amount of solute, n should be as large as possible.

7. Desilverisation of Lead (Parke's process): Desilverisation of lead by Parke's process is based on the distribution law. The lead containing silver (argentiferous lead) is melted and heated to 800°C. Molten zinc is added to this. Zinc and lead form two layers and silver is distributed between them since the distribution ratio is in favour of zinc, the partition coefficient being of the order of 300 at 800° C, most of the silver passes into the zinc layer.

On cooling, the zinc layer containing silver is separated. The Ag–Zn alloy is distilled in a retort when zinc passes over leaving silver behind.

The lead layer which still contains unextracted silver is again treated with molten zinc 2 to 3 times so that most of silver can be extracted.

19.12 ARRHENIUS THEORY OF ELECTROLYTIC DISSOCIATION

The substances, which when dissolved in water or any other solvent are capable of conducting electric current are called electrolytes. The common electrolytes include NaCl, kCl, HCl, NaOH etc. The electrolytes exhibit-anomalous behaviour in solution. For example, the colligative properties of electrolytes are always higher than those of non-electrolytes. (Colligative properties of a dilute solution are those properties which depend entirely upon the number of particles of the solute contained in a known volume of the solvent and not at all upon the nature of the solute. Some of the colligative properties are lowering of vapour pressure, depression of freezing point and elevation of boiling point). To explain this anomalous behaviour of electrolytes, the Swedish chemist, Svante Arrhenius, proposed this theory of electrolytic dissociation in 1887.

The postulates of Arrhenius Theory of Electrolytic Dissociation are:

1. An electrolyte when dissolved in water or any other suitable solvent it spontaneously dissociates to form electrically charged particles called ions.

$$AB \rightleftharpoons A^+ + B^+$$

The positively charged ions are called cations and negatively charged ions are called anions. However, it is assumed that ions are already present in the solid electroyte and are held by strong electrostatic attractions. When dissolved in water, these neutral molecules dissociate to form separate anions and cations. Thus the dissociation can be represented as :

$$A^+B^- \rightleftharpoons A^+ + B^-$$

2. These ions are free to move under the influence of applied electric field. This mobililty of ions towards the oppositely charged electrode constitute the electrolyte current through the electrolytes. In other words conductivity of a solution is due to the concentration of ions present in the solution.
It should be mentioned that electric current does not produce ions but it has only directive effect.
3. Ions present in solution constantly recombine to form neutral molecules and thus there exists a state of equlibrium between dissociated and undissociated molecules,

$$AB \rightleftharpoons A^+ + B^-$$

Applying to law of mass Action to the ionic equilibrum

$$\frac{[A^+]\ [B^-]}{[AB]} = K$$

where K is called the Dissociation constant.

4. The total charge present on cations is equal but opposite to the charge present on anion so that the solution remains electrically neutral.
5. The extent of dissociation is different for different electrolyes and depends on (i) the concentration of electrolyte (ii) temperature and (iii) nature of electrolyte.
6. The degree of dissociation of an electrolyte increases with dilution, *i.e.* the equilibrium of the following reaction shifts to right hand side on dilution,

$$AB \rightleftharpoons A^+ + B^-$$

The fraction of the total number of molecules dissociated is called degree of dissociation and is given by,

$$\alpha = \frac{\text{Number of molecules dissociated}}{\text{Total number of molecules}}$$

7. Depending upon the magnitude of degree of dissociation, electrolytes are of two types:
   (i) **Strong electrolytes:** The elelctrolytes having higher value of degree of dissociation.
   (ii) **Weak electrolytes:** The electrolytes which dissociate very feebly and hence have very low value of degree of dissociation.

Evidence in Favour of Dissociation Theory: Arrhenius theory of electrolytic dissociation successfully explained a number of phenomenon which provide sufficient evidence in its favour. Some of them are discussed below :

1. **Ionic Nature of Reactants:** According to Arrhenius theory, ions are produced in a solution as soon as the electolyte is dissolved in the solvent. The proof of existence of ions is furnished by reactions in inorganic analysis which are infact the reactions of different ions. For example, a metal chloride gives a white precipitate with silver nitrate, irrespective of the nature of the metal radical (Ba, K, Ca, etc). This can be explained by the fact that metal chloride furnish chloride ions which combine with silver ions furnished by $AgNO_3$ to form AgCl

$$KCl \rightleftharpoons K^+ + Cl^-$$

$$AgNO_3 \rightleftharpoons Ag^+ + NO_3^-$$

$$Ag^+ + Cl^- \longrightarrow \underset{\text{white precipitable}}{AgCl}$$

Thus formation of white precipitate is the property of silver ions and chloride ions. On the other hand ethyl chloride on mixing with $AgNO_3$ does not give any white precipitate because it is non-electrolyte and does not furnish any chloride ions.

2. **Heat of Neutralisation:** Arrhenius theory successfully explains the fact that heat of neutralisation of all strong acids and strong bases is similar, i.e. 57.1 kJ mol^{-1}. This is because that in each case the reaction that occurs between an acid and base is the same irrespective of the nature of the acid or base taken. According to Arrhenius, the strong acid and strong base or the salt so formed on their combination, exist in a completely dissociated form, while water molecules remain almost completely unionised. Thus, in any strong acid and strong base reaction the fundamental reaction remains the same. Let us illustrate this by considering the reaction between nitric acid and sodium hydroxide which can be represented as:

$$H^+ + NO_3^- + Na^+ + \overline{O}H \rightarrow H_2O + Na^+ + NO_3^-$$

or

$$H^+ + \overline{O}H \rightarrow H_2O \quad \Delta H = -57.1 \text{ kJ}$$

Similarly, reaction between HCl and NaOH can be represented as:

$$H^+ + Cl^- + Na^+ + \overline{O}H \rightarrow H_2O + Na^+ + Cl^-$$

$$\text{or } H^+ + \overline{O}H \rightarrow H_2O$$

Thus it is evident that the fundamental reaction remains the same and hence heat of neutralisation of strong acids and strong bases is approximately constant.

In case of either the acid or the alkali is weak, the value is comparitively less.

3. **Properties of the common ions:** The theory is further supported by the fact that solutions of electrolytes with a common ion exhibit similar physical and chemical properties. For example, due to common ion Cu^{2+}, the solutions of $CuSO_4$, $CuNO_3$, are all blue in colour because of common cupric ion, which is blue in colour. Similarly, solutions of the electrolytes having CrO^{2-} ions are generally yellow.
4. **Colligative properties of solutions of electrolytes:** One of the remarkable successes of Arrhenius theory is that it provided a suitable explaination for the abnormal colligative properties exhibited by electrolytes. As was discussed earlier that the colligative properties of electrolytes are always higher than those of non-electrolytes of the same molal concentration. For example one gram mole of sugar when dissolved in 1000 grams of water, lowers its freezing point by 1.86°C wheres one gram of an electrolyte such as NaCl, HCl, etc. when dissolved in same volume of water lowers the freezing point by nearly two times. For the electrolytes like H_2SO_4, $BaCl_2$, etc., the value is nearly thrice and for electrolytes like Na_3PO_4, $AlCl_3$, etc., this value is four times approximately as compared to value of non-electrolytes.

According to Arrhenius, since the colligative properties depend upon the numbers of particles, a molecule of electrolytic on dissociation produce ions and each ion will behave as a particle. For example a NaCl molecule will be dissociated into two ions,

$$NaCl \rightarrow Na^+ + Cl^-$$

Thus, the value of colligative property will be two times the value for colligative property for a non-electrolyte. Similarly, H_2SO_4 dissociates into three ions and hence the value of colligative property will be three times the value calculated for non-electrolyte

$$H_2SO_4 \rightarrow 2H^+ + SO_4^{2-}$$

To represent the colligative properties of electrolytes by means of the same relation as for non-electrolytes, Van't Hoff suggested the use of factor 'i' which is known as Van't Hoff factor

$$i = \frac{\textit{Observed Colligative property}}{\textit{Calculated Colligative property}}$$

The value of 'i' can be used to calculate the degree of dissociation at a given concentration. Let $\propto$ be the degree of dissociation of an electrolyte AB.

$$AB \rightleftharpoons A^+ + B^-$$

$$1-\alpha \qquad \alpha \qquad \alpha$$

$$\text{Number of undissociated molecule} = 1 - \alpha$$

$$\text{Number of ions} = 2\alpha$$

$$\text{Total number of particles after dissociation} = 1 - \alpha + 2\alpha = 1 + \alpha$$

Now Van't Hoff factor 'i' will be given by,

$$i = \frac{\textit{Actual number of particles in solution}}{\textit{Number of particles without dissociation}} = \frac{1+\alpha}{1} \quad ...(1)$$

The value of i can be determined by any of the colligative propery and hence ∝ can be calculated. Similarly if the electrolyte dissociates into n ions, the 'i' will be given as:

$$i = \frac{1+(n-1)\alpha}{1}$$

$$\text{and } a = \frac{i-1}{n-1}$$

Ostwald's Dilution Law: According to the Arrhenius Theory of dissociation, an electrolyte dissociates into ions in aqueous solutions into positive and negative ions. These ions are in a state of equilibrium with the unionized molecules. Ostwald's applied the law of chemical equilibrium to such cases.

Let us consider a binary electrolyte AB which dissociates in solution to form ions A^+ and B^-

$$AB \rightleftharpoons A^+ + B^-$$

Let c moles per litre be the concentration of the electrolyte and ∝ be its degree of dissociation

$$AB \rightleftharpoons A^+ + B^-$$

$$\begin{array}{cccl} c & o & o & \text{Initial Concentration} \\ C(1-\alpha) & C\alpha & C\alpha & \text{Equilibrium Concentration} \end{array}$$

According to law of mass action, the equillibrium constant of the above reaction is given by:

$$K = \frac{C\alpha\, C\alpha}{C(1-\alpha)} = \frac{C\alpha^2}{(1-\alpha)} \quad \text{... (3)}$$

Where K is called the **dissociation constant** of the electrolyte. It has a constant value at a constant temperature.

If one mole of an electrolyte is dissolved in V litre of the solutions, then

$$C = \frac{1}{V} \quad \text{... (4)}$$

V is known as the **dilution** for the solution. substituing the value of C from equation (4) in equation (3) we get

$$K = \frac{\alpha^2}{(1-\alpha)V} \quad \text{... (5)}$$

Equation (5) is known as **Ostwald's dilution law** and correlates the variation of degree of dissociation of an electrolyte with dilution. For weak electrolytes, the value of ∝ is very small as compared to unity and hence 1–∝ = 1, thus equation (3) becomes

$$K = C\alpha^2$$

$$\propto = \sqrt{\frac{K}{C}} \quad \text{... (6)}$$

Hence, for weak electrolyte, *degree of dissociation is proportional to the square root of concentration of solution.*

Verification of Ostwald's Dilution law: The Ostwald's Dilution law can be verified if the values of ∝, the degree of dissociation, at different dilutions are known. The degree of dissociation, ∝, is determined by conductance measurement by the relation:

$$\propto = \frac{\wedge_C}{\wedge_O} \quad \text{... (7)}$$

Where A_C and A_0 are the equivalent corductances at dilution V and infinite dilution respectively. The value of ∝ at various dilutions thus determined are inserted in equation (5) and if these values come out to be nearly constant, the law stands verified.

The law was found to hold good in the case of weak electrolytes such as CH_3COOH. NH_4OH etc where in the value of K was found to be fairly constant.

However, when applied to strong electrolytes, the law fails completely. For strong electrolytes, the value of dissociation constant K rapidly falls with dilution

Limitations of Arrhenius Theory: A number of limitations were observed in Arrhenius theory of dissociation, which are as follows:

1. Ostwald's Dilution law which is based on Arrhenius theory fails completely when applied to strong electrolytes.
2. The degree of dissociation obtained from colligative properites and conductance measurement are in complete agreement for uni-univalent electrolytes such as kCl, NaCl, etc. However for bi-univalent or bi-bivalent electrolytes such as $BaCl_2$, $CaCl_2$, etc. the agreement is not at all satisfactory.
3. The strong electrolyte like NaCl, kCl and NaOH are good conductors of electricity in fused state, which implies that these exist in ionized state even when not dissolved in water.
4. This theory assumes the existence of ions but does not take into account the electrostatic attractions between the ions. Though this effect is quite small in dilute solutions but pronouced in concentrated solution.
5. The values of equilibrium constant of strong electrolytes have been found to vary with the concentration of electrolytes. This leads to doubt that whether ionic equilibrium assumed by Arrhenius exist at all in the solution.

Theory of Strong Electrolytes: One of the main weaknesses of Arrhenius theory is that it is applicable only in case of weak electrolytes and fail altogether when applied to strong electrolytes. The failure of Arrhenius method of calulating the degree of dissociation of a strong electrolyte can be explained on the bases of following factors :

1. It is now generally accepted that strong electrolytes such as NaCl, kCl, NaF, Na_2SO_4, etc., consist exclusively of ions even in the crystal lattice. There are no molecules. This fact is supported by X-ray analysis of crystals of strong electrolytes. In such cases the equilibrum between unionized molecules and ions as assumed in the theory of dissociation has no meaning.
2. The ions are formed right at the moment of formation of strong electrolyte. If we recall our knowledge that these compounds are formed by complete transference of electron from one atom to another. For example in the formation of NaCl, sodium transfers its electron to chlorine and as a result sodium acquires positive charge and chlorine acquires negative charge. Thus a crystal of NaCl consists of sodium ions and chloride ions and unionized molecules of NaCl does not exist at all.
3. In crystal lattices the ions are held in position by electrostatic forces and are not free to move. When the crystals are method dissolved in water, the ions become free to move and conduct electricity. When crystals are dissolved in water, which has a high dielectric constant, the electrostatic forces of attraction between the ions are considerably weakened and hence ions begin to move freely and conduct electricity more strongly.
4. The conductance of strong electrolytes increase on dilution and Arrhenius attributed this increase to increase in degree of dissociation. But as the ions are completely ionized at all concentrations this explaination is not correct. It is believed that at high concentration, pairing of ions takes place due to interionic attraction and hence their mobility is decreased resulting in decrease in conductance.

Debye-Huckel Theory of Strong Electrolytes: P. Debye and E. Huckel (1923) put forward the modern theory of strong electrolytes which takes into consideration the interionic attraction. The fundamental ideas of their theory are given below:

(i) Strong electrolytes are completely ionized at all dilutions. These exist as ions even in the solid state.

(ii) If the solution is very dilute, the distance between the ions will be large and hence these are free to move. The electrostatic forces vary inversely as the square of the distance between the ions. Therefore in very dilute solutions the interionic forces will be very less. But if the solution is concentrated, the interionic forces will be appreciable and pairing of ions take place. These pairs of ions are known as ionic doublets. The consequence of this is that the electrolyte is completely ionized but not completely dissociated.

(iii) Increase in molar conductance on dilution was attributed to the increase in degree of dissociation according to Arrhenius theory. For calculating the degree of dissociation of an electrolyte, following equation was suggested,

$$\alpha = \frac{\Lambda_C}{\Lambda_O} \qquad \text{(cf eq. 7)}$$

But this equation is not correct in case of strong electrolyte as there is no question of equilibrium between dissociated and undissociated ions. But then how to explain the increase in equivalent conductance on dilution of solution of strong electrolytes? According to Debye and Huckel theory the increase in equivalent conductance with dilution in case of strong electrolytes is due to increase in the mobilities of ions due to weaker interionic attractions. The solution then behaves as if the electrolyte were completely dissociated as well as completely ionized. Thus interionic attraction and not partial dissociation is the cause of decrease of equivalent conductance with increase in concentration. Fig 19.22 represents schematically the interionic attraction between the ions of NaCl and the effect of diliution.

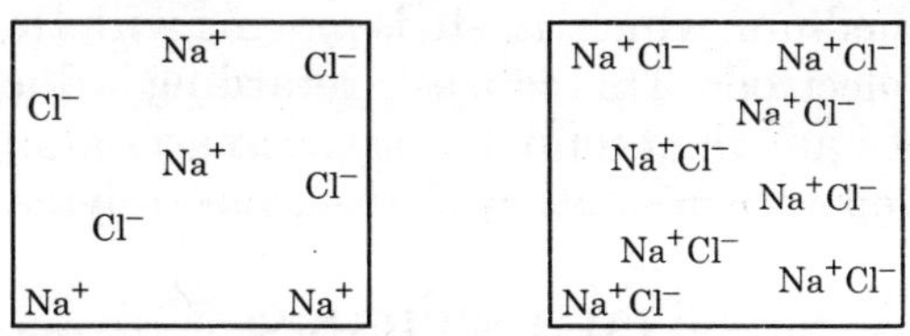

Dilute Solution of Electroyte. The interionic attractions are negligible

Concentrated Solution of Electrolyte. Ions appear as ionic doublets.

Fig. 19.22 Interionic attraction and effect of dilution on solution of NaCl

(iv) **Asymmetry or relaxation effect:** The anions and cations are not uniformly distributed in the solution of an electrolyte but the cations tend to be found in the vicinity of anions and vice-versa. Debye and Huckel derived an equation by which the magnitude of interionic effects can be calculated. The basis of this equation is asymmetry effect. Every ion is regarded as being surrounded by an ionic atmosphere of opposite charge. As long as the system is not exposed to an applied electric field the force of attraction exerted by the atmosphere on the central ion is uniform in all directions. When an electric field is applied, the negative ions move towards the anode and positive ion moves towards the cathode. In otherwords the central ion moves in one direction, and ionic atmosphere in other. The symmetry of the atmosphere about the ion is thus destroyed and the atmosphere becomes distorted Fig 19.23

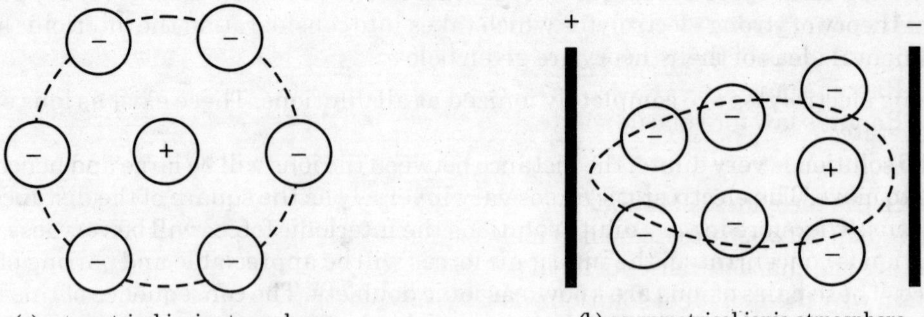

Fig. 19.23 (a) Symmetrical ionic atmosphere and (b) Asymmetrical ionic atmosphere

The force of attraction exerted by the ionic atmosphere on the central ion, which is uniform in all directions before the passage of electicity, becomes greater behind the ion than in front, when electricity is passed. As result the central ion experiences a force which tends to drag it backwards and slows down its movement in the forward direction. This drag on the central ion is known as **asymmetry effect** or **relaxation effect**.

In strong electrolyte at higher concentration, the drag experienced by the central ion is evidently more and hence the conductance is decreased.

(v) **Electrophoretic Effect:** The motion of the ions at higher concentration is also slowed down by another factor called electrophoretic effect. The electrophoretic effect is the tendency of the applied field to move the ionic atmosphere, associated with the molecules of water of hydration, in a direction opposite to that in which the central ion is moving along its ionic atmosphere. For example, a positive ion which migrates towards the cathode has to make its way through the medium, which, itself, is moving with the negative ionic atmosphere towards the positive electrode. This causes a retarding influence on the movement of the ion. Thus the value of equivalent conductance decreases at higher concentration because electrophoretic effect operate in solution of strong electrolytes.

QUESTIONS

Short Answer Questions

Section 19.1, 19.2

1. What is meant by the term: solution?
2. What are the different types of solutions?
3. What is saturated solution?
4. What are the factors affecting solubility?
5. Define the terms solute and solvent?

Section 19.3

6. Define the term molality?
7. How does molality differs from molarity?
8. What is mass fraction?
9. Why in chemistry, molality is preferred to molarity.
10. What is formula weight?
11. Define normality of a solution?

Section 19.4

12. Define vapour pressure.
13. Explain why vapour pressure is lowered by the addition of a non-volatile solute?
14. State Raoult's law.
15. State Raoult's law for volatile solute.

Section 19.5

16. Explain the term ideal solution.
17. Distinguish between ideal and non ideal solution.
18. What are the characteristics of ideal solution.
19. Explain why the non ideal solutions deviate from Raoult's law.
20. Give some examples of solutions exhibiting negative deviations.

Section 19.6, 19.7

21. Construct the vapour pressure — composition diagram for ideal solution showing the composition of vapour and liquid phase.
22. What is lever rule?
23. What is tie line?

Section 19.8, 19.9

24. Construct boiling point composition diagram for an ideal solution.
25. Construct boiling point composition diagram for a solution showing negative deviation from Raoult's law.
26. What is maximum boiling azeotrope?
27. What is minimum boiling azeotrope?
28. Is it possible to separate a mixture of ethyl alcohol and water into pure components?
    Ans. No.
29. Discuss the principle involved in distillation?
30. In which type of solutions, complete separation of the components is possible by distillation?
31. Distinguish between maximum and minimum azeotropic mixtures?

Section 19.10

32. State Henry's Law
33. What is effect of temperature on solubility?
34. What are limitations of Henry's law?
35. Explain: (i) dissolved gases can be removed from water by boiling.
    (ii) on opening a carborated cold drink bottle, bubbles of gas come out.

Section 19.11

36. State Nernst Distribution Law.
37. Under what conditions the law is applicable?
38. State the law which governs the distribution of a solute between two immiscible solvents.
39. What is the modified formula of Nernst distribution law:
    (i) When the solute associates in one of the solvents.
    (ii) When the solute dissociates in one of the solvents.
40. What are the various applications of Distribution Law?
41. How is it used in confirmatory test for bromide and iodide?
42. What is solvent extraction?
43. Discuss the reasons of deviations observed from this law.

Section 19.12

44. What is an electrolyte?
45. Ostwald's Distribution Law.
46. What are the limitations of Arrhenius theory?
47. What is Asymmetry effect?
48. What is electrophoretic effect?

LONG ANSWER QUESTIONS

1. What is meant by the term: solution? Differentiate between a saturated solution and unsaturated solution?
2. State and explain Raoult's law for volatile solutes as well as non-volatile solutes.
3. State 'Henry's Law'. What are the conditions under which it holds good? Show that the volume of a gas that dissolves in a fixed volume of a liquid at a given temperature is independent of pressure.
4. Prove that Raoult's law is a special case of Henry's law.
5. Differentiate between ideal and non-ideal solutions.
6. Explain alongwith diagram why some of the non-ideal solutions show positive deviations while some other show negative deviation from ideal solutions.
7. What type of deviation (positive or negative) from ideal behaviour will be shown by the solution of cyclohexane and ethanol? Give suitable reasons.
8. On the basis of molecular interactions, explain why the non-ideal solutions deviate from Raoult's law.
9. Construct the vapour-pressure composition and boiling point-composition plots for binary liquid pairs showing (i) ideal behaviour and (ii) non ideal behaviour showing both, positive and negative deviations from ideality.
10. Explain, giving reasons, the following :
    (a) Dissolved gases (N_2 or O_2, etc.) from water can be removed by boiling.
    (b) On opening a carbonated cold drink bottle, bubbles of gas come out.
    (c) A solution of benzene and toluene behaves nearly ideal while that of ethanol and water show considerable deviation from ideality.
11. State and explain the Nernst's distribution law. What are its limitations?
12. Deduce the formulae for distribution law in the case of :
    (i) Association of solute in one of the phases.
    (ii) Dissociation of solute in one of the phases.
    (iii) Solute enters into chemical combination with one of the solvents.
13. Discuss the various applications of Nernst distribution law.
14. How is Nernst distribution law is applicable in the process of extraction of solutes. Show that the process of extraction by a given volume of the solvent is more efficient if it is carried out in multiple steps.
15. Write short notes on :
    (a) Parke desilverization process.
    (b) Solvent extraction
    (c) Ideal and non-ideal solutions.
16. Discuss Arrhenius theory of electrolytic dissociation. What are the main evidences in its favour.

17. What are the limitations of Arrhenius theory? Discuss the behaviour of strong electrolytes.
18. (a) State and explain Raoult's law.
    (b) Explain with the help of boiling point composition diagram, the distillation of binary solution exhibiting negative deviation from Raoult's law.

(G.G.S.I.P.U, II Term, B. Tech, Jan. 2000)

19. (a) What is Nernst Distribution Law? Work out formulae which you would apply in case of (i) Association (ii) Dissociation of the solute in one of the solvents.
    (b) Discuss the application of Nernst Distribution law in solvent extraction.

(G.G.S.I.P.U., II Term, B.Tech, Jan. 2000)

20. State Nernst distribution law. Prove that Multi-step extraction is more economical than a single step extraction. **G.G.S.I.P.U. End Term B.Tech, Mar 2000.**
21. (a) What is Raoult's law? State the factors responsible for deviations from this law.
    (b) Explain Henry's law with suitable examples. **G.G.S.I.P.U., End Term, Dec. 2000**
22. State and explain Ostwald's Dilution law. Can this law be applied to the dissociation of HCl in aqueous solution?
23. Discuss Arrhenius Theory of Electrolytic dissociation. What are its limitations?
24. Give evidences in favour of Arrhenius Theory of Electrolytic dissociation.
25. Discuss interionic attraction theory (Debye-Huckel Theory) of strong electrolytes.
26. (a) Explain Binary distillations of solution showing negative and positive deviation from Raoult's law.
    (b) State Nernst distribution Law. How is it modified if the solute undergoes association.

(G.G.S.I.P.U., End Term, Dec. 2001)

27. How Arrhenius theory fails for strong electrolytest? Also explain behaviour of strong electrolytes. **(G.G.S.I.P.U., End Term, Dec. 2001)**
28. (a) Illustrate Raoult's law and its application in fractional distillation
    (b) What are the main postulates of Arrhenius Theory and why it fails in case of strong electrolytes.
    (c) Discuss Nernst Distribution law and its applications.

(G.G.S.I.P.U., End Term, Dec. 2002)

❑❑❑

20

THE PHASE RULE

"It is difficult to say what is impossible, for the dreams of yesterday is the hope of today and the reality of tomorrow"

Robert Goddard

CHAPTER OUTLINE

A substance can exit in various states depending on the external conditions. For example, water can exist in the form of a solid liquid and vapour. Each of the state is referred to as a *phase*. A change in the state of substance is called a *phase transition*. Phase transitions occur when energy (usually in the form of heat) is added or removed from a particular phase. A phase transition, like any other thermodynamic system, proceeds until equilibrium is established. Before discussing the details, let us explain the various terms we would be using in the course of this chapter.

20.1 DEFINITIONS

PHASE

A phase (P) is a homogeneous physically distinct and mechanically separable part of a system which is separated from other parts of the system by definite boundaries.

In otherwords, phase is a state of matter that is uniform throughout, not only in chemical composition but also in physical state.

A system consisting of one phase only is called a homogeneous system and a system consisting two or more phases is called a heterogeneous system.

Examples.

(i) A crystal of a substance is a single phase. But if the crystals of two substances are mixed then there exist two phases.

(ii) A gas or a gaseous mixture consist of a single phase, since there is no demarcation (interface) between one gas and another. Air is in one phase only although it is a mixture of number of gases.

(iii) A solution of two miscible liquids will exist in one phase only as far as *solution* is concerned. But the *system as a whole* will have two phases - one phase will be the solution (liquid) and the other would be the vapour above it.

(iv) If two immiscible liquids are mixed together to form a solution, then the total phases present would be three. Two phases are for the substance in liquid and the third one for vapours, containing the vapours of both the liquids.

(v) Solution of sodium chloride in water is a uniform solution. Therefore, it is a one phase system.

(vi) Saturated solution of water and sugar has two phases (solution of water + sugar and solid sugar settled at the bottom.

(vii) Sulphur exists in two allotropes in nature. As these allotropes are similar in chemical properties but are different in physical properties, therefore a mixture of two allotropes is a two phase system.

(viii) A mixture of two or more chemical substances contains as many phases. Each substance has different physical and chemical properties. Therefore a mixture of $CaCO_3$ and CaO is a two phase system.

(ix) When $CaCO_3$ is heated in a closed vessel, it decomposes to CaO and CO_2. At equilibrium,

$$\underset{\text{(Solid)}}{CaCO_3} \rightleftharpoons \underset{\text{(Solid)}}{CaO} + \underset{\text{(gas)}}{CO_2}$$

At this state there are two solid and one gaseous phase. Therefore, it is a three phase system.

COMPONENTS

The number of components (C) of a system is the smallest number of independently variable constituents at equilibrium, by means of which the *composition of each phase present can be expressed* either directly or in the form of a chemical equation. Zero and negative quantities of the components are acceptable when composition of a phase is expressed in terms of the components.

The components of a system do not express the number of constituents or chemical entities or chemical individuals present in the system.

To understand the meaning of component and to find the number of components of a system following points are noteworthy.

(i) The chemical formula representing the composition of a phase is written on left hand side (LHS).

(ii) The remaining chemical constituents existing independently in the system as represented by chemical formulae are written on right hand side (RHS).

(iii) The quantities of constituents on RHS can be made minus (–) or zero (0) to get the composition of the phase on LHS.

Examples

(i) Consider a water system. At equilibrium there are three phase

$$\underset{\text{(solid)}}{\text{Ice}} \rightleftharpoons \underset{\text{(liquid)}}{\text{Water}} \rightleftharpoons \underset{\text{(gas)}}{\text{Vapour}}$$

It is a one component system because each phase can be expressed in terms of only one component, i.e. water (H_2O).

(ii) In sulphur system there are four phases namely rhombic sulphur, monoclinic sulphur, liquid and vapour.

$$S_R \rightleftharpoons S_M \rightleftharpoons S_L \rightleftharpoons S_V$$

It is a one component system as the composition of each phase can be expressed in terms of sulphur only.

(iii) Decomposition of $CaCO_3$:

$$\underset{\text{(solid)}}{CaCO_3} \rightleftharpoons \underset{\text{(solid)}}{CaO} + \underset{\text{(gas)}}{CO_2}$$

The system consists of three phases – solid $CaCO_3$, solid CaO and gaseous CO_2. Due to existence of equilibrium, composition of each of the three phases can be expressed in terms of any two constituents, $CaCO_3$, CaO and CO_2 as shown below:

If $CaCO_3$ and CaO constituents

Phase	*Component*
$CaCO_3$ (s)	$CaCO_3 + 0CaO$
CaO (s)	$0CaCO_3 + CaO$
CO_2 (g)	$CaCO_3 - CaO$

If $CaCO_3$ and CO_2 constituents are chosen

Phase	*Component*
$CaCO_3$ (s)	$CaCO_3 + 0CO_2$
CaO (s)	$CaCO_3 - CO_2$
CO_2 (g)	$0CaCO_3 + CO_2$

If CaO and CO_2 constituents are chosen

Phase	*Component*
$CaCO_3$ (s)	$CaO + CO_2$
CaO (s)	$CaO + 0CO_2$
CO_2 (g)	$0CaO + CO_2$

Thus in all the cases, only two components are required to express the composition of each phase. Hence, it is a two component system and is also known as binary system.

(iv) $$\underset{\text{(solid)}}{MgCO_3} \rightleftharpoons \underset{\text{(solid)}}{MgO} \rightleftharpoons \underset{\text{(gas)}}{CO_2}$$

It is also a binary system or a two component system because to specify the composition of any phase, we need only two constituents.

(v) $$\underset{\text{(solid)}}{CuSO_4 \cdot 5H_2O} \rightleftharpoons \underset{\text{(solid)}}{CuSO_4 \cdot 3H_2O} + \underset{\text{(gas)}}{2H_2O}$$

It is also a binary system.

(vi) In the dissociation of ammonium chloride in vacuum, equilibrium occurs as follows.

$$\underset{\text{(solid)}}{NH_4Cl} \rightleftharpoons NH_3(g) + HCl\ (g)$$

The system consists of two phases, solid NH_4Cl and a gaseous mixture of NH_3, HCl and NH_4Cl. The composition of either of the two phases can be expressed in terms of *one* component as NH_3 and HCl are formed in fixed stoichiometric proportions by the reaction. Therefore both the phases have formal composition 'NH_4Cl'. However, if NH_3 or HCl is in excess, the system becomes a two component system because decomposition of NH_4Cl would not give the correct composition of the gas phase. Hence NH_3 or HCl would be required to specify the gas composition. Hence it becomes *two component* system.

If there are 'S' substances and 'R' relations (equations representing chemical equilibria) among them, then the number of components (C) can be calculated as:

$$C = S - R$$

For example, consider a solution of glucose in water. There are two substances and hence S = 2, but there is no relation between them so R = 0, therefore.

$$C = S - R$$

or

$$C = 2 - 0 = 2$$

Hence, it is a two component system.

Now consider $NH_4Cl_{(s)}$ in equilibrium with its vapours containing NH_3 and HCl. Here S = 3 and R = 2 because there are two relations as shown,

(a) $NH_4Cl_{(s)} \rightleftharpoons NH_{3\,(g)} + HCl_{(g)}$

(b) Conc. of NH_3 = Conc. of HCl

Therefore C = S – R = 3 – 2 = 1. Thus it is *one* component system.

(vii) Gaseous mixture O_2 and N_2 is a two component system because in the gaseous mixture O_2 and N_2 are two different chemical entities to express the mixture.

Phase	***Component***
Gaseous mixture of O_2 and N_2	$x\,O_2 + y\,N_2$

(viii) The aqueous solution of NaCl constitutes one phase only but to determine the composition of this homogeneous phase (x NaCl y H_2O) two chemical entities are required. Therefore it is a two component system.

Phase	***Component***
Aqueous solution of NaCl	x NaCl + y H_2O

DEGREE OF FREEDOM

The degree of freedom (F) of a system is defined as the number of independent variables, such as temperature, pressure and composition which are to be specified to describe a system completely. It is also termed as variance.*

---

*A system having 1, 2, 3 or 0 degrees of freedom is called univariant, bivariant, trivariant and non variant or invariant respectively.

(i) For a system consisting of Ice, water and vapours,

$$Ice_{(s)} \rightleftharpoons Water_{(l)} \rightleftharpoons Vapours_{(g)}$$

if all the three phases are present in equilibrium, then no condition has to be specified. This is because the three phases of water can exist together in equilibrium only at a particular temperature and pressure (corresponding to the freezing point) and no condition is necessary to be specified to define the system. Thus it is a non-variant system or invariant system, i.e. F = 0.

(ii) For a system consisting of water in contact with its vapour, i.e.

$$Water_{(l)} \rightleftharpoons \text{water vapour}_{(g)}$$

Such a system can be completely defined by specifying either pressure or temperature. This is because on fixing one variable, the other becomes automatically fixed. Hence, either temperature or pressure is required to specify the system. The system is therefore, univariant i.e. F = 1.

(iii) For a system consisting of water vapours only, both temperature and pressure are to be specified in order to describe the system completely. Thus it is a bivariant system i.e. F = 2.

(iv) For a pure gas, F = 2. This is because PV = RT, if the value of pressure (P) and temperature (T) are fixed, volume automatically gets fixed.

(v) A system of saturated NaCl solution has F = 1. Because if only one variable, i.e. temperature is fixed, composition and vapour pressure get automatically fixed.

20.2 PHASE RULE

The phase rule is an important generalisation and a versatile tool which is related to the study of behaviour of heterogeneous systems. In general, it may be stated that with the help of phase rule, it has been possible to predict quantitatively the effect of changing pressure, temperature and concentration on a heterogeneous system in equilibrium by means of a phase diagram.

American physicist Willard Gibbs (1874) deduced this generalisation on the basis of thermodynamics and H.W.B. Roozeboom (1884) explained it further.

GIBBS PHASE RULE

Gibbs phase rule states that if a heterogeneous system is influenced by temperature, pressure and concentration and not by any other action like gravity, electrical, magnetic forces or by surface action, then the sum of number of phases (P) and degrees of freedom (F) is greater than the number of components (C) by two. It is expressed mathematically as follows:

$$P + F = C + 2$$

or

$$F = C - P + 2$$

where P is the number of phases.

C is the number of components and

F is the number of degrees of freedom

Derivation of the Phase Rule

Consider a heterogeneous system in equilibrium consisting of 'C' components distributed in 'P' phases. We have to determine the degrees of freedom of this system.

The number of degrees of freedom (F) of a system in equilibrium is the number of variables (such as temperature, pressure and composition) that must be arbitrarily fixed to define the system completely. Now, let us find out how many of number of variables are there in the above system.

- When the system is in equilibrium, there can be only one temperature and one pressure, hence the total number of the these variables is = 2
- The number of composition variables can be more. For each phase the number of composition variable will be (C – 1) because the composition of remaining component can be found out by difference (i.e. for example, out of 10 components required to describe the composition of a phase, if the mole fraction of 9 are known, remaining one can be found out by the difference)
  Now, the number of phases = P
  Number of composition variable in each phase = (C – 1)
  Total number of composition variables in all phases = P(C – 1)
  Number of other (temperature and pressure) variables = 2
  Total number of variables of the system = P(C – 1) + 2
- Now we have to find the total number of relations among the variables, i.e. number of equations possible at equilibrium.
  When a heterogeneous system is in equilibrium at constant temperature and pressure, the chemical potential of a particular component must be same in all phases in which it appears. For a one component system, having two phases α and β, the equality of chemical potential implies

$$\mu(\alpha) = \mu(\beta)$$

where μ (α) and μ (β) are the chemical potentials of the two phases α and β respectively. Similarly, if there is one component in three phases (say α, β and γ) and one of these phases is referred to as standard phase, then can be expressed in the form of following two equations

$$\mu_\alpha = \mu_\beta$$

$$\mu_\alpha = \mu_\gamma$$

Thus, for each component in equilibrium in *three* phases, *two* equations are possible.
Hence, for each component in 'P' phases (P – 1) equations can be written.
Since, there are 'C' components, *the number of equations or variables possible at* equilibrium are:

$$C(P - 1).$$

Therefore the degrees of freedom or number of independent variables F is given by

$$F = (\text{Total number of variables} - \text{Total number of relations among the variables})$$

$$= [P(C - 1) + 2] - [C(P - 1)]$$

$$\boxed{\therefore \quad F = C - P + 2}$$

Advantages of Phase Rule

(i) It provides a simple method of classifying equilibrium states of systems such as phases components and degrees of freedom.

(ii) It indicates that different systems having the same number of degrees of freedom behave in like manner.

(iii) It explains the behaviour of systems when allowed to changes in the variables such as pressure temperature and concentration.

(iv) The phase rule is applicable to macroscopic systems. Therefore it is not compulsory to take in to account their molecular structures.

(v) It is applicable to physical as well as chemical equlibria.

(vi) Phase rule predicts that under a given set of conditions.

- Whether various substances would exist together in equilibrium.
- Whether some of the substances would be interconverted
- Or whether some of the substances will completely disappear.

Limitations of Phase Rule

(i) As the phase rule is applicable to heterogeneous systems in equilibrium, it is therefore of no use for such systems which are slow in reaching the state of equilibrium.

(ii) As the phase rule is applicable to a single equilibrium state, it never tells about the number of other equilibrium possible in the system.

(iii) In phase rule the different variables are pressure, temperature and concentration or composition. But phase rule does not consider the other factors like electric and magnetic influences.

(iv) All the phases are required to be present under the same temperature, pressure and gravitational force.

(v) Solid or liquid phases should not be in finely divided form, otherwise vapour pressure may differ from the normal values.

(vi) It does not take into consideration the quality of the phases though it takes into account the number of phases.

Application of Phase Rule

All the systems are classified on the basis of number of components present. Thus we may have one, two, three, etc. component systems. The phase rule is perfectly applicable to these systems. The phase diagram is an important medium to clearly indicate the equilibrium conditions between different phases in a system. The phase diagram is helpful for studying and controlling various processes such as phase separation, solidification of metals, the change of structure during heat treatment, etc.

20.3 PHASE DIAGRAM

A phase diagram is a graphical representation of the pressure and temperature under which two or more physical states can exist together in a state of dynamic equilibrium. Phase diagram illustrates the conditions of equilibrium between various phases of a substance. If the temperature is plotted against pressure the diagram is called P – T diagram. The study of phase diagram is vital for proper understanding of the relative stability of equilibrium between phases. With the

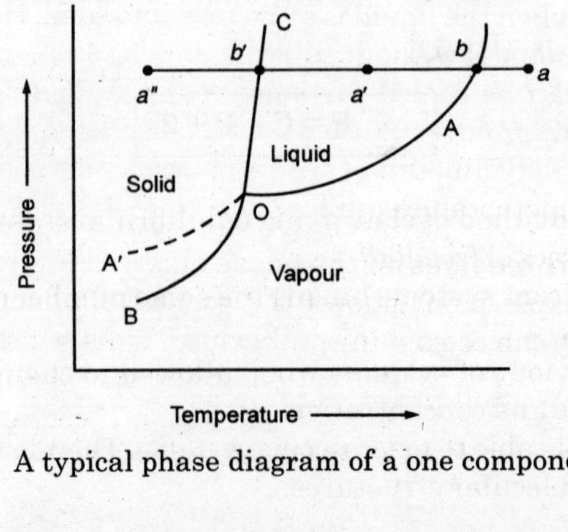

Fig. 20.1 A typical phase diagram of a one component system

help of knowledge of the number of phases and components of the system, phase rule enables one to predict clearly the conditions under which a system can remain in equilibrium.

Fig. 20.1 is a typical case for a *one-component systems*. The diagram consists of:

(a) Areas (b) Curves/Lines

(c) Triple point

(a) **Areas:** The diagram has 3 regions or areas labelled as solid, liquid and vapour, i.e. COB, COA and BOA respectively. Each of the three areas shows the conditions of temperature and pressure under which the respective phase can exit. When only one phase is present,

$$F = 1 - 1 + 2 = 2$$

Thus, each phase has two degrees of freedom; both temperature and pressure must be specified to define the state of the system. Consider a point a in the Fig. 20.1. At this point the gas phase of the substance is thermodynamically the most stable state. If the temperature is reduced at constant pressure, the system moves to point *b* where the liquid and its vapour are in equilibrium. If the temperature is still reduced further it will reach a point a′ where liquid is the stable phase. On reducing the temperature further point b′ is reached, where liquid and solid phase are in equilibrium. On further reducing the temperature upto a″, solid is the stable phase.

(b) **Curves:** There are three lines or curves separating the areas. These show the conditions of equilibrium between any two of the three phases:
These are also known as phase boundaries.

- Curve OA represents the equilibrium,

$$\text{Liquid} \rightleftharpoons \text{Vapour}$$

This is referred to as **vapour pressure curve**. Above the curve liquid is stable phase while below it, vapour is the stable phase.

- Curve OC represents the equilibrium

$$\text{Solid} \rightleftharpoons \text{Liquid}$$

This is referred to as **melting curve** or **fusion curve** or **freezing point curve** and indicates the influence of pressure on melting point of solid.

- Curve OB represents the equilibrium

$$\text{Solid} \rightleftharpoons \text{vapour}$$

This is referred to as **sublimation curve**.

- **Curve OA', Metastable equilibrium:** The vapour pressure curve OA as liquid phase terminates at O, when the liquid is converted into solid. However it is possible to cool the liquid below the point O without solidification as shown by the curve OA'. This means that the liquid can be cooled far below the freezing point or 'supercooled' without solidification. This liquid would be in an unstable condition. On slightest disturbance or by addition of a small amount of seed crystal, the entire liquid solidifies rapidly. Thus OA' represents a metastable equilibrium.

(c) **Triple Point:** The three lines on the phase diagram intersect at a common point called the *Triple Point*. Triple point shows the condition under which all the three phases (solid, liquid, vapour) can coexist in equilibrium. Thus triple point can be represented as,

$$\text{Solid} \rightleftharpoons \text{Liquid} \rightleftharpoons \text{Vapour}$$

$$F = C - P + 2 = 1 - 3 + 2 = 0$$

which predicts that system has no degrees of freedom. At triple point both pressure and temperature are fixed. If we try to change temperature or pressure the three phases would no longer co exist and one of the three phases will disappear.

20.4 ONE COMPONENT SYSTEM

The complexity of one-component systems depend on the number of solid phases that can exist in the system. The simplest case is that in which only single solid phase occurs. Water system (at lower pressure) and carbon dioxide system are the examples of the simplest one component system. For a one component system the phase rule equation can be written as:

$$F = C - P + 2$$
$$= 1 - P + 2$$
$$= 3 - P$$

There may arise three cases.

Case I. When only one phase is present

$$F = 3 - P$$
$$= 3 - 1$$
$$= 2 \quad \text{(Bivariant system)}$$

It can be completely defined by specifying the two variables temperature and pressure or both temperature and pressure can be varied independently. Therefore a single phase is represented by an area on P – T graph.

Case II. When two phases are present in equilibrium, the degree of freedom will be:

$$F = 3 - P$$
$$= 3 - 2$$
$$= 1 \quad \text{(Monovariant or Univariant system)}$$

This means that the pressure can not be altered independently if the temperature is changed. In such a case pressure is fixed automatically for a given temperature. A two phase system is depicted by a line on a P – T graph.

Case III. When three phases are present in equilibrium, the degree of freedom will be:

$$F = 3 - P$$
$$= 3 - 3$$
$$= 0 \quad \text{(Nonvariant or invariant system)}$$

This special condition can only be attained at a define temperature and pressure. A three phase system is depicted by a point on the P – T graph. At this point the three phases are in equilibrium and this point is termed as the Triple Point.

THE WATER SYSTEM

Under normal conditions the system water is a three phase and one component system. The three phases are liquid, ice and vapour. All these are represented by one chemical entity (H_2O), hence it is one component system.

On the basis of experimental data obtained for the water system a plot of relationships between various phases namely solid ice, liquid water and water vapour under different conditions of temperature and pressure are shown in the Fig. 20.2.

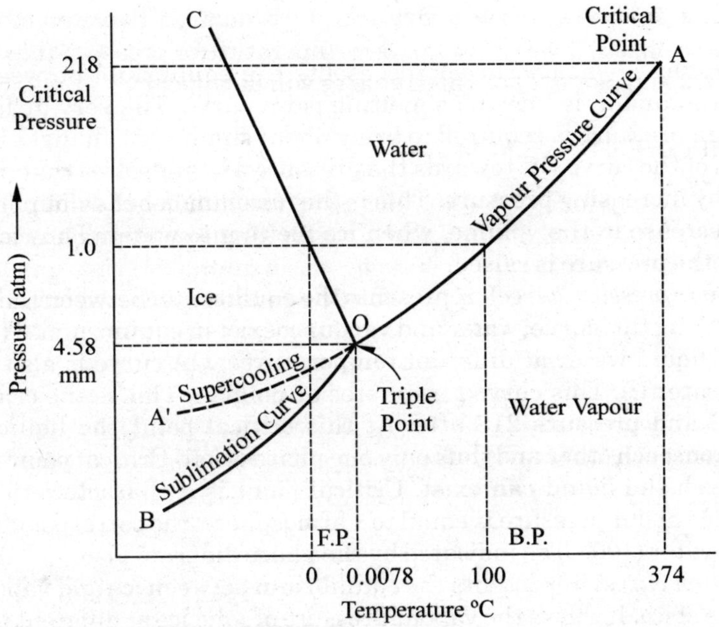

Fig. 20.2 The phase diagram of the system water

The three forms of water (ice, water and vapour) constitute the following equilibria

(i) **Single phase equilibria**

(a) Solid (ice) – represented by area BOC.

(b) Liquid (water) – represented by area AOC.

(c) Gas (vapour) – represented by area AOB.

(ii) **Two phase equilibria**

(a) Solid (ice) $\rightleftharpoons$ Liquid (water) — represented by curve OC.

(b) Liquid (water) $\rightleftharpoons$ Gas (water vapour) — represented by curve OA.

(c) Solid (ice) $\rightleftharpoons$ Gas (water vapour) — represented by curve OB.

(iii) **Three phase equilibria**

Solid (ice) $\rightleftharpoons$ Liquid (water) $\rightleftharpoons$ Gas (vapour) — represented by triple point O.

The phase diagram consists of :

(a) Three areas, BOC, AOC, and AOB.

(b) Three curves, OB, OA and OC.

(c) One triple point, O.

The Areas

Area BOC represents the solid (ice) phase, which is thermodynamically most stable state under these conditions. Area AOC represents the liquid (water) phase in the system and similarly area AOB represents the gas (vapour) phase in the system. Thus these areas represent one phase equilibria.

In order to define the system completely at any point in the area, it is essential to specify both the temerature and pressure. Therefore areas have two degrees of freedom and are called bivariant systems. It can also be concluded by the phase rule equation.

$$F = C - P + 2 = 1 - 1 + 2$$
$$= 2 \quad \text{(Bivariant system)}$$

The Curves

The Curve OC (*melting or fusion curve*) represents the equlibrium betweeen ice and water (two phase equilibria) and it is known as melting point curve. The very little slop of the curve shows that very high pressure is required to bring about significant changes in melting point of ice. The inclination of the curve OC towards the pressure axis indicates that melting point of ice slightly decreases by increasing pressure. This is the uncommon behaviour. The reason for this behaviour is the decrease in the volume, when ice melts into water. Thus ice transforms into denser water when the pressure is raised.

The Curve OA (*Vapour pressure curve*) represents the equilibrium between the two phases liquid water and vapour. Along this curve, water and vapour coexist in equilibrium. The curve shows the vapour pressure of liquid water at different temperatures. OA curve is also known as vapour pressure curve of water (*l*). This curve terminates at point A. This is the critical point having temperature 374°C and pressure 218 atm. At this critical point the liquid and vapour are indistinguishable from each other and thus only one phase is left. Critical point marks the highest temperature at which the liquid can exist. Critical point is a characteristic properly of pure substance. When the vapour pressure is equal to 1 atmosphere, the corresponding temperature is the boiling point of water (100°C) as indicated by the phase diagram also.

Curve OB (*sublimation curve*) represents the equilibrium between ice and vapour and it is called as sublimation curve of ice. It shows the vapour pressure of solid ice at different temperatures. The two phases namely ice and vapour coexist in equilibrium along this curve. At the lower limit OB curve terminates at absolute (– 273°C) where no vapour exists.

Along the curves OC, OA and OB, the number of phases present in equlibrium is two.

$$\begin{aligned} F &= C - P + 2 \\ &= 1 - 2 + 2 \\ &= 1 \end{aligned}$$

Hence, system is univariant along the curve.

Triple Point

The point O where all the three curves OC, OA and OB meet is known as triple point. At the triple point all the three phases of water system namely solid ice, liquid water, and gas vapour are in equilibrium. The equilibrium in three phases is attained at .0076°C temperature and 4.58 mm Hg pressure. Since, there are three phases and one component present in the system, the degree of freedom will be:

$$\begin{aligned} F &= C - P + 2 \\ F &= 1 - 3 + 2 \\ &= 0 \end{aligned}$$

Hence, the system is non variant. If either the temperature or the pressure or both are changed, the three phases would no longer coexist and at least one of them would disappear.

Metastable Curve OA'

The dotted curve OA' is the continuation of vaporisation curve OA. It is vapour pressure curve of supercooled water. As the water does not always freezes at 0°C, therefore if the vessel containing water and vapour is throughly clean and dust free, it is possible to super-cool water several degrees below its freezing point. On slight disturbance, the supercooled water at once changes to ice (solid). This kind of system is a metastable system.

Metastable State

It is the state in which the liquid below its freezing point is in the supercooled state which is not very much stable.

$$\text{Supercooled liquid} \rightleftharpoons \text{Vapour}$$

On adding nucleus of ice (stable state), the system reverts to true stable system

$$\text{Solid} \rightleftharpoons \text{Vapour}$$

THE SULPHUR SYSTEM

It is a one component, four-phase system. The four different phases are:

(i) Rhombic sulphur (S_R)
(ii) Monoclinic Sulphur (S_M)
(iii) Liquid Sulphur (S_L)
(iv) Vapour Sulphur (S_V)

As all the four phases can be represented by only one chemical entity 'Sulphur' (S), it is a one component system.

From the phase rule when C = 1,

$$\begin{aligned} F &= C - P + 2 \\ &= 1 - P + 2 \\ &= 3 - P \end{aligned}$$

As degree of freedom *can not have a minus value*, therefore out of four possible phases, only *three* are present at a time. It is not possible for a single component system to have four phases together at equilibrium. The degree of freedom for different cases will be:

when $P = 1, F = 3 - P = 3 - 1 = 2$ (bivariant system)
$P = 2, F = 3 - P = 3 - 2 = 1$ (monovariant system)
$P = 3, F = 3 - P = 3 - 3 = 0$ (nonvariant system)

The two crystalline forms of sulphur are rhombic sulphur and monoclinic sulphur. S_R and S_M exhibit enantiotropy* with a transition point at 95.6°C. Below 95.6C S_R is stable and above this temperature S_M is the stable variety. At 95.6°C each form can be gradually transformed to another form and the two forms remain in equilibrium. S_M melts at 120°C

Thus, $$S_R \xrightleftharpoons{95.6^\circ C} S_M \xrightleftharpoons{120^\circ C} S_L$$

The number of phases at any instant depends upon the temperataure-pressure conditions at that instant. The P – T diagram of sulphur system showing various equilibriums is given Fig. 20.3.

The phase diagram consists of :

(a) Four areas
   (i) Area ABG (Sulphur rhombic)
   (ii) Area BEC (Sulphur monoclinic)
   (iii) Area GECD (Sulphur liquid)
   (iv) Area ABCD (Sulphur vapour)
(b) Six curves AB, BC, CD, BE, CE, EG
(c) Three triple points B, C and E.

---

*Enantiotropy: In some cases one polymorphic form (or allotrope) can change into another at a definite temperature when the two forms have a common vapour pressure. The temperature is known as the transition temperature. One form is stable above this temperature and the other form below it. When the change of form to the other at the transition temperature is reversible, the phenomenon is called enantiotropy and the polymorphic forms enantitropes. For example,

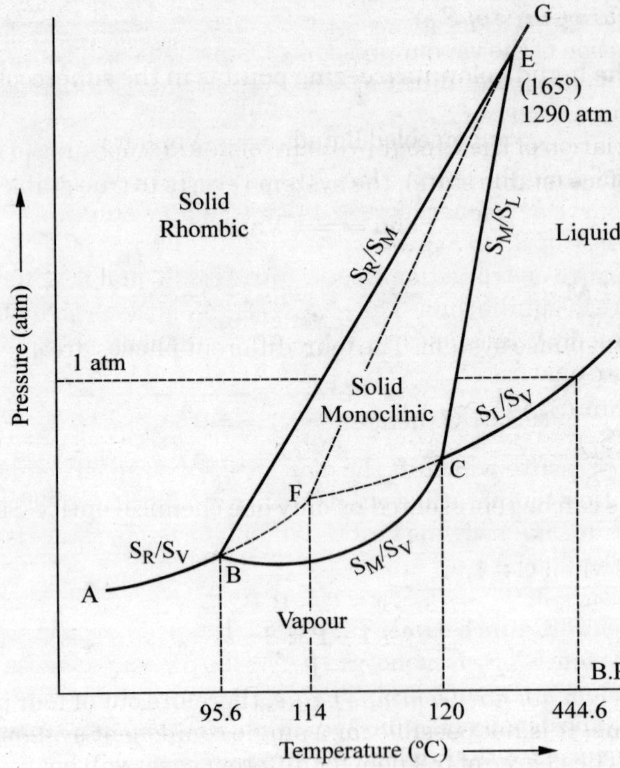

Fig. 20.3 Sulphur system

Areas

The phase diagram of sulphur system consists of four areas or regions. Each area represents the single phase system namely, rhombic sulphur, monoclinic sulphur, liquid sulphur and vapour.

For each area C = 1 and P = 1, and degree of freedom will be:

$$F = C - P + 2$$
$$= 1 - P + 2$$
$$= 2 \quad \text{(Bivariant system)}.$$

Thus it is clear that each of the systems S_R, S_M, S_L and S_V are bivariant systems. It means that in order to locate any point in any of these area, the variables pressure and temperature both are required to be specified.

The Curves

The six curves AB, BC, CD, BE, CE, EG divide the diagram into four areas.

Curve AB (Vapour Pressure Curve of S_R)

Curve AB is the vapour pressure curve of S_R at different temperatures. Along this curve the two phases rhomic sulphur (S_R) and sulphur vapour (S_V) are in equilibrium. The system S_R/S_V has one degree of freedom.

$$F = C - P + 2$$
$$= 1 - 2 + 2$$
$$= 1 \quad \text{Mono variant system.}$$

Curve BC (Vapour Pressure Curve of S_M)

Curve BC shows the variation of the vapour pressure of monoclinic sulphur (S_M) with temperature. Along this curve S_M and S_V are in equilibrium. The system S_M/S_V is monovariant.

Curve CD (Vapour Pressure Curve of S_L)

Curve CD depicts the variation of the vapour pressure of liquid sulphur (S_L) with temperature. S_L and S_V are in equilibrium along this curve. The system S_L/S_V is monovariant. One atmospheric pressure line meets this curve at temperature of 444.6°C(B.P. of sulphur).

Curve BE (Transition Curve of S_R to S_M)

It shows the effect of pressure on transition temperature for S_R and S_M. Along this curve the two solid phases S_R and S_M are in equilibrium. The system S_R/S_M is monovariant. The transformation of S_R to S_M is accompanied by increases in the volume (density of S_R = 2.04 and density of S_M = 1.9) and absorption of heat.

$$S_R + Q \text{ (heat energy)} \rightleftharpoons S_M$$

Thus the increase of pressure will shift the equilibrium to the left hand side (Le Chatelier's Principle) and the transition temperature will therefore be increased. That is why the line BE slopes away from the pressure axis showing that the transition temperature is raised with increases of pressure. The system is monovariant.

Curve CE (Fusion Curve of S_M)

Curve CE represents the equilibrium between S_M and S_L. It represents the effect of pressure on the melting point of S_M. The system S_M/S_L is monovariant. As the melting or fusion of S_M is accompanied by a slight increase of volume, the melting point will be increased by the increase of pressure (Le Chatelier's Principle). The density of liquid sulphur is less than that of the monoclinic solid and hence CE slopes slightly away (to the right) from the pressure axis. The curve ends at point E because monoclinic sulphur (S_M) ceases to exist beyond this point.

Curve EG (Fusion Curve of S_R)

The two phases S_R and S_L coexist in equilibrium along this curve. S_R/S_L is monovariant.

Triple Points

In the sulphur system there are three triple points B, C and E. All the triple points represent the existence of three phases in equlibrium with each other, therefore;

$$\begin{aligned} F &= C + P + 2 \\ &= 1 - 3 + 2 \\ &= 0 \end{aligned}$$

The system corresponding to each point is nonvariant.

(i) *Triple Point B:* At this point the three curves AB, BC and BE meet. Three phases, S_R, S_M and S_V are in equilibrium at this point B. As, P = 3 and C = 1, the system is nonvariant. At this point B (95.6°C) S_R changes to S_M and the process is reversible.

(ii) *Triple Point C:* The three curves BC, CD and CE meet at this point. The three phases S_M, S_L and S_V are in equilibrium at this point. As P = 3 and C = 1, the system is nonvariant. The temperature corresponding to C is 120° C (melting point of S_M)

(iii) *Triple Point E:* The three curves CE, BE and EG join at this point. The two lines CE and BE have different inclinations away from the pressure axis. The system $S_R/S_M/S_L$ is nonvariant at the point E. This point E shows the conditions of existence of the $S_R/S_M/S_L$ at 155°C and 1290 atm pressure.

Metastable Equilibria

The transformation of S_R to S_M is a slow process. If enough time is not given for the change and S_R is heated rapidly, it is possible to pass well above the transition point (95.6°C) without obtaining monoclinic sulphur. In that case, there is existence of three phase only.

$$S_R \rightleftharpoons S_L \rightleftharpoons S_V$$

In this condition, phase diagram like that of water system is obtained which consists of three areas (AFG, GFD and DFA), three curves (BF, CF and FE) and one triple point.

In each of three areas, a single phase exists (S_R, S_L or S_V). As there being P = 1, C = 1 therefore, F = C – P + 2 = 1 – 1 + 2 = 2. In each of the areas, system is bivariant.

The Dashed Curve BF

It is the vapour pressure curve of metastable S_R. This is a continuation of the vapour pressure curve AB to stable S_R. The metastable phases S_R and S_V are in equilibrium along this curve. The system S_R/S_V is monovariant.

The Dashed Curve CF

This is the vapour pressure curve of supercooled S_L. This dashed curve CF is obtained on supercooling liquid sulphur. It is the back prolongation of DC. The curve CF depicts the metastable equilibrium between supercooled S_L and S_V. It meets the curve BF at point F.

The Dashed Curve FE

It is the fusion curve of metastable S_R. Along this curve the two phases S_R and S_L are in equilibrium and the system is monovariant. The curve FE shows that the melting point of S_R is increased with pressure. Beyond E, the curve shows the conditions for the stable equilibrium S_R/S_L as the metastable S_R disappears.

The Metastable Triple Point F

The three metastable phases S_R, S_L and S_V are in equilibrium at this point. The system $S_R/S_L/S_V$ is a non variant one at triple point. The temperature corresponding to triple point is 114°C (melting point of metastable S_R).

20.5 TWO COMPONENT SYSTEM

According to phase rule, the degree of freedom of two component system is given by:

$$F = C - P + 2 = 2 - P + 2$$
$$= 4 - P$$

When a single phase is present in two component system, the degree of freedom is three.

$$F = C - P + 2$$
$$= 2 - 1 + 2$$
$$= 3$$

It means that three variables are required to explain the system completely. Thus in such a system, in addition to pressure and temperature the third variable, i.e. concentration of one of the substance has also to be specified. For representing these three variables graphically, three co-ordinate axis at right angles to each other would be required. Therefore a solid model of phase diagram will be formed. Such a phase diagram cannot be conveniently shown on paper.

But in actual practice, a simple plane diagram with two variables shown at two axis at right angle to each other is considered. It is customary to hold one of them constant. In this way three types of diagrams can be constructed, which are as follows.

(i) Pressure-Temperature diagram when concentration variable is kept constant. (P-T diagram)

(ii) Pressure-Concentration diagram when temperature is kept constant. (P-C diagram)

(iii) Temperature -Concentration diagram when pressure is kept constant. (T-C diagram)

Generally in most of the cases pressure is kept constant at 1 atmosphere and thus it is convenient to use temperature – concentration diagram.

The P – C diagram with constant temperature is known as *isothermal diagram* while the P – T diagram keeping composition constant is known as *isoplethal diagram*.

From the above discussion it is clear that for the sake of having simple plane diagrams we generally consider only two variables, third one being a constant. For example, for a solid-liquid equilibrium, usually the gas phase is absent and effect of pressure on the system is very small which is negligible, hence it is important to take into account the remaining variables like temperature and concentration and the effect of pressure can be discarded. **Such a solid liquid system with gas phase absent is known as a condensed system.**

The experimental measurements of temperature and concentration are usually carried out under atmospheric pressure. Since the degree of freedom in such a system is reduced by one (by keeping pressure constant), the reduced phase rule equation is as follows:

$$F = C - P + 2 \quad \text{(Phase rule equation)}$$

$$F = (C - P + 2) - 1 \quad \text{(Phase rule equation, modified for condensed system)}$$

$$\boxed{F' = C - P + 1} \quad \text{(Reduced phase rule)}$$

where F' shows the remaining degree of freedom of the system. Reduced phase rule is widely applicable to solid-liquid two component reduced systems.

In a two component system the various possible phase equilibria are:

$$\text{Liquid} \rightleftharpoons \text{vapour} \qquad \text{Solid} \rightleftharpoons \text{vapour}$$

$$\text{Solid} \rightleftharpoons \text{liquid} \qquad \text{Solid} \rightleftharpoons \text{solid}$$

$$\text{Liquid} \rightleftharpoons \text{liquid}$$

Here we shall confine our discussion to

$$\text{Solid} \rightleftharpoons \text{Liquid equilibria}$$

20.6 SOLID – LIQUID EQUILIBRIA

As is discussed earlier that measurement on solid-liquid in condensed systems is carried out at atmospheric pressure. Since the degree of freedom in such a case is reduced by one (as pressure is kept constant) the modified phase rule for such systems is

$$F = C - P + 1 \qquad \text{....(ii)}$$

This is know as condensed phase rule. Now for *two component systems* equation (ii) becomes

$$F = 2 - P + 1 = 3 - P$$

When $P = 1$, then $F = 2$

Thus it has two variables (temperature and composition) which have to be specified. Therefore, solid liquid equilibria are best represented by temperature-composition diagram.

CONSTRUCTION OF PHASE DIAGRAMS

To construct the phase diagrams of a two component, solid-liquid equilibria it is necessary to determine the composition of the mixture at different temperatures. Various experimental techniques are there for this purpose but the most generally used one is the *thermal analysis*.

Thermal Analysis

A convenient method to study solid/liquid equilibria is through thermal analysis. In this method, solids of different compositions are separately heated above their melting points. The resulting liquids are cooled slowly and cooling curves are constructed by plotting temperature against time.

From the analysis of cooling curve for any mixture of a definite composition, it is possible to determine its

(i) Freezing point
(ii) Eutectic point

(i) **Analysis of the cooling curves**

(a) **Pure substance**. When a pure substance is cooled slowly and the temperature is noted at definite intervals, a curve of the type shown in Fig. 20.4 is obtained. Point '*b*' denotes the temperature at which the liquid starts solidifying or freezing. After this the temperature remains constant till whole of the liquid is completely solidified (point c). Thereafter, the fall in temperature will again become continuous as is denoted by the curve cd. This curve represents the cooling of the solid mixture.

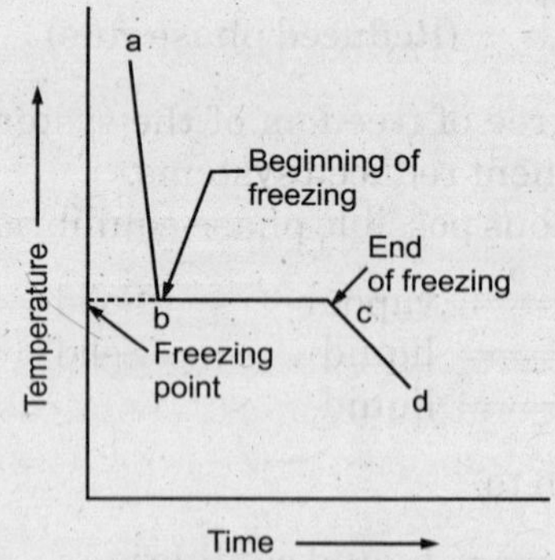

Fig. 20.4 Cooling curve for pure substances

(b) **Mixture of two solids:** When the mixture of two liquids is cooled slowly and the cooling curves is constructed by plotting temperature against time, a cooling curve shown in Fig. 20.5 is obtained. Point b marks the beginning of freezing or appearance of solid. However the temperature does not remain constant. The temperature decreases continuously, but at a different rate (compare with cooling curve of a pure substance). Actually at point b one of the components begins to solidify or crystallize. During this solidification process, heat is liberated and therefore, the temperature falls more slowly. From point b to c as the solidification proceeds, the composition of the mixture changes continuously. At a certain temperature T, known as *eutectic temperature*, the remaining liquid solidifies as a whole, i.e. both the components crystallise out separately. The temperature remains constant from point *c* to *d*. This is also known as *temperature halt*.

(ii) **Eutectic Point:** The term eutectic implies 'easy melting'. A binary system consisting of two substances, which are miscible in all proportions in the liquid phase and do not react chemically, is known as **eutectic system**. The two substances involved have the property of lowering each other's freezing point. Further, solid solution of two or more substances having the **lowest freezing point** of all the possible mixtures of the components is known as **Eutectic mixture**. The minimum freezing point attainable corresponding to the eutectic mixture, is termed as the **eutectic point** (lowest melting point). The eutectic mixture has a definite composition and a sharp melting point, i.e. it melts to a liquid of the same composition. In this respect it resembles a compound. However it is not a compound because the components are not present in stoichiometric proportions.

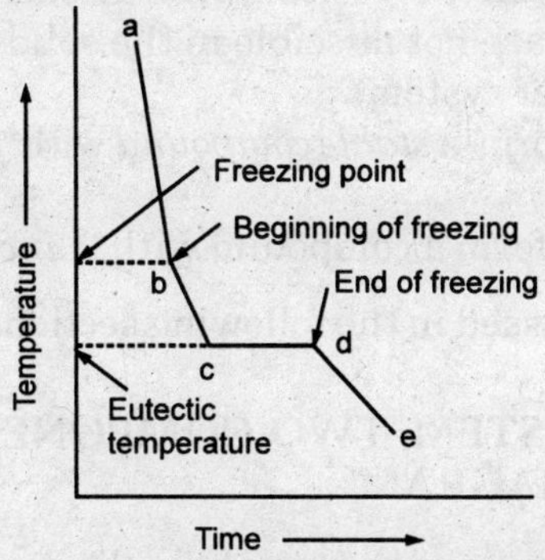

Fig. 20.5 Cooling curve of a binary miscible mixture

A mixture of lead and silver alloy, a mixture of bismuth-cadmium, etc. are few examples of eutectic mixtures. Eutectic mixtures have number of important applications including safety devices where low melting alloys can be used, fire-sprinklers, soldering, etc.

(iii) **Phase diagram for a two component system:** In order to obtain a phase diagram a number of mixtures of two components A and B, having different compositions are prepared. Each mixture is heated to a liquid and the cooling curve is obtained for each mixture. From the cooling curve the freezing point (temperature at which the separation of solid phase just starts) and the temperature of complete halt (eutectic temperature) is determined. A graph is plotted between composition of the mixture and its freezing point. This is the phase diagram for the eutectic system. In (Fig. 20.6) 'C' represents the eutectic temperature.

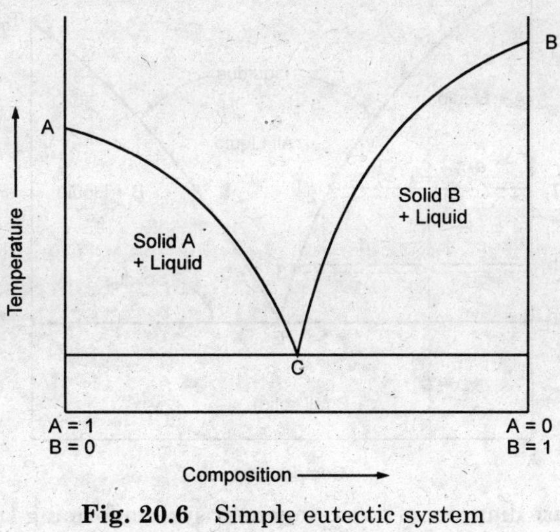

Fig. 20.6 Simple eutectic system

20.7 CLASSIFICATION OF TWO COMPONENT SOLID-LIQUID EQUILIBRIA

For the two component, solid - liquid equilibria we can have following cases, depending upon the miscibility of the two components in the molten state and the nature of the solid phases that separate out during cooling.

(i) The two component are miscible in the liquid state.

(ii) The two components are only partially miscible in the liquid state.

We will restrict our discussion to the case (i) i.e. the two components are miscible in the liquid state. Under this, three categories can be discussed.

(a) The two components are not miscible in the solid state and form a *eutectic mixture* (example: - lead - silver system)

(b) The two components form a *stable compound* with* *congruent melting point* (Example - $FeCl_3 - H_2O$ system)

(c) The two components form a compound with** *incongruent melting point.*

Example These systems are discussed in the following sections.

20.8 SIMPLE EUTECTIC SYSTEM TWO COMPONENT SYSTEM WHICH FORM A GENERAL PHASE DIAGRAM

Figure 20.7 shows the simplest of two component phase diagrams. The components are A and B, and the possible phases are pure crystals of A, pure crystals of B, and liquid with compositions ranging between pure A and pure B. Compositions are plotted across the bottom of the diagram. (Compositions might be expressed as mole fraction of A or B, in which case the total must add up to 1). Temperature is plotted on the vertical axis. The curves separating the fields of A + Liquid from Liquid and B + Liquid from Liquid are termed *liquidus curves*. The horizontal line separating the fields of A + Liquid and B + Liquid from A + B all solid, is termed the *solidus*. The point E, where the liquidus curves and solidus intersect, is termed the *eutectic* point. At the eutectic point in this two component system, all three phases, that is Liquid, crystals of A and crystals of B, all exist in equilibrium. Note that the eutectic is the only point on the diagram where this is true.

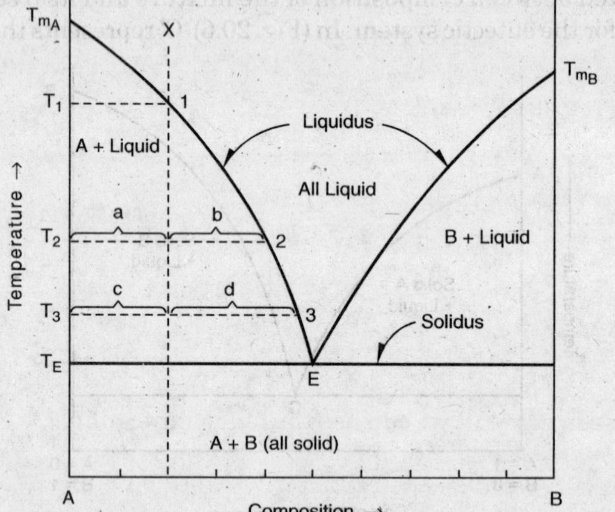

Fig. 20.7 General phase diagram of two component system forming simple eutectic system

---

*Congruent melting point – A compound is said to have congruent melting point if it melts at a constant temperature to give the liquid having the same composition as that of solid compound.

**Incongruent melting point – If the two components undergo chemical combination to form a new compound which is unstable, such a compound decomposes below its melting point to give a new solid phase and a melt which is different from that of the compound, such a compound is said to possess an incongruent melting point.

Since the pressure is kept constant, the phase rule in this case is F = C + 1 – P. The eutectic point is therefore an invariant point. If we change the composition of the liquid or the temperature, the number of phases will be reduced to 2.

If the system contains only pure A, then the system is a one component system and phase A melts at only one temperature, the melting temperature of pure A, T_{ma}. If the system contains only pure B, them it is a one component system and B melts only at the melting temperature of pure B, T_{mB}.

For all compositions between pure A and pure B, the melting temperature is drastically reduced, the minimum melting temperature is at the eutectic point T_E. Note that for all compositions between A and B the melting also occurs over a range of temperatures between the solidus and the liquidus. This is true for all compositions except one, that of the eutectic. The eutectic compositions melts at only one temperature, T_E.

We will now consider the crystallisation of a liquid with composition X in Figure 20.7.

However, it should be mentioned here that in equilibrium crystallisation or melting in a closed system, the final composition of the system will be identical to the initial composition of the system.

Therefore, composition X, which is made up of a mixture of 80% A and 20% B, will have, as its final crystalline product a mixture of 80% crystals of A and 20% crystals of B.

Composition X will be all liquid above the temperature T_1, because it will lie in the field of all Liquid. If the temperature is lowered to T_1, at T_1 crystals of A begin to form.

Further lowering of the temperature causes more crystals of A to form. As a result, the liquid composition must become more enriched in B as more crystals of A form out of liquid. Thus, with lowering of temperature, the liquid composition will change from point 1 to point 2 to point 3 to point E as the temperature is lowered from T_1 to T_2 to T_E respectively. At all temperatures between T_1 and T_E, two phases will be present in the system; liquid and crystals of A. At the eutectic temperature, T_E, crystals of B will begin to from, and three phases will coexist, crystals of A, crystals of B, and liquid. When the liquid crystallises completely, only pure solid A and pure solid B will remain and mixture of these two solid phase will be in the proportions of the original mixture, that is 80% A and 20% B.

The crystallisation of composition X can be written in abbreviated form as follows :

$T > T_1$ -- all liquid

$T_1 - T_E$ -- liquid + A

at T_E -- liquid + A + B

$T < T_E$ -- A + B all solid

If we were to stop the crystallisation process at any point during crystallisation and observe how much of each phase is present, we can use the following example.

For example, at temperature T_2 the amount of crystals of A and liquid (the only two phases present at this temperature) could be determined by measuring the distances *a* and *b* on Figure 20.7. The percentages would then be given by the lever rule :

$$\% \text{ crystals of A} = b/(a + b) \times 100$$

$$\% \text{ liquid} = a/(a + b) \times 100$$

SILVER-LEAD SYSTEM

It is an example of simple eutectic system in which the pure components alone separate as solid phases on cooling a molten mixture. Silver and lead in the solid state do not form intercomponent compounds because these are immiscible in solid state and do not react chemically. When molten silver and molten lead are mixed together in all proportions, a single homogeneous solution is formed. The system consists of four phases :

(i) solid silver (ii) solid lead
(iii) solution of molten silver and lead (iv) vapour

Since, the system is studied at constant pressure, the vapour phase can be ignored and the condensed phase rule is applicable

$$F' = C - P + 1$$

The maximum number of phase (when F = 0) that exist for this system is:

$$P = C - F' + 1$$
$$= 2 - 0 + 1$$
$$= 3$$

The phase diagram of lead silver system is shown in Fig. 20.8.

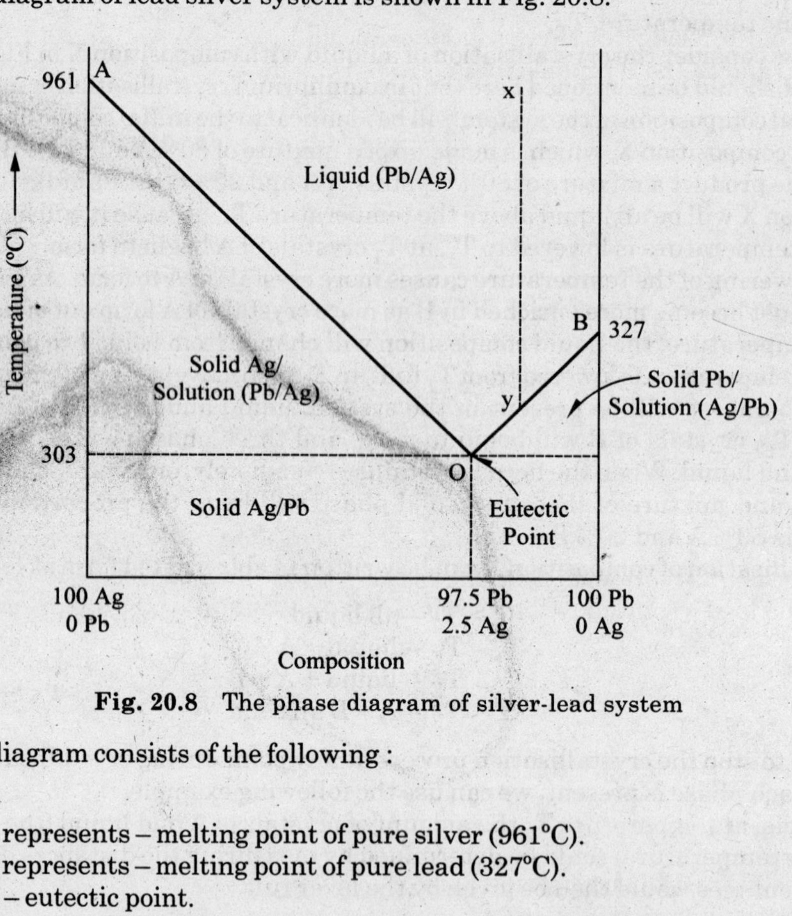

Fig. 20.8 The phase diagram of silver-lead system

The phase diagram consists of the following :

(a) *Points*
Point A represents – melting point of pure silver (961°C).
Point B represents – melting point of pure lead (327°C).
Point O – eutectic point.

(b) *Curves*
Curve AO – melting point curve of silver in presence of lead
Curve BO – melting point of lead in presence of silver.

(c) *Areas*
Area above AOB.
Area below AO.
Area below OB.

Points

In the phase diagram point A represents the melting point of pure silver and b that of pure lead. It is clear from the diagram that adddition of Pb to pure Ag lowers the melting point of Ag. Similarly addition of Ag to pure Pb lowers the melting point of Pb. All the points on the curves AO and BO, represent the melting point of various mixtures of Ag and Pb.

Curves

Curve AO is the melting point curve of silver, it shows the effect of addition of Pb on the melting point of pure Ag. Along this curve solid Ag and solution are in equilibrium.

Curve BO is the melting point curve of lead. It shows the effect of addition of Ag on the melting point of Pb. Along this curve solid Pb and solution are in equilibrium.

Hence, AO and BO both represent univariant systems.

$$\begin{aligned} F' &= C - P + 1 \\ &= 2 - 2 + 1 \\ &= 1 \end{aligned}$$

Areas

In the area above AOB, there is only one phase (liquid solution).On applying reduced phase rule equation, the degree of freeedom comes out to be two, means the system is bivariant in the area above AOB.

$$\begin{aligned} F' &= C - P + 1 \\ &= 2 - 1 + 1 \\ &= 2 \end{aligned}$$

It means that both the temperature and composition are required to be specified to define the system completely.

The area below AO represents the phases solid silver + solution and the area below BO represents the phases solid lead + solution. All these areas have two phases and one degree of freedom.

$$\begin{aligned} F' &= C - P + 1 \\ &= 2 - 2 + 1 \\ &= 1 \end{aligned} \qquad \left[\begin{aligned} &\because C = 2 \text{ (lead and silver)} \\ &P = 2 \text{ (liquid and solid phase)} \end{aligned}\right.$$

The system is univariant in areas below AO and OB.

Eutectic Point O

The two curves AO and BO intersect at point O at a temperature 303°C. The point O is known as eutectic point or eutectic temperature.

At this point three phases solid silver, solid lead and the melt are in equilibrium. On applying phase rule equation to this point we get,

$$\begin{aligned} F' &= C - P + 2 \\ &= 2 - 3 + 1 \\ &= 0 \end{aligned}$$

Thus the system solid silver/solid lead/solution of both Ag and Pb is monvariant at point O. At this point both temperature (303°C) and composition (Ag = 2.6% and Pb = 97.4) are fixed. If the temperature is increased above the eutectic temperature, the solid phase (silver and lead) will disappear and if the temperature is decreased below the eutectic point, solution phase will disappear and only solid phase (silver and lead) will remain in existence.

The mixture (Ag = 2.6% and Pb = 97.4%) at this point is called as eutectic mixture. The eutectic mixture has a definite melting point, but it is not regarded as a compound because

(i) The components are not in stoichiometric proportion.
(ii) The existence of separate crystals of the components can be observed on examining it under microscope.

Application of phase diagram of two component system or Pattinson's process for the desilverisation of argentiferrous lead

The phase diagram of silver-lead system has a special significance in relation to disilverisation of argentiferrous lead. The argentiferrous lead contains very small amount of silver. This argentiferrous lead is first heated to a temperature above its melting point. Thus the system consists of only the liquid phase represented by the point '*a*' in the figure. It is then allowed to cool, the temperature of the melt will fall along the perpendicular line *ab*. As the point '*b*' is reached lead will begin to crystallise and the solution will contain relatively increasing amount of silver. On cooling further, the system will be shifted along the line BO, lead starts separating out continuously and it is constantly removed from the argentiferrous lead. Thus the melt continues to be richer and richer in silver until the point O is reached. At this point O, the percentage of silver increases to 2.6% by mass.

The process of raising the relative proportion of silver in the alloy is called as Pattinson's Process. This is used for the recovery of silver profitably.

Characterstics of Eutectic Point

1. It is an invariant piont and is marked by specific temperature and composition.
2. It is the lowest melting point of any mixture of Ag and Pb.
3. Below the eutectic temperature, no liquid phase of any composition can exist in the system.
4. If the liquid melt of the two components (silver and lead) is cooled below the eutectic point, both the components solidify simultaneously without any change in composition or temperature of the liquid phase.
5. The eutectic mixtures of alloys have greater strength than their components because of their crystal charaterstics.
6. An eutectic system can maintain its temperature contant over long peroids.

Use of Eutectic System

1. The eutectic systems are used in the preparation of low melting alloys such as various types of solders useful in metal joints and electric fuses.
2. An alloy 50% Bi + 25% Pb + 12.5% Sn + 12.5% Cd is known as wood metal alloy. It has a low melting piont of 65°C. This is used for plugs in water sprayers in buildings. In case of fire accident, the plugs melt away and water is released automatically to extinguish fire.
3 Freezing mixutures: A system at the eutectic point can maintain its temperature over long periods and such systems are used as constant temperature baths. The lowest possible temperature attainable with a certain freezing mixture depends upon the eutectic temperature of the system. Some common freezing mixtures are given in the Table 20.1.

Table 20.1 Some Common Freezing Mixtures.

Eutectic system	*% of Salt*	*Eutectic temperature*
$NaCl. 2H_2O$ + Ice	23	–22°C
KI + Ice	52	–23°C
$CaCl_2. 6H_2O$ + Ice	15.2	– 55.9°C
$NaNO_3$ + Ice	33.3	–18.1°C
NH_4Cl + Ice	20.1	– 16°C
NH_4NO_3 + Ice	42.98	– 18°C

Differences among Melting point, Triple point and Eutectic point

- A solid and a liquid having the same composition are in equilibrium at the melting point
- Three phases are in equilibrium at the triple point.
- Two solids and a liquid are in equilibrium at eutectic point.
- All eutectic points are melting points.
- All melting points are not eutectic points.
- All the eutectic points are triple points.
- All triple points are not eutectic points.

20.9 TWO COMPONENT SYSTEM WHICH FORM STABLE COMPOUND WITH CONGRUENT MELTING POINT

If the two components of the system at a certain stage enter into chemical combination with one another forming stable compounds, the compound formed may have congruent or compound melting point.

A compound is said to have congruent melting point if it melts at a constant temperature to give a liquid of same composition as the solid.

GENERAL PHASE DIAGRAM

Let A and B be the two components and AB a stable solid compound formed by their chemical combination. The phase diagram for this system is shown in Fig. 20.9.

The diagram may be regarded as a combination of two simple eutectic diagrams. In this system , there are three different solid phases, A, B and the compound AB. Hence there will be three fusion or freezing point curves, AC, BE and CDE.

Along AC the solid A is in equilibrium with the liquid phase.
Along BE the solid B is in equilibrium with the liquid phase.
Along CDE, the solid compound AB is in equilibrium with the liquid phase.

The point D of the CDE curve is the congruent melting point of the compound because the solid and liquid phases have the same composition, which is indicated by D' on the composition axis. Thus a two component system becomes one component system at point D because both solid and liquid phases contain only the compound AB.

$$F = C - P + 1 = 1 - 2 + 1 = 0$$

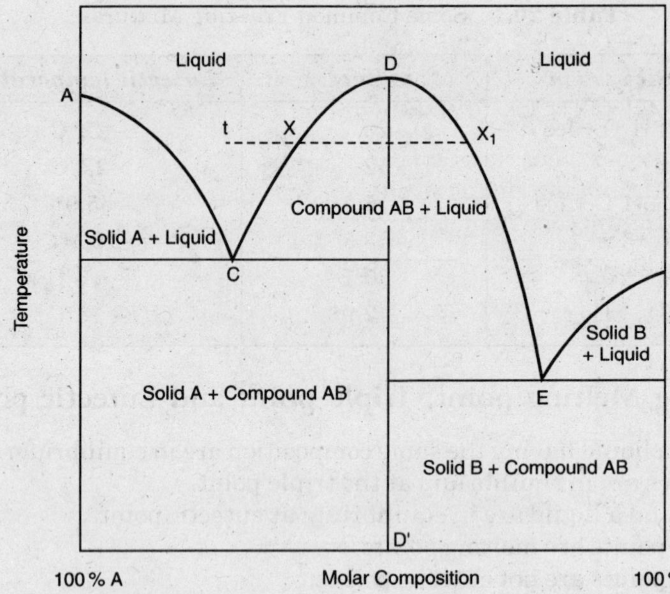

Fig. 20.9 General phase diagram of two component system with congruent melting point

Thus the point D is *non-variant point.*

It is clear from the figure that congruent melting point D of compound AB lies above the melting points of the pure components A and B. But this is not the case always, i.e. congruent melting point may be above, below or in between the melting point of pure components.

There are two eutectic point in such systems. In the figure, these are represented by C and E. At C, the solids A and AB are in equilibrium with the liquid phase, where as at E, the solids B and AB are in equilibrium with the liquid phase. From the figure, one can say that at a certain temperature 't', the liquid phase can have two compositions, x and x_1 in equilibrium with the same solid AB. This means that the compound AB can have two solubilities at the same temperature. This can be explained by dividing the diagram into two halves with DD' as the dividing line. The left half of the diagram is the two component system, A and AB where as right half shows the two the component system B and AB. On left hand side the curve, DC is the usual depression in freezing point of the compound AB on the addition of A. Similarly on the right hand side, the curve DE, represents the usual depression in freezing point of the compound AB on addition of B.

Examples of this type of two component systems which form compound with congruent melting point include : zinc-magnesium, gold-tin, mercury-thallium and salt hydrates, such as $FeCl_3$ – H_2O system.

FERRIC CHLORIDE – WATER SYSTEM:

Ferric chloride - water system provides an example of a two component system, in which a number of compounds (hydrates) with congruent melting points are formed. It is a composite system made of several simple eutectic systems. The phase diagram of the system is shown in Fig. 20.10. It may be mentioned here, when the two components form more than one compound there may be as many humps in the phase diagram as there are compounds. Thus the phase diagram of $FeCl_3$ – H_2O system indicates the existence of four compounds. These four stable hydrates are :

$Fe_2Cl_6 \cdot 12H_2O$	dodecahydrate
$Fe_2Cl_6 \cdot 7H_2O$	heptaydrate
$Fe_2Cl_6 \cdot 5H_2O$	pentahydrate
$Fe_2Cl_6 \cdot 4H_2O$	tetrahydrate

All the hydrates possess congruent melting points.

The sharpness of the maximum in the phase diagram gives an indication of the stability of the compound. A flat maximum corresponds to an unstable compound.

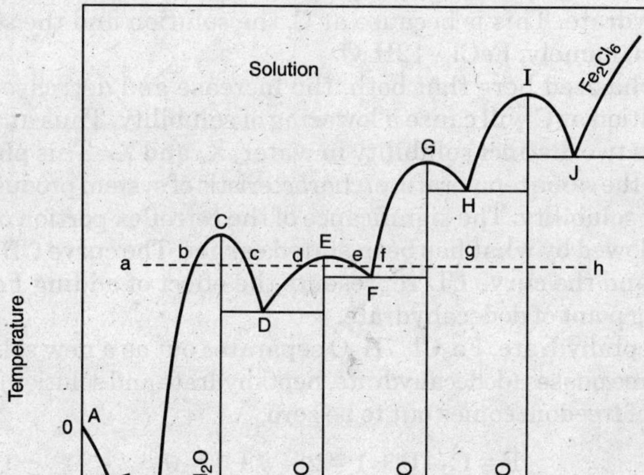

Fig. 20.10 Phase diagram for $FeCl_3$-H_2O system

The point A in the diagram represents the melting point of ice or freezing point of water (0°C) under normal pressure.

- Addition of Fe_2Cl_6 results into the formation of a solution of appropriate composition, and the temperature of the solution falls along the line AB. The system has two components (H_2O and Fe_2Cl_6) and two phases (ice and solution) along this curve. Hence, the degree of freedom by applying condensed phase rule comes out to be 1

$$F = C - P + 1 = 2 - 2 + 1 = 1.$$

  Thus, on adding increasing amount of Fe_2Cl_2, the temperature falls with change in composition, till a eutectic point B is reached. The eutectic point in case of system involving salt and water, is referred to as *cryohydric point*.

- At eutectic point B, the dodecahydrate $FeCl_6 .12H_2O$ separates out as a new solid phase. The system now consists of three phase (ice, solution and solid $FeCl_6$. $12H_2O$) and two components. Applying reduced phase rule equation, we have

$$F = C - P + 1 = 2 - 3 + 1 = 0$$

The system is non-variant at eutectic point B. The temperature at this point is –55°C. As is seen from the figure, this is the lowest temperature that can be attained in this system.

- When the addition of Fe_2Cl_2 is continued further, beyond the point B and the temperature is raised, the composition of the solution will change along the curve BCD. Along this curve the ice disappears and the system becomes two phase (solution and solid Fe_2Cl_6). Thus the degree of freedom by applying condensed phase rule comes out to be 1. The BCD curve is the is the solubility curve of $FeCl_6$. $12H_2O$ because Fe_2Cl_6. $12H_2O$ and the saturated solution are in equilibrium along this curve. The point C on the curve BCD is the congruent melting point of dodecahydrate. This is because at C, the solution and the solid phases have the same composition, namely, $FeCl_6$. $12H_2O$.
- It should be emphasized here that both, the increase and decrease in the proportion of Fe_2Cl_6 in the solution at C will cause a lowering of solubility. Thus at a temperature t°, the $FeCl_6$. $12H_2O$ has two distinct solubility in water, X_1 and X_2. This phenomenon of having two solubilities at the same temperature, characteristic of system producing solid compounds, is called retroflex solubility. The significance of the retroflex portion of the solubility curve can be readily followed by what has been stated earlier. The curve CB represents the effect of adding water and the curve CD represents the effect of adding $FeCl_6$, in lowering the congruent melting point of dodecahydrate.
- At point D, the heptahydrate, Fe_2Cl. $7H_2O$ separates out as a new solid phase. Thus again the system has three phases (dodecahydrate, heptahydrate and solution) and two components. Thus the degree of freedom comes out to be zero,

$$F = C - P + 1 = 2 - 3 + 1 = 0$$

The system again becomes invariant. Point D is thus again eutectic point.

- Similarly, the curve DEF is the solubility curve of Fe_2Cl_6. $7H_2O$ with E as its congruent melting point. F is the eutectic point where the pentahydrate separates out and three phases coexist (heptahydrate, pentahydrate and solution)
- Curve FGH is the solubility curve of pentahydrate Fe_2Cl_6. $5H_2O$ with G as its congruent melting point. On increasing the amount of Fe_2Cl_6, the temperature falls till the eutectic point H is reached. At H terahydrate separates out of system. The three phases present in the system are pentahydrate , tetrahydrate and the solution.
- This is followed by the curve HIJ which is the solubility curve of tetrahydrate- I is the congruent melting point of this hydrate. On further addition of Fe_2Cl_6, the anhydrous Fe_2Cl_6 separates out at J which is the fifth eutectic point. After this JK represents the solubility curve of the anhydrous ferric chloride. No other solid phase separate out along this line.

To summarise — In $Fe_2Cl_6 - H_2O$ system we have,

Curves: The curves AB, BCD, DEF, FGH, HIJ and JK have one solid phase and one liquid (solution) in equilibrium. Therefore applying reduced phase equation we have $F = C - P + 1 = 2 - 2 + 1 = 1$. Hence the systems ice/solution, Fe_2Cl_6. $12H_2O$/solution, Fe_2Cl_6. $7H_2O$/ solution, Fe_2Cl_6. $5H_2O$/ solution, Fe_2Cl_6. $4H_2O$/solution and Fe_2Cl_6./solution represented by various curves are *monovariant*.

Congruent Melting Compounds: At the congruent melting point C, E, G, I, the composition of the solution with respective hydrate is identical. Thus it becomes one component system at these points C, E, G and I has two phases and *one* component. It is therefore nonvariant.

$$F = C - P + 1 = 1 - 2 + 1 = 0$$

Table 20.2 Congruent point of hydrates of Fe_2Cl_6

Point	*Congruent*	*Hydrate*
C	37°	$12H_2O$
E	32.5°	$7H_2O$
G	56°	$5H_2O$
I	73.5°	$4H_2O$

The Eutectic Point: In the phase diagram of $Fe_2Cl_6 - H_2O$ system, we have 5 eutectic points. At each of the eutectic points, the system has three phases and two components

Thus, $$F = C - P + 1 = 2 - 3 + 1 = 0$$

The system is non variant.

Table 20.3 Eutectic points

Eutectic point	*Temperature*	*Phases in Equilibrium*
B	−55°	ice $12H_2O$, solution
D	27.4°	$12H_2O$, $7H_2O$, solution
F	30°	$7H_2O$, $5H_2O$, solution
H	55°	$5H_2O$, $4H_2O$, solution
J	66°	$4H_2O$, $FeCl_6$, solution

20.10 FORMATION OF COMPOUNDS WITH INCONGRUENT MELTING POINTS

In this two component system, the two components undergo chemical combination to form a new compound, which is unstable. Such a compound decomposes below its melting point to give a new solid phase and a melt with a composition different from that of solid phase. Such a compound is said to have incogruent meting point. The temperature at which decomposition takes place is called *transition temperature* or *meritectic* or *peritectic* temperature. The decomposition reaction is represented by the equation,

$$\text{Original solid} \rightleftharpoons \text{New Solid} + \text{Melt.}$$

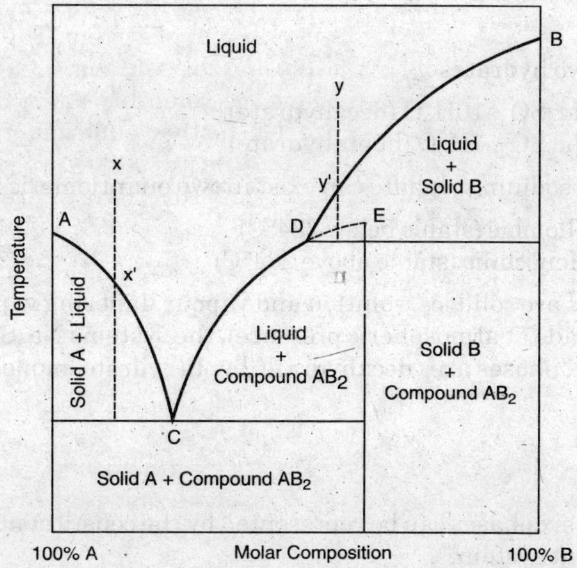

Fig. 20.11 Formation of compound with incongruent melting point

Thus in peritectic reaction, the number of phases is 3 (two solid – one liquid) and degrees of freedom F= 0 (F = 2 + 1 – 3 = 0).

GENERAL PHASE DIAGRAM

The general phase diagram for this type systems is shown in Fig. 20.11.

Let A and B be the components which combine to give the compound AB_2. The point A and B are the respective melting points of the two pure components and the point D represents the incongruent melting point of the compound AB_2. The compound AB_2 decomposes to give a new compound. The point lies somewhere below and may be taken as incongruent melting point of the compound AB_2.

Curves: Curve AC is the fusion curve of A along which the solid A is in equilibrium with the liquid. Curve BD is the fusion curve of B along which the solid B is in equilibrium with the liquid, curve CD is the fusion curve of the compound AB_2 along which solid AB_2 is in equilibrium with the liquid. Let us consider the behaviour of the system when a melt of composition x is cooled. At point x', solid A begins to separate out from melt. On further cooling the proportion of solid A increases and the melt become richer in B, and its composition changes along x'C. At C the compound AB_2 starts forming. Thus the system now consists of two solids and one liquid phase and hence at C the system is non - variant (F = C – P + 1 = 2 – 3 + 1 = 0)

Let us consider another point with the composition y. When the liquid of composition y is cooled down, the solid B separates at y′. The melt becomes richer in A and further cooling will proceed along y'D. At point D, the compound AB_2 separates out with the transition reaction, (meritectic reaction)

$$\text{Solid B} + \text{solution} \rightleftharpoons \text{Solid } AB_2$$

Again, three phases are there and the system is non - variant. The transition of solid B to the compound AB_2 will take place at constant temperature till whole of B disappears. Point D is thus the peritectic point. A number of two components system give rise to compounds with incongruent melting point. We will discuss $Na_2SO_4 - H_2O$ system in detail.

THE SODIUM SULPHATE-WATER SYSTEM

Phases

Sodium sulphate forms two hydrates :

$Na_2SO_4 . 10H_2O$ (decahydrate)
$Na_2SO_4 . 7H_2O$ (heptahydrate)

Mereover, anhydrous sodium sulphate can exist in two enantiomeric forms:

Rhombic (stable below 234°C)
Monoclinic (stable above 234°C)

The remaining phases are solid ice, solution and vapour. Ignoring vapour phase, (because all the determinations are made at atmospheric pressure), the systems Na_2SO_4 -water is a six phase condensed system. The six phases are : decahydrate, heptahydrate, monoclinic, rhombic, ice, and solution.

Components

The composition of all the six phases can be represented by the constituent Na_2SO_4 and water and hence it is a two component system.

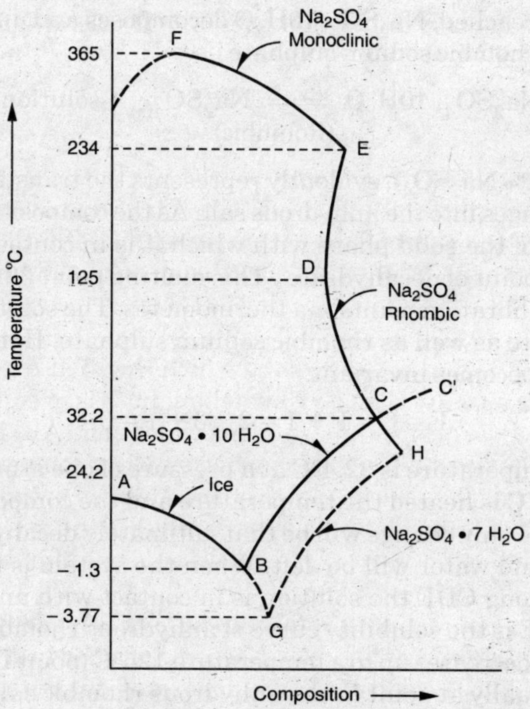

Fig. 20.12 Phase diagram of sodium sulphate-water system

Phase Diagram: The phase diagram of the system is shown in Fig. 20.12. Let us discuss the various portion of this diagram one by one.

- The curve AB is the melting point curve of ice. The point A (0ºC) is the melting point of ice. On addition of Na_2SO_4 the curve AB shows the lowering of melting point of ice. The phases along AB are ice and solution. Thus, applying condensed phase rule equation,

$$F = C - P + 1 = 2 - 2 + 1 = 1$$

  We find that system is univariant. Therefore, with change in composition, there has to be change in temperature. The curve AB can also be regarded as fusion curve of ice.
- At B, a new solid phase, Na_2SO_4, $10H_2O$ separates out (– 1.3°C, 4.6 % Na_2SO_4). This is the lowest temperature attainable or it is the eutectic point. At eutectic point, hence three phases coexist, ice, Na_2SO_4. $10H_2O$ and solution.
  Thus,

$$F = C - P + 1 = 2 - 3 + 1 = 0$$

  The system becomes invariant. The eutectic temperature is referred to as cryohydric temperature.
- The curve BC results with further addition of anhydrous Na_2SO_4 and supplying heat to the system. Ice will melt and the salt will dissolve in the water thus produced. Along the curve BC, solid decahydrate and the solution are in equilibrium and hence BC is the solubility curve of decahydrate. The solubility increases with rise in temperature the system Na_2SO_4. $10H_2O$/solution is monovariant because only two phases (solution and Na_2SO_4. $10H_2O$) are present as ice disappears

$$F = C - P + 1 = 2 - 2 + 1 = 1.$$

- When the point C is reached, $Na_2SO_4 . 10H_2O$ decomposes and undergoes peritectic change yielding anhydrous rhombic sodium sulphate i.e.

$$\underset{}{Na_2SO_4 . 10H_2O} \rightleftharpoons \underset{\text{(Rohmbic)}}{Na_2SO_{4(s)}} + \text{solution}$$

Thus C (32.383°C, 32% Na_2SO_4), evidently represents the transition temperature at which the decahydrate changes into the anhydrous salt. As the composition of the solution phase is different from that of the solid phase with which it is in contact, this is regarded as the *incogruent* melting point of decahydrate . This melting point 32.383°C is so characteristic that it is used or a calibration point for a thermometes. The solution at C is saturated with both, the decahydrate as well as rhombic sodium sulphate. Hence at C, three phases are present and system becomes invariant.

$$F = C - P + 1 = 2 - 3 + 1 = 0$$

This incongruent temperature is 32.4°C at a pressure of one atmosphere.

- When the system at C is heated the temperature and the composition of the solution will remain constant. The only change will be that, ultimately decahydrate disappear and only Na_2SO_4 (rhombic) and water will be left. When the system is further heated, the curve CED is obtained. Along CDE the solution is in contact with anhydrous rhombic sodium sulphate, hence CDE is the solubility curve of anhydrous rhombic salt. It is clear from the curve that solubility decreases up to a temperature 125°C (point D) and then increases up to 234°C (point E). Actually at point E, the anhydrous rhombic salt changes into monoclinic form. Thus E is again a transition point where rhombic Na_2SO_4 changes into monoclinic Na_2SO_4.

$$\underset{\text{Rhombic}}{Na_2SO_4} \rightleftharpoons \underset{\text{Monoclinic}}{Na_2SO_4}$$

At this point, there are three phases, namely, rhombic Na_2SO_4, monoclinic Na_2SO_4 and the solution and hence the solution is invariant

$$F = C - P + 1 = 2 - 3 + 1 = 0.$$

- If heating is continued at E, ultimately, rhombic salt will completely change into monoclinic form. On further heating, the curve the monoclinic Na_2SO_4 and the solution are in equilibrium, hence this is the solubility curve of monoclinic sodium sulphate. As is evident from with figure that the solubility falls with increase in temperature up to 365°C. The temperature 365°C is the critical temperature of the solution. Even at F, the sodium sulphate has slight solubility. Along the curve EF, the system is monovariant as the number of phase that exist are two (monoclinic Na_2SO_4 + Solution)

Metastable Equilibria of Na_2SO_4—Water System

- As is discussed above, at point C in the figure, the decahydrate ($Na_2SO_4 . 10H_2O$) changes into anhydrous rhombic Na_2SO_4 if heating is continued. However it is possible that $Na_2SO_4 . 10H_2O$ can exist as decahydrate even beyond the transition temperature (32.383°C), if enough time is not allowed. The curve CC' represents the curve where decahydrate exists as a metastable phase and CC' is its solubility curve. If is a general rule that the solubility of a metastable state is higher than that of a stable state. This is evident from the figure that the solubility of metastable decahydrate is higher than the stable decahydrate.
- At point C, if the solution is cooled then the rhombic anhydrous salt starts disappearing and crystals of decahydrate separate out as cooling continues along the curve CB. However, it is

possible to prevent crystallisation of decahydrate, if the solution of anhydrous monoclinic rhombic sulphur is cooled rapidly below the transition temperature. When the solution is rapidly cooled the curve CH results, when a metastable rhombic anhydrous Na_2SO_4 exists. The CH being its solubility curve. Again, it can be seen that metastable state is more soluble than the stable anhydrous Na_2SO_4.

- At point B, as was discussed earlier, solid phase of $Na_2SO_4 . 10H_2O$ separates out. However, if the system is cooled below – 1.3°C, the crystallisation of decahydrate is prevented. In such a case at a point G in the figure (at –3.77°C), a second hydrate of formula $Na_2SO_4.7H_2O$ (heptahydrate) separates out. The curve GH represents the solubility curve of heptahydrate. However, the system will be metastable and it has the spontaneous tendency to yield crystals of decahydrate. It can be seen, the point G is the second but metastable eutectic point of the system. At point H the heptahydrate changes into rhombic anhydrous Na_2SO_4.

Other incogruent melting compounds are formed by systems: Gold –Antimony, Gold-Lead, picric acid-benzene, potassium sulphate-cadmium sulphate etc.

PHASE DIAGRAM OF WATER AND ICE SKATING

We can account for the effect of pressure on the melting point of water using the following reasoning. At the melting point, liquid and solid water coexist — they are in dynamic equilibrium, since the rate at which ice is melting is just balanced by the rate at which the water is freezing. What happens if we apply pressure to this system? When subjected to increased pressure, matter reduces its volume. Since a given mass of ice at 0°C has a larger volume than the same mass of liquid water, the system can reduce its volume in response to the increased pressure by changing to liquid. Thus, at 0°C and an external pressure greater than 1 atm, water is liquid. In other words, the freezing point of water is less than 0°C when the pressure is greater than 1 atm.

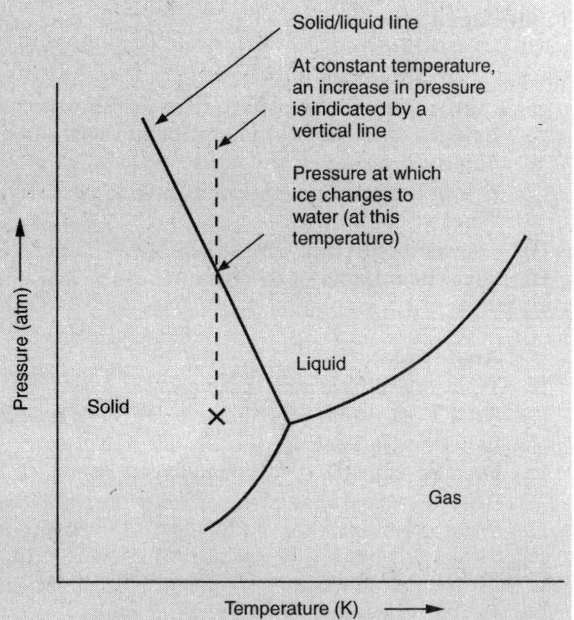

Fig. 20.13 Phase diagram for water

Figure illustrates the effect of pressure on ice. At the point X on the phase diagram, ice is subjected to increased pressure at constant temperature, Note that as the pressure is increased, the solid/liquid line is crossed, indicating that the ice melts, This is exactly what happens in the ice skating. The narrow blade of the skates exerts a large pressure, since the skater's weight is supported by the small area on the blade. The ice under the blade melts because of the pressure, providing lubrication. After the blade passes, the liquid refreezes, as normal pressure returns. Without this lubrication effect due to the thawing ice, ice skating would not be the smooth activity that it can be.

QUESTIONS

Short Answer Questions

(Refer to the respective sections for answers)

Section 20.1

1. How many phases are present in each of the following systems :
   (i) a drop of water placed in a stoppered bottle.
   (*ii*) a piece of molten ice placed in a beaker covered with a watch glass.
   (*iii*) mixture of N_2, and O_2. **[Ans (*i*) 2; (*ii*) 2; (*iii*) 1]**
2. Explain the term component. How many components are present in the following system ?
   (*a*) Water $\rightleftharpoons$ Water-vapour
   (b) KCl + Water $\rightleftharpoons$ kCl hydrate **[Ans (*a*) 1; (*b*) 2]**
3. The number of components present in the following system :
   (*i*) a solution of common salt
   (ii) $MgCO_3(s) \rightleftharpoons MgO(s) + CO_2$ (in a closed vessel)
   (iii) $NH_2Cl(s) \rightleftharpoons NH_3(g) + HCl(g)$ **[Ans (*i*) 2; (*ii*) 3; (*iii*) 3]**
4. Explain the terms
   (i) Phase
   (ii) Component
   (iii) Degree of freedom. How are they related ?
5. Explain that NH_4Cl in equilibrium with its dissociation product is a one component system.
6. Sulphur system at any of its triple point is a non variant system
7. Evaluate the number of phases in a mixture of two allotropes of sulphur
8. What is the number of phases when $CaCO_{3(s)}$ is heated?
9. A gaseous mixture consists of N_2, H_2 and NH_3 in equilibrium. What is the number of phase?
10. Give the number of components of the system

$$Fe_{(s)} + H_2O_{(g)} \rightleftharpoons FeO_{(s)} + H_{2(g)}$$

    Ans: Three
11. What is an invariant system ?
    Ans: The system in which the degree of freedom is zero, i.e. no condition is required to be specified to define the system.
12. Give an example of invariant system.
    Ans: A system consisting of ice, water and water vapour in equilibrium is an invariant system.
13. What is the number of phases and components in a mixture of gases say N_2 and O_2.

Section 20.2, 20.3

14. Derive phase rule
15. What are the advantages and limitations of phase rule
16. Mention some of the applications of phase rule.
17. Explain the following (*i*) Triple point (*ii*) Metastable equilibrium
18. What triple point?

Section 20.4

19. Draw the phase diagram of one component system.
20. For a one component system, triple point is an invariant point. Discuss.
21. In phase diagram of ice, the fusion curve of ice has a negative slope whereas the sublimation curve has positive slope. Explain.
22. Calculate the number of phases, components and degrees of freedom for the following sulphur system

$$S_s \rightleftharpoons S_m \rightleftharpoons S_l \rightleftharpoons S_v$$

23. How many triple points are there in phase diagram of sulphur.
24. In the phase diagram of water, the melting curve is almost a straight line, which slopes backwards. What will happen to the melting point if pressure is increased

Ans: The melting point of ice decreases, i.e. when the pressure is increased ice turns into water at a lower temperature.

25. What is the significance of triple point?

Ans: Triple point is invariant for a given substance, the triple point occurs at a unique set of values of temperature and pressure. If either of these is altered, one of the phase disappears and the system becomes univariant or monovariant.

Section 20.5, 20.6, 20.7, 5,8

26. What is the difference between the phase diagram of one component system and two component system?
27. A eutectic mixture has a definite composition and a sharp melting point, yet it is not a compound.

Ans: This is becomes in eutectic mixture the components are not in stoichiometric proportions as is the case with a compound.

28. Explain the terms eutectic point and a eutectic mixture.

Ans: **Eutectic mixture** is a 'solid solution' of two or more substances having the *lowest freezing* point of all the possible mixture of the components. This is taken advantage of in " alloys of low melting point", which are generally eutectic mixtures.

Eutectic point: Two or more solid substances capable of forming solid solutions with each other have the property of lowering each other's freezing point; and the *minimum freezing* point attainable corresponding to the eutectic mixture, is termed the eutectic point (means lowest **melting point**).

29. Distinguish between congruent melting compound and incongruent melting compound.

Ans: A compound is said to have congruent melting point if it gives the liquid having the same composition, as that of the solid, on melting point. If it decomposes below its melting point to give a new solid and a melt which has different composition than that of the compound then it is in congruent melting compound.

30. Distinguish between triple point and freezing point of a pure substance.

Ans: The triple point is the point where liquid solid and vapour are in equilibrium

The freezing point is the point where solid and liquid are in equilibrium under 1 atm. pressure

31. Distinguish between the triple point and eutectic point.
32. What is eutectic composition of lead-silver system?

Ans: 97.4% Pb and 2.6% Ag

33. Mention an application of Pb - Ag system
34. What is metastable state?

Ans: The state of supercooled or supersaturated solution in which the phase, which is stable under normal condition, does not form.

Section 20.9

35. Give an example of a two component system which form compound with congruent melting point.
36. How many hydrates are depicted by the phase diagram of $FeCl_3 - H_2O$ system.
37. How many eutectic points are there in $FeCl_3 - H_2O$ system.

Section 20.10

38. How many components and phases are present in Na_2SO_4 –- Water system.
39. What is incongruent melting point of decahydrate in Na_2SO_4 – Water system.

Long Answer Questions

1. State phase rule and explain its application to Ice-water-vapour system. What is meant by the triple point.

2. (a) State phase rule and define the terms involved. Explain sulphur system on the basis of phase rule.
   (b) Give the limitations and advantages of phases rule.
3. State and explain the conditions under which the phase rule is represented by the equation $F = C - P + 1$. What is the qualitative difference between the phase diagrams of one component system and two component system?
4. (a) Explain the terms true and metastable equilibrium.
   (b) Write down the number of components, number of phases and calculate the degree of freedom for the following equilibrium.

$$Fe(s) + H_2O(g) \rightleftharpoons FeO(s) + H_2(g)$$

5. Draw and explain the phase diagram of a two-component system of solids, A and B forming a compound AB with a congruent melting.
6. (a) Determine the value of C and P in the following systems
   (i) $NH_4Cl(s) \rightleftharpoons NH_4Cl \rightleftharpoons NH_3(g) + HCl_{(g)}$
   (ii) Ice $\rightleftharpoons$ water $\rightleftharpoons$ vapour
   (b) Draw a neat diagram of sulphur system. Explain the application of phase rule to this system.
7. For one component system, the triple point is an invariant point. Discuss
8. (a) Derive Gibbs phase rule
   (b) Explain why the fusion curve of ice has a negative slope whereas the sublimation curve has positive slope in the phase diagram.
9. (a) Explain the following:
   (i) Eutectic mixture
   (ii) Eutectic point
   (iii) Triple point
   (iv) Condensed phase rule with explanation of terms involved
   (b) Discuss the effect of pressure on
   (i) transition temperature of rhombic sulphur,
   (ii) melting point of ice.
10. Draw and discuss the phase diagram of lead-silver system. Explain the eutectic point and Pattinson's process of desilverisation of lead. (Or discuss practical application of this system.)
11. Draw a neat laballed phase diagram of water system and explain areas, curves and triple point in it.
12. In the phase diagram for water system, explain
    (i) a bivariant system,
    (ii) an univariant system,
    (iii) an invariant system.
13. Describe the application of phase rule in the study of binary system forming compound with congruent melting point.
14. Explain the phase diagram for sodium sulphate water system
    (b) Explain what do you understand by the following :
    (i) Congruent melting point
    (ii) Eutectic point
15. Draw labelled phase diagram of $FeCl_3^{-}H_2O$ system. Mention the various stable hydrates formed in this system.
16. Explain the following terms with the help of the diagrams
    (i) Triple point
    (ii) Invariant system
    (iii) Metastable equilibrium
17. A saturated solution of sodium sulphate with excess of the salt is at equilibrium with its vapour in a closed vessel. How many phases and components are present? What are the degrees of freedom in this system?
18. What do you mean by incongruent melting point. Discuss the phase diagram of a two component system with the formation of a compound having incongruent melting point.

19. Draw a phase diagram representing schematically a temperature composition phase diagram for a binary AB system having a single eutectic.
20. Discuss the application of phase rule in desilverisation of lead.
21. (a) What are the no. of components in the equilibrium reaction of dissociation of ammonium chloride in vacuum.

    (b) An eutectic mixture has a definite composition and a sharp melting point, yet it is not a compound. Comment. **(G.G.S.I.P.U, II-Term, B.Tech., April 2006)**
22. (a) Distinguish between triple point and eutectic point.

    (b) In the phase diagram of ice, the fusion curve of ice has a negative slope. Explain.

    (G.G.S.I.P.U, II-Term, B.Tech., April 2006)
23. Draw and discuss the phase diagram of lead – silver system. Discuss practical application of the system. **(G.G.S.I.P.U, II-Term, B.Tech., April 2006)**
24. Draw and label phase diagram for water system. What is the significance of triple-point.

    (G.G.S.I.P.U, End-Term, B.Tech., May-June 2006)
25. A eutectic mixture has a definite composition and a sharp melting point, yet it is not a compound. Justify. **(G.G.S.I.P.U, II-Term, B.Tech., April 2007)**
26. Draw and discuss the phase diagram of sodium sulphate-water system. How many components and phases are present in it. **(G.G.S.I.P.U, II-Term, B.Tech., April 2007)**
27. In order to get a two metal low melting alloy, what composition is preferred.

    (G.G.S.I.P.U, II-Term, B.Tech., April 2007)
28. State Gibb's phase rule. With the help of suitable examples, explain the terms:

    (i) Components (ii) Degree of freedom (iii) Phase

    (G.G.S.I.P.U, End-Term, B.Tech., Dec. 2007)
29. Give a labelled phase diagram of water system and discuss the importance of triple point.

    (G.G.S.I.P.U, End-Term, B.Tech., Dec. 2007)
30. Draw a labelled diagram of lead-silver system and mention its salient features.

    (G.G.S.I.P.U, End-Term, B.Tech., Dec. 2007)

❑❑❑

21 CHEMICAL KINETICS

Don't go through life, grow through life

— Eric Butterworth

CHAPTER OUTLINE

21.1 INTRODUCTION

Chemical kinetics is the branch of chemistry which deals with the study of the rate with which a chemical reaction occurs and the factors that affect this rate. This information is especially useful for determining *how* a reaction occurs.

What is meant by the *rate* of a reaction? The rate of a reaction is the speed at which the products are formed and the reactants are consumed.

The study of chemical kinetics has been highly useful in determining the factors that influence the rate, maximum yield and conversion in industrial processes. The mechanism or the sequence of steps by which the reaction occurs can be known. It is also useful in selecting the optimum conditions for maximum rate and yield of the chemical process.

Generally, chemical reactions can be put into three categories:

1. Instantaneous Reactions: These reactions take place instantaneously and the period of completion of such reactions is about 10^{-14} to 10^{-16} seconds. The ionic reactions are the examples of such reactions.

(*i*) $AgNO_3 + NaCl \longrightarrow AgCl + NaNO_3$

2. Very Slow Reactions: These reactions take a very long time for completion and it may appear that the reactions have not occurred at all. For example, the rusting of iron

$$Fe(s) + H_2O(g) + CO_2(g) + \frac{1}{2}O_2(g) \longrightarrow \underset{\text{Rust}}{Fe(OH)HCO_3}$$

The reactions proceed so slowly that is becomes difficult to determine the speed of the reactions.

3. Moderate Reactions: There are some reactions which occur in conveniently measurable time and their rates can be measured. These reactions are called moderate reactions. For example:

(*i*) Decomposition of hydrogen peroxide

$$2H_2O_2(l) \longrightarrow 2H_2O(l) + O_2(g)$$

(*ii*) Inversion of canesugar

$$C_{12}H_{22}O_{11}(s) + H_2O(l) \xrightarrow{H^+} \underset{\text{glucose}}{C_6H_{12}O_6(s)} + \underset{\text{fructose}}{C_6H_{12}O_6(s)}$$

(*iii*) Hydrolysis of ethyl acetate

$$CH_3COOC_2H_5(l) + H_2O(l) \xrightarrow{H^+} CH_3COOH(l) + C_2H_5OH(l)$$

These reactions are the main category of reactions for the kinetic study.

21.2 RATE OF REACTION

The quantity of a reactant species consumed or the quantity of a product species formed in a unit time in a chemical reaction is called the rate or velocity of the reaction. Precisely, the *rate* is defined as the change in concentration of any reactant or the product with time.

$$\frac{\text{change in concentration } \Delta c \text{ in time } \Delta t}{\Delta t}$$

Rate of the reaction can be measured by the

- the rate of formation of the product, or
- the rate of consumption of the reactant

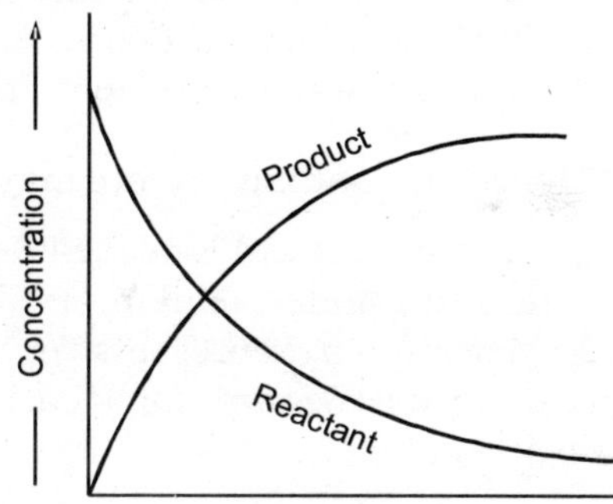

Fig. 21.1

Consider the reaction.

$$A \rightarrow B$$

where A is the reactant and B is the product

With the passage of time, the concentration of A goes on decreasing, whereas the concentration of B goes on increasing (Fig. 21.1).

The rate of reaction at any moment may be expressed as:

$$\text{Rate of reaction } (r) = \frac{\text{Amount of A consumed or decrease in concentration of A}}{\text{Time taken}}$$

$$\text{Rate of reaction } (r) = \frac{\text{Amount of B produced or increase in concentration of B}}{\text{Time taken}}$$

The decrease or increase in the concentration of reactants or products may also be expressed in terms of change in their concentration during the time interval, Δt, as

$$\text{Rate of reaction } (r) = \frac{-\Delta[A]}{\Delta t}$$

$$\text{Rate of reaction } (r) = \frac{+\Delta[B]}{\Delta t}$$

where $\Delta[A]$ gives the **decrease** in concentration of A and $\Delta[B]$ represents **increase** in concentration of B. The square brackets around the substances are used to express molar concentration (mole/litre). This rate is also known as **average rate** of reaction.

During a reaction, the concentration of reactants decreases with the passage of time and therefore, ΔA, (*i.e.*, final concentration – initial concentration) will be negative. The rate of reaction, then will be negative. But reaction rate is **always positive**, as also obtained from the rate of formation of the products. Therefore, the minus sign is put before $\frac{\Delta[A]}{\Delta t}$ so that the rate is positive.

As experiments show that for most reactions, the concentrations of all participants change most rapidly at the beginning of the reaction. That is, the [products] show the greatest rate of increase and the [reactants] show the greatest rate of decrease at the beginning. This means that the *rate of a reaction* **change with** *time,* so it is usual to define the rate over a very small time, Δt. Hence, we can express the rate as the *derivative* of concentration *with respect to time*

$$\text{rate} = \frac{d[\text{concentration}]}{dt}$$

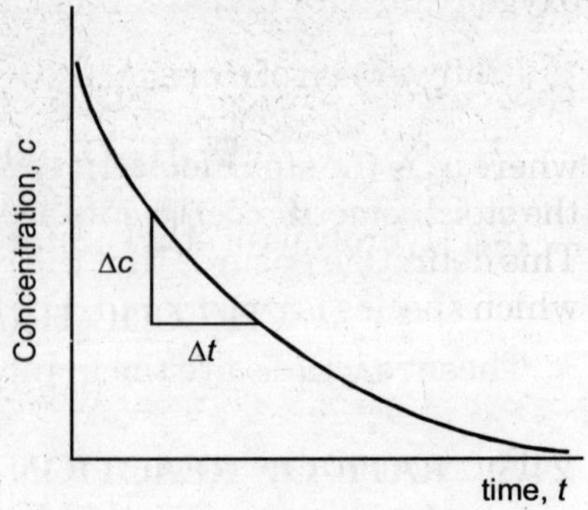

Fig. 21.2

This is known as **instantaneous rate** of reaction. Hence instantaneous rate of reaction is the reaction rate at a particular instant. It is equal to the slope of the line tangent to the concentration versus time graph.

Hence, instantaneous rate may be represented by $-\frac{d[A]}{dt}$ or $\frac{d[B]}{dt}$ instead of $\frac{\Delta[A]}{\Delta t}$ or $\frac{\Delta[B]}{\Delta t}$ respectively. Here $d[A]$ and $d[B]$ represent infinitesimally small changes in concentration of A and B in infinitesimally small interval of time, dt.

In general, if dx represents very small, i.e. infinitesimally small change in concentration of any species during the very small. i.e. infinitesimally small, change of time dt, the rate of reaction may be expressed as:

$$\text{Rate of reaction} = -\frac{dx}{dt}$$

(in terms of reactants; – sign indicates decrease in concentration)

or

$$+\frac{dx}{dt}$$

(in terms of products; + sign indicates increase in concentration)

21.3 RELATION TO STOICHIOMETRIC EQUATIONS

Consider the reaction whose *stoichiometric equation* is

$$O_2 + 2H_2 \rightarrow 2H_2O$$

The stoichiometric equation shows how the number of moles of reactants and products are related. It must be balanced. This equation says that to form two moles of water, one mole of oxygen and two moles of hydrogen must react. It follows that the rate of consumption of hydrogen is twice that of oxygen.

$$\frac{d[H_2]}{dt} = 2\frac{d[O_2]}{dt}$$

Or the rate of formation of water is twice the rate of loss of oxygen, as two moles of water are formed from one mole of oxygen. As water is the product, the rate of change of its concentration is positive, i.e. the concentration is increasing with time; however, the rates of change of the concentrations of both hydrogen and oxygen are negative as these concentrations are decreasing with time. In derivative form:

$$\frac{d[H_2O]}{dt} = -2\frac{d[O_2]}{dt} \quad \text{or equivalently} \quad -\frac{1}{2}\frac{d[H_2O]}{dt} = \frac{d[O_2]}{dt} \qquad ...(1)$$

In order to get unique value of the reaction rate, it becomes necessary to divide each concentration change by its coefficient in the balanced equation of the reaction. These coefficients are the numbers in front of the species in the balanced chemical equation. So the stoichiometric coefficient is 1 for oxygen and 2 for both hydrogen and water. The rate of reaction, r, is defined for any species A as

$$r = \frac{1}{\upsilon_A}\frac{d[A]}{dt} \qquad ...(2)$$

where υ_A is the stoichiometric coefficient of species A in the balanced chemical equation. In addition the stoichiometric coefficients are given negative signs for reactants and positive signs for products. This definition ensures that the rate is always *positive* and the *same* for a given reaction no matter which species is considered. For the above reaction rate is given by:

$$\text{Rate} = -\frac{1}{2}\frac{d[H_2O]}{dt} = \frac{d[O_2]}{dt} \qquad ...(3)$$

For a general reaction,

$$aA + bB \rightarrow cC + dD$$

the reaction rate can be written in a number of different but equivalent ways,

$$\text{Rate} = -\frac{1}{a}\frac{d[A]}{dt} = \frac{1}{b}\frac{d[B]}{dt} = \frac{1}{c}\frac{d[C]}{dt} = \frac{1}{d}\frac{d[D]}{dt} \qquad ...(4)$$

As in our previous example, the negative signs account for material that is being consumed in the reaction and the positive signs account for material that is being formed in the reaction. The stoichiometry is preserved by dividing the rate of change of concentration of each substance by its stoichiometric coefficient.

Example. The rate of the reaction,

$2NO(g) + O_2(g) \rightarrow 2NO_2(g)$, is given by

$$\text{Rate} = \frac{1}{2}\frac{d[NO_2]}{dt} = -\frac{d[O_2]}{dt} = -\frac{1}{2}\frac{d[NO]}{dt}$$

21.4 RATE LAWS (RATE EQUATION)

The *rate law* is an expression indicating how the *rate* depends on the concentrations of the reactants and catalysts (if present).

A rate law of a reaction is a *mathematical expression* relating the rate of a reaction to the concentration of either reactants or products.

Often reactions are found to have rate laws of the form

$$r = k[A]^{\alpha}[B]^{\beta} ...$$

where k is the rate constant, characteristic of a particular reaction.

The rate law of a reaction must be proven *experimentally* by looking at either the appearance of products or disappearance of reactants.

Consider the following overall chemical reaction:

$$2NO(g) + O_2(g) \rightleftharpoons 2NO_2(g)$$

Here is an example of a rate law for the above reaction:

$$\text{Rate} = k[NO]^2[O_2]$$

In this rate law, the rate of disappearance of NO and O_2 is proportional to the concentrations of NO and O_2. Because this equation is an *experimental* law that describes the rate of the reaction, it is called the **rate law** for the reaction. The proportionality constant, *k*, is known as the **rate constant.**

21.5 ORDER AND MOLECULARITY

Law of Mass Action: According to the law of mass action, the rate of a chemical reaction is proportional to the product of the concentrations of the reactants.

For the reaction,

$$aA + bB \longrightarrow \text{Products}$$

$$\text{Rate} \propto [A]^a [B]^b$$

Low of mass Action is simply based upon the stoichiometry of the equation. On the other hand 'Rate Low' is based upon the experimental facts. Order of reaction is determined by Rate Law.

Order of **the reaction is defined as the sum of the powers of the concentration of the reactants in the rate law expression.** Or

Order of the reaction is the total number of reacting species whose concentration determine the rate of the reaction at a given temperature. The *power of the concentration* of a particular reactant in the rate law expression is called the order with respect to that reactant

For any general reaction

$$aA + bB \rightarrow cC + dD$$

If the rate law expression is;

$$\text{Rate} = r[A]^m [B]^n$$

where A and B represent the *molar concentrations* of A and B, m and n are the powers to which the concentrations must be raised, k is a constant of proportionality known as the rate constant. The values of 'm' and 'n' are *not* the stoichiometric numbers obtained from the balanced equation. These are obtained from the *experimental data.*

m is the *order* with respect to A,

n is the order with respect to B.

The overall order of the reaction will be given by *the sum of the powers of the concentration terms of the reactants in the rate law.*

The overall order of the reaction is $(m + n)$.

The rate law for a reaction **cannot** be predicted from the stoichiometry of the reaction; it must be determined **experimentally**. However, sometimes, the rate law is consistent with what we expect from the stoichiometry of the reaction.

$$2HI(g) \rightleftharpoons H_2(g) + I_2(g)$$

$$\text{Rate} = k[HI]^2$$

Often, however, it is not same as the stoichiometry of the reaction. Example:

$$2N_2O_5(g) \longrightarrow NO_2(g) + O_2(g)$$

$$\text{Rate} = k[N_2O_5]$$

Simple Rate Laws

A first order rate law is one in which the rate is proportional to the concentration raised to the power 1 (hence "first")

$$\text{rate} = k[A]^1$$

Here k is called a *first order rate constant*. The rate equation and the reaction it describes, is said to be *first order in A* or it is said that *the order with respect to A is one.*

Units of first order rate constant:

As the units of rate are concentration time^{-1} and the units of $[A]$ are concentration, the units of k are found as:

$$r = \frac{dx}{dt} = k\,[\text{conc.}]^n$$

$$k = \frac{dx}{dt} \times \frac{1}{[\text{conc.}]^n}$$

For first order,

$$k = \frac{r}{[\text{conc.}]^1} = \frac{dx/dt}{\text{conc.}} = \frac{\text{conc.}}{\text{Time}} \times \frac{1}{\text{conc}} = \text{Time}^{-1}$$

Hence first order rate constants have units of time^{-1}.

A second order rate law is the one in which the rate is proportional to the concentration raised to the power of 2.

$$r = k[A]^2$$

Here k is called a *second order rate constant.* The rate equation, and the reaction it describes, is said to be *second order in A,* or *the order with respect to A is two.*

Unit of second order rate constant:

The units of k are found to be conc^{-1} time^{-1}

$$k = \frac{dx}{dt} \times \frac{1}{[\text{conc}]^n}$$

For, second order,

$$k = \frac{dx}{dt} \times \frac{1}{[\text{conc}]^2} = \frac{\text{conc.}}{\text{Time}} \times \frac{1}{\text{conc.}^2} = \text{conc}^{-1}\ \text{time}^{-1}$$

Examples:

The decomposition of N_2O_5 is a **first-order reaction** because the rate of reaction depends on the concentration of N_2O_5 raised to the first power.

$$\text{Rate} = k[N_2O_5]$$

The decomposition of HI is a **second-order reaction** because the rate of reaction depends on the concentration of HI raised to the second power.

$$\text{Rate} = k[\text{HI}]^2$$

When the rate of a reaction depends on more that one reagent, we classify the reaction in terms of the order of each reagent.

For a zero-order reaction, the rate of reaction is a constant and does not depend on concentrations of the reactants. When the limiting reactant is completely consumed, the reaction abruptly stops.

$$\text{Rate} = k$$

Unit of zero order rate constant:

The rate constant, k, has units of conc. time^{-1}

$$k = \frac{dx}{dt} \times \frac{1}{[\text{conc.}]^0} = \frac{\text{conc}}{\text{Time}} = \text{conc. time}^{-1}$$

Example:

$$2\text{HI} \xrightarrow{\text{Au}} H_2 + I_2$$

$$\text{Rate} = k$$

$$H_{2(g)} + Cl_{2(g)} \xrightarrow{h\nu} 2HCl_{(g)}$$

$$\text{Rate} = k$$

Order	*Unit of k*
Zero	mol $litre^{-1}$ $time^{-1}$
First	$time^{-1}$
Second	litre mol^{-1} $time^{-1}$

21.6 THE MOLECULARITY OF A REACTION

Molecularity is defined **as the number of atoms or molecules taking part in an elementary step leading to a chemical reaction.**

The overall chemical reaction may consist of many elementary steps. Each elementary reaction has its own molecularity which is equal to number of atoms or molecules. When a single molecule is consumed, the step is called **unimolecular**. When two molecules are consumed, it is **bimolecular**. When three molecules are colliding simultaneously then it is **ter molecular** reaction, although such reactions are very rare.

Elementary reactions are classified according to their molecularity

- In a **unimolecular Reaction**, a single molecule is involved in the reaction. Example, decomposition of NH_4NO_2 to N_2 and $2H_2O$

$$NH_4NO_2 \longrightarrow N_2 + 2H_2O$$

- **Bimolecular Reactions** involve the collision of pair of molecules. Bimolecular reactions are common. Example, oxidation of nitric oxide by ozone.

$$O_{3(g)} + NO_{(g)} \longrightarrow NO_{2(g)} + O_{2(g)}$$

In **Ter Molecular Reactions**, three molecules collide simultaneously. Molecularity rarely exceeds three. It is so because the occurrence of such reactions will involve the collision of four or more molecules simultaneously. However, chances of collisions of more than three molecules simultaneously, each having energy equal to or greater than the threshold energy, are rare.

$$2NO_{(g)} + O_{2(g)} \longrightarrow 2NO_{2(g)}$$

Thus, possible molecularities can be represented as:

Molecularity	**Elementary step**	**Rate law**
Unimolecular	$A \rightarrow$ product	$k[A]$
Bimolecular	$A + B \rightarrow$ product	$k[A]\,[B]$
	$A + A \rightarrow$ product	$k[A]^2$
Termolecular	$A + B + C \rightarrow$ product	$k[A]\,[B]\,[C]$
	$A + A + A \rightarrow$ product	$k[A]^3$
	$A + A + B \rightarrow$ product	$k[A]^2\,[B]$

Difference between Molecularity and Order

It is most important to distinguish **molecularity** from **reaction order:**

1. The **reaction order** (first order, etc.) is an empirical quantity, and obtained from the experimental rate law.
2. The **molecularity** (unimolecular/bimolecular) refers to an elementary reaction proposed as an individual step in a mechanism. In case of complex reactions, the overall molecularity has no significance because each step in the reaction has its own molecularity.

Molecularity	*Order*
It is the number or molecules of reactants taking part in elementary step of a reaction.	It is the sum of the powers of the concentration terms in the rate law expression.
It is theoretical concept	It is experimentally determined
It has integral values only	It can have fractional values also
It cannot be zero fractional or negative	It can be zero, fractional or negative
It provides no information on reaction mechanism.	The slowest step in the reaction can be judged by the order of the reaction and this gives further information about the mechanism
It is assigned for each elementary step of reaction.	It is assigned for overall reaction.

Let us illustrate the difference between the two with the following example:
Decomposition of nitrogen pentoxide can be represented as:

$$2N_2O_5 \longrightarrow 4NO_2 + O_2$$

From the stoichiometry it seems to be a **second order reaction**, but experimental verification reveals it to be **of first order**. There is however no correlation between the order and molecularity of a reaction, or between the stoichiometric representation and molecularity. The mechanism of the process consists of the following sequence of steps:

$N_2O_5 \rightarrow NO_2 \rightarrow NO_3$ (1) Unimolecular
$NO_2 + NO_3 \rightarrow NO_2 + O_2 + NO$ (2) Bimolecular
$NO + N_2O_5 \rightarrow 3NO_2$ (3) Bimolecular
$NO_2 + NO_3 \rightarrow N_2O_5$ (4) Bimolecular

Step (1) is unimolecular and the other three steps are bimolecular. The slowest step in this reaction is the decomposition of N_2O_5 to NO_2 and NO_3 and is the rate determining step and hence *first order* as only one molecule is involved.

Specific Rate Constant

In the rate equation k represents the rate constant. Rate constant or velocity constant (or) specific reaction rate is defined as the rate of the reaction when the concentration of each of the reactants is unity in the reaction.

When concentration of A and B is unity then, the rate constant is equal to the rate of the reaction.

The generalised rate equation for nth order reaction may be expressed as:

$$\text{Rate, } r = -\frac{dc}{dt} = k[c]^n \qquad ...(1)$$

When the concentration becomes unity, the specific reaction rate and the rate constant becomes equivalent.

$$r = k[1]^a = k = -\frac{dc}{dt}$$

Following are the important differences between rate and rate constant of a reaction.

Rate of Reaction	*Rate Constant of Reaction*
1. It represents the speed at which the reactants are converted into product at any instant.	1. It is the constant of proportionality in the rate law expression.
2. At any instant of time, the rate depends upon the concentration of reactants at that instant.	2. It refers to the rate of a reaction at the specific point when concentration of every reaction species is unity.
3. It decreases as the reaction proceeds.	3. It is constant and does not depend on the progress of the reaction.
4. Rate of rate determining step determines overall rate value.	4. It is an experimental value. It does not depend on the rate determining step.

21.7 PSEUDO-FIRST ORDER REACTION

The reactions which are not truly first order but which show first order kinetics under special conditions are called pseudo-first order reactions. In these reactions, one of the reactants is present in large excess. Since in elementary reactions, molecularity and order are identical, pseudo order reaction are also called pseudo-molecular reaction.

Let us consider a general reaction,

$$A + B \longrightarrow \text{Products}$$

Suppose the rate law for this reaction is,

$$\text{Rate} = k[A]\,[B] \qquad ...(1)$$

Now, if $[B]$ is present in large excess, its concentration remains practically constant in the course of reaction and hence does not affect the rate. Therefore rate equation can be written in the form,

$$\text{Rate} = k'[A] \qquad ...(2)$$

where $k' = k[B]$

Thus the actual order of the reaction is second order according to (1) but in practice it is first order (2). Thus the true law has been **forced into first-order form** by assuming the concentration of one the reactant B, constant, the effective rate law is called pseudo first order.

Many reactions in aqueous solutions that are reported as first or second order are actually pseudo first or second order. The water (solvent) participates in the reaction, but it is in such large excess that its concentration remains constant in the reaction.

Examples of Pseudo-Order Reactions

(1) Hydrolysis of an Ester. For example, ethyl acetate upon hydrolysis in aqueous solution using a mineral acid as catalyst forms acetic acid and ethyl alcohol.

$$\underset{\text{ethyl acetate}}{CH_3\,COOC_2H_5} + \underset{\text{(excess)}}{H_2O} \longrightarrow \underset{\text{acetic acid}}{CH_3COOH} + \underset{\text{ethyl alcohol}}{C_2H_5OH}$$

Here a large excess of water is used and any change in its concentration would be too small to be noticed, the rate law will be written as:

$$\text{Rate} = k[CH_3COOC_2H_5]\,[H_2O] = k'[CH_3COOC_2H_5]$$

The reaction is actually second-order but in practice it is found to be first order. Thus it is a pseudo-first-order reaction, the reaction appears to be dependent upon the concentration of ester only.

(2) Hydrolysis of Sucrose. Sucrose upon hydrolysis in the presence of a dilute mineral acid gives glucose and fructose.

$$\underset{\text{sucrose}}{C_{12}H_{22}O_{11}} + \underset{\text{(excess)}}{H_2O} \longrightarrow \underset{\text{glucose}}{C_6H_{12}O_6} + \underset{\text{fructose}}{C_6H_{12}O_6}$$

If a large excess of water is present, $[H_2O]$ is practically constant and the rate law may be written as:

$$\text{rate} = k[C_{12}H_{22}O_{11}]\,[H_2O] = k'[C_{12}H_{22}O_{11}]$$

The reaction though of second-order is experimentally found to be first-order. Thus it is pseudo-first-order reaction.

21.8 INTEGRATED RATE LAWS

The *differential rate law* describes how the rate of reaction varies with the concentrations of various species, usually reactants, in the system. The rate laws discussed above are differential rate laws and can be *integrated* in order to obtain an **integrated rate equation** that links concentrations of reactants or products with time.

- In other words differential rate law describes how the rate varies with change in concentration of reactants.
- Integrated rate law gives the variation of the concentration of a reactant with time as a chemical reaction proceeds.

Let us deduce the integrated rate laws for zero, first, second order reactions.

(1) Zero Orders Rate Law:

Consider the reaction,

$$A \longrightarrow P$$

For the zero order, the rate law takes the form

$$\text{Rate} = \frac{-d[A]}{dt} = k[A]^\circ = k \qquad ...(i)$$

The minus sign indicates the concentration of the reactant is decreasing with time.
Rearranging (*i*)

$$d[A] = -k\,dt \qquad ...(ii)$$

Integrating both sides, we get,

$$[A] = -kt + I \qquad ...(iii)$$

where I is the constant of integration.

The value of I, integration constant is determined from the initial conditions.

$$\text{At time } t = 0,\ [A] = [A]_0$$

where $[A]_0$ represents the initial concentration of A at time = 0. Therefore equation (*iii*) becomes,

$$[A]_0 = -k \times 0 + I$$

$$[A]_0 = I \qquad ...(iv)$$

Substituting, the value of I from (*iv*) in (*iii*).
We get,

$$[A] = -kt + [A]_0 \qquad ...(v)$$

$$kt = [A]_0 - [A]$$

or $$\boxed{k = \frac{1}{t}\{[A]_0 - [A]\}} \qquad ...(vi)$$

This expression is the rate law expression for zero order reaction.

Half Life Period of Zero Order Reaction

Half life period of a reaction is defined as the time it takes for the concentration of the specified reagent to fall to half its initial value. This implies that, when

$$[A] = [A]_{1/2} \text{ at } t = t_{1/2}$$

Substituting, these values in equation (*vi*), we get,

$$t_{1/2} = \frac{1}{k}\{[A]_0 - [A]_0/2\}$$

$$= \frac{[A]_0}{2k}$$

$$\therefore \boxed{t_{1/2} = \frac{[A]_0}{2k}} \text{ for zero order reaction.}$$

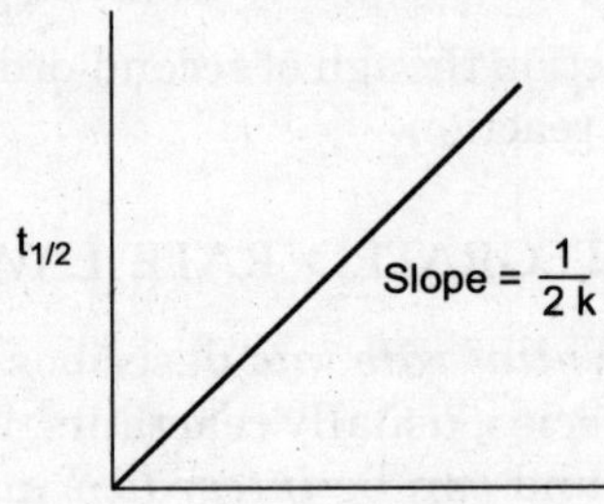

Fig. 21.3 Plot of half life ($t_{1/2}$) versus initial concentration for a zero order reaction

Thus half-life period of a zero order reaction is directly proportional to initial concentration. The plot of $t_{1/2}$ versus $[A]_0$ will be a straight line passing through the origin where slop $= \frac{1}{2k}$

(2) First Order Rate Law:

For a reaction,

$$A \longrightarrow P, \text{ a first order rate law takes the form}$$

$$\text{Rate} = -\frac{d[A]}{dt} = k[A] \qquad ...(i)$$

where $[A]$ represents the concentration of reactant at any time t and k is the rate constant.

Rearranging equation (*i*) we get,

$$\frac{d[A]}{[A]} = -k\,dt \qquad ...(ii)$$

Integrating equation (*ii*) we get,

$$\int \frac{d[A]}{[A]} = -k \int dt$$

$$In[A] = -kt + I \qquad ...(iii)$$

where I is the integration constant. The value of I can be determined as before from initial conditions.

At time $t = 0$, $[A] = [A]_0$, and equation (*iii*) becomes,

$$In[A]_0 = (-k \times 0) + I$$

$$In[A]_0 = I \qquad ...(iv)$$

Substituting, the value of integration constant from (*iv*) into (*iii*) we get,

$$In[A] = -kt + In[A]_0 \qquad ...(v)$$

Rearranging (*v*),

$$In[A] - In[A]_0 = -kt$$

$$In\frac{[A]}{[A]_0} = -kt \quad ...(vi)$$

or
$$In\frac{[A]_0}{[A]} = kt \quad ...(vii)$$

Using logarithm to the base 10, we can write equation (*vii*) as

$$2.303 \log\frac{[A]_0}{[A]} = kt \quad ...(viii)$$

or
$$\boxed{k = \frac{2.303}{t}\log\frac{[A]_0}{[A]}} \quad ...(ix)$$

Equation (*ix*) is the first order rate equation. This equation is sometimes written in other form which is obtained as follows:

Suppose initial concentration of A is represented as 'a' moles per litre at time $t = 0$. After time t, suppose x moles per litre of it is decomposed. Therefore the concentration of A after time $t = (a - x)$ moles per litre.

So,
$$[A]_0 = a \text{ moles/L}$$
$$[A] = a - x \text{ moles/L}$$

Therefore, the equation (*ix*) becomes,

$$\boxed{k = \frac{2.303}{t}\log\frac{a}{a-x}} \quad ...(x)$$

Equation (*vi*) can also be written as exponential form as:

$$In\frac{[A]}{[A]_0} = -kt \quad ...(vi)$$

$$\boxed{[A] = [A]_0\, e^{-kt}} \quad ...(xi)$$

Equation (*xi*) is the exponential form of the first order rate equation.

Characteristics of First Order Rate Equation

(*i*) Any reaction of first order must obey equations (*ix*), (*x*) and (*xi*).

(*ii*) Determination of rate constant of first order reaction:

From equation (*ix*),

$$\frac{k}{2.303}t = \log\frac{[A]_0}{[A]} = \log[A]_0 - \log[A]$$

or
$$\log[A] = \frac{-k}{2.303}t + \log[A]_0 \quad ...(xii)$$

Equation (*xii*) is the equation of straight line ($y = mx + c$). Thus, if log $[A]$ is plotted against time t, the graph obtained should be a straight line. The rate constant k can be calculated from the slope of the graph.

(*iii*) Half life period of a first order reaction:

Equation (*x*) can be written as:

$$t = \frac{2.303}{k}\log\frac{a}{a-x} \quad ...(xiii)$$

When half of the reaction is completed

$$x = \frac{a}{2} \text{ at } t_{1/2}.$$

Hence equation (*xiii*) becomes,

$$t_{1/2} = \frac{2.303}{k} \log \frac{a}{a - a/2}$$

$$= \frac{2.303}{k} \log 2 = \frac{0.693}{k}$$

$$\therefore\ t_{1/2} = \frac{0.693}{k} \qquad ...(xiv)$$

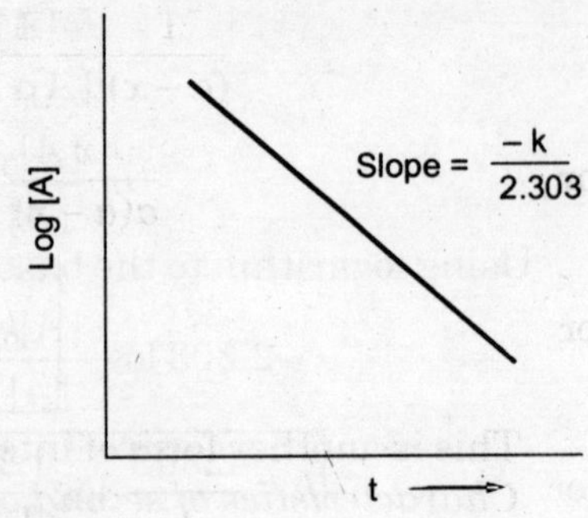

Fig. 21.4 Plot of log [A] vs t

Thus for *the first order reaction the half life period is independent of the initial concentration of the reactant.*

(3) Second Order Rate Law

Second order reactions may be represented as:

Type 1: When there is only one reactant involved

$$2A \longrightarrow \text{Products}$$

$$\frac{dx}{dt} = k[A]^2 \qquad \text{order} = 2$$

Type 2:

$$A + B \longrightarrow \text{Products}$$

$$\frac{dx}{dt} = k[A][B] \qquad \text{order} = 2$$

Now let us discuss the second order rate equation for these two types.

Type 1:

$$2\text{A} \longrightarrow \text{Products}$$

$$\frac{-d[A]}{dt} = k[A]^2 \qquad ...(i)$$

or

$$\frac{-d[A]}{[A]^2} = k\,dt \qquad ...(ii)$$

On integration,

$$\int \frac{-d[A]}{[A]^2} = \int k\,dt \quad \text{or} \quad \frac{1}{[A]} = kt + I \qquad ...(iii)$$

where I is the integration constant. I can be evaluated by putting $[A] = [A]_0$ at $t = 0$ in equation (*ii*)

$$\frac{1}{[A]_0} = k \times 0 + I \qquad ...(iv)$$

$$I = \frac{1}{[A]_0} \qquad ...(v)$$

substituting the value of I from equation (*v*) in equation (*iii*) we get,

$$\frac{1}{[A]} = kt + \frac{1}{[A]_0}$$

or

$$\boxed{\frac{1}{[A]} - \frac{1}{[A]_0} = kt} \qquad ...(vi)$$

Let if initial concentration of the reactant $[A]_0 = a$ at $t = 0$. After time t, let x moles of A have reacted. Then the concentration of A after time t, $[A] = (a - x)$. Putting these in equation (*vi*) we have,

$$\frac{1}{(a-x)} - \frac{1}{a} = kt \qquad ...(vii)$$

or
$$\frac{x}{a(a-x)} = kt$$

or
$$\boxed{k = \frac{1}{t}\cdot\frac{x}{a(a-x)}} \qquad ...(viii)$$

This is another form of integrated rate law for second order.

Characteristics of second order rate law when only one reactant is involved.

(i) Half Life Period

In equation (viii) if

$$x = \frac{1}{2} \text{ at } t = t_{1/2}\text{, then,}$$

$$k = \frac{1}{t_{1/2}}\cdot\frac{a/2}{a(a-a/2)} = \pm\frac{1}{t_{1/2}\cdot a}$$

or
$$\boxed{t_{1/2} = \frac{1}{k_a}} \qquad ...(ix)$$

From equation (*ix*) it is evident that half life period of second order reaction is inversely proportional to the concentration of the reactant.

(ii) Determination of rate constant of second order reaction.

From equation (*viii*), the rate constant of second order is given by,

$$k = \frac{x}{at(a-x)}$$

This equation can be written as,

$$\frac{1}{a-x} = kt + \frac{1}{a} \qquad ...(x)$$

$$y = mx + b$$

This is the equation of a straight line $y = mx + b$, where $y = \frac{1}{a-x}$, hence a plot of $\frac{1}{a-x}$ against time t will be a straight line (Fig. 21.5). The rate constant k = slope.

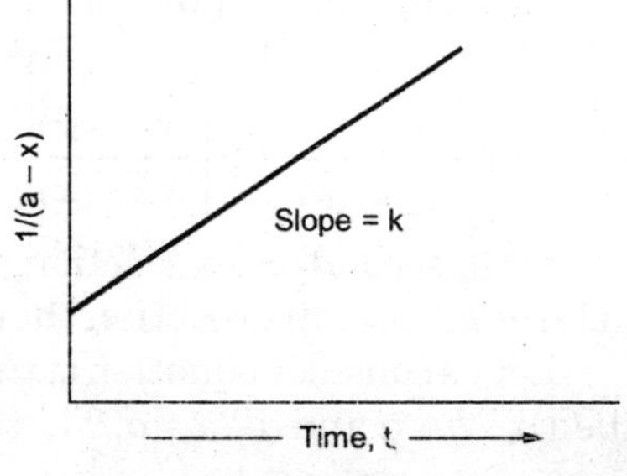

Fig. 21.5 A plot of $1/(a-x)$ time t for second order reaction.

Type 2: $A + B \longrightarrow \text{Products}$

$$\frac{dx}{dt} = \frac{-d[A]}{dt} = \frac{-d[B]}{dt}$$

$$= k[A]\,[B] \qquad ...(xi)$$

Let the initial concentration of A and B be 'a' and 'b' moles respectively. After time 't' let 'x' moles of each been converted into products. The remaining concentration of A and B after time t will be $(a-x)$ and $(b-x)$ respectively

$$\therefore \quad [A] = a - x \quad \text{after time } t$$
$$[B] = b - x \quad \text{after time } t$$

Putting these in (*xi*) we get,

$$\frac{dx}{dt} = k(a-x)(b-x) \qquad ...(xii)$$

Separating the variables,

$$\frac{dx}{(a-x)(b-x)} = k\,dt \qquad ...(xiii)$$

On integration,

$$\int_0^x \frac{dx}{(a-x)(b-x)} = \int_0^t k\,dt \qquad ...(xiv)$$

$$= \frac{-1}{a-b}\int_0^x \left[\frac{1}{(a-x)} - \frac{1}{(b-x)}\right] dx = dt \qquad ...(xv)$$

$$= \frac{1}{(a-b)}[\text{In}(a-x) - \text{In}(b-x)]_0^x = k \cdot t \qquad ...(xvi)$$

$$= \frac{1}{(a-b)}\left[\text{In}\left[\frac{(a-x)}{a} - \text{In}\frac{(b-x)}{b}\right]\right] = k \cdot t \qquad ...(xvii)$$

$$= \frac{1}{(a-b)}\text{In}\frac{b\,(a-x)}{a\,(b-x)} = k \cdot t \qquad ...(xviii)$$

or

$$k = \frac{1}{t(a-b)}\text{In}\frac{b(a-x)}{b(b-x)} \qquad ...(xix)$$

Changing the logarithm to the base 10, we get,

$$\boxed{k = \frac{2.303}{t(a-b)}\log\frac{b(a-x)}{a(b-x)}} \qquad ...(xx)$$

Equation (*xvi*) is the rate order equation for the second order reaction, when there are two different reactants involved.

Characteristics:

(*i*) The above equation is the equation of a straight line

$$\frac{1}{(a-b)}\text{In}\left[\frac{b(a-x)}{a(b-x)}\right] = k \cdot t \qquad ...(xxi)$$
$$y = mx$$

where $y = \frac{1}{(a-b)}\text{In}\left[\frac{b(a-x)}{a(b-x)}\right]$ and $x = t$, with slope $= k$)

(ii) In second order reaction, if one of the reactant is present in large excess, the reaction then becomes first order reaction.

Let us consider equation (*xxi*) once again and assume $a >> b$, then $a - b \approx a$ and $a - x \approx a$. Therefore equation (*xxi*) reduces to

$$k = \frac{2.303}{(a-b)t}\log\frac{b(a-x)}{a(b-x)}$$

$$k = \frac{2.303}{at}\log\frac{b}{(b-x)} \qquad ...(xxii)$$

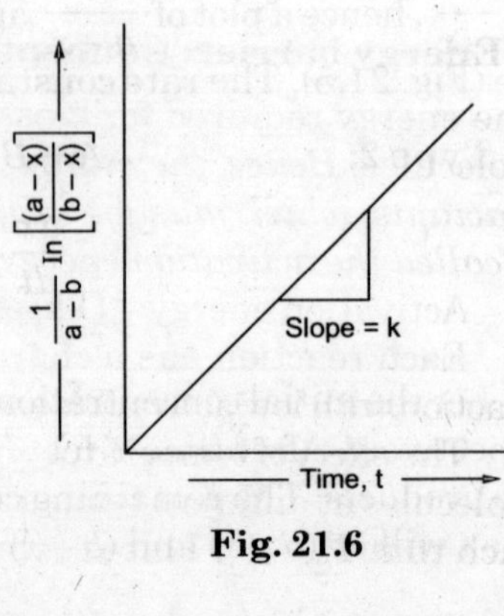

Fig. 21.6

where b = Initial concentration of reactant B

$b - x$ = Concentration of B after time t

a = The concentration of A, present in excess and hence constant.

$\therefore a \times k = k'$ (another constant)

Equation (xvii) becomes

$$k' = \frac{2.303}{t} \log \frac{b}{b-x} \qquad ...(xxiii)$$

The above equation is equivalent to that of first order rate equation.

A summary of the relationships for first and second order reactions is given below:

Order	*Differential rate law*	*Integrated rate law*	*Linear plot*	*Half life*	*Units*
Zero	$\frac{-d[A]}{dt} = k[A]^0$	$[A] = -kt + [A]_0$	$[A]$ vs t	$\frac{[A]_0}{2k}$	mol $L^{-1}s^{-1}$
First order	$\frac{-d[A]}{dt} = k[A]$	$[A] = [A]^0 e^{-kt}$	$\log [A]$ vs t slope = $\frac{-k}{2.303}$	$\frac{0.693}{k}$	s^{-1}
Second	$\frac{-d[A]}{dt} = k[A]^2$	$\frac{1}{[A]} - \frac{1}{[A]_0} = kt$	$\frac{1}{[A]}$ vs t slope = k	$\frac{1}{k[A]}$	L $mol^{-1}s^{-1}$

21.9 THEORIES OF REACTION RATES

(I) Collision Theory

According to this theory, for a reaction to occur, reactant molecules must collide with each other. But not sufficient in condition. The reaction can only takes place when these collisions are *"effective collisions"*. In order for a molecular collision to be effective it must meet two conditions:

- Energy barrier: The reactant molecules must collide with energy greater than some minimum value. This minimum value energy of collision required for two molecules to react is called threshold energy. Collisions with energy greater than this can result in the reaction. This implies that there is an energy barrier between reactants and products which must be crossed before products can be formed.
- Orientation barrier: The collision must occur with proper orientation

1. Energy barrier: Concept of Activation Energy:

The energy required for crossing the energy barrier comes from the kinetic energy of reacting molecules. *Hence, the minimum extra energy over and above the average potential energy of the reactants, which must be supplied so that the reactants are able to cross over the energy barrier is called the activation energy (E_a).*

Activation energy = (Threshold energy) – (Average energy of the reactants)

Each reaction has a characteristic value of energy of activation. At a given temperature, a reaction with low activation energy will proceed faster than a reaction with high energy of activation.

The effect of temperature on reaction rate can be explained from the energy distribution among molecules at different temperatures. Consider a reaction at two different temperatures T_1 and T_2, such that $T_2 > T_1$. This is represented in Fig. 21.8.

It is observed that the energy distribution curve is flatter at higher T and is shifted toward higher energy region. This is because, the kinetic energy of the molecules are higher at higher temperatures. Also, the fraction of molecules with energy treater than the threshold as indicated by the shaded portion is almost *double. It is very clear, that the rate of reaction increases with temperature, mainly due to increase in the number of collisions which occur with sufficient energy, i.e. with energy greater than the threshold.*

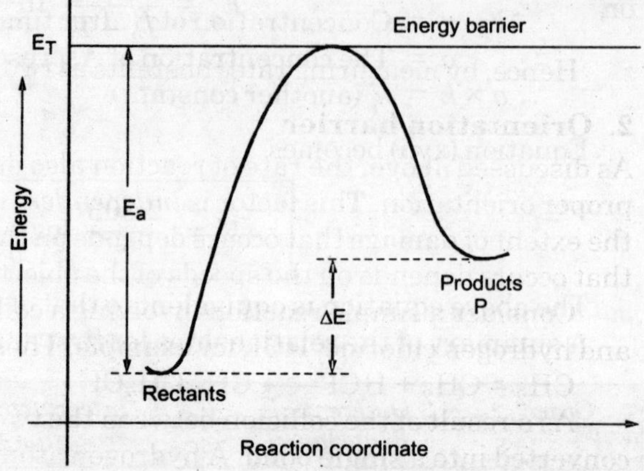

Fig 21.7 Energy profile of a chemical reaction.
E_a = Activation Energy
E_T = Threshold energy.

Quantitatively this relationship between the rate of a reaction proceeds and its temperature is determined by the *Arrhenius Equation*. At higher temperatures, the probability that two molecules will collide is higher. This higher collision rate results in a higher kinetic energy, which has an effect on the activation energy of the reaction.

Arrhenius equation is given by:

$$k = Ae^{-E_a/RT}$$

where k is the rate coefficient, A is a constant (experimentally determined), E_a is the activation energy, R is the universal gas constant, and T is the temperature (in kelvin).
R has the value of 8.314×10^{-3} kJ mol^{-1} K^{-1}

In the logarithm form

$$\text{In } k = \text{In } A^{-E_a/RT}$$

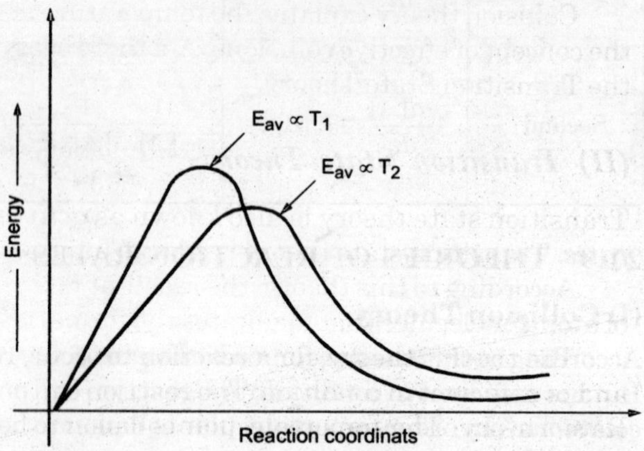

Fig. 21.8 Energy profile of a chemical reaction at two different temperatures

If we plot In k against $\frac{1}{T}$, a straight line having a slope equal to $-E_a/R$ and an intercept In A can be obtained. From the slope and intercept, activation energy and frequency factor can be obtained.

E_a can be calculated by measuring the rate constant of a reaction at two different temperature. If k_1 and k_2 are the rate constants at temperature T_1 and T_2 respectively, then from Arrhenius equation, we have:

$$\text{In } k_1 = \text{In } A - \frac{E_a}{RT_1} \quad ...(i)$$

$$\text{In } k_2 = \text{In } A - \frac{E_a}{RT_2} \quad ....(ii)$$

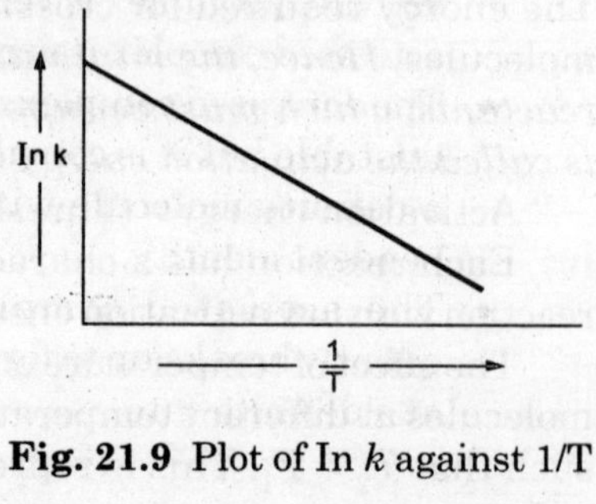

Fig. 21.9 Plot of In k against 1/T

Subtracting eq. (*i*) from eq. (*ii*), we have

$$\text{In } k_2 - \text{In } k_1 = \frac{E_a}{R}\left[\frac{1}{T_1} - \frac{1}{T_2}\right]$$

or,
$$\text{In } \frac{k_2}{k_1} = \frac{E_a}{R}\left[\frac{T_2 - T_1}{T_1 T_2}\right]$$

or, $$E_a = \frac{RT_1T_2}{T_2 - T_1} \ln \frac{k_2}{k_1}$$

Hence, by measuring rate constants at two different temperatures, E_a can easily be found out.

2. Orientation barrier

As discussed above, the rate of reaction also depends on the fraction of collisions that occur with proper orientation. This factor is *independent* of temperature. When two speeding objects collide, the extent of damage that occurs depends on the speeds of the project collide, the extent of damage that occurs depends on the speeds of the object and the position of impact.

Consider a simple reaction involving a collision between two molecules - ethene, $CH_2 = CH_2$, and hydrogen chloride, HCI, for example, These react to give chloroethane.

$CH_2 = CH_2 + HCI \longrightarrow CH_3\,CH_2CI$

As a result of the collision between the two molecules, the double between the two carbons is converted into a single bond. A hydrogen atom gets attached to one of the carbons and a chlorine atom to the other.

The reaction can only happen if the hydrogen end of the H–CI bond approaches the carbon-carbon double bond. Any other collision between the two molecules doesn't work. The two simply bounce off each other.

Collision theory explains the temperature dependence of rate of a reaction based specifically in the concept of effective collisions. Another theory that was proposed a few years later by Eyring, is the Transition State Theory.

(II) Transition State Theory

Transition state theory is also known as *activated-complex theory or theory of absolute reaction rates.* Transition state theory is a large improvement on simple collision theory,

According to this theory, the reactant molecules must come together to form an arrangement of atoms/molecules that facilitates the formation of products.

The theory was developed by Eyring and is also known as Absolute Reaction Rate Theory. The important points can be summarized as:

- Reactant molecules approach each other to form an activated complex by rearrangement of atoms and redistribution of energy.
- The activated complex is in equilibrium with reactants.
- The potential energy of the activated complex is very high and activated complex is very unstable.
- The activated complex although is unstable but it is considered to be a definite molecule with an independent identity.
- The activated complex subsequently breaks up to form products at a definite rate.

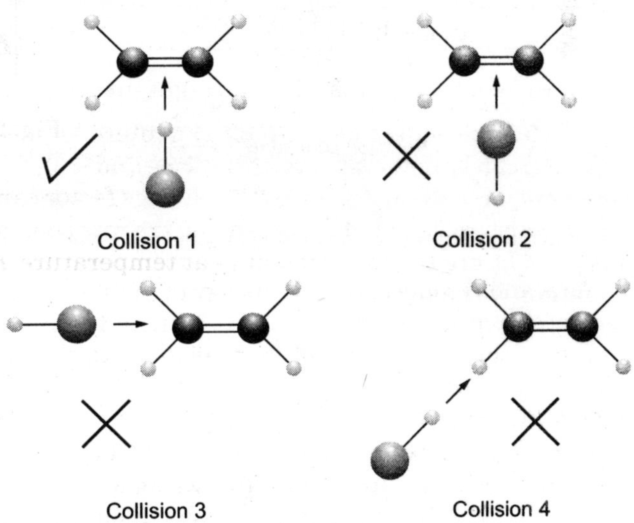

Fig. 21.10 Proper and improper orientations of colliding molecules

Schematically,

$$\text{Reactants} \rightleftharpoons \text{Activated Complex} \rightleftharpoons \text{Products}$$

A simple way of visualising the concept of the transition state is to think of a reaction proceeding along a reaction coordinate. For e.g., in a reaction where two reactants A and B react to form product

$$A + B \longrightarrow \text{Products,}$$

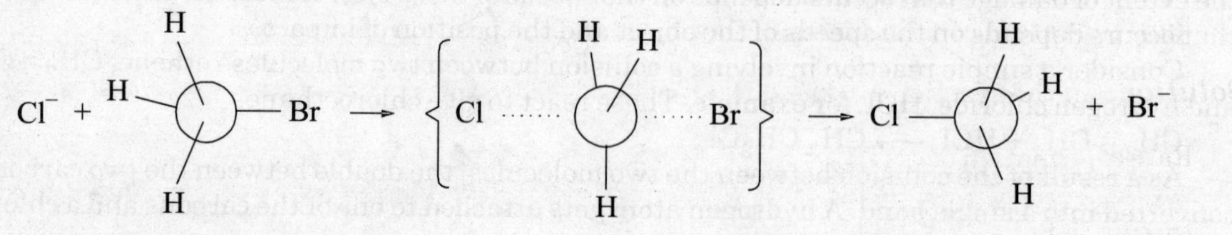

Reactants Transition State Product

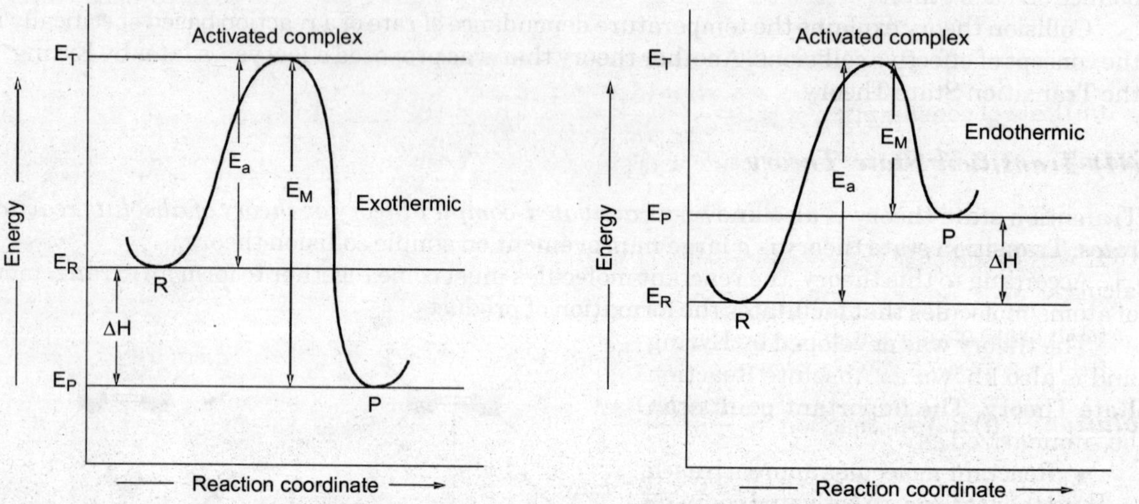

Fig. 21.11 Energy diagram for exothermic reaction (b) Energy diagram for endothermic reaction

As A and B come close, the energy of the system goes up and reaches a maxima. The arrangement of the atoms or molecules corresponding to this energy maxima is the transition state. Therefore, the energy level of the *transition state* is higher than that of the reactants. Therefore, the energy of the system goes down as soon as the transition state breaks and transforms into products. (Fig. 21.11)

Consider, for example, the reaction of CH_3 Br with CI in water at 298K.

SOLVED NUMERICALS

(I) Based on Rate of Reaction

Example 21.1. In the reaction $N_2(g) + 3H_2(g) \longrightarrow 2NH_3(g)$, the rate of formation of NH_3 is 5.0×10^{-7} mol^{-1} L^{-1} s^{-1}. Calculate

(*i*) the general rate of the reaction
(*ii*) the rate of consumption of N_2
(*iii*) the rate of consumption of H_2

Solution: In the reaction $N_2(g) + 3\,H_2(g) \longrightarrow 2\,NH_3(g)$

$$\text{Rate of formation of } NH_3 = \frac{d[NH_3]}{dt} = 5 \times 10^{-7} \text{ mol L}^{-1}\text{ s}^{-1}$$

$$(i)\ \textit{General rate} = \frac{d[N_2]}{dt} = -\frac{1}{3}\frac{d[H_2]}{dt} = \frac{1}{2}\frac{d[NH_3]}{dt}$$

$$(ii)\ \text{Rate of consumption of } N_2 = -\frac{d[N_2]}{dt} = \frac{1}{2}\frac{d[NH_3]}{dt}$$

$$= \frac{1}{2} \times 5 \times 10^{-7} \text{ mol L}^{-1}\text{s}^{-1} = 2.5 \times 10^{-7} \text{ mol L}^{-1}\text{ s}^{-1}$$

$$(iii)\ \text{Rate of consumption of } H_2 = -\frac{d[H_2]}{dt} = \frac{3}{2}\frac{d[NH_3]}{dt} = 7.5 \times 10^{-7} \text{ mol L}^{-1}\text{ s}^{-1}$$

Example 21.2. The following reaction was studied in a closed vessel

$$2\,N_2O_5(g) \rightleftharpoons 4\,NO_2(g) + O_2(g)$$

It was found that the concentration of NO_2 Increases by 2.0×10^{-2} mol L^{-1} in five seconds. Calculate

(*i*) the rate of reaction
(*ii*) the rate of change of concentration of N_2O_5

Solution: (*i*) Rate of reaction $= \dfrac{1}{4}\dfrac{d[NO_2]}{dt}$

$$\text{But} \quad \frac{d[NO_2]}{dt} = \frac{2.0 \times 10^{-2} \text{ mol L}^{-1}}{5s} = 4 \times 10^{-3} \text{ mol L}^{-1}\text{ s}^{-1}$$

$$\Rightarrow \quad \text{Rate of reaction} = \frac{1}{4} \times 4 \times 10^{-3} = 10^{-3} \text{ mol L}^{-1}\text{s}^{-1}$$

(*ii*) Rate of change of concentration of N_2O_5

$$= -\frac{d[N_2O_5]}{dt} = -\frac{1}{2} \times \text{Rate of formation of } NO_2$$

$$= -\frac{1}{2}\frac{d[NO_2]}{dt} = -\frac{1}{2} \times (4 \times 10^{-3} \text{ mol L}^{-1}\text{ s}^{-1})$$

$$= 2 \times 10^{-3} \text{ mol L}^{-1}\text{s}^{-1}$$

Example 21.3. For the decomposition of dinitrogen pentoxide at 200 C,

$$N_2O_3\,(g) \longrightarrow N_2O_4\,(g) + \frac{1}{2}O_2\,(g),$$

if the initial pressure is 114 mm and after 25 minutes of the reaction, total pressure of the reaction mixture is 133 mm, calculate the average rate of reaction in (*a*) atm min^{-1} (*b*) mol L^{-1} s^{-1}.

Solution: (*a*) *If P* is the decrease in pressure of N_2O_3 in 25 minutes, then

$$N_2O_5\,(g) \longrightarrow N_2O_4\,(g) + \frac{1}{2}\,O_2\,(g)$$

Initial pressure	114 mm	0	0
After 25 min	114 – *p*	*p*	$\frac{1}{2}p$

$$\text{Total pressure} = 114 - p + p + \frac{1}{2}\,p = 114 + \frac{1}{2}\,p = 133 \text{ mm (Given)}$$

$$\therefore \quad p = 2 \times (133 - 114) = 38 \text{ mm}$$

i.e., Decrease in pressure in 25 min = 38 mm = $\frac{38}{760}$ atm

$$\therefore \text{ Average rate of reaction} = \frac{38/760\text{atm}}{25 \text{ min}} = 0.002 \text{ atm min}^{-1}$$

(*b*) Applying $pV = nRT$, i.e. $\frac{n}{V} = \frac{P}{RT}$

$$\text{Decrease in molar concentration} = \frac{38/760 \text{ atm}}{0.0821 \text{ L atm K}^{-1} \times 473 \text{ K}} = 0.0012875 \text{ mol L}^{-1}$$

$$\therefore \quad \text{Average rate of reaction} = \frac{0.0012875 \text{ mol L}^{-1}}{25 \times 60\ s} = 8.58 \times 10^{-7} \text{ mol L}^{-1}\text{ s}^{-1}$$

(II) Based on First Order Reactions

Example 21.4. What is the rate of a first order reaction with rate constant $3.0 \times 10^{-10}\ s^{-1}$ when concentration of the reactant is 1.5 M.

Solution: Rate = $-\frac{d[A]}{dt} = k\,[A] = 3.0 \times 10^{-10}\ s^{-1} \times 1.5\ M = 4.5 \times 10^{-10}\ Ms^{-1}$

Example 21.5. The reaction $N_2O_3 \longrightarrow 2NO_2 + \frac{1}{2}O_2$ is of first order in N_2O_5. Its rate constant is $6.2 \times 10^{-6}\ s^{-1}$. If in the beginning $[N_2O_5]$ is 15 mol L^{-1}, calculate the rate of reaction in the beginning.
Solution: Rate of reaction = $k[A] = 6.2 \times 10^{-6}\ s^{-1} \times 15 \text{ mol L}^{-1} = 9.3 \times 10^{-5} \text{ mol L}^{-1}\text{ s}^{-1}$

Example 21.6. The initial concentration of N_2O_5 in the following first order reaction

$$N_2O_5(g) \longrightarrow 2NO_2(g) + \frac{1}{2}\,O_2(g)$$

was 1.24×10^{-2} mol L^{-1} at 45°C. The concentration of N_2O_5 after 60 minutes was 0.20×10^{-2} mol L^{-1}. Calculate the rate constant of the reaction at 45° C.

Solution:
$$\log \frac{[A]_1}{[A]_2} = \frac{k(t_2 - t_1)}{2.303}$$

or
$$k = \frac{2.303}{t_2 - t_1} \log \frac{[A]_1}{[A]_2} = \frac{2.303}{60 \text{ min} - 0 \text{ min}} \log \frac{1.24 \times 10^{-2} \text{ mol L}^{-1}}{0.20 \times 10^{-2} \text{ mol L}^{-1}}$$

$$= \frac{2.303}{60} \log 6.2 \text{ min}^{-1} = \frac{2.303}{60} \times 0.7924 \text{ min}^{-1} = \mathbf{0.0304\ min^{-1}}$$

Example 21.7. The rate law for the decomposition of hydrogen peroxide at 70°C

$$2H_2O_2(aq) \longrightarrow 2H_2O(l) + O_2(g)$$

is first order in H_2O_2. The half-life of the reaction at 70 °C is 20 min. What is the concentration of hydrogen peroxide 60 minutes after the beginning of the reaction in which concentration of hydrogen peroxide is 0.30 mol L^{-1}?

Solution: Rate constant, $k = \dfrac{9.693}{t_{1/2}} = \dfrac{0.693}{20 \text{ min}} = 0.03465 \text{ min}^{-1}$

$$\log \frac{[A]_1}{[A]_2} = \frac{k(t_2 - t_1)}{2.303} = \frac{0.03465 \text{ min}^{-1} \times (60 \text{ min} - 0 \text{ min})}{2.303}$$

$$= \frac{0.03465 \times 60}{2.303} = 0.9027$$

$$\frac{[A]_1}{[A]_2} = \text{Antilog } 0.9027 = 7.993$$

$$[A]_2 = \frac{[A]_1}{7.993} = \frac{0.30 \text{ mol L}^{-1}}{7.993} = 0.0375 \text{ mol L}^{-1}$$

Example 21.8. A solution of H_2O_2 when titrated against $KMnO_4$ solution at different time intervals gave the following results:

t (minutes)	0	10	20
Vol $KMnO_4$ used for 10 ml H_2SO_2	23.8 ml	14.7 ml	9.1 ml

Show that the decomposition of H_2O_2 is a first order reaction.

Solution: The integrated rate equation for first order reaction is:

$$k = \frac{2.303}{t} \log \frac{a}{a - x}$$

Since, volume of $KMnO_4$ used in the titration is a measure of concentration of H_2O_2 in solution,

$$a = 23.8 \text{ ml}$$

$$(a - x) = 14.7 \text{ when } t = 10 \text{ mts}$$

$$(a - x) = 9.1 \text{ when } t = 20 \text{ mts}$$

Substituting, these values in the rate equation above, we have

$$k = \frac{2.303}{10} \log \frac{23.8}{14.7} = 0.2303 (\log 23.8 - \log 14.7)$$

$$= 0.2303(1.3766 - 1.1673) = 0.04794$$

and

$$k = \frac{2.303}{20} \log \frac{23.8}{9.1} = 0.10165 (\log 23.8 - \log 9.1)$$

$$= 0.10165(1.366 - 0.9595) = 0.04806$$

Since, the value of k is constant, it is first order reaction.

Example 21.9. The rate constant of reaction with respect to reactant A is 6 min^{-1}. If we with $[A]$ = 0.8 mol L^{-1}, when would $[A]$ will reach this concentration of 0.08 mol L^{-1}.

Solution:
$$t = \frac{2.303}{k}\log\frac{[A]_o}{[A]_t} = \frac{2.303}{6\ \text{min}^{-1}}\log\frac{0.8\ \text{mol L}^{-1}}{0.08\ \text{mol L}^{-1}}$$
$$= \frac{2.303}{6\ \text{min}^{-1}}\log 10 = \frac{2.303}{6}\times 1\ \text{min} = 0.38\ \text{min}$$

Example 21.10. **The following date was obtained on hydrolysis of methyl acetate at 25 °C in 0.35 N hydrochloric acid. Establish that it is a first order reaction.**

t (secs) ml alkali	0	4500	7140	∞
used	24.36	29.32	31.72	47.15

Solution:

For a first order reaction, $k = \frac{2.303}{t}\log\frac{a}{a-x}$

At any time, the volume of alkali used is needed for the acid present as catalyst and the acid produced by hydrolysis.

The volume of alkali used for total change from t_0 to t_∞ gives the initial concentration of ester. Thus.

$$a = 47.15 - 24.36 = 22.79\ \text{ml}$$
$$(a-x)\ \text{after 4500 sec} = 47.15 - 29.32 = 17.83\ \text{ml}$$
$$(a-x)\ \text{after 7140 sec} = 47.15 - 31.72 = 15.43\ \text{ml}$$

Substituting, these values in the rate equation, we have,

$$k = \frac{2.303}{4500}\log\frac{22.79}{17.83} = 0.000\ 654$$
$$k = \frac{7.303}{7140}\log\frac{22.79}{15.43} = 0.000546$$

The value of '*k*' are fairly constant and hence it is first order reaction

Example 21.11. **The decomposition of hydrogen Peroxide into water and oxygen was studied. The concentration of H_2O_2 was determined by titrating against standard $KMnO_4$ solution and the following data were obtained:**

Time (min)	0	10	20	30	40	50
Conc. of H_2O_2	0.1297	0.1162	0.1045	0.0940	0.0845	0.0760

(*i*) Show that reaction is of first order.
(*ii*) Calculate the rate constant.

Solution: If the reaction is of first order, it must obey the equation.

$$k = \frac{2.303}{t}\log\frac{[A]_0}{[A]}$$

Here $[A]_0 = 0.1297$ and the values of *k* can be calculated by substituting the concentration of *A* at different times.

$$t = 10, [A] = 0.1162, k = \frac{2.303}{10} \log \frac{0.1297}{0.1162} = 0.01099$$

$$t = 20 \text{ min}, [A] = 0.1045, k = \frac{2.303}{20} \log \frac{0.1297}{0.1045} = 0.01080$$

$$t = 30 \text{ min}, [A] = 0.0940, k = \frac{2.303}{30} \log \frac{0.1297}{0.0940} = 0.01073$$

$$t = 40 \text{ min } [A] = 0.08445, k = \frac{2.303}{30} \log \frac{0.1297}{0.0940} = 0.01071$$

$$t = 50 \text{ min}, [A] = 0.0760, k = \frac{2.303}{50} \log \frac{0.1297}{0.0760} = 0.01069$$

Since, the value of k comes out to be almost constant, the reaction, therefore, is of first order. The mean value of $k = 0.01078 \text{ min}^{-1}$.

Example 21.12. A reaction that is of first order with respect to reactant A has a rate constant 6 min^{-1}, If we start with [A] = 5.0 mol L^{-1}, when would [A] reach the value of 0.05 mol L^{-1} ?

Solution: For the first order reaction,

$$t = \frac{2.303}{k} \log \frac{[A]_0}{[A]}$$

$$[A]_0 = 5.0 \text{ mol L}^{-1}, k = 6 \text{ min}^{-1}.$$

$$t = \frac{2.303}{6} \log \frac{5.0}{0.05} = \frac{2.303}{6} \log 100 = 0.768 \text{ min.}$$

Example 21.13. What will be the initial rate of a reaction if its rate constant is 10^{-3} min^{-1} and the concentration of the reactant is 0.2 mol dm^{-3}? How much of the reactant will be converted into the products in 200 minutes?

Solution: Rate constant, $k = 10^{-3} \text{ min}^{-1}$.

$$\text{Initial rate} = k[A]$$

$$= (10^{-3} \text{ min}^{-1}) \times (0.2 \text{ mol dm}^{-3}) = 2 \times 10^{-4} \text{ mol dm}^{-3} \text{ min}^{-1}$$

Now, the integrated rate equation for the first order reaction is:

$$kt = 2.303 \log \frac{a}{a-x}$$

or

$$\log \frac{a}{a-x} = \frac{kt}{2.303}$$

$$t = 200 \text{ min}$$

$$k = 10^{-3} \text{ min}^{-1}$$

so that

$$\log \frac{a}{a-x} = \frac{10^{-3} \times 200}{2.303} = 8.68 \times 10^{-2}$$

or

$$\frac{a}{a-x} = 1.22 \qquad (\text{antilog } 8.68 \times 10^{-2} = 1.22)$$

or

$$a = 1.22a - 1.22x$$

or

$$1.22x = 1.22a - a - a = 0.22a$$

or

$$x = \frac{0.22a}{1.22} = 0.18a$$

$$\% \text{ conversion} = \frac{0.18a \times 100}{a} = 18\%$$

Example 21.14. For a first order reaction prove that the time required for the completion of 99.9% of reaction is about ten times the time taken for the completion of 50% of this reaction.

(*i*) When extent of reaction is 99.9%, then the amount of reactant unconsumed is

$$[A] = [A]_0 - 99.9\% \text{ of } [A]_0 = [A]_0 - 99.9\,[A]_0/100 = 0.1\,[A]_0/100 = 10^{-3}$$

$$\therefore \quad t\,(99.9\%) = \frac{2.303}{k}\log\frac{[A]_0}{[A]} = \frac{2.303}{k}\log\frac{[A]_0}{0.1[A]/100} = \frac{2.303}{k}\log 1000$$

$$= \frac{2.303}{k}\times 3 \qquad \text{...(1)} \because \log 1000 = 3$$

(ii) When extent of reaction is 50% then the amount of reactant unconsumed is

$$[A] = [A]0 - 0.50\% \text{ of } [A]_0 = [A]_0 - 50[A]_0/100 = 50[A]_0/100 = [A]_0/2$$

$$\therefore \quad t\,(50\%) = \frac{2.303}{k}\log\frac{[A]_0}{[A]} = \frac{2.303}{k}\log\frac{[A]_0}{[A]_0/2} = \frac{2.303}{k}\log 2$$

$$= \frac{2.303}{k}\times 0.3010 \qquad \text{...(2)} \because \log 2 = 0.3010$$

$$\text{Now} \quad \frac{t(99.9\%)}{t(50\%)} = \frac{2.303\times 3/k}{2.303\times 0.301/k} = \frac{3}{0.301} = 9.9966 = 10$$

That is, the time required for the completion of 99.9% of a first order reaction is about ten times the time taken for the completion of 50% of the same reaction.

(III) *Based on Half Life Period*

Example 21.15. Calculate the half life period of a first order reaction where the specific rate constant is (a) 200 s^{-1} (b) 2min^{-1} (c) 5 $year^{-1}$.

Solution: For the first order reaction,

$$t_{1/2} = \frac{0.693}{k}$$

(a) $$k = 200\ s^{-1}$$

$$\therefore \quad t_{1/2} = \frac{0.693}{200} = \mathbf{3.465 \times 10^{-3}\ s}$$

(b) $$k = 2\ min^{-1}$$

$$\therefore \quad t_{1/2} = \frac{0.693}{2} = \mathbf{0.3465\ min}$$

(c) $$k = 5\ year^{-1}$$

$$\therefore \quad t_{1/2} = \frac{0.693}{5} = \mathbf{0.1386\ year}$$

Example 21.16. The thermal decomposition of a compound is of first order. If 50% of the compound is decomposed in 120 minutes, how long will it take for 90% of the compound to decompose?

Solution: Let us first calculate, *k* as

$$k = \frac{0.693}{t_{1/2}}$$

Here $$t_{1/2} = 120\ min$$

$$\therefore \quad k = \frac{0.693}{120\ min} = 5.77\times 10^{-3}\ min^{-1}$$

Now, for the first order reaction,

$$t = \frac{2.303}{k}\log\frac{[A]_0}{[A]}$$

If the initial concentration of A i.e. $[A]_0 = a$, then $[A] = a - a \times 0.90 = 0.10a$ because 90% of compound decomposes and only 10% is left.

$$\therefore \quad t = \frac{2.303}{5.77\times10^{-3}}\log\frac{a}{0.1a} = \frac{2.303}{5.77\times10^{-3}}\log 10 = \mathbf{399\ min}$$

Example 21.17. For a first order reaction, calculate the ratio between the time taken to complete three-fourth of the reaction and the time taken to complete half of the reaction.

Solution: For first order reaction,

$$t = \frac{2.303}{k}\log\frac{[A]_0}{[A]}$$

For 3/4 of a reaction to take place, $t = t_{3/4}$, $[A] = [A]_0 - 3/4[A]_0 = 1/4[A]_0$

Thus, $$t_{3/4} = \frac{2.303}{k}\log\frac{[A]_0}{1/4[A]_0} = \frac{2.303}{k}\log 4 \quad \text{...(i)}$$

Now, for half of a reaction to take place,

$$t = t_{1/2};\ [A] = [A]_0 - 1/2[A]_0 = 1/2[A]_0$$

Thus,

$$t_{1/2} = \frac{2.303}{k}\log\frac{[A]_0}{[A]_0/2} = \frac{2.303}{k}\log 2$$

Dividing (i) by (ii)

$$\frac{t_{3/4}}{t_{1/2}} = \frac{\log 4}{\log 2} = \frac{0.6020}{0.3010} = 2$$

Thus, the time required for 3/4th of the reaction to occur is two times that required for half of the reaction.

(IV) Based on Second Order Reactions

Example 21.18. Hydrolysis of ethyl acetate by NaOH was studied by titrating 25 mL of the reaction mixture at different time intervals against standard acid. From the data given below, establish that this is a second order reaction.

t (min)	0	5	15	25
mL of acid used	16.00	10.24	6.13	4.32

Solution: The second order integrated rate equation is

$$k = \frac{1}{at}\cdot\frac{x}{a(a-x)}$$

The volume of acid used at any time is a measure of concentration of the unreacted substance at that them Therefore,

a, initial concentration = 16,00

after 5 mts $(a - x) = 10.24$ and $x = 5.76$

after 15 mts $(a - x) = 6.13$ and $x = 9.85$

after 25 mts $(a - x) = 4.32$ and $x = 11.68$

Substituting values in the rate equation (1), we have

$$k = \frac{1}{16 \times 5} \cdot \frac{5.76}{10.24} = 0.0070$$

$$k = \frac{1}{16 \times 5} \cdot \frac{9.85}{6.13} = 0.0067$$

$$k = \frac{1}{16 \times 25} \times \frac{11.68}{6.13} = 0.00675$$

The value of k being fairly constant, this reaction is of the second order.

(V) Based on Arrhenius Equation

Example 21.19. The rate of a reaction triples when temperature changes from 20° to 50°C. Calculate the energy of activation for such a reaction. ($R = 8.314$ JK^{-1} mol^{-1})

Solution: We know that

$$\log \frac{k_2}{k_1} = \frac{E_a}{2.303\,R}\left[\frac{T_2 - T_1}{T_1 T_2}\right]$$

$$\frac{k_2}{k_1} = 3,\ T_1 = 273 + 20 = 293 \text{ K}$$

$$T_2 = 273 + 50 = 323 \text{ K},\ R = 8.314 \text{ JK}^{-1} \text{ mol}^{-1}$$

$$\log 3 = \frac{E_a}{2.303 \times 8.314}\left[\frac{323 - 293}{323 \times 293}\right]$$

$$0.477 = \frac{E_a}{2.303 \times 8.314} \times \frac{30}{323 \times 293}$$

or

$$E_a = \frac{0.477 \times 2.303 \times 8.314 \times 323 \times 293}{30}$$

$$= 28.8 \times 10^3 \text{ J mol}^{-1} = 28.8 \text{ kJ mol}^{-1}$$

Example 21.20. The reaction,

$$2NO_2 \longrightarrow 2NO + O_2$$

has an activation energy of 110 kJ mol^{-1}. At 400°C, the rate constant is 7.8 mol^{-1}. L^{-1}s^{-1}. What is the value of rate constant at 430°C?

Solution: We know that

$$\log \frac{k_2}{k_1} = \frac{E_a}{2.303\,R}\left[\frac{1}{T_1} - \frac{1}{T_2}\right]$$

$E_a = 110$ kJ mol^{-1} = 110 × 103 J mol^{-1} $\qquad R = 8.314$ J mol^{-1} K^{-1}

$k_1 = 7.8$ mol^{-1} L s^{-1} $\qquad T_1 = 400 + 273 = 673$ K

$k_2 = ?$ $\qquad T_2 = 430 + 273 = 703$ K

Substituting the values,

$$\therefore \quad \log\frac{k_2}{7.8} = \frac{110\times10^3}{2.303\times8.314}\left[\frac{1}{673}-\frac{1}{703}\right]$$

$$\log\frac{k_2}{7.8} = \frac{110\times10^3\times(703-673)}{2.303\times8.314\times673\times703}$$

$$\log\frac{k_2}{7.8} = 0.3643 \text{ or } \frac{k_2}{7.8} = 2.31$$

$$k_2 = 2.31 \times 7.8 = 18.0 \text{ mol}^{-1} \text{ L s}^{-1}.$$

Example 21.21. For a first order reaction the rate constants are $3.20 \times 10^{-4}s^{-1}$ at 600 K and $1.60 \times 10^{-2}s^{-1}$ 1200 K. Calculate the energy of activation of the reaction. Take R = 8.314 JK^{-1} mol^{-1}.

Solution: The energy of activation (E_a) of a reaction is given by Arrhenius equation.

$$E_a = \frac{2.303RT_1T_2}{T_2-T_1}\log\frac{k_2}{k_1}$$

On substituting the given values in equation (i) we get,

$$E_a = \frac{2.303\times8.314 \text{ JK}^{-1}\text{mol}^{-1}\times600 \text{ K}\times1200 \text{ K}}{1200 \text{ K}-600 \text{ K}}\log\frac{1.60\times10^{-2}}{3.20\times10^{-4}}$$

$$= 39036 \text{ J mol}^{-1} = 39.0 \text{ kJ mol}^{-1}$$

Example 21.22. The rate of a reaction is doubled when the temperature is changed from 289 K to 308 K. Calculate the energy of activation of the reaction. R = 8.314 JK^{-1}.

Solution: The energy of activation is given by Arrhenius equation:

$$E_a = \frac{2.303\ RT_1T_2}{T_2-T_1}\log\left(\frac{k_2}{k_1}\right)$$

$$\text{It is given that } \frac{k \text{ at } 308\text{K}}{k \text{ at } 298 \text{ K}} = 2 = \frac{k_2}{k_1}$$

$$\therefore \quad E_a = \frac{2.303\times8.314 \text{ JK}^{-1}\text{mol}^{-1}\times298 \text{ K}\times308 \text{ K} \log 2}{308 \text{ K}-298 \text{ K}}$$

$$= 52897 \text{ J mol}^{-1} = 52.9 \text{ kJ mol}^{-1}.$$

Example 21.23: The rate constant of a reaction is 2×10^{-2} s^{-1} at 300 K and 8×10^{-2} s^{-1} at 340 K. Calculate the energy of activation of the reaction.

Solution: The energy of activation (E_a) is given by:

$$E_a = \frac{2.303\ RT_1T_2}{T_2-T_1}\log\left(\frac{k_2}{k_1}\right)$$

It is given that : when $T_1 = 300$ K, $k_1 = 2 \times 10^{-2}$ s^{-1}
and when $T_2 = 340$ K, $k_2 = 8 \times 10^{-2}$ s^{-1}

$$\text{Now,} \quad \log\frac{k_2}{k_1} = \log\frac{8\times10^{-2}s^{-1}}{2\times10^{-2}s^{-1}} = \log 4 = 2\log 2 = 0.6020$$

$$T_2 - T_1 = 340 \text{ K} - 300 \text{ K} = 40 \text{ K}$$

$$T_2 - T_1 = 340 \text{ K} \times 300 \text{ K} = 102000 \text{ K}^2$$

$$2.303 \text{ R} = 2.303 \times 8.314 \text{ JK}^{-1} \text{ mol}^{-1} = 19.147 \text{ JK}^{-1} \text{ mol}^{-1}$$

On substituting these values into equation (i) we get

$$E_a = \frac{19.147\ \text{JK}^{-1}\text{mol}{-1} \times 102000\ \text{K}^{-2} \times 0.602}{40\ \text{K}}$$

$$= 29392.6\ \text{J mol}^{-1} = 29.39\ \text{kJ mol}^{-1}.$$

QUESTIONS

1. What do you mean by chemical kinetics?
2. What is meant by rate of a reaction? What is the difference between the rate of reaction and rate constant?
3. What is the order of a reaction?
4. Can order of a reaction be fractional? Is it possible for a reaction to have identical values for molecularity and order? Given two examples of first order reaction.
5. A reaction is found to be zero order reaction, will its molecularity be also zero?
6. How is the half-life period of a first order reaction related to its rate constant?
7. What is the difference between order and molecularity of a reaction?
8. Show that the half-life of a first order reaction is independent of the initial concentration of the reactant.
9. What are the units of first order and second order reactions?
10. Given two differences between rate of a reaction and reaction rate constant.
11. What do you determine the order of a reaction by initial rate method?
12. Derive an integrated rate equation for a simple first order reaction
13. What is rate determining step? Show that the slowest step in the mechanism of the reaction largely determines the rate of the reaction.
14. What is meant by order of a chemical reaction? The reaction:

    $$2NO_2 + F_2 \longrightarrow 2NO_2F$$

    has been found to be first order with respect to NO_2 and with respect to F_2.
    (i) Write the rate law equation for the reaction.(ii) What is the overall order of the reaction?
    (iii) Suggest a suitable mechanism for the reaction.
15. Derive the integrated rate equation for rate constant for a first order reaction. What would be the units of (first order) rate constant if the concentration is expressed in moles per litre and time in seconds?
16. Explain with examples the meaning of the terms average rate and instantaneous rate of the reaction.
17. What is the effect of temperature upon the rates of chemical reactions? Explain how does a small rise in temperature appreciably increase the speed of many reactions?
18. How do we know that all collisions between the reactant molecules do not lead to chemical change?
19. What is pseudo unimolecular reaction? Discuss with a simple example.
20. What are the main features of Transition State Theory?
21. Define the rate of a reaction. For the reaction X → Y, the rate of reaction can be denoted by:

    $$-\frac{dX}{dt} \quad \text{or} \quad +\frac{dY}{dt}$$

    State the significance of the plus and minus signs in these expressions.
22. Explain:
    (i) Energy of activation for a reaction. (ii) Half life of areaction.
23. The reaction $2A + B \rightarrow C + D$ has the rate law equation as:

    $$\text{rate} = k[A]^x\,[B]^y$$

    What is the
    (i) order of reaction (ii) apparent molecularity of reaction and
    (iii) order of reaction when B is present in large excess.

24. Write Arrhenius equation showing the effect of temperature in the reaction rates. How does it help to calculate the activation energy of a reaction?
25. Give important differences between
    (i) order and molecularity of a reaction
    (ii) reaction rate and rate constant
    (iii) activation energy and internal energy of a reaction.
26. Define the terms
    (i) Activated complex. (ii) Orientation barrier (iii) Effective Collision
27. Discuss the main points of collision theory of reaction rates.

PROBLEMS

1. In a reaction 5g of ethyl acetate is hydrolysed per litre in presence of dil HCl in 300 min. If the reaction is of first order and initial concentration of ethyl acetate is 22 g/L, calculate the rate constant for this reaction. **(8.73×10^{-4} min^{-1})**
2. A first order reaction has specific reaction rate of 2×10^{-3} s^{-1}. Calculate the half-life of this reaction. **(346.5 s)**
3. Calculate the half-life of a first order reaction when the specific rate constant is
   (i) 100 s^{-1} (ii) 2 min^{-1} **(0.00693 s and 0.3465 min)**
4. Time required to decompose SO_2Cl_2 to half its amount is 60 minutes. If decomposition is a first order reaction, find the rate constant of this reaction. **(0.0115 min^{-1})**
5. The reaction $SO_2Cl_2 \rightarrow SO_2 + Cl_2$ is a first order reaction with $k = 2.2 \times 10^{-5}$ s^{-1} at 320°C. What is its half-life? What percentage of SO_2Cl_2 would be decomposed on heating at 320°C? **(3.15×10^4 s, 11.27%)**
6. For a certain first order reaction, it takes 5 minutes for the initial concentration of 0.6 mol L^{-1} to become 0.2 mol L^{-1} to become 0.2 mol L^{-1}. What is the rate constant for this reaction? **(0.2198 min^{-1})**
7. A first order reaction is 20% complete in 5 minutes. Calculate the time taken for tyhe rectaion to be 60% complete. **(20.53 min)**
8. A first order reaction is found to have a rate constant k = 7.39×10^{-5} sec^{-1}. Find half-life of this reaction. **(0.1875×10^{-5} sec)**
9. The catalytic decomposition of hydrogen peroxide was studied by titrating it at different intervals with $KMnO_4$. Calculate velocity constant from the following data, assuming it to be a first order reaction.

t, sec	0	600	1200
$KMnO_4$ (mls)	22.8	13.8	8.2

(0.0008367 sec^{-1})

10. A reaction that is of first order with respect to reactant A, has rate constant of 6 min^{-1}. If we start with concentration of A equal to 0.5 mol L^{-1}, when would [A] reach the value of 0.05 mol L^{-1}. **(0.38 min)**
11. Show by using rate law, how much the rate of the reaction

$$2NO(g) + O_2(g) \rightleftharpoons 2NO_2(g)$$

will change if volume of the reaction vessel is diminished to 1/3 of its initial volume. The reaction is of first order.

12. From the following data for the decomposition of ammonium nitrate in aqueous solution, show that the reaction is of first order.

Time (minutes) :	10	20	25	∞
Volume of N_2 (ml) :	6.25	11.40	13.65	35.05

(0.0197 min^{-1})

13. A first order reaction is 20% complete in 10 minutes. Calculate:
    (i) the specific rate constant of the reaction and
    (ii) the time taken for the reaction to go to 75% completion. **[(i) 0.02231 min^{-1} (ii) 62.15 min^{-1}]**
14. Show that in case of a first order reaction, the time required for 99.9% of the reaction to take place is ten times that required for half of the reaction.

EXAMINATION QUESTIONS

1. (a) What do you mean by Energy of Activation ? Explain with the help of energy reaction coordinate diagram. Also explain how the nature of activated complex changes in the presence of a catalyst.
   (b) Differentiate between order and molecularity of a reaction.
   [*U.P. Tech., B.Tech., 1st/2nd Sem., 2000-02*]
2. Discuss the differences between order of a reaction and molecularity of a reaction.
   [*U.P. Tech., March 2002*]
3. For a first order reaction, the rate constant is found to be 7.0×10^{-7} at 7°C and 9×10^{-4} at 57°C. Calculate the energy of activation of the reaction.
   Ans. 1.098×105 mol^{-1} ($\log_{10} 7 = 0.8451$, $\log_{10} 9 = 0.9542$) **[*U.P. Tech., 2003-04*]**
4. (i) What is meant by the rate of reaction ? Explain the factors which affect the rate of a reaction.
   (ii) Distinguish between Order and Molecularity of a Reaction. **[*U.P. Tech., 2002-03*]**
5. (i) What meant by order of reaction ?
   (ii) Show that in case of a first order reaction, the time required for 99.9% of the reaction to take place is about 10 times that required for half the reaction. **[*U.P. Tech., 2004-05*]**
6. (i) Define order of a reaction. What is condition that a second order reaction behaves like a first order?
   (ii) With the help of data given show that decomposition of H_2O_2 in aqueous solution is of first order.

Time (min) :	0	10	20	30
Volume of KMnO4 : required to decompose H_2O_2 (ml).	25.0	20.0	15.7	12.5

   [*U.P. Tech., 1st Sem., 2006-07*]
7. (i) The rate constant of a reaction is 1.5×10^7 Sec^{-1} at 50°C and 4.5×10^2 sec^{-1} at 100°C. Evaluate the Arrhenius parameters A and Ea.
   (ii) Derive the equation for half life of first order reaction. **[*U.P. Tech., 2nd-Sem., 2006-07*]**
8. (a) What do you understand by the rate equation and the order of a reaction ?
   (b) Derive the rate Expression for Second-order reaction, when the reactants are different.
   (c) Discuss the effect of temperature on reaction rates. **[*U.P. Tech., 2007-08*]**
9. (i) What is meant by rate of reaction ? Explain the factors which affect the rate of a reaction.
   (ii) What is activation energy ? How is the rate constant of a reaction related to its activation energy? **[*U.P. Tech., 2nd Sem., 2003-04*]**
10. For the first gaseous reaction

$$X_{(g)} \longrightarrow 2Y_{(g)} + Z_{(g)}$$

   The initial pressure is 90 mm Hg. The pressure after 10 minute is 180 mm Hg. Calculate the rate constant of the reaction. **[*U.P. Tech., 1st Sem., (C.O.) 2003*]**
11. Prove that for a second order reaction, the half life period is inversely proportional to its initial concentration. **[*U.P. Tech., 1st Sem., 2007-08*]**
12. (a) Rate of reaction is directly proportional to the ________.
   (b) The reaction : $N_2O_5\,(g) \rightarrow N_2O_4\,(g)\ \frac{1}{2}O_2\,(g)$ is :
   (*i*) Zero order (*ii*) First order
   (*iii*) Second order (*iv*) Fractional order.
   (c) Threshold energy = Activation energy + ________. **[*U.P. Tech., 1st Sem., 2008-09*]**
13. (a) Distinguish between order and molecularity of a reaction, calculate order and molecularity of the following reactions :

$$N_2O_5\,(g) \longrightarrow N_2O_4\,(g) + \frac{1}{2}O_2$$

$$CH_3COOCH_2H_5 + H_2)\text{ (excess)} \xrightarrow{H^+} CH_3COOH + C_2H_5OH$$

$$H_2(g) + I_2\,(g) \rightarrow 2Hi(g)\text{ , and } CH_3CHO \xrightarrow{\Delta} CH_4 + CO$$

(b) What is activation energy ? Discuss its relationship with rate constant of a reaction. **[*U.P. Tech., 1st Sem., 2008-09*]**

14. Derive an equation for half life period of a first order reaction.

A compound decomposes according to the first order rate low with a half life period of 30 min. Calculate the fraction of remaining compound after 120 min. **[*U.P. Tech., 1st Sem., 2008-09*]**

15. (a) Derive the rate equation for a second order chemical reaction. The specific rate constant for the decomposition of formic acid is 5.5×10^{-4} sec^{-1} at 413°K. Calculate the specific rate constant at 458°K if the energy of activation is 2.37×10^4 cal mol^{-1}.

(b) What is half life of a chemical reaction ? Show that time required to complete 99.9% of a first order reaction is about 10 times of its half life. **[*U.P. Tech., B. Tech. 1st Sem., 2008-09*]**

❑❑❑

22 FUNDAMENTALS OF ORGANIC MECHANISM

"The winds and waves are always on the side of the ablest navigators."

Edward Gibbon

CHAPTER OUTLINE

The term 'organic chemistry' was used to describe the compounds obtained from living organisms, while *inorganic chemistry* was used for the study of compounds from non living sources. Today, however we know that the common characteristic of organic compounds from living sources is that, all contain the element 'carbon'. Thus, organic chemistry is now defined as the chemistry of carbon compounds or more precisely hydrocarbons and their derivatives.

Most of the inorganic reactions are very fast and it is extremely difficult to know the detailed pathway of such reactions. However, the reactions of *organic compounds* are generally slow and proceed through well defined steps to form the products. In organic reactions, the reactant change into the *intermediate* substance and these intermediates then change into the final product.

The step-by-step sequence of elementary reactions by which the reactants change into the products is called reaction mechanism. In other words, the mechanism of an organic reaction is a detailed description of all the steps involved in that reaction. The study of mechanism of organic reactions is a complex area.

The general concepts and principles described in the following sections, help better understanding of the organic reaction mechanisms.

22.1 NATURE OF ORGANIC MOLECULES

Why do chemists treat organic chemistry as a separate branch of science? This is because of the ability of carbon atoms to bond together, forming chains or rings. Of all the elements, only carbon has this property and can form compounds from one carbon atom (such as methane) to tens of billions of carbon atoms (such as deoxyribonucleic acid, *DNA*).

Organic molecules have '*covalent bonds*' formed by sharing of electrons between the bonded atoms. Carbon can form multiple covalent bonds also. These covalent bonds are directional and hence the organic molecules have specific three dimensional shapes. Covalent bonding gives organic compounds, properties that are quite different form those of inorganic compounds. The nature of covalent bonds determine the path of the reaction in these compounds.

22.2 CLEAVAGE OF COVALENT BONDS

Organic reactions involve breaking and formation of covalent bonds. The covalent bonds can break by two ways. Different products will be formed depending upon the point of cleavage as discussed below:

(*i*) **Homolytic cleavage (Symmetrical cleavage):** We know that covalent bond is formed by the sharing of electrons between the two atoms. Let us represent the organic molecule as: "*A* : *B*". In homolytic cleavage the bond breaks in such a way that each atom departs with one electron of the shared pair.

$$A \cdot\!\vdots\!\cdot B \longrightarrow \underbrace{A\cdot \;+\; B\cdot}_{\text{Free radicals}}$$

Example:

$$Br : Br \xrightarrow{\text{Homolytic fission}} \cdot Br + \cdot Br$$

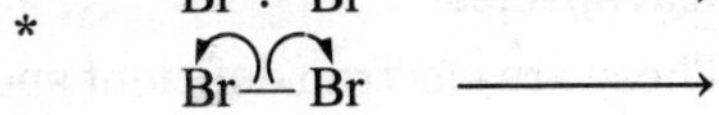

Homolytic cleavage or fission leads to the formation of highly reactive neutral species containing unpaired electrons which are known as *Free Radicals*.

Homolytic cleavage is possible only when the electronegativity ** of both the bonded atoms is same, *i.e.* both the atoms have similar attraction for the shared electron pair.

(*ii*) **Heterolytic cleavage (Unsymmetrical Cleavage):** In this case, the covalent bond is broken in such a way that one of the atoms carries away both the electrons of the shared pair. This result in the formation of a cation and an anion.

Note:* A partial head (fish hook) on the arrow indicates the shift of single electron. A full head arrow indicates the movement of electron pair.

** Electronegativity: It is the measure of the tendency of an atom in a molecule to attract the bonded pair of electrons.

$$A \,\vdots\!\cdot\cdot\, B \longrightarrow A^{\oplus} + B^{\ominus} \quad \text{(When B is more electro negative than A)}$$

or

$$A\cdot\cdot\!\vdots\, B \longrightarrow :A^{\ominus} + B^{\oplus} \quad \text{(When A is more electro negative than B)}$$

or

$$A\!-\!B \longrightarrow A^{+} + B^{-}$$

Heterolytic fission is possible when the electronegativities of the bonded atoms are different. The more electonegative atom retains both the electrons and acquires a negative charge, *i.e.* an anion is formed and the other atom which departs without the electrons acquires the positive charge, *i.e.* a cation is formed.

Example: 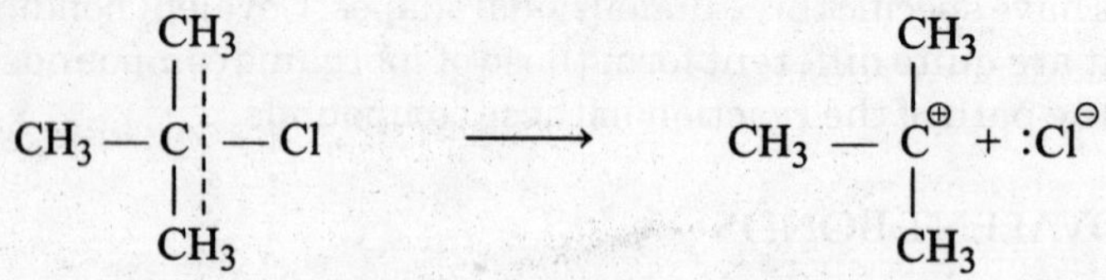

22.3 ATTACKING REAGENTS (ELECTROPHILES AND NUCLEOPHILES)

In a chemical reaction, the substrate (reactant) changes into the product under the influence of attacking reagent.

Substrate + Attacking Reagent → Product

The nature of attacking reagent determines the course of chemical reaction. The attacking reagent are of two types:

(*i*) Electrophiles or Electrophilic reagent
(*ii*) Nucleophiles or Nucleophile reagent.

(*i*) **Electrophiles:** **(Electro = electron, philic = loving) Electrophiles are electron deficient species which can accept a pair of electrons. There are electron loving species and hence always attack at the centre of high electron density. Electrophiles may be either positively charged or electrically neutral chemical species.**

(*a*) ***Positively charged electrophiles:*** **These are electron deficient species and carry a positive charge. Examples,.**

$$H^+, H_3O^+, Cl^+, Br^+, I^+, \overset{+}{N}O_2, \overset{+}{N}O, R^+ \text{ etc.}$$

(*b*) ***Neutral Electrophiles:*** **These are electron deficient species but carry no charge. The central atom in neutral electrophiles has six electrons and can accept two electrons readily.**

Examples: **BF_3, $AlCl_3$, $FeCl_3$, $SnCl_4$, $\dot{R}$. (free radicals), $:CR_2$ (Carbenes), $:NR$ (Nitrencs) etc.**

Since, both the positively charged and neutral electrophiles are short from by a pair of electrons (except free radicals which are short by one electron), they have strong tendency to attract electron from other compounds and hence behave as *Lewis acids*.

(*ii*) **Nucleophiles:** (Nucleo = nucleus, philic = loving) Nucleophiles are electron rich species and have a tendency to donate a pair of electrons. Nucleophiles always attack at electron deficient (low electron density) centres. These may be either negatively charged or neutral chemical species.

(*a*) *Negatively charged nucleophiles:* These are electron rich species having negative charge. Examples: H^- (hydride ion), Cl^-, Br^-, I^-, R^- (Carbanion), $\bar{O}H$, $\bar{O}R$, SR^-, NH_2^-, CN^-, $RCOO^-$ etc.

(*b*) *Neutral Nucleophile:* These are electron rich, having lone pair of electrons but carry no charge.

Examples: $H_2\ddot{\underset{\cdot\cdot}{O}}$, $\ddot{N}H_3$, $R\ddot{N}H_2$, $R\ddot{\underset{\cdot\cdot}{O}}H$, $R\ddot{\underset{\cdot\cdot}{S}}H$, $R\ddot{\underset{\cdot\cdot}{O}}R$ etc.

(c) *Ambident Nucleophile*: Sometimes more than one atom in a nucleophile is electron rich and hence it can attack the substrate through either of the atom. Such nucleophiles are called ambident nucleophile.

Examples: $\bar{C}N$, NO_2^-.

Nucleophiles have strong tendency to donate the electrons to other compounds and hence and behave as Lewis bases.

Table 22.1 gives the difference between electrophiles and Nucleophiles

Table 22.1: Difference between electrophiles and Nucleophiles.

Electrophiles	*Nucleophiles*
1. These are electron deficient species which can accept a pair of electrons.	These are electron rich species which can donate a pair of electrons.
2. They are positively charged or neutral species.	These are negatively charged or neutral species.
3. Attack on electron rich centre.	Attack on electron deficient centre.
4. These are Lewis acids as these can accept an electron pair.	There are Lewis bases as these can donate electron pair.
5. Example NO_2^+, Br^+, BF_3, $AlCl_3$, SO_3 etc.	Example : $\bar{O}H$, $\bar{O}R$, $\bar{C}N$, NH_3, H_2O etc.

22.4 ELECTRON DISPLACEMENTS IN ORGANIC MOLECULES

Covalent bonds in organic molecules undergo electronic displacements, sometimes on their own or under the influence of attacking reagent (nucleophile or electrophile). For a chemical reaction to take place in organic compounds, the substrate must develop some polarity so that charged centres are created. This is made possible by the displacement of electrons of the bonding electrons. The electron displacement influences the mechanism involving such bonds. Now, how do these electron displacements take place? These displacements are due to certain effects, some of which are permanent and others are temporary. Thus, *the permanent or temporary effects involving the displacement of electron in the substrate molecule are termed as electron displacement effects.* Some of the important displacement effects are:

(*i*) Inductive Effect
(*ii*) Mesomeric Effect
] Permanent Effects

(*i*) Electromeric Effect
(*ii*) Inductomeric Effect
] Temporary Effects

Some of these effects are discussed below.

22.4.1 Inductive Effect

A covalent bond is formed by the sharing of electron pair between two atoms. If the two atoms are dissimilar, the shared electrons will be attracted towards more electronegative atom. Thus an atom which is more electronegative acquires partial negative charge (δ^-) where as the other acquires partial positive charge (δ^+). In a carbon chlorine bond, for example, the electron pair between carbon and chlorine is displaced towards more electronegative chlorine atom as compared to carbon. Thus, chlorine atom acquires a small negative charge and the carbon atom a small positive charge.

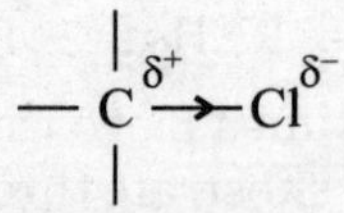

This type of permanent displacement of electrons of a covalent bond towards the more electronegative element or group is called inductive effect.

If this carbon bond forms the terminal of carbon chain, the effect of this electron displacement is passed to other carbon atoms in the chain, *i.e.* the positive charge on C_1 attracts the electrons between C_1 and C_2 and hence C_2 will also acquire slight positive charge. But this positive charge will be less as compared to the positive charge on C_1. Similarly, effect can be passed further to C_3 but as the distance between carbon atom and chlorine increases, the effect decreases rapidly.

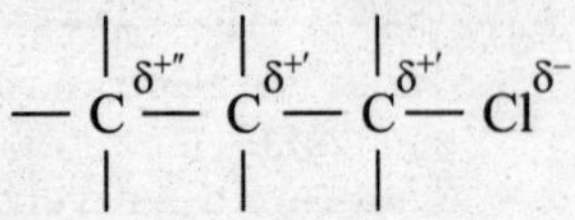

where $\delta'^{+} > \delta^{+''} > \delta^{+'''}$

From the practical point of view, the effect is ignored after the second carbon atom. Hence, inductive effect may be defined as '*the process of electron displacement along a chain of atoms due to the presence of polar covalent bond*'. This is a permanent effect.

Inductive effect is represented by the symbol $\rightarrow$, the arrow pointing towards the more electronegative element or group of atoms.

We can see that the inductive effect causes a certain degree of polarity in the bond which makes the bond much more liable to be attacked by other charged atoms or group.

In the foregoing example, the hetero atom, C_1 is more electronegative than carbon. Inductive effect also takes place if the carbon is bonded to less electronegative atom. In that case, the electron displacement will take place in the opposite direction, *i.e.* the electrons are pushed away form the heteroatom.

(*i*) *– I Effect or electron withdrawing effect:* This effect is exerted by the groups attached to the carbon chain having high electronegativity than carbon. Some of the examples of groups having – *I* effect are:

$$NO_2 > F > Cl > Br > I > OCH_3 > OH$$

(*ii*) *+ I Effect or electron releasing inductive effect:* It is exerted by the atoms or groups attached to the carbon chain having lesser electronegativity than carbon.

Some of the examples of groups having + I effect are:

$$C_6H_5\bar{O} > CO\bar{O} > CR_3 > CHR_2 > CH_2R > CH_3 > H.$$

Characteristics of Inductive Effect

(*i*) Inductive effect does not involve an actual transfer of electrons from one atom to the other. It merely helps in displacing it permanently.

(*ii*) This effect is observed in saturated compounds involving the electrons of sigma bond.

(*iii*) Inductive effect is relayed along the carbon chain but decreases rapidly with increasing distance from the hetero atom.

Application of Inductive Effect

(*i*) **Dipole Moment:** Inductive effect leads to dipolar character in the molecule. More is the inductive effect, greater will be the dipole moment. Example,

$$CH_3,\ CH_3Br,\ CH_3Cl$$

$$\xrightarrow[\text{– I increases}]{}$$

$$\xrightarrow[\text{Dipole Moment increases}]{}$$

(*ii*) **Bond Length:** Usually the bond length of organic molecules decreases with increase in inductive effect. Example,

$$CH_3,\ CH_3Br,\ CH_3Cl$$

$$\xrightarrow[\text{– I increases}]{}$$

$$\xrightarrow[\text{Bond length decreases}]{}$$

(*iii*) **Reactivity of Alkyl halides:** Inductive effect has a great influence an reactivity of alkyl halides. Due to – I effect of halogen atom in the molecule of alkyl halides, the carbon attached to the halogen acquires a partial positive charge and hence is readily attacked by a nucleophile (negatively charged ion).

$$H_3C \rightarrow Cl \quad \text{or} \quad H_3C^{\delta+} — Cl^{\delta-}$$

↑ Attack by Nucleophile

Thus alkyl halides undergo '*nucleophilic substitution reactions*' The influence of inductive effect on alkyl halides is explained in detail under the section 'Nucleophilic Reactions'.

(*iv*) **Relative strengths of the Organic acids:** Carboxylic acids have the tendency to release a proton and change into carboxylate anion.

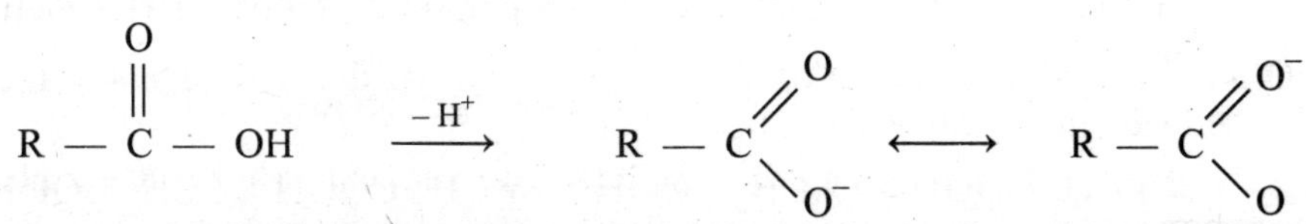

The acidic strength of carboxylic acid depends upon the ease with which it ionizes to give proton and change into carboxylate anion. Any factor which stabilizes the carboxylate anion will increase the acidic strength of the carboxylic acid. Inductive effect influences the strength of carboxylic acids as follows.

(*a*) Effect of +I groups: To explain the effect of +I groups, let us take the example of formic acid (HCOOH) and acetic acid (CH_3COOH)

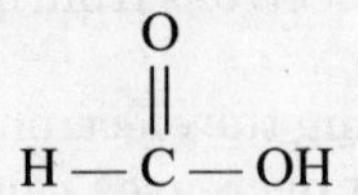

and

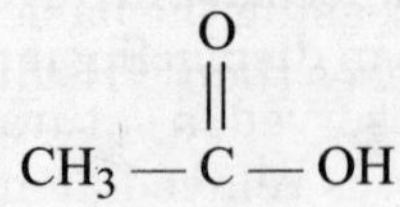

Formic acid (I)

Acetic acid (II)

CH_3 group in (II) has +I effect (electron donating effect) and it will *destabilise* the acetate ion formed by release of proton because it enhances the negative charge on it. Whereas *no such destabilisation* is present in formate ion formed from (I). Hence formic acid is stronger acid than acetic acid.

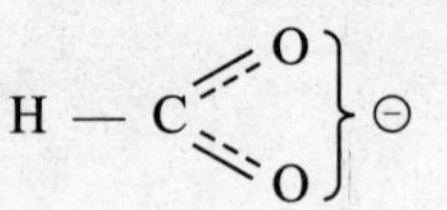

Formate ion

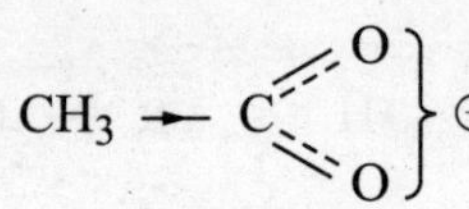

Acetate ion (Destabilised by electron donating CH_3 group)

Further, the +I effect of alkyl groups increases with size,

$$CH_3- < CH_3\,CH_2- < (CH_3)_2\,CH- < (CH_3)_3\,C-$$

Therefore the acidic strength decreases in the same order. For example,

$$HCOOH > CH_3\,COOH > CH_3CH_2COOH > (CH_3)_2\,CHCOOH > (CH_3)_3\,CCOOH$$

(*b*) Effect of –I groups: The groups having –I effect will *stabilise* the carboxylate ion by dispersing the negative charge on it. Example,

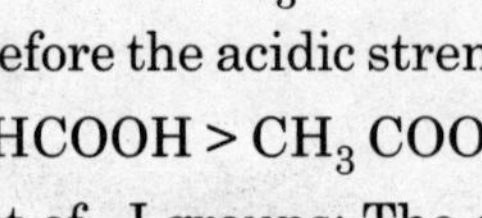

+ I effect of – CH_3 group intensifies the negative charge, destabilise the carboxylate ion

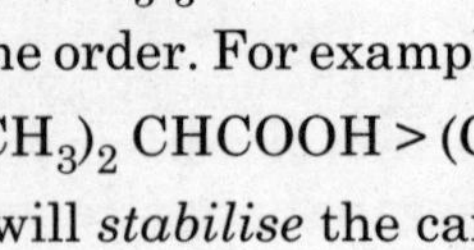

– I effect of halogen atom disperse the negative charge, stabilise the carboxylate ion

the acidic strength of chloroacetic acid is more than the acetic acid because the halogen attached to carbon atom stabilises the carboxylate ion formed after the release of proton as shown above.

Therefore, $Cl-CH_2\,COOH > CH_3\,COOH$ (Acidic strength)

More is the number of halogens on $\alpha-$carbon, stronger is the acid. Example,

$$Cl_3C-COOH > Cl_2\,CH-COOH > ClCH_2-COOH > CH_3COOH.$$

Further, since – I effect of halogens decrease in the order, F > Cl > Br > I, the strength of $\alpha-$halogen acids follow the same order.

$$FCH_2\,COOH > Cl\,CH_2\,COOH > Br\,CH_2\,COOH > I\,CH_2\,COOH$$

(*v*) **Relative basic strength of amines:** Amines contain a lone pair of electrons on the nitrogen atom and therefore have a strong tendency to donate the electron pair. Hence, amines behave as bases. The relative strength of aliphatic amines depends upon inductive effect, steric factors, etc. The inductive effect influences the strength as follows:

(*a*) +I Effect: The electron donating (+I) groups present on the nitrogen in amines increase the electron density on nitrogen. As a result nitrogen release the electron pair more easily, increasing the basic strength. For example, all the three classes of amines are stronger bases then NH_3 due to +I effect of alkyl groups.

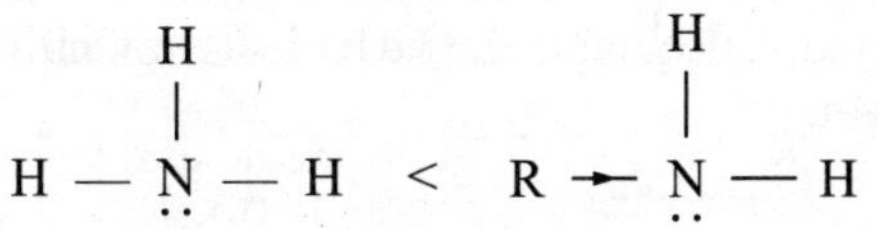

(Basic strength)

For the same reason, one should conclude that more the number of alkyl groups attached to nitrogen, more stronger the base is. On the basis of inductive effect alone, the strength of the amines should follow the order:

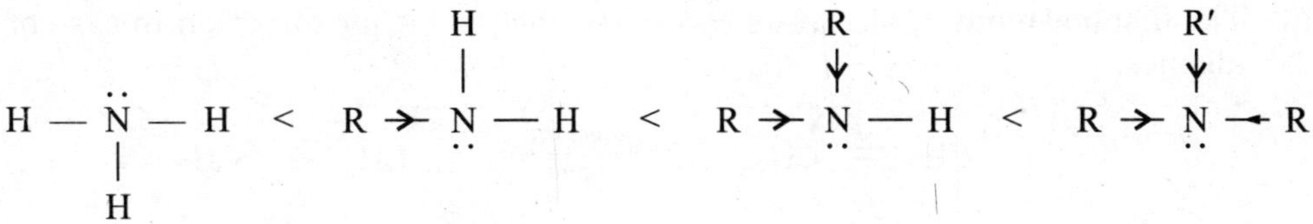

Primary amine   Secondary amine   Tertiary amine

However, the *actual order* of basicity in aqueous solution is:

2° Amine > 1° Amine > 3° Amine.

Thus, apart from inductive effect, other factors such as steric factors, stability of conjugate acid formed also play their role in determining the basicity.

(*b*) – I Effect: The electron withdrawing groups present on the nitrogen in amines decrease the electron density an nitrogen, making the electron release difficult. Hence – I group decrease the basic strength of amines Example:

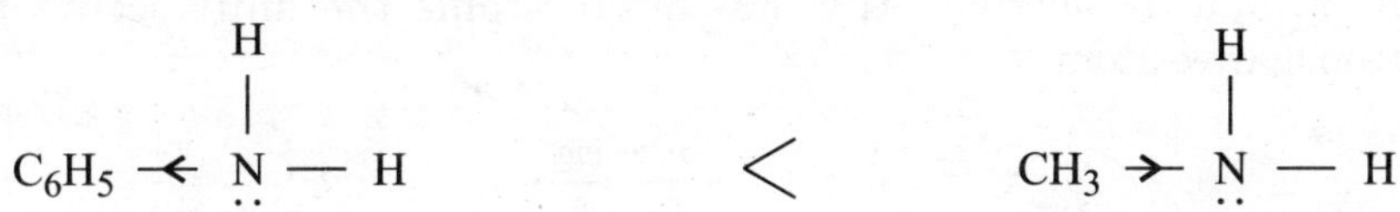

Electron withdrawing phenyl group decreases the electron density on Nitrogen. (Weaker Base)

Electron releasing alkyl group increases the electron density on Nitrogen (Stronger Base)

Similarly, the basic strength of aromatic amines follow the order:

$$C_6H_5NH_2 > (C_6H_5)_2NH > (C_6H_5)_3N$$

(Base strength)

Aniline   Diphenyl amine   Triphenyl amine

22.4.2 Electromeric Effect

Electromeric effect involves the complete transfer of electrons of a multiple bond (double or triple bond) to one of the bonded atoms under the influence of an attacking agent.

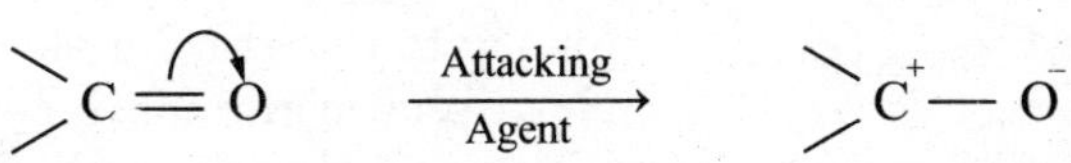

Characteristics:

(*i*) This is a *temporary effect* operating only under the influence of attacking reagent. As soon as the attacking agent is removed the electromeric effect disappears.

(*ii*) It involves complete transfer of π electrons of a multiple bond. For example, during the attack of a carbonyl group by a nucleophile reagent, the π electrons pair bonding the carbon and oxygen are transferred to oxygen.

It causes the polarisation of the molecule leading to the formation positive and negative charges.

(*iii*) The displacement of electrons can take place in either direction in case of symmertical alkenes.

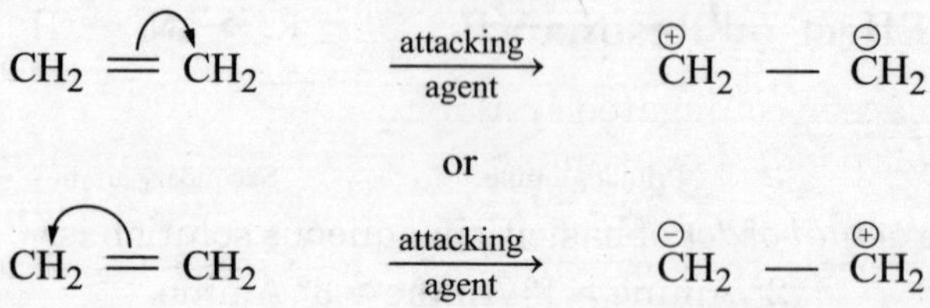

However in case of unsymmetrical alkenes the displacement of electrons is goverened by inductive effect.

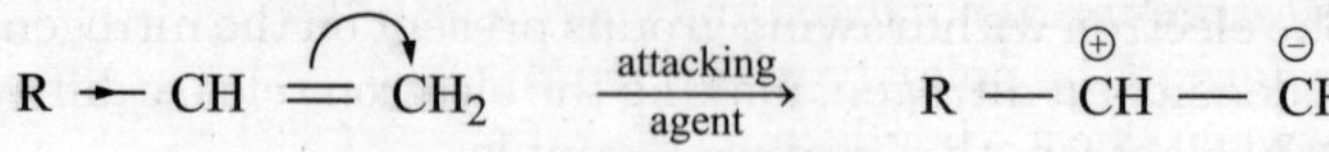

It the π bond is between the dissimilar atoms the shift occurs towards the more electronegative atom.

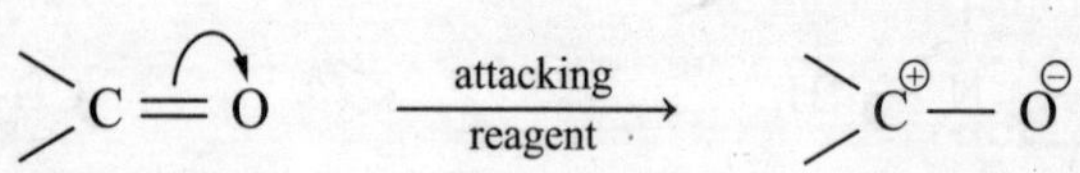

Types of Electromeric Effect

Electromeric effect is of 2 types:

(*i*) + *E* Effect: The electrons of the π bond are transferred to that atom of the double bond to which the reagent gets finally attached, the effect is called + *E* effect. For examples, during the addition of acids to alkenes,

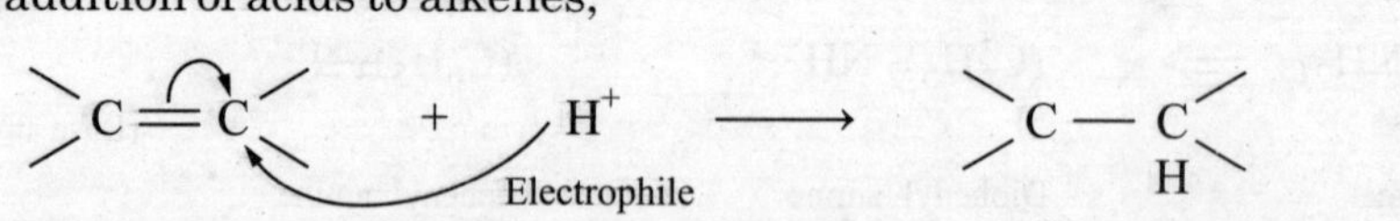

(+E Effect)

(*ii*) – *E* Effect: If the electrons of this double bond are transferred to an atom of the double bond *other than* the one to which the attacking reagent get finally attached, then it is called – E effect. Example,

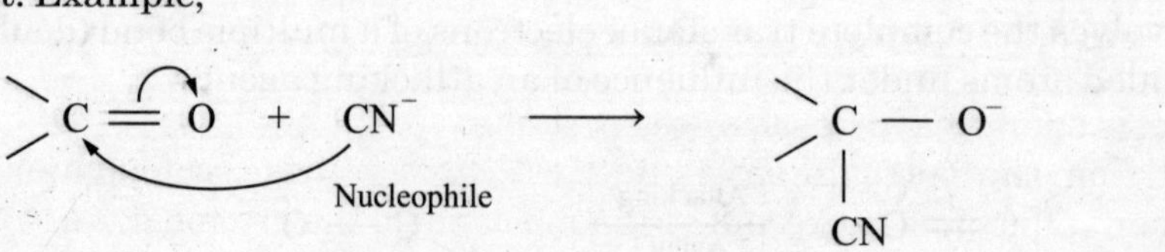

(–E Effect)

Difference between Electromeric and Inductive Effect.

Inductive effect	*Electromeric effect*
(i) Takes place due to the presence of electron donating or electron withdrawing group linked to the carbon chain.	(i) Takes place under the influence of an attacking reagent (electrophile or nucleophile).
(ii) This is a permanent effect	(ii) This is a temporary effect and disappears as soon as the attacking reagent is removed.
(iii) The electron pair is only displaced towards more electronegative atom	(iii) The electron pair is completely transferred to one of the multiple bonded atoms.
(iv) Shown by the substrate usually having single bond.	(iv) Shown by the substrate having multiple bonds.

22.4.3 Mesomeric Effect or Resonance Effect

Mesomeric effect takes place in conjugated systems. *Conjugated systems are the systems having alternate single and double bonds*. For example, butadiene is a conjugated system.

$$CH_2 = CH - CH = CH_2.$$

A group with lone pair of electrons can also be situated in conjugation with double bond. For example,

$$CH_2 = CH - \ddot{\underset{\cdot\cdot}{Cl}}:$$

In such systems, π electrons are delocalised and lead to the formation of polarity in the molecule. The mesomeric effect refers to the polarity which is produced in there conjugated systems as a result of interaction between two π – bonds or a π bond and a lone pair.

Mesomeric effect can be defined as " *the transference of electrons in conjugated systems,*

- from a multiple bond to an atom or
- from a multiple bond to a single covalent bond or
- from an atom having lone pair of electrons to a single covalent bond.

In other words, mesomeric effect is flow of electrons form one part of the molecule to the other creating centres of low and high electron density.

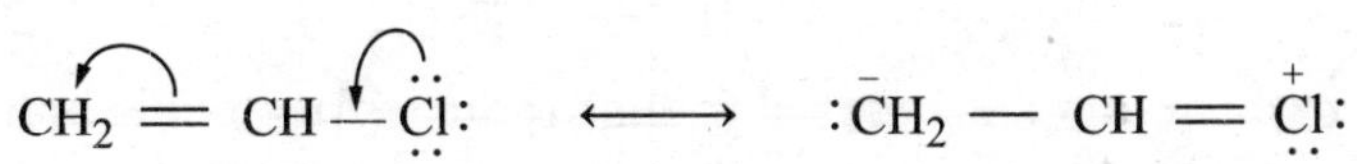

(I) (II)

The movement of the electrons from atom having lone pair to the single covalent bond

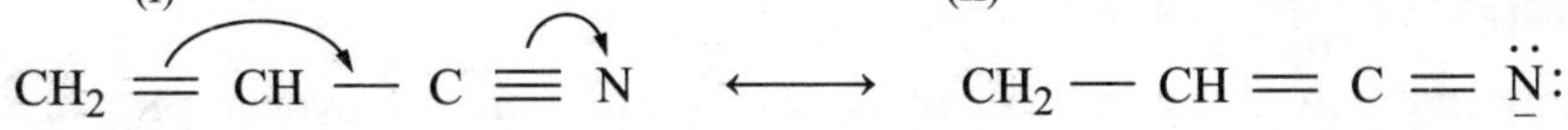

(I) (II)

The movement of the electrons from multiple bond to the single covalent bond and also from multiple bond to the atom

Characteristics:

(i) Mesomeric effect is operative in unsaturated systems.
(ii) It is transmitted from one end to the other end of the chain, provided conjugation is present.
(iii) It is the permanent effect and is present in the molecule in the ground state.
(iv) The actual structure of the molecule lies in between the resonance (mesomeric) or contributing structures.

Types: Mesomeric effect is of two types:

(*i*) + M or + R Effect: A group is said to exhibit + M or + R effect, if it donates the electrons to a conjugated system. In such cases we can say that the transference of electrons is away form the atom or group in conjugated system. Some of the groups having + M effect are:

$$-Cl,\ -Br,\ -I,\ NH_2,\ -NR_2,\ -OH,\ -OCH_3.$$

Example:

(+ M effect)

(I) (II)

(*ii*) – M or – R effect: A group or atom exhibit – M or – R effect if it withdraws electrons from the double bond or from a conjugated system. In such cases, electrons move towards the atom or group in conjugated system. Some of the groups having – M or – R effect are:

Example: $-NO_2$, $-CN$, $-C(=O)-$, $-SO_3H$

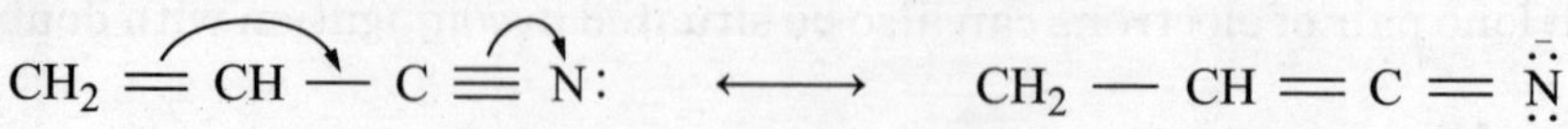

(– M effect)

(I) (II)

22.4.4 Hyperconjugation

Hyperconjugation, also knows as Baker Nathan effect takes place in the organic molecules in which an alkyl group is attached to the unsaturated systems such as multiple bonds or benzene ring.

Baker and Nathan suggested that the alkyl groups with at least one hydrogen atom on the α carbon atom, attached to unsaturated carbon atom are able to transfer electrons by a mechanism similar to electromeric effect. Example,

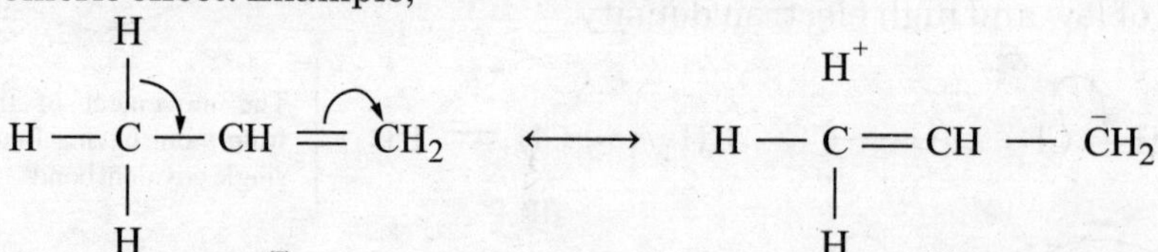

(I) (II)

Thus, hyperconjugation involves the delocalisation of electrons through overlapping of π orbitals of C = C double bond with σ orbital of the adjacent C – H bond in unsaturated system. This is also known as σ – π conjugation.

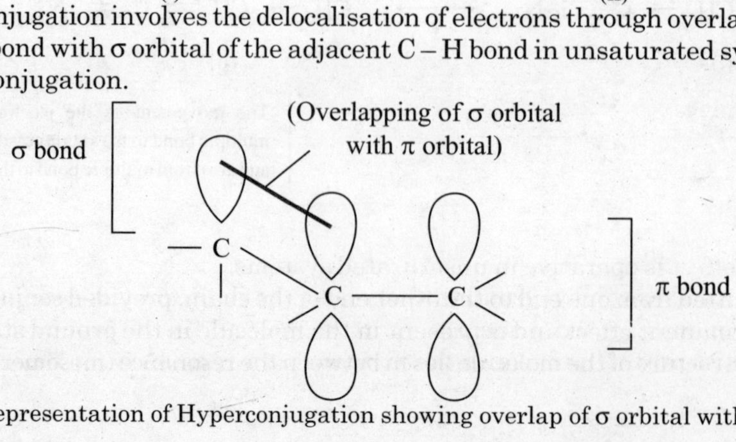

Orbital Representation of Hyperconjugation showing overlap of σ orbital with π orbital.

In fact, hyperconjugation effect is an extension of the mesomeric effect. The difference is that in mesomeric effect the delocalisation of π – electron of two adjacent double bonds or π electrons of a double bond and lone pair of heteroatom takes place, whereas in hyperconjugation the delocalisation of σ electrons takes place through overlapping of *p-orbitals of double bond* with σ – *orbital of the adjacent single bond.*

The polypropylene can have the following hyperconjugated structures in all:

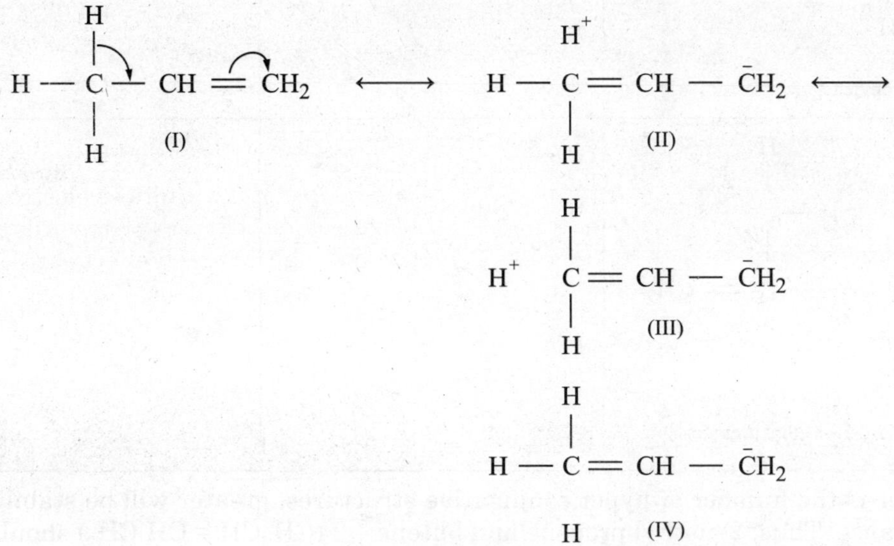

Structures II, III and IV are called contributing hyper conjugative structures. Since there is no real bond between C and one of hydrogens in these structures, hyperconjugation is also called *no bond resonance*. However, this proton is still firmly bond to the π-cloud and is not free to move.

Characteristics

(*i*) Hyperconjugation occurs through the H-atoms present on the carbon atom next to the double bond, *i.e.* α carbon atom.

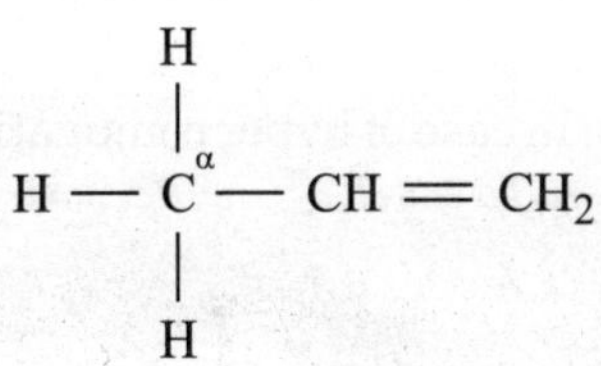

(I)

3 hydrogen atoms on α - carbon atom, 3 possible hyper conjugated structures.

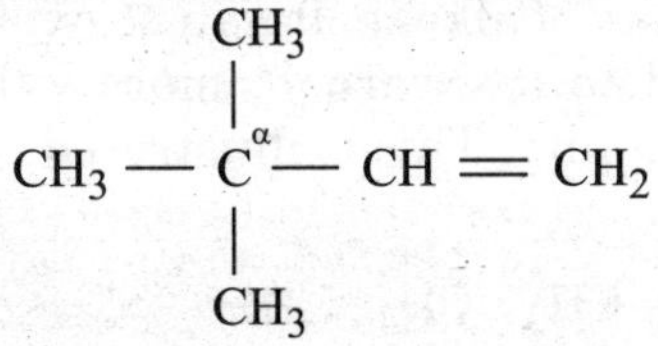

(II)

no hydrogen atoms on α - carbon, hence no hyper conjugation

(*ii*) Greater the number of alkyl groups attached to doubly bonded carbons, greater will be the number of hyperconjugated strictures. Thus, 2 methyl – propene has more hyper conjugated structures than propylene because the former has 6 hyper conjugation structures.

Compound	*No. of H on α carbon*	*Hyper conjugative structure*
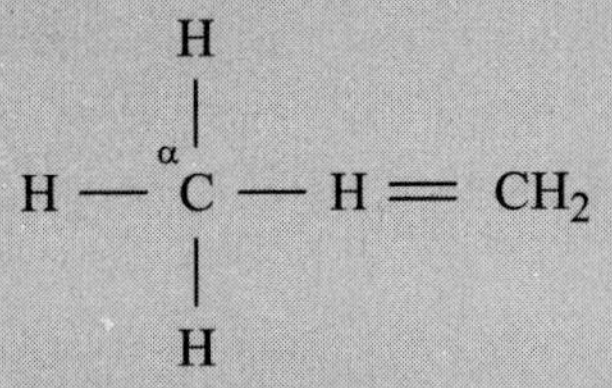 Propylene	3	3
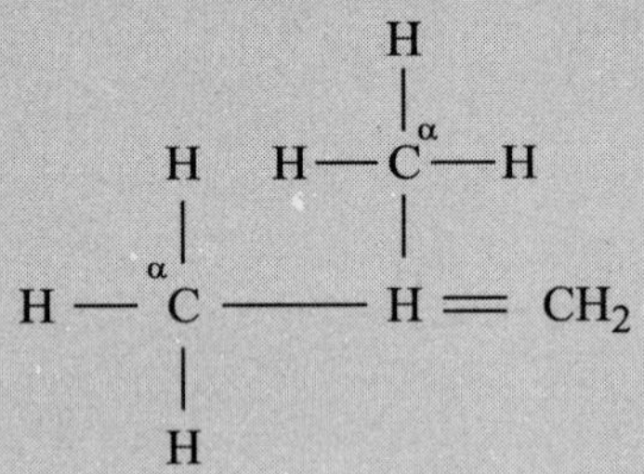 2 - methyl propylene	6	6

(*iii*) Larger is the number of hyper conjugative structures, greater will be stability of that substance. Thus, 2 methyl propene and butene – 2 ($CH_3CH = CH\ CH_3$) should be more stable than propene and this is actually so.

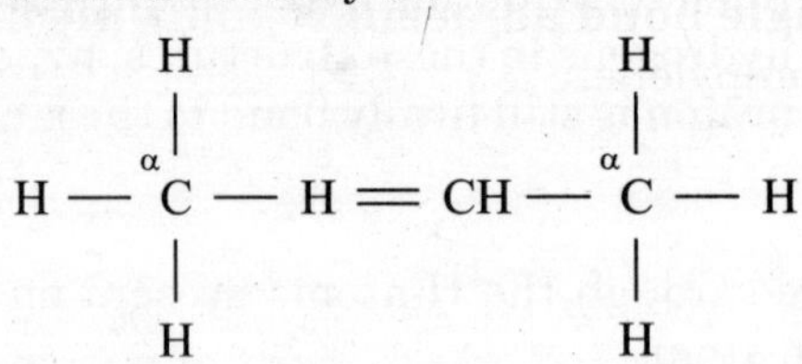

(6 hydrogen atoms on a carbon and hence 6 hyperconjugative structures)

Butene - 2

It may be mentioned here that the stabilising influence in case of hyper conjugative structures is not as large as in case of resonating structures.

Applications of Hyperconjugation

(*i*) *Stability of Alkenes:* Stability of certain alkenes over others can be explained by hyperconjugation. It has been found experimentally that greater is the number of alkyl groups attached to the α – carbon greater is the stability of alkene. It can be explained as follows: More is the number of hydrogen attached to α – carbon, more will be the hyperconjugative structures and more will be the stability. This is illustrated by taking following examples,

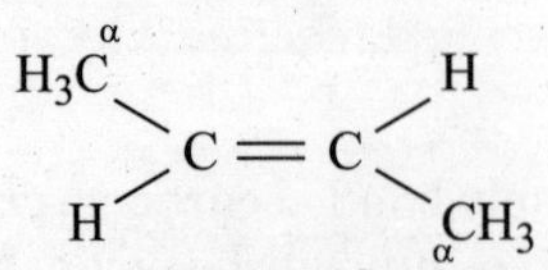

But - 2 - ene
6 α hydrogens
6 hyper conjugatine

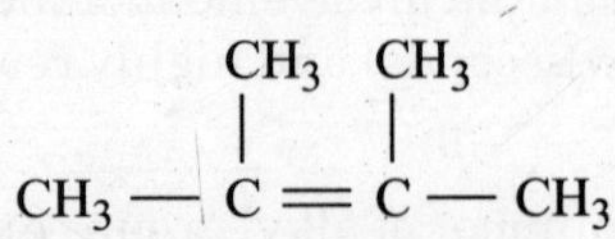

2, 3 Dimethyl but -2-ene
12 α-hydragens
12 hyper conjugatine structure

Thus 2, 3 Dimethyl but – 2 – ene is more stable than But – 2 – ene.
Similarly, propene is more stable than ethene.

$H_2C = CH_2$ $H_3C - CH = CH_2$ (Propene)
No α hydrogen 3 α hydrogen
No hyper conjugative structure 3 hyper conjugative structures (more stable)

(ii) **Directive influence of Alkyl benzenes:** In electrophilic substitution reactions of alkylbenzene, the presence of alkyl group direct the incoming electrophile at ortho and para position. This can be explained by hyper conjugation.

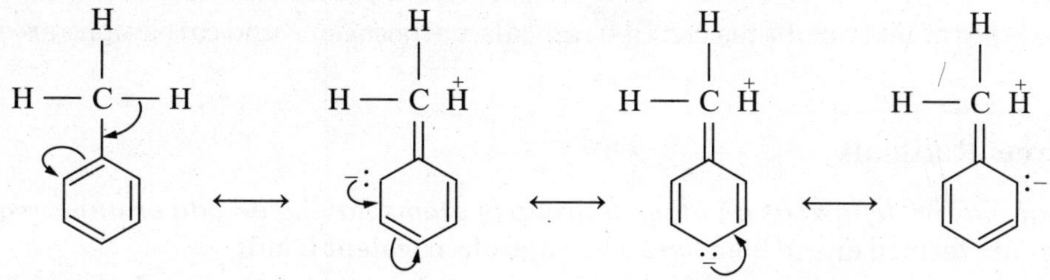

Similarly, six more structures can be drawn as two more α – *H* are present on alkyl group. From all there structures, it is evident that electron density is maximum either in ortho or para position.
Thus at a result of hyper conjugation, the electron density at o and p-position with respect or alkyl group increases and hence electrophile subsitution reactions in alkyl substituted benzene will occur at o-and p-position. *Thus alkyl group are ortho and para directing.*

(iii) **Shortening of C – C single bond adjacent to multiple bonds:**- Let us illustrate this by taking the example of propylene.

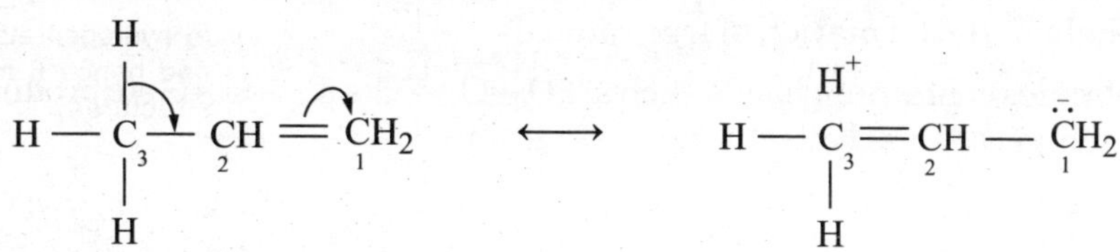

(I) (II)

We can see that the single bond between $C_2 - C_3$ in structure I acquires some double bond character in II and hence is shorter than the normal carbon-carbon bond length in propane.

(iv) **Stability of carbocations:** This is discussed in detail under the section reactive intermediates.

22.5 REACTIVE INTERMEDIATES

As discussed in previous section, homolytic or heterocyclic cleavage bonds leads to the formation of reactive organic species such as free radicals, carbocations, carbanions, etc. Most of the organic reactions occur through the involvement of these chemical species. These are generally short lived (10^{-6} seconds to a few seconds) and highly reactive and are called *reactive intermediate.*

Note: Reactive intermediates may not be confused with transition states. Reaction intermediates are stable enough to be isolated for a short period of time (few second) under appropriate condition whereas transition state refers to an imaginary state and cannot be isolated.

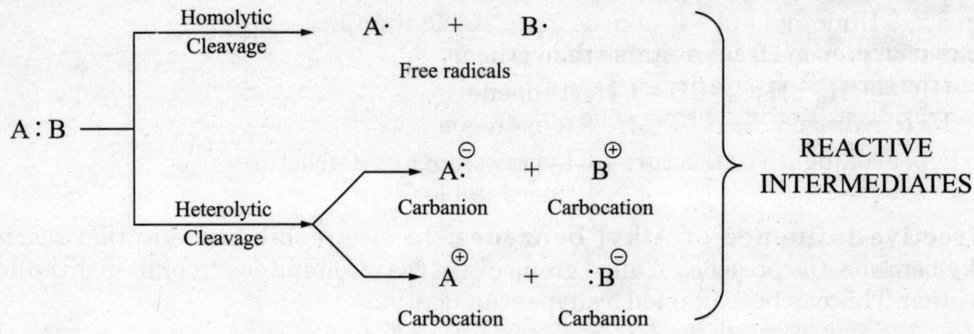

The three types of intermediates *viz*; free radicals, carbocations and carbanions are discussed below.

22.5.1 Free Radicals

A free radical may be defined as an atom or group of atoms having an odd or unpaired electron. Free radicals are formed due to homolytic cleavage of a covalent bond.

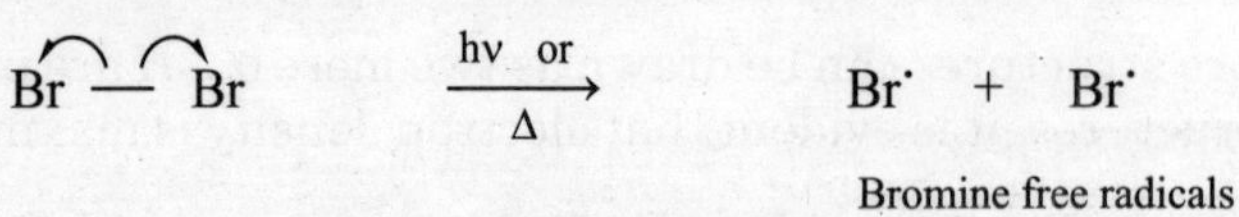

Free radicals are electrically neutral species. Due to the presence of odd electron these are *paramagnetic* in nature. Free radicals are highly reactive specie because of its tendency to gain an additional electron and complete its octet.

(*i*) **Generation:** Free radicals are generally formed in the reactions carried out at high temperature or under the influence of *high energy radiation* Moreover, the presence of weak bond also help in the formation of free radicals.

- The molecules containing bonds such as: O—O, —C—N, —C—S etc produce free radicals at high temperatures. Examples,

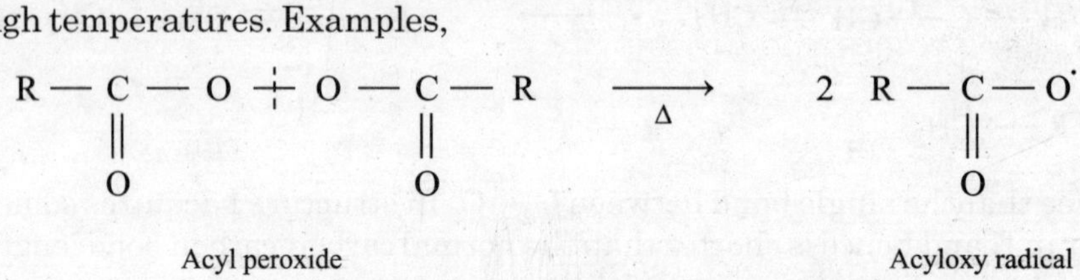

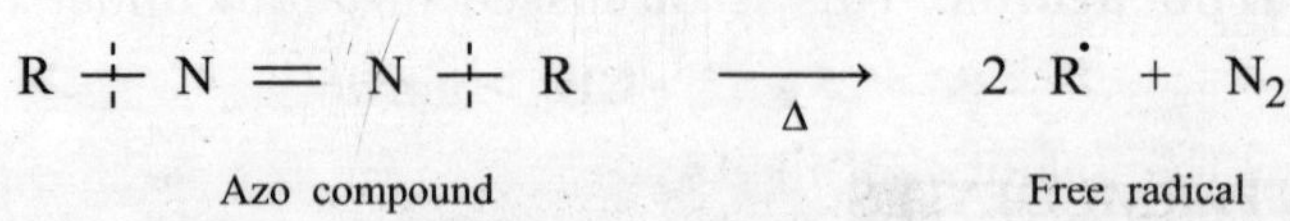

- Free radicals are produced by *high energy radiations* in photo chemical reactions. Example,

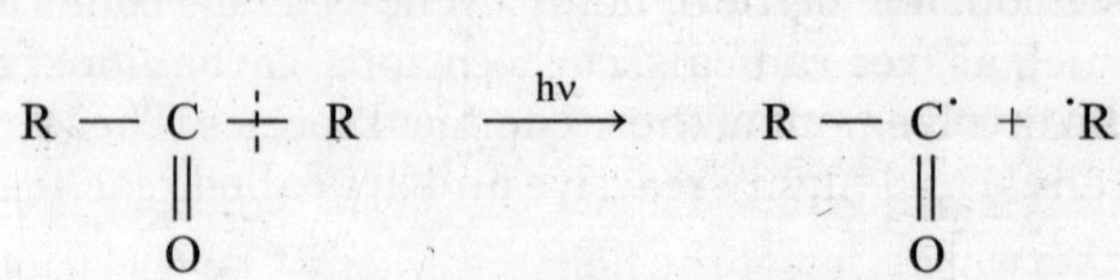

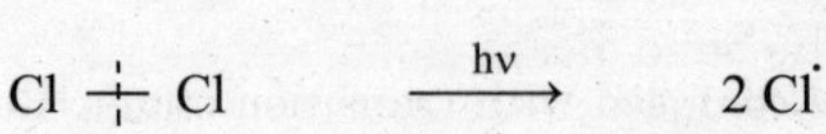

(*ii*) **Structure:** Alkyl free radicals are planar chemical species. The carbon is sp^2 hybridised and the three bonds are directed towards the three corners of planar trigonal geometry. The unhybridised p-orbital lie perpendicular to the plane and contains the odd electron.

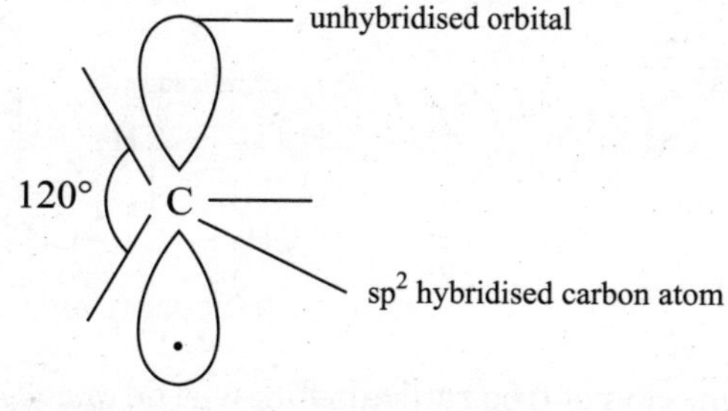

Fig. 22.1 Orbital structure of free radical planar triagonal

Sometimes, free radical may have pyramidal geometry also with sp^3 hybridised carbon.

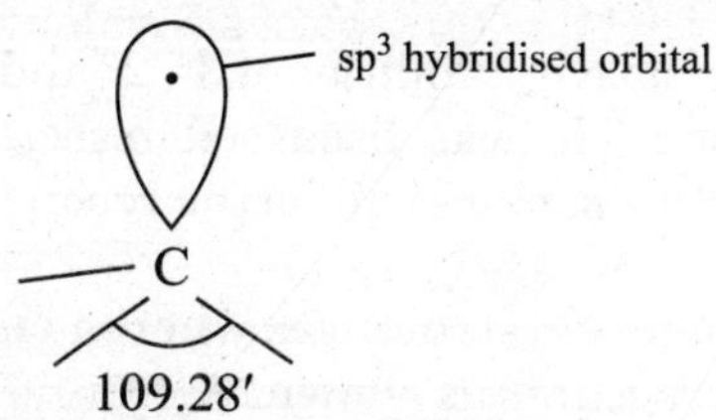

Fig. 22.2 Orbital structure of free radical – pyramidal

However, CH_3 and other simple alkyl radicals predeminantly have planar geometry which is supported by experimental proofs.

(*iii*) **Stability:** Free radicals are classified as primary, secondary or tertiary, according to the carbon atom carrying the free radical is primary, secondary and tertiary respectively.

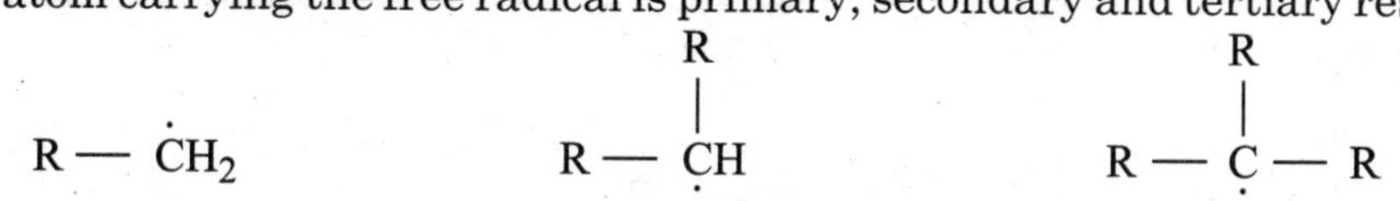

(Primary) 1° (Secondary) 2° (Tertiary) 3°

Relative to alkanes from which the radical is formed the order of stability of alkyl free radicals follows the order:

$$3° > 2° > 1° > CH_3 > \text{vinyl}$$

Note: When we compare the stabilities of alkyl free radicals, we are comparing it with the reference to the alkane form which it is formed, i.e. the difference in energy between methane and methyl radical is more than between ethane and ethyl radical (ref: Morrison and Boyd, Organic chemistry, Pearson).

The above order of stability can be explained on the basis of ease of formation and hyperconjugation.

(*a*) **Ease of formation:** It was discussed above that weak bonds easily cleave to homolytically to form free radicals. In otherwords, lower the homolytic bond * dissociation energy (ΔH) greater is the ease of formation of free radicals. The bond dissociation energies of formation for 1°, 2° and 3° alkyl radicals are:

***Note:** In this case, bond dissociation energy is the amount of energy that must be supplied to convert a mole of alkane into radicals and hydrogen atoms.

Primary $\quad CH_3—H \longrightarrow \dot{C}H_3 + \dot{H} \qquad \Delta H = 104 \text{ Kcal}$

Secondary $\quad CH_3CH—CH_3$ (H) $\longrightarrow CH_3\dot{C}H—CH_3 + \dot{H} \qquad \Delta H = 95 \text{ Kcal}$

Tertiary alkyl radical $CH_3—C(CH_3)(H)—CH_3 \longrightarrow CH_3—\dot{C}(CH_3)—CH_3 + \dot{H} \qquad \Delta H = 92 \text{ Kcal}$

Therefore, the ease of formation of free radicals follows the order,

$$3° > 2° > 1° > CH_3°$$

More stable is the radical, the more rapidly it is formed. Hence the stability of alkyl radicals also follows the same order.

(*b*) **Hyperconjugation:-** The relative stabilities of 3°, 2° and 1° alkyl radicals can be accounted for on the basis of hyper conjugation. The was discussed in section 22.3 that more is the number of hyperconjugative (no-bond) structures, more will be the resonance energy and greater will be the stability of free radicals.

Ethyl radical has 3 hyperconjugative structures whereas isopropyl radical has 6 and t-butyl radical has 9 hyperconjugative structures as shown:

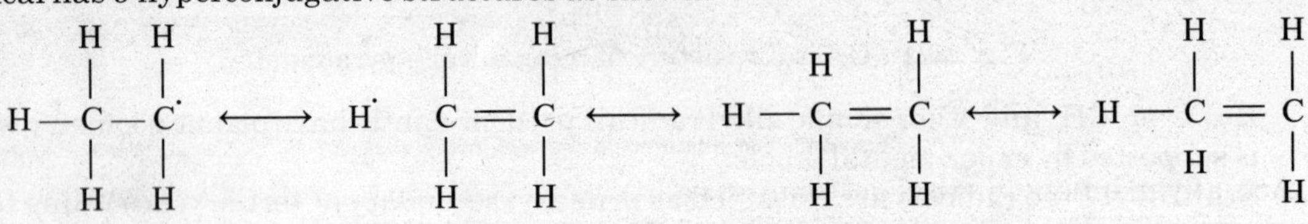

Ethyl radical — Hyperconjugative Structures (three)

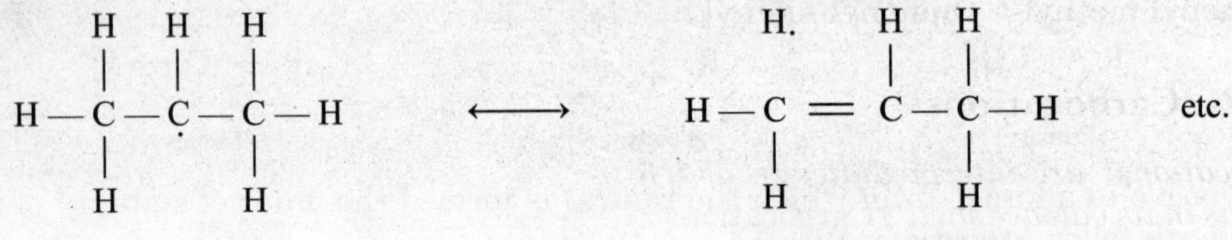

Isopropyl radical — 6 hyper conjugative structures are possible due to 6 – α hydrogen atoms.

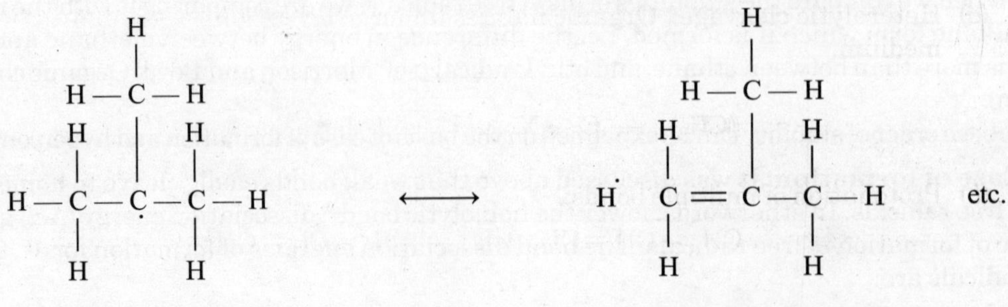

Ter - butyl radical — 9 - hyperconjugative structures possible due to 9 α - hydrogen atoms.

Therefore the order of the stability of alkyl halide free radicals is

$$3° > 2° > 1° > CH_3·$$

(*c*) **Resonance:** Allyl and benzyl radicals are more stable than simple alkyl radicals because there is increased stabilization due to resonance, as shown below:

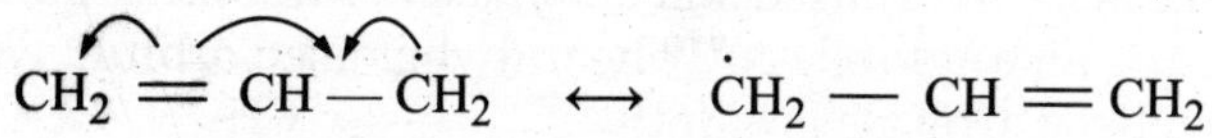

Resonance-stabilisation of allyl radical

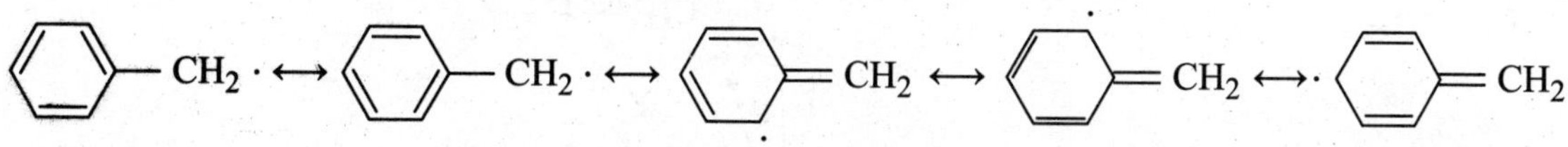

Resonance-stabilisation of benzyl radical

Diphenylmethyl and triphenylmethyl radicals, in which resonance stabilisation is much more marked should show very high stability.

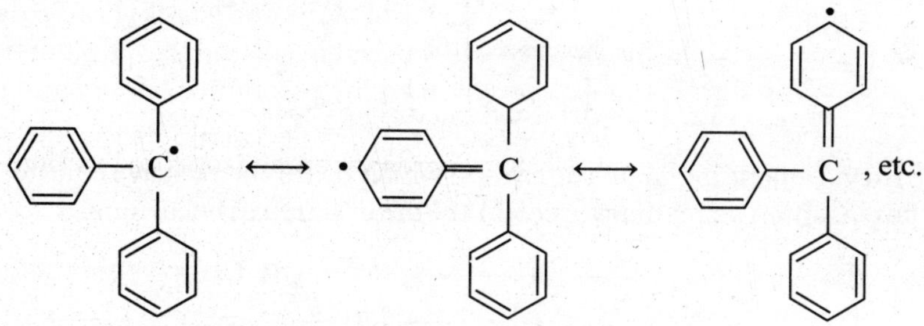

Resonance – stabilisation of triphenyl methyl radical

Actually, triphenylmethyl and similar radicals are stable enough to exist in solution at room temperature and hence can be isolated.

To sum up the decreasing order of stability of radicals mentioned above is:
Triphenyl-methyl > Diphenyl-methyl > Benzyl > Allyl > Tert-alkyl > Sec-alkyl > Prim-alkyl

22.5.2 Carbocations

"Carbocations are intermediates in which the carbon atom bears a positive charge with six electrons in its valence shell"*. Carbocations are formed due to heterolytic cleavage of the covalent bond in which leaving group (more electronegative) takes away with it the shared pair of electrons.

(*i*) **Generation:** Carbocations are produced in one of the following ways:

(*a*) Heterolytic cleavage: Organic halides undergo direct ionisation in presence of polar medium.

$$(CH_3)_3 - C \vdots - X \longrightarrow (CH_3)_3\overset{+}{C} + X^-$$

(*b*) Protonation of multiple bonds:

$$CH_3 - CH = CH_2 + H^+ \rightarrow CH_3 - \overset{+}{C}H - CH_3$$

*Note: Sometimes the alternative names such as carbonium ions or carbenium ion are used sepecially in older literature, for carbocations. But the term carbocation is most appropriate.

Carbocations are very reactive intermediates due to the electron deficient carbon atom which has only six electrons in its valence shell and thus have strong tendency to complete its octet.

(*ii*) **Structure:** Carbocations are planar chemical species. The carbon atom carrying positive charge is sp^2 hybridised and is bonded to three other atoms through its hybridised orbitals forming an angle of 120°. The unhybridised orbital, lying perpendicular to the plane of hybridised orbitals is empty.

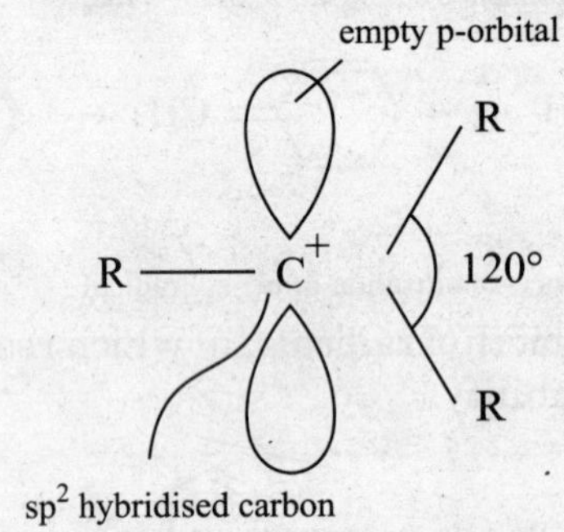

Fig. 22.3 Orbital Structure of Carbocation.

(*iii*) **Stability:** Depending upon the carbon which bears the positive charge, the carbocations can be classified as primary, secondary or tertiary carbocations.

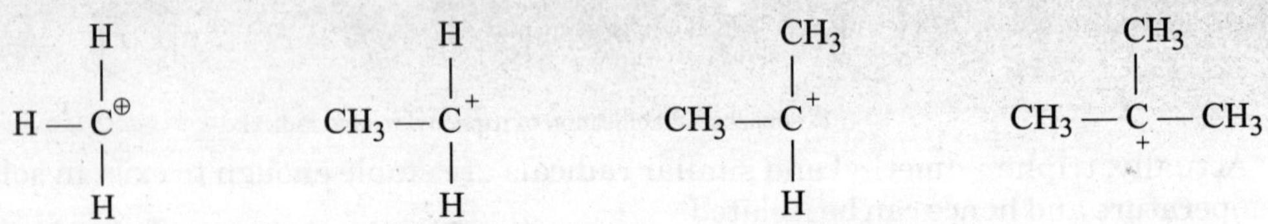

Methyl Carbocation | Primary carbocation | Secondary carbocation | Tertiary carbocation

The stability of carbocation generally follows the order:

$$3° > 2° > 1° > CH_3$$

This order of stability can easily be explained on the basis of inductive effect, hyerconjugation and resonance.

(*a*) *Inductive effect (+I):* Since carbocations are charged species, inductive effect has great influence on their stability The + I effect of alkyl groups decreases the net charge on the carbon and hence increases its stability. In otherwords, the positive charge on carbon is dispersed by alkyl groups. Greater is the number of alkyl groups on carbon atom bearing the positive charge, greater is the dispersal of charge and greater would be the stability.

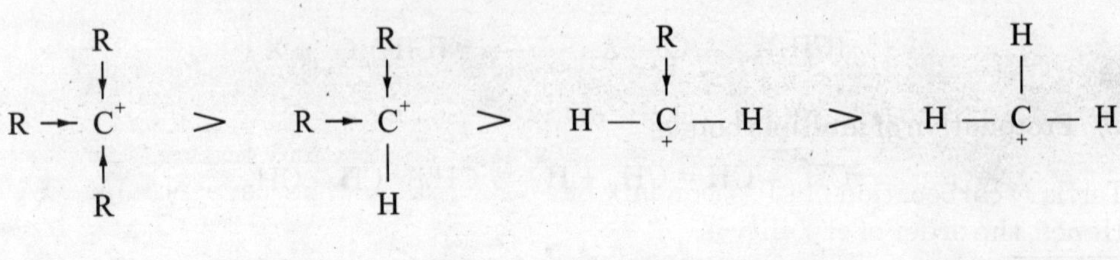

3° 2° 1° Methyl

(*b*) *Hyperconjugation:* Stability of alkyl carbocations can also be explained by hyperconjugation similar to the stability of alkyl free radicals. The + I effect and hyperconjugation in tertiary carbocations are responsible for its greater stability than secondary and primary.

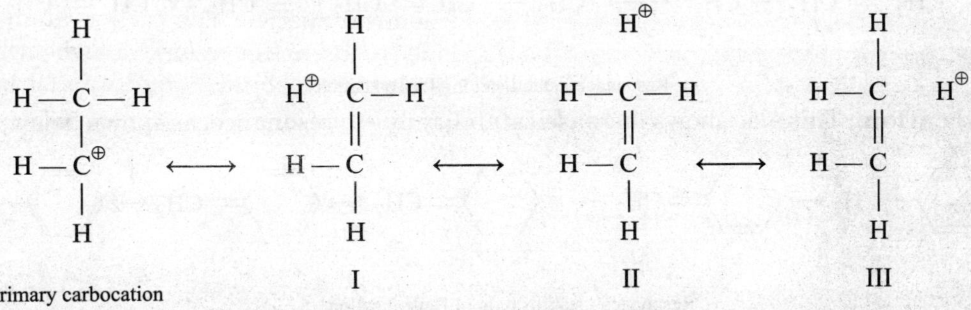

Primary carbocation

Total-3 hyperconjugative structures

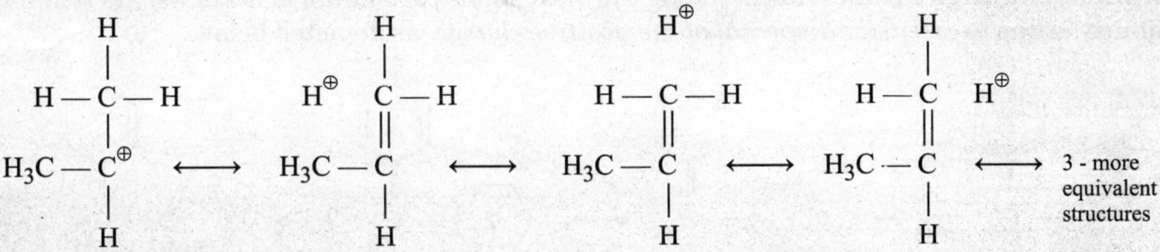

Isopropyl cation
Secondary carbocation

Total 6-hyperconjugative structures (6 α hydrogens)

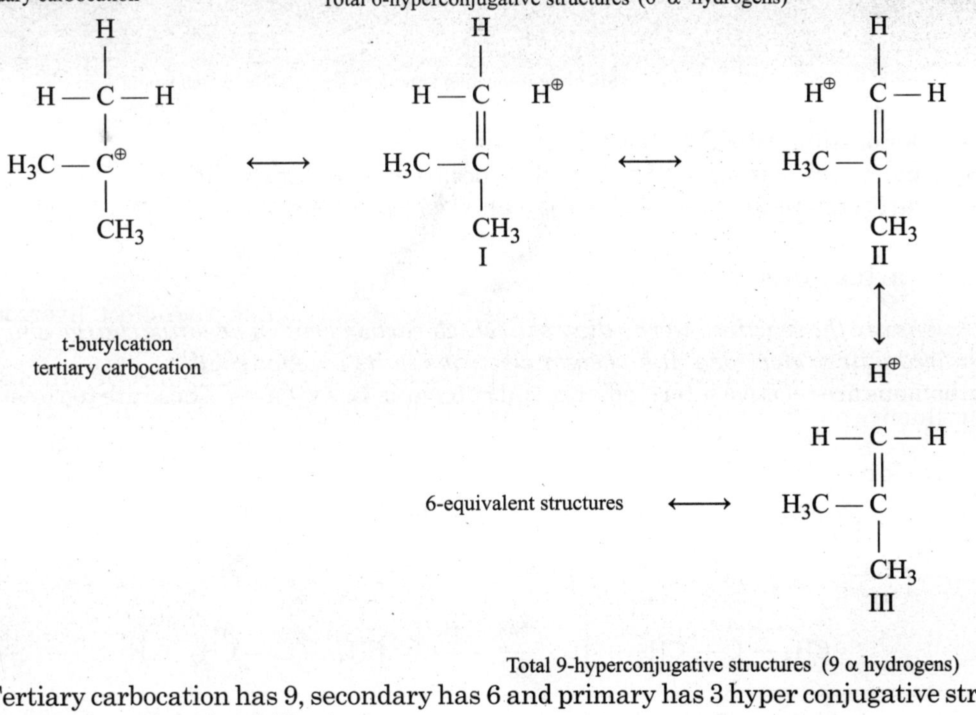

t-butylcation
tertiary carbocation

Total 9-hyperconjugative structures (9 α hydrogens)

Tertiary carbocation has 9, secondary has 6 and primary has 3 hyper conjugative structure. Hence, the order of stability is

$$3° > 2° > 1° > CH_3.$$

(*c*) *Resonance:* The stability of benzyl and allyl carbocation is explained on the basis of resonance as shown below:

Allyl cation: It is stabilised through resonance, as:

$$CH_2 = CH - \overset{\oplus}{C}H_2 \longleftrightarrow \overset{\oplus}{C}H_2 - CH = CH_2 \left(\equiv \overbrace{CH_2 \overset{\cdots}{=} CH \overset{\cdots}{=} CH_2}^{\oplus} \right)$$

Resonance - stabilisation of Allyl cation

Benzyl cation: This also shows increased stability due to resonance, as shown below:

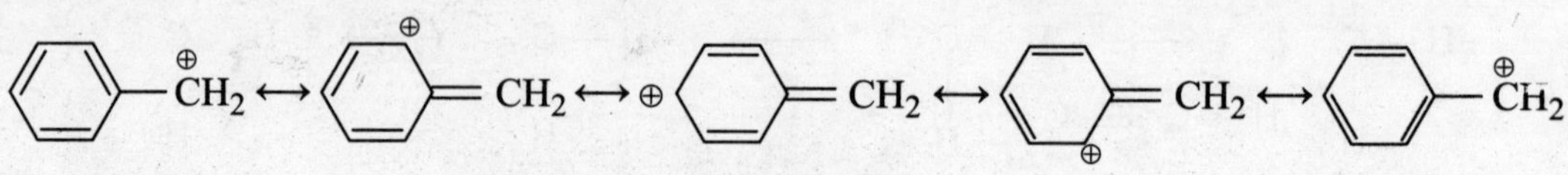

Resonance - stabilisation of Benzyl cation

Triphenyl methylcation: This is among the most stable carbonium ions known. Its remarkable stability is due to *extensive dispersal of the positive charge*, as depicted below:

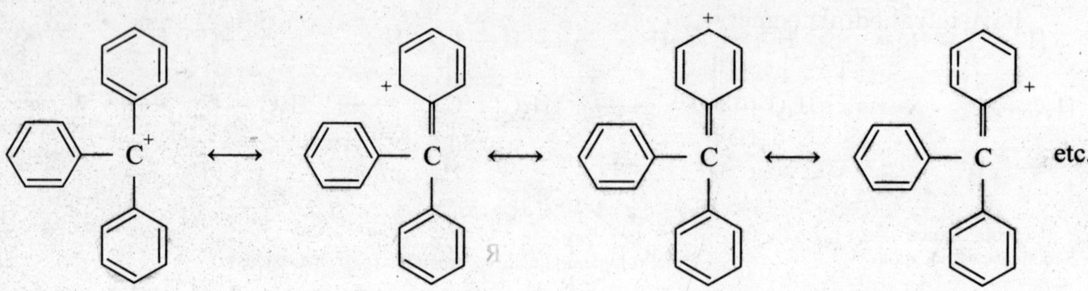

(Six more resonating structures are possible, which would disperse the positive change)

The order of stability of the above three ions is
Triphenyl methyl carbocation > Benzyl carbocation > allyl carbocation
These carbocations are more stable than alkyl carbocations.

22.5.3 Carbanions

"Carbanions are the reactive intermediates in which carbon bears a negative charge and possesses an unshared pair of electrons. It has eight electrons in its valence shell".

Carbanions are reactive intermediates and behave as Lewis bases. These are too unstable to be isolated.

(i) Generation:

- Heterolytic cleavage of covalent bond in which carbon is attached to less electronegative element.

$$CH_3 - \overset{\overset{\displaystyle O}{\|}}{C} - CH_2 - \vdots H \xrightarrow{NaO\,C_2H_5} CH_3 - \overset{\overset{\displaystyle O}{\|}}{C} - \overset{..}{C}H_2 + H^+$$

$$H - \vdots CH_2 - CHO \xrightarrow{\bar{O}H} \overset{..}{C}H_2 - CHO + H_2O$$

Acetaldehyde

- Nucleophilic attack on a carbon multiple bond.

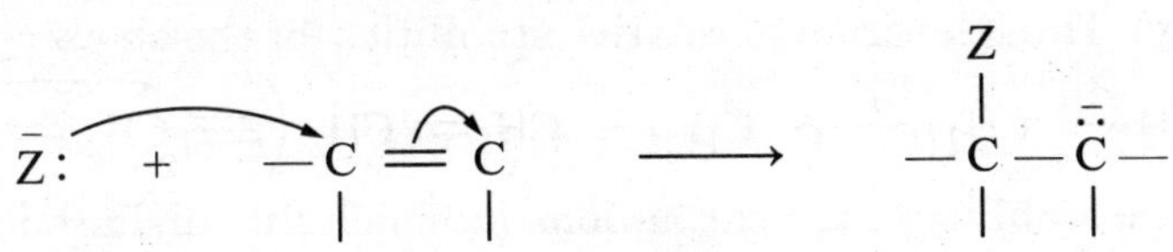

Nucleophile

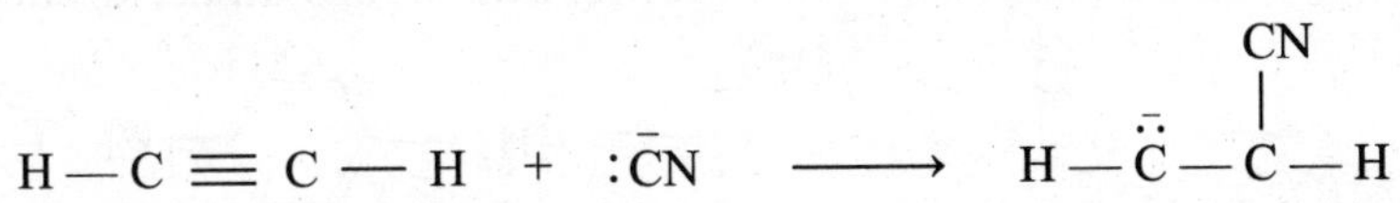

Ethyne

(*ii*) **Structure:** Carbocations have pyramidal structure. The carbon carrying the negative charge is sp^3 hybridised. The three of four sp^3 hybridised orbitals form three σ bonds with other groups whereas the fourth one occupies the lone pair of electrons. Thus carbanions have tetrahedral geometry.

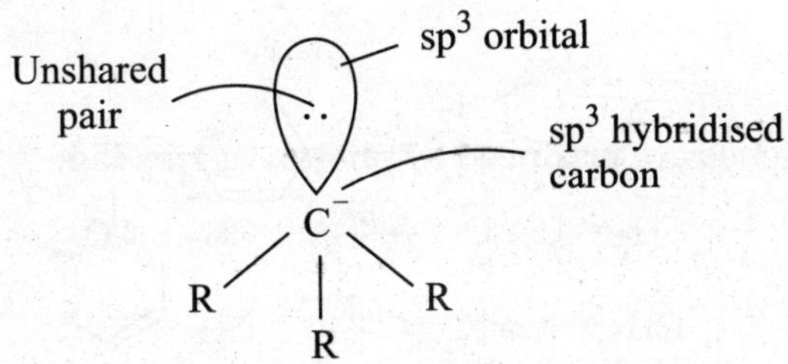

Fig. 22.4 *Orbital Structure of Carbanion*

(*iii*) **Stability:** The order of stability of carbanions is as follows:

$$\bar{C}H_3 > 1° > 2° > 3°.$$

The stability can be explained due to following effects:

(*a*) **Inductive effect:** The order is readily understood if we recall that alkyl groups are electron donating (+I effect) and hence they will intensify the negative charge on carbon, making the anion less stable. Greater the number of alkyl groups, greater would be the negative charge density at central carbon atom of carbanion and lesser would be the stability

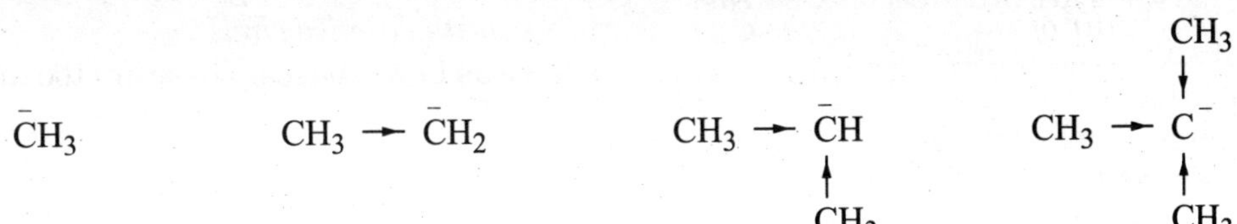

methyl 1° 2° 3°

Hence $\bar{C}H_3$ is most stable followed by 1°, 2° and Tertiary.

(*b*) *s – Character:* The stability of carbanions derived from alkynes, alkenes and alkanes can be explained on the basis of their *s* character.

$HC \equiv \bar{C}$	$H_2C = \bar{C}H$	$H_3C - \bar{C}H_2$
(sp hybridised)	(sp^2 hybridised)	(sp^3 hybridised)
s character 50%	33%	25%

Higher the *s* character, closer the electrons to the nucleus and hence lower the energy, (more stability). Thus the order of relative stabilities for the above carbanions is:

$$HC \equiv \bar{C} > H_2C = \bar{C}H > H_3C - \bar{C}H_2$$

(*c*) *Resonance:* The stability of the carbanions in which the unshared pair is in conjugation with unsaturated bond is due to resonance. Example

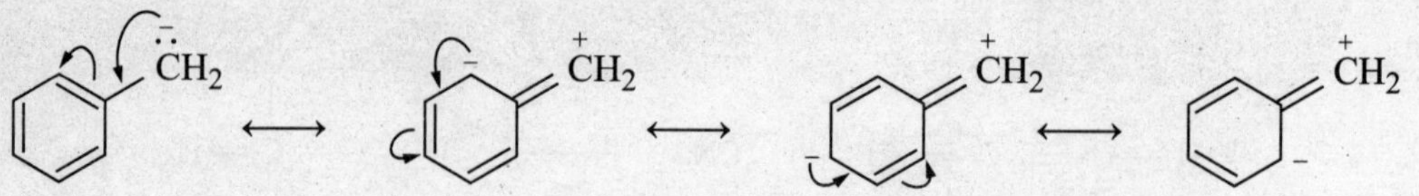

22.5.4 Carbenes

Carbenes are the reactive intermediates in which carbon has six electrons, two of which are free. These are neutral species. Electron deficient specie

The simplest example of carbene is methylene.

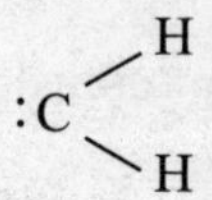

Methylene

(*i*) **Generation:-** Carbenes are produced by photolytic cleavage

$$CH_2{=}C{=}O \xrightarrow{hv} :CH_2 + CO$$

$$CH_2{=}\overset{+}{N}{=}\bar{N} \xrightarrow{hv} :CH_2 + N_2$$

(*ii*) **Forms:** Carbene has two forms:

Singlet carbene: The unshared electron pair in carbene remains paired 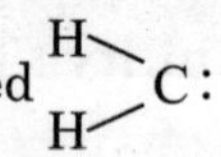

Triplet carbene: The unshared electron pair is unpaired, i.e. it is diradical

$$H - \dot{C} - H\,.$$

22.6 TYPES OF ORGANIC REACTIONS

Organic reactions are broadly classified as:

(*i*) Substitution Reactions
(*ii*) Addition Reactions
(*iii*) Elimination Reactions
(*iv*) Rearrangement Reactions.

This classification is solely based on the structural changes occurring in the reactant molecule as shown below:

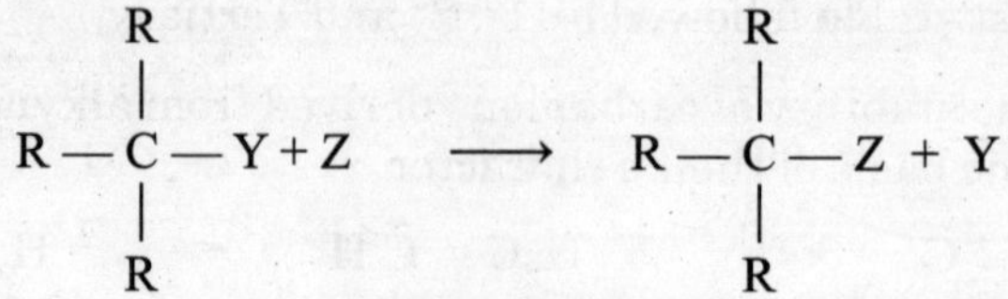

SUBSITUTION
(The group/atom *Y* is substitued by *Z*)

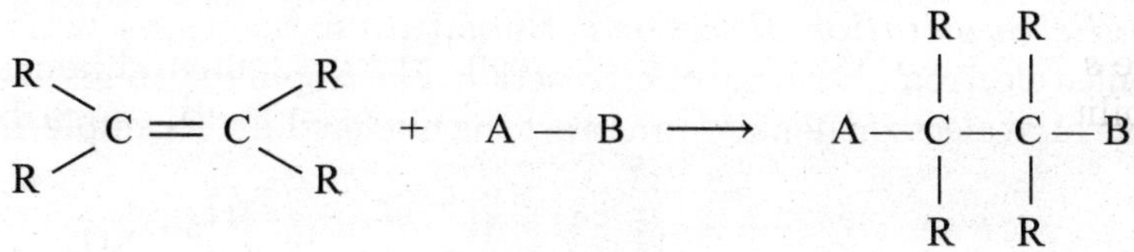

ADDITION

(The molecule A-B is simply added to the reactant)

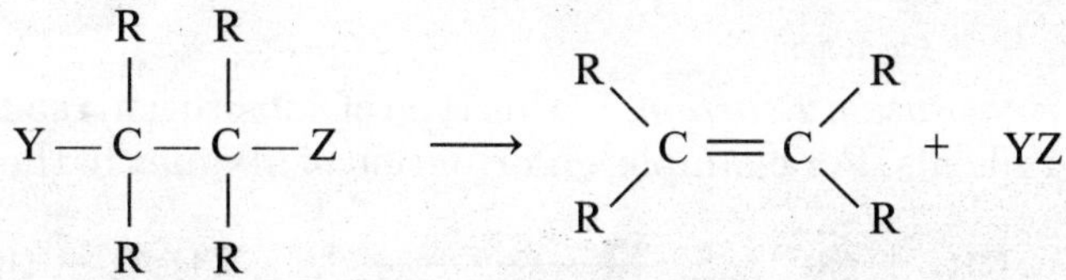

ELIMINATION

(The molecule/group Y-Z is eliminated from the reactant)

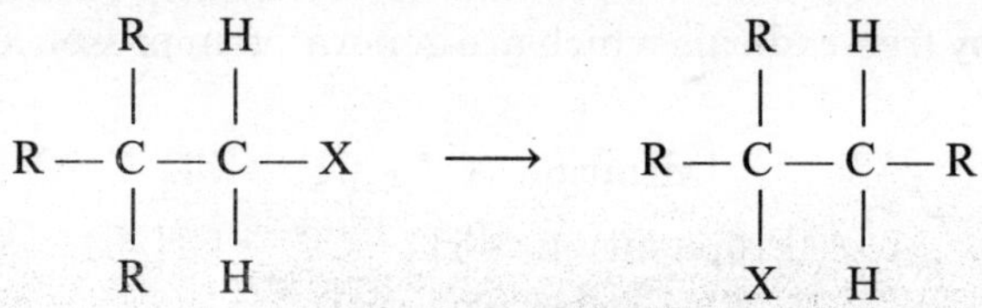

REARRANGEMENT

(The reactant molecule is simple rearranged)

Chemical reactions are commonly represented as equations such as:

$$\text{Reactant} + \text{Reagent} \xrightarrow[\text{Conditions}]{\text{Reaction}} \text{Products}$$

Now, we will discuss the above four classes briefly.

22.6.1 Subsitution Reactions

The subsitution reaction or displacement reactions are those reactions in which atoms or group of atoms attached to a carbon atom in a substrate molecule is replaced by another atoms or group of atoms.

The subsitution reactions can take place through different attacking species (nucleophile, electrophile or a free radical). The subsitution reactions are of the following types:

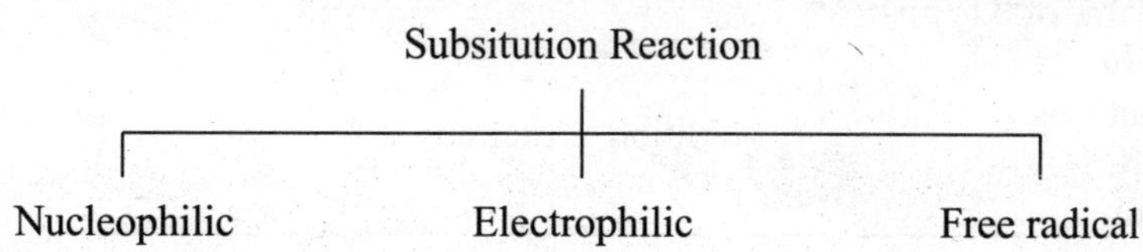

(i) Nucleophilic Subsitution Reaction: Subsitution reactions which are brought about by nucleophiles are called nucleophilic subsitution reactions. There are denoted by SN reaction (S = subsitution, N = Nucleophile). These reactions are typical of alkyl halides. Example

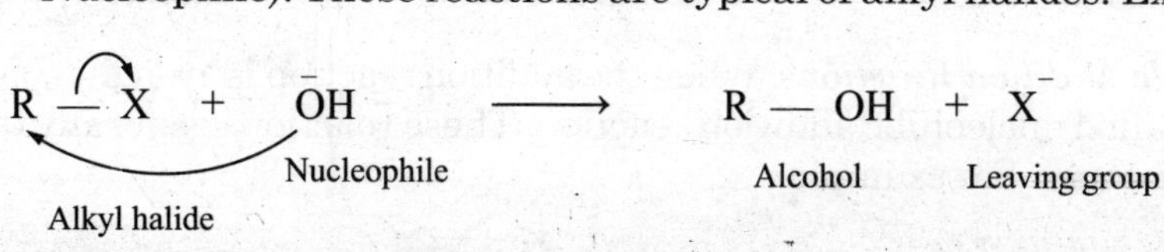

The mechanism of Nucleophilic subsitution reactions is discussed in detail in the next section.

(*ii*) *Electrophilic Subsitution Reactions:* Subsitution reactions which are brought by electrophiles are called electrophilic subsitution reaction. Electrophile systems are typical of aromatic systems which have high electron density to attract the nucleophile. Example, nitration of benzene ring

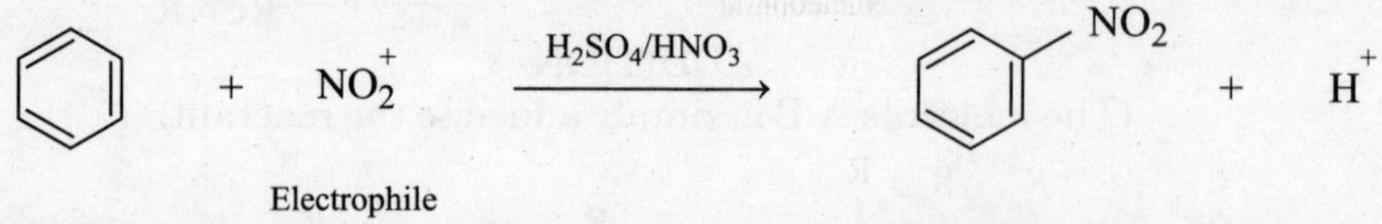

(*iii*) *Free Radical Subsitution Reaction:* In this type of subsitution reaction, the subsitution is brought about by free radicals. For example, chlorination of alkanes in the presence of sunlight.

$$CH_4 + \dot{C}l \longrightarrow CH_3Cl + HCl$$

Free radical — Methyl chloride

The reaction occurs by free-radicals which are generated in presence of sunlight. The reaction involves three steps.

$$\text{Initiation}: Cl^{\bullet} + Cl^{\bullet} \rightarrow Cl_2$$

$$\text{Propagation}: CH_3^{\bullet} + Cl^{\bullet} \rightarrow CH_3Cl$$

$$CH_3^{\bullet} + Cl \rightarrow CH_3Cl$$

Propagation steps keep on producing reactive free radicals by chain propagation until the free radicals are terminated in termination step.

Termination. $Cl^{\bullet} + Cl^{\bullet} \rightarrow Cl_2$

$$CH_3^{\bullet} + Cl^{\bullet} \rightarrow CH_3Cl$$

$$CH_3^{\bullet} + CH^{\bullet} \rightarrow CH_3 - CH_3$$

22.6.2 Addition Reactions

Addition reactions are the reactions in which the attacking reagent adds up to the reactant (substrate) molecule to give the single molecule of the product. Addition reactions are typical of compounds containing double or triple bonds.

Like subsitution reactions, addition reactions can also proceed through different attacking reagents and hence are of 3 types:

Addition Reactions

- Nucleophilic addition reaction
- Electrophilic addition reaction
- Free radical addition reaction

(*i*) *Nucleophile Addition Reactions:* When the addition reaction is brought about by nucleophiles, the addition is called nucleohilic addition reaction. These reactions generally take place in case of aldehydes and ketones. For example,

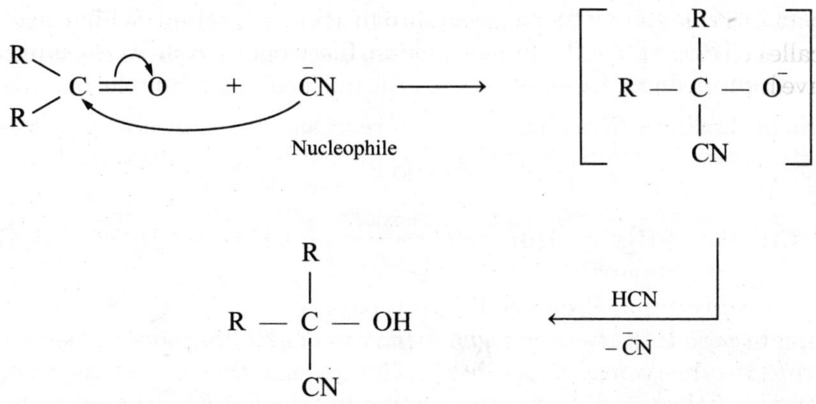

Addition product

(*ii*) ***Electrophilic Addition Reactions:*** Addition reaction brought about by electrophile is called electrophilic addition. Carbocations are formed as reaction intermediates. These reactions are typical of alkenes and alkynes.

$$CH_2 = CH_2 + HBr \longrightarrow CH_3\underset{\displaystyle Br}{\overset{|}{C}H} - CH_3$$

The reaction takes place in two steps:

$$CH_2 = CH_2 + \overset{+}{H} \xrightarrow{\text{slow}} \left[\overset{+}{C}H - CH_3\right]$$

Carbocation

$$\overset{+}{C}H_2 - CH_3 + Br^- \xrightarrow{\text{fast}} \underset{\underset{\displaystyle Br}{|}}{CH} - CH_3$$

Bromoethane

Electrophilic addition to asymmetrical alkenes occurs according to the 'Markovnikav's Rule'.

Markovnikov's Rule:

If the two carbon atoms at the double bond are linked to a different number of hydrogen atoms, the halogen is found preferentially at the carbon with less hydrogen substituents; This observation is known as Markovnikov's rule.

$$CH_3 - CH = CH_2 + HBr \longrightarrow CH_3 - \overset{+}{C}H - CH_3 + Br^- \quad \text{...(i)}$$

$$\underset{\text{Carbocation}}{CH_3 - \overset{+}{C}H - CH_3} + Br^- \longrightarrow CH_3 - \underset{\underset{\displaystyle Br}{|}}{CH} - CH_3 \quad \text{...(ii)}$$

2 - Bromopropane
(Major product)

Markovnikov's rule or Markownikoff's rule is an observation formulated by the Russian chemist Markovnikov in 1870. In chemical reactions found particularly in organic chemisty involving the addition of H—X to the alkene, the rule states that,

"The addition of H-X to an unsymmertrical alkene, the acid hydrogen (H) becomes attached to the carbon with the greatest number of hydrogens, and the halide (X) group becomes attached to the carbon with the fewest hydrogens"

This is due to an intermediate which is produced by the abstraction of a hydrogen atom by the alkene from the acid (HBr) to form the most stable carbocation (i) (relative stability: 3° > 2° > 1° > methyl), as well as generating a Br anion.

(*iii*) *Free Radical Addition Reaction:* Addition reactions brought about by free radicals are called free radical addition reactions. Example, free radical addition of Br_2 to alkene in presence of peroxides.

$$CH_3\,CH = CH_2 + HBr \xrightarrow{\text{Peroxide}} \underset{\text{n-propyl bromide}}{CH_3 - CH_2 - CH_2\,Br}$$

It is interesting to note that the free radical addition of Br_2 to propene takes place by *Anti-Markovnokov's rule*. In other words, the addition of hydrogen halide to the unsymmetrical alkene, in presence of peroxide, takes place in such a way that hydrogen gets attached to the carbon of the double bond having lesser number of hydrogen atom. This is also known as *peroxide effect.*

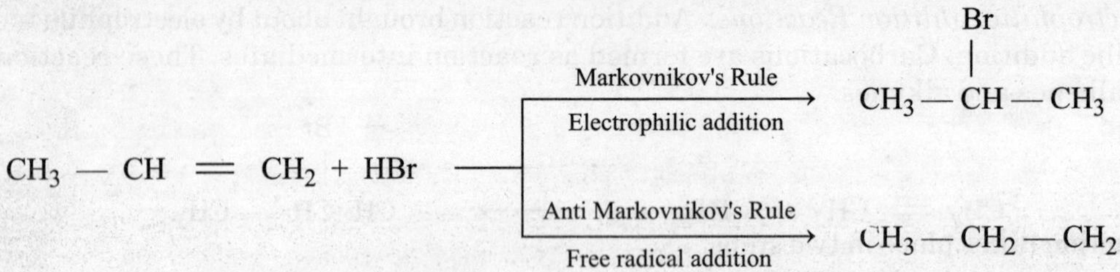

The above selectivity can be explained on the basis of stable intermediates formed. The free radical addition of HBr to unsymmetrical alkenes involves the following steps:

Initiation:

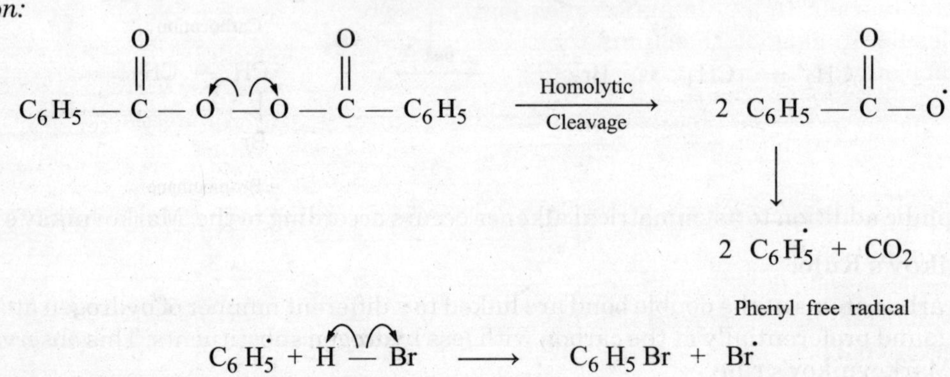

Propagation:

The $\dot{B}r$ free radical then attacks the propene to form primary and secondary free radicals which is chain propagating step.

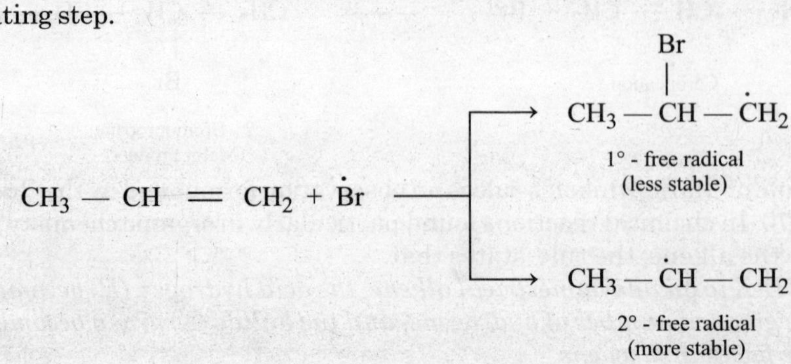

Free radical formed in step II is then attacked by HBr to form the product. Since secondary free radical is more stable than the primary free radical hence the final product is obtained form secondary free radical.

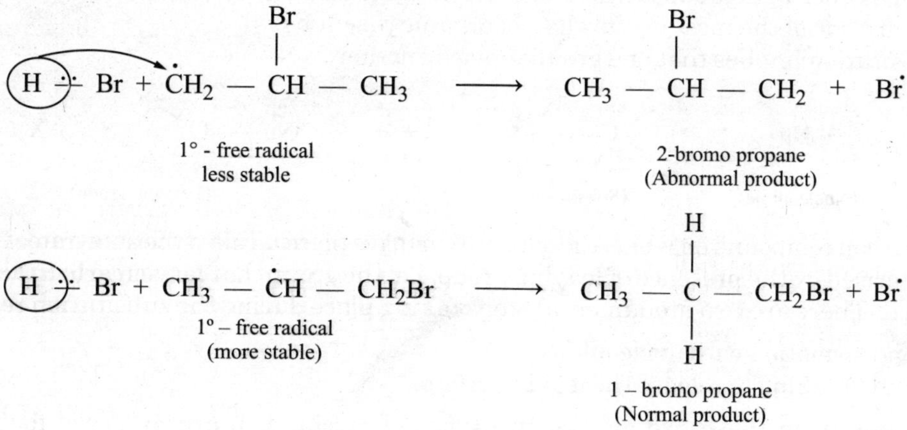

Termination.

$$\dot{Br} + \dot{Br} \rightarrow Br_2$$

22.6.3 Elimination Reactions

The term 'elimination' describes that a small molecule is eliminated during the reaction. These reactions are of two types:

(*i*) β-Elimination: In β elimination reactions, two atoms or groups are lost from the same molecule. In most of the elimination reactions, the atoms groups are lost from adjacent carbon atom thus forming a new double or triple bond. Example,

$$\underset{H}{\overset{\beta}{CH_2}} - \underset{OH}{\overset{\alpha}{CH_2}} \xrightarrow{\text{Conc. } H_2SO_4} CH_2 = CH_2 + H_2O$$

(*ii*) α-Elimination: In these reactions, the loss of two atoms or groups is from the same carbon atom. Example, dehydrohalogenation of chloroform in the presence of base β-elimination reactions are more common. The brief mechanism of elimination reaction is discussed in the section 22.8.

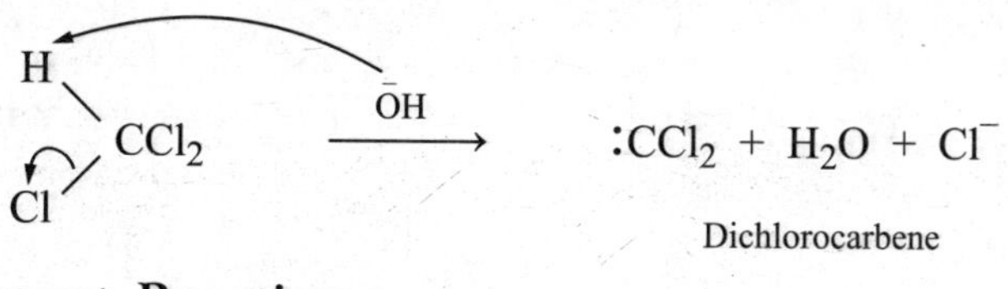

22.6.4 Rearrangement Reactions

These reactions involve the migration of an atom or group within the molecule. For example,

$$CH_3-CH_2-CH_2-CH_2-Br \xrightarrow[575\text{ K}]{\text{Anhyd. } AlCl_3} CH_3-CH_2-\underset{}{\overset{Br}{\overset{|}{CH}}}-CH_3$$

2–Bromobutane

Beckmann's Rearrangement (described later under the chapter Name Reaction)

22.7 NUCLEOPHILIC SUBSITUTION REACTIONS

It was discussed in the previous section that the subsitution reaction brought about by nucleophiles are known as nucleophilic subsitution reactions and are denoted by S_N. Nucleophilic substitutions reactions are one of the most useful class of organic reactions.

A subsitution implies that one group replace another,

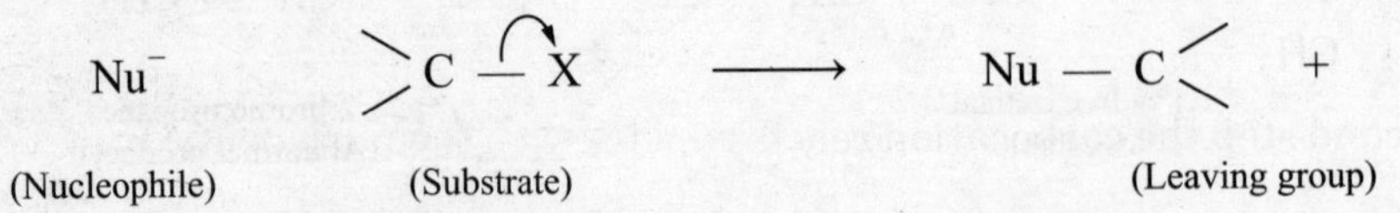

The carbon compound on which the subsitution takes place is called the substrate. The substrate is characterised by the presence of leaving group, i.e. the group that leaves carbon taking electron pair with it. There are two fundamental steps taking place during the subsitution reactions,

- Bond formation with nucleophile
- Bond breaking to release the leaving group.

Depending upon these two events, two types of mechanism are proposed for nucleophilic subsitution reactions:

(*i*) SN_1 Mechanism
(*ii*) SN_2 Mechanism

Let us discues these one by one.

22.7.1 SN_1 Mechanism:

SN_1 indicates a *subsitution, nucleophilic unimolecular reaction.*

The components required for nucleophilic subsitution are: substrate, nucleophile and the solvent. In SN_1 type of reaction, the rate of the reaction depends upon the concentration of the substrate and is independent of the concentration of the nucleophile, i.e. it is unimolecular reaction. It shows first order kinetics and reaction rate is proportional to the concentration of the substrate

$$\text{Rate} \propto [\text{Substrate}].$$

Example: The reaction between tert-butyl bromide and hydroxide ion to yield tertiary butyl alcohol follows first order reaction.

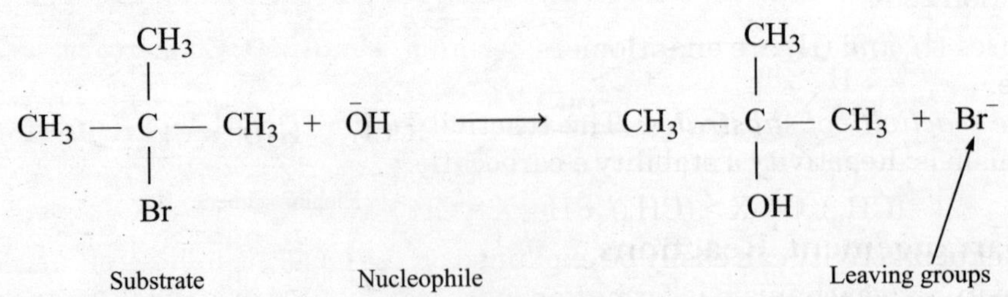

The rate of the reaction depends upon the concentration of tert-butyl bromide only.

$$\text{Rate} = k\,[(CH_3)_3\,C\text{-}Br]$$

Mechanism: The rate law suggest that reaction occurs in two steps:

Step I: In this step, the alkyl halide undergoes heterolytic cleavage to form a bromide ion and a carbocation.

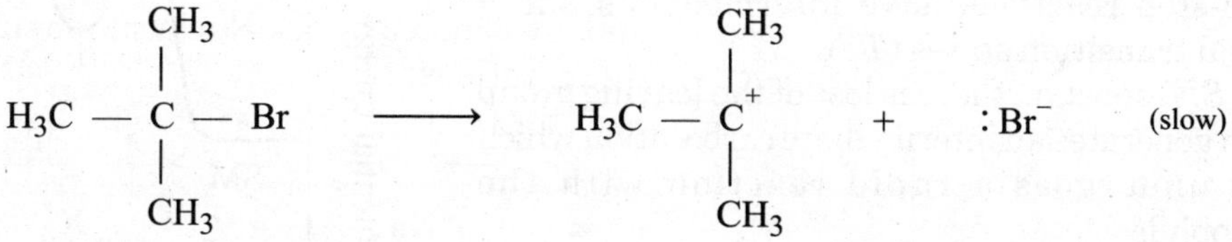

Step II: In the second step the carbocation reacts rapidly with the nucleophile to gtive the product

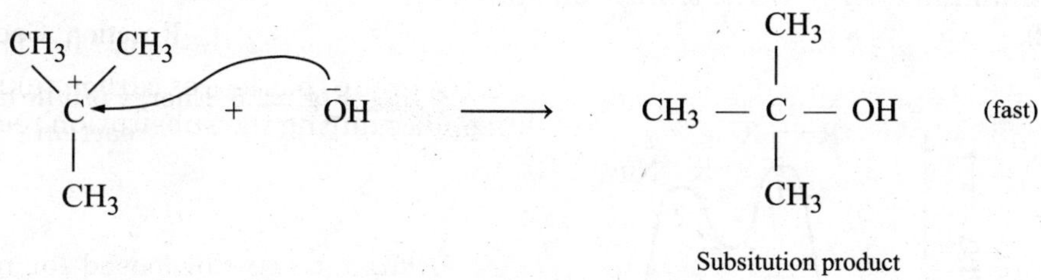

So, the reaction rate is dependent upon the first step, which is the *slow* step. This mechanism is thus called SN_1 (Subsitution, Nucleophilic Unimolecular). The term unimolecular is used here since rate determining step involvers only *one* molecule.

Characteristics of SN_1 Reaction:

(*i*) *Stereochemistry:* We can see that nucleophile can attach from either face of the carbocation, which is a planar molecule. Hence, if the alkyl halide is optically active, then the product will be *racemic mixture.*

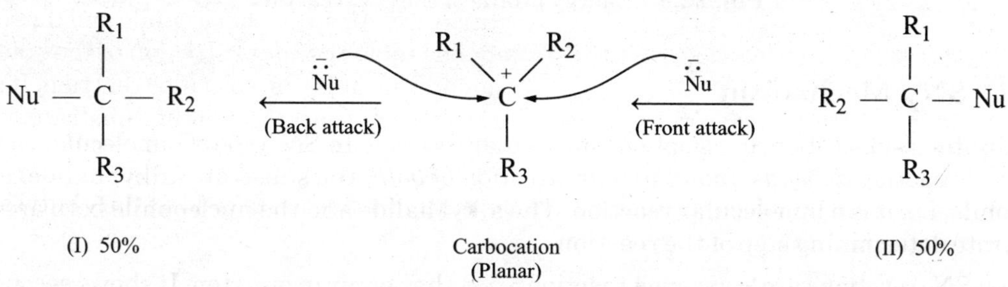

Structures (I) and (II) are enantiomers, forming almost 50 : 50 mixture, i.e. the racemic mixture.

(*ii*) *Relative reactivity of substrates:* The reactivity order of the alkyl halides towards SN_1 mechanism is: Reactivity a stability a carbocation

$$(CH_3)_3C—X > (CH_3)_3CH—X > CH_3CH_2—X > CH_3—X$$

The order of reactivity can easily be explained on the basis of stability of carbocations. The key step in SN_1 mechanism is *formation of intermediate carbocation.* The more stable the carbocation is, more easily it is formed and faster the SN_1 reaction will be. We have studied in the previous section that the stability of carbocation follows the order:

$$3° > 2° > 1° > \overset{+}{C}H_3$$

due to inductive and hyperconjugation. Therefore tert-alkyl halides undergo SN_1 reaction mechanism predominantly.

(*iii*) *Energy diagram for SN_1 reaction:* Energy profile for SN_1 reactions:

Multi-step reactions have intermediates and a several transition states (*TS*).

In an SN_1 reaction there is loss of the leaving group which generates an intermediate carbocation which then undergoes a rapid reaction with the nucleophile

It can be seen that the intermediate carbocation has minimum energy w.r.t. transition states (1) and (2).

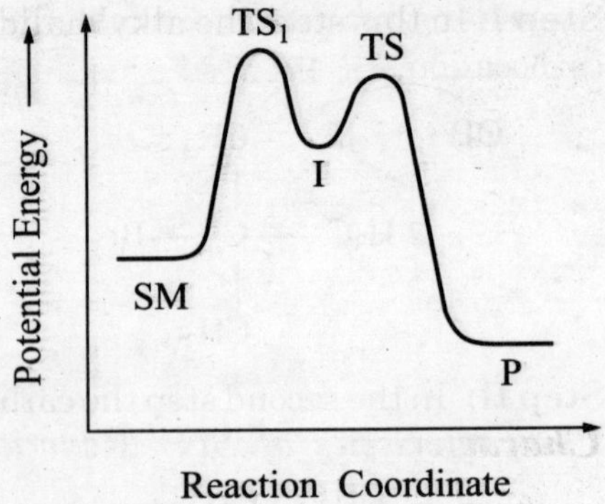

Fig. 22.5 Energy profile of the general reaction

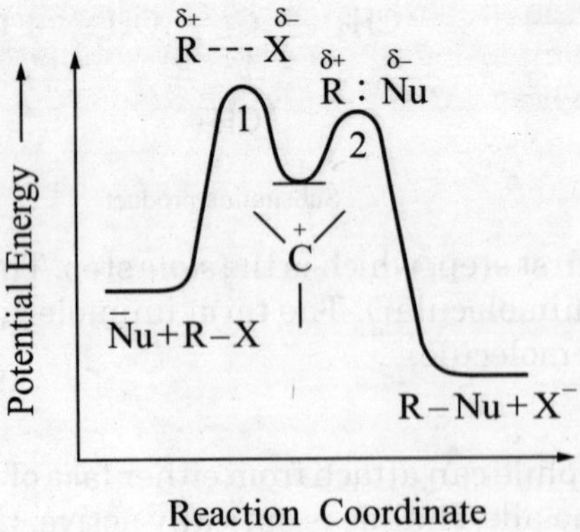

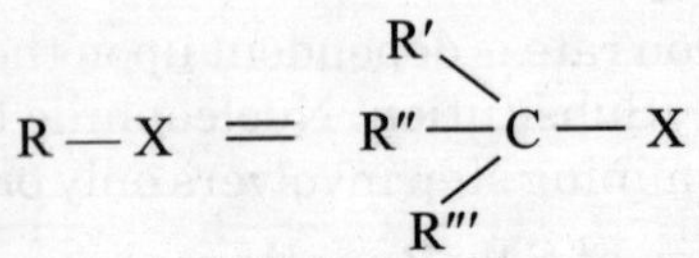

Fig. 22.6 Energy profile of the SN_1 reaction

22.7.2 SN_2 Mechanism

SN_2 indicates a subsitution, nucleophile bimolecular reaction. In SN_2 type of bimolecular reactions, the rate of reaction *depends upon the concentration of both*, the substrate (alkyl halide) and the nucleophile, i.e. it is a bimolecular reaction. The alkyl halide and the nucleophile both are taking part in rate determining step of the reaction.

Thus SN_2 reactions are 'concerted reactions', i.e. they occur in one step. It shows second order i.e. kinetics the rate is proportional to substrate and the nucleophile. Hence,

$$\text{Rate } \alpha \text{ [Substrate] [Nucleophile]}$$

Example: The reaction between CH_3Br and hydroxide ion to yield methanol,

$$CH_3Br + \bar{O}H \longrightarrow CH_3OH + B\bar{r}$$

$$\text{Rate} = k[CH_3Br]\ [\bar{O}H]$$

Mechanism: The rate law suggests that the reaction requires a collision between alkyl halide and nucleophile. The breaking of bond and formation of bond takes place simultaneously.

SN_2 reactions proceed via a transition state in which carbon appear to be pentavalent. The transition state can be pictured as a structure in which carbon is partially bonded to both $\bar{O}H$ and $B\bar{r}$ *i.e.*, the C—OH is not completely formed and C—Br is not completely broken.

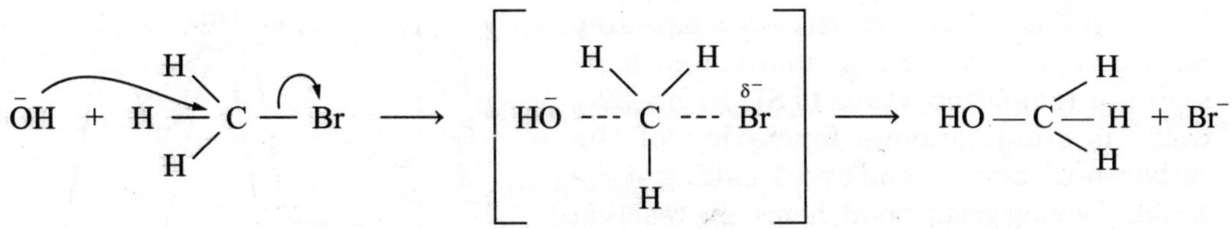

Transition state (I)

Product

Characteristics of SN_2 Reactions

(*i*) *Stereochemistry:* We can see that in SN_2 reaction, the attack of the nucleophile ($\bar{O}H$) occurs from the back side, whereas the leaving group (Br^-) leaves from the front side. As a result, the SN_2 reaction is always accompanied by *inversion* of configuration commonly known as 'Walden Inversion'. This means that if the alkyl halide is optically active, the subsitution product will also be optically active and the product will have opposite configuration to the substrate. If alkyl halide has '*S*' configuration, the product will have '*R*' configuration.

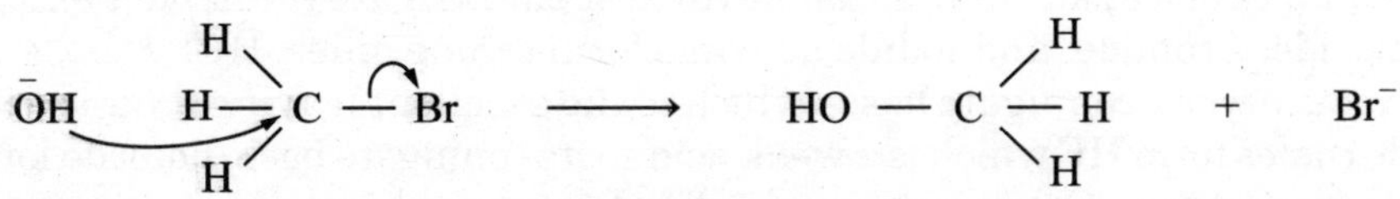

Inversion of configuration

(*ii*) *Relative reactivity of substrate:* The reactivity order of haloalkanes towards SN_2 mechanism is:

$$CH_3X > R–CH_2 – X > R_2\ CH – X > R_3CX$$

or simply, primary halides are more reactive towards SN_2 mechanism as compared to tertiary halides, which are least reactive.

This order of reactivity is explained on the basis of 'steric factors'. We can see from the mechanism that the nucleophile approaches from the opposite side of the leaving halide ion. If bulky groups are attached to the carbon, the approach of nucleophile would be quite difficult. We can say that nucleophile is 'sterically hindered' by the alkyl groups. Hence, as the number of substituents (alkyl groups) attached to the carbon bearing the halogen is *increased*, the reaction towards SN_2 *decreases*. Methyl halides or primary halides have much less sterical hinderance and hence undergo SN_2 type of reactions easily whereas, tertiary halides are sterically hindered and hence does not react through SN_2 mechanism.

(*iii*) *Effect of Nucleophile:* Since the nucleophile is involved in the rate determining step, the nature of the nucleophile is very important in an SN_2 reaction. SN_2 reactions are dependent upon the strength and concentration of the nucleophile. Strong nucleophiles will react more rapidly than nucleophiles that are described as poor or weak. Negative charged nucleophiles tend to be stronger than the neutral nucleophiles with the same central atom. For example, $\bar{O}H$ is stronger than H_2O and $R\bar{O}$ are better nucleophiles than R-O-H. Stronger nucleophiles will enhance the reaction rate of a reaction operating under SN_2 conditions.

Increasing the concentration of the nucleophile will enhance the rate of an SN_2 reaction since kinetically the reaction rate is also dependent upon the nucleophile concentration. This is in contrast to SN_1 where the concentration and nucleophilicity *have no effect* on the rate of an SN_1 reaction.

(*iv*) *Energy profile for SN_2 Reaction:* Single step reactions have no intermediates and have a single transition state (TS). In an SN_2 there is simultaneous formation of the carbon-nucleophile bond and breaking of the carbon-leaving group bond, hence the reaction proceeds via a TS in which the central C is partially bonded to five groups. No intermediate is present in the SN_2 reactions.

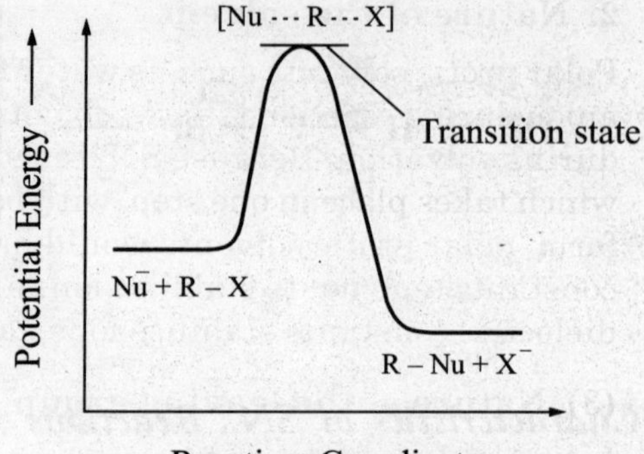

Fig. 22.7 Energy profile of SN_1 Reactions

(*v*) *Leaving Group Ability: Effect on SN_2 Reactions:* The leaving group ability (LGA) is defined as the ability of the leaving group to be displaced and remain stable. Leaving groups that are weak bases (coming form strong acids) are more stable and better leaving groups. After leaving the molecule these do not have any desire to return. However, the more basic the leaving group is (coming form weak acids), the poorer will be its ability to depart

Some examples of leaving groups that have excellent LGA are the halides except for fluoride ion. Chloride, Bromide, and Iodide all come form strong aides, HCl, HBr, and HI. Strong acids produce weak conjugate bases which make excellent leaving groups with high LGA. Flouride comes form HF which is a weak acid so its conjugate base, flouride ion would make a poor leaving group. Other groups with low LGA would include acetate ion, cyanide ion ($\bar{C}N$) and Nitrite ion, $N\bar{O}_2$.

The reaction rate of a reaction occurring via SN_2 path would increase with higher LGA for the leaving group. The same effect is observed for SN_1 reaction.

22.7.3 SN_1 Versus SN_2 Reactions

Whether an alkyl halide will undergo an SN_1 or an SN_2 reaction depends upon a number of factors. Some of the more common factors include the natures of the carbon skeleton, the solvent, the leaving group, and the nature of the nucleophile.

1. Nature of the carbon skeleton

Only those molecules that form extremely stable cations undergo SN_1 mechanisms. Normally, only compounds that yield 3° (tertiary) carbonications (or resonance-stabilised carbocations) undergo SN_1 mechanisms rather than SN_2 mechanisms. Carbocations of tertiary alkyl halides not only exhibit stability due to the inductive effect, but the original molecules exhibit steric hindrance of the rear lobe of the bonding orbital, which inhibits SN_2 mechanisms from occurring. Primary alkyl halides, which have little inductive stability of their cations and exhibit no steric hindrance of the rear lobe of the bonding orbital, generally undergo SN_2 mechanisms. Figure below illustrates the tendencies of alkyl halides toward the two types of substitution mechanisms.

Increasing SN_2 reactions ←

1° alkyl halide 2° alkyl halide 3° alkyl halide

→ Increasing SN_1 reactions

2. Nature of the solvent

Polar protic solvents such as water favor SN_1 reactions, which produce both a cation and an anion during reaction. These solvents are capable of stabilising the charges on the ions formed during solvation. Because SN_2 reactions occur via a **concerted mechanism** (a mechanism which takes place in one step, with bonds breaking and forming at the same time) and no ions form, polar protic solvents would have little effect upon them. Solvents with low dielectric constants tend not to stabilise ions and thus favor SN_2 reactions. Conversely, solvents of high dielectric constants stabilise ions, favoring SN_1 reactions.

(3) Nature of the leaving group

In general, good leaving groups are those capable of forming stable ions or molecules upon displacement from the original molecule. Conversely, poor leaving groups form ions of poor to moderate stability. Strong bases, such as OH^-, NH_2^-, and RO^-, make poor leaving groups. Water, which is less basic than a hydroxide ion, is a better leaving group. Poor bases usually make good leaving groups. A poor base is an ion or group in which the electrons are tightly bound to the molecule due to high electronegativity or resonance.

The following list ranks atoms and molecules in order of their stability as leaving groups, from most to least stable.

$$CH_3SO_3^-, \text{ tosyl}^- > I^- > Br^- > H_2O^+ > Cl^- > F^-$$

22.8 MECHANISM OF ELECTROPHILIC SUBSTITUTION REACTIONS (ELECTROPHILIC AROMATIC SUBSTITUTION REACTIONS)

It was discussed in the previous section that the substitution reactions brought about by the electrophilles are known as the Electrophilic substitution reactions. Electrophilic substitution is characteristic of aromatic compounds. Electrophilic aromatic substitution is an important way of introducing functional groups on benzene rings.

Although aromatic compounds have multiple double bonds, these compounds do not undergo addition reactions. Their lack of reactivity toward addition reactions is due to the great stability of the ring systems that result from complete π electron delocalisation (resonance). Aromatic compounds react by electrophilic aromatic substitution reactions, in which the aromaticity of the ring system is preserved.

Electrophilic Substitution Reactions of Benzene

A two-step mechanism has been proposed for these electrophilic substitution reactions. In the first, slow or rate-determining, step the electrophile forms a sigma-bond to the benzene ring, generating a positively charged **benzenonium intermediate**. In the second, fast step, a proton is removed from this intermediate, yielding a substituted benzene ring.

Every specific electrophilic aromatic substitution will contain these two steps plus whatever steps explain the initial generation of the electrophile required in step 1 above.

The mechanism is depicted in the following reactions , typical of benzene and its derivatives.

(1) Halogenation:

The electrophilic substitution reaction between benzene and chlorine or bromine is termed as halogenation. Benzene reacts with chlorine or bromine in an electrophilic substitution reaction, but only in the presence of a catalyst. The catalyst is either aluminium chloride (or aluminium bromide if benzene is reacted with bromine)

The reaction between benzene and chlorine in the presence of either aluminium chloride gives chlorobenzene.

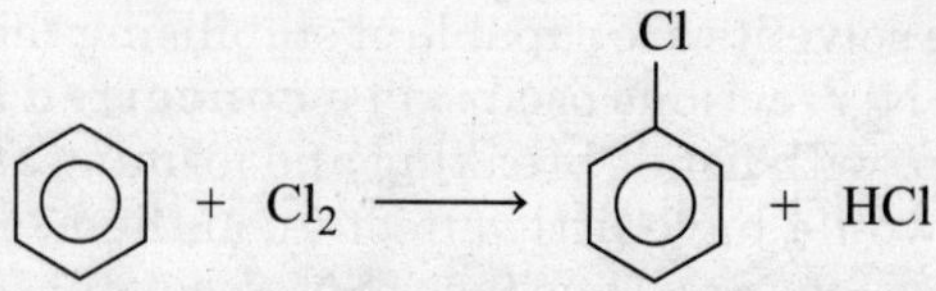

The reaction between benzene and bromine in the presence of either aluminium bromide gives bromobenzene.

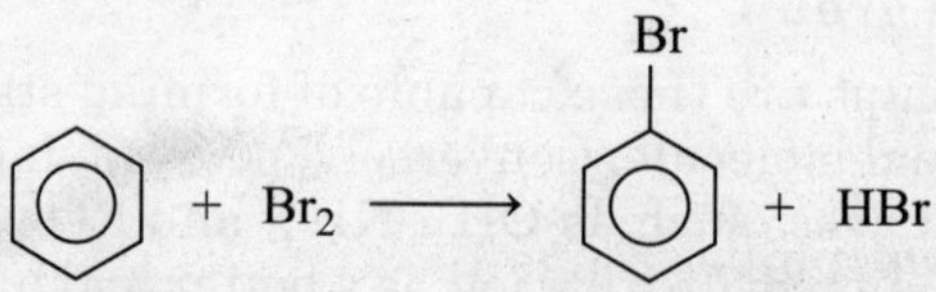

Iron may also be used as a catalyst. Iron is usually used because it is cheaper and more readily available. Strictly speaking iron isn't a catalyst, because it gets permanently changed during the reaction. It reacts with some of the chlorine or bromine to form iron(III) chloride, $FeCl_3$, or iron(III) bromide, $FeBr_3$.

$$2Fe + 3Cl_2 \longrightarrow 2FeCl_3$$

$$2Fe + 3Br_2 \longrightarrow 2FeBr_3$$

These compounds act as the catalyst and behave exactly like aluminium chloride in these reactions. All electrophilic aromatic substitution reactions share a common mechanism.

Mechanism: The bromination of benzene

This mechanism consists of a series of steps.

1. An **electrophile** is generated. For the bromination of benzene reaction, the electrophile is the Br^+ ion generated by the reaction of the bromine molecule with ferric bromide, a Lewis acid.

$$Br—Br + FeBr_3 \longrightarrow Br—Br^+—FeBr_3^- \longrightarrow \overset{+}{Br} + FeBr_4^-$$

2. The electrophile is attracted by the π electron system of the benzene ring to form a nonaromatic carbocation.

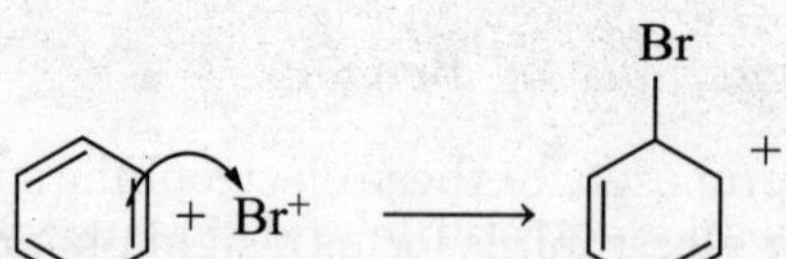

3. The positive charge on the carbocation that is formed is delocalised throughout the molecule, i.e., it is resonance hybridised.

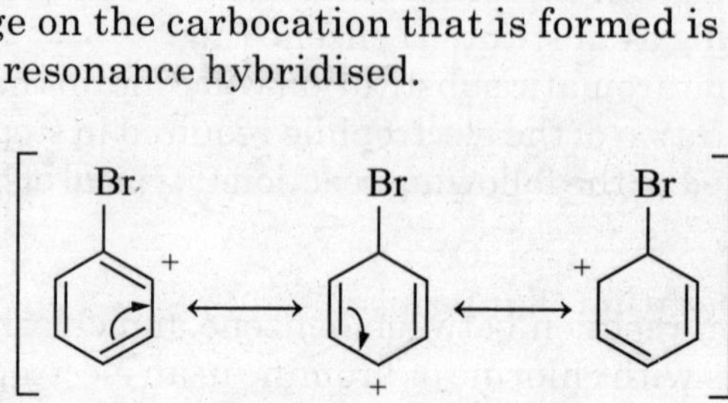

4. The aromaticity is restored by the loss of a proton from the atom to which the bromine atom (the electrophile) has bonded.

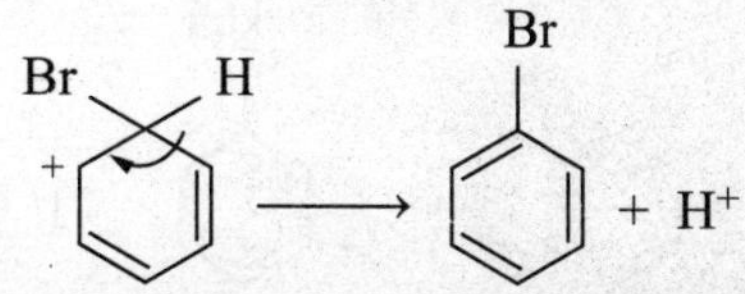

5. Finally, the proton reacts with the $FeBr_4^-$ to regenerate the $FeBr_3$ catalyst and form the product HBr.

$$H^+ + FeBr_4^- \longrightarrow FeBr_3 + HBr$$

2. The nitration of benzene

In another example of an electrophilic aromatic substitution reaction, benzene reacts with a mixture of concentrated nitric and sulfuric acids to create nitrobenzene.

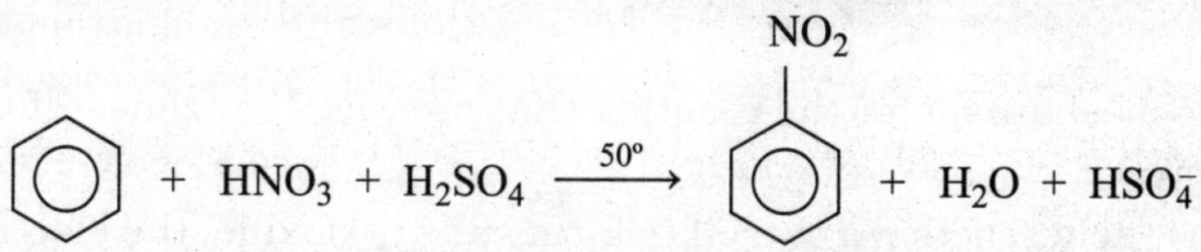

Mechanism: The mechanism for the nitrobenzene reaction is as follows:.

1. Sulfuric acid ionizes to produce a proton.

$$H_2SO_4 \rightleftharpoons H^+ + HSO_4^-$$

2. Nitric acid accepts the proton in an acid-base reaction.

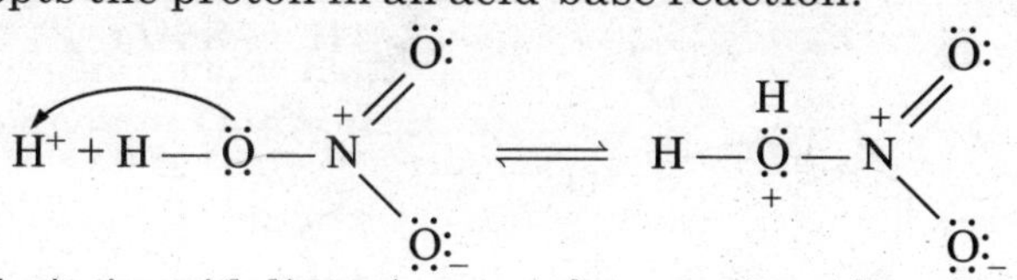

3. The protonated nitric acid dissociates to form a nitronium ion ($^+NO_2$).

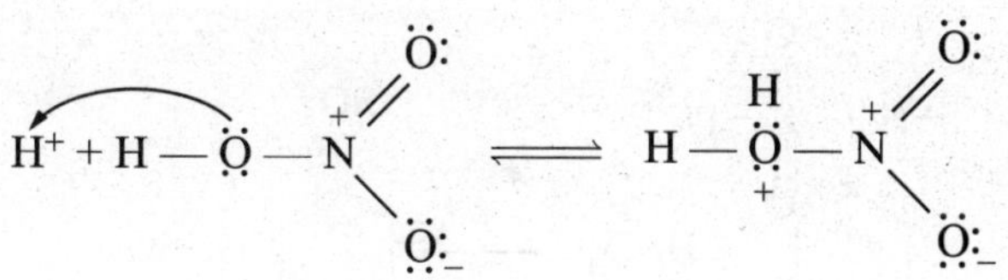

4. The nitronium ion acts as an electrophile and is attracted by the π electron system of the benzene ring.

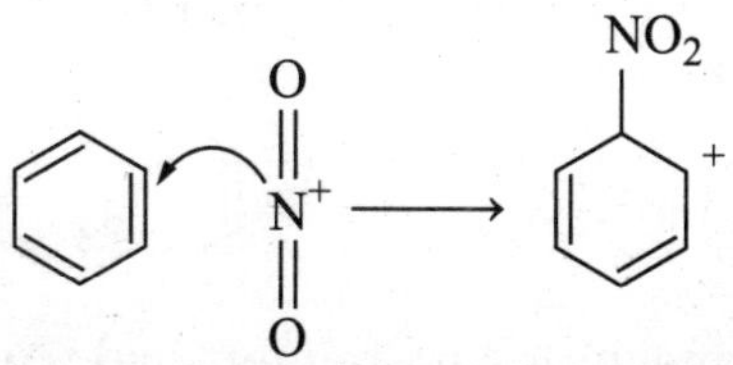

5. The nonaromatic carbocation that forms has resonating structures.

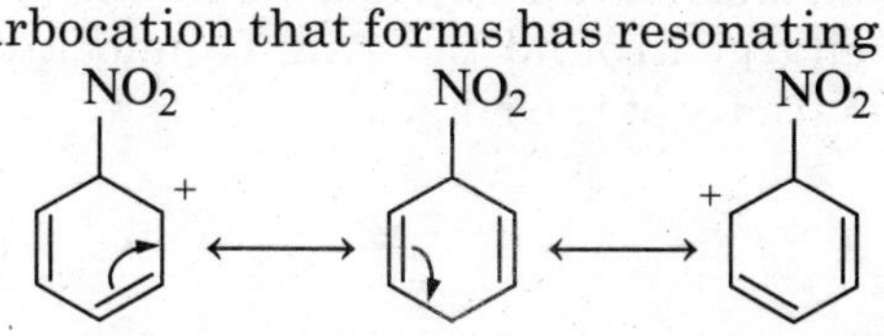

6. The aromaticity of the ring is reestablished by the loss of a proton from the carbon to which the nitro group is attached.

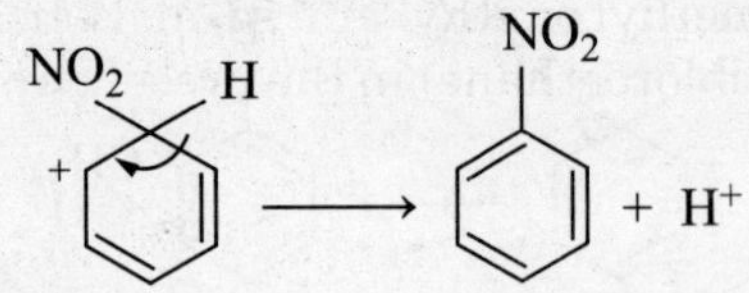

3. The sulfonation of benzene

The reaction of benzene with concentrated sulfuric acid at room temperature produces benzenesulfonic acid.

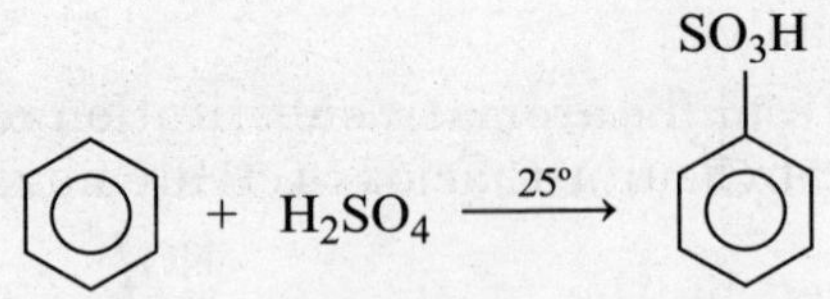

Mechanism: The mechanism for the reaction that produces benzenesulfonic acid occurs in the following steps:

1. The sulfuric acid reacts with itself to form sulfur trioxide, the electrophile.

$$2H_2SO_4 \rightleftharpoons SO_3 + H_3O^+ + HSO_4^-$$

2. The sulfur trioxide is attracted by the π electron system of the benzene molecule.

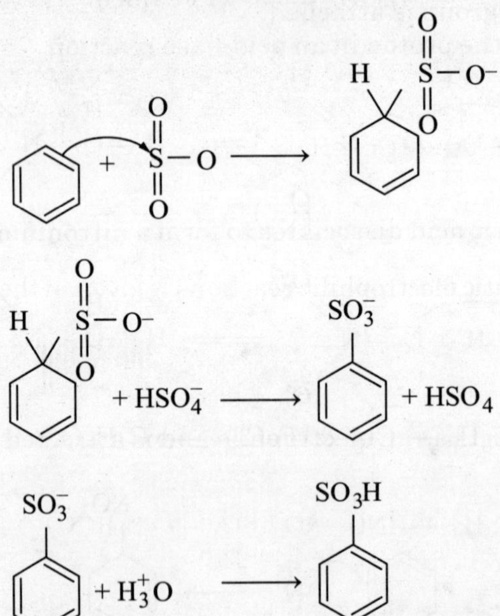

The remaining steps in the mechanism are identical with those in the bromination and nitration mechanisms: The charge around the ring is delocalised, and then the loss of a proton reestablishes the aromaticity of the ring.

3. Alkylation:

Alkylation of benzene means substituting an alkyl group into benzene ring. A hydrogen on the ring is replaced by a group like methyl or ethyl and so on. Benzene is treated with a chloroalkane (for example, chloromethane or chloroethane) in the presence of aluminium chloride as a catalyst

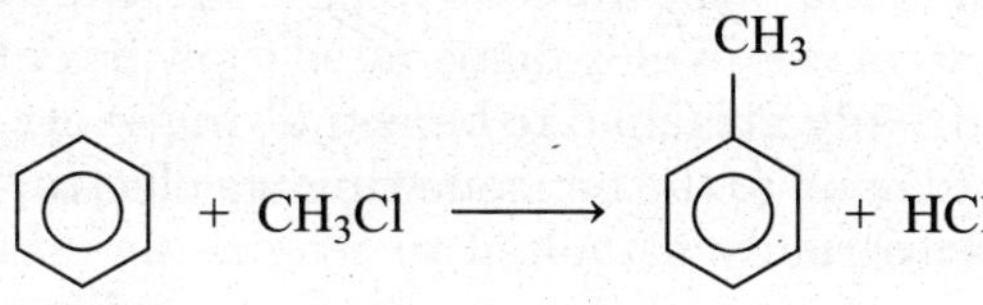

Mechanism:

1. The electrophile is CH_3^+. It is formed by reaction between the chloromethane and the aluminium chloride catalyst.

$$CH_3Cl + AlCl_3 \longrightarrow \overset{+}{C}H_3 + AlCl_4^-$$

2. The CH_3^+ is attracted by the π electron system of the benzene molecule. The nonaromatic carbocation that forms has resonating structures.

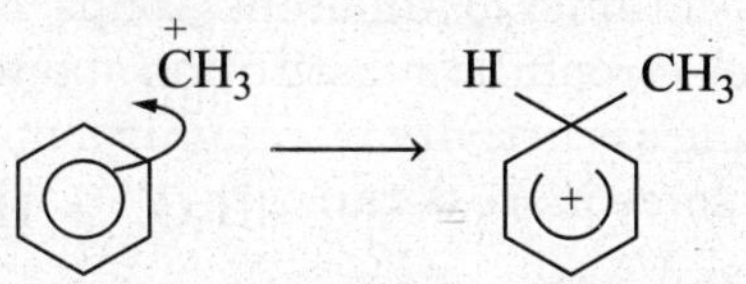

3. The aromaticity of the ring is reestablished by the loss of a proton from the carbon to which the methyl group is attached.

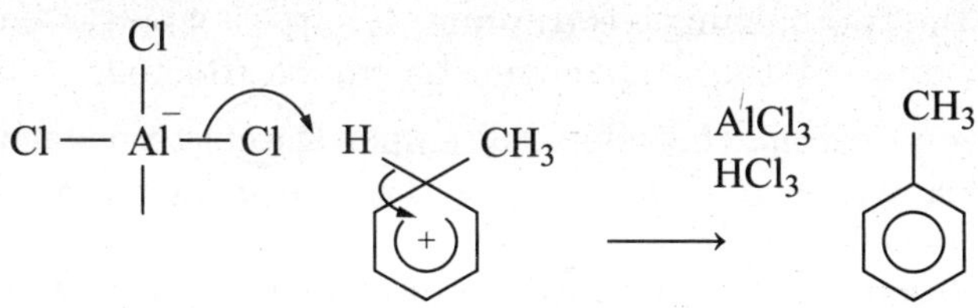

Summary of the aromatic electrophilic reactions is given in the table:

Reaction Type	***Typical Equation***	***Electrophile $E^{(+)}$***
Halogenation:	C_6H_6 + Cl_2 & $FeCl_3$ ⟶ $C_6H_5Cl + HCl$ heat catalyst Chlorobenzene	$Cl^{(+)}$ or $Br^{(+)}$
Nitration:	C_6H_6 + HNO_3 & H_2SO_4 ⟶ $C_6H_5NO_2 + H_2O$ heat catalyst Nitrobenzene	$NO_2^{(+)}$
Sulfonation:	C_6H_6 + $H_2SO_4 + SO_3$ ⟶ $C_6H_5SO_3H + H_2O$ heat Benzenesulfonic acid	$SO_3H^{(+)}$
Alkylation: Friedel-Crafts	C_6H_6 + R-Cl & $AlCl_3$ ⟶ C_6H_5-R + HCl heat catalyst Benzenesulfonic acid	$R^{(+)}$

22.8.1 Effects of Substituents (Directive influence)

We have only considered the substitution of unsubstituted benzene in the above reactions. In discussing these electrophilic aromatic substitutions on benzene, it was apparent that all six carbon atoms were equivalent in their ability to be the center of the substitution since the electron density exhibited by the π electron system was evenly distributed over the entire benzene ring.

However,substituents already attached to benzene ring exert an influence on additional atoms or groups attempting to bond to the benzene ring via electrophilic aromatic substitution reactions. An atom or group already attached to a benzene ring may direct an incoming electrophile to either the *ortho-para* positions or the *meta* position.

- Atoms or groups that make the benzene molecule more reactive by increasing the ring's electron density are called **activating groups.** Activating groups serve as ortho-para directors when they are attached to a benzene ring, meaning that they direct an incoming electrophile to the ortho or para positions.
- An atom or group that makes the benzene molecule less reactive by removing electron density from the ring acts as a **deactivating group.** Deactivating groups direct incoming electrophiles to the meta position.

These kinetic and directive effects differ for different groups. For example, a hydroxy or methoxy substituent increases the rate of electrophilic substitution about ten thousand fold. In contrast, a nitro substituent decreases the ring's reactivity by roughly a million.

Let us discuss this effects in some more detail.

1. Activating Groups: Ortho-Para Directors:

There are some groups which when attached to benzene ring will *activate* the ring toward further substitution. They do this by increasing the negative electron density in the ring thus making the ring more attractive to the approaching electrophile.

There are two ways in which the electron density can be affected:

(a) *Inductive Electronic effects*: This effect relies upon the electronegativity of the atom directly attached to the ring carbon. If the atoms within the group has +I effect than the ring will have an activating effect.

(b) *Resonance Electronic Effect*: This effect is the strongest of the two effects and accounts for some groups being powerful activators. Generally speaking any atom having at least one lone pair that is directly connected to a ring carbon will increase the electron density of the ring by lending that lone pair of electrons to the π electron system making the ring more attractive to the electrophile.

Most ring activators have atoms with unshared electron pairs directly attached to a carbon atom of the benzene ring. For example, the — OH group has two pairs of unshared electrons on the oxygen atom, which will form a bond to a carbon atom of the benzene ring. Thus, the — OH group will be an activating group. The following illustration shows why this group will act as an ortho-para director.

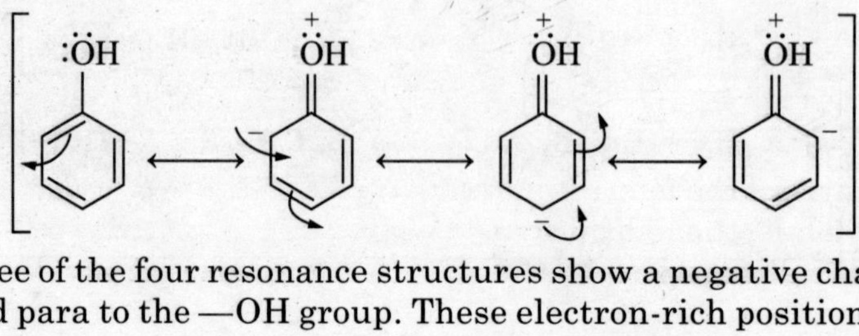

Notice that three of the four resonance structures show a negative charge residing on the positions ortho and para to the —OH group. These electron-rich positions should attract an electrophile more strongly than the less electron-rich meta positions do. Therefore, any group that possesses unshared electron pairs on the atom directly attached to a carbon atom of the benzene ring will be an ortho-para (activating) group. Groups that do not have unshared electron pairs on the atom directly attached to the benzene ring may also supply electrons to the benzene ring. This situation occurs if the atom in a group has weakly bonded π electrons attached to it or if the group has an inductive effect associated with it. The following diagram shows an example of the ð electron movement giving ring activation

The most powerful and strongest activators are shown in Figure below.

Activating Substituents

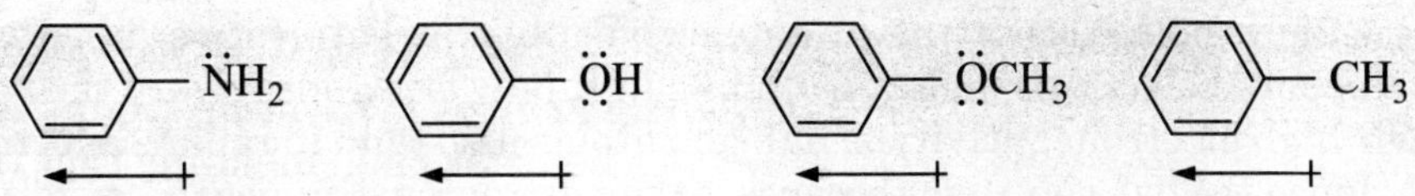

$$-NR_2 > -NHR > -NH_2 > -OH > OR > NH-\overset{O}{\overset{||}{C}}-CH_3 > CH_3 > Ten > C_6H_5$$

2. Deactivating Groups and meta direction

Most of these groups either have atoms attached to a ring carbon with a π bond or highly electronegative atoms attached as in the case of —CF_3 and —CCl_3 groups which are strong deactivators .These deactivators reduce the negative electron density of the ring by withdrawing it out of the ring either by resonance (Those atoms directly attached to the ring that have at least one π bond) or inductive effect as in the case of the —CCl_3, CF_3, and —NR_3^+ groups. Consequently, substituents in which nitrogen, oxygen and halogen atoms form sigma-bonds to the aromatic ring exert an inductive electron withdrawal, which deactivates the ring.

All deactivators with the exception of the weakly deactivating groups spoken of earlier will be meta directors. This is because when the electron density is dimenished in the ring , it occurs more at the ortho and para positions leaving the meta position relatively more negative.Therefore, if the electrophile does manage to be attracted to the deactivated ring, it will find the meta position that will be more negative and more attractive to the incoming electrophile.

Groups capable of doing this usually contain an atom that is directly attached to a carbon atom of the benzene ring and that bears a positive or partially positive charge. A typical example is the nitro group —NO_2. The structure of the nitro group is:

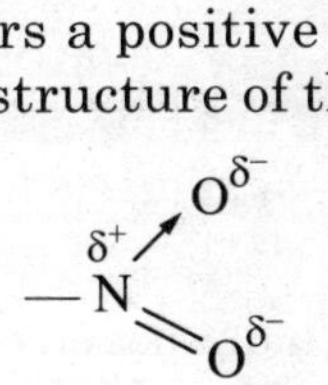

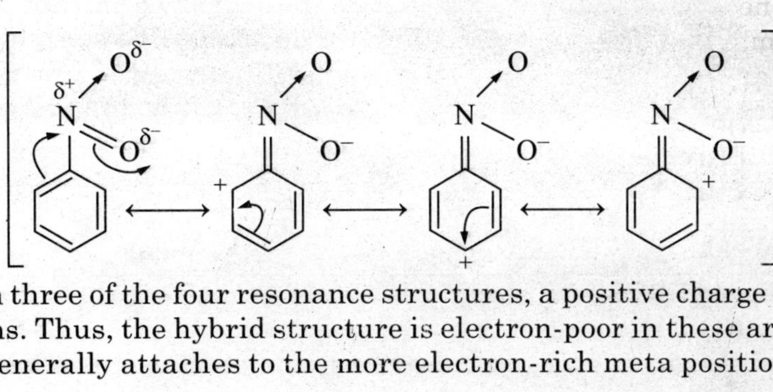

Notice that in three of the four resonance structures, a positive charge exists on the ortho and para positions. Thus, the hybrid structure is electron-poor in these areas, meaning that an electrophile generally attaches to the more electron-rich meta position. Some examples include:

Deactivating Substituents

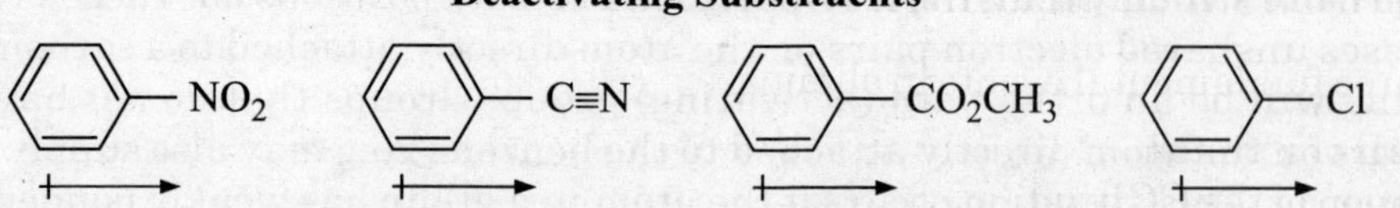

$$-NO_2 > -SO_3H > -C \equiv N - \overset{O}{\overset{||}{C}} - R > -\overset{O}{\overset{||}{C}} - OH$$

3. Halogens Deactivating but ortho and para directing

Halogen atoms show both activating and deactivating characteristics. Because they have three pairs of unshared electrons, halogen atoms can supply electrons toward the ring. However, because of their high electronegativities, halogen atoms also tend to remove electrons from the benzene ring. Hence,the halogen atoms because of their extreme electronegativity will pull electrons out of the ring leaving the ring more positive ,and therefore, less attractive to the incoming electrophile but the three lone pairs of non-bonding electrons found on the halogen atom will provide increased electron density at the ortho and para positions. Thus the benzene ring is deactivated ,but the relative electron density is greater at the ortho and para positions due to the resonance effect of the lone pairs. So, if the substitution occurs it will be slow but occur either at the ortho or para positions

There are three generalisations concerning this effect:

1. all meta-directing groups are deactivating
2. all ortho, para-directing groups are activating except the **halogens**
3. the halogens are deactivating

We can further classify activating and deactivating groups or atoms as strong, moderate, or weak in their directing influence. This table lists some typical activating and deactivating groups by the order of their strength.

The Influence of Various Directing Groups and Atoms

Ortho-Para Directions	Meta Directors
Strong	**Strong**
$—NH_2$	$—NO_2$
$—NHR$	$—NH_3^+$
$—NR_2$	$—NR_3^+$
$—OH$	$—CF_3$
$—O^-$	$—CCl_3$
Moderate	**Moderate**
$—NHCOCH_3$	$—CN\equiv N$
$—NHCOR$	$—SO_3H$
$—OCH_3$	$—COOH$
$—OR$	$—COOR$
$—CHO$	
$—COR$	
Weak	**Weak**
$—CH_3$	
$—CH_2CH_3$	
$—R$	
$—C_6H_5$	

22.9 MECHANISM OF ELIMINATION REACTIONS

There are three fundamental events in elimination reactions:

1. Removal of a proton
2. Formation of the *CC* bond
3. Breaking of the bond to the leaving group

Depending on the relative timing of these events, different mechanisms are possible:

- E_1 Reactions: Loss of the leaving groups to form a carbocation in step1, removal of H^+ and formation of C = C bond in step 2
- E_2 Reactions: Simultaneous removal of H^+, C = C bond formation and loss of the leaving group in single concerted step.

Let us discuss these mechanisms in detail.

22.9.1 E_1 Mechanism (Unimolecular Elimination)

E_1 indicates a *elimination, unimolecular* reaction, where

$$\text{rate} = k\,[\text{substrate}]$$

This implies that rate determining step of the mechanism depends on the decomposition of single molecular species.

Overall, this pathway is a multi-step process with the following two critical steps:

Step I:

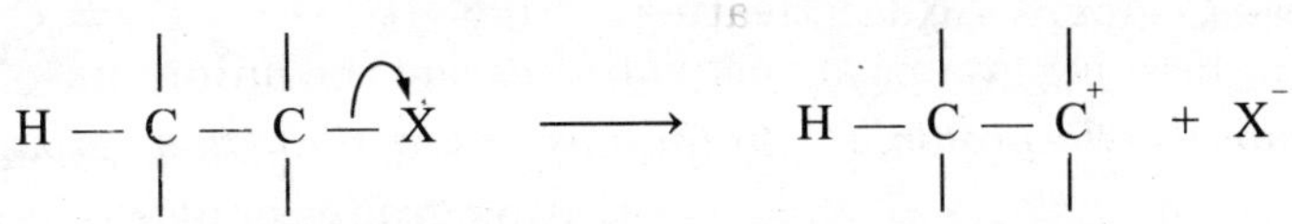

Loss of the leaving group, LG, to generate a carbocation,

Step II:

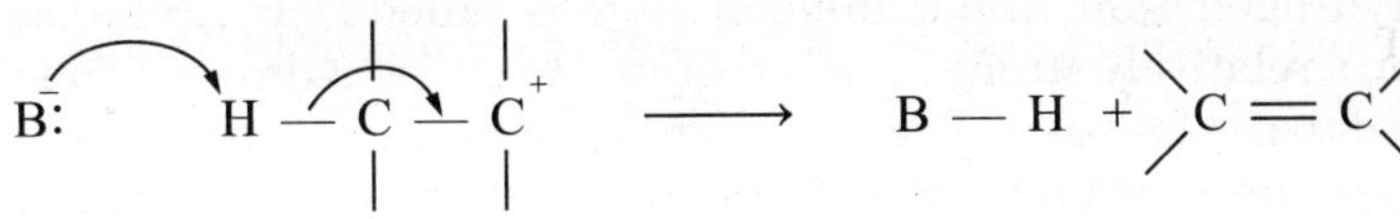

Loss of proton, H^+, from the carbocation to form the π bond

Generally, tertiary halides as well as phenyl alkyl halides and phenyl allyl halides will undergo E1 elimination using a strong base and high temperature

This mechanism is a two step process with the rate determining step being an ionisation of the halide to produce a carbocation. The carbocation will then react with a base that can pull a β hydrogen off to produce the most stable alkene according to the **Say tzeff'e rule.** The mechanism is similar to the SN_1 reaction except here the second step invloves a proton abstraction and formation of π bond.

Note: Saytzeff's rule states that an E_2 reaction will preferntially form the most stable alkene isomer. Alkene stability generally increases with branching at the alkene carbons (details are given in section 22.8.4)

Example: Dehydrohalogenation of a Tertiary Alkyl Halide

Tertiary halides in the presence of a Moderately strong base at high temperature will undergo E1 elimination since the formation of the carbocation in the first rate determining step of the E1 mechanism involves a very stable tertiary carbocation.

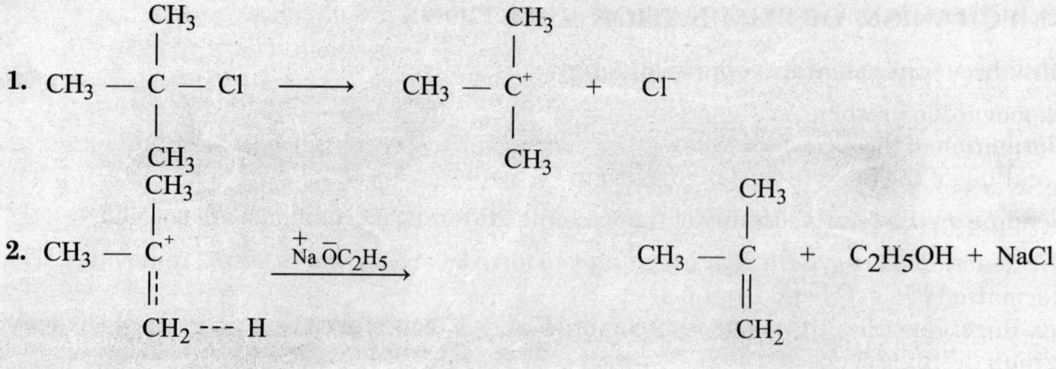

Dehydrohalogenation of Tertiary Butyl Chloride

22.9.2 E_2 Mechanism: (Bimolecular Elimination)

E_2 indicates an elimmination, bimolecular reaction, where

$$\text{rate} = k[\text{substrate}]\,[\text{Base}]$$

This implies that the rate determining step involves an interaction between these two species, the base and the organic substrate.

This pathway is a concerted (single step) process with the following characteristics:

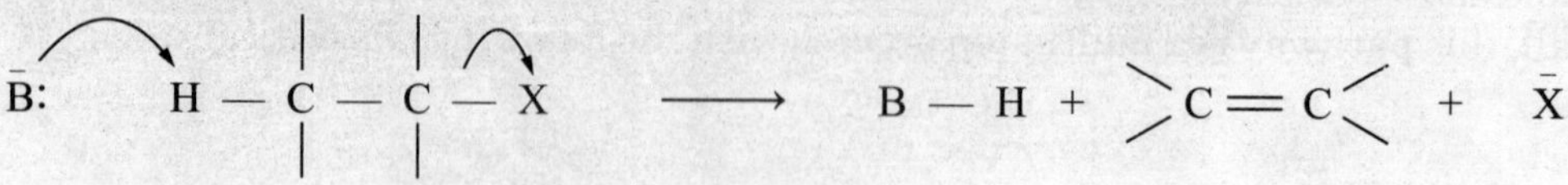

Simultaneous removel of the proton, H^+, by the base, loss of the leaving group, X^-, and formation of the π bond.

Example: Dehydrohalogenation of Primary Halides:

The removal of a Hydrogen atom and a halogen atom is called Dehydrohalogenation. With any dehydrohalogenation a relatively strong base is required to pull a hydrogen from the carbon adjacent to the carbon bonded to the halogen. Such adjacent carbons are referred to as β carbons and the hydrogens attached to these carbons are referred to as β hydrogens. Primary halides, $R—CH_2—CH_2—X$ will require a base to pull a proton from the β carbon as the halogen is leaving. Thus, one step concerted pathway is similar to SN_2 mechaism. Primary halides will only result in elimnation (E_2) if the base is bulky. The rate of such an elimination will depend upon the concentration of the organic alkyl halide as well as concentration of the base used.

Typical strong bases would include alkoxide ions ($R\bar{O}$) prepared from alcohols reacting with active metals like sodium or potassium.

$$CH_3—CH_2—OH + Na \rightarrow CH_3—CH_2—\bar{O}—Na^+ + 1/2\ H_2$$

These bases are strong enough to pull β hydrogens, as the halogen atom is departing from the reactant molecule.

The dechlorination of 1-Chloropropane using sodium ethoxide in ethanol solvent will produce pro-pene

$$CH_3—CH_2—CH_2—Cl + \overset{+}{Na}\ \bar{O}\,C_2H_4 \longrightarrow CH_3—CH{=}{=}CH_2 + NaCl + C_2H_5OH$$

The concerted mechanism can be represented as:

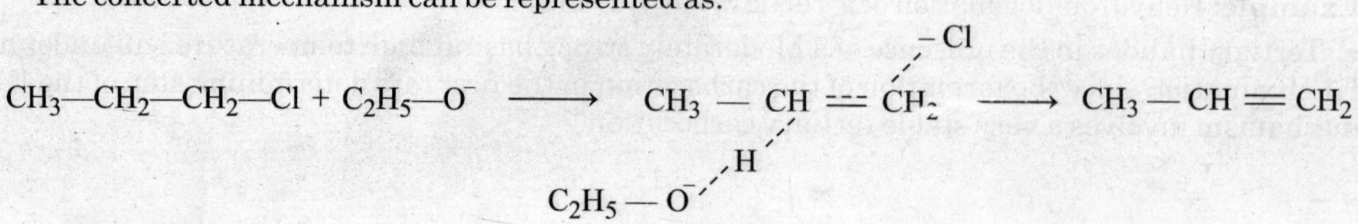

22.9.3 Elimination Versus Substitution

Alkyl Halides undergo two major kinds of reactions:

1. Nucleophilic Substitutions
2. Elimination Reactions

Wherever Nucleophilic reactions occur, often Elimination reations will compete so that it is possible to get two products, one involving a substitution and one involving elimination.

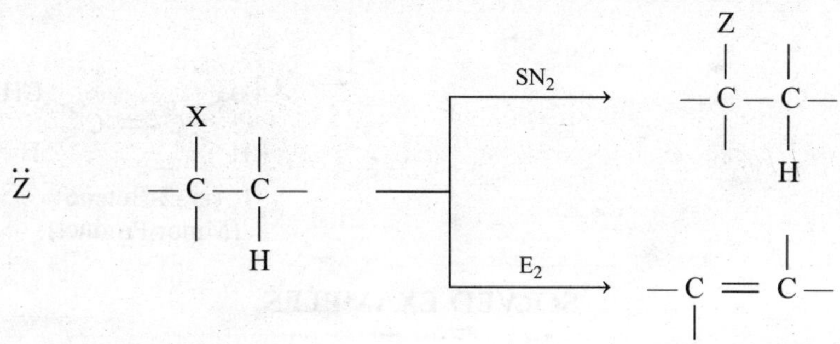

The amount of alkyl substitution on the carbon bonded to the halide will affect this competition. Here are some things to keep in mind:

1. Methyl halides will only undergo SN_2 reactions.
2. Primary halides will always undergo SN_2 except with hindered strong bases in which case E_2 product predominates over SN_2 product
3. Secondary halides will give SN_2 product with weak bases like carboxylate ions, R-COO$^-$, or cyanide ions, CN^-. With strong bases E_2 product will predominate
4. Teriary halides, allylic halides, and benzylic halides will never occur under SN_2. SN_1 product is favoured at low temperatures with weak bases and E_1 product is favoured with high temperatures and weak bases: E_2 will only predominate when the base is strong and the tempreature is high.

22.9.4 Saytzeff Rule

As discussed above that in the dehydrohalogenation of alkyl halides, the most substituted olefin is the main product or the alkene that has the greater number of alkyl groups attached to the doubly bonded carbon atoms is the preferred product in elimination reactions. This genesalization is known as Saytzeff Rule.

Example :

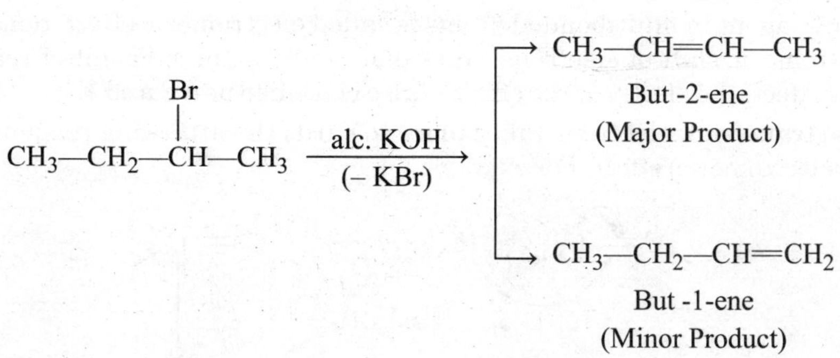

If the substituted alkene formed during elimination reaction is capable of showing cis-trans isomerism, the trans alkene being more stable is formed as the major product.

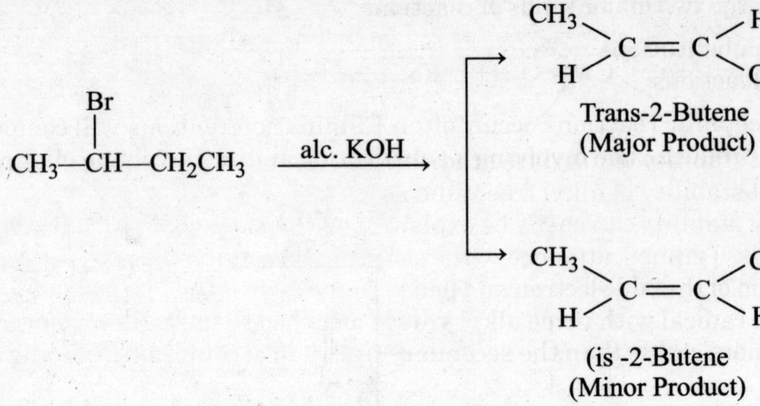

SOLVED EXAMPLES

1. Which of the following molecules and ions are electrophilic?

A	B	C	D	E	F	G	H	
CH_4	H_2O	$Br^{(+)}$	$H_2C{=}CH_2$	BF_3	$NO_2^{(+)}$	NH_3	NH_3	$Br^{(-)}$

Ans: C, E & F

2. Which of the following molecules and ions are nucleophilic?

A	B	C	D	E	F	G	H
CH_4	H_2O	BH_3	$H_2C{=}CH_2$	$OH^{(-)}$	$Br^{(+)}$	NH_3	$Br^{(-)}$

Ans: B, D, E, G, & H

3. What do you mean by +I and –I effect?

Atoms or groups of atoms that have a greater electron-attracting capacity than hydrogen are referred to as having –1 (electron attracting) effect. For example,

$$NO_2 > CN > COOH > F > Cl > Br > I > OH > OCH_3 > C_6H_5 > HI$$

Atoms or groups of atoms, having smaller electron attracting power than hydrogen are referred to as having +I (electron repelling) effect.

$$(CH_3)_3C \longrightarrow (CH_3)_2CH \longrightarrow CH_3CH_2 \longrightarrow CH_3-$$

4. What do you mean by Electromeric effect?

This phenomenon of movement of electrons from one atom to another at the demand of attacking reagent in multibonded atoms is called electromeric effect, denoted as E effect. The electromeric shift of electrons takes place only at the moment of reaction. Like the inductive effect, the electromeric effect is also classified as +E and E:

When the transfer of electrons takes place towards the attacking reagent, it is called +E (positive electromeric) effect. For example,

When the transfer of electrons takes place away from the attacking reagent, it is called, –E (negative electromeric) effect. For example,

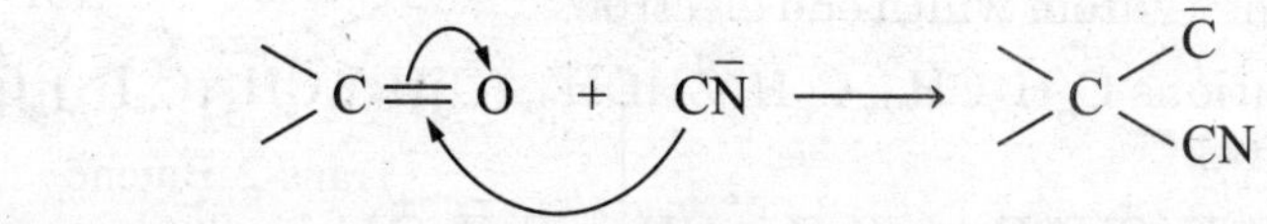

5. The order of stability of alkyl free radicals is $CH_3 < 1° < 2° < 3°$ Comment.
The order of stability of alkyl free radicals is : $CH_3 < 1° < 2° < 3°$
This order of stability can easily be explained on the basis of hyperconjugation, Larger the number of alkyl groups, attached to the carbon tom carrying the odd electron, greater is the delocalisation of the odd electron and hence more stable is the free radicals. Accordingly, the tertiary free radical with three alkyl groups attached to the carbon atom carrying the odd electron is more stable than the secondary free radical containing two alkyl groups and so on.
6. Which of the following is *not* a nucleophile : H_2O, BF_3, NH_3, OH^-?
Ans: BF_3.
7. What type of bond fission is probable in the molecules : Br_2, HBr, H_2O, CCl_4 ?
Ans: Br_2 – homonuclear, HBr – heteronuclear,
H_2O – heteronuclear, CCl_4 – heteronuclear.
8. The transition state in the SN_1 reaction involves a negative ion separating from a positive ion (carbocation). Why is the rate of reaction much faster in a polar solvent rather than a non-polar solvent?
Ans: The carbocations and negative ions can be *solveated* by polar solvent. This solvation *encourages* the formation of the ions, so there are more of them in polar solvent. This results in the rate increasing. On the other and, a non-polar solvent *cannot* solvate the ions, so does nothing to favor the SN_2 reaction.
9. Which among acetic acid, ethanol, p-nitrophenol, p-methoxyhenol are *more acidic* than phenol ?
Ans: *Ethanol and p-methoxyphenol.* This is due to the following *reasons* :
Phenoxide ion ($C_6H_5O^-$) is *stabilised by resonance;* whereas ethoixde ion ($C_2H_5O^-$) is *not* resonance stabilised. Hence, pheonol is *more acidic* than ethanol. Moreover, resonance in p-methoxyphenoxide ion ($CH_3O.C_6H_4.O^-$) is *less* than that in phenol, because electron-release by O– group in phenoxide ion is opposed by electriondonating nature of $-OCH_3$ group. Hence, phenol is *more acidic* than p-methoxyphenol.
10. Why is phenol *less acidic* than carbonic acid ?
Ans: Due to the presence of two O-atoms, the bicarbonate ion $HO-\overset{O}{\overset{\|}{C}}-O^-$ is *more resonance stabilised* than phenoxide ion ($C_6H_5O^-$) having having only one O-atom. Consequently, carbonic acid is a stronger acid than pheonol. Alternatively, phenol is a *weaker* acid than carbonic acid.
11. Arrange CH_3OH, CH_3CH_2OH, $(CH_3)_2CHOH$, $(CH_3)_3COH$ in decreasing order of their *acidity*.
Ans: $CH_3OH > CH_3CH_2OH > (CH_3)_2CHOH > (CH_3)_3COH$, since the electronreleasing methyl groups *decrease* the acidic nature of alcohol.
12. Arrange $\overset{+}{C}H_3$, $R\overset{+}{C}H_2$, $R_2\overset{+}{C}H$, $R_3\overset{+}{C}$ carbonium ions in order of increasing *stability*.
Ans: $\overset{+}{C}H_3 < R\overset{+}{C}H_2 < R_2\overset{+}{C}H < R_3\overset{+}{C}$

13. Which among $\dot{C}H_3$, $CH_3\dot{C}H_2$, $(CH_3)_2\dot{C}H$, $(CH_3)_3\dot{C}$ is *most stable* free radical?

Ans: $(CH_3)_3\dot{C}$, since the stability of carbon free radical *increases* with the number of alkyl groups attached to the C-atom width odd electron.

14. Arrange the carbocations $C_6H_5\overset{+}{C}H_2$, $C_6H_5\overset{+}{C}HCH_3$, $(C_6H_5)_2\overset{+}{C}H$, $(C_6H_5)_3\overset{+}{C}$ in increasing of order of their *stability*.

Ans: $C_6H_5\overset{+}{C}CH_2 < C_6H_5\overset{+}{C}HCH_3 < (C_6H_5)_2\overset{+}{C}H < (C_6H_5\overset{+}{C}$

This is because the resonance effect (+M-effect) is always more stabilising than the inductive effect (+I-effect). Triphenyl carbocation is the *most stable* among these, because +ve charge is dispersed by resonance due to +M-effect of $-C_6H_5$ groups.

15. Which among $C_6H_5CH_2Br$, $(CH_3)_2CHBr$, $(CH_3)_3CBr$, CH_3Br undergoes nucleophilic substitution reaction by SN_2 mechanism ?

Ans: CH_3Br, since SN_2 reactivity of haloalkance in the order : $CH_3 > 1° < 2° > 3°$. The rate determining factor is *steric hindrance.* There is minimum steric hindrance offered by H's of CH_3Br to the approach of nucleophile attacking at the C so C—Br bond.

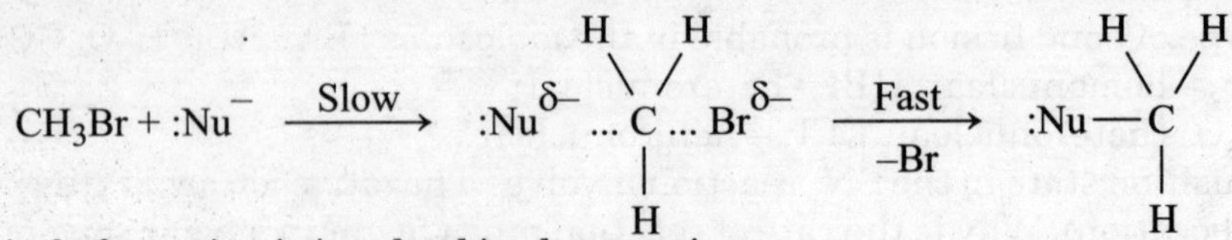

16. What kind of reaction is involved in the reaction :

$$CH_3CH_2BrCH_3 + HOH\ (alc.) \longrightarrow CH_3CH = CH_2 + KBr + H_2O?$$

Ans: *Nucleophilic elimination reaction :*

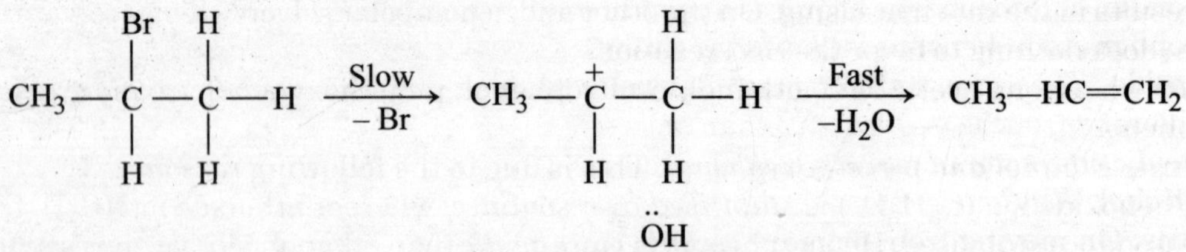

17. Arrange the following in the order of their increasing reactivity in nucleophili substitution reactions : CH_3F, CH_3I, CH_3Br, CH_3Cl.

Ans: $CH_3I < CH_3Cl < CH_3Br < CH_3F$.

Reason : The bond desiccation energices of : C—F > C—Cl > C—Br > C—I. Thus, C—I bond is not readily cleaveraged; while the C—F bond is cleaved with most difficulty. Now ***lesser** there bond dislocation energy, **more** is the reactivity of the haloalkance.* This explains the above order of reactivity.

18. Why is the halogen atom of an alkyl halide *readily substituted* by an other group?

Ans: Alkyl halides (RX) are *partially polar* in nature with halogen atom carrying a fractional negative charge. Thus :

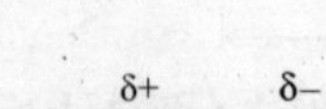

Folar molecule

The electron-releasing inductive effect of an alkyl group facilitates the ionisation of halogen atom and its substitution by a nucleophile. Thus :

19. The treatment of alkyl chloride with aqueous KOH leads to the formation of alcohols; whereas in the presence of alc. KOH, alkences are formed as the major products. Explain.

Ans: *Aqueous KOH* contains *only* OH– *ions,* which act as nucleophile and these brings about *hydrolysis* of alkyl chloride to the corresponding alcohol.

$$RCl + \underset{\text{(Only nucleophile present)}}{OH^-(aq)} \xrightarrow{\text{Hydrolysis}} ROH + Cl^-(aq)$$

On the other hand, *alcoholic KOH* contains *ethoxide ions* (C2H5O–), which are *more basic than* OH– *ions.* Consequently, ethoxide ions preferentially bring about *dehydrohalgenatur* to form alkenes, e.g.

$$\underset{\boxed{H\quad Cl}}{H_2C{-}CH_2} + KOH(alc.) \longrightarrow H_2C{=}CH_2 + H_2O + KBr$$

20. Why are haloakanes more reactive than haloraenes?

Ans: Resonance structure of haloalkance (e.g., chlorobenzene) is :

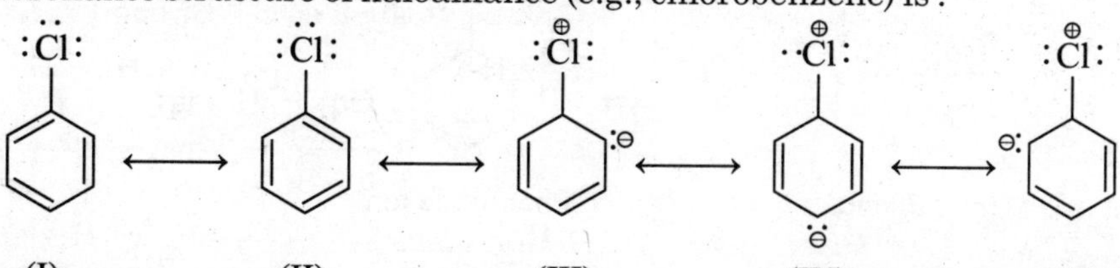

(I) (II) (III) (IV) (V)

Chlorobenzene is *resonance hybrid* of five structures; while haloalkance (e.g., chloromethane) is not resonance stabilised. As a result of contribution of structures III, IV and V, the C—Cl bond in chlorobenzene acquires *a partial double bond character.* Consequently, the C—Cl bond in chlorobenzene is *shortened* (from 1.82Å in CH_3Cl to 1.69Å in C6H5Cl) and is *more, stable. Therefore, chlorobenzene is more stable than chloromethane.* In other words, chlorobenzene is *less reactive* than chloromethane.

21. Acidic character of monohydric alcohols is : primary alcohol > secondary alcohol > tertiary alcohol. Why?

Ans: On moving from primary alcohol through secondary alcohol to tertiary alcohol, the *number of alkyl groups* (having electronreleasing inductive effect) *increases.* Consequently, the electronreleasing inductive effect of three alkyl groups in tertiary alcohols s maximum and it is minimum in primary alcohols. In other words, the primary alcohols are the *strongest* acids; while tertiary alcohols are *weakest* acids. Hence, the acidic character of alcohols is :

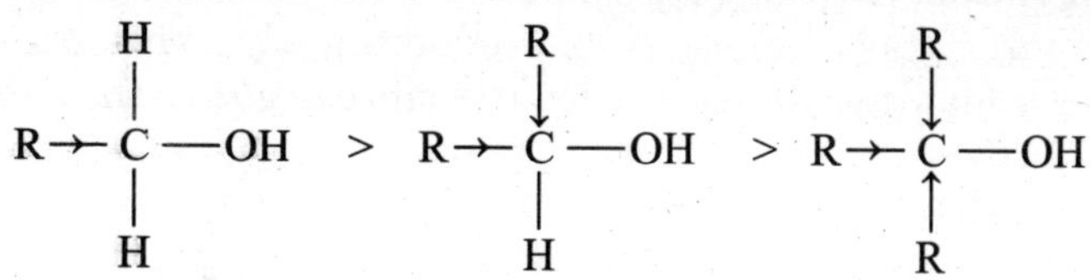

Primary alcohol Secondary alcohol Tertiary alcohol

22. Alcohols are *weaker acids* than water. Why?

Ans: The reaction : $2\,ROH + 2Na \longrightarrow 2RONa + H_2$ suggests that *alcohols are acidic in nature.* This is due to fact that the *highly electronegative* O-atom withdraws partially the electrons of the O—H bond towards itself, thereby the O—H bond becomes *weak* and splits off a proton (H+). However, alcohols are *weaker acids* than water, because *the electronreleasing effect of alkyl group (R–>–O—H) from the O—H bond becomes difficult.* Hence, alcoholds are *weaker acids* than water, due to *electronreleasing inductive effect of the alkyl group.*

23. Why phenol is *more acidic* than alcohols (or ethanol)?

Ans: Phenol is *acid and more acid* than alcohols (or ethanol) because of the following facts:

(i) The phenoxide anion ($C_6H_5O^-$) is ***stabilised dby resonance;*** whereas the alkoxide ion (or ethoxide ion) is *not* resonance stabilised.

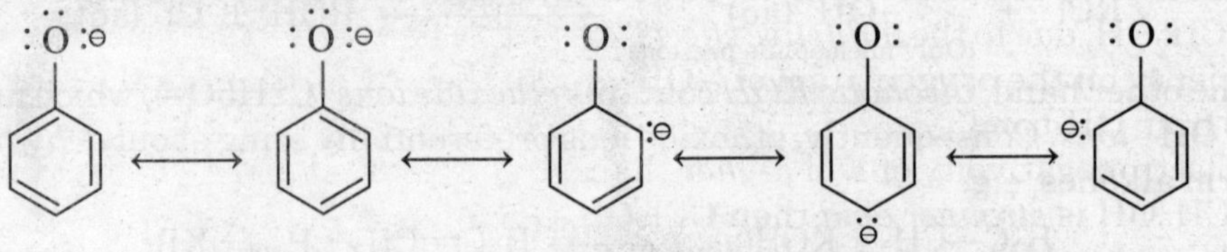

As a result of resonance, the oxygen atom of phenoxide ion acquires a *partial positive charge.* This is-turn *weakens* the O—H bond, thereby facilitating the release of a proton.

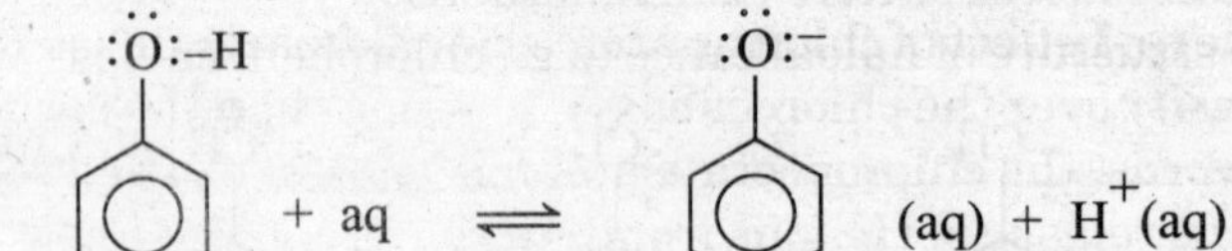

Phenol *(Less stable)* Phenoxide ion *(More stable)*

Thus, phenol has a strong tendency to form more stable phenoxide ion by releasing a proton. *Hence, phenol is acidic in nature.*

On the other hand, the alkoxide (or ethoxide) ion is *less stable* than alcohol (ethanol) and is *not* resonance stabilised. So alcohol (or ethanol) possesses *negligible tendency* to form *less stable* alkoxide (or ethoxide) ion by releasing a proton.

$$R—\ddot{O}—H \rightleftharpoons R—\ddot{O}:^{\ominus} + H^+$$

Alcohol *(More stable)* Alkoxide ion *(Less stable)*

Reason : Due to the aromatic stability imported by the *electron delocalisation,* benzene is *not readily reactive* to electrophilies such as NO_2^+ ion (from conc. HNO_3). On the other hand, in phenol, the phenate ion ($C_6H_5O^-$), due to the presence of 3 lone pairs of electrons on the oxygen, *activates* the aromatic ring, predominantly at the 2 or 6 (*ortho*) and 4 (*para*) positions, thereby making these *significantly more negative* than that in benzene. Thus :

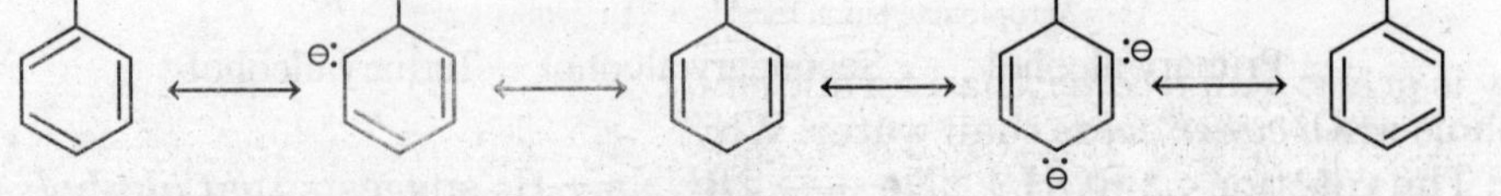

Ortho *Para* *Ortho*

Since, the OH group in phenol activates the 2 or 6 (*ortho*) and 4 (*para*) positions towards electrophile such as NO_2^+, so electrophilic substitution reactions (such as nitration) of phenol occur *more readily* than of benzene.

24. Arrange the following compounds in the *increasing order* of the acid strength. Give reasons for you answers.

CH_3CH_2OH, CF_3CH_2OH, CCl_3CH_2OH

Ans: $CH_3CH_2OH < CCl_3CH_2OH < CF_3CH_2OH$.

Reason : (a) Acid strength of CF_3CH_2OH and CCl_3CH_2OH is *higher* than that of CH_3CH_2OH, due to the *inductive effect* (–I effect) of halogen atoms. This causes the electron-deficiency on the oxygen atom of '–OH' group, thereby *facilitating* the removal of proton (H+) from OH-bond.

(b) Electronegativeity of F of *higher* than Cl, so F exerts *higher* – I effect than Cl. Hence, CF_3CH_2OH is *stronger acid* than CCl_3CH_2OH.

25. Explain why chloroacetic acid is a *stronger acid* than acetic and.

Ans: (1) The chlorine atom in chloroacetic acid has *electron-withdrawing effect* (–I effect), so the chlorine atom withdraws itself the electrons from the O—H bond *decreases.* This in-turn *facilitates the release of H^+ ions.*

(2) Moreover, due to -I effect of chlorine atom, the chloroacetate ion (of chloroacetic acid) has *less electron-density* over the chlorocaboxylate anion than the carboxylate anion (of acetic acid). In other words, the chloroacetate anion is *more stabilised* relative to acetate anion. Consequently, the release of H+ ions from chloroacetic acid is *facilitated.*

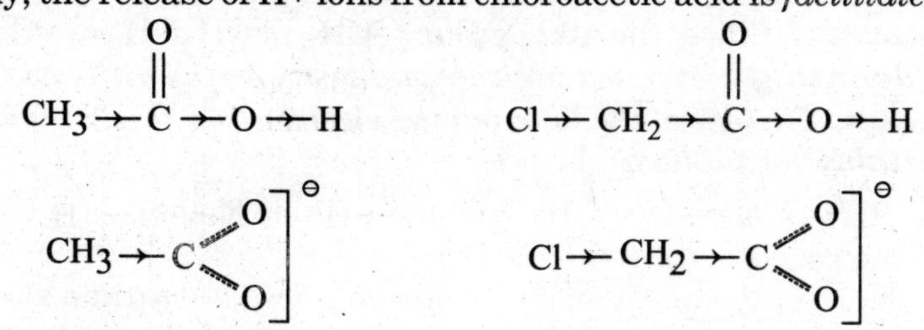

+I effect of the CH_3 group *intensifies* the negative charge on the carboxylate anion, thereby making it *less stable.*

–I effect of the Cl atom *disperses* the negative charge on the carboxylate anion, there making it *more stable.*

As a result of the above two effect, chloracetic acid is a stronger acid than acetic acid, i.e. acetic acid is weaker acid than chloracetic acid.

26. Why is propionic acid *weaker acid* than acetic acid?

Ans: When an *electronreleasing* methyl (CH_3) group is substituted for one H-atom of acetic acid, then we get *propionic acid.* Compared to the acetate ion, the *propionate ion is **less stable***. *So propionic acid is **weaker acid** than acetic acid.*

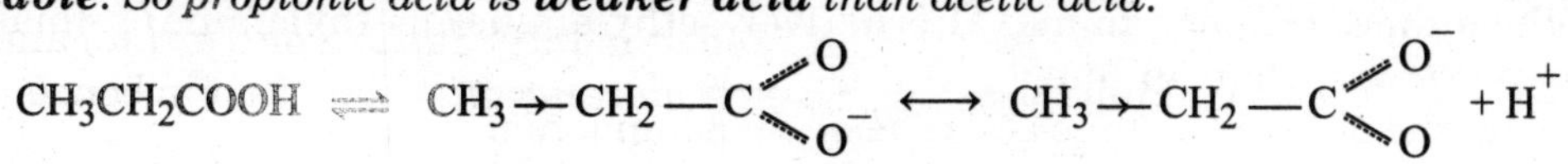

Electron-releasing CH_3 group *destabilises* the propionate anion relative to the acetate anion.

27. Why is acetic acid *weaker* than formic acid ?

Ans: H—C(=O)—O—H; CH_3→H—C(=O)→O→H. The +I effect of methyl group (CH_3–) present in acetic acid *increases* the electron-density on the oxygen atom of the O—H bond. As a result, the release of H^+ ions from acetic acid is *more difficult* as compared to formic acid. Moreover, the +I effect of the methyl group *intensifies* the negative charge on the carboxylate anion, thereby making acetate anion *less stable* than formate anion. As a result, the release of H^+ ions from acetic acid becomes *more difficult* as compared to formic acid. Hence, *acetic acid is a weaker acid than formic acid.*

28. Compared the acid strengths of formic acid, acetic acid and benzoic acid.

Ans: *Acid strength of :* formic acid > benzoic acid > acetic acid.

Reasons : (i) Formic acid does not contain any alkyl or aryl group, *so it is stronger acid than acetic acid as well as benzoic acid.*

(ii) A phenyl group directly attached to carboxylic group (as in benzoic acid) has an overall *electronreleasing effect* (+I effect), which is *weaker* than any alkyl group (e.g., CH_3 group in acetic acid). *Hence benzoic acid is stronger acid than acetic acid.*

29. Methylamine is a *stronger base* than ammonia. Explain why.

Ans: Both ammonia (NH_3) and methylamine (CH_3NH_2) possess *a lone pair of electrons* on their nitrogen atom. Thus:

However, methyl group in CH_3NH_2 is *electron-donating* group. *As a result, the electron-density on N-atom of CH_3NH_2* ***increases*** *and consequently, the lone pair of electrons is* ***more readily available*** *for protonation. Hence, methylamine is more basic than ammonia. In other words, methylamine is a stronger base than ammonia.*

30. Why is aniline *weaker base* than ethylamine / methylamine?

Ans: Any primary amine (e.g., CH_3NH_2 or $C_2H_5NH_2$) possesses *a nitrogen atom having a lone pair of electrons.* Since, the alkyl group (–CH_3 in CH_3NH_2 or –C_2H_5 in $C_2H_2NH_2$) is an ***electron-donating*** *group, so the electron-density on the nitrogen atom of the primary amine* ***increases.*** *Therefore, the lone pair of electrons on the N atom of amine is* ***more readily available*** *for protonation.*

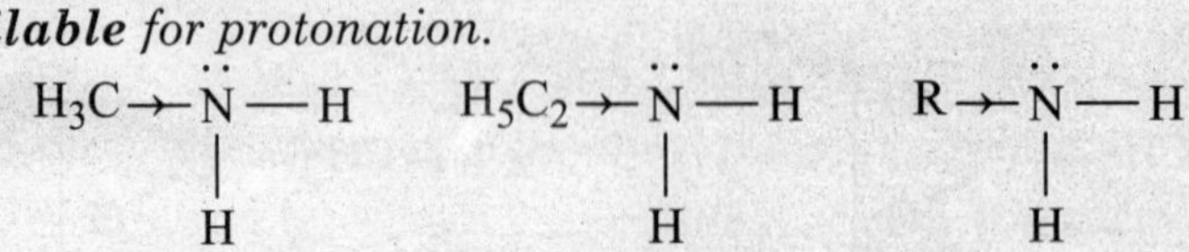

Methylamine (+I effect) Ethylamine (+I effect) Primary amine (+I effect)

On the other hand, *phenyl group* 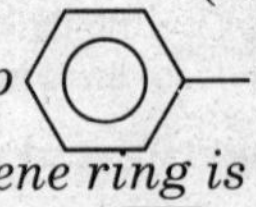 *in aniline*  NH_2 *is electron-withdrawing the nature. Moreover, the benzene ring is* ***resonance stabilised.***

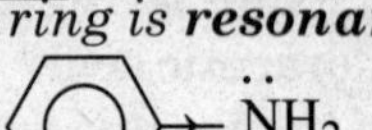

Hence, electrons on the N-atom of amino group of aniline are ***less readily available*** *for protonation than any primary amine.* Hence, aniline is a *weaker base* than ethylamine / methylamine / primary amine. Alternatively, ethylamine / methylamine / primary amine is *stronger base* than aniline.

31. Explain the fact that aniline is *less basic* than ammonia.

Ans: Aniline is a *resonance hybrid* and as a result of this *the lone pair of electrons on the nitrogen atom of amino group in aniline get* ***delocalised*** *over the benzene ring,* thereby this long pair of electrons is ***less readily available*** for protonation. On the other hand, in case of ammonia, delocalisation of the lone pair of electrons on the nitrogen atom is *not possible* through resonance. Therefore, the long pair of electron on nitrogen atom of NH_3 is ***easily available*** for protonation. *Hence, aniline is less basic than ammonia.*

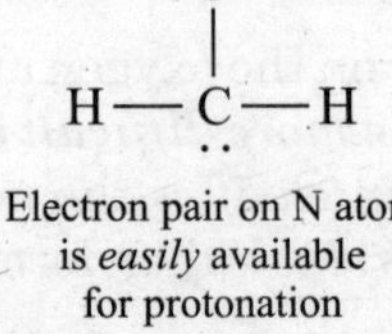

Electron pair on N atom is *easily* available for protonation

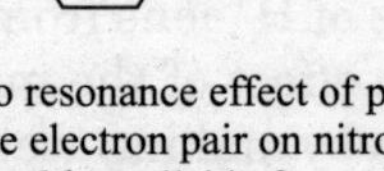

Due to resonance effect of phenyl radical, the electron pair on nitrogen atom is *less readily* available for protonation

QUESTIONS

1. What are the different types of organic reactions? Explain them with one example each.
2. What are: (a) Electrophilic reagents (b) Nucleophilic reagents? Give two examples of each.
3. Write notes on: (a) Electronegativeity (b) Polar and non-polar covalent bonds (c) Inductive effect (d) Electromeric effect
4. Classify the following into electrophilic and nucleophilic reagents.
   (a) H^+ (b) Cl^- (c) NO_2^+ (d) OH^-
5. What is an addition reaction? Explain its reaction mechanism with an example.
6. What is an elimination reaction? Explain its reaction mechanism with an example.
7. Explain SN_1 reactions with examples.
8. Give a detailed account stereochemistry of SN_1 reaction.
9. How do bulky groups affect SN_1 reaction?
   (a) Comment on the stability of carbocations
   (b) Comment on elimination versus substitution.
10. What is the difference between a free radical and an ion?
11. What are different intermediate organic species? Discuss in detail their stabilities.
12. Distinguish between electromeric and inductive effects, giving some suitable examples.

❑❑❑

23 NAME REACTIONS

My god is simple. It is complete understanding of the universe,
why it is as it is and why it exists at all.

—Stephen Hawking

CHAPTER OUTLINE

In the light of the mechanisms discussed in chapter 7, let us discuss some of the name reactions.

23.1 ALDOL CONDENSATION

When aldehydes and ketones having α-hydrogen atoms* are treated with dilute alkali, β-hydroxy aldehyde or β-hydroxy ketone is formed due to the condensation of two molecules of such compounds. This is called aldol condensation. 'Aldol' is an abbreviation of *aldehyde* and *alcohol.*

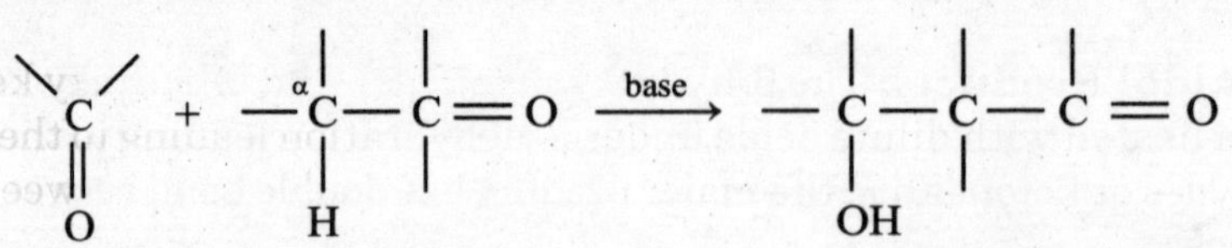

Ketone

An aldol
(A β-hydroxy carbonyl compound)

The product results from the addition of one molecule of aldehyde (or ketone) to a second molecule in such a way that the α-carbon of the one becomes attached to the carbonyl carbon of the other.

---

* α-hydrogen atoms: Hydrogen attached to the carbon next to the carbonyl group

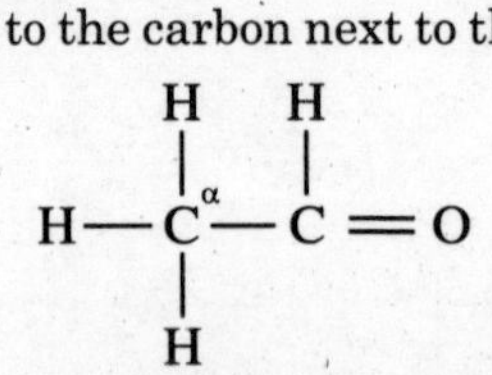

Examples:

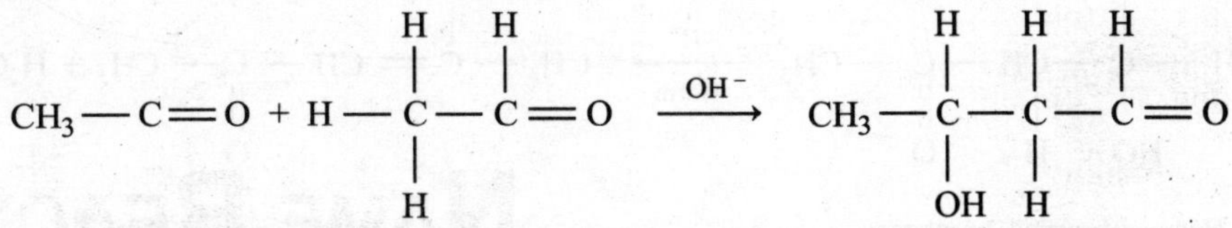

Acetaldehyde 2 moles → β-Hydroxybutyraldehyde (3-Hydroxybutanal)

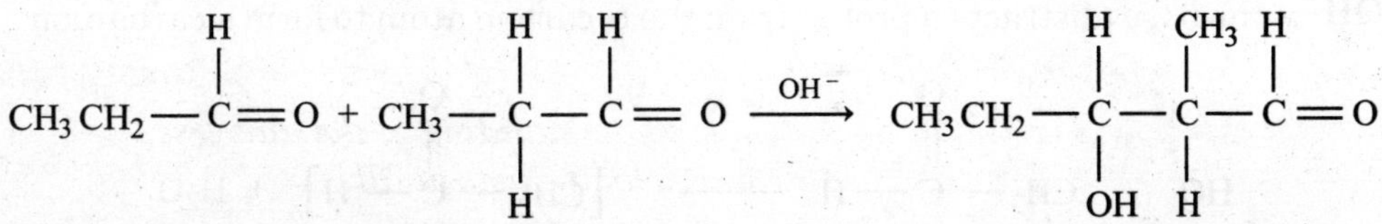

Propionaldehyde 2 moles → β-Hydroxy-α-methylvaleraldehyde (3-Hydroxy-2-methylpentanal)

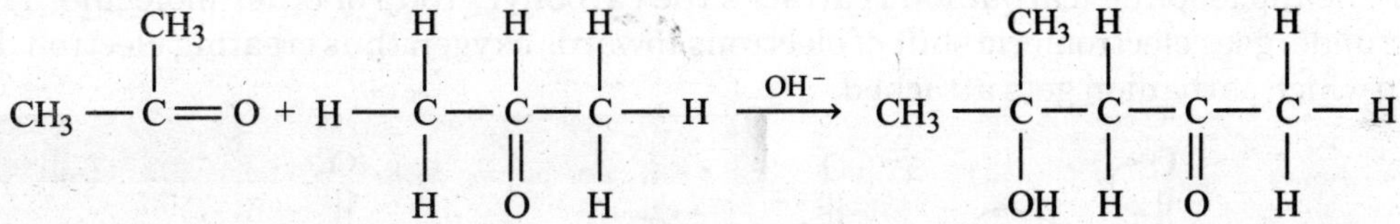

Acetone 2 moles → 4-Hydroxy-4-methyl-2-pentanone

If the aldehyde or ketone does not contain α-hydrogen, a simple aldol condensation cannot take place. For example:

$$\left.\begin{array}{r}\text{ArCHO}\\ \text{HCHO}\\ (CH_3)_3CCHO\\ \text{ArCOAr}\\ \text{ArCOCR}_3\end{array}\right\} \xrightarrow{\text{dilute } OH^-} \text{no reaction}$$

No α-hydrogen atoms

(In concentrated base, however, such aldehydes may undergo the Cannizzaro's reaction, sec. 23.2)

Dehydration of Aldol Products: The β-hydroxy aldehydes and β-hydroxy ketones from aldol condensation when heated with dilute acids undergo dehydration leading to the formation of α–β unsaturated aldehydes or ketones, i.e. the major product has double bond between α- and β-carbon atoms. For examples:

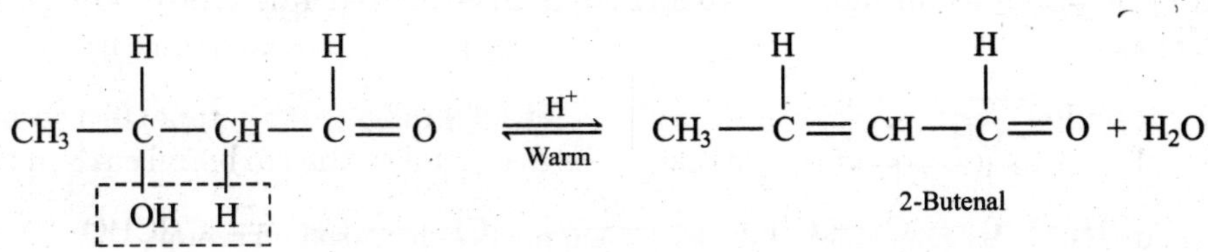

Aldol → 2-Butenal

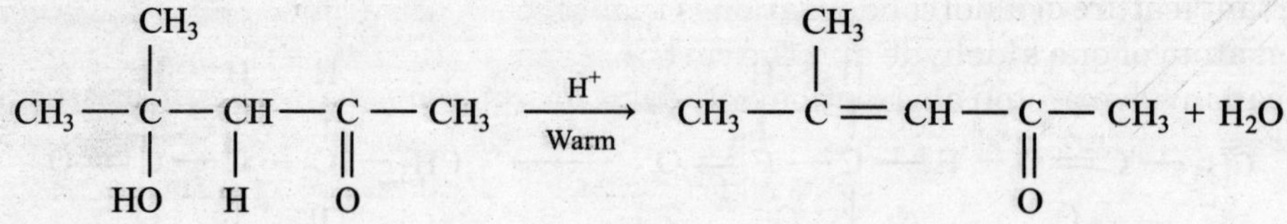

4-Hydroxy-4-methyl-2-pentarone   4-Methyl-3-penten-2-one

Mechanism: The mechanism of aldol condensation involves the following steps:
Step I: $\bar{O}H$ of the base abstracts a proton from the α-carbon atom to form a carbanion

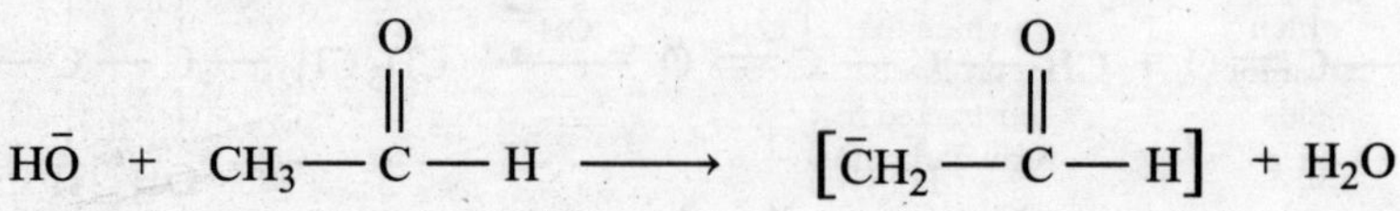

(I)

Step II: The nucleophile (carbanion I) attacks the carbonyl group of other molecule. The other molecule undergoes electromeric shift of electrons towards oxygen thus creating electron deficient carbon to which carbanion gets attached.

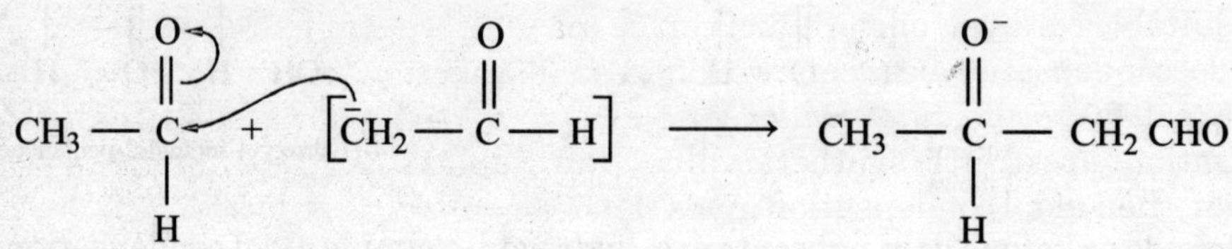

(II)

Step III: The anion (II) abstracts a hydrogen from water to from β-hydroxy aldehyde (aldol) regenerating the $\bar{O}H$ ion.

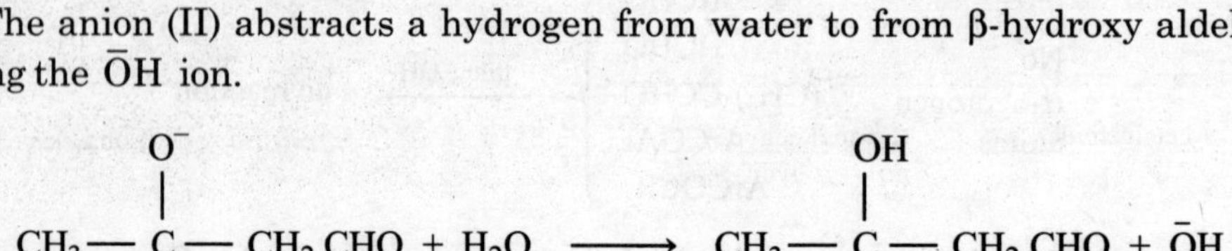

β-hydroxy aldehyde
(aldol)

Aldol may eliminate a water molecule to form α, β unsaturated aldehyde or ketone in presence of acids and bases.

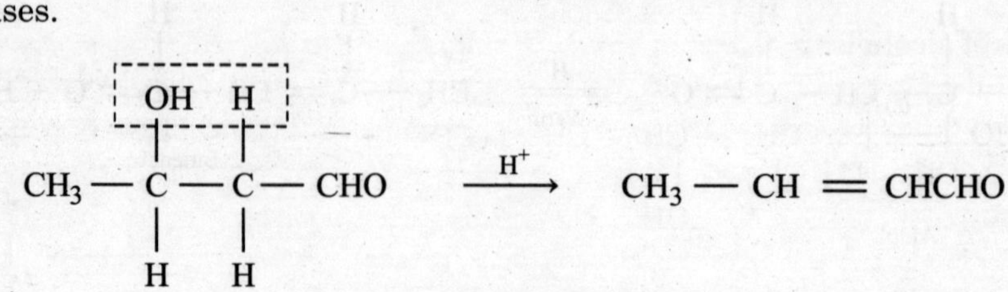

An important feature of aldol condensation is that carbon-carbon bond formation occurs between the α-carbon atom of one aldehyde and the carbonyl group of another. This is because carbanion generation can involve proton abstraction only from the α-carbon atom. Schematically, this can be represented as:

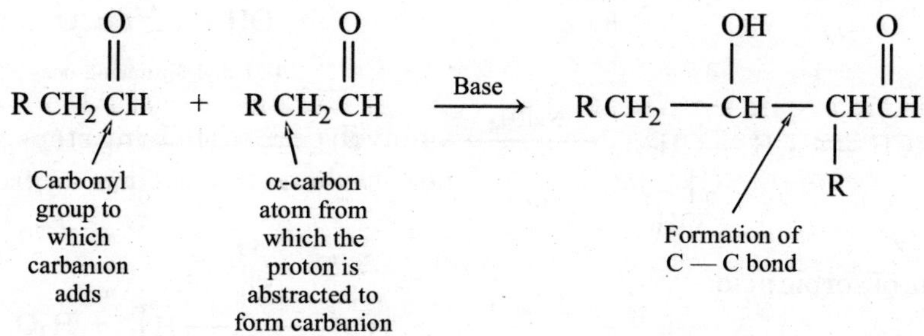

Aldol condensation can occur:

(*i*) between two aldehydes (identical or different)
(*ii*) between two ketones (identical or different)
(*iii*) between an aldehyde and ketone.

Generally with two different aldehydes, all four possible products are obtained. However by the choice of the suitable catalyst, one product can be made to predominate.

Such an aldol condensation between two different aldehydes or ketones or between one aldehyde and ketone are called *crossed aldol condensation*. As mixture of four different products are formed, these condensations are of little synthetic value. However, if one of the carbonyl does not posses α-hydrogen atom, then aldol condensation gives the good yield of useful product.

Example: Crossed Aldol condensation

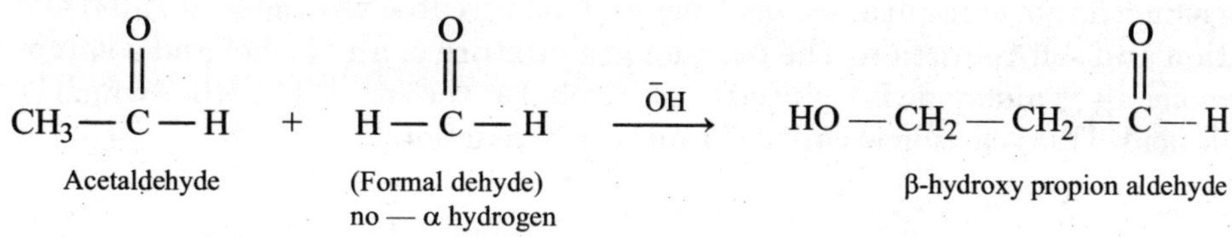

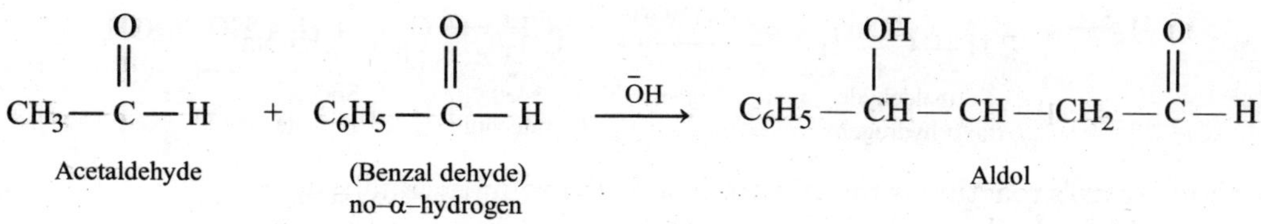

Applications of Aldol Condensation

(*i*) Synthesis of alcohols:

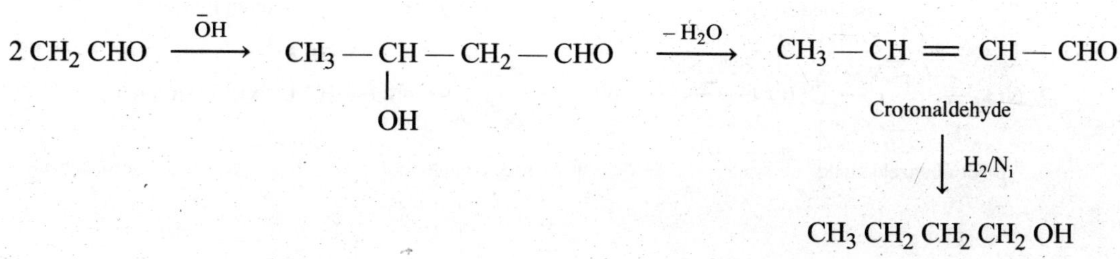

(*ii*) **Synthesis of α–β unsaturated, ketones and alcohols:**

$$RCHO + CH_3-\underset{\underset{O}{\|}}{C}-R \xrightarrow{\bar{O}H} R-\underset{\underset{OH}{|}}{CH}-CH_2-\underset{\underset{O}{\|}}{C}-R$$

$$\xrightarrow{-H_2O} R-CH=CH-\underset{\underset{O}{\|}}{C}-H \xrightarrow{NaBH_4} RCH=CH-\underset{\underset{OH}{|}}{CH_2}$$

(*iii*) **Synthesis of sorbic acid:**

$$\underset{\text{Condensation}}{2\,CH_3\,CHO \xrightarrow[\text{aldol}]{\bar{O}H}} \underset{\text{Crotonaldehyde}}{CH_3\,CH=CH-CHO}$$

$$CH_3\,CH=CH-CHO + CH_3\,CHO \xrightarrow[\text{aldol condensation}]{\bar{O}H} CH_3\,CH=CH-CH=CH-CHO \xrightarrow{[O]} \underset{\text{Sorbic acid}}{CH_3-CH=CH-CH=CH-COOH}$$

23.2 CANNIZZARO'S REACTION

Aldehydes which do not contain any α-hydrogens, when treated with concentrated alkali undergo self-oxidation and self-reduction. The product is a mixture of an alcohol and a salt of carboxylic acid. One molecule of aldehyde is reduced to an alcohol at the cost of the other which is oxidized to a carboxylic acid. This reaction is called Cannizzaro's reaction.

$$\underset{\substack{\text{Formaldehyde}\\ \text{(no } \alpha \text{ hydrogen)}}}{2\,H-\overset{\overset{O}{\|}}{C}-H} \xrightarrow{50\%\ NaOH} \underset{\substack{\text{Methyl}\\ \text{alcohol}}}{CH_3\,OH} + \underset{\substack{\text{Sodium}\\ \text{formate}}}{H\,CO\bar{O}\,Na}$$

The Cannizzaro's reaction is mainly applicable to aromatic aldehydes.

Note that aldehydes having α-hydrogen atom *do not* give Cannizzaro's reaction, instead they undergo aldol condensation.

$$\underset{\text{Benzaldehyde}}{2\,C_6H_5\,CHO} \xrightarrow{50\%\ NaOH} \underset{\text{Benzyl alcohol}}{C_6H_5\,CH_2\,OH} + \underset{\text{Sodium Benzoate}}{C_6H_5\,COONa}$$

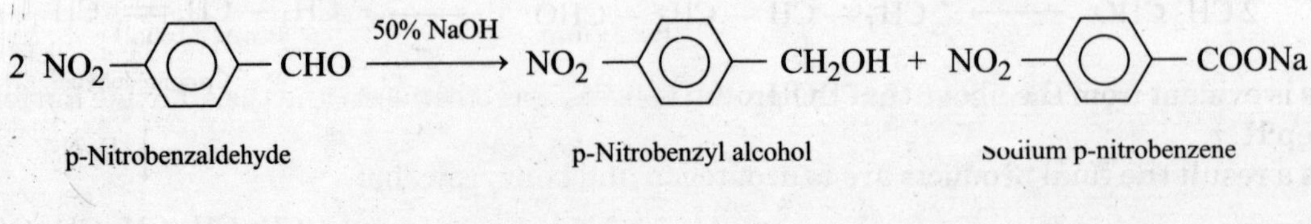

p-Nitrobenzaldehyde p-Nitrobenzyl alcohol Sodium p-nitrobenzene

Crossed Cannizzaro's Reaction

Cannizzaro's reaction can also take place between two different aldehydes. In that case a mixture of all possible products is obtained and the reaction is called Crossed Cannizzaro's Reaction.

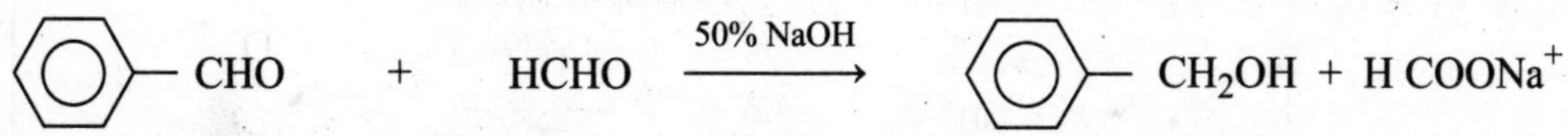

It must be mentioned however that if one of the aldehyde is formaldehyde, the sodium formate is exclusively formed and the other aldehyde is reduced to alcohol.

This is because formaldehyde is most reactive than any other aldehyde and nucleophilic attack of base occurs on formaldehyde which gets oxidised to form formate ion.

Mechanism: The mechanism involves hydride ion (H^-) shift from carbon of one molecule to carbon of another molecule of aldehyde.

Step I: It involves nucleophilic addition of nucleophile ($\bar{O}H$) on carbonyl cation to give intermediate I.

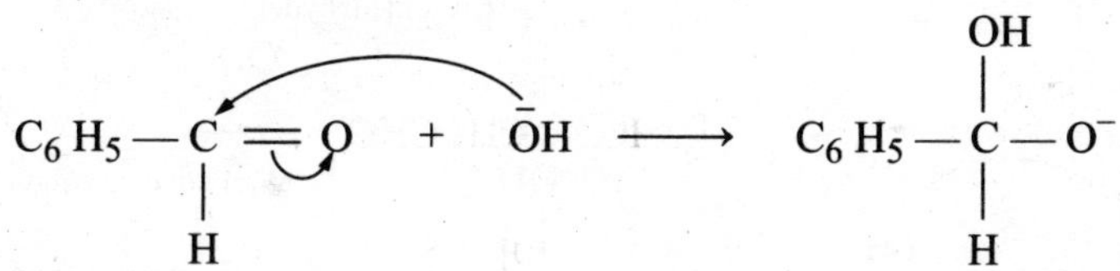

(I)

Step II: The intermediate I act as an hydride donor and transfer the hydride ion to a second molecule of aldehyde. The presence of negative charge on I (hydroxy alkoxide) helps in the loss of hydride ion

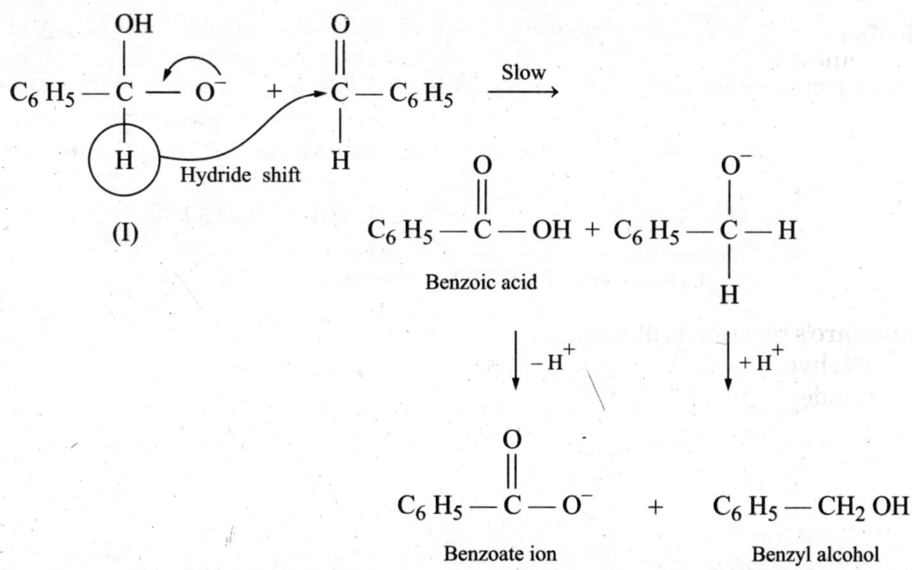

As is evident from the above that the proton is transferred between and the alkoxide ion formed in step II.

As a result the final products are benzoate ion and benzyl alcohol.

Intramolecular Hydride Transfer: It is interesting to note that α-ketoaldehydes and 1, 2 dialdehydes in presence of concentrated bases are capable of hydride transfer and these compound undergo internal Cannizzaro's reaction.

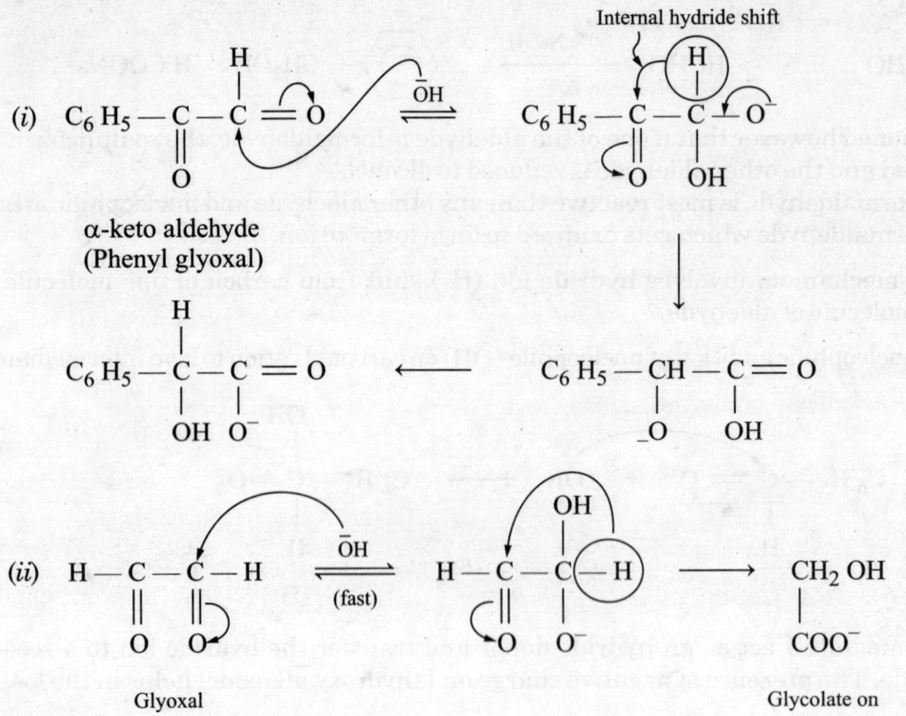

Applications

(i) *Synthesis of pentaerythritol:* A reaction between HCHO and an aldehyde containing one or more α-hydrogen atom first undergo aldol condensation. The aldol condensation continues until all the α-hydrogens are used. Then the product obtained undergoes Cannizzaro's reaction with formaldehyde.

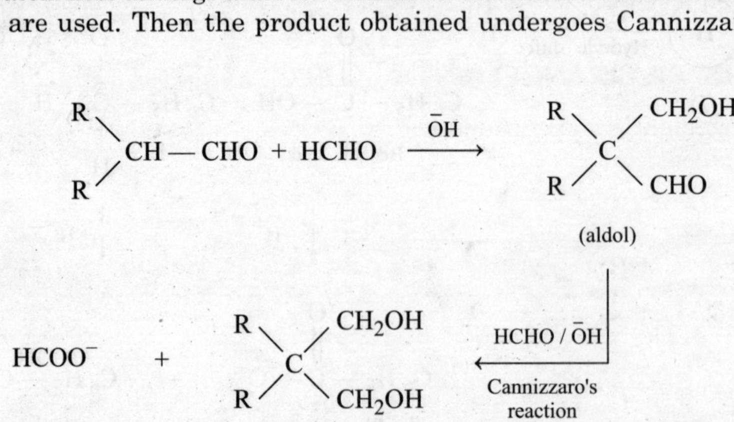

Similarly,

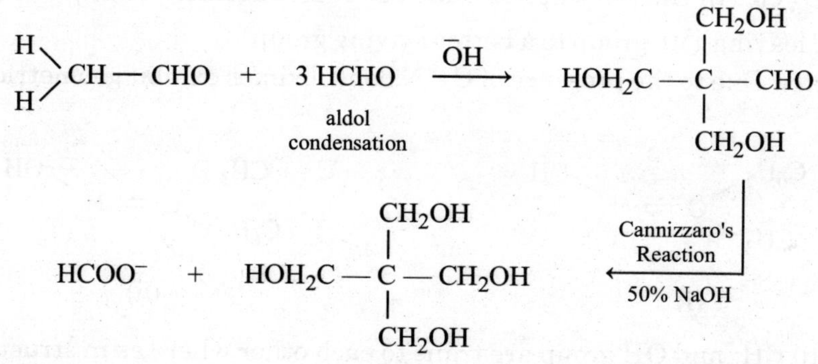

Pentaerythrital

(ii) Synthesis of heterocyclic alcohols and acids: Heterocyclic compounds undergo Cannizzaro's reaction to give heterocyclic alcohols and acids.

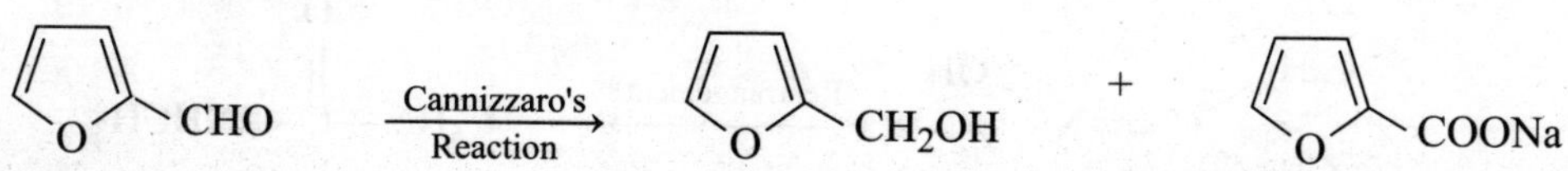

Furfural　　Furfuryl alcohol　　Sodium salt of furoic acid

(iii) Synthesis of Glycollic acid

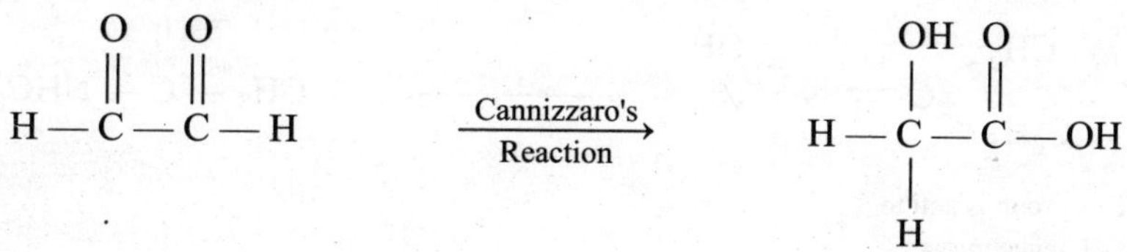

Glyoxal　　Glycollic acid

23.3 BECKMANN REARRANGEMENT

Beckmann Rearrangement involves the conversion of ketoximes to N-substituted acid amides in presence of acidic reagents.

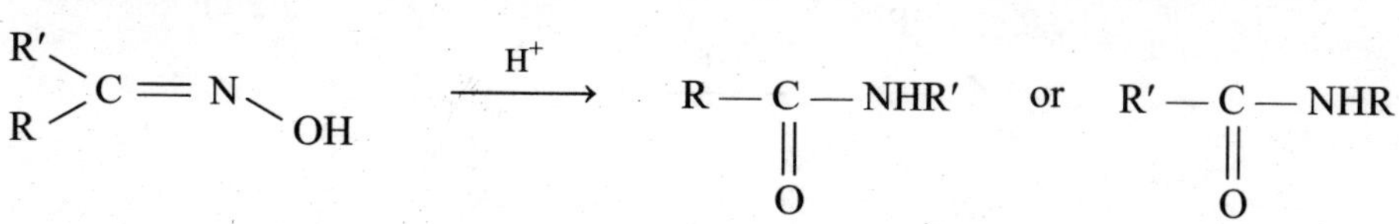

Ketoxime　　N – substituted amide

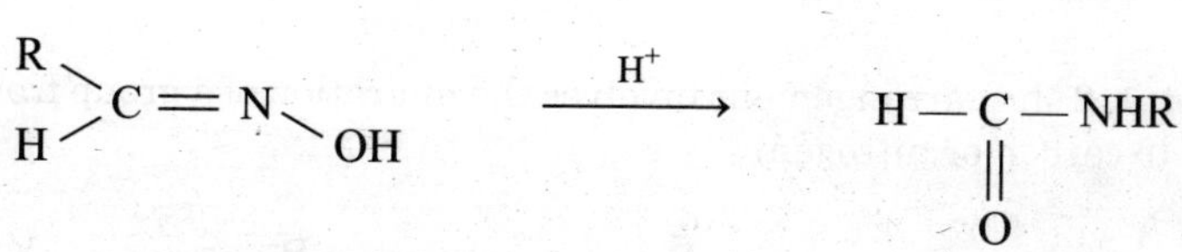

Aldoxime　　N – substituted amide

The acids which are used in Beckmann Rearrangment include conc. H_2SO_4, PCl_5, liquid SO_2, BF_3, $SOCl_2$, P_2O_5 etc. In the presence of acids the rearrangement is important because these convert the poor leaving $\bar{O}H$ group to a better leaving group.

(i) *Stereochemistry:* Due to the presence of C = N bond, oximes exhibit geometrical isomerism as shown below:

(i) (ii)

In structure (i) CH_3 and OH group are trans to each other where as in structure (ii) CH_3 and OH are cis to each other. The Beckmann rearrangement is stereospecific and the group that migrates is generally the one which is trans (anti) to hydroxyl group. The stereospecificity is very significant and this is often used as a method for determining the configuration of oximes.

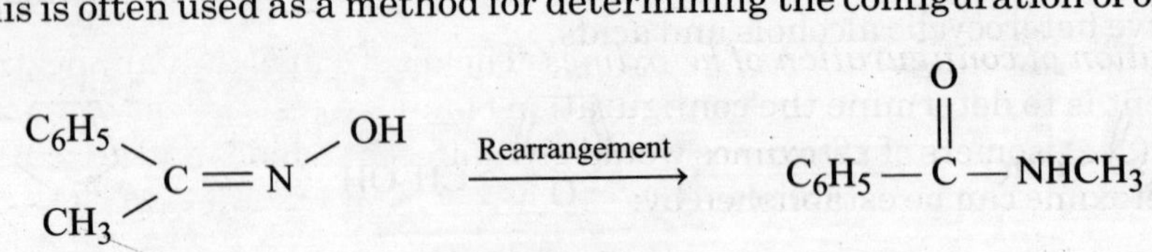

This group is anti to OH and hence migrates

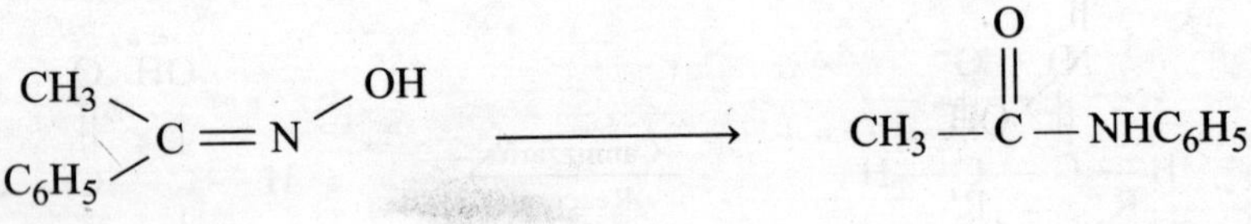

This group is anti to OH hence migrates

The migration of group from carbon to nitrogen is always anti, hence the rearranged products of isomers (i) and (ii) would be:

(ii) *Mechanism:* Beckmann rearrangement can be explained by 1, 2 shift rearrangement.*

Step I: In the first step the OH group of ketoxime is converted by the reagent (acid) to a better leaving group, i.e. proton of the acid converts into OH_2^+

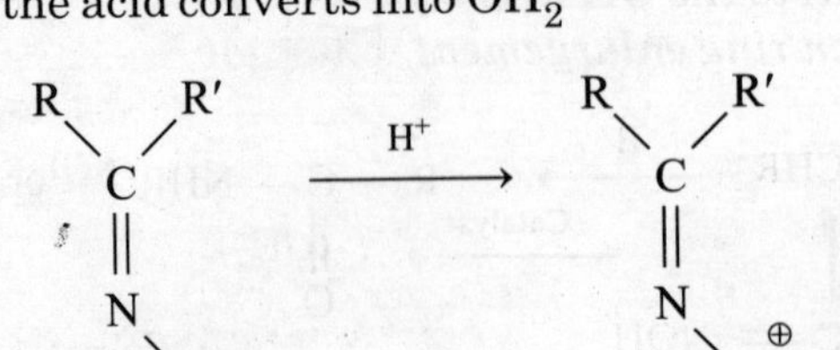

1, 2 SChift arrangement: 1, 2 shift arrangement involves the migration of a group from carbon atom to the adjacent atom (usually to carbon or nitrogen)

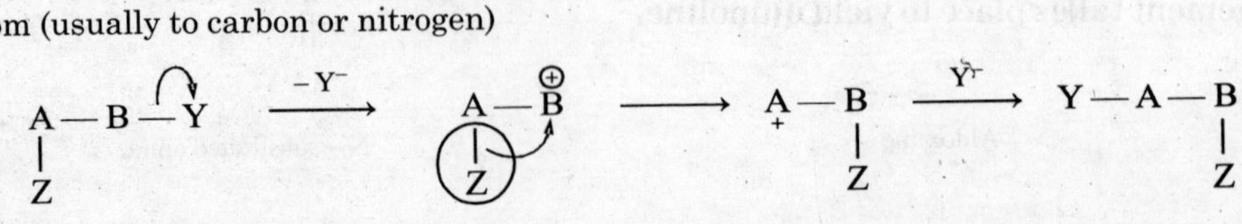

Step II: After this, the group anti to the OH group migrates with its bonding pair to nitrogen. This results in the formation of carbonium ion.

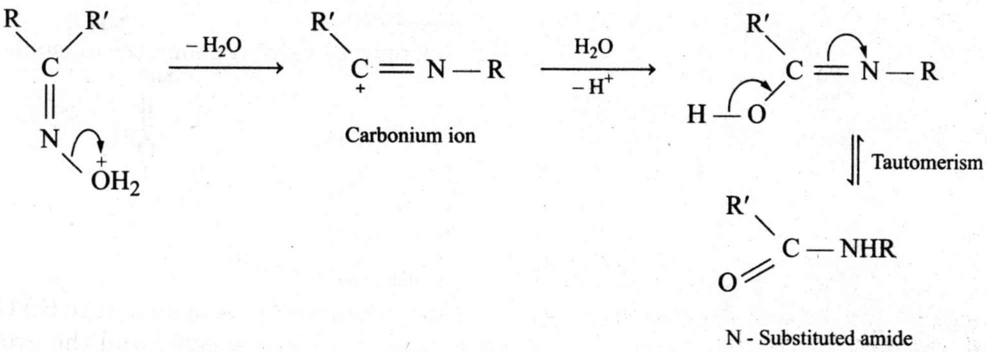

Applications

(i) *Determination of configuration of ketoximes.* The most important application of Beckmann's rearrangement is to determine the configuration of ketoximes. As was discussed that the trans (anti) and cis (syn) isomers of ketoximes would give different amides and hence the configuration of the parent ketoxime can be established by:

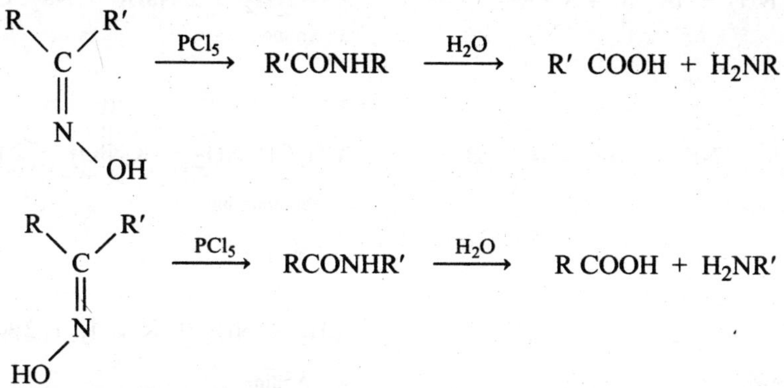

The amides obtained are identified by their hydrolysis product. The amine is formed from the alkyl/aryl group which is anti to the OH in the parent oxime.

(ii) *The cyclic ketoximes given ring enlargement.* Example

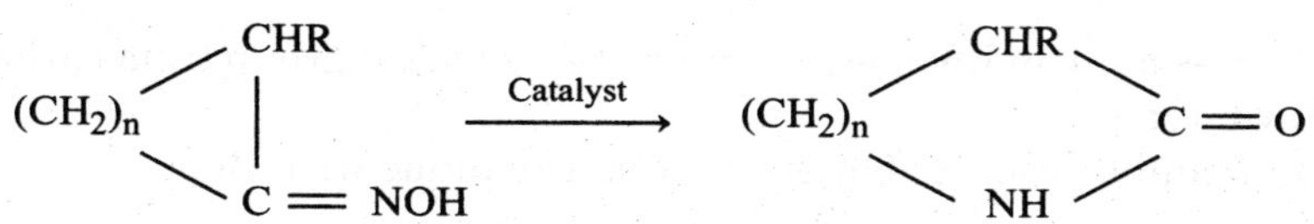

The reaction is useful for preparing rings of all sizes.

(iii) *Synthesis of isoquinolines:* On treating cinnamaldehyde oxime with acid, the Beckmann's rearrangement takes place to yield quinoline.

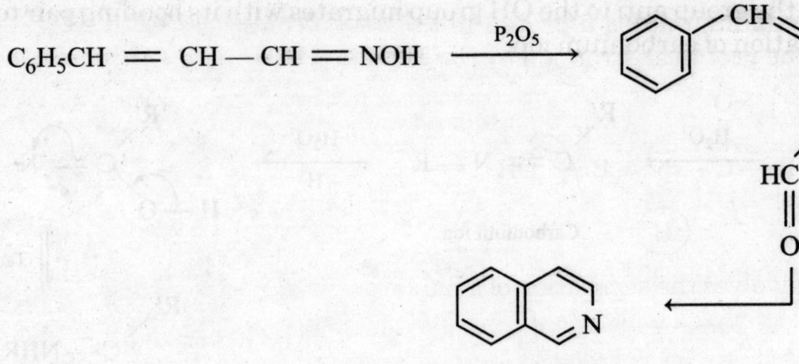

isoquinoline

23.4 HOFFMANN REARRANGEMENT OR HOFFMANN DEGRADATION

Primary amides when heated with bromine in the presence of an alkali, are converted into primary amines with one *carbon atom less than the parent amide*. For example,

$$\underset{1^\circ \text{ Amide}}{R-\overset{O}{\overset{\|}{C}}-NH_2} + Br_2 + 4\,NaOH \xrightarrow{\Delta} \underset{1^\circ \text{ Amine}}{R-NH_2} + 2\,NaBr + Na_2CO_3 + 2\,H_2O$$

$$\underset{\text{Propanamide}}{CH_3CH_2-\overset{O}{\overset{\|}{C}}-NH_2} + Br_2 + 4\,KOH \longrightarrow \underset{\text{Ethanamine}}{CH_3CH_2NH_2} + K_2CO_3 + 2\,KBr + 2\,H_2O$$

$$\underset{\text{Benzamide}}{C_6H_5-\overset{O}{\overset{\|}{C}}-NH_2} + Br_2 + 4\,KOH \longrightarrow \underset{\text{Aniline}}{C_6H_5-NH_2} + K_2CO_3 + 2\,KBr + 2\,H_2O$$

The overall reaction involves the elimination of carbonyl group. It involves a *molecular rearrangement* in which the alkyl or the aryl group migrates from carbon to nitrogen. This reaction is called Hoffmann bromamide reaction and is widely used for descending the homologous series.

Mechanism: The mechanism involves the migration of alkyl group from carbon to nitrogen. It involves the following steps:
Step I: The N-Bromamide is formed by the reaction of amide with Br_2.

$$R-\overset{O}{\overset{\|}{C}}-\ddot{N}H_2 \underset{}{\overset{Br_2}{\rightleftharpoons}} R-\overset{O}{\overset{\|}{C}}-NHR + HBr$$

Step II: The N-Bromamide loses a proton to the base. The presence of electron withdrawing bromine facilitate the proton abstraction by the base as it increases the acidity of amide

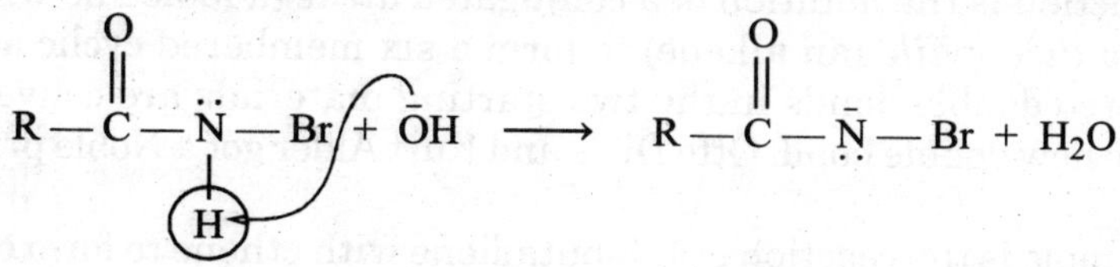

Step III: This step involves the separation of a halide ion, which leaves behind an electron-deficient nitrogen atom.

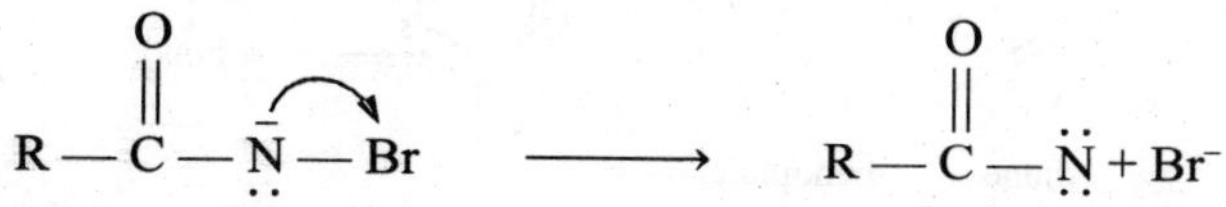

Nitrene

Step IV: In this step the rearrangement occurs, i.e. the alkyl 'R' migrates from carbon to nitrogen along with its group of electrons to from isocyanate.

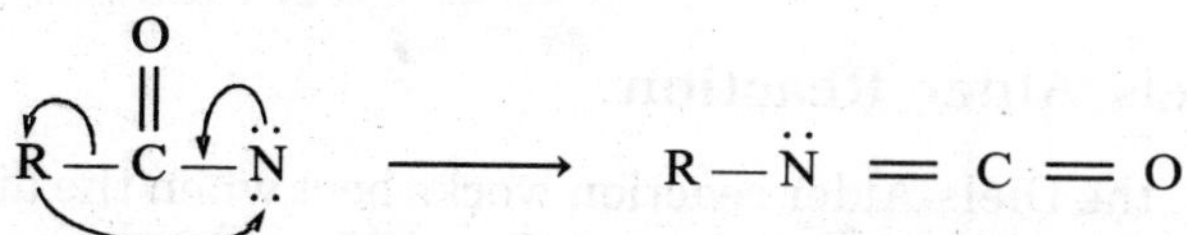

Isocyanate

The steps (iii) and (iv) are believed to occur simultaneously.

Step V: The hydrolysis of isocyanate take place resulting in the formation of a primary amine.

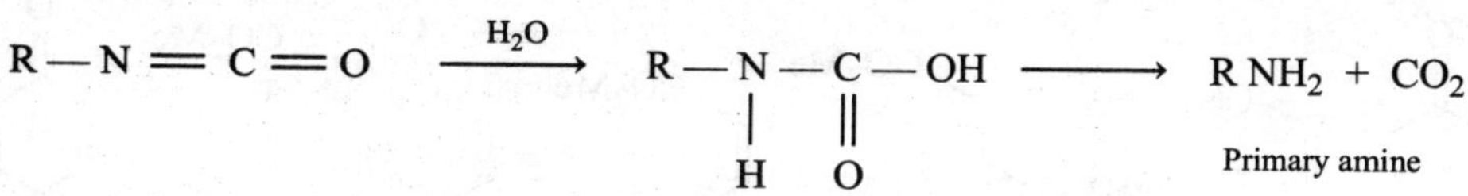

Primary amine

Application

1. The reaction is mainly used in synthetic organic chemistry to decrease the length of carbon chain by one carbon atom.

2. Synthesis of Anthranilic Acid: One of most important industrial applications of this reaction is the synthesis of anthranilic acid:

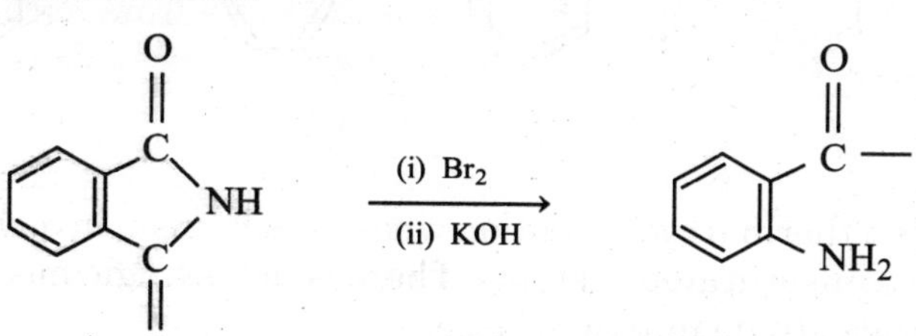

Anthranilic acid

3. Synthesis of Urethane: If this reaction is carried out in alcoholic solution the isocyanate is converted into urethane. The polymerisation of urethane gives polyurethane which is on important industrial polymer used in making adhesives, foams, etc.

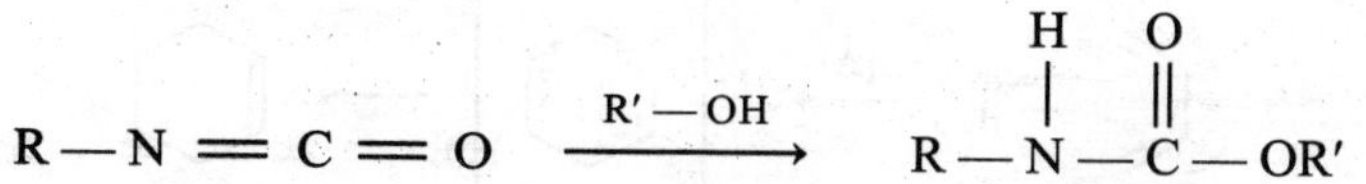

Urethane

23.5 DIELS-ALDER REACTION

The Diels-Alder reaction is the addition of a conjugated *diene* (a molecule with two alternating double bonds) and a *dienophile* (an alkene) to form a six membered cyclic adduct and bicyclic compounds. The three double bonds in the two starting materials are converted into two new single bonds and one new double bond. Otto Diels and Kurt Alder got a Noble prize in chemistry for discovering it.

The simplest example is the reaction of 1, 3-butadiene with ethene to form cyclohexane

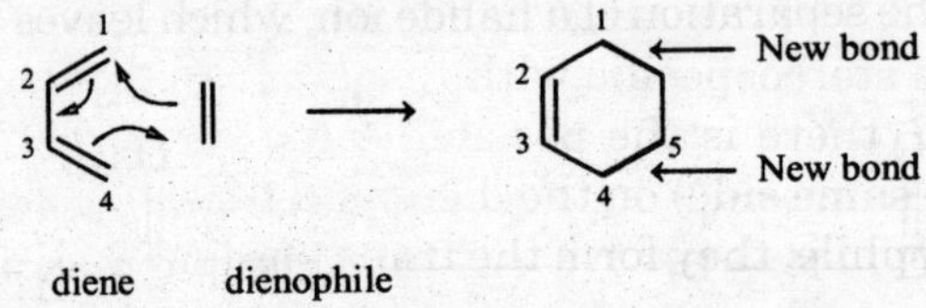

Since, the reaction forms a cyclic product, via a cyclic transition state, it can also be described as a "***cycloaddition***" As is evident from above that Diels-Alder reaction involves a system of 4π electrons (the diene) and a system of 2π electrons, it is known as [4 + 2] cycloaddition.

Components of Diels Alder Reaction

(i) Dienophile: Typically, the Diels-Alder reaction works best when the dienophile is substituted with electron-withdrawing groups (like $-NO_2$, —CN, —COR, etc).

- Some common examples are shown below:

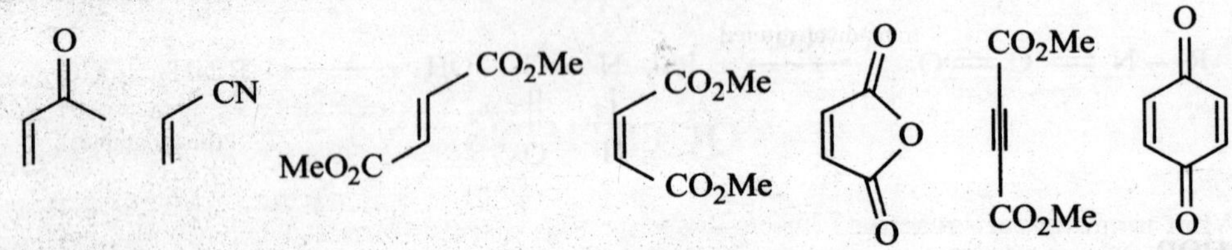

(ii) Diene: The reaction is favoured when the diene is substituted with electron donating groups (like —OR, $—NR_2$, etc).

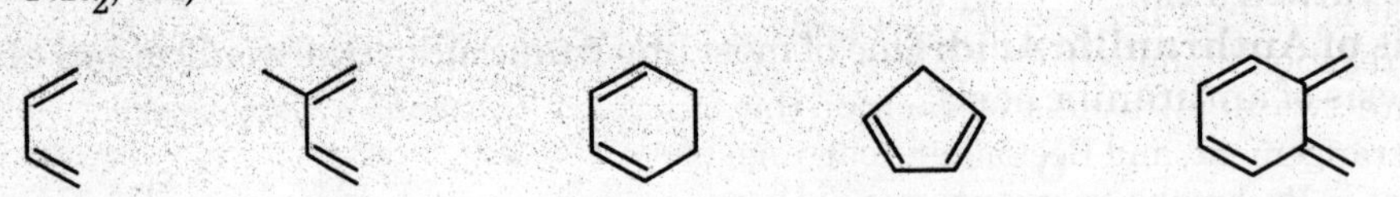

Mechanism

The Diels-Alder reaction is a thermal cycloaddition whose mechanism involves the sigma-overlap of the pi-orbitals of the two unsaturated systems. There is not a single mechanism for all Diels-Alder reaction. We can divided them into two classes:

1. *Symmetrical (concerted) mechanisms* when the two new bonds are formed simultaneously. In the transition state, the two forming bonds have the same lengths. The combination of ethylene and butadiene is one example.

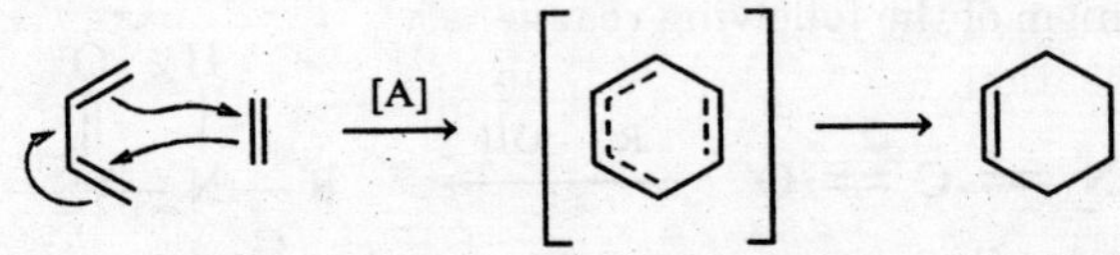

2. The reaction is stepwise, first with the formation of a rate determining single bond, followed by a faster formation of a second bond.

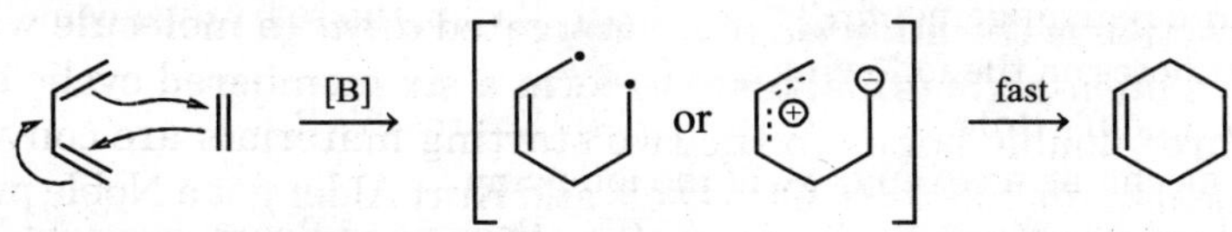

Stereochemistry of Diels-Alder Reaction

The Diels-Alder reaction is stereospecific with respect to both the diene and the dienophile. If dienophile is disubstituted, there is the possibility for stereochemistry in the product. If the substituents are *cis* (on the same side) on the dienophile, they form cis product. If they are trans (opposite sides) on the dienophile, they form the trans product.

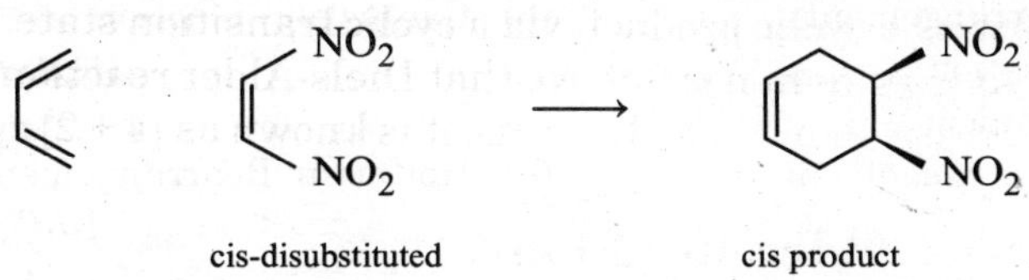

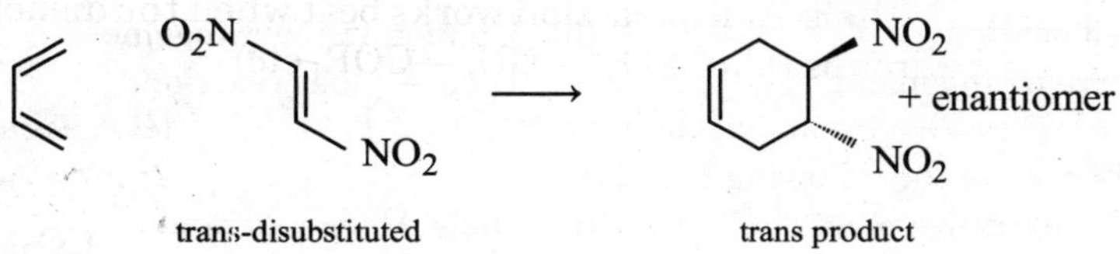

QUESTIONS

1. What is cannizzaro reaction? Discuss mechanism of formaldehyde with NaOH.
2. Write an example of:
   (i) Crossed Cannizzaro reaction (ii) Hoffmann reaction
   (iii) Diel's Alder reaction (iv) Beckmann rearrnagement.
3. What is Diel's Alder reaction? Describe the mechanism of the reaction. What is the importance of the reaction in industry?
4. Write two applications of each of the following reactions : Cannizzaro reaction, Hoffmann rearrangement, and Beckmann rearrangement.
5. What is Beckmann rearrangement? Discuss the mechanism and application of the reaction
6. What is Hoffmann rearrangement? Discuss the mechanism and application of the reaction.

EXAMINATION QUESTIONS

1. Attempt the following
   (i) Write the mechanism of crossed Aldol condensation involving equimolecular amounts of acetaldehyde and formaldehyde. Also indicate the possible side products of the reaction.
   (ii) In the Diels-elder reaction, identify six groups in the dienophile, which promote the reaction.
   (iii) Will N-methylacetamide undergo Hofmann rearrangement ? Explain. **[*U.P. Tech.; 2003-04*]**
2. Write the meachanism of the following reactions:
   (i) Aldol Condensation (ii) Friedel-Crafts reaction. **[*U.P. Tech.; 2002-03*]**

3. Attempt the following :
(a) With the mechanism of the following:
(*i*) Hofmann rearrangement (*ii*) Backmann rearrangement.
(b) Write short notes on the following :
(*i*) Karkownikoff's Rule (*ii*) Kharasch's Rule. ***[U.P. Tech.; 2002-03]***

4. (a) Write the mechanisms of any two of the following :
(i) Aldol condensation (ii) Backmann Rearrangement
(iii) Connizzaro reaction (iv) Diels – Alder reaction
(v) Hafmann Rearrangement.

5. Describe the mechanism of Cannizzaro's reaction. ***[U.P Tech.; 1st Sem., 2006-07]***

6. Give the mechanism of following Reactions :
(i) Hofmann re-arrangement (*ii*) Aldol condensation (*iii*) Canizzaro reaction sessional.
[U.P. Tech.; 2nd Sem., 2006-07]

7. Give the mechanism of the following reactions :
(a) Hoffmann rearrangement (b) Cannizzaro reaction
(c) Aldol condensation. ***[U.P. Tech.; 2nd Sem., 2007-08]***

8. Discuss the mechanism
(a) Beckmann Rearrangement (b) Hoffmann Rearrangement
[U.P. Tech.; 1st Sem., 2008-09]

9. Ketoxime $\xrightarrow{H_2SO_4}$ N-substituted amide. This reaction is known as
(a) Aldol condensation (b) Beckmann rearrangement
(c) Hoffmann rearrangement (d) Diels – Alder reaction.
[U.P. Tech.; 1st Sem., 2008-09]

10. Give the mechanism of the following reaction.
(a) Backmann rearrangement (b) Diels-Alder reaction
[U.P. Tech.; 1st Sem., 2008-09]

❑❑❑

24

STEREO CHEMISTRY

"Imagination is more important than knowledge. Knowledge is limited, imagination encircles the world"

—Albert Einstein

CHAPTER OUTLINE

24.1 INTRODUCTION

One of the most distinctive feature of organic compounds is their ability to show isomerism. The term isomer (Greek 'iso-equal', 'mers'-parts) was coined by Berzelius to different compounds having the same molecular formula. The isomers are defined as—*two or more compounds having the same molecular formula but different chemical and physical properties and the phenomena is termed as isomerism.*

Isomerism is of two types:

1. Structural Isomerism
2. Stereoisomerism

Structural Isomerism: Isomers having the same molecular formula but different structures, i.e. different arrangement of atoms within the molecule are called structural isomers and the phenomena is called structural isomerism. Structural isomers can involve the carbon skeleton or nature and position of functional groups.

A simple example of structural isomers are the isomers ethanol and dimethyl ethers, as shown below:

$$CH_3 - CH_2 - OH \qquad\qquad CH_3 - O - CH_3$$
$$(C_2H_6O) \qquad\qquad\qquad (C_2H_6O)$$

Stereoisomerism: Stereoisomerism is also called space isomerism. It is due to the difference in spatial positions of atoms in a molecule, i.e. *the isomers are different from each other only in the way the atoms are oriented in space.* A small difference in their spatial arrangement may lead to a profound effect on their physical and chemical properties. The space relationship of atoms within organic molecules have assumed such an importance that a new branch of chemistry called "Stereochemistry" has been established. Stereochemistry is the branch of chemistry which deals with the structures of molecules in three dimensions and stereoisomerism is a part of this branch of chemistry.

Stereoisomers have 2 major classes:

1. Configurational isomers.
2. Conformational isomers.

Configurational isomers inturn are of two types:

(*i*) Enantiomers. (*ii*) Diastereomers

All these types of stereoisomers are discussed in the following sections.

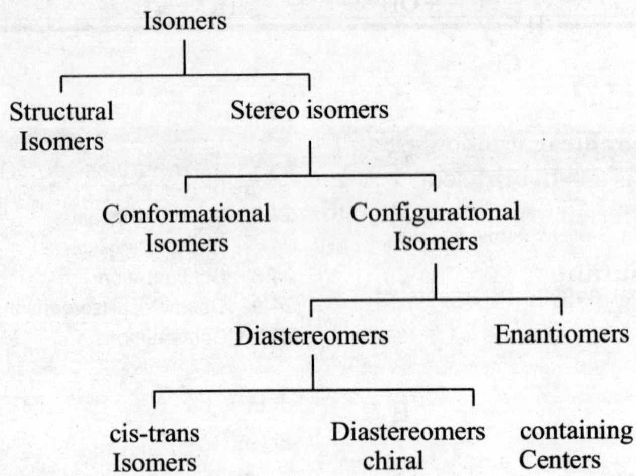

24.2 BASIC CONCEPTS IN STEREOCHEMISTRYS

Before we discuss the details of the stereoisomers, it is worthwhile to be familiar with certain basic concepts involved.

1. Chirality (Stereogenicity)

Chirality (*cheir*, Greek for "hand") refers to objects which are not superimposable on (cannot be made to coincide with) their mirror image. ***A chiral molecule is one that is not superimposable on its mirror image.*** Achiral compounds (objects) can be superimposed on their mirror image. Our right and left hands are very similar, yet they are not identical. They are related to each other as mirror images, and as such they cannot be superimposed on top of each other. Hence, there are is said to be *chiral*. A more contemporary term for "chiral centre" (or chiral carbon) is stereogenic centre (or stereocenter).

Chiral

A molecule is *chiral* if its two mirror image forms are *not* superimposable upon one another.
A molecule is *achiral* if it its two mirror image forms are superimposable

sp^3–hybridised carbon atoms possessing four different substituents display this property due to their tetrahedral geometry. A chiral carbon is one that has four different "group" attached to it.

The groups can be anything from a single H to functional groups to one more other carbons. A chiral carbon is often denote\ by **C***.

An sp^3 hybridised with 1, 2 or 3 different atoms or groups attached can be superimposed on its mirror image and is, therefore *achiral*.

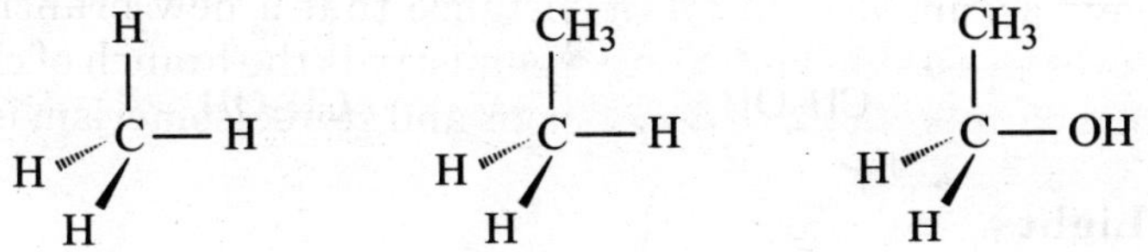

Achiral Carbons

However, if there are 4 different subsituents on carbon, then it is no longer superimposable on its mirror image and we say that carbon is chiral.

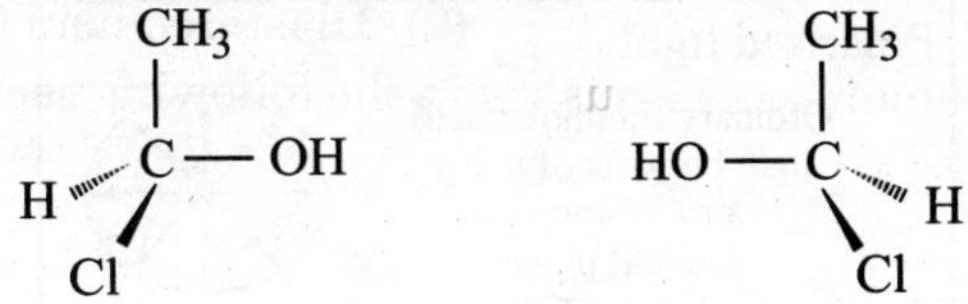

Chiral Carbons

For the carbon to be chiral, it follows that:

- the carbon atom is sp^3-hybridised
- there are four different groups attached to the carbon atom.

Consider 2-bromobutane

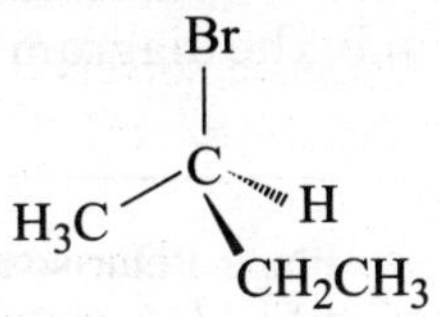

The carbon which is bound to the bromine is a stereogenic centre, it is sp^3 hybridised and it is bonded to four different groups:

1. a hydrogen
2. a bromine
3. a methyl group, and
4. an ethyl group.

2. Fischer Projections

Emil Fischer (late 1800's) introduced formulas depicting the spatial arrangement of groups around chiral carbon atoms. This convention represents molecular structure in a two-dimensional framework of vertical and horizontal bonds.

In Fischer projections, the vertical carbons are always viewed as being positioned behind the plane of the paper (away from the viewer) while the horizontal carbons are imagined as coming out of the plane (toward the viewer).

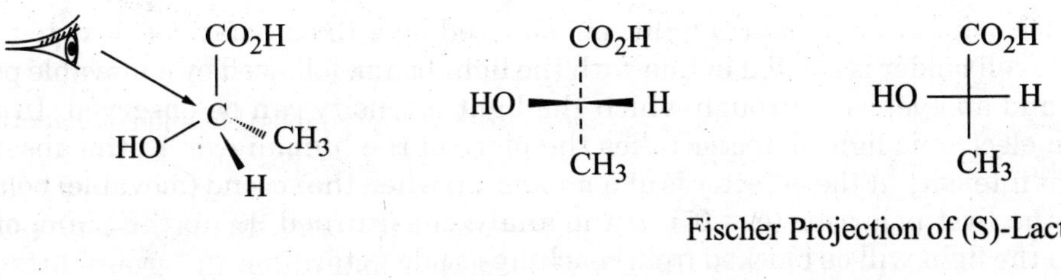

Fischer Projection of (S)-Lactic Acid

Similarly, we represent the isomers of glyceraldehyde as:

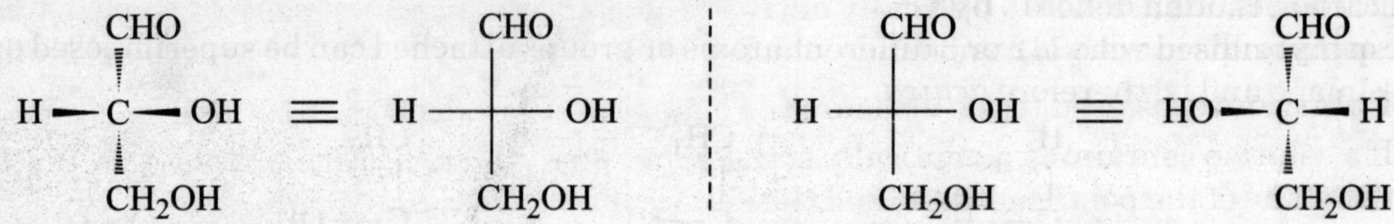

3. Plane Polarised Light

Ordinary light consists of waves vibrating in all planes perpendicular to the direction of propagation. Plane-polarised light is light in which all wave vibrations have been filtered out except for those in one plane.

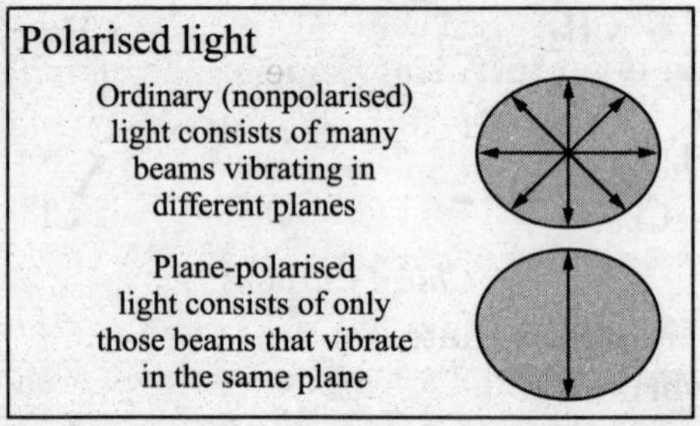

Plane-polarised light is created by passing ordinary light trough a polarising devices. Such devices transmit selectively only that component of a light beam having electrical and magnetic field vectors oscillating in a single plane. The plane of polarisation can be determined by an instrument called a **polarimeter,** shown in the diagram below.

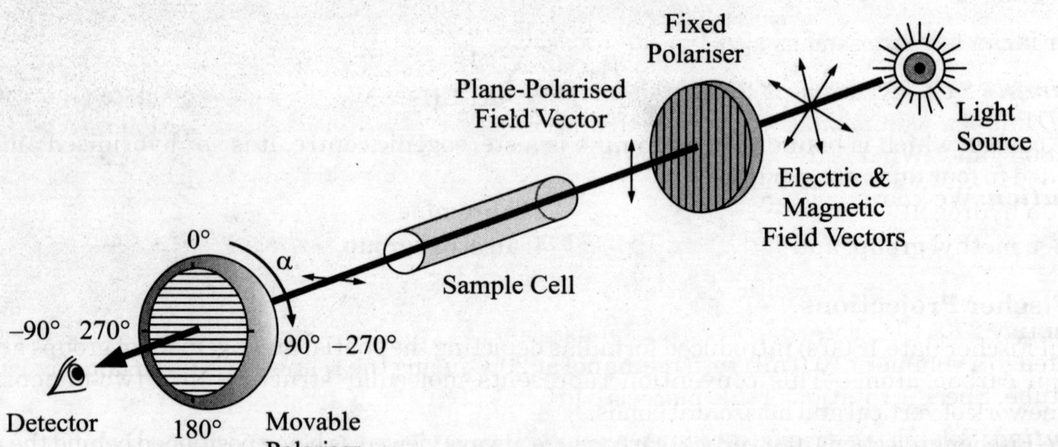

Fig. 24.1 Polarimeter

Monochromatic (single wavelength) light, is polarised by a fixed polariser next to the light source. A sample cell holder is located in line with the light beam, followed by a movable polariser (the analyzer) and an eyepiece through which the light intensity can be observed. In modern instruments an electronic light detector takes the place of the human eye. In the absence of a sample, the light intensity at the detector is at a maximum when the second (movable) polariser is set parallel to the first polariser ($\alpha = 0°$). If the analyzer is turned 90° to the plane of initial polarisation, all the light will be blocked from reaching the detector.

4. Optical Activity

Organic compounds are said to **optically active** because they rotate plane polarised light. Optically active compounds are chiral compounds which possess the ability to rotate plane-polarised light.

- **Polarimeter** is used to the measure the optical activity of compounds.
- If a solution containing an optically active substance is placed in the sample tube, it rotates the plane of the polarised light, and the degree of rotation is measured.
- If it rotates the light to the right (clockwise) it is called *dextrorotary* and is designated with a + sign. (+) Dextrorotatory (Latin *dexter* is "right").
- If it rotates light to the left (counterclockwise) it is called leavorotatory and it is designated – sign (–) Leavorotatory (Latin *laevus* is "left").

5. Specific Rotation

The magnitude of the rotation is an intrinsic property of the molecule. The observed rotation depends:

- on the concentration of the sample compound,
- the length of the sample tube,
- the temperature

Specific Rotation is the degree of rotation by 1.00 gram of compound in 1.00 mL of solution measured in a tube with a 1 decimeter path length. The amount of rotation is also dependent on the wavelength (typically the D line of sodium, 589.3 nm) and solvent.

$[\alpha]_D^{20} = \alpha/lc$ where α = observed rotation at 20°C

l = cell path length in decimeters

c = concentration in g/mL

D = wavelength (typically the D line of sodium, 589.3 nm)

or it can be expressed as : $[\alpha]_\lambda^T = \alpha/lc$

Example 24.1. A sample of pure (S)-2-butanol was placed in a 10.0 cm polarimeter tube. Using the *D* line of a sodium lamp, the observed rotation at 20°C was + 104°. The density of this compound is 0.805 gml^{-1}. What is the specific rotation of (S)-2. butanol?

Solution: We know;

$$[\alpha]_\lambda^7 = \alpha/lc$$

$$[\alpha]_\lambda^T = (+104°)/(1.00\text{ dm})\ (0.805\text{ gm}^{-1}) = +129°$$

$$[\alpha]_\lambda^T = +129°$$

Example 24.2. Calculate the observed rotation of a solution of 0.5245 g of (S)-1-amino-1-phenylethane diluted to a volume of 10.0 mL with methanol at 20°C, using the D line of a sodium lamp and a 1.00 dm tube. Specific rotation of this material: $[\alpha]_\lambda^T = -30.0°$.

Solution: Solving the specific rotation equation for observed rotation, we get $\alpha = [\alpha]_\lambda^T lc$. The sample size is 0.5245 g, but this has been diluted to 10.0 mL, so the sample concentration is 0.05245 g/10.0 mL = 0.05245 g ml^{-1}.

Therefore,

$$\alpha = (-30.0°)\ (1.00\text{ dm})\ (0.05245\text{ g ml}^{-1}).$$

$$\alpha = -1.57°.$$

Example 24.3. A pure enantiomer has an observed optical rotation of – 0.82° when measured in a one dm at a concentration of 0.3 g/10 mL. Calculate the specific rotation for this molecule.

Solution:

The concentration of 0.3 g/10 mL is equivalent to 0.03 g/mL; $c = 0.03$ g/mL

The length of the analysis tube is one dm; 1 = 1.0 dm

The specific rotation is therefore: $[\alpha] = (-0.82°)/(0.03 \text{ g/mL} \times 1.0 \text{ dm})$

$$[\alpha] = -27.3° \text{ g}^{-1} \text{ mL}^{-1} \text{ dm}^{-1}$$

If the specific rotation of a pure enantiomer is known, the observed rotation can also be used to calculate **optical purity,** or the level of contamination of one compound with its enantiomer, using the simple convention:

Optical purity = (% of one enantiomer) – (% of the other enantiomer)

24.3 OPTICAL ISOMERISM

Optical isomers are compounds with the same molecular formula but differ in the way they rotate plane polarised light. These isomers are called "enantiomers" and are said to be optical active. *Enantiomers are stereoisomers which are related to each other as mirror images.* Such isomers will have one carbon that has four non-equivalent groups attached to the "chiral" carbon. The molecule will also be chiral. For example, CHClBrF would be a molecule in which its carbon would be chiral. Its mirror image would be non-superimposible, and the two molecules (the molecule and its mirror image) would be enantiomers.

Mirror-images–Non-superimposable
(enantiomers)

Whereas the molecule CH_2ClBr would not have a chiral carbon therefore the molecule is not chiral or "achiral", and the mirror image would be superimposible and therefore not enantiomers. There would be no optical activity in an achiral molecule.

Mirror-images–Non superimposable
Not Enantiomers

In enantiomers, one molecule will rotate plane polarised light to the left. The other enantiomer will rotate plane polarised light to the right by the same number of degrees. Both enantiomers are said to be optically active because of their ability to rotate plane polarised light. Hence, in a pair of optically active enantiomers, each enantiomer will rotate the plane of polarised light in equal and opposite directions; the enantiomers can therefore be referred to as (+) and (–) enantiomers.

Now, let us consider specific examples of the molecules which show enantiomerism.

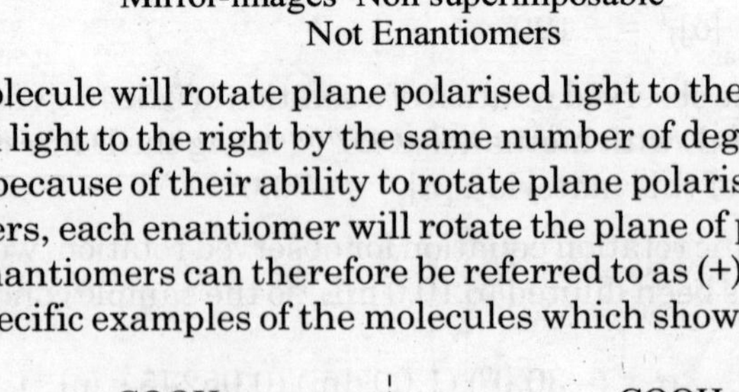

Enantiomers of Lactic acid

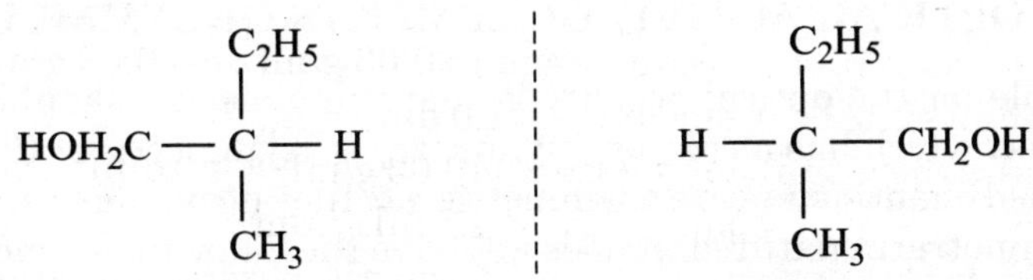

Enantiomers of 2 methyl-1-butanol

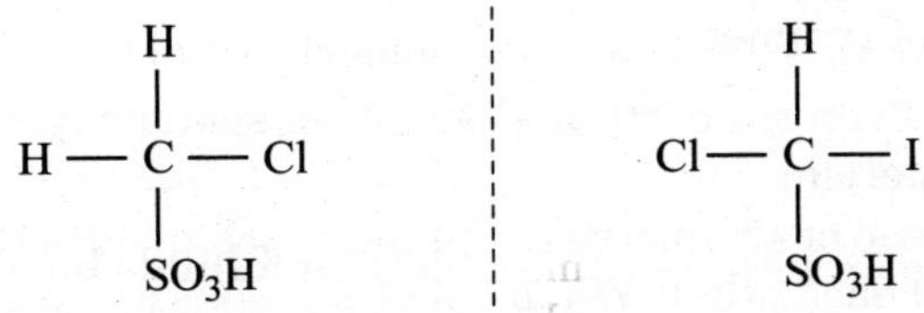

Enantiomers of chloroiodomethane sulf onic acid

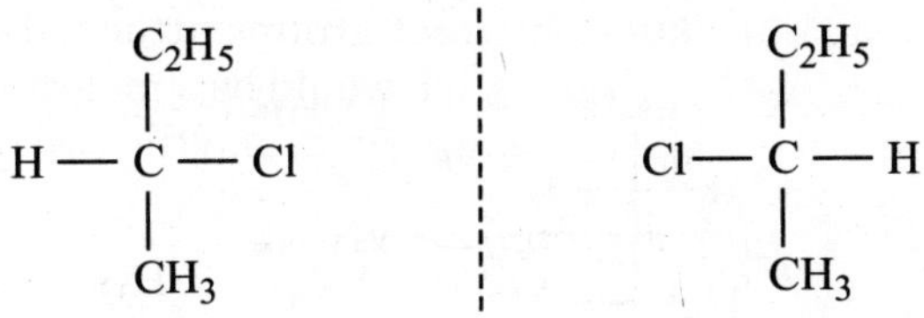

Enantiomers of sec-Butyl chloride

Most of the physical properties of a pair of enantiomers have identical values. For instance, enantiomers have the same boiling point, melting point, solubility, refraction index and many other properties. As mentined above enantiomers differ in only one physical properties, the interaction with plane-polarised light., as shown in the Figure below:

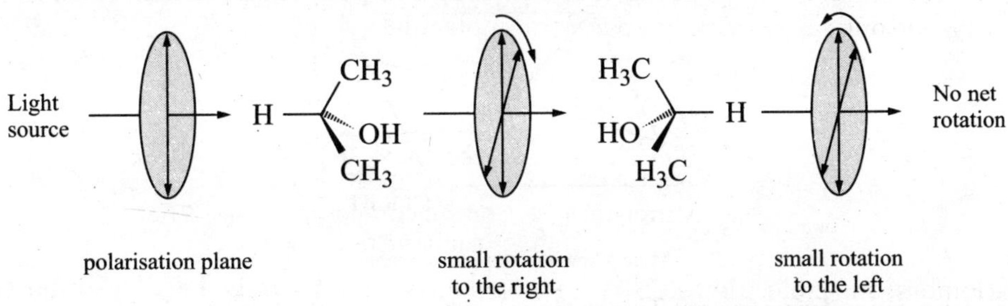

*(a) Interaction of an **achiral** molecule with polarised light.*

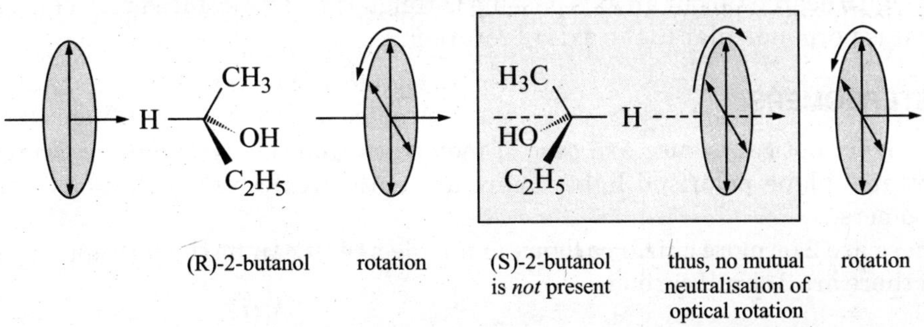

*(b) Interaction of a **chiral** molecule with polarised light.*

Fig. 24.2 Interaction of achiral and chiral molecule with polarised molecule.

24.4 CAUSE OF OPTICAL ACTIVITY (ELEMENTS OF SYMMETRY)

The factor responsible for the optical activity is that the molecule should have asymmetry or disymmetry. "An object which has no plane of symmetry is asymmetric or Chiral". Chiral carbons in the above-mentioned enantiomers are asymmetric, i.e. they do not have any plane of symmetry. A compound is disymmetric in nature if it does not have the following elements of symmetry.

(*i*) a plane of symmetry
(*ii*) a centre of symmetry
(*iii*) an alternating axis of symmetry

If the molecule contains even one of these elements of symmetry, the molecule is symmetrical and will not show enantiomerism.

(*i*) *Plane of symmetry:* A plane of symmetry is a plane which divides the molecule into two halves which are mirror images of each other. When a molecule has a plane of symmetry, it will not be optically active even though it has chiral carbon.

Example:

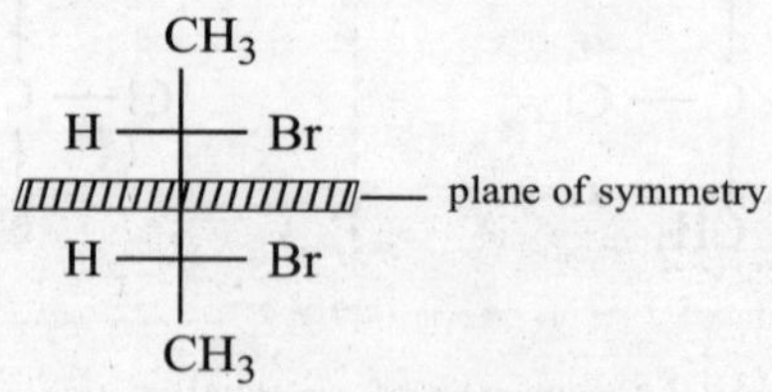

(*ii*) *Centre of symmetry:* Centre of symmetry is an imaginary point in a molecule from which if a line is drawn from any atom group to this point and extended beyond this point, another identical atom/group will be found just at the end of the line.

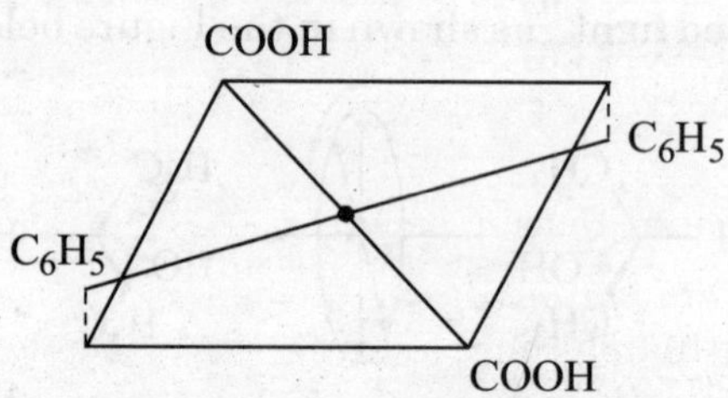

Molecule with centre of symmetry

(*iii*) *Alternating axis of symmetry:* The molecule has alternating axis of symmetry, when it is rotated through θ degrees about an axis passing through it and the rotated molecule is reflected in a mirror that is perpendicular to the axis of rotation.

24.5 DIASTEREOMERS

The other type of optical isomer are pair of molecules that are NOT mirror images but each molecule rotates plane polarised light. There are optically active. Such molecules are called "diastereoisomers".

When there are 2 or more chiral carbons in a molecule, it can have *more* optical isomers. For example, if there are 2 chiral carbons:

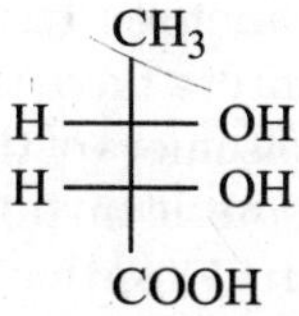

(S, S)
(I)

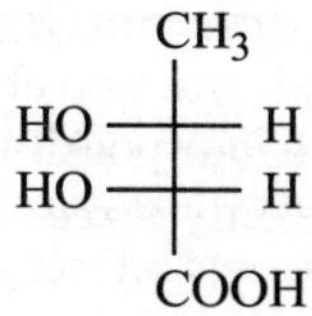

(R, R)
(II)

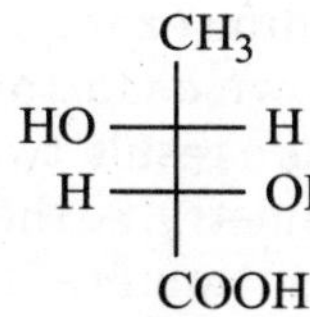

(R, S)
(III)

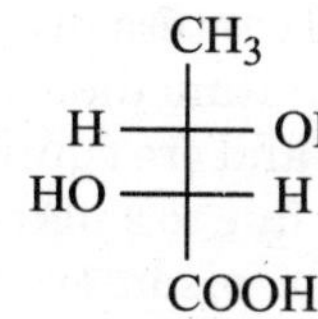

(S, R)
(IV)

This compound has 4 optical isomers and 2 pair of enantiomers. The pair of enantiomers are stereoisomers of one another. Stereoisomers which *are not mirror-image isomers* are called **diastereomers**. The structures I and III are not mirror images but are stereoisomers because they rotate the plane polarised light. Hence, I and III are a pair of diastereoisomers. Similarly, II and IV are diastereoisomers.

A compound with *n* chiral carbon atoms *can have maximum of* 2^n stereoisomers. The acid shown in the previous example has 2 chiral carbons and $2^2 = (= 4)$ stereoisomers.

24.6 MESO COMPOUNDS

Meso compounds are compounds that contain chiral carbons but *can be superimposed* on their mirror-image. Meso compounds are compounds that have chiral carbons (2) within the molecule, but the compound is *optically inactive*. Actually meso compounds have two chiral carbons and there is a plane of symmetry that will have half the molecule on one side of the plane of symmetry while the other half which is the mirror image of the first half on the other side of the plane.

Some compounds with 2 chiral carbons have *less than* 4 stereoisomers. For example,.

Achiral Diastereomers (meso-Compounds)

This molecule has an *internal plane of symmetry;* the top half is the mirror-image of the bottom half of the molecule. Although there are 2 chiral carbons in this molecule, this particular stereoisomers is **achiral**, or optically inactive.

The meso compound is optically inactive because as the plane polarised light passes through the molecule, half the molecule will rotate the light to the right and the other half of the molecule will rotate the light back in the other direction by the same number of degrees. This means that the light exits the sample with no apparent rotation.

The compound above is *meso*-tartaric acid. Since it has 2 chiral carbons, there is another pair of enantiomers possible,

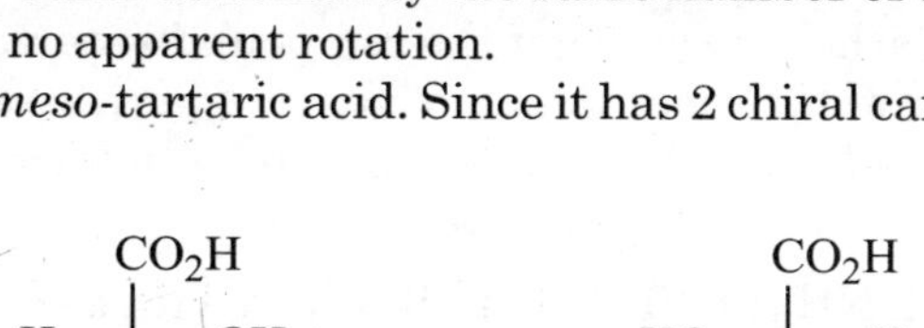

(–) tartaric acid (+) tartaric acid

The chiral centers in the preceding examples are different, from one another. In this case of 2, 3-dihydroxybutanedioic acid, known as tartaric acid, the two chiral centers have the same four substituents and are equivalent. As a result, two of the four possible stereoisomers of this compound are identical due to a plane of symmetry, so there are only *three* stereoisomeric tartaric acids. Two of these stereoisomers are enantiomers and the third is an achiral (optically inactive) diastereoisomers, called a Meso compound. Meso compounds are achiral (optically inactive) diastereomers of chiral stereoisomers. Investigations of isomeric tartaric acid salts were carried out by *Louis Pasteur* in the mid 19th century.

24.7 RACEMIC MIXTURE

A mixture consisting of equal parts of any pair of enantiomers is called a **racemic mixture** (or *racemic modification*) and is designated by (+/–). A racemic mixture does not rotate plane-polarised light because (+)-rotation caused by one enantiomer is cancelled by rotation in the opposite direction by the (–) enantiomer. A solution of a racemic mixture of enantiomers is optically inactive.

Clearly, unequal mixtures of two enantiomers will have a lower optical rotation than a pure enantiomer and the strength of this rotation will depend upon the ***enantiomeric excess (e.e.)*** of the mixture. Mixtures of unequal amounts of enantiomers are referred to as ***scalemic.***

$$\text{Enantiomeric Excess} = (\%\ \text{Enantiomer A} - \%\ \text{Enantiomer B})\%$$

24.8 CONFIGURATION

The arrangement of atoms that characterizes a particular stereoisomer is called its configuration. Although enantiomers may be identified by their characteristic specific rotation, the assignment of a unique configuration to each has not yet been discussed. We have referred to the mirror-image configurations of enantiomers as "right-handed" and "left-handed". There are two conventions used to assign the configuration of enantiomers.

1. Relative Configuration: (D and L Configuration)

In 1960, a New York University chemist, M. A. Rosanoff, proposed the D.L. system. The glyceraldehyde was used as a standard and many thousands of compounds were related to it by chemical reactions that did not change any bond at a stereogenic unit.

In otherwords, this convention assigned a **D** prefix to enantiomers chemically related to a right-handed reference compound (glyceraldehyde) and a **L** prefix to a left-handed reference compound. Glyceraldehyde is chiral itself, and its two isomers are labeled *D* and *L*. In this system, compounds are named by analogy to glyceraldehyde.

The D/L labelling is unrelated to (+)/(–); it does not indicate which enantiomer is dextrorotatory and which is levorotatory. Rather, it says that the compound's stereochemistry is related to that of the dextrorotatory or levorotatory enantiomer of glyceraldehyde—the dextrorotatory isomer of glyceraldehyde is, in fact, the *D* isomer. A rule of thumb for determining the *D*/*L* isomeric form of an amino acid is the "CORN" rule. The groups:

COOH, R, NH_2 and H (where R is a variant carbon chain)

are arranged around the *chiral centre* carbon atom. Sighting with the hydrogen atom away from the viewer, if these groups are arranged clockwise around the carbon atom, then it is *D*-form. If counter-clockwise, it is *L*-form.

Although this notation is still applied to carbohydrates and amino acids, it required chemical transformations to establish group relationships, and proved to be doubtful in its general application.

2. R and S (Absolute) Configuration

In the early sixties, R. S. Cahn, C.K. Ingold, and V. Prelog introduced a new system, the *R*, *S* system, to name the absolute configuration because the *D*, *L* system is not unequivocal in all cases and cannot easily be applied to all families of compounds.

The resulting nomenclature system is sometimes called the **CIP** system or the **R-S** system. For chemists, the *R*/*S* system is the most important nomenclature system for denoting enantiomers, which does not involve a reference molecule such as glyceraldehyde.

In the *RS* system of nomenclature, each chiral centre in molecule is assigned a prefix (*R* or *S*), according to whether its configuration is right-or left-handed. No chemical reactions or interrelationship are required for this assignment. The symbol **R** comes from the Latin rectus for right, and L from the Latin *sinister* for left. The assignment of these prefixes depends on the application of following rules:

1. Rank the 4 atoms (groups) attached to the carbon according to the sequence rules or Cahn-Ingold-Prelog rules. (Given below)
2. Rotate the molecule until the lowest priority (number 4) substituent is in the back
3. Draw a curve from number 1 to number 2 to number 3 substituent.
4. If the curve is clockwise, the stereocenter is of *R* configuration. If the curve is counterclockwise, the stereocenter is of *S* configuration

The sequence Rules for R and S (Absolute) configuration:

(*i*) *Assigned sequence priorities to the four substituents by looking at the atoms attached directly to the chiral centre.* The higher the atomic number of the immediate substituent atom, the higher the priority.

For example, H – < C – < O – < Cl – (Different isotopes of the same element are assigned a priority according to their atomic mass). In example below, iodine would be 1, bromine 2, chlorine 3, and fluorine 4, because iodine has the highest molecular weight (and is therefore highest priority) and fluorine has the lowest molecular weight (and is therefore the lowest priority).

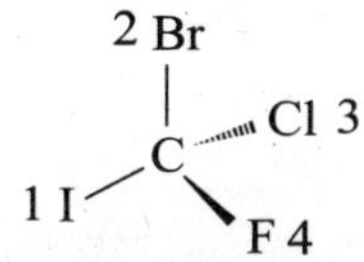

(*ii*) *If the first atom of two substituent happen to be identical in molecular weight, go to the next atom and make the molecular weight comparison* (e.g. and ethyl group would have higher priority over a methyl group). Evaluate atoms progressively further away from the chiral center until a difference is found.

For example, CH_3 – < C_2H_5 – < $ClCH_2$ – < $BrCH_2$ – < CH_3O – .

Thus, the four substituents on 2-bromobutane would be listed in the order. Br > CH_2CH_3 > CH_3 > H.

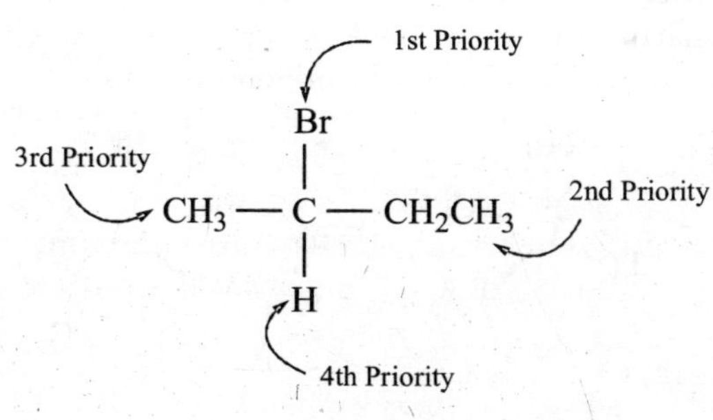

(*iii*) *If double or triple bonded groups are present as substituents, they are treated as an equivalent set of single-bonded atoms, i.e. a carbon with a double bond to another carbon is treated as a carbon singly bounded to two carbons, as shown below.* For example an ethylene substituent, R–CH$=$CH$_2$, will have a higher priority then an ethyl substituent (R–CH$_2$CH$_3$)

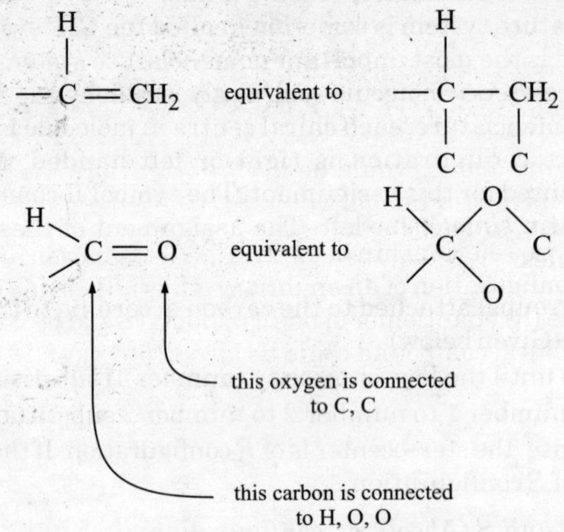

(*i*) Once the relative priorities of the four substituents have been determined, the chiral centre must be viewed from the side opposite the lowest priority group. If we number the substituent groups from 1 to 4, with 1 being the highest and 4 the lowest in priority sequence, the two enantiomeric configurations are shown in the following diagram along with a viewers eye on the side opposite substituent number 4.

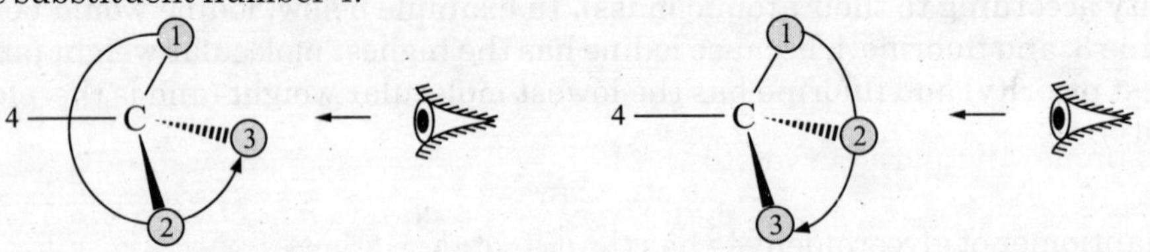

(S) (R)

Example: Let us assign the *R* or *S* configuration to the following molecule of bromochlorofluormethane based on above rules:

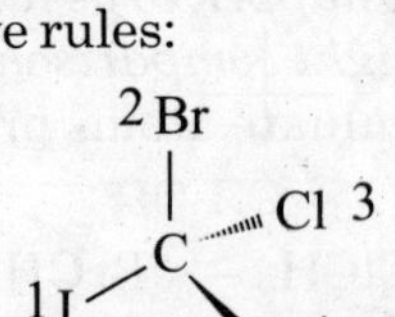

(*ii*) Rotate the molecule so the substituents can be arranged from 1 to 4. Rotate the molecule until the number 4 priority substituent is in the back, as shown below. (Remember that dashed lines mean a bond away from the viewer, and a solid wedged line indicates a bond towards the viewer).

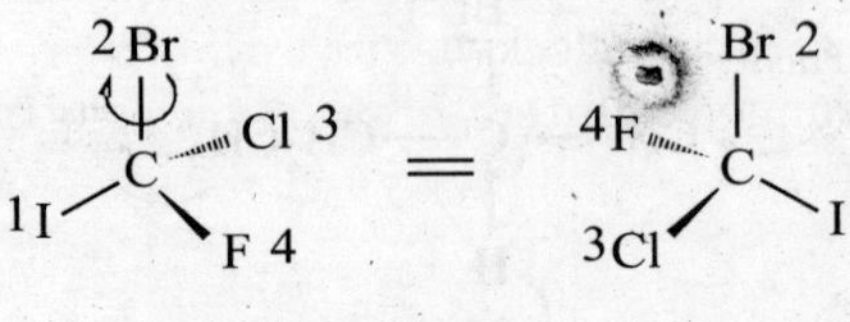

(*iii*) A curve is then drawn from the 1 to 2 to 3 priority substituents, ignoring the 4th priority substituent (as shown below). In the example below, the curve goes countercloskwise so the molecule is of *S* configuration.

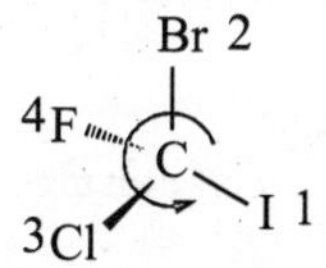

The name of the compound above, would be (S)-bromo, chloro, fluoro, iodomethane, and the name of its enantiomer, or its mirror image, would be (R)-bromo, chloro, fluro, iodomethane.

Note: Please note there is no relationship between R, S designation and the direction of optical rotation, i.e. the absolute configuration of an enantiomer has no bearing on its direction of rotation of plane-polarised light. The (R) enantiomer of one compound could be dextrorotatory (+) while the (R) enantiomer of another chiral compound could be levorotatory (–).

Some examples:

1.

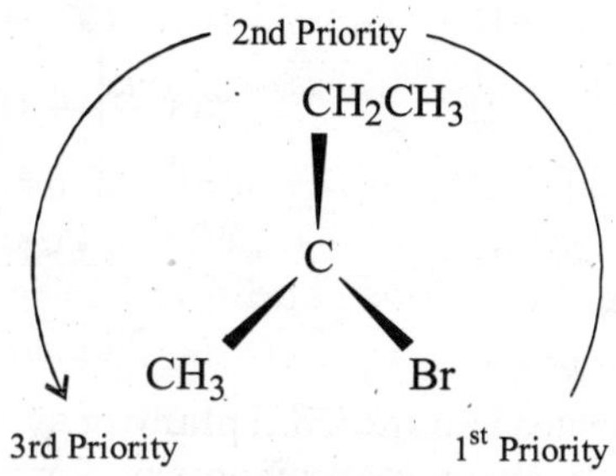

(S) -2-Bromobutane

In this example, the path curves to the left, so this enantiomer is the (*S*)-2-bromobutane stereoisomer

2. This enantiomer of glyceraldehyde has the (*R*) configuration.

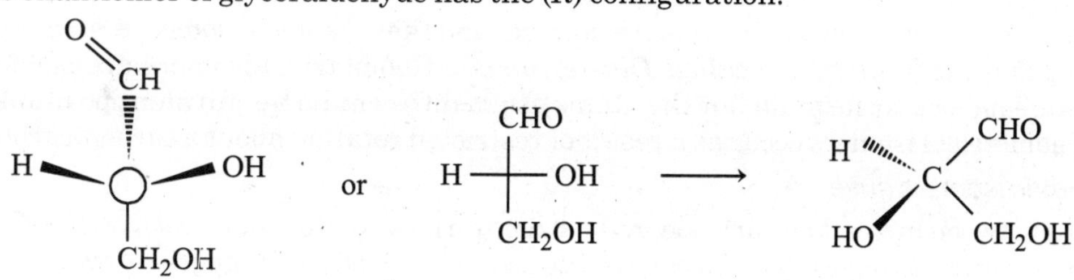

(R)

This naming of absolute configuration with *R*, *S* descriptors can be taken from the Fischer projection.

For this purpose, the structure is illustrated in such a manner in the Fischer projection, so that the substituent with the lowest CIP priority is located at one of the two vertical positions. If the remaining three substituents are arranged clockwise in order of decreasing CIP priority, the absolute configuration is *R*. If they are arranged counterclockwise however, the absolute configuration is *S*.

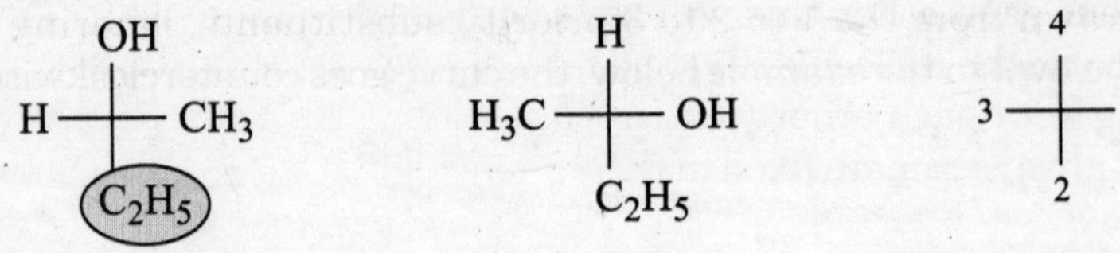

e.g. ethyl is arrested and the remaining substituents are rotated clockwise

thus, the molecule possesses R configuration

Isomers of Tartaric Acid

Fischer projection formulas provide a helpful view of the configurational relationships within the structures of these isomers. In the following illustration a mirror line is drawn between formulas that have a mirror-image relationship. In demonstrating the identity of the two meso-compound formulas, remember that a Fischer projection formula may be rotated 180° in the plane.

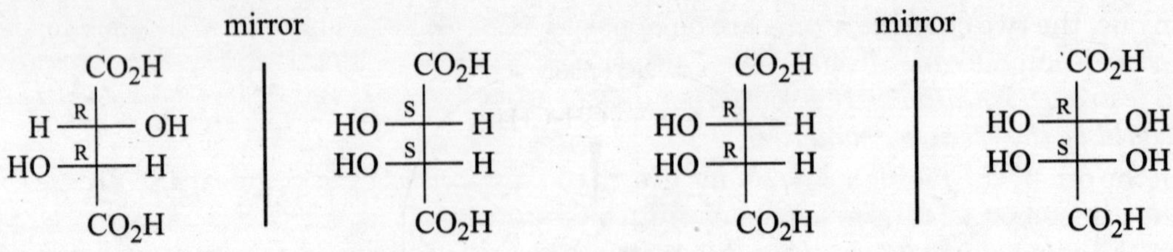

(+)-tartaric acid (–)-tartaric acid meso-tartaric acid

This enantiomeric pair do not possess an internal plane of symmetry and are optically active. Tartaric acid, therefore, has only three stereoisomeric forms: one meso form and a pair of enantiomers. The meso form is a diastereomer of each of the enantiomers.

24.9 GEOMETRICAL ISOMERISM

Geometric isomerism (also known as cis-trans isomerism or E-Z isomerism) is a form of stereoisomerism. It was discussed in the previous section that *the stereoisomers which are not the mirror images of each other, are called Diastereomers.* Geometrical isomers are isomers that have the same molecular formula but the atoms are in different non-equivalent positions to one another. Geometrical isomers occur as a result of restricted rotation about a carbon-carbon bond.

How geometric isomers arise:

Single bonds between tow carbons in a non-cyclic structure may rotate around each other unhindered at room temperature because as the carbons rotate around each other the degree of overlap between the atomic orbitals is not changed and remain at maximum overlap as rotation occurs.

Consider two possible configurations of 1,2-dichloroethane. These two structures represent the same molecule

trans-1, 2-dichloroethene *cis*-1, 2-dichloroethene

However if there is a double bond between two carbons as in the case of alkenes, these sp^2 carbons have a sigma bond and a pi bond between two sp^2 hybrid orbitals. If the two sp^2 carbons attempt to rotate about each other this will reduce the p-p overlap of the Pi bond. This has a destabilising influence on the molecular system and so is resisted. Thus the Pi bond prevents the carbons from rotating abut each other (restricted rotation). As a result, the atoms attached to these sp^2 carbons will always remain fixed in three dimensional space. This gives rise to geometric isomers.

If the two sp^2 carbons have different atoms attached to themselves, then two configurations are possible: either two like groups (one on each sp^2 carbon) being on the SAME side of the double bond or two like groups being on the OPPOSITE sides of the double bond.

Example: as in 1,2-dichloroethene

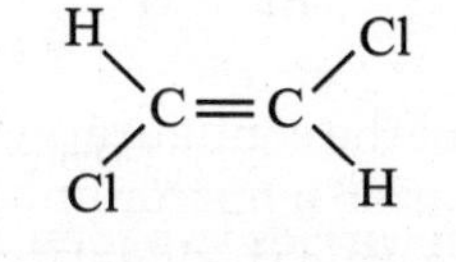

(E)-1, 2-dichloroethene

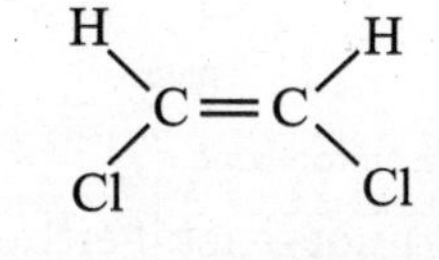

(Z)-1, 2-dichloroethene

- In one, the two chlorine atoms are on opposite sides of the double bond. This is known as the ***trans*** isomer. (*trans* : from latin meaning "across"). *Hence, if the similar groups are fixed due to restricted rotation by the double bond on opposite sides of the double bond, then the isomer would be the "**trans**" isomer.*
- In the other, the two chlorine atoms are on the same side of the double bond. This is know as the ***cis*** isomer. (*cis* : from latin meaning "on this side"). Hence, *if the similar groups are fixed on the same side as a result of this restricted rotation, then the isomer would be classified as "**cis**".*

Cis and trans isomers will differ in their physical and chemical properties. In order to identify them we prefix the cis or trans onto the front of the name of the isomer. For example, cis-1,2-dichloroethene would have two chlorines on the same side of the double bond. Trans-1,2-dichloroethene would have the two chlorines on opposite sides of the double bond.

Examples: The most likely example of geometric isomerism is but-2-ene. In one case, the CH_3 groups are on opposite sides of the double bond, and in the other case they are on the same side.

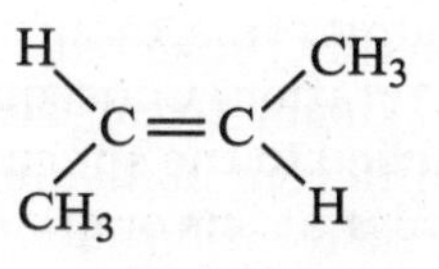

trans-but-2-ene

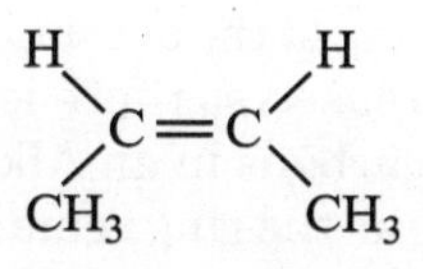

cis-but-2-ene

It must be emphasized that if the two atoms or groups of atoms attached to the same sp^2 carbon are the same, then **three cannot be geometrical isomerism** since the two like groups on the same sp^2 carbon could not differentiate between the two like groups.

On this basis we find that propylene, I-butene, and isobutylene should not show isomerism:

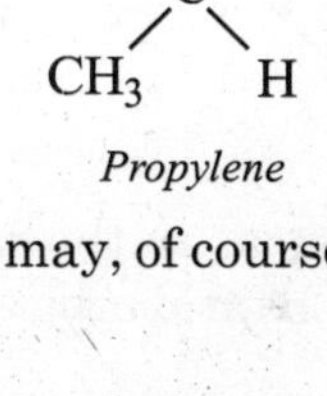

Propylene

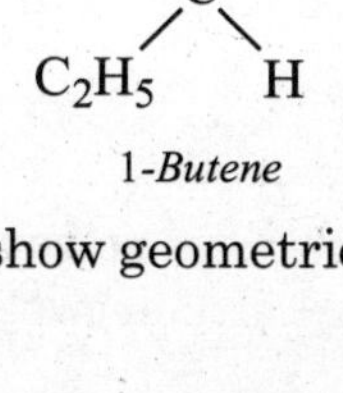

1-*Butene*

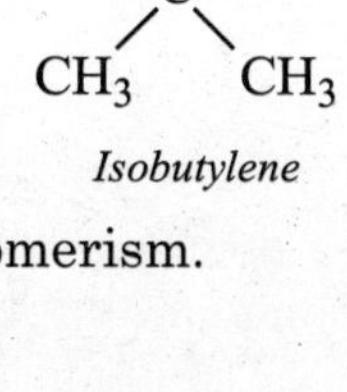

Isobutylene

Many higher alkenes may, of course, show geometric isomerism.

If we consider compounds other than hydrocarbons, we find that 1,1-dichloroethene and 1,1-dibromoethene should not show isomerism, whereas the 1,2-dichloroethenes and 1,2-dibromoethenes should. In every case these predictions have been found correct. Isomers of the following physical properties have been isolated.

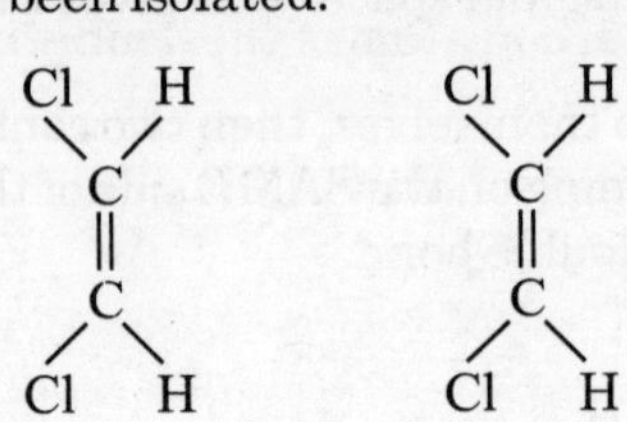

cis- trans-

1, 2-Dichloroethene

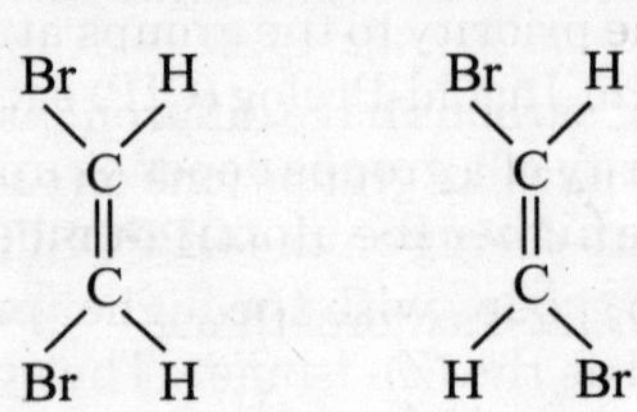

cis- trans-

1, 2-Dibromoethene

Geometric isomerism cannot exist if either carbon carries two identical groups. Some possible combinations are shown below.

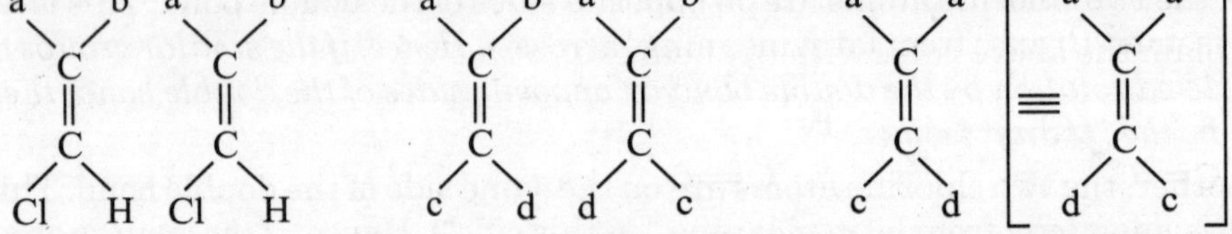

Isomerism *Isomerism* *No isomerism*

One might ask why there is no geometrical isomerism in Alkynes. That is because of the linear geometry around the sp hybrid carbons. The atoms attached to these carbons are in the same line as the carbons, so there is no difference in the geometrical orientation of these groups.

Geometrical Isomerism for Cyclic Hydrocarbon

The carbons found within the ring of a cyclic hydrocarbon also have restricted rotation. The carbons within the ring are sp^3 hybridised. If you attempt to rotate the carbons around each other again there will be a lessening of the overlap between two sp^3 orbitals. This, in turn, will cause instability if the overlap is decreased so there is a resistance to the rotation. Consequently, like the atoms attached to the sp^2 carbons in an Alkene, the atoms attached to the sp^3 carbons within the ring become fixed as long as the ring remains intact. If two similar atoms or groups of atoms find them selves on opposite sides of the ring, then the groups are trans and the isomer is the **trans** form. If the two similar groups are on the same side of the ring structure then we have the **cis** isomer. For example, Cis-1,2-dimethylcyclohexane would have the two methyl groups on the same side of the six-member ring system.

E and Z nomenclature:

The expression cis and trans only applies to alkenes or cycloalkanes if one of the substituents on each of the carbons are the same. However, if there are three or four different substituents, it is not possible to

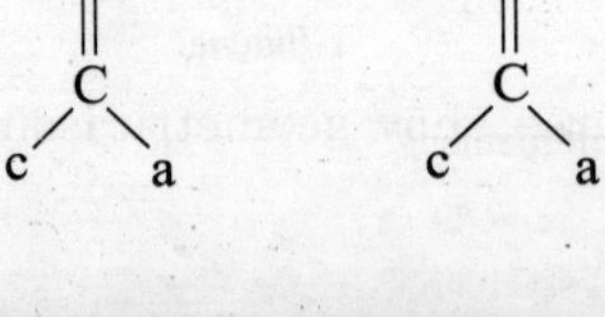

use the cis and trans system of naming the isomers. For such systems Cahn, Ingold and Prelog priority system or commonly known as E,Z nomenclature must be used.

Rules for E and Z nomenclature;

(1) Assign the priority to the groups attached to each carbon atom of the double bond according to the Cahn-Ingold-Prelog (CIP) sequence rules (already discussed under R and S system). (The priority of a groups depends upon the atomic number of the atom attached to the double bond. The higher the atomic number, the higher the priority of the group is)

(2) If the two groups with the higher priorities are on the same side of the double bond, that is described as the (Z)- isomer. The symbol Z comes from a German word (zusammen) which means together

(3) If the two groups with the higher priorities are on opposite sides of the double bond, then this is the (E)- isomer. E comes from the German entgegen which means opposite.

> (E)- Isomer: The higher priority groups are on opposite sides of the double bond.
> (Z)- Isomer: The higher priority groups are on the same side of the double bond

Let us explain the above rules with the help of following examples:

(1) 

- Compare the atomic numbers of the attached atoms to work out the various priorities.

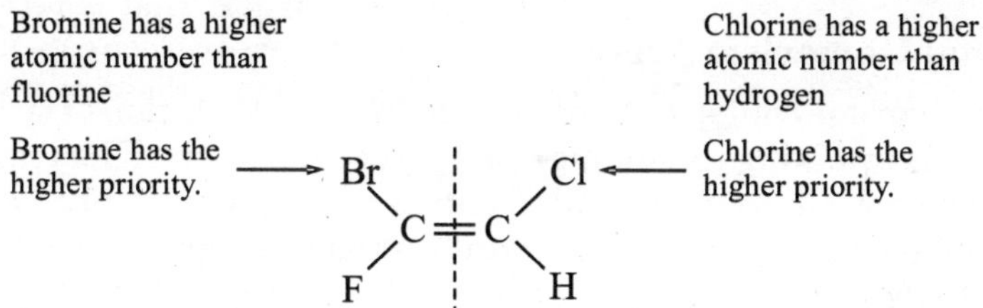

- Notice that the atoms with the higher priorities are both on the same side of the double bond. That counts as the (Z)- isomer.
- The second isomer obviously still has the same atoms at each end, but this time the higher priority atoms are on opposite sides of the double bond. That's the (E)- isomer.

(2) 1,2-dichloroethene.

In the first isomer, the higher priority groups are on opposite sides of the bond. That must be the (E)- isomer. The other one, with the higher priority groups on the same side, is the (Z)- isomer.

(E)-1, 2-dichloroethene (Z)-1, 2-dichloroethene

(3) But-2-ene:

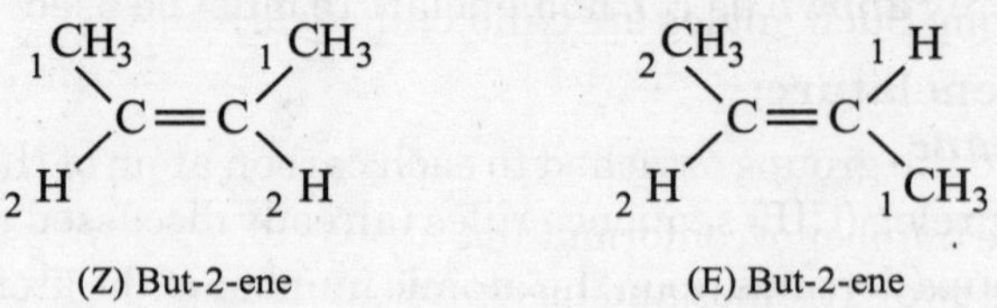

The CH_3 group has the higher priority because its carbon atom has an atomic number of 6 compared with an atomic number of 1 for the hydrogen also attached to the carbon-carbon double bond.

24.10 CONFIRMATIONS

Conformations are different forms of a molecule, related by simple rotation about a single bond. For example alkanes of two or more carbons can be twisted into a number of different three-dimensional arrangements by rotation about a carbon-carbon single bond. The different spatial arrangements that a molecule can adopt due to rotation about σ bonds are called ***conformations*** and hence conformational isomers or ***conformers.***

Such interconversions are usually very rapid, so that a sample of a given molecule may exist in many different conformations.

Conformational isomerism is a form of stereoisomerism involving the phenomenon of molecules with the same structural formula existing as different conformational isomers or conformers due to atoms rotating about a bond. Different conformers can interconvert by rotation around single bonds, without breaking chemical bonds.

How Conformational Isomerism Arise:

The existence of more than one conformation is due to hindered rotation of sp^3 hybridised carbon bonds. Let us explain this by taking the simple example of ethane which has carbon- carbon sigma bond. Sigma bond is cylindrically symmetrical in nature and hence permits free rotation about its axis. In other words sigma bond permits *free rotation* about its axis. Hence, *the energetic barrier to rotation about sigma bonds is generally very low.*

The different spatial arrangements formed by rotations about a single bond are called conformations or conformers.

Visualising Conformations:

A Newman projection can be used to specify the confirmation of a particular bond with clarity and detail. A Newman projection represents the head-on look down the bond of interest. The circle in the Newman projection represents the atom in front of the bond, and the lines radiating from the centre are the bonds of that atom. The bonds of the rear atom emerge from the sides of the circle. For example the Newman projection for ethane can be drawn like:

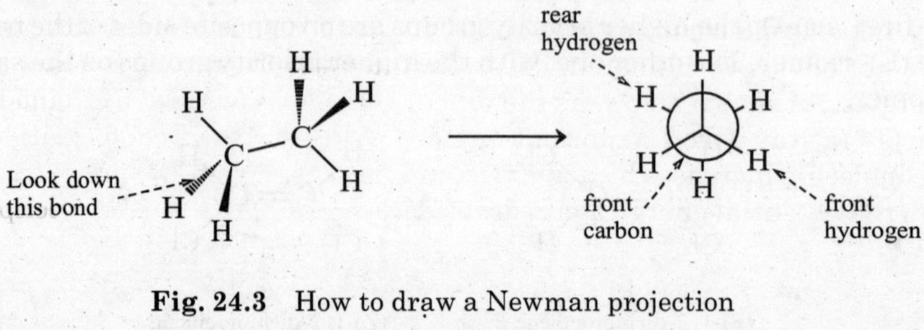

Fig. 24.3 How to draw a Newman projection

Newman projections can be characterised by the angles formed between bonds on the front atom and bonds on the rear atom. Such angles are called dihedral angles.

Conformations of Ethane

While there are an infinite number of conformations about any sigma bond, in ethane two particular conformers represent the two extremes and have special names.

(i) Eclipsed: In the eclipsed conformation, the C-H bonds on the front and back carbons are aligned with each other with dihedral angles of 0 degrees.

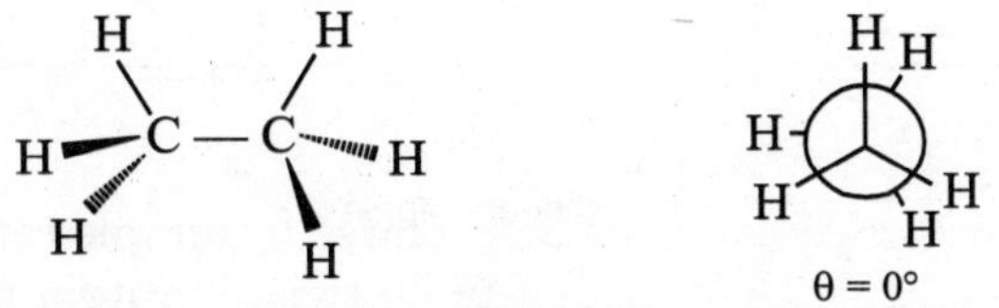

θ = 0°

This conformation of ethane is the **eclipsed** form. The hydrogen atoms have a dihedral angle of 0°.

(ii) **Staggered:** In the staggered conformation, the C-H bonds on the rear carbon lie between those on the front carbon with dihedral angles of 60 degrees.

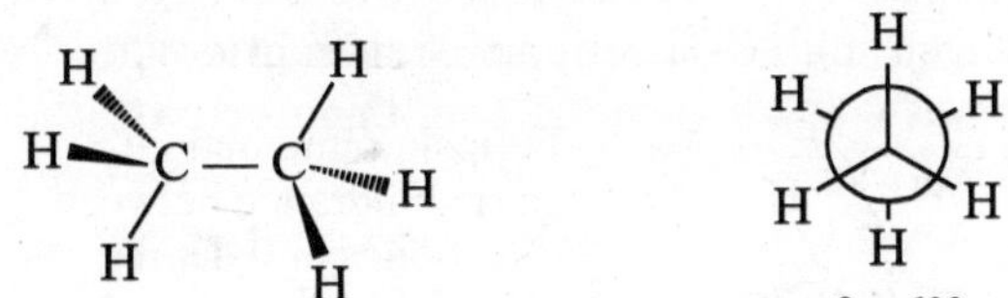

θ = 60°

This conformation of ethane is the **staggered** from. The various hydrogen atoms on the two carbons form a dihedral angle of 60°.

(iii) The infinite number of possible intermediate conformations between these two extremes is referred to **skew** conformations

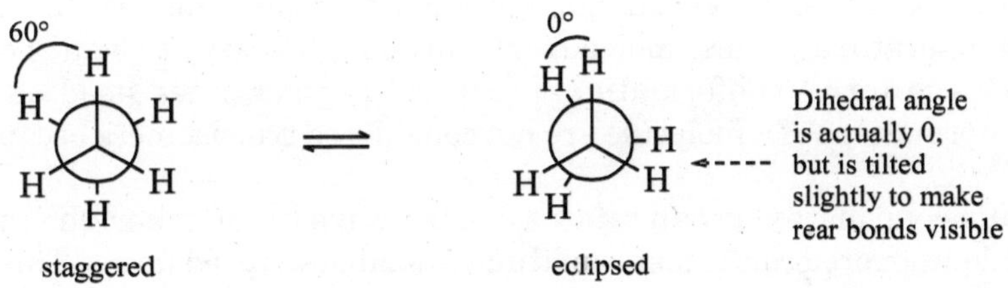

Fig. 24.4 The inter conversion of eclipsed and staggered conformations of ethane.

Energy diagram for rotation about the C-C bond in ethane:

Energetically, not all conformations are equally favoured. The rotation around the single bond in ethane, while not obviously hindered, does generate conformational isomers having different potential energies. As shown below, as the dihedral angle between the ethane hydrogen atoms changes from 60° (a **staggered** conformation) to 120° (an **eclipsed** conformation), the potential energy of the molecule increases by about 3 kCal/mole. As the methyl group continues to rotate towards 180°, the potential energy again drops and rises again as the next eclipsed structure is formed.

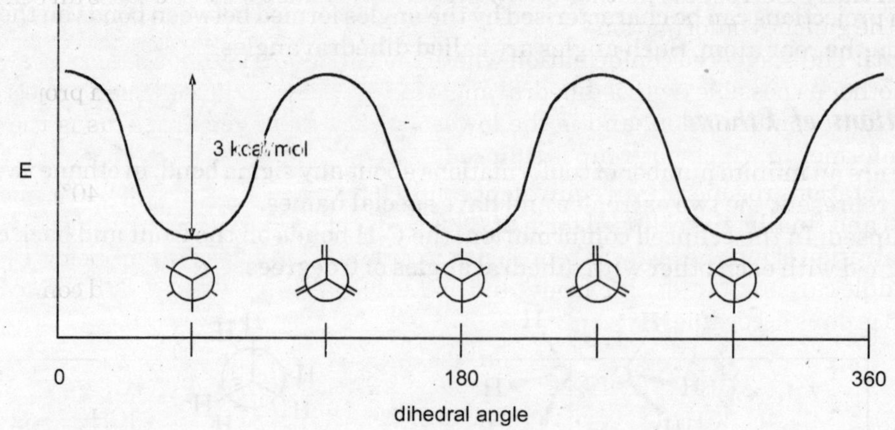

Fig. 24.5 Energy diagram for rotation about the C-C bond in ethane

The staggered conformation is the most stable of all possible conformations of ethane. In the eclipsed form, the electron densities on the C-H bonds are closer together than they are in the staggered form. When two C-H bonds are brought into a dihedral angle of zero degrees, their electron clouds experience repulsion, which raises the energy of the molecule. The eclipsed conformation of ethane has three such C-H eclipsing interactions.

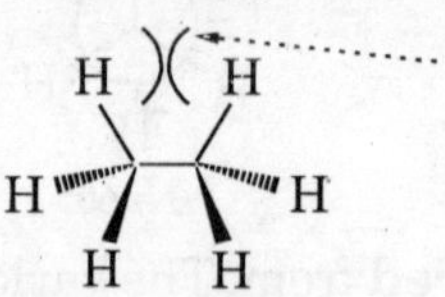

Fig. 24.6 Eclipsing interactions in ethane

Eclipsing interactions, which are due steric hindrance cause resistance to rotation, it is also called *torsional strain.* The energy (3 kCal/mol) needed to overcome this resistance is the torsional energy. At room temperature, ethane molecules have enough energy to be in a constant state of rotation. Although the term "conformational isomer" is sometimes used as a synonym for conformations, conformations of a molecule are not considered true isomers because of their rapid interconversion.

We can say that as one methyl group rotates 360° relative to the other, the compound passes through three stable staggered conformers via three unstable eclipsed forms. This barrier is small enough that at 25°C the compound changes conformation about 50 million times each second.

2. Conformations of Butane

In butane it is the rotation about the C2-C3 bond that is of most interest since the relative position of the two methyl groups is important. Butane has not only eclipsed and staggered conformations, but also forms that vary in the relative orientation of the methyl groups.

(i) **Fully Eclipsed (Syn):** The structure shown at 0° is fully **eclipsed**, that is, both methyl groups are aligned and are interacting.

(ii) **Gauche:** The two staggered forms with the methyl groups in closer proximity (60°) are the *gauche* conformations. As the front methyl group is rotated 60° in structure I, a **gauche** conformation is produced. The two staggered forms with the methyl groups in closer proximity (60°) (II and VI) are the *gauche* conformation

(iii) **Partially Eclipsed:** Another 60° rotation of conformation II produces an eclipsed version of the gauche conformation.

(iv) **Anti:** The staggered conformation where the two methyl groups are as for away from one another as possible (with a dihedral angle of 180°, seen in the Newman projection) is called the *anti* conformation, and is the lowest-energy arrangement. This is the most stable conformer in which no group is eclipsed.

Further rotations regenerate an equivalent eclipsed gauche conformer (at 240°), another gauche form (300°) and finally, the eclipsed form at 360°.

The illustration of butane below is represented by the Newman Projections below by designating the two middle carbons, one as the "front" and one as the "behind" carbon, and connecting the two end methyl groups accordingly.

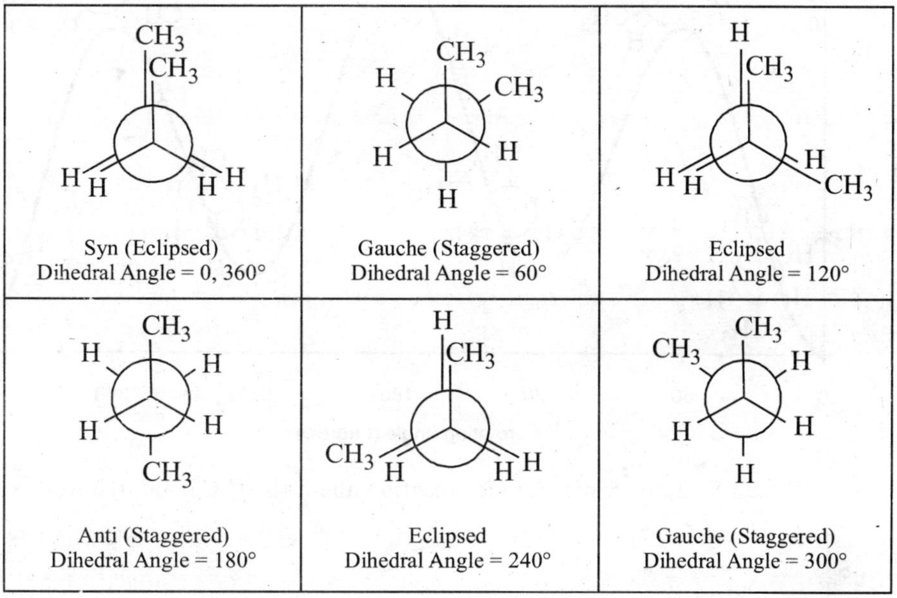

Energy diagram for rotation about the C-C bond in butane: (Conformational Analysis)

- As with ethane, the eclipsed conformations are higher energy than the staggered group are in close proximity with the molecule, therefore the potential energy is at its highest.

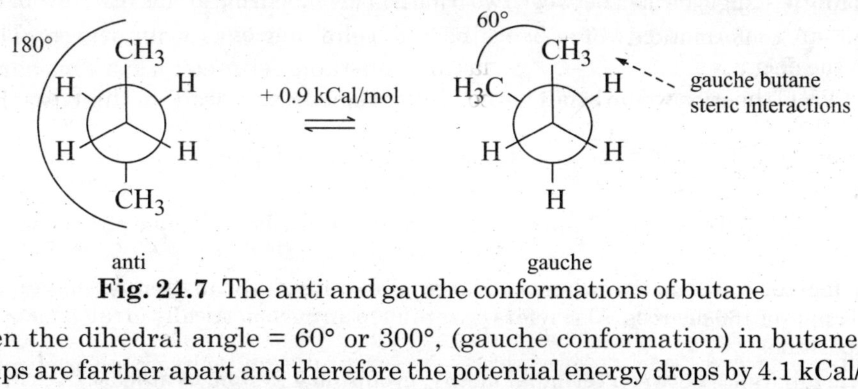

Fig. 24.7 The anti and gauche conformations of butane

- When the dihedral angle = 60° or 300°, (gauche conformation) in butane, the methyl groups are farther apart and therefore the potential energy drops by 4.1 kCal/mol. Gauche form is less stable than the anti form by 0.9 kCal/mol due to steric hindrance between the two methyl groups. Such an interaction is often referred to as a gauche-butane interaction because butane is the first alkane discovered to exhibit such an effect.

- When the dihedral angle = 120° or 240°, (eclipsed conformation) in butane, the methyl groups align themselves with hydrogens on the same plane, resulting in the butane having a higher potential energy than the gauche conformations though still having a lower potential energy than the syn conformation (dihedral angle - 0° or 360°) by 1.5 kCal/mol.
- The staggered conformation (when the dihedral angle = 180°) where the two methyl groups are as far away from one another as possible, i.e. the *anti* conformation, is the lowest-energy arrangement.

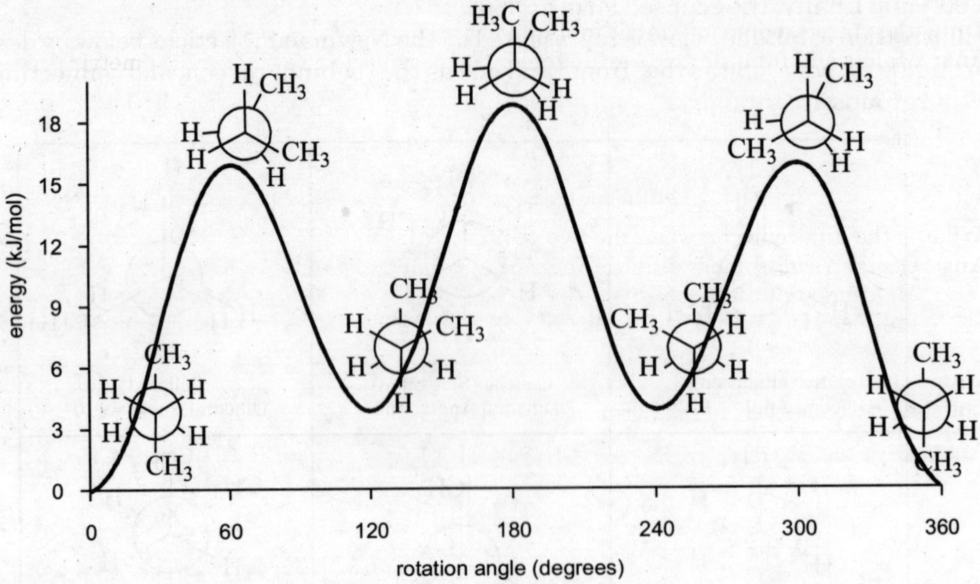

Fig. 24.8 Energy diagram for rotation about the C-C bond in butane

KEY TERMS

Conformation - Different 3-dimensional shapes that can be taken by a molecule by rotating about single bonds.

Newman projection - A depiction of the conformation of a molecule where it is viewed down the atoms of a particular bond.

Dihedral angle - Angle formed between two bonds on neighbouring atoms in a Newman projection.

Staggered - A conformation where bonds have dihedral angles of sixty degrees. This typically minimizes the energy.

Eclipsed - A conformation where bonds have dihedral angles of zero degrees. This typically maximizes the energy.

Steric hindrance - Repulsion between the electron clouds on bulky groups of a molecule or between molecules.

Gauche - A staggered conformation where large groups at a dihedral angle of 60 degrees repel each other.

Anti - A staggered conformation where two large groups are placed in their most stable positions at a dihedral angle of 180 degrees. Also refers generally to any geometrically arrangement of groups in opposing positions

Syn - Steric hindrance between terminal methyl groups in a system of at least five carbon centers. Also a generic term describing groups oriented in the same direction; i.e. the opposite of anti.

Torsional strain - Resistance to rotation about a single bond caused by steric hindrance of eclipsing interactions.

Torsional energy - The energy it takes to overcome torsional strain, or the difference in energy between eclipsed and staggered conformations.

SOLVED EXERCISE

1. Draw the stereoisomers of Maleic and fumaric acids

Ans: Maleic and fumaric acids exhibit geometric isomerism or they are geometric isomers.

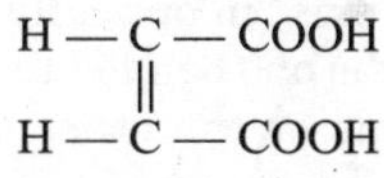

Maleic acid (cis form)

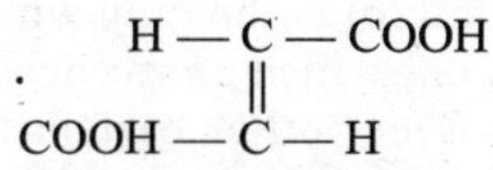

Fumaric acid (trans form)

2. What is the difference between the two geometrical isomers ?

Ans: Geometrical isomers differ in their physical and chemical properties.
Melting point and stability of the cis isomers are lower than those of the trans isomer.
Boiling point, solubility, density, dipole moment, refractive index and heat of combustion are higher for the cis isomer than the trans isomer.

3. Name the isomeric forms of C_6H_{14} according to the IUPAC system. Which one(s), if any, is (are) optically active?

Ans: $CH_3CH_2CH_2CH_2CH_2CH_3$

hexane

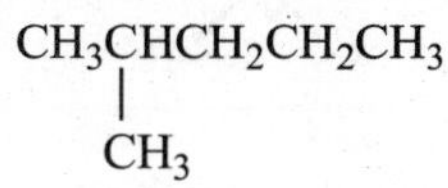

2-methylpentane

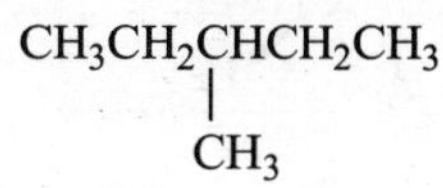

3-methylpentane

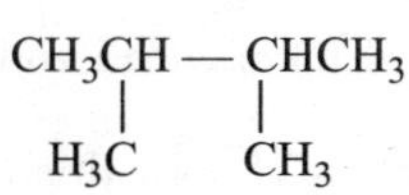

2, 3-dimethylbutane

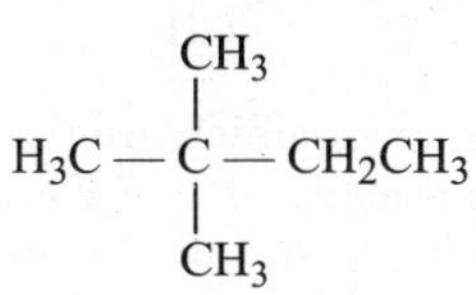

2, 2-dimethylbutane

None of these isomers is optically active.

4. Write formulas for all the structural and geometrical isomers of C_4H_8.

Ans: If the four carbons are not in a ring there must be one double bond to satisfy the tetracovalence of carbon. The double bond occurs either in the center of the molecule or toward an end. In the former case, two geometrical isomers occur with different positions of the terminal carbons relative to the double bond; in the latter case, two structural isomers occur, differing in the extent of branching within the skeleton.

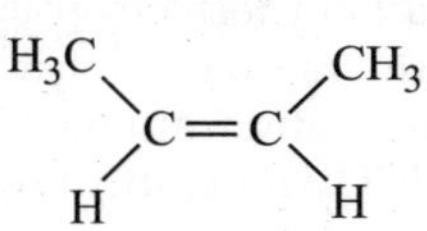

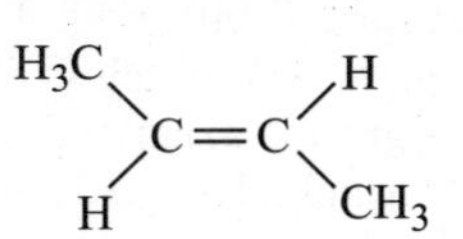

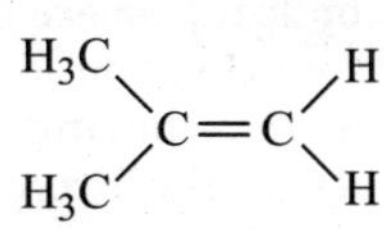

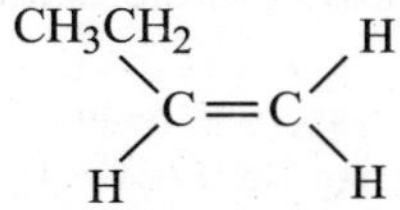

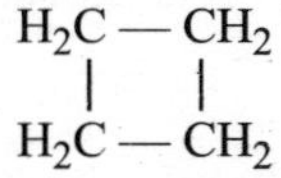

5. Which of the following compounds can exist as geometric isomers? CH_2Cl_2, $CH_2Cl—CH_2Cl—CH_2Cl$, CHBr=CHCl, $CH_2Cl—CH_2Br$

Ans: Only CHBr=CHCl can exist as geometric isomers:

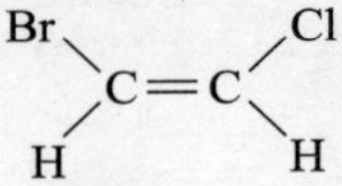 and 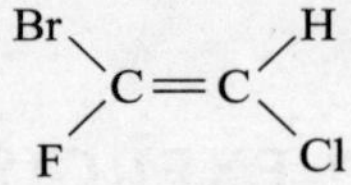

In CH_2ClCH_2Cl and $CH_2Cl—CH_2Br$, the carbon atoms are connected by a single bond about which the groups can rotate relatively freely.

6. Which of the isomeric C_4H_9Cl compounds would you expect to be optically active?

Ans: Compound (*b*) is the only which would exist in optically active isomeric forms, because it is the only one which has a carbon atom (the one bonded to Cl) which is bonded to four different groups. Every other carbon base in this or the other structure is bonded to at least two hydrogen atoms or to two CH_3 groups.

7. Select from the following the pairs of (a) geometric isomers (b) optical isomers (e) structural isomers

(i) 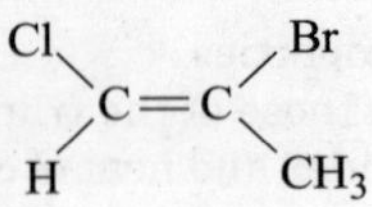 (ii) 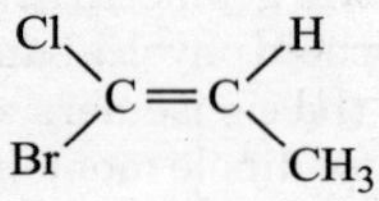 (iii) 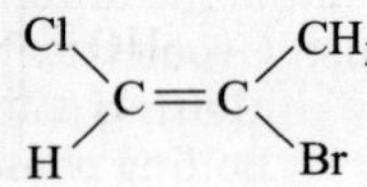

(iv) 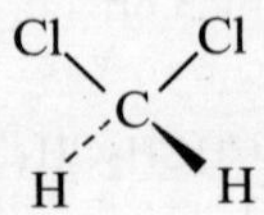 (v) 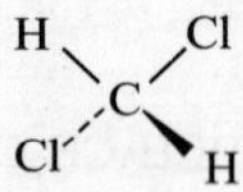 (vi) 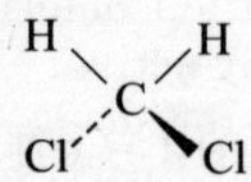

(vii) 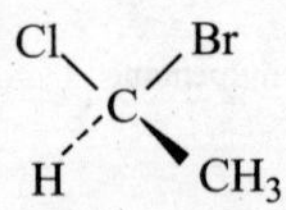 (viii) 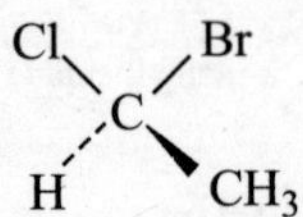 (ix) 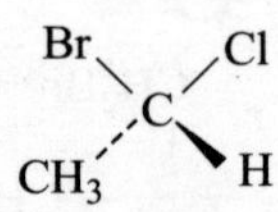

Ans: (*a*) i and iii (*b*) vii with vii or ix, which are the same (c) ii with i and iii

8. Are all optical isomers necessarily optically active? Explain.

Ans: Meso isomers are optically inactive, despite being in the class of optical isomers.

9. Explain the difference between a meso isomer and a racemic mixture. What characteristic(s) do they have in compound.

Ans: A meso isomer is a single compound; a racemic mixture contains an equimolar mixture of two compounds. The meso isomer is inherently inactive because of *internal* compensation; the racemic mixture does not rotate the plane of polarised light because the effect of one of the isomers negates the effect of the other. The racemic mixture can be separated into two optically active compounds by physical means (as Pasteur did). The meso isomer and the racemic mixture have in common their inability to rotate the plane of polarised light. Both have two centres of (opposite) chirality.

10. How many optical isomers can exist for 2,3-butanediol? Would all of these be optically active?

Ans: There are three optical isomers, one of which is inactive (the meso form) because of the two carbon atoms having the same groups arranged in opposite configurations.

11. Tartaric acid is a dihydroxydicarboxylic acid, $HOCOCHOHCHOHCO_2H$. Identify any chiral centers, draw all optical isomers, and explain the total number of such isomers of tartaric acid in terms of the 2^n rule.

Ans: The two middle carbon atoms in each structure are chiral centers. They are starred, to indicate their chirality.

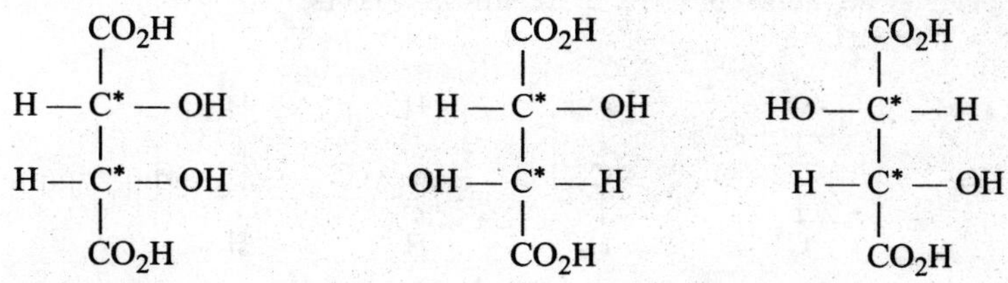

meso

There are 3 isomers instead of $2^2 = 4$, one a meso isomer.

12. (a) Draw all the stereoisomers of 3-chloro-2-pentanol, $CH_{3C}H(OH)CHClCH_2CH_3$.
    (b) Give the stereochemical relationships of the stereoisomers.

Ans: (a) There are four stereoisomers. A and B are enantiomers, as are C and D. (b) A and are diastereomers of C and D.

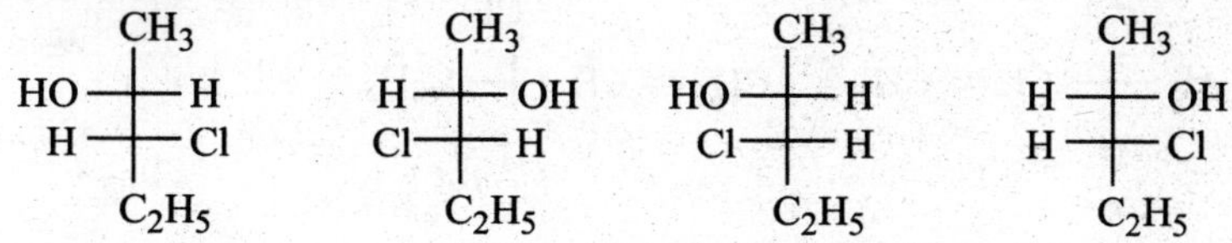

A B C D

13. Give the R/S designation for each of the stereoisomers of 1,3-dibromo-2-methylbutane.

Ans: There are two chiral C's (C^2 and C^3), and the four stereoisomers are shown below with their designation.

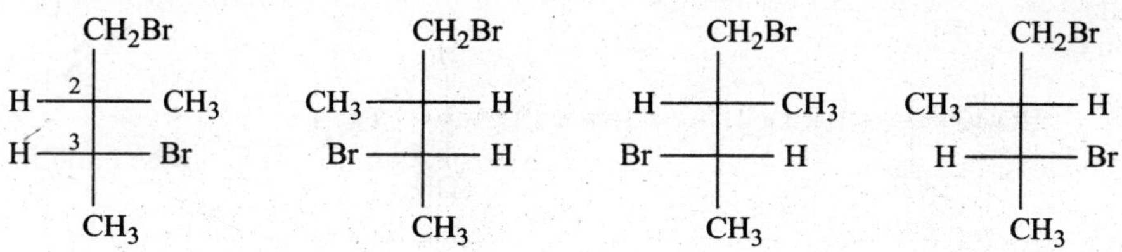

2*S*, 3*R* 2*R*, 3*S* 2*S*, 3*S* 2*R*, 3*R*

14. (a) Theoretically, what is the maximum number of possible stereoisomers of a compound having n dissimilar chiral atoms?
    (b) What is the maximum number of stereoisomers possible for a compound having (i) three (ii) four different chiral centers?

Ans: (a) A compound with n different chiral atoms can have 2^n stereoisomers.
(b) (i) $2^3 = 8$ (four racemic of enantiomers). (ii) $2^4 = 16$ (eight racemic pairs of enantiomers).

15. Draw the enantiomers of (a) chloroiodomethane sulphonic acid and label each as R or S.

Ans:

R *S*

I > Cl > S > H

16. What is the priority sequence of CH_3, 1°, 2° and 3° alkyls?

Ans: 3° > 2° > 1° > CH_3

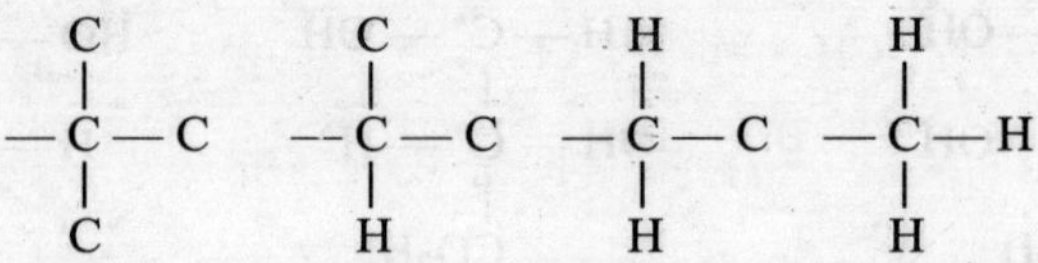

C, C, C > C, C, H > C, H, H > H, H, H

3° 2° 1° CH_3

17. Draw and specify as R or S of the following:

(a) 3-chloro-l-pentene (c) $CH_3CH(NH_2)COOH$ (Alanine)

(c) Malic acid (d) $C_6H_5CH(CH_3)NH_2$

Ans: (a)

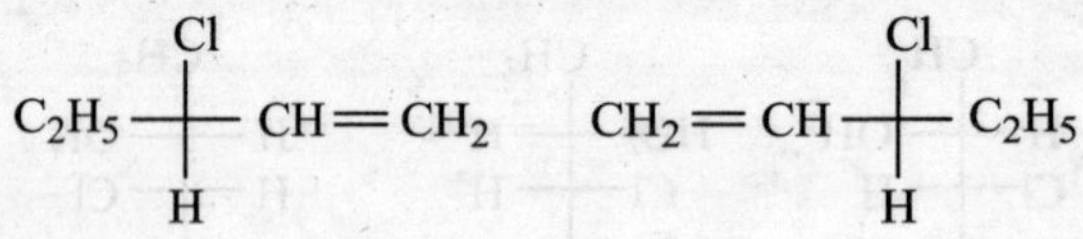

R *S*

Cl > C=C > C—C > H

(b)

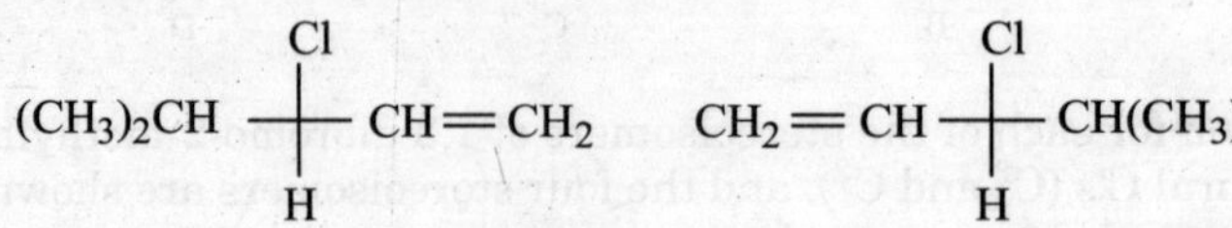

R *S*

Cl > C=C > C—C > H

(c)

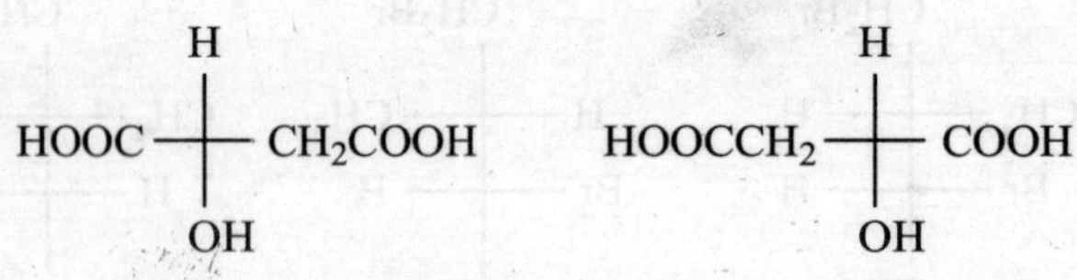

R *S*

O > C—O > C—C > H

(d)

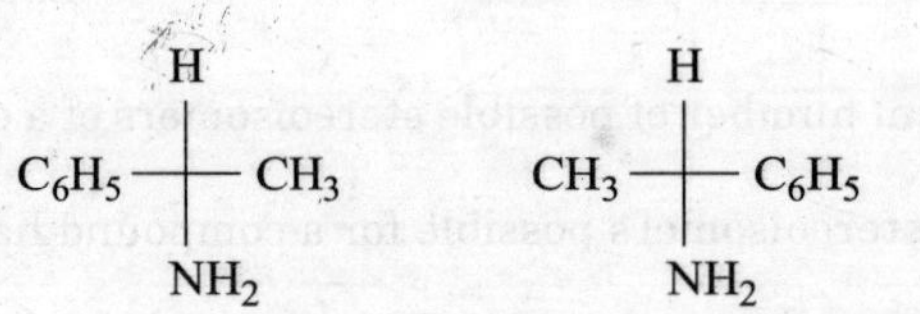

R *S*

N > C_{aryl} > C_{aliph} > H

(e)

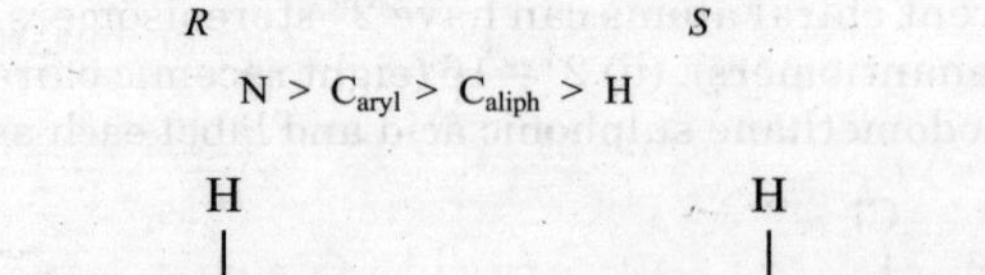

R *S*

N > C—O > C—H > H

18. Draw stereochemical formulas for all the possible stereoisomers of :

(a) 2, 4-dibromopentance (b) 2, 3, 4-tripromohexane

(c) 1, 2, 3, 4-tetrabromobutane (d) 2-bromo-3-chlorobutane

Ans: (a)

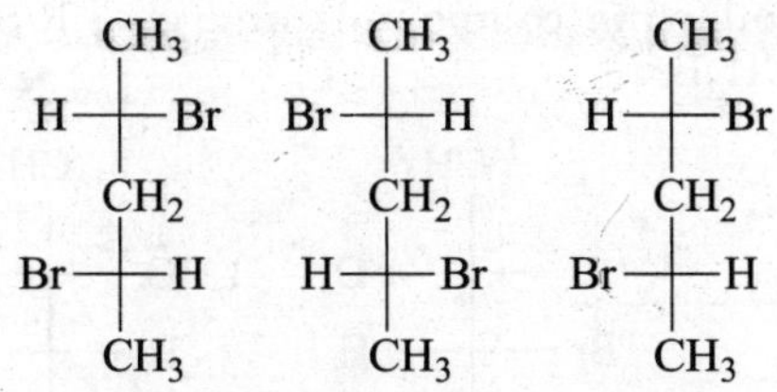

Enantiomers *Meso*

(b)

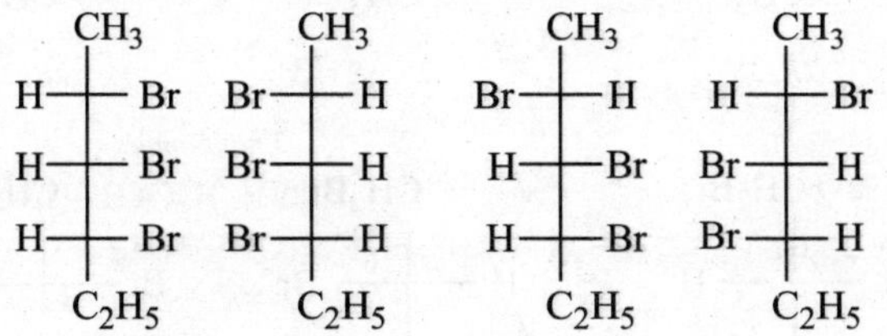

Enantiomers *Enantiomers*

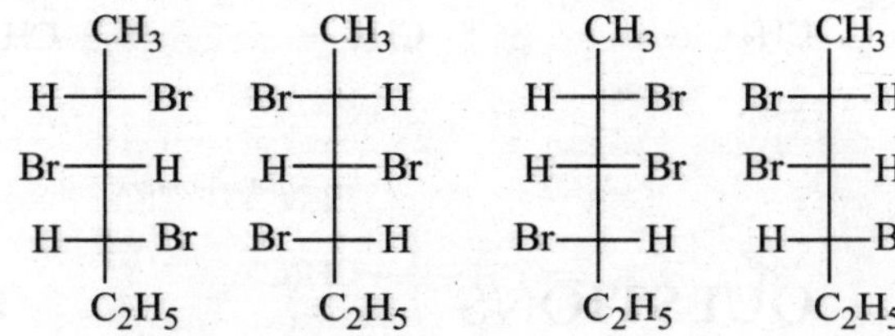

Enantiomers *Enantiomers*

(c)

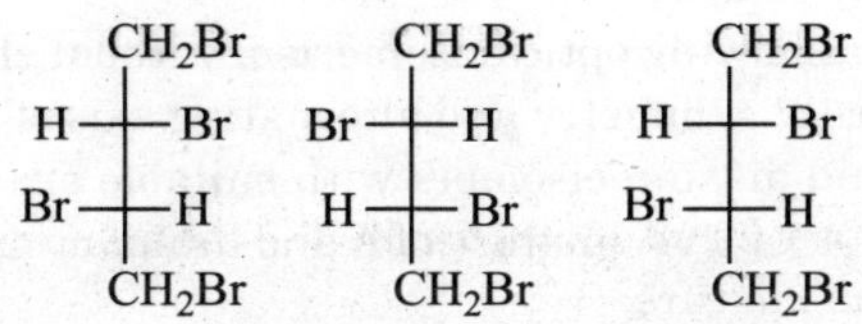

Enantiomers *Meso*

(d)

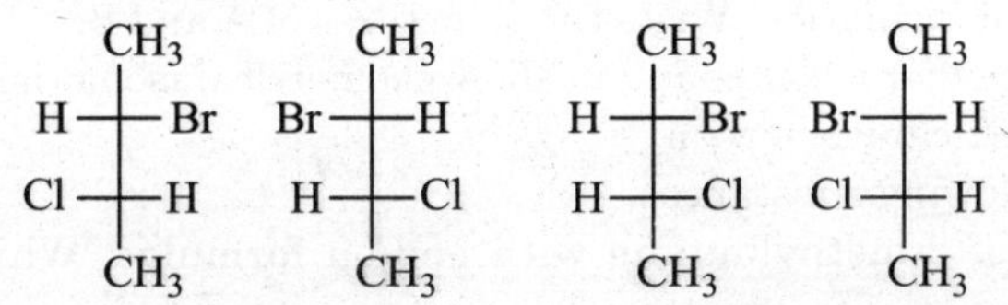

Enantiomers *Enantiomers*

19. Draw and specify as R or S enantiomer (if any) of:

(a) 3-bromohexane (b) 1-deuterio-1-chlorobutane

Ans: (a)

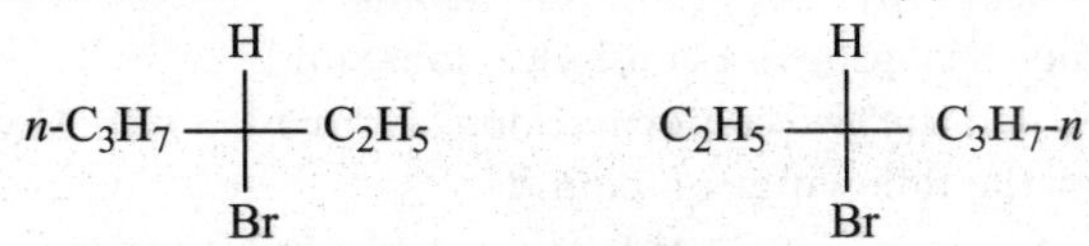

R *S*

Br > C,C,C > C,C,H > H

(b)

R *S*

Cl > C > D > H

20. Draw stereochemical formulas of the following compounds and give R/S designation (a) $CH_3CHBrCHOHCH_3$ (b) $CH_3CHBrCHBrCH_2Br$

Ans: (a)

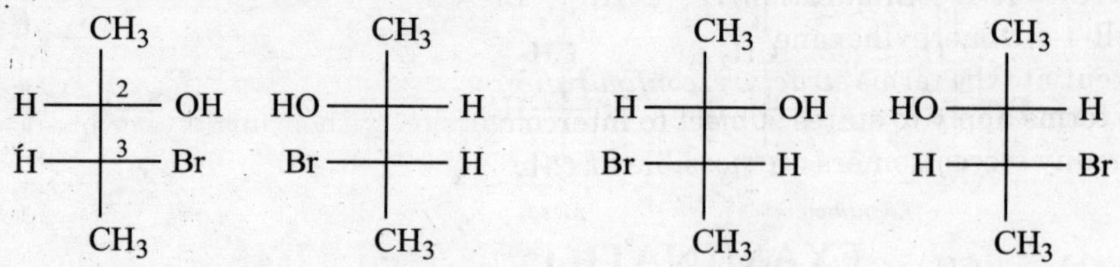

2*S*, 3*R* 2*R*, 3*S* 2*S*, 3*S* 2*R*, 3*R*

Enantiomers *Enantiomers*

(b)

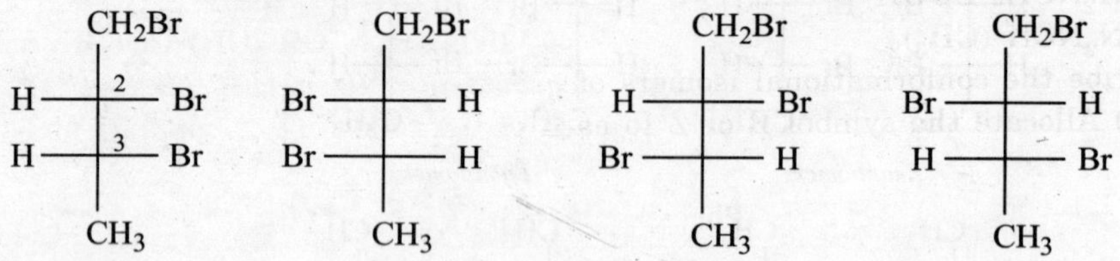

2*S*, 3*R* 2*R*, 3*S* 2*S*, 3*S* 2*R*, 3*R*

Enantiomers *Enantiomers*

QUESTIONS

1. Explain optical isomerism of chiral organic compounds.
2. Write five examples of organic compounds showing optical isomerism without chirality.
3. Define the term; plane of symmetry, centre of symmetry and alternating axis of symmetry.
4. State differences between enantiomers and diastereoisomers with suitable examples.
5. An optical pure compound A, gave an $[\alpha]^{25} = +20°$. A mixture of A and its enantiomers B, gave $[\alpha]^{25} = +10°$. Calculate the ratio of A to B in the mixture.
6. An optically active compound A having molecular formula C_6H_{12} gives inactive compound B of molecular formula C_6H_{14} on hydrogenation. Write the structures of A and B.
7. What are the essential conditions for a compound to show geometrical isomerism?
8. Write three examples of geometrical isomerism.
9. Write out structures for the following substances.
   (a) The three conformations of 2-methylbutane with end-on formulas. Which formulas are superimposable?
   (b) Three conformations of 2-butanone with three-dimensional formulas (the groups attached to any unsaturated carbon atom all lie in one plane).
   (c) The two conformation of *cis*-4-isopropyl-1-methylcyclohexane.
   (d) The two conformations of *trans*-4-isopropyl-1-methylcyclohexane.
   (c) The two conformations of *cis*- and the two conformations of *trans*-1,3-dimethylcyclohexane.
10. Write out structural formulas for the following compounds.
   (a) *cis*-2,3-Dichloro-2-pentene (b) *trans*-4-Methyl-3-octene
   (c) *trans*-2-Phenyl-2-butane (d) *anti*-oxime of a-naphthaldehyde
11. Define the following terms:
   (a) Racemization (b) Mirror plane
   (c) Epimerization (d) Center of symmetry
   (e) Resolution (f) Diastereoisomerism
   (g) Equatorial bond

12. Write three-dimensional formulas for the following compounds:
(a) CHClBrF (R configuration) (b) * (s configuration)
(c) 3-R-4-s-Dimethylhexane

13. Differentiate the terms *structure, configuration,* and *conformation*. In ordinary molecules, which of these terms apply to states subject to interconversion without making or breaking bonds?

14. How many stereoisomers are possible for 2-bromo-4 hydroxycy-clohexanecarboxylic acid?

EXAMINATION QUESTIONS

1. Which of the following compounds would show optical isomerism :
(i) H_2NCH_2 COOH (ii) $(CH_3)_2$ CHCHO
(iii) N_2NCH $(CH_3)_2$ (iv) CH_3 – CH (OH) COOH

2. Describe the conformational isomers of *n*-butane. [*U.P. Tech., $1^{st}/2^{nd}$ Sem., 2000-02*]

3. (a) (i) Allocate the symbol E or Z to each of the following compounds and explain :

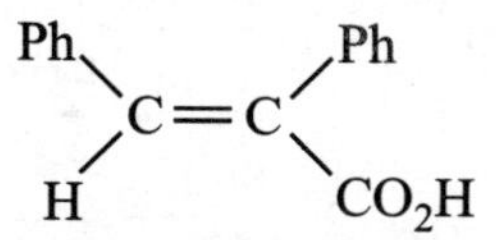 ; 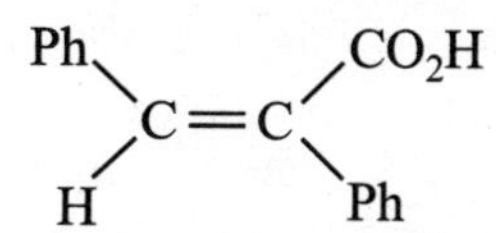

(Ph = Phenyl = C_6H_5)

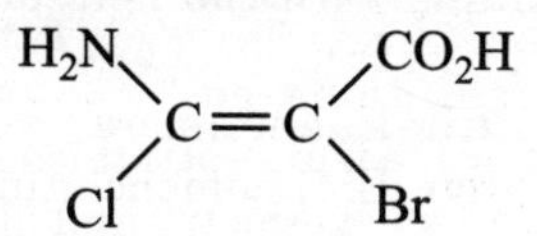 ; 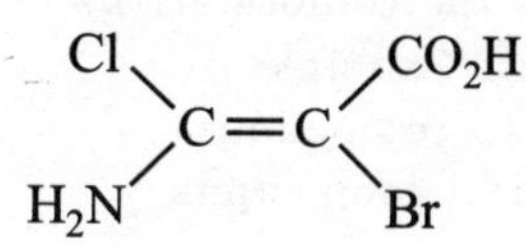

(ii) Under what conditions can the terms cis and trans not be used ?
(b) Discuss, in detail, the case of an organic compound with two Chiral centres.
(c) Draw the potential energy diagram for the various conformations of *n*-butane.
[U.P. Tech., B. Tech : 1st Sem., 2002-03]

4. Draw structures for the equilibrating chair conformes of trans- and cis-1, 2-dimethylcyclohexanes. Show all Me/H1, 3-diaxial interactions. Select the more stable conformation for each diastereoisomer. **[U.P. Tech., 2002-03]**

5. (a) (i) Indicate, with a suitable diagram, the potential energy changes during rotation about $C_{(2)} - C_{(3)}$ bond of *n*-butane. (ii) Comment on the relative stabilities of the conformers.
(b) Name the elements of symmetry. Discuss, with the help of an example, an optically active compound without chirality. **[U.P. Tech., 2003-04]**

6. Attempt the following :
(a) Write the two examples of optically active compounds without chirality.
(b) Write short notes on: (i) Meso compound (ii) Racemisation

7. Write short notes on E, Z Nomenclature, Free Radicals, Electrophilic reagents and Nucleophilic reagents. **[U.P. Tech., 1st Sem., 2004-05]**

8. Define: (i) Enantiomers and Disastereoisoners.
(ii) Which of the following compounds are optically active, and why, *n*-propanol, allenes, n-butanol and 2-chlorobutane.
(iii) What is E-Z system of nomenclature ? In what way it is better than cis, trans-nomenclature? **[U.P. Tech., 1st Sem., 2006-07]**

9. What is optical activity ? Give the stereoisomers of Tartaric acid. How do you account for lack of optical activity in meso-forms and racemic mixtures. **[U.P. Tech., 2nd Sem., 2006-07]**

10. (a) Explain the mechanism of SN^1 and SN^2 organic reactions.

(b) (i) Differentiate between

– Enantiomers and diastereomers

– Racemic mixtures and meso compounds.

(ii) Draw the potential energy diagram for the various conformations of *n*-butane.

[*U.P. Tech., 2nd Sem., 2007-08*]

11. Which of the following compounds shows optical isomerism ?

(i) $CH_3 - CH(OH) - COOH$

(ii) $CH_3 - CHBr - CH_3$

(iii) $CH_3 - CH(CH_3) - CH_2CH_3$

(iv) $CH_3 - CH(OH) - CH_3$ **[*U.P. Tech., 2nd Sem., 2007-08*]**

12. (a) What do you understand by E-Z notations ? Assign E and Z configuration to the following compounds:

(i) 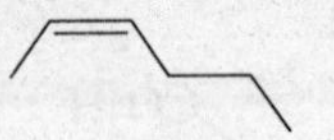

(ii) 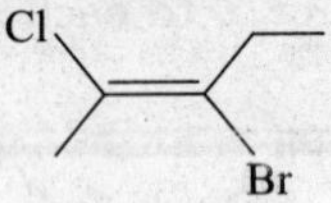

(iii) 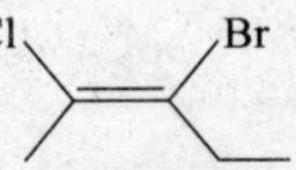

(iv) 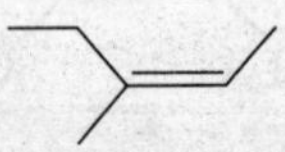

[*U.P. Tech., 2nd Sem., 2008-09*]

(b) Discuss the conformations of n-butane with the help of its potential energy diagram.

13. Explain with example

(i) Meso compounds (ii) Enantiomers

(iii) Chiral compounds (iv) E. Z nomenclature

[*U.P. Tech., 2nd Sem., 2008-09*]

❑❑❑

25

PHOTOCHEMISTRY

Even if you are on the right track, you will get run over if you just sit there.

— ***Will Rogers***

CHAPTER OUTLINE

25.1 INTRODUCTION

Photochemistry is the branch of chemistry which deals with the chemical reactions caused by exposure of reactants to light radiations (photons).

In ordinary thermal reactions, the activation energy to bring about the chemical reaction is provided by the random intermolecular collisions. In photochemical reactions, on the other hand, the necessary activation energy to bring about the molecular collisions is provided by the light radiations.

The term 'radiation' includes electromagnetic radiations ranging from electromagnetic waves of low frequency through ultraviolet, visible and infrared rays to the high frequency X-rays and γ-rays. However, in *photochemical reactions, only the light radiations of the visible and ultraviolet regions lying between 800 nm and 200 nm wave lengths are chiefly involved.* The energies of this range vary from about 90-900 kJ per mole. These energies are comparable with the strengths of the chemical bonds. The radiations of higher energies such as *X*-rays or γ-rays cause ionisation of the molecules and lead to chemical changes through secondary processes. The study of the effects of there ionising radiations form a separate branch of chemistry, *i.e.* radiation chemistry.

To carry out the photochemical reactions, we need a regular source of visible and ultraviolet light. There sources include: sunlight, arc light and mercury vapour lamp.

25.2 DIFFERENCE BETWEEN THERMAL AND PHOTOCHEMICAL REACTIONS

To distinguish the photochemical reactions from the thermal reactions we refer thermal reactions as dark reactions. Hence, chemical reactions can be of two types:

(a) *Dark Reactions or thermal reactions:* The reactions which do not require light radiations to take place. These reactions are induced by temperature, concentration of reactants, presence of catalyst, etc.

Example:

$$N_2 + 3H_2 \rightleftharpoons 2NH_3$$

$$H_2 + I_2 \rightleftharpoons 2HI$$

(b) *Photochemical reactions:* These are the chemical reactions which are brought about directly or indirectly by absorption of visible (400 – 800 nm) and near ultraviolet (200 – 400 nm) radiations.

The main points of difference between these two types of reactions are given below:

Thermal Reactions	*Photochemical Reactions*
1. Thermal reactions do not require light to proceed, *i.e.* these can be carried even in dark.	Photochemical reactions can only take place in presence of light radiations.
2. These reactions involve absorption (endothermic) or evolution of heat (exothermic reactions).	These reactions involve only absorption of light radiations.
3. All thermal reactions, whether exothermic or endothermic proceed with decrease in free energy ($\Delta G = -$ ve)	In photochemical reactions decrease in free energy is not essential. Many photochemical reactions are known which are accompanied by an increase in free energy. Examples: photosynthesis of plants, conversion of oxygen into ozone, decomposition of NH_3.
4. In a thermally activated reaction, energy entering the system in the form of heat is shared among all the molecules in the system.	In a photochemical reaction individual molecule is promoted to highly excited state by absorption of a photon of light without immediate effect on surrounding moloecules and undergoes chemical reactions while still in that excited estate.
5. Temperature has significant effect on the rate of a thermochemical reaction.	Temperature has very little effect on the rate of a photochemical reaction. Instead, the intensity of light has marked effect on the rate of a photochemical reaction.
6. Thermochemical activations are usually not selective. For example exposure of H_2 and Br_2 to heat would cause excitation of both the bromine and hydrogen molecules.	In a photochemical reaction only a single selected species (atom or molecule) can be promoted to an excited state independent of other specie present. For example, when a mixture of H_2 and Br_2 is exposed to a radiation between 450 nm and 550 nm, only Br_2 molecules are excited.

25.3 IMPORTANCE OF PHOTOCHEMISTRY

Photochemistry finds a wide variety of applications, some of these are:

(*a*) Photochemistry is responsible for many of the life processes such as formation of vitamin D, formation of food through photosynthesis of plants, etc.

(*b*) The synthesis of complex molecules such as proteins and nucleic acids occurs through photochemical reactions between simple gaseous molecules such as CO_2, NH_3 and methane.

(*c*) Photochromic materials which have the peculiar property of changing colour when exposed to suitable radiations are prepared through photochemistry. These substances restore to their original colour when the irradiation source is cut off. Spiropyrams, used as photochromic sunglasses belong to this class.

(d) Photochemistry has helped in carrying a large number of typical organic reactions such as preparation of caprolactam (a monomer for nylon-6), several insecticides and synthesis of antioxidants produced by photosulphonation.

(e) Photophysical phenomena such as fluorescence and phosphorescence are another valid applications of photochemistry.

(f) Laser technology is the new field where photochemistry has been used.

(g) The study of photochemical reactions is utilised for the manufacture of solar cells which are replacing traditional sources of fuels.

25.4 LAWS GOVERNING LIGHT ABSORPTION – LAMBERT AND BEER'S LAW

It may be recalled that electromagnetic radiation, which may be considered in terms of an oscillating electric and magnetic field, has a dual nature. It is propagated through space in waveform and obey the relationship,

$$\nu = \frac{c}{\lambda} \qquad ...(1)$$

where ν is the frequency of radiation
λ is the wavelength, and
c is the velocity of light.

The radiation of a given frequency is quantised and is emitted or absorbed in discrete units (photons). The radiation having wavelength 'λ' is associated with energy E, such that,

$$E = h\nu \qquad ...(2)$$

where h is Plank's constant.

Combining (1) and (2),

$$E = h\,\frac{c}{\lambda} \qquad ...(3)$$

The energy value of a photon is sometimes expressed as wave number $\bar{\nu}$, which is defined as number of wavelengths per centimeter.

$$\bar{\nu} = \frac{1}{\lambda} \qquad ...(4)$$

Hence, energy associated with a photon may be written as:

$$E = hc\bar{\nu} \qquad ...(5)$$

The absorbance of radiations is governed by two empirical laws:

(*a*) **Lambert's Law:** When a monochromatic radiation is passed through a medium, the decrease in intensity of radiation with the thickness of the medium is directly proportional to the intensity of the incident radiation. Let,

Intensity of the incident radiation = I

∴ Rate of decrease in intensity of incident radiation with thickness dx of the medium $= \dfrac{-dI}{dx}$ ...(6)

Thus,
$$\frac{-dI}{dx} \propto I$$

or
$$\frac{-dI}{dx} = kI \quad ...(7)$$

where k is the proportionality constant.

Further,
$$\frac{-dI}{I} = k\,dx \quad ...(8)$$

If I_0 is the intensity of the radiation before entering the absorbing medium (*i.e.*, when $x = 0$), then the intensity of radiation I after passing through any finite thickness x, of the medium, can be obtained by integrating equation (8) between these limits.

$$\int_{I_0}^{I} \frac{dI}{I} = -\int_{x=0}^{x=x} k\,dx \quad ...(9)$$

$$\ln \frac{I}{I_0} = -kx \quad ...(10)$$

or
$$\frac{I}{I_0} = e^{-kx} \quad ...(11)$$

$$\boxed{I = I_0\, e^{-kx}} \quad ...(12)$$

The Lambert's law is usually expressed as equation (12). The intensity of the absorbed radiation is given by

$$I_{abs} = I_0 - I = I_0\,(1 - e^{-kx}) \quad ...(13)$$

Equation (10) can also be written as,

$$2.303 \log \frac{I}{I_0} = -kx \quad ...(14)$$

$$\log \frac{I}{I_0} = \frac{-k}{2.303}x \quad ...(15)$$

or
$$\log \frac{I}{I_0} = \frac{k}{2.303}x \quad ...(16)$$

where
$$\log \frac{I}{I_0} = \text{A (Absorbance or optical density)}$$

and
$$\frac{k}{2.303} = \varepsilon \text{ (Absorp co-efficient)}$$

or
$$A = \varepsilon x \quad ...(17)$$

If desired, the natural logarithm can be changed to base 10, in which case equation (15) can be written as

$$I = I_0 10^{-\varepsilon x} \quad ...(18)$$

The absorption co-efficient (ε) also known as extinction co-efficient varies a great deal from substance to substance and also with the wavelength of the light used.

When $$I/I_0 = \frac{1}{10}$$

then $$\varepsilon = \frac{1}{x}$$

Hence, absorption *co-efficient may be defined as the reciprocal of thickness which is required to reduce the intensity of light upto 1/10 of its intensity.*

(*b*) **Lambert-Beer's Law:** If the absorbing substance is present in solution, the decrease in the intensity of radiation with thickness of the solution is given by Beer's law. This law states that when a monochromatic radiation is passed through a solution, the decrease in intensity of radiation with thickness of the absorbing solution is directly proportional to the intensity of incident radiation as well as to the concentration of the solution.

Let a monochromatic light of intensity I pass through a solution of concentration C molar and the thickness dx, then rate of decrease of intensity of radiation is given by:

$$\frac{-dI}{dx} \propto CI \quad ...(19)$$

$$\frac{-dI}{dx} = k'CI \quad ...(20)$$

$$\frac{dI}{dx} = -k'Cdx \quad ...(21)$$

where k' is the proportionality constant.

Let I_0 be the intensity of radiation before entering the absorbing solution (*i.e.* when $x = 0$). Integrating equation (21) between the same limits as before and assuming the concentration C to remain constant, we have,

$$\int_{I_0}^{I} \frac{dI}{I} = -\int_{x=0}^{x=x} k'Cdx \quad ...(22)$$

or $$I_n \frac{I}{I_0} = -k'Cx \quad ...(23)$$

$$\frac{I}{I_0} = e^{-k'Cx} \quad ...(24)$$

$$I = I_0\, e^{-k'Cx} \quad ...(25)$$

or $$2.303 \log \frac{I}{I_0} = -k'Cx \quad ...(26)$$

or $$I = I_0\, 10^{-\varepsilon Cx} \quad ...(27)$$

where ε is called the extinction co-efficient or the absorption co-efficient of the absorbing medium, and is given by,

$$\varepsilon = k'/2.303$$

and its value depends upon units of C and x.

If the concentration 'C' is expressed in moles dm^{-3} and the path length x in cm then ε is referred to as *molar absorption co-efficient* or *molar extinction co-efficient.*

From equation (26),

$$\log \frac{I}{I_0} = \frac{-k}{2.303} Cx \qquad ...(28)$$

or

$$\log \frac{I}{I_0} = \frac{k}{2.303} Cx \qquad ...(29)$$

where $\log \frac{I}{I_0}$ = A (Absorbance or optical density)

$$A = \varepsilon \, Cx \qquad ...(30)$$

Equation (30) implies that the absorbance A is directly proportional to the concentration of the solution and length of the path.

Moreover, transmittance 'T' is defined as:

$$T = \frac{I}{I_0}$$

$$\therefore \quad A = \varepsilon \, Cx = -\log T \quad (\because \; A = \log I_0/I) \qquad ...(31)$$

The plots of T and A versus C are shown in Fig. 25.1. It can be seen that the plot of A versus C is a straight line passing through the origin. Hence, this plot is easier to interpret than the plots of T versus C. The spectrophotometers are, therefore, calibrated to read absorbance, although it is the transmitted light which is actually measured by the instrument.

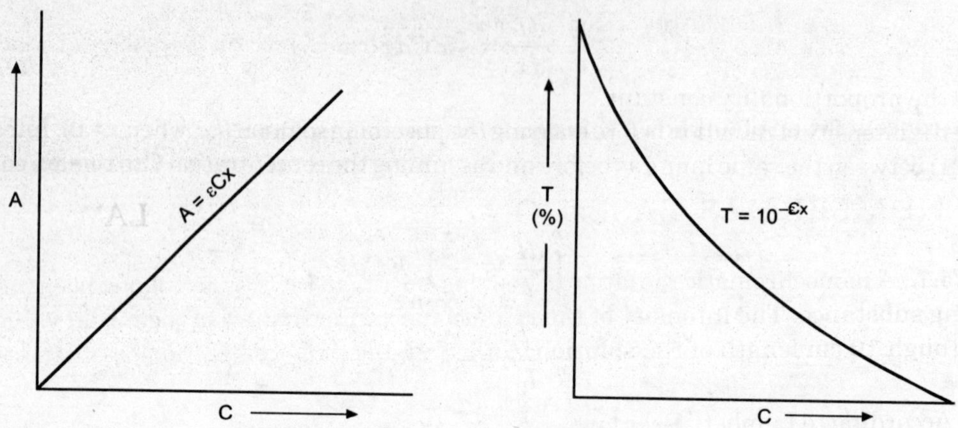

Fig. 25.1 Plots of A and T versus C.

Equation (31) may be put as:

$$\varepsilon = \frac{A}{Cx} \qquad ...(32)$$

If $C = 1$ mol dm^{-3} and $x = 1$cm, equation (32) becomes

$$\varepsilon = A \qquad ...(33)$$

It follows that the molar absorption co-efficient is the specific absorption co-efficient for a concentration of 1 mol dm^{-3} and a path thickness of 1cm.

Limitations of Lambert-Beer Law

1. The Lambert Beer law is not obeyed if the radiation used is not monochromatic.
2. The law is applicable in case of dilute solutions only. In case of high concentrations, the ions

of a solute are close enough to disturb the charge distribution of their neighbours. This inter-ionic interaction drastically alters the ability of the solute to absorb a given wavelength of the incident radiation.

3. The change in temperature also affects the absorption. Increase of temperature shifts the absorption band towards the longer wavelength (bathochromic shift).

Applications of Lambert-Beer's Law

Beer Lambert's law can be used for the determination of an unknown concentration by comparison with a solution of known concentration by using colorimeter or spectrophotometer. According to Lambert-Beer's law,

$$\log I/I_0 = -\varepsilon\, Cx \qquad ...(34)$$

or

$$\log I_0/I = \varepsilon\, Cx = A \qquad ...(35)$$

For determining the concentration of an unknown solution, Absorbance A, of a solution of known concentration C_1 is measured with the help of spectrophotometer and is given as:

$$A_1 = \varepsilon\, C_1\, x \qquad ...(36)$$

The absorbance of solution of unknown concentration C_2 is then measured with the same sample cell so that ε and x are constant

$$A_2 = \varepsilon\, C_2 x \qquad ...(37)$$

From equations (36) and (37), we get,

$$A_2/A_1 = C_2/C_1$$

or

$$C_2 = \frac{A_2}{A_1} \times C_1 \qquad ...(38)$$

Thus the concentration of unknown solution 'C_2' can be determined.

NUMERICALS BASED ON LAMBERT-BEER'S LAW

Problem 25.1. A monochromatic radiation is incident on a solution of 0.05 molar concentration of an absorbing substance. The intensity of the radiation is reduced to 1/4 of the initial value after passing through 10 cm length of the solution. Calculate the molar extinction co-efficient of the substance.

Solution: According to Lambert-Beer law,

$$\log I_0/I = \varepsilon\, Cx$$

$$I/I_0 = .25 = 25\%$$

i.e.

$$I_0/I = 100/25$$

$$\log 100/25 = \varepsilon \times 10 \times 0.05 \text{ mol dm}^{-3}$$

Molar extinction co-efficient, $\varepsilon = 1.204 \text{ dm}^3 \text{ mol}^{-1} \text{ cm}^{-1}$.

Problem 25.2. A solution of thickness 3 cm transmits 30% incident light. Calculate the concentration of the solution, given extinction co-efficient $\varepsilon = 4{,}000 \text{ dm}^3 \text{ mol}^{-1} \text{ cm}^{-1}$.

Solution:

$$\varepsilon = 4{,}000 \text{ dm}^3 \text{ mol}^{-1} \text{ cm}^{-1}$$

$$\frac{I}{I_0} = 0.3$$

$$x = 3 \text{ cm}$$

Absorbance $A = \log I_0/I = \log\left[\frac{1}{0.3}\right] = 0.523$

$\because$ $$A = \varepsilon\, Cx$$

$\therefore$ $$C = \frac{A}{\varepsilon x} = \frac{0.523}{4{,}000\ \text{dm}^3\text{mol}^{-1}\text{cm}^{-1} \times 3\,\text{cm}}$$
$$= 4.35 \times 10^{-5}\ \text{mol dm}^{-3}$$

Problem 25.3. **Calculate the absorbance A of an aqueous solution of unknown compound having percentage transmittance of 20% at 25°C. Also find the molar extinction co-efficient given $x = 2$ cm and $C = 4 \times 10^{-5}$ M.**

Solution: $\frac{I}{I_0} = \frac{20}{100}$

Absorbance $A = \log \frac{I_0}{I} = \log \frac{100}{20} = 0.699$

$$x = 2\ \text{cm}$$
$$C = 4 \times 10^{-5}\ \text{M}$$
$$A = 0.699$$
$$A = \varepsilon\, Cx$$

$\therefore$ Molar extinction co-efficient $\varepsilon = \frac{A}{Cx}$

$\therefore$ $$\varepsilon = \frac{0.699}{4 \times 10^{-5}\text{M} \times 2\,\text{cm}} = 8.7 \times 10^3\ \text{M}^{-1}\ \text{cm}^{-1}$$

25.5 LAWS OF PHOTOCHEMISTRY

There are two basic laws governing the photochemical reactions.

1. **Grotthus Draper Law:** The earliest generalisation related to photochemical reaction was made by Grotthus in 1818 and later by Draper in 1841 as a result of his experimental studies. When light is incident on a system, a part of it may be reflected, a part is absorbed and a part is transmitted through it. It is the absorbed component of light that is capable of bringing about the chemical reaction. Hence Grotthus Draper law may be stated as follows "*only the light which is absorbed by the system can bring about a photochemical change*". This law is also known as the *first law of photochemistry*.
   The law does not mean that a chemical reaction will always take place as soon as light is absorbed by the system. When the conditions are not favourable for the molecules to react, the light energy remains unused. It may be re-emitted as light or heat without producing any chemical change.
2. **Stark-Einstein Law of Photochemical Equivalence:** Stark and Einstein (1905) studied the quantitative aspect of photochemical reactions by applying the concept of quantum theory of light. The law can be stated as: "*In a photochemical reaction, each molecule of a reacting substance absorbs one quantum of the light (photon) causing the reaction.*" In other words, each molecule or atom in a photochemical reaction becomes activated by absorption of one quantum or photon of light. This law is also called *second law of Photochemistry*.

The term, one photon or quantum implies energy equal to $h\nu$, where h is Plank's constant and ν is the frequency of light. Let us consider the following photochemical reaction:

$$A + h\nu \longrightarrow A^*$$
$$A^* \longrightarrow B$$
$$\overline{A + h\nu \longrightarrow B}$$

The molecule A absorbs light and gets activated. The activated molecule A^* then decomposes to yield B.

The number of molecules that are activated is equal to the number of quantas absorbed. In practice, we use molar quantities. That is, one mole of a substance absorbs one mole of photons. The quantity of energy possessed by one mole of photons is called **Einstein.**

$$\text{Energy of 1 photon} = h\nu \qquad ...(39)$$

$$\text{Energy of 1 mole of photon} = Nh\nu \qquad ...(40)$$

where N is Avogadro's number.

or

$$E = Nh\nu \qquad ...(41)$$

where 'E' denotes Einstein

$$E = N h \frac{c}{\lambda} \qquad ...(42)$$

where c is the velocity of light and λ is the wavelength.

We know:

$$N = 6.023 \times 10^{23} \text{ mol}^{-1}$$
$$h = 6.626 \times 10^{-34} \text{ Js}$$
$$c = 3 \times 10^{8} \text{ ms}^{-1}$$

$$\therefore \quad E = \frac{(6.023 \times 10^{23} \text{ mole}^{-1}\ (6.626 \times 10^{-34} \text{ Js})\ (3 \times 10^{8} \text{ ms}^{-1})}{\lambda}$$

$$= \frac{0.1197}{\lambda} \text{ J mol}^{-1} \quad \text{(when } \lambda \text{ is in meter)}$$

$$= \frac{0.1197}{\lambda} \times 10^{10} \text{ J} \quad \text{(when } \lambda \text{ is in } A^{\circ} \because 1\ A^{\circ} = 10^{-10} \text{ m)} \qquad ...(43)$$

In terms of C.G.S. units,

$$N = 6.023 \times 10^{23} \text{ mol}^{-1}$$
$$h = 6.626 \times 10^{-27} \text{ ergs s}$$
$$c = 3 \times 10^{10} \text{ cm/s}$$

$$\therefore \quad E = \frac{(6.023 \times 10^{23} \text{ mole}^{-1})\ (6.626 \times 10^{-27} \text{ ergs s})\ (3 \times 10^{10} \text{ cm/s})}{\lambda}$$

$$= 0.1197 \times 10^{6} \text{ ergs mol}^{-1} \text{ (when } \lambda \text{ is in cm)} \qquad ...(44)$$

Since, 4.184×10^{7} ergs = 1 cal

$$E = \frac{119.7 \times 10^{6}}{4.184 \times 10^{7}\ \lambda} \text{ cal mol}^{-1} \quad (\lambda \text{ in cm})$$

or

$$E = \frac{2.86}{\lambda} \text{ cal mol}^{-1} \quad (\lambda \text{ in cm}) \qquad ...(45)$$

$$E = \frac{2.86 \times 10^{8}}{\lambda} \text{ cal mol}^{-1} \quad \text{(when } \lambda \text{ is in Å)} \qquad ...(46)$$

It is evident from the above expressions that the numerical value of Einstein varies inversely as wavelength of light absorbed. Shorter the wavelength, greater is the energy absorbed.

As most of the photochemical reactions are confined to the visible and ultraviolet regions, i.e. in the wavelength range from 800 to 200 nm.

If $\lambda = 200$ nm, then

$$E = \frac{1.196 \times 10^8}{200} = 5.98 \times 10^5 \text{ J mol}^{-1}$$

If $\lambda = 800$ nm, then

$$E = \frac{1.196 \times 10^8}{800} = 1.495 \times 10^5 \text{ J mol}^{-1}.$$

25.6 QUANTUM EFFICIENCIES OR QUANTUM YIELDS

Quantum efficiency is an important parameter of a photochemical reaction. It is also known as quantum yield. The quantum efficiency (ϕ) of a photochemical process is defined as the number of molecules of reactants consumed or product formed per quantum of light absorbed.

$$\phi = \frac{\text{Number of molecules reacted or product formed}}{\text{Number of photons absorbed}}$$

$$\phi = \frac{\text{Number of moles reacting in given time}}{\text{Number of Einsteins absorbed in the same time}}$$

The law of photochemical equivalence applies to the *primary photochemical processes, i.e.* as a result of primary absorption of one quantum, only one molecule undergoes dissociation and the product does not undergo further reaction, *i.e.* there is 1 : 1 relationship between the number of quantas absorbed and number of reacting molecules. Hence, for these processes $\phi = 1$.

However, the secondary processes alter the quantum yield which is no more equal to 1. The quantum yield deviate from unity because of secondary processes. This can be illustrated by taking the example of formation of hydrogen chloride.

$$Cl_2 \xrightarrow{h\nu} Cl^\bullet + Cl^\bullet$$

$$H_2 + Cl^\bullet \longrightarrow HCl + H^\bullet$$

$$H^\bullet + Cl_2 \longrightarrow HCl + Cl^\bullet$$

Quantum yield of the above photochemical reaction may vary from $10^6 - 10^7$ at room temperature. This is because over a million molecules of HCl are formed for every one quantum of light absorbed by chlorine. A chain mechanism explains this observation. Hence $\phi >> 1$.

The quantum efficiency of a reaction helps in understanding the nature of secondary processes and hence the mechanism of chemical reactions.

Determination of Quantum Yields

Knowledge of quantum efficiency of a photochemical reaction offer valuable information about the mechanism of a photochemical reaction.

In order to determine the value of quantum yield, it is essential to know:

(a) *Number of moles of reacting substance at a given time*: This can be determined by usual analytical methods used in chemical kinetics.

(b) *Number of Einsteins of light absorbed*: This can be determined by a device known 'actinometer' and the process is known as actinometry. For the measurement of quantum yield, a knowledge of incident light (i.e. the number of quanta falling per unit time on the reaction vessel) and the emitted light (the number of quanta emitting per unit time from reaction vessel) is required. The details of Actinometer are, however, beyond the scope of the present syllabus.

Some photochemical reactions and their quantum yields: Quantum yields of a large number of photochemical reactions were studied and it was found that while some reactions followed the law of photochemical equivalence, ($\phi = 1$), others did not. The results obtained in case of some important photochemical reactions together with the effective wavelengths are given in Table 25.2.

Table 25.2 Quantum Yield

Example	*Reaction*	*Quantum yield*
1. Reactions in which the quantum yield is one; $\phi = 1$.		
(i) Formation of Bromocyclohexane:	$Br_2 + C_6H_{12} \rightarrow C_6H_{11}Br + HBr$	
(ii) Formation of Hydrogen peroxide:	$H_2 + O_2 \rightarrow H_2O_2$	1
(iii) Reaction between Chlorine and oxygen	$Cl_2 + O_2 \rightarrow Cl_2O_2$	1
2. Reaction in which the quantum yield $\phi > 1$		
(i) Formation of HBr	$H_2 + Br_2 \rightarrow 2HBr$	2
(ii) Formation of Ozone	$3/2O_2 \rightarrow 2O_3$	2
(iii) Decomposition of HBr	$2HBr \rightarrow H_2 + Br_2$	2
3. Reactions in which the qunatum yeild $\phi < 1$		
(i) Decomposition of Ammonia	$2NH_3 \rightarrow N_2 + 3H_2$	0.14-0.2 at 20°C
(ii) Decomposition of Acetone	$CH_3COCH_3 \rightarrow C_2H_6 + CO + CH_4$	0.2
(iii) Decomposition of Methyliodide	$CH_3I \rightarrow CH_4C_2H_6I_2$	0.01
(iv) Decomposition of Methylbromide	$CH_3Br \rightarrow CH_4 + Br$	4×10^{-3}
4. Reaction in which the quantum yield $\phi >> 1$		
Reaction of chlorine with hydrogen:	$Cl_2 + H_2 \rightarrow 2HCl$	$10^4 - 10^6$
Reaction of Cl_2 with CO:	$Cl_2 + CO \rightarrow COCl_2$	10^3

It is seen from Table 25.2, that the law of photochemical equivalence is strictly valid for a very few reactions only. In many cases the quantum yields are higher and in some cases the quantum yields are lower than expected from the law.

According to the quantum yield, photochemical reactions may be classified into the following three categories:

(a) The reactions in which quantum yield is a small integer such as 1, 2 or 3. Examples are dissociation of hydrogen sulphide, combination of sulphur dioxide and chlorine, oxidation of Fe^{2+} into Fe^{3+} by I_2.

(b) The reactions in which quantum yield is (extremely low) less than 1. Examples are: conversion of maleic into fumaric acid, dissociation of ammonia and dissociation of nitrogen dioxide.

(c) The reactions in which quantum yield is (extremely high) more than 1. Examples are: combination of hydrogen and chlorine and formation of phosgene from CO and Cl_2.

Reasons for High and Low Quantum Yields

In order to explain the variations of the quantum yield from law of chemical equivalence Bodenstein pointed out that photochemical reactions involve two distinct processes.

1. **Primary Processes:** It is the process in which quantum of light is absorbed by a molecule resulting in the formation of an excited molecule. Primary process may involve nothing more than transfer of quantum of energy. For example,

$$Hg_{(vap)} + h\nu \longrightarrow Hg^*$$

The molecule which absorbs light may undergo loss of energy by collisions with other molecules as well as fluorescence and phosphorescence. Law of photochemical equivalence i.e. $\phi = 1$ is applicable only to primary processes.

2. **Secondary Processes:** It is the process in which excited atoms, molecules or free radicals produced in the primary process react further giving rise to higher quantum yields. A secondary process can take place in the dark as well. Due to this process, a number of molecules may undergo reactions by absorbing one quantum only, i.e. the excited atoms/ molecules can initiate a series of chain reactions. In these cases the quantum yield deviates from unity.

Reasons for High Quantum Yield

(i) When the primary step of absorption produces atoms as free radicals which initiate a series of chain reactions. The quantum yield in such cases depends on how far the chain reaction proceeds before termination.

(ii) Formation of an intermediate, which is capable of acting as a catalyst thus giving rise to high quantum yields.

(iii) The reactions may be exothermic and the heat evolved may activate other molecules and thus these molecules react without absorbing additional quantum of energy.

(iv) The active molecules produced after absorption of radiation may collide with other molecules and activate them which in turn activate other reacting molecules.

Example: Combination of hydrogen and chlorine

$$H_{2\,(g)} + Cl_{2\,(g)} \longrightarrow 2HCl_{\,(g)}$$

The quantum yield of this reaction has been found to be exceptionally high of the order of 10^4 to 10^6. It is supposed that the quantum of energy absorbed in the primary process is more than sufficient to bring about the primary excitation process. Hence the molecules carry excess energy which is capable of starting secondary process, resulting into higher quantum yield.

$$Cl_2 + h\nu \longrightarrow Cl^\bullet + Cl^\bullet] \quad \text{Primary process}$$
$$Cl^\bullet + H_2 \longrightarrow HCl + H^\bullet] \quad \text{Secondary Reactions}$$
$$H^\bullet + Cl_2 \longrightarrow HCl + Cl^\bullet]$$

There is possibility of formation of an infinite number of HCl molecules in the chain reaction which will continue unless terminated. Thus a single quantum of energy absorbed in the primary process forms a number of HCl molecules resulting in exceptionally high quantum yield ($\phi >> 1$).

Reasons for Low Quantum Yield

(i) Some processes involve deactivation of the excited molecules before they are able to react to form products.
(ii) Energised molecules may lose their energy by collision with nonexcited molecules.
(iii) Molecules may not receive sufficient energy to enable them to react.
(iv) The primary photochemical reaction may be reversed.

Example: Combination of hydrogen and bromine. The quantum efficiency of this reaction is about 0.01.

$$Br_2 + h\nu \longrightarrow 2Br^\bullet \quad \text{(Primary process)}$$

The expected secondary processes are:

$$Br^\bullet + H_2 \longrightarrow HBr + H^\bullet \quad \text{(i)}$$
$$H^\bullet + Br_2 \longrightarrow HBr + Br^\bullet \quad \text{(ii)}$$
$$H^\bullet + HBr \longrightarrow H_2 + Br^\bullet \quad \text{(iii)}$$
$$Br^\bullet + Br^\bullet \longrightarrow Br_2 \quad \text{(iv)}$$

Reaction (i) is highly endothermic and is, therefore, taking place very slowly at ordinary temperature. At this slow speed, most of the bromine free radicals recombine to give the bromine molecules. Hence, the reactions (ii), (iii) and (iv) cannot occur. Therefore, the quantum yield will be extremely low.

25.7 KINETICS OF PHOTOCHEMICAL REACTIONS

The rate law which photochemical reactions follow is generally more complex than those for thermal reactions because a number of variables are involved.

The rate law expression for a number of photochemical reactions have been derived using the well known 'Study State Approximation' for transient species. The kinetics of some important photochemical reactions are discussed below.

I. REACTIONS WHICH INVOLVE CHAIN REACTIONS

(1) A Typical Example is Combination of Hydrogen and Chlorine

$H_{2(g)} + Cl_{2(g)} \longrightarrow HCl_{(g)}$ when a mixture of H_2 and Cl_2 is exposed to light in the region of chlorine spectrum, i.e. wavelength less than 480 nm, the primary step is photo-decomposition of chlorine molecules.

(i) $Cl_2 + h\nu \xrightarrow{k_1} 2Cl$ (Primary process)

This is followed by the secondary reactions

(ii) $Cl + H_2 \xrightarrow{k_2} HCl + H$ Propagation

(iii) $H + Cl_2 \xrightarrow{k_3} HCl + Cl$ Propagation

(iv) $Cl + Cl \xrightarrow[(wall)]{k_4} \frac{1}{2}Cl_2$ Termination.

The step (iv), the chain termination step might be a recombination of chlorine atoms in the gas phase with the help of a third body (M or wall of the vessel) to carry away the excess of energy. Reactions (ii) and (iii) are responsible for the conversion of a large quantity of H_2 and Cl_2 to HCl. The rate of formation of HCl is given by (ii) and (iii). Therefore, rate is given by:

$$\frac{d\,[HCl]}{dt} = k_2\,[Cl]\,[H_2] + k_3\,[H]\,[Cl_2] \qquad ...(1)$$

For determining the rate from equation (1), we require the concentrations of intermediates [H] and [Cl]. These can be found out by applying steady state approximation, i.e. the concentration of free radical or intermediate is constant at any instant or,

Rate of formation of reactive intermediate = Rate of its decomposition.

The rate of formation of [Cl] in the above mechanisms is given by:

$$\frac{d[Cl]}{dt} = k_1 I_{abs} + k_3 [H][Cl_2] - \{[k_2 [Cl][H_2] + k_4 [Cl]\} \quad ...(2)$$

Applying Steady State Approximation (S.S.A.),

$$\frac{d[Cl]}{dt} = 0$$

or $$k_1 I_{abs} + k_3 [H][Cl_2] = k_2 [Cl][H_2] + k_4 [Cl] \quad ...(3)$$

Similarly for [H], we can write

$$\frac{d[H]}{dt} = 0 = k_2 [Cl][H_2] - k_3 [H][Cl_2] \quad ...(4)$$

or $$k_2 [Cl][H_2] = k_3 [H][Cl_2] \quad ...(5)$$

By adding equations (3) and (5) we get,

$$k_1 I_{abs} - k_4 [Cl] = 0$$

or $$[Cl] = k_1 I_{abs}/k_4 \quad ...(6)$$

Substituting the value of [Cl] from equation (6) in equation (5) we can obtain the value of [H] as:

$$k_2 \frac{k_1 I_{abs}}{k_4} [H_2] = k_3 [H][Cl_2] \quad ...(7)$$

$$[H] = \frac{k_1 k_2 I_{abs} [H_2]}{k_3 k_4 [Cl_2]} \quad ...(8)$$

Substituting, the values of [H] and [Cl] from equations (7) and (8) in (1), we get,

$$\frac{d[HCl]}{dt} = \frac{k_1 k_2 I_{abs} [H_2]}{k_4} + \frac{k_1 k_2 I_{abs} [H_2]}{k_4} \quad ...(9)$$

$$= \frac{2k_1 k_2 I_{abs} [H_2]}{k_4} \quad ...(10)$$

$$= \left(\frac{2 k_1 k_2}{k_4}\right) I_{abs} [H_2]$$

$$\text{Rate} = k' I_{abs} [H_2] \quad ...(11)$$

where $$k' = (2k_1 k_2/k_4)$$

It is seen from equation (11) that rate is directly proportional to the intensity I_{abs} of the absorbed radiation. Equation (11) is in perfect agreement with experimental data. The quantum yield of this reaction is usually of the order of 10^4 to 10^6.

Evidence in Favour of Chain Mechanism

(a) If the reaction is carried out in capillary tubes, the quantum yield decreases to a large extent. The capillary tubes provide large areas of walls where chains might get terminated.

(b) If some chain inhibitor, such as oxygen is present the quantum yield decreases significantly supporting the chain mechanisms. In presence of oxygen the various steps can be written as:

(i) $Cl_2 + h\nu \xrightarrow{k_1} 2Cl$

(ii) $Cl + H_2 \xrightarrow{k_2} HCl + H$

(iii) $H + Cl_2 \xrightarrow{k_3} HCl + Cl$

(iv) $H + O_2 \xrightarrow{k_4} HO_2$

(v) $Cl + O_2 \xrightarrow{k_5} ClO_2$

(vi) $Cl + X \xrightarrow{k_6} ClX$

Steps (iv) and (v) remove reactive H and Cl specie and hence decrease the rate.

(2) Kinetics of Reaction of Hydrogen with Bromine

$$H_{2(g)} + Br_{2(g)} \xrightarrow{h\nu} 2\,HBr_{(g)}$$

When a mixture of H_2 and Br_2 is exposed to light in the continuous spectrum of Br_2 ($\lambda < 510$ nm), Br_2 molecules dissociate into Br atoms.

(i) $Br_2 + h\nu \xrightarrow{k_1} 2Br$ (Primary process)

(ii) $Br + H_2 \xrightarrow{k_2} HBr + H$

(iii) $H + Br_2 \xrightarrow{k_3} HBr + Br$ — Propagation (Secondary Process) [steps (ii)–(iv)]

(iv) $H + HBr \xrightarrow{k_4} H_2 + Br$

(v) $2Br \xrightarrow{k_5} Br_2$ (Termination)

Since, HBr is produced by reactions (ii) and (iii) and removed by reaction (iv), the rate of formation of HBr is given by:

$$\frac{d\,[HBr]}{dt} = k_2\,[Br]\,[H_2] + k_3\,[H]\,[Br_2] - k_4\,[H]\,[HBr] \qquad ...(1)$$

To find the values of [Br] and [H], we apply Steady State Approximation for intermediates, *i.e.*,

$$\frac{d\,[H]}{dt} = 0 = k_2\,[Br]\,[H_2] - k_3\,[H]\,[Br_2] - k_4\,[H]\,[HBr] \qquad ...(2)$$

and
$$\frac{d\,[Br]}{dt} = 0 = k_1\,I_{abs} - k_2\,[Br][H_2] + k_3\,[H]\,[Br_2] + k_4\,[H]\,[HBr] - k_5\,[Br]^2 \qquad ...(3)$$

Addition of equations (2) and (3), yields:

$$k_1\,I_{abs} - k_5\,[Br]^2 = 0 \qquad ...(4)$$

or
$$k_1\,I_{abs} = k_5\,[Br]^2$$

$$[Br] = \left(\frac{k_1}{k_5}\,I_{abs}\right)^{\frac{1}{2}} \qquad ...(5)$$

Similarly, the value of [H] can be obtained from equation (2) which can be rearranged as:

$$k_2\,[Br]\,[H_2] = [H]\;k_3\,[Br_2] + k_4\,[HBr]$$

$$[H] = \frac{k_2\,[Br]\,[H_2]}{k_3\,[Br_2] + k_4\,[HBr]} \qquad ...(6)$$

Putting the value of [Br] from equation (5) into equation (6)

$$[H] = \frac{k_2 [H_2]\left[\frac{k_1}{k_5} I_{abs}\right]^{\frac{1}{2}}}{k_3 [Br_2] + k_4 [HBr]} \quad ...(7)$$

Substituting the values of [H] and [Br] from equations (7) and (5) in equation (1), we get,

$$\text{rate} = \frac{d[HBr]}{dt} = k_2 [H_2]\sqrt{\frac{k_1 I_{abs}}{k_5}} + \{k_3 [Br_2] - k_4 [HBr]\}\frac{k_2 [H_2]\sqrt{\frac{k_1 I_{abs}}{k_5}}}{k_3 [Br_2] + k_4 [HBr]} \quad ...(8)$$

$$= k_2 [H_2]\sqrt{\frac{k_1 I_{abs}}{k_5}}\left\{\frac{1 + k_3 [Br_2] - k_4 [HBr]}{k_3 [Br_2] + k_4 [HBr]}\right\}$$

$$= k_2 [H_2]\sqrt{\frac{k_1 I_{abs}}{k_5}}\left\{\frac{k_3 [Br_2] + k_4 [HBr] + k_3 [Br_2] - k_4 [HBr]}{k_3 [Br_2] + k_4 [HBr]}\right\}$$

$$= k_2 [H_2]\sqrt{\frac{k_1 I_{abs}}{k_5}}\left\{\frac{2k_3 [Br_2]}{k_3 [Br_2] + k_4 [HBr]}\right\}$$

$$= k_2 [H_2]\sqrt{\frac{k_1 I_{abs}}{k_5}}\left\{\frac{2}{1 + k_4 [HBr] / k_3 [Br_2]}\right\}$$

$$= 2k_2 \sqrt{\frac{k_1 I_{abs}}{k_5}}\left\{\frac{H_2}{1 + k_4 [HBr] / k_3 [Br_2]}\right\} \quad ...(9)$$

From equation (9) it is clear that rate varies as the square root of the intensity I_{abs} of the absorbed radiation.

Note: It must be noted that inspite of the chain mechanism, similar to the combination of hydrogen and chlorine, the quantum yield of this photochemical reaction is quite low, being about, 0.01 at ordinary temperatures.

For $H_2 + Cl_2 \longrightarrow 2HCl$ quantum yield $10^4 - 10^6$

For $H_2 + Br_2 \longrightarrow 2HBr$ quantum yield 0.01

The reason lie in the propagation steps of the two reactions.

$$Cl + H_2 \xrightarrow{k_2} HCl + H \quad \text{(Exothermic)}$$

$$Br + H_2 \xrightarrow{k_2} HBr + H \quad \text{(Endothermic)}$$

The propagation step in the mechanism of the formation of HCl is exothermic and hence takes place spontaneously leading to high quantum yield. Whereas, the propagation step in the mechanism of formation of HBr is endothermic and takes place extremely slowly at ordinary temperatures. The quantum yield depends upon the number of times the propagation steps are repeated before the termination step, which is quite less in the latter case. Hence quantum yield is low. Moreover, with increasing time the reversal reaction also becomes predominant and hence rate of formation of HBr decreases.

II. REACTIONS WHICH DO NOT INVOLVE CHAINS

We will discuss the kinetics of the reactions which do not involve chains. Example are: dissociation of HI and HBr.

(1) Dissociation of HI: Decomposition of HI is brought by the radiation in the region 200 – 230 nm. The primary photochemical step is as follows:

(i) $HI + h\nu \xrightarrow{k_1} H + I$

This is followed by the secondary reactions

(ii) $H + HI \xrightarrow{k_2} H_2 + I$

(iii) $I + I \xrightarrow{k_3} I_2$

It must be emphasized here that the possibility of other elementary reactions is ruled out since they involve either high activation energy or a third body collision.

In the above mechanism, HI is consumed in reactions (i) and (ii). Thus rate law for the decomposition of HI is given by

$$\frac{-d\,[HI]}{dt} = k_1\, I_{abs} + k_2\, [H]\,[HI] \qquad \text{...(1)}$$

Also, the rate of formation of [H] is given by,

$$\frac{d\,[H]}{dt} = k_1\, I_{abs} - k_2\, [H]\,[HI] \qquad \text{...(2)}$$

Applying steady state approximation to [H], we have

$$\frac{d\,[H]}{dt} = k_1 I_{abs} - k_2\, [H]\,[HI] = 0 \qquad \text{...(3)}$$

$$\therefore \quad k_2\, [H]\,[HI] = k_1 I_{abs} \qquad \text{...(4)}$$

Substituting the value of k_2 [H] [HI] from equation (4) in equation (1) we have,

$$\frac{-d\,[HI]}{dt} = 2k_1 I_{abs} \qquad \text{...(5)}$$

This shows that the absorption of one photon leads to the decomposition of two molecules of HI. The quantum efficiency, therefore, equals 2 as shown below:

$$\phi = \frac{\text{Rate of appearance of HI}}{\text{Rate of absorption of light}}$$

$$\phi = \frac{2k_1\, I_a}{k_1\, I_a} = 2.$$

However, the quantum yield decreases as the reaction proceeds. This is because as I_2 accumulates, the following thermal reaction takes place,

$$H + I_2 \xrightarrow{k_4} HI + I \qquad \text{...(6)}$$

Hence the steady state approximation for [H] now becomes,

$$\frac{d[H]}{dt} = k_1 I_{abs} - k_2\, [H]\,[HI] - k_4\, [H]\,[I_2] = 0 \qquad \text{...(7)}$$

$$[H] = \frac{k_1\, I_{abs}}{k_2\, [HI] + k_4\, [I_2]} \qquad \text{...(8)}$$

Substituting [H] from equation (8) in (1) we get,

$$\frac{-d\,[\mathrm{HI}]}{dt} = k_1 \mathrm{I_{abs}} + k_2\,[\mathrm{HI}]\,\frac{k_1\,\mathrm{I}_{abs}}{k_2[\mathrm{HI}] + k_4[\mathrm{I_2}]}$$

$$= k_1\,\mathrm{I_{abs}}\left(1 + \frac{1}{1 + \left\{k_4\,[\mathrm{I_2}] \big/ k_2\,[\mathrm{HI}]\right\}}\right) \qquad ...(9)$$

$$\frac{-d\,[\mathrm{HI}]/dt}{\mathrm{I}_{abs}} = k_1 + \frac{k_1}{1 + \left\{k_4\,[\mathrm{I_2}] \big/ k_2\,[\mathrm{HI}]\right\}} \qquad ...(10)$$

As the reaction proceeds, $[I_2]$ increases whereas [HI] decreases. Hence the quantum yield decreases from the original value of 2.

(2) Dissociation of HBr

The quantum yield for the dissociation of HBr is also 2. The mechanism of this reaction is same as discussed for the dissociation of HI.

III. DIMERIZATION OF ANTHRACENE

When anthracene is exposed to uv radiation, it is converted into dianthracene

$$2\,C_{14}H_{10} + h\nu \longrightarrow C_{28}H_{20}$$

The quantum yield should be 2 but actually it is only about 0.5. This is due to the back thermal reaction in which anthracene dissociates to form anthracene again.

$$C_{28}H_{20} \xrightarrow{dark} 2\,C_{14}H_{10}$$

Hence, the dimerization of anthracene can be represented by:

$$2\,C_{14}H_{10} \underset{\text{dark}}{\overset{\text{light}}{\rightleftharpoons}} C_{28}H_{20}$$

The reversible sign indicates that ultimately an equilibrium is attained between the dark and light reactions. Here, we will discuss the kinetics of photochemical reaction of anthracene. The following mechanisms have been suggested:

$$C_{14}H_{10} + h\nu \xrightarrow{k_1} C_{14}H^*_{10} \quad \text{(Excitation)} \qquad ...(1)$$

$$C_{14}H^*_{10} + C_{14}H_{10} \xrightarrow{k_2} C_{28}H_{20} \quad \text{(Dimerization)} \qquad ...(2)$$

$$C_{14}H^*_{10} \xrightarrow{k_3} C_{14}H_{10} + h\nu \quad \text{(Fluorescence)} \qquad ...(3)$$

Rate of formation of dianthracene ($C_{28}H_{20}$) is given by step (2) as

$$\frac{d\,[C_{28}H_{20}]}{dt} = k_2\,[C_{14}H^*_{10}]\,[C_{14}H_{10}] \qquad ...(4)$$

Also, the activated $C_{14}H^*_{10}$ is produced in step (1) and are being removed by step (2), thus according to steady state approximation,

$$\frac{d\,[C_{14}H_{10}]}{dt} = k_1\,\mathrm{I_{abs}} - k_2\,[C_{14}H^*_{10}]\,[C_{14}H_{10}] - k_3\,[C_{14}H^*_{10}] = 0 \qquad ...(5)$$

$$k_1\,\mathrm{I_{abs}} = k_2\,[C_{14}H^*_{10}]\,[C_{14}H_{10}] - k_3\,[C_{14}H^*_{10}] \qquad ...(6)$$

or

$$[C_{14}H^*_{10}] = \frac{k_1\,\mathrm{I_{abs}}}{k_2\,[C_{14}H_{10}] + k_3} \qquad ...(7)$$

Substituting, the value of $[C_{14}\ H^*_{10}]$ from equation (7) in equation (4) we get,

$$\frac{d\,[C_{28}\ H_{20}]}{dt} = k_2 \frac{k_1\ I_{abs}}{k_2\ [C_{14}\ H_{10}] + k_3}\ [C_{14}\ H_{10}] \qquad ...(8)$$

$$= \frac{k_1\ k_2\ I_{abs}\ [C_{14}\ H_{10}]}{k_2\,[C_{14}\ H_{10}] + k_3} \qquad ...(9)$$

The quantum yield for the formation of the dimer is given by:

$$\phi\ (C_{28}\ H_{20}) = \frac{\text{Number of moles of } C_{28}\ H_{20} \text{ formed}}{\text{Number of einsteins absorbed}}$$

$$= \frac{d\,[C_{28}\ H_{20}]/dt}{I_{abs}} = \frac{k_1\ k_2\ [C_{14}\ H_{10}]}{k_2\ [C_{14}\ H_{10}] + k_3} \quad \text{(Using equation (9))} \qquad ...(10)$$

It is clear from equation (10) that quantum yield for dimerization of anthracene is independent of the intensity of the light absorbed. Hence, dimerization of anthracene follow the first order kinetics.

Equation (10) can be written as:

$$\phi\ (C_{28}\ H_{20}) = \frac{k_1 [C_{14}\ H_{10}]}{[C_{14}\ H_{10}] + k_3 / k_2} \qquad ...(11)$$

Two cases can be discussed:

Case 1: $[C_{14}\ H_{10}] << k_3/k_2$, hence from equation (11),

$$\phi\ (C_{28}\ H_{20}) = \frac{k_1\ k_2}{k_3}\ [C_{14}\ H_{10}] \qquad ...(12)$$

Therefore, $\phi\ (C_{28}\ H_{20})$ increases with increase in conc. of $[C_{14}\ H_{10}]$

Case 2: $k_3\ k_2 << [C_{14}\ H_{10}]$, hence from equation (11),

$$\phi\ (C_{28}\ H_{20}) = k_1 \qquad ...(13)$$

Therefore, $\phi\ (C_{28}\ H_{20})$ is independent of $[C_{14}\ H_{10}]$.

25.8 CONSEQUENCES OF PHOTOCHEMICAL EXCITATIONS

The first step in a photochemical process is the absorption of a quantum of radiation by the molecule. As is discussed earlier, this absorption of radiation is called the primary process in photochemistry. In this process, the energy absorbed from the photon in the visible or ultraviolet region raises the molecule from lower energy state to higher electronic state. The excited molecule may then behave in different ways leading to different types of phenomena. Let us discuss the absorption of a photon by a molecule in more detail.

We encounter three kinds of molecular orbitals, *i.e.* bonding (σ and π) , antibonding (σ* and π*) and nonbonding (*n*) molecular orbitals. The absorption in the UV region causes excitation of an electron from an occupied bonding or nonbonding molecular orbital to an antibonding molecular orbital as shown in Fig. 25.2.

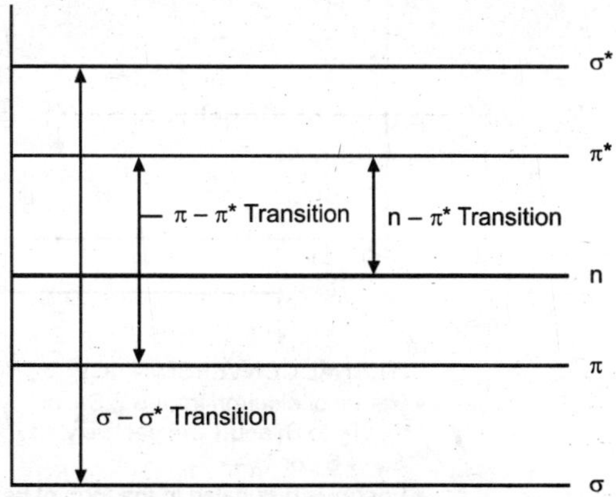

Fig. 25.2 Electronic transitions between molecular orbials.

By far the most important excitation in organic molecules are $n - \pi^*$ and $\pi - \pi^*$ transitions.

Once the photon of light is absorbed, it is not essential that it will bring about photochemical change. The absorption of light may result in a number of other phenomena as well. This is well illustrated with the help of Jablonski diagram.

Jablonski Diagram

There are even number of electrons in typical organic molecules and these electrons are paired in the ground state. In electronic excitation only one electron per molecule undergoes transition to an orbital of higher energy leaving its original orbital half vacant with only one electron. The promoted electron which is in different orbital, may have the same spin (unpaired) as its former partner or the opposite spin (paired). These two states are designated as Triplet (T) and Singlet (S_0) respectively (Fig. 25.3).

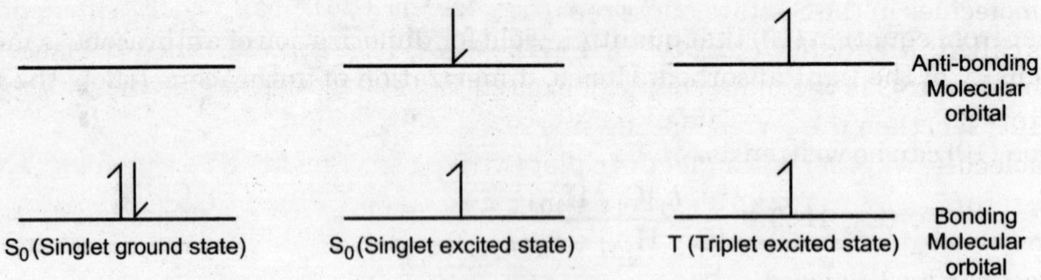

Fig. 25.3

The terms singlet and triplet originate from the fact that singlet state does not split in a magnetic field while the Triplet state splits as it has three possible energy states in an applied magnetic field.

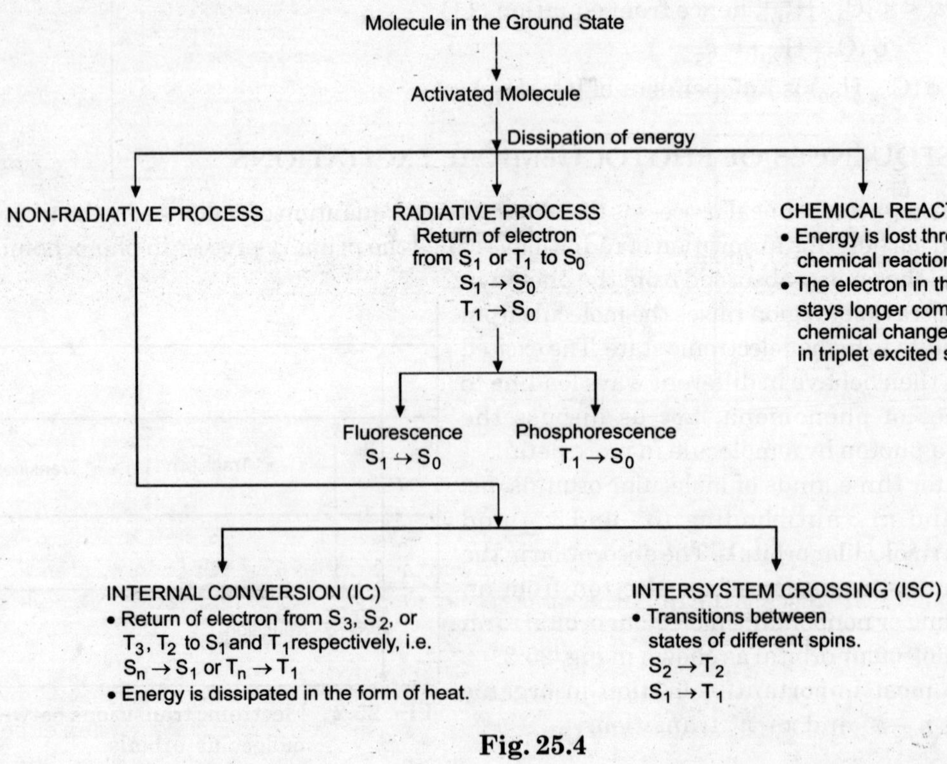

Fig. 25.4

In principle, for every excited singlet state there is a corresponding Triplet state. Singlet state is of higher energy than Triplet state. Promotion of an electron of a molecule takes place without spin, so the electrons remain paired in the excited state. Since, the electron can jump to any of the higher electronic states depending upon the energy of the photon absorbed, we get a series of singlet excited states, S_n where $n = 1, 2, 3, 4, \ldots$. Direct excitation of the ground state (S_0) to the first triplet state (T_1) is not possible as according to selection rules, these transitions are *forbidden*. The triplet state (T_1) is invariably populated *via.*, the first singlet (S_1) state ($S_1 \rightarrow T_1$). Hence, the most promotions are from $S_0 \rightarrow S_1$. The promotions of ($S_1 \longrightarrow S_2$ or S_3, can take place but their life time is usually less than 10^{-11} seconds and they rapidly decay to the lowest level (S_1) by a process called *internal conversion* (IC). Therefore, in most cases, the lowest vibrational level of the S_1 state is the only important singlet state.

Most molecules in the S_1 state may drop to triplet state T_1. The $S_1 \longrightarrow T_1$ conversion occurs by a process called '*intersystem crossing*' *(ISC)*, a non-radiative process. Since, transition from S_0 to T_1 is forbidden, ISC is the main source of excited Triplets. A triplet state has a longer life time (10^{-5} to 10^{-3} sec) then the corresponding singlet state.

The molecule, whether in singlet or triplet excited state is said to be activated. The activated molecule returns to the ground state by dissipating its energy through the following processes as shown in Fig. 25.4.

The various photophysical processes involving transitions between ground and excited states are often summarised on a Jablonski diagram Fig. 25.5.

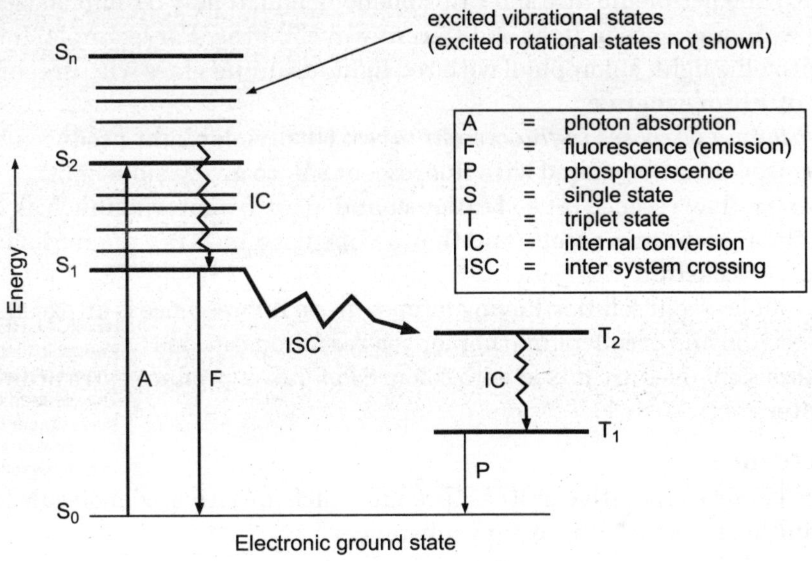

Fig. 25.5 Jablonski diagram

25.9 PHOTOPHYSICAL PROCESSES

In this section, we will discuss the radiative processes (phosphorescence and fluorescence) in detail.

1. Fluorescence

When a beam of light is incident on certain substances, they emit visible light or radiations. This phenomena is known as fluorescence and the substances which emit radiations when exposed to light, are called fluorescent substances. The fluorescent substances stop emitting light as soon as the incident light is cut-off.

Let us explain the phenomena of fluorescence in the light of photochemistry. When atoms of an element or molecules of substance are exposed to radiations, they are excited to higher electronic states as discussed in section 25.7. The process may be represented as:

$$A + h\nu \longrightarrow A^*$$

The excited state can be singlet or triplet (S_n or T). *When the excited singlet (S_1) drop to ground state (S_0) rapidly via emitting excess radiations, the phenomena is called fluorescence.* The excess radiations are emitted within 10^{-6} to 10^{-4} seconds of absorption of radiations.

$$A^* \longrightarrow A_1 + h\nu_1$$

The emitted radiation in fluorescence has a lower energy than the incident radiations. This is due to the fact that molecules which initially are at their lowest vibrational energy states, may acquire higher vibrational energy states. In case of atoms, as they do not have vibrational energy level, the emitted radiations are of exactly same frequency as absorbed radiations. This is known as the *resonance fluorescence.* Example, mercury vapours absorb the radiations of wavelength 253.7 nm and emit the radiations of the same wavelength.

Because of the possibility of fluorescence, any chemical reaction of the S_1 state must take place very rapidly or fluorescence will occur before the reaction takes place.

Examples of fluorescence are solutions of Flouresceine and eosin. When their solutions are placed in light, they show flourescence from green to violet colour. Uranyl sulphate, chlorophyll, flourespar (CaF_2) and petroleum also show this phenomenon. It may be emphasized that different substances show fluorescence in light of different wavelengths. For example, flourespar shows fluorescence with blue light, chlorophyll with red light, uranium glass with green light and so on.

Applications of Fluorescence

(a) *For lightening purposes in fluorescent tubes*: Ultraviolet light produced by mercury arc lamp is used in a tube coated with fluorescent salt to give visible light.

(b) *In industry*, fluorescence is used to detect and identify materials such as rubbers.

(c) Use of fluorescent microscope and fluoroscope used in X-ray diagnosis helps in testing condition of foodstuff.

(d) Aqueous fluorescent solution having intense green fluorescence is an excellent material for leak detection and excellent marker for sea rescue operations.

(e) The fluorescent dyes are mixed with coloured paints to produce extra ordinary brightness and luster.

2. **Phosphorescence**

Phosphorescence is also a radiative process through which an activated molecule loses its energy. When light radiation is incident on certain substances, they emit light continuously even after the incident light is cut-off. This type of delayed fluorescence is called phosphorescence and substances are called phosphorescent substances.

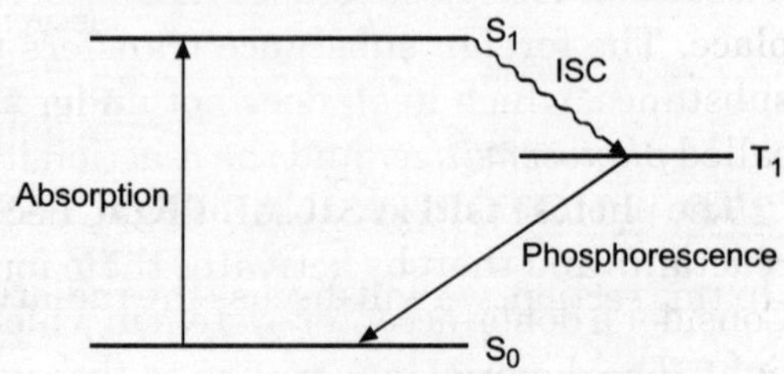

Fig. 25.6

Phosphorescence involves excitation of the electron to the singlet state (S_1) followed by the inter system crossing (ISC) to triplet state (T_1). *Phosphorescence is the radiative emission from triplet excited state to singlet ground state ($T_1 \longrightarrow S_0$) (Fig. 25.6).*

Phosphorescence differs from fluorescence in respect that emission of radiation persists for some time, of the order of seconds or more after the exciting source is removed. This is due to the fact that phosphorescence emissions arise from a triplet state. The transition from the triplet state is "forbidden" transition and this transition is normally long lived. As the triplet state is of lower energy then the singlet state, the phosphorescent emission is almost exclusively of a longer wavelength than the flourescent emission.

The longest phosphorescence lifetimes are around 30 sec (whereas the longest flourescence lifetimes are around 10^{-6} sec). Phosphorescence is more intense in solid samples because molecule gets sufficient time for crossing over from S_1 to T_1 state.

The common substances that exhibit this phenomenon are sulphides of calcium, barium strontium, and zinc. Minerals, such as ruby, emarald are also phosphorescent substances.

Difference between Flourescence and phosphorescence

Fluorescence	*Phosphorescence*
1. Fluorescence is the radiative emission of the absorbed radiation in which the molecule in the excited singlet state returns to singlet ground state, i.e. transition from $S_1 \longrightarrow S_0$.	Phosphorescence is the radiative emission of the absorbed radiation in which the molecule in the excited triplet state returns to triplet ground state, *i.e.* transition from $T_1 \longrightarrow S_0$.
2. The transition from $S_1 \longrightarrow S_0$ is 'allowed' transition and occurs in about 10^{-8} seconds.	The transition from $T_1 \longrightarrow S_0$ is 'forbidden' transition and occurs in about $10^{-3} - 30$ sec.
3. Fluorescence is exhibited by gases, liquids and solids. No flourescence will be observed in gases unless the pressure is low.	Phosphorescence is exhibited mainly by solids.
4. It stops as soon as the incident light is cut-off.	It does not stop as incident light is cut-off because life time of phosphorescence is longer.
5. The extent of fluorescence depends upon the nature of the solvent and presence of certain anions in solution.	The time for which radiation is emitted depend upon the nature of the substance and sometimes on temperature variations.

25.10 PHOTOSENSITIZED REACTIONS (ELECTRONIC ENERGY TRANSFER)

Photosensitized reactions constitute an important class of photochemical reactions. In these reactions, the reactant molecules do not absorb the radiation directly when they are exposed to light. However, if a suitable foreign substance that absorbs the radiation is added to the reactant the reaction takes place. The foreign substance transfers the absorbed radiation to the reactants. This foreign substance, which itself does not undergo any change but initiates a photochemical reaction, is called *photosensitizer* and the reaction it causes is called *photosensitized reaction*.

The photosensitizer, absorbs light, becomes excited and then passes on this energy to one of the reactants and thereby activates them for reaction, without itself undergoing a chemical change. Consider a donor-accepter system in which only the donor D (sensitizer) is capable of absorbing the light. The donor D may return to the ground state by simultaneous transfer of large quanta of energy to other species, the 'accepter' A (reactant) in the system

$$D + h\nu \longrightarrow D^*$$
$$D^* + A \longrightarrow A^* + D$$

Although the photons are absorbed by D, it is A, as the reactant molecule which becomes excited and undergoes chemical reaction. The basic requirement for energy transfer is that the triplet excited state of donor molecule should be higher than the triplet excited state of accepter molecule as shown in Fig. 25.7. Transfer of energy from singlet produces singlet and transfer from triplet produces triplet. In other words, if the acceptor molecule possesses a state of same multiplicity (singlet or triplet) and lower in energy than that of the donor molecule, energy transfer will occur from donor to acceptor.

A typical mechanism of triplet energy transfer involves excitation of donor D to excited singlet state (D'). D^1 gives the triplet excited state (D^3) by intersystem crossing (ISC). The D^3 collides with the acceptor A producing the triplet excited state of acceptor, A^3 and returns to the ground state.

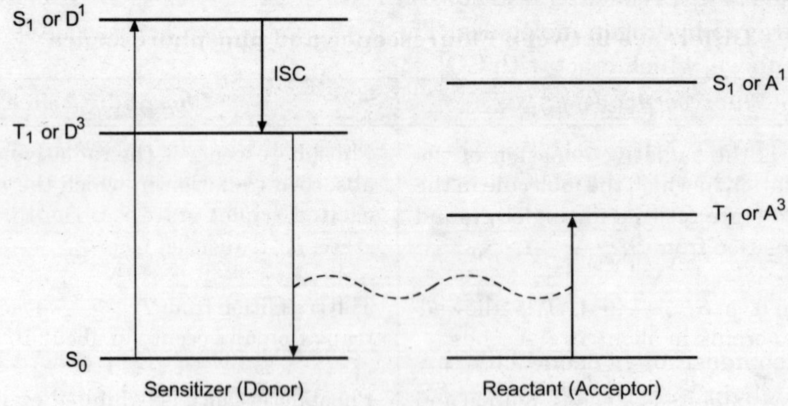

Fig. 25.7 Mechanism of photosensitization

It must be emphasized here that if triplet excited state of reactant A gives the desired product, the process is called photosensitization. If, however, the desired product results from triplet excited state of donor, then A is known as *quencher* and the process is called *quenching*. Schematically, the whole process can be represented as:

$$D + h\nu \longrightarrow D^1 \quad \text{(Singlet excited state of donor)}$$

$$D^1 \xrightarrow{\text{ISC}} D^3 \quad \text{(Triplet excited state of donor)}$$

$$D^3 + A \longrightarrow D + A^3 \quad \text{(Energy transfer)}$$

$$A^3 \longrightarrow \text{Products} \quad \text{(Photosensitization)}$$

or

$$D^3 \longrightarrow \text{Products} \quad \text{(Quenching)}$$

The phenomenon of photosensitization has proved to be of enormous value as it provides a new group of potentially reactive intermediates. A sensitizer should possess high triplet energy (T_1), high intersystem crossing and should be capable of absorbing at lower wavelengths.

The implications of photosensitization are many. The substrates which are unable to absorb light of conveniently longer wavelengths may often be sensitized by additives (photosensitizers) which absorb. In terms of energetics, it implies that compounds whose upper singlet and triplet states are widely separated, thereby making ISC difficult, may be excited to their triplet states by sensitizers having appropriate intermediate energy levels.

Examples of Photosensitized Reactions

(i) Hydrogen is unaffected when exposed to a wavelength of 253.7 nm. However, in presence of mercury vapour, it dissociates into atomic hydrogen with the radiation of the same wavelength.

$$\underset{\text{Sensitizer}}{Hg + hv} \longrightarrow \underset{\text{Excited atom}}{Hg^*}$$

$$Hg^* + H_2 \longrightarrow H_2^* + Hg$$

$$H_2^* \longrightarrow 2H^*$$

(ii) When a mixture of carbon monoxide and hydrogen containing traces of mercury vapours are exposed to ultraviolet light of 253.7 nm, the mercury atoms are excited and transfer their energy to hydrogen molecules. The hydrogen molecule dissociates into hydrogen atoms as above, which react with CO.

$$Hg + h\nu \longrightarrow Hg^* \quad \text{Primary step}$$

$$Hg^* + H_2 \longrightarrow H_2^* + Hg \quad \text{Energy Transfer}$$

$$\left.\begin{aligned} H_2^* &\longrightarrow 2H^\bullet \\ H^\bullet + CO &\longrightarrow HCO^\bullet \\ HCO^\bullet + H_2 &\longrightarrow HCHO + H^\bullet \\ 2HCO^\bullet &\longrightarrow HCHO + CO \end{aligned}\right] \text{Reaction}$$

(iii) In the decomposition of ozone, chlorine acts as a photosensitizer under the influence of ultraviolet light,

$$2O_3 \xrightarrow[h\nu]{Cl_2} 3O_2$$

The mechanism may be written as follows:

$$Cl_2 + h\nu \longrightarrow Cl_2^* \longrightarrow 2Cl$$

$$Cl + O_3 \longrightarrow ClO_3^*$$

$$ClO_3^* \longrightarrow ClO_2 + 2O_2$$

$$ClO_2 + O_3 \longrightarrow ClO_3 + O_2$$

(iv) Isomerization of Butene–2 : The cis-butene–2 undergoes isomerization to yield trans-butene-2 when exposed to radiation of 253.7 nm in the presence of sulphurdioxide as photosensitizer.

$$SO_2 \xrightarrow{h\nu} SO_2^*$$

$$SO_2^* + \text{Cis-Butene–2} \longrightarrow SO_2 + \text{Cis–Butene–2}^*$$

$$\text{Cis – Butene–2}^* \longrightarrow \text{trans–Butene–2}$$

(v) Chlorophyll as photosensitizer in photosynthesis: The photochemical reaction between carbon dioxide and water taking place in plants is photosensitized by chlorophyll. Chlorophyll absorbs strongly in the region between 600 to 700 nm.

$$CO_2 + H_2O \xrightarrow{h\nu} \underset{\text{Formaldehyde}}{CH_2O} + O_2$$

$$6CH_2O \longrightarrow C_6H_{12}O_6$$

25.11 LASERS AND MASERS

LASER stands for **l**ight **a**mplification by **s**timulation **e**mission of **r**adiation.
MASER stands for **m**icrowave **a**mplification by **s**timulated **E**mission of **r**adiation.

Actually, a LASER is a MASER that work with higher frequency photons in the ultraviolet or visible light spectrum.

When the laser was invented, in 1960, it was classified as a ***solution in search of a problem,*** and today laser technology is applied in many different areas such as: medicine, communication, daily use, military and industry. To explain how the laser can be applied in such diverse areas, we need to understand the *basic physical principles of the operation of a laser*.

(a) **Absorption**

Imagine that we have an atom in a lower energy state, and a photon comes along. If this photon has just the right amount of energy, it can be absorbed, causing the atom to make a transition to a higher energy state. This is illustrated in the Fig. 25.8. This will only occur though if the photon has an energy equal to the energy difference between the two levels involved in the transition — if not, the photon will just pass through. This process is called absorption. According to quantum mechanics, "absorption of photons by atoms occurs only if the wavelength of the photon is just the right size (say, of wavelength λ). If it is, the atom will "absorb" it and go to a higher energy state."

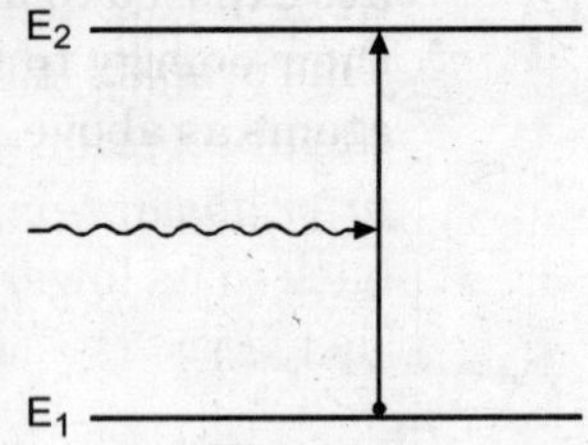

Fig. 25.8

(b) **Spontaneous Emission**

Atoms do not like to stay in high energy states, so after absorbing a photon and going to a high energy state, they will move of their own accord to a lower energy state, emitting a photon in the process. This is called "*spontaneous emission*" because no outside influence triggers the emission.

Normally the average life time for spontaneous emissions by excited atoms is around 10^{-8} seconds (that is, the atom or molecule will usually take around 10^{-8} seconds before emitting the photon).

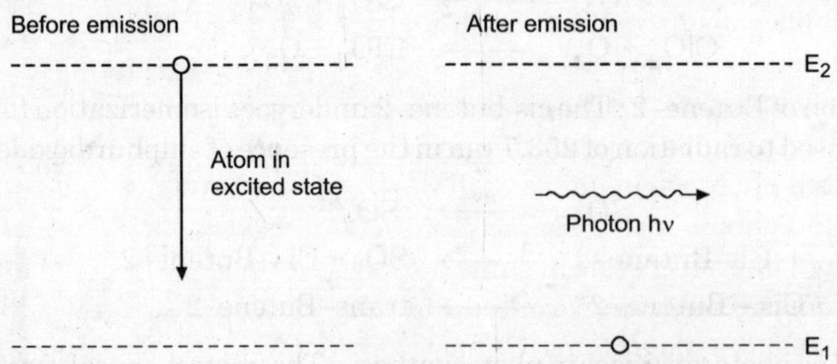

Fig. 25.9

Occasionally, however, there are states for which the life time is much longer, perhaps around 10^{-3} seconds. These states are called *metastable*. Metastable emission levels are essential for a working MASER and will be discussed further.

If the atom is in the excited state with energy E_2, it may spontaneously decay into the **ground state,** with energy E_1, releasing the difference in energies between the two states as a photon. The photon have **frequency** v **and energy** hv, given by:

$$E_2 - E_1 = hv,$$

where **h** is Planck's constant.

(c) **Stimulated Emission**

A third type of photon-related electron transition in an atom is *stimulated emission.* Suppose an atom is in a higher energy state and another photon comes along with an energy equal to the difference between the higher energy state and ground energy state. The photon will stimulate the atom to fall into the lower energy state, thereby emitting another photon from the excited atom.

This process is known as **stimulated emission.** With stimulated emission, a photon of the absorption wavelength λ, is fired at an atom *already in its high energy state from prior absorption.* The atom absorbs this photon, and then quickly emits *two* photons to get back to its lower energy state (Fig. 25.10).

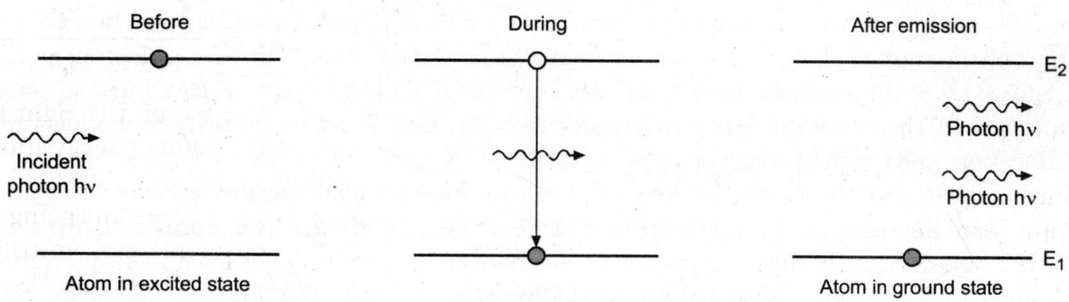

Fig. 25.10 Stimulated emission

Both of these newly emitted photons are of wavelength λ. Stimulated emission is very uncommon process in nature but it is central to the operation of lasers. Furthermore, the two emitted photons are said to be in phase, the crest of the wave associated with one photon occurs at the same time as on the wave associated with the other. Such waves will constructively interfere, leading to a more intense wave.

The stimulated photons have unique properties:

- the emitted photon is in phase with the incident photon.
- the emitted photon has the same wavelength as the incident photon.
- the emitted photon travels in same direction as incident photon.

In a laser atoms are kept in an excited state by "pumping" the laser, and some photons are inserted. This causes some atoms to undergo stimulated emission, and the resulting photons cause other atoms to undergo stimulated emission, leading to a chain reaction. The resultant light is very intense and coherent (composed of one frequency), and can be easily focused.

(d) **Inverted Population or Population Inversion**

For laser action to take place the *stimulated emission* should *predominate* over the *spontaneous emission*. If stimulated emission is to predominate, we must have more atoms in the higher energy state than in the lower one. Hence the atoms or molecules are excited so that more of them are at higher energy levels than are at lower energy levels. This unusual condition, referred to as a *population inversion,* is necessary to create a population inversion for laser action to occur. The process of adding energy to produce an inverted population is called *pumping*. Once the atoms or molecules are in this excited state, they readily emit radiation (Fig. 9).

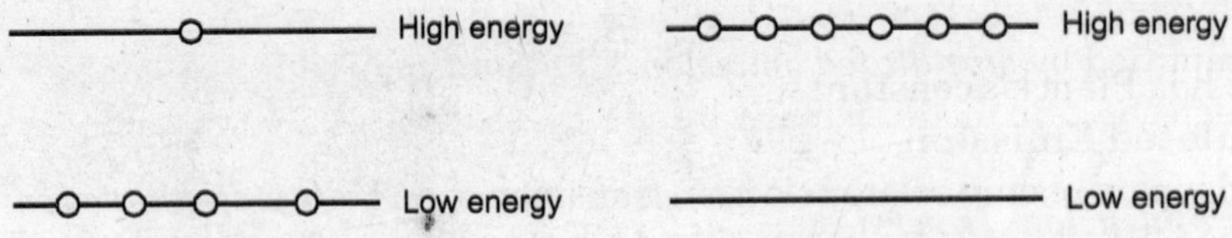

(a) Normal population (b) Inverted population

Fig. 25.11

Amplification of Light: The Laser

Stimulated emission is the process that can give rise to the amplification of light. As is discussed above, if photon whose frequency corresponds to the energy difference between the excited state and the ground state strikes an excited atom, the atom is stimulated to emit a second photon of the same frequency, *in phase with and in the same direction as the bombarding photon.* The bombarding photon and the emitted photon may then each strike other excited atoms, stimulating further emissions of photons, all of the same frequency and all in phase. This produces a sudden burst of *coherent radiation* as all the atoms discharge in a rapid chain reaction. It is this property that allows optical amplification, and the production of a laser system.

During the operation of laser, all three light-matter interactions described above are happening. Initially atoms are energized from the ground state to the excited state by a process called *pumping*, described below. Some of these atoms decay *via* spontaneous emission, releasing incoherent light as photons of frequency υ. These photons are fedback into the laser medium, usually by an **optical resonator**. Some of these photons are absorbed by the atoms in the ground state, and the photons are lost to the laser process. However, some photons cause stimulated emission in excited-state atoms, releasing another coherent photon. In effect, this results in *optical amplification*.

If the number of photons being amplified per unit time is greater than the number of photons being absorbed, then the net result is a continuously increasing number of photons being produced: the laser medium is said to have a gain of greater than unity.

25.11.1 Construction of Lasers

A *laser* system generally consists of three important part:

- An energy source (usually referred to as the *pump* or *pump source*)
  The *pump source* is the part that provides energy to the laser system. Examples of pump source include electrical discharges, flashlamps, arc lamps, light from another laser, chemical reactions and even explosive devices. The type of pump source used depends on the *gain medium*, and this also determines how the energy is transmitted to the medium.
- A *gain medium* or *laser medium*
  The *gain medium* is the major determining factor of the wavelength of operation, and other properties, of the *laser*. The gain medium is excited by the pump source to produce a *population inversion*, and it is in the gain medium that spontaneous and *stimulated emission* of photons takes place, leading to the phenomenon of optical gain, or amplification.

- A mirror, or system or mirrors, forming an *optical resonator*.
  The *optical resonator*, or *optical cavity*, in its simplest form is a system of two parallel mirrors placed around the gain medium which provide *feedback* of the light. Light from the medium, produced by *spontaneous emission*, is reflected by the mirrors back into the medium, where it may be amplified by *stimulated emission*. The light may reflect from the mirrors and thus pass through the gain medium many hundreds of times before exiting the cavity.

Creating a Population Inversion

As described above, a population inversion is required for laser operation, but cannot be achieved in our theoretical group of atoms with two energy-levels when they are in thermal equilibrium. In fact, any method by which the atoms are directly and continuously excited from the ground state to the excited state (such as optical absorption) will eventually reach equilibrium with the de-exciting processes of spontaneous and stimulated emission.

Assume there is a group of N atoms, each of which is capable of being in one of two **energy** states, either

1. The *ground state,* with energy E_1, or
2. The *excited state,* with energy E_2, with $E_2 > E_1$.

The number of those atoms which are in the ground state is given by N_1, and the number in the excited state by N_2. Since, there are N atoms in total.

$$N_1 + N_2 = N$$

At best, an equal population of the two states $N_1 = N_2 = \frac{N}{2}$, can be achieved, resulting in optical transparency but no net optical gain.

To achieve non-equilibrium conditions, an indirect method of populating the excited state must be used, which may include:

(a) Three-Level Lasers

Consider a group of N atoms, with each atom able to exist in any of three energy states, levels 1, 2 and 3, with energies E_1, E_2 and E_3 and population N_1, N_2 and N_3 respectively.

Note that $E_1 < E_2 < E_3$, that is the energy of level 2 lies between that of the ground state and level 3.

Initially, the system of atoms is at thermal equilibrium and the majority of the atoms will be in the ground state. If we now subject the atoms to light of a frequency υ, where $E_3 - E_1 = h\upsilon$, the process of optical absorption will excite the atoms from the ground state to level 3. This process is called *pumping*. Apart from optical absorption other methods of exciting the laser medium, such as electrical discharge or chemical reactions may also be used for exciting the laser medium. The level 3 is sometimes referred to as the *pump level* or *pump band,* and the energy transition E_1 — $\rightarrow E_3$ as the *pump transition*, which is shown as the arrow marked P in Fig. 25.12.

If we continue pumping the atoms, we will excite an appreciable number of them into level 3, such that $N_3 > 0$. In a medium suitable for laser operation, we require these excited atoms to quickly decay to level 2, which is known, as the *metastable state*. The energy released in this transition is non-radiative, with the energy being transferred to vibrational motion **(heat)** of the host material surrounding the atoms.

An atom in level 2 may decay by spontaneous emission to the ground state, releasing a photon of frequency υ_{21} (given by $E_2 - E_1 = h\upsilon_{21}$), which is shown as the transition L, called the *laser transition* in the diagram.

It is to be noted that for an efficient laser action, the transition from level 3 to level 2 must be very rapid compared to the transition from level 2 to level 1 so that a population of excited state atoms will accumulate in level 2.

If over half the N atoms can be accumulated in this state, this will exceed the population of the ground state N_1. A population inversion ($N_2 > N_1$) has thus been achieved between level 1 and 2, and optical amplification at the frequency υ_{21} can be obtained.

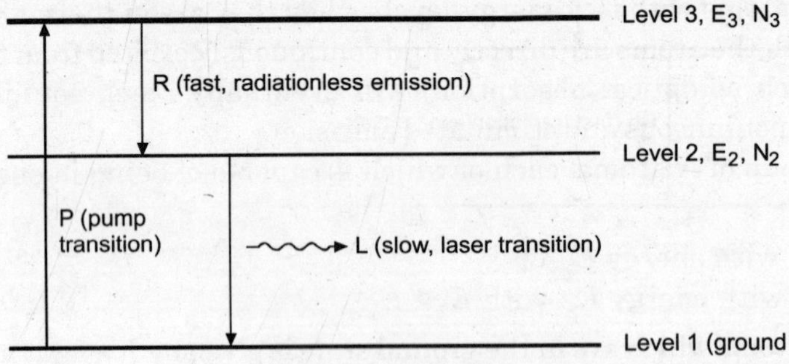

Fig. 25.12 A three-level laser energy diagram

Because at least half the population of atoms must be excited from the ground state to obtain a population inversion, the laser medium must be very strongly pumped. This makes three-level lasers rather inefficient, despite being the first type of laser to be discovered (based on a **ruby** laser medium by Theodore Maiman in 1960. In practice, most lasers are *four-level lasers.*

(b) Four-Level Lasers

Here, there are four energy levels, energies E_1, E_2, E_3, E_4, and populations N_1, N_2, N_3, N_4, respectively. The energies of each level are such that $E_1 < E_2 < E_3 < E_4$.

In this system, the pumping transition P excites the atoms in the ground state (level 1) into the pump band (level 4). From level 4, the atoms again decay by a fast, non-radiative transition Ra into the level 3. Since, the laser transition from level 3 to level 2L is slow as compared to that of Ra, a population accumulates in level 3 (*the upper laser level*), which may relax by spontaneous or stimulated emission into level 2 (*the lower laser level*). This level likewise has a fast, non-radiative decay **Rb** into the ground state. Thus the laser transition terminates in level 2 and not in the ground state (Fig. 25.13).

As before, the presence of a fast, radiationless decay transitions result in population of the pump band being quickly depleted ($N_4 \approx 0$). In a four-level system, any atom in the lower laser level E_2 is also quickly de-excited, leading to a negligible population in that states ($N_2 \approx 0$). This is important, since any appreciable population accumulating in level 3, the upper laser level, will form a population inversion with respect to level 2. That is, as long as $N_3 > 0$, then $N_3 < N_2$ and a population inversion is achieved, the optical amplification, and laser operation can take place.

Since, only a few atoms must be excited into the upper laser level to form a population inversion, a four level laser is much more efficient than a three-level one, and most practical lasers are of this type. In reality, many more than four energy levels may be involved in the laser process with complex excitation and relaxation processes involved between these levels.

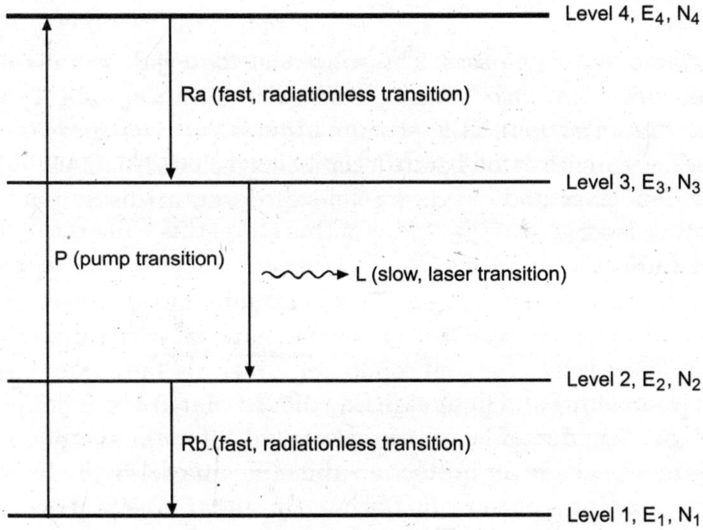

Fig. 25.13 A four-level laser energy diagram

Note that in both three- and four-level lasers, the energy of the pumping transition is greater than that of the laser transition. This means that, if the laser is optically pumped, the frequency of the pumping light must be greater than that of the resulting laser light. In other words, the pump wavelength is shorter than the laser wavelength.

25.11.2 Types of Lasers

On the basis of gain medium, lases can be classified as:

- **Gas Lasers**
  They consist of a laser tube made of quartz, glass, ceramic, or in special cases, even metals. It each end of the tube are the laser mirrors, and the tube is also fitted with electrodes for applying a high voltage to the gas. Gas laser tubes are electrically similar to common fluorescent lights or neon signs. When high voltage is applied, the gas in the tube ionizes and produces a glowing light. In the gas laser, the perfectly aligned mirrors gather and concentrate a specific wavelength of light produced by the gas into an intense laser beam. Gas lasers produce extremely high quality laser beams, many of which are of very high power, suitable for gases, such as *carbon dioxide, argon, krypton* and mixtures such as *helium-neon* are used.
- **Dye Lasers**
  Liquid lasers use fluorescent dyes to produce laser beams of many different colours. The dye is flowed through a glass tube with laser mirrors positioned on each end. Dye lasers can be tuned to produce almost any colour of light by changing the type and concentration of the dye used, and by using special mirror arrangements designed to allow only a specific wavelength of light to circulate within the laser. Dye lasers find widespread use in scientific research and in medicine where different types of living tissue respond to different wavelengths of laser light.

  The liquids are usually *organic* chemical *solvents, such* as methanol, ethanol or ethylene glycol, to which are added chemical dyes such as *coumarin rhodamine* and *fluorescein*. The exact chemical configuration of the dye molecules determines the operation wavelength of the *dye laser*.

- **Solid-State Lasers**
  Solids, such as *crystals* and *glasses*. The solid *host* materials are usually doped with an impurity such as *chromium*, *neodymium*, *erbium* or *titanium ions*. Typical hosts include *YAG* (*yttrium aluminium garnet*), YLF (yttrium *lithium fluoride*), *sapphire* (aluminium oxide) and various glasses. Examples of solid-state laser media include Nd : YAG, *strontium* aluminium fluoride), Er : YLF, Nd : glass, and Er : glass Solid-state lasers are usually pumped by flashlamps or light from another laser.
- **Semiconductor Lasers**
  Today, the most widespread lasers are of the semiconductor type. *Semiconductors*, a type of solid, in which the movement of *electrons* between material with differing dopant levels can cause laser action. These lasers use semiconductor materials similar to those found in transistors, integrated circuit chips and light emitting diodes that are commonplace in today's electronic equipment. Semiconductor lasers are very small, often the size of a grain of salt, so they are mounted in larger package for protection and efficient cooling. In most cases they do not use separate mirrors as larger lasers do. Rather, the ends of the semiconductor material are cut so that the end faces are perfectly parallel, and then coated to enhance reflectivity.
  Semiconductor lasers are typically very small, and can be pumped with a simple electric current, enabling them to be used in consumer devices such as *compact disc* players.

25.11.3 Some Common Lasers

Let's have a brief look at the main type of laser and how they are used as tools of modern technology.

(a) Crystal Lasers

Crystal lasers, also known as "solid state" lasers, use a rod of crystal material positioned between a pair of mirrors. The energy source can be a powerful flash lamp, an arc lamp or even a bank of high power light-emitting diodes that pump energy into the crystal. Maiman's first laser consisted of a rod of synthetic ruby with reflective coatings at each end that served as mirrors. A high power flash lamp pumped energy into the rod. When fired, the device generated a pulsed red laser beam that was powerful enough to pierce a razor blade. Since then, hundreds of exotic crystal types have been successfully used in lasers.

(b) Helium-neon laser

Helium-neon laser, usually called a HeNe laser, is type a type of small *gas laser*. Its usual operation *wavelenth* is 632.8 *nm*, in the *red* portion of the *visible spectrum*.

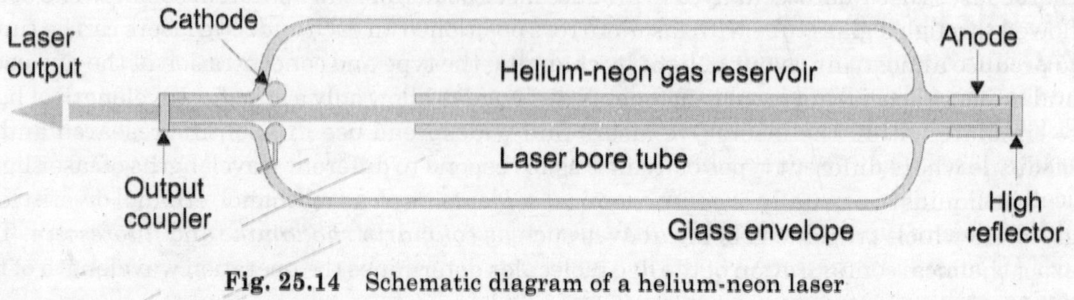

Fig. 25.14 Schematic diagram of a helium-neon laser

The *gain medium* of the laser, as suggested by its name, is a mixture *helium* and *neon* gases, approximately in the ratio 5 : 1, contained at low pressure in glass envelope. The energy or pump source of the laser is provided by an *electrical* discharge of around 1000 *V* through an *anode* and *cathode* at each end of the glass tube. The *optical cavity* of the laser typically consists of a plane, high-reflecting *mirror* at one end of the laser tube, and a concave output coupler mirror of approximately 1% transmission at the other end.

HeNe laser are typically small, with cavity lengths of around 15 cm up to 0.5 m, and optical output *power* ranging from 1 mW to 100 mW.

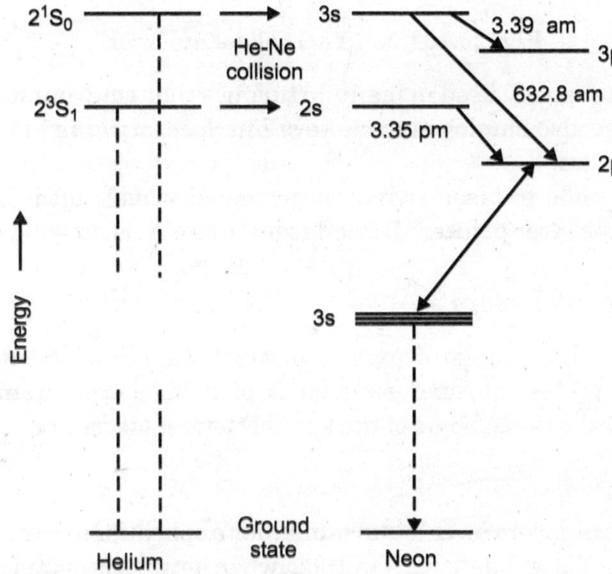

Fig. 25.15 Energy level diagram of a HeNe laser

The laser process in HeNe laser starts with collision of *electrons* form the electrical discharge with the helium atoms in the gas. This excites helium from the *ground state* to the 2^3S_1 and $2^1\,S_0$ long-lived, *metastable* excited states. Collision of the excited helium atoms with the ground-state neon atoms results in transfer of energy to the neon atoms, exciting them into the 2s and 3s states. This is due to a coincidence of *energy levels* between the helium and neon atoms.

HeNe lasers have many industrial and scientific uses, and are often used in *laboratory* demonstrations of *optics*.

(c) Nd:YAG Laser

Nd:YAG is an acronym for *neodymium*-doped *yttrium aluminium garnet* ($Nd : Y_3Al_5O_{12}$), a compound that is used as the *lasing medium* for certain solid-state *lasers*.

Nd:YAG laser are optically pumped using a flashlamp or laser diodes. They are the example of a four level laser. They are one of the most common types of laser, and are used for many different applications.

Nd:YAG lasers typically emit light with a *wavelength* of 1064 *nm* in the *infrared*.

Nd:YAG lasers are commonly used in the medical field and in manufacturing as a means of engraving, etching, or marking a variety of metals and plastics. These lasers are also used extensively in the field of cosmetic medicine for hair removal and the treatment of minor vascular defects such as spider veins on the face and legs.

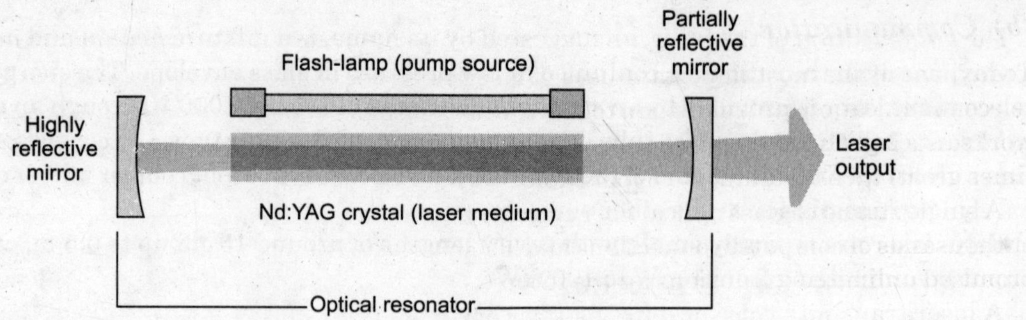

Fig. 25.16 Nd: YAG solid-state laser

Nd:YAG lasers are extensively used in manufacturing for cutting and welding steel and super alloys. Nd:YAG lasers are also employed to make *subsurface marking* in transparent materials such as *glass* or *acrylic glass*.

Nd:YAG can be also made to laser at its non-principal wavelength. The line at 946 nm is typically employed in "blue laser pointer" DPSS lasers, where it is doubled to 473 nm.

25.11.4 Application of Lasers

Lasers are employed over a huge range of applications including basic scientific research, test and measurement, industrial processing, microelectronics, biomedicine, environmental science, entertainment and telecommunications. Some of the applications of lasers are:

(a) In Medicine

Since their first appearance, lasers have found numerous applications in the field of medicine. The first medical applications of lasers date back to 1962, when a ruby laser was used for reattachment of the retina in the eye. Use of lasers as diagnostic tools and sensors is one of the fastest growing branches of biomedical laser development. Lasers have possibilities for seeing through subsurface media, therefore making it thinkable to diagnose cancers before they become visible. They also have the potential for treating skin cancers before they become dangerous.

In medicine, they have been used in surgical operations Surgeons use lasers to remove deceased body figure, with little damage to surrounding area. In addition lasers seal the blood vessels severed during the surgery, thus reducing the amount of bleeding. Laser eye surgery is becoming common for correcting eye defects. Beam from carbon dioxide lasers is heavily absorbed by the water in human tissue, making it an excellent cutting tool for the surgeon. The heat of the beam repairs tissue as it cuts, resulting in almost bloodless surgery. CO_2 lasers can be used to repair cuts from conventional surgical procedures. CO_2 lasers are also used extensively for removing scars and other skin lesions.

Other commonly used lasers in medicine are Excimer, several different harmonic of YAG laser well as carbon dioxide lasers.

One other medical area-dentistry-is begining to benefit from laser technology. Dentists have now started using lasers to remove small cavities. And the technology is beginning to find its way into the cosmetic end of the profession.

In dermatology, lasers are being used to remove tattoos.

(b) Communication

Today, one of the most important uses of lasers is in the field of communications. In the 1980's telecommunication systems began to move away from copper wires in favour or fiber optics. Laser works as a light source in all fiber optics in use. It has greater bandwidth (potentially 100,000 times great) then an ordinary copper cable.

A single strand of glass optical fiber can carry more than half a million telephone conversations, or thousands of computer connections and TV channels. Thinner than a human hair, fiber optics promised unlimited expansions for the future.

A laser transmits voice or data via fiber optic cables at much improved speed and capacity. These lasers are part of the broadband revolution we hear about daily. Other everyday, low power applications for lasers are CD and DVD players.

(c) In Computer Industry

Semiconductor lasers form the heart of every CD and DVD player in the world. Digital information on a commercial CD or DVD is in the form of microscopic pits on an otherwise reflective metallic substrate. When a focussed laser beam scans the disk surface the pattern of pits can be easily be read by an optical sensor and processed to form an audio or video signal.

CD-ROM or DVD-ROM writers operates in a similar way, but instead of a metallic substrate, recordable disks use a light sensitive coating similar to photographic field. To record, the laser is operated at a higher power level to fix a pattern of digital ones and zeros onto the disk. To read the data, the laser is operated at a lower power that is not sufficient to affect the coating, but bright enough to be reflected form the pre-recorded data so an optical sensor can detect it.

A laser printer is a common type of *computer printer* that reproduces high quality text and graphics on plain paper. Laser printers employ a *xerographic* printing process but differ from most *photocopiers* in that the image is produced by the direct scanning of a laser beam across the printer's internal drum.

(d) In Industry

The light beam produced by most lasers is pencil-sized, and maintains its size and direction over very large distances; this sharply focused beam of coherent light it suitable for a wide variety of applications. Lasers have been used in industry for cutting and boring metals and other materials, and for inspecting optical equipment.

In cutting and welding applications, lasers can provide some distinct advantages over other methods. Lasers produce smooth, clean cuts, unlike cutting torches. The heat applied to a metal work-piece is sharply focussed so it affects only a small area around the cut. With a cutting torch, the heat-affected zone is larger. The metal in the heat-affected zone becomes more brittle than the rest of the work-piece, so in many applications such as aircraft parts, the heat affected metal must be removed by grinding

Commonly used lasers in industrial environments are Excimer, Neodymium YAG, and carbon dioxide

(e) Scientific Research

Laser have opened a new field of scientific research, nonlinear optics, which is concerned with the study of such phenomena as the frequency doubling of coherent light by certain crystals. One important result of laser is the development of lasers that can be tuned to emit light over a range of frequencies, instead of producing light of only a single frequency.

(f) Nuclear Fusion

Perhaps the most extravagant use of lasers in science is in the field of fusion research by concentrating the laser beam to a very narrow spot. Temperature may increase to about 10° K, which is the temperature required for nuclear fusion. This technique, known as *inertial confinement fusion* so far has not been able to achieve *breakeven,* that is less power is generated by the fusion reaction than is used to power the lasers. but research continues.

Lasers have been used in several kinds of scientific research. The field of *holography* is based on the fact that actual wave front patterns, captured in a photographic image of an object illuminated with laser light, can be reconstructed to produce a three-dimensional image of the object.

(g) Defense Applications

Recently, some progress has been made in the use of the laser as a directed energy weapon, mostly in defensive applications. One example is a laser system which is designed to destroy missiles in flight. It is mounted in a converted commercial airliner, and could be used, for example, to protect assets or to destroy ballistic missiles.

The *Mobile Tactical High-Energy Laser* (MTHEL) is another defensive laser system under development. This is field-deployed weapon system able to track incoming *artillery* projectiles *cruise missiles* by *radar* and destroy them with a powerful *deuterium fluoride laser*.

25.11.5 Masers

As discussed previously, a maser is any optical amplifier that produces coherent microwaves due to stimulated emission. The difference between a maser and a laser is that the laser utilises visible light.

As in lasers, amplification of radiation in masers is obtained by *stimulated emission*. Masers make use of those transitions in *molecules* or crystals that correspond to the energies of microwave or radio frequency.

On Earth masers and lasers are relatively difficult things to produce requiring special conditions to produce them. However, the physical conditions of interstellar space (very low density of gas and very low temperature) are very different to those on Earth and in some environments masers occur naturally. Helium emission from these naturally occurring masers is very bright, which makes them relatively easy to find. The masers are also usually very small and have a size less than the distance from the Earth to the sun.

In 1963 Harold Weaver and his radio astronomers at Berkeley were studying the radio waves emitted by hydroxyl (OH) groups in the Orion Nebula, the nearest region in the galaxy where stars are still being formed. Hydroxyl emits some of the brightest radiation of all the molecules. Weaver and his colleagues were surprised to find that one part of the nebula showed a strange emission spectrum. Its line intensities were all wrong, and one was so strong that Weaver suspected it of belonging to a different molecule altogether. They called it Mysterium.

Before long, however the evidence was clear Mysterium was simply OH. But the OH had formed a *natural maser*. The molecules were being pumped by infrared light from nearby gas clouds that were collapsing to form stars. Since then, hundreds of interstellar OH masers have been discovered. The first maser oscillator was developed by the American physicists Charles Hard Townes, James P. Gordon, and Herbert J. Zeiger in 1954, and made use of the *frequency of the ammonia molecule*.

Applications of Masers

- Because of the high stability of the generated frequencies, masers serve as time standards in atomic clocks.
- Masers are also used as low-noise radio frequency amplifiers in satellite communication and *radio astronomy*.

Some Interesting Techniques Involving LASERS

LASIK

Today, a surgical technique that is gaining popularity is laser vision correction, or LASIK. In less than 15 minuets this procedure can greatly reduce a patient's reliance on corrective eyeglasses. In some cases after surgery, eyeglasses will no longer be required. The surgeon first uses a high-precision instrument to separate the layers of corneal tissue and create a thin flap on the surface on the surface of the eye.

This flap is folded back, and the ultraviolet beam from an eximer laser is guided to gently reshape the underlying tissue. The cornea's new shape is what improves vision. After reshaping is complete, the flap is replaced and the patient can resume most normal activities within 24 hours.

LITHOTRIPSY

Optical fibers can deliver laser beams inside the body to reduce the need for more invasive surgery. An interesting example is the destruction of painful kidney or gall stones in a process called lithotripsy. An optical fiber is first passed through the urethra or bile ducts to the location of the stone. High power pulses from a tunable liquid dye laser cause intense shock waves within the stones that break them into tiny fragments, allowing them to pass painlessly from the body.

SCANNERS

Almost every product you buy been tagged with a universal product code. or UPC. At the checkout A clerk scans the bar code to tally up your purchases. The scanners use a small semiconductor laser that is repeatedly scanned across the bar code to illuminate it. An optical sensor reads the series of black and white stripes of the bar code and converts it to digital data. Which is passed to the store's computer system. Not only does this system tell you how much you must pay, but it also helps the store track and control its inventory in a fraction of the time of past methods.

25.12 PHOTOSYNTHESIS

Photosynthesis is the most important biological process on earth. Nearly all living things depend on energy produced from photosynthesis for their nourishment, making it vital to life on *Earth*. It is also responsible for producing the *oxygen* that makes up a large portion of the *Earth's atmosphere*.

Photosynthesis is an important *biochemical* process in which *plants, algae* some *bacteria,* and some *protists,* which contain the green pigment chlorophyll, convert the energy of *sunlight* to chemical energy.

In photosynthesis, solar energy is converted to chemical energy. The chemical energy is stored in the form of glucose (sugar). Carbon dioxide, water and sunlight are used to produced glucose, oxygen, and water. The chemical equation of this process is:

$$6CO_2 + 12H_2O + light \rightarrow C_6H_{12}O_6 + 6O_2 + 6H_2O$$

6 molecules of carbon dioxide ($6CO_2$) and 12 molecules of water ($12H_2O$) are consumed in the process, while glucose ($C_6H_{12}O_6$), six molecules of oxygen ($6O_2$), and six molecules of water ($6H_2O$) are produced.

Since photosynthesis requires carbon dioxide, water, and sunlight, all of these substance must be obtained by or transported to the leaves. Carbon dioxide is obtained through tiny pores in plant leaves called stomata. Oxygen is also released through the stomata. Water is obtained by the plant through the roots and delivered to the leaves through *vascular plant tissue systems.* Sunlight is absorbed by chlorophyll, a green pigment located in plant cell structures called *chloroplasts. Chloroplasts* are the sites of photosynthesis.

Chloroplasts contain several structures, each having specific functions:

Outer and Inner Membranes: Protective coverings that keep chloroplast structures enclosed.

Stroma: Dense fluid within the chloroplast. Site of conversion of carbon dioxide to sugar.

Thylakoid: Innermost membrane of the chloroplast. It is flattened sac-like membrane structure. Sites of conversion of light energy to chemical energy.

Grana: Dense layered stacks of thylakoid sacs. Sites of conversion of light energy to chemical energy.

Chlorophyll: A green pigment within the chloroplast. Absorbs light energy.
The thylakoid is the structural unit of photosynthesis.

Thylakoids are stacked like pancakes in stacks known collectively as grana. The areas between grana are referred to as stroma.

Chlorophyll looks green because it absorbs red and blue light.

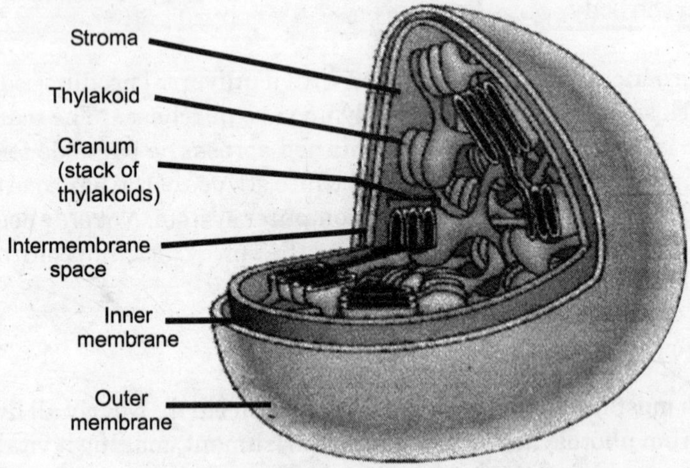

Fig. 25.17 Structure of chloroplast

25.12.1 Stages of Photosynthesis

Photosynthesis occurs in two stages. These stages are called the light reactions and the dark reactions. The light take place in the presence of light. The dark reaction do not require direct light, however dark reactions in most plants occur during the day.

Light Reactions

Light reaction occur mostly in the thylakoid stacks of the grana. Here, sunlight is converted to chemical energy. The energy harvested *via* the light reaction is stored by forming a chemical called *ATP* (*adenosine triphosphate*), a compound used by cells for energy storage. This chemical is made of the nucleotide adenine bonded to a ribose sugar, and that is bonded to three phosphate groups.

The light reactions are a series of reactions which occur in the grana, and require the direct energy of light to make energy-carrier molecules that are used in the second process:

- light energy is trapped by chlorophyll to make ATP (photophosphorylation)
- at the same time water is split into oxygen, hydrogen ions and free electrons:

$$2H_2O \rightarrow 4H^+ + O_2 + 4e^- \text{ (photolysis)}$$

- the electrons then react with a carrier molecule nicotinamide adenine dinucleotide phosphate (NADP), changing it from it oxidised state ($NADP^+$) to its reduced state (NADPH):

$$NADP^+ + 2e^- + 2H^+ \rightarrow NADPH\ ^+\ H^+$$

Dark Reactions

Dark reactions are the light-independent reactions, which occur in the stroma of the chloroplasts. During dark reactions, the products of the light reaction, ATP and NADPH, are used to make carbohydrates from carbon dioxide (reduction). This process is known as *carbon fixation* or the Calvin cycle. Carbon dioxide is combined with a 5-carbon sugar creating a 6-carbon sugar. The 6-carbon sugar is eventually broken-down into two molecules, glucose and fructose. These two molecules make sucrose or sugar.

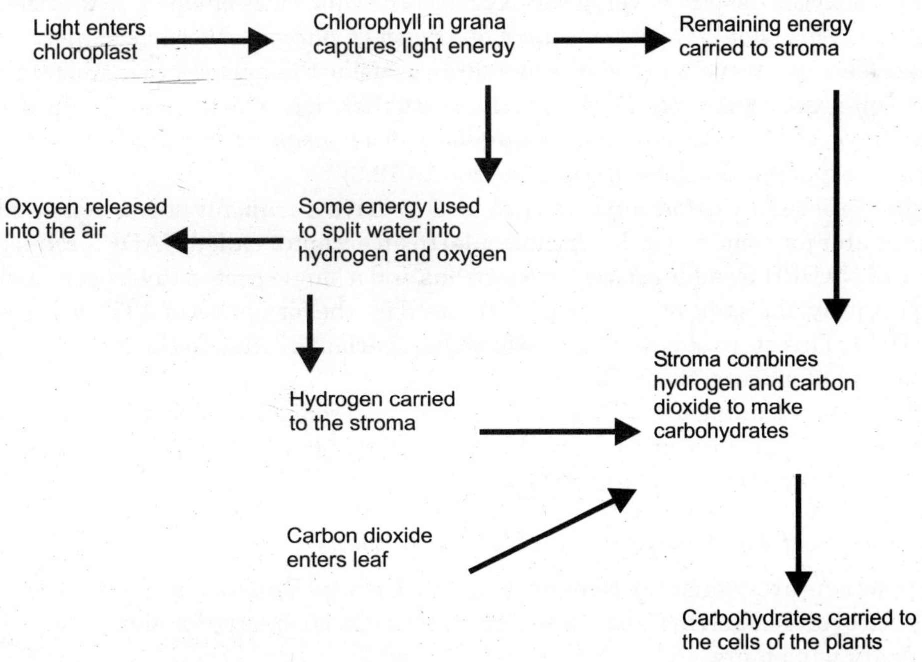

Fig. 25.18 Summary of stage of photosynthesis

25.12.2 Mechanism of Photosynthesis

In view to the above steps, we can discuss the mechanism of photosynthesis in detail. As is evident from the above section, the photosynthesis primarily involve two chemical stages:

(1) Oxygen liberation (Light reactions)
(2) Consumption of carbon dioxide (Dark reactions)

1. Oxygen Liberation (Light Reactions)

The purpose of the light reactions is to change light energy into chemical energy of ATP and NADPH along with the liberation of oxygen.

- When sunlight strikes a chlorophyll complex (cluster of chlorophyll), the molecule is excited, and an electron is transferred to a higher energy level of the molecule.

  The central part of the chemical structure of a chlorophyll molecule is a *porphyrin ring*, which consists of several fused rings of carbon and nitrogen with a magnesium ion in the centre.

  Only chlorophyll of the reaction centre is capable of transferring an electron to an electron acceptor

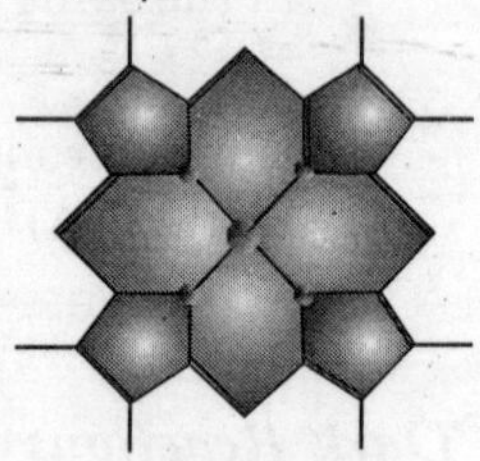

Fig. 25.19

- Light energy entering the plant splits the water into hydrogen and oxygen;

$$H_2O + \text{light energy} \rightarrow 1/2\ O_2 + 2H^+ + 2 \text{ electrons}$$

  Early in the 20th century, researchers took advantages of the use of isotopes to better understand the basic equation of photosynthesis. It was discovered that when carbon dioxide was labelled with a heavy isotope of oxygen, only the lighter isotope was emitted from the plant as oxygen gas. However, if the oxygen of the water was labelled, so was the oxygen gas emitted. *This showed that the source of oxygen in photosynthesis is water.*

- These electrons travel through the membrance, using their energy to pump protons through the membrane. The proton gradient thus established can be used to synthesize ATP. The generation of ATP is by powering the addition of a *phosphate* group to *ADP*, a process called photophosphorylation (discussed in section 25.12.3).
- Light absorbed by *chlorophyll* is also used to drive a transfer of *electrons* and hydrogen from water (or some other donor molecule) to an acceptor called $NADP^+$, reducing it to the form of NADPH by adding a pair of electrons and a single proton (hydrogen nucleus). More importantly, the same electron which are used for the formation of ATP, reduces $NADP^+$ to NADPH. This store of power serves to reduce carbon dioxide to the more complex carbon structure of glucose, the building block of life.
- The reactions leading to the production of ATP and reduction of $NADP^+$ are called the *light reaction* because they are initiated by the splitting of water by light energy. ATP and NADPH carry energy to the dark reactions.

2. Consumption of Carbon Dioxide (Dark Reactions)

The dark reactions are commonly referred to as the **Calvin-Benson cycle** after the pioneering work of its discoverers. Melvin Calvin won a Nobel prize in chemistry for elucidating the pathway of carbon fixation in plants.

The purpose of the dark reactions is to change CO_2 into sugar by adding energy & the H+ from ATP and NADPH.

- Can occur in light or dark.
- $6CO_2$ and lots of ATP and NADPH are needed to make one sugar molecule.

The reactions of the Calvin cycle can be thought of as occurring in 3 stages: (Fig. 25.20)

1. **Carboxylation** - Fixation of CO_2 into a stable organic intermediate.
2. **Reduction** - Reduction of this intermediate to the level of carbohydrate
3. **Regeneration** - Regeneration of the CO_2 acceptor

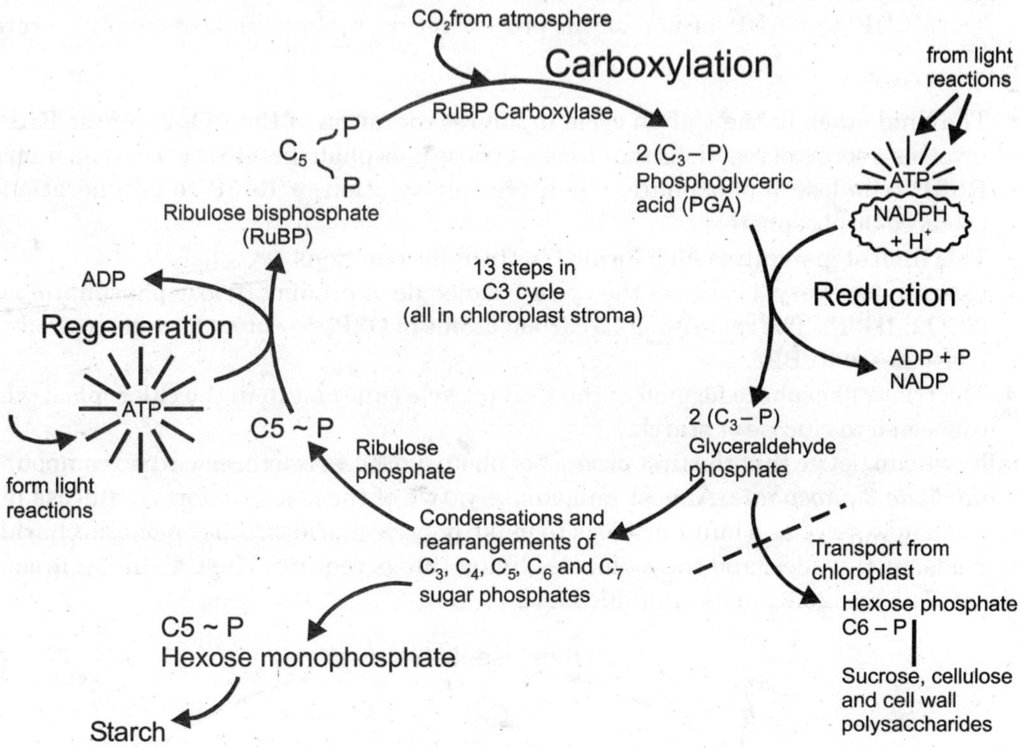

Fig. 25.20

1. **Carboxylation**
   - Carboxylation involves the addition of one molecule of $\mathbf{CO_2}$ to a 5-carbon "acceptor" molecule, **ribulose bis-phosphate (RuBP)**. This reaction is catalyzed by the enzyme **RUBISCO**. Plants invest a huge amount of their available nitrogen into making this one protein. As a result, Rubisco is the most abundant protein in the biosphere.
   - The resulting 6-carbon product splits into 2 identical 3-carbon products. This product is **3-phosphoglyceric acid** or simply **PGA**.
   - At this point in the cycle, $\mathbf{CO_2}$ has been "fixed" into an organic product but no energy has been added to the molecule.

2. **Reduction**
   - The second step in the Calvin cycle is the reduction of PGA to the level of carbohydrate.
   - This reaction occurs in two steps: (1) phosphorylation of PGA by ATP to form a "bisphosphate", (2) reduction of di-PGA by NADPH to form **triose phosphate, a simple 3-carbon carbohydrate**.

- This reaction requires both, ATP and NADPH, the high energy chemical intermediates formed in the light reactions.
- The overall equation can be written as follows:

$$PGA + ATP + NADPH \rightarrow G3P + NADP + ADP + P1$$

where PGA represents 3-phosphoglycerate and G3P represents glyceraldehyde-3-phosphate.
- Note that the basis for the term "C3 photosynthesis" comes from the fact that the initial product is the 3-carbon acid PGA.
- The $NADP^+$ and ADP formed in this process return to the thylakoids to regenerate.

3. Regeneration

- The final stage in the Calvin cycle is the regeneration of the CO_2 acceptor **RuBP.** This involves a series of reactions that convert triose phosphate first to the 5-carbon intermediate **Ru5P** (ribulose 5-phosphate), then phosphorylation of Ru5P to regeneration **RuBP** (ribulose-bisphosphate)
- This final step requires ATP formed in the light reactions.
- Overall, for every 3 turns of the cycle 1 molecule of product (triose phosphate) is formed (3CO2; 1G3P). The remaining 15 carbon atoms (5 G3P) re-enter the cycle to produce three molecules of RuBP.
- The triose phosphate formed in the Calvin cycle can remain in the chloroplast where it is converted to glucose or starch.

Actually, we can notice that the first product of photosynthesis is a three-carbon compound called *glyceraldehyde 3-phosphate*. Almost immediately, two of these join to form a glucose molecule. Glucose can now serve as a building block to make polysaccharides, other monosaccharides, fats, amino acids, nucleotides, and the molecules living things require. The Calvin-Cycle can also be represented as Fig. 25.21 in its simplified form.

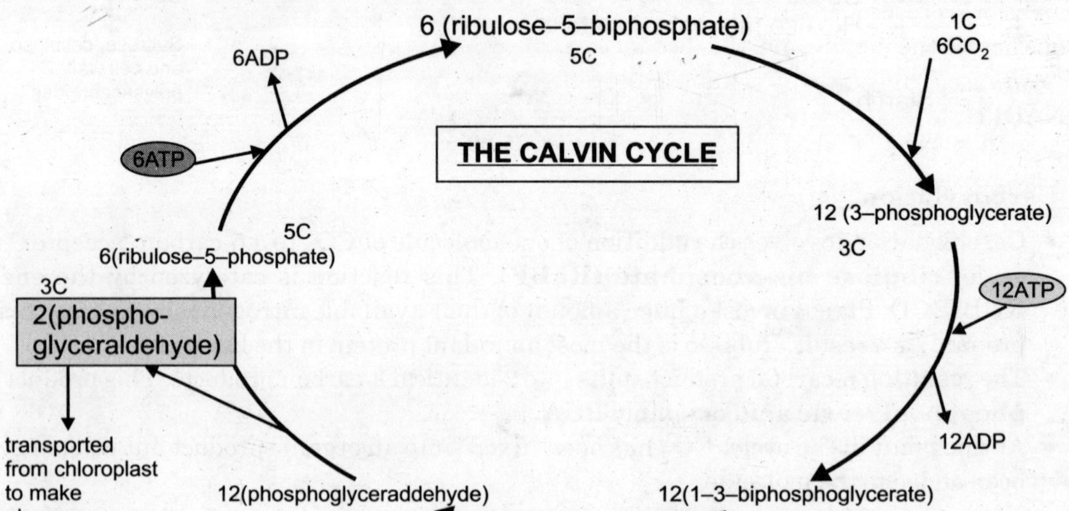

Fig. 25.21

25.12.3 Electron Transfer

The process of synthesizing ATP and NADPH are accomplished via the mechanism of an *electron transport chain*. This is a series of proteins embedded in a *biological membrane* that transfer high energy electrons from one to another. When sunlight strikes a chlorophyll complex (cluster of chlorophyll), the molecule is excited, and an electron is transferred to a higher energy level of the molecule. Only chlorophyll of the reaction is capable of transferring an electron to an electron acceptor.

An electron transfer system (a series of chemical reactions) carries the two electrons to and from across the thylakoid membrane. The form of photosynthesis with which we are most familiar is photophosphorylation. It consists of two sets of pigments to excite. They are called PSI, or photosystem I, and PS II, or photosystem 2, PS I better excited by light at about 700 nm, and is thus sometimes called P-700, PS II cannot use photons of wavelength longer than 680 nm, and is thus sometimes called P-680

- Photosystem II (PSII) (P680)
- Photosystem I (PSI) (P700)

It may seem confusing but PSII occurs before PSI. It is named because it was the second to be discovered and hence named second.

The energy changes accompanying the two sets of changes make a Z shape when drawn out. This is why the electron transfer process is sometimes called the Z scheme.

The chlorophyll's electron can follow either of two different pathways, cyclic or non-cyclic.

(a) Cyclic Photophosphorylation

In cycle electron flow, the electron begins in a pigment complex called photosystem I passes from the primary acceptor to *ferredoxin* (fd), then to a complex of two *cytochromes* and then to *plastocyanin* before returning to chlorophyll. This transport chain produces a proton-motive force, pumping H^+ ions across the membrane, this produces a concentration gradient which can be used to power *ATP synthase*. This pathway is known as cyclic photophosphorylation, and it produces neither O_2 nor NADPH.

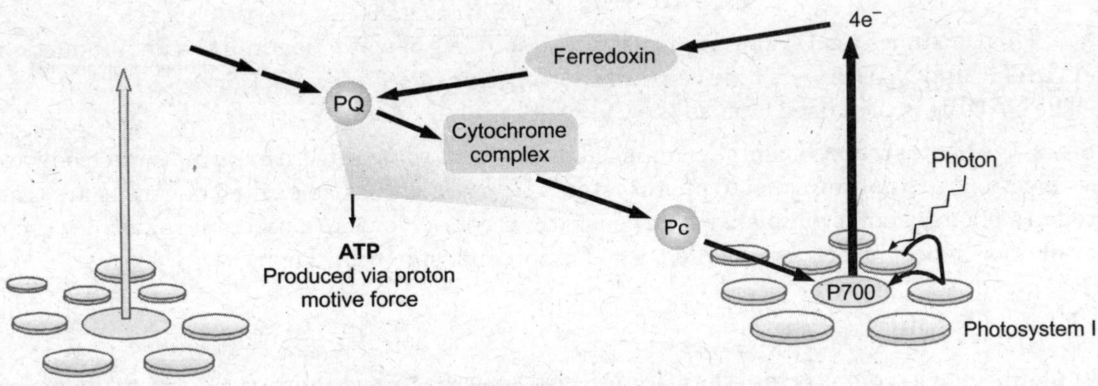

Fig. 25.22 Cyclic Phosphorylation

The electron ejected from the chlorophyll molecule thus returns back to the same molecule and hence the process is called the cyclic photophosphorylation.

In bacterial photosynthesis a single photosystem is used, and therefore is involved in cyclic photophosphorylation.

(b) Non-cyclic photophosphorylation

The other pathway, noncyclic photophosphorylation, is a two-stage process involving two different chlorophyll photosystems (Fig. 25.23).

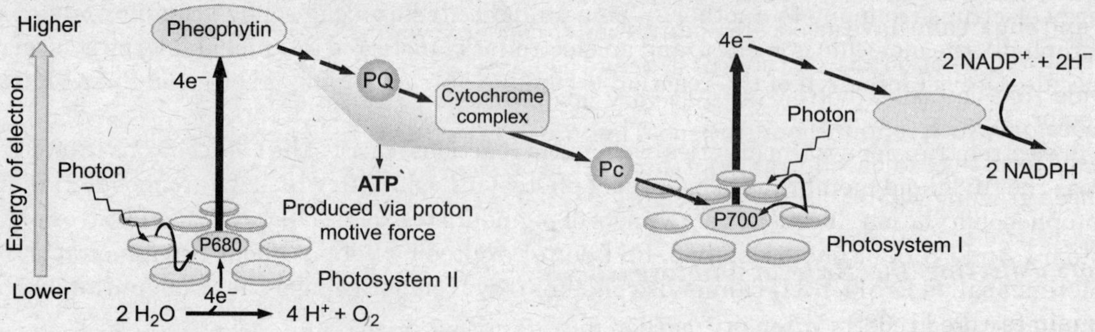

Fig. 25.23 Non-cyclic phosphorylation

It is important to note that both photosystems are almost simultaneously; thus both photosystems begin functioning at almost the same time.

- Light strikes photosystem II and the energy is absorbed and passed along until it reaches P680 chlorophyll.
- The excited electron is passed to the primary electron acceptor. Photolysis in the thylakoid takes the electrons from water replace the P680 electrons that were passed to the primary electron acceptor. (O_2 is released as a product).
- The electrons are passed to photosystem I via the electron transport chain (ETC) and in the process used to pump protons across the thylakoid membrane.
- The stored energy in the proton gradient is used to produce ATP which is used later in the Calvin-Benson Cycle.
- P700 chlorophyll then uses light to excite the electron to its second primary acceptor.
- The electron is sent down another ETC and used to reduce $NADP^+$ to NADPH.

$$NADP^+ + H^+ 2e^- \rightarrow NADPH$$

This consumes the H^+ ions produced by the splitting of water, leading to a net production of O_2, ATP, and NADPH with the consumption of solar photons and water.

- The NADPH is then used later in the Calvin-Benson Cycle.

Since, the photosystem II complex replaced its lost electrons from an external source however, these electrons are not returned to photosystems II Instead, the still excited electrons are transferred to a photosystem I complex, which boosts their energy level to a higher level using a second solar photon, hence the process is called nocyclic photophosphorylation.

25.12.4 C_3 and C_4 Plants

Most plants put CO_2 directly into the Calvin cycle. Thus the first stable organic compound formed is the glyceraldehyde 3-phosphate. Since, the molecule contains three carbon atoms, these plants are called C_3 plants. For all plants, hot summer weather increases the amount of water that evaporates from the plant. Plants lessen the amount of water that evaporates by keeping their stomates closed during hot, dry weather. Unfortunately, this means that once the CO_2 in their leaves reaches a low level, they must stop doing photosynthesis. Even if there is a tiny bit of CO_2 left, the enzymes used to grab it and put it into the Calvin cycle just don't have enough CO_2 to use. Typically the grass in our yards just turns brown and goes dormant. Some plants like crabgrass, corn and sugarcane have a special modification to conserve water. These plants capture CO_2 in a different way. They do an extra step first, before doing the Calvin cycle. These plants have a special enzyme that can work better, even at very low CO_2 level, to grab CO_2 and turn it first into *oxaloacetate* which contains four carbons. Thus, these plants are called C_4 plants. The CO_2 is then released form the oxaloacetate and put into the Calvin cycle. This is why crabgrass can stay green and keep growing when all the rest of your grass is dried up and brown.

Factors Affecting the Rate of Photosynthesis

The main factors are light intensity, carbon dioxide concentration and temperature, known as limiting factors.

- As light intensity increases, the rate of the light-dependent reaction, and therefore photosynthesis generally, increases proportionatly. As light intensity is increased however, the rate of photosynthesis is eventually limited by some other factor. Chlorophyll a is used in both photosystems. The wavelength of light is also important. PSI absorbs energy most efficiently at 700 nm and PSII at 680 nm. Light with high proportion of energy concentrated in these wavelengths will produce a high rate of photosynthesis.
- An increase in the carbon dioxide concentration increases the rate at which carbon is incorporated into carbohydrate in the light-independent reaction and so the rate of photosynthesis generally increases until limited by another factor.
- Photosynthesis is dependent on temperature. It is a reaction catalysed by enzymes. As the enzymes approach their optimum temperatures the overall rate increases. Above the optimum temperature the rate begins to decrease until it stops.

QUESTIONS

1. How do photochemical reactions differ from thermal reactions?
2. Explain Lambert's and Beer's Laws.
3. State and Explain
   (i) Grotthuss-Draper's law
   (ii) Einstein's law of Photochemical equivalence.
4. What is meant by primary and secondary process in Photochemistry?
5. What do you understand by the term "quantum yield" or "quantum efficiency"? What does it signify?
6. Explain why in certain cases, quantum yield is lower than that expected from law of photochemical equivalence? Give examples.
7. Discuss the photochemical reaction: $H_2(g) + Ci_2(g) \rightarrow 2HCl(g)$
8. Give one example of photochemical reaction in which quantum yield is very high. Briefly explain the reasons for same.
9. Give photosynthesis of HBr. What is the quantum yield of this process?
10. Write short notes on "fluorescence" and "Phosphorescence". What is the difference between them?
11. How do you explain that in fluorescence the wavelength of the light emitted is usually greater than that of the light absorbed?

12. Write short notes on
    (i) Photosensitization
    (ii) Molecular exictation
    (iii) Intersystem crossing.
13. Why the wavelength of light emitted in fluorescence is usually greater than that of light absorbed.
14. (a) Explain the processes of internal and intersystem crossing.
    (b) Discuss the photolysis of acetone.
15. Define the terms : extinction coefficient and molar extinction coefficient.
16. Distinguish between fluorescence and phosphorescence. Draw energy level diagrams to explain these processes and discuss them briefly.
17. (a) Write a short note on "photosensitization". Give its use in photography.
    (b) What are photo inhibitors? How do they work?
18. Give the mechanism of
    (i) Fluorescence
    (ii) Phosphorescence and
    (iii) Photosensitisation.
19. (a) Write note on Lasers and its applications.
    (b) Explain the following:
        (i) Three phase Laser system
        (ii) Four phase laser system
20. Write short note on Masers.
21. What do you understand by population inversion.
22. Explain
    (i) Spontaneous Emission
    (ii) Stimulated Emission.
23. Explain the terms
    (i) Internal Crossing
    (ii) Intersystem crossing.
24. Explain
    (i) Molar Extinction coefficient
    (ii) Optical density **(PTU, June 2004)**
25. Write short notes on
    (i) Photochemical reaction
    (ii) Phosphorescence **(PTU, June 2004)**
26. Explain the terms Photosynthesis and bring out its importance.
27. Discuss the mechanism of consumption of carbon dioxide in photosynthesis.
28. Explain the two photo systems involved in photosynthesis and the process of cyclic and non-cyclic photophosphorylation.
29. Give evidence for the presence of types of reactions (light and dark) involved in photosynthesis.
30. Describe the path of carbon in photosynthesis by Calvin cycle (C_3 cycle).
31. Explain C_4 cycle for the consumption of carbon dioxide in photosynthesis.
32. What is meant by photosensitization? Give one example. **(PTU, June 2003)**
33. Explain what is meant by optical density of a substance. **(PTU, Dec. 2002)**
34. Explain photo-chemical reaction. **(PTU, Jun. 2002)**
35. Distinguish between Phosphorescence and Flourescence **(PTU, May, 2005)**
36. (a) Explain the following:
        (i) Intensities and line width of spectra
        (ii) Charge transfer transitions

(b) Explain Beer-Lambert Law. **(PTU, Dec. 2005)**

37. What type of information is obtained by studying the UV, IR and H-NMR spectra of an organic compound. **(PTU, Dec. 2005)**
38. (a) What is Franck Condon principle?
    (b) Calculate the force constant for HCl^{35} when its fundamental vibrational frequency is 8.667×10^{13}/sec. How does the force constant of HCl^{37} compare with it?
    (c) How can you establish whether a carbonyl is part of aldehyde, ester or ketone from U.V. spectroscopy? **(PTU, May 2006)**
39. What is the importance of IR spectroscopy in finger point region? **(PTU, Dec. 2006/May 2007)**
40. What is UV spectrum? Give various regions associated with UV spectrum. **(PTU, Dec. 2006/May 2007)**
41. (a) Define the teams bathochromic shift and hypsochromic shift. What structural feature may produce bathochromic or hypsochromic shift in an organic compound.
    (b) In an absorption cell, the transmittance of 0.1 M solution of a substance X is 80% and that of 0.1 M solution of another substance Y is 60% at a given wavelength. What is the transmittance of solution that is simultaneously 0.1 M in X and 0.1 M in Y?
    (c) Using IR spectrocopy, how will you determine whether the oxygen in an organic compound is present as carbonyl or hydroxyl group. **(PTU, Dec. 2006/May 2007)**
42. Distinguish between 1-hexyne, 1-hexene and hexane on the basis of IR spectrocopy. **(PTU, May 2007 (C))**
43. (a) Discuss various types of electronic transitions and explain the effect of the polarity of the solvent each type of transition.
    (b) How will you detect the presence of carbonyl group in aldehydes and ketones.
    (c) Distinguish between chromophores and auxochromes. **(PTU, May 2007(C))**
44. What is importance of IR spectrocopy in fingerprint region? **(PTU, Jan. 2008)**
45. Why are electronic bands broad as compared to infrared. **(PTU, Jan. 2008)**
46. (a) State and illustrate with suitable potential energy curve the Frank-Condon principle in the vibronic spectrum of a diatomic molecule.
    (b) A substance when dissolved in water at 10^{-3} M concentration absorbs 10% of an incident radiation in a path of 1 cm length. What should be concentration of solution in order to absorb 90% of the same radiation?
    (c) Define wave number? What are its units? **(PTU, Jan. 2008)**

PROBLEMS

1. A 2×10^{-3} thick glass sheet transmits 10% of the incident light of the wavelength 300 mm. What percentage of light of the same wavelength will absorbed by 1×10^{-2} m thick glass sheet.
2. In a Beer's law, the absorbance of 0.001 M aqueous solution of the substance is 3. The path length of the cell is 1 cm. What is the molar extinction coefficient of the substance ?
3. In a photochemical reaction A $\rightarrow$ B, 1×10^{-5} moles of B are formed by absorption of 6.00×10^{7} ergs at 3600 Å. Calculate the quantum yield of the reaction.
4. A certain system absorbs 3.0×10^{16} quanta of light per second. When it was irradiated for 10 minutes, it was found that 0.0002 mole of the reactant has reached. What is the quantum yield of the reaction?
5. For the photochemical formation of ethylene from di-n-propyl ketone using a radiation of wavelength 313 nm, the quantum yield is 0.21. Calculate the number of moles of ethylene formed when the sample is irradiated with 50 watt of this radiation assuming that all the radiation is absorbed by the sample. **(Ans. 2.75×10^{-5} mol) (PTU, Dec. 2002)**
6. A monochromatic radiation is incident on a solution of 0.05 molar concentration of an absorbing substance. The intensity of radiation is reduced to one forth of the initial value after passing through 10 cm length of the solution. Calculate the value of Molar extinction coefficient of the substance. **(Ans. 1.204 $dm^3\ mol^{-1}\ cm^{-1}$) (PTU, July 2002)**

7. The molar extinction coefficient of phenanthroline complex of iron (II) is 12000 $dm^3\ mol^{-1}\ cm^{-1}$ and the minimum detectable absorbance is 0.01. Calculate the minimum concentration of the complex that can be detected in a Lambert Beer Law call of path length 1.00 cm. **(Ans. 1.20 × 10^{-6} M)**
8. Photobromination of cinnamic acid to dibromocinnamic acid was carried out in a light of wavelength 440 nm. Using light intensity of $1.5 \times 10^{-3} js^{-1}$, and exposure of 20 min. produced a decrease of 0.075 m mole of Br_2. The solution absorbed 80% of the light passing through it. Calculate the quantum yield of the reaction. **(Ans. 14.15) (PTU, June 2003)**

26 SOLID STATE CHEMISTRY

Knowledge is a treasure, but practice is the key to it.

— ***Thomas Fullen***

CHAPTER OUTLINE

Matter is anything occupying space and having mass; it is the material of the universe.

There are three main phases of matter: gases, liquids, and solids. There is also a fourth phase, plasma, but it exists at very high temperatures and molecules in that state can behave very differently.

Out of these three different states of matter, *solids are characterised by definite shape and definite volume. These are highly incompressible and rigid.* The constituent particles (atoms, molecules, ions) in the solids are closely packed as they are held together by strong intermolecular forces. These occupy fixed positions and can only oscillate about their mean positions.

26.1 CLASSIFICATION OF SOLIDS

Classification of solids can be based on atomic arrangement, binding energy, physical and chemical properties, or the geometrical aspects of the crystalline structure.

Broadly the solids can be divided into following two classes:

Crystalline

Crystalline solids are based upon a simple pattern of arrangement of atoms that is repeated many times throughout the molecule. The atoms or group of atoms in the solid are arranged in a regular order. The order is so regular that knowing the arrangement at one point, the arrangement at other site can be predicted. This is called *long range order*. These solids are referred to as the crystalline solids. The regular repeating pattern that the atoms arrange in is called the *crystalline lattice*. Examples of crystalline solids are ice, methanol, and sodium chloride .

Crystalline substances have the following properties:

- They have orderly arranged units and are practically incompressible. Crystalline solids also show a definite melting point and so they pass rather sharply from solid to liquid state.
- **Anisotropy:** These solids have different physical properties in different directions. This phenomenon is known as anisotropy.
- Crystalline solids have *definite* geometries.
- The crystalline solids can be further divided into two categories: the single-crystalline and the polycrystalline solids. In a single-crystalline solid, the regular order extends over the entire crystal. In a polycrystalline solid, however, the regular order exists only over a small region of the crystal, with grain size ranging from a few hundred angstroms to a few centimeters.
- A polycrystalline solid contains many of these small single-crystalline regions surrounded by the grain boundaries.

Amorphous

Amorphous solids have no regular arrangement of their atoms or molecules, in other words, in amorphous solids, the atoms are set in an irregular manner, without any short or long-range order in their atomic arrangement. This class of solids is commonly known as non crystalline or amorphous materials. Examples of amorphous solids are glass and plastic

Amorphous substances have the following properties:

- They are sometimes described as supercooled liquids because their molecules are arranged in a random manner some what as in the liquid state. For example, glass is commonly made from silicon dioxide or quartz sand, which has a crystalline structure. When the sand is melted and the liquid is cooled rapidly enough to avoid crystallisation, an amorphous solid called a glass is formed.
- Amorphous solids do not show a sharp phase change from solid to liquid at a definite melting point, but rather soften gradually when they are heated.
- The physical properties of amorphous solids are identical in all directions along any axis so they are said to have **isotropic** properties,
- These *do not have definite* geometrical shapes.

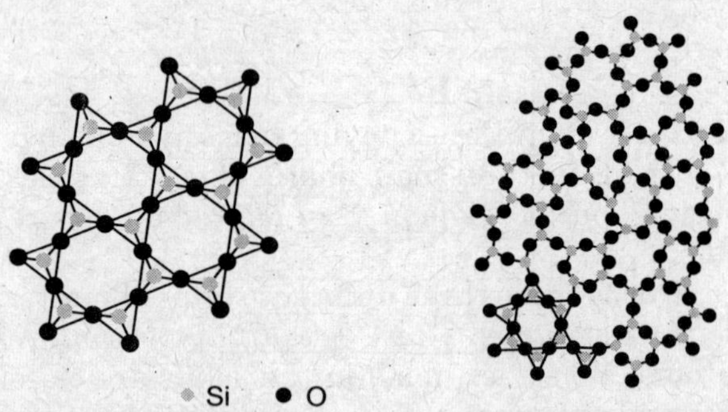

(*a*) *Crystalline* SiO_2 (*Quartz*)   (*b*) *Amorphous* SiO_2 (*Glass*)

Fig. 26.1

Distinction between these two classes of solids—amorphous and crystalline—can be made through the use of X-ray or electron diffraction techniques

Depending on the nature of the acting interatomic forces, all solids may be subdivided into:

(a) Ionic solids (NaCl)
(b) Covalent solids (Diamond)
(c) Metallic solids (Fe, Ni, etc.)
(d) Molecular solids (ice, solid He)

A brief description of these types is given in the following table.

Table 26.1

Types of Solids				
Types of Solid	*Form of Unit Partides*	*Forces Between Partides*	*Properties*	*Examples*
Molecular	Atoms or molecules	London dispersions, dipole-dipole forces, hydrogen bonds	Fairly soft, low to moderately high melting point, poor thermal and electrical conduction	Argon, Ar; Methane, CH_4; sucrose, $C_{12}H_{22}O_{11}$; Dry Ice, CO_2
Covalent Network	Atoms connected in a network of Covalent bonds	Covalent bonds	Very hard, very high melting point, often poor thermal and electric conduction	Diamond, C: quartz, SiO_2
Ionic	Positive and negative ions	Electrostatic attractions	Hard and brittle, high melting point, poor thermal and electrical conduction	Typical salts — for example, NaCl, $Ca(NO_3)_2$
Metallic	Atoms	Metallic bonds	Soft to very hard, low to very high melting point, excellent thermal and electrical conduction, malleable and Ductile	All metallic elements —for example, Cu, Fe. Al, W

26.2 SPACE LATTICE OR CRYSTAL LATTICE

The constituent particles of a crystalline solid are arranged in a definite fashion in the three dimensional space. A lattice is a regular array of (imaginary) points showing how molecules, atoms and ions are arranged in space.

One such arrangement by representing the particles with points is shown below (Fig. 26.2).

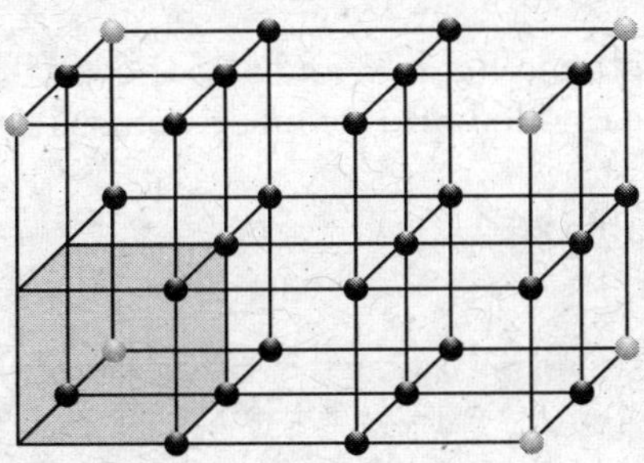

Fig. 26.2(a) Representation of space lattice and unit cell

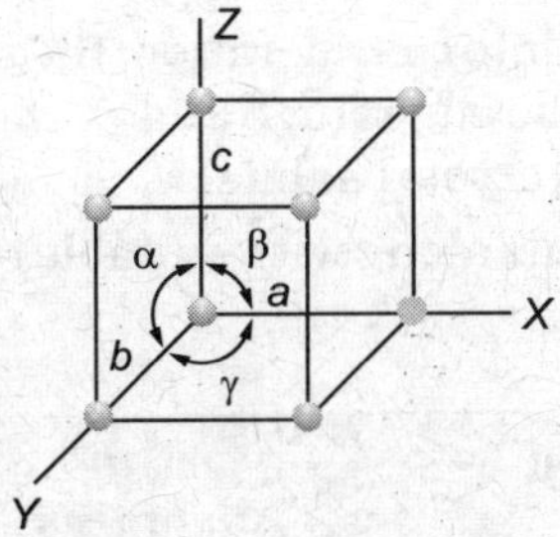

Fig. 26.2(b) Representation of dimensions of a unit cell

Such a regular arrangement of the constituent particles of a crystal in a three dimensional space is called crystal lattice or space lattice.

From the complete space lattice, it is possible to select a smallest three dimensional portion which repeats itself in different directions to generate the complete space lattice. This is called a Unit Cell.

Unit Cell

Unit cell is the smallest three-dimensional portion of a complete space lattice, which when repeated over and again in different directions produces the complete space lattice Fig. 26.2 (b).

Parameters of the unit cell:

The size and shape of a unit cell is determined by the lengths of the edges of the unit cell (a, b and c) and by the angles α, β and γ between the edges b and c, c and a, and a and b respectively. Thus a unit cell is characterised by six parameters, i.e. axial length (a, b, c) and also the axial angles between the edges (α, β and γ), refer to Fig. 26.2 (b).

26.3 CRYSTAL STRUCTURE

A crystal is a regular, ordered arrangement of atoms over a large scale. The atoms may be of a single type or the repetition of a complex arrangement of many different types of atoms; The geometrical arrangement of atoms in an ideal crystal (one that extends to infinity) can be described using 2 elements – a lattice and a basis.

Crystal structure = lattice + basis

The lattice is an ordered arrangement *of points* in space, while the basis consists of the simplest *arrangement of atoms* which is repeated at every point in the lattice to build up the crystal structure.

A good analogy is patterned wallpaper. The basis is like a motif on the wallpaper and the lattice would be periodic pattern of points on which the motif is repeated. In Figure 26.3, the lattice points correspond with the centre of the basis, but this does not have to be the case.

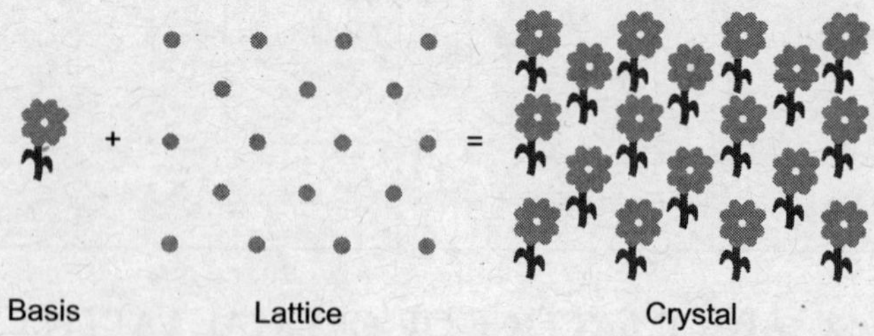

Fig. 26.3 A crystal can be thought of as being like wallpaper. The motif is analogous to the basis and the arrangement of the motif over the surface is like the lattice.

26.4 TYPES OF UNIT CELL

If we take into consideration, the symmetry of the axial distances (a, b, c) and also the axial angles between the edges (α, β and γ), the various crystals can be divided into seven systems. These are also called crystal habits.

Crystals along with suitable examples are listed in Table 26.2.

Table 26.2

System	*Axial distances*	*Axial angles*	*Examples*
Cubic	$a = b = c$	$\alpha = \beta = \gamma = 90°$	NaCl, kCl Diamond, Cu, Ag
Tetragonal	$a = b \neq c$	$\alpha = \beta = \gamma = 90°$	White tin, SnO_2
Orthorhombic	$a \neq b \neq c$	$\alpha = \beta = \gamma = 90°$	Rhombic Sulphur, KNO_3, K_2SO_4, $BaSO_4$, $PbCO_3$
Monoclinic	$a \neq b \neq c$	$\alpha = \gamma = 90°$, $\beta \neq 90°$	Monoclinic sulphur, Na_2SO_4, $10H_2O$
Hexagonal	$a = b \neq c$	$\alpha = \beta = 90°$, $\gamma = 120°$	Graphite, ZnO, CdS
Rhombohedral	$a = b = c$	$\alpha = \beta = \gamma \neq 90°$	Calcite, quartz, $NaNO_3$
Triclinic	$a \neq b \neq c$	$\alpha \neq \beta \neq \gamma \neq 90°$	$CuSO_4$, $5H_2O$, $K_2Cr_2O_7$

But it has been observed that the particles may also be present at some other special positions within the unit cell. Hence unit cells can further be classified as:

(a) Primitive Unit Cell

The primitive cell is the simplest unit cell which has lattice points only at the corners-It is denoted by P.

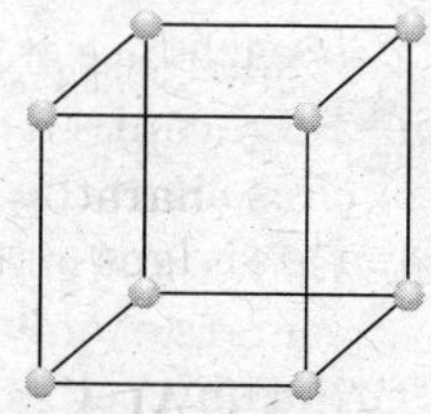

Fig. 26.4 Simple/Primitive Unit Cell.

(b) Non primitive Cell:

The unit cell in which the lattice points are present not only at the corners of the unit cell but also at some other positions are known as the non primitive unit cell. It is divided into following categories:

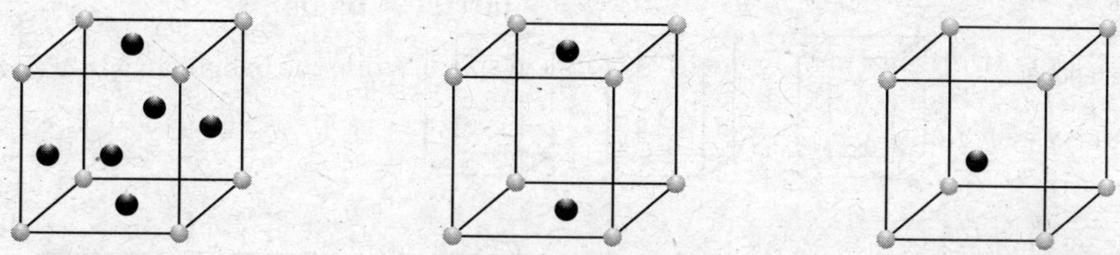

(a) Face-centred (b) End face-centred (c) Body-centred

Fig. 26.5 Non-primitive Unit cells

Table 26.3 Seven crystal systems make up fourteen Bravais lattice types in three dimensions.

System	*Number of Lattices*	*Lattice Symbol*	*Restriction on Crystal cell angle*
Cubic	3	P or sc, l or bcc, F or fcc	$a = b = c$ $\alpha = \beta = \gamma = 90°$
Tetragonal	2	P, I	$a = b \neq c$ $\alpha = \beta = \gamma = 90°$
Orthorhombic	4	P, C, I, F	$a \neq b \neq c$ $\alpha = \beta = \gamma = 90°$
Monoclinic	2	F, C	$a \neq b \neq c$ $\alpha = \beta = 90° \neq \beta$
Triclinic	1	P	$a \neq b \neq c$ $\alpha \neq \beta \neq \gamma$
Trigonal	1	R	$a = b = c$ $\alpha = \beta = \gamma < 120°, \neq 90°$
Hexagonal	1	P	$a = b \neq c$ $\alpha = \beta = 90°$ $\gamma = 120°$

P – Primitive: simple unit cell

F – Face-centred: additional point in the centre of each face

I – Body-centred: additional point in the centre of the cell

C – Centred: additional point in the centre of each end

R – Rhombohedral: Hexagonal class only

(i) Face centred cubic: The particles are present not only at the comers but also at the centre of each face of the unit cell

(ii) Body centred cubic : In addition to the particles at the comer of the cube, there, is one present at the body centre position of the unit cell

(iii) End centered Unit cell: In an end centered unit cell, there are lattice points in the face centers of only one set of opposite faces, in addition to the lattice points at the corners of the unit cell.

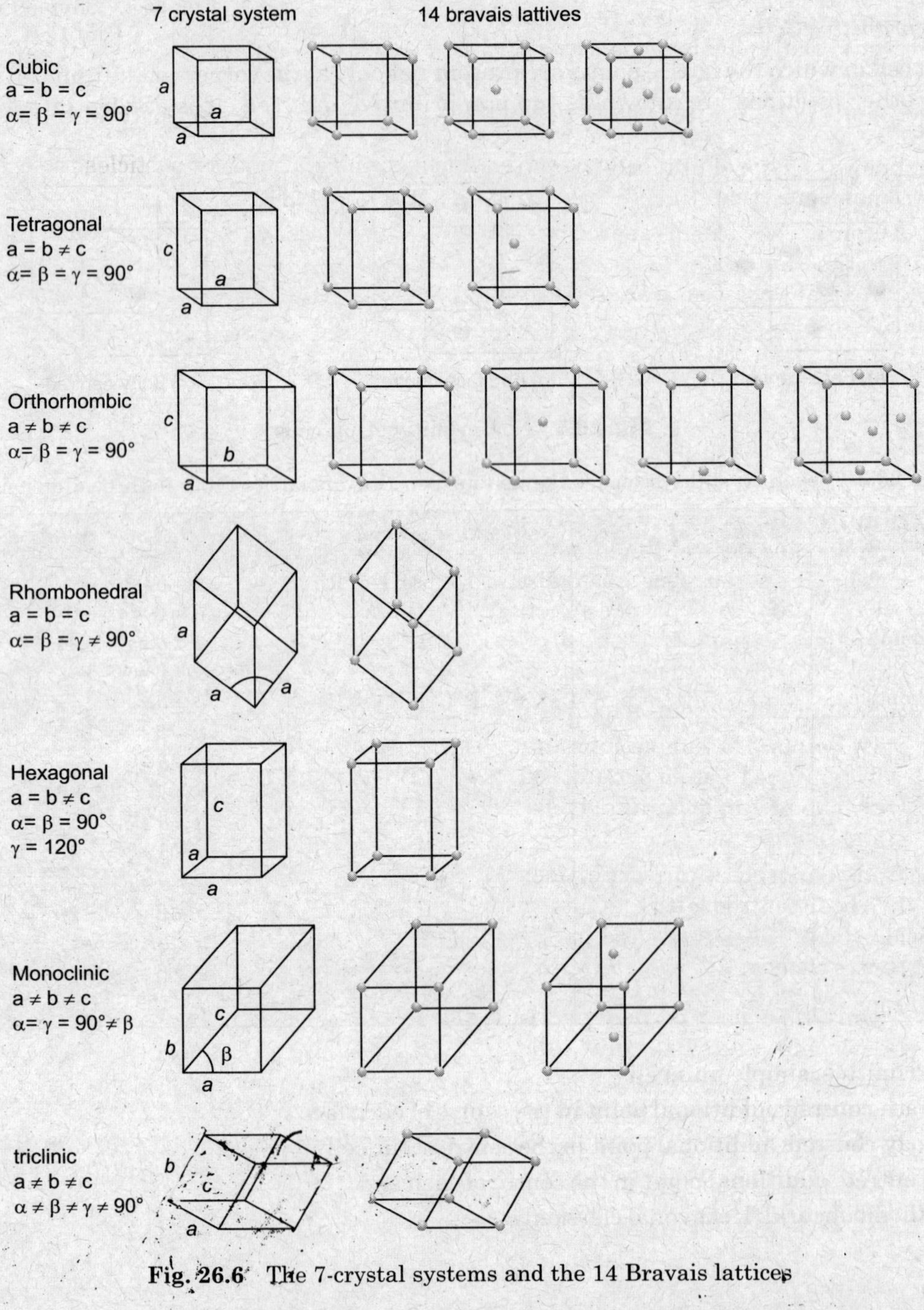

Fig. 26.6 The 7-crystal systems and the 14 Bravais lattices

Although all the seven crystal systems discussed above are expected to have all the four different unit cells (simple, face centered, body centered, end centered unit cell), but actually all of them cannot exist in each case. Auguste Bravais demonstrated that there are in fact ***only fourteen*** possible point lattices and no more. For his efforts, the term ***Bravais lattice*** is often used in place of point lattice.

The fourteen possible crystal lattices are as in Table 26.3

Thus, there are seven crystal types for which fourteen different types of lattices (7 primitive or simple, 3 body centred, 2 face centred and 2 end centred are possible).

The seven crystal systems and fourteen Bravias Lattices are represented in the following Fig. 26.6.

26.5 NUMBER OF PARTICLES PER UNIT CELL OF A CUBIC CRYSTAL SYSTEM

Once we know the type of unit cell, we can count the number of atoms /particles per unit cell. It must be remembered that:

- An atom at the corner is shared by eight unit cells. Hence, an atom at the corner contributes 1/8 to the unit cell
- An atom at the face is a shared by two unit cells. Hence, contribution of each atom on the face is 1/2 to the unit cell.
- An atom within the body of the unit cell is shared by no other unit cell. Hence, contribution of each atom within the body is 1 to the unit cell
- An atom present on the edge is shared by four unit cells. Contribution of each atom on the edge is 1/4 to the unit cell.

Now let us calculate the number of atoms in different cubic cells.

(i) Simple cubic structure: In a simple cubic lattice, the atoms or ions lie only at the corners of a cube.

Only 1/8 of each corner sphere lies whin any single unit cell. Since there are 8 corner atoms, the simple cubic unit cell contains a total of $\frac{1}{8} \times 8 = 1$ atom. The simple cubic unit cell is assigned 1 atom per unit cell. Only one metal, polonium, has a simple cubic unit cell.

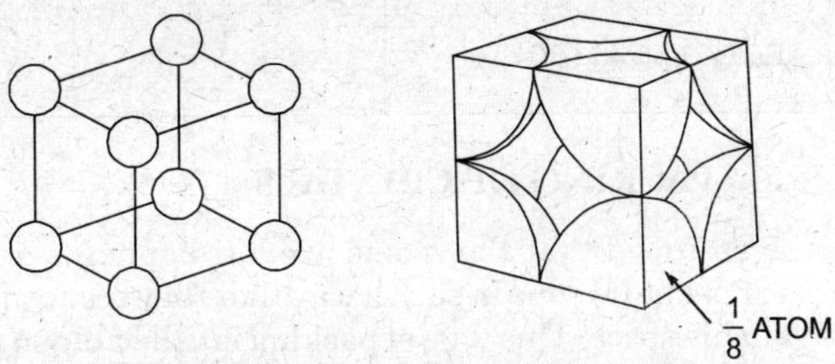

Fig. 26.7 The simple cubic structure

(ii) Body centred cubic structure: In a body centred lattice, atoms or ions lie at the corners and in the centre of cube.

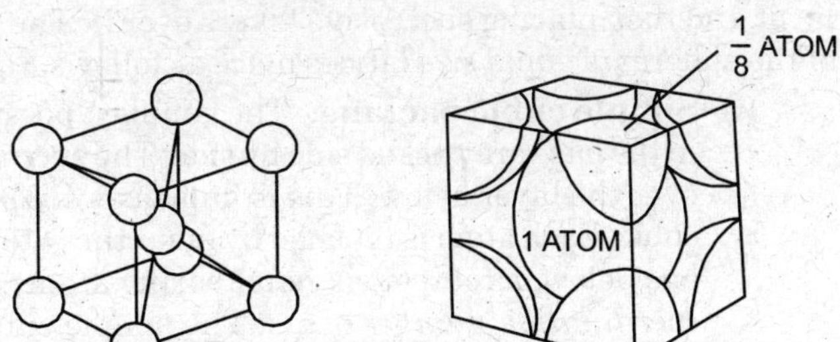

Fig. 26.8 Body centred cubic lattice.

Only 1/8 of each corner sphere lies within any single unit cell. Since, there are 8 corner atoms, the number of atoms contributed by them is $\frac{1}{8} \times 8 = 1$ atom. As a body centred cubic lattice has a unshared atom in the centre, the total number of atoms per unit cell is:

$$= \frac{1}{8} \times 8 \text{ atoms} + 1 \text{ unshared atom} = 1 \text{ atom} + 1 \text{ atom} = 2 \text{ atoms}.$$

(iii) Face centred cubic lattice: In a face centred cubic lattice, the atom lie at the corners of a cube and one atom lies at each face of the cube.

The face centred cubic unit cell is assigned 4 atoms per unit cell. Only 1/8 of each corner sphere and 1/2 of each face-sphere lie within any single unit cell. Since there are 8 corner atoms and 6 face atoms in a face centred cubic unit cell, the total number of atoms in its unit cell is

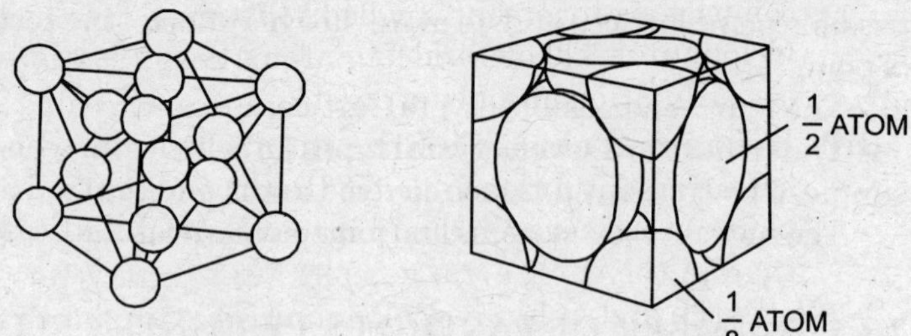

Fig. 26.9 The face centred cubic lattice.

$$= 8 \times \frac{1}{8} \text{ atoms} + 6 \times \frac{1}{2} \text{ atoms} = 1 \text{ atom} + 3 \text{ atom} = 4 \text{ atoms.}$$

Table 26.4 Number of atoms per unit cell in cubic close packed structure of atoms.

Unit cell	*Number of atoms at*			*No. of atoms per unit cell*
	Corners	*Faces*	*Centre*	
Primitive (simple cube)	$8 \times \frac{1}{8} = 1$	0	0	1
Face centred (FCC)	$8 \times \frac{1}{8} = 1$	$6 \times \frac{1}{2} = 3$	0	4
Body centred (BCC)	$8 \times \frac{1}{8} = 1$	0	1	2

26.6 PACKING OF CRYSTALS

It is assumed that the atoms are hard spheres of identical size. Packing is done in such a way that they occupy maximum available space. This type of packing is called **close packing.**

It is easiest to picture the atomic arrays by arranging one layer and then placing successive layers over it. The packing of the spheres of equal size latices place as follows Fig. 26.10.

Fig. 26.10 Close packing of spheres in one dimension

(i) Simple cubic packing: The simplest possible arrangement is that in which the atoms in the base are packed side by side. The successive layers of atoms are then placed directly over the layer below. This is known as *simple cubic packing (sc)* Fig. 26.11 or primitive cubic. Each atom is touched by four other atoms in its own plane plus one atom above and one below, a total of six neighboring atoms. As a result, it is said that each atom has a *coordination number* of six. The simple cubic arrangement is not very compact and is known only for polonium in metal structures.

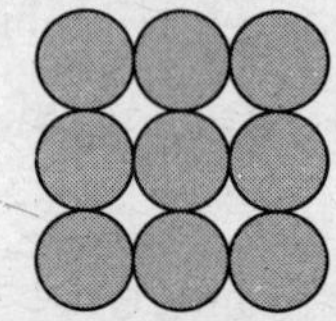

Fig. 26.11 Simple cubic packing. Successive layers are superimposed over the first array.

Fig. 26.12 Body-centred cubic packing. The second layer (shaded) in placed over holes in the first layer and the third over the holes in the second.

(ii) Body-centered cubic (bcc): An alternative cubic packing arrangement is to place the second layer of atoms over the holes in the first layer. The third layer then fits over the holes in the second layer—which happens to be exactly over the first layer. This more compact arrangement is called *body-centered cubic (bcc)* (Fig. 26.12). Each atom is touched by four atoms above and four atoms below its plane. Thus, body-centered cubic results in a coordination number of eight.

(iii) Hexagonal close packing (hcp): The other two possibilities are based on a hexagon arrangement for each layer; that is each atom is surrounded by six neighbours in the plane. In the hexagonal arrangement, the holes between the atoms are much closer together than in the cubic arrangement Fig. 26.13. When the second hexagonal layer is placed over the first, it is physically impossible to place atoms over all the holes in the first layer. In fact, only half the holes can be covered. If the third layer is placed over holes in the second layers so that it is superimposed over the first layer, then a *hexagonal close-packed (hcp)* arrangement is obtained. The fourth layer is then superimposed over the second layer. Hence, it is also known as an *abab* packing arrangement Fig. 26.14.

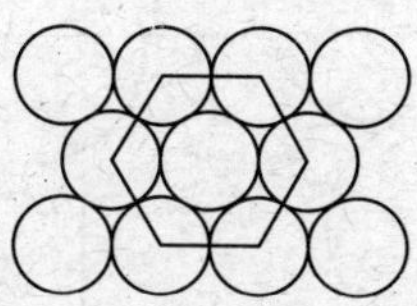

Fig. 26.13 The first layer of the hexagonal arrangement

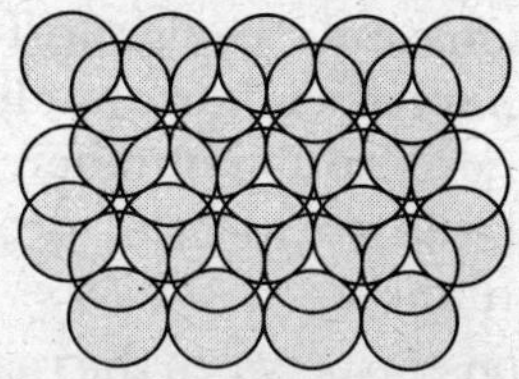

Fig. 26.14 In hexagonal packing, the second layer (shaded) fits over alternate holes in the first layer. The hexagonal close-packed arrangement involves placing the third layer over the top of the first layer

(iv) Cubic close packing (ccp) or (fcc): A alternative hexagonal packing arrangement involves placement of the third layer over the holes in the first and second layers. It is the fourth layer, then, that repeats the alignment of the first layer. This *abcabc* packing arrangement is known as *cubic close-packed (ccp)* or *face-centered cubic (fcc)* Fig. 26.15. Both packings based on the hexagonal arrangement are 12-coordinate.

Fig. 26.15 The cubic close-packed arrangement has the third layer placed over voids in both the first and second layers

Figure 26.16 gives the expanded view of the *hcp* and *ccp* arrangements.

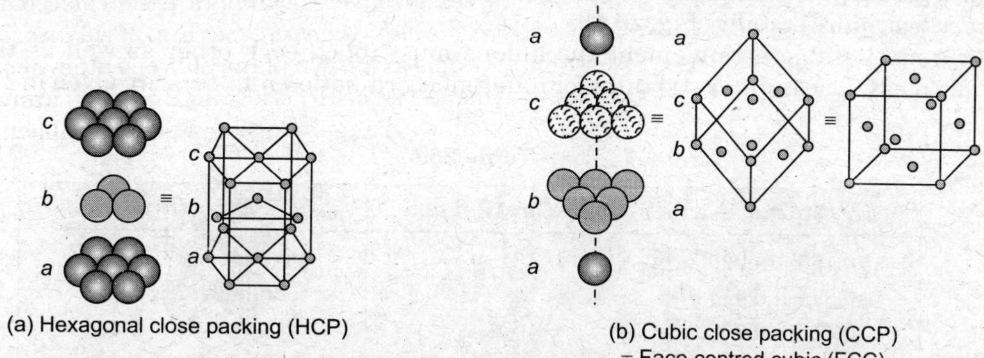

Fig. 26.16 (a) abab ... type is hexagonal close packing
(b) abc abc ... type is cubic close packing

Most metals adopt one of the three more compact arrangement (*bcc, hcp, fcc*);

The types of packing, the coordination numbers, and the percentage of occupancy (filling) of the total volume are shown in Table 26.5

Table 26.5 Properties of the different packing types

Packing type	*Coordination number*	*Occupancy (%)*
Simple cubic (sc)	6	52
Body-centered cubic (bcc)	8	68
Hexagonal close-packed (hcp)	12	74
Cubic close-packed (ccp/fcc)	12	74

26.7 RADIUS RATIO RULE

For the stability of an ionic compound, each cation should be surrounded by maximum number of anions and vice-versa. The number of oppositely charged ions surrounding each ion is called its co-ordination number. The larger is the size of the cation the greater is its co-ordination number.

The relationship between the radius and the co-ordination number and the structural arrangement are called radius ratio rules. The radius ratio is given by:

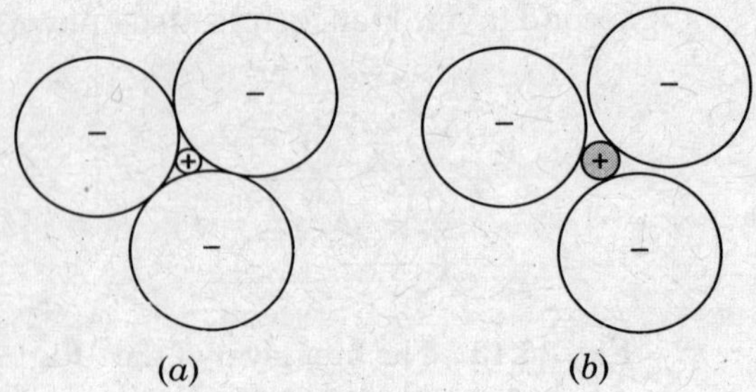

Fig. 26.17

$$\text{Radius ratio} = \frac{\text{Radius of cation } (r^+)}{\text{Radius of anion } (r^-)} = \frac{r^+}{r^-}$$

For example, in a planar triangular structure, the ideal arrangement is represented in the Fig. 26.17 (a).

But if the size of the cation increases (Keeping the size of anions same) then the anions will no longer be touching each others as shown in the Fig. 26.17 (b). So, readjust in such a way that they touch each other as well as the cation. Under such situation four anions may be touching the cation. The leads to a *tetrahedral* arrangement.

Similarly, in case of an *octahedral arrangement,* if the size of the cation decreases, i.e. the radius ratio decreases, the arrangement changes to tetrahedral to become stable. But if the size of the cation increases, i.e. the radius ratio increases the arrangement changes to body centred cubic structure to acquire stability Fig. 26.18.

For a most stable arrangement the anions must touch each other as well as the cation simultaneously. The radius ratio rules for different coordination numbers are given in Table 26.6 below:

Table 26.6

Radius	*Co-ordination no.*	*Structure*
0.155 – .0225	3	Plane triangular
0.225 – 0.414	4	tetrahedral
0.414 – 0.732	6	octahedral
0.732 – 1	8	Body-centered cubic

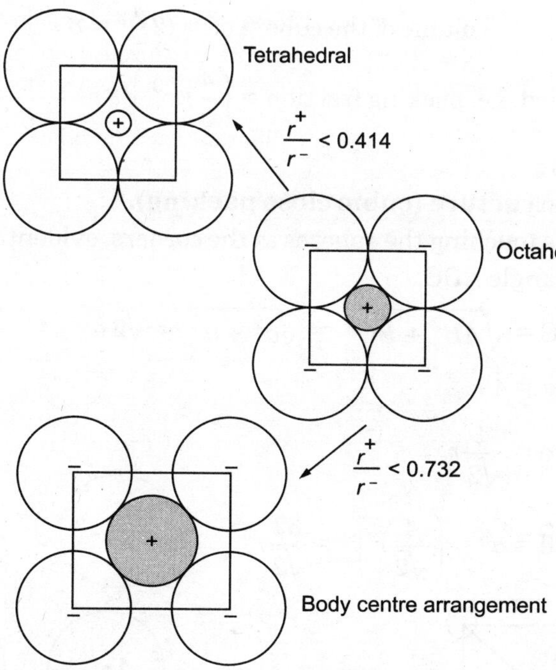

Fig. 26.18

26.8 PACKING EFFICIENCY AND ATOMIC RADIUS

We have seen above that in any type of packing, some voids (empty spaces) are always present.

The percentage of the total space filled by the particles is called **packing efficiency** *or the fraction the total space filled is called* **packing fraction.**

The calculations for different types of structures are described below:

(1) In a simple cubic unit cell

Suppose the edge length of the unit cell = a and radius of the sphere = r.

As spheres are touching each other, evidently, $a = 2r$

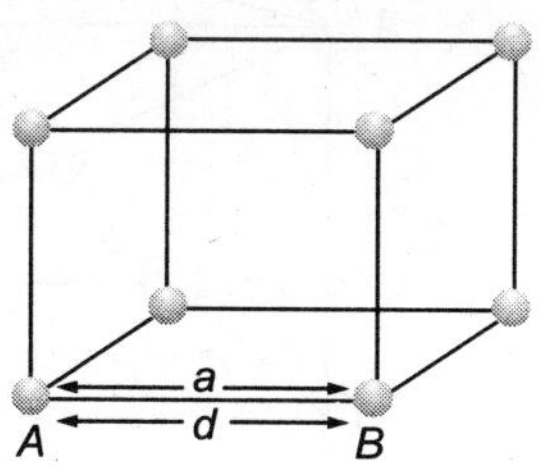

(a) Simple cubic unit cell

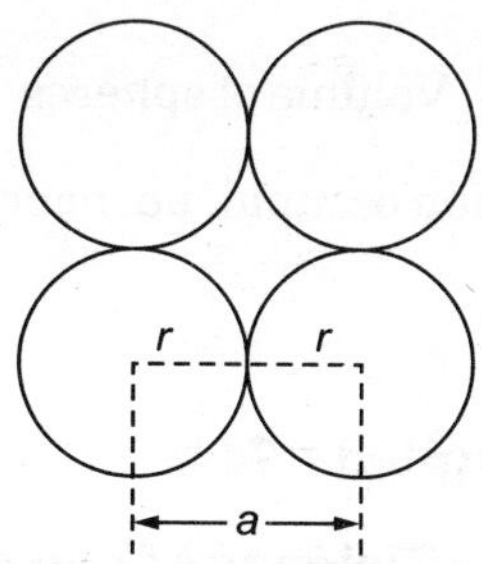

(b) Top/Bottom/Side view of simple cubic unit cell

Fig. 26.19

$$\text{No. of spheres per unit cell} = \frac{1}{8} \times 8 = 1$$

$$\text{Volume of the sphere} = \frac{4}{3}\pi r^3$$

$$\text{Volume of the cube} = a^3 = (2r)^3 = 8r^3$$

$$\therefore \quad \text{Fraction occupied, i.e. packing fraction} = \left(\frac{4}{3}\pi r^3\right)/8r^3 = \frac{\pi}{6} = 0.524$$

or **% occupied = 52.4%**

(2) In face-centred cubic structure (cubic close packing).

As sphere on the face centre is touching the spheres at the corners, evidently $AC = 4r$ (Fig. 26.20)

But from right angled triangle ABC,

$$AC = \sqrt{AB^2 + BC^2} = \sqrt{a^2 + a^2} = \sqrt{2}\,a$$

$$\therefore \quad \sqrt{2}\,a = 4r$$

$$\text{or} \quad a = \frac{4}{\sqrt{2}} r$$

$$\therefore \quad \text{Volume of the unit cell} = a^3 = \left(\frac{4}{\sqrt{2}} r\right)^3 = \frac{32}{\sqrt{2}} r^3$$

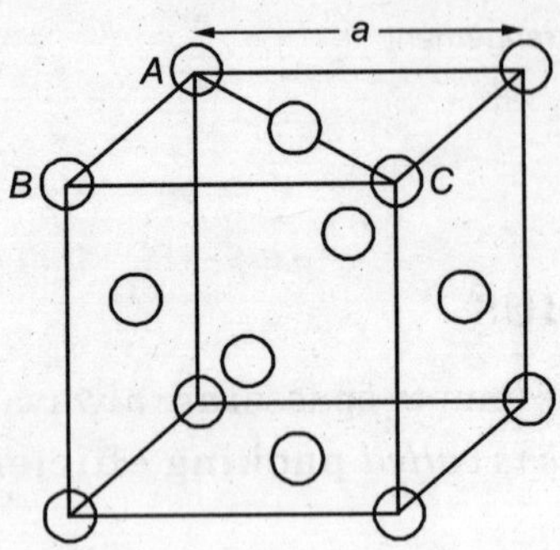

(*a*) Face-centred cubic unit cell.

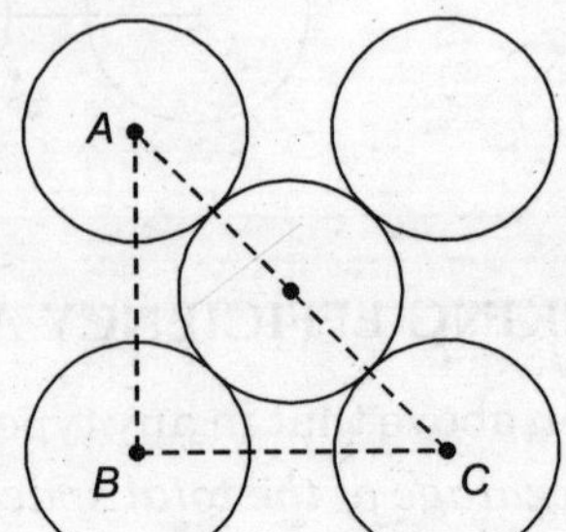

(*b*) top view of face-centred cubic unit cell.

Fig. 26.20

$$\text{No. of spheres in the unit cell} = 8 \times \frac{1}{8} + 6 \times \frac{1}{2} = 4$$

$$\text{Volume of spheres} = 4 \times \frac{4}{3}\pi r^3 = \frac{16}{3}\pi r^3$$

$\therefore$ Fraction occupid, i.e. packing fraction

$$= \frac{16\pi r^3/3}{32 r^3/\sqrt{2}} = \frac{\pi\sqrt{2}}{6} = 0.74$$

or **% occupied = 74 %**

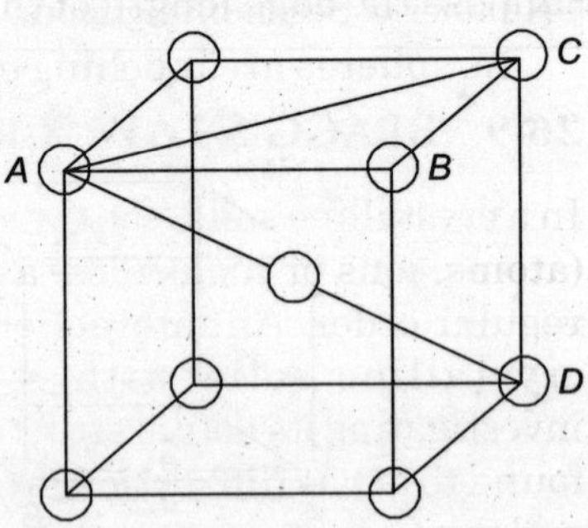

Fig. 26.21 Body-centred cubic unit cell.

(3) Packing efficiency of hcp and ccp structures.

Cubic close packing (ccp) is same as face-centred cubic packing.

Hence, packing efficiency is 74%. Further, ccp and hcp are equally possible. Hence, hcp also has a packing efficiency of 74%.

(4) In body centred cubic structure.

As the sphere at the body centre touches the spheres at the corners, therefore, AD = 4 *r* (Fig. 26.21). AD is the body diagonal

Further, face diagonal, AC $= \sqrt{AB^2 + BC^2} = \sqrt{a^2 + a^2} = \sqrt{2}\,a$

and body diagonal, AD $= \sqrt{AC^2 + CD^2} = \sqrt{2a^2 + a^2} = \sqrt{3}\,a$

$$\therefore \qquad \sqrt{3}\,a = 4\,r \quad \text{or} \quad a = \frac{4r}{\sqrt{3}}$$

$$\therefore \qquad \text{Volume of the unit cell} = a^3 = \left(\frac{4r}{\sqrt{3}}\right)^3 = \frac{64\,r^3}{3\sqrt{3}}$$

$$\text{No. of sphereas per unit cell} = 8 \times \frac{1}{8} + 1 = 2$$

$$\text{Volume of 2 spheres} = 2 \times \frac{4}{3}\pi r^3 = \frac{8}{3}\pi r^3$$

∴ Fraction occupied, i.e. packing fraction

$$= \frac{\frac{8}{3}\pi r^3}{\frac{64\,r^3}{3\sqrt{3}}} = \frac{\pi\sqrt{3}}{8} = 0.68 \text{ or } \% \text{ occupied} = 68\%.$$

Table 26.7 Characteristics of Cubic Lattices

	Simple	*Body-Centered*	*Face-Centered*
Unit Cell Volume	a^3	a^3	a^3
Lattice Points Per Cell	1	2	4
Nearest Neighbour Distance	a	$\frac{a\sqrt{3}}{2}$	$\frac{a}{\sqrt{2}}$
Number of Nearest Neighbours	6	8	12
Second Nearest Neighbour Distance	$a\sqrt{2}$	a	a
Number of Second Neighbour	12	6	6

26.9 BRAGG'S LAW: X-RAY STUDY OF CRYSTALS

In a crystalline solid, the constituent particles (atoms, ions or molecules) are arranged in a regular order. An interaction of a particular crystalline solid with X-rays helps in investigating its actual structure. Crystals are found to act as diffraction gratings for X-rays and this indicates that the constituent particles in the crystals are arranged in planes at close distances in repeating patterns. The phenomenon of diffraction of X-rays by crystals was studied by W.L. Bragg and his father W.H. Bragg in 1913.

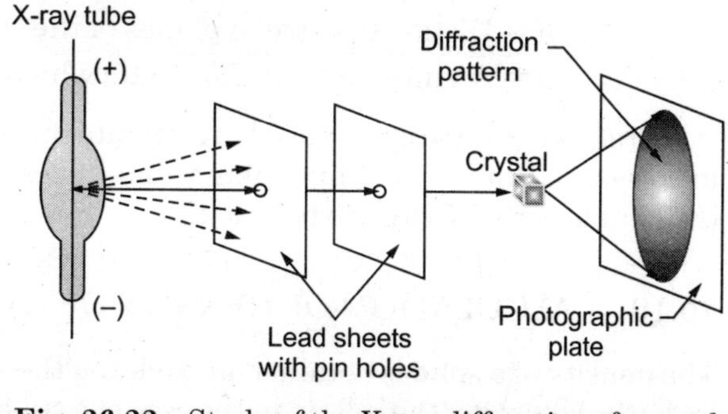

Fig. 26.22 Study of the X-ray diffraction of crystal

They used crystals of zinc sulphide (ZnS) for this purpose. The experimental setup is shown in the Fig. 26.22.

The diffraction patterns were analysed to determine the interplanar distances in a particular crystal. A simple representation of the X-ray diffraction is shown in Fig. 26.23.

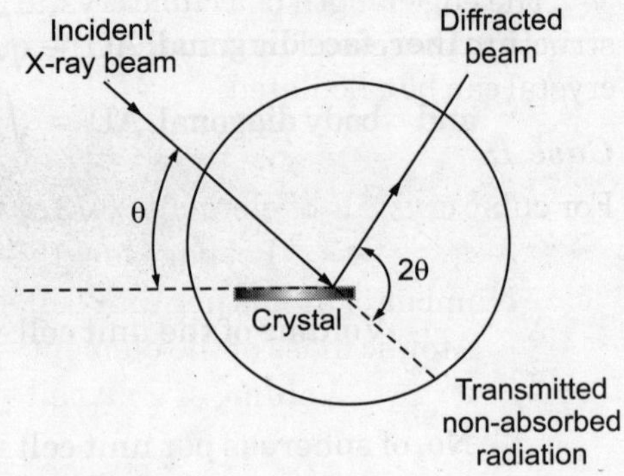

Fig. 26.23 A simple representation of X-ray diffraction

The process was based upon the principle that a crystal may be considered to be made up of a number of parallel equidistant atomic planes, as represented by lines AB, CD and EF in the Fig. 26.24.

Suppose two waves (Y and Z) of X-ray beams, which are in phase falls on the surface of the crystal. If the ray Y gets reflected from the first layer, i.e. AB line and the ray Z is reflected from the second layer of atoms, i.e. CD line, then it is evident that as compared to the ray Y, ray Z has to travel a longer distance, equal to QRS in order to emerge out of the crystal. If the waves Y and Z are in phase after reflection, the difference in distance travelled by the two rays must be equal to an integral number of wavelength (nX), for constructive interference.

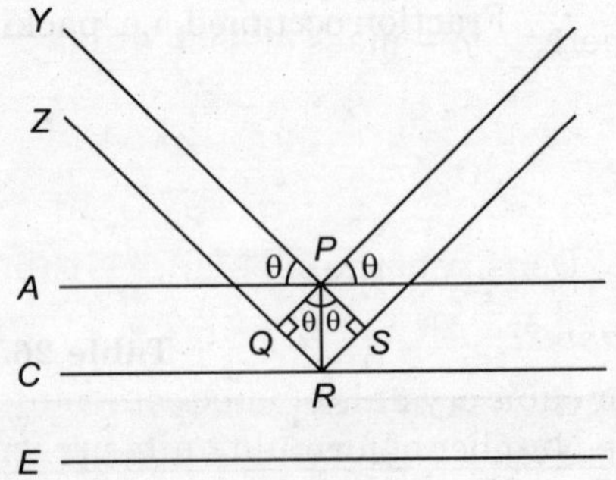

Fig. 26.24

Thus,

$$\text{Distance } QRS = n\lambda \quad \text{...(i)}$$

It is obvious from the figure that

$$QR = RS = PR \sin\theta$$

Therefore $$QRS = 2\,PR \sin\theta \quad \text{...(ii)}$$

Then, $$PR = d \quad \text{...(iii)}$$

So, from equations (i), (ii) and (iii)

$$\boxed{n\lambda = 2d\sin\theta}$$

This is called Bragg's Equations

Thus, Bragg gave a mathematical equation to establish a relationship between wave length of the incident *X*-ray, the distance between the layers and the angle of diffraction.

Here, λ = wavelength of x-ray used

θ = Angle between incident x-rays and plane of the crystal. The diffracted beam makes an angle 2θ.

d = Distance between planes of the constituent particles in a crystal.

n = An integer (1, 2, 3,4, ...etc) which represents the serial order of diffracted beams.

Bragg's equation can be used to calculate the distances between repeating planes of the particles in a crystal. Similarly, if interplanar distances are given, the corresponding wavelengths of the incident beam of X-ray can be calculated.

26.10 CALCULATION OF DENSITY OF A CUBIC CRYSTAL FROM ITS EDGE LENGTH

The density of a solid is obtained by dividing the mass of the atoms (*n atoms* × M_{atom}) and dividing by V_c the volume of the cell (a^3 in the case of a cube). If the mass of the atom is given in amu (*A*), then we have to divide it by the Avogadro number to get M_{atom}. Thus, the formula for the density is:

$$\text{Density } (\rho) = \frac{nA}{V_c N_A}$$

The edge length of a cubic crystal can be obtained from X-ray studies and knowing the crystal structure possessed by it so that the number of particles per unit cell are known, the density of the crystal can be calculated.

Case 1:

For cubic crystals of elements (e.g., Fe, Cu)

$$\text{Let edge length} = a \text{ pm}$$

$$\text{Number of atoms per unit cell} = Z$$

$$\text{Atomic mass of the element} = M$$

$$\text{Volume of Unit cell} = (a \text{ pm})^3 = a^3 \times 10^{-30} \text{ cm}$$

$$\text{Density} = \frac{\text{Mass of unit cell}}{\text{Volume of unit cell}}$$

$$\text{Mass of unit cell} = \text{Number of atoms per unit cell} \times \text{Mass of each atom} = Z \times m$$

where, $m = \text{mass of each atom} = \dfrac{\text{Atomic mass}}{\text{Avogadro's number}} = \dfrac{M}{N_0}$

$$\text{Density} = \frac{Z \times M}{a^3 \times N_0 \times 10^{-30}} \text{ gm/cc}$$

Density of the substance is same as the density of the unit cell.

Case 2:

For cubic crystals of ionic compounds (NaCl, ZnS). The formula is the same except that now, Z is the Number of formula units present in one unit cell and M is Formula mass.

$$\therefore \qquad \rho = \frac{Z \times m}{a^3 \times N_0 \times 10^{-30}} \text{ g/cm}^3$$

The above equation involves five parameters, Knowing any four, the fifth can be calculated.

With the help of the above formula and the relation between a and r of ions, the ionic radius can be calculated for ionic crystals.

SOLVED NUMERICALS

(I) Calculation of density of a cubic crystal from the edge.

Example 26.1. Chromium has mono atomic body-centred cubic structure. Its cell edge is 300 pm, what is its density? (Molar mass of $Cr = 52$ g mol^{-1}, Avogadro number $N = 6.023 \times 10^{23}$ mol),

Solution: Density $(\rho) = \dfrac{zM}{a^3 N_A}$

Given: $z = 2$ since structure is body-centred cubic

$M = 52 \text{ g mol}^{-1}$

$a = 300 \text{ pm} = 300 \times 10^{-10} \text{ cm}$

$N = 6.023 \times 10^{23} \text{ mol}^{-1}$

$$\rho = \frac{2 \times 52 \text{ g mol}^{-1}}{(300 \times 10^{-10})^3 \text{ cm}^3 \times 6.023 \times 10^{23} \text{ mol}^{-1}}$$

$$= \frac{2 \times 52}{2.7 \times 10^{-23} \times 6.023 \times 10^{23}} \text{ g cm}^{-3} = \frac{2 \times 52}{2.7 \times 6.023} \text{ g cm}^{-3} = 6.40 \text{ g cm}^{-3}$$

Example 26.2. Aluminium forms a face-centred cubic lattice whose unit cell edge length is 404 pm. Calculate the density of aluminium.

Solution: Length of side of the unit cell $= 404 \text{ pm} = 404 \text{ pm} \times \dfrac{1 \text{ cm}}{10^{10} \text{ pm}} = 404 \times 10^{-10}$ cm

$$\text{Volume of the unit cell} = (404 \times 10^{-10} \text{ cm})^3 = 6.6 \times 10^{-23} \text{ cm}^3.$$

Number of atoms per unit cell is 4 because aluminium forms a face-centred cubic lattice,

$\therefore$ $Z = 4$ atoms/unit cell.

Atomic mass of aluminium = 27 g /mol

$$\text{Density of Al} = \frac{\text{Atomic mass} \times Z}{\text{Unit cell volume (cm}^3) \times N_{Avog}}$$

$$= \frac{27 \text{ g/mol} \times 4 \text{ atoms/unit cell}}{6.6 \times 10^{-23} \text{cm}^3 \times 6.023 \times 10^{23} \text{ atoms/mol}} = 2.72 \text{ g/cm}^3.$$

Example 26.3. Sodium has a *bcc* structure with nearest neighbour distance 365.9 pm. Calculate its density (Atomic mass of sodium = 23).

Solution: For the *bcc* structure, nearest neighbour distance (*d*) is related to edge (*a*) as:

$$d = \frac{\sqrt{3}}{2} a$$

or $$a = \frac{2}{\sqrt{3}} d = \frac{2}{1.732} \times 365.9 = 422.5 \text{ pm}$$

For bcc structure, Z = 2

For sodium, M = 23

$$\therefore \quad r = \frac{Z \times M}{a^3 \times N_0} = \frac{2 \times 23 \, g \text{ mol}^{-1}}{(422.5 \times 10^{-10} \text{ cm})^3 \times (6.02 \times 10^{23} \text{ mol}^{-1})} = 1.51 \text{ g/cm}^3$$

(II) Calculation of edge and interionic distances from density.

Example 26.4. CsCl has cubic structure. Its density is 3.99 g cm^{-1}. What is the distance between Cs^+ and Cl^- ions? (At. mass of Cs = 133).

Solution: CsCl has BCC structure. It has one formula unit the unit cell. So $Z = 1$.

$$\rho = \frac{Z \times M}{a^3 \times N_0} \quad \text{or} \quad a^3 = \frac{Z \times M}{\rho \times N_0}$$

$$= \frac{1 \times (133 + 35.5) \, g \text{ mol}^{-1}}{3.99 \, g \text{ cm}^{-3} \times 6.02 \times 10^{23} \text{ mol}^{-1}} = 70.15 \times 10^{-24} \text{ cm}^3$$

$$a = (70.15)^{1/3} \times 10^{-8} \text{ cm} = (70.15)^{1/3} \times 10^2 \text{ pm } (1 \text{ pm} = 10^{-10} \text{ cm})$$

$$= 4.124 \times 10^2 \text{ pm} = 412.4 \text{ pm}$$

$$\text{Interionic distance} = \frac{\sqrt{3} a}{2} = \frac{1.732}{2} \times 412.4 = 357 \text{ pm}$$

Example 26.5. The length of the unit edge of a body centered cubic metal crystal is 352 pm. Calculate the radius of an atom of the metal.

Solution. In body-centered cubic unit cell, the radius *r* of the atom is related to the edge of the unit cell by the relation.

$$r = \frac{\sqrt{3}}{4} a = \frac{\sqrt{3} \times 354}{4} = 153.3 \text{ pm}$$

Example 26.6. Sodium chloride crystal has FCC structure. Its density is 2.163 × 103 kg m^{-3}. Calculate the edge of the unit cell cube.

Solution: ($M = 58.45 \times 10^{-3}$ kg mol^{-1}; $N_0 = 6.023 \times 10^{23}$)

$$a = \frac{zM}{pN_0}$$

z = 4 for a FCC structures

$$a = \frac{4 \times 58.45 \times 10^{-3}\ \text{kg mol}^{-1}}{2.163 \times 10^3\ \text{kg m}^{-3} \times 6.023 \times 10^{23}\ \text{mol}^{-1}}$$

$$a^3 = 179.5 \times 10^{-30}\ \text{m}^3$$

or $$a = 5.64 \times 10^{-10}\ \text{m} = 564\ \text{pm}.$$

(III) Calculation of Avogadro's Number

Example 26.7. Calculate the value of Avogadro's number from the following data: Density of NaCl = 2.165 g cm^{-3}. Distance between Na^+ and Cl^- in NaCl = 281 pm.

Solution: A Unit cell of NaCl contains 4 NaCl units, therefore, $Z = 4$, $M = 23 + 35.5 = 58.5$ g mol^{-1}

$$\rho = 2.165\ \text{g cm}^{-3} \quad \text{(Given)}$$

As distance between Na^+ and Cl^- = 281 pm

∴ Edge of the unit cell = 2 × 281 = 562 pm

Substituting these values in the expression

$$\rho = \frac{Z \times M}{a^3 \times N_0}$$

$$2.165\ \text{g cm}^{-3} = \frac{4 \times 58.5\text{g mol}^{-1}}{(562 \times 10^{-10}\ \text{cm})^3 \times N_0}$$

$$N_0 = 6.09 \times 10^{23}\ \text{mol}^{-1}.$$

Example 26.8. An element of atomic mass 98.0 g mol^{-1} occurs in FCC structure. If the unit cell edge length is 500 pm and its density is 5.22 g cm^{-3}. What is the value of the Avogadro's constant?

Solution: $N_0 = \dfrac{zM}{a^3 P}$

Given: $z = 4$

Since the structure is FCC: $M = 98.0$ g mol^{-1}

$$A = 500\ \text{pm} = 500 \times 10^{-10}\ \text{cm} = 5 \times 10^{-8}\ \text{cm}$$

$$\rho = 5.22\ \text{g cm}^{-3}$$

∴ $$N_0 = \frac{4 \times 98.0\text{g mol}^{-3}}{(5 \times 10^{-8})^3\ \text{cm}^3 \times 5.22\text{g cm}^{-3}} = 6.0 \times 10^{23}\ \text{mol}^{-1}$$

(IV) Mixed

Example 26.9. Calculate the density of Mo which forms body centered cubic crystal in which the distance between the centers of closest atoms is 274 pm. Atomic mass of Mo is 95.94.

Solution: In a body centered cubic crystal, the closest distance between touching atom is

$$d_B = 2r = \frac{\sqrt{3}}{2} a = 274\ \text{pm}$$

$$\therefore \qquad a = \frac{274 \times 2}{\sqrt{3}} = 316.4 \text{ pm} = 3.164 \times 10^{-8} \text{ cm}$$

$$\rho = \frac{zM}{N_0 a^3} = \frac{2 \times 95.94}{6.023 \times 10^{23} \times (3.164 \times 10^{-8})^3}$$

or $$\rho = 10.06 \text{ g cm}^{-3}.$$

Example 26.10. **NH_4Cl crystallises in a body centered cubic lattice with a, edge length of 400 pm. Calculate the**

(a) Distance between oppositely charged ions in the lattice.

(b) Radius of Cl^- ions if radius of that of NH_4^+ ion is 160 pm.

Solution: (a) $r_{NH_4^+} + r_{Cl^-} = \frac{1}{2}$ body diagonal $= \frac{1}{2} \times \sqrt{3}\,a$

or $$r_{NH_4^+} + r_{Cl^-} = \frac{\sqrt{3} \times a}{2} = \frac{\sqrt{3} \times 400}{2} = 346.4 \text{ pm}$$

(b) $r_{NH_4^+} = 160$ pm $\qquad r_{Cl^-} = 346.4 - 160 = 186.4$ pm

Example 26.11. **Copper which crystallises as a face-centred cubic lattice has a density of 8.930 g/ cm^3 at 20°C.**

(i) Calculate the length of the unit cell

(ii) Calculate the radius of a copper atom.

Solution: The number of atoms per unit cell is 4 because copper crystallises as face centred cubic lattice.

$$Z = 4 \text{ atoms/unit cell}$$

$$\text{Atomic mass of cooper} = 63.5 \text{ g/mol}$$

$$\text{Density} = \frac{\text{Atomic mass} \times Z}{\text{Unit cell volume (cm}^3) \times N_0}$$

$$\text{Unit cell volume} = \frac{\text{Atomic mass} \times Z}{\text{Density} \times N_0}$$

$$= \frac{63.5\, g/\text{mol} \times 4 \text{ atoms/unit cell}}{8.930\,\text{g/cm}^3 \times 6.023 \times 10^{23} \text{ atoms/mol}}$$

$$= 4.727 \times 10^{-23} \text{ cm}^3\text{/unit cell}$$

$$\therefore \quad \text{Length of side of the unit cell} = (V \text{ cm}^3)^{1/3} = [4.722 \times 10^{-23} \text{ cm}^3\text{/unit cell}]^{1/3}$$

$$= 3.616 \times 10^{-8} \text{ cm}$$

$$= 3.616 \times 10^{-8} \text{ cm} \times \frac{10^{10} \text{ pm}}{1 \text{ cm}} = 361.6 \text{ pm}$$

For a face centred cubic lattice $l = \sqrt{8}.r$

or $$r = \frac{1}{\sqrt{8}} = \frac{361.6 \text{ pm}}{\sqrt{8}} = \mathbf{128 \text{ pm.}}$$

Example 26.12. **Chromium crystallizes in a body centred cubic lattice with a density of 7.20 g cm^{-3}. Given that the length of an edge of a unit cell is 288.4 pm, calculate Avogadro's number.**

Solution: The number of atoms per unit cell is 2 because chromium crystallises as a body centred cubic lattice.

$$Z = 2 \text{ atoms/unit cell}$$

$$\text{Atomic mass of chromium} = 52 \text{ g/mol}$$

$$\text{Length of edge of the unit cell} = 288.4 \text{ pm}$$

$$= 288.4 \text{ pm} \times \frac{1 \text{ cm}}{10^{10} \text{ pm}}$$

$$= 288.4 \times 10^{-10} \text{ cm}$$

$$\text{Volume of the unit cell} = (288.4 \times 10^{-10} \text{ cm})^3$$

$$\text{Density} = \frac{\text{Atomic mass} \times Z}{\text{Unit cell volume cm}^3 \times N_{Avog}}$$

$$N_0 = \frac{\text{Atomic mass} \times Z}{\text{Unit cell voluem cm}^3 \times \text{Density}}$$

$$= \frac{52 \text{ g/mol} \times 2 \text{ atoms/unit cell}}{2.4 \times 10^{-23} \text{ cm}^3 \times 7.20 \text{ g/cm}^3} = 6.02 \times 10^{23} \text{ atoms/mol.}$$

Example 26.13. The length of a side of a unit cell of cesium chloride as determined by X-ray diffraction pattern is 412.1 pm. What is the density of cesium chloride? Cesium chloride crystallises as a body centred cubic lattice.

Solution: Length of edge of the unit cell $= 412.1 \text{ pm} \times \dfrac{1 \text{ cm}}{10^{10} \text{ pm}} = 412.1 \times 10^{-10} \text{ cm}$

$$\text{Volume of the unit cell} = (412.1 \times 10^{-10} \text{ cm})^3 = 7.00 \times 10^{-23} \text{ cm}^3$$

$$\text{Formula mass of CsCl} = 133 \text{ a.m.u.} + 35.5 \text{ a.m.u.} = 168.5 \text{ a.m.u.}$$

The number of formula units (ion-pairs) per unit cell of cesium chloride is 1 as it has a body centred cubic lattice.

$\therefore \quad Z = 1$ Formula unit/Unit cell

$$\text{Density} = \frac{\text{Formula mass} \times Z}{\text{Unit cell volume cm}^3 \times N_0} = \frac{168.5 \text{g/mol} \times 1 \text{ Formula unit/unit cell}}{7.00 \times 10^{-23} \text{ cm}^3 \times 6.023 \times 10^{23} \text{ Formula units/mol}}$$

$$= 4.0 \text{ g/cm}^3.$$

Note: The body centred cubic unit cell of a metal has 2 atoms per unit cell whereas the body centred unit cell of an ionic compounds of the type AB has 1 formula unit per unit cell.

Example 26.14. An element 'A' (atomic mass 100) having b.c.c. structure is unit cell edge 400 pm. Calculate the density of A and the number of unit its for 10 g of A.

Solution: Length of the unit cell edge $= 400 \text{ pm} = 400 \text{ pm} \times \dfrac{1 \text{ cm}}{10^{10} \text{ pm}} = 400 \times 10^{-10} \text{ cm}$

$$\text{Volume of the unit cell} = (400 \times 10^{-10} \text{ cm})^3 = 6.4 \times 10^{-23} \text{ cm}^3$$

Number of atoms per unit cell is 2 because the element forms a body centred lattice.

$$Z = 2 \text{ atoms/unit cell}$$

Atomic mass of the element = 100 g/mol

$$\text{Density of element} = \frac{\text{Atomic mass} \times Z}{\text{Unit cell volume (cm}^3) \times N_0}$$

$$= \frac{100 \text{ g/mol} \times 2 \text{ atoms/unit cell}}{6.4 \times 10^{-23} \text{ cm}^3 \times 6.023 \times 10^{23} \text{ atoms/mol}} = 5.188 \text{ g/cm}^3$$

$$\text{Volume of 10 g of } A = \frac{\text{Mass}}{\text{Density}} = \frac{10\text{g}}{5.188\text{g cm}^{-13}} = 1.9275 \text{ cm}^{-1}$$

Number of unit-cells is 1.9275 cm^3 volume

$$= \frac{\text{Volume of the substance}}{\text{Unit cell volume}} = \frac{1.9275 \text{ cm}^3}{6.4 \times 10^{-23} \text{ cm}^3}$$

$= 3.0 \times 10^{22}$ unit cells.

Example 26.15. The density of KBr is 2.75 g cm^{-3}. The length of edge of the unit cell is 654 pm. Predict the type of cubic lattice to which unit cell of KBr belong.

(N_0 = 6.023 × 1023 mol^{-1}, At. mass: K = 39, Br = 80)

Solution: For cubic crystals, $\rho = \dfrac{Z \times M}{a^3 \times N_0}$

or

$$Z = \frac{\rho \times a^3 \times N_0}{M} = \frac{(2.75\text{g cm}^{-3})(654 \times 10^{-10} \text{ cm})^3 \times (6.023 \times 10^{23} \text{ mol}^{-1})}{(39 + 80)\text{g mol}^{-1}}$$

$= 3.89 = 4$

Thus, there are four formula units of KBr present per unit cell. Hence, it has face-centred cubic lattice (similar to that of NaCl).

26.11 STRUCTURES OF SIMPLE IONIC COMPOUNDS

(a) Rock Salt (or NaCl) Type of Structure

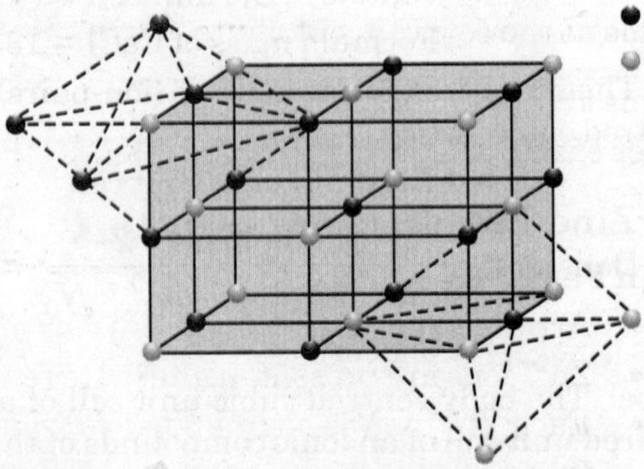

Fig. 26.25

Main Features

- It has a face centred arrangement (or CCP). Cl^- ions occupy the corners and face centres, Na^+ occupy body centre and edge centers.
- Each Na^+ is surrounded by six Cl^- and each Cl^- is surrounded by 6 Na^+. Therefore it has 6 : 6 coordination.
- Number of NaCl Units per unit cell.

Cl^- ions

8 at the corrners contribute 1/8 per unit cell $8 \times \dfrac{1}{8} = 1$

6 at the face center contribute $\dfrac{1}{2}$ per unit cell $6 \times \dfrac{1}{2} = 3$

total number of Cl^- ions = 4

Na^+ ions

12 at the edge centers contribute $\dfrac{1}{4}$ per unit cell $12 \times \dfrac{1}{4} = 3$

One at the body center contributes fully = 1 × 1 = 1

Total number of Na^+ = 3 + 1 = 4

Therefore total NaCl units in one unit cell = 4

(b) Caesium Chloride (CsCl) Type of Structure

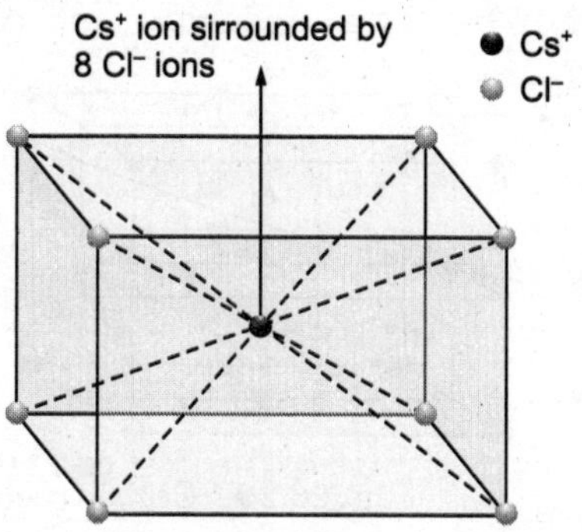

Fig. 26.26

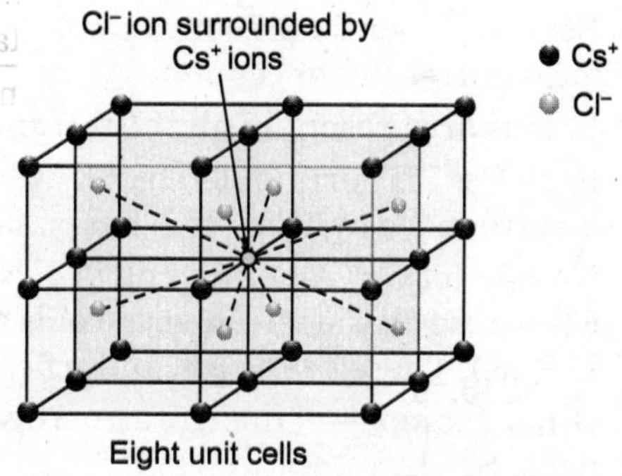

Fig. 26.27

Main Features

- It has body centred cubic (bcc) arrangement, with Cs^+ at the centre and Cl^- at the corners.
- Each Cs^+ ion is surrounded by 8 Cl^- ions and each Cl^- ions is surrounded by 8 Cs^+ ions. Therefore the crystal has 8.8 co-ordination.
- It has one CsCl per unit cell because.

Atoms at the corners contribute $\frac{1}{8}$ so $8 \times \frac{1}{8} = 1$

Atoms at the centre contribute fully, i.e. 1.

Other examples: CsBr, Csl, CsCN, TiCl, TiBr, TiCN, etc.

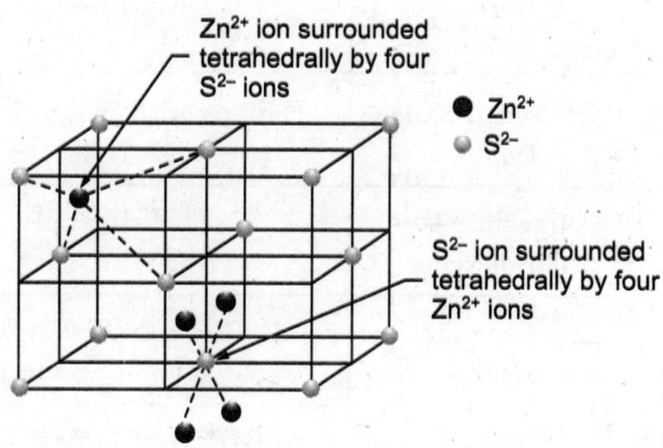

Fig. 26.28 ZnS structure

(c) Zinc Blende (ZnS) Structure

Main Features

- Arrangement is fcc on ccp. S^{2-} ions are present at the corners and face centres.
- Zn^{2+} ions are present in alternate tetrahedral voids.
- Each Zn^{2+} ion is surrounded tetrahedrally by S^{-2} ions each and S^{-2} ion is surrounded tetrahedrally by Zn^{2+} ions. Therefore, this structure has 4 : 4 coordination.
- Number of S^{2-} ions per unit cell 8 at corners contributed

$$\frac{1}{8} \text{ per unit cell} = 8 \times \frac{1}{8} = 1$$

6 at the face centre contribute

$$\frac{1}{2} \text{ per unit cell} = 6 \times \frac{1}{2} = 3$$

Total number of S^{-2} per unit cell = 4.

Since Zn^{2+} ions occupy half the tetrahedral sites, the number of Zn^{2+} ions per unit cell will be 4.

Total number of ZnS per unit cell = 4.

Other examples : CuCl, CuBr, Cul, Agl, BeS.

(d) Fluorite Structure (CaF_2)

Main Features

- Each Ca^{2+} ion is present at the corners and the face centres (fcc or ccp).
- F^- ions are present in all the tetrahedral voids.
- Each Ca^{2+} is surrounded by $8F^-$ where as each F^- is surrounded by $4Ca^{2+}$ so it has 8 : 4 co-ordianation.
- Number of Ca^{2+} per unit cell is 4. Number of F^- is 8 (because all the tetrahedral voids are occupied by F^- ions). Therefore $4CaF_2$ units per unit cell.
- Other examples of fluorite structure: BaF_2, $BaCl_2$, SrF_2, $SrCl_2$.

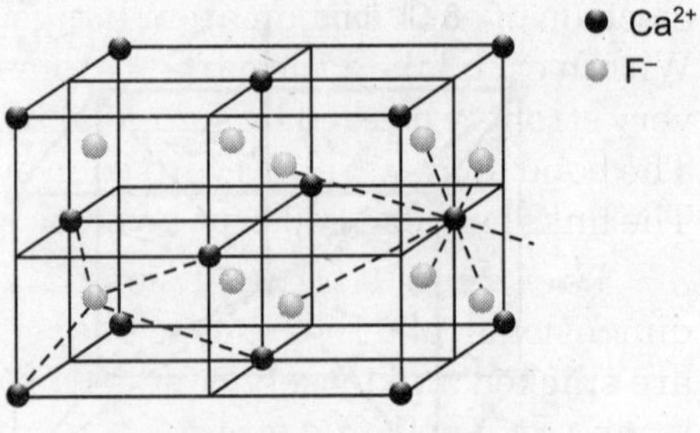

Fig. 26.29 CaF_2 structure

Summary of Main Characteristics of Ionic Solids

Crystal Structure	*Examples*	*Co-ordination number*	*Number of formula units per unit cell*
Rock Salt (NaCl) Type	Halides of Li, Na, K, Rb, AgF, AgBr	$Na^+ = 6$ $Cl^- = 6$	4
Caesium Chloride type	CsCl, CsBr, CsI TiCl, TiBr, Til	$Cs^+ = 8$ $Cl^- = 8$	1
Zinc Chloride type (ZnS)	CuCl, CuBr, CuI, AgI	$Zn^{+2} = 4$ $S^{2-} = 4$	4
Fluorite (CaF_2) Type	BaF_2, $BaCl_2$, SrF_2, $SrCl_2$, CaF_2, PBF_2	$Ca^{2+} = 8$ $F^- = 4$	4
Anti Fluorite (Na_2O) type	Na_2O	$Na^+ = 4$ $O^{2-} = 8$	4

Effect of Temperature and Pressure on the Crystal Structure

At ordinary temperatures and pressure, chlorides, bromides and iodides of lithium, sodium and rubidium and some halides of silver possess the NaCl structure with 6 : 6 co-ordination.

On application of high pressure they transform to the CsCl structure with 8 : 8 co-ordination. On the other hand CsCl on heating transforms to the NaCl Structure with 6 : 6 co-ordination.

26.12 GRAPHITE

Graphite is a second allotrope of carbon. As is the case with diamond, pure graphite contains only carbon atoms. However, in the case of graphite, the atoms are arranged in layers

Graphite itself comes in two allotropic forms.

- Naturally occurring graphite is called beta-graphite and it comes in a hexagonal form.
- Synthetically produced graphite is called alpha-graphite and it comes in a rhombohedral form.

The hexagonal form of graphite has carbon atoms arranged in a hexagon; the hexagons form a plane. Each carbon atom in the hexagon is attached to only three other carbon atoms. The fourth electron of each carbon atom hangs loosely and is not bound. Graphite shows sp^2 hybridisation. Within each layer, the carbon atoms are very strongly bonded by sigma (a) bonds. The bond angles are 120 with each other. The links between adjacent layers is weak

The resulting network is 2-dimensional, and the resulting flat sheets are stacked and loosely bonded through weak *van der Waals forces*.

The sheet like structure with weak inter-layer forces makes graphite a very soft substance. The melting point of graphite is relatively high (3700 °C). This can be explained because of the strong overlap of the sigma (σ) bonds within the carbon atoms in the planes. Sigma bonds are known to be strong and they need a large amount of energy to break. Hence, just like diamond, graphite too has a high melting point.

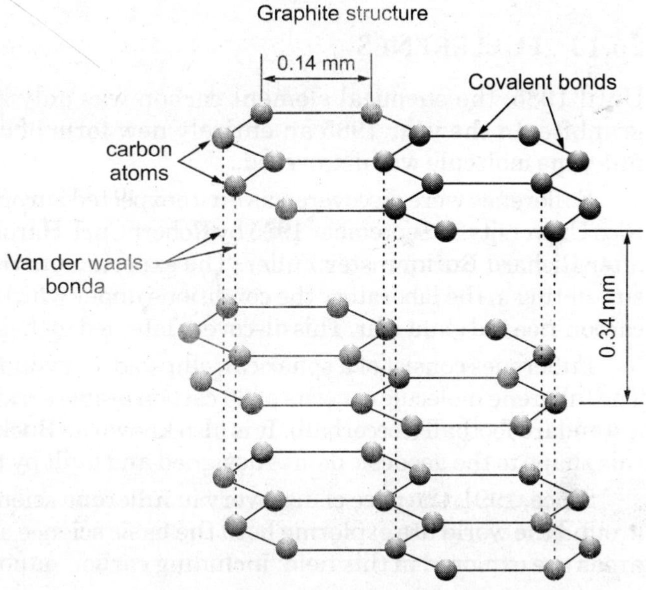

Fig. 26.30 Structure of Graphite

Bond Length

In graphite, the C-C bonds are all equal in length and intermediate between normal single and double bonds. This "delocalisation" can be explained by noting that graphite has many different but equivalent "resonance structures". (The resonance structures differ only in the placement of the double bonds.) Two of these are drawn for a portion of graphite in Fig. 26.31.

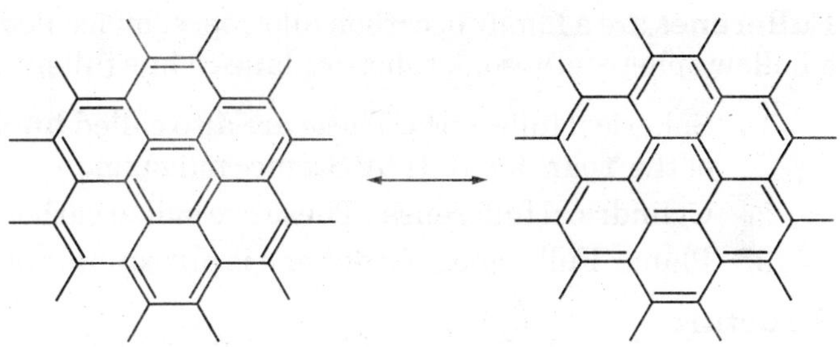

Fig. 26.31: Two resonance structures of a portion of graphite

Of the three bonds, two are single bonds and one is a double bond. Therefore the overall bond order is 1.33. Graphite is a soft, slippery substance, attributed to the fact that the layers are not chemically bonded and can slide over one another.

The most stable is *graphite* with which we write every time you use a pencil. It writes because graphite solidifies as infinite sheets of hexagons which glide over one another and onto your paper easily. (Pencils are called "lead pencils" because the soft metal lead will leave a similar dark mark, but there's no lead in a pencil...only graphite.)

Conductivity:

Graphite is a good conductor of heat and electricity. This can be understood by looking at the loosely bound fourth electron of each carbon atom. Because of the delocalisation of one of the outer electrons of each atom to form a *π-cloud*, graphite conducts *electricity*, but only in the plane of each *covalently bonded* sheet. The electron is relatively free to travel and move about in the crystal. The electrons are carriers of heat and electricity in a crystal. Hence, it is clear why graphite is a good conductor of heat and electricity.

Density of graphite is 2.2 gm/cm^3. From the structure of graphite it is clear that the distance between two hexagonal layers is large; thus the carbon atoms are not rigidly bound. This gives rise to low density of graphite as observed.

26.13 FULLERENES

Until 1985, the chemical element carbon was only known to exist in two forms -diamond and graphite. In the year 1985 an entirely new form of carbon, which became known as C_{60} or the fullerene molecule was discovered.

Fullerenes were discovered as an unexpected surprise during laser spectroscopy experiments at Rice University in September 1985 by Robert Curl, Harold Kroto and Richard Smalley, and are named after Richard Buckminster Fuller. The experiments that led to the discovery of C_{60} were aimed at simulating in the laboratory the conditions under which carbon nucleates in the atmosphere of a cool carbon-rich red giant star. This discovery later led to their award of the 1996 Nobel Prize in Chemistry.

Fullerenes consist of a spherical, ellipsoid, or cylindrical arrangement of dozens of carbon atoms The fullerene molecule consists of 60 carbon atoms arranged in pentagons and hexagons, very like in a standard football (soccerball). It is also known as Buckminster Fullerene due to the resemblance of this shape to the geodesic domes designed and built by the architect R Buckminster Fuller.

Since, 1991, the pace of discovery in fullerene science has continued to accelerate. Researchers around the world are exploring both the basic science and potential applications of fullerenes. New areas are explored in this field, including carbon nanotubes and nanowires.

Types

Fullerenes are a family of carbon allotropes, molecules composed entirely of carbon, in the form of a hollow sphere, ellipsoid, tube, or plane . Thus fullerenes are of the following types:

1. Spherical fullerenes: These are also called **buckyballs.** Buckyball was named "Molecule of the Year" for 1991 by *Science* magazine
2. Cylindrical fullerenes : These are called carbon nanotubes or **buckytubes.**
3. Planar Fullerenes : Graphene is an example of a planar fullerene sheet.

Structure

Fullerenes are similar in structure to graphite, which is composed of stacked sheets of linked hexagonal rings, but may also contain pentagonal (or sometimes heptagonal) rings that would prevent a sheet from being planar.

The structure of fullerene (spherical) is Truncated icosahedron. The **truncated icosahedron** is an Archimedean solid. It comprises 12 regular pentagonal faces, 20 regular hexagonal faces, 60 vertices and 90 edges

The Truncated Icosahedron:

The proposed structure for C_{60} "truncated icosahedron", is derived from an icosahedron by truncating or "snipping off each of the twelve vertices (see Fig. 26.32). Hence, each vertex is replaced by a five-membered ring - a pentagon. This snipping process also converts each of the twenty former triangular faces into six-membered rings-hexagons.

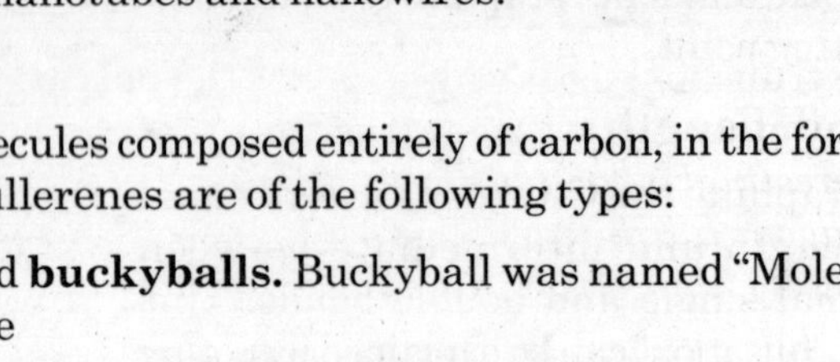

Fig. 26.32 Icosahedron (left) and "truncated icosahedron" (right)

Bonding in C_{60}

Fullerenes consist of 20 hexagonal and 12 pentagonal rings as the basis of an icosohedral symmetry closed cage structure. In the proposed structure for C_{60}, each vertex of the truncated icosahedron is occupied by a carbon atom Each carbon atom is bonded to three others and is sp^2 hybridised. The sp^2 hybridised orbitals form the three sigma bonds with the adjacent carbon atoms . The remaining 2*p* unhybridised orbital (2pz) is responsible for the pi-bond.

Clearly, the proposed structure for C_{60} is not planar. The angle θ between a p_x axis and a C-C bond vector, is 101.6° (as compared to 90° in planar graphite). The bowl-shape or concavity at each sp^2 carbon center introduces some strain into the molecule. However, the high symmetry distributes that strain evenly across the entire structure.

The C_{60} molecule has two bond lengths - the hexagonal ring bonds can be considered "double bonds" and are shorter than the pentagonal bonds. C_{60} is not "superaromatic" as it tends to avoid double bonds in the pentagonal rings, resulting in poor electron delocalisation. As a result, C_{60} behaves like an electron deficient alkene, and reacts readily with electron rich species. The electronic bonding factors in the structure account for the stability of the molecule.

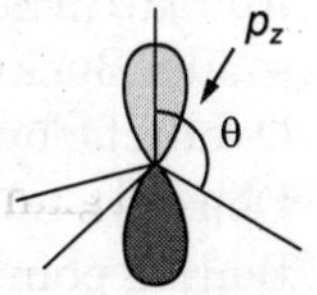

Fig. 26.33 Theta (θ) is defined as the angle between the pi axis and a carbon-carbon bond vector

Other Fullerenes

The truncated icosahedron is not the only possible fullerene-type structure. There are, in fact, many other hollow cage structures that can be constructed using only pentagons and hexagons . *Interestingly, each of these structures contains exactly twelve pentagons, while the number of hexagons is arbitrary.* The pentagons are necessary for closure. The number of vertices in closed fullerene-type structures is necessarily even.

"Buckytubes" & "Buckyonions": Many fullerenes have now been discovered in carbon soot, including tubes of carbon many thousands of times long as they are wide, with the same icosahedral structure as the carbon nanotube.

Synthesis:

The first method of production of fullerenes used laser vaporization of carbon in an inert atmosphere, but this produced microscopic amounts of fullerenes. In 1990, a new type of apparatus using an arc to vaporise graphite was developed in Germany by Kratschmer and Huffmann

1. Kraitchmer synthesis: In 1990 physicists W. Kratschmer and D.R. Huffman for the first time produced isolable quantities of C60 in graphitic carbon "soot", produced by evaporating graphite electrodes *via* resistive heating in an atmosphere of ~ 100 torr helium (1 atm = 760 torr)

Although the soot contained only a few per cent by weight of C_{60}, it could be conveniently extracted using benzene as solvent. The red-brown benzene solution could be decanted from the black insoluble soot and then dried using gentle heat, leaving a residue of dark brown to black crystalline material.

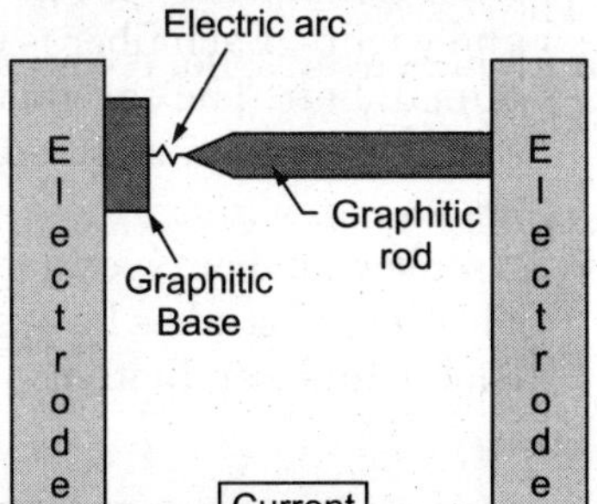

Fig. 26.34 Schematic diagram of the contact-arc apparatus used to generate macroscopic quantities of C_{60}

2. Contact Arc Apparatus: Smalley modified the above method. In the Smalley apparatus (Fig. 26.34), an electric arc is maintained between two nearly contacting graphite electrodes. Hence, most of the power is dissipated in the arc *and not in resistive heating of the rod.* The entire electrode assembly is enclosed in a reaction kettle that is filled with ~ 100 torr pressure of helium. Black soot, like that observed by Kraitschmer, is produced, and extraction with organic solvents yields fullerenes.

Physical Properties Of C_{60}

Some of the physical properties of fullerenes are:

1. Appearance: While Buckyball soot is a very finely divided black powder, the C_{60} is black solid.
2. Odour: Odourless
3. Density: 1.65 g cm^{-3}
4. Boiling point: Sublimes at 800K
5. Solubility: The fullerenes are soluble in common solvents such as benzene, toluene or chloroform. If one shakes up some fullerene soot with toluene and filter the mixture, one obtains a red solution. As the solvent evaporates, crystals of pure carbon appear
6. Resistivity: 1014 ohms m^{-1}
7. Crystal form : Hexagonal cubic

5. Applications of Fullerenes

Because of their unique structures and properties, fullerenes draw lots of attention physicists, chemists and engineers, who are trying to find some potential applications new carbon structure

- A fullerene nanotube has tensile strength about 20 times that of high-strength steel alloys, and a density half that of aluminum.
- Carbon nanotubes demonstrate superconductive properties, and single nanotubes up to 4 centimeters in length have been synthesized. A range of companies exists to develop nanotubes for commercial applications, including computer memory, electronic wires, and materials science. One day nanotubes could be used to create futuristic computers not possible with conventional lithographic techniques.
- One potentially important application for these carbon nanotubes is in the area of composites. Carbon fibers, made from organic polymers, are used to strengthen lightweight high-tech materials such as the carbon/epoxy resins used in golf clubs, tennis racquets, bicycle frames, and yachts. Each fiber is 6,000-10,000 nm in diameter or about five times thinner than a human hair. The carbon nanotubes are substantially thinner, 4-30 nm in diameter, and could be an even more effective strengthening agent than carbon fibers in carbon/epoxy resins.
- A somewhat more speculative use for carbon nanotubes is as molecular wires (of "nanowires")- Theoretical studies have suggested that the extremely small diameters of the tubes could lead to high conductivity, comparable to that of metals at room temperature.

MOOD RINGS

Mood rings were first seen as an extremely popular fad in the late 1970s .The idea behind a mood ring is simple. Wear it on your finger and it will reflect the state of your emotions. The ring's stone should be dark blue if you're happy, and it supposedly turns black if you are anxious or stressed. While mood rings cannot reflect your mood with any real scientific accuracy, they actually are indicators of your body's involuntary physical reaction to your

The stone in a mood ring is either a hollow glass shell filled with *thermotropic liquid crystals*, or a clear glass stone sitting on top of a thin sheet of liquid crystals. These liquid crystal molecules are very sensitive: they change position, or twist, according to changes in temperature. This change in molecular structure affects the wavelengths of Hsihi that are absorbed or reflected by the liquid crystals, resulting in an apparent change in the colour of the stone. For example, as the temperature increases, the liquid crystal molecules twist slightly in one direction. This twist causes the liquid crystal substance to absorb more of the red and green portions of the visible light, and reflect the blue part. This causes the stone to appear dark blue. When the temperature decreases, the molecules begin to twist in the other direction, and reflect a different portion of the spectrum.

The inside of the ring conducts heat from your finger to the liquid crystals in the "stone." The colour **green,** which signifies "average" on the mood ring colour scale, is calibrated to the **surface temperature** of a typical person, approximately 82 degrees Fahrenheit (28 degrees Celsius). If your surface temperature varies far enough from the norm, then the liquid crystals in the stone alter enough to cause a change in the colour reflected. And if you take a mood ring off, it will normally change to black unless the ambient temperature is very high.

Take a look at the Mood Ring colours listed below, and what "mood" they represent. The colours are listed according to the change in temperature they represent, with dark blue being the warmest and black the coolest.

- Dark blue: Happy, romantic or passionate
- Blue: Calm or relaxed
- Blue-green: Somewhat relaxed
- Green: Normal or average
- Amber: A little nervous or anxious
- Gray: Very nervous or anxious

Black: Stressed, tense or feeling harried

QUESTIONS

1. What is the number of atoms present per unit cell in a face centred cubic and a body centred cubic arrangement ?
2. Define a unit cell. How many atoms are present in a simple cubic unit cell ?
3. What do you understand by the terms 'space lattice' and 'unit cell'?
4. Define coordination number in crystals? Is the coordination number of a sphere in *hcp* and *ccp* arrangements same or different?
5. Discuss the following types of cubic structure:
   (i) Simple cubic
   (ii) Body-centred cubic
   (iii) Face-centred cubic.
6. Calculate the number of atoms in different cubic unit cells of monoatomic substances in the above three types of cubic structures.
7. How can you calculate the density of a crystal whose length of the the edge of the unit cell is known?
8. Derive Bragg's Equation for diffraction of X-rays by crystals.

EXAMINATION QUESTIONS

1. Derive Bragg's Equation. How is it used to determine the internal crystal structure ?
[U.P. Tech., 1st/2nd Sem., 2001-02]

2. (a) Derive Bragg's equation for diffraction of X-rays by crystals.
   (b) Derive an expression for the density (ρ) of a cubic crystal.
   (c) Write short note on fullerenes, indicating the method of preparation, assuming that four molecules of NaCl are associated per unit cell. *[U.P. Tech., 1st Sem., 2004-05]*

3. (a) Write down a brief note on fullerences and its applications.
   (b) Calculate density of a BCC crystal. Solid of cube is 4Å and M = 60
   (c) Derive Brag's equation. *[U.P. Tech., 1st Sem., 2006-07]*

4. Explain the structure of graphite. Also explain the reasons for its electrical and lubricating properties. *[U.P. Tech., 2nd Sem., 2006-07]*

5. What is Bragg's law ? How do you calculate the density of unit cell ?
[U.P. Tech., B.Tech., 1st Sem., 2007-08]

6. (a) An FÇC structure of an element has density 10.3 gm/cm^3 and has an edge length 310 pm. Calculate the atomic mass of the element. *[U.P. Tech, B. Tech, Ist Sem., 2007-08]*
   (b) Derive Bragg's equation.
   In Bragg's reflection of X-ray, a reflection was found at 30° with lattice plane of spacing 1.87Å. If this is a second order reflection, calculate the wavelength of *X*-rays.
   [U.P. Tech, B. Tech, Ist Sem., 2007-08]
   (c) Calculate the radius ratio for trigonal backing of spheres is solids.
   [U.P. Tech, B. Tech, Ist Sem., 2007-08]

7. Bragg's equation is __________. *[U.P. Tech, B. Tech, Ist Sem., 2008-09]*

8. An edge of cubic cell of NaCl crystal is 6.5 × 10^{-8} cm. Assuming that four molecules of NaCl are associated per unit cell, calculate its density.
   Given: Avogadro's number = 6.023 × 10^{23} *[U.P. Tech, B. Tech, Ist Sem., 2008-09]*

9. (a) What do you understand by liquid crystalline state ? Discuss the classification of liquid crystals and write their applications.
   (b) Discuss properties and applications of fullerene. *[U.P. Tech, B. Tech, Ist Sem., 2008-09]*

10. Derive Bragg's equation for diffraction of X-rays by crystals.
[U.P. Tech, B. Tech, Ist Sem., 2008-09]

11. Write short note on fullerene. *[U.P. Tech, B. Tech, Ist Sem., 2008-09]*

❑❑❑

27

LIQUID CRYSTALS

Knowledge is a treasure, but practice is the key to it.

— ***Thomas Fullen***

CHAPTER OUTLINE

27.1 INTRODUCTION

Normally, we consider matter to have three distinct states: solid, liquid, and gas. Solids, namely crystals, are highly ordered, in which the constituting atoms or molecules possess positional as well as orientational order. On contrary, simple liquids like water or gases exhibit no order at all. However, there are states of matter which do not meet the necessary requirements of any of these categories. For example, a substances such as mayonnaise is somewhere between a liquid and a solid.

Liquid crystals *are substances that exhibit a phase of matter that has properties between those of a conventional liquid, and those of a solid crystal.* For instance, a liquid crystal (LC) may flow like a liquid, but have the ***molecules*** in the liquid arranged and/or oriented in a crystal-like way. Liquid crystals exhibit intermediate thermodynamic phases between the crystalline solid and simple liquid state These phases are called *liquid crystalline* phases, or *mesomorphic* phases (*mesophases).*

Molecules which form Liquid Crystals:

What makes a liquid crystal different from ordinary liquids is the shape of its molecules. Molecules of substances with a liquid crystal state are generally oblong and rigid, that is, rod-shaped. They are composed of moderate size organic molecules which tend to be elongated and shaped like a cigar. Following are the *typical Liquid Crystalline Molecules:*

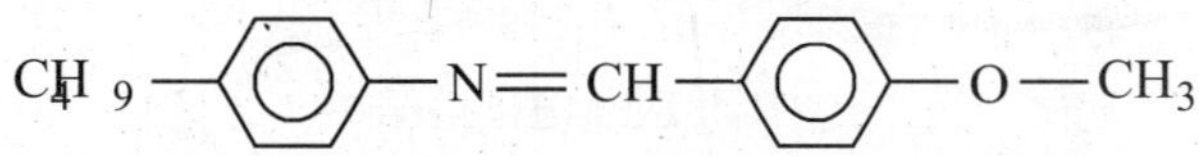

Methoxybenzilidene Butylanaline ("MBBA")

Other examples include:

- cholesterol ester
- surfactants
- glyco lipids
- phenyl benzoates
- paraffines
- cellulose derivatives

Mesogen: Mesogen is the fundamental unit of a *liquid crystal* that induces structural order in the crystals. Typically, a liquid-crystalline molecule consists of a rigid *moiety* and one or more flexible parts. The rigid part aligns molecules in one direction, whereas the flexible parts induce fluidity in the liquid crystal. This rigid part is referred to as mesogen, and it plays a crucial role in the molecule. The optimum balance of these two parts is essential to form liquid-crystalline materials.

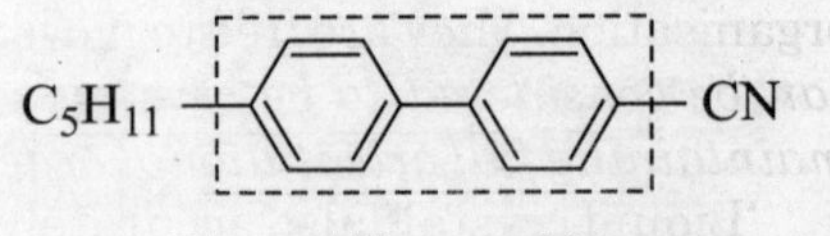

(*a*) Rigid rod-like part = Mesogen

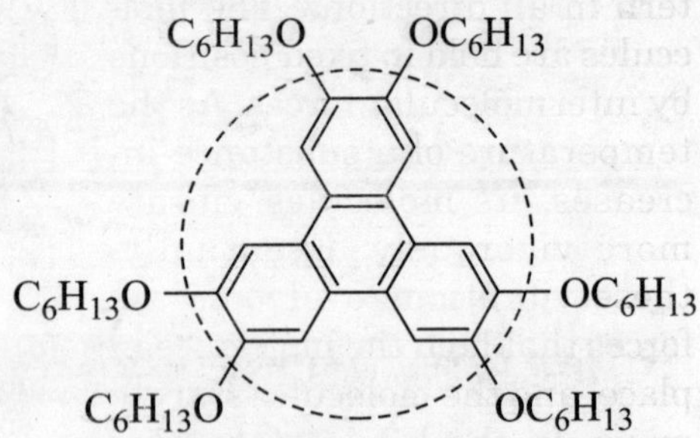

(*b*) Rigid disk-like part = Mesogen

Fig. 27.1 Typical building blocks of liquid crystals

Calamitic and Discotic Liquid Crystals

Liquid cystals usually consist of steric "rod like-" (*calamitic*) or "disk-like" (*discotic*) organic molecules which tend to align themselves with a long range order due to anisotropic intermolecular forces.

- In a *calamitic liquid, crystal,* the mesogen is a rod-like structure composed of two or more *aromatic* and *aliphatic* rings connected in one direction. [Fig. 27.1(a)]
- In a *discotic liquid crystal,* the flat-shaped aromatic core that makes molecules stack in one direction is defined as the mesogen. [Fig. 27.1(b)]

Other examples include:

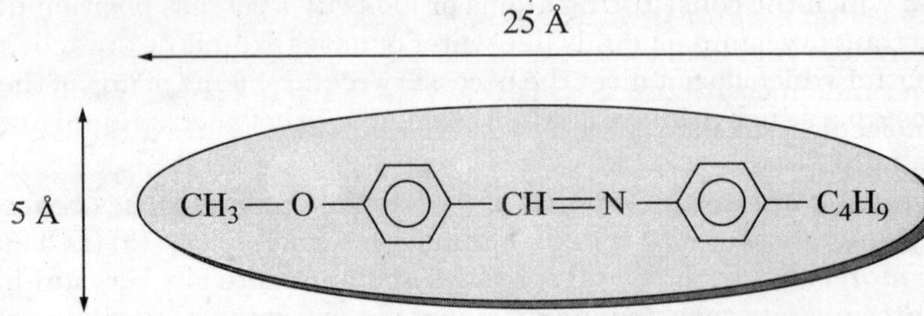

MIBBA
(4-methoxybenzyli den-4' -butyl anilin)

Rod like (calamitie)

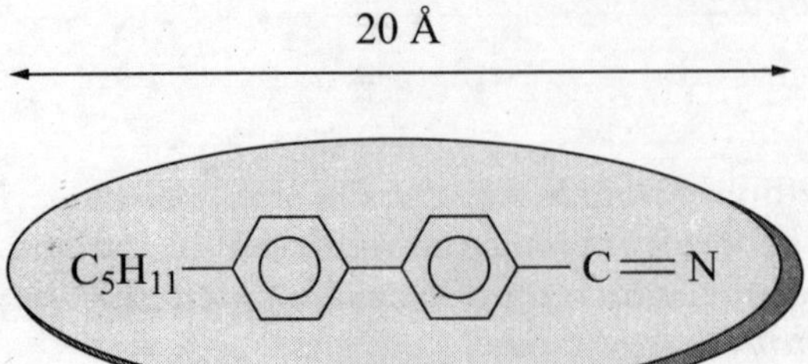

5CB
(4-pentyl-4'-cyanobiphenyl)

Disk like (discotic)

Typical Behavior of Liquid Crystals:

Even if the molecules' *positions* are random, their *orientations* can be aligned with one another in a regular pattern—and that's what creates the ordered structure of a liquid crystal. For example, imagine a large number of toothpicks put into a rectangular box and shaken. When you open the box, the toothpicks will be facing in about the same direction, but will have no definite spatial organisation. They are free to move, but like to line up in about the same direction. *Liquid crystals can be considered to be crystals which have lost some or all of their positional order, while maintaining full orientational order.*

'Liquid crystal" also accurately describes the arrangement of molecules in this state. In the crystalline solid state, as represented in Fig. 27.2(*a*), the arrangement of molecules is regular, with a regularly repeating pattern in all directions. The molecules are held in fixed positions by intermolecular forces. As the temperature of a substance increases, its molecules vibrate more vigorously. Eventually, these vibrations overcome the forces that hold the molecules in place, and the molecules start to move. In the liquid state, this motion overcomes the intermolecular forces that maintain a crystalline state, and the molecules move into random positions, without pattern in location or orientation, as represented in Fig. 27.2(*b*).

(*a*) Arrangement of molecules in a solid crystal (b) Arrangement of molecules in a liquid (c) Arrangement of molecules in a liquid crystal

Fig. 27.2

In materials that form liquid crystals, the intermolecular forces in the crystalline solid are not the same in all directions; in some directions the forces are weaker than in other directions. As such a material is heated, the increased molecular motion overcomes the weaker forces first, but its molecules remain bound by the stronger forces. This produces a molecular arrangement that is random in some directions and regular in others. The arrangement of molecules in one type of liquid crystal is represented in Fig. 27.2(*c*). The molecules are still in layers, but within each layer, they are arranged in random positions, although they remain more or less parallel to each other. Within layers, the molecules can slide around each other, and the layers can slide over one another. This molecular mobility produces the fluidity characteristic of a liquid crystals are also not quite liquid and not quite solid. Physically, they are observed to flow like liquids, but they have some properties of crystalline solids.

27.2 CLASSIFICATION OF LIQUID CRYSTALS

Liquid crystals can be divided into *thermotropic* and *lyotropic* LCs.

(i) The transitions between the liquid crystals are given by the change in temperature and hence these are called *thermotropic.*

(ii) The phase transitions may also depend on concentration of the mesogen in the solvent, and these liquid crystals are called *lyotropic.*

While thermotropic are presently mostly used for technical applications, *lyotropic* are important for biological systems, e.g. membranes

Thermotropic Liquid Crystals are classified as:

(i) Isotropic Phase

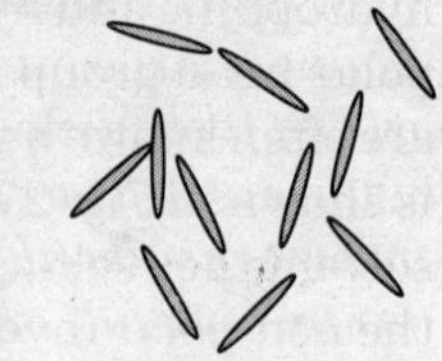

In the isotropic phase the molecules are randomly aligned and exhibit no long range order.

The isotropic phase has a low viscosity and will often appear to be crystal clear. There is no long range positional or orientational order of the molecules, although this sort of order may exist on very short length scales of order tens of Angstroms, corresponding to a few molecular distances. For all practical purposes, the isotropic phase macroscopically appears to be like any other isotropic liquid such as water.

(ii) Nematic Phase

The simplest liquid crystal phase is called the nematic phase (N). It is characterised by a high degree of long range orientation order but no translational order. Molecules in a particular direction. Nematics are polarisable rod-like organic molecules on the order of 20 Angstroms in length. Nematics have fluidity similar to that of ordinary (isotropic) liquids but they can be easily aligned by an external magnetic or electric field. An twisted (aligned) nematic has the optical properties of a uniaxial crystal and this makes them extremely useful in *liquid, crystal displays* (LCD).

The digital watch you used to wear back in the 80's functioned using nematic liquid crystals. Today, many more useful and interesting properties of nematics are known and exploited. This is the phase which is used in many *liquid crystal devices* (e.g., the "twisted nematic" cell), because the average orientation may be manipulated with an electric field.

(iii) Smectic Phase

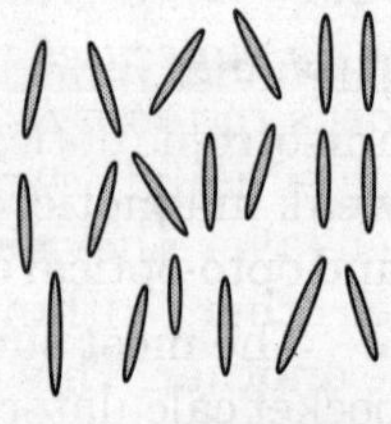

The smectic phases, which are found at lower temperatures than the nematic, form well-defined layers that can slide over one another like soap. The smectics are thus positionally ordered along one direction.

Smectic Liquid crystals are different from nematics in that they have one more degree of orientational order than do the nematics. There are several different categories to describe smectics. The two best known of these are :

Smectic A, in which the molecules are aligned perpendicular to the layer planes, and

Smectic C, in which the alignment of the molecules is at *some arbitrary angle* to the normal.

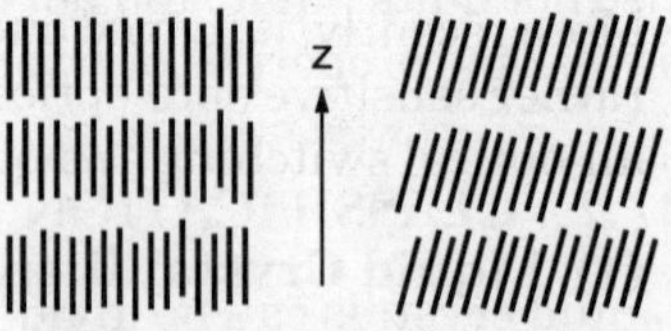

The smectic phase, like the nematic, exhibits long range orientational order of the molecules. The viscosity is rather high, and this phase is generally not useful for devices.

(vi) Chiral Nematic:

Another common phase is cholesteric, also known as chiral nematic. In this phase, the molecules twist slightly from one layer to the next, resulting in a spiral formation.

This phase shows nematic ordering but the preferred direction rotates throughout the sample. An example of this is shown in Fig 27.3.

Lyotropic Liquid Crystals:

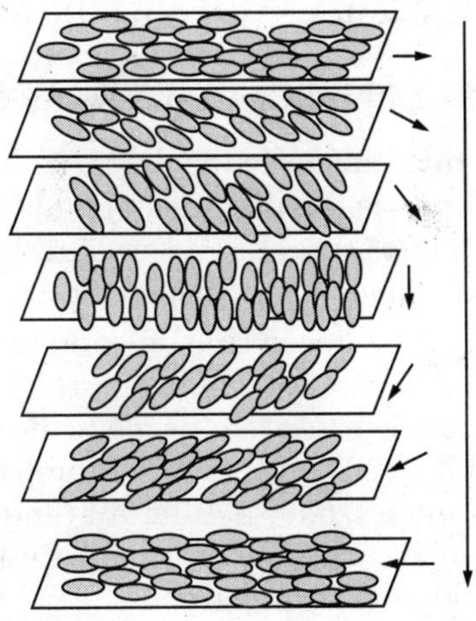

Fig. 27.3 Schematic diagram of a chiral nematic liquid

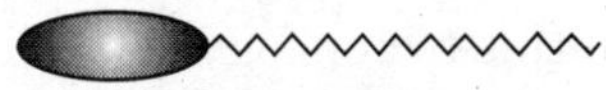

Fig. 27.4 Schematic diagram of an amphiphilic molecule

Lyotropic liquid crystal phases are formed by amphiphilic (both hydrophilic and hydrophobic) molecules .These often consist of a polar head group attached to one or more non-polar chains and are often known as *surfactants* (surface active agents). A schematic is shown in Fig. 27.4. When these are dissolved in an appropriate solvent they self-assemble so the polar (hydrophilic) heads protect the non-polar (hydrophobic) tails. These structures are known as micelles.

Lyotropic liquid-crystalline phases are abundant in living systems, the study of which is referred to as *polymorphism*. In particular, *biological membranes* and *cell membranes* are a form of liquid crystal. Their constituent molecules (e.g., *phospholipids*) are perpendicular to the membrane surface, yet the membrane is capable of a range of elastic stress. These lipids vary in shape. The constituent molecules can inter-mingle easily, but tend not to leave the membrane due to the high energy requirement of this process. Lipid molecules can flip from one side of the membrane to the other, this process being catalysed by *flippases* and floppases (depending on the direction of movement).

Many other biological structures exhibit LC behaviour. For instance, the concentrated protein solution that is extruded by a spider to generate *silk* is, in fact, a liquid crystal phase. The precise ordering of molecules in silk is critical to its renowned strength.

27.3 APPLICATION OF LIQUID CRYSTALS

The most remarkable features of liquid crystals with respect to applications are due to their anisotropic optical properties. Orientation order can be changed easily, e.g. with the help of rather weak magnetic, electric or optical fields, leading to a number of magneto-optical, electro-optical and opto-optical effects.

The most successful application are *liquid crystal displays* well-known from wrist watches, pocket calculators or flat screens of laptop computer which take advantage of electro-optical effects. More recently, it turned out that orientational order can be also affected by optical fields leading to rather sensitive opto-optical effects and *nonlinear optical properties*, which are important, e.g. for all-optical switching and other photonic devices in future optical information technologies.

(1) Liquid Crystal Displays

The most common application of liquid crystal technology is liquid crystal displays (LCDs.) This field has grown into a multi-billion dollar industry, and many significant scientific and engineering discoveries have been made.

Liquid crystals rely on the *optical* properties of certain liquid crystalline substances in the presence or absence of an *electric field.*

A *liquid crystal* **display (LCD)** is a thin, flat *display device* made up of any number of colour or *monochrome pixels* arrayed in front of a *light source* or *reflector*. It is often utilised in battery-powered *electronic* devices because it uses very small amounts of *electric power*. LCD (liquid crystal display) is the technology used for displays in notebook and other smaller computers. Like light-emitting diode (LED) and gas-plasma technologies, LCDs allow displays to be much thinner than cathode ray tube (CRT) technology. LCDs consume much less power than LED and gas-display displays because they work on the principle of blocking light rather than emitting it.

(2) Liquid Crystal Thermometers

Chiral nematic (cholesteric) liquid crystals reflect light with a wavelength equal to the pitch. Because the pitch is dependent upon temperature, the colour reflected also is dependent upon temperature. Liquid crystals make it possible to accurately gauge temperature just by looking at the colour of the thermometer.

The "mood ring", a popular novelty a few years ago, took advantage of the unique ability of the chiral nematic liquid crystal. More important and practical applications have been developed in such diverse areas as medicine and electronics. Special liquid crystal devices can be attached to the skin to show a "map" of temperatures. This is useful because often physical problems, such as tumours, have a different temperature than the surrounding tissue. Liquid crystal temperature sensors can also be used to find bad connections on a circuit board by detecting the characteristic higher temperature.

(3) Optical Imaging

An application of liquid crystals that is only now being explored is optical imaging and recording. In this technology, a liquid crystal cell is placed between two layers of photoconductor. Light is applied to the photoconductor, which increases the material's conductivity. This causes an electric field to develop in the liquid crystal corresponding to the intensity of the light. The electric pattern can be transmitted by an electrode, which enables the image to be recorded. This technology is still being developed and is one of the most promising areas of liquid crystal research.

(4) Other Liquid Crystal Applications

Liquid crystals have a multitude of other uses. They are used for nondestructive mechanical testing of materials under stress. This technique is also used for the visualisation of RF (radio frequency) waves in waveguides. They are used in medical applications where, for example, transient pressure transmitted by a walking foot on the ground is measured. Low molar mass (LMM) liquid crystals have applications including erasable optical disks, full colour "electronic slides" for computer-aided drawing (CAD), and light modulators for colour electronic imaging.

Liquid crystal technology has had a major effect many areas of science and engineering, as well as device technology. Applications for this special kind of material are still being discovered and continue to provide effective solutions to many different problems.

QUESTIONS

1. Write short notes on: (a) Liquid crystals (b) Graphite.
2. What are the different types of liquid crystals?
3. Discuss fullerenes and their applications.
4. Discuss the applications of liquid crystals.

❑❑❑

28 ORGANOMETALLICS COMPOUNDS

"Every great advance in science has issued from a new audacity of imagination".

— *John Dewey*

CHAPTER OUTLINE

28.1 INTRODUCTION

An organometallic compound is generally a compound which possesses a metal-carbon bond. Organometallic chemistry can be viewed as a bridge between organic and inorganic chemistry. The importance of organometallic compounds lies in the fact that these are excellent catalysts which play a major role in the production of pharmaceuticals agrichemicals, flavours and semiconductors, etc. Organometallic chemistry is rapidly growing field of chemistry. The discovery of Grinard reagents in 1900 not gave versatile intermediates for a variety of organic preparations.

It may be noted that not all the compounds containing carbon and a metal atom are organometallic. The term 'organometallic' is reserved to the compounds which contain atleast one Metal-Carbon bond. An alkoxide such as $(C_3H_7O_4)$ Ti is not considered as organometallic compound because here the organic group is bonded to Ti through oxygen and not through carbon. Whereas the compound $C_6H_5Ti\ (OC_3H_7)_3$ is an organometallic compound because in this compound Ti is bonded to carbon of the organic group

$$(C_6H_5 — Ti)$$

$$Ti — (O_3C_4H_7)$$

Not an organometallic compound

$Ti\ (OC_3H_7)_3$

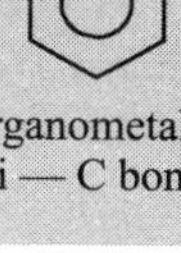

Organometallic (Ti — C bond)

The common examples of organometallic compounds are those $Pb(C_2H_5)_4$ (tetracthyl lead), $(C_5H_5)_2$ Fe (ferrocene), $K[PtCl_3 (C_2H_4)]$ (Zeise salt), etc.

Despite the rigorous definition, the boderline that distinguishes organometallic chemistry from other branches is unclear. For example metal cyanides (M–CN) and metal carbides (CaC_2, Al_4C_3 etc.) contain Metal–Carbon bond but are *not* considered organometallics because their properties are closely related to those of many other compounds of inorganic chemistry. On the other hand, $Ni(CO)_4$ (nickel tetracarbonyl) and other complexes having the ligand CD are considered organometallic because of the resemblence of their properties to organometallics.

Organometallics are also known as organo-inorganics, metallo-organics and metal-organics. Organometallic compounds are named by prefixing the metal with organo-, e.g. organopalladium compounds, organophosphorus compounds, organomagnesium compounds.

Reactivity

Organometallics are very reactive compounds and hence are kept in solutions in organic solvents (especially with H_2O, O_2, etc.).

The organometallic compounds form electron rich or anionic carbon atoms, i.e. carbanians during their reactions. This means, they will function as bases or nucleophiles. Thus it is reasonable to write the general formula of these compounds as R^-M^+.

$CH_3 \rightarrow Cl$	$CH_3 — Li$
alkyl halide	**Organometallic alkyl lithium**
Methyl group has lower electron density and hence is an electrophile	The methyl group has higher electron density and hence is a nucleophile

The alkyl group in alkyl halides acts as electrophile whereas the alkyl group in organometallics acts as nucleophile. In otherwords organolithium and organomagnesium compounds are strong bases since the negative charge is on carbon.

It is the carbonionic nature of organometallics that is responsible for their usefulness as synthetic reagents.

28.2 CLASSIFICATION

The organometallic compounds can be classified on the basis of the nature of metal-carbon bond.

(1) Ionic compounds of electropositive metals:

The organosodium and organopotassium compounds are essentially ionic organometallic compounds. In these type of compounds the bonding between the metal and the carbon is ionic. The carbon of the organic group (hydrocarbon) carry a negative charge which is strongly attracted by the positively charged metal ion by electrostatic forces of attraction.

These organometallic compounds are formed by the most electropositive metals of group IA and IIA. The formation of these compounds is favoured if the negative charge on the carbon atom is stabilised, i.e. it is delocalised over several carbon atoms.

Example: $K^+C_5H_5^-$ (potassium cyclopentadienyl). The negative charge is stabilised due to delocalisation over the five carbon atoms of cyclopentadienyl ring.

Other examples include $Na^+C_6H_5^-$, $Na^+C_4H_9^-$ etc. These compounds contain a very reactive C_5H_5 group. These compounds are referred to as metal cyclopentadienides. These organometallic compounds show behaviour typical of ionic compounds and are insoluble in organic or non-polar solvents. The reactivity of these compounds depend upon the stability of the anion. However, there are certain metal cyclopentadienides which have substantial amount of covalent character present.

(2) Covalent organometallic compounds

These compounds have organic past bonded to the metal atom by a normal sigma covalent bond. These compounds are very common and are generally formed by Zn, Cd, Hg and representative elements of group III, IV and V. The covalent organometallic compounds are generally metal-alkyl compounds. Examples are $Pb(C_2H_5)_2$, $(C_6H_5)_2Zn$, $(CH_3)_2Cd$, C_4H_9Li, etc. The polarity of the covalent bond between metal and alkyl groups depends upon the difference in electronegativity between metal atom and carbon atom.

(3) Electron deficient organometallic compounds

These type of compounds include the compounds having bridging alkyl groups. For example, dimeric trialkyl aluminium which have bridging alkyl groups.

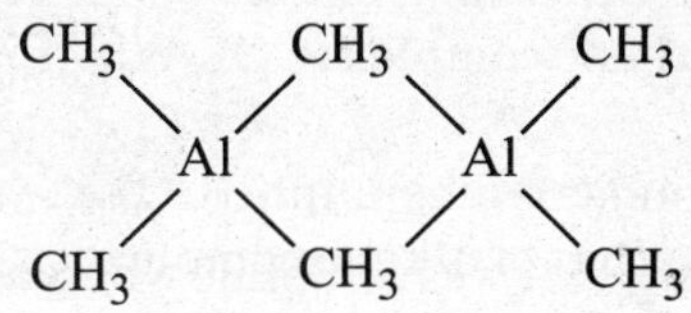

Dimeric dialkyl aluminium

(4) Organometallics involving p bonds

These constitute an important class of organometallics having metal-ligand bond between a metal and the π orbitals of organic ligands such as alkenes, benzene ring, etc. Ferrocene was the first of many π complexes which came to be known as metallocenes. In these complexes the metal ion is sandwitched between two parallel carbocyclic rings became known as "sandwich" compounds. Example,

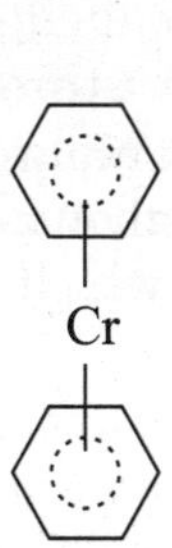

$(C_6H_5)_2$ Cr

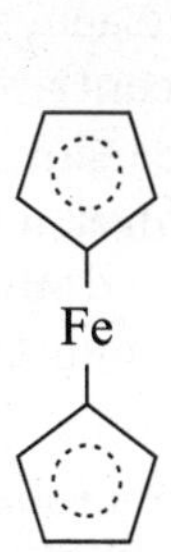

$(C_5H_5)_2$ Fe

common example of ligands forming such type of organometallic compounds are :

Complexes of alkenes are well known for transition metals like Pt(II), Pd(II), Cu(II), Ag(I), Hg(II) etc. The first compound of alkene with the metal ion was reported by Zeise (a Danish Chemist) and is known as Zeise salt. $K[PtCl_3.(C_2H_4)]$.

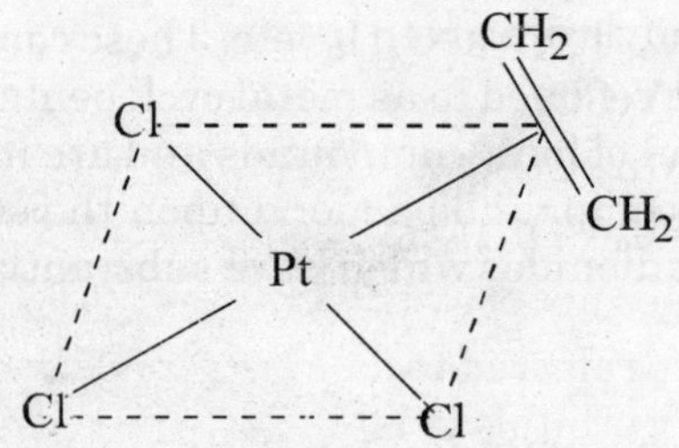

Zeise salt
(πcomplex)

(5) Carbene/Carbyne complexes

The chemistry of metal-carbon single bond dates back to 19th century (e.g. Grignard's reagent) Transition metal compounds containing metal carbon double bond or triple bond have come to be understood more recently.

$$M—CR_3, \quad M{=}CR_2, \quad M{\equiv}CR$$

Metal Carbones Metal Carbynes

If these occurs multiple bonding between metal and carbon, the bond strength would be increased. It has been observed that in carbene compounds, the metal-carbon bond is shorter than metal-alkyl sigma bond but larger than metal-carbon double bond. Thus metal carbenes are generally stable organometallics. The first carbene complex $(CO)_5$ W[:CRX] was synthesized by Noble Laureate in 1964 by E.O. Fisher.

Metal Carbynes have triple bonds. For example $(CO)_4M$ ($\equiv$CR)X. The chemistry of alkyne complexes is more complicated than that of alkene complexes.

(6) Transition-metal Organometallic Compounds

A large number of organometallics are based on transition metals. Examples include organometallics containing iron, nickel, chromium, platinum and rhodium.

A transition-metal complex consists of a transition-metal to which are attached groups called *ligands*. Essentially, anything attached to a metal is a ligand. A ligand can be an element (O_2, N_2), a compound (NO), or an ion (CN^-); it can be inorganic as in the examples just cited or it can be an organic ligand. Ligands differ in the number of electrons that they share with the transition metal to which they are attached. Carbon monoxide is a frequently encountered ligand in transition-metal complexes and contributes two electrons; it is best through of in terms of the Lewis structure $:\bar{C}\equiv\overset{+}{O}:$ in which carbon is the reactive site. An example of a carbonyl complex of a transition metal is nickel carbonyl, a very toxic substance, which was first prepared over a hundred years ago and is an intermediate in the purification of nickel. It forms spontaneously when carbon monoxide is passed over elemental nickel.

$$\underset{\text{Nickel}}{Ni} + \underset{\text{Carbon monoxide}}{4CO} \rightarrow \underset{\text{Nickel carbonyl}}{Ni(CO)_4}$$

Almost all the transition metals forms compounds in which CO act as a ligand. The 18 electron rule is obeyed by these complexes remaskably well (99% time). In these compounds the metal atom are always in their lowest oxidation states, mostly zero, low positive or negative oxidation state. These complexes are of 3 types:

(i) Mononuclear Carbonyls
(ii) Binuclear Carbonyls
(iii) Polynuclear Carbonyls

Example of the stable carbonyls are given below:

Mononuclear—	*$V(CO)_6$	$Cr(CO)_6$	$Ni(CO)_4$
Binuclear—	$Mn_2(CO)_{10}$	$Fe_2(CO)_9$	$Co_2(CO)_8$
Polynuclear—	$Fe_3(CO)_{12}$	$Co_4(CO)_{12}$	

The mononuclear $V(CO)_6$ does not obey the 18 electron rule.

The metal carbonyls are also referred to as complexes of π-*accepter* ligands. This is because metal carbonyls posses & vacant π orbitals in addition to lone pairs. Due to the excessive negative charge on the central metal atom (by the donation of electrons from the ligands), the *back donation* of electrons from the metal atom to the vacant π-orbitals of CO takes place. Hence, the CO is known as π- accepter.

18 Electron Rule (Sidgwick, 1927)

Many transition metal complexes containing organic, carbonyl or hydride ligands, which are isolable at room temperature, can be regarded as having 18 electrons in the valence shell of the metal. In this way, the metal formally attain the electron configuration of the next higher noble gas [18 valence electron rule is also known as the inert gas rule, **effective atomic number** (EAN) rule].

To determine the electron count for a metal complex:

(1) Determine the oxidation state of the transition metal centre(s) and the metal centres resulting *d*-electron count. To do this one must:

(2) Add up the electron counts for the metal centre and ligands

Complexes with 18 e- counts are referred to as *saturated*, because there are no empty low-lying orbitals to which another incoming ligand can coordinate. Complexes with counts lower than 18e- are called *unsaturated* and can electronically bind additional ligands.

Example:

(i) Tetracarbonyl nickel (0), $Ni(CO)_4$

Atomic number of Ni = 28

Ni(0) in $Ni(CO)_4$ = 28

Four CO groups give two electrons each (4 × 2) = 8

EAN of $[Ni(CO)_4]$ = 36

Thus, the complex $[Ni(CO)_4]$ obeys EAN rule.

(ii) Hexacarbonyl chromium (0), $Cr(CO)_6$

Atomic number of Cr = 24

Cr(0) = 24

Six CO groups (6 × 2) = 12

EAN = 36

Thus, EAN rule is obeyed.

(iii) Pentacarbonyl iron (0), $Fe(CO)_8$

Atomic number of Fe = 26

Fe(0) = 26

Five CO groups (5 × 2) = 10

EAN = 36

Thus, EAN rule is obeyed.

(iv) **Pentacarbonyl manganate (—I) ion, $[Mn(CO)_5]^-$**

Atomic number of Mn = 25

Mn(–1)	= 26
Five CO groups (5 × 2)	= 10
EAN	= 36

Thus, $[Mn(CO)_5]^-$ obeys EAN rule.

General Methods of Synthesis of Organometallics

(1) BY REDUCTIVE SUBSTITUTION OF ALKYL HALIDES: (PREPARATION OF METAL ALKYLS)

The first reported organometallic compounds were prepared by the reductive substitution of alkyl halides. The following equations illustrate these reactions for the commonly used metals lithium and magnesium (R may be hydrogen or alkyl groups in any combination).

$R_3C\text{-}X + 2Li \rightarrow R_3C\text{-}Li + LiX$ An Alkyl Lithium Reagent

$R_3C\text{-}X + Mg \rightarrow R_3C\text{-}MgX$ A Grignard Reagent

The alkyl magnesium halides described in the second reaction are called *Grignard Reagents* after the French chemist, Victor Grignard, who discovered them. The solvent used in this reaction is anhydrous diethyl ether. Typical solvents are normally anhydrous diethyl ether but pentane or hexane can also be used.

The other metals react in a similar manner, but the two shown here are the most widely used.

$R\text{-}X + Zn \rightarrow R\text{-}Zn\text{-}X$	An Alkyl Zinc Reagent	1849 E. Frankland
$R\text{-}X + Mg \rightarrow R\text{-}Mg\text{-}X$	A Grignard Reagent	1900 V. Grignard
$R\text{-}X + 2Li \rightarrow R\text{-}Li + LiX$	An Alkyl Lithium Reagent	1930 K. Ziegler

Reactions of organolithium and Grignard reagents reflect the nucleophilic (and basic) character of the functional carbon in these compounds

$$\% \text{ Ionic Character of } H_3C - \text{Metal}$$

$$(CH_3)_2Hg < (CH_3)_2Cd < (CH_3)_2Zn < (CH_3)_2Mg < CH_3Li$$

These reactions are obviously substitution reactions, *but they cannot be classified as nucleophilic substitutions*, as were the earlier reactions of alkyl halides. Because the functional carbon atom has been reduced, the polarity of the resulting functional group is inverted (an originally electrophilic carbon becomes nucleophilic). This change, shown below, makes alkyl lithium and Grignard reagents unique and useful reactants in synthesis.

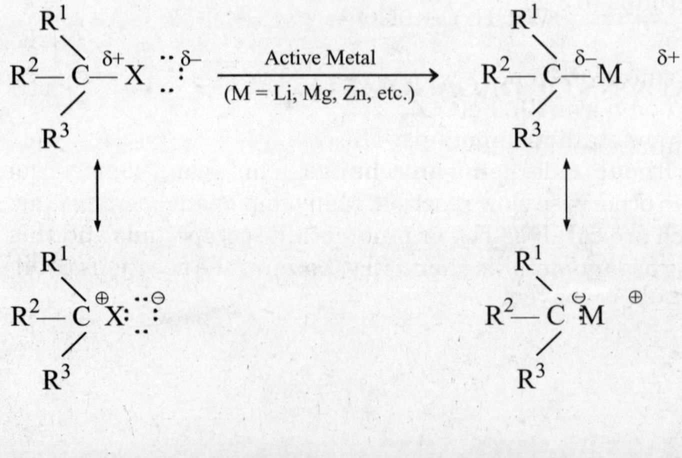

(2) METAL EXCHANGE REACTIONS

Alternative methods of preparing a wide variety of organometallic compounds generally involve an exchange reaction in which a given metal is either moved to a new location or replaced by a new metal, which may include B, Al, Ti, V, Fe, Ni, Cu, Mo, Ru, Pd, Sn, Pt, Hg & Pb. Some of these reactions are:

Type 1: $R–M + R'–H \rightleftharpoons R–H + R'–M$

Examples:

$$R—C \equiv R—C + C_2H_5MgBr \longrightarrow R—C \equiv R—MgBr + C_2H_6$$

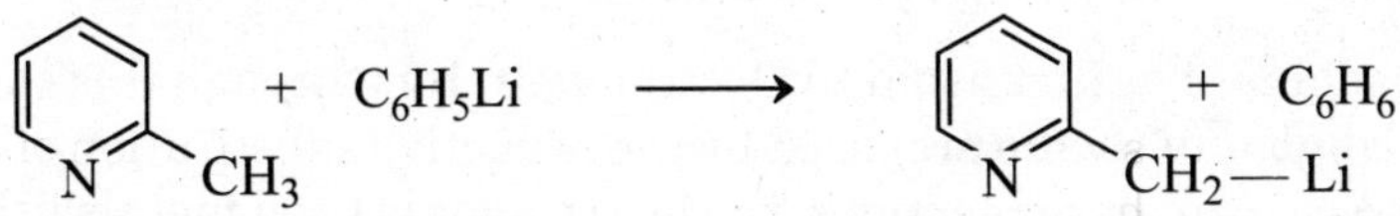

Type 2: $R–M + M \rightleftharpoons R–M' + M$

R-M is often a dialkylmercury compound.

Examples:

$$3(C_6H_5)_2Hg + 2Al \rightarrow 2(C_6H_5)_3Al + 3Hg$$

$$(C_2H_5)_2Hg + Mg \rightarrow 2(C_2H_5)_2Mg + Hg$$

Type 3: $R–M + M'X \rightleftharpoons R–M' + MX$

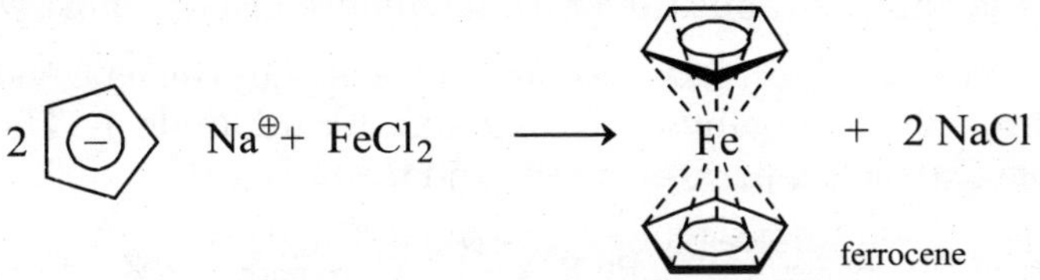

(3) SYNTHESES OF METAL CARBONYLS

Metal carbonyls can beprepared in a variety of ways:

- For Ni and Fe, binary metal carbonyls can be made by the direction interaction with the metal

$$Ni + 4CO \xrightarrow{25°C} Ni(CO_4)$$

- In other cases, a reduction of a metal precursor in the presence of CO (or using CO as the reductant) is used

$$W(CH_3)_6 + 9CO \rightarrow W(CO)_6 + 3(CH_3)_2CO$$

28.3 APPLICATIONS OF ORGANOMETALLICS

Catalysis by organometallic compounds: The catalyst is defined as a substance, which alters the rate of reaction without undergoing any change in mass and composition. Generally, catalyst is used to speed up an otherwise slow reaction. Many important chemicals are produced commercially by reactions which are catalysed by organometallic compounds and this is one of the important factor for studying organometallic chemistry. Some of the reactions catalysed by organometallics are given below:

(1) **Alkene Hydrogenation** (Wilkinson's Catalysis) Although the reaction of hydrogen gas with ethylene is thermodynamically favourable, it does *not* take place at room temperature and pressure.

$$H_2C=CH_2 + H_2 \rightarrow H_3C-CH_3$$

However, in the presence of the organometallic complex chlorotris (tri-phenylphosphine) rhodium(I), $(Ph_3P)_3$ RhCl, which is known as wilkinson's Catalyst, the reaction is very fact.

Wilkinson's catalyst is most widely used for homogeneous hydrogenation. This catalyst was discovered by Sir Geoffrey Wilkinson. It is a complex of a transition metal rhodium called tris (triphenylphosphine) chloro rhodium (I) having the formula $RhCl\ (PPh_3)_3$ where Ph stands for phenyl group.

The phosphorus atom of phosphines has an unshared electron pair. It is through this electron pair, that triphenylphosphine is bonded to rhodium.

Rhodium can exhibit variable oxidation state of I and III and hence forms unstable intermediate with the reactant molecules. The mechanism can be explained as follows–

- It is believed that in solution the complex RhCl $(PPh_3)_3$ exchanges one of its PPh_3 for a solvent molecule, reversibly.

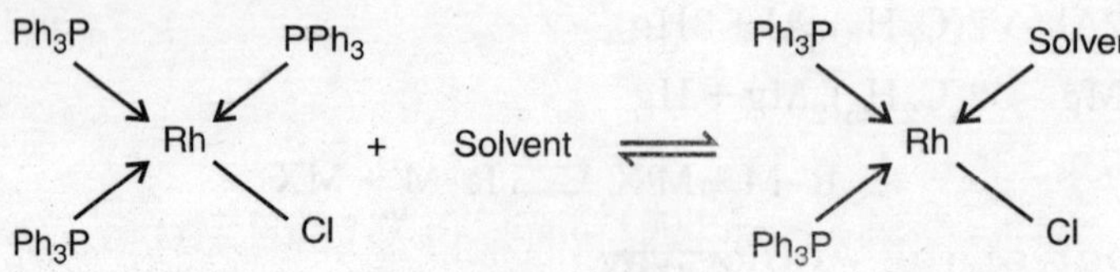

- The reactants, alkene and molecular hydrogen are brought close to the catalyst. The catalyst reacts with the hydrogen to form dihydrido complex RhH_2 Cl $(PPh_3)_2$. The H–H bond breaks and each hydrogen attach seperately to the rhodium.

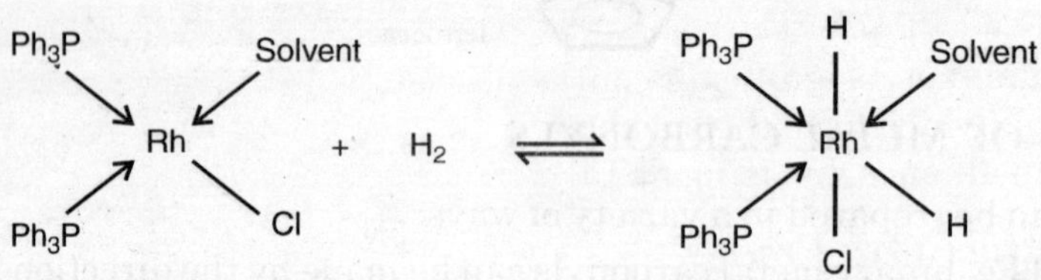

Thus in this step rhodium changes its oxidation state from (I) to (III).

- In the next step, alkene reacts with the catalyst by replacing a solvent molecule. The bond between the alkene and the metal atom rhodium is formed by the overlap of an empty orbital of the metal with the π cloud of alkene resulting in π complex.

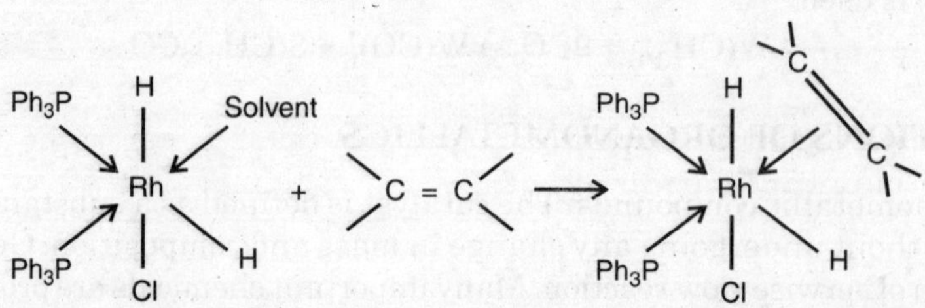

At this point both the reactants are bounded to rhodium and are in close proximity to react.

- A hydrogen atom migrates from the metal atom to one of the carbon atom of alkene whereas the other carbon attaches itself to the metal by a σ bond forming metal alkyl.

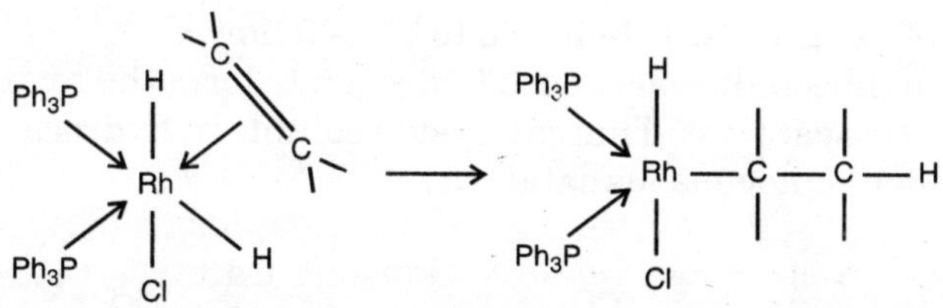

- The migration of alkene in between the metal hydrogen bond as in the above case is often referred to as insertion. Insertion can be defined as a process in which any atom or group is inserted between two other atoms initially bound together.
- Now, the second hydrogen migrates from the metal to carbon, which is still bonded to the metal atom, forming the hydrogenated product. The product detaches itself from the metal atom regenerating the catalyst. The rhodium atom regains its oxidation state of (I).

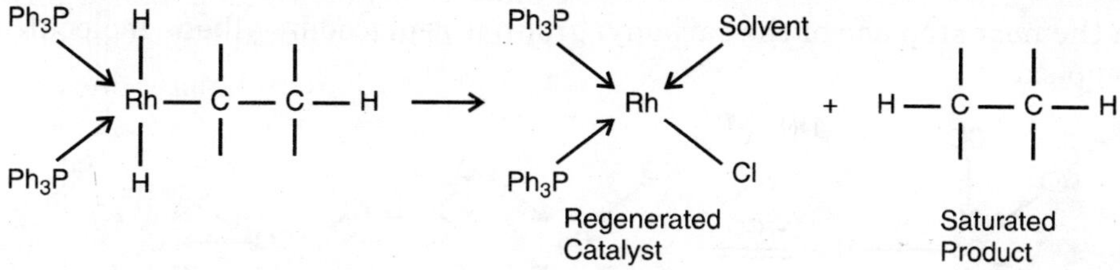

Evidence in Favour of the Above Mechanism

The most significant evidence comes from the fact that the above mentioned intermediates have been detected in solution and in some cases, isolated. Their structures have been well established by using spectroscopic techniques and have found to be in close agreement with the postulated structures.

Metal catalyst makes a significant contribution in industrial applications of catalysts. For example, the alcohols which form the backbone of aliphatic organic synthesis, can be obtained by the oxo process from alkenes, by using metal carbonyl, as catalyst.

(2) Hydroformylation Oxo Process

In presence of a metal catalyst, alkenes react with carbon monoxide and hydrogen to yield aldehydes. The aldehydes are again catalytically reduced to form alcohols.

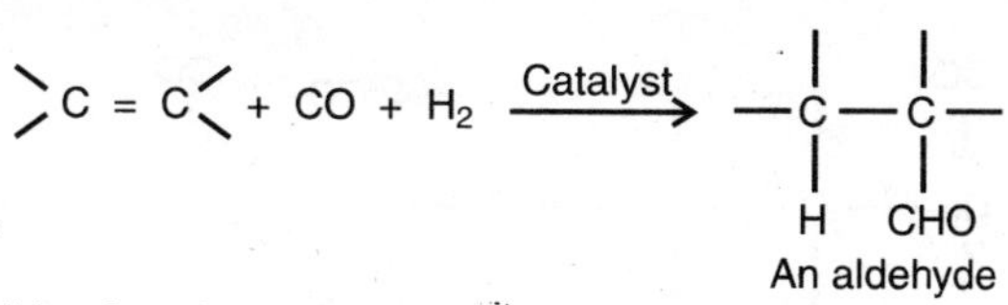

The catalyst employed in the above process is octacarbonyldicobalt, $CO_2\,(CO)_8$, **This process is the first industrial application of catalysis by a transition metal complex.**

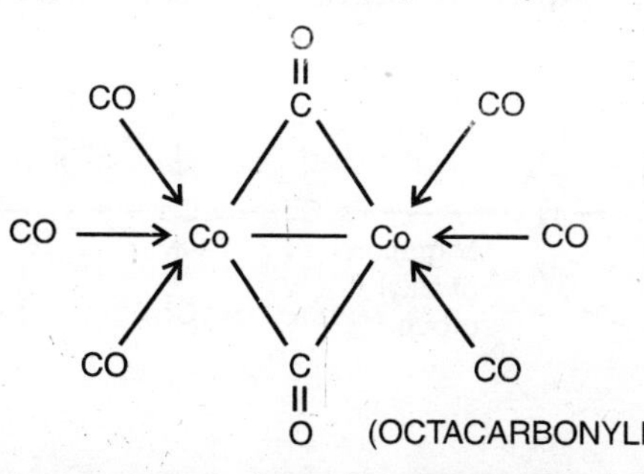

(OCTACARBONYLDICOBALT)

The mechanism of oxo process is believed to be as follows:

- The octacarbonyldicobalt reacts with hydrogen to form the hydrido complex COH $(CO)_4$ which is the active catalyst. This catalyst is soluble in hydrocarbons, so oxo process can be grouped under homogeneous catalysis.

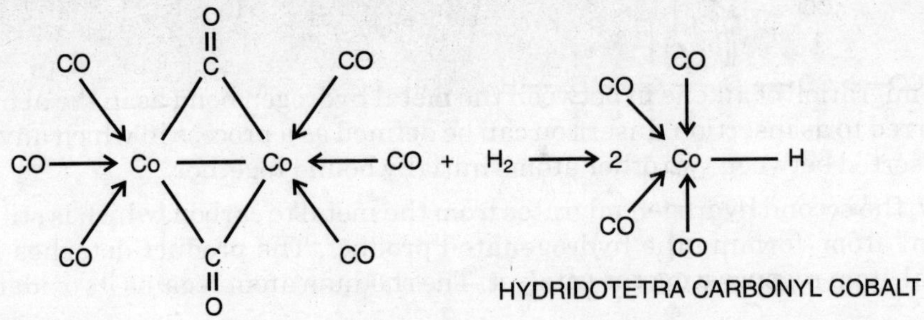

- In the next step one of the carbonyl group is replaced by alkene molecule to form a π-complex.

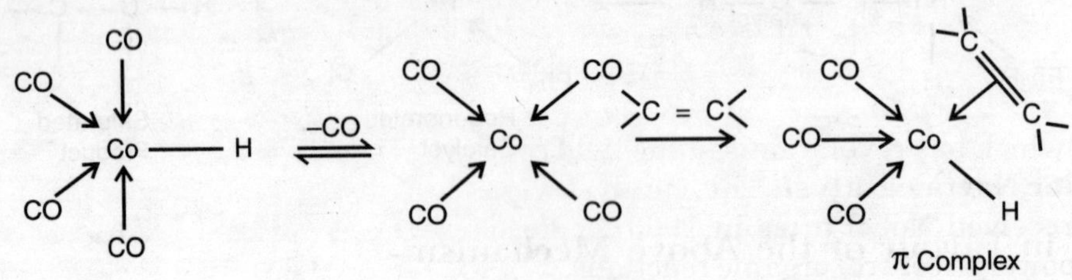

Thus all the three reactants, viz CO, H_2 and alkene are in close proximity.

- The hydrogen migrates from CO to one of the carbon atom of alkene and the other carbon atom of alkene attaches itself to cobalt forming metal alkyl. In other words the alkene is inserted inbetween cobalt and hydrogen atom.

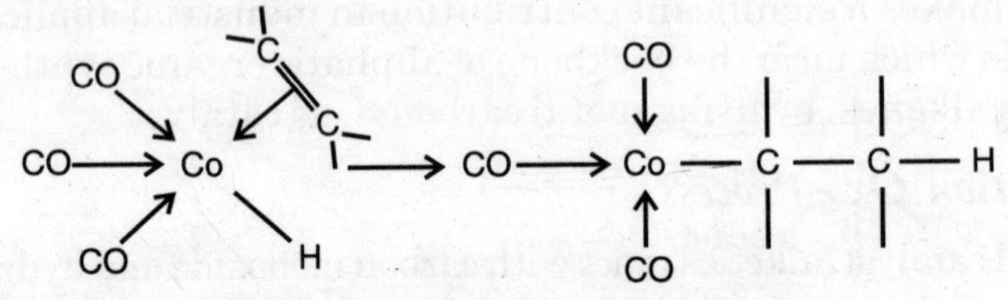

One molecule of CO is further added to the metal alkyl.

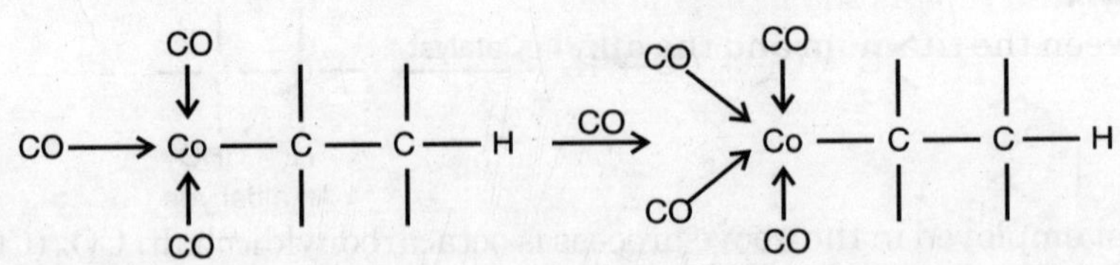

- The newly formed alkyl group then migrates to the carbonyl group, forming carbon-carbon bond.

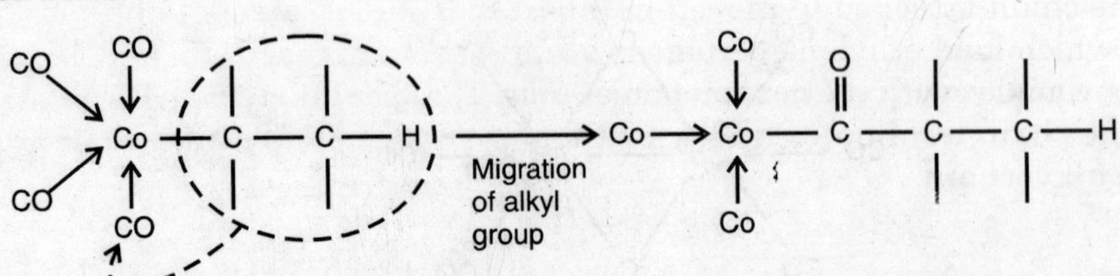

- At this stage hydrogen is absorbed to form dihydrido complex. One of these hydrogens migrates to carbon of the C = O group and aldehyde molecule is formed which detaches itself from the metal complex, which becomes ready for fresh action.

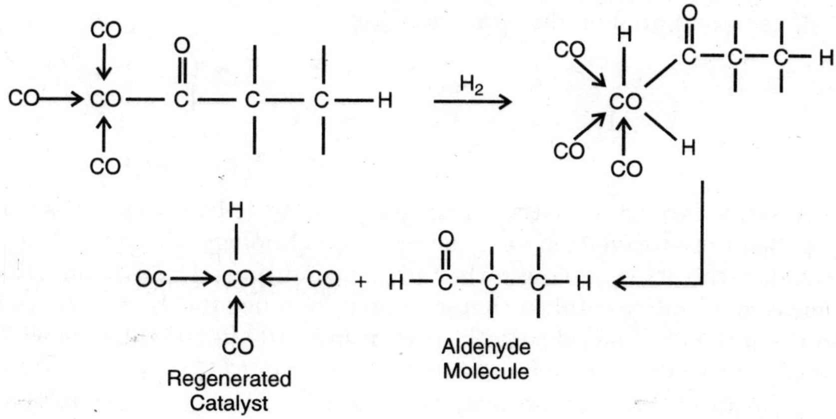

(3) Polymerization

The catalyst which has revolutionised the field of polymer chemistry also belongs to metal salts i.e. the **Ziegler Natta catalyst**. This catalyst was discovered by Karl Ziegler and Giulio Natta who jointly received Nobel prize in 1963 for developing this catalyst. This catalyst exhibits remarkable power to control organic reactions.

Ziegler Natta catalyst comprises of a transition metal salt like titanium trichloride in combination with a metal alkyl like triethyl aluminium. These two react to form a complex having an ethyl group in situ.

The alkene which has to undergo polymerisation, attaches itself to titanium by π bond forming π complex.

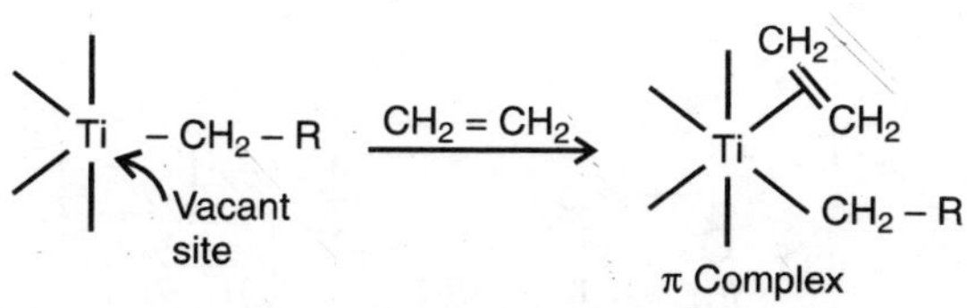

Now the alkyl and the alkene molecule are both held by titanium atom. The ethylene unit inserts itself between the titanium and the alkyl group.

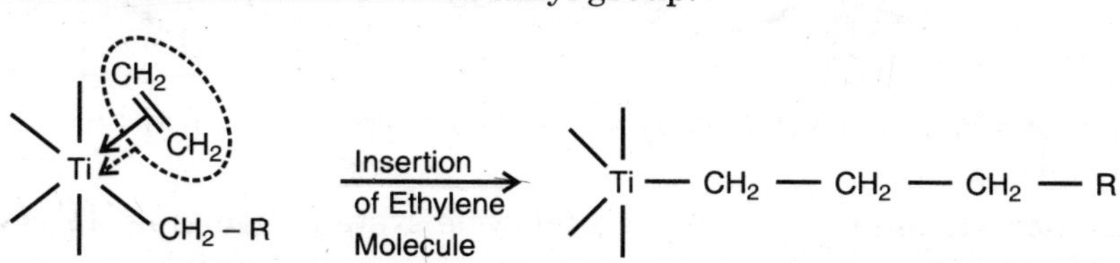

Thus, the chain attached to the titanium extends further undergoing similar steps. The bonding site which held ethylene becomes vacant with each insertion, and is followed by the formation of π complex with the next monomer unit. Thus the alkyl group keeps on growing with each cycle. The chain terminates finally by the insertion of hydrogen atom between titanium atom and the alkyl chain.

Ziegler Natta catalyst thus gives linear polymer molecules a great degree of stereochemical control.

(4) Monsanto Acetic Acid Process (Manufacture of Acetic Acid):

The carbonylation of methanol produces acetic acid:

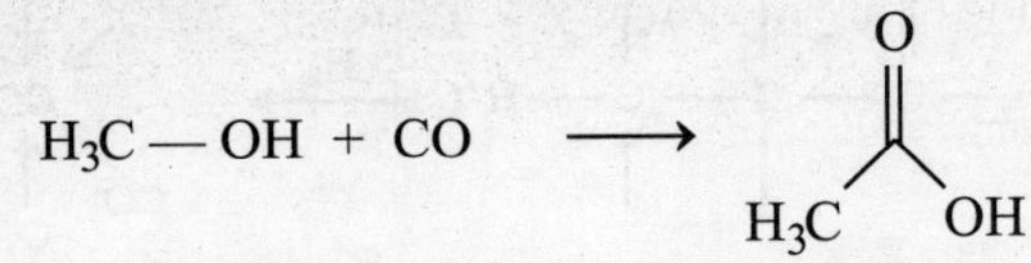

This is the second largest industrial homogeneous carbonlyation process with over 7 billion pounds of acetic acid produced each year using this technology. Prior to 1970, acetic acid was made using cobalt catalysts requiring rather severe conditions. In 1970 Monsanto commercialized a rhodiumcarbonyl iodide catalyst that is commonly called the *Mansanto Acetic Acid Process* (developed in the late 60's by James Roth. Rhodium is 10^3 to 10^4 times more active than the corresponding cobalt catalyst, which means that much lower CO pressures and moderately lower temperatures are required. Most importantly, the rhodium catlyst gives extremely high selectivities to acetic acid:

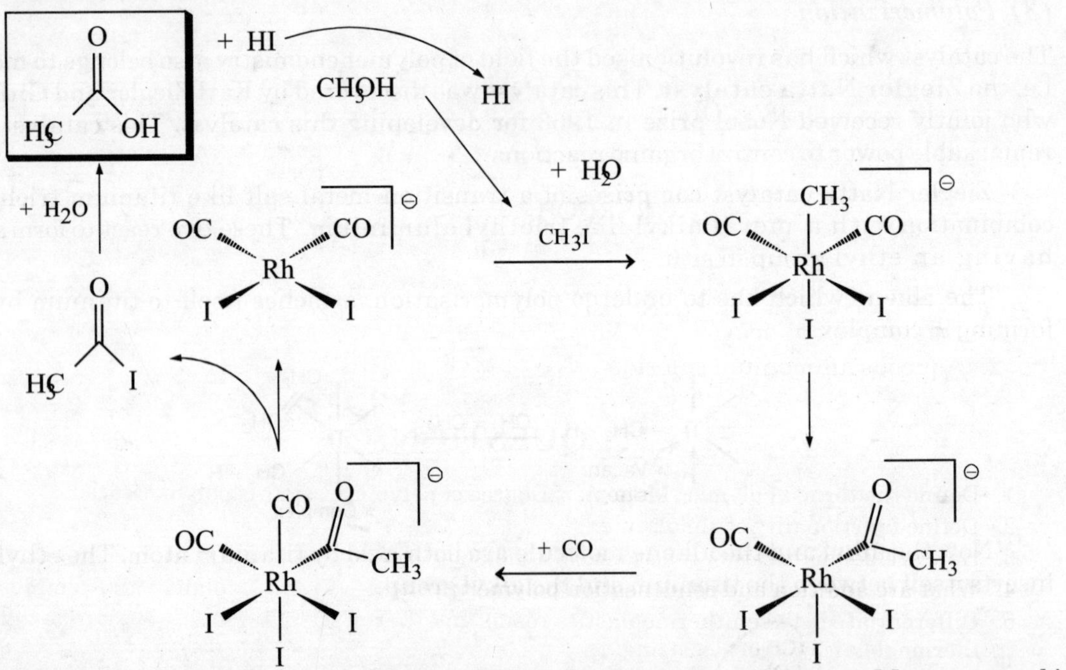

The mechanism has been extensively studied by Forster and coworkers at Monsanto and is shown below:

This catalytic reaction is an unusual dual cycle system involving HI as one catalyst and $[RhI_2(CO)_2]$ as the transition metal component. HI catalyzes the conversion of MeOH to MeI and H_2O at the beginning of the Rh-catalysed carbonylation reaction, followed by regeneration of HI at the end of the Rh-cycle by hydrolysis of the acyl-iodide. The Rh catalyst carbonylates the MeI to produce the acyl-iodide.

(5) Synthesis of organic compounds using Grignad Reactions

Insert Organometallic compounds are very useful as catalysts or reagents in the **synthesis** of organic compounds, such as pharmaceutical products. One of the major advantages of organometallic compounds, as compared with organic or inorganic compounds, is their high reactivity. Reactions that cannot be carried out with the usual types of organic reagent can sometimes be easily carried out using one of a wide variety of available organometallics

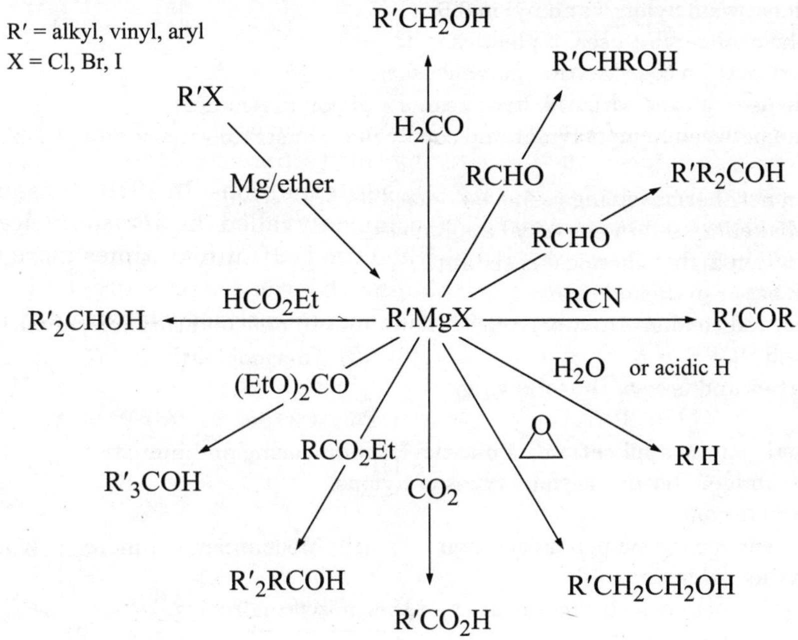

Typical work-up for these reactions:
1. Dilute aqueous acid or
2. Aqueous ammonium chloride

QUESTIONS

1. Define the terms: Polymer; Monomer; Degree of polymerization; copolymerization.
2. Define functionality of a molecule.
3. How are polymerization reactions classified on the basis of mechanism?
4. What are addition and condensation polymerization processes? Explain with example.
5. Differentiate between thermoplastics resins and thermosetting resins. Classify the following as thermoplastic or thermosetting:
   polyethylene, nylon, polyester, urea-formaldehyde resin, teflon, polystyrene, bakelite.
6. What do you mean by tacticity? Explain an isotactic, a syndiotactic and an atactic polymer.
7. Explain (a) cationic polymerization, (b) anionic polymerization, (c) coordination polymerization.
8. Explain step growth and chain growth polymerization with examples.
9. What are the chemical and structural differences between thermosetting and thermoplastic polymers. Name three important examples of each class.
10. What are the monomers used for each of the following polymers. Give commercial application of each:
    (a) Bakelite (b) Nylon-6 (c) Nylon 6-6 (d) Terylene (e) Teflon (f) PVC

11. (a) List out the difference between LDPE and HDPE.
    (b) What are the materials from which an epoxy adhesive is prepared.
    (c) Name any flourine containing polymer. State whether it is thermoplastic or thermosetting.
12. Give properties and uses of the following polymers:
    (a) Nylon 6-6 (b) PMMA (c) PVC (d) Teflon (e) Polyester
13. Give main properties and uses of:
    (a) HDPE (b) Plexiglass (c) Epoxy resins (d) Polypropylene.
14. Distinguish between nylon-6 and nylon 6,6.
15. What are the commercial uses of plastics.
16. Write a short note on coordination polymerization.
17. Mention the uses of urea-formaldehyde resins and polyurethane.
18. Differentiate between homopolymer and copolymer. Illustrate your point with examples.
19. Explain:
    (a) Why can not thermosetting resins be reused and reshaped?
    (b) Why is bakelite used in electrical appliances.
    (c) Why is teflon highly chemical resistant. (d) Why is PVC generally plasticised?
20. Distinguish between the following:
    (a) Addition and condensation polymerization (b) Anionic and cationic polymerization.
    (c) LDPE and HDPE (d) Thermoplastic and thermosetting resins.
21. Give properties and uses of the following:
    (a) PVC (b) PVA (c) Polystyrene (d) PMMA
22. Outline briefly on the applications of plastics in engineering and industry.
23. What is polyamide? Discuss various types of nylons.
24. Write short notes on:
    (i) Mechanism of cationic polymerization (ii) Mechanism of anionic polymerization
    (iii) Conducting polymers
25. (a) Distinguish between thermoplastic and thermosetting resins.
    (b) Give the properties and uses of
        (i) Phenolic resins (ii) Nylon 6:6 (iii) Urea-formaldehyde.
26. (a) Differentiate between thermoplastic and thermosetting resins.
    (b) Write the structural unit and two important applications of each of the following:
        (i) PE (ii) PVC (iii) Nylon 6,6 (iv) PMMA (v) Bakelite
27. (a) Classify the following polymers on the basis of action of heat on them
        (i) Bakelite (ii) Nylon (iii) Dacron (iv) Polyethylene
    (b) Write a short note on conducting polymers.
28. What is PET? Discuss its properties and applications. What are the advantages of blending PET with wool and cotton.
29. What is an epoxy resin? Give its properties and uses.
30. Name the polymers used for the following:
    (i) Squeeze bottles (ii) Domestic gas piping
    (iii) Overhead tanks for water storage. (iv) Baby bottles
    (v) Wrapping confectionary items (vi) Bobbins, Spools, Spindles, etc.
    (vii) Radiator grills (viii) Radio and television cabinets
    (ix) Ion exchange resins (x) Inner liners for refrigerators
    (xi) Loudspeaker housings (xii) Safety helmets
    (xiii) Containers for detergents, cosmetics, mineral water
    (xiv) Ladies handbags
    (xv) Adhesives (xvi) Laminating safety glass

(xvii) Alternative to glass
(xviii) Housing for rear lamps,
(xix) Gaskets and reals
(xx) Non-stick coating
(xxi) Moulding for computers, calculating machines
(xxii) Microwave cookware
(xxiii) Substitute for metals in gears and bearings.
(xxiv) Bullet resistant vests
(xxv) Switch board, heater handles.
(xxvi) Tanning material for leather.
(xxvii) Araldite (adhesive)
(xxviii) Skid resistant surface for highways.

31. What are elastomers?
32. Give the structure and uses of:
    (i) SBR
    (ii) Butyl Rubber
    (iii) Buna N
    (iv) Styrene acrylonitrile rubber.
33. Compare the structure of elastomers and plastics.
34. Explain vulcanisation of rubber.
35. Explain number average and weight average molecular mass.
36. Weight average molecular weight is higher than number average molecular weight. Explain
37. What do you mean by the term copolymerisation. What are its advantages?
38. How structure of a polymer is related to its (i) Mechanical property (ii) Electrical property.
39. (a) What are resins? Describe some distinguished characteristics of 'Thermoplastic' and 'Ther mosetting' Resins.
    (b) Write down the preparation and uses of any two industrially important "Vinyl" based Resins.
40. Define resins. Describe thermoplastic and thermosetting resins, their properties and applications.
41. Give the preparation, propert ies and uses of following:
    (a) Polyvinyl chloride (b) Bakelite (c) Polyester
42. (a) Distinguish between thermosetting and thermoplastics resins.
    (b) What are nylons? How are they distinguished from one another. Explain with the help of suitable example.
    (c) What is meant by functionality of monomer.
43. Write notes on
    (a) Conducting Polymers (b) Fiber reinforced composites.
44. (a) Give the preparation, properties and applications of any one thermoplastic polymer.
    (b) Define functionality of a monomer; classify the following on the basis of their functionality.
    (i) Ethylene Glycol (ii) Vinyl Chloride (iii) Lactic Acid
    (c) Discuss a technique by which a polymer can be made into a conducting polymer, Mention few applications of conducting polymers.
45. What are organometallic compounds?
46. Name different categories of organometallic compounds?
47. What is EAN rule? Give two examples of organometallic compounds in which EAN rule is obeyed.
48. What are metal carbonyls?
49. Discuss catalysis by organometallics?
50. How organometallics are being used for polymerization reaction?

EXAMINATION QUESTIONS

1. What is meant by functional of monomers? Explain its significance with suitable examples.
   [U.P. Tech., B. Tech., 2nd Sem., 2003-04]
2. What are the characteristic of polymer? Why do polymers have an average molecules weight?
   [U.P. Tech., 2nd Sem., 2003-04]

3. (a) What are the characteristics of polymers?
   (b) Classify the following polymers into natural and synthetic polymers:
   (i) polyamides (ii) polyesters (iii) Nucleic acid (iv) Starch

   (v) 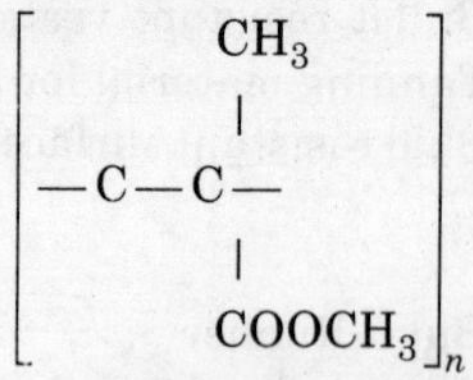

   (c) If average degree of polystyrene is 105, calculate its average molecular weight.
   [U.P., Tech., (C.O.) Exam. 2003]
4. (a) What are the characteristics of polymer ? Why do polymers have an average molecular weight?
   (b) What are the limitations of raw rubber ? Explain the process of Vulcanisation of rubber.
   [U.P., Tech., 2nd Sem., 2004-04]
5. (a) Suggest a process that would cause the break up of condensation polymer (say, Nylon 6,6) by essentially reversing the condensation process.
   (b) Why low density and high density polythene differ in density ?
   (c) Why is Teflon highly chemical resistant ?
   (d) What is copolymerisation ? Give one example.
   (e) Name the type of polymerisation and the method by which conducting polymers can be synthesized. Give examples of such polymers, including their applications.
   [U.P., Tech., Ist Sem., 2004-04]
6. (a) With suitable example differentiate isotactic, syndiotactic and atactic polymers.
   (b) What are Copolymers ? How does Buna-S differ from Buna-N?
   (c) Show the chemical reactions for the formation of Nylon-6 and Nylon-66.
   (d) Describe in brief about conducting polymers with their applications.
   [U.P., Tech., 1st Sem., 2006-07]
7. (a) Explain valcanisation and stage the improvement in properties of rubber after vulcanisation is carried out.
   (b) Write the structure (only) of the following polymer: (i) Polyacrylonitrile (ii) Buna-S (iii) Dacron (iv) Nylon-5.
   (c) Give the mechanism of addition and condensation polymerisation.
   [U.P., Tech., 2nd Sem., 2006-07]
8. (a) Enumerate the difference between thermoplastic & thermosetting resins.
   (b) Give the preparation of Buna-S and Buna-N.
   (c) How do you prepare the following polymers (i) Bakelite (ii) Perspex.
   [U.P., Tech., Ist Sem., 2007-08]
9. (a) Write the method of preparation for the following polymers:
   (i) Nylon-6 (ii) Buna-S
   (iii) Polyacrylonitrile.
   (b) (i) Differentiate between addition and condensation polymerisation with suitable examples.
   (ii) Writ applications of conducting polymers. **[U.P., Tech., B. Tech., 2nd Sem., 2007-08]**
10. Which stereoisomeric polymer of styrene is formed when polymerisation reaction is carried out in the presence of Ziegler Natta catalyst **[U.P., Tech., B. Tech., 1st Sem., 2008-09]**
11. (a) Write down the structure and application of following polymers :
   (i) Starch (ii) Cellulose
   (iii) Natural rubber (iv) polyacrylonitrile
   (v) BUNA-S

(b) Identify the monomers in the following polymers:

(i) $-\overset{\overset{\large O}{||}}{C}-(CH_2)_6-\overset{\overset{\large O}{||}}{C}-NH-(CH_2)_8-Nh-$

(ii) $-CH_2-\underset{\underset{\large Cl}{|}}{CH}-CH_2-\underset{\underset{\large Cl}{|}}{CH}-$ **[*U.P., Tech., B. Tech., 1st Sem., 2008-09*]**

12. (a) The degree of polymerisation represents the __________.

(b) The polymerisation which is accompanied by elimination of small molecules is called.

(i) Addition (ii) Copolymerization

(iii) Condensation (iv) Crosslinking polymerisation.

(c) Polystyrene is prepared from styrene in presence of ___________.

[*U.P., Tech., B. Tech., 1st Sem., 2008-09*]

13. Discuss the thermoplastic resins. Write the synthesis and applications of polystyrene and Polyvinyl chloride. **[*U.P., Tech., B. Tech., 1st Sem., 2008-09*]**

❑❑❑

29 TRANSITION ELEMENTS

"Our age of anxiety is, in great part, the result of trying to do today's jobs with yesterday's tools."
Marshall Mcluhan

"Chemistry leads man into that domain of latent forces whose power controls the whole material world."
Leipzig

CHAPTER OUTLINE

Descriptive chemistry in the broadest sense, includes all we know about matter. But this has been possible only due to the attempts of various scientists who organised and systematised the chemistry of the elements by distributing these into groups and periods. Extensive tabulation of elements based on the regular, periodic reoccurance of properties, proposed by Mendeleev and Meyer in 1870's was a major breakthrough in the study of the elements. It helped in assessment of facts, prediction and verification of various chemical properties. In otherwords, chemistry of elements becomes a manageable subject after the periodic table was proposed.

Long form of periodic table has already been discussed in your previous classes in detail. In brief, the elements are divided into different blocks on the basis of their electronic configuration as:

s-block elements : Elements in which the last-electron enters the s-orbital of their respective outermost shells are called s-block elements. These are the group1 (alkali metals) and group 2(alkaline earth metals) with the outermost electronic configuration as ns^1 and ns^2 respectively.

p block elements : The elements in which the last electron enter any one of the three p-orbitals of their respective outermost shells are called p-block elements. Since a p-subshell has three degenerate p-orbitals, each of which can accommodate two electrons, therefore in all, there are six groups of p-block elements. These are the elements in group 13 to 18 with the outermost electronic configuration varying between $ns^2\ np^1$ and $ns^2\ np^6$. Elements in the zero group with $ns^2\ np^6$ configuration come at the end of this block.

Elements of s and p blocks are often called **representative elements.**

d block elements : Elements in which the last electron enters any one of the five d-orbitals of their respective penultimate shells are called d block elements. Since a d subshell has five d-orbitals, each one of which can accommodate two electrons, therefore in all there are ten groups of d block elements. These are the elements in groups 3 to 12 with the outermost electronic configuration varying between $(n\text{-}1)d^{1\text{-}10}\ ns^{0\text{-}2}$.

f block elements : Elements in which the last electron enters any one of the seven f-orbitals of their respective ante-penultimate shells are called f-block elements. In all these elements the s-orbital of the last shell (n) is completely filled, the d-orbitals of the penultimate (n-1) shell invariably contains zero or one electron but the f-orbitals of the antepenultimate (n-2) shell gets progressively filled in. There are two series of f-block elements each containing 14 elements. Therefore, in all there are 28 f-block elements in the periodic table with outer electronic configuration varying between $(n\text{-}2)f^{1\text{-}14}\ (n\text{-}1)d^{0\text{-}1}\ ns^2$.

The d block elements are called **transition** elements and f block elements are called **inner transition elements.**

In the present chapter we will confine our discussion to elements of d block.

29.1 d-BLOCK ELEMENTS

As is discussed, transition elements may be strictly defined as those elements which have partly filled d or f sub shells. But the electronic configuration of transition elements reveal that the elements Cu, Ag and Au, i.e. the coinage metals and also the metals Zn, Cd and Hg have filled d subshells and hence cannot be treated as transition elements. But if we broaden our definition and include the elements which have partly filled d orbitals when they form compounds, then certainly Cu, Ag and Au are treated as transition elements. This is because the ions of the Cu^{2+} Ag^{2+} and Au^{3+} have partly filled d orbitals as shown in Table 29.1. Moreover, their chemical behaviour is quite similar to that of other transition elements.

Table 29.1 Electronic Configuration of Free Atoms and Ions

Element	*Electronic configuration*	*Ion*	*Electronic configuration*
Cu	$[Ar]3d^{10}4s^1$	Cu^{2+}	$[Ar]3d^9$
Ag	$[Kr]4d^{10}\ 5s^1$	Ag^{2+}	$[Kr]4d^9$
Au	$[Xe]4f^{14}\ 5d^{10}6s^1$	Au^{3+}	$[Xe]4f^{14}\ 5d^8$

Elements of group 12 viz. Zn, Cd and Hg cannot be regarded as transition elements because these do not have partly filled d-orbitals either in their atoms or ions. The d-orbitals in these elements do not take part in chemical bonding.

On the same basis, Cu^+, Ag^+ and Au^+ are also not considered to be transition metal ions. Hence, d block elements are those which have:

- incomplete (n-1) d subshells or
- form ions having d electrons varying between 1 to 9 electrons.

The transition elements are placed in between those of s block elements and p block elements in the periodic table. The word **transition** is used because these represent a change of state from highly electropositive elements (metals) of s-block to highly electronegative elements (non-metals) of p block. Moreover, their properties are intermediate between those of s and p block elements.

29.2 DIVISION OF d BLOCK ELEMENTS

Whereas s and p block elements are usually discussed in groups, d block elements are discussed by classifying them into horizontal rows or series. d block elements consist of four series (Table 29.2).

(i) ***First Transition Series:*** The elements from Sc (Z=21) to Cu (Z=29) constitute the first transition series or 3 d series. The elements of first transition series have partly filled 3 d shells, either in the ground state of the free atom or in one or more of their ions.

(ii) ***Second Transition Series:*** The elements from Y (Z= 39) to Ag (Z=47) constitute the second transition series or 4 d series. The elements of second transition series have partially filled 4 d shells, either in the ground state or in one or more of their ions.

(iii) ***Third Transition Series:*** The elements from La (Z=57), Hf (Z=72) to Au (Z=80) constitute the third transition series or 5 d series. These elements have partially filled 5 d shells either in the ground state or in one or more of their ions.

(iv) ***Fourth Transition Series:*** The fourth transition series starts from Ac (Z=89) and is incomplete.

Table 29.2 Transition Series

Period number	Group → 3	4	5	6	7	8	9	10	11	12
4	Scandium $_{21}Sc$ [$_{18}Ar$] $4s^2 3d^1$	Titanium $_{22}T$ [$_{18}Ar$] $4s^2 3d^2$	Vanadium $_{23}V$ [$_{18}Ar$] $4s^2 3d^3$	Chromium $_{23}Cr$ [$_{18}Ar$] $4s^1 3d^5$	Manganese $_{25}Mn$ [$_{18}Ar$] $4s^2 3d^5$	Iron $_{26}Fe$ [$_{18}Ar$] $4s^2 3d^6$	Cobalt $_{27}Co$ [$_{18}Ar$] $4s^2 3d^7$	Nickel $_{28}Ni$ [$_{18}Ar$] $4s^2 3d^8$	Copper $_{29}Cu$ [$_{18}Ar$] $4s^1 3d^{10}$	Zinc $_{30}Zn$ [^{28}Ar] $4s^2 3d^{10}$
5	Yttrium $_{39}Y$ [$_{36}Kr$] $5s^2 4d^1$	Zirconium $_{40}Zr$ [$_{36}Kr$] $5s^2 4d^2$	Niobium $_{41}Nb$ [$_{36}Kr$] $5s^1 4d^4$	Molybdenum $_{42}Mo$ [$_{36}Kr$] $5s^1 4d^5$	Technetium $_{43}Tc$ [$_{36}Kr$] $5s^1 4d^6$	Ruthenium $_{44}Ru$ [$_{36}Kr$] $5s^1 4d^7$	Rhodium $_{45}Rh$ [$_{36}Kr$] $5s^1 4d^8$	Palladium $_{46}Pd$ [$_{36}Kr$] $5s^2 4d^{10}$	Silver $_{47}Ag$ [$_{36}Kr$] $5s^1 4d^{10}$	Cadmium $_{48}Cd$ [$_{36}Kr$] $5s^2 4d^{10}$
6	Lanthanum $_{57}La$ [$_{54}Xe$] $6s^2 5d^1$	Hafnium $_{72}Hf$ [$_{54}Xe$] $4f^{14} 5d^2 6s^2$	Tantalum $_{73}Ta$ [$_{54}Xe$] $4f^{14} 5d^3 6s^2$	Tungsten $_{74}W$ [$_{54}X$] $4f^{14} 5d^4 6s^2$	Rhenium $_{75}Re$ [$_{54}Xe$] $4f^{14} 5d^5 6s^2$	Osmium $_{76}Os$ [$_{54}Xe$] $4f^{14} 5d^6 6s^2$	Iridium $_{77}Ir$ [$_{54}Xe$] $4f^{14} 5d^7 6s^2$	Platinum $_{78}Pt$ [$_{54}Xe$] $4f^{14} 5d^8 6s^2$	Gold $_{79}Au$ [$_{54}Xe$] $4f^{14} 5d^{10} 6s^1$	Mercury $_{80}Hg$ [$_{54}Xe$] $f^{14} 5d^{10} 6s^2$
7	Actinium $_{89}Ac$ [$_{86}Rn$] $6d^1 7s^2$	Rutherfordium $_{104}Rf$ [$_{86}Rn$] $5f^{14} 6d^2 7s^?$	Unnilpentium $_{105}Unp$ [$_{86}Rn$] $5f^{14} 6d^3 7s^2$	Unnilhexium Unh	Unnilseptium $_{107}Uns$	Unniloctium $_{108}Uno$	Unnilennium Une			

29.3 ELECTRONIC CONFIGURATION OF d BLOCK ELEMENTS

The general outer electronic confugration of d block elements is $ns^{1\text{-}2}$ $(n\text{-}1)d^{1\text{-}10}$ where n is the outermost orbit. Sc has electronic configuration $4s^2$, $3d^1$. The next element Ti has one more d electron with the configuration $3d^2$, $4s^2$. This addition of one electron in the penultmate shells continues expanding d subshell from d^1 to d^9 electron, step by step from Sc to Cu on moving towards right along the first transition series. The element Zn having outer electronic configuration $3d^{10}$, $4s^2$ though not a transition element is studied along the d block elements as it marks the completion of the d sub shell.

The second series starts with the element yttrium with the outer electronic configuration $5s^2$ $4d^1$. The filling up of 4 d orbitals continues on moving through the element Y to Ag, adding one electron as one moves from one group to the next. The element Cd having outer electronic configuration $5s^2$ $4d^{10}$ is also studied along the transition elements.

The first element of the third transition series La has outer electronic configuration $5d^1\ 6s^2$. Now if we repeat the pattern we have observed in the first two series, then we may expect the next eight elements in which filling of d orbital take place step-by-step, i.e. from d^1 to d^9. However this does not happen. The 4f subshell becomes slightly more stable than 5d subshell and the next electron enters into the 4f subshell and not d orbital. This continues through the next 14 elements untill at Lu the 4f subshell becomes filled. Lutetium thus has the outer-electronic configuration of $4f^{14}\ 5d^1\ 6s^2$. These 14 elements which follow lanthanum are called *lanthanides*. It is only after the f orbital is completely filled, the next electron in Hf enters the 5d orbital. The outer electronic configuration of Hf being $4f^{14}\ 5d^2\ 6s^2$ and thereafter the electrons are added in the penultimate 5 d subshell untill at Hg the d orbital is completely filled. The outer electronic configuration of Hg being $4f^{14}\ 5d^{10}\ 6s^2$ and hence Hg is not included in the transition elements but studied along with them.

The fourth transition series starts with Ac having electronic configuration $7s^2\ 6d^1$. Similar to what happened after Lanthanum in third transition series, the electron in the next element enters the 5f orbital. The electrons in succeeding elements enter the 5f orbital which becomes complete at Lr with the electronic configuration $5f^{14}\ 6d^1\ 7s^2$. These 14 elements which follow actinium are called *actinides*. Thus the next d block element after lanthanum, is Ku with the electronic configuration $5f^{14}\ 6d^2\ 7s^2$. The fourth transition series is incomplete and attempts are on to isolate the missing members. Some of them have already been isolated and have been assigned their respective places on the basis of their electronic configuration.

Anomalous Configurations of Cr and Cu in the first series and Mo and Ag in the second series can be explained on the bases of extra stability attributed to them as a result of exactly half filled or completely filled orbitals.

Table 29.3 Anomalous Configuration of Some Elements

Element	*Electronic Configuration*	
	Expected	*Observed*
Chromium (Cr)	$3d^4\ 4s^2$	$3d^5\ 4s^1$
Copper (Cu)	$3d^9\ 4s^2$	$3d^{10}\ 4s^1$
Molybdenum (Mo)	$4d^4\ 5s^2$	$4d^5\ 5s^1$
Silver (Ag)	$4d^9\ 5s^2$	$4d^{10}\ 5s^1$

It is evident from the table 29.3 that the configuration actually observed in the case of Cr, Cu, Mo and Ag is such that the s and d orbitals are either exactly half filled or completely filled, which gives extra stability to the atom. For example the 3d level in case of chromium, gets exactly half filled with configuration $3d^5\ 4s^1$ and that in case of Cu it gets completely filled with configuration $3d^{10}\ 4s^1$. The extra stability of exactly half filled or completely filled orbitals can be explained on the basis of exchange energy (explained in the box).

Exceptional Configurations- Stability of completely Filled and Exactly Half Filled Orbitals: In certain elements when two subshells differ slightly in their energies, an electron may shift from a subshell of lower energy to a subshell of higher energy to form either completely filled or exactly half filled orbitals, which confers extra stability to the atom. The extra stability is attained due to the following reasons:

(i) **Exchange Energy:** If on exchanging the position in space of two electrons with parallel spins, there is no change in the electronic arrangement, it would lead to decrease in energy. The pair of electrons is called the exchange pair. The larger the number of exchange pairs of electrons, greater would be the decrease of energy. The decrease per exchange pair of electrons is termed as exchange energy.

Exchange energy for d^4 arrangement would be given as follows:

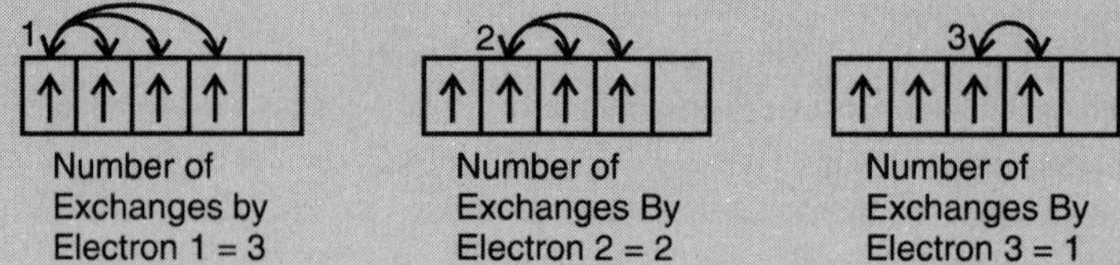

∴ Total number of exchange pairs=3+2+1=6.

If E is the exchange energy per exchange pair then the energy for the above arrangement for d^4 electrons energy would decrease by 6E.

Similarly exchange energy for d^5 arrangement would be given as:

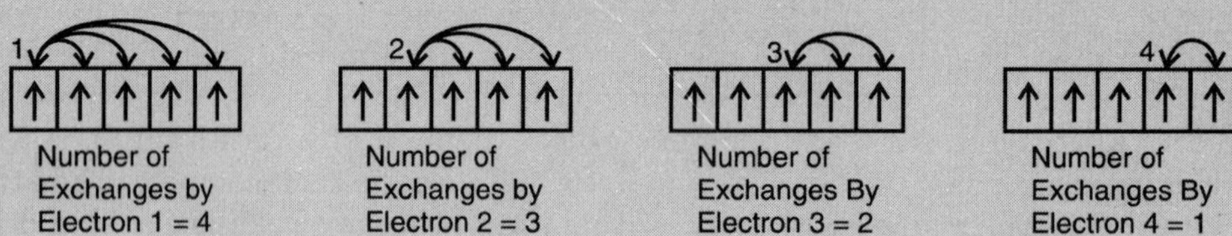

∴ Total number of exchanges = 4+3+2+1 = 10.

Decrease in electron arrangement energy = 10E.

Thus we see that arrangement d^5 is associated with more exchange energy than d^4 and is hence more stable.

(ii) **Symmetrical Distribution:** The other factor which explains the extra stability of the exactly half filled or fully filled orbitals is the symmetrical distribution of charge in the orbitals. Configurations with even or uniform or symmetrical distribution of charge in all directions would evidently be associated with lower energy and hence leads to higher stability. Of all the d configurations, the configurations having exactly half filled or fully filled orbitals would have a uniform or symmetrical distribution of charge in space and hence would be more stable. For example the electronic configuration of Cr $3d^5\ 4d^1$ is more uniform and hence more stable than $3d^4\ 4s^2$ as shown:

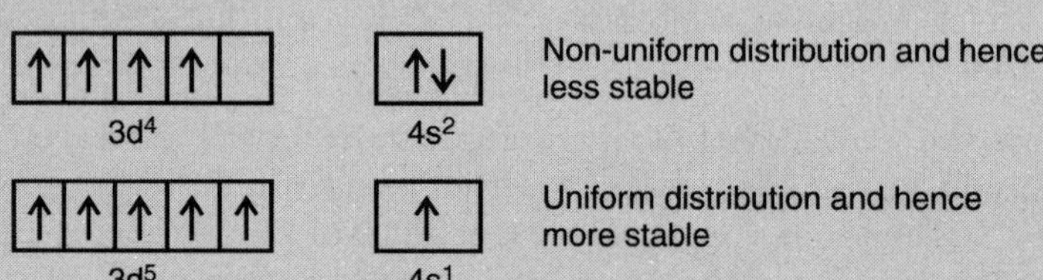

The concept of exchange energy however cannot explain the anomalous configuration of Nb, Ru, Rh and Pd. Nuclear-electron and electron-electron interactions play significant role in determining the configuration of these elements.

Naming of elements beyond atomic number 100 :

As is mentioned above the fourth series is incomplete series. The electrons in elements with atomic numbers 103 to 112 enter the inner 6d orbitals. Some of these elements have been prepared artificially through nuclear reactions. The d subshell gets completely filled with the element with atomic number 112. Though these elements which have been isolated are known by their common names but the elements beyond atomic number 100 are now named in accordance with IUPAC system. According to which:

— the elements beyond atomic number 100 will have the ending -ium.

— the name will be constructed from the following numerical roots.

0 = nil, 1 = un, 2 = bi, 3 = tri, 4 = quad
5 = pent, 6 = hex, 7 = sept 8 = oct, 9 = enn

— the symbols of these elements will consist of three letters- the first letters from the roots which make up the name.

For example element with atomic number 106 will be named as:

1 0 6
un nil hex ie unnil hexium.

Table 29.4 gives the common and IUPAC names of some of these elements.

Table 29.4 Common and IUPAC name of some of the elements belonging to fourth transition series.

Atomic Number	*Common Name*	*IUPAC Name*
101	Mendelevlum	Unnilunium
102	Nobelium	Un nil bium
103	Lawrencium	Un nil trium
104	Rutherfordium	Un nil quadium
105	Dubnium	Un nil pentium
106	Seaborgium	Un nil hexium
107	Bohrium	Un nil septium
108	Hassium	Un nil octium
109	Meitnerium	Un nil ennium

The IUPAC names should be written as one word without leaving any spaces in between. Here the spaces are left for clarity.

29.4 GENERAL PROPERTIES OF TRANSITION ELEMENTS

The d block elements have similar ns^2 orbital electronic configuration in the outermost shell and they differ only in the number of unpaired electrons in d orbitals of the penultimate shell. Because of the same number of electrons in the outermost shell, the transition elements have resemblance in their physical and chemical properties. Some of the general characteristics of transition elements are discussed below:

1. Atomic RADDII:

The atomic radii of the transition metals lie in between those of s and p block elements. The following trends are observed in atomic raddii of the elements:

(i) **The atomic radii of d block elements in a series decrease with increase in atomic number as is expected but the decrease in atomic size is small after midway.** Table 29.5 gives the atomic radii of the elements of the transition elements. The atomic radii decrease gradually from Sc to Cr but thereafter from Cr to Cu, it remains practically constant.

Table 29.5 Atomic Radii of d-Block elements in pm

Sc 144	Ti 132	V 122	Cr 117	Mn 117	Fe 117	Co 116	Ni 115	Cu 117	Zn 125
y 162	Zr 145	Nb 134	Mo 129	Te -	Ru 124	Rh 125	Pd 128	Ag 134	Cd 141
Ca 169	Hf 144	Ta 134	W 130	Re 128	Os 126	Ir 126	Pt 129	Au 134	Hg 144

Reason: The decrease in the atomic radius in the beginning is due to the increase in the nuclear charge on moving along the series horizontally. The increased nuclear charge attracts the electron towards itself more strongly thus lowering the atomic radius.

The electron is added in the penultimate shell on moving from left to right in the periodic table. This electron shields the outermost ns electrons from the nuclear pull and hence counterbalances the increased nuclear charge due to increase in atomic number. Thus the atomic radii decrease slightly after chromium. Similar, trend has been observed for the second and third transition series.

(ii) **There is a slight increase in the atomic radii towards the end of the period.** From table 29.5, it is seen that atomic radii of the elements towards the end of the transition series increase slightly.

Reason: Towards the end of the series the factors causing the change in atomic radii counterbalance each other. After chromium the pairing of electrons takes place causing electron-electron repulsions. This increases the atomic radius. Now the increased nuclear charge contracts the electron cloud, whereas electron-electron repulsions cause it to expand. In Fe, the two opposing effects balance each other and therefore no change of size is observed on moving from Mn to Fe. However towards the end of the series, the repulsive forces exceeds the attractive forces and hence the size increases. In the last elements of each series viz Zn, Cd, Hg the electron-electron repulsions become predominant and hence the size of the atom increases.

(iii) **Atomic radii increases down the group.** This trend is expected but it is seen that whereas the atomic radii of the second transition series are larger than their group members of first transition series, the atomic radii of the third transition series have nearly the same radii as their group members in second transition series.

Reason : The member of shells increase on moving down the group and hence their atomic radii also increases accordingly.

The atomic radii of the respective group elements of second and third transition series are similar due to lanthanide contraction discussed later in the chapter.

Ionic radii of the transition elements follow the same trend as followed by the atomic radii.

2. Ionization Energies:

Ionisation energies of transition elements are fairly high due to the small size of atoms.

(i) **Ionisation energies increase with increase in atomic number along a given transition series**. The trend is however not regular (Table 29.6).

Table 29.6: First Ionization Energies of d Block Elements

First Transition Seris :										
Elements	Sc	Ti	V	Cr	Mn	Fe	Co	Ni	Cu	Zn
Ionization Energy ($kJmol^{-1}$)	631	656	650	652	717	762	758	736	745	906
Second Transition Series :										
Elements	Y	Zr	Nb	Mo	Tc	Ru	Rh	Pd	Ag	Cd
Ionization Energy (kJ mol-)	616	674	664	685	703	711	720	804	731	876
Third Transition Series :										
Elements	La	Hf	Ta	W	Re	Os	Ir	Pt	Au	Hg
Ionization Energy ($kJ\ mol^{-1}$)	541	760	760	770	759	840	900	870	889	1007

Reason: The increase in ionization energies along a given period is expected. The increased nuclear charge with increase in atomic number reduces the size of the atom, making the removal of outer electron difficult. However, addition of electron takes place in the penultimate shell on moving along the period. The shielding effect due to these electrons also increases along the period. Thus the effect of increased nuclear charge is opposed by the additional screening effect of the electrons and hence ionization energy increases but quite slowly and irregularly along the period in d-block elements.

(ii) **First Ionisation energies of third transition series is higher than those of first and second transition series.**

Reason: In 5d elements the shielding of the nucleus by of electrons is very weak. Thus due to the greater effective nuclear charge acting on the outer valence electrons the ionization energies are higher.

3. Standard Electrode Potentials:

The standard electrode potential of a transition element is generally higher than that of standard hydrogen electrode, which is arbitrarily given a value of zero as is evident from (Table 29.7).

Standard oxidation potentials of the elements of first transition series except copper are higher than that of hydrogen (zero).

Table 29.7: Standard Oxidation Electrode Potentials of the Elements of First Transition Series

Elements	*Electrode Reactions*	*Standard Oxidation (volts) Potential*
Sc	$Sc \rightarrow Sc^{3+} + 3e^-$	2.10
Ti	$Ti \rightarrow Ti^{2+} + 2e^-$	1.60
V	$V \rightarrow V^{2+} + 2e^-$	1.20
Cr	$Cr \rightarrow Cr^{3+} + 3e^-$	0.74
Mn	$Mn \rightarrow Mn^{2+} + 2e^-$	1.18
Fe	$Fe \rightarrow Fe^{2+} + 2e^-$	0.44
Co	$Co \rightarrow Co^{2+} + 2e^-$	0.28
Ni	$Ni \rightarrow Ni^{2+} + 2e^-$	0.25
Cu	$Cu \rightarrow Cu^{2+} + 2e^-$	– 0.34
Zn	$Zn \rightarrow Zn^{2+} + 2e^-$	0.76

These metals, except Cu, would evolve hydrogen gas when react with acids, i.e.

$$M + 2H^+_{(aq)} \longrightarrow M^{2+}(aq) + H_2(g)$$

Thus it implies that transition metals should be more reactive than hydrogen, according to their electrode potentials. But it has been observed that **most of those do not displace hydrogen on treatment with dilute acids**.

Reason: This is because the surfaces of the transition metals get covered with oxide films which are insoluble and inert in nature. These films protect the bulk of the metal from further action. So these metals cannot react with dilute acids once the protective oxide film is formed on the surface. For example chromium, in spite of its high electrode potential is so inert to the attack of dilute acids that it is used as a protective non-oxidising metal. This is because it gets coated with a non-reactive Cr_2O_3 layer.

4. Metallic Character

All the transition elements are metals having hcp, ccp or bcc lattices:

(i) **These exhibit all the characteristics of metals, e.g. They**

— are hard, lustrous
— are malleable and ductile.
— have high melting and boiling points.
— have high thermal and electrical conductivity.
— have high tensile strength.

Reason: The metallic character of transition elements is due to their relatively low ionization energies because of which the ns electrons can easily be lost.

(ii) **These elements are harder and more brittle than s block elements.**

Reason: Hardness and brittleness are associated with covalent bonding between the metal atoms. The unpaired d-electrons in these metals overlap with unpaired electrons of the same metal atom to form a covalent bond. Greater the number of unpaired d-electrons, greater the number of covalent bonds. The number of unpaired electrons increase up to group 6 (Cr, Mo, W) and then decreases. That is why Cr, Mo and W are very hard. The elements Zn, Cd, Hg where d orbitals are all paired are not hard and infact Hg is a liquid.

It may be emphasized here that the existence of metallic bonding cannot be ruled out because the other metallic properties like malleability, ductility, electrical and thermal conductivity can be explained on the basis of metallic bonding only. Thus one may conclude that whereas overlapping of orbitals, having unpaired electrons, causes covalent bonding, the valence s electrons cause metallic bonding. Thus transition elements have both, covalent as well as metallic bonding.

Table 29.8: Densities, Melting and Boiling point of Transition Elements

	Symbol	*atomic number*	*density (grams per cubic centimeter, 20°C*	*melting point °C*	*boiling point °C*
First Series	Ti	22	4.507	1,668	3,260
	V	23	6.11	1,890	3,000
	Cr	24	7.19	1,875	2,199
	Mn	25	7.44	1,244	2,097
	Fe	26	7.873	1,536	3,000
	Co	27	8.90	1,493	3,100
	Ni	28	8.908	1,453	2,730
	Cu	29	8.94	1,083	2,582
Second Series	Zr	40	6.506	1,850	4,377
	Nb	41	8.58	2,468	4,927
	Mo	42	10.22	2,610	5,560
	Tc	43	11.49	2,170	5,030
	Ru	44	12.45	2,310	4,080
	Rh	45	12.41	1,960	3,700
	Pd	46	12.02	1,552	2,020
	Ag	47	10.5	960.8	2,210
Third Series	Hf	72	13.29	2,230	5,200
	Ta	73	16.65	2,996	5,427
	W	74	19.3	3,410	5,930
	Re	75	21.04	3,170	5,630
	Os	76	22.61	3,050	5,020
	Ir	77	22.65	2,443	4,500
	Pt	78	21.45	2,769.3	4,100
	Au	79	19.32	1,063	2,808

5. Densities

Transition elements have high densities, As we move along a period, the densities increase.

Reason: The atomic volumes of the transition elements are low. This is beacuse the electrons are added in the penultimate shell and the increased nuclear charge is poorly shielded by the d electrons. Thus, the outer electrons are strongly attracted by the nucleus. Consequently, the densities of transition metals are quite high as a result of decrease in molar volume. Most of these elements have densities greater than 5g cm^{-3}. Sc, Y and Ti have densities 3.0 g cm^{-3}, 4.51 g cm^{-3} and 4.47g cm^{-3} respectively. Osmium and iridium are the elements having highest densities of 22.61g cm^{-3} and 22.65g m^{-3} respectively. (Table 29.8).

6. Melting and Boiling Points

These elements have high melting and boiling points: From the Fig. 29.2 it is clear that melting points increase upto the elements with maximum unpaired electrons and then decrease.

Reason: The melting and boiling points of a metal depend upon the strength of forces acting betwen its atoms. Transition metals are assumed to have two types of bonding as discussed earlier, i.e typical metallic bonding and second is the covalent bonding which makes transition elements hard and brittle. The incomplete d orbitals overlap with each other to form covalent bonds. The presence of this covalent bonding due to d-d overbonding along with the strong metallic bonds results in higher melting and boiling points.

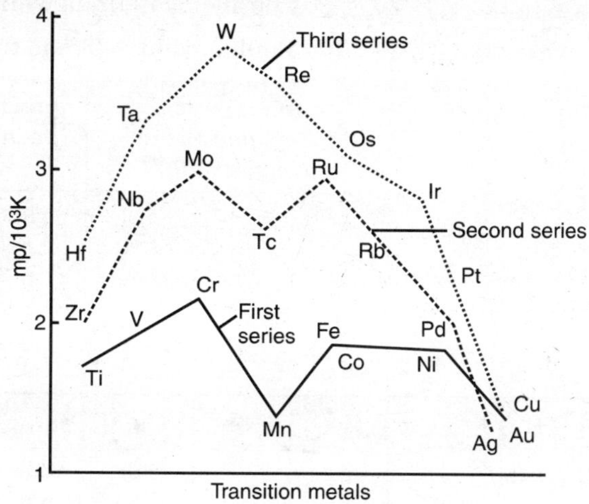

Fig. 29.1 Melting and boiling points of transition metals.

In general greater the number of valence electrons, stronger the bonding will be, hence melting and boiling point increase us we move along the period because of increase in number of unpaired electrons uptill the middle. After that pairing starts and melting and boiling point decrease due to the decreasing availability of unpaired electrons.

The alkali and alkaline earth metals have only one or two unpaired electrons and hence have low melting and boiling points as compared to transition elements. Similarly, the elements of group 12, excluded from transition elements but studied along them, have low melting and boiling points. This is because there are no unpaired electrons available for covalent bonding and hence their boiling and melting points are low being 419.5°C, 320.9°C and -38.4°C for Zn, Cd and Hg respectively. The melting and boiling points for first, second and third transition series are given in Table 29.8.

7. Oxidation States

One of the remarkable feature of the transition elements is that they exhibit variable oxidation states. These oxidation states are related to their electronic configurations. The oxidation states exhibited by the transition elements are shown in Table 29.9.

Table 29.9: Oxidation States of Transition Series Different oxidation states of elements of first transition series

Elements	*Sc*	*Ti*	*V*	*Cr*	*Mn*	*Fe*	*Co*	*Ni*	*Cu*	*Zn*
Oxidation states				+1					+1	
	+2	+2	+2	+2	+2	+2	+2	+2	+2	+2
	+3	+3	+3	+3	+3	+3	+3	+3	+3	
		+4	+4	+4	+4	+4	+4	+4	+4	
			+5	+5	+5	+5	+5			
				+6	+6	+6				
					+7					

Different oxidation states of elements of second transition elements

Elements	Y	Zr	Nb	Mo	Tc	Ru	Rh	Pd	Ag	Cd
Oxidation states	+3	+4	+3	+3	+4	+2	+3	+2	+1	+2
			+5	+4	(+6)	+3	+4	(+3)	(+2)	
				+5	+7	+4,	+6	+4	(+3)	
				+6		+5				
						+6				
						(+7),				
						(+8)				

Different oxidation states of elements of third transition series

Elements	La	Hf	Ta	W	Re	Os	Ir	Pt	Au	Hg
Oxidation states	+3	(+4)	(+4),	(+2)	(+3)	(+2)	+2,	+1	+1	
			+5	(+3),	+4	(+3)	+3	(+3)	+3	+2
				+4,	+5,	+4,	+4,	+4		
				+5,	+6,	+5,	+6,			
				+6,	+7,	+6				
						+8,				

Reason: The existence of the transition elements in the various oxidation states is due to the tendency of the penultimate d-orbitals to enter into chemical bond formation because of comparable energies of ns and (n-1) d orbitals.

The outer electronic configuration of scandium Sc, the first element of transition series is $3d^1$, $4s^2$. If both the s elements are used for bonding then Sc exhibits an oxidation state of +2. When two s and one d electrons are used for bonding Sc has an oxidation state of +3.

Titanium Ti, having outer electronic configuration $3d^2$, $4s^2$ exhibits the oxidation state of +2 when it uses both the 4s electrons. When it uses one of its d electron along with 4s electrons for bonding, it shows oxidation state of +3 and oxidation state of +4 is shown when both the d electrons along with two 4s electrons are used for bonding.

Similarly, the highest oxidation state upto manganese, in the first transition series, is equal to the sum total of ns and (n-1) d electrons. Upto manganese the d orbitals are singly filled, the minimum oxidation state is given by the number of electrons in the outer s subshell and the maximum oxidation state is given by the sum of the outer s and d electrons.

In the last five elements of the first transition series, the tendency for all the d electrons to participate in bonding decreases. This may be due to increased nuclear charge which binds the d electrons more strongly than the ns orbitals. Thus Fe has oxidation state of +4. The study of oxidation states of transition elements yield the following important conclusions:

(i) The common oxidation state of transition elements is +2 (except Scandium).

(ii) The transition elements usually show their highest oxidation states in compounds with high electronegative elements like chlorides, oxides and flourides. The bonds formed by these elements in higher oxidation states are covalent because the bonds are formed by the sharing of electrons.

(iii) In the +2 and +3 oxidation states, the bonds formed are mostly ionic.

(iv) The maximum oxidation state shown by a transition element increases down the group. For example, common oxidation states of Fe are +2 and +3 but of Ru and Os belonging to the same group are +4, +6 and +8. This difference between Fe and the other two elements Ru and Os is due to the increased size of Ru and Os.

(v) The transition elements also exhibit the oxidation states of +1 and 0. Many stable compounds of Chromium, Copper and other elements in +1 oxidation state are known e.g. CuCl. The oxidation state of the transition elements in carbonyls and nitrosyls is 0, e.g. $[Ni(CO)_4]_3$ $[Fe(CO)_5]$ in which Ni and Fe are in zero oxidation states.

(vi) Relative stability of different oxidation states: Elements at the left of transition elements prefer higher oxidation states. Towards the right of a given transition series, lower oxidation state become stable. Higher oxidation state also becomes more stable as one moves down the group.

The stable oxidation states are found in oxides and halides because oxygen and halogens are the most electronegative elements. The anions F^-,and O_2^{2-} have high tendency to be oxidised and transition metal cation in higher oxidation state has higher ability to act as oxidising agent. This leads to the formation of a strong M-F and M-O chemical bond in metal flourides and oxides respectively.

(vii) The stability of oxidation states can be compared by their electrode potential values. For e.g.

$$Cu^{2+}(aq) + 2e^- \longrightarrow Cu(s) \quad E^\circ red = 0.34V$$

$$Cu^{+}(aq) + e^- \longrightarrow Cu(s) \quad E^\circ red = 0.52V.$$

Thus Cu^+ has strong tendency to reduce as compared to Cu^{2+} and hence in Cu the +2 oxidation state is more stable than Cu^+.

8. Catalytic Properties

Transition elements and their compounds are well known for their catalytic properties. Iron, Cobalt, nickel, chromium, platinum, vanadium and manganese are very good catalysts for various reactions (refer to chapter 9) some of the transition metal catalyst are shown in Table 29.10.

Table 29.10: Catalytic Properties of Transition Elements

Transition Element	*Substance Used As Catalyst*	*Reaction Catalysed*
Fe	Fe or Fe_2O_3	Haber Process for manufacturer of NH_3.
Ti	$TiCl_3/Al\ (C_2H_5)_3$	Polymerization of Ethene
V	V_2O_5	Contact Process for the manufacturer of H_2SO_4.
Ni	Ni	Hydrogenation of Vegetable Oils.
Co	Co-Th	Fischer Tropsch Process for the synthesis of gasoline.
Pt	Spongy Pt	Ostwald's Process-Oxidation of ammonia to nitric acid by ostwald's process.
Mn	MnO_2	Decomposition of $KClO_3$.
Ru	$[Ru(bipy)_3]^{2+}$	Photo decomposition of water to produce H_2 and O_2.
Co	Co^{2+} (salts)	Decomposition of bleaching powder.
Cu	Cu	Oxidation of ethnol to ethanal.

Reasons: The main reasons for the catalytic activity of transition metals are:

(a) Tendency to exhibit variable oxidation states. The transtion elements thus can form reaction intermediates. (Intermediate Compound formation theory refer to Chap. 9). These intermediates provides a new path with lower activation energy.

$$\underset{\text{Reactant}}{A} + \underset{\text{Catalyst}}{K} \longrightarrow \underset{\text{Intermediate}}{AK}$$

$$AK + \underset{\text{reactant}}{B} \longrightarrow \underset{\text{(Product)}}{AB} + \underset{\text{(Catalyst)}}{K}$$

For example reaction between SO_2 and O_2 to form SO_3 in presence of V_2O_5.

$$V_2O_5 + SO_2 \longrightarrow V_2O_4 + SO_3$$

$$2V_2O_4 + O_2 \longrightarrow 2V_2O_5$$

(b) The transition elements can also act as catalysts by providing suitable large surface area with free valencies on which the reactants are adsorbed and the close proximity of the reactants facilitates the reaction. (Adsorption Theory).

For example-hydrogenation of vegetable oils in the presence of finaly divided nickel.

$$R{-}CH = CH{-}R_1 + H_2 \xrightarrow{Ni} R{-}CH_2{-}CH_2{-}R$$

9. Complex Formation

An important feature of transition elements is their tendency to form complexes. Certain molecules having lone pair of electrons or anions, attack transition metals cations through coordinate bonds to form complex compounds. The neutral molecules having lone pair of electrons or anions are called ligands. The common examples of complexes are $[Ni\ (NH_3)_6]^{2+}$, $[Co(NH_3)_6]^{3+}$, $[Fe\ (CN)_6]^{3-}$ etc.

Reason: Transition elements are capable of forming complexes due to the following:

(a) Transition metal cations, because of small size and high effective nuclear charge have high charge density. Therefore, these can accept lone pairs of electrons from the ligands.

(b) Moreover, these have vacant d orbitals to accommodate lone pair of electrons from the ligands.

10. Interstitial Compounds

The transition metals form interstitial compounds in which small atoms like hydrogen, carbon, boron and nitrogen occupy the empty spaces called interstitial sites in their lattices. As a result, the transition metals become rigid and hard. Such compounds are known as interstitial compounds. These interstitial compounds have similar chemical properties but differ significantly in their physical properties. For example, cast iron (containing 2 to 4.5% carbon) and steel (containing .6 to 1.5% carbon) are hard because of trapped carbon in their interstitial sites.

11. Formation of Alloys

Transition metal alloys have a very significant role to play in engineering chemistry. A wide range of alloys with the required properties are manufactured from transition elements. These transition metals have similar atomic sizes due to which atoms of one metal can easily take up position in the crystal lattice of the other. Transition metals are thus miscible with each other in the molten state. Actually, an alloy is a metallic solid mixture of two or more different elements. The physical properties of the elements forming the alloys are different from their constituent elements. Some of the important alloys with their composition and uses are given in Table 29.11.

Table 29.11: Some Important Alloys

Metal	*Alloy*	*Composition*
Aluminium	Duralumin	95% Al+Cu, Mg, Mn
	Magnalium	90% Al+10% Mg
Copper	Brass	Cu and Zn
	Bronze	Cu and Sn
	Gun metal	Cu (88%), Sn (10%), Zn (2%)
	Bell metal	Cu (80%), Sn (20%)
	German silver	Cu, Zn and Ni
Lead	Solder	Pb (67%) & Sn (33%)
	Type metal	Pb, Sb and Sn
	Wood's metal	Bi, Pb, Sn, Cd
Magnesium	Electron	Mg and Zn
Nickel	Monel metal	Ni and Cu
	Nichrome wire	Ni and Cr
Steel	Stainless steel	Steel, Cr + Ni
	Invar	Steel and NI
	Tool steel	Steel, Ti, V, W
Tin	Babbit metal	Sn, Pb and Cu
	White metal	Sn, Sb, Cu

The other two important characteristics of Transition elements are Colour and Magnetic properties which can be expalined on the bases of Crystal Field of Theory. Hence, it is worthwhile to discuss this theory in brief before taking up these two properties of transition elements.

29.5 CRYSTAL FIELD THEORY

This theory was developed by H.Bethe and V.Bleck (1935). According to this theory, when the compounds of transition metal atoms, for e.g. complex compounds are formed the d orbitals of the metal atoms do not remain degenerate. The main points of crystal field splitting can be summarised as follows:

(i) Crystal field theory considers the central metal atom (transition metal atom) to be point charge, i.e. a charge concentrated into a very small volume. Similarly, surrounding negative ligands or neutral molecules which have a lone pair of electrons are also considered to be point masses.

(ii) The attraction between central metal and ligands in a complex is purely electrostatic.

(iii) The d orbitals in the metal atom in the free state are degenerate, i.e. they have the same energy. However when a complex is formed the degeneracy of these orbitals is lost. In otherwords, d orbitals split into different energy levels under the influence of approaching ions towards the transition metal ion.

In an free metal atom the five d orbitals viz, dxy, dyz, dzx, $dx^2–y^2$, dz^2 all have same energy and are termed as degenerate.

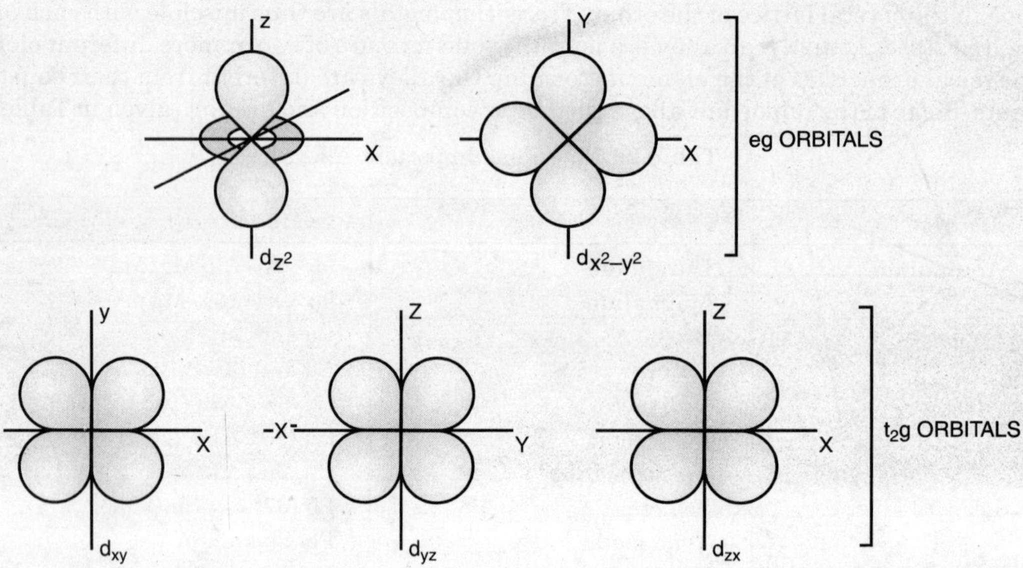

Fig. 29.2 Five d-orbitals of central metal atom having similar energies in an isolated atom

As is seen from the Fig. 29.3, the three orbitals dxy, dyz, dzx are oriented in between the axes. These are termed as t_2g orbitals. The orbitals dx^2-y^2, dz^2 are oriented along the axes and are termed as e_g orbitals.

Now, if a spherically symmetrical field of negative charge approach the central metal atom, the energy of the d orbitals of central metal atom is raised, because of the repulsion between the approaching negative field and the electrons of the central metal. Thus if the field is symmetrical about the central metal atom, the d orbitals are affected equally and the energy of these is raised to the same level, i.e. they may remain degenerate. But this is not the case in reality. In most transition metal atoms, either 6 or 4 ligands, considered as point charges, are coordinated to the central metal atom forming octahedral or tetrahedral complexes. The approaching field thus, is not symmetrical around the central metal atom and hence d orbitals are raised to different energy levels depending upon the direction of the ligand. Let us illustrate the above point by considering octahedral and tetrahedral complexes separately.

Crystal Field Splitting in Octahedral Complexes:

Consider the case of transition metal ion surrounded octahedrally by six ligands as shown in Fig. 29.4.

The ligands lie along x, y and z axes. When these are brought towards the positively charged central metal ion, there will be net electrostatic attraction. But repulsion between the electrons of the ligands and electrons present in the d orbital of the metal atom, increases the energy of the d orbitals. As the orientation of the five d orbitals is different relative to the x, y, and z axes, the d orbitals of the central metal atom will be affected differently.

In the above case of octahedral complex, the d orbitals which are oriented along the direction of the ligands will experience greater repulsion and their energy will be raised more than the d orbitals, which are oriented in between the direction of the ligands. Thus, dx^2-y^2 and dz^2 orbitals will be raised to higher energy levels than dxy dzx and dxz orbitals. In simple words, the degenerate d orbitals split up into two sets having different energies:

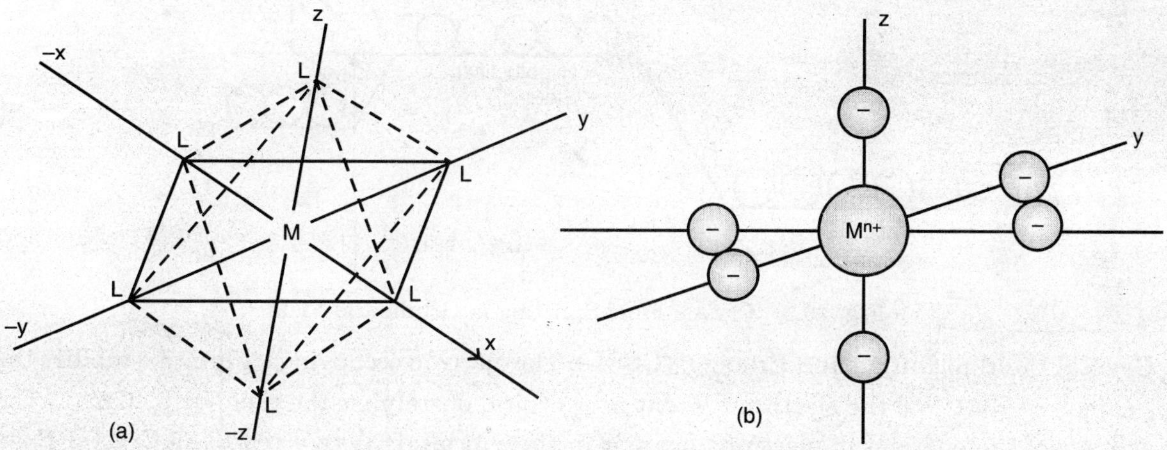

Fig. 29.3 An octahedrally arranged set of ligands in relation to a set of Cartesian coordinates with a positively charged metal ion, M

one of the sets comprises of dxy, dzx dyz or t_2g orbitals-the other set comprises of $dx^2–y^2$ and dz^2 or e_g orbitals which are raised to higher energy level than t_2g orbital.

This splitting of degenerate d orbitals of an isolated central metal atom into two sets of different energies, under the influence of approaching ligands is called **Crystal Field Splitting**.

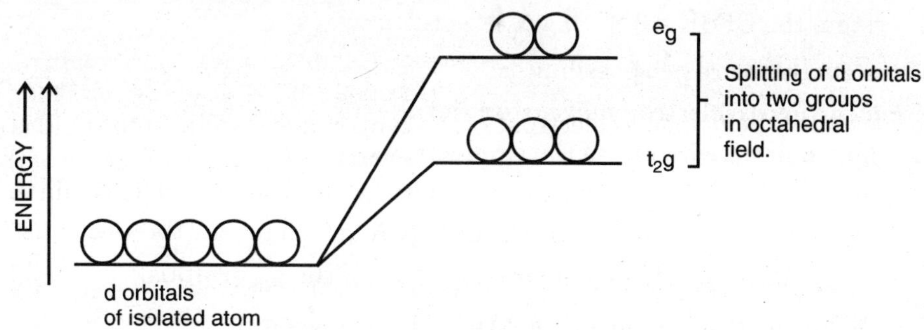

Fig. 29.4 Crystal Field Splitting in Octahedral Field

The average increase in energies of the d orbital, if the approaching field were symmetrical is called **Bari Centre**. It is the hypothetical state which represents the mean of the variations of the energies of the d orbitals, which is taken as zero. This means that the energy of e_g orbitals must be raised by the same amount, by which the energy of t_2g orbitals are lowered as compared to the hypothetical state. The difference in energy between the two d levels is measured in terms of Δ_0 in octahedral complexes. Sometimes a symbol Dq is also used for representing the crystal field splitting. $\Delta_0 = 10$ Dq by convention.

e_g orbitals are $+0.6\,\Delta_0$ above the average level (hypothetical state) whereas t_2g orbitals are $-0.4\ \Delta_0$ below the average.

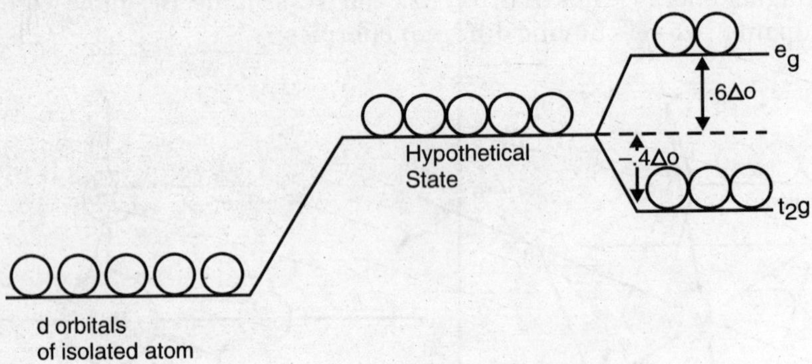

Fig. 29.5 Crystal Field Splitting in Octahedral Field

Crystal Field Stabilization Energy: (CFSE): The electron occupying t_2g orbital stabilize the complex by $.4\,\Delta_0$ and the electron present in e_g orbital destabilise the complex by $0.6\,\Delta_0$. The average of the energies of electrons present in these orbitals is referred to as Crystal Field Stabilisation Energies.

In an octahedral complex, if the central metal atom has one electron, then it will occupy t_2g orbital and stablise the complex by $0.4\,\Delta_0$, i.e. it is $0.4\ \Delta_0$ more stable than predicted by electrostatic theory. The CFSE is given by- $0.4\,\Delta_0$. Crystal Field Stabilisation Energy (CFSE) is thus lowering in the energy of a transition metal ion, during complex formation, due to crystal field effects. Similarly:

for d^2 system the CFSE $= 2 \times -0.4\,\Delta_0 = -0.8\ \Delta_0$.

for d^3 system the CFSE $= 3 \times -0.4\,\Delta_0 = -1.2\,\Delta_0$.

for d^4 system there are two possibilites:

- either the fourth electron may go to e_g level
- or it may pair up with one of the t_{2g} electrons.

The actual state depends upon the magnitude of Δ_0, i.e. the crystal field splitting and pairing energy. Pairing energy (P) is defined as the energy required to pair the electrons. Thus,

If $\Delta_0 > P$, the fourth electron will pair up in one of the t_{2g} orbitals.

If $\Delta_0 < P$, the fourth electron will go to one of the e_g orbital.

Both the cases are shown in Fig. 29.7.

Similar, conclusions can be derived for the systems having d^5 to d^{10} electrons.

When Δ_0, i.e. crystal field splitting is large, the electrons pair up in t_{2g} level and thus the complex would have less number of unpaired electrons. Such type of complexes are called **low**

spin complexes. On the other hand, if the magnitude of Δ_0 is less, the electrons will tend to move to e_g level and thus the complex would have more number of unpaired electrons and the complex would be termed as **high spin complex.**

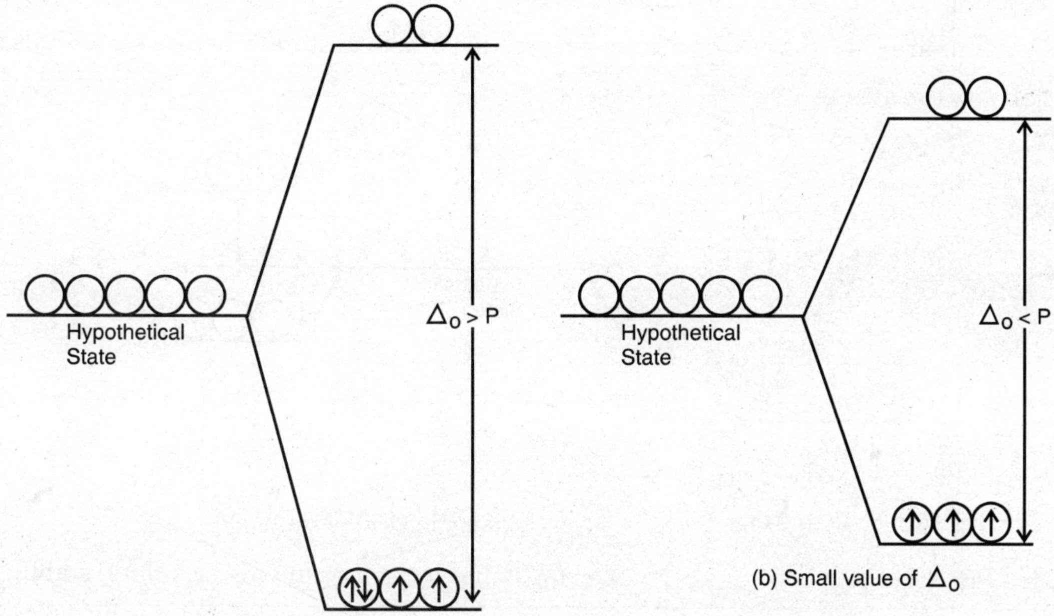

(a) Large value of Δ_o

(b) Small value of Δ_o

Fig. 29.6 Electronic Configuration for d^4 System

The crystal field stabilisation energies for metal ions having different numbers of d electrons in octahedral complexes are given in Table 29.12.

Effects of Crystal Field splitting.

Table 29.12: CFSE and electrons arrangements in octahedral complexes.

Number of d trons	*Arrangement in weak ligand field* t_{2g}	e_g	*CFSE* Δ_0	*Arrangement in strong ligand field* t_{2g}	e_g	*CFSE e l e c -* Δ_0
d^1	(↑)( )( )	( )( )	–0.4	(↑)( )( )	( )( )	–0.4
d^2	(↑)(↑)( )	( )( )	–0.8	(↑)(↑)( )	( )( )	–0.8
d^3	(↑)(↑)(↑)	( )( )	–1.2	(↑)(↑)(↑)	( )( )	–1.2
d^4	(↑)(↑)(↑)	(↑)( )	–1.2 +0.6 = –0.6	(↑↓)(↑)(↑)	( )( )	–1.6
d^5	(↑)(↑)(↑)	(↑)(↑)	–1.2 +1.2 = –0.0	(↑↓)(↑↓)(↑)	( )( )	–2.0

Number of d trons	*Arrangement in weak ligand field*			*Arrangement in strong ligand field*		
	t_{2g}	e_g	*CFSE* Δ_0	t_{2g}	e_g	*CFSE elec-* Δ_0
d^6	(↑↓)(↑)(↑)	(↑)(↑)	−1.6 +1.2 =−0.4	(↑↓)(↑↓)(↑↓)	( )( )	−2.4
d^7	(↑↓)(↑↓)(↑)	(↑)(↑)	−2.0 +1.2 = − 0.8	(↑↓)(↑↓)(↑↓)	(↑)( )	−2.4 +0.6 = − 1.8
d^8	(↑↓)(↑↓)(↑↓)	(↑)(↑)	−2.4 + 1.2 =−1.2	(↑↓)(↑↓)(↑↓)	(↑)(↑)	−2.4 +1.2 =−1.2
d^9	(↑↓)(↑↓)(↑↓)	(↑↓)(↑)	−2.4 +1.8 =−0.6	(↑↓)(↑↓)(↑↓)	(↑↓)(↑)	−2.4 +1.8 =−0.6
d^{10}	(↑↓)(↑↓)(↑↓)	(↑↓)(↑↓)	−2.4 +2.4 =−0.0	(↑↓)(↑↓)(↑↓)	(↑↓)(↑↓)	−2.4 +2.4 = − 0.0

Spectro Chemical Series

The magnitude of Δ_0 depends on

(i) The nature of the ligands

(ii) Charge on the metal ion.

Nature of the ligands: The Δ_0 is approximately same for ions of a given charge with the same ligand as is evident from the Table 29.13

Table 29.13: Crystal Field Splitting, D_0 for the Metal ions having same charge.

Metal Ion	*Δ_0 Values (K cal/mol)*		
	H_2O As Ligand	*NH_3 as Ligand*	*CN^- as Ligand*
Ti (III)	58		
Cr (III)	50	62	75
Co (III)	53	65	97
Co (II)	28	29	
Ni (II)	24	31	
Cu (II)	36	43	

From table 29.13 it is also observed that changing the ligand, changes the value of Δ_0. For example Δ_0 for Co (III) with H_2O as ligand is 53 k Cal/mol whereas with NH_3 and CN^- it is 65 k Cal/mol and 97k Cal/mole respectively.

Ligands which cause only a small degree of crystal field splitting are referred to as **weak field ligands**. The ligands which cause a large splitting are called **strong field ligands.** It is possible to arrange the common ligands in ascending order of crystal field splitting Δ_0 from the measurement of spectrochemical data. More or less this order remains constant for different metals and this order is known as **spectro chemical series.**

$$I^- < Br^- < Cl^- < \bar{O}H < C_2H_5\,OH < C_2O_4^2 < H_2O < EDTA < Py = NH_3$$
$$< ethylendia\min e < NO_2^- < CN^- < CO$$

Spectrochemical Series

This series is experimentally determined series. The above order for solid ions can be explained on the basis of their different sizes. Smaller the ion, smaller is the ligand-central ion separation and larger is the splitting, e.g. F^- cause more splitting than I^-.

However the order of the entire spectro chemical series is difficult to explain as it incorporates both the effects of σ and π bonding.

Charge on the metal ion: The charge on the metal ion also affects the magnitude of crystal field splitting. The value of Δ_0 increases as the charge on the metal ion increases. It has been observed that the values of Δ_0 for the metals of the first transition series, having +3 oxidation states are approximately 50% higher than the values for the same metals having +2 oxidation state. (Table 29.14).

Table 29.14: Crystal Field Splitting for Transition Metals in Octahedral Complexes in Different Oxidation States with H_2O as Ligand.

Metal Atom	*Crystal Field Splitting Δ_0 in $kJmol^{-1}$*	
	When the Oxidation State of metal atom is +3	*When the Oxidation State of metal atom +2*
V	151	226
Cr	166	213
Fe	124	164
Co	111	222

So, far our discussion is restricted to crystal field splitting in octahedral complexes. Other common type of geometry which is exhibited by transition metal complexes in regular tetrahedral.

Crystal Field Splitting in Tetrahedral Complexes

The tetrahedral arrangement of four ligands can be visualised by placing ligands at the alternate corners of a cube surrounding the central metal atom as shown in Fig. 29.7.

From the Fig. 29.8 it is clear that none of the d orbitals points exactly in the direction of the ligands. The three t_{2g} orbitals viz., dxy, dyz and dzx are nearer to the direction of the ligands as compared to e_g orbitals ie $dx^2–y^2$ and dz^2 which lie in between the ligands. The approach of the ligands, treated as point charges, raises the energy of both the sets. But the

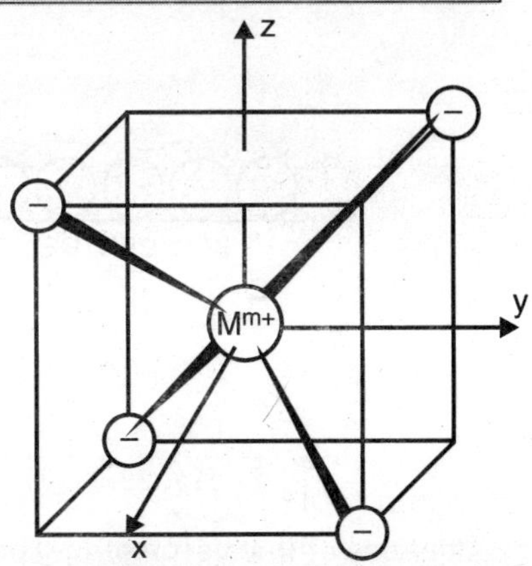

Fig. 29.7: Tetrahedral Arrangement of Four Ligands around the metal atom.

energy of t_{2g} orbitals will be raised more as compared to the energy of e_g orbitals. Thus the d orbitals are split into two groups with the approach of the ligands- t_{2g} orbitals having higher energy as compared to the hypothetical state or bari centre. The splitting in regular tetrahedral is inverted compared with that for a regular octahedral structure as shown in the Fig. 29.9.

The t_{2g} orbitals are 0.4 Δ_t above the weighed average energy (bari centre) of the two groups whereas e_g orbitals are 0.6 Δ_t below the average. (Fig. 29.9)

The symbol Δ_t denotes the splitting in tetrahedral complexes. As is evident from the Fig. 29.9, the magnitude of crystal field splitting in tetrahedral complexes is less than in octahedral field. This is due to the reason that the approaching ligands do not coincide exactly with the direction of the ligands which reduces the repulsion and hence crystal field splitting. Moreover, the number of ligands is just four as compared to six in octahedral complexes and hence the repulsion is again less which reduces the crystal field splitting further.

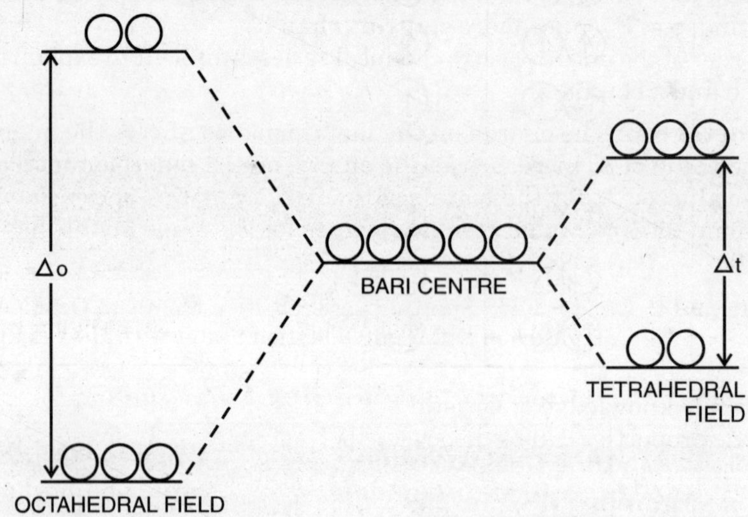

Fig. 29.8 Crystal field splitting diagram for octahedral and tetrahedral fields.

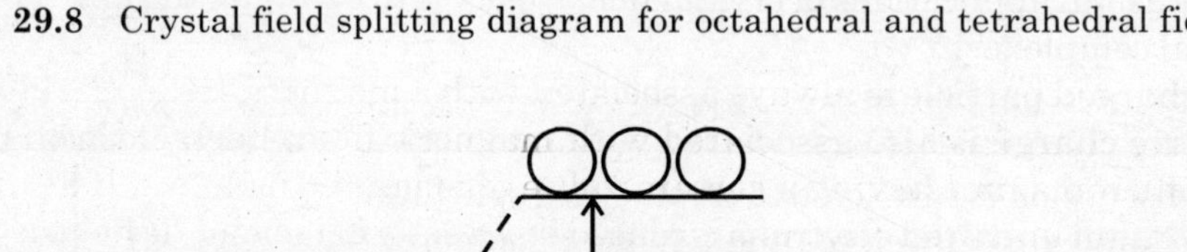

Fig. 29.9 Energy levels for d orbitals in a tetrahedral field.

Because of low value of Δ_t, the pairing only takes place after the t_{2g} orbitals are also singly occupied. Hence the tetrahedral complexes are high spin complexes.

The CFSE values for d^1 to d^{10} complexes are given in Table 29.14.

Table 29.14: CFSE and electronic arragements in tetrahedral complexes.

Number of d electrons	*Arrangement of electrons* e_g	t_{2g}	*Tetrahedral CFSE* Δ_t
d^1	(↑)( )	( )( )( )	–0.6
d^2	(↑)(↑)	( )( )( )	– 1.2
d^3	(↑)(↑)	(↑)( )( )	–1.2 + 0.4 = – 0.8
d^4	(↑)(↑)	(↑)(↑)( )	– 1.2 + 0.8 = –0.4
d^5	(↑)(↑)	(↑)(↑)(↑)	– 1.2 + 1.2 = 00
d^6	(↑↓)(↑)	(↑)(↑)(↑)	– 1.8 – 1.2 = – 0.6
d^7	(↑↓)(↑↓)	(↑)(↑)(↑)	– 2.4 + 1.2 = – 1.2
d^8	(↑↓)(↑↓)	(↑↓)(↑)(↑)	– 2.4 + 1.6 = – 0.8
d^9	(↑↓)(↑↓)	(↑↓)(↑↓)(↑)	– 2.4 + 2.0 = – 0.4
d^{10}	(↑↓)(↑↓)	(↑↓)(↑↓)(↑↓)	–2.4 + 2.4 = 0.0

After having a fair knowledge of Crystal Field Theory, the most important properties, i.e. colour and magnetic properties can be discussed in the light of the above.

29.6 MAGNETIC PROPERTIES OF TRANSITION METAL COMPOUNDS

One of the most useful applications of crystal field theory is to correlate the magnetic properties of transition metal complexes.

A moving charged particle is always associated with a magnetic field. The electron carrying one unit negative charge is also associated with magnetic field. Each electron may, infact, be regarded as a micromagnet having a certain value of magnetic moment. If a substance whose atoms or ions contain unpaired electrons is placed in an external magnetic field, is attracted by it, it is termed as **paramagnetic substance**. The magnetic lines of force travel through a paramagnetic substance more easily as compared to that of vacuum. The intensity of the magnetic field of the paramagnetic substance when placed in the external magnetic field, will be greater than the magnetic field strength in which it is placed. On the other hand, a substance whose atoms or ions contain paired electrons is repelled by the external magnetic field in which it is placed. Such type of substance is termed as **diamagnetic substance**. Diamagnetic substances tend to repel lines of force. The magnetic lines of force travel through a diamagnetic substance with much more difficulty compared to that of vacuum. The intensity of the magnetic field of the diamagnetic substance is much less as compared to the strength of magnetic field in which it is placed.

The total magnetic moment of a substance is the resultant of the magnetic moments of all the individual electrons. In diamagnetic substances, the electrons in the atoms are completely paired and as a result the effects of individual electrons are mutually compensated. While in the case of paramagnetic substances, containing unpaired electrons in their atoms, the magnetic moments due to spin and orbital motion of such electrons get aligned, parallel to the direction of external magnetic field applied. Such a substance therefore experiences attraction in a magnetic field. To summarize:

Substance	Due to the presence of	When placed in the external magnetic field	Intensity of the magnetic field in the substance when placed in external magnetic field
Paramagnetic	Unpaired electrons	Attracted by the external magnetic field	Higher
Diamagnetic	Paired electrons	Repelled by the external magnetic field	Lower

From the above discussion it follows that paramagnetism of a substance is directly related to the number of unpaired electrons. The magnetic moment of an unpaired electron is considered to be the sum of, magnetic moment due to spin motion (μ spin) and magnetic moment due to orbital motion (μ orbital).

μ effective = μ orbital + μ spin

In many compounds of the first row transition elements, the orbital contribution is quenched by the electric-fields of the surrounding ligands. Thus the μ orbital can be ignored and the magnetic moment in such cases will be approximately given by the μ spin. The **spin only** magnetic moment, μ s is given by the formula:

$$\mu_s = \sqrt{n(n+2)}$$

Where n is the number of unpaired electrons. μ_s is expressed in Bohr Magnetons (BM)

BM = eh/4πmc.

where h is the Plank's constant, e is the electron charge, c is the velocity of light and m is the mass of electron.

Thus, the magnetic moment of substance depends mainly on the number of unpaired electrons present and can be calculated by using the above expression for μ spin. The formula gives good agreement with many complexes of the first row transition elements. The μ_s for different number of unpaired electrons as calculated from the above expression is given in Table 29.15 for easy reference.

Table 29.15: Magnetic moment calculated from number of unpaired electrons.

No. of unpaired electrons μ spin (B.M)	*Calculated Magnetic Moment*
1	1.73
2	2.84
3	3.87
4	4.90
5	5.92

To ascertain the magnetic properties of transition metal complexes, the basic requirement is to know the number of unpaired electrons present in the complex. This can be worked out with the help of Crystal Field Theory. Let us illustrate this by taking the example of cobalt (III) complexes.

From the experiments it has been observed that some of the complexes of cobalt like [CO $(NH_3)_6$] are not attracted by the magnet, i.e. they are diamagnetic, whereas some other complexes like [CoF_6] are paramagnetic. This observation can be understood on the bases of crystal field theory. The outer electronic configuration of Co^{3+} is d^6. Now the d orbitals in octahedral complexes split up into two sets t_{2g} and e_g. In this case, the first three atoms occupy the t_{2g} orbitals. Now the remainging three may (i) either pair up with electrons in t_{2g} orbitals or (ii) two of these occupy e_g orbitals with the last one pairing with t_{2g} depending upon the strength of Δ_0 and pairing energy.

The two cases are represented in Fig. 29.11

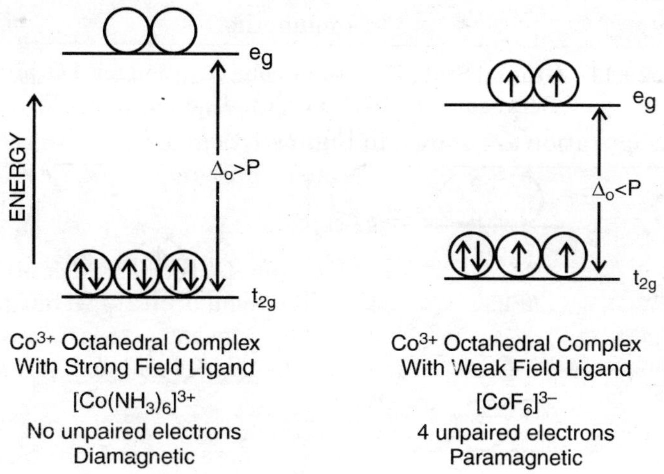

Fig. 29.10 Complexes of Co^{3+} with strong and weak field ligands

NH_3 is a strong field ligand and hence produce larger splitting resulting in diamagnetic or Low Spin Complex. Flouride ion is weak field ligand and hence produce less splitting and the electrons can be placed in higher orbitals producing paramagnetic or High Spin Complex In general- strong **field ligands produce Low Spin Complexes. Weak field ligands produce high spin complexes.**

An arguement as above can be applied to determine the magnetic properties of octahedral complexes containing d^1 to d^{10} electrons. It is seen from Fig. 29.12, the d^1, d^2 and d^3 electrons have only one type of configuration. These will follow Hund's rule and enter the most stable t_{2g} orbitals with their spins all parallel, regardless of the magnitude of crystal field splitting. Similarly for the ions with 8,9 and 10 d electrons, again there is only one way in which these can be arranged as shown in Fig. 29.12.

For the configuration d^4, d^5, d^6 and d^7 there can be two possible arrangements, depending upon the magnitude of crystal field splitting Δ_0. The arrangement having maximum possible number of unpaired electrons are called the high spin complexes and those with minimum number of unpaired electrons are called low spin complexes.

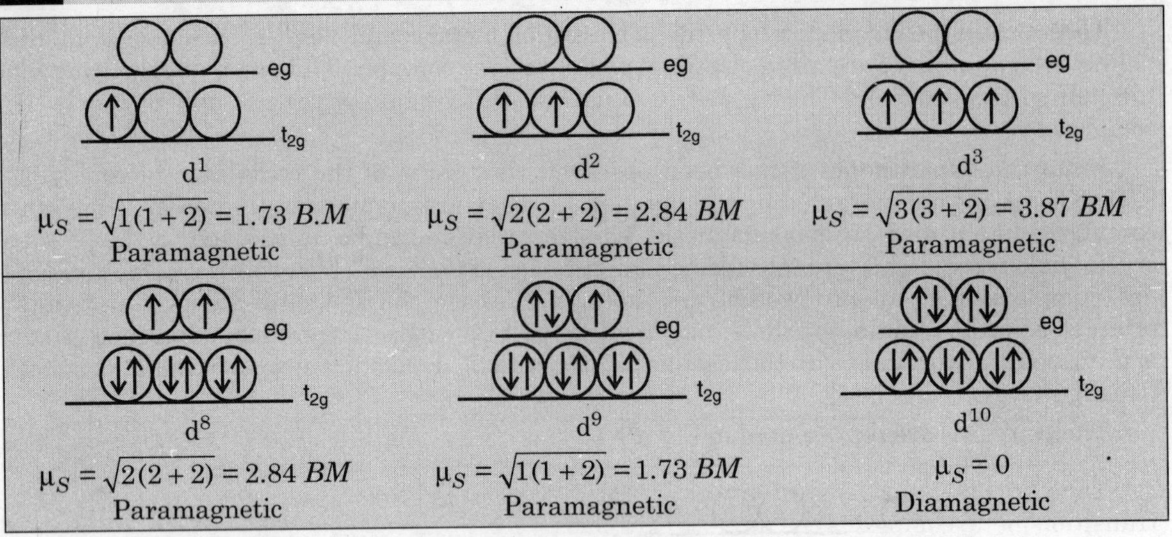

Fig. 29.11 Ground State Electronic configuration for d Orbitals in Octahedral Fields for d^1, d^2, d^3, d^8, d^9 and d^{10}

Both types of configuration are shown in Fig. 29.13.

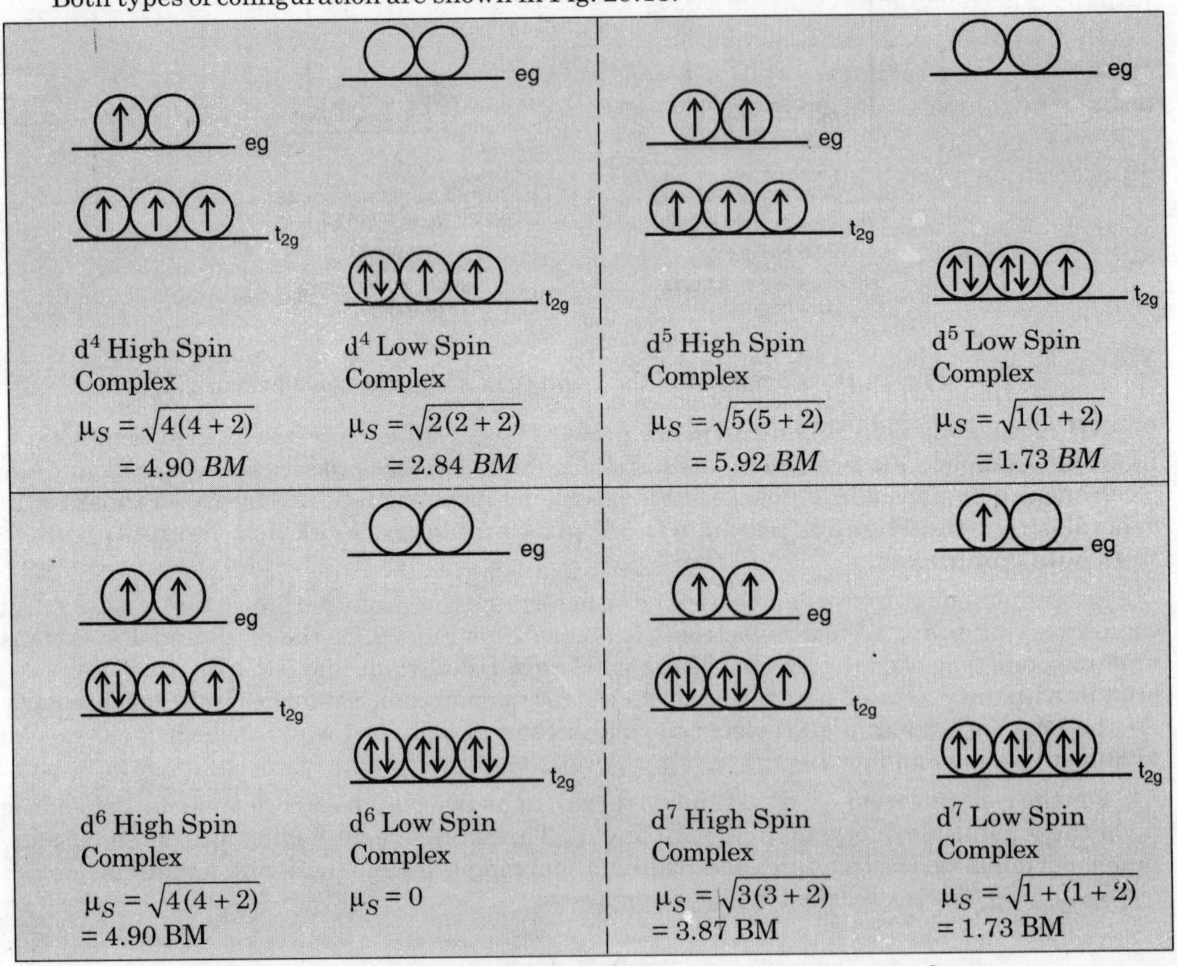

Fig. 29.12 High Spin and Low Spin arrangements for the ground state configurations of d^4 d^5 d^6 and d^7 ions in Octahedral fields

The configurations can also be written out in the form of the notations shown in Table 29.16.

Table 29.16: Notation for electronic configuration after crystal field splitting.

Electronic Configuration Before splitting	*Electronic Configuration After splitting*	
	Weak field (High Spin Complex)	Strong field (Low Spin Complex)
d^4	$t_{2g}^3\ e_g^1$	$t_{2g}^4\ e_g^o$
d^5	$t_{2g}^3\ e_g^2$	$t_{2g}^5\ e_g^o$
d^6	$t_{2g}^3\ e_g^3$	$t_{2g}^6\ e_g^o$
d^7	$t_{2g}^4\ e_g^3$	$t_{2g}^6\ e_g^1$

Thus from the knowledge of the configuration of the d electrons after crystal field splitting in transition metal complexes the number of unpaired electrons and hence their magnetic moment can be calculated.

The magnetic moments calculated on the basis of spin only moment are in close agreement with the experimental values.

In the case of tetrahedral complexes, the magnitudes of crystal field splitting Δ_t is much less and hence pairing takes place only after all the orbitals t_{2g} and e_g are singly filled. Thus all the tetrahedral complexes are high spin complexes.

29.7 COLOUR IN TRANSITION METAL COMPOUNDS

Most of the transition metal compounds are coloured in their solid or solution form. For example Ni^{2+} compounds are green in colour whereas Ti^{3+} compounds are purple. Once again, this property of transition elements can be explained on the basis of crystal field theory. We know that-

A substance looks coloured if it absorbs the visible light in one region and transmits the rest. For e.g. if a compound absorbs light in red region (6500-7000A°) of the spectrum and transmits the rest it will appear blue in colour. A relationship between absorbed colour and the colour observed is given in Table 29.17.

- The compound which do not absorb light radiations in the visible region will appear white. The electrons get excited to higher energy levels by absorbing in infra red or ultraviolet region.
- A compound appears black, if it absorbs all the visible light that strikes it.
- A compound also appears coloured if it absorbs all the radiations except the one which corresponds to the colour which is reflected. For e.g. it will appear green if it absorb all the light but reflects the green component.

The colour of transition metal ion is readily explained in terms of crystal field theory. The energy difference between t_{2g} and e_g orbitals is usually small and can be provided by the visible light. The electrons from lower energy level are excited to higher energy level (from t_{2g} to e_g in octahedral complexes) by absorbing some low energy radiations of visible light corresponding to the energy gap. The other remaining radiations are transmitted and hence the substance appears coloured.

Table 29.17: Relation between absorbed and observed colour

Colour Absorsed	*Colour Observed*	*Wavelenth Observed (cm^{-1})*
Yellow Green	Violet	24000–26000
Yellow	Indigo	23000–24000
Orange	Blue	21000-23000
Purple	Green	18000–20,000
Indigo	Yellow	16400–17300
Blue	Orange	15300–16400
Blue Green	Red	12800–15300

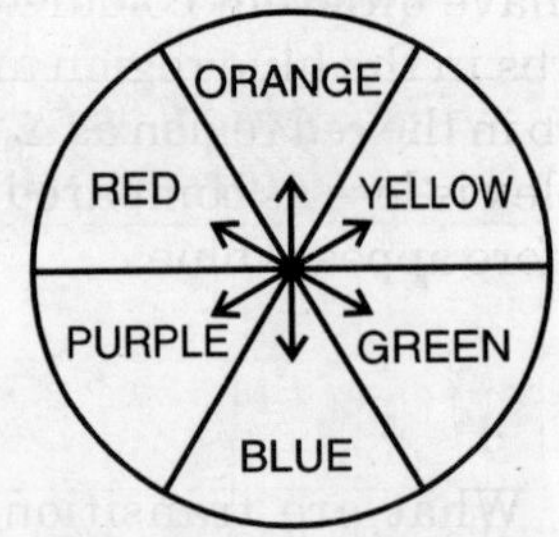

A colour Wheel Showing Complementary Colours (For Quick Reference)

If λ is the wave length of radiations absorbed, the energy associated with it will be given by E = h_v, where h is the Plank's constant.

Consider for example the complex $[Ti(H_2O)_6]^{3+}$ where the metal Ti^{3+} has the configuration d^1. In the ground state, this solitary electron occupies one of the t_{2g} orbitals. The e_g level in this complex lies at an energy corresponding to 20,000 cm^{-1} or 5000 A which is the energy possessed by the greenish-yellow region of the visible spectrum. When white light is impinged on this complex, the greenish-yellow light is absorbed and the rest transmitted. Hence, the complex looks purple which is the complementary colour of the absorbed radiation. Fig. 29.14.

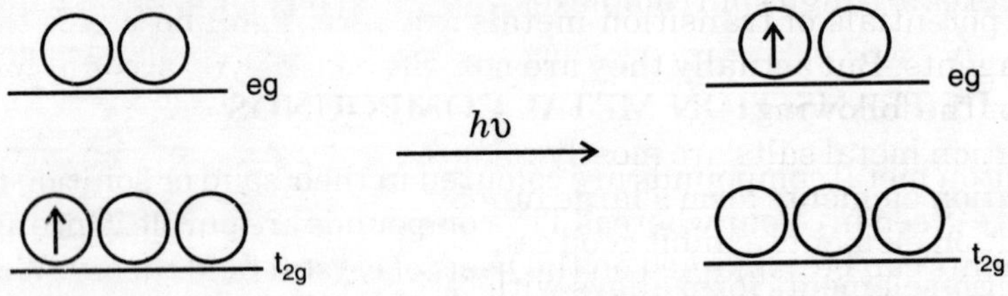

Fig. 29.13 Excitation of Electron in the Complex $[T_i(H_2O)_6]^{3+}$

Crystal Field Theory thus attributes the colour of this transition metal complex to d— d transitions of the electrons. In the absence of ligands, crystal field splitting does not take place and hence the free metal ion is not coloured. If $[Ti(H_2O)_6]Cl_3$ is heated, H_2O molecules evaporate and resulting compound $TiCl_3$ is white. Another example is $CuSO_4 . 5H_2O$ which is blue in colour and in its anhydrous form $CuSO_4$ is white.

It can be thus concluded that stronger is the ligand, larger will be the magnitude of the splitting, shorter will be the absorption wavelengh or higher is the energy required to excite the electron from low energy level to high energy level in d-d transition. Since, NH_3 is a stronger ligand and hence the complex $[Co(NH_3)_6]^{3+}$ requires radiation corresponding to high energy for excitation of its t_{2g} electron. It absorbs in blue region and hence appear orange in solution. The ligand CN^- ion is stronger than NH_3 and in this case the energy required is still higher for excitation and hence it absorbs in violet region and therefore appears yellow.

From the similar grounds it can be said that the transition metals having completely filled or completely empty d orbitals are colourless. Thus the complexes of Cu^{2+}, Zn^{2+}, Ag^+, Ti^{4+} are all colourless.

Moreover, the crystal field splitting also depends upon the geometry of the complex formed. The radiations absorbed for different geometries for the same metal ion are different and hence they have different colours. For example, octahedral hydrated cobalt II complex $[Co(H_2O)_6]^{2+}$ absorbs in the blue region and appears red whereas tetrahedral complex of cobalt (II) $[Co X_4]^{2-}$ absorb in the red region as $\Delta_0 > \Delta_t$ and therefore energy required to excite the electron in tetrahedral complex is less as compared to octahedral geometry. The tetrahedral complex mentioned above therefore appears blue.

QUESTIONS

1. What are transition elements? Which of the d-block elements are not regarded as transitionelements?
2. Write the electronic configuration of Cu^{2+} and Cr^{3+} ions, given atomic numbers of Cu and Cr are 29 and 24 respectively.
3. Why do transition elements exhibit variable oxidation states?
4. Write the IUPAC name of each element from atomic number 104 to atomic number 107.
5. The 3-rd orbital of Cu is completely filled ($3d^{10}$ $4s^1$) still it is considered as transition metal but Zn ($3d^{10}$ $4s^2$) is not. Give reason.
6. The ionization energies of the elements of first transition series do not vary much with increasing atomic number. Explain why?
7. Although the number of electrons in the outermost level in transition elements is the same (i.e. 1 or 2) as in alkali or alkaline earth metals, yet the transition elements are much less reactive. Give reason.
8. Electrode potentials of transition metals are very high and hence they should be good reducing agents. But actually they are not. How would you account for it?
9. Account for the following :
   (a) Transition metal salts are mostly coloured.
   (b) Transition elements form a large number of complexes.
   (c) Transition elements exhibit good catalytic properties.
   (d) Transition elements form alloys with other transition metals easily.
10. Why transition elements have high melting and boiling points? Why the melting points of the elements of the first transition series rise to a maximum at chromium and then decrease?
11. Most of the transition metals do not displace hydrogen from dilute acids. Give reason.
12. Transition elements are harder and brittle than s block elements. Explain why?
13. Discuss concisely the splitting of d orbitals in case of
    (i) Octahedral complexes
    (ii) Tetrahedral complexes
14. How does crystal field theory accounts for the fact that $[CoF_6]^{3-}$ is paramagnetic but $[Co(NH_3)_6]^{3+}$ is diamagnetic, though both are octahedral.
15. Which complex has larger crystal field splitting :
    (i) $[Co(CN)_6]^{3+}$ or $[Co(NH_3)_6]^{3+}$
    (ii) $[Co(H_2O)_6]^{3+}$ or $[Co(H_2O)_6]^{3+}$
    Give reasons for your answer.
16. What do you mean by low-spin and high-spin complexes.
17. Define CFSE. Calculate its value for the following systems :
    (i) d^5 low spin octahedral
    (ii) d^5 high spin octahedral
    (iii) d^4 tetrahedral

18. How crystal field theory explains satisfactorily the magnetic properties of complexes of transition metals.
19. Account for the fact that $CuSO_4 . 5H_2O$ is blue in colour while anhydrous $CuSO_4$ is white.
20. Cu^{2+}, Ti^{3+}, Co^{2+} and Fe^{3+} ions are coloured. Explain why?
21. Sc^{3+} ion is colourless while Cr^{3+} is coloured. Give reason.
22. What is the difference in interstitial solid and an alloy of transition metal.
23. The paramagnetic property in the 3d transition series first increases upto manganese, then decreases. Explain why?
24. How is magnetic moment of a species related to the number of unpaired electrons. Calculate the magnetic moment of Fe^{3+} ions (At. No. of Fe = 26).
25. Why are Zn, Cd, Hg are volatile metals.
26. Why does Mn^{2+} ion show maximum paramagnetic character amongst the bivalent ions of the first transition series?
27. Explain the following in relation to transition metals :
    (a) Metallic character
    (b) Ionization energies
    (c) Oxidation states
    (d) Ionic radii
28. What are inner transition elements?
29. Discuss electronic configurations of lanthanides and actinides.
30. In what way observed oxidation states of the lanthanides related to their electronic configurations.
31. What is lanthanide contraction and what are its consequences?
32. Explain the following:
    (a) Why do Zr and Hf exhibit similar properties?
    (b) Why lanthanides have similar chemical properties?
33. What are actinides? Give their general configuration.
34. Give important uses of lanthanides and actinides.
35. Describe giving reason, which one of the following has the property indicated:
    (a) Fe or Cu has higher melting point.
    (b) Co^{2+} or Ni^{2+} has lower magnetic moment.
    (c) Sc^{3+} or Cr^{3+} exhibits paramagnetism.
    (d) V or Mn exhibits more number of oxidation states (Atomic numbers : Sc = 21, Cr = 24, V = 23, Mn = 25).
36. Write the main points of similarity between lanthanides and actinides. What are the main differences?
37. Account for the colour exhibited by lanthanide compounds.
38. (a) Describe the magnetic properties of $[Fe(CN)_6]^{4-}$, $[Fe(CN)_6]^{3-}$ and $[FeF_6]^{3-}$ on the basis of crystal field theory.
    (b) Why Ti^{4+} salts are colourless while Ti^{3+} are coloured.

G.G.S.I.P.U., I Term, B.Tech., Dec. 1999

39. (a) Why do transition metals :
    (i) form coloured ions.
    (ii) exhibit variable oxidation states.
    (iii) exhibit paramagnetism.
    (b) Second ionization energies of Cr and Cu are exceptionally high. Give reasons.
    (c) The complex ion, $[Co(NH_3)_6]^{3+}$ is octahedral, but diamagnetic; while complex ion, $[CoF_6]^{3-}$ is also octahedral, but paramagnetic. Explain on the basis of Crystal Field Theory.

G.G.S.I.P.U, End Term, B.Tech, March, 2000.

40. Explain, why?
 (a) Transition metals form complexes.
 (b) Complexes of transition metals are generally coloured.
 (c) Transition metals exhibit variable valency.
 (d) Octahedral complexes are most stable and common than tetrahedral complexes.

G.G.S.I.P.U, End Term, B.Tech, Dec., 2000.

41. (a) Explain the crystal field splitting of d-orbitals in an octahedral and tetrahedral fields in complexes.
 (b) Show the hybridisation and magnetic moments of the following complex ions:
 (i) $[Fe(CN)_6]^{3-}$ (ii) $[CoF_6]^{3-}$
 (iii) $Ni(CO)_4$ (iv) $[CuCl_4]^{2-}$

G.G.S.I.P.U, End Term, B.Tech, Dec., 2000.

42. (a) What are transition elements? Why do second and third rows of transition elements resemble each other much more closely than they resemble the first row?
 (b) How do the following properties vary in transition elements?
 (i) Ionic character
 (ii) Basic properties
 (iii) Ability to form complexes

G.G.S.I.P.U, End Term, B.Tech, II Batch, March, 2000.

43. (a) Explain giving reasons why?
 (i) Most of the transition metals are paramagnetic.
 (ii) Compounds of transition metals are generally coloured.
 (b) Which complex has larger crystal field splitting? Explain.
 (i) $[Co(CN)_6]^{3+}$ or $[Co(NH_3)_6]^{3+}$
 (ii) $[Co\ (H_2O)_6]^{2+}$ or $[Co(H_2O)_6]^{3+}$

G.G.S.I.P.U, II Term, B.Tech, Nov. 2001.

44. Why is density of Tungsten about twice as great as that of Molebdenum whereas the difference in the densities between Cr and Mo is much smaller?

G.G.S.I.P.U, End Term, Dec. 2001.

45. Explain on the basis of crystal field theory the magnetic behaviour of $[CO(NH_3)_6]^{3+}$ and $[CoF_6]^{3-}$

G.G.S.I.P.U, End Term, Dec. 2001.

46. Give differences between 3d and other two transition series, i.e. 4d and 5d series.

G.G.S.I.P.U, End Term, Dec. 2001.

❑❑❑

30 ENVIRONMENTAL POLLUTION-I

"When the quality of life goes down for environment, the quality of life goes down for humans".

*— **George Holland***

CHAPTER OUTLINE

The environmental pollution is one of the most challenging problem today. Inspite of continuous efforts of our environmental engineers, scientists and an conscious society, our problem of environmental pollution remains the same. It may be a case of polluted water at some industrial discharge point, the chloroflouro carbons (CFCs) that are attacking the earth's ozone layer, the carcinogens that are contaminating our ground water or the carbon dioxide emissions leading to climate change. The endless list of environmental pollution is a cause of great concern. Moreover radiation hazards and hazardous waste management join the array of environmental pollution problems.

Human beings depend upon the quality of environment they are living in and they also exert a tremendous impact on the environment through:

- extraction of resources
- modification and manipulation of the environment
- adding to the pollution through the process of industrialisation.

All the three basic amenities of human existence viz. air, water and land are experiencing the serious ecological imbalance. *Water* is polluted by pollutants such as sewage, solid wastes and industrial effluents. *Air* is polluted by automobile exhausts, exhaust gases from industries. *Land* is polluted by hazardous waste disposal. These environmental problems are becoming a serious threat to our existence. Furthermore, the effects of these problems seem to be far more longer than we anticipate. For example, CFC's emitted today will be around for *decades*, carbon dioxide from our power plants will be in the atmosphere for *centuries*, and our radioactive wastes will be problematic for *thousands of years*.

Hence there is a serious need to address these issues for harmonious existence of industry and the world at large.

Environmental science is defined as "Applying knowledge from many disciplines to the study and management of the environment. It deals with the analysis of the conditions, circumstances and influences affecting life and how life in turn responds".

Before proceeding to the environmental pollution let us have a brief idea of our environment.

30.1 ENVIRONMENT

The human environment encompasses all physical, chemical, biological and social processes which exert a significant influence, directly or indirectly on human beings. In turn, human beings also exert influence on quality of environment. Thus, the subject of environment sciences has its roots in ecology which is the study of how living and non-living things in nature relate to one anothers.

The environment is not a simple distribution of gases but is highly complex and structured. It comprises of following four segments:

1. Lithosphere
2. Hydrosphere
3. Atmosphere
4. Biosphere

The names of the four sphere are derived from the Greek words for stone (litho), air (atom), water (hydro) and life (bio.).

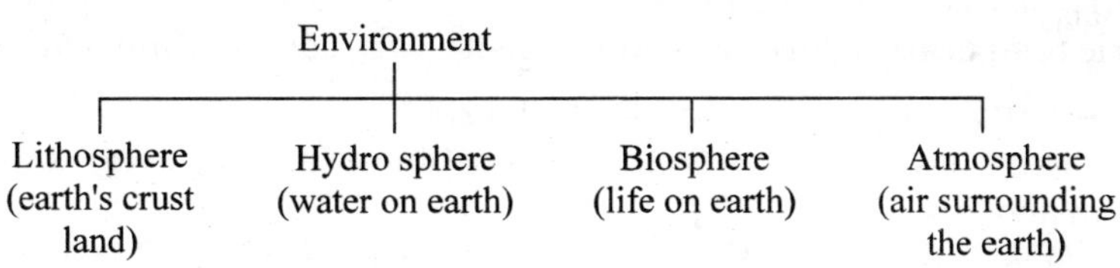

Lithosphere

The lithosphere contains all of the hard solid land of the planet's crust (surface) and the semi-solid land underneath the crust. The surface of the lithosphere is very uneven. There are high mountain ranges, huge plains or flat areas and deep valleys along the ocean flow.

The solid and semi-solid land of the lithosphere form layers that are physically and chemically different. The outermost layer of the lithosphere consists of lose soil rich in nutrients, oxygen, and silicon. Beneath that layer lies a very thin, solid crust of oxygen and silicon. Next is a thick, semi-solid mantle of oxygen, silicon, iron, and magnesium. Below that is a liquid outer core of nickel and iron. At the center of earth is a solid inner core of nickel and iron.

Generally, the word "lithosphere" means only the cold, hard surface of earth, not the entire inside of the planet.

Hydrosphere

The water on earth's surface constitute hydrosphere. It ranges from 10 to 20 kilometers in thickness. The hydrosphere extends from earth's surface downwards, several kilometers into the lithosphere and upwards about 12 kilometers into the atmosphere.

A small portion of the water in the hydrosphere is fresh (non-salty). This water flows as precipitation from the atmosphere down to earth's surface, as rivers and streams along earth's surface, and as groundwater beneath earth's surface. Most of earth's fresh water, however, is frozen.

Ninety-seven percent of earth's water is salty. The salty water collects in deep valleys along earth's surface. These large collections of salty water are called oceans. On the other hand, water near the poles is very cold, while water near the equator is very warm. The differences in temperature cause water to change physical states. Extremely low temperatures like those found at the poles cause water to freeze into a solid such as a polar ice-cap, a glacier, or an iceberg. Extremely high temperatures, like those found at the equator cause water to evaporate into a gas.

Frozen water (glaciers, icecaps, and icebergs) in its own sphere is called the "cryosphere".

Biosphere

The biosphere is composed of all living organisms. This sphere includes all of the microorganisms, plants, and animals of Earth. In fact, biosphere involves the interaction of the living beings with other segments. Most of the planet's life is found from three meters below the ground to thirty meters above it and in the top 200 meters of the oceans and seas.

Atmosphere

The atmosphere is the body of air which surrounds the earth. It extends from less than 1 m below the earth's surface to more than 348 km above the earth's surface. The upper portion of the atmosphere protects the organisms of the biosphere from the sun's ultraviolet radiation. It also absorbs and emits heat. When air temperature in the lower portion of this sphere changes, weather occurs. As air in the lower atmosphere is heated or cooled, it moves around the planet. The result can be as simple as breeze or as complex as a tornado endangering the survival of life. There is a constant exchange of matter taking place between the atmosphere, biosphere and hydrosphere. The atmospheric temperature, pressure and density vary considerably with altitude.

- The temperature varies from $-100°C$ to $+1200°C$.
- Atmospheric pressure varies from 1 atmosphere (at sea level) to 3×10^{-7} atmosphere (at 100 km above sea level)
- The atmospheric density at the surface of the earth is about 0.0013 g/m^3 which decreases sharply with increasing altitude.

It is interesting to note that all four spheres can be and often are present in the single location. For example a piece of soil will have mineral material from the lithosphere, there will be elements of the hydrosphere present as moisture within the soil, the biosphere as insects and plants, and even the atmosphere as pockets of air between soil pieces. A detailed description of the structure of atmosphere is given in the next section.

30.2 ATMOSPHERE-COMPOSITION AND STRUCTURE

Composition

The proportions of the various constituents of the atmosphere remain more or less constant upto a height of about 16 km from the surface of earth. This region is sometimes referred to as homosphere. Above this the composition starts showing variability with altitude. In this region, called the heterosphere, gravitational separation takes place as the lightest and fastest atoms and molecules move farther from the earth.

Finally in the upper heterosphere hydrogen and heliums are the most typical gases found. The atmospheric composition of clean, dry air, near the sea-level is given in Table 30.1.

Table 30.1 Atmospheric constituents at sea-level [%]

N_2	78.1
O_2	20.9
Ar	0.93
CO_2	0.035
Ne	0.0018
He	0.00052
CH_2	0.0002
Kr	0.00011
H_2	0.00005
Xe	0.00001

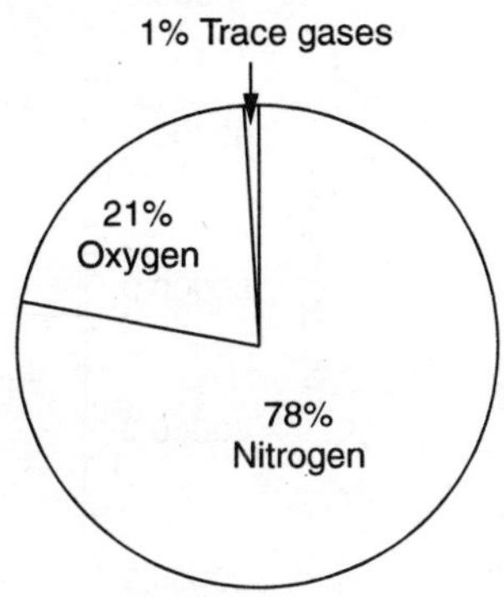

Fig. 30.1 Composition of atmosphere

The Layers of the Atmosphere

The earth's atmosphere is highly structured. Because of earth's gravitational pull, the atmosphere is horizontally stratified. It can be divided into four zones depending on the temperature profile. (Fig. 30.2)

(*i*) Troposphere
(*ii*) Stratosphere
(*iii*) Mesosphere
(*iv*) Thermosphere

(*i*) **Troposphere:** The word troposphere comes from tropein, meaning to turn or change. All of the weather occurs in the troposphere.
It has the following characterstics:

- It is nearest to the earth's surface and extends from the earth's surface to an average of 12 km.
- The temperature generally decreases with increasing height upto the top of the troposphere (tropopause). It drops from about 17°C to – 52°C.
- The troposphere ends at a point where temperature no longer varies with height. This area is known as tropopause. Tropopause marks the transition to the stratosphere.
- The density of air, in this region, decrease exponentially with increasing altitude. The lower density of molecules higher up would not give us enough oxygen to survive.

- The moisture content also decreases with height upto tropopause.
- Towards the top of the troposphere speed of the winds increases.
- The tropopause and troposphere are known as lower atmosphere.
- The top of the troposphere is extremely cold and windy place.

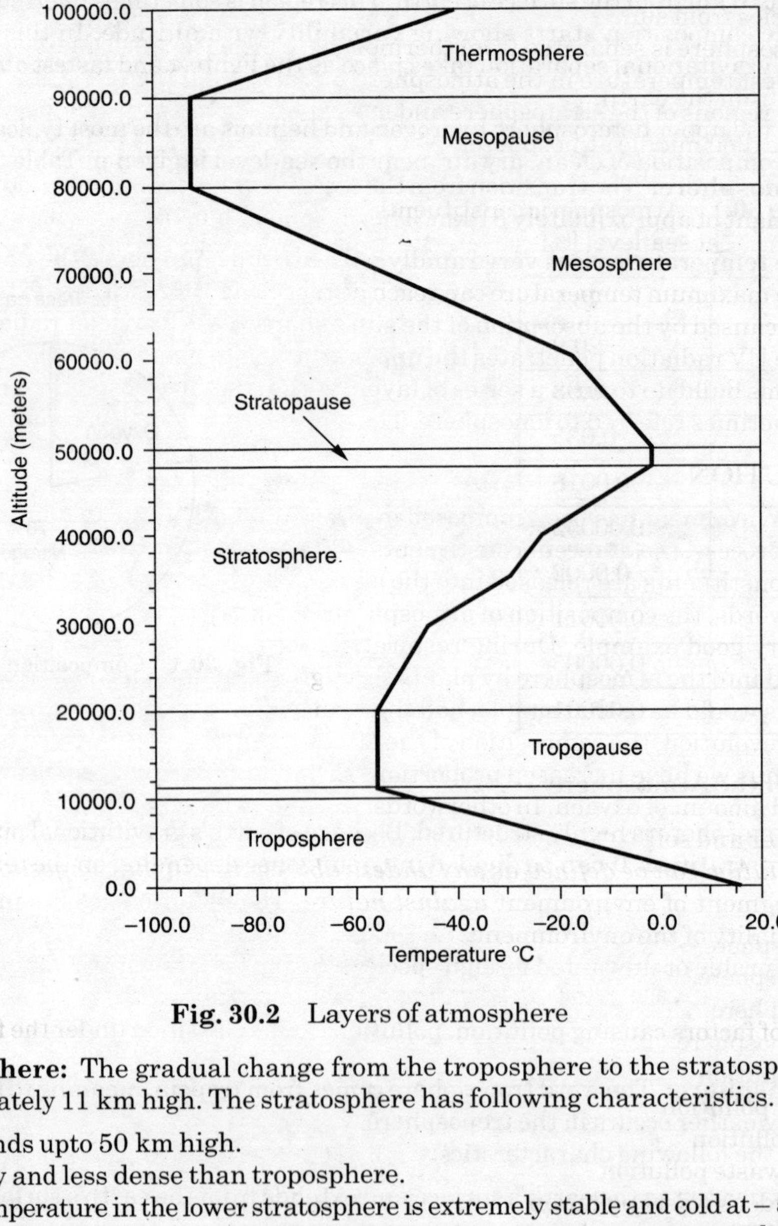

Fig. 30.2 Layers of atmosphere

(*ii*) **Stratosphere:** The gradual change from the troposphere to the stratosphere begins at approximately 11 km high. The stratosphere has following characteristics.

- It extends upto 50 km high.
- It is dry and less dense than troposphere.
- The temperature in the lower stratosphere is extremely stable and cold at – 57°C. However, from the middle of the stratosphere, and up, the temperature pattern changes, sharply increasing with height. This is due to increasing levels of ozone concentration which absorbs ultra violet radiation. The temperature reaches to 18°C in the upper atmosphere at 40 km altitude.
- The ozone layer is in the stratosphere.
- The stratopause separates the stratosphere from the next layer.

(*iii*) **Mesosphere:** The mesosphere has the following characteristics:

- It starts just above the stratosphere and extends to 85 km high.
- The temperature again falls in this region. It reaches to as low as – 93°C with increase in altitude. This is due to low levels of ozone and other species that can absorb ultra violet species from sun.
- Mesosphere is separated from thermosphere by mesopause. Mesopause is the region of coldest temperature in the atmosphere (i.e. about – 100°C).
- The regions of the stratosphere and mesosphere along with stratopause and mesopause are called middle atmosphere.

(*iv*) **Thermosphere:** The transition from the mesosphere to the final thermosphere layer begins at a height of approximately 81 km. Thermosphere has the following characteristics:

- The temperature rises very rapidly with attitude and hence the name thermosphere. The maximum temperature can reach a staggering 1982°C. These extreme temperatures are caused by the absorption of the sun's shortwave ultraviolet radiation.
- The UV radiation penetrates the upper atmosphere and ionizes the atoms. The charged atoms build up to form a series of layers within the thermosphere. This charged layer is sometimes referred to ionosphere. The ionosphere can deflect some radio signals.

30.3 POLLUTION

The earth's environment has been composed in a very unique way for sustaining the life on earth. Through one process, the different constituents of the atmosphere are consumed and through the other, these constituents are released into the atmosphere.

In other words, the composition of atmosphere remains constant. The oxygen cycle in the air provides a very good example. During respiration, oxygen is consumed by living beings and is again released into the atmosphere by plants through a process called photosynthesis.

The things would have that simple, had the man not interfered with the nature. But in the era of industrial revolution, the composition of the atmosphere has been changed undesirably against the nature. Thus we have increased proportions of different harmful constituents in atmosphere and decreased amount of oxygen. In otherwords, we have polluted the atmosphere. The same is the case with water and soil.

Thus, pollution can be defined as any undesirable change in the compostion of air, water and soil or any segment of environment *against nature*. The pollution results in deteriorating the nature and quality of the environment.

The undesirable or unwanted foreign species causing pollution in air water and soil are called *pollutants*.

On the basis of factors causing pollution, pollution can be classified under the following heads:

1. Air pollution
2. Water pollution
3. Soil pollution
4. Solid waste pollution
5. Hazardous waste pollution

We will discuss each one of these under the following sections.

30.4 AIR POLLUTION

Air pollution can be defined as the presence of chemicals in the atmosphere in quantities and duration that are harmful to human health and the environment. It occurs when the concentration of certain substances become high enough to cause the atmospheric environment to become toxic.

Air pollution is the most crucial from the point of human health. Every human being inhales about 15 - 22 kg of air daily and if that air is polluted, one can easily imagine the damage it can cause to human health.

Source of Air Pollution: The sources of air pollution can be natural or man made.

(i) *Natural sources include*

- Volcanic eruptions emitting poisonous gases like H_2S, SO_2, etc.
- Decay of vegetagion,
- Marsh gases
- Pollen grains
- Forest fires

(ii) *Man made sources include*

- Increase in population
- Deforestation
- Fossil fuel combustion
- Vehicular emissions
- Industrialisation
- Use of pesticides, insecticides in agriculture.
- Explosives used in wars.

Types of air pollutants: The undesirable component in the atmosphere, causing pollution is referred to as pollutant.

Pollutants can be classified on the basis of origin and physical state.

On the basis of origin, pollutants are of two types:

(i) *Primary pollutant*: Primary pollutants are the products of natural events (like fires, volcanic eruptions) and human activities added directly to air. Examples include, CO, NO_2, SO_2 and hydrocarbons.

(ii) *Secondary pollutants*: These are formed by the interaction of primary pollutants with each other or with normal components of the air. Examples include ozone, photochemical smog, etc.

On the basis of physical state, pollutants are of two types:

(i) *Gaseous pollutants*: These are the gases which mix with air without settling down. Gaseous pollutants include: CO, CO_2, SO_2, SO_3, NO_x, and hydrocarbons.

(ii) *Particulate pollutants*: Particulate pollutants comprise of finely divided solids or liquids. These include dust, smoke, smog, lead, mercury, cadmium and asbestos.

Classification of Air Pollutants

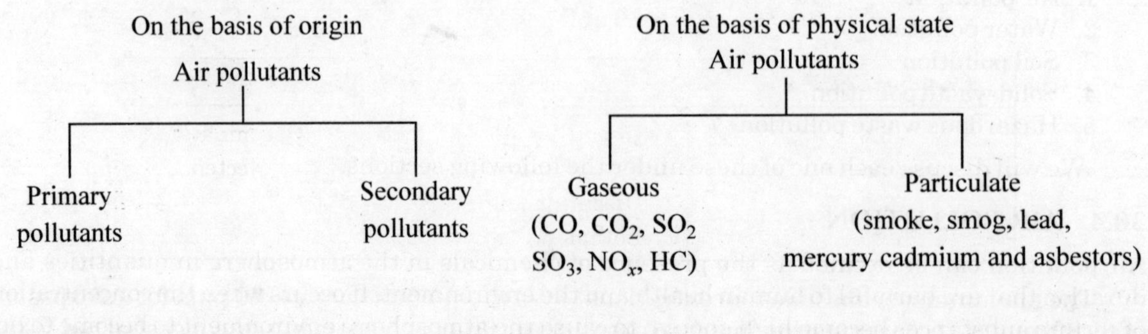

30.5 SOME COMMON AIR POLLUTANTS

1. Carbon Monoxide (CO)

Carbon monoxide is a colourless, odourless, tasteless gas that is by far the most abundant of the air pollutants.

Source: CO is produced due to:

- incomplete combustion of fuels.
- automobile exhausts. (accounts for 70% of CO in the atmosphere)
- industrial operation

CO is also produced through natural processes such as volcanic activity, natural gas and marsh gas emissions.

Sink for CO: Sink is a system which absorbs/store the pollutant, thereby, nullifying its harmful effect. A large mass of CO is generated and also there is continued increase in CO emissions within recent years. So it is obvious that the amount of CO should also increase in the atmosphere. But it has been found that the amount of CO in the atmosphere remains relatively constant suggesting that a sink or scavenging process also exists in the atmosphere. The micro organisms present in the soil act as a major sink for CO.

A significant amount of CO is converted into CO_2 by these micro organisms.

Effects: The levels of CO present in the urban air do not effect significantly the plants and materials. However, these levels adversely affect human health. In urban areas, the soil available is insufficient to act as a sink thereby increasing the level of CO beyond permissible limits.

Carbon monoxide interferes with the blood's ability to carry oxygen to different parts of the body. The oxygen combines with haemoglobin, which is also known as oxygen carrier, to form oxyhaemoglobin. This oxyhaemoglobin travels to the different body cells where it gives oxygen to the cell and takes up CO_2 through the lungs.

$$Hb + O_2 \longrightarrow HbO_2 \longrightarrow Hb + O_2$$

However, if large amount of CO is present in the surroundings, is inhaled, it readily binds to haemoglobin to form carboxyhaemoglobin. Carbon monoxide, in fact, has a much greater affinity for haemoglobin (about 250 times) than oxygen because it forms a stronger complex with haemoglobin.

$$Hb + CO \rightleftharpoons HbCO$$

Thus the blood (haemoglobin) would carry less oxygen to various parts which inturn affects the brain function. The symptoms include laziness, exhaustion, headache. Acute starvation of oxygen (asyphyxiation) can lead to coma and death.

Physiological effects can be noted even at small percentage of COHb. As the concentration increases, the severity also increases as is evident from the following Table 30.2

Table 30.2

Conc. of COHb	*Effect*
2.5%	Impairment in time-interval discrimination.
5%	Psychomotor response times are affected.
10%	Dizziness and headache
50%	Lethal

The maximum permissible concentrations of CO in the ambient air is 40 ppm for an exposure of 6 - 8 hours. CO concentration in urban areas range from 5 - 50 ppm but on congested red lights the concentration can increase to more than 100 ppm. Cigarette smoke contains more than 400 ppm of CO and smokers frequently have COHb levels between 5 to 10 per cent.

Fortunately CO is removed from the blood stream when clean air is inhaled.

Control of CO : Control of the CO pollution can be achieved through the following techniques:

(a) *Modification of engine design:* A low fuel-air ratio reduces NO_x emissions but increase CO emissions. So engine design should be modified so that right proportion of oxygen is maintained for complete oxidation of carbon and hydrogen to CO_2 and H_2O respectively.

(b) *Fuel modification:* The fuels which release lesser amounts of CO are recommended to be used. These fuels include natural gas, methane and blends of light hydrocarbons.

(c) *Treatment of exhaust gases:* Two stage catalytic converters are in use to lower the pollution from exhaust gases. In the first stage NO_x are reduced to N_2 and NH_3 in presence of catalyst such as Pt, Pd and Ruthenium in the presence of reducing gas such as CO. In second stage, oxidizing catalysts of noble metals (Pt, Ru) supported on ceramic materials are used, which ensure oxidation of CO to CO_2.

2. Oxides of Nitrogen (NO_x)

A number of oxides of nitrogen such as NO, N_2O, NO_2, N_2O_3 and N_2O_5 are introduced into the atmosphere due to natural as well as human activity. Out of these, the two oxides NO and NO_2 are responsible for pollution and are considered as pollutants and are represented by NO_x.

NO_2 is reddish brown in colour having pungent smell and is suffocating, whereas NO is colourless and odourless gas.

Source: The source of NO_x include:

(i) *Natural*: During lightening discharge, N_2 and O_2 in air combine to form NO

$$N_2 + O_2 \longrightarrow NO$$

NO_x is also formed by the fixation of nitrogen from atmosphere or from nitrogen compounds present in organic matter.

(ii) *Man-made sources*:

- NO is formed from N_2 and O_2 at very high temperatures (~ 1210 C to 1765°C). This temperature is usually attained during combustion of fossil fuels in air

$$N_2 + O_2 \xrightleftharpoons{1200-1750°C} 2NO \quad ...(i)$$

$$2NO + O_2 \xrightarrow[1100°C]{} 2NO_2 \quad ...(ii)$$

Though the (*i*) reaction is reversible but rapid cooling of combustion product prevents the dissociation of NO.

The formation of NO_2 by (*ii*) reaction is quite slow and is not more than 0.5% of the total NO_x present. However, the main reaction through which NO_2 is formed is by reacting with ozone.

$$NO + O_3 \longrightarrow NO_2 + O_2 \quad ...(iii)$$

- NO_x is also produced in chemical industries as by products such as in coal based power plants, sulphuric acid and nitric acid manufacturing industries.

Sink for NO_x: The NO and NO_2 undergo various photochemical and chemical reactions in the atmosphere leading to the formation of HNO_3. Thus NO_2, NO, pollutants get precipitated as nitrates during rainfall. Ozone plays a significant role in these photochemical reactions.

$$NO + O_3 \longrightarrow NO_2 + O_2 \quad ...(i)$$

It should be noted that NO remove ozone from atmosphere and this occurs naturally in unpolluted atmosphere.

$$NO_2 + \text{Sunlight } (h\nu) \longrightarrow NO + O \quad ...(ii)$$

The nascent oxygen formed in the above reaction then reacts with one of the oxygen molecules producing ozone.

$$O + O_2 \longrightarrow O_3 \quad ...(iii)$$

Thus the destruction of NO_2 (reaction *ii*) and regeneration of NO_2 reaction (*i*) is a cyclic process. Moreover the amount of ozone should also remain constant as it is used in reaction (*i*) and regenerated in reaction (*iii*). However, in the presence of volatile organic compounds, NO_2 follows an alternate path. It reacts with water in presence of oxygen resulting in the formation of nitric acid.

$$4NO_2 + O_2 + 2H_2O \longrightarrow 4HNO_3 \quad ...(iv)$$

$$2NO_2 + O_3 \longrightarrow N_2O_5 + O_2$$

$$N_2O_5 + H_2O \longrightarrow 2HNO_3 \quad ...(v)$$

The HNO_3 comes down from the atmosphere to the surface of earth in the form of acid rain. Here it reacts with bases such as ammonia, lime, etc. to form nitrates.

Effects

- Almost all the NO_x emissions are in the form of NO, which has no known adverse health effects at the concentrations found in the atmosphere.
- However NO can oxidise to NO_2, which in turn may react with hydrocarbons in the presence of sunlight to form *photochemical smog* (discussed later). Photochemical smog is injurious to health.
- The HNO_3 formed by NO_2 causes *acid rain* (discussed later) which has corroding effect on marble and the metallic structures. Moreover, acid rain decreases the pH of the soil, affecting its fertility.
- At higher concentrations than are normally found in the atmosphere, it is an acute irritant. Prolonged exposure to relatively low concentration, has been found to cause bronchitis in children.

Control of NO_x: The control measures for NO_x emissions include the following:

(a) *Modifying the engine design:* Reducing the amount of excess air for combustion in air helps in controlling NO_x emissions. For instance NO_x decreases from about 600 ppm at 25% excess air to 175 ppm at 1.5% excess air. The burners are so modified that the fuel and air mix more slowly reducing the intensity and temperature of combustion. Two staged combustion process is employed for the purpose.

(b) *Scrubbing the flue gases:* The flue gases are scrubbed with H_2SO_4 in a scrubber

$$NO_2 + NO \longrightarrow N_2O_3$$

$$N_2O_3 + 2H_2SO_4 \longrightarrow 2NOHSO_4 + H_2O$$

The reaction product obtained from the scrubber is then decomposed.

$$2NOHSO_4 + \frac{1}{2}O_2 + H_2O \longrightarrow 2H_2SO_4 + 2NO_2$$

$$2NO_2 + H_2O \longrightarrow 2HNO_3 + NO$$

Excess NO and NO_2 are again circulated though the scrubber.

(c) *Selective catalytic reduction:* The selective catalytic reduction can be achieved through CH_4, NH_3 and CO. The NO_x is added to the exhaust gases and mixture is passed over a fixed bed catalyst such as copper oxide.

$$6NO + 4NH_3 \longrightarrow 5N_2 + 6H_2O$$

or

$$NO + 2CO \longrightarrow N_2 + 2CO_2$$

It is interesting to note that the air pollution control methods also require modification of combustion methods. But the combustion process that improve CO emissions tend to make the emissions of NO_x problem worse, and vice-versa. For example, to control CO, the air supply should be increased and temperature should be raised so that complete combustion takes place. This however is just the reverse if we want to control the emissions of NO_x which requires just opposite conditions.

3. Oxides of Sulphur

Oxides of sulphur, SO_2, SO_3 are represented as SO_x. SO_2 is the second most important contributer to air pollution. Though a small concentration of sulphur oxides is desirable for plants and animals but at higher concentrations, it is injurious. Both SO_2 and SO_3 are colourless gases with pungent smell.

Sources

- Natural sources such as volcanic eruptions constitute to about 67% of SO_x pollution.
- Man made sources the sulphur emissions from petroleum refining, copper smelting and cement manufacture plants.

$$Cu_2S + O_2 \longrightarrow 2Cu + SO_2$$

- The burning of fossil fuels in thermal power plants, electric power plants are also one of the main sources of SO_x emissions.

$$S + O_2 \longrightarrow SO_2$$

  SO_x produced (from the fuel) from natural sources are uniformly distributed all over the globe but SO_x made from man-made sources is more in urban areas.
- A small amount of SO_2 undergoes photolytic oxidation in the air to form SO_3

$$2SO_2 + O_2 \longrightarrow 2SO_3$$

Sinks for SO_x: The sulphur dioxide released in the atmosphere is converted to sulphur SO_3 by photolytic and catalytic processes involving ozone.

$$SO_2 + O_3 \longrightarrow SO_3 + O_2$$

Sulphur oxides combine with water vapour to produce H_2SO_4 aerosol which give rise to the so called "acid-rain" droplets

$$SO_3 + H_2O \longrightarrow H_2SO_4$$

H_2SO_4 combines with the lime-stone minerals and thus removed from atmosphere. In other words limestone minerals act as a sink for sulphur dioxide.

Effects

- Sulphur dioxide is perhaps the most damaging among the various gaseous air pollutants.
- One of the main causes of 'acid rain' are the oxides of sulphur.
- SO_2 and SO_3 are strong irritants to the respiratory tract. SO_2 is highly water soluble, much more so than any of the other criteria pollutants. As a result, when it is inhaled it is absorbed in the upper respiratory tract.
- If along with sulphur dioxide, the particulate matter in air is also present, the combination becomes detrimental to health. In fact, in every major air pollution episode, this combination has been found to be the major cause leading to death. This combination is suggested as one of the reasons of chronic bronchitis.
- Sulphur dioxide has a very damaging effect on plants. If plants are exposed to SO_3 even at a low concentration (0.03 ppm) for a few weeks, the formation of chlorophyll is slowed down resulting in loss of green colour of leaf. This is called chlorosis.
- Prolonged exposure to sulphur oxide which gets converted to sulphates causes serious damage to buildings, marble, lime stone. The carbonates in these materials are replaced by sulphates.

$$CaCO_3 + H_2SO_4 \longrightarrow CaSO_4 + CO_2 + H_2O$$

The $CaSO_4$ so formed is water soluble and easily washes away, leaving a pitted surface. Many of the world's historic buildings and statues are rapidly being degraded due to this exposure.

- Wool, cotton and leather materials are rotted as a result of the absorption of SO_2 and its oxidation to sulphuric acid, which attacks the proteins in these materials.
- Paper also absorbs SO_2 which is oxidised to H_2SO_4 causing paper to become brittle and fragile.

Control of SO_x: SO_x can be reduced and controlled by number of ways.

- Use of fuels having low content of sulphur.
- By removing SO_2 from the exhaust gases by using chemical scrubbers. The chemical scrubbers which are used include :

(a) *Lime stone*:

$$2CaCO_3 + 2SO_2 + O_2 \longrightarrow 2CaSO_4 + CO_2$$
$$CaCO_3 + SO_3 \longrightarrow CaSO_4 + CO_2$$

The flue gases (exhaust gases) are passed through the slurry of $CaCO_3$ which reacts with SO_2 and SO_3 to form $CaSO_4$. Though this method is economical but huge amounts of $CaSO_4$ are produced, the disposal of which is a great problem.

(b) *Citric acid*: Another method employs the use of citric acid which absorbs the SO_2 from the flue gases. The flue gases are passed through a solution containing citrate ions. The HSO_3^- ions, produced from SO_2 and H_2O, combine with citrate ions to form citrate complex.

$$SO_2 + H_2O \rightleftharpoons HSO_3^- + H^+$$
$$HSO_3^- + H_2Cit^- \rightleftharpoons [HSO_3 . H_2Cit]^{2-}$$

The solution containing citrate ion is transferred in a closed vessel and H_2S is passed through it. Sulphur precipitates out and citrate ions are regenerated which are used again.

4. Particulates

Atmospheric particulate matter is defined to be any dispersed matter, solid or liquid, in air. The important characteristics of a particulate are:

(*a*) **Size:** The size of the particulate matter ranges from a diameter of 0.002 μm to 500 μm.

(*b*) **Concentration:** The density of particulate material vary from few hundred per cm^3 in clean air to 10^5 per cm^3 in polluted air.

A number of terms are used to categorise particulates depending on their size and phase (liquid or solid). Some of these are:

Aerosol: Aerosol is any tiny particle, liquid or solid dispersed in atmosphere.

Dust: Solid particles are called dust if they are caused by grinding or crushing operations.

Fumes: Solid particles are called fumes if they are formed when vapours condense.

Mist: Liquid particles are called mist, or more loosly, fog.

Smoke or Soot: These are the terms used to describe particles composed primarily of carbon that results from incomplete combustion.

Fly Ash: These are smaller ash particles coming alongwith the furnace flue gases and are produced due to combustion of high-ash fossil fuels as in the thermal power plants.

Source: Natural sources include the particulates released from volcanic eruptions, wind and dust storms, etc. Man-made sources include the activities such as burning of wood, coal, oil and gaseous fuels, industrial processes, smelting and mining operations. The fly ash emissions from power plants, forest fires etc. release about 450 million tonnes of particulates per year.

Composition: Particulates include organic and inorganic matter. It is estimated that particulate contaminants contain nearly 22 elements. The particulate matter can be organic or inorganic.

Organic Particulates	*Inorganic Particulates*
Organic particulates arise from automobile exhausts, combustion of fuels. Organic particulates contain polycyclic aromatic hydrocarbons (PAH) which are carcinogenic.	Inorganic particulates include iron oxide, calcium oxide which result from combustion of coal and metallurgical operations. Lead halides are produced in automobile exhausts. Asbestos dust, a fibrous silicate material is introduced by industrial processes. Aerosol mists are generated from oxidations of SO_2 to SO_3 which form H_2SO_4 in presence of water. Cadmium, arsenic, mercury also contribute to the particulate matter.

Effects: The size and chemical characteristics of the particles in atmosphere are more significant than their concentration.

Harmful Effects on Human Life:

The ability of the human respiratory system to defend itself against particulate matter is determined by the size of the particulate matter to a large extent. Large particles that enter the respiratory system can be trapped by lining and hair of the nose. Smaller particles which move to the throat can be captured by mucus membrane and can be removed by spitting. However, smaller particles traverse right upto the lungs and may deposit there. The harmful effects of particulate matter on human life include the following:

- The smaller particles (smaller than 10 μm) which are capable of reaching the lungs, accumulate there and act as excellent sites for adsorption of carcinogenic compounds and may cause lung cancer, bronchital asthma, chronic bronchitis, etc. Some of the pollutants, even in small doses, lead to cumulative poisoning, i.e. these pollutants are not excreted once they enter the body. The health effects caused by different types of particulates is summarised in Table 30.3.

Harmful effects on materials

- Particulates such as fumes, dust, soot, mists and aerosols can bring serve damage to soil, buildings, sculpture and monuments.
- Corrosive particulates accelerates corrosion of the materials.
- Smoke particles stick on every surface of material, forming a film of soot.

Effects on climate

- Particulates play a significant role in maintaining radiation balance and heat balance of earth.
- They provide nuclei for condensation of water vapour leading to the formation of cloud and smoke.
- Particles can absorb free radicals thereby acting as inhibitors for free radical chain reactions.

Control of Particulate Emissions: Many techniques for the control of particulate emissions have been developed. These include:

- Gravity Settling Chamber
- Cyclone Collector
- Filters
- Scrubbers

The details are discussed in section 30.7.

Table 12.3

Pollutant	*Source*	*Effect on health*
Asbestos (fibrous silicate material)	Industry, used as insulation materials in floor and ceilings, fire-proofing, asbestos cement sheets.	Lung disease namely asbestosis, lung cancer known as mesothelioma and other respiratory problems.
Lead	Exhaust from leaded gasoline combustion, lead paint, contaminated soil, plumbing.	Cumulative poisoning impaired mental and physical development, loss of memory, convulsions, coma, inhibits the action of enzymes, causes anaemia, impairs kidney functions.
Silica	Dust containing free SiO_2	Silicosis, exerts a noxious or fibrotic local action in lungs.
Mercury	Spillage from thermometers, thermosets, etc. mining and refining of mercury, organic mercurials used in pesticides.	• In vapour form it adversely effect the neurological behaviour in humans, mercury poisoning cause brain damage.
Cadmium	Tobacco products, fungicides, cadmium-nickle batteries, fertilizer industries.	• Can remain in the body for almost ten years. Lung cancer and kidney damage, bone marrow disorder and cancer.
Arsenic	Smoking, pesticides, rodent poisons chemical wastes.	Toxic, carcinogen conjuctivitis, coagulates proteins.
Fly Ash	Industrial combustion processes	Respiratory deseases, silicosis.

5. Hydrocarbons

Many different hydrocarbons are present in air. Most of these are low molecular weight and gases and volatile liquids at ordinary temperatures.

Source

- Natural sources include trees which emit large quantities of hydrocarbons into the air. Methane is the major naturally occuring hydrocarbon emitted in the atmosphere which is produced by aneorobic bacteria through decomposition of organic matter in the soil, water and sediments.

$$\text{Organic Matter} \xrightarrow{\text{Bacteria}} CH_4 + CO_2$$

- Man made sources include motor fuels which are mixtures of hydrocarbons. Automobile exhausts emit maximum amount of hydrocarbons in the atmosphere.
- Evaporation of organic solvents used in paints, such as benzene, carbontetrachloride, etc. also emit hydrocarbons in air.

Sinks: Hydrocarbons undergo a number of chemical reactions and photochemical reactions through a number of steps and form water and CO_2 in the end.

$$HC \xrightarrow{O_2} H_2O + CO_2$$

Effects

- At high concentrations hydrocarbons have carcinogenic effect on lungs.
- Inhalation of benzene, toluene, etc. cause irritation in the mucus membrane.
- They are responsible for the of photochemical smog (discussed later)
- Methane causes narcotic effect on human beings.
- Acetylene and propylene at 50-500 ppm show toxicity towards plants, damaging their growth.
- Benzpyrine present in tobacco, charcoal, etc. induces cancer.

Control

- Hydrocarbons, being thermodynamically unstable tend to get oxidized in the atmosphere by a series of chemical and photochemical reactions. This gives rise to the formation of various end-products such as CO_2, solid organic particulates and water soluble acids. The solid organic particulates settle down and water soluble acids are washed down by rain.
- The hydrocarbons from vehicular traffic are controlled by the techniques such as incineration, adsorption and absorption.
  Adsorption is carried on a bed of carbon or by passing exhaust gases through a liquid in which hydrocarbons will dissolve or become suspended. Incineration or 'after burning' completes the oxidation of hydrocarbon to CO_2 and H_2O.
- Loss of hydrocarbons by evaporation from fuel tanks and carburettor is reduced by installation of a connection system which eventually returns them to the fuel induction system.

30.6 ENVIRONMENTAL EFFECTS OF AIR POLLUTION

The air pollutants discussed above have a great impact on our environment. They are posing a threat to the Earth's general environment. The major environmental effects of air pollution are summarised in Fig. 30.3

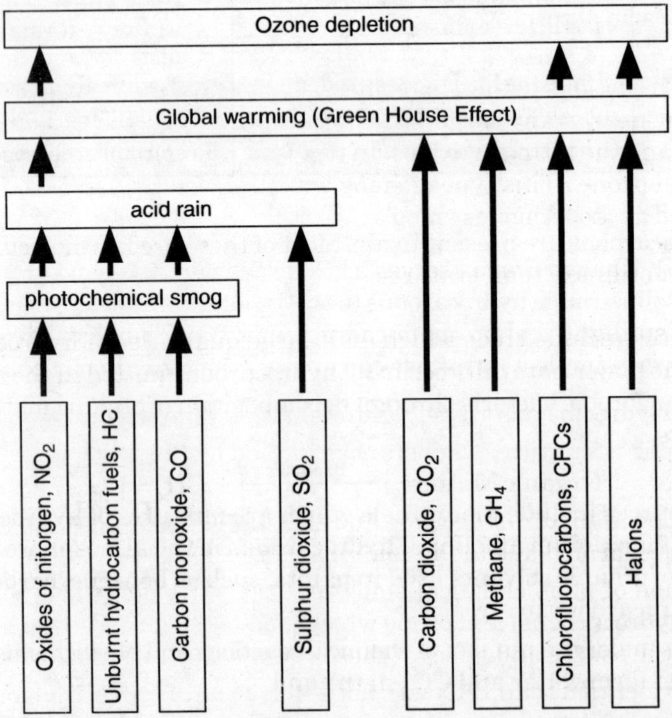

Fig. 30.3 A summary of various environmental problems caused by air pollution

As is evident from the figure that these effects are interelated and one leads to another. A detailed description of these effects is given below.

1. PHOTOCHEMICAL SMOG

'Smog' is the combination of two words, 'smoke' and 'fog'. The name was given because fog in the atmosphere condense on the carbon particles of smoke to form smog. However, the smog can have a number of other compositions and methods of formation. Depending upon these factors, smog is of two types:

(i) *London Smog*: The smog formed from SO_x particulates (soot, ammonium sulphate) and humidity is known as London smog. This type of smog caused the death of around 4000 people in London in 1952, hence the name. Five days of calm foggy weather created a toxic atmosphere which caused bronchits and acute respiratory problems claiming 4000 human lives. This mixture of smoke, SO_x and fog is chemically a reducing mixture, so it is also termed as *reducing smog*.

SO_2 in the atmosphere is converted into SO_3 by a number of chemicals present in the particulates. SO_3 combines with water in the atmosphere forming a fog of sulphuric acid droplets. These droplets condense on the carbon particles of soot and are drawn into the respiratory tract during breathing.

$$2SO_2 + O_2 \longrightarrow 2SO_3 \qquad\qquad SO_2 + H_2O \longrightarrow H_2SO_3$$

$$SO_3 + H_2O \longrightarrow H_2SO_4 \quad \text{or} \quad H_2SO_3 + \frac{1}{2}O \xrightarrow[\text{salts}]{\text{Metal}} H_2SO_4$$

London smog is formed in the early morning hours of winter months. After sunrise it increases due to the oxidation of SO_2 to SO_3 and subsequent combination with water to form sulphuric acid aerosol.

(ii) *Los Angeles Fog (Photochemical Smog)*: Photochemical smog results from the photochemical reactions of the atmosphere. Photochemical smog is the atmospheric haze that is formed near many large cities and it is due to the action of sunlight on the hydrocarbons and the nitrogen oxides. In this type of smog, the main unhealthy ingredient is 'ground level ozone'. This type of smog was first observed in Los Angeles in 1950 and hence is named as 'Los Angeles smog'.

Composition: In 1951, Haagen-Smit observed a key mechanism involving an atmospheric reaction between unburned gasoline-range hydrocarbons (from unburned automobile emission) and nitrogen dioxide in presence of sunlight, a photochemical reaction. The products were peroxyacyl nitrates (PANs) and ozone. In very simplest terms, we can express the formation of photochemical smog as

$$\text{Hydrocarbons} + NO_x + \text{Sunlight} \longrightarrow \text{Photochemical smog}$$

Photochemical smog is mixture of substances formed by the reaction of hydrocarbons, NO_x and sunlight.

Photochemical smog is an *oxidising smog* and should be clearly distinguished from the usual reducing 'smog' which is formed due to the combination of smoke and fog.

Thus, the formation of photochemical smog requires nitrogen oxides ($NO + NO_2$), volalite organic compounds (hydrocarbons) and ozone which lead to the formation of peroxyacetyl nitrate (PAN). An account of the source and effect of these chemical pollutants is given in Table 30.4

It is worthwhile to mention than PAN was not detected until recognized in smog. It has higher toxicity to plants than ozone.

Table 30.4: Major chemical pollutants in photochemical smog: Source and environmental effects.

Toxic Chemical	*Source*	*Environmental effects*
Nitrogen oxides (NO and NO_2)	• combustion of oil, coal, gas in both automobiles and industry • bacterial action in soil • forest fires • volcanic action • lightning	• decreased visibility due to yellowish colour of NO_2 • NO_2 contributes to heart and lung problems • NO_2 can supress plant growth • decreased resistance to infection • may encourage the spread of cancer
Volatile organic compounds (VOCs)	• evaporation of solvents • evaporation of fuels • incomplete combustion of fossil fuels • naturally occurring compounds like terpenes from trees	• eye irritation • respiratory irritation • some are carcinogenic • decreased visibility to due to blue-brown haze
Ozone (O_3)	• formed from photolysis of NO_2 • sometimes results from stratospheric ozone intrusions	• bronchial construction • coughing wheezing • respiratory irritation • eye irritation • decreased crop yields • retards plant growth • damages plastics • break down rubber • harsh odour
Peroxyacetyl Nitrates (PAN)	• formed by the reaction of NO_2 with VOCs (can be formed naturally in some environments)	• eye irritation • high toxicity to plants respiratory • damaging to proteinsirritation

Development of Photochemical Smog (Conditions required):

For photochemical smog to occur there are four necessary conditions:

- presence of nitrogen oxides
- sunlight
- hydrocarbons
- temperature above 18°C

In cities, early morning traffic increases emission of NO_x and hydrocarbons. These NO_x and hydrocarbons (volatile organic compounds) react and form NO_2. NO_2 breaks down forming by products and atomic oxygen. Atomic oxygen combines with the molecular oxygen to form ozone. NO_2 also reacts with VOC to produce toxic chemicals such as PAN. In the evening hours, when the sun goes down, the production of ozone is halted. The ozone in the atmosphere is consumed by different reactions.

Chemistry of Photochemical Smog

Photochemical smog formation proceeds through a sequence of reactions, all involving a free radical mechanism. The main steps can be outlined as follows:

- Nitrogen dioxide is formed as a result of one of the following reactions. At high temperature of the combustion chamber of the vehicle N_2 and O_2 combine to form NO

$$N_2 + O_2 \rightleftharpoons 2NO \qquad ...(i)$$

$$2NO + O_2 \rightleftharpoons 2NO_2 \qquad ...(ii)$$

The (*i*) reaction is endothermic so the formation of NO is favoured at high temperatures.

- NO_2 so formed is an efficient absorber of ultra-violet (UV) radiations. Thus UV rays in the sunlight cause the photodissociation of NO_2 to form NO and atomic oxygen

$$NO_2 + h\nu \xrightarrow{\text{Sunlight}} NO + O \qquad ...(iii)$$

- The atomic oxygen formed in (*iii*) then reacts with abundant oxygen molecules producing ozone

$$O + O_2 \longrightarrow O_3 \qquad ...(iv)$$

Ozone can convert NO back to NO_2

$$NO + O_3 \longrightarrow NO_2 + O_2$$

Note that nitric oxide (NO) acts to remove ozone (O_3) from the atmosphere.

This set of reaction creates a cycle as represented in Fig. 30.4

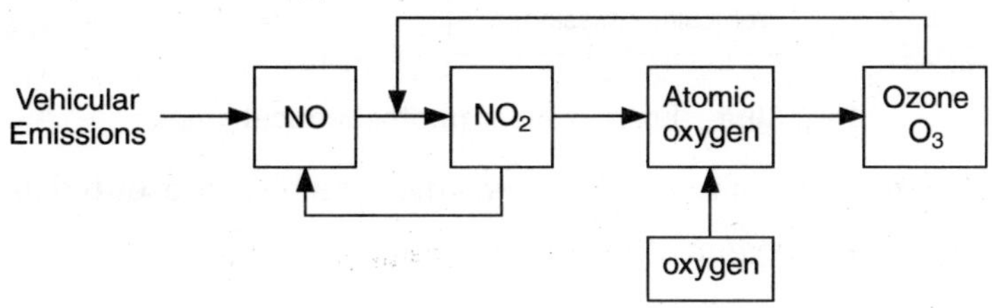

Fig. 30.4

Thus NO and O_3 produced are used up and no extra NO_2 is added up to the atmosphere. But when hydrocarbons are present, then the chain propagation reactions take place.

- The hydrocarbons combine with the oxygen atom produced by the photolysis of NO_2, to form highly reactive intermediates (free radicals)

$$O + RH \longrightarrow R^{\bullet} + \text{other products}$$
$$R^{\bullet} + O_2 \longrightarrow RO_2^{\bullet}$$
$$RO_2^{\bullet} + NO \longrightarrow RO^{\bullet} + NO_2$$

Oxygen atom can also react with H_2O to form hydroxyl radicals.

$$H_2O + O \longrightarrow H_2O_2 \xrightarrow[\text{dissociation}]{\text{Photo}} \dot{O}H$$

- The free radicals then react with O_2 or NO to produce a number of by products which are highly reactive organic oxidants, i.e. secondary pollutants such as acrolein, formaldehyde, peroxyacetyl nitrate (PAN). These secondary pollutants collectively form photochemical smog.

1. $R\dot{O} + O_2 \longrightarrow RCHO + H\dot{O}_2$
2. $RCHO + {}^{\bullet}OH \longrightarrow RCO^{\bullet} + H_2O$
3. $RCO^{\bullet} + O_2 \longrightarrow ROC^{\bullet}$
4. $R-C(=O)-O-O^{\bullet} + NO_2 \longrightarrow CH_3-C(=O)-O-O-ONO_2$ (PAN)

(Acetyl proxy radical)

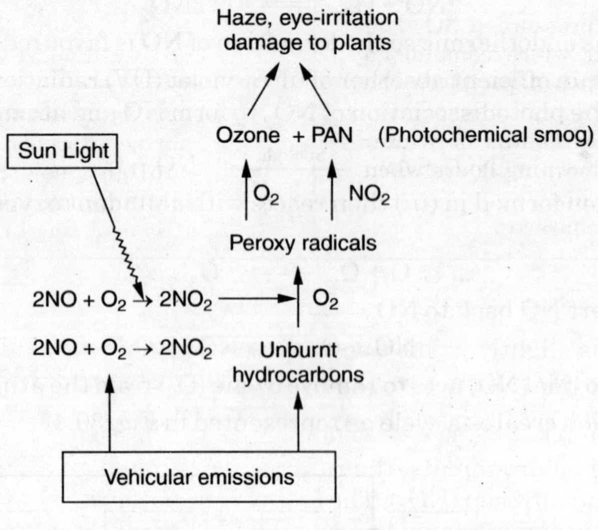

Fig. 30.5 Formation of photochemical smog.

To summarise, the sequence of photochemical smog can be represented as:

NO_x generate oxygen atoms

↓

Oxygen atoms generate free radicals

↓

Free radicals generate hydrocarbon radicals

↓

Hydrocarbon radicals form hydrocarbon peroxides

↓

Peroxides form aldehydes or aldehyde peroxides

↓

Aldehyde peroxide form peroxy acyl nitrate (PAN)

↓

O_3, PAN, NO_x constitute photochemical smog.

All these compounds (PAN, aldehydes, ketones, ozone) produce a number of respiratory problems and irritation in the eyes as is discussed in Table 30.4.

The word 'smog' is misnomer here because smog is a combination of smoke and fog and neither of these are present in photochemical smog which is a mixture of NO_2, O_3, PAN, aldehydes, and hydrocarbons. The difference between London smog and photochemical smog are tabulated in Table 30.5.

Table 30.5

Classical smog (London smog)	*Photochemical smog (Los Angeles smog)*
1. It involves smoke and fog.	1. This type of smog was observed in Los Angeles in 1950.
2. This type of smog was first observed in London in 1952.	2. It is formed due to photochemical reaction taking place when air contains NO_2 and hydrocarbons.
3. It is formed due to pressure of SO_2 and humidity in the air which combine to form H_2SO_4 fog which deposits on the particulates.	3. It does not involve any smoke or fog. The word smog is a misnomer here.
4. It is formed in the months of winter particularly in the morning hours when the temperature is low.	4. It is formed in the months of summer during afternoon when there is bright sunlight so that photochemical reactions can take place.
5. It is a reducing in character.	5. It is oxidizing in character.

2. Acid Rain

[Unpolluted rain water is slightly acidic due to the presence of carbondioxide in air. The CO_2 combines with water droplets to form a weak solution of carbonic acid with a pH of about 5.6

$$CO_{2(g)} + H_2O_{(l)} \longrightarrow H^+_{(aq)} + HCO^-_{3(aq)}$$

However, in polluted environments, the rain passes through an atmosphere polluted with oxides of sulphur (SO_x) and nitrogen (NO_x). The falling rain water reacts with these oxides to form a mixture of sulphuric acid and nitric acid and water. This is known as acid rain. Therefore rain water with pH value lower than 5.7 is called acid rain. In some parts of the world, a pH as low as 2.5 in rainwater has been recorded.

The term acid precipitation is often used to specifically describe wet forms of acid pollution that is found in rain, snow, fog, cloud and vapour. The term 'acid rain' was used by Angus Smith.

It is more correct to refer to acid rain as *acid deposition* because it may be exhibited in more physical states than rain alone.

Actually acid rain is the one phase of acid deposition which can either be wet or dry.

Wet deposition: Occurs in the atmosphere and falls on earth as rain, snow, dew, etc.

Dry deposition: Dust particles containing sulphates and nitrates, settled on earth, is called dry deposition.

Acid Rain Formation: It is well known that oxides of nitrogen and sulphur are present in the atmosphere. The sources of these oxides have been already discussed earlier in the chapter:

- The oxides of sulphur react with moisture to form H_2SO_4 in sequence of steps:

$$SO_2 + O_2 \xrightarrow[\text{Soot or metal ions}]{\text{NO}} 2SO_3$$

This reaction is catalysed by aerosol containing metal ions of Cu(II), Fe(II) and oxides of Pb, etc. Even soot particles are also known to catalyze the reaction.

The SO_3 formed reacts with water vapour in air to form H_2SO_4.

$$SO_3 + H_2O \longrightarrow H_2SO_4$$

- Similarly, the oxides of nitrogen also undergo a series of photochemical and chemical reactions to form HNO_3.

$$NO + O_3 \longrightarrow NO_2 + O_2$$
$$NO_2 + O_3 \longrightarrow NO_3 + O_2$$
$$NO_2 + NO_3 \longrightarrow N_2O_5$$
$$N_2O_5 + H_2O \longrightarrow 2HNO_3$$

These acid droplets are partly neutralised with bases such as particulate lime, NH_3, etc.

$$NH_3 + H^+ \longrightarrow NH_4^+$$

- These salts and remaining H_2SO_4, HNO_3 droplets alongwith HCl, which is released into the atmosphere by man made and natural HCl emissions, give rise to acid precipitation, known as 'Acid Rain'. H_2SO_4 contributes maximum followed by HNO_3 which in turn is followed by HCl. Fig. 30.6 illustrates the formation of acid rain.

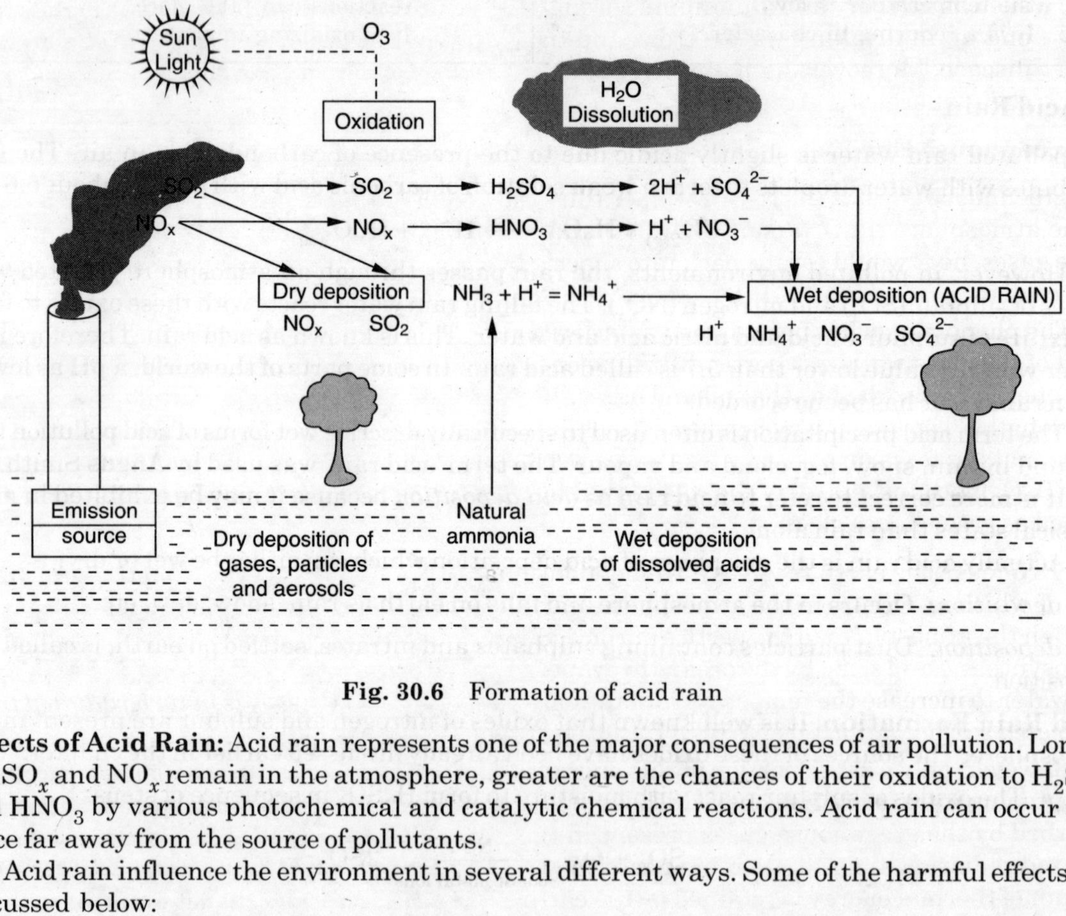

Fig. 30.6 Formation of acid rain

Effects of Acid Rain: Acid rain represents one of the major consequences of air pollution. Longer the SO_x and NO_x remain in the atmosphere, greater are the chances of their oxidation to H_2SO_4 and HNO_3 by various photochemical and catalytic chemical reactions. Acid rain can occur at a place far away from the source of pollutants.

Acid rain influence the environment in several different ways. Some of the harmful effects are discussed below:

- *Acidification of lakes and streams:* Acid rain increases the acidity of water in lakes and streams which is detrimental to fish life. At pH 5, most fish cannot hatch and at lower pH some fish die. In some lakes, acid rain has completely eliminated fish species. Moreover as acid rain flows through soils, aluminium is released from soils into the lakes. Both low pH and increased aluminium levels are toxic to fish.

- *Damage to the vegetation:* Acid rain does not usually kill trees directly. It causes damage due to:
  - —damaging their leaves
  - —leaching the nutrients available to the plant
  - —exposing the plants to toxic substances slowly released from soil.
  - —alteration of seed germination characteristics.
- *Damage to building and sculptual material:* The buildings and materials made of marble, lime-stone, mortar etc. suffer extensive damage due to acid rains. The reaction with marble takes place as

$$CaCO_3 + H_2SO_4 \longrightarrow CaSO_4 + CO_2 + H_2O$$

  The fumes from Mathura refinary are responsible for causing acid rain in Agra, attacking the Taj Mahal. Paints on car can react with acid rain causing fading.
- *Effect on human:* Acid deposition can effect human being life by releasing toxic metals such as mercury and aluminium through the acidification of soil. These toxic metals then end up in drinking water, crops and fish. Aluminium is believed to cause Alzheimer's disease. Moreover, high amount of acidic pollution show increased frequencies of chest colds, allergies and coughs.

3. Greenhouse Effect

The greenhouse effect is the rise in temperature that the earth experiences because certain gases in the atmosphere (H_2O vapour, CO_2, NO and methane) absorb energy from the sun. Without these gases, heat would escape back into space and earth's average temperature would be about 60°C colder.

The phenomenon has been named as green house effect, because it is similar to heat trapping effect of the glass walls in a horticultural green house. The sun's energy enter through the glass. This energy is absorbed by the soil and other structures of green house and is remitted in longer wavelengths. The longer wavelengths cannot pass through the glass and are retained in the green house making it warm and cosy for the growth of the plants. Similarly, in the atmosphere the CO_2, H_2O, methane, etc. play the role played by glass in green house and hence these gases are termed as green house gases.

The Process: The earth receives a tremendous quantity of radiant energy from the sun, about 30% of which is reflected into space by the earth's atmosphere. The remaining energy passes through the atmosphere to the earth's surface. Some of this energy is absorbed by plants to drive photosynthesis and some by the oceans to evaporate water, but most of it is absorbed by soil, rock, and water to increase the temperature of the earth's surface. *This energy is in turn radiated from the heated surface mainly as infrared radiation,* often called heat radiation.

However, only a small portion of this energy actually makes it back to space. The majority of the outgoing infrared radiations, which have longer wavelengths than the solar radiations, are absorbed by the greenhouse gases present in the atmosphere. Thus CO_2, H_2O strongly absorb infrared radiation and radiate it back into the atmosphere towards the earth. (Fig. 30.7). A net amount of thermal energy is retained by the earth and makes it warmer than it would be without these gases in the atmosphere. The average temperature of the earth's surface is 288 K. It would be – 255 K without the greenhouse gases. Without the greenhouse effect, life on this planet would probably not exist as the average temperature would be a chilly – 18°C.

Scientists use the term *albedo* to define the percentage of solar energy reflected back by a surface.

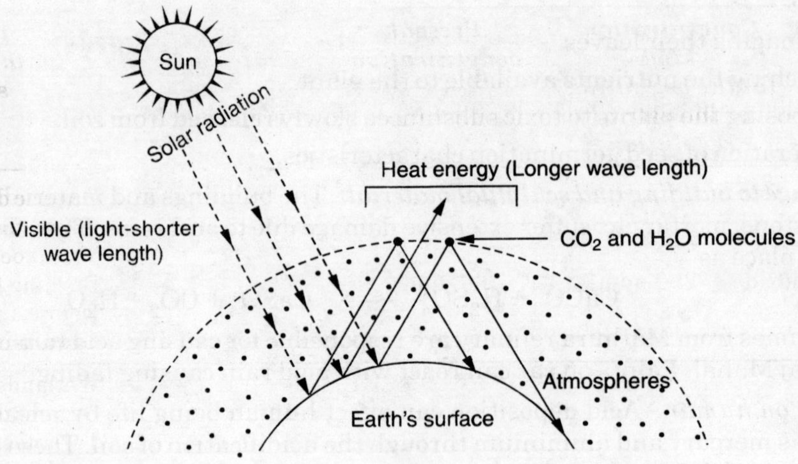

Fig. 30.7 The Visible Light Strikes the Earth and part of it is transforms into Infrared Radiation. Infrared Radiation is strongly absorbed by CO_2, H_2O, CH_4 and N_2O in the Atmosphere keeping the Earth warmer.

Green House Gases: The four major green house gases are carbondioxide (CO_2), methane (CH_4), nitrous oxide (N_2O) and chloroflourocarbons (CFCs). The CO_2 is the major contributer to green house effect as in shown in Fig. 30.8

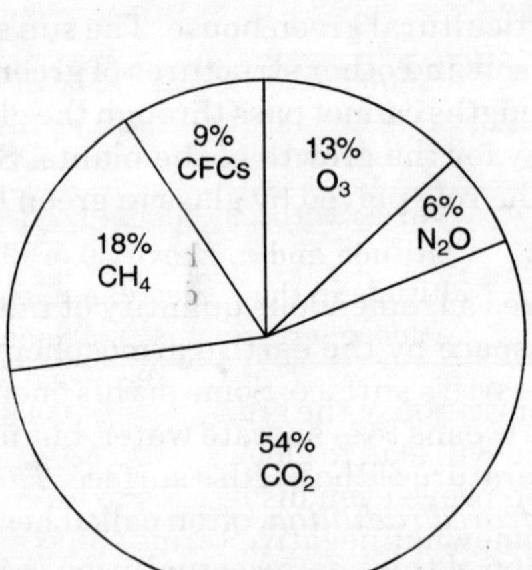

Fig. 30.8 Contribution of atmospheric gases to greenhouse effect

Enhanced Green House Effect (Global Warming)

The green house has been described as a natural phenomenon that is responsible for life on earth. Natural greenhouse effect is good for human life. However, due to industrial revolution, the concentration of green house gases is increasing constantly as is apparent from Table 30.6.

Table 30.6: Past and present concentration and sources

Greenhouse Gas	*Concentration 1750 (at the beginning of industrial revolution*	*Present concentration*	*Pecent increase*	*Natural and anthropogenic source*
Carbon Dioxide	280 ppm	360 ppm	29%	Organic decay; Forest fires; Volcanoes; Burning fossil fuels; Deforestation; Land-use change
Methane	0.70 ppm	1.70 ppm	143%	Wetlands; Organic decay; Termites; Natural gas & oil extraction; Biomass burning; Rice cultivation; Cattle; Refuse landfills
Nitrous Oxide	280 ppb	310 ppb	11%	Forests; Grasslands; Oceans; Soils; Soil cultivation; Fertilizers; Biomass burning; Burning of fossil fuels
Chlorofluorocarbons (CFCs)	0	900 ppt	Not applicable	Refrigerators; Aerosol spray propellants; Cleaning solvents
Ozone	Unknown	Varies with latitude and altitude in the atmosphere	Global levels have generally decreased in the stratosphere	Created naturally by the action of sunlight on molecular oxygen

As a result of higher concentration of the green house gases, scientists predict that the earth's atmosphere will be warmer and will lead to global warming. The global warming will have drastic effect on climate and may even prove even disastrous to human life. That is why, we often hear about green house effect in somewhat negative terms. But the negative concerns are related to possible impacts of an *enhanced* green house effect. Some scientists fear that if proper precautions are not taken, the concentration of the green house gases in the atmosphere may double in next 50 years leading to increase in global temperature by 4 to 5°C.

Effects of Global Warming: Global warming is one of the greatest challenges of 21st century:

- Predicting the amount of warming is accomplished by computer modelling. Computer model suggest that doubling of CO_2 may raise the temperature between 1 and 3°C. Some scientists estimate that the earth's average temperature has already increased by 0.3 to 0.6°C.
- The increased temperature would lead to melting of glaciers and polar ice caps and the level of sea level may rise thereby flooding the coastal lands and change the climate radically.
- Even 1.5°C raise in surface temperature can adversely affect the food production in the world.
- Changing regional climates could alter forests, crop yields and water supplies.

Remedical Measures for Global Warming

- Reducing consumption of fossil fuels such as coal and petroleum.
- Recovering green house gases from the atmosphere.
- Reduction in CFC production
- Reforestation and conservation of forests, so that plants can take up CO_2.
- Development of environmentally compatible technologies.

4. Ozone Layer and Its Depletion

Ozone is a very small part of our atmosphere but its presence is nevertheless vital to human well being.

Most ozone resides in the upper part of the atmosphere, called stratosphere extending from 16 km to 40 km. About 90% of ozone is contained in this 'ozone-layer' and remaining 10% ozone is present in troposhere. The high concentration of ozone (about 10 ppm) in statosphere shields us from harmful ultra-violet light from sun and hence is referred to as earth's protective umbrella.

Ozone is naturally present in our atmosphere. It is formed in the atmosphere in multistep chemical processes that require sunlight. Oxygen which make up 21% of the atmosphere undergo photolytic decomposition in presence of sunlight.

$$O_2 + hv \longrightarrow O + O$$

The atomic oxygen rapidly reacts with molecular oxygen to form ozone

$$O + O_2 + M \longrightarrow O_3 + M$$

where *M* represents a third body (N_2 or O_2) necessary to carry away the energy released in the reaction.

Stratospheric ozone is considered good ozone because it absorbs ultra-violet (*UV–B*) radiation, which are harmful for human beings.

$$O_3 + \underset{\substack{(UV\text{ - }B \\ 200\text{ - }320\text{ nm})}}{hv} \longrightarrow O_2 + O$$

The above combination of reactions form a long chain in which oxygen atoms are constantly being shuttled back and forth between the various molecular forms.

Other *UV* radiation, *UV-A* is not absorbed significantly by ozone, which reaches the surface of earth. However *UV-A* is not carcinogenic at usual exposure level of earth. Fig. 30.9 shows the two bands designated as *UV-A* and *UV-B*.

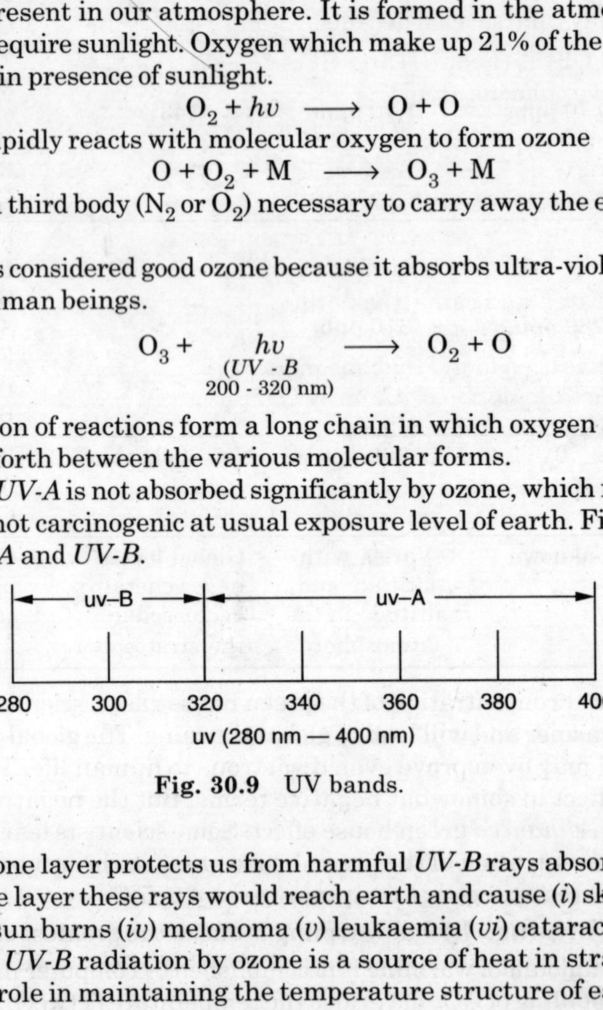

Fig. 30.9 UV bands.

Effect of Ozone Layer

- As is discussed ozone layer protects us from harmful *UV-B* rays absorbing these radiation. In absence of ozone layer these rays would reach earth and cause (*i*) skin-cancer (*ii*) damage to the plants (*iii*) sun burns (*iv*) melonoma (*v*) leukaemia (*vi*) cataracts, etc.
- The absorption of *UV-B* radiation by ozone is a source of heat in stratosphere As a result ozone plays a key role in maintaining the temperature structure of earth's atmosphere.

Bad ozone (Tropospheric ozone): Ozone is also formed near earth's surface by chemical reactions caused by presence of pollutants. This ozone is considered bad because more ozone comes in contact with humans, plants and animals. Ozone strongly reacts to destroy or alter many other molecules. In humans excessive exposure to ozone can lead to chest pains, throat irritation and cough. It also contributes to global warming. It also aids in the formation of photochemical smog.

Depletion of Stratospheric Ozone: It was discovered in the mid-1970s that some human activities could destroy ozone and deplete the ozone layer. The emission of ozone depletion gases is a cause of great concern. Ozone depletion has been found to occur through following types of compounds:

- Chloroflouro carbons (freons)
- Nitric oxide
- Reactive hydroxy radicals
- Atomic oxygen

Chloroflouro carbons (CFCs): These are the compounds containing chlorine and flourine. These are introduced into the atmosphere by aerosol sprays in which they act as propellants, refrigerating equipments in which they act as coolants, and in plastic foams such as 'thermocole' or styrofoam.

CFCs are insolubel in water and are not washed out through rain. Slowly they diffuse into stratosphere, which may take several decades.

The CFCs such as $CFCl_3$ (Feon-II) and CF_2Cl_2 (Freon-12), absorb *UV* radiation resulting in photodissociation to give chlorine atoms.

$$CFCl_3 + h\nu \longrightarrow CFCl_2 + Cl$$
$$CF_2Cl_2 + h\nu \longrightarrow CF_2Cl + Cl$$

The free Cl now reacts with ozone in a catalytic manner,

$$Cl + O_3 \longrightarrow ClO + O_2$$
$$ClO + O \longrightarrow Cl + O_2$$

Thus Cl is regenerated and can attack other ozone molecules. ClO can react with ozone *six times faster* than any of the nitrogen species, it becomes the dominant mechanism to destroy stratospheric ozone. It has been found that one molecule of CFC can destroy one lakh of O_3 molecules in the stratosphere Due to depletion of ozone by CFCs it was found in eighties that a large hole has been created in the ozone layer. That is why the use of CFCs has been completely banned.

Nitric Oxide: The main source of nitric oxide in the atmosphere is nitrons oxide N_2O produced through bacterial action of the surface of the earth. It gradually crosses to the stratosphere and reacts with ozone

$$NO + O_3 \longrightarrow NO_2 + O_2$$
$$NO_2 + O \longrightarrow NO + O_2$$
$$O_3 + h\nu \longrightarrow O_2 + O$$

Net reaction

$$2O_3 \longrightarrow 3O_2$$

Thus in each cycle two molecules of ozone are destroyed to yield three molecules of oxygen.

Reactive hydroxy radicals: The hydroxy radicals enhance the reactions of CFCs and NO with O_3. $\dot{O}H$ reacts with ozone as follows:

$$O_3 + \dot{O}H \longrightarrow O_2 + \dot{O}OH$$

$$\dot{O}OH + O \longrightarrow \dot{O}H + O_2$$

Effect of Ozone Depletion

- The existing ozone layer screens out more than 99% of incoming *UV* radiations. The small fraction that gets through is known to cause sunburn, skin cancer, and various kinds of damage to animals and plants.

- The genetic material DNA has the capacity to absorb *UV* radiations which may cause various disruptive effects.
- A 5% depletion in ozone produces 10% increase in radiations reaching the earth according to an estimate. That is, it would produce 5 times additional skin cancer per year worldwide.
- Increased *UV* radiations impairs the growth of certain crop plants and adversely affects a wide variety of organisms.
- Exposure to *UV* radiations damages the cornea and lens of eye causing blindness.
- Ozone depletion has a strong effect on climate. Ozone depletion itself is not the principal cause of climate change. However because ozone is a greenhouse gas, ozone changes and climate change are linked in important ways.

Substitute for CFCs: CFCs are found to be the major culprit in ozone depletion hence the use of CFCs has been totally banned worldwide.

Montreal Protocol, an international agreement on substances that deplete the ozone layer, established legally binding controls for about 185 nations on the production and consumption of ozone depleting gases.

Thus substitutes for CFCs were worked out and these days hydrochloroflouro carbons (HCFC) are being substituted for CFCs in air conditioning and refrigerating units. Because HCFC are chemically removed primarily in the troposphere and hence do not cause ozone depletion inspite of chlorine atom present. Hydroflouro carbons (HFCs) are also used as substitute compounds for CFCs. HFCs contain no chlorine atom they do not contribute to ozone depletion. However, these may contribute to global warming.

The ozone layer is expected to recover by the middle of 21st century if nations comply with Montreal Protocol.

30.7 AIR POLLUTION CONTROL

The most effective method of dealing with air pollution problem is to prevent the formation of pollutant or to reduce their emissions at the source itself. Some of the industrial processes may be suitably modified so as to minimise the extent of pollution.

Main routes to cutting pollutant emissions are:

(*i*) Reducing the consumption of fossil fuels.

(*ii*) Better designing of the equipment to control emissions.

(*iii*) Removing pollutants from exhaust and gases (source correction methods)

Based on the above points various pollution control methods have been adopted. Some of which are discussed below;

1. Source Correction Methods: Air pollution can be minimised at source by taking the following measures:

(*a*) *Substitution of the raw material:* A simple control of air pollution is to use pure-grade raw material for combustion. For example, use of low-sulphur oil and coal is recommended.

(*b*) *Modification in the process:* It involves changes and modification in industrial technologies aimed at reducing the atmospheric emissions of pollutants. For example, the use of exhaust hoods and ducts over several types of industrial ovens allow the recovery of various solvents that could have become air pollutants. H_2S emissions can be recycled and used to recover sulphur.

(*c*) *Alternations in equipment:* By suitably modifying the carburation and ignition systems in an automobile engine, air pollution can be reduced. For example, the basic oxygen furnace which are replacing open hearth furnace in steel industry pose much less air pollution problems. Evaporation from petroleum refineries can be minimised by suitable design modification of the tanks.

2. **Air pollution control devices:** Even if the above measures are taken, in certain cases emissons of pollutants cannot be prevented and pollution control devices are employed depending upon the type of pollutant (gaseous or particulate).

(*i*) **Devices for controlling Particulate Emissions:** Particulate matter in gaseous effluents consists of discrete and minute particles of sizes ranging from 0.1 μm to 100 μm. The device selection is based upon physical and chemical characteristics of particulates, size, temperature and humidity. The equipments presently available use one of the following techniques:

(*a*) *Gravitational Settling:* This technique is generally employed to remove large particulates (> 50 μm) from the gaseous effluents. In this the velocity of the horizontal carrier gas is reduced adequately so that the particles settle by gravitational force. The collector efficiency is improved by dividing the setting chambers by a number of trays. (Fig. 30.10)

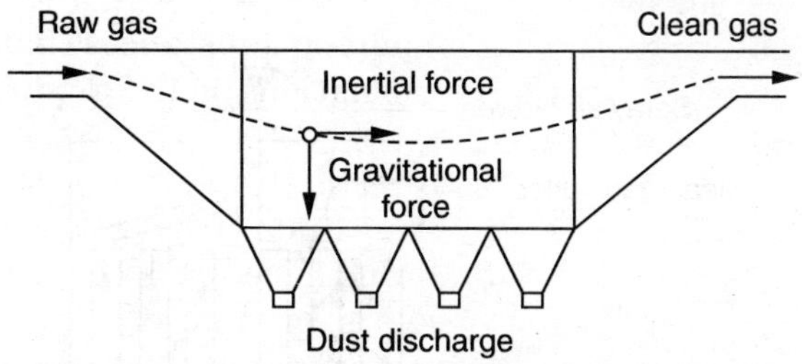

Fig. 30.10 Gravitational settling chamber

(*b*) *Cyclonic separators:* In this method, gas containing particulates is allowed to flow into a tight circular spiral-fitted chamber. Now the particulates present in the gaseous stream possess greater inertia than the gas molecules and experience greater centrifugal force, which drive these particulates towards the wall of the chamber. From there they settle down due to force of gravity. These type of cyclone separators one used in power plants and industries dealing with rock products. (Fig. 30.11)

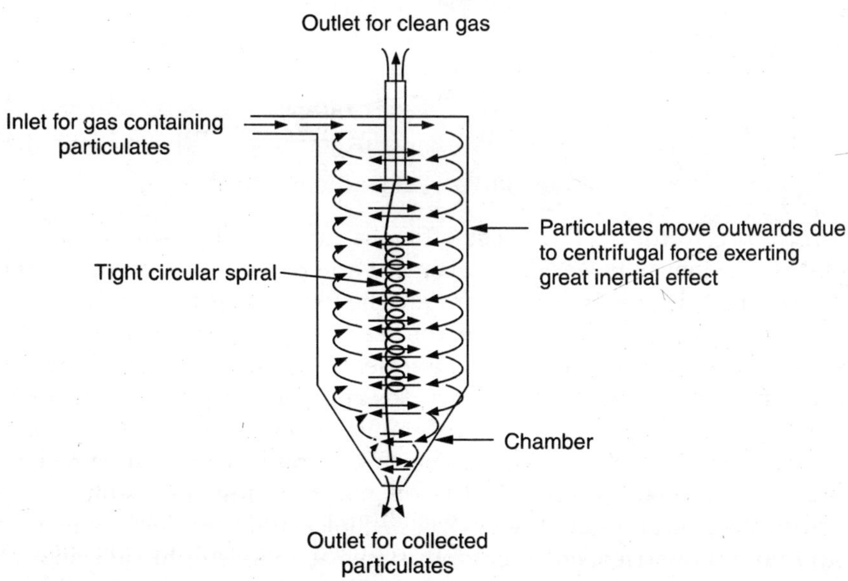

Fig. 30.11 A cyclone collector

The cyclone collector is most efficient for the removal of large particles of size ranging from 5 - 20 μm (smoke, dust etc.). This technique is more efficient than gravitational setting because centrifugal force on particles in a spinning gas stream is much stronger than the force of gravity.

(c) *Fabric filters:* Small solid particles are removed effectively by the use of fabric filters. The exhaust gases are forced through bags of very fine cloth-like mesh which are capable of retaining small particles. Fabric filters generally consist of cotton, wool, nylon, dacron or coated glass cloth. Fabric filters are used in industries dealing with rock products, pigments etc. (Fig. 30.12).

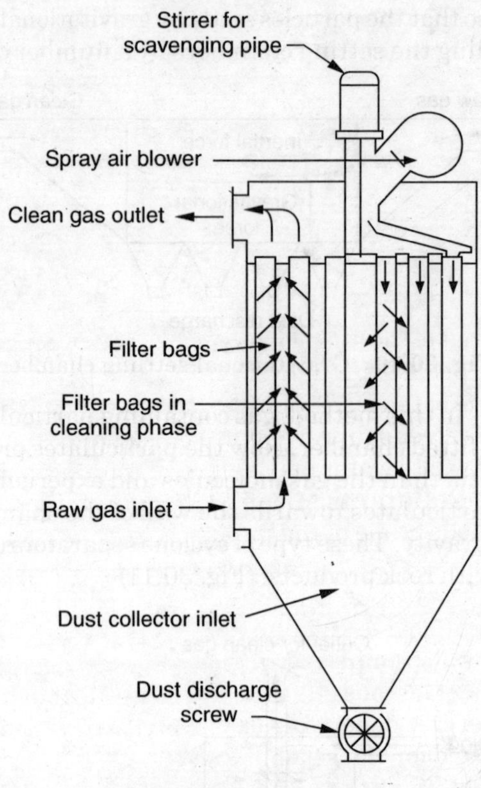

Fig. 30.12 Fabric filters

(d) *Electrostatic precipitator:* The most efficient and widely used device for controlling particulate emissions in power plants, cement and paper industry and oil refineries is the electrostatic precipitator. In this technique the particles present in gaseous exhaust are charged electrically and separated from the gas stream under the influence of electric field. The electrostatic precipitator consists of a series of grounded thin plates which are charged to high voltages, alternately + and –. Particles approaching these plates (around 50,000 V) become charged and are attracted towards opposite plates. From these plates they fall into a collector (hopper) below (Fig. 30.13). The efficiency of electrostatic precipitator is 97.9% and it is capable of cleaning 150,000 litres of gas per minute at a temperature of 600°C. Electrostatic precipitators are the devices of choice and are widely used in power plants, paper and pulp industries, chemical industries such as sulphuric acid plants, iron and steel plants.

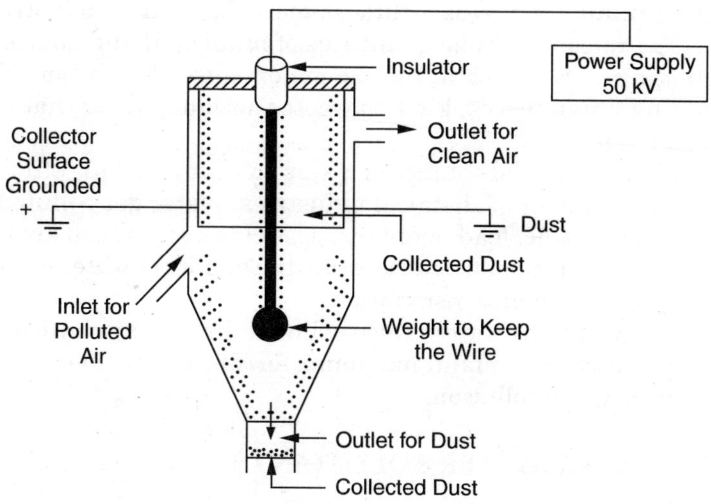

Fig. 30.13 Electrostatic Precipitator

(*e*) *Wet Scrubbers:* A natural cleansing mechanism is wet precipitation through which some pollutants are washed out of the atmosphere. The same concept is employed in wet scrubbers. The exhaust gases are passed through a fire spring of water so that the effective size of the particle is increased due to the formation of water particulate agglomerates. These are then separated from the gas stream by any of the collection mechanism.
However the collected waste water may become a source of water pollution.

(*ii*) **Controlling Gaseous Pollutants:** Control of gaseous pollutants can be achieved through source correction methods Some of the methods adopted for control of gaseous pollutants are:

(*a*) *Combustion:* When the pollutant contains gases or vapours, which are organic in nature, flame combustion is employed. Combustion converts them to water vapours and CO_2. Catalytic combustion is also employed where low operating temperatures are desired.

(*b*) *Absorption/Adsorption:* Gaseous effluents are passed through scrubbers or absorbers. The technique can be dry or wet depending on the phase in which the main reactions occurs. Some of the solvents used for wet scrubbing are:

$$NaOH, Na_2SO_3 \text{ (for } SO_x) \quad \text{and} \quad NH_4HCO_3\ Ca(OH)_2 \text{ (for } NO_x)$$

For dry scrubbing, activated carbon is used since it attracts thin layers of gases and vapours to its surface.

30.8 WATER POLLUTION

When toxic substances enter lakes, streams, rivers, oceans and other water bodies, they get dissolved or lie suspended in water. This results in the pollution of water where by the quality of water deteriorates.

The word polluted water is defined as the deterioration of physical, chemical and biological properties of water brought about mainly by human activities and natural resources and which cause harmful effects on human and aquatic life.

Sources of Water Pollution: Water pollution is caused by natural as well as human activities.

(i) Pollution from **natural sources** include the decomposed vegetable, animal and weathered products which are brought into main water resources.

(ii) Pollution from **human activities** include domestic as well as industrial pollution.

(a) *Domestic:* The release of huge quantities of municipal and domestic wastes through drains into the rivers and canals is the major cause of pollution. The domestic waste water contain human faeces, kitchen wastes, organic water that provide nutrition to bacteria and fungi.

(b) *Industrial:* The industrial sources of water pollution include effluents from factories, refineries and a number of chemical industries. Water get polluted by acids, alkalies, detergents, copper, zinc, lead, mercury, pesticides, etc. which are constantly added to water by industrial effluents, deteriorating the quality of water to such an extent that it becomes unsafe for human consumption.

(c) Apart from these two, the water is also polluted through **agricultural discharge** such as pesticides, insecticides plant, nutrients, fertilisers, etc. Beside this, bacteria, algae, virus also cause water pollution.

30.9 CLASSIFICATION OF WATER POLLUTANTS

The problem of water pollution due to discharge of domestic and industrial water into water has already become a serious problem in the country. Nearly 75 to 80% of India's population is exposed, to unsafe drinking water as a result, enteric diseases often reach epidemic proportions.

To aid in a systematic discussion of water pollutants, they have been classified into nine categories as described below:

1. Oxygen demanding wastes
2. Pathogens (Disease causing agents)
3. Synthetic organic compounds
4. Plant nutrients
5. Inorganic chemicals and minerals
6. Sediments
7. Radioactive substances
8. Thermal discharges
9. Oil

Water pollution is often caused by the combination of the above categories, increasing the severity of the problem. The categories are discussed briefly in the following sections:

1. Oxygen Demanding Wastes

Dissolved oxygen (DO) is essential for sustaining the plant and animal life in any aquatic system. Warm-water fish requires a minimum DO level of at least 5mg/L (5 ppm). If DO level drops below the level necessary to sustain normal life, then water is classified as polluted.

The amount of DO in water is reduced because of oxygen demanding wastes. Oxygen demanding wastes are substances that oxidise in water, reducing the amount of DO. These substances include the organic substances which are formed due to death and decay of the algae and related species. Oxygen demanding wastes also include organic substances contained in municipal waste water or in effluents from certain industries, such as food processing and paper production. In addition, the oxidation of certain inorganic compounds may also contribute to the oxygen demand. Even naturally occuring organic matter, such as leaves and animal droppings, that find their way into surface water, add to the DO depletion.

The oxygen demanding wastes are oxidised by bacteria or micro organisms consuming dissolved oxygen in water to CO_2 and water. Thus DO levels drop. These substances produce undesirable odour, tastes and reduce the acceptability of water as a domestic supply.

There are several measures of oxygen demand commonly used.

(a) Biological Oxygen Demand (BOD) BOD is the amount of oxygen required by microorganisms to biologically degrade the wastes. It is most important measure of the strength of organic pollution. It is discussed in more details in section—

(b) Chemical Oxygen demand (COD) COD is the amount of oxygen needed to chemically oxidise the waste. It is discussed in more details in section —

2. Pathogens

Pathogens are disease causing organisms that grow and multiply within the host. Water is a potential carrier of pathogenic micro-organisms.. These pathogens are carried into the water bodies by sewage and wastes from farm and various industries, specially tanning and wheat packaging industries.

Contaminated water caused by poor sanitation can lead to both, water borne and water contact diseases.

Water borne diseases are those acquired by ingestion of pathogens not only in drinking water, but also from water that makes it into a person's mouth from washing food, utensils and hands.

Examples of pathogens associated with water borne diseases include bacteria, viruses and protozoa. A brief account of the diseases caused by these organisms given in Table 30.7

Table 30.7: Diseases Commonly found in polluted water

Disease	*Agent Type of organism*	*Symptoms*
Cholera	Vibriocholerae (Bacterium)	Severe diarrhoea, vomiting fluid loss, cramps and coiiapse.
Dysentery	Shigella dysenteriae (Bacterium)	Infection of the colon causes painful diarrhoea with mucus and blood in the stools
Enteritis	Clostridium perfringens (Bacterium)	Inflammation of the small intestine cause general dicomfort loss of appetite abdominal cramps and diarrhoea
Typhoid	Salmonella typhi (Bacterium)	Early symptoms include headache, loss of energy, fever later a pink rash appears along with hemorrhage in the intestine
Infectious Hepatitis	Hepatitis virus A (Virus)	Inflammation of liver causes jaundice, fever, haeadache, nausea, vomiting sever loss of appetite;
Poliomyelitis	Poliovirus (Virus)	Early symptoms include sore throat, fever, diarrhoea and aching in limbs Spreads to spinal cord, paralysis and atrophy of musles.
Amoebic Dysentery	Entamoeba histolytica (Amoeba)	Infection of the colon causes painful stools; abdominal pain
Schistosomiasis	Schistosoma (Fluke)	Disorder of the liver and bladder, causes blood in urine, diarrhoea, weakness, lack of energy, repeated abdominal pain.
Ancylostomiasis	Ancylostoma (Hookworm)	Severe anemia sometimes symptoms of bronchitis.

Water-Contact diseases do not even require that individuals ingest the water. Just contact with contaminated water may cause disease. For example, Schistosomiasis (bilharzia) is the most common water-contact disease in the world, affecting approximately 200 million people, spread by free swimming larva in water, called cercaria. These larvae attach themselves to human skin, penetrate it and enter the blood stream.

According to WHO report approximately 80 per cent of all sickness in the world is attributed to contaminated water.

The coliform group of bacteria are considered reliable indicators of the hygienic quality of water. **Escherichia coli** (*E* coli), belonging to the coliform group is harmless bacteria found in large number in human faeces. *A* large concentration of *E.* Coli in water indicates faecal contamination and evidence of the presence of pathogens.

3. Synthetic Organic Compounds

These include pesticides, synthetic organic chemicals and detergents. In contrast to organic wastes, these compounds are not biodegradable and persist for longer periods. Most of these are accumulative toxic poisons and ultimately reach objectionable levels in water. Out of various synthetic organic compounds, the pesticides and detergents have been subjected to intensive study because of their liberal use resulting in water pollution.

Pesticides The term pesticide is used to cover a range of chemicals that kill organisms that humans consider undesirable. Pesticides can be classified into three main classes of insecticides, herbicides and fungicides. Other types include rodenticides, molluscicides and acaricides.

Pesticides enter the water body from surface run off from agricultural lands, waste discharge by pesticide manufacturers and by other means. Because of their world-wide usage, nearly all the rivers and oceans of the world contain pesticide residues.

There are three main groups under which the pesticides can be categorized — chlorinated hydrocarbons, organophosphorus compounds and carbonate pesticides. The common pesticides and their effects are tabulated in Table 30.8

Table 30.8

Pesticide	*Effects*
1. **Chlorinated hydrocarbons** - DDT (Dichloro diphenyl trichloro ethane)	• Very persistent (last a long time in the envi ronment) • Soluble in lipids and hence can accumulate in fatty tissues • Affect the calcium metabolism in predatory birds, resulting in birds, laying eggs with thin shells and consequent reproduc tive failure. • Tends to become more concentrated from one trophic level to another due to the process of *biological magnification.
Chlordane	• Produce lives cancers
Heptachlor	
Aldrin	• Cause birth defects
Dieldrin Kepone	• Severe neurological damage.
Chlorophenoxy compounds (2,4,5,T, 2 – 4 D)	• Contains the impurity dioxin, one of the most potent toxins known.

Contd....

* Biological magnification implies that successively higher trophic levels in food chain are consuming food that has successively higher concentration of pesticides. For example, the concentration of DDT increases as we move from water → plankton → Fresh water fish → Predatory birds.

Pesticide	*Effects*
2. Organophosphorus compounds Parathion Malathion Diazinon	• Non-persistent but damage nervous system Acute exposure result in slurred speech, muscle twitching and convulsions.
3. Carbonate Pesticides Propoxur Carbaryl Aldicarb	• Acute exposure results in nausea, vomiting, blurred vision and convulsions.

It may be emphasized that India is one of the few world's producers of DDT, which is banned in US in 1972 for environmental reasons.

Detergents: The term detergent means cleansing agent. The basic active ingredient in detergents is the surfactant or surface-active agent which contains hydrophobic (hydrocarbon) and hydrophilic (polar) groups.

Surfactants decrease the surface tension of water so that they can penetrate the surface and interstices of the object being cleaned. Although reducing the surface tension is an essential feature of cleaning agents, detergents usually contain 10 – 30 % surfactant. The remainder comprises of polysulphate salts (the builder) and a number of other ingredients. Both the surfactants and builders of detergents create serious pollution problems in water.

Surfactant concentrations as low as 1 ppm produce foam in rivers and in sewage treatment plants. Although these concentrations are non-toxic to humans, the presence of surfactants gives an off taste to drinking water. The presence of surfactants also reduce the rate of oxygen absorption in water and may lead to the symptoms similar to asphyxia in certain kind of fish (trout).Till the early 1960s the surfactants present in synthetic detergent was alkylbenzene sulphonate (ABS). ABS is non biodegradable and cause rivers and sewage treatment plants to become covered with huge amounts of foams. ABS have been now replaced by linear alkyl sulphonate (LAS) which is rapidly biodegradable.

The detergent builders pose greater problem to day. The polyphosphates builders are released into water and act as plant nutrients. The extensive growth of algae consumes most of dissolved oxygen from water.

Nitrilotriacetate (NTA) was considered a replacement but it proved to be hazardous to human health

4. Plant Nutrients (Sewage and agricultural runoff)

Nutrients are chemicals, such as nitrogen, phosphorus, carbon, sulphur, calcium, potassium, iron, manganese etc. that are essential to growth of living things. However, in terms of water, these nutrients are considered as pollutants, when their concentrations are sufficient to allow excessive growth of aquatic plants, particularly algae. When these algae die and decompose they add undesirable odour and objectionable taste to water. Further, the decaying organic matter oxidizes and leads to reduced DO levels. The gradual accumulation of silt and organic matter in the lake is known as **eutrophication.** A eutrophic lake gets shallower and warmer with further growth and decay of algae and eventually lake turns into a marsh or bog.

The plant nutrients enter in the water through municipal waste water, industrial wastes and runoff from fertilised lands. The green slimy layer seen on the surface of some water bodies is an example of eutrophication of the lake. The slimy layer reduces light penetration and restricts atmospheric reoxygenation of water.

High concentration of nitrogen not only poses the problem of eutrophication but when found in drinking water, it is serious health hazard. Certain bacteria found in the intestine, reduce the nitrate in water to highly toxic nitrites (NO_2). Nitrites have a greater affinity for haemoglobin as compared to oxygen. Thus nitrite attaches to haemoglobin forming methaemoglobin. As a result, the oxygen carrying capacity of the blood is reduced producing a condition called methaemoglobin anemia commonly referred to *blue baby syndrome*. In young babies, it is extremely fatal. Further the nitrates can be converted into nitrosoamines in the body leading to gastric cancer.

5. Inorganic Chemicals and Minerals

This category of water pollutants includes inorganic salts, mineral acids, finely divided metals or metal compounds, trace elements, cyanides, organometallic compounds, etc.

These pollutants are added to the water bodies through municipal and industrial waste waters and mine runoff. Most of these substances are toxic and are a health hazard to human life. Some of these pollutants are discussed below in brief.

Acid Mine Drainage: Mine drainage is an important source of increased acidity in natural water. Acid rain also adds to the acidity. The mining of sulphur bearing ores containing lead, zinc and copper lead to acid drainage. Coal mines discharge containing varying amounts of iron sulphide (pyrite) is also a major cause of acid mine drainage. These discharges release considerable quantities of sulphuric acid and ferric hydroxide, (which are formed as result of reactions between air, water and pyrite) into local streams through seepage. Acids cause corrosion of metals and concrete and is fatal to fish.

Soluble Salts: Water naturally accumulates variety of dissolved solids or salts as it passes through soils and rocks on its way to sea. These salts include cations such sodium, calcium, magnesium and potassium and anions such as chloride, sulphate and bicarbonate. The measure of salinity is to determine the concentration of total dissolved solids (TDS). Water with less than 500 mg/L TDS is considered safe for drinking purpose. For irrigation purposes, the TDS above 500 mg/L is also considered harmful. Salts also cause 'hardness' of water leading to the formation of scale and sludges. High chloride concentration are fatal to fish. One way to approach salt problem is through desalination (chapter I).

Heavy Metals: The term heavy metals' is less precisely defined. In chemical terms it refers to metals with specific gravity greater than about 4 or 5 metals. But more often, the term is simply used to denote metals that are toxic. The toxic properties of numerous metals have been known for years. The metals of particular concern in industrial waste waters are cadmium, chromium lead, mercury, silver, arsenic, aluminium, copper, cobalt, manganese, etc. Some of these, such as chromium and iron, are extremely toxic in higher doses.

Metals may be inhaled or may be digested. These have a range of adverse impacts on the body, including nervous system and kidney damage, creation of mutations and induction of tumours. The kidneys contain millions of excretory units called nephrons and the chemicals that are toxic to kidneys are called nephrotoxins. Cadmium, lead and mercury are examples of nephrotoxic metals.

Table 30.9 summarises the effects of some of the toxic metals.

Table 30.9: Effects of trace metals present in air and water

Pollutant	*Source*	*Effect*
Lead	Automobile emissions, paints, storage batteries, lead piping, effluents from printing and dyeing industries	• Impairment of intellectual development and cognitive functioning • Absorbed through gastointestinal and respira tory tract and cause liver and kidney damage. • Permissible limit in drinking water 0.05 mg/L
Cadmium	Effluents from textile, electro plating and chemical plants, Fertilizer industry cadmium nickel Satteries (Nicad)	• Inhalation of vapours causes kidney damage, bronchitis • Once absorbed by the organism remains resident for many years in the body. • High dose has been linked to lung cancers. • May produce bone defects. • Permissible limits in drinking water is 0.01mg/L
Chromium	Metallurgical and chemical industries, pigments for paints cement, paper, ceramic and glass industry.	• Low level exposure can irritate the skin and cause ulceration • Long term exposure can cause kidney and liver damage. • Accumulates in the body of fish and adds to the health effects of eating the fish exposed to high levels of chromium. • Permissible limit in water 0.0.5mg/L
Mercury	Various industrial efflments such as those from paint and paper manufacturing units. Highly prevalent in medical waste, lamps, thermometers, used in dentistry as amalgum for fillings.	• Builds up in the tissues of fish and other creatures living in contaminated water and may be carried up the food chain to humans. • It is proved to be the most toxic aquatic pollutant because of its rapid methylation in aquatic enviro nment by some bacteria. Monomethyl mercury and dimethyl mercury and dimethyl mercury are highly toxic and cause neuro-toxicological disorders.
Arsenic	Introduced as a by-product of fossil fuel combustion, irrigation water containing pesticides, chemical wastes	• Carcinogenic • Can cause mild bronchitis • Attack–SH group of enzymes and can coagulate proteins.
Silver	Electroplating and photographic industries	• Fish and lower organisms are susceptible to silver poisoning • It is cumulative poison and chronic injestion can cause permanent discolouration. Permissible limit in drinking water is 0.05 mg/L.

6. Sediments

Sediments include soil, sand and mineral particles washed into the aquatic environment by storms and flood waters. Sediments are the sources of organic and inorganic matter in the streams, fresh water, rivers, and other water bodies. Soil particles eroded by running water ultimately find their way into water reservoirs and such a process is called 'siltation'. Reservoirs and dams are filled with soil particles and other solid materials because of siltation. This reduces their storage capacity. Further presence of suspended solid particles present in the water bodies block the sunlight and hence decrease the process of photosynthesis for aquatic plants. As a result, evolution of oxygen is decreased and it is not enough as required by aquatic animals.

- Sediments increase the cost of water treatment used for culinary purposes.
- Sediments enter pumping equipment and power turbines and increase the turbidity.
- In suspension, the solids may cause thickening of fish gills which may lead to eventual asphyxiation of the fish.

7. Radioactive substances

Radioactive pollutants enter into the aquatic system through a number of human activities involving the use of naturally occuring or artificially produced radioactive materials. The radioactive pollutants enter into the water system through a variety of sources such as:

- Mining and processing of ores to produce radioactive ores, e.g, uranium and thorium.
- From nuclear power plants and from industrial use of radioactive materials
- Leakage from underground nuclear detonations
- Use of radio isotopes in medicine, industry, agriculture and research operations

The refining of uranium ore is an important source of radioactive waste, producing radionuclides of radium, bismuth, etc. Radium is considered to be a hazard in drinking water. It causes various genetic disasters in human beings. Radioactive substances can enter humans with food and water and get accumulated in blood and certain vital organs like thyroid gland, the liver, bone and muscular tissues. These may cause cancers, leukemia and eye cataract.

8. Thermal Discharges

Industry and power plants use large quantities of water for cooling purposes. Used coolant water is usually discharged directly into water bodies. This results in increase in temperature of the water bodies which is called thermal pollution. Coal fired power plants, electric powers, steel and chemical industries as well as atomic energy plants discharge their heated effluents into nearby lakes or rivers leading to thermal pollution.

Rise in temperature of water decreases dissolved oxygen content of water which effects the equatic life. At 32°F the DO content of water is 15 ppm which is decreased to 6.5 ppm at 64° F Fishes are killed due its the action of heat on nervous system, inactivation of enzymes and coagulation of cell protoplasm.

An increase in temperature also increase the toxicity of some chemical pollutants.

9. Oil

Oil is an important commodity used in one way or the other. Oil and oil wastes are added to the water bodies from industries as effluents, oil refineries, storage tanks, automobile waste oil and petrochemical plants. Oil is insoluble in water and hence floats over it as a thin layer. This oil may penetrate the feather of the birds and effect their insulation and buoyancy. Thus birds experience difficulty in floating and flying. The birds may ingest oil while they dive to feed. This ingestion may produce toxic effects.

Moreover the oil slick formed on the surface prevents the diffusion of oxygen into water resulting in decreased concentration of DO.

10. Volatile Organic Compounds

These are one of the most commonly found contaminants in water. Volatile organic compounds are often used as solvents in industrial processes. They are volatile and hence their concentrations remain as low as few micrograms per litre in surface water. But in ground water their concentrations can be hundreds or thousands of times higher. Volatile organic compounds are toxic and their presence in drinking water is cause of great concern. The following VOCs are tabulated in Table 30.10.

Table 30.10: Volatile Organic Compounds

Vinyl Chloride	Production of polyvinyl chloride resins	Carcinogenic
Tetrachloroethylene	Used as a solvent, heat transfer medium, manufacture of CFCs.	Causes tumours in animals.
Trichloroethylene	solvent, used to clean electronic parts	Carcinogenic
1,2 Dichloroethane	Metal degreaser, used in manufacture of vinyl chloride, varnish removers and soap compounds.	Cause injury to central nervous system liver and kidneys
Carbon tetrachloride	Common household cleaning agent, used in fire extinguisher, solvent	very toxic, only a few millilitres can result in to death

30.10 WASTE WATER TREATMENT

Until 1900's human wastes were simply dumped as raw sewage into the nearest stream or river. But later on when a number of diseases were linked to the waste waters, the treatment of waste water was taken up seriously.

The purpose of waste water treatment is to remove the contaminants from water so that the treated water can meet the acceptable quality standards. The quality standards usually depend upon whether water will be reused or discharged in to a receiving stream..

The waste water is often referred to as *sewage*. Sewage is the liquid water which includes human and domestic waste waters, industrial wastes, ground wastes and street washings. Sewage contains 99.9 per cent water and rest being the organic, inorganic compounds in dissolved colloidal or suspended states. It is this small percentage of pollutants that brings about polluted conditions.

Constituents of Sewage

There are various types of chemicals founds in waste water, depending upon the source from where it is being discharged. Depending upon the nature of the sewage it can be grouped as:

(i) *Domestic sewage:* Domestic sewage includes human excreta as well as discharges from kitchens, baths, lavatories, etc. from public and private buildings.

(ii) *Industrial Sewage:* Industrial sewage or waste water may be regarded as the liquid waste water in which industrial effluents are present as main wastes

(iii) *Storm water:* This is the rain water which flows as run-off from streets, open yards, etc.

(iv) *Combined Sewage:* Combined sewage is the combination of domestic sewage, industrial waste and storm water.

Characterisation of Waste Water (Strength of the sewage)

The strength or impurity in a particular sewage is usually measured on the following catogories.

1. Dissolved Oxygen

Oxygen is poorly soluble in water. The amount of oxygen in water depends on physical, chemical and biological activities taking place in water. The solubility of dissolved oxygen in water at saturation at any temperature and pressure is given by Henry's law.

For determination of dissolved oxygen refer to chapter 1.

Biochemical Oxygen Demand

When biodegradable organic matter is released into water, the micro-organisms feed on the wastes and break it down into simpler organic or inorganic substances. If the decomposition takes place in the **presence of oxygen**, i.e. aerobically, **the non objectionable**, stable end products are formed.

$$\text{Organic Matter} + O_2 \xrightarrow{\text{MICRO ORGANISMS}} CO_2 + H_2O + \text{Other Stable Products.}$$

On the other hand if **insufficient oxygen** is available, the decomposition takes place anaerobically. The micro organisms causing the decomposition of the organic matter without the presence of oxygen are entirely different from aerobic bacteria and produce **highly objectionable** end products including H_2S, NH_3 and CH_4.

$$\text{Organic Matter} \xrightarrow{\text{MICRO ORGANISMS}} CO_2 + CH_4 + \text{Unstable Products}$$

The amount of oxygen required by micro-organisms to oxidize organic wastes aerobically is called biochemical oxygen demand (BOD). The BOD of raw water will indicate the extent of organic matter present, thus indicating the extent of treatment required to make it safe for use. The BOD of treated water should be nil, so as to make it free from any organic matter. If BOD is high, the dissolved oxygen becomes low and this results in greater pollution.

Determination: Polluted waters will continue to absorb oxygen for many months, till the oxidation gets completed and hence it is not practical to determine this ultimate oxygen demand which would require an extended period of time. As a result, it has become standard practice simply to measure the oxygen over a shorter period of 5 days. This is known as 5-day BOD test. **BOD_5 is the total amount of oxygen consumed by micro organisms during the first five days of biodegradation.**

In its simplest form BOD_5 would involve diluting a known volume of sample water with a known volume of pure water, whose oxygen content is already known. This is kept for 5 days at 20°C in a stoppered bottle, away from light. The dissolved oxygen is measured after the period of incubation. The difference between the original oxygen content and residual oxygen content will indicate the oxygen consumed by the water sample in five days in oxidising the organic matter present in water sample. Thus BOD_5 is given by:

$$BOD_5 = \text{Loss of oxygen in mg/L} \times \text{Dilution factor}$$

Chemical Oxygen Demand

Although BOD test is applicable to organic wastes, there are many drawbacks. Certain organic materials are not biodegredable and hence can give wrong conclusions that less organic matter is present because BOD of such water samples will be low. In this case COD, Chemical Oxygen demand reveals the real organic content present. Here the oxidation of organic substances present in water is done chemically.

COD, is the amount of oxygen required by organic matter in a sample of water for its oxidation by a strong chemical oxidizing agent such as $K_2Cr_2O_7$.

Determination: A known amount of $K_2Cr_2O_7$ is added to a measured amount of the sample and the mixture is boiled with concentrated H_2SO_4. After boiling, the amount of unreacted $K_2Cr_2O_7$ is determined by titrating against a standard Mohr's salt solution using ferroin as indicator.

The difference between the dichromate originally present and the dichromate remaining unreacted gives the amount of dichromate used for oxidation of organic matter.

2. Dissolved CO_2

Varying amounts of CO_2 are present in water. CO_2 can be picked up from the atmosphere or it is generated at the bottom of the lake due to the decay or organic matter. Rivers receiving acid wastes may also show high CO_2 content.

Determination: CO_2 is determined by titrating the sample water with N/10 sodium carbonate solution using phenolphthalein as indicator. The end point is appearance of pink colour.

3. Free Chlorine

Chlorine determination is necessary in those samples of water that have been treated with chlorine or hypochlorites.

Determination: The estimation of free chlorine is based on oxidation of KI by free chlorine, i.e. by iodometric titration. (Refer to chapter 1 for details)

4. Dissolved Chlorides

The presence of chlorides in natural waters is due to the discharges of effluents from the industries, irrigation drainage, presence of marine sedimentary deposits, etc.

Determination: Chlorides ions can be determined by titration of the water sample against a standard solution of $AgNO_3$ using potassium chromate as indicator. (Argentometric titration). (Refer to chapter 1 for details.)

5. Total Dissolved Solids (TDS)

Dissolved solids denote mainly the various kinds of minerals present in water. However, some organic substances present in water also contribute to the dissolved solids.

Determination: The amount of total solids in a water sample can be determined very easily as the residue left after evaporation of the filtered sample. Refer to chapter 1 for details.

30.11 WASTE WATER TREATMENT PROCESSES

Available waste water treatment processes can be broadly classified as:

- Physical processes
- Chemical processes
- Biological processes

These processes consists of a series of unit operations. These are applied in different combinations composition and specifications of the waste water.

Further, the waste waters treatment processes are generally grouped according to the water quality they are expected to produce. These processes are (physical, chemical and biological) grouped as:

(i) *Primary Treatment:* Removes suspended solids and floating matter

(ii) *Secondary Treatment:* Involves biological process and organic matter is removed.

(iii) *Tertiary or Advanced treatment:* Physical, chemical or biological processes depending upon the impurities to be removed.

We would be discussing these in detail in the following section.

1. Primary Treatment

Primary treatment plant utilises physical processes such as screening or sedimentation to remove a portion of pollutants that will settle, float or that are too large to pass through simple screening devices.

Primary treatment comprises two steps, Pretreatment and Sedimentation.

(a) *Pretreatment:* Pretreatment consists of screening and grit removal. Screening removes large floating objects such as rags, sticks, wood and other large floating and suspended solids. A typical screen consists of a parallel steel bars spaced anywhere from 2 to 7 cm apart. This may be followed by a wire mesh screen. The material collected on the screens is disposed off. One way to dispose off the material is to use a device called comminuter, which grinds the coarse material in to small pieces which flow along with water and handled in sedimentation tank. After screening, the waste water passes into a grit chamber where the velocity of water is reduced or it is detained for a few minutes. Here sand, grit and other heavy materials settle down. The grit settling chambers are periodically disconnected from the main system to remove grit manually, for possible use in land-filling, road making and on sludge drying beds. Grit also is a food manure for growing crops.

(b) *Sedimentation:* From the grit chamber, the sewage passes to a primary settling tank known as sedimentation basin. Here the velocity of the water is reduced considerably to allow most of the suspended solids to settle out by gravity. The most common equipment used include horizontal flow sedimentation tanks, similar to the one described under domestic water. The water is detained in the horizontal flow tanks for 2-3 hours resulting in removal of 50% of the suspended solid matter. An efficient sedimentation tank or clarifier removes about 80-90% of the suspended solids and 40% of organic matter. The solids that settle are called primary sludge or raw sludge.

Sedimentation aids: The removal of finely divided solids is difficult by simple sedimentation technique. In such cases, mechanical flocculation or chemical coagulation is employed.

In *mechanical flocculation*, the waste water is passed through a sedimentation tank which is fitted with rotating paddles which rotate slowly at a speed of 0.43 m/s. This slow mechanical stirring allows the finely divided solid particles to coalesce into larger particles and settle out.

In *chemical coagulation*, certain chemical substances, known as coagulants are added. Coagulation used for sewage treatment are similar to the ones used in water treatment. The coagulants react with colloidal matter in the sewage to form floc. The floc entraps the smaller particles and eventually settles down as sludge. Common coagulants used in sewage treatment are:

Alum $Al_2(SO_4)_3 \cdot 18H_2O$, Copperas $FeSO_4 \cdot 7H_2O$, hydrated lime, ferric chloride, and chlorinated copperas, $FeSO_4.Cl$ (mixture of ferric sulphate and chloride).

Apart from using coagulants, coagulant aids are also used sometimes for efficient removal of finely divided particles. The commonly used coagulant aids are, activated silica polyelectrolytes (polymers of cynamide or methacrylic acids), and hydrolysed high molecular weight polymers. The coagulant aids draw a number of suspended particles together because of their long chain molecular structure. They also neutralise the charge on the particles by them bringing them closure to form large particles.

In *industrial waste water*, sometimes the primary treatment also includes equalisation and neutralisation

Equalisation: Sometimes different types of wastes are produced by some industries. To apply uniform treatment, different effluents are held in big tanks for certain periods and are mixed thoroughly to produce homogeneous equalised effluent.

Neutralisation: Acidic wastes are neutralised with lime stone and alkaline wastes are neutralised by treatment with sulphuric acid or CO_2 or waste boiler flue gas. If both acid and alkaline wastes are produced in the nearby plants then mutual neutralisation by mixing there is them cheapest method of neutralisation.

2. Secondary Treatment

In primary treatment, only those materials are taken care of, which can be removed by simple physical or mechanical action. However, in waste water much of the organic material is dissolved or in colloidal form which is not removed by primary treatment. This organic material is oxygen demanding and must be removed further. Thus the main purpose of the secondary treatment is to remove the organic matter. This is achieved through biological processes which are quite similar in concept to the natural biodegradation of organic matter by aerobic bacteria. The various changes which are brought by secondary treatment include:

— Coagulation of the finely divided or colloidal matter.
— Oxidation of the organic matter to CO_2
— Conversion of nitrogenous organic matter to ammonia, which is eventually converted into nitrite and nitrate.
— Anoerobic digestion of the sludge so obtained.

There are three commonly used approaches, all of which take advantage of the biological processes. These processes are:

(i) Trickling filters.
(ii) Activated Sludge Process
(iii) Oxidation Ponds (Lagoons).

(i) Trickling Filters (Aerobic filteration)

A trickling filter consists of a rotating distribution arm that sprays the liquid over a circular bed of rocks or other coarse material (Fig. 30.14)

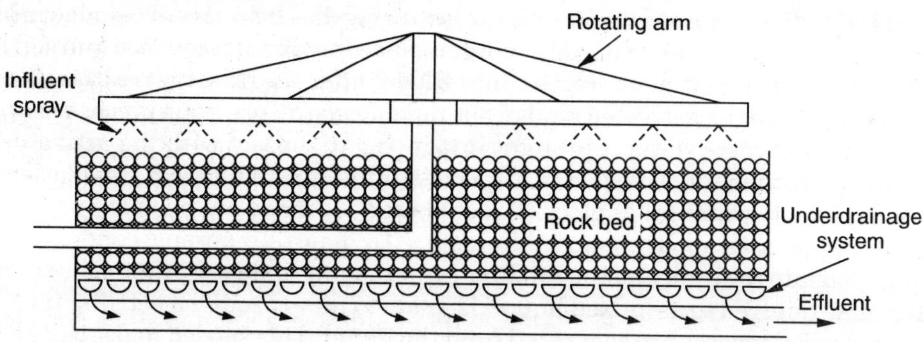

Fig. 30.14 Sketch of the cross section of a trickling filter.

The rock bed is 1 to 3 meters deep with enough open spaces between rocks to allow air to circulate easily. The individual rocks becomes coated with a layer of biological slime or a gelatinous film that absorbs and consumes the wastes trickling through the bed. The slime consists mainly of aerobic micro-organisms (bacteria, algae, protozoa, etc.) which are collectively called **zooglea** and thrive on the nutrients supplied by the waste water. The in fluent (waste water) is sprinkled over this bed packing. As the liquid trickles over the packing, oxygen and the dissolved organic matter diffuses into the slimy layer and are oxidised by the micro-organisms to form CO_2, NO_3, etc., which diffuse back out of film and appear in the filter effluent. As the micro-organisms utilise the organic matter, the thickness of the slime layer increases and it gets detached from the surface of the rock and is carried away along with the effluent, which is referred to as *Humus*. The effluent along with detached slimy layer and treated water is collected at the bottom of the filter and passed on to settling tank, where the detached slimy layer settles down and is removed. The slimy layer (sludge) is then pumped to the sludge digestion unit.

More recently, plastic filters have been used for trickling filters. The plastic filters are light and can be stacked many times higher than the conventional bed rock which crumbles under its own weight if stacked high. These filters, made with the plastic media are sometimes referred to as 'biological towers'.

Advantages of the Process

- Trickling filters are simple to operate and can produce BOD removal to the extent of 65 to 85%
- Constant monitoring is not required.
- Effluents so produced are of better quality.

Limitations

- The microbial film formed is sensitive to temperature changes in winters.
- Efficiency of the filter is dependent upon the composition of waste, pH, size uniformity of the filtering medium and the supply of air.
- Cost of construction is high
- Trickling filters are used for treating industrial waste water from dairy, brewery, food processing, pulp and paper mills, pharmaceuticals, petrochemicals, etc.

(ii) Activated Sludge Process

This is the most versatile biological oxidation method, employed for the treatment of waste water containing organic matter. It is a biological waste water treatment process in which a mixture of waste water and activated sludge is agitated and aerated. *The activated sludge is the sludge obtained by settling the sewage in presence of excess of oxygen.* The activated sludge is biologically active because it is heavily laden with micro-organisms which are in active state of growth.

The activated sludge treatment process, the effluent after the primary treatment, is passed to an aeration tank where the activated sludge obtained from the previous process is present. The organic matter in the waste water is brought into intimate contact with the activated sludge in presence of adequate amount of oxygen or air. During this aeration process, oxidation of organic matter present is waste water takes place. The oxidation is carried out by micro-organisms which are aggregated in masses about .1mm in diameter. To maintain aerobic conditions, the air or oxygen in pumped into the tank and the mixture is thoroughly agitated. After about 6-8 hours, the waste water, now referred to as 'mixed liquor' is made to flow through a settling tank where the solid, mostly bacterial masses are separated from the liquid. This settled mass, called the sludge, is partly reintroduced in the aeration tank to provide an effective microbial population for a fresh treatment cycle and is therefore termed as activated sludge. Rest of the sludge is digested in a sludge digester, along with the primary sludge obtained from primary treatment. Fig. 30.15.

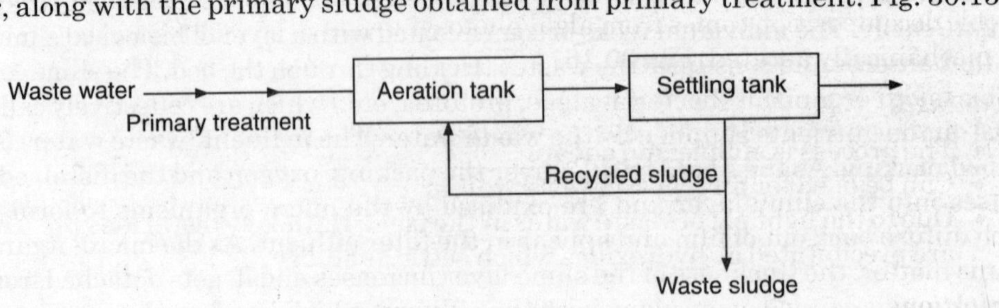

Fig. 30.15 Activated sludge process

Advantages

- The primary advantage of the activated sludge process is, good effluent quality. The effluent after going through activated sludge has little BOD (< 20mg/L)
- It takes less area as compared to trickling water filters.
- The activated sludge process equipment is less expensive.

Limitation

- For the process to be efficient, at least 0.5 ppm oxygen must be present all time.
- The optimum pH range 6.5 to 9.0 has to be maintained throughout.
- Low temperature slows down the rate of metabolism, while high temperature increases the metabolic activity to such an extent that the oxygen is consumed fast, leading to anaerobic conditions.
- The presence of detergents (which are not biodegradable) lead to the formation of foam, making the process difficult.
- The disadvantage of activated sludge process is the production of a huge amount of sludge, which should be digested and disposed off.
- Table 30.11 gives a comparison between trickling filters and activated sludge process for water treatment.

Table 30.11: Comparison of trickling filters with activated sludge systems

Trickling filters	*Activated sludge systems*
1. Bacterial growth is fixed on the media	1. Bacterial growth is suspended as a dispersed floc
2. All solids from the settler are wasted	2. Solids from the settler are partially recycled
3. Less sensitive to shock loading, more stable	3. More sensitive to shock loadings, required closer process control
4. Produce odours	4. Produce spray clouds
5. Less effective in removing disease-causing organisms	5. More effective in removing pathogens than trickling filters
6. Low operating costs	6. High operating costs

(iii) Oxidation Ponds

Oxidation ponds are large shallow ponds, typically 1-2 m deep, where the partially treated waste water is decomposed by micro-organisms. The waste water enter the pond at one and of the treated waste water is collected at the other end. The organic matter is oxidized by the micro-organisms present in the pond (such as pseudmonas, flavo bacterium, etc). The decomposition of the organic matter near the surface is aerobic, whereas near the bottom, it is anaerobic. Such ponds having a mix of aerobic and anaerobic conditions are called facultative ponds. The oxygen required for aerobic decomposition comes from algal photosynthesis. However deeper ponds known as lagoons are mechanically aerated. Fig. 30.16.

Advantages

- The process is simple and cheap.
- Can be used for all types of waste waters
- Due to the light pH of waste water in the pond, the heavy metal ions present in waste water are precipitated as hydroxides which settle as sludge.

Limitations

- The oxidation ponds require larger space.
- Anaerobic conditions may lead to release of bad odours.

The main drawback of the above secondary treatment processes is the formation of sludge. The collection, processing and disposal of sludge can be the most costly and complex aspect of waste water treatment.

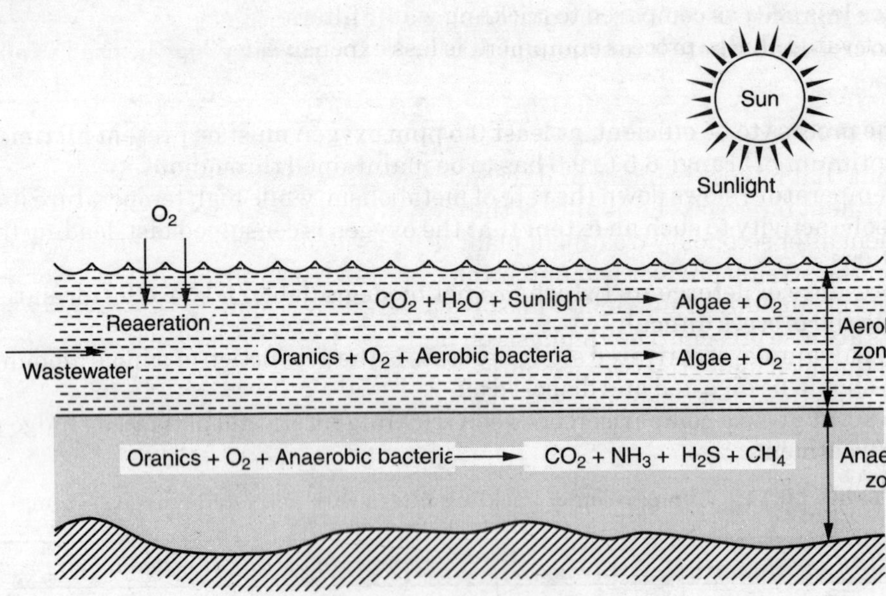

Fig. 30.16 Schematic diagram of an oxidation pond.

Sludge Treatment and Disposal

Sludge is the watery residue from the primary sedimentation tank and humus tank from secondary treatment. The quantity of sludge produced may be as high as 2 per cent of the original volume of waste water, depending upon the treatment process used. One of the primary goals of sludge treatment is to separate as much of the water from the solids as possible. In addition to reducing the water content of the sludge, it may be stabilised so that its biological activity and tendency towards putrefaction is reduced.

The traditional method of sludge digestion is **anaerobic digestion.** It involves the microorganisms that thrive in absence of oxygen. The organic material in the sludge is digested by these micro-organisms under anaerobic conditions to give carbon dioxide and methane gas. The components of the sewage which can be converted into gases are called volatile solids. Though the sludge digestion is complex, it can be summarised in two steps.

- First the complex organics such as fats, proteins, and carbohydrates are converted to organic fatty acids by acid forming bacteria.
- Then these acids are converted into methane, CO_2 etc by methane forming bacteria.

Most treatment plants use a sludge digester in which the sludge is maintained at 35° for 30 days at a pH of 7.0 to 8.0. CH_4, CO_2 and NH_3 are liberated as the end products (Fig. 30.17)

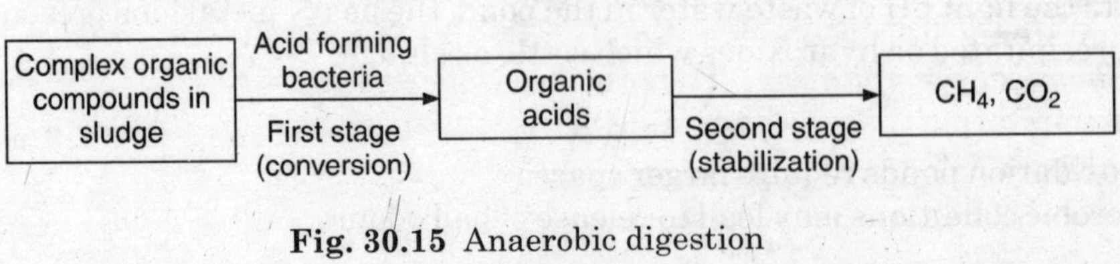

Fig. 30.15 Anaerobic digestion

Digested sludge is removed from the anaerobic digester. This sludge contains 90 to 93% water and is *dewatered.* Dewatering is accomplished by mechanical methods, the most common being centrifugation and filteration, which includes pressure filteration and vacuum filteration. Drying beds are also commonly used. The dewatered sludge is sent for ultimate disposal. Wet sludge is sprayed on to crop land where it functions as fertiliser. Dried sludge may be used as a landfill or a soil conditioner.

Tertiary Treatment

Any treatment that follows primary and secondary treatment is considered tertiary treatment. Tertiary treatment improves the quality of the effluent further.

The effluent after secondary treatment plant still contains suspended solids (20-40 mg/L) which may settle on the stream or river bed and inhibit certain forms of aquatic life. Some amount of BOD, significant amount of nutrients, dissolved solids, traces of organic chemicals and other contaminants are also present. The emphasis on recovery of valuables from industrial wastewaters have created the need for tertiary treatment.

The type of tertiary treatment depends upon the specific goal which include removal of:

- suspended solids
- bacteria
- dissolved organic solids
- toxic substances
- nutrients (phosphorus and nitrogen)

Depending upon the required quality of final effluent a number of techniques can be used. Some of which are described in brief as follows:

(i) **Removal of suspended solids (Microstraining):** This can be achieved by **microstraining.** The filter media consists of finely woven stainless steel fabric. The treated waste is allowed to pass through it. The solids which are retained on the fabric are washed into a trough, which recycles the solids to the sedimentation tank.

(ii) **Removal of dissolved solids:**

(a) *Adsorption:* Dissolved solids can be organics or inorganic. The soluble organics are removed by **adsorption** on activated carbon. Special adsorbents are commercially available for the removal of toxic heavy metals from industrial waste water.

(b) *Solvent Extraction:* It is sometimes used to recover phenolic materials from waste waters of refineries and coke plants. The waste water is intimately brought in contact with a solvent having high affinity for the solute. The layers (aqueous and the solvent layer containing solute).

(c) *Ion Exchange:* Ion exchange is extensively used to remove hardness and iron and manganese salts from drinking water as described in chapter 1. The same technique has been extended to waste water treatment for the removal and recovery of waste water treatment.

(d) *Reverse Osmosis:* When waste water containing dissolved solids is allowed to pass through a semi-permeable membrane at a pressure, which is more than osmotic pressure, the water from the waste passes through the membrane. Hence a highly concentrated solution containing dissolved solids is left behind.

(e) *Chemical precipitating:* The precipitating agents like lime etc remove heavy ions by precipitating these as hydroxides. Precipitating agents include $FeSO_4$, alum and ferric chloride.

(iii) **Removal of Nutrients**

(a) *Nitrogen Removal:* All forms of nitrogen in wastewater are harmful because plants can utilize the inorganic forms as nutrients, NH_3 can be utilized by bacteria resulting in reduced oxygen in water.

One of the method of nitrogen removal is **ammonia stripping**. Ammonia is present in natural water as ammonium in (NH_4^+). This NH_4 is changed to ammonia gas by raising the pH (the $\bar{O}H$ concentration) of the waste water by adding quick lime.

$$NH_4^+ + \bar{O}H \rightarrow NH_3\uparrow + H_2O$$

The ammonia gas is liberated

Another approach of nitrogen removal is nitrification, i.e. to convert NH_4^+ to NO_3^-, followed by anaerobic stage in which micro-organisms convert nitrates to nitrogen gas (N_2)

$$NH_4^+ + 2O_2 \xrightarrow{bacteria} NO_3^- + 2H^+ + H_2O$$

$$2NO_3^- + \text{organic matter} \rightarrow N_2\uparrow + CO_2 + H_2O$$

(b) *Phosphorus Removal (Chemical precipitation):* Phosphorus is present in the form of orthophosphates ($H_2PO_4^-$, HPO_4^{2-} and PO^{3-}) Phosphates are removed by adding coagulants usually alum [$Al_2(SO_4)_3$] or lime [$Ca(OH)_2$].

$$Al_2(SO_4) + 2PO_4^{3-} \rightarrow 2AlPO_4\downarrow + 3SO_4^{2-}$$

(iv) **Removal of bacteria:** Bacteria are removed by retaining the effluents in maturation ponds or lagoons for specified period of times.

30.12 WASTE WATERS FROM SOME TYPICAL INDUSTRIES

1. Textile Industries

Introduction

The textile industry uses vegetable fibers such as cotton; animal fibers such as wool and silk; and a wide range of synthetic materials such as nylon, polyester, and acrylics.

The stages of textile production are fiber production, fiber processing and spinning, yarn preparation, fabric production, bleaching, dyeing and printing, and finishing. Each stage generates wastes that require proper management.

Waste Characteristics

Taste production involves a number of wet processes that may use solvent.

- Emissions of volatile organic compounds (VOCs) mainly arise from textile finishing, drying processes, and solvent use.
- Process wastewater is a major source of pollutants. It is typically alkaline and has high BOD — from 700 to 2,000 milligrams per liter (mg/1) — and high chemical oxygen demand (COD) in the range 2 to 5 times BOD.
- Wastewater also contains solids, oil, and possibly toxic organics, including phenols from dyeing and finishing and halogenated organics from processes such as bleaching.
- Dye wastewaters are frequently highly coloured and may contain heavy metals such as copper and chromium.
- Wool processing may release bacteria and other pathogens.

- Pesticides are sometimes used for the preservation of natural fibers, and these are transferred to wastewater during washing operations.
- Pesticides are used for mothproofing, brominated flame retardants are used for synthetic fabrics, and isocyanates are used for lamination.
- Air emissions include dust, oil mists, acid vapours, odours, and boiler exhausts.
- Cleaning and production changes result in sludges from tanks and spent process chemicals, which may contain toxic organics and metals.

Treatment Technologies

- VOC abatement measures include using scrubbers, employing activated carbon adsorbers, and routing the vapors through a combustion system.
- A common approach to wastewater treatment consists of screening, flow equalisation, and settling to remove suspended solids, followed by biological treatment.
- Physical-chemical treatment is also practiced: careful control of pH, followed by the addition of a coagulant such as alum before settling, can achieve good first-stage treatment. Further treatment to reduce BOD, if required, can be carried out using oxidation ponds or another aerobic process; up to 95% removal of BOD can be achieved.
- Anaerobic treatment systems are not widely used for textile wastes.
- Carbon adsorption is sometimes used to enhance removal. In some case, precipitation and filtration may also be required.
- Residues and sludges often contain toxic organic chemicals and metals. These should be properly managed, with final disposal in an approved, secure landfill.
- Sludges containing halogenated organics and other toxic organics should be effectively treated by, for example, incineration before disposal of the residue in a secure landfill.

2. Tanning and Leather Finishing Industries

Introduction

In the tanning process, animal hides and skins are treated to remove hair and nonstructured proteins and fats, leaving and essentially pure collagen matrix. The hides are then preserved by impregnation with tanning agents. Leather production usually involves three distinct phases: preparation (in the beamhouse); tanning (in the tanyard); and finishing, including dyeing and surface treatment. A wide range of processes and chemicals, including chrome salts, used in the tanning and finishing processes.

The tanning and finishing process generally consists of :

- Soaking and washing to remove salt, restore the moisture content of the hides, and remove any foreign material such as dirt and manure
- Liming to open up the collagen structure by removing interstitial material
- Fleshing to remove excess tissue from the interior of the hide
- Dehairing or dewooling to remove hair or wool by mechanical or chemical means
- Bating and pickling to delime the skins and condition the hides to receive the tanning agents.
- Tanning to stabilise the hide material and impart basic properties of the hides
- Retanning, dyeing, and fat-liquoring to impart special properties to the leather, increase penetration of tanning solution, replenish oils in the hides, and impart color to the leather
- Finishing to attain final product specifications.

Waste Characteristics

- Composite untreated wastewater, amounting to 20—80 cubic meters per metric ton (m^3/t) of hide or skin, is turbid, clouted, and foul smelling. It consist of acidic and alkaline liquors, with chromium levels of 100—400 milligrams per liter (mg/1) sulfide levels of 200—800mg/1 – nitrogen levels of 200—1,000mg/1 – biochemical oxygen demand (BOD) levels of 900—6,000mg/1, usually ranging from 160 to 24,000mg/1 – chemical oxygen demand (COD) ranging from 2,400 to 14,000mg/1 – chloride ranging from to 27,000mg/1 and high levels of fat. Suspended solids are usually half of chloride levels.

- Wastewater may also contain residues of pesticides used to preserve hides during transport, as well as significant levels of pathogens.
- Significant volumes of solid wastes are produced, including trimmings, degraded hide, and hair from the beamhouse processes.
- The solid wastes can represent up to 70% of the wet weigh of the original hides. In addition, large quantities of sludges are generated.
- Decaying organic material produces strong odours. Hydrogen sulfide is released during dehairing, and ammonia is released in deliming.
- Air quality may be further degraded by release of solvent vapours from spray application, decreasing, and finishing (for example, dye application).

Treatment Technologies

- Treatment of tannery wastewaters is always required. Some streams, such as slaking liquor (which has high salinity), sulfide-right lime liquor, and chrome wastewaters should be segregated.
- Preliminary screening of wastewaters is required because of the large quantities of solids present.
- Recovery of hair from the dehairing and liming prices reduces the BOD of the process effluent.
- Physical-chemical treatment precipitates metals and removes a large portion of solids, BOD, and COD.
- Biological treatment is usually required to reduce the remaining organic loads to acceptable levels (0.3kg BOD, 2kg COD, and 0.004 kg chromium per metric ton of raw hide).
- Good ventilation and minimisation of solvent release can avoid the need to collect and treat vapours in carbon adsorption beds, VOC emissions from finishing are approximately 30 kg/t if pollution prevention measures are not adopted.
- Maximum upstream pollutant reduction is essential for tanneries, but treatment is also required.

3. Electroplating Industries

Introduction

Electroplating involves the deposition of a thin protective layer (usually metallic) onto a prepared metal surface, using electrochemical processes. The process involves pretreatment (cleaning, degreasing, and other preparation steps), plating, rinsing, passivating, and drying. The cleaning and pretreatment stages involve a variety of solvents (often chlorinated hydrocarbons, whose use is discouraged) and surface stripping agents, including caustic soda and a range of strong acids, depending on the metal surface to be plated.

In the plating process, the object to be plated is usually used as the cathode in an electrolytic bath. Plating solutions are acid or alkaline and may contain complexion agents such as cyanides.

Waste Characteristics

- Any or all of the substances used in electroplating (such as acidic solutions, toxic metals, solvents, and cyanides) can be found in the wastewater, either via rinsing of the product or from spillage and dumping of prices baths.

- The solvents and vapours from hot plating baths result in elevated levels of volatile metal compounds, which may contain chromates.
- The mixing of cyanide and acidic wastewaters can generate lethal hydrogen cyanide gas, and this must be avoided. The overall wastewater stream is usually high in heavy metals, including cadmium, chrome, lead copper, zinc, and nickel, and in cyanides, fluorides, and oil and grease., all of which are process dependent
- Air emissions may contain toxic organics such as trichloroethylene and trichloroethane.
- Cleaning or changing of process tanks and treatment of wastewaters can generate substantial quantities of wet sludges containing high levels of toxic organics or metals.

Treatment Technologies

- Segregation of waste streams is essential because of the dangerous reactions that can occur. Strong acid and caustic reactions can generate boiling and splashing of corrosive liquids; acids can react with cyanides and generate lethal hydrogen cyanide gas. In addition, segregated streams that are concentrated are easier to treat.
- Exhaust hoods and good ventilation systems protect the working environment, but the exhaust streams should be treated to reduce VOCs and heavy metals to acceptable levels before venting to the atmosphere.
- Acid mists and vapors should be scrubbed with water before venting. In some cases, VOC levels of the vapors are reduced by use of carbon filters, which allow the reuse of solvents, or by combustion (and energy recovery) after scrubbing, adsorption, or other treatment methods.
- Cyanide destruction, flow equalisation and neutralisation, and metals removal are required, as a minimum, for electroplating plants.
- Individual design is necessary to address the characteristics of the specific plant, but there are a number of common treatment steps.
- Cyanide destruction must be carried out upstream of the other treatment processes.
- If hexavalent chrome (Cr^{+6}) occurs in the wastewaters, the wastewater is usually pretreated to reduce the chromium to a trivalent form using reducing agent, such as a sulfide.
- The main treatment processes are equalization, pH adjustment for precipitation, flocculation, and sedimentation/ filtration.
- The optimum pH for metal precipitation is usually in the range 8.5-11, but this depends on the mixture of metals present.
- Flocculation agents are sometimes used to facilitate the filteration of suspend solids.
- Modern wastewater treatment systems use ion exchange, membrane filtration, and evaporation to reduce the release of toxins and the quantity of effluent that needs to be discharged.
- Treatment sludges contain high levels of metals, and these should normally be managed as hazardous waste or sent for metals recovery.
- Electrolytical methods may be used to recover metals. Sludges are usually thickened, dewatered, and stabilized using chemical agents (such as lime) before disposal, which must be in an approved and controlled landfill.

4. Pulp and Paper Industries

Introduction

Pulp and paper are manufactured from raw materials containing cellulose fibers, generally wood, recycled paper, and agricultural residues.

The main steps in pulp and paper manufacturing are raw material preparation, such as wood debarking and chip making; paper manufacturing; and fiber recycling. Pulp mills and paper mills may exist separately or as integrated operations. Manufactured pulp is used as a source of cellulose for fiber manufacture and for conversion into paper or cardboard.

Chemical pulps are made by cooking (digesting) the raw materials, using the kraft (sulfate) and sulfite processes. Kraft processes produce a variety of pulps used mainly for packaging and high-strength papers and board.

Mechanical pulp can be used without bleaching to make printing papers for applications in which low brightness is acceptable-primarily, newsprint. However, for most printing, for copying, and some packaging grades, the pulp has to be bleached.

The soluble organic substances removed from the pulp in bleaching stages that use chlorine or chlorine compounds, as well as the substances removed in the subsequent alkaline stages, are chlorinated. Some of these chlorinated organic substances are toxic; they include dioxins, chlorinated phenols, and many other chemicals.

The finished pulp may be dried for shipment (market pulp) or may be used to manufacture paper on site.

Waste Characteristics

- The significant environmental impacts of the manufacture of pulp and paper result from the pulping and bleaching processes.
- In some processes, sulfur compounds and nitrogen oxides are emitted to the air, and chlorinated and organic compounds, nutrients, and metals are chlorinated to the waste waters.
- In the pulping process, emissions of reduced sulfur compounds, measured as total reduced sulfur (TRS) and including hydrogen sulfide, methyl mercaptan, demethy1 sulfide, and dimethyl disulfide, are emitted, other pollutants are particulate matter, sulfur oxides, nitrogen oxides, and volatile organic compounds (VOCs).
- Waste waters are discharged are high in biochemical oxygen demand (BOD), total suspended solids, chemical oxygen demand (COD), and chlorinated organic compounds, which may include dioxins, furans, and other adsorbable organic halides,
- Phosphorus and nitrogen are also released into wastewater. The main source of nutrients, nitrogen, and phosphorous compounds is raw material such as wood. The use of peroxide, ozone, and other chemicals in bleaching makes it necessary to use a complexing agent for heavy metals such as manganese.

Treatment Technologies

- Sulfur oxide emissions are scrubbed with slightly alkaline solutions. The reduced sulfur-com-pounds gases are collected using headers, hoods, and venting equipment. Condensates from the digester relief condenser and evaporation of black liquor are stripped of reduced sulfur compound. The stripper overhead and noncondensable are incinerated in a lime kiln or a dedicated combustion unit.
- Electrostatic precipitators are used to control the release of particulate matter into the atmosphere.
- Wastewater treatment typically includes (a) neutralisation screening, sedimentation, and floatation/hydrocycloning to remove suspended solids and (b) biological/secondary treatment to reduce the organic content in waste water and destroy toxic organics.
- Chemical precipitation is also used to remove certain cations. Fibers collected in primary treatment should be recovered and recycled.
- Biological treatment systems, such as activated sludge, aerated lagoons, and anaerobic fermentation, can reduce BOD by over 99%.
- Tertiary treatment may be performed to reduce toxicity, suspended solids, and color.
- Solid waste treatment steps include dewatering of sludge and combustion in an incinerator, bark boiler, or fossil-fuel-fired boiler.

5. Food Industries

(i) **Dairy : Sources of Waste Waters :** Spillage, washing of cans, water from butter and cheese producing units.

Characteristics : High organic content, high BOD, Foul odour easily putrescible

Treatment : Equalisation, Skimming, Grease removal, Biological treatment

(ii) **Distilleries :**

Sources of waste waters : Washings, Spillage

Characteristics : Low pH, High organic matter, very high COD High suspended ad dissolved organic matter.

Treatment : Equalization, Biological treatment

(iii) **Cane Sugar Industry :**

Sources : Cane wash water, effluents from evaporators, spray overflow, molasses effluents

Characteristics : Coloured, disagreeable odour, contains oil and grease, high BOD, high carbonaceous matter

Treatment : Biological treatment methods are not effective due to high concentration of carbohydrates.

Lagooning, oxidation ponds have been found to be the most practical treatment which is proceeded by equalisation and anaerobic digestion.

WEATHER

By absorbing infra-red (as well as by its contact with the hot ground), air heats up. As hot air expands, each cubic meter of it weighs less than before heating. Where the heating is most pronounced, the warm air is more buoyant than the cooler air surrounding it, and tends to float upwards; soaring birds and glider pilots seek such "thermal currents" and allow themselves to be carried upwards by them. This buoyancy is the basic process responsible for weather.

Rising air expands, and expansion of a gas cools it down, which is why mountain tops are cooler. Ultimately, a height is reached where not enough air remains on top to stop the IR radiation from escaping to space. The air then cools by radiation and stops rising, producing a relatively stable layer of the atmosphere known as the **stratosphere**. As one goes up in the stratosphere, air actually gets warmer from the absorption of ultra-violet light by ozone, and this extra heating keeps it even more stable.

Just below the boundary of the stratosphere ("tropopause"), air which has cooled is forced down again by warmer air rising from below. The result is a **circulation** of air, rising hot and returning cold, going around again and again, a motion known as **convection**. On a cold winter day such convection also occurs in homes: near poorly insulated windows the air cools and descends while further inside the room it rises again. The region between the ground and the stratosphere where convection and weather take place is known as the **troposphere**.

Sunlight also evaporates water-from the oceans, from lakes and rivers and from green plants. Energy is invested in turning liquid water into vapour, and therefore humid air has **more** energy stored in it than dry air.

The capacity of air to hold water vapour depends strongly on temperature, and is smaller in cold air (just as less sugar can dissolve in cold water). As warm humid air rises, it expands and cools, and since it then cannot hold as much water as before, the excess is forced out: Initially into the tiny droplets of clouds, then if the cooling is more drastic, into raindrops.

The remaining air is drier and **warmer** - warmed by water vapor turning back to liquid and returning energy to its surroundings - and warmer air is better able to radiate its heat into space. That is how water, clouds and rain play a major role in the transport of solar heat from the ground back to space and help create the complex patterns of weather and climate.

QUESTIONS

Short Answer Questions

Section 30.1, 30.2, 30.3

1. What are hydrosphere, lithosphere and biosphere?
2. What are primary and secondary pollutants?
3. What are biodegradable and non-biodegradable pollutions?

Section 30.4

4. What is the compound formed when CO combines with blood?
5. How are NO and NO_2 formed in the atmosphere?
6. What are the reactions involved in removing SO_2 from the atmosphere by passing it through a solution containing citrate ions?
7. What type of aromatic compounds are present as particulates in the air?

Section 30.5

8. What is the composition of pure dry air? What do you mean by the term 'sink' with respect to pollution.
9. What are the main air pollutants? Write about the sources and harmful effects of any one of them.
10. How carbon monoxide acts as a poison for human beings?
11. What remedial steps should be taken to save a person suffering from CO poisoning?
12. What are sources and sinks of nitrogen oxides as air pollutant? What are their effects?
13. How can pollution due to nitrogen oxides be controlled?
14. What are sources and sinks of sulphur oxide pollution.
15. What are particulates? What are their sources?
16. What are the harmful effects of particulate pollutants? List atleast five of them.

Section 30.6, 30.7

17. List the main points of different between 'London smog' and 'Phyotochemical smog'.
18. What is 'Acid Rain'? What are harmful effects?
19. Brief explain 'Electrostatic precipitation' method for controlling particulate pollution.
20. What is 'Greenhouse effect'? How does it affect the global climate?
21. What are the factors responsible for depletion of ozone layer? What are the chemical reactions involved?

Section 30.9

22. How synthetic detergents present as water pollutant create problems?
23. What are oxygen-demanding wastes? Define 'BOD_5'. How is it determined?
24. How plant nutrient and pesticides act as water pollutants?
25. What are thermal pollutants? What are their sources? What damage is done by them?

LONG ANSWER QUESTIONS

1. What are the main components of our environment. Explain each of them briefly.
2. What is Air Pollution? What are the main sources of air pollution.
3. What are major air pollutants. Describe sources of pollution, sinks, harmful effects and methods to control the following pollutants: CO, NO_x, and SO_x.
4. What are the various chemical species and particulates present in the atmosphere?

5. Write informative notes on:
   (a) Photochemical smog
   (b) Green House Effect.
6. What are the major gases responsible for causing green house effect and how they are released into the atmosphere.
7. (a) What are the possible consequences of uncontrolled global warming.
   (b) Write the chemical reactions involved in the depletion of ozone by nitric oxide.
8. List the auto-exhaust gases which contribute in polluting our environment. How could we control such pollution?
9. Mention the sources of particulate matter that pollute air. Explain the method for removal of particulate by Cottrell precipitator.
10. What is photochemical smog? Discuss the mechanism of photochemical smog formation with relevant reactions.
11. Mention the sources of water pollution.
12. Explain the various steps involved in primary treatment of sewage. Discuss the active sludge process.
13. Explain the terms BOD and COD in industrial waste water or distinguish between BOD and COD.
14. What is the difference between aerobic and anaerobic oxidation of sewage?
15. Write notes no: (i) sewage, (ii) primary and secondary air pollutants, (iii) trickling filter process, (iv) biological oxygen demand, (v) activated sludge process.
16. (a) What is BOD? Explain.
    (b) Explain the primary treatment methods adopted for domestic sewage.
17. Name the principal sources of water pollution. Discuss briefly treatment of domestic sewage.
18. Write and explanatory note on "Acid Rain."
19. What is "Greenhouse effect"? Why is it so called? What are its con sequences?
20. How can you classify water pollution into different types? Briefly explain each one of them.

❑❑❑

31 ENVIRONMENTAL POLLUTION-II

"It is difficult to say what is impossible, for the dream of yesterday is the hope of today and the reality of tomorrow."

— Roberi Goddard

CHAPTER OUTLINE

31.1 Solid Waste Pollution
- Classification

31.2 Treatment and Disposal Methods
- Dumping
- Sanitary Landfills
- Incineration
- Compositing
- Recycling and Reuse
- Segregation of Waste

31.3 Soil Pollution
- Composition
- Source of Soil Pollution
- Effects of Soil Pollution
- Control of Soil Pollution

31.4 Hazardous Wastes
- Radiochemical Waste
- Biomedical Waste
- Chemical Waste

31.5 Hazardous Waste Treatment Technologies
- Physical Processes: Sedmen-tation, Aeration, Reverse Osmosis Ion exchange process
- Chemical Processes: Neutralization, Chemical Precipitation, Oxidation and reduction,
- Biological Processes: in Situ Biodegration
- Incineration
- Land Disposal

Landfills, surface impoundments, underground injections.

31.1 SOLID WASTE POLLUTION

Solid waste is that material which arises from various human activities and is normally discarded as useless or unwanted.

It consists of discarded materials from the urban community as well as accumulation of agriculture, industrial, and mining wastes. The quality of solid wastes generated in some major cities of India is very high (300 to 600 g per person per day). Thus crores of rupees are being spent on solid waste management. Moreover, in some cases such as cans, bottles, plastic containers, tyres, etc. are considered cheaper to throw away than to reclaim. This has increased the generation of solid waste many folds.

Classification

Solid wastes can be classified on the basis of content or on moisture basis or on the basis of heating value.

A typical classification is given in Fig. 31.1

(i) *Garbage:* This is the solid waste produced during the preparation or storage of meat, fruit, vegetables, etc. This putrescible waste have a moisture content of about 70% and a heating value of around 6×10^6 J/kg. It may contain valuable minerals and nutrients.

(ii) *Rubbish:* Rubbish is non — putrescible solid waste. It can be either combustible (such as paper, wood, scrap, rubber, etc.) or non combustible (such as metals, glass, ceramics, etc.) The water content of the waste is about 25% and the heating value of the waste is around 15×10^6 J/kg.

(iii) *Pathological Waste:* Dead animals, humans, etc. constitute pathological waste. It has a moisture content of 85% and a heating value of around 2.5×10^6 J/kg. The waste way also include hospital waste which comprises of disposable syringes, swabs, bandages, body fluids etc. This kind of waste is highly infectious.

(iv) *Industrial Waste:* The industrial waste include the waste from industries. The industrial wastes include waste from coal or ore mines, electroplating works, textile industries, paper industries, dairies, chemical industries, oil refineries, etc. Industrial wastes contain inorganic or organic pollutants which may or may not be soluble.

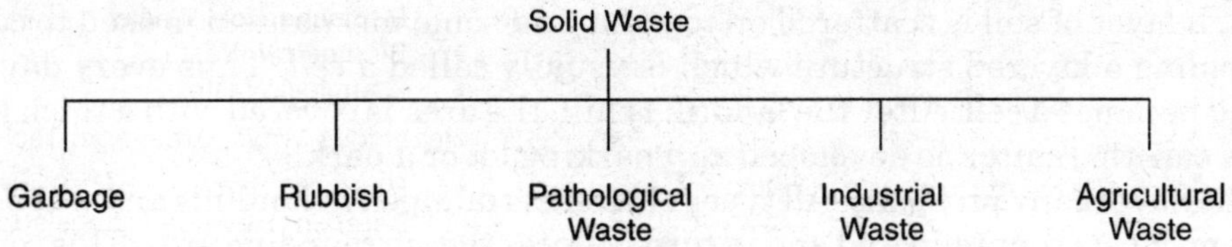

Fig. 31.1

(v) *Agricultural Wastes:* Agricultural wastes include animal manure, crop residues, etc. Herbicides, fungicides, etc. also contribute to agricultural waste

(vi) *Ashes:* These are residues of the combustion of solid fuels.

The principal sources of solid wastes are domestic, commercial and industrial and agricultural activities. Sometimes domestic and commercial wastes are considered together and called *urban wastes*. The composition of the urban wastes remain similar throughout the world but the amount generated, density and the ratio of the constituents vary widely from place to place. Urban wastes amount to between 0.3 to 0.5 kg per person per day.

On the other hand, agricultural wastes amount to around 2 kg per person per day in India. Waste is also referred to as trash, junk, debris, refuse.

31.2 TREATMENT AND DISPOSAL METHODS

The composition of solid waste is not the same everywhere, so the process of selection of the right solid waste disposal method is a complex one. In rural areas, agricultural waste forms the major portion of solid waste, whereas in urban areas, plastic, paper, glass, metal, chemical, food waste etc. are found in large quantities.

Several disposal methods are being used in various parts of the world and the most prominent of these are discussed below:

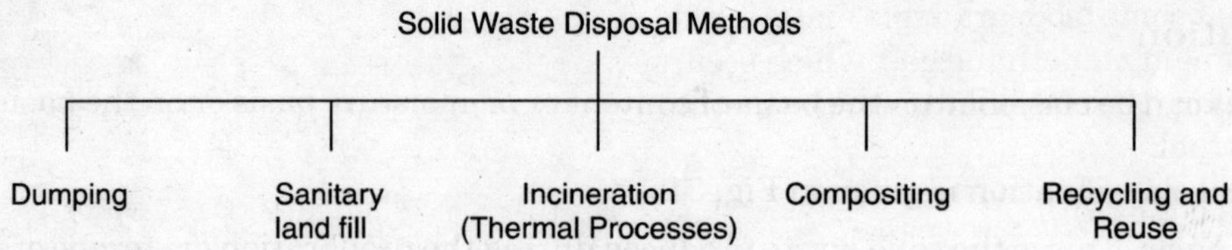

Fig. 31.2

1. Dumping

Open dumping of solid wastes is practized extensively in India because it requires no planning and is cheap. Open dumps refer to uncovered areas that are used to dump solid waste of all kinds. The waste is untreated, uncovered and not segregated. It is the breeding ground for flies, rats, and other insects that breed diseases. The rainwater run-off from these dumps contaminates nearby land and water thereby spreading disease. This system of open dumping in no more used in some countries.

2. Sanitary Landfills

Historically landfills have been the main method used to dispose of solid wastes because of their convenience and low maintenance costs

Landfills are generally located in urban areas where a large amount of waste is generated and has to be dumped in common place. Unlike an open dump, it is pit that is dug in the ground. The garbage is dumped and the pit is covered thus preventing the breeding of flies and rats. At the end of each day, a layer of soil is scattered on top of it and some mechanism is used to compress the garbage forming a layered structure which is usually called a *cell.* Thus every day, garbage is dumped and becomes a cell. After the landfill is full, the area is covered with a thick layer of mud and the site can thereafter be developed as a parking lot or a park.

Landfills have many problems. All types of waste is dumped in landfills and when water seeps through them, it gets contaminated and in turn pollutes the surrounding area. This contamination of groundwater and soil through landfills is known as leaching.

An alternative to landfills which will solve the problem of leaching to some extent is, a **sanitary landfill** which is more hygienic and built in methodical manner. These are lined with materials that are impermeable such as plastics and clay, and are also built over impermeable soil. The lining materials are called liners.

Daily collections of waste are spread evenly across the depression and are then covered by a layer of soil. Drainage systems are set up around and beneath the liner to collect liquids leaching from garbage and waste. Collected liquids are treated to make them more environmentally safe.

The landfill operation is essentially a biological method of waste treatment. The whole process takes place in five phases

- In the first phase aerobic bacteria deplete the available oxygen as a result of aerobic respiration.
- In the second phase anearobic conditions prevail and H_2 and CO_2 are evolved
- In phase three and four methane forming bacteria decompose the matter to methane and CO_2. That is why escape vents for gases are provided in the landfills.
- In phase five the methanogenic activity decreases, representing the depletion of organic matter and ultimately the system returns to aerobic conditions within the land fill.

Constructing sanitary landfills is very costly and they are have their own problems. Some authorities claim that often the plastic liner develops cracks as it reacts with various chemical solvents present is the waste.

The rate of decomposition in sanitary landfills is also extremely variable. This can be due to the fact that less oxygen is available as the garbage is compressed very tightly. It has also been observed that some biodegradable materials do not decompose in landfill. Another major problem is the development of methane gas, which occurs when little oxygen is present, i.e. during anaerobic decomposition. In some countries, the methane being produced from sanitary landfills is tapped and sold as fuel.

Advantages of Landfill

- Infectious diseases are minimised because flies, rats and other pests are unable to breed in covered pits.
- No air pollution from burning
- Fire hazards are minimal.

3. Incineration

The process of burning waste in large furnaces is known as incineration. In these plants the recyclable material is segregated and the rest of the material is burnt. At the end of the process all that is left behind is ash. During the process some of the ash floats out with the hot air. This is called fly ash. Both the fly ash and the ash that is left in the furnace after burning have high concentration of dangerous toxins such as dioxins and heavy metals. Disposing of this ash is a problem. The ash that is buried at the landfills leaches the area and cause severe contamination.

Burning garbage is not a clean process as it produces tonnes of toxic ash and pollutes the air and water. A large amount of the waste that is burnt here can be recovered and recycled. In fact, at present, incineration is kept as the last resort and is used mainly for treating the infectious waste.

To make the process more effective, governments or private operators can implement properly equipped incinerators which can convert water into steam or generate electricity Burning solid waste can generate energy, while reducing the amount of waste by up to 90 per cent in volume and 75 per cent in weight

In addition to heat, the products of incineration include the normal primary products of combustion, (carbon dioxide and water) as well as oxides of sulfur and nitrogen the other gaseous pollutants. Non gaseous products are fly ash and unburnt solid residue.

A variety of pollution control technologies reduce the toxic materials emitted in combustion smoke. Among these are scrubbers (a device that uses a liquid spray to neutralize acid gases in smoke) — and filters, which remove tiny ash particles from the smoke. Burning waste at extremely high temperatures also destroys harmful chemical compounds and disease causing bacteria. Regular testing ensures that residual ash is nonhazardous before being landfilled.

Advantages

1. Useful material and energy can be recovered, if incineration is done in properly equipped incinerators.
2. Varying solid wastes loads are converted into small volumes of ash which can be handled.

4. Composting

Organic matter constitutes 35–40% of the municipal solid waste generated in India. This waste can be recycled by the method of composting, one of the oldest forms of disposal. In contrast to a sanitary landfill, compositing of waste is an *aerobic* method of decomposing. It is the natural process of decomposition of organic waste that yields manure or compost, which is very rich in nutrients. Composting is a biological process in which micro organisms, mainly fungi and bacteria, convert degradable organic waste into humus like substance.

Initially the process starts with oxidization of organic matter in the waste to produce CO and liberate heat. The temperature rises to 45°C and at this point, the thermophilic bacteria take over and continue decomposition. After about three weeks composite is stabilized. The compost should have an earthy smell and a dark brown colour. This finished product, which looks like soil, is high in carbon and nitrogen and is an excellent medium for growing plants.

For an optimum compositing operation, the following conditions should be satisfied:

Temperature	40 to 50°C	Above this biological activity is reduced
pH	4.5 to 9.5	Above this N_2 is lost as NH_3
Moisture	40 – 70%	
Air	0.5 to 0.8 m^3/day/kg of volatile compost	
Carbon: Nitrogen ratio	35 to 50: 1	
Carbon: Phosphorus ratio	100: 1	

Vermi-composting has become very popular in the last few years. In this method, worms are added to the compost. These help to break the waste and the added excreta of the worms makes the compost very rich in nutrients.

The compositing process is generally carried out in three steps:

(i) *Waste Preparation:* The solid waste is placed on slow moving conveyer belts. Materials like corrugated paper are hand picked and iron material is removed by magnetic separation. The waste is grounded to the desired size. A nutrient source is mixed in it along with water to provide 50% moisture.

(ii) *Digestion:* The mixture is subjected to digestion by micro-organisms for 4/6 weeks.

(iii) *Product Upgradation:* Sometimes the resulting compost is upgraded by operations such as curing, grinding, etc. for better market prospects.

Advantages

1. The process of composting ensures the waste that is produced in the kitchens is not carelessly thrown and left to rot.
2. It recycles the nutrients and returns them to the soil as nutrients.
3. Apart from being clean, cheap and safe, composting can significantly reduce the amount of disposable garbage.
4. This organic fertiliser can be used instead of chemical fertilisers and is better specially when used for vegetables. It helps reduce the adverse effects of excessive alkalinity, acidity or the excessive use of chemical fertiliser.
5. It increases the soil's ability to hold water and makes the soil easier to cultivate. It helps the soil to retain more of the plant nutrient over a longer period.
6. It helps to keep soil cool in summer and warm in winter

5. Recycling and Reuse

Solid wastes contain significant amounts of valuable materials like steel, aluminium, copper and other metals which can be recovered or reused. This will reduce the volumes of the wastes to be collected and at the same time would yield significant resale value. Recycling involves the collection of used and discarded materials, *processing* these materials and *making them* into new products. It reduces the amount of waste that is thrown into the community dustbins thereby making the environment cleaner and the air more fresh to breathe.

Surveys carried out by Government and non-government agencies in the country have all recognised the importance of recycling wastes. However, the methodology for safe recycling of waste has not been standardised. Studies have revealed that 7-15% of the waste is recycled. If recycling is done in a proper manner, it will solve the problems of waste or garbage.

The steps involved in the process prior to recycling include:

(a) Collection of waste from doorsteps, commercial places, etc.
(b) Collection of waste from community dumps.
(c) Collection/packing up of waste from final disposal sites.

The four 'R's to be followed for managing the problem of solid waste management are:

1. *Refuse:* Instead of buying new containers from the market, use the ones that are in the house. Refuse to buy new items.
2. *Reuse:* Do not throw away the soft drink cans or the bottles; cover them with homemade paper or paint on them and use them as pencil stands or small vases. Use shopping bags made of cloth or jute, which can be used over and over again.
3. *Recycle:* Segregate the waste to make sure that it is collected and taken for recycling.
4. *Reduce:* Reduce the generation of unnecessary waste, e.g. carry your own shopping bag when you go to the market and put all your purchases directly into it.

Some Items that Can be Recycled or Reused	
Paper	Old copies Old books Paper bags News paper Old greeting cards Cardboard box
Plastic	Containers Bottles Bags Sheets
Glass and ceramics	Bottles Plates Cup Bowls
Miscellaneous	Old cans Utensils Clothes Furniture

Most of the garbage generated in the household can be recycled and reused. Organic kitchen waste such as leftover foodstuff, vegetable peels, and spoilt or dried fruits and vegetables can be recycled by putting them in the compost pits.

The schematic diagram below depicts recycling of wastes.

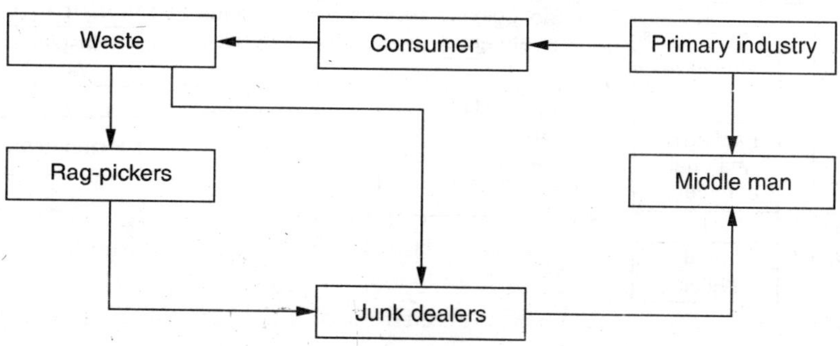

Fig. 31.3

Waste recycling has some significant advantages. It –

- leads to less utilisation of raw materials.
- reduces environmental impacts arising from waste treatment and disposal.
- makes the surroundings cleaner and healthier.
- saves money.
- reduces the amount of energy required to manufacture new products.
- prevent the creation of waste at the source.

Segregation of Waste

Prior to recycling the waste should be segregated . Separating our waste is essential as the amount of waste being generated today causes immense problem. Certain items are not biodegradable but can be reused or recycled as is discussed above. In fact, it is believed that a larger portion can be recycled, a part of it can be converted to compost, and only a smaller portion of it is real waste that has no use and has to be discarded.

Waste can be segregated as:

1. Biodegradable and
2. Nonbiodegradable.

Biodegradable waste: Include organic waste, e.g. kitchen waste vegetable, fruits, flowers leaves from the garden, and paper.

Nonbiodegradable waste: Can be further segregated into

(a) Recyclable waste – plastics, paper, glass, metal, etc.
(b) Toxic waste – old medicines, paints, chemicals, bulbs, spray cans, fertiliser and pesticide containers, batteries, shoe polish.
(c) Soiled – hospital waste such as cloth soiled with blood and other body fluids.

Toxic and soiled waste must be disposed of with utmost care.

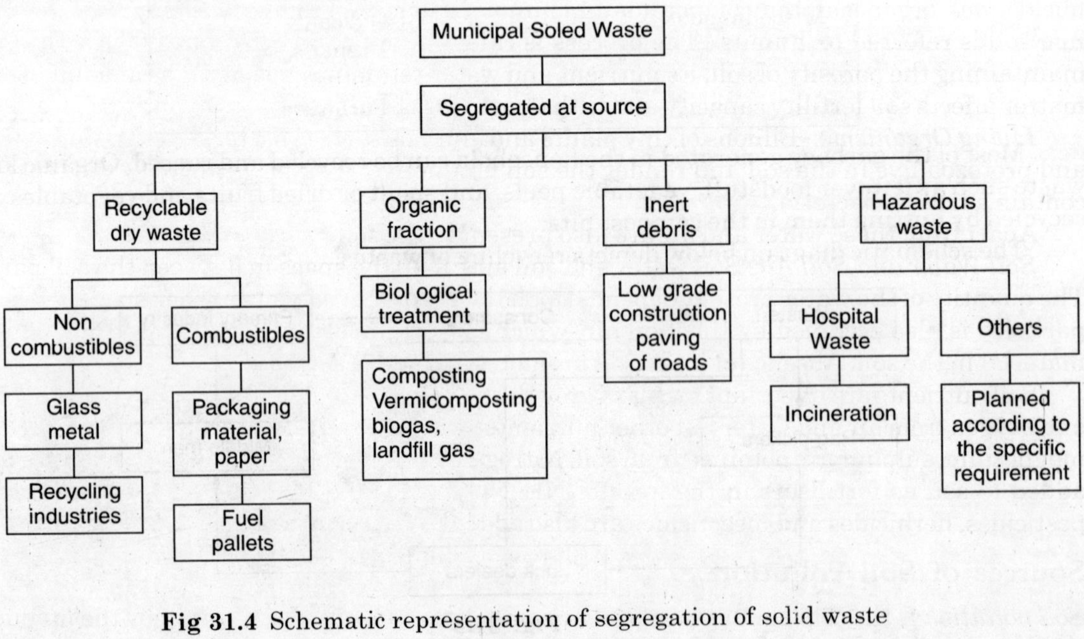

Fig 31.4 Schematic representation of segregation of solid waste

31.3 SOIL POLLUTION

Soil is the thin layer of organic and *inorganic* materials that covers the earth's rocky surface. It plays a vital role in the production of food for the survival of life on earth. Soil can be defined as the weathered layer of the earth's crust.

Composition

Soil comprises of a mixture of organic and inorganic matter including water and gases in variable proportions. Depending upon the particle size it can be called:

Soil	*Particle size (diameter in mm)*
• Stone and gravel	2 and above
• Coarse sand	0.2 – 2.0
• Fine sand	0.02 – 0.2
• Silt	0.002 – 0.0
• Clay	less than 0.002

Inorganic matter: The inorganic portion is made up of rock fragments and was formed over thousands of year by physical and chemical weathering of rocks. The inorganic mineral constituents of soil include complex mixture of the following inorganic compounds.

- Silicates of Na, K, Ca, Al and Fe
- Oxides of Fe, Mn and Ti
- Carbonates of Ca and Mg

Primary minerals come from rocks such as quartz (SiO_2), feldspars and micas. Secondary formed from primary minerals by weathering. Clay are essentially hydrates, and silicates of aluminium and iron. The clay particles bind Na^+, K^+, Ca^{2+}, and Mg^{2+} which serve as plant nutrients.

Organic Matter: Organic matter constitute 50% of soil, which determines the productivity of soil. The organic matter consists of biologically active compounds such as polysaccharides, nucleotides, organosulphur compounds and humus. Partially decomposed organic matter present in a soil is referred to 'humus'. The process is called humification. Humus plays a vital role in maintaining the porosity of soil, its nutrient and water retaining capacity. The amount of organic matter affects soil fertility capacity and tilth (looseness of soil)

Living Organisms: Billions of tiny plants and animals such as bacteria, fungi, worms, insects and protozoa live in the soil and render the soil productive. A single gram of fertile topsoil may contain a trillion bacteria.

Apart from these water and air are also present in the soil.

Soil Water and Soil Air: Soil water and soil air fill up the space in between the soil particles. The quantity of the water present depends upon the pH of the soil water. Some space between soil particles is also occupied by the 'Soil air' which is very important for the oxidation of organic material in the soil. Absence of air in soil promotes anaerobic activity.

Soil nutrients are important for plant growth and development. Plants obtain carbon, hydrogen and oxygen from air and water but other nutrients like nitrogen, phosphorus, potassium, calcium, magnesium, sulphur are obtained from soil. Nitrogen, Phosphorus, Potassium, (N,P,K) are usually added to soil as fertilisers in the form of NH_4NO_3, P_2O_5 and K_2O etc. Apart from fertilisers, pesticides, herbicides and insecticides are also added to the soil to kill unwanted organisms

Sources of Soil Pollution

Soil pollution is the contamination of soil by some chemical substances, whereby the productivity and fertility of soil decreases. It the degradation of earth's land through human misuse.

The human influences on soil include:

- Industrial Waste
- Disposal of Soil Wastes in urban areas.
- Agricultural practices
- Biological Agents
- Soil Erosion

Industrial Wastes: Disposal of industrial waste is the main cause of soil pollution. These discharges are from pulp and paper mills, chemical , steel, coal and mining industries, etc. The pollutants added affect and alter the chemical and biological properties of soil. Industrial sludges containing calcium salts and several toxic volatile elements such as arsenic, selenium, mercury, etc. pose a detrimental threat to environment. Moreover fallout of gaseous and particulate air pollutants from mining and smelting industry also add to soil pollution.

Soil pollution by urban wastes: Accumulation of lead particulates from automobile exhausts, garbage containing plastic, glasses, metallic cans, fibers, papers etc constitute soil pollution by urban wastes.

Agricultural Practices: Excessive use of fertilisers such as nitrates and phosphates may lead to excessive release of these nutrients in soils which may leach out nearby water body and may cause water pollution.

Use of fertilisers, pesticides, herbicides to increase crop yield may also pose soil pollution when used excessively. For example, soil become acidic, when $(NH_4) SO_4$ is used as a fertiliser excessively, because $NH^+{}_4$ is absorbed by the plants and $SO^-{}_4$ remains in the soil making it acidic. Similarly, use of KNO_3 make the soil alkaline as $NO^-{}_3$ ions are used by plants and K^+ remains in the soil. This alkaline soil becomes unfit for crops.

Pesticides used to kill the unwanted species such as DDT, BHC, aldrin, heptachlor, etc. are absorbed by the soil particles and contaminate root crops grown in soil.

Animal wastes (cow dung, etc.) contain several pathogenic bacteria and viruses which enter into plant metabolism and ultimately to man. Though cow dung is richest source of nitrogen still when dumped into heaps, it becomes the breeding ground for insects.

Biological Agents: Soil gets large quantities of human, animals and birds excreta which constitute the major source of land pollution by biological agents. In addition to these excreta, faulty sanitation, municipal garbage waste water also add to soil pollution.

Soil Erosion: Soil erosion is when the topsoil of the land is stripped of essential nutrients needed for plant and vegetative growth. When the soil erodes, the rich top layer called humus is removed. The result of soil erosion over a long period of time includes, the decaying of vegetation and negative microbial effects. This man-make disaster leaves large areas of land destroyed and unfit for regrowth for many years, and thus, thousands of acres of land are wasted each year.

Coal based Thermal plants: Disposal of large quantities of fly ash and bottom ash from coal based thermal plants also contribute to the soil pollution. It is estimated that two lakh tons of ash is being deposited per day by thermal plants in our country. This renders large areas of agriculture unfit for cultivation. Moreover the toxic metals also leach out of the soil and pollute ground waters.

Effect of Soil Pollution

Soil nutrients are essential for plant growth and development which inturn have a great impact on human life.

The prosperity of the nation depends upon the quality of its soil. However, soil pollution due to various sources listed above cause detrimental effects on environment and human life. Some of the effects of soil pollution are discussed below:

1. Soil pollution through industrial wastes adds a number of chemicals to the soil which are extremely toxic to living beings. This toxic metals are transferred to human beings in their food chain through plants and cause undesirable effects.
2. Metallic contaminants (Hg, Pb, Zn, As, Cr, K, Na, Cu, etc) destroy the beneficial microorganisms in the soil. These metals act as indestructible poisons to the soil and are highly fatal to living organisms in the soil.
3. Excessive use of fertilisers renders the soil alkaline or acidic, depending upon the fertiliser used. The acidic or alkaline soil is unfit for the growth of many crops. An estimate shows that soil fertility between 30 – 80% of world's irrigated land has been degraded due to increased salinity. Nitrates and phosphates added as fertilizers may be leached out of soil by run off waters and pollute the nearby surface water or ground water sources. High concentration of nitrates is objectionable in drinking water because they reduce to nitrites and cause a disease methenoglobinemia.
4. Use of pesticides contaminate the soil and are persistent pollutants. They enter the food chain and pose serious health hazards. The degradation products of these pesticides is more dangerous than their parent compounds and some of there are carcinogenic. These pesticides residues in food cause hazards to human health. Some arsenic pesticides if used indiscriminately, may render the soil permanently infertile. The most threatening pesticide DDT concentrates and accumulates in food chain. Polychlorinated biphenyls (PCBs) are more dangerous having half life of 25 years in soil. PBCs when enter in human bodies cause deformities in foetus, nervous disorders, liver and stomach cancer in animals.

Control of Soil Pollution

Control measures of soil pollution include primarily the methods for the solid wastes to be reduced and to dispose it safely. Some of the control measures are:

(i) Proper dumping of solid waste: Dumping of solid waste in open dumps should be discouraged. Instead, land fill should be the possible way of disposal.

(ii) Highly loxic and persistent synthetic chemical pesticides should be banned. Biopesticides should be used instead of toxic chemical pesticides.

(iii) Proper awareness: Anti litter campaigns should be carried on to educate people. People should be trained regarding sanitary habits.

(iv) Crop rotation should be encouraged to revive the depleted nutrients.

(v) High concentration of lead and cadmium can be controlled by growing a special type of grass, the vetiver grass. The contaminants concentrates in its roots and reduce the concentration of lead and cadmium in soil by 38 - 60% and 35 - 40% respectively.

(vi) Recycling of waste such as ferrous materials, metal, paper, glass, plastics is another method of reducing the amount of pollutants added to soil.

(vii) Plantation of trees can help to check soil erosion. Deterrent measures against deforestation should be implemented.

31.4 HAZARDOUS WASTES

Although the legal definition of Hazardous Waste is complex, the term generally refers to the waste that could pose a threat to human health and the environment, if managed improperly. In other words, hazardous waste can be defined as *any waste which because of its quantity, concentration, or physical, chemical or infectious characteristics, may cause significant hazard to human health and environment when improperly treated, stored, transported or disposed of.* Hazardous waste will significantly contribute to:

- An increases in mortality
- An increase in serious, irreversible or in capacitating reversible illness
- Pose a substantial potential hazard to human health.

More specifically, a substance is defined as hazardous if it posses any of the following characteristic attributes:

Reactivity: Any waste which is not stable at normal conditions and hence can cause explosive reactions or liberate toxic fumes, gases and vapours, reacts violently when mixed with water. Example of reactive waste include, sodium and potassium metal, dry picric acid, compounds that form explosive peroxides, cyanide plating operations, etc.

Ignitability: Wastes which are easily ignited and burn vigorously at or below 60°C. Examples include volatile liquids, such as solvents (acetone, toluene, methanol, ethers), rubber, cement glue, paint thinners, etc.)

Corrosivity: Wastes which include liquids with pH less than 2 or greater than 12.5, and those that are capable of corroding metal containers. These include H_2SO_4, HCl, NaOH, OH, lime, battery acid, drain openers and other products that contain strong acids or bases.

Toxicity: Toxic substances are harmful or fatal when ingested or absorbed. The wastes which release toxic materials on leaching in excess of the permissible concentration pose a substantial hazard to human health and are termed as toxic. These are measured by Toxicity Characteristics Leaching Procedure (TCLP). Examples are, products that contain benzene (many petroleum based products), cadmium (nickel cadmium batteries), lead (lead batteries, paints) silver (spent photofixer, silver mercury), chromium, mercury (mercury batteries), rat poison and antifreeze, etc.

Table 31.1 gives maximum concentration of contaminants for environment Protection Agency (EPA) test

Maximum Concentration of Contaminants For EPA Toxicity Test

EPA hazardous waste number	Contaminant	Maximum concentration (mg/L)
D004	Arsenic	5.0
D005	Barium	100
D006	Cadmium	1.0
D007	Chromium	5.0
D008	Lead	5.0
D009	Mercury	0.2
D010	Selenium	1.0
D011	Silver	5.0
D012	Endrin	0.02
D013	Lindane	0.4
D014	Methoxychlor	10.0
D015	Toxaphene	0.5
D016	2, 4-D (2, 4-dichloropheno-xyacetic acid)	10.5
D017	2, 4, 5-TP Silver (2, 4, 5-trichl-orophenoxypropionic acid)	1.0

Toxicity can be acute toxicity (one massive dose kills you) or chronic toxicity (long exposure to small dose resulting in chronic disease).

While the four characteristic attributes given above are sufficient to consider a substance to be hazardous, EPA has given a list of specific hazardous wastes. The hazardous wastes have been classified into three categories as given in the schematic representation (Fig. 31.4)

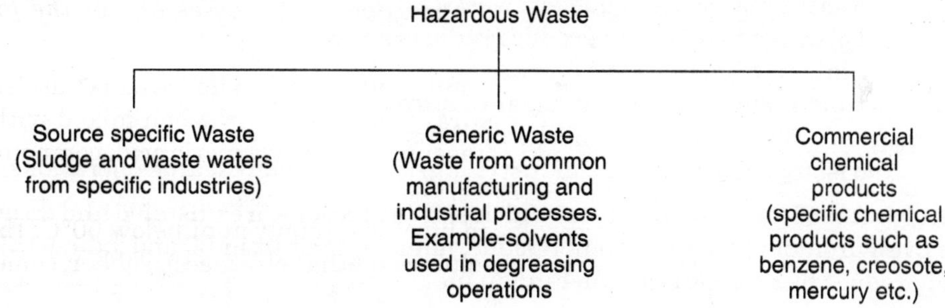

Fig. 31.5 Schematic representation of classification of hazardous waste by EPA

Hazardous Waste is generated through various industrial processes and can be named accordingly. The prominent among them include:

(i) Radiochemical Wastes
(ii) Biomedical Wastes
(iii) Chemical Wastes

A brief description of these is given below:

1. Radiochemical Waste

Radioactivity arises naturally from the decay of particular forms of some elements, called isotopes. Some isotopes are radioactive, most are not, though in this discussion we concentrate on the former.

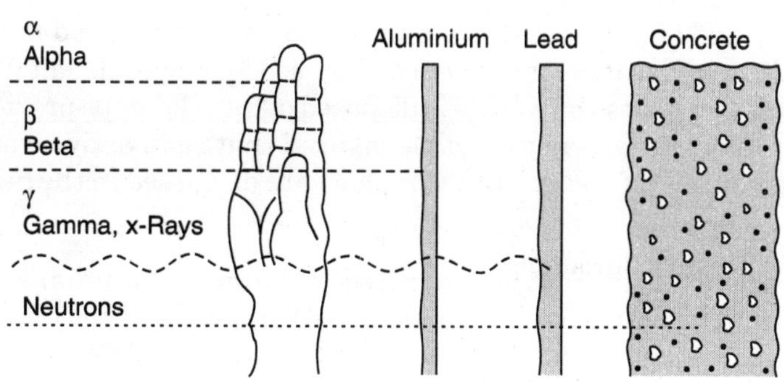

Fig. 31.6

There are three kinds of radiation to consider: **alpha, beta and gamma**. A fourth kind, neutron radiation, generally only occurs inside a nuclear reactor (31.6).

Different types of radiation require different forms of protection:

- Alpha radiation cannot penetrate the skin and can be blocked out by a sheet of paper, but is dangerous in the lung.
- Beta radiation can penetrate into the body but can be blocked out by a sheet of aluminium foil.
- Gamma radiation can penetrate into the body, aluminium foil, but can be blocked by lead.

All of these kinds of radiation are, at low levels, naturally part of our environment. Any of them may be present in any classification on waste.

Source: The radioactive waste is generated in:

(i) Discharges from nuclear reactions contain a number of radio-nuclides. These nuclides emit radiations harmful to living organisms.
(ii) Use of radioactive materials in nuclear weapons
(iii) Mining and processing of ores to produce radio isotopes.
(iv) Radioactive fallout from nuclear bombs.
(v) Emissions from industrial use of nuclear energy
(vi) Leakage from underground nuclear detonations.
(vii) Use of radio isotopes in medicine, industry, agriculture and research operations.

Fortunately most of the radio-nuclides, which emit radiations are unstable and do not persist for long time in environment. But two radionuclide, namely, strontium-90 and iodine-137 are slow decaying and produce hazardous effect on human life.

Effects: Radiation damages the living organisms. The damage can be classified as somatic damage or genetic damage

Somatic damage: It is the damage to the organism itself, resulting in sickness and death. The effects may appear almost immediately if a massive dose of radiation is received. For smaller doses, damage, may appear years later, usually in the form of cancer.

Genetic Damage: It is the damage which passes from one generation to another. The radiations affect the chromosomes causing mutations in the genes.

Method of Disposel of Radioactive Wastes

The radioactive wastes must be disposed of in such a manner that they create no hazard to the living organism. The wastes can be disposed of in ground, air or ocean.

Disposal in ground: Soil absorbs the radioactive materials easily that is why disposal in ground is an easy method for disposal of radioactive wastes. The radioactive waste containing calcium and strontium is first separated and stored in the tanks, which are buried in ground. This is done to allow the radioactivity of the material to decay so that it is much easier to handle. The burial of the radioactive waste is around 500 metres deep. The wastes are kept buried for about 13 years and finally disposed off in the sea. Vacated coal mines can be used for disposal. Salt is a powerful absorber of radioactive radiations and hence the wastes is disposed off in salt heaps provided in mines. Before disposing the buried tanks in the sea, it must be ensured that radioactive waste should be below harmful levels. Further details of treatment technology are discussed in the next section.

2. Biomedical Waste

"Biomedical waste" means waste, that is generated by human or animal health care facilities, medical or veterinary and teaching establishments, health care teaching establishments, clinical testing or research laboratories or facilities involved in the production or testing of vaccines. Biomedical wastes include:

(i) Human anatomical waste, that consists of tissues, organs or body parts, but does not include teeth, hair and nails.
(ii) Animal waste, that consists of tissues, organs, body parts, fluid blood and blood products, items saturated or dripping with blood, body fluids contaminated with blood, and body fluids removed for diagnosis or removed during surgery.
(iii) Microbiology laboratory waste, that consists of laboratory cultures, stocks for specimens of microorganisms, live or attenuated vaccines, human or animal cell cultures used in research
(iv) Waste sharps, that are clinical and laboratory materials consisting of needles, syringes, blades or laboratory glass capable of causing punctures or cuts.

3. Chemical Wastes

Chemical wastes include the effluents from various industries, solvents, etc. and have been discussed earlier.

31.5 HAZARDOUS WASTE TREATMENT TECHNOLOGIES

In the past the disposal of hazardous waste was most often on Land. But RCRA (Resource Conservation and Recovery Act) emphasize the development and use of alternative and innovative treatment technologies that result in permanent destruction of wastes or a reduction in toxicity.

The following treatment technologies are available for the treatment of hazardous wastes:

- Physical Processes
- Chemical Processes
- Biological Processes

1. Physical Processes

Physical treatment processes include gravity separation, phase change systems, (air and steam coming out of votaliles from liquid wastes) and various filtering operations such as carbon adsorption, reverse osmosis, etc.

Some of the methods have been discussed below:

(i) *Sedimentation:* This is the simplest physical treatment system that separate solids from liquids. It involves gravity settling and natural floating. The solids in the hazardous wastes are made to settle in the specially designed sedimentation tanks, where they are collected as sludge. Some solids naturally float to the surface and are removed by skimming device. The floatation is encouraged by introducing finely divided bubbles into the waste stream. When these bubble rise they collect the particles and the combination is skimmed out.

(ii) *Adsorption:* Dissolved hazardous substances that would never settle out can be removed by this technique of adsorption. The chemicals are adsorbed onto the adsorbent. The most commonly used adsorbent is granular activated carbon (GAC), which has large surface area of about 1000 m^2/g. A single handful of GAC has an internal surface area of about one acre.

The process consists of a series of large vessels partially filled with the adsorbent. Contaminated water enters the top of each vessel, trickles down the adsorbent and is released at the bottom. After sometime the GAC filter gets clogged by contaminants and is regenerated, which is quite expensive process. The regeneration is usually done by burning the contaminants from the surface of the adsorbent granules or using a solvent. Carbon filters which cannot be regenerated due to their contaminant composition must be disposed off.

(iii) *Aeration:* This process typically use the process of aeration to drive the contaminants out of solution. The air or steam is used in the process. Contaminated wastes is sprayed downward through a packing material in a tower, through which air is blown upward carrying away the chemicals that are relatively volatile. This process can remove 95% of VOCs (volatile organic compounds) such as trichloro ethylene, tetrachloro ethylene, benzene, toluene, etc.

Sometimes the adsorption is combined with aeration and many volatile and nonvolatile compounds are removed from water to non detectable levels. First the waste water is passed through the air stripper which removes volatile substances before the waste water is subjected to adsorption process. This extends the life of carbon because no clogging with volatile compounds take place.

(iv) *Reverse Osmosis:* The contaminated water is forced against a semi-permeable membrane, which acts as a filter. The water passes through the pores, and hazardous chemical substances are left behind.

(v) *Ion exchange process:* This is similar to the process, discussed under the chapter 'Water' for removing hardness causing ions. Special exchange resins are used to remove toxic substances from waste water.

(vi) *Electrodialysis:* This technique which is also discussed in detail in chapter 'Water' for purifying brackish water, is also used to remove hazardous substances. Metal salts from electroplating rinses are removed this way.

2. Chemical Processes

In chemical treatment methods the hazardous waste is converted into less hazardous form and sometimes useful products are also formed. The various chemical techniques are:

(i) *Neutralisation:* As per the definition of hazardous material, one of the characteristic is that hazardous waste comprises of corrosive material having pH less than 2 or more that 12.5. Neutralisation converts these materials to less hazardous by changing their pH.
Acidic waste waters are neutralised with [$Ca(OH)_2$]. The waste water is continuously stirred with $Ca(OH)_2$ in a chemical reactor. The rate of addition of lime is controlled with a feedback control system that monitors pH and adjust the feed.
Alkaline waste waters are neutralised by adding acid directly or by bubbling CO_2. Simultaneous alkaline and acidic neutralisation can be accomplished in the same vessel as shown in Fig. 31.7

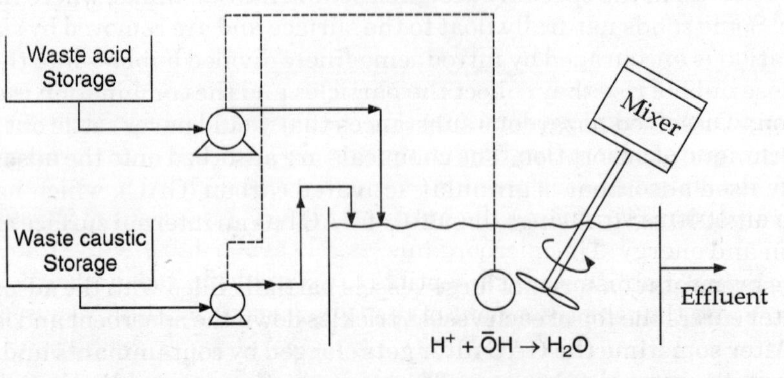

Fig. 31.7 Simultaneous neutralisation of acid and base

(ii) *Chemical Precipitation*: This is a common method for removing heavy metals from a liquid waste. This method is again pH dependent. The pH of waste waters is so adjusted that the solubility of toxic metals is decreased, leading to the formation of a precipitate that can be removed by settling and filteration. For example, the ion M^{2+} can be removed by converting it into its insoluble hydroxide, using lime

$$M^{2+} + Ca(OH)_2 \longrightarrow M(OH)_2 + Ca^{2+}$$

The metal hydroxides in general are relatively insoluble in basic solutions. Each metal has its own pH at which its solubility its minimum. If the waste contains several metals then it becomes difficult to control the precipitation of a mix of different metals in the same waste. In such cases more than one stage of precipitation is required to allow different values of pH to control the removal of different hazardous substances.

If the metals are present in very low concentrations in waste water then the metals are precipitated as sulphides. Metal sulphides have lower solubilities than metal hydroxides and hence can be precipitated out. The only disadvantage of precipitating metals as sulphides is the potential formation of toxic H_2S gas.

(iii) *Oxidation – Reduction Process:* Certain hazardous substances have variable oxidation states. All the oxidation states do not have the same toxicity. Therefore, the substances can be converted to the oxidation state which is less toxic by redox reactions One of the most important redox treatment process is the reduction of hexavalent chromium (Cr^{6+}) to trivalent (Cr^{3+}). The trivalent chromium is less toxic and can be precipitated easily as compared to Cr^{6+}. In this process SO_2 is often used as reducing agent.

$$3SO_2 + 3H_2O \longrightarrow 3H_2SO_3$$

$$2CrO_3 + 3H_2SO_3 \longrightarrow Cr_2(SO_4)_3 + 3H_2O$$

Similarly, cyanide wastes are treated using redox process. These hazardous substances which are common in effluents from metal finishing industry, are converted to less toxic cyanate using alkaline chlorination.

$$NaCN + Cl_2 + NaOH \longrightarrow NaCNO + NaCl + H_2O$$

If chlorination is continued further, the cyanate ion oxidizes to CO_2 and N_2 resulting in complete destruction of cyanide

$$2NaCNO + 3Cl_2 + 4NaOH \longrightarrow 2CO_2 + N_2 + 6NaCl + 2H_2O$$

The wastes that can be treated using redox *oxidation* include the waters which contain benzene, toluene, phenols, cyanide, arsenic ions, iron, and manganese. The wastes which can be treated using redox *reduction* include the ones containing chromium (IV), mercury, lead, silver, and chlorinated organics.

3. Biological Processes

Mostly the wastewater treatment plants rely on biological processes to decompose the organic water. As was discussed earlier in the chapter, the biological treatment systems use micro-organisms (bacteria, fungi, etc.) to oxidise organic matter to CO_2 and H_2O.

Now, living organisms survive on the nutrients such as nitrogen and phosphorus as well as a source of carbon and energy. The microorganisms get carbon and energy from the organics they consume. These living organisms are susceptible to toxic substances in hazardous wastes present in organics water and that the treatment of these hazardous wastes by these organisms seems unlikely.

But, surprisingly, most hazardous organics can be treated with these micro-organisms. This is because, for a given organic hazardous substance, there may be one type of organisms which find it toxic whereas the other type may find the same substance to be an acceptable food, thus decomposing it. Moreover micro-organisms are sensitive to concentration of these substances. These many flourish in one concentration and may die when concentration is increased beyond a critical level.

The following biological processes are common for the treatment of hazardous wastes. It may be emphasized here that *biological treatment stage utilises processes already described for municipal waste water treatment plants.*

In situ biodegradation:

In, in situ biodegradation the bacteria are used to degrade organic compounds in the soil and ground water on site itself. The advantage of in situ biodegradation is that soil and ground water do not have to be removed and hence there is less risk associated with hazardous waste transportation.

In situ biodegradation can be carried out by using two approaches:

- In one approach the population of already existing micro-organisms is increased by supplying necessary nutrients such as ammonium sulphate, magnesium sulphate, sodium carbonate etc, to the contaminated waste. Oxygen can be supplied by injecting an oxidant such as H_2O_2 or by forcing air.
- In another approach, the underground population of micro-organism is altered by seeding with new micro-organisms, which have proven to be effective to decompose the hazardous wastes under consideration, based on laboratory studies.

In situ biodegradation have been used to treat wastes containing gasoline and diesel. If also can be used to treat chlorinated solvents such as trichloroethylene (TCE), tetrachloroethylene (PCE) and 1, 2 dichloroethylene (DCE), etc.

4. Incineration

Incineration was discussed in the section 'sewage disposal'. The same technology is particularly effective with organic containing hazardous substances. Carcinogens (cancer causing), mutagens (causing mutation in genes), tetragens (causing birth defects) can all be completely detoxified in a properly operated incinerator. However, inorganic compounds our not destroyed in incinerators and are only concentrated in ash.

Incinerators are however have a disadvantage that the substance which volatilize at temperatures below 2000°F are added in the air and cause air pollution. They are difficult to remove using conventional air pollution control equipment.

The measure of incinerator's performance is known as the destruction and removal efficiency DRE.; A DRE of 99.99%, for example, means that one moleule of an organic compound is released to the air for every 10,000 moleules entering the incinerator.

A number of hazardous waste incinerators are there, but the two most commonly used are:

— liquid injection incinerator and
— rotary kiln incinerator.

Liquid injection Incinerator: In liquid injection incinerators the waste containing gases, liquids and slurries are pumped through an atomizing nozzle. The nozzle emits tiny droplets of waste that are mixed with air and fuel gas. The resulting mixture is burned at high temperature.

Rotary kiln Incinerator: In rotary kiln incinerator all sorts of wastes can be treated. Wastes and fuel gas are introduced in the rotary kiln from one end and combustion takes place while the cylinder slowly rotates. Rotation helps to increase turbulence which improves the combustion efficiencey.

The emissions from these incinerators include unburned organic compounds, products of incomplete combustions, CO_2, N_2, SO_X, and particulates. The unburned ash, which may contain hazardous substance, should be treated with care.

5. Land Disposal

The purpose of a hazardous waste landfill is to store the waste which are not yet recoverable. Land disposal techniques include landfills, surface impoundments, underground injection wells and waste piles.

Conventionally, land disposal methods were used to dispose off hazardous wastes because this was the convenient and inexpensive method. Unfortunately, the poorly monitored disposal sites have resulted in tragic incidences. Therefore RCRA (Resources Conservation and Recovery Act) has put ban on some of the hazardous substances from land disposal. A few of the restrictions are:

- Banning liquids from landfills
- Banning underground disposal of hazardous waste within 1/4 mile of a drinking water well.
- Requiring more stringent structural designs for landfills, impounds
- Requiring corrective action, if hazardous substance leaks from a facility
- Requiring disposable facilities to be constructed only in suitable hydrogeolic settings.

Some of the Land disposal Techniques used are:

(i) *Landfills:* The sanitary landfill has already been discussed in the section 'sewage disposal'. The similar type of treatment is applicable to hazardous water treatment. However, for disposing of hazardous waste more stringent RCRA requirements are there. The hazardous waste landfill is designed as a modular series of three-dimensional control cells. The wastes are segregated, and only compatible wastes are disposed of together. After dumping the wastes in the appropriate cells, these cells are covered at the end of each working day with a layer of soil.

A double-liner system is placed beneath the hazardous wastes to stop the leakage of liquids form the waste. The leakage is often referred to as leachate. The leachate may enter the soil and ground water beneath the site. The double-liner system prevents this. The upper liner of this double liner must be a flexible-membrane lining and is made of sheets of plastic or rubber. The plastics used are polyvinyl chloride (PVC), high density polyethylene (HDPE). The rubbers used for the purpose are chlorosulphonated polyethylene (CSPE) and ethylene propylene diene monomer (EPDM). The lower lining can be recompacted clay having a thickness of atleast 3 feet. Flexible plastic membranes can also be used in lower linings. The leachate that collects over each double liner is collected in a series of perforated drainage pipes and pumped to the surface for further treatment. A low permeability cap is placed over completed cell to minimise the leakage of leachate the landfill must also include monitoring facilities to check the possibility of any contamination from the site (Fig. 31.8).

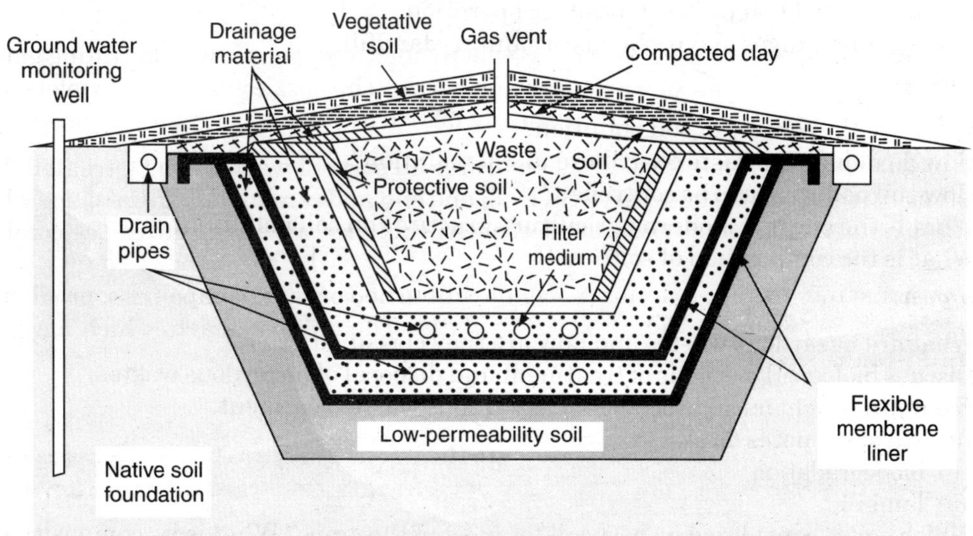

Fig. 31.8 Schematic of a hazardous waste landfill.

(ii) *Surface Impoundments:* Surface impoundments are excavated areas used to store liquid hazardous wastes. Usually storage is temporary unless the impoundment has been designed to eventually be closed as a landfill. In these impoundments the waste remains accessible for further treatment during storage. The treatments include neutralisation, precipitation, settling and biodegradation.
Conventional surface impoundments have typically been poorly constructed and monitored. More than half were located over thin or very permeable soils that would allow easy transport of leachate to groundwater. Recent regulations require new surface impoundments with two or more liners, leachate collection system and a monitoring system, similar to those required in landfills.

(iii) *Underground Injections:* Underground injection is to force the waste underground through deep injection walls. This is the most popular way of disposing hazardous waste. Typical injection depths are more than 700 m below the surface. However, the main concern with underground injection is the potential for contaminating underground drinking water supplies. Thus injection wells are required to extend below the lowest formation containing underground sources of drinking water.

QUESTIONS

Section 31.1, 31.2

1. How are solid wastes classified?
2. Why dumping is disadvantageous?
3. What are advantages of Sanitary Landfills?
4. What do you mean by incineration?
5. What are the advantages of incineration?
6. What do you understand by compositing?
7. Give an appraisal on recycling and reuse of solid wastes.
8. Discuss different steps in compositing operation.
9. Discuss advantages and problems of sanitary landfill.

Section 31.3

10. What are the sources of soil pollution?
11. How do industrial waste contribute to soil pollution?
12. How soil pollution can be controlled?
13. What is the effect of modern agriculture practices on soil pollution?
14. What is the composition of soil?

Section 31.4

15. What are hazardous wastes and how are they classified?
16. Discuss biological methods of treatment and disposal of hazardous wastes.
17. What are the chemical processes of hazardous waste treatment.
18. Write a short notes on
    (i) biodegradation
    (ii) landfill
19. Which soil is considered as best soil for most of the crops? What is its composition?
20. What do you mean by hazardous substance?
21. What are the advantages of sanitary landfill?
22. What is underground injection technique?

❑❑❑

32 GREEN CHEMISTRY AND ITS APPLICATIONS

Modern technology, Owes Ecology, An apology

— ***Alan M.Eddison***

CHAPTER OUTLINE

32.1 INTRODUCTORY OVERVIEW

Green chemistry, also known as sustainable chemistry, is an umbrella concept that has grown substantially since it fully emerged a decade ago. *By definition, green chemistry is the design, development, and implementation of chemical products and processes to reduce or eliminate the use and generation of substances hazardous to human health and the environment*

Chemistry has played, and continues to play, a fundamental role in almost every aspect of modern society .The range of chemical products in today's society is enormous and these products make an invaluable contribution to the quality of our lives. In medicine, the design and manufacture of pharmaceutical products has enabled us to cure diseases that have ravaged humankind throughout history. Crop protection and growth enhancement chemicals have enabled us to increase our food yields dramatically. Petroleum products have revolutionized the transportation and other industries and semiconductors that have made possible computers and other electronic devices. Infact, chemistry changed for ever the way we live.

In recent times, however, the chemical industry is often viewed as causing more harm than good. There can be no denying that in years past, and even at present, chemistry has been misused in many respects, such as the release of pollutants and toxic substances and the production of nonbiodegradable materials, resulting in harm to the environment and living things, including humans.

The biggest challenge for the chemical industry in the twenty first century is to continue to provide the benefits it has provided but without the adverse environmental side effects

Fortunately, the practice of chemical science is moving steadily in the direction of environmental friendliness and resource sustainability. ***The practice of chemistry in a manner that maximizes its benefits while eliminating or at least greatly reducing its adverse impacts has come to be known as green chemistry***

32.2 DEFINITION AND CONCEPTS OF GREEN CHEMISTRY

During the early 1990s the US Environmental Protection Agency (EPA) coined the phrase "Green Chemistry" given by Paul Anastas to promote innovative chemical technologies that reduce or eliminate the use or generation of hazardous substances in the manufacture and use of chemical products. Over the last 10 years Green Chemistry has gradually become recognized asboth a culture and a methodology for achieving sustainability.

Green chemistry is the design of chemical products and processes that reduce or eliminate the use and generation of hazardous substances.

Green chemistry is defined by IUPAC as follows:

> "The invention, design, and application of chemical products and processes to reduce or to eliminate the use and generation of hazardous substances" (*Pure and Applied Chemistry*, 2000, Vol. 72, No. 7, pp. 1207–1228).

Green and sustainable chemistry concerns the development of processes and technologies that result in more efficient chemical reactions that generate little waste and fewer environmental emissions than "traditional" chemical reactions do. Green chemistry encompasses all aspects and types of chemical processes that reduce negative impacts to human health and the environment relative to the current state-of-the-art practices.

By reducing or eliminating the use or generation of hazardous substances associated with a particular synthesis or process, chemists can greatly reduce risks to both human health and the environment.

32.3 EMERGENCE OF GREEN CHEMISTRY

The adoption of green chemistry as one of the primary methods of pollution prevention is a fairly recent phenomenon. Green and sustainable chemistry, a new concept that arose in the early 1990s, gained wider interest and support only at the turn of the millennium. Since the early 1960s, environmental statutes and regulations have proliferated at an exponential rate. With these regulations came costs, restrictions on the use of chemicals, and increased testing of chemical substances to determine their hazards. This provided powerful incentives for industry to find replacements, substitutes, or alternatives. The term green chemistry was first used in 1991 by P.T. Anastas in a special program launched by the US Environmental Protection Agency (EPA) to implement sustainable development in chemistry and chemical technology by industry, academia and government.

Green Chemistry has been described as having three major time periods of development by Linthorst (An overview: origins and development of green chemistry *J. A. Linthorst Foundations of Chemistry Volume 12, Number 1*, 55-68, DOI: 10.1007/s10698-009-9079-4)

1. **Period before 1993:** In this time frame the awareness about pollution and its control started to originate. Major events include the formation of USEPA in 1970, Rachael Carson's *Silent Spring*, and the first Pollution Prevention rules in 1990. In this time EPA evolved as an active organization to promote the efforts to prevent pollution
2. **Time period is between 1993 and 1998** The next time period is between 1993 and 1998. In this time frame a chemical philosophy evolved which is a combination of several chemical concepts, a conceptual framework that can be used in the design of chemical processes achieving environmental and economic goals by way of preventing pollution. This comprises, for instance, ambient reaction conditions, renewable feed stocks and minimization of reaction steps.

3. **From 1999 onwards** The third time period is from 1999 to now which has seen an explosion of Green Chemistry based scientific articles. Networking played a key role in the dissemination of Green Chemistry. Successful networking originated from contextual political, economic and environmental influences. The observed political support for US EPA contributed significantly to the growth of green chemistry. This is especially true for the phase transition around 1998.

32.4 TWELVE PRINCIPLES OF GREEN CHEMISTRY

Green chemistry is commonly presented as a set of twelve principles proposed by Anastas and Warner (P.C. Anastas and J.C. Warner 'Green Chemistry Theory and Practice', Oxford University Press, New York, 1998)

The principles comprise instructions for professional chemists to implement new chemical compounds, new syntheses and new technological processes.

The 12 principles of green chemistry as given by Anastas and Clark et al are listed below.

The 12 principles of Green Chemistry:

1. **Prevention** It is better to prevent waste than to treat or clean up waste after it has been created.
2. **Atom Economy** Synthetic methods should be designed to maximize the incorporation of all materials used in the process into the final product.
3. **Less Hazardous Chemical Synthesis** Wherever practicable, synthetic methods should be designed to use and generate substances that possess little or no toxicity to people or the environment.
4. **Designing Safer Chemicals** Chemical products should be designed to effect their desired function while minimizing their toxicity.
5. **Safer Solvents and Auxiliaries** The use of auxiliary substances (*e.g.,* solvents or separation agents) should be made unnecessary whenever possible and innocuous when used.
6. **Design for Energy Efficiency** Energy requirements of chemical processes should be recognised for their environmental and economic impacts and should be minimized. If possible, synthetic methods should be conducted at ambient temperature and pressure.
7. **Use of Renewable Feedstocks** A raw material or feedstock should be renewable rather than depleting whenever technically and economically practicable.
8. **Reduce Derivatives** Unnecessary derivatization (use of blocking groups, protectiodde-protection, and temporary modification of physicalkhemical processes) should be minimized or avoided if possible, because such steps require additional reagents and can generate waste.
9. **Catalysis** Catalytic reagents (as selective as possible) are superior to stoichiometric reagents.
10. **Design for Degradation** Chemical products should be designed so that at the end of their function they break down into innocuous degradation products and do not persist in the environment.
11. **Real-time Analysis for Pollution Prevention** Analytical methodologies need to be further developed to allow for real-time, in-process monitoring and control prior to the formation of hazardous substances.
12. **Inherently Safer Chemistry for Accident Prevention** Substances and the form of a substance used in a chemical process should be chosen to minimize the potential for chemical accidents, including releases, explosions, and fires.

It is evident from these principles that green chemistry encompasses much more of the concepts of sustainability than simply preventing pollution. Some of the important aspects which encompass these principles are discussed in the following sections

32.5 USE OF ALTERNATIVE FEEDSTOCKS (BIO FUELS)

The era of society's almost total dependence on fossil resources will come to an end during the twenty-first century. Other than longer-term supply issues the main driver for moving away from fossil resources is pollution. There is therefore a growing need for alternative feedstocks to replace fossil resources.

For energy use there are several options including wind, water, solar and nuclear as well as biomass. For organic chemical production the only non-fossil option is biomass.

Green chemistry tries to utilize benign, renewable feedstocks as raw materials. For example, many vehicles around the world are fueled with diesel oil. The main problems with petroleum based diesel include:

- depletion of natural resources
- CO_2 emission
- contribution to smog
- SO, emission

The production of **biodiesel oil** is thus a promising possibility. As the name indicates, biodiesel oil is produced from cultivated plants oil. Most recent research into biodiesel has focused on vegetable oils such as soybean, sunflower, palm and rapeseed.

The main constituents of vegetable oils are triglycerides, which are esters of glycerol with long-chain saturated and unsaturated fatty acids. In order to convert the raw oils into useful material, trans esterification technology is used . The oil is reacted with a low molecular weight alcohol, commonly methanol, in the presence of a catalyst to form the fatty acid ester and glycerol. The ester is subsequently separated from the glycerol and used as biodiesel, the glycerol being used as a raw material for fine chemicals' production. Biodiesel oil also can be obtained from wasted plant oils, e.g. oils used in restaurants.

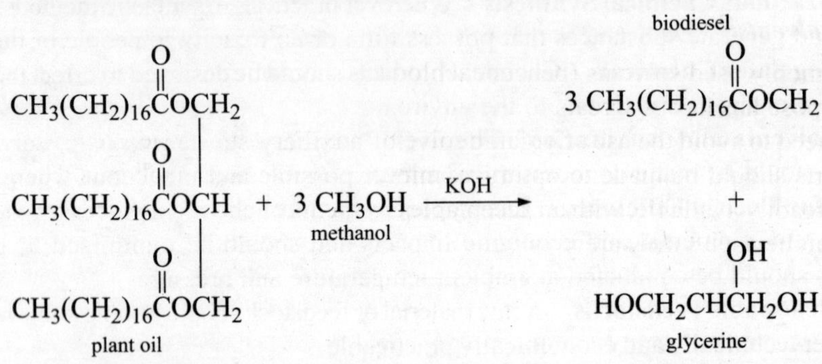

The advantages of using biodiesel oil are many.

- It is a fuel from renewable resources
- contrary to normal diesel oil, the combustion of biodiesel does not generate sulphur compounds
- and generally does not increase the amount of carbon dioxide in the atmosphere.

32.6 USE OF ALTERNATIVE SOLVENTS

Solvents are used in almost all areas of chemistry including synthetic chemistry, analytical chemistry, pharmaceutical production and processing, the food and flavour industry and the materials and coatings sectors. In chemical manufacture, organic solvents are widely used in a variety of unit operations including extraction, recrystallization and the dissolution of solids for ease of handling.

Many of the commonly used solvents are volatile organic compounds (VOCs), hazardous air pollutants (HAPs), flammable,and/or toxic. They also pose serious environmental, health, and safety (EHS) concerns, including human and eco-toxicity issues, process safety hazards, and waste management issues. However, the principles of green chemistry guide us to use less of them, or to use safer, more environmentally friendly solvents only if they are essential.

The main environmental issue concerned with VOCs is their ability to form low-level ozone and smog through free radical air oxidation processes. The EPA has published a list of a number of adverse health effects, which are now thought to originate from the presence of VOCs in the environment, including:

- conjunctival irritation
- nose and throat discomfort
- headache
- allergic skin reaction
- dyspnea
- declines in serum cholinesterase levels
- nausea
- fatigue
- dizziness

Hence alternatives to organic solvents are needed to decrease the negative environmental impact of these substances.

Some of the more common alternatives to using organic solvents are:

- use of safer/green solvents,
- use of water as solvent,
- solvent-free processes
- use of supercritical liquids (supercritical carbon dioxide (31.1°C, 73 atm),and supercritical water (374°C, 218 atm),
- Room-temperature ionic liquids.

(a) Use of safer solvents:

Many of the commonly used solvents (benzene, chlorinated organic solvents, etc.) are known carcinogens, and many others pose hazardous threats to the environment.

There is the need to avoid the use of organic solvents where ever possible. However, when elimination is not feasible, efforts should be made to optimize, minimize, and recycle solvents. Solvents that are stable, inexpensive, and readily available, with an acceptable environment impact, are the most suitable. Some of the solvents which can be used safely are:

• Isoamyl alcohol	• 2-Ethylhexanol
• 2-Butanol	• Ethylene glycol
• 1-Butanol	• Diethylene glycol butyl ether
• *t*-Butyl acetate	• Butyl acetate
• *n*-Propyl acetate	• Isopropyl acetate
• Dimethylpropylene urea	• Propionic acid
• Ethyl acetate	• Methyl isobutyl ketone

Another class of safer solvents is the *Green Solvents. Green solvents* are environmentally friendly solvents, or biosolvents, which are derived from the processing of agricultural crops.

Ethyl lactate, for example, is a green solvent derived from processing corn. Ethyl lactate is the ester of lactic acid. Lactate ester solvents are commonly used solvents in the paints and coatings industry and have numerous attractive advantages such as 100% biodegradable, easy to recycle, noncorrosive, noncarcinogenic, and nonozone-depleting. Ethyl lactate has replaced solvents such as toluene, acetone, and xylene, resulting in a much safer workplace.

(b) Use of supercritical liquids

The term "supercritical fluids" comprises the liquid and gases at temperatures and pressures higher than their critical temperatures and pressures (Fig. 32.1). Above the critical point the liquid-vapour phase boundary disappears and the supercritical liquid exhibits properties between those of gas and liquid. High compressibility of supercritical fluids in the vicinity of the critical point makes it easy to adjust density and solution ability by a small change of temperature or pressure. Due to this; the supercritical fluids are able to dissolve many compounds with different polarity and molecular mass.

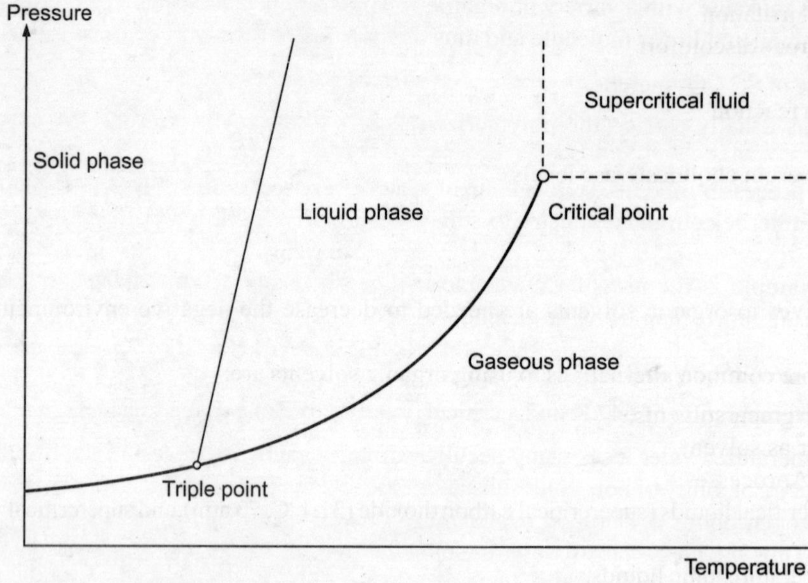

Fig 32.1 Phase diagram showing supercritical fluid region

Two supercritical fluids are of particular interest fulfilling the green chemistry demands as the reaction media, *carbon dioxide and water*.

(i) Supercritical carbondioxide (SC CO_2)

Carbon dioxide has a low critical temperature of 304 K and a moderate critical pressure of 73 bar. Carbon dioxide as a supercritical fluid is most frequently used as medium for reactions.

- It is inflammable, easily available and cheap.
- Its application gives considerable energy savings because the critical point is easy to reach due to a low evaporation heat of CO_2.
- Carbon dioxide as a supercritical fluid dissolves non-polar compounds and some polar (e.g. methanol, acetone) like fluorocarbon solvents.
- The non-flammability of SC CO_2 makes it an attractive medium for carrying out oxidation reactions outside the explosive limit.

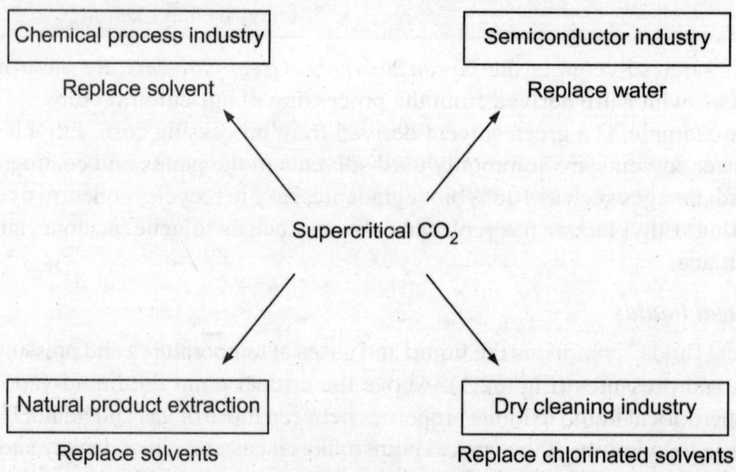

Fig. 32.2 Uses of supercritical CO_2.

- Further, it is miscible with a variety of organic solvents and is readily recovered after processing.
- It is also a small and linear molecule and thus diffuses faster than conventional liquid solvents.

Examples of use of SC CO_2 include

- the reactions usually carried the polymerization of tetrafluoroethylene in CO_2 using fluorinated initiators ,
- CO_2-based processes used for dry cleaning, metal cleaning, and textile processing. Liquid CO_2 is also used in the microelectronics industry to spin-coat photoresists instead of using traditional organic solvents.
- Another example is the use of CO_2 to clean integrated circuits and flat-panel displays during manufacturing rather than using large amounts of water and organic solvents.

(ii) Supercritical water (SC H_2O):

Water has a critical temperature of 647 K and a critical pressure of 220 bar due to its high polarity.

- At this temperature, water loses many peculiar properties it has in its liquid state, such as hydrogen bonding and insolubility of non-polar substances in it. At the critical point, they disappear completely.
- When hydrogen bonds are broken, water molecules can dissolve chemicals that were previously insoluble. Thus the character of water at supercritical conditions changes from one that dissolves only ionic species at normal conditions to one that dissolves paraffins, aromatics, gases and salts.
- Due to this unique property, research has been carried out on supercritical water for reaction and separation processes to treat toxic wastewater.
- Both near-critical and supercritical water (SCH_2O) have increased acidity, reduced density, and lower polarity, greatly extending the possible range of chemistry that could be carried out in water.
- A key advantage of chemistry with supercritical water is the possibility of varying the properties of the reaction medium over a wide range solely by changing the pressure and temperature and of optimizing the reaction in this way without changing solvent.

Supercritical fluids such as water and carbon dioxide are substances that are compatible with the earth's environment. However, several other supercritical fluids can be used, but the final choice would depend on the specific application and additional factors such as safety, flammability, phase behavior and solubility at the operating conditions and the cost of the fluid.

(c) Room-temperature ionic liquids.

- Most of the common liquids (e.g., water, ethanol, benzene, etc.) are molecular and are basically made up of molecules. However, since the early 1980s an exciting new class of room-temperature liquids has become available. These are the ambient-temperature **ionic liquids**. Ionic liquids are composed entirely of ions.
- An ionic liquid is a salt in which the ions are poorly coordinated, which results in these solvents being liquid below 100°C, or even at room temperature. In broad terms they can be viewed like common ionic materials such as sodium chloride, the difference being that they are liquid at low temperatures, this being due to poor packing of the respective ions.
- Ionic liquids possess a unique array of physico-chemical properties that make them excellent solvents such as high thermal stability and high electrical conductivity.
- These simple liquid salts (single anion and cation) can be mixed with other salts (including inorganic salts) to form multicomponent ionic liquids.
- The first room-temperature ionic liquid ($EtNH_3$)(NO_3) was discovered in 1914, and later on binary ionic liquids from mixtures of aluminium (III) chloride and N-alkylpyridinium or 1,3-dialkylimidazolium chloride were discovered.
- There are many thousands of possible ionic liquids of which only a few have been studied in detail; Figure 32.3 shows some of the more common ones.

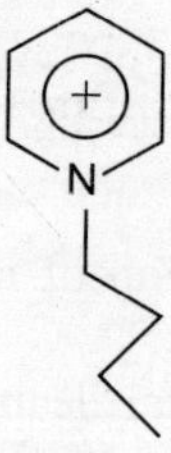

N-butylpyridinium-aluminium(III) chlorlde ([NBupy]$AlCl_4$])

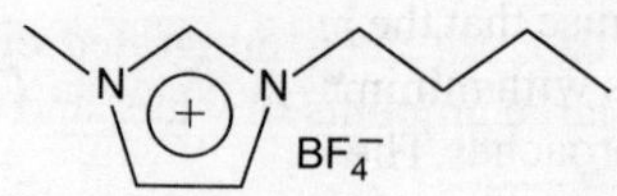

1-butyl-3-methylimidazolium fluoride-boron trifluoride ([bmim]BF_4)

Fig. 32.3 Some common room temperature ionic liquids

- The absence of volatility is one of the most important benefits of ionic liquids, offering a much lower toxicity as compared to low-boiling-point solvents.
- Ionic liquids have been found useful for a wide range of chemical reactions and processes, including hydrogenation reactions, biocatalysis reactions such as transesterification and hydrolysis, and electrochemical applications such as battery electrolytes.

(d) Water as Solvent

The use of water as solvent for organic reactions is one of the finest solutions to the problem of solvent toxicity and disposal. The chemistry in natural systems (biochemical reactions) is based on water. The use of water as solvent for synthetic chemistry holds great promise for the future in terms of the cheaper and less hazardous production of chemicals.

(e) Solvent-Free Conditions

In general, many reactions involving miscible or partially miscible reagents can proceed under solvent-free, often mild, conditions Several advantages are associated with the use of a solvent-free system over the use of organic solvent. These include

- There is no reaction media to collect, dispose of, or purify and recycle.
- On a laboratory's preparative scale, there is often no need for specialized equipment.
- Extensive and expensive purification procedures such as chromatography can often be avoided due to the formation of sufficiently pure compounds.
- Greater selectivity is often observed.
- Reaction times can be rapid, often with increased yields and lower energy usage.

Not surprisingly, solvent-free synthesis has recently drawn attention from the wider synthetic community. Reactions achieved by solvent-free systems include aldol condensations, sequential aldol and Michael additions, Stobbe condensations etc.

The aldol reaction is carried out by grinding together solid sodium hydroxide with a benzaldehyde and acetyl pyridine (both usually liquids). The product of the reaction is, however, a solid which forms after a few minutes.

32.7 DESIGN OF SAFER CHEMICALS

Historically there have been two factors which governed our approach to toxic materials. First it is only since the 1960s that our knowledge of the toxicity of chemicals has been handled scientifically. Before then we tended to know about the toxic effects only after the event, usually entailing ill health or death amongst those working with the substance. One early example of this was the early production of nickel from its carbonyl; it was only after several strange deaths amongst workers that it was finally realized that nickel carbonyl is highly toxic. This is only one of many examples where we have learned through 'trial and error' of the toxic nature of chemicals.

Kenneth Geiser in his book Materials and Matter, defines safer chemicals and materials as being: *inherently safer, physically benign, biodegradable, generated on demand, and manufactured in contained systems under ambient conditions*

Green chemistry is based on the premise that the most desirable and efficient way of pollution prevention is to design commercially used chemicals with minimal or no toxicity. This area is one of the most difficult to achieve, and requires many different approaches.The key considerations are summarized below:

- Design alternative synthetic pathways for the new and existing chemicals so that they utilize neither toxic reagents or solvents, nor produce toxic by-products or coproducts
- Use of molecular structure design to minimize the intrinsic toxicity of the product while maintaining its efficacy of function
- Elimination of 'toxic' functional groups
- Innovative and ingeneous molecular design

***Example*:** Carbon-carbon bond formation *via* free radical chemistry is a very versatile method of producing a range of materials of interest to pharmaceutical companies. Unfortunately in the traditional way free radicals are generated using highly neurotoxic organotin compounds, especially tributyl tin hydride, an excellent radical chain carrier. Whilst handling this compound is a significant hazard there is also the possibility that trace amounts of organotin compounds may be left in the product at the end of the reaction and purification stages. This is completely unacceptable for any product destined for ingestion by humans. Replacement of organotin compounds has therefore been the subject of much recent research. One possible alternative is to use an easily oxidized sulfide which could transfer an electron to an electrophile to give a radical anion .This in turn could fragment to give an organic radical.

$$Ar_2S + RX \longrightarrow RX^- + Ar_2S^+ \longrightarrow R + Ar_2^+X^-$$

32.8 DESIGN OF ALTERNATIVE REACTIVE METHODOLOGY

This aspect overlaps with some previously described ones such as use of safer solvents, safer chemicals etc. Industries have resorted to alternative methodologies such as heterogeneous catalytic, biocatalytic, and fermentation routes to improve process efficiency, reduce energy usage, and decrease raw material usage. The key points are:

- Replacement of the chemical step with a biocatalytic step , which is followed by replacing the entire multiple-step process with a single fermentation process
- Developing atom-economical processes
- Use of enzyme technology has become the focus of many pharmaceutical, food, and specialty industries.

***Example*:** In the pharmaceutical industry, the largest scale biocatalytic process is the conversion of the fermentation product penicillin G into 6-aminopenicillanic acid by the enzyme penicillin acylase

- Also, scientists have realized that biotechnology routes are inherently safe and mild and do not adversely impact the environment.
- Heterogeneous catalysis also appears to offer several advantages since the reactions are clean and selective. Also, the spent catalyst can be filtered and recovered for later use.

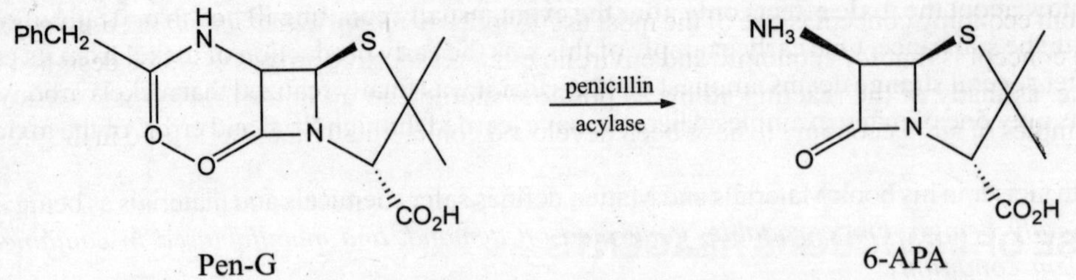

Fig. 32.4 Conversion of penicillin G (Pen-G) to 6-aminopenicillanic acid (6- APA) by penicillin acylase.

- The semiconductor, tannery, dyes, paper and pulp, sugar and distillery, and textile industries are reeling from large amounts of toxic effluents. Pressures from government and NGOs are pushing these industries to the corner. They need to embrace the concepts of green chemistry and develop innovative processes that would eliminate use of toxic chemicals, use of large amounts of pure water, and generation of large quantities of VOC.
- The leather industry is developing processes that use lipase instead of harsh chemicals for dehairing, while the semiconductor industry is using SO_3 and SC-CO_2 for cleaning printed circuit boards.

Atom Economy One of the major areas of reactive methodology is "Atom Economy": Atom economy is a method of determining the efficiency with which raw materials (reactants or feedstocks) are used *regardless of* the percentage yield obtained in the reaction.

Consider the reaction in which ethyl chloride (CH_3CH_2Cl) is to be produced in the reaction among ethyl alcohol (CH_3CH_2OH), sodium chloride (NaCl), and sulfuric acid (H_2SO_4)

$$CH_3CH_2OH + NaCl + H_2SO_4 \longrightarrow CH_3CH_2Cl + NaHSO_4 + H_2O$$

The atoms that are *not* directly involved in the formation of the product are "wasted" atoms. In this particular example, the final by-products, sodium bisulfate and water, are wasted. Many chemical reactions result in hazardous chemicals that do pose a threat to the health of plants and animals and that do, therefore, become factors in air and water pollution and waste management issues. The measure of the atom economy of the reaction is the mass percentage of reactants used in the final product.By dividing that number by the total mass of reactants, a numerical measure of atom economy can be obtained. That is:

$$\text{Atom economy} = \frac{\text{Formula weight of all reactants utilize}}{\text{Formula weight of all reactants}} \times 100\%$$

For the above example, atom economy,

$$= \frac{64}{202} = 32\%$$

One goal of the green chemist, then, is to find a way of making the same product with the waste of fewer reactant atoms. One well-known option is to chlorinate ethane (CH_3CH_3) directly with chlorine gas, as shown in the following equation:

$$CH_3CH_3 + Cl_2 \longrightarrow CH_3CH_2Cl + HCl$$

Notice in this case that the atom economy of the reaction is about double that of the first reaction:

$$\text{Atom economy} = \frac{64}{101} = 63\%$$

The atom economy concept is one of the most useful tools available for design of reactions with minimum waste. The concept is that for economic and environmental reasons reactions should be designed to be atom efficient, i.e. as many of the reacting atoms as possible should end up in useful products. One of the best-known examples of atom economy to have been developed commercially thus far is used in the preparation of ibuprofen.

32.9 USE OF INNOCUOUS REAGENTS

Principles of Green Chemistry guide us to make use of reagents that are inherently less hazardous and are catalytic whenever feasible. Frequently, many chemical reactions involve the use of reagents such as protecting groups and so-called catalysts that do not end up in the useful product. Organic solvents, often thought to be essential but sometimes not actually required at all, fall into this category. Some of these materials end up as waste and some are recovered, but in all cases valuable resources and energy are consumed that do not form part of the required product.

Example: Taking Friedel–Crafts reactions as an example, alkylation and acylation reactions often are referred to as being catalysed by lewis acids such as aluminium chloride. In both cases, reactions usually are quenched with water, leading to copious amounts of aluminous waste and releasing three equivalents of HCl. Now a days active solid catalysts such as zeolites are used that are highly selective and avoid waste raw materials and byproducts.

Carboxylation is the process by which carbon dioxide is added to compounds. The bacterium *Bacillus megaterium* in supercritical carbon dioxide at temperatures of about 100°F (40°C) and 100 atmospheres of pressure is being to bring about carboxylation. The process illustrates yet one more way of using more than one green chemistry principle in a process: the use of a safer catalyst as well as a safer solvent.

32.10 MICROWAVE AND ULTRASONIC RADIATION IN GREEN TECHNOLOGY

32.10.1 Mininizing Energy Consumption

Most chemical processes use thermal sources of energy originating from fossil (or nuclear) fuels. The energy input to the process is non-specific, *i.e.* it is not directly targeted at the chemical bond or even the molecules undergoing reaction. Much of the energy is 'wasted' in heating up reactors, solvent and even the general environment.

In order to minimize energy and control reactions with a view to *green* chemistry, attempts are being made to make the energy input in chemical systems as efficient as possible. Approaches are being taken to use other forms of energy, so-called nonclassical energy forms, in order to optimize the duration and product yield and avoid undesired side products.

The *nonclassical methods*, include

- Ultrasonication,
- Microwave radiations

(a) Ultrasonic radiation:

Ultrasound refers to sound waves with frequencies higher than those detectable by the human ear, *i.e.* around 18 kHz. The terminology has become common knowledge owing to the widespread use of ultrasound scanning equipment in medical applications. The ultrasound frequencies of interest for chemical reactions (typically 20-100 kHz) are much lower than used for medical applications, but the power used is higher; the two should not be confused.

Ultrasound, an efficient means of activation in synthetic chemistry, has been employed for decades. Not only can this high-energy input enhance mechanical effects in heterogeneous processes, but it is also known to induce new reactions, leading to the formation of unexpected chemical species. The term sonochemistry is used to denote reactions initiated by ultrasound.

Theory When a sound wave, propagated by a series of *compression and rarefaction* cycles, passes through a liquid medium it causes the molecules to oscillate around their mean position. During the compression cycle the average distance between molecules is reduced and, conversely, it is increased during rarefaction. Under appropriate conditions in the rarefaction cycle the attractive forces of the molecules of the liquid may be overcome, causing bubbles to form. If the internal forces are great enough to cause collapse of these bubbles very high local temperatures (around 5000 °C) and pressures (over 1000 bar) may be created. *It is these very high temperatures and pressures that initiate chemical reaction.* Ultrasound is known to enhance the reaction rate, thus minimizing the duration of a reaction. The real benefit of using ultrasound lies in its unique selectivity and reactivity enhancement.

The range of reactions carried out under sonochemical conditions is large and growing rapidly; in many cases some green benefits are obvious.

Typical reaction types assessed include

- *oxidation,* which can often be carried out more rapidly at lower temperatures.
- *radical reactions,* with the radicals being generated under mild conditions,
- *synthesis of nanopartides* Here the high temperatures generated, followed by rapid cooling are favourable to the formation of these particles.
- Apart from the use of ultrasound in enhancing the reactivity in organic reactions, ultrasound has varied uses in industry, such as welding, cutting, emulsification, solvent degassing, powder dispersion, cell disruption, and atomization

(b) Use of Microwaves of Synthesis

Microwaves have wavelengths between 1 mm and 1 m and hence have similar frequencies to radar and telecommunication devices. In synthetic chemistry, 1986 was an important year for the use of microwave devices. Since that year, countless syntheses initiated by microwaves have been carried out on a laboratory scale.The result is often a drastic reduction in the reaction time with comparable product yields, if microwaves are used instead of classical methods of energy input. Unwanted side reactions can often be suppressed and solvents dispensed with. Numerous reactions, such as esterifications, Diels Alder reactions,hydrolyses, or the production of inorganic pigments, have been investigated in recent years. Apart from the obvious advantages of the use of microwaves in chemical syntheses, microwave technologies are being tested as energy- and cost-saving alternatives.

Theory The overall mechanism of how energy is imparted to a substance under microwave irradiation is complex, consisting of several different aspects.

One process, which explains why microwaves heat certain substances and not others, is termed dipolar polarization. When a substance possessing a dipole moment (water is the obvious example) is subject to electromagnetic radiation it will attempt to align itself to the electromagnetic field by rotation. In liquids this rotation causes friction between adjacent molecules, which in turn causes a temperature rise. In the microwave region of the spectrum, rotation rates are high enough to produce rapid temperature rise in dipolar substances. Liquids which do not have a dipole moment (or in which one cannot be induced) are not directly heated by microwaves. Because of the large distances between molecules, gases are not heated by microwave irradiation.

Examples

(a) Water-based microwave reactions have been relatively well studied for hydrolysis and hydrogen peroxide oxidation reactions.In the hydrolysis of benzamide with sulfuric acid, quantitative conversion was achieved in 7 rnin under microwave irradiation at 140 °C, compared to a 90% conversion after a 1-h reflux using a conventional heating source. The difference in reaction time can be accounted for by the difference in reaction temperature coupled with the very rapid heating of the microwave reactor.

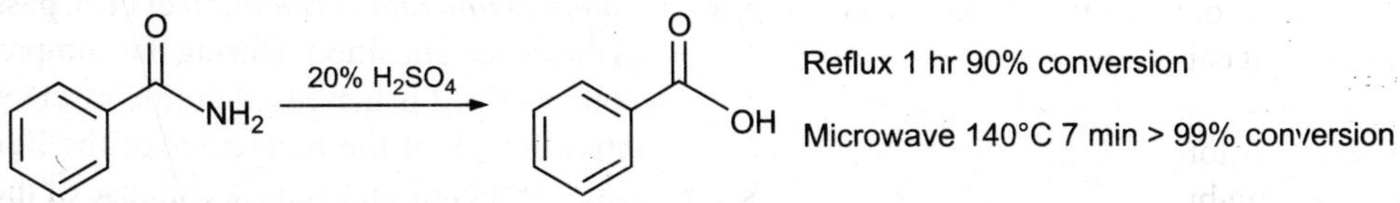

(b) A range of primary alcohols have been oxidized to the corresponding carboxylic acids using sodium tungstate as catalyst in 30% aqueous hydrogen peroxide. Yields, although variable, of up to 85% have been achieved in a rapid, clean, safe and atom efficient reaction

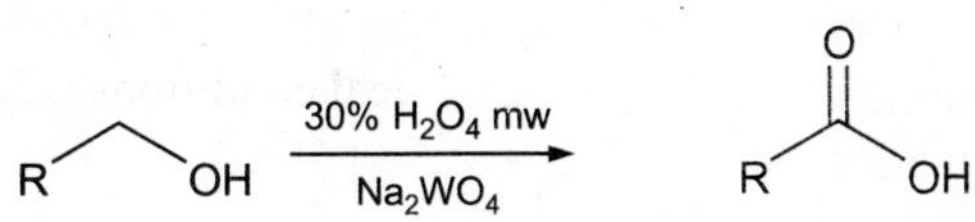

QUESTIONS

1. What is the requirement of Green Chemistry in today's scientific world?
2. Fist twelve Principles of Green Chemistry.
3. Write short notes:
   (i) Use of Alternative Feedstocks
   (ii) Use of Innocuous reagents.
4. What are the alternatives to using organic solvents with reference to green chemistry.
5. How supercritical liquids help in synthesis using green approach.
6. What are the advantages of room temperature ionic liquids
7. What do you mean by Atom Economy.
8. Discuss the role of Microwaves in Green technology.
9. How ultrasonic radiations help in green synthesis. Explain with the help of examples.

❑❑❑

INDEX

D

E

Q

R

S

T

❑❑❑

Win Prizes!

ATTENTION READERS

We request you for your frank assessment regarding some of the aspects of the book given as under:

A TEXTBOOK OF ENGINEERING CHEMISTRY

By

DR. SUNITA RATTAN

Please fill up the given space in neat capital letters.

(i) What topics of your syllabus that are important from you examination point of view are not covered in the book ?

....

....

....

....

(ii) What are the chapters wherein the treatment of the subject matter is not systematic or organised or updated ?

....

....

....

....

(iii) Have you come across misprints/mistakes/factual inaccuracies in the book? Please specify the chapters and the page numbers.

....

....

....

....

(iv) What other books on the same subject have you found/heard better than the present book? Please specify in terms of price and quality.

....

....

....

....

(v) Further suggestions and comments for the improvement of the book:

....

....

....

....

(PS: If need be please attach a separate sheet to write your views.)

Other Details :

(i) Who recommended you the book ? (Please tick in the box near the option relevant to you.)

☐ Teachers ☐ Friends ☐ Bookseller

(ii) Name of the recommending teacher, his designation and address.

....

....

....

....

(iii) Name and address of the bookseller you purchased the book from:

....

....

....

....

(iv) Name and address of your institution (Please mention the University or Board, as the case may be)

....

....

....

....

(v) Your name and complete postal address:

....

....

....

....

(vi) Course you are enrolled in:

....

....

....

....

The best assessment will be awarded each month. The award will be in the form of our publications as decided by the Editorial Board.

Please mail the filled up coupon at your earliest to:

S.K. Kataria & Sons®

4885/109, Prakash Mahal, Dr. Subhash Bhargav Lane,
Opposite Delhi Medical Association, Daryaganj, New Delhi-110002
Phones: +91-11-43551243, Mobile: +91-9871775858
e-mail: katariabook@gmail.com, katariabooks@yahoo.com
Website: www.skkatariaandsons.com